Concept Check exercises focus on mathematical thinking and conceptual understanding. Mathematics involves more than notation, calculations, and formulas; pausing to think about the underlying concepts will help you remember and apply them correctly when the time comes.

Concept Check *Give an equation of the form $f(x) = a^x$ to define the exponential function whose graph contains the given point.*

83. $(3, 8)$

84. $(-3, 64)$

Concept Check *Use properties of exponents to write each function in the form $f(t) = ka^t$, where k is a constant. (Hint: Recall that $a^{x+y} = a^x \cdot a^y$.)*

85. $f(t) = 3^{2t+3}$

86. $f(t) = \left(\frac{1}{3}\right)^{1-2t}$

Additional complete, step-by-step solutions are now provided at the back of the text for selected exercises. These are solutions to exercises that extend the skills and concepts presented in the examples—actually providing you with a pool of examples to different and/or more challenging problems.

55. Solve $\left(\frac{1}{e}\right)^{-x} = \left(\frac{1}{e^2}\right)^{x+1}$.

55. $\left(\frac{1}{e}\right)^{-x} = \left(\frac{1}{e^2}\right)^{x+1}$

$(e^{-1})^{-x} = (e^{-2})^{x+1}$	Definition of negative exponent
$e^x = e^{-2(x+1)}$	$(a^m)^n = a^{mn}$
$e^x = e^{-2x-2}$	Distributive property
$x = -2x - 2$	Property (b)
$3x = -2$	Add $2x$.
$x = -\frac{2}{3}$	Divide by 3.

Solution set: $\left\{-\frac{2}{3}\right\}$

Chapter 4 Summary

KEY TERMS

4.1 one-to-one function inverse function 4.2 exponential function exponential equation compound interest	future value present value compound amount continuous compounding	4.3 logarithm logarithmic equation logarithmic function 4.4 common logarithm pH	natural logarithm 4.6 doubling time half-life

NEW SYMBOLS

$f^{-1}(x)$ the inverse of $f(x)$

e a constant, approximately 2.718281828

$\log_a x$ the logarithm of x to the base a

$\log x$ common (base 10) logarithm of x

$\ln x$ natural (base e) logarithm of x

QUICK REVIEW

CONCEPTS | **EXAMPLES**

4.2 Exponential Functions

Additional Properties of Exponents

For any real number $a > 0$, $a \neq 1$,

(a) a^x is a unique real number for all real numbers x.
(b) $a^b = a^c$ if and only if $b = c$.
(c) If $a > 1$ and $m < n$, then $a^m < a^n$.
(d) If $0 < a < 1$ and $m < n$, then $a^m > a^n$.

(a) 2^x is a unique real number for all real numbers x.
(b) $2^x = 2^3$ if and only if $x = 3$.
(c) $2^5 < 2^{10}$, because $2 > 1$ and $5 < 10$.
(d) $\left(\frac{1}{2}\right)^5 > \left(\frac{1}{2}\right)^{10}$ because $0 < \frac{1}{2} < 1$ and $5 < 10$.

Exponential Function

If $a > 0$ and $a \neq 1$, then $f(x) = a^x$ defines the exponential function with base a.

$f(x) = 3^x$ defines the exponential function with base 3.

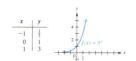

Graph of $f(x) = a^x$

1. The graph contains the points $\left(-1, \frac{1}{a}\right)$, $(0, 1)$, and $(1, a)$.
2. If $a > 1$, then f is an increasing function; if $0 < a < 1$, then f is a decreasing function.
3. The x-axis is a horizontal asymptote.
4. The domain is $(-\infty, \infty)$; the range is $(0, \infty)$.

x	y
-1	$\frac{1}{3}$
0	1
1	3

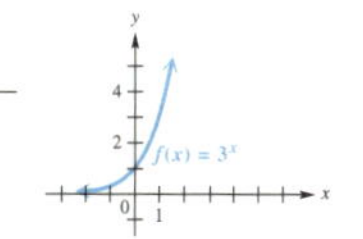

Extensive mid-chapter and end-of-chapter summaries help you prepare for quizzes and tests. Each end-of-chapter Summary includes a section-by-section list of Key Terms, New Symbols, and a Quick Review of Concepts illustrated by a multitude of Examples. This comprehensive review is a hallmark feature of the Lial/Hornsby/Schneider systematic approach to student success.

Precalculus

Third Edition

Margaret L. Lial
American River College

John Hornsby
University of New Orleans

David I. Schneider
University of Maryland

Boston San Francisco New York
London Toronto Sydney Tokyo Singapore Madrid
Mexico City Munich Paris Cape Town Hong Kong Montreal

Publisher: Greg Tobin
Senior Acquisitions Editor: Anne Kelly
Editorial Assistant: Cecilia Fleming
Associate Project Editor: Joanne Ha
Senior Production Supervisor: Karen Wernholm
Production Coordination and Text Design: Elm Street Publishing Services, Inc.
Senior Marketing Manager: Becky Anderson
Marketing Coordinator: Carolyn Buddeke
Senior Author Support/Technology Specialist: Joe Vetere
Associate Media Producer: Sara Anderson
Software Development: Kathleen Bowler and Malcolm Litowitz
Senior Manufacturing Buyer: Evelyn Beaton
Rights and Permissions Advisor: Dana Weightman
Cover Design: Barbara T. Atkinson
Composition: Beacon Publishing Services
Illustrations: Techsetters, Inc.
Cover Photo: © Getty/National Geographics

Photo Credits: **p. 1, 7, 470** AFP/Corbis; **p. 13, 372** Reuters New Media Inc./Corbis; **p. 19, 280** Brand X Pictures; **p. 63, 292, 389, 413, 653, 664, 684, 960, 810** PhotoDisc Blue; **p. 74** Duomo/Corbis; **p. 80** Comstock; **p. 84, 85, 99, 386, 531, 556, 597, 631, 788, 912** Corbis RF; **p. 103, 158** Kobal Collection; **p. 131, 186, 388, 434, 995, 1012** Beth Anderson; **p. 180, 181, 197, 225, 248, 293, 302, 361, 472, 473, 518, 550, 605, 699, 728 (T) (B), 902** PhotoDisc; **p. 129, 938, 969** St. Andrew's University MacTutor Archives; **p. 232, 238, 934** Digital Vision; **p. 100, 328** SAU; **p. 306** Jeff Greenberg/PhotoEdit; **p. 311** ThinkStock; **p. 459** Archivo Iconografico, S.A./Corbis ; **p. 533** NASA; **p. 573** Courtesy Joseph A. Dellinger; **p. 603** Thinkstock; **p. 617** Jutta Klee/Corbis; **p. 685** University of Waterloo; **p. 779** © Susan Ragan/Reuters New Media Inc./Corbis; **p. 781** Artville; **p. 783** Frank W. Olsen, Kanab Field Office, Bureau of Land Management; **p. 891, 909** Michael Freeman/Corbis; **p.935, 946** Getty Images/Taxi/David Seed; **p. 990** Courtesy of Margaret Westmoreland

Library of Congress Cataloging-in-Publication Data

Lial, Margaret L.
Precalculus/Margaret L. Lial, John Hornsby, David I. Schneider.—3rd ed.
p. cm.
Includes index.
ISBN 0-321-22762-X
1. Algebra. 2. Trigonometry. I. Hornsby, E. John. II. Schneider, David I. III. Title.
QA154.3.L56 2005
512—dc22

2003055617

Student Edition: 4 5 6 7 8 9 10—QWT—070605

Contents

Preface

In the new edition of this text, we continue our ongoing commitment to providing the best possible text to help instructors teach and students succeed. To help students develop both the conceptual understanding and the analytical skills necessary to experience success in mathematics, we present each mathematical topic in the text using a systematic approach designed to actively engage students in the learning process. We have tried to address the diverse needs of today's students through a more open design, updated figures and graphs, helpful features, careful explanations of topics, and a comprehensive package of supplements and study aids. Students planning to continue their study of mathematics in calculus, statistics, or other disciplines, as well as those taking this as their final mathematics course, will benefit from the text's student-oriented approach. We believe instructors will particularly welcome the new *Annotated Instructor's Edition,* which provides answers in the margins to almost all exercises, plus helpful Teaching Tips.

New or Enhanced Features

We are pleased to offer the following new or enhanced features.

Chapter Openers These have been updated and streamlined and provide a quick preview of the chapter topics, plus a reference to a motivating application within the chapter.

Examples We have added even more examples in this edition. Step-by-step solutions to examples are now easily identified using a **Solution** head. We have carefully polished all solutions and incorporated more side comments and explanations, including helpful section references to previously covered material. Selected examples continue to provide graphing calculator solutions alongside traditional algebraic solutions. ***The graphing calculator solutions can be easily omitted if desired.***

Now try Exercises To actively engage students in the learning process, each example now concludes with a reference to one or more parallel odd-numbered exercises from the corresponding exercise set. In this way, students are able to immediately apply and reinforce the concepts and skills presented in the examples.

Exercise Sets We have taken special care to respond to the suggestions of users and reviewers and have added many new exercises to this edition based on their feedback. As a result, the text includes more problems than ever to provide students with ample opportunities to practice, apply, connect, and extend concepts and skills. We have included writing exercises and optional graphing calculator problems, as well as multiple-choice, matching, true/

false, and completion problems. *Concept Check* problems, which focus on mathematical thinking and conceptual understanding, were well received in the previous edition and have been expanded in this edition.

Solutions to Selected Exercises Exercise numbers enclosed in a blue circle, such as 11., indicate that a complete solution for the problem is included at the back of the text. These new solutions are given for selected exercises that extend the skills and concepts presented in the section examples—actually providing students with a pool of examples to different and/or more challenging problems.

Summary Exercises These new sets of in-chapter exercises give students the all-important *mixed* review problems they need to synthesize concepts and select appropriate solution methods.

Function Boxes Beginning in Chapter 2, functions are a unifying theme throughout the remainder of the text. Special function boxes offer a comprehensive, visual introduction to each class of function and also serve as an excellent resource for student reference and review throughout the course. Each function box includes a table of values alongside traditional and calculator graphs, as well as the domain, range, and other specific information about the function.

Figures and Photos Today's students are more visually oriented than ever. As a result, we have made a concerted effort to include mathematical figures, diagrams, tables, and graphs whenever possible. Drawings of famous mathematicians accompany historical exposition, and photos accompany selected applications in examples and exercises.

Chapter Reviews Each chapter ends with an expanded Summary, featuring a section-by-section list of Key Terms, New Symbols, and a Quick Review of important Concepts, presented alongside corresponding all-new Examples. A comprehensive set of Review Exercises and a Chapter Test are also provided.

Quantitative Reasoning Now appearing at the end of each chapter, these problems enable students to apply algebraic and trigonometric concepts to life situations, such as financial planning for retirement or determining the value of a college education. A photo highlights each problem.

Glossary As an additional student study aid, a comprehensive glossary of key terms from throughout the text is provided at the back of the book.

Continuing Features

We have retained the popular features of previous editions of the text, some of which follow.

Real-Life Applications We are always on the lookout for interesting data to use in real-life applications. As a result, we have incorporated many new or updated applied examples and exercises from fields such as business, pop culture, sports, the life sciences, and environmental studies that show the relevance of algebra and trigonometry to daily life. Applications that feature mathematical modeling are labeled with a *Modeling* head. All applications are titled, and a comprehensive Index of Applications is included at the back of the text.

Use of Technology As in the previous edition, we have integrated the use of graphing calculators where appropriate, although graphing technology is not a central feature of this text. We continue to stress that graphing calculators are an aid to understanding and that students must master the underlying mathematical concepts. We have included graphing calculator solutions for selected examples and continue to mark all graphing calculator notes and exercises that use graphing calculators with an icon for easy identification and added flexibility. ***This graphing calculator material is optional and can be omitted without loss of continuity.***

Cautions and Notes We often give students warnings of common errors and emphasize important ideas in **CAUTION** and **NOTE** comments that appear throughout the exposition.

Looking Ahead to Calculus These margin notes offer glimpses of how the algebraic or trigonometric topics currently being studied are used in calculus.

Connections Connections boxes continue to provide connections to the real world or to other mathematical concepts, historical background, and thought-provoking questions for writing, class discussion, or group work.

Relating Concepts Exercises Appearing in selected exercise sets, these sets of problems help students tie together topics and develop problem-solving skills as they compare and contrast ideas, identify and describe patterns, and extend concepts to new situations. These exercises make great collaborative activities for pairs or small groups of students.

Content Changes

A primary focus of this revision of the text was to polish and enhance individual presentations of topics, and we have worked hard to do this throughout the book. Additional content changes you may notice include the following:

- A thorough review of algebraic prerequisites is included in Chapter R. This chapter is former Chapter 1, and all successive chapters have been renumbered accordingly.
- The presentation on relations, functions, and graphs has undergone extensive reorganization and rewriting and is now covered in Chapter 2. Among the many specific revisions in this key chapter, you will find new material on graphing equations in Section 2.1. Functions and relations are now introduced together in Section 2.2, which includes twice as many examples as in the previous edition. Coverage of slope/average rate of change has been expanded in Section 2.3, and new material on even and odd functions is included in Section 2.6.
- The presentation of rational functions in Section 3.5 has been expanded and now includes function boxes to highlight this important class of function.
- Continuous compounding is presented along with the material on compound interest in Section 4.2, and several new examples are provided based on reviewer feedback. Among other enhancements, new function boxes have been included for exponential and logarithmic functions in Sections 4.2 and 4.3.

- Former Chapter 6 on Trigonometric Functions has been split into two chapters. Chapter 5 covers right triangles and the trigonometric functions, while Chapter 6 defines the circular functions in terms of the unit circle. In Sections 6.3 and 6.4, we include function boxes to highlight important information about the circular functions. The section on harmonic motion now concludes this chapter.
- Function boxes have been included in Section 7.5 for the inverse trigonometric functions.
- Section 8.7 features increased coverage of polar forms of circles and lines.
- Chapter 10 incorporates over a dozen new figures to help students visualize and understand the terminology, definitions, and equations of conic sections. The material in former Sections 10.4 and 10.5 on rectangular and polar forms of conics and rotation of axes has been moved to two appendices.
- The binomial theorem is now covered in Chapter 11.
- Appendix C includes a helpful list of geometry formulas for student reference.

Supplements

For a comprehensive list of the supplements and study aids that accompany *Precalculus,* Third Edition, see pages xv and xvi.

Acknowledgments

Previous editions of this text were published after thousands of hours of work, not only by the authors, but also by reviewers, instructors, students, answer checkers, and editors. To these individuals and all those who have worked in some way on this text over the years, we are most grateful for your contributions. We could not have done it without you. We especially wish to thank the following individuals who provided valuable input into this edition of the text.

Richard Andrews, *Florida A&M University*
Sandra Arman, *Motlow State Community College*
Melissa Berta, *Saddleback College*
Steven Bogart, *Shoreline Community College*
Larry Bouldin, *Roane State Community College*
Brian Carter, *San Diego City College*
Susan Duke, *Meridian Community College*
Diane Ellis, *Mississippi State University*
Sam Evers, *University of Alabama Tuscaloosa*
Jon Freedman, *Skyline College*
Jeffrey Hughes, *Hinds Community College*
Lynne Kendall, *Metropolitan State College of Denver*
Maria Maspons, *Miami-Dade Community College*
Amy McLanahan, *Foothill College—Los Altos Hills*
Charles Odion, *Houston Community College*
Jane Roads, *Moberly Area Community College*
Alex Rolon, *Northampton Community College*
Mahmoud Shagroni, *Houston Community College*
Virginia Starkenburg, *San Diego City College*
Lewis Walston, *Methodist College*

Over the years, we have come to rely on an extensive team of experienced professionals. Our sincere thanks go to these dedicated individuals at Addison-Wesley, who worked long and hard to make this revision a success: Greg Tobin, Anne Kelly, Becky Anderson, Karen Guardino, Karen Wernholm, Barbara Atkinson, Joanne Ha, Cecilia Fleming, and Joe Vetere.

Terry McGinnis continues to provide invaluable behind-the-scenes guidance—we have come to rely on her expertise during all phases of the revision process. Thanks are due Gina Linko and Elm Street Publishing Services for their excellent production work. Perian Herring, Kitty Pellissier, and Abby Tanenbaum did an outstanding job checking the answers to all exercises, and Abby Tanenbaum also wrote the new solutions to selected exercises that appear at the back of the book. We especially thank Paul Van Erden for the many indexes he has so diligently prepared for us over the years. Joanne Still prepared the index for this text, and Becky Troutman compiled the comprehensive Index of Applications. Special thanks to Perian Herring, Deana Richmond, Lauri Semarne, and Margaret Westmoreland for accuracy checking page proofs.

As an author team, we are committed to the goal stated earlier in the Preface—to provide the best possible text to help instructors teach and students succeed. As we continue to work toward it, we would welcome any comments or suggestions you might have via e-mail to *math@aw.com.*

Margaret L. Lial
John Hornsby
David I. Schneider

Student Supplements

Student's Solutions Manual

- By Heidi Howard, *Florida Community College at Jacksonville*
- Provides detailed solutions to all odd-numbered text exercises
  ISBN: 0-321-22770-0

Graphing Calculator Manual

- By Darryl Nester, *Bluffton College*
- Provides instructions and keystroke operations for the TI-83/83 Plus, TI-84 Plus, TI-85, TI-86, and TI-89
  ISBN: 0-321-22771-9

Videotape Series

- Features an engaging team of lecturers
- Provides comprehensive coverage of each section and topic in the text
  ISBN: 0-321-22769-7

Digital Video Tutor

- Complete set of digitized videos for student use at home or on campus
- Ideal for distance learning or supplemental instruction
  ISBN: 0-321-23733-1

New! Additional Skill & Drill Manual

- By Cathy Ferrer, *Valencia Community College*
- Provides additional practice and test preparation for students
  ISBN: 0-321-23829-X

A Review of Algebra

- By Heidi Howard, *Florida Community College at Jacksonville*
- Provides additional support for students needing further algebra review
  ISBN: 0-201-77347-3

Addison-Wesley Math Tutor Center

- Provides tutoring through a registration number packaged with a new textbook or purchased separately
- Staffed by college mathematics instructors
- Accessible via toll-free telephone, toll-free fax, e-mail, and the Internet
  www.aw-bc.com/tutorcenter

Instructor Supplements

New! Annotated Instructor's Edition

- Special edition of the text
- Provides answers in the margins to almost all text exercises, plus helpful Teaching Tips
  ISBN: 0-321-25769-3

Instructor's Solutions Manual

- By Heidi Howard, *Florida Community College at Jacksonville*
- Provides complete solutions to all text exercises
  ISBN: 0-321-22766-2

Instructor's Testing Manual

- By Rodney Lynch, *Indiana University-Purdue University Indianapolis*
- Includes diagnostic pretests, chapter tests, and additional test items, grouped by section, with answers provided
  ISBN: 0-321-22767-0

TestGen-EQ with Quizmaster-EQ

- Enables instructors to build, edit, print, and administer tests
- Features a computerized bank of questions developed to cover all text topics
- Available on a dual-platform Windows/Macintosh CD-ROM
  ISBN: 0-321-22768-9

New! Adjunct Support Manual

- Includes resources to help new and adjunct faculty with course preparation and classroom management
- Provides helpful teaching tips correlated to the sections of the text
  ISBN: 0-321-23826-5

PowerPoint Lecture Presentation

- Classroom presentation software geared specifically to this textbook sequence
- Available within MyMathLab or on the Supplements Central Web site at http://suppscentral.aw.com

New! Adjunct Support Center

- Offers consultation on suggested syllabi, helpful tips on using the textbook support package, assistance with content, and advice on classroom strategies
- Available Sunday–Thursday evenings from 5 P.M. to midnight EST; telephone: 1-800-435-4084; e-mail: AdjunctSupport@aw.com; fax: 1-877-262-9774

MathXL®

MathXL is a powerful online homework, tutorial, and assessment system that accompanies your Addison-Wesley textbook in mathematics or statistics. With MathXL, instructors can create, edit, and assign online homework and tests using algorithmically generated exercises correlated to your textbook. All student work is tracked in MathXL's online gradebook. Students can take chapter tests in MathXL and receive personalized study plans based on their test results. The study plan diagnoses weaknesses and links students directly to tutorial exercises for the topics they need to study and retest. Students can also access supplemental animations and video clips directly from selected exercises. MathXL is available to qualified adopters. For more information, visit our Web site at www.mathxl.com.

MathXL Tutorials on CD® (ISBN 0-321-26803-2)

This interactive tutorial CD-ROM provides algorithmically generated practice exercises that are correlated to the exercises in the textbook. Every practice exercise is accompanied by an example and a guided solution designed to involve students in the solution process. Selected exercises may also include a video clip to help students visualize concepts. The software tracks student activity and scores and can generate printed summaries of students' progress.

MyMathLab®

MyMathLab is a series of text-specific, easily customizable online courses for Addison-Wesley textbooks in mathematics and statistics. MyMathLab is powered by CourseCompass™—Pearson Education's online teaching and learning environment—and by MathXL®—our online homework, tutorial, and assessment system. MyMathLab gives you the tools you need to deliver your course online, whether your students are in a lab setting or working from home. MyMathLab provides a rich and flexible set of course materials, featuring free-response exercises that are algorithmically generated for unlimited practice and mastery. Students can also use online tools, such as video lectures, animations, and a multimedia textbook, to independently improve their understanding and performance. Instructors can use MyMathLab's homework and test managers to select and assign online exercises correlated directly to the textbook, and they can import TestGen tests into MyMathLab for added flexibility. MyMathLab's online gradebook—designed specifically for mathematics and statistics—automatically tracks students' homework and test results and gives the instructor control over how to calculate final grades. MyMathLab is available to qualified adopters. For more information, visit our Web site at www.mymathlab.com.

InterAct Math® Tutorial Web Site www.interactmath.com

Get practice and tutorial help online! This interactive tutorial Web site provides algorithmically generated practice exercises that correlate directly to the exercises in the text. A detailed worked-out example and guided solution accompany each practice exercise. The Web site recognizes student errors and provides feedback.

R

Review of Basic Concepts

With the availability of computerized statistics, fans of professional football can now use mathematics to analyze the performances of their favorite players. The National Football League (NFL) began keeping statistics in 1932. Ever since, the passing effectiveness of NFL quarterbacks has been rated by several different methods. In Section R.1, Example 5, we give one such rating formula that is used again in Exercises 47–50. (*Source:* www.NFL.com)

R.1 Real Numbers and Their Properties

Sets of Numbers and the Number Line ▪ Exponents ▪ Order of Operations ▪ Properties of Real Numbers

Sets of Numbers and the Number Line The idea of counting goes back to the early days of civilization. When people first counted they used only the **natural numbers,** written in set notation as

$$\{1, 2, 3, 4, \ldots\}.$$

More recent is the idea of counting *no* object—that is, the idea of the number 0. As early as A.D. 150 the Greeks used the symbol o or $\bar{o}$ to represent 0. Including 0 with the set of natural numbers gives the set of **whole numbers,**

$$\{0, 1, 2, 3, 4, \ldots\}.$$

About 500 years ago, people came up with the idea of counting backward, from 4 to 3 to 2 to 1 to 0, and continuing this process, called the new numbers -1, -2, -3, and so on. Including these numbers with the set of whole numbers gives the set of **integers,**

$$\{\ldots, -3, -2, -1, 0, 1, 2, 3, \ldots\}.$$

Integers can be shown pictorially with a **number line.** The elements of the set $\{-3, -1, 0, 1, 3, 5\}$ are located on the number line in Figure 1.

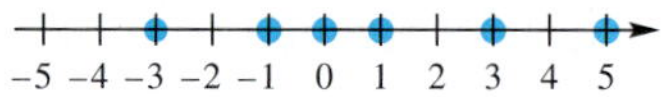

Figure 1

The result of dividing two integers (with a nonzero divisor) is called a *rational number* or *fraction.* A **rational number** is an element of the set

$$\left\{\frac{p}{q} \,\middle|\, p \text{ and } q \text{ are integers and } q \neq 0\right\}.$$

Rational numbers include the natural numbers, whole numbers, and integers. For example, the integer -3 is a rational number because it can be written as $\frac{-3}{1}$. Numbers that can be written as repeating or terminating decimals are also rational numbers. For example, $.\overline{6} = .66666\ldots$ represents a rational number that can be expressed as the fraction $\frac{2}{3}$.

The set of all numbers that correspond to points on a number line is called the **real numbers,** shown in Figure 2.

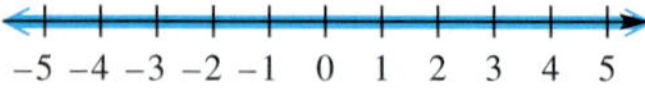

Figure 2

Real numbers can be represented by decimals. Since every fraction has a decimal form—for example, $\frac{1}{4} = .25$—real numbers include rational numbers. However, some real numbers cannot be represented by fractions. These numbers are called **irrational numbers.** The set of irrational numbers includes $\sqrt{3}$ and $\sqrt{5}$, but not $\sqrt{1}$, $\sqrt{4}$, $\sqrt{9}, \ldots$, which equal $1, 2, 3, \ldots$, and hence are rational numbers. Another irrational number is π, which is approximately equal to 3.14159. The numbers in the set $\left\{-\frac{2}{3}, 0, \sqrt{2}, \sqrt{5}, \pi, 4\right\}$ can be located on a number line, as shown in Figure 3. (Only $\sqrt{2}$, $\sqrt{5}$, and π are irrational here. The others are rational.) Since $\sqrt{2}$ is approximately equal to 1.41, it is located between 1 and 2, slightly closer to 1.

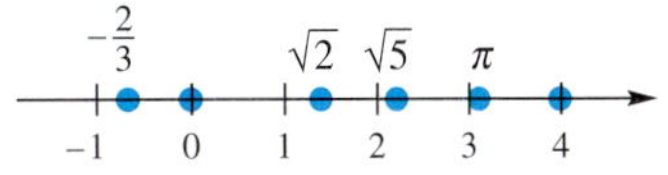

Figure 3

The sets of numbers discussed so far are summarized as follows.

Sets of Numbers

Set	Description
Natural Numbers	$\{1, 2, 3, 4, \ldots\}$
Whole Numbers	$\{0, 1, 2, 3, 4, \ldots\}$
Integers	$\{\ldots, -3, -2, -1, 0, 1, 2, 3, \ldots\}$
Rational Numbers	$\left\{\frac{p}{q} \,\middle|\, p \text{ and } q \text{ are integers and } q \neq 0\right\}$
Irrational Numbers	$\{x \mid x \text{ is real but not rational}\}$
Real Numbers	$\{x \mid x \text{ corresponds to a point on a number line}\}$

EXAMPLE 1 Identifying Elements of Subsets of the Real Numbers

Let set $A = \left\{-8, -6, -\frac{12}{4}, -\frac{3}{4}, 0, \frac{3}{8}, \frac{1}{2}, 1, \sqrt{2}, \sqrt{5}, 6, \frac{9}{0}\right\}$. List the elements from set A that belong to each set.

(a) natural numbers **(b)** whole numbers **(c)** integers

(d) rational numbers **(e)** irrational numbers **(f)** real numbers

Solution

(a) The natural numbers in set A are 1 and 6.

(b) The whole numbers are 0, 1, and 6.

(c) The integers are -8, -6, $-\frac{12}{4}$ (or -3), 0, 1, and 6.

(d) The rational numbers are -8, -6, $-\frac{12}{4}$ (or -3), $-\frac{3}{4}$, 0, $\frac{3}{8}$, $\frac{1}{2}$, 1, and 6.

(e) The irrational numbers are $\sqrt{2}$ and $\sqrt{5}$.

(f) All elements of A are real numbers except $\frac{9}{0}$. Division by 0 is not defined, so $\frac{9}{0}$ is not a number.

Now try Exercises 1, 3, 5, 11, and 13.

The relationships among the subsets of the real numbers are shown in Figure 4.

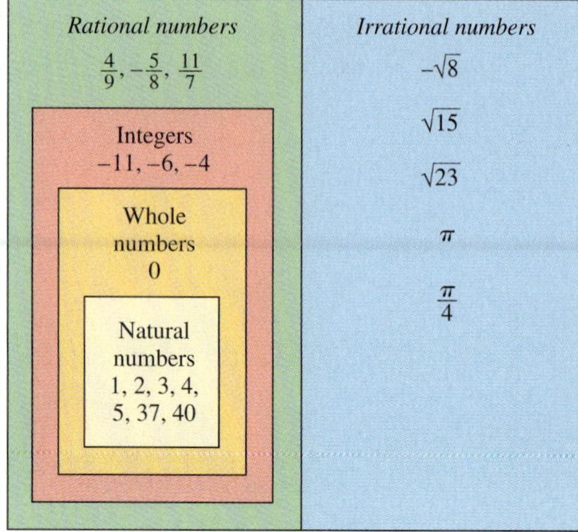

The Real Numbers

Figure 4

Exponents Exponential notation is used to write the products of repeated *factors*. When two numbers are multiplied, each number is called a **factor** of the product. For example, the product $2 \cdot 2 \cdot 2$ can be written as 2^3, where the 3 shows that three factors of 2 appear in the product. The notation a^n is defined as follows.

Exponential Notation

If n is any positive integer and a is any real number, then the nth power of a is

$$a^n = \underbrace{a \cdot a \cdot a \cdots a}_{n \text{ factors of } a}.$$

That is, a^n means the product of n factors of a. The integer n is the **exponent,** a is the **base,** and a^n is a **power** or an **exponential expression** (or simply an **exponential**). Read a^n as "a to the nth power," or just "a to the nth."

EXAMPLE 2 Evaluating Exponential Expressions

Evaluate each exponential expression, and identify the base and the exponent.

(a) 4^3 **(b)** $(-6)^2$ **(c)** -6^2 **(d)** $4 \cdot 3^2$ **(e)** $(4 \cdot 3)^2$

Solution

(a) $4^3 = \underbrace{4 \cdot 4 \cdot 4}_{3 \text{ factors of } 4} = 64$ The base is 4 and the exponent is 3.

(b) $(-6)^2 = (-6)(-6) = 36$
The base is -6 and the exponent is 2.

(c) $-6^2 = -(6 \cdot 6) = -36$
The base is 6 and the exponent is 2.

(d) $4 \cdot 3^2 = 4 \cdot 3 \cdot 3 = 36$
The base is 3 and the exponent is 2.

(e) $(4 \cdot 3)^2 = 12^2 = 144$
The base is $4 \cdot 3$ or 12 and the exponent is 2.

Now try Exercises 15, 17, 19, and 21.

CAUTION Notice in Examples 2(d) and (e) that $4 \cdot 3^2 \neq (4 \cdot 3)^2$.

Order of Operations When a problem involves more than one operation symbol, we use the following order of operations.

Order of Operations

If grouping symbols such as parentheses, square brackets, or fraction bars are present:

Step 1 Work separately above and below each **fraction bar.**

Step 2 Use the rules below within each set of **parentheses** or **square brackets.** Start with the innermost set and work outward.

If no grouping symbols are present:

Step 1 Simplify all **powers** and **roots,** working from left to right.

Step 2 Do any **multiplications** or **divisions** in order, working from left to right.

Step 3 Do any **negations, additions,** or **subtractions** in order, working from left to right.

EXAMPLE 3 Using Order of Operations

Evaluate each of the following.

(a) $6 \div 3 + 2^3 \cdot 5$

(b) $(8 + 6) \div 7 \cdot 3 - 6$

(c) $\dfrac{4 + 3^2}{6 - 5 \cdot 3}$

(d) $\dfrac{-(-3)^3 + (-5)}{2(-8) - 5(3)}$

Solution

(a)
$$\begin{aligned} 6 \div 3 + 2^3 \cdot 5 &= 6 \div 3 + 8 \cdot 5 && \text{Evaluate the exponential.} \\ &= 2 + 8 \cdot 5 && \text{Divide.} \\ &= 2 + 40 && \text{Multiply.} \\ &= 42 && \text{Add.} \end{aligned}$$

(b)
$$\begin{aligned} (8 + 6) \div 7 \cdot 3 - 6 &= 14 \div 7 \cdot 3 - 6 && \text{Work within the parentheses.} \\ &= 2 \cdot 3 - 6 && \text{Divide.} \\ &= 6 - 6 && \text{Multiply.} \\ &= 0 && \text{Subtract.} \end{aligned}$$

(c) $\dfrac{4 + 3^2}{6 - 5 \cdot 3} = \dfrac{4 + 9}{6 - 15}$ Evaluate the exponential and multiply.

$= \dfrac{13}{-9}$ or $-\dfrac{13}{9}$ Add and subtract; $\frac{a}{-b} = -\frac{a}{b}$.

(d) $\dfrac{-(-3)^3 + (-5)}{2(-8) - 5(3)} = \dfrac{-(-27) + (-5)}{2(-8) - 5(3)}$ Evaluate the exponential.

$= \dfrac{27 + (-5)}{-16 - 15}$ Multiply.

$= \dfrac{22}{-31}$ or $-\dfrac{22}{31}$ Add and subtract; $\frac{a}{-b} = -\frac{a}{b}$.

Now try Exercises 27, 29, and 35.

EXAMPLE 4 Using Order of Operations

Evaluate each expression if $x = -2$, $y = 5$, and $z = -3$.

(a) $-4x^2 - 7y + 4z$ **(b)** $\dfrac{2(x - 5)^2 + 4y}{z + 4}$

Solution

(a) $-4x^2 - 7y + 4z = -4(-2)^2 - 7(5) + 4(-3)$ Substitute: $x = -2$, $y = 5$, and $z = -3$.

$= -4(4) - 7(5) + 4(-3)$ Evaluate the exponential.

$= -16 - 35 - 12$ Multiply.

$= -63$ Subtract.

(b) $\dfrac{2(x - 5)^2 + 4y}{z + 4} = \dfrac{2(-2 - 5)^2 + 4(5)}{-3 + 4}$ Substitute: $x = -2$, $y = 5$, and $z = -3$.

$= \dfrac{2(-7)^2 + 20}{1}$ Work inside parentheses; multiply; add.

$= 2(49) + 20$ Evaluate the exponential.

$= 98 + 20$ Multiply.

$= 118$ Add.

Now try Exercises 39 and 45.

EXAMPLE 5 Calculating Passing Rating for an NFL Quarterback

A formula derived from recent quarterback ratings provided by the NFL is

$$\text{Passing Rating} \approx 85.68\left(\frac{C}{A}\right) + 4.31\left(\frac{Y}{A}\right) + 326.42\left(\frac{T}{A}\right) - 419.07\left(\frac{I}{A}\right),$$

where A = number of passes attempted, C = number of passes completed, Y = total number of yards gained passing, T = number of touchdown passes, and I = number of interceptions.

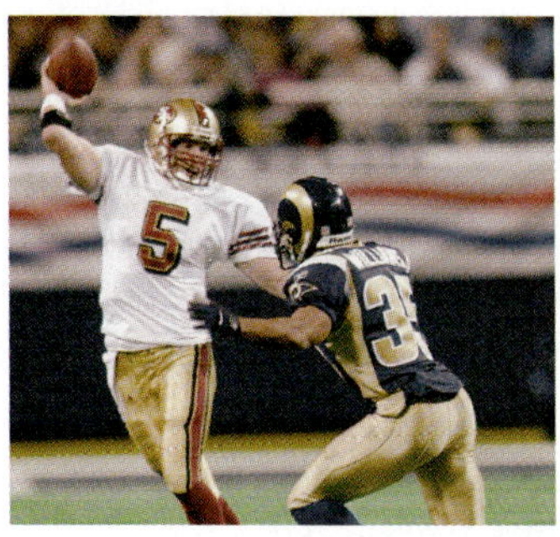

This information for Randall Cunningham of the Minnesota Vikings in a recent year is

$$A = 425, \qquad C = 259, \qquad Y = 3704, \qquad T = 34, \qquad \text{and} \qquad I = 10.$$

Find Cunningham's rating. (*Source:* www.NFL.com)

Solution Substitute the given numbers for Randall Cunningham in the formula for quarterback passing rating.

$$\begin{aligned}\text{Rating} &\approx 85.68\left(\frac{259}{425}\right) + 4.31\left(\frac{3704}{425}\right) + 326.42\left(\frac{34}{425}\right) - 419.07\left(\frac{10}{425}\right)\\ &\approx 106.0\end{aligned}$$

Using the order of operations, we calculated each quotient first (because of the parentheses), then found the products, and finally added the products, working from left to right.

Now try Exercise 47.

NOTE Because of the properties of real numbers that we discuss next, we could have added the products in any order in the final step of the calculations in Example 5.

Properties of Real Numbers The following basic properties express results that occur consistently in adding and multiplying numbers, so we generalize them to apply to expressions with variables.

The **closure properties** assure us that the sum or the product of two real numbers is also a real number. That is, for all real numbers a and b,

$$a + b \text{ is a real number} \qquad \text{and} \qquad ab \text{ is a real number.}$$

For example, since $\sqrt{2}$ and π are both real numbers, $\sqrt{2} + \pi$ and $\sqrt{2}\pi$ are also real numbers.

The **commutative properties** state that we can add or multiply two numbers in any order:

$$4 + (-12) = -12 + 4 \qquad \text{and} \qquad 4(-12) = -12(4).$$

Generalizing, for all real numbers a and b,

$$a + b = b + a \qquad \text{and} \qquad ab = ba.$$

By the **associative properties,** if we add or multiply three numbers, either the first two numbers or the last two numbers may be "associated" (or grouped). For example, the sum of the three numbers -9, 8, and 7 may be found in either of two ways:

$$-9 + (8 + 7) = -9 + 15 = 6,$$

or

$$(-9 + 8) + 7 = -1 + 7 = 6.$$

Also,

$$5(-3 \cdot 2) = 5(-6) = -30$$

or

$$(5 \cdot -3)2 = (-15)2 = -30.$$

In summary, for all real numbers a, b, and c, the associative properties state that

$$(a + b) + c = a + (b + c) \quad \text{and} \quad (ab)c = a(bc).$$

CAUTION To avoid confusing the associative and commutative properties, check the order of the terms: with the commutative properties the order changes from one side of the equals sign to the other; with the associative properties the order does not change, but the grouping does.

Commutative Properties	Associative Properties
$(x + 4) + 9 = (4 + x) + 9$	$(x + 4) + 9 = x + (4 + 9)$
$7 \cdot (5 \cdot 2) = (5 \cdot 2) \cdot 7$	$7 \cdot (5 \cdot 2) = (7 \cdot 5) \cdot 2$

EXAMPLE 6 Using the Commutative and Associative Properties to Simplify Expressions

Simplify each expression.

(a) $6 + (9 + x)$ **(b)** $\frac{5}{8}(16y)$ **(c)** $-10p\left(\frac{6}{5}\right)$

Solution

(a) $6 + (9 + x) = (6 + 9) + x = 15 + x$ Associative property

(b) $\frac{5}{8}(16y) = \left(\frac{5}{8} \cdot 16\right)y = 10y$ Associative property

(c) $-10p\left(\frac{6}{5}\right) = \frac{6}{5}(-10p)$ Commutative property

$= \left[\frac{6}{5}(-10)\right]p$ Associative property

$= -12p$ Multiply.

Now try Exercises 67 and 69.

The **identity properties** state special properties of the numbers 0 and 1. The sum of 0 and any real number a is a itself. For example,

$$0 + 4 = 4 \quad \text{and} \quad -5 + 0 = -5.$$

The number 0 preserves the identity of a number under addition, so 0 is the **identity element for addition** (or **additive identity**).

The number 1, the **identity element for multiplication** (or **multiplicative identity**), preserves the identity of a number under multiplication, since the product of 1 and any number a is a. For example,

$$5 \cdot 1 = 5 \quad \text{and} \quad 1\left(-\frac{2}{3}\right) = -\frac{2}{3}.$$

In summary, the identity properties say that for every real number a, there exist unique real numbers 0 and 1 such that

$$a + 0 = a \quad \text{and} \quad 0 + a = a;$$
$$a \cdot 1 = a \quad \text{and} \quad 1 \cdot a = a.$$

The **inverse properties** ensure that each real number has a negative and each nonzero real number has a reciprocal. For example,

$$5 + (-5) = 0 \quad \text{and} \quad 5\left(\frac{1}{5}\right) = 1.$$

For each real number a, we call the number $-a$ the **additive inverse** or **negative** of a. Similarly, for each *nonzero* real number a, we call the number $\frac{1}{a}$ the **multiplicative inverse** or **reciprocal** of a. In general, for every real number a,

$$a + (-a) = 0 \quad \text{and} \quad -a + a = 0,$$

and for every nonzero real number a,

$$a \cdot \frac{1}{a} = 1 \quad \text{and} \quad \frac{1}{a} \cdot a = 1.$$

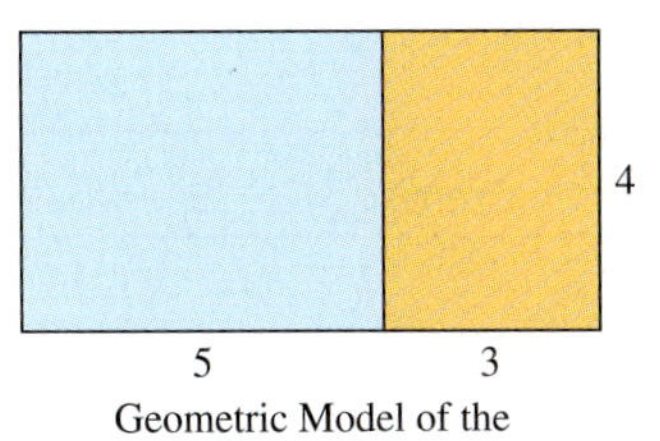

Geometric Model of the Distributive Property

Figure 5

Figure 5 helps to explain a key property of the real numbers. The area of the entire region shown can be found in two ways. We can multiply the length of the base of the entire region, $5 + 3 = 8$, by the width of the region.

$$4(5 + 3) = 4(8) = 32$$

Or, we can add the areas of the smaller rectangles, $4(5) = 20$ and $4(3) = 12$.

$$4(5) + 4(3) = 20 + 12 = 32$$

The result is the same. This means that

$$4(5 + 3) = 4(5) + 4(3),$$

which illustrates the **distributive property.** (This property also applies to subtraction.) In summary, for all real numbers a, b, and c,

$$a(b + c) = ab + ac \quad \text{and} \quad a(b - c) = ab - ac.$$

Using a commutative property, the distributive property can be rewritten as

$$(b + c)a = ba + ca.$$

Also, the distributive property can be extended to include more than two terms in the sum. For example,

$$\begin{aligned} 9(5x + y + 4z) &= 9(5x) + 9y + 9(4z) \\ &= 45x + 9y + 36z. \end{aligned}$$

NOTE The distributive property is a key property of real numbers because it is used to change products to sums and sums to products.

EXAMPLE 7 Using the Distributive Property

Rewrite each expression using the distributive property and simplify if possible.

(a) $3(x + y)$ **(b)** $-(m - 4n)$

(c) $\frac{1}{3}\left(\frac{4}{5}m - \frac{3}{2}n - 27\right)$ **(d)** $7p + 21$

Solution

(a) $3(x + y) = 3x + 3y$

(b)
$$\begin{aligned}-(m - 4n) &= -1(m - 4n)\\ &= -1(m) + (-1)(-4n)\\ &= -m + 4n\end{aligned}$$

(c)
$$\begin{aligned}\frac{1}{3}\left(\frac{4}{5}m - \frac{3}{2}n - 27\right) &= \frac{1}{3}\left(\frac{4}{5}m\right) + \frac{1}{3}\left(-\frac{3}{2}n\right) + \frac{1}{3}(-27)\\ &= \frac{4}{15}m - \frac{1}{2}n - 9\end{aligned}$$

(d)
$$\begin{aligned}7p + 21 &= 7p + 7 \cdot 3\\ &= 7(p + 3)\end{aligned}$$

Now try Exercises 63, 65, and 71.

A summary of the properties of real numbers follows.

Properties of Real Numbers

For all real numbers a, b, and c:

Property	Description
Closure Properties	
$\boldsymbol{a + b}$ is a real number. $\boldsymbol{ab}$ is a real number.	The sum or product of two real numbers is a real number.
Commutative Properties $\boldsymbol{a + b = b + a}$ $\boldsymbol{ab = ba}$	The sum or product of two real numbers is the same regardless of their order.
Associative Properties $\boldsymbol{(a + b) + c = a + (b + c)}$ $\boldsymbol{(ab)c = a(bc)}$	The sum or product of three real numbers is the same no matter which two are added or multiplied first.
Identity Properties	
There exists a unique real number 0 such that $\boldsymbol{a + 0 = a}$ and $\boldsymbol{0 + a = a.}$ There exists a unique real number 1 such that $\boldsymbol{a \cdot 1 = a}$ and $\boldsymbol{1 \cdot a = a.}$	The sum of a real number and 0 is that real number, and the product of a real number and 1 is that real number.

(continued)

Inverse Properties There exists a unique real number $-a$ such that $\mathbf{a + (-a) = 0}$ and $\mathbf{-a + a = 0.}$ If $a \neq 0$, there exists a unique real number $\frac{1}{a}$ such that $\mathbf{a \cdot \frac{1}{a} = 1}$ and $\mathbf{\frac{1}{a} \cdot a = 1.}$	The sum of any real number and its negative is 0, and the product of any nonzero real number and its reciprocal is 1.
Distributive Properties $\mathbf{a(b + c) = ab + ac}$ $\mathbf{a(b - c) = ab - ac}$	The product of a real number and the sum (or difference) of two real numbers equals the sum (or difference) of the products of the first number and each of the other numbers.

R.1 Exercises

Concept Check *Match each number from Column I with the letter or letters of the sets of numbers from Column II to which the number belongs. There may be more than one choice, so give all choices. See Example 1.*

I	II
1. 0	**A.** Natural numbers
2. 34	**B.** Whole numbers
3. $-\frac{9}{4}$	**C.** Integers
4. $\sqrt{36}$	**D.** Rational numbers
5. $\sqrt{13}$	**E.** Irrational numbers
6. 2.16	**F.** Real numbers

7. Explain why no answer in Exercises 1–6 can contain both D and E as choices.

8. The number π is irrational. Yet 3.14 and $\frac{22}{7}$ are often used as values for π. The first is a terminating decimal and the second is a quotient of integers, so both are rational. How is this possible?

9. *Concept Check* Give three examples of rational numbers that are not integers.

10. *Concept Check* Give three examples of integers that are not natural numbers.

Let set $B = \left\{-6, -\frac{12}{4}, -\frac{5}{8}, -\sqrt{3}, 0, \frac{1}{4}, 1, 2\pi, 3, \sqrt{12}\right\}$. *List all the elements of B that belong to each set. See Example 1.*

11. Natural numbers
12. Whole numbers
13. Integers
14. Rational numbers

Evaluate each expression. See Example 2.

15. -3^4
16. -3^5
17. $(-3)^4$
18. -2^6
19. $(-3)^5$
20. $(-2)^5$
21. $-2 \cdot 3^4$
22. $-4(-5)^3$

23. Why does $-5^2 = -25$, not 25? Is $(-5)^2$ positive or negative? What about $-(-5)^2$? Explain your answers.

Evaluate each expression. See Example 3.

24. $8^2 - (-4) + 11$

25. $16(-9) - 4$

26. $-2 \cdot 5 + 12 \div 3$

27. $9 \cdot 3 - 16 \div 4$

28. $-4(9 - 8) + (-7)(2)^3$

29. $6(-5) - (-3)(2)^4$

30. $-(-5)^3 - (-5)^2$

31. $(4 - 2^3)(-2 + \sqrt{25})$

32. $[-3^2 - (-2)][\sqrt{16} - 2^3]$

33. $\left(-\frac{2}{9} - \frac{1}{4}\right) - \left[-\frac{5}{18} - \left(-\frac{1}{2}\right)\right]$

34. $\left[-\frac{5}{8} - \left(-\frac{2}{5}\right)\right] - \left(\frac{3}{2} - \frac{11}{10}\right)$

35. $\dfrac{-8 + (-4)(-6) \div 12}{4 - (-3)}$

36. $\dfrac{15 \div 5 \cdot 4 \div 6 - 8}{-6 - (-5) - 8 \div 2}$

Evaluate each expression if $p = -4$, $q = 8$, and $r = -10$. See Example 4.

37. $2(q - r)$

38. $\dfrac{p}{q} + \dfrac{3}{r}$

39. $2p - 7q + r^2$

40. $-p^3 - 2q + r$

41. $\dfrac{q + r}{q + p}$

42. $\dfrac{3q}{3p - 2r}$

43. $\dfrac{3q}{r} - \dfrac{5}{p}$

44. $\dfrac{\frac{q}{4} - \frac{r}{5}}{\frac{p}{2} + \frac{q}{2}}$

45. $\dfrac{-(p + 2)^2 - 3r}{2 - q}$

46. $\dfrac{5q + 2(1 + p)^3}{r + 3}$

Passing Rating for NFL Quarterbacks *Use the formula*

$$\text{Passing Rating} \approx 85.68\left(\frac{C}{A}\right) + 4.31\left(\frac{Y}{A}\right) + 326.42\left(\frac{T}{A}\right) - 419.07\left(\frac{I}{A}\right),$$

where A = number of passes attempted, C = number of passes completed, Y = total number of yards gained passing, T = number of touchdown passes, and I = number of interceptions, to approximate the passing rating for each NFL quarterback. (The formula is exact to one decimal place in Exercises 47–49 and in Exercise 50 differs by only .1.) See Example 5. (Source: www.NFL.com*)*

NFL Quarterback/Team	*A*	*C*	*Y*	*T*	*I*
47. Brad Johnson/Buccaneers	451	281	3049	22	6
48. Trent Green/Chiefs	470	287	3690	26	13
49. Drew Bledsoe/Bills	610	375	4359	24	15
50. Peyton Manning/Colts	591	392	4200	27	19

Blood Alcohol Concentration *The Blood Alcohol Concentration (BAC) of a person who has been drinking is given by the expression*

number of oz × % alcohol × .075 ÷ body weight in lb − hr of drinking × .015.

(*Source:* Lawlor, J., *Auto Math Handbook: Mathematical Calculations, Theory, and Formulas for Automotive Enthusiasts,* HP Books, 1991.)

51. Suppose a policeman stops a 190-lb man who, in 2 hr, has ingested four 12-oz beers (48 oz), each having a 3.2% alcohol content. Calculate the man's BAC to the nearest thousandth. Follow the order of operations.

52. Find the BAC to the nearest thousandth for a 135-lb woman who, in 3 hr, has drunk three 12-oz beers (36 oz), each having a 4.0% alcohol content.

53. Calculate the BACs in Exercises 51 and 52 if each person weighs 25 lb more and the rest of the variables stay the same. How does increased weight affect a person's BAC?

54. Predict how decreased weight would affect the BAC of each person in Exercises 51 and 52. Calculate the BACs if each person weighs 25 lb less and the rest of the variables stay the same.

Identify the property illustrated in each statement. Assume all variables represent real numbers. See Examples 6 and 7.

55. $6 \cdot 12 + 6 \cdot 15 = 6(12 + 15)$

56. $8(m + 4) = (m + 4) \cdot 8$

57. $(x + 6) \cdot \left(\frac{1}{x + 6}\right) = 1, \quad \text{if } x + 6 \neq 0$

58. $\frac{2 + m}{2 - m} \cdot \frac{2 - m}{2 + m} = 1, \quad \text{if } m \neq 2 \text{ or } -2$

59. $(7 + y) + 0 = 7 + y$

60. $5 + \pi$ is a real number.

61. Is there a commutative property for subtraction? That is, in general, is $a - b$ equal to $b - a$? Support your answer with examples.

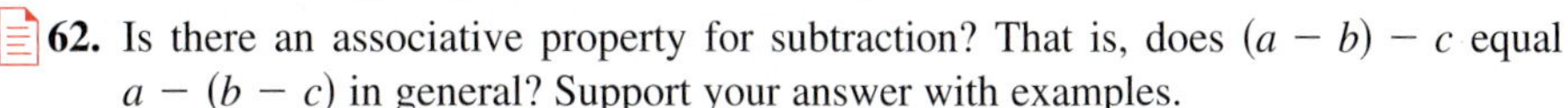

62. Is there an associative property for subtraction? That is, does $(a - b) - c$ equal $a - (b - c)$ in general? Support your answer with examples.

Use the distributive property to rewrite sums as products and products as sums. See Example 7.

63. $8p - 14p$

64. $15x - 10x$

65. $-3(z - y)$

66. $-2(m + n)$

Simplify each expression. See Examples 6 and 7.

67. $\frac{10}{11}(22z)$

68. $\left(\frac{3}{4}r\right)(-12)$

69. $(m + 5) + 3$

70. $2 + (a + 7)$

71. $\frac{3}{8}\left(\frac{16}{9}y + \frac{32}{27}z - \frac{40}{9}\right)$

72. $-\frac{1}{4}(20m + 8y - 32z)$

Solve each problem.

73. ***Average Golf Score*** To find the average of n real numbers, we add the numbers and then divide the sum by n. Ernie Els broke the record for the lowest score on the PGA Tour by winning the 2003 Mercedes Open with a score of 31 under par. His scores for the four rounds were 64, 65, 65, and 67. What was his average score per round? (*Source:* www.kapaluamaui.com/golf)

74. ***Average Golf Score*** Tiger Woods dominated the 1997 Masters, winning by 12 strokes over his nearest competitor. Tiger's scores for the four rounds were 70, 66, 65, and 69. What was his average score per round? (*Source: The Sports Illustrated 1998 Sports Almanac,* 1998.)

75. ***Average Speed of Earth*** The average distance from the center of Earth to the center of the sun is 92,960,000 mi. There are approximately 365.26 days per year. Estimate the average speed (in miles per hour) that Earth is moving around the sun if it is assumed that Earth's orbit is circular. Use $\pi \approx 3.14$ and speed = distance/time. (*Source:* Wright, J. (editor), *The Universal Almanac,* Universal Press Syndicate Company, 1998.)

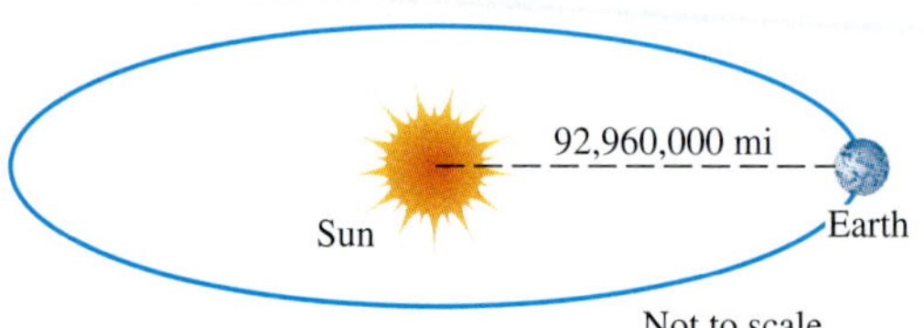

76. ***Average Velocity*** The deepest place in the ocean is the Mariana Trench near Japan with a depth of 35,840 ft. If a person dropped a 2.2-lb steel ball there, it would take 1 hr 4 min for it to reach the bottom. What would be the ball's average velocity in miles per hour during this time period? (*Source: The Guinness Book of Records 1998.*)

Exercises 77–82 provide practice in reading and interpreting tables.

Talking to ATMs *In this day of Automated Teller Machines (ATMs), people often find themselves doing what they have done for years when faced with a soft drink machine that won't respond: they talk to it. According to one report, the following are percentages of people in the United States, the United Kingdom (UK), and Germany who talk to ATMs and what they say.*

	United States	UK	Germany
Thanking the ATM	22%	24%	14%
Cursing the ATM	31%	41%	53%
Telling the ATM to Hurry Up	47%	36%	33%

Source: BMRB International for NCR.

In a random sample of 3000 *people from the specified country, how many would there be in each category?*

77. People in the United States who curse the ATM

78. People in the UK who thank the ATM

79. People in Germany who tell the ATM to hurry up

80. How many more German cursers would there be than United States thankers in random samples of 3000 from each country?

Wave Heights *The table lists the wave heights produced in the ocean for various wind speeds and durations.*

Wind Speed	Duration of the Wind			
mph	**10 hr**	**20 hr**	**30 hr**	**40 hr**
11.5	2 ft	2 ft	2 ft	2 ft
17.3	4 ft	5 ft	5 ft	5 ft
23.0	7 ft	8 ft	9 ft	9 ft
34.5	13 ft	17 ft	18 ft	19 ft
46.0	21 ft	28 ft	31 ft	33 ft
57.5	29 ft	40 ft	45 ft	48 ft

Source: Navarra, J., *Atmosphere, Weather and Climate,* International Thomson Publishers, 1979.

81. What is the expected wave height if a 46-mph wind blows for 30 hr?

82. If the wave height is 5 ft, is it possible to determine the speed of the wind and its duration? Explain.

83. This exercise outlines a proof that $\sqrt{2}$ is irrational. Give a reason for each of steps (a)–(h).

There are two possibilities:

(1) A rational number $\frac{a}{b}$ exists such that $\left(\frac{a}{b}\right)^2 = 2$.

(2) There is no such rational number.

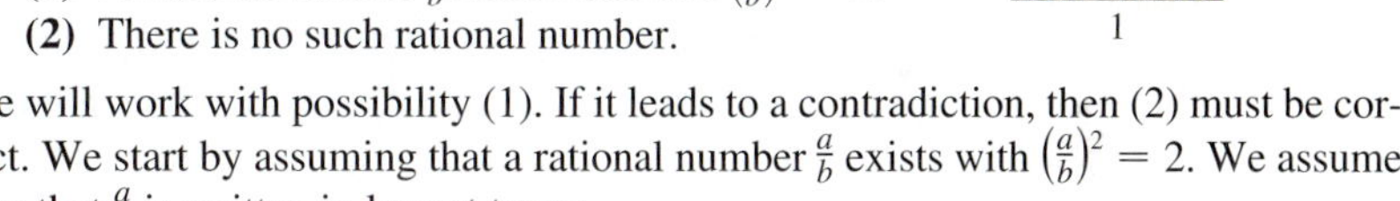

We will work with possibility (1). If it leads to a contradiction, then (2) must be correct. We start by assuming that a rational number $\frac{a}{b}$ exists with $\left(\frac{a}{b}\right)^2 = 2$. We assume also that $\frac{a}{b}$ is written in lowest terms.

(a) Since $\left(\frac{a}{b}\right)^2 = 2$, we must have $\frac{a^2}{b^2} = 2$ or $a^2 = 2b^2$.

(b) $2b^2$ is an even number.

(c) Therefore, a^2, and a itself, must be even numbers.

(d) Since a is an even number, it must be a multiple of 2. That is, we can find a natural number c such that $a = 2c$. This changes $a^2 = 2b^2$ into $(2c)^2 = 2b^2$.

(e) Therefore, $4c^2 = 2b^2$ or $2c^2 = b^2$.

(f) $2c^2$ is an even number.

(g) This makes b^2 an even number, so b must be even.

(h) We have reached a contradiction. Show where the contradiction occurs.

(i) Since assuming possibility (1) leads to a contradiction, we are forced to accept possibility (2), which says that $\sqrt{2}$ is irrational.

84. Do you think an argument similar to that of Exercise 83 could be used to "prove" that $\sqrt{36}$ is irrational?

Concept Check *Use the distributive property to calculate each value mentally.*

85. $72 \cdot 17 + 28 \cdot 17$

86. $32 \cdot 80 + 32 \cdot 20$

87. $123\frac{5}{8} \cdot 1\frac{1}{2} - 23\frac{5}{8} \cdot 1\frac{1}{2}$

88. $17\frac{2}{5} \cdot 14\frac{3}{4} - 17\frac{2}{5} \cdot 4\frac{3}{4}$

R.2 Order and Absolute Value

Order on the Number Line ▪ **Absolute Value** ▪ **Properties of Absolute Value**

Order on the Number Line

Figure 6 shows a number line with the points corresponding to several different numbers marked on the line. A number that corresponds to a particular point on a line is called the **coordinate** of the point. For example, the leftmost marked point in Figure 6 has coordinate -4. The correspondence between points on a line and the real numbers is called a **coordinate system** for the line. (From now on, the phrase "the point on a number line with coordinate a" will be abbreviated as "the point with coordinate a," or simply "the point a.")

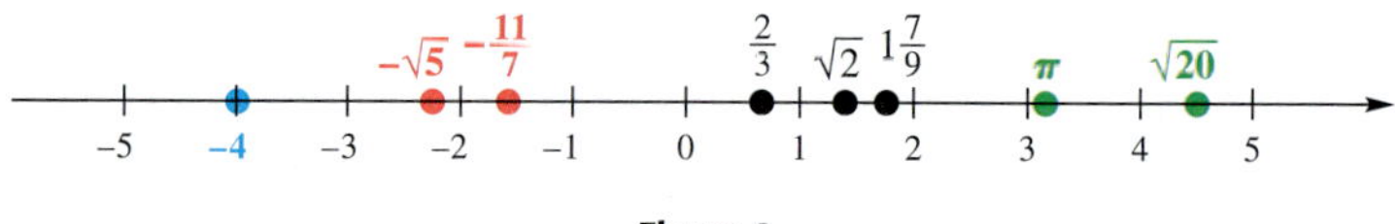

Figure 6

If the real number a is to the left of the real number b on a number line, then

$$a \text{ is less than } b, \text{ written } \quad a < b.$$

If a is to the right of b, then

$$a \text{ is greater than } b, \text{ written } \quad a > b.$$

For example, in Figure 6, $-\sqrt{5}$ is to the left of $-\frac{11}{7}$ on the number line, so $-\sqrt{5} < -\frac{11}{7}$, and $\sqrt{20}$ is to the right of π, indicating $\sqrt{20} > \pi$.

NOTE Remember that the inequality symbol points toward the lesser number.

EXAMPLE 1 Determining the Order of a List of Numbers

Write the following numbers in numerical order from least to greatest:

$$-\frac{11}{7},\ \pi,\ \sqrt{20},\ -4.2.$$

Solution Writing each number in decimal form to the nearest tenth gives, respectively,

$$-1.6,\ 3.1,\ 4.5,\ -4.2,$$

so the correct order is

$$-4.2,\ -\frac{11}{7},\ \pi,\ \sqrt{20}.$$

Now try Exercise 3.

We often use the following variations on $<$ and $>$.

Symbol	Meaning (Reading Left to Right)
$\leq$	is less than or equal to
$\geq$	is greater than or equal to
$\nless$	is not less than
$\ngtr$	is not greater than

Statement	Reason
$8 \leq 10$	$8 < 10$
$8 \leq 8$	$8 = 8$
$-9 \geq -14$	$-9 > -14$
$-8 \ngtr -2$	$-8 < -2$
$4 \nless 2$	$4 > 2$

Statements involving these symbols, as well as the symbols $<$ and $>$, are called *inequalities*. The table in the margin shows several such statements and the reason each is true.

The inequality $a < b < c$ says that b is *between* a and c since

$$a < b < c$$

means $\quad a < b \quad$ and $\quad b < c.$

In the same way, $\quad a \leq b \leq c$

means $\quad a \leq b \quad$ and $\quad b \leq c.$

CAUTION When writing "between" statements, make sure that both inequality symbols point in the same direction, toward the smallest number. For example,

$$2 < 7 < 11 \quad \text{and} \quad 5 > 4 > -1$$

are true statements, but $3 < 5 > 8$ is false. Generally, it is best to rewrite statements such as $5 > 4 > -1$ as $-1 < 4 < 5$, which is the order of these numbers on a number line when read from left to right.

Now try Exercise 7.

Absolute Value The distance on the number line from a number to 0 is called the **absolute value** of that number. The absolute value of the number a is written $|a|$. For example, the distance on the number line from 9 to 0 is 9, as is the distance from -9 to 0. (See Figure 7.) Therefore,

$$|9| = 9 \quad \text{and} \quad |-9| = 9.$$

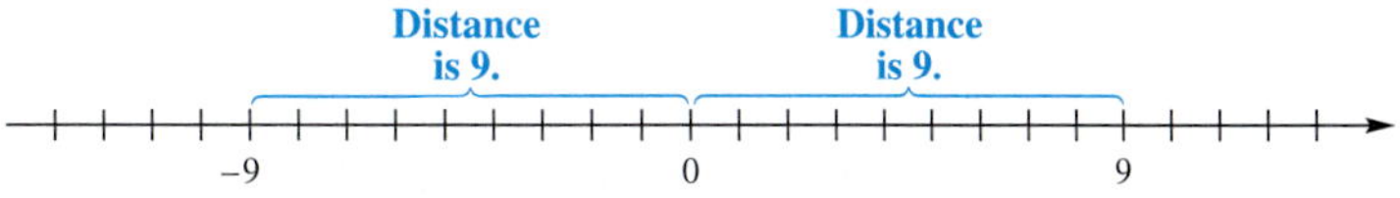

Figure 7

NOTE Since distance cannot be negative, ***the absolute value of a number is always nonnegative.***

Looking Ahead to Calculus

One of the most important definitions in calculus is that of the *limit*. While you are not expected to understand the definition that follows, look at the last few lines and notice how absolute value is used.

Suppose that a function f is defined at every number in an open interval I containing a, except perhaps at a itself. Then the limit of $f(x)$ as x approaches a is L, written

$$\lim_{x \to a} f(x) = L,$$

if for every $\epsilon > 0$ there exists a $\delta > 0$ such that $|f(x) - L| < \epsilon$ whenever $0 < |x - a| < \delta$.

The algebraic definition of absolute value follows.

Absolute Value

For all real numbers a,

$$|a| = \begin{cases} a & \text{if } a \geq 0 \\ -a & \text{if } a < 0. \end{cases}$$

That is, the absolute value of a positive number or 0 equals that number; the absolute value of a negative number equals its negative (or opposite).

The second part of this definition requires some thought. If a is a negative number (that is, if $a < 0$), then $-a$ is positive. Thus, for a *negative* number a,

$$|a| = -a,$$

or the negative of a. For example,

$$\text{if } a = -5, \quad \text{then} \quad |a| = |-5| = -(-5) = 5.$$

Think of $-a$ as the "opposite" of a.

EXAMPLE 2 Evaluating Absolute Values

Evaluate each expression.

(a) $\left|-\frac{5}{8}\right|$ **(b)** $-|8|$ **(c)** $-|-2|$ **(d)** $|2x|$, if $x = \pi$

Solution

(a) $\left|-\frac{5}{8}\right| = \frac{5}{8}$

(b) $-|8| = -(8) = -8$

(c) $-|-2| = -(2) = -2$

(d) $|2\pi| = 2\pi$

Now try Exercises 15 and 17.

EXAMPLE 3 Finding Absolute Values of Sums or Differences

Write each expression without absolute value bars.

(a) $|-8 + 2|$ **(b)** $|\sqrt{5} - 2|$ **(c)** $|\pi - 4|$ **(d)** $|m - 2|$, if $m < 2$

Solution

(a) Work inside the absolute value bars first. Since $-8 + 2 = -6$,

$$|-8 + 2| = |-6| = 6.$$

(b) Since $\sqrt{5} > 2$ $\left(\text{and } \sqrt{5} \approx 2.24\right)$, $\sqrt{5} - 2$ is positive (greater than 0).

$$|\sqrt{5} - 2| = \sqrt{5} - 2$$

(c) Here, $\pi < 4$, so $\pi - 4$ is negative (less than 0).

$$|\pi - 4| = -(\pi - 4)$$
$$= -\pi + 4 \quad \text{or} \quad 4 - \pi$$

(d) If $m < 2$, then $m - 2 < 0$, so

$$|m - 2| = -(m - 2)$$
$$= -m + 2 \quad \text{or} \quad 2 - m.$$

Now try Exercises 29 and 31.

Absolute value is useful in applications where only the *size* (or magnitude), not the *sign*, of the difference between two numbers is important.

EXAMPLE 4 Measuring Blood Pressure Difference

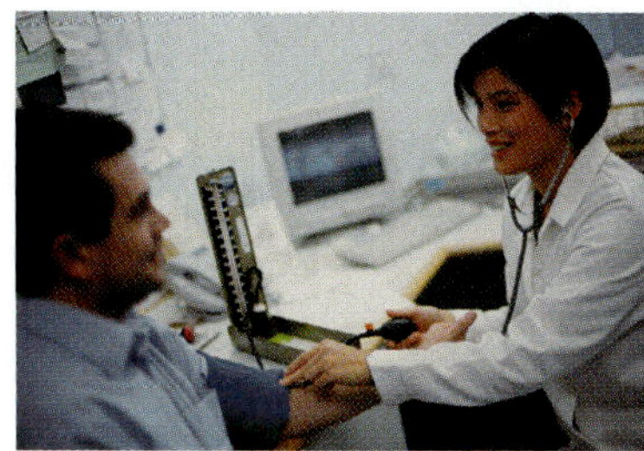

Systolic blood pressure is the maximum pressure produced by each heartbeat. Both low blood pressure and high blood pressure may be cause for medical concern. Therefore, health care professionals are interested in a patient's "pressure difference from normal," or P_d. If 120 is considered a normal systolic pressure, $P_d = |P - 120|$, where P is the patient's recorded systolic pressure. Find P_d for a patient with a systolic pressure, P, of 113.

Solution

$$P_d = |P - 120|$$
$$= |113 - 120|$$
$$= |-7|$$
$$= 7$$

Now try Exercise 49.

Properties of Absolute Value The definition of absolute value can be used to prove the following.

Properties of Absolute Value

For all real numbers a and b:

Property	Description
1. $\lvert a \rvert \geq 0$	The absolute value of a real number is positive or 0.
2. $\lvert -a \rvert = \lvert a \rvert$	The absolute values of a real number and its opposite are equal.
3. $\lvert a \rvert \cdot \lvert b \rvert = \lvert ab \rvert$	The product of the absolute values of two real numbers equals the absolute value of their product.
4. $\dfrac{\lvert a \rvert}{\lvert b \rvert} = \left\lvert \dfrac{a}{b} \right\rvert \quad (b \neq 0)$	The quotient of the absolute values of two real numbers equals the absolute value of their quotient.
5. $\lvert a + b \rvert \leq \lvert a \rvert + \lvert b \rvert$ (the triangle inequality)	The absolute value of the sum of two real numbers is less than or equal to the sum of their absolute values.

EXAMPLE 5 Illustrating the Properties of Absolute Value

In parts (a)–(d), use a property of absolute value to rewrite each expression.

(a) $|-15|$ **(b)** $|-10|$ **(c)** $|5x|$ **(d)** $\left|\frac{2}{y}\right|$

(e) For $a = 3$ and $b = -7$, find $|a + b|$. Show how the triangle inequality applies.

(f) For $a = 2$ and $b = 12$, find $|a + b|$. Show how the triangle inequality applies.

Solution

(a) $|-15| = 15 \geq 0$ Property 1

(b) $|-10| = 10$ and $|10| = 10$, so $|-10| = |10|$. Property 2

(c) $|5x| = |5| \cdot |x| = 5|x|$ since 5 is positive. Property 3

(d) $\left|\frac{2}{y}\right| = \frac{|2|}{|y|} = \frac{2}{|y|}$, $y \neq 0$ Property 4

(e) For $a = 3$ and $b = -7$,

$$|a + b| = |3 + (-7)| = |-4| = 4$$

$$|a| + |b| = |3| + |-7| = 3 + 7 = 10.$$

Thus, $|a + b| < |a| + |b|$. Property 5

(f) For $a = 2$ and $b = 12$,

$$|a + b| = |2 + 12| = |14| = 14$$

$$|a| + |b| = |2| + |12| = 2 + 12 = 14.$$

Thus, $|a + b| = |a| + |b|$. Property 5

Now try Exercises 39, 41, 43, and 45.

EXAMPLE 6 Evaluating Absolute Value Expressions

Let $x = -6$ and $y = 10$. Evaluate each expression.

(a) $|2x - 3y|$ **(b)** $\frac{2|x| - |3y|}{|xy|}$

Solution

(a) $|2x - 3y| = |2(-6) - 3(10)|$ Substitute.

$= |-12 - 30|$ Work inside absolute value bars; multiply.

$= |-42|$ Subtract.

$= 42$

(b)
$$\frac{2|x| - |3y|}{|xy|} = \frac{2|-6| - |3(10)|}{|-6(10)|} \quad \text{Substitute.}$$
$$= \frac{2 \cdot 6 - |30|}{|-60|} \quad |-6| = 6\text{; multiply.}$$
$$= \frac{12 - 30}{60} \quad \text{Multiply; } |30| = 30\text{; } |-60| = 60.$$
$$= \frac{-18}{60} \quad \text{Subtract.}$$
$$= -\frac{3}{10} \quad \text{Lowest terms; } \frac{-a}{b} = -\frac{a}{b}$$

Now try Exercises 23 and 25.

Absolute value is used to find the distance between two points on a number line.

Distance between Points on a Number Line

If P and Q are points on the number line with coordinates a and b, respectively, then the distance $d(P, Q)$ between them is

$$d(P,Q) = |b - a| \quad \text{or} \quad d(P,Q) = |a - b|.$$

That is, the distance between two points on a number line is the absolute value of the difference between their coordinates in either order. See Figure 8.

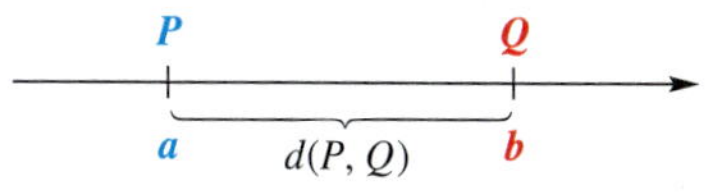

Figure 8

EXAMPLE 7 Finding the Distance between Two Points

Find the distance between -5 and 8.

Solution The distance is given by

$$|8 - (-5)| = |8 + 5| = |13| = 13.$$

Alternatively,

$$|(-5) - 8| = |-13| = 13.$$

Now try Exercise 55.

R.2 Exercises

Write the numbers in each list in numerical order, from least to greatest. Use a calculator as necessary. See Examples 1 and 2.

1. $\sqrt{8}, -4, -\sqrt{3}, -2, -5, \sqrt{6}, 3$

2. $\sqrt{2}, -1, 4, 3, \sqrt{8}, -\sqrt{6}, \sqrt{7}$

3. $\frac{3}{4}, \sqrt{2}, \frac{7}{5}, \frac{8}{5}, \frac{22}{15}$

4. $-\frac{9}{8}, -3, -\sqrt{3}, -\sqrt{5}, -\frac{9}{5}, -\frac{8}{5}$

5. $|-8|, -|9|, -|-6|$

6. $-|7|, -|-2|, -|-9|$

7. What is wrong with writing the statement "$x < 2$ or $x > 5$" as $5 < x < 2$?

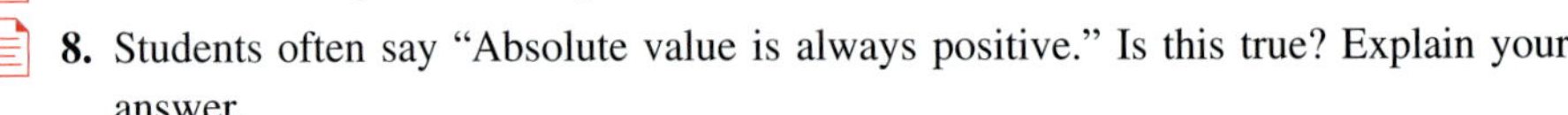
8. Students often say "Absolute value is always positive." Is this true? Explain your answer.

Concept Check *Decide whether each statement is* true *or* false. *If false, correct the statement so it is true.*

9. $|5 - 7| = |5| - |7|$

10. $|(-3)^3| = -|3^3|$

11. $|-5| \cdot |4| = |-5 \cdot 4|$

12. $\frac{|-8|}{|2|} = \left|\frac{-8}{2}\right|$

13. $|a - b| = |a| - |b|$, if $b > a > 0$.

14. If a is negative, then $|a| = -a$.

Evaluate each expression. See Example 2.

15. $|-9|$

16. $|-12|$

17. $-\left|\frac{4}{5}\right|$

18. $-\left|\frac{27}{2}\right|$

Let $x = -4$ and $y = 2$. Evaluate each expression. See Examples 3, 5, and 6.

19. $|2x|$

20. $|-3y|$

21. $|x - y|$

22. $|2x + 5y|$

23. $|3x + 4y|$

24. $|-5y + x|$

25. $\frac{2|y| - 3|x|}{|xy|}$

26. $\frac{4|x| + 4|y|}{|x|}$

27. $\frac{|-8y + x|}{-|x|}$

28. $\frac{|x| + 2|y|}{5 + x}$

Write each expression without absolute value bars. See Example 3.

29. $|\pi - 3|$

30. $|\pi - 5|$

31. $|y - 3|$, if $y < 3$

32. $|x - 4|$, if $x > 4$

33. $|2k - 8|$, if $k < 4$

34. $|3r - 15|$, if $r > 5$

35. $|x - y|$, if $x < y$

36. $|x - y|$, if $x > y$

37. $|3 + x^2|$

38. $|x^2 + 4|$

Justify each statement by giving the correct property of absolute value. Assume all variables represent real numbers. See Example 5.

39. $|m| = |-m|$

40. $|-k| \geq 0$

41. $|-3| \cdot |-5| = |15|$

42. $|8| \cdot |-4| = |-32|$

43. $|k - m| \leq |k| + |-m|$

44. $|5 + x| \leq |5| + |x|$

45. $|12 + 11r| \geq 0$

46. $\left|\frac{-12}{5}\right| = \frac{|-12|}{|5|}$

Solve each problem.

47. *Total Football Yardage* During his 16 yr in the NFL, Marcus Allen gained 12,243 yd rushing, 5411 yd receiving, and −6 yd returning fumbles. Find his total yardage (called *all-purpose yards*). Is this the same as the sum of the absolute values of the three categories? Why or why not? (*Source: The Sports Illustrated 2003 Sports Almanac,* 2003.)

48. *Golf Scores* In the 2002 U.S. Women's Open golf tournament, Juli Inkster won with a score that was 4 under par, while Shani Waugh finished third with a score that was 3 over par. Using −4 to represent 4 under par and +3 to represent 3 over par, find the difference between these scores (in either order) and take the absolute value of this difference. What does this final number represent? (*Source: The Sports Illustrated 2003 Sports Almanac,* 2003.)

49. *Blood Pressure Difference* Calculate the P_d value for a woman whose actual systolic pressure is 116 and whose normal value should be 125. (See Example 4.)

50. *Systolic Blood Pressure* If a patient's P_d value is 17 and the normal pressure for his gender and age should be 130, what are the two possible values for his systolic blood pressure? (See Example 4.)

Windchill *The windchill factor is a measure of the cooling effect that the wind has on a person's skin. It calculates the equivalent cooling temperature if there were no wind. The chart gives the windchill factor for various wind speeds and temperatures at which frostbite is a risk, and how quickly it may occur.*

Legend: 30 minutes; 10 minutes; 5 minutes

Wind speed (mph)		Temperature (°F)								
Calm		**40**	**30**	**20**	**10**	**0**	**−10**	**−20**	**−30**	**−40**
	5	36	25	13	1	−11	−22	−34	−46	−57
	10	34	21	9	−4	−16	−28	−41	−53	−66
	15	32	19	6	−7	−19	−32	−45	−58	−71
	20	30	17	4	−9	−22	−35	−48	−61	−74
	25	29	16	3	−11	−24	−37	−51	−64	−78
	30	28	15	1	−12	−26	−39	−53	−67	−80
	35	28	14	0	−14	−27	−41	−55	−69	−82
	40	27	13	−1	−15	−29	−43	−57	−71	−84

Source: National Oceanic and Atmospheric Administration, National Weather Service

If we are interested only in the magnitude of the difference between two of these entries, then we subtract the two entries and find the absolute value. Find the magnitude of the difference of each pair of windchill factors.

51. wind at 15 mph with a 30°F temperature and wind at 10 mph with a −10°F temperature

52. wind at 20 mph with a −20°F temperature and wind at 5 mph with a 30°F temperature

53. wind at 30 mph with a −30°F temperature and wind at 15 mph with a −20°F temperature

54. wind at 40 mph with a 40°F temperature and wind at 25 mph with a −30°F temperature

Find the given distances between points P, Q, R, and S on a number line, with coordinates −4, −1, 8, and 12, respectively. See Example 7.

55. $d(P, Q)$ **56.** $d(P, R)$ **57.** $d(Q, R)$

58. $d(P, S)$ **59.** $d(Q, S)$ **60.** $d(R, S)$

Concept Check *Determine what signs on values of x and y would make each statement true. Assume that x and y are not 0. (You should be able to work mentally.)*

61. $xy > 0$ **62.** $x^2y > 0$ **63.** $\frac{x}{y} < 0$

64. $\frac{y^2}{x} < 0$ **65.** $\frac{x^3}{y} > 0$ **66.** $-\frac{x}{y} > 0$

R.3 Polynomials

Rules for Exponents ▪ Polynomials ▪ Addition and Subtraction ▪ Multiplication ▪ Division

Rules for Exponents Work with exponents is simplified by using rules for exponents. From Section R.1, the notation a^m (where m is a positive integer and a is a real number) means that a appears as a factor m times. In the same way, a^n (where n is a positive integer) means that a appears as a factor n times. In the product $a^m \cdot a^n$, the number a would appear $m + n$ times, so the **product rule** states that

$$a^m \cdot a^n = a^{m+n}.$$

EXAMPLE 1 Using the Product Rule

Find each product.

(a) $y^4 \cdot y^7$ **(b)** $(6z^5)(9z^3)(2z^2)$

Solution

(a) $y^4 \cdot y^7 = y^{4+7} = y^{11}$ Product rule

(b) $(6z^5)(9z^3)(2z^2) = (6 \cdot 9 \cdot 2) \cdot (z^5z^3z^2)$ Commutative and associative properties (Section R.1)

$= 108z^{5+3+2}$ Product rule

$= 108z^{10}$

Now try Exercises 5 and 7.

The expression $(2^5)^3$ can be written as

$$(2^5)^3 = 2^5 \cdot 2^5 \cdot 2^5.$$

By a generalization of the product rule for exponents, this product is

$$(2^5)^3 = 2^{5+5+5} = 2^{15}.$$

The same exponent could have been obtained by multiplying 3 and 5. This example suggests the first of the **power rules** below. The others are found in a similar way. For positive integers m and n and all real numbers a and b,

1. $(a^m)^n = a^{mn}$ **2.** $(ab)^m = a^m b^m$ **3.** $\left(\frac{a}{b}\right)^m = \frac{a^m}{b^m}$ $(b \neq 0)$.

EXAMPLE 2 Using the Power Rules

Simplify.

(a) $(5^3)^2$ **(b)** $(3^4x^2)^3$ **(c)** $\left(\frac{2^5}{b^4}\right)^3$ **(d)** $\left(\frac{-2m^6}{t^2z}\right)^5$

Solution

(a) $(5^3)^2 = 5^{3(2)} = 5^6$ Power rule 1

(b) $(3^4x^2)^3 = (3^4)^3(x^2)^3$ Power rule 2

$= 3^{4(3)}x^{2(3)}$ Power rule 1

$= 3^{12}x^6$

(c) $\left(\frac{2^5}{b^4}\right)^3 = \frac{(2^5)^3}{(b^4)^3}$ Power rule 3

$= \frac{2^{15}}{b^{12}}, \quad b \neq 0$ Power rule 1

(d) $\left(\frac{-2m^6}{t^2z}\right)^5 = \frac{(-2m^6)^5}{(t^2z)^5}$ Power rule 3

$= \frac{(-2)^5(m^6)^5}{(t^2)^5z^5}$ Power rule 2

$= \frac{-32m^{30}}{t^{10}z^5}$ or $-\frac{32m^{30}}{t^{10}z^5}$ Evaluate $(-2)^5$; power rule 1

Now try Exercises 9 and 13.

CAUTION Do not confuse exponentials like mn^2 and $(mn)^2$. The two expressions are *not* equal. The second power rule given above can be used only with the second expression: $(mn)^2 = m^2n^2$.

These rules for exponents are summarized here.

Rules for Exponents

For all positive integers m and n and all real numbers a and b:

Rule	Description
Product Rule $a^m \cdot a^n = a^{m+n}$	When multiplying powers of like bases, keep the base and add the exponents.
Power Rule 1 $(a^m)^n = a^{mn}$	To raise a power to a power, multiply exponents.
Power Rule 2 $(ab)^m = a^m b^m$	To raise a product to a power, raise each factor to that power.
Power Rule 3 $\left(\frac{a}{b}\right)^m = \frac{a^m}{b^m} \quad (b \neq 0)$	To raise a quotient to a power, raise the numerator and the denominator to that power.

A zero exponent is defined as follows.

Zero Exponent

For any nonzero real number a, $\quad a^0 = 1.$

That is, any nonzero number with a zero exponent equals 1. We show why a^0 is defined this way in Section R.6. The symbol **0^0 is undefined.**

EXAMPLE 3 Using the Definition of a^0

Evaluate each power.

(a) 4^0 **(b)** $(-4)^0$ **(c)** -4^0 **(d)** $-(-4)^0$ **(e)** $(7r)^0$

Solution

(a) $4^0 = 1$ Base is 4.

(b) $(-4)^0 = 1$ Base is -4.

(c) $-4^0 = -(4^0) = -1$ Base is 4.

(d) $-(-4)^0 = -(1) = -1$ Base is -4.

(e) $(7r)^0 = 1, \quad r \neq 0$ Base is $7r$.

Now try Exercise 15.

Polynomials An **algebraic expression** is the result of adding, subtracting, multiplying, dividing (except by 0), raising to powers, or taking roots on any combination of variables, such as x, y, m, a, and b, or constants, such as -2, 3, 15, and 64.

$$-2x^2 + 3x, \quad \frac{15y}{2y-3}, \quad \sqrt{m^3 - 64}, \quad (3a+b)^4 \quad \text{Algebraic expressions}$$

The simplest algebraic expressions, *polynomials,* are discussed in this section.

The product of a real number and one or more variables raised to powers is called a **term.** The real number is called the **numerical coefficient,** or just the **coefficient.** The coefficient in $-3m^4$ is -3, while the coefficient in $-p^2$ is -1. **Like terms** are terms with the same variables each raised to the same powers.

$$-13x^3, \quad 4x^3, \quad -x^3 \qquad \text{Like terms}$$

$$6y, \quad 6y^2, \quad 4y^3 \qquad \text{Unlike terms}$$

A **polynomial** is defined as a term or a finite sum of terms, with only positive or zero integer exponents permitted on the variables. If the terms of a polynomial contain only the variable x, then the polynomial is called a **polynomial in** $\boldsymbol{x}$. (Polynomials in other variables are defined similarly.)

$$5x^3 - 8x^2 + 7x - 4, \quad 9p^5 - 3, \quad 8r^2, \quad 6 \qquad \text{Polynomials}$$

The terms of a polynomial cannot have variables in a denominator.

$$9x^2 - 4x + \frac{6}{x} \qquad \text{Not a polynomial}$$

The **degree of a term** with one variable is the exponent on the variable. For example, the degree of $2x^3$ is 3, the degree of $-x^4$ is 4, and the degree of $17x$ (that is, $17x^1$) is 1. The greatest degree of any term in a polynomial is called the **degree of the polynomial.** For example,

$$4x^3 - 2x^2 - 3x + 7 \text{ has degree } 3,$$

because the greatest degree of any term is 3 (the degree of $4x^3$). A nonzero constant such as -6, which can be written as $-6x^0$, has degree 0. (The polynomial 0 has no degree.)

A polynomial can have more than one variable. A term containing more than one variable has degree equal to the sum of all the exponents appearing on the variables in the term. For example, $-3x^4y^3z^5$ has degree $4 + 3 + 5 = 12$. The degree of a polynomial in more than one variable is equal to the greatest degree of any term appearing in the polynomial. By this definition, the polynomial

$$2x^4y^3 - 3x^5y + x^6y^2 \text{ has degree } 8$$

because of the x^6y^2 term.

A polynomial containing exactly three terms is called a **trinomial;** one containing exactly two terms is a **binomial;** and a single-term polynomial is called a **monomial.** The table shows several examples.

Polynomial	Degree	Type
$9p^7 - 4p^3 + 8p^2$	7	Trinomial
$29x^{11} + 8x^{15}$	15	Binomial
$-10r^6s^8$	14	Monomial
$5a^3b^7 - 3a^5b^5 + 4a^2b^9 - a^{10}$	11	None of these

Now try Exercises 19, 21, and 25.

Addition and Subtraction Since the variables used in polynomials represent real numbers, a polynomial represents a real number. This means that all the properties of the real numbers mentioned in Section R.1 hold for polynomials. In particular, the distributive property holds, so

$$3m^5 - 7m^5 = (3 - 7)m^5 = -4m^5.$$

Thus, polynomials are added by adding coefficients of like terms; polynomials are subtracted by subtracting coefficients of like terms.

EXAMPLE 4 Adding and Subtracting Polynomials

Add or subtract, as indicated.

(a) $(2y^4 - 3y^2 + y) + (4y^4 + 7y^2 + 6y)$

(b) $(-3m^3 - 8m^2 + 4) - (m^3 + 7m^2 - 3)$

(c) $(8m^4p^5 - 9m^3p^5) + (11m^4p^5 + 15m^3p^5)$

(d) $4(x^2 - 3x + 7) - 5(2x^2 - 8x - 4)$

Solution

(a) $(2y^4 - 3y^2 + y) + (4y^4 + 7y^2 + 6y)$

$= (2 + 4)y^4 + (-3 + 7)y^2 + (1 + 6)y$ Add coefficients of like terms.

$= 6y^4 + 4y^2 + 7y$

(b) $(-3m^3 - 8m^2 + 4) - (m^3 + 7m^2 - 3)$

$= (-3 - 1)m^3 + (-8 - 7)m^2 + [4 - (-3)]$ Subtract coefficients of like terms.

$= -4m^3 - 15m^2 + 7$

(c) $(8m^4p^5 - 9m^3p^5) + (11m^4p^5 + 15m^3p^5) = 19m^4p^5 + 6m^3p^5$

(d) $4(x^2 - 3x + 7) - 5(2x^2 - 8x - 4)$

$= 4x^2 - 4(3x) + 4(7) - 5(2x^2) - 5(-8x) - 5(-4)$ Distributive property (Section R.1)

$= 4x^2 - 12x + 28 - 10x^2 + 40x + 20$

$= -6x^2 + 28x + 48$ Add like terms.

Now try Exercises 29 and 31.

As shown in parts (a), (b), and (d) of Example 4, polynomials in one variable are often written with their terms in *descending order,* so the term of greatest degree is first, the one with the next greatest degree is next, and so on.

Multiplication We also use the associative and distributive properties, together with the properties of exponents, to find the product of two polynomials. To find the product of $3x - 4$ and $2x^2 - 3x + 5$, we treat $3x - 4$ as a single expression and use the distributive property.

$$(3x - 4)(2x^2 - 3x + 5) = (3x - 4)(2x^2) - (3x - 4)(3x) + (3x - 4)(5)$$

Now we use the distributive property three separate times.

$$\begin{aligned} &= 3x(2x^2) - 4(2x^2) - 3x(3x) - (-4)(3x) + 3x(5) - 4(5) \\ &= 6x^3 - 8x^2 - 9x^2 + 12x + 15x - 20 \\ &= 6x^3 - 17x^2 + 27x - 20 \end{aligned}$$

It is sometimes more convenient to write such a product vertically.

$$\begin{array}{rrrrl} & 2x^2 & - 3x & + 5 & \\ & & 3x & - 4 & \\ \hline & -8x^2 & + 12x & - 20 & \leftarrow -4(2x^2 - 3x + 5) \\ 6x^3 & - 9x^2 & + 15x & & \leftarrow 3x(2x^2 - 3x + 5) \\ \hline 6x^3 & - 17x^2 & + 27x & - 20 & \text{Add in columns.} \end{array}$$

EXAMPLE 5 Multiplying Polynomials

Multiply $(3p^2 - 4p + 1)(p^3 + 2p - 8)$.

Solution

$$\begin{array}{rrrrrrl} & & & 3p^2 & - 4p & + 1 & \\ & & p^3 & & + 2p & - 8 & \\ \hline & & & -24p^2 & + 32p & - 8 & \leftarrow -8(3p^2 - 4p + 1) \\ & & 6p^3 & - 8p^2 & + 2p & & \leftarrow 2p(3p^2 - 4p + 1) \\ 3p^5 & - 4p^4 & + p^3 & & & & \leftarrow p^3(3p^2 - 4p + 1) \\ \hline 3p^5 & - 4p^4 & + 7p^3 & - 32p^2 & + 34p & - 8 & \text{Add in columns.} \end{array}$$

Now try Exercise 43.

The FOIL method is a convenient way to find the product of two binomials. The memory aid **FOIL** (for **F**irst, **O**utside, **I**nside, **L**ast) gives the pairs of terms to be multiplied to get the product, as shown in the next example.

EXAMPLE 6 Using FOIL to Multiply Two Binomials

Find each product.

(a) $(6m + 1)(4m - 3)$ **(b)** $(2x + 7)(2x - 7)$ **(c)** $r^2(3r + 2)(3r - 2)$

Solution

(a) $$(6m + 1)(4m - 3) = \overset{\text{F}}{6m(4m)} + \overset{\text{O}}{6m(-3)} + \overset{\text{I}}{1(4m)} + \overset{\text{L}}{1(-3)}$$

$$= 24m^2 - 14m - 3 \qquad -18m + 4m = -14m$$

(b) $$\begin{aligned} (2x + 7)(2x - 7) &= 4x^2 - 14x + 14x - 49 && \text{FOIL} \\ &= 4x^2 - 49 \end{aligned}$$

(c) $$\begin{aligned} r^2(3r + 2)(3r - 2) &= r^2(9r^2 - 6r + 6r - 4) && \text{FOIL} \\ &= r^2(9r^2 - 4) \\ &= 9r^4 - 4r^2 \end{aligned}$$

Now try Exercises 35 and 37.

In part (a) of Example 6, the product of two binomials was a trinomial, while in parts (b) and (c), the product of two binomials was a binomial. The product of two binomials of the forms $x + y$ and $x - y$ is always a binomial. The squares of binomials, $(x + y)^2$ and $(x - y)^2$, are also special products.

Special Products

Product of the Sum and Difference of Two Terms $\quad (x + y)(x - y) = x^2 - y^2$

Square of a Binomial $\quad (x + y)^2 = x^2 + 2xy + y^2$

$(x - y)^2 = x^2 - 2xy + y^2$

In Section R.4 on factoring polynomials, you must be able to recognize and apply these special products.

EXAMPLE 7 Using the Special Products

Find each product.

(a) $(3p + 11)(3p - 11)$

(b) $(5m^3 - 3)(5m^3 + 3)$

(c) $(9k - 11r^3)(9k + 11r^3)$

(d) $(2m + 5)^2$

(e) $(3x - 7y^4)^2$

Solution

(a) $(3p + 11)(3p - 11) = (3p)^2 - 11^2 \quad (x + y)(x - y) = x^2 - y^2$

$= 9p^2 - 121$

(b) $(5m^3 - 3)(5m^3 + 3) = (5m^3)^2 - 3^2$

$= 25m^6 - 9$

(c) $(9k - 11r^3)(9k + 11r^3) = (9k)^2 - (11r^3)^2$

$= 81k^2 - 121r^6$

(d) $(2m + 5)^2 = (2m)^2 + 2(2m)(5) + 5^2 \quad (x + y)^2 = x^2 + 2xy + y^2$

$= 4m^2 + 20m + 25$

(e) $(3x - 7y^4)^2 = (3x)^2 - 2(3x)(7y^4) + (7y^4)^2 \quad (x - y)^2 = x^2 - 2xy + y^2$

$= 9x^2 - 42xy^4 + 49y^8$

Now try Exercises 45, 47, 49, and 51.

CAUTION As shown in Examples 7(d) and (e), the square of a binomial has *three* terms. Do not give $x^2 + y^2$ as the result of expanding $(x + y)^2$.

$$(x + y)^2 = x^2 + 2xy + y^2 \quad \text{Remember the middle term.}$$

Also, $\quad (x - y)^2 = x^2 - 2xy + y^2$

EXAMPLE 8 Multiplying More Complicated Binomials

Find each product.

(a) $[(3p - 2) + 5q][(3p - 2) - 5q]$ **(b)** $(x + y)^3$

(c) $(2a + b)^4$

Solution

(a) $[(3p - 2) + 5q][(3p - 2) - 5q]$

$= (3p - 2)^2 - (5q)^2$ Product of the sum and difference of terms

$= 9p^2 - 12p + 4 - 25q^2$ Square both quantities.

(b) $(x + y)^3 = (x + y)^2(x + y)$

$= (x^2 + 2xy + y^2)(x + y)$ Square $x + y$.

$= x^3 + 2x^2y + xy^2 + x^2y + 2xy^2 + y^3$ Multiply.

$= x^3 + 3x^2y + 3xy^2 + y^3$ Combine like terms.

(c) $(2a + b)^4 = (2a + b)^2(2a + b)^2$

$= (4a^2 + 4ab + b^2)(4a^2 + 4ab + b^2)$ Square each $2a + b$.

$= 16a^4 + 16a^3b + 4a^2b^2 + 16a^3b + 16a^2b^2$
$+ 4ab^3 + 4a^2b^2 + 4ab^3 + b^4$

$= 16a^4 + 32a^3b + 24a^2b^2 + 8ab^3 + b^4$

Now try Exercises 55, 59, and 61.

Division The quotient of two polynomials can be found with an algorithm for long division similar to that used for dividing whole numbers. (An *algorithm* is a step-by-step procedure [or "recipe"] for working a problem.) This algorithm requires that both polynomials be written in descending order.

EXAMPLE 9 Dividing Polynomials

Divide $4m^3 - 8m^2 + 4m + 6$ by $2m - 1$.

Solution

$4m^3$ divided by $2m$ is $2m^2$.

$-6m^2$ divided by $2m$ is $-3m$.

m divided by $2m$ is $\frac{1}{2}$.

$$
\begin{array}{r|l}
 & 2m^2 - 3m + \frac{1}{2} \\
2m - 1 & 4m^3 - 8m^2 + 4m + 6 \\
 & 4m^3 - 2m^2 \quad \leftarrow 2m^2(2m - 1) = 4m^3 - 2m^2 \\
 & -6m^2 + 4m \quad \leftarrow \text{Subtract; bring down the next term.} \\
 & -6m^2 + 3m \quad \leftarrow -3m(2m - 1) = -6m^2 + 3m \\
 & m + 6 \quad \leftarrow \text{Subtract; bring down the next term.} \\
 & m - \frac{1}{2} \quad \leftarrow \frac{1}{2}(2m - 1) = m - \frac{1}{2} \\
 & \frac{13}{2} \quad \leftarrow \text{Subtract; the remainder is } \frac{13}{2}.
\end{array}
$$

Thus, $$\frac{4m^3 - 8m^2 + 4m + 6}{2m - 1} = 2m^2 - 3m + \frac{1}{2} + \frac{\frac{13}{2}}{2m - 1}.$$

In the division on the previous page, $4m^3 - 2m^2$ is subtracted from $4m^3 - 8m^2 + 4m + 6$. The complete result, $-6m^2 + 4m + 6$, should be written under the line. However, to save work we "bring down" just the $4m$, the only term needed for the next step.

Now try Exercise 77.

The polynomial $3x^3 - 2x^2 - 150$ has a missing term, the term in which the power of x is 1. When a polynomial has a missing term, we allow for that term by inserting a term with a 0 coefficient for it.

EXAMPLE 10 Dividing Polynomials with Missing Terms

Divide $3x^3 - 2x^2 - 150$ by $x^2 - 4$.

Solution Both polynomials have missing first-degree terms. Insert each missing term with a 0 coefficient.

$$\begin{array}{r} 3x - 2 \\ x^2 + 0x - 4 \overline{)3x^3 - 2x^2 + 0x - 150} \\ \underline{3x^3 + 0x^2 - 12x} \\ -2x^2 + 12x - 150 \\ \underline{-2x^2 + 0x + 8} \\ 12x - 158 \end{array}$$

Missing term (the $0x$ in the dividend); Missing term (the $0x$ in the divisor); $12x - 158 \leftarrow$ Remainder

The division process ends when the remainder is 0 or the degree of the remainder is less than that of the divisor. Since $12x - 158$ has lesser degree than the divisor, it is the remainder. Thus,

$$\frac{3x^3 - 2x^2 - 150}{x^2 - 4} = 3x - 2 + \frac{12x - 158}{x^2 - 4}.$$

Now try Exercise 79.

R.3 Exercises

Concept Check *Decide whether each expression has been simplified correctly. If not, correct it.*

1. $(mn)^2 = mn^2$ **2.** $y^2 \cdot y^5 = y^7$ **3.** $\left(\frac{k}{5}\right)^3 = \frac{k^3}{5}$ **4.** $3^0y = 0$

Simplify each expression. See Example 1.

5. $9^3 \cdot 9^5$ **6.** $4^2 \cdot 4^8$ **7.** $(-4x^5)(4x^2)$ **8.** $(3y^4)(-6y^3)$

Use the properties of exponents to simplify each expression. See Examples 1–3.

9. $(2^2)^5$
10. $(6^4)^3$
11. $-(4m^3n^0)^2$
12. $(2x^0y^4)^3$
13. $\left(\frac{r^8}{s^2}\right)^3$
14. $-\left(\frac{p^4}{q}\right)^2$

Match each expression in Column I with its equivalent in Column II. See Example 3.

	I	II
15.	**(a)** 6^0	**A.** 0
	(b) -6^0	**B.** 1
	(c) $(-6)^0$	**C.** -1
	(d) $-(-6)^0$	**D.** 6
		E. -6

	I	II
16.	**(a)** $3p^0$	**A.** 0
	(b) $-3p^0$	**B.** 1
	(c) $(3p)^0$	**C.** -1
	(d) $(-3p)^0$	**D.** 3
		E. -3

17. Explain why $x^2 + x^2 \neq x^4$.
18. Explain why $(x + y)^2 \neq x^2 + y^2$.

Identify each expression as a polynomial *or* not a polynomial. *For each polynomial, give the degree and identify it as a* monomial, binomial, trinomial, *or* none of these.

19. $-5x^{11}$
20. $9y^{12} + y^2$
21. $18p^5q + 6pq$
22. $2a^6 + 5a^2 + 4a$
23. $\sqrt{2}x^2 + \sqrt{3}x^6$
24. $-\sqrt{7}m^5n^2 + 2\sqrt{3}m^3n^2$
25. $\frac{1}{3}r^2s^2 - \frac{3}{5}r^4s^2 + rs^3$
26. $\frac{13}{10}p^7 - \frac{2}{7}p^5$
27. $\frac{5}{p} + \frac{2}{p^2} + \frac{5}{p^3}$
28. $-5\sqrt{z} + 2\sqrt{z^3} - 5\sqrt{z^5}$

Find each sum or difference. See Example 4.

29. $(3x^2 - 4x + 5) + (-2x^2 + 3x - 2)$
30. $(4m^3 - 3m^2 + 5) + (-3m^3 - m^2 + 5)$
31. $2(12y^2 - 8y + 6) - 4(3y^2 - 4y + 2)$
32. $3(8p^2 - 5p) - 5(3p^2 - 2p + 4)$
33. $(6m^4 - 3m^2 + m) - (2m^3 + 5m^2 + 4m) + (m^2 - m)$
34. $-(8x^3 + x - 3) + (2x^3 + x^2) - (4x^2 + 3x - 1)$

Find each product. See Examples 5 and 6.

35. $(4r - 1)(7r + 2)$
36. $(5m - 6)(3m + 4)$
37. $x^2\left(3x - \frac{2}{3}\right)\left(5x + \frac{1}{3}\right)$
38. $\left(2m - \frac{1}{4}\right)\left(3m + \frac{1}{2}\right)$
39. $4x^2(3x^3 + 2x^2 - 5x + 1)$
40. $2b^3(b^2 - 4b + 3)$
41. $(2z - 1)(-z^2 + 3z - 4)$
42. $(k + 2)(12k^3 - 3k^2 + k + 1)$
43. $(m - n + k)(m + 2n - 3k)$
44. $(r - 3s + t)(2r - s + t)$

Find each product. See Examples 7 and 8.

45. $(2m + 3)(2m - 3)$
46. $(8s - 3t)(8s + 3t)$
47. $(4x^2 - 5y)(4x^2 + 5y)$
48. $(2m^3 + n)(2m^3 - n)$
49. $(4m + 2n)^2$
50. $(a - 6b)^2$
51. $(5r - 3t^2)^2$
52. $(2z^4 - 3y)^2$
53. $[(2p - 3) + q]^2$
54. $[(4y - 1) + z]^2$

55. $[(3q + 5) - p][(3q + 5) + p]$

56. $[(9r - s) + 2][(9r - s) - 2]$

57. $[(3a + b) - 1]^2$

58. $[(2m + 7) - n]^2$

59. $(y + 2)^3$

60. $(z - 3)^3$

61. $(q - 2)^4$

62. $(r + 3)^4$

Perform the indicated operations. See Examples 4–8.

63. $(p^3 - 4p^2 + p) - (3p^2 + 2p + 7)$

64. $(2z + y)(3z - 4y)$

65. $(7m + 2n)(7m - 2n)$

66. $(3p + 5)^2$

67. $-3(4q^2 - 3q + 2) + 2(-q^2 + q - 4)$

68. $2(3r^2 + 4r + 2) - 3(-r^2 + 4r - 5)$

69. $p(4p - 6) + 2(3p - 8)$

70. $m(5m - 2) + 9(5 - m)$

71. $-y(y^2 - 4) + 6y^2(2y - 3)$

72. $-z^3(9 - z) + 4z(2 + 3z)$

Perform each division. See Examples 9 and 10.

73. $\dfrac{-4x^7 - 14x^6 + 10x^4 - 14x^2}{-2x^2}$

74. $\dfrac{-8r^3s - 12r^2s^2 + 20rs^3}{4rs}$

75. $\dfrac{10x^8 - 16x^6 - 4x^4}{-2x^6}$

76. $\dfrac{3x^3 - 2x + 5}{x - 3}$

77. $\dfrac{6m^3 + 7m^2 - 4m + 2}{3m + 2}$

78. $\dfrac{6x^4 + 9x^3 + 2x^2 - 8x + 7}{3x^2 - 2}$

79. $\dfrac{3x^4 - 6x^2 + 9x - 5}{3x + 3}$

80. $\dfrac{k^4 - 4k^2 + 2k + 5}{k^2 + 1}$

Relating Concepts*

For individual or collaborative investigation

(Exercises 81–84)

The special product $(a + b)(a - b) = a^2 - b^2$ *can be used to perform some multiplication problems. For example,*

$$\begin{aligned} 51 \times 49 &= (50 + 1)(50 - 1) \\ &= 50^2 - 1^2 \\ &= 2500 - 1 \\ &= 2499. \end{aligned}$$

Similarly, the perfect square pattern gives

$$\begin{aligned} 47^2 &= (50 - 3)^2 \\ &= 50^2 - 2(50)(3) + 3^2 \\ &= 2500 - 300 + 9 \\ &= 2209. \end{aligned}$$

Use special products to evaluate each expression.

81. 99×101

82. 63×57

83. 102^2

84. 71^2

*Many exercise sets will contain groups of exercises under the heading Relating Concepts. These exercises are provided to illustrate how the concepts currently being studied relate to previously learned concepts. In most cases, they should be worked sequentially. We provide the answers to all such exercises, both even- and odd-numbered, in the Answer Section at the back of the book.

Solve each problem.

85. ***Geometric Modeling*** Consider the figure, which is a square divided into two squares and two rectangles.

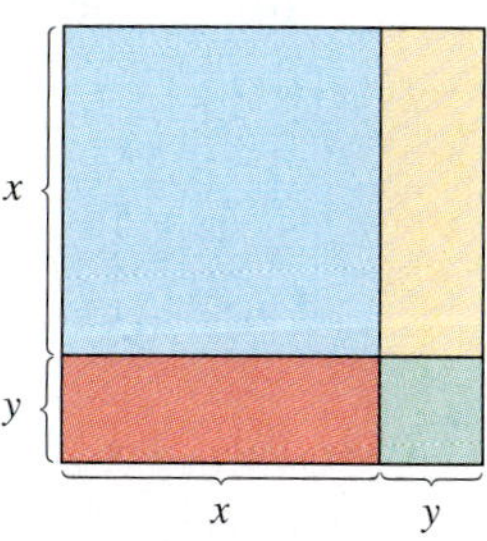

(a) The length of each side of the largest square is $x + y$. Use the formula for the area of a square to write the area of the largest square as a power.

(b) Use the formulas for the area of a square and the area of a rectangle to write the area of the largest square as a trinomial that represents the sum of the areas of the four figures that comprise it.

(c) Explain why the expressions in parts (a) and (b) must be equivalent.

(d) What special product from this section does this exercise reinforce geometrically?

86. ***Geometric Modeling*** Use the reasoning process of Exercise 85 and the accompanying figure to geometrically support the distributive property. Write a short paragraph explaining this process.

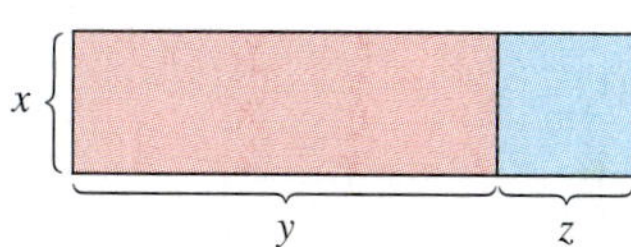

87. ***Volume of the Great Pyramid*** One of the most amazing formulas in all of ancient mathematics is the formula discovered by the Egyptians to find the volume of the frustum of a square pyramid shown in the figure. Its volume is given by

$$V = \frac{1}{3}h(a^2 + ab + b^2),$$

where b is the length of the base, a is the length of the top, and h is the height. (*Source:* Freebury, H. A., *A History of Mathematics,* MacMillan Company, New York, 1968.)

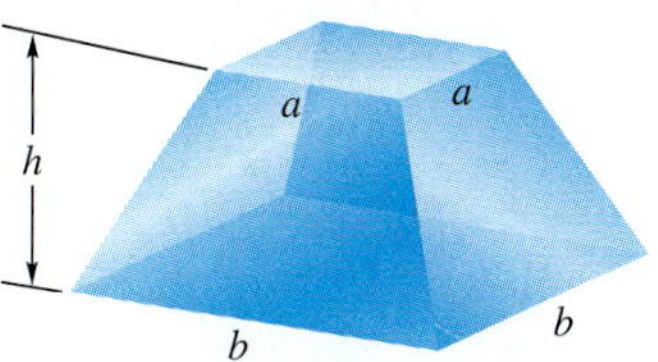

(a) When the Great Pyramid in Egypt was partially completed to a height h of 200 ft, b was 756 ft, and a was 314 ft. Calculate its volume at this stage of construction.

(b) Try to visualize the figure if $a = b$. What is the resulting shape? Find its volume.

(c) Let $a = b$ in the Egyptian formula and simplify. Are the results the same?

88. *Volume of the Great Pyramid* Refer to the formula and the discussion in Exercise 87.

(a) Use $V = \frac{1}{3}h(a^2 + ab + b^2)$ to determine a formula for the volume of a pyramid with square base b and height h by letting $a = 0$.

(b) The Great Pyramid in Egypt had a square base of 756 ft and a height of 481 ft. Find the volume of the Great Pyramid. Compare it with the 273-ft-tall Superdome in New Orleans, which has an approximate volume of 100 million ft^3. (*Source: The Guinness Book of Records 1998.*)

(c) The Superdome covers an area of 13 acres. How many acres does the Great Pyramid cover? (*Hint:* 1 acre $= 43{,}560\ \text{ft}^2$)

(Modeling) Number of Farms in the United States *A statistician from the U.S. Department of Agriculture, asked for the most significant trend in food production, mentioned the decline in the number of farms. The graph shows the number of farms in the United States since 1935 for selected years. The polynomial*

$$.000020591075x^3 - .1201456829x^2 + 233.5530856x - 151{,}249.8184$$

provides a good approximation of the number of farms for these years by substituting the year for x and evaluating the polynomial. For example, if $x = 1987$, *the value of the polynomial is approximately* 1.9, *which differs from the data in the bar graph by only* .2.

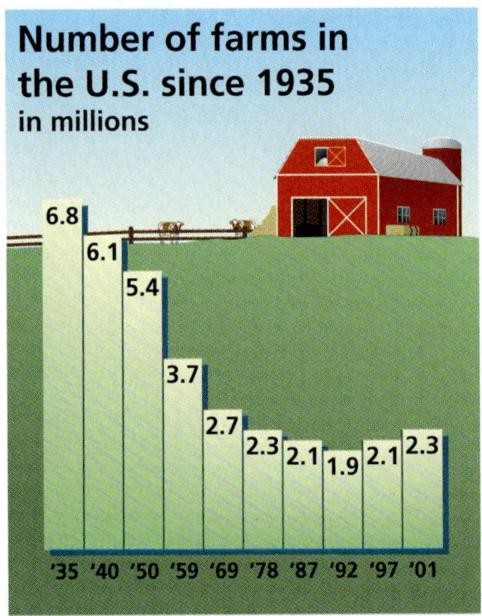

Source: U.S. Bureau of the Census.

Evaluate the polynomial for each year and then give the difference from the value in the graph.

89. 1940 **90.** 1959 **91.** 1978 **92.** 1997

Concept Check *Perform each operation mentally.*

93. $(.25^3)(400^3)$ **94.** $(24^2)(.5^2)$ **95.** $\frac{4.2^5}{2.1^5}$ **96.** $\frac{15^4}{5^4}$

97. Show that $(y - x)^3 = -(x - y)^3$. **98.** Show that $(y - x)^2 = (x - y)^2$.

R.4 Factoring Polynomials

Factoring Out the Greatest Common Factor ▪ Factoring by Grouping ▪ Factoring Trinomials ▪ Factoring Binomials ▪ Factoring by Substitution

The process of finding polynomials whose product equals a given polynomial is called **factoring.** For example, since $4x + 12 = 4(x + 3)$, both 4 and $x + 3$ are called *factors* of $4x + 12$. Also, $4(x + 3)$ is called the **factored form** of $4x + 12$. A polynomial with variable terms that cannot be written as a product of two polynomials of lower degree is a **prime polynomial.** A polynomial is **factored completely** when it is written as a product of prime polynomials.

Factoring Out the Greatest Common Factor

Polynomials are factored by using the distributive property. For example, to factor $6x^2y^3 + 9xy^4 + 18y^5$, we look for a monomial that is the greatest common factor (GCF) of each term.

$$6x^2y^3 + 9xy^4 + 18y^5 = 3y^3(2x^2) + 3y^3(3xy) + 3y^3(6y^2) \quad \text{GCF} = 3y^3$$
$$= 3y^3(2x^2 + 3xy + 6y^2) \quad \text{Distributive property (Section R.1)}$$

EXAMPLE 1 Factoring Out the Greatest Common Factor

Factor out the greatest common factor from each polynomial.

(a) $9y^5 + y^2$ **(b)** $6x^2t + 8xt + 12t$

(c) $14(m + 1)^3 - 28(m + 1)^2 - 7(m + 1)$

Solution

(a) $9y^5 + y^2 = y^2(9y^3) + y^2(1) \quad \text{GCF} = y^2$

$= y^2(9y^3 + 1) \quad$ Distributive property

To *check,* multiply out the factored form: $y^2(9y^3 + 1) = \underbrace{9y^5 + y^2}_{\text{Original polynomial}}$.

(b) $6x^2t + 8xt + 12t = 2t(3x^2 + 4x + 6) \quad \text{GCF} = 2t$

Check: $2t(3x^2 + 4x + 6) = 6x^2t + 8xt + 12t$

(c) $14(m + 1)^3 - 28(m + 1)^2 - 7(m + 1)$

$= 7(m + 1)[2(m + 1)^2 - 4(m + 1) - 1] \quad \text{GCF} = 7(m + 1)$

$= 7(m + 1)[2(m^2 + 2m + 1) - 4m - 4 - 1] \quad$ Square $m + 1$; distributive property

$= 7(m + 1)(2m^2 + 4m + 2 - 4m - 4 - 1) \quad$ Distributive property

$= 7(m + 1)(2m^2 - 3) \quad$ Combine like terms.

Now try Exercises 3, 9, and 15.

CAUTION In Example 1(a), remember to include the 1. Since $y^2(9y^3) \neq 9y^5 + y^2$, the 1 is essential in the answer. ***Factoring can always be checked by multiplying.***

Factoring by Grouping When a polynomial has more than three terms, it can sometimes be factored using **factoring by grouping.** For example, to factor

$$ax + ay + 6x + 6y,$$

group the terms so that each group has a common factor.

$$ax + ay + 6x + 6y = \underbrace{(ax + ay)}_{\text{Terms with common factor } a} + \underbrace{(6x + 6y)}_{\text{Terms with common factor 6}}$$

$$= a(x + y) + 6(x + y) \quad \text{Factor each group.}$$

$$= (x + y)(a + 6) \quad \text{Factor out } x + y.$$

It is not always obvious which terms should be grouped. Experience and repeated trials are the most reliable tools when factoring.

EXAMPLE 2 Factoring by Grouping

Factor each polynomial by grouping.

(a) $mp^2 + 7m + 3p^2 + 21$ **(b)** $2y^2 + az - 2z - ay^2$

(c) $4x^3 + 2x^2 - 2x - 1$

Solution

(a) $mp^2 + 7m + 3p^2 + 21 = (mp^2 + 7m) + (3p^2 + 21)$ Group the terms.

$= m(p^2 + 7) + 3(p^2 + 7)$ Factor each group.

$= (p^2 + 7)(m + 3)$ $p^2 + 7$ is a common factor.

Check: $(p^2 + 7)(m + 3) = mp^2 + 3p^2 + 7m + 21$ FOIL (Section R.3)

$= mp^2 + 7m + 3p^2 + 21$ Commutative property (Section R.1)

(b) $2y^2 + az - 2z - ay^2 = 2y^2 - 2z - ay^2 + az$ Rearrange the terms.

$= (2y^2 - 2z) + (-ay^2 + az)$ Group the terms.

$= 2(y^2 - z) + a(-y^2 + z)$ Factor each group.

The expression $-y^2 + z$ is the negative of $y^2 - z$, so factor out $-a$ instead of a.

$= 2(y^2 - z) - a(y^2 - z)$

$= (y^2 - z)(2 - a)$ Factor out $y^2 - z$.

Check by multiplying.

(c) $4x^3 + 2x^2 - 2x - 1 = 2x^2(2x + 1) - 1(2x + 1)$ Factor each group.

$= (2x + 1)(2x^2 - 1)$ Factor out $2x + 1$.

Check by multiplying.

Now try Exercises 17 and 21.

Factoring Trinomials As shown here, factoring is the opposite of multiplication.

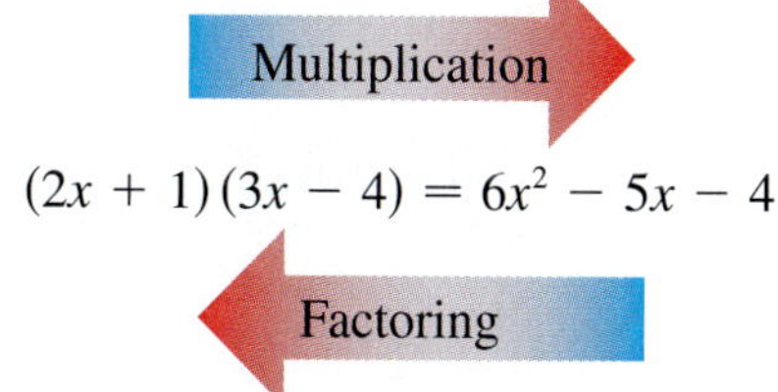

$$(2x + 1)(3x - 4) = 6x^2 - 5x - 4$$

Since the product of two binomials is usually a trinomial, we can expect factorable trinomials (that have terms with no common factor) to have two binomial factors. Thus, factoring trinomials requires using FOIL in reverse.

EXAMPLE 3 Factoring Trinomials

Factor each trinomial.

(a) $4y^2 - 11y + 6$ **(b)** $6p^2 - 7p - 5$

Solution

(a) To factor this polynomial, we must find integers a, b, c, and d such that

$$4y^2 - 11y + 6 = (ay + b)(cy + d). \quad \text{FOIL}$$

Using FOIL, we see that $ac = 4$ and $bd = 6$. The positive factors of 4 are 4 and 1 or 2 and 2. Since the middle term is negative, we consider only negative factors of 6. The possibilities are -2 and -3 or -1 and -6. Now we try various arrangements of these factors until we find one that gives the correct coefficient of y.

$$(2y - 1)(2y - 6) = 4y^2 - 14y + 6 \quad \text{Incorrect}$$
$$(2y - 2)(2y - 3) = 4y^2 - 10y + 6 \quad \text{Incorrect}$$
$$(y - 2)(4y - 3) = 4y^2 - 11y + 6 \quad \text{Correct}$$

Therefore, $4y^2 - 11y + 6 = (y - 2)(4y - 3)$

Check: $(y - 2)(4y - 3) = 4y^2 - 3y - 8y + 6$ FOIL
$= 4y^2 - 11y + 6$ Original polynomial

(b) Again, we try various possibilities to factor $6p^2 - 7p - 5$. The positive factors of 6 could be 2 and 3 or 1 and 6. As factors of -5 we have only -1 and 5 or -5 and 1.

$$(2p - 5)(3p + 1) = 6p^2 - 13p - 5 \quad \text{Incorrect}$$
$$(3p - 5)(2p + 1) = 6p^2 - 7p - 5 \quad \text{Correct}$$

Therefore, $6p^2 - 7p - 5 = (3p - 5)(2p + 1)$. *Check.*

Now try Exercises 23 and 25.

NOTE In Example 3, we chose positive factors of the positive first term. We could have used two negative factors, but the work is easier if positive factors are used.

Each of the special patterns for multiplication given in Section R.3 can be used in reverse to get a pattern for factoring. Perfect square trinomials can be factored as follows.

Factoring Perfect Square Trinomials

$$x^2 + 2xy + y^2 = (x + y)^2$$
$$x^2 - 2xy + y^2 = (x - y)^2$$

EXAMPLE 4 Factoring Perfect Square Trinomials

Factor each polynomial.

(a) $16p^2 - 40pq + 25q^2$ **(b)** $36x^2y^2 + 84xy + 49$

Solution

(a) Since $16p^2 = (4p)^2$ and $25q^2 = (5q)^2$, we use the second pattern shown in the box with $4p$ replacing x and $5q$ replacing y.

$$16p^2 - 40pq + 25q^2 = (4p)^2 - 2(4p)(5q) + (5q)^2$$
$$= (4p - 5q)^2$$

Make sure that the middle term of the trinomial being factored, $-40pq$ here, is twice the product of the two terms in the binomial $4p - 5q$.

$$-40pq = 2(4p)(-5q)$$

Thus, $16p^2 - 40pq + 25q^2 = (4p - 5q)^2$.

Check by squaring $4p - 5q$: $(4p - 5q)^2 = 16p^2 - 40pq + 25q^2$.

(b) $36x^2y^2 + 84xy + 49 = (6xy + 7)^2$, since $2(6xy)(7) = 84xy$.

Check by squaring $6xy + 7$: $(6xy + 7)^2 = 36x^2y^2 + 84xy + 49$.

Now try Exercises 35 and 39.

Factoring Binomials Check first to see whether the terms of a binomial have a common factor. If so, factor it out. The binomial may also fit one of the following patterns. Each can be verified by multiplying on the right side of the equation.

Factoring Binomials

Difference of Squares	$x^2 - y^2 = (x + y)(x - y)$
Difference of Cubes	$x^3 - y^3 = (x - y)(x^2 + xy + y^2)$
Sum of Cubes	$x^3 + y^3 = (x + y)(x^2 - xy + y^2)$

CAUTION There is no factoring pattern for a sum of squares. That is, $x^2 + y^2 \neq (x + y)^2$.

EXAMPLE 5 Factoring Differences of Squares

Factor each polynomial.

(a) $4m^2 - 9$ **(b)** $256k^4 - 625m^4$ **(c)** $(a + 2b)^2 - 4c^2$

(d) $x^2 - 6x + 9 - y^4$ **(e)** $y^2 - x^2 + 6x - 9$

Solution

(a) $4m^2 - 9 = (2m)^2 - 3^2$ Write as the difference of squares.

$= (2m + 3)(2m - 3)$

Check by multiplying.

(b) $256k^4 - 625m^4 = (16k^2)^2 - (25m^2)^2$ Write as the difference of squares.

$= (16k^2 + 25m^2)(16k^2 - 25m^2)$ Factor.

$= (16k^2 + 25m^2)(4k + 5m)(4k - 5m)$ Factor $16k^2 - 25m^2$.

Check: $(16k^2 + 25m^2)(4k + 5m)(4k - 5m)$

$= (16k^2 + 25m^2)(16k^2 - 25m^2)$ Multiply the last two factors.

$= 256k^4 - 625m^4$ Original polynomial

(c) $(a + 2b)^2 - 4c^2 = (a + 2b)^2 - (2c)^2$ Write as the difference of squares.

$= [(a + 2b) + 2c][(a + 2b) - 2c]$ Factor.

$= (a + 2b + 2c)(a + 2b - 2c)$

Check by multiplying.

(d) $x^2 - 6x + 9 - y^4 = (x^2 - 6x + 9) - y^4$ Group terms.

$= (x - 3)^2 - y^4$ Factor the trinomial.

$= (x - 3)^2 - (y^2)^2$ Write as the difference of squares.

$= [(x - 3) + y^2][(x - 3) - y^2]$ Factor.

$= (x - 3 + y^2)(x - 3 - y^2)$

Check by multiplying.

(e) $y^2 - x^2 + 6x - 9 = y^2 - (x^2 - 6x + 9)$ Factor out the negative and group the last three terms.

$= y^2 - (x - 3)^2$ Write as the difference of squares.

$= [y - (x - 3)][y + (x - 3)]$ Factor.

$= (y - x + 3)(y + x - 3)$ Distributive property

Check by multiplying.

Now try Exercises 45, 47, 49, and 51.

EXAMPLE 6 Factoring Sums or Differences of Cubes

Factor each polynomial.

(a) $x^3 + 27$ **(b)** $m^3 - 64n^3$ **(c)** $8q^6 + 125p^9$

Solution

(a) $x^3 + 27 = x^3 + 3^3$ Write as a sum of cubes.

$= (x + 3)(x^2 - 3x + 9)$ Factor.

(b) $m^3 - 64n^3 = m^3 - (4n)^3$ Write as a difference of cubes.

$= (m - 4n)[m^2 + m(4n) + (4n)^2]$ Factor.

$= (m - 4n)(m^2 + 4mn + 16n^2)$ Simplify.

(c) $8q^6 + 125p^9 = (2q^2)^3 + (5p^3)^3$ Write as a sum of cubes.

$= (2q^2 + 5p^3)[(2q^2)^2 - 2q^2(5p^3) + (5p^3)^2]$ Factor.

$= (2q^2 + 5p^3)(4q^4 - 10q^2p^3 + 25p^6)$ Simplify.

Now try Exercises 53, 55, and 57.

Factoring by Substitution Sometimes a polynomial can be factored by substituting one expression for another.

EXAMPLE 7 Factoring by Substitution

Factor each polynomial.

(a) $6z^4 - 13z^2 - 5$ **(b)** $10(2a - 1)^2 - 19(2a - 1) - 15$

(c) $(2a - 1)^3 + 8$

Solution

(a) Replace z^2 with y, so $y^2 = (z^2)^2 = z^4$.

$6z^4 - 13z^2 - 5 = 6y^2 - 13y - 5$

$= (2y - 5)(3y + 1)$ Use FOIL to factor.

$= (2z^2 - 5)(3z^2 + 1)$ Replace y with z^2.

(Some students prefer to factor this type of trinomial directly using trial and error with FOIL.)

(b) $10(2a - 1)^2 - 19(2a - 1) - 15$

$= 10m^2 - 19m - 15$ Replace $2a - 1$ with m.

$= (5m + 3)(2m - 5)$ Factor.

$= [5(2a - 1) + 3][2(2a - 1) - 5]$ Let $m = 2a - 1$.

$= (10a - 5 + 3)(4a - 2 - 5)$ Distributive property

$= (10a - 2)(4a - 7)$ Add.

$= 2(5a - 1)(4a - 7)$ Factor out the common factor.

(c) $(2a-1)^3+8=t^3+8$ Let $2a-1=t$.

$=t^3+2^3$ Write as a sum of cubes.

$=(t+2)(t^2-2t+4)$ Factor.

$=[(2a-1)+2][(2a-1)^2-2(2a-1)+4]$ Let $t=2a-1$.

$=(2a+1)(4a^2-4a+1-4a+2+4)$ Add; multiply.

$=(2a+1)(4a^2-8a+7)$ Combine like terms.

Now try Exercises 71, 73, and 89.

R.4 Exercises

Factor out the greatest common factor from each polynomial. See Examples 1 and 2.

1. $12m+60$
2. $15r-27$
3. $8k^3+24k$
4. $9z^4+81z$
5. $xy-5xy^2$
6. $5h^2j+hj$
7. $-4p^3q^4-2p^2q^5$
8. $-3z^5w^2-18z^3w^4$
9. $4k^2m^3+8k^4m^3-12k^2m^4$
10. $28r^4s^2+7r^3s-35r^4s^3$
11. $2(a+b)+4m(a+b)$
12. $4(y-2)^2+3(y-2)$
13. $(5r-6)(r+3)-(2r-1)(r+3)$
14. $(3z+2)(z+4)-(z+6)(z+4)$
15. $2(m-1)-3(m-1)^2+2(m-1)^3$
16. $5(a+3)^3-2(a+3)+(a+3)^2$

Factor each polynomial by grouping. See Example 2.

17. $6st+9t-10s-15$
18. $10ab-6b+35a-21$
19. $2m^4+6-am^4-3a$
20. $15-5m^2-3r^2+m^2r^2$
21. $20z^2-8x+5pz^2-2px$
22. ***Concept Check*** Layla factored $16a^2-40a-6a+15$ by grouping and obtained $(8a-3)(2a-5)$. Jamal factored the same polynomial and gave an answer of $(3-8a)(5-2a)$. Which answer is correct?

Factor each trinomial. See Examples 3 and 4.

23. $6a^2-11a+4$
24. $8h^2-2h-21$
25. $3m^2+14m+8$
26. $9y^2-18y+8$
27. $6k^2+5kp-6p^2$
28. $14m^2+11mr-15r^2$
29. $5a^2-7ab-6b^2$
30. $12s^2+11st-5t^2$
31. $9x^2-6x^3+x^4$
32. $30a^2+am-m^2$
33. $24a^4+10a^3b-4a^2b^2$
34. $18x^5+15x^4z-75x^3z^2$
35. $9m^2-12m+4$
36. $16p^2-40p+25$
37. $32a^2+48ab+18b^2$
38. $20p^2-100pq+125q^2$
39. $4x^2y^2+28xy+49$
40. $9m^2n^2+12mn+4$
41. $(a-3b)^2-6(a-3b)+9$
42. $(2p+q)^2-10(2p+q)+25$

43. *Concept Check* Match each polynomial in Column I with its factored form in Column II.

I	II
(a) $x^2 + 10xy + 25y^2$	**A.** $(x + 5y)(x - 5y)$
(b) $x^2 - 10xy + 25y^2$	**B.** $(x + 5y)^2$
(c) $x^2 - 25y^2$	**C.** $(x - 5y)^2$
(d) $25y^2 - x^2$	**D.** $(5y + x)(5y - x)$

44. *Concept Check* Match each polynomial in Column I with its factored form in Column II.

I	II
(a) $8x^3 - 27$	**A.** $(3 - 2x)(9 + 6x + 4x^2)$
(b) $8x^3 + 27$	**B.** $(2x - 3)(4x^2 + 6x + 9)$
(c) $27 - 8x^3$	**C.** $(2x + 3)(4x^2 - 6x + 9)$

Factor each polynomial. See Examples 5 and 6.

45. $9a^2 - 16$
46. $16q^2 - 25$
47. $25s^4 - 9t^2$
48. $36z^2 - 81y^4$
49. $(a + b)^2 - 16$
50. $(p - 2q)^2 - 100$
51. $p^4 - 625$
52. $m^4 - 81$
53. $8 - a^3$
54. $r^3 + 27$
55. $125x^3 - 27$
56. $8m^3 - 27n^3$
57. $27y^9 + 125z^6$
58. $27z^3 + 729y^3$
59. $(r + 6)^3 - 216$
60. $(b + 3)^3 - 27$
61. $27 - (m + 2n)^3$
62. $125 - (4a - b)^3$

63. *Concept Check* Which of the following is the correct complete factorization of $x^4 - 1$?

A. $(x^2 - 1)(x^2 + 1)$
B. $(x^2 + 1)(x + 1)(x - 1)$
C. $(x^2 - 1)^2$
D. $(x - 1)^2(x + 1)^2$

64. *Concept Check* Which of the following is the correct factorization of $x^3 + 8$?

A. $(x + 2)^3$
B. $(x + 2)(x^2 + 2x + 4)$
C. $(x + 2)(x^2 - 2x + 4)$
D. $(x + 2)(x^2 - 4x + 4)$

Relating Concepts

For individual or collaborative investigation
(Exercises 65–70)

*The polynomial $x^6 - 1$ can be considered either a difference of squares or a difference of cubes. **Work Exercises 65–70 in order,** to connect the results obtained when two different methods of factoring are used.*

65. Factor $x^6 - 1$ by first factoring as the difference of squares, and then factor further by using the patterns for the sum of cubes and the difference of cubes.

66. Factor $x^6 - 1$ by first factoring as the difference of cubes, and then factor further by using the pattern for the difference of squares.

67. Compare your answers in Exercises 65 and 66. Based on these results, what is the factorization of $x^4 + x^2 + 1$?

68. The polynomial $x^4 + x^2 + 1$ cannot be factored using the methods described in this section. However, there is a technique that allows us to factor it, as shown on the next page.

Supply the reason that each step is valid.

$$\begin{aligned} x^4 + x^2 + 1 &= x^4 + 2x^2 + 1 - x^2 \\ &= (x^4 + 2x^2 + 1) - x^2 \\ &= (x^2 + 1)^2 - x^2 \\ &= (x^2 + 1 - x)(x^2 + 1 + x) \\ &= (x^2 - x + 1)(x^2 + x + 1) \end{aligned}$$

69. Compare your answer in Exercise 67 with the final line in Exercise 68. What do you notice?

70. Factor $x^8 + x^4 + 1$ using the technique outlined in Exercise 68.

Factor each polynomial by substitution. See Example 7.

71. $m^4 - 3m^2 - 10$

72. $a^4 - 2a^2 - 48$

73. $7(3k - 1)^2 + 26(3k - 1) - 8$

74. $6(4z - 3)^2 + 7(4z - 3) - 3$

75. $9(a - 4)^2 + 30(a - 4) + 25$

76. $20(4 - p)^2 - 3(4 - p) - 2$

Factor by any method. See Examples 1–7.

77. $4b^2 + 4bc + c^2 - 16$

78. $(2y - 1)^2 - 4(2y - 1) + 4$

79. $x^2 + xy - 5x - 5y$

80. $8r^2 - 3rs + 10s^2$

81. $p^4(m - 2n) + q(m - 2n)$

82. $36a^2 + 60a + 25$

83. $4z^2 + 28z + 49$

84. $6p^4 + 7p^2 - 3$

85. $1000x^3 + 343y^3$

86. $b^2 + 8b + 16 - a^2$

87. $125m^6 - 216$

88. $q^2 + 6q + 9 - p^2$

89. $64 + (3x + 2)^3$

90. $216p^3 + 125q^3$

91. $4p^2 + 3p - 1$

92. $100r^2 - 169s^2$

93. $144z^2 + 121$

94. $(3a + 5)^2 - 18(3a + 5) + 81$

95. $(x + y)^2 - (x - y)^2$

96. $4z^4 - 7z^2 - 15$

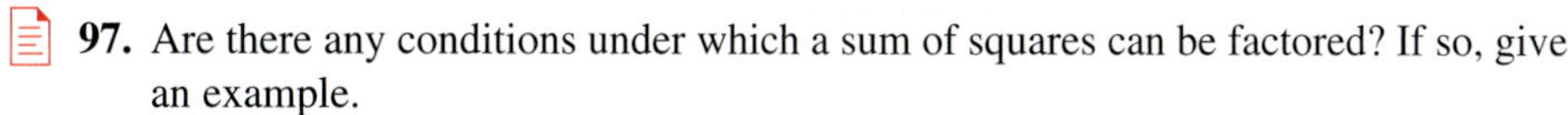

97. Are there any conditions under which a sum of squares can be factored? If so, give an example.

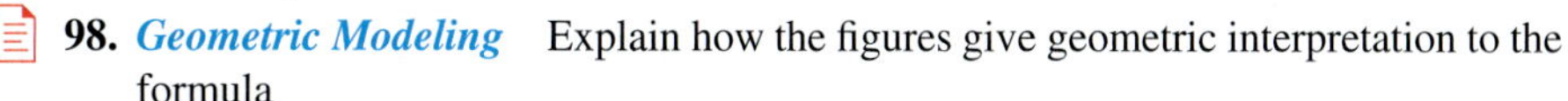

98. ***Geometric Modeling*** Explain how the figures give geometric interpretation to the formula

$$x^2 + 2xy + y^2 = (x + y)^2.$$

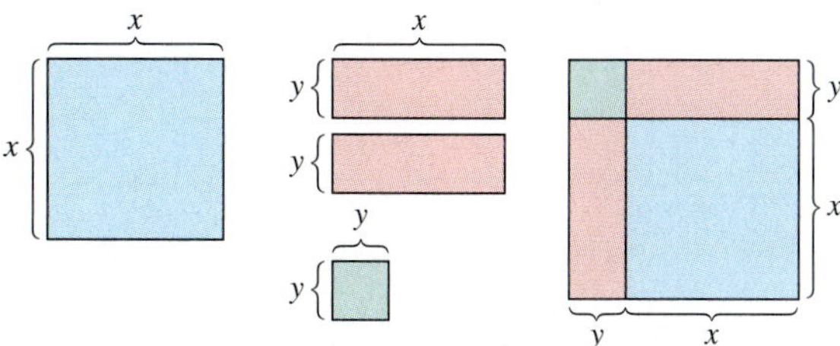

Concept Check *Find all values of b or c that will make the polynomial a perfect square trinomial.*

99. $4z^2 + bz + 81$

100. $9p^2 + bp + 25$

101. $100r^2 - 60r + c$

102. $49x^2 + 70x + c$

R.5 Rational Expressions

Rational Expressions ▪ Lowest Terms of a Rational Expression ▪ Multiplication and Division ▪ Addition and Subtraction ▪ Complex Fractions

Rational Expressions The quotient of two polynomials P and Q, with $Q \neq 0$, is called a **rational expression.**

$$\frac{x+6}{x+2}, \quad \frac{(x+6)(x+4)}{(x+2)(x+4)}, \quad \frac{2p^2+7p-4}{5p^2+20p} \qquad \text{Rational expressions}$$

The **domain** of a rational expression is the set of real numbers for which the expression is defined. Because the denominator of a fraction cannot be 0, the domain consists of all real numbers except those that make the denominator 0. We find these numbers by setting the denominator equal to 0 and solving the resulting equation. For example, in the rational expression

$$\frac{x+6}{x+2},$$

the solution to the equation $x + 2 = 0$ is excluded from the domain. Since this solution is -2, the domain is the set of all real numbers $x \neq -2$, written $\{x \mid x \neq -2\}$.

If the denominator of a rational expression contains a product, we determine the domain with the *zero-factor property* (covered in more detail in Section 1.4), which states that $ab = 0$ if and only if $a = 0$ or $b = 0$. For example, to find the domain of

$$\frac{(x+6)(x+4)}{(x+2)(x+4)},$$

we solve as follows.

$$(x+2)(x+4) = 0$$

$$x + 2 = 0 \quad \text{or} \quad x + 4 = 0 \qquad \text{Set each factor equal to 0.}$$

$$x = -2 \quad \text{or} \quad x = -4 \qquad \text{Solve each equation.}$$

The domain is the set of real numbers not equal to -2 or -4, written $\{x \mid x \neq -2, -4\}$.

Now try Exercises 1 and 3.

Lowest Terms of a Rational Expression A rational expression is written in **lowest terms** when the greatest common factor of its numerator and its denominator is 1. We use the following basic principle of fractions to write a rational expression in lowest terms.

$$\frac{ac}{bc} = \frac{a}{b} \quad (b \neq 0, c \neq 0)$$

Looking Ahead to Calculus

A standard problem in calculus is investigating what value an expression such as $\frac{x^2-1}{x-1}$ approaches as x approaches 1. We cannot do this by simply substituting 1 for x in the expression since the result is the indeterminate form $\frac{0}{0}$. By factoring the numerator and writing the expression in lowest terms, it becomes $x + 1$. Then by substituting 1 for x, we get $1 + 1 = 2$, which is called the *limit* of $\frac{x^2-1}{x-1}$ as x approaches 1.

EXAMPLE 1 Writing Rational Expressions in Lowest Terms

Write each rational expression in lowest terms.

(a) $\dfrac{2p^2 + 7p - 4}{5p^2 + 20p}$ **(b)** $\dfrac{6 - 3k}{k^2 - 4}$

Solution

(a) $\dfrac{2p^2 + 7p - 4}{5p^2 + 20p} = \dfrac{(2p - 1)(p + 4)}{5p(p + 4)}$ Factor. (Section R.4)

$= \dfrac{2p - 1}{5p}$ Divide out the common factor.

To determine the domain, we find values of p that make the original denominator $5p^2 + 20p$ equal to 0.

$$5p^2 + 20p = 0$$

$$5p(p + 4) = 0 \quad \text{Factor.}$$

$$5p = 0 \quad \text{or} \quad p + 4 = 0 \quad \text{Set each factor equal to 0.}$$

$$p = 0 \quad \text{or} \quad p = -4 \quad \text{Solve.}$$

The domain is $\{p \mid p \neq 0, -4\}$. From now on, we will assume such restrictions when writing rational expressions in lowest terms.

(b) $\dfrac{6 - 3k}{k^2 - 4} = \dfrac{3(2 - k)}{(k + 2)(k - 2)}$ Factor.

$= \dfrac{3(2 - k)(-1)}{(k + 2)(k - 2)(-1)}$ $2 - k$ and $k - 2$ are opposites; multiply numerator and denominator by -1.

$= \dfrac{3(2 - k)(-1)}{(k + 2)(2 - k)}$ $(k - 2)(-1) = -k + 2 = 2 - k$

$= \dfrac{-3}{k + 2}$ Lowest terms

Working in an alternative way would lead to the equivalent result

$$\frac{3}{-k - 2}.$$

Now try Exercises 13 and 17.

CAUTION One of the most common errors made in algebra is the incorrect use of the basic principle to write a fraction in lowest terms. Remember, the basic principle requires a pair of common *factors*, one in the numerator and one in the denominator. *Only after a rational expression has been factored can any common factors be divided out.* For example,

$$\frac{2x + 4}{6} = \frac{2(x + 2)}{6} = \frac{2(x + 2)}{2 \cdot 3} = \frac{x + 2}{3}. \quad \text{Factor first, then divide.}$$

Multiplication and Division We multiply and divide rational expressions using definitions from earlier work with fractions.

Multiplication and Division

For fractions $\frac{a}{b}$ and $\frac{c}{d}$ $(b \neq 0, d \neq 0)$,

$$\frac{a}{b} \cdot \frac{c}{d} = \frac{ac}{bd} \quad \text{and} \quad \frac{a}{b} \div \frac{c}{d} = \frac{a}{b} \cdot \frac{d}{c} \quad (c \neq 0).$$

That is, to find the product of two fractions, multiply their numerators to get the numerator of the product; multiply their denominators to get the denominator of the product. To divide two fractions, multiply the first fraction by the reciprocal of the second fraction.

EXAMPLE 2 Multiplying or Dividing Rational Expressions

Multiply or divide, as indicated.

(a) $\frac{2y^2}{9} \cdot \frac{27}{8y^5}$

(b) $\frac{3m^2 - 2m - 8}{3m^2 + 14m + 8} \cdot \frac{3m + 2}{3m + 4}$

(c) $\frac{3p^2 + 11p - 4}{24p^3 - 8p^2} \div \frac{9p + 36}{24p^4 - 36p^3}$

(d) $\frac{x^3 - y^3}{x^2 - y^2} \cdot \frac{2x + 2y + xz + yz}{2x^2 + 2y^2 + zx^2 + zy^2}$

Solution

(a) $\frac{2y^2}{9} \cdot \frac{27}{8y^5} = \frac{2y^2 \cdot 27}{9 \cdot 8y^5}$ Multiply fractions.

$= \frac{2 \cdot 9 \cdot 3 \cdot y^2}{9 \cdot 2 \cdot 4 \cdot y^2 \cdot y^3}$ Factor.

$= \frac{3}{4y^3}$ Lowest terms

While we usually factor first and then multiply the fractions (see parts (b)–(d)), we did the opposite here. Either is acceptable.

(b) $\frac{3m^2 - 2m - 8}{3m^2 + 14m + 8} \cdot \frac{3m + 2}{3m + 4} = \frac{(m - 2)(3m + 4)}{(m + 4)(3m + 2)} \cdot \frac{3m + 2}{3m + 4}$ Factor.

$= \frac{(m - 2)(3m + 4)(3m + 2)}{(m + 4)(3m + 2)(3m + 4)}$ Multiply fractions.

$= \frac{m - 2}{m + 4}$ Lowest terms

(c) $\dfrac{3p^2 + 11p - 4}{24p^3 - 8p^2} \div \dfrac{9p + 36}{24p^4 - 36p^3} = \dfrac{(p + 4)(3p - 1)}{8p^2(3p - 1)} \div \dfrac{9(p + 4)}{12p^3(2p - 3)}$ Factor.

$= \dfrac{(p + 4)(3p - 1)}{8p^2(3p - 1)} \cdot \dfrac{12p^3(2p - 3)}{9(p + 4)}$ Multiply by the reciprocal of the divisor.

$= \dfrac{12p^3(2p - 3)}{9 \cdot 8p^2}$ Divide out common factors; multiply fractions.

$= \dfrac{p(2p - 3)}{6}$ Lowest terms

(d) $\dfrac{x^3 - y^3}{x^2 - y^2} \cdot \dfrac{2x + 2y + xz + yz}{2x^2 + 2y^2 + zx^2 + zy^2}$

$= \dfrac{(x - y)(x^2 + xy + y^2)}{(x + y)(x - y)} \cdot \dfrac{2(x + y) + z(x + y)}{2(x^2 + y^2) + z(x^2 + y^2)}$ Factor; group terms.

$= \dfrac{(x - y)(x^2 + xy + y^2)}{(x + y)(x - y)} \cdot \dfrac{(2 + z)(x + y)}{(2 + z)(x^2 + y^2)}$ Factor by grouping. **(Section R.4)**

$= \dfrac{x^2 + xy + y^2}{x^2 + y^2}$ Lowest terms

Now try Exercises 21, 31, and 33.

Addition and Subtraction Adding and subtracting rational expressions also depends on definitions from earlier work with fractions.

Addition and Subtraction

For fractions $\dfrac{a}{b}$ and $\dfrac{c}{d}$ $(b \neq 0, d \neq 0)$,

$$\frac{a}{b} + \frac{c}{d} = \frac{ad + bc}{bd} \quad \text{and} \quad \frac{a}{b} - \frac{c}{d} = \frac{ad - bc}{bd}.$$

That is, to add (or subtract) two fractions, find their least common denominator (LCD) and change each fraction to one with the LCD as denominator. The sum (or difference) of their numerators is the numerator of their sum (or difference), and the LCD is the denominator of their sum (or difference).

Finding the Least Common Denominator (LCD)

Step 1 Write each denominator as a product of prime factors.

Step 2 Form a product of all the different prime factors. Each factor should have as exponent the *greatest* exponent that appears on that factor.

EXAMPLE 3 Adding or Subtracting Rational Expressions

Add or subtract, as indicated.

(a) $\frac{5}{9x^2} + \frac{1}{6x}$ **(b)** $\frac{y}{y-2} + \frac{8}{2-y}$

(c) $\frac{3}{(x-1)(x+2)} - \frac{1}{(x+3)(x-4)}$

Solution

(a) $\frac{5}{9x^2} + \frac{1}{6x}$

Step 1 Write each denominator as a product of prime factors.

$$9x^2 = 3^2 \cdot x^2$$
$$6x = 2^1 \cdot 3^1 \cdot x^1$$

Step 2 For the LCD, form the product of all the prime factors, with each factor having the greatest exponent that appears on it.

Greatest exponent on 3 is 2. Greatest exponent on x is 2.

$$\begin{aligned} \text{LCD} &= 2^1 \cdot 3^2 \cdot x^2 \\ &= 18x^2 \end{aligned}$$

Now write both of the given expressions with this denominator, then add.

$$\begin{aligned} \frac{5}{9x^2} + \frac{1}{6x} &= \frac{5 \cdot 2}{9x^2 \cdot 2} + \frac{1 \cdot 3x}{6x \cdot 3x} \\ &= \frac{10}{18x^2} + \frac{3x}{18x^2} \\ &= \frac{10 + 3x}{18x^2} \qquad \text{Add the numerators.} \end{aligned}$$

Always check to see that the answer is in lowest terms.

(b) $\frac{y}{y-2} + \frac{8}{2-y}$

To get a common denominator of $y - 2$, multiply the second expression by -1 in both the numerator and the denominator.

$$\begin{aligned} \frac{y}{y-2} + \frac{8}{2-y} &= \frac{y}{y-2} + \frac{8(-1)}{(2-y)(-1)} \\ &= \frac{y}{y-2} + \frac{-8}{y-2} \\ &= \frac{y-8}{y-2} \qquad \text{Add the numerators.} \end{aligned}$$

If $2 - y$ is used as the common denominator instead of $y - 2$, the equivalent expression $\frac{8-y}{2-y}$ results. (Verify this.)

(c) $\dfrac{3}{(x-1)(x+2)} - \dfrac{1}{(x+3)(x-4)}$

The LCD here is $(x-1)(x+2)(x+3)(x-4)$. Write each fraction with this denominator, then subtract.

$$\frac{3}{(x-1)(x+2)} - \frac{1}{(x+3)(x-4)}$$

$$= \frac{3(x+3)(x-4)}{(x-1)(x+2)(x+3)(x-4)} - \frac{(x-1)(x+2)}{(x+3)(x-4)(x-1)(x+2)}$$

$$= \frac{3(x^2-x-12)-(x^2+x-2)}{(x-1)(x+2)(x+3)(x-4)}$$ Multiply in the numerators, then subtract them.

$$= \frac{3x^2-3x-36-x^2-x+2}{(x-1)(x+2)(x+3)(x-4)}$$ Be careful with signs.

$$= \frac{2x^2-4x-34}{(x-1)(x+2)(x+3)(x-4)}$$ Combine terms in the numerator.

Now try Exercises 45, 49, and 57.

CAUTION When subtracting fractions where the second fraction has more than one term in the numerator, as in Example 3(c), **be sure to distribute the negative sign to *each* term.** Use parentheses as in the second step to avoid an error.

Complex Fractions We call the quotient of two rational expressions a **complex fraction.**

EXAMPLE 4 Simplifying Complex Fractions

Simplify each complex fraction.

(a) $\dfrac{6-\frac{5}{k}}{1+\frac{5}{k}}$ **(b)** $\dfrac{\frac{a}{a+1}+\frac{1}{a}}{\frac{1}{a}+\frac{1}{a+1}}$

Solution

(a) Multiply both numerator and denominator by the LCD of all the fractions, k.

$$\frac{6-\frac{5}{k}}{1+\frac{5}{k}} = \frac{k\left(6-\frac{5}{k}\right)}{k\left(1+\frac{5}{k}\right)} = \frac{6k-k\left(\frac{5}{k}\right)}{k+k\left(\frac{5}{k}\right)} = \frac{6k-5}{k+5}$$

(b) Multiply both numerator and denominator by the LCD of all the fractions, $a(a + 1)$.

$$\frac{\frac{a}{a+1}+\frac{1}{a}}{\frac{1}{a}+\frac{1}{a+1}} = \frac{\left(\frac{a}{a+1}+\frac{1}{a}\right)a(a+1)}{\left(\frac{1}{a}+\frac{1}{a+1}\right)a(a+1)}$$

$$= \frac{\frac{a}{a+1}(a)(a+1)+\frac{1}{a}(a)(a+1)}{\frac{1}{a}(a)(a+1)+\frac{1}{a+1}(a)(a+1)} \quad \text{Distributive property (Section R.1)}$$

$$= \frac{a^2+(a+1)}{(a+1)+a}$$

$$= \frac{a^2+a+1}{2a+1} \quad \text{Combine terms.}$$

As an alternative method of solution, first perform the indicated additions in the numerator and denominator, and then divide.

$$\frac{\frac{a}{a+1}+\frac{1}{a}}{\frac{1}{a}+\frac{1}{a+1}} = \frac{\frac{a^2+1(a+1)}{a(a+1)}}{\frac{1(a+1)+1(a)}{a(a+1)}} \quad \text{Get the LCD; add terms in the numerator and denominator of the complex fraction.}$$

$$= \frac{\frac{a^2+a+1}{a(a+1)}}{\frac{2a+1}{a(a+1)}} \quad \text{Combine terms in the numerator and denominator.}$$

$$= \frac{a^2+a+1}{a(a+1)} \cdot \frac{a(a+1)}{2a+1} \quad \text{Definition of division}$$

$$= \frac{a^2+a+1}{2a+1} \quad \text{Multiply fractions; write in lowest terms.}$$

Now try Exercises 59 and 63.

R.5 Exercises

Find the domain of each rational expression.

1. $\frac{x+3}{x-6}$ **2.** $\frac{2x-4}{x+7}$ **3.** $\frac{3x+7}{(4x+2)(x-1)}$

4. $\frac{9x+12}{(2x+3)(x-5)}$ **5.** $\frac{12}{x^2+5x+6}$ **6.** $\frac{3}{x^2-5x-6}$

Relating Concepts

For individual or collaborative investigation

(Exercises 7–10)

Work Exercises 7–10 in order.

7. Let $x = 4$ and $y = 2$. Evaluate **(a)** $\frac{1}{x} + \frac{1}{y}$ and **(b)** $\frac{1}{x + y}$.

8. Are the answers for Exercises 7(a) and (b) the same? What can you conclude?

9. Let $x = 3$ and $y = 5$. Evaluate **(a)** $\frac{1}{x} - \frac{1}{y}$ and **(b)** $\frac{1}{x - y}$.

10. Are the answers for Exercises 9(a) and (b) the same? What can you conclude?

Write each rational expression in lowest terms. See Example 1.

11. $\dfrac{8k + 16}{9k + 18}$ **12.** $\dfrac{20r + 10}{30r + 15}$ **13.** $\dfrac{3(3 - t)}{(t + 5)(t - 3)}$ **14.** $\dfrac{-8(4 - y)}{(y + 2)(y - 4)}$

15. $\dfrac{8x^2 + 16x}{4x^2}$ **16.** $\dfrac{36y^2 + 72y}{9y}$ **17.** $\dfrac{m^2 - 4m + 4}{m^2 + m - 6}$ **18.** $\dfrac{r^2 - r - 6}{r^2 + r - 12}$

19. $\dfrac{8m^2 + 6m - 9}{16m^2 - 9}$ **20.** $\dfrac{6y^2 + 11y + 4}{3y^2 + 7y + 4}$

Find each product or quotient. See Example 2.

21. $\dfrac{15p^3}{9p^2} \div \dfrac{6p}{10p^2}$ **22.** $\dfrac{3r^2}{9r^3} \div \dfrac{8r^3}{6r}$

23. $\dfrac{2k + 8}{6} \div \dfrac{3k + 12}{2}$ **24.** $\dfrac{5m + 25}{10} \cdot \dfrac{12}{6m + 30}$

25. $\dfrac{x^2 + x}{5} \cdot \dfrac{25}{xy + y}$ **26.** $\dfrac{3m - 15}{4m - 20} \cdot \dfrac{m^2 - 10m + 25}{12m - 60}$

27. $\dfrac{4a + 12}{2a - 10} \div \dfrac{a^2 - 9}{a^2 - a - 20}$ **28.** $\dfrac{6r - 18}{9r^2 + 6r - 24} \cdot \dfrac{12r - 16}{4r - 12}$

29. $\dfrac{p^2 - p - 12}{p^2 - 2p - 15} \cdot \dfrac{p^2 - 9p + 20}{p^2 - 8p + 16}$ **30.** $\dfrac{x^2 + 2x - 15}{x^2 + 11x + 30} \cdot \dfrac{x^2 + 2x - 24}{x^2 - 8x + 15}$

31. $\dfrac{m^2 + 3m + 2}{m^2 + 5m + 4} \div \dfrac{m^2 + 5m + 6}{m^2 + 10m + 24}$ **32.** $\dfrac{y^2 + y - 2}{y^2 + 3y - 4} \div \dfrac{y^2 + 3y + 2}{y^2 + 4y + 3}$

33. $\dfrac{xz - xw + 2yz - 2yw}{z^2 - w^2} \cdot \dfrac{4z + 4w + xz + wx}{16 - x^2}$

34. $\dfrac{ac + ad + bc + bd}{a^2 - b^2} \cdot \dfrac{a^3 - b^3}{2a^2 + 2ab + 2b^2}$

35. $\dfrac{x^3 + y^3}{x^3 - y^3} \cdot \dfrac{x^2 - y^2}{x^2 + 2xy + y^2}$ **36.** $\dfrac{x^2 - y^2}{(x - y)^2} \cdot \dfrac{x^2 - xy + y^2}{x^2 - 2xy + y^2} \div \dfrac{x^3 + y^3}{(x - y)^4}$

37. *Concept Check* Which of the following rational expressions equals -1? (In parts A, B, and D, $x \neq -4$, and in part C, $x \neq 4$.) (*Hint:* There may be more than one answer.)

A. $\dfrac{x - 4}{x + 4}$ **B.** $\dfrac{-x - 4}{x + 4}$ **C.** $\dfrac{x - 4}{4 - x}$ **D.** $\dfrac{x - 4}{-x - 4}$

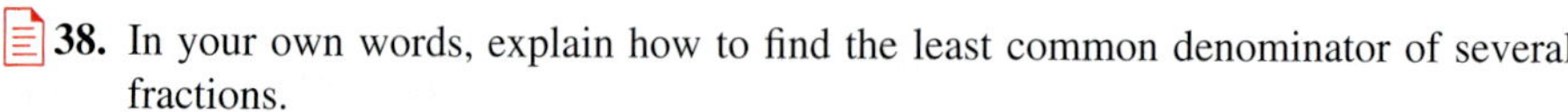

38. In your own words, explain how to find the least common denominator of several fractions.

Perform each addition or subtraction. See Example 3.

39. $\dfrac{3}{2k} + \dfrac{5}{3k}$

40. $\dfrac{8}{5p} + \dfrac{3}{4p}$

41. $\dfrac{1}{6m} + \dfrac{2}{5m} + \dfrac{4}{m}$

42. $\dfrac{8}{3p} + \dfrac{5}{4p} + \dfrac{9}{2p}$

43. $\dfrac{1}{a} - \dfrac{b}{a^2}$

44. $\dfrac{3}{z} + \dfrac{x}{z^2}$

45. $\dfrac{5}{12x^2y} - \dfrac{11}{6xy}$

46. $\dfrac{7}{18a^3b^2} - \dfrac{2}{9ab}$

47. $\dfrac{17y + 3}{9y + 7} - \dfrac{-10y - 18}{9y + 7}$

48. $\dfrac{7x + 8}{3x + 2} - \dfrac{x + 4}{3x + 2}$

49. $\dfrac{1}{x + z} + \dfrac{1}{x - z}$

50. $\dfrac{m + 1}{m - 1} + \dfrac{m - 1}{m + 1}$

51. $\dfrac{3}{a - 2} - \dfrac{1}{2 - a}$

52. $\dfrac{q}{p - q} - \dfrac{q}{q - p}$

53. $\dfrac{x + y}{2x - y} - \dfrac{2x}{y - 2x}$

54. $\dfrac{m - 4}{3m - 4} + \dfrac{3m + 2}{4 - 3m}$

55. $\dfrac{4}{x + 1} + \dfrac{1}{x^2 - x + 1} - \dfrac{12}{x^3 + 1}$

56. $\dfrac{5}{x + 2} + \dfrac{2}{x^2 - 2x + 4} - \dfrac{60}{x^3 + 8}$

57. $\dfrac{3x}{x^2 + x - 12} - \dfrac{x}{x^2 - 16}$

58. $\dfrac{p}{2p^2 - 9p - 5} - \dfrac{2p}{6p^2 - p - 2}$

Simplify each expression. See Example 4.

59. $\dfrac{1 + \dfrac{1}{x}}{1 - \dfrac{1}{x}}$

60. $\dfrac{2 - \dfrac{2}{y}}{2 + \dfrac{2}{y}}$

61. $\dfrac{\dfrac{1}{x + 1} - \dfrac{1}{x}}{\dfrac{1}{x}}$

62. $\dfrac{\dfrac{1}{y + 3} - \dfrac{1}{y}}{\dfrac{1}{y}}$

63. $\dfrac{1 + \dfrac{1}{1 - b}}{1 - \dfrac{1}{1 + b}}$

64. $m - \dfrac{m}{m + \dfrac{1}{2}}$

65. $\dfrac{m - \dfrac{1}{m^2 - 4}}{\dfrac{1}{m + 2}}$

66. $\dfrac{\dfrac{3}{p^2 - 16} + p}{\dfrac{1}{p - 4}}$

67. $\dfrac{\dfrac{1}{x + h} - \dfrac{1}{x}}{h}$

68. $\dfrac{\dfrac{1}{(x + h)^2 + 9} - \dfrac{1}{x^2 + 9}}{h}$

(Modeling) Distance from the Origin of the Nile River *The Nile River in Africa is about* 4000 mi *long. The Nile begins as an outlet of Lake Victoria at an altitude of* 7000 ft *above sea level and empties into the Mediterranean Sea at sea level* (0 ft)*. The distance from its origin in thousands of miles is related to its height above sea level in thousands of feet (x) by the rational expression*

$$\frac{7 - x}{.639x + 1.75}.$$

For example, when the river is at an altitude of 600 ft, $x = .6$ *(thousand), and the distance from the origin is*

$$\frac{7 - .6}{.639(.6) + 1.75} \approx 3,$$

which represents 3000 mi. *(Source: The World Almanac and Book of Facts,* 2001.*)*

69. What is the distance from the origin of the Nile when the river has an altitude of 7000 ft?

70. Find the distance from the origin of the Nile when the river is 1200 ft high.

(Modeling) Cost-Benefit Model for a Pollutant *In situations involving environmental pollution, a cost-benefit model expresses cost in terms of the percentage of pollutant removed from the environment. Suppose a cost-benefit model is expressed as*

$$y = \frac{6.7x}{100 - x},$$

where y is the cost in thousands of dollars of removing x percent of a certain pollutant. Find the value of y for each given value of x.

71. $x = 75$ (75%)

72. $x = 95$ (95%)

R.6 Rational Exponents

Negative Exponents and the Quotient Rule ▪ Rational Exponents ▪ Complex Fractions Revisited

In this section we complete our review of exponents.

Negative Exponents and the Quotient Rule In the product rule, $a^m \cdot a^n = a^{m+n}$, the exponents are *added.* By the definition of exponent in Section R.1, if $a \neq 0$,

$$\frac{a^3}{a^7} = \frac{a \cdot a \cdot a}{a \cdot a \cdot a \cdot a \cdot a \cdot a \cdot a} = \frac{1}{a \cdot a \cdot a \cdot a} = \frac{1}{a^4}.$$

This example suggests that we should *subtract* exponents when dividing. Subtracting exponents gives

$$\frac{a^3}{a^7} = a^{3-7} = a^{-4}.$$

The only way to keep these results consistent is to define a^{-4} as $\frac{1}{a^4}$. This example suggests the following definition.

Negative Exponent

If a is a nonzero real number and n is any integer, then

$$a^{-n} = \frac{1}{a^n}.$$

EXAMPLE 1 Using the Definition of a Negative Exponent

Evaluate each expression. In parts (d) and (e), write the expression without negative exponents. Assume all variables represent nonzero real numbers.

(a) 4^{-2} **(b)** $\left(\frac{2}{5}\right)^{-3}$ **(c)** -4^{-2} **(d)** $(xy)^{-3}$ **(e)** xy^{-3}

Solution

(a) $4^{-2} = \frac{1}{4^2} = \frac{1}{16}$

(b) $\left(\frac{2}{5}\right)^{-3} = \frac{1}{\left(\frac{2}{5}\right)^3} = \frac{1}{\frac{8}{125}} = 1 \div \frac{8}{125} = 1 \cdot \frac{125}{8} = \frac{125}{8}$

Multiply by the reciprocal.

(c) $-4^{-2} = -\frac{1}{4^2} = -\frac{1}{16}$

(d) $(xy)^{-3} = \frac{1}{(xy)^3}$

Base is xy.

(e) $xy^{-3} = x \cdot \frac{1}{y^3} = \frac{x}{y^3}$

Base is y.

Now try Exercises 3, 5, 7, 9, and 11.

CAUTION A negative exponent indicates a reciprocal, *not* a negative expression.

Part (b) of Example 1 showed that

$$\left(\frac{2}{5}\right)^{-3} = \frac{125}{8} = \left(\frac{5}{2}\right)^3.$$

We can generalize this result. If $a \neq 0$ and $b \neq 0$, then for any integer n,

$$\left(\frac{a}{b}\right)^{-n} = \left(\frac{b}{a}\right)^n.$$

The **quotient rule** for exponents follows from the definition of exponents, as shown at the beginning of this section.

Quotient Rule

For all integers m and n and all nonzero real numbers a,

$$\frac{a^m}{a^n} = a^{m-n}.$$

That is, when dividing powers of like bases, keep the same base and subtract the exponent of the denominator from the exponent of the numerator.

By the quotient rule, if $a \neq 0$, then

$$\frac{a^m}{a^m} = a^{m-m} = a^0.$$

On the other hand, any nonzero quantity divided by itself equals 1. This is why we defined $a^0 = 1$ in Section R.3.

EXAMPLE 2 Using the Quotient Rule

Simplify each expression. Assume all variables represent nonzero real numbers.

(a) $\frac{12^5}{12^2}$ **(b)** $\frac{a^5}{a^{-8}}$ **(c)** $\frac{16m^{-9}}{12m^{11}}$ **(d)** $\frac{25r^7z^5}{10r^9z}$

Solution

(a) $\frac{12^5}{12^2} = 12^{5-2} = 12^3$ **(b)** $\frac{a^5}{a^{-8}} = a^{5-(-8)} = a^{13}$

(c) $\frac{16m^{-9}}{12m^{11}} = \frac{16}{12} \cdot m^{-9-11} = \frac{4}{3}m^{-20} = \frac{4}{3} \cdot \frac{1}{m^{20}} = \frac{4}{3m^{20}}$

(d) $\frac{25r^7z^5}{10r^9z} = \frac{25}{10} \cdot \frac{r^7}{r^9} \cdot \frac{z^5}{z^1} = \frac{5}{2}r^{-2}z^4 = \frac{5z^4}{2r^2}$

Now try Exercises 15, 21, 23, and 25.

The rules for exponents from Section R.3 also apply to negative exponents.

EXAMPLE 3 Using the Rules for Exponents

Simplify each expression. Write answers without negative exponents. Assume all variables represent nonzero real numbers.

(a) $3x^{-2}(4^{-1}x^{-5})^2$ **(b)** $\frac{12p^3q^{-1}}{8p^{-2}q}$ **(c)** $\frac{(3x^2)^{-1}(3x^5)^{-2}}{(3^{-1}x^{-2})^2}$

Solution

(a)
$$\begin{aligned} 3x^{-2}(4^{-1}x^{-5})^2 &= 3x^{-2}(4^{-2}x^{-10}) && \text{Power rules (Section R.3)} \\ &= 3 \cdot 4^{-2} \cdot x^{-2+(-10)} && \text{Rearrange factors; product rule (Section R.3)} \\ &= 3 \cdot 4^{-2} \cdot x^{-12} \\ &= \frac{3}{16x^{12}} && \text{Write with positive exponents.} \end{aligned}$$

(b)
$$\begin{aligned} \frac{12p^3q^{-1}}{8p^{-2}q} &= \frac{12}{8} \cdot \frac{p^3}{p^{-2}} \cdot \frac{q^{-1}}{q^1} \\ &= \frac{3}{2} \cdot p^{3-(-2)}q^{-1-1} && \text{Quotient rule} \\ &= \frac{3}{2}p^5q^{-2} \\ &= \frac{3p^5}{2q^2} && \text{Write with positive exponents.} \end{aligned}$$

(c) $\dfrac{(3x^2)^{-1}(3x^5)^{-2}}{(3^{-1}x^{-2})^2} = \dfrac{3^{-1}x^{-2}3^{-2}x^{-10}}{3^{-2}x^{-4}}$ Power rules

$= \dfrac{3^{-1+(-2)}x^{-2+(-10)}}{3^{-2}x^{-4}} = \dfrac{3^{-3}x^{-12}}{3^{-2}x^{-4}}$ Product rule

$= 3^{-3-(-2)}x^{-12-(-4)} = 3^{-1}x^{-8}$ Quotient rule

$= \dfrac{1}{3x^8}$ Write with positive exponents.

Now try Exercises 29, 33, and 35.

CAUTION Notice the use of the power rule $(ab)^n = a^nb^n$ in Example 3(c): $(3x^2)^{-1} = 3^{-1}(x^2)^{-1} = 3^{-1}x^{-2}$. Remember to apply the exponent to a numerical coefficient.

Rational Exponents The definition of a^n can be extended to rational values of n by defining $a^{1/n}$ to be the nth root of a. By one of the power rules of exponents (extended to a rational exponent),

$$(a^{1/n})^n = a^{(1/n)n} = a^1 = a,$$

which suggests that $a^{1/n}$ is a number whose nth power is a.

$a^{1/n}$, n Even If n is an *even* positive integer, and if $a > 0$, then $a^{1/n}$ is the positive real number whose nth power is a. That is, $(a^{1/n})^n = a$. In this case, $a^{1/n}$ is the principal nth root of a.

$a^{1/n}$, n Odd If n is an *odd* positive integer, and a is *any real number,* then $a^{1/n}$ is the positive or negative real number whose nth power is a. That is, $(a^{1/n})^n = a$.

EXAMPLE 4 Using the Definition of $a^{1/n}$

Evaluate each expression.

(a) $36^{1/2}$ **(b)** $-100^{1/2}$ **(c)** $625^{1/4}$

(d) $(-1296)^{1/4}$ **(e)** $(-27)^{1/3}$ **(f)** $-32^{1/5}$

Solution

(a) $36^{1/2} = 6$ because $6^2 = 36$. **(b)** $-100^{1/2} = -10$

(c) $625^{1/4} = 5$

(d) $(-1296)^{1/4}$ is not a real number. (But $-1296^{1/4} = -6$.)

(e) $(-27)^{1/3} = -3$ **(f)** $-32^{1/5} = -2$

Now try Exercises 37, 39, and 43.

We now consider more general rational exponents. The notation $a^{m/n}$ should be defined so that all the previous rules for exponents still hold. For the power rule to hold, $(a^{1/n})^m$ must equal $a^{m/n}$. Therefore, $a^{m/n}$ is defined as follows.

Rational Exponent

For all integers m, all positive integers n, and all real numbers a for which $a^{1/n}$ is a real number,

$$a^{m/n} = (a^{1/n})^m.$$

EXAMPLE 5 Using the Definition of $a^{m/n}$

Evaluate each expression.

(a) $125^{2/3}$ **(b)** $32^{7/5}$ **(c)** $-81^{3/2}$

(d) $(-27)^{2/3}$ **(e)** $16^{-3/4}$ **(f)** $(-4)^{5/2}$

Solution

(a) $125^{2/3} = (125^{1/3})^2 = 5^2 = 25$

(b) $32^{7/5} = (32^{1/5})^7 = 2^7 = 128$

(c) $-81^{3/2} = -(81^{1/2})^3 = -9^3 = -729$

(d) $(-27)^{2/3} = [(-27)^{1/3}]^2 = (-3)^2 = 9$

(e) $16^{-3/4} = \dfrac{1}{16^{3/4}} = \dfrac{1}{(16^{1/4})^3} = \dfrac{1}{2^3} = \dfrac{1}{8}$

(f) $(-4)^{5/2}$ is not a real number because $(-4)^{1/2}$ is not a real number.

Now try Exercises 47, 51, and 53.

NOTE By starting with $(a^{1/n})^m$ and $(a^m)^{1/n}$ and raising each expression to the nth power, we can show that $(a^{1/n})^m$ is equal to $(a^m)^{1/n}$. This means that $a^{m/n}$ could be defined in either of the following ways.

For all real numbers a, integers m, and positive integers n for which $a^{1/n}$ is a real number,

$$a^{m/n} = (a^{1/n})^m \quad \text{or} \quad a^{m/n} = (a^m)^{1/n}.$$

Now $a^{m/n}$ can be evaluated in either of two ways: as $(a^{1/n})^m$ or as $(a^m)^{1/n}$. It is usually easier to find $(a^{1/n})^m$. For example, $27^{4/3}$ can be evaluated as

$$27^{4/3} = (27^{1/3})^4 = 3^4 = 81$$

or

$$27^{4/3} = (27^4)^{1/3} = 531{,}441^{1/3} = 81.$$

The form $(27^{1/3})^4$ is easier to evaluate.

All the earlier results concerning integer exponents also apply to rational exponents. These definitions and rules are summarized here.

Definitions and Rules for Exponents

Let r and s be rational numbers. The results here are valid for all positive numbers a and b.

Product rule	$a^r \cdot a^s = a^{r+s}$	**Power rules**	$(a^r)^s = a^{rs}$
Quotient rule	$\frac{a^r}{a^s} = a^{r-s}$		$(ab)^r = a^r b^r$
Negative exponent	$a^{-r} = \frac{1}{a^r}$		$\left(\frac{a}{b}\right)^r = \frac{a^r}{b^r}$

EXAMPLE 6 Combining the Definitions and Rules for Exponents

Simplify each expression.

(a) $\frac{27^{1/3} \cdot 27^{5/3}}{27^3}$ **(b)** $81^{5/4} \cdot 4^{-3/2}$ **(c)** $6y^{2/3} \cdot 2y^{1/2}$

(d) $\left(\frac{3m^{5/6}}{y^{3/4}}\right)^2\left(\frac{8y^3}{m^6}\right)^{2/3}$ **(e)** $m^{2/3}(m^{7/3} + 2m^{1/3})$

Solution

(a) $\frac{27^{1/3} \cdot 27^{5/3}}{27^3} = \frac{27^{1/3+5/3}}{27^3}$ Product rule

$= \frac{27^2}{27^3} = 27^{2-3}$ Quotient rule

$= 27^{-1} = \frac{1}{27}$ Negative exponent

(b) $81^{5/4} \cdot 4^{-3/2} = (81^{1/4})^5(4^{1/2})^{-3} = 3^5 \cdot 2^{-3} = \frac{3^5}{2^3}$ or $\frac{243}{8}$

(c) $6y^{2/3} \cdot 2y^{1/2} = 12y^{2/3+1/2} = 12y^{7/6}, \quad y \geq 0$

(d) $\left(\frac{3m^{5/6}}{y^{3/4}}\right)^2\left(\frac{8y^3}{m^6}\right)^{2/3} = \frac{9m^{5/3}}{y^{3/2}} \cdot \frac{4y^2}{m^4}$

$= 36m^{5/3-4}y^{2-3/2}$

$= \frac{36y^{1/2}}{m^{7/3}}, \quad m > 0, y > 0$

(e) $m^{2/3}(m^{7/3} + 2m^{1/3}) = (m^{2/3+7/3} + 2m^{2/3+1/3})$

$= m^3 + 2m$

Now try Exercises 59, 61, 65, 67, and 73.

EXAMPLE 7 Factoring Expressions with Negative or Rational Exponents

Factor out the least power of the variable or variable expression. Assume all variables represent positive real numbers.

(a) $12x^{-2} - 8x^{-3}$ **(b)** $4m^{1/2} + 3m^{3/2}$ **(c)** $(y - 2)^{-1/3} + (y - 2)^{2/3}$

Solution

(a) The least exponent on $12x^{-2} - 8x^{-3}$ is -3. Since 4 is a common numerical factor, factor out $4x^{-3}$.

$$12x^{-2} - 8x^{-3} = 4x^{-3}(3x^{-2-(-3)} - 2x^{-3-(-3)})$$
$$= 4x^{-3}(3x - 2)$$

Check by multiplying on the right.

(b) $4m^{1/2} + 3m^{3/2} = m^{1/2}(4 + 3m)$

To *check,* multiply $m^{1/2}$ by $4 + 3m$.

(c) $(y - 2)^{-1/3} + (y - 2)^{2/3} = (y - 2)^{-1/3}[1 + (y - 2)]$
$= (y - 2)^{-1/3}(y - 1)$

Looking Ahead to Calculus

The technique of Example 7(c) is used often in calculus.

Now try Exercises 81, 85, and 87.

Complex Fractions Revisited Negative exponents are sometimes used to write complex fractions. Recall that complex fractions are simplified either by first multiplying the numerator and denominator by the LCD of all the denominators or by performing any indicated operations in the numerator and the denominator and then using the definition of division for fractions.

EXAMPLE 8 Simplifying a Fraction with Negative Exponents

Simplify $\dfrac{(x + y)^{-1}}{x^{-1} + y^{-1}}$. Write the result with only positive exponents.

Solution

$$\frac{(x + y)^{-1}}{x^{-1} + y^{-1}} = \frac{\frac{1}{x + y}}{\frac{1}{x} + \frac{1}{y}} \quad \text{Definition of negative exponent}$$

$$= \frac{\frac{1}{x + y}}{\frac{y + x}{xy}} \quad \text{Add fractions in the denominator. (Section R.5)}$$

$$= \frac{1}{x + y} \cdot \frac{xy}{x + y} \quad \text{Multiply by the reciprocal.}$$

$$= \frac{xy}{(x + y)^2} \quad \text{Multiply fractions.}$$

Now try Exercise 89.

CAUTION Remember that if $r \neq 1$, $(x + y)^r \neq x^r + y^r$. In particular, this means that $(x + y)^{-1} \neq x^{-1} + y^{-1}$.

R.6 Exercises

Concept Check *In Exercises 1 and 2, match each expression in Column I with its equivalent expression in Column II. Choices may be used once, more than once, or not at all.*

	I	II
1.	**(a)** 4^{-2}	**A.** 16
	(b) -4^{-2}	**B.** $\frac{1}{16}$
	(c) $(-4)^{-2}$	**C.** -16
	(d) $-(-4)^{-2}$	**D.** $-\frac{1}{16}$

	I	II
2.	**(a)** 5^{-3}	**A.** 125
	(b) -5^{-3}	**B.** -125
	(c) $(-5)^{-3}$	**C.** $\frac{1}{125}$
	(d) $-(-5)^{-3}$	**D.** $-\frac{1}{125}$

Write each expression with only positive exponents and evaluate if possible. Assume all variables represent nonzero real numbers. See Example 1.

3. $(-4)^{-3}$ **4.** $(-5)^{-2}$ **5.** -5^{-4} **6.** -7^{-2}

7. $\left(\frac{1}{3}\right)^{-2}$ **8.** $\left(\frac{4}{3}\right)^{-3}$ **9.** $(4x)^{-2}$ **10.** $(5t)^{-3}$

11. $4x^{-2}$ **12.** $5t^{-3}$ **13.** $-a^{-3}$ **14.** $-b^{-4}$

Perform the indicated operations. Write each answer using only positive exponents. Assume all variables represent nonzero real numbers. See Examples 2 and 3.

15. $\frac{4^8}{4^6}$ **16.** $\frac{5^9}{5^7}$ **17.** $\frac{x^{12}}{x^8}$ **18.** $\frac{y^{14}}{y^{10}}$

19. $\frac{r^7}{r^{10}}$ **20.** $\frac{y^8}{y^{12}}$ **21.** $\frac{6^4}{6^{-2}}$ **22.** $\frac{7^5}{7^{-3}}$

23. $\frac{4r^{-3}}{6r^{-6}}$ **24.** $\frac{15s^{-4}}{5s^{-8}}$ **25.** $\frac{16m^{-5}n^4}{12m^2n^{-3}}$ **26.** $\frac{15a^{-5}b^{-1}}{25a^{-2}b^4}$

27. $-4r^{-2}(r^4)^2$ **28.** $-2m^{-1}(m^3)^2$ **29.** $(5a^{-1})^4(a^2)^{-3}$ **30.** $(3p^{-4})^2(p^3)^{-1}$

31. $\frac{(p^{-2})^0}{5p^{-4}}$ **32.** $\frac{(m^4)^0}{9m^{-3}}$ **33.** $\frac{(3pq)q^2}{6p^2q^4}$ **34.** $\frac{(-8xy)y^3}{4x^5y^4}$

35. $\frac{4a^5(a^{-1})^3}{(a^{-2})^{-2}}$ **36.** $\frac{12k^{-2}(k^{-3})^{-4}}{6k^5}$

Simplify each expression. See Example 4.

37. $169^{1/2}$ **38.** $121^{1/2}$ **39.** $16^{1/4}$ **40.** $625^{1/4}$

41. $\left(-\frac{64}{27}\right)^{1/3}$ **42.** $\left(\frac{8}{27}\right)^{1/3}$ **43.** $(-4)^{1/2}$ **44.** $(-64)^{1/4}$

Concept Check *In Exercises 45 and 46, match each expression from Column I with its equivalent expression from Column II. Choices may be used once, more than once, or not at all.*

I		II	
45. **(a)** $\left(\frac{4}{9}\right)^{3/2}$	**46.** **(a)** $\left(\frac{8}{27}\right)^{2/3}$	**A.** $\frac{9}{4}$	**B.** $-\frac{9}{4}$
(b) $\left(\frac{4}{9}\right)^{-3/2}$	**(b)** $\left(\frac{8}{27}\right)^{-2/3}$	**C.** $-\frac{4}{9}$	**D.** $\frac{4}{9}$
(c) $-\left(\frac{9}{4}\right)^{3/2}$	**(c)** $-\left(\frac{27}{8}\right)^{2/3}$	**E.** $\frac{8}{27}$	**F.** $-\frac{27}{8}$
(d) $-\left(\frac{4}{9}\right)^{-3/2}$	**(d)** $-\left(\frac{27}{8}\right)^{-2/3}$	**G.** $\frac{27}{8}$	**H.** $-\frac{8}{27}$

Perform the indicated operations. Write each answer using only positive exponents. Assume all variables represent positive real numbers. See Examples 5 and 6.

47. $8^{2/3}$ **48.** $27^{4/3}$ **49.** $100^{3/2}$

50. $64^{3/2}$ **51.** $-81^{3/4}$ **52.** $(-32)^{-4/5}$

53. $\left(\frac{27}{64}\right)^{-4/3}$ **54.** $\left(\frac{121}{100}\right)^{-3/2}$ **55.** $3^{1/2} \cdot 3^{3/2}$

56. $6^{4/3} \cdot 6^{2/3}$ **57.** $\frac{64^{5/3}}{64^{4/3}}$ **58.** $\frac{125^{7/3}}{125^{5/3}}$

59. $y^{7/3} \cdot y^{-4/3}$ **60.** $r^{-8/9} \cdot r^{17/9}$ **61.** $\frac{k^{1/3}}{k^{2/3} \cdot k^{-1}}$

62. $\frac{z^{3/4}}{z^{5/4} \cdot z^{-2}}$ **63.** $\frac{(x^{1/4}y^{2/5})^{20}}{x^2}$ **64.** $\frac{(r^{1/5}s^{2/3})^{15}}{r^2}$

65. $\frac{(x^{2/3})^2}{(x^2)^{7/3}}$ **66.** $\frac{(p^3)^{1/4}}{(p^{5/4})^2}$ **67.** $\left(\frac{16m^3}{n}\right)^{1/4}\left(\frac{9n^{-1}}{m^2}\right)^{1/2}$

68. $\left(\frac{25^4a^3}{b^2}\right)^{1/8}\left(\frac{4^2b^{-5}}{a^2}\right)^{1/4}$ **69.** $\frac{p^{1/5}p^{7/10}p^{1/2}}{(p^3)^{-1/5}}$ **70.** $\frac{z^{1/3}z^{-2/3}z^{1/6}}{(z^{-1/6})^3}$

Solve each applied problem.

71. *(Modeling) Holding Time of Athletes* A group of ten athletes were tested for isometric endurance by measuring the length of time they could resist a load pulling on their legs while seated. The approximate amount of time (called the *holding time*) that they could resist the load was given by the formula

$$t = \frac{31{,}293}{w^{1.5}},$$

where w is the weight of the load in pounds and the holding time t is measured in seconds. (*Source:* Townend, M. Stewart, *Mathematics in Sport,* Chichester, Ellis Horwood Limited, 1984.)

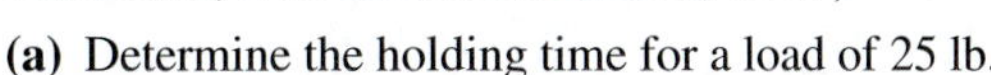

(a) Determine the holding time for a load of 25 lb.

(b) When the weight of the load is doubled, by what factor is the *holding time* changed?

72. *Duration of a Storm* Meteorologists can approximate the duration of a storm by using the formula

$$T = .07D^{3/2},$$

where T is the time in hours that a storm of diameter D (in miles) lasts.

(a) The National Weather Service reports that a storm 4 mi in diameter is headed toward New Haven. How long can the residents expect the storm to last?

(b) After weeks of dry weather, a thunderstorm is predicted for the farming community of Apple Valley. The crops need at least 1.5 hr of rain. Local radar shows that the storm is 7 mi in diameter. Will it rain long enough to meet the farmers' need?

Find each product. Assume all variables represent positive real numbers. See Example 6(e) in this section and Example 7 in Section R.3.

73. $y^{5/8}(y^{3/8} - 10y^{11/8})$

74. $p^{11/5}(3p^{4/5} + 9p^{19/5})$

75. $-4k(k^{7/3} - 6k^{1/3})$

76. $-5y(3y^{9/10} + 4y^{3/10})$

77. $(x + x^{1/2})(x - x^{1/2})$

78. $(2z^{1/2} + z)(z^{1/2} - z)$

79. $(r^{1/2} - r^{-1/2})^2$

80. $(p^{1/2} - p^{-1/2})(p^{1/2} + p^{-1/2})$

Factor, using the given common factor. Assume all variables represent positive real numbers. See Example 7.

81. $4k^{-1} + k^{-2}$; k^{-2}

82. $y^{-5} - 3y^{-3}$; y^{-5}

83. $9z^{-1/2} + 2z^{1/2}$; $z^{-1/2}$

84. $3m^{2/3} - 4m^{-1/3}$; $m^{-1/3}$

85. $p^{-3/4} - 2p^{-7/4}$; $p^{-7/4}$

86. $6r^{-2/3} - 5r^{-5/3}$; $r^{-5/3}$

87. $(p + 4)^{-3/2} + (p + 4)^{-1/2} + (p + 4)^{1/2}$; $(p + 4)^{-3/2}$

88. $(3r + 1)^{-2/3} + (3r + 1)^{1/3} + (3r + 1)^{4/3}$; $(3r + 1)^{-2/3}$

Perform all indicated operations and write each answer with positive integer exponents. See Example 8.

89. $\dfrac{a^{-1} + b^{-1}}{(ab)^{-1}}$

90. $\dfrac{p^{-1} - q^{-1}}{(pq)^{-1}}$

91. $\dfrac{r^{-1} + q^{-1}}{r^{-1} - q^{-1}} \cdot \dfrac{r - q}{r + q}$

92. $\dfrac{xy^{-1} + yx^{-1}}{x^2 + y^2}$

93. $\dfrac{x - 9y^{-1}}{(x - 3y^{-1})(x + 3y^{-1})}$

94. $\dfrac{(m + n)^{-1}}{m^{-2} - n^{-2}}$

Concept Check *Answer each question.*

95. If $a^7 = 30$, what is a^{21}?

96. If $a^{-3} = .2$, what is a^6?

97. If the lengths of the sides of a cube are tripled, by what factor will the volume change?

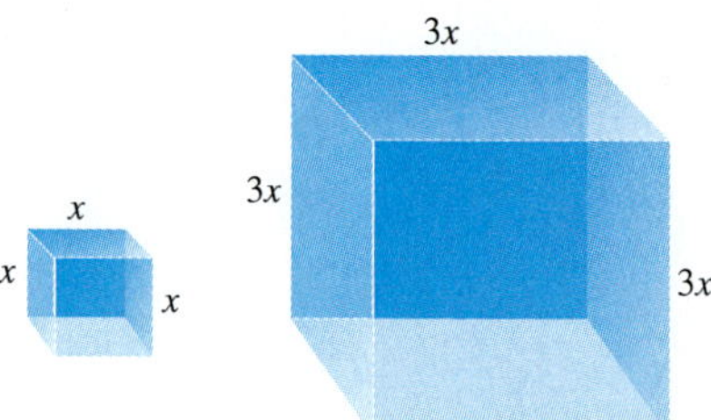

98. If the radius of a circle is doubled, by what factor will the area change?

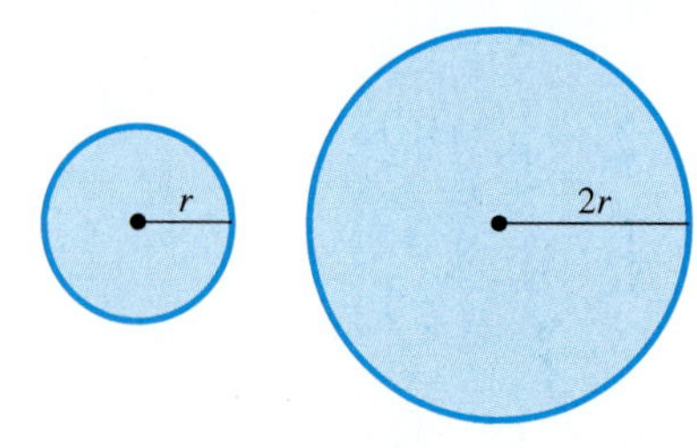

Concept Check *Calculate each value mentally.*

99. $.2^{2/3} \cdot 40^{2/3}$ **100.** $.1^{3/2} \cdot 90^{3/2}$ **101.** $\frac{2^{2/3}}{2000^{2/3}}$ **102.** $\frac{20^{3/2}}{5^{3/2}}$

R.7 Radical Expressions

Radical Notation ▪ Simplified Radicals ▪ Operations with Radicals ▪ Rationalizing Denominators

Radical Notation In the previous section we used rational exponents to express roots. An alternative (and more familiar) notation for roots is *radical notation.*

Radical Notation for $a^{1/n}$

If a is a real number, n is a positive integer, and $a^{1/n}$ is a real number, then

$$\sqrt[n]{a} = a^{1/n}.$$

Radical Notation for $a^{m/n}$

If a is a real number, m is an integer, n is a positive integer, and $\sqrt[n]{a}$ is a real number, then

$$a^{m/n} = \left(\sqrt[n]{a}\right)^m = \sqrt[n]{a^m}.$$

In the radical $\sqrt[n]{a}$, the symbol $\sqrt[n]{\ }$ is a **radical sign,** the number a is the **radicand,** and n is the **index.** We usually use the familiar notation $\sqrt{a}$ instead of $\sqrt[2]{a}$ for the square root.

For even values of n (square roots, fourth roots, and so on), when a is positive, there are two nth roots, one positive and one negative. In such cases, the notation $\sqrt[n]{a}$ represents the positive root, the **principal nth root.** We write the negative root as $-\sqrt[n]{a}$.

Looking Ahead to Calculus

In calculus, the "power rule" for derivatives requires converting radicals to rational exponents.

EXAMPLE 1 Evaluating Roots

Write each root using exponents and evaluate.

(a) $\sqrt[4]{16}$ **(b)** $-\sqrt[4]{16}$ **(c)** $\sqrt[5]{-32}$

(d) $\sqrt[3]{1000}$ **(e)** $\sqrt[6]{\frac{64}{729}}$ **(f)** $\sqrt[4]{-16}$

Solution

(a) $\sqrt[4]{16} = 16^{1/4} = 2$

(b) $-\sqrt[4]{16} = -16^{1/4} = -2$

(c) $\sqrt[5]{-32} = (-32)^{1/5} = -2$

(d) $\sqrt[3]{1000} = 1000^{1/3} = 10$

(e) $\sqrt[6]{\frac{64}{729}} = \left(\frac{64}{729}\right)^{1/6} = \frac{2}{3}$

(f) $\sqrt[4]{-16}$ is not a real number.

Now try Exercises 21, 23, 25, and 27.

EXAMPLE 2 Converting from Rational Exponents to Radicals

Write in radical form and simplify.

(a) $8^{2/3}$ **(b)** $(-32)^{4/5}$ **(c)** $-16^{3/4}$ **(d)** $x^{5/6}$

(e) $3x^{2/3}$ **(f)** $2p^{-1/2}$ **(g)** $(3a + b)^{1/4}$

Solution

(a) $8^{2/3} = (\sqrt[3]{8})^2 = 2^2 = 4$

(b) $(-32)^{4/5} = (\sqrt[5]{-32})^4 = (-2)^4 = 16$

(c) $-16^{3/4} = -(\sqrt[4]{16})^3 = -(2)^3 = -8$

(d) $x^{5/6} = \sqrt[6]{x^5}, \quad x \geq 0$

(e) $3x^{2/3} = 3\sqrt[3]{x^2}$

(f) $2p^{-1/2} = \dfrac{2}{p^{1/2}} = \dfrac{2}{\sqrt{p}}, \quad p > 0$

(g) $(3a + b)^{1/4} = \sqrt[4]{3a + b}, \quad 3a + b \geq 0$

Now try Exercises 1, 3, and 5.

CAUTION It is not possible to "distribute" exponents over a sum, so in Example 2(g), $(3a + b)^{1/4}$ *cannot be written as* $(3a)^{1/4} + b^{1/4}$. More generally,

$$\sqrt[n]{x^n + y^n} \textit{ is not equal to } x + y.$$

(For example, let $n = 2$, $x = 3$, and $y = 4$ to see this.)

EXAMPLE 3 Converting from Radicals to Rational Exponents

Write in exponential form.

(a) $\sqrt[4]{x^5}$ **(b)** $\sqrt{3y}$ **(c)** $10(\sqrt[5]{z})^2$

(d) $5\sqrt[3]{(2x^4)^7}$ **(e)** $\sqrt{p^2 + q}$

Solution

(a) $\sqrt[4]{x^5} = x^{5/4}, \quad x \geq 0$

(b) $\sqrt{3y} = (3y)^{1/2}, \quad y \geq 0$

(c) $10(\sqrt[5]{z})^2 = 10z^{2/5}$

(d) $5\sqrt[3]{(2x^4)^7} = 5(2x^4)^{7/3} = 5 \cdot 2^{7/3}x^{28/3}$

(e) $\sqrt{p^2 + q} = (p^2 + q)^{1/2}, \quad p^2 + q \geq 0$

Now try Exercises 7 and 9.

We *cannot* simply write $\sqrt{x^2} = x$ for all real numbers x. For example, if $x = -5$, then

$$\sqrt{x^2} = \sqrt{(-5)^2} = \sqrt{25} = 5 \neq x.$$

To take care of the fact that a negative value of x can produce a positive result, we use absolute value. For any real number a,

$$\sqrt{a^2} = |a|.$$

For example, $\sqrt{(-9)^2} = |-9| = 9$ and $\sqrt{13^2} = |13| = 13.$

We can generalize this result to any *even* nth root.

$\sqrt[n]{a^n}$

If n is an *even* positive integer, then $\sqrt[n]{a^n} = |a|$.

If n is an *odd* positive integer, then $\sqrt[n]{a^n} = a$.

EXAMPLE 4 Using Absolute Value to Simplify Roots

Simplify each expression.

(a) $\sqrt{p^4}$ **(b)** $\sqrt[4]{p^4}$ **(c)** $\sqrt{16m^8r^6}$ **(d)** $\sqrt[6]{(-2)^6}$

(e) $\sqrt[5]{m^5}$ **(f)** $\sqrt{(2k + 3)^2}$ **(g)** $\sqrt{x^2 - 4x + 4}$

Solution

(a) $\sqrt{p^4} = \sqrt{(p^2)^2} = |p^2| = p^2$ **(b)** $\sqrt[4]{p^4} = |p|$

(c) $\sqrt{16m^8r^6} = |4m^4r^3| = 4m^4|r^3|$ **(d)** $\sqrt[6]{(-2)^6} = |-2| = 2$

(e) $\sqrt[5]{m^5} = m$ **(f)** $\sqrt{(2k + 3)^2} = |2k + 3|$

(g) $\sqrt{x^2 - 4x + 4} = \sqrt{(x - 2)^2} = |x - 2|$

Now try Exercises 15, 17, and 19.

NOTE When working with variable radicands, we usually assume that all variables in radicands represent only nonnegative real numbers.

Three key rules for working with radicals are given below. These rules are just the power rules for exponents written in radical notation.

Rules for Radicals

For all real numbers a and b, and positive integers m and n for which the indicated roots are real numbers:

Rule	Description
Product rule $\sqrt[n]{a} \cdot \sqrt[n]{b} = \sqrt[n]{ab}$	The product of two radicals is the radical of the product.
Quotient rule $\sqrt[n]{\frac{a}{b}} = \frac{\sqrt[n]{a}}{\sqrt[n]{b}}$ $(b \neq 0)$	The radical of a quotient is the quotient of the radicals.
Power rule $\sqrt[m]{\sqrt[n]{a}} = \sqrt[mn]{a}$.	The index of the radical of a radical is the product of their indexes.

EXAMPLE 5 Using the Rules for Radicals to Simplify Radical Expressions

Apply the rules for radicals to the following.

(a) $\sqrt{6} \cdot \sqrt{54}$ **(b)** $\sqrt[3]{m} \cdot \sqrt[3]{m^2}$ **(c)** $\sqrt{\frac{7}{64}}$

(d) $\sqrt[4]{\frac{a}{b^4}}$ **(e)** $\sqrt[7]{\sqrt[3]{2}}$ **(f)** $\sqrt[4]{\sqrt{3}}$

Solution

(a) $\sqrt{6} \cdot \sqrt{54} = \sqrt{6 \cdot 54}$ Product rule

$= \sqrt{324} = 18$

(b) $\sqrt[3]{m} \cdot \sqrt[3]{m^2} = \sqrt[3]{m^3} = m$

(c) $\sqrt{\frac{7}{64}} = \frac{\sqrt{7}}{\sqrt{64}} = \frac{\sqrt{7}}{8}$ Quotient rule

(d) $\sqrt[4]{\frac{a}{b^4}} = \frac{\sqrt[4]{a}}{\sqrt[4]{b^4}} = \frac{\sqrt[4]{a}}{b}, \quad a \geq 0, b > 0$

(e) $\sqrt[7]{\sqrt[3]{2}} = \sqrt[21]{2}$ Power rule

(f) $\sqrt[4]{\sqrt{3}} = \sqrt[4 \cdot 2]{3} = \sqrt[8]{3}$

Now try Exercises 29, 33, 37, and 57.

NOTE In Example 5, converting to rational exponents would show why these rules work. For example, in part (e)

$$\sqrt[7]{\sqrt[3]{2}} = (2^{1/3})^{1/7} = 2^{(1/3)(1/7)} = 2^{1/21} = \sqrt[21]{2}.$$

Simplified Radicals In working with numbers, we prefer to write a number in its simplest form. For example, $\frac{10}{2}$ is written as 5 and $-\frac{9}{6}$ is written as $-\frac{3}{2}$. Similarly, expressions with radicals should be written in their simplest forms.

Simplified Radicals

An expression with radicals is simplified when all of the following conditions are satisfied.

1. The radicand has no factor raised to a power greater than or equal to the index.
2. The radicand has no fractions.
3. No denominator contains a radical.
4. Exponents in the radicand and the index of the radical have no common factor.
5. All indicated operations have been performed (if possible).

EXAMPLE 6 Simplifying Radicals

Simplify each radical.

(a) $\sqrt{175}$ **(b)** $-3\sqrt[5]{32}$ **(c)** $\sqrt[3]{81x^5y^7z^6}$

Solution

(a) $\sqrt{175} = \sqrt{25 \cdot 7} = \sqrt{25} \cdot \sqrt{7} = 5\sqrt{7}$

(b) $-3\sqrt[5]{32} = -3\sqrt[5]{2^5} = -3 \cdot 2 = -6$

(c) $\sqrt[3]{81x^5y^7z^6} = \sqrt[3]{27 \cdot 3 \cdot x^3 \cdot x^2 \cdot y^6 \cdot y \cdot z^6}$ Factor. (Section R.4)

$= \sqrt[3]{(27x^3y^6z^6)(3x^2y)}$ Group all perfect cubes.

$= 3xy^2z^2\sqrt[3]{3x^2y}$ Remove all perfect cubes from the radical.

Now try Exercise 43.

Operations with Radicals Radicals with the same radicand and the same index, such as $3\sqrt[4]{11pq}$ and $-7\sqrt[4]{11pq}$, are called **like radicals.** On the other hand, examples of *unlike radicals* are

$2\sqrt{5}$ and $2\sqrt{3}$, Radicands are different.

as well as $2\sqrt{3}$ and $2\sqrt[3]{3}$. Indexes are different.

We add or subtract like radicals by using the distributive property. Only like radicals can be combined. Sometimes we need to simplify radicals before adding or subtracting.

EXAMPLE 7 Adding and Subtracting Like Radicals

Add or subtract as indicated. Assume all variables represent positive real numbers.

(a) $3\sqrt[4]{11pq} + \left(-7\sqrt[4]{11pq}\right)$ **(b)** $\sqrt{98x^3y} + 3x\sqrt{32xy}$

(c) $\sqrt[3]{64m^4n^5} - \sqrt[3]{-27m^{10}n^{14}}$

Solution

(a) $3\sqrt[4]{11pq} + \left(-7\sqrt[4]{11pq}\right) = -4\sqrt[4]{11pq}$

(b) $\sqrt{98x^3y} + 3x\sqrt{32xy} = \sqrt{49 \cdot 2 \cdot x^2 \cdot x \cdot y} + 3x\sqrt{16 \cdot 2 \cdot x \cdot y}$ Factor.

$= 7x\sqrt{2xy} + 3x(4)\sqrt{2xy}$ Remove all perfect squares from the radicals.

$= 7x\sqrt{2xy} + 12x\sqrt{2xy}$

$= 19x\sqrt{2xy}$ Distributive property (Section R.1)

(c) $\sqrt[3]{64m^4n^5} - \sqrt[3]{-27m^{10}n^{14}} = \sqrt[3]{(64m^3n^3)(mn^2)} - \sqrt[3]{(-27m^9n^{12})(mn^2)}$

$= 4mn\sqrt[3]{mn^2} - (-3)m^3n^4\sqrt[3]{mn^2}$

$= 4mn\sqrt[3]{mn^2} + 3m^3n^4\sqrt[3]{mn^2}$

Now try Exercises 59, 61, and 65.

If the index of the radical and an exponent in the radicand have a common factor, we can simplify the radical by writing it in exponential form, simplifying the rational exponent, then writing the result as a radical again.

EXAMPLE 8 Simplifying Radicals by Writing Them with Rational Exponents

Simplify each radical.

(a) $\sqrt[6]{3^2}$ **(b)** $\sqrt[6]{x^{12}y^3}$ **(c)** $\sqrt[9]{\sqrt{6^3}}$

Solution

(a) $\sqrt[6]{3^2} = 3^{2/6} = 3^{1/3} = \sqrt[3]{3}$

(b) $\sqrt[6]{x^{12}y^3} = (x^{12}y^3)^{1/6} = x^2y^{3/6} = x^2y^{1/2} = x^2\sqrt{y}, \quad y \geq 0$

(c) $\sqrt[9]{\sqrt{6^3}} = \sqrt[9]{6^{3/2}} = (6^{3/2})^{1/9} = 6^{1/6} = \sqrt[6]{6}$

Now try Exercises 55 and 57.

In Example 8(a), we simplified $\sqrt[6]{3^2}$ as $\sqrt[3]{3}$. However, to simplify $(\sqrt[6]{x})^2$, the variable x must be nonnegative. For example, consider the statement

$$(-8)^{2/6} = [(-8)^{1/6}]^2.$$

This result is not a real number, since $(-8)^{1/6}$ is not defined. On the other hand,

$$(-8)^{1/3} = -2.$$

Here, even though $\frac{2}{6} = \frac{1}{3}$,

$$(\sqrt[6]{x})^2 \neq \sqrt[3]{x}.$$

If a is nonnegative, then it is always true that $a^{m/n} = a^{mp/(np)}$. Simplifying rational exponents on negative bases must be considered case by case.

Multiplying radical expressions is much like multiplying polynomials.

EXAMPLE 9 Multiplying Radical Expressions

Find each product.

(a) $(\sqrt{7} - \sqrt{10})(\sqrt{7} + \sqrt{10})$ **(b)** $(\sqrt{2} + 3)(\sqrt{8} - 5)$

Solution

(a) $(\sqrt{7} - \sqrt{10})(\sqrt{7} + \sqrt{10}) = (\sqrt{7})^2 - (\sqrt{10})^2$ Product of the sum and difference of two terms (Section R.3)

$$= 7 - 10$$

$$= -3$$

(b) $\left(\sqrt{2}+3\right)\left(\sqrt{8}-5\right) = \sqrt{2}\left(\sqrt{8}\right) - \sqrt{2}(5) + 3\sqrt{8} - 3(5)$ FOIL (Section R.3)

$= \sqrt{16} - 5\sqrt{2} + 3\left(2\sqrt{2}\right) - 15$ Multiply; $\sqrt{8} = 2\sqrt{2}$.

$= 4 - 5\sqrt{2} + 6\sqrt{2} - 15$

$= -11 + \sqrt{2}$ Combine terms.

Now try Exercises 67 and 73.

Rationalizing Denominators The third condition for a simplified radical requires that no denominator contain a radical. We achieve this by **rationalizing the denominator,** that is, multiplying by a form of 1, as explained in Example 10.

EXAMPLE 10 Rationalizing Denominators

Rationalize each denominator.

(a) $\dfrac{4}{\sqrt{3}}$ **(b)** $\sqrt[4]{\dfrac{3}{5}}$

Solution

(a) To rationalize the denominator in $\dfrac{4}{\sqrt{3}}$, multiply by $\dfrac{\sqrt{3}}{\sqrt{3}}$ (which equals 1) so that the denominator of the product is a rational number.

$$\frac{4}{\sqrt{3}} \cdot \frac{\sqrt{3}}{\sqrt{3}} = \frac{4\sqrt{3}}{3}$$

(b) $\sqrt[4]{\dfrac{3}{5}} = \dfrac{\sqrt[4]{3}}{\sqrt[4]{5}}$ Quotient rule

The denominator will be a rational number if it equals $\sqrt[4]{5^4}$. That is, four factors of 5 are needed under the radical. Since $\sqrt[4]{5}$ has just one factor of 5, three additional factors are needed, so multiply by $\dfrac{\sqrt[4]{5^3}}{\sqrt[4]{5^3}}$.

$$\frac{\sqrt[4]{3}}{\sqrt[4]{5}} = \frac{\sqrt[4]{3} \cdot \sqrt[4]{5^3}}{\sqrt[4]{5} \cdot \sqrt[4]{5^3}} = \frac{\sqrt[4]{3 \cdot 5^3}}{\sqrt[4]{5^4}} = \frac{\sqrt[4]{375}}{5}$$

Now try Exercises 47 and 51.

EXAMPLE 11 Simplifying Radical Expressions with Fractions

Simplify each expression. Assume all variables represent positive real numbers.

(a) $\dfrac{\sqrt[4]{xy^3}}{\sqrt[4]{x^3y^2}}$ **(b)** $\sqrt[3]{\dfrac{5}{x^6}} - \sqrt[3]{\dfrac{4}{x^9}}$

Solution

(a)
$$\frac{\sqrt[4]{xy^3}}{\sqrt[4]{x^3y^2}} = \sqrt[4]{\frac{xy^3}{x^3y^2}} \quad \text{Quotient rule}$$
$$= \sqrt[4]{\frac{y}{x^2}} \quad \text{Simplify radicand.}$$
$$= \frac{\sqrt[4]{y}}{\sqrt[4]{x^2}} \quad \text{Quotient rule}$$
$$= \frac{\sqrt[4]{y}}{\sqrt[4]{x^2}} \cdot \frac{\sqrt[4]{x^2}}{\sqrt[4]{x^2}} \quad \sqrt[4]{x^2} \cdot \sqrt[4]{x^2} = \sqrt[4]{x^4} = x$$
$$= \frac{\sqrt[4]{x^2y}}{x} \quad \text{Product rule}$$

(b)
$$\sqrt[3]{\frac{5}{x^6}} - \sqrt[3]{\frac{4}{x^9}} = \frac{\sqrt[3]{5}}{\sqrt[3]{x^6}} - \frac{\sqrt[3]{4}}{\sqrt[3]{x^9}} \quad \text{Quotient rule}$$
$$= \frac{\sqrt[3]{5}}{x^2} - \frac{\sqrt[3]{4}}{x^3} \quad \text{Simplify the denominators.}$$
$$= \frac{x\sqrt[3]{5}}{x^3} - \frac{\sqrt[3]{4}}{x^3} \quad \text{Write with a common denominator. (Section R.5)}$$
$$= \frac{x\sqrt[3]{5} - \sqrt[3]{4}}{x^3} \quad \text{Subtract the numerators.}$$

Now try Exercises 75 and 77.

Looking Ahead to Calculus

Another standard problem in calculus is investigating the value that an expression such as $\frac{\sqrt{x^2+9}-3}{x^2}$ approaches as x approaches 0. This cannot be done by simply substituting 0 for x, since the result is $\frac{0}{0}$. However, by rationalizing the *numerator,* we can show that for $x \neq 0$ the expression is equivalent to $\frac{1}{\sqrt{x^2+9}+3}$. Then, by substituting 0 for x, we find that the original expression approaches $\frac{1}{6}$ as x approaches 0.

In Example 9(a), we saw that
$$\left(\sqrt{7} - \sqrt{10}\right)\left(\sqrt{7} + \sqrt{10}\right) = -3,$$
a rational number. This suggests a way to rationalize a denominator that is a binomial in which one or both terms is a radical. The expressions $a - b$ and $a + b$ are called **conjugates.**

EXAMPLE 12 Rationalizing a Binomial Denominator

Rationalize the denominator of $\frac{1}{1-\sqrt{2}}$.

Solution

Multiply both the numerator and the denominator by the conjugate of the denominator, $1 + \sqrt{2}$.
$$\frac{1}{1-\sqrt{2}} = \frac{1\left(1+\sqrt{2}\right)}{\left(1-\sqrt{2}\right)\left(1+\sqrt{2}\right)} = \frac{1+\sqrt{2}}{1-2} = -1-\sqrt{2}$$

Now try Exercise 83.

R.7 Exercises

Concept Check *Match the rational exponent expression in Column I for Exercises 1 and 2 with the equivalent radical expression in Column II. Assume that x is not 0. See Example 2.*

I		II	
1. (a) $(-3x)^{1/3}$	**2. (a)** $-3x^{1/3}$	**A.** $\frac{3}{\sqrt[3]{x}}$	**B.** $-3\sqrt[3]{x}$
(b) $(-3x)^{-1/3}$	**(b)** $-3x^{-1/3}$	**C.** $\frac{1}{\sqrt[3]{3x}}$	**D.** $\frac{-3}{\sqrt[3]{x}}$
(c) $(3x)^{1/3}$	**(c)** $3x^{-1/3}$	**E.** $3\sqrt[3]{x}$	**F.** $\sqrt[3]{-3x}$
(d) $(3x)^{-1/3}$	**(d)** $3x^{1/3}$	**G.** $\sqrt[3]{3x}$	**H.** $\frac{1}{\sqrt[3]{-3x}}$

Write in radical form. Assume all variables represent positive real numbers. See Example 2.

3. $m^{2/3}$ **4.** $p^{5/4}$ **5.** $(2m + p)^{2/3}$ **6.** $(5r + 3t)^{4/7}$

Write in exponential form. Assume all variables represent positive real numbers. See Example 3.

7. $\sqrt[5]{k^2}$ **8.** $-\sqrt[4]{z^5}$ **9.** $-3\sqrt{5p^3}$ **10.** $m\sqrt{2y^5}$

Concept Check *Answer each question.*

11. For which of the following cases is $\sqrt{ab} = \sqrt{a} \cdot \sqrt{b}$ a true statement?

A. a and b both positive **B.** a and b both negative

12. For what positive integers n greater than or equal to 2 is $\sqrt[n]{a^n} = a$ always a true statement?

13. For what values of x is $\sqrt{9ax^2} = 3x\sqrt{a}$ a true statement? Assume $a \geq 0$.

14. Which of the following expressions is *not* simplified? Give the simplified form.

A. $\sqrt[3]{2y}$ **B.** $\frac{\sqrt{5}}{2}$ **C.** $\sqrt[4]{m^3}$ **D.** $\sqrt{\frac{3}{4}}$

Simplify each expression. See Example 4.

15. $\sqrt{(-5)^2}$ **16.** $\sqrt[6]{x^6}$ **17.** $\sqrt{25k^4m^2}$

18. $\sqrt[4]{81p^{12}q^4}$ **19.** $\sqrt{(4x - y)^2}$ **20.** $\sqrt[4]{(5 + 2m)^4}$

Simplify each expression. Assume all variables represent positive real numbers. See Examples 1, 4–6, and 8–11.

21. $\sqrt[3]{125}$ **22.** $\sqrt[4]{81}$ **23.** $\sqrt[4]{-64}$ **24.** $\sqrt[6]{-64}$

25. $\sqrt[3]{81}$ **26.** $\sqrt[3]{250}$ **27.** $-\sqrt[4]{32}$ **28.** $-\sqrt[4]{243}$

29. $\sqrt{14} \cdot \sqrt{3pqr}$ **30.** $\sqrt{7} \cdot \sqrt{5xt}$ **31.** $\sqrt[3]{7x} \cdot \sqrt[3]{2y}$ **32.** $\sqrt[3]{9x} \cdot \sqrt[3]{4y}$

33. $-\sqrt{\frac{9}{25}}$ **34.** $-\sqrt{\frac{12}{49}}$ **35.** $-\sqrt[3]{\frac{5}{8}}$ **36.** $\sqrt[4]{\frac{3}{16}}$

37. $\sqrt[4]{\frac{m}{n^4}}$ **38.** $\sqrt[6]{\frac{r}{s^6}}$ **39.** $3\sqrt[5]{-3125}$ **40.** $5\sqrt[3]{343}$

41. $\sqrt[3]{16(-2)^4(2)^8}$ **42.** $\sqrt[3]{25(3)^4(5)^3}$ **43.** $\sqrt{8x^5z^8}$ **44.** $\sqrt{24m^6n^5}$

45. $\sqrt[4]{x^4+y^4}$ **46.** $\sqrt[3]{27+a^3}$ **47.** $\sqrt{\frac{2}{3x}}$ **48.** $\sqrt{\frac{5}{3p}}$

49. $\sqrt{\frac{x^5y^3}{z^2}}$ **50.** $\sqrt{\frac{g^3h^5}{r^3}}$ **51.** $\sqrt[3]{\frac{8}{x^2}}$ **52.** $\sqrt[3]{\frac{9}{16p^4}}$

53. $\sqrt[4]{\frac{g^3h^5}{9r^6}}$ **54.** $\sqrt[4]{\frac{32x^5}{y^5}}$ **55.** $\sqrt[8]{3^4}$ **56.** $\sqrt[9]{5^3}$

57. $\sqrt[3]{\sqrt{4}}$ **58.** $\sqrt[4]{\sqrt[3]{2}}$

Simplify each expression, assuming all variables represent positive real numbers. See Examples 7, 9, and 11.

59. $5\sqrt{6}+2\sqrt{10}$ **60.** $3\sqrt{11}-5\sqrt{13}$

61. $8\sqrt{2x}-\sqrt{8x}+\sqrt{72x}$ **62.** $4\sqrt{18k}-\sqrt{72k}+\sqrt{50k}$

63. $2\sqrt[3]{3}+4\sqrt[3]{24}-\sqrt[3]{81}$ **64.** $\sqrt[3]{32}-5\sqrt[3]{4}+2\sqrt[3]{108}$

65. $\sqrt[4]{81x^6y^3}-\sqrt[4]{16x^{10}y^3}$ **66.** $\sqrt[4]{256x^5y^6}+\sqrt[4]{625x^9y^2}$

67. $(\sqrt{2}+3)(\sqrt{2}-3)$ **68.** $(\sqrt{5}+\sqrt{2})(\sqrt{5}-\sqrt{2})$

69. $(\sqrt[3]{11}-1)(\sqrt[3]{11^2}+\sqrt[3]{11}+1)$ **70.** $(\sqrt[3]{7}+3)(\sqrt[3]{7^2}-3\sqrt[3]{7}+9)$

71. $(\sqrt{3}+\sqrt{8})^2$ **72.** $(\sqrt{2}-1)^2$

73. $(3\sqrt{2}+\sqrt{3})(2\sqrt{3}-\sqrt{2})$ **74.** $(4\sqrt{5}-1)(3\sqrt{5}+2)$

75. $\frac{\sqrt[3]{mn}\cdot\sqrt[3]{m^2}}{\sqrt[3]{n^2}}$ **76.** $\frac{\sqrt[4]{8m^2n^3}\cdot\sqrt[4]{2m^2}}{\sqrt[4]{32m^4n^3}}$ **77.** $\sqrt[3]{\frac{2}{x^6}}-\sqrt[3]{\frac{5}{x^9}}$

78. $\sqrt[4]{\frac{7}{t^{12}}}+\sqrt[4]{\frac{9}{t^4}}$ **79.** $\frac{1}{\sqrt{2}}+\frac{3}{\sqrt{8}}+\frac{1}{\sqrt{32}}$ **80.** $\frac{5}{\sqrt[3]{2}}-\frac{2}{\sqrt[3]{16}}+\frac{1}{\sqrt[3]{54}}$

81. $\frac{-4}{\sqrt[3]{3}}+\frac{1}{\sqrt[3]{24}}-\frac{2}{\sqrt[3]{81}}$

Rationalize the denominator of each radical expression. Assume all variables represent nonnegative numbers and that no denominators are 0. *See Example 12.*

82. $\frac{\sqrt{3}}{\sqrt{5}+\sqrt{3}}$ **83.** $\frac{\sqrt{7}}{\sqrt{3}-\sqrt{7}}$ **84.** $\frac{1+\sqrt{3}}{3\sqrt{5}+2\sqrt{3}}$

85. $\frac{\sqrt{7}-1}{2\sqrt{7}+4\sqrt{2}}$ **86.** $\frac{p}{\sqrt{p}+2}$ **87.** $\frac{\sqrt{r}}{3-\sqrt{r}}$

88. $\frac{a}{\sqrt{a+b}-1}$ **89.** $\frac{3m}{2+\sqrt{m+n}}$

Solve each applied problem.

90. *(Modeling)* ***Rowing Speed*** Olympic rowing events have one-, two-, four-, or eight-person crews, with each person pulling a single oar. Increasing the size of the crew increases the speed of the boat. An analysis of 1980 Olympic rowing events concluded that the approximate speed, s, of the boat (in feet per second) was given by the formula

$$s = 15.18\sqrt[9]{n},$$

where n is the number of oarsmen. Estimate the speed of a boat with a four-person crew. (*Source:* Townend, M. Stewart, *Mathematics in Sport,* Chichester, Ellis Horwood Limited, 1984.)

(Modeling) Windchill *In Section R.2, Exercises 51–54, we used a table to give windchill for various combinations of temperature and wind speed. The National Weather Service uses the formula*

$$\text{Windchill temperature} = 35.74 + .6215T - 35.75V^{.16} + .4275TV^{.16},$$

where T is the temperature in °F and V is the wind speed in miles per hour, to calculate windchill. (Source: National Oceanic and Atmospheric Administration, National Weather Service.)

Use the formula to calculate the windchill to the nearest tenth of a degree given the following conditions. Compare your answers with the appropriate entries in the table in Section R.2.

91. 30°F, 15 mph wind

92. 10°F, 30 mph wind

Concept Check *Simplify each expression mentally.*

93. $\dfrac{\sqrt[3]{54}}{\sqrt[3]{2}}$

94. $\sqrt[4]{8} \cdot \sqrt[4]{2}$

95. $\sqrt{.1} \cdot \sqrt{40}$

96. $\dfrac{\sqrt[5]{320}}{\sqrt[5]{10}}$

97. $\sqrt[6]{2} \cdot \sqrt[6]{4} \cdot \sqrt[6]{8}$

Calculators are a wonderful tool for approximating radicals; however, we must use caution interpreting the results. For example, the screen in Figure A seems to indicate that π and $\sqrt[4]{\frac{2143}{22}}$ are exactly equal, since the eight decimal values given by the calculator agree. However, as shown in Figure B using one more decimal place in the display, they differ in the ninth decimal place. The radical expression is a very good approximation for π, but it is still only an approximation.

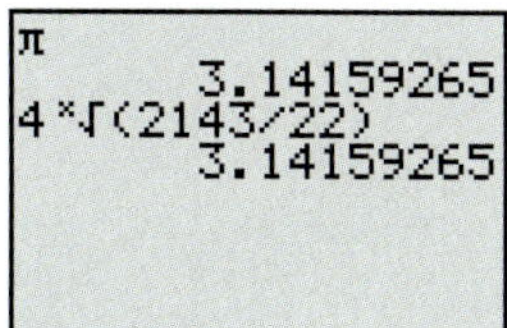

Figure A

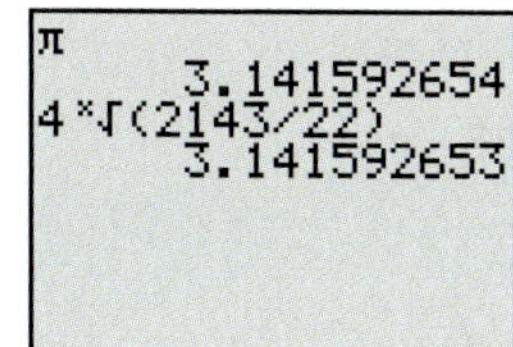

Figure B

Use your calculator to answer each question. Refer to the display for π in Figure B. (*Source:* Eves, H., *An Introduction to the History of Mathematics,* © 1990 Brooks/Cole Publishing. Used with permission of Thomson Learning.)

98. A value for π that the Greeks used circa A.D. 150 is equivalent to $\frac{377}{120}$. In which decimal place does this value first differ from π?

99. The Chinese of the fifth century used $\frac{355}{113}$ as an approximation for π. How many decimal places of accuracy does this fraction give?

100. The Hindu mathematician Bhaskara used $\frac{3927}{1250}$ as an approximation for π circa A.D. 1150. In which decimal place does this value first differ from π?

Chapter R Summary

KEY TERMS

R.1 factor
exponent
base
power or exponential expression (exponential)
identity element for addition (additive identity)
identity element for multiplication (multiplicative identity)
additive inverse (negative)
multiplicative inverse (reciprocal)

R.2 coordinate
coordinate system
absolute value

R.3 algebraic expression
term
coefficient
like terms
polynomial
polynomial in x
degree of a term
degree of a polynomial
trinomial
binomial
monomial

R.4 factoring
factored form
prime polynomial
factored completely
factoring by grouping

R.5 rational expression
domain of a rational expression
lowest terms
complex fraction

R.7 radicand
index of a radical
principal nth root
like radicals
rationalizing the denominator
conjugates

NEW SYMBOLS

$\{x \mid x$ **has property** $p\}$	set-builder notation
a^n	n factors of a
$<$	is less than
$>$	is greater than
$\leq$	is less than or equal to
$\geq$	is greater than or equal to
$\lvert a \rvert$	absolute value of a
$\sqrt{}$	radical sign

QUICK REVIEW

CONCEPTS	EXAMPLES
R.1 Real Numbers and Their Properties	
SETS OF NUMBERS	
Natural Numbers $\{1, 2, 3, 4, \ldots\}$	5, 17, 142
Whole Numbers $\{0, 1, 2, 3, 4, \ldots\}$	0, 27, 96
Integers $\{\ldots, -3, -2, -1, 0, 1, 2, 3, \ldots\}$	$-24, 0, 19$
Rational Numbers $\left\{\frac{p}{q} \,\middle\vert\, p \text{ and } q \text{ are integers and } q \neq 0\right\}$	$-\frac{3}{4}, -.28, 0, 7, \frac{9}{16}, .6\overline{6}$
Irrational Numbers $\{x \mid x \text{ is real but not rational}\}$	$-\sqrt{15}, .101101110\ldots, \sqrt{2}, \pi$
Real Numbers $\{x \mid x$ corresponds to a point on a number line$\}$	$-46, .7, \pi, \sqrt{19}, \frac{8}{5}$
PROPERTIES OF REAL NUMBERS	
For all real numbers a, b, and c:	
Closure Properties	
$a + b$ is a real number.	$1 + \sqrt{2}$ is a real number.
ab is a real number.	$3\sqrt{7}$ is a real number.

CONCEPTS	EXAMPLES
Commutative Properties	
$a + b = b + a$	$5 + 18 = 18 + 5$
$ab = ba$	$-4 \cdot 8 = 8 \cdot (-4)$
Associative Properties	
$(a + b) + c = a + (b + c)$	$[6 + (-3)] + 5 = 6 + (-3 + 5)$
$(ab)c = a(bc)$	$(7 \cdot 6)20 = 7(6 \cdot 20)$
Identity Properties	
There exists a unique real number 0 such that	
$a + 0 = a$ and $0 + a = a$.	$145 + 0 = 145$ and $0 + 145 = 145$
There exists a unique real number 1 such that	
$a \cdot 1 = a$ and $1 \cdot a = a$.	$-60 \cdot 1 = -60$ and $1 \cdot (-60) = -60$
Inverse Properties	
There exists a unique real number $-a$ such that	
$a + (-a) = 0$ and $-a + a = 0$.	$17 + (-17) = 0$ and $-17 + 17 = 0$
If $a \neq 0$, there exists a unique real number $\frac{1}{a}$ such that	
$a \cdot \frac{1}{a} = 1$ and $\frac{1}{a} \cdot a = 1$.	$22 \cdot \frac{1}{22} = 1$ and $\frac{1}{22} \cdot 22 = 1$
Distributive Properties	
$a(b + c) = ab + ac$	$3(5 + 8) = 3 \cdot 5 + 3 \cdot 8$
$a(b - c) = ab - ac$	$6(4 - 2) = 6 \cdot 4 - 6 \cdot 2$

R.2 Order and Absolute Value

$a > b$ if a is to the right of b on a number line.	$7 > -5$
$a < b$ if a is to the left of b on a number line.	$0 < 15$
$\lvert a \rvert = \begin{cases} a & \text{if } a \geq 0 \\ -a & \text{if } a < 0 \end{cases}$	$\lvert 3 \rvert = 3$ and $\lvert -3 \rvert = 3$

R.3 Polynomials

SPECIAL PRODUCTS	
Product of the Sum and Difference of Two Terms $(x + y)(x - y) = x^2 - y^2$	$(7 - x)(7 + x) = 7^2 - x^2 = 49 - x^2$
Square of a Binomial $(x + y)^2 = x^2 + 2xy + y^2$	$(3a + b)^2 = (3a)^2 + 2(3a)(b) + b^2$ $= 9a^2 + 6ab + b^2$
$(x - y)^2 = x^2 - 2xy + y^2$	$(2m - 5)^2 = (2m)^2 - 2(2m)(5) + 5^2$ $= 4m^2 - 20m + 25$

CONCEPTS	EXAMPLES

R.4 Factoring Polynomials

FACTORING PATTERNS

Concepts	Examples
Difference of Squares $x^2 - y^2 = (x+y)(x-y)$	$4t^2 - 9 = (2t+3)(2t-3)$
Perfect Square Trinomial $x^2 + 2xy + y^2 = (x+y)^2$	$p^2 + 4pq + 4q^2 = (p+2q)^2$
$x^2 - 2xy + y^2 = (x-y)^2$	$9m^2 - 12mn + 4n^2 = (3m-2n)^2$
Difference of Cubes $x^3 - y^3 = (x-y)(x^2 + xy + y^2)$	$r^3 - 8 = (r-2)(r^2 + 2r + 4)$
Sum of Cubes $x^3 + y^3 = (x+y)(x^2 - xy + y^2)$	$27c^3 + 64 = (3c+4)(9c^2 - 12c + 16)$

R.5 Rational Expressions

Operations

For fractions $\frac{a}{b}$ and $\frac{c}{d}$ $(b \neq 0, d \neq 0)$,

$$\frac{a}{b} \pm \frac{c}{d} = \frac{ad \pm bc}{bd}$$

$$\frac{2}{x} + \frac{5}{y} = \frac{2y + 5x}{xy} \qquad \frac{x}{6} - \frac{2y}{5} = \frac{5x - 12y}{30}$$

$$\frac{a}{b} \cdot \frac{c}{d} = \frac{ac}{bd} \quad \text{and} \quad \frac{a}{b} \div \frac{c}{d} = \frac{ad}{bc} \quad (c \neq 0).$$

$$\frac{3}{q} \cdot \frac{3}{2p} = \frac{9}{2pq} \qquad \frac{z}{4} \div \frac{z}{2t} = \frac{z}{4} \cdot \frac{2t}{z} = \frac{2zt}{4z} = \frac{t}{2}$$

R.6 Rational Exponents

Rules for Exponents

Let r and s be rational numbers. The following results are valid for all positive numbers a and b.

$$a^r \cdot a^s = a^{r+s} \qquad (ab)^r = a^r \cdot b^r \qquad (a^r)^s = a^{rs}$$

$$6^2 \cdot 6^3 = 6^5 \qquad (3x)^4 = 3^4x^4 \qquad (m^2)^3 = m^6$$

$$a^{-r} = \frac{1}{a^r} \qquad \left(\frac{a}{b}\right)^r = \frac{a^r}{b^r} \qquad a^{-r} = \frac{1}{a^r}$$

$$\frac{p^5}{p^2} = p^3 \qquad \left(\frac{x}{3}\right)^2 = \frac{x^2}{3^2} \qquad 4^{-3} = \frac{1}{4^3}$$

R.7 Radical Expressions

Radical Notation

If a is a real number, n is a positive integer, and $a^{1/n}$ is defined, then

$$\sqrt[n]{a} = a^{1/n}.$$

$$\sqrt[4]{16} = 16^{1/4} = 2$$

If m is an integer, n is a positive integer, and a is a real number for which $\sqrt[n]{a}$ is defined, then

$$a^{m/n} = \left(\sqrt[n]{a}\right)^m = \sqrt[n]{a^m}.$$

$$8^{2/3} = \left(\sqrt[3]{8}\right)^2 = \sqrt[3]{8^2} = 4$$

Operations

Operations with radical expressions are much like operations with polynomials.

$$\sqrt{8x} + \sqrt{32x} = 2\sqrt{2x} + 4\sqrt{2x} = 6\sqrt{2x}$$

$$\left(\sqrt{5} - \sqrt{3}\right)\left(\sqrt{5} + \sqrt{3}\right) = 5 - 3 = 2$$

$$\left(\sqrt{2} + \sqrt{7}\right)\left(\sqrt{3} - \sqrt{6}\right)$$
$$= \sqrt{6} - 2\sqrt{3} + \sqrt{21} - \sqrt{42} \quad \text{FOIL; } \sqrt{12} = 2\sqrt{3}$$

Rationalize the denominator by multiplying numerator and denominator by a form of 1.

$$\frac{\sqrt{7y}}{\sqrt{5}} = \frac{\sqrt{7y}}{\sqrt{5}} \cdot \frac{\sqrt{5}}{\sqrt{5}} = \frac{\sqrt{35y}}{5}$$

Chapter R Review Exercises

Let set $K = \left\{-12, -6, -.9, -\sqrt{7}, -\sqrt{4}, 0, \frac{1}{8}, \frac{\pi}{4}, 6, \sqrt{11}\right\}$. *List all elements of K that belong to each set.*

1. Integers

2. Rational numbers

For Exercises 3–5, choose all words from the following list that apply.

natural number, whole number, integer, rational number, irrational number, real number

3. 0

4. $-\sqrt{36}$

5. $\dfrac{4\pi}{5}$

Write each algebraic identity (true statement) as a complete English sentence without using the names of the variables. For instance, $z(x + y) = zx + zy$ *can be stated as "The multiple of a sum is the sum of the multiples."*

6. $a(b - c) = ab - ac$

7. $\dfrac{1}{xy} = \dfrac{1}{x} \cdot \dfrac{1}{y}$

8. $a^2 - b^2 = (a + b)(a - b)$

9. $(ab)^n = a^n b^n$

10. $|st| = |s| \cdot |t|$

Identify by name each property illustrated.

11. $8(5 + 9) = (5 + 9)8$

12. $4 \cdot 6 + 4 \cdot 12 = 4(6 + 12)$

13. $3 \cdot (4 \cdot 2) = (3 \cdot 4) \cdot 2$

14. $-8 + 8 = 0$

15. $(9 + p) + 0 = 9 + p$

16. Use the distributive property to write the product in part (a) as a sum and to write the sum in part (b) as a product: **(a)** $3x^2(4y + 5)$ **(b)** $4m^2n + xm^2n$.

17. ***Evaporative Cooling*** The table shows what the indoor temperature would likely be in a house using evaporative cooling when the temperature and humidity are at various levels.

Outdoor Temperature	% Relative Humidity							
°F	**10**	**20**	**30**	**40**	**50**	**60**	**70**	**80**
75	57	59	62	64	66	68	70	72
80	60	63	66	68	71	73	76	77
85	63	67	70	72	74	76	79	
90	67	70	74	77	79	82	84	
95	70	74	78	81	84	87		
100	73	78	82	85	88			
105	77	81	86	89				
110	80	85	90					

Source: Phillips Arizona Almanac.

Find each indoor temperature given the following conditions.

(a) outdoor temperature: 90°F; relative humidity: 30%

(b) outdoor temperature: 90°F; relative humidity: 70%

18. *Ages of College Undergraduates* The following table shows the age distribution of college students in October 2000. In a random sample of 5000 such students, how many would you expect to be over 19?

Age	Percent
15–19	25
20–24	38
25–34	21
35 and older	16

Source: U.S. Bureau of the Census.

Simplify each expression.

19. $(-4-1)(-3-5)-2^3$

20. $(6-9)(-2-7)\div(-4)$

21. $\left(-\frac{5}{9}-\frac{2}{3}\right)-\frac{5}{6}$

22. $\left(-\frac{2^3}{5}-\frac{3}{4}\right)-\left(-\frac{1}{2}\right)$

23. $\dfrac{6(-4)-3^2(-2)^3}{-5[-2-(-6)]}$

24. $\dfrac{(-7)(-3)-(-2^3)(-5)}{(-2^2-2)(-1-6)}$

Evaluate each expression if $a=-1$, $b=-2$, and $c=4$.

25. $-c(2a-5b)$

26. $(a-2)\div 5\cdot b+c$

27. $\dfrac{9a+2b}{a+b+c}$

28. $\dfrac{3|b|-4|c|}{|ac|}$

Write the numbers in each list in numerical order, from smallest to largest.

29. $|6-4|, -|-2|, |8+1|, -|3-(-2)|$

30. $\sqrt{7}, -\sqrt{8}, -|\sqrt{16}|, |-\sqrt{12}|$

Write without absolute value bars.

31. $-|-6|+|3|$

32. $7-|-8|$

33. $|\sqrt{8}-8|$

34. $|m-3|$, if $m>3$

Perform the indicated operations.

35. $(3q^3-9q^2+6)+(4q^3-8q+3)$

36. $2(3y^6-9y^2+2y)-(5y^6-10y^2-4y)$

37. $(8y-7)(2y+7)$

38. $(2r+11s)(4r-9s)$

39. $(3k-5m)^2$

40. $(4a-3b)^2$

(Modeling) Internet Users *The bar graph indicates the number of Internet users in North America (in millions).*

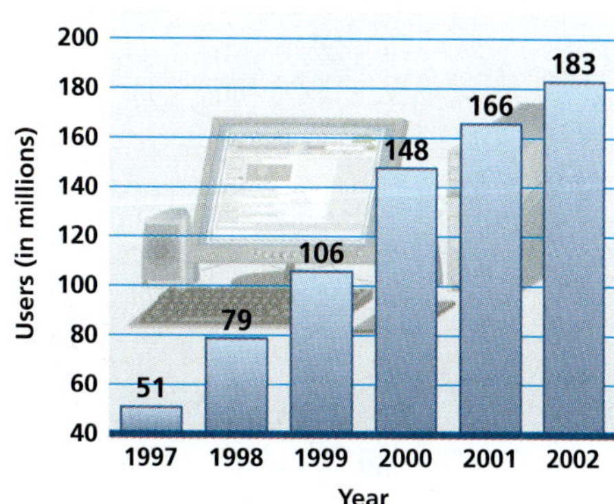

Source: NUA Internet Surveys.

The polynomial

$$.146x^4 - 2.54x^3 + 11.0x^2 + 16.6x + 51.5$$

models the number of users in year x, where $x = 0$ corresponds to 1997, $x = 1$ corresponds to 1998, and so on. For each given year,

(a) *use the bar graph to determine the number of users, and then*
(b) *use the polynomial to determine the number of users.*
(c) *How closely does the polynomial approximate the number of users compared with the data?*

41. 1997 **42.** 1999 **43.** 2002

Perform each division.

44. $\dfrac{72r^2 + 59r + 12}{8r + 3}$

45. $\dfrac{30m^3 - 9m^2 + 22m + 5}{5m + 1}$

46. $\dfrac{5m^3 - 7m^2 + 14}{m^2 - 2}$

47. $\dfrac{3b^3 - 8b^2 + 12b - 30}{b^2 + 4}$

Factor as completely as possible.

48. $7z^2 - 9z^3 + z$

49. $3(z - 4)^2 + 9(z - 4)^3$

50. $r^2 + rp - 42p^2$

51. $z^2 - 6zk - 16k^2$

52. $6m^2 - 13m - 5$

53. $48a^8 - 12a^7b - 90a^6b^2$

54. $169y^4 - 1$

55. $49m^8 - 9n^2$

56. $8y^3 - 1000z^6$

57. $6(3r - 1)^2 + (3r - 1) - 35$

58. $15mp + 9mq - 10np - 6nq$

Factor each expression. (These expressions arise in calculus from a technique called the product rule that is used to determine the shape of a curve.)

59. $(3x - 4)^2 + (x - 5)(2)(3x - 4)(3)$

60. $(5 - 2x)(3)(7x - 8)^2(7) + (7x - 8)^3(-2)$

Perform the indicated operations.

61. $\dfrac{k^2 + k}{8k^3} \cdot \dfrac{4}{k^2 - 1}$

62. $\dfrac{3r^3 - 9r^2}{r^2 - 9} \div \dfrac{8r^3}{r + 3}$

63. $\dfrac{x^2 + x - 2}{x^2 + 5x + 6} \div \dfrac{x^2 + 3x - 4}{x^2 + 4x + 3}$

64. $\dfrac{27m^3 - n^3}{3m - n} \div \dfrac{9m^2 + 3mn + n^2}{9m^2 - n^2}$

65. $\dfrac{p^2 - 36q^2}{(p - 6q)^2} \cdot \dfrac{p^2 - 5pq - 6q^2}{p^2 - 6pq + 36q^2} \div \dfrac{5p}{p^3 + 216q^3}$

66. $\dfrac{1}{4y} + \dfrac{8}{5y}$

67. $\dfrac{m}{4 - m} + \dfrac{3m}{m - 4}$

68. $\dfrac{3}{x^2 - 4x + 3} - \dfrac{2}{x^2 - 1}$

69. $\dfrac{\frac{1}{p} + \frac{1}{q}}{1 - \frac{1}{pq}}$

70. $\dfrac{3 + \frac{2m}{m^2 - 4}}{\frac{5}{m - 2}}$

Simplify each expression. Write the answer with only positive exponents. Assume all variables represent positive real numbers.

71. 2^{-6}

72. -3^{-2}

73. $\left(-\dfrac{5}{4}\right)^{-2}$

74. $3^{-1} - 4^{-1}$

75. $(5z^3)(-2z^5)$

76. $(8p^2q^3)(-2p^5q^{-4})$

77. $(-6p^5w^4m^{12})^0$

78. $(-6x^2y^{-3}z^2)^{-2}$

79. $\dfrac{-8y^7p^{-2}}{y^{-4}p^{-3}}$

80. $\dfrac{a^{-6}(a^{-8})}{a^{-2}(a^{11})}$

81. $\dfrac{(p + q)^4(p + q)^{-3}}{(p + q)^6}$

82. $\dfrac{[p^2(m + n)^3]^{-2}}{p^{-2}(m + n)^{-5}}$

83. $(7r^{1/2})(2r^{3/4})(-r^{1/6})$

84. $(a^{3/4}b^{2/3})(a^{5/8}b^{-5/6})$

85. $\dfrac{y^{5/3} \cdot y^{-2}}{y^{-5/6}}$

86. $\left(\dfrac{25m^3n^5}{m^{-2}n^6}\right)^{-1/2}$

Find each product. Assume all variables represent positive real numbers.

87. $2z^{1/3}(5z^2 - 2)$

88. $-m^{3/4}(8m^{1/2} + 4m^{-3/2})$

Simplify. Assume all variables represent positive real numbers.

89. $\sqrt{200}$

90. $\sqrt[3]{16}$

91. $\sqrt[4]{1250}$

92. $-\sqrt{\dfrac{16}{3}}$

93. $-\sqrt[3]{\dfrac{2}{5p^2}}$

94. $\sqrt{\dfrac{2^7y^8}{m^3}}$

95. $\sqrt[4]{\sqrt[3]{m}}$

96. $\dfrac{\sqrt[4]{8p^2q^5} \cdot \sqrt[4]{2p^3q}}{\sqrt[4]{p^5q^2}}$

97. $\left(\sqrt[3]{2} + 4\right)\left(\sqrt[3]{2^2} - 4\sqrt[3]{2} + 16\right)$

98. $\dfrac{3}{\sqrt{5}} - \dfrac{2}{\sqrt{45}} + \dfrac{6}{\sqrt{80}}$

99. $\sqrt{18m^3} - 3m\sqrt{32m} + 5\sqrt{m^3}$

100. $\dfrac{2}{7 - \sqrt{3}}$

101. $\dfrac{6}{3 - \sqrt{2}}$

102. $\dfrac{k}{\sqrt{k} - 3}$

Concept Check *Correct each INCORRECT statement by changing the right side.*

103. $x(x^2 + 5) = x^3 + 5$ **104.** $-3^2 = 9$ **105.** $(m^2)^3 = m^5$

106. $(3x)(3y) = 3xy$ **107.** $\dfrac{\left(\frac{a}{b}\right)}{2} = \dfrac{2a}{b}$ **108.** $\dfrac{m}{r} \cdot \dfrac{n}{r} = \dfrac{mn}{r}$

109. $\dfrac{1}{\sqrt{a} + \sqrt{b}} = \dfrac{1}{\sqrt{a}} + \dfrac{1}{\sqrt{b}}$ **110.** $\dfrac{(2x)^3}{2y} = \dfrac{x^3}{y}$

111. $4 - (t + 1) = 4 - t + 1$ **112.** $\dfrac{1}{(-2)^3} = 2^{-3}$

113. $(-5)^2 = -5^2$ **114.** $\left(\dfrac{8}{7} + \dfrac{a}{b}\right)^{-1} = \dfrac{7}{8} + \dfrac{b}{a}$

Chapter R Test

1. Let $A = \left\{-13, -\frac{12}{4}, 0, \frac{3}{5}, \frac{\pi}{4}, 5.9, \sqrt{49}\right\}$. List the elements of A that belong to the given set.

(a) Integers **(b)** Rational numbers **(c)** Real numbers

2. Evaluate the expression if $x = -2$, $y = -4$, and $z = 5$: $\left|\dfrac{x^2 + 2yz}{3(x + z)}\right|$.

3. Identify each property illustrated. Let a, b, and c represent any real numbers.

(a) $a + (b + c) = (a + b) + c$ **(b)** $a + (c + b) = a + (b + c)$

(c) $a(b + c) = ab + ac$ **(d)** $a + [b + (-b)] = a + 0$

4. *Passing Rating for an NFL Quarterback* Use the formula in Section R.1, Example 5 (page 6) to approximate the quarterback rating of Matt Hasselbeck of the Seattle Seahawks in 2002. He attempted 419 passes, completed 267, had 3075 total yards, threw for 15 touchdowns, and had 10 interceptions. (*Source:* www.NFL.com)

Perform the indicated operations.

5. $(x^2 - 3x + 2) - (x - 4x^2) + 3x(2x + 1)$ **6.** $(6r - 5)^2$

7. $(t + 2)(3t^2 - t + 4)$ **8.** $\dfrac{2x^3 - 11x^2 + 28}{x - 5}$

(Modeling) Adjusted Poverty Threshold *The adjusted poverty threshold for a single person between the years 1997 and 2003 can be approximated by the polynomial*

$$5.476x^2 + 154.3x + 7889,$$

where $x = 0$ corresponds to 1997, $x = 1$ corresponds to 1998, and so on, and the amount is in dollars. According to this model, what was the adjusted poverty threshold in each given year? (Source: U.S. Department of Health and Human Services.)

9. 2000 **10.** 2002

Factor completely.

11. $6x^2 - 17x + 7$

12. $x^4 - 16$

13. $24m^3 - 14m^2 - 24m$

14. $x^3y^2 - 9x^3 - 8y^2 + 72$

Perform the indicated operations.

15. $\dfrac{5x^2 - 9x - 2}{30x^3 + 6x^2} \cdot \dfrac{2x^8 + 6x^7 + 4x^6}{x^4 - 3x^2 - 4}$

16. $\dfrac{x}{x^2 + 3x + 2} + \dfrac{2x}{2x^2 - x - 3}$

17. $\dfrac{a + b}{2a - 3} - \dfrac{a - b}{3 - 2a}$

18. $\dfrac{y - 2}{y - \dfrac{4}{y}}$

19. Simplify $\left(\dfrac{x^{-2}y^{-1/3}}{x^{-5/3}y^{-2/3}}\right)^3$ so there are no negative exponents. Assume all variables represent positive real numbers.

20. Evaluate $\left(-\dfrac{64}{27}\right)^{-2/3}$.

Simplify. Assume all variables represent positive real numbers.

21. $\sqrt{18x^5y^8}$

22. $\sqrt{32x} + \sqrt{2x} - \sqrt{18x}$

23. $(\sqrt{x} - \sqrt{y})(\sqrt{x} + \sqrt{y})$

24. Rationalize the denominator of $\dfrac{14}{\sqrt{11} - \sqrt{7}}$ and simplify.

25. ***(Modeling) Period of a Pendulum*** The period t in seconds of the swing of a pendulum is given by the equation

$$t = 2\pi\sqrt{\frac{L}{32}},$$

where L is the length of the pendulum in feet. Find the period of a pendulum 3.5 ft long. Use a calculator.

Chapter R Quantitative Reasoning

Are you paying too much for a large pizza?

Pizza is one of the most popular foods available today, and the take-out pizza has become a staple in today's hurried world. But are you paying too much for that large pizza you ordered? Pizza sizes are typically designated by their diameters. A pizza of diameter d inches has area $\pi\left(\frac{d}{2}\right)^2$.

Let's assume that the cost of a pizza is determined by its area. Suppose a pizza parlor charges \$4.00 for a 10-in. pizza and \$9.25 for a 15-in. pizza. Evaluate the area of each pizza to show the owner that he is overcharging you by \$.25 for the larger pizza.

1 Equations and Inequalities

Indoor air quality has become a major health concern during the past decade, as people spend 80 to 90 percent of their time in tightly sealed, energy-efficient buildings. Many contaminants, such as tobacco smoke, formaldehyde, radon, lead, and carbon monoxide, are allowed to increase to unsafe levels. Other air pollutants occur in such low levels that until recently scientists were unable to detect or measure them. (*Source:* Boubel, R., D. Fox, D. Turner, and A. Stern, *Fundamentals of Air Pollution,* Academic Press, 1994; Ghosh, T., *Indoor Air Pollution Control,* Lewis Publishers, 1989.)

In Example 6 of Section 1.2, we use a linear equation to model how indoor contaminants are controlled using vented hoods for gas ranges.

1.1 Linear Equations

Basic Terminology of Equations ▪ Solving Linear Equations ▪ Identities, Conditional Equations, and Contradictions ▪ Solving for a Specified Variable (Literal Equations)

Basic Terminology of Equations An **equation** is a statement that two expressions are equal.

$$x + 2 = 9, \qquad 11x = 5x + 6x, \qquad x^2 - 2x - 1 = 0 \qquad \text{Equations}$$

To *solve* an equation means to find all numbers that make the equation a true statement. These numbers are called **solutions** or **roots** of the equation. A number that is a solution of an equation is said to *satisfy* the equation, and the solutions of an equation make up its **solution set.** Equations with the same solution set are **equivalent equations.** For example, $x = 4$, $x + 1 = 5$, and $6x + 3 = 27$ are equivalent equations because they have the same solution set, $\{4\}$. However, the equations $x^2 = 9$ and $x = 3$ are not equivalent, since the first has solution set $\{-3, 3\}$ while the solution set of the second is $\{3\}$.

One way to solve an equation is to rewrite it as a series of simpler equivalent equations using the *addition and multiplication properties of equality.*

Addition and Multiplication Properties of Equality

For real numbers a, b, and c:

$$a = b \text{ and } a + c = b + c \text{ are equivalent.}$$

That is, the same number may be added to both sides of an equation without changing the solution set.

$$\text{If } c \neq 0, \text{ then } a = b \text{ and } ac = bc \text{ are equivalent.}$$

That is, both sides of an equation may be multiplied by the same nonzero number without changing the solution set.

These properties can be extended: The same number may be subtracted from both sides of an equation, and both sides may be divided by the same nonzero number, without changing the solution set.

Solving Linear Equations We use the properties of equality to solve *linear equations.*

Linear Equation in One Variable

A **linear equation in one variable** is an equation that can be written in the form

$$ax + b = 0,$$

where a and b are real numbers with $a \neq 0$.

A linear equation is also called a *first-degree* equation since the greatest degree of the variable is one.

$$3x + \sqrt{2} = 0, \quad \frac{3}{4}x = 12, \quad .5(x + 3) = 2x - 6 \quad \text{Linear equations}$$

$$\sqrt{x} + 2 = 5, \quad \frac{1}{x} = -8, \quad x^2 + 3x + .2 = 0 \quad \text{Nonlinear equations}$$

EXAMPLE 1 Solving a Linear Equation

Solve $3(2x - 4) = 7 - (x + 5)$.

Solution

$$\begin{aligned}
3(2x - 4) &= 7 - (x + 5) \\
6x - 12 &= 7 - x - 5 && \text{Distributive property (Section R.1)} \\
6x - 12 &= 2 - x && \text{Combine terms. (Section R.3)} \\
6x - 12 + x &= 2 - x + x && \text{Add } x \text{ to each side.} \\
7x - 12 &= 2 && \text{Combine terms.} \\
7x - 12 + 12 &= 2 + 12 && \text{Add 12 to each side.} \\
7x &= 14 && \text{Combine terms.} \\
\frac{7x}{7} &= \frac{14}{7} && \text{Divide each side by 7.} \\
x &= 2
\end{aligned}$$

Check:

$$\begin{aligned}
3(2x - 4) &= 7 - (x + 5) && \text{Original equation} \\
3(2 \cdot 2 - 4) &\stackrel{?}{=} 7 - (2 + 5) && \text{Let } x = 2. \\
3(4 - 4) &\stackrel{?}{=} 7 - (7) \\
0 &= 0 && \text{True}
\end{aligned}$$

Since replacing x with 2 results in a true statement, 2 is a solution of the given equation. The solution set is $\{2\}$.

Now try Exercise 11.

If an equation has fractions as coefficients, we can begin by multiplying both sides by the least common denominator to clear the fractions.

EXAMPLE 2 Clearing Fractions before Solving a Linear Equation

Solve $\frac{2t + 4}{3} + \frac{1}{2}t = \frac{1}{4}t - \frac{7}{3}$.

Solution

$$\begin{aligned}
\frac{2t + 4}{3} + \frac{1}{2}t &= \frac{1}{4}t - \frac{7}{3} \\
12\left(\frac{2t + 4}{3} + \frac{1}{2}t\right) &= 12\left(\frac{1}{4}t - \frac{7}{3}\right) && \text{Multiply by 12, the LCD of the fractions. (Section R.5)} \\
4(2t + 4) + 6t &= 3t - 28 && \text{Distributive property} \\
8t + 16 + 6t &= 3t - 28 && \text{Distributive property} \\
14t + 16 &= 3t - 28 && \text{Combine terms.} \\
11t &= -44 && \text{Subtract } 3t\text{; subtract 16.} \\
t &= -4 && \text{Divide by 11.}
\end{aligned}$$

Check: $\frac{2(-4)+4}{3} + \frac{1}{2}(-4) = \frac{1}{4}(-4) - \frac{7}{3}$? Let $t = -4$.

$$\frac{-4}{3} + (-2) = -1 - \frac{7}{3} \quad ?$$

$$-\frac{10}{3} = -\frac{10}{3} \quad \text{True}$$

The solution set is $\{-4\}$.

Now try Exercise 19.

Identities, Conditional Equations, and Contradictions An equation satisfied by every number that is a meaningful replacement for the variable is called an **identity.** The equation $3(x + 1) = 3x + 3$ is an example of an identity. An equation that is satisfied by some numbers but not others, such as $2x = 4$, is called a **conditional equation.** The equations in Examples 1 and 2 are conditional equations. An equation that has no solution, such as $x = x + 1$, is called a **contradiction.**

EXAMPLE 3 Identifying Types of Equations

Decide whether each equation is an identity, a conditional equation, or a contradiction. Give the solution set.

(a) $-2(x + 4) + 3x = x - 8$ **(b)** $5x - 4 = 11$ **(c)** $3(3x - 1) = 9x + 7$

Solution

(a) $-2(x + 4) + 3x = x - 8$

$-2x - 8 + 3x = x - 8$ Distributive property

$x - 8 = x - 8$ Combine terms.

$0 = 0$ Subtract x; add 8.

When a *true* statement such as $0 = 0$ results, the equation is an identity, and the solution set is {all real numbers}.

(b) $5x - 4 = 11$

$5x = 15$ Add 4.

$x = 3$ Divide by 5.

This is a conditional equation, and its solution set is $\{3\}$.

(c) $3(3x - 1) = 9x + 7$

$9x - 3 = 9x + 7$ Distributive property

$-3 = 7$ Subtract $9x$.

When a *false* statement such as $-3 = 7$ results, the equation is a contradiction, and the solution set is the **empty set** or **null set,** symbolized $\emptyset$.

Now try Exercises 29, 31, and 35.

Identifying Linear Equations as Identities, Conditional Equations, or Contradictions

1. If solving a linear equation leads to a true statement such as $0 = 0$, the equation is an **identity.** Its solution set is **{all real numbers}.** (See Example 3(a).)
2. If solving a linear equation leads to a single solution such as $x = 3$, the equation is **conditional.** Its solution set consists of a single element. (See Example 3(b).)
3. If solving a linear equation leads to a false statement such as $-3 = 7$, the equation is a **contradiction.** Its solution set is $\emptyset$. (See Example 3(c).)

Solving for a Specified Variable (Literal Equations) The solution of a problem sometimes requires the use of a formula that relates several variables. One such formula is the one for *simple interest.* The **simple interest** I on P dollars at an annual interest rate r for t years is $\boldsymbol{I = Prt}$**.** A formula is an example of a **literal equation.** The methods used to solve linear equations can be used to solve some literal equations for a specified variable.

EXAMPLE 4 Solving for a Specified Variable

Solve for the specified variable.

(a) $I = Prt$, for t

(b) $A = P(1 + rt)$, for r

(c) $3(2x - 5a) + 4b = 4x - 2$, for x

Solution

(a) Treat t as if it were the only variable, and the other variables as if they were constants.

$$I = Prt \qquad \text{Goal: Isolate } t \text{ on one side of the equation.}$$

$$\frac{I}{Pr} = \frac{Prt}{Pr} \qquad \text{Divide both sides by } Pr.$$

$$\frac{I}{Pr} = t \quad \text{or} \quad t = \frac{I}{Pr}$$

(b) The formula $A = P(1 + rt)$ gives the *future* or *maturity value* A of P dollars invested for t years at annual simple interest rate r.

$$A = P(1 + rt) \qquad \text{Goal: Isolate } r\text{, the specified variable.}$$

$$\frac{A}{P} = 1 + rt \qquad \text{Divide by } P.$$

$$\frac{A}{P} - 1 = rt \qquad \text{Subtract 1.}$$

$$\frac{1}{t}\left(\frac{A}{P} - 1\right) = r \quad \text{or} \quad r = \frac{1}{t}\left(\frac{A}{P} - 1\right) \qquad \text{Divide by } t.$$

(c)
$$\begin{aligned} 3(2x - 5a) + 4b &= 4x - 2 && \text{Solve for } x. \\ 6x - 15a + 4b &= 4x - 2 && \text{Distributive property} \\ 6x - 4x &= 15a - 4b - 2 && \text{Isolate the } x\text{-terms on one side.} \\ 2x &= 15a - 4b - 2 && \text{Combine terms.} \\ x &= \frac{15a - 4b - 2}{2} && \text{Divide by 2.} \end{aligned}$$

Now try Exercises 39, 43, and 49.

EXAMPLE 5 Applying the Simple Interest Formula

Laquanda Nelson borrowed \$5240 for new furniture. She will pay it off in 11 months at an annual interest rate of 4.5%. How much interest will she pay?

Solution Here, $r = .045$, $P = 5240$, and $t = \frac{11}{12}$ (year). Using the formula,

$$I = Prt = 5240(.045)\left(\frac{11}{12}\right) = \$216.15.$$

She will pay \$216.15 interest on her purchase.

Now try Exercise 59.

1.1 Exercises

Concept Check *In Exercises 1–4, decide whether each statement is* true *or* false.

1. The solution set of $2x + 3 = x - 5$ is $\{-8\}$.

2. The equation $5(x - 9) = 5x - 45$ is an example of an identity.

3. The equations $x^2 = 4$ and $x + 1 = 3$ are equivalent equations.

4. It is possible for a linear equation to have exactly two solutions.

5. Explain the difference between an identity and a conditional equation.

6. Make a complete list of the steps needed to solve a linear equation. (Some equations will not require every step.)

7. *Concept Check* Which one is not a linear equation?

A. $5x + 7(x - 1) = -3x$ **B.** $8x^2 - 4x + 3 = 0$
C. $7x + 8x = 13x$ **D.** $.04x - .08x = .40$

8. In solving the equation $3(2x - 4) = 6x - 12$, a student obtains the result $0 = 0$ and gives the solution set $\{0\}$. Is this correct? Explain.

Solve each equation. See Examples 1 and 2.

9. $5x + 2 = 3x - 6$

10. $9x + 1 = 7x - 9$

11. $6(3x - 1) = 8 - (10x - 14)$

12. $4(-2x + 1) = 6 - (2x - 4)$

13. $\frac{5}{6}x - 2x + \frac{1}{3} = \frac{2}{3}$

14. $\frac{3}{4} + \frac{1}{5}x - \frac{1}{2} = \frac{4}{5}x$

15. $3x + 2 - 5(x + 1) = 6x + 4$

16. $5(x + 3) + 4x - 5 = -(2x - 4)$

17. $2[x - (4 + 2x) + 3] = 2x + 2$

18. $4[2x - (3 - x) + 5] = -7x - 2$

19. $\frac{1}{7}(3x - 2) = \frac{x + 10}{5}$

20. $\frac{1}{5}(2x + 5) = \frac{x + 2}{3}$

21. $.2x - .5 = .1x + 7$

22. $.01x + 3.1 = 2.03x - 2.96$

23. $-4(2x - 6) + 7x = 5x + 24$

24. $-8(3x + 4) + 2x = 4(x - 8)$

25. $.5x + \frac{4}{3}x = x + 10$

26. $\frac{2}{3}x + .25x = x + 2$

27. $.08x + .06(x + 12) = 7.72$

28. $.04(x - 12) + .06x = 1.52$

Decide whether each equation is an identity, *a* conditional equation, *or a* contradiction. *Give the solution set. See Example 3.*

29. $4(2x + 7) = 2x + 25 + 3(2x + 1)$

30. $\frac{1}{2}(6x + 14) = x + 1 + 2(x + 3)$

31. $2(x - 7) = 3x - 14$

32. $-8(x + 3) = -8x - 5(x + 1)$

33. $.3(x + 2) - .5(x + 2) = -.2x - .4$

34. $-.3(x - 5) + .4(x - 6) = .1x - .9$

35. $8(x + 7) = 4(x + 12) + 4(x + 1)$

36. $-6(2x + 1) - 3(x - 4) = -15x + 1$

37. A student claims that the equation $5x = 4x$ is a contradiction, since dividing both sides by x leads to $5 = 4$, a false statement. Explain why the student is incorrect.

38. If $k \neq 0$, is the equation $x + k = x$ a contradiction, a conditional equation, or an identity? Explain.

Solve each formula for the indicated variable. Assume that the denominator is not 0 if variables appear in the denominator. See Examples 4(a) and (b).

39. $V = lwh$ for l (volume of a rectangular box)

40. $I = Prt$ for P (simple interest)

41. $P = a + b + c$ for c (perimeter of a triangle)

42. $P = 2l + 2w$ for w (perimeter of a rectangle)

43. $A = \frac{1}{2}(B + b)h$ for B (area of a trapezoid)

44. $A = \frac{1}{2}(B + b)h$ for h (area of a trapezoid)

45. $S = 2\pi rh + 2\pi r^2$ for h (surface area of a right circular cylinder)

46. $s = \frac{1}{2}gt^2$ for g (distance traveled by a falling object)

47. $S = 2lw + 2wh + 2hl$ for h (surface area of a rectangular box)

48. Refer to Exercise 45. Why is it not possible to solve this formula for r using the methods of this section?

Solve each equation for x. See Example 4(c).

49. $2(x - a) + b = 3x + a$

50. $5x - (2a + c) = a(x + 1)$

51. $ax + b = 3(x - a)$

52. $4a - ax = 3b + bx$

53. $\frac{x}{a - 1} = ax + 3$

54. $\frac{2a}{x - 1} = a - b$

55. $a^2x + 3x = 2a^2$

56. $ax + b^2 = bx - a^2$

57. $3x = (2x - 1)(m + 4)$

58. $-x = (5x + 3)(3k + 1)$

Work each problem. See Example 5.

59. *Simple Interest* Miguel Rodriguez borrowed \$1575 from his brother Julio to pay for books and tuition. He agreed to repay Julio in 6 months with simple interest at 8%.

(a) How much will the interest amount to?

(b) What amount must Miguel pay Julio at the end of the 6 months?

60. *Simple Interest* Jennifer Kerber borrows \$10,450 from her bank to open a florist shop. She agrees to repay the money in 18 months with simple interest of 10.4%.

(a) How much must she pay the bank in 18 months?

(b) How much of the amount in part (a) is interest?

Celsius and Fahrenheit Temperatures *In the metric system of weights and measures, temperature is measured in degrees Celsius (°C) instead of degrees Fahrenheit (°F). To convert between the two systems, we use the equations*

$$C = \frac{5}{9}(F - 32) \quad \text{and} \quad F = \frac{9}{5}C + 32.$$

In each exercise, convert to the other system. Round answers to the nearest tenth of a degree if necessary.

61. 20°C **62.** 100°C **63.** 59°F **64.** 86°F **65.** 100°F **66.** 350°F

Work each problem.

67. *Temperature of Venus* Venus is the hottest planet with a surface temperature of 867°F. What is this temperature in Celsius? (*Source: The World Almanac,* 2003.)

68. *Temperature at Soviet Antarctica Station* A record low temperature of −89.4°C was recorded at the Soviet Antarctica Station of Vostok on July 21, 1983. Find the corresponding Fahrenheit temperature. (*Source: The World Almanac,* 2003.)

69. *Temperature in Montreal* The average daily low temperature for January in Montreal, Canada (based on a 30-year study from 1961–1990) was 5.2°F. What is the corresponding Celsius temperature? (*Source: The World Almanac,* 2003.)

70. *Temperature in Dublin* The average daily high temperature for July in Dublin, Ireland (based on a 30-year study from 1961–1990) was 18.9°C. What is the corresponding Fahrenheit temperature? (*Source: The World Almanac,* 2003.)

1.2 Applications and Modeling with Linear Equations

Solving Applied Problems ▪ Geometry Problems ▪ Motion Problems ▪ Work Rate Problems ▪ Mixture Problems ▪ Modeling with Linear Equations

One of the main reasons for learning mathematics is to be able to solve practical problems. In this section, we give a few hints that may help.

Solving Applied Problems While there is no one method that allows us to solve all types of applied problems, the following six steps are helpful.

Solving an Applied Problem

Step 1 **Read** the problem carefully until you understand what is given and what is to be found.

Step 2 **Assign a variable** to represent the unknown value, using diagrams or tables as needed. Write down what the variable represents. If necessary, express any other unknown values in terms of the variable.

Step 3 **Write an equation** using the variable expression(s).

Step 4 **Solve** the equation.

Step 5 **State the answer** to the problem. Does it seem reasonable?

Step 6 **Check** the answer in the words of the original problem.

Geometry Problems

EXAMPLE 1 Finding the Dimensions of a Square

If the length of each side of a square is increased by 3 cm, the perimeter of the new square is 40 cm more than twice the length of each side of the original square. Find the dimensions of the original square.

Solution

Step 1 **Read** the problem. We must find the length of each side of the original square.

Step 2 **Assign a variable.** Since the length of a side of the original square is to be found, let the variable represent this length.

$$x = \text{length of side of the original square in centimeters}$$

The length of a side of the new square is 3 cm more than the length of a side of the old square, so

$$x + 3 = \text{length of side of the new square.}$$

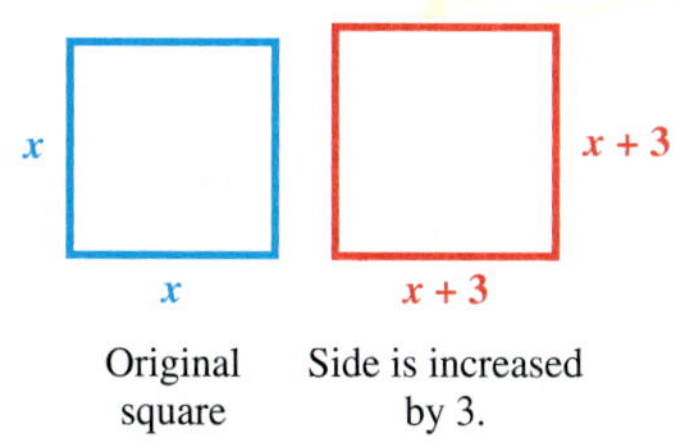

Figure 1

See Figure 1. Now write a variable expression for the perimeter of the new square. Since the perimeter of a square is 4 times the length of a side,

$$4(x + 3) = \text{perimeter of the new square.}$$

Step 3 **Write an equation.**

The new perimeter	is	40	more than	twice the length of the side of the original square.
$4(x + 3)$	$=$	40	$+$	$2x$

Step 4 **Solve** the equation.

$$4x + 12 = 40 + 2x \quad \text{Distributive property (Section R.1)}$$
$$2x = 28 \quad \text{Subtract } 2x \text{ and } 12. \text{ (Section 1.1)}$$
$$x = 14 \quad \text{Divide by 2. (Section 1.1)}$$

Step 5 **State the answer.** Each side of the original square measures 14 cm.

Step 6 **Check.** Go back to the words of the original problem to see that all necessary conditions are satisfied. The length of a side of the new square would be $14 + 3 = 17$ cm; its perimeter would be $4(17) = 68$ cm. Twice the length of a side of the original square is $2(14) = 28$ cm. Since $40 + 28 = 68$, the answer checks.

Now try Exercise 13.

Motion Problems

Looking Ahead to Calculus

In calculus the concept of the definite integral is used to find the distance traveled by an object traveling at a *non-constant* velocity.

PROBLEM SOLVING In a motion problem, the components *distance, rate,* and *time* are denoted by the letters d, r, and t, respectively. (The *rate* is also called the *speed* or *velocity.* Here, rate is understood to be constant.) These variables are related by the equation

$$d = rt, \quad \text{and its related forms} \quad r = \frac{d}{t} \quad \text{and} \quad t = \frac{d}{r}.$$

EXAMPLE 2 Solving a Motion Problem

Maria and Eduardo are traveling to a business conference. The trip takes 2 hr for Maria and 2.5 hr for Eduardo, since he lives 40 mi farther away. Eduardo travels 5 mph faster than Maria. Find their average rates.

Solution

Step 1 **Read** the problem. We must find both Maria's and Eduardo's average rates.

Step 2 **Assign a variable.** Let x = Maria's rate. Then $x + 5$ = Eduardo's rate. Summarize the given information in a table.

	r	t	d
Maria	x	2	$2x$
Eduardo	$x + 5$	2.5	$2.5(x + 5)$

Use $d = rt$.

Step 3 **Write an equation.** Use the fact that Eduardo's distance traveled exceeds Maria's distance by 40 mi.

Eduardo's distance is 40 more than Maria's.

$$2.5(x + 5) = 2x + 40$$

Step 4 **Solve.**

$$2.5x + 12.5 = 2x + 40 \quad \text{Distributive property}$$

$$.5x = 27.5 \quad \text{Subtract } 2x\text{; subtract 12.5.}$$

$$x = 55 \quad \text{Divide by .5.}$$

Step 5 **State the answer.** Maria's rate of travel is 55 mph, and Eduardo's rate is $55 + 5 = 60$ mph.

Step 6 **Check.** The diagram shows that the conditions of the problem are satisfied.

Distance traveled by Maria: $2(55) = 110$ mi

Distance traveled by Eduardo: $2.5(60) = 150$ mi

$150 - 110 = 40$

Now try Exercise 19.

Work Rate Problems In Example 2 (a motion problem), we used the formula relating rate, time, and distance. In problems involving rate of work, we use a similar idea.

PROBLEM SOLVING If a job can be done in t units of time, then the rate of work is $\frac{1}{t}$ of the job per time unit. Therefore,

rate × time = portion of the job completed.

If the letters r, t, and A represent the rate at which work is done, the time, and the amount of work accomplished, respectively, then

$$A = rt.$$

Amounts of work are often measured in terms of the number of jobs accomplished. For instance, if one job is accomplished in t time units, then $A = 1$ and

$$r = \frac{1}{t}.$$

EXAMPLE 3 Solving a Work Rate Problem

One computer can do a job twice as fast as another. Working together, both computers can do the job in 2 hr. How long would it take each computer, working alone, to do the job?

Solution

Step 1 **Read** the problem. We must find the time it would take each computer working alone to do the job.

Step 2 **Assign a variable.** Let x represent the number of hours it would take the faster computer, working alone, to do the job. The time for the slower computer to do the job alone is then $2x$ hours. Therefore,

$$\frac{1}{x} = \text{rate of faster computer (job per hour)}$$

and $$\frac{1}{2x} = \text{rate of slower computer (job per hour).}$$

The time for the computers to do the job together is 2 hr. Multiplying each rate by the time will give the fractional part of the job accomplished by each.

	Rate	Time	Part of the Job Accomplished
Faster Computer	$\frac{1}{x}$	2	$2\left(\frac{1}{x}\right) = \frac{2}{x}$
Slower Computer	$\frac{1}{2x}$	2	$2\left(\frac{1}{2x}\right) = \frac{1}{x}$

$A = rt$

Step 3 **Write an equation.** The sum of the two parts of the job accomplished is 1, since one whole job is done.

$$\underbrace{\text{Part of the job done by the faster computer}}_{\frac{2}{x}} + \underbrace{\text{Part of the job done by the slower computer}}_{\frac{1}{x}} = \underbrace{\text{One whole job}}_{1}$$

Step 4 **Solve.**

$$x\left(\frac{2}{x} + \frac{1}{x}\right) = x(1) \quad \text{Multiply both sides by } x. \text{ (Section 1.1)}$$

$$2 + 1 = x$$

$$3 = x$$

Step 5 **State the answer.** The faster computer would take 3 hr to do the job alone, while the slower computer would take $2(3) = 6$ hr.

Step 6 **Check.** The answer is reasonable, since the time working together is less than the time it would take the faster computer working alone.

Now try Exercise 31.

NOTE Example 3 can also be solved by using the fact that the sum of the rates of the individual computers is equal to their rate working together:

$$\frac{1}{x} + \frac{1}{2x} = \frac{1}{2}$$

$$2 + 1 = x \quad \text{Multiply both sides by } 2x.$$

$$x = 3. \quad \text{Same solution found earlier}$$

Mixture Problems Problems involving mixtures of two types of the same substance, salt solution, candy, and so on, often involve percent.

PROBLEM SOLVING In mixture problems, the rate (percent) of concentration is multiplied by the quantity to get the amount of pure substance present. Also, the concentration of the final mixture must be *between* the concentrations of the two solutions making up the mixture.

EXAMPLE 4 Solving a Mixture Problem

Charlotte Besch is a chemist. She needs a 20% solution of alcohol. She has a 15% solution on hand, as well as a 30% solution. How many liters of the 15% solution should she add to 3 L of the 30% solution to obtain her 20% solution?

Solution

Step 1 **Read** the problem. We must find the required number of liters of 15% alcohol solution.

Step 2 **Assign a variable.**

Let x = number of liters of 15% solution to be added.

Figure 2 and the table show what is happening in the problem. The numbers in the last column were found by multiplying the strengths and the numbers of liters. The number of liters of pure alcohol in the 15% solution plus the number of liters in the 30% solution must equal the number of liters in the 20% solution.

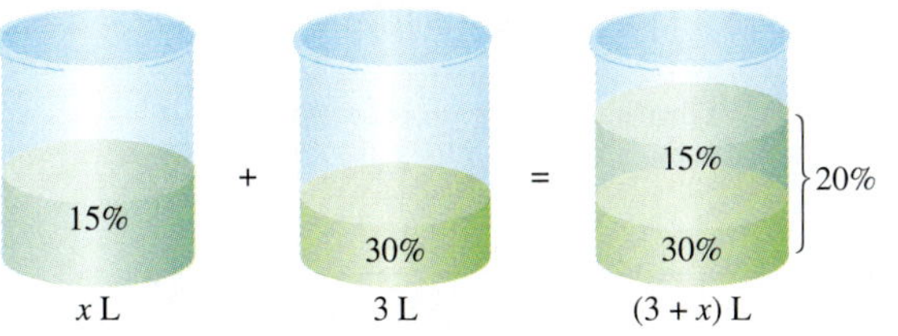

Figure 2

Strength	Liters of Solution	Liters of Pure Alcohol
15%	x	$.15x$
30%	3	$.30(3)$
20%	$3 + x$	$.20(3 + x)$

(Sum must equal: $.15x$ + $.30(3)$ must equal $.20(3 + x)$.)

Step 3 **Write an equation.** Since the number of liters of pure alcohol in the 15% solution plus the number of liters in the 30% solution must equal the number of liters in the final 20% solution,

$$\underbrace{.15x}_{\text{Liters in 15\%}} + \underbrace{.30(3)}_{\text{Liters in 30\%}} = \underbrace{.20(3 + x)}_{\text{Liters in 20\%}}.$$

Step 4 **Solve.**

$$.15x + .90 = .60 + .20x \quad \text{Distributive property}$$

$$.30 = .05x \quad \text{Subtract .60; subtract } .15x.$$

$$6 = x \quad \text{Divide by .05.}$$

Step 5 **State the answer.** Thus, 6 L of 15% solution should be mixed with 3 L of 30% solution, giving $6 + 3 = 9$ L of 20% solution.

Step 6 **Check.** Since

$$.15(6) + .9 = .9 + .9 = 1.8$$

and

$$.20(3 + 6) = .20(9) = 1.8,$$

the answer checks.

Now try Exercise 35.

PROBLEM SOLVING In mixed investment problems, multiply each principal by the interest rate to find the amount of interest earned.

EXAMPLE 5 Solving an Investment Problem

An artist has sold a painting on eBay for \$410,000. He needs some of the money in 6 months and the rest in 1 yr. He can get a treasury bond for 6 months at 4.65% and one for a year at 4.91%. His broker tells him the two investments will earn a total of \$14,961. How much should be invested at each rate to obtain that amount of interest?

Solution

Step 1 **Read** the problem. We must find the amount to be invested at each rate.

Step 2 **Assign a variable.**

Let x = the dollar amount to be invested for 6 months at 4.65%;

$410{,}000 - x$ = the dollar amount to be invested for 1 yr at 4.91%.

Summarize this information in a table using the formula $I = Prt$.

Amount Invested	Interest Rate (%)	Time (in years)	Interest Earned
x	4.65	.5	$x(.0465)(.5)$
$410{,}000 - x$	4.91	1	$(410{,}000 - x)(.0491)(1)$

Step 3 **Write an equation.**

$$\underbrace{.5x(.0465)}_{\text{Interest from 4.65\% investment}} + \underbrace{.0491(410{,}000 - x)}_{\text{Interest from 4.91\% investment}} = \underbrace{14{,}961}_{\text{Total interest}}$$

Step 4 **Solve.**

$$.02325x + 20{,}131 - .0491x = 14{,}961$$
$$20{,}131 - .02585x = 14{,}961$$
$$-.02585x = -5170$$
$$x = 200{,}000$$

Step 5 **State the answer.** The artist should invest \$200,000 at 4.65% for 6 months and \$410,000 − \$200,000 = \$210,000 at 4.91% for 1 yr to earn \$14,961 in interest.

Step 6 **Check.** The 6-month investment earns

$$.5(\$200{,}000)(.0465) = \$4650,$$

while the 1-yr investment earns

$$1(\$210{,}000)(.0491) = \$10{,}311.$$

The total amount of interest earned is

$$\$4650 + \$10{,}311 = \$14{,}961, \quad \text{as required.}$$

Now try Exercise 41.

Modeling with Linear Equations A **mathematical model** is an equation (or inequality) that describes the relationship between two quantities. A *linear model* is a linear equation; the next example shows how a linear model is applied. In Chapter 2 we actually determine linear models.

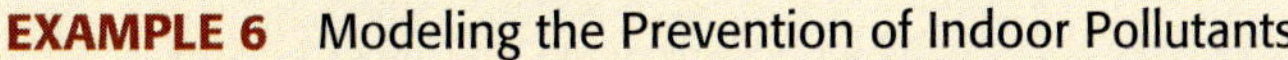

EXAMPLE 6 Modeling the Prevention of Indoor Pollutants

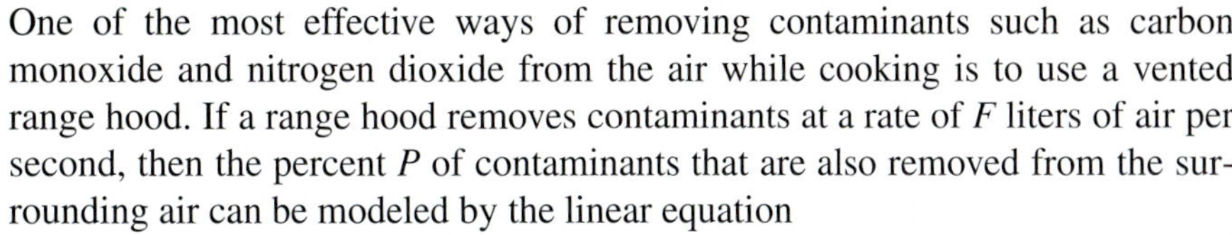

One of the most effective ways of removing contaminants such as carbon monoxide and nitrogen dioxide from the air while cooking is to use a vented range hood. If a range hood removes contaminants at a rate of F liters of air per second, then the percent P of contaminants that are also removed from the surrounding air can be modeled by the linear equation

$$P = 1.06F + 7.18,$$

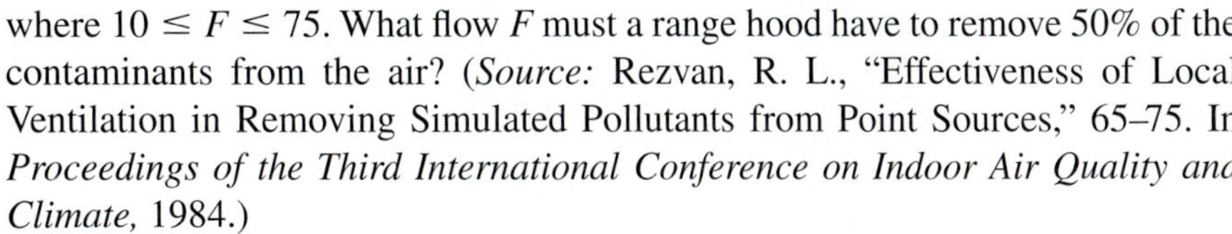

where $10 \leq F \leq 75$. What flow F must a range hood have to remove 50% of the contaminants from the air? (*Source:* Rezvan, R. L., "Effectiveness of Local Ventilation in Removing Simulated Pollutants from Point Sources," 65–75. In *Proceedings of the Third International Conference on Indoor Air Quality and Climate,* 1984.)

Solution Since $P = 50$, the equation becomes

$$\begin{aligned} 50 &= 1.06F + 7.18 \\ 42.82 &= 1.06F && \text{Subtract 7.18.} \\ F &\approx 40.40. && \text{Divide by 1.06.} \end{aligned}$$

Therefore, to remove 50% of the contaminants, the flow rate must be approximately 40.40 L of air per second.

Now try Exercise 47.

EXAMPLE 7 Modeling Oven Temperatures during Cleaning Cycles

To the nearest Fahrenheit degree, the temperature T during oven heating and cooling cycles can be modeled by linear equations, where x represents time in hours after START.

Heating cycle: $T = 1033.3x + 100$

Cooling cycle: $T = -1033.3x + 3975$

(*Source: Whirlpool Use and Care Guide, Self-Cleaning Electric Range.*)

(a) The heating cycle begins after about $\frac{3}{4}$ hr. To the nearest degree, what is the temperature at that time?

(b) Using the answer to part (a), find the time when the cooling cycle begins.

Solution

(a) Let $x = \frac{3}{4}$ and find the value of T in the first equation, to determine how hot the oven has become.

$$T = 1033.3\left(\frac{3}{4}\right) + 100 \approx 875.$$

After $\frac{3}{4}$ hr, the temperature is approximately 875°F.

(b) Let $T = 875$ and find the value of x in the second equation.

$$875 = -1033.3x + 3975$$
$$-3100 = -1033.3x$$
$$x \approx 3$$

The cooling cycle begins after 3 hr.

Now try Exercise 51.

George Polya (1887–1985)

CONNECTIONS George Polya proposed an excellent general outline for solving applied problems in his classic book *How to Solve It.*

1. Understand the problem. **2.** Devise a plan.
3. Carry out the plan. **4.** Look back.

Polya, a native of Budapest, Hungary, wrote more than 250 papers in many languages, as well as a number of books. He was a brilliant lecturer and teacher, and numerous mathematical properties and theorems bear his name. He once was asked why so many good mathematicians came out of Hungary at the turn of the century. He theorized that it was because mathematics was the cheapest science, requiring no expensive equipment, only pencil and paper.

For Discussion or Writing

Look back at the six problem-solving steps given at the beginning of this section, and compare them to Polya's four steps.

1.2 Exercises

Concept Check *Exercises 1–8 should be done mentally. They will prepare you for some of the applications found in this exercise set.*

1. If a train travels at 80 mph for 15 min, what is the distance traveled?

2. If 40 L of an acid solution is 75% acid, how much pure acid is there in the mixture?

3. If a person invests \$100 at 4% simple interest for 2 yr, how much interest is earned?

4. If a jar of coins contains 30 half-dollars and 100 quarters, what is the monetary value of the coins?

5. *Acid Mixture* Suppose two acid solutions are mixed. One is 26% acid and the other is 32% acid. Which one of the following concentrations cannot possibly be the concentration of the mixture?

A. 36% **B.** 28% **C.** 30% **D.** 31%

6. *Sale Price* Suppose that a computer that originally sold for x dollars has been discounted 30%. Which one of the following expressions does not represent its sale price?

A. $x - .30x$ **B.** $.70x$ **C.** $\frac{7}{10}x$ **D.** $x - .30$

7. ***Unknown Numbers*** Consider the following problem.

One number is 3 less than 6 times a second number. Their sum is 32. Find the numbers.

If x represents the second number, which equation is correct for solving this problem?

A. $32 - (x + 3) = 6x$ **B.** $(3 - 6x) + x = 32$
C. $32 - (3 - 6x) = x$ **D.** $(6x - 3) + x = 32$

8. ***Unknown Numbers*** Consider the following problem.

The difference between six times a number and 9 is equal to five times the sum of the number and 2. Find the number.

If x represents the number, which equation is correct for solving this problem?

A. $6x - 9 = 5(x + 2)$ **B.** $9 - 6x = 5(x + 2)$
C. $6x - 9 = 5x + 2$ **D.** $9 - 6x = 5x + 2$

Solve each problem. See Example 1.

9. ***Perimeter of a Rectangle*** The perimeter of a rectangle is 98 cm. The width is 19 cm. Find the length.

10. ***Perimeter of a Storage Shed*** Gunner Van Erden must build a rectangular storage shed. He wants the length to be 3 ft greater than the width, and the perimeter must be 22 ft. Find the length and the width of the shed.

11. ***Dimensions of a Label*** The length of a rectangular label is 2.5 cm less than twice the width. The perimeter is 40.6 cm. Find the width.

Side lengths are in centimeters.

12. ***Dimensions of a Puzzle Piece*** A puzzle piece in the shape of a triangle has perimeter 30 cm. Two sides of the triangle are each twice as long as the shortest side. Find the length of the shortest side.

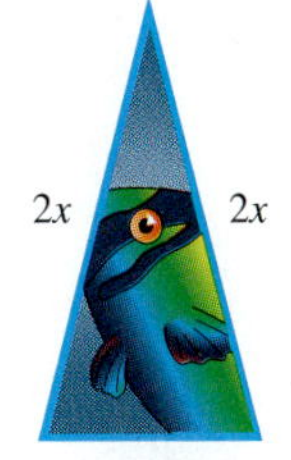

Side lengths are in centimeters.

13. ***Perimeter of a Plot of Land*** The perimeter of a triangular plot of land is 2400 ft. The longest side is 200 ft less than twice the shortest. The middle side is 200 ft less than the longest side. Find the lengths of the three sides of the triangular plot.

14. ***World's Largest Tablecloth*** The world's largest rectangular tablecloth has perimeter 44,252 in. It was manufactured by Döhler, S.A. of Joinville, Santa Catarina, Brazil on December 7, 1999. Its length is 22,000 in. more than its width. What is its length, to the nearest hundredth, in yards? (*Source:* www.guinnessworldrecords.com/index.asp)

15. ***Smallest Ticket Size*** The smallest ticket ever produced was for admission to the Asian-Pacific Exposition Fukuoka '89 in Japan. It was rectangular, with length 4 mm more than its width. If the width had been increased by 1 mm and the length had been increased by 4 mm, the ticket would have had perimeter equal to 30 mm. What were the actual dimensions of the ticket? (*Source: The Guinness Book of World Records 1995.*)

16. *Cylinder Dimensions* A right circular cylinder has radius 6 in. and volume 144π in.^{3}. What is its height? (In the figure, h = height.)

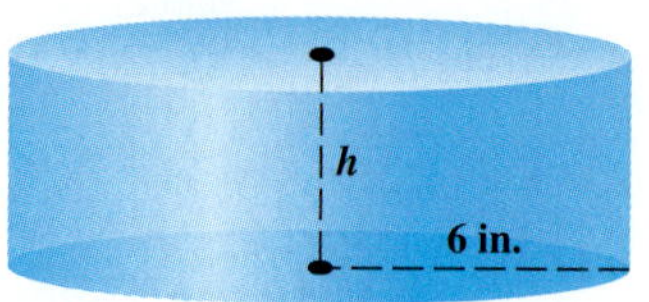

h is in inches.

17. *Recycling Bin Dimensions* A recycling bin is in the shape of a rectangular box. Find the height of the box if its length is 18 ft, its width is 8 ft, and its surface area is 496 ft^2. (In the figure, h = height. Assume that the given surface area includes that of the top lid of the box.)

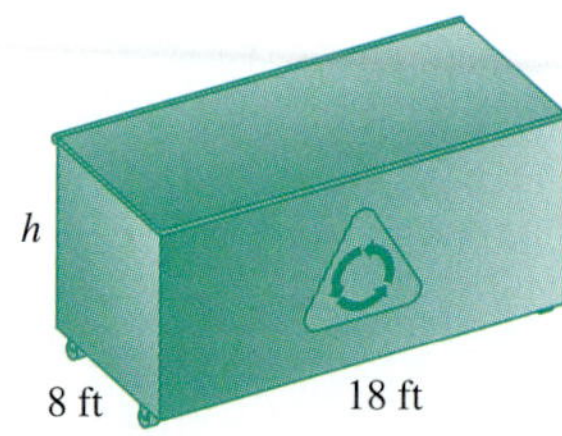

h is in feet.

18. *Concept Check* Which one or more of the following cannot be a correct equation to solve a geometry problem, if x represents the length of a rectangle? (*Hint:* Solve each equation and consider the solution.)

A. $2x + 2(x - 1) = 14$ **B.** $-2x + 7(5 - x) = 62$
C. $4(x + 2) + 4x = 8$ **D.** $2x + 2(x - 3) = 22$

Solve each problem. See Example 2.

19. *Distance to an Appointment* In the morning, Marge drove to a business appointment at 50 mph. Her average speed on the return trip in the afternoon was 40 mph. The return trip took $\frac{1}{4}$ hr longer because of heavy traffic. How far did she travel to the appointment?

	r	t	d
Morning	50	x	
Afternoon	40	$x + \frac{1}{4}$	

20. *Distance between Cities* On a vacation, Johnny averaged 50 mph traveling from Denver to Minneapolis. Returning by a different route that covered the same number of miles, he averaged 55 mph. What is the distance between the two cities if his total traveling time was 32 hr?

	r	t	d
Going	50	x	
Returning	55	$32 - x$	

21. ***Distance to Work*** David gets to work in 20 min when he drives his car. Riding his bike (by the same route) takes him 45 min. His average driving speed is 4.5 mph greater than his average speed on his bike. How far does he travel to work?

22. ***Speed of a Plane*** Two planes leave Los Angeles at the same time. One heads south to San Diego; the other heads north to San Francisco. The San Francisco plane flies 50 mph faster. In $\frac{1}{2}$ hr, the planes are 275 mi apart. What are their speeds?

23. ***Running Times*** Russ and Janet are running in the Apple Hill Fun Run. Russ runs at 7 mph, Janet at 5 mph. If they start at the same time, how long will it be before they are $\frac{1}{2}$ mi apart?

24. ***Running Times*** If the run in Exercise 23 has a staggered start, and Janet starts first, with Russ starting 10 min later, how long will it be before he catches up with her?

25. ***Track Event Speeds*** On September 14, 2002, Tim Montgomery (USA) set a world record in the 100-m dash with a time of 9.78 sec. If this pace could be maintained for an entire 26-mi marathon, what would his time be? How would this time compare to the fastest time for a marathon of 2 hr, 5 min, 38 sec? (*Hint:* 1 m $\approx$ 3.281 ft.) (*Source: The World Almanac,* 2003.)

26. ***Track Event Speeds*** At the 1996 Olympics, Donovan Bailey (Canada) set what was then the world record in the 100-m dash with a time of 9.84 sec. Refer to Exercise 25, and answer the questions using Bailey's time. (*Source: The World Almanac,* 2000.)

27. ***Boat Speed*** Joann took 20 min to drive her boat upstream to water-ski at her favorite spot. Coming back later in the day, at the same boat speed, took her 15 min. If the current in that part of the river is 5 km per hr, what was her boat speed?

28. ***Wind Speed*** Joe traveled against the wind in a small plane for 3 hr. The return trip with the wind took 2.8 hr. Find the speed of the wind if the speed of the plane in still air is 180 mph.

Solve each problem. See Example 3.

29. ***Painting a House*** (This problem appears in the 1994 movie *Little Big League.*) If Joe can paint a house in 3 hr, and Sam can paint the same house in 5 hr, how long does it take them to do it together?

30. *Painting a House* Repeat Exercise 29, but assume that Joe takes 6 hr working alone, and Sam takes 8 hr working alone.

31. *Pollution in a River* Two chemical plants are polluting a river. If plant A produces a predetermined maximum amount of pollutant twice as fast as plant B, and together they produce the maximum pollutant in 26 hr, how long will it take plant B alone?

	Rate	Time	Part of Job Accomplished
Pollution from A	$\frac{1}{x}$	26	$\frac{1}{x}(26)$
Pollution from B		26	

32. *Filling a Settling Pond* A sewage treatment plant has two inlet pipes to its settling pond. One can fill the pond in 10 hr, the other in 12 hr. If the first pipe is open for 5 hr and then the second pipe is opened, how long will it take to fill the pond?

33. *Filling a Pool* An inlet pipe can fill Reynaldo's pool in 5 hr, while an outlet pipe can empty it in 8 hr. In his haste to surf the Internet, Reynaldo left both pipes open. How long did it take to fill the pool?

34. *Filling a Pool* Suppose Reynaldo discovered his error (see Exercise 33) after an hour-long surf. If he then closed the outlet pipe, how much more time would be needed to fill the pool?

Solve each problem. See Example 4.

35. *Acid Mixture* How many gallons of a 5% acid solution must be mixed with 5 gal of a 10% solution to obtain a 7% solution?

Strength	Gallons of Solution	Gallons of Pure Acid
5%	x	
10%	5	
7%	$x + 5$	

36. *Alcohol Mixture* How many gallons of pure alcohol should be mixed with 20 gal of a 15% alcohol solution to obtain a mixture that is 25% alcohol?

37. *Alcohol Mixture* Bailey Eckles wishes to strengthen a mixture from 10% alcohol to 30% alcohol. How much pure alcohol should be added to 7 L of the 10% mixture?

38. *Acid Mixture* Travis Hayes needs 10% hydrochloric acid for a chemistry experiment. How much 5% acid should be mixed with 60 mL of 20% acid to get a 10% solution?

39. *Saline Solution* How much water should be added to 8 mL of 6% saline solution to reduce the concentration to 4%?

40. *Acid Mixture* How much pure acid should be added to 6 L of 30% acid to increase the concentration to 50% acid?

Solve each problem. See Example 5.

41. *Real Estate Financing* Cody Westmoreland wishes to sell a piece of property for \$240,000. He wants the money to be paid off in two ways—a short-term note at 6% interest and a long-term note at 5%. Find the amount of each note if the total annual interest paid is \$13,000.

42. *Buying and Selling Land* Anoa bought two plots of land for a total of \$120,000. When she sold the first plot, she made a profit of 15%. When she sold the second, she lost 10%. Her total profit was \$5500. How much did she pay for each piece of land?

43. *Retirement Planning* In planning her retirement, Karen Guardino deposits some money at 2.5% interest with twice as much deposited at 3%. Find the amount deposited at each rate if the total annual interest income is \$850.

44. *Investing a Building Fund* A church building fund has invested some money in two ways: part of the money at 4% interest and four times as much at 3.5%. Find the amount invested at each rate if the total annual income from interest is \$3600.

45. *Lottery Winnings* Marietta won \$200,000 in a state lottery. She first paid income tax of 30% on the winnings. Of the rest she invested some at 1.5% and some at 4%, earning \$4350 interest per year. How much did she invest at each rate?

46. *Cookbook Royalties* Latasha Williams earned \$48,000 from royalties on her cookbook. She paid a 28% income tax on these royalties. The balance was invested in two ways, some of it at 3.25% interest and some at 1.75%. The investments produced \$904.80 interest per year. Find the amount invested at each rate.

(Modeling) Solve each problem. See Examples 6 and 7.

47. *Indoor Air Quality and Control* The excess lifetime cancer risk R is a measure of the likelihood that an individual will develop cancer from a particular pollutant. For example, if $R = .01$ then a person has a 1% increased chance of developing cancer during a lifetime. (This would translate into 1 case of cancer for every 100 people during an average lifetime.) The value of R for formaldehyde, a highly toxic indoor air pollutant, can be calculated using the linear model $R = kd$, where k is a constant and d is the daily dose in parts per million. The constant k for formaldehyde can be calculated using the formula

$$k = \frac{.132B}{W},$$

where B is the total number of cubic meters of air a person breathes in one day and W is a person's weight in kilograms. (*Source:* Hines, A., T. Ghosh, S. Loyalka, and R. Warder, *Indoor Air: Quality & Control,* Prentice-Hall, 1993; Ritchie, I., and R. Lehnen, "An Analysis of Formaldehyde Concentration in Mobile and Conventional Homes," *J. Env. Health* 47: 300–305.)

(a) Find k for a person who breathes in 20 m^3 of air per day and weighs 75 kg.

(b) Mobile homes in Minnesota were found to have a mean daily dose d of .42 part per million. Calculate R using the value of k found in part (a).

(c) For every 5000 people, how many cases of cancer could be expected each year from these levels of formaldehyde? Assume an average life expectancy of 72 yr.

48. *Indoor Air Quality and Control* (See Exercise 47.) For nonsmokers exposed to environmental tobacco smoke (passive smokers), $R = 1.5 \times 10^{-3}$. (*Source:* Hines, A., T. Ghosh, S. Loyalka, and R. Warder, *Indoor Air: Quality & Control,* Prentice-Hall, 1993.)

(a) If the average life expectancy is 72 yr, what is the excess lifetime cancer risk from second-hand tobacco smoke per year?

(b) Write a linear equation that will model the expected number of cancer cases C per year if there are x passive smokers.

(c) Estimate the number of cancer cases each year per 100,000 passive smokers.

(d) The excess lifetime risk of death from smoking is $R = .44$. Currently 26% of the U.S. population smoke. If the U.S. population is 260 million, approximate the excess number of deaths caused by smoking each year.

49. *Eye Irritation from Formaldehyde* When concentrations of formaldehyde in the air exceed 33 $\mu g/ft^3$ (1 $\mu g = 1$ microgram $= 10^{-6}$ gram), a strong odor and irritation to the eyes often occurs. One square foot of hardwood plywood paneling can emit 3365 μg of formaldehyde per day. (*Source:* Hines, A., T. Ghosh, S. Loyalka, and R. Warder, *Indoor Air: Quality & Control,* Prentice-Hall, 1993.) A 4-ft by 8-ft sheet of this paneling is attached to an 8-ft wall in a room having floor dimensions of 10 ft by 10 ft.

(a) How many cubic feet of air are there in the room?

(b) Find the total number of micrograms of formaldehyde that are released into the air by the paneling each day.

(c) If there is no ventilation in the room, write a linear equation that models the amount of formaldehyde F in the room after x days.

(d) How long will it take before a person's eyes become irritated in the room?

50. *Classroom Ventilation* According to the American Society of Heating, Refrigerating and Air-Conditioning Engineers, Inc. (ASHRAE), a nonsmoking classroom should have a ventilation rate of 15 ft^3 per min for each person in the classroom. (*Source: ASHRAE,* 1989.)

(a) Write an equation that models the total ventilation V (in cubic feet per hour) necessary for a classroom with x students.

(b) A common unit of ventilation is an air change per hour (ach). 1 ach is equivalent to exchanging all of the air in a room every hour. If x students are in a classroom having volume 15,000 ft^3, determine how many air exchanges per hour (A) are necessary to keep the room properly ventilated.

(c) Find the necessary number of ach A if the classroom has 40 students in it.

(d) In areas like bars and lounges that allow smoking, the ventilation rate should be increased to 50 ft^3 per min per person. Compared to classrooms, ventilation should be increased by what factor in heavy smoking areas?

51. *College Enrollments* The graph on the next page shows the projections in total enrollment at degree-granting institutions from Fall 2003 to Fall 2012. The linear model

$$y = .2145x + 15.69$$

provides the approximate enrollment, in millions, between the years 2003 and 2012, where $x = 0$ corresponds to 2003, $x = 1$ to 2004, and so on, and y is in millions of students.

(a) Use the model to determine projected enrollment for Fall 2008.

(b) Use the model to determine the year in which enrollment is projected to reach 17 million.

(c) How do your answers to parts (a) and (b) compare to the corresponding values shown in the graph?

(d) The enrollment in Fall 1993 was 14.3 million. The model here is based on data from 2003 to 2012. If you were to use the model for 1993, what would the enrollment be?

(e) Why do you think there is such a discrepancy between the actual value and the value based on the model in part (d)? Discuss the pitfalls of using the model to predict for years preceding 2003.

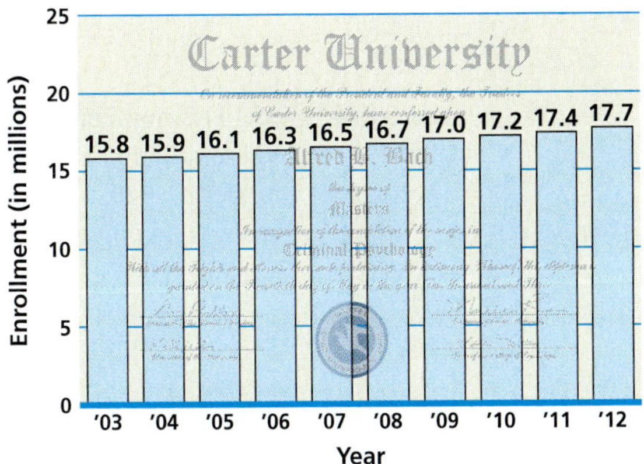

Source: U.S. Department of Education, National Center for Educational Statistics.

52. *Mobility of Americans* The bar graph shows the percent of Americans moving each year in the decades since the 1950s. The linear model

$$y = -.9x + 21.2,$$

where y represents the percent moving each year in decade x, approximates the data fairly well. The decades are coded, so $x = 1$ represents the 1950s, $x = 2$ represents the 1960s, and so on. Use this model to answer each question.

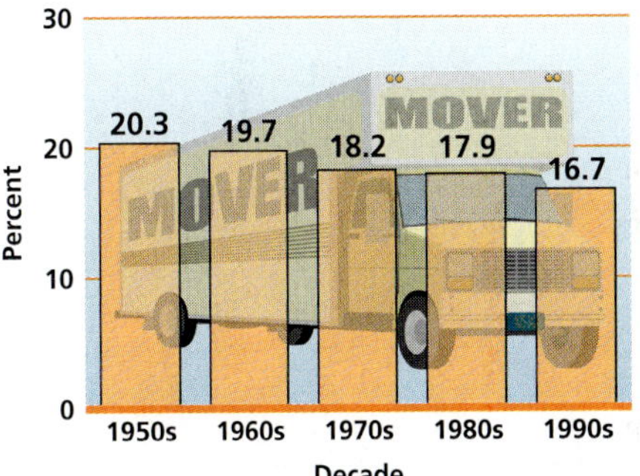

Source: U.S. Bureau of the Census.

(a) What was the approximate percent of Americans moving in the 1960s? Compare your answer to the data in the graph.

(b) What was the approximate percent of Americans moving in the 1980s? How does the approximation compare to the data in the graph?

(c) The 1997 U.S. population was about 268 million. From the data given in the graph, how many Americans moved that year?

1.3 Complex Numbers

Basic Concepts of Complex Numbers ▪ Operations on Complex Numbers

Basic Concepts of Complex Numbers So far, we have worked only with real numbers. The set of real numbers, however, does not include all numbers needed in algebra. For example, there is no real number solution of the equation

$$x^2 = -1,$$

since -1 has no real square root. Square roots of negative numbers were not incorporated into an integrated number system until the 16th century. They were then used as solutions of equations and, later in the 18th century, in surveying. Today, such numbers are used extensively in science and engineering.

To extend the real number system to include numbers such as $\sqrt{-1}$, the number i is defined to have the following property.

$$i^2 = -1$$

Thus, $\boldsymbol{i = \sqrt{-1}}$. The number i is called the **imaginary unit.** Numbers of the form $a + bi$, where a and b are real numbers, are called **complex numbers.** In the complex number $a + bi$, a is the **real part** and b is the **imaginary part.***

The relationships among the various sets of numbers are shown in Figure 3.

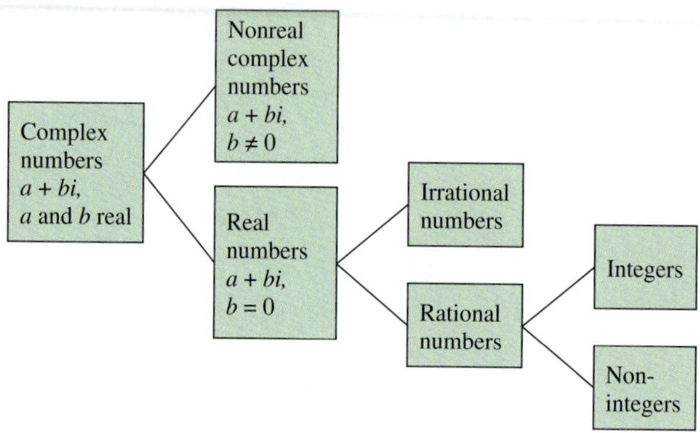

Figure 3

Two complex numbers $a + bi$ and $c + di$ are equal provided that their real parts are equal and their imaginary parts are equal; that is,

$$\boldsymbol{a + bi = c + di} \quad \textbf{if and only if} \quad \boldsymbol{a = c} \quad \textbf{and} \quad \boldsymbol{b = d}.$$

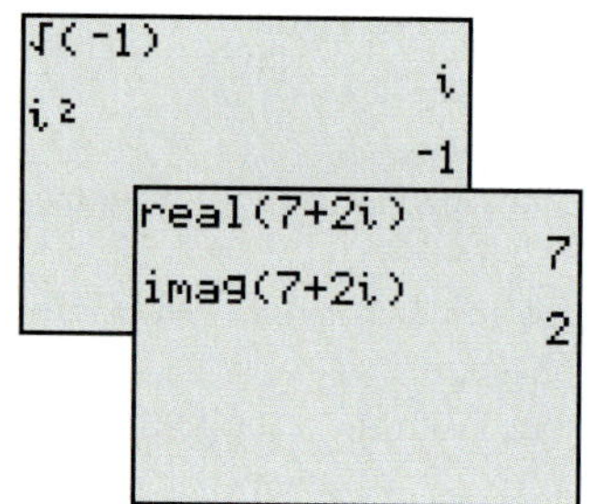

The calculator is in complex number mode.

Figure 4

For a complex number $a + bi$, if $b = 0$, then $a + bi = a$, which is a real number. Thus, the set of real numbers is a subset of the set of complex numbers. If $a = 0$ and $b \neq 0$, the complex number is said to be a **pure imaginary number.** For example, $3i$ is a pure imaginary number. A number such as $7 + 2i$ is a nonreal complex number. A complex number written in the form $a + bi$ (or $a + ib$) is in **standard form.** (The form $a + ib$ is used to write expressions such as $i\sqrt{5}$, since $\sqrt{5}i$ could be mistaken for $\sqrt{5i}$.)

Now try Exercises 9, 11, and 13.

Some graphing calculators, such as the TI-83 Plus, are capable of working with complex numbers, as seen in Figure 4. The top screen supports the definition of i. The bottom screen shows how the calculator returns the real and imaginary parts of $7 + 2i$. ■

For a positive real number a, $\sqrt{-a}$ is defined as follows.

$\sqrt{-a}$

If $a > 0$, then $\boldsymbol{\sqrt{-a} = i\sqrt{a}}$.

*In some texts, the term bi is defined to be the imaginary part.

EXAMPLE 1 Writing $\sqrt{-a}$ as $i\sqrt{a}$

Write as the product of a real number and i, using the definition of $\sqrt{-a}$.

(a) $\sqrt{-16}$ **(b)** $\sqrt{-70}$ **(c)** $\sqrt{-48}$

Solution

(a) $\sqrt{-16} = i\sqrt{16} = 4i$ **(b)** $\sqrt{-70} = i\sqrt{70}$

(c) $\sqrt{-48} = i\sqrt{48} = i\sqrt{16 \cdot 3} = 4i\sqrt{3}$ Product rule for radicals (Section R.7)

Now try Exercises 17, 19, and 21.

Operations on Complex Numbers Products or quotients with negative radicands are simplified by first rewriting $\sqrt{-a}$ as $i\sqrt{a}$ for a positive number a. Then the properties of real numbers are applied, together with the fact that $i^2 = -1$.

CAUTION When working with negative radicands, *use the definition* $\sqrt{-a} = i\sqrt{a}$ *before using any of the other rules for radicals.* In particular, the rule $\sqrt{c} \cdot \sqrt{d} = \sqrt{cd}$ is valid only when c and d are *not* both negative. For example,

$$\sqrt{(-4)(-9)} = \sqrt{36} = 6,$$

while

$$\sqrt{-4} \cdot \sqrt{-9} = 2i(3i) = 6i^2 = -6,$$

so

$$\sqrt{-4} \cdot \sqrt{-9} \neq \sqrt{(-4)(-9)}.$$

EXAMPLE 2 Finding Products and Quotients Involving Negative Radicands

Multiply or divide as indicated. Simplify each answer.

(a) $\sqrt{-7} \cdot \sqrt{-7}$ **(b)** $\sqrt{-6} \cdot \sqrt{-10}$ **(c)** $\dfrac{\sqrt{-20}}{\sqrt{-2}}$ **(d)** $\dfrac{\sqrt{-48}}{\sqrt{24}}$

Solution

(a)
$$\begin{aligned} \sqrt{-7} \cdot \sqrt{-7} &= i\sqrt{7} \cdot i\sqrt{7} \\ &= i^2 \cdot (\sqrt{7})^2 \\ &= -1 \cdot 7 \quad i^2 = -1 \\ &= -7 \end{aligned}$$

(b)
$$\begin{aligned} \sqrt{-6} \cdot \sqrt{-10} &= i\sqrt{6} \cdot i\sqrt{10} \\ &= i^2 \cdot \sqrt{60} \\ &= -1\sqrt{4 \cdot 15} \\ &= -1 \cdot 2\sqrt{15} \\ &= -2\sqrt{15} \end{aligned}$$

(c) $\dfrac{\sqrt{-20}}{\sqrt{-2}} = \dfrac{i\sqrt{20}}{i\sqrt{2}} = \sqrt{\dfrac{20}{2}} = \sqrt{10}$ Quotient rule for radicals (Section R.7)

(d) $\dfrac{\sqrt{-48}}{\sqrt{24}} = \dfrac{i\sqrt{48}}{\sqrt{24}} = i\sqrt{\dfrac{48}{24}} = i\sqrt{2}$

Now try Exercises 25, 27, 29, and 31.

EXAMPLE 3 Simplifying a Quotient Involving a Negative Radicand

Write $\frac{-8 + \sqrt{-128}}{4}$ in standard form $a + bi$.

Solution

$$\frac{-8 + \sqrt{-128}}{4} = \frac{-8 + \sqrt{-64 \cdot 2}}{4}$$

$$= \frac{-8 + 8i\sqrt{2}}{4} \qquad \sqrt{-64} = 8i$$

$$= \frac{4\left(-2 + 2i\sqrt{2}\right)}{4} \qquad \text{Factor. (Section R.4)}$$

$$= -2 + 2i\sqrt{2} \qquad \text{Lowest terms (Section R.5)}$$

Now try Exercise 37.

With the definitions $i^2 = -1$ and $\sqrt{-a} = i\sqrt{a}$ for $a > 0$, all properties of real numbers are extended to complex numbers. As a result, complex numbers are added, subtracted, multiplied, and divided using the following definitions and real number properties.

Addition and Subtraction of Complex Numbers

For complex numbers $a + bi$ and $c + di$,

$$(a + bi) + (c + di) = (a + c) + (b + d)i$$

and
$$(a + bi) - (c + di) = (a - c) + (b - d)i.$$

That is, to add or subtract complex numbers, add or subtract the real parts and add or subtract the imaginary parts.

EXAMPLE 4 Adding and Subtracting Complex Numbers

Find each sum or difference.

(a) $(3 - 4i) + (-2 + 6i)$

(b) $(-9 + 7i) + (3 - 15i)$

(c) $(-4 + 3i) - (6 - 7i)$

(d) $(12 - 5i) - (8 - 3i)$

Solution

(a) $(3 - 4i) + (-2 + 6i) = [3 + (-2)] + [-4 + 6]i$ Commutative, associative, distributive properties (Section R.1)

$$= 1 + 2i$$

(b) $(-9 + 7i) + (3 - 15i) = -6 - 8i$

(c) $(-4 + 3i) - (6 - 7i) = (-4 - 6) + [3 - (-7)]i$
$= -10 + 10i$

(d) $(12 - 5i) - (8 - 3i) = 4 - 2i$

Now try Exercises 43 and 45.

The product of two complex numbers is found by multiplying as if the numbers were binomials and using the fact that $i^2 = -1$, as follows.

$$(a + bi)(c + di) = ac + adi + bic + bidi \quad \text{FOIL (Section R.3)}$$
$$= ac + adi + bci + bdi^2$$
$$= ac + (ad + bc)i + bd(-1) \quad \text{Distributive property; } i^2 = -1$$
$$= (ac - bd) + (ad + bc)i$$

Multiplication of Complex Numbers

For complex numbers $a + bi$ and $c + di$,

$$(a + bi)(c + di) = (ac - bd) + (ad + bc)i.$$

This definition is not practical in routine calculations. To find a given product, it is easier just to multiply as with binomials.

EXAMPLE 5 Multiplying Complex Numbers

Find each product.

(a) $(2 - 3i)(3 + 4i)$ **(b)** $(4 + 3i)^2$ **(c)** $(6 + 5i)(6 - 5i)$

Solution

(a) $(2 - 3i)(3 + 4i) = 2(3) + 2(4i) - 3i(3) - 3i(4i)$ FOIL
$= 6 + 8i - 9i - 12i^2$
$= 6 - i - 12(-1)$ $i^2 = -1$
$= 18 - i$

```
(2-3i)(3+4i)
            18-i
(4+3i)²
           7+24i
(6+5i)(6-5i)
              61
```

This screen shows how the TI–83 Plus displays the results found in Example 5.

(b) $(4 + 3i)^2 = 4^2 + 2(4)(3i) + (3i)^2$ Square of a binomial (Section R.3)
$= 16 + 24i + 9i^2$
$= 16 + 24i + 9(-1)$ $i^2 = -1$
$= 7 + 24i$

(c) $(6 + 5i)(6 - 5i) = 6^2 - (5i)^2$ Product of the sum and difference of two terms (Section R.3)
$= 36 - 25(-1)$ $i^2 = -1$
$= 36 + 25$
$= 61$ or $61 + 0i$ Standard form

Now try Exercises 51, 55, and 59.

Powers of i can be simplified using the facts

$$i^2 = -1 \qquad \text{and} \qquad i^4 = (i^2)^2 = (-1)^2 = 1.$$

EXAMPLE 6 Simplifying Powers of *i*

Simplify each power of i.

(a) i^{15} **(b)** i^{-3}

Solution

(a) Since $i^2 = -1$ and $i^4 = 1$, write the given power as a product involving i^2 or i^4. For example,

$$i^3 = i^2 \cdot i = (-1) \cdot i = -i.$$

Alternatively, using i^4 and i^3 to rewrite i^{15} gives

$$i^{15} = i^{12} \cdot i^3 = (i^4)^3 \cdot i^3 = 1^3(-i) = -i.$$

(b) $i^{-3} = i^{-4} \cdot i = (i^4)^{-1} \cdot i = (1)^{-1} \cdot i = i$

Now try Exercises 69 and 77.

We can use the method of Example 6 to construct the following list of powers of i.

```
i²
          -1
i³
          -i
i^4
           1
```

Powers of i on the TI–83 Plus calculator

Powers of *i*

$i^1 = i$	$i^5 = i$	$i^9 = i$	
$i^2 = -1$	$i^6 = -1$	$i^{10} = -1$	
$i^3 = -i$	$i^7 = -i$	$i^{11} = -i$	
$i^4 = 1$	$i^8 = 1$	$i^{12} = 1,$	and so on.

Example 5(c) showed that $(6 + 5i)(6 - 5i) = 61$. The numbers $6 + 5i$ and $6 - 5i$ differ only in the sign of their imaginary parts, and are called **complex conjugates.** The product of a complex number and its conjugate is always a real number.

Property of Complex Conjugates

For real numbers a and b,

$$(a + bi)(a - bi) = a^2 + b^2.$$

The complex conjugate of the divisor is used to find the quotient of two complex numbers in standard form. Multiplying both the numerator and the denominator by the complex conjugate of the denominator gives the quotient, as shown in the next example.

EXAMPLE 7 Dividing Complex Numbers

Find each quotient.

(a) $\dfrac{3+2i}{5-i}$ **(b)** $\dfrac{3}{i}$

Solution

```
(3+2i)/(5-i)▶Fra
c
        1/2+1/2i
3/i
            -3i
```

This screen supports the results in Example 7. Notice that the answer to part (a) must be interpreted $\frac{1}{2}+\frac{1}{2}i$, *not* $\frac{1}{2}+\frac{1}{2i}$.

(a) $\dfrac{3+2i}{5-i} = \dfrac{(3+2i)(5+i)}{(5-i)(5+i)}$ Multiply by the complex conjugate of the denominator in both the numerator and the denominator.

$= \dfrac{15+3i+10i+2i^2}{25-i^2}$ Multiply.

$= \dfrac{13+13i}{26}$ $i^2=-1$

$= \dfrac{13}{26}+\dfrac{13i}{26}$ $\dfrac{a+bi}{c}=\dfrac{a}{c}+\dfrac{bi}{c}$ (Section R.5)

$= \dfrac{1}{2}+\dfrac{1}{2}i$ Lowest terms; standard form

To *check* this answer, show that

$$\left(\frac{1}{2}+\frac{1}{2}i\right)(5-i)=3+2i. \quad \text{Quotient} \times \text{Divisor} = \text{Dividend}$$

(b) $\dfrac{3}{i} = \dfrac{3(-i)}{i(-i)}$ $-i$ is the conjugate of i.

$= \dfrac{-3i}{-i^2}$

$= \dfrac{-3i}{1}$ $-i^2=-(-1)=1$

$= -3i$ or $0-3i$ Standard form

Now try Exercises 83 and 89.

1.3 Exercises

Concept Check Determine whether each statement is true *or* false. *If it is false, tell why.*

1. Every real number is a complex number.
2. No real number is a pure imaginary number.
3. Every pure imaginary number is a complex number.
4. A number can be both real and complex.
5. There is no real number that is a complex number.
6. A complex number might not be a pure imaginary number.

Identify each number as real, complex, *or* pure imaginary. *(More than one of these descriptions may apply.)*

7. -6 **8.** 0 **9.** $10i$ **10.** $-8i$ **11.** $2 + i$

12. $-5 - 2i$ **13.** π **14.** $\sqrt{8}$ **15.** $\sqrt{-9}$ **16.** $\sqrt{-16}$

Write each number as the product of a real number and i. See Example 1.

17. $\sqrt{-25}$ **18.** $\sqrt{-36}$ **19.** $\sqrt{-10}$ **20.** $\sqrt{-15}$

21. $\sqrt{-288}$ **22.** $\sqrt{-500}$ **23.** $-\sqrt{-18}$ **24.** $-\sqrt{-80}$

Multiply or divide as indicated. Simplify each answer. See Example 2.

25. $\sqrt{-13} \cdot \sqrt{-13}$ **26.** $\sqrt{-17} \cdot \sqrt{-17}$ **27.** $\sqrt{-3} \cdot \sqrt{-8}$

28. $\sqrt{-5} \cdot \sqrt{-15}$ **29.** $\dfrac{\sqrt{-30}}{\sqrt{-10}}$ **30.** $\dfrac{\sqrt{-70}}{\sqrt{-7}}$

31. $\dfrac{\sqrt{-24}}{\sqrt{8}}$ **32.** $\dfrac{\sqrt{-54}}{\sqrt{27}}$ **33.** $\dfrac{\sqrt{-10}}{\sqrt{-40}}$

34. $\dfrac{\sqrt{-40}}{\sqrt{20}}$ **35.** $\dfrac{\sqrt{-6} \cdot \sqrt{-2}}{\sqrt{3}}$ **36.** $\dfrac{\sqrt{-12} \cdot \sqrt{-6}}{\sqrt{8}}$

Write each number in standard form a + bi. See Example 3.

37. $\dfrac{-6 + \sqrt{-24}}{2}$ **38.** $\dfrac{-9 + \sqrt{-18}}{3}$ **39.** $\dfrac{10 - \sqrt{-200}}{5}$

40. $\dfrac{20 - \sqrt{-8}}{2}$ **41.** $\dfrac{3 + \sqrt{-18}}{24}$ **42.** $\dfrac{5 + \sqrt{-50}}{10}$

Find each sum or difference. Write the answer in standard form. See Example 4.

43. $(3 + 2i) + (4 - 3i)$ **44.** $(4 - i) + (2 + 5i)$

45. $(-2 + 3i) - (-4 + 3i)$ **46.** $(-3 + 5i) - (-4 + 5i)$

47. $(2 - 5i) - (3 + 4i) - (-2 + i)$ **48.** $(-4 - i) - (2 + 3i) + (-4 + 5i)$

49. $-i - 2 - (6 - 4i) - (5 - 2i)$ **50.** $3 - (4 - i) - 4i + (-2 + 5i)$

Find each product. Write the answer in standard form. See Example 5.

51. $(2 + i)(3 - 2i)$ **52.** $(-2 + 3i)(4 - 2i)$ **53.** $(2 + 4i)(-1 + 3i)$

54. $(1 + 3i)(2 - 5i)$ **55.** $(-3 + 2i)^2$ **56.** $(2 + i)^2$

57. $(3 + i)(-3 - i)$ **58.** $(-5 - i)(5 + i)$ **59.** $(2 + 3i)(2 - 3i)$

60. $(6 - 4i)(6 + 4i)$ **61.** $(\sqrt{6} + i)(\sqrt{6} - i)$ **62.** $(\sqrt{2} - 4i)(\sqrt{2} + 4i)$

63. $i(3 - 4i)(3 + 4i)$ **64.** $i(2 + 7i)(2 - 7i)$ **65.** $3i(2 - i)^2$

66. $-5i(4 - 3i)^2$ **67.** $(2 + i)(2 - i)(4 + 3i)$ **68.** $(3 - i)(3 + i)(2 - 6i)$

Simplify each power of i. See Example 6.

69. i^{21} **70.** i^{25} **71.** i^{22} **72.** i^{26}

73. i^{23} **74.** i^{27} **75.** i^{24} **76.** i^{32}

77. i^{-9} **78.** i^{-10} **79.** $\dfrac{1}{i^{-11}}$ **80.** $\dfrac{1}{i^{12}}$

81. Suppose that your friend, Ceci Stashwick, tells you that she has discovered a method of simplifying a positive power of i. "Just divide the exponent by 4," she says, "and then look at the remainder. Then refer to the table of powers of i in this section. The large power of i is equal to the power indicated by the remainder. And if the remainder is 0, the result is $i^0 = 1$." Explain why her method works.

82. Explain why the following method of simplifying i^{-46} works.

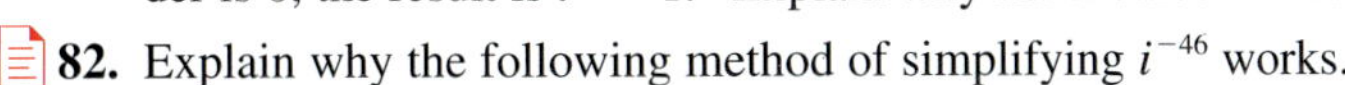

$$i^{-46} = i^{-46} \cdot i^{48} = i^{-46+48} = i^2 = -1$$

Find each quotient. Write the answer in standard form. See Example 7.

83. $\dfrac{6 + 2i}{1 + 2i}$ **84.** $\dfrac{14 + 5i}{3 + 2i}$ **85.** $\dfrac{2 - i}{2 + i}$

86. $\dfrac{4 - 3i}{4 + 3i}$ **87.** $\dfrac{1 - 3i}{1 + i}$ **88.** $\dfrac{-3 + 4i}{2 - i}$

89. $\dfrac{5}{i}$ **90.** $\dfrac{6}{i}$ **91.** $\dfrac{-8}{-i}$

92. $\dfrac{-12}{-i}$ **93.** $\dfrac{2}{3i}$ **94.** $\dfrac{5}{9i}$

95. Show that $\dfrac{\sqrt{2}}{2} + \dfrac{\sqrt{2}}{2}i$ is a square root of i.

96. Show that $\dfrac{\sqrt{3}}{2} + \dfrac{1}{2}i$ is a cube root of i.

97. Evaluate $3z - z^2$ if $z = 3 - 2i$.

98. Evaluate $-2z + z^3$ if $z = -6i$.

1.4 Quadratic Equations

Solving a Quadratic Equation ▪ Completing the Square ▪ The Quadratic Formula ▪ Solving for a Specified Variable ▪ The Discriminant

A **quadratic equation** is defined as follows.

Quadratic Equation in One Variable

An equation that can be written in the form

$$ax^2 + bx + c = 0,$$

where a, b, and c are real numbers with $a \neq 0$, is a **quadratic equation.**

A quadratic equation is a *second-degree* equation—that is, an equation with a squared term and no terms of greater degree.

$$x^2 = 25, \qquad 4x^2 + 4x - 5 = 0, \qquad 3x^2 = 4x - 8 \quad \text{Quadratic equations}$$

A quadratic equation written in the form $ax^2 + bx + c = 0$ is in *standard form.*

Solving a Quadratic Equation Factoring is the simplest method of solving a quadratic equation (but one that is not always easily applied). This method depends on the following **zero-factor property.**

Zero-Factor Property

If a and b are complex numbers with $ab = 0$, then $a = 0$ or $b = 0$ or both.

EXAMPLE 1 Using the Zero-Factor Property

Solve $6x^2 + 7x = 3$.

Solution

$$6x^2 + 7x = 3$$

$$6x^2 + 7x - 3 = 0 \qquad \text{Standard form}$$

$$(3x - 1)(2x + 3) = 0 \qquad \text{Factor. (Section R.4)}$$

$$3x - 1 = 0 \quad \text{or} \quad 2x + 3 = 0 \qquad \text{Zero-factor property}$$

$$3x = 1 \quad \text{or} \quad 2x = -3 \qquad \text{Solve each equation. (Section 1.1)}$$

$$x = \frac{1}{3} \quad \text{or} \quad x = -\frac{3}{2}$$

Check: $6x^2 + 7x = 3$ Original equation

$$6\left(\frac{1}{3}\right)^2 + 7\left(\frac{1}{3}\right) = 3 \qquad \text{Let } x = \tfrac{1}{3}.$$

$$\frac{6}{9} + \frac{7}{3} = 3 \qquad ?$$

$$3 = 3 \qquad \text{True}$$

$$6\left(-\frac{3}{2}\right)^2 + 7\left(-\frac{3}{2}\right) = 3 \qquad \text{Let } x = -\tfrac{3}{2}.$$

$$\frac{54}{4} - \frac{21}{2} = 3 \qquad ?$$

$$3 = 3 \qquad \text{True}$$

Both values check, and the solution set is $\left\{\frac{1}{3}, -\frac{3}{2}\right\}$.

Now try Exercise 15.

A quadratic equation of the form $x^2 = k$ can also be solved by factoring.

$$x^2 = k$$

$$x^2 - k = 0 \qquad \text{Subtract } k.$$

$$\left(x - \sqrt{k}\right)\left(x + \sqrt{k}\right) = 0 \qquad \text{Factor.}$$

$$x - \sqrt{k} = 0 \quad \text{or} \quad x + \sqrt{k} = 0 \qquad \text{Zero-factor property}$$

$$x = \sqrt{k} \quad \text{or} \quad x = -\sqrt{k} \qquad \text{Solve each equation.}$$

This proves the **square root property.**

Square Root Property

If $x^2 = k$, then

$$x = \sqrt{k} \quad \text{or} \quad x = -\sqrt{k}.$$

That is, the solution set of $x^2 = k$ is $\{\sqrt{k}, -\sqrt{k}\}$, which may be abbreviated $\{\pm\sqrt{k}\}$. Both solutions are real if $k > 0$, and both are pure imaginary if $k < 0$. If $k < 0$, we write the solution set as $\{\pm i\sqrt{|k|}\}$. (If $k = 0$, there is only one distinct solution, 0, sometimes called a *double* solution.)

EXAMPLE 2 Using the Square Root Property

Solve each quadratic equation.

(a) $x^2 = 17$ **(b)** $x^2 = -25$ **(c)** $(x - 4)^2 = 12$

Solution

(a) By the square root property, the solution set of $x^2 = 17$ is $\{\pm\sqrt{17}\}$.

(b) Since $\sqrt{-1} = i$, the solution set of $x^2 = -25$ is $\{\pm 5i\}$.

(c) Use a generalization of the square root property.

$$(x - 4)^2 = 12$$

$$x - 4 = \pm\sqrt{12} \qquad \text{Generalized square root property}$$

$$x = 4 \pm \sqrt{12} \qquad \text{Add 4.}$$

$$x = 4 \pm 2\sqrt{3} \qquad \sqrt{12} = \sqrt{4 \cdot 3} = 2\sqrt{3} \text{ (Section R.7)}$$

The solution set is $\{4 \pm 2\sqrt{3}\}$.

Now try Exercises 21, 23, and 25.

Completing the Square Any quadratic equation can be solved by the method of **completing the square,** as summarized here.

Solving a Quadratic Equation by Completing the Square

To solve $ax^2 + bx + c = 0$, $a \neq 0$, by completing the square:

Step 1 If $a \neq 1$, multiply both sides of the equation by $\frac{1}{a}$.

Step 2 Rewrite the equation so that the constant term is alone on one side of the equal sign.

Step 3 Square half the coefficient of x, and add this square to both sides of the equation.

Step 4 Factor the resulting trinomial as a perfect square and combine terms on the other side.

Step 5 Use the square root property to complete the solution.

EXAMPLE 3 Using the Method of Completing the Square, $a = 1$

Solve $x^2 - 4x - 14 = 0$ by completing the square.

Solution

Step 1 This step is not necessary since $a = 1$.

Step 2 $x^2 - 4x = 14$ Add 14 to both sides.

Step 3 $x^2 - 4x + 4 = 14 + 4$ $\left[\frac{1}{2}(-4)\right]^2 = 4$; add 4 to both sides.

Step 4 $(x - 2)^2 = 18$ Factor **(Section R.4)**; combine terms.

Step 5 $x - 2 = \pm\sqrt{18}$ Square root property

$x = 2 \pm \sqrt{18}$ Add 2.

$x = 2 \pm 3\sqrt{2}$ Simplify the radical.

The solution set is $\{2 \pm 3\sqrt{2}\}$.

Now try Exercise 35.

EXAMPLE 4 Using the Method of Completing the Square, $a \neq 1$

Solve $9x^2 - 12x - 1 = 0$ by completing the square.

Solution

$$9x^2 - 12x - 1 = 0$$

$$x^2 - \frac{4}{3}x - \frac{1}{9} = 0 \quad \text{Multiply by } \tfrac{1}{9}. \text{ (Step 1)}$$

$$x^2 - \frac{4}{3}x = \frac{1}{9} \quad \text{Add } \tfrac{1}{9}. \text{ (Step 2)}$$

$$x^2 - \frac{4}{3}x + \frac{4}{9} = \frac{1}{9} + \frac{4}{9} \quad \left[\tfrac{1}{2}\left(-\tfrac{4}{3}\right)\right]^2 = \tfrac{4}{9}; \text{ add } \tfrac{4}{9}. \text{ (Step 3)}$$

$$\left(x - \frac{2}{3}\right)^2 = \frac{5}{9} \quad \text{Factor; combine terms. (Step 4)}$$

$$x - \frac{2}{3} = \pm\sqrt{\frac{5}{9}} \quad \text{Square root property (Step 5)}$$

$$x - \frac{2}{3} = \pm\frac{\sqrt{5}}{3} \quad \text{Quotient rule for radicals } \textbf{(Section R.7)}$$

$$x = \frac{2}{3} \pm \frac{\sqrt{5}}{3} \quad \text{Add } \tfrac{2}{3}.$$

The solution set is $\left\{\frac{2 \pm \sqrt{5}}{3}\right\}$.

Now try Exercise 39.

The Quadratic Formula The method of completing the square can be used to solve any quadratic equation. However, in the long run it is better to start with the general quadratic equation, $ax^2 + bx + c = 0$, $a \neq 0$, and use the method of completing the square to solve this equation for x in terms of the constants a, b, and c. The result is a general formula for solving any quadratic equation. For now, assume that $a > 0$.

$$ax^2 + bx + c = 0$$

$$x^2 + \frac{b}{a}x + \frac{c}{a} = 0 \qquad \text{Multiply by } \tfrac{1}{a}. \text{ (Step 1)}$$

$$x^2 + \frac{b}{a}x = -\frac{c}{a} \qquad \text{Subtract } \tfrac{c}{a}. \text{ (Step 2)}$$

Square half the coefficient of x: $\left[\frac{1}{2}\left(\frac{b}{a}\right)\right]^2 = \left(\frac{b}{2a}\right)^2 = \frac{b^2}{4a^2}$.

$$x^2 + \frac{b}{a}x + \frac{b^2}{4a^2} = -\frac{c}{a} + \frac{b^2}{4a^2} \qquad \text{Add } \tfrac{b^2}{4a^2} \text{ to each side. (Step 3)}$$

$$\left(x + \frac{b}{2a}\right)^2 = \frac{b^2}{4a^2} + \frac{-c}{a} \qquad \text{Factor; commutative property (Step 4)}$$

$$\left(x + \frac{b}{2a}\right)^2 = \frac{b^2}{4a^2} + \frac{-4ac}{4a^2} \qquad \text{Write fractions with a common denominator. (Section R.5)}$$

$$\left(x + \frac{b}{2a}\right)^2 = \frac{b^2 - 4ac}{4a^2} \qquad \text{Add fractions. (Section R.5)}$$

$$x + \frac{b}{2a} = \pm\sqrt{\frac{b^2 - 4ac}{4a^2}} \qquad \text{Square root property (Step 5)}$$

$$x + \frac{b}{2a} = \frac{\pm\sqrt{b^2 - 4ac}}{2a} \qquad \text{Since } a > 0, \sqrt{4a^2} = 2a. \text{ (Section R.7)}$$

$$x = \frac{-b}{2a} \pm \frac{\sqrt{b^2 - 4ac}}{2a} \qquad \text{Add } \tfrac{-b}{2a}.$$

Combining terms on the right side leads to the **quadratic formula.** (A slight modification of the derivation produces the same result for $a < 0$.)

Quadratic Formula

The solutions of the quadratic equation $ax^2 + bx + c = 0$, where $a \neq 0$, are

$$x = \frac{-b \pm \sqrt{b^2 - 4ac}}{2a}.$$

CAUTION Notice that the fraction bar in the quadratic formula extends under the $-b$ term in the numerator.

EXAMPLE 5 Using the Quadratic Formula (Real Solutions)

Solve $x^2 - 4x = -2$.

Solution $x^2 - 4x + 2 = 0$ Write in standard form.

Here $a = 1$, $b = -4$, and $c = 2$. Substitute these values into the quadratic formula and solve for x.

$$x = \frac{-b \pm \sqrt{b^2 - 4ac}}{2a} \quad \text{Quadratic formula}$$

$$= \frac{-(-4) \pm \sqrt{(-4)^2 - 4(1)2}}{2(1)} \quad a = 1, b = -4, c = 2$$

$$= \frac{4 \pm \sqrt{16 - 8}}{2}$$

$$= \frac{4 \pm 2\sqrt{2}}{2} \quad \sqrt{16 - 8} = \sqrt{8} = \sqrt{4 \cdot 2} = 2\sqrt{2} \text{ (Section R.7)}$$

$$= \frac{2(2 \pm \sqrt{2})}{2} \quad \text{Factor out 2 in the numerator. (Section R.5)}$$

$$= 2 \pm \sqrt{2} \quad \text{Lowest terms}$$

The solution set is $\{2 \pm \sqrt{2}\}$.

Now try Exercise 47.

EXAMPLE 6 Using the Quadratic Formula (Nonreal Complex Solutions)

Solve $2x^2 = x - 4$.

Solution

$$2x^2 - x + 4 = 0 \quad \text{Standard form}$$

$$x = \frac{-(-1) \pm \sqrt{(-1)^2 - 4(2)(4)}}{2(2)} \quad \text{Quadratic formula; } a = 2, b = -1, c = 4$$

$$= \frac{1 \pm \sqrt{1 - 32}}{4}$$

$$= \frac{1 \pm \sqrt{-31}}{4}$$

$$= \frac{1 \pm i\sqrt{31}}{4} \quad \sqrt{-1} = i \text{ (Section 1.3)}$$

The solution set is $\left\{\frac{1}{4} \pm \frac{i\sqrt{31}}{4}\right\}$.

Now try Exercise 51.

The equation $x^3 + 8 = 0$ in Example 7 is called a *cubic equation* because of the degree 3 term. Some higher-degree equations can be solved using factoring and the quadratic formula.

EXAMPLE 7 Solving a Cubic Equation

Solve $x^3 + 8 = 0$.

Solution

$$x^3 + 8 = 0$$

$$(x + 2)(x^2 - 2x + 4) = 0 \quad \text{Factor as a sum of cubes. (Section R.4)}$$

$$x + 2 = 0 \quad \text{or} \quad x^2 - 2x + 4 = 0 \quad \text{Zero-factor property}$$

$$x = -2 \quad \text{or} \quad x = \frac{2 \pm \sqrt{4 - 16}}{2} \quad \text{Quadratic formula; } a = 1, b = -2, c = 4$$

$$x = \frac{2 \pm \sqrt{-12}}{2}$$

$$x = \frac{2 \pm 2i\sqrt{3}}{2} \quad \text{Simplify the radical.}$$

$$x = \frac{2(1 \pm i\sqrt{3})}{2} \quad \text{Factor out 2 in the numerator.}$$

$$x = 1 \pm i\sqrt{3} \quad \text{Lowest terms}$$

The solution set is $\{-2, 1 \pm i\sqrt{3}\}$.

Now try Exercise 59.

Solving for a Specified Variable To solve a quadratic equation for a specified variable, we usually apply the square root property or the quadratic formula.

EXAMPLE 8 Solving for a Variable That Is Squared

Solve for the specified variable.

(a) $A = \frac{\pi d^2}{4}$, for d **(b)** $rt^2 - st = k \ (r \neq 0)$, for t

Solution

(a)

$$A = \frac{\pi d^2}{4} \quad \text{Goal: Isolate } d\text{, the specified variable.}$$

$$4A = \pi d^2 \quad \text{Multiply by 4.}$$

$$\frac{4A}{\pi} = d^2 \quad \text{Divide by } \pi.$$

$$d = \pm\sqrt{\frac{4A}{\pi}} \quad \text{Square root property}$$

$$d = \frac{\pm\sqrt{4A}}{\sqrt{\pi}} \cdot \frac{\sqrt{\pi}}{\sqrt{\pi}} \quad \text{Rationalize the denominator. (Section R.7)}$$

$$d = \frac{\pm\sqrt{4A\pi}}{\pi} \quad \text{Multiply numerators; multiply denominators.}$$

$$d = \frac{\pm 2\sqrt{A\pi}}{\pi} \quad \text{Simplify.}$$

(b) Because $rt^2 - st = k$ has terms with t^2 and t, use the quadratic formula. Subtract k from both sides.

$$rt^2 - st - k = 0 \quad \text{Standard form}$$

Now use the quadratic formula to find t, with $a = r$, $b = -s$, and $c = -k$.

$$t = \frac{-b \pm \sqrt{b^2 - 4ac}}{2a}$$

$$t = \frac{-(-s) \pm \sqrt{(-s)^2 - 4(r)(-k)}}{2(r)}$$

$$t = \frac{s \pm \sqrt{s^2 + 4rk}}{2r}$$

Now try Exercises 63 and 67.

The Discriminant The quantity under the radical in the quadratic formula, $b^2 - 4ac$, is called the **discriminant.**

$$x = \frac{-b \pm \sqrt{b^2 - 4ac}}{2a} \longleftarrow \text{Discriminant}$$

When the numbers a, b, and c are *integers* (but not necessarily otherwise), the value of the discriminant can be used to determine whether the solutions of a quadratic equation are rational, irrational, or nonreal complex numbers, as shown in the following table. If the discriminant is 0, there will be only one distinct solution. (Why?)

Discriminant	Number of Solutions	Kind of solutions
Positive, perfect square	Two	Rational
Positive, but not a perfect square	Two	Irrational
Zero	One (a double solution)	Rational
Negative	Two	Nonreal complex

CAUTION The restriction on a, b, and c is important. For example, for the equation

$$x^2 - \sqrt{5}x - 1 = 0,$$

the discriminant is $b^2 - 4ac = 5 + 4 = 9$, which would indicate two rational solutions *if the coefficients were integers.* By the quadratic formula, however, the two solutions

$$x = \frac{\sqrt{5} \pm 3}{2}$$

are *irrational* numbers.

EXAMPLE 9 Using the Discriminant

Determine the number of distinct solutions and tell whether they are rational, irrational, or nonreal complex numbers.

(a) $5x^2 + 2x - 4 = 0$ **(b)** $x^2 - 10x = -25$ **(c)** $2x^2 - x + 1 = 0$

Solution

(a) For $5x^2 + 2x - 4 = 0$, $a = 5$, $b = 2$, and $c = -4$. The discriminant is

$$b^2 - 4ac = 2^2 - 4(5)(-4) = 84.$$

Since 84 is positive and not a perfect square, there are two distinct irrational solutions.

(b) First, write the equation in standard form as $x^2 - 10x + 25 = 0$. Thus, $a = 1$, $b = -10$, and $c = 25$, and

$$b^2 - 4ac = (-10)^2 - 4(1)(25) = 0.$$

There is one distinct rational solution, a "double solution."

(c) For $2x^2 - x + 1 = 0$, $a = 2$, $b = -1$, and $c = 1$, so

$$b^2 - 4ac = (-1)^2 - 4(2)(1) = -7.$$

There are two distinct nonreal complex solutions. (They are complex conjugates.)

Now try Exercises 71, 73, and 75.

1.4 Exercises

Concept Check *Match the equation in Column I with its solution(s) in Column II.*

I		II	
1. $x^2 = 4$	**2.** $x^2 = -4$	**A.** $\pm 2i$	**B.** $\pm 2\sqrt{2}$
3. $x^2 + 2 = 0$	**4** $x^2 - 2 = 0$	**C.** $\pm i\sqrt{2}$	**D.** 2
5. $x^2 = -8$	**6.** $x^2 = 8$	**E.** $\pm\sqrt{2}$	**F.** -2
7. $x - 2 = 0$	**8.** $x + 2 = 0$	**G.** ± 2	**H.** $\pm 2i\sqrt{2}$

Concept Check *Answer each question.*

9. Which one of the following equations is set up for direct use of the zero-factor property? Solve it.

A. $3x^2 - 17x - 6 = 0$ **B.** $(2x + 5)^2 = 7$
C. $x^2 + x = 12$ **D.** $(3x + 1)(x - 7) = 0$

10. Which one of the following equations is set up for direct use of the square root property? Solve it.

A. $3x^2 - 17x - 6 = 0$ **B.** $(2x + 5)^2 = 7$
C. $x^2 + x = 12$ **D.** $(3x + 1)(x - 7) = 0$

11. Only one of the following equations does not require Step 1 of the method for completing the square described in this section. Which one is it? Solve it.

A. $3x^2 - 17x - 6 = 0$ **B.** $(2x + 5)^2 = 7$
C. $x^2 + x = 12$ **D.** $(3x + 1)(x - 7) = 0$

12. Only one of the following equations is set up so that the values of a, b, and c can be determined immediately. Which one is it? Solve it.

A. $3x^2 - 17x - 6 = 0$ **B.** $(2x + 5)^2 = 7$
C. $x^2 + x = 12$ **D.** $(3x + 1)(x - 7) = 0$

Solve each equation by the zero-factor property. See Example 1.

13. $x^2 - 5x + 6 = 0$ **14.** $x^2 + 2x - 8 = 0$ **15.** $5x^2 - 3x - 2 = 0$
16. $2x^2 - x - 15 = 0$ **17.** $-4x^2 + x = -3$ **18.** $-6x^2 + 7x = -10$

Solve each equation by the square root property. See Example 2.

19. $x^2 = 16$ **20.** $x^2 = 25$ **21.** $x^2 = 27$
22. $x^2 = 48$ **23.** $x^2 = -16$ **24.** $x^2 = -100$
25. $(3x - 1)^2 = 12$ **26.** $(4x + 1)^2 = 20$ **27.** $(x + 5)^2 = -3$
28. $(x - 4)^2 = -5$ **29.** $(5x - 3)^2 = -3$ **30.** $(-2x + 5)^2 = -8$

Solve each equation by completing the square. See Examples 3 and 4.

31. $x^2 + 4x + 3 = 0$ **32.** $x^2 + 7x + 12 = 0$ **33.** $2x^2 - x - 28 = 0$
34. $4x^2 - 3x - 10 = 0$ **35.** $x^2 - 2x - 2 = 0$ **36.** $x^2 - 3x - 6 = 0$
37. $2x^2 + x - 10 = 0$ **38.** $3x^2 + 2x - 5 = 0$ **39.** $2x^2 - 4x - 3 = 0$
40. $-3x^2 + 6x + 5 = 0$ **41.** $-4x^2 + 8x = 7$ **42.** $3x^2 - 9x = -7$

43. ***Concept Check*** Francisco claimed that the equation $x^2 - 4x = 0$ cannot be solved by the quadratic formula since there is no value for c. Is he correct?

44. ***Concept Check*** Francesca, Francisco's twin sister, claimed that the equation $x^2 - 17 = 0$ cannot be solved by the quadratic formula since there is no value for b. Is she correct?

Solve each equation using the quadratic formula. See Examples 5 and 6.

45. $x^2 - x - 1 = 0$ **46.** $x^2 - 3x - 2 = 0$ **47.** $x^2 - 6x = -7$
48. $x^2 - 4x = -1$ **49.** $x^2 = 2x - 5$ **50.** $x^2 = 2x - 10$
51. $4x^2 = 12x - 11$ **52.** $6x^2 = -3x - 2$ **53.** $\frac{1}{2}x^2 + \frac{1}{4}x - 3 = 0$
54. $\frac{2}{3}x^2 + \frac{1}{4}x = 3$ **55.** $.2x^2 + .4x - .3 = 0$ **56.** $.1x^2 - .1x = .3$
57. $4 + \frac{3}{x} - \frac{2}{x^2} = 0$ **58.** $3 - \frac{4}{x} - \frac{2}{x^2} = 0$

Solve each cubic equation. See Example 7.

59. $x^3 - 8 = 0$ **60.** $x^3 - 27 = 0$ **61.** $x^3 + 27 = 0$ **62.** $x^3 + 64 = 0$

Solve each equation for the indicated variable. Assume no denominators are 0. See Example 8.

63. $s = \frac{1}{2}gt^2$ for t **64.** $A = \pi r^2$ for r
65. $F = \frac{kMv^2}{r}$ for v **66.** $s = s_0 + gt^2 + k$ for t
67. $h = -16t^2 + v_0 t + s_0$ for t **68.** $S = 2\pi rh + 2\pi r^2$ for r

*For each equation, **(a)** solve for x in terms of y, and **(b)** solve for y in terms of x. See Example 8.*

69. $4x^2 - 2xy + 3y^2 = 2$

70. $3y^2 + 4xy - 9x^2 = -1$

Evaluate the discriminant for each equation. Then use it to predict the number of distinct solutions, and whether they are rational, irrational, *or* nonreal complex. *Do not solve the equation. See Example 9.*

71. $x^2 + 8x + 16 = 0$

72. $x^2 - 4x + 4 = 0$

73. $3x^2 - 5x + 2 = 0$

74. $8x^2 = 14x - 3$

75. $4x^2 = 6x + 3$

76. $2x^2 - 4x + 1 = 0$

77. $9x^2 + 11x + 4 = 0$

78. $3x^2 = 4x - 5$

79. $8x^2 - 72 = 0$

80. Show that the discriminant for the equation $\sqrt{2}x^2 + 5x - 3\sqrt{2} = 0$ is 49. If this equation is completely solved, it can be shown that the solution set is $\left\{-3\sqrt{2}, \frac{\sqrt{2}}{2}\right\}$. Here we have a discriminant that is positive and a perfect square, yet the two solutions are irrational. Does this contradict the discussion in this section? Explain.

81. Is it possible for the solution set of a quadratic equation with integer coefficients to consist of a single irrational number? Explain.

82. Is it possible for the solution set of a quadratic equation with real coefficients to consist of one real number and one nonreal complex number? Explain.

Find the values of a, b, and c for which the quadratic equation $ax^2 + bx + c = 0$ has the given numbers as solutions. (Hint: Use the zero-factor property in reverse.)

83. 4, 5

84. $-3, 2$

85. $1 + \sqrt{2}, 1 - \sqrt{2}$

86. $i, -i$

1.5 Applications and Modeling with Quadratic Equations

Geometry Problems ▪ Using the Pythagorean Theorem ▪ Height of a Propelled Object ▪ Modeling with Quadratic Equations

Looking Ahead to Calculus

In calculus, you will need to be able to write an algebraic expression from the description in a problem like those in this section. Using calculus techniques, you will be asked to find the value of the variable that produces an optimum (a maximum or minimum) value of the expression.

In this section we give several examples of problems whose solutions require solving quadratic equations.

PROBLEM SOLVING When solving problems that lead to quadratic equations, we may get a solution that does not satisfy the physical constraints of the problem. For example, if x represents a width and the two solutions of the quadratic equation are -9 and 1, the value -9 must be rejected since a width must be a positive number.

Geometry Problems Problems involving the area or volume of a geometric object often lead to quadratic equations. We continue to use the problem-solving strategy discussed in Section 1.2.

EXAMPLE 1 Solving a Problem Involving the Volume of a Box

A piece of machinery is capable of producing rectangular sheets of metal such that the length is three times the width. Furthermore, equal-sized squares measuring 5 in. on a side can be cut from the corners so that the resulting piece of metal can be shaped into an open box by folding up the flaps. If specifications call for the volume of the box to be 1435 in.^{3}, what should the dimensions of the original piece of metal be?

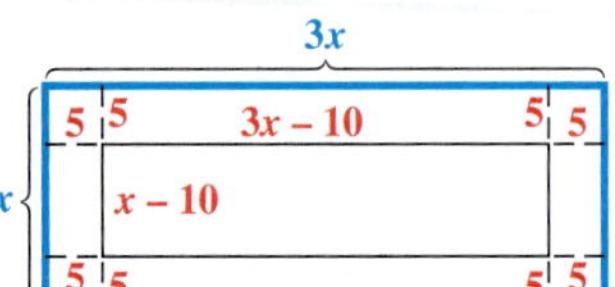

Figure 5

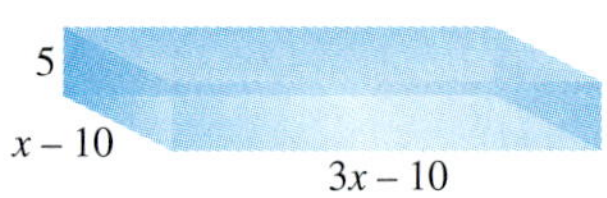

Figure 6

Solution

Step 1 **Read** the problem. We must find the dimensions of the original piece of metal.

Step 2 **Assign a variable.** We know that the length is three times the width, so let $x =$ the width (in inches) and thus $3x =$ the length.

The box is formed by cutting $5 + 5 = 10$ in. from both the length and the width. See Figure 5. Figure 6 indicates that the width of the bottom of the box is $x - 10$, the length of the bottom of the box is $3x - 10$, and the height is 5 in. (the length of the side of each cutout square).

Step 3 **Write an equation.** The formula for volume of a box is $V = lwh$.

$$\begin{array}{ccccccc} \text{Volume} & = & \text{length} & \times & \text{width} & \times & \text{height} \\ \downarrow & & \downarrow & & \downarrow & & \downarrow \\ 1435 & = & (3x - 10) & & (x - 10) & & (5) \end{array}$$

(Note that the dimensions of the box must be positive numbers, so $3x - 10$ and $x - 10$ must be greater than 0, which implies $x > \frac{10}{3}$ and $x > 10$. These are both satisfied when $x > 10$.)

Step 4 **Solve** the equation.

$$1435 = 15x^2 - 200x + 500 \qquad \text{Multiply. (Section R.3)}$$

$$0 = 15x^2 - 200x - 935 \qquad \text{Subtract 1435.}$$

$$0 = 3x^2 - 40x - 187 \qquad \text{Divide by 5.}$$

$$0 = (3x + 11)(x - 17) \qquad \text{Factor. (Section R.4)}$$

$$3x + 11 = 0 \quad \text{or} \quad x - 17 = 0 \qquad \text{Zero-factor property (Section 1.4)}$$

$$x = -\frac{11}{3} \quad \text{or} \quad x = 17 \qquad \text{Solve. (Section 1.1)}$$

Step 5 **State the answer.** Only 17 satisfies the restriction that $x > 10$. Thus, the dimensions of the original piece of metal should be 17 in. by $3(17) = 51$ in.

Step 6 **Check.** The length of the bottom of the box is $51 - 2(5) = 41$ in. The width is $17 - 2(5) = 7$ in. The height is 5 in. (the amount cut on each corner), so the volume of the box is

$$V = lwh = 41 \times 7 \times 5 = 1435 \text{ in.}^3, \quad \text{as required.}$$

Now try Exercise 11.

Using the Pythagorean Theorem Example 2 requires the use of the **Pythagorean theorem** from geometry.

Pythagorean Theorem

In a right triangle, the sum of the squares of the lengths of the legs is equal to the square of the length of the hypotenuse.

$$a^2 + b^2 = c^2$$

Hypotenuse c
Leg a
Leg b

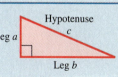

EXAMPLE 2 Solving a Problem Involving the Pythagorean Theorem

Erik Van Erden finds a piece of property in the shape of a right triangle. To get some idea of its dimensions, he measures the three sides, starting with the shortest side. He finds that the longer leg is 20 m longer than twice the length of the shorter leg. The hypotenuse is 10 m longer than the length of the longer leg. Find the lengths of the sides of the triangular lot.

Solution

Step 1 **Read** the problem. We must find the lengths of the three sides.

Step 2 **Assign a variable.**

Let s = the length of the shorter leg (in meters).

Then $2s + 20$ = the length of the longer leg, and

$(2s + 20) + 10$ or $2s + 30$ = the length of the hypotenuse.

See Figure 7.

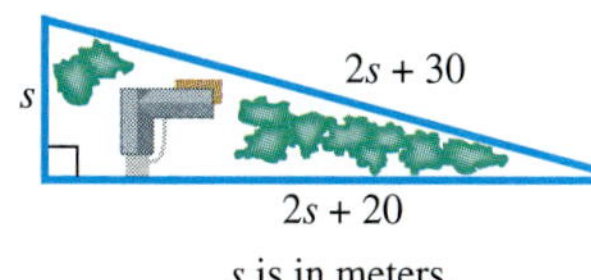

s is in meters.

Figure 7

Step 3 **Write an equation.**

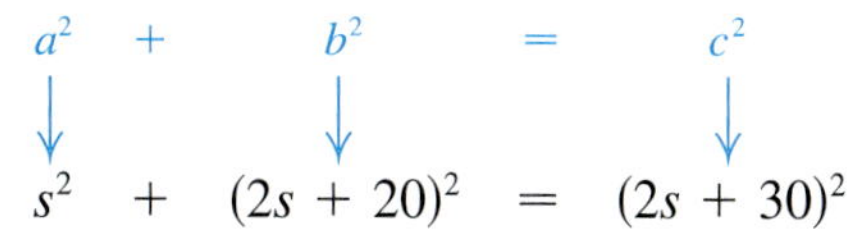

$$s^2 + (2s + 20)^2 = (2s + 30)^2 \quad \text{Pythagorean theorem}$$

Step 4 **Solve** the equation.

$$s^2 + 4s^2 + 80s + 400 = 4s^2 + 120s + 900 \quad \text{Square the binomials. (Section R.3)}$$

$$s^2 - 40s - 500 = 0 \quad \text{Standard form}$$

$$(s - 50)(s + 10) = 0 \quad \text{Factor.}$$

$$s - 50 = 0 \quad \text{or} \quad s + 10 = 0 \quad \text{Zero-factor property}$$

$$s = 50 \quad \text{or} \quad s = -10 \quad \text{Solve.}$$

Step 5 **State the answer.** Since s represents a length, -10 is not reasonable. The lengths of the sides of the triangular lot are 50 m, $2(50) + 20 =$ 120 m, and $2(50) + 30 = 130$ m.

Step 6 **Check.** The lengths 50, 120, and 130 satisfy the words of the problem, and also satisfy the Pythagorean theorem.

Now try Exercise 17.

Height of a Propelled Object If air resistance is neglected, the height s (in feet) of an object propelled directly upward from an initial height of s_0 feet, with initial velocity v_0 feet per second, is

$$s = -16t^2 + v_0 t + s_0,$$

where t is the number of seconds after the object is propelled. The coefficient of t^2, -16, is a constant based on the gravitational force of Earth. This constant varies on other surfaces, such as the moon and other planets.

EXAMPLE 3 Solving a Problem Involving the Height of a Projectile

If a projectile is shot vertically upward from the ground with an initial velocity of 100 ft per sec, neglecting air resistance, its height s (in feet) above the ground t seconds after projection is given by

$$s = -16t^2 + 100t.$$

(a) After how many seconds will it be 50 ft above the ground?

(b) How long will it take for the projectile to return to the ground?

Solution

(a) We must find value(s) of t so that height s is 50 ft. Let $s = 50$ in the given equation.

$$50 = -16t^2 + 100t$$

$$16t^2 - 100t + 50 = 0 \quad \text{Standard form}$$

$$8t^2 - 50t + 25 = 0 \quad \text{Divide by 2.}$$

$$t = \frac{-(-50) \pm \sqrt{(-50)^2 - 4(8)(25)}}{2(8)} \quad \text{Quadratic formula (Section 1.4)}$$

$$t = \frac{50 \pm \sqrt{1700}}{16}$$

$$t \approx .55 \quad \text{or} \quad t \approx 5.70 \quad \text{Use a calculator.}$$

Both solutions are acceptable, since the projectile reaches 50 ft twice: once on its way up (after .55 sec) and once on its way down (after 5.70 sec).

(b) When the projectile returns to the ground, the height s will be 0 ft, so let $s = 0$ in the given equation.

$$0 = -16t^2 + 100t$$

$$0 = -4t(4t - 25) \quad \text{Factor.}$$

$$-4t = 0 \quad \text{or} \quad 4t - 25 = 0 \quad \text{Zero-factor property}$$

$$t = 0 \quad \text{or} \quad 4t = 25 \quad \text{Solve.}$$

$$t = 6.25$$

The first solution, 0, represents the time at which the projectile was on the ground prior to being launched, so it does not answer the question. The projectile will return to the ground 6.25 sec after it is launched.

Now try Exercise 29.

Galileo Galilei (1564–1642)

CONNECTIONS Galileo Galilei did more than construct theories to explain physical phenomena—he set up experiments to test his ideas. According to legend, Galileo dropped objects of different weights from the Leaning Tower of Pisa to disprove the Aristotelian view that heavier objects fall faster than lighter objects. He developed a formula for freely falling objects that stated

$$d = 16t^2,$$

where d is the distance in feet that an object falls (neglecting air resistance) in t seconds, regardless of weight.

For Discussion or Writing

1. Use Galileo's formula to find the distance a freely falling object will fall in **(a)** 5 sec and **(b)** 10 sec. Is the second answer twice the first? If not, how are the two answers related? Explain.
2. Compare Galileo's formula with the one we gave for the height of a propelled object. How are they similar? How do they differ?

Modeling with Quadratic Equations

EXAMPLE 4 Analyzing Sport Utility Vehicle (SUV) Sales

The bar graph in Figure 8 shows sales of SUVs (sport utility vehicles) in the United States, in millions. The quadratic equation

$$S = .016x^2 + .124x + .787$$

models sales of SUVs from 1990 to 2001, where S represents sales in millions, and $x = 0$ represents 1990, $x = 1$ represents 1991, and so on.

Sales of SUVs in the United States (in millions)

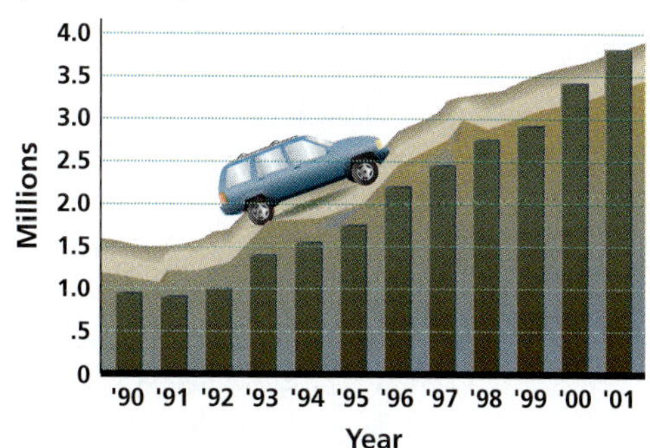

Source: CNW Marketing Research of Bandon, OR, based on automakers' reported sales.

Figure 8

(a) Use the model to determine sales in 2000 and 2001. Compare the results to the actual figures of 3.4 million and 3.8 million from the graph.

(b) According to the model, in what year do sales reach 3 million? (Round down to the nearest year.) Is the result accurate?

Solution

(a)

$$S = .016x^2 + .124x + .787 \qquad \text{Original model}$$

$$S = .016(10)^2 + .124(10) + .787 \qquad \text{For 2000, } x = 10.$$

$$\approx 3.6 \text{ million}$$

$$S = .016(11)^2 + .124(11) + .787 \qquad \text{For 2001, } x = 11.$$

$$\approx 4.1 \text{ million}$$

The predictions are greater than the actual figures of 3.4 and 3.8 million.

(b)

$$3 = .016x^2 + .124x + .787 \qquad \text{Let } S = 3 \text{ in the original model.}$$

$$0 = .016x^2 + .124x - 2.213 \qquad \text{Standard form}$$

$$x = \frac{-.124 \pm \sqrt{(.124)^2 - 4(.016)(-2.213)}}{2(.016)} \qquad \text{Quadratic formula}$$

$x \approx -16.3$ or $x \approx 8.5$

Reject the negative solution and round 8.5 down to 8. The year 1998 corresponds to $x = 8$. The model is a bit misleading, since SUV sales did not reach 3 million until 2000.

Now try Exercise 31.

1.5 Exercises

Concept Check *Answer each question.*

1. *Area of a Parking Lot* For the rectangular parking area of the shopping center shown, which one of the following equations says that the area is 40,000 yd²?

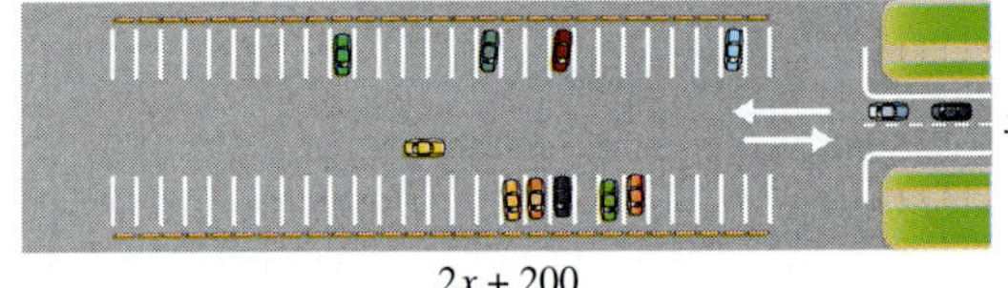

A. $x(2x + 200) = 40{,}000$
B. $2x + 2(2x + 200) = 40{,}000$
C. $x + (2x + 200) = 40{,}000$
D. none of the above

2. *Diagonal of a Rectangle* If a rectangle is r feet long and s feet wide, which one of the following expressions is the length of its diagonal in terms of r and s?

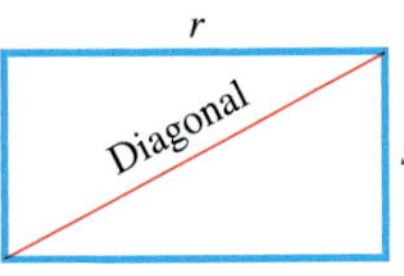

A. $\sqrt{rs}$
B. $r + s$
C. $\sqrt{r^2 + s^2}$
D. $r^2 + s^2$

3. *Sides of a Right Triangle* To solve for the lengths of the right triangle sides, which equation is correct?

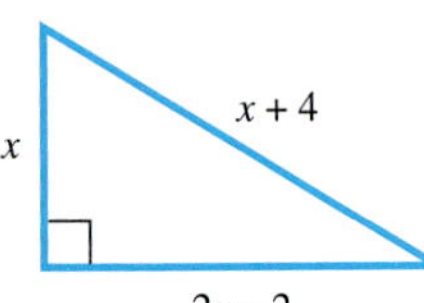

A. $x^2 = (2x - 2)^2 + (x + 4)^2$
B. $x^2 + (x + 4)^2 = (2x - 2)^2$
C. $x^2 = (2x - 2)^2 - (x + 4)^2$
D. $x^2 + (2x - 2)^2 = (x + 4)^2$

4. *Area of a Picture* The mat around the picture shown measures x inches across. Which one of the following equations says that the area of the picture itself is 600 in.^{2}?

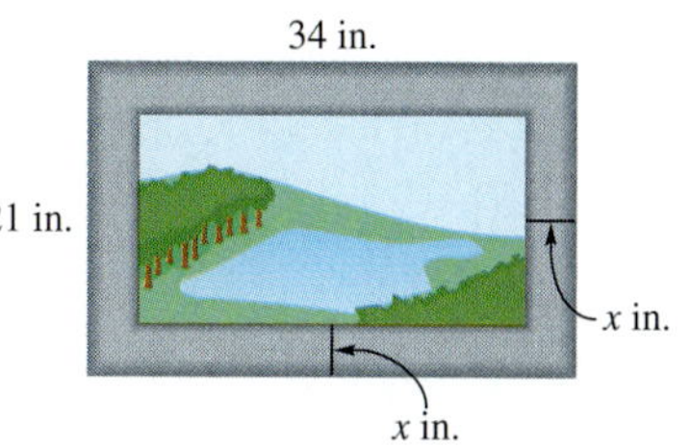

A. $2(34 - 2x) + 2(21 - 2x) = 600$
B. $(34 - 2x)(21 - 2x) = 600$
C. $(34 - x)(21 - x) = 600$
D. $x(34)(21) = 600$

Solve each problem. See Example 1.

5. *Dimensions of a Parking Lot* A shopping center has a rectangular area of 40,000 yd^2 enclosed on three sides for a parking lot. The length is 200 yd more than twice the width. What are the dimensions of the lot? (See Exercise 1.)

6. *Dimensions of a Garden* An ecology center wants to set up an experimental garden using 300 m of fencing to enclose a rectangular area of 5000 m^2. Find the dimensions of the garden.

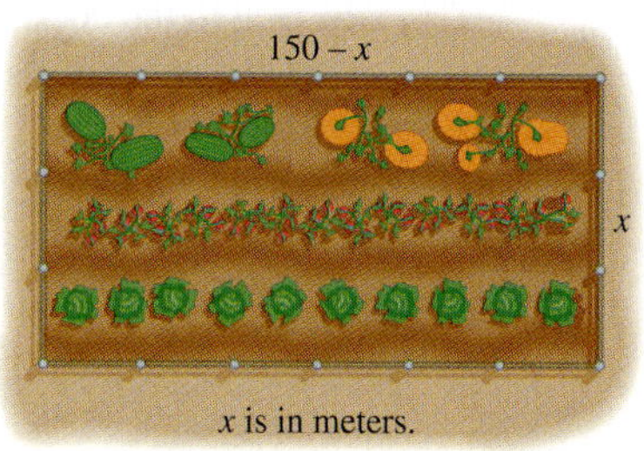

7. *Dimensions of a Rug* Cynthia Herring wants to buy a rug for a room that is 12 ft wide and 15 ft long. She wants to leave a uniform strip of floor around the rug. She can afford to buy 108 ft^2 of carpeting. What dimensions should the rug have?

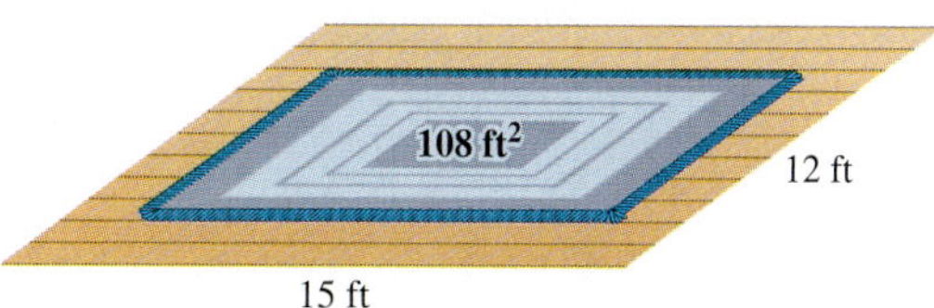

8. *Width of a Flower Border* A landscape architect has included a rectangular flower bed measuring 9 ft by 5 ft in her plans for a new building. She wants to use two colors of flowers in the bed, one in the center and the other for a border of the same width on all four sides. If she has enough plants to cover 24 ft^2 for the border, how wide can the border be?

9. *Volume of a Box* A rectangular piece of metal is 10 in. longer than it is wide. Squares with sides 2 in. long are cut from the four corners, and the flaps are folded upward to form an open box. If the volume of the box is 832 in.^{3}, what were the original dimensions of the piece of metal?

10. *Volume of a Box* In Exercise 9, suppose that the piece of metal has length twice the width, and 4-in. squares are cut from the corners. If the volume of the box is 1536 in.^{3}, what were the original dimensions of the piece of metal?

11. *Radius of a Can* A can of Blue Runner Red Kidney Beans has surface area 371 cm^2. Its height is 12 cm. What is the radius of the circular top? Round to the nearest hundredth. (*Hint:* The surface area consists of the circular top and bottom and a rectangle that represents the side cut open vertically and unrolled.)

12. *Dimensions of a Cereal Box* The volume of a 10-oz box of Cheerios cereal is 182.742 in.^{3}. The width of the box is 3.1875 in. less than the length, and its depth is 2.3125 in. Find the length and width of the box to the nearest thousandth.

13. *Manufacturing to Specifications* A manufacturing firm wants to package its product in a cylindrical container 3 ft high with surface area 8π ft^2. What should the radius of the circular top and bottom of the container be?

14. *Manufacturing to Specifications* In Exercise 13, what radius would produce a container with a volume of π times the radius?

15. *Dimensions of a Square* What is the length of the side of a square if its area and perimeter are numerically equal?

16. *Dimensions of a Rectangle* A rectangle has an area that is numerically twice its perimeter. If the length is twice the width, what are its dimensions?

Solve each problem. See Example 2.

17. *Height of a Dock* A boat is being pulled into a dock with a rope attached to the boat at water level. When the boat is 12 ft from the dock, the length of the rope from the boat to the dock is 3 ft longer than twice the height of the dock above the water. Find the height of the dock.

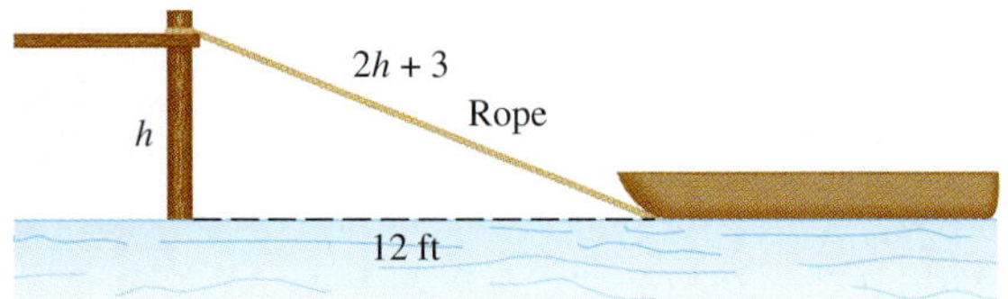

18. *Height of a Kite* A kite is flying on 50 ft of string. Its vertical distance from the ground is 10 ft more than its horizontal distance from the person flying it. Assuming that the string is being held at ground level, find its horizontal distance from the person and its vertical distance from the ground.

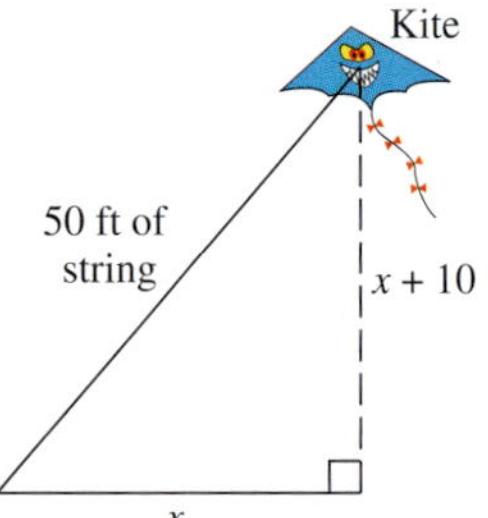

19. *Radius Covered by a Circular Lawn Sprinkler* A square lawn has area 800 ft^2. A sprinkler placed at the center of the lawn sprays water in a circular pattern that just covers the lawn. What is the radius of the circle?

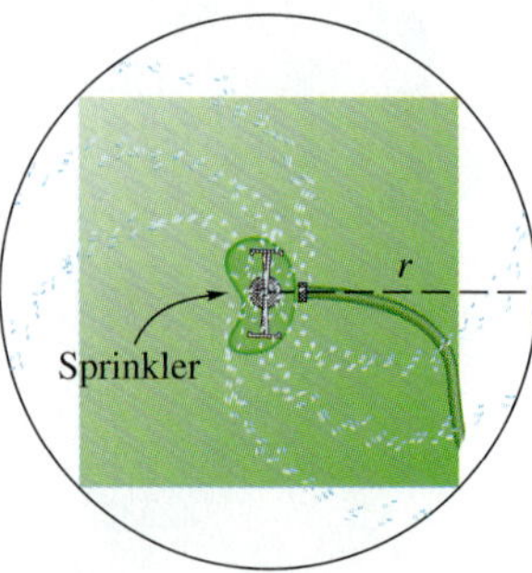

20. ***Dimensions of a Solar Panel Frame*** Christine has a solar panel with a width of 26 in. To get the proper inclination for her climate, she needs a right triangular support frame that has one leg twice as long as the other. To the nearest tenth of an inch, what dimensions should the frame have?

21. ***Length of a Ladder*** A building is 2 ft from a 9-ft fence that surrounds the property. A worker wants to wash a window in the building 13 ft from the ground. He plans to place a ladder over the fence so it rests against the building. (See the figure.) He decides he should place the ladder 8 ft from the fence for stability. To the nearest tenth of a foot, how long a ladder will he need?

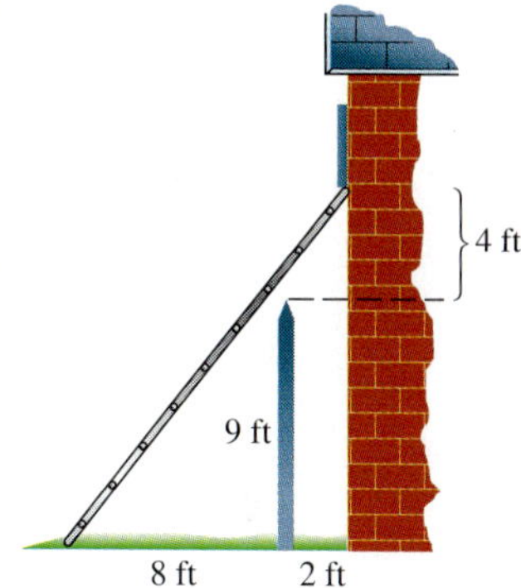

22. ***Range of Walkie-Talkies*** Corbin Eckles and Connor Hayes have received walkie-talkies for Christmas. If they leave from the same point at the same time, Corbin walking north at 2.5 mph and Connor walking east at 3 mph, how long will they be able to talk to each other if the range of the walkie-talkies is 4 mi? Round your answer to the nearest minute.

23. ***Length of a Walkway*** A nature conservancy group decides to construct a raised wooden walkway through a wetland area. To enclose the most interesting part of the wetlands, the walkway will have the shape of a right triangle with one leg 700 yd longer than the other and the hypotenuse 100 yd longer than the longer leg. Find the total length of the walkway.

24. ***Broken Bamboo*** Problems involving the Pythagorean theorem have appeared in mathematics for thousands of years. This one is taken from the ancient Chinese work, *Arithmetic in Nine Sections:*

> *There is a bamboo 10 ft high, the upper end of which, being broken, reaches the ground 3 ft from the stem. Find the height of the break.*

(Modeling) *Solve each problem. See Examples 3 and 4.*

Height of a Projectile *A projectile is launched from ground level with an initial velocity of v_0 feet per second. Neglecting air resistance, its height in feet t seconds after launch is given by*

$$s = -16t^2 + v_0 t.$$

In Exercises 25–28, find the time(s) that the projectile will ***(a)*** *reach a height of* 80 ft *and* ***(b)*** *return to the ground for the given value of v_0. Round answers to the nearest hundredth if necessary.*

25. $v_0 = 96$

26. $v_0 = 128$

27. $v_0 = 32$

28. $v_0 = 16$

29. *Height of a Propelled Ball* An astronaut on the moon throws a baseball upward. The astronaut is 6 ft, 6 in. tall, and the initial velocity of the ball is 30 ft per sec. The height s of the ball in feet is given by the equation

$$s = -2.7t^2 + 30t + 6.5,$$

where t is the number of seconds after the ball was thrown.

(a) After how many seconds is the ball 12 ft above the moon's surface? Round to the nearest hundredth.

(b) How many seconds will it take for the ball to return to the surface? Round to the nearest hundredth.

30. The ball in Exercise 29 will never reach a height of 100 ft. How can this be determined algebraically?

31. *Carbon Monoxide Exposure* Carbon monoxide (CO) combines with the hemoglobin of the blood to form carboxyhemoglobin (COHb), which reduces transport of oxygen to tissues. Smokers routinely have a 4% to 6% COHb level in their blood, which can cause symptoms such as blood flow alterations, visual impairment, and poorer vigilance ability. The quadratic model

$$T = .00787x^2 - 1.528x + 75.89$$

approximates the exposure time in hours necessary to reach this 4% to 6% level, where $50 \leq x \leq 100$ is the amount of carbon monoxide present in the air in parts per million (ppm). (*Source: Indoor Air Quality Environmental Information Handbook: Combustion Sources,* Report No. DOE/EV/10450-1, U.S. Department of Energy, 1985.)

(a) A kerosene heater or a room full of smokers is capable of producing 50 ppm of carbon monoxide. How long would it take for a nonsmoking person to start feeling the above symptoms?

(b) Find the carbon monoxide concentration necessary for a person to reach the 4% to 6% COHb level in 3 hr. Round to the nearest tenth.

32. *Lead Emissions* The table gives lead emissions from all sources in the United States, in thousands of tons. The quadratic model

$$L = -10.64x^2 + 93.93x + 3758$$

approximates these emissions. In the model, x represents the number of years since 1990, so $x = 3$ represents 1993, and so on.

(a) Find the year that this model predicts the emissions reached 3000 thousand tons.

(b) In what year will emissions reach 2500 thousand tons?

Year	Thousands of Tons
1993	3911
1994	4043
1995	3924
1996	3910
1997	3915

Source: U.S. Environmental Protection Agency, *National Air Pollution Emission Trends.*

33. *Carbon Monoxide Exposure* High concentrations of carbon monoxide CO can cause coma and possible death. The time required for a person to reach a COHb level capable of causing a coma can be approximated by the quadratic model

$$T = .0002x^2 - .316x + 127.9,$$

where T is the exposure time in hours necessary to reach this level and $500 \leq x \leq 800$ is the amount of carbon monoxide present in the air in parts per million (ppm). (*Source: Indoor Air Quality Environmental Information Handbook: Combustion Sources,* Report No. DOE/EV/10450-1, U.S. Department of Energy, 1985.)

(a) What is the exposure time when $x = 600$ ppm?

(b) Estimate the concentration of CO necessary to produce a coma in 4 hr.

34. *Cost of Public Colleges* The average cost for tuition and fees at public colleges from 1990–2002, in dollars, can be modeled by the equation

$$y = 18.34x^2 + 512.9x + 9314,$$

where $x = 0$ corresponds to 1990, $x = 1$ to 1991, and so on. Based on this model, for what year was the cost \$11,702?

35. *Online Bill Paying* The number of U.S. households estimated to see and pay at least one bill online each month during the years 2000 through 2006 can be modeled by the equation

$$y = .808x^2 + 2.625x + .502,$$

where $x = 0$ corresponds to the year 2000, $x = 1$ corresponds to 2001, and so on, and y is in millions. Approximate the number of households expected to pay at least one bill online each month in 2004.

36. *Growth of HDTV* It is projected that for the period 2002 through 2007, the number of households having high-definition television (HDTV) can be modeled by the equation

$$y = 1.318x^2 - 3.526x + 2.189,$$

where $x = 2$ corresponds to 2002, $x = 3$ corresponds to 2003, and so on, and y is in millions. Based on this model, how many households will have HDTV in 2007? (*Source:* The Yankee Group, May 2003.)

Relating Concepts

For individual or collaborative investigation

(Exercises 37–41)

If p units of an item are sold for x dollars per unit, the revenue is px. Use this idea to analyze the following problem, **working Exercises 37–41 in order.**

Number of Apartments Rented The manager of an 80-unit apartment complex knows from experience that at a rent of \$300, all the units will be full. On the average, one additional unit will remain vacant for each \$20 increase in rent over \$300. Furthermore, the manager must keep at least 30 units rented due to other financial considerations. Currently, the revenue from the complex is \$35,000. How many apartments are rented?

37. Suppose that x represents the number of \$20 increases over \$300. Represent the number of apartment units that will be rented in terms of x.

38. Represent the rent per unit in terms of x.

39. Use the answers in Exercises 37 and 38 to write an expression that defines the revenue generated when there are x increases of \$20 over \$300.

40. According to the problem, the revenue currently generated is \$35,000. Write a quadratic equation in standard form using your expression from Exercise 39.

41. Solve the equation from Exercise 40 and answer the question in the problem.

Solve each problem. (See Relating Concepts Exercises 37–41.)

42. *Harvesting a Cherry Orchard* The manager of a cherry orchard is trying to decide when to schedule the annual harvest. If the cherries are picked now, the average yield per tree will be 100 lb, and the cherries can be sold for 40 cents per pound. Past experience shows that the yield per tree will increase about 5 lb per week, while the price will decrease about 2 cents per pound per week. How many weeks should the manager wait to get an average revenue of \$38.40 per tree?

43. *Number of Airline Passengers* A local club is arranging a charter flight to Miami. The cost of the trip is \$225 each for 75 passengers, with a refund of \$5 per passenger for each passenger in excess of 75. How many passengers must take the flight to produce a revenue of \$16,000?

44. *Recycling Aluminum Cans* A local group of scouts has been collecting old aluminum cans for recycling. The group has already collected 12,000 lb of cans, for which they could currently receive \$4 per hundred pounds. The group can continue to collect cans at the rate of 400 lb per day. However, a glut in the old-can market has caused the recycling company to announce that it will lower its price, starting immediately, by \$.10 per hundred pounds per day. The scouts can make only one trip to the recycling center. How many days should they wait in order to get \$490 for their cans?

1.6 Other Types of Equations

Rational Equations ▪ Equations with Radicals ▪ Equations Quadratic in Form

Rational Equations

A **rational equation** is an equation that has a rational expression for one or more terms. Since a rational expression is not defined when its denominator is 0, values of the variable for which any denominator equals 0 cannot be solutions of the equation. To solve a rational equation, begin by multiplying both sides by the least common denominator (LCD) of the terms of the equation.

EXAMPLE 1 Solving Rational Equations That Lead to Linear Equations

Solve each equation.

(a) $\dfrac{3x-1}{3} - \dfrac{2x}{x-1} = x$ **(b)** $\dfrac{x}{x-2} = \dfrac{2}{x-2} + 2$

Solution

(a) The least common denominator is $3(x-1)$, which is equal to 0 if $x = 1$. Therefore, 1 cannot possibly be a solution of this equation.

$$3(x-1)\left(\frac{3x-1}{3}\right) - 3(x-1)\left(\frac{2x}{x-1}\right) = 3(x-1)x$$ Multiply by the LCD, $3(x-1)$, where $x \neq 1$. (Section R.5)

$$(x-1)(3x-1) - 3(2x) = 3x(x-1)$$ Simplify on both sides.

$$3x^2 - 4x + 1 - 6x = 3x^2 - 3x$$ Multiply. (Section R.3)

$$1 - 10x = -3x$$ Subtract $3x^2$; combine terms.

$$1 = 7x$$ Add $10x$.

$$x = \frac{1}{7}$$ Divide by 7.

The restriction $x \neq 1$ does not affect this result. To check for correct algebra, substitute $\frac{1}{7}$ for x in the original equation.

Check:

$$\frac{3x-1}{3} - \frac{2x}{x-1} = x \quad \text{Original equation}$$

$$\frac{3\left(\frac{1}{7}\right)-1}{3} - \frac{2\left(\frac{1}{7}\right)}{\frac{1}{7}-1} \stackrel{?}{=} \frac{1}{7} \quad \text{Let } x = \tfrac{1}{7}.$$

$$-\frac{4}{21} - \left(-\frac{1}{3}\right) \stackrel{?}{=} \frac{1}{7}$$

$$\frac{1}{7} = \frac{1}{7} \quad \text{True}$$

The solution set is $\left\{\frac{1}{7}\right\}$.

(b) $(x-2)\left(\dfrac{x}{x-2}\right) = (x-2)\left(\dfrac{2}{x-2}\right) + (x-2)2$ Multiply by the LCD, $x-2$, where $x \neq 2$.

$$x = 2 + 2(x-2)$$

$$x = 2 + 2x - 4 \quad \text{Distributive property (Section R.1)}$$

$$-x = -2 \quad \text{Subtract } 2x\text{; combine terms.}$$

$$x = 2 \quad \text{Multiply by } -1.$$

The only possible solution is 2. However, the variable is restricted to real numbers except 2; if $x = 2$, then multiplying by $x - 2$ in the first step is multiplying both sides by 0, which is not valid. Thus, the solution set is $\emptyset$.

Now try Exercises 7 and 9.

EXAMPLE 2 Solving Rational Equations That Lead to Quadratic Equations

Solve each equation.

(a) $\dfrac{3x+2}{x-2} + \dfrac{1}{x} = \dfrac{-2}{x^2-2x}$ **(b)** $\dfrac{-4x}{x-1} + \dfrac{4}{x+1} = \dfrac{-8}{x^2-1}$

Solution

(a)

$$\frac{3x+2}{x-2} + \frac{1}{x} = \frac{-2}{x(x-2)} \quad \text{Factor the last denominator. (Section R.4)}$$

$$x(x-2)\left(\frac{3x+2}{x-2}\right) + x(x-2)\left(\frac{1}{x}\right) = x(x-2)\left(\frac{-2}{x(x-2)}\right) \quad \text{Multiply by } x(x-2),\ x \neq 0, 2.$$

$$x(3x+2) + (x-2) = -2$$

$$3x^2 + 2x + x - 2 = -2 \quad \text{Distributive property}$$

$$3x^2 + 3x = 0 \quad \text{Standard form}$$

$$3x(x+1) = 0 \quad \text{Factor.}$$

$$3x = 0 \quad \text{or} \quad x + 1 = 0 \quad \text{Zero-factor property (Section 1.4)}$$

$$x = 0 \quad \text{or} \quad x = -1 \quad \text{Proposed solutions}$$

Because of the restriction $x \neq 0$, the only valid solution is -1. Check by substituting -1 for x in the original equation. (What happens if you substitute 0 for x?) The solution set is $\{-1\}$.

(b)

$$\frac{-4x}{x-1} + \frac{4}{x+1} = \frac{-8}{x^2-1}$$

$$\frac{-4x}{x-1} + \frac{4}{x+1} = \frac{-8}{(x+1)(x-1)} \quad \text{Factor.}$$

The restrictions on x are $x \neq \pm 1$. Multiply by the LCD, $(x+1)(x-1)$.

$$(x+1)(x-1)\left(\frac{-4x}{x-1}\right) + (x+1)(x-1)\left(\frac{4}{x+1}\right) = (x+1)(x-1)\left(\frac{-8}{(x+1)(x-1)}\right)$$

$$-4x(x+1) + 4(x-1) = -8$$

$$-4x^2 - 4x + 4x - 4 = -8 \quad \text{Distributive property}$$

$$-4x^2 + 4 = 0 \quad \text{Standard form}$$

$$x^2 - 1 = 0 \quad \text{Divide by } -4.$$

$$(x+1)(x-1) = 0 \quad \text{Factor.}$$

$$x + 1 = 0 \quad \text{or} \quad x - 1 = 0 \quad \text{Zero-factor property}$$

$$x = -1 \quad \text{or} \quad x = 1 \quad \text{Proposed solutions}$$

Neither proposed solution is valid, so the solution set is $\emptyset$.

Now try Exercises 15 and 17.

Equations with Radicals To solve an equation such as

$$x - \sqrt{15 - 2x} = 0,$$

in which the variable appears in a radicand, we use the following **power property** to eliminate the radical.

Power Property

If P and Q are algebraic expressions, then every solution of the equation $P = Q$ is also a solution of the equation $P^n = Q^n$, for any positive integer n.

CAUTION ***Be very careful when using the power property.*** It does *not* say that the equations $P = Q$ and $P^n = Q^n$ are equivalent; it says only that each solution of the original equation $P = Q$ is also a solution of the new equation $P^n = Q^n$.

We also use the power property to solve equations such as

$$(4x+3)^{1/3} = (2x-1)^{1/3},$$

where the variable appears in an expression that is the base of a term with a rational exponent.

When using the power property to solve equations, the new equation may have *more* solutions than the original equation. For example, the solution set of the equation $x = -2$ is $\{-2\}$. If we square both sides of the equation $x = -2$, we obtain the new equation $x^2 = 4$, which has solution set $\{-2, 2\}$. Since the solution sets are not equal, the equations are not equivalent. Because of this, when an equation contains radicals (or rational exponents), ***it is essential to check all proposed solutions in the original equation.***

To solve an equation containing radicals, follow these steps.

Solving an Equation Involving Radicals

Step 1 Isolate the radical on one side of the equation.

Step 2 Raise each side of the equation to a power that is the same as the index of the radical so that the radical is eliminated.

If the equation still contains a radical, repeat Steps 1 and 2.

Step 3 Solve the resulting equation.

Step 4 Check each proposed solution in the *original* equation.

EXAMPLE 3 Solving an Equation Containing a Radical (Square Root)

Solve $x - \sqrt{15 - 2x} = 0$.

Solution

$$x = \sqrt{15 - 2x} \quad \text{Isolate the radical. (Step 1)}$$

$$x^2 = \left(\sqrt{15 - 2x}\right)^2 \quad \text{Square both sides. (Step 2)}$$

$$x^2 = 15 - 2x$$

$$x^2 + 2x - 15 = 0 \quad \text{Solve the quadratic equation. (Step 3)}$$

$$(x + 5)(x - 3) = 0$$

$$x + 5 = 0 \quad \text{or} \quad x - 3 = 0$$

$$x = -5 \quad \text{or} \quad x = 3$$

Check: $x - \sqrt{15 - 2x} = 0$ Original equation (Step 4)

If $x = -5$, then

$$-5 - \sqrt{15 - 2(-5)} = 0 \quad ?$$

$$-5 - \sqrt{25} = 0 \quad ?$$

$$-5 - 5 = 0 \quad ?$$

$$-10 = 0. \quad \text{False}$$

If $x = 3$, then

$$3 - \sqrt{15 - 2(3)} = 0 \quad ?$$

$$3 - \sqrt{9} = 0 \quad ?$$

$$3 - 3 = 0 \quad ?$$

$$0 = 0. \quad \text{True}$$

As the check shows, only 3 is a solution, giving the solution set $\{3\}$.

Now try Exercise 27.

EXAMPLE 4 Solving an Equation Containing Two Radicals

Solve $\sqrt{2x+3} - \sqrt{x+1} = 1$.

Solution When an equation contains two radicals, begin by isolating one of the radicals on one side of the equation.

$$\sqrt{2x+3} - \sqrt{x+1} = 1$$

$$\sqrt{2x+3} = 1 + \sqrt{x+1} \quad \text{Isolate } \sqrt{2x+3}. \text{ (Step 1)}$$

$$\left(\sqrt{2x+3}\right)^2 = \left(1 + \sqrt{x+1}\right)^2 \quad \text{Square both sides. (Step 2)}$$

$$2x + 3 = 1 + 2\sqrt{x+1} + (x+1)$$

Be careful: $(a+b)^2 = a^2 + 2ab + b^2$. **(Section R.3)**

$$x + 1 = 2\sqrt{x+1} \quad \text{Isolate the remaining radical. (Step 1)}$$

$$(x+1)^2 = \left(2\sqrt{x+1}\right)^2 \quad \text{Square again. (Step 2)}$$

$$x^2 + 2x + 1 = 4(x+1)$$

$$x^2 + 2x + 1 = 4x + 4$$

$$x^2 - 2x - 3 = 0 \quad \text{Solve the quadratic equation. (Step 3)}$$

$$(x-3)(x+1) = 0$$

$$x - 3 = 0 \quad \text{or} \quad x + 1 = 0$$

$$x = 3 \quad \text{or} \quad x = -1$$

Check: $\sqrt{2x+3} - \sqrt{x+1} = 1$ Original equation (Step 4)

If $x = 3$, then

$$\sqrt{2(3)+3} - \sqrt{3+1} = 1 \quad ?$$

$$\sqrt{9} - \sqrt{4} = 1 \quad ?$$

$$3 - 2 = 1 \quad ?$$

$$1 = 1. \quad \text{True}$$

If $x = -1$, then

$$\sqrt{2(-1)+3} - \sqrt{-1+1} = 1 \quad ?$$

$$\sqrt{1} - \sqrt{0} = 1 \quad ?$$

$$1 - 0 = 1 \quad ?$$

$$1 = 1. \quad \text{True}$$

Both 3 and -1 are solutions of the original equation, so $\{3, -1\}$ is the solution set.

Now try Exercise 41.

CAUTION Avoid the common error of neglecting to isolate a radical in Step 1. It would be incorrect to square each term individually as the first step in Example 4.

EXAMPLE 5 Solving an Equation Containing a Radical (Cube Root)

Solve $\sqrt[3]{4x^2 - 4x + 1} - \sqrt[3]{x} = 0$.

Solution

$$\sqrt[3]{4x^2 - 4x + 1} = \sqrt[3]{x} \quad \text{Isolate a radical. (Step 1)}$$

$$\left(\sqrt[3]{4x^2 - 4x + 1}\right)^3 = \left(\sqrt[3]{x}\right)^3 \quad \text{Cube both sides. (Step 2)}$$

$$4x^2 - 4x + 1 = x$$

$$4x^2 - 5x + 1 = 0 \quad \text{Solve the quadratic equation. (Step 3)}$$

$$(4x - 1)(x - 1) = 0$$

$$4x - 1 = 0 \quad \text{or} \quad x - 1 = 0$$

$$x = \frac{1}{4} \quad \text{or} \quad x = 1$$

Check: $\sqrt[3]{4x^2 - 4x + 1} - \sqrt[3]{x} = 0$ Original equation (Step 4)

If $x = \frac{1}{4}$, then

$$\sqrt[3]{4\left(\tfrac{1}{4}\right)^2 - 4\left(\tfrac{1}{4}\right) + 1} - \sqrt[3]{\tfrac{1}{4}} = 0 \quad ?$$

$$\sqrt[3]{\tfrac{1}{4}} - \sqrt[3]{\tfrac{1}{4}} = 0 \quad ?$$

$$0 = 0. \quad \text{True}$$

If $x = 1$, then

$$\sqrt[3]{4(1)^2 - 4(1) + 1} - \sqrt[3]{1} = 0 \quad ?$$

$$\sqrt[3]{1} - \sqrt[3]{1} = 0 \quad ?$$

$$0 = 0. \quad \text{True}$$

Both are valid solutions, and the solution set is $\left\{\frac{1}{4}, 1\right\}$.

Now try Exercise 51.

Equations Quadratic in Form Many equations that are not quadratic equations can be solved by the methods discussed in Section 1.4. The equation

$$12x^4 - 11x^2 + 2 = 0$$

is not a quadratic equation because of the x^4 term. However, with the substitutions

$$u = x^2 \quad \text{and} \quad u^2 = (x^2)^2 = x^4$$

the equation becomes

$$12u^2 - 11u + 2 = 0,$$

which is a quadratic equation in u. This quadratic equation can be solved to find u, and then $u = x^2$ can be used to find the values of x, the solutions to the original equation.

Equation Quadratic in Form

An equation is said to be **quadratic in form** if it can be written as

$$au^2 + bu + c = 0,$$

where $a \neq 0$ and u is some algebraic expression.

EXAMPLE 6 Solving an Equation Quadratic in Form

Solve $12x^4 - 11x^2 + 2 = 0$.

Solution

$$12(x^2)^2 - 11x^2 + 2 = 0 \qquad x^4 = (x^2)^2$$

$$12u^2 - 11u + 2 = 0 \qquad \text{Let } u = x^2\text{; thus } u^2 = x^4.$$

$$(3u - 2)(4u - 1) = 0 \qquad \text{Solve the quadratic equation.}$$

$$3u - 2 = 0 \quad \text{or} \quad 4u - 1 = 0 \qquad \text{Zero-factor property}$$

$$u = \frac{2}{3} \quad \text{or} \quad u = \frac{1}{4}$$

To find x, replace u with x^2.

$$x^2 = \frac{2}{3} \quad \text{or} \quad x^2 = \frac{1}{4} \qquad u = x^2$$

$$x = \pm\sqrt{\frac{2}{3}} \quad \text{or} \quad x = \pm\sqrt{\frac{1}{4}} \qquad \text{Square root property (Section 1.4)}$$

$$x = \frac{\pm\sqrt{2}}{\sqrt{3}} \cdot \frac{\sqrt{3}}{\sqrt{3}} \quad \text{or} \quad x = \pm\frac{1}{2} \qquad \text{Simplify radicals. (Section R.7)}$$

$$x = \pm\frac{\sqrt{6}}{3}$$

Check that the solution set is $\left\{\pm\frac{\sqrt{6}}{3}, \pm\frac{1}{2}\right\}$.

Now try Exercise 61.

EXAMPLE 7 Solving Equations Quadratic in Form

Solve each equation.

(a) $(x + 1)^{2/3} - (x + 1)^{1/3} - 2 = 0$ **(b)** $6x^{-2} + x^{-1} = 2$

Solution

(a) Since $(x + 1)^{2/3} = [(x + 1)^{1/3}]^2$, let $u = (x + 1)^{1/3}$.

$$u^2 - u - 2 = 0 \qquad \text{Substitute.}$$

$$(u - 2)(u + 1) = 0 \qquad \text{Factor.}$$

$$u - 2 = 0 \quad \text{or} \quad u + 1 = 0 \qquad \text{Zero-factor property}$$

$$u = 2 \quad \text{or} \quad u = -1$$

Now, replace u with $(x + 1)^{1/3}$.

$$(x + 1)^{1/3} = 2 \quad \text{or} \quad (x + 1)^{1/3} = -1$$

$$[(x + 1)^{1/3}]^3 = 2^3 \quad \text{or} \quad [(x + 1)^{1/3}]^3 = (-1)^3 \qquad \text{Cube each side.}$$

$$x + 1 = 8 \quad \text{or} \quad x + 1 = -1$$

$$x = 7 \quad \text{or} \quad x = -2$$

Check: $(x+1)^{2/3}-(x+1)^{1/3}-2=0$ Original equation

If $x=7$, then

$$(7+1)^{2/3}-(7+1)^{1/3}-2=0 \quad ?$$
$$8^{2/3}-8^{1/3}-2=0 \quad ?$$
$$4-2-2=0 \quad ?$$
$$0=0. \quad \text{True}$$

If $x=-2$, then

$$(-2+1)^{2/3}-(-2+1)^{1/3}-2=0 \quad ?$$
$$(-1)^{2/3}-(-1)^{1/3}-2=0 \quad ?$$
$$1+1-2=0 \quad ?$$
$$0=0. \quad \text{True}$$

Both check, so the solution set is $\{-2,7\}$.

(b) $6x^{-2}+x^{-1}-2=0$ Subtract 2.

$6u^2+u-2=0$ Let $u=x^{-1}$; thus $u^2=x^{-2}$.

$(3u+2)(2u-1)=0$ Factor.

$3u+2=0$ or $2u-1=0$ Zero-factor property

$u=-\frac{2}{3}$ or $u=\frac{1}{2}$

$x^{-1}=-\frac{2}{3}$ or $x^{-1}=\frac{1}{2}$ Substitute again.

$x=-\frac{3}{2}$ or $x=2$ x^{-1} is the reciprocal of x. **(Section R.6)**

Both check, so the solution set is $\left\{-\frac{3}{2},2\right\}$.

Now try Exercises 71 and 75.

EXAMPLE 8 Solving an Equation That Leads to One That is Quadratic in Form

Solve $(5x^2-6)^{1/4}=x$.

Solution

$[(5x^2-6)^{1/4}]^4=x^4$ Raise both sides to the fourth power.

$5x^2-6=x^4$ Power rule for exponents **(Section R.3)**

$x^4-5x^2+6=0$

$u^2-5u+6=0$ Let $u=x^2$; thus $u^2=x^4$.

$(u-3)(u-2)=0$ Factor.

$u-3=0$ or $u-2=0$ Zero-factor property

$u=3$ or $u=2$

$x^2=3$ or $x^2=2$ $u=x^2$

$x=\pm\sqrt{3}$ or $x=\pm\sqrt{2}$ Square root property

Checking the four proposed solutions in the original equation shows that only $\sqrt{3}$ and $\sqrt{2}$ are solutions, since the left side of the equation cannot represent a negative number. The solution set is $\{\sqrt{2},\sqrt{3}\}$.

Now try Exercise 59.

CAUTION If a substitution variable is used when solving an equation that is quadratic in form, do not forget the step that gives the solution in terms of the *original* variable.

1.6 Exercises

Decide what values of the variable cannot possibly be solutions for each equation. Do not solve. See Examples 1 and 2.

1. $\frac{5}{2x+3} + \frac{1}{x-6} = 0$

2. $\frac{2}{x+1} - \frac{3}{5x+5} = 0$

3. $\frac{3}{x-2} + \frac{1}{x+1} = \frac{1}{x^2-x-2}$

4. $\frac{2}{x+3} - \frac{5}{x-1} = \frac{-1}{x^2+2x-3}$

5. $\frac{1}{4x} + \frac{2}{x} = 3$

6. $\frac{5}{2x} - \frac{2}{x} = 6$

Solve each equation. See Example 1.

7. $\frac{2x+5}{2} - \frac{3x}{x-2} = x$

8. $\frac{4x+3}{4} - \frac{2x}{x+1} = x$

9. $\frac{x}{x-3} = \frac{3}{x-3} + 3$

10. $\frac{x}{x-4} = \frac{4}{x-4} + 4$

11. $\frac{2}{x-3} - \frac{3}{x+3} = \frac{12}{x^2-9}$

12. $\frac{3}{x-2} + \frac{1}{x+2} = \frac{12}{x^2-4}$

13. $\frac{4}{x^2+x-6} - \frac{1}{x^2-4} = \frac{2}{x^2+5x+6}$

14. $\frac{3}{x^2+x-2} - \frac{1}{x^2-1} = \frac{7}{2x^2+6x+4}$

Solve each equation. See Example 2.

15. $\frac{2x+1}{x-2} + \frac{3}{x} = \frac{-6}{x^2-2x}$

16. $\frac{4x+3}{x+1} + \frac{2}{x} = \frac{1}{x^2+x}$

17. $\frac{-x}{x-1} + \frac{1}{x+1} = \frac{-2}{x^2-1}$

18. $\frac{x}{x+1} + \frac{1}{x-1} = \frac{2}{x^2-1}$

19. $\frac{5}{x^2} - \frac{43}{x} = 18$

20. $\frac{7}{x^2} + \frac{19}{x} = 6$

21. $2 = \frac{3}{2x-1} + \frac{-1}{(2x-1)^2}$

22. $6 = \frac{7}{2x-3} + \frac{3}{(2x-3)^2}$

23. $\frac{2x-5}{x} = \frac{x-2}{3}$

24. $\frac{x+4}{2x} = \frac{x-1}{3}$

25. $\frac{2x}{x-2} = 5 + \frac{4x^2}{x-2}$

26. $\frac{-3x}{2} + \frac{9x-5}{3} = \frac{11x+8}{6x}$

Solve each equation. See Examples 3–5 and 8.

27. $x - \sqrt{2x+3} = 0$

28. $x - \sqrt{3x+18} = 0$

29. $\sqrt{3x+7} = 3x + 5$

30. $\sqrt{4x+13} = 2x - 1$

31. $\sqrt{4x+5} - 2 = 2x - 7$

32. $\sqrt{6x+7} - 1 = x + 1$

33. $\sqrt{4x} - x + 3 = 0$

34. $\sqrt{2x} - x + 4 = 0$

35. $\sqrt{x} - \sqrt{x-5} = 1$

36. $\sqrt{x} - \sqrt{x-12} = 2$

37. $\sqrt{x+7} + 3 = \sqrt{x-4}$

38. $\sqrt{x+5} - 2 = \sqrt{x-1}$

39. $\sqrt{x+2} = \sqrt{2x+5} - 1$

40. $\sqrt{4x+1} = \sqrt{x-1} + 2$

41. $\sqrt{3x} = \sqrt{5x+1} - 1$

42. $\sqrt{2x} = \sqrt{3x+12} - 2$

43. $\sqrt{x+2} = 1 - \sqrt{3x+7}$

44. $\sqrt{2x-5} = 2 + \sqrt{x-2}$

45. $\sqrt{2\sqrt{7x+2}} = \sqrt{3x+2}$

46. $\sqrt{3\sqrt{2x+3}} = \sqrt{5x-6}$

47. $3 - \sqrt{x} = \sqrt{2\sqrt{x} - 3}$

48. $\sqrt{x} + 2 = \sqrt{4 + 7\sqrt{x}}$

49. $\sqrt[3]{4x+3} = \sqrt[3]{2x-1}$

50. $\sqrt[3]{2x} = \sqrt[3]{5x+2}$

51. $\sqrt[3]{5x^2 - 6x + 2} - \sqrt[3]{x} = 0$

52. $\sqrt[3]{3x^2 - 9x + 8} = \sqrt[3]{x}$

53. $(2x+5)^{1/3} - (6x-1)^{1/3} = 0$

54. $(3x+7)^{1/3} - (4x+2)^{1/3} = 0$

55. $\sqrt[4]{x-15} = 2$

56. $\sqrt[4]{3x+1} = 1$

57. $\sqrt[4]{x^2+2x} = \sqrt[4]{3}$

58. $\sqrt[4]{x^2+6x} = 2$

59. $(x^2 + 24x)^{1/4} = 3$

60. $(3x^2 + 52x)^{1/4} = 4$

Solve each equation. See Examples 6 and 7.

61. $2x^4 - 7x^2 + 5 = 0$

62. $4x^4 - 8x^2 + 3 = 0$

63. $x^4 + 2x^2 - 15 = 0$

64. $3x^4 + 10x^2 - 25 = 0$

65. $(2x-1)^{2/3} = x^{1/3}$

66. $(x-3)^{2/5} = (4x)^{1/5}$

67. $x^{2/3} = 2x^{1/3}$

68. $3x^{3/4} = x^{1/2}$

69. $(x-1)^{2/3} + (x-1)^{1/3} - 12 = 0$

70. $(2x-1)^{2/3} + 2(2x-1)^{1/3} - 3 = 0$

71. $(x+1)^{2/5} - 3(x+1)^{1/5} + 2 = 0$

72. $(x+5)^{4/3} + (x+5)^{2/3} - 20 = 0$

73. $6(x+2)^4 - 11(x+2)^2 = -4$

74. $8(x-4)^4 - 10(x-4)^2 = -3$

75. $10x^{-2} + 33x^{-1} - 7 = 0$

76. $7x^{-2} - 10x^{-1} - 8 = 0$

77. $x^{-2/3} + x^{-1/3} - 6 = 0$

78. $2x^{-4/3} - x^{-2/3} - 1 = 0$

79. Refer to the equation in Exercise 39. A student attempted to solve the equation by "squaring both sides" to get $x + 2 = (2x + 5) + 1$. What was incorrect about the student's method?

80. Refer to the equation in Exercise 48. What should be the first step in solving the equation using the method described in this section?

Relating Concepts

For individual or collaborative investigation

(Exercises 81–84)

In this section we introduced methods of solving equations quadratic in form by substitution, and solving equations involving radicals by raising both sides of the equation to a power. Suppose we wish to solve

$$x - \sqrt{x} - 12 = 0.$$

We can solve this equation using either of the two methods. ***Work Exercises 81–84 in order,*** *to see how both methods apply.*

81. Let $u = \sqrt{x}$ and solve the equation by substitution. What is the value of u that does not lead to a solution of the equation?

82. Solve the equation by isolating $\sqrt{x}$ on one side and then squaring. What is the value of x that does not satisfy the equation?

83. Which one of the methods used in Exercises 81 and 82 do you prefer? Why?

84. Solve $3x - 2\sqrt{x} - 8 = 0$ using one of the two methods described.

Solve each equation for the indicated variable. Assume all denominators are nonzero.

85. $d = k\sqrt{h}$ for h

86. $x^{2/3} + y^{2/3} = a^{2/3}$ for y

87. $m^{3/4} + n^{3/4} = 1$ for m

88. $\frac{1}{R} = \frac{1}{r_1} + \frac{1}{r_2}$ for R

89. $\frac{E}{e} = \frac{R + r}{r}$ for e

Summary Exercises on Solving Equations

This section of miscellaneous equations provides practice in solving all the types introduced in this chapter. Solve each equation.

1. $4x - 5 = 2x + 1$

2. $5 - (6x + 3) = 2(2 - 2x)$

3. $x(x + 6) = 9$

4. $x^2 - 8x + 12 = 0$

5. $\sqrt{x + 2} + 5 = \sqrt{x + 15}$

6. $\frac{5}{x + 3} - \frac{6}{x - 2} = \frac{3}{x^2 + x - 6}$

7. $\frac{3x + 4}{3} - \frac{2x}{x - 3} = x$

8. $x + \frac{8}{3}x = 2x + 10$

9. $5 - \frac{2}{x} + \frac{1}{x^2} = 0$

10. $(2x + 1)^2 = 9$

11. $x^{-2/5} - 2x^{-1/5} - 15 = 0$

12. $\sqrt{x + 2} = \sqrt{2x + 6} - 1$

13. $x^4 - 3x^2 - 4 = 0$

14. $1.2x + .3 = .7x - .9$

15. $\sqrt[4]{2x + 1} = \sqrt[4]{9}$

16. $3x^2 - 2x + 1 = 0$

17. $3[2x - (6 - 2x) + 1] = 5x$

18. $\sqrt{x} + 1 = \sqrt{11 - \sqrt{x}}$

19. $(14 - 2x)^{2/3} = 4$

20. $2x^{-1} - x^{-2} = 1$

1.7 Inequalities

Linear Inequalities ▪ Three-Part Inequalities ▪ Quadratic Inequalities ▪ Rational Inequalities

An **inequality** says that one expression is greater than, greater than or equal to, less than, or less than or equal to, another. As with equations, a value of the variable for which the inequality is true is a solution of the inequality; the set of all solutions is the solution set of the inequality. Two inequalities with the same solution set are equivalent.

Inequalities are solved with the properties of inequality, which are similar to the properties of equality in Section 1.1.

Properties of Inequality

For real numbers a, b, and c:

1. **If $a < b$, then $a + c < b + c$,**
2. **If $a < b$ and if $c > 0$, then $ac < bc$,**
3. **If $a < b$ and if $c < 0$, then $ac > bc$.**

Replacing $<$ with $>$, $\leq$, or $\geq$ results in similar properties. (Restrictions on c remain the same.)

NOTE Multiplication may be replaced by division in properties 2 and 3. ***Always remember to reverse the direction of the inequality symbol when multiplying or dividing by a negative number.***

Linear Inequalities The definition of a linear inequality is similar to the definition of a linear equation.

Linear Inequality in One Variable

A **linear inequality in one variable** is an inequality that can be written in the form

$$ax + b > 0,$$

where a and b are real numbers with $a \neq 0$. (Any of the symbols $\geq$, $<$, or $\leq$ may also be used.)

EXAMPLE 1 Solving a Linear Inequality

Solve $-3x + 5 > -7$.

Solution

$$-3x + 5 > -7$$

$$-3x + 5 - 5 > -7 - 5 \quad \text{Subtract 5.}$$

$$-3x > -12$$

$$\frac{-3x}{-3} < \frac{-12}{-3} \quad \text{Divide by } -3\text{; reverse the direction of the inequality symbol.}$$

$$x < 4$$

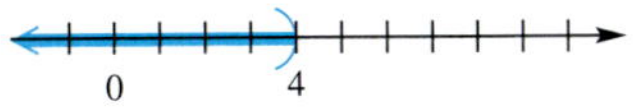

0 4

Figure 9

The original inequality is satisfied by any real number less than 4. The solution set can be written $\{x \mid x < 4\}$. A graph of the solution set is shown in Figure 9, where the parenthesis is used to show that 4 itself does not belong to the solution set.

The solution set for the inequality in Example 1, $\{x \mid x < 4\}$, is an example of an **interval.** We use a simplified notation, called **interval notation,** to write intervals. With this notation, we write the interval in Example 1 as $(-\infty, 4)$. The symbol $-\infty$ does not represent an actual number; it is used to show that the interval includes all real numbers less than 4. The interval $(-\infty, 4)$ is an example of an **open interval,** since the endpoint, 4, is not part of the interval. A **closed interval** includes both endpoints. A square bracket is used to show that a number *is* part of the graph, and a parenthesis is used to indicate that a number *is not* part of the graph. In the table that follows, we assume that $a < b$.

Type of Interval	Set	Interval Notation	Graph
Open interval	$\{x \mid x > a\}$	(a, ∞)	
	$\{x \mid a < x < b\}$	(a, b)	
	$\{x \mid x < b\}$	$(-\infty, b)$	
Half-open interval	$\{x \mid x \geq a\}$	$[a, \infty)$	
	$\{x \mid a < x \leq b\}$	$(a, b]$	
	$\{x \mid a \leq x < b\}$	$[a, b)$	
	$\{x \mid x \leq b\}$	$(-\infty, b]$	
Closed interval	$\{x \mid a \leq x \leq b\}$	$[a, b]$	
All real numbers	$\{x \mid x \text{ is a real number}\}$	$(-\infty, \infty)$	

Now try Exercise 15.

EXAMPLE 2 Solving a Linear Inequality

Solve $4 - 3x \leq 7 + 2x$. Give the solution set in interval notation and graph it.

Solution

$$4 - 3x \leq 7 + 2x$$

$$4 - 3x - 4 \leq 7 + 2x - 4 \quad \text{Subtract 4.}$$

$$-3x \leq 3 + 2x$$

$$-3x - 2x \leq 3 + 2x - 2x \quad \text{Subtract } 2x.$$

$$-5x \leq 3$$

$$\frac{-5x}{-5} \geq \frac{3}{-5} \quad \text{Divide by } -5\text{; reverse the direction of the inequality symbol.}$$

$$x \geq -\frac{3}{5}$$

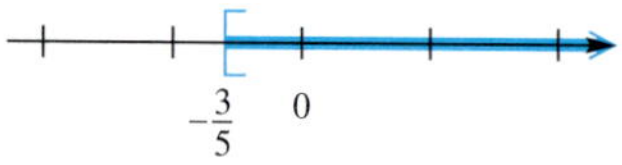

$-\frac{3}{5}$ 0

Figure 10

In interval notation the solution set is $\left[-\frac{3}{5}, \infty\right)$. See Figure 10 for the graph.

Now try Exercise 17.

From now on, we write solution sets of inequalities in interval notation.

Three-Part Inequalities The inequality $-2 < 5 + 3x < 20$ in the next example says that $5 + 3x$ is *between* -2 and 20. This inequality is solved using an extension of the properties of inequality given earlier, working with all three expressions at the same time.

EXAMPLE 3 Solving a Three-Part Inequality

Solve $-2 < 5 + 3x < 20$.

Solution

$$-2 < 5 + 3x < 20$$

$$-2 - 5 < 5 + 3x - 5 < 20 - 5 \quad \text{Subtract 5 from each part.}$$

$$-7 < 3x < 15$$

$$\frac{-7}{3} < \frac{3x}{3} < \frac{15}{3} \quad \text{Divide each part by 3.}$$

$$-\frac{7}{3} < x < 5$$

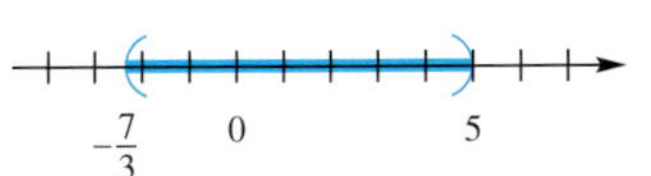

Figure 11

The solution set, graphed in Figure 11, is the interval $\left(-\frac{7}{3}, 5\right)$.

Now try Exercise 25.

A product will break even, or begin to produce a profit, only if the revenue from selling the product at least equals the cost of producing it. If R represents revenue and C is cost, then the **break-even point** is the point where $R = C$.

EXAMPLE 4 Finding the Break-Even Point

If the revenue and cost of a certain product are given by

$$R = 4x \qquad \text{and} \qquad C = 2x + 1000,$$

where x is the number of units produced and sold, at what production level does R at least equal C?

Solution Set $R \geq C$ and solve for x.

$$R \geq C$$

$$4x \geq 2x + 1000 \quad \text{Substitute.}$$

$$2x \geq 1000 \quad \text{Subtract } 2x.$$

$$x \geq 500 \quad \text{Divide by 2.}$$

The break-even point is at $x = 500$. This product will at least break even only if the number of units produced and sold is in the interval $[500, \infty)$.

Now try Exercise 35.

Quadratic Inequalities The solution of *quadratic inequalities* depends on the solution of quadratic equations, introduced in Section 1.4.

Quadratic Inequality

A **quadratic inequality** is an inequality that can be written in the form

$$ax^2 + bx + c < 0$$

for real numbers a, b, and c with $a \neq 0$. (The symbol $<$ can be replaced with $>$, $\leq$, or $\geq$.)

One method of solving a quadratic inequality involves finding the solutions of the corresponding quadratic equation, and then testing values in the intervals on a number line determined by those solutions.

Solving a Quadratic Inequality

Step 1 Solve the corresponding quadratic equation.

Step 2 Identify the intervals determined by the solutions of the equation.

Step 3 Use a test value from each interval to determine which intervals form the solution set.

EXAMPLE 5 Solving a Quadratic Inequality

Solve $x^2 - x - 12 < 0$.

Solution

Step 1 Find the values of x that satisfy $x^2 - x - 12 = 0$.

$$x^2 - x - 12 = 0 \quad \text{Corresponding quadratic equation}$$

$$(x + 3)(x - 4) = 0 \quad \text{Factor. (Section R.4)}$$

$$x + 3 = 0 \quad \text{or} \quad x - 4 = 0 \quad \text{Zero-factor property (Section 1.4)}$$

$$x = -3 \quad \text{or} \quad x = 4$$

Step 2 The two numbers -3 and 4 divide a number line into the three intervals shown in Figure 12. If a value in Interval A, for example, makes the polynomial $x^2 - x - 12$ negative, then all values in Interval A will make that polynomial negative.

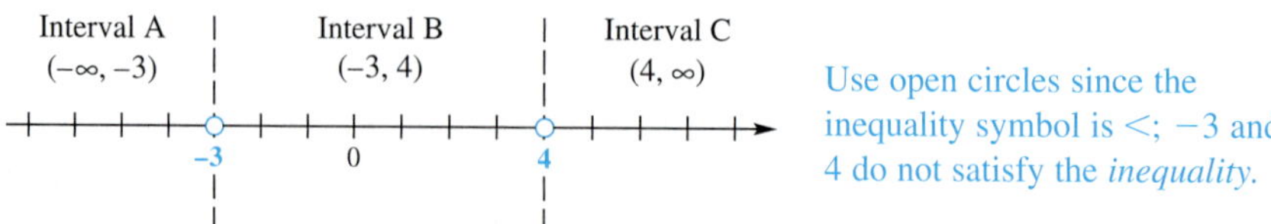

Figure 12

Step 3 Choose a test value in each interval to see if it satisfies the original inequality, $x^2 - x - 12 < 0$. If the test value makes the statement true, then the entire interval belongs to the solution set.

Interval	Test Value	Is $x^2 - x - 12 < 0$ True or False?
A: $(-\infty, -3)$	-4	$(-4)^2 - (-4) - 12 < 0$? $8 < 0$ False
B: $(-3, 4)$	0	$0^2 - 0 - 12 < 0$? $-12 < 0$ True
C: $(4, \infty)$	5	$5^2 - 5 - 12 < 0$? $8 < 0$ False

Since the values in Interval B make the inequality true, the solution set is $(-3, 4)$. See Figure 13.

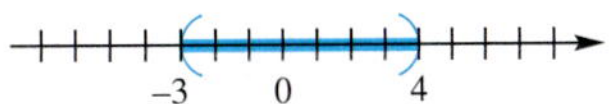

Figure 13

Now try Exercise 39.

EXAMPLE 6 Solving a Quadratic Inequality

Solve $2x^2 + 5x - 12 \geq 0$.

Solution

Step 1 Find the values of x that satisfy $2x^2 + 5x - 12 = 0$.

$$2x^2 + 5x - 12 = 0 \quad \text{Corresponding quadratic equation}$$
$$(2x - 3)(x + 4) = 0 \quad \text{Factor.}$$
$$2x - 3 = 0 \quad \text{or} \quad x + 4 = 0 \quad \text{Zero-factor property}$$
$$x = \frac{3}{2} \quad \text{or} \quad x = -4$$

Step 2 The values $\frac{3}{2}$ and -4 form the intervals $(-\infty, -4)$, $\left(-4, \frac{3}{2}\right)$, and $\left(\frac{3}{2}, \infty\right)$ on the number line, as seen in Figure 14.

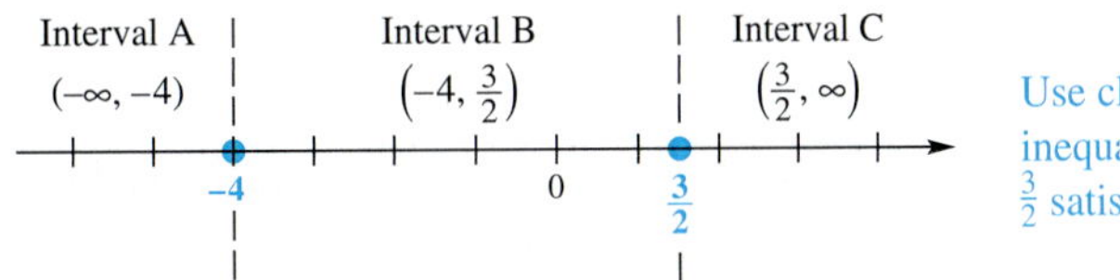

Use closed circles since the inequality symbol is $\geq$; -4 and $\frac{3}{2}$ satisfy the *inequality*.

Figure 14

Step 3 Choose a test value from each interval.

Interval	Test Value	Is $2x^2 + 5x - 12 \geq 0$ True or False?
A: $(-\infty, -4)$	-5	$2(-5)^2 + 5(-5) - 12 \geq 0$? $13 \geq 0$ True
B: $\left(-4, \frac{3}{2}\right)$	0	$2(0)^2 + 5(0) - 12 \geq 0$? $-12 \geq 0$ False
C: $\left(\frac{3}{2}, \infty\right)$	2	$2(2)^2 + 5(2) - 12 \geq 0$? $6 \geq 0$ True

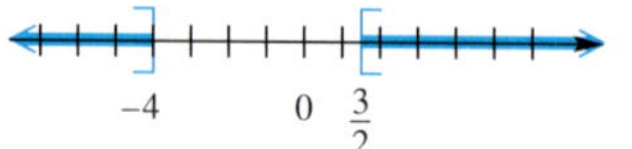

Figure 15

The values in Intervals A and C make the inequality true, so the solution set is the *union** of the intervals, written $(-\infty, -4] \cup \left[\frac{3}{2}, \infty\right)$. The graph of the solution set is shown in Figure 15.

Now try Exercise 41.

NOTE Inequalities that use the symbols $<$ and $>$ are called **strict inequalities;** $\leq$ and $\geq$ are used in **nonstrict inequalities.** The solutions of the equation in Example 5 were not included in the solution set since the inequality was a *strict* inequality. In Example 6, the solutions of the equation *were* included in the solution set because of the nonstrict inequality.

EXAMPLE 7 Solving a Problem Involving the Height of a Projectile

If a projectile is launched from ground level with an initial velocity of 96 ft per sec, its height in feet t seconds after launching is s feet, where

$$s = -16t^2 + 96t.$$

When will the projectile be greater than 80 ft above ground level?

Solution We want $s > 80$, so we must solve

$$-16t^2 + 96t > 80.$$

$$-16t^2 + 96t - 80 > 0 \quad \text{Subtract 80.}$$

$$t^2 - 6t + 5 < 0 \quad \text{Divide by } -16\text{; reverse the direction of the inequality symbol.}$$

Now solve the corresponding *equation.*

$$t^2 - 6t + 5 = 0$$

$$(t - 1)(t - 5) = 0 \quad \text{Factor.}$$

$$t = 1 \quad \text{or} \quad t = 5 \quad \text{Zero-factor property}$$

*The **union** of sets A and B, written $A \cup B$, is defined as

$$A \cup B = \{x \mid x \text{ is an element of } A \text{ or } x \text{ is an element of } B\}.$$

The **intersection** of sets A and B, written $A \cap B$, is defined as

$$A \cap B = \{x \mid x \text{ is an element of } A \text{ and } x \text{ is an element of } B.\}$$

Use these values to determine intervals. See Figure 16.

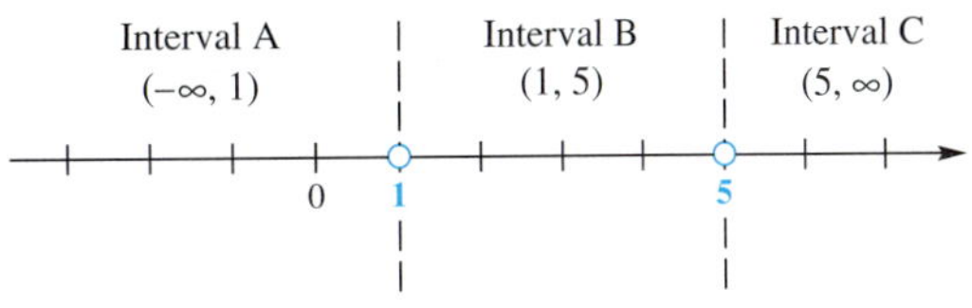

Figure 16

Use the procedure of Examples 5 and 6 to determine that values in Interval B, $(1, 5)$, satisfy the inequality. The projectile is greater than 80 ft above ground level between 1 and 5 sec after it is launched.

Now try Exercise 95.

Rational Inequalities Inequalities involving rational expressions such as

$$\frac{5}{x+4} \geq 1 \quad \text{and} \quad \frac{2x-1}{3x+4} < 5 \qquad \text{Rational inequalities}$$

are called **rational inequalities,** and are solved in a manner similar to the procedure for solving quadratic inequalities.

Solving a Rational Inequality

Step 1 Rewrite the inequality, if necessary, so that 0 is on one side and there is a single fraction on the other side.

Step 2 Determine the values that will cause either the numerator or the denominator of the rational expression to equal 0. These values determine the intervals on the number line to consider.

Step 3 Use a test value from each interval to determine which intervals form the solution set.

CAUTION Solving a rational inequality such as

$$\frac{5}{x+4} \geq 1$$

by multiplying both sides by $x + 4$ to obtain $5 \geq x + 4$, requires considering *two cases* since the sign of $x + 4$ depends on the value of x. If $x + 4$ were negative, then the inequality symbol must be reversed. The procedure described in the preceding box and used in the next two examples eliminates the need for considering separate cases.

EXAMPLE 8 Solving a Rational Inequality

Solve $\frac{5}{x+4} \geq 1.$

Solution

Step 1

$$\frac{5}{x+4} - 1 \geq 0$$ Subtract 1 so that 0 is on one side.

$$\frac{5}{x+4} - \frac{x+4}{x+4} \geq 0$$ Use $x + 4$ as the common denominator.

$$\frac{5-(x+4)}{x+4} \geq 0$$ Write as a single fraction. (Section R.5)

$$\frac{1-x}{x+4} \geq 0$$ Combine terms in the numerator; be careful with signs.

Step 2 The quotient possibly changes sign only where x-values make the numerator or denominator 0. This occurs at

$$1 - x = 0 \quad \text{or} \quad x + 4 = 0$$
$$x = 1 \quad \text{or} \quad x = -4.$$

These values form the intervals $(-\infty, -4)$, $(-4, 1)$, and $(1, \infty)$ on the number line, as seen in Figure 17.

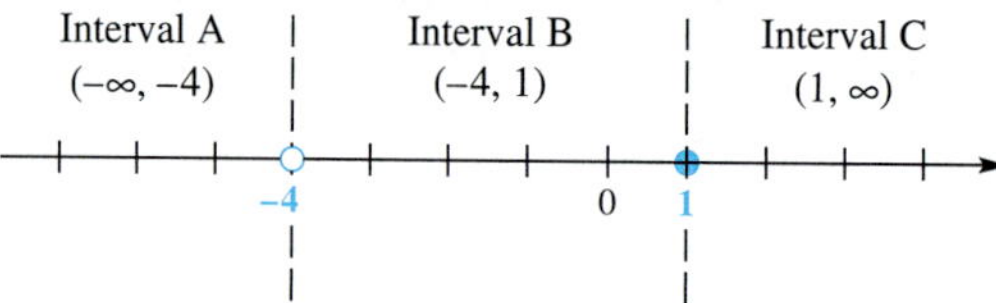

Use a solid circle on 1, since the symbol is $\geq$. The value -4 cannot be in the solution set since it causes the denominator to equal 0. Use an open circle on -4.

Figure 17

Step 3 Choose test values.

Interval	Test Value	Is $\frac{5}{x+4} \geq 1$ True or False?
A: $(-\infty, -4)$	-5	$\frac{5}{-5+4} \geq 1$? $-5 \geq 1$ False
B: $(-4, 1)$	0	$\frac{5}{0+4} \geq 1$? $\frac{5}{4} \geq 1$ True
C: $(1, \infty)$	2	$\frac{5}{2+4} \geq 1$? $\frac{5}{6} \geq 1$ False

The values in the interval $(-4, 1)$ satisfy the original inequality. The value 1 makes the nonstrict inequality true, so it must be included in the solution set. Since -4 makes the denominator 0, it must be excluded. The solution set is $(-4, 1]$.

Now try Exercise 73.

CAUTION As suggested by Example 8, be careful with the endpoints of the intervals when solving rational inequalities.

EXAMPLE 9 Solving a Rational Inequality

Solve $\frac{2x - 1}{3x + 4} < 5$.

Solution

$$\frac{2x - 1}{3x + 4} < 5$$

$$\frac{2x - 1}{3x + 4} - 5 < 0 \quad \text{Subtract 5.}$$

$$\frac{2x - 1 - 5(3x + 4)}{3x + 4} < 0 \quad \text{Common denominator is } 3x + 4.$$

$$\frac{-13x - 21}{3x + 4} < 0 \quad \text{Multiply and combine terms in the numerator.}$$

Set the numerator and denominator equal to 0 and solve the resulting equations to get the values of x where sign changes may occur.

$$-13x - 21 = 0 \quad \text{or} \quad 3x + 4 = 0$$

$$x = -\frac{21}{13} \quad \text{or} \quad x = -\frac{4}{3}$$

Use these values to form intervals on the number line, as seen in Figure 18.

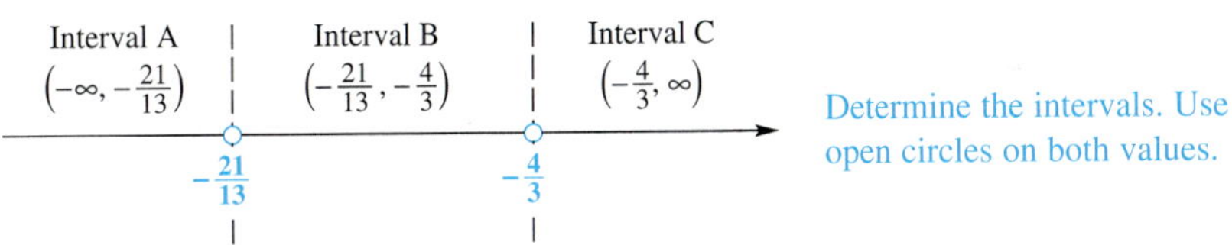

Figure 18

Now choose test values from the intervals in Figure 18. Verify that

-2 from Interval A makes the inequality true;

-1.5 from Interval B makes the inequality false;

0 from Interval C makes the inequality true.

Because of the $<$ symbol, neither endpoint satisfies the inequality, so the solution set is $\left(-\infty, -\frac{21}{13}\right) \cup \left(-\frac{4}{3}, \infty\right)$.

Now try Exercise 85.

1.7 Exercises

Concept Check *Match the inequality in each exercise in Column I with its equivalent interval notation in Column II.*

I	II
1. $x < -4$	**A.** $(-2, 6]$
2. $x \le 4$	**B.** $[-2, 6)$
3. $-2 < x \le 6$	**C.** $(-\infty, -4]$
4. $0 \le x \le 8$	**D.** $[4, \infty)$
5. $x \ge -3$	**E.** $(3, \infty)$
6. $4 \le x$	**F.** $(-\infty, -4)$
7. 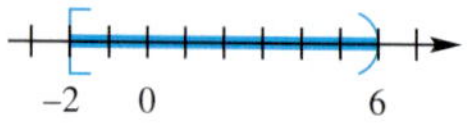	**G.** $(0, 8)$
8. 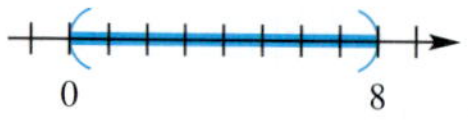	**H.** $[0, 8]$
9. 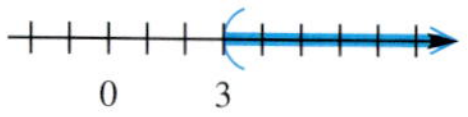	**I.** $[-3, \infty)$
10. 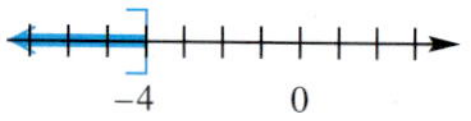	**J.** $(-\infty, 4]$

11. Explain how to determine whether to use a parenthesis or a square bracket when graphing the solution set of a linear inequality.

12. *Concept Check* The three-part inequality $a < x < b$ means "a is less than x and x is less than b." Which one of the following inequalities is not satisfied by some real number x?

A. $-3 < x < 5$ **B.** $0 < x < 4$
C. $-3 < x < -2$ **D.** $-7 < x < -10$

Solve each inequality. Write each solution set in interval notation, and graph it. See Examples 1 and 2.

13. $2x + 1 \le 9$

14. $3x - 2 \le 13$

15. $-3x - 2 \le 1$

16. $-5x + 3 \ge -2$

17. $2(x + 5) + 1 \ge 5 + 3x$

18. $6x - (2x + 3) \ge 4x - 5$

19. $8x - 3x + 2 < 2(x + 7)$

20. $2 - 4x + 5(x - 1) < -6(x - 2)$

21. $\frac{4x + 7}{-3} \le 2x + 5$

22. $\frac{2x - 5}{-8} \le 1 - x$

23. $\frac{1}{3}x + \frac{2}{5}x - \frac{1}{2}(x + 3) \le \frac{1}{10}$

24. $-\frac{2}{3}x - \frac{1}{6}x + \frac{2}{3}(x + 1) \le \frac{4}{3}$

Solve each inequality. Write each solution set in interval notation, and graph it. See Example 3.

25. $-3 < 7 + 2x < 13$

26. $-4 < 5 + 3x < 8$

27. $10 \le 2x + 4 \le 16$

28. $-6 \le 6x + 3 \le 21$

29. $-10 > -3x + 2 > -16$

30. $4 > -6x + 5 > -1$

31. $-4 \le \frac{x + 1}{2} \le 5$

32. $-5 \le \frac{x - 3}{3} \le 1$

33. $-3 \leq \frac{x-4}{-5} < 4$

34. $1 \leq \frac{4x-5}{-2} < 9$

Break-Even Interval *Find all intervals where each product will at least break even. See Example 4.*

35. The cost to produce x units of picture frames is $C = 50x + 5000$, while the revenue is $R = 60x$.

36. The cost to produce x units of baseball caps is $C = 100x + 6000$, while the revenue is $R = 500x$.

37. The cost to produce x units of coffee cups is $C = 85x + 900$, while the revenue is $R = 105x$.

38. The cost to produce x units of briefcases is $C = 70x + 500$, while the revenue is $R = 60x$.

Solve each quadratic inequality. Write each solution set in interval notation. See Examples 5 and 6.

39. $x^2 - x - 6 > 0$

40. $x^2 - 7x + 10 > 0$

41. $2x^2 - 9x - 18 \leq 0$

42. $3x^2 + x - 4 \leq 0$

43. $x^2 + 4x + 6 \geq 3$

44. $x^2 + 6x + 16 < 8$

45. $x(x - 1) \leq 6$

46. $x(x + 1) < 12$

47. $x^2 \leq 9$

48. $x^2 > 16$

49. $x^2 + 5x - 2 < 0$

50. $4x^2 + 3x + 1 \leq 0$

51. $x^2 - 2x \leq 1$

52. $x^2 + 4x > -1$

53. ***Concept Check*** Which one of the following inequalities has solution set $(-\infty, \infty)$?

A. $(x + 3)^2 \geq 0$

B. $(5x - 6)^2 \leq 0$

C. $(6x + 4)^2 > 0$

D. $(8x - 7)^2 < 0$

54. ***Concept Check*** Which one of the inequalities in Exercise 53 has solution set $\emptyset$?

Relating Concepts

For individual or collaborative investigation

(Exercises 55–58)

Inequalities that involve more than two factors, such as

$$(3x - 4)(x + 2)(x + 6) \leq 0,$$

can be solved using an extension of the method shown in Examples 5 and 6. ***Work Exercises 55–58 in order,*** *to see how the method is extended.*

55. Use the zero-factor property to solve $(3x - 4)(x + 2)(x + 6) = 0$.

56. Plot the three solutions in Exercise 55 on a number line.

57. The number line from Exercise 56 should show four intervals formed by the three points. For each interval, choose a number from the interval and decide whether it satisfies the original inequality.

58. On a single number line, graph the intervals that satisfy the inequality, including endpoints. This is the graph of the solution set of the inequality. Write the solution set in interval notation.

Use the technique described in Relating Concepts Exercises 55–58 to solve each inequality. Write each solution set in interval notation.

59. $(2x - 3)(x + 2)(x - 3) \geq 0$

60. $(x + 5)(3x - 4)(x + 2) \geq 0$

61. $4x - x^3 \geq 0$

62. $16x - x^3 \geq 0$

63. $(x + 1)^2(x - 3) < 0$

64. $(x - 5)^2(x + 1) < 0$

65. $x^3 + 4x^2 - 9x - 36 \geq 0$

66. $x^3 + 3x^2 - 16x - 48 \leq 0$

67. $x^2(x + 4)^2 \geq 0$

68. $x^2(2x - 3)^2 < 0$

Solve each rational inequality. Write each solution set in interval notation. See Examples 8 and 9.

69. $\dfrac{x - 3}{x + 5} \leq 0$

70. $\dfrac{x + 1}{x - 4} > 0$

71. $\dfrac{x - 1}{x + 2} > 1$

72. $\dfrac{x - 6}{x + 2} < -1$

73. $\dfrac{3}{x - 6} \leq 2$

74. $\dfrac{3}{x - 2} < 1$

75. $\dfrac{4}{x - 1} < 5$

76. $\dfrac{6}{5 - 3x} \leq 2$

77. $\dfrac{10}{3 + 2x} \leq 5$

78. $\dfrac{1}{x + 2} \geq 3$

79. $\dfrac{7}{x + 2} \geq \dfrac{1}{x + 2}$

80. $\dfrac{5}{x + 1} > \dfrac{12}{x + 1}$

81. $\dfrac{3}{2x - 1} > \dfrac{-4}{x}$

82. $\dfrac{-5}{3x + 2} \geq \dfrac{5}{x}$

83. $\dfrac{4}{x - 2} \leq \dfrac{3}{x - 1}$

84. $\dfrac{4}{x + 1} < \dfrac{2}{x + 3}$

85. $\dfrac{x + 3}{x - 5} \leq 1$

86. $\dfrac{x + 2}{3 + 2x} \leq 5$

Solve each rational inequality. Write each solution set in interval notation.

87. $\dfrac{2x - 3}{x^2 + 1} \geq 0$

88. $\dfrac{9x - 8}{4x^2 + 25} < 0$

89. $\dfrac{(3x - 5)^2}{(2x - 5)^3} > 0$

90. $\dfrac{(5x - 3)^3}{(8x - 25)^2} \leq 0$

91. $\dfrac{(2x - 3)(3x + 8)}{(x - 6)^3} \geq 0$

92. $\dfrac{(9x - 11)(2x + 7)}{(3x - 8)^3} > 0$

(Modeling) Solve each problem. For Exercises 95 and 96, see Example 7.

93. ***Box Office Receipts*** U.S. box office receipts, in billions of dollars, for films in the years 1998 through 2002 are shown in the table.

Year	Receipts
1998	6.9
1999	7.4
2000	7.7
2001	8.4
2002	9.2

Source: Motion Picture Association of America.

The receipts R are approximated reasonably well by the linear model

$$R = .56x + 6.8$$

where $x = 0$ corresponds to 1998, $x = 1$ corresponds to 1999, and so on. From the model, in what years was the revenue below \$7.5 billion? In what years did it exceed \$8.5 billion? Compare your answers with the data from the table.

94. *Recovery of Solid Waste* The percent W of municipal solid waste recovered is shown in the bar graph. The linear model

$$W = .737x + 25.7,$$

where $x = 0$ represents 1995, $x = 1$ represents 1996, and so on, fits the data reasonably well. Based on this model, when did the percent of waste recovered first exceed 26%? In what years was it between 26% and 28%?

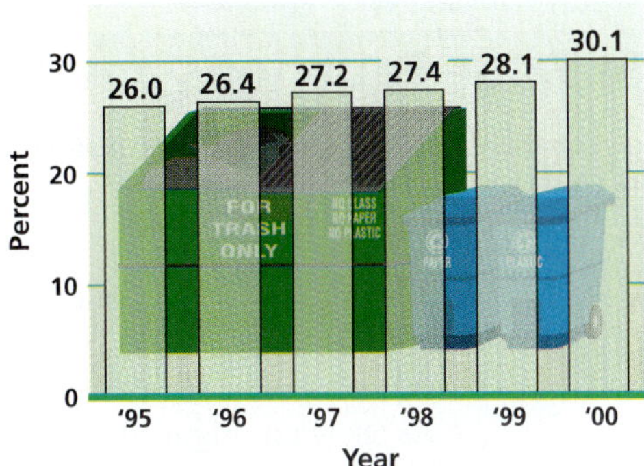

Source: Municipal Solid Waste in the United States: 2000 Facts and Figures; U.S. Environmental Protection Agency.

95. *Height of a Projectile* A projectile is fired straight up from ground level. After t seconds, its height above the ground is s feet, where

$$s = -16t^2 + 220t.$$

For what time period is the projectile at least 624 ft above the ground?

96. *Velocity of an Object* Suppose the velocity of an object is given by

$$v = 2t^2 - 5t - 12,$$

where t is time in seconds. (Here t can be positive or negative.) Find the intervals where the velocity is negative.

97. *Cancer Risk from Radon Gas Exposure* Radon gas occurs naturally in homes and is produced when uranium radioactively decays into lead. Exposure to radon gas is a known lung cancer risk. According to the Environmental Protection Agency (EPA) the individual lifetime excess cancer risk R for radon exposure is between

$$1.5 \times 10^{-3} \quad \text{and} \quad 6.0 \times 10^{-3},$$

where $R = .01$ represents a 1% increase in risk of developing lung cancer.

(a) Calculate the range of individual annual risk by dividing R by an average life expectancy of 72 yr.

(b) Approximate the range of new cases of lung cancer each year (to the nearest hundred) caused by radon if the population of the United States is 260 million.

(Source: Indoor-Air Assessment: A Review of Indoor Air Quality Risk Characterization Studies. Report No. EPA/600/8-90/044, Environmental Protection Agency, 1991.)

98. A student attempted to solve the inequality

$$\frac{2x-1}{x+2} \le 0$$

by multiplying both sides by $x + 2$ to get

$$2x - 1 \le 0$$

$$x \le \frac{1}{2}.$$

He wrote the solution set as $\left(-\infty, \frac{1}{2}\right]$. Is his solution correct? Explain.

99. A student solved the inequality $x^2 \le 16$ by taking the square root of both sides to get $x \le 4$. She wrote the solution set as $(-\infty, 4]$. Is her solution correct? Explain.

100. *Concept Check* Use the discriminant to find the values of k for which $x^2 - kx + 8 = 0$ has two real solutions.

1.8 Absolute Value Equations and Inequalities

Absolute Value Equations ▪ Absolute Value Inequalities ▪ Special Cases ▪ Absolute Value Models for Distance and Tolerance

Recall from Chapter R that the absolute value of a number a, written $|a|$, gives the distance from a to 0 on a number line. By this definition, the equation $|x| = 3$ can be solved by finding all real numbers at a distance of 3 units from 0. As shown in Figure 19, two numbers satisfy this equation, 3 and -3, so the solution set is $\{-3, 3\}$.

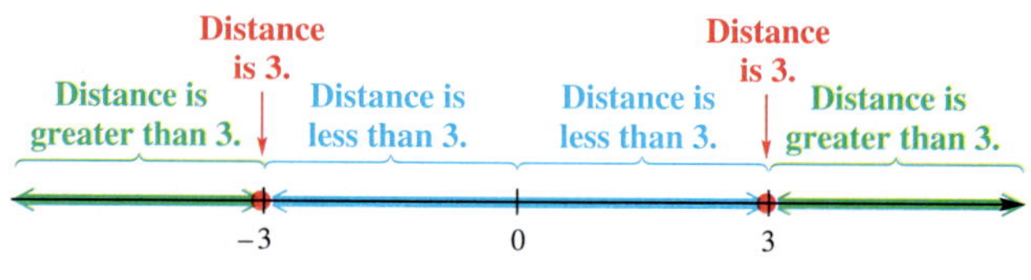

Figure 19

Similarly, $|x| < 3$ is satisfied by all real numbers whose distances from 0 are less than 3, that is, the interval $-3 < x < 3$ or $(-3, 3)$. See Figure 19. Finally, $|x| > 3$ is satisfied by all real numbers whose distances from 0 are greater than 3. As Figure 19 shows, these numbers are less than -3 or greater than 3, so the solution set is $(-\infty, -3) \cup (3, \infty)$. Notice in Figure 19 that the union of the solution sets of $|x| = 3$, $|x| < 3$, and $|x| > 3$ is the set of real numbers.

These observations support the following properties of absolute value.

Properties of Absolute Value

1. For $b > 0$, $|a| = b$ **if and only if** $a = b$ **or** $a = -b$.
2. $|a| = |b|$ **if and only if** $a = b$ **or** $a = -b$.

For any positive number b:

3. $|a| < b$ **if and only if** $-b < a < b$.
4. $|a| > b$ **if and only if** $a < -b$ **or** $a > b$.

Absolute Value Equations We use Properties 1 and 2 to solve absolute value equations.

EXAMPLE 1 Solving Absolute Value Equations

Solve each equation.

(a) $|5 - 3x| = 12$ **(b)** $|4x - 3| = |x + 6|$

Solution

(a) Use Property 1, with $a = 5 - 3x$ and $b = 12$.

$$|5 - 3x| = 12$$

$$5 - 3x = 12 \quad \text{or} \quad 5 - 3x = -12 \qquad \text{Property 1}$$

$$-3x = 7 \quad \text{or} \quad -3x = -17 \qquad \text{Subtract 5.}$$

$$x = -\frac{7}{3} \quad \text{or} \quad x = \frac{17}{3} \qquad \text{Divide by } -3.$$

Check the solutions $-\frac{7}{3}$ and $\frac{17}{3}$ by substituting them in the original absolute value equation. The solution set is $\left\{-\frac{7}{3}, \frac{17}{3}\right\}$.

(b)

$$|4x - 3| = |x + 6|$$

$$4x - 3 = x + 6 \quad \text{or} \quad 4x - 3 = -(x + 6) \qquad \text{Property 2}$$

$$3x = 9 \quad \text{or} \quad 4x - 3 = -x - 6$$

$$x = 3 \quad \text{or} \quad 5x = -3$$

$$x = -\frac{3}{5}$$

Check these solutions. The solution set is $\left\{-\frac{3}{5}, 3\right\}$.

Now try Exercises 9 and 19.

Absolute Value Inequalities We use Properties 3 and 4 to solve absolute value inequalities.

EXAMPLE 2 Solving Absolute Value Inequalities

Solve each inequality.

(a) $|2x + 1| < 7$ **(b)** $|2x + 1| > 7$

Solution

(a) Use Property 3, replacing a with $2x + 1$ and b with 7.

$$|2x + 1| < 7$$

$$-7 < 2x + 1 < 7 \qquad \text{Property 3}$$

$$-8 < 2x < 6 \qquad \text{Subtract 1 from each part. (Section 1.7)}$$

$$-4 < x < 3 \qquad \text{Divide each part by 2.}$$

The final inequality gives the solution set $(-4, 3)$.

(b) $|2x + 1| > 7$

$2x + 1 < -7$	or	$2x + 1 > 7$	Property 4
$2x < -8$	or	$2x > 6$	Subtract 1 from each side.
$x < -4$	or	$x > 3$	Divide each side by 2.

The solution set of the final compound inequality is $(-\infty, -4) \cup (3, \infty)$.

Now try Exercises 27 and 29.

Properties 1, 3, and 4 require that the absolute value expression be *alone* on one side of the equation or inequality.

EXAMPLE 3 Solving an Absolute Value Inequality Requiring a Transformation

Solve $|2 - 7x| - 1 > 4$.

Solution $|2 - 7x| - 1 > 4$

$|2 - 7x| > 5$ Add 1 to each side.

$2 - 7x < -5$	or	$2 - 7x > 5$	Property 4
$-7x < -7$	or	$-7x > 3$	Subtract 2.
$x > 1$	or	$x < -\frac{3}{7}$	Divide by -7; reverse the direction of each inequality. (Section 1.7)

The solution set is $\left(-\infty, -\frac{3}{7}\right) \cup (1, \infty)$.

Now try Exercise 45.

Special Cases Three of the four properties given in this section require the constant, b, to be positive. When $b \leq 0$, use the fact that the absolute value of any expression must be nonnegative and consider the truth of the statement.

EXAMPLE 4 Solving Special Cases of Absolute Value Equations and Inequalities

Solve each equation or inequality.

(a) $|2 - 5x| \geq -4$ **(b)** $|4x - 7| < -3$ **(c)** $|5x + 15| = 0$

Solution

(a) Since the absolute value of a number is always nonnegative, the inequality $|2 - 5x| \geq -4$ is always true. The solution set includes all real numbers, written $(-\infty, \infty)$.

(b) There is no number whose absolute value is less than -3 (or less than *any* negative number). The solution set of $|4x - 7| < -3$ is $\emptyset$.

(c) The absolute value of a number will be 0 only if that number is 0. Therefore, $|5x + 15| = 0$ is equivalent to $5x + 15 = 0$, which has solution set $\{-3\}$.

Now try Exercises 51, 53, and 55.

Absolute Value Models for Distance and Tolerance Recall from Section R.2 that if a and b represent two real numbers, then the absolute value of their difference, either $|a - b|$ or $|b - a|$, represents the distance between them. This fact is used to write absolute value equations or inequalities to express distances.

EXAMPLE 5 Using Absolute Value Inequalities to Describe Distances

Write each statement using an absolute value inequality.

(a) k is no less than 5 units from 8. **(b)** n is within .001 unit of 6.

Solution

(a) Since the distance from k to 8, written $|k - 8|$ or $|8 - k|$, is no less than 5, the distance is greater than or equal to 5. This can be written as

$$|k - 8| \geq 5, \qquad \text{or equivalently} \qquad |8 - k| \geq 5.$$

Either form is acceptable.

(b) This statement indicates that the distance between n and 6 is less than .001, written

$$|n - 6| < .001, \qquad \text{or equivalently} \qquad |6 - n| < .001.$$

Now try Exercises 77 and 79.

Looking Ahead to Calculus

The precise definition of a limit in calculus requires writing absolute value inequalities as in Examples 5 and 6.

In quality control and other applications, as well as in more advanced mathematics, we often wish to keep the difference between two quantities within some predetermined amount, called the *tolerance*.

EXAMPLE 6 Using Absolute Value to Model Tolerance

Suppose $y = 2x + 1$ and we want y to be within .01 unit of 4. For what values of x will this be true?

Solution

$\lvert y - 4\rvert < .01$	Write an absolute value inequality.
$\lvert 2x + 1 - 4\rvert < .01$	Substitute $2x + 1$ for y.
$\lvert 2x - 3\rvert < .01$	
$-.01 < 2x - 3 < .01$	Property 3
$2.99 < 2x < 3.01$	Add 3 to each part.
$1.495 < x < 1.505$	Divide each part by 2.

Reversing these steps shows that keeping x in the interval $(1.495, 1.505)$ ensures that the difference between y and 4 is within .01 unit.

Now try Exercise 83.

Looking Ahead to Calculus

A standard problem in calculus is to find the "interval of convergence" of something called a *power series,* by solving an inequality of the form

$$|x - a| < r.$$

This inequality says that x can be any number within r units of a on the number line, so its solution set is indeed an interval—namely the interval $(a - r, a + r)$.

1.8 Exercises

Concept Check *Match each equation or inequality in Column I with the graph of its solution set in Column II.*

I	II
1. $\lvert x\rvert = 4$	A. −4 0 4
2. $\lvert x\rvert = -4$	B. −4 0 4 $\emptyset$
3. $\lvert x\rvert > -4$	C. −4 0 4
4. $\lvert x\rvert > 4$	D. −4 0 4
5. $\lvert x\rvert < 4$	E. −4 0 4
6. $\lvert x\rvert \geq 4$	F. −4 0 4
7. $\lvert x\rvert \leq 4$	G. −4 0 4
8. $\lvert x\rvert \neq 4$	H. −4 0 4

Solve each equation. See Example 1.

9. $\lvert 3x - 1\rvert = 2$
10. $\lvert 4x + 2\rvert = 5$
11. $\lvert 5 - 3x\rvert = 3$

12. $\lvert 7 - 3x\rvert = 3$
13. $\left\lvert \frac{x-4}{2}\right\rvert = 5$
14. $\left\lvert \frac{x+2}{2}\right\rvert = 7$

15. $\left\lvert \frac{5}{x-3}\right\rvert = 10$
16. $\left\lvert \frac{3}{2x-1}\right\rvert = 4$
17. $\left\lvert \frac{6x+1}{x-1}\right\rvert = 3$

18. $\left\lvert \frac{3x-4}{2x+3}\right\rvert = 1$
19. $\lvert 2x - 3\rvert = \lvert 5x + 4\rvert$
20. $\lvert x + 1\rvert = \lvert 3x - 1\rvert$

21. $\lvert 4 - 3x\rvert = \lvert 2 - 3x\rvert$
22. $\lvert 3 - 2x\rvert = \lvert 5 - 2x\rvert$
23. $\lvert 5x - 2\rvert = \lvert 2 - 5x\rvert$

24. The equation $\lvert 5x - 6\rvert = 6x$ cannot have a negative solution. Why?

25. The equation $\lvert 7x + 3\rvert = -7x$ cannot have a positive solution. Why?

26. *Concept Check* Determine the solution set of each equation by inspection.

(a) $-\lvert x\rvert = \lvert x\rvert$ **(b)** $\lvert -x\rvert = \lvert x\rvert$ **(c)** $\lvert x^2\rvert = \lvert x\rvert$ **(d)** $-\lvert x\rvert = 3$

Solve each inequality. Give the solution set using interval notation. See Example 2.

27. $\lvert 2x + 5\rvert < 3$
28. $\lvert 3x - 4\rvert < 2$
29. $\lvert 2x + 5\rvert \geq 3$

30. $\lvert 3x - 4\rvert \geq 2$
31. $\left\lvert x - \frac{1}{2}\right\rvert < 2$
32. $\left\lvert x + \frac{3}{5}\right\rvert < 1$

33. $4\lvert x - 3\rvert > 12$
34. $3\lvert x + 1\rvert > 6$
35. $\lvert 5 - 3x\rvert > 7$

36. $\lvert 7 - 3x\rvert > 4$
37. $\lvert 5 - 3x\rvert \leq 7$
38. $\lvert 7 - 3x\rvert \leq 4$

39. $\left\lvert \frac{2}{3}x + \frac{1}{2}\right\rvert \leq \frac{1}{6}$
40. $\left\lvert \frac{5}{3} - \frac{1}{2}x\right\rvert > \frac{2}{9}$

Solve each equation or inequality. See Example 3.

41. $|4x + 3| - 2 = -1$ **42.** $|8 - 3x| - 3 = -2$ **43.** $|6 - 2x| + 1 = 3$

44. $|4 - 4x| + 2 = 4$ **45.** $|3x + 1| - 1 < 2$ **46.** $|5x + 2| - 2 < 3$

47. $\left|5x + \frac{1}{2}\right| - 2 < 5$ **48.** $\left|2x + \frac{1}{3}\right| + 1 < 4$ **49.** $|10 - 4x| + 1 \geq 5$

50. $|12 - 6x| + 3 \geq 9$

Solve each equation or inequality. See Example 4.

51. $|10 - 4x| \geq -3$ **52.** $|12 - 9x| \geq -6$ **53.** $|6 - 3x| < -5$

54. $|18 - 3x| < -12$ **55.** $|8x + 5| = 0$ **56.** $|7 + 2x| = 0$

57. $|4.3x + 8.6| < 0$ **58.** $|1.5x - 3| < 0$ **59.** $|2x + 1| \leq 0$

60. $|3x + 2| \leq 0$ **61.** $|3x + 2| > 0$ **62.** $|4x + 3| > 0$

Relating Concepts

For individual or collaborative investigation

(Exercises 63–66)

To see how to solve an equation that involves the absolute value of a quadratic polynomial, such as $|x^2 - x| = 6$, ***work Exercises 63–66 in order.***

63. For $x^2 - x$ to have an absolute value equal to 6, what are the two possible values that it may be? (*Hint:* One is positive and the other is negative.)

64. Write an equation stating that $x^2 - x$ is equal to the positive value you found in Exercise 63, and solve it using factoring.

65. Write an equation stating that $x^2 - x$ is equal to the negative value you found in Exercise 63, and solve it using the quadratic formula. (*Hint:* The solutions are not real numbers.)

66. Give the complete solution set of $|x^2 - x| = 6$, using the results from Exercises 64 and 65.

Use the method described in Relating Concepts Exercises 63–66 to solve each equation or inequality.

67. $|4x^2 - 23x - 6| = 0$ **68.** $|6x^3 + 23x^2 + 7x| = 0$

69. $|x^2 + 1| = |2x|$ **70.** $\left|\frac{x^2 + 2}{x}\right| = \frac{11}{3}$

71. $|x^4 + 2x^2 + 1| < 0$ **72.** $|x^4 + 2x^2 + 1| \geq 0$

73. $\left|\frac{3x + 1}{x - 4}\right| \geq 0$ **74.** $\left|\frac{8x + 7}{x - 9}\right| \geq 0$

75. *Concept Check* Write an equation involving absolute value that says the distance between p and q is 5 units.

76. *Concept Check* Write an equation involving absolute value that says the distance between r and s is 9 units.

Write each statement as an absolute value equation or inequality. See Example 5.

77. m is no more than 8 units from 9.

78. z is no less than 2 units from 12.

79. p is within .0001 unit of 9.

80. k is within .0002 unit of 3.

81. r is no less than 1 unit from 19.

82. q is no more than 4 units from 12.

83. *Tolerance* Suppose that $y = 5x + 1$ and we want y to be within .002 unit of 6. For what values of x will this be true?

84. *Tolerance* Repeat Exercise 83, but let $y = 10x + 2$.

(Modeling) Solve each problem. See Example 6.

85. *Weights of Babies* Dr. Tydings has found that, over the years, 95% of the babies he has delivered weighed y pounds, where

$$|y - 8.0| \leq 1.5.$$

What range of weights corresponds to this inequality?

86. *Temperatures on Mars* The temperatures on the surface of Mars in degrees Celsius approximately satisfy the inequality

$$|C + 84| \leq 56.$$

What range of temperatures corresponds to this inequality?

87. *Conversion of Methanol to Gasoline* The industrial process that is used to convert methanol to gasoline is carried out at a temperature range of 680°F to 780°F. Using F as the variable, write an absolute value inequality that corresponds to this range.

88. *Wind Power Extraction Tests* When a model kite was flown in crosswinds in tests to determine its limits of power extraction, it attained speeds of 98 to 148 ft per sec in winds of 16 to 26 ft per sec. Using x as the variable in each case, write absolute value inequalities that correspond to these ranges.

(Modeling) Carbon Dioxide Emissions When humans breathe, carbon dioxide is emitted. In one study, the emission rates of carbon dioxide by college students were measured during both lectures and exams. The average individual rate R_L (in grams per hour) during a lecture class satisfied the inequality

$$|R_L - 26.75| \leq 1.42,$$

whereas during an exam the rate R_E satisfied the inequality

$$|R_E - 38.75| \leq 2.17.$$

(*Source:* Wang, T. C., *ASHRAE Trans.*, 81 (Part 1), 32, 1975.)

Use this information in Exercises 89–91.

89. Find the range of values for R_L and R_E.

90. The class had 225 students. If T_L and T_E represent the total amounts of carbon dioxide in grams emitted during a one-hour lecture and exam, respectively, write inequalities that model the ranges for T_L and T_E.

91. Discuss any reasons that might account for these differences between the rates during lectures and the rates during exams.

92. Is $|a - b|^2$ always equal to $(b - a)^2$? Explain your answer.

Chapter 1 Summary

KEY TERMS

1.1 equation
solution or root
solution set
equivalent equations
linear equation in one variable
identity
conditional equation
contradiction
simple interest
literal equation

1.2 mathematical model

1.3 imaginary unit
complex number
real part
imaginary part
pure imaginary number
standard form
complex conjugate

1.4 quadratic equation
discriminant

1.6 rational equation
quadratic in form

1.7 inequality
linear inequality in one variable
interval
interval notation
open interval
closed interval
break-even point
quadratic inequality
strict inequality
nonstrict inequality
rational inequality

NEW SYMBOLS

$\emptyset$ empty or null set
i imaginary unit
∞ infinity
(a, b), $(-\infty, a]$, $[a, b)$ interval notation
$\cup$ union of sets
$\cap$ intersection of sets

QUICK REVIEW

CONCEPTS	EXAMPLES

1.1 Linear Equations

Addition and Multiplication Properties of Equality
For real numbers a, b, and c:

$a = b$ and $a + c = b + c$ are equivalent.

If $c \neq 0$, then $a = b$ and $ac = bc$ are equivalent.

Solve. $5(x + 3) = 3x + 7$

$5x + 15 = 3x + 7$ Distributive property

$2x = -8$ Subtract $3x$; subtract 15.

$x = -4$ Divide by 2.

Solution set: $\{-4\}$

1.2 Applications and Modeling with Linear Equations

Problem-Solving Steps

Step 1 Read the problem.

How many liters of 30% alcohol solution and 80% alcohol solution must be mixed to obtain 50 L of 50% alcohol solution?

Step 2 Assign a variable.

Let x = number of liters of 30% solution needed;
then $50 - x$ = number of liters of 80% solution needed.

Summarize the information of the problem in table.

Strength	Liters of Solution	Liters of Pure Alcohol
30%	x	$.30x$
80%	$50 - x$	$.80(50 - x)$
50%	50	$.50(50)$

(continued)

CONCEPTS	EXAMPLES
Step 3 Write an equation.	The equation is $.30x + .80(50 - x) = .50(50)$.
Step 4 Solve the equation.	Solve the equation to obtain $x = 30$.
Step 5 State the answer.	Therefore, 30 L of the 30% solution and $50 - 30 = 20$ L of the 80% solution must be mixed.
Step 6 Check.	*Check:* $.30(30) + .80(50 - 30) = .50(50)$? $25 = 25$ True

1.3 Complex Numbers

Definition of *i*

$$i^2 = -1 \quad \text{or} \quad i = \sqrt{-1}$$

Definition of $\sqrt{-a}$

For $a > 0$,

$$\sqrt{-a} = i\sqrt{a}.$$

$$\sqrt{-4} = 2i$$

$$\sqrt{-12} = i\sqrt{12} = 2i\sqrt{3}$$

Adding and Subtracting Complex Numbers

Add or subtract the real parts and add or subtract the imaginary parts.

$$\begin{aligned}(2 + 3i) &+ (3 + i) - (2 - i) \\ &= (2 + 3 - 2) + (3 + 1 + 1)i \\ &= 3 + 5i\end{aligned}$$

Multiplying and Dividing Complex Numbers

Multiply complex numbers as with binomials, and use the fact that $i^2 = -1$.

$$\begin{aligned}(6 + i)(3 - 2i) &= 18 - 12i + 3i - 2i^2 && \text{FOIL} \\ &= (18 + 2) + (-12 + 3)i && i^2 = -1 \\ &= 20 - 9i\end{aligned}$$

Divide complex numbers by multiplying the numerator and denominator by the complex conjugate of the denominator.

$$\frac{3 + i}{1 + i} = \frac{(3 + i)(1 - i)}{(1 + i)(1 - i)} = \frac{3 - 3i + i - i^2}{1 - i^2}$$

$$= \frac{4 - 2i}{2} = \frac{2(2 - i)}{2} = 2 - i$$

1.4 Quadratic Equations

Zero-Factor Property

If a and b are complex numbers with $ab = 0$, then $a = 0$ or $b = 0$ or both.

Solve. $6x^2 + x - 1 = 0$

$$(3x - 1)(2x + 1) = 0 \quad \text{Factor.}$$

$$3x - 1 = 0 \quad \text{or} \quad 2x + 1 = 0$$

$$x = \frac{1}{3} \quad \text{or} \quad x = -\frac{1}{2}$$

Solution set: $\left\{-\frac{1}{2}, \frac{1}{3}\right\}$

Square Root Property

The solution set of $x^2 = k$ is $\{\sqrt{k}, -\sqrt{k}\}$, abbreviated $\{\pm\sqrt{k}\}$.

Solve. $x^2 = 12$

$$x = \pm\sqrt{12} = \pm 2\sqrt{3}$$

Solution set: $\{\pm 2\sqrt{3}\}$

CONCEPTS	EXAMPLES
Quadratic Formula The solutions of the quadratic equation $ax^2 + bx + c = 0$, where $a \neq 0$, are given by $$x = \frac{-b \pm \sqrt{b^2 - 4ac}}{2a}.$$	Solve. $x^2 + 2x + 3 = 0$ $$x = \frac{-2 \pm \sqrt{2^2 - 4(1)(3)}}{2(1)} \quad a = 1, b = 2, c = 3$$ $$= \frac{-2 \pm \sqrt{-8}}{2} = \frac{-2 \pm 2i\sqrt{2}}{2} = \frac{2(-1 \pm i\sqrt{2})}{2}$$ $$= -1 \pm i\sqrt{2}$$ Solution set: $\{-1 \pm i\sqrt{2}\}$

1.5 Applications and Modeling with Quadratic Equations

Pythagorean Theorem In a right triangle, the sum of the squares of the lengths of legs a and b is equal to the square of the length of hypotenuse c: $$a^2 + b^2 = c^2.$$	In a right triangle, the shorter leg is 7 in. less than the longer leg, and the hypotenuse is 2 in. longer than the longer leg. What are the lengths of the sides? Let x represent the length of the longer leg. 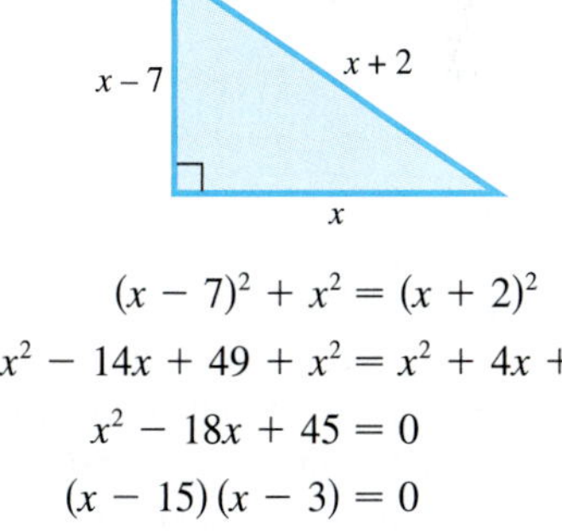 $$(x - 7)^2 + x^2 = (x + 2)^2$$ $$x^2 - 14x + 49 + x^2 = x^2 + 4x + 4$$ $$x^2 - 18x + 45 = 0$$ $$(x - 15)(x - 3) = 0$$ $$x = 15 \quad \text{or} \quad x = 3$$
Height of a Propelled Object The height s (in feet) of an object propelled directly upward from an initial height of s_0 feet, with initial velocity v_0 feet per second, is $$s = -16t^2 + v_0 t + s_0,$$ where t is the number of seconds after the object is propelled.	The value 3 must be rejected. The lengths of the sides are 15 in., 8 in., and 17 in. Check to see that the conditions of the problem are satisfied.

1.6 Other Types of Equations

Power Property If P and Q are algebraic expressions, then every solution of the equation $P = Q$ is also a solution of the equation $P^n = Q^n$, for any positive integer n. **Quadratic in Form** An equation in the form $au^2 + bu + c = 0$, where u is an algebraic expression in x, can often be solved by using a substitution variable. If the power property is applied, or if both sides of an equation are multiplied by a variable expression, ***check all proposed solutions.***	Solve. $(x + 1)^{2/3} + (x + 1)^{1/3} - 6 = 0$ $$u^2 + u - 6 = 0 \quad \text{Let } u = (x + 1)^{1/3}.$$ $$(u + 3)(u - 2) = 0$$ $$u = -3 \quad \text{or} \quad u = 2$$ $$(x + 1)^{1/3} = -3 \quad \text{or} \quad (x + 1)^{1/3} = 2$$ $$x + 1 = -27 \quad \text{or} \quad x + 1 = 8$$ $$x = -28 \quad \text{or} \quad x = 7$$ Both solutions check; the solution set is $\{-28, 7\}$.

CONCEPTS	EXAMPLES

1.7 Inequalities

Properties of Inequality

For real numbers a, b, and c:

1. If $a < b$, then $a + c < b + c$.
2. If $a < b$ and if $c > 0$, then $ac < bc$.
3. If $a < b$ and if $c < 0$, then $ac > bc$.

Solve. $-3(x + 4) + 2x < 6$

$$-3x - 12 + 2x < 6$$

$$-x < 18$$

$$x > -18 \quad \text{Multiply by } -1; \text{ change } < \text{ to } >.$$

Solution set: $(-18, \infty)$

Solving Quadratic Inequalities

Step 1 Solve the corresponding quadratic equation.

Step 2 Identify the intervals determined by the solutions of the equation.

Step 3 Use a test value from each interval to determine which intervals form the solution set.

Solve. $x^2 + 6x \le 7$

$$x^2 + 6x - 7 = 0 \quad \text{Corresponding equation}$$

$$(x + 7)(x - 1) = 0 \quad \text{Factor.}$$

$$x = -7 \quad \text{or} \quad x = 1 \quad \text{Zero-factor property}$$

The intervals formed by these solutions are $(-\infty, -7)$, $(-7, 1)$, and $(1, \infty)$.

Test values show that values in the intervals $(-\infty, -7)$ and $(1, \infty)$ do not satisfy the original inequality, while those in $(-7, 1)$ do. Since the symbol is $\le$, the endpoints are included.

Solution set: $[-7, 1]$

Solving Rational Inequalities

To solve a rational inequality, rewrite the inequality so that 0 is on one side and a single fraction is on the other. Find the values that make either the numerator or denominator 0. Then follow Step 3 in solving quadratic inequalities.

Solve. $\dfrac{x}{x + 3} \ge \dfrac{5}{x + 3}$

$$\frac{x}{x + 3} - \frac{5}{x + 3} \ge 0$$

$$\frac{x - 5}{x + 3} \ge 0$$

The values -3 and 5 make either the numerator or denominator 0. The intervals formed are

$$(-\infty, -3), (-3, 5), \text{ and } (5, \infty).$$

The value -3 must be excluded and 5 must be included. Test values show that values in the intervals $(-\infty, -3)$ and $(5, \infty)$ yield true statements.

Solution set: $(-\infty, -3) \cup [5, \infty)$

CONCEPTS	EXAMPLES
1.8 Absolute Value Equations and Inequalities	
Properties of Absolute Value **1.** For $b > 0$, $\lvert a \rvert = b$ if and only if $a = b$ or $a = -b$. **2.** $\lvert a \rvert = \lvert b \rvert$ if and only if $a = b$ or $a = -b$. If b is a positive number: **3.** $\lvert a \rvert < b$ if and only if $-b < a < b$. **4.** $\lvert a \rvert > b$ if and only if $a < -b$ or $a > b$.	Solve. $\lvert 5x - 2 \rvert = 3$ $5x - 2 = 3$ or $5x - 2 = -3$ $5x = 5$ or $5x = -1$ $x = 1$ or $x = -\frac{1}{5}$ Solution set: $\left\{-\frac{1}{5}, 1\right\}$ Solve. $\lvert 5x - 2 \rvert < 3$ $-3 < 5x - 2 < 3$ $-1 < 5x < 5$ $-\frac{1}{5} < x < 1$ Solution set: $\left(-\frac{1}{5}, 1\right)$ Solve. $\lvert 5x - 2 \rvert \geq 3$ $5x - 2 \geq 3$ or $5x - 2 \leq -3$ $5x \geq 5$ or $5x \leq -1$ $x \geq 1$ or $x \leq -\frac{1}{5}$ Solution set: $\left(-\infty, -\frac{1}{5}\right] \cup [1, \infty)$

Chapter 1 Review Exercises

Solve each equation.

1. $2x + 7 = 3x + 1$

2. $4x - 2(x - 1) = 12$

3. $5x - 2(x + 4) = 3(2x + 1)$

4. $9x - 11(k + p) = x(a - 1)$ for x

5. $A = \dfrac{24f}{B(p + 1)}$ for f (approximate annual interest rate)

6. *Concept Check* Which of the following cannot be a correct equation to solve a geometry problem, if x represents the length of a rectangle? (*Hint:* Solve the equations and consider the solutions.)

A. $2x + 2(x + 2) = 20$

B. $2x + 2(5 + x) = -2$

C. $8(x + 2) + 4x = 16$

D. $2x + 2(x - 3) = 10$

7. *Concept Check* If x represents the number of pennies in a jar in an applied problem, which of the following equations cannot be a correct equation for finding x? (*Hint:* Solve the equations and consider the solutions.)

   A. $5x + 3 = 9$ **B.** $12x + 3 = -4$
   C. $100x = 50(x + 3)$ **D.** $6(x + 4) = x + 24$

8. *Airline Carry-On Baggage Size* Carry-on rules for domestic economy-class travel differ from one airline to another, as shown in the table.

Airline	Size (linear inches)
American	45
Continental	45
Delta	45
Northwest	45
Southwest	50
United	45
USAirways	50
America West	45

Source: Individual airlines.

To determine the number of linear inches for a carry-on, add the length, width, and height of the bag.

   (a) One Samsonite rolling bag measures 9 in. by 12 in. by 21 in. Are there any airlines that would not allow it as a carry-on?
   (b) A Lark wheeled bag measures 10 in. by 14 in. by 22 in. On which airlines does it qualify as a carry-on?

Solve each problem.

9. *Dimensions of a Square* If the length of each side of a square is decreased by 4 in., the perimeter of the new square is 10 in. more than half the perimeter of the original square. What are the dimensions of the original square?

10. *Distance from a Library* Alison Romike can ride her bike to the university library in 20 min. The trip home, which is all uphill, takes her 30 min. If her rate is 8 mph slower on the return trip, how far does she live from the library?

11. *Alcohol Mixture* A chemist wishes to strengthen a mixture that is 10% alcohol to one that is 30% alcohol. How much pure alcohol should be added to 12 L of the 10% mixture?

12. *Loan Interest Rates* A realtor borrowed $90,000 to develop some property. He was able to borrow part of the money at 11.5% interest and the rest at 12%. The annual interest on the two loans amounts to $10,525. How much was borrowed at each rate?

13. *Speed of an Excursion Boat* An excursion boat travels upriver to a landing and then returns to its starting point. The trip upriver takes 1.2 hr, and the trip back takes .9 hr. If the average speed on the return trip is 5 mph faster than on the trip upriver, what is the boat's speed upriver?

49. *Concept Check* Which one of the equations in Exercise 47 has two nonreal complex solutions?

Evaluate the discriminant for each equation, and then use it to predict the number and type of solutions.

50. $8x^2 = 2x - 6$ **51.** $6x^2 - 2x = 3$ **52.** $16x^2 + 3 = 26x$

53. $-8x^2 - 10x = 7$ **54.** $25x^2 - 110x + 121 = 0$ **55.** $x(9x - 6) = -1$

56. Explain how the discriminant of $ax^2 + bx + c = 0$ $(a \neq 0)$ is used to determine the number and type of solutions.

Solve each problem.

57. *(Modeling) Height of a Projectile* A projectile is fired straight up from ground level. After t seconds its height s, in feet above the ground, is given by

$$s = 220t - 16t^2.$$

At what times is the projectile exactly 750 ft above the ground?

58. *Dimensions of a Picture Frame* Mitchel Levy went into a frame-it-yourself shop. He wanted a frame 3 in. longer than it was wide. The frame he chose extended 1.5 in. beyond the picture on each side. Find the outside dimensions of the frame if the area of the unframed picture is 70 in.^{2}.

59. *Kitchen Flooring* Paula Story plans to replace the vinyl floor covering in her 10-ft by 12-ft kitchen. She wants to have a border of even width of a special material. She can afford only 21 ft^2 of this material. How wide a border can she have?

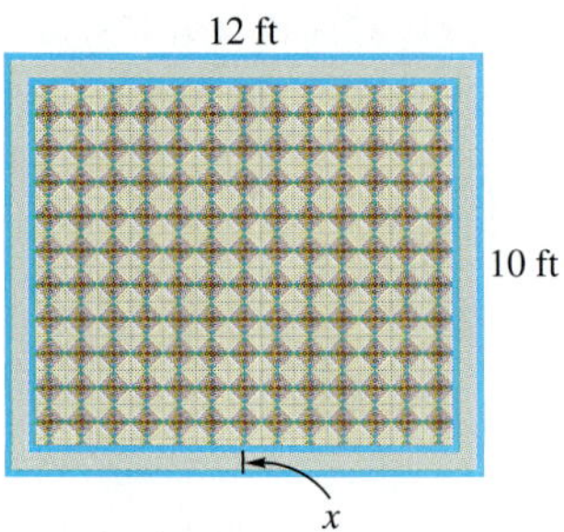

60. *(Modeling) Airplane Landing Speed* To determine the appropriate landing speed of a small airplane, the formula

$$D = .1s^2 - 3s + 22$$

is used, where s is the initial landing speed in feet per second and D is the length of the runway in feet. If the landing speed is too fast, the pilot may run out of runway; if the speed is too slow, the plane may stall. If the runway is 800 ft long, what is the appropriate landing speed? Round to the nearest tenth.

61. *(Modeling) Number of Airports in the U.S.* The number of airports in the United States during the period 1970–1997 can be approximated by the equation

$$y = -6.77x^2 + 445.34x + 11{,}279.82,$$

where $x = 0$ corresponds to 1970, $x = 1$ to 1971, and so on. According to this model, how many airports were there in 1980?

62. *Dimensions of a Right Triangle* The lengths of the sides of a right triangle are such that the shortest side is 7 in. shorter than the middle side, while the longest side (the hypotenuse) is 1 in. longer than the middle side. Find the lengths of the sides.

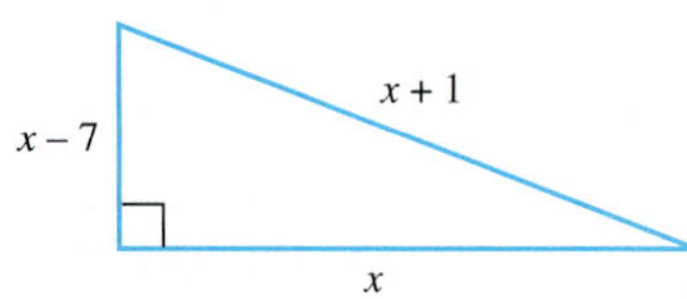

Solve each equation.

63. $4x^4 + 3x^2 - 1 = 0$

64. $2x^4 - x^2 = 0$

65. $\frac{2}{x} - \frac{4}{3x} = 8 + \frac{3}{x}$

66. $2 - \frac{5}{x} = \frac{3}{x^2}$

67. $\frac{10}{4x - 4} = \frac{1}{1 - x}$

68. $\frac{13}{x^2 + 10} = \frac{2}{x}$

69. $\frac{x}{x + 2} + \frac{1}{x} + 3 = \frac{2}{x^2 + 2x}$

70. $\frac{2}{x + 2} + \frac{1}{x + 4} = \frac{4}{x^2 + 6x + 8}$

71. $(2x + 3)^{2/3} + (2x + 3)^{1/3} = 6$

72. $(x + 3)^{-2/3} - 2(x + 3)^{-1/3} - 3 = 0$

73. $\sqrt{4x - 2} = \sqrt{3x + 1}$

74. $\sqrt{2x + 3} = x + 2$

75. $\sqrt{x + 2} = 2 + x$

76. $\sqrt{x} - \sqrt{x + 3} = -1$

77. $\sqrt{x + 3} - \sqrt{3x + 10} = 1$

78. $\sqrt{5x - 15} - \sqrt{x + 1} = 2$

79. $\sqrt{x^2 + 3x} - 2 = 0$

80. $\sqrt[3]{2x} = \sqrt[3]{3x + 2}$

81. $\sqrt[3]{6x + 2} - \sqrt[3]{4x} = 0$

82. $(x - 2)^{2/3} = x^{1/3}$

Solve each inequality. Write each solution set using interval notation.

83. $-9x < 4x + 7$

84. $11x \geq 2(x - 4)$

85. $-5x - 4 \geq 3(2x - 5)$

86. $7x - 2(x - 3) \leq 5(2 - x)$

87. $5 \leq 2x - 3 \leq 7$

88. $-8 > 3x - 5 > -12$

89. $x^2 + 3x - 4 \leq 0$

90. $x^2 + 4x > 21$

91. $6x^2 - 11x - 10 < 0$

92. $x^2 - 3x - 5 \geq 0$

93. $x^3 - 16x \leq 0$

94. $2x^3 - 3x^2 - 5x < 0$

95. $\frac{3x + 6}{x - 5} > 0$

96. $\frac{x + 7}{2x + 1} \leq 1$

97. $\frac{3x - 2}{x} > 4$

98. $\frac{5x + 2}{x} < -1$

99. $\frac{3}{x - 1} \leq \frac{5}{x + 3}$

100. $\frac{3}{x + 2} > \frac{2}{x - 4}$

(Modeling) Solve each problem.

101. ***Ozone Concentration*** Automobiles are a major source of tropospheric ozone (ground-level ozone). Ozone in outdoor air can enter buildings through ventilation systems. Guideline levels for indoor ozone are less than 50 parts per billion (ppb). In a scientific study, a purafil air filter was used to reduce an initial ozone concentration of 140 ppb. The filter removed 43% of the ozone. (*Source:* Parmar and Grosjean, *Removal of Air Pollutants from Museum Display Cases,* Getty Conservation Institute, Marina del Rey, CA, 1989.)

(a) Determine whether this type of filter reduced the ozone concentration to acceptable levels. Explain your answer.

(b) What is the maximum initial concentration of ozone that this filter will reduce to an acceptable level?

102. ***Break-Even Interval*** A company produces videotapes. The revenue from the sale of x units of tapes is $R = 8x$. The cost to produce x units of tapes is $C = 3x + 1500$. In what interval will the company at least break even?

103. ***Height of a Projectile*** A projectile is launched upward from the ground. Its height s in feet above the ground after t seconds is given by

$$s = 320t - 16t^2.$$

(a) After how many seconds in the air will it hit the ground?

(b) During what time interval is the projectile more than 576 ft above the ground?

104. *Social Security Benefits* The total benefits paid by the Social Security Administration during the period from 1990 to 2002 can be approximated by the linear model

$$y = 28.07x + 363.6,$$

where $x = 0$ corresponds to 1990, $x = 1$ corresponds to 1991, and so on, and y is in billions of dollars. Based on this model, when did this amount first exceed \$500 billion? Round to the nearest year. Compare your answer to the graph.

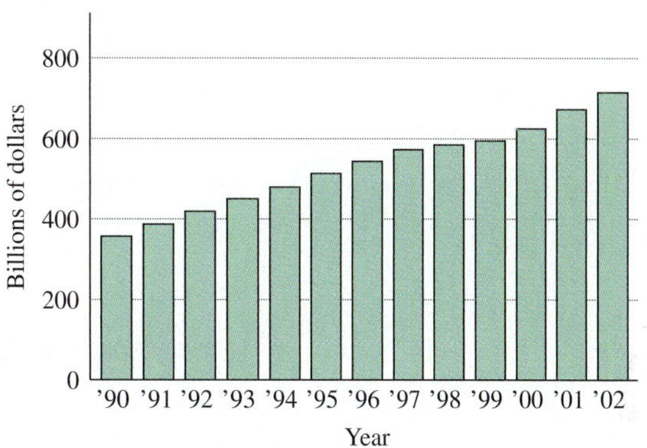

105. Without actually solving the inequality, explain why 3 cannot be in the solution set of $\frac{2x + 5}{x - 3} < 0$.

106. Without actually solving the inequality, explain why -4 must be in the solution set of $\frac{x + 4}{x - 3} \geq 0$.

Rewrite the numerical part of each statement from a newspaper or magazine article as an inequality using the indicated variable.

107. *Whale Corpses* The corpses of at least 65 whales were reported washed up on the Pacific beaches of the Baja peninsula. (*Source: The Sacramento Bee,* June 9, 1999, page A12.) Let W represent the number of whale corpses.

108. *Whales* These enormous mammals (whales), up to 35 tons and 45 ft of bumpy and mottled gray, lead remarkable lives. (*Source: The Sacramento Bee,* June 9, 1999, page A12.) Let w represent weight and L represent length.

109. *Archaeological Sites* Grand Staircase-Escalante National Monument holds as many as 100,000 archaeological sites, most unsurveyed. (*Source: National Geographic,* July 1999, page 106.) Let a represent the number of archaeological sites.

110. *Humpback Whale Population* The North Pacific (humpback whale) population was thought to have tumbled to fewer than 2000. (*Source: National Geographic,* July 1999, page 115.) Let p represent the population.

Solve each equation or inequality.

111. $|x + 4| = 7$

112. $|2 - x| = 3$

113. $\left|\frac{7}{2 - 3x}\right| = 9$

114. $\left|\frac{8x - 1}{3x + 2}\right| = 7$

115. $|5x - 1| = |2x + 3|$

116. $|x + 7| = |x - 8|$

117. $|2x + 9| \le 3$ **118.** $|5x - 8| \ge 2$ **119.** $|7x - 3| > 4$

120. $\left|\frac{1}{2}x + \frac{2}{3}\right| < 3$ **121.** $|3x + 7| - 5 = 0$ **122.** $|7x + 8| - 2 > 1$

123. $|4x - 12| \ge -2$ **124.** $|7 - 2x| \le -2$

125. $|x^2 + 4x| \le 0$ **126.** $|x^2 + 4x| > 0$

Write as an absolute value equation or inequality.

127. k is 12 units from 3 on the number line.

128. p is at least 5 units from 1 on the number line.

129. t is no less than .01 unit from 4.

130. s is no more than .001 unit from 10.

Chapter 1 Test

Solve each equation.

1. $3(x - 4) - 5(x + 2) = 2 - (x + 24)$ **2.** $\frac{2}{3}x + \frac{1}{2}(x - 4) = x - 4$

3. $6x^2 - 11x = 7$ **4.** $(3x + 1)^2 = 8$

5. $3x^2 + 2x + 2 = 0$ **6.** $\frac{12}{x^2 - 9} + \frac{3}{x + 3} = \frac{2}{x - 3}$

7. $\frac{4x}{x - 2} + \frac{3}{x} = \frac{-6}{x^2 - 2x}$ **8.** $\sqrt{3x + 4} + 4 = 2x$

9. $\sqrt{-2x + 3} + \sqrt{x + 3} = 3$ **10.** $\sqrt[3]{3x - 8} = \sqrt[3]{9x + 4}$

11. $x^4 - 17x^2 + 16 = 0$ **12.** $(x + 3)^{2/3} + (x + 3)^{1/3} - 6 = 0$

13. $|4x + 3| = 7$ **14.** $|2x + 1| = |x - 5|$

15. ***Surface Area of a Rectangular Solid*** The formula for the surface area of a rectangular solid is

$$S = 2HW + 2LW + 2LH,$$

where S, H, W, and L represent surface area, height, width, and length, respectively. Solve this formula for W.

16. Perform each operation. Give the answer in standard form.

(a) $(7 - 3i) - (2 + 5i)$ **(b)** $(4 + 3i)(-5 + 3i)$

(c) $(8 + 3i)^2$ **(d)** $\frac{3 + 19i}{1 + 3i}$

17. Simplify each power of i.

(a) i^{18} **(b)** i^{-27} **(c)** $\frac{1}{i^{15}}$

Solve each problem.

18. ***(Modeling) Water Consumption for Snowmaking*** Ski resorts require large amounts of water in order to make snow. Snowmass Ski Area in Colorado plans to pump between 1120 and 1900 gal of water per minute at least 12 hr per day from Snowmass Creek between mid-October and late December. (*Source:* York Snow Incorporated.)

(a) Determine an equation that will calculate the *minimum* amount of water A (in gallons) pumped after x days during mid-October to late December.
(b) Find the minimum amount of water pumped in 30 days.
(c) Suppose the water being pumped from Snowmass Creek was used to fill swimming pools. The average backyard swimming pool holds 20,000 gal of water. Determine an equation that will give the minimum number of pools P that could be filled after x days. How many pools could be filled each day?
(d) In how many days could a minimum of 1000 pools be filled?

19. ***Dimensions of a Rectangle*** The perimeter of a rectangle is 310 m. The length is 10 m less than twice the width. What are the length and width?

20. ***Nut Mixture*** To make a special mix, the owner of a fruit and nut stand wants to combine cashews that sell for \$7.00 per lb with walnuts that sell for \$5.50 per lb to obtain 35 lb of a mixture that sells for \$6.50 per lb. How many pounds of each type of nut should be used in the mixture?

21. ***Speed of a Plane*** Sheryl left by plane to visit her mother in Hartford, 420 km away. Fifteen minutes later, her mother left to meet her at the airport. She drove the 20 km to the airport at 40 km per hr, arriving just as the plane taxied in. What was the speed of the plane?

22. ***(Modeling) Height of a Projectile*** A projectile is launched straight up from ground level with an initial velocity of 96 ft per sec. Its height in feet, s, after t seconds is given by the equation $s = -16t^2 + 96t$.

(a) At what time(s) will it reach a height of 80 ft?
(b) After how many seconds will it return to the ground?

23. ***(Modeling) Airline Passenger Growth*** The number of fliers on 10- to 30-seat commuter aircraft was 1.4 million in 1975 and 3.1 million in 1994. It is estimated that there will be 9.3 million fliers in the year 2006. Here are three possible models for these data, where $x = 0$ corresponds to the year 1975.

Year	A	B	C
0	.60	1.40	1.40
19	5.16	3.11	2.25
31	8.04	9.33	7.43

Source: Federal Aviation Administration (*USA Today* 3/27/95, page 1B).

A. $y = .24x + .6$
B. $y = .0138x^2 - .172x + 1.4$
C. $y = .0125x^2 - .193x + 1.4$

The table shows each equation evaluated at the years 1975, 1994, and 2006. Decide which equation most closely models the data for these years.

Solve each inequality. Give the answer using interval notation.

24. $-2(x - 1) - 10 < 2(x + 2)$

25. $-2 \le \frac{1}{2}x + 3 \le 4$

26. $2x^2 - x - 3 \ge 0$

27. $\frac{x + 1}{x - 3} < 5$

28. $|2x - 5| < 9$

29. $|2x + 1| \ge 11$

30. $|3x + 7| < 0$

Chapter 1 Quantitative Reasoning

How many new shares will double the percent of ownership?

Many situations in everyday life involve percent. A useful measure of change in business reports, advertisements, sports data, and government reports, for example, is the percent increase or decrease of a statistic. Remember that percent times the base gives the percentage, a part of the base. To calculate with a percent, change it to decimal form.

An acquaintance of one of the authors was about to take his startup Internet company public. He owned 5% of the 900,000 shares of stock that were issued when the company was first formed. The board of directors decided to reward him by issuing additional stock so that he would then own 10% of the company. How many new shares of stock should be issued and then rewarded to him? (*Be careful.* Doubling the number of shares he owned would not double the percent.)

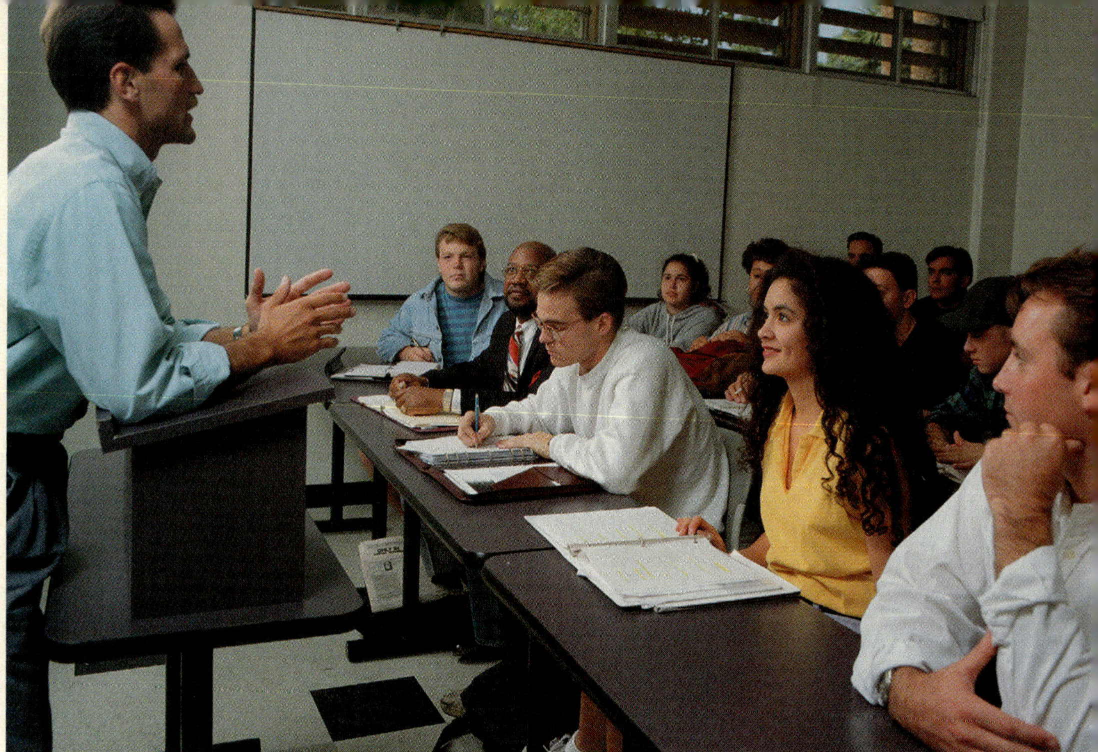

2

Graphs and Functions

Nearly 75 years ago, Calvin Coolidge, the 30th president of the United States, said "The chief business of America is business." Clearly, this is still true today. Business, management, accounting, and marketing degrees have consistently been the most popular choices of entering students for their college major. (*Source:* Higher Education Research Institute.)

In this chapter we introduce the concept of a *function,* one of the most important mathematical concepts in business and other applications. Another important concept is rate of change. In Section 2.3, Exercise 73, you are asked to find the average rate of change in the percent of college freshmen majoring in business.

2.1 Graphs of Equations

Ordered Pairs ▪ The Rectangular Coordinate System ▪ The Distance Formula ▪ The Midpoint Formula ▪ Graphing Equations ▪ Circles

Type of Entertainment	Amount Spent
VHS rentals/ sales	\$47
DVD rentals/ sales	\$25
CDs	\$47
theme parks	\$34
sports tickets	\$33
movie tickets	\$30

Source: PricewaterhouseCoopers; Pollstar.

Ordered Pairs

The idea of pairing one quantity with another is often encountered in everyday life. For example, a numerical grade in a mathematics course is paired with a corresponding letter grade. The number of gallons of gasoline pumped into a tank is paired with the amount of money needed to purchase it. Another example is shown in the table, which gives the dollars the average American spent in 2001 on entertainment. For each type of entertainment, there is a corresponding number of dollars spent.

Pairs of related quantities, such as a 96 determining a grade of A, 3 gallons of gasoline costing \$5.25, and 2001 spending on CDs of \$47, can be expressed as *ordered pairs:* (96, A), (3, \$5.25), (CDs, \$47). An **ordered pair** consists of two components, written inside parentheses, in which the order of the components is important.

EXAMPLE 1 Writing Ordered Pairs

Use the table to write ordered pairs to express the relationship between each type of entertainment and the amount spent on it.

(a) DVD rentals/sales **(b)** movie tickets

Solution

(a) Use the data in the second row: (DVD rentals/sales, \$25).

(b) Use the data in the last row: (movie tickets, \$30).

Now try Exercise 7.

In mathematics, we are most often interested in ordered pairs whose components are numbers. Note that $(4, 2)$ and $(2, 4)$ are different ordered pairs because the order of the numbers is different.

The Rectangular Coordinate System

As mentioned in Chapter R, each real number corresponds to a point on a number line. This idea is extended to ordered pairs of real numbers by using two perpendicular number lines, one horizontal and one vertical, that intersect at their zero-points. This point of intersection is called the **origin.** The horizontal line is called the ***x*-axis,** and the vertical line is called the ***y*-axis.** Starting at the origin, on the x-axis the positive numbers go to the right and the negative numbers go to the left. The y-axis has positive numbers going up and negative numbers going down.

The x-axis and y-axis together make up a **rectangular coordinate system,** or **Cartesian coordinate system** (named for one of its coinventors, René Descartes; the other coinventor was Pierre de Fermat). The plane into which the coordinate system is introduced is the **coordinate plane,** or ***xy*-plane.** The x-axis and y-axis divide the plane into four regions, or **quadrants,** labeled as shown in Figure 1. The points on the x-axis and y-axis belong to no quadrant.

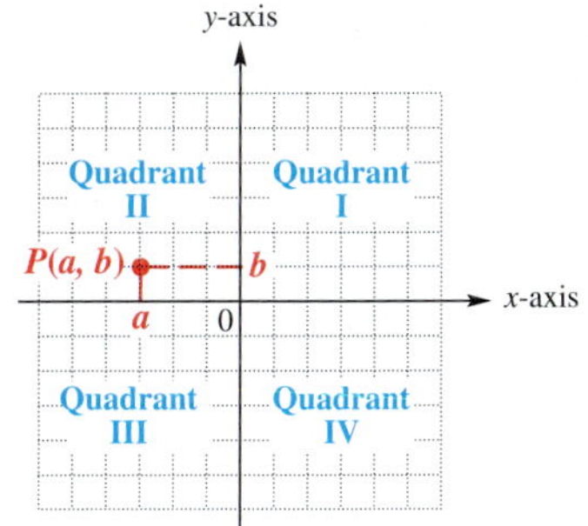

Figure 1

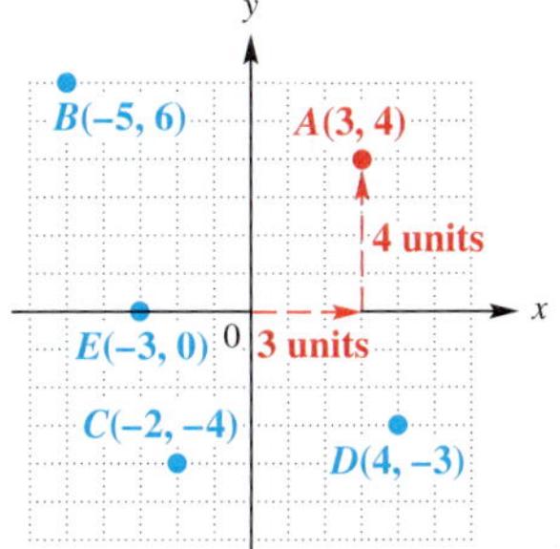

Figure 2

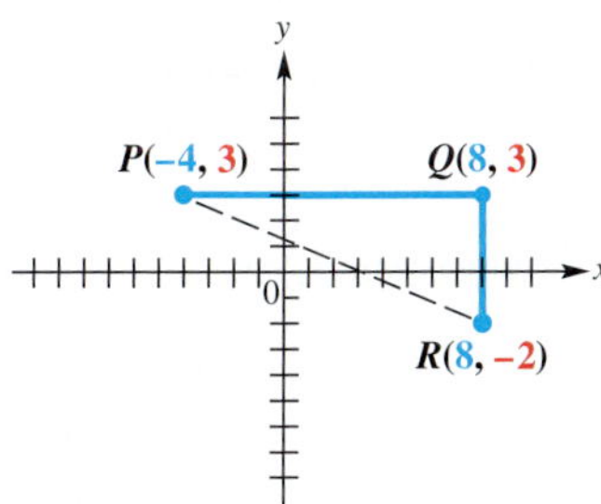

Figure 3

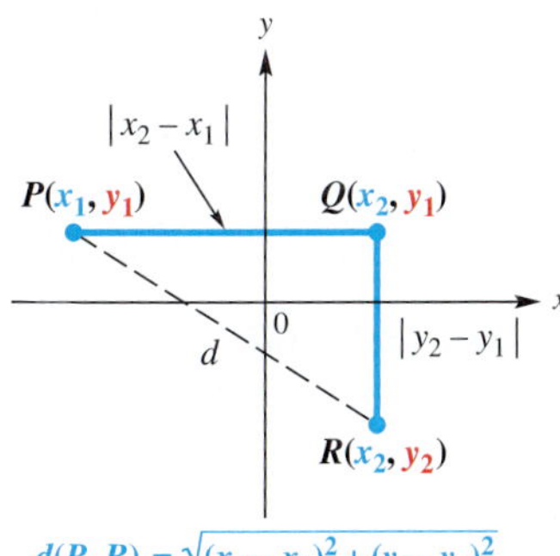

Figure 4

Each point P in the xy-plane corresponds to a unique ordered pair (a, b) of real numbers. The numbers a and b are the **coordinates** of point P. (See Figure 1.) To locate on the xy-plane the point corresponding to the ordered pair $(3, 4)$, for example, start at the origin, move 3 units in the positive x-direction, and then move 4 units in the positive y-direction. (See Figure 2.) Point A corresponds to the ordered pair $(3, 4)$. Also in Figure 2, B corresponds to the ordered pair $(-5, 6)$, C to $(-2, -4)$, D to $(4, -3)$, and E to $(-3, 0)$. The point P corresponding to the ordered pair (a, b) often is written $P(a, b)$ as in Figure 1 and referred to as "the point (a, b)."

The Distance Formula Recall that the distance on a number line between points P and Q with coordinates x_1 and x_2 is

$$d(P, Q) = |x_1 - x_2| = |x_2 - x_1|. \quad \text{(Section R.2)}$$

By using the coordinates of their ordered pairs, we can extend this idea to find the distance between any two points in a plane.

Figure 3 shows the points $P(-4, 3)$ and $R(8, -2)$. To find the distance between these points, we complete a right triangle as in the figure. This right triangle has its 90° angle at $Q(8, 3)$. The horizontal side of the triangle has length

$$d(P, Q) = |8 - (-4)| = 12. \quad \text{Definition of distance}$$

The vertical side of the triangle has length

$$d(Q, R) = |3 - (-2)| = 5.$$

By the Pythagorean theorem, the length of the remaining side of the triangle is

$$\sqrt{12^2 + 5^2} = \sqrt{144 + 25} = \sqrt{169} = 13. \quad \text{(Section 1.5)}$$

Thus, the distance between $(-4, 3)$ and $(8, -2)$ is 13.

To obtain a general formula for the distance between two points in a coordinate plane, let $P(x_1, y_1)$ and $R(x_2, y_2)$ be any two distinct points in a plane, as shown in Figure 4. Complete a triangle by locating point Q with coordinates (x_2, y_1). The Pythagorean theorem gives the distance between P and R as

$$d(P, R) = \sqrt{(x_2 - x_1)^2 + (y_2 - y_1)^2}.$$

NOTE Absolute value bars are not necessary in this formula, since for all real numbers a and b, $|a - b|^2 = (a - b)^2$.

The *distance formula* can be summarized as follows.

Distance Formula

Suppose that $P(x_1, y_1)$ and $R(x_2, y_2)$ are two points in a coordinate plane. Then the distance between P and R, written $d(P, R)$, is given by the **distance formula,**

$$\mathbf{d(P, R) = \sqrt{(x_2 - x_1)^2 + (y_2 - y_1)^2}.}$$

That is, the distance between two points in a coordinate plane is the square root of the sum of the square of the difference between their x-coordinates and the square of the difference between their y-coordinates.

Looking Ahead to Calculus

The distance formula is used to find the distance between two points in the plane. In analytic geometry and calculus, the formula is extended to two points in space. Points in space can be represented by *ordered triples.* The *distance between the two points* (x_1, y_1, z_1) *and* (x_2, y_2, z_2) is given by the expression

$$\sqrt{(x_2 - x_1)^2 + (y_2 - y_1)^2 + (z_2 - z_1)^2}.$$

Although our derivation of the distance formula assumed that P and R are not on a horizontal or vertical line, the result is true for any two points.

EXAMPLE 2 Using the Distance Formula

Find the distance between $P(-8, 4)$ and $Q(3, -2)$.

Solution According to the distance formula,

$$\begin{aligned} d(P,Q) &= \sqrt{[3 - (-8)]^2 + (-2 - 4)^2} \quad x_1 = -8, y_1 = 4, x_2 = 3, y_2 = -2 \\ &= \sqrt{11^2 + (-6)^2} \\ &= \sqrt{121 + 36} \\ &= \sqrt{157}. \end{aligned}$$

Now try Exercise 9(a).

A statement of the form "If p, then q" is called a *conditional statement.* The related statement "If q, then p" is called its *converse.* In Chapter 1 we studied the Pythagorean theorem. The converse of the Pythagorean theorem is also a true statement:

> If the sides a, b, and c of a triangle satisfy $a^2 + b^2 = c^2$, then the triangle is a right triangle with legs having lengths a and b and hypotenuse having length c.

We can use this fact to determine whether three points are the vertices of a right triangle.

EXAMPLE 3 Determining Whether Three Points Are the Vertices of a Right Triangle

Are points $M(-2, 5)$, $N(12, 3)$, and $Q(10, -11)$ the vertices of a right triangle?

Solution A triangle with the three given points as vertices is shown in Figure 5. This triangle is a right triangle if the square of the length of the longest side equals the sum of the squares of the lengths of the other two sides. Use the distance formula to find the length of each side of the triangle.

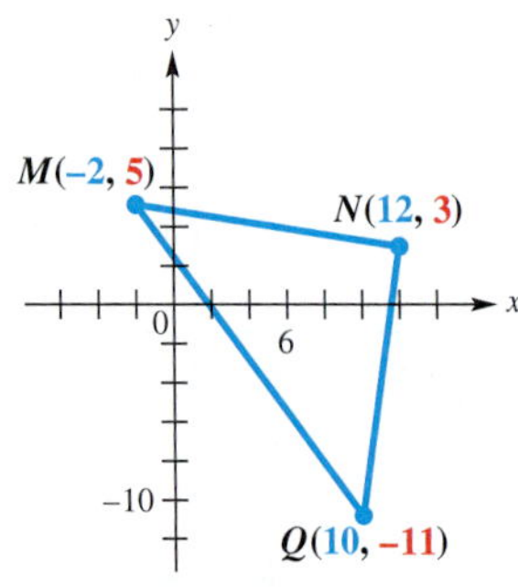

Figure 5

$$d(M,N) = \sqrt{[12 - (-2)]^2 + (3 - 5)^2} = \sqrt{196 + 4} = \sqrt{200}$$

$$d(M,Q) = \sqrt{[10 - (-2)]^2 + (-11 - 5)^2} = \sqrt{144 + 256} = \sqrt{400} = 20$$

$$d(N,Q) = \sqrt{(10 - 12)^2 + (-11 - 3)^2} = \sqrt{4 + 196} = \sqrt{200}$$

The longest side has length 20 units. Since

$$\left(\sqrt{200}\right)^2 + \left(\sqrt{200}\right)^2 = 400 = 20^2,$$

the triangle is a right triangle with hypotenuse joining M and Q.

Now try Exercise 17.

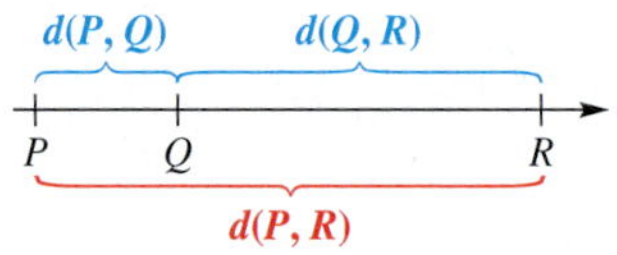

d(P, Q) d(Q, R)
P Q R
d(P, R)

$d(P, Q) + d(Q, R) = d(P, R)$

Figure 6

Using a similar procedure, we can tell whether three points are **collinear,** that is, lie on a straight line. Three points are collinear if the sum of the distances between two pairs of the points is equal to the distance between the remaining pair of points. See Figure 6.

EXAMPLE 4 Determining Whether Three Points are Collinear

Are the points $(-1, 5)$, $(2, -4)$, and $(4, -10)$ collinear?

Solution The distance between $(-1, 5)$ and $(2, -4)$ is

$$\sqrt{(-1-2)^2 + [5-(-4)]^2} = \sqrt{9+81} = \sqrt{90} = 3\sqrt{10}.$$

(Section R.7)

The distance between $(2, -4)$ and $(4, -10)$ is

$$\sqrt{(2-4)^2 + [-4-(-10)]^2} = \sqrt{4+36} = \sqrt{40} = 2\sqrt{10}.$$

The distance between the remaining pair of points $(-1, 5)$ and $(4, -10)$ is

$$\sqrt{(-1-4)^2 + [5-(-10)]^2} = \sqrt{25+225} = \sqrt{250} = 5\sqrt{10}.$$

Because $3\sqrt{10} + 2\sqrt{10} = 5\sqrt{10}$, the three points are collinear.

Now try Exercise 23.

NOTE In Exercises 67–70 of Section 2.4, we examine another method of determining whether three points are collinear.

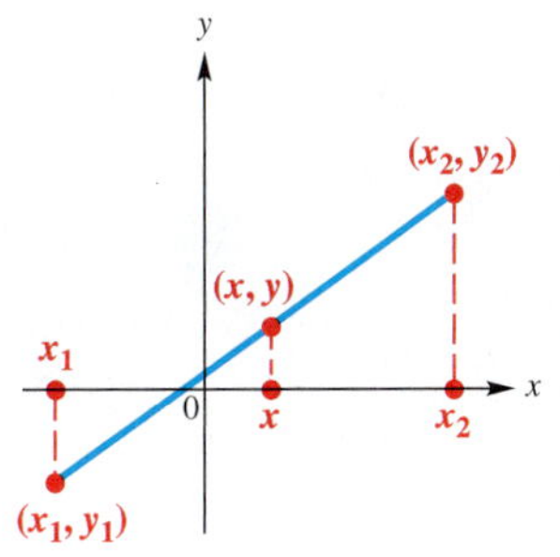

Figure 7

The Midpoint Formula The *midpoint formula* is used to find the coordinates of the midpoint of a line segment. (Recall that the midpoint of a line segment is equidistant from the endpoints of the segment.) To develop the midpoint formula, let (x_1, y_1) and (x_2, y_2) be any two distinct points in a plane. (Although Figure 7 shows $x_1 < x_2$, no particular order is required.) Let (x, y) be the midpoint of the segment connecting (x_1, y_1) and (x_2, y_2). Draw vertical lines from each of the three points to the x-axis, as shown in Figure 7.

Since (x, y) is the midpoint of the line segment connecting (x_1, y_1) and (x_2, y_2), the distance between x and x_1 equals the distance between x and x_2, so

$$x_2 - x = x - x_1$$

$$x_2 + x_1 = 2x \qquad \text{Add } x\text{; add } x_1.$$

$$x = \frac{x_1 + x_2}{2}. \qquad \text{Divide by 2; rewrite.}$$

Similarly, the y-coordinate of the midpoint is $\frac{y_1 + y_2}{2}$, yielding the following formula.

Midpoint Formula

The midpoint of the line segment with endpoints (x_1, y_1) and (x_2, y_2) is

$$\left(\frac{x_1 + x_2}{2}, \frac{y_1 + y_2}{2}\right).$$

That is, the x-coordinate of the midpoint of a line segment is the average of the x-coordinates of the segment's endpoints, and the y-coordinate is the average of the y-coordinates of the segment's endpoints.

EXAMPLE 5 Using the Midpoint Formula

Find the midpoint M of the segment with endpoints $(8, -4)$ and $(-6, 1)$.

Solution Use the midpoint formula to find that the coordinates of M are

$$\left(\frac{8 + (-6)}{2}, \frac{-4 + 1}{2}\right) = \left(1, -\frac{3}{2}\right).$$

Now try Exercise 9(b).

EXAMPLE 6 Applying the Midpoint Formula to Data

Figure 8 depicts how the number of McDonald's restaurants worldwide increased from 1995 through 2001. Use the midpoint formula and the two given points to estimate the number of restaurants in 1998, and compare it to the actual (rounded) figure of 24,000.

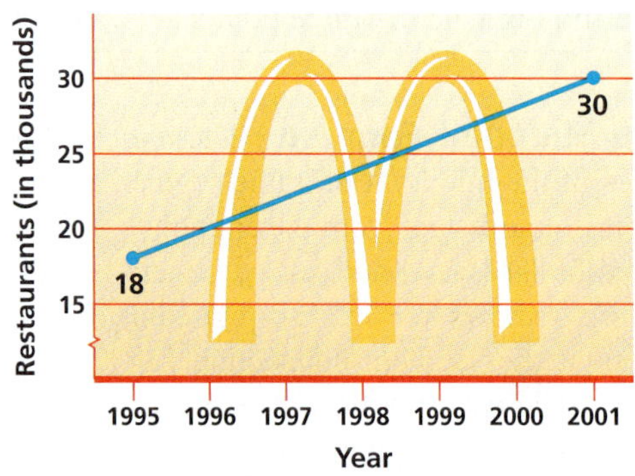

Source: McDonald's Corp.; Yahoo.com.

Figure 8

Solution The year 1998 lies halfway between 1995 and 2001, so we must find the coordinates of the midpoint of the segment that has endpoints $(1995, 18)$ and $(2001, 30)$. (Here, y is in thousands.) By the midpoint formula, this is

$$\left(\frac{1995 + 2001}{2}, \frac{18 + 30}{2}\right) = (1998, 24).$$

Thus, our estimate is 24,000 restaurants in 1998, which matches the actual (rounded) figure.

Now try Exercise 29.

Graphing Equations Ordered pairs are used to express the solutions of equations in two variables. When an ordered pair represents the solution of an equation with the variables x and y, the x-value is written first. For example, we say that $(1, 2)$ is a solution of $2x - y = 0$, since substituting 1 for x and 2 for y in the equation gives a true statement.

$$2(1) - 2 = 0$$
$$0 = 0 \quad \text{True}$$

EXAMPLE 7 Finding Ordered Pairs That Are Solutions of Equations

For each equation, find three ordered pairs that are solutions.

(a) $y = 4x - 1$ **(b)** $x = \sqrt{y - 1}$ **(c)** $x^2 + y^2 = 9$

Solution

(a) Choose any real number for x or y and substitute in the equation to get the corresponding value of the other variable. For example, let $x = -2$ and then let $y = 3$.

$y = 4x - 1$	$y = 4x - 1$
$y = 4(-2) - 1$ Let $x = -2$.	$3 = 4x - 1$ Let $y = 3$.
$y = -8 - 1$	$4 = 4x$
$y = -9$	$1 = x$

This gives the ordered pairs $(-2, -9)$ and $(1, 3)$. Verify that the ordered pair $(0, -1)$ is also a solution.

(b)

$x = \sqrt{y - 1}$ Given equation

$1 = \sqrt{y - 1}$ Let $x = 1$.

$1 = y - 1$ Square both sides. **(Section 1.6)**

$2 = y$

One ordered pair is $(1, 2)$. Verify that the ordered pairs $(0, 1)$ and $(2, 5)$ are also solutions of the equation.

(c)

$x^2 + y^2 = 9$ Given equation

$2^2 + y^2 = 9$ Let $x = 2$.

$y^2 = 5$

$y = \sqrt{5}$ or $y = -\sqrt{5}$ Square root property **(Section 1.4)**

We get two ordered pairs here, $(2, \sqrt{5})$ and $(2, -\sqrt{5})$. Each choice of a real number for x such that $-3 < x < 3$ will give two ordered pairs. Verify that $(0, 3)$ and $(0, -3)$ are also solutions.

Now try Exercises 33(a), 37(a), and 39(a).

The **graph** of an equation is found by plotting ordered pairs that are solutions of the equation. The *intercepts* of the graph are good points to plot first. An ***x*-intercept** is an x-value where the graph intersects the x-axis. A ***y*-intercept** is a y-value where the graph intersects the y-axis.* In other words, the x-intercept is the x-coordinate of an ordered pair where $y = 0$, and the y-intercept is the y-coordinate of an ordered pair where $x = 0$.

*The intercepts are sometimes defined as ordered pairs, such as $(3, 0)$ and $(0, -4)$, instead of numbers, like x-intercept 3 and y-intercept -4. At this level, however, they are usually defined as numbers.

A general algebraic approach for graphing an equation follows.

Graphing An Equation by Point Plotting

Step 1 Find the intercepts.

Step 2 Find as many additional ordered pairs as needed.

Step 3 Plot the ordered pairs from Steps 1 and 2.

Step 4 Connect the points from Step 3 with a smooth line or curve.

EXAMPLE 8 Graphing Equations

Graph each equation from Example 7.

(a) $y = 4x - 1$ **(b)** $x = \sqrt{y - 1}$ **(c)** $x^2 + y^2 = 9$

Solution

(a) *Step 1* Let $y = 0$ to find the x-intercept, and let $x = 0$ to find the y-intercept.

$$y = 4x - 1 \qquad\qquad y = 4x - 1$$
$$0 = 4x - 1 \qquad\qquad y = 4(0) - 1$$
$$1 = 4x \qquad\qquad y = 0 - 1$$
$$\frac{1}{4} = x \quad x\text{-intercept} \qquad\qquad y = -1 \quad y\text{-intercept}$$

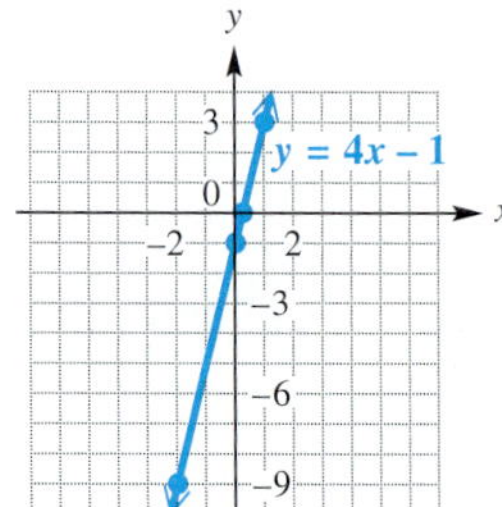

Figure 9

These intercepts lead to the ordered pairs $\left(\frac{1}{4}, 0\right)$ and $(0, -1)$. Note that the y-intercept yields one of the ordered pairs we found in Example 7(a).

Step 2 We use the other ordered pairs found in Example 7(a): $(-2, -9)$, $(1, 3)$.

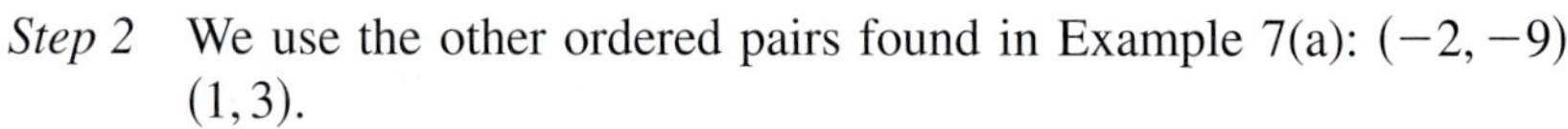

Step 3 Plot the four ordered pairs from Steps 1 and 2 as shown in Figure 9.

Step 4 Connect the points plotted in Step 3 with a straight line. This line, also shown in Figure 9, is the graph of the equation $y = 4x - 1$.

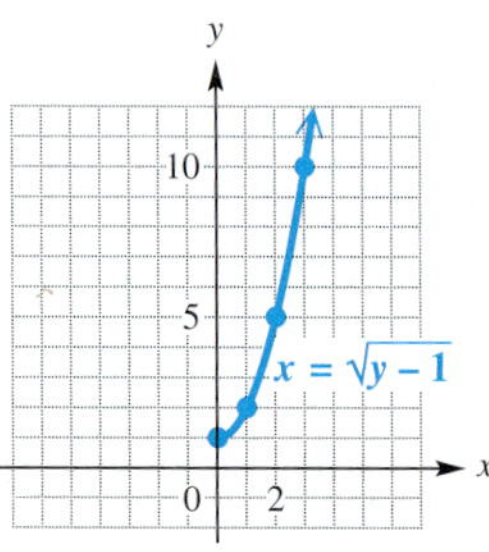

Figure 10

(b) For $x = \sqrt{y - 1}$, the y-intercept 1 was found in Example 7(b). Solve $x = \sqrt{0 - 1}$ for the x-intercept. Since the quantity under the radical is negative, there is no x-intercept. In fact, $y - 1$ must be greater than or equal to 0, so y must be greater than or equal to 1. We start by plotting the ordered pairs from Example 7(b), then connect the points with a smooth curve as in Figure 10. To confirm the direction the curve will take as x increases, we find another solution, $(3, 10)$.

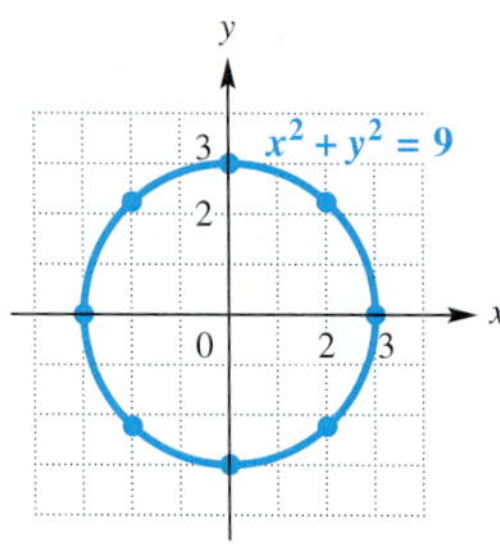

Figure 11

(c) For $x^2 + y^2 = 9$, we found the y-intercepts 3 and -3 in Example 7(c). Verify that the x-intercepts are 3 and -3. We also found the solutions $\left(2, -\sqrt{5}\right)$ and $\left(2, \sqrt{5}\right)$ in Example 7(c). Verify that $\left(-2, -\sqrt{5}\right)$ and $\left(-2, \sqrt{5}\right)$ also lie on the graph. Plotting these ordered pairs and connecting the corresponding points with a smooth curve gives the circle in Figure 11.

Now try Exercises 33(b), 37(b), and 39(b).

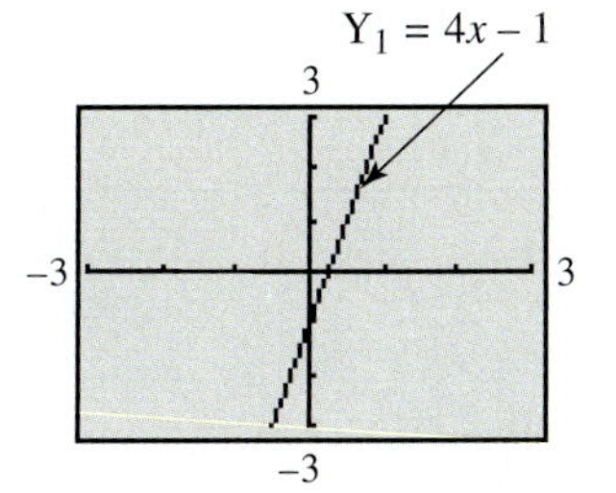

Figure 12

To graph an equation such as $y = 4x - 1$ on a calculator, we must first solve it for y (if necessary). Here the equation is already in the correct form, $y = 4x - 1$, so we enter $4x - 1$ for Y_1. The intercepts can help determine an appropriate window, since we want them to appear in the graph. Usually, a good choice is the *standard viewing window,* which has X minimum $= -10$, X maximum $= 10$, Y minimum $= -10$, Y maximum $= 10$, with X scale $= 1$ and Y scale $= 1$. (The X and Y scales determine the spacing of the tick marks.) Since the intercepts here are very close to the origin, we might choose the X and Y minimum and maximum to be -3 and 3 instead. See Figure 12. ■

Looking Ahead to Calculus

The circle $x^2 + y^2 = 1$ is called the **unit circle.** It is important in interpreting the *trigonometric* or *circular* functions that appear in the study of calculus.

Circles In Examples 7(c) and 8(c) we worked with the equation $x^2 + y^2 = 9$, which we saw is the equation of a circle. By definition, a **circle** is the set of all points in a plane that lie a given distance from a given point. The given distance is the **radius** of the circle, and the given point is the **center.** We can find the equation of a circle from its definition by using the distance formula.

In Figure 13 we graphed a circle of radius 3 with center at the origin. To find the equation of this circle, we let (x, y) be any point on the circle. The distance between (x, y) and the center of the circle, $(0, 0)$, is given by

$$\sqrt{(x - 0)^2 + (y - 0)^2}.$$

Since this distance equals the radius, 3,

$$\sqrt{(x - 0)^2 + (y - 0)^2} = 3$$

$$\sqrt{x^2 + y^2} = 3$$

$$x^2 + y^2 = 9. \quad \text{Square both sides.}$$

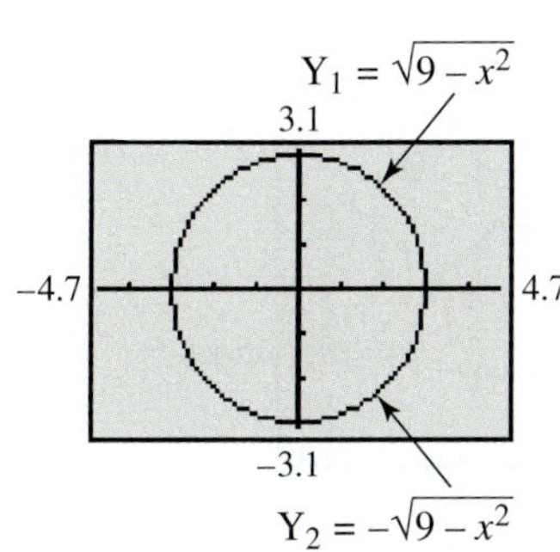

The graph of this circle has equation $x^2 + y^2 = 9$.

Figure 13

To graph the circle $x^2 + y^2 = 9$ using a graphing calculator, we must first solve the equation for y.

$$x^2 + y^2 = 9$$

$$y^2 = 9 - x^2 \quad \text{Subtract } x^2.$$

$$y = \pm\sqrt{9 - x^2} \quad \text{Square root property}$$

Now we let $Y_1 = \sqrt{9 - x^2}$ and $Y_2 = -\sqrt{9 - x^2}$. These are the equations of the two semicircles indicated in Figure 13. Because of the design of the viewing window, it is necessary to use a *square viewing window* to avoid distortion when graphing circles. Refer to your owner's manual to see how to do this. ■

EXAMPLE 9 Finding the Equation of a Circle

Find an equation for the circle having radius 6 and center at $(-3, 4)$.

Solution Use the distance formula. Let (x, y) be any point on the circle. The distance from (x, y) to $(-3, 4)$ is given by

$$\sqrt{[x - (-3)]^2 + (y - 4)^2} = \sqrt{(x + 3)^2 + (y - 4)^2}.$$

This distance equals the radius, 6. Therefore,

$$\sqrt{(x + 3)^2 + (y - 4)^2} = 6$$

or

$$(x + 3)^2 + (y - 4)^2 = 36. \quad \text{Square both sides.}$$

Generalizing from the work in Example 9 gives the following result.

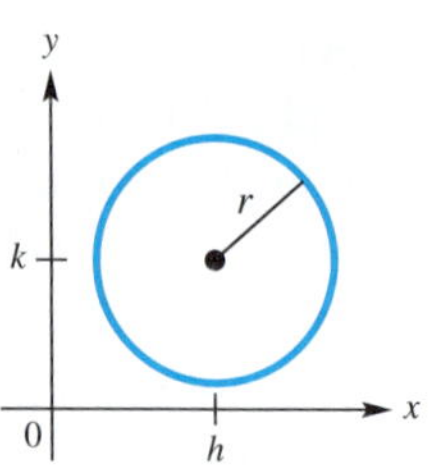

Figure 14

Center-Radius Form of the Equation of a Circle

The circle with center (h, k) and radius r has equation

$$(x - h)^2 + (y - k)^2 = r^2,$$

the **center-radius form** of the equation of a circle. (See Figure 14.) A circle with center $(0, 0)$ and radius r has equation

$$x^2 + y^2 = r^2.$$

Now try Exercise 49(a).

EXAMPLE 10 Graphing a Circle

Graph the circle with equation $(x + 3)^2 + (y - 4)^2 = 36$ from Example 9.

Algebraic Solution

Writing the equation as

$$[x - (-3)]^2 + (y - 4)^2 = 6^2$$

gives $(-3, 4)$ as the center and 6 as the radius. The graph is shown in Figure 15.

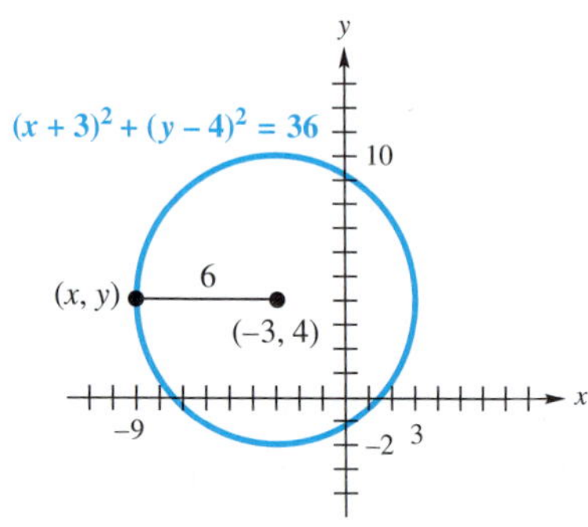

Figure 15

Graphing Calculator Solution

Solve $(x + 3)^2 + (y - 4)^2 = 36$ for y to obtain

$$y = 4 \pm \sqrt{36 - (x + 3)^2}.$$

Entering these *two* equations as Y_1 and Y_2 gives the graph shown in Figure 16.

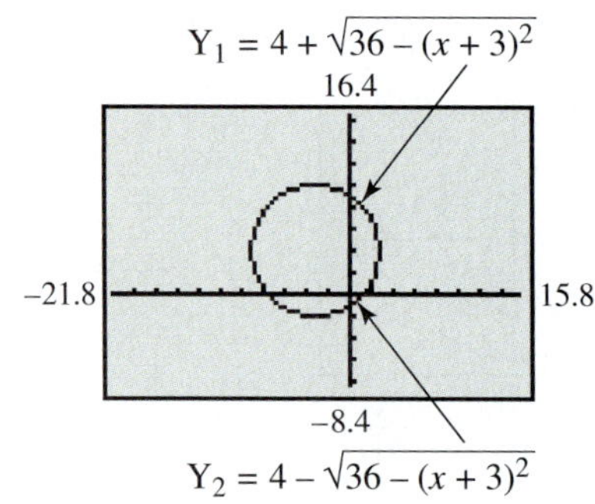

This is a square window with X scale = Y scale = 2.

Figure 16

Now try Exercise 49(b).

Starting with $(x - h)^2 + (y - k)^2 = r^2$ and squaring $x - h$ and $y - k$ gives the **general form of the equation of a circle,**

$$x^2 + y^2 + cx + dy + e = 0, \quad (1)$$

where c, d, and e are real numbers. Starting with an equation in this form, we can complete the square to get an equation of the form

$$(x - h)^2 + (y - k)^2 = m, \qquad \text{for some number } m.$$

There are three possibilities for the graph based on the value of m.

1. If $m > 0$, then $r^2 = m$, and the equation represents a circle with radius $\sqrt{m}$.
2. If $m = 0$, then the equation represents the single point (h, k).
3. If $m < 0$, then no points satisfy the equation.

EXAMPLE 11 Finding the Center and Radius by Completing the Square

Decide whether or not each equation has a circle as its graph.

(a) $x^2 - 6x + y^2 + 10y + 25 = 0$ **(b)** $x^2 + 10x + y^2 - 4y + 33 = 0$

(c) $2x^2 + 2y^2 - 6x + 10y = 1$

Solution

(a) Since $x^2 - 6x + y^2 + 10y + 25 = 0$ has the form of equation (1), it either represents a circle, a single point, or no points at all. To decide, complete the square on x and y separately, as explained in Section 1.4. Start with

$$(x^2 - 6x \qquad) + (y^2 + 10y \qquad) = -25.$$

$$\left[\frac{1}{2}(-6)\right]^2 = (-3)^2 = 9 \quad \text{and} \quad \left[\frac{1}{2}(10)\right]^2 = 5^2 = 25$$

Add 9 and 25 on the left to complete the two squares, and to compensate, add 9 and 25 on the right.

$$(x^2 - 6x + 9) + (y^2 + 10y + 25) = -25 + 9 + 25 \quad \text{Complete the square. (Section 1.4)}$$

$$(x - 3)^2 + (y + 5)^2 = 9 \quad \text{Factor. (Section R.4)}$$

Since $9 > 0$, the equation represents a circle with center at $(3, -5)$ and radius 3.

(b)

$$x^2 + 10x + y^2 - 4y + 33 = 0$$

$$\left[\frac{1}{2}(10)\right]^2 = 25 \quad \text{and} \quad \left[\frac{1}{2}(-4)\right]^2 = 4$$

$$(x^2 + 10x + 25) + (y^2 - 4y + 4) = -33 + 25 + 4 \quad \text{Complete the square.}$$

$$(x + 5)^2 + (y - 2)^2 = -4 \quad \text{Factor.}$$

Since $-4 < 0$, there are no ordered pairs (x, y), with x and y both real numbers, satisfying the equation. The graph of the given equation contains no points.

(c) $2x^2 + 2y^2 - 6x + 10y = 1$

To complete the square, the coefficients of the x^2- and y^2-terms must be 1.

$$2(x^2 - 3x) + 2(y^2 + 5y) = 1 \quad \text{Group the terms; factor out 2.}$$

$$2\left(x^2 - 3x + \frac{9}{4}\right) + 2\left(y^2 + 5y + \frac{25}{4}\right) = 1 + 2\left(\frac{9}{4}\right) + 2\left(\frac{25}{4}\right) \quad \text{Complete the square.}$$

$$2\left(x - \frac{3}{2}\right)^2 + 2\left(y + \frac{5}{2}\right)^2 = 18 \quad \text{Factor.}$$

$$\left(x - \frac{3}{2}\right)^2 + \left(y + \frac{5}{2}\right)^2 = 9 \quad \text{Divide both sides by 2.}$$

The equation has a circle with center at $\left(\frac{3}{2}, -\frac{5}{2}\right)$ and radius 3 as its graph.

Now try Exercises 57, 61, and 63.

René Descartes (1596–1650)

CONNECTIONS The rectangular coordinate system introduced in this section is credited to the French mathematician and philosopher René Descartes. The concept is so common today that we often use rectangular coordinates without realizing it. For example, to locate a city on a detailed state map, you may use a key that designates its location by a letter and a number. You would read across from left to right to find the letter of the grid in which the city lies, and then would read up and down to find the number. This is the same idea used to locate a point in the Cartesian plane.

In *An Introduction to the History of Mathematics,* Sixth Edition, Howard Eves relates a story suggesting how Descartes may have come up with the idea of using rectangular coordinates in developing the branch of mathematics known as *analytic geometry:*

> Another story...says that the initial flash of analytic geometry came to Descartes when watching a fly crawling about on the ceiling near a corner of his room. It struck him that the path of the fly on the ceiling could be described if only one knew the relation connecting the fly's distances from two adjacent walls.

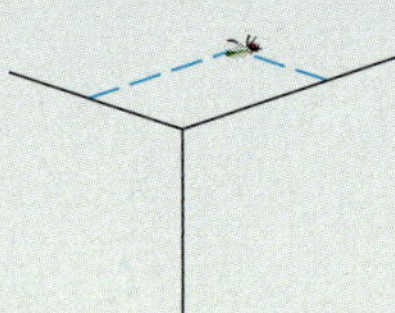

For Discussion or Writing

1. What are some other situations where objects are located by using the concept of rectangular coordinates?
2. How are latitudes and longitudes similar to the Cartesian coordinate system?

2.1 Exercises

Concept Check *Decide whether each statement in Exercises 1–3 is* true *or* false. *If the statement is false, tell why.*

1. The distance from the origin to the point (a, b) is $\sqrt{a^2 + b^2}$.
2. The midpoint of the segment joining (a, b) and $(3a, -3b)$ has coordinates $(2a, -b)$.
3. The equation defined by $x^2 - y^2 = 4$ is a circle with its center at the origin and radius 2.
4. In your own words, list the steps for graphing an equation.

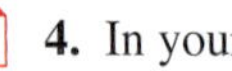

In Exercises 5–8, give three ordered pairs from each table.

5.

x	y
2	−5
−1	7
3	−9
5	−17
6	−21

6.

x	y
3	3
−5	−21
8	18
4	6
0	−6

7. *Percent of High School Students Who Smoke*

Year	Percent
1993	31
1995	35
1997	37
1999	35
2001	28

Source: Centers for Disease Control and Prevention.

8. *Number of Viewers of the Super Bowl*

Year	Viewers (millions)
1997	87.8
1998	90.0
1999	83.7
2000	88.5
2001	84.3

Source: Advertising Age.

*For the points P and Q, find **(a)** the distance $d(P, Q)$ and **(b)** the coordinates of the midpoint of the segment PQ. See Examples 2 and 5.*

9. $P(-5, -7), Q(-13, 1)$

10. $P(-4, 3), Q(2, -5)$

11. $P(8, 2), Q(3, 5)$

12. $P(-6, -5), Q(6, 10)$

13. $P(-8, 4), Q(3, -5)$

14. $P(6, -2), Q(4, 6)$

15. $P(3\sqrt{2}, 4\sqrt{5}), Q(\sqrt{2}, -\sqrt{5})$

16. $P(-\sqrt{7}, 8\sqrt{3}), Q(5\sqrt{7}, -\sqrt{3})$

Determine whether the three points are the vertices of a right triangle. See Example 3.

17. $(-6, -4), (0, -2), (-10, 8)$

18. $(-2, -8), (0, -4), (-4, -7)$

19. $(-4, 1), (1, 4), (-6, -1)$

20. $(-2, -5), (1, 7), (3, 15)$

21. $(-4, 3), (2, 5), (-1, -6)$

22. $(-7, 4), (6, -2), (0, -15)$

Determine whether the three points are collinear. See Example 4.

23. $(0, -7), (-3, 5), (2, -15)$

24. $(-1, 4), (-2, -1), (1, 14)$

25. $(0, 9), (-3, -7), (2, 19)$

26. $(-1, -3), (-5, 12), (1, -11)$

27. $(-7, 4), (6, -2), (-1, 1)$

28. $(-4, 3), (2, 5), (-1, 4)$

Solve each problem. See Example 6.

29. *Aging of College Freshmen* The graph shows a straight line that approximates the results from an annual survey of college freshmen. Use the midpoint formula and the two given points to estimate the percent in 1992. Compare your answer with the actual percent of 75.3.

Percent of College Freshmen Age 18 or Younger on December 31

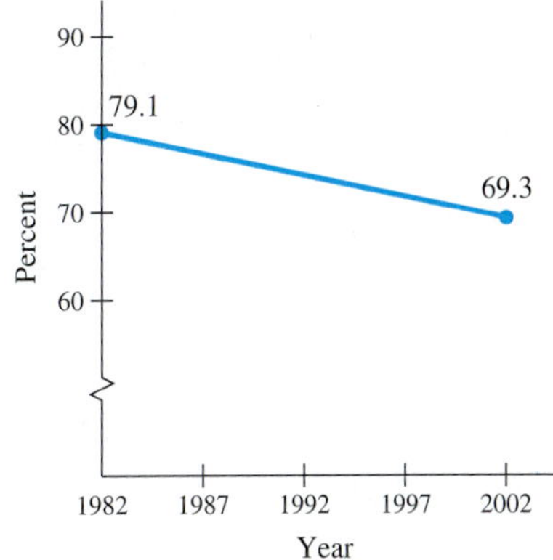

Source: Astin, A., L. Oseguera, L. Sax, and W. Korn, *The American Freshmen: Thirty-Five Year Trends*; Higher Education Research Institute, UCLA, 2002.

30. *Payment to Families with Dependent Children* The graph shows an idealized linear relationship for the average monthly family payment to families with dependent children. Based on this information, what was the average payment in 1998?

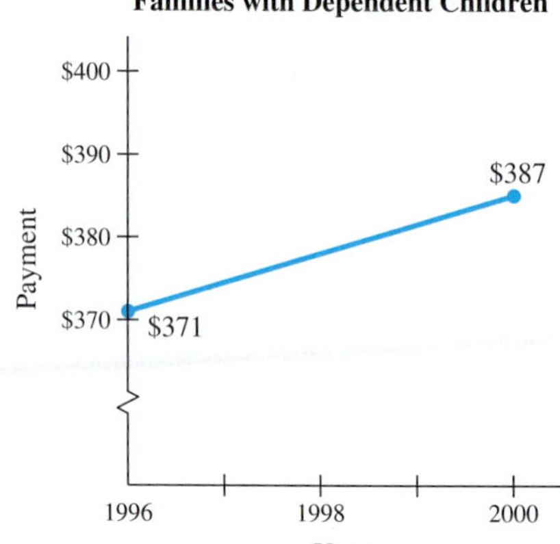

Source: U.S. Administration for Children and Families.

31. *Poverty Level Income Cutoffs* The table lists how poverty level income cutoffs (in dollars) for a family of four have changed over time. Use the midpoint formula to approximate the poverty level cutoff in 1995.

Year	Income (in dollars)
1960	3022
1970	3968
1980	8414
1990	13,359
2000	17,603

Source: U.S. Bureau of the Census.

32. *Two-Year College Enrollment* Enrollments in two-year colleges for recent years are shown in the table. Assuming a linear relationship, estimate the enrollments for 1985 and 1995.

Year	Enrollment (in millions)
1980	4.5
1990	5.2
2000	5.8

Source: Statistical Abstract of the United States.

For each equation, **(a)** *give a table with three ordered pairs that are solutions, and* **(b)** *graph the equation. (Hint: You will need more than three points for Exercises 37–44.) See Examples 7 and 8.*

33. $6y = 3x - 12$ **34.** $6y = -6x + 18$ **35.** $2x + 3y = 5$

36. $3x - 2y = 6$ **37.** $y = x^2$ **38.** $y = x^2 + 2$

39. $y = \sqrt{x - 3}$ **40.** $y = \sqrt{x} - 3$ **41.** $y = |x - 2|$

42. $y = -|x + 4|$ **43.** $y = x^3$ **44.** $y = -x^3$

In Exercises 45–52, **(a)** *find the center-radius form of the equation of each circle, and* **(b)** *graph it. See Examples 9 and 10.*

45. center $(0, 0)$, radius 6 **46.** center $(0, 0)$, radius 9 **47.** center $(2, 0)$, radius 6

48. center $(0, -3)$, radius 7 **49.** center $(-2, 5)$, radius 4 **50.** center $(4, 3)$, radius 5

51. center $(5, -4)$, radius 7 **52.** center $(-3, -2)$, radius 6

53. Find the center-radius form of the equation of a circle with center $(3, 2)$ and tangent to the x-axis. (*Hint:* A line *tangent to* a circle means touching it at exactly one point.)

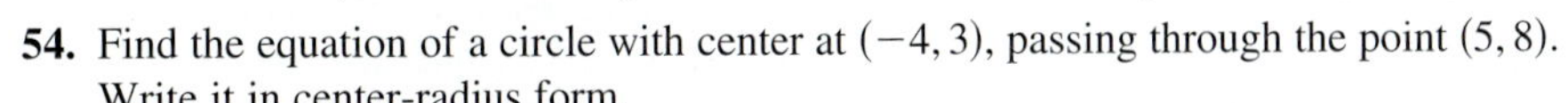

54. Find the equation of a circle with center at $(-4, 3)$, passing through the point $(5, 8)$. Write it in center-radius form.

55. When the equation of a circle is written in the form $(x - h)^2 + (y - k)^2 = m$, how does the value of m indicate whether the graph is a circle, a point, or does not exist?

56. Which one of the two screens is the correct graph of the circle with center $(-3, 5)$ and radius 4?

A.

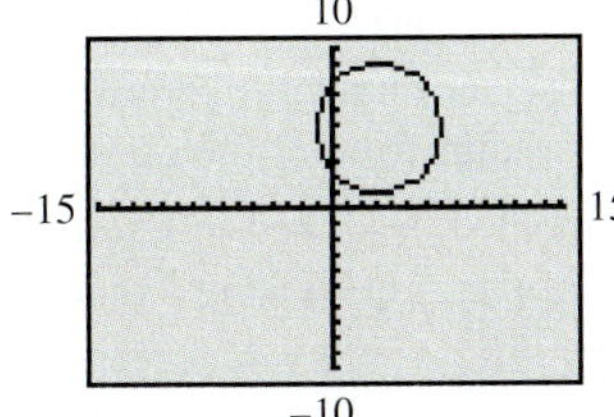

B.

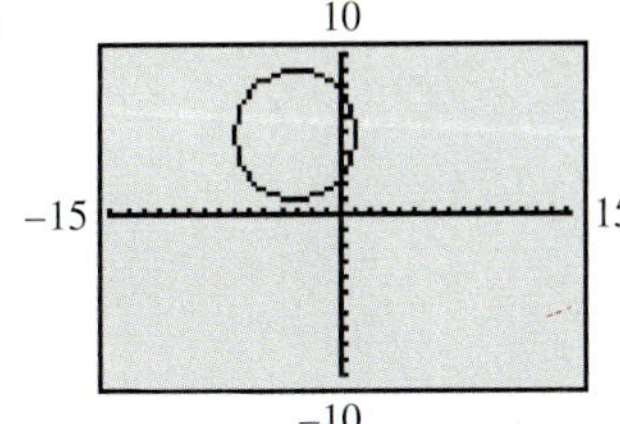

Decide whether or not each equation has a circle as its graph. If it does, give the center and the radius. See Example 11.

57. $x^2 + 6x + y^2 + 8y + 9 = 0$

58. $x^2 + 8x + y^2 - 6y + 16 = 0$

59. $x^2 - 4x + y^2 + 12y = -4$

60. $x^2 - 12x + y^2 + 10y = -25$

61. $4x^2 + 4x + 4y^2 - 16y - 19 = 0$

62. $9x^2 + 12x + 9y^2 - 18y - 23 = 0$

63. $x^2 + 2x + y^2 - 6y + 14 = 0$

64. $x^2 + 4x + y^2 - 8y + 32 = 0$

Relating Concepts

For individual or collaborative investigation

(Exercises 65–70)

The distance formula, the midpoint formula, and the center-radius form of the equation of a circle are closely related in the following problem.

A circle has a diameter with endpoints $(-1, 3)$ *and* $(5, -9)$. *Find the center-radius form of the equation of this circle.*

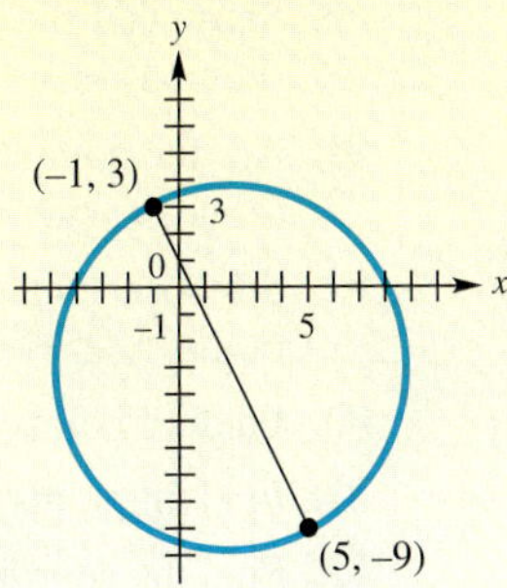

Work Exercises 65–70 in order, *to see the relationships among these concepts.*

65. To find the center-radius form, we must find both the radius and the coordinates of the center. Find the coordinates of the center using the midpoint formula. (The center of the circle must be the midpoint of the diameter.)

(continued)

66. There are several ways to find the radius of the circle. One way is to find the distance between the center and the point $(-1, 3)$. Use your result from Exercise 65 and the distance formula to find the radius.

67. Another way to find the radius is to repeat Exercise 66, but use the point $(5, -9)$ rather than $(-1, 3)$. Do this to obtain the same answer you found in Exercise 66.

68. There is yet another way to find the radius. Because the radius is half the diameter, it can be found by finding half the length of the diameter. Using the endpoints of the diameter given in the problem, find the radius in this manner. You should once again obtain the same answer you found in Exercise 66.

69. Using the center found in Exercise 65 and the radius found in Exercises 66–68, give the center-radius form of the equation of the circle.

70. Use the method described in Exercises 65–69 to find the center-radius form of the equation of the circle having a diameter whose endpoints are $(3, -5)$ and $(-7, 3)$.

Concept Check *Find the coordinates of the other endpoint of each segment, given its midpoint and one endpoint. (Hint: Let (x, y) be the unknown endpoint. Apply the midpoint formula, and solve the two equations for x and y.)*

71. midpoint $(5, 8)$, endpoint $(13, 10)$

72. midpoint $(-7, 6)$, endpoint $(-9, 9)$

73. midpoint $(12, 6)$, endpoint $(19, 16)$

74. midpoint $(-9, 8)$, endpoint $(-16, 9)$

75. Show that if M is the midpoint of the segment with endpoints $P(x_1, y_1)$ and $Q(x_2, y_2)$, then $d(P, M) + d(M, Q) = d(P, Q)$ and $d(P, M) = d(M, Q)$.

76. The distance formula as given in the text involves a square root radical. Write the distance formula using a rational exponent.

Concept Check *Answer the following.*

77. If a vertical line is drawn through the point $(4, 3)$, where will it intersect the x-axis?

78. If a horizontal line is drawn through the point $(4, 3)$, where will it intersect the y-axis?

79. If the point (a, b) is in the second quadrant, in what quadrant is $(a, -b)$? $(-a, b)$? $(-a, -b)$? (b, a)?

80. Show that the points $(-2, 2)$, $(13, 10)$, $(21, -5)$, and $(6, -13)$ are the vertices of a rhombus (all sides equal in length).

81. Are the points $A(1, 1)$, $B(5, 2)$, $C(3, 4)$, and $D(-1, 3)$ the vertices of a parallelogram (opposite sides equal in length)? of a rhombus (all sides equal in length)?

82. Suppose that a circle is tangent to both axes, is in the third quadrant, and has radius $\sqrt{2}$. Find the center-radius form of its equation.

83. Find all points (x, y) with $x = y$ that are 4 units from $(1, 3)$.

84. Find all points satisfying $x + y = 0$ that are 8 units from $(-2, 3)$.

85. Find the coordinates of all points whose distance from $(1, 0)$ is $\sqrt{10}$ and whose distance from $(5, 4)$ is $\sqrt{10}$.

86. Find the equation of the circle of smallest radius that contains the points $(1, 4)$ and $(-3, 2)$ within or on its boundary.

87. Find all values of y such that the distance between $(3, y)$ and $(-2, 9)$ is 12.

88. Find the coordinates of the points that divide the line segment joining $(4, 5)$ and $(10, 14)$ into three equal parts.

2.2 Functions

Relations and Functions ▪ Domain and Range ▪ Determining Functions from Graphs or Equations ▪ Function Notation ▪ Increasing, Decreasing, and Constant Functions

Relations and Functions

Recall from Section 2.1 how we described one quantity in terms of another.

- The letter grade you receive in a mathematics course depends on your numerical scores.
- The amount you pay (in dollars) for gas at the gas station depends on the number of gallons pumped.
- The dollars spent on entertainment depends on the type of entertainment.

We used ordered pairs to represent these corresponding quantities. For example, (3, \$5.25) indicates that you pay \$5.25 for 3 gallons of gas. Since the amount you pay *depends* on the number of gallons pumped, the amount (in dollars) is called the *dependent variable,* and the number of gallons pumped is called the *independent variable.* Generalizing, if the value of the variable y depends on the value of the variable x, then y is the **dependent variable** and x is the **independent variable.**

Independent variable → x, Dependent variable → y

$$(x, y)$$

Because we can write related quantities using ordered pairs, a set of ordered pairs such as $\{(3, 5.25), (8, 10), (10, 12.50)\}$ is called a *relation.*

Relation

A **relation** is a set of ordered pairs.

A special kind of relation called a *function* is very important in mathematics and its applications.

Function

A **function** is a relation in which, for each value of the first component of the ordered pairs, there is *exactly one* value of the second component.

NOTE The relation from the beginning of this section representing the number of gallons of gasoline and the corresponding cost is a function since each x-value is paired with exactly one y-value. You would not be happy, for example, if you and a friend each pumped 20 gal of regular gasoline at the same station and your bill was \$32 while his bill was \$28.

EXAMPLE 1 Deciding Whether Relations Define Functions

Decide whether each relation defines a function.

$$F = \{(1, 2), (-2, 4), (3, -1)\}$$
$$G = \{(1, 1), (1, 2), (1, 3), (2, 3)\}$$
$$H = \{(-4, 1), (-2, 1), (-2, 0)\}$$

Solution Relation F is a function, because for each different x-value there is exactly one y-value. We can show this correspondence as follows.

$\{1, -2, 3\}$ x-values of F
↓ ↓ ↓
$\{2, 4, -1\}$ y-values of F

As the correspondence below shows, relation G is not a function because one first component corresponds to *more than one* second component.

$\{1, 2\}$ x-values of G
$\{1, 2, 3\}$ y-values of G

In relation H the last two ordered pairs have the same x-value paired with two different y-values (-2 is paired with both 1 and 0), so H is a relation but not a function. ***In a function, no two ordered pairs can have the same first component and different second components.***

Different y-values
$H = \{(-4, 1), (-2, 1), (-2, 0)\}$ Not a function
Same x-value

Now try Exercises 5 and 7.

In a function, there is *exactly one* value of the dependent variable, the second component, for each value of the independent variable, the first component. This is what makes functions so important in applications.

Relations and functions can also be expressed as a correspondence or *mapping* from one set to another, as shown in Figure 17 for function F and relation H from Example 1. The arrow from 1 to 2 indicates that the ordered pair $(1, 2)$ belongs to F—each first component is paired with exactly one second component. In the mapping for relation H, which is not a function, the first component -2 is paired with two different second components, 1 and 0.

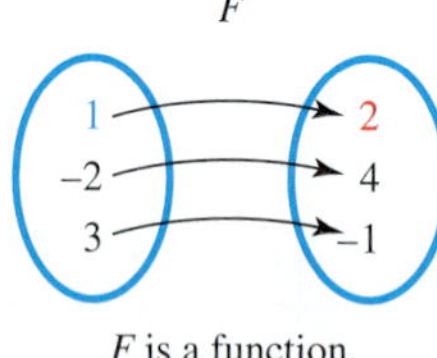

F is a function.

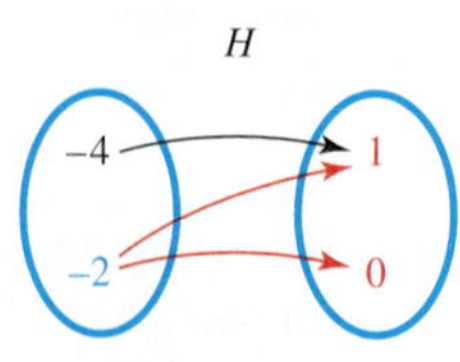

H is not a function.

Figure 17

Since relations and functions are sets of ordered pairs, we can represent them using tables and graphs. A table and graph for function F is shown in Figure 18.

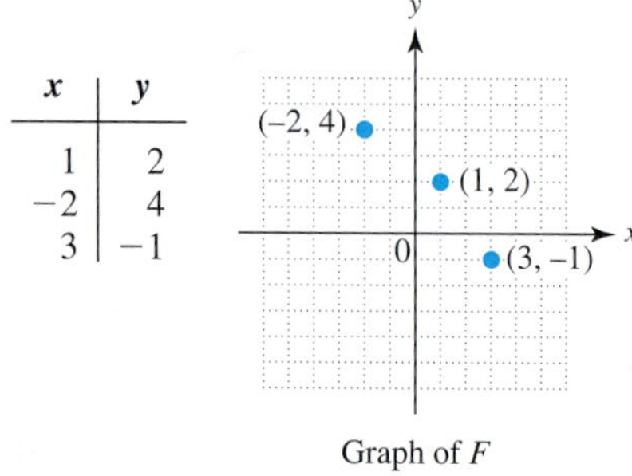

x	y
1	2
-2	4
3	-1

Graph of F

Figure 18

Finally, we can describe a relation or function using a rule that tells how to determine the dependent variable for a specific value of the independent variable. The rule may be given in words:

for instance, "the dependent variable is twice the independent variable." Usually the rule is an equation:

$$y = 2x.$$

↑ Dependent variable ↑ Independent variable

This is the most efficient way to define a relation or function.

NOTE Another way to think of a function relationship is to think of the independent variable as an input and the dependent variable as an output. This is illustrated by the input-output (function) machine for the function defined by $y = 2x$.

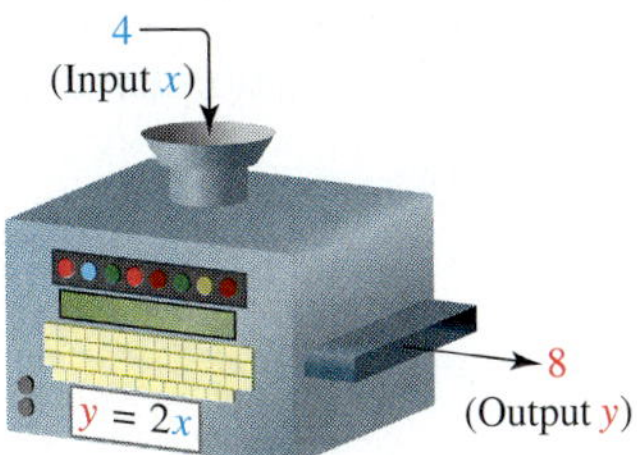

Function machine

Domain and Range For every relation there are two important sets of elements called the *domain* and *range*.

Domain and Range

In a relation, the set of all values of the independent variable (x) is the **domain;** the set of all values of the dependent variable (y) is the **range.**

EXAMPLE 2 Finding Domains and Ranges of Relations

Give the domain and range of each relation. Tell whether the relation defines a function.

(a) $\{(3, -1), (4, 2), (4, 5), (6, 8)\}$

(b)

(c)

x	y
−5	2
0	2
5	2

Solution

(a) The domain, the set of x-values, is $\{3, 4, 6\}$; the range, the set of y-values, is $\{-1, 2, 5, 8\}$. This relation is not a function because the same x-value, 4, is paired with two different y-values, 2 and 5.

(b) The domain of this relation is $\{4, 6, 7, -3\}$; the range is $\{A, B, C\}$. This mapping defines a function—each x-value corresponds to exactly one y-value.

x	y
-5	2
0	2
5	2

(c) This relation is a set of ordered pairs, so the domain is the set of x-values $\{-5, 0, 5\}$ and the range is the set of y-values $\{2\}$. The table defines a function because each different x-value corresponds to exactly one y-value (even though it is the same y-value).

Now try Exercises 11, 13, and 15.

As mentioned previously, the graph of a relation is the graph of its ordered pairs. The graph gives a picture of the relation, which can be used to determine its domain and range.

EXAMPLE 3 Finding Domains and Ranges from Graphs

Give the domain and range of each relation.

(a)

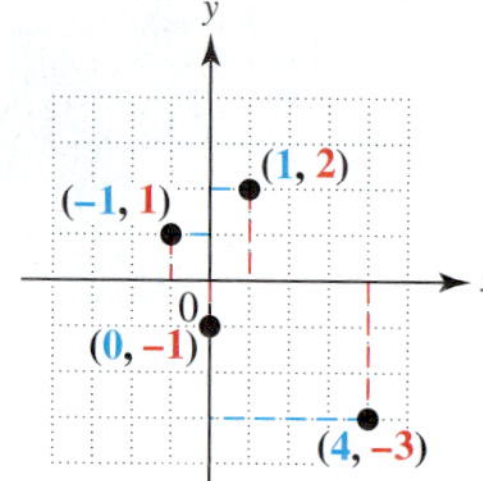

(b)

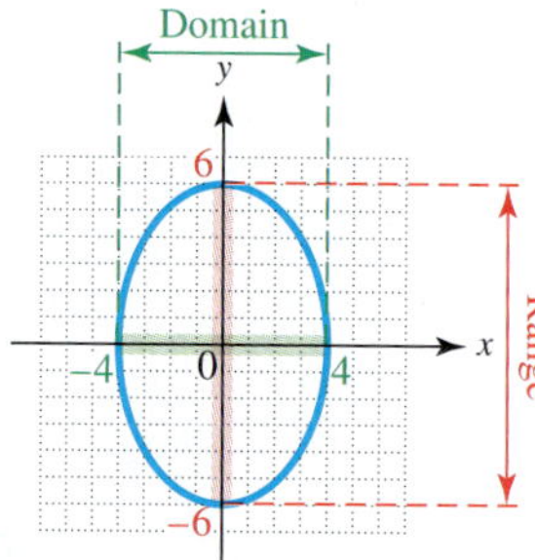

(c)

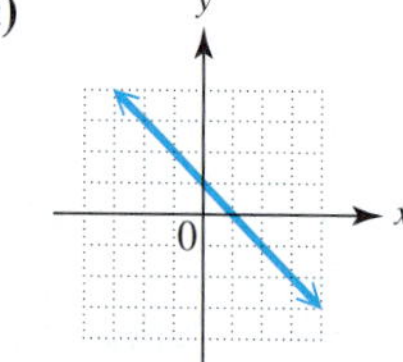

(d)

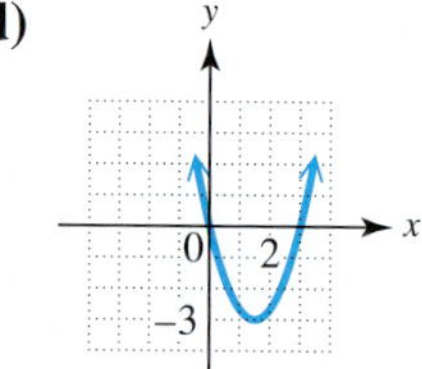

Solution

(a) The domain is the set of x-values, $\{-1, 0, 1, 4\}$. The range is the set of y-values, $\{-3, -1, 1, 2\}$.

(b) The x-values of the points on the graph include all numbers between -4 and 4, inclusive. The y-values include all numbers between -6 and 6, inclusive. Using interval notation,

the domain is $[-4, 4]$ and the range is $[-6, 6]$.

(c) The arrowheads indicate that the line extends indefinitely left and right, as well as up and down. Therefore, both the domain and the range include all real numbers, written $(-\infty, \infty)$.

(d) The arrowheads indicate that the graph extends indefinitely left and right, as well as upward. The domain is $(-\infty, \infty)$. Because there is a least y-value, -3, the range includes all numbers greater than or equal to -3, written $[-3, \infty)$.

Now try Exercises 17 and 19.

Since relations are often defined by equations, such as $y = 2x + 3$ and $y^2 = x$, we must sometimes determine the domain of a relation from its equation. In this book, we assume the following agreement on the domain of a relation.

Agreement on Domain

Unless specified otherwise, the domain of a relation is assumed to be all real numbers that produce real numbers when substituted for the independent variable.

To illustrate this agreement, since any real number can be used as a replacement for x in $y = 2x + 3$, the domain of this function is the set of all real numbers. As another example, the function defined by $y = \frac{1}{x}$ has all real numbers except 0 as domain, since y is undefined if $x = 0$. In general, the domain of a function defined by an algebraic expression is all real numbers, except those numbers that lead to division by 0 or an even root of a negative number.

Determining Functions from Graphs or Equations Most of the relations we have seen in the examples are functions—that is, each x-value corresponds to exactly one y-value. Since each value of x leads to only one value of y in a function, any vertical line drawn through the graph of a function must intersect the graph in at most one point. This is the *vertical line test* for a function.

Vertical Line Test

If each vertical line intersects a graph in at most one point, then the graph is that of a function.

The graph in Figure 19(a) represents a function—each vertical line intersects the graph in at most one point. The graph in Figure 19(b) is not the graph of a function since a vertical line intersects the graph in more than one point.

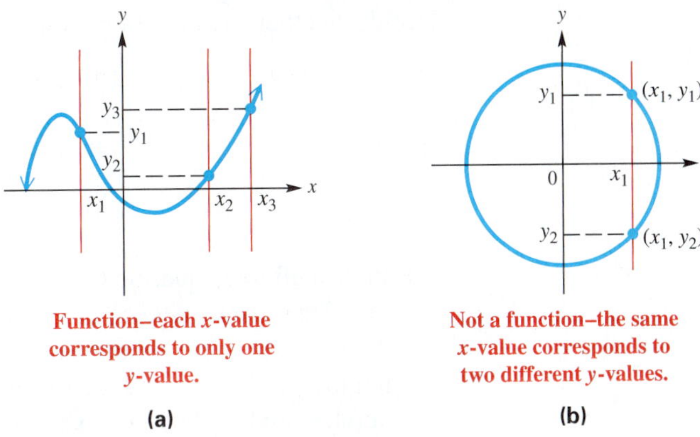

Function–each x-value corresponds to only one y-value.

(a)

Not a function–the same x-value corresponds to two different y-values.

(b)

Figure 19

EXAMPLE 4 Using the Vertical Line Test

Use the vertical line test to determine whether each relation graphed in Example 3 is a function.

(a)

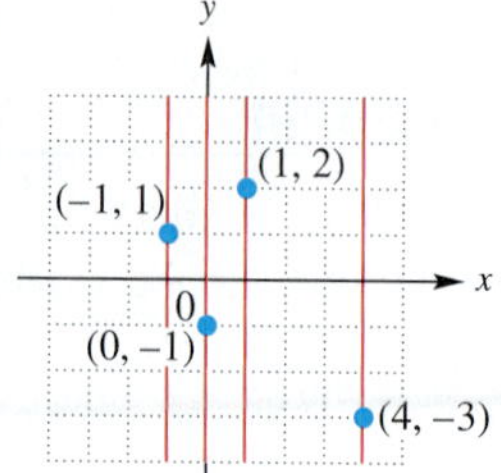

(b)

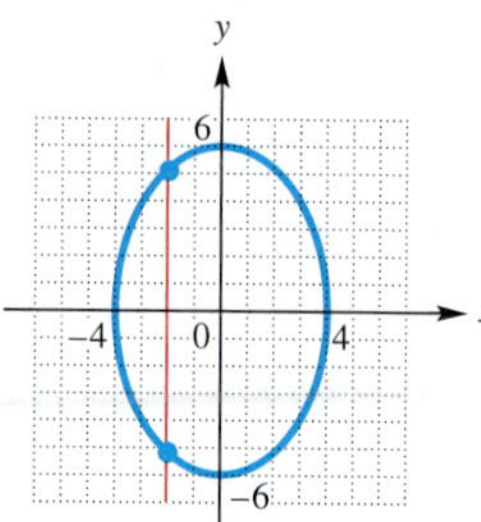

(c)

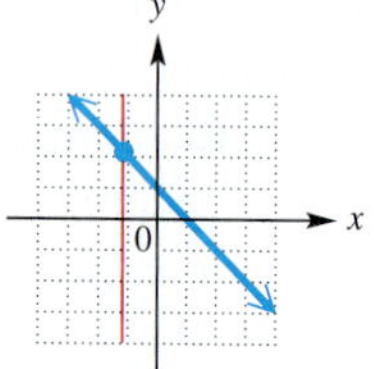

(d)

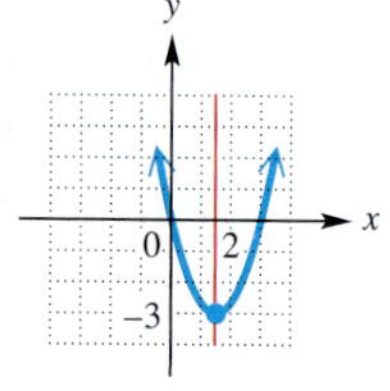

Solution The graphs in (a), (c), and (d) represent functions. The graph of the relation in (b) fails the vertical line test, since the same x-value corresponds to two different y-values; therefore, it is not the graph of a function.

Now try Exercise 21.

NOTE Graphs that do not represent functions are still relations. Remember that all equations and graphs represent relations and that all relations have a domain and range.

The vertical line test is a simple method for identifying a function defined by a graph. It is more difficult deciding whether a relation defined by an equation or an inequality is a function, as well as determining the domain and range. The next example gives some hints that may help.

EXAMPLE 5 Identifying Functions, Domains, and Ranges from Equations

Decide whether each relation defines a function and give the domain and range.

(a) $y = x + 4$ **(b)** $y = \sqrt{2x - 1}$ **(c)** $y^2 = x$

(d) $y \leq x - 1$ **(e)** $y = \dfrac{5}{x - 1}$

Solution

(a) In the defining equation (or rule), $y = x + 4$, y is always found by adding 4 to x. Thus, each value of x corresponds to just one value of y and the relation defines a function; x can be any real number, so the domain is $\{x \mid x \text{ is a real number}\}$ or $(-\infty, \infty)$. Since y is always 4 more than x, y also may be any real number, and so the range is $(-\infty, \infty)$.

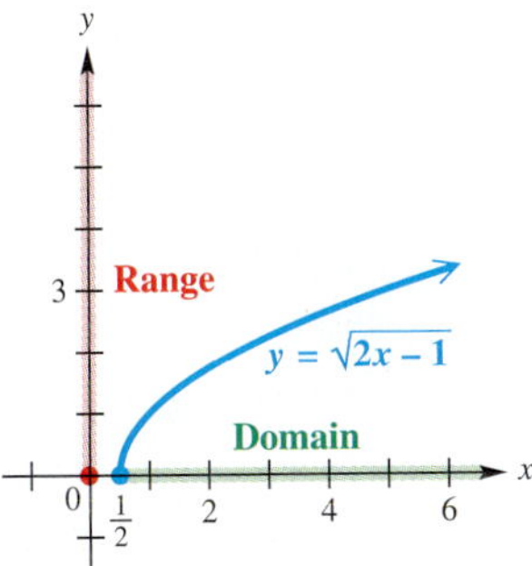

Figure 20

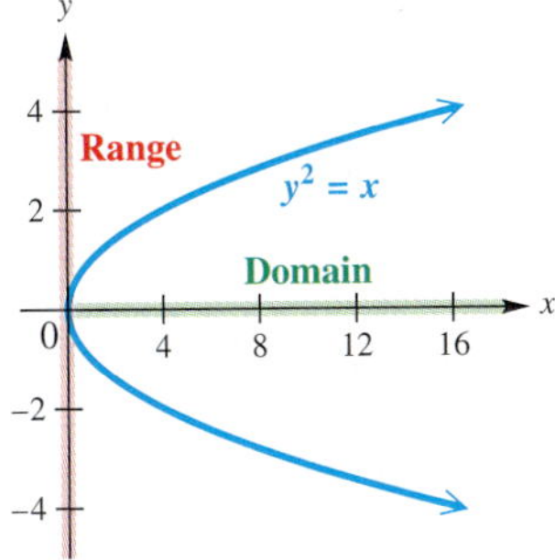

Figure 21

(b) For any choice of x in the domain of $y = \sqrt{2x - 1}$, there is exactly one corresponding value for y (the radical is a nonnegative number), so this equation defines a function. Refer to the agreement on domain stated previously. Since the equation involves a square root, the quantity under the radical sign cannot be negative. Thus,

$$2x - 1 \geq 0 \qquad \text{Solve the inequality. (Section 1.7)}$$

$$2x \geq 1$$

$$x \geq \frac{1}{2},$$

and the domain of the function is $\left[\frac{1}{2}, \infty\right)$. Because the radical is a nonnegative number, as x takes values greater than or equal to $\frac{1}{2}$, the range is $y \geq 0$, that is, $[0, \infty)$. See Figure 20.

(c) The ordered pairs $(16, 4)$ and $(16, -4)$ both satisfy the equation $y^2 = x$. Since one value of x, 16, corresponds to two values of y, 4 and -4, *this equation does not define a function.* Because x is equal to the square of y, the values of x must always be nonnegative. The domain of the relation is $[0, \infty)$. Any real number can be squared, so the range of the relation is $(-\infty, \infty)$. See Figure 21.

(d) By definition, y is a function of x if every value of x leads to exactly one value of y. Substituting a particular value of x, say 1, into $y \leq x - 1$, corresponds to many values of y. The ordered pairs $(1, 0)$, $(1, -1)$, $(1, -2)$, $(1, -3)$, and so on, all satisfy the inequality. For this reason, *an inequality rarely defines a function.* Any number can be used for x or for y, so the domain and the range of this relation are both the set of real numbers, $(-\infty, \infty)$.

(e) Given any value of x in the domain of

$$y = \frac{5}{x - 1},$$

we find y by subtracting 1, then dividing the result into 5. This process produces exactly one value of y for each value in the domain, so this equation defines a function. The domain includes all real numbers except those that make the denominator 0. We find these numbers by setting the denominator equal to 0 and solving for x.

$$x - 1 = 0$$

$$x = 1$$

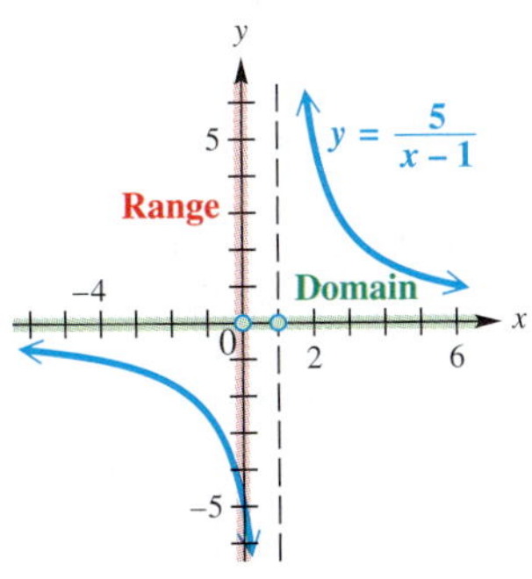

Figure 22

Thus, the domain includes all real numbers except 1, written as the interval $(-\infty, 1) \cup (1, \infty)$. Values of y can be positive or negative, but never 0, because a fraction cannot equal 0 unless its numerator is 0. Therefore, the range is the interval $(-\infty, 0) \cup (0, \infty)$, as shown in Figure 22.

Now try Exercises 27, 29, and 35.

In summary, we give three variations of the definition of function.

Variations of the Definition of Function

1. A **function** is a relation in which, for each value of the first component of the ordered pairs, there is exactly one value of the second component.
2. A **function** is a set of ordered pairs in which no first component is repeated.
3. A **function** is a rule or correspondence that assigns exactly one range value to each domain value.

Looking Ahead to Calculus

One of the most important concepts in calculus, that of the *limit of a function,* is defined using function notation: $\lim_{x \to a} f(x) = L$ (read "the limit of $f(x)$ as x approaches a is equal to L") means that the values of $f(x)$ become as close as we wish to L when we choose values of x sufficiently close to a.

Function Notation When a function f is defined with a rule or an equation using x and y for the independent and dependent variables, we say "y is a function of x" to emphasize that y *depends on* x. We use the notation

$$y = f(x),$$

called **function notation,** to express this and read $f(x)$ as "f of x." (In this notation the parentheses do not indicate multiplication.) The letter f stands for *function.* For example, if $y = 9x - 5$, we can name this function f and write

$$f(x) = 9x - 5.$$

Note that ***$f(x)$ is just another name for the dependent variable y.*** For example, if $y = f(x) = 9x - 5$ and $x = 2$, then we find y, or $f(2)$, by replacing x with 2.

$$\begin{aligned} y &= f(2) \\ &= 9 \cdot 2 - 5 \\ &= 18 - 5 \\ &= 13 \end{aligned}$$

The statement "if $x = 2$, then $y = 13$" represents the ordered pair (2, 13) and is abbreviated with function notation as

$$f(2) = 13.$$

Read $f(2)$ as "f of 2" or "f at 2." Also,

$$f(0) = 9 \cdot 0 - 5 = -5 \quad \text{and} \quad f(-3) = 9(-3) - 5 = -32.$$

These ideas and the symbols used to represent them can be illustrated as follows.

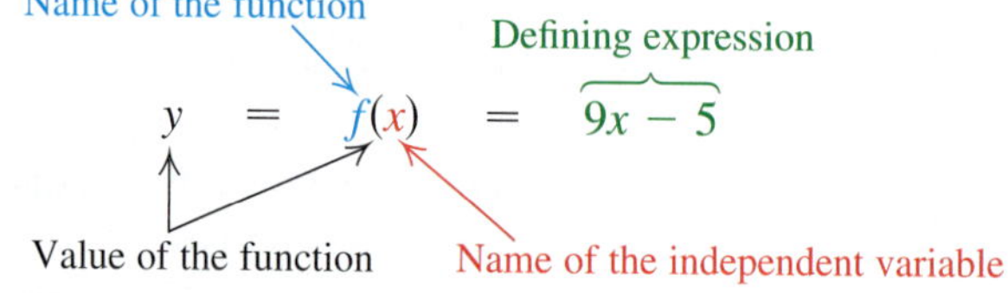

CAUTION The symbol $f(x)$ *does not* indicate "f times x," but represents the y-value for the indicated x-value. As just shown, $f(2)$ is the y-value that corresponds to the x-value 2.

EXAMPLE 6 Using Function Notation

Let $f(x) = -x^2 + 5x - 3$. Find the following.

(a) $f(2)$ **(b)** $f(q)$

Solution

(a) $f(x) = -x^2 + 5x - 3$

$f(2) = -2^2 + 5 \cdot 2 - 3$ Replace x with 2.

$= -4 + 10 - 3$

$= 3$

Thus, $f(2) = 3$; the ordered pair $(2, 3)$ belongs to f.

(b) $f(x) = -x^2 + 5x - 3$

$f(q) = -q^2 + 5q - 3$ Replace x with q.

The replacement of one variable with another is important in later courses.

Now try Exercises 41 and 45.

Sometimes letters other than f, such as g, h, or capital letters F, G, and H are used to name functions.

EXAMPLE 7 Using Function Notation

Let $g(x) = 2x + 3$. Find and simplify $g(a + 1)$.

Solution

$g(x) = 2x + 3$

$g(a + 1) = 2(a + 1) + 3$ Replace x with $a + 1$.

$= 2a + 2 + 3$

$= 2a + 5$

Now try Exercise 49.

Functions can be evaluated in a variety of ways, as shown in Example 8.

EXAMPLE 8 Using Function Notation

For each function, find $f(3)$.

(a) $f(x) = 3x - 7$

(b) $f = \{(-3, 5), (0, 3), (3, 1), (6, -1)\}$

(c)

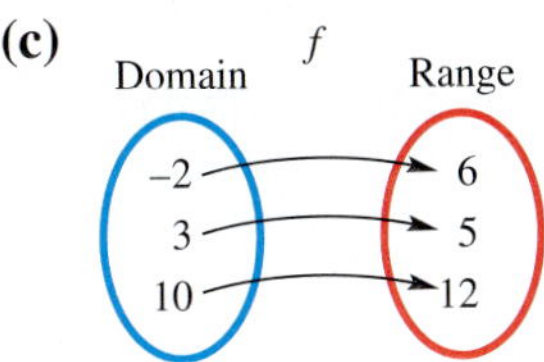

(d)

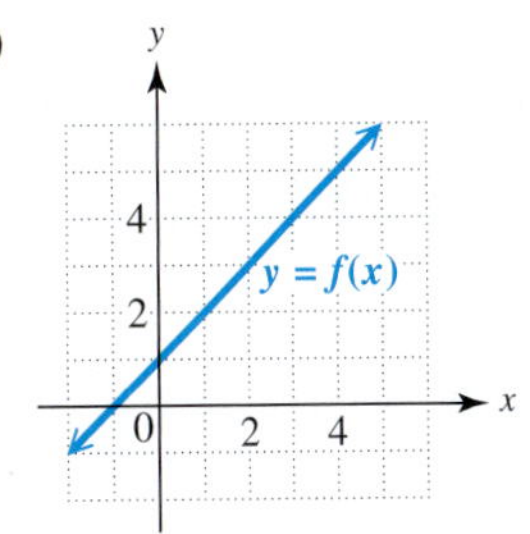

Solution

(a) $f(x) = 3x - 7$

$f(3) = 3(3) - 7$ Replace x with 3.

$f(3) = 2$

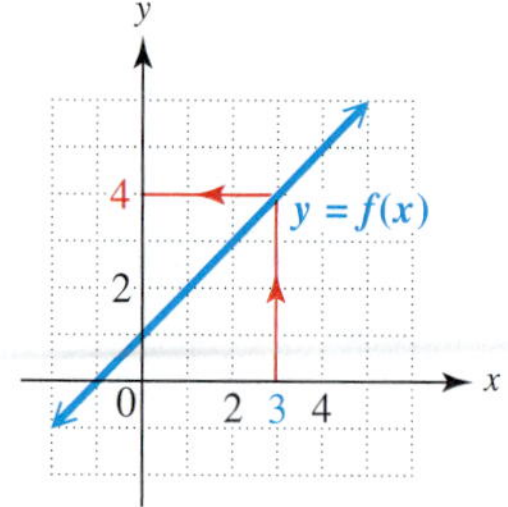

Figure 23

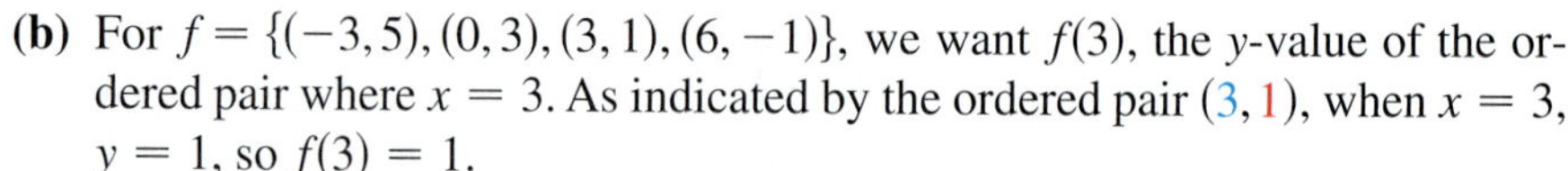

(b) For $f = \{(-3, 5), (0, 3), (3, 1), (6, -1)\}$, we want $f(3)$, the y-value of the ordered pair where $x = 3$. As indicated by the ordered pair $(3, 1)$, when $x = 3$, $y = 1$, so $f(3) = 1$.

(c) In the mapping on the previous page, the domain element 3 is paired with 5 in the range, so $f(3) = 5$.

(d) To evaluate $f(3)$, find 3 on the x-axis. See Figure 23. Then move up until the graph of f is reached. Moving horizontally to the y-axis gives 4 for the corresponding y-value. Thus, $f(3) = 4$.

Now try Exercises 53, 55, and 57.

If a function f is defined by an equation with x and y, not with function notation, use the following steps to find $f(x)$.

Finding an Expression for $f(x)$

Step 1 Solve the equation for y.

Step 2 Replace y with $f(x)$.

EXAMPLE 9 Writing Equations Using Function Notation

Rewrite each equation using function notation. Then find $f(-2)$ and $f(a)$.

(a) $y = x^2 + 1$ **(b)** $x - 4y = 5$

Solution

(a) This equation is already solved for y. Since $y = f(x)$,

$$f(x) = x^2 + 1.$$

To find $f(-2)$, let $x = -2$.

$$\begin{aligned} f(-2) &= (-2)^2 + 1 \\ &= 4 + 1 \\ &= 5 \end{aligned}$$

Find $f(a)$ by letting $x = a$: $f(a) = a^2 + 1$.

(b) First solve $x - 4y = 5$ for y. Then replace y with $f(x)$.

$$x - 4y = 5$$

$$x - 5 = 4y$$

$$y = \frac{x-5}{4} = \frac{x}{4} - \frac{5}{4}, \quad \text{so} \quad f(x) = \frac{1}{4}x - \frac{5}{4}.$$

Now find $f(-2)$ and $f(a)$.

$$f(-2) = \frac{1}{4}(-2) - \frac{5}{4} = -\frac{7}{4} \quad \text{Let } x = -2.$$

$$f(a) = \frac{1}{4}a - \frac{5}{4} \quad \text{Let } x = a.$$

Now try Exercises 59 and 61.

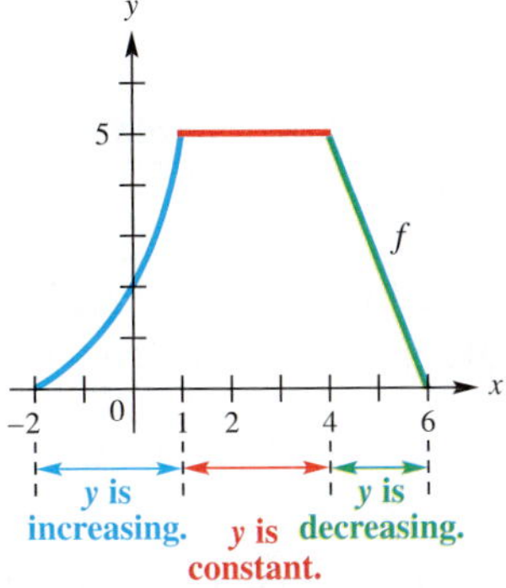

Figure 24

Increasing, Decreasing, and Constant Functions Informally speaking, a function *increases* on an interval of its domain if its graph rises from left to right on the interval. It *decreases* on an interval of its domain if its graph falls from left to right on the interval. It is *constant* on an interval of its domain if its graph is horizontal on the interval.

For example, look at Figure 24. The function increases on the interval $[-2, 1]$ because the y-values continue to get larger for x-values in that interval. Similarly, the function is constant on the interval $[1, 4]$ because the y-values are always 5 for all x-values there. Finally, the function decreases on the interval $[4, 6]$ because there the y-values continuously get smaller. Notice that the y-values produce the shape of the graph. *The intervals refer to the x-values where the y-values either increase, decrease, or are constant.*

The formal definitions of these concepts follow.

Increasing, Decreasing, and Constant Functions

Suppose that a function f is defined over an interval I. If x_1 and x_2 are in I,

(a) f **increases** on I if, whenever $x_1 < x_2$, $f(x_1) < f(x_2)$;

(b) f **decreases** on I if, whenever $x_1 < x_2$, $f(x_1) > f(x_2)$;

(c) f is **constant** on I if, for every x_1 and x_2, $f(x_1) = f(x_2)$.

Figure 25 illustrates these ideas.

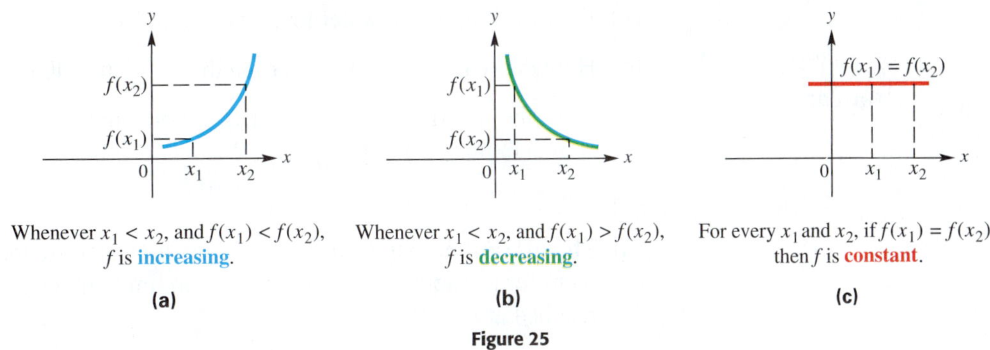

Whenever $x_1 < x_2$, and $f(x_1) < f(x_2)$, f is increasing.

(a)

Whenever $x_1 < x_2$, and $f(x_1) > f(x_2)$, f is decreasing.

(b)

For every x_1 and x_2, if $f(x_1) = f(x_2)$, then f is constant.

(c)

Figure 25

NOTE To decide whether a function is increasing, decreasing, or constant on an interval, ask yourself **"What does *y* do as *x* goes from left to right?"**

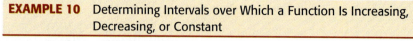

EXAMPLE 10 Determining Intervals over Which a Function Is Increasing, Decreasing, or Constant

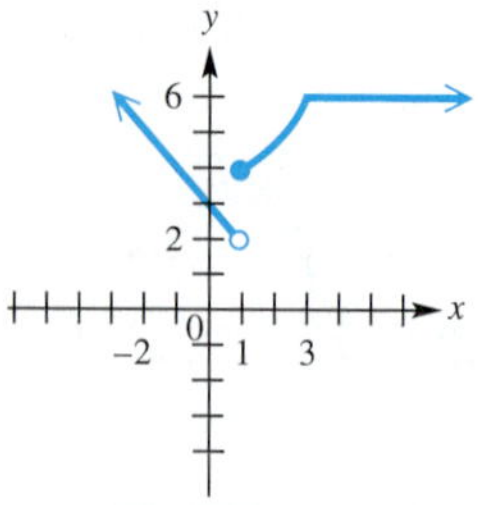

Figure 26

Figure 26 shows the graph of a function. Determine the intervals over which the function is increasing, decreasing, or constant.

Solution We should ask, "What is happening to the y-values as the x-values are getting larger?" Moving from left to right on the graph, we see that on the interval $(-\infty, 1)$, the y-values are *decreasing;* on the interval $[1, 3]$, the y-values are *increasing;* and on the interval $[3, \infty)$, the y-values are *constant* (and equal to 6). Therefore, the function is decreasing on $(-\infty, 1)$, increasing on $[1, 3]$, and constant on $[3, \infty)$.

Now try Exercise 73.

CAUTION When specifying intervals over which a function is increasing, decreasing, or constant, use *domain* values. Range values are not involved when writing the intervals.

One advantage of using a graph to display information is that it can convey that information more efficiently than words.

EXAMPLE 11 Interpreting a Graph

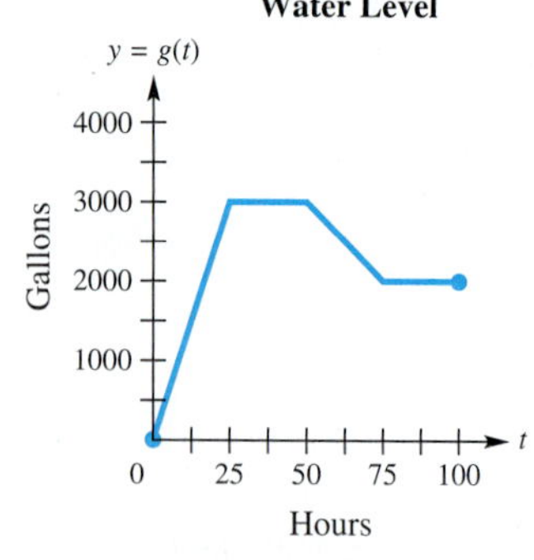

Figure 27

Figure 27 shows the relationship between the number of gallons of water in a small swimming pool and time in hours. By looking at this graph of the function, we can answer questions about the water level in the pool at various times. For example, at time 0 the pool is empty. The water level then increases, stays constant for a while, decreases, then becomes constant again. Use the graph to respond to the following.

(a) What is the maximum number of gallons of water in the pool? When is the maximum water level first reached?

(b) For how long is the water level increasing? decreasing? constant?

(c) How many gallons of water are in the pool after 90 hr?

(d) Describe a series of events that could account for the water level changes shown in the graph.

Solution

(a) The maximum range value is 3000, as indicated by the horizontal line segment for the hours 25 to 50. This maximum number of gallons, 3000, is first reached at $t = 25$ hr.

(b) The water level is increasing for $25 - 0 = 25$ hr and is decreasing for $75 - 50 = 25$ hr. It is constant for

$$(50 - 25) + (100 - 75) = 25 + 25 = 50 \text{ hr.}$$

(c) When $t = 90$, $y = g(90) = 2000$. There are 2000 gal after 90 hr.

(d) The pool is empty at the beginning and then is filled to a level of 3000 gal during the first 25 hr. For the next 25 hr, the water level remains the same. At 50 hr, the pool starts to be drained, and this draining lasts for 25 hr, until only 2000 gal remain. For the next 25 hr, the water level is unchanged.

Now try Exercise 79.

2.2 Exercises

1. In your own words, define a function and give an example.

2. In your own words, define the domain of a function and give an example.

3. *Concept Check* In an ordered pair of a relation, is the first element the independent or the dependent variable?

4. *Concept Check* Give an example of a relation that is not a function, having domain $\{-3, 2, 6\}$ and range $\{4, 6\}$. (There are many possible correct answers.)

Decide whether each relation defines a function. See Example 1.

5. $\{(5, 1), (3, 2), (4, 9), (7, 6)\}$

6. $\{(8, 0), (5, 4), (9, 3), (3, 8)\}$

7. $\{(2, 4), (0, 2), (2, 5)\}$

8. $\{(9, -2), (-3, 5), (9, 2)\}$

9. $\{(-3, 1), (4, 1), (-2, 7)\}$

10. $\{(-12, 5), (-10, 3), (8, 3)\}$

Decide whether each relation defines a function and give the domain and range. See Examples 1–4.

11. $\{(1, 1), (1, -1), (0, 0), (2, 4), (2, -4)\}$

12. $\{(2, 5), (3, 7), (4, 9), (5, 11)\}$

13.

2
5
11
17
3

1
7
20

14.

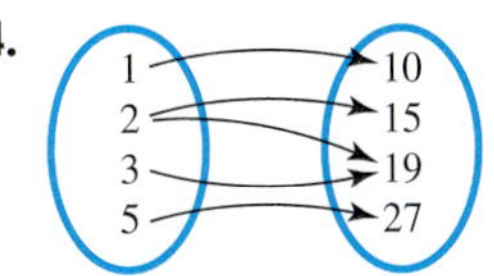

15. *Number of Rounds of Golf Played in the U.S.*

Year (x)	Number of Rounds (y)
1997	547,200,000
1998	528,500,000
1999	564,100,000
2000	587,100,000

Source: National Golf Foundation.

16. *Attendance at NCAA Women's College Basketball Games*

Year (x)	Attendance
1998	7,387,000
1999	8,010,000
2000	8,698,000

Source: National Collegiate Athletic Association.

17.

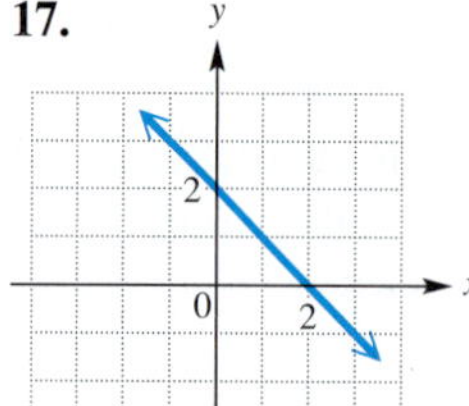

18.

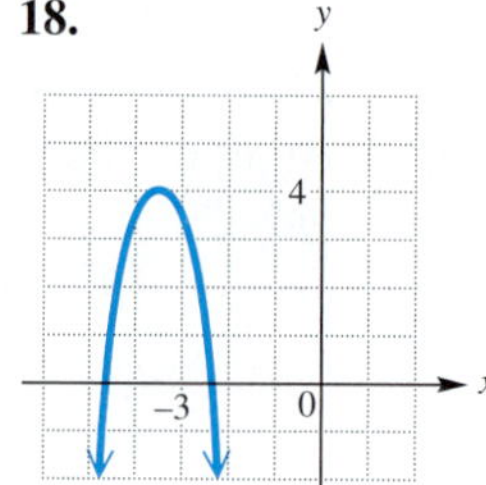

19.

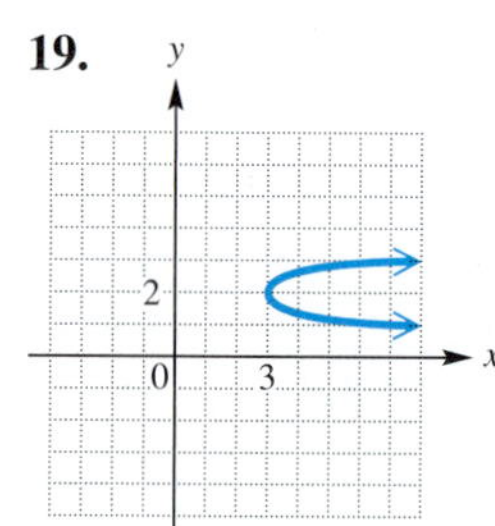

20.

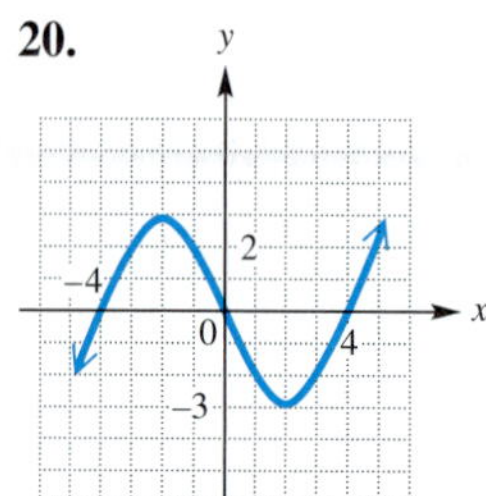

21.

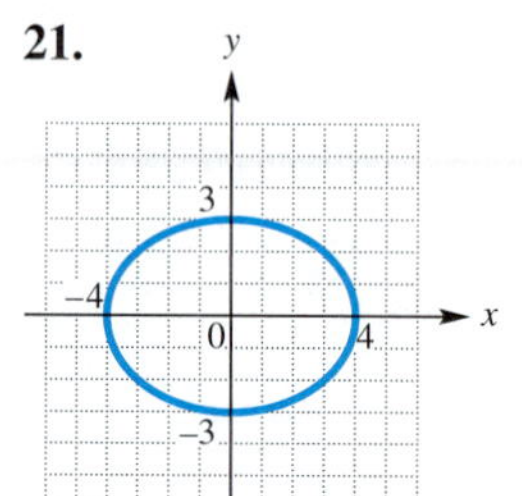

22.

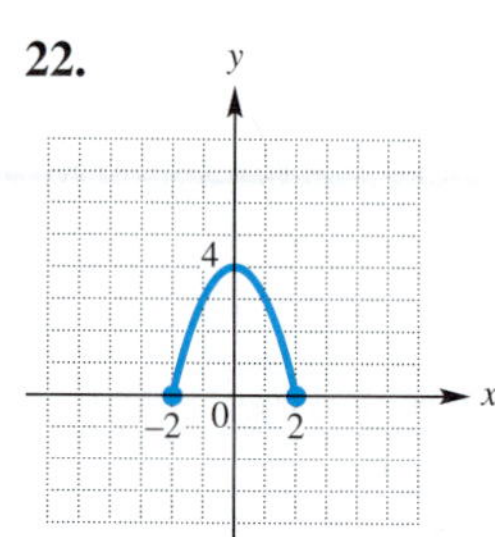

Decide whether each relation defines y as a function of x. Give the domain and range. See Example 5.

23. $y = x^2$ **24.** $y = x^3$ **25.** $x = y^6$ **26.** $x = y^4$

27. $y = 2x - 6$ **28.** $y = -6x + 8$ **29.** $x + y < 4$ **30.** $x - y < 3$

31. $y = \sqrt{x}$ **32.** $y = -\sqrt{x}$ **33.** $xy = 1$ **34.** $xy = -3$

35. $y = \sqrt{4x + 2}$ **36.** $y = \sqrt{9 - 2x}$ **37.** $y = \dfrac{2}{x - 9}$ **38.** $y = \dfrac{-7}{x - 16}$

39. ***Concept Check*** Choose the correct response: The notation $f(3)$ means

A. the variable f times 3 or $3f$.
B. the value of the dependent variable when the independent variable is 3.
C. the value of the independent variable when the dependent variable is 3.
D. f equals 3.

40. ***Concept Check*** Give an example of a function from everyday life. (*Hint:* Fill in the blanks: ______ depends on ______, so ______ is a function of ______.)

Let $f(x) = -3x + 4$ *and* $g(x) = -x^2 + 4x + 1$. *Find the following. See Examples 6 and 7.*

41. $f(0)$ **42.** $f(-3)$ **43.** $g(-2)$ **44.** $g(10)$

45. $f(p)$ **46.** $g(k)$ **47.** $f(-x)$ **48.** $g(-x)$

49. $f(x + 2)$ **50.** $f(a + 4)$ **51.** $f(2m - 3)$ **52.** $f(3t - 2)$

For each function, find ***(a)*** $f(2)$ *and* ***(b)*** $f(-1)$. *See Example 8.*

53. $f = \{(-1, 3), (4, 7), (0, 6), (2, 2)\}$ **54.** $f = \{(2, 5), (3, 9), (-1, 11), (5, 3)\}$

55.

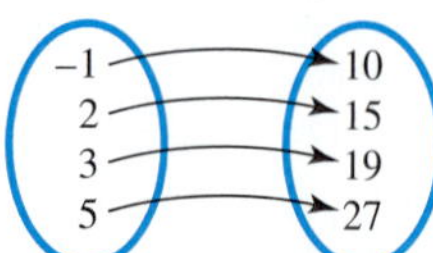

56.

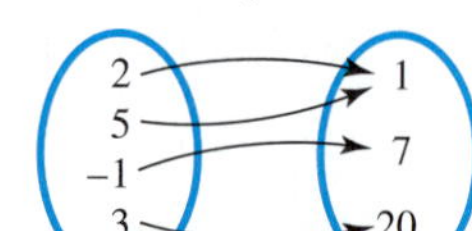

57.

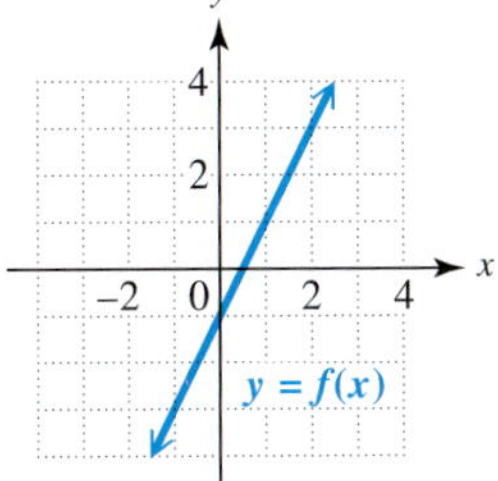

58.

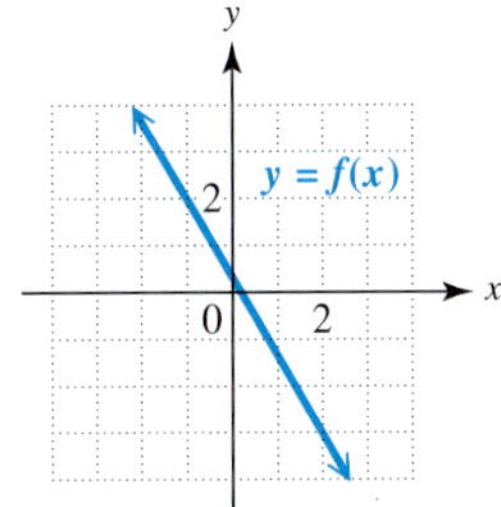

An equation that defines y as a function of x is given. ***(a)*** *Solve for y in terms of x and replace y with the function notation $f(x)$.* ***(b)*** *Find $f(3)$. See Example 9.*

59. $x + 3y = 12$

60. $x - 4y = 8$

61. $y + 2x^2 = 3$

62. $y - 3x^2 = 2$

63. $4x - 3y = 8$

64. $-2x + 5y = 9$

65. If $(3, 4)$ is on the graph of $y = f(x)$, which one of the following must be true: $f(3) = 4$ or $f(4) = 3$? Explain your answer.

Concept Check *Answer each question.*

66. The figure shows a portion of the graph of $f(x) = x^2 + 3x + 1$ and a rectangle with its base on the x-axis and a vertex on the graph. What is the area of the rectangle? (*Hint:* $f(.2)$ is the height.)

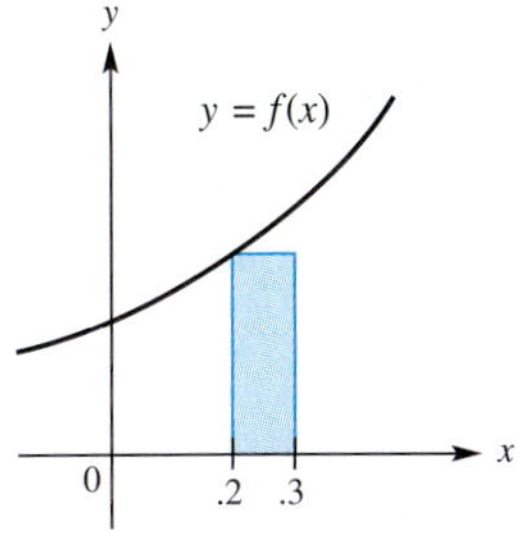

67. The graph of $Y_1 = f(X)$ is shown with a display at the bottom. What is $f(3)$?

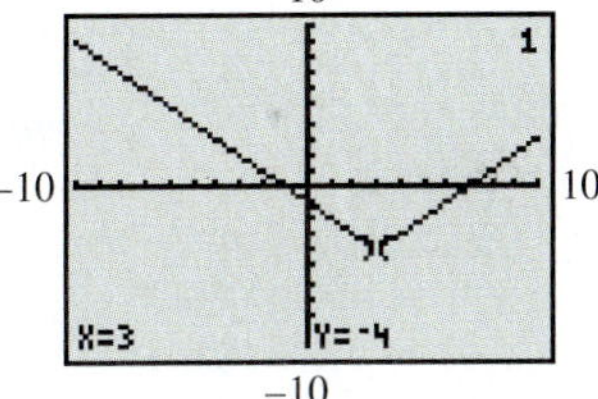

68. The graph of $Y_1 = f(X)$ is shown with a display at the bottom. What is $f(-2)$?

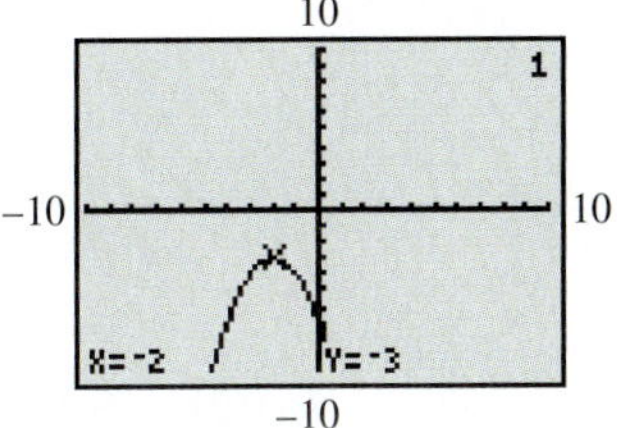

Concept Check *Use the graph of $y = f(x)$ to find each function value:* ***(a)*** *$f(-2)$,* ***(b)*** *$f(0)$,* ***(c)*** *$f(1)$, and* ***(d)*** *$f(4)$.*

69.

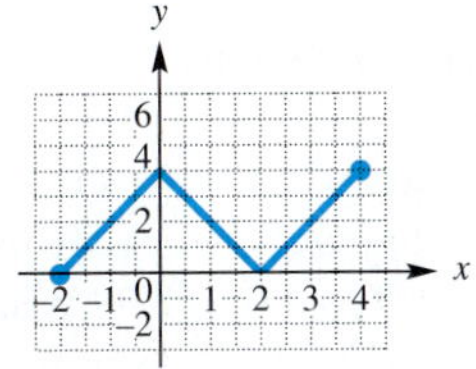

70.

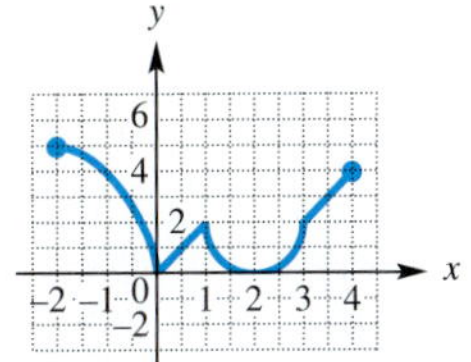

71.

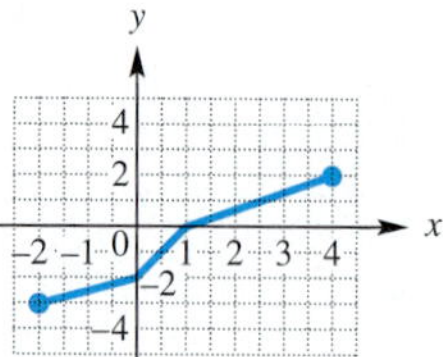

72.

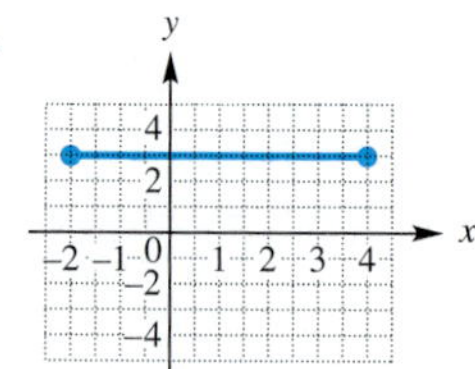

*Determine the intervals of the domain for which each function is **(a)** increasing, **(b)** decreasing, and **(c)** constant. See Example 10.*

73.

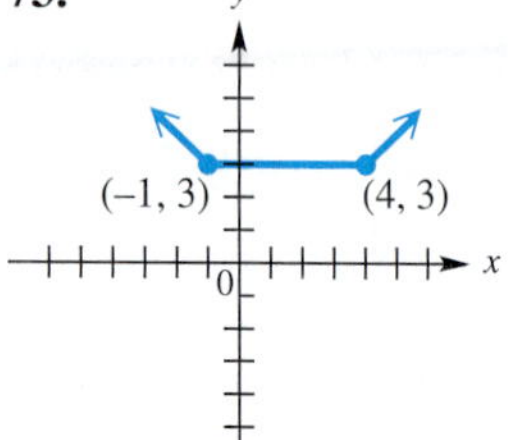

74.

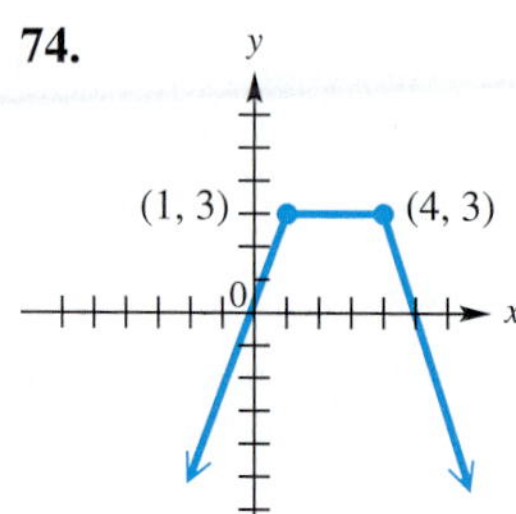

75.

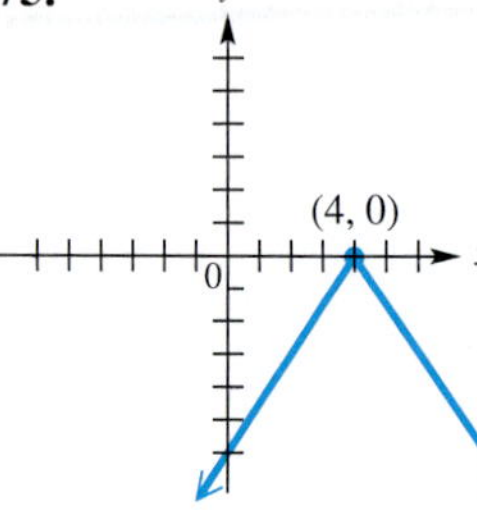

76.

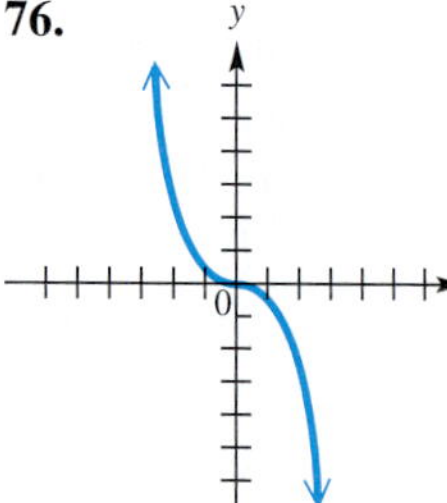

77.

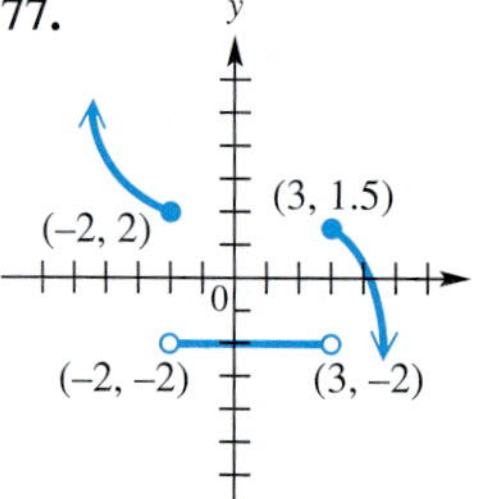

78.

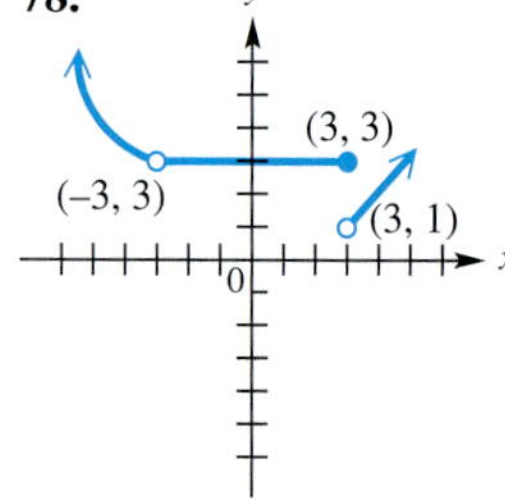

Solve each problem by obtaining information from the associated graph. See Example 11.

79. *Electricity Usage* The graph shows the daily megawatts of electricity used on a record-breaking summer day in Sacramento, California.

Electricity Use

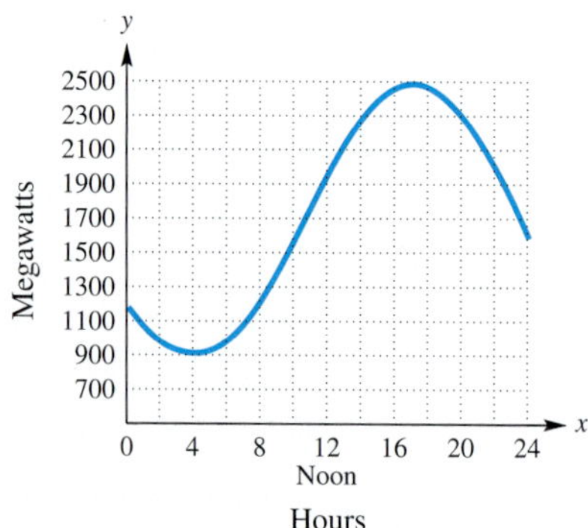

Source: Sacramento Municipal Utility District.

(a) Is this the graph of a function?
(b) What is the domain?
(c) Estimate the number of megawatts used at 8 A.M.
(d) At what time was the most electricity used? the least electricity?
(e) Call this function f. What is $f(12)$? What does it mean?
(f) During what time intervals is electricity usage increasing? decreasing?

80. ***Height of a Ball*** A ball is thrown straight up into the air. The function defined by $y = h(t)$ in the graph gives the height of the ball (in feet) at t seconds. (*Note:* The graph does *not* show the path of the ball. The ball is rising straight up and then falling straight down.)

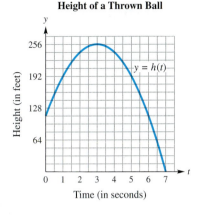

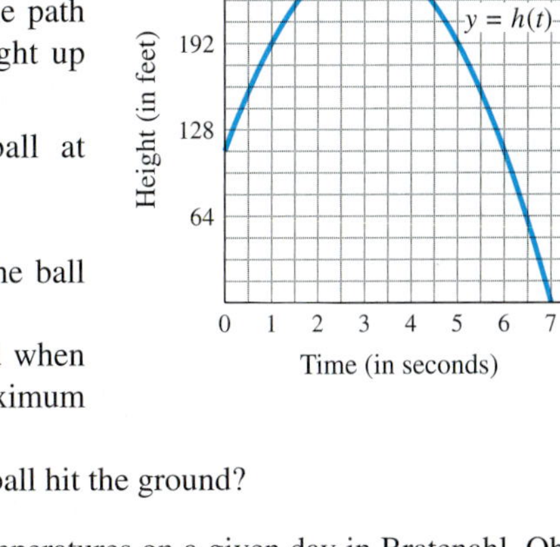

(a) What is the height of the ball at 2 sec?
(b) When will the height be 192 ft?
(c) During what time intervals is the ball going up? down?
(d) How high does the ball go, and when does the ball reach its maximum height?
(e) At how many seconds does the ball hit the ground?

81. ***Temperature*** The graph shows temperatures on a given day in Bratenahl, Ohio.

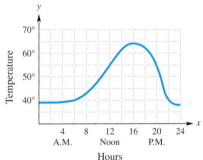

Temperature in Bratenahl, Ohio

(a) At what times during the day was the temperature over 55°?
(b) When was the temperature below 40°?
(c) Greenville, South Carolina, is 500 mi south of Bratenahl, Ohio, and its temperature is 7° higher all day long. At what time was the temperature in Greenville the same as the temperature at noon in Bratenahl?

82. ***Drug Levels in the Bloodstream*** When a drug is taken orally, the amount of the drug in the bloodstream at t hours is given by the function defined by $y = f(t)$, as shown in the graph.

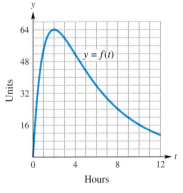

Drug Levels in the Bloodstream

(a) How many units of the drug are in the bloodstream at 8 hr?
(b) During what time interval is the drug level in the bloodstream increasing? decreasing?
(c) When does the level of the drug in the bloodstream reach its maximum value, and how many units are in the bloodstream at that time?
(d) When the drug reaches its maximum level in the bloodstream, how many additional hours are required for the level to drop to 16 units?

2.3 Linear Functions

Graphing Linear Functions ▪ Standard Form $Ax + By = C$ ▪ Slope ▪ Average Rate of Change ▪ Linear Models

Graphing Linear Functions We begin our study of specific functions by looking at *linear functions.* The name "*line*ar" comes from the fact that the graph of every linear function is a straight line.

Linear Function

A function f is a **linear function** if, for real numbers a and b,

$$f(x) = ax + b.$$

If $a \neq 0$, the domain and the range of a linear function are both $(-\infty, \infty)$. If $a = 0$, then the equation becomes $f(x) = b$. In this case, the domain is $(-\infty, \infty)$ and the range is $\{b\}$.

In Section 2.1, we graphed lines by finding ordered pairs and plotting them. Although only two points are necessary to graph a linear function, we usually plot a third point as a check. The intercepts are often good points to choose for graphing lines.

EXAMPLE 1 Graphing a Linear Function Using Intercepts

Graph $f(x) = -2x + 6$. Give the domain and range.

Solution The x-intercept is found by letting $f(x) = 0$ and solving for x.

$$0 = -2x + 6$$
$$3 = x$$

The x-intercept is 3, so we plot $(3, 0)$. The y-intercept is

$$f(0) = -2(0) + 6 = 6.$$

Therefore another point on the graph is $(0, 6)$. We plot this point, and join the two points with a straight line to get the graph. We use the point $(2, 2)$ as a check. See Figure 28. The domain and the range are both $(-\infty, \infty)$.

The corresponding calculator graph is shown in Figure 29.

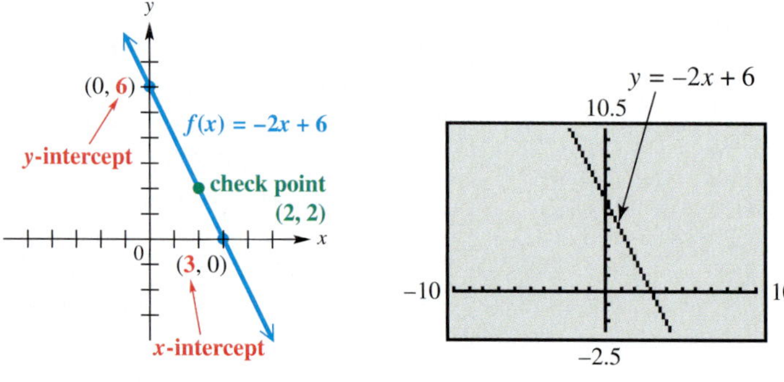

Figure 28

Figure 29

Now try Exercise 9.

A function of the form $f(x) = b$ is called a **constant function,** and its graph is a horizontal line.

EXAMPLE 2 Graphing a Horizontal Line

Graph $f(x) = -3$. Give the domain and range.

Solution Since $f(x)$, or y, always equals -3, the value of y can never be 0. This means that the graph has no x-intercept. The only way a straight line can have no x-intercept is for it to be parallel to the x-axis, as shown in Figure 30. Notice that the domain of this linear function is $(-\infty, \infty)$; the range is $\{-3\}$.

Figure 31 shows the calculator graph.

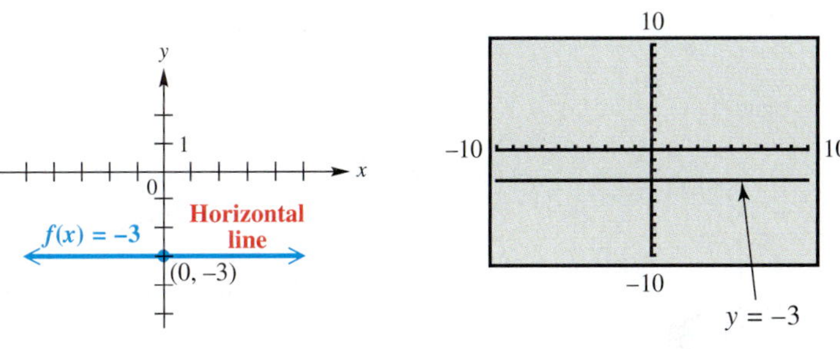

Figure 30 Figure 31

Now try Exercise 17.

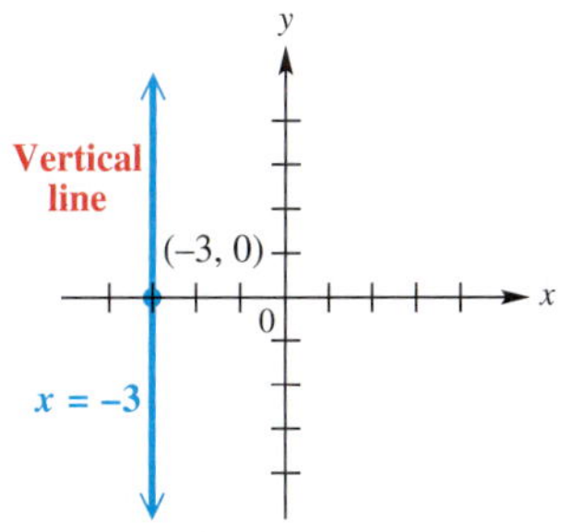

Figure 32

EXAMPLE 3 Graphing a Vertical Line

Graph $x = -3$. Give the domain and range of this relation.

Solution Since x always equals -3, the value of x can never be 0, and the graph has no y-intercept. Using reasoning similar to that of Example 2, we find that this graph is parallel to the y-axis, as shown in Figure 32. The domain of this relation, which is *not* a function, is $\{-3\}$, while the range is $(-\infty, \infty)$.

Now try Exercise 19.

Examples 2 and 3 illustrate that a linear function of the form $y = b$ has as its graph a horizontal line through $(0, b)$, and a relation of the form $x = a$ has as its graph a vertical line through $(a, 0)$.

Standard Form $Ax + By = C$ Equations of lines are often written in the form $Ax + By = C$, called **standard form.**

NOTE The definition of "standard form" is not standard from one text to another. Any linear equation can be written in many different but equivalent forms. For example, the equation $2x + 3y = 8$ can be written as

$$2x = 8 - 3y, \qquad 3y = 8 - 2x, \qquad x + \frac{3}{2}y = 4, \qquad 4x + 6y = 16,$$

and so on. In addition to writing it in the form $Ax + By = C$, with A, B, and C integers and $A \geq 0$, let us agree that the form $2x + 3y = 8$ is preferred over any multiples of both sides, such as $4x + 6y = 16$.

EXAMPLE 4 Graphing $Ax + By = C$ with $C = 0$

Graph $4x - 5y = 0$. Give the domain and range.

Solution Find the intercepts.

$4(0) - 5y = 0$ Let $x = 0$.	$4x - 5(0) = 0$ Let $y = 0$.
$y = 0$ y-intercept	$x = 0$ x-intercept

The graph of this function has just one intercept—at the origin $(0, 0)$. We need to find an additional point to graph the function by choosing a different value for x (or y). Choosing $x = 5$ gives

$$4(5) - 5y = 0$$
$$20 - 5y = 0$$
$$4 = y,$$

which leads to the ordered pair $(5, 4)$. Complete the graph using the two points $(0, 0)$ and $(5, 4)$, with a third point as a check. The domain and range are both $(-\infty, \infty)$. See Figure 33.

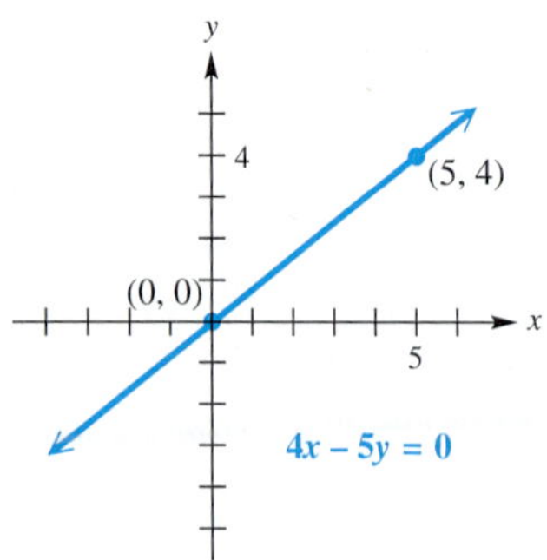

Figure 33

Now try Exercise 13.

To use a graphing calculator to graph a linear function given in standard form, we must first solve the defining equation for y. Solving $4x - 5y = 0$ for y, we get $y = \frac{4}{5}x$, so we graph $Y_1 = \frac{4}{5}X$. See Figure 34. ■

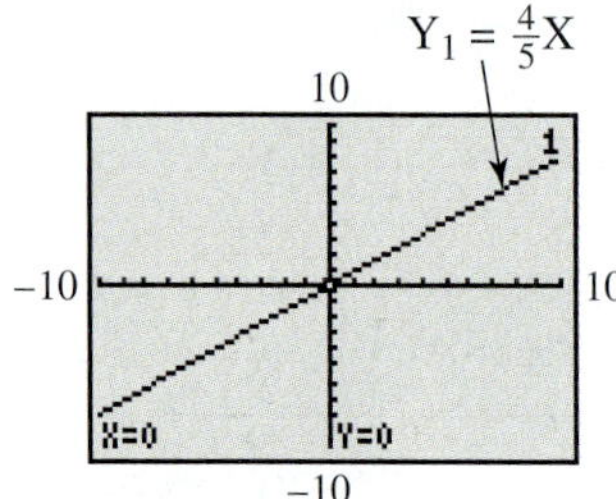

Figure 34

Slope An important characteristic of a straight line is its *slope,* a numerical measure of the steepness of the line. (Geometrically, this may be interpreted as the ratio of *rise* to *run.*) The slope of a highway (sometimes called the *grade*) is often given as a percent. For example, a 10% $\left(\text{or } \frac{10}{100} = \frac{1}{10}\right)$ slope means the highway rises 1 unit for every 10 horizontal units. Roofs have slopes, as shown in the figure.

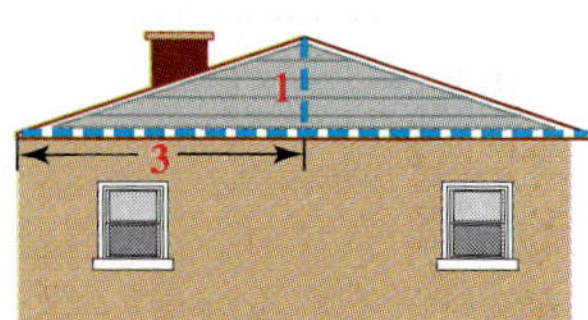

Slope (or pitch) is $\frac{1}{3}$.

To find the slope of a line, start with two distinct points (x_1, y_1) and (x_2, y_2) on the line, as shown in Figure 35, where $x_1 \neq x_2$. As we move along the line from (x_1, y_1) to (x_2, y_2), the horizontal difference

$$x_2 - x_1$$

is called the **change in *x*,** denoted by Δx (read "delta x"), where Δ is the Greek letter *delta.* In the same way, the vertical difference, called the **change in *y*,** can be written

$$\Delta y = y_2 - y_1.$$

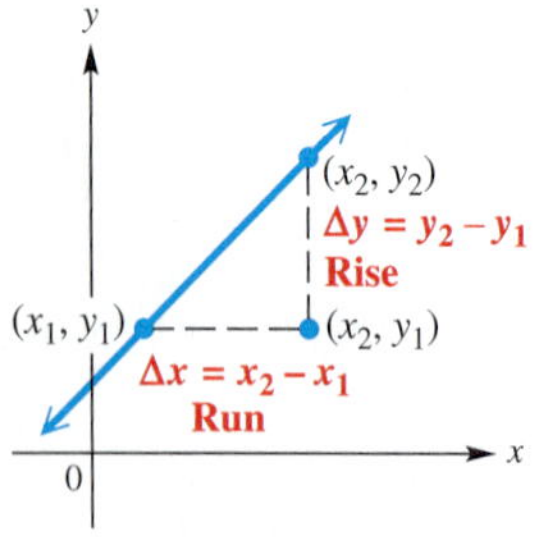

Figure 35

The *slope* of a nonvertical line is defined as the quotient (ratio) of the change in y and the change in x, as follows.

Slope

The **slope** m of the line through the points (x_1, y_1) and (x_2, y_2) is

$$m = \frac{\text{rise}}{\text{run}} = \frac{\Delta y}{\Delta x} = \frac{y_2 - y_1}{x_2 - x_1}, \quad \text{where } \Delta x \neq 0.$$

Looking Ahead to Calculus

The concept of slope of a line is extended in calculus to general curves. The *slope of a curve at a point* is understood to mean the slope of the line tangent to the curve at that point.

The line in the figure is tangent to the curve at point P.

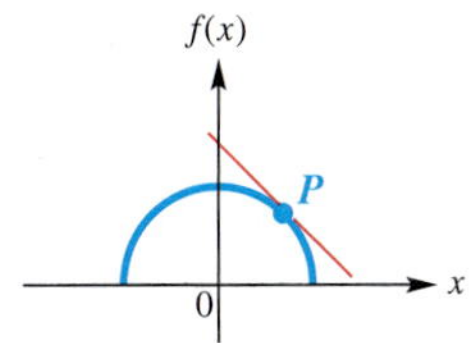

That is, the slope of a line is the change in y divided by the corresponding change in x, where the change in x is not 0.

CAUTION When using the slope formula, it makes no difference which point is (x_1, y_1) or (x_2, y_2); however, it is important to be consistent. Start with the x- and y-values of *one* point (either one) and subtract the corresponding values of the *other* point.

$$\frac{y_2 - y_1}{x_2 - x_1} \quad \text{or} \quad \frac{y_1 - y_2}{x_1 - x_2} \qquad \textbf{not} \qquad \frac{y_2 - y_1}{x_1 - x_2} \quad \text{or} \quad \frac{y_1 - y_2}{x_2 - x_1}$$

Be sure to put the difference of the y-values in the *numerator* and the difference of the x-values in the *denominator.*

The slope of a line can be found only if the line is nonvertical. This guarantees that $x_2 \neq x_1$ so that the denominator $x_2 - x_1 \neq 0$. It is not possible to define the slope of a vertical line.

> The slope of a vertical line is undefined.

EXAMPLE 5 Finding Slopes with the Slope Formula

Find the slope of the line through the given points.

(a) $(-4, 8), (2, -3)$ **(b)** $(2, 7), (2, -4)$ **(c)** $(5, -3), (-2, -3)$

Solution

(a) Let $x_1 = -4$, $y_1 = 8$, and $x_2 = 2$, $y_2 = -3$. Then

$$\text{rise} = \Delta y = -3 - 8 = -11 \quad \text{and} \quad \text{run} = \Delta x = 2 - (-4) = 6.$$

The slope is

$$m = \frac{\text{rise}}{\text{run}} = \frac{\Delta y}{\Delta x} = \frac{-11}{6} = -\frac{11}{6}.$$

Note that we could subtract in the opposite order, letting $x_1 = 2$, $y_1 = -3$, and $x_2 = -4$, $y_2 = 8$.

$$m = \frac{8 - (-3)}{-4 - 2} = \frac{11}{-6} = -\frac{11}{6}$$

The same slope results.

(b) If we use the formula, we get

$$m = \frac{-4 - 7}{2 - 2} = \frac{-11}{0}. \quad \text{Undefined}$$

The formula is not valid here because $\Delta x = x_2 - x_1 = 2 - 2 = 0$. A sketch would show that the line through $(2, 7)$ and $(2, -4)$ is vertical. As mentioned above, the slope of a vertical line is undefined.

(c) For $(5, -3)$ and $(-2, -3)$,

$$m = \frac{-3 - (-3)}{-2 - 5} = \frac{0}{-7} = 0.$$

Now try Exercises 35, 39, and 41.

Looking Ahead to Calculus

The *derivative* of a function provides a formula for determining the slope of a line tangent to a curve. If the slope is positive for a given domain value, then the function is increasing at that point; if it is negative, then the function is decreasing, and if it is 0 on an interval, then the function is constant there.

Drawing a graph through the points in Example 5(c) would produce a horizontal line, which suggests the following generalization.

> The slope of a horizontal line is 0.

Theorems for similar triangles can be used to show that a line's slope is independent of the choice of points on the line. That is, the slope of a line is the same no matter which pair of distinct points on the line are used to find it.

EXAMPLE 6 Finding the Slope from an Equation

Find the slope of the line $y = -4x - 3$.

Solution Find any two ordered pairs that are solutions of the equation.

If $x = -2$, then $y = -4(-2) - 3 = 5$,

and if $x = 0$, then $y = -4(0) - 3 = -3$,

so two ordered pairs are $(-2, 5)$ and $(0, -3)$. The slope is

$$m = \frac{\text{rise}}{\text{run}} = \frac{-3 - 5}{0 - (-2)} = \frac{-8}{2} = -4.$$

Notice that the slope is the same as the coefficient of x in the equation. In Section 2.4, we show that this always happens if the equation is solved for y.

Now try Exercise 45.

Since the slope of a line is the ratio of vertical change (rise) to horizontal change (run), if we know the slope of a line and the coordinates of a point on the line, we can draw the graph of the line.

EXAMPLE 7 Graphing a Line Using a Point and the Slope

Graph the line passing through $(-1, 5)$ and having slope $-\frac{5}{3}$.

Solution First locate the point $(-1, 5)$ as shown in Figure 36. Since the slope of this line is $\frac{-5}{3}$, a change of -5 units vertically (that is, 5 units down) corresponds to a change of 3 units horizontally (that is, 3 units to the right). This gives a second point, $(2, 0)$, which can then be used to complete the graph.

Because $\frac{-5}{3} = \frac{5}{-3}$, another point could be obtained by starting at $(-1, 5)$ and moving 5 units *up* and 3 units to the *left*. We would reach a different second point, but the graph would be the same. Confirm this in Figure 36.

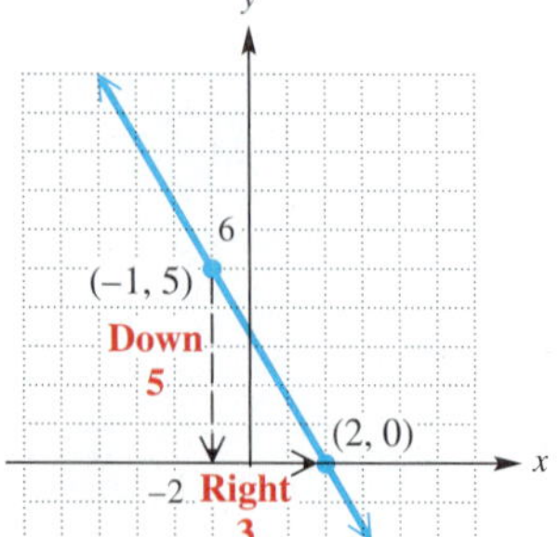

Figure 36

Now try Exercise 53.

Figure 37 shows lines with various slopes. Notice that a line with a positive slope rises from left to right, but a line with a negative slope falls from left to right. When the slope is positive, the function is increasing, and when the slope is negative, the function is decreasing.

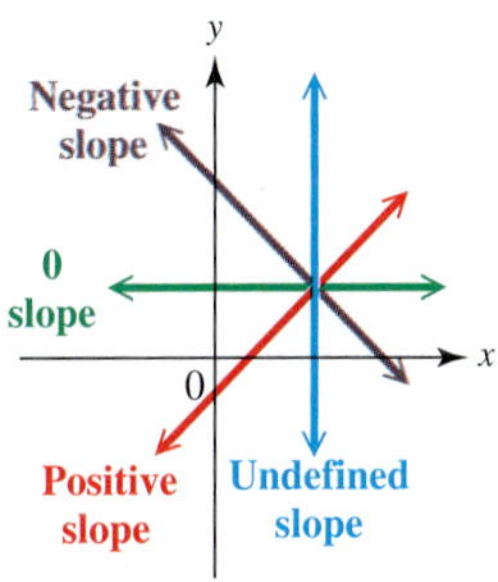

Figure 37

Average Rate of Change We know that the slope of a line is the ratio of the vertical change in y to the horizontal change in x. Thus, slope gives the *average rate of change* in y per unit of change in x, where the value of y depends on the value of x. The next example illustrates this idea. We assume a linear relationship between x and y.

EXAMPLE 8 Interpreting Slope as Average Rate of Change

In 1997, sales of VCRs numbered 16.7 million. In 2002, estimated sales of VCRs were 13.3 million. Find the average rate of change in VCR sales, in millions, per year. (*Source: The Gazette,* June 22, 2002.)

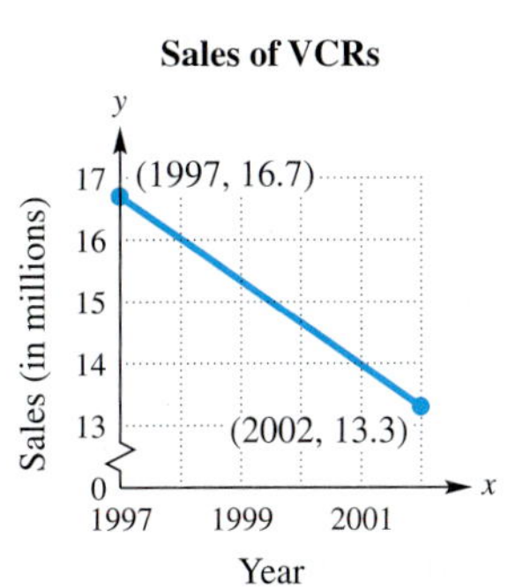

Figure 38

Solution To use the slope formula, we need two ordered pairs. Here, if $x = 1997$, then $y = 16.7$ and if $x = 2002$, then $y = 13.3$, which gives the ordered pairs (1997, 16.7) and (2002, 13.3). (Note that y is in millions.)

$$\text{average rate of change} = \frac{13.3 - 16.7}{2002 - 1997} = \frac{-3.4}{5} = -.68$$

The graph in Figure 38 confirms that the line through the ordered pairs falls from left to right and therefore has negative slope. Thus, sales of VCRs *decreased* by an average of .68 million each year from 1997 to 2002.

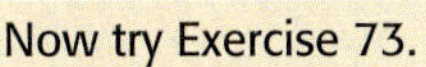

Linear Models In Example 8, we used the graph of a line to approximate real data. Recall from Section 1.2 that this is called *mathematical modeling.* Points on the straight line graph model (approximate) the actual points that correspond to the data.

A *linear cost function* has the form

$$C(x) = mx + b,$$

where x represents the number of items produced, m represents the *variable cost* per item, and b represents the *fixed cost.* The fixed cost is constant for a particular product and does not change as more items are made. The variable cost per item, which increases as more items are produced, covers labor, materials, packaging, shipping, and so on. The *revenue function* for selling a product depends on the price per item p and the number of items sold x, and is given by

$$R(x) = px.$$

Profit is described by the *profit function* defined as

$$P(x) = R(x) - C(x).$$

EXAMPLE 9 Writing Linear Cost, Revenue, and Profit Functions

Assume that the cost to produce an item is a linear function and all items produced are sold. The fixed cost is \$1500, the variable cost per item is \$100, and the item sells for \$125. Write linear functions to model

(a) cost, **(b)** revenue, and **(c)** profit.

(d) How many items must be sold for the company to make a profit?

Solution

(a) Since the cost function is linear, it will have the form $C(x) = mx + b$, with $m = 100$ and $b = 1500$. That is,

$$C(x) = 100x + 1500.$$

(b) The revenue function is

$$R(x) = px = 125x. \quad \text{Let } p = 125.$$

(c) Using the cost and revenue functions found in parts (a) and (b), the profit function is given by

$$\begin{aligned} P(x) &= R(x) - C(x) \\ &= 125x - (100x + 1500) \\ &= 125x - 100x - 1500 \\ &= 25x - 1500. \end{aligned}$$

Algebraic Solution

(d) To make a profit, $P(x)$ must be positive. From part (c),

$$P(x) = 25x - 1500.$$

Thus,

$$\begin{aligned} P(x) &> 0 \\ 25x - 1500 &> 0 \quad && P(x) = 25x - 1500 \\ 25x &> 1500 \quad && \text{Add 1500 to each side. (Section 1.7)} \\ x &> 60. \quad && \text{Divide by 25.} \end{aligned}$$

Since the number of items must be a whole number, at least 61 items must be sold for the company to make a profit.

Graphing Calculator Solution

(d) Define Y_1 as $25X - 1500$ and graph the line. Use the capability of your calculator to locate the x-intercept. See Figure 39. As the graph shows, y-values for x less than 60 are negative, and y-values for x greater than 60 are positive, so at least 61 items must be sold for the company to make a profit.

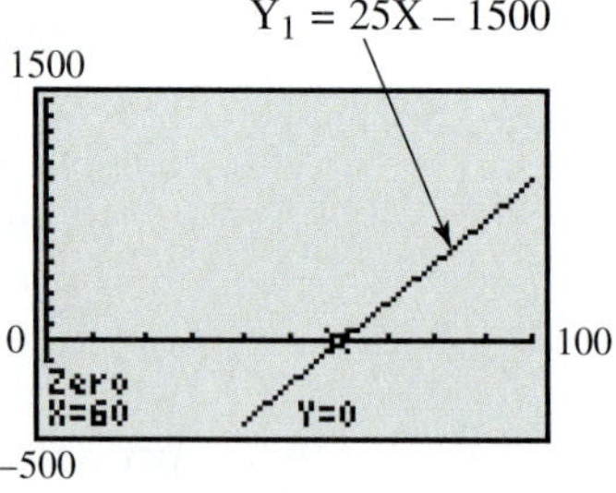

Figure 39

Now try Exercise 87.

2.3 Exercises

Concept Check *Match the description in Column I with the correct response in Column II. Some choices may not be used.*

I	II
1. a linear function whose graph has y-intercept 6	**A.** $f(x) = 2x$
2. a vertical line	**B.** $f(x) = 2x + 6$
3. a constant function	**C.** $f(x) = 7$
4. a linear function whose graph has x-intercept -2 and y-intercept 4	**D.** $f(x) = x^2$
5. a linear function whose graph passes through the origin	**E.** $x + y = 4$
6. a function that is not linear	**F.** $f(x) = 3x + 7$
	G. $2x - y = -4$
	H. $x = 3$

Graph each linear function. Identify any constant functions. Give the domain and range. See Examples 1, 2, and 4.

7. $f(x) = x - 4$
8. $f(x) = -x + 4$
9. $f(x) = \frac{1}{2}x - 6$
10. $f(x) = \frac{2}{3}x + 2$
11. $-4x + 3y = 9$
12. $2x + 5y = 10$
13. $3y - 4x = 0$
14. $3x + 2y = 0$
15. $f(x) = 3x$
16. $f(x) = -2x$
17. $f(x) = -4$
18. $f(x) = 3$

Graph each vertical line. Give the domain and range of the relation. See Example 3.

19. $x = 3$
20. $x = -4$
21. $2x + 4 = 0$
22. $-3x + 6 = 0$
23. $-x + 5 = 0$
24. $3 + x = 0$

Match each equation with the sketch that most closely resembles its graph. See Examples 2 and 3.

25. $y = 2$
26. $y = -2$
27. $x = 2$
28. $x = -2$

A.

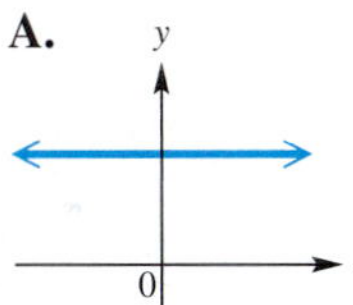

B.

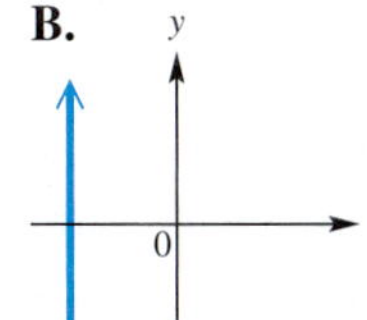

C.

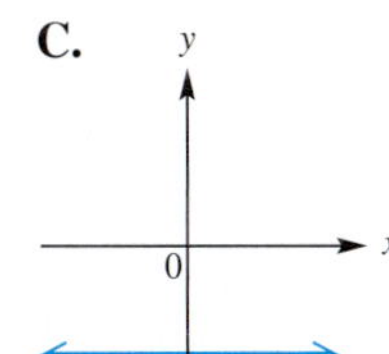

D.

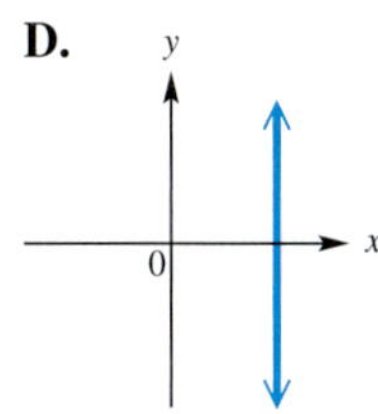

Use a graphing calculator to graph each equation in the standard viewing window. See Examples 1–4.

29. $y = 3x + 4$
30. $y = -2x + 3$
31. $3x + 4y = 6$
32. $-2x + 5y = 10$

33. *Concept Check* If a walkway rises 2.5 ft for every 10 ft on the horizontal, which of the following express its slope (or grade)? (There are several correct choices.)

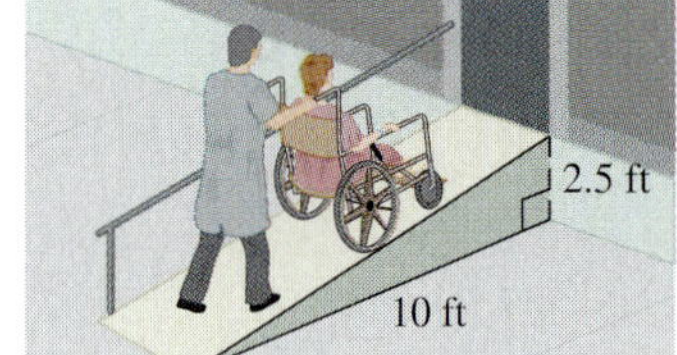

A. .25 B. 4 C. $\frac{2.5}{10}$ D. 25%

E. $\frac{1}{4}$ F. $\frac{10}{2.5}$ G. 400% H. 2.5%

34. *Concept Check* If the pitch of a roof is $\frac{1}{4}$, how many feet in the horizontal direction correspond to a rise of 4 ft?

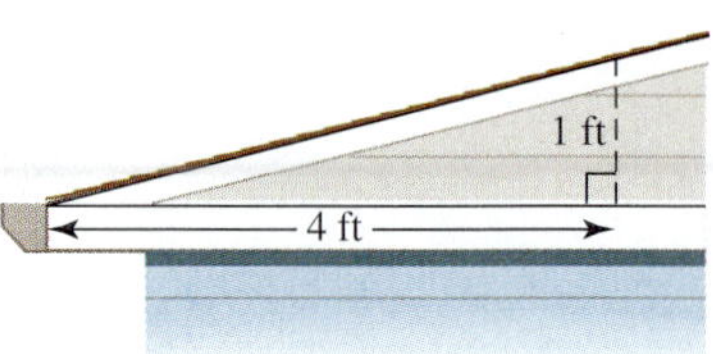

Find the slope of the line satisfying the given conditions. See Example 5.

35. through $(2, -1)$ and $(-3, -3)$
36. through $(5, -3)$ and $(1, -7)$
37. through $(5, 9)$ and $(-2, 9)$
38. through $(-2, 4)$ and $(6, 4)$
39. horizontal, through $(3, -7)$
40. horizontal, through $(-6, 5)$
41. vertical, through $(3, -7)$
42. vertical, through $(-6, 5)$

43. Which of the following forms of the slope formula are correct? Explain.

A. $m = \frac{y_1 - y_2}{x_2 - x_1}$ B. $m = \frac{y_1 - y_2}{x_1 - x_2}$ C. $m = \frac{y_2 - y_1}{x_2 - x_1}$ D. $m = \frac{x_2 - x_1}{y_2 - y_1}$

44. Can the graph of a linear function have undefined slope? Explain.

Find the slope of each line and sketch the graph. See Example 6.

45. $y = 3x + 5$
46. $y = 2x - 4$
47. $2y = -3x$
48. $-4y = 5x$
49. $5x - 2y = 10$
50. $4x + 3y = 12$

51. Explain in your own words what is meant by the slope of a line.

52. Explain how to graph a line using a point on the line and the slope of the line.

Graph the line passing through the given point and having the indicated slope. Plot two points on the line. See Example 6.

53. through $(-1, 3)$, $m = \frac{3}{2}$
54. through $(-2, 8)$, $m = -1$
55. through $(3, -4)$, $m = -\frac{1}{3}$
56. through $(-2, -3)$, $m = -\frac{3}{4}$
57. through $\left(-\frac{1}{2}, 4\right)$, $m = 0$
58. through $\left(\frac{9}{4}, 2\right)$, undefined slope

Concept Check *For each given slope in Exercises 59–64, identify the line in A–F at the top of the next page having that slope.*

59. $\frac{1}{2}$
60. -2
61. 0
62. $-\frac{1}{2}$
63. 2
64. undefined

A.

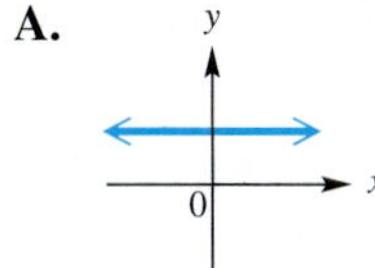

B.

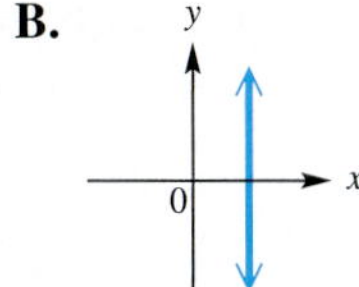

C.

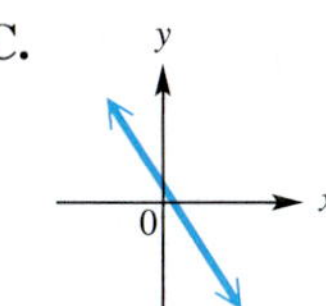

D.

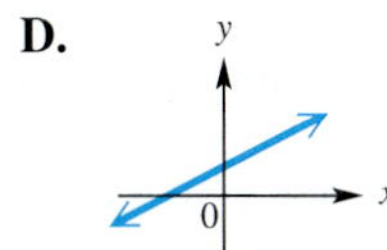

E.

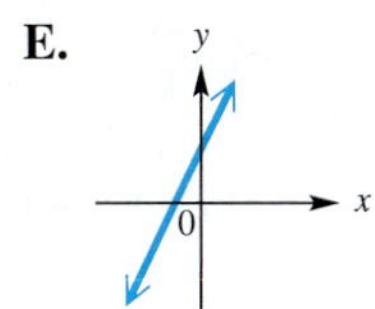

F.

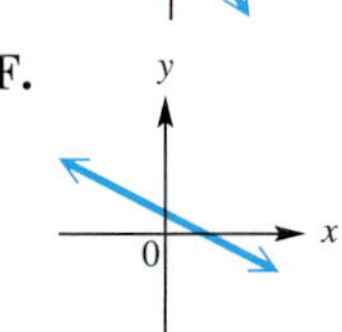

Concept Check *Find and interpret the average rate of change illustrated in each graph.*

65.

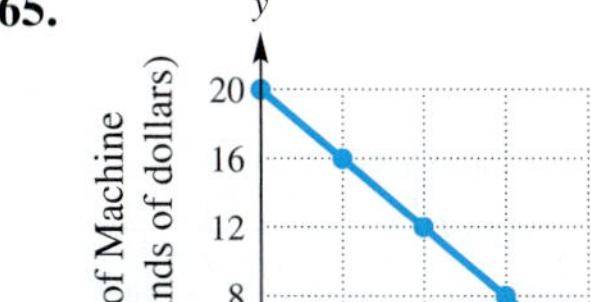

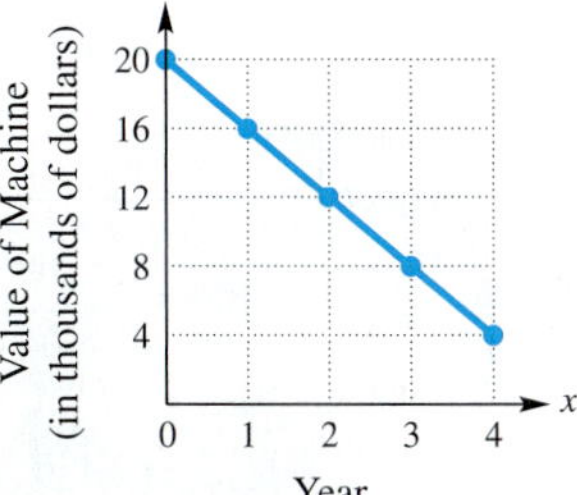

66.

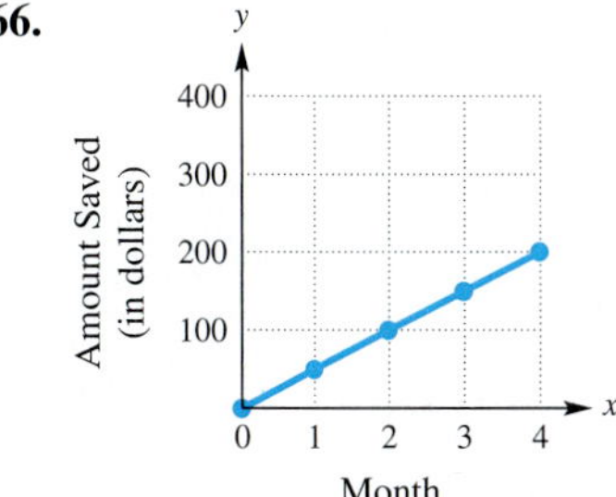

67.

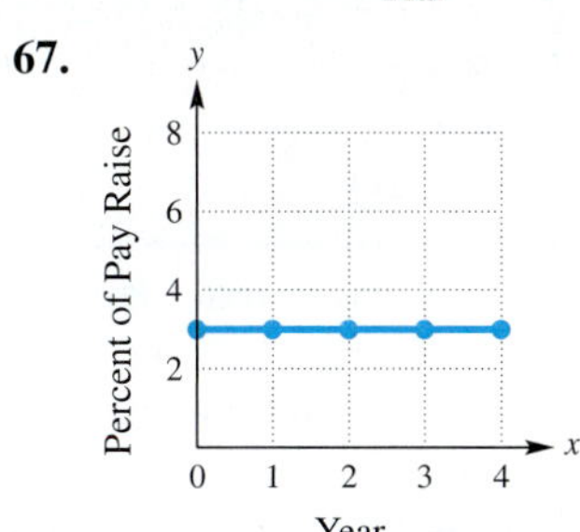

Solve each problem. See Example 8.

68. ***(Modeling) Olympic Times for 5000 Meter Run*** The graph shows the winning times (in minutes) at the Olympic Games for the men's 5000 m run together with a linear approximation of the data.

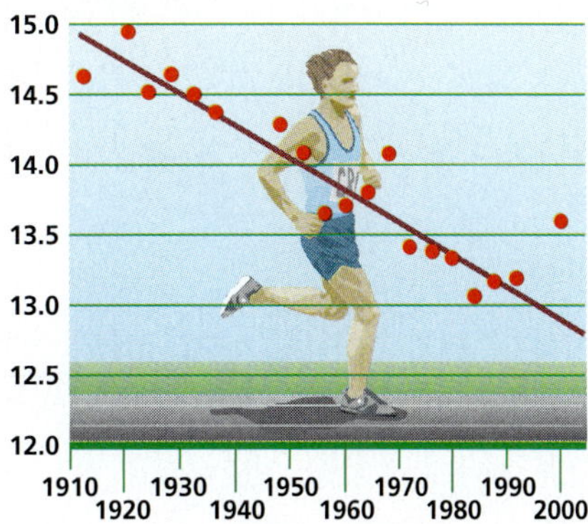

Source: United States Olympic Committee.

(continued)

(a) An equation for the linear model, based on data from 1912–2000 (where x represents the year), is

$$y = -.0187x + 50.60.$$

Determine the slope. (See Example 6.) What does the slope of this line represent? Why is the slope negative?

(b) Can you think of any reason why there are no data points for the years 1940 and 1944?

(c) The winning time for the 1996 Olympic Games was 13.13 min. What does the model predict? How far is the prediction from the actual value?

69. *(Modeling) U.S. Radio Stations* The graph shows the number of U.S. radio stations on the air along with the graph of a linear function that models the data.

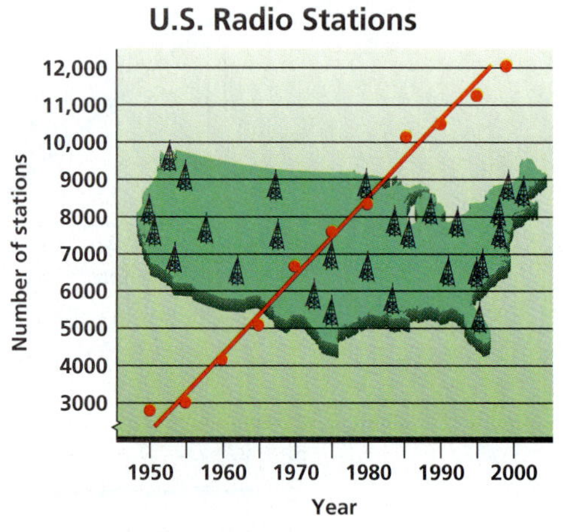

Source: National Association of Broadcasters.

(a) Discuss the predictive accuracy of the linear function.

(b) Use the two data points (1950, 2773) and (1999, 12,057) to find the approximate slope of the line shown. Interpret this number.

70. *Cellular Telephone Subscribers* The table gives the number of cellular telephone subscribers (in thousands) from 1994 through 1999.

(a) Find the change in subscribers for 1994–1995, 1995–1996, and so on.

(b) Is the change in successive years approximately the same? If the ordered pairs in the table were plotted, could an approximately straight line be drawn through them?

Year	Subscribers (in thousands)
1994	24,134
1995	33,786
1996	44,043
1997	55,312
1998	69,209
1999	86,047

Source: Cellular Telecommunications Industry Association, Washington, D.C., *State of the Cellular Industry* (Annual).

71. *Food Stamp Recipients* The graph provides a good approximation of the number of food stamp recipients (in millions) from 1994 through 1998.

(a) Use the given ordered pairs to find the average rate of change in food stamp recipients per year during this period.

(b) Interpret what a negative slope means in this situation.

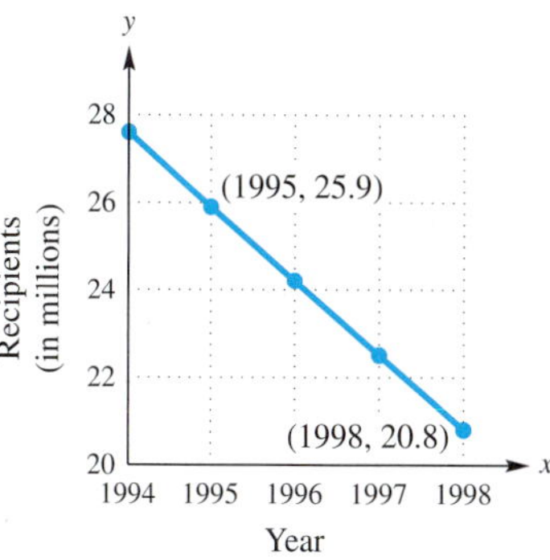

Source: U.S. Bureau of the Census.

72. *Olympic Times for 5000 Meter Run* Exercise 68 showed the winning times for the Olympic men's 5000 m run. Find and interpret the average rate of change for the following periods.

(a) The winning time in 1912 was 14.6 min; in 1996 it was 13.1 min.

(b) The winning time in 1912 was 14.6 min; in 2000 it was 13.6 min.

73. *College Freshmen Studying Business* In 1995, 15% of all U.S. college freshmen listed business as their probable field of study. By 2000, this figure had increased to 16.7%. Find and interpret the average rate of change in the percent per year of freshmen listing business as their probable field of study. (*Source:* Higher Education Research Institute.)

74. *Price of DVD Players* When introduced in 1997, a DVD player sold for about \$500. In 2002, the average price was \$155. Find and interpret the average rate of change in price per year. (*Source: The Gazette,* June 22, 2002.)

75. *Sales of DVD Players* In 1997, .349 million (that is, 349,000) DVD players were sold. In 2002, sales of DVD players reached 15.5 million. Find and interpret the average rate of change in sales, in millions, per year. Round your answer to the nearest hundredth. (*Source: The Gazette,* June 22, 2002.)

Relating Concepts

For individual or collaborative investigation

(Exercises 76–85)

The table shows several points on the graph of a linear function. ***Work Exercises 76–85 in order,*** *to see connections between the slope formula, the distance formula, the midpoint formula, and linear functions.*

x	y
0	−6
1	−3
2	0
3	3
4	6
5	9
6	12

76. Use the first two points in the table to find the slope of the line.

77. Use the second and third points in the table to find the slope of the line.

78. Make a conjecture by filling in the blank: If we use any two points on a line to find its slope, we find that the slope is ______ in all cases.

(continued)

x	y
0	−6
1	−3
2	0
3	3
4	6
5	9
6	12

79. Use the distance formula to find the distance between the first two points in the table.

80. Use the distance formula to find the distance between the second and fourth points in the table.

81. Use the distance formula to find the distance between the first and fourth points in the table.

82. Add the results in Exercises 79 and 80, and compare the sum to the answer you found in Exercise 81. What do you notice?

83. Fill in the blanks, basing your answers on your observations in Exercises 79–82: If points A, B, and C lie on a line in that order, then the distance between A and B added to the distance between _____ and _____ is equal to the distance between _____ and _____.

84. Use the midpoint formula to find the midpoint of the segment joining $(0, -6)$ and $(6, 12)$. Compare your answer to the middle entry in the table. What do you notice?

85. If the table were set up to show an x-value of 4.5, what would be the corresponding y-value?

(Modeling) Cost, Revenue, and Profit Analysis *A firm will break even (no profit and no loss) as long as revenue just equals cost. The value of x (the number of items produced and sold) where $C(x) = R(x)$ is called the* ***break-even point.*** *Assume that each of the following can be expressed as a linear function. Find*

(a) *the cost function,* ***(b)*** *the revenue function, and* ***(c)*** *the profit function.*

(d) *Find the break-even point and decide whether the product should be produced based on the restrictions on sales.*

See Example 9.

	Fixed Cost	Variable Cost	Price of Item	
86.	\$500	\$10	\$35	No more than 18 units can be sold.
87.	\$180	\$11	\$20	No more than 30 units can be sold.
88.	\$2700	\$150	\$280	No more than 25 units can be sold.
89.	\$1650	\$400	\$305	All units produced can be sold.

90. ***(Modeling) Break-Even Point*** The manager of a small company that produces roof tile has determined that the total cost in dollars, $C(x)$, of producing x units of tile is given by

$$C(x) = 200x + 1000,$$

while the revenue in dollars, $R(x)$, from the sale of x units of tile is given by

$$R(x) = 240x.$$

(a) Find the break-even point.
(b) What are the cost and revenue at the break-even point?
(c) Suppose the manager finds he has miscalculated his variable cost, which is actually \$220 per unit, instead of \$200. How does this affect the break-even point? Is he better off or not?

2.4 Equations of Lines; Curve Fitting

Point-Slope Form ▪ **Slope-Intercept Form** ▪ **Vertical and Horizontal Lines** ▪ **Parallel and Perpendicular Lines** ▪ **Modeling Data** ▪ **Solving Linear Equations in One Variable by Graphing**

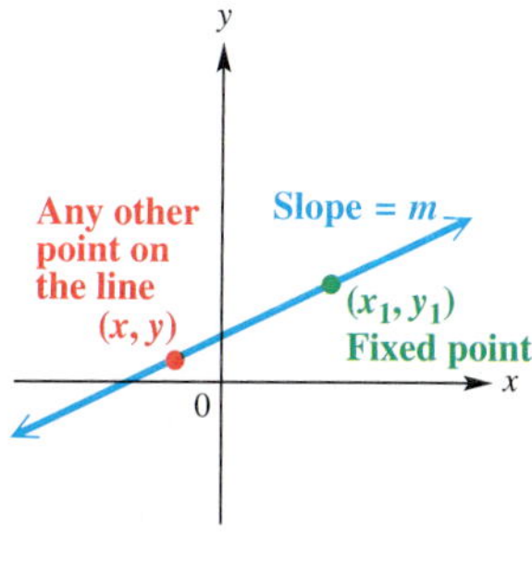

Figure 40

Point-Slope Form

In the previous section we saw that the graph of a linear function is a straight line. In this section we develop various forms for the equation of a line. Figure 40 shows the line passing through the fixed point (x_1, y_1) having slope m. (Assuming that the line has a slope guarantees that it is not vertical.) Let (x, y) be any other point on the line. Since the line is not vertical, $x - x_1 \neq 0$. By the definition of slope, the slope of this line is

$$m = \frac{y - y_1}{x - x_1}$$

$$m(x - x_1) = y - y_1 \qquad \text{Multiply both sides by } x - x_1.$$

or

$$y - y_1 = m(x - x_1).$$

This result is called the *point-slope form* of the equation of a line.

Looking Ahead to Calculus

A standard problem in calculus is to find the equation of the line tangent to a curve at a given point. The derivative (see *Looking Ahead to Calculus* on page 218) is used to find the slope of the desired line, and then the slope and the given point are used in the point-slope form to solve the problem.

Point-Slope Form

The line with slope m passing through the point (x_1, y_1) has an equation

$$\mathbf{y - y_1 = m(x - x_1),}$$

the **point-slope form** of the equation of a line.

EXAMPLE 1 Using the Point-Slope Form (Given a Point and the Slope)

Write an equation of the line through $(-4, 1)$ having slope -3.

Solution Here $x_1 = -4$, $y_1 = 1$, and $m = -3$.

$$\begin{aligned} y - y_1 &= m(x - x_1) && \text{Point-slope form} \\ y - 1 &= -3[x - (-4)] && x_1 = -4,\ y_1 = 1,\ m = -3 \\ y - 1 &= -3(x + 4) && \\ y - 1 &= -3x - 12 && \text{Distributive property (Section R.1)} \\ y &= -3x - 11 && \text{Add 1. (Section 1.1)} \end{aligned}$$

Now try Exercise 5.

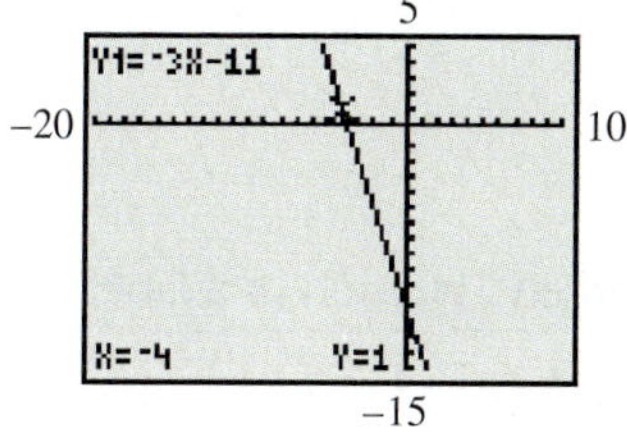

Figure 41

Figure 41 shows how a graphing calculator supports the result of Example 1. The display at the bottom of the screen indicates that the point $(-4, 1)$ lies on the graph of $y = -3x - 11$. We can verify that the slope is -3 using the discussion that follows Example 2. ■

EXAMPLE 2 Using the Point-Slope Form (Given Two Points)

Find an equation of the line through $(-3, 2)$ and $(2, -4)$.

Solution Find the slope first.

$$m = \frac{-4 - 2}{2 - (-3)} = -\frac{6}{5} \qquad \text{Definition of slope}$$

The slope m is $-\frac{6}{5}$. Either $(-3, 2)$ or $(2, -4)$ can be used for (x_1, y_1). We choose $(-3, 2)$.

$$y - y_1 = m(x - x_1) \quad \text{Point-slope form}$$

$$y - 2 = -\frac{6}{5}[x - (-3)] \quad x_1 = -3,\ y_1 = 2,\ m = -\tfrac{6}{5}$$

$$5(y - 2) = -6(x + 3) \quad \text{Multiply by 5.}$$

$$5y - 10 = -6x - 18 \quad \text{Distributive property}$$

$$y = -\frac{6}{5}x - \frac{8}{5} \quad \text{Add 10; divide by 5.}$$

Verify that we get the same equation if we use $(2, -4)$ instead of $(-3, 2)$ in the point-slope form.

Now try Exercise 11.

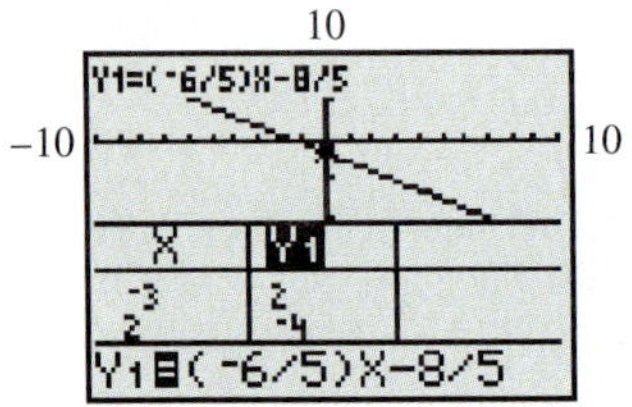

Figure 42

The screen in Figure 42 supports the result of Example 2. ■

Slope-Intercept Form As a special case of the point-slope form of the equation of a line, suppose that a line passes through the point $(0, b)$, so the line has y-intercept b. If the line has slope m, then using the point-slope form with $x_1 = 0$ and $y_1 = b$ gives

$$y - y_1 = m(x - x_1)$$

$$y - b = m(x - 0)$$

$$y = mx + b.$$

Slope ↑ (m) ↑ (b) y-intercept

Since this result shows the slope of the line and the y-intercept, it is called the *slope-intercept form* of the equation of the line. This is an important form, since linear functions are written this way.

Slope-Intercept Form

The line with slope m and y-intercept b has an equation

$$y = mx + b,$$

the **slope-intercept form** of the equation of a line.

EXAMPLE 3 Using the Slope-Intercept Form to Find an Equation of a Line

Find an equation of the line with slope $-\frac{4}{5}$ and y-intercept -2.

Solution Here $m = -\frac{4}{5}$ and $b = -2$. Substitute these values into the slope-intercept form.

$$y = mx + b \quad \text{Slope-intercept form}$$

$$y = -\frac{4}{5}x - 2 \quad m = -\tfrac{4}{5},\ b = -2$$

Now try Exercise 19.

Look again at Figure 41 on page 227. The line has equation $y = -3x - 11$, and thus we can conclude that its slope is -3. ■

EXAMPLE 4 Using the Slope-Intercept Form to Graph a Line

Find the slope and y-intercept of $3x - y = 2$. Then graph the line.

Solution First write $3x - y = 2$ in slope-intercept form, $y = mx + b$, by solving for y.

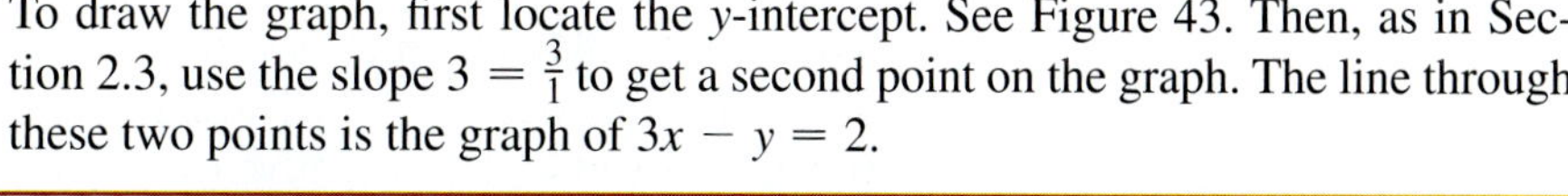

To draw the graph, first locate the y-intercept. See Figure 43. Then, as in Section 2.3, use the slope $3 = \frac{3}{1}$ to get a second point on the graph. The line through these two points is the graph of $3x - y = 2$.

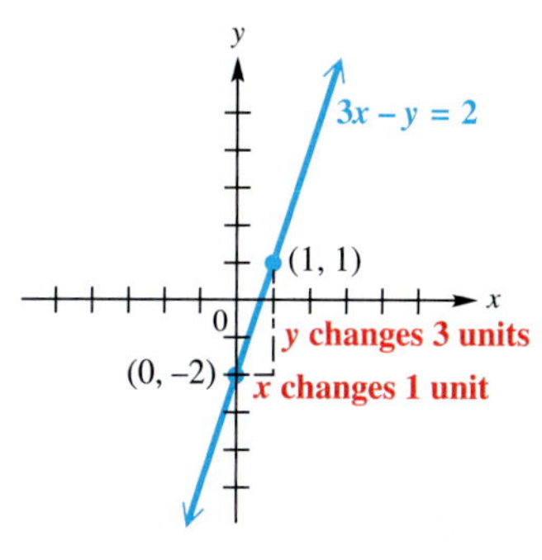

Figure 43

Now try Exercise 29.

Vertical and Horizontal Lines In the preceding discussion, we assumed that the given line had slope. The only lines having undefined slope are vertical lines. The vertical line through the point (a, b) passes through all points of the form (a, y), for any value of y. Consequently, the equation of a vertical line through (a, b) is $x = a$.

For example, the vertical line through $(-3, 9)$ has equation $x = -3$. See Figure 32 in Section 2.3 on page 215 for the graph of this equation. Since each point on the y-axis has x-coordinate 0, the equation of the y-axis is $x = 0$.

The horizontal line through the point (a, b) passes through all points of the form (x, b), for any value of x. Therefore, the equation of a horizontal line through (a, b) is $y = b$.

For example, the horizontal line through $(1, -3)$ has equation $y = -3$. See Figure 30 in Section 2.3 on page 215 for the graph of this equation. Since each point on the x-axis has y-coordinate 0, the equation of the x-axis is $y = 0$.

Equations of Vertical and Horizontal Lines

An equation of the vertical line through the point (a, b) is $\boldsymbol{x = a}$.
An equation of the horizontal line through the point (a, b) is $\boldsymbol{y = b}$.

Now try Exercises 15 and 21.

NOTE All the lines discussed above have equations that can be written in standard form $Ax + By = C$ for real numbers A, B, and C, where A and B are not both 0.

Parallel and Perpendicular Lines We use slope to decide whether or not two lines are parallel. Since two parallel lines are equally "steep," they should have the same slope. Also, two distinct lines with the same "steepness" are parallel. The following result summarizes this discussion.

Parallel Lines

Two distinct nonvertical lines are parallel if and only if they have the same slope.

Slopes are also used to determine whether two lines are perpendicular. Whenever two lines have slopes with a product of -1, the lines are perpendicular.

Perpendicular Lines

Two lines, neither of which is vertical, are perpendicular if and only if their slopes have a product of -1. Thus, the slopes of perpendicular lines, neither of which is vertical, are negative reciprocals.

For example, if the slope of a line is $-\frac{3}{4}$, the slope of any line perpendicular to it is $\frac{4}{3}$, since $-\frac{3}{4}\left(\frac{4}{3}\right) = -1$. (Numbers like $-\frac{3}{4}$ and $\frac{4}{3}$ are called *negative reciprocals* of each other.) A proof of this result is outlined in Exercises 58–65.

EXAMPLE 5 Finding Equations of Parallel and Perpendicular Lines

Find the equation in slope-intercept form of the line that passes through the point $(3, 5)$ and satisfies the given condition.

(a) parallel to the line $2x + 5y = 4$ **(b)** perpendicular to the line $2x + 5y = 4$

Solution

(a) Since we know that the point $(3, 5)$ is on the line, we need only find the slope to use the point-slope form. We find the slope by writing the equation of the given line in slope-intercept form. (That is, solve for y.)

$$2x + 5y = 4$$
$$y = -\frac{2}{5}x + \frac{4}{5}$$

The slope is $-\frac{2}{5}$. Since the lines are parallel, $-\frac{2}{5}$ is also the slope of the line whose equation is to be found. Substituting $m = -\frac{2}{5}$, $x_1 = 3$, and $y_1 = 5$ into the point-slope form gives

$$y - y_1 = m(x - x_1)$$
$$y - 5 = -\frac{2}{5}(x - 3)$$
$$y - 5 = -\frac{2}{5}x + \frac{6}{5} \quad \text{Distributive property}$$
$$y = -\frac{2}{5}x + \frac{31}{5}. \quad \text{Add } 5 = \tfrac{25}{5}.$$

(b) In part (a) we found that the slope of the line $2x + 5y = 4$ is $-\frac{2}{5}$, so the slope of any line perpendicular to it is $\frac{5}{2}$. Therefore, use $m = \frac{5}{2}$, $x_1 = 3$, and $y_1 = 5$ in the point-slope form.

$$y - y_1 = m(x - x_1)$$

$$y - 5 = \frac{5}{2}(x - 3)$$

$$y - 5 = \frac{5}{2}x - \frac{15}{2} \quad \text{Distributive property}$$

$$y = \frac{5}{2}x - \frac{5}{2} \quad \text{Add } 5 = \tfrac{10}{2}.$$

Now try Exercises 35 and 37.

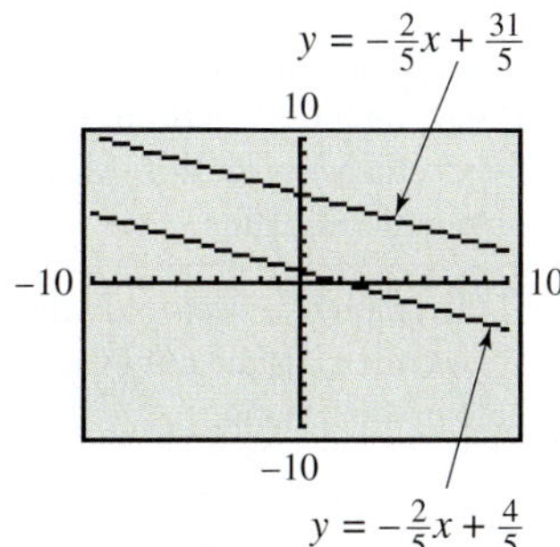

These lines are parallel.

(a)

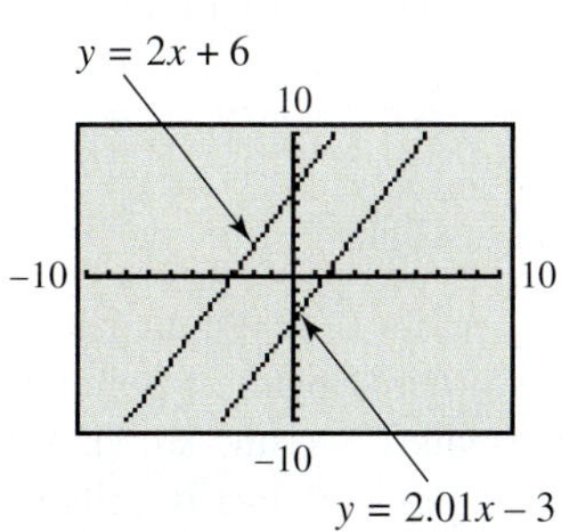

These lines are not parallel, but appear to be.

(b)

Figure 44

We can use a graphing calculator to support the results of Example 5. In Figure 44(a), we graph

$$y = -\frac{2}{5}x + \frac{4}{5} \quad \text{and} \quad y = -\frac{2}{5}x + \frac{31}{5}$$

from part (a). The lines appear to be parallel, giving visual support for our result. We must use caution, however, when viewing such graphs, as the limited resolution of a graphing calculator screen may cause two lines to *appear* to be parallel even when they are not. For example, Figure 44(b) shows the graphs of

$$y = 2x + 6 \quad \text{and} \quad y = 2.01x - 3$$

in the standard viewing window, and they appear to be parallel. This is not the case, however, because their slopes, 2 and 2.01, are different.

To support the result of part (b), we graph

$$y = -\frac{2}{5}x + \frac{4}{5} \quad \text{and} \quad y = \frac{5}{2}x - \frac{5}{2}.$$

If we use the standard viewing window, the lines do not appear to be perpendicular. See Figure 45(a). To obtain the correct perspective, we must use a square viewing window, as in Figure 45(b).

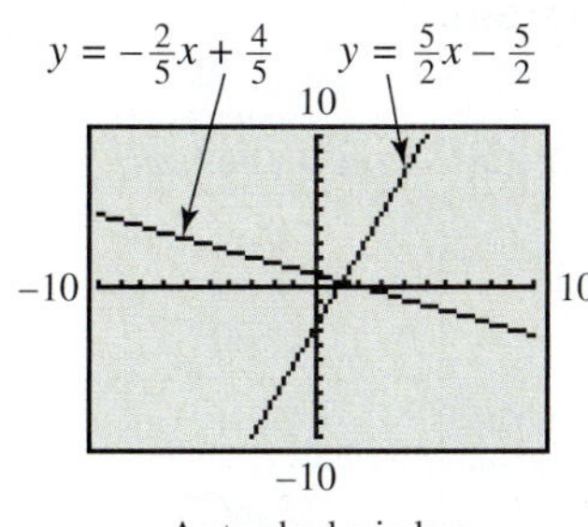

A standard window

(a)

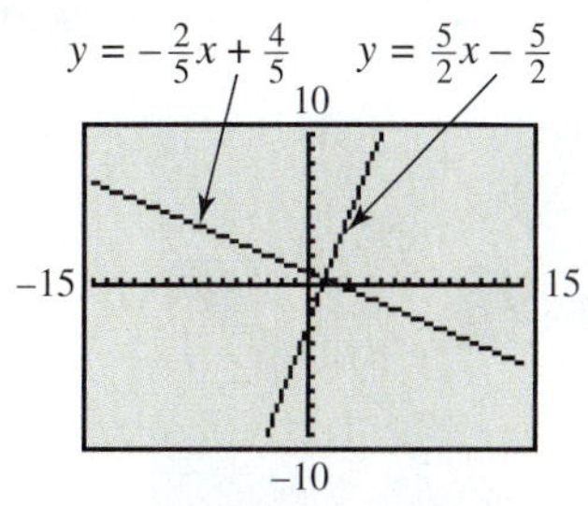

A square window

(b)

Figure 45

A summary of the various forms of linear equations follows.

Equation	Description	When to Use
$y = mx + b$	**Slope-Intercept Form** Slope is m. y-intercept is b.	The slope and y-intercept can be easily identified and used to quickly graph the equation.
$y - y_1 = m(x - x_1)$	**Point-Slope Form** Slope is m. Line passes through (x_1, y_1).	This form is ideal for finding the equation of a line if the slope and a point on the line or two points on the line are known.
$Ax + By = C$	**Standard Form** (A, B, and C integers, $A \geq 0$) Slope is $-\frac{A}{B}$ $(B \neq 0)$. x-intercept is $\frac{C}{A}$ $(A \neq 0)$. y-intercept is $\frac{C}{B}$ $(B \neq 0)$.	The x- and y-intercepts can be found quickly and used to graph the equation. Slope must be calculated.
$y = b$	**Horizontal Line** Slope is 0. y-intercept is b.	If the graph intersects only the y-axis, then y is the only variable in the equation.
$x = a$	**Vertical Line** Slope is undefined. x-intercept is a.	If the graph intersects only the x-axis, then x is the only variable in the equation.

Modeling Data We can use the information presented in this section to write equations of lines that mathematically describe, or *model,* real data if the given set of data changes at a fairly constant rate. In this case, the data fit a linear pattern, and the rate of change is the slope of the line.

EXAMPLE 6 Finding an Equation of a Line That Models Data

Average annual tuition and fees for in-state students at public 4-year colleges are shown in the table for selected years and graphed as ordered pairs of points in Figure 46, where $x = 0$ represents 1990, $x = 4$ represents 1994, and so on, and y represents the cost in dollars. This graph of ordered pairs of data is called a **scatter diagram.**

Year	Cost (in dollars)
1990	2035
1994	2820
1996	3151
1998	3486
2000	3774

Source: U.S. National Center for Education Statistics.

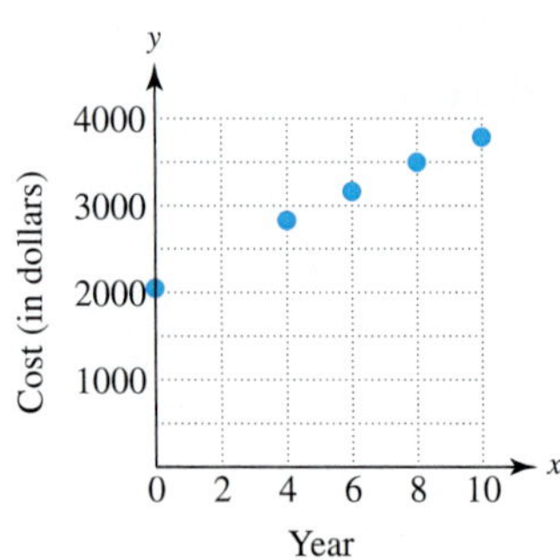

Figure 46

(a) Find an equation that models the data.

(b) Use the equation from part (a) to approximate the cost of tuition and fees at public 4-year colleges in 2002.

Solution

(a) Since the points in Figure 46 lie approximately on a straight line, we can write a linear equation that models the relationship between year x and cost y. We choose two data points, $(0, 2035)$ and $(10, 3774)$, to find the slope of the line.

$$m = \frac{3774 - 2035}{10 - 0} = \frac{1739}{10} = 173.9$$

The slope 173.9 indicates that the cost of tuition and fees for in-state students at public 4-year colleges increased by about \$174 per year from 1990 to 2000. We use this slope, the y-intercept 2035, and the slope-intercept form to write an equation of the line,

$$y = mx + b \qquad \text{Slope-intercept form}$$

$$y = 173.9x + 2035.$$

(b) The value $x = 12$ corresponds to the year 2002, so we substitute 12 for x.

$$y = 173.9x + 2035 \qquad \text{Model from part (a)}$$

$$y = 173.9(12) + 2035 \qquad \text{Let } x = 12.$$

$$y = 4121.8$$

According to the model, average tuition and fees for in-state students at public 4-year colleges in 2002 were about \$4122.

Now try Exercise 47(a) and (b).

NOTE In Example 6, if we had chosen different data points, we would have gotten a slightly different equation. However, all such equations should be similar.

EXAMPLE 7 Finding a Linear Equation that Models Data

The table below and graph in Figure 47 on the next page illustrate how the percent of women in the civilian labor force has changed from 1955 to 2000.

Year	1955	1960	1965	1970	1975	1980	1985	1990	1995	2000
% Women	35.7	37.7	39.3	43.3	46.3	51.5	54.5	57.5	58.9	60.0

Source: U.S. Bureau of Labor Statistics.

(a) Use the points $(1955, 35.7)$ and $(1995, 58.9)$ to find a linear equation that models the data.

(b) Use the equation to estimate the percent for 2000. How does the result compare to the actual figure of 60.0%?

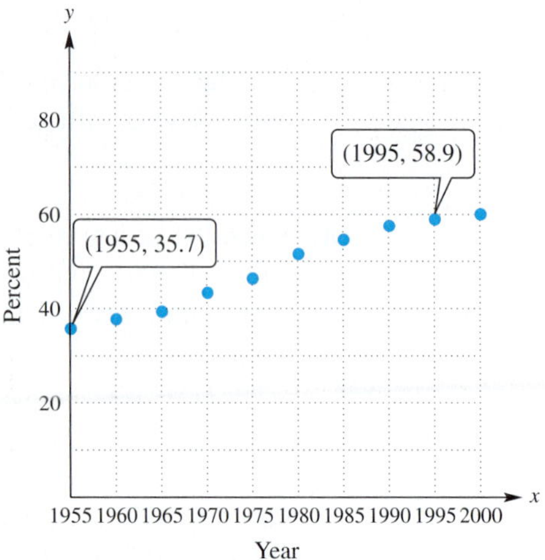

Source: U.S. Bureau of Labor Statistics.

Figure 47

Solution

(a) The slope must be positive, since the scatter diagram in Figure 47 rises from left to right. Using the two given points, the slope is

$$m = \frac{58.9 - 35.7}{1995 - 1955} = \frac{23.2}{40} = .58.$$

Now use either point, say (1955, 35.7), and the point-slope form to find an equation.

$$
\begin{aligned}
y - y_1 &= m(x - x_1) && \text{Point-slope form} \\
y - 35.7 &= .58(x - 1955) && m = .58,\ x_1 = 1955,\ y_1 = 35.7 \\
y - 35.7 &= .58x - 1133.9 && \text{Distributive property} \\
y &= .58x - 1098.2 && \text{Add 35.7.}
\end{aligned}
$$

(b) To use this equation to estimate the percent in 2000, let $x = 2000$ and solve for y.

$$
\begin{aligned}
y &= .58(2000) - 1098.2 \\
&\approx 61.8
\end{aligned}
$$

This figure of 61.8% is 1.8% greater than the actual figure of 60.0%.

Now try Exercise 45.

The steps for fitting a curve to a set of data are summarized here.

Curve Fitting

Step 1 Make a scatter diagram of the data.

Step 2 Find an equation that models the data. For a line, this involves selecting two data points and finding the equation of the line through them.

In Example 7, choosing a different pair of points would yield a slightly different linear equation. (See Exercises 45 and 46.) A technique from statistics called *linear regression* provides the line of "best fit." Figure 48 shows how a TI-83 Plus calculator can accept the data points, calculate the equation of this line of best fit (in this case, $y = .60230x - 1142.58424$), and plot both the data points and the line on the same screen.

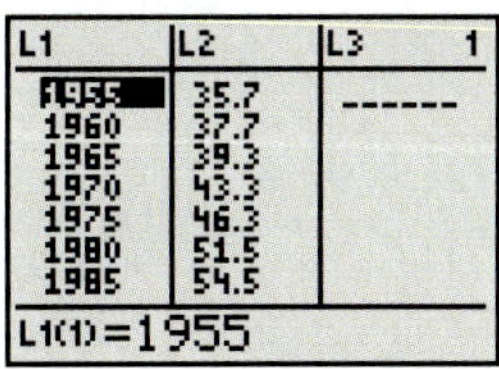

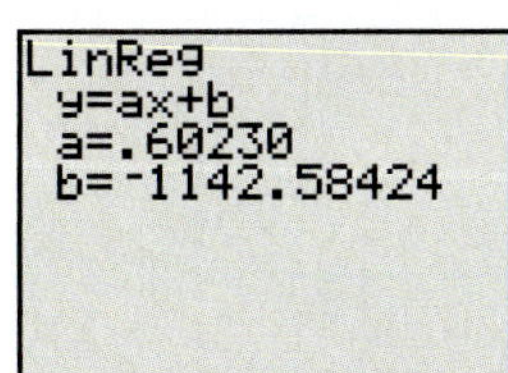

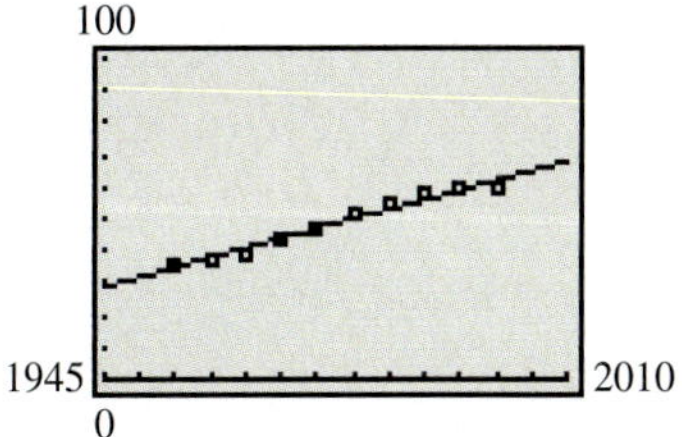

Figure 48

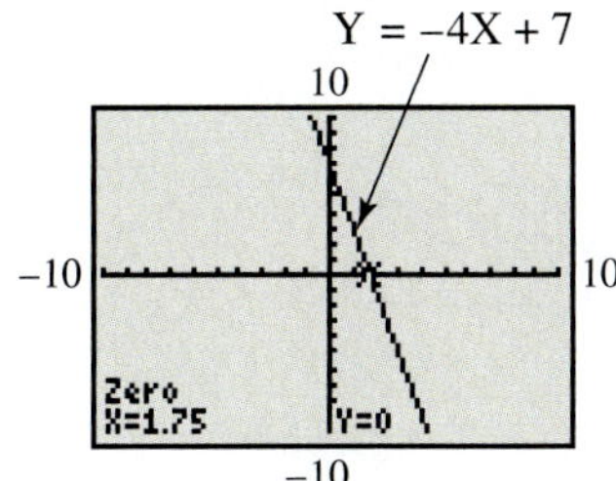

Figure 49

Solving Linear Equations in One Variable by Graphing

Figure 49 shows the graph of $Y = -4X + 7$. From the values at the bottom of the screen, we see that when $X = 1.75$, $Y = 0$. This means that $X = 1.75$ satisfies the equation $-4X + 7 = 0$, a linear equation in one variable. Therefore, the solution set of $-4X + 7 = 0$ is $\{1.75\}$. We can verify this algebraically by substitution. (The word "Zero" indicates that the x-intercept has been located.)

EXAMPLE 8 Solving an Equation with a Graphing Calculator

Use a graphing calculator to solve $-2x - 4(2 - x) = 3x + 4$.

Solution We must write the equation as an equivalent equation with 0 on one side.

$$-2x - 4(2 - x) - 3x - 4 = 0 \quad \text{Subtract } 3x \text{ and } 4.$$

Then we graph $Y = -2X - 4(2 - X) - 3X - 4$ to find the x-intercept. The standard viewing window cannot be used because the x-intercept does not lie in the interval $[-10, 10]$. As seen in Figure 50, the x-intercept of the graph is -12, and thus the solution (or zero) of the equation is -12. The solution set is $\{-12\}$.

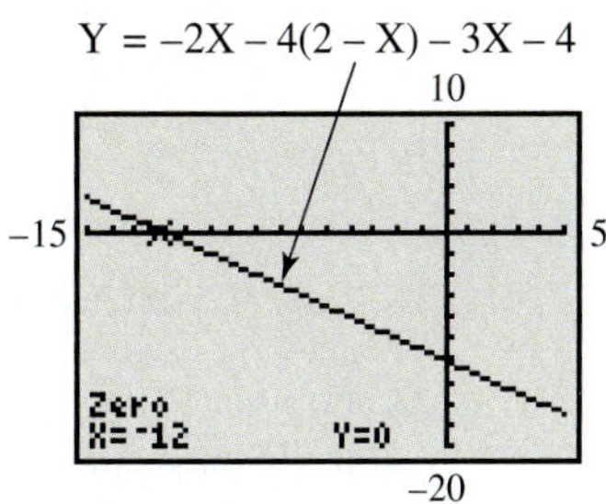

Figure 50

Now try Exercise 53.

2.4 Exercises

Concept Check *Match each equation in Exercises 1–4 to the correct graph in A–D.*

1. $y = \frac{1}{4}x + 2$

2. $4x + 3y = 12$

3. $y - (-1) = \frac{3}{2}(x - 1)$

4. $y = 4$

A.

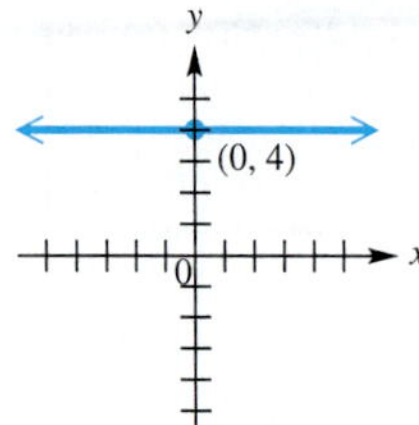

B.

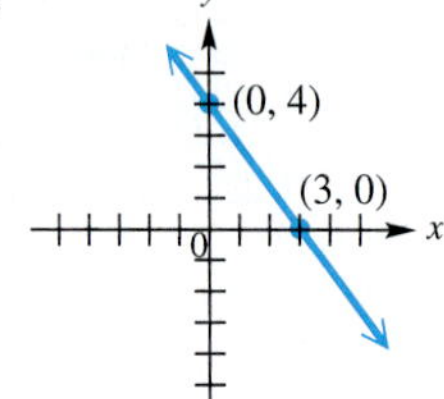

C.

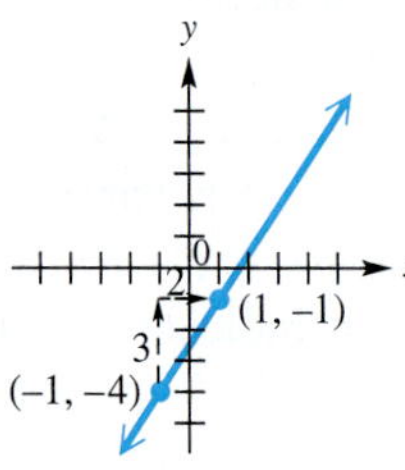

D.

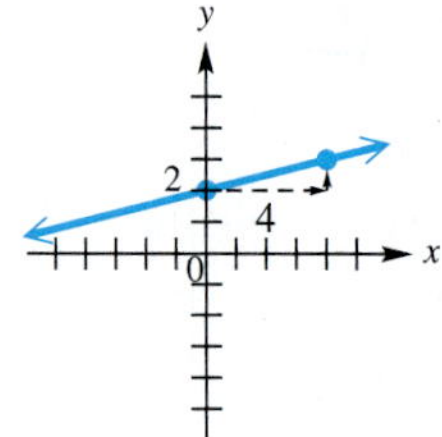

In Exercises 5–22, write an equation for the line described. Give answers in standard form for Exercises 5–10 and in slope-intercept form (if possible) for Exercises 11–22. See Examples 1–3.

5. through $(1, 3), m = -2$

6. through $(2, 4)$, $m = -1$

7. through $(-5, 4), m = -\frac{3}{2}$

8. through $(-4, 3)$, $m = \frac{3}{4}$

9. through $(-8, 4)$, undefined slope

10. through $(5, 1)$, $m - 0$

11. through $(-1, 3)$ and $(3, 4)$

12. through $(8, -1)$ and $(4, 3)$

13. x-intercept 3, y-intercept -2

14. x-intercept -2, y-intercept 4

15. vertical, through $(-6, 4)$

16. horizontal, through $(2, 7)$

17. $m = 5, b = 15$

18. $m = -2, b = 12$

19. $m = -\frac{2}{3}, b = -\frac{4}{5}$

20. $m = -\frac{5}{8}, b = -\frac{1}{3}$

21. slope 0, y-intercept $\frac{3}{2}$

22. undefined slope, x-intercept $-\frac{5}{4}$

23. *Concept Check* Fill in each blank with the appropriate response: The line $x + 2 = 0$ has x-intercept ______. It ______ (does/does not) have a y-intercept. The slope of this line is ______ (0/undefined). The line $4y = 2$ has y-intercept ______. It ______ (does/does not) have an x-intercept. The slope of this line is ______ (0/undefined).

24. *Concept Check* Match each equation with the line that would most closely resemble its graph. (*Hint:* Consider the signs of m and b in the slope-intercept form.)

(a) $y = 3x + 2$ **(b)** $y = -3x + 2$ **(c)** $y = 3x - 2$ **(d)** $y = -3x - 2$

A. **B.**

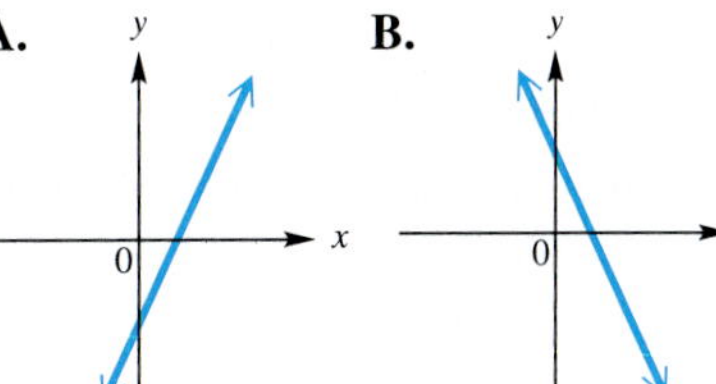

C. **D.**

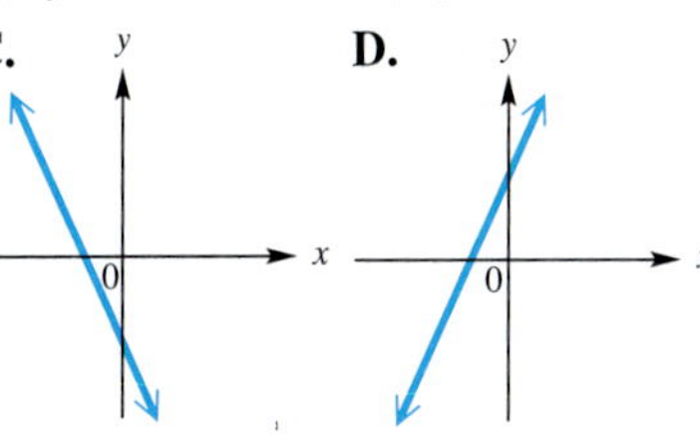

25. *Concept Check* Match each equation with its calculator graph. The standard viewing window is used in each case, but no tick marks are shown.

(a) $y = 2x + 3$ **(b)** $y = -2x + 3$ **(c)** $y = 2x - 3$ **(d)** $y = -2x - 3$

A.

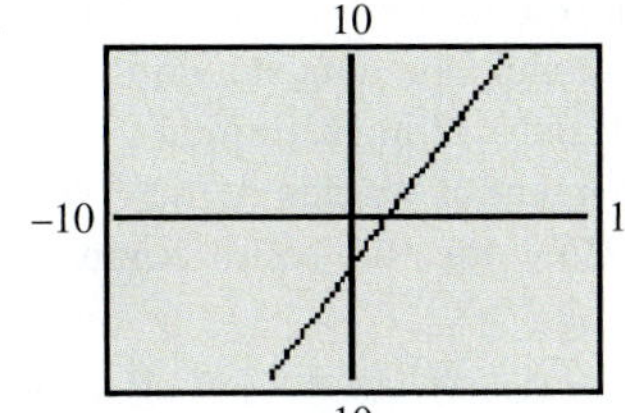

B.

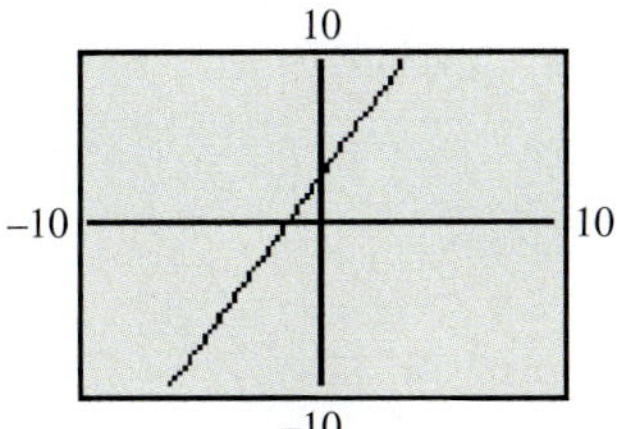

C.

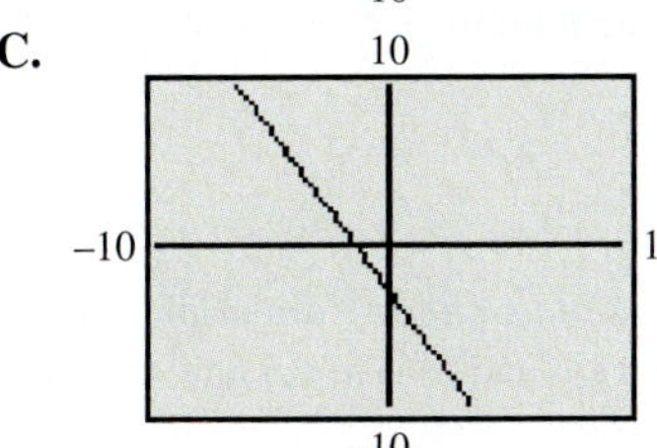

D.

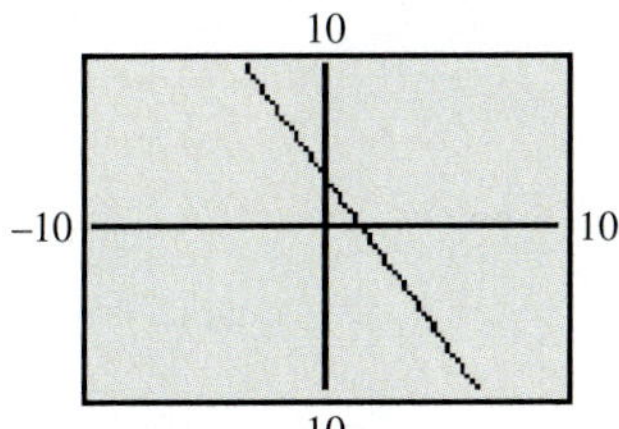

26. *Concept Check* The table represents a linear function f.

(a) Find the slope of the line defined by $y = f(x)$.
(b) Find the y-intercept of the line.
(c) Find the equation for this line in slope-intercept form.

x	y
−2	−11
−1	−8
0	−5
1	−2
2	1
3	4

Give the slope and y-intercept of each line and graph it. See Example 4.

27. $y = 3x - 1$ **28.** $y = -2x + 7$ **29.** $4x - y = 7$

30. $2x + 3y = 16$ **31.** $4y = -3x$ **32.** $2y - x = 0$

33. $x + 2y = -4$ **34.** $x + 3y = -9$

In Exercises 35–42, write an equation **(a)** *in standard form and* **(b)** *in slope-intercept form for the line described. See Example 5.*

35. through $(-1, 4)$, parallel to $x + 3y = 5$

36. through $(3, -2)$, parallel to $2x - y = 5$

37. through $(1, 6)$, perpendicular to $3x + 5y = 1$

38. through $(-2, 0)$, perpendicular to $8x - 3y = 7$

39. through $(4, 1)$, parallel to $y = -5$

40. through $(-2, -2)$, parallel to $y = 3$

41. through $(-5, 6)$, perpendicular to $x = -2$

42. through $(4, -4)$, perpendicular to $x = 4$

43. Find k so that the line through $(4, -1)$ and $(k, 2)$ is

(a) parallel to $3y + 2x = 6$;
(b) perpendicular to $2y - 5x = 1$.

44. Find r so that the line through $(2, 6)$ and $(-4, r)$ is

(a) parallel to $2x - 3y = 4$;
(b) perpendicular to $x + 2y = 1$.

(Modeling) Solve each problem. See Examples 6 and 7.

45. *Women in the Work Force* Use the data points $(1970, 43.3)$ and $(1995, 58.9)$ to find a linear equation that models the data shown in the table accompanying Figure 47 in Example 7. Then use it to predict the percent of women in the civilian labor force in 1996. How does the result compare to the actual figure of 59.3%?

46. *Women in the Work Force* Repeat Exercise 45 using the data points for the years 1975 and 2000.

47. *Cost of Private College Education* The table lists the average annual cost (in dollars) of tuition and fees at private 4-year colleges for selected years.

Year	Tuition and Fees (in dollars)
1990	9,340
1992	10,448
1994	11,719
1996	12,994
1998	14,709
2000	16,233
2002	18,116

Source: The College Board.

(a) Determine a linear function defined by $f(x) = mx + b$ that models the data, where $x = 0$ represents 1990, $x = 1$ represents 1991, and so on. Use the points $(0, 9340)$ and $(12, 18{,}116)$. Graph f and a scatter diagram of the data on the same coordinate axes. (You may wish to use a graphing calculator.) What does the slope of the graph of f indicate?

(b) Use this function to approximate tuition and fees in 1995. Compare your approximation to the actual value of \$12,143.

(c) Use the linear regression feature of a graphing calculator to find the equation of the line of best fit.

48. *Distances and Velocities of Galaxies* The table lists the distances (in megaparsecs; 1 megaparsec $= 3.085 \times 10^{24}$ cm and 1 megaparsec $= 3.26$ million light-years) and velocities (in kilometers per second) of four galaxies moving rapidly away from Earth.

Galaxy	Distance	Velocity
Virgo	15	1600
Ursa Minor	200	15,000
Corona Borealis	290	24,000
Bootes	520	40,000

Source: Acker, A., and C. Jaschek, *Astronomical Methods and Calculations,* John Wiley & Sons, 1986. Karttunen, H. (editor), *Fundamental Astronomy,* Springer-Verlag, 2003.

(a) Plot the data using distances for the x-values and velocities for the y-values. What type of relationship seems to hold between the data?

(b) Find a linear equation in the form $y = mx$ that models these data using the points $(520, 40{,}000)$ and $(0, 0)$. Graph your equation with the data on the same coordinate axes.

(c) The galaxy Hydra has a velocity of 60,000 km per sec. How far away is it according to the model in part (b)?

(d) The value of m is called the *Hubble constant.* The Hubble constant can be used to estimate the age of the universe A (in years) using the formula

$$A = \frac{9.5 \times 10^{11}}{m}.$$

Approximate A using your value of m.

(e) Astronomers currently place the value of the Hubble constant between 50 and 100. What is the range for the age of the universe A?

49. *Celsius and Fahrenheit Temperatures* When the Celsius temperature is 0°, the corresponding Fahrenheit temperature is 32°. When the Celsius temperature is 100°, the corresponding Fahrenheit temperature is 212°. Let C represent the Celsius temperature and F the Fahrenheit temperature.

(a) Express F as an exact linear function of C.

(b) Solve the equation in part (a) for C, thus expressing C as a function of F.

(c) For what temperature is $F = C$?

50. *Water Pressure on a Diver* The pressure p of water on a diver's body is a linear function of the diver's depth, x. At the water's surface, the pressure is 1 atmosphere. At a depth of 100 ft, the pressure is about 3.92 atmospheres.

(a) Find the linear function that relates p to x.

(b) Compute the pressure at a depth of 10 fathoms (60 ft).

51. *Consumption Expenditures* Economists frequently use linear models as approximations for more complicated models. In Keynesian macroeconomic theory, total consumption expenditure on goods and services, C, is assumed to be a linear function of national income, I. The table gives the values of C and I for 1990 and 1997 in the United States.

Year	1990	1997
Total consumption (C)	3839	5494
National income (I)	6650	4215

Source: The Wall Street Journal Almanac; New York Times Almanac.

(a) Find the formula for C as a function of I.

(b) The slope of the linear function is called the *marginal propensity to consume.* What is the marginal propensity to consume for the United States from 1990–1997?

In Exercises 52–55, do the following.

(a) *Simplify and rewrite the equation so that the right side is* 0. *Then replace* 0 *with y.*

(b) *The graph of the equation for y is shown with each exercise. Use the graph to determine the solution of the given equation. See Example 8.*

(c) *Solve the equation using the methods of Chapter 1.*

52. $2x + 7 - x = 4x - 2$

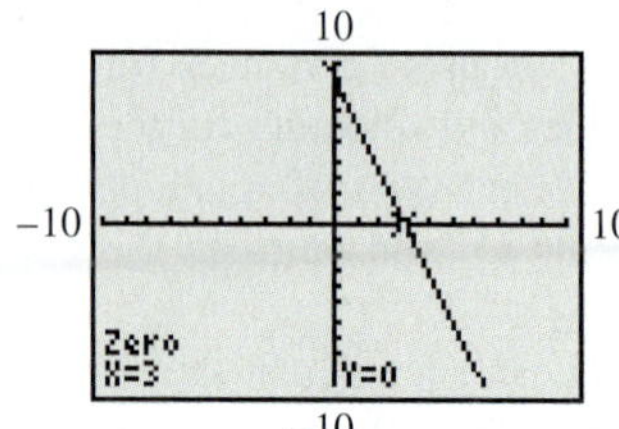

53. $7x - 2x + 4 - 5 = 3x + 1$

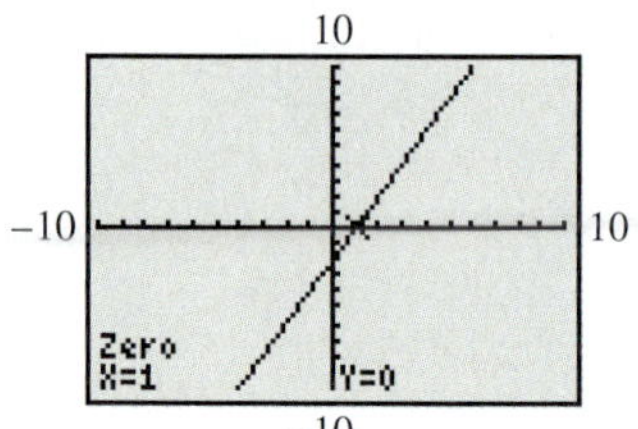

54. $3(2x + 1) - 2(x - 2) = 5$

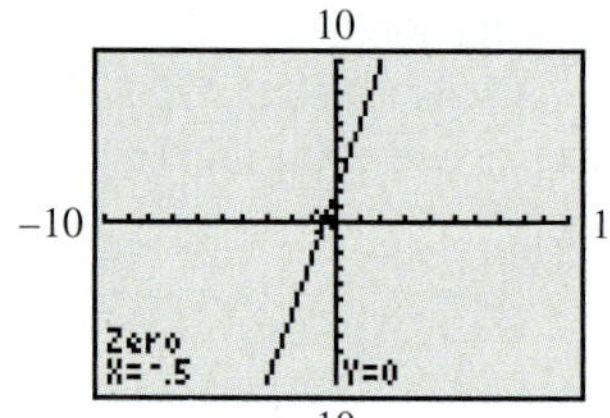

55. $4x - 3(4 - 2x) = 2(x - 3) + 6x + 2$

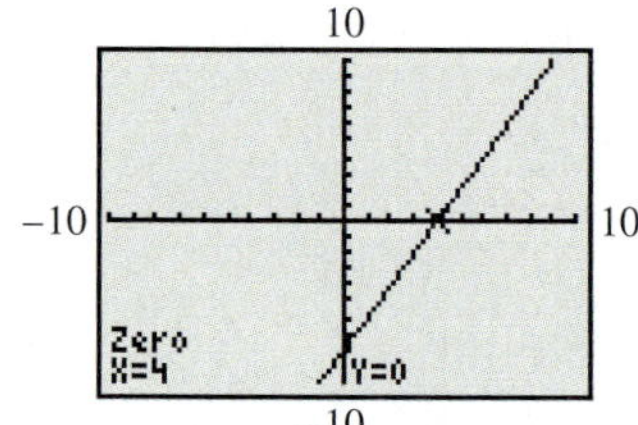

56. The graph of y_1 is shown in the standard viewing window. Which is the only value of x that could possibly be the solution of the equation $y_1 = 0$?

A. −15 **B.** 0 **C.** 5 **D.** 15

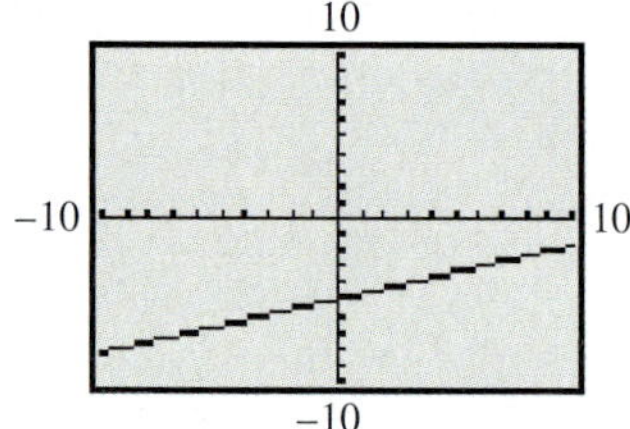

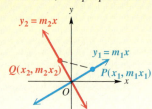

57. **(a)** Solve $-2(x - 5) = -x - 2$ using the methods of Chapter 1.

(b) Explain why the standard viewing window of a graphing calculator cannot graphically support the solution found in part (a). What minimum and maximum x-values would make it possible for the solution to be seen?

Relating Concepts

For individual or collaborative investigation

(Exercises 58–65)

In this section we state that two lines, neither of which is vertical, are perpendicular if and only if their slopes have a product of −1. *In Exercises 58–65, we outline a partial proof of this for the case where the two lines intersect at the origin.* ***Work these exercises in order,*** *and refer to the figure as needed.*

y
$y_2 = m_2x$
$y_1 = m_1x$
$Q(x_2, m_2x_2)$
$P(x_1, m_1x_1)$
O
x

58. In triangle OPQ, angle POQ is a right angle if and only if

$$[d(O, P)]^2 + [d(O, Q)]^2 = [d(P, Q)]^2.$$

What theorem from geometry assures us of this?

59. Find an expression for the distance $d(O, P)$.

60. Find an expression for the distance $d(O, Q)$.

61. Find an expression for the distance $d(P, Q)$.

62. Use your results from Exercises 59–61, and substitute into the equation in Exercise 58. Simplify to show that this leads to the equation $-2m_1m_2x_1x_2 - 2x_1x_2 = 0$.

63. Factor $-2x_1x_2$ from the final form of the equation in Exercise 62.

64. Use the property that if $ab = 0$ then $a = 0$ or $b = 0$ to solve the equation in Exercise 63, showing that $m_1m_2 = -1$.

65. State your conclusion based on Exercises 58–64.

66. Show that the line $y = x$ is the perpendicular bisector of the segment with endpoints (a, b) and (b, a), where $a \neq b$. (*Hint:* Use the midpoint formula and the slope formula.)

67. Refer to Example 4 in Section 2.1, and prove that the three points are collinear by taking them two at a time showing that in all three cases, the slope is the same.

Determine whether the three points are collinear by using slopes as in Exercise 67. (Note: These problems were first seen in Exercises 23–25 in Section 2.1.)

68. $(0, -7), (-3, 5), (2, -15)$

69. $(-1, 4), (-2, -1), (1, 14)$

70. $(0, 9), (-3, -7), (2, 19)$

Summary Exercises on Graphs, Functions, and Equations

These summary exercises provide practice with some of the concepts from Sections 2.1–2.4.

For the points P and Q, find **(a)** *the distance d(P, Q),* **(b)** *the coordinates of the midpoint of the segment PQ, and* **(c)** *an equation for the line through the two points. Write the equation in slope-intercept form if possible.*

1. $P(3, 5), Q(2, -3)$

2. $P(-1, 0), Q(4, -2)$

3. $P(-2, 2), Q(3, 2)$

4. $P(2\sqrt{2}, \sqrt{2}), Q(\sqrt{2}, 3\sqrt{2})$

5. $P(5, -1), Q(5, 1)$

6. $P(1, 1), Q(-3, -3)$

7. $P(2\sqrt{3}, 3\sqrt{5}), Q(6\sqrt{3}, 3\sqrt{5})$

8. $P(0, -4), Q(3, 1)$

Write an equation for each of the following, and sketch the graph.

9. the line through $(-2, 1)$ and $(4, -1)$

10. the horizontal line through $(2, 3)$

11. the circle with center $(2, -1)$ and radius 3

12. the circle with center $(0, 2)$ and tangent to the x-axis

13. the line through $(3, -5)$ with slope $-\frac{5}{6}$

14. the line through the origin and perpendicular to the line $3x - 4y = 2$

15. the line through $(-3, 2)$ and parallel to the line $2x + 3y = 6$

16. the vertical line through $(-4, 3)$

Decide whether or not each equation has a circle as its graph. If it does, give the center and the radius.

17. $x^2 - 4x + y^2 + 2y = 4$

18. $x^2 + 6x + y^2 + 10y + 36 = 0$

19. $x^2 - 12x + y^2 + 20 = 0$

20. $x^2 + 2x + y^2 + 16y = -61$

21. $x^2 - 2x + y^2 + 10 = 0$

22. $x^2 + y^2 - 8y - 9 = 0$

For each of the following relations, ***(a)*** *find the domain and range, and* ***(b)*** *if the relation defines y as a function of x, rewrite the relation using $f(x)$ notation and find $f(-2)$.*

23. $3x - 2y = 1$

24. $y + 3x^2 = -1$

25. $x - 4y = -6$

26. $y^2 - x = 5$

27. $(x + 2)^2 + y^2 = 25$

28. $x^2 - 2y = 3$

2.5 Graphs of Basic Functions

Continuity ▪ **The Identity, Squaring, and Cubing Functions** ▪ **The Square Root and Cube Root Functions** ▪ **The Absolute Value Function** ▪ **Piecewise-Defined Functions** ▪ **The Relation $x = y^2$**

Continuity Earlier in this chapter we graphed linear functions. The graph of a linear function, a straight line, may be drawn by hand over any interval of its domain without picking the pencil up from the paper. In mathematics we say that a function with this property is *continuous* over any interval. The formal definition of continuity requires concepts from calculus, but we can give an informal definition at the college algebra level.

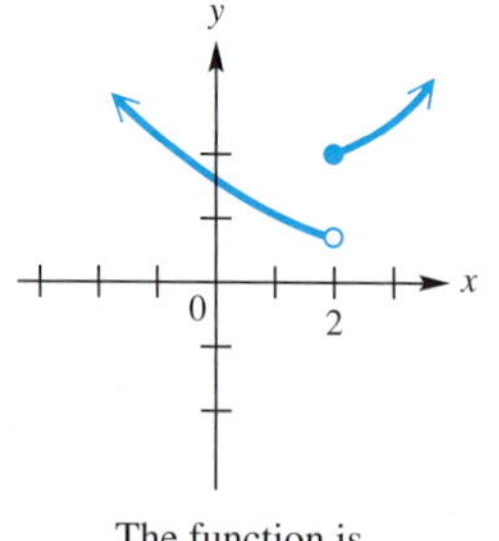

The function is discontinuous at $x = 2$.

Figure 51

Continuity

A function is **continuous** over an interval of its domain if its hand-drawn graph over that interval can be sketched without lifting the pencil from the paper.

If a function is not continuous at a *point,* then it has a *discontinuity* there. Figure 51 shows the graph of a function with a discontinuity at the point where $x = 2$.

Looking Ahead to Calculus

Many calculus theorems apply only to continuous functions.

EXAMPLE 1 Determining Intervals of Continuity

Describe the intervals of continuity for each function in Figure 52.

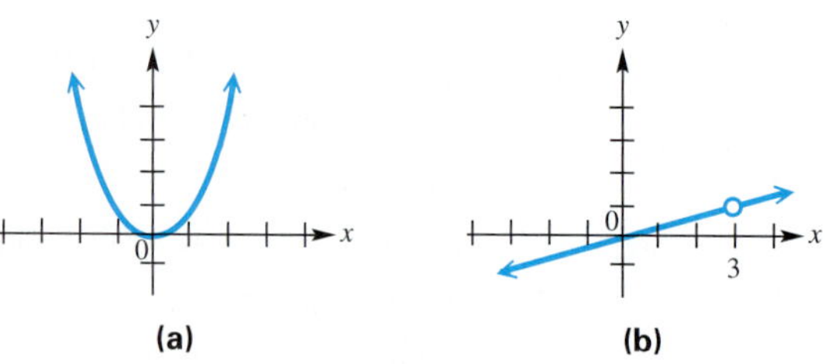

Figure 52

Solution The function in Figure 52(a) is continuous over its entire domain, $(-\infty, \infty)$. The function in Figure 52(b) has a point of discontinuity at $x = 3$. Thus, it is continuous over the intervals $(-\infty, 3)$ and $(3, \infty)$.

Now try Exercises 1 and 5.

Graphs of the basic functions studied in college algebra can be sketched by careful point plotting or generated by a graphing calculator. As you become more familiar with these graphs, you should be able to provide quick rough sketches of them.

The Identity, Squaring, and Cubing Functions The **identity function** defined by $f(x) = x$ pairs every real number with itself. See Figure 53.

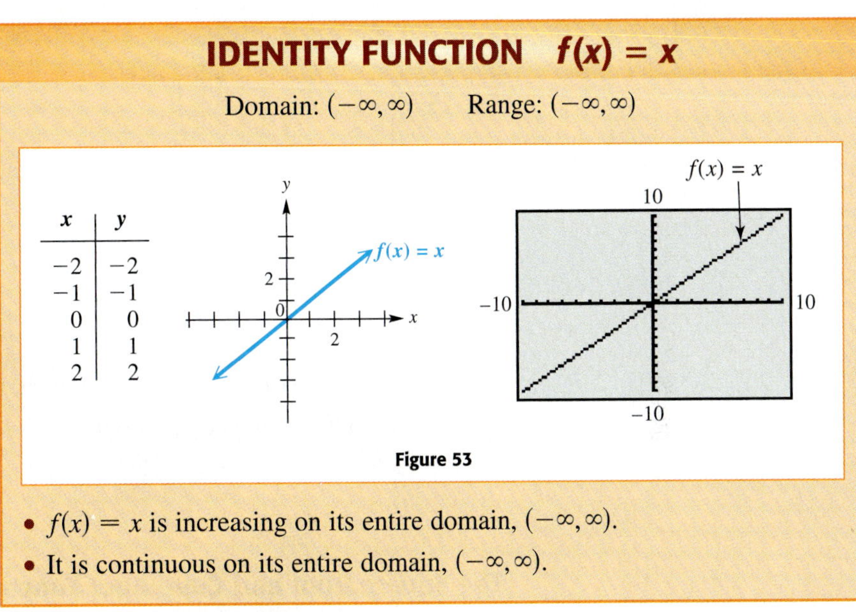

x	y
−2	−2
−1	−1
0	0
1	1
2	2

Figure 53

- $f(x) = x$ is increasing on its entire domain, $(-\infty, \infty)$.
- It is continuous on its entire domain, $(-\infty, \infty)$.

The **squaring function,** $f(x) = x^2$, pairs each real number with its square; its graph is called a **parabola.** The point $(0, 0)$ at which the graph changes from decreasing to increasing is called the **vertex** of the parabola. See Figure 54.

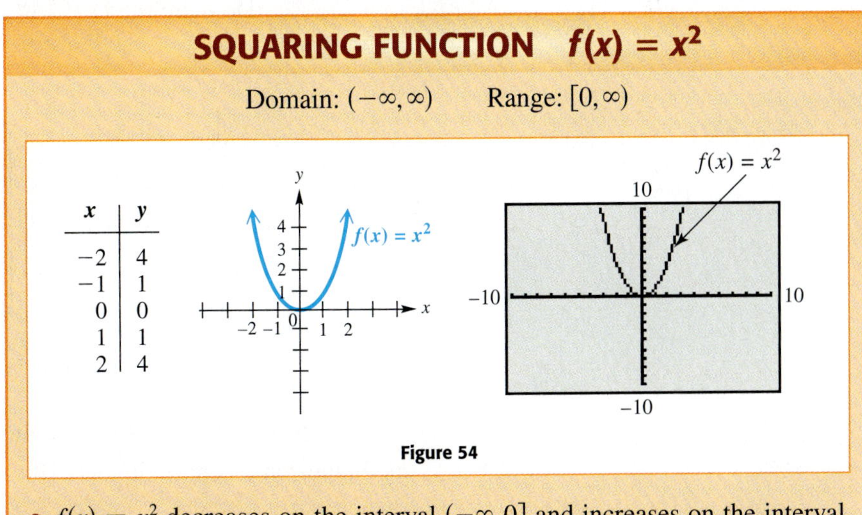

x	y
−2	4
−1	1
0	0
1	1
2	4

Figure 54

- $f(x) = x^2$ decreases on the interval $(-\infty, 0]$ and increases on the interval $[0, \infty)$.
- It is continuous on its entire domain, $(-\infty, \infty)$.

The function defined by $f(x) = x^3$ is called the **cubing function.** It pairs with each real number the cube of the number. See Figure 55.

CUBING FUNCTION $f(x) = x^3$

Domain: $(-\infty, \infty)$ Range: $(-\infty, \infty)$

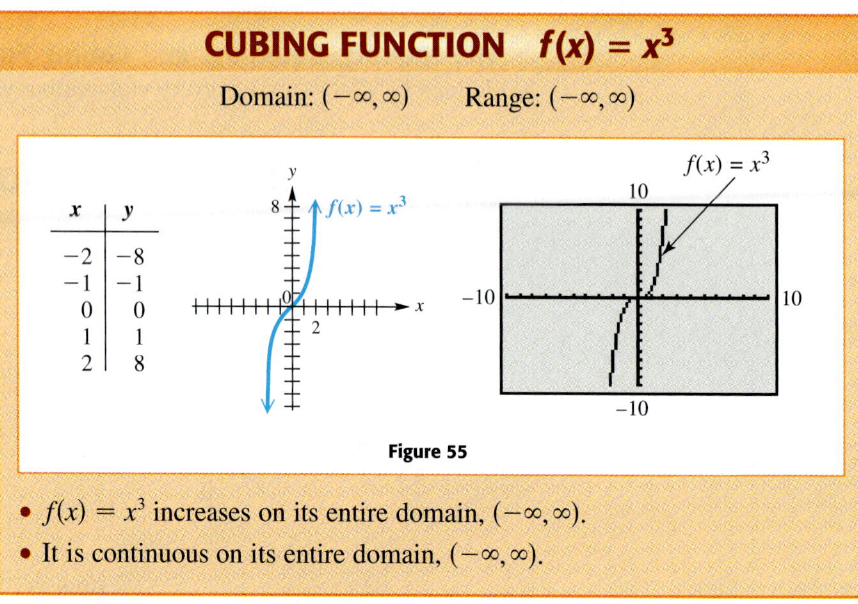

Figure 55

- $f(x) = x^3$ increases on its entire domain, $(-\infty, \infty)$.
- It is continuous on its entire domain, $(-\infty, \infty)$.

The Square Root and Cube Root Functions The **square root function,** $f(x) = \sqrt{x}$, pairs each real number with its principal square root. See Figure 56.

SQUARE ROOT FUNCTION $f(x) = \sqrt{x}$

Domain: $[0, \infty)$ Range: $[0, \infty)$

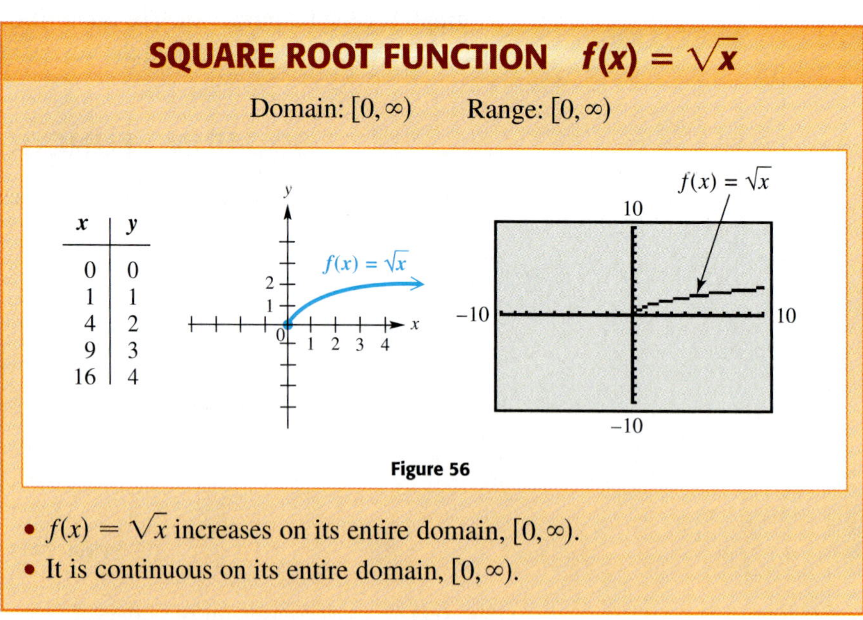

Figure 56

- $f(x) = \sqrt{x}$ increases on its entire domain, $[0, \infty)$.
- It is continuous on its entire domain, $[0, \infty)$.

The **cube root function,** $f(x) = \sqrt[3]{x}$, pairs each real number with its cube root. See Figure 57.

CUBE ROOT FUNCTION $f(x) = \sqrt[3]{x}$

Domain: $(-\infty, \infty)$ Range: $(-\infty, \infty)$

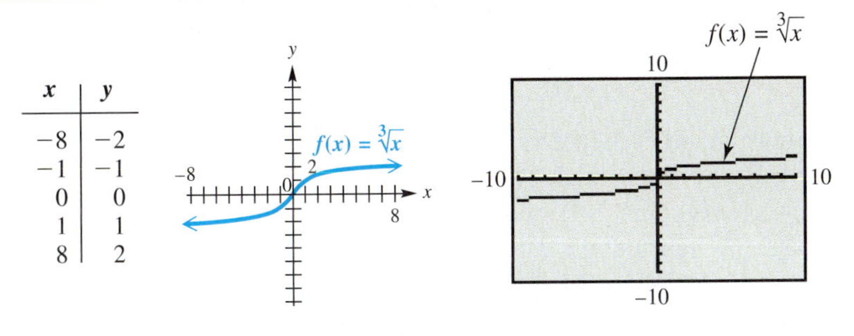

x	y
−8	−2
−1	−1
0	0
1	1
8	2

Figure 57

- $f(x) = \sqrt[3]{x}$ increases on its entire domain, $(-\infty, \infty)$.
- It is continuous on its entire domain, $(-\infty, \infty)$.

The Absolute Value Function The **absolute value function,** $f(x) = |x|$, which pairs every real number with its absolute value, is graphed in Figure 58 and defined as follows.

$$f(x) = |x| = \begin{cases} x & \text{if } x \geq 0 \\ -x & \text{if } x < 0 \end{cases}$$

That is, we use $|x| = x$ if x is positive or 0, and we use $|x| = -x$ if x is negative.

ABSOLUTE VALUE FUNCTION $f(x) = |x|$

Domain: $(-\infty, \infty)$ Range: $[0, \infty)$

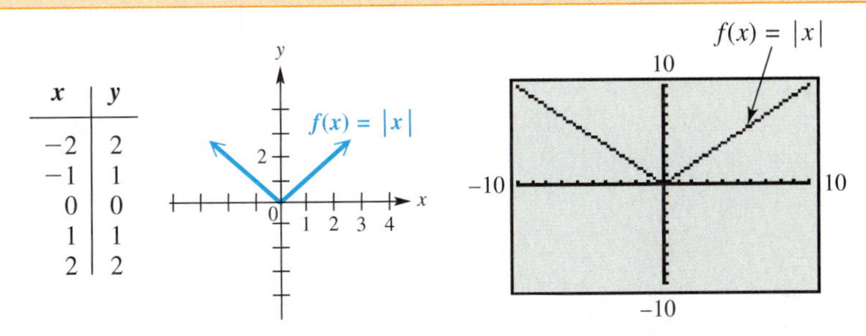

x	y
−2	2
−1	1
0	0
1	1
2	2

Figure 58

- $f(x) = |x|$ decreases on the interval $(-\infty, 0]$ and increases on $[0, \infty)$.
- It is continuous on its entire domain, $(-\infty, \infty)$.

Piecewise-Defined Functions The absolute value function is defined by different rules over different intervals of its domain. Such functions are called **piecewise-defined functions.**

EXAMPLE 2 Graphing Piecewise-Defined Functions

Graph each function.

(a) $f(x) = \begin{cases} -2x + 5 & \text{if } x \leq 2 \\ x + 1 & \text{if } x > 2 \end{cases}$ **(b)** $f(x) = \begin{cases} 2x + 3 & \text{if } x \leq 1 \\ -x + 6 & \text{if } x > 1 \end{cases}$

Algebraic Solution

(a) We must graph each interval of the domain separately. If $x \leq 2$, the graph of $f(x) = -2x + 5$ has an endpoint at $x = 2$. We find the corresponding y-value by substituting 2 for x in $-2x + 5$ to get $y = 1$. To get another point on this part of the graph, we choose $x = 0$, so $y = 5$. Draw the graph through $(2, 1)$ and $(0, 5)$ as a partial line with endpoint $(2, 1)$.

Graph the function for $x > 2$ similarly, using $f(x) = x + 1$. This partial line has an open endpoint at $(2, 3)$. Use $y = x + 1$ to find another point with x-value greater than 2 to complete the graph. See Figure 59.

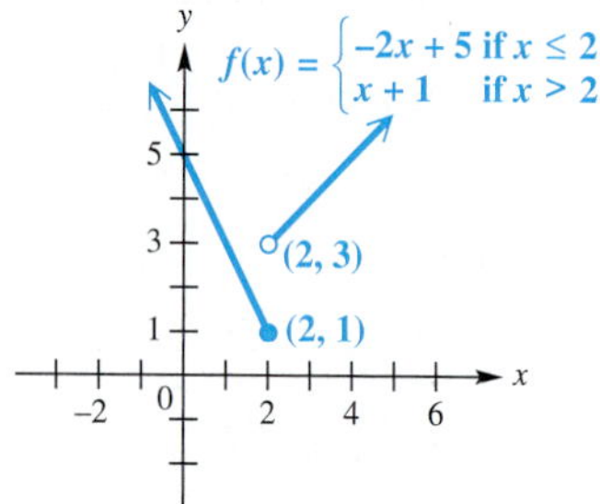

Figure 59

(b) Graph $f(x) = 2x + 3$ for $x \leq 1$. For $x > 1$, graph $f(x) = -x + 6$. The graph consists of the two pieces shown in Figure 61. The two partial lines meet at the point $(1, 5)$.

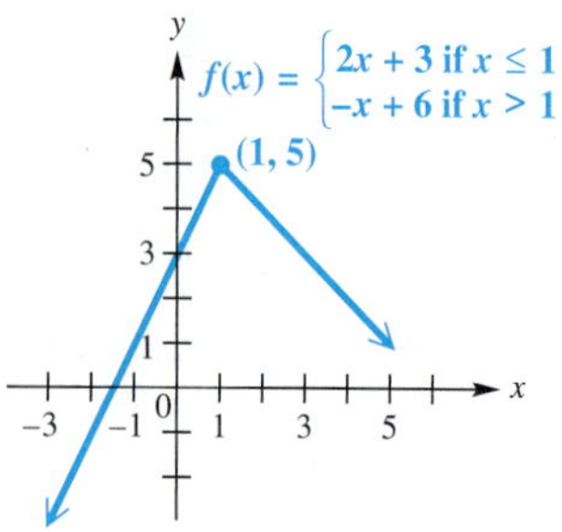

Figure 61

Graphing Calculator Solution

(a) By defining

$$Y_1 = (-2x + 5)(x \leq 2)$$

and

$$Y_2 = (x + 1)(x > 2)$$

and using *dot mode,* we obtain the graph of f shown in Figure 60. Remember that inclusion or exclusion of endpoints is not readily apparent when observing a calculator graph. This decision must be made using your knowledge of inequalities.

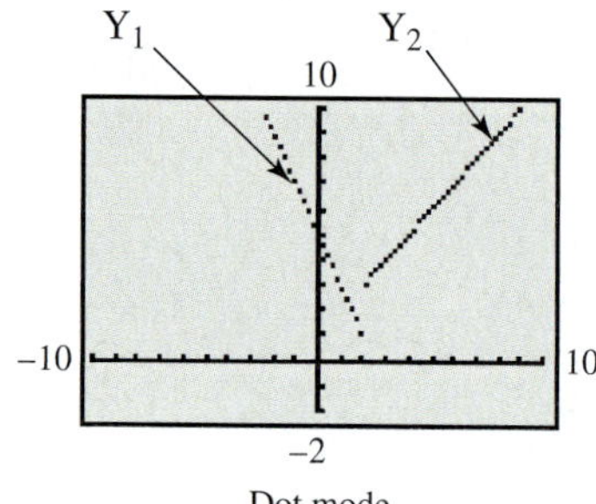

Dot mode

Figure 60

(b) Figure 62 shows an alternative method that can be used to obtain a calculator graph of this function. Here we entered the function as one expression

$$Y_1 = (2X + 3)(X \leq 1) + (-X + 6)(X > 1).$$

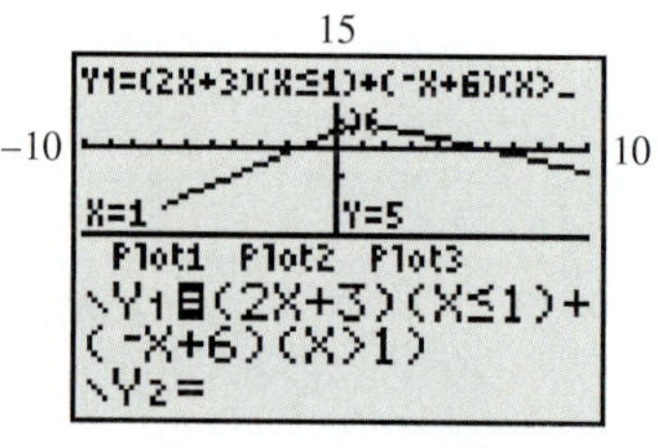

Figure 62

Now try Exercise 23.

Another piecewise-defined function is the *greatest integer function.*

> The **greatest integer function,** $f(x) = [\![x]\!]$, pairs every real number x with the greatest integer less than or equal to x.

For example, $[\![8.4]\!] = 8$, $[\![-5]\!] = -5$, $[\![\pi]\!] = 3$, and $[\![-6.9]\!] = -7$. In general, if $f(x) = [\![x]\!]$, then

$$\begin{array}{ll} \text{for } -2 \le x < -1, & f(x) = -2, \\ \text{for } -1 \le x < 0 & f(x) = -1, \\ \text{for } 0 \le x < 1, & f(x) = 0, \\ \text{for } 1 \le x < 2, & f(x) = 1, \\ \text{for } 2 \le x < 3, & f(x) = 2, \end{array}$$

and so on. The graph of the greatest integer function is shown in Figure 63.

GREATEST INTEGER FUNCTION $f(x) = [\![x]\!]$

Domain: $(-\infty, \infty)$

Range: $\{y \mid y \text{ is an integer}\} = \{\ldots, -3, -2, -1, 0, 1, 2, 3, \ldots\}$

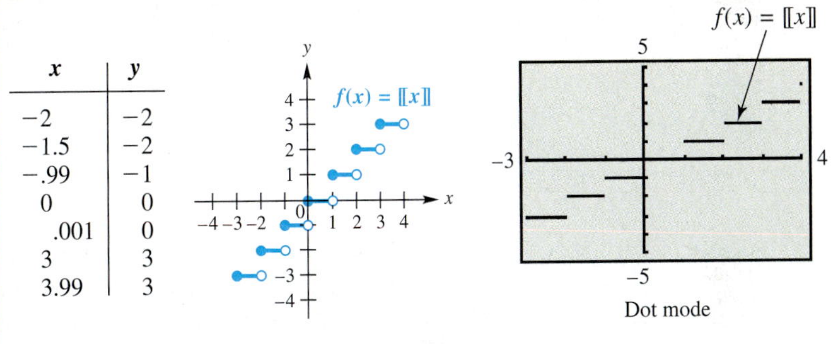

x	y
-2	-2
-1.5	-2
$-.99$	-1
0	0
$.001$	0
3	3
3.99	3

Figure 63

- $f(x) = [\![x]\!]$ is constant on the intervals $\ldots, [-2, -1), [-1, 0), [0, 1), [1, 2), [2, 3), \ldots$.
- It is discontinuous at all integer values in its domain, $(-\infty, \infty)$.

Looking Ahead to Calculus

The *greatest integer function* is used in calculus as a classic example of how the limit of a function may not exist at a particular value in its domain. For a limit to exist, both the left- and right-hand limits must be equal. We can see from the graph of the greatest integer function that for an integer value such as 3, as x approaches 3 from the left, function values are all 2, while as x approaches 3 from the right, function values are all 3. Since the left- and right-hand limits are different, the limit as x approaches 3 does not exist.

EXAMPLE 3 Graphing a Greatest Integer Function

Graph $f(x) = [\![\frac{1}{2}x + 1]\!]$.

Solution If x is in the interval $[0, 2)$, then $y = 1$. For x in $[2, 4)$, $y = 2$, and so on. Some sample ordered pairs are given here.

x	0	$\frac{1}{2}$	1	$\frac{3}{2}$	2	3	4	-1	-2	-3
y	1	1	1	1	2	2	3	0	0	-1

The ordered pairs in the table suggest the graph shown in Figure 64. The domain is $(-\infty, \infty)$. The range is $\{\ldots, -2, -1, 0, 1, 2, \ldots\}$.

Now try Exercise 35.

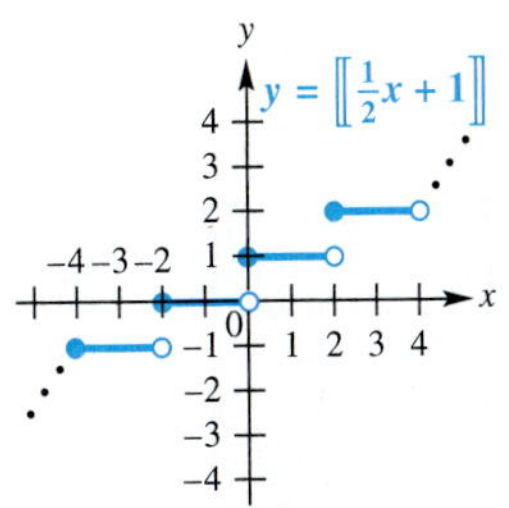

The dots indicate that the graph continues indefinitely in the same pattern.

Figure 64

The greatest integer function is an example of a **step function,** a function with a graph that looks like a series of steps.

EXAMPLE 4 Applying a Greatest Integer Function

An express mail company charges \$25 for a package weighing up to 2 lb. For each additional pound or fraction of a pound there is an additional charge of \$3. Let $D(x)$ represent the cost to send a package weighing x pounds. Graph $D(x)$ for x in the interval $(0, 6]$.

Solution For x in the interval $(0, 2]$, $y = 25$. For x in $(2, 3]$, $y = 25 + 3 = 28$. For x in $(3, 4]$, $y = 28 + 3 = 31$, and so on. The graph, which is that of a step function, is shown in Figure 65. In this case, the first step has a different width.

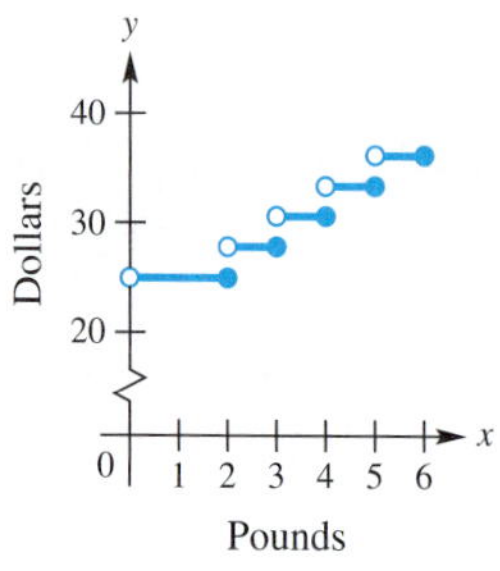

Figure 65

Now try Exercise 37.

The Relation $x = y^2$

Recall that a function is a relation where every domain value is paired with one and only one range value. However, there are cases where we are interested in graphing relations that are not functions, and one of the simplest of these is the relation defined by the equation $x = y^2$. Notice from the table of selected ordered pairs in the margin that this relation has two different y-values for each positive value of x. If we plot these points and join them with a smooth curve, we find that the graph of $x = y^2$ is a parabola opening to the right with vertex $(0, 0)$. See Figure 66. As the graph indicates, the domain is $[0, \infty)$ and the range is $(-\infty, \infty)$.

Selected Ordered Pairs for $x = y^2$

x	y
0	0
1	± 1
4	± 2
9	± 3

Two different y-values for the same x-value

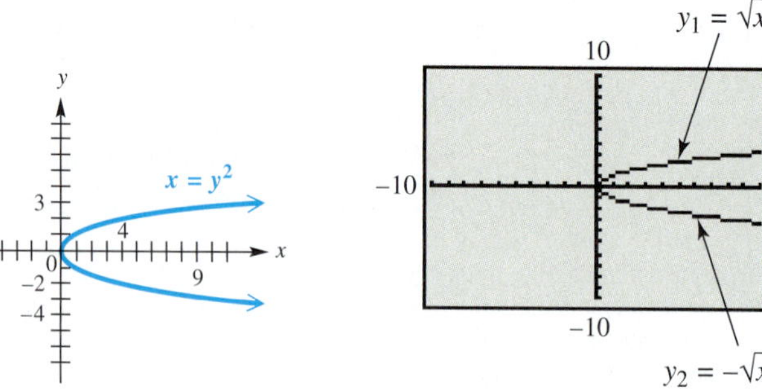

Figure 66

To use a calculator in function mode to graph the relation $x = y^2$, we graph the two functions $y_1 = \sqrt{x}$ and $y_2 = -\sqrt{x}$, as shown in Figure 66. ■

2.5 Exercises

Determine the intervals of the domain over which each function is continuous. See Example 1.

1.
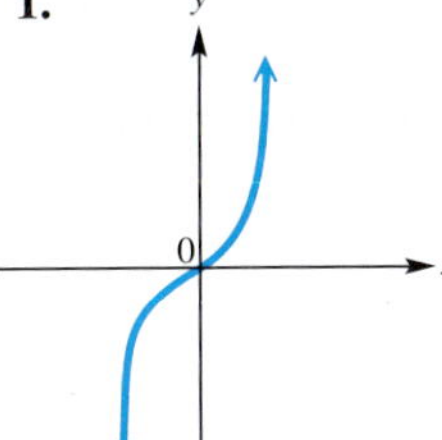

2.
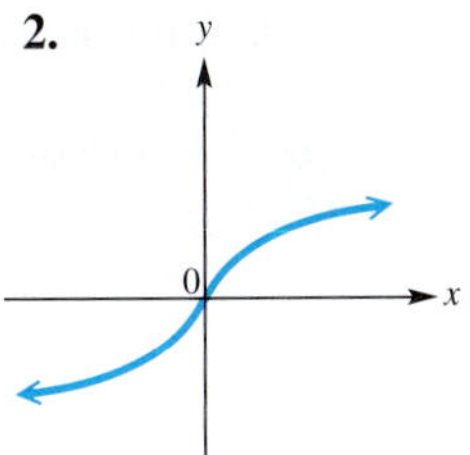

3.
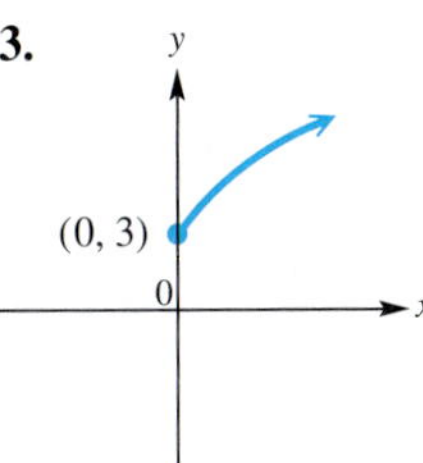

4.
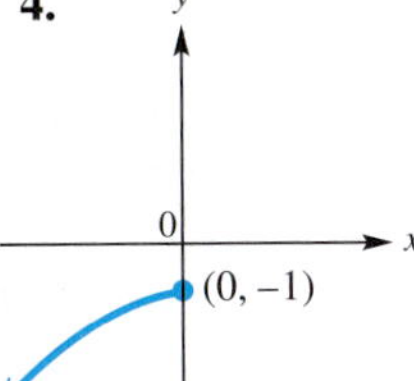

5.
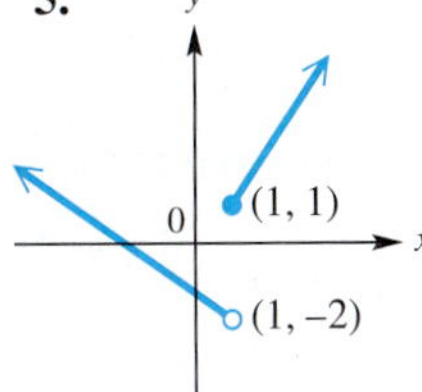

6.
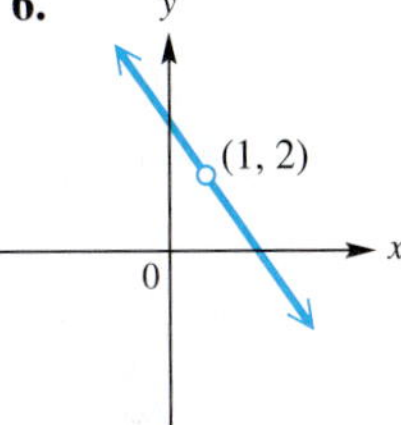

Concept Check *For Exercises 7–16, refer to the following basic graphs.*

A.
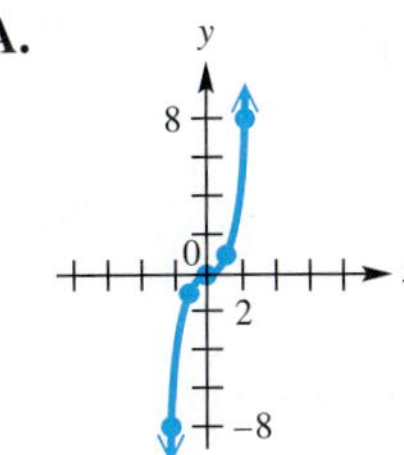

B.
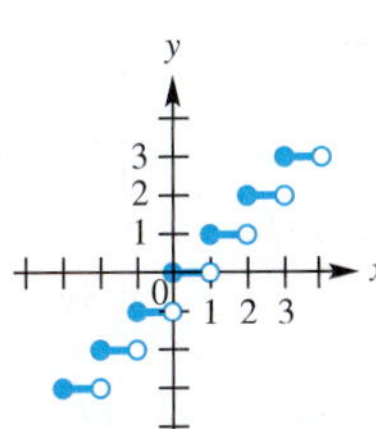

C.
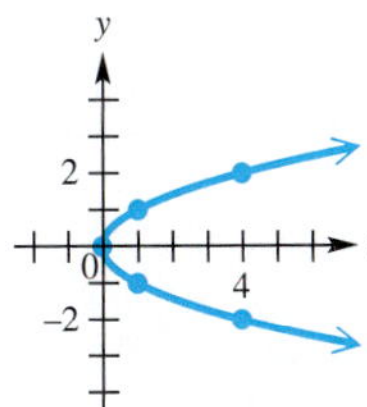

D.
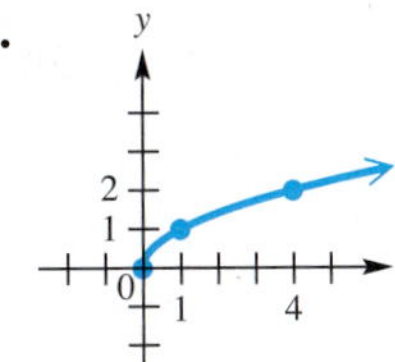

E.
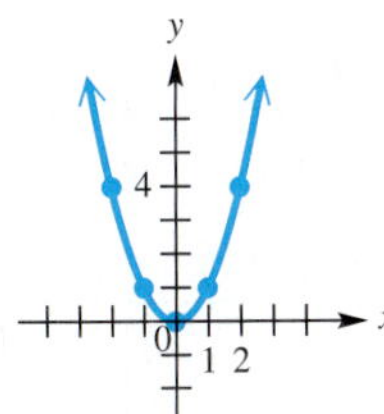

F.
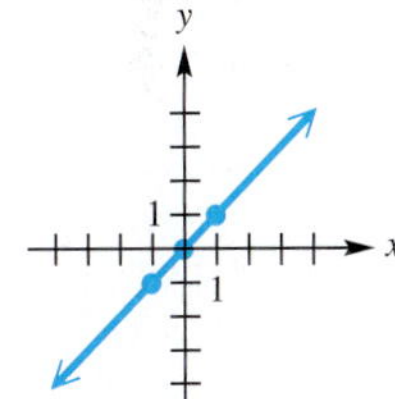

G.
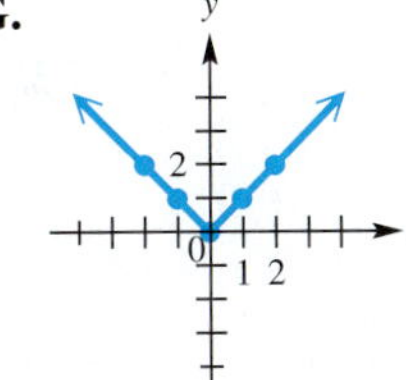

H.
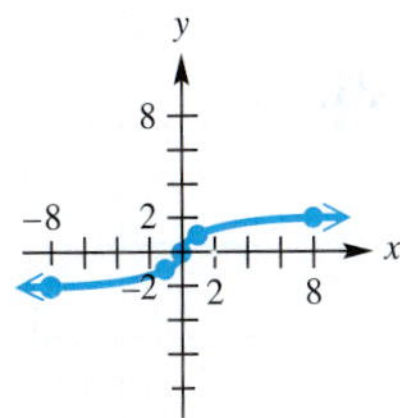

7. Which one is the graph of $y = x^2$? What is its domain?
8. Which one is the graph of $y = |x|$? On what interval is it increasing?
9. Which one is the graph of $y = x^3$? What is its range?
10. Which one is not the graph of a function? What is its equation?
11. Which one is the identity function? What is its equation?
12. Which one is the graph of $y = [\![x]\!]$? What is the value of y when $x = 1.5$?

13. Which one is the graph of $y = \sqrt[3]{x}$? Is there any interval over which the function is decreasing?

14. Which one is the graph of $y = \sqrt{x}$? What is its domain?

15. Which one is discontinuous at many points? What is its range?

16. Which graphs of functions decrease over part of the domain and increase over the rest of the domain? On what intervals do they increase? decrease?

For each piecewise-defined function, find ***(a)*** $f(-5)$, ***(b)*** $f(-1)$, ***(c)*** $f(0)$, *and* ***(d)*** $f(3)$. *See Example 2.*

17. $f(x) = \begin{cases} 2x & \text{if } x \leq -1 \\ x - 1 & \text{if } x > -1 \end{cases}$

18. $f(x) = \begin{cases} x - 2 & \text{if } x < 3 \\ 5 - x & \text{if } x \geq 3 \end{cases}$

19. $f(x) = \begin{cases} 2 + x & \text{if } x < -4 \\ -x & \text{if } -4 \leq x \leq 2 \\ 3x & \text{if } x > 2 \end{cases}$

20. $f(x) = \begin{cases} -2x & \text{if } x < -3 \\ 3x - 1 & \text{if } -3 \leq x \leq 2 \\ -4x & \text{if } x > 2 \end{cases}$

Graph each piecewise-defined function. See Example 2.

21. $f(x) = \begin{cases} x - 1 & \text{if } x \leq 3 \\ 2 & \text{if } x > 3 \end{cases}$

22. $f(x) = \begin{cases} 6 - x & \text{if } x \leq 3 \\ 3x - 6 & \text{if } x > 3 \end{cases}$

23. $f(x) = \begin{cases} 4 - x & \text{if } x < 2 \\ 1 + 2x & \text{if } x \geq 2 \end{cases}$

24. $f(x) = \begin{cases} 2x + 1 & \text{if } x \geq 0 \\ x & \text{if } x < 0 \end{cases}$

25. $f(x) = \begin{cases} 5x - 4 & \text{if } x \leq 1 \\ x & \text{if } x > 1 \end{cases}$

26. $f(x) = \begin{cases} -2 & \text{if } x \leq 1 \\ 2 & \text{if } x > 1 \end{cases}$

27. $f(x) = \begin{cases} 2 + x & \text{if } x < -4 \\ -x & \text{if } -4 \leq x \leq 5 \\ 3x & \text{if } x > 5 \end{cases}$

28. $f(x) = \begin{cases} -2x & \text{if } x < -3 \\ 3x - 1 & \text{if } -3 \leq x \leq 2 \\ -4x & \text{if } x > 2 \end{cases}$

Concept Check *Give a rule for each piecewise-defined function. Also give the domain and range.*

29.

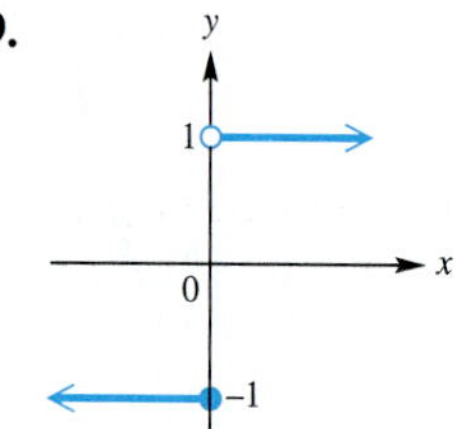

30.

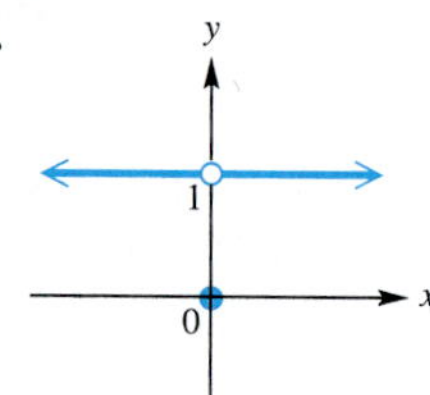

31.

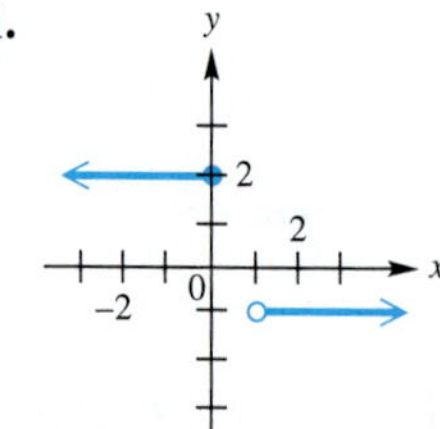

32.

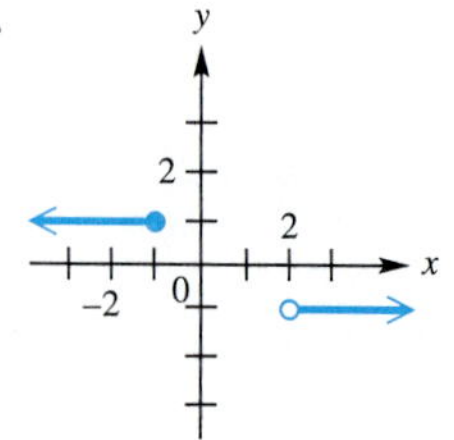

Graph each function. Give the domain and range. See Example 3.

33. $f(x) = [\![-x]\!]$

34. $f(x) = [\![2x]\!]$

35. $g(x) = [\![2x - 1]\!]$

36. *Concept Check* If x is an even integer and $f(x) = [\![\frac{1}{2}x]\!]$, how would you describe the function value?

(Modeling) Solve each problem. See Example 4.

37. *Postage Charges* Assume that postage rates are 37¢ for the first ounce, plus 23¢ for each additional ounce, and that each letter carries one 37¢ stamp and as many 23¢ stamps as necessary. Graph the function f that models the number of stamps on a letter weighing x ounces over the interval $(0, 5]$.

38. *Airport Parking Charges* The cost of parking a car at an airport hourly parking lot is \$3 for the first half-hour and \$2 for each additional half-hour or fraction of a half-hour. Graph the function f that models the cost of parking a car for x hours over the interval $(0, 2]$.

Match each piecewise-defined function with its calculator graph. (All graphs are shown in dot mode.)

39. $f(x) = \begin{cases} x^2 - 4 & \text{if } x \geq 0 \\ -x + 5 & \text{if } x < 0 \end{cases}$

40. $g(x) = \begin{cases} |x - 4| & \text{if } x \geq -1 \\ -x^2 & \text{if } x < -1 \end{cases}$

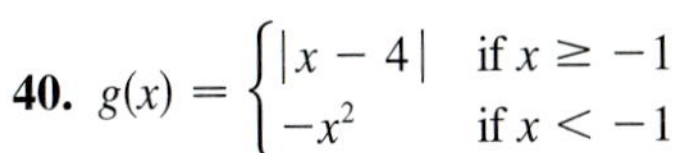

41. $h(x) = \begin{cases} 6 & \text{if } x \geq 0 \\ -6 & \text{if } x < 0 \end{cases}$

42. $k(x) = \begin{cases} \sqrt{x} & \text{if } x \geq 0 \\ -x^2 & \text{if } x < 0 \end{cases}$

A.

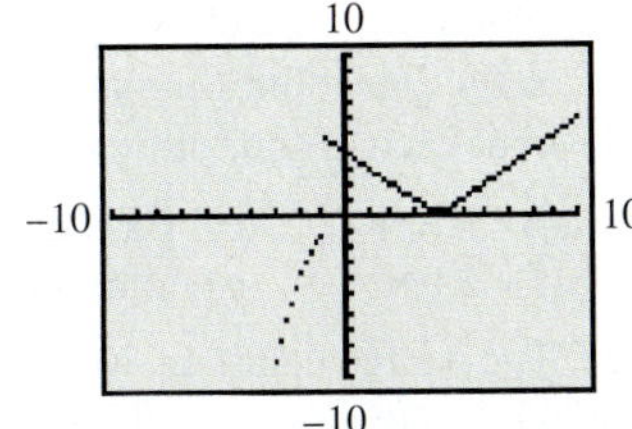

B.

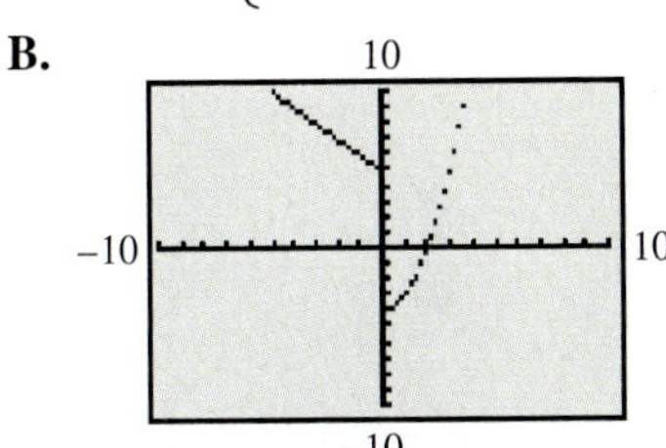

C.

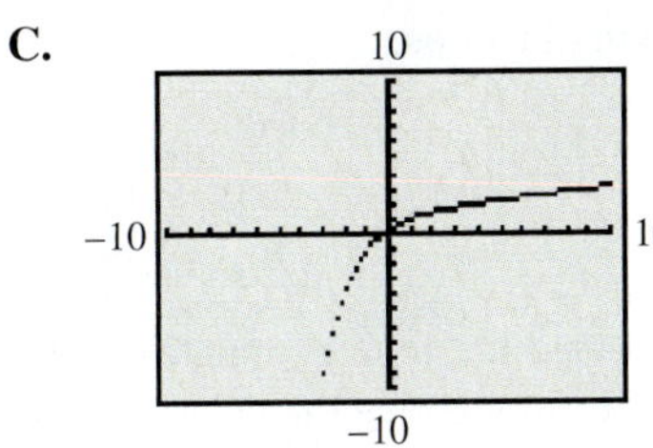

D.

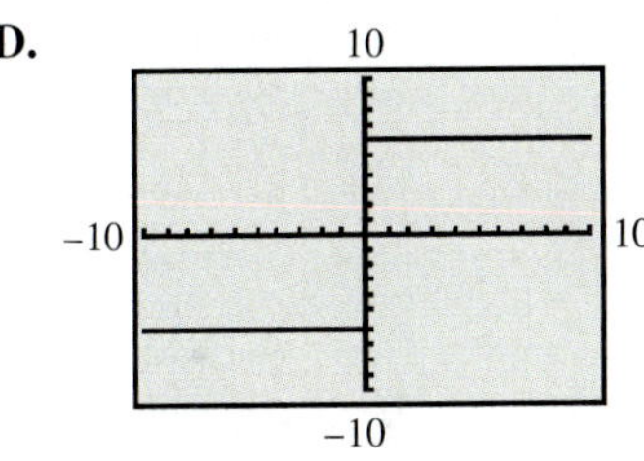

(Modeling) Solve each problem.

43. *Light Vehicle Market Share* The light vehicle market share (in percent) in the U.S. for domestic cars is shown in the graph. Let $x = 3$ represent 1993, $x = 6$ represent 1996, and so on.

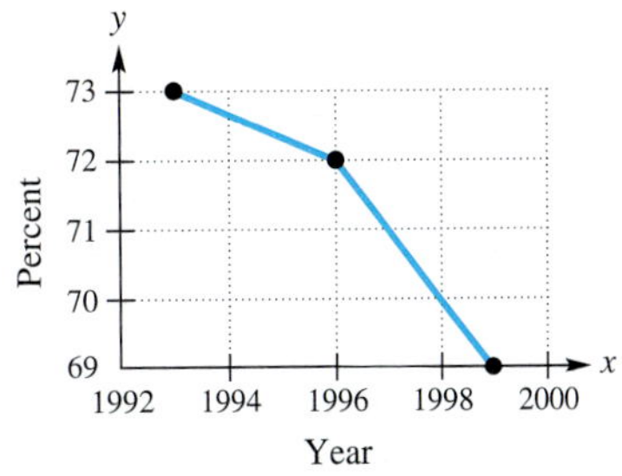

Source: J.D. Power & Associates.

(a) Use the points on the graph to write equations for the line segments in the intervals $[3, 6]$ and $(6, 9]$.

(b) Define $f(x)$ for the piecewise-defined function.

44. *Car Rental* To rent a midsized car from Avis costs \$30 per day or fraction of a day. If you pick up the car in Lansing and drop it off in West Lafayette, there is a fixed \$50 dropoff charge. Let $C(x)$ represent the cost of renting the car for x days, taking it from Lansing to West Lafayette. Find each of the following.

(a) $C\left(\frac{3}{4}\right)$ **(b)** $C\left(\frac{9}{10}\right)$ **(c)** $C(1)$ **(d)** $C\left(1\frac{5}{8}\right)$ **(e)** $C(2.4)$

(f) Graph $y = C(x)$.

45. *Insulin Level* When a diabetic takes long-acting insulin, the insulin reaches its peak effect on the blood sugar level in about 3 hr. This effect remains fairly constant for 5 hr, then declines, and is very low until the next injection. In a typical patient, the level of insulin might be modeled by the following function.

$$i(t) = \begin{cases} 40t + 100 & \text{if } 0 \le t \le 3 \\ 220 & \text{if } 3 < t \le 8 \\ -80t + 860 & \text{if } 8 < t \le 10 \\ 60 & \text{if } 10 < t \le 24 \end{cases}$$

Here $i(t)$ is the blood sugar level, in appropriate units, at time t measured in hours from the time of the injection. Suppose a patient takes insulin at 6 A.M. Find the blood sugar level at each of the following times.

(a) 7 A.M. **(b)** 9 A.M. **(c)** 10 A.M. **(d)** noon
(e) 3 P.M. **(f)** 5 P.M. **(g)** midnight **(h)** Graph $y = i(t)$.

46. *Snow Depth* The snow depth in Michigan's Isle Royale National Park varies throughout the winter. In a typical winter, the snow depth in inches is approximated by the following function.

$$f(x) = \begin{cases} 6.5x & \text{if } 0 \le x \le 4 \\ -5.5x + 48 & \text{if } 4 < x \le 6 \\ -30x + 195 & \text{if } 6 < x \le 6.5 \end{cases}$$

Here, x represents the time in months with $x = 0$ representing the beginning of October, $x = 1$ representing the beginning of November, and so on.

(a) Graph $y = f(x)$.
(b) In what month is the snow deepest? What is the deepest snow depth?
(c) In what months does the snow begin and end?

Graph each equation in the standard viewing window of a graphing calculator. First solve for y_1 and y_2, and give the equations for both.

47. $x = y^2$

48. $x = |y|$

2.6 Graphing Techniques

Stretching and Shrinking ▪ **Reflecting** ▪ **Symmetry** ▪ **Even and Odd Functions** ▪ **Translations**

One of the main objectives of this course is to recognize and learn to graph various functions. Graphing techniques presented in this section show how to graph functions that are defined by altering the equation of a basic function.

Stretching and Shrinking We begin by considering how the graph of $y = a \cdot f(x)$ or $y = f(ax)$ compares to the graph of $y = f(x)$.

EXAMPLE 1 Stretching or Shrinking a Graph

Graph each function.

(a) $g(x) = 2|x|$ **(b)** $h(x) = \frac{1}{2}|x|$ **(c)** $k(x) = |2x|$

Solution

(a) Comparing the tables of values for $f(x) = |x|$ and $g(x) = 2|x|$ in Figure 67, we see that for corresponding x-values, the y-values of g are each twice those of f. Thus the graph of $g(x)$, shown in blue in Figure 67, is narrower than that of $f(x)$, shown in red for comparison.

x	$f(x)$ $\lvert x \rvert$	$g(x)$ $2\lvert x \rvert$
−2	2	4
−1	1	2
0	0	0
1	1	2
2	2	4

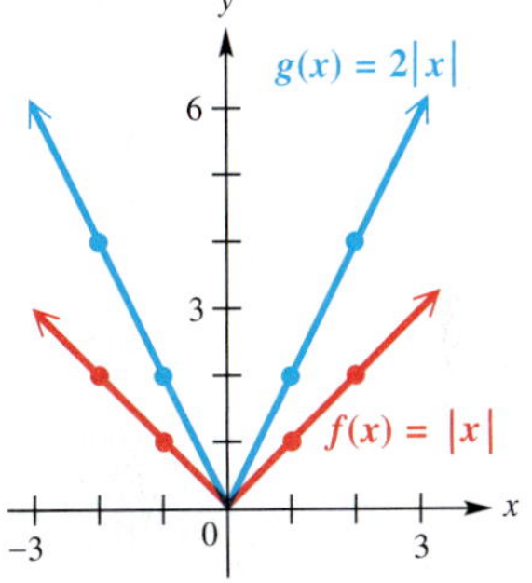

Figure 67

(b) The graph of $h(x)$ is also the same general shape as that of $f(x)$, but here the coefficient $\frac{1}{2}$ causes the graph of $h(x)$ to be wider than the graph of $f(x)$, as we see by comparing the tables of values. See Figure 68.

x	$f(x)$ $\lvert x \rvert$	$h(x)$ $\frac{1}{2}\lvert x \rvert$
−2	2	1
−1	1	$\frac{1}{2}$
0	0	0
1	1	$\frac{1}{2}$
2	2	1

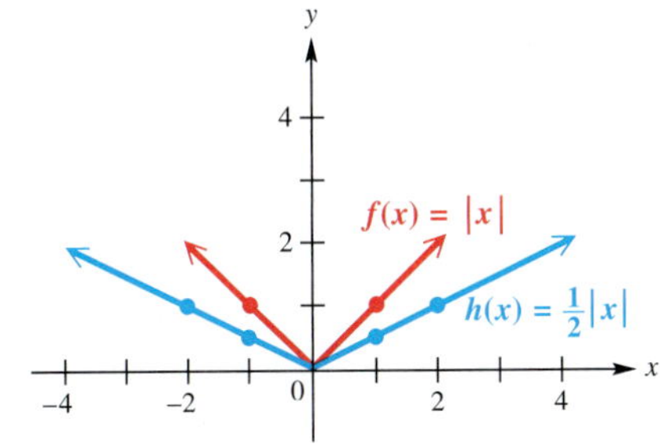

Figure 68

(c) Use Property 2 of absolute value ($|ab| = |a| \cdot |b|$) to rewrite $|2x|$.

$$k(x) = |2x| = |2| \cdot |x| = 2|x| \quad \text{Property 2 (Section R.2)}$$

Therefore, the graph of $k(x) = |2x|$ is the same as the graph of $g(x) = 2|x|$ in part (a). See Figure 67 on the previous page.

Now try Exercises 5 and 7.

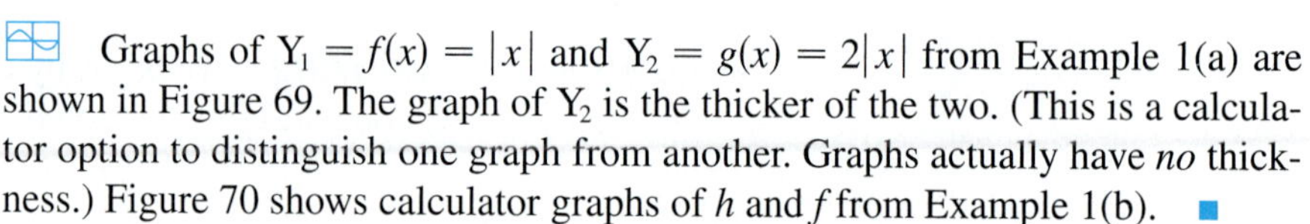

Graphs of $Y_1 = f(x) = |x|$ and $Y_2 = g(x) = 2|x|$ from Example 1(a) are shown in Figure 69. The graph of Y_2 is the thicker of the two. (This is a calculator option to distinguish one graph from another. Graphs actually have *no* thickness.) Figure 70 shows calculator graphs of h and f from Example 1(b). ■

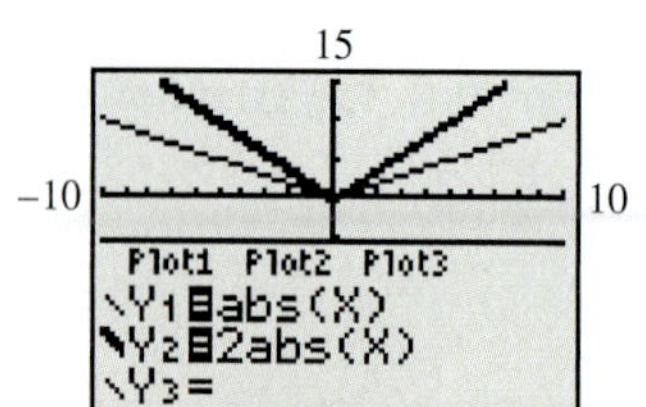

Figure 69

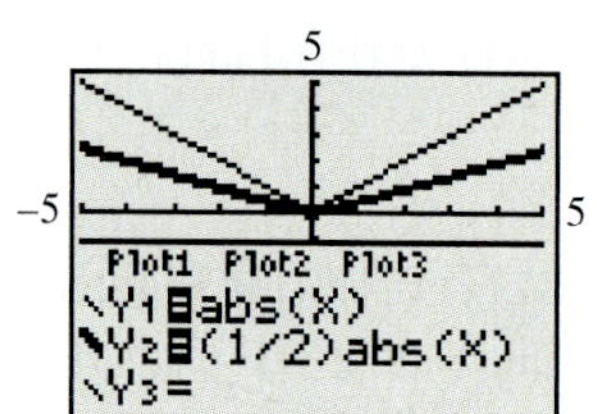

Figure 70

The graphs in Example 1 suggest the following generalizations.

Stretching and Shrinking

The graph of $\mathbf{g(x) = a \cdot f(x)}$ has the same general shape as the graph of $f(x)$. If $|a| > 1$, then the graph is stretched vertically (narrower) compared to the graph of $f(x)$.

If $0 < |a| < 1$, then the graph is shrunken vertically (wider) compared to the graph of $f(x)$.

In general, the larger the value of $|a|$, the greater the stretch. The smaller the value of $|a|$, the greater the shrink.

Reflecting Forming the mirror image of a graph across a line is called *reflecting the graph across the line.*

EXAMPLE 2 Reflecting a Graph Across an Axis

Graph each function.

(a) $g(x) = -\sqrt{x}$ **(b)** $h(x) = \sqrt{-x}$

Solution

(a) The tables of values for $g(x) = -\sqrt{x}$ and $f(x) = \sqrt{x}$ are shown with their graphs in Figure 71. As the tables suggest, every y-value of the graph of $g(x) = -\sqrt{x}$ is the negative of the corresponding y-value of $f(x) = \sqrt{x}$. This has the effect of reflecting the graph across the x-axis.

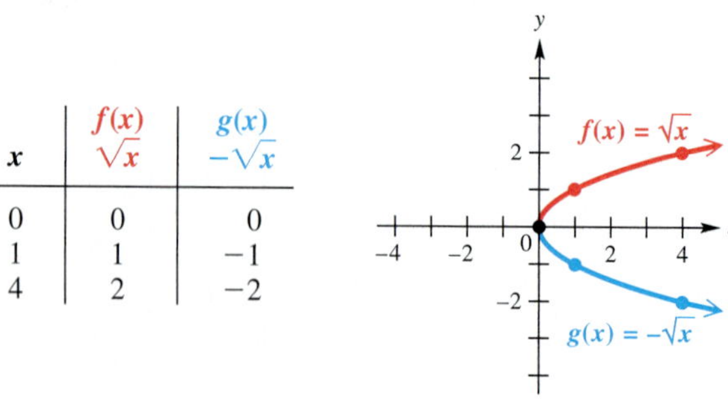

x	$f(x)$ $\sqrt{x}$	$g(x)$ $-\sqrt{x}$
0	0	0
1	1	−1
4	2	−2

Figure 71

(b) The domain of $h(x) = \sqrt{-x}$ is $x \le 0$, while the domain of $f(x) = \sqrt{x}$ is $x \ge 0$. If we choose x-values for $h(x)$ that are the negatives of those we use for $f(x)$, we see that the corresponding y-values are the same. Thus, the graph of h is a reflection of the graph of f across the y-axis. See Figure 72.

x	$f(x)$ $\sqrt{x}$	$h(x)$ $\sqrt{-x}$
-4	undefined	2
-1	undefined	1
0	0	0
1	1	undefined
4	2	undefined

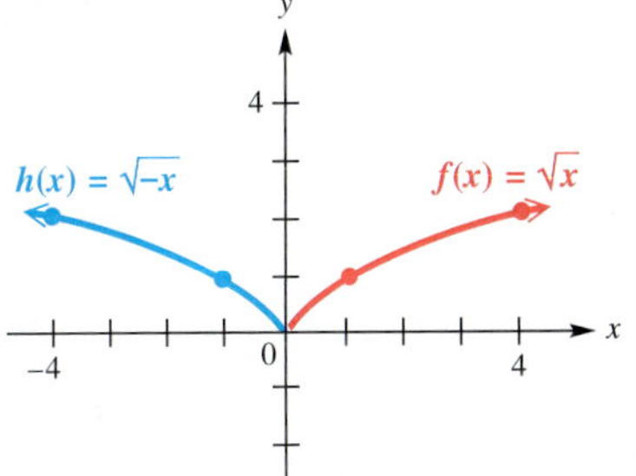

Figure 72

Now try Exercises 9 and 11.

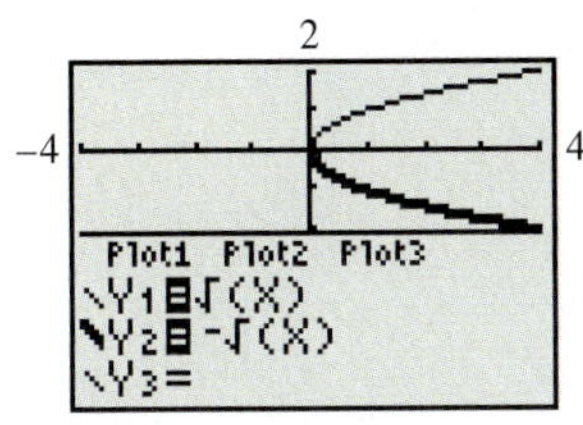

Figure 73

See Figure 73 for calculator graphs of g and f from Example 2(a). Calculator graphs for Example 2(b) are in Figure 74. ■

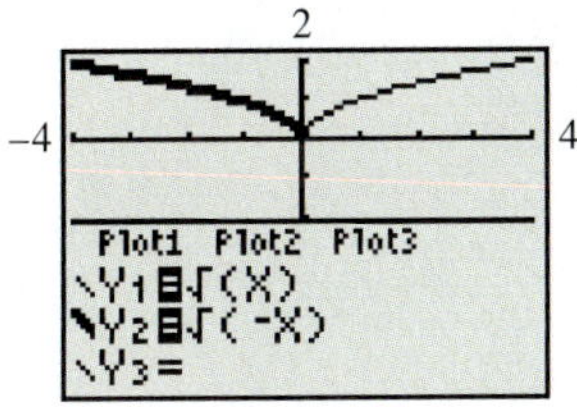

Figure 74

The graphs in Example 2 suggest the following generalizations.

Reflecting Across an Axis

The graph of $\boldsymbol{y = -f(x)}$ is the same as the graph of $y = f(x)$ reflected across the x-axis.

The graph of $\boldsymbol{y = f(-x)}$ is the same as the graph of $y = f(x)$ reflected across the y-axis.

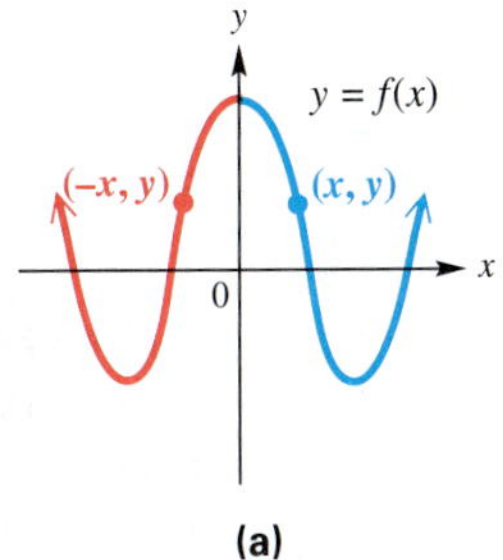

(a)

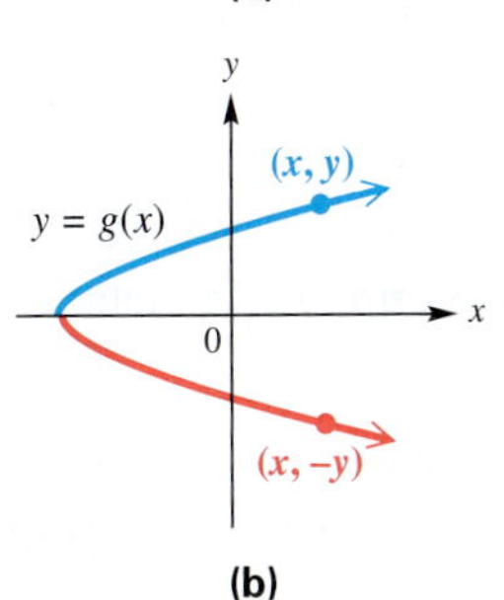

(b)

Figure 75

Notice that if Figure 71 were folded on the x-axis, the graph of $f(x)$ would exactly match the graph of $g(x)$. A similar match occurs if Figure 72 is folded on the y-axis.

Symmetry The graph of f shown in Figure 75(a) is cut in half by the y-axis with each half the mirror image of the other half. A graph with this property is said to be *symmetric with respect to the y-axis.* As this graph suggests, a graph is symmetric with respect to the y-axis if the point $(-x, y)$ is on the graph whenever the point (x, y) is on the graph.

Similarly, if the graph of g in Figure 75(b) were folded in half along the x-axis, the portion at the top would exactly match the portion at the bottom. Such a graph is *symmetric with respect to the x-axis:* the point $(x, -y)$ is on the graph whenever the point (x, y) is on the graph.

Symmetry with Respect to an Axis

The graph of an equation is **symmetric with respect to the y-axis** if the replacement of x with $-x$ results in an equivalent equation.

The graph of an equation is **symmetric with respect to the x-axis** if the replacement of y with $-y$ results in an equivalent equation.

EXAMPLE 3 Testing for Symmetry with Respect to an Axis

Test for symmetry with respect to the x-axis and the y-axis.

(a) $y = x^2 + 4$ **(b)** $x = y^2 - 3$ **(c)** $x^2 + y^2 = 16$ **(d)** $2x + y = 4$

Solution

(a) In $y = x^2 + 4$, replace x with $-x$.

$$y = x^2 + 4$$
$$y = (-x)^2 + 4 \quad \text{Equivalent}$$
$$y = x^2 + 4$$

The result is the same as the original equation, so the graph, shown in Figure 76, is symmetric with respect to the y-axis. Check algebraically that the graph is *not* symmetric with respect to the x-axis.

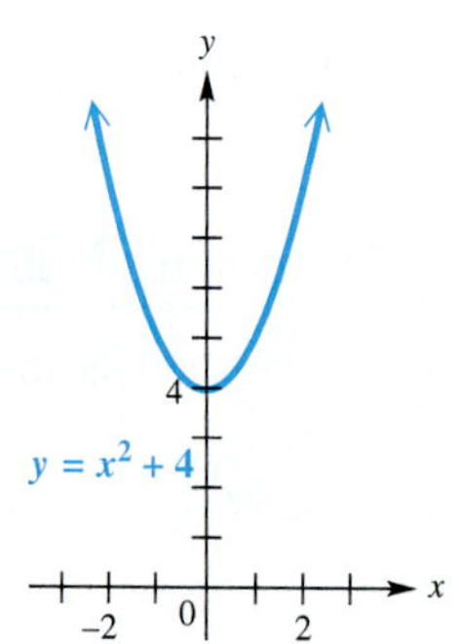

Figure 76

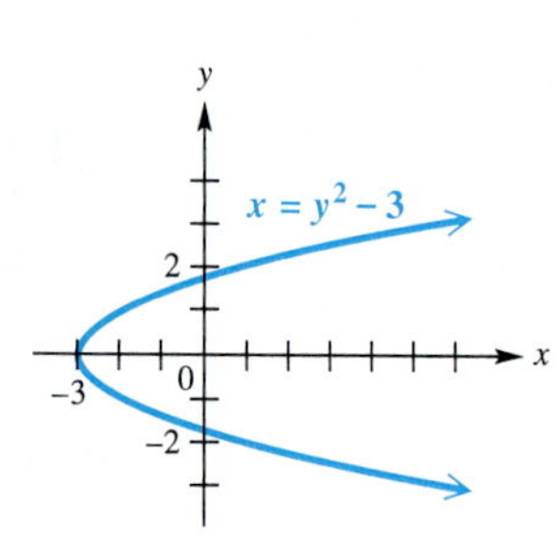

Figure 77

(b) In $x = y^2 - 3$, replace y with $-y$ to get $x = (-y)^2 - 3 = y^2 - 3$, the same as the original equation. The graph is symmetric with respect to the x-axis, as shown in Figure 77. It is not symmetric with respect to the y-axis.

(c) Substituting $-x$ for x and $-y$ for y in $x^2 + y^2 = 16$, we get

$$(-x)^2 + y^2 = 16 \quad \text{and} \quad x^2 + (-y)^2 = 16.$$

Both simplify to $\quad x^2 + y^2 = 16.$

Thus the graph, a circle of radius 4 centered at the origin, is symmetric with respect to both axes.

(d) In $2x + y = 4$, replace x with $-x$, and then replace y with $-y$; neither case produces an equivalent equation. This graph is not symmetric with respect to either axis.

Looking Ahead to Calculus

The tools of calculus allow us to find areas of regions in the plane. To find the area of the region below the graph of $y = x^2$, above the x-axis, bounded on the left by the line $x = -2$, and on the right by $x = 2$, draw a sketch of this region. Notice that due to the symmetry of the graph of $y = x^2$, the desired area is twice that of the area to the right of the y-axis. Thus, symmetry can be used to reduce the original problem to an easier one by simply finding the area to the right of the y-axis and then doubling the answer.

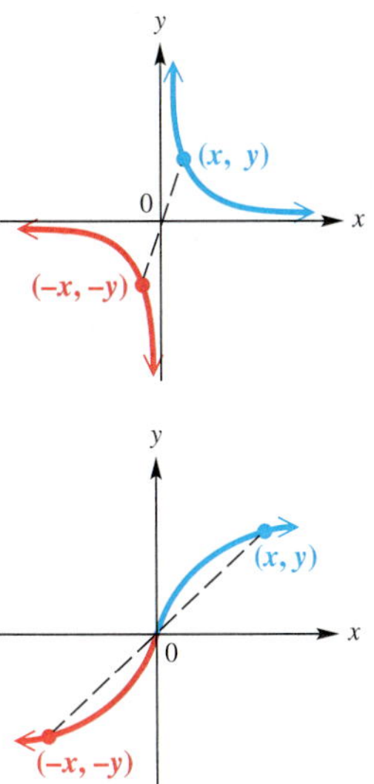

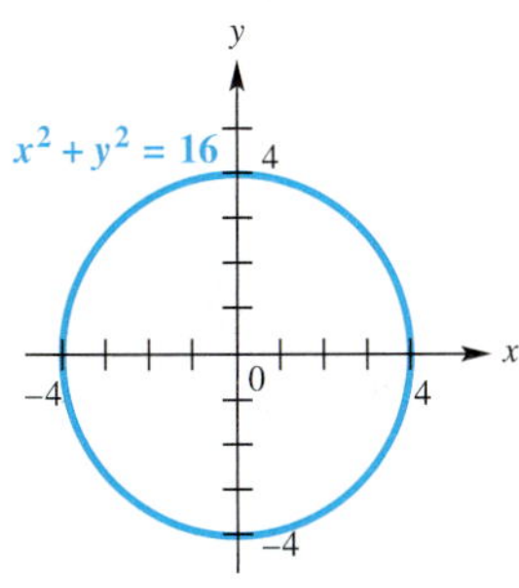

Figure 78

Another kind of symmetry occurs when a graph can be rotated 180° about the origin, with the result coinciding exactly with the original graph. Symmetry of this type is called *symmetry with respect to the origin.* A graph is symmetric with respect to the origin if the point $(-x, -y)$ is on the graph whenever the point (x, y) is on the graph. Figure 78 shows two graphs that are symmetric with respect to the origin.

Symmetry with Respect to the Origin

The graph of an equation is **symmetric with respect to the origin** if the replacement of both x with $-x$ and y with $-y$ results in an equivalent equation.

EXAMPLE 4 Testing for Symmetry with Respect to the Origin

Are the following graphs symmetric with respect to the origin?

(a) $x^2 + y^2 = 16$ **(b)** $y = x^3$

Solution

(a) Replace x with $-x$ and y with $-y$.

$$x^2 + y^2 = 16$$
$$(-x)^2 + (-y)^2 = 16$$
$$x^2 + y^2 = 16$$
Equivalent

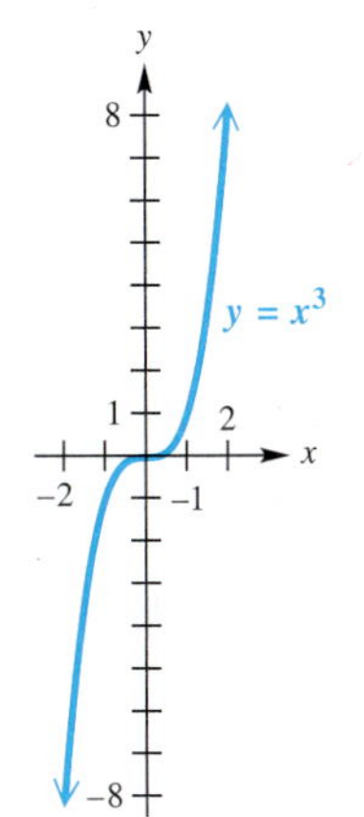

Figure 79

The graph, shown in Figure 79, is symmetric with respect to the origin.

(b) Replace x with $-x$ and y with $-y$.

$$y = x^3$$
$$-y = (-x)^3$$
$$-y = -x^3$$
$$y = x^3$$
Equivalent

The graph, symmetric with respect to the origin, is shown in Figure 80.

Figure 80

Now try Exercise 23.

A graph symmetric with respect to both the x- and y-axes is automatically symmetric with respect to the origin. However, a graph symmetric with respect to the origin need not be symmetric with respect to either axis. (See Figure 80.) Of the three types of symmetry—with respect to the x-axis, the y-axis, and the origin—a graph possessing any two must also exhibit the third type.

The various tests for symmetry are summarized below.

Tests for Symmetry

	Symmetry with Respect to:		
	x-Axis	**y-Axis**	**Origin**
Equation is unchanged if:	y is replaced with $-y$	x is replaced with $-x$	x is replaced with $-x$ and y is replaced with $-y$
Example:			

Now try Exercises 19, 21, and 25.

Even and Odd Functions The concepts of symmetry with respect to the y-axis and symmetry with respect to the origin are closely associated with the concepts of *even* and *odd functions.*

Even and Odd Functions

A function f is called an **even function** if $f(-x) = f(x)$ for all x in the domain of f. (Its graph is symmetric with respect to the y-axis.)

A function f is called an **odd function** if $f(-x) = -f(x)$ for all x in the domain of f. (Its graph is symmetric with respect to the origin.)

EXAMPLE 5 Determining Whether Functions Are Even, Odd, or Neither

Decide whether each function defined is even, odd, or neither.

(a) $f(x) = 8x^4 - 3x^2$ **(b)** $f(x) = 6x^3 - 9x$ **(c)** $f(x) = 3x^2 + 5x$

Solution

(a) Replacing x in $f(x) = 8x^4 - 3x^2$ with $-x$ gives

$$f(-x) = 8(-x)^4 - 3(-x)^2 = 8x^4 - 3x^2 = f(x).$$

Since $f(-x) = f(x)$ for each x in the domain of the function, f is even.

(b) Here

$$f(-x) = 6(-x)^3 - 9(-x) = -6x^3 + 9x = -f(x).$$

The function f is odd because $f(-x) = -f(x)$.

(c) $f(x) = 3x^2 + 5x$

$f(-x) = 3(-x)^2 + 5(-x)$ Replace x with $-x$.

$= 3x^2 - 5x$

Since $f(-x) \neq f(x)$ and $f(-x) \neq -f(x)$, f is neither even nor odd.

Now try Exercises 27, 29, and 31.

Translations The next examples show the results of horizontal and vertical shifts, called **translations,** of the graph of $f(x) = |x|$.

EXAMPLE 6 Translating a Graph Vertically

Graph $g(x) = |x| - 4$.

Solution By comparing the table of values for $g(x) = |x| - 4$ and $f(x) = |x|$ shown with Figure 81, we see that for corresponding x-values, the y-values of g are each 4 less than those for f. Thus, the graph of $g(x) = |x| - 4$ is the same as that of $f(x) = |x|$, but translated 4 units down. See Figures 81 and 82. The lowest point is at $(0, -4)$. The graph is symmetric with respect to the y-axis.

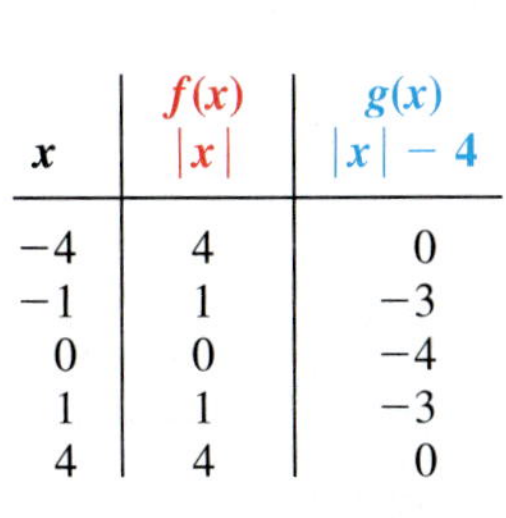

x	$f(x)$ $\lvert x \rvert$	$g(x)$ $\lvert x \rvert - 4$
−4	4	0
−1	1	−3
0	0	−4
1	1	−3
4	4	0

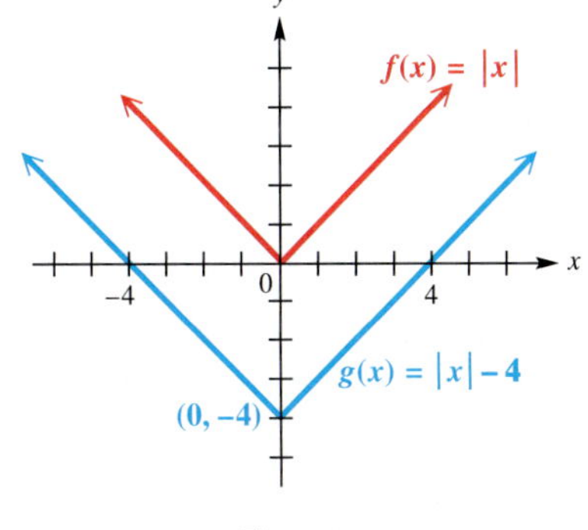

Figure 81

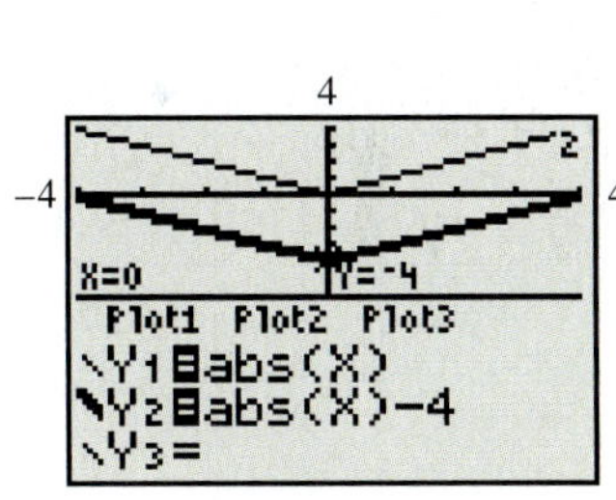

Figure 82

Now try Exercise 33.

The graphs in Example 6 suggest the following generalization.

Vertical Translations

If a function g is defined by $\mathbf{g(x) = f(x) + c}$, where c is a real number, then for every point (x, y) on the graph of f, there will be a corresponding point $(x, y + c)$ on the graph of g. The graph of g will be the same as the graph of f, but translated c units up if c is positive or $|c|$ units down if c is negative. The graph of g is called a **vertical translation** of the graph of f.

Figure 83 shows a graph of a function f and two vertical translations of f. Figure 84 shows two vertical translations of $Y_1 = x^2$.

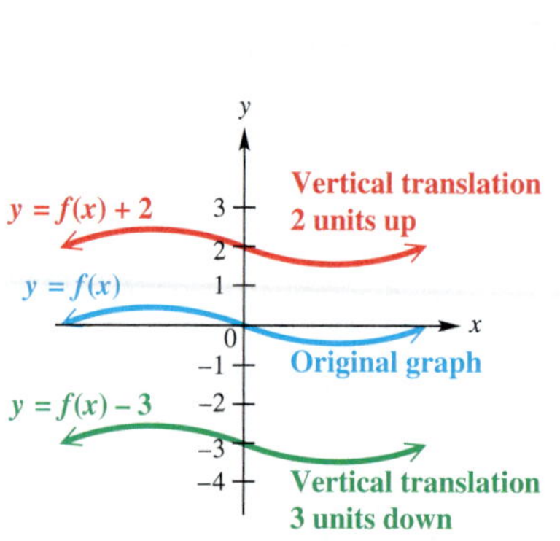

Figure 83

Y_2 is the graph of Y_1 translated 2 units *up*. Y_3 is that of Y_1 translated 6 units *down*.

Figure 84

EXAMPLE 7 Translating a Graph Horizontally

Graph $g(x) = |x - 4|$.

Solution Comparing the tables of values given with Figure 85 shows that for corresponding y-values, the x-values of g are each 4 *more* than those for f. The graph of $g(x) = |x - 4|$ is the same as that of $f(x) = |x|$, but translated 4 units to the right. The lowest point is at $(4, 0)$. As suggested by the graphs in Figures 85 and 86, this graph is symmetric with respect to the line $x = 4$.

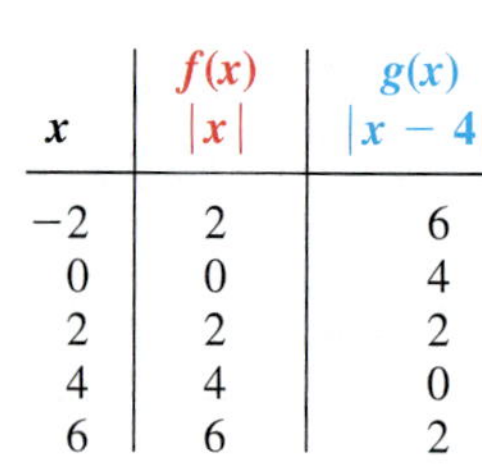

x	$f(x)$ $\lvert x \rvert$	$g(x)$ $\lvert x - 4 \rvert$
−2	2	6
0	0	4
2	2	2
4	4	0
6	6	2

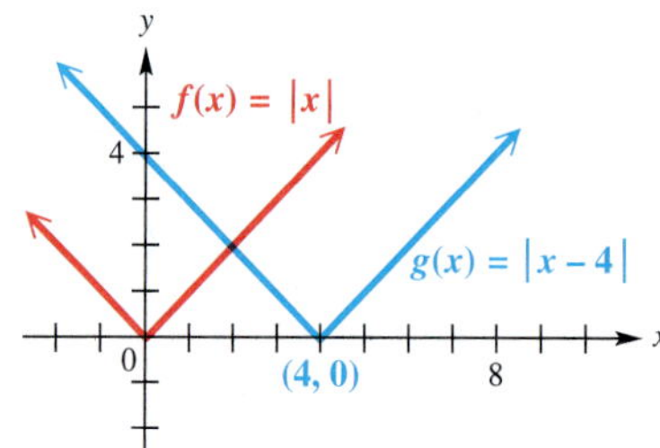

Figure 85

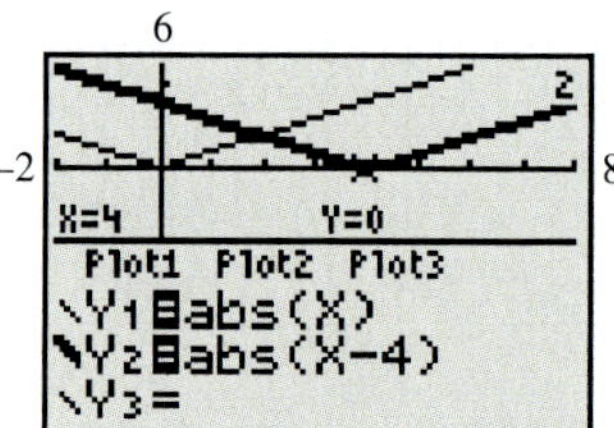

Figure 86

Now try Exercise 37.

The graphs in Example 7 suggest the following generalization.

Horizontal Translations

If a function g is defined by $g(x) = f(x - c)$, where c is a real number, then for every point (x, y) on the graph of f, there will be a corresponding point $(x + c, y)$ on the graph of g. The graph of g will be the same as the graph of f, but translated c units to the right if c is positive or $|c|$ units to the left if c is negative. The graph of g is called a **horizontal translation** of the graph of f.

Figure 87 shows a graph of a function f and two horizontal translations of f. Figure 88 shows two horizontal translations of $Y_1 = x^2$.

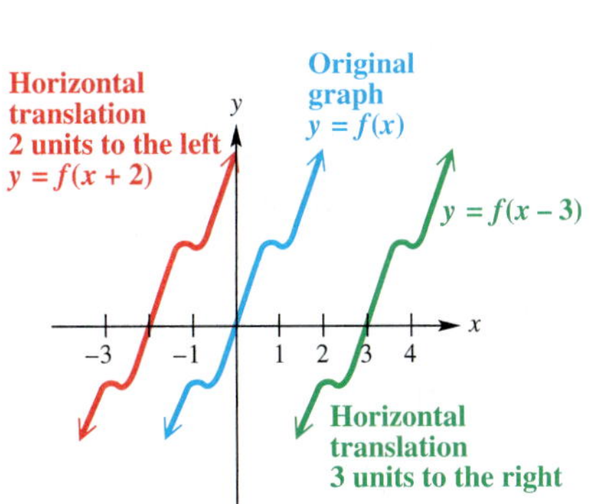

Figure 87

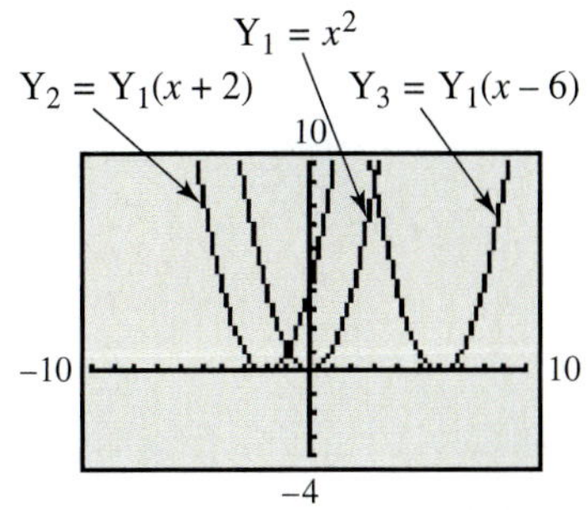

Y_2 is the graph of Y_1 translated 2 units to the *left*. Y_3 is that of Y_1 translated 6 units to the *right*.

Figure 88

Vertical and horizontal translations are summarized in the table, where f is a function, and c is a positive number.

To Graph:	Shift the Graph of $y = f(x)$ by c Units:
$y = f(x) + c$	up
$y = f(x) - c$	down
$y = f(x + c)$	left
$y = f(x - c)$	right

CAUTION Be careful when translating graphs horizontally. To determine the direction and magnitude of horizontal translations, find the value that would cause the expression in parentheses to equal 0. For example, the graph of $y = (x - 5)^2$ would be translated 5 units to the *right* of $y = x^2$, because $x = +5$ would cause $x - 5$ to equal 0. On the other hand, the graph of $y = (x + 5)^2$ would be translated 5 units to the *left* of $y = x^2$, because $x = -5$ would cause $x + 5$ to equal 0.

EXAMPLE 8 Using More Than One Transformation on Graphs

Graph each function.

(a) $f(x) = -|x + 3| + 1$ **(b)** $h(x) = |2x - 4|$ **(c)** $g(x) = -\frac{1}{2}x^2 + 4$

Solution

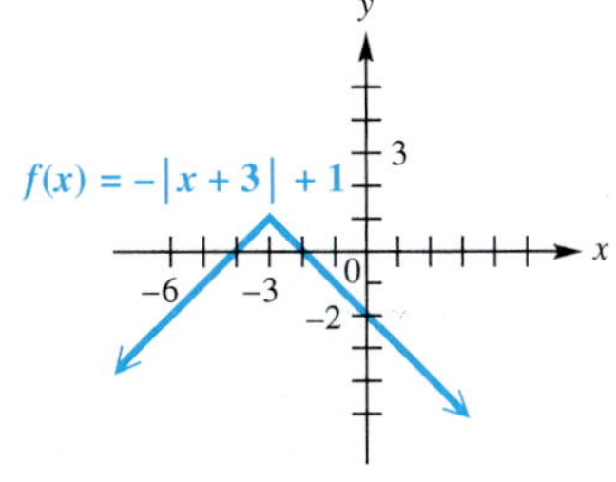

Figure 89

(a) The *lowest* point on the graph of $y = |x|$ is translated 3 units to the left and 1 unit up. The graph opens down because of the negative sign in front of the absolute value expression, making the lowest point now the highest point on the graph, as shown in Figure 89. The graph is symmetric with respect to the line $x = -3$.

(b) To determine the horizontal translation, factor out 2.

$$\begin{aligned} h(x) &= |2x - 4| \\ &= |2(x - 2)| && \text{Factor out 2.} \\ &= |2| \cdot |x - 2| && |ab| = |a| \cdot |b| \\ &= 2|x - 2| && |2| = 2 \end{aligned}$$

The graph of h is the graph of $y = |x|$ translated 2 units to the right, and stretched by a factor of 2. See Figure 90.

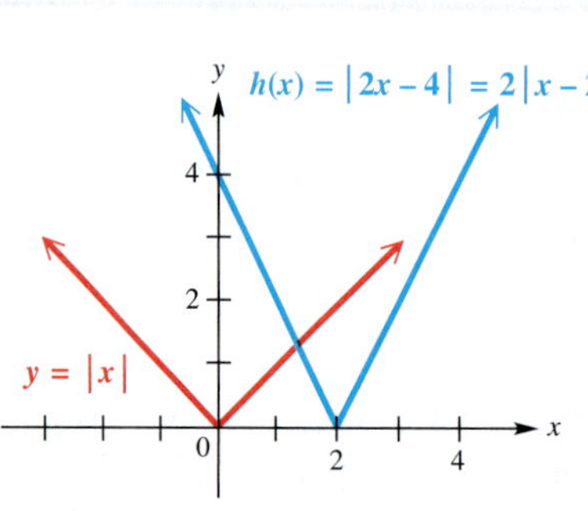

Figure 90

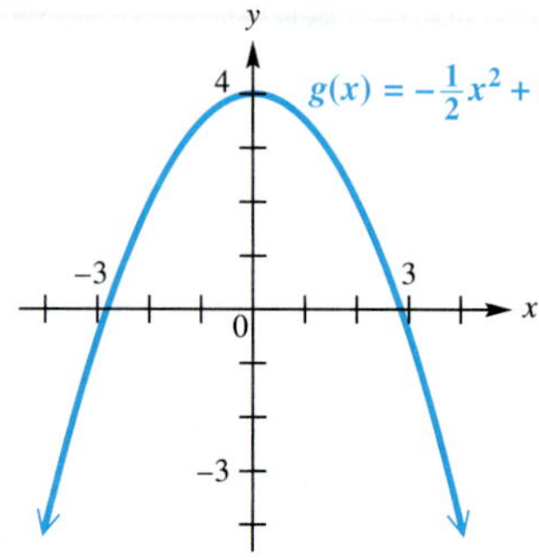

Figure 91

(c) The graph of $g(x) = -\frac{1}{2}x^2 + 4$ will have the same shape as that of $y = x^2$, but is wider (that is, shrunken vertically) and reflected across the x-axis because of the coefficient $-\frac{1}{2}$, and then translated 4 units up. See Figure 91.

Now try Exercises 41, 43, and 45.

EXAMPLE 9 Graphing Translations Given the Graph of $y = f(x)$

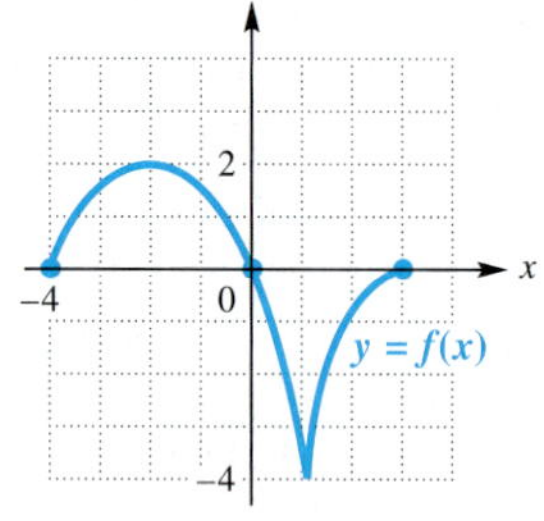

Figure 92

A graph of a function defined by $y = f(x)$ is shown in Figure 92. Use this graph to sketch each of the following graphs.

(a) $g(x) = f(x) + 3$ **(b)** $h(x) = f(x + 3)$ **(c)** $k(x) = f(x - 2) + 3$

Solution

(a) The graph of $g(x) = f(x) + 3$ is the same as the graph in Figure 92, translated 3 units up. See Figure 93(a).

(b) To get the graph of $h(x) = f(x + 3)$, translate the graph of $y = f(x)$ 3 units to the left since $x + 3 = 0$ if $x = -3$. See Figure 93(b).

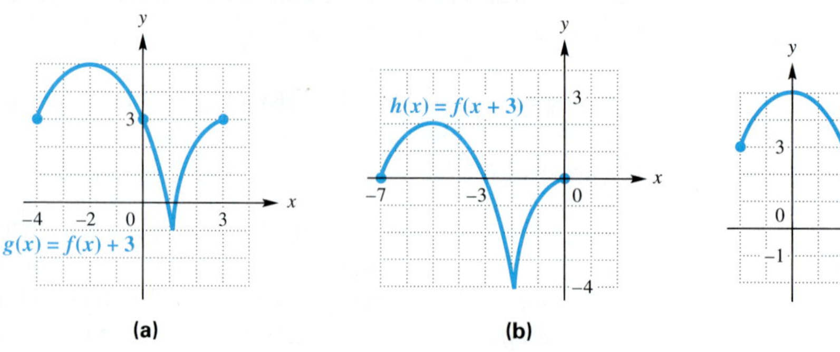

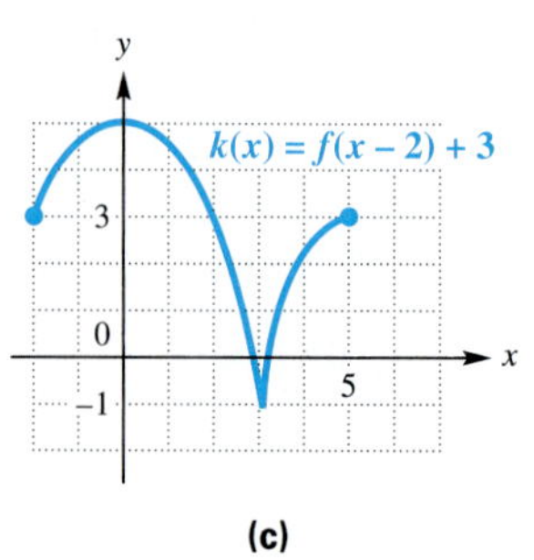

(a) (b) (c)

Figure 93

(c) The graph of $k(x) = f(x - 2) + 3$ will look like the graph of $f(x)$ translated 2 units to the right and 3 units up, as shown in Figure 93(c).

Now try Exercise 49.

Summary of Graphing Techniques

In the descriptions that follow, assume that $a > 0$, $h > 0$, and $k > 0$. In comparison with the graph of $y = f(x)$:

1. The graph of $y = f(x) + k$ is translated k units up.

2. The graph of $y = f(x) - k$ is translated k units down.

3. The graph of $y = f(x + h)$ is translated h units to the left.

4. The graph of $y = f(x - h)$ is translated h units to the right.

4. The graph of $y = a \cdot f(x)$ is stretched vertically by a factor of a, if $a > 1$.

6. The graph of $y = a \cdot f(x)$ is shrunken vertically by a factor of a, if $0 < a < 1$.

7. The graph of $y = -f(x)$ is reflected across the x-axis.

8. The graph of $y = f(-x)$ is reflected across the y-axis.

CONNECTIONS

A figure has *rotational symmetry* around a point P if it coincides with itself by all rotations about P. Because of their complete rotational symmetry, the circle in the plane and the sphere in space were considered by the early Greeks to be the most perfect geometric figures. Aristotle assumed a spherical shape for the celestial bodies because any other would detract from their heavenly perfection.

Symmetry is found throughout nature, from the hexagons of snowflakes to the diatom, a microscopic sea plant. Perhaps the most striking examples of symmetry in nature are crystals.

A diatom

A cross section of tourmaline

Source: Mathematics, Life Science Library, Time Inc., New York, 1963.

For Discussion or Writing

Discuss other examples of symmetry in art and nature.

2.6 Exercises

1. *Concept Check* Match each equation in Column I with a description of its graph from Column II as it relates to the graph of $y = x^2$.

I	II
(a) $y = (x - 7)^2$	**A.** a translation 7 units to the left
(b) $y = x^2 - 7$	**B.** a translation 7 units to the right
(c) $y = 7x^2$	**C.** a translation 7 units up
(d) $y = (x + 7)^2$	**D.** a translation 7 units down
(e) $y = x^2 + 7$	**E.** a vertical stretch by a factor of 7

2. *Concept Check* Match each equation in Column I with a description of its graph from Column II as it relates to the graph of $y = \sqrt[3]{x}$.

I	II
(a) $y = 4\sqrt[3]{x}$	**A.** a translation 4 units to the right
(b) $y = -\sqrt[3]{x}$	**B.** a translation 4 units down
(c) $y = \sqrt[3]{-x}$	**C.** a reflection across the x-axis
(d) $y = \sqrt[3]{x - 4}$	**D.** a reflection across the y-axis
(e) $y = \sqrt[3]{x} - 4$	**E.** a vertical stretch by a factor of 4

3. *Concept Check* Match each equation in parts (a)–(h) with the sketch of its graph.

(a) $y = x^2 + 2$ **(b)** $y = x^2 - 2$ **(c)** $y = (x + 2)^2$
(d) $y = (x - 2)^2$ **(e)** $y = 2x^2$ **(f)** $y = -x^2$
(g) $y = (x - 2)^2 + 1$ **(h)** $y = (x + 2)^2 + 1$

A.
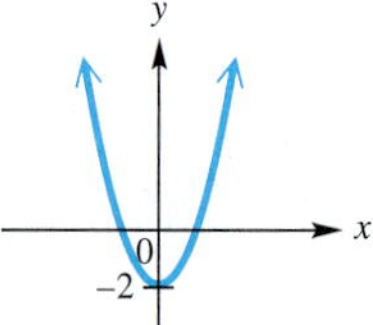

B.
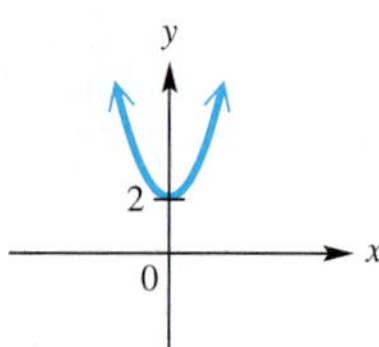

C.
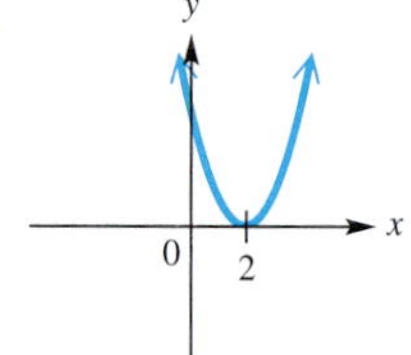

D.
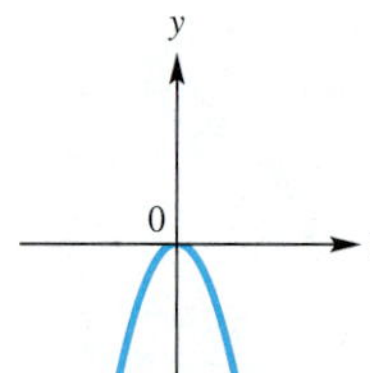

E.
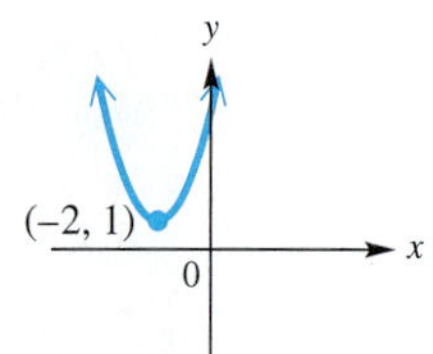

F.
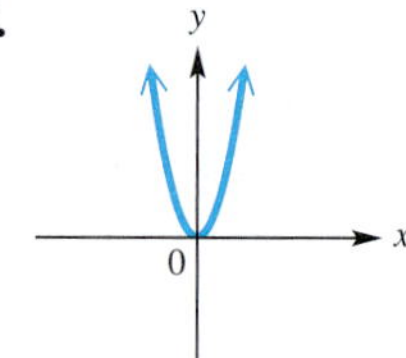

G.
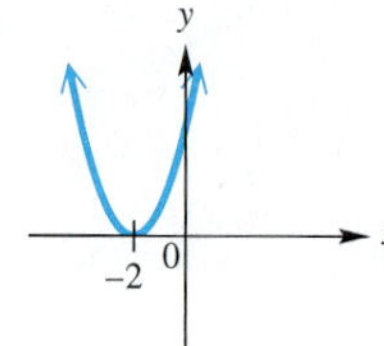

H.
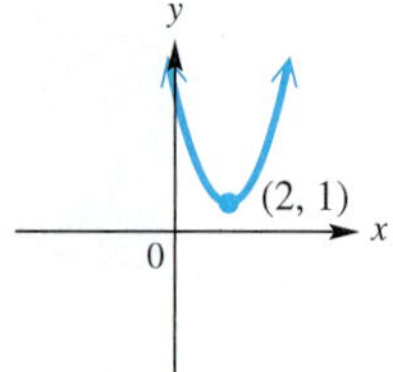

Graph each function. See Examples 1 and 2.

4. $y = 2x^2$ 5. $y = 3|x|$ 6. $y = \frac{1}{3}x^2$ 7. $y = \frac{2}{3}|x|$

8. $y = -\frac{1}{2}x^2$ **9.** $y = -3|x|$ **10.** $y = (-2x)^2$ **11.** $y = \left|-\frac{1}{2}x\right|$

Concept Check Suppose the point $(8, 12)$ *is on the graph of* $y = f(x)$. *Find a point on the graph of each function.*

12. (a) $y = f(x + 4)$
(b) $y = f(x) + 4$

13. (a) $y = \frac{1}{4}f(x)$
(b) $y = 4f(x)$

14. (a) the reflection of the graph of $y = f(x)$ across the x-axis
(b) the reflection of the graph of $y = f(x)$ across the y-axis

Concept Check Plot each point, and then plot the points that are symmetric to the given point with respect to the **(a)** *x-axis,* **(b)** *y-axis, and* **(c)** *origin.*

15. $(5, -3)$ **16.** $(-6, 1)$ **17.** $(-4, -2)$ **18.** $(-8, 0)$

Without graphing, determine whether each equation has a graph that is symmetric with respect to the x-axis, the y-axis, the origin, or none of these. See Examples 3 and 4.

19. $y = x^2 + 2$ **20.** $y = 2x^4 - 1$ **21.** $x^2 + y^2 = 10$ **22.** $y^2 = \frac{-5}{x^2}$

23. $y = -3x^3$ **24.** $y = x^3 - x$ **25.** $y = x^2 - x + 7$ **26.** $y = x + 12$

Decide whether each function is even, odd, or neither. See Example 5.

27. $f(x) = -x^3 + 2x$ **28.** $f(x) = x^5 - 2x^3$

29. $f(x) = .5x^4 - 2x^2 + 1$ **30.** $f(x) = .75x^2 + |x| + 1$

31. $f(x) = x^3 - x + 3$ **32.** $f(x) = x^4 - 5x + 2$

Graph each function. See Examples 6–8.

33. $y = x^2 - 1$ **34.** $y = x^2 + 3$ **35.** $y = x^2 + 2$

36. $y = x^2 - 2$ **37.** $y = (x - 4)^2$ **38.** $y = (x - 2)^2$

39. $y = (x + 2)^2$ **40.** $y = (x + 3)^2$ **41.** $y = |x| - 1$

42. $y = |x + 3| + 2$ **43.** $y = -(x + 1)^3$ **44.** $y = (-x + 1)^3$

45. $y = 2x^2 - 1$ **46.** $y = \frac{2}{3}(x - 2)^2$ **47.** $f(x) = 2(x - 2)^2 - 4$

48. $f(x) = -3(x - 2)^2 + 1$

For Exercises 49 and 50, see Example 9.

49. Given the graph of $y = g(x)$ in the figure, sketch the graph of each function, and explain how it is obtained from the graph of $y = g(x)$.

(a) $y = g(-x)$ **(b)** $y = g(x - 2)$
(c) $y = -g(x) + 2$

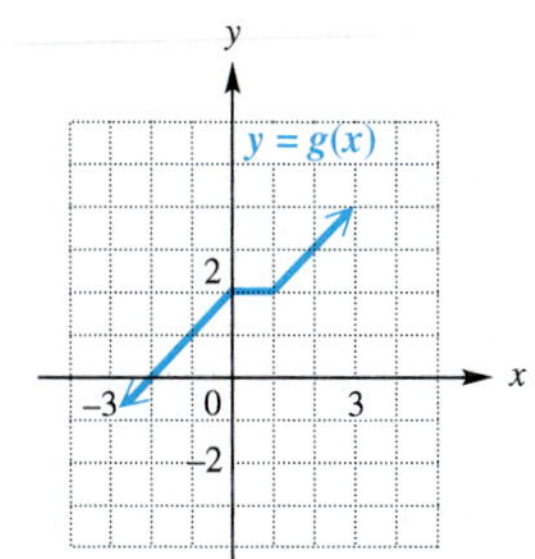

50. Given the graph of $y = f(x)$ in the figure, sketch the graph of each function, and explain how it is obtained from the graph of $y = f(x)$.

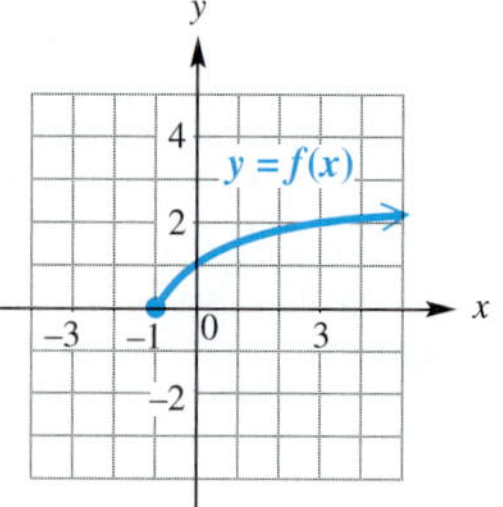

(a) $y = -f(x)$ **(b)** $y = 2f(x)$

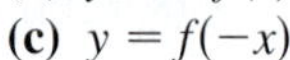
(c) $y = f(-x)$

Concept Check *Each of the following graphs is obtained from the graph of* $f(x) = |x|$ *or* $g(x) = \sqrt{x}$ *by applying several of the transformations discussed in this section. Describe the transformations and give the equation for the graph.*

51.

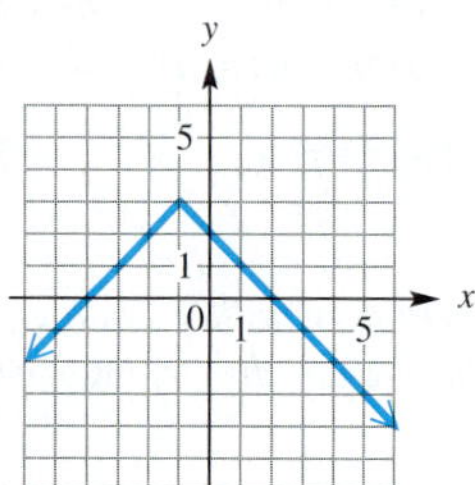

52.

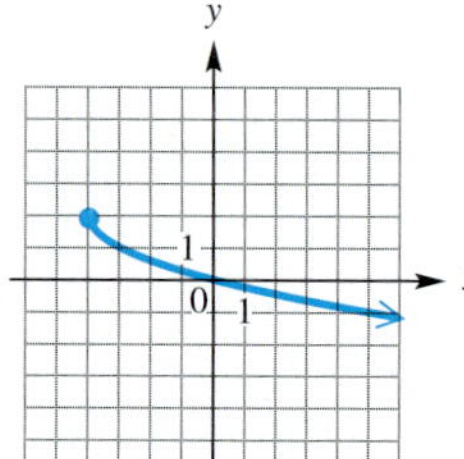

53.

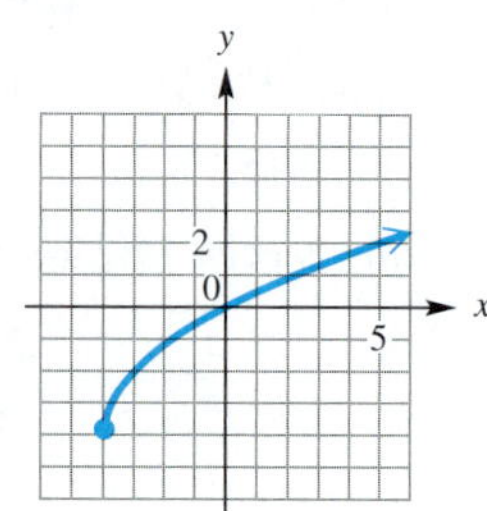

54.

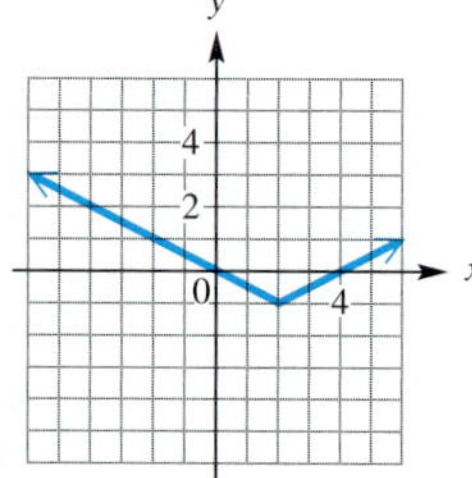

Concept Check *Suppose* $f(3) = 6$. *For the given assumptions in Exercises 55–60, find another function value.*

55. The graph of $y = f(x)$ is symmetric with respect to the origin.

56. The graph of $y = f(x)$ is symmetric with respect to the y-axis.

57. The graph of $y = f(x)$ is symmetric with respect to the line $x = 6$.

58. For all x, $f(-x) = f(x)$.

59. For all x, $f(-x) = -f(x)$.

60. f is an odd function.

61. Find the function g whose graph can be obtained by translating the graph of $f(x) = 2x + 5$ up 2 units and to the left 3 units.

62. Find the function g whose graph can be obtained by translating the graph of $f(x) = 3 - x$ down 2 units and to the right 3 units.

63. Complete the left half of the graph of $y = f(x)$ in the figure for each condition.

(a) $f(-x) = f(x)$ **(b)** $f(-x) = -f(x)$

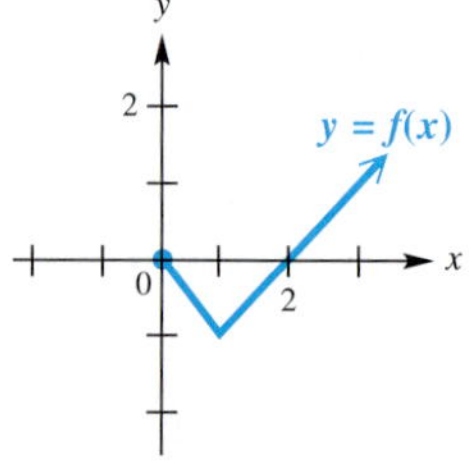

64. Complete the right half of the graph of $y = f(x)$ in the figure for each condition.

(a) f is odd. (b) f is even.

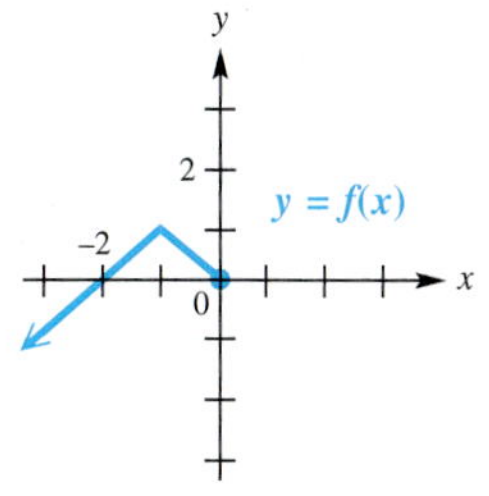

65. Suppose the equation $y = F(x)$ is changed to $y = c \cdot F(x)$, for some constant c. What is the effect on the graph of $y = F(x)$? Discuss the effect depending on whether $c > 0$ or $c < 0$, and $|c| > 1$ or $|c| < 1$.

66. Suppose $y = F(x)$ is changed to $y = F(x + h)$. How are the graphs of these equations related? Is the graph of $y = F(x) + h$ the same as the graph of $y = F(x + h)$? If not, how do they differ?

Relating Concepts

For individual or collaborative investigation
(Exercises 67–73)

In Section 2.3 we introduced linear functions of the form $g(x) = ax + b$. Consider the graph of the simplest linear function defined by $g(x) = x$, shown here. ***Work Exercises 67–73 in order.***

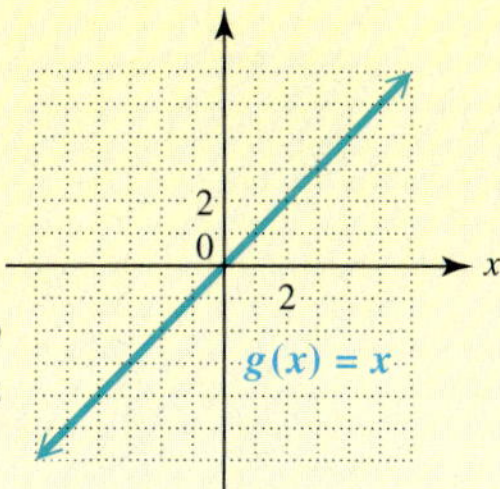

67. How does the graph of $F(x) = x^2 + 6$ compare to the graph of $f(x) = x^2$ if a *vertical* translation is considered?

68. Graph the linear function defined by $G(x) = x + 6$.

69. How does the graph of $G(x) = x + 6$ compare to the graph of $g(x) = x$ if a vertical translation is considered? (*Hint:* Look at the y-intercept.)

70. How does the graph of $F(x) = (x - 6)^2$ compare to the graph of $f(x) = x^2$ if a *horizontal* translation is considered?

71. Graph the linear function defined by $G(x) = x - 6$.

72. How does the graph of $G(x) = x - 6$ compare to the graph of $g(x) = x$ if a horizontal translation is considered? (*Hint:* Look at the x-intercept.)

73. Consider the two functions in the figure.

(a) Find a value of c for which $g(x) = f(x) + c$.
(b) Find a value of c for which $g(x) = f(x + c)$.

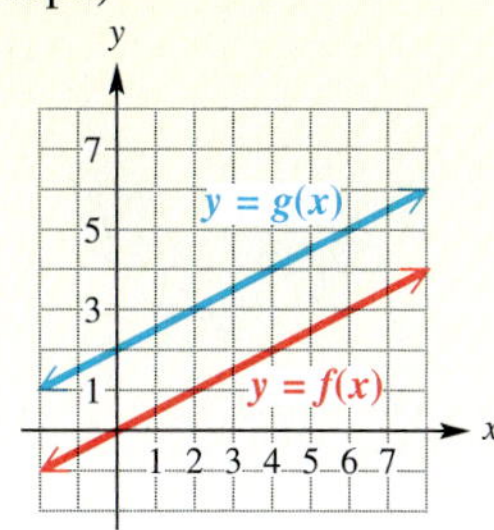

2.7 Function Operations and Composition

Arithmetic Operations on Functions ▪ The Difference Quotient ▪ Composition of Functions

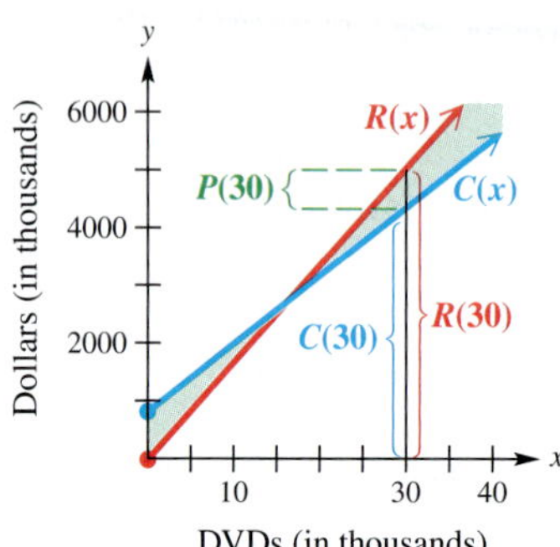

Figure 94

Arithmetic Operations on Functions As mentioned near the end of Section 2.3, economists frequently use the equation "profit equals revenue minus cost," or $P(x) = R(x) - C(x)$, where x is the number of items produced and sold. That is, the profit function is found by subtracting the cost function from the revenue function. Figure 94 shows the situation for a company that manufactures DVDs. The two lines are the graphs of the linear functions for revenue $R(x) = 168x$ and cost $C(x) = 118x + 800$, with x, $R(x)$, and $C(x)$ given in thousands. When 30,000 (that is, 30 thousand) DVDs are produced and sold, profit is

$$\begin{aligned} P(30) &= R(30) - C(30) \\ &= 5040 - 4340 \qquad R(30) = 168(30);\ C(30) = 118(30) + 800 \\ &= 700. \end{aligned}$$

Thus, the profit from the sale of 30,000 DVDs is \$700,000.

New functions can be formed by using other operations as well.

Operations on Functions

Given two functions f and g, then for all values of x for which both $f(x)$ and $g(x)$ are defined, the functions $f + g$, $f - g$, fg, and $\frac{f}{g}$ are defined as follows.

$$(f + g)(x) = f(x) + g(x) \qquad \text{Sum}$$

$$(f - g)(x) = f(x) - g(x) \qquad \text{Difference}$$

$$(fg)(x) = f(x) \cdot g(x) \qquad \text{Product}$$

$$\left(\frac{f}{g}\right)(x) = \frac{f(x)}{g(x)}, \quad g(x) \neq 0 \qquad \text{Quotient}$$

NOTE The condition $g(x) \neq 0$ in the definition of the quotient means that the domain of $\left(\frac{f}{g}\right)(x)$ is restricted to all values of x for which $g(x)$ is not 0. The condition does not mean that $g(x)$ is a function that is never 0.

EXAMPLE 1 Using Operations on Functions

Let $f(x) = x^2 + 1$ and $g(x) = 3x + 5$. Find each of the following.

(a) $(f + g)(1)$ **(b)** $(f - g)(-3)$ **(c)** $(fg)(5)$ **(d)** $\left(\frac{f}{g}\right)(0)$

Solution

(a) Since $f(1) = 2$ and $g(1) = 8$, use the definition above to get

$$\begin{aligned} (f + g)(1) &= f(1) + g(1) \qquad (f + g)(x) = f(x) + g(x) \\ &= 2 + 8 \\ &= 10. \end{aligned}$$

(b) $(f - g)(-3) = f(-3) - g(-3)$ $\quad (f - g)(x) = f(x) - g(x)$
$= 10 - (-4)$
$= 14$

(c) $(fg)(5) = f(5) \cdot g(5)$
$= 26 \cdot 20$
$= 520$

(d) $\left(\frac{f}{g}\right)(0) = \frac{f(0)}{g(0)} = \frac{1}{5}$

Now try Exercises 1, 3, and 5.

Domains

For functions f and g, the domains of $f + g$, $f - g$, and fg include all real numbers in the intersection of the domains of f and g, while the domain of $\frac{f}{g}$ includes those real numbers in the intersection of the domains of f and g for which $g(x) \neq 0$.

EXAMPLE 2 Using Operations on Functions and Determining Domains

Let $f(x) = 8x - 9$ and $g(x) = \sqrt{2x - 1}$. Find each of the following.

(a) $(f + g)(x)$ **(b)** $(f - g)(x)$ **(c)** $(fg)(x)$ **(d)** $\left(\frac{f}{g}\right)(x)$

(e) Give the domains of the functions in parts (a)–(d).

Solution

(a) $(f + g)(x) = f(x) + g(x) = 8x - 9 + \sqrt{2x - 1}$

(b) $(f - g)(x) = f(x) - g(x) = 8x - 9 - \sqrt{2x - 1}$

(c) $(fg)(x) = f(x) \cdot g(x) = (8x - 9)\sqrt{2x - 1}$

(d) $\left(\frac{f}{g}\right)(x) = \frac{f(x)}{g(x)} = \frac{8x - 9}{\sqrt{2x - 1}}$

(e) To find the domains of the functions in parts (a)–(d), we first find the domains of f and g as in Section 2.2. The domain of f is the set of all real numbers $(-\infty, \infty)$, and the domain of g, since $g(x) = \sqrt{2x - 1}$, includes just the real numbers that make $2x - 1$ nonnegative. That is, $2x - 1 \geq 0$, so $x \geq \frac{1}{2}$. The domain of g is $\left[\frac{1}{2}, \infty\right)$.

The domains of $f + g$, $f - g$, and fg are the intersection of the domains of f and g, which is

$$(-\infty, \infty) \cap \left[\frac{1}{2}, \infty\right) = \left[\frac{1}{2}, \infty\right).$$

The domain of $\frac{f}{g}$ includes those real numbers in the intersection above for which $g(x) = \sqrt{2x - 1} \neq 0$; that is, the domain of $\frac{f}{g}$ is $\left(\frac{1}{2}, \infty\right)$.

Now try Exercise 9.

EXAMPLE 3 Evaluating Combinations of Functions

If possible, use the given representations of functions f and g to evaluate

$$(f + g)(4), \qquad (f - g)(-2), \qquad (fg)(1), \qquad \text{and} \qquad \left(\frac{f}{g}\right)(0).$$

(a)

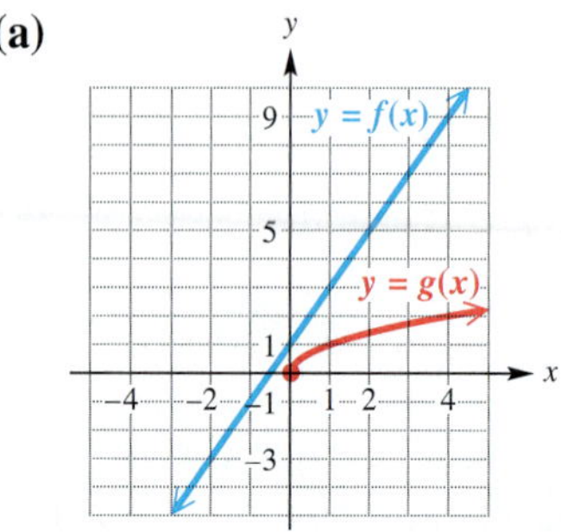

Figure 95

(b)

x	$f(x)$	$g(x)$
−2	−3	undefined
0	1	0
1	3	1
4	9	2

(c) $f(x) = 2x + 1$, $g(x) = \sqrt{x}$

Solution

(a) In Figure 95, $f(4) = 9$ and $g(4) = 2$. Thus

$$(f + g)(4) = f(4) + g(4) = 9 + 2 = 11.$$

For $(f - g)(-2)$, although $f(-2) = -3$, $g(-2)$ is undefined because -2 is not in the domain of g. Thus $(f - g)(-2)$ is undefined. The domains of f and g include 1, so

$$(fg)(1) = f(1) \cdot g(1) = 3(1) = 3.$$

The graph of g includes the origin, so $g(0) = 0$. Thus $\left(\frac{f}{g}\right)(0)$ is undefined.

(b) From the table, $f(4) = 9$ and $g(4) = 2$. As in part (a),

$$(f + g)(4) = f(4) + g(4) = 9 + 2 = 11.$$

In the table $g(-2)$ is undefined, so $(f - g)(-2)$ is also undefined. Similarly,

$$(fg)(1) = f(1) \cdot g(1) = 3(1) = 3,$$

and $$\left(\frac{f}{g}\right)(0) = \frac{f(0)}{g(0)} \text{ is undefined since } g(0) = 0.$$

(c) Use the formulas $f(x) = 2x + 1$ and $g(x) = \sqrt{x}$.

$$(f + g)(4) = f(4) + g(4) = (2 \cdot 4 + 1) + \sqrt{4} = 9 + 2 = 11$$

$$(f - g)(-2) = f(-2) - g(-2) = [2(-2) + 1] - \sqrt{-2} \text{ is undefined.}$$

$$(fg)(1) = f(1) \cdot g(1) = (2 \cdot 1 + 1)\sqrt{1} = 3(1) = 3$$

$$\left(\frac{f}{g}\right)(0) = \frac{f(0)}{g(0)} \text{ is undefined since } g(0) = 0.$$

Now try Exercises 23 and 27.

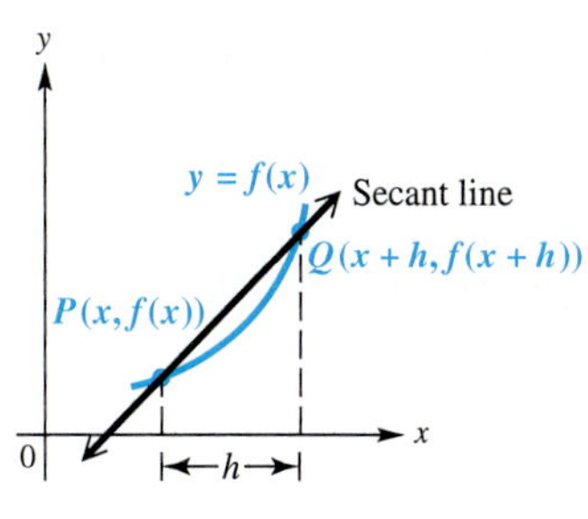

Figure 96

The Difference Quotient Suppose the point P lies on the graph of $y = f(x)$, and h is a positive number. If we let $(x, f(x))$ denote the coordinates of P and $(x + h, f(x + h))$ denote the coordinates of Q, then the line joining P and Q has slope

$$m = \frac{f(x + h) - f(x)}{(x + h) - x} = \frac{f(x + h) - f(x)}{h}, \quad h \neq 0.$$

This expression is called the **difference quotient.**

Figure 96 shows the graph of the line PQ (called a *secant line*). As h approaches 0, the slope of this secant line approaches the slope of the line tangent to the curve at P. Important applications of this idea are developed in calculus.

The next example illustrates a three-step process for finding the difference quotient of a function.

EXAMPLE 4 Finding the Difference Quotient

Let $f(x) = 2x^2 - 3x$. Find the difference quotient and simplify the expression.

Solution

Step 1 Find $f(x + h)$.
Replace x in $f(x)$ with $x + h$.

$$f(x + h) = 2(x + h)^2 - 3(x + h)$$

Step 2 Find $f(x + h) - f(x)$.

$$\begin{aligned} f(x + h) - f(x) &= [2(x + h)^2 - 3(x + h)] - (2x^2 - 3x) && \text{Substitute.} \\ &= 2(x^2 + 2xh + h^2) - 3(x + h) - (2x^2 - 3x) && \text{Square } x + h. \text{ (Section R.3)} \\ &= 2x^2 + 4xh + 2h^2 - 3x - 3h - 2x^2 + 3x && \text{Distributive property (Section R.1)} \\ &= 4xh + 2h^2 - 3h && \text{Combine terms.} \end{aligned}$$

Step 3 Find the difference quotient.

$$\begin{aligned} \frac{f(x + h) - f(x)}{h} &= \frac{4xh + 2h^2 - 3h}{h} && \text{Substitute.} \\ &= \frac{h(4x + 2h - 3)}{h} && \text{Factor out } h. \text{ (Section R.4)} \\ &= 4x + 2h - 3 && \text{Divide. (Section R.5)} \end{aligned}$$

Now try Exercise 35.

Looking Ahead to Calculus

The difference quotient is essential in the definition of the *derivative of a function* in calculus. The derivative provides a formula, in function form, for finding the slope of the tangent line to the graph of the function at a given point.

To illustrate, it is shown in calculus that the derivative of $f(x) = x^2 + 3$ is given by the function $f'(x) = 2x$. Now, $f'(0) = 2(0) = 0$, meaning that the slope of the tangent line to $f(x) = x^2 + 3$ at $x = 0$ is 0, implying that the tangent line is horizontal. If you draw this tangent line, you will see that it is the line $y = 3$, which is indeed a horizontal line.

CAUTION Notice that $f(x + h)$ is not the same as $f(x) + f(h)$. For $f(x) = 2x^2 - 3x$ in Example 4,

$$f(x + h) = 2(x + h)^2 - 3(x + h) = 2x^2 + 4xh + 2h^2 - 3x - 3h$$

but

$$f(x) + f(h) = (2x^2 - 3x) + (2h^2 - 3h) = 2x^2 - 3x + 2h^2 - 3h.$$

These expressions differ by $4xh$.

Composition of Functions The diagram in Figure 97 shows a function f that assigns to each x in its domain a value $f(x)$. Then another function g assigns to each $f(x)$ in its domain a value $g[f(x)]$. This two-step process takes an element x and produces a corresponding element $g[f(x)]$.

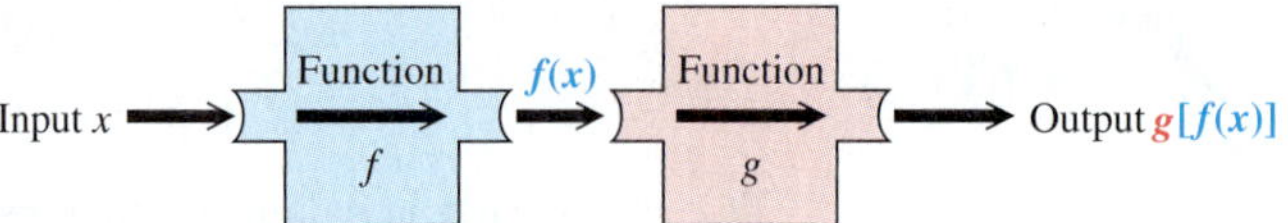

Figure 97

The function with y-values $g[f(x)]$ is called the *composition* of functions g and f, written $g \circ f$.

Composition of Functions

If f and g are functions, then the **composite function,** or **composition,** of g and f is defined by

$$(g \circ f)(x) = g[f(x)].$$

The domain of $g \circ f$ is the set of all numbers x in the domain of f such that $f(x)$ is in the domain of g.

t	$r(t)$	$A[r(t)]$
1	5	$\pi(5)^2 = 25\pi$
2	10	$\pi(10)^2 = 100\pi$
3	15	$\pi(15)^2 = 225\pi$
t	$5t$	$\pi(5t)^2 = 25\pi t^2$

As a real-life example of function composition, suppose an oil well off the California coast is leaking, with the leak spreading oil in a circular layer over the water's surface. (See the figure.) At any time t, in minutes, after the beginning of the leak, the radius of the circular oil slick is $r(t) = 5t$ feet. Since $A(r) = \pi r^2$ gives the area of a circle of radius r, the area can be expressed as a function of time by substituting $5t$ for r in $A(r) = \pi r^2$ to get

$$A[r(t)] = \pi(5t)^2 = 25\pi t^2.$$

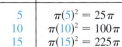

See the table in the margin. The function $(A \circ r)(t) = A[r(t)]$ is a composite function of the functions A and r.

EXAMPLE 5 Evaluating Composite Functions

Let $f(x) = 2x - 1$ and $g(x) = \dfrac{4}{x - 1}$. Find each composition.

(a) $(f \circ g)(2)$ **(b)** $(g \circ f)(-3)$

(c) Find the domain of $g \circ f$.

Solution

(a) First find $g(2)$. Since $g(x) = \dfrac{4}{x - 1}$,

$$g(2) = \frac{4}{2 - 1} = \frac{4}{1} = 4.$$

Now find $(f \circ g)(2) = f[g(2)] = f(4)$:

$$f[g(2)] = f(4) = 2(4) - 1 = 7.$$

Plot1 Plot2 Plot3
\Y1 = 2X−1
\Y2 = 4/(X−1)
\Y3=
\Y4=
\Y5=
\Y6=
\Y7=

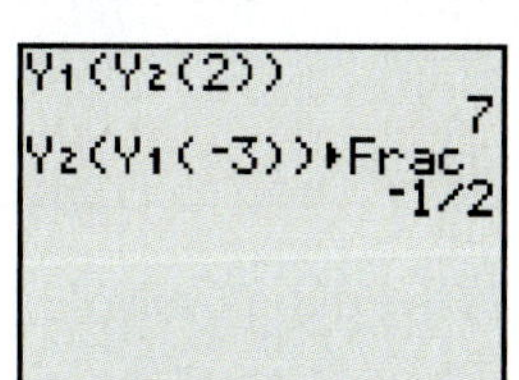

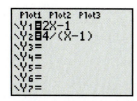

The screens show how a graphing calculator evaluates the expressions in Example 5(a) and (b).

(b) Since $f(-3) = 2(-3) - 1 = -7$,

$$\begin{aligned}(g \circ f)(-3) &= g[f(-3)] = g(-7)\\ &= \frac{4}{-7-1} = \frac{4}{-8}\\ &= -\frac{1}{2}.\end{aligned}$$

(c) Note that $(g \circ f)(x) = \frac{4}{f(x)-1}$. To find the domain of $g \circ f$, we consider a two-step process. We use each x in the domain of f to get a corresponding $f(x)$. These range values then become the domain of g. Here, $f(x) = 2x - 1$, so $f(x)$ can be any real number. However, the *denominator* of $g[f(x)]$, which is $f(x) - 1$, cannot equal 0. This happens for $x = 1$, since $f(1) = 2(1) - 1 = 1$, which makes $g[f(x)]$ undefined. Thus, the domain of $g \circ f$ is $\{x \mid x \neq 1\}$, or the interval

$$(-\infty, 1) \cup (1, \infty).$$

Now try Exercise 41.

EXAMPLE 6 Finding Composite Functions

Let $f(x) = 4x + 1$ and $g(x) = 2x^2 + 5x$. Find each composition.

(a) $(g \circ f)(x)$ **(b)** $(f \circ g)(x)$

Solution

(a) By definition, $(g \circ f)(x) = g[f(x)]$. Using the given functions,

$$\begin{aligned}(g \circ f)(x) = g[f(x)] &= g(4x + 1) && f(x) = 4x + 1\\ &= 2(4x + 1)^2 + 5(4x + 1) && g(x) = 2x^2 + 5x\\ &= 2(16x^2 + 8x + 1) + 20x + 5 && \text{Square } 4x + 1; \text{ distributive property.}\\ &= 32x^2 + 16x + 2 + 20x + 5 && \text{Distributive property}\\ &= 32x^2 + 36x + 7. && \text{Combine terms.}\end{aligned}$$

(b) If we use the definition above with f and g interchanged, $(f \circ g)(x)$ becomes $f[g(x)]$.

$$\begin{aligned}(f \circ g)(x) &= f[g(x)]\\ &= f(2x^2 + 5x) && g(x) = 2x^2 + 5x\\ &= 4(2x^2 + 5x) + 1 && f(x) = 4x + 1\\ &= 8x^2 + 20x + 1 && \text{Distributive property}\end{aligned}$$

Now try Exercise 59.

As Example 6 shows, it is not always true that $f \circ g = g \circ f$. In fact, the composite functions $f \circ g$ and $g \circ f$ are equal only for a special class of functions, discussed in Section 4.1. In Example 6, the domain of both composite functions is $(-\infty, \infty)$.

CAUTION In general, the composite function $f \circ g$ is not the same as the product fg. For example, with f and g defined as in Example 6,

$$(f \circ g)(x) = 8x^2 + 20x + 1$$

but $$(fg)(x) = (4x + 1)(2x^2 + 5x) = 8x^3 + 22x^2 + 5x.$$

EXAMPLE 7 Finding Composite Functions and Their Domains

Let $f(x) = \frac{1}{x}$ and $g(x) = \sqrt{3 - x}$. Find $f \circ g$ and $g \circ f$. Give the domain of each.

Solution First find $f \circ g$.

$$\begin{aligned}(f \circ g)(x) &= f[g(x)] \\ &= f\left(\sqrt{3 - x}\right) && g(x) = \sqrt{3 - x} \\ &= \frac{1}{\sqrt{3 - x}} && f(x) = \tfrac{1}{x}\end{aligned}$$

The radical $\sqrt{3 - x}$ is a nonzero real number only when $3 - x > 0$ or $x < 3$, so the domain of $f \circ g$ is the interval $(-\infty, 3)$.

Use the same functions to find $g \circ f$, as follows.

$$\begin{aligned}(g \circ f)(x) &= g[f(x)] \\ &= g\left(\frac{1}{x}\right) && f(x) = \tfrac{1}{x} \\ &= \sqrt{3 - \frac{1}{x}} && g(x) = \sqrt{3 - x} \\ &= \sqrt{\frac{3x - 1}{x}} && \text{Write as a single fraction. (Section R.5)}\end{aligned}$$

The domain of $g \circ f$ is the set of all real numbers x such that $x \neq 0$ and $3 - f(x) \geq 0$. As shown above,

$$3 - f(x) = \frac{3x - 1}{x}.$$

We need to solve the inequality

$$\frac{3x - 1}{x} \geq 0. \quad \text{(Section 1.7)}$$

To find the numbers where the quotient changes sign, recall from Section 1.7 that to solve an inequality in this form, we need to find the x-values where the numerator or denominator equals 0.

$$3x - 1 = 0 \quad \text{or} \quad x = 0$$

$$x = \frac{1}{3}$$

The numbers 0 and $\frac{1}{3}$ divide the number line into the intervals $(-\infty, 0)$, $\left(0, \frac{1}{3}\right)$, and $\left(\frac{1}{3}, \infty\right)$. Use a test value from each interval to decide which intervals include the solutions of the inequality $\frac{3x - 1}{x} \geq 0$.

Interval	**Test Value**	**Is $\frac{3x-1}{x} \geq 0$ True or False?**
$(-\infty, 0)$	-1	$\frac{3(-1) - 1}{-1} \geq 0$? $4 \geq 0$ True
$\left(0, \frac{1}{3}\right)$	$\frac{1}{4}$	$\frac{3\left(\frac{1}{4}\right) - 1}{\frac{1}{4}} \geq 0$? $-1 \geq 0$ False
$\left(\frac{1}{3}, \infty\right)$	1	$\frac{3(1) - 1}{1} \geq 0$? $2 \geq 0$ True

Since $\frac{1}{3}$ satisfies the inequality, but 0 does not, include $\frac{1}{3}$ in the domain and exclude 0. Therefore, the domain of $g \circ f$ is

$$(-\infty, 0) \cup \left[\frac{1}{3}, \infty\right).$$

Now try Exercise 61.

Looking Ahead to Calculus

Finding the derivative of a function in calculus is called *differentiation.* To differentiate a composite function such as $h(x) = (3x + 2)^4$, we interpret $h(x)$ as $(f \circ g)(x)$, where $g(x) = 3x + 2$ and $f(x) = x^4$. *The chain rule* allows us to differentiate composite functions. Notice the use of the composition symbol and function notation in the following, which comes from the chain rule.

If $h(x) = (f \circ g)(x)$, then
$h'(x) = f'[g(x)] \cdot g'(x)$.

In calculus it is sometimes necessary to treat a function as a composition of two functions. The next example shows how this can be done.

EXAMPLE 8 Finding Functions That Form a Given Composite

Find functions f and g such that

$$(f \circ g)(x) = (x^2 - 5)^3 - 4(x^2 - 5) + 3.$$

Solution Note the repeated quantity $x^2 - 5$. If we choose

$$g(x) = x^2 - 5 \quad \text{and} \quad f(x) = x^3 - 4x + 3,$$

then

$$\begin{aligned}(f \circ g)(x) &= f[g(x)] \\ &= f(x^2 - 5) \\ &= (x^2 - 5)^3 - 4(x^2 - 5) + 3.\end{aligned}$$

There are other pairs of functions f and g that also work. For instance, let

$$f(x) = (x - 5)^3 - 4(x - 5) + 3 \quad \text{and} \quad g(x) = x^2.$$

Now try Exercise 75.

2.7 Exercises

Let $f(x) = 5x^2 - 2x$ and $g(x) = 6x + 4$. Find each of the following. See Example 1.

1. $(f + g)(3)$ **2.** $(f - g)(-5)$ **3.** $(fg)(4)$ **4.** $(fg)(-3)$

5. $\left(\frac{f}{g}\right)(-1)$ **6.** $\left(\frac{f}{g}\right)(4)$ **7.** $(f - g)(m)$ **8.** $(f + g)(2k)$

For the pair of functions defined, find $f + g$, $f - g$, fg, and $\frac{f}{g}$. Give the domain of each. See Example 2.

9. $f(x) = 3x + 4,\ g(x) = 2x - 5$ **10.** $f(x) = 6 - 3x,\ g(x) = -4x + 1$

11. $f(x) = 2x^2 - 3x,\ g(x) = x^2 - x + 3$

12. $f(x) = 4x^2 + 2x - 3,\ g(x) = x^2 - 3x + 2$

13. $f(x) = \sqrt{4x - 1},\ g(x) = \frac{1}{x}$ **14.** $f(x) = \sqrt{5x - 4},\ g(x) = -\frac{1}{x}$

Sodas Consumed by Adolescents *The graph shows the number of sodas (soft drinks) adolescents (age 12–19) drank per week from 1978–1996. $G(x)$ gives the number of sodas for girls, $B(x)$ gives the number of sodas for boys, and $T(x)$ gives the total number for both groups. Use the graph to do the following.*

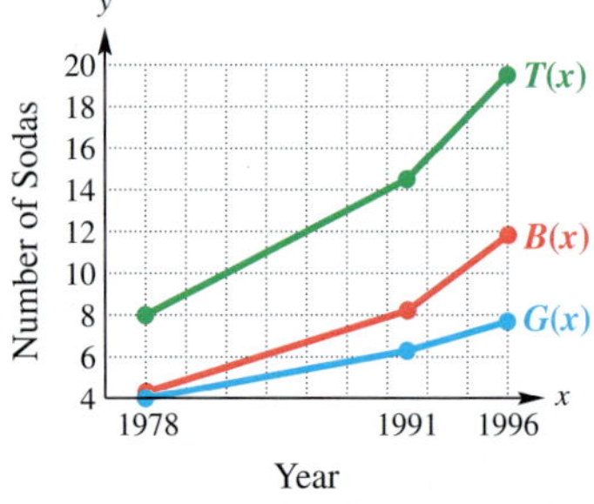

Source: U.S. Department of Agriculture.

15. Estimate $G(1996)$ and $B(1996)$ and use your results to estimate $T(1996)$.

16. Estimate $G(1991)$ and $B(1991)$ and use your results to estimate $T(1991)$.

17. Use the slopes of the line segments to decide in which period (1978–1991 or 1991–1996) the number of sodas per week increased more rapidly.

18. Give a reason that might explain why girls drank fewer sodas in each of the three periods.

Science and Space/Technology Spending *The graph shows dollars (in billions) spent for general science and for space/other technologies in selected years. $G(x)$ represents the dollars spent for general science, and $S(x)$ represents the dollars spent for space and other technologies. $T(x)$ represents the total expenditures for these two categories. Use the graph to answer the following.*

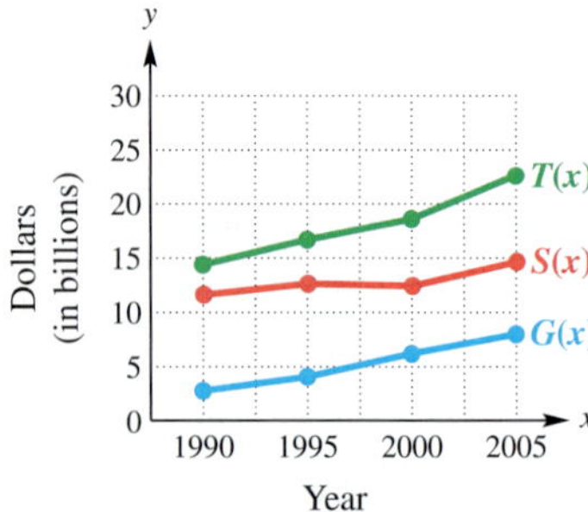

Source: U.S. Office of Management and Budget.

19. Estimate $(T - S)(2000)$. What does this function represent?

20. Estimate $(T - G)(2005)$. What does this function represent?

21. In which of the categories was spending almost static for several years? In which years did this occur?

22. In which period and which category does spending for $G(x)$ or $S(x)$ increase most?

Use the graph to evaluate each expression. See Example 3(a).

23. **(a)** $(f+g)(2)$ **(b)** $(f-g)(1)$

(c) $(fg)(0)$ **(d)** $\left(\frac{f}{g}\right)(1)$

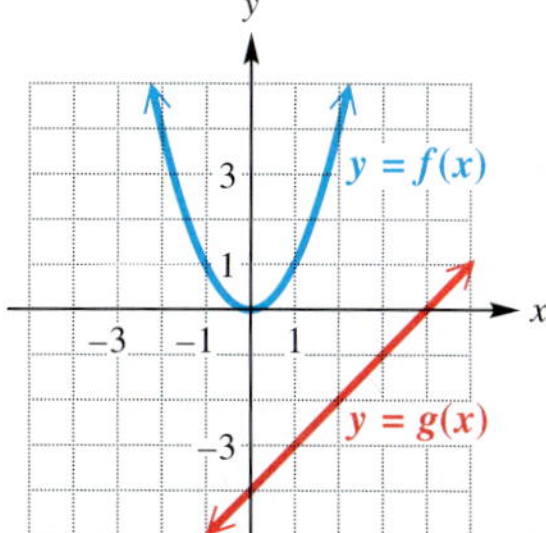

24. **(a)** $(f+g)(0)$ **(b)** $(f-g)(-1)$

(c) $(fg)(1)$ **(d)** $\left(\frac{f}{g}\right)(2)$

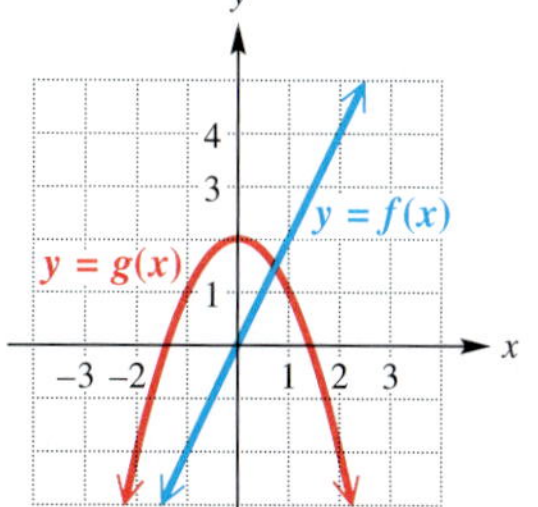

25. **(a)** $(f+g)(-1)$ **(b)** $(f-g)(-2)$

(c) $(fg)(0)$ **(d)** $\left(\frac{f}{g}\right)(2)$

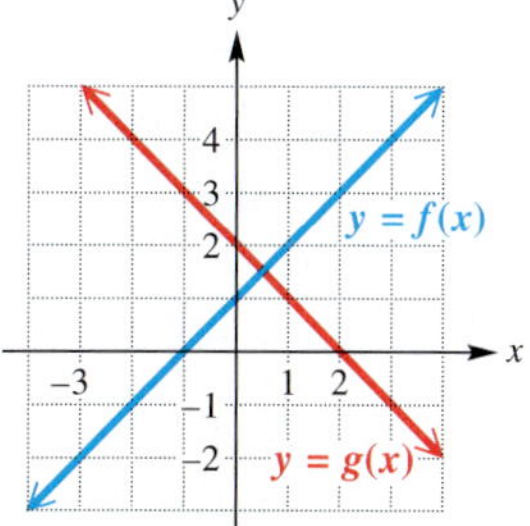

26. **(a)** $(f+g)(1)$ **(b)** $(f-g)(0)$

(c) $(fg)(-1)$ **(d)** $\left(\frac{f}{g}\right)(1)$

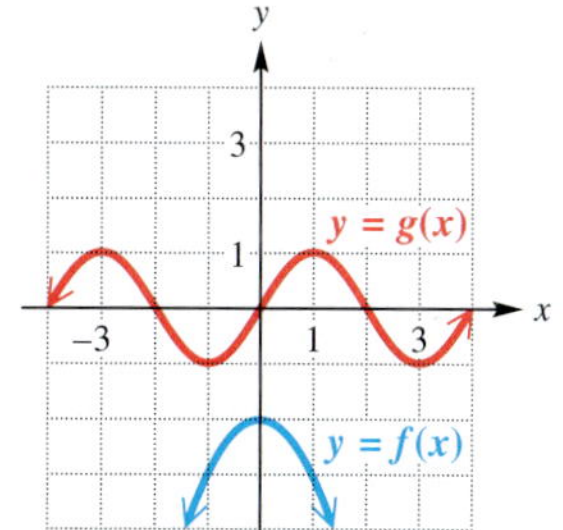

In Exercises 27 and 28, use the table to evaluate each expression in parts (a)–(d), if possible. See Example 3(b).

(a) $(f+g)(2)$ **(b)** $(f-g)(4)$ **(c)** $(fg)(-2)$ **(d)** $\left(\frac{f}{g}\right)(0)$

27.

x	$f(x)$	$g(x)$
−2	0	6
0	5	0
2	7	−2
4	10	5

28.

x	$f(x)$	$g(x)$
−2	−4	2
0	8	−1
2	5	4
4	0	0

29. Use the table in Exercise 27 to complete the following table.

x	$(f+g)(x)$	$(f-g)(x)$	$(fg)(x)$	$\left(\frac{f}{g}\right)(x)$
−2				
0				
2				
4				

30. Use the table in Exercise 28 to complete the following table.

x	$(f+g)(x)$	$(f-g)(x)$	$(fg)(x)$	$\left(\frac{f}{g}\right)(x)$
-2				
0				
2				
4				

31. How is the difference quotient related to slope?

32. Refer to Figure 96. How is the secant line PQ related to the tangent line to a curve at point P?

For each of the functions defined as follows, find **(a)** $f(x+h)$, **(b)** $f(x+h)-f(x)$, *and* **(c)** $\frac{f(x+h)-f(x)}{h}$. *See Example 4.*

33. $f(x) = 2 - x$

34. $f(x) = 1 - x$

35. $f(x) = 6x + 2$

36. $f(x) = 4x + 11$

37. $f(x) = -2x + 5$

38. $f(x) = 1 - x^2$

39. $f(x) = x^2 - 4$

40. $f(x) = 8 - 3x^2$

Let $f(x) = 2x - 3$ *and* $g(x) = -x + 3$. *Find each composite function. See Example 5.*

41. $(f \circ g)(4)$

42. $(f \circ g)(2)$

43. $(f \circ g)(-2)$

44. $(g \circ f)(3)$

45. $(g \circ f)(0)$

46. $(g \circ f)(-2)$

47. $(f \circ f)(2)$

48. $(g \circ g)(-2)$

Concept Check *The tables give some selected ordered pairs for functions f and g.*

x	3	4	6
$f(x)$	1	3	9

x	2	7	1	9
$g(x)$	3	6	9	12

Find each of the following.

49. $(f \circ g)(2)$

50. $(f \circ g)(7)$

51. $(g \circ f)(3)$

52. $(g \circ f)(6)$

53. $(f \circ f)(4)$

54. $(g \circ g)(1)$

55. Why can you not determine $(f \circ g)(1)$ given the information in the tables for Exercises 49–54?

56. Extend the concept of composition of functions to evaluate $[g \circ (f \circ g)](7)$ using the tables for Exercises 49–54.

Find $(f \circ g)(x)$ *and* $(g \circ f)(x)$ *for each pair of functions. Give the domains in Exercises 61 and 62. See Examples 6 and 7.*

57. $f(x) = -6x + 9,\ g(x) = 5x + 7$

58. $f(x) = 8x + 12,\ g(x) = 3x - 1$

59. $f(x) = 4x^2 + 2x + 8,\ g(x) = x + 5$

60. $f(x) = 5x + 3,\ g(x) = -x^2 + 4x + 3$

61. $f(x) = \frac{2}{x^4},\ g(x) = 2 - x$

62. $f(x) = \frac{1}{x},\ g(x) = x^2$

63. $f(x) = 9x^2 - 11x,\ g(x) = 2\sqrt{x+2}$

64. $f(x) = \sqrt{x+2},\ g(x) = 8x^2 - 6$

65. *Concept Check* Fill in the missing entries in the table.

x	$f(x)$	$g(x)$	$g[f(x)]$
1	3	2	7
2	1	5	
3	2		

66. *Concept Check* Suppose $f(x)$ is an odd function and $g(x)$ is an even function. Fill in the missing entries in the table.

x	-2	-1	0	1	2
$f(x)$			0	-2	
$g(x)$	0	2	1		
$(f \circ g)(x)$		1	-2		

67. Composition is an operation that is unique to functions. Is composition of functions commutative? That is, does $f \circ g = g \circ f$ for all functions f and g? Explain.

68. Describe the steps required to find the composite function $f \circ g$, given $f(x) = 2x - 5$ and $g(x) = x^2 + 3$.

For certain pairs of functions f and g, $(f \circ g)(x) = x$ and $(g \circ f)(x) = x$. Show that this is true for each pair in Exercises 69–72.

69. $f(x) = 4x + 2,\ g(x) = \frac{1}{4}(x - 2)$

70. $f(x) = -3x,\ g(x) = -\frac{1}{3}x$

71. $f(x) = \sqrt[3]{5x + 4},\ g(x) = \frac{1}{5}x^3 - \frac{4}{5}$

72. $f(x) = \sqrt[3]{x + 1},\ g(x) = x^3 - 1$

Find functions f and g such that $(f \circ g)(x) = h(x)$. (There are many possible ways to do this.) See Example 8.

73. $h(x) = (6x - 2)^2$

74. $h(x) = (11x^2 + 12x)^2$

75. $h(x) = \sqrt{x^2 - 1}$

76. $h(x) = (2x - 3)^3$

77. $h(x) = \sqrt{6x} + 12$

78. $h(x) = \sqrt[3]{2x + 3} - 4$

Solve each problem.

79. *Relationship of Measurement Units* The function defined by $f(x) = 12x$ computes the number of inches in x feet, and the function defined by $g(x) = 5280x$ computes the number of feet in x miles. What does $(f \circ g)(x)$ compute?

80. *Perimeter of a Square* The perimeter x of a square with side of length s is given by the formula $x = 4s$.

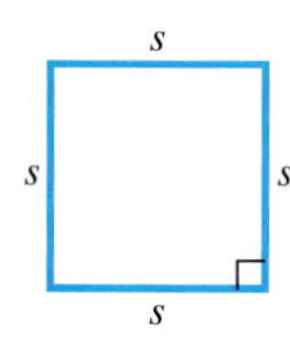

(a) Solve for s in terms of x.

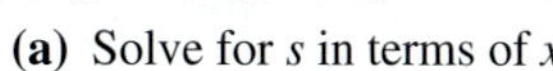

(b) If y represents the area of this square, write y as a function of the perimeter x.

(c) Use the composite function of part (b) to find the area of a square with perimeter 6.

81. *Area of an Equilateral Triangle* The area of an equilateral triangle with sides of length x is given by the function defined by $A(x) = \frac{\sqrt{3}}{4}x^2$.

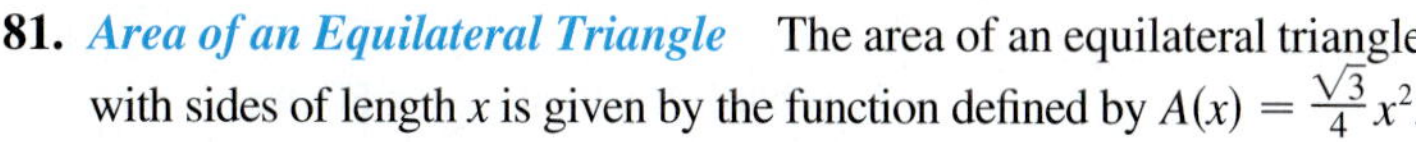

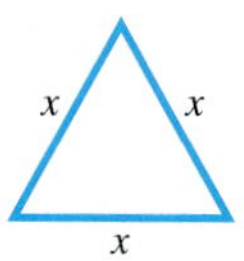

(a) Find $A(2x)$, the function representing the area of an equilateral triangle with sides of length twice the original length.

(b) Find the area of an equilateral triangle with side length 16. Use the formula $A(2x)$ found in part (a).

82. *Software Author Royalties* A software author invests his royalties in two accounts for 1 yr.

(a) The first account pays 4% simple interest. If he invests x dollars in this account, write an expression for y_1 in terms of x, where y_1 represents the amount of interest earned.

(b) He invests in a second account \$500 more than he invested in the first account. This second account pays 2.5% simple interest. Write an expression for y_2, where y_2 represents the amount of interest earned.

(c) What does $y_1 + y_2$ represent?

(d) How much interest will he receive if \$250 is invested in the first account?

83. *Oil Leak* An oil well off the Gulf Coast is leaking, with the leak spreading oil over the water's surface as a circle. At any time t, in minutes, after the beginning of the leak, the radius of the circular oil slick on the surface is $r(t) = 4t$ feet. Let $A(r) = \pi r^2$ represent the area of a circle of radius r.

(a) Find $(A \circ r)(t)$. **(b)** Interpret $(A \circ r)(t)$.

(c) What is the area of the oil slick after 3 min?

84. *Emission of Pollutants* When a thermal inversion layer is over a city (as happens in Los Angeles), pollutants cannot rise vertically but are trapped below the layer and must disperse horizontally. Assume that a factory smokestack begins emitting a pollutant at 8 A.M. Assume that the pollutant disperses horizontally over a circular area. If t represents the time, in hours, since the factory began emitting pollutants ($t = 0$ represents 8 A.M.), assume that the radius of the circle of pollutants is $r(t) = 2t$ miles. Let $A(r) = \pi r^2$ represent the area of a circle of radius r.

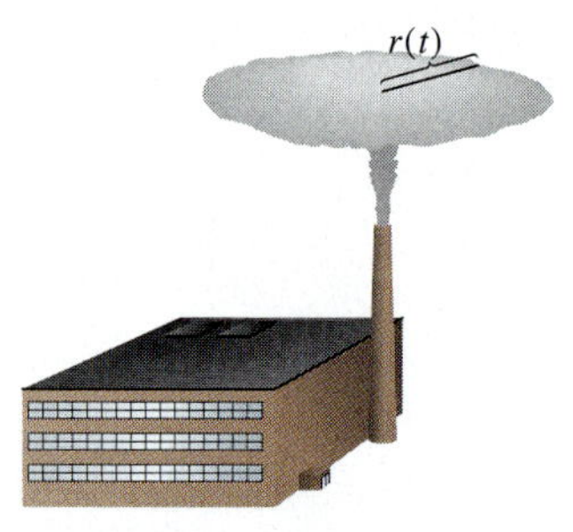

(a) Find $(A \circ r)(t)$. **(b)** Interpret $(A \circ r)(t)$.

(c) What is the area of the circular region covered by the layer at noon?

85. *(Modeling) Catering Cost* A couple planning their wedding has found that the cost to hire a caterer for the reception depends on the number of guests attending. If 100 people attend, the cost per person will be \$20. For each person less than 100, the cost will increase by \$5. Assume that no more than 100 people will attend. Let x represent the number less than 100 who do not attend. For example, if 95 attend, $x = 5$.

(a) Write a function defined by $N(x)$ giving the number of guests.

(b) Write a function defined by $G(x)$ giving the cost per guest.

(c) Write a function defined by $N(x) \cdot G(x)$ for the total cost, $C(x)$.

(d) What is the total cost if 80 people attend?

86. *Area of a Square* The area of a square is x^2 square inches. Suppose that 3 in. is added to one dimension and 1 in. is subtracted from the other dimension. Express the area $A(x)$ of the resulting rectangle as a product of two functions.

Chapter 2 Summary

KEY TERMS

2.1	ordered pair origin x-axis y-axis rectangular (Cartesian) coordinate system coordinate plane (xy-plane) quadrants coordinates collinear graph of an equation x-intercept y-intercept circle radius center of a circle	**2.2**	dependent variable independent variable relation function domain range function notation increasing function decreasing function constant function	**2.3**	linear function standard form change in x change in y slope average rate of change	**2.4**	point-slope form slope-intercept form scatter diagram
2.5	continuous function parabola vertex piecewise-defined function step function	**2.6**	symmetry even function odd function vertical translation horizontal translation	**2.7**	difference quotient composite function (composition)		

NEW SYMBOLS

(a, b)	ordered pair
$f(x)$	function of x (read "f of x")
Δx	change in x
Δy	change in y
m	slope
$[\![x]\!]$	greatest integer less than or equal to x
$g \circ f$	composite function

QUICK REVIEW

CONCEPTS | **EXAMPLES**

2.1 Graphs of Equations

Distance Formula

Suppose $P(x_1, y_1)$ and $R(x_2, y_2)$ are two points in a coordinate plane. Then the distance between P and R, written $d(P, R)$, is

$$d(P, R) = \sqrt{(x_2 - x_1)^2 + (y_2 - y_1)^2}.$$

Example: The distance between the points $(-1, 4)$ and $(6, -3)$ is

$$\sqrt{[6 - (-1)]^2 + (-3 - 4)^2} = \sqrt{49 + 49} = 7\sqrt{2}.$$

Midpoint Formula

The midpoint of the line segment with endpoints (x_1, y_1) and (x_2, y_2) is

$$\left(\frac{x_1 + x_2}{2}, \frac{y_1 + y_2}{2}\right).$$

Example: The midpoint of the line segment with endpoints $(-1, 4)$ and $(6, -3)$ is

$$\left(\frac{-1 + 6}{2}, \frac{4 + (-3)}{2}\right) = \left(\frac{5}{2}, \frac{1}{2}\right).$$

Center-Radius Form of the Equation of a Circle

$$(x - h)^2 + (y - k)^2 = r^2$$

is the equation of a circle with center at (h, k) and radius r.

Example: The center-radius form of the equation of the circle with center at $(-2, 3)$ and radius 4 is

$$[x - (-2)]^2 + (y - 3)^2 = 4^2.$$

(continued)

CONCEPTS	EXAMPLES

General Form of the Equation of a Circle

$$x^2 + y^2 + cx + dy + e = 0$$

The general form of the equation of the preceding circle is

$$x^2 + y^2 + 4x - 6y - 3 = 0.$$

2.2 Functions

A **relation** is a set of ordered pairs. A **function** is a relation in which, for each value of the first component of the ordered pairs, there is *exactly one* value of the second component. The set of first components is called the **domain**, and the set of second components is called the **range.**

The relation $y = x^2$ defines a function, because each choice of a number for x corresponds to one and only one number for y. The domain is $(-\infty, \infty)$, and the range is $[0, \infty)$.

The relation $x = y^2$ does *not* define a function because a number x may correspond to two numbers for y. The domain is $[0, \infty)$, and the range is $(-\infty, \infty)$.

Vertical Line Test

If each vertical line intersects a graph in at most one point, then the graph is that of a function.

A.

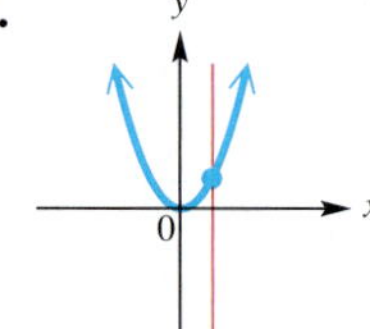

B.

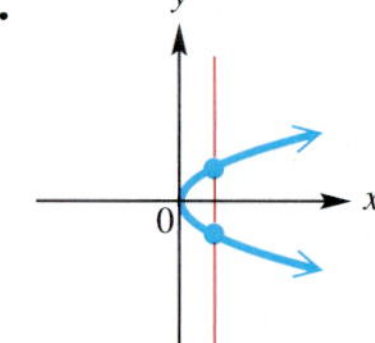

By the vertical line test, graph A is the graph of a function, but graph B is not.

Increasing, Decreasing, and Constant Functions

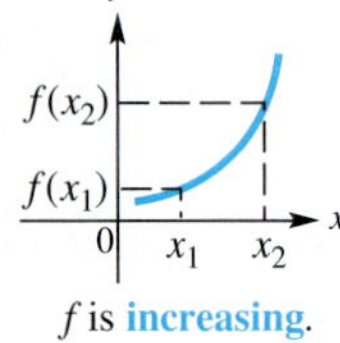

f is increasing.

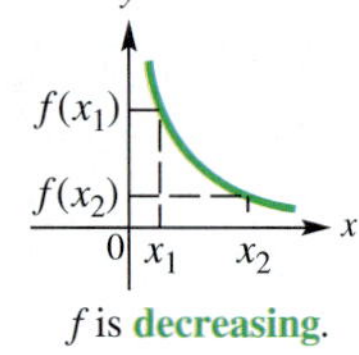

f is decreasing.

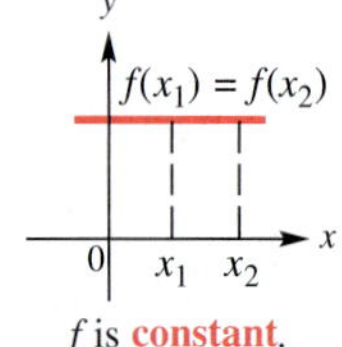

f is constant.

The function in graph A is decreasing on the interval $(-\infty, 0]$ and increasing on the interval $[0, \infty)$.

2.3 Linear Functions

A function f is a **linear function** if, for real numbers a and b,

$$f(x) = ax + b.$$

The graph of a linear function is a line.

The equation

$$y = \frac{1}{2}x - 4$$

defines a linear function.

Definition of Slope

The slope m of the line through the points (x_1, y_1) and (x_2, y_2) is

$$m = \frac{\text{rise}}{\text{run}} = \frac{\Delta y}{\Delta x} = \frac{y_2 - y_1}{x_2 - x_1}, \quad \text{where } \Delta x \neq 0.$$

The slope of the line through the points $(2, 4)$ and $(-1, 7)$ is

$$m = \frac{7 - 4}{-1 - 2} = \frac{3}{-3} = -1.$$

CONCEPTS	EXAMPLES

2.4 Equations of Lines; Curve Fitting

Forms of Linear Equations

Equation	Description	Examples
$y = mx + b$	**Slope-Intercept Form** Slope is m. y-intercept is b.	$y = x + \frac{2}{3}$ The slope is 1; the y-intercept is $\frac{2}{3}$.
$y + y_1 = m(x - x_1)$	**Point-Slope Form** Slope is m. Line passes through (x_1, y_1).	$y - 3 = -2(x + 5)$ The slope is -2; the line passes through the point $(-5, 3)$.
$Ax + By = C$	**Standard Form** (A, B, and C integers, $A \geq 0$) Slope is $-\frac{A}{B}$ $(B \neq 0)$. x-intercept is $\frac{C}{A}$ $(A \neq 0)$. y-intercept is $\frac{C}{B}$ $(B \neq 0)$.	$4x + 5y = 12$ The slope is $-\frac{4}{5}$; the x-intercept is 3; the y-intercept is $\frac{12}{5}$.
$y = b$	**Horizontal Line** Slope is 0. y-intercept is b.	$y = -6$ The slope is 0; the y-intercept is -6.
$x = a$	**Vertical Line** Slope is undefined. x-intercept is a.	$x = 3$ The slope is undefined; the x-intercept is 3.

2.5 Graphs of Basic Functions

Basic Functions

Identity Function $f(x) = x$

Squaring Function $f(x) = x^2$

Cubing Function $f(x) = x^3$

Square Root Function $f(x) = \sqrt{x}$

Cube Root Function $f(x) = \sqrt[3]{x}$

Absolute Value Function $f(x) = |x|$

Greatest Integer Function $f(x) = [\![x]\!]$

Refer to the function boxes on pages 243–245 and page 247. Graphs of the basic functions are also shown on the back inside covers.

2.6 Graphing Techniques

Stretching and Shrinking

The graph of $g(x) = a \cdot f(x)$ has the same shape as the graph of $f(x)$. The graph is

narrower if $|a| > 1$ and wider if $0 < |a| < 1$.

Reflection Across an Axis

The graph of $y = -f(x)$ is the same as the graph of $y = f(x)$ reflected across the x-axis.

The graph of $y = f(-x)$ is the same as the graph of $y = f(x)$ reflected across the y-axis.

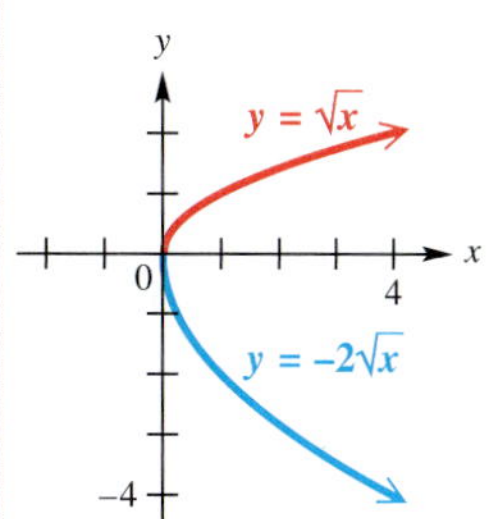

The graph of $y = -2\sqrt{x}$ is the graph of $y = \sqrt{x}$ stretched vertically by a factor of 2 and reflected across the x-axis.

(continued)

CONCEPTS	EXAMPLES

Symmetry

The graph of an equation is **symmetric with respect to the y-axis** if the replacement of x with $-x$ results in an equivalent equation.

The graph of an equation is **symmetric with respect to the x-axis** if the replacement of y with $-y$ results in an equivalent equation.

The graph of an equation is **symmetric with respect to the origin** if the replacement of both x with $-x$ and y with $-y$ results in an equivalent equation.

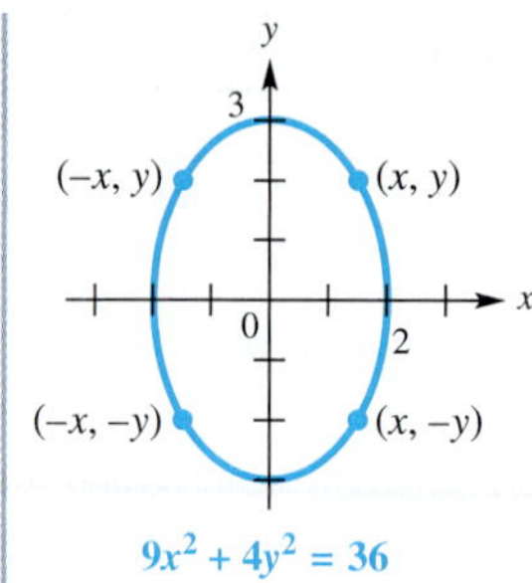

The graph of $9x^2 + 4y^2 = 36$ is symmetric with respect to the y-axis, the x-axis, and the origin.

Translations

Let f be a function and c be a positive number.

To Graph:	Shift the Graph of $y = f(x)$ by c Units:
$y = f(x) + c$	up
$y = f(x) - c$	down
$y = f(x + c)$	left
$y = f(x - c)$	right

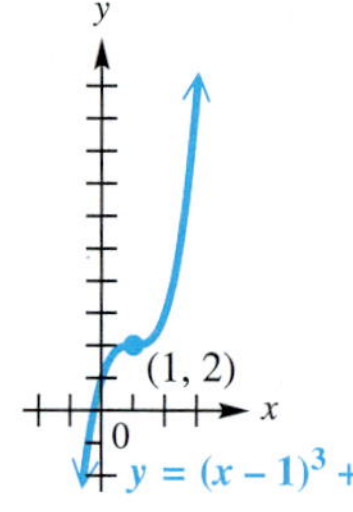

The graph of $y = (x - 1)^3 + 2$ is the graph of $y = x^3$ translated 1 unit to the right and 2 units up.

2.7 Function Operations and Composition

Operations on Functions

Given two functions f and g, then for all values of x for which both $f(x)$ and $g(x)$ are defined, the following operations are defined.

$$(f + g)(x) = f(x) + g(x) \quad \text{Sum}$$
$$(f - g)(x) = f(x) - g(x) \quad \text{Difference}$$
$$(fg)(x) = f(x) \cdot g(x) \quad \text{Product}$$
$$\left(\frac{f}{g}\right)(x) = \frac{f(x)}{g(x)}, \quad g(x) \neq 0 \quad \text{Quotient}$$

Let $f(x) = 2x - 4$ and $g(x) = \sqrt{x}$.

$$\left.\begin{aligned} (f + g)(x) &= 2x - 4 + \sqrt{x} \\ (f - g)(x) &= 2x - 4 - \sqrt{x} \\ (fg)(x) &= (2x - 4)\sqrt{x} \end{aligned}\right\} \text{The domain is } [0, \infty).$$

$$\left.\left(\frac{f}{g}\right)(x) = \frac{2x - 4}{\sqrt{x}}\right\} \text{The domain is } (0, \infty).$$

Difference Quotient

The line joining $P(x, f(x))$ and $Q(x + h, f(x + h))$ has slope

$$m = \frac{f(x + h) - f(x)}{h}, \quad h \neq 0.$$

Refer to Example 4 in Section 2.7 on page 271.

Composition of Functions

If f and g are functions, then the composite function, or composition, of g and f is defined by

$$(g \circ f)(x) = g[f(x)].$$

The domain of $g \circ f$ is the set of all x in the domain of f such that $f(x)$ is in the domain of g.

Using f and g as defined above,

$$(g \circ f)(x) = \sqrt{2x - 4}.$$

The domain is all x such that $2x - 4 \geq 0$; that is, the interval $[2, \infty)$.

Chapter 2 Review Exercises

Find the distance between each pair of points, and give the coordinates of the midpoint of the segment joining them.

1. $P(3, -1), Q(-4, 5)$ **2.** $M(-8, 2), N(3, -7)$ **3.** $A(-6, 3), B(-6, 8)$

4. Are the points $(5, 7)$, $(3, 9)$, and $(6, 8)$ the vertices of a right triangle?

5. Find all possible values of k so that $(-1, 2)$, $(-10, 5)$, and $(-4, k)$ are the vertices of a right triangle.

6. Use the distance formula to determine whether the points $(-2, -5)$, $(1, 7)$, and $(3, 15)$ are collinear.

Find an equation for each circle satisfying the given conditions.

7. center $(-2, 3)$, radius 15

8. center $(\sqrt{5}, -\sqrt{7})$, radius $\sqrt{3}$

9. center $(-8, 1)$, passing through $(0, 16)$

10. center $(3, -6)$, tangent to the x-axis

Find the center and radius of each circle.

11. $x^2 - 4x + y^2 + 6y + 12 = 0$

12. $x^2 - 6x + y^2 - 10y + 30 = 0$

13. $2x^2 + 14x + 2y^2 + 6y + 2 = 0$

14. $3x^2 + 33x + 3y^2 - 15y = 0$

15. Find all possible values of x so that the distance between $(x, -9)$ and $(3, -5)$ is 6.

16. Find all points (x, y) with $x = 6$ so that (x, y) is 4 units from $(1, 3)$.

17. Describe the graph of $(x - 4)^2 + (y + 5)^2 = 0$.

Decide whether each graph is that of a function of x. Give the domain and range of each relation.

18.

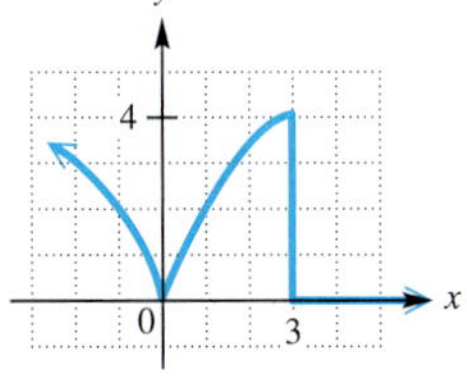

19.

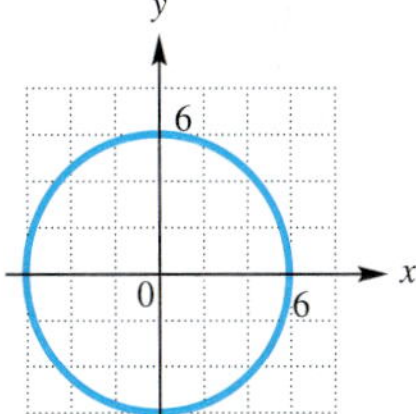

20.

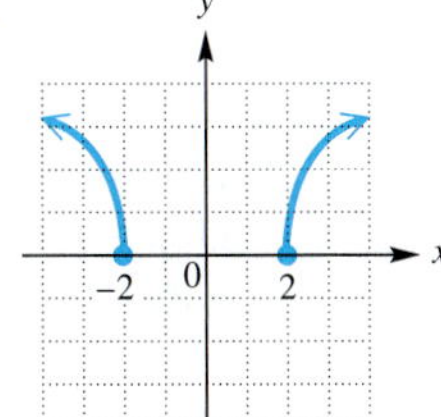

21.

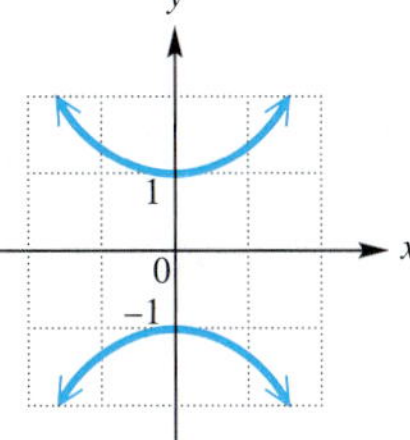

22.

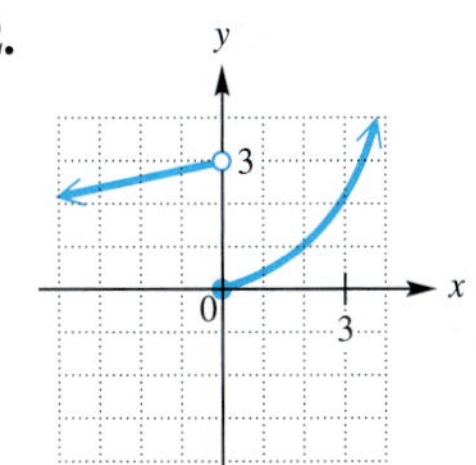

23.

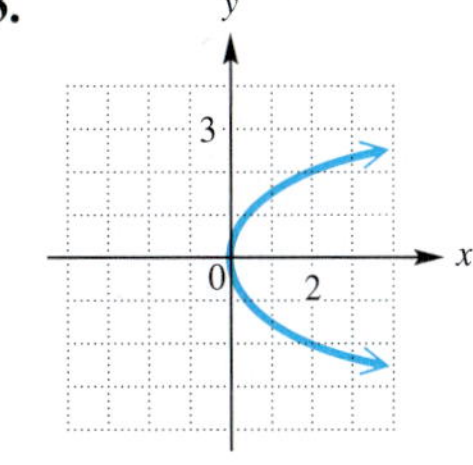

Determine whether each equation defines y as a function of x.

24. $x = \frac{1}{2}y^2$ **25.** $y = 3 - x^2$ **26.** $y = -\frac{8}{x}$ **27.** $y = \sqrt{x - 7}$

Give the domain of each function defined.

28. $y = -4 + |x|$ **29.** $y = \frac{8 + x}{8 - x}$

30. $y = -\sqrt{\frac{5}{x^2 + 9}}$ **31.** $y = \sqrt{49 - x^2}$

32. For the function graphed in Exercise 20, give the interval over which it is **(a)** increasing and **(b)** decreasing.

33. Suppose the graph in Exercise 22 is that of $y = f(x)$. What is $f(0)$?

Given $f(x) = -2x^2 + 3x - 6$, find each function value or expression.

34. $f(0)$ **35.** $f(2.1)$ **36.** $f\left(-\frac{1}{2}\right)$ **37.** $f(k)$

Graph each equation.

38. $3x + 7y = 14$ **39.** $2x - 5y = 5$ **40.** $3y = x$

41. $2x + 5y = 20$ **42.** $x - 4y = 8$ **43.** $f(x) = x$

44. $f(x) = 3$ **45.** $x = -5$ **46.** $x - 4 = 0$

47. $y + 2 = 0$

Graph the line satisfying the given conditions.

48. through $(2, -4)$, $m = \frac{3}{4}$ **49.** through $(0, 5)$, $m = -\frac{2}{3}$

Find the slope for each line, provided that it has a slope.

50. through $(8, 7)$ and $\left(\frac{1}{2}, -2\right)$ **51.** through $(2, -2)$ and $(3, -4)$

52. through $(5, 6)$ and $(5, -2)$ **53.** through $(0, -7)$ and $(3, -7)$

54. $9x - 4y = 2$ **55.** $11x + 2y = 3$

56. $x - 5y = 0$ **57.** $x - 2 = 0$

58. *(Modeling) Job Market* The figure shows the number of jobs gained or lost in a recent period from September to May.

(a) Is this the graph of a function?

(b) In what month were the most jobs lost? the most gained?

(c) What was the largest number of jobs lost? of jobs gained?

(d) Do these data show an upward or downward trend? If so, which is it?

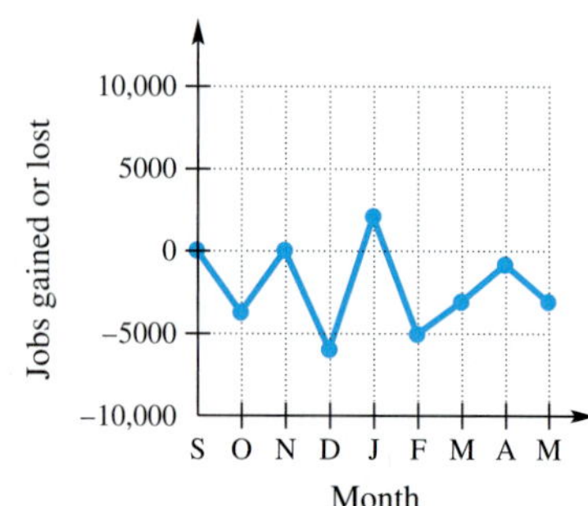

59. *(Modeling) Distance from Home* The graph depicts the distance y that a person driving a car on a straight road is from home after x hours. Interpret the graph. What speeds did the car travel?

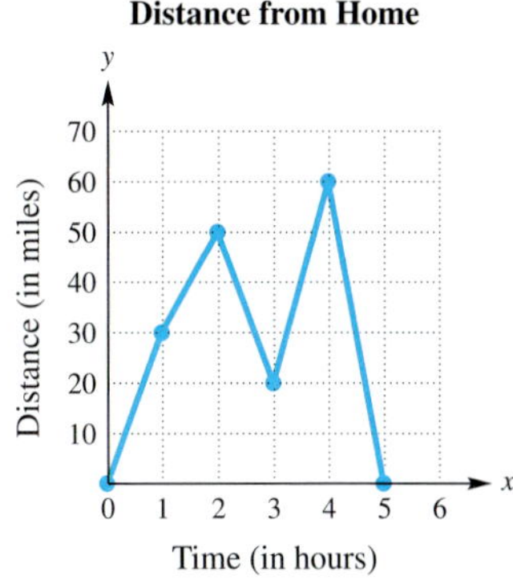

60. *Family Income* Family income in the United States has steadily increased for many years (primarily due to inflation). In 1970 the median family income was about \$10,000 per year. In 1999 it was about \$49,000 per year. Find the average rate of change of median family income to the nearest dollar over that period. (*Source:* U.S. Bureau of the Census.)

61. *(Modeling) E-Filing Tax Returns* The percent of tax returns filed electronically for the years 1996–2001 is shown in the graph.

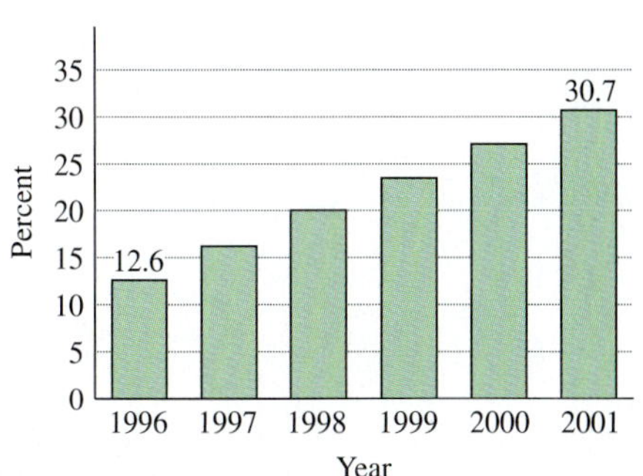

Source: Internal Revenue Service.

(a) Use the information given for the years 1996 and 2001, letting $x = 6$ represent 1996, $x = 11$ represent 2001, and y represent the percent of returns filed electronically to find a linear equation that models the data. Write the equation in slope-intercept form. Interpret the slope of this equation.

(b) Use your equation from part (a) to predict the percent of tax returns that will be filed electronically in 2005. (Assume a constant rate of change.)

For each line described, write the equation in slope-intercept form, if possible.

62. through $(-2, 4)$ and $(1, 3)$

63. through $(3, -5)$ with slope -2

64. x-intercept -3, y-intercept 5

65. through $(2, -1)$, parallel to $3x - y = 1$

66. through $(0, 5)$, perpendicular to $8x + 5y = 3$

67. through $(2, -10)$, perpendicular to a line with undefined slope

68. through $(3, -5)$, parallel to $y = 4$

69. through $(-7, 4)$, perpendicular to $y = 8$

Graph each function.

70. $f(x) = -|x|$

71. $f(x) = |x| - 3$

72. $f(x) = -|x| - 2$

73. $f(x) = -|x + 1| + 3$

74. $f(x) = 2|x - 3| - 4$

75. $f(x) = [\![x - 3]\!]$

76. $f(x) = \left[\!\!\left[\frac{1}{2}x - 2\right]\!\!\right]$

77. $f(x) = \begin{cases} -4x + 2 & \text{if } x \leq 1 \\ 3x - 5 & \text{if } x > 1 \end{cases}$

78. $f(x) = \begin{cases} 3x + 1 & \text{if } x < 2 \\ -x + 4 & \text{if } x \geq 2 \end{cases}$

79. $f(x) = \begin{cases} |x| & \text{if } x < 3 \\ 6 - x & \text{if } x \geq 3 \end{cases}$

Concept Check *Decide whether each statement is* true *or* false. *If false, tell why.*

80. The graph of a nonzero function cannot be symmetric with respect to the x-axis.

81. The graph of an even function is symmetric with respect to the y-axis.

82. The graph of an odd function is symmetric with respect to the origin.

83. If (a, b) is on the graph of an even function, so is $(a, -b)$.

84. If (a, b) is on the graph of an odd function, so is $(-a, b)$.

85. The constant function $f(x) = 0$ is both even and odd.

Decide whether each equation has a graph that is symmetric with respect to the x-axis, the y-axis, the origin, or none of these.

86. $3y^2 - 5x^2 = 15$ **87.** $x + y^2 = 8$ **88.** $y^3 = x + 1$

89. $x^2 = y^3$ **90.** $|y| = -x$ **91.** $|x + 2| = |y - 3|$

92. $|x| = |y|$

Describe how the graph of each function can be obtained from the graph of $f(x) = |x|$.

93. $g(x) = -|x|$ **94.** $h(x) = |x| - 2$ **95.** $k(x) = 2|x - 4|$

Let $f(x) = 3x - 4$. *Find an equation for each reflection of the graph of* $f(x)$.

96. across the x-axis **97.** across the y-axis **98.** across the origin

99. **Concept Check** The graph of a function f is shown in the figure. Sketch the graph of each function defined as follows.

(a) $y = f(x) + 3$
(b) $y = f(x - 2)$
(c) $y = f(x + 3) - 2$
(d) $y = |f(x)|$

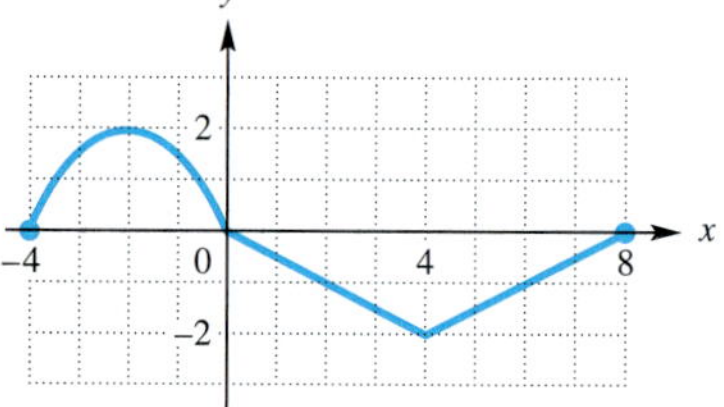

Let $f(x) = 3x^2 - 4$ *and* $g(x) = x^2 - 3x - 4$. *Find each of the following.*

100. $(f + g)(x)$ **101.** $(fg)(x)$ **102.** $(f - g)(4)$

103. $(f + g)(-4)$ **104.** $(f + g)(2k)$ **105.** $\left(\frac{f}{g}\right)(3)$

106. $\left(\frac{f}{g}\right)(-1)$ **107.** the domain of $(fg)(x)$ **108.** the domain of $\left(\frac{f}{g}\right)(x)$

109. Which of the following is *not* equal to $(f \circ g)(x)$ for $f(x) = \frac{1}{x}$ and $g(x) = x^2 + 1$? (*Hint:* There may be more than one.)

A. $f[g(x)]$ **B.** $\frac{1}{x^2 + 1}$ **C.** $\frac{1}{x^2}$ **D.** $(g \circ f)(x)$

For each function, find and simplify $\frac{f(x + h) - f(x)}{h}$.

110. $f(x) = 2x + 9$ **111.** $f(x) = x^2 - 5x + 3$

Let $f(x) = \sqrt{x - 2}$ and $g(x) = x^2$. Find each of the following.

112. $(f \circ g)(x)$ **113.** $(g \circ f)(x)$ **114.** $(f \circ g)(-6)$

115. $(g \circ f)(3)$ **116.** the domain of $f \circ g$

Use the table to evaluate each expression, if possible.

117. $(f + g)(1)$

118. $(f - g)(3)$

119. $(fg)(-1)$

120. $\left(\dfrac{f}{g}\right)(0)$

x	$f(x)$	$g(x)$
−1	3	−2
0	5	0
1	7	1
3	9	9

Tables for f and g are given. Use them to evaluate the expressions in Exercises 121 and 122.

121. $(g \circ f)(-2)$

122. $(f \circ g)(3)$

x	$f(x)$
−2	1
0	4
2	3
4	2

x	$g(x)$
1	2
2	4
3	−2
4	0

Concept Check *The graphs of two functions f and g are shown in the figures.*

123. Find $(f \circ g)(2)$.

124. Find $(g \circ f)(3)$.

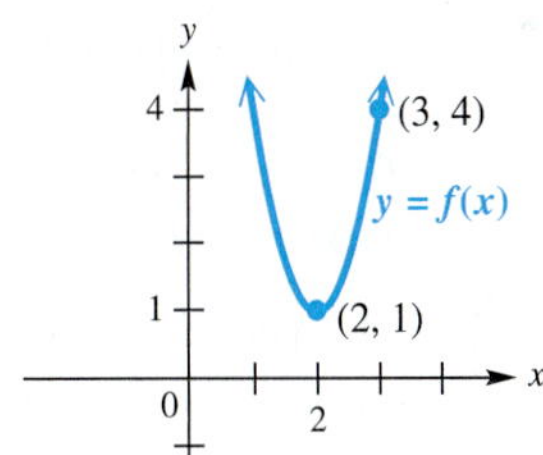

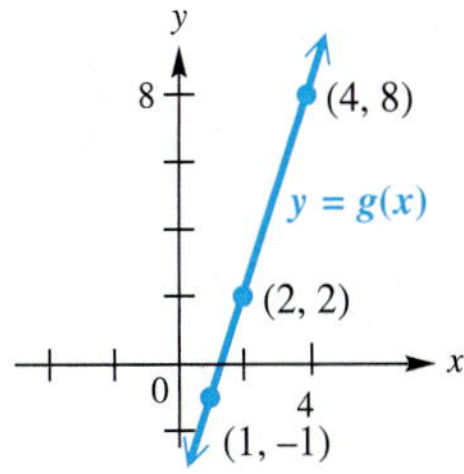

Solve each problem.

125. Relationship of Measurement Units There are 36 in. in 1 yd, and there are 1760 yd in 1 mi. Express the number of inches x in 1 mi by forming two functions and then considering their composition.

126. (Modeling) Perimeter of a Rectangle Suppose the length of a rectangle is twice its width. Let x represent the width of the rectangle. Write a formula for the perimeter P of the rectangle in terms of x alone. Then use $P(x)$ notation to describe it as a function. What type of function is this?

127. (Modeling) Volume of a Sphere The formula for the volume of a sphere is $V(r) = \frac{4}{3}\pi r^3$, where r represents the radius of the sphere. Construct a model function V representing the amount of volume gained when the radius r (in inches) of a sphere is increased by 3 in.

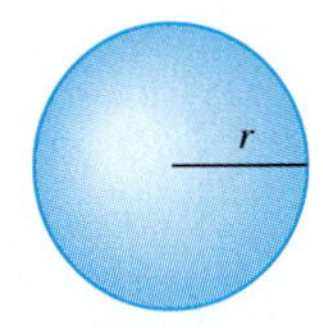

128. *(Modeling) Dimensions of a Cylinder* A cylindrical can makes the most efficient use of materials when its height is the same as the diameter of its top.

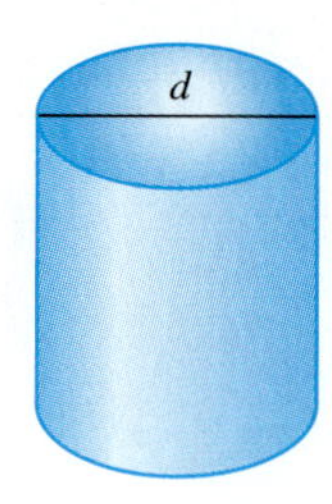

(a) Express the volume V of such a can as a function of the diameter d of its top.

(b) Express the surface area S of such a can as a function of the diameter d of its top. (*Hint:* The curved side is made from a rectangle whose length is the circumference of the top of the can.)

Chapter 2 Test

1. Match the set described in Column I with the correct interval notation from Column II. Choices in Column II may be used once, more than once, or not at all.

I	II
(a) Domain of $f(x) = \sqrt{x} + 3$	**A.** $[-3, \infty)$
(b) Range of $f(x) = \sqrt{x - 3}$	**B.** $[3, \infty)$
(c) Domain of $f(x) = x^2 - 3$	**C.** $(-\infty, \infty)$
(d) Range of $f(x) = x^2 + 3$	**D.** $[0, \infty)$
(e) Domain of $f(x) = \sqrt[3]{x - 3}$	**E.** $(-\infty, 3)$
(f) Range of $f(x) = \sqrt[3]{x} + 3$	**F.** $(-\infty, 3]$
(g) Domain of $f(x) = \lvert x \rvert - 3$	**G.** $(3, \infty)$
(h) Range of $f(x) = \lvert x + 3 \rvert$	**H.** $(-\infty, 0]$
(i) Domain of $x = y^2$	
(j) Range of $x = y^2$	

The graph shows the line that passes through the points $(-2, 1)$ *and* $(3, 4)$. *Refer to it to answer Exercises 2–6.*

2. What is the slope of the line?

3. What is the distance between the two points shown?

4. What are the coordinates of the midpoint of the segment joining the two points?

5. Find the standard form of the equation of the line.

6. Write the linear function defined by $f(x) = ax + b$ that has this line as its graph.

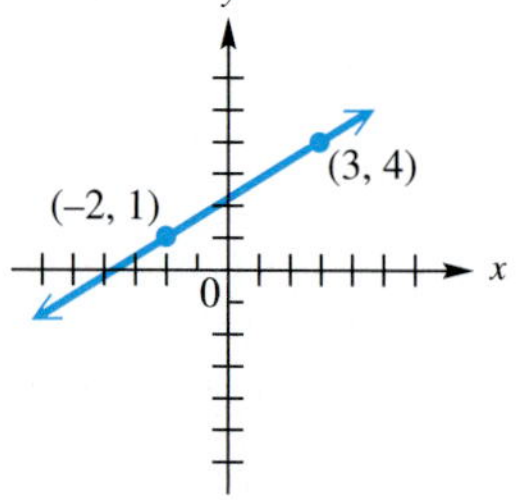

7. Tell whether each graph is that of a function. Give the domain and range. If it is a function, give the intervals where it is increasing, decreasing, or constant.

(a)

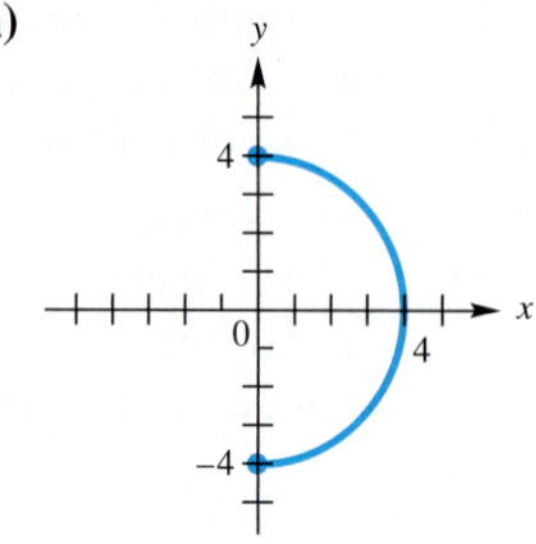

(b)

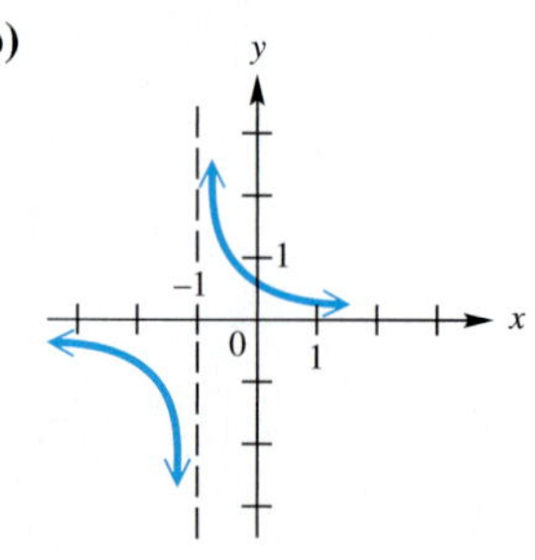

8. Suppose point A has coordinates $(5, -3)$.

(a) What is the equation of the vertical line through A?
(b) What is the equation of the horizontal line through A?

9. Find the slope-intercept form of the equation of the line passing through $(2, 3)$ and

(a) parallel to the graph of $y = -3x + 2$;
(b) perpendicular to the graph of $y = -3x + 2$.

10. Consider the graph of the function shown here.

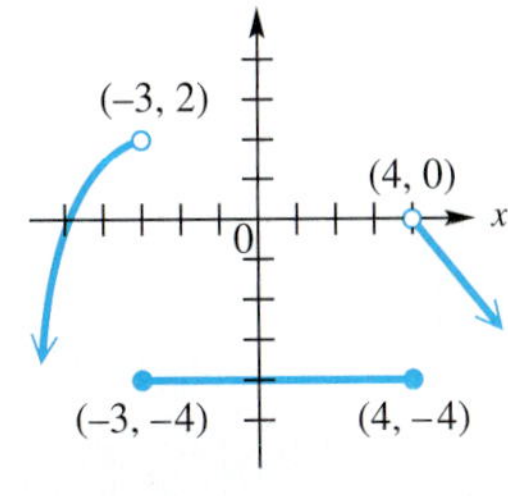

(a) Give the interval over which the function is increasing.
(b) Give the interval over which the function is decreasing.
(c) Give the interval over which the function is constant.
(d) Give the intervals over which the function is continuous.
(e) What is the domain of this function?
(f) What is the range of this function?

Graph each function.

11. $y = |x - 2| - 1$

12. $f(x) = [\![x + 1]\!]$

13. $f(x) = \begin{cases} 3 & \text{if } x < -2 \\ 2 - \frac{1}{2}x & \text{if } x \geq -2 \end{cases}$

14. The graph of $y = f(x)$ is shown here. Sketch the graph of each of the following. Use ordered pairs to indicate three points on the graph.

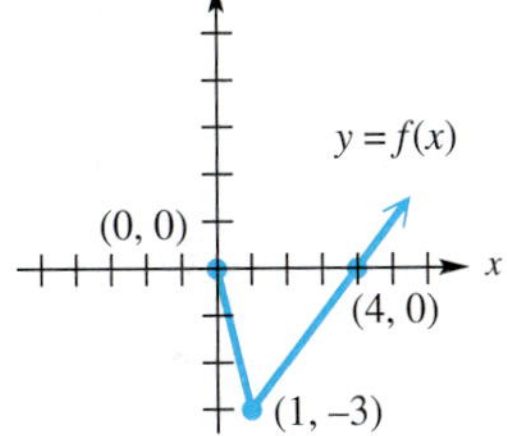

(a) $y = f(x) + 2$
(b) $y = f(x + 2)$
(c) $y = -f(x)$
(d) $y = f(-x)$
(e) $y = 2 \cdot f(x)$

15. Explain how the graph of $y = -2\sqrt{x + 2} - 3$ can be obtained from the graph of $y = \sqrt{x}$.

16. Determine whether the graph of $3x^2 - 2y^2 = 3$ is symmetric with respect to

(a) the x-axis, **(b)** the y-axis, and **(c)** the origin.

Given $f(x) = 2x^2 - 3x + 2$ and $g(x) = -2x + 1$, find each of the following. Simplify the expressions when possible.

17. (a) $(f - g)(x)$

(b) $\left(\frac{f}{g}\right)(x)$

(c) the domain of $\frac{f}{g}$

(d) $\frac{f(x + h) - f(x)}{h} \quad (h \neq 0)$

18. (a) $(f + g)(1)$

(b) $(fg)(2)$

(c) $(f \circ g)(0)$

19. *(Modeling) Long-Distance Call Charges* A certain long-distance carrier provides service between Podunk and Nowheresville. If x represents the number of minutes for the call, where $x > 0$, then the function f defined by

$$f(x) = .40[\![x]\!] + .75$$

gives the total cost of the call in dollars. Find the cost of a 5.5-min call.

20. *(Modeling) Cost, Revenue, and Profit Analysis* Tyler McGinnis starts up a small business manufacturing bobble-head figures of famous baseball players. His initial cost is \$3300. Each figure costs \$4.50 to manufacture.

(a) Write a cost function C, where x represents the number of figures manufactured.
(b) Find the revenue function R, if each figure in part (a) sells for \$10.50.
(c) Give the profit function P.
(d) How many figures must be produced and sold before Tyler earns a profit?

Chapter 2 Quantitative Reasoning

How can you decide whether to buy a municipal bond or a treasury bond?

Instead of a traditional pension plan, many employers now match an employee's savings in a retirement account set up by the employee. It is up to the employee to decide what kind of investments to make with these funds. Suppose you have earned such an account. Two possible investments are municipal bonds and treasury bonds.

Interest earned on a municipal bond is free of both federal and state taxes, whereas interest earned on a treasury bond is free only of state taxes. However, treasury bonds normally pay higher interest rates than municipal bonds. Deciding on the better choice depends not only on the interest rates but also on your *federal tax bracket*. (The federal tax bracket, also known as the *marginal tax bracket,* is the percent of tax paid on the last dollar earned.) If t is the tax-free rate and x is your tax bracket (as a decimal), then the formula

$$r = \frac{t}{1 - x}$$

gives the taxable rate r that is equivalent to a given tax-free rate.

1. Suppose a 1-yr municipal bond pays 4% interest. Graph the function defined by

$$r(x) = \frac{.04}{1 - x}$$

for $0 < x \leq .5$. If you graph by hand, use x-values .1, .2, .3, .4, and .5. If you use a graphing calculator, use the viewing window $[0, .5]$ by $[0, .08]$. Is this a linear function? Why or why not?

2. What is the equivalent rate for a 1-yr treasury bond if your tax bracket is 31%?

3. If a 1-yr treasury bond pays 6.26%, in how high a tax bracket must you be before the municipal bond is more attractive?

3 Polynomial and Rational Functions

Traffic congestion is one of the biggest hassles in modern life. Although the population of the United States has only increased by about 20% since 1982, the time spent waiting in traffic has increased by 236%. Today, the average driver spends over 10 hr a year stuck in traffic. Traffic congestion costs Americans $78 billion a year due to wasted fuel and lost time from work.

Traffic is subject to a *nonlinear effect.* According to Joe Sussman, an engineer from Massachusetts Institute of Technology, you can put more cars on the road up to a point. Then, if traffic intensity increases even slightly beyond this point, congestion and waiting time increase dramatically. (*Source:* Longman, Phillip J., "American Gridlock," *U.S. News & World Report,* May 28, 2001.) In Example 10 of Section 3.5, we use a rational function to model this phenomenon.

3.1 Quadratic Functions and Models

Quadratic Functions ▪ Graphing Techniques ▪ Completing the Square ▪ The Vertex Formula ▪ Quadratic Models and Curve Fitting

A *polynomial function* is defined as follows.

Polynomial Function

A **polynomial function** of degree n, where n is a nonnegative integer, is a function defined by an expression of the form

$$f(x) = a_n x^n + a_{n-1}x^{n-1} + \cdots + a_1 x + a_0,$$

where $a_n, a_{n-1}, \ldots, a_1$, and a_0 are real numbers, with $a_n \neq 0$.

$$f(x) = 5x - 1, \qquad f(x) = x^4 + \sqrt{2}x^3 - 4x^2, \qquad f(x) = 4x^2 - x + 2$$

Polynomial functions

For the polynomial function defined by

$$f(x) = 2x^3 - \frac{1}{2}x + 5,$$

n is 3 and the polynomial has the form

$$a_3x^3 + a_2x^2 + a_1x + a_0,$$

where a_3 is 2, a_2 is 0, a_1 is $-\frac{1}{2}$, and a_0 is 5. The polynomial functions defined by $f(x) = x^4 + \sqrt{2}x^3 - 4x^2$ and $f(x) = 4x^2 - x + 2$ have degrees 4 and 2, respectively. The number a_n is the **leading coefficient** of $f(x)$. The function defined by $f(x) = 0$ is called the **zero polynomial.** The zero polynomial has no degree. However, a polynomial function defined by $f(x) = a_0$ for a nonzero number a_0 has degree 0.

Looking Ahead to Calculus

In calculus, polynomial functions are used to approximate more complicated functions, such as trigonometric, exponential, or logarithmic functions. For example, the trigonometric function $\sin x$ is approximated by the polynomial $x - \frac{x^3}{3} + \frac{x^5}{5} - \frac{x^7}{7}$.

Quadratic Functions In Sections 2.3 and 2.4, we discussed first-degree (*linear*) polynomial functions, in which the greatest power of the variable is 1. Now we look at polynomial functions of degree 2, called *quadratic functions.*

Quadratic Function

A function f is a **quadratic function** if

$$f(x) = ax^2 + bx + c,$$

where a, b, and c are real numbers, with $a \neq 0$.

The simplest quadratic function is given by $f(x) = x^2$ with $a = 1$, $b = 0$, and $c = 0$. To find some points on the graph of this function, choose some values for x and find the corresponding values for $f(x)$, as in the table with Figure 1. Then plot these points, and draw a smooth curve through them. This graph is called a **parabola.** Every quadratic function has a graph that is a parabola.

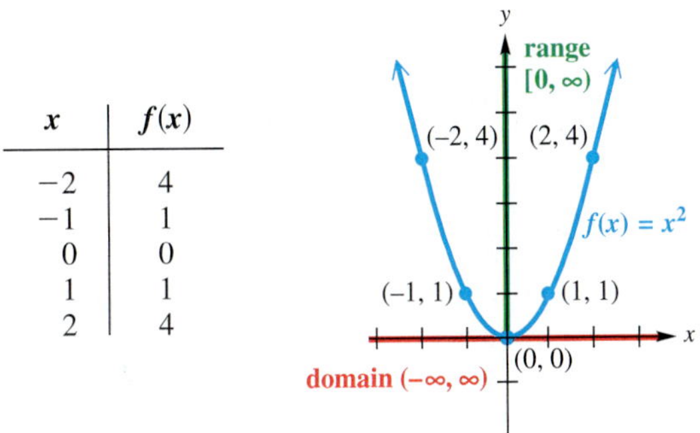

x	$f(x)$
−2	4
−1	1
0	0
1	1
2	4

Figure 1

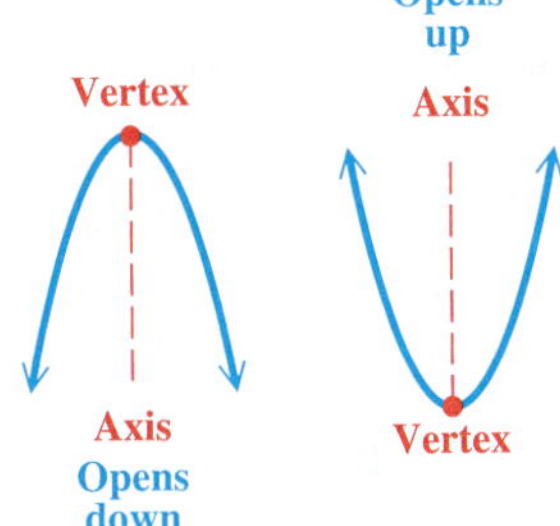

Figure 2

We can determine the domain and the range of a quadratic function whose graph is a parabola, such as the one in Figure 1, from its graph. Since the graph extends indefinitely to the right and to the left, the domain is $(-\infty, \infty)$. Since the lowest point is $(0, 0)$, the minimum range value (y-value) is 0. The graph extends upward indefinitely; there is no maximum y-value, so the range is $[0, \infty)$.

Parabolas are symmetric with respect to a line (the y-axis in Figure 1). The line of symmetry for a parabola is called the **axis** of the parabola. The point where the axis intersects the parabola is the **vertex** of the parabola. As Figure 2 shows, the vertex of a parabola that opens down is the highest point of the graph and the vertex of a parabola that opens up is the lowest point of the graph.

Graphing Techniques The graphing techniques of Section 2.6 applied to the graph of $f(x) = x^2$ give the graph of *any* quadratic function. The graph of $g(x) = ax^2$ is a parabola with vertex at the origin that opens up if a is positive and down if a is negative. The width of the graph of $g(x)$ is determined by the magnitude of a. That is, the graph of $g(x)$ is narrower than that of $f(x) = x^2$ if $|a| > 1$ and is broader (wider) than that of $f(x) = x^2$ if $|a| < 1$. By completing the square, any quadratic function can be written in the form

$$F(x) = a(x - h)^2 + k.$$

The graph of $F(x)$ is the same as the graph of $g(x) = ax^2$ translated $|h|$ units horizontally (to the right if h is positive and to the left if h is negative) and translated $|k|$ units vertically (up if k is positive and down if k is negative).

EXAMPLE 1 Graphing Quadratic Functions

Graph each function. Give the domain and range.

(a) $f(x) = x^2 - 4x - 2$ (by plotting points as in Section 2.1)

(b) $g(x) = -\dfrac{1}{2}x^2$ (and compare to $y = x^2$ and $y = \frac{1}{2}x^2$)

(c) $F(x) = -\dfrac{1}{2}(x - 4)^2 + 3$ (and compare to the graph in part (b))

Solution

(a) A table of selected points is shown accompanying the graph of $f(x) = x^2 - 4x - 2$. See Figure 3 on the next page. The domain is $(-\infty, \infty)$, the range is $[-6, \infty)$, the vertex is $(2, -6)$, and the axis has equation $x = 2$. Figure 4 shows how a graphing calculator displays this graph.

x	$f(x)$
-1	3
0	-2
1	-5
2	-6
3	-5
4	-2
5	3

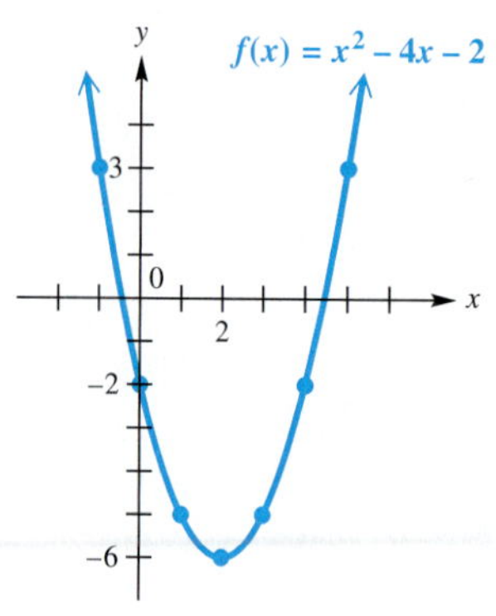

Figure 3

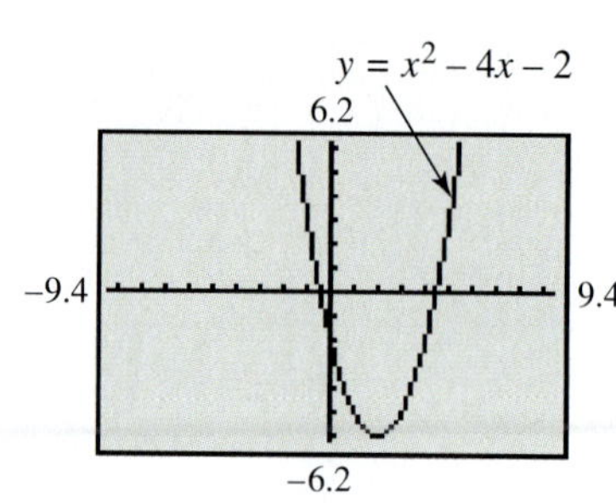

Figure 4

(b) Think of $g(x) = -\frac{1}{2}x^2$ as $g(x) = -\left(\frac{1}{2}x^2\right)$. The graph of $y = \frac{1}{2}x^2$ is a broader version of the graph of $y = x^2$, and the graph of $g(x) = -\left(\frac{1}{2}x^2\right)$ is a reflection of the graph of $y = \frac{1}{2}x^2$ across the x-axis. See Figure 5. The vertex is $(0, 0)$, and the axis of the parabola is the line $x = 0$ (the y-axis). The domain is $(-\infty, \infty)$, and the range is $(-\infty, 0]$. Calculator graphs for $y = x^2$, $y = \frac{1}{2}x^2$, and $g(x) = -\frac{1}{2}x^2$ are shown in Figure 6.

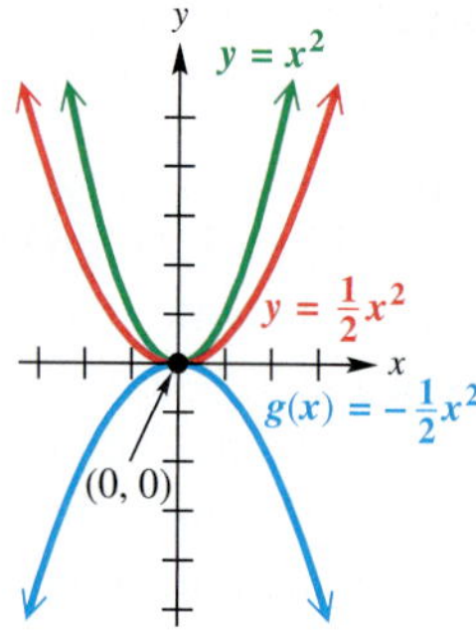

Figure 5

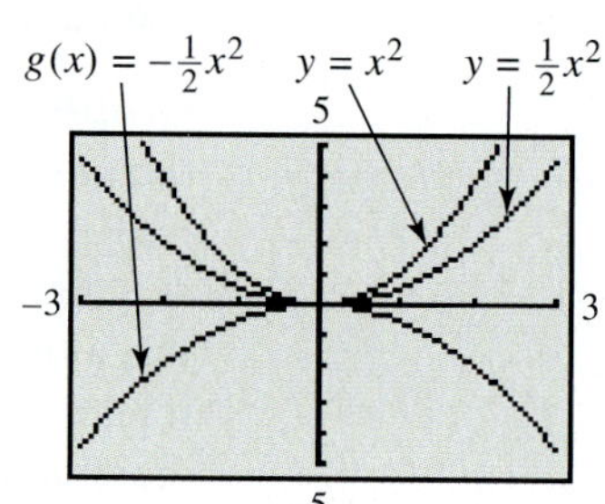

Figure 6

(c) We can write $F(x) = -\frac{1}{2}(x - 4)^2 + 3$ as $F(x) = g(x - h) + k$, where $g(x)$ is the function of part (b), h is 4, and k is 3. Therefore, the graph of $F(x)$ is the graph of $g(x)$ translated 4 units to the right and 3 units up. See Figure 7. The vertex is $(4, 3)$, which is also shown in the calculator graph in Figure 8, and the axis of the parabola is the line $x = 4$. The domain is $(-\infty, \infty)$, and the range is $(-\infty, 3]$.

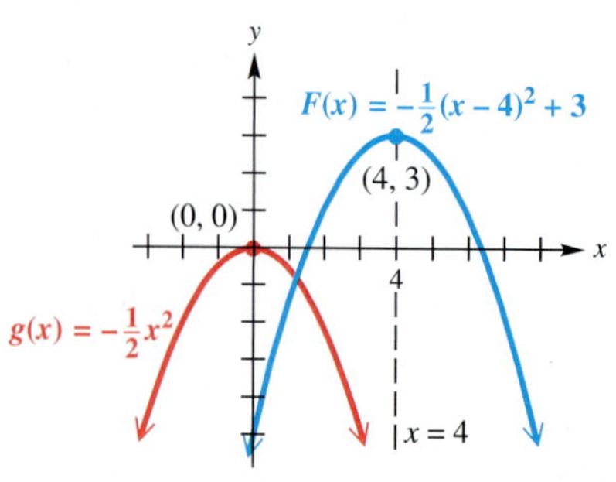

Figure 7

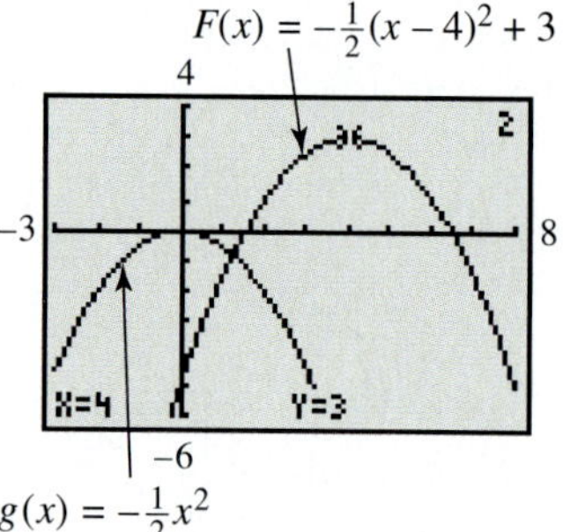

Figure 8

Now try Exercises 9, 17, and 19.

Completing the Square In general, the graph of the quadratic function defined by

$$f(x) = a(x - h)^2 + k$$

is a parabola with vertex (h, k) and axis $x = h$. The parabola opens up if a is positive and down if a is negative. With these facts in mind, we *complete the square* to graph a quadratic function defined by $f(x) = ax^2 + bx + c$.

EXAMPLE 2 Graphing a Parabola by Completing the Square

Graph $f(x) = x^2 - 6x + 7$ by completing the square and locating the vertex.

Solution We express $x^2 - 6x + 7$ in the form $(x - h)^2 + k$ by completing the square. First, write

$$f(x) = (x^2 - 6x \qquad) + 7. \quad \text{Complete the square. (Section 1.4)}$$

We must add a number inside the parentheses to get a perfect square trinomial. Find this number by taking half the coefficient of x and squaring the result:

$$\left[\frac{1}{2}(-6)\right]^2 = (-3)^2 = 9.$$

Add and subtract 9 inside the parentheses. (This is the same as adding 0.)

$$f(x) = (x^2 - 6x + 9 - 9) + 7 \quad \text{Add and subtract 9.}$$

$$f(x) = (x^2 - 6x + 9) - 9 + 7 \quad \text{Regroup terms.}$$

$$f(x) = (x - 3)^2 - 2 \quad \text{Factor; simplify. (Section R.4)}$$

This form shows that the vertex of the parabola is $(3, -2)$ and the axis is the line $x = 3$.

Now find additional ordered pairs that satisfy the equation. Since the y-intercept is 7, $(0, 7)$ is on the graph. Verify that $(1, 2)$ is also on the graph. Use symmetry about the axis of the parabola to find the ordered pairs $(5, 2)$ and $(6, 7)$. Plot and connect these points to obtain the graph in Figure 9. The domain of this function is $(-\infty, \infty)$, and the range is $[-2, \infty)$.

$f(x) = x^2 - 6x + 7$
$= (x - 3)^2 - 2$

10, −3, 9, −4
Minimum X=3 Y=-2

This screen shows that the vertex of the graph in Figure 9 is $(3, -2)$. Because it is the *lowest* point on the graph, we direct the calculator to find the *minimum*.

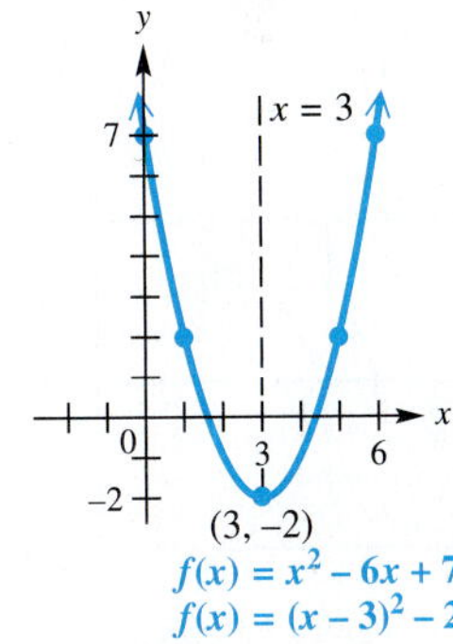

Figure 9

Now try Exercise 21.

NOTE In Example 2 we added and subtracted 9 *on the same side* of the equation to complete the square. This differs from adding the same number to *each side of the equation,* as when we completed the square in Chapter 1. Since we want $f(x)$ (or y) alone on one side of the equation, we adjusted that step in the process of completing the square slightly.

EXAMPLE 3 Graphing a Parabola by Completing the Square

Graph $f(x) = -3x^2 - 2x + 1$ by completing the square and locating the vertex.

Solution To complete the square, the coefficient of x^2 must be 1.

$$f(x) = -3\left(x^2 + \frac{2}{3}x \qquad\right) + 1 \qquad \text{Factor } -3 \text{ from the first two terms.}$$

$$f(x) = -3\left(x^2 + \frac{2}{3}x + \frac{1}{9} - \frac{1}{9}\right) + 1 \qquad \left[\tfrac{1}{2}\left(\tfrac{2}{3}\right)\right]^2 = \left(\tfrac{1}{3}\right)^2 = \tfrac{1}{9}\text{; add and subtract } \tfrac{1}{9}.$$

$$= -3\left(x^2 + \frac{2}{3}x + \frac{1}{9}\right) - 3\left(-\frac{1}{9}\right) + 1 \qquad \text{Distributive property (Section R.1)}$$

$$= -3\left(x + \frac{1}{3}\right)^2 + \frac{4}{3} \qquad \text{Factor; simplify.}$$

The vertex is $\left(-\frac{1}{3}, \frac{4}{3}\right)$. Find additional points by substituting x-values into the original equation. For example, $\left(\frac{1}{2}, -\frac{3}{4}\right)$ is on the graph shown in Figure 10. The intercepts are often good additional points to find. Here, the y-intercept is

$$y = -3(0)^2 - 2(0) + 1 = 1, \qquad \text{Let } x = 0.$$

giving the point (0, 1). The x-intercepts are found by setting $f(x)$ equal to 0 in the original equation.

$$0 = -3x^2 - 2x + 1 \qquad \text{Let } f(x) = 0.$$

$$3x^2 + 2x - 1 = 0 \qquad \text{Multiply by } -1.$$

$$(3x - 1)(x + 1) = 0 \qquad \text{Factor.}$$

$$x = \frac{1}{3} \quad \text{or} \quad x = -1 \qquad \text{Zero-factor property (Section 1.4)}$$

Therefore, the x-intercepts are $\frac{1}{3}$ and -1.

$f(x) = -3x^2 - 2x + 1$
$= -3\left(x + \frac{1}{3}\right)^2 + \frac{4}{3}$

3.1

−4.7 4.7

Maximum X=-.333333 Y=1.3333333

−3.1

This screen gives the vertex of the graph in Figure 10 as $(-.\bar{3}, 1.\bar{3}) = \left(-\frac{1}{3}, \frac{4}{3}\right)$. We want the highest point on the graph, so we direct the calculator to find the *maximum*.

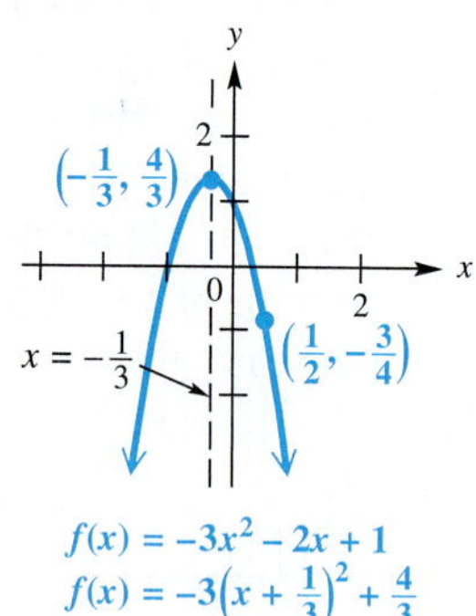

Figure 10

Now try Exercise 23.

NOTE The square root property or the quadratic formula can be used to find the x-intercepts if $f(x)$ in the equation $f(x) = 0$ is not readily factorable. This would be the case in Example 2.

Looking Ahead to Calculus

An important concept in calculus is the *definite integral.* If the graph of f lies above the x-axis, the symbol

$$\int_a^b f(x)\,dx$$

represents the area of the region above the x-axis and below the graph of f from $x = a$ to $x = b$. For example, in Figure 10 with

$$f(x) = -3x^2 - 2x + 1,$$

$a = -1$, and $b = \frac{1}{3}$, calculus provides the tools for determining that the area enclosed by the parabola and the x-axis is $\frac{32}{27}$ (square units).

The Vertex Formula We can generalize the earlier work to obtain a formula for the vertex of a parabola. Starting with the general quadratic form $f(x) = ax^2 + bx + c$ and completing the square will change the form to $f(x) = a(x - h)^2 + k$.

$$f(x) = ax^2 + bx + c$$

$$= a\left(x^2 + \frac{b}{a}x \qquad\right) + c \qquad \text{Factor } a \text{ from the first two terms.}$$

$$= a\left(x^2 + \frac{b}{a}x + \frac{b^2}{4a^2}\right) + c - a\left(\frac{b^2}{4a^2}\right) \qquad \text{Add } \left[\tfrac{1}{2}\left(\tfrac{b}{a}\right)\right]^2 = \tfrac{b^2}{4a^2} \text{ inside the parentheses; subtract } a\left(\tfrac{b^2}{4a^2}\right) \text{ outside the parentheses.}$$

$$f(x) = a\left(x + \frac{b}{2a}\right)^2 + c - \frac{b^2}{4a} \qquad \text{Factor; simplify.}$$

Comparing the last result with $f(x) = a(x - h)^2 + k$ shows that

$$h = -\frac{b}{2a} \quad \text{and} \quad k = c - \frac{b^2}{4a}.$$

Letting $x = h$ in $f(x) = a(x - h)^2 + k$ gives $f(h) = a(h - h)^2 + k = k$, so $k = f(h)$, or $k = f\left(-\frac{b}{2a}\right)$.

The following statement summarizes this discussion.

Graph of a Quadratic Function

The quadratic function defined by $f(x) = ax^2 + bx + c$ can be written in the form

$$\mathbf{y = f(x) = a(x - h)^2 + k}, \qquad a \neq 0,$$

where

$$h = -\frac{b}{2a} \quad \text{and} \quad k = f(h).$$

The graph of f has the following characteristics.

1. It is a parabola with vertex (h, k) and the vertical line $x = h$ as axis.
2. It opens up if $a > 0$ and down if $a < 0$.
3. It is broader than the graph of $y = x^2$ if $|a| < 1$ and narrower if $|a| > 1$.
4. The y-intercept is $f(0) = c$.
5. If $b^2 - 4ac \geq 0$, the x-intercepts are

$$x = \frac{-b \pm \sqrt{b^2 - 4ac}}{2a}.$$

If $b^2 - 4ac < 0$, there are no x-intercepts.

We can find the vertex and axis of a parabola from its equation either by completing the square or by remembering that $h = -\frac{b}{2a}$ and letting $k = f(h)$.

EXAMPLE 4 Finding the Axis and the Vertex of a Parabola Using the Formula

Find the axis and vertex of the parabola having equation $f(x) = 2x^2 + 4x + 5$ using the formula.

Solution Here $a = 2$, $b = 4$, and $c = 5$. The axis of the parabola is the vertical line

$$x = h = -\frac{b}{2a} = -\frac{4}{2(2)} = -1.$$

The vertex is $(-1, f(-1)) = (-1, 3)$.

Now try Exercise 25.

Looking Ahead to Calculus

The derivative of a function provides a formula for finding the slope of a line tangent to the graph of a function. Using the methods of calculus, the function of Example 3, $f(x) = -3x^2 - 2x + 1$, has derivative $f'(x) = -6x - 2$. If we solve $f'(x) = 0$, we find the x-coordinate for which the graph of f has a horizontal tangent. Solve this equation and show that its solution gives the x-coordinate of the vertex. Notice that if you draw a tangent line at the vertex, it is a horizontal line with slope 0.

Quadratic Models and Curve Fitting From the graphs in this section, we see that quadratic functions make good models for data sets where the data either increases, levels off, and then decreases or decreases, levels off, and then increases.

Since the vertex of a vertical parabola is the highest or lowest point on the graph, equations of the form $y = ax^2 + bx + c$ are important in problems where we must find the maximum or minimum value of some quantity. When $a < 0$, the y-coordinate of the vertex gives the maximum value of y and the x-value tells where it occurs. Similarly, when $a > 0$, the y-coordinate of the vertex gives the minimum y-value.

An application of quadratic functions models the height of a propelled object as a function of the time elapsed after it is propelled. Recall that if air resistance is neglected, the height s (in feet) of an object propelled directly upward from an initial height s_0 feet with initial velocity v_0 feet per second is

$$s(t) = -16t^2 + v_0 t + s_0, \quad \text{(Section 1.5)}$$

where t is the number of seconds after the object is propelled. The coefficient of t^2 (that is, -16) is a constant based on the gravitational force of Earth. This constant varies on other surfaces, such as the moon or the other planets.

EXAMPLE 5 Solving a Problem Involving Projectile Motion

A ball is thrown directly upward from an initial height of 100 ft with an initial velocity of 80 ft per sec.

(a) Give the function that describes the height of the ball in terms of time t.

(b) Graph this function on a graphing calculator so that the y-intercept, the positive x-intercept, and the vertex are visible.

(c) Figure 11 shows that the point (4.8, 115.36) lies on the graph of the function. What does this mean for this particular situation?

(d) After how many seconds does the projectile reach its maximum height? What is this maximum height?

(e) For what interval of time is the height of the ball greater than 160 ft?

(f) After how many seconds will the ball hit the ground?

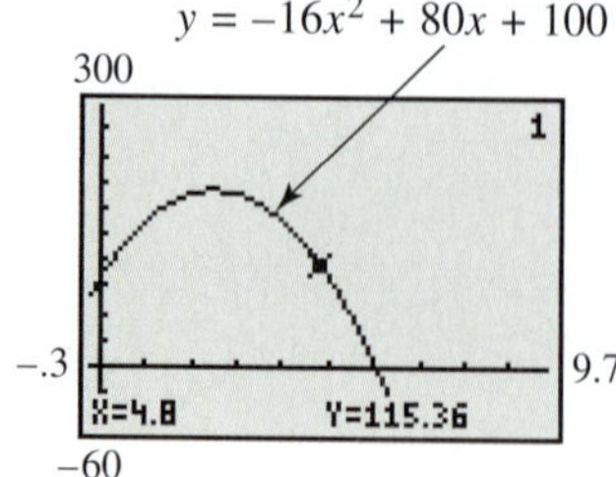

Figure 11

Solution

(a) Use the projectile height function with $v_0 = 80$ and $s_0 = 100$:

$$s(t) = -16t^2 + 80t + 100.$$

(b) There are many suitable choices for such a window. One choice is $[-.3, 9.7]$ by $[-60, 300]$, as in Figure 11, which shows the graph of $y = -16x^2 + 80x + 100$. (Here, $x = t$.) It is easy to misinterpret the graph in Figure 11. *The graph does not show the path followed by the ball; it defines height as a function of time.*

(c) In Figure 11, when $x = 4.8$, $y = 115.36$. Therefore, when 4.8 sec have elapsed, the projectile is at a height of 115.36 ft.

Algebraic Solution

(d) Find the coordinates of the vertex of the parabola. Using the vertex formula with $a = -16$ and $b = 80$,

$$x = -\frac{b}{2a} = -\frac{80}{2(-16)} = 2.5$$

and

$$y = -16(2.5)^2 + 80(2.5) + 100$$
$$= 200.$$

Therefore, after 2.5 sec the ball reaches its maximum height of 200 ft.

(e) We must solve the quadratic *inequality*

$$-16x^2 + 80x + 100 > 160.$$

$$-16x^2 + 80x - 60 > 0 \quad \text{Subtract 160.}$$

$$4x^2 - 20x + 15 < 0$$

Divide by -4; reverse the inequality sign. (Section 1.7)

By the quadratic formula, the solutions of $4x^2 - 20x + 15 = 0$ are

$$\frac{5 - \sqrt{10}}{2} \approx .92 \text{ and } \frac{5 + \sqrt{10}}{2} \approx 4.08.$$

These numbers divide the number line into three intervals:

$$(-\infty, .92), \quad (.92, 4.08), \quad \text{and} \quad (4.08, \infty).$$

Using a test value from each interval shows that $(.92, 4.08)$ satisfies the *inequality*. The ball is more than 160 ft above the ground between .92 sec and 4.08 sec.

Graphing Calculator Solution

(d) Using the capabilities of the calculator, we see in Figure 12 that the vertex coordinates are indeed (2.5, 200).

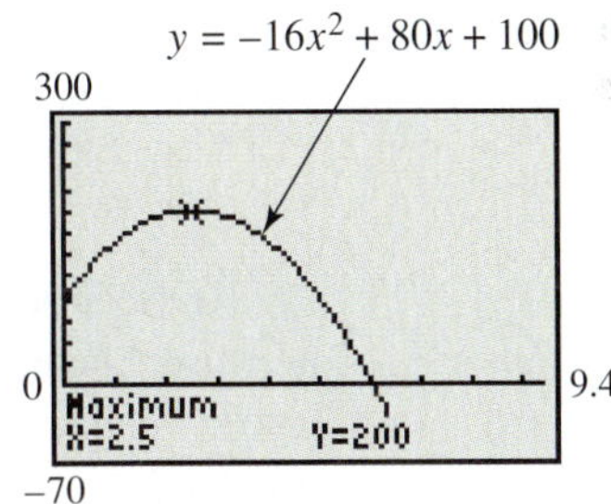

Figure 12

(e) If we graph

$$y_1 = -16x^2 + 80x + 100 \quad \text{and} \quad y_2 = 160,$$

as shown in Figure 13, and locate the two points of intersection, we find that the x-coordinates for these points are approximately .92 and 4.08. Therefore, between .92 sec and 4.08 sec, the ball is more than 160 ft above the ground, that is, $y_1 > y_2$.

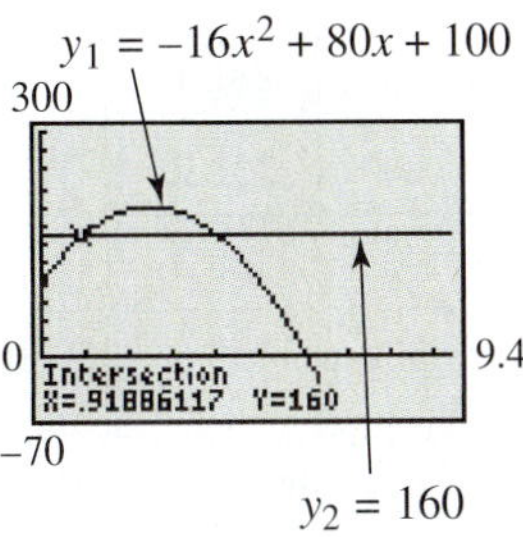

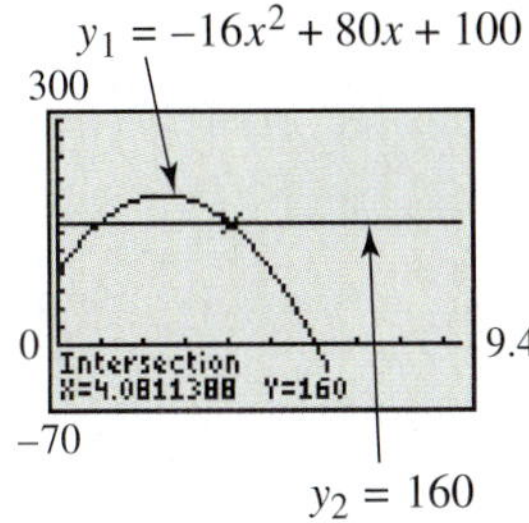

Figure 13

(continued)

(f) The height is 0 when the ball hits the ground. We use the quadratic formula to find the *positive* solution of

$$-16x^2 + 80x + 100 = 0.$$

Here, $a = -16$, $b = 80$, and $c = 100$.

$$x = \frac{-80 \pm \sqrt{80^2 - 4(-16)(100)}}{2(-16)}$$

$x \approx -1.04$ or $x \approx 6.04$
Reject

The ball hits the ground after about 6.04 sec.

(f) Figure 14 shows that the positive x-intercept of the graph of

$$y = -16x^2 + 80x + 100$$

is approximately 6.04, which means that the ball hits the ground after about 6.04 sec.

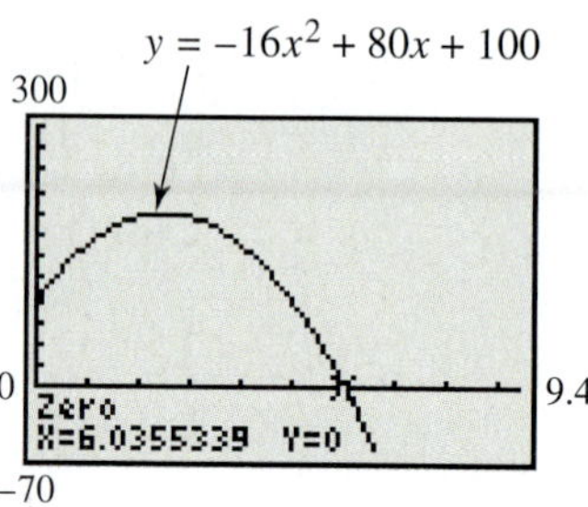

Figure 14

Now try Exercise 47.

In Section 2.4 we introduced curve fitting and used linear regression to determine linear equations that modeled data. With a graphing calculator, we can use a technique called *quadratic regression* to find quadratic equations that model data.

EXAMPLE 6 Modeling the Number of Hospital Outpatient Visits

The number of hospital outpatient visits (in millions) for selected years is shown in the table.

Year	Visits	Year	Visits
75	254.9	95	488.2
80	263.0	96	505.5
85	282.1	97	520.6
88	336.2	98	545.5
90	368.2	99	573.5
93	435.6	100	592.7

Source: American Hospital Association.

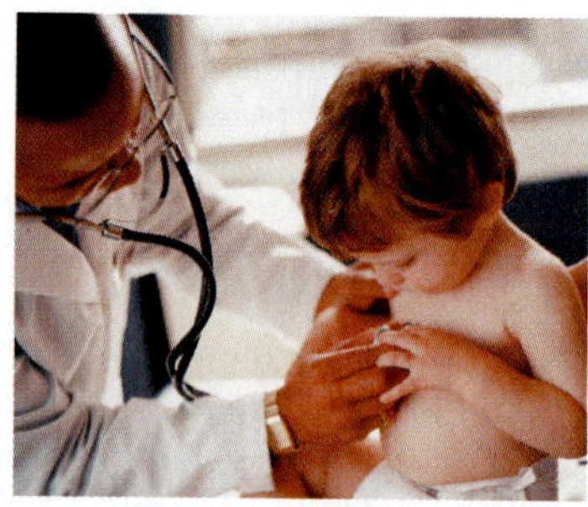

In the table, 75 represents 1975, 85 represents 1985, 100 represents 2000, and so on, and the number of outpatient visits is given in millions.

(a) Prepare a scatter diagram, and determine a quadratic model for these data.

(b) Use the model from part (a) to predict the number of visits in 2005.

Solution

(a) The scatter diagram in Figure 15(a) suggests that a quadratic function with a positive value of a (so the graph opens up) would be a reasonable model for the data. Using quadratic regression, the quadratic function defined by $f(x) = .6161x^2 - 93.67x + 3810$ approximates the data well. See Figure 15(b). Figure 15(c) displays the quadratic regression values of a, b, and c.

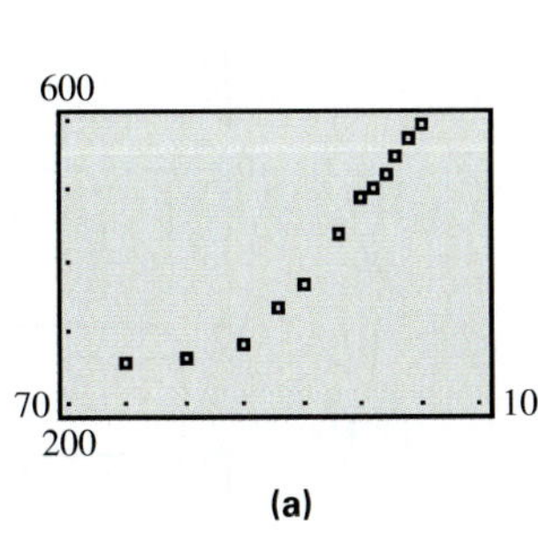

(a)

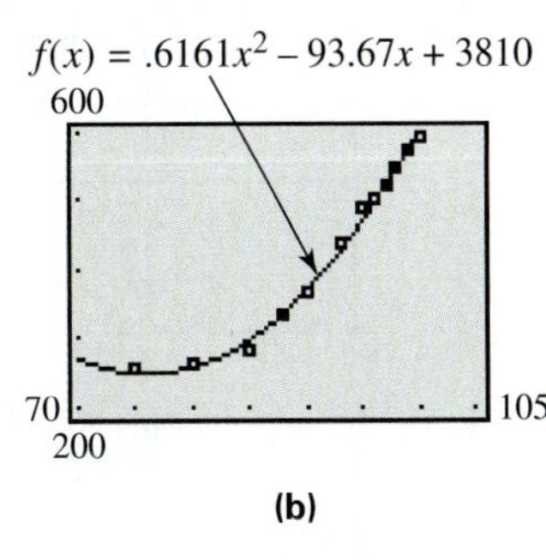

(b)

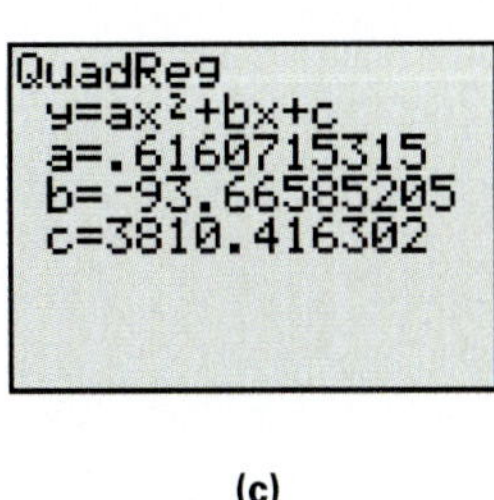

(c)

Figure 15

(b) Since 2005 corresponds to $x = 105$, the model predicts that in 2005 the number of visits will be

$$f(105) = .6161(105)^2 - 93.67(105) + 3810 \approx 767 \text{ million.}$$

Now try Exercise 61.

3.1 Exercises

In Exercises 1–4, you are given an equation and the graph of a quadratic function. Do each of the following. See Examples 1(c) and 2–4.

(a) *Give the domain and range.*
(b) *Give the coordinates of the vertex.*
(c) *Give the equation of the axis.*
(d) *Find the y-intercept.*
(e) *Find the x-intercepts.*

1. $f(x) = (x + 3)^2 - 4$

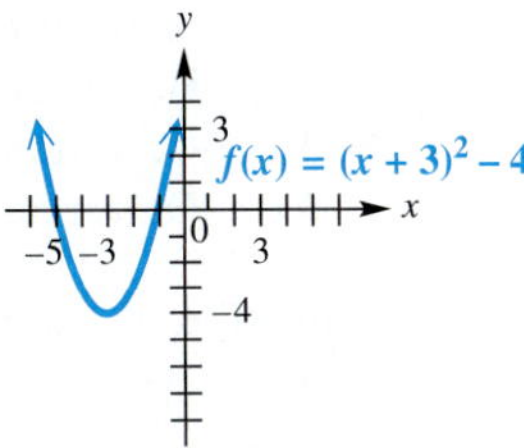

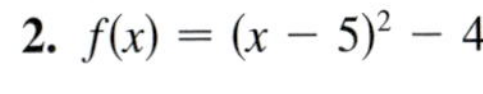
2. $f(x) = (x - 5)^2 - 4$

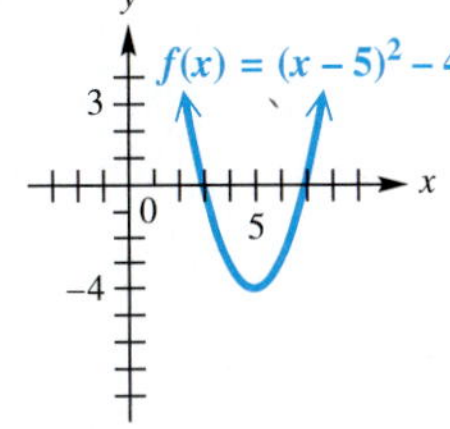

3. $f(x) = -2(x + 3)^2 + 2$

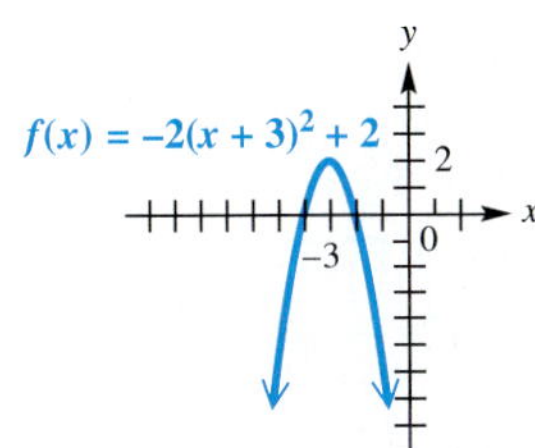

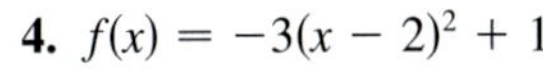
4. $f(x) = -3(x - 2)^2 + 1$

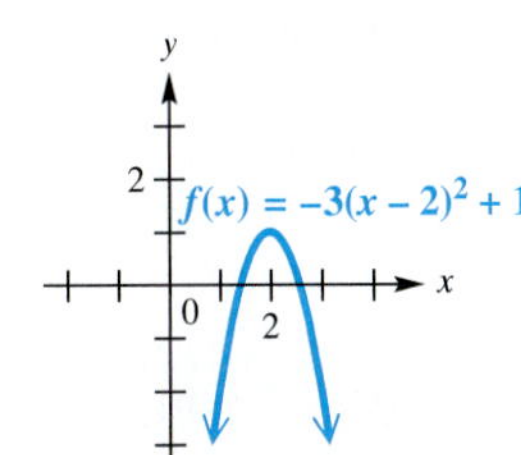

Concept Check *Calculator graphs of the functions in Exercises 5–8 are shown in Figures A–D. Match each function with its graph without actually entering it into your calculator. Then, after you have completed the exercises, check your answers with your calculator. Use the standard viewing window.*

5. $f(x) = (x - 4)^2 - 3$ **6.** $f(x) = -(x - 4)^2 + 3$

7. $f(x) = (x + 4)^2 - 3$ **8.** $f(x) = -(x + 4)^2 + 3$

A.

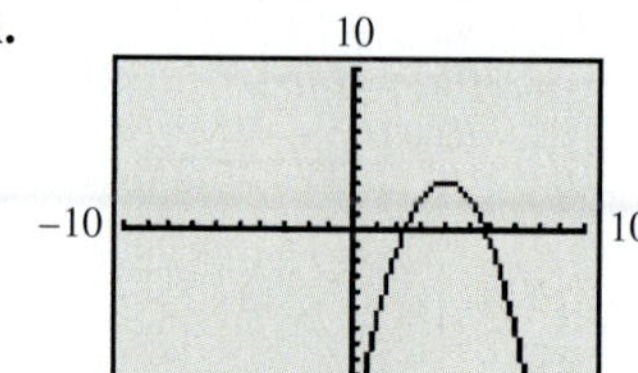

B.

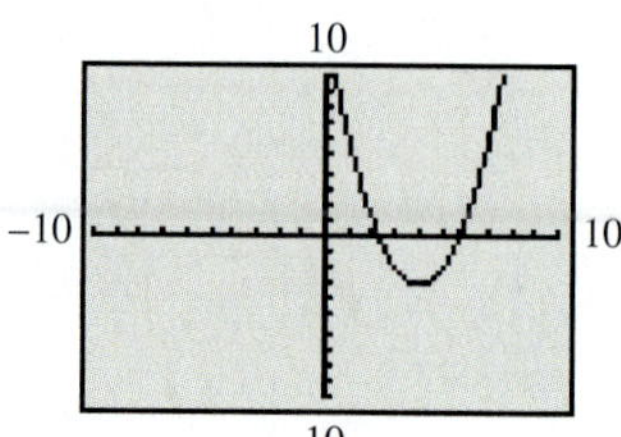

C.

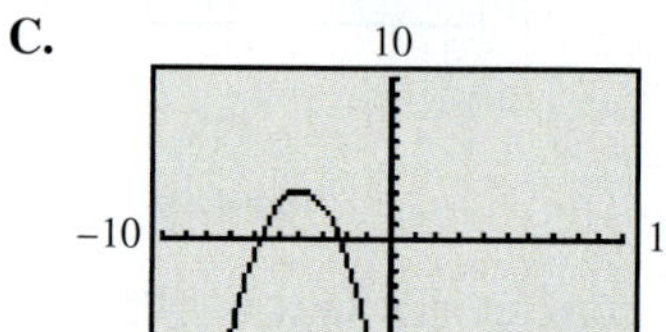

D.

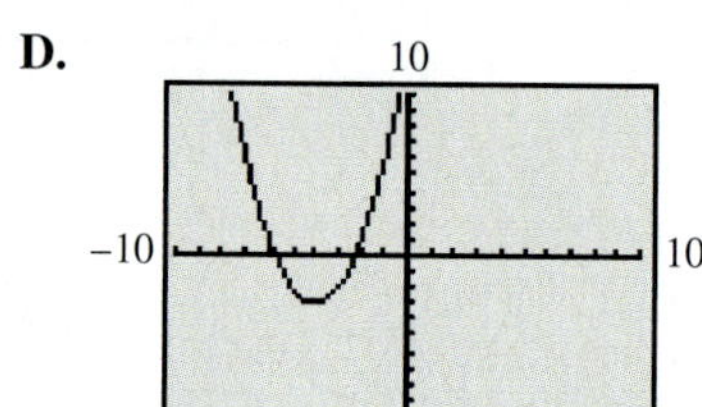

9. Graph the following on the same coordinate system.

(a) $y = 2x^2$ **(b)** $y = 3x^2$ **(c)** $y = \frac{1}{2}x^2$ **(d)** $y = \frac{1}{3}x^2$

(e) How does the coefficient of x^2 affect the shape of the graph?

10. Graph the following on the same coordinate system.

(a) $y = x^2 + 2$ **(b)** $y = x^2 - 1$ **(c)** $y = x^2 + 1$ **(d)** $y = x^2 - 2$

(e) How do these graphs differ from the graph of $y = x^2$?

11. Graph the following on the same coordinate system.

(a) $y = (x - 2)^2$ **(b)** $y = (x + 1)^2$ **(c)** $y = (x + 3)^2$ **(d)** $y = (x - 4)^2$

(e) How do these graphs differ from the graph of $y = x^2$?

12. *Concept Check* Match each equation with the description of the parabola that is its graph.

(a) $y = (x - 4)^2 - 2$ **A.** vertex $(2, -4)$, opens down

(b) $y = (x - 2)^2 - 4$ **B.** vertex $(2, -4)$, opens up

(c) $y = -(x - 4)^2 - 2$ **C.** vertex $(4, -2)$, opens down

(d) $y = -(x - 2)^2 - 4$ **D.** vertex $(4, -2)$, opens up

Graph each quadratic function. Give the vertex, axis, domain, and range. See Examples 1–4.

13. $f(x) = (x - 2)^2$ **14.** $f(x) = (x + 4)^2$

15. $f(x) = (x + 3)^2 - 4$ **16.** $f(x) = (x - 5)^2 - 4$

17. $f(x) = -\frac{1}{2}(x + 1)^2 - 3$ **18.** $f(x) = -3(x - 2)^2 + 1$

19. $f(x) = x^2 - 2x + 3$ **20.** $f(x) = x^2 + 6x + 5$

21. $f(x) = x^2 - 10x + 21$ **22.** $f(x) = 2x^2 - 4x + 5$

23. $f(x) = -2x^2 - 12x - 16$ **24.** $f(x) = -3x^2 + 24x - 46$

25. $f(x) = -x^2 - 6x - 5$ **26.** $f(x) = \frac{2}{3}x^2 - \frac{8}{3}x + \frac{5}{3}$

Concept Check *The figure shows the graph of a quadratic function $y = f(x)$. Use it to work Exercises 27–30.*

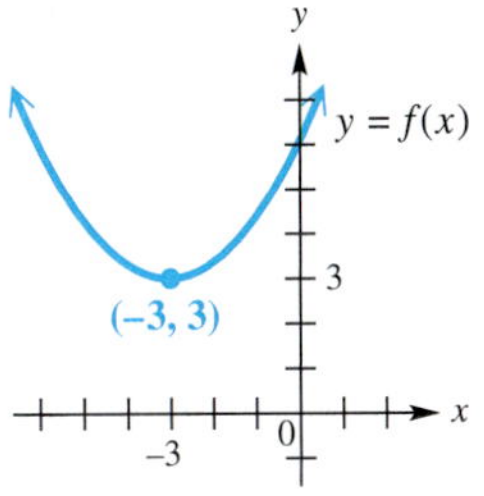

27. What is the minimum value of $f(x)$?

28. For what value of x is $f(x)$ as small as possible?

29. How many real solutions are there to the equation $f(x) = 1$?

30. How many real solutions are there to the equation $f(x) = 4$?

31. In Chapter 2, we saw how certain changes to an equation cause the graph of the equation to be stretched, shrunken, reflected across an axis, or translated vertically or horizontally. The order in which these changes are done affects the final graph. For example, stretching and then shifting vertically produces a graph that differs from the one produced by shifting vertically, then stretching. To see this, use a graphing calculator to graph

$$y = 3x^2 - 2 \quad \text{and} \quad y = 3(x^2 - 2),$$

and then compare the results. Are the two expressions equivalent algebraically?

32. *Concept Check* Suppose that a quadratic function with $a > 0$ is written in the form $f(x) = a(x - h)^2 + k$. Match each statement in Column I with one of the choices A, B, or C in Column II.

I	II
(a) k is positive.	**A.** The graph of $f(x)$ intersects the x-axis at only one point.
(b) k is negative.	**B.** The graph of $f(x)$ does not intersect the x-axis.
(c) k is zero.	**C.** The graph of $f(x)$ intersects the x-axis twice.

Concept Check *The following figures show several possible graphs of $f(x) = ax^2 + bx + c$. For the restrictions on a, b, and c given in Exercises 33–38, select the corresponding graph from choices A–F. (Hint: Use the discriminant.)*

33. $a < 0$; $b^2 - 4ac = 0$ **34.** $a > 0$; $b^2 - 4ac < 0$ **35.** $a < 0$; $b^2 - 4ac < 0$

36. $a < 0$; $b^2 - 4ac > 0$ **37.** $a > 0$; $b^2 - 4ac > 0$ **38.** $a > 0$; $b^2 - 4ac = 0$

A.

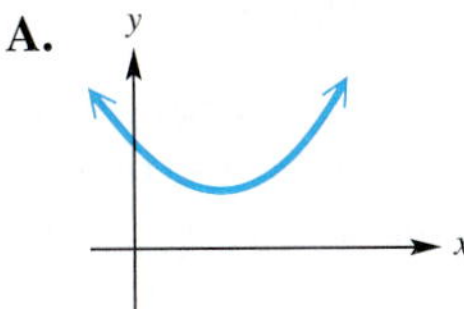

B.

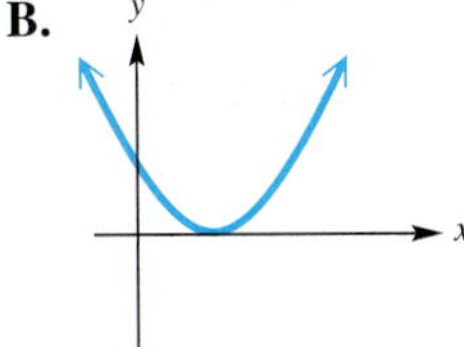

C.

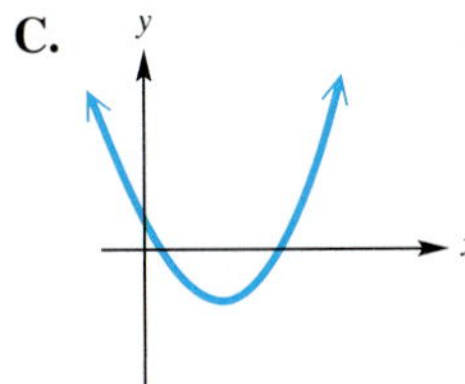

D.

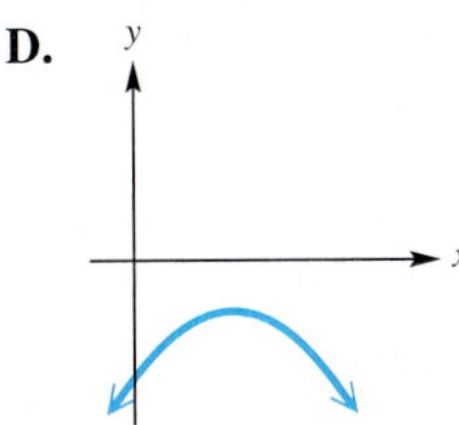

E.

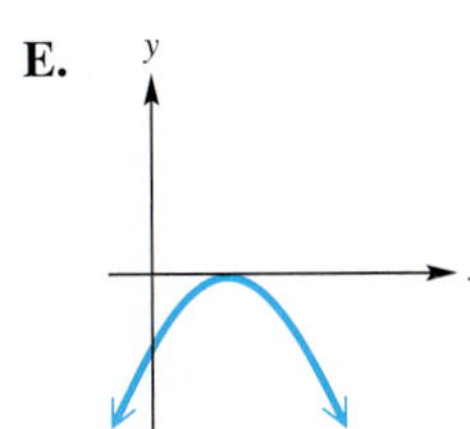

F.

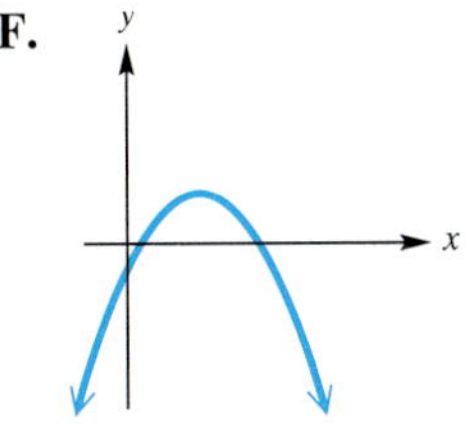

Concept Check *In Exercises 39 and 40, find a polynomial function f whose graph matches the one in the figure. Then use a graphing calculator to graph the function and verify your result.*

39.

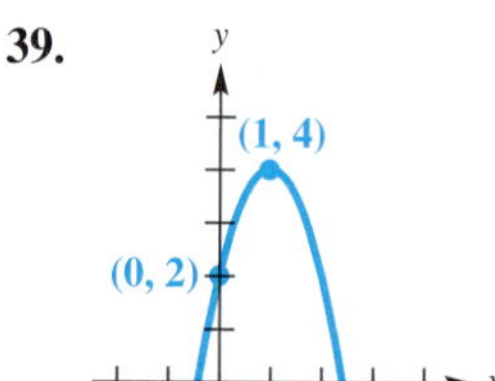

40.

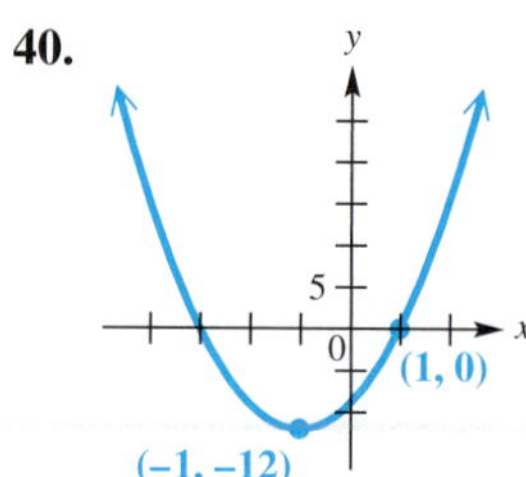

Curve Fitting *Exercises 41–46 show scatter diagrams of sets of data. In each case, tell whether a linear or quadratic model is appropriate for the data. If linear, tell whether the slope should be positive or negative. If quadratic, decide whether the leading coefficient of x^2 should be positive or negative.*

41. Social Security assets as a function of time

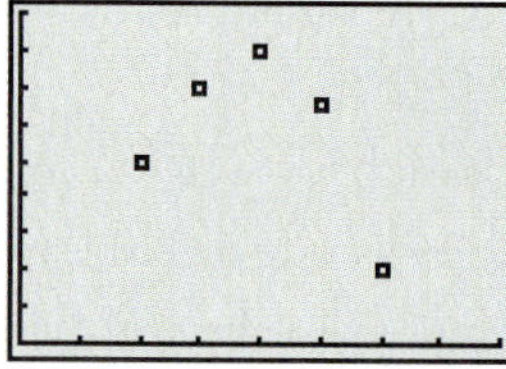

42. growth in science centers/museums as a function of time

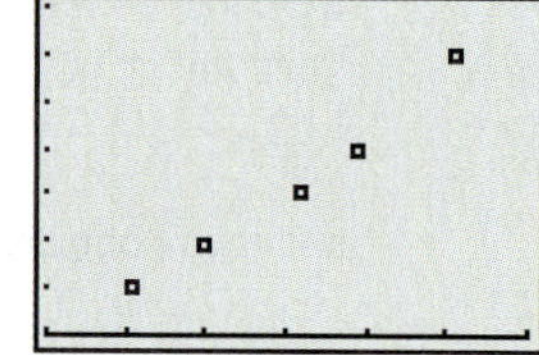

43. value of U.S. salmon catch as a function of time

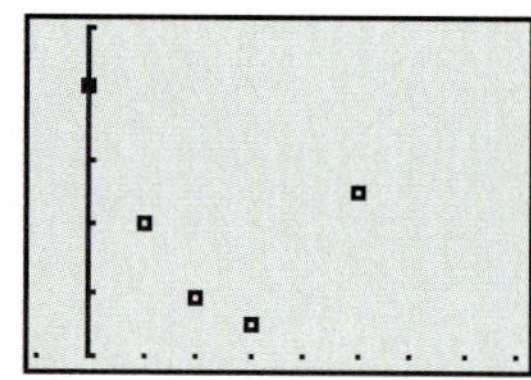

44. height of an object thrown upward from a building as a function of time

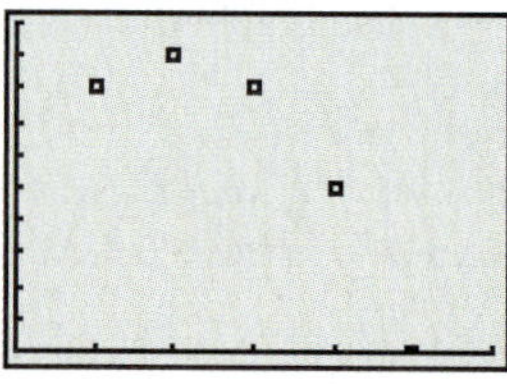

45. number of shopping centers as a function of time

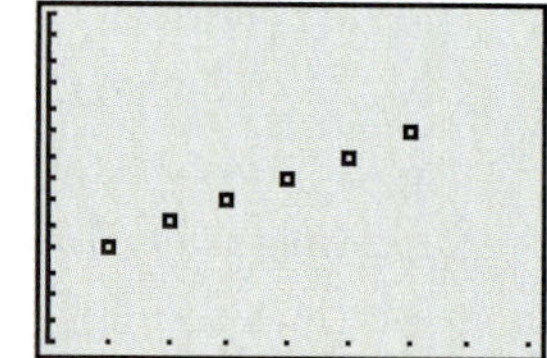

46. newborns with AIDS as a function of time

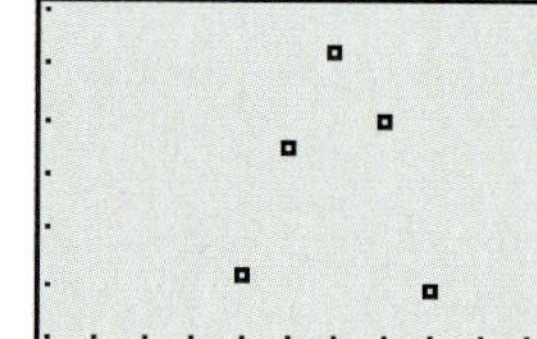

(Modeling) *Solve each problem. See Example 5.*

47. Height of a Toy Rocket A toy rocket is launched straight up from the top of a building 50 ft tall at an initial velocity of 200 ft per sec.

(a) Give the function that describes the height of the rocket in terms of time t.

(b) Determine the time at which the rocket reaches its maximum height, and the maximum height in feet.

(c) For what time interval will the rocket be more than 300 ft above ground level?

(d) After how many seconds will it hit the ground?

48. Height of a Propelled Rock A rock is propelled directly upward from ground level with an initial velocity of 90 ft per sec.

(a) Give the function that describes the height of the rock in terms of time t.

(b) Determine the time at which the rock reaches its maximum height, and the maximum height in feet.

(c) For what time interval will the rock be more than 120 ft above ground level?

(d) After how many seconds will it return to the ground?

49. *Height of a Propelled Ball* Determine whether a ball propelled straight up from ground level with initial velocity of 150 ft per sec will reach a height of 400 ft. If it will, determine the time(s) at which this happens. If it will not, explain why.

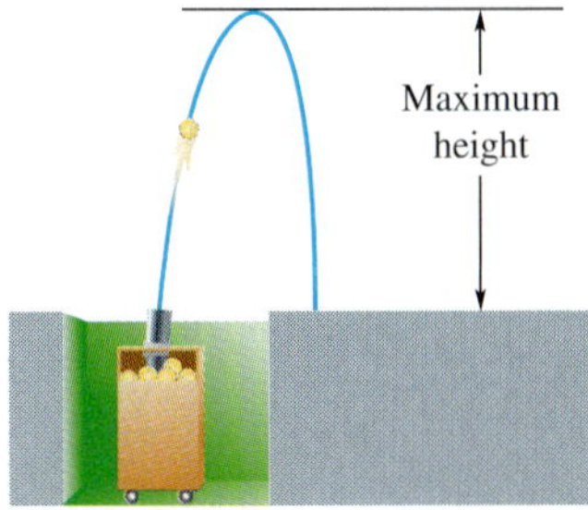

50. *Height of a Propelled Ball* Repeat Exercise 49 for initial velocity of 128 ft per sec and height 256 ft.

51. *Sum and Product of Two Numbers* Suppose that x represents one of two *positive* numbers whose sum is 30.

(a) Represent the other of the two numbers in terms of x.

(b) What are the restrictions on x?

(c) Determine a function f that represents the product of these two numbers.

(d) What are the two such numbers that yield the maximum product? What is their product?

(e) For what two such numbers is the product equal to 104?

52. *Sum and Product of Two Numbers* Suppose that x represents one of two *positive* numbers whose sum is 45.

(a) Represent the other of the two numbers in terms of x.

(b) What are the restrictions on x?

(c) Determine a function f that represents the product of these two numbers.

(d) What are the two such numbers that yield the maximum product? What is their product?

(e) For what two such numbers is the product equal to 504?

53. *Area of a Parking Lot* One campus of Houston Community College has plans to construct a rectangular parking lot on land bordered on one side by a highway. There are 640 ft of fencing available to fence the other three sides. Let x represent the length of each of the two parallel sides of fencing.

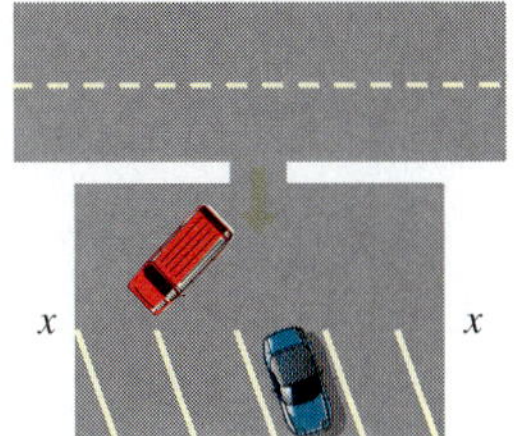

(a) Express the length of the remaining side to be fenced in terms of x.

(b) What are the restrictions on x?

(c) Determine a function A that represents the area of the parking lot in terms of x.

(d) Determine the values of x that will give an area between 30,000 and 40,000 ft^2.

(e) What dimensions will give a maximum area, and what will this area be?

54. *Area of a Rectangular Region* A farmer wishes to enclose a rectangular region bordering a river with fencing, as shown in the diagram. Suppose that x represents the length of each of the three parallel pieces of fencing. She has 600 ft of fencing available.

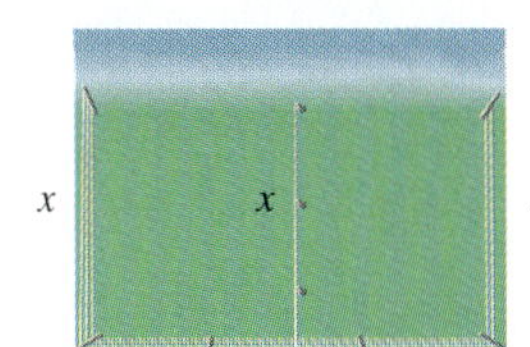

(a) What is the length of the remaining piece of fencing in terms of x?

(b) Determine a function A that represents the total area of the enclosed region. Give any restrictions on x.

(c) What dimensions for the total enclosed region would give an area of 22,500 ft^2?

(d) What is the maximum area that can be enclosed?

55. *Volume of a Box* A piece of cardboard is twice as long as it is wide. It is to be made into a box with an open top by cutting 2-in. squares from each corner and folding up the sides. Let x represent the width of the original piece of cardboard.

(a) Represent the length of the original piece of cardboard in terms of x.

(b) What will be the dimensions of the bottom rectangular base of the box? Give the restrictions on x.

(c) Determine a function V that represents the volume of the box in terms of x.

(d) For what dimensions of the bottom of the box will the volume be 320 in.^{3}?

(e) Find the values of x (to the nearest tenth of an inch) if such a box is to have a volume between 400 and 500 in.^{3}.

56. *Volume of a Box* A piece of sheet metal is 2.5 times as long as it is wide. It is to be made into a box with an open top by cutting 3-in. squares from each corner and folding up the sides. Let x represent the width of the original piece of sheet metal.

(a) Represent the length of the original piece of sheet metal in terms of x.

(b) What are the restrictions on x?

(c) Determine a function V that represents the volume of the box in terms of x.

(d) For what values of x (that is, original widths) will the volume of the box be between 600 and 800 in.^{3}? Give values to the nearest tenth of an inch.

57. *Path of a Frog's Leap* A frog leaps from a stump 3 ft high and lands 4 ft from the base of the stump. We can consider the initial position of the frog to be at $(0, 3)$ and its landing position to be at $(4, 0)$. See the figure. It is determined that the height of the frog as a function of its horizontal distance x from the base of the stump is given by

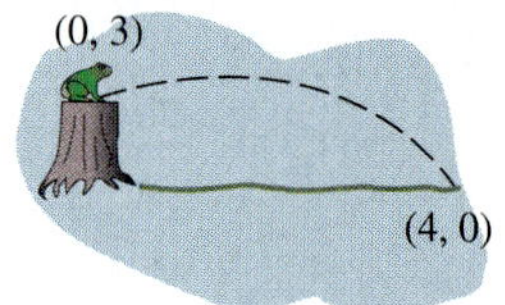

$$h(x) = -.5x^2 + 1.25x + 3,$$

where x and $h(x)$ are both in feet.

(a) How high was the frog when its horizontal distance from the base of the stump was 2 ft?

(b) At what two horizontal distances from the base of the stump was the frog 3.25 ft above the ground?

(c) At what horizontal distance from the base of the stump did the frog reach its highest point?

(d) What was the maximum height reached by the frog?

58. *Path of a Frog's Leap* Refer to Exercise 57. Suppose that the initial position of the frog is $(0, 4)$ and its landing position is $(6, 0)$. The height of the frog is given by

$$h(x) = -\frac{1}{3}x^2 + \frac{4}{3}x + 4.$$

(a) After how many feet did it reach its maximum height?

(b) What was the maximum height?

59. ***Shooting a Foul Shot*** To make a foul shot in basketball, the ball must follow a parabolic arc. This arc depends on both the angle and velocity with which the basketball is released. If a person shoots the basketball overhand from a position 8 ft above the floor, then the path can sometimes be modeled by the parabola

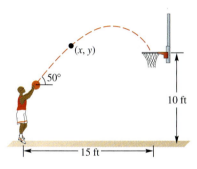

$$y = \frac{-16x^2}{.434v^2} + 1.15x + 8,$$

where v is the initial velocity of the ball in feet per second, as illustrated in the figure. (*Source:* Rist, C., "The Physics of Foul Shots," *Discover,* October 2000.)

(a) If the basketball hoop is 10 ft high and located 15 ft away, what initial velocity v should the basketball have?

(b) What is the maximum height of the basketball?

60. ***Shooting a Foul Shot*** Refer to Exercise 59. If a person releases a basketball underhand from a position 3 ft above the floor, it often has a steeper arc than if it is released overhand and its path can sometimes be modeled by

$$y = \frac{-16x^2}{.117v^2} + 2.75x + 3.$$

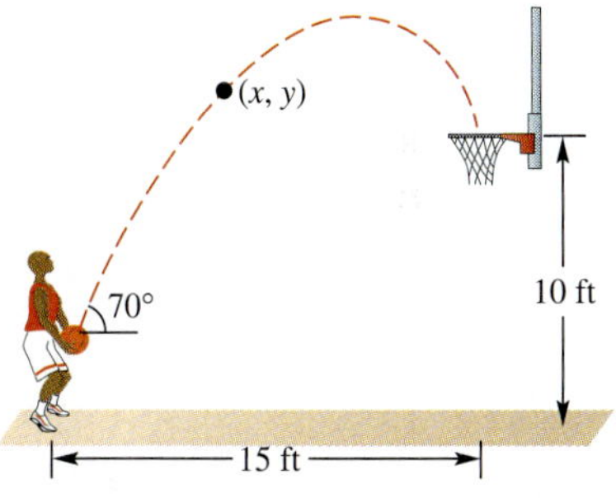

Repeat parts (a) and (b) from Exercise 59. Then compare the paths for the overhand shot and the underhand shot.

(Modeling) Solve each problem. See Example 6.

61. ***Births to Unmarried Women*** The percent of births to unmarried women from 1990–2000 are shown in the table. The data is modeled by the quadratic function defined by

$$f(x) = -.0661x^2 + 1.118x + 28.3,$$

where $x = 0$ corresponds to 1990 and $f(x)$ is the percent.

Year	Percent	Year	Percent
1990	28	1996	32.4
1991	29.6	1997	32.4
1992	30.1	1998	32.8
1993	31.0	1999	33
1994	32.6	2000	33.2
1995	32.2		

Source: National Center for Health Statistics.

(a) If this model continues to apply, what would it predict for the percent of these births in 2005?

(b) There are pitfalls in using models to predict very far into the future. How does your answer in part (a) support this?

62. ***College Freshmen Pursuing Medical Degrees*** Between 1996 and 2002, the percent of college freshmen who planned to get a professional degree in a medical field (that is, M.D., D.O., D.D.S., or D.V.M.) can be modeled by

$$f(x) = .07726x^2 - .8004x + 10.84,$$

where $x = 0$ represents 1996. (*Source:* Higher Education Research Institute, UCLA, 2002.) Based on this model, in what year did the percent of freshmen planning to get a medical degree reach its minimum?

63. ***Household Income*** Between 1989 and 1997, the percent of households with incomes of \$100,000 or more (in 1997 dollars) can be modeled by

$$f(x) = .071x^2 - .426x + 8.05,$$

where $x = 0$ represents 1989. (*Source: American Demographics,* January 1999.) Based on this model, in what year did the percent of affluent households reach its minimum?

64. ***Accident Rate*** According to data from the National Highway Traffic Safety Administration, the accident rate as a function of the age of the driver in years x can be approximated by the function defined by

$$f(x) = .0232x^2 - 2.28x + 60.0$$

for $16 \leq x \leq 85$. Find both the age at which the accident rate is a minimum and the minimum rate.

65. ***AIDS Cases in the United States*** The table lists the total (cumulative) number of AIDS cases diagnosed in the United States up to 1996. For example, a total of 22,620 AIDS cases were diagnosed through 1985.

Year	AIDS Cases	Year	AIDS Cases
1982	1563	1990	193,245
1983	4647	1991	248,023
1984	10,845	1992	315,329
1985	22,620	1993	361,509
1986	41,662	1994	441,406
1987	70,222	1995	515,586
1988	105,489	1996	584,394
1989	147,170		

Source: "Facts and Figures," Joint United Nations Programme on HIV/AIDS (UNAIDS), February 1999.

(a) Plot the data. Let $x = 0$ correspond to the year 1980.

(b) Would a linear or quadratic function model the data better? Explain.

(c) Find a quadratic function defined by $f(x) = a(x - h)^2 + k$ that models the data. (*Hint:* Use $(2, 1563)$ as the vertex and then choose a second point such as $(13, 361{,}509)$ to determine a.)

(d) Plot the data together with f on the same coordinate plane. How well does f model the number of AIDS cases?

(e) Use f to predict the total number of AIDS cases diagnosed by the years 1999 and 2000.

(f) According to the model, how many new cases were diagnosed in the year 2000?

66. ***AIDS Deaths in the United States*** The table lists the total (cumulative) number of known deaths caused by AIDS in the United States up to 1996.

Year	Deaths	Year	Deaths
1982	620	1990	119,821
1983	2122	1991	154,567
1984	5600	1992	191,508
1985	12,529	1993	220,592
1986	24,550	1994	269,992
1987	40,820	1995	320,692
1988	61,723	1996	359,892
1989	89,172		

Source: "Facts and Figures," Joint United Nations Programme on HIV/AIDS (UNAIDS), February 1999.

(a) Plot the data. Let $x = 0$ correspond to the year 1980.

(b) Would a linear or quadratic function model the data better? Explain.

(c) Find a quadratic function defined by $g(x) = a(x - h)^2 + k$ that models the data. Use $(2, 620)$ as the vertex and $(13, 220{,}592)$ as the other point to determine a.

(d) Plot the data together with g on the same coordinate plane. How well does g model the number of deaths caused by AIDS?

(e) Use g to predict the number of new deaths caused by AIDS during the year 2000.

(f) Both f (from the previous exercise) and g are quadratic functions. Discuss why it is reasonable to expect that the number of AIDS cases and AIDS-related deaths could be modeled using the same type of function.

67. ***Americans Over 100 Years of Age*** The table lists the number of Americans (in thousands) who were over 100 yr old for selected years.

Year	Number (in thousands)
1994	50
1996	56
1998	65
2000	75
2002	94
2004	110

Source: U.S. Bureau of the Census.

(a) Plot the data. Let $x = 4$ correspond to the year 1994, $x = 6$ correspond to 1996, and so on.

(b) Find a quadratic function defined by $f(x) = a(x - h)^2 + k$ that models the data. Use $(4, 50)$ as the vertex and $(14, 110)$ as the other point to determine a.

(c) Plot the data together with f in the same window. How well does f model the number of Americans (in thousands) who are expected to be over 100 yr old?

(d) Use the quadratic regression feature of a graphing calculator to determine the quadratic function g that provides the best fit for the data.

(e) Use the functions f and g to predict the number of Americans, in thousands, who will be over 100 yr old in the year 2006.

68. ***Automobile Stopping Distance*** Selected values of the stopping distance y in feet of a car traveling x miles per hour are given in the table.

Speed (in mph)	Stopping Distance (in feet)
20	46
30	87
40	140
50	240
60	282
70	371

Source: National Safety Institute Student Workbook, 1993, p. 7.

(a) Plot the data.

(b) The quadratic function defined by $f(x) = .056057x^2 + 1.06657x$ is one model that has been used to approximate stopping distances. Find and interpret $f(45)$.

(c) How well does f model the stopping distance?

69. *Coast-Down Time* The coast-down time y for a typical car as it drops 10 mph from an initial speed x depends on several factors, such as average drag, tire pressure, and whether the transmission is in neutral. The table gives coast-down time in seconds for a car under standard conditions for selected speeds in miles per hour.

(a) Plot the data.

(b) Use the quadratic regression feature of a graphing calculator to find the quadratic function g that best fits the data. Graph this function in the same window as the data. Is g a good model for the data?

(c) Use g to predict the coast-down time at an initial speed of 70 mph.

(d) Use the graph to find the speed that corresponds to a coast-down time of 24 sec.

Initial Speed (in mph)	Coast-Down Time (in seconds)
30	30
35	27
40	23
45	21
50	18
55	16
60	15
65	13

Source: Scientific American, December 1994.

70. *Concentration of Atmospheric CO_2* The International Panel on Climate Change (IPCC) has published data indicating that if current trends continue, the quadratic function defined by

$$f(x) = .0167x^2 - 65.37x + 64{,}440$$

models future amounts of atmospheric carbon dioxide in parts per million (ppm), where x represents the year. According to this model, what would be the concentration in 2010?

Concept Check *Work Exercises 71–80.*

71. Find a value of c so that $y = x^2 - 10x + c$ has exactly one x-intercept.

72. For what values of a does $y = ax^2 - 8x + 4$ have no x-intercepts?

73. Define the quadratic function f having x-intercepts 2 and 5, and y-intercept 5.

74. Define the quadratic function f having x-intercepts 1 and -2, and y-intercept 4.

75. Find the largest possible value of y if $y = -(x-2)^2 + 9$. Then find the following.

(a) the largest possible value of $\sqrt{-(x-2)^2 + 9}$

(b) the smallest possible positive value of $\dfrac{1}{-(x-2)^2 + 9}$

76. Find the smallest possible value of y if $y = 3 + (x+5)^2$. Then find the following.

(a) the smallest possible value of $\sqrt{3 + (x+5)^2}$

(b) the largest possible value of $\dfrac{1}{3 + (x+5)^2}$

77. From the distance formula in Section 2.1, the distance between the two points $P(x_1, y_1)$ and $R(x_2, y_2)$ is

$$d(P, R) = \sqrt{(x_1 - x_2)^2 + (y_1 - y_2)^2}.$$

Using the result of Exercise 75, find the closest point on the line $y = 2x$ to the point $(1, 7)$. (*Hint:* Every point on $y = 2x$ has the form $(x, 2x)$, and the closest point has the minimum distance.)

78. A quadratic equation $f(x) = 0$ has a solution $x = 2$. Its graph has vertex $(5, 3)$. What is the other solution of the equation?

79. Let $f(x) = 3x^2 + 9x + 5$. Without doing any arithmetic, explain why

$$f\left(\frac{-9 + \sqrt{9^2 - 4 \cdot 3 \cdot 5}}{2 \cdot 3}\right) = 0.$$

(*Hint:* Consider the quadratic formula.)

80. What is the other value of x for which $f(x) = 0$ for the function in Exercise 79?

Relating Concepts

For individual or collaborative investigation
(Exercises 81–86)

The solution set of $f(x) = 0$ consists of all x-values where the graph of $y = f(x)$ intersects the x-axis (i.e., the x-intercepts). The solution set of $f(x) < 0$ consists of all x-values where the graph lies below *the x-axis, while the solution set of $f(x) > 0$ consists of all x-values where the graph lies* above *the x-axis.*

In Chapter 1 we used intervals on a number line to solve a quadratic inequality. ***Work Exercises 81–86 in order,*** *to see why we must reverse the direction of the inequality sign when multiplying or dividing an inequality by a negative number.*

81. Graph $f(x) = x^2 + 2x - 8$. This function has a graph with two x-intercepts. What are they?

82. Use the graph from Exercise 81 to determine the solution set of $x^2 + 2x - 8 < 0$.

83. Graph $g(x) = -f(x) = -x^2 - 2x + 8$. Using the terminology of Chapter 2, how is the graph of g obtained by a transformation of the graph of f?

84. Use the graph from Exercise 83 to determine the solution set of $-x^2 - 2x + 8 > 0$.

85. Compare the two solution sets of the inequalities in Exercises 82 and 84.

86. Write a short paragraph explaining how Exercises 81–85 illustrate the property for multiplying an inequality by a negative number.

3.2 Synthetic Division

Synthetic Division ▪ Evaluating Polynomial Functions Using the Remainder Theorem ▪ Testing Potential Zeros

The quotient of two polynomials was found in Chapter R with an algorithm for long division similar to that used to divide whole numbers. More formally, we define the *division algorithm* as follows.

Division Algorithm

Let $f(x)$ and $g(x)$ be polynomials with $g(x)$ of lower degree than $f(x)$ and $g(x)$ of degree one or more. There exist unique polynomials $q(x)$ and $r(x)$ such that

$$\mathbf{f(x) = g(x) \cdot q(x) + r(x),}$$

where either $r(x) = 0$ or the degree of $r(x)$ is less than the degree of $g(x)$.

For instance, we saw in Example 10 of Section R.3 that

$$\frac{3x^3 - 2x^2 - 150}{x^2 - 4} = 3x - 2 + \frac{12x - 158}{x^2 - 4}. \quad \text{(Section R.3)}$$

We can express this result using the division algorithm:

$$\underbrace{3x^3 - 2x^2 - 150}_{\substack{f(x) \\ \text{Dividend} \\ \text{(original polynomial)}}} \; = \; \underbrace{(x^2 - 4)}_{\substack{g(x) \\ \text{Divisor}}} \cdot \underbrace{(3x - 2)}_{\substack{q(x) \\ \text{Quotient}}} + \underbrace{12x - 158}_{\substack{r(x) \\ \text{Remainder}}}.$$

Synthetic Division A shortcut method of performing long division with certain polynomials, called **synthetic division,** is used only when a polynomial is divided by a first-degree binomial of the form $x - k$, where the coefficient of x is 1. To illustrate, notice the example worked on the left below. On the right, the division process is simplified by omitting all variables and writing only coefficients, with 0 used to represent the coefficient of any missing terms. Since the coefficient of x in the divisor is always 1 in these divisions, it too can be omitted. These omissions simplify the problem, as shown on the right.

$$\begin{array}{r|l}
 & 3x^2 + 10x + 40 \\ \hline
x - 4 & 3x^3 - 2x^2 \qquad\qquad - 150 \\
 & \underline{3x^3 - 12x^2} \\
 & \qquad 10x^2 \\
 & \qquad \underline{10x^2 - 40x} \\
 & \qquad\qquad 40x - 150 \\
 & \qquad\qquad \underline{40x - 160} \\
 & \qquad\qquad\qquad 10
\end{array}
\qquad\qquad
\begin{array}{r|l}
 & 3 \quad 10 \quad 40 \\ \hline
-4 & 3 - 2 + 0 - 150 \\
 & \underline{3 - 12} \\
 & \quad 10 \\
 & \quad \underline{10 - 40} \\
 & \qquad 40 - 150 \\
 & \qquad \underline{40 - 160} \\
 & \qquad\qquad 10
\end{array}$$

The numbers in color that are repetitions of the numbers directly above them can also be omitted.

$$\begin{array}{r|l}
 & 3 \quad 10 \quad 40 \\ \hline
-4 & 3 - 2 + 0 - 150 \\
 & \underline{\;\; - 12} \\
 & \quad 10 \\
 & \quad \underline{\;\; - 40} \\
 & \qquad 40 - 150 \\
 & \qquad \underline{\;\;\; - 160} \\
 & \qquad\qquad 10
\end{array}$$

The entire problem can now be condensed vertically, and the top row of numbers can be omitted since it duplicates the bottom row if the 3 is brought down.

$$\begin{array}{r|rrrr}
-4 & 3 & -2 & 0 & -150 \\
 & & -12 & -40 & -160 \\ \hline
 & 3 & 10 & 40 & 10
\end{array}$$

The rest of the bottom row is obtained by subtracting -12, -40, and -160 from the corresponding terms above them.

With synthetic division it is useful to change the sign of the divisor, so the -4 at the left is changed to 4, which also changes the sign of the numbers in the second row. To compensate for this change, subtraction is changed to addition. Doing this gives the following result.

$$\begin{array}{r|rrrr} 4 & 3 & -2 & 0 & -150 \\ & & 12 & 40 & 160 \\ \hline & 3 & 10 & 40 & 10 \end{array}$$

Additive inverse → 4; 160 ← Signs changed.; 10 ← Remainder

$$\text{Quotient} \rightarrow 3x^2 + 10x + 40 + \frac{10}{x - 4}$$

In summary, to use synthetic division to divide a polynomial by a binomial of the form $x - k$, begin by writing the coefficients of the polynomial in decreasing powers of the variable, using 0 as the coefficient of any missing powers. The number k is written to the left in the same row. In the example above, $x - k$ is $x - 4$, so k is 4. Next bring down the leading coefficient of the polynomial, 3 in the previous example, as the first number in the last row. Multiply the 3 by 4 to get the first number in the second row, 12. Add 12 to -2; this gives 10, the second number in the third row. Multiply 10 by 4 to get 40, the next number in the second row. Add 40 to 0 to get the third number in the third row, and so on. This process of multiplying each result in the third row by k and adding the product to the number in the next column is repeated until there is a number in the last row for each coefficient in the first row.

CAUTION To avoid errors, use 0 as coefficient for any missing terms, including a missing constant, when setting up the division.

EXAMPLE 1 Using Synthetic Division

Use synthetic division to divide

$$\frac{5x^3 - 6x^2 - 28x - 2}{x + 2}.$$

Solution Express $x + 2$ in the form $x - k$ by writing it as $x - (-2)$. Use this and the coefficients of the polynomial to obtain

$$x + 2 \text{ leads to } -2. \rightarrow \begin{array}{r|rrrr} -2 & 5 & -6 & -28 & -2. \end{array} \leftarrow \text{Coefficients}$$

Bring down the 5, and multiply: $-2(5) = -10$.

$$\begin{array}{r|rrrr} -2 & 5 & -6 & -28 & -2 \\ & \downarrow & -10 & & \\ \hline & 5 & & & \end{array}$$

Add -6 and -10 to obtain -16. Multiply: $-2(-16) = 32$.

$$\begin{array}{r|rrrr} -2 & 5 & -6 & -28 & -2 \\ & & -10 & 32 & \\ \hline & 5 & -16 & & \end{array}$$

Add -28 and 32, obtaining 4. Finally, $-2(4) = -8$.

$$\begin{array}{r|rrrr} -2 & 5 & -6 & -28 & -2 \\ & & -10 & 32 & -8 \\ \hline & 5 & -16 & 4 & \end{array}$$

Add -2 and -8 to obtain -10.

$$\begin{array}{r|rrrr} -2 & 5 & -6 & -28 & -2 \\ & & -10 & 32 & -8 \\ \hline & \underbrace{5 \quad -16 \quad 4}_{\text{Quotient}} & & & -10 \leftarrow \text{Remainder} \end{array}$$

Since the divisor $x - k$ has degree 1, the degree of the quotient will always be one less than the degree of the polynomial to be divided. Thus,

$$\frac{5x^3 - 6x^2 - 28x - 2}{x + 2} = 5x^2 - 16x + 4 + \frac{-10}{x + 2}.$$

Now try Exercise 1.

The result of the division in Example 1 can be written as

$$5x^3 - 6x^2 - 28x - 2 = (x + 2)(5x^2 - 16x + 4) + (-10)$$

by multiplying both sides by the denominator $x + 2$. The following theorem is a generalization of the division process illustrated above.

For any polynomial $f(x)$ and any complex number k, there exists a unique polynomial $q(x)$ and number r such that

$$\boldsymbol{f(x) = (x - k)q(x) + r.}$$

For example, in the synthetic division above,

$$\underbrace{5x^3 - 6x^2 - 28x - 2}_{f(x)} = \underbrace{(x + 2)}_{(x - k)\,\cdot}\underbrace{(5x^2 - 16x + 4)}_{q(x)} + \underbrace{(-10)}_{r}.$$

This theorem is a special case of the division algorithm given earlier. Here $g(x)$ is the first-degree polynomial $x - k$.

Evaluating Polynomial Functions Using the Remainder Theorem

Suppose that $f(x)$ is written as $f(x) = (x - k)q(x) + r$. This equality is true for all complex values of x, so it is true for $x = k$. Replacing x with k gives

$$f(k) = (k - k)q(k) + r \qquad \text{or} \qquad f(k) = r.$$

This proves the following **remainder theorem,** which gives a new method of evaluating polynomial functions.

Remainder Theorem

If the polynomial $f(x)$ is divided by $x - k$, then the remainder is equal to $f(k)$.

For example, in the synthetic division problem given in Example 1, when $f(x) = 5x^3 - 6x^2 - 28x - 2$ was divided by $x + 2$ or $x - (-2)$, the remainder was -10. Substituting -2 for x in $f(x)$ gives

$$\begin{aligned} f(-2) &= 5(-2)^3 - 6(-2)^2 - 28(-2) - 2 \\ &= -40 - 24 + 56 - 2 \\ &= -10. \end{aligned}$$

As shown here, a simpler way to find the value of a polynomial is often by using synthetic division. By the remainder theorem, instead of replacing x by -2 to find $f(-2)$, divide $f(x)$ by $x + 2$ using synthetic division as in Example 1. Then $f(-2)$ is the remainder, -10.

$$\begin{array}{r|rrrr} -2 & 5 & -6 & -28 & -2 \\ & & -10 & 32 & -8 \\ \hline & 5 & -16 & 4 & -10 \leftarrow f(-2) \end{array}$$

EXAMPLE 2 Applying the Remainder Theorem

Let $f(x) = -x^4 + 3x^2 - 4x - 5$. Use the remainder theorem to find $f(-3)$.

Solution Use synthetic division with $k = -3$.

$$\begin{array}{r|rrrrr} -3 & -1 & 0 & 3 & -4 & -5 \\ & & 3 & -9 & 18 & -42 \\ \hline & -1 & 3 & -6 & 14 & -47 \leftarrow \text{Remainder} \end{array}$$

By this result, $f(-3) = -47$.

Now try Exercise 31.

Testing Potential Zeros A **zero** of a polynomial function f is a number k such that $f(k) = 0$. The zeros are the x-intercepts of the graph of the function.

The remainder theorem gives a quick way to decide if a number k is a zero of a polynomial function defined by $f(x)$. Use synthetic division to find $f(k)$; if the remainder is 0, then $f(k) = 0$ and k is a zero of $f(x)$. A zero of $f(x)$ is called a **root** or **solution** of the equation $f(x) = 0$.

EXAMPLE 3 Deciding Whether a Number Is a Zero

Decide whether the given number k is a zero of $f(x)$.

(a) $f(x) = x^3 - 4x^2 + 9x - 6$; $k = 1$

(b) $f(x) = x^4 + x^2 - 3x + 1$; $k = -4$

(c) $f(x) = x^4 - 2x^3 + 4x^2 + 2x - 5$; $k = 1 + 2i$

Solution

(a)

$$\begin{array}{r|rrrr} \text{Proposed zero} \rightarrow 1 & 1 & -4 & 9 & -6 \leftarrow f(x) = x^3 - 4x^2 + 9x - 6 \\ & & 1 & -3 & 6 \\ \hline & 1 & -3 & 6 & 0 \leftarrow \text{Remainder} \end{array}$$

Since the remainder is 0, $f(1) = 0$, and 1 is a zero of the polynomial function defined by $f(x) = x^3 - 4x^2 + 9x - 6$.

(b) For $f(x) = x^4 + x^2 - 3x + 1$, remember to use 0 as coefficient for the missing x^3-term in the synthetic division.

$$\text{Proposed zero} \rightarrow -4\overline{)\begin{array}{rrrrr} 1 & 0 & 1 & -3 & 1 \\ & -4 & 16 & -68 & 284 \\ \hline 1 & -4 & 17 & -71 & 285 \end{array}} \leftarrow \text{Remainder}$$

The remainder is not 0, so -4 is not a zero of $f(x) = x^4 + x^2 - 3x + 1$. In fact, $f(-4) = 285$.

(c) Use synthetic division and operations with complex numbers to determine whether $1 + 2i$ is a zero of $f(x) = x^4 - 2x^3 + 4x^2 + 2x - 5$.

$$1 + 2i\overline{)\begin{array}{rrrrr} 1 & -2 & 4 & 2 & -5 \\ & 1 + 2i & -5 & -1 - 2i & 5 \\ \hline 1 & -1 + 2i & -1 & 1 - 2i & 0 \end{array}} \leftarrow \text{Remainder}$$

Since the remainder is 0, $1 + 2i$ is a zero of the given polynomial function.

Now try Exercises 41 and 55.

CONNECTIONS In Example 3(c) we found that $f(x) = x^4 - 2x^3 + 4x^2 + 2x - 5$ has the nonreal complex zero $1 + 2i$. At the beginning of this chapter, we defined a polynomial with *real* coefficients. The theorems and definitions of this chapter also apply to polynomials with complex coefficients. For example, we can show that

$$f(x) = 3x^3 + (-1 + 3i)x^2 + (-12 + 5i)x + 4 - 2i$$

has $2 - i$ as a zero. We must show that $f(2 - i) = 0$.

$$2 - i\overline{)\begin{array}{rrrr} 3 & -1 + 3i & -12 + 5i & 4 - 2i \\ & 6 - 3i & 10 - 5i & -4 + 2i \\ \hline 3 & 5 & -2 & 0 \end{array}} \leftarrow \text{Remainder}$$

By the division algorithm,

$$\begin{aligned} f(x) &= (x - 2 + i)(3x^2 + 5x - 2) \\ &= (x - 2 + i)(3x - 1)(x + 2). \quad \text{Factor. (Section R.4)} \end{aligned}$$

Thus, the quotient function q has zeros $\frac{1}{3}$ and -2, which are also zeros of f. Even though f has nonreal complex coefficients, it has two real zeros, as well as the one nonreal complex zero we tested for above.

For Discussion or Writing

1. Find $f(-2 + i)$ if $f(x) = x^3 - 4x^2 + 2x - 29i$.
2. Is i a zero of $f(x) = x^3 + 2ix^2 + 2x + i$? Is $-i$?
3. Give a simple function with real coefficients that has at least one nonreal complex zero.

3.2 Exercises

Use synthetic division to perform each division. See Example 1.

1. $\dfrac{x^3 + 2x^2 - 17x - 12}{x + 5}$
2. $\dfrac{x^3 + 4x^2 - 5x + 44}{x + 6}$
3. $\dfrac{4x^3 - 3x - 2}{x + 1}$
4. $\dfrac{3x^3 - 4x + 2}{x - 1}$
5. $\dfrac{x^4 + 4x^3 + 2x^2 + 9x + 4}{x + 4}$
6. $\dfrac{x^4 + 5x^3 + 4x^2 - 3x + 9}{x + 3}$
7. $\dfrac{x^5 + 3x^4 + 2x^3 + 2x^2 + 3x + 1}{x + 2}$
8. $\dfrac{x^6 - 3x^4 + 2x^3 - 6x^2 - 5x + 3}{x + 2}$
9. $\dfrac{-9x^3 + 8x^2 - 7x + 2}{x - 2}$
10. $\dfrac{-11x^4 + 2x^3 - 8x^2 - 4}{x + 1}$
11. $\dfrac{\frac{1}{3}x^3 - \frac{2}{9}x^2 + \frac{1}{27}x + 1}{x - \frac{1}{3}}$
12. $\dfrac{x^3 + x^2 + \frac{1}{2}x + \frac{1}{8}}{x + \frac{1}{2}}$
13. $\dfrac{x^4 - 3x^3 - 4x^2 + 12x}{x - 2}$
14. $\dfrac{x^4 + 5x^3 - 6x^2 + 2x}{x + 1}$
15. $\dfrac{x^3 - 1}{x - 1}$
16. $\dfrac{x^4 - 1}{x - 1}$
17. $\dfrac{x^5 + 1}{x + 1}$
18. $\dfrac{x^7 + 1}{x + 1}$

Express $f(x)$ in the form $f(x) = (x - k)q(x) + r$ for the given value of k.

19. $f(x) = 2x^3 + x^2 + x - 8;\ k = -1$
20. $f(x) = 2x^3 + 3x^2 - 16x + 10;\ k = -4$
21. $f(x) = -x^3 + 2x^2 + 4;\ k = -2$
22. $f(x) = -4x^3 + 2x^2 - 3x - 10;\ k = 2$
23. $f(x) = 4x^4 - 3x^3 - 20x^2 - x;\ k = 3$
24. $f(x) = 2x^4 + x^3 - 15x^2 + 3x;\ k = -3$
25. $f(x) = 3x^4 + 4x^3 - 10x^2 + 15;\ k = -1$
26. $f(x) = -5x^4 + x^3 + 2x^2 + 3x + 1;\ k = 1$

For each polynomial function, use the remainder theorem and synthetic division to find $f(k)$. See Example 2.

27. $f(x) = x^2 + 5x + 6;\ k = -2$
28. $f(x) = x^2 - 4x + 5;\ k = 3$
29. $f(x) = 2x^2 - 3x - 3;\ k = 2$
30. $f(x) = -x^3 + 8x^2 + 63;\ k = 4$
31. $f(x) = x^3 - 4x^2 + 2x + 1;\ k = -1$
32. $f(x) = 2x^3 - 3x^2 - 5x + 4;\ k = 2$
33. $f(x) = 2x^5 - 10x^3 - 19x^2 - 50;\ k = 3$
34. $f(x) = x^4 + 6x^3 + 9x^2 + 3x - 3;\ k = 4$
35. $f(x) = 6x^4 + x^3 - 8x^2 + 5x + 6;\ k = \dfrac{1}{2}$
36. $f(x) = 6x^3 - 31x^2 - 15x;\ k = -\dfrac{1}{2}$
37. $f(x) = x^2 - 5x + 1;\ k = 2 + i$
38. $f(x) = x^2 - x + 3;\ k = 3 - 2i$

Use synthetic division to decide whether the given number k is a zero of the given polynomial function. If it is not, give the value of $f(k)$. See Examples 2 and 3.

39. $f(x) = x^2 + 2x - 8;\ k = 2$

40. $f(x) = x^2 + 4x - 5;\ k = -5$

41. $f(x) = x^3 - 3x^2 + 4x - 4;\ k = 2$

42. $f(x) = x^3 + 2x^2 - x + 6;\ k = -3$

43. $f(x) = 2x^3 - 6x^2 - 9x + 4;\ k = 1$

44. $f(x) = 2x^3 + 9x^2 - 16x + 12;\ k = 1$

45. $f(x) = x^3 + 7x^2 + 10x;\ k = 0$

46. $f(x) = 2x^3 - 3x^2 - 5x;\ k = 0$

47. $f(x) = 2x^4 + 3x^3 - 8x^2 - 2x + 15;\ k = -\frac{3}{2}$

48. $f(x) = 3x^4 + 2x^3 - 5x + 10;\ k = -\frac{4}{3}$

49. $f(x) = 5x^4 + 2x^3 - x + 3;\ k = \frac{2}{5}$

50. $f(x) = 16x^4 + 4x^2 - 2;\ k = \frac{1}{2}$

51. $f(x) = x^2 - 2x + 2;\ k = 1 - i$

52. $f(x) = x^2 - 4x + 5;\ k = 2 - i$

53. $f(x) = x^2 + 3x + 4;\ k = 2 + i$

54. $f(x) = x^2 - 3x + 5;\ k = 1 - 2i$

55. $f(x) = x^3 + 3x^2 - x + 1;\ k = 1 + i$

56. $f(x) = 2x^3 - x^2 + 3x - 5;\ k = 2 - i$

Relating Concepts

For individual or collaborative investigation

(Exercises 57–62)

We used synthetic division to evaluate $f(1)$ for $f(x) = x^3 - 4x^2 + 9x - 6$ in Example 3(a). There is an even quicker method for evaluating a polynomial $f(x)$ for $x = 1$. ***Work Exercises 57–62 in order,*** *to discover this method and another for $x = -1$.*

57. If 1 is raised to *any* power, what is the result?

58. If we multiply the result found in Exercise 57 by a real number, how does the product compare with the real number?

59. Based on your answer to Exercise 58, how can we evaluate $f(1)$ for $f(x) = x^3 - 4x^2 + 9x - 6$ without using direct substitution or synthetic division?

60. Support your answer in Exercise 59 by actually applying the method. Does your answer agree with the one found in Example 3(a)?

61. Find $f(-x)$ and $f(-1)$.

62. Add the coefficients of $f(-x)$ in Exercise 61, and compare to $f(-1)$. Make a conjecture about how to find $f(-1)$.

3.3 Zeros of Polynomial Functions

Factor Theorem ▪ Rational Zeros Theorem ▪ Number of Zeros ▪ Conjugate Zeros Theorem ▪ Finding Zeros of a Polynomial Function ▪ Descartes' Rule of Signs

Factor Theorem By the remainder theorem, if $f(k) = 0$, then the remainder when $f(x)$ is divided by $x - k$ is 0. This means that $x - k$ is a factor of $f(x)$. Conversely, if $x - k$ is a factor of $f(x)$, then $f(k)$ must equal 0. This is summarized in the following **factor theorem.**

Factor Theorem

The polynomial $x - k$ is a factor of the polynomial $f(x)$ if and only if $f(k) = 0$.

EXAMPLE 1 Deciding Whether $x - k$ Is a Factor of $f(x)$

Determine whether $x - 1$ is a factor of $f(x)$.

(a) $f(x) = 2x^4 + 3x^2 - 5x + 7$

(b) $f(x) = 3x^5 - 2x^4 + x^3 - 8x^2 + 5x + 1$

Solution

(a) By the factor theorem, $x - 1$ will be a factor of $f(x)$ only if $f(1) = 0$. Use synthetic division and the remainder theorem to decide.

$$\begin{array}{r|rrrrr} 1 & 2 & 0 & 3 & -5 & 7 \\ & & 2 & 2 & 5 & 0 \\ \hline & 2 & 2 & 5 & 0 & 7 \end{array} \leftarrow f(1) = 7$$

(Section 3.2)

Since the remainder is 7 and not 0, $x - 1$ is not a factor of $f(x)$.

(b)

$$\begin{array}{r|rrrrrr} 1 & 3 & -2 & 1 & -8 & 5 & 1 \\ & & 3 & 1 & 2 & -6 & -1 \\ \hline & 3 & 1 & 2 & -6 & -1 & 0 \end{array} \leftarrow f(1) = 0$$

Because the remainder is 0, $x - 1$ is a factor. Additionally, we can determine from the coefficients in the bottom row that the other factor is

$$3x^4 + x^3 + 2x^2 - 6x - 1,$$

and $$f(x) = (x - 1)(3x^4 + x^3 + 2x^2 - 6x - 1).$$

Now try Exercises 5 and 7.

We can use the factor theorem to factor a polynomial of higher degree into linear factors of the form $ax - b$.

EXAMPLE 2 Factoring a Polynomial Given a Zero

Factor $f(x) = 6x^3 + 19x^2 + 2x - 3$ into linear factors if -3 is a zero of f.

Solution Since -3 is a zero of f, $x - (-3) = x + 3$ is a factor.

$$\begin{array}{r|rrrr} -3 & 6 & 19 & 2 & -3 \\ & & -18 & -3 & 3 \\ \hline & 6 & 1 & -1 & 0 \end{array}$$

Use synthetic division to divide $f(x)$ by $x + 3$.

The quotient is $6x^2 + x - 1$, so

$$f(x) = (x + 3)(6x^2 + x - 1)$$

$$f(x) = (x + 3)(2x + 1)(3x - 1).$$ Factor $6x^2 + x - 1$. (Section R.4)

These factors are all linear.

Now try Exercise 17.

Looking Ahead to Calculus

Finding the derivative of a polynomial function is one of the basic skills required in a first calculus course. For the functions defined by

$$f(x) = x^4 - x^2 + 5x - 4,$$
$$g(x) = -x^6 + x^2 - 3x + 4,$$
$$h(x) = 3x^3 - x^2 + 2x - 4,$$
and $k(x) = -x^7 + x - 4,$

the derivatives are

$$f'(x) = 4x^3 - 2x + 5,$$
$$g'(x) = -6x^5 + 2x - 3,$$
$$h'(x) = 9x^2 - 2x + 2,$$
and $k'(x) = -7x^6 + 1.$

Notice the use of the "prime" notation: for example, the derivative of $f(x)$ is denoted $f'(x)$.

Do you see the pattern among the exponents and the coefficients? What do you think is the derivative of $F(x) = 4x^4 - 3x^3 + 6x - 4$? See the answer at the bottom of the next page.

Rational Zeros Theorem The **rational zeros theorem** gives a method to determine all possible candidates for rational zeros of a polynomial function with integer coefficients.

Rational Zeros Theorem

If $\frac{p}{q}$ is a rational number written in lowest terms, and if $\frac{p}{q}$ is a zero of f, a polynomial function with integer coefficients, then p is a factor of the constant term and q is a factor of the leading coefficient.

Proof $f\left(\frac{p}{q}\right) = 0$ since $\frac{p}{q}$ is a zero of $f(x)$, so

$$a_n\left(\frac{p}{q}\right)^n + a_{n-1}\left(\frac{p}{q}\right)^{n-1} + \cdots + a_1\left(\frac{p}{q}\right) + a_0 = 0.$$

This also can be written as

$$a_n\left(\frac{p^n}{q^n}\right) + a_{n-1}\left(\frac{p^{n-1}}{q^{n-1}}\right) + \cdots + a_1\left(\frac{p}{q}\right) + a_0 = 0.$$

Multiply both sides of this result by q^n and add $-a_0q^n$ to both sides.

$$a_np^n + a_{n-1}p^{n-1}q + \cdots + a_1pq^{n-1} = -a_0q^n$$

Factoring out p gives

$$p(a_np^{n-1} + a_{n-1}p^{n-2}q + \cdots + a_1q^{n-1}) = -a_0q^n.$$

This result shows that $-a_0q^n$ equals the product of the two factors p and $(a_np^{n-1} + \cdots + a_1q^{n-1})$. For this reason, p must be a factor of $-a_0q^n$. Since it was assumed that $\frac{p}{q}$ is written in lowest terms, p and q have no common factor other than 1, so p is not a factor of q^n. Thus, p must be a factor of a_0. In a similar way, it can be shown that q is a factor of a_n.

EXAMPLE 3 Using the Rational Zeros Theorem

Do each of the following for the polynomial function defined by

$$f(x) = 6x^4 + 7x^3 - 12x^2 - 3x + 2.$$

(a) List all possible rational zeros.

(b) Find all rational zeros and factor $f(x)$ into linear factors.

Solution

(a) For a rational number $\frac{p}{q}$ to be a zero, p must be a factor of $a_0 = 2$ and q must be a factor of $a_4 = 6$. Thus, p can be ± 1 or ± 2, and q can be ± 1, ± 2, ± 3, or ± 6. The possible rational zeros, $\frac{p}{q}$, are

$$\pm 1, \quad \pm 2, \quad \pm\frac{1}{2}, \quad \pm\frac{1}{3}, \quad \pm\frac{1}{6}, \quad \pm\frac{2}{3}.$$

(b) Use the remainder theorem to show that 1 and -2 are zeros.

$$\begin{array}{r|rrrrr} 1 & 6 & 7 & -12 & -3 & 2 \\ & & 6 & 13 & 1 & -2 \\ \hline & 6 & 13 & 1 & -2 & 0 \end{array}$$

The 0 remainder shows that 1 is a zero. Now, use the quotient polynomial $6x^3 + 13x^2 + x - 2$ and synthetic division to find that -2 is also a zero.

$$\begin{array}{r|rrrr} -2 & 6 & 13 & 1 & -2 \\ & & -12 & -2 & 2 \\ \hline & 6 & 1 & -1 & 0 \end{array}$$

The new quotient polynomial is $6x^2 + x - 1$, which is easily factored:

$$6x^2 + x - 1 = (3x - 1)(2x + 1).$$

Setting $3x - 1 = 0$ and $2x + 1 = 0$ yields the zeros $\frac{1}{3}$ and $-\frac{1}{2}$. Thus, the rational zeros are $1, -2, \frac{1}{3}$, and $-\frac{1}{2}$, and the linear factors of $f(x)$ are

$$x - 1, \quad x + 2, \quad 3x - 1, \quad \text{and} \quad 2x + 1.$$

Therefore,

$$\begin{aligned} f(x) &= 6x^4 + 7x^3 - 12x^2 - 3x + 2 \\ &= (x - 1)(x + 2)(3x - 1)(2x + 1). \end{aligned}$$

Now try Exercise 35.

NOTE In Example 3, once we obtained the quadratic factor $6x^2 + x - 1$, we were able to complete the work by factoring it directly. Had it not been easily factorable, we could have used the quadratic formula to find the other two zeros (and factors).

CAUTION The rational zeros theorem gives only *possible* rational zeros; it does not tell us whether these rational numbers are *actual* zeros. We must rely on other methods to determine whether or not they are indeed zeros. Furthermore, the function must have integer coefficients. To apply the rational zeros theorem to a polynomial with fractional coefficients, multiply through by the least common denominator of all the fractions. For example, any rational zeros of

$$p(x) = x^4 - \frac{1}{6}x^3 + \frac{2}{3}x^2 - \frac{1}{6}x - \frac{1}{3}$$

will also be rational zeros of

$$q(x) = 6x^4 - x^3 + 4x^2 - x - 2.$$

The polynomial $q(x)$ was obtained by multiplying the terms of $p(x)$ by 6.

Answer: $F'(x) = 16x^3 - 9x^2 + 6$

Number of Zeros The next theorem says that every function defined by a polynomial of degree 1 or more has a zero, which means that every such polynomial can be factored.

Fundamental Theorem of Algebra

Every function defined by a polynomial of degree 1 or more has at least one complex zero.

From the fundamental theorem, if $f(x)$ is of degree 1 or more then there is some number k_1 such that $f(k_1) = 0$. By the factor theorem,

$$f(x) = (x - k_1)q_1(x)$$

for some polynomial $q_1(x)$. If $q_1(x)$ is of degree 1 or more, the fundamental theorem and the factor theorem can be used to factor $q_1(x)$ in the same way. There is some number k_2 such that $q_1(k_2) = 0$, so

$$q_1(x) = (x - k_2)q_2(x)$$

and $$f(x) = (x - k_1)(x - k_2)q_2(x).$$

Assuming that $f(x)$ has degree n and repeating this process n times gives

$$f(x) = a(x - k_1)(x - k_2) \cdots (x - k_n),$$

where a is the leading coefficient of $f(x)$. Each of these factors leads to a zero of $f(x)$, so $f(x)$ has the n zeros $k_1, k_2, k_3, \ldots, k_n$. This result suggests the next theorem.

Number of Zeros Theorem

A function defined by a polynomial of degree n has at most n distinct zeros.

This theorem says that there exist *at most n* distinct zeros. For example, the polynomial function defined by

$$f(x) = x^3 + 3x^2 + 3x + 1 = (x + 1)^3$$

is of degree 3 but has only one zero, -1. Actually, the zero -1 occurs three times, since there are three factors of $x + 1$; this zero is called a **zero of multiplicity** 3.

EXAMPLE 4 Finding a Polynomial Function That Satisfies Given Conditions (Real Zeros)

Find a function f defined by a polynomial of degree 3 that satisfies the given conditions.

(a) Zeros of -1, 2, and 4; $f(1) = 3$

(b) -2 is a zero of multiplicity 3; $f(-1) = 4$

Solution

(a) These three zeros give $x - (-1) = x + 1$, $x - 2$, and $x - 4$ as factors of $f(x)$. Since $f(x)$ is to be of degree 3, these are the only possible factors by the number of zeros theorem. Therefore, $f(x)$ has the form

$$f(x) = a(x + 1)(x - 2)(x - 4)$$

for some real number a. To find a, use the fact that $f(1) = 3$.

$$\begin{aligned} f(1) = a(1 + 1)(1 - 2)(1 - 4) &= 3 \\ a(2)(-1)(-3) &= 3 \\ 6a &= 3 \\ a &= \frac{1}{2} \end{aligned}$$

Thus, $$f(x) = \frac{1}{2}(x + 1)(x - 2)(x - 4),$$

or, $$f(x) = \frac{1}{2}x^3 - \frac{5}{2}x^2 + x + 4. \quad \text{Multiply.}$$

(b) The polynomial function defined by $f(x)$ has the form

$$\begin{aligned} f(x) &= a(x + 2)(x + 2)(x + 2) \\ &= a(x + 2)^3. \end{aligned}$$

Since $f(-1) = 4$,

$$\begin{aligned} f(-1) = a(-1 + 2)^3 &= 4 \\ a(1)^3 &= 4 \\ a &= 4, \end{aligned}$$

and $$f(x) = 4(x + 2)^3 = 4x^3 + 24x^2 + 48x + 32.$$

Now try Exercises 49 and 53.

NOTE In Example 4(a), we cannot clear the denominators in $f(x)$ by multiplying both sides by 2 because the result would equal $2 \cdot f(x)$, not $f(x)$.

Conjugate Zeros Theorem The following properties of complex conjugates are needed to prove the *conjugate zeros theorem.* We use a simplified notation for conjugates here. If $z = a + bi$, then the conjugate of z is written $\overline{z}$, where $\overline{z} = a - bi$. For example, if $z = -5 + 2i$, then $\overline{z} = -5 - 2i$. The proofs of these properties are left for the exercises. (See Exercises 79–82.)

Properties of Conjugates

For any complex numbers c and d,

$$\overline{c + d} = \overline{c} + \overline{d}, \quad \overline{c \cdot d} = \overline{c} \cdot \overline{d}, \quad \text{and} \quad \overline{c^n} = (\overline{c})^n.$$

The remainder theorem can be used to show, for example, that both $2 + i$ and $2 - i$ are zeros of $f(x) = x^3 - x^2 - 7x + 15$. In general, if z is a zero of a polynomial function with *real* coefficients, then so is $\overline{z}$.

Conjugate Zeros Theorem

If $f(x)$ defines a polynomial function *having only real coefficients* and if $z = a + bi$ is a zero of $f(x)$, where a and b are real numbers, then $\overline{z} = a - bi$ is also a zero of $f(x)$.

Proof Start with the polynomial function defined by

$$f(x) = a_n x^n + a_{n-1} x^{n-1} + \cdots + a_1 x + a_0,$$

where all coefficients are real numbers. If the complex number z is a zero of $f(x)$, then

$$f(z) = a_n z^n + a_{n-1} z^{n-1} + \cdots + a_1 z + a_0 = 0.$$

Taking the conjugate of both sides of this last equation gives

$$\overline{a_n z^n + a_{n-1} z^{n-1} + \cdots + a_1 z + a_0} = \overline{0}.$$

Using generalizations of the properties $\overline{c + d} = \overline{c} + \overline{d}$ and $\overline{c \cdot d} = \overline{c} \cdot \overline{d}$ gives

$$\overline{a_n z^n} + \overline{a_{n-1} z^{n-1}} + \cdots + \overline{a_1 z} + \overline{a_0} = \overline{0}$$

or

$$\overline{a_n}\,\overline{z^n} + \overline{a_{n-1}}\,\overline{z^{n-1}} + \cdots + \overline{a_1}\,\overline{z} + \overline{a_0} = \overline{0}.$$

Now use the property $\overline{c^n} = (\overline{c})^n$ and the fact that for any real number a, $\overline{a} = a$, to obtain

$$a_n(\overline{z})^n + a_{n-1}(\overline{z})^{n-1} + \cdots + a_1(\overline{z}) + a_0 = 0$$
$$f(\overline{z}) = 0.$$

Hence $\overline{z}$ is also a zero of $f(x)$, which completes the proof.

CAUTION It is essential that the polynomial have only real coefficients. For example, $f(x) = x - (1 + i)$ has $1 + i$ as a zero, but the conjugate $1 - i$ is not a zero.

EXAMPLE 5 Finding a Polynomial Function That Satisfies Given Conditions (Complex Zeros)

Find a polynomial function of least degree having only real coefficients and zeros 3 and $2 + i$.

Solution The complex number $2 - i$ also must be a zero, so the polynomial has at least three zeros, 3, $2 + i$, and $2 - i$. For the polynomial to be of least degree, these must be the only zeros. By the factor theorem there must be three factors, $x - 3$, $x - (2 + i)$, and $x - (2 - i)$, so

$$\begin{aligned} f(x) &= (x - 3)[x - (2 + i)][x - (2 - i)] \\ &= (x - 3)(x - 2 - i)(x - 2 + i) \\ &= x^3 - 7x^2 + 17x - 15. \end{aligned}$$

Any nonzero multiple of $x^3 - 7x^2 + 17x - 15$ also satisfies the given conditions on zeros. The information on zeros given in the problem is not enough to give a specific value for the leading coefficient.

Now try Exercise 57.

Finding Zeros of a Polynomial Function The theorem on conjugate zeros helps predict the number of real zeros of polynomial functions with real coefficients. A polynomial function with real coefficients of odd degree n, where $n \geq 1$, must have at least one real zero (since zeros of the form $a + bi$, where $b \neq 0$, occur in conjugate pairs). On the other hand, a polynomial function with real coefficients of even degree n may have no real zeros.

EXAMPLE 6 Finding All Zeros of a Polynomial Function Given One Zero

Find all zeros of $f(x) = x^4 - 7x^3 + 18x^2 - 22x + 12$, given that $1 - i$ is a zero.

Solution Since the polynomial function has only real coefficients and since $1 - i$ is a zero, by the conjugate zeros theorem $1 + i$ is also a zero. To find the remaining zeros, first use synthetic division to divide the original polynomial by $x - (1 - i)$.

$$\begin{array}{r|rrrrr} 1 - i & 1 & -7 & 18 & -22 & 12 \\ & & 1 - i & -7 + 5i & 16 - 6i & -12 \\ \hline & 1 & -6 - i & 11 + 5i & -6 - 6i & 0 \end{array}$$

By the factor theorem, since $x = 1 - i$ is a zero of $f(x)$, $x - (1 - i)$ is a factor, and $f(x)$ can be written as

$$f(x) = [x - (1 - i)][x^3 + (-6 - i)x^2 + (11 + 5i)x + (-6 - 6i)].$$

We know that $x = 1 + i$ is also a zero of $f(x)$, so

$$f(x) = [x - (1 - i)][x - (1 + i)]q(x),$$

for some polynomial $q(x)$. Thus,

$$x^3 + (-6 - i)x^2 + (11 + 5i)x + (-6 - 6i) = [x - (1 + i)]q(x).$$

Use synthetic division to find $q(x)$.

$$\begin{array}{r|rrrr} 1 + i & 1 & -6 - i & 11 + 5i & -6 - 6i \\ & & 1 + i & -5 - 5i & 6 + 6i \\ \hline & 1 & -5 & 6 & 0 \end{array}$$

Since $q(x) = x^2 - 5x + 6$, $f(x)$ can be written as

$$f(x) = [x - (1 - i)][x - (1 + i)](x^2 - 5x + 6).$$

Factoring $x^2 - 5x + 6$ as $(x - 2)(x - 3)$, we see that the remaining zeros are 2 and 3. The four zeros of $f(x)$ are $1 - i$, $1 + i$, 2, and 3.

Now try Exercise 31.

Descartes' Rule of Signs The following rule helps to determine the number of positive and negative real zeros of a polynomial function.

Descartes' Rule of Signs

Let $f(x)$ define a polynomial function with real coefficients and a nonzero constant term, with terms in descending powers of x.

(a) The number of positive real zeros of f either equals the number of variations in sign occurring in the coefficients of $f(x)$, or is less than the number of variations by a positive even integer.

(b) The number of negative real zeros of f either equals the number of variations in sign occurring in the coefficients of $f(-x)$, or is less than the number of variations by a positive even integer.

In the theorem, a *variation in sign* is a change from positive to negative or negative to positive in successive terms of the polynomial when written in descending powers of the variable. Missing terms (those with 0 coefficients) are counted as no change in sign and can be ignored.

EXAMPLE 7 Applying Descartes' Rule of Signs

Determine the possible number of positive real zeros and negative real zeros of $f(x) = x^4 - 6x^3 + 8x^2 + 2x - 1$.

Solution We first consider the possible number of positive zeros by observing that $f(x)$ has three variations in signs:

$$+x^4 - 6x^3 + 8x^2 + 2x - 1.$$

1 2 3

Thus, by Descartes' rule of signs, f has either 3 or $3 - 2 = 1$ positive real zeros.

For negative zeros, consider the variations in signs for $f(-x)$:

$$\begin{aligned} f(-x) &= (-x)^4 - 6(-x)^3 + 8(-x)^2 + 2(-x) - 1 \\ &= x^4 + 6x^3 + 8x^2 - 2x - 1. \end{aligned}$$

1

Since there is only one variation in sign, $f(x)$ has only 1 negative real zero.

Now try Exercise 73.

Carl Friedrich Gauss (1777–1855)

CONNECTIONS The fundamental theorem of algebra was first proved by the German mathematician Carl Friedrich Gauss in 1797 as part of his doctoral dissertation completed in 1799. This theorem had challenged the world's finest mathematicians for at least 200 years. Gauss's proof used advanced mathematical concepts outside the field of algebra. To this day, no purely algebraic proof has been discovered. Gauss returned to the theorem many times and in 1849 published his fourth and last proof, in which he extended the coefficients of the unknown quantities to include complex numbers.

(continued)

Although methods of solving linear and quadratic equations were known since the Babylonians, mathematicians struggled for centuries to find a formula that solved cubic equations to find the zeros of cubic functions. In 1545, a method of solving a cubic equation of the form $x^3 + mx = n$, developed by Niccolo Tartaglia, was published in the *Ars Magna,* a work by Girolamo Cardano. The formula for finding the one real solution of the equation is

$$x = \sqrt[3]{\frac{n}{2} + \sqrt{\left(\frac{n}{2}\right)^2 + \left(\frac{m}{3}\right)^3}} - \sqrt[3]{\frac{-n}{2} + \sqrt{\left(\frac{n}{2}\right)^2 + \left(\frac{m}{3}\right)^3}}.$$

(*Source:* Gullberg, J., *Mathematics from the Birth of Numbers,* W. W. Norton & Company, 1997.)

For Discussion or Writing

Use the formula to solve the equation $x^3 + 9x = 26$ for the one real solution.

3.3 Exercises

Concept Check *Decide whether each statement is* true *or* false. *If false, tell why.*

1. Since $x - 1$ is a factor of $f(x) = x^6 - x^4 + 2x^2 - 2$, we can conclude that $f(1) = 0$.
2. Since $f(1) = 0$ for $f(x) = x^6 - x^4 + 2x^2 - 2$, we can conclude that $x - 1$ is a factor of $f(x)$.
3. For $f(x) = (x + 2)^4(x - 3)$, 2 is a zero of multiplicity 4.
4. Since $2 + 3i$ is a zero of $f(x) = x^2 - 4x + 13$, we can conclude that $2 - 3i$ is also a zero.

Use the factor theorem and synthetic division to decide whether the second polynomial is a factor of the first. See Example 1.

5. $x^3 - 5x^2 + 3x + 1;\ x - 1$
6. $x^3 + 6x^2 - 2x - 7;\ x + 1$
7. $2x^4 + 5x^3 - 8x^2 + 3x + 13;\ x + 1$
8. $-3x^4 + x^3 - 5x^2 + 2x + 4;\ x - 1$
9. $-x^3 + 3x - 2;\ x + 2$
10. $-2x^3 + x^2 - 63;\ x + 3$
11. $4x^2 + 2x + 54;\ x - 4$
12. $5x^2 - 14x + 10;\ x + 2$
13. $x^3 + 2x^2 - 3;\ x - 1$
14. $2x^3 + x + 2;\ x + 1$
15. $2x^4 + 5x^3 - 2x^2 + 5x + 6;\ x + 3$
16. $5x^4 + 16x^3 - 15x^2 + 8x + 16;\ x + 4$

Factor $f(x)$ into linear factors given that k is a zero of $f(x)$. See Example 2.

17. $f(x) = 2x^3 - 3x^2 - 17x + 30;\ k = 2$
18. $f(x) = 2x^3 - 3x^2 - 5x + 6;\ k = 1$
19. $f(x) = 6x^3 + 13x^2 - 14x + 3;\ k = -3$
20. $f(x) = 6x^3 + 17x^2 - 63x + 10;\ k = -5$
21. $f(x) = 6x^3 + 25x^2 + 3x - 4;\ k = -4$
22. $f(x) = 8x^3 + 50x^2 + 47x - 15;\ k = -5$

23. $f(x) = x^3 + (7 - 3i)x^2 + (12 - 21i)x - 36i;\ k = 3i$

24. $f(x) = 2x^3 + (3 + 2i)x^2 + (1 + 3i)x + i;\ k = -i$

25. $f(x) = 2x^3 + (3 - 2i)x^2 + (-8 - 5i)x + (3 + 3i);\ k = 1 + i$

26. $f(x) = 6x^3 + (19 - 6i)x^2 + (16 - 7i)x + (4 - 2i);\ k = -2 + i$

27. $f(x) = x^4 + 2x^3 - 7x^2 - 20x - 12;\ k = -2$ (multiplicity 2)

28. $f(x) = 2x^4 + x^3 - 9x^2 - 13x - 5;\ k = -1$ (multiplicity 3)

For each polynomial function, one zero is given. Find all others. See Examples 2 and 6.

29. $f(x) = x^3 - x^2 - 4x - 6;\ 3$

30. $f(x) = x^3 + 4x^2 - 5;\ 1$

31. $f(x) = x^3 - 7x^2 + 17x - 15;\ 2 - i$

32. $f(x) = 4x^3 + 6x^2 - 2x - 1;\ \frac{1}{2}$

33. $f(x) = x^4 + 5x^2 + 4;\ -i$

34. $f(x) = x^4 + 10x^3 + 27x^2 + 10x + 26;\ i$

For each polynomial function, ***(a)*** *list all possible rational zeros,* ***(b)*** *find all rational zeros, and* ***(c)*** *factor $f(x)$. See Example 3.*

35. $f(x) = x^3 - 2x^2 - 13x - 10$

36. $f(x) = x^3 + 5x^2 + 2x - 8$

37. $f(x) = x^3 + 6x^2 - x - 30$

38. $f(x) = x^3 - x^2 - 10x - 8$

39. $f(x) = 6x^3 + 17x^2 - 31x - 12$

40. $f(x) = 15x^3 + 61x^2 + 2x - 8$

41. $f(x) = 24x^3 + 40x^2 - 2x - 12$

42. $f(x) = 24x^3 + 80x^2 + 82x + 24$

For each polynomial function, find all zeros and their multiplicities.

43. $f(x) = 7x^3 + x$

44. $f(x) = (x + 1)^2(x - 1)^3(x^2 - 10)$

45. $f(x) = 3(x - 2)(x + 3)(x^2 - 1)$

46. $f(x) = 5x^2(x + 1 - \sqrt{2})(2x + 5)$

47. $f(x) = (x^2 + x - 2)^5(x - 1 + \sqrt{3})^2$

48. $f(x) = (7x - 2)^3(x^2 + 9)^2$

Find a polynomial function of degree 3 with real coefficients that satisfies the given conditions. See Example 4.

49. Zeros of -3, 1, and 4; $f(2) = 30$

50. Zeros of 1, -1, and 0; $f(2) = 3$

51. Zeros of -2, 1, and 0; $f(-1) = -1$

52. Zeros of 2, -3, and 5; $f(3) = 6$

53. Zero of -3 having multiplicity 3; $f(3) = 36$

54. Zero of 4 having multiplicity 2 and zero of 2 having multiplicity 1; $f(1) = -18$

Find a polynomial function of least degree having only real coefficients with zeros as given. See Examples 4–6.

55. $5 + i$ and $5 - i$

56. $7 - 2i$ and $7 + 2i$

57. 2 and $1 + i$

58. $-3, 2, -i$, and $2 + i$

59. $1 + \sqrt{2}, 1 - \sqrt{2}$, and 1

60. $1 - \sqrt{3}, 1 + \sqrt{3}$, and 1

61. $2 + i, 2 - i, 3$, and -1

62. $3 + 2i, -1$, and 2

63. 2 and $3 + i$

64. -1 and $4 - 2i$

65. $1 - \sqrt{2}, 1 + \sqrt{2}$, and $1 - i$

66. $2 + \sqrt{3}, 2 - \sqrt{3}$, and $2 + 3i$

67. $2 - i$ and $6 - 3i$

68. $5 + i$ and $4 - i$

69. $4, 1 - 2i$, and $3 + 4i$

70. $-1, 1 + \sqrt{2}, 1 - \sqrt{2}$, and $1 + 4i$

71. $1 + 2i$ and 2 (multiplicity 2)

72. $2 + i$ and -3 (multiplicity 2)

Use Descartes' rule of signs to determine the possible number of positive real zeros and negative real zeros for each function. See Example 7.

73. $f(x) = 2x^3 - 4x^2 + 2x + 7$

74. $f(x) = x^3 + 2x^2 + x - 10$

75. $f(x) = 5x^4 + 3x^2 + 2x - 9$

76. $f(x) = 3x^4 + 2x^3 - 8x^2 - 10x - 1$

77. $f(x) = x^5 + 3x^4 - x^3 + 2x + 3$

78. $f(x) = 2x^5 - x^4 + x^3 - x^2 + x + 5$

If c and d are complex numbers, prove each statement. (Hint: Let $c = a + bi$ and $d = m + ni$ and form all the conjugates, the sums, and the products.)

79. $\overline{c + d} = \overline{c} + \overline{d}$

80. $\overline{cd} = \overline{c} \cdot \overline{d}$

81. $\overline{a} = a$ for any real number a

82. $\overline{c^2} = (\overline{c})^2$

3.4 Polynomial Functions: Graphs, Applications, and Models

Graphs of $f(x) = ax^n$ ▪ Graphs of General Polynomial Functions ▪ Turning Points and End Behavior ▪ Graphing Techniques ▪ Intermediate Value and Boundedness Theorems ▪ Approximating Real Zeros ▪ Polynomial Models and Curve Fitting

Graphs of $f(x) = ax^n$ We can now graph polynomial functions of degree 3 or more with real number domains (since we will be graphing in the real number plane).

EXAMPLE 1 Graphing Functions of the Form $f(x) = ax^n$ $(a = 1)$

Graph each function.

(a) $f(x) = x^3$ **(b)** $g(x) = x^5$ **(c)** $f(x) = x^4$, $g(x) = x^6$

Solution

(a) Choose several values for x, and find the corresponding values of $f(x)$, or y, as shown in the left table beside Figure 16. Plot the resulting ordered pairs and connect the points with a smooth curve. The graph of $f(x) = x^3$ is shown in blue in Figure 16.

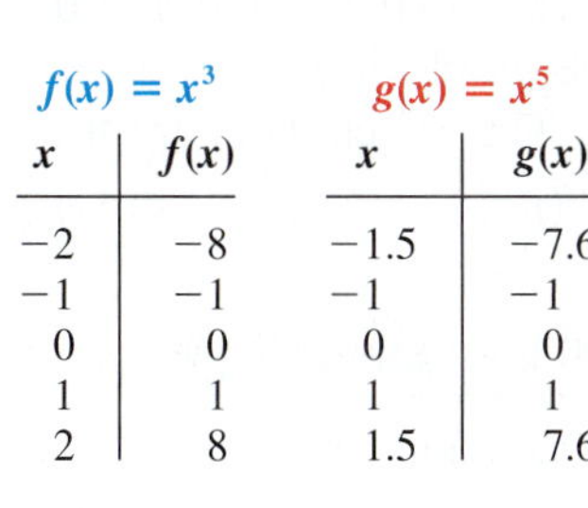

$f(x) = x^3$

x	$f(x)$
−2	−8
−1	−1
0	0
1	1
2	8

$g(x) = x^5$

x	$g(x)$
−1.5	−7.6
−1	−1
0	0
1	1
1.5	7.6

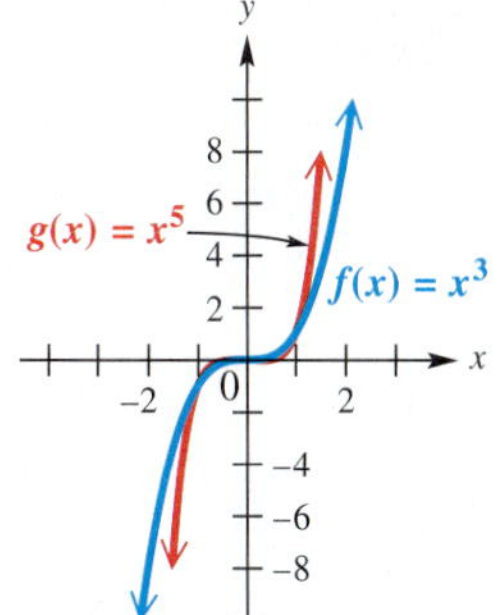

Figure 16

(b) Work as in part (a) to obtain the graph shown in red in Figure 16. Notice that the graphs of $f(x) = x^3$ and $g(x) = x^5$ are both symmetric with respect to the origin.

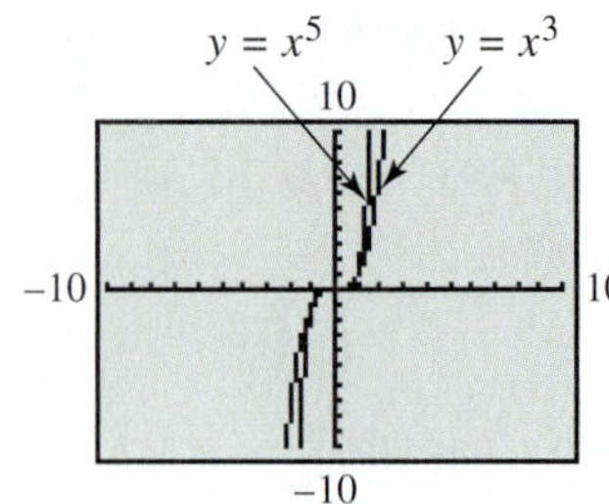

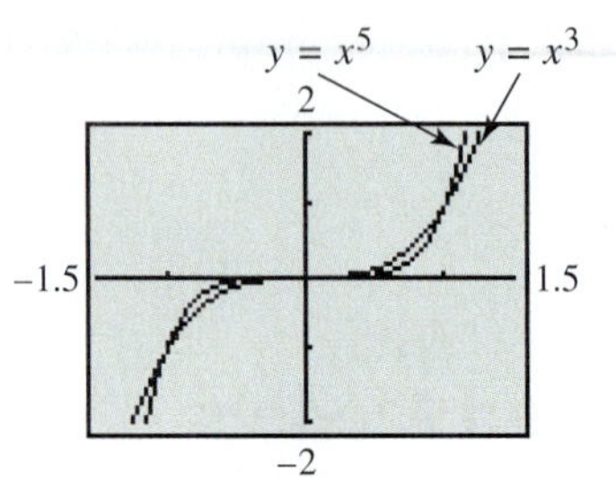

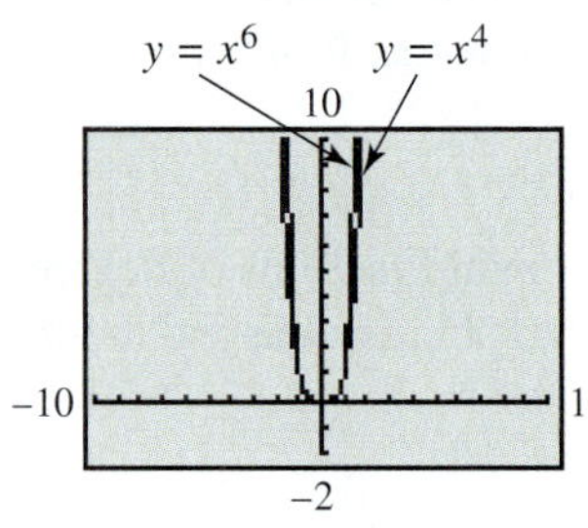

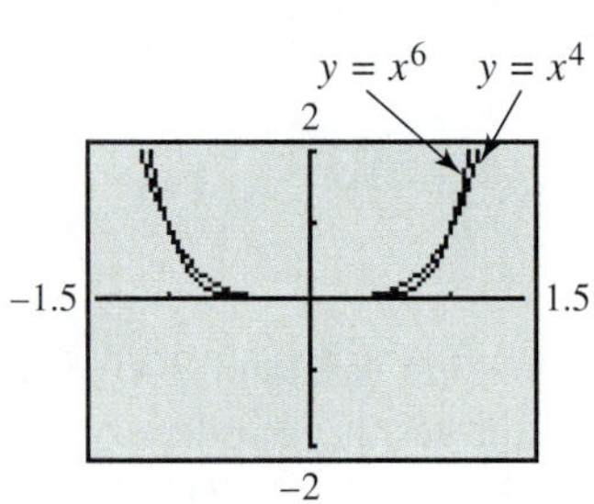

Figure 18

(c) Some typical ordered pairs for the graphs of $f(x) = x^4$ and $g(x) = x^6$ are given in the tables beside Figure 17. These graphs are symmetric with respect to the y-axis, as is the graph of $f(x) = ax^2$ for any nonzero real number a.

$f(x) = x^4$

x	$f(x)$
−2	16
−1	1
0	0
1	1
2	16

$g(x) = x^6$

x	$g(x)$
−1.5	11.4
−1	1
0	0
1	1
1.5	11.4

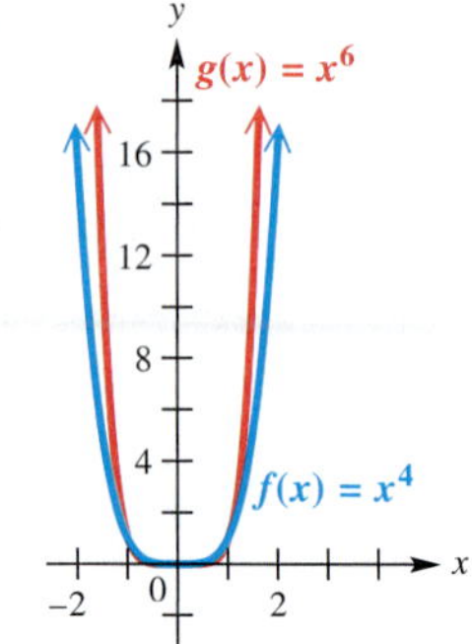

Figure 17

The ZOOM feature of a graphing calculator is useful with graphs like those in Example 1 to show the difference between the graphs of $y = x^3$ and $y = x^5$ and between $y = x^4$ and $y = x^6$ for values of x in the interval $[-1.5, 1.5]$. See Figure 18. In each case the first window shows x in $[-10, 10]$ and the second window shows x in $[-1.5, 1.5]$. ■

Graphs of General Polynomial Functions As with quadratic functions, the value of a in $f(x) = ax^n$ determines the width of the graph. When $|a| > 1$, the graph is stretched vertically, making it narrower, while when $0 < |a| < 1$, the graph is shrunk or compressed vertically, so the graph is broader. The graph of $f(x) = -ax^n$ is reflected across the x-axis compared to the graph of $f(x) = ax^n$.

Now try Exercises 9 and 11.

Compared with the graph of $f(x) = ax^n$, the graph of $f(x) = ax^n + k$ is translated (shifted) k units up if $k > 0$ and $|k|$ units down if $k < 0$. Also, when compared with the graph of $f(x) = ax^n$, the graph of $f(x) = a(x - h)^n$ is translated h units to the right if $h > 0$ and $|h|$ units to the left if $h < 0$.

The graph of $f(x) = a(x - h)^n + k$ shows a combination of these translations. The effects here are the same as those we saw earlier with quadratic functions.

EXAMPLE 2 Examining Vertical and Horizontal Translations

Graph each function.

(a) $f(x) = x^5 - 2$ **(b)** $f(x) = (x + 1)^6$ **(c)** $f(x) = -2(x - 1)^3 + 3$

Solution

(a) The graph of $f(x) = x^5 - 2$ will be the same as that of $f(x) = x^5$, but translated 2 units down. See Figure 19.

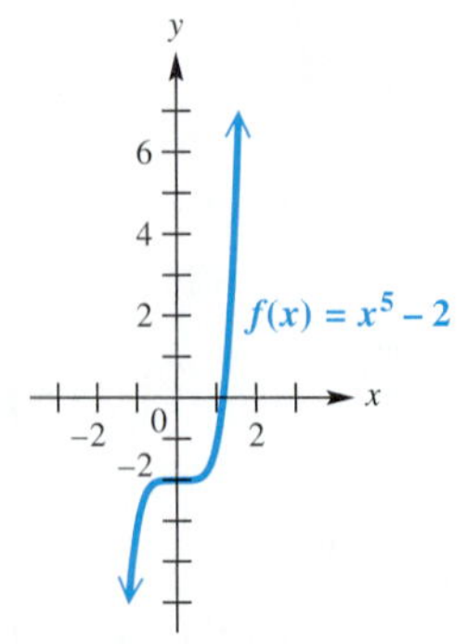

Figure 19

(b) In $f(x) = (x + 1)^6$, function f has a graph like that of $f(x) = x^6$, but since $x + 1 = x - (-1)$, it is translated 1 unit to the left as shown in Figure 20.

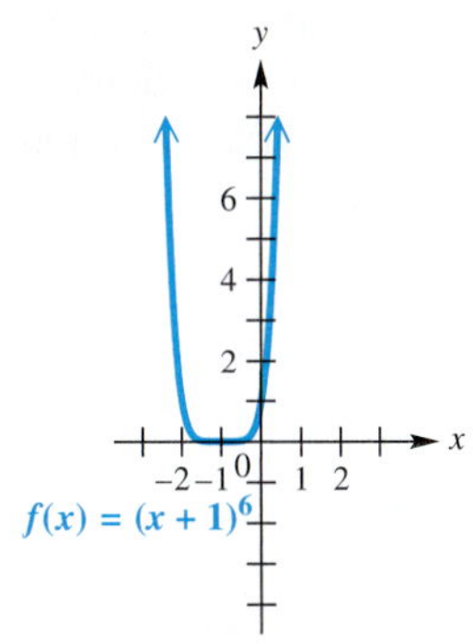

Figure 20

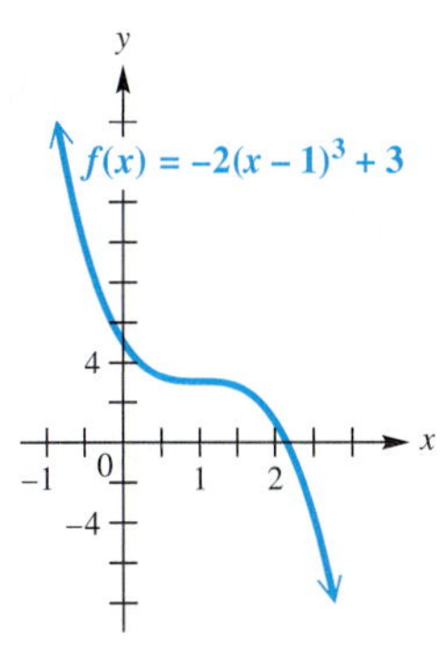

Figure 21

(c) The negative sign in -2 causes the graph of $f(x) = -2(x - 1)^3 + 3$ to be reflected across the x-axis when compared with the graph of $f(x) = x^3$. Because $|-2| > 1$, the graph is stretched vertically as compared to the graph of $f(x) = x^3$. As shown in Figure 21, the graph is also translated 1 unit to the right and 3 units up.

Now try Exercises 13, 15, and 19.

The domain of every polynomial function is the set of all real numbers; thus polynomial functions are continuous on the interval $(-\infty, \infty)$. The range of a polynomial function of odd degree is also the set of all real numbers. Typical graphs of polynomial functions of odd degree are shown in Figure 22. These graphs suggest that for every polynomial function f of odd degree there is at least one real value of x that makes $f(x) = 0$. The zeros are the x-intercepts of the graph.

Even Degree

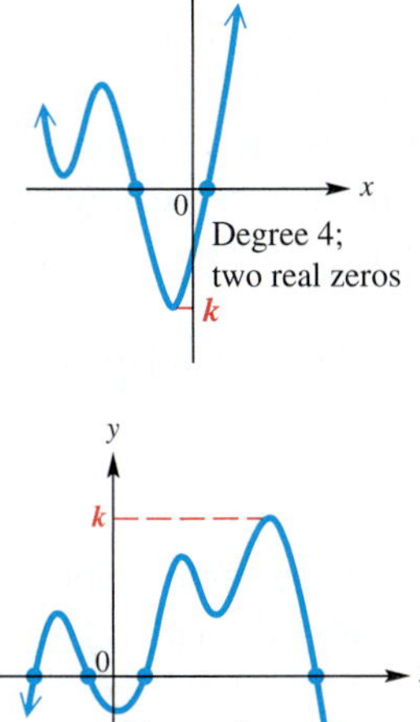

Figure 23

Odd Degree

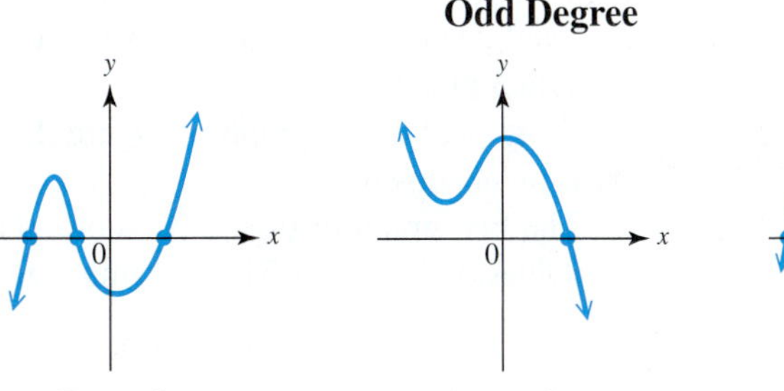

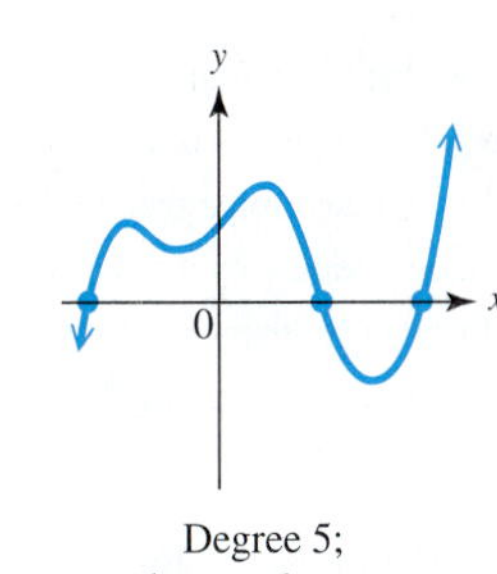

Figure 22

A polynomial function of even degree has a range of the form $(-\infty, k]$ or $[k, \infty)$ for some real number k. Figure 23 shows two typical graphs of polynomial functions of even degree.

Recall that a zero k of a polynomial function has as multiplicity the exponent of the factor $x - k$. Determining whether a zero has even or odd multiplicity aids in sketching the graph near that zero. If the zero has odd multiplicity, the graph crosses the x-axis at the corresponding x-intercept as seen in Figure 24(a). If the zero has even multiplicity, the graph is tangent to the x-axis at the corresponding x-intercept (that is, it touches but does not cross the x-axis there). See Figure 24(b).

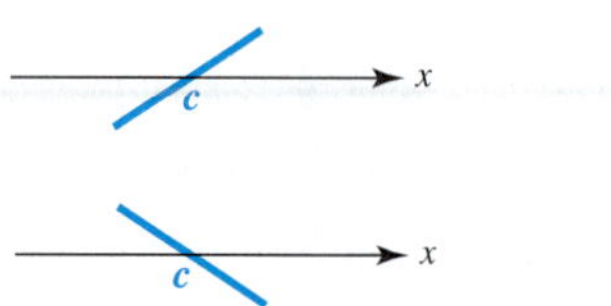

The graph crosses the x-axis at $(c, 0)$ if c is a zero of odd multiplicity.

(a)

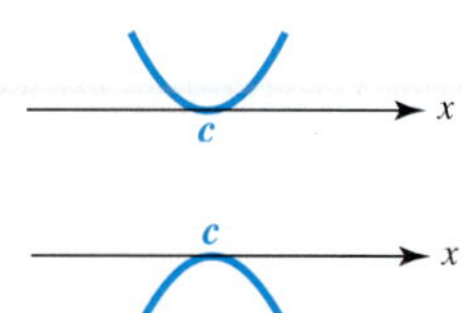

The graph is tangent to the x-axis at $(c, 0)$ if c is a zero of even multiplicity.

(b)

Figure 24

Turning Points and End Behavior The graphs in Figures 22 and 23 on the previous page show that polynomial functions often have **turning points** where the function changes from increasing to decreasing or from decreasing to increasing.

Turning Points

A polynomial function of degree n has at most $n - 1$ turning points, with at least one turning point between each pair of successive zeros.

Looking Ahead to Calculus

To find the x-coordinates of the two turning points of the graph of

$$f(x) = 2x^3 - 8x^2 + 9,$$

we can use the "maximum" and "minimum" capabilities of a graphing calculator and determine that, to the nearest thousandth, they are 0 and 2.667. In calculus, their exact values can be found by determining the zeros of the derivative function of $f(x)$,

$$f'(x) = 6x^2 - 16x,$$

because the turning points occur precisely where the tangent line has 0 slope. Factoring would show that the zeros are 0 and $\frac{8}{3}$, which agree with the calculator approximations.

The **end behavior** of a polynomial graph is determined by the **dominating term,** that is, the term of greatest degree. A polynomial of the form $f(x) = a_nx^n + a_{n-1}x^{n-1} + \cdots + a_0$ has the same end behavior as $f(x) = a_nx^n$. For instance, $f(x) = 2x^3 - 8x^2 + 9$ has the same end behavior as $f(x) = 2x^3$. It is large and positive for large positive values of x and large and negative for negative values of x with large absolute value. The arrows at the ends of the graph look like those of the first graph in Figure 22; the right arrow points up and the left arrow points down.

The first graph in Figure 22 shows that as x takes on larger and larger positive values, y does also. This is symbolized

$$\text{as} \quad x \to \infty, \quad y \to \infty,$$

read "as x approaches infinity, y approaches infinity." For the same graph, as x takes on negative values of larger and larger absolute value, y does also:

$$\text{as} \quad x \to -\infty, \quad y \to -\infty.$$

For the middle graph in Figure 22, we have

$$\text{as} \quad x \to \infty, \quad y \to -\infty$$

and

$$\text{as} \quad x \to -\infty, \quad y \to \infty.$$

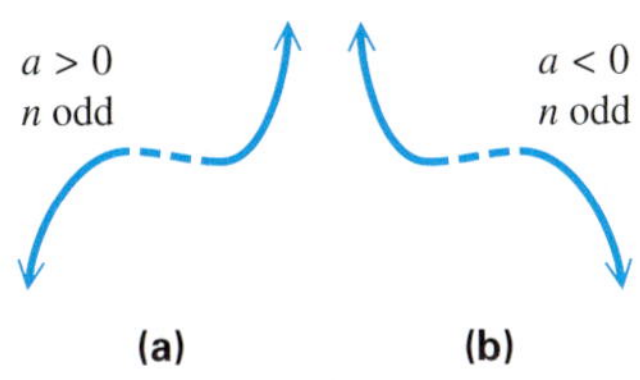

Figure 25

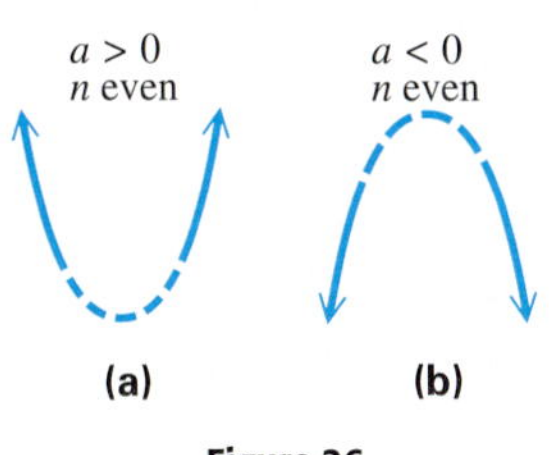

Figure 26

End Behavior of Graphs of Polynomial Functions

Suppose that ax^n is the dominating term of a polynomial function f of *odd degree.*

1. If $a > 0$, then as $x \to \infty$, $f(x) \to \infty$, and as $x \to -\infty$, $f(x) \to -\infty$. Therefore, the end behavior of the graph is of the type shown in Figure 25(a). We symbolize it as .
2. If $a < 0$, then as $x \to \infty$, $f(x) \to -\infty$, and as $x \to -\infty$, $f(x) \to \infty$. Therefore, the end behavior of the graph is of the type shown in Figure 25(b). We symbolize it as .

Suppose that ax^n is the dominating term of a polynomial function f of *even degree.*

1. If $a > 0$, then as $|x| \to \infty$, $f(x) \to \infty$. Therefore, the end behavior of the graph is of the type shown in Figure 26(a). We symbolize it as .
2. If $a < 0$, then as $|x| \to \infty$, $f(x) \to -\infty$. Therefore, the end behavior of the graph is of the type shown in Figure 26(b). We symbolize it as .

EXAMPLE 3 Determining End Behavior Given the Defining Polynomial

The graphs of the functions defined as follows are shown in A–D.

$$f(x) = x^4 - x^2 + 5x - 4, \qquad g(x) = -x^6 + x^2 - 3x - 4,$$

$$h(x) = 3x^3 - x^2 + 2x - 4, \quad \text{and} \quad k(x) = -x^7 + x - 4.$$

Based on the discussion in the preceding box, match each function with its graph.

A.

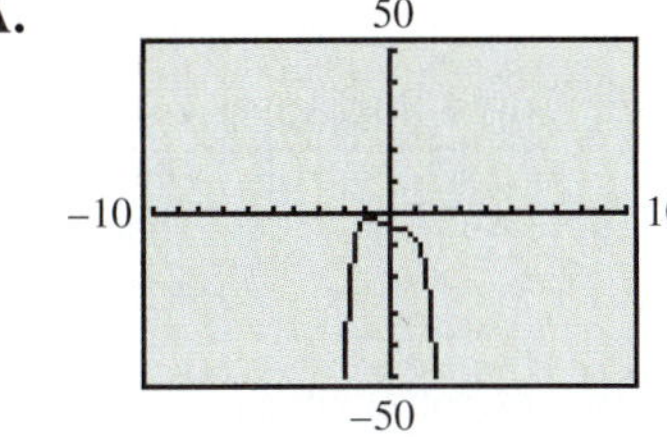

B.

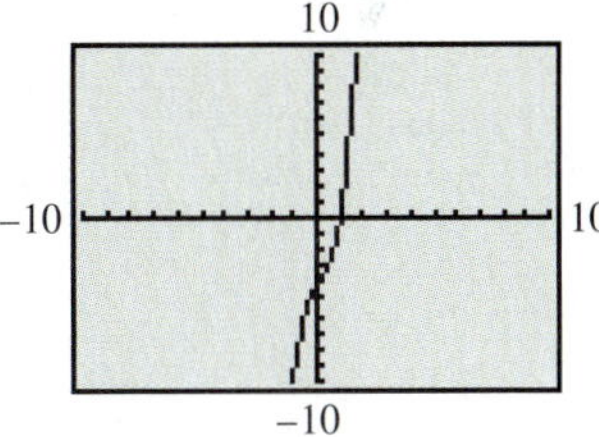

C.

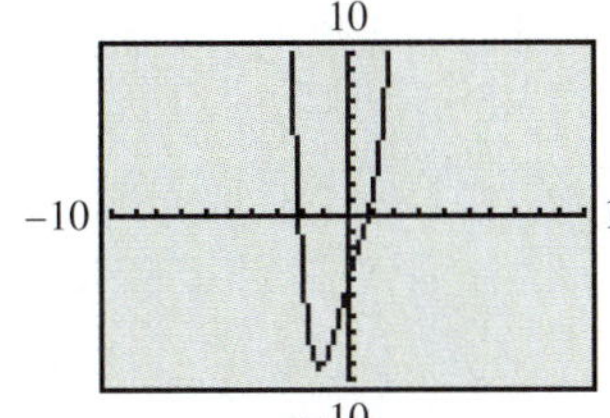

D.

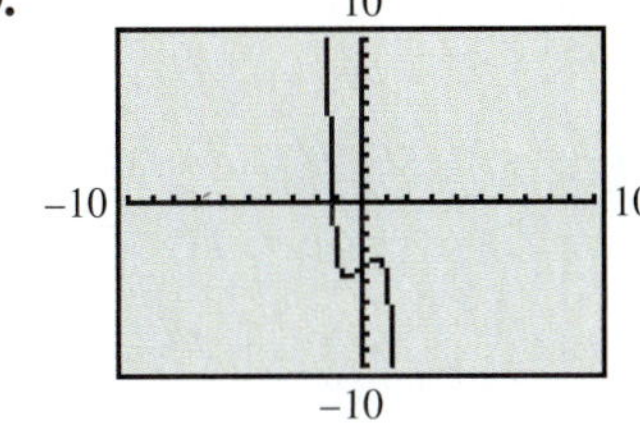

Solution Because f is of even degree with positive leading coefficient, its graph is in C. Because g is of even degree with negative leading coefficient, its graph is in A. Function h has odd degree and the dominating term is positive, so its graph is in B. Because function k has odd degree and a negative dominating term, its graph is in D.

Now try Exercises 21, 23, 25, and 27.

Graphing Techniques We have discussed several characteristics of the graphs of polynomial functions that are useful for graphing the function by hand. A *comprehensive graph* of a polynomial function will show the following characteristics.

1. all x-intercepts (zeros)
2. the y-intercept
3. all turning points
4. enough of the domain to show the end behavior

In Example 4, we sketch the graph of a polynomial function by hand. While there are several ways to approach this, here are some guidelines.

Graphing a Polynomial Function

Let $f(x) = a_n x^n + a_{n-1}x^{n-1} + \cdots + a_1 x + a_0$, $a_n \neq 0$, be a polynomial function of degree n. To sketch its graph, follow these steps.

Step 1 Find the real zeros of f. Plot them as x-intercepts.

Step 2 Find $f(0) = a_0$. Plot this as the y-intercept.

Step 3 Use test points within the intervals formed by the x-intercepts to determine the sign of $f(x)$ in the interval. This will determine whether the graph is above or below the x-axis in that interval.

Use end behavior, whether the graph crosses or is tangent to the x-axis at the x-intercepts, and selected points as necessary to complete the graph.

EXAMPLE 4 Graphing a Polynomial Function

Graph $f(x) = 2x^3 + 5x^2 - x - 6$.

Solution

Step 1 The possible rational zeros are

$$\pm 1, \quad \pm 2, \quad \pm 3, \quad \pm 6, \quad \pm \frac{1}{2}, \quad \text{and} \quad \pm \frac{3}{2}.$$

Use synthetic division to show that 1 is a zero.

$$\begin{array}{r|rrrr} 1 & 2 & 5 & -1 & -6 \\ & & 2 & 7 & 6 \\ \hline & 2 & 7 & 6 & 0 \end{array} \quad \leftarrow f(1) = 0$$

(Section 3.2)

Thus,

$$\begin{aligned} f(x) &= (x-1)(2x^2+7x+6) \\ &= (x-1)(2x+3)(x+2) \end{aligned}$$

Factor $2x^2 + 7x + 6$. (Section R.4)

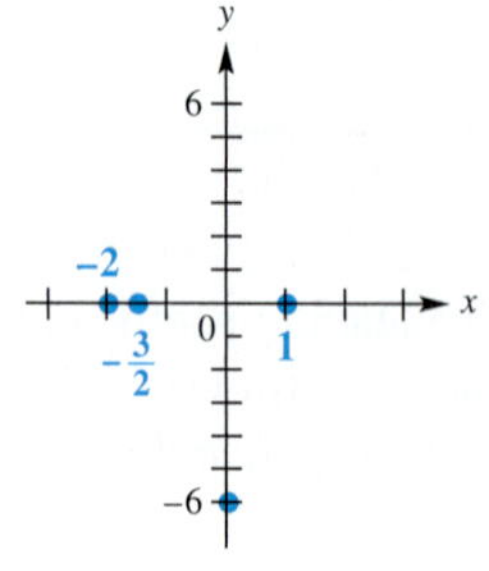

Figure 27

The three zeros of f are 1, $-\frac{3}{2}$, and -2. See Figure 27.

Step 2 $f(0) = -6$, so plot $(0, -6)$. See Figure 27.

Step 3 The x-intercepts divide the x-axis into four intervals:

$$(-\infty, -2), \quad \left(-2, -\frac{3}{2}\right), \quad \left(-\frac{3}{2}, 1\right), \quad \text{and} \quad (1, \infty).$$

Because the graph of a polynomial function has no breaks, gaps, or sudden jumps, the values of $f(x)$ are either always positive or always negative in any given interval. To find the sign of $f(x)$ in each interval, select an x-value in each interval and substitute it into the equation for $f(x)$ to determine whether the values of the function are positive or negative in that interval. When the values of $f(x)$ are negative, the graph is below the x-axis, and when $f(x)$ has positive values, the graph is above the x-axis.

A typical selection of test points and the results of the tests are shown below. (As a bonus, this procedure also locates points that lie on the graph.)

Interval	Test Point	Value of $f(x)$	Sign of $f(x)$	Graph Above or Below x-Axis
$(-\infty, -2)$	-3	-12	Negative	Below
$\left(-2, -\frac{3}{2}\right)$	$-\frac{7}{4}$	$\frac{11}{32}$	Positive	Above
$\left(-\frac{3}{2}, 1\right)$	0	-6	Negative	Below
$(1, \infty)$	2	28	Positive	Above

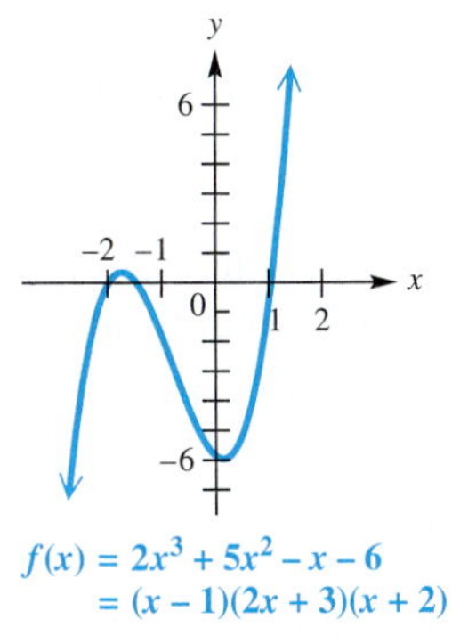

$f(x) = 2x^3 + 5x^2 - x - 6$
$= (x - 1)(2x + 3)(x + 2)$

Figure 28

Plot the test points and join the x-intercepts, y-intercept, and test points with a smooth curve to get the graph. Because each zero has odd multiplicity (1), the graph crosses the x-axis each time. The graph in Figure 28 shows that this function has two turning points, the maximum number for a third-degree polynomial function. The sketch could be improved by plotting the points found in each interval in the table. Notice that the left arrow points down and the right arrow points up. This end behavior is correct since the dominating term of the polynomial is $2x^3$.

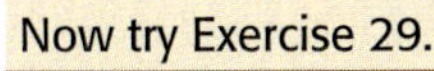
Now try Exercise 29.

10

Y1=2X³+5X²-X-6

−4.7 4.7

−10

Here is a calculator graph of the function in Example 4.

NOTE If a polynomial function is given in factored form, such as

$$f(x) = (-2x + 5)(x - 3)(x + 4)^2,$$

Step 1 of the guidelines is easier to perform, since real zeros can be determined by inspection. For this function, $\frac{5}{2}$ and 3 are zeros of multiplicity 1, and -4 is a zero of multiplicity 2. Since the dominating term is $-2x(x)(x^2) = -2x^4$, the end behavior is ↙ ↘. The y-intercept is $f(0) = 5(-3)(4)^2 = -240$.

We emphasize the important relationships among the following concepts.

1. the x-intercepts of the graph of $y = f(x)$
2. the zeros of the function f
3. the solutions of the equation $f(x) = 0$
4. the factors of $f(x)$

For example, the graph of the function in Example 4, defined by

$$f(x) = 2x^3 + 5x^2 - x - 6$$
$$= (x - 1)(2x + 3)(x + 2), \quad \text{Factored form}$$

has x-intercepts 1, $-\frac{3}{2}$, and -2, as shown in Figure 28. Since 1, $-\frac{3}{2}$, and -2 are the x-values where the function is 0, they are the zeros of f. Also, 1, $-\frac{3}{2}$, and -2 are the solutions of the polynomial equation $2x^3 + 5x^2 - x - 6 = 0$. This discussion is summarized as follows.

x-Intercepts, Zeros, Solutions, and Factors

If a is an x-intercept of the graph of $y = f(x)$, then a is a zero of f, a is a solution of $f(x) = 0$, and $x - a$ is a factor of $f(x)$.

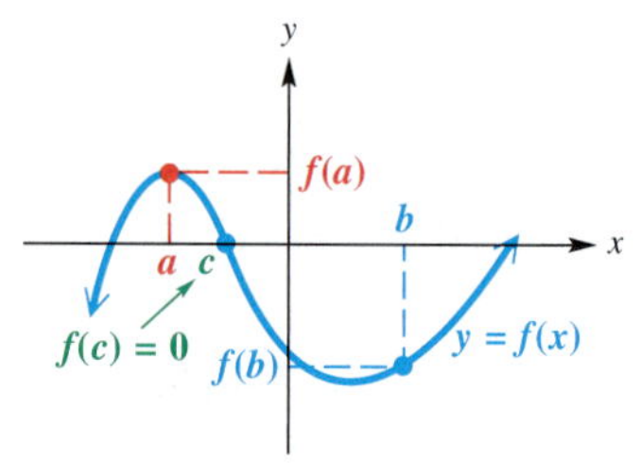

Figure 29

Intermediate Value and Boundedness Theorems As Example 4 shows, one key to graphing a polynomial function is locating its zeros. In the special case where the potential zeros are rational numbers, the zeros are found by the rational zeros theorem (Section 3.3). Occasionally, irrational zeros can be found by inspection. For instance, $f(x) = x^3 - 2$ has the irrational zero $\sqrt[3]{2}$. The next two theorems presented in this section apply to the zeros of every polynomial function with real coefficients. The first theorem uses the fact that graphs of polynomial functions are continuous curves. The proof requires advanced methods, so it is not given here. Figure 29 illustrates the theorem.

Intermediate Value Theorem for Polynomials

If $f(x)$ defines a polynomial function with only real coefficients, and if for real numbers a and b, the values $f(a)$ and $f(b)$ are opposite in sign, then there exists at least one real zero between a and b.

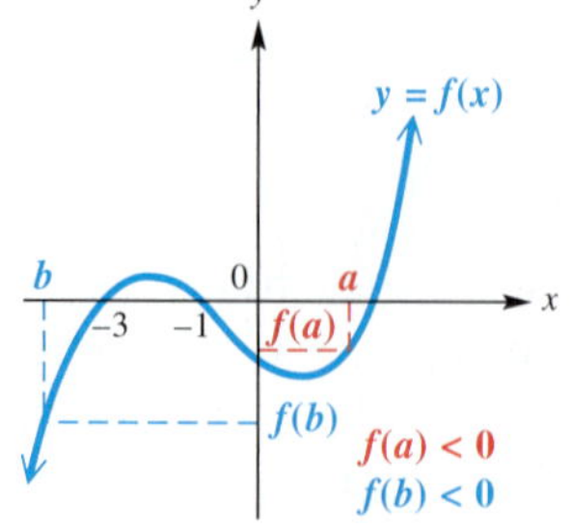

Figure 30

This theorem helps identify intervals where zeros of polynomial functions are located. If $f(a)$ and $f(b)$ are opposite in sign, then 0 is between $f(a)$ and $f(b)$, and so there must be a number c between a and b where $f(c) = 0$.

CAUTION Be careful how you interpret the intermediate value theorem. If $f(a)$ and $f(b)$ are *not* opposite in sign, it does not necessarily mean that there is no zero between a and b. For example, in Figure 30, $f(a)$ and $f(b)$ are both negative, but -3 and -1, which are between a and b, are zeros of $f(x)$.

EXAMPLE 5 Locating a Zero

Use synthetic division and a graph to show that $f(x) = x^3 - 2x^2 - x + 1$ has a real zero between 2 and 3.

Algebraic Solution

Use synthetic division to find $f(2)$ and $f(3)$.

$$\begin{array}{r|rrrl} 2 & 1 & -2 & -1 & 1 \\ & & 2 & 0 & -2 \\ \hline & 1 & 0 & -1 & -1 = f(2) \end{array}$$

$$\begin{array}{r|rrrl} 3 & 1 & -2 & -1 & 1 \\ & & 3 & 3 & 6 \\ \hline & 1 & 1 & 2 & 7 = f(3) \end{array}$$

Since $f(2)$ is negative and $f(3)$ is positive, by the intermediate value theorem there must be a real zero between 2 and 3.

Graphing Calculator Solution

The graphing calculator screen in Figure 31 indicates that this zero is approximately 2.2469796. (Notice that there are two other zeros as well.)

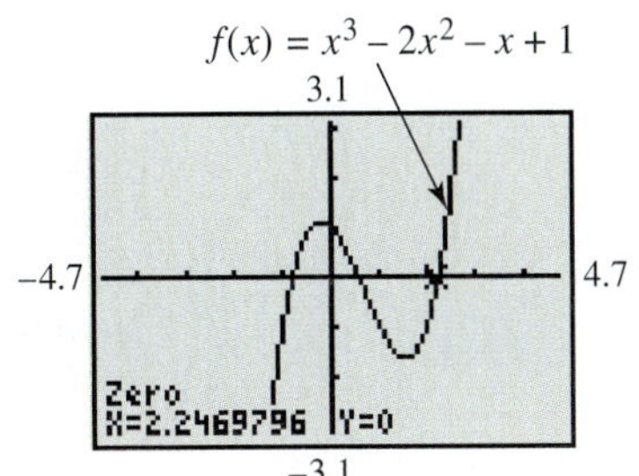

Figure 31

Now try Exercise 45.

The intermediate value theorem for polynomials helps limit the search for real zeros to smaller and smaller intervals. In Example 5, we used the theorem to verify that there is a real zero between 2 and 3. We could use the theorem repeatedly to locate the zero more accurately. The next theorem, the **boundedness theorem,** shows how the bottom row of a synthetic division is used to place upper and lower bounds on possible real zeros of a polynomial function.

Boundedness Theorem

Let $f(x)$ be a polynomial function of degree $n \geq 1$ with real coefficients and with a positive leading coefficient. If $f(x)$ is divided synthetically by $x - c$, and

(a) if $c > 0$ and all numbers in the bottom row of the synthetic division are nonnegative, then $f(x)$ has no zero greater than c;

(b) if $c < 0$ and the numbers in the bottom row of the synthetic division alternate in sign (with 0 considered positive or negative, as needed), then $f(x)$ has no zero less than c.

Proof We outline the proof of part (a). The proof for part (b) is similar. By the division algorithm, if $f(x)$ is divided by $x - c$, then for some $q(x)$ and r,

$$f(x) = (x - c)q(x) + r,$$

where all coefficients of $q(x)$ are nonnegative, $r \geq 0$, and $c > 0$. If $x > c$, then $x - c > 0$. Since $q(x) > 0$ and $r \geq 0$,

$$f(x) = (x - c)q(x) + r > 0.$$

This means that $f(x)$ will never be 0 for $x > c$.

EXAMPLE 6 Using the Boundedness Theorem

Show that the real zeros of $f(x) = 2x^4 - 5x^3 + 3x + 1$ satisfy the following conditions.

(a) No real zero is greater than 3. **(b)** No real zero is less than -1.

Solution

(a) Since $f(x)$ has real coefficients and the leading coefficient, 2, is positive, use the boundedness theorem. Divide $f(x)$ synthetically by $x - 3$.

$$\begin{array}{r|rrrrr} 3 & 2 & -5 & 0 & 3 & 1 \\ & & 6 & 3 & 9 & 36 \\ \hline & 2 & 1 & 3 & 12 & 37 \end{array}$$

Since $3 > 0$ and all numbers in the last row of the synthetic division are nonnegative, $f(x)$ has no real zero greater than 3.

(b)

$$\begin{array}{r|rrrrr} -1 & 2 & -5 & 0 & 3 & 1 \\ & & -2 & 7 & -7 & 4 \\ \hline & 2 & -7 & 7 & -4 & 5 \end{array}$$ Divide $f(x)$ by $x + 1$.

Here $-1 < 0$ and the numbers in the last row alternate in sign, so $f(x)$ has no zero less than -1.

Now try Exercises 53 and 55.

Approximating Real Zeros We can approximate the irrational real zeros of a polynomial function using a graphing calculator.

EXAMPLE 7 Approximating Real Zeros of a Polynomial Function

Approximate the real zeros of $f(x) = x^4 - 6x^3 + 8x^2 + 2x - 1$.

Solution The greatest degree term is x^4, so the graph will have end behavior similar to the graph of $f(x) = x^4$, which is positive for all values of x with large absolute values. That is, the end behavior is up at the left and the right, ↖↗. There are at most four real zeros, since the polynomial is fourth-degree.

Since $f(0) = -1$, the y-intercept is -1. Because the end behavior is positive on the left and the right, by the intermediate value theorem f has at least one zero on either side of $x = 0$. To approximate the zeros, we use a graphing calculator. The graph in Figure 32 shows that there are four real zeros, and the table indicates that they are between -1 and 0, 0 and 1, 2 and 3, and 3 and 4 because there is a sign change in $f(x)$ in each case.

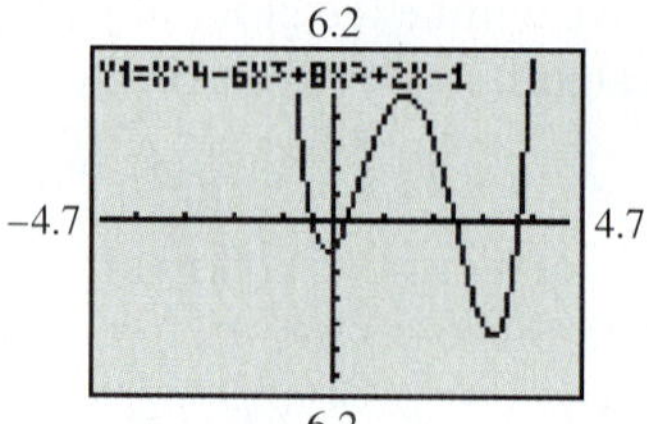

X	Y1
-1	12
0	-1
1	4
2	3
3	-4
4	7
5	84

Y1■X^4-6X^3+8X^2+...

Figure 32

$f(x) = x^4 - 6x^3 + 8x^2 + 2x - 1$

6.2

−4.7 4.7

Zero X=−.4142136 Y=0

−6.2

Figure 33

Using the capability of the calculator, we can find the zeros to a great degree of accuracy. Figure 33 shows that the negative zero is approximately $-.4142136$. Similarly, we find that the other three zeros are approximately .26794919, 2.4142136, and 3.7320508.

Now try Exercise 69.

Polynomial Models and Curve Fitting Polynomial functions are often good choices to model real data. Graphing calculators provide several types of polynomial regression equations to model data: linear, quadratic, cubic (degree 3), and quartic (degree 4).

EXAMPLE 8 Examining a Polynomial Model for Debit Card Use

Year	Transactions (in millions)
1990	127
1992	204
1995	829
1998	3765
2000	6655

Source: Statistical Abstract of the United States, 2000.

The table shows the number of transactions, in millions, by users of bank debit cards for selected years.

(a) Using $x = 0$ to represent 1990, $x = 2$ to represent 1992, and so on, use the regression feature of a calculator to determine the quadratic function that best fits the data. Plot the data and the graph.

(b) Repeat part (a) for a cubic function.

(c) Repeat part (a) for a quartic function.

(d) The *correlation coefficient, R,* is a measure of how good a fit a regression equation is for a function. The values of R and R^2 are used to determine how reliable a predictor a regression equation is. The closer the value of R^2 is to 1, the better the fit. Compare R^2 for the three functions found in parts (a)–(c) to decide which function best fits the data.

Solution

(a) The best-fitting quadratic function for the data is defined by

$$y = 98.08x^2 - 343.7x + 248.6.$$

The regression coordinates screen and the data points with the graph are shown in Figure 34.

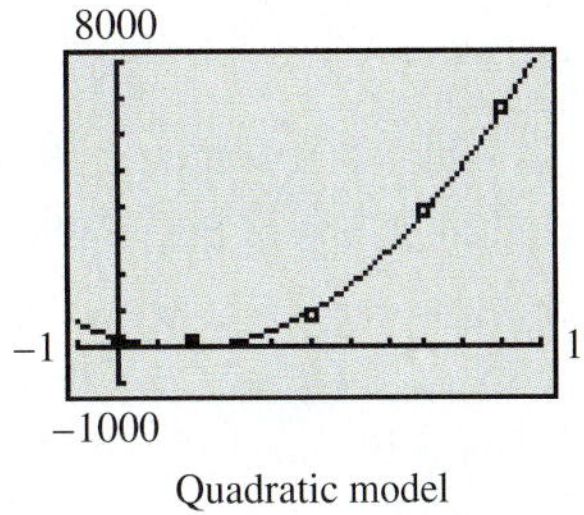

Quadratic model

Figure 34

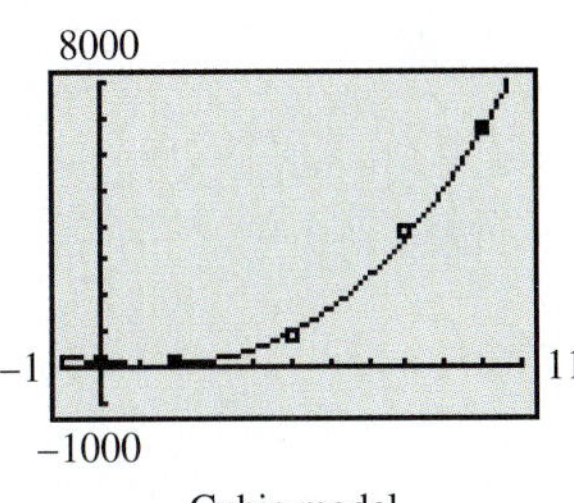

Cubic model

Figure 35

(b) The best-fitting cubic function is defined by

$$y = 3.706x^3 + 42.48x^2 - 142.7x + 170.$$

See Figure 35.

Quartic model

Figure 36

(c) The best-fitting quartic function is defined by

$$y = -1.619x^4 + 36.09x^3 - 155.5x^2 + 218.1x + 127.$$

See Figure 36.

(d) Find the correlation coefficient values R^2. Figure 34 shows R^2 for the quadratic function. The others are .9982771996 for the cubic function and 1 for the quartic function. Therefore, the quartic function provides the best fit.

Now try Exercise 97.

3.4 Exercises

Concept Check *Comprehensive graphs of four polynomial functions are shown in A–D. They represent the graphs of functions defined by these four equations, but not necessarily in the order listed.*

$$y = x^3 - 3x^2 - 6x + 8 \qquad y = x^4 + 7x^3 - 5x^2 - 75x$$

$$y = -x^3 + 9x^2 - 27x + 17 \qquad y = -x^5 + 36x^3 - 22x^2 - 147x - 90$$

Apply the concepts of this section to answer each question.

A.

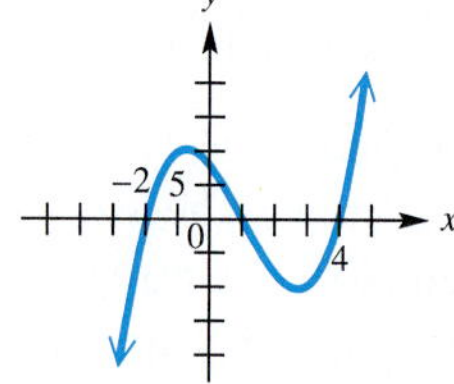

B.

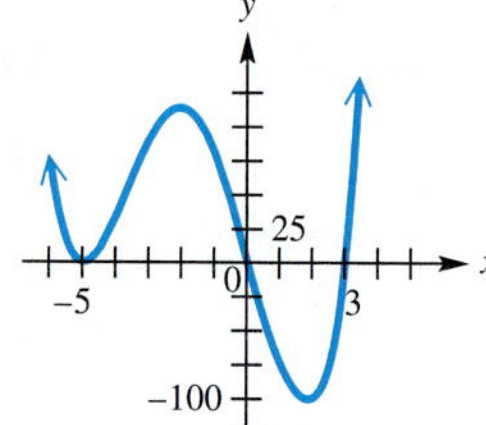

C.

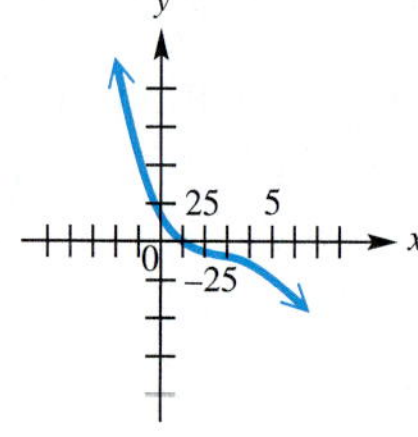

D.

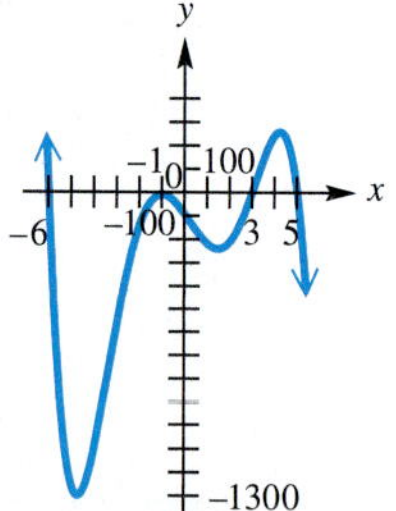

1. Which one of the graphs is that of $y = x^3 - 3x^2 - 6x + 8$?
2. Which one of the graphs is that of $y = x^4 + 7x^3 - 5x^2 - 75x$?
3. How many real zeros does the graph in C have?
4. Which one of C and D is the graph of $y = -x^3 + 9x^2 - 27x + 17$? (*Hint:* Look at the y-intercept.)
5. Which of the graphs cannot be that of a cubic polynomial function?
6. Which one of the graphs is that of a function whose range is *not* $(-\infty, \infty)$?
7. The function defined by $f(x) = x^4 + 7x^3 - 5x^2 - 75x$ has the graph shown in B. Use the graph to factor the polynomial.
8. The function defined by $f(x) = -x^5 + 36x^3 - 22x^2 - 147x - 90$ has the graph shown in D. Use the graph to factor the polynomial.

Sketch the graph of each polynomial function. See Examples 1 and 2.

9. $f(x) = 2x^4$
10. $f(x) = \frac{1}{4}x^6$
11. $f(x) = -\frac{2}{3}x^5$
12. $f(x) = -\frac{5}{4}x^5$
13. $f(x) = \frac{1}{2}x^3 + 1$
14. $f(x) = -x^4 + 2$
15. $f(x) = -(x + 1)^3$
16. $f(x) = (x + 2)^3 - 1$
17. $f(x) = (x - 1)^4 + 2$
18. $f(x) = \frac{1}{3}(x + 3)^4$
19. $f(x) = \frac{1}{2}(x - 2)^2 + 4$
20. $f(x) = \frac{1}{3}(x + 1)^3 - 3$

Use an end behavior diagram ↖ ↗, ↙ ↘, ↖ ↘, ↙ ↗, *to describe the end behavior of the graph of each polynomial function. See Example 3.*

21. $f(x) = 5x^3 + 2x^2 - 3x + 4$
22. $f(x) = -6x^3 - 4x^2 + 2x - 1$
23. $f(x) = -4x^5 + 3x^2 - 1$
24. $f(x) = 8x^7 - x^5 + x - 1$
25. $f(x) = 9x^4 - 3x^2 + x - 2$
26. $f(x) = 12x^6 - x^5 + 2x - 2$
27. $f(x) = 3 + 2x - 4x^2 - 5x^8$
28. $f(x) = 8 + 2x - 5x^2 - 10x^4$

Graph each polynomial function. Factor first if the expression is not in factored form. See Example 4.

29. $f(x) = x^3 + 5x^2 + 2x - 8$
30. $f(x) = x^3 + 3x^2 - 13x - 15$
31. $f(x) = 2x(x - 3)(x + 2)$
32. $f(x) = x^2(x + 1)(x - 1)$
33. $f(x) = x^2(x - 2)(x + 3)^2$
34. $f(x) = x^2(x - 5)(x + 3)(x - 1)$
35. $f(x) = (3x - 1)(x + 2)^2$
36. $f(x) = (4x + 3)(x + 2)^2$
37. $f(x) = x^3 + 5x^2 - x - 5$
38. $f(x) = x^3 + x^2 - 36x - 36$
39. $f(x) = x^3 - x^2 - 2x$
40. $f(x) = 3x^4 + 5x^3 - 2x^2$
41. $f(x) = 2x^3(x^2 - 4)(x - 1)$
42. $f(x) = x^2(x - 3)^3(x + 1)$

Use the intermediate value theorem for polynomials to show that each polynomial function has a real zero between the numbers given. See Example 5.

43. $f(x) = 2x^2 - 7x + 4$; 2 and 3
44. $f(x) = 3x^2 - x - 4$; 1 and 2
45. $f(x) = 2x^3 - 5x^2 - 5x + 7$; 0 and 1
46. $f(x) = 2x^3 - 9x^2 + x + 20$; 2 and 2.5
47. $f(x) = 2x^4 - 4x^2 + 4x - 8$; 1 and 2
48. $f(x) = x^4 - 4x^3 - x + 3$; .5 and 1
49. $f(x) = x^4 + x^3 - 6x^2 - 20x - 16$; 3.2 and 3.3
50. $f(x) = x^4 - 2x^3 - 2x^2 - 18x + 5$; 3.7 and 3.8
51. $f(x) = x^4 - 4x^3 - 20x^2 + 32x + 12$; -1 and 0
52. $f(x) = x^5 + 2x^4 + x^3 + 3$; -1.8 and -1.7

Show that the real zeros of each polynomial function satisfy the given conditions. See Example 6.

53. $f(x) = x^4 - x^3 + 3x^2 - 8x + 8$; no real zero greater than 2
54. $f(x) = 2x^5 - x^4 + 2x^3 - 2x^2 + 4x - 4$; no real zero greater than 1
55. $f(x) = x^4 + x^3 - x^2 + 3$; no real zero less than -2

56. $f(x) = x^5 + 2x^3 - 2x^2 + 5x + 5$; no real zero less than -1

57. $f(x) = 3x^4 + 2x^3 - 4x^2 + x - 1$; no real zero greater than 1

58. $f(x) = 3x^4 + 2x^3 - 4x^2 + x - 1$; no real zero less than -2

59. $f(x) = x^5 - 3x^3 + x + 2$; no real zero greater than 2

60. $f(x) = x^5 - 3x^3 + x + 2$; no real zero less than -3

Concept Check *In Exercises 61 and 62, find a cubic polynomial having the graph shown.*

61.

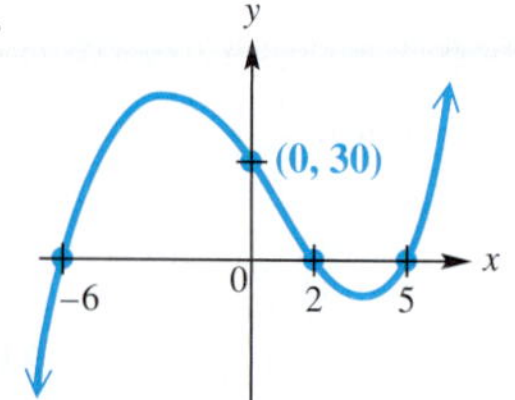

62.

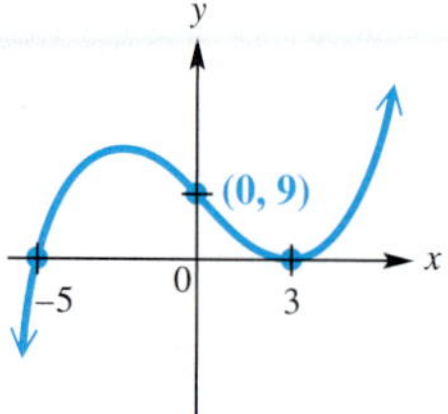

Use a graphing calculator to graph the function defined by $f(x)$ in the viewing window specified. Compare the graph to the one shown in the answer section of this text. Then use the graph to find $f(1.25)$.

63. $f(x) = 2x(x - 3)(x + 2)$; window: $[-3, 4]$ by $[-20, 12]$
Compare to Exercise 31.

64. $f(x) = x^2(x - 2)(x + 3)^2$; window: $[-4, 3]$ by $[-24, 4]$
Compare to Exercise 33.

65. $f(x) = (3x - 1)(x + 2)^2$; window: $[-4, 2]$ by $[-15, 15]$
Compare to Exercise 35.

66. $f(x) = x^3 + 5x^2 - x - 5$; window: $[-6, 2]$ by $[-30, 30]$
Compare to Exercise 37.

Use a graphing calculator to approximate the real zero discussed in each specified exercise. See Example 7.

67. Exercise 43 **68.** Exercise 45 **69.** Exercise 47 **70.** Exercise 46

For the given polynomial function, approximate each zero as a decimal to the nearest tenth. See Example 7.

71. $f(x) = x^3 + 3x^2 - 2x - 6$

72. $f(x) = x^3 - 3x + 3$

73. $f(x) = -2x^4 - x^2 + x + 5$

74. $f(x) = -x^4 + 2x^3 + 3x^2 + 6$

Use a graphing calculator to find the coordinates of the turning points of the graph of each polynomial function in the given domain interval. Give answers to the nearest hundredth.

75. $f(x) = x^3 + 4x^2 - 8x - 8$; $[-3.8, -3]$

76. $f(x) = x^3 + 4x^2 - 8x - 8$; $[.3, 1]$

77. $f(x) = 2x^3 - 5x^2 - x + 1$; $[-1, 0]$

78. $f(x) = 2x^3 - 5x^2 - x + 1$; $[1.4, 2]$

79. $f(x) = x^4 - 7x^3 + 13x^2 + 6x - 28$; $[-1, 0]$

80. $f(x) = x^3 - x + 3$; $[-1, 0]$

81. *(Modeling) Social Security Numbers* Your Social Security number (SSN) is unique, and with it you can construct your own personal Social Security polynomial. Let the polynomial function be defined as follows, where a_i represents the ith digit in your SSN:

$$\text{SSN}(x) = (x - a_1)(x + a_2)(x - a_3)(x + a_4)(x - a_5) \cdot (x + a_6)(x - a_7)(x + a_8)(x - a_9).$$

For example, if the SSN is 539-58-0954, the polynomial function is

$$\text{SSN}(x) = (x - 5)(x + 3)(x - 9)(x + 5)(x - 8)(x + 0)(x - 9)(x + 5)(x - 4).$$

A comprehensive graph of this function is shown in Figure A. In Figure B, we show a screen obtained by zooming in on the positive zeros, as the comprehensive graph does not show the local behavior well in this region. Use a graphing calculator to graph your own "personal polynomial."

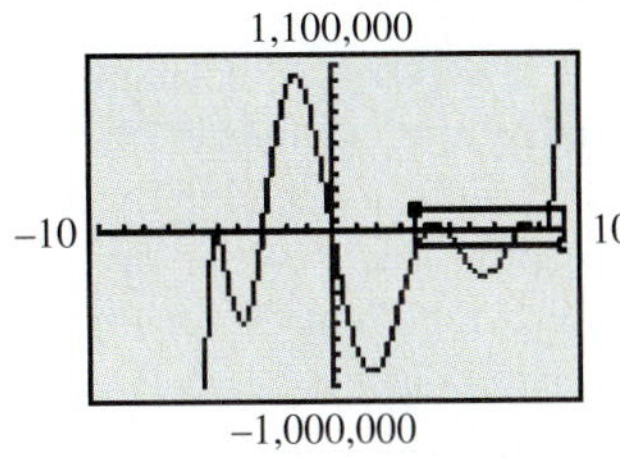

Figure A

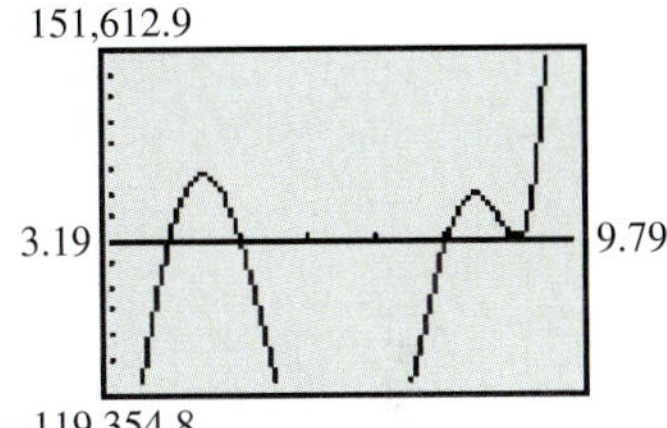

Figure B

82. A comprehensive graph of $f(x) = x^4 - 7x^3 + 18x^2 - 22x + 12$ is shown in the two screens, along with displays of the two real zeros. Find the two remaining nonreal complex zeros.

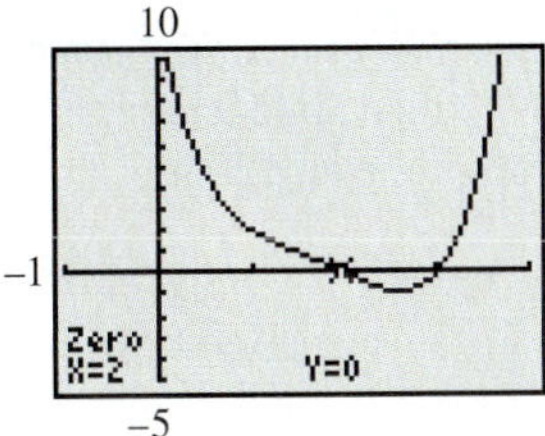

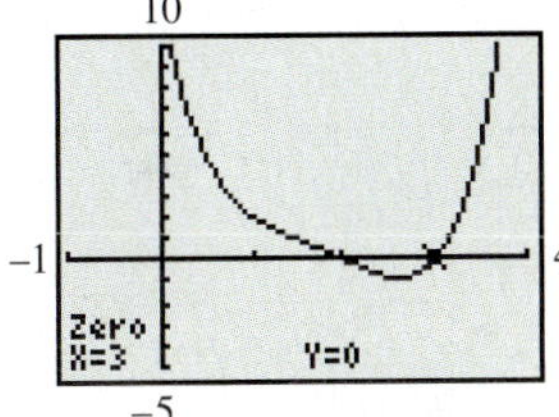

Relating Concepts

For individual or collaborative investigation

(Exercises 83–88)

For any function $y = f(x)$,

(a) the real solutions of $f(x) = 0$ are the x-intercepts of the graph;

(b) the real solutions of $f(x) < 0$ are the x-values for which the graph lies *below* the x-axis; and

(c) the real solutions of $f(x) > 0$ are the x-values for which the graph lies *above* the x-axis.

In Exercises 83–88, a polynomial function defined by $f(x)$ is given in both expanded and factored forms. Graph the function, and solve the equations and inequalities. Give multiplicities of solutions when applicable.

83. $f(x) = x^3 - 3x^2 - 6x + 8$
$= (x - 4)(x - 1)(x + 2)$

(a) $f(x) = 0$ **(b)** $f(x) < 0$
(c) $f(x) > 0$

84. $f(x) = x^3 + 4x^2 - 11x - 30$
$= (x - 3)(x + 2)(x + 5)$

(a) $f(x) = 0$ **(b)** $f(x) < 0$
(c) $f(x) > 0$

85. $f(x) = 2x^4 - 9x^3 - 5x^2 + 57x - 45$
$= (x-3)^2(2x+5)(x-1)$

(a) $f(x) = 0$ **(b)** $f(x) < 0$
(c) $f(x) > 0$

86. $f(x) = 4x^4 + 27x^3 - 42x^2$
$- 445x - 300$
$= (x+5)^2(4x+3)(x-4)$

(a) $f(x) = 0$ **(b)** $f(x) < 0$
(c) $f(x) > 0$

87. $f(x) = -x^4 - 4x^3 + 3x^2 + 18x$
$= x(2-x)(x+3)^2$

(a) $f(x) = 0$ **(b)** $f(x) \geq 0$
(c) $f(x) \leq 0$

88. $f(x) = -x^4 + 2x^3 + 8x^2$
$= x^2(4-x)(x+2)$

(a) $f(x) = 0$ **(b)** $f(x) \geq 0$
(c) $f(x) \leq 0$

(Modeling) Exercises 89–94 are geometric in nature, and lead to polynomial models. Solve each problem.

89. *Volume of a Box* A rectangular piece of cardboard measuring 12 in. by 18 in. is to be made into a box with an open top by cutting equal size squares from each corner and folding up the sides. Let x represent the length of a side of each such square in inches.

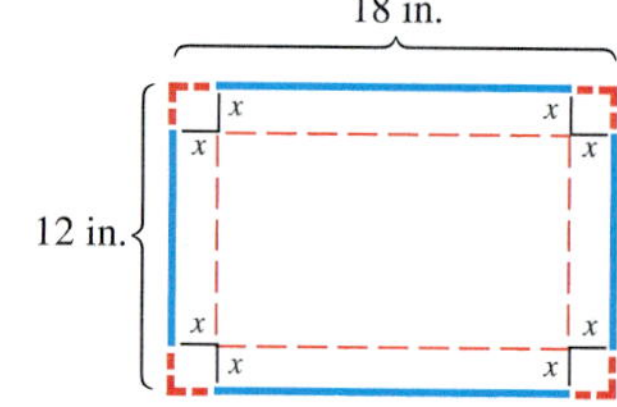

(a) Give the restrictions on x.
(b) Determine a function V that gives the volume of the box as a function of x.
(c) For what value of x will the volume be a maximum? What is this maximum volume? (*Hint:* Use the function of a graphing calculator that allows you to determine a maximum point within a given interval.)
(d) For what values of x will the volume be greater than 80 in.^{3}?

90. *Construction of a Rain Gutter* A piece of rectangular sheet metal is 20 in. wide. It is to be made into a rain gutter by turning up the edges to form parallel sides. Let x represent the length of each of the parallel sides.

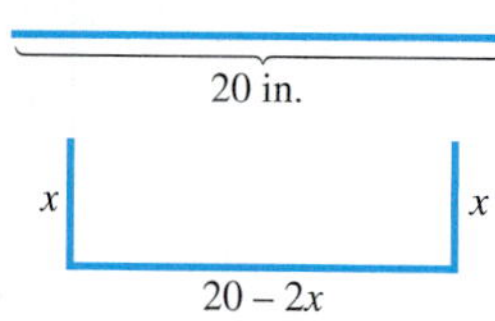

(a) Give the restrictions on x.
(b) Determine a function A that gives the area of a cross section of the gutter.
(c) For what value of x will A be a maximum (and thus maximize the amount of water that the gutter will hold)? What is this maximum area?
(d) For what values of x will the area of a cross section be less than 40 in.^{2}?

91. *Sides of a Right Triangle* A certain right triangle has area 84 in.^{2}. One leg of the triangle measures 1 in. less than the hypotenuse. Let x represent the length of the hypotenuse.

(a) Express the length of the leg mentioned above in terms of x. Give the domain of x.
(b) Express the length of the other leg in terms of x.
(c) Write an equation based on the information determined thus far. Square both sides and then write the equation with one side as a polynomial with integer coefficients, in descending powers, and the other side equal to 0.
(d) Solve the equation in part (c) graphically. Find the lengths of the three sides of the triangle.

92. ***Area of a Rectangle*** Find the value of x in the figure that will maximize the area of rectangle $ABCD$.

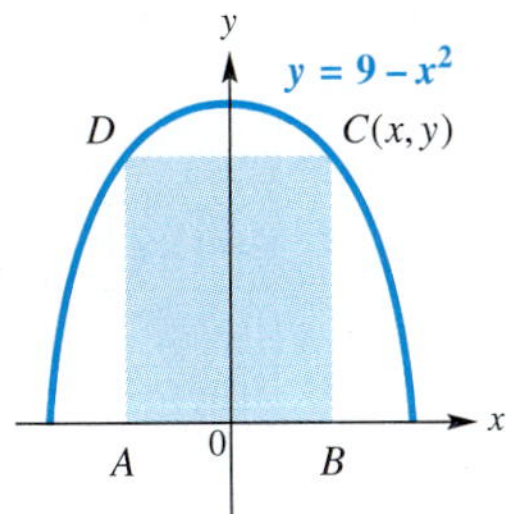

93. ***Butane Gas Storage*** A storage tank for butane gas is to be built in the shape of a right circular cylinder of altitude 12 ft, with a half sphere attached to each end. If x represents the radius of each half sphere, what radius should be used to cause the volume of the tank to be 144π ft^3?

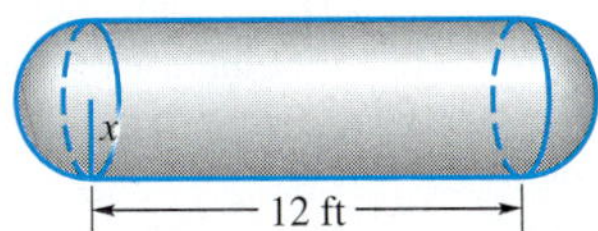

94. ***Volume of a Box*** A standard piece of notebook paper measuring 8.5 in. by 11 in. is to be made into a box with an open top by cutting equal-size squares from each corner and folding up the sides. Let x represent the length of a side of each such square in inches.

(a) Use the table feature of your graphing calculator to find the maximum volume of the box.

(b) Use the table feature to determine when the volume of the box will be greater than 40 in.^{3}.

95. ***Floating Ball*** The polynomial function defined by

$$f(x) = \frac{\pi}{3}x^3 - 5\pi x^2 + \frac{500\pi d}{3}$$

can be used to find the depth that a ball 10 cm in diameter sinks in water. The constant d is the density of the ball, where the density of water is 1. The smallest *positive* zero of $f(x)$ equals the depth that the ball sinks. Approximate this depth for each material and interpret the results.

(a) A wooden ball with $d = .8$

(b) A solid aluminum ball with $d = 2.7$

(c) A spherical water balloon with $d = 1$

96. ***Floating Ball*** Refer to Exercise 95. If a ball has a 20-cm diameter, then the function becomes

$$f(x) = \frac{\pi}{3}x^3 - 10\pi x^2 + \frac{4000\pi d}{3}.$$

This function can be used to determine the depth that the ball sinks in water. Find the depth that this size ball sinks when $d = .6$.

(Modeling) *Solve each problem involving a polynomial function model. See Example 8.*

97. ***Highway Design*** To allow enough distance for cars to pass on two-lane highways, engineers calculate minimum sight distances between curves and hills. The table shows the minimum sight distance y in feet for a car traveling at x miles per hour.

x (in mph)	20	30	40	50	60	65	70
y (in feet)	810	1090	1480	1840	2140	2310	2490

Source: Haefner, L., *Introduction to Transportation Systems,* Holt, Rinehart and Winston, 1986.

(a) Make a scatter diagram of the data.
(b) Use the regression feature of a calculator to find the best-fitting linear function for the data. Graph the function with the data.
(c) Repeat part (b) for a cubic function.
(d) Estimate the minimum sight distance for a car traveling 43 mph using the functions from parts (b) and (c).
(e) Which function models the data better? Why?

98. ***Water Pollution*** In one study, freshwater mussels were used to monitor copper discharge into a river from an electroplating works. Copper in high doses can be lethal to aquatic life. The table lists copper concentrations in mussels after 45 days at various distances downstream from the plant. The concentration C is measured in micrograms of copper per gram of mussel x kilometers downstream.

x	5	21	37	53	59
C	20	13	9	6	5

Source: Foster, R., and J. Bates, "Use of mussels to monitor point source industrial discharges," *Eviron. Sci. Technol.*; Mason, C., *Biology of Freshwater Pollution,* John Wiley and Sons, 1991.

(a) Make a scatter diagram of the data.
(b) Use the regression feature of a calculator to find the best-fitting quadratic function for the data. Graph the function with the data.
(c) Repeat part (b) for a cubic function.
(d) By comparing graphs of the functions in parts (b) and (c) with the data, decide which function best fits the given data.
(e) Concentrations above 10 are lethal to mussels. Find the values of x (using the cubic function) for which this is the case.

99. ***Government Spending on Research*** The table lists the annual amount (in billions of dollars) spent by the federal government on research programs at universities and related institutions. Which one of the following provides the best model for these data, where x represents the year?

Year	Amount	Year	Amount
1995	15.7	1999	20.0
1996	16.3	2000	21.7
1997	17.3	2001	25.2
1998	18.5	2002	25.9

Source: National Center for Educational Statistics.

A. $f(x) = .5(x - 1995)^2 + 15$
B. $g(x) = 1.6(x - 1995) + 15$
C. $h(x) = 4\sqrt{x - 1995} + 15$

100. *Swing of a Pendulum* A simple pendulum will swing back and forth in regular time intervals. Grandfather clocks use pendulums to keep accurate time. The relationship between the length of a pendulum L and the time T for one complete oscillation can be expressed by the equation $L = kT^n$, where k is a constant and n is a positive integer to be determined. The data in the table were taken for different lengths of pendulums.

L (ft)	T (sec)	L (ft)	T (sec)
1.0	1.11	3.0	1.92
1.5	1.36	3.5	2.08
2.0	1.57	4.0	2.22
2.5	1.76		

(a) As the length of the pendulum increases, what happens to T?
(b) Discuss how n and k could be found.
(c) Use the data to approximate k and determine the best value for n.
(d) Using the values of k and n from part (c), predict T for a pendulum having length 5 ft.
(e) If the length L of a pendulum doubles, what happens to the period T?

Summary Exercises on Polynomial Functions, Zeros, and Graphs

In this chapter we have studied many characteristics of polynomial functions. This set of review exercises is designed to put these ideas together.

For each polynomial function, do the following in order.

(a) Use Descartes' rule of signs to find the possible number of positive and negative real zeros.
(b) Use the rational zeros theorem to determine the possible rational zeros of the function.
(c) Find the rational zeros, if any.
(d) Find all other real zeros, if any.
(e) Find any other complex zeros (that is, zeros that are not real), if any.
(f) Find the x-intercepts of the graph, if any.
(g) Find the y-intercept of the graph.
(h) Use synthetic division to find $f(4)$, and give the coordinates of the corresponding point on the graph.
(i) Determine the end behavior of the graph.
(j) Sketch the graph.

1. $f(x) = x^4 + 3x^3 - 3x^2 - 11x - 6$

2. $f(x) = -2x^5 + 5x^4 + 34x^3 - 30x^2 - 84x + 45$

3. $f(x) = 2x^5 - 10x^4 + x^3 - 5x^2 - x + 5$

4. $f(x) = 3x^4 - 4x^3 - 22x^2 + 15x + 18$

5. $f(x) = -2x^4 - x^3 + x + 2$

6. $f(x) = 4x^5 + 8x^4 + 9x^3 + 27x^2 + 27x$ (*Hint:* Factor out x first.)

7. $f(x) = 3x^4 - 14x^2 - 5$ (*Hint:* Factor the polynomial.)

8. $f(x) = -x^5 - x^4 + 10x^3 + 10x^2 - 9x - 9$

9. $f(x) = -3x^4 + 22x^3 - 55x^2 + 52x - 12$

10. For the polynomial functions in Exercises 1–9 that have irrational zeros, find approximations to the nearest thousandth.

3.5 Rational Functions: Graphs, Applications, and Models

The Reciprocal Function $f(x) = \frac{1}{x}$ ▪ The Function $f(x) = \frac{1}{x^2}$ ▪ Asymptotes ▪ Steps for Graphing Rational Functions ▪ Rational Function Models

X	Y1
-1	-1
-.1	-10
-.01	-100
-.001	-1000
-1E-4	-10000
-1E-5	-1E5
-1E-6	-1E6

Y1=1/X

As X approaches 0 from the left, $Y_1 = \frac{1}{X}$ approaches $-\infty$. (–1E–6 means -1×10^{-6}, and –1E6 means -1×10^{6}.)

X	Y1
1	1
.1	10
.01	100
.001	1000
1E-4	10000
1E-5	100000
1E-6	1E6

Y1=1/X

As X approaches 0 from the right, $Y_1 = \frac{1}{X}$ approaches ∞.

Figure 37

A rational expression is a fraction that is the quotient of two polynomials. A function defined by a rational expression is called a *rational function.*

Rational Function

A function f of the form

$$f(x) = \frac{p(x)}{q(x)},$$

where $p(x)$ and $q(x)$ are polynomials, with $q(x) \neq 0$, is called a **rational function.**

$$f(x) = \frac{1}{x}, \qquad f(x) = \frac{x + 1}{2x^2 + 5x - 3}, \qquad f(x) = \frac{3x^2 - 3x - 6}{x^2 + 8x + 16}$$

Rational functions

Since any values of x such that $q(x) = 0$ are excluded from the domain of a rational function, this type of function often has a *discontinuous* graph, that is, a graph that has one or more breaks in it. (See Chapter 2.)

The Reciprocal Function $f(x) = \frac{1}{x}$ The simplest rational function with a variable denominator is the **reciprocal function,** defined by

$$f(x) = \frac{1}{x}.$$

The domain of this function is the set of all real numbers except 0. The number 0 cannot be used as a value of x, but it is helpful to find values of $f(x)$ for some values of x close to 0. We use the table feature of a graphing calculator to do this. The tables in Figure 37 suggest that $|f(x)|$ gets larger and larger as x gets closer and closer to 0, which is written in symbols as

$$|f(x)| \to \infty \quad \text{as} \quad x \to 0.$$

(The symbol $x \to 0$ means that x approaches 0, without necessarily ever being equal to 0.) Since x cannot equal 0, the graph of $f(x) = \frac{1}{x}$ will never intersect the vertical line $x = 0$. This line is called a **vertical asymptote.**

On the other hand, as $|x|$ gets larger and larger, the values of $f(x) = \frac{1}{x}$ get closer and closer to 0, as shown in the tables in Figure 38. Letting $|x|$ get larger and larger without bound (written $|x| \to \infty$) causes the graph of $f(x) = \frac{1}{x}$ to move closer and closer to the horizontal line $y = 0$. This line is called a **horizontal asymptote.**

X	Y1
1	1
10	.1
100	.01
1000	.001
10000	1E-4
100000	1E-5
1E6	1E-6

Y1=1/X

As X approaches ∞, $Y_1 = \frac{1}{X}$ approaches 0 through positive values.

X	Y1
-1	-1
-10	-.1
-100	-.01
-1000	-.001
-10000	-1E-4
-1E5	-1E-5
-1E6	-1E-6

Y1=1/X

As X approaches $-\infty$, $Y_1 = \frac{1}{X}$ approaches 0 through negative values.

Figure 38

The graph and important features of $f(x) = \frac{1}{x}$ are summarized in the following box and shown in Figure 39.

RECIPROCAL FUNCTION $f(x) = \dfrac{1}{x}$

Domain: $(-\infty, 0) \cup (0, \infty)$ Range: $(-\infty, 0) \cup (0, \infty)$

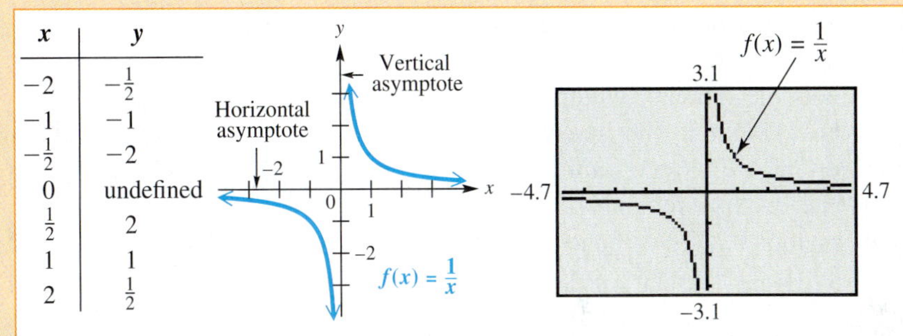

x	y
-2	$-\frac{1}{2}$
-1	-1
$-\frac{1}{2}$	-2
0	undefined
$\frac{1}{2}$	2
1	1
2	$\frac{1}{2}$

Figure 39

- $f(x) = \frac{1}{x}$ decreases on the intervals $(-\infty, 0)$ and $(0, \infty)$.
- It is discontinuous at $x = 0$.
- The y-axis is a vertical asymptote, and the x-axis is a horizontal asymptote.
- It is an odd function, and its graph is symmetric with respect to the origin.

The graph of $y = \frac{1}{x}$ can be translated and reflected in the same way as other basic graphs in Chapter 2.

EXAMPLE 1 Graphing a Rational Function

Graph $y = -\dfrac{2}{x}$. Give the domain and range.

Solution The expression $-\frac{2}{x}$ can be written as $-2\left(\frac{1}{x}\right)$ or $2\left(\frac{1}{-x}\right)$, indicating that the graph may be obtained by stretching the graph of $y = \frac{1}{x}$ vertically by a factor of 2 and reflecting it across either the y-axis or x-axis. The x- and y-axes remain the horizontal and vertical asymptotes. The domain and range are both still $(-\infty, 0) \cup (0, \infty)$. See Figure 40.

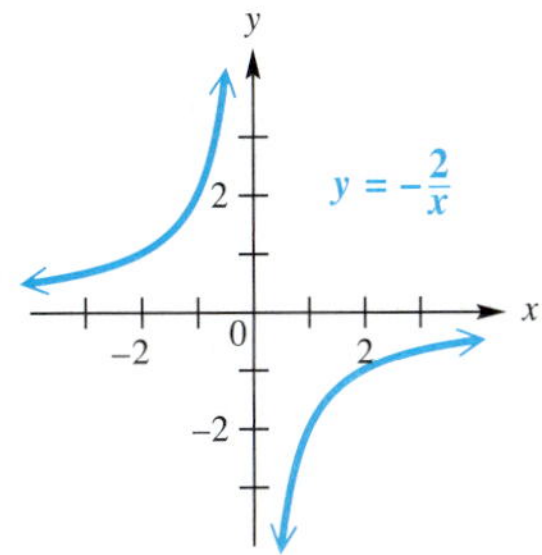

Figure 40

Now try Exercise 17.

$y = -\frac{2}{x}$ 3.1 −4.7 4.7 −3.1

The graph in Figure 40 is shown here using a *decimal window*. Using a nondecimal *window may* produce an extraneous vertical line that is not part of the graph.

EXAMPLE 2 Graphing a Rational Function

Graph $f(x) = \dfrac{2}{x+1}$. Give the domain and range.

Algebraic Solution

The expression $\frac{2}{x+1}$ can be written as $2\left(\frac{1}{x+1}\right)$, indicating that the graph may be obtained by shifting the graph of $y = \frac{1}{x}$ to the left 1 unit and stretching it vertically by a factor of 2. The graph is shown in Figure 41. The horizontal shift affects the domain, which is now $(-\infty, -1) \cup (-1, \infty)$. The line $x = -1$ is the vertical asymptote, and the line $y = 0$ (the x-axis) remains the horizontal asymptote. The range is still $(-\infty, 0) \cup (0, \infty)$.

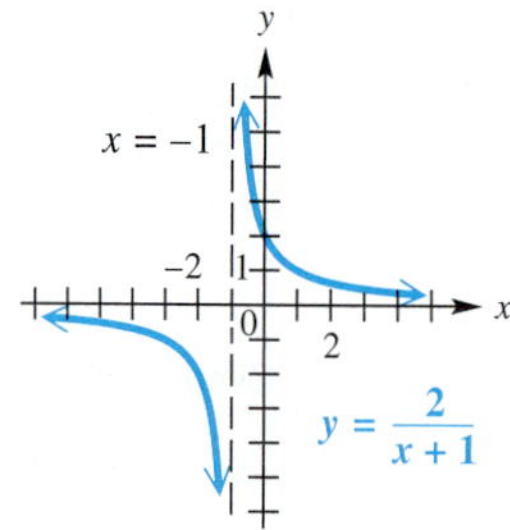

Figure 41

Graphing Calculator Solution

If a calculator is in *connected mode,* the graph it generates of a rational function may show a vertical line for each vertical asymptote. While this may be interpreted as a graph of the asymptote, *dot mode* produces a more realistic graph. See Figure 42.

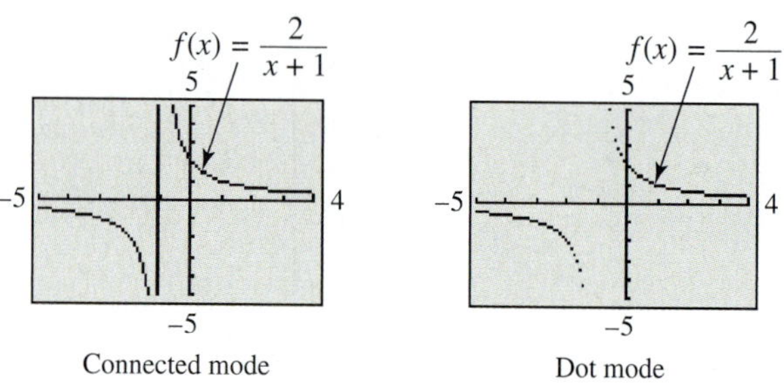

Connected mode Dot mode

Figure 42

We can eliminate the vertical line in the connected mode graph by changing the viewing window for x slightly to $[-5, 3]$. This works because it places the asymptote exactly in the center of the domain. You may need to experiment with various modes and windows to obtain an accurate calculator graph of a rational function.

Now try Exercise 19.

The Function $f(x) = \frac{1}{x^2}$

The rational function defined by

$$f(x) = \frac{1}{x^2}$$

also has domain $(-\infty, 0) \cup (0, \infty)$. We can use the table feature of a graphing calculator to examine values of $f(x)$ for some x-values close to 0. See Figure 43.

X	Y1
-1	1
-.1	100
-.01	10000
-.001	1E6
-1E-4	1E8
-1E-5	1E10
-1E-6	1E12

Y1 = 1/X²

As X approaches 0 from the left, $Y_1 = \frac{1}{X^2}$ approaches ∞.

X	Y1
1	1
.1	100
.01	10000
.001	1E6
1E-4	1E8
1E-5	1E10
1E-6	1E12

Y1 = 1/X²

As X approaches 0 from the right, $Y_1 = \frac{1}{X^2}$ approaches ∞.

Figure 43

The tables suggest that $f(x)$ gets larger and larger as x gets closer and closer to 0. Notice that as x approaches 0 from *either* side, function values are all positive and there is symmetry with respect to the y-axis. Thus, $f(x) \to \infty$ as $x \to 0$. The y-axis ($x = 0$) is the vertical asymptote.

X	Y1
1	1
10	.01
100	1E-4
1000	1E-6
10000	1E-8
100000	1E-10
1E6	1E-12

Y1=1/X²

As X approaches ∞, $Y_1 = \frac{1}{X^2}$ approaches 0 through positive values.

X	Y1
-1	1
-10	.01
-100	1E-4
-1000	1E-6
-10000	1E-8
-1E5	1E-10
-1E6	1E-12

Y1=1/X²

As X approaches $-\infty$, $Y_1 = \frac{1}{X^2}$ approaches 0 through positive values.

Figure 44

As $|x|$ gets larger and larger, $f(x)$ approaches 0, as suggested by the tables in Figure 44. Again, function values are all positive. The x-axis is the horizontal asymptote of the graph.

The graph and important features of $f(x) = \frac{1}{x^2}$ are summarized in the following box and shown in Figure 45.

RATIONAL FUNCTION $f(x) = \frac{1}{x^2}$

Domain: $(-\infty, 0) \cup (0, \infty)$ Range: $(0, \infty)$

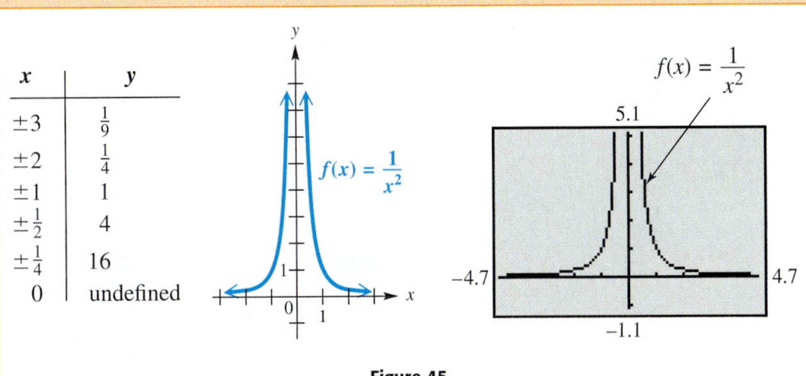

x	y
± 3	$\frac{1}{9}$
± 2	$\frac{1}{4}$
± 1	1
$\pm \frac{1}{2}$	4
$\pm \frac{1}{4}$	16
0	undefined

Figure 45

- $f(x) = \frac{1}{x^2}$ increases on the interval $(-\infty, 0)$ and decreases on the interval $(0, \infty)$.
- It is discontinuous at $x = 0$.
- The y-axis is a vertical asymptote, and the x-axis is a horizontal asymptote.
- It is an even function, and its graph is symmetric with respect to the y-axis.

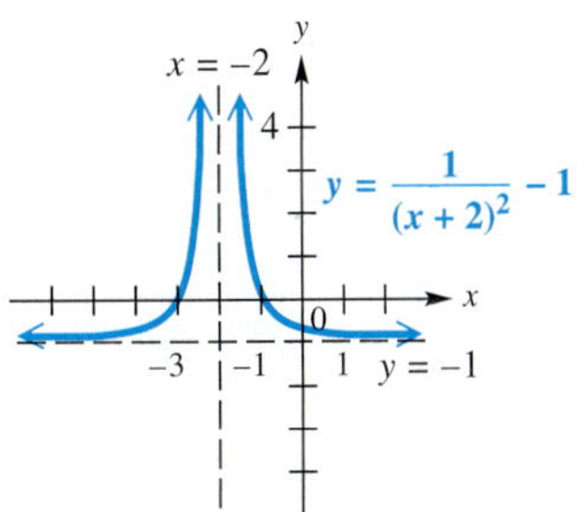

Figure 46

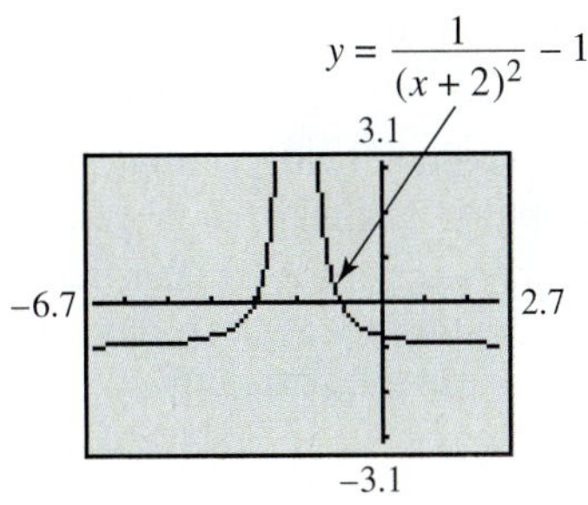

A choice of window other than the one here *may* produce an undesired vertical line at $x = -2$.

Figure 47

EXAMPLE 3 Graphing a Rational Function

Graph $y = \frac{1}{(x+2)^2} - 1$. Give the domain and range.

Solution The equation $y = \frac{1}{(x+2)^2} - 1$ is equivalent to

$$y = f(x + 2) - 1,$$

where $f(x) = \frac{1}{x^2}$. This indicates that the graph will be shifted 2 units to the left and 1 unit down. The horizontal shift affects the domain, which is now $(-\infty, -2) \cup (-2, \infty)$, while the vertical shift affects the range, now $(-1, \infty)$. The vertical asymptote has equation $x = -2$, and the horizontal asymptote has equation $y = -1$. A traditional graph is shown in Figure 46, and a calculator graph is shown in Figure 47.

Now try Exercise 27.

Looking Ahead to Calculus

The rational function defined by

$$f(x) = \frac{2}{x+1}$$

in Example 2 has a vertical asymptote at $x = -1$. In calculus, the behavior of the graph of this function for values close to -1 is described using *one-sided limits.* As x approaches -1 from the *left,* the function values decrease without bound: This is written

$$\lim_{x \to -1^-} f(x) = -\infty.$$

As x approaches -1 from the *right* the function values increase without bound: This is written

$$\lim_{x \to -1^+} f(x) = \infty.$$

Asymptotes The preceding examples suggest the following definitions of vertical and horizontal asymptotes.

Asymptotes

Let $p(x)$ and $q(x)$ define polynomials. For the rational function defined by $f(x) = \frac{p(x)}{q(x)}$, written in lowest terms, and for real numbers a and b:

1. If $|f(x)| \to \infty$ as $x \to a$, then the line $x = a$ is a **vertical asymptote.**

2. If $f(x) \to b$ as $|x| \to \infty$, then the line $y = b$ is a **horizontal asymptote.**

Locating asymptotes is important when graphing rational functions. We find vertical asymptotes by determining the values of x that make the denominator equal to 0. To find horizontal asymptotes (and, in some cases, *oblique asymptotes*), we must consider what happens to $f(x)$ as $|x| \to \infty$. These asymptotes determine the end behavior of the graph.

Determining Asymptotes

To find the asymptotes of a rational function defined by a rational expression in *lowest terms,* use the following procedures.

1. Vertical Asymptotes
Find any vertical asymptotes by setting the denominator equal to 0 and solving for x. If a is a zero of the denominator, then the line $x = a$ is a vertical asymptote.

2. Other Asymptotes
Determine any other asymptotes. Consider three possibilities:

(a) If the numerator has lower degree than the denominator, then there is a horizontal asymptote $y = 0$ (the x-axis).

(b) If the numerator and denominator have the same degree, and the function is of the form

$$f(x) = \frac{a_n x^n + \cdots + a_0}{b_n x^n + \cdots + b_0}, \qquad \text{where } a_n, b_n \neq 0,$$

then the horizontal asymptote has equation $y = \frac{a_n}{b_n}$.

(c) If the numerator is of degree exactly one more than the denominator, then there will be an oblique (slanted) asymptote. To find it, divide the numerator by the denominator and disregard the remainder. Set the rest of the quotient equal to y to obtain the equation of the asymptote.

NOTE The graph of a rational function may have more than one vertical asymptote, or it may have none at all. The graph cannot intersect any vertical asymptote. There can be at most one other (nonvertical) asymptote, and the graph *may* intersect that asymptote as we shall see in Example 7.

EXAMPLE 4 Finding Asymptotes of Graphs of Rational Functions

For each rational function f, find all asymptotes.

(a) $f(x) = \dfrac{x + 1}{(2x - 1)(x + 3)}$ **(b)** $f(x) = \dfrac{2x + 1}{x - 3}$ **(c)** $f(x) = \dfrac{x^2 + 1}{x - 2}$

Solution

(a) To find the vertical asymptotes, set the denominator equal to 0 and solve.

$$(2x - 1)(x + 3) = 0$$

$$2x - 1 = 0 \quad \text{or} \quad x + 3 = 0 \qquad \text{Zero-factor property (Section 1.4)}$$

$$x = \frac{1}{2} \quad \text{or} \quad x = -3 \qquad \text{Solve each equation. (Section 1.1)}$$

The equations of the vertical asymptotes are $x = \frac{1}{2}$ and $x = -3$.

To find the equation of the horizontal asymptote, divide each term by the largest power of x in the expression. First, multiply the factors in the denominator.

$$f(x) = \frac{x + 1}{(2x - 1)(x + 3)} = \frac{x + 1}{2x^2 + 5x - 3}$$

Now divide each term in the numerator and denominator by x^2 since 2 is the greatest power of x.

$$f(x) = \frac{\frac{x}{x^2} + \frac{1}{x^2}}{\frac{2x^2}{x^2} + \frac{5x}{x^2} - \frac{3}{x^2}} = \frac{\frac{1}{x} + \frac{1}{x^2}}{2 + \frac{5}{x} - \frac{3}{x^2}}$$

As $|x|$ gets larger and larger, the quotients $\frac{1}{x}$, $\frac{1}{x^2}$, $\frac{5}{x}$, and $\frac{3}{x^2}$ all approach 0, and the value of $f(x)$ approaches

$$\frac{0 + 0}{2 + 0 - 0} = \frac{0}{2} = 0.$$

The line $y = 0$ (that is, the x-axis) is therefore the horizontal asymptote.

(b) Set the denominator $x - 3$ equal to 0 to find that the vertical asymptote has equation $x = 3$. To find the horizontal asymptote, divide each term in the rational expression by x since the greatest power of x in the expression is 1.

$$f(x) = \frac{2x + 1}{x - 3} = \frac{\frac{2x}{x} + \frac{1}{x}}{\frac{x}{x} - \frac{3}{x}} = \frac{2 + \frac{1}{x}}{1 - \frac{3}{x}}$$

As $|x|$ gets larger and larger, both $\frac{1}{x}$ and $\frac{3}{x}$ approach 0, and $f(x)$ approaches

$$\frac{2 + 0}{1 - 0} = \frac{2}{1} = 2,$$

so the line $y = 2$ is the horizontal asymptote.

(c) Setting the denominator $x - 2$ equal to 0 shows that the vertical asymptote has equation $x = 2$. If we divide by the greatest power of x as before (x^2 in this case), we see that there is no horizontal asymptote because

$$f(x) = \frac{x^2 + 1}{x - 2} = \frac{\frac{x^2}{x^2} + \frac{1}{x^2}}{\frac{x}{x^2} - \frac{2}{x^2}} = \frac{1 + \frac{1}{x^2}}{\frac{1}{x} - \frac{2}{x^2}}$$

does not approach any real number as $|x| \to \infty$ since $\frac{1}{0}$ is undefined. This happens whenever the degree of the numerator is greater than the degree of the denominator. In such cases, divide the denominator into the numerator to write the expression in another form. We use synthetic division, as shown in the margin. The result allows us to write the function as

$$\begin{array}{r|rrr} 2 & 1 & 0 & 1 \\ & & 2 & 4 \\ \hline & 1 & 2 & 5 \end{array}$$ (Section 3.2)

$$f(x) = x + 2 + \frac{5}{x - 2}.$$

For very large values of $|x|$, $\frac{5}{x-2}$ is close to 0, and the graph approaches the line $y = x + 2$. This line is an **oblique asymptote** (slanted, neither vertical nor horizontal) for the graph of the function.

Now try Exercises 37, 39, and 41.

Steps for Graphing Rational Functions A comprehensive graph of a rational function exhibits these features:

1. all x- and y-intercepts;
2. all asymptotes: vertical, horizontal, and/or oblique;
3. the point at which the graph intersects its nonvertical asymptote (if there is any such point);
4. enough of the graph to exhibit the correct end behavior.

Graphing a Rational Function

Let $f(x) = \frac{p(x)}{q(x)}$ define a function where $p(x)$ and $q(x)$ are polynomials and the rational expression is written in lowest terms. To sketch its graph, follow these steps.

Step 1 Find any vertical asymptotes.

Step 2 Find any horizontal or oblique asymptotes.

Step 3 Find the y-intercept by evaluating $f(0)$.

Step 4 Find the x-intercepts, if any, by solving $f(x) = 0$. (These will be the zeros of the numerator, $p(x)$.)

Step 5 Determine whether the graph will intersect its nonvertical asymptote $y = b$ or $y = mx + b$ by solving $f(x) = b$ or $f(x) = mx + b$.

Step 6 Plot selected points, as necessary. Choose an x-value in each domain interval determined by the vertical asymptotes and x-intercepts.

Step 7 Complete the sketch.

EXAMPLE 5 Graphing a Rational Function with the x-Axis as Horizontal Asymptote

Graph $f(x) = \dfrac{x + 1}{2x^2 + 5x - 3}$.

Solution

Step 1 Since $2x^2 + 5x - 3 = (2x - 1)(x + 3)$, from Example 4(a), the vertical asymptotes have equations $x = \frac{1}{2}$ and $x = -3$.

Step 2 Again, as shown in Example 4(a), the horizontal asymptote is the x-axis.

Step 3 The y-intercept is $-\frac{1}{3}$, since

$$f(0) = \frac{0 + 1}{2(0)^2 + 5(0) - 3} = -\frac{1}{3}.$$

Step 4 The x-intercept is found by solving $f(x) = 0$.

$$\frac{x + 1}{2x^2 + 5x - 3} = 0$$

$$x + 1 = 0 \quad \text{If a rational expression is equal to 0, then its numerator must equal 0.}$$

$$x = -1 \quad \text{The } x\text{-intercept is } -1.$$

Step 5 To determine whether the graph intersects its horizontal asymptote, solve

$$f(x) = 0. \leftarrow y\text{-value of horizontal asymptote}$$

Since the horizontal asymptote is the x-axis, the solution of this equation was found in Step 4. The graph intersects its horizontal asymptote at $(-1, 0)$.

Step 6 Plot a point in each of the intervals determined by the x-intercepts and vertical asymptotes, $(-\infty, -3)$, $(-3, -1)$, $\left(-1, \frac{1}{2}\right)$ and $\left(\frac{1}{2}, \infty\right)$, to get an idea of how the graph behaves in each interval.

Interval	Test Point	Value of $f(x)$	Sign of $f(x)$	Graph Above or Below x-Axis
$(-\infty, -3)$	-4	$-\frac{1}{3}$	Negative	Below
$(-3, -1)$	-2	$\frac{1}{5}$	Positive	Above
$\left(-1, \frac{1}{2}\right)$	0	$-\frac{1}{3}$	Negative	Below
$\left(\frac{1}{2}, \infty\right)$	2	$\frac{1}{5}$	Positive	Above

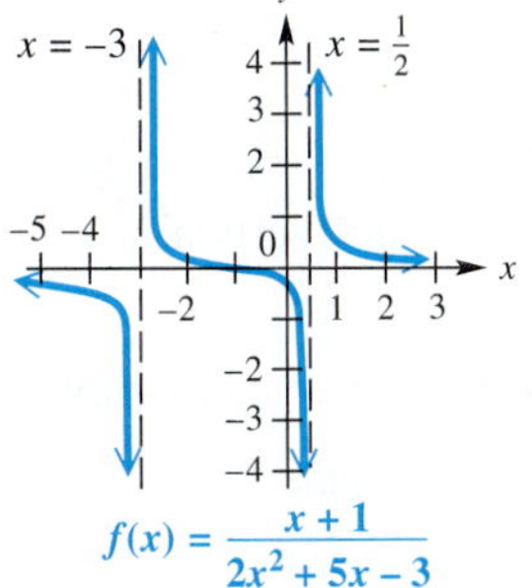

Figure 48

Step 7 Complete the sketch as shown in Figure 48.

Now try Exercise 53.

EXAMPLE 6 Graphing a Rational Function That Does Not Intersect Its Horizontal Asymptote

Graph $f(x) = \dfrac{2x + 1}{x - 3}$.

Solution

Steps 1 and 2 As determined in Example 4(b), the equation of the vertical asymptote is $x = 3$. The horizontal asymptote has equation $y = 2$.

Step 3 $f(0) = -\frac{1}{3}$, so the y-intercept is $-\frac{1}{3}$.

Step 4 Solve $f(x) = 0$ to find any x-intercepts.

$$\frac{2x + 1}{x - 3} = 0$$

$$2x + 1 = 0$$

$$x = -\frac{1}{2} \quad x\text{-intercept}$$

Step 5 The graph does not intersect its horizontal asymptote since $f(x) = 2$ has no solution. (Verify this.)

Steps 6 and 7 The points $(-4, 1)$, $\left(1, -\frac{3}{2}\right)$, and $\left(6, \frac{13}{3}\right)$ are on the graph and can be used to complete the sketch, seen in Figure 49.

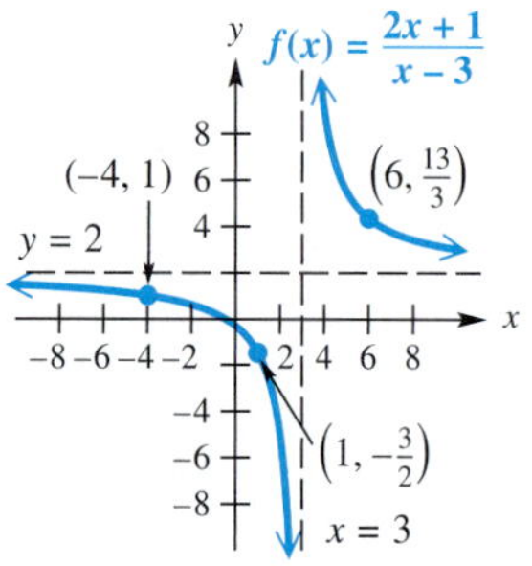

Figure 49

Now try Exercise 51.

Looking Ahead to Calculus

The rational function defined by

$$f(x) = \frac{2x + 1}{x - 3},$$

seen in Example 6, has horizontal asymptote $y = 2$. In calculus, the behavior of the graph of this function as x approaches $-\infty$ and as x approaches ∞ is described using *limits at infinity.* As x approaches $-\infty$, $f(x)$ approaches 2. This is written

$$\lim_{x \to -\infty} f(x) = 2.$$

As x approaches ∞, $f(x)$ approaches 2. This is written

$$\lim_{x \to \infty} f(x) = 2.$$

EXAMPLE 7 Graphing a Rational Function That Intersects Its Horizontal Asymptote

Graph $f(x) = \dfrac{3x^2 - 3x - 6}{x^2 + 8x + 16}$.

Solution

Step 1 To find the vertical asymptote(s), solve $x^2 + 8x + 16 = 0$.

$$x^2 + 8x + 16 = 0$$

$$(x + 4)^2 = 0$$

$$x = -4$$

Since the numerator is not 0 when $x = -4$, the only vertical asymptote has equation $x = -4$.

Step 2 We divide all terms by x^2 to get the equation of the horizontal asymptote,

$$y = \frac{3}{1}, \quad \begin{matrix} \leftarrow \text{Leading coefficient of numerator} \\ \leftarrow \text{Leading coefficient of denominator} \end{matrix}$$

or $y = 3$.

Step 3 The y-intercept is $f(0) = -\frac{3}{8}$.

Step 4 To find the x-intercept(s), if any, we solve $f(x) = 0$.

$$\frac{3x^2 - 3x - 6}{x^2 + 8x + 16} = 0$$

$$3x^2 - 3x - 6 = 0 \quad \text{Set the numerator equal to 0.}$$

$$x^2 - x - 2 = 0 \quad \text{Divide by 3.}$$

$$(x - 2)(x + 1) = 0 \quad \text{Factor. (Section R.4)}$$

$$x = 2 \quad \text{or} \quad x = -1 \quad \text{Zero-factor property}$$

The x-intercepts are -1 and 2.

Step 5 We set $f(x) = 3$ and solve to locate the point where the graph intersects the horizontal asymptote.

$$\frac{3x^2 - 3x - 6}{x^2 + 8x + 16} = 3$$

$$3x^2 - 3x - 6 = 3x^2 + 24x + 48 \quad \text{Multiply by } x^2 + 8x + 16.$$

$$-3x - 6 = 24x + 48 \quad \text{Subtract } 3x^2.$$

$$-27x = 54 \quad \text{Subtract } 24x\text{; add 6.}$$

$$x = -2 \quad \text{Divide by } -27.$$

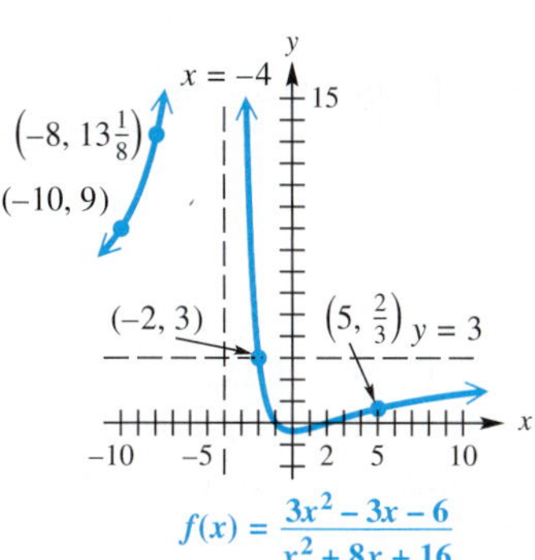

Figure 50

The graph intersects its horizontal asymptote at $(-2, 3)$.

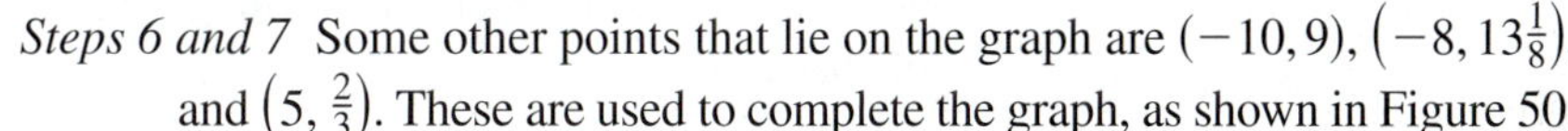

Steps 6 and 7 Some other points that lie on the graph are $(-10, 9)$, $\left(-8, 13\frac{1}{8}\right)$, and $\left(5, \frac{2}{3}\right)$. These are used to complete the graph, as shown in Figure 50.

Now try Exercise 57.

Notice the behavior of the graph of the function in Example 7 in Figure 50 near the line $x = -4$. As $x \to -4$ from either side, $f(x) \to \infty$. On the other hand, if we examine the behavior of the graph of the function in Example 6 in Figure 49 near the line $x = 3$, $f(x) \to -\infty$ as x approaches 3 from the left, while $f(x) \to \infty$ as x approaches 3 from the right. The behavior of the graph of a rational function near a vertical asymptote $x = a$ will partially depend on the exponent on $x - a$ in the denominator.

Behavior of Graphs of Rational Functions Near Vertical Asymptotes

Suppose that $f(x)$ is defined by a rational expression in lowest terms. If n is the largest positive integer such that $(x - a)^n$ is a factor of the denominator of $f(x)$, the graph will behave in the manner illustrated.

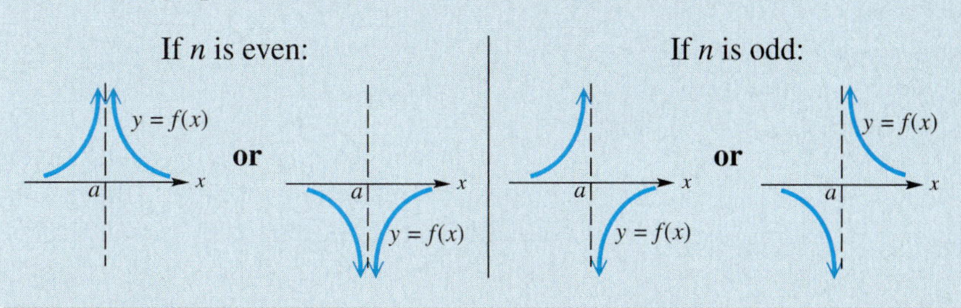

In Section 3.4 we observed that the behavior of the graph of a polynomial function near its zeros is dependent on whether the multiplicity of the zero is even or odd. The same statement can be made for rational functions. Suppose that $f(x)$ is defined by a rational expression in lowest terms. If n is the largest positive integer such that $(x - c)^n$ is a factor of the numerator of $f(x)$, the graph will behave in the manner illustrated.

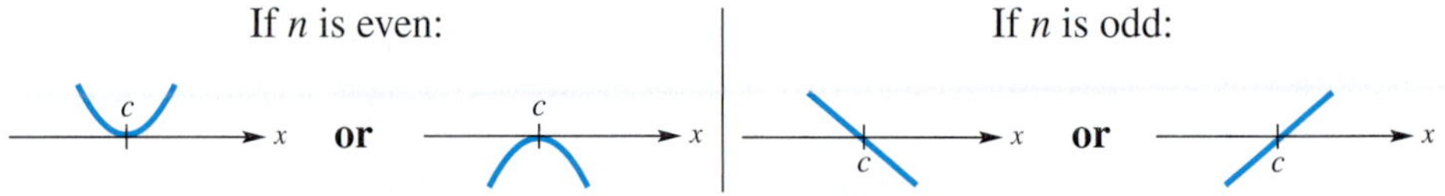

EXAMPLE 8 Graphing a Rational Function with an Oblique Asymptote

Graph $f(x) = \dfrac{x^2 + 1}{x - 2}$.

Solution As shown in Example 4, the vertical asymptote has equation $x = 2$, and the graph has an oblique asymptote with equation $y = x + 2$. Refer to the preceding discussion to determine the behavior near the vertical asymptote $x = 2$. The y-intercept is $-\frac{1}{2}$, and the graph has no x-intercepts since the numerator, $x^2 + 1$, has no real zeros. The graph does not intersect its oblique asymptote because

$$\frac{x^2 + 1}{x - 2} = x + 2$$

has no solution. Using the y-intercept, asymptotes, the points $\left(4, \frac{17}{2}\right)$ and $\left(-1, -\frac{2}{3}\right)$, and the general behavior of the graph near its asymptotes leads to the graph in Figure 51.

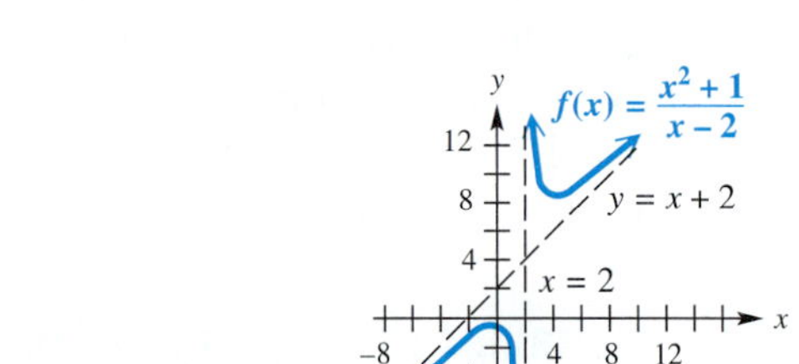

Figure 51

Now try Exercise 63.

Looking Ahead to Calculus

Different types of discontinuity are discussed in calculus. The function in Example 9,

$$f(x) = \frac{x^2 - 4}{x - 2},$$

is said to have a *removeable* discontinuity at $x = 2$, since the discontinuity can be removed by redefining f at 2. The function of Example 8,

$$f(x) = \frac{x^2 + 1}{x - 2},$$

has *infinite* discontinuity at $x = 2$, as indicated by the vertical asymptote there. The greatest integer function, discussed in Section 2.5, has *jump* discontinuities because the function values "jump" from one value to another for integer domain values.

As mentioned earlier, a rational function must be defined by an expression in lowest terms before we can use the methods discussed in this section to determine the graph. A rational function that is not in lowest terms usually has a "hole," or point of discontinuity, in its graph.

EXAMPLE 9 Graphing a Rational Function Defined by an Expression That Is Not in Lowest Terms

$$\text{Graph } f(x) = \frac{x^2 - 4}{x - 2}.$$

Algebraic Solution

The domain of this function cannot include 2. The expression $\frac{x^2 - 4}{x - 2}$ should be written in lowest terms.

$$\begin{aligned} f(x) &= \frac{x^2 - 4}{x - 2} \\ &= \frac{(x + 2)(x - 2)}{x - 2} \quad \text{Factor.} \\ &= x + 2, \quad x \neq 2 \end{aligned}$$

Therefore, the graph of this function will be the same as the graph of $y = x + 2$ (a straight line), with the exception of the point with x-value 2. A "hole" appears in the graph at $(2, 4)$. See Figure 52.

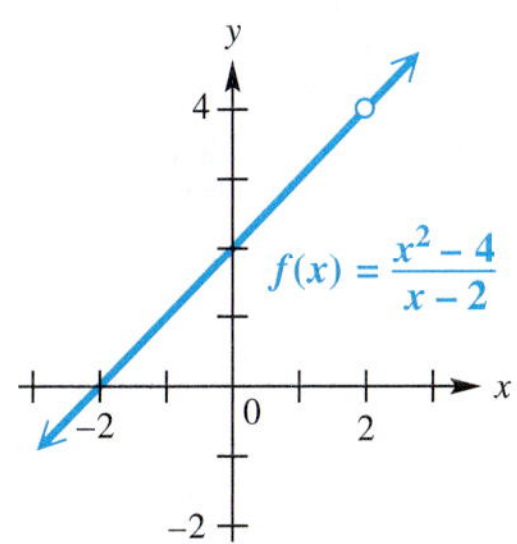

Figure 52

Graphing Calculator Solution

If we set the window of a graphing calculator so that an x-value of 2 is displayed, then we can see that the calculator cannot determine a value for y. We define

$$Y_1 = \frac{X^2 - 4}{X - 2},$$

and graph it in such a window, as in Figure 53. The error message in the table further supports the existence of a discontinuity at $X = 2$. (For the table, $Y_2 = X + 2$.)

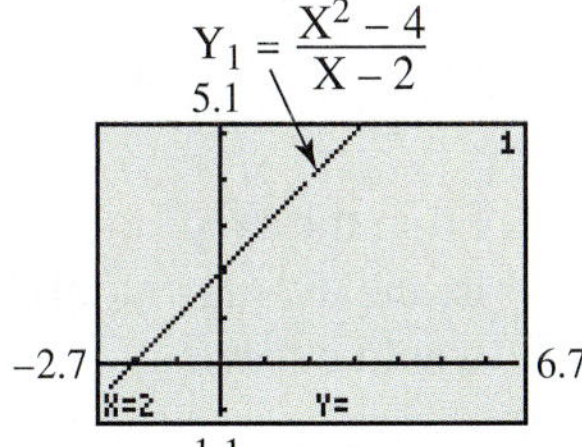

X	Y1	Y2
-1	1	1
0	2	2
1	3	3
2	ERROR	4
3	5	5
4	6	6
5	7	7

Y1=(X²-4)/(X-2)

Figure 53

Notice the visible discontinuity at $X = 2$ in the graph. The window was chosen so the "hole" would be visible. This requires a decimal viewing window or a window with x-values centered at 2. Other window choices may not show this discontinuity.

Now try Exercise 67.

Rational Function Models

Rational functions have a variety of applications.

EXAMPLE 10 Modeling Traffic Intensity with a Rational Function

Vehicles arrive randomly at a parking ramp at an average rate of 2.6 vehicles per minute. The parking attendant can admit 3.2 vehicles per minute. However, since arrivals are random, lines form at various times. (*Source:* Mannering, F. and W. Kilareski, *Principles of Highway Engineering and Traffic Analysis,* 2nd ed., John Wiley & Sons, 1998.)

(a) The *traffic intensity* x is defined as the ratio of the average arrival rate to the average admittance rate. Determine x for this parking ramp.

(b) The average number of vehicles waiting in line to enter the ramp is given by

$$f(x) = \frac{x^2}{2(1 - x)},$$

where $0 \le x < 1$ is the traffic intensity. Graph $f(x)$ and compute $f(.8125)$ for this parking ramp.

(c) What happens to the number of vehicles waiting as the traffic intensity approaches 1?

Solution

(a) The average arrival rate is 2.6 vehicles and the average admittance rate is 3.2 vehicles, so

$$x = \frac{2.6}{3.2} = .8125.$$

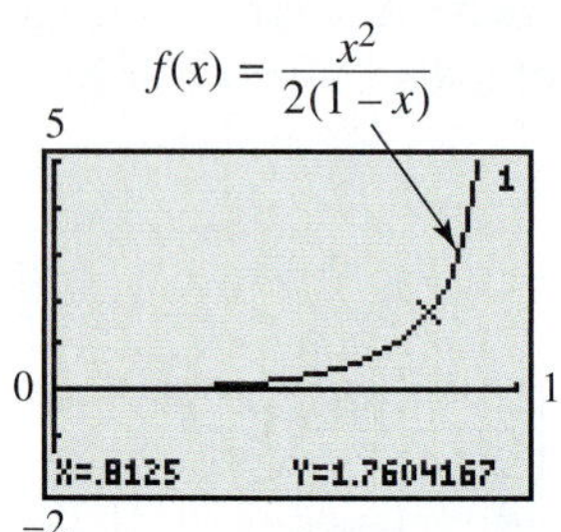

Figure 54

(b) A calculator graph of f is shown in Figure 54.

$$f(.8125) = \frac{.8125^2}{2(1 - .8125)} \approx 1.76 \text{ vehicles}$$

(c) From the graph we see that as x approaches 1, $y = f(x)$ gets very large; that is, the average number of waiting vehicles gets very large. This is what we would expect.

Now try Exercise 79.

3.5 Exercises

Concept Check *Provide a short answer to each question.*

1. What is the domain of $f(x) = \frac{1}{x}$? What is its range?

2. What is the domain of $f(x) = \frac{1}{x^2}$? What is its range?

3. What is the interval over which $f(x) = \frac{1}{x}$ increases? decreases? is constant?

4. What is the interval over which $f(x) = \frac{1}{x^2}$ increases? decreases? is constant?

5. What is the equation of the vertical asymptote of the graph of $y = \frac{1}{x - 3} + 2$? of the horizontal asymptote?

6. What is the equation of the vertical asymptote of the graph of $y = \frac{1}{(x + 2)^2} - 4$? of the horizontal asymptote?

7. Is $f(x) = \frac{1}{x^2}$ an even or odd function? What symmetry does its graph exhibit?

8. Is $f(x) = \frac{1}{x}$ an even or odd function? What symmetry does its graph exhibit?

Concept Check *Use the graphs of the rational functions in A–D to answer each question. Give all possible answers, as there may be more than one correct choice.*

9. Which choices have domain $(-\infty, 3) \cup (3, \infty)$?

10. Which choices have range $(-\infty, 3) \cup (3, \infty)$?

11. Which choices have range $(-\infty, 0) \cup (0, \infty)$?

12. Which choices have range $(0, \infty)$?

13. If f represents the function, only one choice has a single solution to the equation $f(x) = 3$. Which one is it?

14. What is the range of the function in B?

15. Which choices have the x-axis as a horizontal asymptote?

16. Which choices are symmetric with respect to a vertical line?

A.

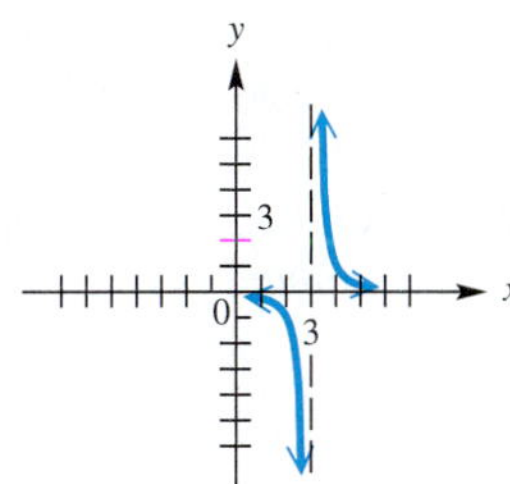

B.

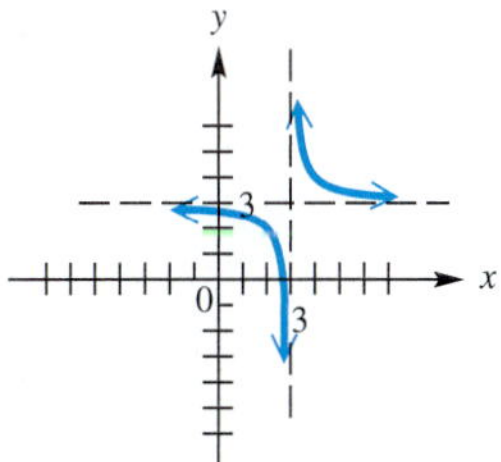

C.

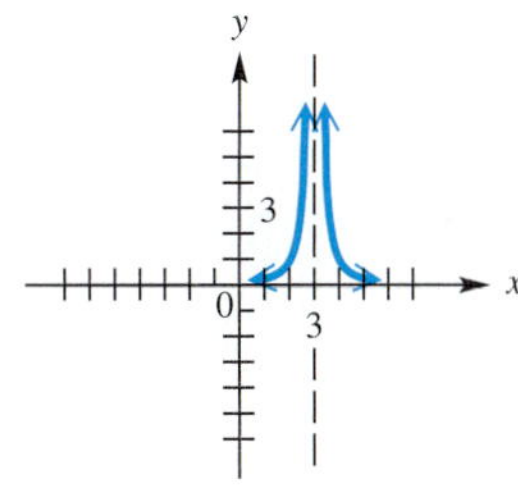

D.

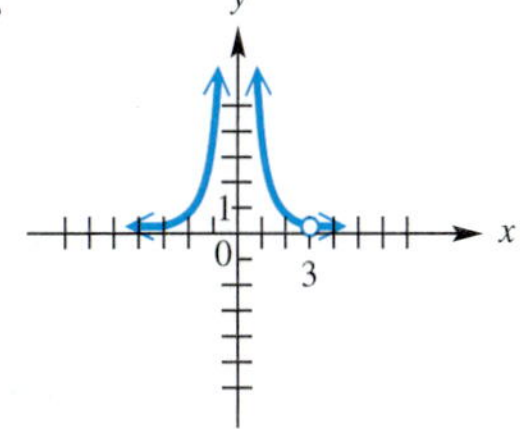

Explain how the graph of each function can be obtained from the graph of $y = \frac{1}{x}$ or $y = \frac{1}{x^2}$. Then graph f and give the domain and range. See Examples 1–3.

17. $f(x) = \frac{2}{x}$

18. $f(x) = -\frac{3}{x}$

19. $f(x) = \frac{1}{x + 2}$

20. $f(x) = \frac{1}{x - 3}$

21. $f(x) = \frac{1}{x} + 1$

22. $f(x) = \frac{1}{x} - 2$

23. $f(x) = -\frac{2}{x^2}$

24. $f(x) = \frac{1}{x^2} + 3$

25. $f(x) = \frac{1}{(x - 3)^2}$

26. $f(x) = \frac{-2}{(x - 3)^2}$

27. $f(x) = \frac{-1}{(x + 2)^2} - 3$

28. $f(x) = \frac{-1}{(x - 4)^2} + 2$

Concept Check *Match the rational function in Column I with the appropriate description in Column II. Choices in Column II can be used only once.*

I	II
29. $f(x) = \frac{x+7}{x+1}$	**A.** The x-intercept is -3.
30. $f(x) = \frac{x+10}{x+2}$	**B.** The y-intercept is 5.
31. $f(x) = \frac{1}{x+4}$	**C.** The horizontal asymptote is $y = 4$.
32. $f(x) = \frac{-3}{x^2}$	**D.** The vertical asymptote is $x = -1$.
33. $f(x) = \frac{x^2-16}{x+4}$	**E.** There is a "hole" in its graph at $x = -4$.
34. $f(x) = \frac{4x+3}{x-7}$	**F.** The graph has an oblique asymptote.
35. $f(x) = \frac{x^2+3x+4}{x-5}$	**G.** The x-axis is its horizontal asymptote, and the y-axis is not its vertical asymptote.
36. $f(x) = \frac{x+3}{x-6}$	**H.** The x-axis is its horizontal asymptote, and the y-axis is its vertical asymptote.

Give the equations of any vertical, horizontal, or oblique asymptotes for the graph of each rational function. See Example 4.

37. $f(x) = \frac{3}{x-5}$

38. $f(x) = \frac{-6}{x+9}$

39. $f(x) = \frac{4-3x}{2x+1}$

40. $f(x) = \frac{2x+6}{x-4}$

41. $f(x) = \frac{x^2-1}{x+3}$

42. $f(x) = \frac{x^2+4}{x-1}$

43. $f(x) = \frac{x^2-2x-3}{2x^2-x-10}$

44. $f(x) = \frac{3x^2-6x-24}{5x^2-26x+5}$

45. $f(x) = \frac{x^2+1}{x^2+9}$

46. $f(x) = \frac{4x^2+25}{x^2+9}$

47. *Concept Check* Let f be the function whose graph is obtained by translating the graph of $y = \frac{1}{x}$ to the right 3 units and up 2 units.

(a) Write an equation for $f(x)$ as a quotient of two polynomials.
(b) Determine the zero(s) of f.
(c) Identify the asymptotes of the graph of $f(x)$.

48. *Concept Check* Repeat Exercise 47 with f the function whose graph is obtained by translating the graph of $y = -\frac{1}{x^2}$ to the left 3 units and up 1 unit.

49. *Concept Check* After the numerator is divided by the denominator,

$$f(x) = \frac{x^5 + x^4 + x^2 + 1}{x^4 + 1} \quad \text{becomes} \quad f(x) = x + 1 + \frac{x^2 - x}{x^4 + 1}.$$

(a) What is the oblique asymptote of the graph of the function?
(b) Where does the graph of the function intersect its asymptote?
(c) As $x \to \infty$, does the graph of the function approach its asymptote from above or below?

50. *Concept Check* The figures below show the four ways that the graph of a rational function can approach the vertical line $x = 2$ as an asymptote. Identify the graph of each rational function defined as follows.

(a) $f(x) = \dfrac{1}{(x-2)^2}$ **(b)** $f(x) = \dfrac{1}{x-2}$

(c) $f(x) = \dfrac{-1}{x-2}$ **(d)** $f(x) = \dfrac{-1}{(x-2)^2}$

A.

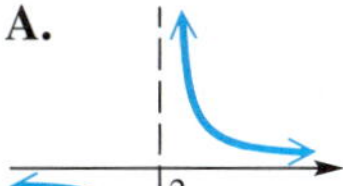

B.

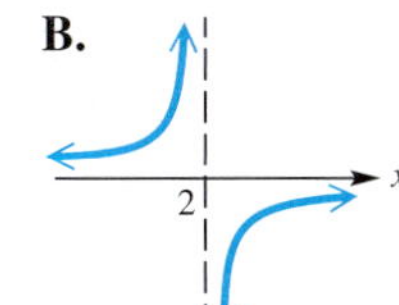

C.

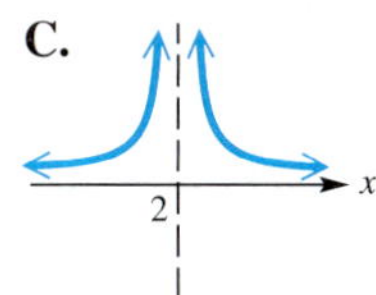

D.

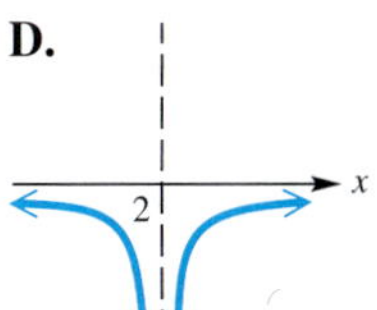

Sketch the graph of each rational function. See Examples 5–9.

51. $f(x) = \dfrac{x+1}{x-4}$ **52.** $f(x) = \dfrac{x-5}{x+3}$ **53.** $f(x) = \dfrac{3x}{x^2-x-2}$

54. $f(x) = \dfrac{2x+1}{x^2+6x+8}$ **55.** $f(x) = \dfrac{5x}{x^2-1}$ **56.** $f(x) = \dfrac{x}{4-x^2}$

57. $f(x) = \dfrac{x^2-2x}{x^2+6x+9}$ **58.** $f(x) = \dfrac{x^2-2x-3}{x^2-2x+1}$ **59.** $f(x) = \dfrac{x}{x^2-9}$

60. $f(x) = \dfrac{-5}{2x+4}$ **61.** $f(x) = \dfrac{1}{x^2+1}$ **62.** $f(x) = \dfrac{x^2-7x+10}{x^2+9}$

63. $f(x) = \dfrac{x^2+1}{x+3}$ **64.** $f(x) = \dfrac{2x^2+3}{x-4}$ **65.** $f(x) = \dfrac{x^2+2x}{2x-1}$

66. $f(x) = \dfrac{x^2-x}{x+2}$ **67.** $f(x) = \dfrac{x^2-9}{x+3}$ **68.** $f(x) = \dfrac{x^2-16}{x+4}$

Concept Check *Find an equation for each rational function graph.*

69.

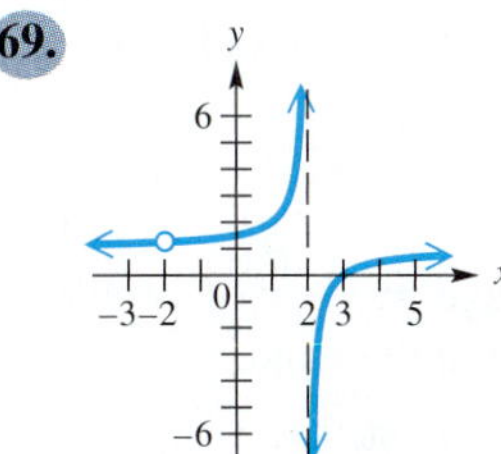

70.

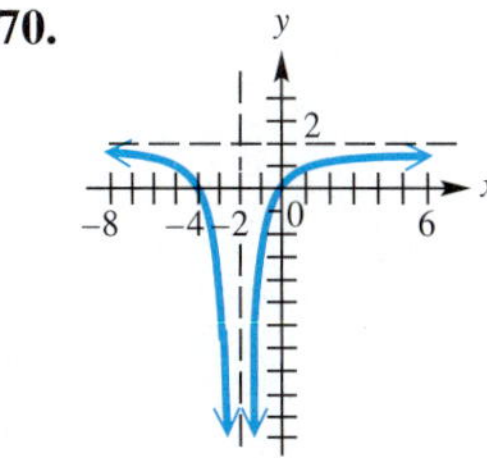

71.

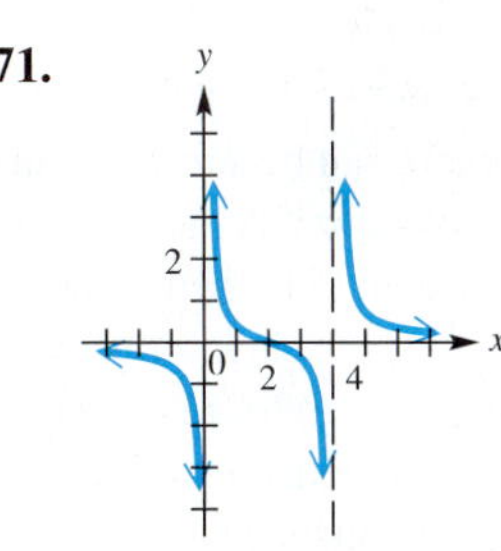

72.

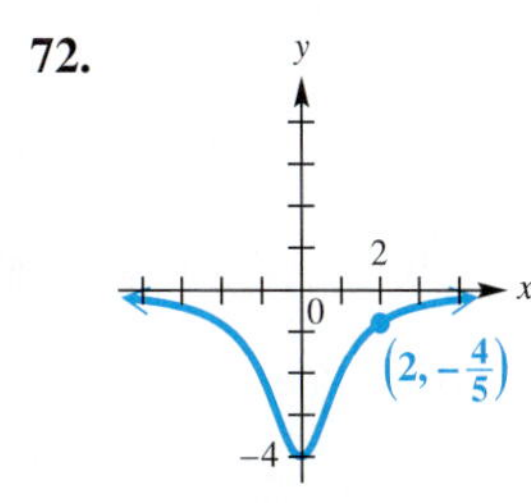

Concept Check *In Exercises 73 and 74, find a possible equation for the function with a graph having the given features.*

73. x-intercepts: -1 and 3, y-intercept: -3, vertical asymptote: $x = 1$, horizontal asymptote: $y = 1$

74. x-intercepts: 1 and 3, y-intercept: none, vertical asymptotes: $x = 0$ and $x = 2$, horizontal asymptote: $y = 1$

Use a graphing calculator to graph the rational function in the specified exercise. Then use the graph to find $f(1.25)$.

75. Exercise 51 **76.** Exercise 53 **77.** Exercise 65 **78.** Exercise 67

(Modeling) *Solve each problem. See Example 10.*

79. ***Traffic Intensity*** Let the average number of vehicles arriving at the gate of an amusement park per minute be equal to k, and let the average number of vehicles admitted by the park attendants be equal to r. Then, the average waiting time T (in minutes) for each vehicle arriving at the park is given by the rational function defined by

$$T(r) = \frac{2r - k}{2r^2 - 2kr},$$

where $r > k$. (*Source:* Mannering, F., and W. Kilareski, *Principles of Highway Engineering and Traffic Analysis,* 2nd ed., John Wiley & Sons, 1998.)

(a) It is known from experience that on Saturday afternoon $k = 25$. Use graphing to estimate the admittance rate r that is necessary to keep the average waiting time T for each vehicle to 30 sec.

(b) If one park attendant can serve 5.3 vehicles per minute, how many park attendants will be needed to keep the average wait to 30 sec?

80. ***Waiting in Line*** *Queuing theory* (also known as *waiting-line theory*) investigates the problem of providing adequate service economically to customers waiting in line. Suppose customers arrive at a fast-food service window at the rate of 9 people per hour. With reasonable assumptions, the average time (in hours) a customer will wait in line before being served is modeled by

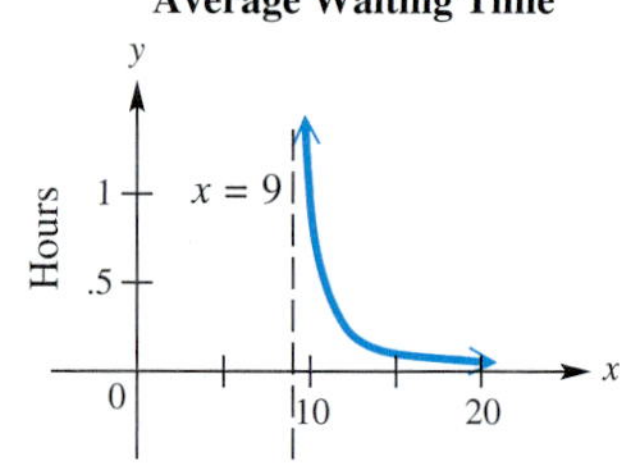

$$f(x) = \frac{9}{x(x - 9)},$$

where x is the average number of people served per hour. A graph of $f(x)$ for $9 < x \leq 20$ is shown in the figure.

(a) Why is the function meaningless if the average number of people served per hour is less than 9?

Suppose the average time to serve a customer is 5 min.

(b) How many customers can be served in an hour?

(c) How long will a customer have to wait in line (on the average)?

(d) Suppose you were managing the business, and you decided to halve the average waiting time to 7.5 min $\left(\frac{1}{8}\text{ hr}\right)$. How fast must your employee work to serve a customer (on the average)? (*Hint:* Let $f(x) = \frac{1}{8}$ and solve the equation for x. Remember to convert your answer to minutes.) How might this reduction in serving time be accomplished?

81. *Braking Distance* The rational function defined by

$$d(x) = \frac{8710x^2 - 69{,}400x + 470{,}000}{1.08x^2 - 324x + 82{,}200}$$

can be used to accurately model the braking distance for automobiles traveling at x miles per hour, where $20 \le x \le 70$. (*Source:* Mannering, F., and W. Kilareski, *Principles of Highway Engineering and Traffic Analysis,* 2nd ed., John Wiley & Sons, 1998.)

(a) Use graphing to estimate x when $d(x) = 300$.
(b) Complete the table for each value of x.
(c) If a car doubles its speed, does the stopping distance double or more than double? Explain.
(d) Suppose the stopping distance doubled whenever the speed doubled. What type of relationship would exist between the stopping distance and the speed?

x	$d(x)$	x	$d(x)$
20		50	
25		55	
30		60	
35		65	
40		70	
45			

82. *Deaths Due to AIDS* Refer to Exercises 65 and 66 in Section 3.1.

(a) Make a table listing the ratios of total deaths caused by AIDS to total cases of AIDS in the United States for each year from 1982 to 1996. (For example, in 1982 there were 620 deaths and 1563 cases, so the ratio is $\frac{620}{1563} \approx .397$.)
(b) As time progresses, what happens to the values of the ratio?
(c) Using the polynomial functions f and g that were found in the exercises cited above, define the rational function h, where $h(x) = \frac{g(x)}{f(x)}$. Graph $h(x)$ on the interval $[2, 20]$. Compare $h(x)$ to the values for the ratio found in the table.
(d) Use $h(x)$ to write an equation that models the relationship between the functions defined by $f(x)$ and $g(x)$ as x increases.
(e) The ratio of AIDS deaths to AIDS cases can be used to estimate the total number of AIDS deaths. According to the World Health Organization, by the end of 1998 there had been 13.9 million AIDS cases diagnosed worldwide since the disease began. Predict the total number of deaths caused by AIDS by the end of 1998.

83. *Tax Revenue* Economist Arthur Laffer has been a center of controversy because of his **Laffer curve,** an idealized version of which is shown here. According to this curve, increasing a tax rate, say from x_1 percent to x_2 percent on the graph, can actually lead to a decrease in government revenue. All economists agree on the endpoints, 0 revenue at tax rates of both 0% and 100%, but there is much disagreement on the location of the rate x_1 that produces maximum revenue. Suppose an economist studying the Laffer curve produces the rational function defined by

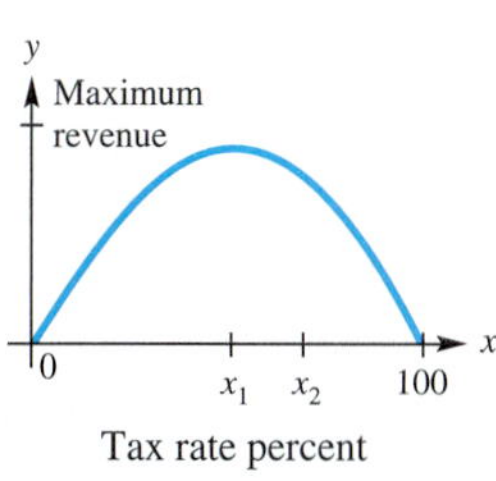

$$R(x) = \frac{80x - 8000}{x - 110},$$

where $R(x)$ is government revenue in tens of millions of dollars for a tax rate of x percent, with the function valid for $55 \le x \le 100$. Find the revenue for the following tax rates.

(a) 55% **(b)** 60% **(c)** 70% **(d)** 90% **(e)** 100%
(f) Graph R in the window $[0, 100]$ by $[0, 80]$.

84. *Tax Revenue* See Exercise 83. Suppose an economist determines that

$$R(x) = \frac{60x - 6000}{x - 120},$$

where $y = R(x)$ is government revenue in tens of millions of dollars for a tax rate of x percent, with $y = R(x)$ valid for $50 \le x \le 100$. Find the revenue for each tax rate.

(a) 50% **(b)** 60% **(c)** 80% **(d)** 100%

(e) Graph R in the window $[0, 100]$ by $[0, 50]$.

Relating Concepts

For individual or collaborative investigation

(Exercises 85–94)

Consider the following "monster" rational function.

$$f(x) = \frac{x^4 - 3x^3 - 21x^2 + 43x + 60}{x^4 - 6x^3 + x^2 + 24x - 20}$$

Analyzing this function will synthesize many of the concepts of this and earlier chapters. ***Work Exercises 85–94 in order.***

85. Find the equation of the horizontal asymptote.

86. Given that -4 and -1 are zeros of the numerator, factor the numerator completely.

87. **(a)** Given that 1 and 2 are zeros of the denominator, factor the denominator completely.

(b) Write the entire quotient for f so that the numerator and the denominator are in factored form.

88. **(a)** What is the common factor in the numerator and the denominator?

(b) For what value of x will there be a point of discontinuity (i.e., a "hole")?

89. What are the x-intercepts of the graph of f?

90. What is the y-intercept of the graph of f?

91. Find the equations of the vertical asymptotes.

92. Determine the point or points of intersection of the graph of f with its horizontal asymptote.

93. Sketch the graph of f.

94. Use the graph of f to solve the inequalities **(a)** $f(x) < 0$ and **(b)** $f(x) > 0$.

3.6 Variation

Direct Variation ▪ **Inverse Variation** ▪ **Combined and Joint Variation**

To apply mathematics we often need to express relationships between quantities. For example, in chemistry, the ideal gas law describes how temperature, pressure, and volume are related. In physics, various formulas in optics describe the relationship between focal length of a lens and the size of an image. This section introduces some special applications of polynomial and rational functions.

Direct Variation When one quantity is a constant multiple of another quantity, the two quantities are said to *vary directly.* For example, if you work for an hourly wage of \$6, then [pay] = 6[hours worked]. Doubling the hours doubles the pay. Tripling the hours triples the pay, and so on. This is stated more precisely as follows.

Direct Variation

y **varies directly** as x, or y is **directly proportional** to x, if there exists a nonzero real number k, called the **constant of variation,** such that

$$y = kx.$$

The phrase "directly proportional" is sometimes abbreviated to just "proportional."

The steps involved in solving a variation problem are summarized here.

Solving Variation Problems

Step 1 Write the general relationship among the variables as an equation. Use the constant k.

Step 2 Substitute given values of the variables and find the value of k.

Step 3 Substitute this value of k into the equation from Step 1, obtaining a specific formula.

Step 4 Substitute the remaining values and solve for the required unknown.

EXAMPLE 1 Solving a Direct Variation Problem

The area of a rectangle varies directly as its length. If the area is 50 m^2 when the length is 10 m, find the area when the length is 25 m.

Solution

Step 1 Since the area varies directly as the length,

$$A = kL,$$

where A represents the area of the rectangle, L is the length, and k is a nonzero constant.

$A = 50$ m^2

10 m

Step 2 Since $A = 50$ when $L = 10$, the equation $A = kL$ becomes

$$50 = 10k$$
$$k = 5.$$

Step 3 Using this value of k, we can express the relationship between the area and the length as

$$A = 5L. \quad \text{Direct variation equation}$$

A = ?

25 m

Step 4 To find the area when the length is 25, we replace L with 25.

$$A = 5L = 5(25) = 125$$

The area of the rectangle is 125 m^2 when the length is 25 m.

Now try Exercise 21.

Sometimes y varies as a power of x. If n is a positive integer greater than or equal to 2, then y is a greater power polynomial function of x.

Direct Variation as *n*th Power

Let n be a positive real number. Then y **varies directly as the *n*th power** of x, or y is **directly proportional to the *n*th power** of x, if there exists a nonzero real number k such that

$$y = kx^n.$$

For example, the area of a square of side x is given by the formula $A = x^2$, so the area varies directly as the square of the length of a side. Here $k = 1$.

Inverse Variation The case where y increases as x decreases is an example of *inverse variation.* In this case, the product of the variables is constant, and this relationship can be expressed as a rational function.

Inverse Variation

Let n be a positive real number. Then y **varies inversely as the *n*th power** of x, or y is **inversely proportional to the *n*th power** of x, if there exists a nonzero real number k such that

$$y = \frac{k}{x^n}.$$

If $n = 1$, then $y = \frac{k}{x}$, and y **varies inversely** as x.

EXAMPLE 2 Solving an Inverse Variation Problem

In a certain manufacturing process, the cost of producing a single item varies inversely as the square of the number of items produced. If 100 items are produced, each costs \$2. Find the cost per item if 400 items are produced.

Solution

Step 1 Let x represent the number of items produced and y represent the cost per item. Then for some nonzero constant k,

$$y = \frac{k}{x^2}.$$

Step 2 $$2 = \frac{k}{100^2} \quad \text{Substitute; } y = 2 \text{ when } x = 100.$$

$$k = 20{,}000 \quad \text{Solve for } k.$$

Step 3 The relationship between x and y is $y = \frac{20{,}000}{x^2}$.

Step 4 When 400 items are produced, the cost per item is

$$y = \frac{20{,}000}{400^2} = .125, \quad \text{or} \quad 12.5¢.$$

Now try Exercise 31.

Combined and Joint Variation One variable may depend on more than one other variable. Such variation is called **combined variation.** More specifically, when a variable depends on the *product* of two or more other variables, it is referred to as *joint variation.*

Joint Variation

Let m and n be real numbers. Then y **varies jointly** as the nth power of x and the mth power of z if there exists a nonzero real number k such that

$$\mathbf{y = kx^n z^m.}$$

CAUTION Note that *and* in the expression "y varies jointly as x and z" translates as the product $y = kxz$. The word *and* does not indicate addition here.

EXAMPLE 3 Solving a Joint Variation Problem

The area of a triangle varies jointly as the lengths of the base and the height. A triangle with base 10 ft and height 4 ft has area 20 ft^2. Find the area of a triangle with base 3 ft and height 8 ft.

Solution

Step 1 Let A represent the area, b the base, and h the height of the triangle. Then for some number k,

$$A = kbh.$$

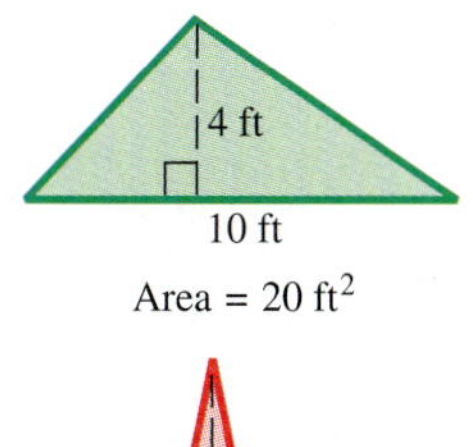

Step 2 Since A is 20 when b is 10 and h is 4,

$$20 = k(10)(4)$$

$$\frac{1}{2} = k.$$

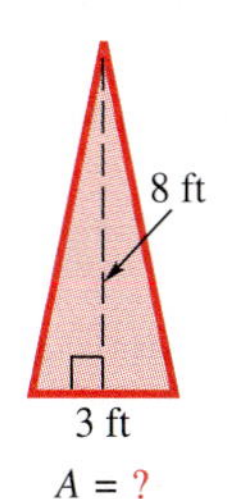

Step 3 The relationship among the variables is the familiar formula for the area of a triangle,

$$A = \frac{1}{2}bh.$$

Step 4 When $b = 3$ ft and $h = 8$ ft,

$$A = \frac{1}{2}(3)(8) = 12 \text{ ft}^2.$$

Now try Exercise 33.

EXAMPLE 4 Solving a Combined Variation Problem

The number of vibrations per second (the pitch) of a steel guitar string varies directly as the square root of the tension and inversely as the length of the string. If the number of vibrations per second is 5 when the tension is 225 kg and the length is .60 m, find the number of vibrations per second when the tension is 196 kg and the length is .65 m.

Solution Let n represent the number of vibrations per second, T represent the tension, and L represent the length of the string. Then, from the information in the problem, write the variation equation. (Step 1)

$$n = \frac{k\sqrt{T}}{L}$$

Substitute the given values for n, T, and L to find k. (Step 2)

$$5 = \frac{k\sqrt{225}}{.60} \quad \text{Let } n = 5, T = 225, L = .60.$$

$$3 = k\sqrt{225} \quad \text{Multiply by .60.}$$

$$3 = 15k \quad \sqrt{225} = 15$$

$$k = \frac{1}{5} = .2 \quad \text{Divide by 15.}$$

Substitute for k to find the relationship among the variables (Step 3).

$$n = \frac{.2\sqrt{T}}{L}$$

Now use the second set of values for T and L to find n. (Step 4)

$$n = \frac{.2\sqrt{196}}{.65} \approx 4.3 \quad \text{Let } T = 196, L = .65.$$

The number of vibrations per second is approximately 4.3.

Now try Exercise 37.

3.6 Exercises

Concept Check *Write each formula as an English phrase using the word* varies *or* proportional.

1. $C = 2\pi r$, where C is the circumference of a circle of radius r
2. $d = \frac{1}{5}s$, where d is the approximate distance (in miles) from a storm and s is the number of seconds between seeing lightning and hearing thunder
3. $v = \frac{d}{t}$, where v is the average speed when traveling d miles in t hours
4. $d = \frac{1}{4\pi nr^2}$, where d is the average distance a gas atom of radius r travels between collisions and n is the number of atoms per unit volume

5. $s = kx^3$, where s is the strength of a muscle that has length x

6. $f = \dfrac{mv^2}{r}$, where f is the centripetal force of an object of mass m moving along a circle of radius r at velocity v

Concept Check *Match each statement with its corresponding graph. In each case,* $k > 0$.

7. y varies directly as x. $(y = kx)$

8. y varies inversely as x. $\left(y = \frac{k}{x}\right)$

9. y varies directly as the second power of x. $(y = kx^2)$

10. x varies directly as the second power of y. $(x = ky^2)$

A.

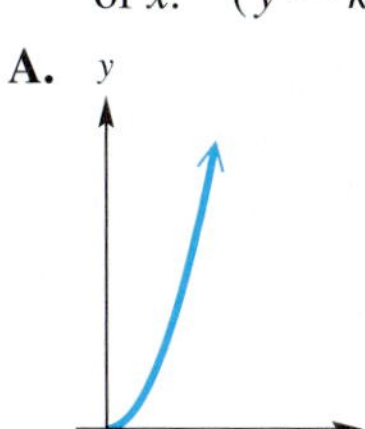

B.

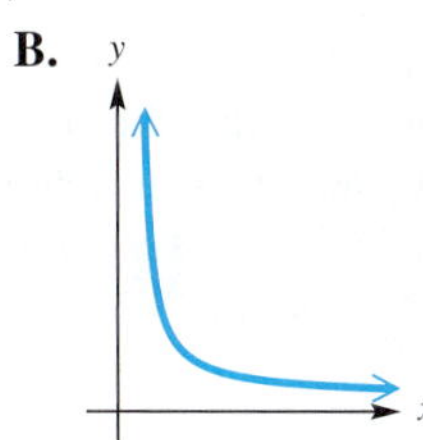

C.

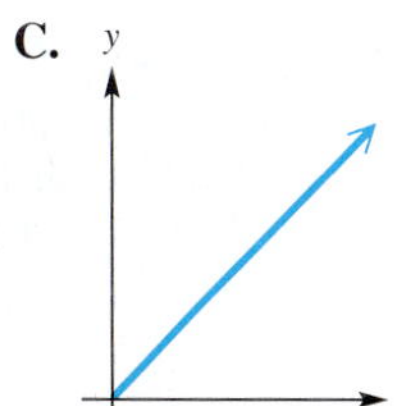

D.

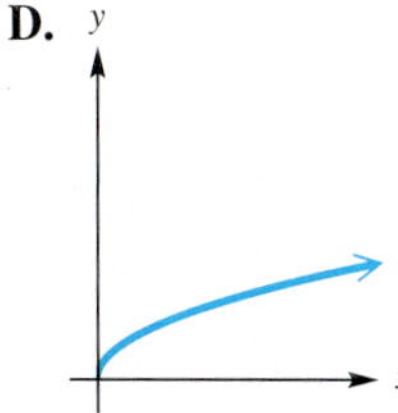

Solve each variation problem. See Examples 1–4.

11. If y varies directly as x, and $y = 10$ when $x = 2$, find y when $x = -6$.

12. If y varies directly as x, and $y = 3$ when $x = 10$, find y when $x = 40$.

13. If m varies jointly as x and y, and $m = 10$ when $x = 4$ and $y = 7$, find m when $x = 11$ and $y = 8$.

14. If m varies jointly as z and p, and $m = 10$ when $z = 3$ and $p = 5$, find m when $z = 5$ and $p = 7$.

15. If y varies inversely as x, and $y = 10$ when $x = 3$, find y when $x = 12$.

16. If y varies inversely as x, and $y = 20$ when $x = \frac{1}{4}$, find y when $x = 20$.

17. Suppose r varies directly as the square of m, and inversely as s. If $r = 12$ when $m = 6$ and $s = 4$, find r when $m = 4$ and $s = 10$.

18. Suppose p varies directly as the square of z, and inversely as r. If $p = \frac{32}{5}$ when $z = 4$ and $r = 10$, find p when $z = 2$ and $r = 16$.

19. Let a be directly proportional to m and n^2, and inversely proportional to y^3. If $a = 9$ when $m = 4$, $n = 9$, and $y = 3$, find a when $m = 6$, $n = 2$, and $y = 5$.

20. If y varies directly as x, and inversely as m^2 and r^2, and $y = \frac{5}{3}$ when $x = 1$, $m = 2$, and $r = 3$, find y when $x = 3$, $m = 1$, and $r = 8$.

Solve each problem. See Examples 1–4.

21. ***Circumference of a Circle*** The circumference of a circle varies directly as the radius. A circle with radius 7 in. has circumference 43.96 in. Find the circumference of the circle if the radius changes to 11 in.

22. ***Pressure Exerted by a Liquid*** The pressure exerted by a certain liquid at a given point varies directly as the depth of the point beneath the surface of the liquid. The pressure at 10 ft is 50 pounds per square inch (psi). What is the pressure at 15 ft?

23. ***Resistance of a Wire*** The resistance in ohms of a platinum wire temperature sensor varies directly as the temperature in degrees Kelvin (K). If the resistance is 646 ohms at a temperature of 190 K, find the resistance at a temperature of 250 K.

24. *Distance to the Horizon* The distance that a person can see to the horizon on a clear day from a point above the surface of Earth varies directly as the square root of the height at that point. If a person 144 m above the surface of Earth can see 18 km to the horizon, how far can a person see to the horizon from a point 64 m above the surface?

25. *Weight on the Moon* The weight of an object on Earth is directly proportional to the weight of that same object on the moon. A 200-lb astronaut would weigh 32 lb on the moon. How much would a 50-lb dog weigh on the moon?

26. *Water Emptied by a Pipe* The amount of water emptied by a pipe varies directly as the square of the diameter of the pipe. For a certain constant water flow, a pipe emptying into a canal will allow 200 gal of water to escape in an hour. The diameter of the pipe is 6 in. How much water would a 12-in. pipe empty into the canal in an hour, assuming the same water flow?

27. *Hooke's Law for a Spring* Hooke's law for an elastic spring states that the distance a spring stretches varies directly as the force applied. If a force of 15 lb stretches a certain spring 8 in., how much will a force of 30 lb stretch the spring?

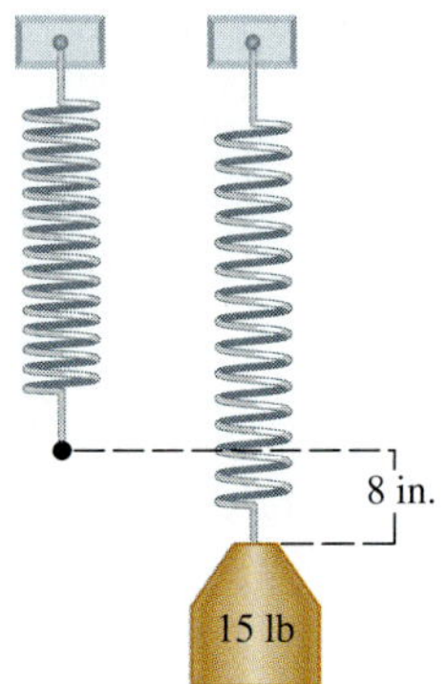

28. *Current in a Circuit* The current in a simple electrical circuit varies inversely as the resistance. If the current is 50 amps when the resistance is 10 ohms, find the current if the resistance is 5 ohms.

29. *Speed of a Pulley* The speed of a pulley varies inversely as its diameter. One kind of pulley, with diameter 3 in., turns at 150 revolutions per minute. Find the speed of a similar pulley with diameter 5 in.

30. *Weight of an Object* The weight of an object varies inversely as the square of its distance from the center of Earth. If an object 8000 mi from the center of Earth weighs 90 lb, find its weight when it is 12,000 mi from the center of Earth.

31. *Current Flow* In electric current flow, it is found that the resistance (measured in units called ohms) offered by a fixed length of wire of a given material varies inversely as the square of the diameter of the wire. If a wire .01 in. in diameter has a resistance of .4 ohm, what is the resistance of a wire of the same length and material with diameter .03 in. to the nearest ten-thousandth?

32. *Illumination* The illumination produced by a light source varies inversely as the square of the distance from the source. The illumination of a light source at 5 m is 70 candela. What is the illumination 12 m from the source?

33. *Simple Interest* Simple interest varies jointly as principal and time. If $1000 left at interest for 2 yr earned $110, find the amount of interest earned by $5000 for 5 yr.

34. *Volume of a Gas* Natural gas provides 35.8% of U.S. energy. (*Source:* U.S. Energy Department.) The volume of a gas varies inversely as the pressure and directly as the temperature in degrees Kelvin (K). If a certain gas occupies a volume of 1.3 L at 300 K and a pressure of 18 newtons per square centimeter, find the volume at 340 K and a pressure of 24 newtons per square centimeter.

35. *Force of Wind* The force of the wind blowing on a vertical surface varies jointly as the area of the surface and the square of the velocity. If a wind of 40 mph exerts a force of 50 lb on a surface of $\frac{1}{2}$ ft^2, how much force will a wind of 80 mph place on a surface of 2 ft^2?

36. *Volume of a Cylinder* The volume of a right circular cylinder is jointly proportional to the square of the radius of the circular base and to the height. If the volume is 300 cm^3 when the height is 10.62 cm and the radius is 3 cm, find the volume to the nearest tenth of a cylinder with radius 4 cm and height 15.92 cm.

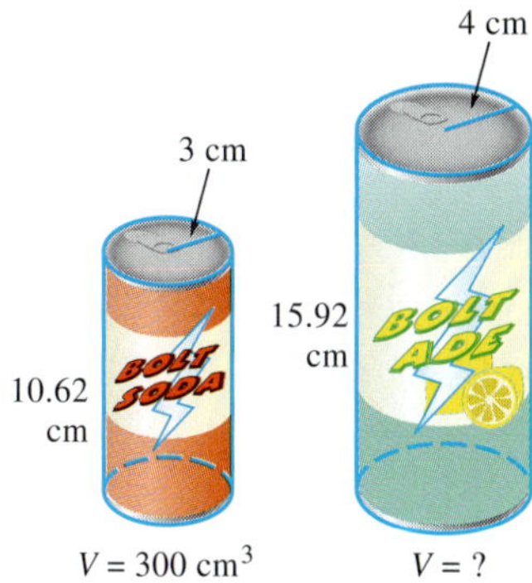

37. *Sports Arena Construction* The roof of a new sports arena rests on round concrete pillars. The maximum load a cylindrical column of circular cross section can hold varies directly as the fourth power of the diameter and inversely as the square of the height. The arena has 9-m tall columns that are 1 m in diameter and will support a load of 8 metric tons. How many metric tons will be supported by a column 12 m high and $\frac{2}{3}$ m in diameter?

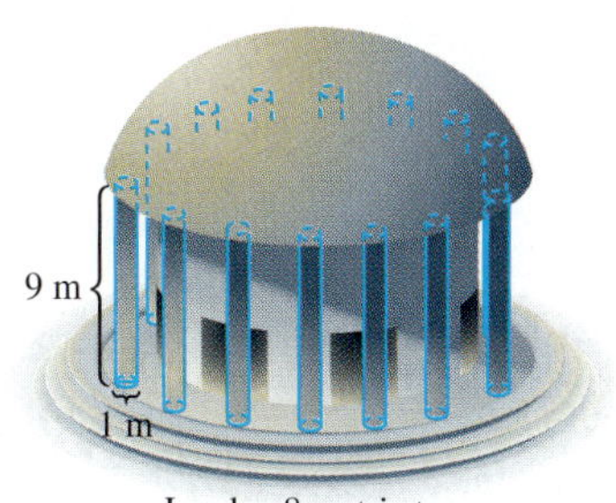

Load = 8 metric tons

38. *Sports Arena Construction* The sports arena in Exercise 37 requires a beam 16 m long, 24 cm wide, and 8 cm high. The maximum load of a horizontal beam that is supported at both ends varies directly as the width and square of the height and inversely as the length between supports. If a beam of the same material 8 m long, 12 cm wide, and 15 cm high can support a maximum of 400 kg, what is the maximum load the beam in the arena will support?

39. *Period of a Pendulum* The period of a pendulum varies directly as the square root of the length of the pendulum and inversely as the square root of the acceleration due to gravity. Find the period when the length is 121 cm and the acceleration due to gravity is 980 cm per second squared, if the period is 6π seconds when the length is 289 cm and the acceleration due to gravity is 980 cm per second squared.

40. *Long-Distance Phone Calls* The number of long-distance phone calls between two cities in a certain time period varies directly as the populations p_1 and p_2 of the cities, and inversely as the distance between them. If 10,000 calls are made between two cities 500 mi apart, having populations of 50,000 and 125,000, find the number of calls between two cities 800 mi apart, having populations of 20,000 and 80,000.

41. *Body Mass Index* The federal government has developed the *body mass index* (BMI) to determine ideal weights. A person's BMI is directly proportional to his or her weight in pounds and inversely proportional to the square of his or her height in inches. (A BMI of 19 to 25 corresponds to a healthy weight.) A 6-foot-tall person weighing 177 lb has BMI 24. Find the BMI (to the nearest whole number) of a person whose weight is 130 lb and whose height is 66 in. (*Source: Washington Post.*)

42. *Poiseuille's Law* According to Poiseuille's law, the resistance to flow of a blood vessel, R, is directly proportional to the length, l, and inversely proportional to the fourth power of the radius, r. If $R = 25$ when $l = 12$ and $r = .2$, find R to the nearest hundredth as r increases to .3, while l is unchanged.

43. *Stefan-Boltzmann Law* The Stefan-Boltzmann law says that the radiation of heat R from an object is directly proportional to the fourth power of the Kelvin temperature of the object. For a certain object, $R = 213.73$ at room temperature (293 K). Find R to the nearest hundredth if the temperature increases to 335 K.

44. *Nuclear Bomb Detonation* Suppose a nuclear bomb is detonated at a certain site. The effects of the bomb will be felt over a distance from the point of detonation that is directly proportional to the cube root of the yield of the bomb. Suppose a 100-kiloton bomb has certain effects to a radius of 3 km from the point of detonation. Find the distance to the nearest tenth that the effects would be felt for a 1500-kiloton bomb.

45. *Malnutrition Measure* A measure of malnutrition, called the *pelidisi,* varies directly as the cube root of a person's weight in grams and inversely as the person's sitting height in centimeters. A person with a pelidisi below 100 is considered to be undernourished, while a pelidisi greater than 100 indicates overfeeding. A person who weighs 48,820 g with a sitting height of 78.7 cm has a pelidisi of 100. Find the pelidisi (to the nearest whole number) of a person whose weight is 54,430 g and whose sitting height is 88.9 cm. Is this individual undernourished or overfed?

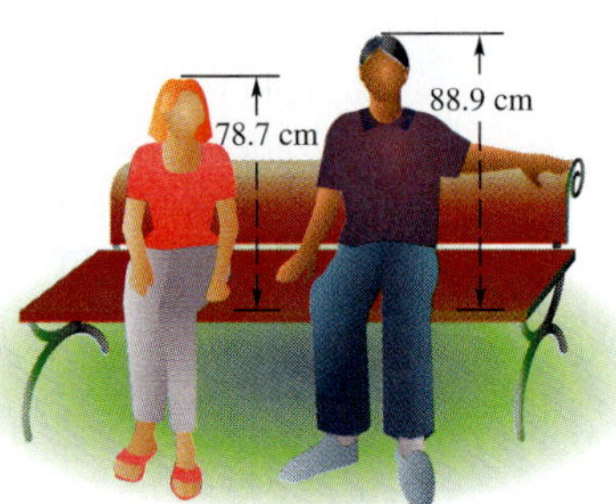

46. *Photography* Variation occurs in a formula from photography. In the formula

$$L = \frac{25F^2}{st},$$

the luminance, L, varies directly as the square of the F-stop, F, and inversely as the product of the film ASA number, s, and the shutter speed, t.

(a) What would an appropriate F-stop be for 200 ASA film and a shutter speed of $\frac{1}{250}$ sec when 500 footcandles of light are available?

(b) If 125 footcandles of light are available and an F-stop of 2 is used with 200 ASA film, what shutter speed should be used?

Concept Check *Work each problem.*

47. For $k > 0$, if y varies directly as x, then when x increases, y __________, and when x decreases, y __________.

48. For $k > 0$, if y varies inversely as x, then when x increases, y __________, and when x decreases, y __________.

49. What happens to y if y varies inversely as x, and x is doubled?

50. If y varies directly as x, and x is halved, how is y changed?

51. Suppose y is directly proportional to x, and x is replaced by $\frac{1}{3}x$. What happens to y?

52. What happens to y if y is inversely proportional to x, and x is tripled?

53. Suppose p varies directly as r^3 and inversely as t^2. If r is halved and t is doubled, what happens to p?

54. If m varies directly as p^2 and q^4, and p doubles while q triples, what happens to m?

Chapter 3 Summary

KEY TERMS

3.1 polynomial function, leading coefficient, zero polynomial, quadratic function, parabola, axis, vertex

3.2 synthetic division, zero of a polynomial function, root (or solution) of an equation

3.3 zero of multiplicity n

3.4 turning points, end behavior, dominating term

3.5 rational function, vertical asymptote, horizontal asymptote, oblique asymptote

3.6 varies directly (directly proportional to), constant of variation, varies inversely (inversely proportional to), combined variation, varies jointly

NEW SYMBOLS

$\bar{z}$ conjugate of $z = a + bi$

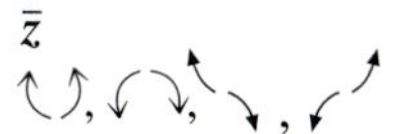
end behavior diagrams

$|f(x)| \to \infty$ absolute value of $f(x)$ approaches infinity

$x \to a$ x approaches a

QUICK REVIEW

CONCEPTS	EXAMPLES

3.1 Quadratic Functions and Models

1. The graph of
$$f(x) = a(x - h)^2 + k, \quad a \neq 0,$$
is a parabola with vertex at (h, k) and the vertical line $x = h$ as axis.

2. The graph opens up if a is positive and down if a is negative.

3. The graph is wider than the graph of $f(x) = x^2$ if $|a| < 1$ and narrower if $|a| > 1$.

Graph $f(x) = -(x + 3)^2 + 1$.

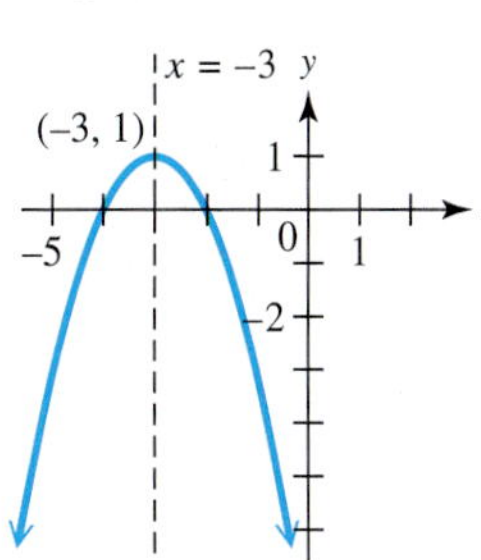

The graph opens down since $a < 0$. It is the graph of $y = x^2$ shifted 3 units left and 1 unit up, so the vertex is $(-3, 1)$, with axis $x = -3$. The domain is $(-\infty, \infty)$; the range is $(-\infty, 1]$.

Vertex Formula

The vertex of the graph of $f(x) = ax^2 + bx + c$, $a \neq 0$, may be found by completing the square. The vertex has coordinates
$$\left(-\frac{b}{2a}, f\left(-\frac{b}{2a}\right)\right).$$

Graph $f(x) = x^2 + 4x + 3$.
The vertex of the graph is
$$\left(-\frac{b}{2a}, f\left(-\frac{b}{2a}\right)\right) = (-2, -1). \quad a = 1, b = 4, c = 3$$

Graphing a Quadratic Function

Step 1 Find the vertex either by using the vertex formula or by completing the square.
Step 2 Find the y-intercept by evaluating $f(0)$.
Step 3 Find any x-intercepts by solving $f(x) = 0$.
Step 4 Find and plot any additional points as needed, using symmetry about the axis.

The graph opens up (if $a > 0$) or down (if $a < 0$).

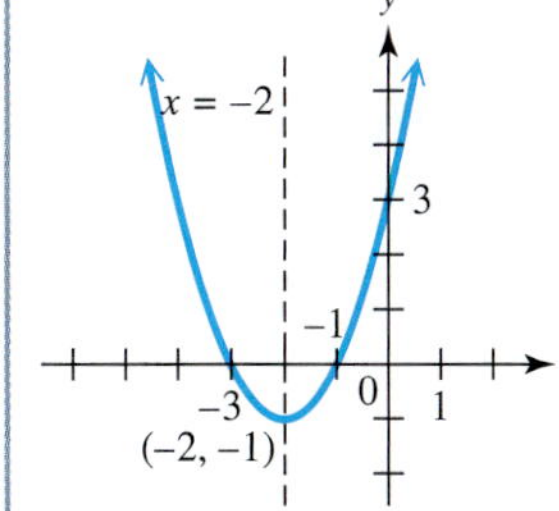

The graph opens up since $a > 0$. Since $f(0) = 3$, the y-intercept is 3. The solutions of $x^2 + 4x + 3 = 0$ are -1 and -3, which are the x-intercepts. The domain is $(-\infty, \infty)$, and the range is $[-1, \infty)$.

CONCEPTS	EXAMPLES
3.2 Synthetic Division	
Synthetic division is a shortcut method for dividing a polynomial by a binomial of the form $x - k$.	Use synthetic division to divide $$f(x) = 2x^3 - 3x + 2$$ by $x - 1$, and write the result in the form $f(x) = g(x) \cdot q(x) + r(x)$. $$\begin{array}{r\|rrrr} 1 & 2 & 0 & -3 & 2 \\ & & 2 & 2 & -1 \\ \hline & 2 & 2 & -1 & 1 \end{array}$$ Coefficients of the quotient; Remainder. $2x^3 - 3x + 2 = (x - 1)(2x^2 + 2x - 1) + 1$
Remainder Theorem If the polynomial $f(x)$ is divided by $x - k$, the remainder is $f(k)$.	By the result above, for $f(x) = 2x^3 - 3x + 2$, $f(1) = 1$.
3.3 Zeros of Polynomial Functions	
Factor Theorem The polynomial $x - k$ is a factor of the polynomial $f(x)$ if and only if $f(k) = 0$.	For the polynomial functions defined by $$f(x) = x^3 + x + 2 \quad \text{and} \quad g(x) = x^3 - 1,$$ $f(-1) = 0$. Therefore, $x - (-1)$, or $x + 1$, is a factor of $f(x)$. Also, since $x - 1$ is a factor of $g(x)$, $g(1) = 0$.
Rational Zeros Theorem If $\frac{p}{q}$ is a rational number written in lowest terms, and if $\frac{p}{q}$ is a zero of f, a polynomial function with integer coefficients, then p is a factor of the constant term and q is a factor of the leading coefficient.	The only rational numbers that can possibly be zeros of $$f(x) = 2x^3 - 9x^2 - 4x - 5$$ are ± 1, $\pm\frac{1}{2}$, ± 5, and $\pm\frac{5}{2}$. By synthetic division, it can be shown that the only rational zero of $f(x)$ is 5. $$\begin{array}{r\|rrrr} 5 & 2 & -9 & -4 & -5 \\ & & 10 & 5 & 5 \\ \hline & 2 & 1 & 1 & 0 \leftarrow f(5) \end{array}$$
Fundamental Theorem of Algebra Every function defined by a polynomial of degree 1 or more has at least one complex zero.	$f(x) = x^3 + x + 2$ has at least 1 and at most 3 zeros.
Number of Zeros Theorem A function defined by a polynomial of degree n has at most n distinct zeros.	

CONCEPTS	EXAMPLES
Conjugate Zeros Theorem If $f(x)$ defines a polynomial function having only real coefficients and if $a + bi$ is a zero of $f(x)$, where a and b are real numbers, then the conjugate $a - bi$ is also a zero of $f(x)$.	Since $1 + 2i$ is a zero of $$f(x) = x^3 - 5x^2 + 11x - 15,$$ its conjugate $1 - 2i$ is a zero as well.
Descartes' Rule of Signs See page 328.	For $f(x) = 3x^3 - 2x^2 + x - 4$, there are three sign changes, so there will be 3 or 1 positive real zeros. Since $f(-x) = -3x^3 - 2x^2 - x - 4$ has no sign changes, there will be no negative real zeros.

3.4 Polynomial Functions: Graphs, Applications, and Models

Graphing Using Translations The graph of the function $$f(x) = a(x - h)^n + k$$ can be found by considering the effects of the constants a, h, and k on the graph of $y = x^n$.	Graph $f(x) = -(x + 2)^4 + 1$. 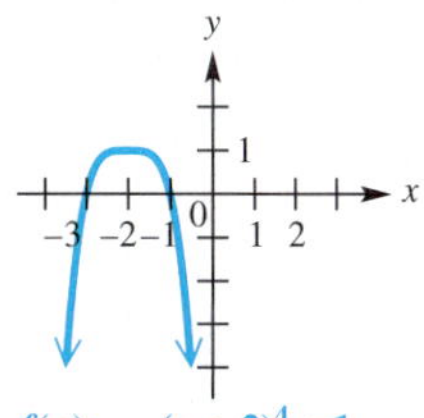 $f(x) = -(x + 2)^4 + 1$ The negative sign causes the graph to be reflected across the x-axis compared to the graph of $y = x^4$. The graph is translated 2 units to the left and 1 unit up.
Turning Points A polynomial function of degree n has at most $n - 1$ turning points.	The graph of $$f(x) = 4x^5 - 2x^3 + 3x^2 + x - 10$$ has at most $5 - 1 = 4$ turning points.
Graphing Polynomial Functions To graph a polynomial function f, where $f(x)$ is factorable, find x-intercepts and y-intercepts. Choose a value in each interval determined by the x-intercepts to decide whether the graph is above or below the x-axis there. If f is not factorable, use the procedure in the Summary Exercises that follow Section 3.4 to graph $f(x)$.	Graph $f(x) = (x + 2)(x - 1)(x + 3)$. 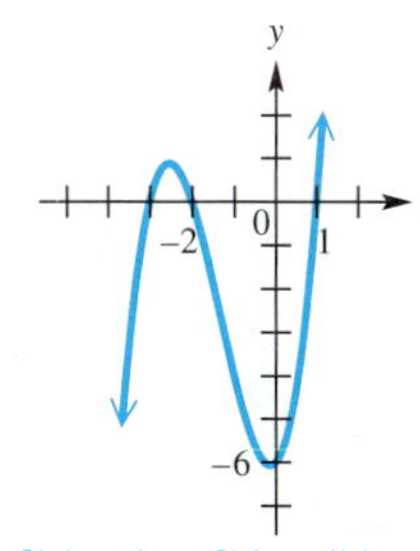 $f(x) = (x + 2)(x - 1)(x + 3)$ The zeros of f are -2, 1, and -3. Since $f(0) = 2(-1)(3) = -6$, the y-intercept is -6. Plot the intercepts and test points in the intervals determined by the x-intercepts. The dominating term is $x(x)(x)$ or x^3, so the end behavior is ↙ ↗.
Intermediate Value Theorem for Polynomials If $f(x)$ defines a polynomial function with *real coefficients,* and if for real numbers a and b the values of $f(a)$ and $f(b)$ are opposite in sign, then there exists at least one real zero between a and b.	For the polynomial function $$f(x) = -x^4 + 2x^3 + 3x^2 + 6,$$ $$f(3.1) = 2.0599 \text{ and } f(3.2) = -2.6016.$$ Since $f(3.1) > 0$ and $f(3.2) < 0$, there exists at least one real zero between 3.1 and 3.2.

(continued)

CONCEPTS	EXAMPLES

Boundedness Theorem

Let $f(x)$ be a polynomial function with *real coefficients* and with a *positive* leading coefficient. If $f(x)$ is divided synthetically by $x - c$, and

(a) if $c > 0$ and all numbers in the bottom row of the synthetic division are nonnegative, then $f(x)$ has no zero greater than c;

(b) if $c < 0$ and the numbers in the bottom row of the synthetic division alternate in sign (with 0 considered positive or negative, as needed), then $f(x)$ has no zero less than c.

Show that $f(x) = x^3 - x^2 - 8x + 12$ has no zero greater than 4 and no zero less than -4.

$$\begin{array}{r|rrrr} 4 & 1 & -1 & -8 & 12 \\ & & 4 & 12 & 16 \\ \hline & 1 & 3 & 4 & 28 \end{array} \leftarrow \text{All positive}$$

$$\begin{array}{r|rrrr} -4 & 1 & -1 & -8 & 12 \\ & & -4 & 20 & -48 \\ \hline & 1 & -5 & 12 & -36 \end{array} \leftarrow \text{Alternating signs}$$

3.5 Rational Functions: Graphs, Applications, and Models

Graphing Rational Functions

To graph a rational function in lowest terms, find asymptotes and intercepts. Determine whether the graph intersects a nonvertical asymptote. Plot a few points, as necessary, to complete the sketch.

Graph $f(x) = \dfrac{x^2 - 1}{(x + 3)(x - 2)}$.

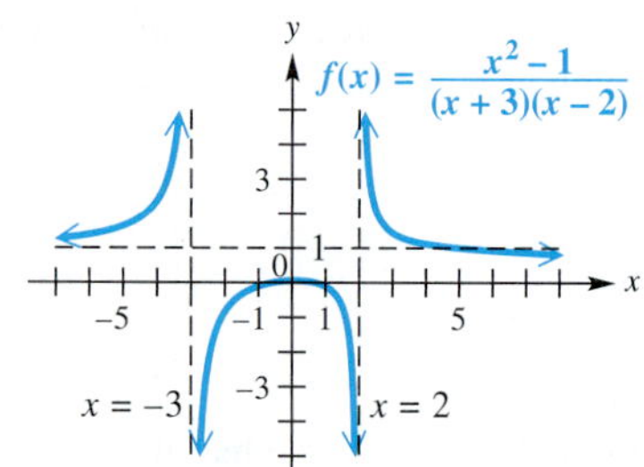

Point of Discontinuity

If a rational function is not written in lowest terms, there may be a "hole" in the graph instead of an asymptote.

Graph $f(x) = \dfrac{x^2 - 1}{x + 1}$.

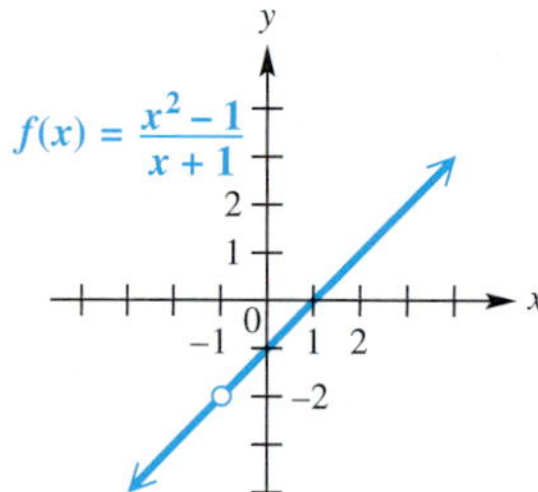

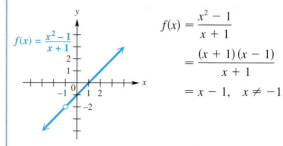

$$f(x) = \frac{x^2 - 1}{x + 1} = \frac{(x + 1)(x - 1)}{x + 1} = x - 1, \quad x \neq -1$$

3.6 Variation

Direct Variation

y varies directly as the nth power of x if there exists a nonzero real number k such that $\boldsymbol{y = kx^n}$.

The area of a circle varies directly as the square of the radius.

$$A = kr^2$$

Inverse Variation

y varies inversely as the nth power of x if there exists a nonzero real number k such that $\boldsymbol{y = \dfrac{k}{x^n}}$.

Pressure of a gas varies inversely as volume.

$$P = \frac{k}{V}$$

CONCEPTS	EXAMPLES
Joint Variation For real numbers m and n, y varies jointly as the nth power of x and the mth power of z if there exists a nonzero real number k such that $y = kx^n z^m$.	The area of a triangle varies jointly as its base and its height. $A = \frac{1}{2}bh$

Chapter 3 Review Exercises

Graph each quadratic function. Give the vertex, axis, x-intercepts, y-intercept, domain, and range of each graph.

1. $f(x) = 3(x + 4)^2 - 5$

2. $f(x) = -\frac{2}{3}(x - 6)^2 + 7$

3. $f(x) = -3x^2 - 12x - 1$

4. $f(x) = 4x^2 - 4x + 3$

Concept Check In Exercises 5–8, consider the function defined by $f(x) = a(x - h)^2 + k$, with $a > 0$.

5. What are the coordinates of the lowest point of its graph?

6. What is the y-intercept of its graph?

7. Under what conditions will its graph have one or more x-intercepts? For these conditions, express the x-intercept(s) in terms of a, h, and k.

8. If a is positive, what is the smallest value of $ax^2 + bx + c$ in terms of a, b, and c?

9. *(Modeling) Area of a Rectangle* Use a quadratic function to find the dimensions of the rectangular region of maximum area that can be enclosed with 180 m of fencing, if no fencing is needed along one side of the region.

10. *(Modeling) Height of a Projectile* A projectile is fired vertically upward, and its height $s(t)$ in feet after t seconds is given by the function defined by

$$s(t) = -16t^2 + 800t + 600.$$

(a) From what height was the projectile fired?
(b) After how many seconds will it reach its maximum height?
(c) What is the maximum height it will reach?
(d) Between what two times (in seconds, to the nearest tenth) will it be more than 5000 ft above the ground?
(e) How long to the nearest tenth of a second will the projectile be in the air?

11. *(Modeling) Food Bank Volunteers* During the course of a year, the number of volunteers available to run a food bank each month is modeled by $V(x)$, where

$$V(x) = 2x^2 - 32x + 150$$

between the months of January and August. Here x is time in months, with $x = 1$ representing January. From August to December, $V(x)$ is modeled by

$$V(x) = 31x - 226.$$

Find the number of volunteers in each of the following months.

(a) January **(b)** May **(c)** August **(d)** October **(e)** December
(f) Sketch a graph of $y = V(x)$ for January through December. In what month are the fewest volunteers available?

12. *(Modeling) Concentration of Atmospheric CO_2* In 1990, a publication of the International Panel on Climate Change (IPCC) stated that if current trends of burning fossil fuel and deforestation were to continue, then future amounts of atmospheric carbon dioxide in parts per million (ppm) would increase, as shown in the table.

Year	Carbon Dioxide
1990	353
2000	375
2075	590
2175	1090
2275	2000

Source: IPCC.

(a) Let $x = 0$ represent 1990, $x = 10$ represent 2000, and so on. Find a function of the form $f(x) = a(x - h)^2 + k$ that models the data. Use $(0, 353)$ as the vertex and $(285, 2000)$ as another point to determine a.

(b) Use the function to predict the amount of carbon dioxide in 2300.

Consider the function defined by $f(x) = -2.64x^2 + 5.47x + 3.54$ for Exercises 13–16.

13. Use the discriminant to explain how you can determine the number of x-intercepts the graph of f will have even before graphing it on your calculator. (See Section 1.4.)

14. Graph the function in the standard viewing window of your calculator, and use the calculator to solve the equation $f(x) = 0$. Express solutions as approximations to the nearest hundredth.

15. Use your answer to Exercise 14 and the graph of f to solve

(a) $f(x) > 0$, and **(b)** $f(x) < 0$.

16. Use the capabilities of your calculator to find the coordinates of the vertex of the graph. Express coordinates to the nearest hundredth.

Use synthetic division to perform each division.

17. $\dfrac{x^3 + x^2 - 11x - 10}{x - 3}$

18. $\dfrac{3x^3 + 8x^2 + 5x + 10}{x + 2}$

19. $\dfrac{2x^3 - x + 6}{x + 4}$

20. $\dfrac{\frac{1}{3}x^3 - 2x^2 - 9x + 4}{x + 3}$

Express $f(x)$ in the form $f(x) = (x - k)q(x) + r$ for the given value of k.

21. $5x^3 - 3x^2 + 2x - 6; \quad k = 2$

22. $-3x^3 + 5x - 6; \quad k = -1$

Use synthetic division to find $f(2)$.

23. $f(x) = -x^3 + 5x^2 - 7x + 1$

24. $f(x) = 2x^3 - 3x^2 + 7x - 12$

25. $f(x) = 5x^4 - 12x^2 + 2x - 8$

26. $f(x) = x^5 + 4x^2 - 2x - 4$

Use synthetic division to determine whether k is a zero of the function.

27. $f(x) = -2x^3 + x^2 - 2x - 69; \quad k = -3$

28. $f(x) = \dfrac{1}{4}x^3 + \dfrac{3}{4}x^2 - \dfrac{1}{2}x + 6; \quad k = -4$

29. *Concept Check* If $f(x)$ is defined by a polynomial with real coefficients, and if $7 + 2i$ is a zero of the function, then what other complex number must also be a zero?

30. *Concept Check* Suppose the polynomial function f has a zero at $x = 3$. Which of the following statements must be true?

A. 3 is an x-intercept of the graph of f. **B.** 3 is a y-intercept of the graph of f.
C. $x - 3$ is a factor of $f(x)$. **D.** $f(-3) = 0$

Find a polynomial function with real coefficients and least degree having the given zeros.

31. $-1, 4, 7$

32. $8, 2, 3$

33. $\sqrt{3}, -\sqrt{3}, 2, 3$

34. $-2 + \sqrt{5}, -2 - \sqrt{5}, -2, 1$

Find all rational zeros of each function.

35. $f(x) = 2x^3 - 9x^2 - 6x + 5$

36. $f(x) = 8x^4 - 14x^3 - 29x^2 - 4x + 3$

In Exercises 37–39, show that the polynomial function has a real zero as described.

37. $f(x) = 3x^3 - 8x^2 + x + 2$

(a) between -1 and 0 **(b)** between 2 and 3

38. $f(x) = 4x^3 - 37x^2 + 50x + 60$

(a) between 2 and 3 **(b)** between 7 and 8

39. $f(x) = 6x^4 + 13x^3 - 11x^2 - 3x + 5$

(a) no zero greater than 1 **(b)** no zero less than -3

40. Use Descartes' rule of signs to determine the possible number of positive zeros and negative zeros of $f(x) = x^3 + 3x^2 - 4x - 2$.

41. Is $x + 1$ a factor of $f(x) = x^3 + 2x^2 + 3x + 2$?

42. Find a polynomial function with real coefficients of degree 4 with 3, 1, and $-1 - 3i$ as zeros, and $f(2) = -36$.

43. Find a polynomial function of degree 3 with -2, 1, and 4 as zeros, and $f(2) = 16$.

44. Find all zeros of $f(x) = x^4 - 3x^3 - 8x^2 + 22x - 24$, given that $1 - i$ is a zero.

45. Find all zeros of $f(x) = 2x^4 - x^3 + 7x^2 - 4x - 4$, given that 1 and $2i$ are zeros.

46. Find a value of s such that $x - 4$ is a factor of $f(x) = x^3 - 2x^2 + sx + 4$.

47. Find a value of s such that when the polynomial $x^3 - 3x^2 + sx - 4$ is divided by $x - 2$, the remainder is 5.

48. *Concept Check* Give an example of a fourth-degree polynomial function having exactly two distinct real zeros, and then sketch its graph.

49. *Concept Check* Give an example of a cubic polynomial function having exactly one real zero, and then sketch its graph.

50. Give the maximum number of turning points of the graph of each function.

(a) $f(x) = x^3 - 9x$ **(b)** $f(x) = 4x^4 - 6x^2 + 2$

51. *Concept Check* Suppose the leading term of a polynomial function is $6x^7$. What can you conclude about each of the following features of the graph of the function?

(a) domain **(b)** range **(c)** end behavior **(d)** number of zeros

(e) number of turning points

52. *Concept Check* Repeat Exercise 51 for a polynomial function with leading term $-7x^6$.

Concept Check *For each polynomial function, identify its graph from choices A–F.*

53. $f(x) = (x - 2)^2(x - 5)$

54. $f(x) = -(x - 2)^2(x - 5)$

55. $f(x) = (x - 2)^2(x - 5)^2$

56. $f(x) = (x - 2)(x - 5)$

57. $f(x) = -(x - 2)(x - 5)$

58. $f(x) = -(x - 2)^2(x - 5)^2$

A.

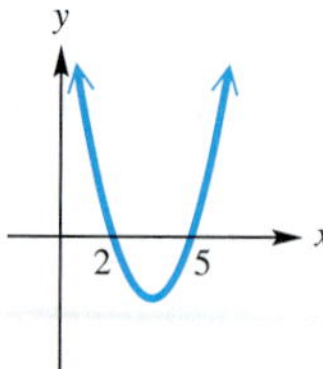

B.

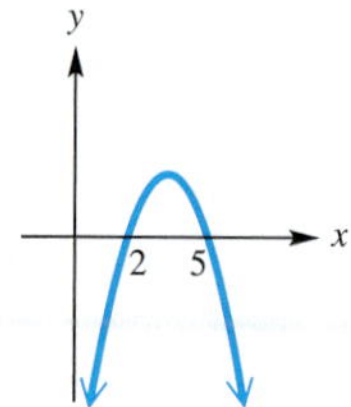

C.

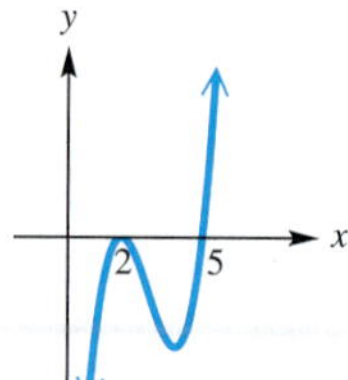

D.

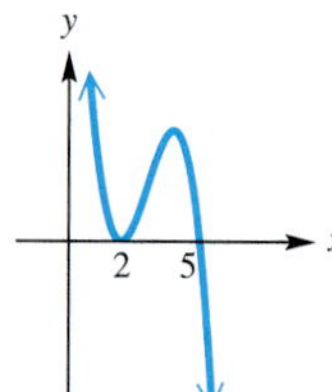

E.

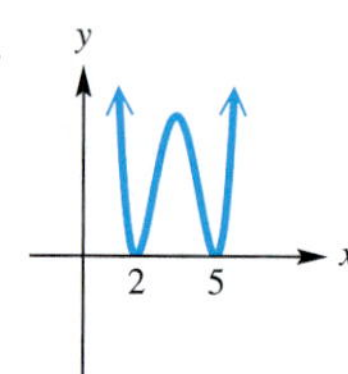

F.

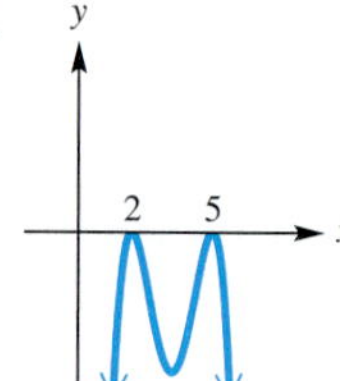

Sketch the graph of each polynomial function.

59. $f(x) = (x - 2)^2(x + 3)$

60. $f(x) = -2x^3 + 7x^2 - 2x - 3$

61. $f(x) = 2x^3 + x^2 - x$

62. $f(x) = x^4 - 3x^2 + 2$

Use a graphing calculator to graph each polynomial function in the viewing window specified. Then determine the real zeros to as many decimal places as the calculator will provide.

63. $f(x) = x^3 - 8x^2 + 2x + 5$; window: $[-10, 10]$ by $[-60, 60]$

64. $f(x) = x^4 - 4x^3 - 5x^2 + 14x - 15$; window: $[-10, 10]$ by $[-60, 60]$

65. ***(Modeling) Medicare Beneficiary Spending*** Out-of-pocket spending projections for a typical Medicare beneficiary as a share of his or her income are given in the table. Let $x = 0$ represent 1990, so $x = 8$ represents 1998. Use a graphing calculator to do the following.

(a) Graph the data points.

(b) Find a quadratic function to model the data.

(c) Find a cubic function to model the data.

(d) Graph each function in the same viewing window as the data points.

(e) Compare the two functions. Which is a better fit for the data?

Year	Percent of Income
1998	18.6
2000	19.3
2005	21.7
2010	24.7
2015	27.5
2020	28.3
2025	28.6

Source: Urban Institute's Analysis of 1998 Medicare Trustees' Report.

Solve each problem.

66. ***Dimensions of a Cube*** After a 2-in. slice is cut off the top of a cube, the resulting solid has a volume of 32 in.^{3}. Find the dimensions of the original cube.

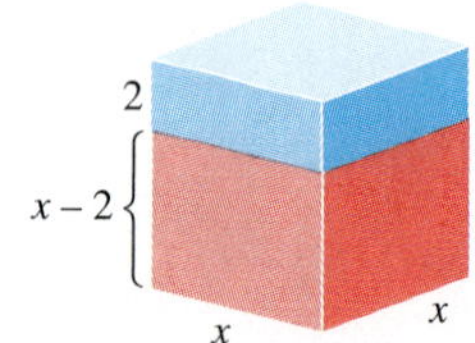

67. *Dimensions of a Box* The width of a rectangular box is three times its height, and its length is 11 in. more than its height. Find the dimensions of the box if its volume is 720 in.^{3}.

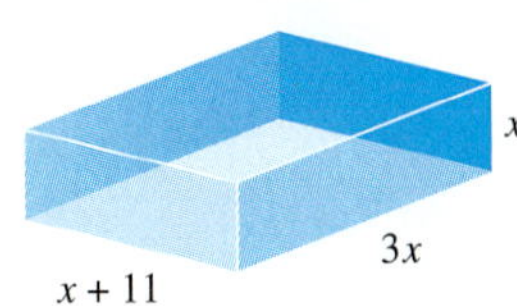

68. The function defined by $f(x) = \frac{1}{x}$ is negative at $x = -1$ and positive at $x = 1$, but has no zero between -1 and 1. Explain why this does not contradict the intermediate value theorem.

Graph each rational function.

69. $f(x) = \dfrac{4}{x - 1}$

70. $f(x) = \dfrac{4x - 2}{3x + 1}$

71. $f(x) = \dfrac{6x}{x^2 + x - 2}$

72. $f(x) = \dfrac{2x}{x^2 - 1}$

73. $f(x) = \dfrac{x^2 + 4}{x + 2}$

74. $f(x) = \dfrac{x^2 - 1}{x}$

75. $f(x) = \dfrac{-2}{x^2 + 1}$

76. $f(x) = \dfrac{4x^2 - 9}{2x + 3}$

77. *Concept Check*

(a) Sketch the graph of a function that has the line $x = 3$ as a vertical asymptote, the line $y = 1$ as a horizontal asymptote, and x-intercepts 2 and 4.

(b) Find an equation for a possible corresponding rational function.

78. *Concept Check*

(a) Sketch the graph of a function that is never negative and has the lines $x = -1$ and $x = 1$ as vertical asymptotes, the x-axis as a horizontal asymptote, and 0 as an x-intercept.

(b) Find an equation for a possible corresponding rational function.

79. *Concept Check* Find an equation for the rational function graphed here.

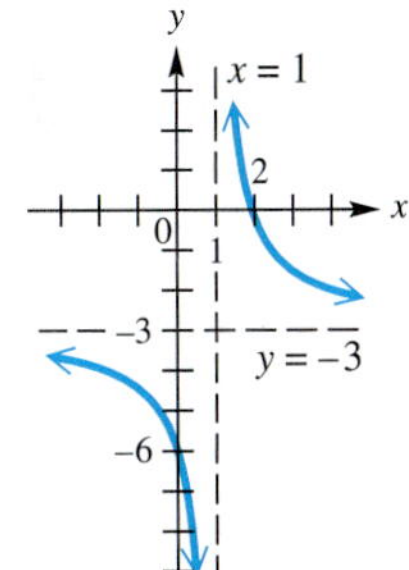

(Modeling) Solve each problem.

80. *Antique-Car Competition* Antique-car owners often enter their cars in a *concours d'elegance* in which a maximum of 100 points can be awarded to a particular car. Points are awarded for the general attractiveness of the car. The function defined by

$$C(x) = \frac{10x}{49(101 - x)}$$

models the cost, in thousands of dollars, of restoring a car so that it will win x points.

(a) Graph the function in the window [0, 101] by [0, 10].

(b) How much would an owner expect to pay to restore a car in order to earn 95 points?

81. *Environmental Pollution* In situations involving environmental pollution, a cost-benefit model expresses cost as a function of the percentage of pollutant removed from the environment. Suppose a cost-benefit model is expressed as

$$C(x) = \frac{6.7x}{100 - x},$$

where $C(x)$ is cost in thousands of dollars of removing x percent of a pollutant.

(a) Graph the function in the window [0, 100] by [0, 100].

(b) How much would it cost to remove 95% of the pollutant?

Solve each variation problem.

82. If x varies directly as y, and $x = 9$ when $y = 3$, find x when $y = 12$.

83. If x varies directly as y, and $x = 10$ when $y = 7$, find y when $x = 50$.

84. If z varies inversely as w, and $z = 10$ when $w = \frac{1}{2}$, find z when $w = 8$.

85. If t varies inversely as s, and $t = 3$ when $s = 5$, find s when $t = 5$.

86. p varies jointly as q and r^2, and $p = 200$ when $q = 2$ and $r = 3$. Find p when $q = 5$ and $r = 2$.

87. f varies jointly as g^2 and h, and $f = 50$ when $g = 4$ and $h = 2$. Find f when $g = 3$ and $h = 6$.

88. *Pressure in a Liquid* The pressure on a point in a liquid is directly proportional to the distance from the surface to the point. In a certain liquid, the pressure at a depth of 4 m is 60 kg per m^2. Find the pressure at a depth of 10 m.

89. *Skidding Car* The force needed to keep a car from skidding on a curve varies inversely as the radius r of the curve and jointly as the weight of the car and the square of the speed. It takes 3000 lb of force to keep a 2000-lb car from skidding on a curve of radius 500 ft at 30 mph. What force will keep the same car from skidding on a curve of radius 800 ft at 60 mph?

90. *Power of a Windmill* The power a windmill obtains from the wind varies directly as the cube of the wind velocity. If a 10 km per hr wind produces 10,000 units of power, how much power is produced by a wind of 15 km per hr?

Chapter 3 Test

1. Sketch the graph of the quadratic function defined by $f(x) = -2x^2 + 6x - 3$. Give the intercepts, vertex, axis, domain, and range.

2. *(Modeling) Rock Concert Attendance* The number of tickets sold at the top 50 rock concerts from 1998–2002 is modeled by the quadratic function defined by

$$f(x) = -.3857x^2 + 1.2829x + 11.329,$$

where $x = 0$ corresponds to 1998, $x = 1$ to 1999, and so on, and $f(x)$ is in millions. (*Source:* Pollstar Online.) If the model continued to apply, what would have been the ticket sales in 2003?

Use synthetic division to perform each division.

3. $\dfrac{3x^3 + 4x^2 - 9x + 6}{x + 2}$

4. $\dfrac{2x^3 - 11x^2 + 28}{x - 5}$

5. Use synthetic division to determine $f(5)$, if $f(x) = 2x^3 - 9x^2 + 4x + 8$.

6. Use the factor theorem to determine whether the polynomial $x - 3$ is a factor of $6x^4 - 11x^3 - 35x^2 + 34x + 24$. If it is, what is the other factor? If it is not, explain why.

7. Find all zeros of $f(x)$, given that $f(x) = x^3 + 8x^2 + 25x + 26$ and -2 is one zero.

8. Find a fourth degree polynomial function $f(x)$ having only real coefficients, -1, 2, and i as zeros, and $f(3) = 80$.

9. Why can't the polynomial function defined by $f(x) = x^4 + 8x^2 + 12$ have any real zeros?

10. Consider the function defined by $f(x) = x^3 - 5x^2 + 2x + 7$.

(a) Use the intermediate value theorem to show that f has a zero between 1 and 2.

(b) Use Descartes' rule of signs to determine the possible number of positive zeros and negative zeros.

(c) Use a graphing calculator to find all real zeros to as many decimal places as the calculator will give.

11. Graph the functions defined by $f_1(x) = x^4$ and $f_2(x) = -2(x + 5)^4 + 3$ on the same axes. How can the graph of f_2 be obtained by a transformation of the graph of f_1?

12. Use end behavior to determine which one of the following graphs is that of $f(x) = -x^7 + x - 4$.

A.

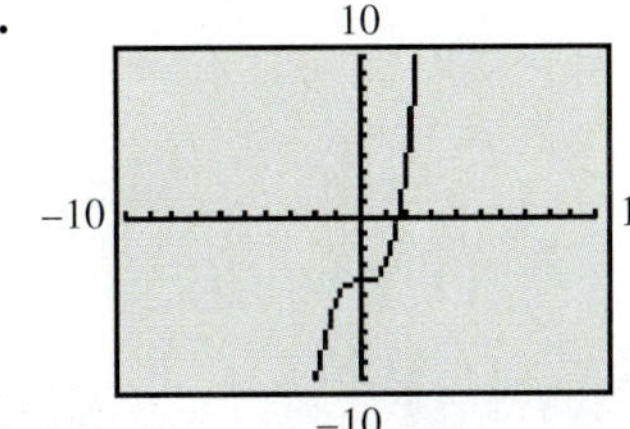

B.

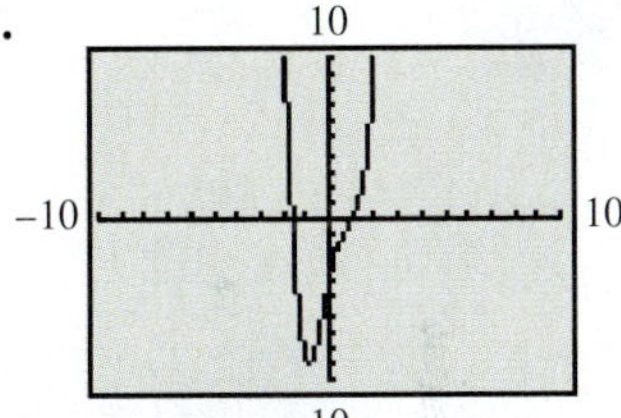

C.

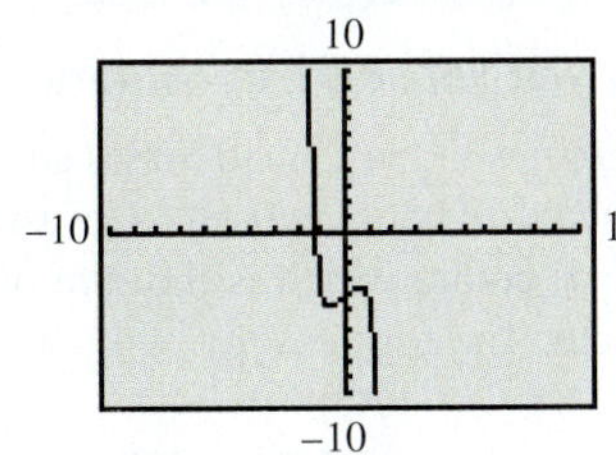

Graph each polynomial function.

13. $f(x) = x^3 - 5x^2 + 3x + 9$

14. $f(x) = 2x^2(x - 2)^2$

15. $f(x) = -x^3 - 4x^2 + 11x + 30$

16. Find a cubic polynomial function having the graph shown.

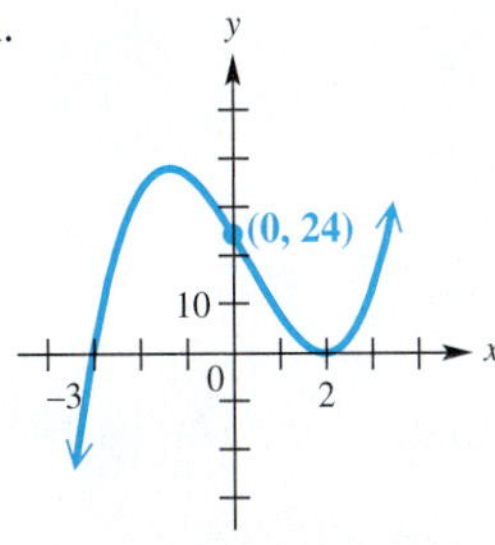

17. *(Modeling) Oil Pressure* The pressure of oil in a reservoir tends to drop with time. Engineers found that the change in pressure is modeled by

$$f(t) = 1.06t^3 - 24.6t^2 + 180t,$$

for t (in years) in the interval $[0, 15]$.

(a) What was the change after 2 yr?

(b) For what time periods is the amount of change in pressure increasing? decreasing? Use a graph to decide.

Graph each rational function.

18. $f(x) = \dfrac{3x - 1}{x - 2}$

19. $f(x) = \dfrac{x^2 - 1}{x^2 - 9}$

20. For the rational function defined by $f(x) = \dfrac{2x^2 + x - 6}{x - 1}$,

(a) determine the equation of the oblique asymptote;
(b) determine the x-intercepts; (c) determine the y-intercept; (d) determine the equation of the vertical asymptote; (e) sketch the graph.

21. If y varies directly as the square root of x, and $y = 12$ when $x = 4$, find y when $x = 100$.

22. *Weight On and Above Earth* The weight w of an object varies inversely as the square of the distance d between the object and the center of Earth. If a man weighs 90 kg on the surface of Earth, how much would he weigh 800 km above the surface? (*Hint:* The radius of Earth is about 6400 km.)

Chapter 3 Quantitative Reasoning

How would you vote on a fluoridation referendum?

Many communities in the United States add fluoride to their drinking water to promote dental health. The scale used to measure dental health is the DMF (Decayed, Missing, Filled) count, which is the sum of overtly decayed, filled, and extracted permanent teeth. The formula

$$\text{DMF} = 200 + \frac{100}{x}, \quad x > .25,$$

gives the DMF count per 100 examinees for fluoride content x (in parts per million, ppm). This formula was developed with data from the World Health Organization.

Suppose your community will vote on whether to fluoridate the local water. To vote intelligently, you need to know how much fluoride must be used to get desirable DMF counts, and what levels of fluoride are considered safe. A further concern is cost. Does cost influence the amount of fluoride that can be added? The formula will provide an answer to the first consideration. To decide on a safe level, you may need to seek information on the Internet or from experts. The usual fluoride level added to drinking water is 1 ppm. What DMF count does that correspond to? What level of fluoride would provide a DMF count of 250?

4 Exponential and Logarithmic Functions

In 1990 the International Panel on Climate Change (IPCC) reported that if current trends of burning fossil fuels and deforestation continue, then future amounts of atmospheric carbon dioxide in parts per million (ppm) would increase exponentially. Atmospheric carbon dioxide accumulation is irreversible and its effect requires hundreds of years to disappear.

In Example 11 of Section 4.2 and other examples and exercises in this chapter, we learn about exponential growth.

4.1 Inverse Functions

Inverse Operations ▪ One-to-One Functions ▪ Inverse Functions ▪ Equations of Inverses ▪ An Application of Inverse Functions to Cryptography

In this chapter we study two important types of functions, *exponential* and *logarithmic*. These functions are related in a special way: They are *inverses* of one another.

Inverse Operations Addition and subtraction are inverse operations: starting with a number x, adding 5, and subtracting 5 gives x back as the result. Similarly, some functions are *inverses* of each other. For example, the functions defined by

$$f(x) = 8x \quad \text{and} \quad g(x) = \frac{1}{8}x$$

are inverses of each other with respect to function composition. This means that if a value of x such as $x = 12$ is chosen, then

$$f(12) = 8 \cdot 12 = 96.$$

Calculating $g(96)$ gives

$$g(96) = \frac{1}{8} \cdot 96 = 12.$$

Thus, $$g[f(12)] = 12.$$

Also, $f[g(12)] = 12$. For these functions f and g, it can be shown that

$$f[g(x)] = x \quad \text{and} \quad g[f(x)] = x$$

for any value of x.

One-to-One Functions Suppose we define the function

$$F = \{(-2, 2), (-1, 1), (0, 0), (1, 3), (2, 5)\}.$$

We can form another set of ordered pairs from F by interchanging the x- and y-values of each pair in F. We call this set G, so

$$G = \{(2, -2), (1, -1), (0, 0), (3, 1), (5, 2)\}.$$

To show that these two sets are related, G is called the *inverse* of F. For a function f to have an inverse, f must be a *one-to-one function*.

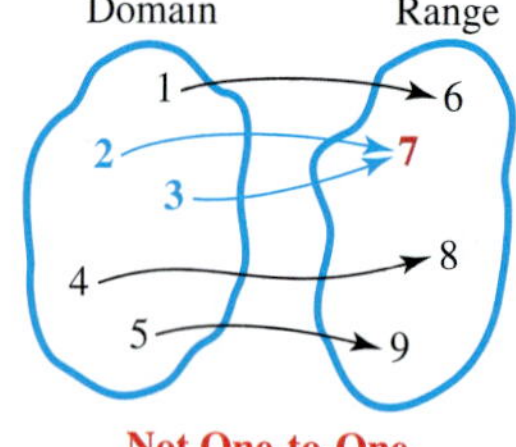

Not One-to-One

Figure 1

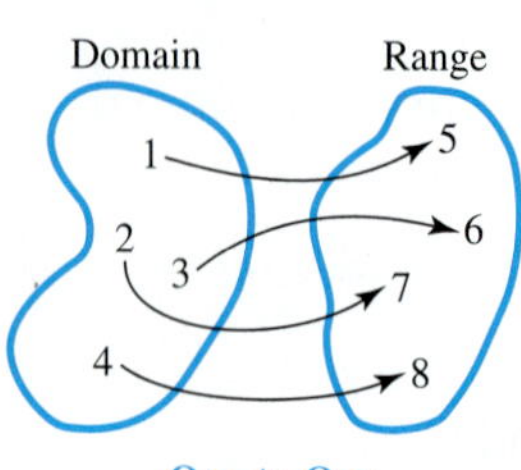

One-to-One

Figure 2

One-to-One Function

In a **one-to-one function,** each x-value corresponds to only one y-value, and each y-value corresponds to only one x-value.

The function shown in Figure 1 is not one-to-one because the y-value 7 corresponds to *two* x-values, 2 and 3. That is, the ordered pairs $(2, 7)$ and $(3, 7)$ both belong to the function. The function in Figure 2 is one-to-one.

EXAMPLE 1 Deciding Whether a Function is One-to-One

Determine whether the function defined by $f(x) = \sqrt{25 - x^2}$ is one-to-one.

Solution If any y-value in a function corresponds to more than one x-value, then the function is not one-to-one. Here, a given y-value may correspond to more than one x-value. For example, $x = 3$ and $x = -3$ correspond to $y = 4$.

$$f(3) = \sqrt{25 - 3^2} = \sqrt{25 - 9} = \sqrt{16} = 4$$

$$f(-3) = \sqrt{25 - (-3)^2} = \sqrt{25 - 9} = 4.$$

Thus, by definition, f is not a one-to-one function.

Now try Exercises 11 and 13.

As shown in Example 1, a way to show that a function is *not* one-to-one is to produce a pair of different numbers that lead to the same function value. There is also a useful graphical test, the *horizontal line test,* that tells whether or not a function is one-to-one.

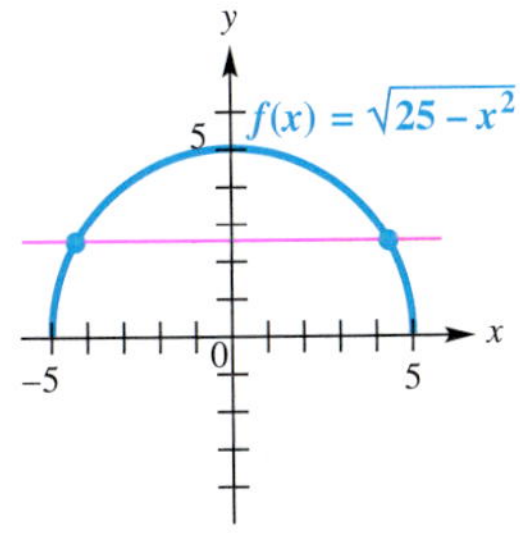

Figure 3

Horizontal Line Test

If any horizontal line intersects the graph of a function in no more than one point, then the function is one-to-one.

NOTE In Example 1, the graph of the function is a semicircle, as shown in Figure 3. There are infinitely many horizontal lines that intersect the graph in two points, so the horizontal line test shows that the function is not one-to-one.

EXAMPLE 2 Using the Horizontal Line Test

Determine whether each graph is the graph of a one-to-one function.

(a)

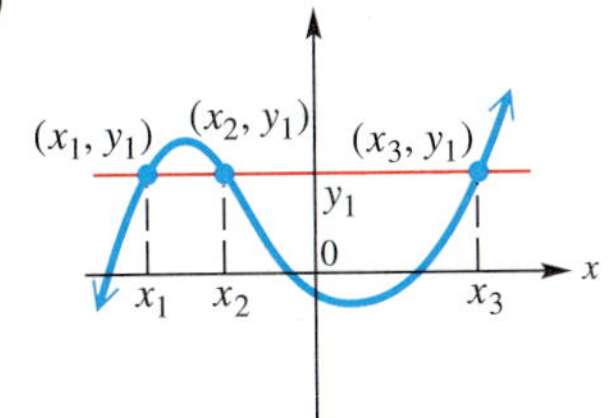

(b)

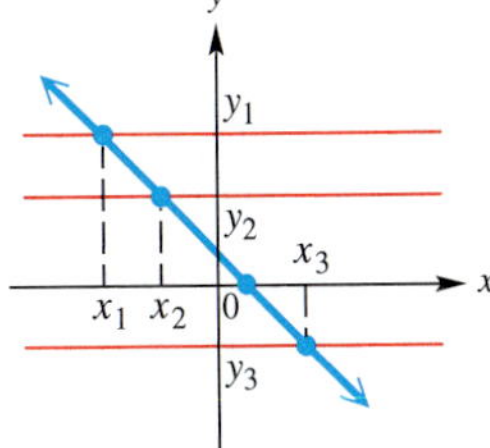

Solution

(a) Each point where the horizontal line intersects the graph has the same value of y but a different value of x. Since more than one (here three) different values of x lead to the same value of y, the function is not one-to-one.

(b) Since every horizontal line will intersect the graph at exactly one point, this function is one-to-one.

Now try Exercises 3 and 7.

Notice that the function graphed in Example 2(b) decreases on its entire domain. In general, a function that is either increasing or decreasing on its entire domain, such as $f(x) = -x$, $f(x) = x^3$, and $g(x) = \sqrt{x}$, must be one-to-one.

In summary, there are three ways to decide whether a function is one-to-one.

Tests for One-to-One Functions

1. In a one-to-one function every y-value corresponds to no more than one x-value. To show that a function is not one-to-one, find at least two x-values that produce the same y-value. (Example 1)
2. Sketch the graph and use the horizontal line test. (Example 2)
3. If the function either increases or decreases on its entire domain, then it is one-to-one. A sketch is helpful here, too. (Figure 3(b))

Inverse Functions As mentioned earlier, certain pairs of one-to-one functions "undo" one another. For example, if

$$f(x) = 8x + 5 \quad \text{and} \quad g(x) = \frac{x-5}{8},$$

then $f(10) = 8 \cdot 10 + 5 = 85$ and $g(85) = \dfrac{85-5}{8} = 10.$

Starting with 10, we "applied" function f and then "applied" function g to the result, which returned the number 10. See Figure 4.

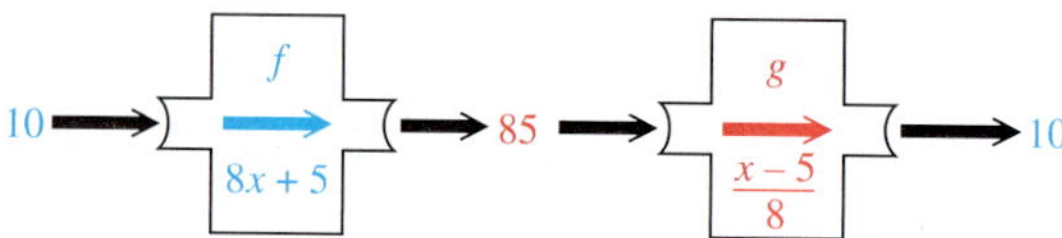

Figure 4

As further examples, check that

$$f(3) = 29 \quad \text{and} \quad g(29) = 3,$$
$$f(-5) = -35 \quad \text{and} \quad g(-35) = -5,$$
$$g(2) = -\frac{3}{8} \quad \text{and} \quad f\left(-\frac{3}{8}\right) = 2.$$

In particular, for this pair of functions,

$$f[g(2)] = 2 \quad \text{and} \quad g[f(2)] = 2.$$

In fact, for *any* value of x,

$$f[g(x)] = x \quad \text{and} \quad g[f(x)] = x,$$

or

$$(f \circ g)(x) = x \quad \text{and} \quad (g \circ f)(x) = x.$$

Because of this property, g is called the *inverse* of f.

Inverse Function

Let f be a one-to-one function. Then g is the **inverse function** of f if

$$(f \circ g)(x) = x \quad \text{for every } x \text{ in the domain of } g,$$

and

$$(g \circ f)(x) = x \quad \text{for every } x \text{ in the domain of } f.$$

EXAMPLE 3 Deciding Whether Two Functions Are Inverses

Let functions f and g be defined by $f(x) = x^3 - 1$ and $g(x) = \sqrt[3]{x + 1}$, respectively. Is g the inverse function of f?

Solution As shown in Figure 5, the horizontal line test applied to the graph indicates that f is one-to-one, so the function does have an inverse. Since it is one-to-one, we now find $(f \circ g)(x)$ and $(g \circ f)(x)$.

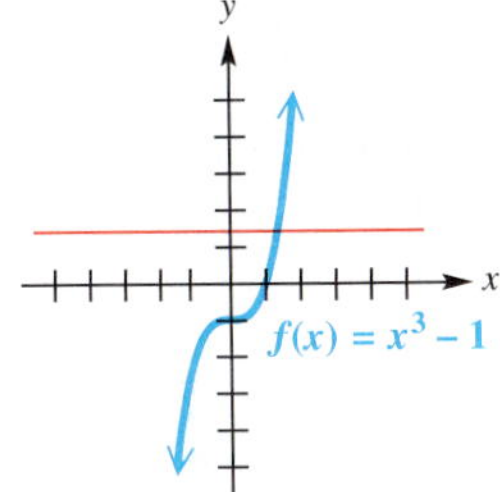

Figure 5

$$\begin{aligned}(f \circ g)(x) = f[g(x)] &= \left(\sqrt[3]{x + 1}\right)^3 - 1 \\ &= x + 1 - 1 \\ &= x\end{aligned}$$

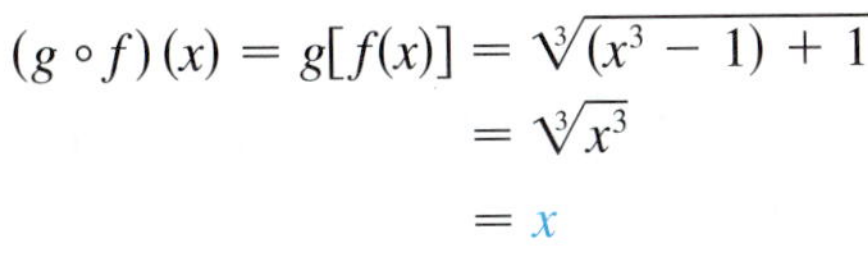

$$\begin{aligned}(g \circ f)(x) = g[f(x)] &= \sqrt[3]{(x^3 - 1) + 1} \\ &= \sqrt[3]{x^3} \\ &= x\end{aligned}$$

Since $(f \circ g)(x) = x$ and $(g \circ f)(x) = x$, function g is the inverse of function f.

Now try Exercise 39.

A special notation is often used for inverse functions: If g is the inverse of a function f, then g is written as f^{-1} (read "f-inverse"). In Example 3,

$$f^{-1}(x) = \sqrt[3]{x + 1}.$$

CAUTION Do not confuse the -1 in f^{-1} with a negative exponent. The symbol $f^{-1}(x)$ does not represent $\frac{1}{f(x)}$; it represents the inverse function of f.

By the definition of inverse function, the domain of f equals the range of f^{-1}, and the range of f equals the domain of f^{-1}. See Figure 6.

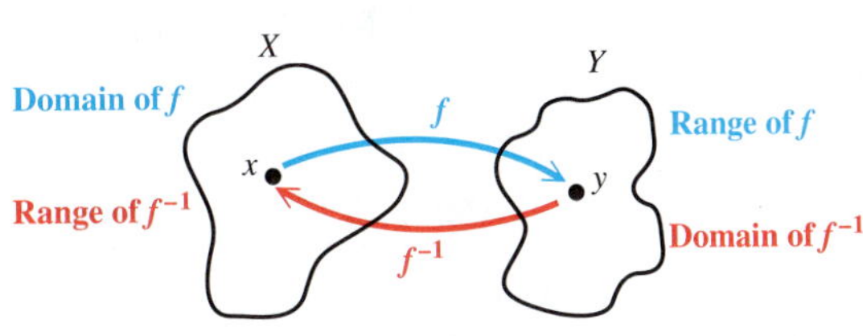

Figure 6

EXAMPLE 4 Finding Inverses of One-to-One Functions

Find the inverse of each function that is one-to-one.

(a) $F = \{(-2, 1), (-1, 0), (0, 1), (1, 2), (2, 2)\}$

(b) $G = \{(3, 1), (0, 2), (2, 3), (4, 0)\}$

(c) If the Pollutant Standard Index (PSI), an indicator of air quality, exceeds 100 on a particular day, then that day is classified as unhealthy. The table in the margin shows the number of unhealthy days in Chicago for selected years. Let f be the function defined in the table, with the years forming the domain and the numbers of unhealthy days forming the range.

Year	Number of Unhealthy Days
1991	21
1992	4
1993	3
1994	8
1995	21
1996	6
1997	9

Source: U.S. Environmental Protection Agency.

Solution

(a) Each x-value in F corresponds to just one y-value. However, the y-value 2 corresponds to two x-values, 1 and 2. Also, the y-value 1 corresponds to both -2 and 0. Because some y-values correspond to more than one x-value, F is not one-to-one and does not have an inverse.

(b) Every x-value in G corresponds to only one y-value, and every y-value corresponds to only one x-value, so G is a one-to-one function. The inverse function is found by interchanging the x- and y-values in each ordered pair.

$$G^{-1} = \{(1, 3), (2, 0), (3, 2), (0, 4)\}$$

Notice how the domain and range of G becomes the range and domain, respectively, of G^{-1}.

(c) Here, f is not one-to-one, because in two different years (1991 and 1995), the number of unhealthy days was the same, 21.

Now try Exercises 33, 45, and 47.

Equations of Inverses By definition, the inverse of a one-to-one function is found by interchanging the x- and y-values of each of its ordered pairs. The equation of the inverse of a function defined by $y = f(x)$ is found in the same way.

Finding the Equation of the Inverse of $y = f(x)$

For a one-to-one function f defined by an equation $y = f(x)$, find the defining equation of the inverse as follows.

Step 1 Interchange x and y.

Step 2 Solve for y.

Step 3 Replace y with $f^{-1}(x)$.

EXAMPLE 5 Finding Equations of Inverses

Decide whether each equation defines a one-to-one function. If so, find the equation of the inverse.

(a) $f(x) = 2x + 5$ **(b)** $y = x^2 + 2$ **(c)** $f(x) = (x - 2)^3$

Solution

(a) The graph of $y = 2x + 5$ is a nonhorizontal line, so by the horizontal line test, f is a one-to-one function. To find the equation of the inverse, follow the steps in the preceding box, first replacing $f(x)$ with y.

$$y = 2x + 5 \quad y = f(x)$$
$$x = 2y + 5 \quad \text{Interchange } x \text{ and } y. \text{ (Step 1)}$$
$$2y = x - 5 \quad \text{Solve for } y. \text{ (Step 2) } \textbf{(Section 1.1)}$$
$$y = \frac{x - 5}{2}$$
$$f^{-1}(x) = \frac{x - 5}{2} \quad \text{Replace } y \text{ with } f^{-1}(x). \text{ (Step 3)}$$

Thus, f^{-1} is a linear function. In the function defined by $y = 2x + 5$, the value of y is found by starting with a value of x, multiplying by 2, and adding 5. The equation for the inverse has us *subtract* 5, and then *divide* by 2. This shows how an inverse is used to "undo" what a function does to the variable x.

(b) The equation $y = x^2 + 2$ has a parabola opening up as its graph, so some horizontal lines will intersect the graph at two points. For example, both $x = 3$ and $x = -3$ correspond to $y = 11$. Because of the x^2-term, there are many pairs of x-values that correspond to the same y-value. This means that the function defined by $y = x^2 + 2$ is not one-to-one and does not have an inverse.

If we did not notice this, then following the steps for finding the equation of an inverse leads to

$$y = x^2 + 2$$
$$x = y^2 + 2 \quad \text{Interchange } x \text{ and } y.$$
$$x - 2 = y^2 \quad \text{Solve for } y.$$
$$\pm\sqrt{x - 2} = y. \quad \text{Square root property } \textbf{(Section 1.4)}$$

The last step shows that there are two y-values for each choice of $x > 2$, so the given function is not one-to-one and cannot have an inverse.

(c) Refer to Sections 2.5 and 2.6 to see that translations of the graph of the cubing function are one-to-one.

$$y = (x - 2)^3 \quad \text{Replace } f(x) \text{ with } y.$$
$$x = (y - 2)^3 \quad \text{Interchange } x \text{ and } y.$$
$$\sqrt[3]{x} = \sqrt[3]{(y - 2)^3} \quad \text{Take the cube root on each side. } \textbf{(Section 1.6)}$$
$$\sqrt[3]{x} = y - 2$$
$$\sqrt[3]{x} + 2 = y \quad \text{Solve for } y.$$
$$f^{-1}(x) = \sqrt[3]{x} + 2 \quad \text{Replace } y \text{ with } f^{-1}(x).$$

Now try Exercises 49(a), 53(a), and 55(a).

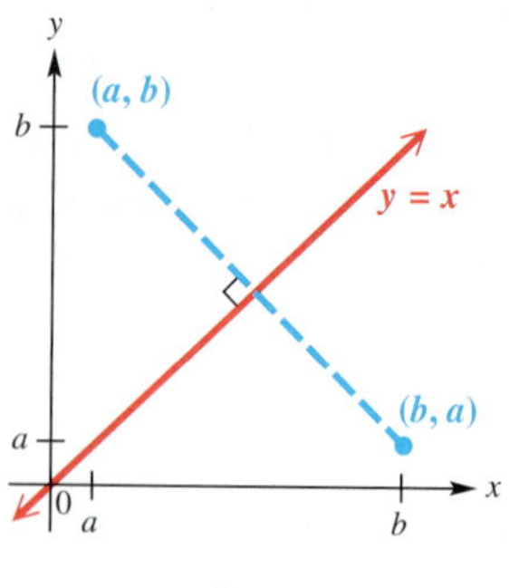

Figure 7

One way to graph the inverse of a function f whose equation is known is to find some ordered pairs that are on the graph of f, interchange x and y to get ordered pairs that are on the graph of f^{-1}, plot those points, and sketch the graph of f^{-1} through the points. A simpler way is to select points on the graph of f and use symmetry to find corresponding points on the graph of f^{-1}.

For example, suppose the point (a, b) shown in Figure 7 is on the graph of a one-to-one function f. Then the point (b, a) is on the graph of f^{-1}. The line segment connecting (a, b) and (b, a) is perpendicular to, and cut in half by, the line $y = x$. The points (a, b) and (b, a) are "mirror images" of each other with respect to $y = x$. For this reason we can find the graph of f^{-1} from the graph of f by locating the mirror image of each point in f with respect to the line $y = x$.

EXAMPLE 6 Graphing the Inverse

Graph the inverses of the functions f (shown in blue) in Figure 8.

Solution In Figure 8 the graphs of two functions f are shown in blue. Their inverses are shown in red. In each case, the graph of f^{-1} is a reflection of the graph of f with respect to the line $y = x$.

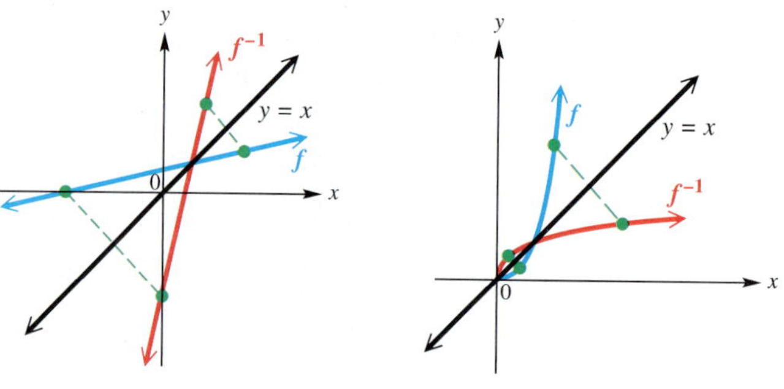

Figure 8

Now try Exercise 49(b).

EXAMPLE 7 Finding the Inverse of a Function with a Restricted Domain

Let $f(x) = \sqrt{x + 5}$. Find $f^{-1}(x)$.

Solution First, notice that the domain of f is restricted to the interval $[-5, \infty)$. Function f is one-to-one because it is increasing on its entire domain, and thus has an inverse function. Now we find the equation of the inverse.

$$y = \sqrt{x + 5}, \quad x \geq -5 \qquad y = f(x)$$

$$x = \sqrt{y + 5}, \quad y \geq -5 \qquad \text{Interchange } x \text{ and } y.$$

$$x^2 = y + 5 \qquad \text{Square both sides. (Section 1.6)}$$

$$y = x^2 - 5 \qquad \text{Solve for } y.$$

However, we cannot define $f^{-1}(x)$ as $x^2 - 5$. The domain of f is $[-5, \infty)$, and its range is $[0, \infty)$. The range of f is the domain of f^{-1}, so f^{-1} must be defined as

$$f^{-1}(x) = x^2 - 5, \qquad x \geq 0.$$

As a check, the range of f^{-1}, $[-5, \infty)$, is the domain of f. Graphs of f and f^{-1} are shown in Figures 9 and 10. The line $y = x$ is included on the graphs to show that the graphs of f and f^{-1} are mirror images with respect to this line.

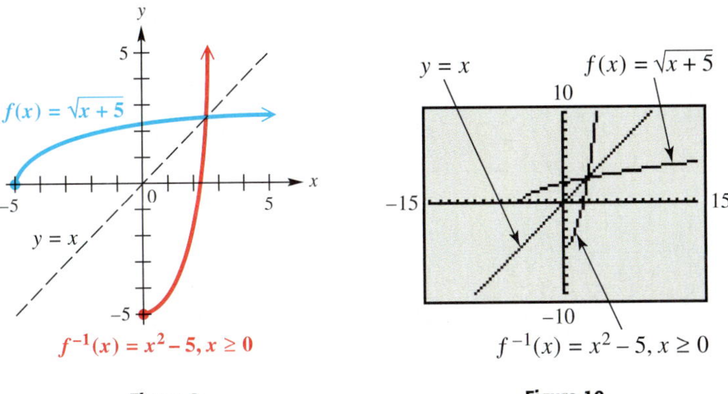

Figure 9 **Figure 10**

Now try Exercise 57.

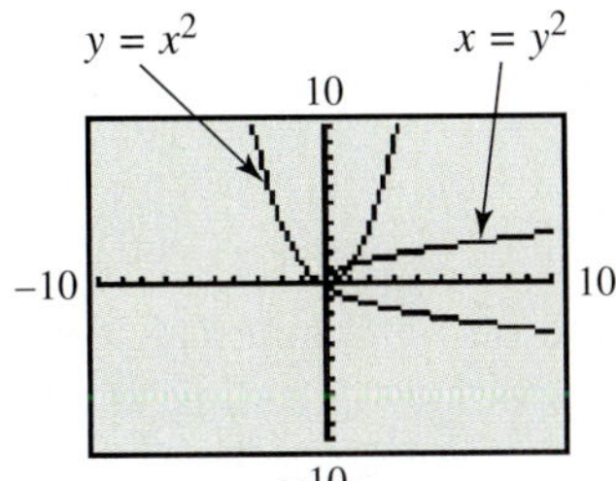

Despite the fact that $y = x^2$ is not one-to-one, the calculator will draw its "inverse," $x = y^2$.

Figure 11

Some graphing calculators have the capability of "drawing" the reflection of a graph across the line $y = x$. This feature does not require that the function be one-to-one, however, so the resulting figure may not be the graph of a function. See Figure 11. Again, it is necessary to understand the mathematics to interpret results correctly. ■

Important Facts about Inverses

1. If f is one-to-one, then f^{-1} exists.
2. The domain of f is equal to the range of f^{-1}, and the range of f is equal to the domain of f^{-1}.
3. If the point (a, b) lies on the graph of f, then (b, a) lies on the graph of f^{-1}, so the graphs of f and f^{-1} are reflections of each other across the line $y = x$.
4. To find the equation for f^{-1}, replace $f(x)$ with y, interchange x and y, and solve for y. This gives $f^{-1}(x)$.

An Application of Inverse Functions to Cryptography A one-to-one function and its inverse can be used to make information secure. The function is used to encode a message, and its inverse is used to decode the coded message. In practice, complicated functions are used. We illustrate the process with the simple function defined by $f(x) = 2x + 5$. Each letter of the alphabet is assigned a numerical value according to its position in the alphabet, as follows.

A	1	H	8	O	15	V	22
B	2	I	9	P	16	W	23
C	3	J	10	Q	17	X	24
D	4	K	11	R	18	Y	25
E	5	L	12	S	19	Z	26
F	6	M	13	T	20		
G	7	N	14	U	21		

A	1	N	14
B	2	O	15
C	3	P	16
D	4	Q	17
E	5	R	18
F	6	S	19
G	7	T	20
H	8	U	21
I	9	V	22
J	10	W	23
K	11	X	24
L	12	Y	25
M	13	Z	26

EXAMPLE 8 Using a Function to Encode a Message

Use the one-to-one function defined by $f(x) = 2x + 5$ and the numerical values repeated in the margin to encode the word ALGEBRA.

Solution The word ALGEBRA would be encoded as

$$7 \quad 29 \quad 19 \quad 15 \quad 9 \quad 41 \quad 7,$$

because A corresponds to 1, and

$$f(1) = 2(1) + 5 = 7,$$

L corresponds to 12, and

$$f(12) = 2(12) + 5 = 29,$$

and so on. Using the inverse $f^{-1}(x) = \frac{x-5}{2}$ to decode,

$$f^{-1}(7) = \frac{7-5}{2} = 1,$$

which corresponds to A, and

$$f^{-1}(29) = \frac{29-5}{2} = 12,$$

which corresponds to L, and so on. The table feature of a graphing calculator can be very useful for this procedure. Figure 12 shows how decoding can be accomplished for the word ALGEBRA.

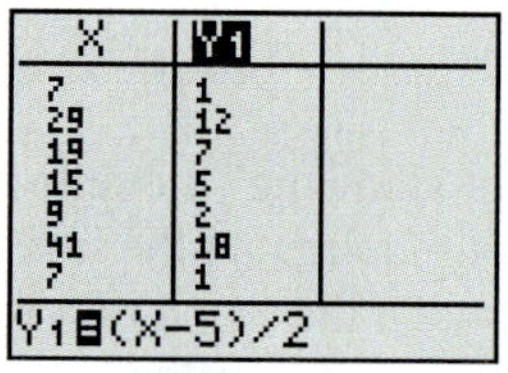

X	Y1
7	1
29	12
19	7
15	5
9	2
41	18
7	1

Y1=(X−5)/2

Figure 12

Now try Exercise 85.

4.1 Exercises

Concept Check *In Exercises 1 and 2, answer the question and then write a short explanation.*

1. The table shows the number of uncontrolled hazardous waste sites that require further investigation to determine whether remedies are needed under the Superfund program. The seven states listed are ranked in the top ten in the United States.

If this correspondence is considered to be a function that pairs each state with its number of uncontrolled waste sites, is it one-to-one? If not, explain why.

State	Number of Sites
New Jersey	108
Pennsylvania	101
California	94
New York	79
Florida	53
Illinois	40
Wisconsin	40

Source: U.S. Environmental Protection Agency.

2. The table shows emissions of a major air pollutant, carbon monoxide, in the United States for the years 1992 through 1998.

If this correspondence is considered to be a function that pairs each year with its emissions amount, is it one-to-one? If not, explain why.

Year	Amount of Emissions (in thousands of tons)
1992	97,630
1993	98,160
1994	102,643
1995	93,353
1996	95,479
1997	94,410
1998	89,454

Source: U.S. Environmental Protection Agency.

Decide whether each function as defined or graphed is one-to-one. See Examples 1 and 2.

3.

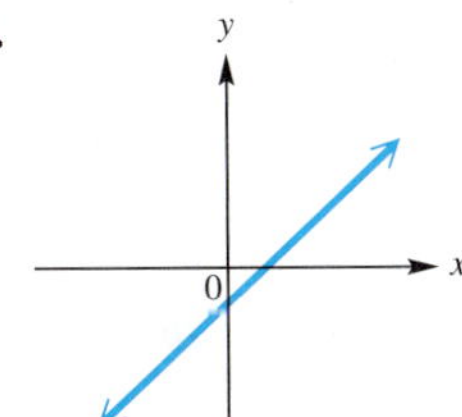

4.

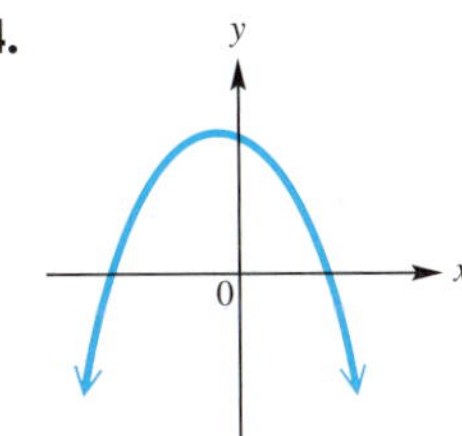

5.

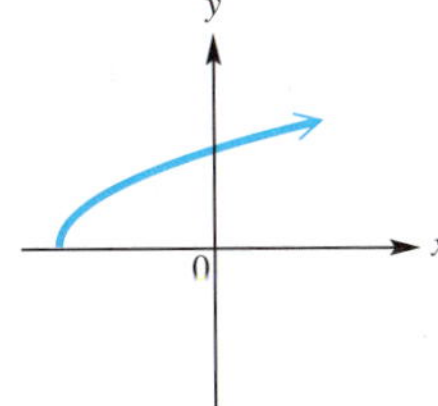

6.

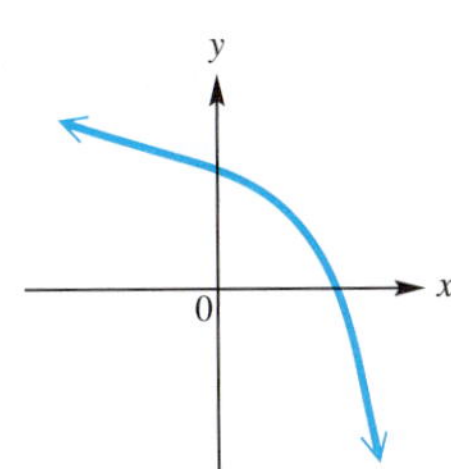

7.

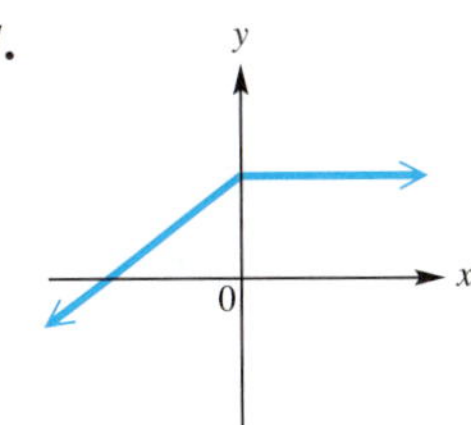

8.

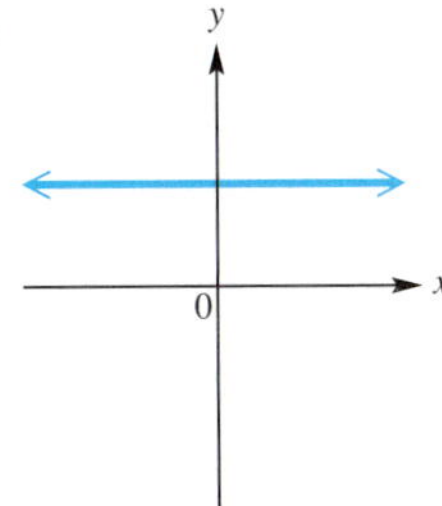

9. $y = (x - 2)^2$

10. $y = -(x + 3)^2 - 8$

11. $y = \sqrt{36 - x^2}$

12. $y = -\sqrt{100 - x^2}$

13. $y = 2x^3 + 1$

14. $y = -\sqrt[3]{x + 5}$

15. $y = \dfrac{1}{x + 2}$

16. $y = \dfrac{-4}{x - 8}$

Concept Check *Answer each of the following.*

17. For a function to have an inverse, it must be ____________.

18. If two functions f and g are inverses, then $(f \circ g)(x) =$ ____________, and ____________ $= x$.

19. The domain of f is equal to the ____________ of f^{-1}, and the range of f is equal to the ____________ of f^{-1}.

20. If the point (a, b) lies on the graph of f, and f has an inverse, then the point ____________ lies on the graph of f^{-1}.

21. True or false: If $f(x) = x^2$, then $f^{-1}(x) = \sqrt{x}$.

22. If a function f has an inverse, then the graph of f^{-1} may be obtained by reflecting the graph of f across the line with equation ____________.

23. If a function f has an inverse and $f(-3) = 6$, then $f^{-1}(6) =$ ________.

24. If $f(-4) = 16$ and $f(4) = 16$, then f ________ (does/does not) have an inverse because ________.

25. Explain why a polynomial function of even degree cannot have an inverse.

26. Explain why a polynomial function of odd degree *may* not be one-to-one.

Concept Check *In Exercises 27–32, an everyday activity is described. Keeping in mind that an inverse operation "undoes" what an operation does, describe each inverse activity.*

27. tying your shoelaces **28.** starting a car **29.** entering a room

30. climbing the stairs **31.** fast-forwarding a tape **32.** filling a cup

Decide whether the given functions are inverses. See Example 4.

33.

x	$f(x)$
3	−4
2	−6
5	8
1	9
4	3

x	$g(x)$
−4	3
−6	2
8	5
9	1
3	4

34.

x	$f(x)$
−2	−8
−1	−1
0	0
1	1
2	8

x	$g(x)$
8	−2
1	−1
0	0
−1	1
−8	2

35. $f = \{(2, 5), (3, 5), (4, 5)\}$; $g = \{(5, 2)\}$

Determine whether each pair of functions as graphed or defined are inverses of each other. See Examples 3 and 6.

36.

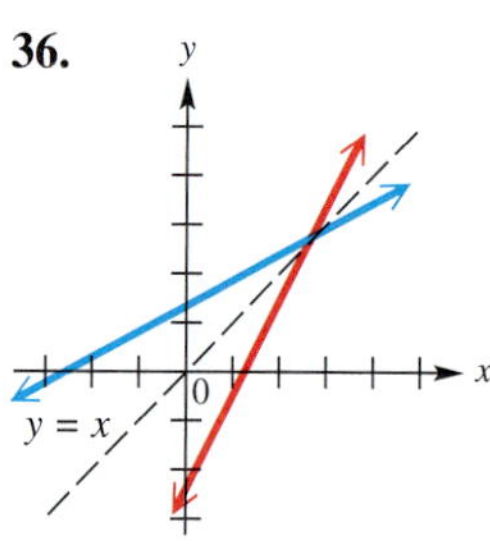

37.

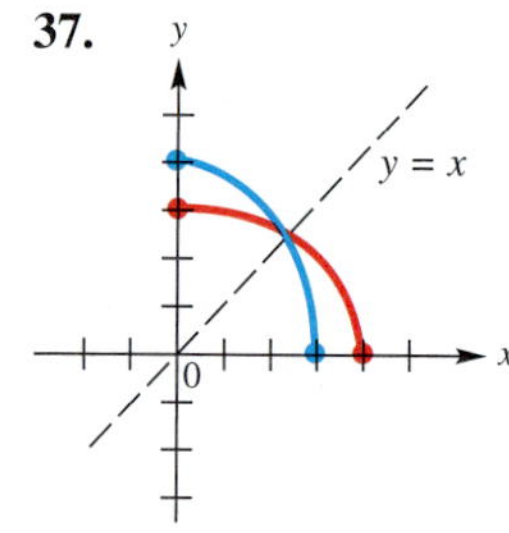

38.

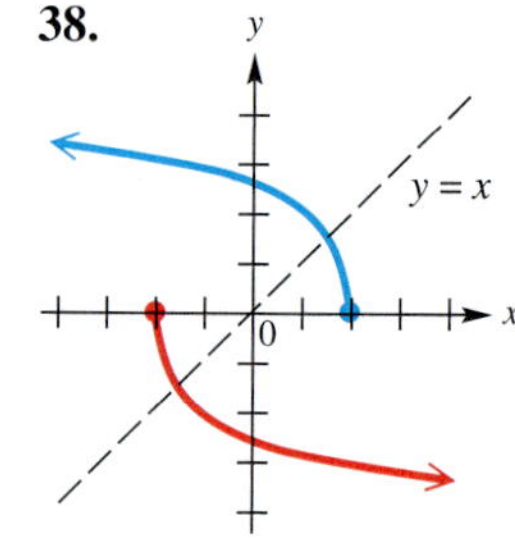

39. $f(x) = 2x + 4, \quad g(x) = \dfrac{1}{2}x - 2$

40. $f(x) = 8x - 7, \quad g(x) = \dfrac{x + 8}{7}$

41. $f(x) = \dfrac{2}{x + 6}, \quad g(x) = \dfrac{6x + 2}{x}$

42. $f(x) = \dfrac{1}{x + 1}, \quad g(x) = \dfrac{1 - x}{x}$

43. $f(x) = x^2 + 3$, domain $[0, \infty)$; $g(x) = \sqrt{x - 3}$, domain $[3, \infty)$

44. $f(x) = \sqrt{x + 8}$, domain $[-8, \infty)$; $g(x) = x^2 - 8$, domain $[0, \infty)$

If the function is one-to-one, find its inverse. See Example 4.

45. $\{(-3, 6), (2, 1), (5, 8)\}$

46. $\left\{(3, -1), (5, 0), (0, 5), \left(4, \dfrac{2}{3}\right)\right\}$

47. $\{(1, -3), (2, -7), (4, -3), (5, -5)\}$

48. $\{(6, -8), (3, -4), (0, -8), (5, -4)\}$

*For each function as defined that is one-to-one, **(a)** write an equation for the inverse function in the form $y = f^{-1}(x)$, **(b)** graph f and f^{-1} on the same axes, and **(c)** give the domain and the range of f and f^{-1}. If the function is not one-to-one, say so.*

49. $y = 3x - 4$ **50.** $y = 4x - 5$ **51.** $f(x) = -4x + 3$

52. $f(x) = -6x - 8$

53. $f(x) = x^3 + 1$

54. $f(x) = -x^3 - 2$

55. $y = x^2$

56. $y = -x^2 + 2$

57. $y = \frac{1}{x}$

58. $y = \frac{4}{x}$

59. $f(x) = \frac{1}{x - 3}$

60. $f(x) = \frac{1}{x + 2}$

61. $f(x) = \sqrt{6 + x},\ x \geq -6$

62. $f(x) = -\sqrt{x^2 - 16},\ x \geq 4$

Graph the inverse of each one-to-one function. See Example 6.

63.

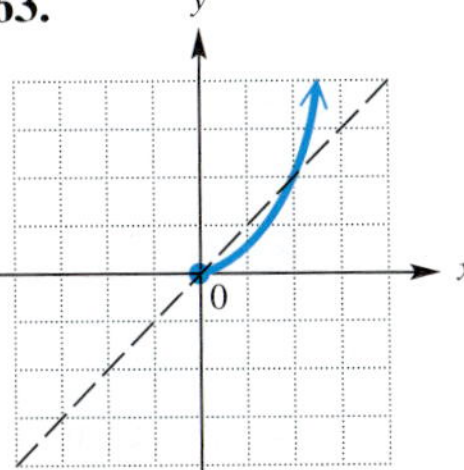

64.

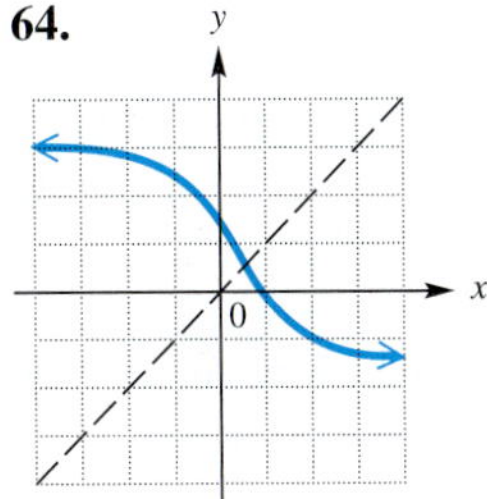

65.

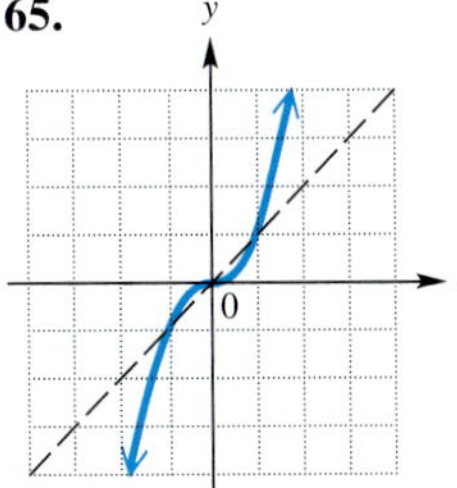

66.

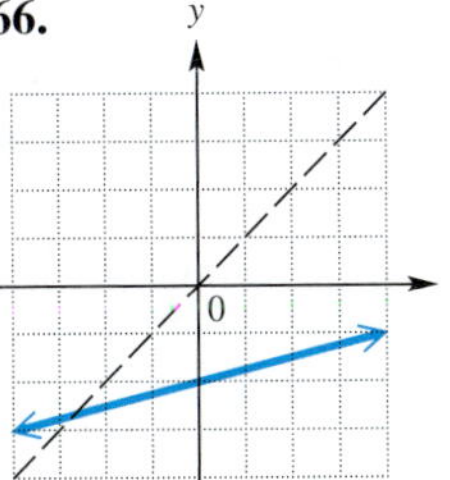

67.

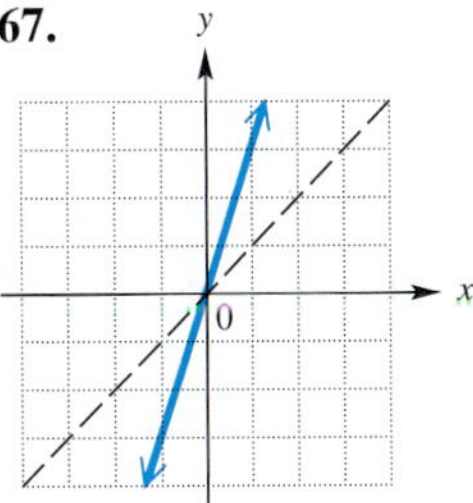

68.

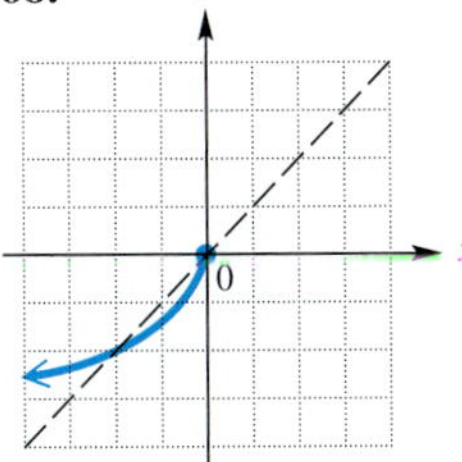

Concept Check *The graph of a function f is shown in the figure. Use the graph to find each value.*

69. $f^{-1}(4)$

70. $f^{-1}(2)$

71. $f^{-1}(0)$

72. $f^{-1}(-2)$

73. $f^{-1}(-3)$

74. $f^{-1}(-4)$

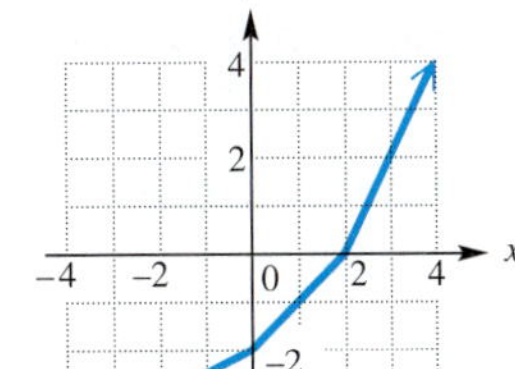

Concept Check *Answer each of the following.*

75. Suppose $f(x)$ is the number of cars that can be built for x dollars. What does $f^{-1}(1000)$ represent?

76. Suppose $f(r)$ is the volume (in cubic inches) of a sphere of radius r inches. What does $f^{-1}(5)$ represent?

77. If a line has slope a, what is the slope of its reflection across the line $y = x$?

78. Find $f^{-1}(f(2))$, where $f(2) = 3$.

Use a graphing calculator to graph each function defined as follows, using the given viewing window. Use the graph to decide which functions are one-to-one. If a function is one-to-one, give the equation of its inverse.

79. $f(x) = 6x^3 + 11x^2 - x - 6$; $[-3, 2]$ by $[-10, 10]$

80. $f(x) = x^4 - 5x^2 + 6$; $[-3, 3]$ by $[-1, 8]$

81. $f(x) = \dfrac{x - 5}{x + 3}$; $[-8, 8]$ by $[-6, 8]$

82. $f(x) = \dfrac{-x}{x - 4}$; $[-1, 8]$ by $[-6, 6]$

Use the alphabet coding assignment given for Example 8 for Exercises 83–86.

83. The function defined by $f(x) = 3x - 2$ was used to encode a message as

37 25 19 61 13 34 22 1 55 1 52 52 25 64 13 10.

Find the inverse function and determine the message.

84. The function defined by $f(x) = 2x - 9$ was used to encode a message as

−5 9 5 5 9 27 15 29 −1 21 19 31 −3 27 41.

Find the inverse function and determine the message.

85. Encode the message SEND HELP, using the one-to-one function defined by $f(x) = x^3 - 1$. Give the inverse function that the decoder would need when the message is received.

86. Encode the message SAILOR BEWARE, using the one-to-one function defined by $f(x) = (x + 1)^3$. Give the inverse function that the decoder would need when the message is received.

87. Why is a one-to-one function essential in this encoding/decoding process?

4.2 Exponential Functions

Exponents and Properties ▪ Exponential Functions ▪ Exponential Equations ▪ Compound Interest ▪ The Number *e* and Continuous Compounding ▪ Exponential Models and Curve Fitting

Exponents and Properties Recall the definition of a^r, where r is a rational number: if $r = \frac{m}{n}$, then for appropriate values of m and n,

$$a^{m/n} = \left(\sqrt[n]{a}\right)^m. \quad \text{(Section R.7)}$$

For example,
$$16^{3/4} = \left(\sqrt[4]{16}\right)^3 = 2^3 = 8,$$

$$27^{-1/3} = \frac{1}{27^{1/3}} = \frac{1}{\sqrt[3]{27}} = \frac{1}{3},$$

and
$$64^{-1/2} = \frac{1}{64^{1/2}} = \frac{1}{\sqrt{64}} = \frac{1}{8}.$$

In this section we extend the definition of a^r to include all real (not just rational) values of the exponent r. For example, $2^{\sqrt{3}}$ might be evaluated by approximating the exponent $\sqrt{3}$ with the rational numbers 1.7, 1.73, 1.732, and so on. Since these decimals approach the value of $\sqrt{3}$ more and more closely, it seems reasonable that $2^{\sqrt{3}}$ should be approximated more and more closely by the numbers $2^{1.7}$, $2^{1.73}$, $2^{1.732}$, and so on. (Recall, for example, that $2^{1.7} = 2^{17/10} = \left(\sqrt[10]{2}\right)^{17}$. In fact, this is exactly how $2^{\sqrt{3}}$ is defined (in a more advanced course). To show that this assumption is reasonable, Figure 13 gives graphs of the function $f(x) = 2^x$ with three different domains.

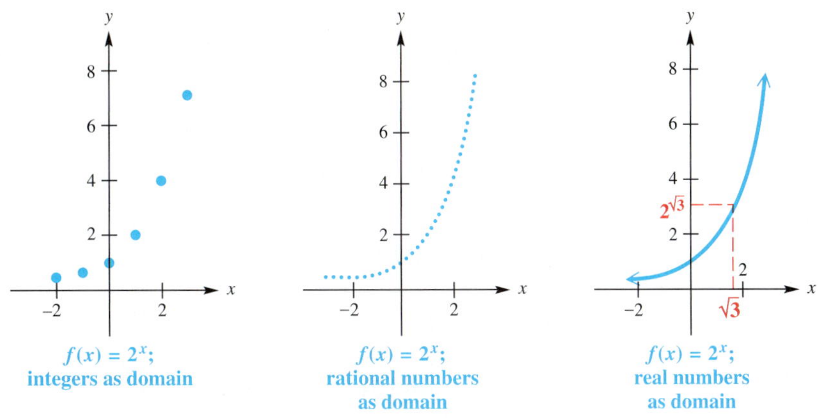

Figure 13

Using this interpretation of real exponents, all rules and theorems for exponents are valid for all real number exponents, not just rational ones. In addition to the rules for exponents presented earlier, we use several new properties in this chapter. For example, if $y = 2^x$, then each real value of x leads to exactly one value of y, and therefore, $y = 2^x$ defines a function. Furthermore,

$$\text{if} \quad 2^x = 2^4, \quad \text{then} \quad x = 4,$$

$$\text{and} \quad \text{if} \quad x = 4, \quad \text{then} \quad 2^x = 2^4.$$

$$\text{Also,} \quad 4^2 < 4^3 \quad \text{but} \quad \left(\frac{1}{2}\right)^2 > \left(\frac{1}{2}\right)^3.$$

In general, when $a > 1$, increasing the exponent on a leads to a *larger* number, but when $0 < a < 1$, increasing the exponent on a leads to a *smaller* number.

These properties are generalized below. Proofs of the properties are not given here, as they require more advanced mathematics.

Additional Properties of Exponents

For any real number $a > 0$, $a \neq 1$, the following statements are true.

(a) a^x **is a unique real number for all real numbers** x.

(b) $a^b = a^c$ **if and only if** $b = c$.

(c) If $a > 1$ **and** $m < n$, **then** $a^m < a^n$.

(d) If $0 < a < 1$ **and** $m < n$, **then** $a^m > a^n$.

Properties (a) and (b) require $a > 0$ so that a^x is always defined. For example, $(-6)^x$ is not a real number if $x = \frac{1}{2}$. This means that a^x will always be positive, since a must be positive. In property (a), a cannot equal 1 because $1^x = 1$ for every real number value of x, so each value of x leads to the same real number, 1. For Property (b) to hold, a must not equal 1 since, for example, $1^4 = 1^5$, even though $4 \neq 5$.

EXAMPLE 1 Evaluating an Exponential Expression

If $f(x) = 2^x$, find each of the following.

(a) $f(-1)$ **(b)** $f(3)$ **(c)** $f\left(\frac{5}{2}\right)$ **(d)** $f(4.92)$

Solution

(a) $f(-1) = 2^{-1} = \frac{1}{2}$ Replace x with -1.

(b) $f(3) = 2^3 = 8$

(c) $f\left(\frac{5}{2}\right) = 2^{5/2} = (2^5)^{1/2} = 32^{1/2} = \sqrt{32} = 4\sqrt{2}$ (Section R.7)

(d) $f(4.92) = 2^{4.92} \approx 30.2738447$ Use a calculator.

Now try Exercises 3, 9, and 11.

Exponential Functions We can now define a function $f(x) = a^x$ whose domain is the set of all real numbers.

Exponential Function

If $a > 0$ and $a \neq 1$, then

$$f(x) = a^x$$

defines the **exponential function** with base a.

NOTE If $a = 1$, the function becomes the constant function defined by $f(x) = 1$, not an exponential function.

Figure 13 showed the graph of $f(x) = 2^x$ with three different domains. We repeat the final graph (with real numbers as domain) here. The y-intercept is $y = 2^0 = 1$. Since $2^x > 0$ for all x and $2^x \to 0$ as $x \to -\infty$, the x-axis is a horizontal asymptote. As the graph suggests, the domain of the function is $(-\infty, \infty)$ and the range is $(0, \infty)$. The function is increasing on its entire domain, and therefore is one-to-one. Our observations from Figure 13 lead to the following generalizations about the graphs of exponential functions.

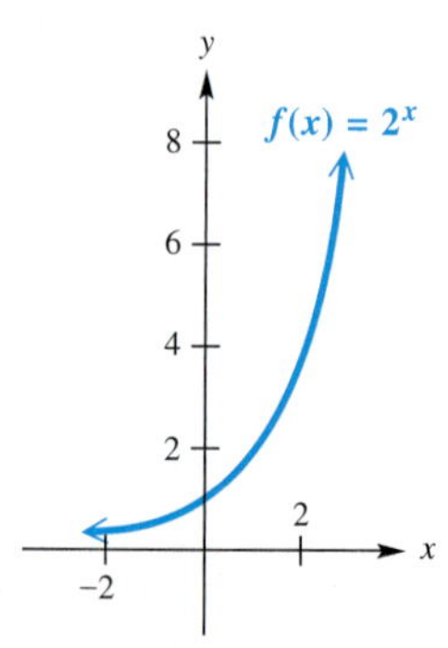

EXPONENTIAL FUNCTION $f(x) = a^x$

Domain: $(-\infty, \infty)$ Range: $(0, \infty)$

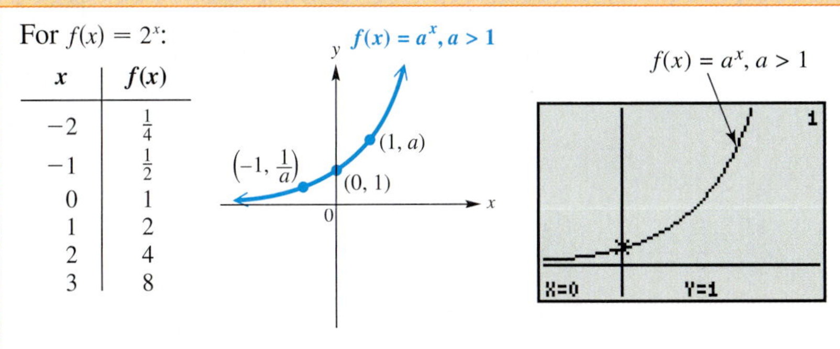

For $f(x) = 2^x$:

x	$f(x)$
-2	$\frac{1}{4}$
-1	$\frac{1}{2}$
0	1
1	2
2	4
3	8

Figure 14

- $f(x) = a^x$, $a > 1$, is increasing and continuous on its entire domain, $(-\infty, \infty)$.
- The x-axis is a horizontal asymptote as $x \to -\infty$.
- The graph goes through the points $\left(-1, \frac{1}{a}\right)$, $(0, 1)$, and $(1, a)$.

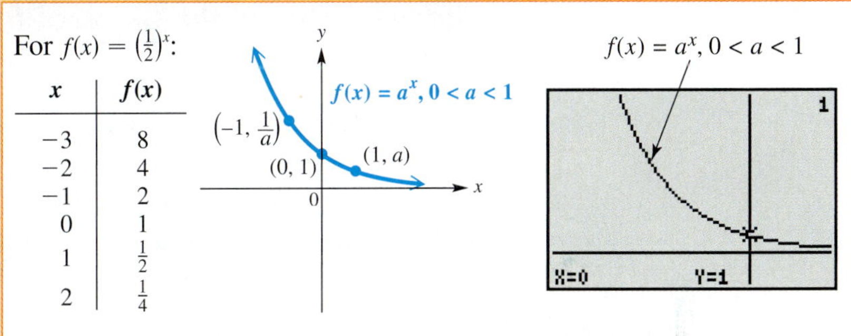

For $f(x) = \left(\frac{1}{2}\right)^x$:

x	$f(x)$
-3	8
-2	4
-1	2
0	1
1	$\frac{1}{2}$
2	$\frac{1}{4}$

Figure 15

- $f(x) = a^x$, $0 < a < 1$, is decreasing and continuous on its entire domain, $(-\infty, \infty)$.
- The x-axis is a horizontal asymptote as $x \to \infty$.
- The graph goes through the points $\left(-1, \frac{1}{a}\right)$, $(0, 1)$, and $(1, a)$.

Starting with $f(x) = 2^x$ and replacing x with $-x$ gives $f(-x) = 2^{-x} = (2^{-1})^x = \left(\frac{1}{2}\right)^x$. For this reason, the graphs of $f(x) = 2^x$ and $f(x) = \left(\frac{1}{2}\right)^x$ are reflections of each other across the y-axis. This is supported by the graphs in Figures 14 and 15.

The graph of $f(x) = 2^x$ is typical of graphs of $f(x) = a^x$ where $a > 1$. For larger values of a, the graphs rise more steeply, but the general shape is similar to the graph in Figure 14. When $0 < a < 1$, the graph decreases in a manner similar to the graph of $f(x) = \left(\frac{1}{2}\right)^x$. In Figure 16 on the next page, the graphs of several typical exponential functions illustrate these facts.

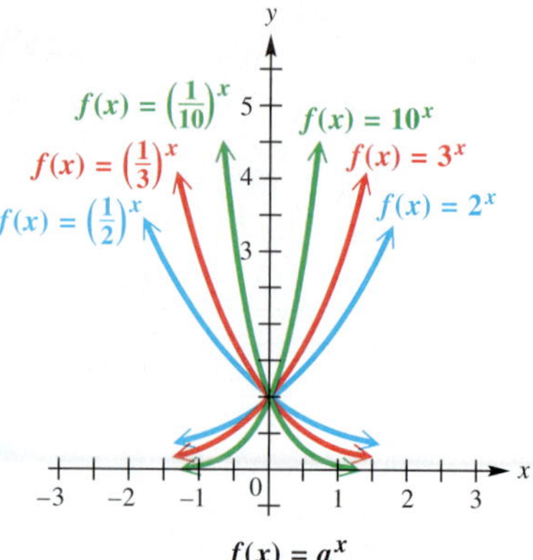

$f(x) = a^x$

Domain: $(-\infty, \infty)$; Range: $(0, \infty)$

When $a > 1$, the function is increasing.

When $0 < a < 1$, the function is decreasing.

Figure 16

In summary, the graph of a function of the form $f(x) = a^x$ has the following features.

Characteristics of the Graph of $f(x) = a^x$

1. The points $\left(-1, \frac{1}{a}\right)$, $(0, 1)$, and $(1, a)$ are on the graph.
2. If $a > 1$, then f is an increasing function; if $0 < a < 1$, then f is a decreasing function.
3. The x-axis is a horizontal asymptote.
4. The domain is $(-\infty, \infty)$, and the range is $(0, \infty)$.

EXAMPLE 2 Graphing an Exponential Function

Graph $f(x) = 5^x$.

Solution The y-intercept is 1, and the x-axis is a horizontal asymptote. Plot a few ordered pairs, and draw a smooth curve through them as shown in Figure 17. Like the function $f(x) = 2^x$, this function also has domain $(-\infty, \infty)$ and range $(0, \infty)$ and is one-to-one. The graph is increasing on its entire domain.

x	$f(x)$
-1	.2
0	1
.5	2.2
1	5
1.5	11.2
2	25

Figure 17

Now try Exercise 13.

EXAMPLE 3 Graphing Reflections and Translations

Graph each function.

(a) $f(x) = -2^x$ **(b)** $f(x) = 2^{x+3}$ **(c)** $f(x) = 2^x + 3$

Solution

(a) The graph of $f(x) = -2^x$ is that of $f(x) = 2^x$ reflected across the x-axis. The domain is $(-\infty, \infty)$, and the range is $(-\infty, 0)$. See Figure 18.

(b) The graph of $f(x) = 2^{x+3}$ is the graph of $f(x) = 2^x$ translated 3 units to the left, as shown in Figure 19.

(c) The graph of $f(x) = 2^x + 3$ is that of $f(x) = 2^x$ translated 3 units up. See Figure 20.

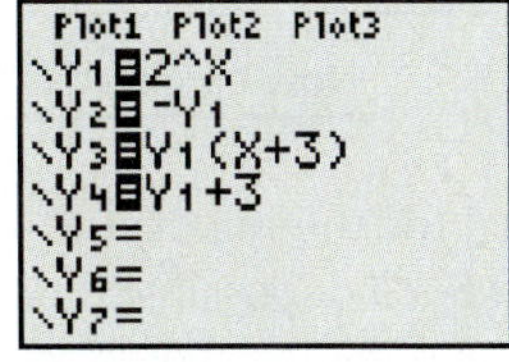

The screen shows how a graphing calculator can be directed to graph the three functions in Example 3. Y_1 is defined as 2^X, and Y_2, Y_3, and Y_4 are defined as reflections and/or translations of Y_1.

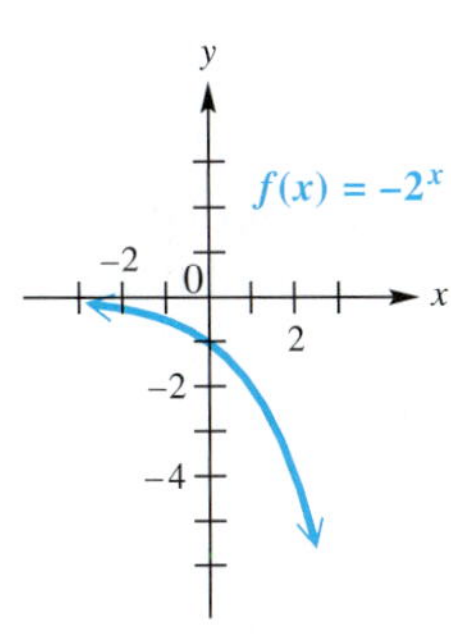

Figure 18

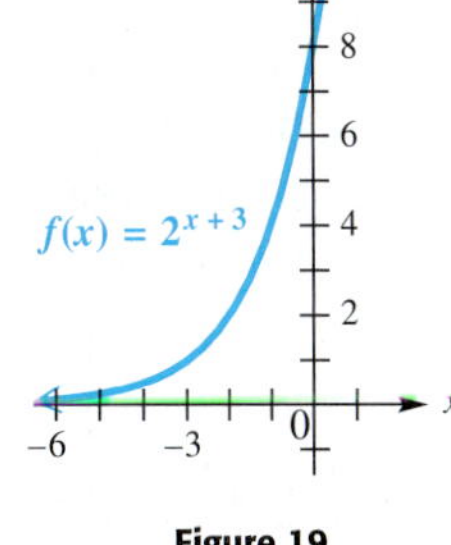

Figure 19

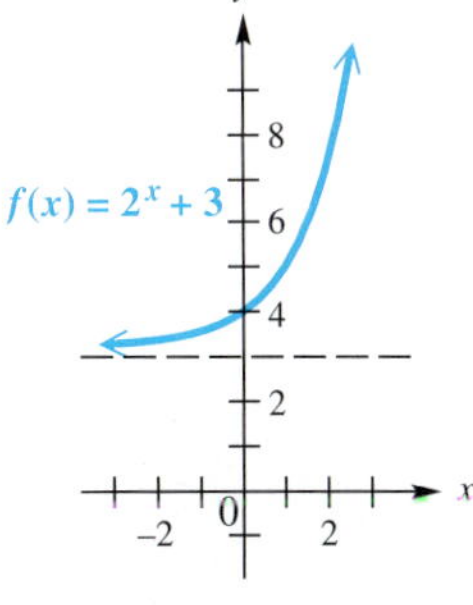

Figure 20

Now try Exercises 25 and 27.

Exponential Equations

Property (b) given at the beginning of this section is used to solve **exponential equations,** equations with variables as exponents.

EXAMPLE 4 Using a Property of Exponents to Solve an Equation

Solve $\left(\frac{1}{3}\right)^x = 81$.

Solution Write each side of the equation using a common base.

$$\left(\frac{1}{3}\right)^x = 81$$

$$(3^{-1})^x = 81 \quad \text{Definition of negative exponent (Section R.6)}$$

$$3^{-x} = 81 \quad (a^m)^n = a^{mn} \text{ (Section R.3)}$$

$$3^{-x} = 3^4 \quad \text{Write 81 as a power of 3. (Section R.1)}$$

$$-x = 4 \quad \text{Property (b)}$$

$$x = -4 \quad \text{Multiply by } -1.$$

The solution set of the original equation is $\{-4\}$.

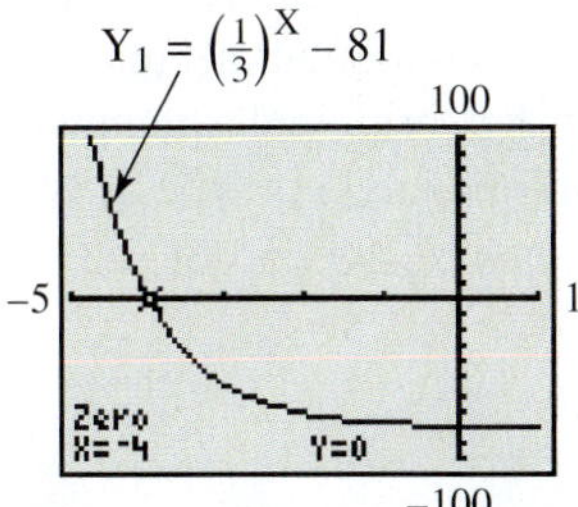

The screen supports the algebraic result in Example 4, using the x-intercept method of solution.

Now try Exercise 45.

EXAMPLE 5 Using a Property of Exponents to Solve an Equation

Solve $2^{x+4} = 8^{x-6}$.

Solution Write each side of the equation using a common base.

$$\begin{aligned} 2^{x+4} &= 8^{x-6} & & \\ 2^{x+4} &= (2^3)^{x-6} & & \text{Write 8 as a power of 2.} \\ 2^{x+4} &= 2^{3x-18} & & (a^m)^n = a^{mn} \\ x + 4 &= 3x - 18 & & \text{Property (b)} \\ -2x &= -22 & & \text{Subtract } 3x \text{ and 4.} \\ x &= 11 & & \text{Divide by } -2. \end{aligned}$$

Check by substituting 11 for x in the original equation. The solution set is $\{11\}$.

Now try Exercise 53.

Later in this chapter, we describe a general method for solving exponential equations where the approach used in Examples 4 and 5 is not possible. For instance, the above method could not be used to solve an equation like $7^x = 12$, since it is not easy to express both sides as exponential expressions with the same base.

EXAMPLE 6 Using a Property of Exponents to Solve an Equation

Solve $81 = b^{4/3}$.

Solution Begin by writing $b^{4/3}$ as $\left(\sqrt[3]{b}\right)^4$.

$$\begin{aligned} 81 &= \left(\sqrt[3]{b}\right)^4 & & \text{Radical notation for } a^{m/n} \textbf{ (Section R.7)} \\ \pm 3 &= \sqrt[3]{b} & & \text{Take fourth roots on both sides. } \textbf{(Section 1.6)} \\ \pm 27 &= b & & \text{Cube both sides.} \end{aligned}$$

Check *both* solutions in the original equation. Both check, so the solution set is $\{-27, 27\}$.

Now try Exercise 61.

Compound Interest The formula for *compound interest* (interest paid on both principal and interest) is an important application of exponential functions. Recall the formula for simple interest, $I = Prt$, where P is principal (amount deposited), r is annual rate of interest expressed as a decimal, and t is time in years that the principal earns interest. Suppose $t = 1$ yr. Then at the end of the year the amount has grown to

$$P + Pr = P(1 + r),$$

the original principal plus interest. If this balance earns interest at the same interest rate for another year, the balance at the end of that year will be

$$\begin{aligned} [P(1 + r)] + [P(1 + r)]r &= [P(1 + r)](1 + r) \\ &= P(1 + r)^2. \end{aligned}$$

After the third year, this will grow to

$$\begin{aligned}[P(1+r)^2] + [P(1+r)^2]r &= [P(1+r)^2](1+r) \\ &= P(1+r)^3.\end{aligned}$$

Continuing in this way produces the formula

$$A = P(1+r)^t$$

for interest compounded annually.

The following general formula for compound interest can be derived in the same way as the formula given above.

Compound Interest

If P dollars are deposited in an account paying an annual rate of interest r compounded (paid) m times per year, then after t years the account will contain A dollars, where

$$A = P\left(1 + \frac{r}{m}\right)^{tm}.$$

In the formula for compound interest, A is sometimes called the **future value** and P the **present value.** A is also called the **compound amount** and is the balance *after* interest has been earned.

EXAMPLE 7 Using the Compound Interest Formula

Suppose \$1000 is deposited in an account paying 4% interest per year compounded quarterly (four times per year).

(a) Find the amount in the account after 10 yr with no withdrawals.

(b) How much interest is earned over the 10-yr period?

Solution

(a)

$$A = P\left(1 + \frac{r}{m}\right)^{tm} \quad \text{Compound interest formula}$$

$$A = 1000\left(1 + \frac{.04}{4}\right)^{10(4)} \quad \text{Let } P = 1000,\ r = .04,\ m = 4, \text{ and } t = 10.$$

$$A = 1000(1 + .01)^{40}$$

$$A = 1488.86 \quad \text{Round to the nearest cent.}$$

Thus, \$1488.86 is in the account after 10 yr.

(b) The interest earned for that period is

$$\$1488.86 - \$1000 = \$488.86.$$

Now try Exercise 63(a).

EXAMPLE 8 Finding Present Value

Becky Anderson must pay a lump sum of \$6000 in 5 yr.

(a) What amount deposited today at 3.1% compounded annually will grow to \$6000 in 5 yr?

(b) If only \$5000 is available to deposit now, what annual interest rate is necessary for the money to increase to \$6000 in 5 yr?

Solution

(a)

$$A = P\left(1 + \frac{r}{m}\right)^{tm} \quad \text{Compound interest formula}$$

$$6000 = P\left(1 + \frac{.031}{1}\right)^{5(1)} \quad \text{Let } A = 6000,\ r = .031,\ m = 1, \text{ and } t = 5.$$

$$6000 = P(1.031)^5 \quad \text{Solve for } P.$$

$$P = \frac{6000}{(1.031)^5} \quad \text{Divide by } (1.031)^5.$$

$$P \approx 5150.60 \quad \text{Use a calculator.}$$

If Becky leaves \$5150.60 for 5 yr in an account paying 3.1% compounded annually, she will have \$6000 when she needs it. We say that \$5150.60 is the present value of \$6000 if interest of 3.1% is compounded annually for 5 yr.

(b)

$$A = P\left(1 + \frac{r}{m}\right)^{tm}$$

$$6000 = 5000(1 + r)^5 \quad \text{Let } A = 6000,\ P = 5000,\ m = 1, \text{ and } t = 5.$$

$$\frac{6}{5} = (1 + r)^5 \quad \text{Divide by 5000.}$$

$$\left(\frac{6}{5}\right)^{1/5} = 1 + r \quad \text{Take the fifth root.}$$

$$\left(\frac{6}{5}\right)^{1/5} - 1 = r \quad \text{Subtract 1.}$$

$$r \approx .0371 \quad \text{Use a calculator.}$$

An interest rate of 3.71% will produce enough interest to increase the \$5000 to \$6000 by the end of 5 yr.

Now try Exercises 65 and 69.

The Number e and Continuous Compounding The more often interest is compounded within a given time period, the more interest will be earned. Surprisingly, however, there is a limit on the amount of interest, no matter how often it is compounded. To see this, suppose that \$1 is invested at 100% interest per year, compounded m times per year. Then the interest rate (in decimal form)

m	$\left(1+\frac{1}{m}\right)^m$ (rounded)
1	2
2	2.25
5	2.48832
10	2.59374
25	2.66584
50	2.69159
100	2.70481
500	2.71557
1000	2.71692
10,000	2.71815
1,000,000	2.71828

is 1.00 and the interest rate per period is $\frac{1}{m}$. According to the formula (with $P = 1$), the compound amount at the end of 1 yr will be

$$A = \left(1 + \frac{1}{m}\right)^m.$$

A calculator gives the results in the margin for various values of m. The table suggests that as m increases, the value of $\left(1 + \frac{1}{m}\right)^m$ gets closer and closer to some fixed number. This is indeed the case. This fixed number is called e.

Value of e

To nine decimal places, $\quad e \approx \mathbf{2.718281828}.$

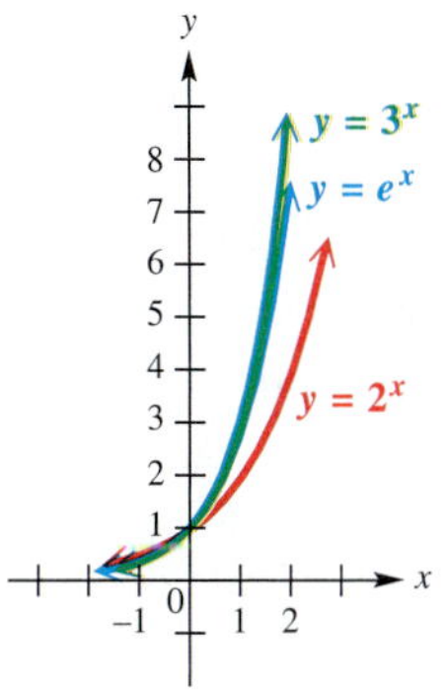

Figure 21

Figure 21 shows the functions defined by $y = 2^x$, $y = 3^x$, and $y = e^x$. Notice that because $2 < e < 3$, the graph of $y = e^x$ lies "between" the other two graphs.

As mentioned above, the amount of interest earned increases with the frequency of compounding, but the value of the expression $\left(1 + \frac{1}{m}\right)^m$ approaches e as m gets larger. Consequently, the formula for compound interest approaches a limit as well, called the compound amount from *continuous compounding.*

Continuous Compounding

If P dollars are deposited at a rate of interest r compounded continuously for t years, the compound amount in dollars on deposit is

$$A = Pe^{rt}.$$

EXAMPLE 9 Solving a Continuous Compounding Problem

Suppose \$5000 is deposited in an account paying 3% interest compounded continuously for 5 yr. Find the total amount on deposit at the end of 5 yr.

Solution

$A = Pe^{rt}$ — Continuous compounding formula

$= 5000e^{.03(5)}$ — Let $P = 5000$, $t = 5$, and $r = .03$.

$= 5000e^{.15}$

$\approx 5000(1.161834)$ — Use a calculator.

$= 5809.17$

or \$5809.17. Check that daily compounding would have produced a compound amount about 3¢ less.

Now try Exercise 63(b).

Looking Ahead to Calculus

In calculus, the derivative allows us to determine the slope of a tangent line to the graph of a function. For the function $f(x) = e^x$, the derivative is the function f itself: $f'(x) = e^x$. Therefore, in calculus the exponential function with base e is much easier to work with than exponential functions having other bases.

EXAMPLE 10 Comparing Interest Earned as Compounding Is More Frequent

In Example 7, we found that \$1000 invested at 4% compounded quarterly for 10 yr grew to \$1488.86. Compare this same investment compounded annually, semiannually, monthly, daily, and continuously.

Solution Substituting into the compound interest formula and the formula for continuous compounding gives the following results for amounts of \$1 and \$1000.

Compounded	\$1	\$1000
annually	$(1 + .04)^{10} \approx 1.48024$	\$1480.24
semiannually	$\left(1 + \frac{.04}{2}\right)^{10(2)} \approx 1.48595$	\$1485.95
quarterly	$\left(1 + \frac{.04}{4}\right)^{10(4)} \approx 1.48886$	\$1488.86
monthly	$\left(1 + \frac{.04}{12}\right)^{10(12)} \approx 1.49083$	\$1490.83
daily	$\left(1 + \frac{.04}{365}\right)^{10(365)} \approx 1.49179$	\$1491.79
continuously	$e^{10(.04)} \approx 1.49182$	\$1491.82

Compounding semiannually rather than annually increases the value of the account after 10 yr by \$5.71. Quarterly compounding grows to \$2.91 more than semiannual compounding after 10 yr. Each increase in the frequency of compounding earns less and less additional interest, until going from daily to continuous compounding increases the value by only \$.03.

Now try Exercise 71.

Exponential Models and Curve Fitting The number e is important as the base of an exponential function in many practical applications. For example, in situations involving growth or decay of a quantity, the amount or number present at time t often can be closely modeled by a function defined by

$$y = y_0 e^{kt},$$

where y_0 is the amount or number present at time $t = 0$ and k is a constant.

The next example, which refers to the problem stated at the beginning of this chapter, illustrates exponential growth.

EXAMPLE 11 Using Data to Model Exponential Growth

If current trends of burning fossil fuels and deforestation continue, then future amounts of atmospheric carbon dioxide in parts per million (ppm) will increase as shown in the table on the next page.

Year	Carbon Dioxide (ppm)
1990	353
2000	375
2075	590
2175	1090
2275	2000

Source: International Panel on Climate Change (IPCC), 1990.

(a) Make a scatter diagram of the data. Do the carbon dioxide levels appear to grow exponentially?

(b) The function defined by

$$y = 353e^{.0060857(t-1990)}$$

is a good model for the data.* Use a graph of this model to estimate when future levels of carbon dioxide will double and triple over the preindustrial level of 280 ppm.

2100
1975 2300
300

(a)

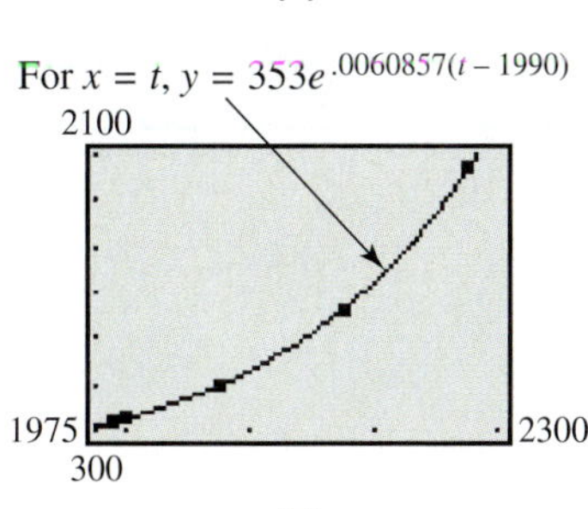

(b)

Figure 22

Solution

(a) We show a calculator graph for the data in Figure 22(a). The data appear to have the shape of the graph of an increasing exponential function.

(b) A graph of $y = 353e^{.0060857(t-1990)}$ in Figure 22(b) shows that it is very close to the data points. We graph $y = 2 \cdot 280 = 560$ in Figure 23(a) and $y = 3 \cdot 280 = 840$ in Figure 23(b) on the same coordinate axes as the given function, and use the calculator to find the intersection points.

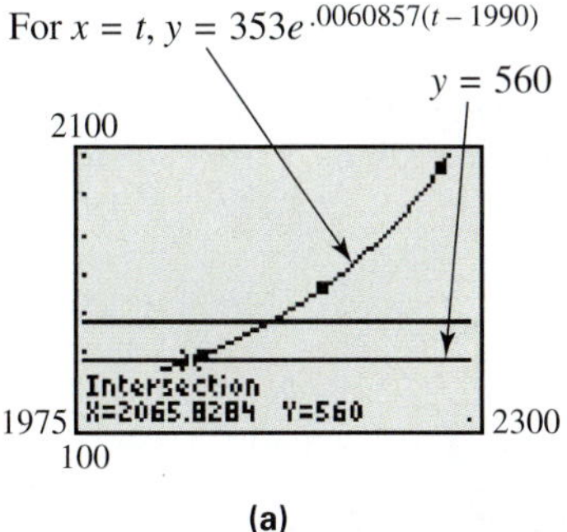

(a)

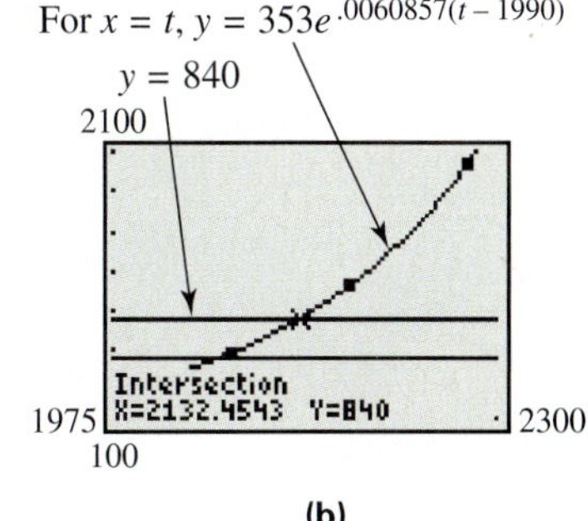

(b)

Figure 23

The graph of the function intersects the horizontal lines at approximately 2065.8 and 2132.5. According to this model, carbon dioxide levels will double by 2065 and triple by 2132.

Now try Exercise 73.

*In Section 4.6, we show how this expression for y was found.

ExpReg
y=a*b^x
a=.0019230377
b=1.006108701

(a)

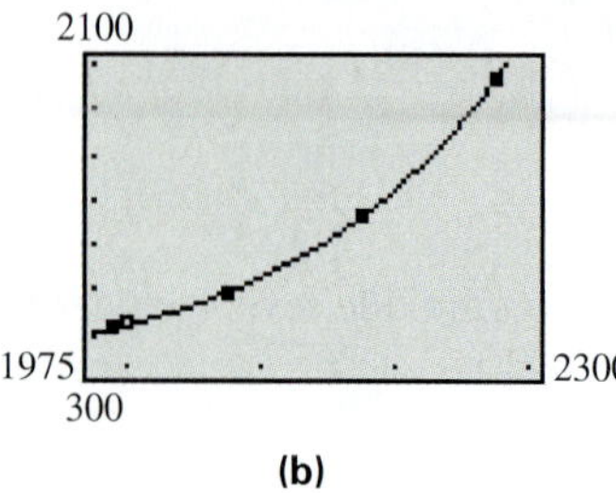

(b)

Figure 24

Graphing calculators are capable of fitting exponential curves to scatter diagrams like the one found in Example 11. Figure 24(a) shows how the TI-83 Plus displays another (different) equation for the atmospheric carbon dioxide example: $y = .0019 \cdot 1.0061^x$. (Coefficients are rounded here.) Notice that this calculator form differs from the model in Example 11. Figure 24(b) shows the data points and the graph of this exponential regression equation. ■

Further examples of exponential growth and decay are given in Section 4.6.

CONNECTIONS In calculus, it is shown that

$$e^x = 1 + x + \frac{x^2}{2 \cdot 1} + \frac{x^3}{3 \cdot 2 \cdot 1} + \frac{x^4}{4 \cdot 3 \cdot 2 \cdot 1} + \frac{x^5}{5 \cdot 4 \cdot 3 \cdot 2 \cdot 1} + \cdots.$$

By using more and more terms, a more and more accurate approximation may be obtained for e^x.

For Discussion or Writing

1. Use the terms shown here and replace x with 1 to approximate $e^1 = e$ to three decimal places. Check your results with a calculator.
2. Use the terms shown here and replace x with $-.05$ to approximate $e^{-.05}$ to four decimal places. Check your results with a calculator.
3. Give the next term in the sum for e^x.

4.2 Exercises

If $f(x) = 3^x$ and $g(x) = \left(\frac{1}{4}\right)^x$, find each of the following. If a result is irrational, round the answer to the nearest thousandth. See Example 1.

1. $f(2)$ **2.** $f(3)$ **3.** $f(-2)$ **4.** $f(-3)$

5. $g(2)$ **6.** $g(3)$ **7.** $g(-2)$ **8.** $g(-3)$

9. $f\left(\frac{3}{2}\right)$ **10.** $g\left(\frac{3}{2}\right)$ **11.** $g(2.34)$ **12.** $f(1.68)$

Graph each function. See Example 2.

13. $f(x) = 3^x$ **14.** $f(x) = 4^x$ **15.** $f(x) = \left(\frac{1}{3}\right)^x$ **16.** $f(x) = \left(\frac{1}{4}\right)^x$

17. $f(x) = \left(\frac{3}{2}\right)^x$ **18.** $f(x) = \left(\frac{2}{3}\right)^x$ **19.** $f(x) = 10^x$ **20.** $f(x) = 10^{-x}$

21. $f(x) = 4^{-x}$ **22.** $f(x) = 6^{-x}$ **23.** $f(x) = 2^{|x|}$ **24.** $f(x) = 2^{-|x|}$

Sketch the graph of $f(x) = 2^x$. Then refer to it and use the techniques of Chapter 2 to graph each function as defined. See Example 3.

25. $f(x) = 2^x + 1$ **26.** $f(x) = 2^x - 4$ **27.** $f(x) = 2^{x+1}$ **28.** $f(x) = 2^{x-4}$

Sketch the graph of $f(x) = \left(\frac{1}{3}\right)^x$. Then refer to it and use the techniques of Chapter 2 to graph each function as defined. See Example 3.

29. $f(x) = \left(\dfrac{1}{3}\right)^x - 2$

30. $f(x) = \left(\dfrac{1}{3}\right)^x + 4$

31. $f(x) = \left(\dfrac{1}{3}\right)^{x+2}$

32. $f(x) = \left(\dfrac{1}{3}\right)^{x-4}$

Concept Check *The graphs of $y = a^x$ for $a = 1.8, 2.3, 3.2, .4, .75,$ and $.31$ are given in the figure. They are identified by letter, but not necessarily in the same order as the values just given. Use your knowledge of how the exponential function behaves for various values of the base to identify each lettered graph.*

33. A

34. B

35. C

36. D

37. E

38. F

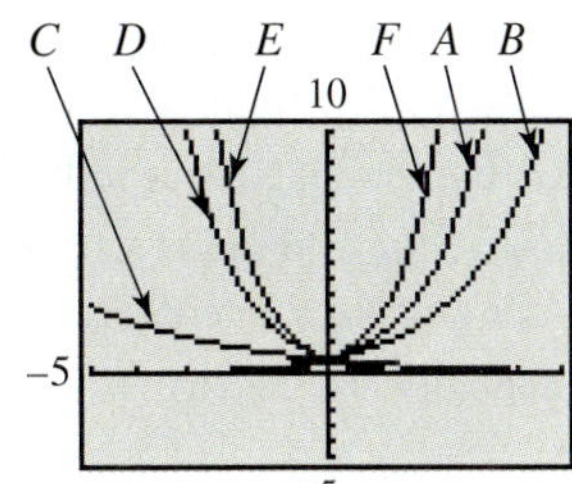

Use a graphing calculator to graph each function as defined.

39. $f(x) = \dfrac{e^x - e^{-x}}{2}$

40. $f(x) = \dfrac{e^x + e^{-x}}{2}$

41. $f(x) = x \cdot 2^x$

42. $f(x) = x^2 \cdot 2^{-x}$

Solve each equation. See Examples 4–6.

43. $4^x = 2$

44. $125^r = 5$

45. $\left(\dfrac{1}{2}\right)^k = 4$

46. $\left(\dfrac{2}{3}\right)^x = \dfrac{9}{4}$

47. $2^{3-y} = 8$

48. $5^{2p+1} = 25$

49. $e^{4x-1} = (e^2)^x$

50. $e^{3-x} = (e^3)^{-x}$

51. $27^{4z} = 9^{z+1}$

52. $32^t = 16^{1-t}$

53. $4^{x-2} = 2^{3x+3}$

54. $\left(\dfrac{1}{2}\right)^{3x-6} = 8^{x+1}$

55. $\left(\dfrac{1}{e}\right)^{-x} = \left(\dfrac{1}{e^2}\right)^{x+1}$

56. $e^{k-1} = \left(\dfrac{1}{e^4}\right)^{k+1}$

57. $(\sqrt{2})^{x+4} = 4^x$

58. $(\sqrt[3]{5})^{-x} = \left(\dfrac{1}{5}\right)^{x+2}$

59. $\dfrac{1}{27} = b^{-3}$

60. $\dfrac{1}{81} = k^{-4}$

61. $4 = r^{2/3}$

62. $z^{5/2} = 32$

Solve each problem involving compound interest. See Examples 7–9.

63. Future Value Find the future value and interest earned if \$8906.54 is invested for 9 yr at 5% compounded
(a) semiannually **(b)** continuously.

64. Future Value Find the future value and interest earned if \$56,780 is invested at 5.3% compounded
(a) quarterly for 23 quarters **(b)** continuously for 15 yr.

65. Present Value Find the present value of \$25,000 if interest is 6% compounded quarterly for 11 quarters.

66. *Present Value* Find the present value of \$45,000 if interest is 3.6% compounded monthly for 1 yr.

67. *Present Value* Find the present value of \$5000 if interest is 3.5% compounded quarterly for 10 yr.

68. *Interest Rate* Find the required annual interest rate to the nearest tenth of a percent for \$65,000 to grow to \$65,325 if interest is compounded monthly for 6 months.

69. *Interest Rate* Find the required annual interest rate to the nearest tenth of a percent for \$1200 to grow to \$1500 if interest is compounded quarterly for 5 yr.

70. *Interest Rate* Find the required annual interest rate to the nearest tenth of a percent for \$5000 to grow to \$8400 if interest is compounded quarterly for 8 yr.

Solve each problem. See Example 10.

71. *Comparing Loans* Bank A is lending money at 6.4% interest compounded annually. The rate at Bank B is 6.3% compounded monthly, and the rate at Bank C is 6.35% compounded quarterly. Which bank will you pay the *least* interest?

72. *Future Value* Suppose \$10,000 is invested at an annual rate of 5% for 10 yr. Find the future value if interest is compounded as follows.

(a) annually **(b)** quarterly **(c)** monthly **(d)** daily (365 days)

(Modeling) Solve each problem. See Example 11.

73. *Atmospheric Pressure* The atmospheric pressure (in millibars) at a given altitude (in meters) is shown in the table.

Altitude	Pressure	Altitude	Pressure
0	1013	6000	472
1000	899	7000	411
2000	795	8000	357
3000	701	9000	308
4000	617	10,000	265
5000	541		

Source: Miller, A. and J. Thompson, *Elements of Meteorology,* Fourth Edition, Charles E. Merrill Publishing Company, Columbus, Ohio, 1993.

(a) Use a graphing calculator to make a scatter diagram of the data for atmospheric pressure P at altitude x.

(b) Would a linear or exponential function fit the data better?

(c) The function defined by

$$P(x) = 1013e^{-.0001341x}$$

approximates the data. Use a graphing calculator to graph P and the data on the same coordinate axes.

(d) Use P to predict the pressures at 1500 m and 11,000 m, and compare them to the actual values of 846 millibars and 227 millibars, respectively.

74. *World Population Growth* Since 1990, world population in millions closely fits the exponential function defined by

$$y = 6073e^{.0137x},$$

where x is the number of years since 1990.

(a) The world population was about 6079 million in 2000. How closely does the function approximate this value?

(b) Use this model to approximate the population in 2005.

(c) Use this model to predict the population in 2015.

(d) Explain why this model may not be accurate for 2015.

75. ***Deer Population*** The exponential growth of the deer population in Massachusetts can be calculated using the model

$$T = 50{,}000(1 + .06)^n,$$

where 50,000 is the initial deer population and .06 is the rate of growth. T is the total population after n years have passed.

(a) Predict the total population after 4 yr.

(b) If the initial population was 30,000 and the growth rate was .12, approximately how many deer would be present after 3 yr?

(c) How many additional deer can we expect in 5 yr if the initial population is 45,000 and the current growth rate is .08?

76. ***Employee Training*** A person learning certain skills involving repetition tends to learn quickly at first. Then learning tapers off and approaches some upper limit. Suppose the number of symbols per minute that a person using a word processor can type is given by

$$p(t) = 250 - 120(2.8)^{-.5t},$$

where t is the number of months the operator has been in training. Find each value.

(a) $p(2)$ **(b)** $p(4)$ **(c)** $p(10)$

(d) What happens to the number of symbols per minute after several months of training?

Use a graphing calculator to find the solution set of each equation. Estimate the solution(s) to the nearest tenth.

77. $5e^{3x} = 75$ **78.** $6^{-x} = 1 - x$ **79.** $3x + 2 = 4^x$ **80.** $x = 2^x$

81. A function of the form $f(x) = x^r$, where r is a constant, is called a **power function.** Discuss the difference between an exponential function and a power function.

82. ***Concept Check*** If $f(x) = a^x$ and $f(3) = 27$, find each value of $f(x)$.

(a) $f(1)$ **(b)** $f(-1)$ **(c)** $f(2)$ **(d)** $f(0)$

Concept Check *Give an equation of the form $f(x) = a^x$ to define the exponential function whose graph contains the given point.*

83. $(3, 8)$ **84.** $(-3, 64)$

Concept Check *Use properties of exponents to write each function in the form $f(t) = ka^t$, where k is a constant. (Hint: Recall that $a^{x+y} = a^x \cdot a^y$.)*

85. $f(t) = 3^{2t+3}$ **86.** $f(t) = \left(\frac{1}{3}\right)^{1-2t}$

87. Explain why the exponential equation $3^x = 12$ cannot be solved by using the properties of exponents given in this section.

88. The graph of $y = e^{x-3}$ can be obtained by translating the graph of $y = e^x$ to the right 3 units. Find a constant C such that the graph of $y = Ce^x$ is the same as the graph of $y = e^{x-3}$. Verify your result by graphing both functions.

Relating Concepts

For individual or collaborative investigation
(Exercises 89–94)

These exercises are designed to prepare you for the next section, on logarithmic functions. Assume $f(x) = a^x$, where $a > 1$. ***Work these exercises in order.***

89. Is f a one-to-one function? If so, based on Section 4.1, what kind of related function exists for f?

90. If f has an inverse function f^{-1}, sketch f and f^{-1} on the same set of axes.

91. If f^{-1} exists, find an equation for $y = f^{-1}(x)$ using the method described in Section 4.1. You need not solve for y.

92. If $a = 10$, what is the equation for $y = f^{-1}(x)$? (You need not solve for y.)

93. If $a = e$, what is the equation for $y = f^{-1}(x)$? (You need not solve for y.)

94. If the point (p, q) is on the graph of f, then the point ________ is on the graph of f^{-1}.

4.3 Logarithmic Functions

Logarithms ▪ **Logarithmic Equations** ▪ **Logarithmic Functions** ▪ **Properties of Logarithms**

Logarithms The previous section dealt with exponential functions of the form $y = a^x$ for all positive values of a, where $a \neq 1$. The horizontal line test shows that exponential functions are one-to-one, and thus have inverse functions. The equation defining the inverse of a function is found by interchanging x and y in the equation that defines the function. Doing so with

$$y = a^x \qquad \text{gives} \qquad x = a^y$$

as the equation of the inverse function of the exponential function defined by $y = a^x$. This equation can be solved for y by using the following definition.

Logarithm

For all real numbers y, and all positive numbers a and x, where $a \neq 1$:

$$\mathbf{y = \log_a x} \qquad \text{if and only if} \qquad \mathbf{x = a^y}.$$

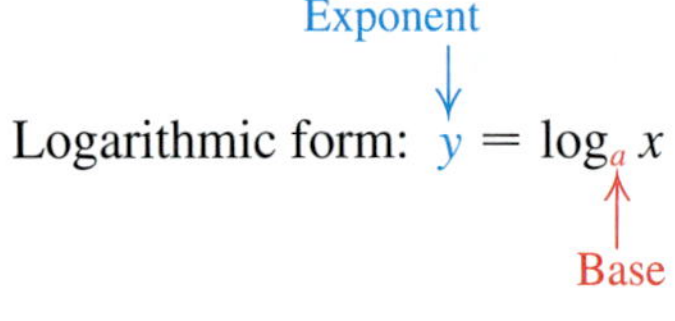

The "log" in the definition above is an abbreviation for **logarithm.** Read $\log_a x$ as "the logarithm to the base a of x."

By the definition of logarithm, if $y = \log_a x$, then the power to which a must be raised to obtain x is y, or $x = a^y$. To remember the location of the base and the exponent in each form, refer to the diagrams in the margin.

Exponent
Exponential form: $a^y = x$
Base

Meaning of $\log_a x$

A logarithm is an exponent; $\log_a x$ is the exponent to which the base a must be raised to obtain x.

The table shows several pairs of equivalent statements, written in both logarithmic and exponential forms.

Logarithmic Form	Exponential Form
$\log_2 8 = 3$	$2^3 = 8$
$\log_{1/2} 16 = -4$	$\left(\frac{1}{2}\right)^{-4} = 16$
$\log_{10} 100{,}000 = 5$	$10^5 = 100{,}000$
$\log_3 \frac{1}{81} = -4$	$3^{-4} = \frac{1}{81}$
$\log_5 5 = 1$	$5^1 = 5$
$\log_{3/4} 1 = 0$	$\left(\frac{3}{4}\right)^0 = 1$

Now try Exercises 3, 5, 7, and 9.

Logarithmic Equations The definition of logarithm can be used to solve a **logarithmic equation,** an equation with a logarithm in at least one term, by first writing the equation in exponential form.

EXAMPLE 1 Solving Logarithmic Equations

Solve each equation.

(a) $\log_x \frac{8}{27} = 3$ **(b)** $\log_4 x = \frac{5}{2}$ **(c)** $\log_{49} \sqrt[3]{7} = x$

Solution

(a)

$$\log_x \frac{8}{27} = 3$$

$$x^3 = \frac{8}{27} \quad \text{Write in exponential form.}$$

$$x^3 = \left(\frac{2}{3}\right)^3 \quad \tfrac{8}{27} = \left(\tfrac{2}{3}\right)^3 \text{ (Section R.1)}$$

$$x = \frac{2}{3} \quad \text{Take cube roots. (Section 1.6)}$$

The solution set is $\left\{\frac{2}{3}\right\}$.

(b)

$$\log_4 x = \frac{5}{2}$$

$$4^{5/2} = x \quad \text{Write in exponential form.}$$

$$(4^{1/2})^5 = x \quad a^{mn} = (a^m)^n \text{ (Section R.3)}$$

$$2^5 = x \quad 4^{1/2} = (2^2)^{1/2} = 2$$

$$32 = x$$

The solution set is $\{32\}$.

(c)

$$\log_{49} \sqrt[3]{7} = x$$

$$49^x = \sqrt[3]{7} \quad \text{Write in exponential form.}$$

$$(7^2)^x = 7^{1/3} \quad \text{Write with the same base.}$$

$$7^{2x} = 7^{1/3} \quad \text{Power rule for exponents}$$

$$2x = \frac{1}{3} \quad \text{Set exponents equal.}$$

$$x = \frac{1}{6} \quad \text{Divide by 2.}$$

The solution set is $\left\{\frac{1}{6}\right\}$.

Now try Exercises 13, 23, and 27.

Logarithmic Functions We define the logarithmic function with base a as follows.

Logarithmic Function

If $a > 0$, $a \neq 1$, and $x > 0$, then

$$f(x) = \log_a x$$

defines the **logarithmic function** with base a.

Exponential and logarithmic functions are inverses of each other. The graph of $y = 2^x$ is shown in red in Figure 25(a). The graph of its inverse is found by reflecting the graph of $y = 2^x$ across the line $y = x$. The graph of the inverse function, defined by $y = \log_2 x$, shown in blue, has the y-axis as a vertical asymptote. Figure 25(b) shows a calculator graph of the two functions.

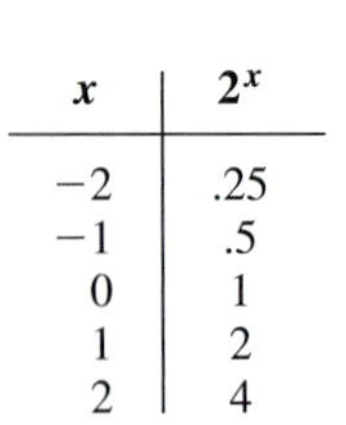

x	2^x
−2	.25
−1	.5
0	1
1	2
2	4

x	$\log_2 x$
.25	−2
.5	−1
1	0
2	1
4	2

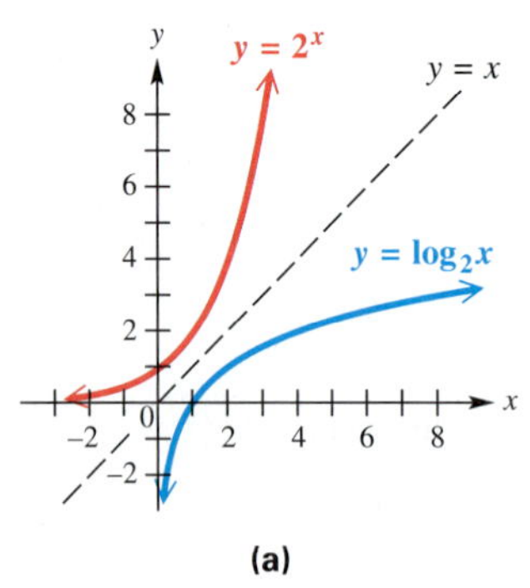

(a)

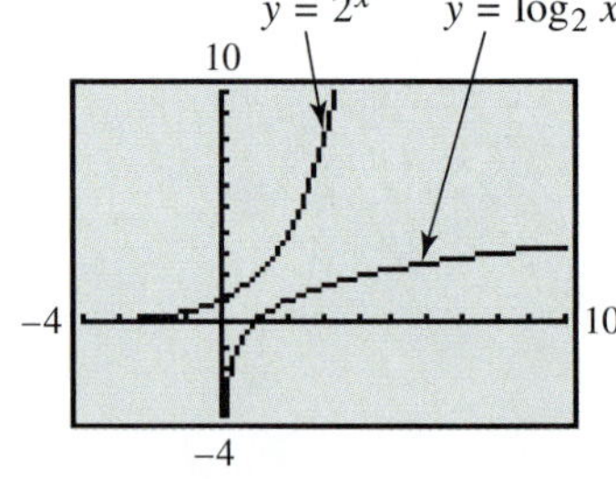

The graph of $y = \log_2 x$ can be obtained by drawing the inverse of $y = 2^x$.

(b)

Figure 25

Since the domain of an exponential function is the set of all real numbers, the range of a logarithmic function also will be the set of all real numbers. In the same way, both the range of an exponential function and the domain of a logarithmic function are the set of all positive real numbers, so ***logarithms can be found for positive numbers only.***

We now summarize information about the graphs of logarithmic functions.

LOGARITHMIC FUNCTION $f(x) = \log_a x$

Domain: $(0, \infty)$ Range: $(-\infty, \infty)$

For $f(x) = \log_2 x$:

x	$f(x)$
$\frac{1}{4}$	-2
$\frac{1}{2}$	-1
1	0
2	1
4	2
8	3

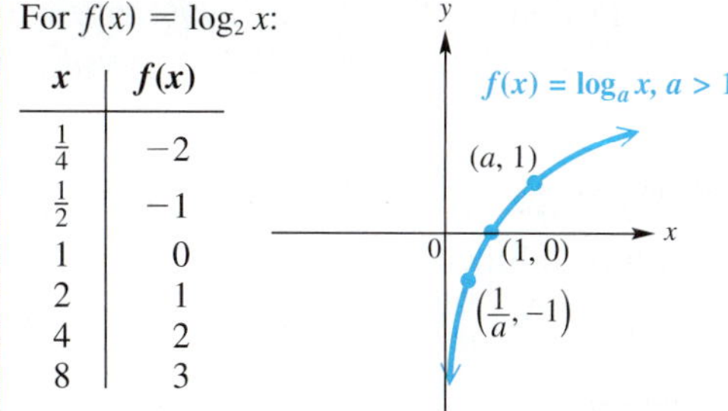

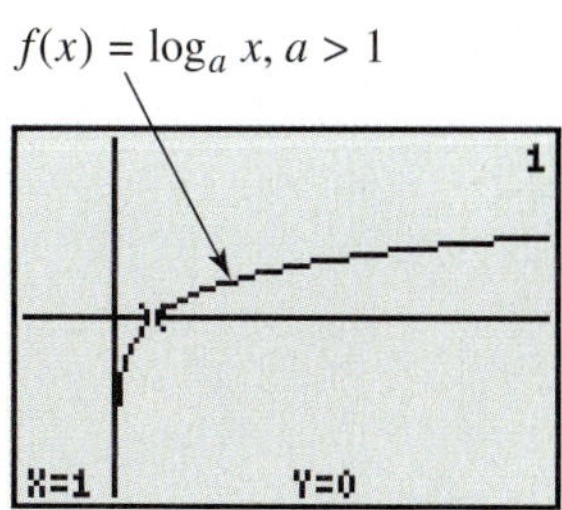

Figure 26

- $f(x) = \log_a x$, $a > 1$, is increasing and continuous on its entire domain, $(0, \infty)$.
- The y-axis is a vertical asymptote as $x \to 0$ from the right.
- The graph goes through the points $\left(\frac{1}{a}, -1\right)$, $(1, 0)$, and $(a, 1)$.

For $f(x) = \log_{1/2} x$:

x	$f(x)$
$\frac{1}{4}$	2
$\frac{1}{2}$	1
1	0
2	-1
4	-2
8	-3

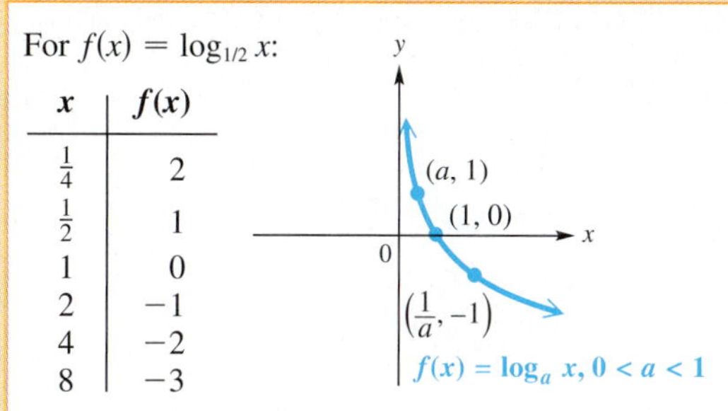

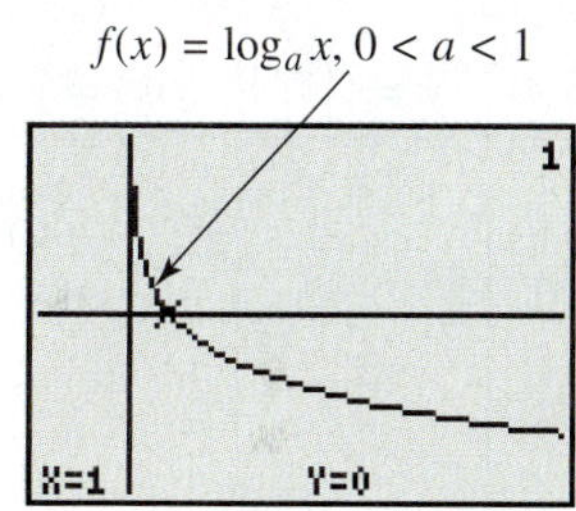

Figure 27

- $f(x) = \log_a x$, $0 < a < 1$, is decreasing and continuous on its entire domain, $(0, \infty)$.
- The y-axis is a vertical asymptote as $x \to 0$ from the right.
- The graph goes through the points $\left(\frac{1}{a}, -1\right)$, $(1, 0)$, and $(a, 1)$.

Calculator graphs of logarithmic functions do not, in general, give an accurate picture of the behavior of the graphs near the vertical asymptotes. While it may seem as if the graph has an endpoint, this is not the case. The resolution of the calculator screen is not precise enough to indicate that the graph approaches the vertical asymptote as the value of x gets closer to it. Do not draw incorrect conclusions just because the calculator does not show this behavior. ■

The graphs in Figures 26 and 27 and the information with them suggest the following generalizations about the graphs of logarithmic functions of the form $f(x) = \log_a x$.

Characteristics of the Graph of $f(x) = \log_a x$

1. The points $\left(\frac{1}{a}, -1\right)$, $(1, 0)$, and $(a, 1)$ are on the graph.
2. If $a > 1$, then f is an increasing function; if $0 < a < 1$, then f is a decreasing function.
3. The y-axis is a vertical asymptote.
4. The domain is $(0, \infty)$, and the range is $(-\infty, \infty)$.

EXAMPLE 2 Graphing Logarithmic Functions

Graph each function.

(a) $f(x) = \log_{1/2} x$ **(b)** $f(x) = \log_3 x$

Solution

(a) First graph $y = \left(\frac{1}{2}\right)^x$, which defines the inverse function of f, by plotting points. Some ordered pairs are given in the table with the graph shown in red in Figure 28. The graph of $f(x) = \log_{1/2} x$ is the reflection of the graph of $y = \left(\frac{1}{2}\right)^x$ across the line $y = x$. The ordered pairs for $y = \log_{1/2} x$ are found by interchanging the x- and y-values in the ordered pairs for $y = \left(\frac{1}{2}\right)^x$. See the graph in blue in Figure 28.

x	$y = \left(\frac{1}{2}\right)^x$	$f(x) = y = \log_{1/2} x$
-2	4	undefined
-1	2	undefined
0	1	undefined
1	$\frac{1}{2}$	0
2	$\frac{1}{4}$	-1
4	$\frac{1}{16}$	-2

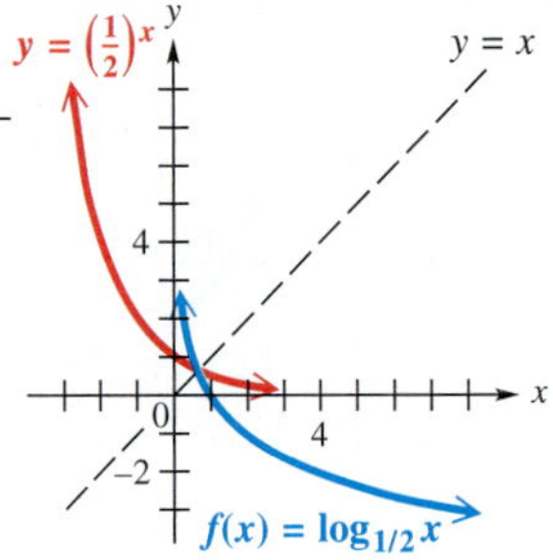

Figure 28

x	$y = \log_3 x$
$\frac{1}{3}$	-1
1	0
3	1
9	2

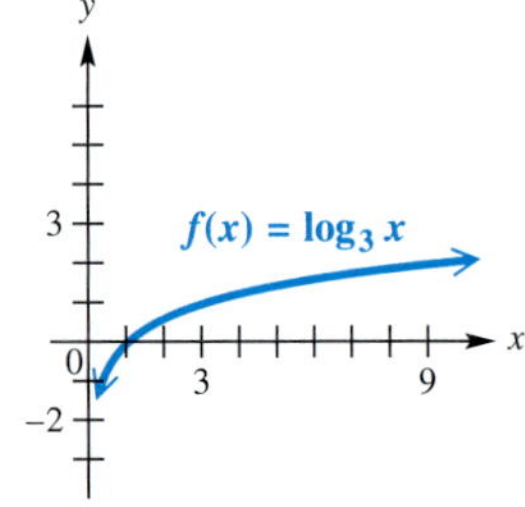

Figure 29

(b) Another way to graph a logarithmic function is to write $f(x) = y = \log_3 x$ in exponential form as $x = 3^y$. Now find some ordered pairs; several are shown in the table with the graph in Figure 29.

Now try Exercise 45.

CAUTION If you write a logarithmic function in exponential form, choosing y-values to calculate x-values as we did in Example 2(b), be careful to get the ordered pairs in the correct order.

More general logarithmic functions can be obtained by forming the composition of $h(x) = \log_a x$ with a function defined by $g(x)$ to get

$$f(x) = h[g(x)] = \log_a[g(x)].$$

EXAMPLE 3 Graphing Translated Logarithmic Functions

Graph each function.

(a) $f(x) = \log_2(x - 1)$ **(b)** $f(x) = (\log_3 x) - 1$

Solution

(a) The graph of $f(x) = \log_2(x - 1)$ is the graph of $f(x) = \log_2 x$ translated 1 unit to the right. The vertical asymptote is $x = 1$. The domain of this function is $(1, \infty)$ since logarithms can be found only for positive numbers. To find some ordered pairs to plot, use the equivalent exponential form of the equation $y = \log_2(x - 1)$.

$$y = \log_2(x - 1)$$

$$x - 1 = 2^y \quad \text{Write in exponential form.}$$

$$x = 2^y + 1 \quad \text{Add 1.}$$

We choose values for y and then calculate each of the corresponding x-values. See Figure 30.

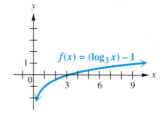

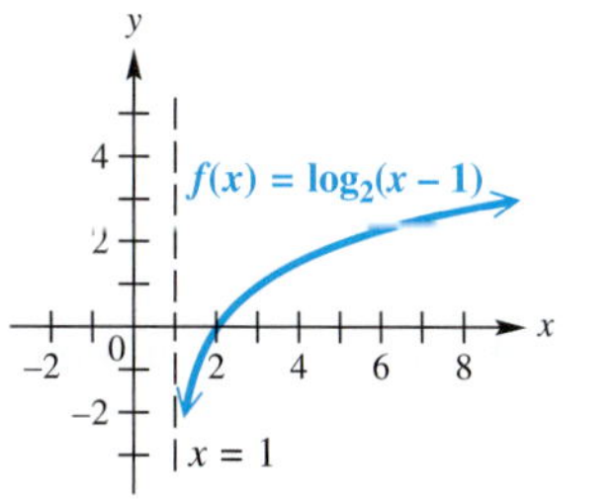

Figure 30

Figure 31

(b) The function defined by $f(x) = (\log_3 x) - 1$ has the same graph as $g(x) = \log_3 x$ translated 1 unit down. We find ordered pairs to plot by writing $y = (\log_3 x) - 1$ in exponential form.

$$y = (\log_3 x) - 1$$

$$y + 1 = \log_3 x \quad \text{Add 1.}$$

$$x = 3^{y+1} \quad \text{Write in exponential form.}$$

Again, choose y-values and calculate the corresponding x-values. The graph is shown in Figure 31.

Now try Exercises 33 and 37.

Properties of Logarithms Since a logarithmic statement can be written as an exponential statement, it is not surprising that the properties of logarithms are based on the properties of exponents. The properties of logarithms allow us to change the form of logarithmic statements so that products can be converted to sums, quotients can be converted to differences, and powers can be converted to products.

Properties of Logarithms

For $x > 0$, $y > 0$, $a > 0$, $a \neq 1$, and any real number r:

Property	Description
Product Property $\log_a xy = \log_a x + \log_a y$	The logarithm of the product of two numbers is equal to the sum of the logarithms of the numbers.
Quotient Property $\log_a \frac{x}{y} = \log_a x - \log_a y$	The logarithm of the quotient of two numbers is equal to the difference between the logarithms of the numbers.
Power Property $\log_a x^r = r \log_a x$	The logarithm of a number raised to a power is equal to the exponent multiplied by the logarithm of the number.

Looking Ahead to Calculus

A technique called *logarithmic differentiation,* which uses the properties of logarithms, can often be used to differentiate complicated functions.

Two additional properties of logarithms follow directly from the definition of $\log_a x$ since $a^0 = 1$ and $a^1 = a$.

$$\log_a 1 = 0 \quad \text{and} \quad \log_a a = 1$$

Proof To prove the product property, let $m = \log_b x$ and $n = \log_b y$, and recall that

$$\log_b x = m \quad \text{means} \quad b^m = x.$$

$$\log_b y = n \quad \text{means} \quad b^n = y.$$

Now consider the product xy.

$$xy = b^m \cdot b^n \qquad \text{Substitute.}$$

$$xy = b^{m+n} \qquad \text{Product rule for exponents (Section R.3)}$$

$$\log_b xy = m + n \qquad \text{Write in logarithmic form.}$$

$$\log_b xy = \log_b x + \log_b y \qquad \text{Substitute.}$$

The last statement is the result we wished to prove. The quotient and power properties are proved similarly. (See Exercises 87 and 88.)

EXAMPLE 4 Using the Properties of Logarithms

Rewrite each expression. Assume all variables represent positive real numbers, with $a \neq 1$ and $b \neq 1$.

(a) $\log_6(7 \cdot 9)$ **(b)** $\log_9 \frac{15}{7}$ **(c)** $\log_5 \sqrt{8}$

(d) $\log_a \frac{mnq}{p^2}$ **(e)** $\log_a \sqrt[3]{m^2}$ **(f)** $\log_b \sqrt[n]{\frac{x^3y^5}{z^m}}$

Solution

(a) $\log_6(7 \cdot 9) = \log_6 7 + \log_6 9$ — Product property

(b) $\log_9 \dfrac{15}{7} = \log_9 15 - \log_9 7$ — Quotient property

(c) $\log_5 \sqrt{8} = \log_5(8^{1/2}) = \dfrac{1}{2} \log_5 8$ — Power property

(d)
$$\log_a \frac{mnq}{p^2} = \log_a m + \log_a n + \log_a q - \log_a p^2$$
$$= \log_a m + \log_a n + \log_a q - 2 \log_a p$$

(e) $\log_a \sqrt[3]{m^2} = \log_a m^{2/3} = \dfrac{2}{3} \log_a m$

(f)
$$\log_b \sqrt[n]{\frac{x^3y^5}{z^m}} = \log_b \left(\frac{x^3y^5}{z^m}\right)^{1/n}$$ — $\sqrt[n]{a} = a^{1/n}$ (Section R.7)

$$= \frac{1}{n} \log_b \frac{x^3y^5}{z^m}$$ — Power property

$$= \frac{1}{n}(\log_b x^3 + \log_b y^5 - \log_b z^m)$$ — Product and quotient properties

$$= \frac{1}{n}(3 \log_b x + 5 \log_b y - m \log_b z)$$ — Power property

$$= \frac{3}{n} \log_b x + \frac{5}{n} \log_b y - \frac{m}{n} \log_b z$$ — Distributive property (Section R.1)

Notice the use of parentheses in the second and third steps. The factor $\frac{1}{n}$ applies to each term.

Now try Exercises 57, 59, and 63.

EXAMPLE 5 Using the Properties of Logarithms

Write each expression as a single logarithm with coefficient 1. Assume all variables represent positive real numbers, with $a \neq 1$ and $b \neq 1$.

(a) $\log_3(x + 2) + \log_3 x - \log_3 2$ **(b)** $2 \log_a m - 3 \log_a n$

(c) $\dfrac{1}{2} \log_b m + \dfrac{3}{2} \log_b 2n - \log_b m^2 n$

Solution

(a) $\log_3(x + 2) + \log_3 x - \log_3 2 = \log_3 \dfrac{(x + 2)x}{2}$ — Product and quotient properties

(b)
$$2 \log_a m - 3 \log_a n = \log_a m^2 - \log_a n^3$$ — Power property

$$= \log_a \frac{m^2}{n^3}$$ — Quotient property

(c) $\frac{1}{2}\log_b m + \frac{3}{2}\log_b 2n - \log_b m^2 n$

$= \log_b m^{1/2} + \log_b(2n)^{3/2} - \log_b m^2 n$ Power property

$= \log_b \frac{m^{1/2}(2n)^{3/2}}{m^2 n}$ Product and quotient properties

$= \log_b \frac{2^{3/2} n^{1/2}}{m^{3/2}}$ Rules for exponents (Sections R.3 and R.6)

$= \log_b \left(\frac{2^3 n}{m^3}\right)^{1/2}$ Rules for exponents

$= \log_b \sqrt{\frac{8n}{m^3}}$ Definition of $a^{1/n}$

Now try Exercises 65, 67, and 69.

CAUTION There is no property of logarithms to rewrite a logarithm of a *sum* or *difference*. That is why, in Example 5(a), $\log_3(x + 2)$ was not written as $\log_3 x + \log_3 2$. Remember, $\log_3 x + \log_3 2 = \log_3(x \cdot 2)$.

The distributive property does not apply in a situation like this because $\log_3(x + y)$ is one term; "log" is a function name, not a factor.

EXAMPLE 6 Using the Properties of Logarithms with Numerical Values

Assume that $\log_{10} 2 = .3010$. Find each logarithm.

(a) $\log_{10} 4$ **(b)** $\log_{10} 5$

Solution

(a) $\log_{10} 4 = \log_{10} 2^2 = 2\log_{10} 2 = 2(.3010) = .6020$

(b) $\log_{10} 5 = \log_{10} \frac{10}{2} = \log_{10} 10 - \log_{10} 2 = 1 - .3010 = .6990.$

Now try Exercise 73.

Compositions of the exponential and logarithmic functions can be used to get two more useful properties. If $f(x) = a^x$ and $g(x) = \log_a x$, then

$$f[g(x)] = a^{\log_a x} \quad \text{and} \quad g[f(x)] = \log_a(a^x).$$

Theorem on Inverses

For $a > 0, a \neq 1$:

$$\mathbf{a^{\log_a x} = x} \quad \text{and} \quad \mathbf{\log_a a^x = x.}$$

By the results of this theorem,

$$7^{\log_7 10} = 10, \quad \log_5 5^3 = 3, \quad \text{and} \quad \log_r r^{k+1} = k + 1.$$

The second statement in the theorem will be useful in Sections 4.5 and 4.6 when we solve other logarithmic and exponential equations.

Napier's Rods
Source: IBM Corporate Archives

CONNECTIONS The search for making calculations easier has been a long, ongoing process. Machines built by Charles Babbage and Blaise Pascal, a system of "rods" used by John Napier, and slide rules were the forerunners of today's calculators and computers. The invention of logarithms by John Napier in the sixteenth century was a great breakthrough in the search for easier calculation methods.

Since logarithms are exponents, their properties allowed users of tables of common logarithms to multiply by adding, divide by subtracting, raise to powers by multiplying, and take roots by dividing. Although logarithms are no longer used for computations, they still play an important role in higher mathematics.

For Discussion or Writing

1. To multiply 458.3 by 294.6 using logarithms, we add $\log_{10} 458.3$ and $\log_{10} 294.6$, then find 10 raised to the sum. Perform this multiplication using the log key and the 10^x key on your calculator.* Check your answer by multiplying directly with your calculator.
2. Try division, raising to a power, and taking a root by this method.

4.3 Exercises

Concept Check *In Exercises 1 and 2, match the logarithm in Column I with its value in Column II. Remember that* $\log_a x$ *is the exponent to which a must be raised in order to obtain x.*

	I	II
1.	**(a)** $\log_2 16$	**A.** 0
	(b) $\log_3 1$	**B.** $\frac{1}{2}$
	(c) $\log_{10} .1$	**C.** 4
	(d) $\log_2 \sqrt{2}$	**D.** -3
	(e) $\log_e \frac{1}{e^2}$	**E.** -1
	(f) $\log_{1/2} 8$	**F.** -2

	I	II
2.	**(a)** $\log_3 81$	**A.** -2
	(b) $\log_3 \frac{1}{3}$	**B.** -1
	(c) $\log_{10} .01$	**C.** 0
	(d) $\log_6 \sqrt{6}$	**D.** $\frac{1}{2}$
	(e) $\log_e 1$	**E.** $\frac{9}{2}$
	(f) $\log_3 27^{3/2}$	**F.** 4

For each statement, write an equivalent statement in logarithmic form.

3. $3^4 = 81$ **4.** $2^5 = 32$ **5.** $\left(\frac{2}{3}\right)^{-3} = \frac{27}{8}$ **6.** $10^{-4} = .0001$

For each statement, write an equivalent statement in exponential form.

7. $\log_6 36 = 2$ **8.** $\log_5 5 = 1$ **9.** $\log_{\sqrt{3}} 81 = 8$ **10.** $\log_4 \frac{1}{64} = -3$

*In this text, the notation $\log x$ is used to mean $\log_{10} x$. This is also the meaning of the log key on calculators.

11. Explain why logarithms of negative numbers are not defined.

12. ***Concept Check*** Why does $\log_a 1$ always equal 0 for any valid base a?

Solve each logarithmic equation. See Example 1.

13. $x = \log_5 \dfrac{1}{625}$ **14.** $x = \log_3 \dfrac{1}{81}$ **15.** $x = \log_{10} .001$

16. $x = \log_6 \dfrac{1}{216}$ **17.** $x = \log_8 \sqrt[4]{8}$ **18.** $x = 8 \log_{100} 10$

19. $x = 3^{\log_3 8}$ **20.** $x = 12^{\log_{12} 5}$ **21.** $x = 2^{\log_2 9}$

22. $x = 8^{\log_8 11}$ **23.** $\log_x 25 = -2$ **24.** $\log_x \dfrac{1}{16} = -2$

25. $\log_4 x = 3$ **26.** $\log_2 x = -1$ **27.** $x = \log_4 \sqrt[3]{16}$

28. $x = \log_5 \sqrt[4]{25}$ **29.** $\log_x 3 = -1$ **30.** $\log_x 1 = 0$

31. Compare the summary of characteristics of the graph of $f(x) = \log_a x$ with the similar summary about the graph of $f(x) = a^x$ in Section 4.2. Make a list of characteristics that reinforce the idea that these are inverse functions.

32. The calculator graph of $y = \log_2 x$ shows the values of the ordered pair with $x = 5$. What does the value of y represent?

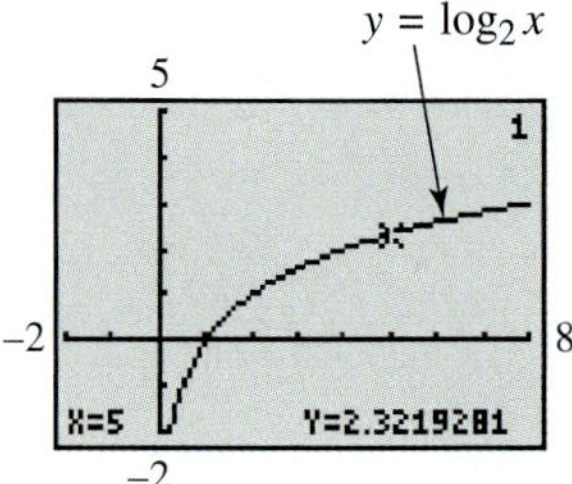

Sketch the graph of $f(x) = \log_2 x$. Then refer to it and use the techniques of Chapter 2 to graph each function. See Example 3.

33. $f(x) = (\log_2 x) + 3$ **34.** $f(x) = \log_2(x + 3)$ **35.** $f(x) = |\log_2(x + 3)|$

Sketch the graph of $f(x) = \log_{1/2} x$. Then refer to it and use the techniques of Chapter 2 to graph each function. See Example 3.

36. $f(x) = (\log_{1/2} x) - 2$ **37.** $f(x) = \log_{1/2}(x - 2)$ **38.** $f(x) = |\log_{1/2}(x - 2)|$

Concept Check *In Exercises 39–44, match the function with its graph from choices A–F.*

39. $f(x) = \log_2 x$ **40.** $f(x) = \log_2 2x$ **41.** $f(x) = \log_2 \dfrac{1}{x}$

42. $f(x) = \log_2 \dfrac{x}{2}$ **43.** $f(x) = \log_2(x - 1)$ **44.** $f(x) = \log_2(-x)$

A.

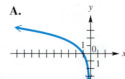

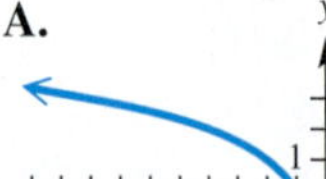

B.

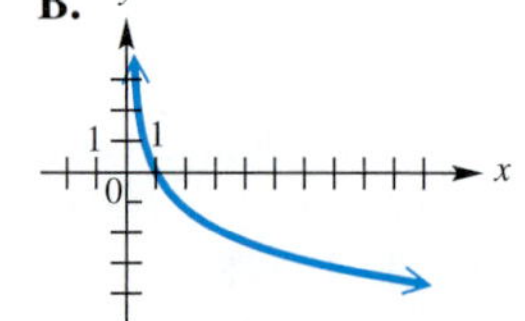

C.

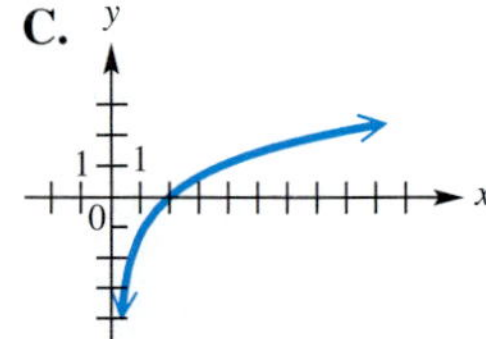

D.

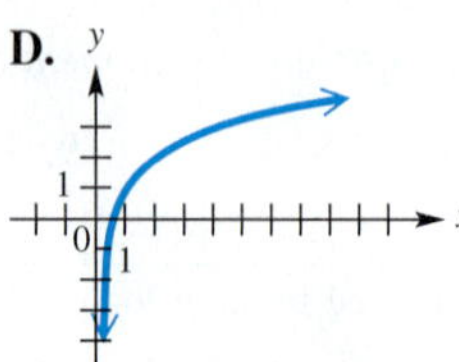

E.

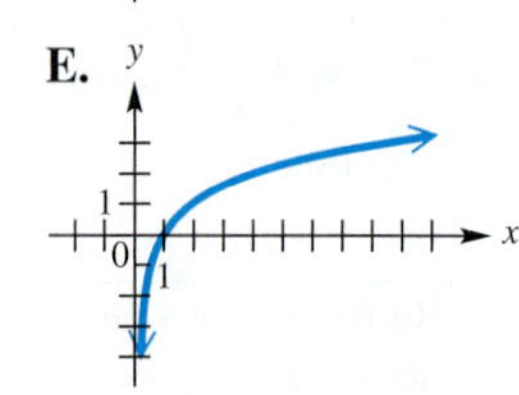

F.

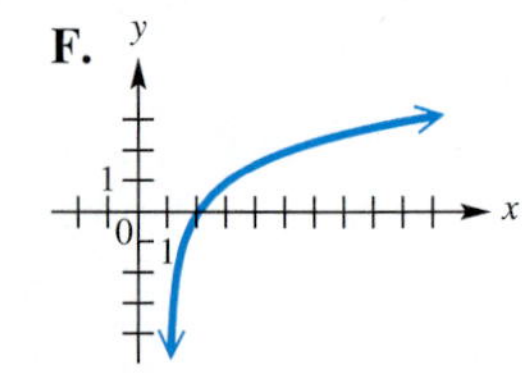

Graph each function. See Examples 2 and 3.

45. $f(x) = \log_5 x$ **46.** $f(x) = \log_{10} x$ **47.** $f(x) = \log_{1/2}(1 - x)$

48. $f(x) = \log_{1/3}(3 - x)$ **49.** $f(x) = \log_3(x - 1)$ **50.** $f(x) = \log_2(x^2)$

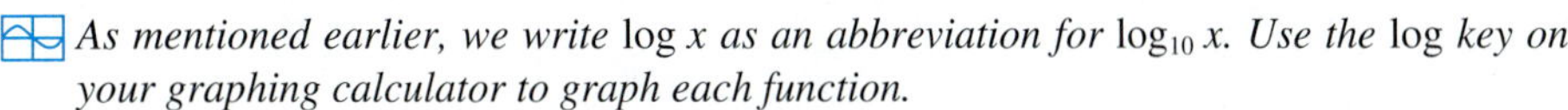

As mentioned earlier, we write $\log x$ *as an abbreviation for* $\log_{10} x$. *Use the* log *key on your graphing calculator to graph each function.*

51. $f(x) = x \log_{10} x$ **52.** $f(x) = x^2 \log_{10} x$

Relating Concepts

For individual or collaborative investigation

(Exercises 53–56)

Exercises 53–56 show how the quotient property for logarithms is related to the vertical translation of graphs. ***Work these exercises in order.***

53. Complete the following statement of the quotient property for logarithms: If x and y are positive numbers, then $\log_a \frac{x}{y} =$ ____________.

54. Use the quotient property to explain how the graph of $f(x) = \log_2 \frac{x}{4}$ can be obtained from the graph of $g(x) = \log_2 x$ by a vertical translation.

55. Graph f and g on the same axes and explain how these graphs support your answer in Exercise 54.

56. If $x = 4$, then $\log_2 \frac{x}{4} =$ ______; since $\log_2 x =$ ______ and $\log_2 4 =$ ______, $\log_2 x - \log_2 4 =$ ______. How does this support the quotient property in Exercise 53?

Use the properties of logarithms to rewrite each expression. Simplify the result if possible. Assume all variables represent positive real numbers. See Example 4.

57. $\log_2 \frac{6x}{y}$ **58.** $\log_3 \frac{4p}{q}$ **59.** $\log_5 \frac{5\sqrt{7}}{3}$

60. $\log_2 \frac{2\sqrt{3}}{5}$ **61.** $\log_4(2x + 5y)$ **62.** $\log_6(7m + 3q)$

63. $\log_m \sqrt{\frac{5r^3}{z^5}}$ **64.** $\log_p \sqrt[3]{\frac{m^5 n^4}{t^2}}$

Write each expression as a single logarithm with coefficient 1. Assume all variables represent positive real numbers. See Example 5.

65. $\log_a x + \log_a y - \log_a m$ **66.** $(\log_b k - \log_b m) - \log_b a$

67. $2 \log_m a - 3 \log_m b^2$ **68.** $\frac{1}{2} \log_y p^3 q^4 - \frac{2}{3} \log_y p^4 q^3$

69. $2 \log_a(z - 1) + \log_a(3z + 2), \quad z > 1$ **70.** $\log_b(2y + 5) - \frac{1}{2} \log_b(y + 3)$

71. $-\frac{2}{3} \log_5 5m^2 + \frac{1}{2} \log_5 25m^2$ **72.** $-\frac{3}{4} \log_3 16p^4 - \frac{2}{3} \log_3 8p^3$

Given $\log_{10} 2 = .3010$ and $\log_{10} 3 = .4771$, find each logarithm without using a calculator. See Example 6.

73. $\log_{10} 6$

74. $\log_{10} 12$

75. $\log_{10} \frac{9}{4}$

76. $\log_{10} \frac{20}{27}$

77. $\log_{10} \sqrt{30}$

78. $\log_{10} 36^{1/3}$

Solve each problem.

79. *(Modeling) Interest Rates of Treasury Securities* The table gives interest rates for various U.S. Treasury Securities on June 27, 2003.

(a) Make a scatter diagram of the data.

(b) Discuss which type of function will model these data best: linear, exponential, or logarithmic.

Time	Yield
3-month	.83%
6-month	.91%
2-year	1.35%
5-year	2.46%
10-year	3.54%
30-year	4.58%

Source: Charles Schwab.

80. *Concept Check* Use the graph to estimate each logarithm.

(a) $\log_3 .3$ (b) $\log_3 .8$

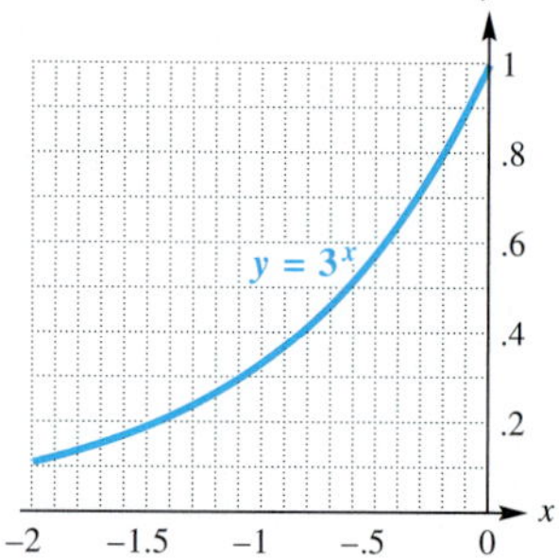

81. *Concept Check* Suppose $f(x) = \log_a x$ and $f(3) = 2$. Determine each function value.

(a) $f\left(\frac{1}{9}\right)$ (b) $f(27)$

82. Use properties of logarithms to evaluate each expression.

(a) $100^{\log_{10} 3}$ (b) $\log_{10} .01^3$

83. Refer to the compound interest formula from Section 4.2. Show that the amount of time required for a deposit to double is $\dfrac{1}{\log_2\left(1 + \frac{r}{m}\right)^m}$.

84. *Concept Check* If $(5, 4)$ is on the graph of the logarithmic function of base a, does $5 = \log_a 4$ or does $4 = \log_a 5$?

Use a graphing calculator to find the solution set of each equation. Give solutions to the nearest hundredth.

85. $\log_{10} x = x - 2$

86. $2^{-x} = \log_{10} x$

87. Prove the quotient property of logarithms: $\log_a \frac{x}{y} = \log_a x - \log_a y$.

88. Prove the power property of logarithms: $\log_a x^r = r \log_a x$.

Summary Exercises on Inverse, Exponential, and Logarithmic Functions

The following exercises are designed to help solidify your understanding of inverse, exponential, and logarithmic functions from Sections 4.1–4.3.

Determine whether the functions in each pair are inverses of each other.

1. $f(x) = 3x - 4, \quad g(x) = \dfrac{x + 4}{3}$

2. $f(x) = 8 - 5x, \quad g(x) = 8 + \dfrac{1}{5}x$

3. $f(x) = 1 + \log_2 x, \quad g(x) = 2^{x-1}$

4. $f(x) = 3^{x/5} - 2, \quad g(x) = 5 \log_3 (x + 2)$

Determine whether each function is one-to-one. If it is, then sketch the graph of its inverse function.

5.

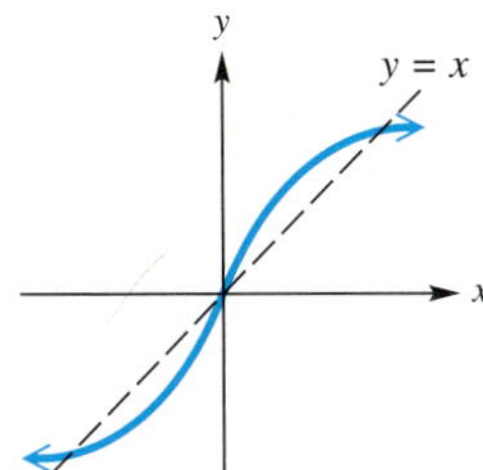

6.

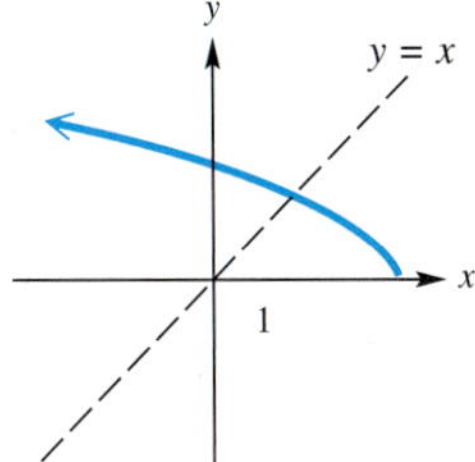

7.

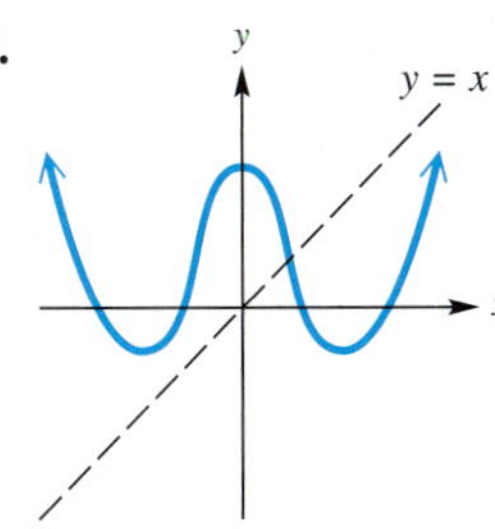

8.

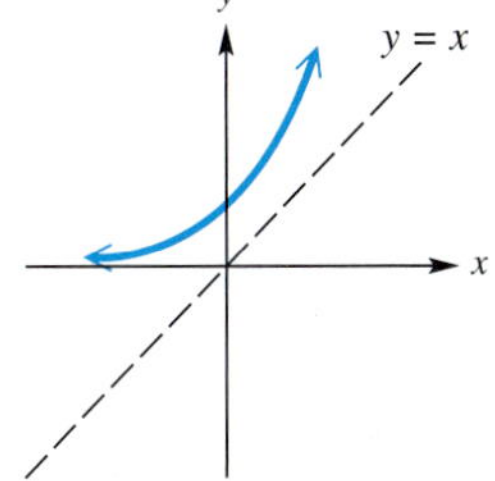

In Exercises 9–12, match the function with its graph from choices A–D.

9. $y = \log_3(x + 2)$

10. $y = 5 - 2^x$

11. $y = \log_2(5 - x)$

12. $y = 3^x - 2$

A.

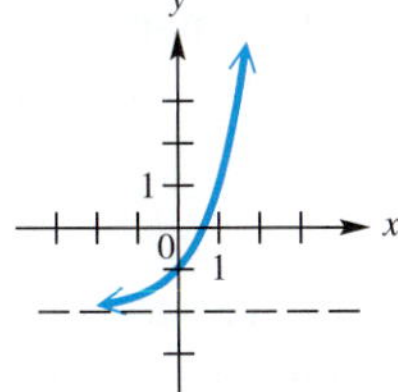

B.

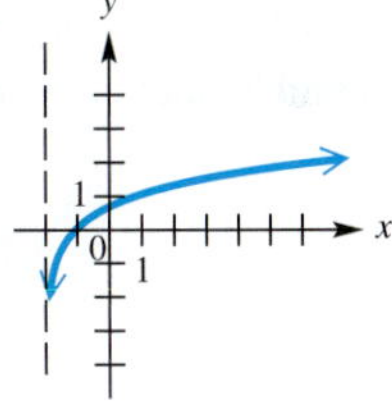

C.

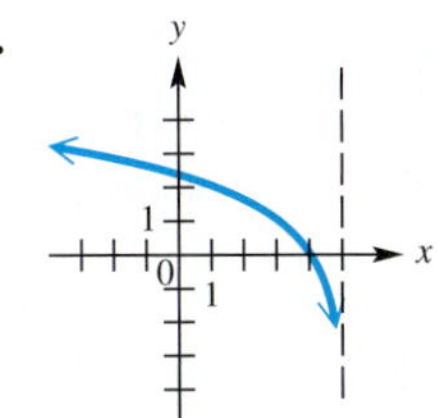

D.

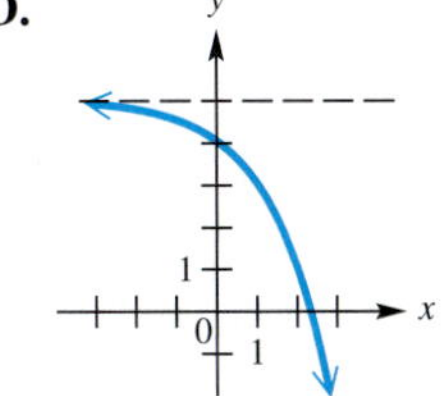

13. The functions in Exercises 9–12 form two pairs of inverse functions. Determine which functions are inverses of each other.

14. Determine the inverse of the function defined by $f(x) = \log_5 x$. (*Hint:* Replace $f(x)$ with y, and write in exponential form.)

For each function that is one-to-one, write an equation for the inverse function. Give the domain and range of f and f^{-1}. If the function is not one-to-one, say so.

15. $f(x) = 3x - 6$

16. $f(x) = 2(x + 1)^3$

17. $f(x) = 3x^2$

18. $f(x) = \dfrac{2x - 1}{5 - 3x}$

19. $f(x) = \sqrt[3]{5 - x^4}$

20. $f(x) = \sqrt{x^2 - 9}, x \geq 3$

Write an equivalent statement in logarithmic form.

21. $\left(\dfrac{1}{10}\right)^{-3} = 1000$

22. $a^b = c$

23. $(\sqrt{3})^4 = 9$

24. $4^{-3/2} = \dfrac{1}{8}$

25. $2^x = 32$

26. $27^{4/3} = 81$

Solve each equation.

27. $3x = 7^{\log_7 6}$

28. $\log_x 5 = \dfrac{1}{2}$

29. $\log_{10} .01 = x$

30. $x = \log_2 \sqrt{8}$

31. $\log_x \sqrt[3]{5} = \dfrac{1}{3}$

32. $x = \log_{4/5} \dfrac{25}{16}$

33. $2x - 1 = \log_6 6^x$

34. $\log_3 x = -2$

35. $\log_2 6 + \log_2 x = 5$ (*Hint:* Write the left side as a single logarithm, and then convert to exponential form.)

4.4 Evaluating Logarithms and the Change-of-Base Theorem

Common Logarithms ▪ Applications and Modeling with Common Logarithms ▪ Natural Logarithms ▪ Applications and Modeling with Natural Logarithms ▪ Logarithms to Other Bases

Common Logarithms The two most important bases for logarithms are 10 and e. Base 10 logarithms are called **common logarithms.** The common logarithm of x is written $\log x$, where the base is understood to be 10.

Common Logarithm

For all positive numbers x, $\quad \mathbf{\log x = \log_{10} x.}$

A calculator with a log key can be used to find the base 10 logarithm of any positive number. Consult your owner's manual for the keystrokes needed to find common logarithms.

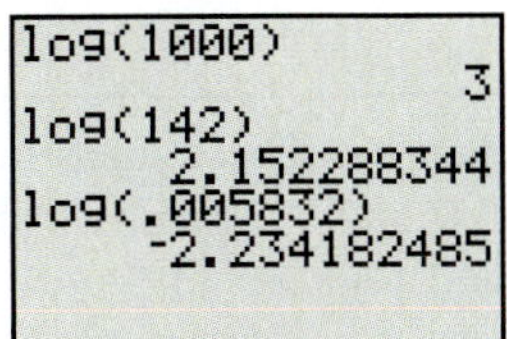

Figure 32

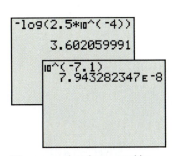

Figure 33

Figure 32 shows how a graphing calculator displays common logarithms. A common logarithm of a power of 10, such as 1000, is an integer (in this case, 3). Most common logarithms used in applications, such as log 142 and log .005832, are irrational numbers.

Figure 33 reinforces the concept presented in the previous section: log x is the exponent to which 10 must be raised in order to obtain x. ■

NOTE Base a, $a > 1$, logarithms of numbers less than 1 are always negative, as suggested by the graphs in Section 4.3.

Applications and Modeling with Common Logarithms

In chemistry, the **pH** of a solution is defined as

$$\text{pH} = -\log[H_3O^+],$$

where $[H_3O^+]$ is the hydronium ion concentration in moles* per liter. The pH value is a measure of the acidity or alkalinity of a solution. Pure water has pH 7.0, substances with pH values greater than 7.0 are alkaline, and substances with pH values less than 7.0 are acidic. It is customary to round pH values to the nearest tenth.

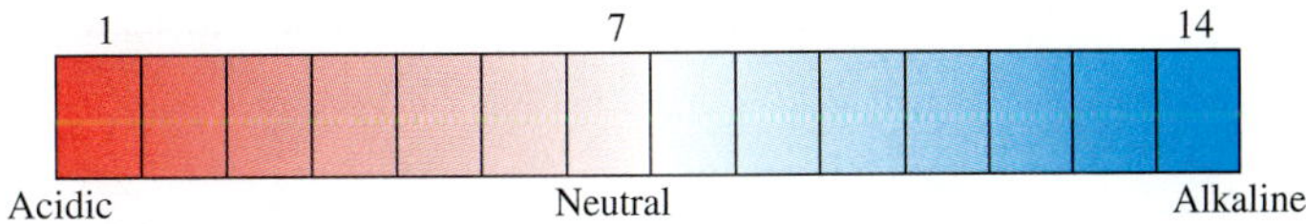

EXAMPLE 1 Finding pH

(a) Find the pH of a solution with $[H_3O^+] = 2.5 \times 10^{-4}$.

(b) Find the hydronium ion concentration of a solution with pH = 7.1.

Solution

(a)

$$\begin{aligned} \text{pH} &= -\log[H_3O^+] \\ &= -\log(2.5 \times 10^{-4}) && \text{Substitute.} \\ &= -(\log 2.5 + \log 10^{-4}) && \text{Product property (Section 4.3)} \\ &= -(.3979 - 4) && \log 10^{-4} = -4 \text{ (Section 4.3)} \\ &= -.3979 + 4 && \text{Distributive property (Section R.1)} \\ \text{pH} &\approx 3.6 \end{aligned}$$

(b)

$$\begin{aligned} \text{pH} &= -\log[H_3O^+] \\ 7.1 &= -\log[H_3O^+] && \text{Substitute.} \\ -7.1 &= \log[H_3O^+] && \text{Multiply by } -1. \\ [H_3O^+] &= 10^{-7.1} && \text{Write in exponential form. (Section 4.3)} \\ [H_3O^+] &\approx 7.9 \times 10^{-8} && \text{Evaluate } 10^{-7.1} \text{ with a calculator.} \end{aligned}$$

The screens show how a graphing calculator evaluates the pH and $[H_3O^+]$ in Example 1.

Now try Exercises 23 and 27.

---

*A *mole* is the amount of a substance that contains the same number of molecules as the number of atoms in exactly 12 grams of carbon 12.

EXAMPLE 2 Using pH in an Application

Wetlands are classified as *bogs*, *fens*, *marshes*, and *swamps*. These classifications are based on pH values. A pH value between 6.0 and 7.5, such as that of Summerby Swamp in Michigan's Hiawatha National Forest, indicates that the wetland is a "rich fen." When the pH is between 4.0 and 6.0, it is a "poor fen," and if the pH falls to 3.0 or less, the wetland is a "bog." (*Source:* R. Mohlenbrock, "Summerby Swamp, Michigan," *Natural History,* March 1994.)

Suppose that the hydronium ion concentration of a sample of water from a wetland is 6.3×10^{-5}. How would this wetland be classified?

Solution

$$\begin{aligned} \text{pH} &= -\log[\text{H}_3\text{O}^+] && \text{Definition of pH} \\ &= -\log(6.3 \times 10^{-5}) && \text{Substitute.} \\ &= -(\log 6.3 + \log 10^{-5}) && \text{Product property} \\ &= -\log 6.3 - (-5) && \text{Distributive property; } \log 10^n = n \\ &= -\log 6.3 + 5 \\ \text{pH} &\approx 4.2 && \text{Use a calculator.} \end{aligned}$$

Since the pH is between 4.0 and 6.0, the wetland is a poor fen.

Now try Exercise 31.

EXAMPLE 3 Measuring the Loudness of Sound

The loudness of sounds is measured in a unit called a *decibel.* To measure with this unit, we first assign an intensity of I_0 to a very faint sound, called the *threshold sound.* If a particular sound has intensity I, then the decibel rating of this louder sound is

$$d = 10 \log \frac{I}{I_0}.$$

Find the decibel rating of a sound with intensity $10{,}000I_0$.

Solution

$$\begin{aligned} d &= 10 \log \frac{10{,}000I_0}{I_0} && \text{Let } I = 10{,}000I_0. \\ &= 10 \log 10{,}000 \\ &= 10(4) && \log 10{,}000 = \log 10^4 = 4 \text{ (Section 4.3)} \\ &= 40 \end{aligned}$$

The sound has a decibel rating of 40.

Now try Exercise 59.

Looking Ahead to Calculus

The natural logarithmic function defined by $f(x) = \ln x$ and the reciprocal function defined by $g(x) = \frac{1}{x}$ have an important relationship in calculus. The derivative of the natural logarithmic function is the reciprocal function. Using *Leibniz notation* (named after one of the co-inventors of calculus), this fact is written $\frac{d}{dx}(\ln x) = \frac{1}{x}$.

Natural Logarithms In Section 4.2, we introduced the irrational number e. In most practical applications of logarithms, e is used as base. Logarithms to base e are called **natural logarithms,** since they occur in the life sciences and economics in natural situations that involve growth and decay. The base e logarithm of x is written $\ln x$ (read "el-en x").

Natural Logarithm

For all positive numbers x, $\quad \ln x = \log_e x.$

A graph of the natural logarithmic function defined by $f(x) = \ln x$ is given in Figure 34.

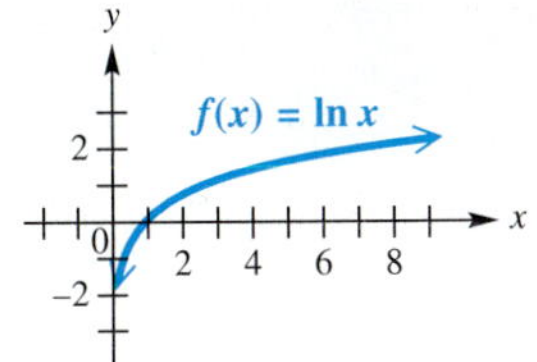

Figure 34

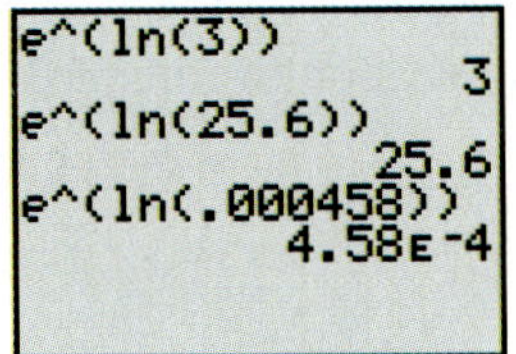

Figure 35

e^(ln(3))
3
e^(ln(25.6))
25.6
e^(ln(.000458))
4.58E-4

Figure 36

Natural logarithms can be found using a calculator. (Consult your owner's manual.) As in the case of common logarithms, when used in applications natural logarithms are usually irrational numbers. Figure 35 shows how three natural logarithms are evaluated with a graphing calculator.

Figure 36 reinforces the fact that $\ln x$ is the exponent to which e must be raised in order to obtain x. ■

Applications and Modeling with Natural Logarithms

EXAMPLE 4 Measuring the Age of Rocks

Geologists sometimes measure the age of rocks by using "atomic clocks." By measuring the amounts of potassium 40 and argon 40 in a rock, the age t of the specimen in years is found with the formula

$$t = (1.26 \times 10^9)\frac{\ln\left[1 + 8.33\left(\frac{A}{K}\right)\right]}{\ln 2},$$

where A and K are the numbers of atoms of argon 40 and potassium 40, respectively, in the specimen.

(a) How old is a rock in which $A = 0$ and $K > 0$?

(b) The ratio $\frac{A}{K}$ for a sample of granite from New Hampshire is .212. How old is the sample?

Solution

(a) If $A = 0$, $\frac{A}{K} = 0$ and the equation becomes

$$t = (1.26 \times 10^9)\frac{\ln 1}{\ln 2} = (1.26 \times 10^9)(0) = 0.$$

The rock is new (0 yr old).

(b) Since $\frac{A}{K} = .212$, we have

$$t = (1.26 \times 10^9)\frac{\ln[1 + 8.33(.212)]}{\ln 2} \approx 1.85 \times 10^9.$$

The granite is about 1.85 billion yr old.

Now try Exercise 65.

EXAMPLE 5 Modeling Global Temperature Increase

Carbon dioxide in the atmosphere traps heat from the sun. The additional solar radiation trapped by carbon dioxide is called *radiative forcing*. It is measured in watts per square meter (w/m^2). In 1896 the Swedish scientist Svante Arrhenius modeled radiative forcing R caused by additional atmospheric carbon dioxide using the logarithmic equation

$$R = k \ln \frac{C}{C_0},$$

where C_0 is the preindustrial amount of carbon dioxide, C is the current carbon dioxide level, and k is a constant. Arrhenius determined that $10 \leq k \leq 16$ when $C = 2C_0$. (*Source:* Clime, W., *The Economics of Global Warming,* Institute for International Economics, Washington, D.C., 1992.)

(a) Let $C = 2C_0$. Is the relationship between R and k linear or logarithmic?

(b) The average global temperature increase T (in °F) is given by $T(R) = 1.03R$. Write T as a function of k.

Solution

(a) If $C = 2C_0$, then $\frac{C}{C_0} = 2$, so $R = k \ln 2$ is a linear relation, because $\ln 2$ is a constant.

(b)

$$T(R) = 1.03R$$

$$T(k) = 1.03k \ln \frac{C}{C_0} \qquad \text{Use the given expression for } R.$$

Now try Exercise 71.

Logarithms to Other Bases We can use a calculator to find the values of either natural logarithms (base e) or common logarithms (base 10). However, sometimes we must use logarithms to other bases. The following theorem can be used to convert logarithms from one base to another.

Looking Ahead to Calculus

In calculus, natural logarithms are more convenient to work with than logarithms to other bases. The change-of-base theorem enables us to convert any logarithmic function to a *natural* logarithmic function.

Change-of-Base Theorem

For any positive real numbers x, a, and b, where $a \neq 1$ and $b \neq 1$:

$$\log_a x = \frac{\log_b x}{\log_b a}.$$

Proof Let

$$y = \log_a x.$$

$$a^y = x \qquad \text{Change to exponential form.}$$

$$\log_b a^y = \log_b x \qquad \text{Take logarithms on both sides.}$$

$$y \log_b a = \log_b x \qquad \text{Power property (Section 4.3)}$$

$$y = \frac{\log_b x}{\log_b a} \qquad \text{Divide both sides by } \log_b a.$$

$$\log_a x = \frac{\log_b x}{\log_b a} \qquad \text{Substitute } \log_a x \text{ for } y.$$

Any positive number other than 1 can be used for base b in the change-of-base theorem, but usually the only practical bases are e and 10 since calculators give logarithms only for these two bases.

With the change-of-base theorem, we can now graph the equation $y = \log_2 x$, for example, by directing the calculator to graph $y = \frac{\log x}{\log 2}$, or equivalently, $y = \frac{\ln x}{\ln 2}$. ■

EXAMPLE 6 Using the Change-of-Base Theorem

Use the change-of-base theorem to find an approximation to four decimal places for each logarithm.

(a) $\log_5 17$ **(b)** $\log_2 .1$

```
log(17)/log(5)
          1.7604
ln(.1)/ln(2)
         -3.3219
```

The screen shows how the result of Example 6(a) can be found using *common* logarithms, and how the result of Example 6(b) can be found using *natural* logarithms. The results are the same as those in Example 6.

Solution

(a) We use natural logarithms.

$$\log_5 17 = \frac{\ln 17}{\ln 5} \approx \frac{2.8332}{1.6094} \approx 1.7604$$

(b) Here, we use common logarithms.

$$\log_2 .1 = \frac{\log .1}{\log 2} \approx \frac{-1.0000}{.3010} \approx -3.3219$$

Now try Exercises 35 and 37.

NOTE In Example 6, logarithms evaluated in the intermediate steps, such as ln 17 and ln 5, were shown to four decimal places. However, the final answers were obtained *without* rounding these intermediate values, using all the digits obtained with the calculator. In general, it is best to wait until the final step to round off the answer; otherwise, a build-up of round-off errors may cause the final answer to have an incorrect final decimal place digit.

EXAMPLE 7 Modeling Diversity of Species

One measure of the diversity of the species in an ecological community is modeled by the formula

$$H = -[P_1 \log_2 P_1 + P_2 \log_2 P_2 + \cdots + P_n \log_2 P_n],$$

where $P_1, P_2, \ldots, P_n$ are the proportions of a sample that belong to each of n species found in the sample. (*Source:* Ludwig, J., and J. Reynolds, *Statistical Ecology: A Primer on Methods and Computing,* New York, Wiley, 1988, p. 92.)

Find the measure of diversity in a community with two species where there are 90 of one species and 10 of the other.

Solution Since there are 100 members in the community, $P_1 = \frac{90}{100} = .9$ and $P_2 = \frac{10}{100} = .1$, so

$$H = -[.9 \log_2 .9 + .1 \log_2 .1].$$

In Example 6(b), we found that $\log_2 .1 \approx -3.32$. Now we find $\log_2 .9$.

$$\log_2 .9 = \frac{\log .9}{\log 2} \approx \frac{-.0458}{.3010} \approx -.152$$

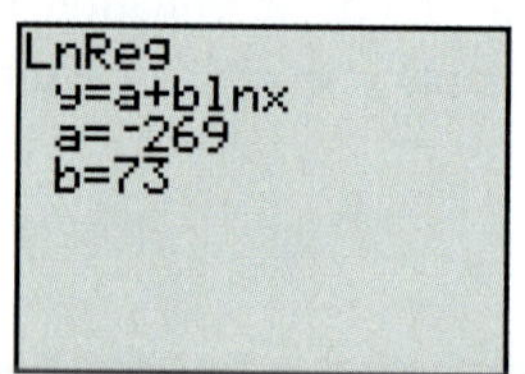

Therefore,

$$H = -[.9 \log_2 .9 + .1 \log_2 .1]$$
$$\approx -[.9(-.152) + .1(-3.32)] \approx .469.$$

Verify that $H \approx .971$ if there are 60 of one species and 40 of the other. As the proportions of n species get closer to $\frac{1}{n}$ each, the measure of diversity increases to a maximum of $\log_2 n$.

Now try Exercise 69.

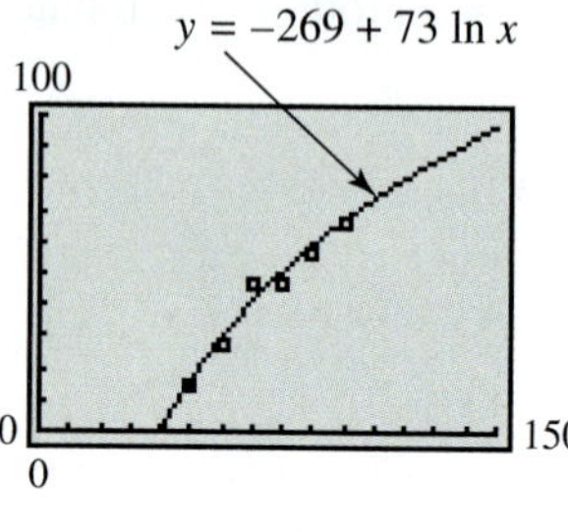

Figure 37

At the end of Section 4.2, we saw that graphing calculators are capable of fitting exponential curves to data that suggest such behavior. The same is true for logarithmic curves. Figure 37 shows how a calculator gives the best-fitting natural logarithmic curve for the data in Exercise 65 in this section: $y = -269 + 73 \ln x$. ■

4.4 Exercises

Concept Check *Answer each of the following.*

1. For the exponential function defined by $f(x) = a^x$, where $a > 1$, is the function increasing or is it decreasing over its entire domain?
2. For the logarithmic function defined by $g(x) = \log_a x$, where $a > 1$, is the function increasing or is it decreasing over its entire domain?
3. If $f(x) = 5^x$, what is the rule for $f^{-1}(x)$?
4. What is the name given to the exponent to which 4 must be raised in order to obtain 11?
5. A base e logarithm is called a(n) ____________ logarithm; a base 10 logarithm is called a(n) ____________ logarithm.
6. How is $\log_3 12$ written in terms of natural logarithms?
7. Why is $\log_2 0$ undefined?
8. Between what two consecutive integers must $\log_2 12$ lie?
9. The graph of $y = \log x$ shows a point on the graph. Write the logarithmic equation associated with that point.

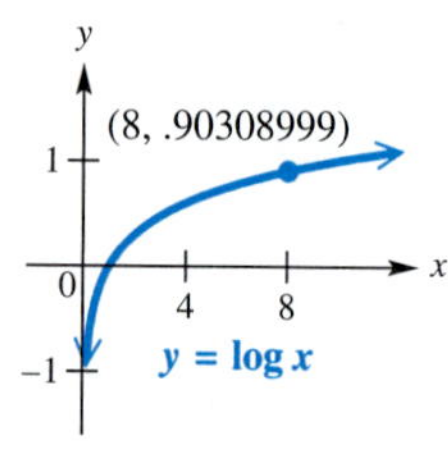

10. The graph of $y = \ln x$ shows a point on the graph. Write the logarithmic equation associated with that point.

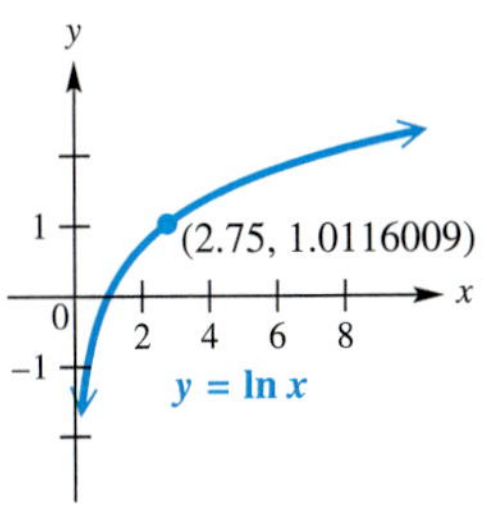

Use a calculator with logarithm keys to find an approximation to four decimal places for each expression.

11. $\log 36$ **12.** $\log 72$ **13.** $\log .042$ **14.** $\log .319$

15. $\log(2 \times 10^4)$ **16.** $\log(2 \times 10^{-4})$ **17.** $\ln 36$ **18.** $\ln 72$

19. $\ln .042$ **20.** $\ln .319$ **21.** $\ln(2 \times e^4)$ **22.** $\ln(2 \times e^{-4})$

For each substance, find the pH from the given hydronium ion concentration. See Example 1(a).

23. grapefruit, 6.3×10^{-4} **24.** crackers, 3.9×10^{-9}

25. limes, 1.6×10^{-2} **26.** sodium hydroxide (lye), 3.2×10^{-14}

Find the $[H_3O^+]$ *for each substance with the given pH. See Example 1(b).*

27. soda pop, 2.7 **28.** wine, 3.4

29. beer, 4.8 **30.** drinking water, 6.5

In Exercises 31–33, suppose that water from a wetland area is sampled and found to have the given hydronium ion concentration. Determine whether the wetland is a rich fen, poor fen, or bog. See Example 2.

31. 2.49×10^{-5} **32.** 2.49×10^{-2} **33.** 2.49×10^{-7}

34. Use your calculator to find an approximation for each logarithm.

(a) $\log 398.4$ **(b)** $\log 39.84$ **(c)** $\log 3.984$

(d) From your answers to parts (a)–(c), make a conjecture concerning the decimal values in the approximations of common logarithms of numbers greater than 1 that have the same digits.

Use the change-of-base theorem to find an approximation for four decimal places for each logarithm. See Example 6.

35. $\log_2 5$ **36.** $\log_2 9$ **37.** $\log_8 .59$ **38.** $\log_8 .71$

39. $\log_{\sqrt{13}} 12$ **40.** $\log_{\sqrt{19}} 5$ **41.** $\log_{.32} 5$ **42.** $\log_{.91} 8$

43. **Concept Check** Which of the following is the same as $2\ln(3x)$ for $x > 0$?

A. $\ln 9 + \ln x$ **B.** $\ln(6x)$ **C.** $\ln 6 + \ln x$ **D.** $\ln(9x^2)$

44. **Concept Check** Which of the following is the same as $\ln(4x) - \ln(2x)$ for $x > 0$?

A. $2\ln x$ **B.** $\ln(2x)$ **C.** $\dfrac{\ln(4x)}{\ln(2x)}$ **D.** $\ln 2$

Let $u = \ln a$ *and* $v = \ln b$. *Write each expression in terms of u and v without using the* ln *function.*

45. $\ln(b^4\sqrt{a})$ **46.** $\ln \dfrac{a^3}{b^2}$ **47.** $\ln \sqrt{\dfrac{a^3}{b^5}}$ **48.** $\ln(\sqrt[3]{a} \cdot b^4)$

49. Given $g(x) = e^x$, evaluate **(a)** $g(\ln 3)$ **(b)** $g[\ln(5^2)]$ **(c)** $g\left[\ln\left(\frac{1}{e}\right)\right]$.

50. Given $f(x) = 3^x$, evaluate **(a)** $f(\log_3 7)$ **(b)** $f[\log_3(\ln 3)]$ **(c)** $f[\log_3(2 \ln 3)]$.

51. Given $f(x) = \ln x$, evaluate **(a)** $f(e^5)$ **(b)** $f(e^{\ln 3})$ **(c)** $f(e^{2 \ln 3})$.

52. Given $f(x) = \log_2 x$, evaluate **(a)** $f(2^3)$ **(b)** $f(2^{\log_2 2})$ **(c)** $f(2^{2 \log_2 2})$.

53. The function defined by $f(x) = \ln|x|$ plays a prominent role in calculus. Find its domain, range, and the symmetries of its graph.

54. Consider the function defined by $f(x) = \log_3|x|$.

(a) What is the domain of this function?

(b) Use a graphing calculator to graph $f(x) = \log_3|x|$ in the window $[-4, 4]$ by $[-4, 4]$.

(c) How might one easily misinterpret the domain of the function simply by observing the calculator graph?

55. The table is for $Y_1 = \log_3(4 - X)$. Why do the values of Y_1 show ERROR for $X \geq 4$?

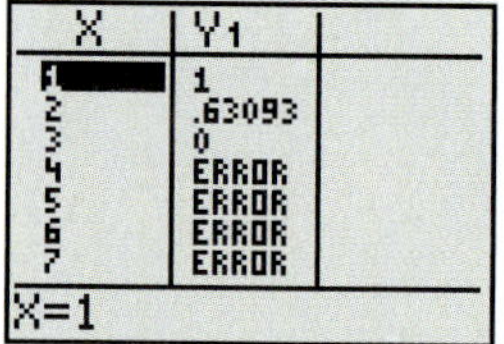

X	Y1
1	1
2	.63093
3	0
4	ERROR
5	ERROR
6	ERROR
7	ERROR

X=1

56. The function defined by Y_1 is of the form $Y_1 = \log_a X$. What is the value of a?

X	Y1
1	0
4	1
8	1.5
16	2
32	2.5
64	3
128	3.5

X=1

Use the properties of logarithms and the terminology of Chapter 2 to describe how the graph of the given function compares to the graph of $f(x) = \ln x$.

57. $f(x) = \ln e^2x$

58. $f(x) = \ln \dfrac{x}{e}$

Solve each application of logarithms. See Examples 3–5.

59. ***Decibel Levels*** Find the decibel ratings of sounds having the following intensities.

(a) $100I_0$ **(b)** $1000I_0$ **(c)** $100{,}000I_0$ **(d)** $1{,}000{,}000I_0$

(e) If the intensity of a sound is doubled, by how much is the decibel rating increased?

60. ***Decibel Levels*** Find the decibel ratings of the following sounds, having intensities as given. Round each answer to the nearest whole number.

(a) whisper, $115I_0$ **(b)** busy street, $9{,}500{,}000I_0$

(c) heavy truck, 20 m away, $1{,}200{,}000{,}000I_0$

(d) rock music, $895{,}000{,}000{,}000I_0$

(e) jetliner at takeoff, $109{,}000{,}000{,}000{,}000I_0$

61. ***Earthquake Intensity*** The magnitude of an earthquake, measured on the Richter scale, is $\log_{10} \frac{I}{I_0}$, where I is the amplitude registered on a seismograph 100 km from the epicenter of the earthquake, and I_0 is the amplitude of an earthquake of a certain (small) size. Find the Richter scale ratings for earthquakes having the following amplitudes.

(a) $1000I_0$ **(b)** $1{,}000{,}000I_0$ **(c)** $100{,}000{,}000I_0$

62. *Earthquake Intensity* On June 16, 1999, the city of Puebla in central Mexico was shaken by an earthquake that measured 6.7 on the Richter scale. Express this reading in terms of I_0. See Exercise 61. *(Source: Times Picayune.)*

63. *Earthquake Intensity* On September 19, 1985, Mexico's largest recent earthquake, measuring 8.1 on the Richter scale, killed about 9500 people. Express the magnitude of an 8.1 reading in terms of I_0. *(Source: Times Picayune.)*

64. Compare your answers to Exercises 62 and 63. How much greater was the force of the 1985 earthquake than the 1999 earthquake?

65. *(Modeling) Visitors to U.S. National Parks* The heights of the bars in the graph represent the number of visitors (in millions) to U.S. National Parks from 1950–2000. Suppose x represents the number of years since 1900—thus, 1950 is represented by 50, 1960 is represented by 60, and so on. The logarithmic function defined by

$$f(x) = -269 + 73 \ln x$$

closely models the data. Use this function to estimate the number of visitors in the year 2004. What assumption must we make to estimate the number of visitors in years beyond 2000?

Visitors to National Parks
(in millions)

1950	1960	1970	1980	1990	2000
14	28	46	47	57	66

Source: Statistical Abstract of the United States, 2000.

66. *(Modeling) Volunteerism among College Freshmen* The growth in the percentage of college freshmen who reported involvement in volunteer work during their last year of high school is shown in the bar graph. Connecting the tops of the bars with a continuous curve would give a graph that indicates logarithmic growth. The function defined by

$$f(t) = 74.61 + 3.84 \ln t, \quad t \geq 1,$$

where t represents the number of years since 1994 and $f(t)$ is the percent, approximates the curve reasonably well.

Volunteerism among College Freshman

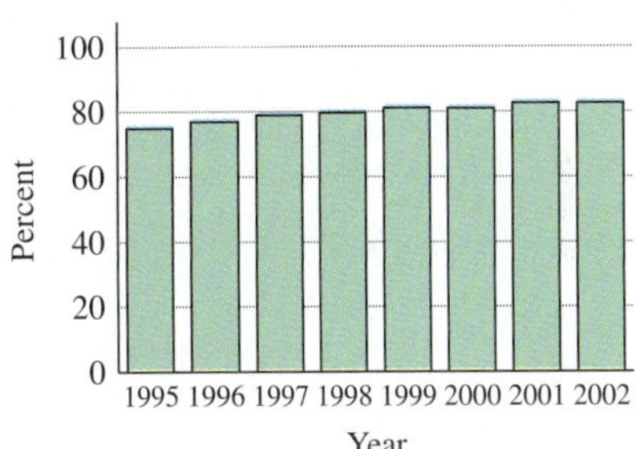

Source: Higher Education Research Institute, UCLA.

(a) What does the function predict for the percent of freshmen entering college in 2002 who performed volunteer work during their last year of high school? How does this compare to the actual percent of 82.6?

(b) Explain why an exponential function would *not* provide a good model for these data.

67. *(Modeling) Diversity of Species* The number of species in a sample is given by

$$S(n) = a \ln\left(1 + \frac{n}{a}\right).$$

Here n is the number of individuals in the sample, and a is a constant that indicates the diversity of species in the community. If $a = .36$, find $S(n)$ for each value of n. *(Hint: $S(n)$ must be a whole number.)*

(a) 100 **(b)** 200 **(c)** 150 **(d)** 10

68. ***(Modeling) Diversity of Species*** In Exercise 67, find $S(n)$ if a changes to .88. Use the following values of n.

(a) 50 **(b)** 100 **(c)** 250

69. ***(Modeling) Diversity of Species*** Suppose a sample of a small community shows two species with 50 individuals each. Find the measure of diversity H. (See Example 7.)

70. ***(Modeling) Diversity of Species*** A virgin forest in northwestern Pennsylvania has 4 species of large trees with the following proportions of each: hemlock, .521; beech, .324; birch, .081; maple, .074. Find the measure of diversity H. (See Example 7.)

71. ***(Modeling) Global Temperature Increase*** In Example 5, we expressed the average global temperature increase T (in °F) as

$$T(k) = 1.03k \ln \frac{C}{C_0},$$

where C_0 is the preindustrial amount of carbon dioxide, C is the current carbon dioxide level, and k is a constant. Arrhenius determined that $10 \leq k \leq 16$ when C was double the value C_0. Use $T(k)$ to find the range of the rise in global temperature T (rounded to the nearest degree) that Arrhenius predicted. (*Source:* Clime, W., *The Economics of Global Warming,* Institute for International Economics, Washington, D.C., 1992.)

72. ***(Modeling) Global Temperature Increase*** (Refer to Exercise 71.) According to the IPCC, if present trends continue, future increases in average global temperatures (in °F) can be modeled by

$$T(C) = 6.489 \ln \frac{C}{280},$$

where C is the concentration of atmospheric carbon dioxide (in ppm). C can be modeled by the function defined by

$$C(x) = 353(1.006)^{x-1990},$$

where x is the year. (*Source:* International Panel on Climate Change (IPCC), 1990.)

(a) Write T as a function of x.

(b) Using a graphing calculator, graph $C(x)$ and $T(x)$ on the interval [1990, 2275] using different coordinate axes. Describe the graph of each function. How are C and T related?

(c) Approximate the slope of the graph of T. What does this slope represent?

(d) Use graphing to estimate x and $C(x)$ when $T(x) = 10$°F.

73. ***(Modeling) Planets' Distances from the Sun and Periods of Revolution*** The table contains the planets' average distances D from the sun and their periods P of revolution around the sun in years. The distances have been normalized so that Earth is one unit away from the sun. For example, since Jupiter's distance is 5.2, its distance from the sun is 5.2 times farther than Earth's.

Planet	D	P
Mercury	.39	.24
Venus	.72	.62
Earth	1	1
Mars	1.52	1.89
Jupiter	5.2	11.9
Saturn	9.54	29.5
Uranus	19.2	84.0
Neptune	30.1	164.8

Source: Ronan, C., *The Natural History of the Universe,* MacMillan Publishing Co., New York, 1991.

(a) Make a scatter diagram by plotting the point ($\ln D$, $\ln P$) for each planet on the xy-coordinate axes using a graphing calculator. Do the data points appear to be linear?

(b) Determine a linear equation that models the data points. Graph your line and the data on the same coordinate axes.

(c) Use this linear model to predict the period of the planet Pluto if its distance is 39.5. Compare your answer to the actual value of 248.5 yr.

74. Explain the error in the following "proof" that $2 < 1$.

$$\frac{1}{9} < \frac{1}{3}$$

$$\left(\frac{1}{3}\right)^2 < \frac{1}{3} \qquad \text{Rewrite the left side.}$$

$$\log\left(\frac{1}{3}\right)^2 < \log\frac{1}{3} \qquad \text{Take logarithms on both sides.}$$

$$2\log\frac{1}{3} < 1\log\frac{1}{3} \qquad \text{Property of logarithms}$$

$$2 < 1 \qquad \text{Divide both sides by } \log\tfrac{1}{3}.$$

4.5 Exponential and Logarithmic Equations

Exponential Equations ▪ **Logarithmic Equations** ▪ **Applications and Modeling**

Exponential Equations We solved exponential equations in earlier sections. General methods for solving these equations depend on the property below, which follows from the fact that logarithmic functions are one-to-one.

Property of Logarithms

If $x > 0$, $y > 0$, $a > 0$, and $a \neq 1$, then

$$x = y \quad \text{if and only if} \quad \log_a x = \log_a y.$$

EXAMPLE 1 Solving an Exponential Equation

Solve $7^x = 12$. Give the solution to four decimal places.

Solution The properties of exponents given in Section 4.2 cannot be used to solve this equation, so we apply the preceding property of logarithms. While any appropriate base b can be used, the best practical base is base 10 or base e. We choose base e (natural) logarithms here.

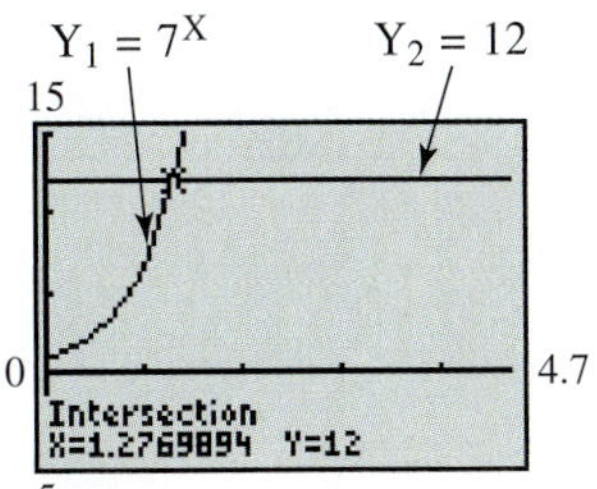

As seen in the display at the bottom of the screen, when rounded to four decimal places, the solution agrees with that found in Example 1.

$$7^x = 12$$

$$\ln 7^x = \ln 12 \qquad \text{Property of logarithms}$$

$$x \ln 7 = \ln 12 \qquad \text{Power property (Section 4.3)}$$

$$x = \frac{\ln 12}{\ln 7} \qquad \text{Divide by } \ln 7.$$

$$x \approx 1.2770 \qquad \text{Use a calculator.}$$

The solution set is $\{1.2770\}$.

Now try Exercise 5.

CAUTION Be careful when evaluating a quotient like $\frac{\ln 12}{\ln 7}$ in Example 1. Do not confuse this quotient with $\ln \frac{12}{7}$, which can be written as $\ln 12 - \ln 7$. You *cannot* change the quotient of *two logarithms* to a difference of logarithms.

$$\frac{\ln 12}{\ln 7} \neq \ln \frac{12}{7}$$

EXAMPLE 2 Solving an Exponential Equation

Solve $3^{2x-1} = .4^{x+2}$. Give the solution to four decimal places.

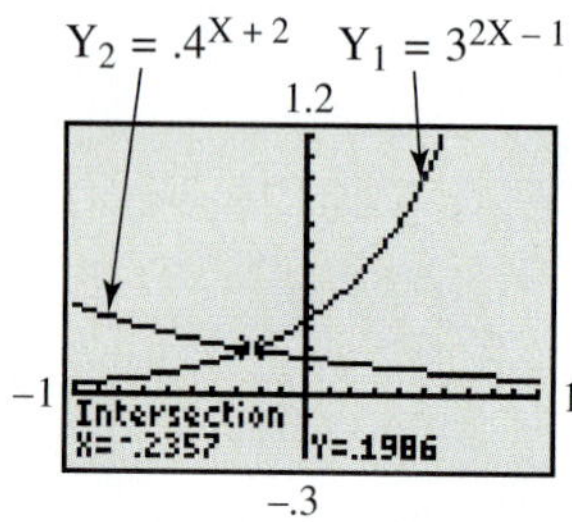

This screen supports the solution found in Example 2.

Solution

$$3^{2x-1} = .4^{x+2}$$

$$\ln 3^{2x-1} = \ln .4^{x+2}$$ Take natural logarithms on both sides.

$$(2x - 1)\ln 3 = (x + 2)\ln .4$$ Power property

$$2x \ln 3 - \ln 3 = x \ln .4 + 2 \ln .4$$ Distributive property (Section R.1)

$$2x \ln 3 - x \ln .4 = 2 \ln .4 + \ln 3$$ Write the terms with x on one side.

$$x(2 \ln 3 - \ln .4) = 2 \ln .4 + \ln 3$$ Factor out x. (Section R.4)

$$x = \frac{2 \ln .4 + \ln 3}{2 \ln 3 - \ln .4}$$ Divide by $2 \ln 3 - \ln .4$.

$$x = \frac{\ln .16 + \ln 3}{\ln 9 - \ln .4}$$ Power property

$$x = \frac{\ln .48}{\ln \frac{9}{.4}}$$ Product property, Quotient property (Section 4.3)

$$x \approx -.2357$$ Use a calculator.

The solution set is $\{-.2357\}$.

Now try Exercise 9.

EXAMPLE 3 Solving Base e Exponential Equations

Solve each equation. Give solutions to four decimal places.

(a) $e^{x^2} = 200$ **(b)** $e^{2x+1} \cdot e^{-4x} = 3e$

Solution

(a)

$$e^{x^2} = 200$$

$$\ln e^{x^2} = \ln 200$$ Take natural logarithms on both sides.

$$x^2 = \ln 200$$ $\ln e^{x^2} = x^2$ (Section 4.3)

$$x = \pm\sqrt{\ln 200}$$ Square root property (Section 1.4)

$$x \approx \pm 2.3018$$ Use a calculator.

The solution set is $\{\pm 2.3018\}$.

(b)

$$e^{2x+1} \cdot e^{-4x} = 3e$$

$$e^{-2x+1} = 3e \qquad a^m \cdot a^n = a^{m+n} \text{ (Section R.3)}$$

$$e^{-2x} = 3 \qquad \text{Divide by } e;\ \frac{a^m}{a^n} = a^{m-n}. \text{ (Section R.3)}$$

$$\ln e^{-2x} = \ln 3 \qquad \text{Take natural logarithms on both sides.}$$

$$-2x \ln e = \ln 3 \qquad \text{Power property}$$

$$-2x = \ln 3 \qquad \ln e = 1$$

$$x = -\frac{1}{2} \ln 3 \qquad \text{Multiply by } -\tfrac{1}{2}.$$

$$x \approx -.5493$$

The solution set is $\{-.5493\}$.

Now try Exercises 11 and 17.

Logarithmic Equations The next examples show some ways to solve logarithmic equations.

EXAMPLE 4 Solving a Logarithmic Equation

Solve $\log_a(x + 6) - \log_a(x + 2) = \log_a x$.

Solution Rewrite the equation as

$$\log_a \frac{x+6}{x+2} = \log_a x. \qquad \text{Quotient property}$$

$$\frac{x+6}{x+2} = x \qquad \text{Property of logarithms}$$

$$x + 6 = x(x + 2) \qquad \text{Multiply by } x + 2.$$

$$x + 6 = x^2 + 2x \qquad \text{Distributive property}$$

$$x^2 + x - 6 = 0 \qquad \text{Standard form}$$

$$(x + 3)(x - 2) = 0 \qquad \text{Factor.}$$

$$x = -3 \quad \text{or} \quad x = 2 \qquad \text{Zero-factor property (Section 1.4)}$$

The negative solution ($x = -3$) is not in the domain of $\log_a x$ in the original equation, so the only valid solution is the positive number 2, giving the solution set $\{2\}$.

Now try Exercise 25.

CAUTION Recall that the domain of $y = \log_b x$ is $(0, \infty)$. For this reason, ***it is always necessary to check that apparent solutions of a logarithmic equation result in logarithms of positive numbers in the original equation.***

EXAMPLE 5 Solving a Logarithmic Equation

Solve $\log(3x+2) + \log(x-1) = 1$.

Solution The notation $\log x$ is an abbreviation for $\log_{10} x$, and $1 = \log_{10} 10$.

$$\log(3x+2) + \log(x-1) = 1$$
$$\log(3x+2) + \log(x-1) = \log 10 \qquad \text{Substitute.}$$
$$\log[(3x+2)(x-1)] = \log 10 \qquad \text{Product property}$$
$$(3x+2)(x-1) = 10 \qquad \text{Property of logarithms}$$
$$3x^2 - x - 2 = 10 \qquad \text{Multiply. (Section R.3)}$$
$$3x^2 - x - 12 = 0 \qquad \text{Subtract 10.}$$
$$x = \frac{1 \pm \sqrt{1+144}}{6} \qquad \text{Quadratic formula (Section 1.4)}$$

The number $\frac{1-\sqrt{145}}{6}$ is negative, so $x-1$ is negative. Therefore, $\log(x-1)$ is not defined and this proposed solution must be discarded. Since $\frac{1+\sqrt{145}}{6} > 1$, both $3x+2$ and $x-1$ are positive and the solution set is $\left\{\frac{1+\sqrt{145}}{6}\right\}$.

Now try Exercise 29.

NOTE The definition of logarithm could have been used in Example 5 by first writing

$$\log(3x+2) + \log(x-1) = 1$$
$$\log_{10}[(3x+2)(x-1)] = 1 \qquad \text{Product property}$$
$$(3x+2)(x-1) = 10^1, \qquad \text{Definition of logarithm (Section 4.3)}$$

then continuing as shown above.

EXAMPLE 6 Solving a Base e Logarithmic Equation

Solve $\ln e^{\ln x} - \ln(x-3) = \ln 2$.

Solution

$$\ln e^{\ln x} - \ln(x-3) = \ln 2$$
$$\ln x - \ln(x-3) = \ln 2 \qquad e^{\ln x} = x \text{ (Section 4.3)}$$
$$\ln \frac{x}{x-3} = \ln 2 \qquad \text{Quotient property}$$
$$\frac{x}{x-3} = 2 \qquad \text{Property of logarithms}$$
$$x = 2x - 6 \qquad \text{Multiply by } x-3.$$
$$6 = x \qquad \text{Solve for } x. \text{ (Section 1.1)}$$

Verify that the solution set is $\{6\}$.

Now try Exercise 37.

A summary of the methods used for solving equations in this section follows.

Solving Exponential or Logarithmic Equations

To solve an exponential or logarithmic equation, change the given equation into one of the following forms, where a and b are real numbers, $a > 0$ and $a \neq 1$.

1. $\boldsymbol{a^{f(x)} = b}$
   Solve by taking logarithms on both sides.
2. $\boldsymbol{\log_a f(x) = b}$
   Solve by changing to exponential form $a^b = f(x)$.
3. $\boldsymbol{\log_a f(x) = \log_a g(x)}$
   The given equation is equivalent to the equation $f(x) = g(x)$. Solve algebraically.
4. In a more complicated equation, such as the one in Example 3(b), it may be necessary to first solve for $a^{f(x)}$ or $\log_a f(x)$ and then solve the resulting equation using one of the methods given above.

Applications and Modeling

EXAMPLE 7 Applying an Exponential Equation to the Strength of a Habit

The strength of a habit is a function of the number of times the habit is repeated. If N is the number of repetitions and H is the strength of the habit, then, according to psychologist C. L. Hull,

$$H = 1000(1 - e^{-kN}),$$

where k is a constant. Solve this equation for k.

Solution First solve the equation for e^{-kN}.

$$\frac{H}{1000} = 1 - e^{-kN} \quad \text{Divide by 1000.}$$

$$\frac{H}{1000} - 1 = -e^{-kN} \quad \text{Subtract 1.}$$

$$e^{-kN} = 1 - \frac{H}{1000} \quad \text{Multiply by } -1\text{; rewrite.}$$

Now we can solve for k.

$$\ln e^{-kN} = \ln\left(1 - \frac{H}{1000}\right) \quad \text{Take natural logarithms on both sides.}$$

$$-kN = \ln\left(1 - \frac{H}{1000}\right) \quad \ln e^x = x$$

$$k = -\frac{1}{N}\ln\left(1 - \frac{H}{1000}\right) \quad \text{Multiply by } -\tfrac{1}{N}.$$

With the final equation, if one pair of values for H and N is known, k can be found, and the equation can then be used to find either H or N for given values of the other variable.

Now try Exercise 45.

The equation used in the final example was determined by curve fitting, which was described at the end of Section 4.4.

EXAMPLE 8 Modeling Coal Consumption in the U.S.

Year	Coal Consumption (in quads)
1980	15.42
1985	17.48
1990	19.25
1995	20.03
2000	22.41

Source: Statistical Abstract of the United States, 2002.

The table gives U.S. coal consumption (in quadrillions of British thermal units, or *quads*) for several years. The data can be modeled with the function defined by

$$f(t) = 29.64 \ln t - 114.36, \qquad t \geq 80,$$

where t is the number of years after 1900, and $f(t)$ is in quads.

(a) Approximately what amount of coal was consumed in the United States in 1993?

(b) If this trend continues, approximately when will annual consumption reach 25 quads?

Solution

(a) The year 1993 is represented by $t = 1993 - 1900 = 93$.

$$\begin{aligned} f(93) &= 29.64 \ln 93 - 114.36 \\ &\approx 19.99 && \text{Use a calculator.} \end{aligned}$$

Based on this model, 19.99 quads were used in 1993.

(b) Let $f(t) = 25$, and solve for t.

$$\begin{aligned} 25 &= 29.64 \ln t - 114.36 \\ 139.36 &= 29.64 \ln t && \text{Add 114.36.} \\ \ln t &= \frac{139.36}{29.64} && \text{Divide by 29.64; rewrite.} \\ t &= e^{139.36/29.64} && \text{Write in exponential form.} \\ t &\approx 110 && \text{Use a calculator.} \end{aligned}$$

Add 110 to 1900 to get 2010. Annual consumption will reach 25 quads in approximately 2010.

Now try Exercise 67.

4.5 Exercises

Concept Check *An exponential equation such as* $5^x = 9$ *can be solved for its exact solution using the meaning of logarithm and the change-of-base theorem. Since x is the exponent to which 5 must be raised in order to obtain 9, the exact solution is* $\log_5 9$, *or* $\frac{\log 9}{\log 5}$ *or* $\frac{\ln 9}{\ln 5}$. *For the following equations, give the exact solution in three forms similar to the forms explained here.*

1. $7^x = 19$ **2.** $3^x = 10$ **3.** $\left(\frac{1}{2}\right)^x = 12$ **4.** $\left(\frac{1}{3}\right)^x = 4$

Solve each equation. When solutions are irrational, give them as decimals correct to four decimal places. See Examples 1–6.

5. $3^x = 6$

6. $4^x = 12$

7. $6^{1-2x} = 8$

8. $3^{2x-5} = 13$

9. $2^{x+3} = 5^x$

10. $6^{x+3} = 4^x$

11. $e^{x-1} = 4$

12. $e^{2-x} = 12$

13. $2e^{5x+2} = 8$

14. $10e^{3x-7} = 5$

15. $2^x = -3$

16. $3^x = -6$

17. $e^{8x} \cdot e^{2x} = e^{20}$

18. $e^{6x} \cdot e^x = e^{21}$

19. $100(1.02)^{x/4} = 200$

20. $500(1.05)^{x/4} = 200$

21. $\ln(6x + 1) = \ln 3$

22. $\ln(7 - x) = \ln 12$

23. $\log 4x - \log(x - 3) = \log 2$

24. $\ln(-x) + \ln 3 = \ln(2x - 15)$

25. $\log(2x - 1) + \log 10x = \log 10$

26. $\ln 5x - \ln(2x - 1) = \ln 4$

27. $\log(x + 25) = 1 + \log(2x - 7)$

28. $\log(11x + 9) = 3 + \log(x + 3)$

29. $\log x + \log(x - 21) = 2$

30. $\log x + \log(3x - 13) = 1$

31. $\ln(5 + 4x) - \ln(3 + x) = \ln 3$

32. $\ln(2x + 5) + \ln x = \ln 7$

33. $\log_6 4x - \log_6(x - 3) = \log_6 12$

34. $\log_2 3x + \log_2 3 = \log_2(2x + 15)$

35. $5^{x+2} = 2^{2x-1}$

36. $6^{x-3} = 3^{4x+1}$

37. $\ln e^x - \ln e^3 = \ln e^5$

38. $\ln e^x - 2 \ln e = \ln e^4$

39. $\log_2(\log_2 x) = 1$

40. $\log x = \sqrt{\log x}$

41. $\log x^2 = (\log x)^2$

42. $\log_2 \sqrt{2x^2} = \frac{3}{2}$

43. Suppose you overhear the following statement: "I must reject any negative answer when I solve an equation involving logarithms." Is this correct? Write an explanation of why it is or is not correct.

44. What values of x could not possibly be solutions of the following equation?

$$\log_a(4x - 7) + \log_a(x^2 + 4) = 0$$

Solve each equation for the indicated variable. Use logarithms to the appropriate bases. See Example 7.

45. $I = \frac{E}{R}(1 - e^{-Rt/2})$ for t

46. $r = p - k \ln t$ for t

47. $p = a + \frac{k}{\ln x}$ for x

48. $T = T_0 + (T_1 - T_0)10^{-kt}$ for t

Relating Concepts

For individual or collaborative investigation

(Exercises 49–54)

Earlier, we introduced methods of solving quadratic equations and showed how they can be applied to equations that are not actually quadratic, but are quadratic in form. Consider the following equation and ***work Exercises 49–54 in order.***

$$e^{2x} - 4e^x + 3 = 0$$

49. The expression e^{2x} is equivalent to $(e^x)^2$. Why is this so?

50. The given equation is equivalent to $(e^x)^2 - 4e^x + 3 = 0$. Factor the left side of this equation.

51. Solve the equation in Exercise 50 by using the zero-factor property. Give exact solutions.

(continued)

52. Support your solution(s) in Exercise 51 by graphing $y = e^{2x} - 4e^x + 3$ with a calculator.

53. Use the graph from Exercise 52 to identify the x-intervals where $y > 0$. These intervals give the solutions of $e^{2x} - 4e^x + 3 > 0$.

54. Use the graph from Exercise 52 and your answer to Exercise 53 to give the intervals where $e^{2x} - 4e^x + 3 < 0$.

Find $f^{-1}(x)$, and give the domain and range.

55. $f(x) = e^{x+1} - 4$

56. $f(x) = 2 \ln 3x$

Solve each inequality with a graphing calculator by rearranging terms so that one side is 0, then graphing the expression on the other side as y and observing from the graph where y is positive or negative as applicable. These inequalities are studied in calculus to determine where certain functions are increasing.

57. $\log_3 x > 3$

58. $\log_x .2 < -1$

Use a graphing calculator to solve each equation. Give irrational solutions correct to the nearest hundredth.

59. $e^x + \ln x = 5$

60. $e^x - \ln(x + 1) = 3$

61. $2e^x + 1 = 3e^{-x}$

62. $e^x + 6e^{-x} = 5$

63. $\log x = x^2 - 8x + 14$

64. $\ln x = -\sqrt[3]{x + 3}$

(Modeling) Solve each application. See Example 8.

65. ***Average Annual Public University Costs*** The table shows the cost of a year's tuition, room and board, and fees at a public university from 2000–2008. (*Note:* The amounts for 2002–2008 are projections.) Letting y represent the cost and x represent the number of years since 2000, we find that the function defined by

$$f(x) = 8160(1.06)^x$$

models the data quite well. According to this function, when will the cost in 2000 be doubled?

Year	Average Annual Cost
2000	\$7,990
2001	\$8,470
2002	\$9,338
2003	\$9,805
2004	\$10,295
2005	\$10,810
2006	\$11,351
2007	\$11,918
2008	\$12,514

Source: www.princetonreview.com

66. ***Race Speed*** At the World Championship races held at Rome's Olympic Stadium in 1987, American sprinter Carl Lewis ran the 100-m race in 9.86 sec. His speed in meters per second after t seconds is closely modeled by the function defined by

$$f(t) = 11.65(1 - e^{-t/1.27}).$$

(*Source:* Banks, Robert B., *Towing Icebergs, Falling Dominoes, and Other Adventures in Applied Mathematics,* Princeton University Press, 1998.)

(a) How fast was he running as he crossed the finish line?

(b) After how many seconds was he running at the rate of 10 m per sec?

67. *Fatherless Children* The percent of U.S. children growing up without a father has increased rapidly since 1950. If x represents the number of years since 1900, the function defined by

$$f(x) = \frac{25}{1 + 1364.3e^{-x/9.316}}$$

models the percent fairly well. (*Source:* National Longitudinal Survey of Youth; U.S. Department of Commerce; U.S. Bureau of the Census.)

(a) What percent of U.S. children lived in a home without a father in 1997?
(b) In what year were 10% of these children living in a home without a father?

68. *Height of the Eiffel Tower* Paris's Eiffel Tower was constructed in 1889 to commemorate the one hundredth anniversary of the French Revolution. The right side of the Eiffel Tower has a shape that can be approximated by the graph of the function defined by

$$f(x) = -301 \ln \frac{x}{207}.$$

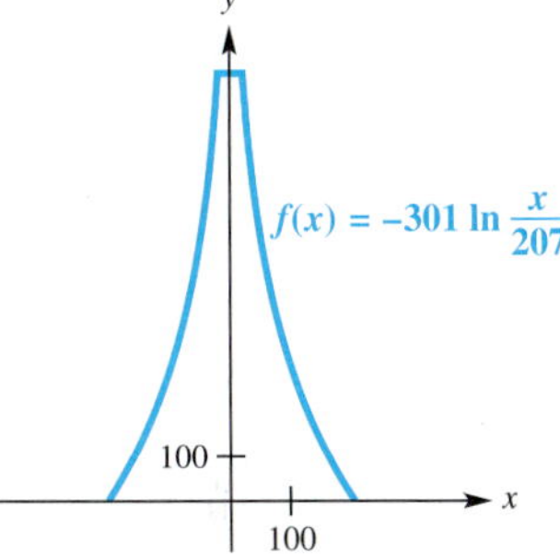

See the figure. (*Source:* Banks, Robert B., *Towing Icebergs, Falling Dominoes, and Other Adventures in Applied Mathematics,* Princeton University Press, 1998.)

(a) Explain why the shape of the left side of the Eiffel Tower has the formula given by $f(-x)$.
(b) The short horizontal line at the top of the figure has length 15.7488 ft. Approximately how tall is the Eiffel Tower?
(c) Approximately how far from the center of the tower is the point on the right side that is 500 ft above the ground?

69. *CO_2 Emissions Tax* One action that government could take to reduce carbon emissions into the atmosphere is to place a tax on fossil fuel. This tax would be based on the amount of carbon dioxide emitted into the air when the fuel is burned. The *cost-benefit* equation

$$\ln(1 - P) = -.0034 - .0053T$$

models the approximate relationship between a tax of T dollars per ton of carbon and the corresponding percent reduction P (in decimal form) of emissions of carbon dioxide. (*Source:* Nordhause, W., "To Slow or Not to Slow: The Economics of the Greenhouse Effect," Yale University, New Haven, Connecticut.)

(a) Write P as a function of T.
(b) Graph P for $0 \leq T \leq 1000$. Discuss the benefit of continuing to raise taxes on carbon.
(c) Determine P when $T = \$60$, and interpret this result.
(d) What value of T will give a 50% reduction in carbon emissions?

70. *Radiative Forcing* (Refer to Example 5 in Section 4.4.) Using computer models, the International Panel on Climate Change (IPCC) in 1990 estimated k to be 6.3 in the radiative forcing equation

$$R = k \ln \frac{C}{C_0},$$

where C_0 is the preindustrial amount of carbon dioxide and C is the current level. (*Source:* Clime, W., *The Economics of Global Warming,* Institute for International Economics, Washington, D.C., 1992.)

(a) Use the equation $R = 6.3 \ln \frac{C}{C_0}$ to determine the radiative forcing R (in watts per square meter) expected by the IPCC if the carbon dioxide level in the atmosphere doubles from its preindustrial level.

(b) Determine the global temperature increase T predicted by the IPCC if the carbon dioxide levels were to double. (*Hint:* $T(R) = 1.03R$.)

For Exercises 71–74, refer to the formula for compound interest given in Section 4.2.

$$A = P\left(1 + \frac{r}{m}\right)^{tm}$$

71. ***Interest on an Account*** Tom Tupper wants to buy a \$30,000 car. He has saved \$27,000. Find the number of years (to the nearest tenth) it will take for his \$27,000 to grow to \$30,000 at 4% interest compounded quarterly.

72. ***Investment Time*** Find t to the nearest hundredth if \$1786 becomes \$2063 at 2.6%, with interest compounded monthly.

73. ***Interest Rate*** Find the interest rate that will produce \$2500 if \$2000 is left at interest compounded semiannually for 3.5 yr.

74. ***Interest Rate*** At what interest rate will \$16,000 grow to \$20,000 if invested for 5.25 yr and interest is compounded quarterly?

4.6 Applications and Models of Exponential Growth and Decay

The Exponential Growth or Decay Function ▪ Growth Function Models ▪ Decay Function Models

The Exponential Growth or Decay Function In many situations that occur in ecology, biology, economics, and the social sciences, a quantity changes at a rate proportional to the amount present. In such cases the amount present at time t is a special function of t called an **exponential growth or decay function.**

Looking Ahead to Calculus

The exponential growth and decay function formulas are studied in calculus in conjunction with the topic known as *differential equations.*

Exponential Growth or Decay Function

Let y_0 be the amount or number present at time $t = 0$. Then, under certain conditions, the amount present at any time t is modeled by

$$y = y_0 e^{kt},$$

where k is a constant.

When $k > 0$, the function describes growth; in Section 4.2, we saw examples of exponential growth: compound interest and atmospheric carbon dioxide, for example. When $k < 0$, the function describes decay; one example of exponential decay is radioactivity.

Growth Function Models The amount of time it takes for a quantity that grows exponentially to become twice its initial amount is called its **doubling time.** The first two examples involve doubling time.

Year	Carbon Dioxide (ppm)
1990	353
2000	375
2075	590
2175	1090
2275	2000

Source: International Panel on Climate Change (IPCC), 1990.

EXAMPLE 1 Determining an Exponential Function to Model the Increase of Carbon Dioxide

In Example 11, Section 4.2, we discussed the growth of atmospheric carbon dioxide over time. A function based on the data from the table was given in that example. Now we can see how to determine such a function from the data.

(a) Find an exponential function that gives the amount of carbon dioxide y in year t.

(b) Estimate the year when future levels of carbon dioxide will double the 1951 level of 280 ppm.

Solution

(a) Recall that the graph of the data points showed exponential growth, so the equation will take the form $y = y_0e^{kt}$. We must find the values of y_0 and k. The data begin with the year 1990, so to simplify our work we let 1990 correspond to $t = 0$, 1991 correspond to $t = 1$, and so on. Since y_0 is the initial amount, $y_0 = 353$ in 1990, that is, when $t = 0$. Thus the equation is

$$y = 353e^{kt}.$$

From the last pair of values in the table, we know that in 2275 the carbon dioxide level is expected to be 2000 ppm. The year 2275 corresponds to $2275 - 1990 = 285$. Substitute 2000 for y and 285 for t and solve for k.

$$2000 = 353e^{k(285)}$$

$$\frac{2000}{353} = e^{285k} \qquad \text{Divide by 353.}$$

$$\ln\left(\frac{2000}{353}\right) = \ln(e^{285k}) \qquad \text{Take logarithms on both sides. (Section 4.5)}$$

$$\ln\left(\frac{2000}{353}\right) = 285k \qquad \ln e^x = x \text{ (Section 4.3)}$$

$$k = \frac{1}{285} \cdot \ln\left(\frac{2000}{353}\right) \qquad \text{Multiply by } \tfrac{1}{285}\text{; rewrite.}$$

$$k \approx .00609 \qquad \text{Use a calculator.}$$

A function that models the data is

$$y = 353e^{.00609t}.$$

(Note that this is a different function than the one given in Section 4.2, Example 11, because we used $t = 0$ to represent 1990 here.)

(b) Let $y = 2(280) = 560$ and find t.

$$560 = 353e^{.00609t}$$

$$\frac{560}{353} = e^{.00609t} \qquad \text{Divide by 353.}$$

$$\ln \frac{560}{353} = \ln e^{.00609t} \qquad \text{Take logarithms on both sides.}$$

$$\ln \frac{560}{353} = .00609t \qquad \ln e^x = x$$

$$t = \frac{1}{.00609} \cdot \ln\left(\frac{560}{353}\right) \quad \text{Multiply by } \tfrac{1}{.00609}\text{; rewrite.}$$

$$t \approx 75.8 \quad \text{Use a calculator.}$$

Since $t = 0$ corresponds to 1990, the 1951 carbon dioxide level will double in the 75th year after 1990, or during 2065.

Now try Exercise 17.

EXAMPLE 2 Finding Doubling Time for Money

How long will it take for the money in an account that is compounded continuously at 3% interest to double?

Solution

$$A = Pe^{rt} \quad \text{Continuous compounding formula (Section 4.2)}$$

$$2P = Pe^{.03t} \quad \text{Let } A = 2P \text{ and } r = .03.$$

$$2 = e^{.03t} \quad \text{Divide by } P.$$

$$\ln 2 = \ln e^{.03t} \quad \text{Take logarithms on both sides.}$$

$$\ln 2 = .03t \quad \ln e^x = x$$

$$\frac{\ln 2}{.03} = t \quad \text{Divide by .03.}$$

$$23.10 \approx t \quad \text{Use a calculator.}$$

It will take about 23 yr for the amount to double.

Now try Exercise 21.

EXAMPLE 3 Determining an Exponential Function to Model Population Growth

According to the U.S. Bureau of the Census, the world population reached 6 billion people on July 18, 1999, and was growing exponentially. The projected world population (in billions of people) t years after July 18, 1999, is given by the function defined by

$$f(t) = 6e^{.0121t}.$$

(a) Based on this model, what will the world population be on July 18, 2008?

(b) In what year will the world population reach 7 billion?

Solution

(a) Since $t = 0$ on July 18, 1999, on July 18, 2008, t would be $2008 - 1999 = 9$ yr. We must find $f(t)$ when $t = 9$.

$$f(t) = 6e^{.0121t}$$

$$f(9) = 6e^{(.0121)9} \quad \text{Let } t = 9.$$

$$\approx 6.69$$

According to the model, the population will be 6.69 billion on July 18, 2008.

(b)

$$f(t) = 6e^{.0121t}$$

$$7 = 6e^{.0121t} \quad \text{Let } f(t) = 7.$$

$$\frac{7}{6} = e^{.0121t} \quad \text{Divide by 6.}$$

$$\ln \frac{7}{6} = \ln e^{.0121t} \quad \text{Take logarithms on both sides.}$$

$$\ln \frac{7}{6} = .0121t \quad \ln e^x = x$$

$$t = \frac{\ln \frac{7}{6}}{.0121} \quad \text{Divide by .0121; rewrite.}$$

$$t \approx 12.7 \quad \text{Use a calculator.}$$

World population will reach 7 billion 12.7 yr after July 18, 1999, that is, during the year 2012.

Now try Exercise 25.

Decay Function Models

EXAMPLE 4 Determining an Exponential Function to Model Radioactive Decay

If 600 g of a radioactive substance are present initially and 3 yr later only 300 g remain, how much of the substance will be present after 6 yr?

Solution To express the situation as an exponential equation $y = y_0e^{kt}$, we use the given values to first find y_0 and then find k.

$$600 = y_0e^{k(0)} \quad \text{Let } y = 600 \text{ and } t = 0.$$

$$600 = y_0 \quad e^0 = 1 \text{ (Section R.3)}$$

This gives

$$y = 600e^{kt} \quad \text{Substitute into } y = y_0e^{kt}.$$

$$300 = 600e^{3k} \quad \text{Let } y = 300 \text{ and } t = 3.$$

$$.5 = e^{3k} \quad \text{Divide by 600.}$$

$$\ln .5 = \ln e^{3k} \quad \text{Take logarithms on both sides.}$$

$$\ln .5 = 3k \quad \ln e^x = x$$

$$\frac{\ln .5}{3} = k \quad \text{Divide by 3.}$$

$$k \approx -.231. \quad \text{Use a calculator.}$$

Thus, the exponential decay equation is $y = 600e^{-.231t}$. To find the amount present after 6 yr, let $t = 6$.

$$y = 600e^{-.231(6)} \approx 600e^{-1.386} \approx 150$$

After 6 yr, about 150 g of the substance will remain.

Now try Exercise 5(a).

Analogous to the idea of doubling time is **half-life,** the amount of time that it takes for a quantity that decays exponentially to become half its initial amount.

EXAMPLE 5 Solving a Carbon Dating Problem

Carbon 14, also known as radiocarbon, is a radioactive form of carbon that is found in all living plants and animals. After a plant or animal dies, the radiocarbon disintegrates. Scientists can determine the age of the remains by comparing the amount of radiocarbon with the amount present in living plants and animals. This technique is called *carbon dating.* The amount of radiocarbon present after t years is given by

$$y = y_0 e^{-(\ln 2)(1/5700)t},$$

where y_0 is the amount present in living plants and animals.

(a) Find the half-life.

(b) Charcoal from an ancient fire pit on Java contained $\frac{1}{4}$ the carbon 14 of a living sample of the same size. Estimate the age of the charcoal.

Solution

(a) If y_0 is the amount of radiocarbon present in a living thing, then $\frac{1}{2}y_0$ is half this initial amount. Thus, we substitute and solve the given equation for t.

$$y = y_0 e^{-(\ln 2)(1/5700)t}$$

$$\frac{1}{2}y_0 = y_0 e^{-(\ln 2)(1/5700)t} \quad \text{Let } y = \tfrac{1}{2}y_0.$$

$$\frac{1}{2} = e^{-(\ln 2)(1/5700)t} \quad \text{Divide by } y_0.$$

$$\ln\frac{1}{2} = \ln e^{-(\ln 2)(1/5700)t} \quad \text{Take logarithms on both sides.}$$

$$\ln\frac{1}{2} = -\frac{\ln 2}{5700}t \quad \ln e^x = x$$

$$-\frac{5700}{\ln 2}\ln\frac{1}{2} = t \quad \text{Multiply by } -\tfrac{5700}{\ln 2}.$$

$$-\frac{5700}{\ln 2}(\ln 1 - \ln 2) = t \quad \text{Quotient property (Section 4.3)}$$

$$-\frac{5700}{\ln 2}(-\ln 2) = t \quad \ln 1 = 0 \text{ (Section 4.3)}$$

$$5700 = t \quad \text{Simplify.}$$

The half-life is 5700 yr.

(b) Solve again for t, this time letting the amount $y = \frac{1}{4}y_0$.

$$y = y_0 e^{-(\ln 2)(1/5700)t} \quad \text{Given equation}$$

$$\frac{1}{4}y_0 = y_0 e^{-(\ln 2)(1/5700)t} \quad \text{Let } y = \tfrac{1}{4}y_0.$$

$$\frac{1}{4} = e^{-(\ln 2)(1/5700)t} \quad \text{Divide by } y_0.$$

$$\ln \frac{1}{4} = \ln e^{-(\ln 2)(1/5700)t} \quad \text{Take logarithms on both sides.}$$

$$\ln \frac{1}{4} = -\frac{\ln 2}{5700}t \quad \ln e^x = x$$

$$-\frac{5700}{\ln 2}\ln \frac{1}{4} = t \quad \text{Multiply by } -\tfrac{5700}{\ln 2}.$$

$$t = 11{,}400 \quad \text{Use a calculator.}$$

The charcoal is 11,400 yr old.

Now try Exercise 11.

EXAMPLE 6 Modeling Newton's Law of Cooling

Newton's law of cooling says that the rate at which a body cools is proportional to the difference C in temperature between the body and the environment around it. The temperature $f(t)$ of the body at time t in appropriate units after being introduced into an environment having constant temperature T_0 is

$$f(t) = T_0 + Ce^{-kt},$$

where C and k are constants.

A pot of coffee with a temperature of 100°C is set down in a room with a temperature of 20°C. The coffee cools to 60°C after 1 hr.

(a) Write an equation to model the data.

(b) Find the temperature after half an hour.

(c) How long will it take for the coffee to cool to 50°C?

Solution

(a) We must find values for C and k. From the given information, when $t = 0$, $T_0 = 20$, and the temperature of the coffee is $f(0) = 100$. Also, when $t = 1$, $f(1) = 60$. Substitute the first pair of values into the equation along with $T_0 = 20$.

$$f(t) = T_0 + Ce^{-kt} \quad \text{Given formula}$$

$$100 = 20 + Ce^{-0k} \quad \text{Let } t = 0,\ f(0) = 100, \text{ and } T_0 = 20.$$

$$100 = 20 + C \quad e^0 = 1$$

$$80 = C \quad \text{Subtract 20.}$$

Thus,

$$f(t) = 20 + 80e^{-kt}. \quad \text{Let } T_0 = 20 \text{ and } C = 80.$$

Now use the remaining pair of values in this equation to find k.

$$f(t) = 20 + 80e^{-kt}$$

$$60 = 20 + 80e^{-1k} \quad \text{Let } t = 1 \text{ and } f(1) = 60.$$

$$40 = 80e^{-k} \quad \text{Subtract 20.}$$

$$\frac{1}{2} = e^{-k} \quad \text{Divide by 80.}$$

$$\ln\frac{1}{2} = \ln e^{-k} \quad \text{Take logarithms on both sides.}$$

$$\ln\frac{1}{2} = -k \quad \ln e^x = x$$

$$k = -\ln\frac{1}{2} \approx .693$$

Thus, the model is $f(t) = 20 + 80e^{-.693t}$.

(b) To find the temperature after $\frac{1}{2}$ hr, let $t = \frac{1}{2}$ in the model from part (a).

$$f(t) = 20 + 80e^{-.693t} \quad \text{Model from part (a)}$$

$$f\left(\frac{1}{2}\right) = 20 + 80e^{(-.693)(1/2)} \approx 76.6°\text{C} \quad \text{Let } t = \tfrac{1}{2}.$$

(c) To find how long it will take for the coffee to cool to 50°C, let $f(t) = 50$.

$$50 = 20 + 80e^{-.693t} \quad \text{Let } f(t) = 50.$$

$$30 = 80e^{-.693t} \quad \text{Subtract 20.}$$

$$\frac{3}{8} = e^{-.693t} \quad \text{Divide by 80.}$$

$$\ln\frac{3}{8} = \ln e^{-.693t} \quad \text{Take logarithms on both sides.}$$

$$\ln\frac{3}{8} = -.693t \quad \ln e^x = x$$

$$t = \frac{\ln\frac{3}{8}}{-.693} \approx 1.415 \text{ hr} \quad \text{or} \quad \text{about 1 hr 25 min}$$

Now try Exercise 15.

4.6 Exercises

Population Growth *A population is increasing according to the exponential function defined by $y = 2e^{.02x}$, where y is in millions and x is the number of years. Match each question in Column I with the correct procedure in Column II, to answer the question.*

I	II
1. How long will it take for the population to triple?	**A.** Evaluate $y = 2e^{.02(1/3)}$.
2. When will the population reach 3 million?	**B.** Solve $2e^{.02x} = 6$.
3. How large will the population be in 3 yr?	**C.** Evaluate $y = 2e^{.02(3)}$.
4. How large will the population be in 4 months?	**D.** Solve $2e^{.02x} = 3$.

(Modeling) *The exercises in this set are grouped according to discipline; they involve exponential or logarithmic models. Solve them by referring to Examples 1–6.*

Physical Sciences *(Exercises 5–18)*

5. ***Decay of Lead*** A sample of 500 g of radioactive lead 210 decays to polonium 210 according to the function defined by

$$A(t) = 500e^{-.032t},$$

where t is time in years. Find the amount of the sample remaining after

(a) 4 yr, **(b)** 8 yr, **(c)** 20 yr. **(d)** Find the half-life.

6. ***Decay of Plutonium*** Repeat Exercise 5 for 500 g of plutonium 241, which decays according to the function defined as follows, where t is time in years.

$$A(t) = A_0e^{-.053t}$$

7. ***Decay of Radium*** Find the half-life of radium 226, which decays according to the function defined as follows, where t is time in years.

$$A(t) = A_0e^{-.00043t}$$

8. ***Decay of Iodine*** How long will it take any quantity of iodine 131 to decay to 25% of its initial amount, knowing that it decays according to the function defined as follows, where t is time in days?

$$A(t) = A_0e^{-.087t}$$

9. ***Snow Levels*** In the central Sierra Nevada mountains of California, the percent of moisture that falls as snow rather than rain is modeled reasonably well by the function defined by

$$p(h) = 86.3 \ln h - 680,$$

where h is the altitude in feet, and $p(h)$ is the percent of snow. (This model is valid for $h \geq 3000$.) Find the percent of snow that falls at the following altitudes.

(a) 3000 ft **(b)** 4000 ft **(c)** 7000 ft

10. ***Magnitude of a Star*** The magnitude M of a star is modeled by

$$M = 6 - \frac{5}{2} \log \frac{I}{I_0},$$

where I_0 is the intensity of a just-visible star and I is the actual intensity of the star being measured. The dimmest stars are of magnitude 6, and the brightest are of magnitude 1. Determine the ratio of light intensities between a star of magnitude 1 and a star of magnitude 3.

11. ***Carbon 14 Dating*** Suppose an Egyptian mummy is discovered in which the amount of carbon 14 present is only about one-third the amount found in living human beings. About how long ago did the Egyptian die?

12. ***Carbon 14 Dating*** A sample from a refuse deposit near the Strait of Magellan had 60% of the carbon 14 of a contemporary living sample. How old was the sample?

13. ***Carbon 14 Dating*** Paint from the Lascaux caves of France contains 15% of the normal amount of carbon 14. Estimate the age of the paintings.

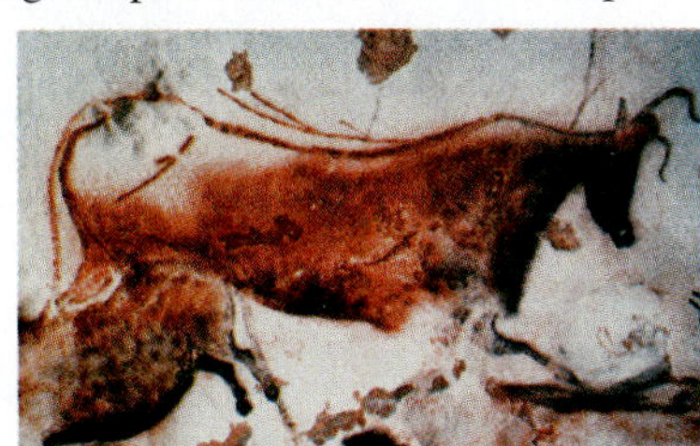

14. *Dissolving a Chemical* The amount of a chemical that will dissolve in a solution increases exponentially as the (Celsius) temperature t is increased according to the model

$$A(t) = 10e^{.0095t}.$$

At what temperature will 15 g dissolve?

15. *Newton's Law of Cooling* Boiling water, at 100°C, is placed in a freezer at 0°C. The temperature of the water is 50°C after 24 min. Find the temperature of the water after 96 min. (*Hint:* Change minutes to hours.)

16. *Newton's Law of Cooling* A piece of metal is heated to 300°C and then placed in a cooling liquid at 50°C. After 4 min, the metal has cooled to 175°C. Find its temperature after 12 min. (*Hint:* Change minutes to hours.)

17. *CFC-11 Levels* Chlorofluorocarbons (CFCs) found in refrigeration units, foaming agents, and aerosols have great potential for destroying the ozone layer, and as a result, governments have phased out their production. The graph displays approximate concentrations of atmospheric CFC-11 in parts per billion (ppb) since 1950.

(a) Use the graph to find an exponential function defined by

$$f(x) = A_0 a^{x-1950}$$

that models the concentration of CFC-11 in the atmosphere from 1950–2000, where x is the year.

(b) Approximate the average annual percent increase of CFC-11 during this time period.

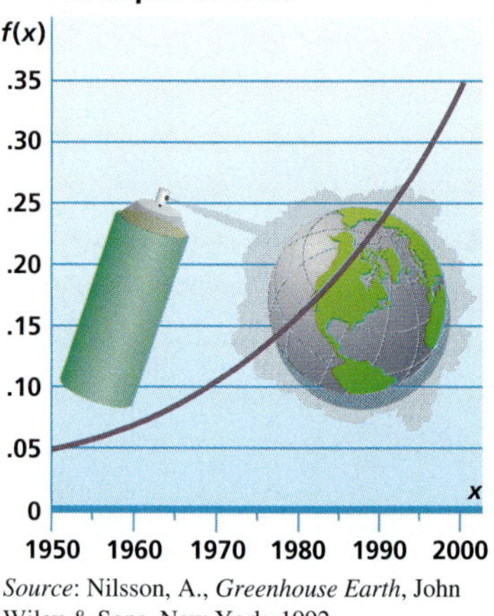

Source: Nilsson, A., *Greenhouse Earth*, John Wiley & Sons, New York, 1992.

18. *Rock Sample Age* Use the function defined by

$$t = T\frac{\ln\left[1 + 8.33\left(\frac{A}{K}\right)\right]}{\ln 2}$$

to estimate the age of a rock sample, if tests show that $\frac{A}{K}$ is .103 for the sample. Let $T = 1.26 \times 10^9$.

Finance *(Exercises 19–24)*

19. *Comparing Investments* Debbie Blanchard, who is self-employed, wants to invest \$60,000 in a pension plan. One investment offers 7% compounded quarterly. Another offers 6.75% compounded continuously.

(a) Which investment will earn more interest in 5 yr?

(b) How much more will the better plan earn?

20. *Growth of an Account* If Mrs. Blanchard (see Exercise 19) chooses the plan with continuous compounding, how long will it take for her \$60,000 to grow to \$80,000?

21. *Doubling Time* Find the doubling time of an investment earning 2.5% interest if interest is compounded continuously.

22. *Doubling Time* If interest is compounded continuously and the interest rate is tripled, what effect will this have on the time required for an investment to double?

23. *Growth of an Account* How long will it take an investment to triple, if interest is compounded continuously at 5%?

24. *Growth of an Account* Use the Table feature of your graphing calculator to find how long it will take \$1500 invested at 5.75% compounded daily to triple in value. Zoom in on the solution by systematically decreasing the increment for x. Find the answer to the nearest day. (Find the answer to the nearest day by eventually letting the increment of x equal $\frac{1}{365}$. The decimal part of the solution can be multiplied by 365 to determine the number of days greater than the nearest year. For example, if the solution is determined to be 16.2027 yr, then multiply .2027 by 365 to get 73.9855. The solution is then, to the nearest day, 16 yr, 74 days.) Confirm your answer algebraically.

Social Sciences *(Exercises 25–30)*

25. *Population Growth* In 2000 India's population reached 1 billion, and in 2025 it is projected to be 1.4 billion. (*Source:* U.S. Bureau of the Census.)

(a) Find values for P_0 and a so that $f(x) = P_0 a^{x-2000}$ models the population of India in year x.

(b) Estimate India's population in 2010.

(c) Use f to determine the year when India's population might reach 1.5 billion.

26. *Population Decline* A midwestern city finds its residents moving to the suburbs. Its population is declining according to the function defined by

$$P(t) = P_0 e^{-.04t},$$

where t is time measured in years and P_0 is the population at time $t = 0$. Assume that $P_0 = 1{,}000{,}000$.

(a) Find the population at time $t = 1$.

(b) Estimate the time it will take for the population to decline to 750,000.

(c) How long will it take for the population to decline to half the initial number?

27. *Gambling Revenues* Revenues in the United States from all forms of legal gambling increased between 1991 and 1995. The function represented by

$$f(x) = 26.6e^{.131x}$$

models these revenues in billions of dollars. In this model x represents the year, where $x = 0$ corresponds to 1991. (*Source: International Gaming & Wagering Business.*)

(a) Estimate gambling revenues in 1995.

(b) Determine the year when these revenues reached \$30 billion. (Round to the nearest whole number.)

28. *Recreation Expenditures* Personal consumption expenditures for recreation in billions of dollars in the United States during the years 1990–2000 can be approximated by the function defined by

$$A(t) = 283.84e^{.069t},$$

where $t = 0$ corresponds to the year 1990. (*Source:* U.S. Bureau of Economic Analysis.) Based on this model, how much were personal consumption expenditures in 1996?

29. *Living Standards* One measure of living standards in the United States is given by

$$L = 9 + 2e^{.15t},$$

where t is the number of years since 1982. Find L for the following years.

(a) 1982 **(b)** 1986 **(c)** 1992

(d) Graph L in the window $[0, 10]$ by $[0, 30]$.

(e) What can be said about the growth of living standards in the United States according to this model?

30. ***Evolution of Language*** The number of years, n, since two independently evolving languages split off from a common ancestral language is approximated by

$$n \approx -7600 \log r,$$

where r is the proportion of words from the ancestral language common to both languages.

(a) Find n if $r = .9$. **(b)** Find n if $r = .3$.

(c) How many years have elapsed since the split if half of the words of the ancestral language are common to both languages?

Life Sciences *(Exercises 31 and 32)*

31. ***Medication Effectiveness*** When physicians prescribe medication, they must consider how the drug's effectiveness decreases over time. If, each hour, a drug is only 90% as effective as the previous hour, at some point the patient will not be receiving enough medication and must receive another dose. This situation can be modeled with a geometric sequence. (See the section on Geometric Sequences and Series.) If the initial dose was 200 mg and the drug was administered 3 hr ago, the expression $200(.90)^2$ represents the amount of effective medication still available. Thus, $200(.90)^2 = 162$ mg are still in the system. (The exponent is equal to the number of hours since the drug was administered, less one.) How long will it take for this initial dose to reach the dangerously low level of 50 mg?

Population Size *Many environmental situations place effective limits on the growth of the number of an organism in an area. Many such limited growth situations are described by the* ***logistic function*** *defined by*

$$G(t) = \frac{MG_0}{G_0 + (M - G_0)e^{-kMt}},$$

where G_0 is the initial number present, M is the maximum possible size of the population, and k is a positive constant. The screens shown here illustrate a typical logistic function calculation and graph.

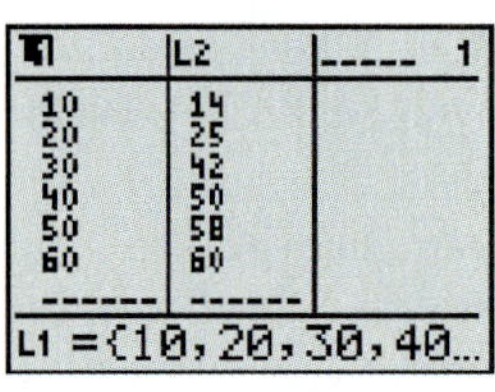

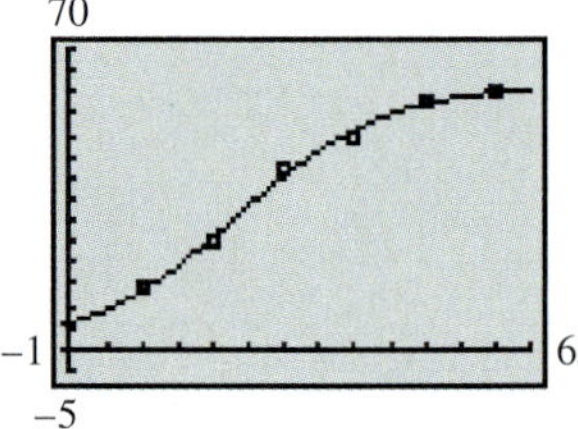

32. Assume that $G_0 = 100$, $M = 2500$, $k = .0004$, and $t =$ time in decades (10-yr periods).

(a) Use a calculator to graph the S-shaped function, using $0 \le t \le 8$, $0 \le y \le 2500$.

(b) Estimate the value of $G(2)$ from the graph. Then, evaluate $G(2)$ algebraically to find the population after 20 yr.

(c) Find the t-coordinate of the intersection of the curve with the horizontal line $y = 1000$ to estimate the number of decades required for the population to reach 1000. Then, solve $G(t) = 1000$ algebraically to obtain the exact value of t.

33. ***Grade Inflation*** During the years 1994 to 2002, the percent of college freshmen whose high school grade average was A or A+ can be approximated by

$$P(t) = 17.77e^{.0305t},$$

where $t = 0$ corresponds to the year 1994, $t = 1$ corresponds to the year 1995, and so on. Based on this model, what percent of the freshman class in 2002 had an A or A+ high school grade average?

Economics (Exercises 34–39)

34. *Consumer Price Index* The U.S. Consumer Price Index is approximated by

$$A(t) = 100e^{.024t},$$

where t represents the number of years after 1990. (For instance, since $A(12)$ is about 133, the amount of goods that could be purchased for \$100 in 1990 cost about \$133 in 2002.) Use the function to determine the year in which costs were 25% higher than in 1990.

35. *Product Sales* Sales of a product, under relatively stable market conditions but in the absence of promotional activities such as advertising, tend to decline at a constant yearly rate. This rate of sales decline varies considerably from product to product, but seems to remain the same for any particular product. The sales decline can be expressed by the function defined by

$$S(t) = S_0e^{-at},$$

where $S(t)$ is the rate of sales at time t measured in years, S_0 is the rate of sales at time $t = 0$, and a is the sales decay constant.

(a) Suppose the sales decay constant for a particular product is $a = .10$. Let $S_0 = 50{,}000$ and find $S(1)$ and $S(3)$.
(b) Find $S(2)$ and $S(10)$ if $S_0 = 80{,}000$ and $a = .05$.

36. *Product Sales* Use the sales decline function given in Exercise 35. If $a = .1$, $S_0 = 50{,}000$, and t is time measured in years, find the number of years it will take for sales to fall to half the initial sales.

37. *Cost of Bread* Assume the cost of a loaf of bread is \$2. With continuous compounding, find the time it would take for the cost to triple at an annual inflation rate of 6%.

38. *Electricity Consumption* Suppose that in a certain area the consumption of electricity has increased at a continuous rate of 6% per year. If it continued to increase at this rate, find the number of years before twice as much electricity would be needed.

39. *Electricity Consumption* Suppose a conservation campaign together with higher rates caused demand for electricity to increase at only 2% per year. (See Exercise 38.) Find the number of years before twice as much electricity would be needed.

(Modeling) Exercises 40 and 41 use another type of logistic function.

40. *Heart Disease* As age increases, so does the likelihood of coronary heart disease (CHD). The fraction of people x years old with some CHD is modeled by

$$f(x) = \frac{.9}{1 + 271e^{-.122x}}.$$

(*Source:* Hosmer, D., and S. Lemeshow, *Applied Logistic Regression,* John Wiley and Sons, 1989.)

(a) Evaluate $f(25)$ and $f(65)$. Interpret the results.
(b) At what age does this likelihood equal 50%?

41. *Tree Growth* The height of a certain tree in feet after x years is modeled by

$$f(x) = \frac{50}{1 + 47.5e^{-.22x}}.$$

(a) Make a table for f starting at $x = 10$, incrementing by 10. What appears to be the maximum height of the tree?
(b) Graph f and identify the horizontal asymptote. Explain its significance.
(c) After how long was the tree 30 ft tall?

Chapter 4 Summary

KEY TERMS

4.1 one-to-one function
inverse function

4.2 exponential function
exponential equation
compound interest
future value
present value
compound amount
continuous compounding

4.3 logarithm
logarithmic equation
logarithmic function

4.4 common logarithm
pH
natural logarithm

4.6 doubling time
half-life

NEW SYMBOLS

$f^{-1}(x)$	the inverse of $f(x)$
e	a constant, approximately 2.718281828
$\log_a x$	the logarithm of x to the base a
$\log x$	common (base 10) logarithm of x
$\ln x$	natural (base e) logarithm of x

QUICK REVIEW

CONCEPTS	EXAMPLES

4.1 Inverse Functions

One-to-One Function

In a one-to-one function, each x value corresponds to only one y-value, and each y-value corresponds to only one x-value.

The function defined by $y = x^2$ is not one-to-one, because $y = 16$, for example, corresponds to both $x = 4$ and $x = -4$.

Horizontal Line Test

If any horizontal line intersects the graph of a function in no more than one point, then the function is one-to-one.

The graph of $f(x) = 2x - 1$ is a straight line with slope 2, so f is a one-to-one function by the horizontal line test.

Inverse Functions

Let f be a one-to-one function. Then g is the inverse function of f if

$$(f \circ g)(x) = x \quad \text{for every } x \text{ in the domain of } g,$$

and

$$(g \circ f)(x) = x \quad \text{for every } x \text{ in the domain of } f.$$

To find $g(x)$, interchange x and y in $y = f(x)$, solve for y, and replace y with $g(x)$ or $f^{-1}(x)$.

To find $f^{-1}(x)$, interchange x and y in the equation $y = 2x - 1$ that defines f.

$$x = 2y - 1$$

$$y = \frac{x + 1}{2} \quad \text{Solve for } y.$$

$$f^{-1}(x) = \frac{x + 1}{2} \quad \text{Replace } y \text{ with } f^{-1}(x).$$

CONCEPTS | **EXAMPLES**

4.2 Exponential Functions

Additional Properties of Exponents

For any real number $a > 0$, $a \neq 1$,

(a) a^x is a unique real number for all real numbers x.

(b) $a^b = a^c$ if and only if $b = c$.

(c) If $a > 1$ and $m < n$, then $a^m < a^n$.

(d) If $0 < a < 1$ and $m < n$, then $a^m > a^n$.

(a) 2^x is a unique real number for all real numbers x.

(b) $2^x = 2^3$ if and only if $x = 3$.

(c) $2^5 < 2^{10}$, because $2 > 1$ and $5 < 10$.

(d) $\left(\frac{1}{2}\right)^5 > \left(\frac{1}{2}\right)^{10}$ because $0 < \frac{1}{2} < 1$ and $5 < 10$.

Exponential Function

If $a > 0$ and $a \neq 1$, then $\boldsymbol{f(x) = a^x}$ defines the exponential function with base a.

$f(x) = 3^x$ defines the exponential function with base 3.

Graph of $f(x) = a^x$

1. The graph contains the points $\left(-1, \frac{1}{a}\right)$, $(0, 1)$, and $(1, a)$.
2. If $a > 1$, then f is an increasing function; if $0 < a < 1$, then f is a decreasing function.
3. The x-axis is a horizontal asymptote.
4. The domain is $(-\infty, \infty)$; the range is $(0, \infty)$.

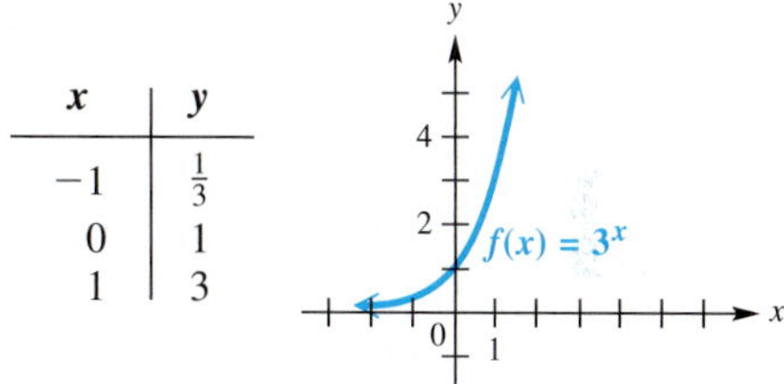

x	y
-1	$\frac{1}{3}$
0	1
1	3

4.3 Logarithmic Functions

Logarithmic Function

If $a > 0$, $a \neq 1$, and $x > 0$, then $\boldsymbol{f(x) = \log_a x}$ defines the logarithmic function with base a.

$f(x) = \log_3 x$ defines the logarithmic function with base 3.

Graph of $f(x) = \log_a x$

1. The graph contains the points $\left(\frac{1}{a}, -1\right)$, $(1, 0)$, and $(a, 1)$.
2. If $a > 1$, then f is an increasing function; if $0 < a < 1$, then f is a decreasing function.
3. The y-axis is a vertical asymptote.
4. The domain is $(0, \infty)$; the range is $(-\infty, \infty)$.

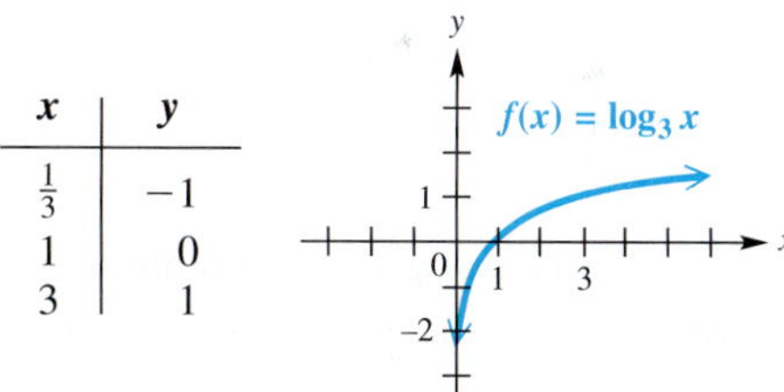

x	y
$\frac{1}{3}$	-1
1	0
3	1

Properties of Logarithms

For any positive real numbers x and y, real number r, and positive real number a, $a \neq 1$:

$\log_a xy = \log_a x + \log_a y$ Product property

$\log_a \frac{x}{y} = \log_a x - \log_a y$ Quotient property

$\log_a x^r = r \log_a x.$ Power property

$\log_2 (3 \cdot 5) = \log_2 3 + \log_2 5$

$\log_2 \frac{3}{5} = \log_2 3 - \log_2 5$

$\log_6 3^5 = 5 \log_6 3$

Also,

$\log_a 1 = 0$ and $\log_a a = 1.$

$\log_{10} 1 = 0$ and $\log_{10} 10 = 1$

Theorem on Inverses

$a^{\log_a x} = x$ and $\log_a (a^x) = x$

$e^{\ln 4} = 4$ and $\ln e^2 = 2$

CONCEPTS	EXAMPLES

4.4 Evaluating Logarithms and the Change-of-Base Theorem

Common and Natural Logarithms

For all positive numbers x,

$$\log x = \log_{10} x \quad \text{Common logarithm}$$
$$\ln x = \log_e x. \quad \text{Natural logarithm}$$

$\log .045 \approx -1.3468$

$\ln 247.1 \approx 5.5098$

Change-of-Base Theorem

For any positive real numbers x, a, and b, where $a \neq 1$ and $b \neq 1$:

$$\log_a x = \frac{\log_b x}{\log_b a}.$$

$$\log_8 7 = \frac{\log 7}{\log 8} = \frac{\ln 7}{\ln 8} \approx .9358$$

4.5 Exponential and Logarithmic Equations

Property of Logarithms

If $x > 0$, $y > 0$, $a > 0$, and $a \neq 1$, then

$$x = y \quad \text{if and only if} \quad \log_a x = \log_a y.$$

If $e^{5x} = 10$, then $\ln e^{5x} = \ln 10$.

If $\log_a(x + 1) = \log_a 7$, then $x + 1 = 7$.

4.6 Applications and Models of Exponential Growth and Decay

Exponential Growth or Decay Function

The exponential growth or decay function is defined by

$$y = y_0 e^{kt},$$

for constant k, where y_0 is the amount or number present at time $t = 0$.

The formula for continuous compounding,

$$A = Pe^{rt},$$

is an example of exponential growth. Here, A is the compound amount if P is invested at an annual interest rate r for t years.

If $P = \$200$, $r = 3\%$, and $t = 5$ yr, then

$$A = 200e^{.03(5)} \approx \$232.37.$$

Chapter 4 Review Exercises

Decide whether each function as defined or graphed is one-to-one.

1.

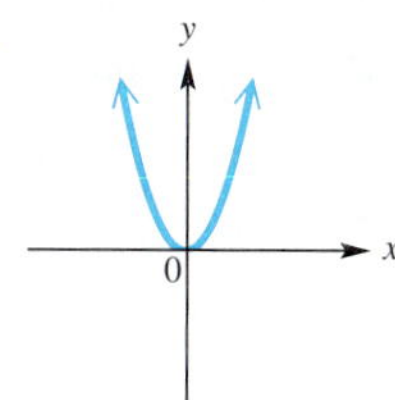

2.

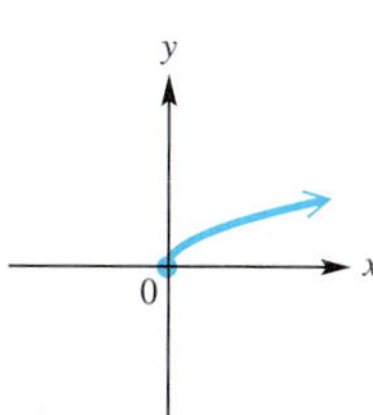

3. $y = 5x - 4$

4. $y = x^3 + 1$

5. $y = (x + 3)^2$

6. $y = \sqrt{3x^2 + 2}$

Write an equation, if possible, for each inverse function in the form $y = f^{-1}(x)$.

7. $f(x) = x^3 - 3$

8. $f(x) = \sqrt{25 - x^2}$

9. *Concept Check* Suppose $f(t)$ is the amount an investment will grow to t years after 2004. What does $f^{-1}(\$50{,}000)$ represent?

10. *Concept Check* The graphs of two functions are shown. Based on their graphs, are these functions inverses?

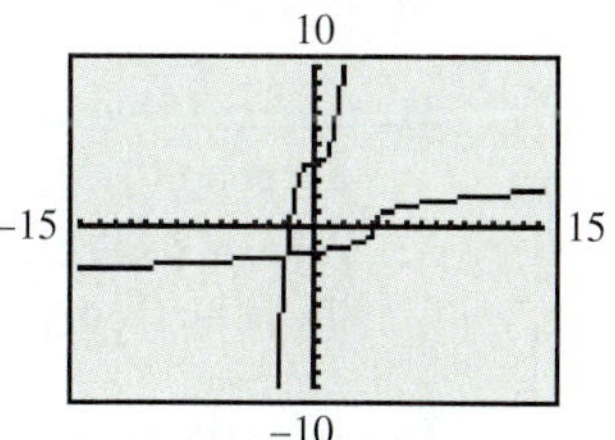

11. *Concept Check* To have an inverse, a function must be a(n) ________ function.

12. *Concept Check* Assuming that f has an inverse, is it true that every x-intercept of the graph of $y = f(x)$ is a y-intercept of the graph of $y = f^{-1}(x)$?

Match each equation with the figure that most closely resembles its graph.

13. $y = \log_{.3} x$

14. $y = e^x$

15. $y = \ln x$

16. $y = (.3)^x$

A.

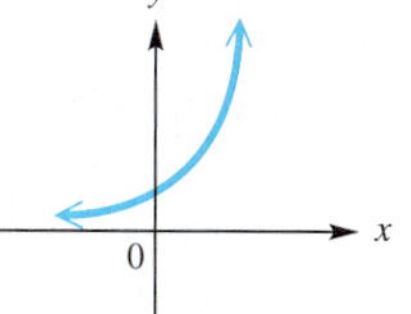

B.

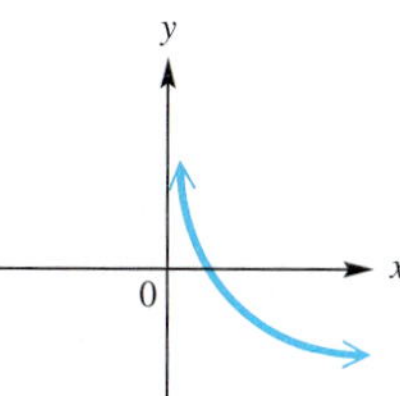

C.

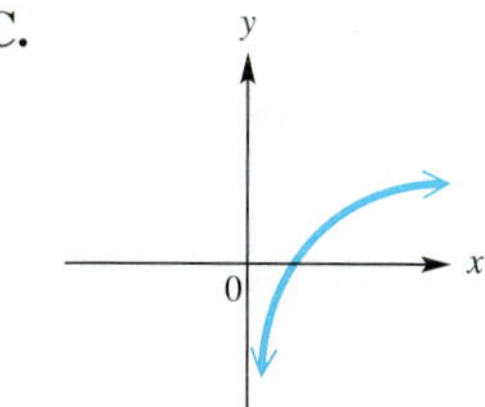

D.

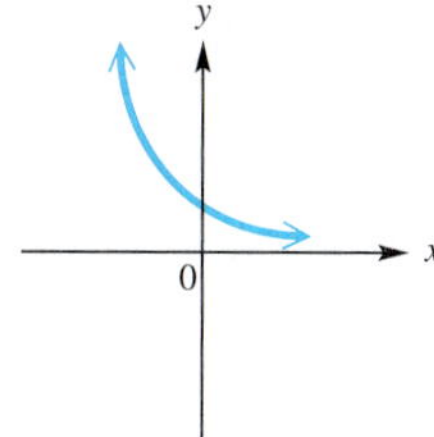

Write each equation in logarithmic form.

17. $2^5 = 32$

18. $100^{1/2} = 10$

19. $\left(\dfrac{3}{4}\right)^{-1} = \dfrac{4}{3}$

20. Graph $y = (1.5)^{x+2}$.

21. *Concept Check* Is the logarithm to the base 3 of 4 written as $\log_4 3$ or $\log_3 4$?

Relating Concepts

For individual or collaborative investigation
(Exercises 22–27)

Work Exercises 22–27 in order.

22. What is the exact value of $\log_3 9$?

23. What is the exact value of $\log_3 27$?

24. Between what two consecutive integers must $\log_3 16$ lie?

25. Use the change-of-base theorem to support your answer for Exercise 24.

26. Repeat Exercises 22 and 23 for $\log_5 \frac{1}{5}$ and $\log_5 1$.

27. Repeat Exercises 24 and 25 for $\log_5 .68$.

Write each equation in exponential form.

28. $\log_9 27 = \dfrac{3}{2}$ **29.** $\log 3.45 \approx .5378$ **30.** $\ln 45 \approx 3.8067$

31. *Concept Check* What is the base of the logarithmic function whose graph contains the point $(81, 4)$?

32. *Concept Check* What is the base of the exponential function whose graph contains the point $\left(-4, \frac{1}{16}\right)$?

Use properties of logarithms to rewrite each expression. Simplify the result, if possible. Assume all variables represent positive numbers.

33. $\log_3 \dfrac{mn}{5r}$ **34.** $\log_5\left(x^2y^4\sqrt[5]{m^3p}\right)$ **35.** $\log_7(7k + 5r^2)$

Find each logarithm. Round to four decimal places.

36. $\log 45.6$ **37.** $\log .0411$ **38.** $\ln 470$

39. $\ln 144{,}000$ **40.** $\log_3 769$ **41.** $\log_{2/3} \dfrac{5}{8}$

Solve each equation. Round irrational solutions to four decimal places.

42. $8^x = 32$ **43.** $x^{-3} = \dfrac{8}{27}$ **44.** $10^{2x-3} = 17$

45. $4^{x+3} = 5^{2-x}$ **46.** $e^{x+1} = 10$ **47.** $\log_{64} x = \dfrac{1}{3}$

48. $\ln(6x) - \ln(x + 1) = \ln 4$ **49.** $\log_{16}\sqrt{x + 1} = \dfrac{1}{4}$

50. $\ln x + 3 \ln 2 = \ln \dfrac{2}{x}$ **51.** $\log x + \log(x - 3) = 1$

52. $\ln[\ln(e^{-x})] = \ln 3$ **53.** $S = a \ln\left(1 + \dfrac{n}{a}\right)$ for n

Solve each problem.

54. ***Earthquake Intensity*** On July 14, 1991, Peshawar, Pakistan, was shaken by an earthquake that measured 6.6 on the Richter scale.

(a) Express this reading in terms of I_0. (See Section 4.4 Exercises.)
(b) In February of the same year a quake measuring 6.5 on the Richter scale killed about 900 people in the mountains of Pakistan and Afghanistan. Express the magnitude of a 6.5 reading in terms of I_0.
(c) How much greater was the force of the earthquake with measure 6.6?

55. ***Earthquake Intensity***

(a) The San Francisco earthquake of 1906 had a Richter scale rating of 8.3. Express the magnitude of this earthquake as a multiple of I_0.
(b) In 1989, the San Francisco region experienced an earthquake with a Richter scale rating of 7.1. Express the magnitude of this earthquake as a multiple of I_0.
(c) Compare the magnitudes of the two San Francisco earthquakes discussed in parts (a) and (b).

56. ***(Modeling) Decibel Levels*** Recall from Section 4.4 that the model for the decibel rating of the loudness of a sound is

$$d = 10 \log \frac{I}{I_0}.$$

A few years ago, there was a controversy about a proposed government limit on factory noise. One group wanted a maximum of 89 decibels, while another group wanted 86. This difference seemed very small to many people. Find the percent by which the 89-decibel intensity exceeds that for 86 decibels.

57. ***Interest Rate*** What annual interest rate, to the nearest tenth, will produce \$5760 if \$3500 is left at interest (compounded annually) for 10 yr?

58. ***Growth of an Account*** Find the number of years (to the nearest tenth) needed for \$48,000 to become \$58,344 at 5% interest compounded semiannually.

59. ***Growth of an Account*** Manuel deposits \$10,000 for 12 yr in an account paying 8% compounded annually. He then puts this total amount on deposit in another account paying 10% compounded semiannually for another 9 yr. Find the total amount on deposit after the entire 21-yr period.

60. ***Growth of an Account*** Anne Kelly deposits \$12,000 for 8 yr in an account paying 5% compounded annually. She then leaves the money alone with no further deposits at 6% compounded annually for an additional 6 yr. Find the total amount on deposit after the entire 14-yr period.

61. ***Cost from Inflation*** If the inflation rate were 4%, use the formula for continuous compounding to find the number of years, to the nearest tenth, for a \$1 item to cost \$2.

62. ***(Modeling) Drug Level in the Bloodstream*** After a medical drug is injected directly into the bloodstream it is gradually eliminated from the body. Graph the following functions on the interval $[0, 10]$. Use $[0, 500]$ for the range of $A(t)$. Determine the function that best models the amount $A(t)$ (in milligrams) of a drug remaining in the body after t hours if 350 mg were initially injected.

(a) $A(t) = t^2 - t + 350$
(b) $A(t) = 350 \log(t + 1)$
(c) $A(t) = 350(.75)^t$
(d) $A(t) = 100(.95)^t$

63. *(Modeling) New York Yankees' Payroll* The table shows the total payroll (in millions of dollars) of the New York Yankees baseball team for the years 2000–2003.

Year	Total Payroll (millions of dollars)
2000	92.9
2001	112.3
2002	125.9
2003	152.7

Source: *USA Today.*

Letting y represent the total payroll and x represent the number of years since 2000, we find that the function defined by

$$f(x) = 93.54e^{.16x}$$

models the data quite well. According to this function, when would the total payroll double its 2003 value?

64. *(Modeling) Transistors on Computer Chips* Computing power of personal computers has increased dramatically as a result of the ability to place an increasing number of transistors on a single processor chip. The table lists the number of transistors on some popular computer chips made by Intel.

Year	Chip	Transistors
1971	4004	2300
1986	386DX	275,000
1989	486DX	1,200,000
1993	Pentium	3,300,000
1995	P6	5,500,000
1997	Pentium 3	9,500,000
2000	Pentium 4	42,000,000

Source: Intel.

(a) Make a scatter diagram of the data. Let the x-axis represent the year, where $x = 0$ corresponds to 1971, and let the y-axis represent the number of transistors.

(b) Decide whether a linear, a logarithmic, or an exponential function describes the data best.

(c) Determine a function f that approximates these data. Plot f and the data on the same coordinate axes.

(d) Assuming that the present trend continues, use f to predict the number of transistors on a chip in the year 2005.

65. Consider $f(x) = \log_4(2x^2 - x)$.

(a) Use the change-of-base theorem with base e to write $\log_4(2x^2 - x)$ in a suitable form to graph with a calculator.

(b) Graph the function using a graphing calculator. Use the window $[-2.5, 2.5]$ by $[-5, 2.5]$.

(c) What are the x-intercepts?

(d) Give the equations of the vertical asymptotes.

(e) Explain why there is no y-intercept.

Chapter 4 Test

1. Consider the function defined by $f(x) = \sqrt[3]{2x - 7}$.
   (a) What are the domain and range of f?
   (b) Explain why f^{-1} exists.
   (c) Find the rule for $f^{-1}(x)$.
   (d) What are the domain and range of f^{-1}?
   (e) Graph both f and f^{-1}. How are the two graphs related to the line $y = x$?

2. Match each equation with its graph.

   (a) $y = \log_{1/3} x$ (b) $y = e^x$ (c) $y = \ln x$ (d) $y = \left(\frac{1}{3}\right)^x$

A.

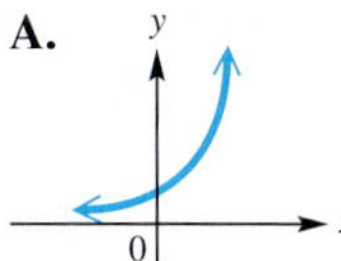

B.

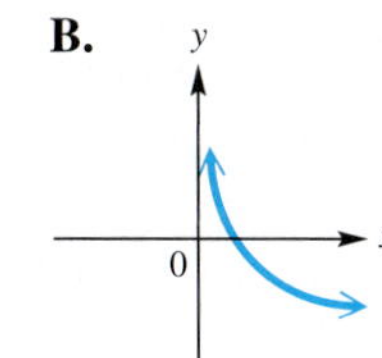

C.

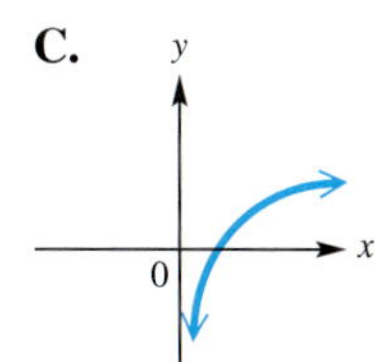

D.

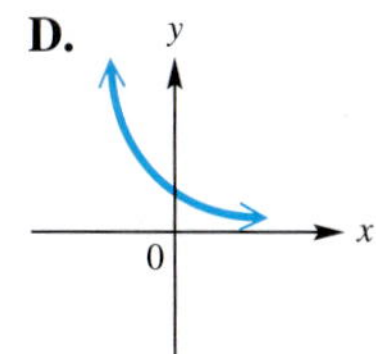

3. Solve $\left(\frac{1}{8}\right)^{2x-3} = 16^{x+1}$.

4. (a) Write $4^{3/2} = 8$ in logarithmic form.
   (b) Write $\log_8 4 = \frac{2}{3}$ in exponential form.

5. Graph $f(x) = \left(\frac{1}{2}\right)^x$ and $g(x) = \log_{1/2} x$ on the same axes. What is their relationship?

6. Use properties of logarithms to write $\log_7 \frac{x^2\sqrt[4]{y}}{z^3}$ as a sum, difference, or product of logarithms. Assume all variables represent positive numbers.

Use a calculator to find an approximation for each logarithm. Express answers to four decimal places.

7. $\log 237.4$

8. $\ln .0467$

9. $\log_9 13$

10. $\log(2.49 \times 10^{-3})$

Use properties of logarithms to solve each equation. Express the solutions as directed.

11. $\log_x 25 = 2$ (exact)

12. $\log_4 32 = x$ (exact)

13. $\log_2 x + \log_2(x + 2) = 3$ (exact)

14. $5^{x+1} = 7^x$ (approximate, to four decimal places)

15. $\ln x - 4 \ln 3 = \ln \frac{5}{x}$ (approximate, to four decimal places)

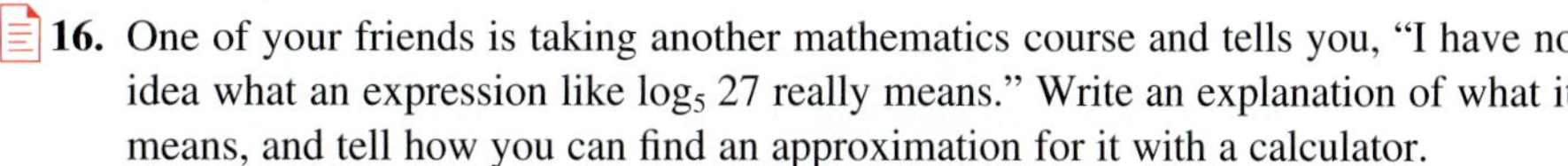

16. One of your friends is taking another mathematics course and tells you, "I have no idea what an expression like $\log_5 27$ really means." Write an explanation of what it means, and tell how you can find an approximation for it with a calculator.

Solve each problem.

17. *(Modeling) Skydiver Fall Speed* A skydiver in free fall travels at a speed modeled by

$$v(t) = 176(1 - e^{-.18t})$$

feet per second after t seconds. How long will it take for the skydiver to attain a speed of 147 ft per sec (100 mph)?

18. *Growth of an Account* How many years, to the nearest tenth, will be needed for \$5000 to increase to \$18,000 at 6.8% compounded **(a)** monthly **(b)** continuously?

19. *(Modeling) Radioactive Decay* The amount of radioactive material, in grams, present after t days is modeled by

$$A(t) = 600e^{-.05t}.$$

(a) Find the amount present after 12 days.

(b) Find the half-life of the material.

20. *(Modeling) Population Growth* In the year 2000, the population of New York state was 18.15 million and increasing exponentially with growth constant .0021. The population of Florida was 15.23 million and increasing exponentially with growth constant .0131. Assuming these trends were to continue, in what year would the population of Florida equal the population of New York?

Chapter 4 Quantitative Reasoning

Financial Planning for Retirement—What are your options?

The traditional IRA (Individual Retirement Account) is a common tax-deferred savings plan in the United States. Earned income deposited into an IRA is not taxed in the current year, and no taxes are incurred on the interest paid in subsequent years. However, when you withdraw the money from the account after age $59\frac{1}{2}$, you pay taxes on the entire amount.

Suppose you deposited \$3000 of earned income into an IRA, you can earn an annual interest rate of 8%, and you are in a 40% tax bracket. (*Note:* Interest rates and tax brackets are subject to change over time, but some assumptions must be made to evaluate the investment.) Also, suppose you deposit the \$3000 at age 25 and withdraw it at age 60.

1. How much money will remain after you pay the taxes at age 60?
2. Suppose that instead of depositing the money into an IRA, you pay taxes on the money and the annual interest. How much money will you have at age 60? (*Note:* You effectively start with \$1800 (60% of \$3000), and the money earns 4.8% (60% of 8%) interest after taxes.)
3. To the nearest dollar, how much additional money will you earn with the IRA?
4. Suppose you pay taxes on the original \$3000, but are then able to earn 8% in a tax-free investment. Compare your balance at age 60 with the IRA balance.

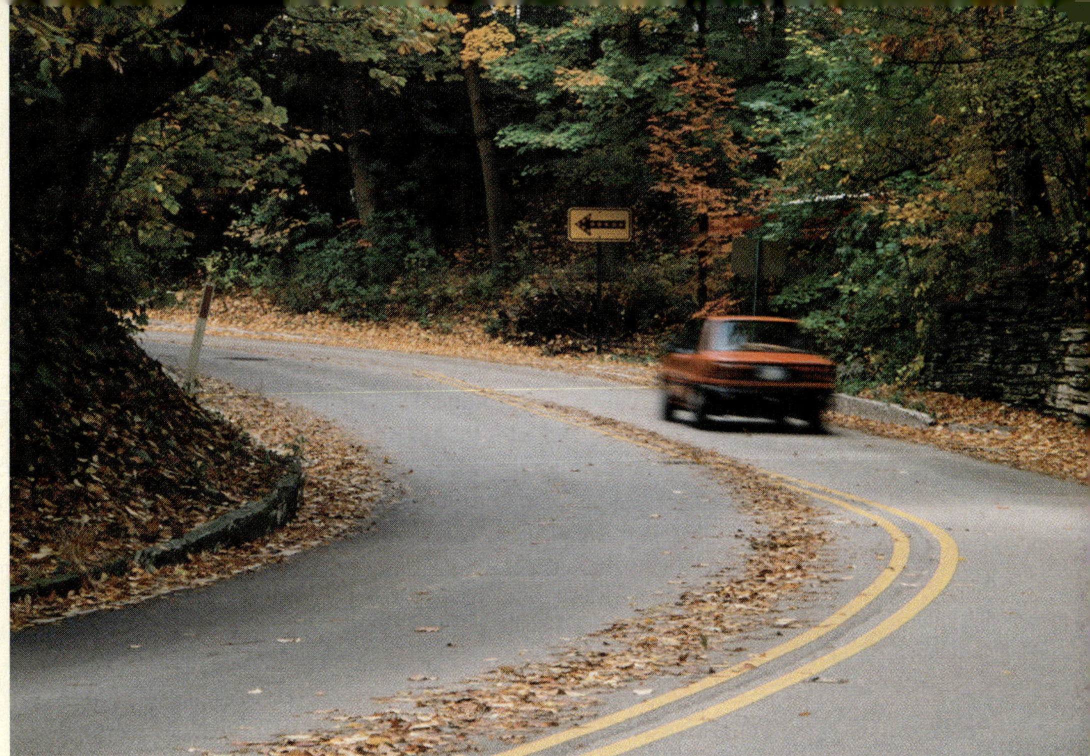

5 Trigonometric Functions

Highway transportation is critical to the economy of the United States. In 1970 there were 1150 billion miles traveled, and by the year 2000 this increased to approximately 2500 billion miles. When an automobile travels around a curve, objects like trees, buildings, and fences situated on the curve may obstruct a driver's vision. Trigonometry is used to determine how far inside the curve land must be cleared to provide visibility for a safe stopping distance. (*Source:* Mannering, F. and W. Kilareski, *Principles of Highway Engineering and Traffic Analysis,* 2nd Edition, John Wiley & Sons, 1998.)

A problem like this is presented in Exercise 61 of Section 5.4.

5.1 Angles

Basic Terminology ▪ Degree Measure ▪ Standard Position ▪ Coterminal Angles

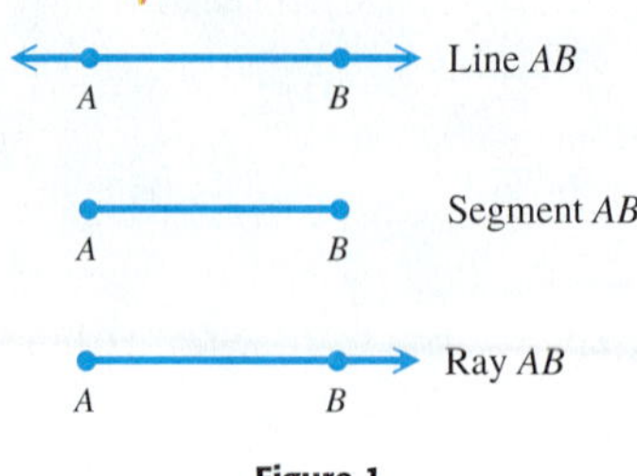

Figure 1

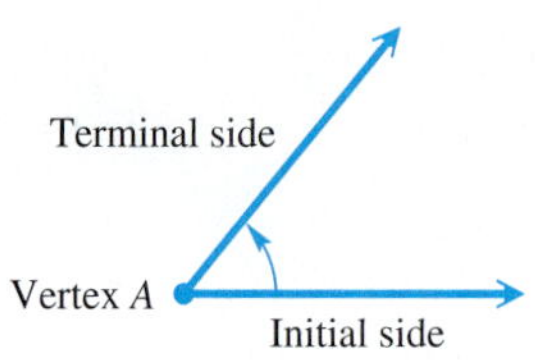

Figure 2

Basic Terminology

Two distinct points A and B determine a line called **line AB.** The portion of the line between A and B, including points A and B themselves, is **line segment AB,** or simply **segment AB.** The portion of line AB that starts at A and continues through B, and on past B, is called **ray AB.** Point A is the *endpoint of the ray.* (See Figure 1.)

An **angle** is formed by rotating a ray around its endpoint. The ray in its initial position is called the **initial side** of the angle, while the ray in its location after the rotation is the **terminal side** of the angle. The endpoint of the ray is the **vertex** of the angle. Figure 2 shows the initial and terminal sides of an angle with vertex A. If the rotation of the terminal side is counterclockwise, the angle is **positive.** If the rotation is clockwise, the angle is **negative.** Figure 3 shows two angles, one positive and one negative.

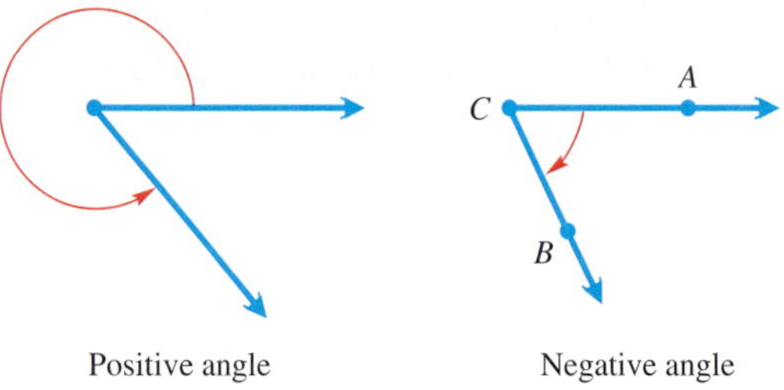

Figure 3

An angle can be named by using the name of its vertex. For example, the angle on the right in Figure 3 can be called angle C. Alternatively, an angle can be named using three letters, with the vertex letter in the middle. Thus, the angle on the right also could be named angle ACB or angle BCA.

Degree Measure

The most common unit for measuring angles is the **degree.** (The other common unit of measure, called the *radian*, is discussed in Section 6.1.) Degree measure was developed by the Babylonians, 4000 years ago. To use degree measure, we assign 360 degrees to a complete rotation of a ray.* In Figure 4, notice that the terminal side of the angle corresponds to its initial side when it makes a complete rotation. One degree, written $1°$, represents $\frac{1}{360}$ of a rotation. Therefore, $90°$ represents $\frac{90}{360} = \frac{1}{4}$ of a complete rotation, and $180°$ represents $\frac{180}{360} = \frac{1}{2}$ of a complete rotation. An angle measuring between $0°$ and $90°$ is called an **acute angle.** An angle measuring exactly $90°$ is a **right angle.** An angle measuring more than $90°$ but less than $180°$ is an **obtuse angle,** and an angle of exactly $180°$ is a **straight angle.** See Figure 5, where we use the Greek letter θ (theta)** to name each angle.

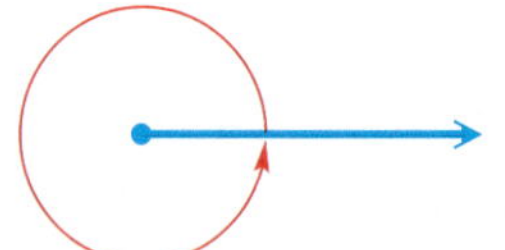

A complete rotation of a ray gives an angle whose measure is 360°.

Figure 4

*The Babylonians were the first to subdivide the circumference of a circle into 360 parts. There are various theories as to why the number 360 was chosen. One is that it is approximately the number of days in a year, and it has many divisors, which makes it convenient to work with. Another involves a roundabout theory dealing with the length of a Babylonian mile.

**In addition to θ (theta), other Greek letters such as α (alpha) and β (beta) are sometimes used to name angles.

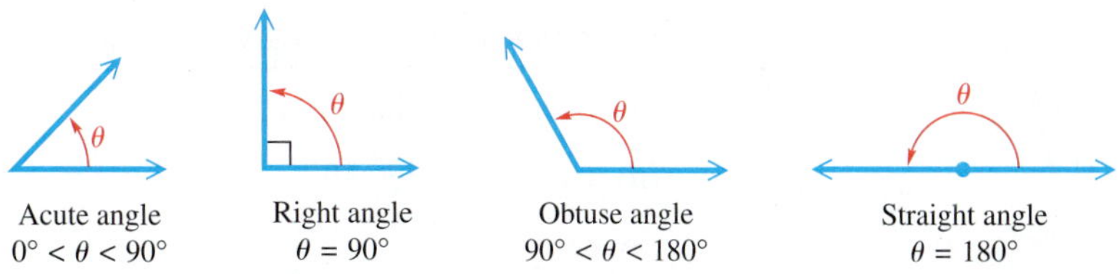

Figure 5

If the sum of the measures of two positive angles is 90°, the angles are called **complementary.** Two positive angles with measures whose sum is 180° are **supplementary.**

EXAMPLE 1 Finding Measures of Complementary and Supplementary Angles

Find the measure of each angle in Figure 6.

Solution

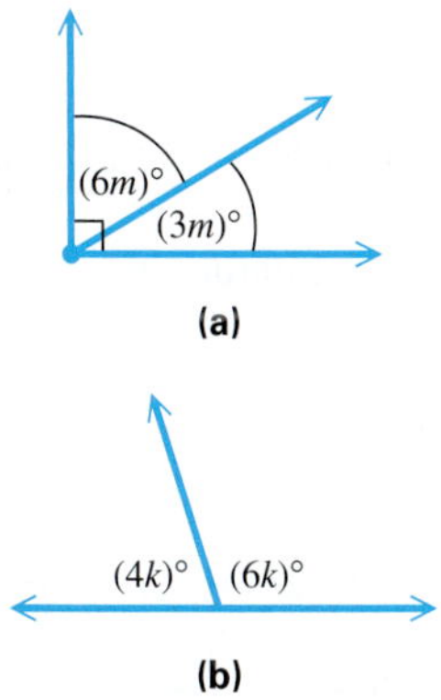

(a) In Figure 6(a), since the two angles form a right angle (as indicated by the ⌉ symbol), they are complementary angles. Thus,

$$6m + 3m = 90$$
$$9m = 90$$
$$m = 10.$$

The two angles have measures of $6(10) = 60°$ and $3(10) = 30°$.

(b) The angles in Figure 6(b) are supplementary, so

$$4k + 6k = 180$$
$$10k = 180$$
$$k = 18.$$

These angle measures are $4(18) = 72°$ and $6(18) = 108°$.

(a)

(b)

Figure 6

Now try Exercises 13 and 15.

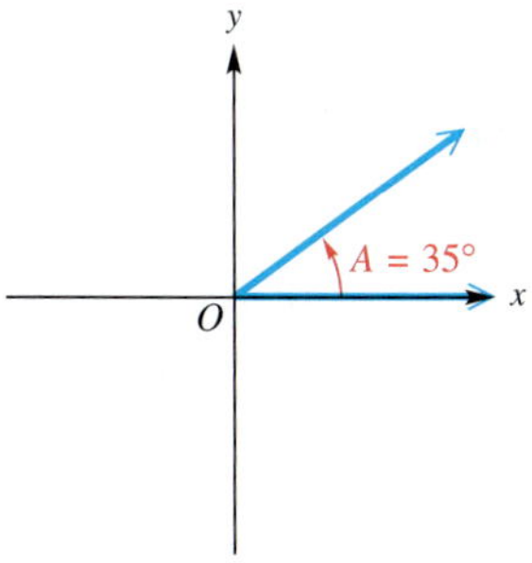

Do not confuse an angle with its measure. Angle A of Figure 7 is a rotation; the measure of the rotation is 35°. This measure is often expressed by saying that $m(\text{angle } A)$ is 35°, where $m(\text{angle } A)$ is read "the measure of angle A." It is convenient, however, to abbreviate $m(\text{angle } A) = 35°$ as $A = 35°$.

Traditionally, portions of a degree have been measured with minutes and seconds. One **minute,** written $1'$, is $\frac{1}{60}$ of a degree.

$$1' = \frac{1}{60}^{\circ} \quad \text{or} \quad 60' = 1°$$

One **second,** $1''$, is $\frac{1}{60}$ of a minute.

$$1'' = \frac{1}{60}' = \frac{1}{3600}^{\circ} \quad \text{or} \quad 60'' = 1'$$

The measure 12° 42′ 38″ represents 12 degrees, 42 minutes, 38 seconds.

Figure 7

EXAMPLE 2 Calculating with Degrees, Minutes, and Seconds

Perform each calculation.

(a) $51° 29' + 32° 46'$ **(b)** $90° - 73° 12'$

Solution

(a) Add the degrees and the minutes separately.

$$\begin{array}{r} 51° 29' \\ +\ 32° 46' \\ \hline 83° 75' \end{array}$$

Since $75' = 60' + 15' = 1° 15'$, the sum is written

$$\begin{array}{r} 83° \\ +\quad 1° 15' \\ \hline 84° 15'. \end{array}$$

(b) $$\begin{array}{r} 89° 60' \\ -\ 73° 12' \\ \hline 16° 48' \end{array}$$ Write 90° as 89° 60'.

Now try Exercises 23 and 27.

Because calculators are now so prevalent, angles are commonly measured in decimal degrees. For example, 12.4238° represents

$$12.4238° = 12\frac{4238}{10,000}°.$$

EXAMPLE 3 Converting Between Decimal Degrees and Degrees, Minutes, and Seconds

(a) Convert 74° 8′ 14″ to decimal degrees.

(b) Convert 34.817° to degrees, minutes, and seconds.

Solution

```
74°8'14"
        74.13722222
74°8'14"
             74.137
34.817▸DMS
        34°49'1.2"
```

A graphing calculator performs the conversions in Example 3 as shown above.

(a) $74° 8' 14'' = 74° + \left(\frac{8}{60}\right)° + \left(\frac{14}{3600}\right)°$ $\quad 1' = \left(\frac{1}{60}\right)°$ and $1'' = \left(\frac{1}{3600}\right)°$

$\approx 74° + .1333° + .0039°$

$\approx 74.137°$ Add; round to the nearest thousandth.

(b) $34.817° = 34° + .817°$

$= 34° + .817(60')$ $\quad 1° = 60'$

$= 34° + 49.02'$

$= 34° + 49' + .02'$

$= 34° + 49' + .02(60'')$ $\quad 1' = 60''$

$= 34° + 49' + 1.2''$

$= 34° 49' 1.2''$

Now try Exercises 33 and 37.

Standard Position An angle is in **standard position** if its vertex is at the origin and its initial side is along the positive x-axis. The angles in Figures 8(a) and 8(b) are in standard position. An angle in standard position is said to lie in the quadrant in which its terminal side lies. An acute angle is in quadrant I (Figure 8(a)) and an obtuse angle is in quadrant II (Figure 8(b)). Figure 8(c) shows ranges of angle measures for each quadrant when $0° < \theta < 360°$. Angles in standard position having their terminal sides along the x-axis or y-axis, such as angles with measures 90°, 180°, 270°, and so on, are called **quadrantal angles.**

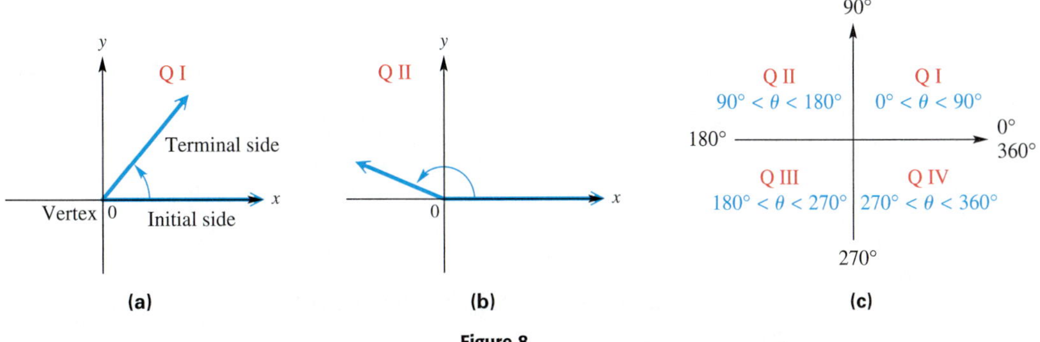

Figure 8

Coterminal Angles A complete rotation of a ray results in an angle measuring 360°. By continuing the rotation, angles of measure larger than 360° can be produced. The angles in Figure 9 with measures 60° and 420° have the same initial side and the same terminal side, but different amounts of rotation. Such angles are called **coterminal angles;** their measures differ by a multiple of 360°. As shown in Figure 10, angles with measures 110° and 830° are coterminal.

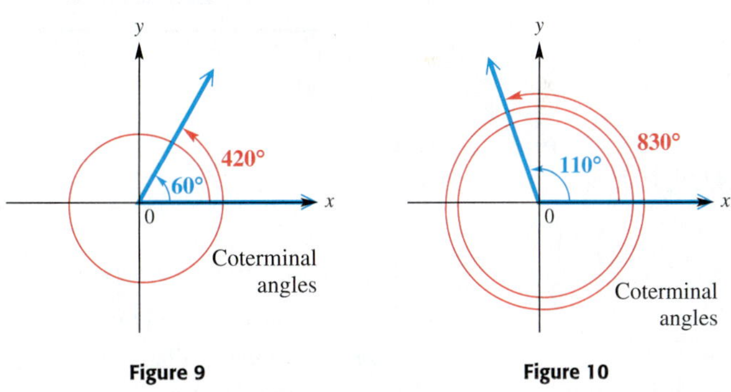

Figure 9 **Figure 10**

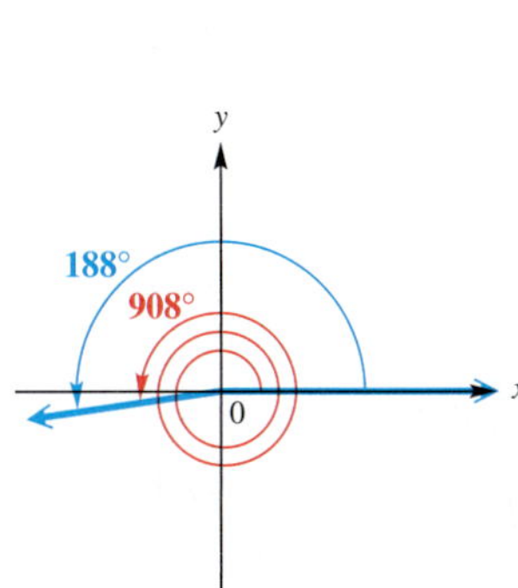

Figure 11

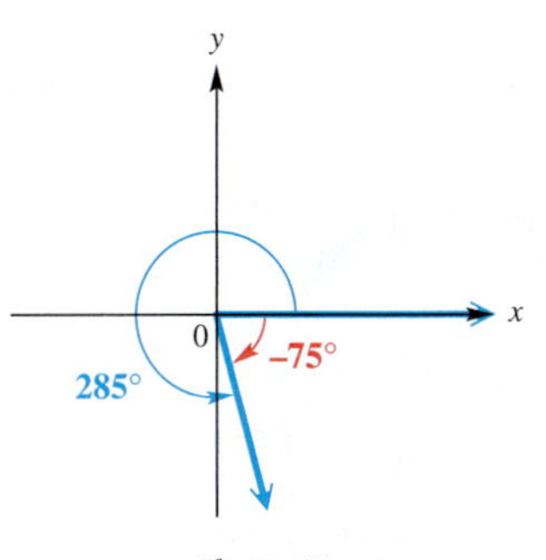

Figure 12

EXAMPLE 4 Finding Measures of Coterminal Angles

Find the angles of smallest possible positive measure coterminal with each angle.

(a) 908° **(b)** $-75°$

Solution

(a) Add or subtract 360° as many times as needed to obtain an angle with measure greater than 0° but less than 360°. Since $908° - 2 \cdot 360° = 908° - 720° = 188°$, an angle of 188° is coterminal with an angle of 908°. See Figure 11.

(b) Use a rotation of $360° + (-75°) = 285°$. See Figure 12.

Now try Exercises 45 and 49.

Sometimes it is necessary to find an expression that will generate all angles coterminal with a given angle. For example, we can obtain any angle coterminal with 60° by adding an appropriate integer multiple of 360° to 60°. Let n represent any integer; then the expression

$$60^\circ + n \cdot 360^\circ$$

represents all such coterminal angles. The table shows a few possibilities.

Value of n	Angle Coterminal with 60°
2	$60^\circ + 2 \cdot 360^\circ = 780^\circ$
1	$60^\circ + 1 \cdot 360^\circ = 420^\circ$
0	$60^\circ + 0 \cdot 360^\circ = 60^\circ$ (the angle itself)
-1	$60^\circ + (-1) \cdot 360^\circ = -300^\circ$

EXAMPLE 5 Analyzing the Revolutions of a CD Player

CAV (Constant Angular Velocity) CD players always spin at the same speed. Suppose a CAV player makes 480 revolutions per min. Through how many degrees will a point on the edge of a CD move in 2 sec?

Solution The player revolves 480 times in 1 min or $\frac{480}{60}$ times $= 8$ times per sec (since 60 sec $=$ 1 min). In 2 sec, the player will revolve $2 \cdot 8 = 16$ times. Each revolution is 360°, so a point on the edge of the CD will revolve $16 \cdot 360^\circ = 5760^\circ$ in 2 sec.

Now try Exercise 75.

5.1 Exercises

1. Explain the difference between a segment and a ray.

2. What part of a complete revolution is an angle of 45°?

3. *Concept Check* What angle is its own complement?

4. *Concept Check* What angle is its own supplement?

Find **(a)** *the complement and* **(b)** *the supplement of each angle.*

5. 30° **6.** 60° **7.** 45° **8.** 18° **9.** 54° **10.** 89°

Find the measure of the smaller angle formed by the hands of a clock at the following times.

11.

12.

Find the measure of each angle in Exercises 13–18. See Example 1.

13. **14.** **15.**

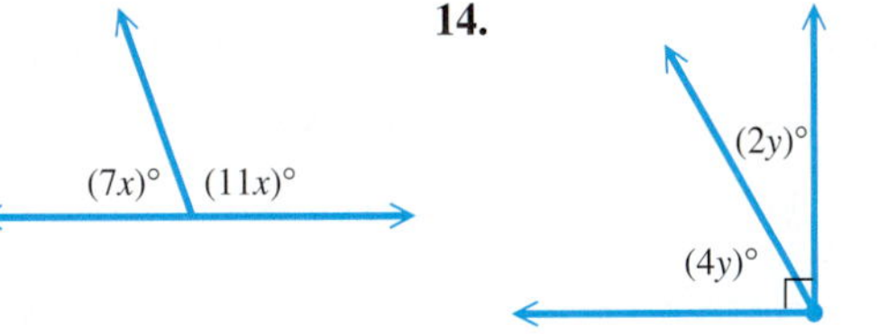

16. supplementary angles with measures $10m + 7$ and $7m + 3$ degrees

17. supplementary angles with measures $6x - 4$ and $8x - 12$ degrees

18. complementary angles with measures $9z + 6$ and $3z$ degrees

Concept Check *Answer each question.*

19. If an angle measures $x°$, how can we represent its complement?

20. If an angle measures $x°$, how can we represent its supplement?

21. If a positive angle has measure $x°$ between 0° and 60°, how can we represent the first negative angle coterminal with it?

22. If a negative angle has measure $x°$ between 0° and −60°, how can we represent the first positive angle coterminal with it?

Perform each calculation. See Example 2.

23. 62° 18′ + 21° 41′ **24.** 75° 15′ + 83° 32′ **25.** 71° 18′ − 47° 29′

26. 47° 23′ − 73° 48′ **27.** 90° − 51° 28′ **28.** 180° − 124° 51′

29. 90° − 72° 58′ 11″ **30.** 90° − 36° 18′ 47″

Convert each angle measure to decimal degrees. Round to the nearest thousandth of a degree. See Example 3.

31. 20° 54′ **32.** 38° 42′ **33.** 91° 35′ 54″

34. 34° 51′ 35″ **35.** 274° 18′ 59″ **36.** 165° 51′ 9″

Convert each angle measure to degrees, minutes, and seconds. See Example 3.

37. 31.4296° **38.** 59.0854° **39.** 89.9004°

40. 102.3771° **41.** 178.5994° **42.** 122.6853°

43. Read about the degree symbol (°) in the manual for your graphing calculator. How is it used?

44. Show that 1.21 hr is the same as 1 hr, 12 min, 36 sec. Discuss the similarity between converting hours, minutes, and seconds to decimal hours and converting degrees, minutes, and seconds to decimal degrees.

Find the angle of smallest positive measure coterminal with each angle. See Example 4.

45. −40° **46.** −98° **47.** −125° **48.** −203°

49. 539° **50.** 699° **51.** 850° **52.** 1000°

Give an expression that generates all angles coterminal with each angle. Let n represent any integer.

53. 30° **54.** 45° **55.** 135° **56.** 270° **57.** −90° **58.** −135°

59. Explain why the answers to Exercises 56 and 57 give the same set of angles.

60. ***Concept Check*** Which two of the following are not coterminal with $r°$?

A. $360° + r°$ **B.** $r° - 360°$ **C.** $360° - r°$ **D.** $r° + 180°$

Concept Check *Sketch each angle in standard position. Draw an arrow representing the correct amount of rotation. Find the measure of two other angles, one positive and one negative, that are coterminal with the given angle. Give the quadrant of each angle.*

61. $75°$ **62.** $89°$ **63.** $174°$ **64.** $234°$

65. $300°$ **66.** $512°$ **67.** $-61°$ **68.** $-159°$

Concept Check *Locate each point in a coordinate system. Draw a ray from the origin through the given point. Indicate with an arrow the angle in standard position having smallest positive measure. Then find the distance r from the origin to the point, using the distance formula of Section 2.1.*

69. $(-3, -3)$ **70.** $(-5, 2)$ **71.** $(-3, -5)$

72. $(\sqrt{3}, 1)$ **73.** $(-2, 2\sqrt{3})$ **74.** $(4\sqrt{3}, -4)$

Solve each problem. See Example 5.

75. *Revolutions of a Turntable* A turntable in a shop makes 45 revolutions per min. How many revolutions does it make per second?

76. *Revolutions of a Windmill* A windmill makes 90 revolutions per min. How many revolutions does it make per second?

77. *Rotating Tire* A tire is rotating 600 times per min. Through how many degrees does a point on the edge of the tire move in $\frac{1}{2}$ sec?

78. *Rotating Airplane Propeller* An airplane propeller rotates 1000 times per min. Find the number of degrees that a point on the edge of the propeller will rotate in 1 sec.

79. *Rotating Pulley* A pulley rotates through 75° in 1 min. How many rotations does the pulley make in an hour?

80. *Surveying* One student in a surveying class measures an angle as 74.25°, while another student measures the same angle as 74° 20′. Find the difference between these measurements, both to the nearest minute and to the nearest hundredth of a degree.

81. *Viewing Field of a Telescope* Due to Earth's rotation, celestial objects like the moon and the stars appear to move across the sky, rising in the east and setting in the west. As a result, if a telescope on Earth remains stationary while viewing a celestial object, the object will slowly move outside the viewing field of the telescope. For this reason, a motor is often attached to telescopes so that the telescope rotates at the same rate as Earth. Determine how long it should take the motor to turn the telescope through an angle of 1 min in a direction perpendicular to Earth's axis.

82. *Angle Measure of a Star on the American Flag* Determine the measure of the angle in each point of the five-pointed star appearing on the American flag. (*Hint:* Inscribe the star in a circle, and use the following theorem from geometry: *An angle whose vertex lies on the circumference of a circle is equal to half the central angle that cuts off the same arc.* See the figure.)

5.2 Trigonometric Functions

Trigonometric Functions ▪ Quadrantal Angles ▪ Reciprocal Identities ▪ Signs and Ranges of Function Values ▪ Pythagorean Identities ▪ Quotient Identities

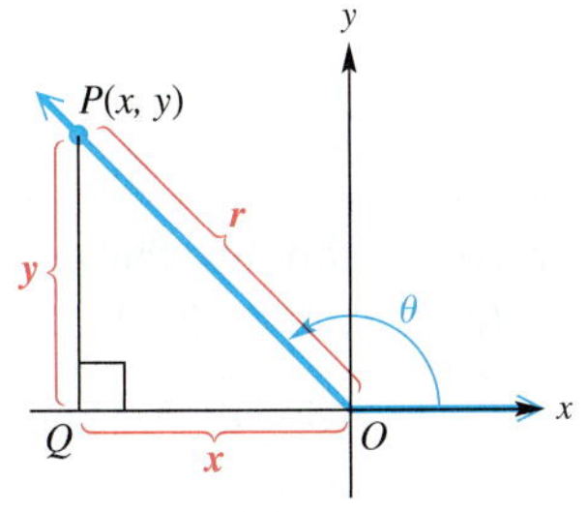

Figure 13

Trigonometric Functions

To define the six *trigonometric functions,* we start with an angle θ in standard position, and choose any point P having coordinates (x, y) on the terminal side of angle θ. (The point P must not be the vertex of the angle.) See Figure 13. A perpendicular from P to the x-axis at point Q determines a right triangle, having vertices at O, P, and Q. We find the distance r from $P(x, y)$ to the origin, $(0, 0)$, using the distance formula.

$$r = \sqrt{(x-0)^2 + (y-0)^2} = \sqrt{x^2 + y^2} \quad \text{(Section 2.1)}$$

Notice that $r > 0$ since distance is never negative.

The six trigonometric functions of angle θ are **sine, cosine, tangent, cotangent, secant,** and **cosecant.** In the following definitions, we use the customary abbreviations for the names of these functions.

Trigonometric Functions

Let (x, y) be a point other than the origin on the terminal side of an angle θ in standard position. The distance from the point to the origin is $r = \sqrt{x^2 + y^2}$. The six trigonometric functions of θ are defined as follows.

$$\sin\theta = \frac{y}{r} \qquad \cos\theta = \frac{x}{r} \qquad \tan\theta = \frac{y}{x} \quad (x \neq 0)$$

$$\csc\theta = \frac{r}{y} \quad (y \neq 0) \qquad \sec\theta = \frac{r}{x} \quad (x \neq 0) \qquad \cot\theta = \frac{x}{y} \quad (y \neq 0)$$

NOTE Although Figure 13 shows a second quadrant angle, these definitions apply to any angle θ. Because of the restrictions on the denominators in the definitions of tangent, cotangent, secant, and cosecant, some angles will have undefined function values.

EXAMPLE 1 Finding Function Values of an Angle

The terminal side of an angle θ in standard position passes through the point $(8, 15)$. Find the values of the six trigonometric functions of angle θ.

Solution Figure 14 shows angle θ and the triangle formed by dropping a perpendicular from the point $(8, 15)$ to the x-axis. The point $(8, 15)$ is 8 units to the right of the y-axis and 15 units above the x-axis, so $x = 8$ and $y = 15$. Since $r = \sqrt{x^2 + y^2}$,

$$r = \sqrt{8^2 + 15^2} = \sqrt{64 + 225} = \sqrt{289} = 17.$$

Figure 14

We can now find the values of the six trigonometric functions of angle θ.

$$\sin \theta = \frac{y}{r} = \frac{15}{17} \qquad \cos \theta = \frac{x}{r} = \frac{8}{17} \qquad \tan \theta = \frac{y}{x} = \frac{15}{8}$$

$$\csc \theta = \frac{r}{y} = \frac{17}{15} \qquad \sec \theta = \frac{r}{x} = \frac{17}{8} \qquad \cot \theta = \frac{x}{y} = \frac{8}{15}$$

Now try Exercise 7.

EXAMPLE 2 Finding Function Values of an Angle

The terminal side of an angle θ in standard position passes through the point $(-3, -4)$. Find the values of the six trigonometric functions of angle θ.

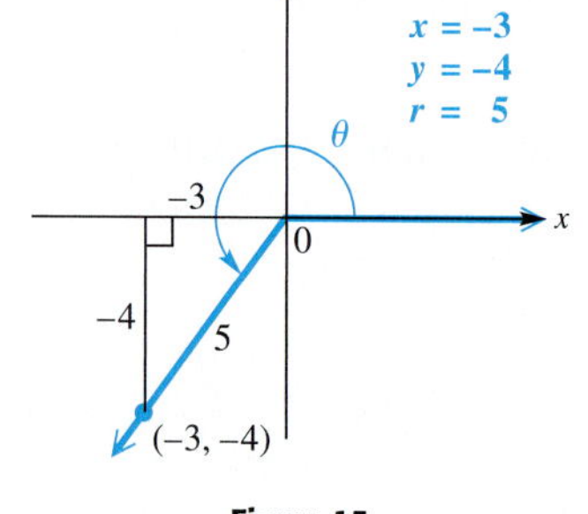

Figure 15

Solution As shown in Figure 15, $x = -3$ and $y = -4$. The value of r is

$$r = \sqrt{(-3)^2 + (-4)^2} = \sqrt{25} = 5. \qquad \text{Remember that } r > 0.$$

Then by the definitions of the trigonometric functions,

$$\sin \theta = \frac{-4}{5} = -\frac{4}{5} \qquad \cos \theta = \frac{-3}{5} = -\frac{3}{5} \qquad \tan \theta = \frac{-4}{-3} = \frac{4}{3}$$

$$\csc \theta = \frac{5}{-4} = -\frac{5}{4} \qquad \sec \theta = \frac{5}{-3} = -\frac{5}{3} \qquad \cot \theta = \frac{-3}{-4} = \frac{3}{4}.$$

Now try Exercise 3.

We can find the six trigonometric functions using *any* point other than the origin on the terminal side of an angle. To see why any point may be used, refer to Figure 16, which shows an angle θ and two distinct points on its terminal side. Point P has coordinates (x, y), and point P' (read "P-prime") has coordinates (x', y'). Let r be the length of the hypotenuse of triangle OPQ, and let r' be the length of the hypotenuse of triangle $OP'Q'$. Since corresponding sides of similar triangles are proportional,

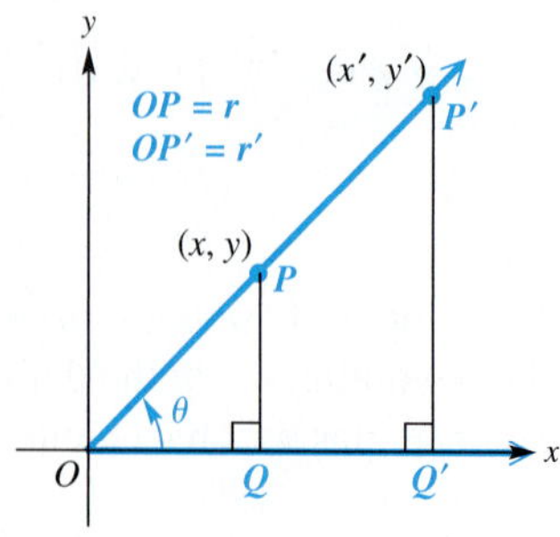

Figure 16

$$\frac{y}{r} = \frac{y'}{r'},$$

so $\sin \theta = \frac{y}{r}$ is the same no matter which point is used to find it. A similar result holds for the other five trigonometric functions.

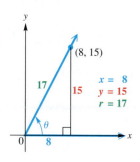

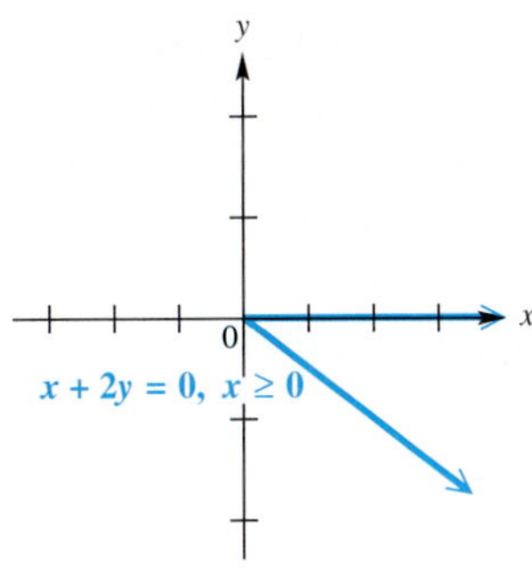

Figure 17

We can also find the trigonometric function values of an angle if we know the equation of the line coinciding with the terminal ray. Recall from algebra that the graph of the equation

$$Ax + By = 0 \quad \text{(Section 2.3)}$$

is a line that passes through the origin. If we restrict x to have only nonpositive or only nonnegative values, we obtain as the graph a ray with endpoint at the origin. For example, the graph of $x + 2y = 0$, $x \geq 0$, shown in Figure 17, is a ray that can serve as the terminal side of an angle in standard position. By choosing a point on the ray, we can find the trigonometric function values of the angle.

EXAMPLE 3 Finding Function Values of an Angle

Find the six trigonometric function values of the angle θ in standard position, if the terminal side of θ is defined by $x + 2y = 0$, $x \geq 0$.

Solution The angle is shown in Figure 18. We can use *any* point except $(0, 0)$ on the terminal side of θ to find the trigonometric function values. We choose $x = 2$ and find the corresponding y-value.

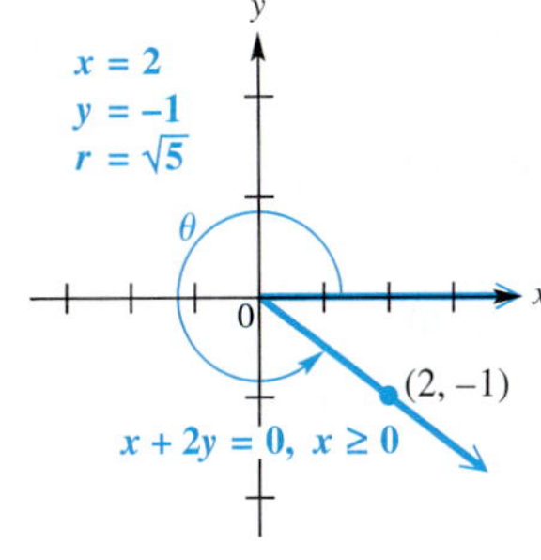

Figure 18

$$x + 2y = 0, x \geq 0$$
$$2 + 2y = 0 \quad \text{Let } x = 2.$$
$$2y = -2 \quad \text{Subtract 2.}$$
$$y = -1 \quad \text{Divide by 2.}$$

The point $(2, -1)$ lies on the terminal side, and the corresponding value of r is $r = \sqrt{2^2 + (-1)^2} = \sqrt{5}$. Now we use the definitions of the trigonometric functions.

$$\sin\theta = \frac{y}{r} = \frac{-1}{\sqrt{5}} = \frac{-1}{\sqrt{5}} \cdot \frac{\sqrt{5}}{\sqrt{5}} = -\frac{\sqrt{5}}{5}$$

$$\cos\theta = \frac{x}{r} = \frac{2}{\sqrt{5}} = \frac{2}{\sqrt{5}} \cdot \frac{\sqrt{5}}{\sqrt{5}} = \frac{2\sqrt{5}}{5} \quad \text{Rationalize denominators. (Section R.7)}$$

$$\tan\theta = \frac{y}{x} = -\frac{1}{2}$$

$$\csc\theta = \frac{r}{y} = -\sqrt{5} \qquad \sec\theta = \frac{r}{x} = \frac{\sqrt{5}}{2} \qquad \cot\theta = \frac{x}{y} = -2$$

Now try Exercise 17.

Recall that when the equation of a line is written in the form $y = mx + b$, the coefficient of x is the slope of the line. In Example 3, $x + 2y = 0$ can be written as $y = -\frac{1}{2}x$, so the slope is $-\frac{1}{2}$. Notice that $\tan\theta = -\frac{1}{2}$. ***In general, it is true that $m = \tan\theta$.***

NOTE The trigonometric function values we found in Examples 1–3 are *exact.* If we were to use a calculator to approximate these values, the decimal results would not be acceptable if exact values were required.

Quadrantal Angles If the terminal side of an angle in standard position lies along the y-axis, any point on this terminal side has x-coordinate 0. Similarly, an angle with terminal side on the x-axis has y-coordinate 0 for any point on the terminal side. Since the values of x and y appear in the denominators of some trigonometric functions, and since a fraction is undefined if its denominator is 0, some trigonometric function values of quadrantal angles (i.e., those with terminal side on an axis) are undefined.

EXAMPLE 4 Finding Function Values of Quadrantal Angles

Find the values of the six trigonometric functions for each angle.

(a) an angle of 90°

(b) an angle θ in standard position with terminal side through $(-3, 0)$

Solution

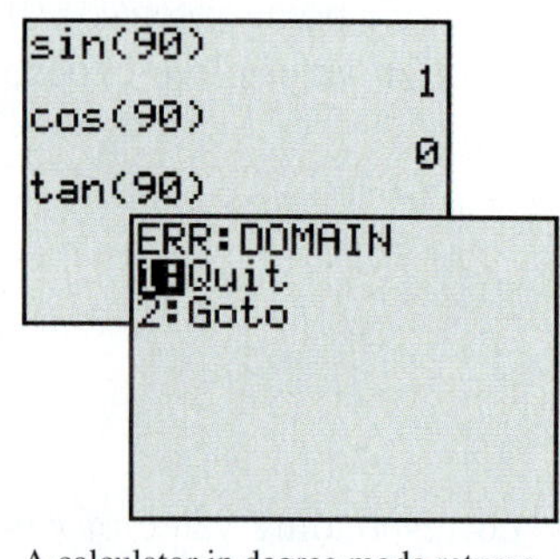

A calculator in degree mode returns the correct values for sin 90° and cos 90°. The second screen shows an ERROR message for tan 90°, because 90° is not in the domain of the tangent function.

(a) First, we select any point on the terminal side of a 90° angle. We choose the point $(0, 1)$, as shown in Figure 19. Here $x = 0$ and $y = 1$, so $r = 1$. Then,

$$\sin 90° = \frac{1}{1} = 1 \qquad \cos 90° = \frac{0}{1} = 0 \qquad \tan 90° = \frac{1}{0} \text{ (undefined)}$$

$$\csc 90° = \frac{1}{1} = 1 \qquad \sec 90° = \frac{1}{0} \text{ (undefined)} \qquad \cot 90° = \frac{0}{1} = 0.$$

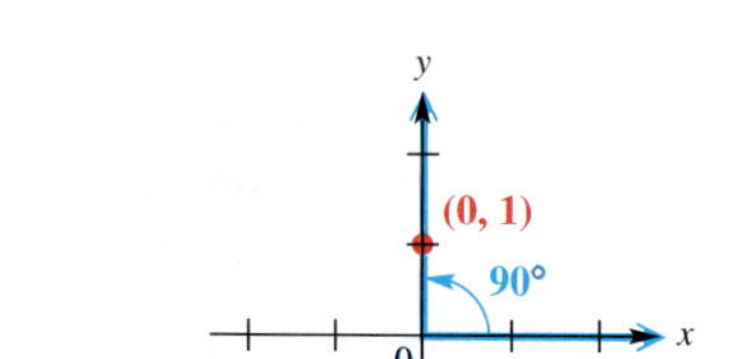

Figure 19

Figure 20

(b) Figure 20 shows the angle. Here, $x = -3$, $y = 0$, and $r = 3$, so the trigonometric functions have the following values.

$$\sin \theta = \frac{0}{3} = 0 \qquad \cos \theta = \frac{-3}{3} = -1 \qquad \tan \theta = \frac{0}{-3} = 0$$

$$\csc \theta = \frac{3}{0} \text{ (undefined)} \qquad \sec \theta = \frac{3}{-3} = -1 \qquad \cot \theta = \frac{-3}{0} \text{ (undefined)}$$

Now try Exercises 5 and 9.

The conditions under which the trigonometric function values of quadrantal angles are undefined are summarized here.

Undefined Function Values

If the terminal side of a quadrantal angle lies along the y-axis, then the tangent and secant functions are undefined. If it lies along the x-axis, then the cotangent and cosecant functions are undefined.

The function values of the most commonly used quadrantal angles, 0°, 90°, 180°, 270°, and 360°, are summarized in the following table.

θ	$\sin \theta$	$\cos \theta$	$\tan \theta$	$\cot \theta$	$\sec \theta$	$\csc \theta$
0°	0	1	0	Undefined	1	Undefined
90°	1	0	Undefined	0	Undefined	1
180°	0	−1	0	Undefined	−1	Undefined
270°	−1	0	Undefined	0	Undefined	−1
360°	0	1	0	Undefined	1	Undefined

The values given in this table can be found with a calculator that has trigonometric function keys. Make sure the calculator is set in *degree mode.*

CAUTION One of the most common errors involving calculators in trigonometry occurs when the calculator is set for *radian measure,* rather than *degree measure.* (Radian measure of angles is discussed in Chapter 6.) Be sure you know how to set your calculator in *degree mode.*

Reciprocal Identities Identities are equations that are true for all values of the variables for which all expressions are defined.

$$(x + y)^2 = x^2 + 2xy + y^2 \qquad 2(x + 3) = 2x + 6 \qquad \text{Identities}$$

The definitions of the trigonometric functions at the beginning of this section were written so that functions in the same column are reciprocals of each other. Since $\sin \theta = \frac{y}{r}$ and $\csc \theta = \frac{r}{y}$,

$$\sin \theta = \frac{1}{\csc \theta} \quad \text{and} \quad \csc \theta = \frac{1}{\sin \theta},$$

provided $\sin \theta \neq 0$. Also, $\cos \theta$ and $\sec \theta$ are reciprocals, as are $\tan \theta$ and $\cot \theta$. In summary, we have the **reciprocal identities** that hold for any angle θ that does not lead to a 0 denominator.

```
1/sin(90)
                1
1/cos(180)
               -1
(sin(-270))-1
                1
1/cos(90)
```

(a)

```
ERR:DIVIDE BY 0
1:Quit
2:Goto
```

(b)

Figure 21

Reciprocal Identities

$$\sin \theta = \frac{1}{\csc \theta} \qquad \cos \theta = \frac{1}{\sec \theta} \qquad \tan \theta = \frac{1}{\cot \theta}$$

$$\csc \theta = \frac{1}{\sin \theta} \qquad \sec \theta = \frac{1}{\cos \theta} \qquad \cot \theta = \frac{1}{\tan \theta}$$

The screen in Figure 21(a) shows how to find csc 90°, sec 180°, and csc(−270°), using the appropriate reciprocal identities and the reciprocal key of a graphing calculator in degree mode. Be sure *not* to use the *inverse trigonometric function* keys to find the reciprocal function values. Attempting to find sec 90° by entering 1/cos 90° produces an ERROR message, indicating the reciprocal is undefined. See Figure 21(b). Compare these results with the ones found in the table of quadrantal angle function values. ■

NOTE Identities can be written in different forms. For example,

$$\sin \theta = \frac{1}{\csc \theta}$$

can be written $\csc \theta = \frac{1}{\sin \theta}$ and $(\sin \theta)(\csc \theta) = 1.$

EXAMPLE 5 Using the Reciprocal Identities

Find each function value.

(a) $\cos \theta$, if $\sec \theta = \frac{5}{3}$ **(b)** $\sin \theta$, if $\csc \theta = -\frac{\sqrt{12}}{2}$

Solution

(a) Since $\cos \theta$ is the reciprocal of $\sec \theta$,

$$\cos \theta = \frac{1}{\sec \theta} = \frac{1}{\frac{5}{3}} = \frac{3}{5}.$$

Simplify the complex fraction. (Section R.5)

(b) $\sin \theta = \frac{1}{-\frac{\sqrt{12}}{2}}$ $\quad \sin \theta = \frac{1}{\csc \theta}$

$= -\frac{2}{\sqrt{12}}$

$= -\frac{2}{2\sqrt{3}}$ $\quad \sqrt{12} = \sqrt{4 \cdot 3} = 2\sqrt{3}$ (Section R.7)

$= -\frac{1}{\sqrt{3}}$ $\quad$ Simplify.

$= -\frac{\sqrt{3}}{3}$ $\quad$ Multiply by $\frac{\sqrt{3}}{\sqrt{3}}$ to rationalize the denominator.

Now try Exercises 45 and 47.

Signs and Ranges of Function Values In the definitions of the trigonometric functions, r is the distance from the origin to the point (x, y). Distance is never negative, so $r > 0$. If we choose a point (x, y) in quadrant I, then both x and y will be positive. Thus, the values of all six functions will be positive in quadrant I.

A point (x, y) in quadrant II has $x < 0$ and $y > 0$. This makes the values of sine and cosecant positive for quadrant II angles, while the other four functions take on negative values. Similar results can be obtained for the other quadrants, as summarized on the next page.

Signs of Function Values

θ in Quadrant	$\sin\theta$	$\cos\theta$	$\tan\theta$	$\cot\theta$	$\sec\theta$	$\csc\theta$
I	+	+	+	+	+	+
II	+	−	−	−	−	+
III	−	−	+	+	−	−
IV	−	+	−	−	+	−

$x < 0, y > 0, r > 0$
II
Sine and cosecant positive

$x > 0, y > 0, r > 0$
I
All functions positive

$x < 0, y < 0, r > 0$
III
Tangent and cotangent positive

$x > 0, y < 0, r > 0$
IV
Cosine and secant positive

EXAMPLE 6 Identifying the Quadrant of an Angle

Identify the quadrant (or quadrants) of any angle θ that satisfies $\sin\theta > 0$, $\tan\theta < 0$.

Solution Since $\sin\theta > 0$ in quadrants I and II, while $\tan\theta < 0$ in quadrants II and IV, both conditions are met only in quadrant II.

Now try Exercise 57.

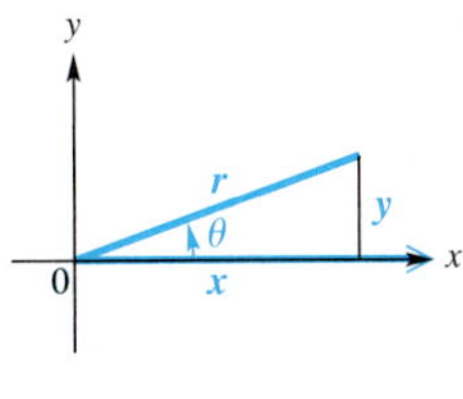

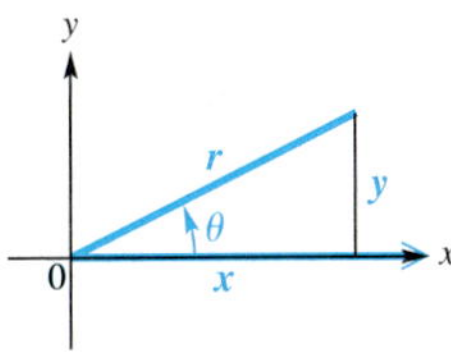

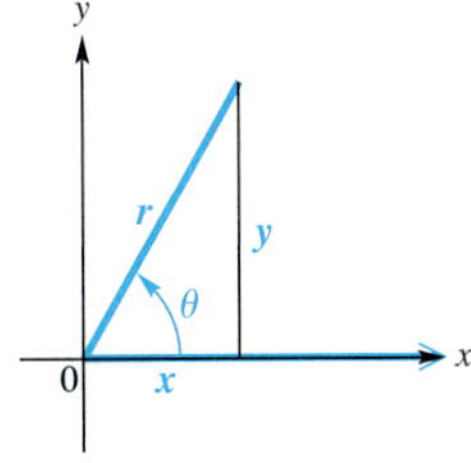

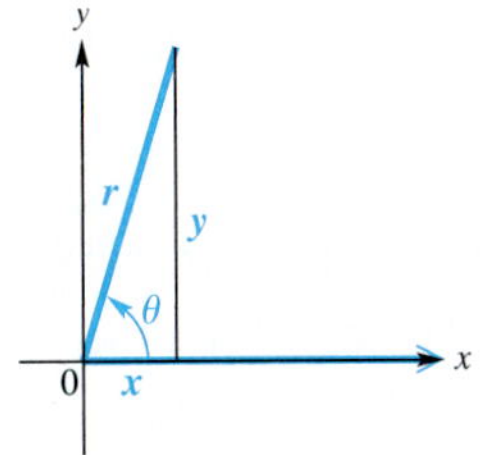

Figure 22

Figure 22 shows an angle θ as it increases in measure from near $0°$ toward $90°$. In each case, the value of r is the same. As the measure of the angle increases, y increases but never exceeds r, so $y \le r$. Dividing both sides by the positive number r gives $\frac{y}{r} \le 1$.

In a similar way, angles in quadrant IV suggest that

$$-1 \le \frac{y}{r},$$

so $$-1 \le \frac{y}{r} \le 1$$

and $$-1 \le \sin\theta \le 1. \quad \frac{y}{r} = \sin\theta \text{ for any angle } \theta.$$

Similarly, $$-1 \le \cos\theta \le 1.$$

The tangent of an angle is defined as $\frac{y}{x}$. It is possible that $x < y$, $x = y$, or $x > y$. Thus, $\frac{y}{x}$ can take any value, so $\tan\theta$ can be any real number, as can $\cot\theta$.

The functions $\sec\theta$ and $\csc\theta$ are reciprocals of the functions $\cos\theta$ and $\sin\theta$, respectively, making

$$\sec\theta \le -1 \text{ or } \sec\theta \ge 1 \qquad \text{and} \qquad \csc\theta \le -1 \text{ or } \csc\theta \ge 1.$$

In summary, the ranges of the trigonometric functions are as follows.

Ranges of Trigonometric Functions

For any angle θ for which the indicated functions exist:

1. $-1 \le \sin\theta \le 1$ and $-1 \le \cos\theta \le 1$;
2. $\tan\theta$ and $\cot\theta$ can equal any real number;
3. $\sec\theta \le -1$ or $\sec\theta \ge 1$ and $\csc\theta \le -1$ or $\csc\theta \ge 1$.

(Notice that $\sec\theta$ and $\csc\theta$ are *never* between -1 and 1.)

EXAMPLE 7 Deciding Whether a Value Is in the Range of a Trigonometric Function

Decide whether each statement is *possible* or *impossible.*

(a) $\sin \theta = \sqrt{8}$ **(b)** $\tan \theta = 110.47$ **(c)** $\sec \theta = .6$

Solution

(a) For any value of θ, $-1 \leq \sin \theta \leq 1$. Since $\sqrt{8} > 1$, it is impossible to find a value of θ with $\sin \theta = \sqrt{8}$.

(b) Tangent can equal any value. Thus, $\tan \theta = 110.47$ is possible.

(c) Since $\sec \theta \leq -1$ or $\sec \theta \geq 1$, the statement $\sec \theta = .6$ is impossible.

Now try Exercises 71 and 73.

Pythagorean Identities We derive three new identities from the relationship $x^2 + y^2 = r^2$.

$$\frac{x^2}{r^2} + \frac{y^2}{r^2} = \frac{r^2}{r^2} \quad \text{Divide by } r^2.$$

$$\left(\frac{x}{r}\right)^2 + \left(\frac{y}{r}\right)^2 = 1 \quad \text{Power rule for exponents (Section R.3)}$$

$$(\cos \theta)^2 + (\sin \theta)^2 = 1 \quad \cos \theta = \frac{x}{r},\ \sin \theta = \frac{y}{r}$$

or $$\sin^2 \theta + \cos^2 \theta = 1.$$

Starting again with $x^2 + y^2 = r^2$ and dividing through by x^2 gives

$$\frac{x^2}{x^2} + \frac{y^2}{x^2} = \frac{r^2}{x^2} \quad \text{Divide by } x^2.$$

$$1 + \left(\frac{y}{x}\right)^2 = \left(\frac{r}{x}\right)^2 \quad \text{Power rule for exponents}$$

$$1 + (\tan \theta)^2 = (\sec \theta)^2 \quad \tan \theta = \frac{y}{x},\ \sec \theta = \frac{r}{x}$$

or $$\tan^2 \theta + 1 = \sec^2 \theta.$$

On the other hand, dividing through by y^2 leads to

$$1 + \cot^2 \theta = \csc^2 \theta.$$

These three identities are called the **Pythagorean identities** since the original equation that led to them, $x^2 + y^2 = r^2$, comes from the Pythagorean theorem.

Pythagorean Identities

$$\sin^2 \theta + \cos^2 \theta = 1 \qquad \tan^2 \theta + 1 = \sec^2 \theta \qquad 1 + \cot^2 \theta = \csc^2 \theta$$

Although we usually write $\sin^2 \theta$, for example, it should be entered as $(\sin \theta)^2$ in your calculator. To test this, verify that in degree mode, $\sin^2 30° = .25 = \frac{1}{4}$. ■

As before, we have given only one form of each identity. However, algebraic transformations produce equivalent identities. For example, by subtracting $\sin^2\theta$ from both sides of $\sin^2\theta + \cos^2\theta = 1$, we get the equivalent identity

$$\cos^2\theta = 1 - \sin^2\theta.$$

You should be able to transform these identities quickly and also recognize their equivalent forms.

Looking Ahead to Calculus

The reciprocal, Pythagorean, and quotient identities are used in calculus to find derivatives and integrals of trigonometric functions. A standard technique of integration called *trigonometric substitution* relies on the Pythagorean identities.

Quotient Identities Recall that $\sin\theta = \frac{y}{r}$ and $\cos\theta = \frac{x}{r}$. Consider the quotient of $\sin\theta$ and $\cos\theta$, where $\cos\theta \neq 0$.

$$\frac{\sin\theta}{\cos\theta} = \frac{\frac{y}{r}}{\frac{x}{r}} = \frac{y}{r} \div \frac{x}{r} = \frac{y}{r} \cdot \frac{r}{x} = \frac{y}{x} = \tan\theta$$

Similarly, $\frac{\cos\theta}{\sin\theta} = \cot\theta$, for $\sin\theta \neq 0$. Thus, we have the **quotient identities.**

Quotient Identities

$$\frac{\sin\theta}{\cos\theta} = \tan\theta \qquad \frac{\cos\theta}{\sin\theta} = \cot\theta$$

EXAMPLE 8 Finding Other Function Values Given One Value and the Quadrant

Find $\sin\theta$ and $\cos\theta$, if $\tan\theta = \frac{4}{3}$ and θ is in quadrant III.

Solution Since θ is in quadrant III, $\sin\theta$ and $\cos\theta$ will both be negative. It is tempting to say that since $\tan\theta = \frac{\sin\theta}{\cos\theta}$ and $\tan\theta = \frac{4}{3}$, then $\sin\theta = -4$ and $\cos\theta = -3$. This is *incorrect,* however, since both $\sin\theta$ and $\cos\theta$ must be in the interval $[-1, 1]$.

We use the Pythagorean identity $\tan^2\theta + 1 = \sec^2\theta$ to find $\sec\theta$, and then the reciprocal identity $\cos\theta = \frac{1}{\sec\theta}$ to find $\cos\theta$.

$$\tan^2\theta + 1 = \sec^2\theta$$

$$\left(\frac{4}{3}\right)^2 + 1 = \sec^2\theta \qquad \tan\theta = \frac{4}{3}$$

$$\frac{16}{9} + 1 = \sec^2\theta$$

$$\frac{25}{9} = \sec^2\theta$$

$$-\frac{5}{3} = \sec\theta$$ Choose the negative square root since $\sec\theta$ is negative when θ is in quadrant III.

$$-\frac{3}{5} = \cos\theta$$ Secant and cosine are reciprocals.

Since $\sin^2 \theta = 1 - \cos^2 \theta$,

$$\sin^2 \theta = 1 - \left(-\frac{3}{5}\right)^2 \quad \cos \theta = -\tfrac{3}{5}$$

$$\sin^2 \theta = 1 - \frac{9}{25}$$

$$\sin^2 \theta = \frac{16}{25}$$

$$\sin \theta = -\frac{4}{5}. \quad \text{Choose the negative square root.}$$

Therefore, we have $\sin \theta = -\frac{4}{5}$ and $\cos \theta = -\frac{3}{5}$.

Now try Exercise 79.

NOTE Example 8 can also be worked by drawing θ in standard position in quadrant III, finding r to be 5, and then using the definitions of $\sin \theta$ and $\cos \theta$ in terms of x, y, and r.

5.2 Exercises

Concept Check *Sketch an angle θ in standard position such that θ has the smallest possible positive measure, and the given point is on the terminal side of θ.*

1. $(5, -12)$ **2.** $(-12, -5)$

Find the values of the six trigonometric functions for each angle in standard position having the given point on its terminal side. Rationalize denominators when applicable. See Examples 1, 2, and 4.

3. $(-3, 4)$ **4.** $(-4, -3)$ **5.** $(0, 2)$ **6.** $(-4, 0)$

7. $(1, \sqrt{3})$ **8.** $(-2\sqrt{3}, -2)$ **9.** $(-2, 0)$ **10.** $(3, -4)$

11. For any nonquadrantal angle θ, $\sin \theta$ and $\csc \theta$ will have the same sign. Explain why.

12. *Concept Check* If the terminal side of an angle θ is in quadrant III, what is the sign of each of the trigonometric function values of θ?

Concept Check *Suppose that the point (x, y) is in the indicated quadrant. Decide whether the given ratio is positive or negative. (Hint: Drawing a sketch may help.)*

13. II, $\frac{x}{r}$ **14.** III, $\frac{y}{r}$ **15.** IV, $\frac{y}{x}$ **16.** IV, $\frac{x}{y}$

In Exercises 17–20, an equation of the terminal side of an angle θ in standard position is given with a restriction on x. Sketch the smallest positive such angle θ, and find the values of the six trigonometric functions of θ. See Example 3.

17. $2x + y = 0, x \geq 0$ **18.** $3x + 5y = 0, x \geq 0$

19. $-6x - y = 0, x \leq 0$ **20.** $-5x - 3y = 0, x \leq 0$

21. Find the six trigonometric function values of the quadrantal angle 450°.

Use the trigonometric function values of quadrantal angles given in this section to evaluate each expression. An expression such as $\cot^2 90°$ *means* $(\cot 90°)^2$, *which is equal to* $0^2 = 0$.

22. $3 \sec 180° - 5 \tan 360°$

23. $4 \csc 270° + 3 \cos 180°$

24. $\tan 360° + 4 \sin 180° + 5 \cos^2 180°$

25. $2 \sec 0° + 4 \cot^2 90° + \cos 360°$

26. $\sin^2 180° + \cos^2 180°$

27. $\sin^2 360° + \cos^2 360°$

If n is an integer, $n \cdot 180°$ *represents an integer multiple of* 180°, *and* $(2n + 1) \cdot 90°$ *represents an odd integer multiple of* 90°. *Decide whether each expression is equal to* 0, 1, −1, *or is* undefined.

28. $\cos[(2n + 1) \cdot 90°]$

29. $\sin[n \cdot 180°]$

30. $\tan[n \cdot 180°]$

31. $\tan[(2n + 1) \cdot 90°]$

Provide conjectures in Exercises 32–35.

32. The angles 15° and 75° are complementary. With your calculator determine sin 15° and cos 75°. Make a conjecture about the sines and cosines of complementary angles, and test your hypothesis with other pairs of complementary angles. (*Note:* This relationship will be discussed in detail in the next section.)

33. The angles 25° and 65° are complementary. With your calculator determine tan 25° and cot 65°. Make a conjecture about the tangents and cotangents of complementary angles, and test your hypothesis with other pairs of complementary angles. (*Note:* This relationship will be discussed in detail in the next section.)

34. With your calculator determine sin 10° and sin(−10°). Make a conjecture about the sines of an angle and its negative, and test your hypothesis with other angles. Also, use a geometry argument with the definition of $\sin \theta$ to justify your hypothesis. (*Note:* This relationship will be discussed in detail in Section 7.1.)

35. With your calculator determine cos 20° and cos(−20°). Make a conjecture about the cosines of an angle and its negative, and test your hypothesis with other angles. Also, use a geometry argument with the definition of $\cos \theta$ to justify your hypothesis. (*Note:* This relationship will be discussed in detail in Section 7.1.)

In Exercises 36–41, set your graphing calculator in parametric and degree modes. Set the window and functions (see the third screen) as shown here, and graph. A circle of radius 1 *will appear on the screen. Trace to move a short distance around the circle. In the screen, the point on the circle corresponds to an angle* T = 25°. *Since* $r = 1$, cos 25° *is* X = .90630779, *and* sin 25° *is* Y = .42261826.

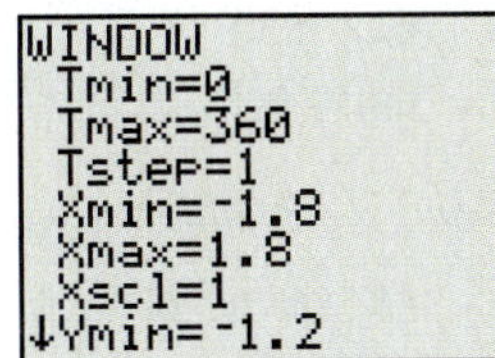

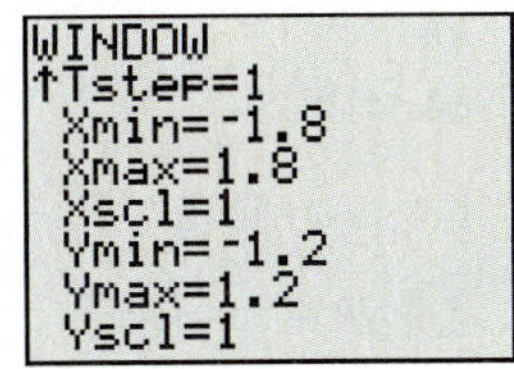

This screen is a continuation of the previous one.

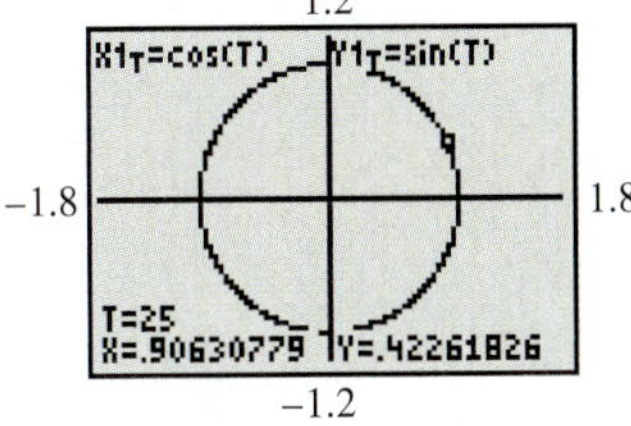

36. Use the right- and left-arrow keys to move to the point corresponding to 20°. What are cos 20° and sin 20°?

37. For what angle T, 0° ≤ T ≤ 90°, is cos T ≈ .766?

38. For what angle T, 0° ≤ T ≤ 90°, is sin T ≈ .574?

39. For what angle T, 0° ≤ T ≤ 90°, does cos T = sin T?

40. As T increases from 0° to 90°, does the cosine increase or decrease? What about the sine?

41. As T increases from 90° to 180°, does the cosine increase or decrease? What about the sine?

42. *Concept Check* What positive number a is its own reciprocal? Find a value of θ for which $\sin \theta = \csc \theta = a$.

43. *Concept Check* What negative number a is its own reciprocal? Find a value of θ for which $\cos \theta = \sec \theta = a$.

Use the appropriate reciprocal identity to find each function value. Rationalize denominators when applicable. In Exercises 48 and 49, use a calculator. See Example 5.

44. $\cos \theta$, if $\sec \theta = -2.5$

45. $\cot \theta$, if $\tan \theta = -\frac{1}{5}$

46. $\sin \theta$, if $\csc \theta = \sqrt{15}$

47. $\tan \theta$, if $\cot \theta = -\frac{\sqrt{5}}{3}$

48. $\sin \theta$, if $\csc \theta = 1.42716321$

49. $\cos \theta$, if $\sec \theta = 9.80425133$

50. Can a given angle θ satisfy both $\sin \theta > 0$ and $\csc \theta < 0$? Explain.

51. *Concept Check* Explain what is wrong with the following item that appears on a trigonometry test:

$$\text{Find } \sec \theta, \text{ given that } \cos \theta = \frac{3}{2}.$$

Find the tangent of each angle. See Example 5.

52. $\cot \theta = -3$

53. $\cot \theta = \frac{\sqrt{3}}{3}$

54. $\cot \theta = .4$

Find a value of each variable.

55. $\tan(3\theta - 4°) = \dfrac{1}{\cot(5\theta - 8°)}$

56. $\sec(2\theta + 6°)\cos(5\theta + 3°) = 1$

Identify the quadrant or quadrants for the angle satisfying the given conditions. See Example 6.

57. $\sin \theta > 0, \cos \theta < 0$

58. $\cos \theta > 0, \tan \theta > 0$

59. $\tan \theta > 0, \cot \theta > 0$

60. $\tan \theta < 0, \cot \theta < 0$

Concept Check Give the signs of the sine, cosine, and tangent functions for each angle.

61. 129°

62. 183°

63. 298°

64. 412°

65. −82°

66. −121°

Concept Check Without using a calculator, decide which is greater.

67. sin 30° or tan 30°

68. sin 20° or sin 21°

69. sin 33° or sec 33°

Decide whether each statement is possible *or* impossible *for an angle θ. See Example 7.*

70. $\sin \theta = 2$

71. $\cos \theta = -1.001$

72. $\tan \theta = .92$

73. $\cot \theta = -12.1$

74. $\sec \theta = 1$

75. $\tan \theta = 1$

76. $\sin \theta = \frac{1}{2}$ and $\csc \theta = 2$

77. $\tan \theta = 2$ and $\cot \theta = -2$

Use identities to find each function value. Use a calculator in Exercises 84 and 85. See Example 8.

78. $\tan \theta$, if $\sec \theta = 3$, with θ in quadrant IV

79. $\sin \theta$, if $\cos \theta = -\frac{1}{4}$, with θ in quadrant II

80. $\csc \theta$, if $\cot \theta = -\frac{1}{2}$, with θ in quadrant IV

81. $\sec \theta$, if $\tan \theta = \frac{\sqrt{7}}{3}$, with θ in quadrant III

82. $\cos \theta$, if $\csc \theta = -4$, with θ in quadrant III

83. $\sin \theta$, if $\sec \theta = 2$, with θ in quadrant IV

84. $\cot \theta$, if $\csc \theta = -3.5891420$, with θ in quadrant III

85. $\tan \theta$, if $\sin \theta = .49268329$, with θ in quadrant II

86. *Concept Check* Does there exist an angle θ with $\cos \theta = -.6$ and $\sin \theta = .8$?

Find all trigonometric function values for each angle. Use a calculator in Exercises 91 and 92. See Example 8.

87. $\tan \theta = -\frac{15}{8}$, with θ in quadrant II

88. $\cos \theta = -\frac{3}{5}$, with θ in quadrant III

89. $\tan \theta = \sqrt{3}$, with θ in quadrant III

90. $\sin \theta = \frac{\sqrt{5}}{7}$, with θ in quadrant I

91. $\cot \theta = -1.49586$, with θ in quadrant IV

92. $\sin \theta = .164215$, with θ in quadrant II

Work each problem.

93. Derive the identity $1 + \cot^2 \theta = \csc^2 \theta$ by dividing $x^2 + y^2 = r^2$ by y^2.

94. Using a method similar to the one given in this section showing that $\frac{\sin \theta}{\cos \theta} = \tan \theta$, show that $\frac{\cos \theta}{\sin \theta} = \cot \theta$.

95. *Concept Check* True or false: For all angles θ, $\sin \theta + \cos \theta = 1$. If false, give an example showing why it is false.

96. *Concept Check* True or false: Since $\cot \theta = \frac{\cos \theta}{\sin \theta}$, if $\cot \theta = \frac{1}{2}$ with θ in quadrant I, then $\cos \theta = 1$ and $\sin \theta = 2$. If false, explain why.

Use a trigonometric function ratio to solve each problem. (Source for Exercises 97–98: Parker, M., Editor, *She Does Math,* Mathematical Association of America, 1995.)

97. *Height of a Tree* A civil engineer must determine the height of the tree shown in the figure. The given angle was measured with a *clinometer*. She knows that $\sin 70° \approx .9397$, $\cos 70° \approx .3420$, and $\tan 70° \approx 2.747$. Use the pertinent trigonometric function and the measurement given in the figure to find the height of the tree to the nearest whole number.

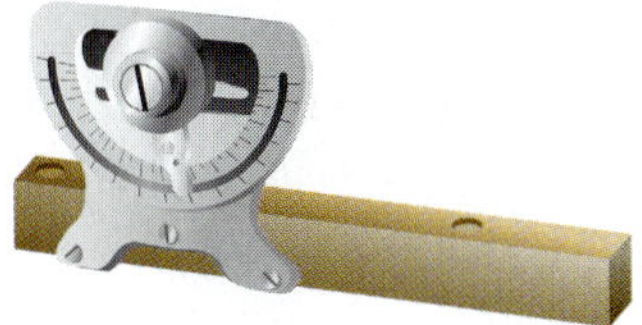

This is a picture of one type of clinometer, called an Abney hand level and clinometer. The picture is courtesy of Keuffel & Esser Co.

98. ***(Modeling) Double Vision*** To correct mild double vision, a small amount of prism is added to a patient's eyeglasses. The amount of light shift this causes is measured in *prism diopters*. A patient needs 12 prism diopters horizontally and 5 prism diopters vertically. A prism that corrects for both requirements should have length r and be set at angle θ. See the figure.

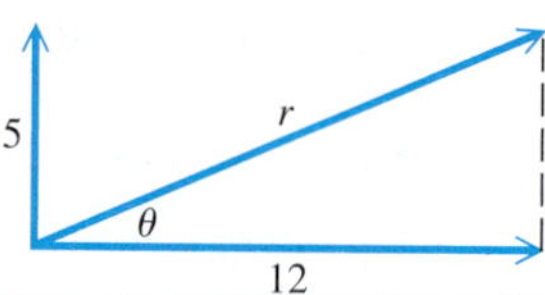

(a) Use the Pythagorean theorem to find r.

(b) Write an equation involving a trigonometric function of θ and the known prism measurements 5 and 12.

99. ***(Modeling) Distance Between the Sun and a Star*** Suppose that a star forms an angle θ with respect to Earth and the sun. Let the coordinates of Earth be (x, y), those of the star $(0, 0)$, and those of the sun $(x, 0)$. See the figure. Find an equation for x, the distance between the sun and the star, as follows.

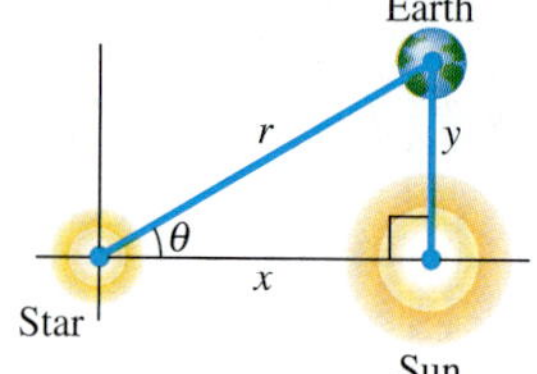

(a) Write an equation involving a trigonometric function that relates x, y, and θ.

(b) Solve your equation for x.

100. ***Area of a Solar Cell*** A solar cell converts the energy of sunlight directly into electrical energy. The amount of energy a cell produces depends on its area. Suppose a solar cell is hexagonal, as shown in the figure. Express its area in terms of $\sin\theta$ and any side x. (*Hint:* Consider one of the six equilateral triangles from the hexagon. See the figure.) (*Source:* Kastner, B., *Space Mathematics,* NASA, 1985.)

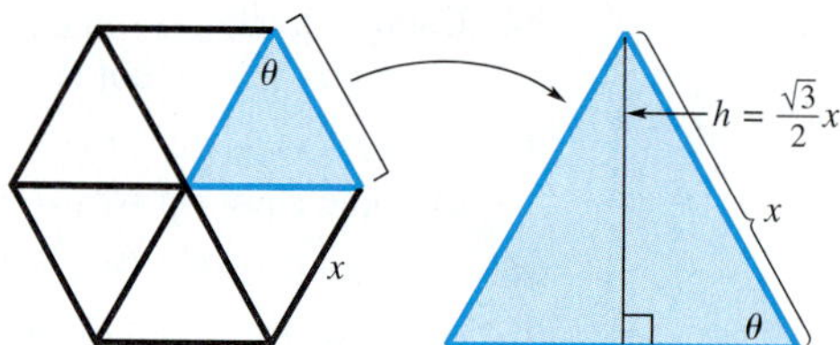

101. The straight line in the figure determines both angle α (alpha) and angle β (beta) with the positive x-axis. Explain why $\tan\alpha = \tan\beta$.

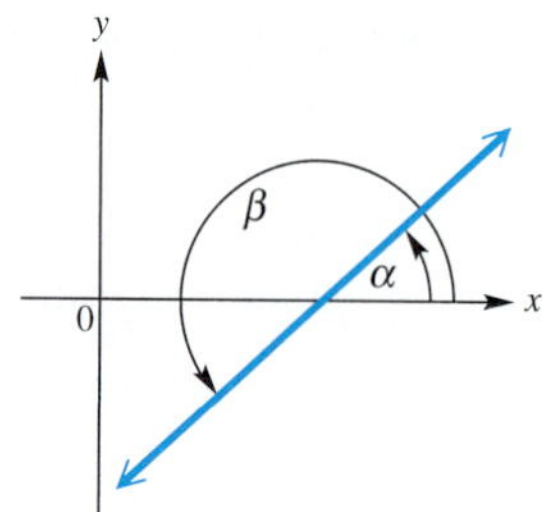

5.3 Evaluating Trigonometric Functions

Definitions of the Trigonometric Functions ▪ Trigonometric Function Values of Special Angles ▪ Reference Angles ▪ Special Angles as Reference Angles ▪ Finding Function Values with a Calculator ▪ Finding Angle Measures

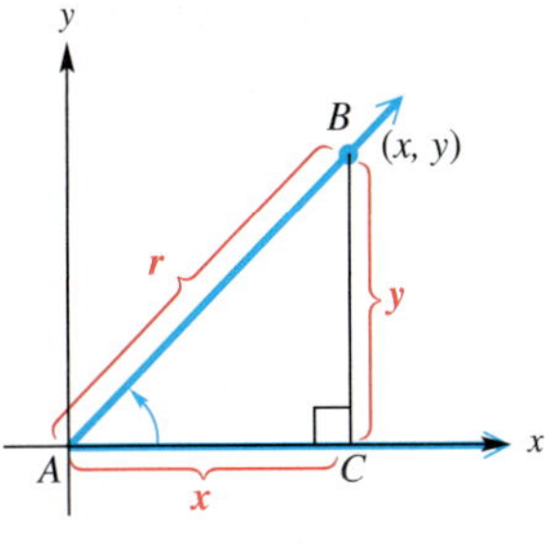

Figure 23

Definitions of the Trigonometric Functions

In Section 5.2 we used angles in standard position to define the trigonometric functions. There is another way to approach them: as ratios of the lengths of the sides of right triangles. Figure 23 shows an acute angle A in standard position. The definitions of the trigonometric function values of angle A require x, y, and r. As drawn in Figure 23, x and y are the lengths of the two legs of the right triangle ABC, and r is the length of the hypotenuse.

The side of length y is called the **side opposite** angle A, and the side of length x is called the **side adjacent** to angle A. We use the lengths of these sides to replace x and y in the definitions of the trigonometric functions, and the length of the hypotenuse to replace r, to get the following right-triangle-based definitions.

Right-Triangle-Based Definitions of Trigonometric Functions

For any acute angle A in standard position,

$$\sin A = \frac{y}{r} = \frac{\text{side opposite}}{\text{hypotenuse}} \qquad \csc A = \frac{r}{y} = \frac{\text{hypotenuse}}{\text{side opposite}}$$

$$\cos A = \frac{x}{r} = \frac{\text{side adjacent}}{\text{hypotenuse}} \qquad \sec A = \frac{r}{x} = \frac{\text{hypotenuse}}{\text{side adjacent}}$$

$$\tan A = \frac{y}{x} = \frac{\text{side opposite}}{\text{side adjacent}} \qquad \cot A = \frac{x}{y} = \frac{\text{side adjacent}}{\text{side opposite}}.$$

EXAMPLE 1 Finding Trigonometric Function Values of an Acute Angle

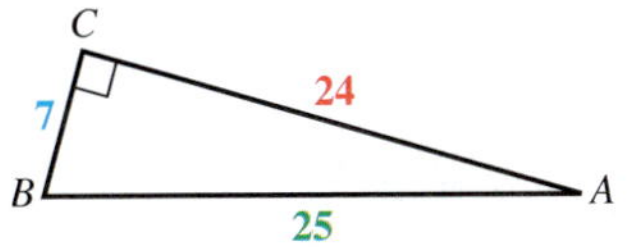

Find the values of $\sin A$, $\cos A$, and $\tan A$ in the right triangle in Figure 24.

C, 7, 24, B, 25, A

Figure 24

Solution The length of the side opposite angle A is 7, the length of the side adjacent to angle A is 24, and the length of the hypotenuse is 25. Use the relationships given in the box.

$$\sin A = \frac{\text{side opposite}}{\text{hypotenuse}} = \frac{7}{25} \qquad \cos A = \frac{\text{side adjacent}}{\text{hypotenuse}} = \frac{24}{25} \qquad \tan A = \frac{\text{side opposite}}{\text{side adjacent}} = \frac{7}{24}$$

Now try Exercise 1.

NOTE Because the cosecant, secant, and cotangent ratios are the reciprocals of the sine, cosine, and tangent values, respectively, in Example 1 we can conclude that $\csc A = \frac{25}{7}$, $\sec A = \frac{25}{24}$, and $\cot A = \frac{24}{7}$.

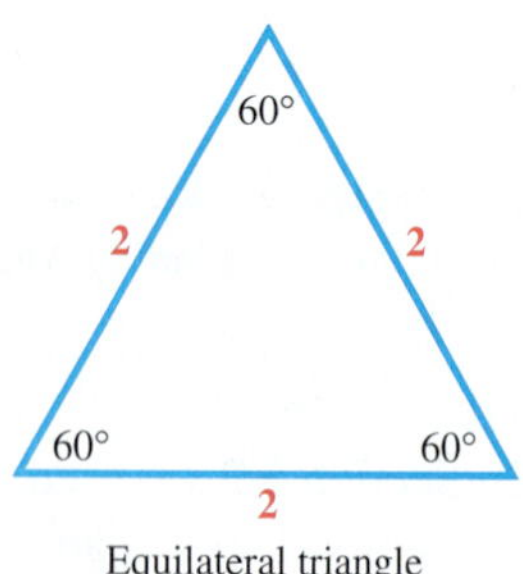

Equilateral triangle

(a)

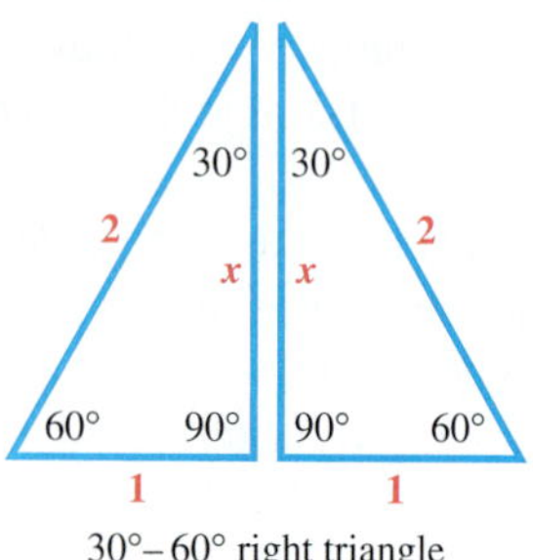

30°–60° right triangle

(b)

Figure 25

Trigonometric Function Values of Special Angles Certain special angles, such as 30°, 45°, and 60°, occur so often in trigonometry and in more advanced mathematics that they deserve special study. We start with an equilateral triangle, a triangle with all sides of equal length. Each angle of such a triangle measures 60°. While the results we will obtain are independent of the length, for convenience we choose the length of each side to be 2 units. See Figure 25(a).

Bisecting one angle of this equilateral triangle leads to two right triangles, each of which has angles of 30°, 60°, and 90°, as shown in Figure 25(b). Since the hypotenuse of one of these right triangles has length 2, the shortest side will have length 1. (Why?) If x represents the length of the medium side, then,

$$\begin{aligned} 2^2 &= 1^2 + x^2 && \text{Pythagorean theorem (Section 1.5)} \\ 4 &= 1 + x^2 \\ 3 &= x^2 && \text{Subtract 1.} \\ \sqrt{3} &= x. && \text{Choose the positive root. (Section 1.4)} \end{aligned}$$

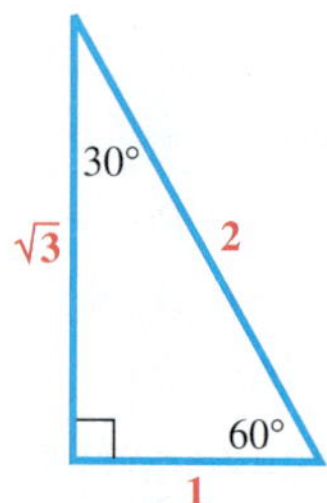

Figure 26

Figure 26 summarizes our results using a 30°–60° right triangle. As shown in the figure, the side opposite the 30° angle has length 1; that is, for the 30° angle,

$$\text{hypotenuse} = 2, \quad \text{side opposite} = 1, \quad \text{side adjacent} = \sqrt{3}.$$

Now we use the definitions of the trigonometric functions.

$$\sin 30° = \frac{\text{side opposite}}{\text{hypotenuse}} = \frac{1}{2} \qquad \csc 30° = \frac{2}{1} = 2$$

$$\cos 30° = \frac{\text{side adjacent}}{\text{hypotenuse}} = \frac{\sqrt{3}}{2} \qquad \sec 30° = \frac{2}{\sqrt{3}} = \frac{2\sqrt{3}}{3}$$

$$\tan 30° = \frac{\text{side opposite}}{\text{side adjacent}} = \frac{1}{\sqrt{3}} = \frac{\sqrt{3}}{3} \qquad \cot 30° = \frac{\sqrt{3}}{1} = \sqrt{3}$$

EXAMPLE 2 Finding Trigonometric Function Values for 60°

Find the six trigonometric function values for a 60° angle.

Solution Refer to Figure 26 to find the following ratios.

$$\sin 60° = \frac{\sqrt{3}}{2} \qquad \cos 60° = \frac{1}{2} \qquad \tan 60° = \sqrt{3}$$

$$\csc 60° = \frac{2\sqrt{3}}{3} \qquad \sec 60° = 2 \qquad \cot 60° = \frac{\sqrt{3}}{3}$$

Now try Exercises 11, 13, and 15.

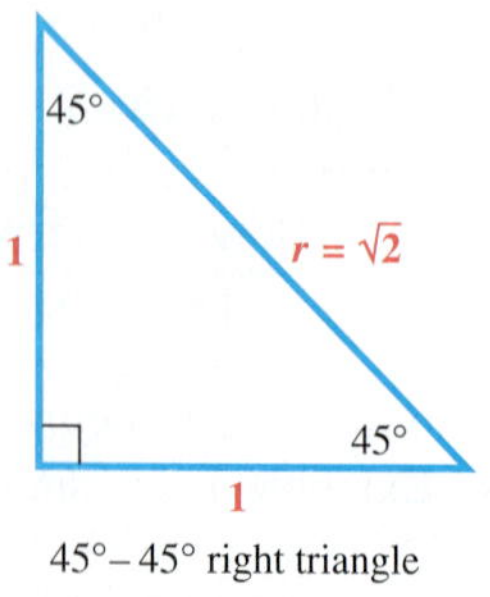

45°–45° right triangle

Figure 27

We find the values of the trigonometric functions for 45° by starting with a 45°–45° right triangle, as shown in Figure 27. This triangle is isosceles; we choose the lengths of the equal sides to be 1 unit. (As before, the results are independent of the length of the equal sides.) Since the shorter sides each have length 1, if r represents the length of the hypotenuse, then

$$\begin{aligned} 1^2 + 1^2 &= r^2 && \text{Pythagorean theorem} \\ 2 &= r^2 \\ \sqrt{2} &= r. && \text{Choose the positive square root.} \end{aligned}$$

Now we use the measures indicated on the 45°–45° right triangle in Figure 27.

$$\sin 45° = \frac{1}{\sqrt{2}} = \frac{\sqrt{2}}{2} \qquad \cos 45° = \frac{1}{\sqrt{2}} = \frac{\sqrt{2}}{2} \qquad \tan 45° = \frac{1}{1} = 1$$

$$\csc 45° = \frac{\sqrt{2}}{1} = \sqrt{2} \qquad \sec 45° = \frac{\sqrt{2}}{1} = \sqrt{2} \qquad \cot 45° = \frac{1}{1} = 1$$

Function values for 30°, 45°, and 60° are summarized in the table that follows.

Function Values of Special Angles

θ	$\sin\theta$	$\cos\theta$	$\tan\theta$	$\cot\theta$	$\sec\theta$	$\csc\theta$
30°	$\frac{1}{2}$	$\frac{\sqrt{3}}{2}$	$\frac{\sqrt{3}}{3}$	$\sqrt{3}$	$\frac{2\sqrt{3}}{3}$	2
45°	$\frac{\sqrt{2}}{2}$	$\frac{\sqrt{2}}{2}$	1	1	$\sqrt{2}$	$\sqrt{2}$
60°	$\frac{\sqrt{3}}{2}$	$\frac{1}{2}$	$\sqrt{3}$	$\frac{\sqrt{3}}{3}$	2	$\frac{2\sqrt{3}}{3}$

Reference Angles Associated with every nonquadrantal angle in standard position is a positive acute angle called its *reference angle.* A **reference angle** for an angle θ, written θ', is the positive acute angle made by the terminal side of angle θ and the x-axis. Figure 28 shows several angles θ (each less than one complete counterclockwise revolution) in quadrants II, III, and IV, respectively, with the reference angle θ' also shown. In quadrant I, θ and θ' are the same. If an angle θ is negative or has measure greater than 360°, its reference angle is found by first finding its coterminal angle that is between 0° and 360°, and then using the diagrams in Figure 28.

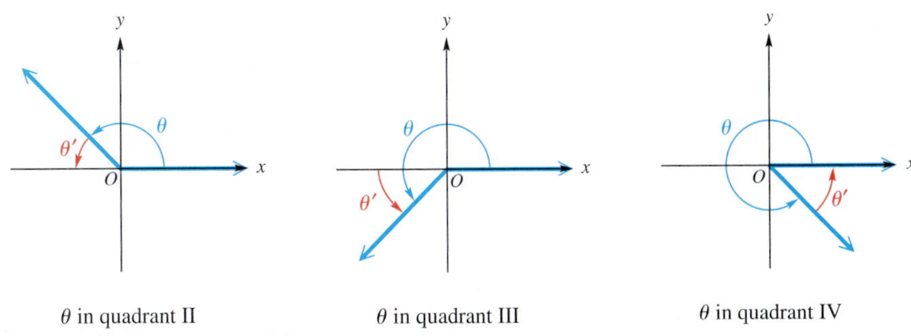

Figure 28

CAUTION A common error is to find the reference angle by using the terminal side of θ and the y-axis. ***The reference angle is always found with reference to the x-axis.***

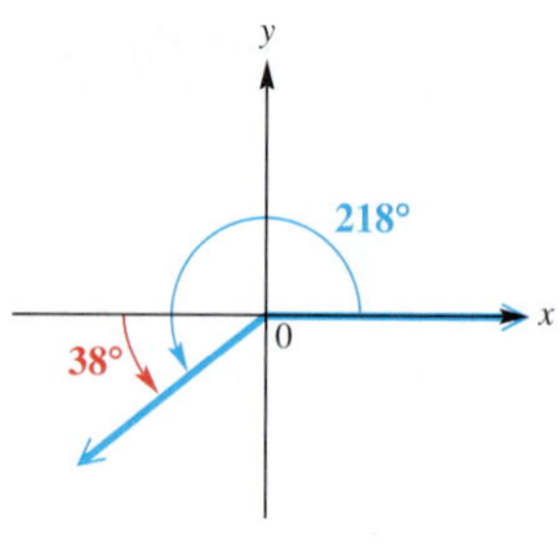

Figure 29

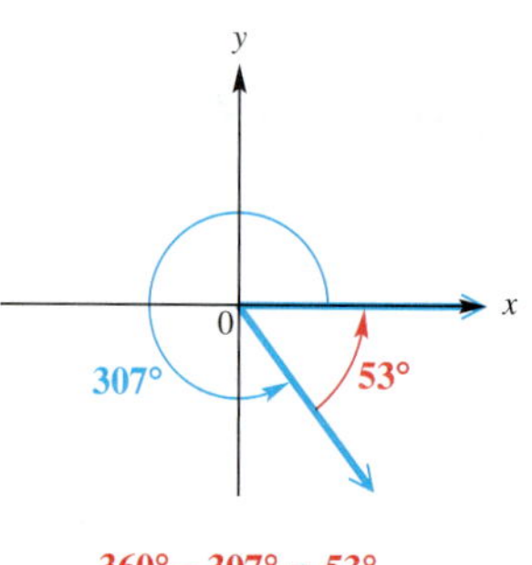

Figure 30

EXAMPLE 3 Finding Reference Angles

Find the reference angle for each angle.

(a) 218° **(b)** 1387°

Solution

(a) As shown in Figure 29, the positive acute angle made by the terminal side of this angle and the x-axis is $218° - 180° = 38°$. For $\theta = 218°$, the reference angle $\theta' = 38°$.

(b) First find a coterminal angle between 0° and 360°. Divide 1387° by 360° to get a quotient of about 3.9. Begin by subtracting 360° three times (because of the 3 in 3.9):

$$1387° - 3 \cdot 360° = 307°.$$

The reference angle for 307° (and thus for 1387°) is $360° - 307° = 53°$. See Figure 30.

Now try Exercises 41 and 45.

Special Angles as Reference Angles We can now find exact trigonometric function values of angles with reference angles of 30°, 45°, or 60°.

EXAMPLE 4 Finding Trigonometric Function Values of a Quadrant III Angle

Find the values of the trigonometric functions for 210°.

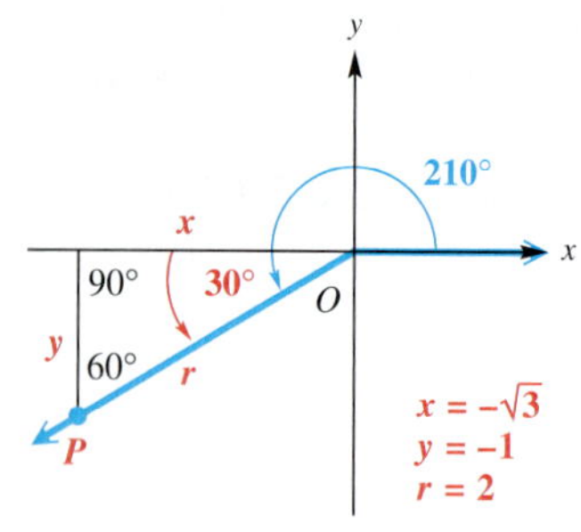

Figure 31

Solution An angle of 210° is shown in Figure 31. The reference angle is $210° - 180° = 30°$. To find the trigonometric function values of 210°, choose point P on the terminal side of the angle so that the distance from the origin O to P is 2. By the results from 30°–60° right triangles, the coordinates of point P become $(-\sqrt{3}, -1)$, with $x = -\sqrt{3}$, $y = -1$, and $r = 2$. Then, by the definitions of the trigonometric functions,

$$\sin 210° = -\frac{1}{2} \qquad \cos 210° = -\frac{\sqrt{3}}{2} \qquad \tan 210° = \frac{\sqrt{3}}{3}$$

$$\csc 210° = -2 \qquad \sec 210° = -\frac{2\sqrt{3}}{3} \qquad \cot 210° = \sqrt{3}.$$

Now try Exercise 59.

Notice in Example 4 that the trigonometric function values of 210° correspond in absolute value to those of its reference angle 30°. The signs are different for the sine, cosine, secant, and cosecant functions because 210° is a quadrant III angle. These results suggest a shortcut for finding the trigonometric function values of a nonacute angle, using the reference angle. In Example 4, the reference angle for 210° is 30°. Using the trigonometric function values of 30°, and choosing the correct signs for a quadrant III angle, we obtain the results found in Example 4.

Similarly, we determine the values of the trigonometric functions for any nonquadrantal angle θ by finding the function values for its reference angle between 0° and 90°, and choosing the appropriate signs.

Finding Trigonometric Function Values for Any Nonquadrantal Angle θ

Step 1 If $\theta > 360°$, or if $\theta < 0°$, then find a coterminal angle by adding or subtracting 360° as many times as needed to get an angle greater than 0° but less than 360°.

Step 2 Find the reference angle θ'.

Step 3 Find the trigonometric function values for reference angle θ'.

Step 4 Determine the correct signs for the values found in Step 3. (Use the table of signs in Section 5.2, if necessary.) This gives the values of the trigonometric functions for angle θ.

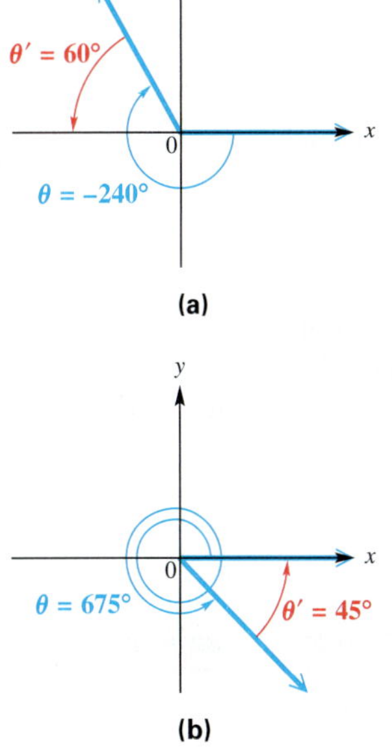

Figure 32

EXAMPLE 5 Finding Trigonometric Function Values Using Reference Angles

Find the exact value of each expression.

(a) $\cos(-240°)$ **(b)** $\tan 675°$

Solution

(a) Since an angle of $-240°$ is coterminal with an angle of

$$-240° + 360° = 120°,$$

the reference angle is $180° - 120° = 60°$, as shown in Figure 32(a). Since the cosine is negative in quadrant II,

$$\cos(-240°) = \cos 120° = -\cos 60° = -\frac{1}{2}.$$

(b) Begin by subtracting 360° to get a coterminal angle between 0° and 360°.

$$675° - 360° = 315°$$

As shown in Figure 32(b), the reference angle is $360° - 315° = 45°$. An angle of 315° is in quadrant IV, so the tangent will be negative, and

$$\tan 675° = \tan 315° = -\tan 45° = -1.$$

Now try Exercises 65 and 67.

```
cos(-240)
           -.5
tan(675)
           -1
```

Degree mode

Figure 33

Finding Function Values with a Calculator Calculators are capable of finding trigonometric function values. For example, the values of $\cos(-240°)$ and $\tan 675°$, found in Example 5, are found with a calculator as shown in Figure 33.

CAUTION We have studied only degree measure of angles; radian measure will be introduced in Chapter 6. When evaluating trigonometric functions of angles given in degrees, remember that the calculator must be set in *degree mode.* Get in the habit of always starting work by entering sin 90. If the displayed answer is 1, then the calculator is set for degree measure.

```
sin(49°12')
      .7569950557
cos(97.977)
     -.1387755707
Ans-1
     -7.205879213
```

```
tan(51.4283)
      1.253948151
Ans-1
       .797481139
sin(-246)
      .9135454576
```

These screens support the results of Example 6. We entered the angle measure in degrees and minutes for part (a). In the fifth line of the first screen, Ans^{-1} tells the calculator to find the reciprocal of the answer given in the previous line.

EXAMPLE 6 Finding Function Values with a Calculator

Approximate the value of each expression.

(a) $\sin 49^\circ 12'$ **(b)** $\sec 97.977^\circ$ **(c)** $\cot 51.4283^\circ$ **(d)** $\sin(-246^\circ)$

Solution

(a) $49^\circ\, 12' = 49\frac{12}{60}^\circ = 49.2^\circ$ Convert $49^\circ\, 12'$ to decimal degrees. (Section 5.1)

$\sin 49^\circ\, 12' = \sin 49.2^\circ \approx .75699506$ To eight decimal places

(b) Calculators do not have secant keys. However, $\sec \theta = \frac{1}{\cos \theta}$ for all angles θ where $\cos \theta \neq 0$. First find $\cos 97.977^\circ$, and then take the reciprocal to get

$$\sec 97.977^\circ \approx -7.205879213.$$

(c) $\cot 51.4283^\circ \approx .79748114$ Use the identity $\cot \theta = \frac{1}{\tan \theta}$.

(d) $\sin(-246^\circ) \approx .91354546$

Now try Exercises 75, 79, 81, and 85.

Finding Angle Measures Sometimes we need to find the measure of an angle having a certain trigonometric function value. Graphing calculators have three *inverse functions* (denoted $\sin^{-1}$, $\cos^{-1}$, and $\tan^{-1}$) that do just that. If x is an appropriate number, then $\sin^{-1}x$, $\cos^{-1}x$, or $\tan^{-1}x$ give the measure of an angle whose sine, cosine, or tangent is x. For the applications in this section, these functions will return values of x in quadrant I.

```
sin-1(.9677091705
)
             75.4
```

Degree mode

Figure 34

EXAMPLE 7 Using an Inverse Trigonometric Function to Find an Angle

Use a calculator to find an angle θ in the interval $[0^\circ, 90^\circ]$ that satisfies $\sin \theta \approx .9677091705$.

Solution With the calculator in degree mode, we find that an angle θ having sine value .9677091705 is 75.4°. (While there are infinitely many such angles, the calculator gives only this one.) We write this result as

$$\sin^{-1} .9677091705 \approx 75.4^\circ.$$

See Figure 34.

Now try Exercise 91.

EXAMPLE 8 Finding Angle Measures

Find all values of θ, if θ is in the interval $[0°, 360°)$ and $\cos \theta = -\frac{\sqrt{2}}{2}$.

Algebraic Solution

Since $\cos \theta$ is negative, θ must lie in quadrant II or III. Since the absolute value of $\cos \theta$ is $\frac{\sqrt{2}}{2}$, the reference angle θ' must be 45°. The two possible angles θ are sketched in Figure 35. The quadrant II angle θ equals $180° - 45° = 135°$, and the quadrant III angle θ equals $180° + 45° = 225°$.

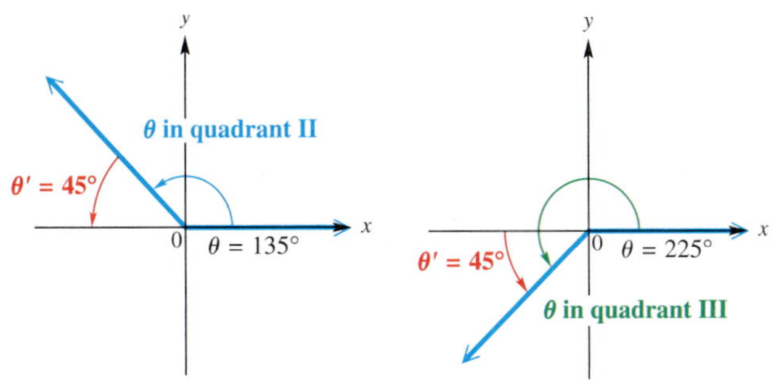

Figure 35

Graphing Calculator Solution

The screen in Figure 36 shows how the inverse cosine function is used to find the two values in $[0°, 360°)$ for which $\cos \theta = -\frac{\sqrt{2}}{2}$. Notice that $\cos^{-1}\left(-\frac{\sqrt{2}}{2}\right)$ yields only one value, 135°; to find the other value, we use the reference angle.

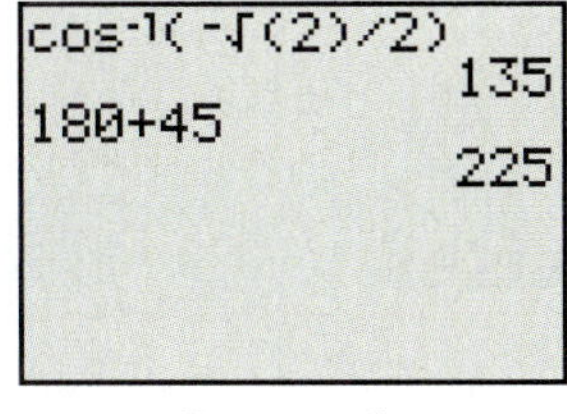

Degree mode

Figure 36

Now try Exercise 95.

EXAMPLE 9 Finding Grade Resistance

When an automobile travels uphill or downhill on a highway, it experiences a force due to gravity. This force F in pounds is called *grade resistance* and is modeled by the equation $F = W \sin \theta$, where θ is the grade and W is the weight of the automobile. If the automobile is moving uphill, then $\theta > 0°$; if downhill, then $\theta < 0°$. See Figure 37. (*Source:* Mannering, F. and W. Kilareski, *Principles of Highway Engineering and Traffic Analysis,* 2nd Edition, John Wiley & Sons, 1998.)

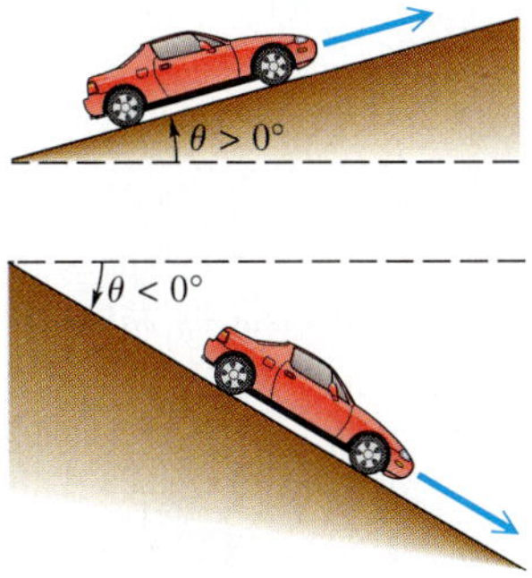

$\theta > 0°$

$\theta < 0°$

Figure 37

(a) Calculate F to the nearest 10 lb for a 2500-lb car traveling an uphill grade with $\theta = 2.5°$.

(b) Calculate F to the nearest 10 lb for a 5000-lb truck traveling a downhill grade with $\theta = -6.1°$.

(c) Calculate F for $\theta = 0°$ and $\theta = 90°$. Do these answers agree with your intuition?

Solution

(a) $F = W \sin \theta = 2500 \sin 2.5° \approx 110$ lb

(b) $F = W \sin \theta = 5000 \sin(-6.1°) \approx -530$ lb

F is negative because the truck is moving downhill.

(c) $F = W \sin \theta = W \sin 0° = W(0) = 0$ lb

$F = W \sin \theta = W \sin 90° = W(1) = W$ lb

This agrees with intuition because if $\theta = 0°$, then there is level ground and gravity does not cause the vehicle to roll. If $\theta = 90°$, the road would be vertical and the full weight of the vehicle would be pulled downward by gravity, so $F = W$.

Now try Exercises 109 and 111.

5.3 Exercises

Find exact values or expressions for $\sin A$, $\cos A$, *and* $\tan A$. *See Example 1.*

1.

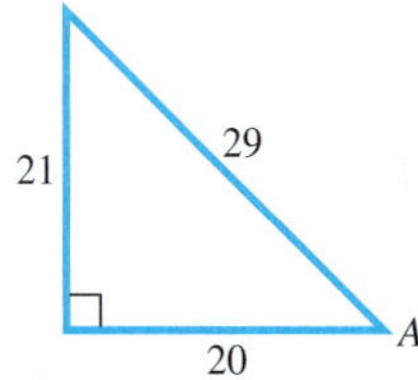

2.

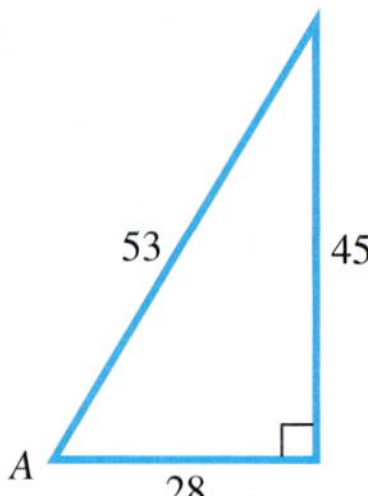

3.

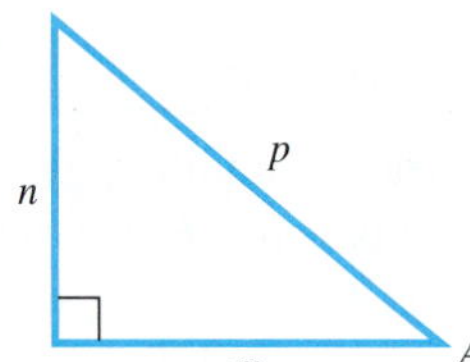

4.

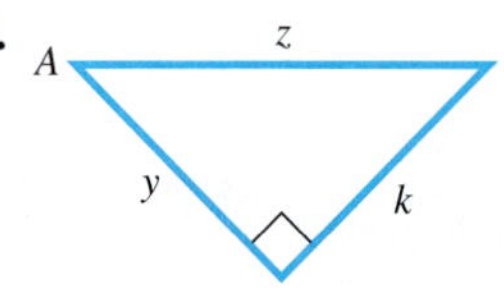

Concept Check *For each trigonometric function in Column I, choose its value from Column II.*

I		II		
5. $\sin 30°$	6. $\cos 45°$	A. $\sqrt{3}$	B. 1	C. $\frac{1}{2}$
7. $\tan 45°$	8. $\sec 60°$	D. $\frac{\sqrt{3}}{2}$	E. $\frac{2\sqrt{3}}{3}$	F. $\frac{\sqrt{3}}{3}$
9. $\csc 60°$	10. $\cot 30°$	G. 2	H. $\frac{\sqrt{2}}{2}$	I. $\sqrt{2}$

For each expression, give the exact value. See Example 2.

11. $\tan 30°$ **12.** $\cot 30°$ **13.** $\sin 30°$ **14.** $\cos 30°$

15. $\sec 30°$ **16.** $\csc 30°$ **17.** $\csc 45°$ **18.** $\sec 45°$

19. $\cos 45°$ **20.** $\cot 45°$

Relating Concepts

For individual or collaborative investigation

(Exercises 21–24)

The figure shows a 45° central angle in a circle with radius 4 units. To find the coordinates of point P on the circle, ***work Exercises 21–24 in order.***

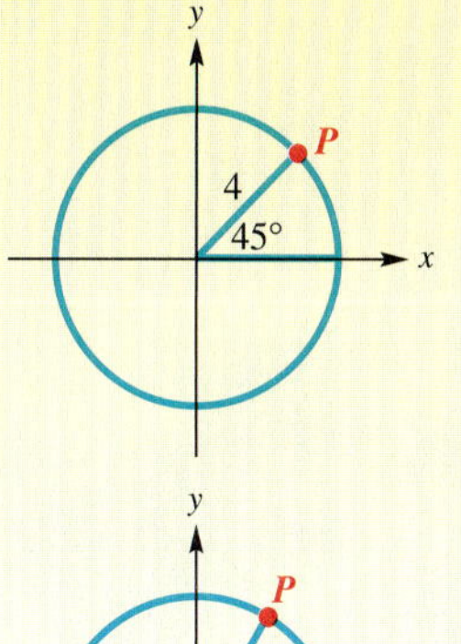

21. Add a line from point P perpendicular to the x-axis.

22. Use the trigonometric ratios for a 45° angle to label the sides of the right triangle you sketched in Exercise 21.

23. Which sides of the right triangle give the coordinates of point P? What are the coordinates of P?

24. Follow the same procedure to find the coordinates of P in the figure given here.

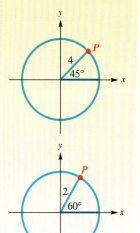

25. Refer to the table. What trigonometric functions are y_1 and y_2?

$x°$	y_1	y_2
0	0	0
15	.25882	.26795
30	.5	.57735
45	.70711	1
60	.86603	1.7321
75	.96593	3.7321
90	1	undefined

26. Refer to the table. What trigonometric functions are y_1 and y_2?

$x°$	y_1	y_2
0	1	undefined
15	.96593	3.8637
30	.86603	2
45	.70711	1.4142
60	.5	1.1547
75	.25882	1.0353
90	0	1

27. *Concept Check* What value of A between 0° and 90° will produce the output shown on the graphing calculator screen?

```
√(3)/2
          .8660254038
sin(A)
          .8660254038
```

28. A student was asked to give the exact value of sin 45°. Using a calculator, he gave the answer .7071067812. The teacher did not give him credit. What was the teacher's reason for this?

29. With a graphing calculator, find the coordinates of the point of intersection of $y = x$ and $y = \sqrt{1 - x^2}$. These coordinates are the cosine and sine of what angle between 0° and 90°?

Concept Check *Work each problem.*

30. Find the equation of the line passing through the origin and making a 60° angle with the x-axis.

31. Find the equation of the line passing through the origin and making a 30° angle with the x-axis.

32. What angle does the line $y = \frac{\sqrt{3}}{3}x$ make with the positive x-axis?

33. What angle does the line $y = \sqrt{3}x$ make with the positive x-axis?

Find the exact value of each part labeled with a variable in each figure.

34.

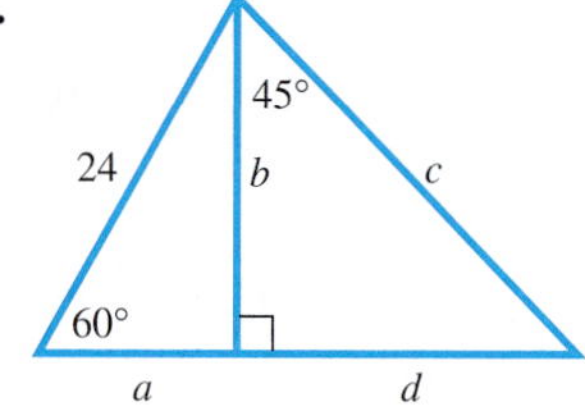

35.

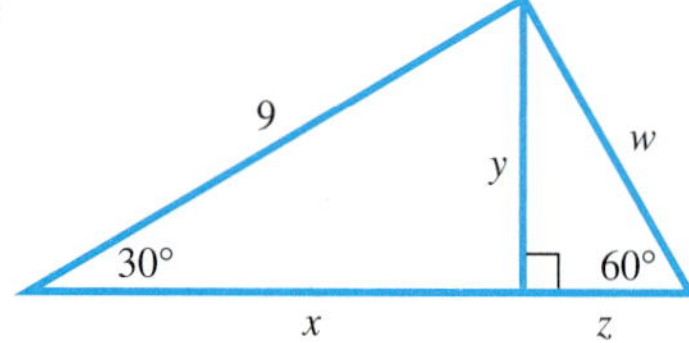

36.

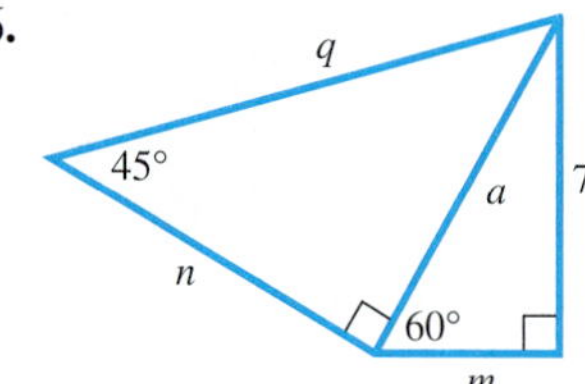

37.

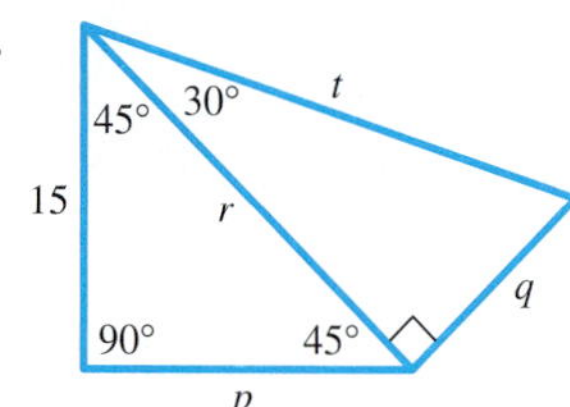

Find a formula for the area of each figure in terms of s.

38.

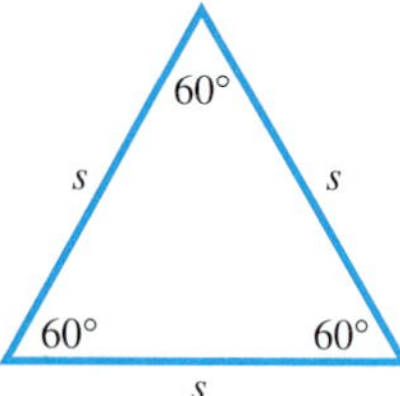

39.

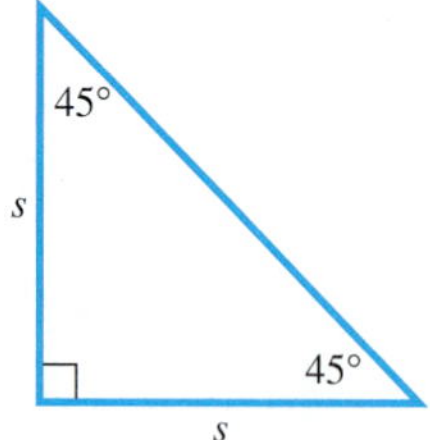

Match each angle in Column I with its reference angle in Column II. Choices may be used once, more than once, or not at all. See Example 3.

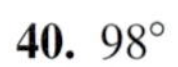

I		II	
40. 98°	**41.** 212°	**A.** 45°	**B.** 60°
42. −135°	**43.** −60°	**C.** 82°	**D.** 30°
44. 750°	**45.** 480°	**E.** 38°	**F.** 32°

Give a short explanation in Exercises 46–49.

46. In Example 4, why was 2 a good choice for r? Could any other positive number have been used?

47. Explain how the reference angle is used to find values of the trigonometric functions for an angle in quadrant III.

48. Explain why two coterminal angles have the same values for their trigonometric functions.

49. If two angles have the same values for each of the six trigonometric functions, must the angles be coterminal? Explain your reasoning.

Complete the table with exact trigonometric function values. Do not use a calculator. See Examples 2, 4, and 5.

	θ	$\sin\theta$	$\cos\theta$	$\tan\theta$	$\cot\theta$	$\sec\theta$	$\csc\theta$
50.	30°	$\frac{1}{2}$	$\frac{\sqrt{3}}{2}$			$\frac{2\sqrt{3}}{3}$	2
51.	45°			1	1		
52.	60°		$\frac{1}{2}$	$\sqrt{3}$		2	
53.	120°	$\frac{\sqrt{3}}{2}$		$-\sqrt{3}$			$\frac{2\sqrt{3}}{3}$
54.	135°	$\frac{\sqrt{2}}{2}$	$-\frac{\sqrt{2}}{2}$			$-\sqrt{2}$	$\sqrt{2}$
55.	150°		$-\frac{\sqrt{3}}{2}$	$-\frac{\sqrt{3}}{3}$			2
56.	210°	$-\frac{1}{2}$		$\frac{\sqrt{3}}{3}$	$\sqrt{3}$		-2
57.	240°	$-\frac{\sqrt{3}}{2}$	$-\frac{1}{2}$			-2	$-\frac{2\sqrt{3}}{3}$

Find exact values of the six trigonometric functions for each angle. Rationalize denominators when applicable. See Examples 2, 4, and 5.

58. 300° **59.** 315° **60.** 405° **61.** −300° **62.** −510° **63.** 750°

Find the exact value of each expression. See Example 5.

64. sin 1305° **65.** cos(−510°) **66.** tan(−1020°) **67.** sin 1500°

Tell whether each statement is true *or* false. *If false, tell why.*

68. $\sin 30° + \sin 60° = \sin(30° + 60°)$

69. $\sin(30° + 60°) = \sin 30° \cdot \cos 60° + \sin 60° \cdot \cos 30°$

70. $\cos 60° = 2\cos^2 30° - 1$

71. $\cos 60° = 2\cos 30°$

72. $\sin 120° = \sin 150° - \sin 30°$

73. $\sin 120° = \sin 180° \cdot \cos 60° - \sin 60° \cdot \cos 180°$

Use a calculator to find a decimal approximation for each value. Give as many digits as your calculator displays. In Exercises 86–89, simplify the expression before using the calculator. See Example 6.

74. $\tan 29° 30'$

75. $\sin 38° 42'$

76. $\cot 41° 24'$

77. $\sec 13° 15'$

78. $\csc 145° 45'$

79. $\cot 183° 48'$

80. $\cos 421° 30'$

81. $\sec 312° 12'$

82. $\tan(-80° 6')$

83. $\sin(-317° 36')$

84. $\cot(-512° 20')$

85. $\cos(-15')$

86. $\dfrac{1}{\sec 14.8°}$

87. $\dfrac{1}{\cot 23.4°}$

88. $\dfrac{\sin 33°}{\cos 33°}$

89. $\dfrac{\cos 77°}{\sin 77°}$

Find a values of θ in $[0°, 90°]$ that satisfies each statement. Leave answers in decimal degrees. See Example 7.

90. $\sin \theta = .84802194$

91. $\tan \theta = 1.4739716$

92. $\sec \theta = 1.1606249$

93. $\cot \theta = 1.2575516$

Find all values of θ, if θ is in the interval $[0°, 360°)$ and has the given function value. See Example 8.

94. $\sin \theta = \dfrac{1}{2}$

95. $\cos \theta = \dfrac{\sqrt{3}}{2}$

96. $\tan \theta = -\sqrt{3}$

97. $\sec \theta = -\sqrt{2}$

98. $\cot \theta = -\dfrac{\sqrt{3}}{3}$

99. $\cos \theta = \dfrac{\sqrt{2}}{2}$

Work each problem.

100. ***Measuring Speed by Radar*** Any offset between a stationary radar gun and a moving target creates a "cosine effect" that reduces the radar mileage reading by the cosine of the angle between the gun and the vehicle. That is, the radar speed reading is the product of the actual reading and the cosine of the angle. Find the radar readings for Auto A and Auto B shown in the figure. (*Source:* "Working Knowledge," Fischetti, M., *Scientific American,* March 2001.)

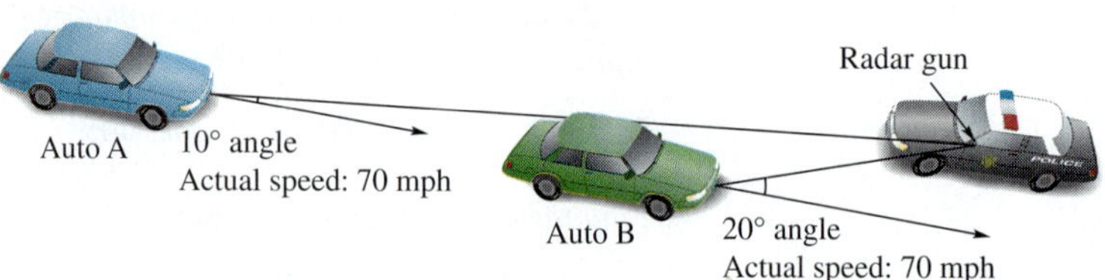

101. ***Measuring Speed by Radar*** In Exercise 100, we saw that the mileage reported by a radar gun is reduced by the cosine of angle θ, shown in the figure. In the figure, r represents reduced speed and a represents the actual speed. Use the figure to show why this "cosine effect" occurs.

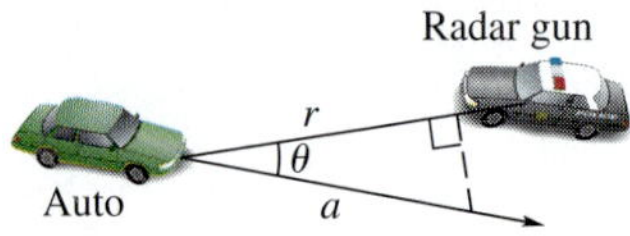

(Modeling) Speed of Light *When a light ray travels from one medium, such as air, to another medium, such as water or glass, the speed of the light changes, and the direction in which the ray is traveling changes. (This is why a fish under water is in a different position than it appears to be.) These changes are given by Snell's law*

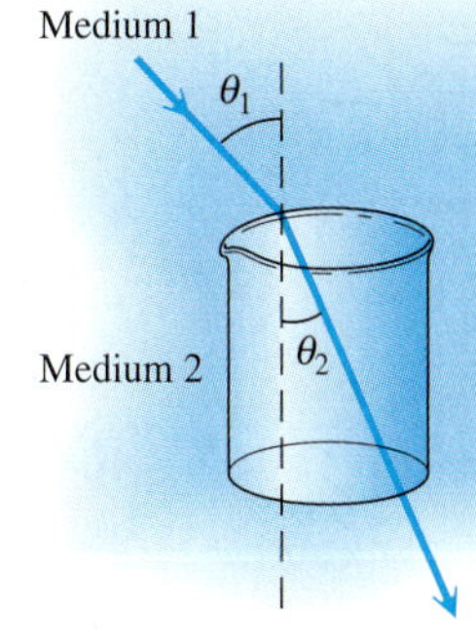

$$\frac{c_1}{c_2} = \frac{\sin \theta_1}{\sin \theta_2},$$

where c_1 is the speed of light in the first medium, c_2 is the speed of light in the second medium, and θ_1 and θ_2 are the angles shown in the figure. (Source: The Physics Classroom, www.glenbrook.k12.il.us) *In Exercises 102 and 103, assume that $c_1 = 3 \times 10^8$ m per sec.*

102. Find the speed of light in the second medium for each of the following.

(a) $\theta_1 = 46°, \theta_2 = 31°$ **(b)** $\theta_1 = 39°, \theta_2 = 28°$

103. Find θ_2 for each of the following values of θ_1 and c_2. Round to the nearest degree.

(a) $\theta_1 = 40°, c_2 = 1.5 \times 10^8$ m per sec **(b)** $\theta_1 = 62°, c_2 = 2.6 \times 10^8$ m per sec

(Modeling) Fish's View of the World *The figure shows a fish's view of the world above the surface of the water. (Source:* Walker, J., "The Amateur Scientist," *Scientific American,* March 1984.) *Suppose that a light ray comes from the horizon, enters the water, and strikes the fish's eye.*

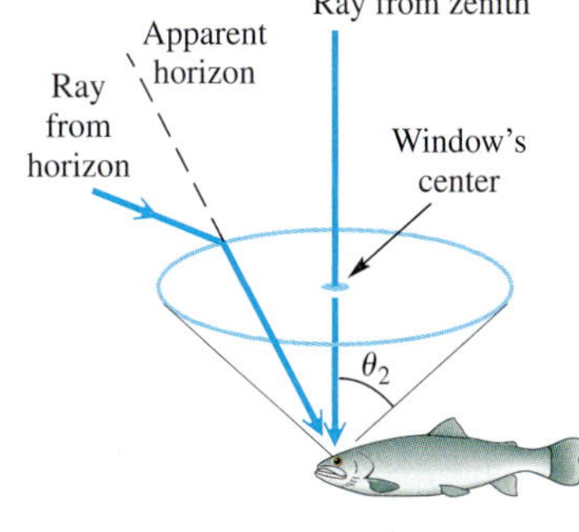

104. Assume that this ray gives a value of 90° for angle θ_1 in the formula for Snell's law. (In a practical situation, this angle would probably be a little less than 90°.) The speed of light in water is about 2.254×10^8 m per sec. Find angle θ_2.

105. Suppose an object is located at a true angle of 29.6° above the horizon. Find the apparent angle above the horizon to a fish.

106. ***(Modeling) Braking Distance*** If aerodynamic resistance is ignored, the braking distance D (in feet) for an automobile to change its velocity from V_1 to V_2 (feet per second) can be modeled using the equation

$$D = \frac{1.05(V_1^2 - V_2^2)}{64.4(K_1 + K_2 + \sin \theta)}.$$

K_1 is a constant determined by the efficiency of the brakes and tires, K_2 is a constant determined by the rolling resistance of the automobile, and θ is the grade of the highway. (*Source:* Mannering, F. and W. Kilareski, *Principles of Highway Engineering and Traffic Analysis,* 2nd Edition, John Wiley & Sons, 1998.)

(a) Compute the number of feet required to slow a car from 55 mph to 30 mph while traveling uphill with a grade of $\theta = 3.5°$. Let $K_1 = .4$ and $K_2 = .02$. (*Hint:* Change miles per hour to feet per second.)

(b) Repeat part (a) with $\theta = -2°$.

(c) How is braking distance affected by grade θ? Does this agree with your driving experience?

107. *(Modeling) Car's Speed at Collision* Refer to Exercise 106. An automobile is traveling at 90 mph on a highway with a downhill grade of $\theta = -3.5°$. The driver sees a stalled truck in the road 200 ft away and immediately applies the brakes. Assuming that a collision cannot be avoided, how fast (in miles per hour) is the car traveling when it hits the truck? (Use the same values for K_1 and K_2 as in Exercise 106.)

(Modeling) Grade Resistance *See Example 9 to work Exercises 108–112.*

108. What is the grade resistance of a 2400-lb car traveling on a $-2.4°$ downhill grade?

109. What is the grade resistance of a 2100-lb car traveling on a 1.8° uphill grade?

110. A car traveling on a $-3°$ downhill grade has a grade resistance of -145 lb. What is the weight of the car?

111. A 2600-lb car traveling downhill has a grade resistance of -130 lb. What is the angle of the grade?

112. Which has the greater grade resistance: a 2200-lb car on a 2° uphill grade or a 2000-lb car on a 2.2° uphill grade?

113. *(Modeling) Design of Highway Curves* When highway curves are designed, the outside of the curve is often slightly elevated or inclined above the inside of the curve. See the figure. This inclination is called *superelevation.* For safety reasons, it is important that both the curve's radius and superelevation are correct for a given speed limit.

If an automobile is traveling at velocity V (in feet per second), the safe radius R for a curve with superelevation θ is modeled by the formula

$$R = \frac{V^2}{g(f + \tan \theta)},$$

where f and g are constants. (*Source:* Mannering, F. and W. Kilareski, *Principles of Highway Engineering and Traffic Analysis,* 2nd Edition, John Wiley & Sons, 1998.)

(a) A roadway is being designed for automobiles traveling at 45 mph. If $\theta = 3°$, $g = 32.2$, and $f = .14$, calculate R.

(b) What should the radius of the curve be if the speed in part (a) is increased to 70 mph?

(c) How would increasing the angle θ affect the results? Verify your answer by repeating parts (a) and (b) with $\theta = 4°$.

114. *(Modeling) Speed Limit on a Curve* Refer to Exercise 113 and use the same values for f and g. A highway curve has radius $R = 1150$ ft and a superelevation of $\theta = 2.1°$. What should the speed limit (in miles per hour) be for this curve?

5.4 Solving Right Triangles

Significant Digits ▪ Solving Triangles ▪ Angles of Elevation or Depression ▪ Bearing ▪ Further Applications

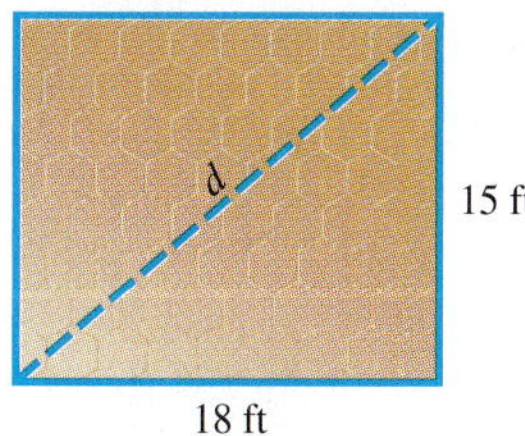

Figure 38

Significant Digits Suppose we quickly measure a room as 15 ft by 18 ft. See Figure 38. To calculate the length of a diagonal of the room, we can use the Pythagorean theorem.

$$d^2 = 15^2 + 18^2 \quad \text{(Section 1.5)}$$
$$d^2 = 549$$
$$d = \sqrt{549} \approx 23.430749$$

Should this answer be given as the length of the diagonal of the room? Of course not. The number 23.430749 contains six decimal places, while the original data of 15 ft and 18 ft are only accurate to the nearest foot. Since the results of a problem can be no more accurate than the least accurate number in any calculation, we really should say that the diagonal of the 15-by-18-ft room is about 23 ft.

If a wall measured to the nearest foot is 18 ft long, this actually means that the wall has length between 17.5 ft and 18.5 ft. If the wall is measured more accurately as 18.3 ft long, then its length is really between 18.25 ft and 18.35 ft. A measurement of 18.00 ft would indicate that the length of the wall is between 17.995 ft and 18.005 ft. The measurement 18 ft is said to have two *significant digits* of accuracy; 18.0 has three significant digits, and 18.00 has four.

A **significant digit** is a digit obtained by actual measurement. A number that represents the result of counting, or a number that results from theoretical work and is not the result of a measurement, is an **exact number.** There are 50 states in the United States, so in that statement, 50 is an exact number.

Most values obtained for trigonometric functions are approximations, and virtually all measurements are approximations. When performing calculations involving approximate numbers, start by determining the number that has the least number of significant digits. Round your final answer to the same number of significant digits as this number. Remember that *your answer is no more accurate than the least accurate number in your calculation.*

Use the following table to determine the significant digits in angle measure.

Significant Digits for Angles

Number of Significant Digits	Angle Measure to Nearest:
2	Degree
3	Ten minutes, or nearest tenth of a degree
4	Minute, or nearest hundredth of a degree
5	Tenth of a minute, or nearest thousandth of a degree

For example, an angle measuring $52° \, 30'$ has three significant digits (assuming that $30'$ is measured to the nearest ten minutes).

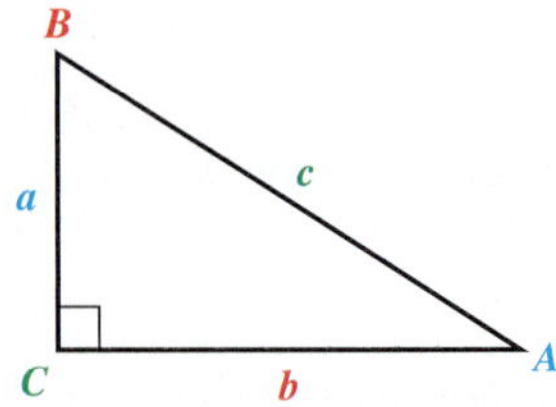

When solving triangles, a labeled sketch is an important aid.

Figure 39

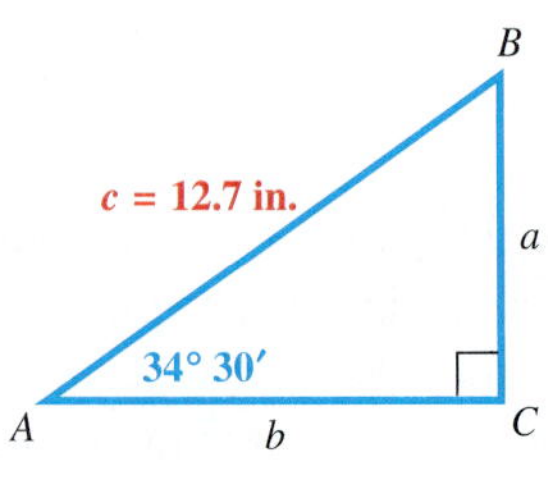

Figure 40

Solving Triangles To *solve a triangle* means to find the measures of all the angles and sides of the triangle. As shown in Figure 39, we use a to represent the length of the side opposite angle A, b for the length of the side opposite angle B, and so on. In a right triangle, the letter c is reserved for the hypotenuse.

EXAMPLE 1 Solving a Right Triangle Given an Angle and a Side

Solve right triangle ABC, if $A = 34° 30'$ and $c = 12.7$ in. See Figure 40.

Solution To solve the triangle, find the measures of the remaining sides and angles. To find the value of a, use a trigonometric function involving the known values of angle A and side c. Since the sine of angle A is given by the quotient of the side opposite A and the hypotenuse, use sin A.

$$\sin A = \frac{a}{c} \qquad \sin A = \frac{\text{side opposite}}{\text{hypotenuse}} \text{ (Section 5.3)}$$

$$\sin 34° 30' = \frac{a}{12.7} \qquad A = 34° 30', c = 12.7$$

$$a = 12.7 \sin 34° 30' \qquad \text{Multiply by 12.7; rewrite.}$$

$$a = 12.7 \sin 34.5° \qquad \text{Convert to decimal degrees.}$$

$$a = 12.7(.56640624) \qquad \text{Use a calculator.}$$

$$a \approx 7.19 \text{ in.} \qquad \text{Three significant digits}$$

To find the value of b, we could use the Pythagorean theorem. It is better, however, to use the information given in the problem rather than a result just calculated. If a mistake is made in finding a, then b also would be incorrect. Also, rounding more than once may cause the result to be less accurate. Use cos A.

$$\cos A = \frac{b}{c} \qquad \cos A = \frac{\text{side adjacent}}{\text{hypotenuse}} \text{ (Section 5.3)}$$

$$\cos 34° 30' = \frac{b}{12.7}$$

$$b = 12.7 \cos 34° 30'$$

$$b \approx 10.5 \text{ in.}$$

Once b is found, the Pythagorean theorem can be used as a check. All that remains to solve triangle ABC is to find the measure of angle B. Since $A + B = 90°$,

$$B = 90° - A$$

$$B = 89° 60' - 34° 30' \qquad A = 34° 30'$$

$$B = 55° 30'.$$

Now try Exercise 17.

NOTE In Example 1, we could have found the measure of angle B first, and then used the trigonometric function values of B to find the unknown sides. The process of solving a right triangle can usually be done in several ways, each producing the correct answer. To maintain accuracy, always use given information as much as possible, and avoid rounding off in intermediate steps.

Looking Ahead to Calculus

The derivatives of the *parametric equations* $x = f(t)$ and $y = g(t)$ often represent the rate of change of physical quantities, such as velocities. When x and y are related by an equation, the derivatives are called *related rates* because a change in one causes a related change in the other. Determining these rates in calculus often requires solving a right triangle. Some problems that ask for the maximum or minimum value of a quantity also involve solving a right triangle.

EXAMPLE 2 Solving a Right Triangle Given Two Sides

Solve right triangle ABC if $a = 29.43$ cm and $c = 53.58$ cm.

Solution We draw a sketch showing the given information, as in Figure 41. One way to begin is to find angle A by using the sine function.

Figure 41

$$\sin A = \frac{\text{side opposite}}{\text{hypotenuse}} = \frac{29.43}{53.58} \approx .5492721165$$

Using the $\sin^{-1}$ function on a calculator, we find that $A \approx 33.32°$. The measure of B is approximately

$$90° - 33.32° = 56.68°.$$

We now find b from the Pythagorean theorem, using $a^2 + b^2 = c^2$, or $b^2 = c^2 - a^2$. Since $c = 53.58$ and $a = 29.43$,

$$b^2 = c^2 - a^2 \qquad \text{Pythagorean theorem (Section 1.5)}$$

$$b^2 = 53.58^2 - 29.43^2$$

$$b \approx 44.77 \text{ cm.}$$

Now try Exercise 21.

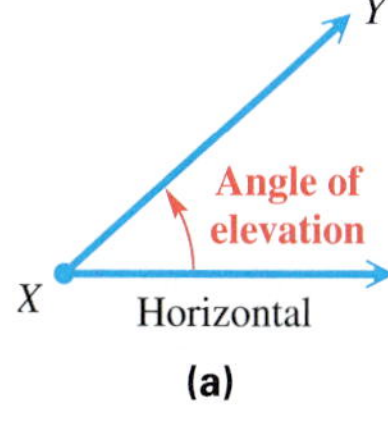

(a)

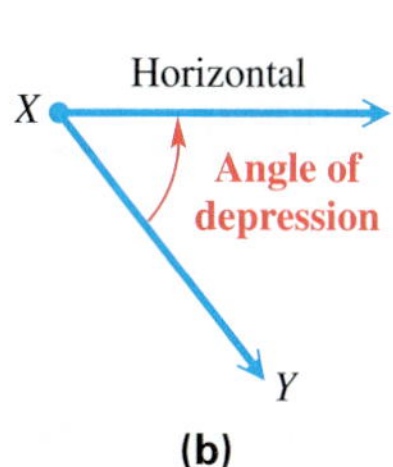

(b)

Figure 42

Angles of Elevation or Depression Many applications of right triangles involve angles of elevation or depression. The **angle of elevation** from point X to point Y (above X) is the acute angle formed by ray XY and a horizontal ray with endpoint at X. See Figure 42(a). The **angle of depression** from point X to point Y (below X) is the acute angle formed by ray XY and a horizontal ray with endpoint X. See Figure 42(b).

CAUTION Be careful when interpreting the angle of depression. Both the angle of elevation and the angle of depression are measured between the line of sight and a *horizontal* line.

To solve applied trigonometry problems, follow the same procedure as solving a triangle.

Solving an Applied Trigonometry Problem

Step 1 Draw a sketch, and label it with the given information. Label the quantity to be found with a variable.

Step 2 Use the sketch to write an equation relating the given quantities to the variable.

Step 3 Solve the equation, and check that your answer makes sense.

EXAMPLE 3 Finding the Angle of Elevation When Lengths Are Known

The length of the shadow of a building 34.09 m tall is 37.62 m. Find the angle of elevation of the sun.

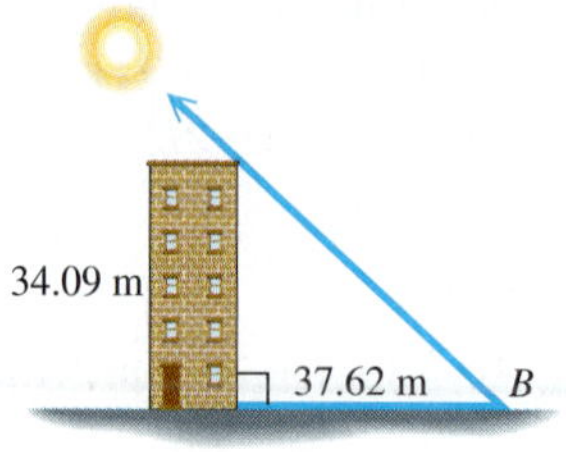

Solution As shown in Figure 43, the angle of elevation of the sun is angle B. Since the side opposite B and the side adjacent to B are known, use the tangent ratio to find B.

$$\tan B = \frac{34.09}{37.62}, \quad \text{so} \quad B = \tan^{-1}\frac{34.09}{37.62} \approx 42.18^\circ. \quad \text{(Section 5.3)}$$

The angle of elevation of the sun is 42.18°.

Figure 43

Now try Exercise 29.

Bearing Other applications of right triangles involve **bearing,** an important idea in navigation. There are two methods for expressing bearing. When a single angle is given, such as 164°, it is understood that the bearing is measured in a clockwise direction from due north. Several sample bearings using this first method are shown in Figure 44.

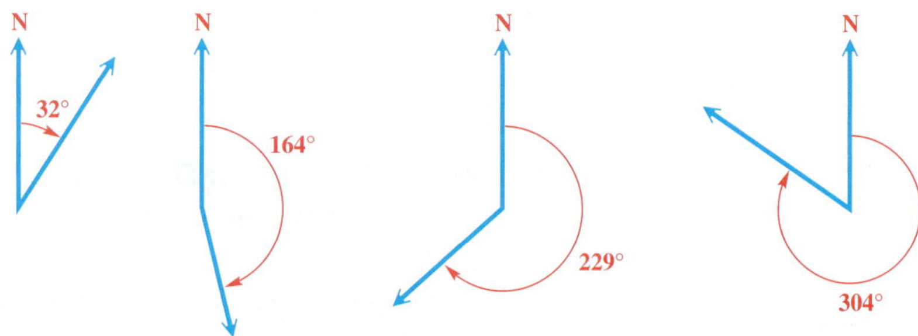

Figure 44

EXAMPLE 4 Solving a Problem Involving Bearing (First Method)

Radar stations A and B are on an east-west line, 3.7 km apart. Station A detects a plane at C, on a bearing of 61°. Station B simultaneously detects the same plane, on a bearing of 331°. Find the distance from A to C.

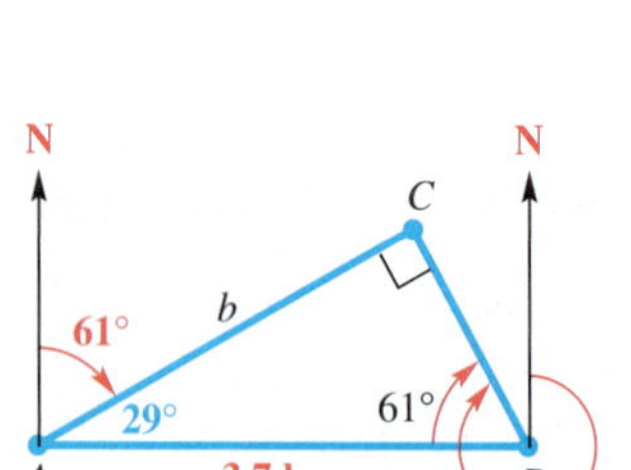

Figure 45

Solution Draw a sketch showing the given information, as in Figure 45. Since a line drawn due north is perpendicular to an east-west line, right angles are formed at A and B, so angles CAB and CBA can be found as shown in Figure 45. Angle C is a right angle because angles CAB and CBA are complementary. Find distance b by using the cosine function.

$$\cos 29^\circ = \frac{b}{3.7}$$

$$3.7 \cos 29^\circ = b$$

$$b \approx 3.2 \text{ km} \quad \text{Use a calculator; round to the nearest tenth.}$$

Now try Exercise 45.

CAUTION The importance of a correctly labeled sketch when solving applications like that in Example 4 cannot be overemphasized. Some of the necessary information is often not directly stated in the problem and can be determined only from the sketch.

The second method for expressing bearing starts with a north-south line and uses an acute angle to show the direction, either east or west, from this line. Figure 46 shows several sample bearings using this system. Either N or S always comes first, followed by an acute angle, and then E or W.

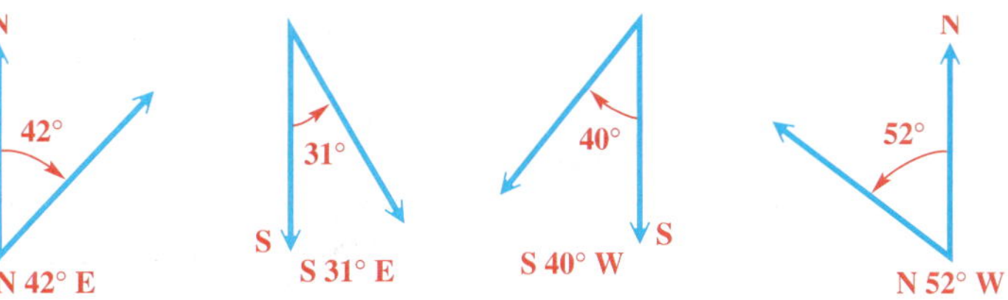

Figure 46

EXAMPLE 5 Solving a Problem Involving Bearing (Second Method)

The bearing from A to C is S 52° E. The bearing from A to B is N 84° E. The bearing from B to C is S 38° W. A plane flying at 250 mph takes 2.4 hr to go from A to B. Find the distance from A to C.

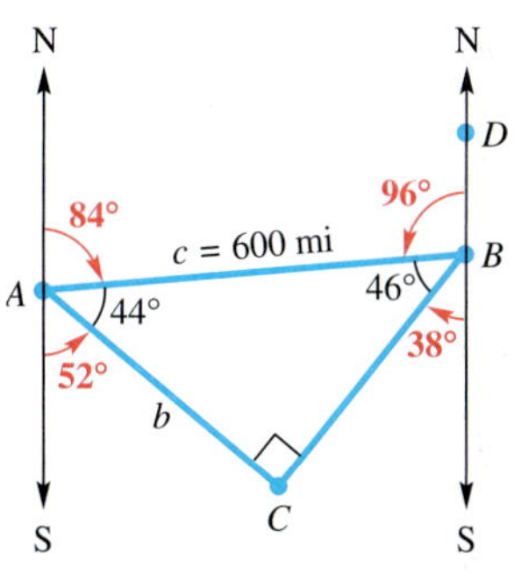

Figure 47

Solution Make a sketch. First draw the two bearings from point A. Choose a point B on the bearing N 84° E from A, and draw the bearing to C. Point C will be located where the bearing lines from A and B intersect, as shown in Figure 47.

Since the bearing from A to B is N 84° E, angle ABD is $180° - 84° = 96°$. Thus, angle ABC is 46°. Also, angle BAC is $180° - (84° + 52°) = 44°$. Angle C is $180° - (44° + 46°) = 90°$. From the statement of the problem, a plane flying at 250 mph takes 2.4 hr to go from A to B. The distance from A to B is the product of rate and time, or

$$c = \text{rate} \times \text{time} = 250(2.4) = 600 \text{ mi.}$$ (Section 1.2)

To find b, the distance from A to C, use the sine. (The cosine could also be used.)

$$\sin 46° = \frac{b}{c}$$

$$\sin 46° = \frac{b}{600}$$

$$600 \sin 46° = b$$

$$b \approx 430 \text{ mi}$$

Now try Exercise 47.

Further Applications

EXAMPLE 6 Using Trigonometry to Measure a Distance

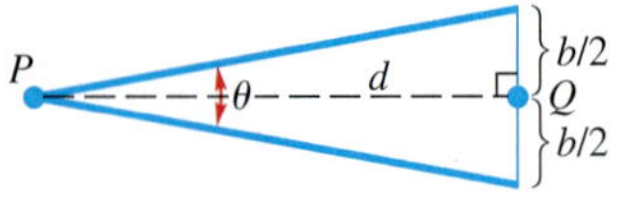

Figure 48

A method that surveyors use to determine a small distance d between two points P and Q is called the *subtense bar method.* The subtense bar with length b is centered at Q and situated perpendicular to the line of sight between P and Q. See Figure 48. Angle θ is measured, and then the distance d can be determined.

(a) Find d when $\theta = 1° \, 23' \, 12''$ and $b = 2.0000$ m.

(b) Angle θ usually cannot be measured more accurately than to the nearest $1''$. How much change would there be in the value of d if θ were measured $1''$ larger?

Solution

(a) From Figure 48, we see that

$$\cot \frac{\theta}{2} = \frac{d}{\frac{b}{2}}$$

$$d = \frac{b}{2} \cot \frac{\theta}{2}. \quad \text{Multiply; rewrite.}$$

Let $b = 2$. To evaluate $\frac{\theta}{2}$, we change θ to decimal degrees.

$$1° \, 23' \, 12'' = 1.386667° \quad \text{(Section 5.1)}$$

Then $$d = \frac{2}{2} \cot \frac{1.386667°}{2} \approx 82.6341 \text{ m.}$$

(b) Since θ is $1''$ larger, use $\theta = 1° \, 23' \, 13'' \approx 1.386944°$.

$$d = \frac{2}{2} \cot \frac{1.386944°}{2} \approx 82.6176 \text{ m}$$

The difference is $82.6341 - 82.6176 \approx .0170$ m.

Now try Exercise 59.

EXAMPLE 7 Solving a Problem Involving Angles of Elevation

Francisco needs to know the height of a tree. From a given point on the ground, he finds that the angle of elevation to the top of the tree is 36.7°. He then moves back 50 ft. From the second point, the angle of elevation to the top of the tree is 22.2°. See Figure 49. Find the height of the tree.

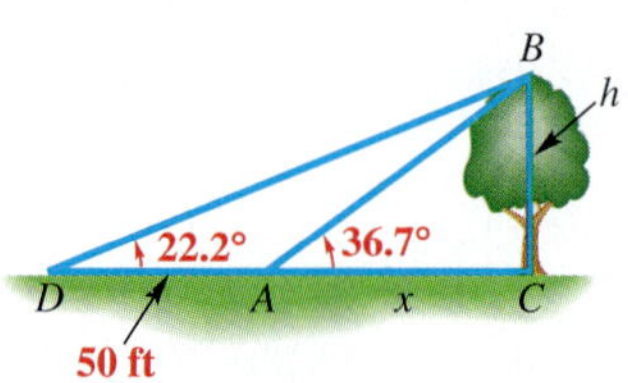

Figure 49

Algebraic Solution

Figure 49 shows two unknowns: x, the distance from the center of the trunk of the tree to the point where the first observation was made, and h, the height of the tree. Since nothing is given about the length of the hypotenuse of either triangle ABC or triangle BCD, use a ratio that does not involve the hypotenuse—namely the tangent. See Figure 50 in the Graphing Calculator Solution.

In triangle ABC, $\tan 36.7° = \frac{h}{x}$ or $h = x \tan 36.7°$.

In triangle BCD, $\tan 22.2° = \frac{h}{50 + x}$ or $h = (50 + x) \tan 22.2°$.

Each expression equals h, so the expressions must be equal.

$$x \tan 36.7° = (50 + x) \tan 22.2°$$
Solve for x.

$$x \tan 36.7° = 50 \tan 22.2° + x \tan 22.2°$$
Distributive property **(Section R.1)**

$$x \tan 36.7° - x \tan 22.2° = 50 \tan 22.2°$$
Get x-terms on one side. **(Section 1.1)**

$$x(\tan 36.7° - \tan 22.2°) = 50 \tan 22.2°$$
Factor out x. **(Section R.4)**

$$x = \frac{50 \tan 22.2°}{\tan 36.7° - \tan 22.2°}$$
Divide by the coefficient of x.

We saw above that $h = x \tan 36.7°$. Substituting for x,

$$h = \left(\frac{50 \tan 22.2°}{\tan 36.7° - \tan 22.2°}\right) \tan 36.7°.$$

From a calculator,

$$\tan 36.7° = .74537703 \quad \text{and} \quad \tan 22.2° = .40809244$$

so $\tan 36.7° - \tan 22.2° = .74537703 - .40809244 = .33728459$

and $$h = \left(\frac{50(.40809244)}{.33728459}\right).74537703 \approx 45.$$

The height of the tree is approximately 45 ft.

Graphing Calculator Solution*

In Figure 50, we superimposed Figure 49 on coordinate axes with the origin at D. By definition, the tangent of the angle between the x-axis and the graph of a line with equation $y = mx + b$ is the slope of the line, m. For line DB, $m = \tan 22.2°$. Since b equals 0, the equation of line DB is $y_1 = (\tan 22.2°)x$. The equation of line AB is $y_2 = (\tan 36.7°)x + b$. Since $b \neq 0$ here, we use the point $A(50, 0)$ and the point-slope form to find the equation.

$$y_2 - y_1 = m(x - x_1) \quad \textbf{(Section 2.4)}$$
$$y_2 - 0 = m(x - 50) \quad x_1 = 50, y_1 = 0$$
$$y_2 = \tan 36.7°(x - 50)$$

Lines y_1 and y_2 are graphed in Figure 51. The y-coordinate of the point of intersection of the graphs gives the length of BC, or h. Thus, $h \approx 45$.

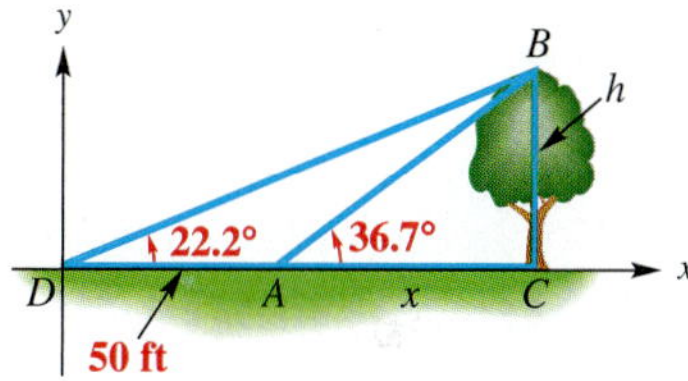

Figure 50

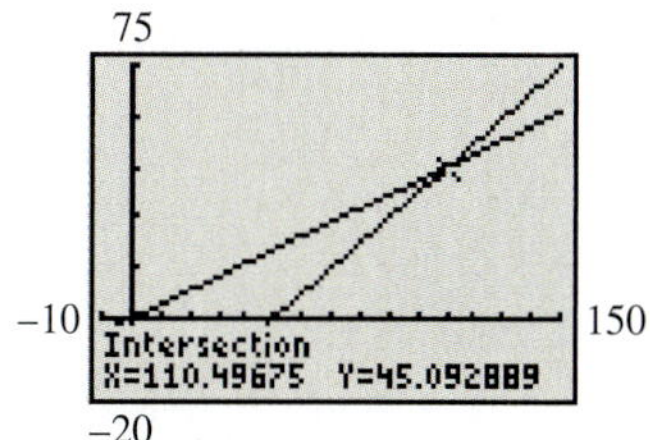

Figure 51

Now try Exercise 53.

NOTE In practice, we usually do not write down intermediate calculator approximation steps. We did in Example 7 so you could follow the steps more easily.

*Source: Adapted with permission from "Letter to the Editor," by Robert Ruzich (*Mathematics Teacher*, Volume 88, Number 1). Copyright © 1995 by the National Council of Teachers of Mathematics.

5.4 Exercises

Concept Check *Refer to the discussion of accuracy and significant digits in this section to work Exercises 1–8.*

1. ***Leading NFL Receiver*** At the end of the 2001 National Football League season, Oakland Raider Jerry Rice was the leading career receiver with 20,386 yd. State the range represented by this number. (*Source: The World Almanac and Book of Facts,* 2003.)

2. ***Height of Mt. Everest*** When Mt. Everest was first surveyed, the surveyors obtained a height of 29,000 ft to the nearest foot. State the range represented by this number. (The surveyors thought no one would believe a measurement of 29,000 ft, so they reported it as 29,002.) (*Source:* Dunham, W., *The Mathematical Universe,* John Wiley & Sons, 1994.)

3. ***Longest Vehicular Tunnel*** The E. Johnson Memorial Tunnel in Colorado, which measures 8959 ft, is one of the longest land vehicular tunnels in the United States. What is the range of this number? (*Source: The World Almanac and Book of Facts,* 2003.)

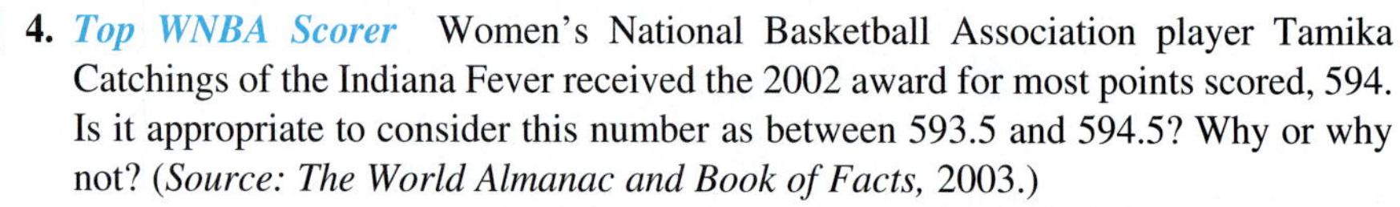

4. ***Top WNBA Scorer*** Women's National Basketball Association player Tamika Catchings of the Indiana Fever received the 2002 award for most points scored, 594. Is it appropriate to consider this number as between 593.5 and 594.5? Why or why not? (*Source: The World Almanac and Book of Facts,* 2003.)

5. ***Circumference of a Circle*** The formula for the circumference of a circle is $C = 2\pi r$. Suppose you use the $\boxed{\pi}$ key on your calculator to find the circumference of a circle with radius 54.98 cm, getting 345.44953. Since 2 has only one significant digit, the answer should be given as 3×10^2, or 300 cm. Is this conclusion correct? If not, explain how the answer should be given.

6. Explain the difference between a measurement of 23.0 ft and a measurement of 23.00 ft.

7. If h is the actual height of a building and the height is measured as 58.6 ft, then $|h - 58.6| \leq$ _____.

8. If w is the actual weight of a car and the weight is measured as 15.00×10^2 lb, then $|w - 1500| \leq$ _____.

Solve each right triangle. See Example 1.

9.
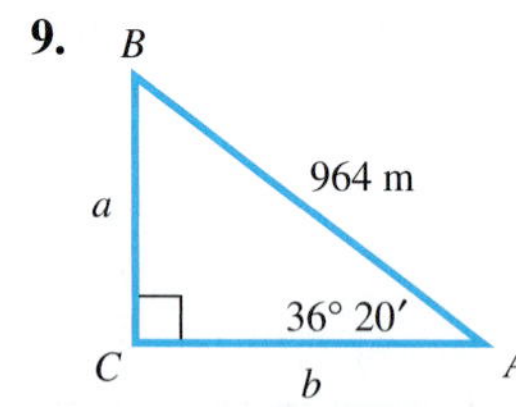

10.
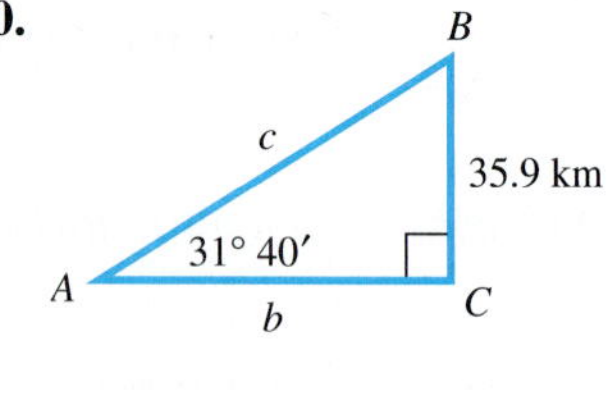

11.
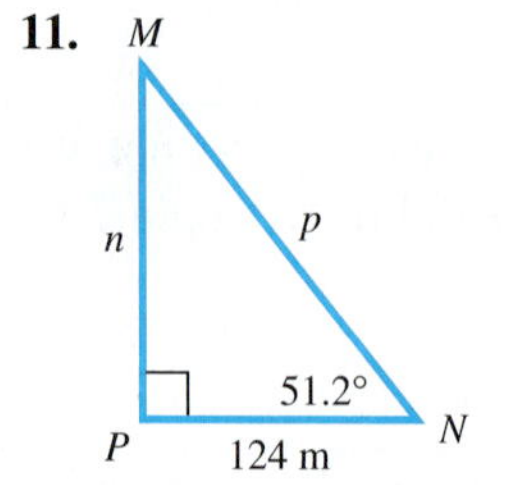

12.
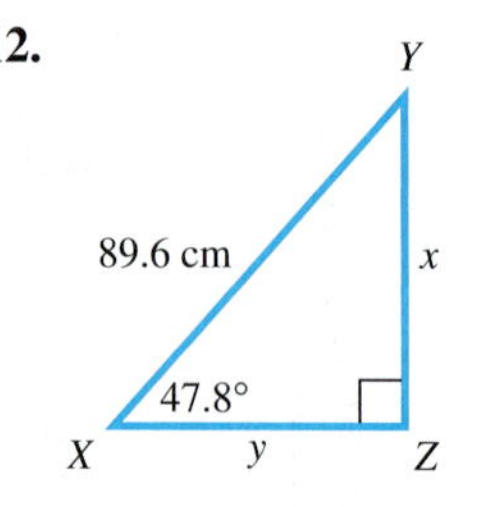

13. Can a right triangle be solved if we are given measures of its two acute angles and no side lengths? Explain.

14. *Concept Check* If we are given an acute angle and a side in a right triangle, what unknown part of the triangle requires the least work to find?

15. Explain why you can always solve a right triangle if you know the measures of one side and one acute angle.

16. Explain why you can always solve a right triangle if you know the lengths of two sides.

Solve each right triangle. In each case, $C = 90°$. If angle information is given in degrees and minutes, give answers in the same way. If given in decimal degrees, do likewise in answers. When two sides are given, give angles in degrees and minutes. See Examples 1 and 2.

17. $A = 28.00°$, $c = 17.4$ ft
18. $B = 46.00°$, $c = 29.7$ m
19. $B = 73.00°$, $b = 128$ in.
20. $A = 61° \, 00'$, $b = 39.2$ cm
21. $a = 76.4$ yd, $b = 39.3$ yd
22. $a = 958$ m, $b = 489$ m
23. *Concept Check* When is an angle of elevation equal to 90°?
24. *Concept Check* Can an angle of elevation be more than 90°?
25. Explain why the angle of depression *DAB* has the same measure as the angle of elevation *ABC* in the figure.

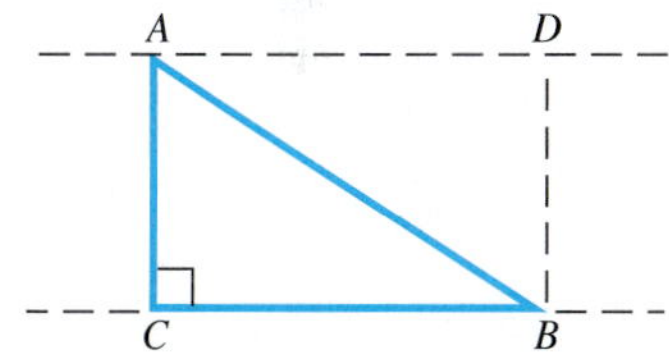

AD is parallel to *BC*.

26. Why is angle *CAB* *not* an angle of depression in the figure for Exercise 25?

27. *Concept Check* When bearing is given as a single angle measure, how is the angle represented in a sketch?

28. *Concept Check* When bearing is given as N (or S), the angle measure, then E (or W), how is the angle represented in a sketch?

Solve each problem involving triangles.

29. *Height of a Ladder on a Wall* A 13.5-m fire truck ladder is leaning against a wall. Find the distance *d* the ladder goes up the wall (above the top of the fire truck) if the ladder makes an angle of 43° 50′ with the horizontal.

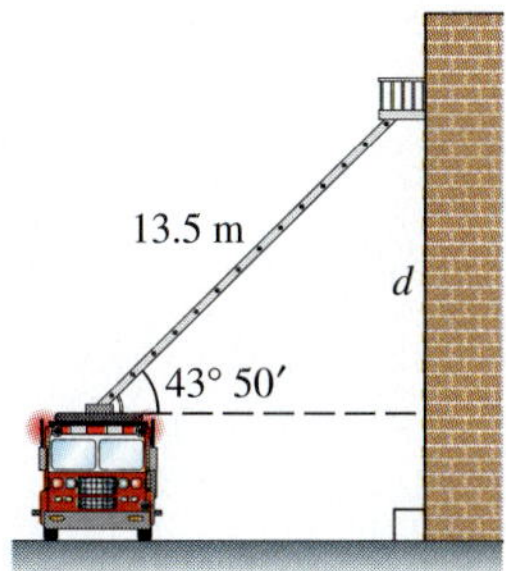

30. *Distance Across a Lake* To find the distance *RS* across a lake, a surveyor lays off $RT = 53.1$ m, with angle $T = 32° \, 10'$ and angle $S = 57° \, 50'$. Find length *RS*.

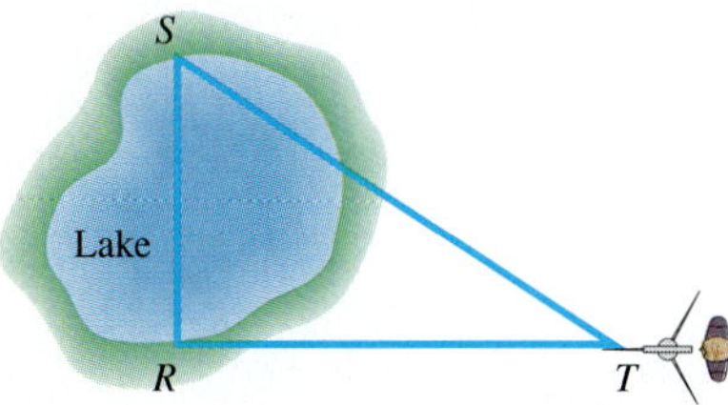

31. *Height of a Building* From a window 30 ft above the street, the angle of elevation to the top of the building across the street is 50.0° and the angle of depression to the base of this building is 20.0°. Find the height of the building across the street.

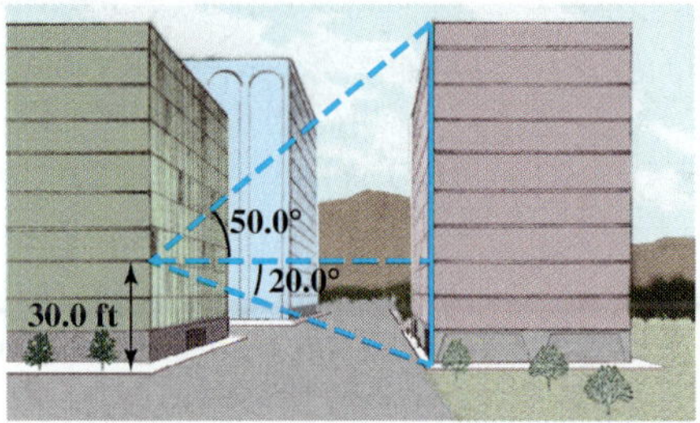

32. *Diameter of the Sun* To determine the diameter of the sun, an astronomer might sight with a *transit* (a device used by surveyors for measuring angles) first to one edge of the sun and then to the other, finding that the included angle equals 1° 4′. Assuming that the distance from Earth to the sun is 92,919,800 mi, calculate the diameter of the sun.

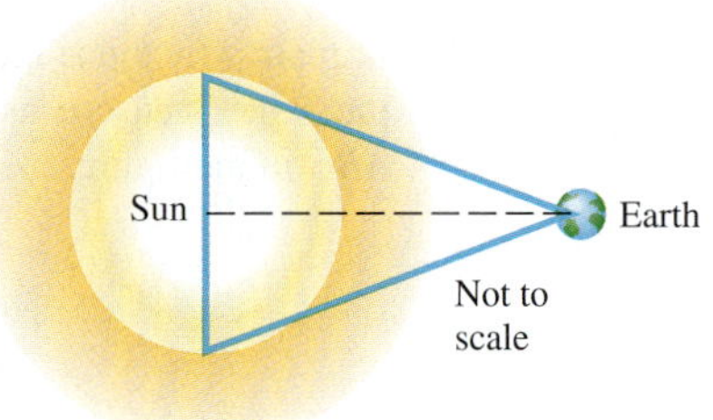

33. *Side Lengths of a Triangle* The length of the base of an isosceles triangle is 42.36 in. Each base angle is 38.12°. Find the length of each of the two equal sides of the triangle. (*Hint:* Divide the triangle into two right triangles.)

34. *Altitude of a Triangle* Find the altitude of an isosceles triangle having base 184.2 cm if the angle opposite the base is 68° 44′.

Solve each problem involving an angle of elevation or depression. See Example 3.

35. *Angle of Elevation of Pyramid of the Sun* The Pyramid of the Sun in the ancient Mexican city of Teotihuacan was the largest and most important structure in the city. The base is a square with sides 700 ft long, and the height of the pyramid is 200 ft. Find the angle of elevation of the edge indicated in the figure to two significant digits. (*Hint:* The base of the triangle in the figure is half the diagonal of the square base of the pyramid.)

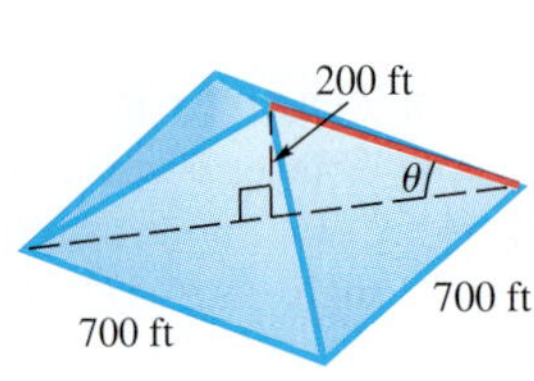

36. *Cloud Ceiling* The U.S. Weather Bureau defines a *cloud ceiling* as the altitude of the lowest clouds that cover more than half the sky. To determine a cloud ceiling, a powerful searchlight projects a circle of light vertically on the bottom of the cloud. An observer sights the circle of light in the crosshairs of a tube called a *clinometer*. A pendant hanging vertically from the tube and resting on a protractor gives the angle of elevation. Find the cloud ceiling if the searchlight is located 1000 ft from the observer and the angle of elevation is 30.0° as measured with a clinometer at eye-height 6 ft. (Assume three significant digits.)

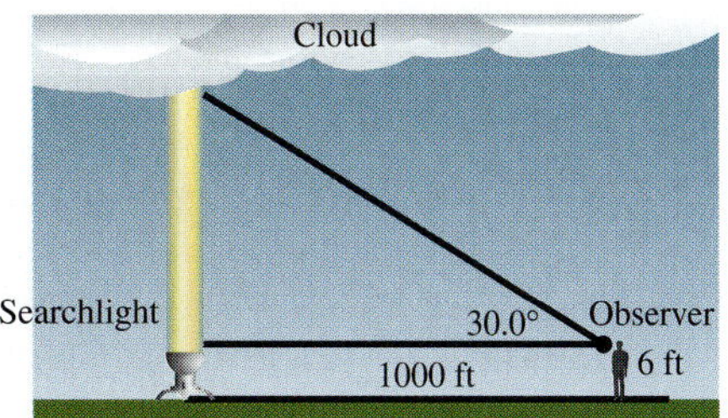

37. *Height of a Tower* The shadow of a vertical tower is 40.6 m long when the angle of elevation of the sun is 34.6°. Find the height of the tower.

38. *Distance from the Ground to the Top of a Building* The angle of depression from the top of a building to a point on the ground is 32° 30′. How far is the point on the ground from the top of the building if the building is 252 m high?

39. *Length of a Shadow* Suppose that the angle of elevation of the sun is 23.4°. Find the length of the shadow cast by Diane Carr, who is 5.75 ft tall.

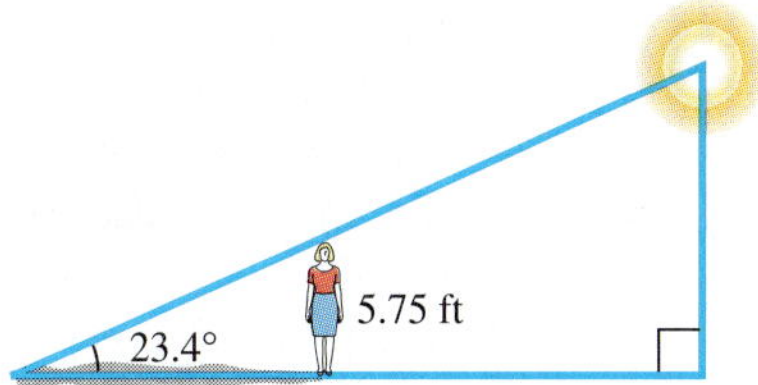

40. *Airplane Distance* An airplane is flying 10,500 ft above the level ground. The angle of depression from the plane to the base of a tree is 13° 50′. How far horizontally must the plane fly to be directly over the tree?

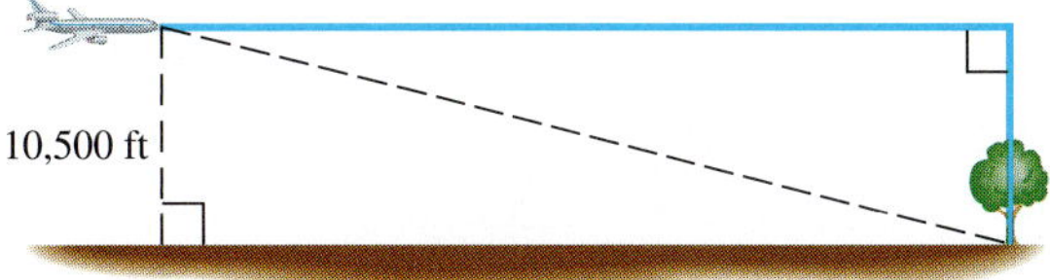

41. *Height of a Building* The angle of elevation from the top of a small building to the top of a nearby taller building is 46° 40′, while the angle of depression to the bottom is 14° 10′. If the smaller building is 28.0 m high, find the height of the taller building.

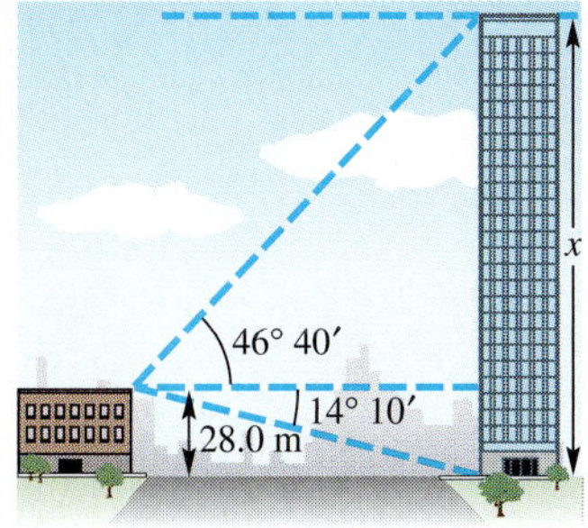

42. *Angle of Depression of a Light* A company safety committee has recommended that a floodlight be mounted in a parking lot so as to illuminate the employee exit. Find the angle of depression of the light.

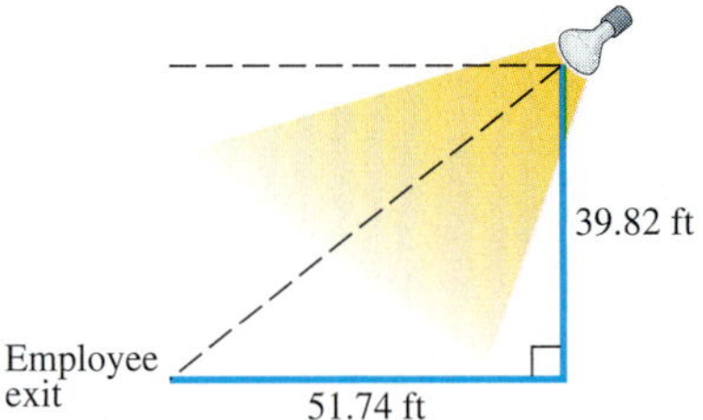

43. *Height of Mt. Everest* The highest mountain peak in the world is Mt. Everest, located in the Himalayas. The height of this enormous mountain was determined in 1856 by surveyors using trigonometry long before it was first climbed in 1953. This difficult measurement had to be done from a great distance. At an altitude of 14,545 ft on a different mountain, the straight line distance to the peak of Mt. Everest is 27.0134 mi and its angle of elevation is $\theta = 5.82°$. (*Source:* Dunham, W., *The Mathematical Universe,* John Wiley & Sons, 1994.)

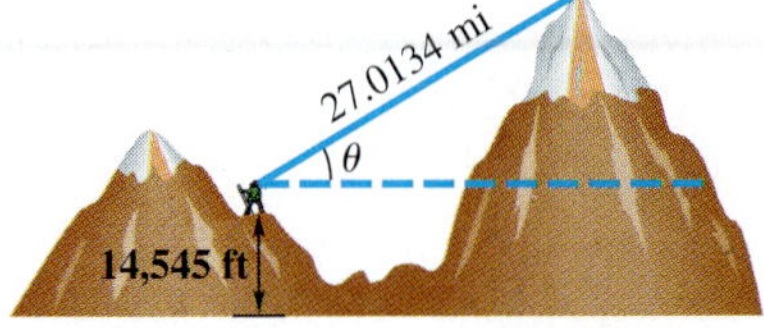

(a) Approximate the height (in feet) of Mt. Everest.

(b) In the actual measurement, Mt. Everest was over 100 mi away and the curvature of Earth had to be taken into account. Would the curvature of Earth make the peak appear taller or shorter than it actually is?

44. *Error in Measurement* A degree may seem like a very small unit, but an error of one degree in measuring an angle may be very significant. For example, suppose a laser beam directed toward the visible center of the moon misses its assigned target by 30 sec. How far is it (in miles) from its assigned target? Take the distance from the surface of Earth to that of the moon to be 234,000 mi. (*Source: A Sourcebook of Applications of School Mathematics* by Donald Bushaw et al. Copyright © 1980 by The Mathematical Association of America.)

Work each problem. Assume the course of a plane or ship is on the indicated bearing. See Examples 4 and 5.

45. *Distance Flown by a Plane* A plane flies 1.3 hr at 110 mph on a bearing of 40°. It then turns and flies 1.5 hr at the same speed on a bearing of 130°. How far is the plane from its starting point?

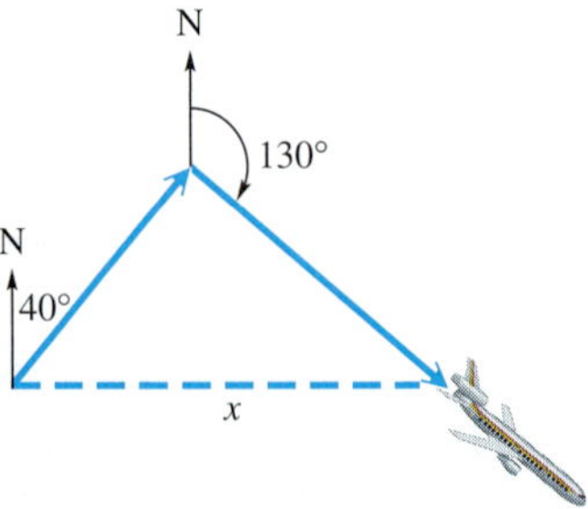

46. *Distance Traveled by a Ship* A ship travels 50 km on a bearing of 27°, then travels on a bearing of 117° for 140 km. Find the distance traveled from the starting point to the ending point.

47. *Distance Between Two Ships* Two ships leave a port at the same time. The first ship sails on a bearing of 40° at 18 knots (nautical miles per hour) and the second at a bearing of 130° at 26 knots. How far apart are they after 1.5 hr?

48. *Distance Between Two Cities* The bearing from Winston-Salem, North Carolina, to Danville, Virginia, is N 42° E. The bearing from Danville to Goldsboro, North Carolina, is S 48° E. A car driven by Mark Ferrari, traveling at 60 mph, takes 1 hr to go from Winston-Salem to Danville and 1.8 hr to go from Danville to Goldsboro. Find the distance from Winston-Salem to Goldsboro.

49. *Distance Between Two Cities* The bearing from Atlanta to Macon is S 27° E, and the bearing from Macon to Augusta is N 63° E. An automobile traveling at 60 mph needs $1\frac{1}{4}$ hr to go from Atlanta to Macon and $1\frac{3}{4}$ hr to go from Macon to Augusta. Find the distance from Atlanta to Augusta.

50. *Distance Between Two Ships* A cruise ship leaves port and sails on a bearing of N 28° 10′ E. Another ship leaves the same port at the same time and sails on a bearing of S 61° 50′ E. If the first ship sails at 24.0 mph and the second sails at 28.0 mph, find the distance between the two ships after 4 hr.

51. *Distance Between Transmitters* Radio direction finders are set up at two points A and B, which are 2.50 mi apart on an east-west line. From A, it is found that the bearing of a signal from a radio transmitter is N 36° 20′ E, while from B the bearing of the same signal is N 53° 40′ W. Find the distance of the transmitter from B.

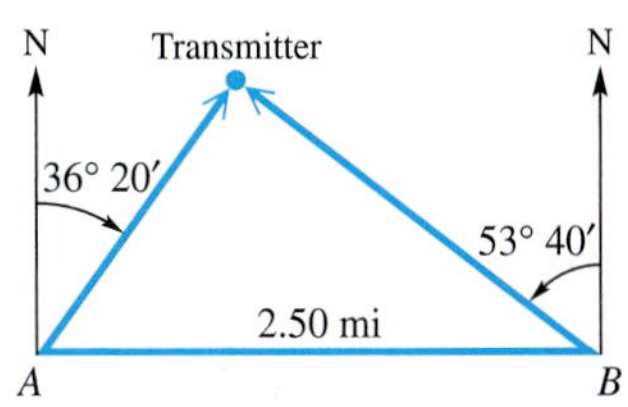

In Exercises 52–55, use the method of Example 7. Drawing a sketch for the problems where one is not given may be helpful.

52. *Height of a Pyramid* The angle of elevation from a point on the ground to the top of a pyramid is 35° 30′. The angle of elevation from a point 135 ft farther back to the top of the pyramid is 21° 10′. Find the height of the pyramid.

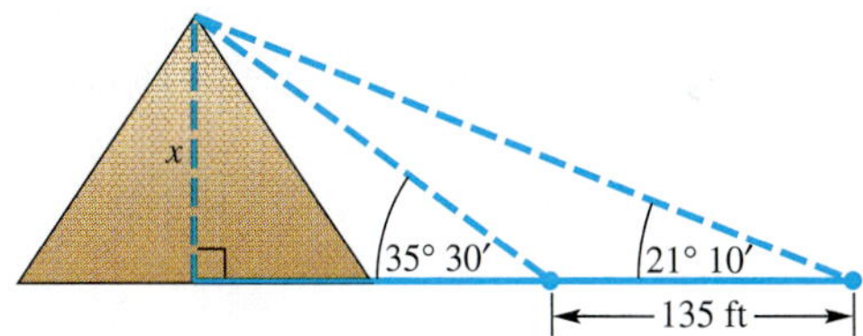

53. *Distance Between a Whale and a Lighthouse* Debbie Glockner-Ferrari, a whale researcher, is watching a whale approach directly toward her as she observes from the top of a lighthouse. When she first begins watching the whale, the angle of depression of the whale is 15° 50′. Just as the whale turns away from the lighthouse, the angle of depression is 35° 40′. If the height of the lighthouse is 68.7 m, find the distance traveled by the whale as it approaches the lighthouse.

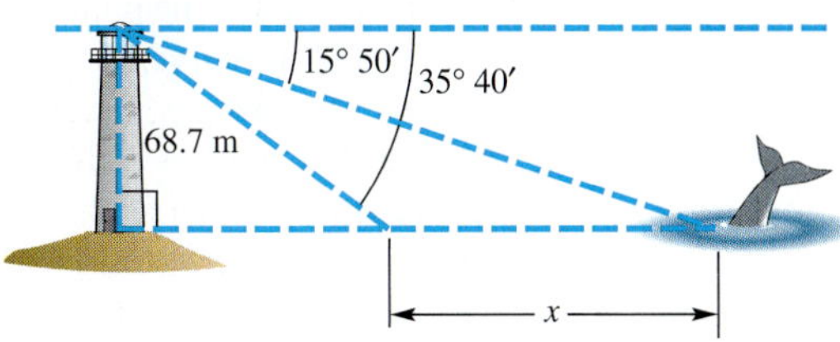

54. *Height of an Antenna* A scanner antenna is on top of the center of a house. The angle of elevation from a point 28.0 m from the center of the house to the top of the antenna is 27° 10′, and the angle of elevation to the bottom of the antenna is 18° 10′. Find the height of the antenna.

55. *Height of Mt. Whitney* The angle of elevation from Lone Pine to the top of Mt. Whitney is $10° 50'$. Van Dong Le, traveling 7.00 km from Lone Pine along a straight, level road toward Mt. Whitney, finds the angle of elevation to be $22° 40'$. Find the height of the top of Mt. Whitney above the level of the road.

Solve each problem.

56. *Height of a Plane Above Earth* Find the minimum height h above the surface of Earth so that a pilot at point A in the figure can see an object on the horizon at C, 125 mi away. Assume that the radius of Earth is 4.00×10^3 mi.

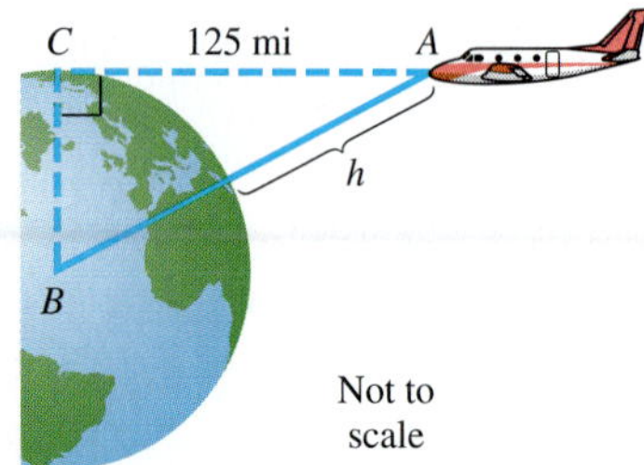

57. *Distance of a Plant from a Fence* In one area, the lowest angle of elevation of the sun in winter is $23° 20'$. Find the minimum distance x that a plant needing full sun can be placed from a fence 4.65 ft high.

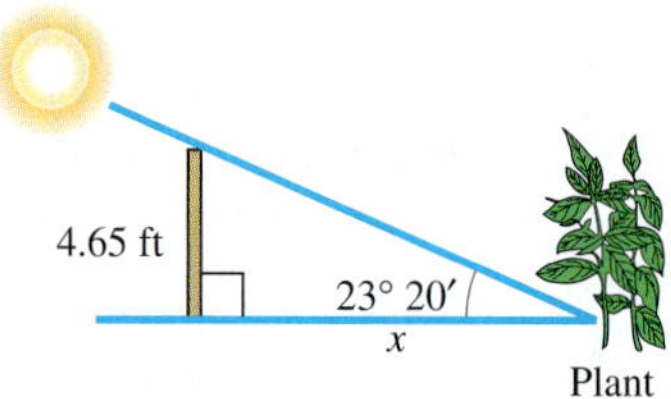

58. *Distance Through a Tunnel* A tunnel is to be dug from A to B. Both A and B are visible from C. If AC is 1.4923 mi and BC is 1.0837 mi, and if C is 90°, find the measures of angles A and B.

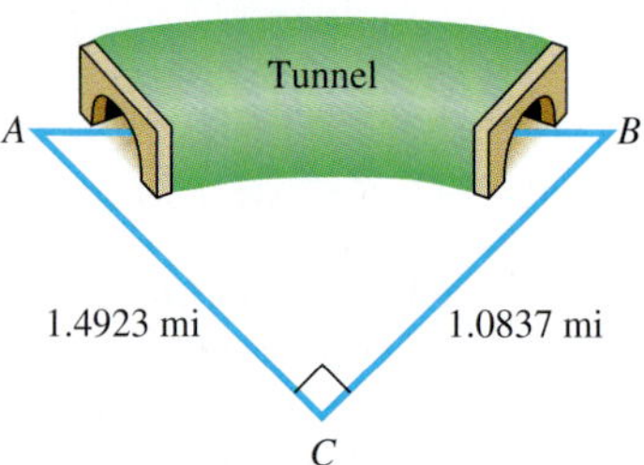

59. *(Modeling) Highway Curves* A basic highway curve connecting two straight sections of road is often circular. In the figure, the points P and S mark the beginning and end of the curve. Let Q be the point of intersection where the two straight sections of highway leading into the curve would meet if extended. The radius of the curve is R, and the central angle θ denotes how many degrees the curve turns. (*Source:* Mannering, F. and W. Kilareski, *Principles of Highway Engineering and Traffic Analysis,* 2nd Edition, John Wiley & Sons, 1998.)

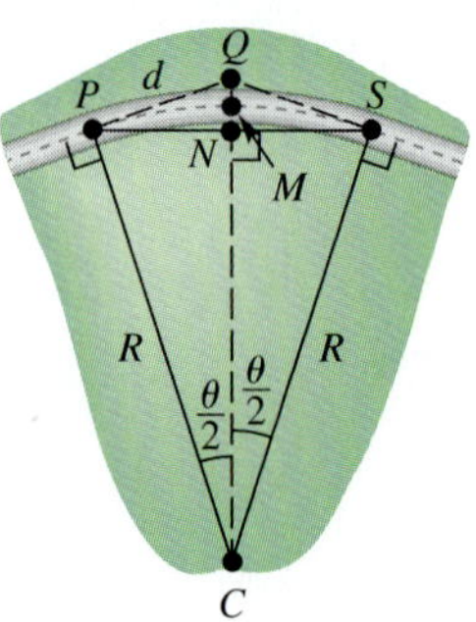

(a) If $R = 965$ ft and $\theta = 37°$, find the distance d between P and Q.

(b) Find an expression in terms of R and θ for the distance between points M and N.

60. *Length of a Side of a Piece of Land* A piece of land has the shape shown in the figure. Find x.

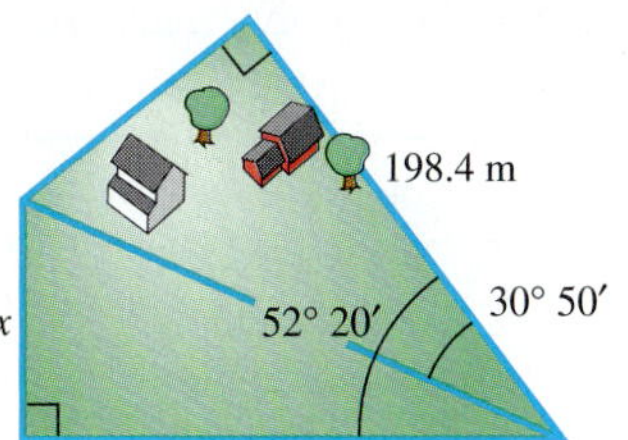

61. *(Modeling) Stopping Distance on a Curve* Refer to Exercise 59. When an automobile travels along a circular curve, objects like trees and buildings situated on the inside of the curve can obstruct a driver's vision. These obstructions prevent the driver from seeing sufficiently far down the highway to ensure a safe stopping distance. In the figure, the *minimum* distance d that should be cleared on the inside of the highway is modeled by the equation

$$d = R\left(1 - \cos\frac{\theta}{2}\right).$$

(*Source:* Mannering, F. and W. Kilareski, *Principles of Highway Engineering and Traffic Analysis,* 2nd Edition, John Wiley & Sons, 1998.)

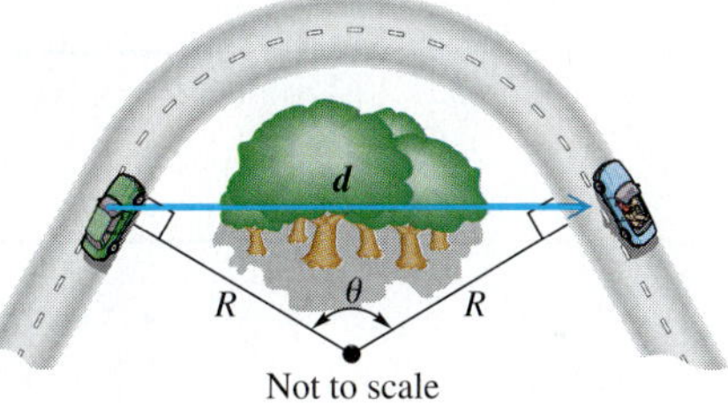

(a) It can be shown that if θ is measured in degrees, then $\theta \approx \frac{57.3S}{R}$, where S is safe stopping distance for the given speed limit. Compute d for a 55 mph speed limit if $S = 336$ ft and $R = 600$ ft.

(b) Compute d for a 65 mph speed limit if $S = 485$ ft and $R = 600$ ft.

(c) How does the speed limit affect the amount of land that should be cleared on the inside of the curve?

Chapter 5 Summary

KEY TERMS

5.1 line
line segment (or segment)
ray
angle
initial side
terminal side
vertex of an angle
positive angle
negative angle
degree
acute angle
right angle
obtuse angle
straight angle
complementary angles
supplementary angles
minute
second
angle in standard position
quadrantal angle
coterminal angles

5.2 sine
cosine
tangent
cotangent
secant
cosecant

5.3 side opposite
side adjacent
reference angle

5.4 significant digit
exact number
angle of elevation
angle of depression
bearing

NEW SYMBOLS

θ Greek letter theta
$^\circ$ degree
$'$ minute
$''$ second

QUICK REVIEW

CONCEPTS	EXAMPLES

5.1 Angles

Types of Angles

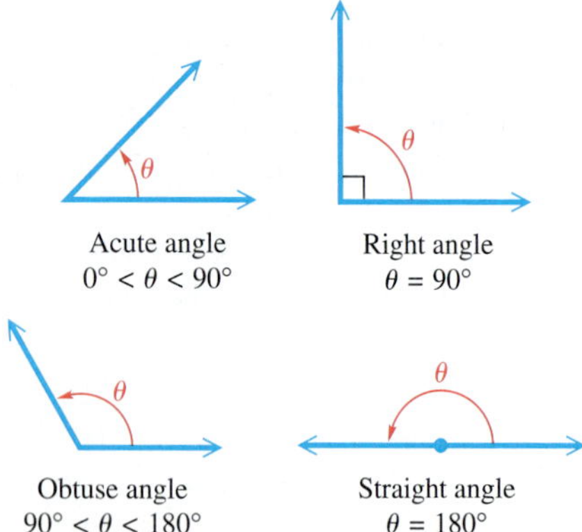

If $\theta = 32^\circ$, then angle θ is an acute angle.

If $\theta = 90^\circ$, then angle θ is a right angle.

If $\theta = 148^\circ$, then angle θ is an obtuse angle.

If $\theta = 180^\circ$, then angle θ is a straight angle.

5.2 Trigonometric Functions

Definitions of the Trigonometric Functions

Let (x, y) be a point other than the origin on the terminal side of an angle θ in standard position. Let $r = \sqrt{x^2 + y^2}$, the distance from the origin to (x, y). Then

$$\sin\theta = \frac{y}{r} \qquad \cos\theta = \frac{x}{r} \qquad \tan\theta = \frac{y}{x}\ (x \neq 0)$$

$$\csc\theta = \frac{r}{y}\ (y \neq 0) \quad \sec\theta = \frac{r}{x}\ (x \neq 0) \quad \cot\theta = \frac{x}{y}\ (y \neq 0).$$

If a point $(-2, 3)$ is on the terminal side of angle θ in standard position, then $x = -2$, $y = 3$, $r = \sqrt{4 + 9} = \sqrt{13}$, and

$$\sin\theta = \frac{3\sqrt{13}}{13}, \qquad \cos\theta = -\frac{2\sqrt{13}}{13}, \qquad \tan\theta = -\frac{3}{2}$$

$$\csc\theta = \frac{\sqrt{13}}{3}, \qquad \sec\theta = -\frac{\sqrt{13}}{2}, \qquad \cot\theta = -\frac{2}{3}.$$

CONCEPTS	EXAMPLES

Reciprocal Identities

$$\sin\theta = \frac{1}{\csc\theta} \qquad \cos\theta = \frac{1}{\sec\theta} \qquad \tan\theta = \frac{1}{\cot\theta}$$

$$\csc\theta = \frac{1}{\sin\theta} \qquad \sec\theta = \frac{1}{\cos\theta} \qquad \cot\theta = \frac{1}{\tan\theta}$$

The preceding examples satisfy these identities.

Pythagorean Identities

$$\sin^2\theta + \cos^2\theta = 1 \qquad \tan^2\theta + 1 = \sec^2\theta$$

$$1 + \cot^2\theta = \csc^2\theta$$

Using the preceding trigonometric values,

$$\left(\frac{3\sqrt{13}}{13}\right)^2 + \left(-\frac{2\sqrt{13}}{13}\right)^2 = \frac{9}{13} + \frac{4}{13} = 1,$$

$$\left(-\frac{3}{2}\right)^2 + 1 = \frac{9}{4} + \frac{4}{4} = \frac{13}{4} = \left(-\frac{\sqrt{13}}{2}\right)^2,$$

$$1 + \left(-\frac{2}{3}\right)^2 = 1 + \frac{4}{9} = \frac{13}{9} = \left(\frac{\sqrt{13}}{3}\right)^2.$$

Quotient Identities

$$\frac{\sin\theta}{\cos\theta} = \tan\theta \qquad \frac{\cos\theta}{\sin\theta} = \cot\theta$$

From the preceding trigonometric values,

$$\frac{\sin\theta}{\cos\theta} = \frac{\frac{3\sqrt{13}}{13}}{-\frac{2\sqrt{13}}{13}} = \frac{3\sqrt{13}}{13}\left(-\frac{13}{2\sqrt{13}}\right) = -\frac{3}{2} = \tan\theta.$$

5.3 Evaluating Trigonometric Functions

Right-Triangle-Based Definitions of the Trigonometric Functions

For any acute angle A in standard position,

$$\sin A = \frac{y}{r} = \frac{\text{side opposite}}{\text{hypotenuse}} \qquad \csc A = \frac{r}{y} = \frac{\text{hypotenuse}}{\text{side opposite}}$$

$$\cos A = \frac{x}{r} = \frac{\text{side adjacent}}{\text{hypotenuse}} \qquad \sec A = \frac{r}{x} = \frac{\text{hypotenuse}}{\text{side adjacent}}$$

$$\tan A = \frac{y}{x} = \frac{\text{side opposite}}{\text{side adjacent}} \qquad \cot A = \frac{x}{y} = \frac{\text{side adjacent}}{\text{side opposite}}.$$

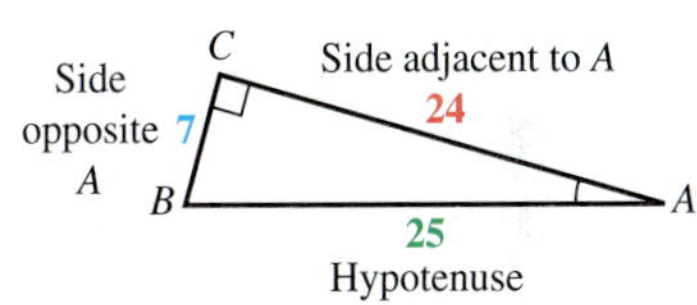

$$\sin A = \frac{7}{25} \qquad \cos A = \frac{24}{25} \qquad \tan A = \frac{7}{24}$$

$$\csc A = \frac{25}{7} \qquad \sec A = \frac{25}{24} \qquad \cot A = \frac{24}{7}$$

Function Values of Special Angles

θ	$\sin\theta$	$\cos\theta$	$\tan\theta$	$\cot\theta$	$\sec\theta$	$\csc\theta$
30°	$\frac{1}{2}$	$\frac{\sqrt{3}}{2}$	$\frac{\sqrt{3}}{3}$	$\sqrt{3}$	$\frac{2\sqrt{3}}{3}$	2
45°	$\frac{\sqrt{2}}{2}$	$\frac{\sqrt{2}}{2}$	1	1	$\sqrt{2}$	$\sqrt{2}$
60°	$\frac{\sqrt{3}}{2}$	$\frac{1}{2}$	$\sqrt{3}$	$\frac{\sqrt{3}}{3}$	2	$\frac{2\sqrt{3}}{3}$

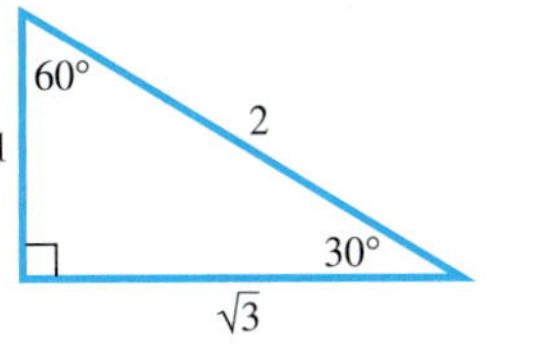

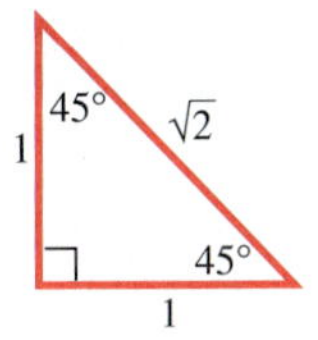

(continued)

CONCEPTS	EXAMPLES

Reference Angle θ' for θ in $(0°, 360°)$

θ in Quadrant	I	II	III	IV
θ' is	θ	$180° - \theta$	$\theta - 180°$	$360° - \theta$

Quadrant I: For $\theta = 25°$, $\theta' = 25°$
Quadrant II: For $\theta = 152°$, $\theta' = 28°$
Quadrant III: For $\theta = 200°$, $\theta' = 20°$
Quadrant IV: For $\theta = 320°$, $\theta' = 40°$

Finding Trigonometric Function Values for Any Angle

Step 1 Add or subtract 360° as many times as needed to get an angle greater than 0° but less than 360°.
Step 2 Find the reference angle θ'.
Step 3 Find the trigonometric function values for θ'.
Step 4 Determine the correct signs for the values found in Step 3.

Find sin 1050°.

$$1050° - 2(360°) = 330°$$

Thus, $\theta' = 30°$.

$$\sin 1050° = -\sin 30° = -\frac{1}{2}$$

5.4 Solving Right Triangles

Solving an Applied Trigonometry Problem

Step 1 Draw a sketch, and label it with the given information. Label the quantity to be found with a variable.
Step 2 Use the sketch to write an equation relating the given quantities to the variable.
Step 3 Solve the equation, and check that your answer makes sense.

Figures 44 and 46 on pages 512 and 513 illustrate the two methods for expressing bearing.

Find the angle of elevation of the sun if a 48.6-ft flagpole casts a shadow 63.1 ft long.

Step 1 See the sketch. We must find θ.

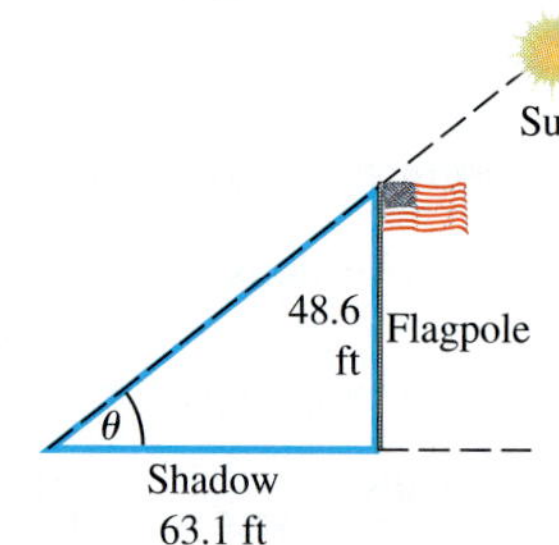

Step 2 $\tan \theta = \dfrac{48.6}{63.1} \approx .770206$

Step 3 $\theta = \tan^{-1} .770206 \approx 37.6°$

The angle of elevation is 37.6°.

Chapter 5 Review Exercises

1. Give the complement and the supplement of an angle of 35°.
2. Convert 32.25° to degrees, minutes, and seconds.
3. Convert 59° 35′ 30″ to decimal degrees.
4. Find the angle of smallest possible positive measure coterminal with each angle.
   (a) $-174°$ **(b)** 560°
5. Let n represent any integer, and write an expression for all angles coterminal with an angle of 270°.

Find the six trigonometric function values for each angle. If a value is undefined, say so.

6.

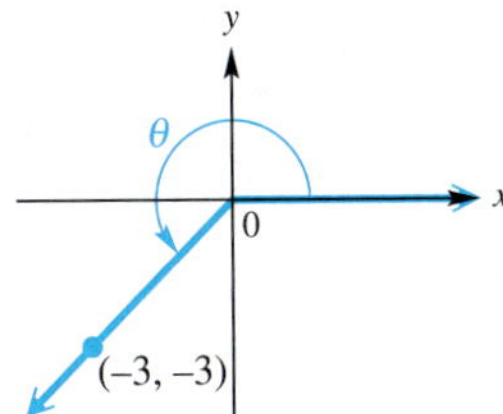

7.

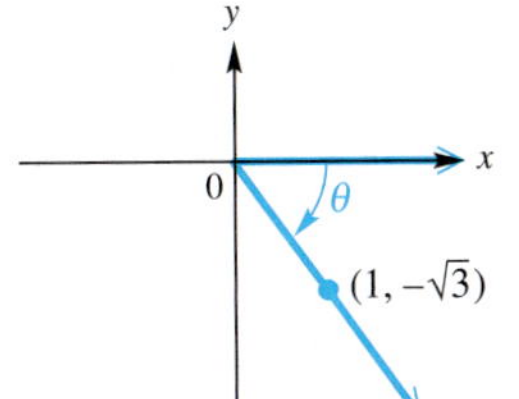

8. 180°

Find the values of the six trigonometric functions for an angle in standard position having each point on its terminal side.

9. $(3, -4)$
10. $(9, -2)$
11. $\left(-2\sqrt{2}, 2\sqrt{2}\right)$
12. *Concept Check* If the terminal side of a quadrantal angle lies along the y-axis, which of its trigonometric functions are undefined?

In Exercises 13 and 14, consider an angle θ in standard position whose terminal side has the equation $y = -5x$, with $x \leq 0$.

13. Sketch θ and use an arrow to show the rotation if $0° \leq \theta < 360°$.
14. Find the exact values of $\sin \theta$ and $\cos \theta$.
15. Decide whether each statement is *possible* or *impossible.*
    (a) $\sec \theta = -\dfrac{2}{3}$ **(b)** $\tan \theta = 1.4$ **(c)** $\csc \theta = 5$

Find all six trigonometric function values for each angle. Rationalize denominators when applicable.

16. $\sin \theta = \dfrac{\sqrt{3}}{5}$ and $\cos \theta < 0$
17. $\cos \theta = -\dfrac{5}{8}$, with θ in quadrant III
18. *Concept Check* If, for some particular angle θ, $\sin \theta < 0$ and $\cos \theta > 0$, in what quadrant must θ lie? What is the sign of $\tan \theta$?
19. Explain how you would find the cotangent of an angle θ whose tangent is 1.6778490 using a calculator. Then find $\cot \theta$.

Find the values of the six trigonometric functions for each angle A.

20.

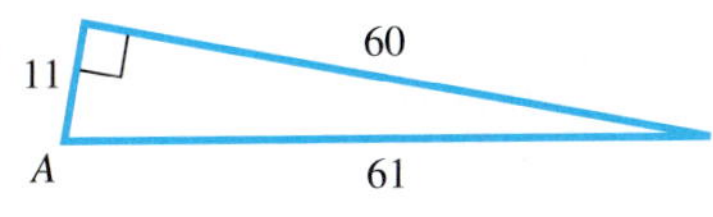

21.

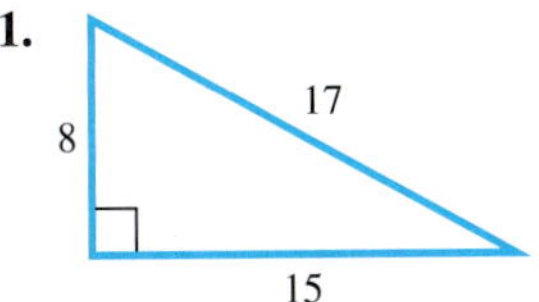

22. Explain why, in the figure, the cosine of angle A is equal to the sine of angle B.

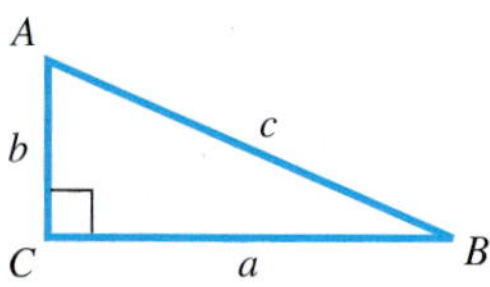

23. Complete the following table of exact function values.

θ	$\sin\theta$	$\cos\theta$	$\tan\theta$	$\cot\theta$	$\sec\theta$	$\csc\theta$
30°						
45°						
60°						

24. Give the reference angle for each angle measure.

(a) 100° **(b)** −365°

Find exact values of the six trigonometric functions for each angle. Do not use a calculator. Rationalize denominators when applicable.

25. 300° **26.** −225° **27.** −390°

Use a calculator to find each value.

28. $\sin 72° 30'$ **29.** $\sec 222° 30'$ **30.** $\tan(-100°)$ **31.** $\tan 11.7689°$

32. Find all values of θ, if θ is in the interval $[0°, 360°)$ and $\tan\theta = -\frac{\sqrt{3}}{3}$.

33. *Concept Check* Which one of the following cannot be *exactly* determined using the methods of this chapter?

A. $\cos 135°$ **B.** $\cot(-45°)$ **C.** $\sin 300°$ **D.** $\tan 140°$

Use a calculator to find each value of θ, where θ is in the interval $[0°, 90°)$. Give answers in decimal degrees.

34. $\sin\theta = .82584121$ **35.** $\cot\theta = 1.1249386$

36. Find two angles in the interval $[0, 360°)$ that satisfy $\sin\theta = .68163876$.

37. A student wants to use a calculator to find the value of $\cot 25°$. However, instead of entering $\frac{1}{\tan 25}$, he enters $\tan^{-1} 25$. Assuming the calculator is in degree mode, will this produce the correct answer? Explain.

38. For $\theta = 1997°$, use a calculator to find $\cos\theta$ and $\sin\theta$. Use your results to decide in which quadrant the angle lies.

Solve each right triangle.

39.

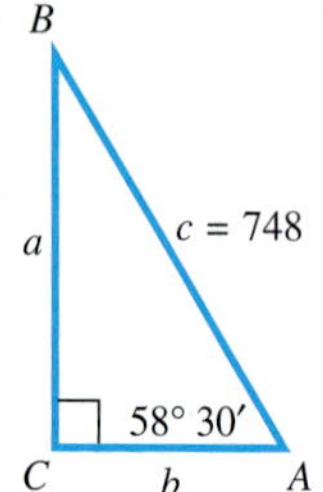

40. $A = 39.72°, b = 38.97$ m

Solve each problem.

41. ***Height of a Tower*** The angle of elevation from a point 93.2 ft from the base of a tower to the top of the tower is $38° 20'$. Find the height of the tower.

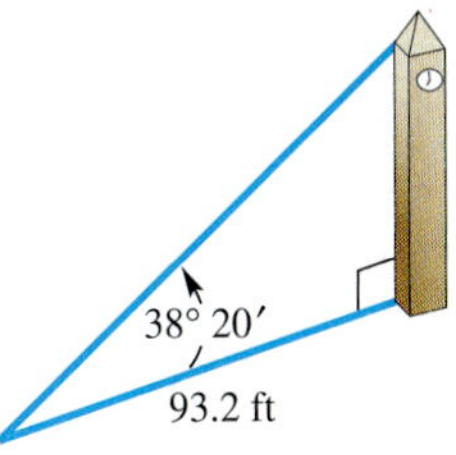

42. ***Length of a Shadow*** The length of a building's shadow is 48 ft when the angle of elevation of the sun is 35.3°. Find the height of the building.

43. ***Height of a Tower*** The angle of depression of a television tower to a point on the ground 36.0 m from the bottom of the tower is 29.5°. Find the height of the tower.

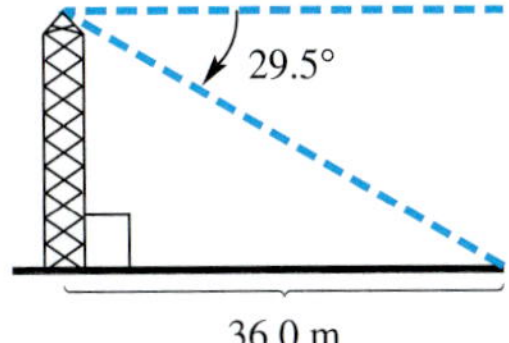

44. Find h as indicated in the figure.

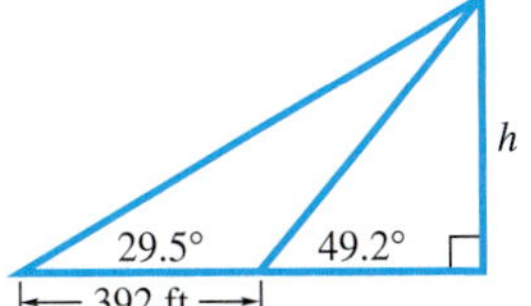

45. ***Height of Mt. Kilimanjaro*** From a point A the angle of elevation of Mt. Kilimanjaro in Africa is 13.7°, and from a point B directly behind A, the angle of elevation is 10.4°. If the distance between A and B is 5 mi, approximate the height of Mt. Kilimanjaro to the nearest hundred feet.

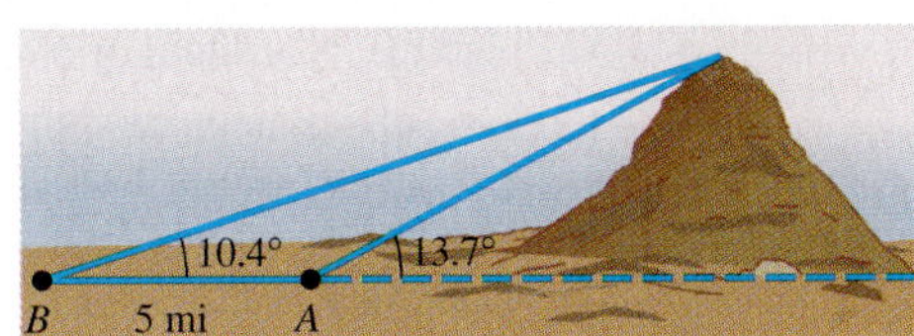

46. ***Distance Between Two Points*** The bearing of B from C is 254°. The bearing of A from C is 344°. The bearing of A from B is 32°. The distance from A to C is 780 m. Find the distance from A to B.

47. ***Distance a Ship Sails*** The bearing from point A to point B is S 55° E and from point B to point C is N 35° E. If a ship sails from A to B, a distance of 80 km, and then from B to C, a distance of 74 km, how far is it from A to C?

48. ***(Modeling) Height of a Satellite*** Artificial satellites that orbit Earth often use VHF signals to communicate with the ground. VHF signals travel in straight lines. The height h of the satellite above Earth and the time T that the satellite can communicate with a fixed location on the ground are related by the model

$$h = R\left(\frac{1}{\cos \frac{180T}{P}} - 1\right),$$

where $R = 3955$ mi is the radius of Earth and P is the period for the satellite to orbit Earth. (*Source:* Schlosser, W., T. Schmidt-Kaler, and E. Milone, *Challenges of Astronomy,* Springer-Verlag, 1991.)

(a) Find h when $T = 25$ min and $P = 140$ min. (Evaluate the cosine function in degree mode.)

(b) What is the value of h if T is increased to 30?

Chapter 5 Test

1. Find the angle of smallest positive measure coterminal with $-157°$.
2. Suppose θ is in the interval $(90°, 180°)$. Find the sign of each value.
   (a) $\cos \dfrac{\theta}{2}$ (b) $\cot(\theta + 180°)$
3. If $\cos \theta < 0$ and $\cot \theta > 0$, in what quadrant does θ lie?
4. If $(2, -5)$ is on the terminal side of an angle θ in standard position, find $\sin \theta$, $\cos \theta$, and $\tan \theta$.
5. What angle does the line $y = -\sqrt{3}x$ make with the positive x-axis?
6. If $\cos \theta = \frac{4}{5}$ and θ is in quadrant IV, find the values of the other trigonometric functions of θ.
7. Find the exact values of each part of the triangle labeled with a letter.

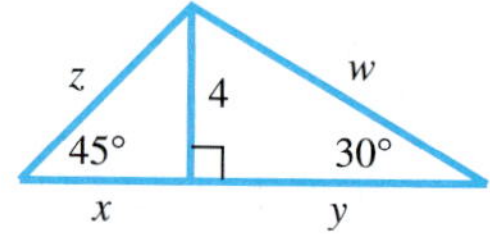

8. Find the exact value of $\cot(-750°)$.
9. Decide whether each statement is *true* or *false*.
   (a) $\cos 40° = 2 \cos 20°$ (b) $\sin 10° + \sin 10° = \sin 20°$
10. Use a calculator to approximate each value.
    (a) $\sin 78° 21'$ (b) $\tan 117.689°$ (c) $\sec 58.9041°$
11. Find the two values of θ to the nearest tenth in $[0°, 360°)$, if $\sin \theta = -\frac{2}{3}$.
12. Solve the triangle.

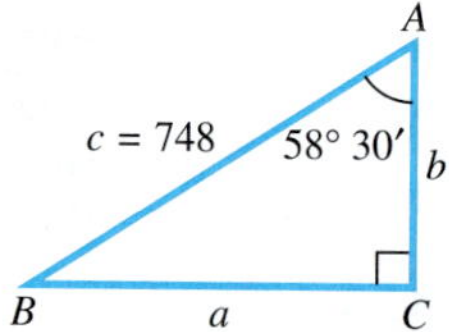

13. *Height of a Flagpole* To measure the height of a flagpole, Amado Carillo found that the angle of elevation from a point 24.7 ft from the base to the top is $32° 10'$. What is the height of the flagpole?
14. *Distance of a Ship from a Pier* A ship leaves a pier on a bearing of S 62° E and travels for 75 km. It then turns and continues on a bearing of N 28° E for 53 km. How far is the ship from the pier?
15. Find h as indicated in the figure.

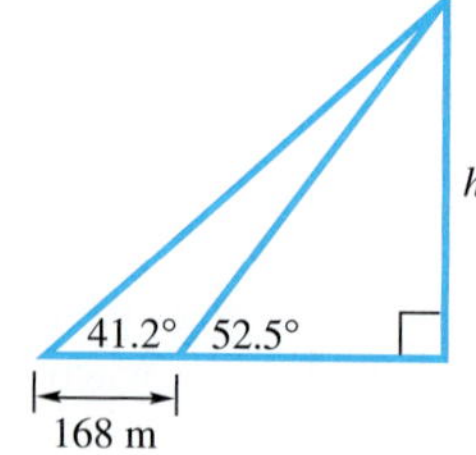

Chapter 5 Quantitative Reasoning

Can trigonometry be used to win an Olympic medal?

A shot-putter trying to improve performance may wonder: Is there an optimal angle to aim for, or is the velocity (speed) at which the ball is thrown more important? The figure shows the path of a steel ball thrown by a shot-putter. The distance D depends on initial velocity v, height h, and angle θ.

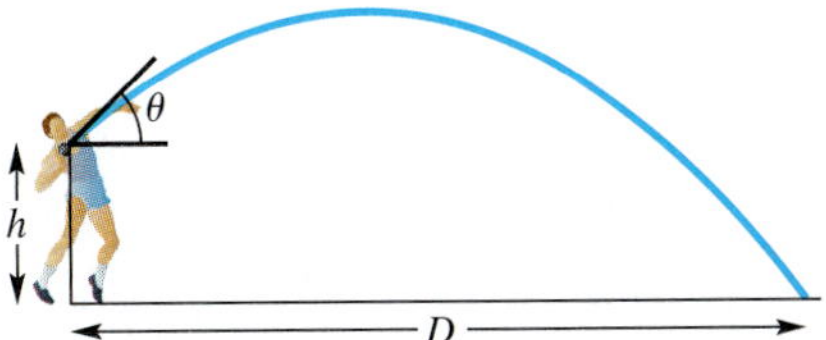

One model developed for this situation gives D as

$$D = \frac{v^2 \sin\theta\cos\theta + v\cos\theta\sqrt{(v\sin\theta)^2 + 64h}}{32}.$$

Typical ranges for the variables are v: 33–46 ft per sec; h: 6–8 ft; and θ: 40°–45°. (*Source:* Kreighbaum, E. and K. Barthels, *Biomechanics*, Allyn & Bacon, 1996.)

1. To see how angle θ affects distance D, let $v = 44$ ft per sec and $h = 7$ ft. Calculate D for $\theta = 40°$, 42°, and 45°. How does distance D change as θ increases?
2. To see how velocity v affects distance D, let $h = 7$ and $\theta = 42°$. Calculate D for $v = 43$, 44, and 45 ft per sec. How does distance D change as v increases?
3. Which affects distance D more, v or θ? What should the shot-putter do to improve performance?

6 The Circular Functions and Their Graphs

In August 2003, the planet Mars passed closer to Earth than it had in almost 60,000 years. Like Earth, Mars rotates on its axis and thus has days and nights. The photos here were taken by the Hubble telescope and show two nearly opposite sides of Mars. (*Source:* www.hubblesite.org) In Exercise 92 of Section 6.2, we examine the length of a Martian day.

Phenomena such as rotation of a planet on its axis, high and low tides, and changing of the seasons of the year are modeled by *periodic functions.* In this chapter, we see how the trigonometric functions of the previous chapter, introduced there in the context of ratios of the sides of a right triangle, can also be viewed from the perspective of motion around a circle.

6.1 Radian Measure

Radian Measure ▪ Converting Between Degrees and Radians ▪ Arc Length of a Circle ▪ Area of a Sector of a Circle

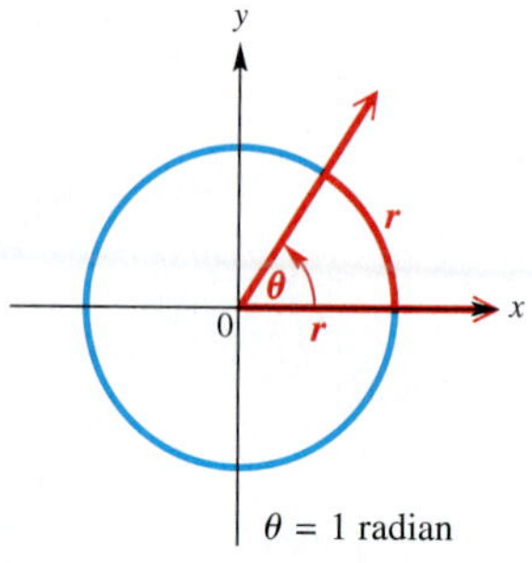

Figure 1

Radian Measure In most applications of trigonometry, angles are measured in degrees. In more advanced work in mathematics, *radian measure* of angles is preferred. Radian measure allows us to treat the trigonometric functions as functions with domains of *real numbers,* rather than angles.

Figure 1 shows an angle θ in standard position along with a circle of radius r. The vertex of θ is at the center of the circle. Because angle θ intercepts an arc on the circle equal in length to the radius of the circle, we say that angle θ has a measure of 1 radian.

Radian

An angle with its vertex at the center of a circle that intercepts an arc on the circle equal in length to the radius of the circle has a measure of **1 radian.**

It follows that an angle of measure 2 radians intercepts an arc equal in length to twice the radius of the circle, an angle of measure $\frac{1}{2}$ radian intercepts an arc equal in length to half the radius of the circle, and so on. In general, if θ is a central angle of a circle of radius r and θ intercepts an arc of length s, then the radian measure of θ is $\frac{s}{r}$.

Converting Between Degrees and Radians The circumference of a circle—the distance around the circle—is given by $C = 2\pi r$, where r is the radius of the circle. The formula $C = 2\pi r$ shows that the radius can be laid off 2π times around a circle. Therefore, an angle of 360°, which corresponds to a complete circle, intercepts an arc equal in length to 2π times the radius of the circle. Thus, an angle of 360° has a measure of 2π radians:

$$360° = 2\pi \text{ radians.}$$

An angle of 180° is half the size of an angle of 360°, so an angle of 180° has half the radian measure of an angle of 360°.

$$180° = \frac{1}{2}(2\pi) \text{ radians} = \pi \text{ radians} \quad \text{Degree/radian relationship}$$

We can use the relationship $180° = \pi$ radians to develop a method for converting between degrees and radians as follows.

$$180° = \pi \text{ radians}$$

$$1° = \frac{\pi}{180} \text{ radian} \quad \text{Divide by 180.} \qquad \text{or} \qquad 1 \text{ radian} = \frac{180°}{\pi} \quad \text{Divide by } \pi.$$

Converting Between Degrees and Radians

1. Multiply a degree measure by $\frac{\pi}{180}$ radian and simplify to convert to radians.
2. Multiply a radian measure by $\frac{180°}{\pi}$ and simplify to convert to degrees.

```
45°
      .7853981634
π/4
      .7853981634
249.8°
      4.359832471
```

Some calculators (in radian mode) have the capability to convert directly between decimal degrees and radians. This screen shows the conversions for Example 1. Note that when *exact* values involving π are required, such as $\frac{\pi}{4}$ in part (a), calculator approximations are not acceptable.

EXAMPLE 1 Converting Degrees to Radians

Convert each degree measure to radians.

(a) 45° **(b)** 249.8°

Solution

(a) $45° = 45\left(\frac{\pi}{180}\text{ radian}\right) = \frac{\pi}{4}\text{ radian}$ Multiply by $\frac{\pi}{180}$ radian.

(b) $249.8° = 249.8\left(\frac{\pi}{180}\text{ radian}\right) \approx 4.360\text{ radians}$ Nearest thousandth

Now try Exercises 1 and 13.

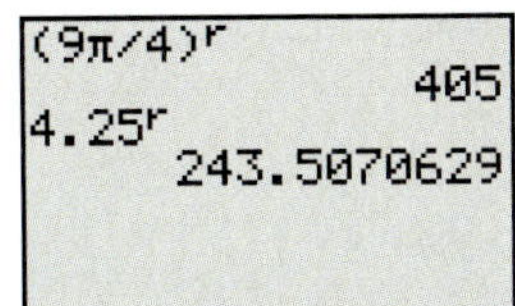

This screen shows how a calculator converts the radian measures in Example 2 to degree measures.

EXAMPLE 2 Converting Radians to Degrees

Convert each radian measure to degrees.

(a) $\frac{9\pi}{4}$ **(b)** 4.25 (Give the answer in decimal degrees.)

Solution

(a) $\frac{9\pi}{4} = \frac{9\pi}{4}\left(\frac{180°}{\pi}\right) = 405°$ Multiply by $\frac{180°}{\pi}$.

(b) $4.25 = 4.25\left(\frac{180°}{\pi}\right) \approx 243.5°$ Use a calculator.

Now try Exercises 21 and 31.

If no unit of angle measure is specified, then radian measure is understood.

CAUTION Figure 2 shows angles measuring 30 radians and 30°. Be careful not to confuse them.

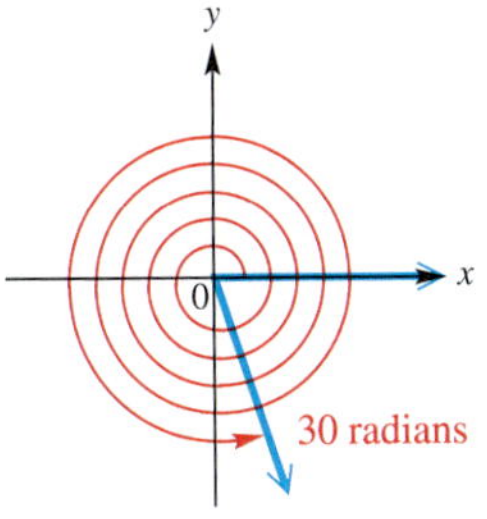

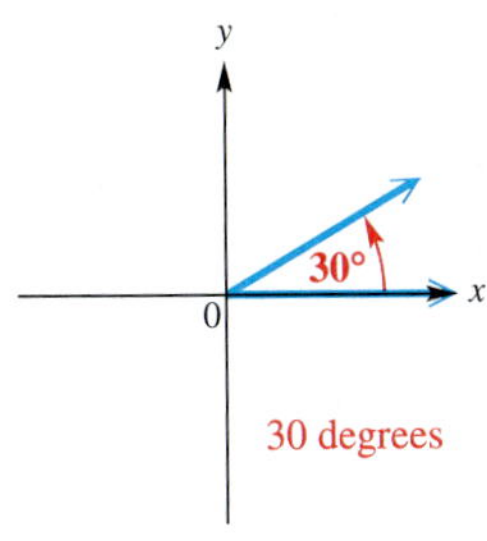

Figure 2

The following table and Figure 3 on the next page give some equivalent angle measures in degrees and radians. Keep in mind that **180° = π radians.**

Degrees	Radians		Degrees	Radians	
	Exact	Approximate		Exact	Approximate
0°	0	0	90°	$\frac{\pi}{2}$	1.57
30°	$\frac{\pi}{6}$	.52	180°	π	3.14
45°	$\frac{\pi}{4}$	.79	270°	$\frac{3\pi}{2}$	4.71
60°	$\frac{\pi}{3}$	1.05	360°	2π	6.28

Looking Ahead to Calculus

In calculus, radian measure is much easier to work with than degree measure. If x is measured in radians, then the derivative of $f(x) = \sin x$ is

$$f'(x) = \cos x.$$

However, if x is measured in degrees, then the derivative of $f(x) = \sin x$ is

$$f'(x) = \frac{\pi}{180} \cos x.$$

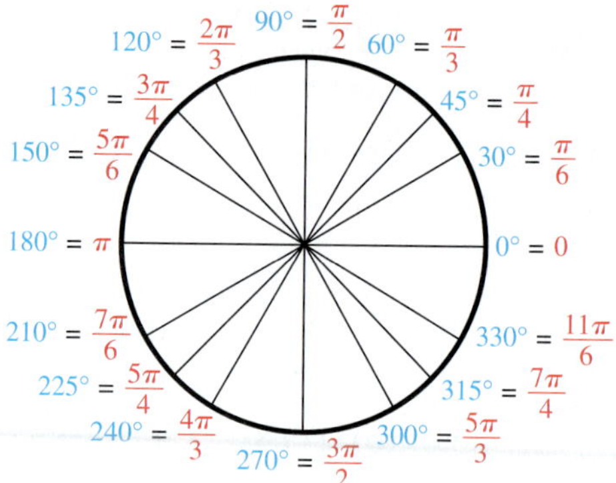

Figure 3

We use radian measure to simplify certain formulas, two of which follow. Each would be more complicated if expressed in degrees.

Arc Length of a Circle We use the first formula to find the length of an arc of a circle. This formula is derived from the fact (proven in geometry) that the length of an arc is proportional to the measure of its central angle.

In Figure 4, angle QOP has measure 1 radian and intercepts an arc of length r on the circle. Angle ROT has measure θ radians and intercepts an arc of length s on the circle. Since the lengths of the arcs are proportional to the measures of their central angles,

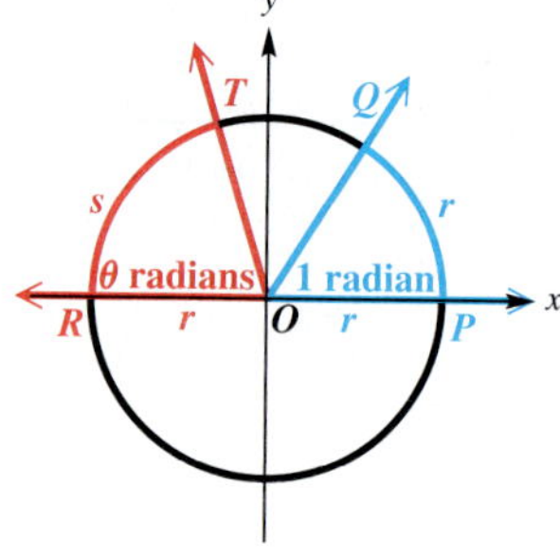

Figure 4

$$\frac{s}{r} = \frac{\theta}{1}.$$

Multiplying both sides by r gives the following result.

Arc Length

The length s of the arc intercepted on a circle of radius r by a central angle of measure θ radians is given by the product of the radius and the radian measure of the angle, or

$$s = r\theta, \qquad \theta \text{ in radians.}$$

CAUTION When applying the formula $s = r\theta$, the value of θ ***must be expressed in radians.***

EXAMPLE 3 Finding Arc Length Using $s = r\theta$

A circle has radius 18.2 cm. Find the length of the arc intercepted by a central angle having each of the following measures.

(a) $\frac{3\pi}{8}$ radians **(b)** $144°$

Solution

(a) As shown in the figure, $r = 18.2$ cm and $\theta = \frac{3\pi}{8}$.

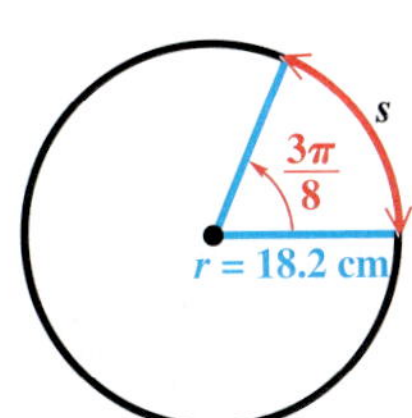

$$s = r\theta \qquad \text{Arc length formula}$$

$$s = 18.2\left(\frac{3\pi}{8}\right) \text{ cm} \qquad \text{Substitute for } r \text{ and } \theta.$$

$$s = \frac{54.6\pi}{8} \text{ cm} \approx 21.4 \text{ cm}$$

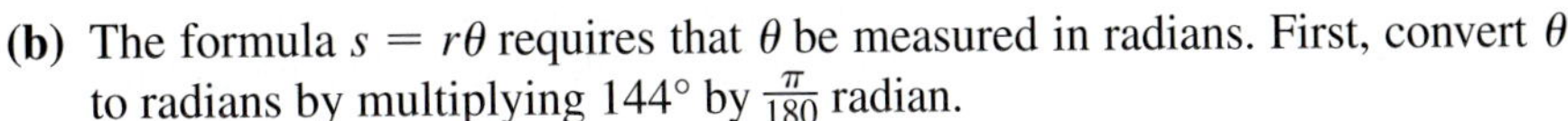

(b) The formula $s = r\theta$ requires that θ be measured in radians. First, convert θ to radians by multiplying $144°$ by $\frac{\pi}{180}$ radian.

$$144° = 144\left(\frac{\pi}{180}\right) = \frac{4\pi}{5} \text{ radians} \qquad \text{Convert from degrees to radians.}$$

The length s is given by

$$s = r\theta = 18.2\left(\frac{4\pi}{5}\right) = \frac{72.8\pi}{5} \approx 45.7 \text{ cm.}$$

Now try Exercises 49 and 51.

EXAMPLE 4 Using Latitudes to Find the Distance Between Two Cities

Reno, Nevada, is approximately due north of Los Angeles. The latitude of Reno is 40° N, while that of Los Angeles is 34° N. (The N in 34° N means *north* of the equator.) The radius of Earth is 6400 km. Find the north-south distance between the two cities.

Solution Latitude gives the measure of a central angle with vertex at Earth's center whose initial side goes through the equator and whose terminal side goes through the given location. As shown in Figure 5, the central angle between Reno and Los Angeles is $40° - 34° = 6°$. The distance between the two cities can be found by the formula $s = r\theta$, after 6° is first converted to radians.

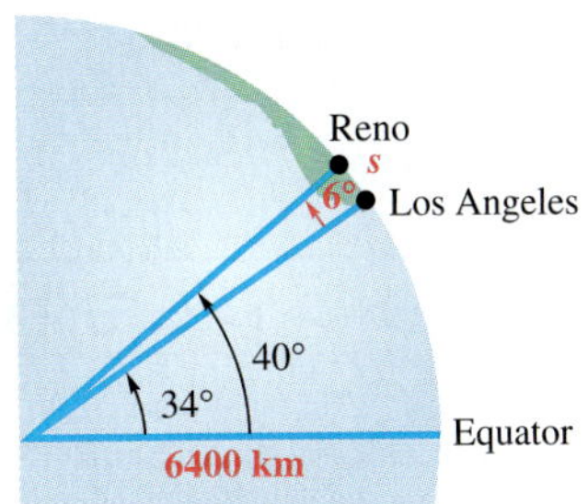

Figure 5

$$6° = 6\left(\frac{\pi}{180}\right) = \frac{\pi}{30} \text{ radian}$$

The distance between the two cities is

$$s = r\theta = 6400\left(\frac{\pi}{30}\right) \approx 670 \text{ km.} \qquad \text{Let } r = 6400 \text{ and } \theta = \frac{\pi}{30}.$$

Now try Exercise 55.

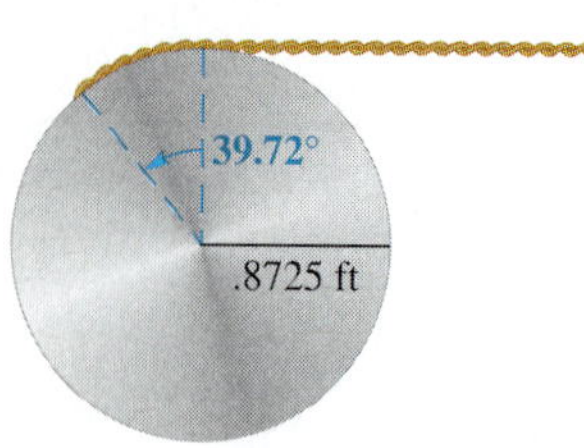

Figure 6

EXAMPLE 5 Finding a Length Using $s = r\theta$

A rope is being wound around a drum with radius .8725 ft. (See Figure 6.) How much rope will be wound around the drum if the drum is rotated through an angle of 39.72°?

Solution The length of rope wound around the drum is the arc length for a circle of radius .8725 ft and a central angle of 39.72°. Use the formula $s = r\theta$, with the angle converted to radian measure. The length of the rope wound around the drum is approximately

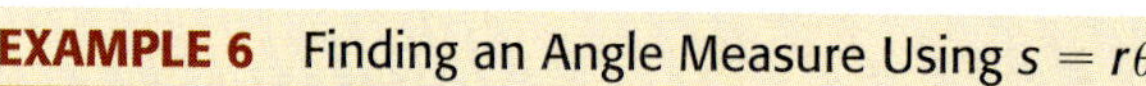

$$s = r\theta = .8725\left[39.72\left(\frac{\pi}{180}\right)\right] \approx .6049 \text{ ft.}$$

Now try Exercise 61(a).

EXAMPLE 6 Finding an Angle Measure Using $s = r\theta$

Figure 7

Two gears are adjusted so that the smaller gear drives the larger one, as shown in Figure 7. If the smaller gear rotates through 225°, through how many degrees will the larger gear rotate?

Solution First find the radian measure of the angle, and then find the arc length on the smaller gear that determines the motion of the larger gear. Since $225° = \frac{5\pi}{4}$ radians, for the smaller gear,

$$s = r\theta = 2.5\left(\frac{5\pi}{4}\right) = \frac{12.5\pi}{4} = \frac{25\pi}{8} \text{ cm.}$$

An arc with this length on the larger gear corresponds to an angle measure θ, in radians, where

$$s = r\theta$$

$$\frac{25\pi}{8} = 4.8\theta \quad \text{Substitute } \tfrac{25\pi}{8} \text{ for } s \text{ and 4.8 for } r.$$

$$\frac{125\pi}{192} \approx \theta. \quad 4.8 = \tfrac{48}{10} = \tfrac{24}{5}\text{; multiply by } \tfrac{5}{24} \text{ to solve for } \theta.$$

Converting θ back to degrees shows that the larger gear rotates through

$$\frac{125\pi}{192}\left(\frac{180°}{\pi}\right) \approx 117°. \quad \text{Convert } \theta = \tfrac{125\pi}{192} \text{ to degrees.}$$

Now try Exercise 63.

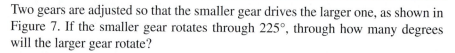

Area of a Sector of a Circle

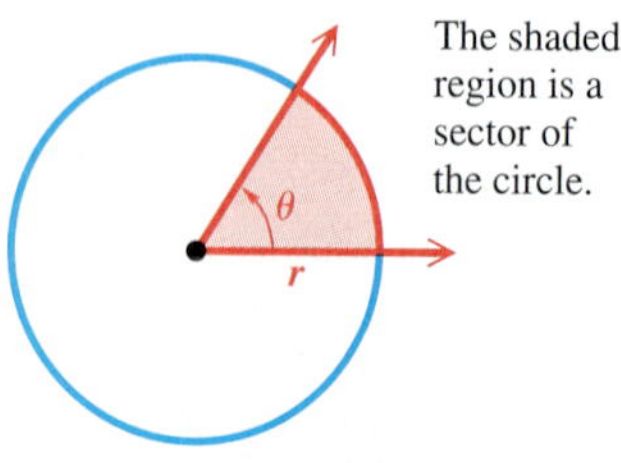

The shaded region is a sector of the circle.

Figure 8

A **sector of a circle** is the portion of the interior of a circle intercepted by a central angle. Think of it as a "piece of pie." See Figure 8. A complete circle can be thought of as an angle with measure 2π radians. If a central angle for a sector has measure θ radians, then the sector makes up the fraction $\frac{\theta}{2\pi}$ of a complete circle. The area of a complete circle with radius r is $A = \pi r^2$. Therefore,

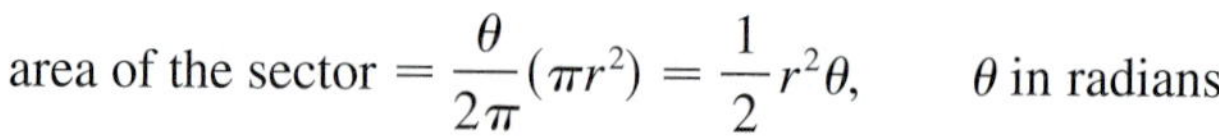

$$\text{area of the sector} = \frac{\theta}{2\pi}(\pi r^2) = \frac{1}{2}r^2\theta, \qquad \theta \text{ in radians.}$$

This discussion is summarized as follows.

Area of a Sector

The area of a sector of a circle of radius r and central angle θ is given by

$$A = \frac{1}{2}r^2\theta, \qquad \theta \text{ in radians.}$$

CAUTION As in the formula for arc length, the value of θ *must be in radians* when using this formula for the area of a sector.

EXAMPLE 7 Finding the Area of a Sector-Shaped Field

Find the area of the sector-shaped field shown in Figure 9.

Figure 9

Solution First, convert 15° to radians.

$$15° = 15\left(\frac{\pi}{180}\right) = \frac{\pi}{12} \text{ radian}$$

Now use the formula to find the area of a sector of a circle with radius $r = 321$.

$$A = \frac{1}{2}r^2\theta = \frac{1}{2}(321)^2\left(\frac{\pi}{12}\right) \approx 13{,}500 \text{ m}^2$$

Now try Exercise 77.

6.1 Exercises

Convert each degree measure to radians. Leave answers as multiples of π. See Example 1(a).

1. 60° **2.** 90° **3.** 150° **4.** 270°

5. 315° **6.** 480° **7.** −45° **8.** −210°

Convert each degree measure to radians. See Example 1(b).

9. 39° **10.** 74° **11.** 139° 10′

12. 174° 50′ **13.** 64.29° **14.** 122.62°

Concept Check *In Exercises 15–18, each angle θ is an integer when measured in radians. Give the radian measure of the angle.*

15.

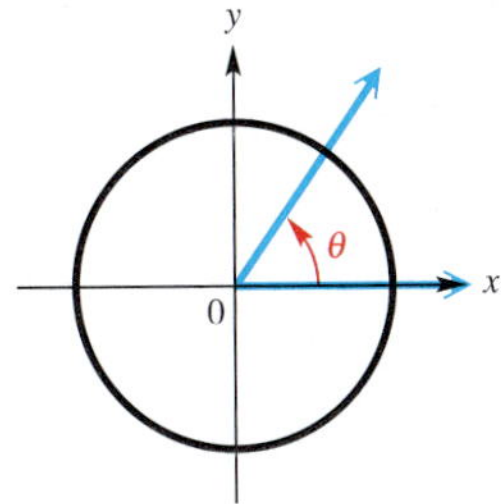

16.

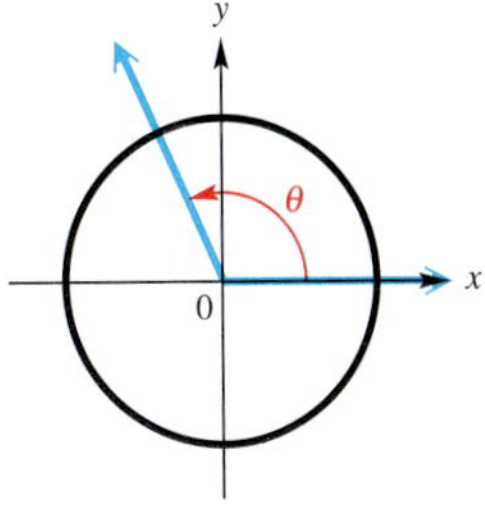

17.

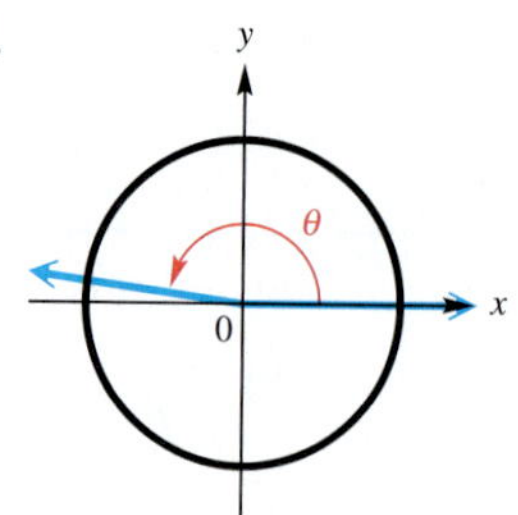

18.

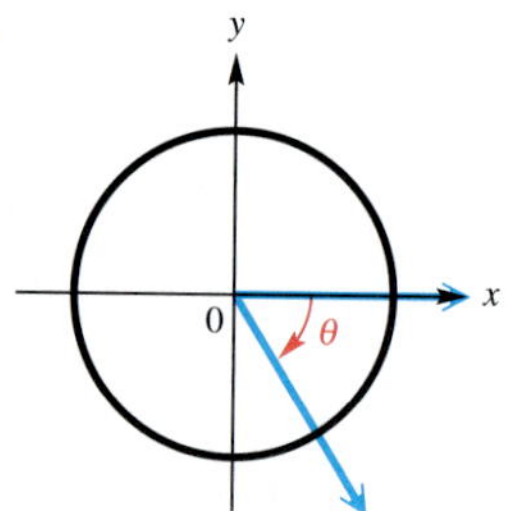

19. In your own words, explain how to convert
(a) degree measure to radian measure; (b) radian measure to degree measure.

20. Explain the difference between degree measure and radian measure.

Convert each radian measure to degrees. See Example 2(a).

21. $\frac{\pi}{3}$ 22. $\frac{8\pi}{3}$ 23. $\frac{7\pi}{4}$ 24. $\frac{2\pi}{3}$

25. $\frac{11\pi}{6}$ 26. $\frac{15\pi}{4}$ 27. $-\frac{\pi}{6}$ 28. $-\frac{7\pi}{20}$

Convert each radian measure to degrees. Give answers using decimal degrees to the nearest tenth. See Example 2(b).

29. 2 30. 5 31. 1.74

32. .3417 33. −9.84763 34. −3.47189

Relating Concepts

For individual or collaborative investigation
(Exercises 35–42)

In anticipation of the material in the next section, we show how to find the trigonometric function values of radian-measured angles. Suppose we want to find $\sin \frac{5\pi}{6}$*. One way to do this is to convert* $\frac{5\pi}{6}$ *radians to 150°, and then use the methods of Chapter 5 to evaluate:*

$$\sin \frac{5\pi}{6} = \sin 150^\circ = + \sin 30^\circ = \frac{1}{2}. \quad \text{(Section 5.3)}$$

Sine is positive in quadrant II.

Reference angle for 150°

Use this technique to find each function value. Give exact values.

35. $\tan \frac{\pi}{4}$ 36. $\csc \frac{\pi}{4}$ 37. $\cot \frac{2\pi}{3}$ 38. $\cos \frac{\pi}{3}$

39. $\sec \pi$ 40. $\sin\left(-\frac{7\pi}{6}\right)$ 41. $\cos\left(-\frac{\pi}{6}\right)$ 42. $\tan\left(-\frac{9\pi}{4}\right)$

Concept Check Find the exact length of each arc intercepted by the given central angle.

43.

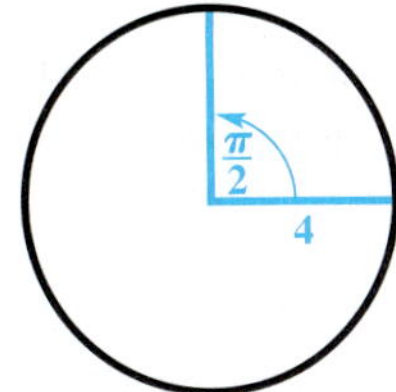

44.

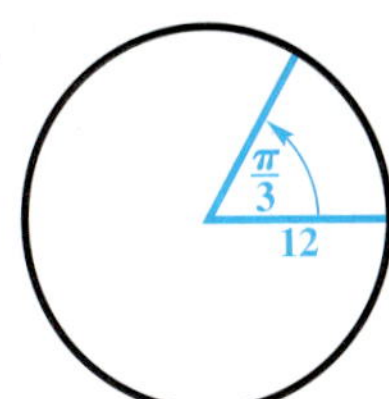

Concept Check Find the radius of each circle.

45.

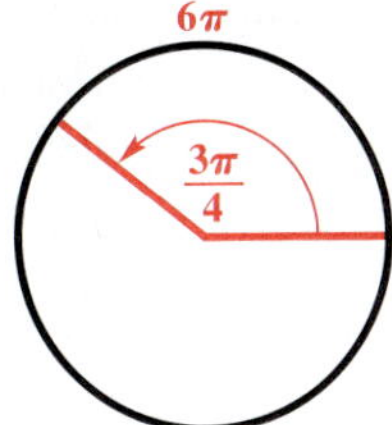

46.

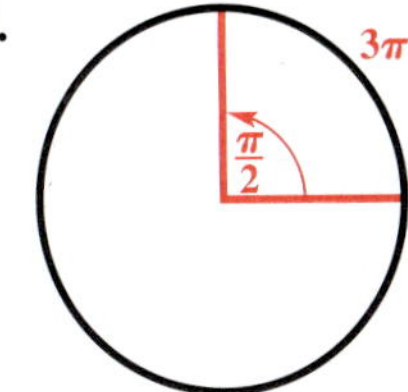

Concept Check Find the measure of each central angle (in radians).

47.

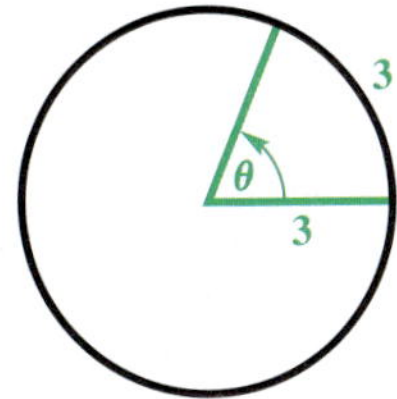

48.

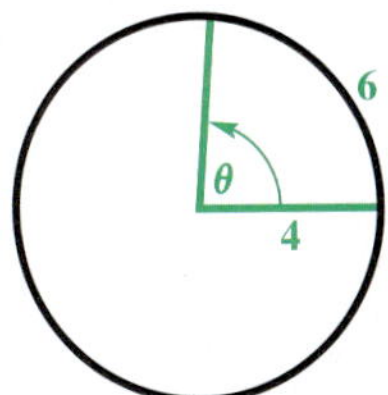

Unless otherwise directed, give calculator approximations in your answers in the rest of this exercise set.

Find the length of each arc intercepted by a central angle θ in a circle of radius r. See Example 3.

49. $r = 12.3$ cm, $\theta = \frac{2\pi}{3}$ radians

50. $r = .892$ cm, $\theta = \frac{11\pi}{10}$ radians

51. $r = 4.82$ m, $\theta = 60°$

52. $r = 71.9$ cm, $\theta = 135°$

53. *Concept Check* If the radius of a circle is doubled, how is the length of the arc intercepted by a fixed central angle changed?

54. *Concept Check* Radian measure simplifies many formulas, such as the formula for arc length, $s = r\theta$. Give the corresponding formula when θ is measured in degrees instead of radians.

Distance Between Cities Find the distance in kilometers between each pair of cities, assuming they lie on the same north-south line. See Example 4.

55. Panama City, Panama, 9° N, and Pittsburgh, Pennsylvania, 40° N

56. Farmersville, California, 36° N, and Penticton, British Columbia, 49° N

57. New York City, New York, 41° N, and Lima, Peru, 12° S

58. Halifax, Nova Scotia, 45° N, and Buenos Aires, Argentina, 34° S

59. *Latitude of Madison* Madison, South Dakota, and Dallas, Texas, are 1200 km apart and lie on the same north-south line. The latitude of Dallas is 33° N. What is the latitude of Madison?

60. *Latitude of Toronto* Charleston, South Carolina, and Toronto, Canada, are 1100 km apart and lie on the same north-south line. The latitude of Charleston is 33° N. What is the latitude of Toronto?

Work each problem. See Examples 5 and 6.

61. *Pulley Raising a Weight*

(a) How many inches will the weight in the figure rise if the pulley is rotated through an angle of 71° 50′?

(b) Through what angle, to the nearest minute, must the pulley be rotated to raise the weight 6 in.?

62. *Pulley Raising a Weight* Find the radius of the pulley in the figure if a rotation of 51.6° raises the weight 11.4 cm.

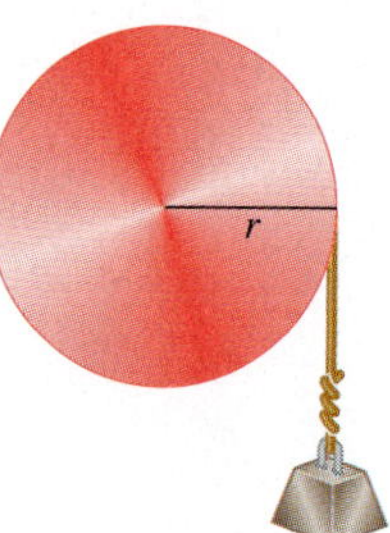

63. *Rotating Wheels* The rotation of the smaller wheel in the figure causes the larger wheel to rotate. Through how many degrees will the larger wheel rotate if the smaller one rotates through 60.0°?

64. *Rotating Wheels* Find the radius of the larger wheel in the figure if the smaller wheel rotates 80.0° when the larger wheel rotates 50.0°.

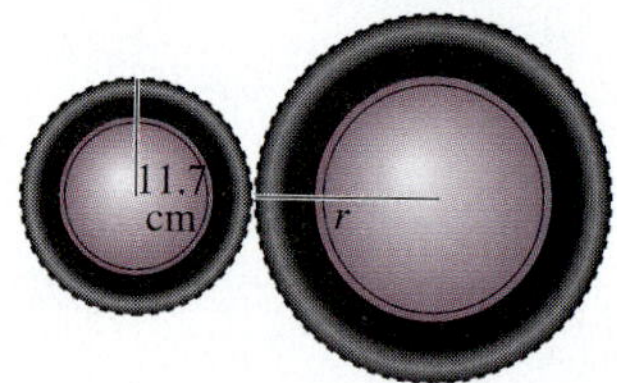

65. *Bicycle Chain Drive* The figure shows the chain drive of a bicycle. How far will the bicycle move if the pedals are rotated through 180°? Assume the radius of the bicycle wheel is 13.6 in.

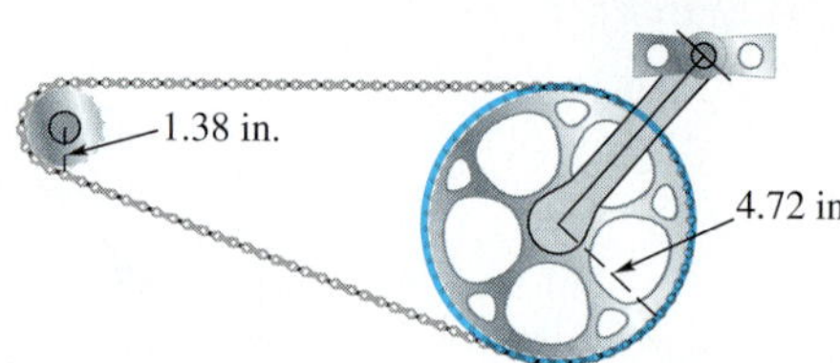

66. *Pickup Truck Speedometer* The speedometer of Terry's small pickup truck is designed to be accurate with tires of radius 14 in.

(a) Find the number of rotations of a tire in 1 hr if the truck is driven at 55 mph.

(b) Suppose that oversize tires of radius 16 in. are placed on the truck. If the truck is now driven for 1 hr with the speedometer reading 55 mph, how far has the truck gone? If the speed limit is 55 mph, does Terry deserve a speeding ticket?

If a central angle is very small, there is little difference in length between an arc and the inscribed chord. See the figure. Approximate each of the following lengths by finding the necessary arc length. (Note: When a central angle intercepts an arc, the arc is said to **subtend** *the angle.)*

Arc length ≈ length of inscribed chord

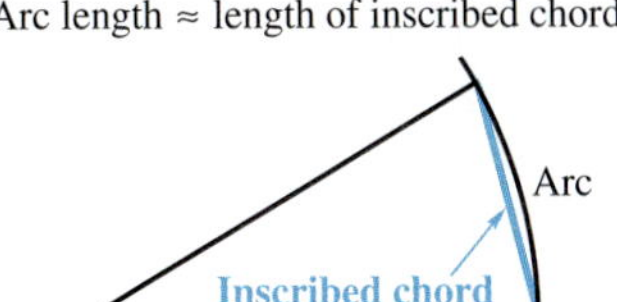

67. *Length of a Train* A railroad track in the desert is 3.5 km away. A train on the track subtends (horizontally) an angle of $3^\circ 20'$. Find the length of the train.

68. *Distance to a Boat* The mast of Brent Simon's boat is 32 ft high. If it subtends an angle of $2^\circ 10'$, how far away is it?

Concept Check *Find the area of each sector.*

69.

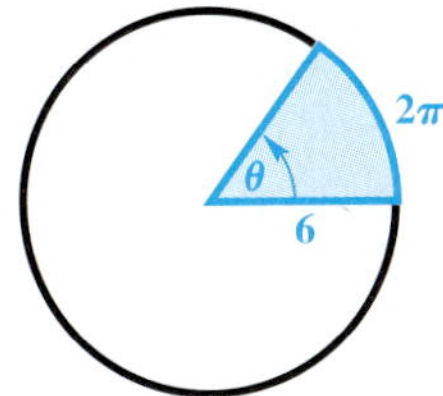

70.

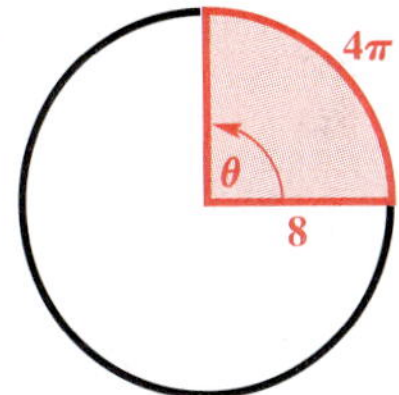

Concept Check *Find the measure (in radians) of each central angle. The number inside the sector is the area.*

71.

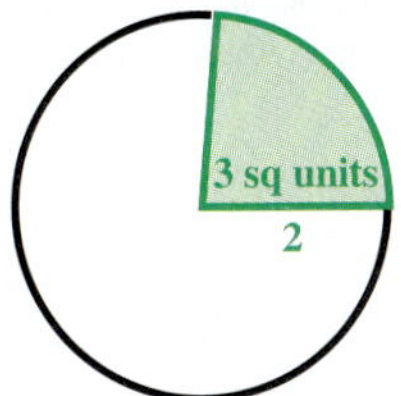

72.

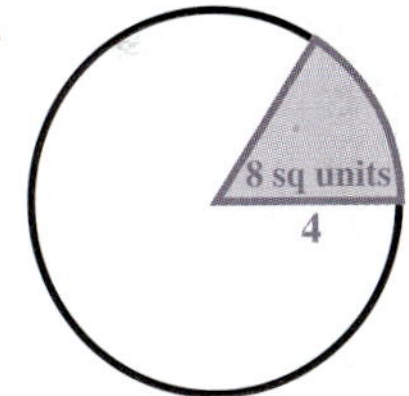

Find the area of a sector of a circle having radius r and central angle θ. See Example 7.

73. $r = 29.2$ m, $\theta = \dfrac{5\pi}{6}$ radians

74. $r = 59.8$ km, $\theta = \dfrac{2\pi}{3}$ radians

75. $r = 30.0$ ft, $\theta = \dfrac{\pi}{2}$ radians

76. $r = 90.0$ yd, $\theta = \dfrac{5\pi}{6}$ radians

77. $r = 12.7$ cm, $\theta = 81^\circ$

78. $r = 18.3$ m, $\theta = 125^\circ$

79. $r = 40.0$ mi, $\theta = 135^\circ$

80. $r = 90.0$ km, $\theta = 270^\circ$

Work each problem.

81. Find the measure (in radians) of a central angle of a sector of area 16 in.^{2} in a circle of radius 3.0 in.

82. Find the radius of a circle in which a central angle of $\frac{\pi}{6}$ radian determines a sector of area 64 m^2.

83. ***Measures of a Structure*** The figure shows Medicine Wheel, a Native American structure in northern Wyoming. This circular structure is perhaps 2500 yr old. There are 27 aboriginal spokes in the wheel, all equally spaced.

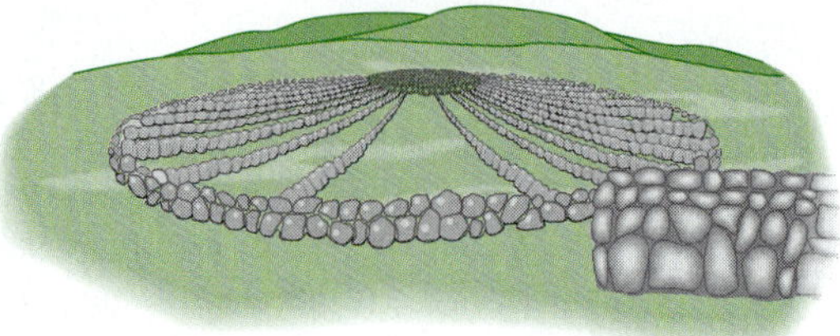

(a) Find the measure of each central angle in degrees and in radians.
(b) If the radius of the wheel is 76 ft, find the circumference.
(c) Find the length of each arc intercepted by consecutive pairs of spokes.
(d) Find the area of each sector formed by consecutive spokes.

84. ***Area Cleaned by a Windshield Wiper*** The Ford Model A, built from 1928 to 1931, had a single windshield wiper on the driver's side. The total arm and blade was 10 in. long and rotated back and forth through an angle of 95°. The shaded region in the figure is the portion of the windshield cleaned by the 7-in. wiper blade. What is the area of the region cleaned?

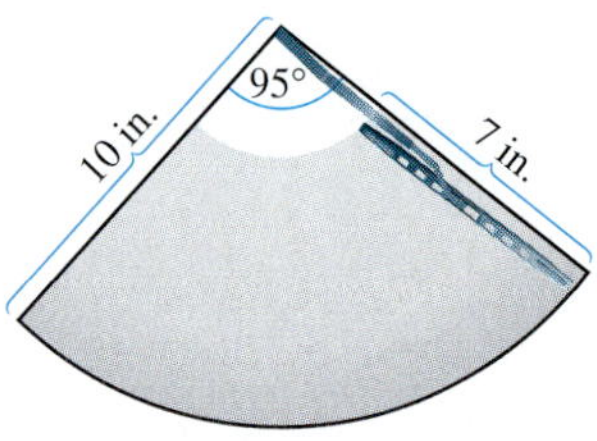

85. ***Circular Railroad Curves*** In the United States, circular railroad curves are designated by the *degree of curvature,* the central angle subtended by a chord of 100 ft. Suppose a portion of track has curvature 42°. (*Source:* Hay, W., *Railroad Engineering,* John Wiley & Sons, 1982.)

(a) What is the radius of the curve?
(b) What is the length of the arc determined by the 100-ft chord?
(c) What is the area of the portion of the circle bounded by the arc and the 100-ft chord?

86. ***Land Required for a Solar-Power Plant*** A 300-megawatt solar-power plant requires approximately 950,000 m^2 of land area in order to collect the required amount of energy from sunlight.

(a) If this land area is circular, what is its radius?
(b) If this land area is a 35° sector of a circle, what is its radius?

87. ***Area of a Lot*** A frequent problem in surveying city lots and rural lands adjacent to curves of highways and railways is that of finding the area when one or more of the boundary lines is the arc of a circle. Find the area of the lot shown in the figure. (*Source:* Anderson, J. and E. Michael, *Introduction to Surveying,* McGraw-Hill, 1985.)

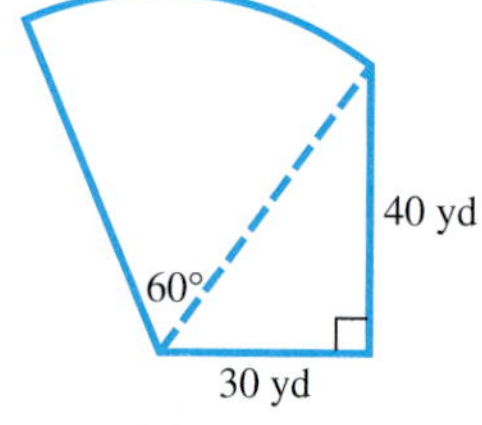

88. ***Nautical Miles*** *Nautical miles* are used by ships and airplanes. They are different from *statute miles,* which equal 5280 ft. A nautical mile is defined to be the arc length along the equator intercepted by a central angle AOB of 1 min, as illustrated in the figure. If the equatorial radius of Earth is 3963 mi, use the arc length formula to approximate the number of statute miles in 1 nautical mile. Round your answer to two decimal places.

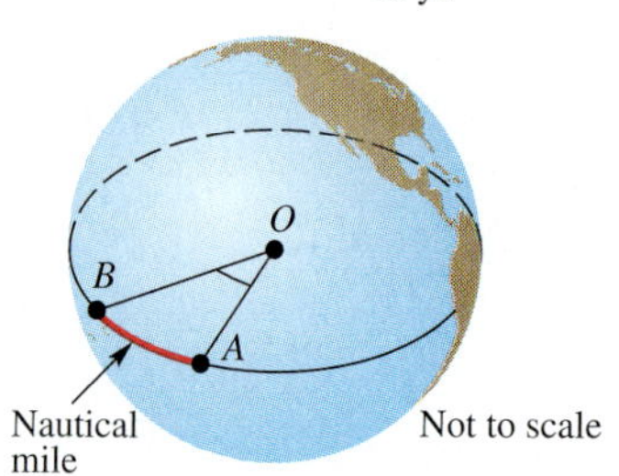

89. *Circumference of Earth* The first accurate estimate of the distance around Earth was done by the Greek astronomer Eratosthenes (276–195 B.C.), who noted that the noontime position of the sun at the summer solstice differed by 7° 12′ from the city of Syene to the city of Alexandria. (See the figure.) The distance between these two cities is 496 mi. Use the arc length formula to estimate the radius of Earth. Then find the circumference of Earth. (*Source:* Zeilik, M., *Introductory Astronomy and Astrophysics,* Third Edition, Saunders College Publishers, 1992.)

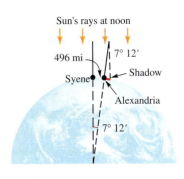

90. *Diameter of the Moon* The distance to the moon is approximately 238,900 mi. Use the arc length formula to estimate the diameter d of the moon if angle θ in the figure is measured to be .517°.

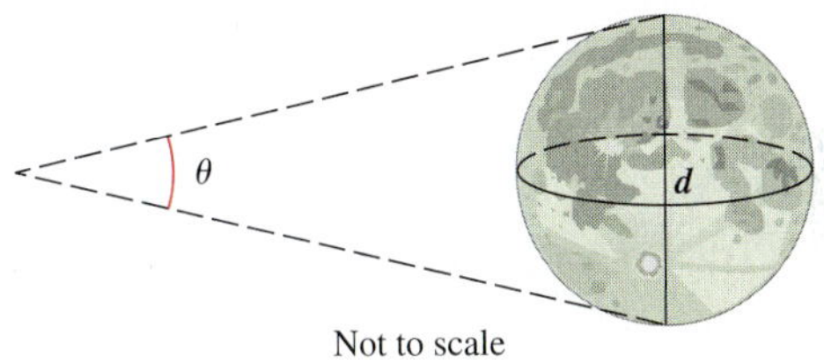

91. *Concept Check* If the radius of a circle is doubled and the central angle of a sector is unchanged, how is the area of the sector changed?

92. *Concept Check* Give the corresponding formula for the area of a sector when the angle is measured in degrees.

6.2 The Unit Circle and Circular Functions

Circular Functions ▪ Finding Values of Circular Functions ▪ Determining a Number with a Given Circular Function Value ▪ Angular and Linear Speed

In Section 5.2, we defined the six trigonometric functions in such a way that the domain of each function was a set of *angles* in standard position. These angles can be measured in degrees or in radians. In advanced courses, such as calculus, it is necessary to modify the trigonometric functions so that their domains consist of *real numbers* rather than angles. We do this by using the relationship between an angle θ and an arc of length s on a circle.

Unit circle $x^2 + y^2 = 1$

Figure 10

Circular Functions

In Figure 10, we start at the point $(1, 0)$ and measure an arc of length s along the circle. If $s > 0$, then the arc is measured in a counterclockwise direction, and if $s < 0$, then the direction is clockwise. (If $s = 0$, then no arc is measured.) Let the endpoint of this arc be at the point (x, y). The circle in Figure 10 is a **unit circle**—it has center at the origin and radius 1 unit (hence the name *unit circle*). Recall from algebra that the equation of this circle is

$$x^2 + y^2 = 1. \quad \text{(Section 2.1)}$$

We saw in the previous section that the radian measure of θ is related to the arc length s. In fact, for θ measured in radians, we know that $s = r\theta$. Here, $r = 1$, so s, which is measured in linear units such as inches or centimeters, is numerically equal to θ, measured in radians. Thus, the trigonometric functions of angle θ in radians found by choosing a point (x, y) on the unit circle can be rewritten as functions of the arc length s, a real number. When interpreted this way, they are called **circular functions.**

Looking Ahead to Calculus

If you plan to study calculus, you must become very familiar with radian measure. In calculus, the trigonometric or circular functions are always understood to have real number domains.

Circular Functions

$$\sin s = y \qquad \cos s = x \qquad \tan s = \frac{y}{x} \quad (x \neq 0)$$

$$\csc s = \frac{1}{y} \quad (y \neq 0) \qquad \sec s = \frac{1}{x} \quad (x \neq 0) \qquad \cot s = \frac{x}{y} \quad (y \neq 0)$$

Since x represents the cosine of s and y represents the sine of s, and because of the discussion in Section 6.1 on converting between degrees and radians, we can summarize a great deal of information in a concise manner, as seen in Figure 11.*

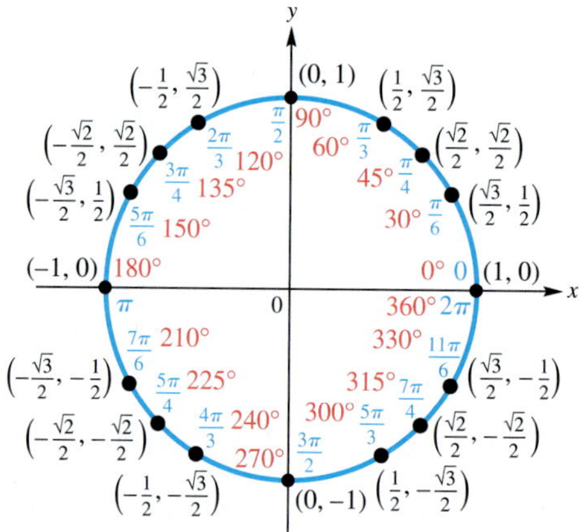

Figure 11

NOTE Since $\sin s = y$ and $\cos s = x$, we can replace x and y in the equation $x^2 + y^2 = 1$ and obtain the Pythagorean identity

$$\cos^2 s + \sin^2 s = 1.$$

The ordered pair (x, y) represents a point on the unit circle, and therefore

$$-1 \le x \le 1 \quad \text{and} \quad -1 \le y \le 1,$$

so

$$-1 \le \cos s \le 1 \quad \text{and} \quad -1 \le \sin s \le 1.$$

*The authors thank Professor Marvel Townsend of the University of Florida for her suggestion to include this figure.

For any value of s, both $\sin s$ and $\cos s$ exist, so the domain of these functions is the set of all real numbers. For $\tan s$, defined as $\frac{y}{x}$, x must not equal 0. The only way x can equal 0 is when the arc length s is $\frac{\pi}{2}$, $-\frac{\pi}{2}$, $\frac{3\pi}{2}$, $-\frac{3\pi}{2}$, and so on. To avoid a 0 denominator, the domain of the tangent function must be restricted to those values of s satisfying

$$s \neq (2n + 1)\frac{\pi}{2}, \qquad n \text{ any integer.}$$

The definition of secant also has x in the denominator, so the domain of secant is the same as the domain of tangent. Both cotangent and cosecant are defined with a denominator of y. To guarantee that $y \neq 0$, the domain of these functions must be the set of all values of s satisfying

$$s \neq n\pi, \qquad n \text{ any integer.}$$

In summary, the domains of the circular functions are as follows.

Domains of the Circular Functions

Assume that n is any integer and s is a real number.

Sine and Cosine Functions: $(-\infty, \infty)$

Tangent and Secant Functions: $\left\{s \,\middle|\, s \neq (2n + 1)\frac{\pi}{2}\right\}$

Cotangent and Cosecant Functions: $\{s \mid s \neq n\pi\}$

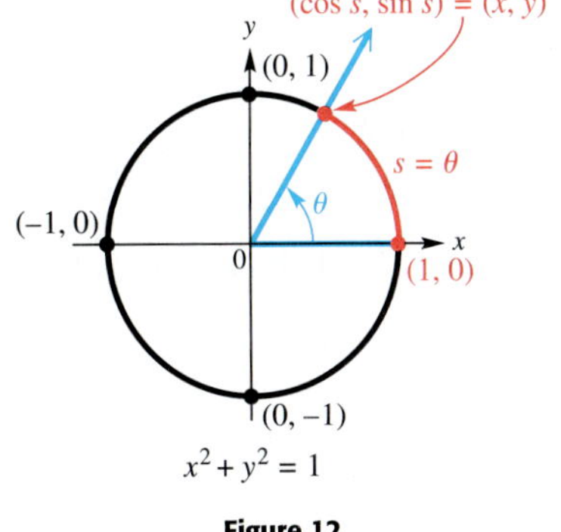

Figure 12

Finding Values of Circular Functions The circular functions (functions of real numbers) are closely related to the trigonometric functions of angles measured in radians. To see this, let us assume that angle θ is in standard position, superimposed on the unit circle, as shown in Figure 12. Suppose further that θ is the *radian* measure of this angle. Using the arc length formula $s = r\theta$ with $r = 1$, we have $s = \theta$. Thus, the length of the intercepted arc is the real number that corresponds to the radian measure of θ. Using the definitions of the trigonometric functions, we have

$$\sin \theta = \frac{y}{r} = \frac{y}{1} = y = \sin s, \qquad \text{and} \qquad \cos \theta = \frac{x}{r} = \frac{x}{1} = x = \cos s,$$

and so on. As shown here, the trigonometric functions and the circular functions lead to the same function values, provided we think of the angles as being in radian measure. This leads to the following important result concerning evaluation of circular functions.

Evaluating a Circular Function

Circular function values of real numbers are obtained in the same manner as trigonometric function values of angles measured in radians. This applies both to methods of finding exact values (such as reference angle analysis) and to calculator approximations. Calculators must be in radian mode when finding circular function values.

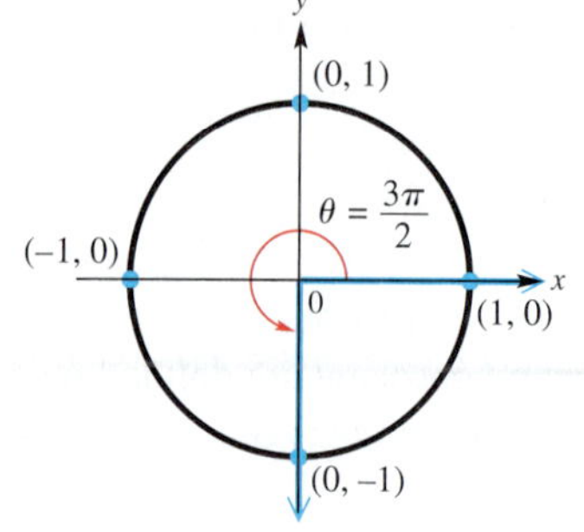

Figure 13

EXAMPLE 1 Finding Exact Circular Function Values

Find the exact values of $\sin \frac{3\pi}{2}$, $\cos \frac{3\pi}{2}$, and $\tan \frac{3\pi}{2}$.

Solution Evaluating a circular function at the real number $\frac{3\pi}{2}$ is equivalent to evaluating it at $\frac{3\pi}{2}$ radians. An angle of $\frac{3\pi}{2}$ radians intersects the unit circle at the point $(0, -1)$, as shown in Figure 13. Since

$$\sin s = y, \qquad \cos s = x, \qquad \text{and} \qquad \tan s = \frac{y}{x},$$

it follows that

$$\sin \frac{3\pi}{2} = -1, \qquad \cos \frac{3\pi}{2} = 0, \qquad \text{and} \qquad \tan \frac{3\pi}{2} \text{ is undefined.}$$

Now try Exercise 1.

EXAMPLE 2 Finding Exact Circular Function Values

(a) Use Figure 11 to find the exact values of $\cos \frac{7\pi}{4}$ and $\sin \frac{7\pi}{4}$.

(b) Use Figure 11 to find the exact value of $\tan\left(-\frac{5\pi}{3}\right)$.

(c) Use reference angles and degree/radian conversion to find the exact value of $\cos \frac{2\pi}{3}$.

Solution

(a) In Figure 11, we see that the terminal side of $\frac{7\pi}{4}$ radians intersects the unit circle at $\left(\frac{\sqrt{2}}{2}, -\frac{\sqrt{2}}{2}\right)$. Thus,

$$\cos \frac{7\pi}{4} = \frac{\sqrt{2}}{2} \qquad \text{and} \qquad \sin \frac{7\pi}{4} = -\frac{\sqrt{2}}{2}.$$

(b) Angles of $-\frac{5\pi}{3}$ radians and $\frac{\pi}{3}$ radians are coterminal. Their terminal sides intersect the unit circle at $\left(\frac{1}{2}, \frac{\sqrt{3}}{2}\right)$, so

$$\tan\left(-\frac{5\pi}{3}\right) = \tan \frac{\pi}{3} = \frac{\frac{\sqrt{3}}{2}}{\frac{1}{2}} = \sqrt{3}.$$

(c) An angle of $\frac{2\pi}{3}$ radians corresponds to an angle of 120°. In standard position, 120° lies in quadrant II with a reference angle of 60°, so

Cosine is negative in quadrant II.

$$\cos \frac{2\pi}{3} = \cos 120^\circ = -\cos 60^\circ = -\frac{1}{2}.$$

Reference angle (Section 5.3)

Now try Exercises 7, 17, and 21.

NOTE Examples 1 and 2 illustrate that there are several methods of finding exact circular function values.

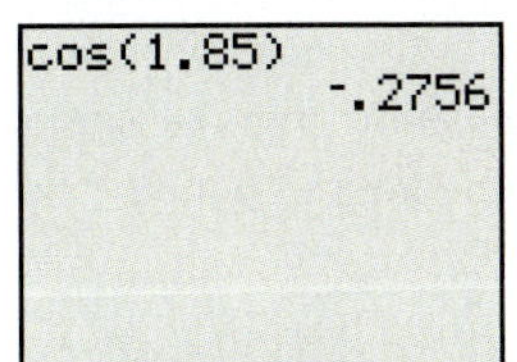

Radian mode

This is how a calculator displays the result of Example 3(a), correct to four decimal digits.

EXAMPLE 3 Approximating Circular Function Values

Find a calculator approximation to four decimal places for each circular function value.

(a) cos 1.85 **(b)** cos .5149 **(c)** cot 1.3209 **(d)** sec(−2.9234)

Solution

(a) With a calculator in radian mode, we find $\cos 1.85 \approx -.2756$.

(b) $\cos .5149 \approx .8703$ Use a calculator in radian mode.

(c) As before, to find cotangent, secant, and cosecant function values, we must use the appropriate reciprocal functions. To find cot 1.3209, first find tan 1.3209 and then find the reciprocal.

$$\cot 1.3209 = \frac{1}{\tan 1.3209} \approx .2552$$

(d) $$\sec(-2.9234) = \frac{1}{\cos(-2.9234)} \approx -1.0243$$

Now try Exercises 23, 29, and 33.

CAUTION A common error in trigonometry is using calculators in degree mode when radian mode should be used. Remember, ***if you are finding a circular function value of a real number, the calculator must be in radian mode.***

Determining a Number with a Given Circular Function Value Recall from Section 5.3 how we used a calculator to determine an angle measure, given a trigonometric function value of the angle.

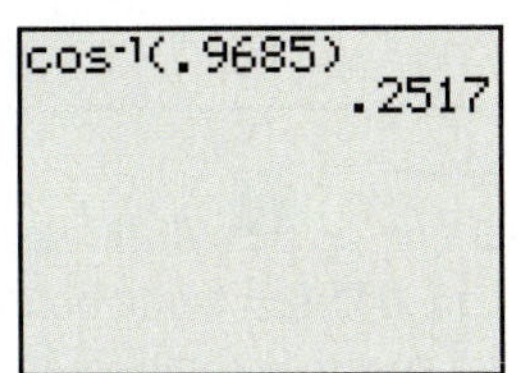

The calculator is set to show four decimal digits in the answer.

Figure 14

tan⁻¹(1)
.7853981634
Ans+π
3.926990817
tan(Ans)
1

This screen supports the result of Example 4(b). The calculator is in radian mode.

Figure 15

EXAMPLE 4 Finding a Number Given Its Circular Function Value

(a) Approximate the value of s in the interval $\left[0, \frac{\pi}{2}\right]$, if $\cos s = .9685$.

(b) Find the exact value of s in the interval $\left[\pi, \frac{3\pi}{2}\right]$, if $\tan s = 1$.

Solution

(a) Since we are given a cosine value and want to determine the real number in $\left[0, \frac{\pi}{2}\right]$ having this cosine value, we use the *inverse cosine* function of a calculator. With the calculator in radian mode, we find

$$\cos^{-1}(.9685) \approx .2517. \quad \text{(Section 5.3)}$$

See Figure 14. (Refer to your owner's manual to determine how to evaluate the $\sin^{-1}$, $\cos^{-1}$, and $\tan^{-1}$ functions with your calculator.)

(b) Recall that $\tan \frac{\pi}{4} = 1$, and in quadrant III tan s is positive. Therefore,

$$\tan\left(\pi + \frac{\pi}{4}\right) = \tan \frac{5\pi}{4} = 1,$$

and $s = \frac{5\pi}{4}$. Figure 15 supports this result.

Now try Exercises 49 and 55.

CONNECTIONS A convenient way to see the sine, cosine, and tangent trigonometric ratios geometrically is shown in Figure 16 for θ in quadrants I and II. The circle shown is the unit circle, which has radius 1. By remembering this figure and the segments that represent the sine, cosine, and tangent functions, you can quickly recall properties of the trigonometric functions. Horizontal line segments to the left of the origin and vertical line segments below the x-axis represent negative values. Note that the tangent line must be tangent to the circle at $(1, 0)$, for any quadrant in which θ lies.

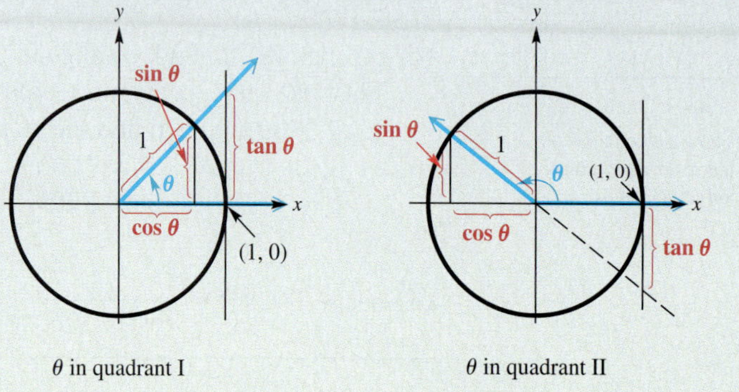

Figure 16

For Discussion or Writing

See Figure 17. Use the definition of the trigonometric functions and similar triangles to show that $PQ = \sin\theta$, $OQ = \cos\theta$, and $AB = \tan\theta$.

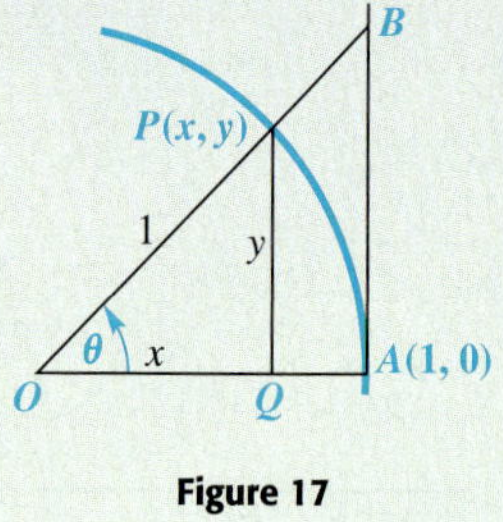

Figure 17

Angular and Linear Speed

The human joint that can be flexed the fastest is the wrist, which can rotate through 90°, or $\frac{\pi}{2}$ radians, in .045 sec while holding a tennis racket. **Angular speed** ω (omega) measures the speed of rotation and is defined by

$$\omega = \frac{\theta}{t},$$

where θ is the angle of rotation in radians and t is time. The angular speed of a human wrist swinging a tennis racket is

$$\omega = \frac{\theta}{t} = \frac{\frac{\pi}{2}}{.045} \approx 35 \text{ radians per sec.}$$

The **linear speed** v at which the tip of the racket travels as a result of flexing the wrist is given by

$$v = r\omega,$$

where r is the radius (distance) from the tip of the racket to the wrist joint. If $r = 2$ ft, then the speed at the tip of the racket is

$$v = r\omega \approx 2(35) = 70 \text{ ft per sec,} \qquad \text{or about 48 mph.}$$

In a tennis serve the arm rotates at the shoulder, so the final speed of the racket is considerably faster. (*Source:* Cooper, J. and R. Glassow, *Kinesiology,* Second Edition, C.V. Mosby, 1968.)

EXAMPLE 5 Finding Angular Speed of a Pulley and Linear Speed of a Belt

A belt runs a pulley of radius 6 cm at 80 revolutions per min.

(a) Find the angular speed of the pulley in radians per second.

(b) Find the linear speed of the belt in centimeters per second.

Solution

(a) In 1 min, the pulley makes 80 revolutions. Each revolution is 2π radians, for a total of

$$80(2\pi) = 160\pi \text{ radians per min.}$$

Since there are 60 sec in 1 min, we find ω, the angular speed in radians per second, by dividing 160π by 60.

$$\omega = \frac{160\pi}{60} = \frac{8\pi}{3} \text{ radians per sec}$$

(b) The linear speed of the belt will be the same as that of a point on the circumference of the pulley. Thus,

$$v = r\omega = 6\left(\frac{8\pi}{3}\right) = 16\pi \approx 50.3 \text{ cm per sec.}$$

Now try Exercise 95.

Suppose that an object is moving at a constant speed. If it travels a distance s in time t, then its linear speed v is given by

$$v = \frac{s}{t}.$$

If the object is traveling in a circle, then $s = r\theta$, where r is the radius of the circle and θ is the angle of rotation. Thus, we can write

$$v = \frac{r\theta}{t}.$$

The formulas for angular and linear speed are summarized in the table.

Angular Speed	Linear Speed
$\omega = \frac{\theta}{t}$ (ω in radians per unit time, θ in radians)	$v = \frac{s}{t}$ $v = \frac{r\theta}{t}$ $v = r\omega$

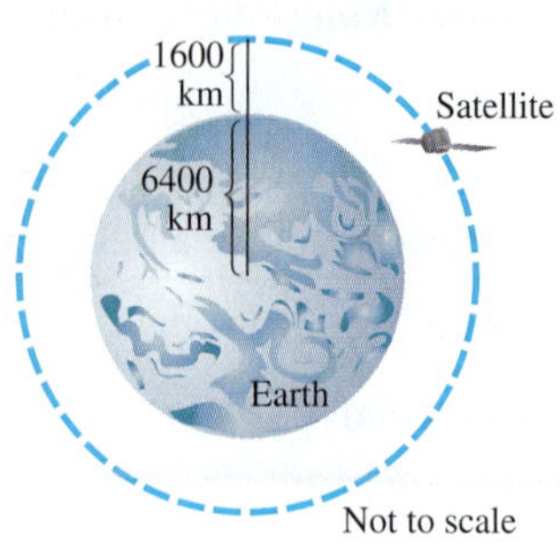

Figure 18

EXAMPLE 6 Finding Linear Speed and Distance Traveled by a Satellite

A satellite traveling in a circular orbit 1600 km above the surface of Earth takes 2 hr to make an orbit. The radius of Earth is 6400 km. See Figure 18.

(a) Find the linear speed of the satellite.

(b) Find the distance the satellite travels in 4.5 hr.

Solution

(a) The distance of the satellite from the center of Earth is

$$r = 1600 + 6400 = 8000 \text{ km}.$$

For one orbit, $\theta = 2\pi$, and

$$s = r\theta = 8000(2\pi) \text{ km.} \quad \text{(Section 6.1)}$$

Since it takes 2 hr to complete an orbit, the linear speed is

$$v = \frac{s}{t} = \frac{8000(2\pi)}{2} = 8000\pi \approx 25{,}000 \text{ km per hr.}$$

(b) $s = vt = 8000\pi(4.5) = 36{,}000\pi \approx 110{,}000$ km

Now try Exercise 93.

6.2 Exercises

For each value of θ, find **(a)** $\sin\theta$, **(b)** $\cos\theta$, *and* **(c)** $\tan\theta$. *See Example 1.*

1. $\theta = \frac{\pi}{2}$ **2.** $\theta = \pi$ **3.** $\theta = 2\pi$

4. $\theta = 3\pi$ **5.** $\theta = -\pi$ **6.** $\theta = -\frac{3\pi}{2}$

Find the exact circular function value for each of the following. See Example 2.

7. $\sin\frac{7\pi}{6}$ **8.** $\cos\frac{5\pi}{3}$ **9.** $\tan\frac{3\pi}{4}$ **10.** $\sec\frac{2\pi}{3}$

11. $\csc\frac{11\pi}{6}$ **12.** $\cot\frac{5\pi}{6}$ **13.** $\cos\left(-\frac{4\pi}{3}\right)$ **14.** $\tan\frac{17\pi}{3}$

15. $\cos\frac{7\pi}{4}$ **16.** $\sec\frac{5\pi}{4}$ **17.** $\sin\left(-\frac{4\pi}{3}\right)$ **18.** $\sin\left(-\frac{5\pi}{6}\right)$

19. $\sec\frac{23\pi}{6}$ **20.** $\csc\frac{13\pi}{3}$ **21.** $\tan\frac{5\pi}{6}$ **22.** $\cos\frac{3\pi}{4}$

Find a calculator approximation for each circular function value. See Example 3.

23. sin .6109 **24.** sin .8203 **25.** cos(−1.1519)

26. cos(−5.2825) **27.** tan 4.0203 **28.** tan 6.4752

29. csc(−9.4946) **30.** csc 1.3875 **31.** sec 2.8440

32. sec(−8.3429) **33.** cot 6.0301 **34.** cot 3.8426

Concept Check *The figure displays a unit circle and an angle of 1 radian. The tick marks on the circle are spaced at every two-tenths radian. Use the figure to estimate each value.*

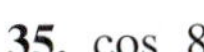

35. cos .8

36. sin 4

37. an angle whose cosine is $-.65$

38. an angle whose sine is $-.95$

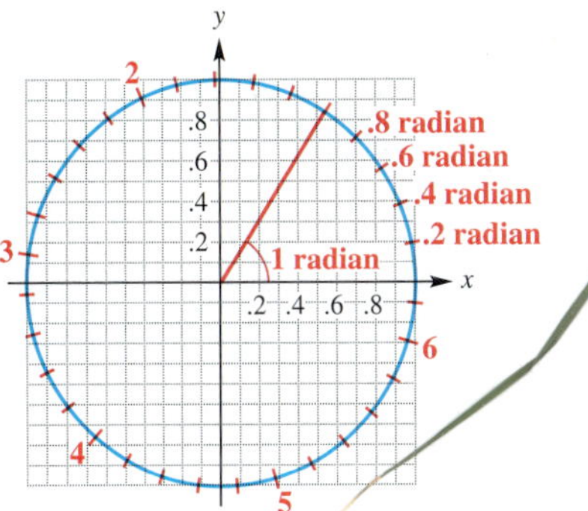

Concept Check *Without using a calculator, decide whether each function value is positive or negative. (Hint: Consider the radian measures of the quadrantal angles.)*

39. cos 2 **40.** $\sin(-1)$ **41.** sin 5

42. cos 6 **43.** tan 6.29 **44.** $\tan(-6.29)$

Concept Check *Each figure in Exercises 45–48 shows angle θ in standard position with its terminal side intersecting the unit circle. Evaluate the six circular function values of θ.*

45.

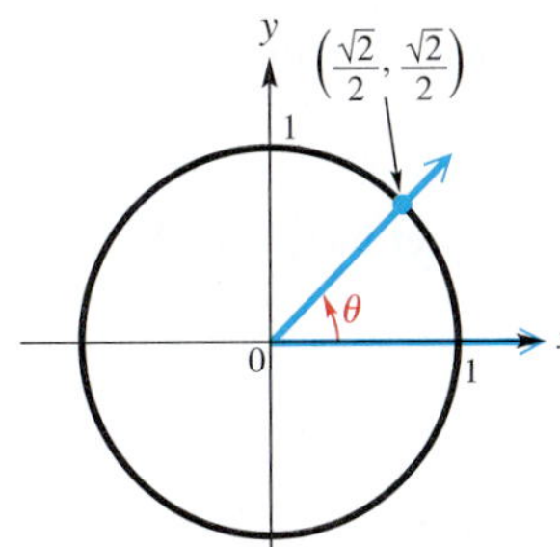

46.

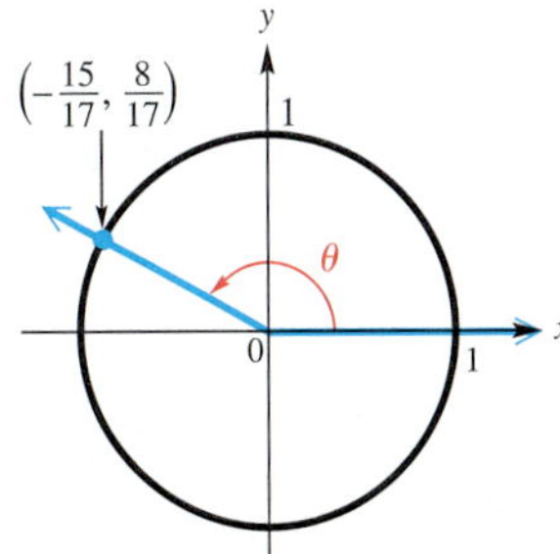

47.

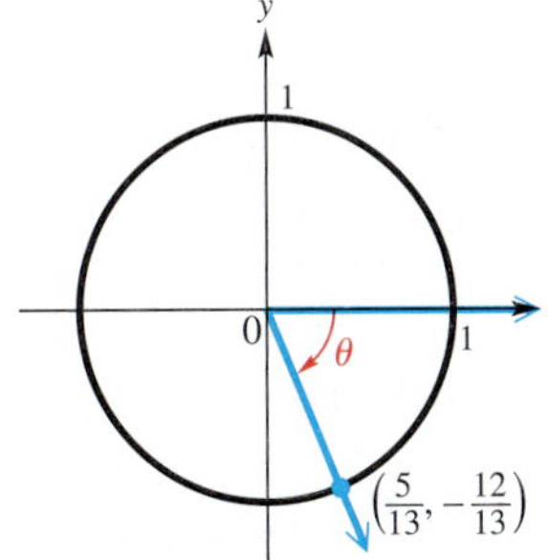

48.

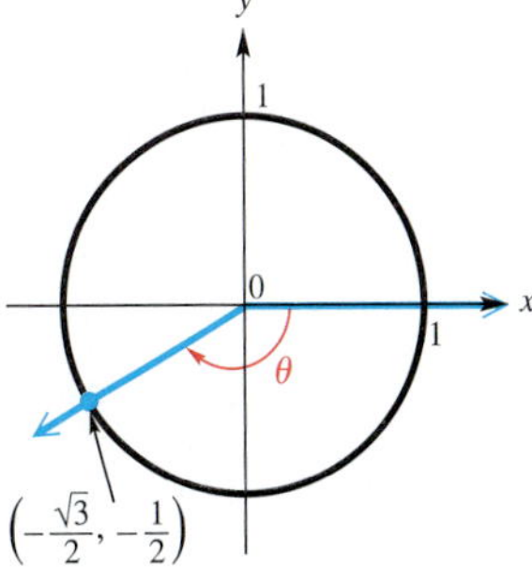

Find the value of s in the interval $\left[0, \frac{\pi}{2}\right]$ that makes each statement true. See Example 4(a).

49. $\tan s = .2126$ **50.** $\cos s = .7826$ **51.** $\sin s = .9918$

52. $\cot s = .2994$ **53.** $\sec s = 1.0806$ **54.** $\csc s = 1.0219$

Find the exact value of s in the given interval that has the given circular function value. Do not use a calculator. See Example 4(b).

55. $\left[\frac{\pi}{2}, \pi\right]$; $\sin s = \frac{1}{2}$ **56.** $\left[\frac{\pi}{2}, \pi\right]$; $\cos s = -\frac{1}{2}$

57. $\left[\pi, \frac{3\pi}{2}\right]$; $\tan s = \sqrt{3}$

58. $\left[\pi, \frac{3\pi}{2}\right]$; $\sin s = -\frac{1}{2}$

59. $\left[\frac{3\pi}{2}, 2\pi\right]$; $\tan s = -1$

60. $\left[\frac{3\pi}{2}, 2\pi\right]$; $\cos s = \frac{\sqrt{3}}{2}$

Suppose an arc of length s lies on the unit circle $x^2 + y^2 = 1$, starting at the point $(1, 0)$ and terminating at the point (x, y). (See Figure 10.) Use a calculator to find the approximate coordinates for (x, y). (Hint: $x = \cos s$ and $y = \sin s$.)

61. $s = 2.5$ **62.** $s = 3.4$ **63.** $s = -7.4$ **64.** $s = -3.9$

Concept Check *For each value of s, use a calculator to find* $\sin s$ *and* $\cos s$ *and then use the results to decide in which quadrant an angle of s radians lies.*

65. $s = 51$ **66.** $s = 49$ **67.** $s = 65$ **68.** $s = 79$

Use the formula $\omega = \frac{\theta}{t}$ to find the value of the missing variable.

69. $\theta = \frac{3\pi}{4}$ radians, $t = 8$ sec

70. $\theta = \frac{2\pi}{5}$ radians, $t = 10$ sec

71. $\theta = \frac{2\pi}{9}$ radian, $\omega = \frac{5\pi}{27}$ radian per min

72. $\theta = \frac{3\pi}{8}$ radians, $\omega = \frac{\pi}{24}$ radian per min

73. $\theta = 3.871142$ radians, $t = 21.4693$ sec

74. $\omega = .90674$ radian per min, $t = 11.876$ min

Use the formula $v = r\omega$ to find the value of the missing variable.

75. $v = 9$ m per sec, $r = 5$ m

76. $v = 18$ ft per sec, $r = 3$ ft

77. $v = 107.692$ m per sec, $r = 58.7413$ m

78. $r = 24.93215$ cm, $\omega = .372914$ radian per sec

The formula $\omega = \frac{\theta}{t}$ can be rewritten as $\theta = \omega t$. Using ωt for θ changes $s = r\theta$ to $s = r\omega t$. Use the formula $s = r\omega t$ to find the value of the missing variable.

79. $r = 6$ cm, $\omega = \frac{\pi}{3}$ radians per sec, $t = 9$ sec

80. $r = 9$ yd, $\omega = \frac{2\pi}{5}$ radians per sec, $t = 12$ sec

81. $s = 6\pi$ cm, $r = 2$ cm, $\omega = \frac{\pi}{4}$ radian per sec

82. $s = \frac{3\pi}{4}$ km, $r = 2$ km, $t = 4$ sec

Find ω for each of the following.

83. the hour hand of a clock

84. a line from the center to the edge of a CD revolving 300 times per min

Find v for each of the following.

85. the tip of the minute hand of a clock, if the hand is 7 cm long

86. a point on the tread of a tire of radius 18 cm, rotating 35 times per min

87. the tip of an airplane propeller 3 m long, rotating 500 times per min (*Hint:* $r = 1.5$ m.)

88. a point on the edge of a gyroscope of radius 83 cm, rotating 680 times per min

89. *Concept Check* If a point moves around the circumference of the unit circle at the speed of 1 unit per sec, how long will it take for the point to move around the entire circle?

90. What is the difference between linear velocity and angular velocity?

Solve each problem. See Examples 5 and 6.

91. *Speed of a Bicycle* The tires of a bicycle have radius 13 in. and are turning at the rate of 200 revolutions per min. See the figure. How fast is the bicycle traveling in miles per hour? (*Hint:* 5280 ft = 1 mi.)

92. *Hours in a Martian Day* Mars rotates on its axis at the rate of about .2552 radian per hr. Approximately how many hours are in a Martian day? (*Source:* Wright, John W. (General Editor), *The Universal Almanac,* Andrews and McMeel, 1997.)

93. *Angular and Linear Speeds of Earth* Earth travels about the sun in an orbit that is almost circular. Assume that the orbit is a circle with radius 93,000,000 mi. Its angular and linear speeds are used in designing solar-power facilities.

(a) Assume that a year is 365 days, and find the angle formed by Earth's movement in one day.
(b) Give the angular speed in radians per hour.
(c) Find the linear speed of Earth in miles per hour.

94. *Angular and Linear Speeds of Earth* Earth revolves on its axis once every 24 hr. Assuming that Earth's radius is 6400 km, find the following.

(a) angular speed of Earth in radians per day and radians per hour
(b) linear speed at the North Pole or South Pole
(c) linear speed at Quito, Ecuador, a city on the equator
(d) linear speed at Salem, Oregon (halfway from the equator to the North Pole)

95. *Speeds of a Pulley and a Belt* The pulley shown has a radius of 12.96 cm. Suppose it takes 18 sec for 56 cm of belt to go around the pulley.

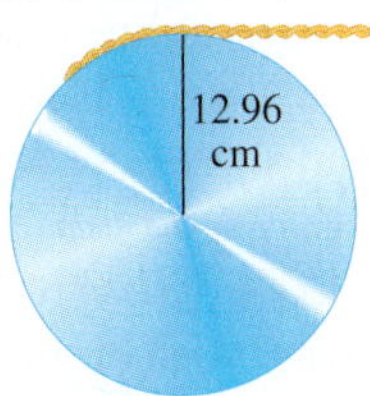

(a) Find the angular speed of the pulley in radians per second.
(b) Find the linear speed of the belt in centimeters per second.

96. *Angular Speed of Pulleys* The two pulleys in the figure have radii of 15 cm and 8 cm, respectively. The larger pulley rotates 25 times in 36 sec. Find the angular speed of each pulley in radians per second.

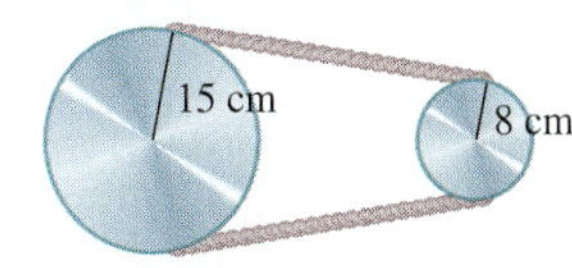

97. *Radius of a Spool of Thread* A thread is being pulled off a spool at the rate of 59.4 cm per sec. Find the radius of the spool if it makes 152 revolutions per min.

98. *Time to Move Along a Railroad Track* A railroad track is laid along the arc of a circle of radius 1800 ft. The circular part of the track subtends a central angle of 40°. How long (in seconds) will it take a point on the front of a train traveling 30 mph to go around this portion of the track?

99. *Angular Speed of a Motor Propeller* A 90-horsepower outboard motor at full throttle will rotate its propeller at 5000 revolutions per min. Find the angular speed of the propeller in radians per second.

100. *Linear Speed of a Golf Club* The shoulder joint can rotate at about 25 radians per sec. If a golfer's arm is straight and the distance from the shoulder to the club head is 5 ft, estimate the linear speed of the club head from shoulder rotation. (*Source:* Cooper, J. and R. Glassow, *Kinesiology,* Second Edition, C.V. Mosby, 1968.)

6.3 Graphs of the Sine and Cosine Functions

Periodic Functions ▪ Graph of the Sine Function ▪ Graph of the Cosine Function ▪ Graphing Techniques, Amplitude, and Period ▪ Translations ▪ Combinations of Translations ▪ Determining a Trigonometric Model Using Curve Fitting

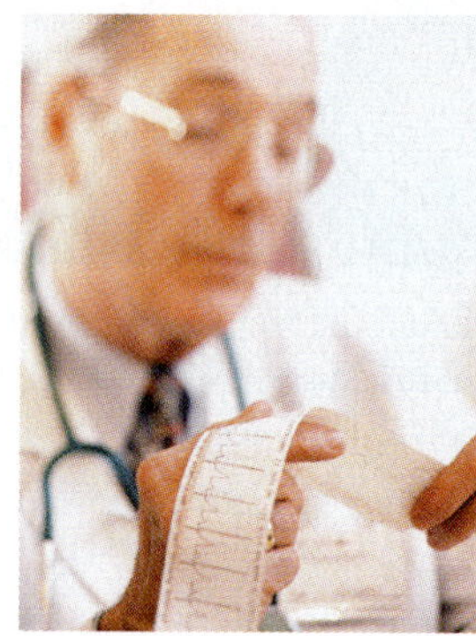

Periodic Functions Many things in daily life repeat with a predictable pattern: in warm areas electricity use goes up in summer and down in winter, the price of fresh fruit goes down in summer and up in winter, and attendance at amusement parks increases in spring and declines in autumn. Because the sine and cosine functions repeat their values in a regular pattern, they are **periodic functions.** Figure 19 shows a periodic graph that represents a normal heartbeat.

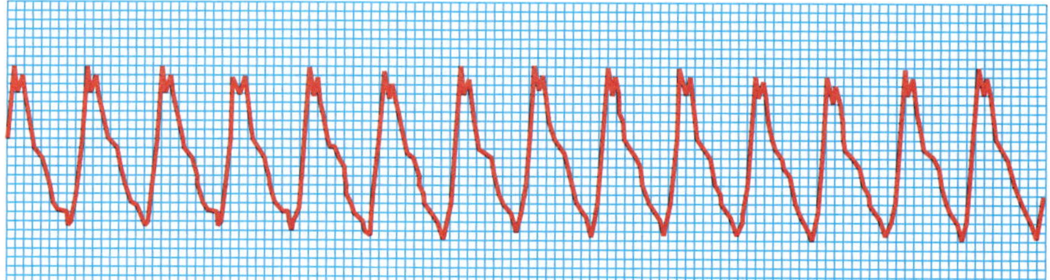

Figure 19

Looking Ahead to Calculus

Periodic functions are used throughout calculus, so you will need to know their characteristics. One use of these functions is to describe the location of a point in the plane using *polar coordinates,* an alternative to rectangular coordinates. (See Chapter 8).

Periodic Function

A **periodic function** is a function f such that

$$f(x) = f(x + np),$$

for every real number x in the domain of f, every integer n, and some positive real number p. The smallest possible positive value of p is the **period** of the function.

The circumference of the unit circle is 2π, so the smallest value of p for which the sine and cosine functions repeat is 2π. Therefore, the sine and cosine functions are periodic functions with period 2π.

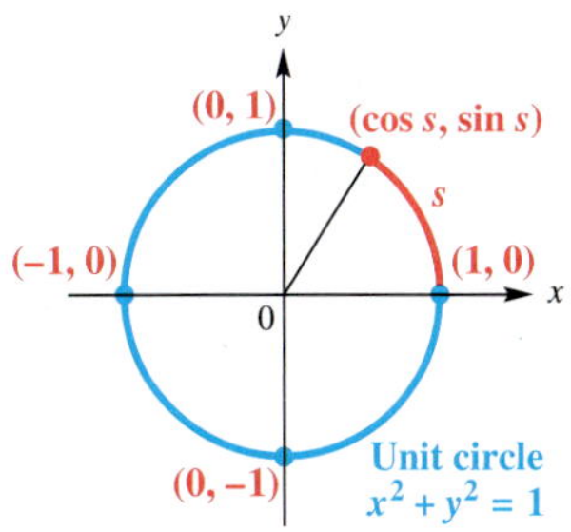

Figure 20

Graph of the Sine Function In Section 6.1 we saw that for a real number s, the point on the unit circle corresponding to s has coordinates $(\cos s, \sin s)$. See Figure 20. Trace along the circle to verify the results shown in the table.

As s Increases from	$\sin s$	$\cos s$
0 to $\frac{\pi}{2}$	Increases from 0 to 1	Decreases from 1 to 0
$\frac{\pi}{2}$ to π	Decreases from 1 to 0	Decreases from 0 to -1
π to $\frac{3\pi}{2}$	Decreases from 0 to -1	Increases from -1 to 0
$\frac{3\pi}{2}$ to 2π	Increases from -1 to 0	Increases from 0 to 1

To avoid confusion when graphing the sine function, we use x rather than s; this corresponds to the letters in the xy-coordinate system. Selecting key values of x and finding the corresponding values of $\sin x$ leads to the table in Figure 21. To obtain the traditional graph in Figure 21, we plot the points from the table, use symmetry, and join them with a smooth curve. Since $y = \sin x$ is periodic with period 2π and has domain $(-\infty, \infty)$, the graph continues in the same pattern in both directions. This graph is called a **sine wave** or **sinusoid.**

SINE FUNCTION $f(x) = \sin x$

Domain: $(-\infty, \infty)$ Range: $[-1, 1]$

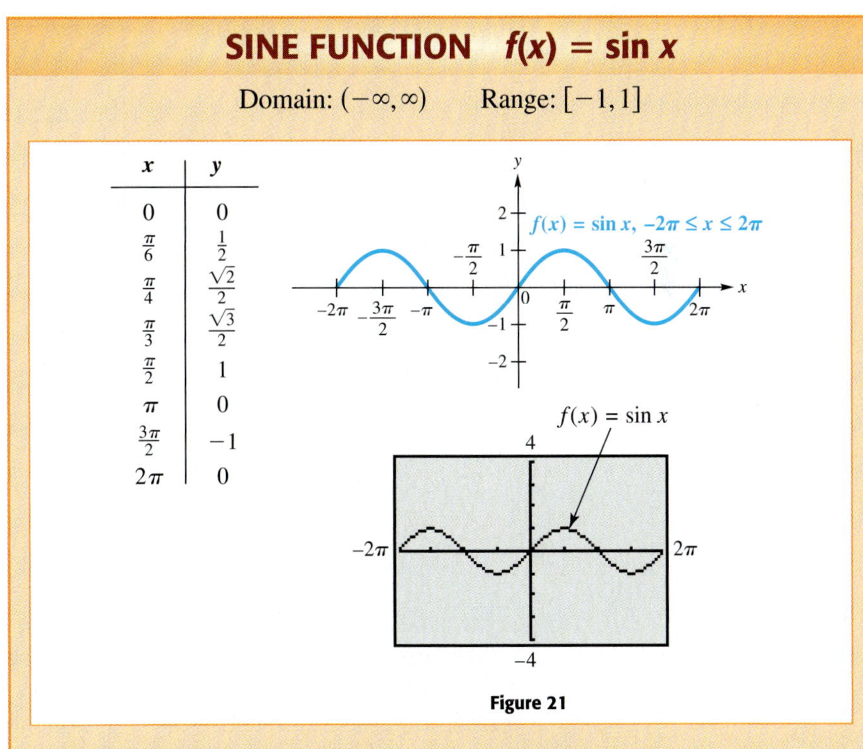

x	y
0	0
$\frac{\pi}{6}$	$\frac{1}{2}$
$\frac{\pi}{4}$	$\frac{\sqrt{2}}{2}$
$\frac{\pi}{3}$	$\frac{\sqrt{3}}{2}$
$\frac{\pi}{2}$	1
π	0
$\frac{3\pi}{2}$	-1
2π	0

Figure 21

- The graph is continuous over its entire domain, $(-\infty, \infty)$.
- Its x-intercepts are of the form $n\pi$, where n is an integer.
- Its period is 2π.
- The graph is symmetric with respect to the origin, so the function is an odd function. For all x in the domain, $\sin(-x) = -\sin x$.

Looking Ahead to Calculus

The discussion of the derivative of a function in calculus shows that for the sine function, the slope of the tangent line at any point x is given by $\cos x$. For example, look at the graph of $y = \sin x$ and notice that a tangent line at $x = \pm\frac{\pi}{2}, \pm\frac{3\pi}{2}, \pm\frac{5\pi}{2}, \ldots$ will be horizontal and thus have slope 0. Now look at the graph of $y = \cos x$ and see that for these values, $\cos x = 0$.

Graph of the Cosine Function We find the graph of $y = \cos x$ in much the same way as the graph of $y = \sin x$. In the table of values shown with Figure 22 for $y = \cos x$, we use the same values for x as we did for the graph of $y = \sin x$. Notice that the graph of $y = \cos x$ in Figure 22 has the same shape as the graph of $y = \sin x$. It is, in fact, the graph of the sine function shifted, or translated, $\frac{\pi}{2}$ units to the left.

COSINE FUNCTION $f(x) = \cos x$

Domain: $(-\infty, \infty)$ Range: $[-1, 1]$

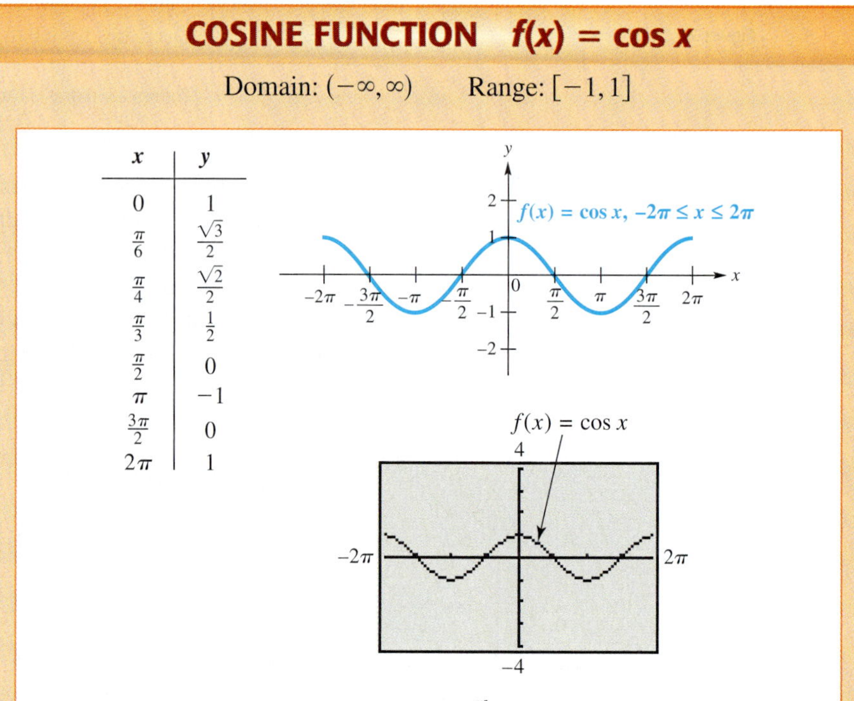

x	y
0	1
$\frac{\pi}{6}$	$\frac{\sqrt{3}}{2}$
$\frac{\pi}{4}$	$\frac{\sqrt{2}}{2}$
$\frac{\pi}{3}$	$\frac{1}{2}$
$\frac{\pi}{2}$	0
π	-1
$\frac{3\pi}{2}$	0
2π	1

Figure 22

- The graph is continuous over its entire domain, $(-\infty, \infty)$.
- Its x-intercepts are of the form $(2n + 1)\frac{\pi}{2}$, where n is an integer.
- Its period is 2π.
- The graph is symmetric with respect to the y-axis, so the function is an even function. For all x in the domain, $\cos(-x) = \cos x$.

Notice that the calculator graphs of $f(x) = \sin x$ in Figure 21 and $f(x) = \cos x$ in Figure 22 are graphed in the window $[-2\pi, 2\pi]$ by $[-4, 4]$, with Xscl $= \frac{\pi}{2}$ and Yscl $= 1$. This is called the **trig viewing window.** (Your model may use a different "standard" trigonometric viewing window. Consult your owner's manual.) ■

Graphing Techniques, Amplitude, and Period The examples that follow show graphs that are "stretched" or "compressed" either vertically, horizontally, or both when compared with the graphs of $y = \sin x$ or $y = \cos x$.

EXAMPLE 1 Graphing $y = a \sin x$

Graph $y = 2 \sin x$, and compare to the graph of $y = \sin x$.

Solution For a given value of x, the value of y is twice as large as it would be for $y = \sin x$, as shown in the table of values. The only change in the graph is the range, which becomes $[-2, 2]$. See Figure 23, which includes a graph of $y = \sin x$ for comparison.

x	0	$\frac{\pi}{2}$	π	$\frac{3\pi}{2}$	2π
$\sin x$	0	1	0	-1	0
$2 \sin x$	0	2	0	-2	0

4

Y1=2sin(X)

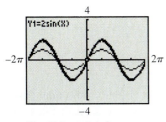

-2π 2π

-4

The thick graph style represents the function in Example 1.

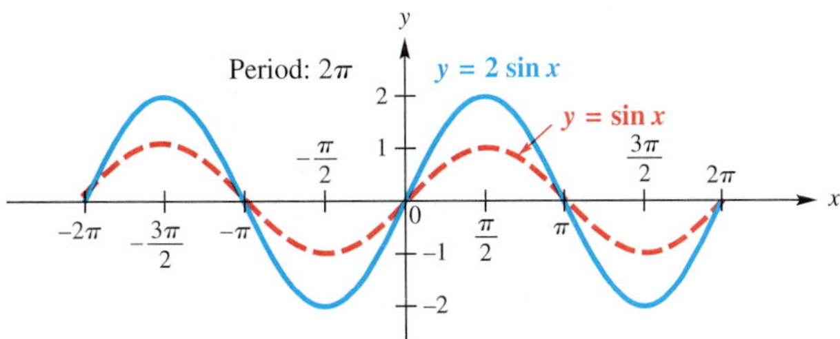

Figure 23

The **amplitude** of a periodic function is half the difference between the maximum and minimum values. Thus, for both the basic sine and cosine functions, the amplitude is

$$\frac{1}{2}[1 - (-1)] = \frac{1}{2}(2) = 1.$$

Generalizing from Example 1 gives the following.

Amplitude

The graph of $\boldsymbol{y = a \sin x}$ or $\boldsymbol{y = a \cos x}$, with $a \neq 0$, will have the same shape as the graph of $y = \sin x$ or $y = \cos x$, respectively, except with range $[-|a|, |a|]$. The amplitude is $|a|$.

Now try Exercise 7.

No matter what the value of the amplitude, the periods of $y = a \sin x$ and $y = a \cos x$ are still 2π. Consider $y = \sin 2x$. We can complete a table of values for the interval $[0, 2\pi]$.

x	0	$\frac{\pi}{4}$	$\frac{\pi}{2}$	$\frac{3\pi}{4}$	π	$\frac{5\pi}{4}$	$\frac{3\pi}{2}$	$\frac{7\pi}{4}$	2π
$\sin 2x$	0	1	0	-1	0	1	0	-1	0

Note that one complete cycle occurs in π units, not 2π units. Therefore, the period here is π, which equals $\frac{2\pi}{2}$. Now consider $y = \sin 4x$. Look at the next table.

x	0	$\frac{\pi}{8}$	$\frac{\pi}{4}$	$\frac{3\pi}{8}$	$\frac{\pi}{2}$	$\frac{5\pi}{8}$	$\frac{3\pi}{4}$	$\frac{7\pi}{8}$	π
$\sin 4x$	0	1	0	-1	0	1	0	-1	0

These values suggest that a complete cycle is achieved in $\frac{\pi}{2}$ or $\frac{2\pi}{4}$ units, which is reasonable since

$$\sin\left(4 \cdot \frac{\pi}{2}\right) = \sin 2\pi = 0.$$

In general, the graph of a function of the form $y = \sin bx$ or $y = \cos bx$, for $b > 0$, will have a period different from 2π when $b \neq 1$. To see why this is so, remember that the values of $\sin bx$ or $\cos bx$ will take on all possible values as bx ranges from 0 to 2π. Therefore, to find the period of either of these functions, we must solve the three-part inequality

$$0 \leq bx \leq 2\pi \quad \text{(Section 1.7)}$$

$$0 \leq x \leq \frac{2\pi}{b}. \quad \text{Divide by the positive number } b.$$

Thus, the period is $\frac{2\pi}{b}$. By dividing the interval $\left[0, \frac{2\pi}{b}\right]$ into four equal parts, we obtain the values for which $\sin bx$ or $\cos bx$ is -1, 0, or 1. These values will give minimum points, x-intercepts, and maximum points on the graph. Once these points are determined, we can sketch the graph by joining the points with a smooth sinusoidal curve. (If a function has $b < 0$, then the identities of the next chapter can be used to rewrite the function so that $b > 0$.)

NOTE One method to divide an interval into four equal parts is as follows.

Step 1 Find the midpoint of the interval by adding the x-values of the endpoints and dividing by 2.

Step 2 Find the midpoints of the two intervals found in Step 1, using the same procedure.

EXAMPLE 2 Graphing $y = \sin bx$

Graph $y = \sin 2x$, and compare to the graph of $y = \sin x$.

Solution In this function the coefficient of x is 2, so the period is $\frac{2\pi}{2} = \pi$. Therefore, the graph will complete one period over the interval $[0, \pi]$.

The endpoints are 0 and π, and the three middle points are

$$\frac{1}{4}(0 + \pi), \quad \frac{1}{2}(0 + \pi), \quad \text{and} \quad \frac{3}{4}(0 + \pi),$$

which give the following x-values.

$0,$	$\frac{\pi}{4},$	$\frac{\pi}{2},$	$\frac{3\pi}{4},$	π
↑	↑	↑	↑	↑
Left endpoint	First-quarter point	Midpoint	Third-quarter point	Right endpoint

We plot the points from the table of values given on page 559, and join them with a smooth sinusoidal curve. More of the graph can be sketched by repeating this cycle, as shown in Figure 24. The amplitude is not changed. The graph of $y = \sin x$ is included for comparison.

Figure 24

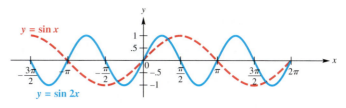

Now try Exercise 15.

Generalizing from Example 2 leads to the following result.

Period

For $b > 0$, the graph of $\mathbf{y = \sin bx}$ will resemble that of $y = \sin x$, but with period $\frac{2\pi}{b}$. Also, the graph of $\mathbf{y = \cos bx}$ will resemble that of $y = \cos x$, but with period $\frac{2\pi}{b}$.

EXAMPLE 3 Graphing $y = \cos bx$

Graph $y = \cos \dfrac{2}{3}x$ over one period.

Solution The period is $\dfrac{2\pi}{\frac{2}{3}} = 3\pi$. We divide the interval $[0, 3\pi]$ into four equal parts to get the following x-values that yield minimum points, maximum points, and x-intercepts.

$$0, \quad \frac{3\pi}{4}, \quad \frac{3\pi}{2}, \quad \frac{9\pi}{4}, \quad 3\pi$$

We use these values to obtain a table of key points for one period.

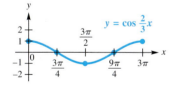

x	0	$\frac{3\pi}{4}$	$\frac{3\pi}{2}$	$\frac{9\pi}{4}$	3π
$\frac{2}{3}x$	0	$\frac{\pi}{2}$	π	$\frac{3\pi}{2}$	2π
$\cos \frac{2}{3}x$	1	0	-1	0	1

Figure 25

The amplitude is 1 because the maximum value is 1, the minimum value is -1, and half of $1 - (-1)$ is $\frac{1}{2}(2) = 1$. We plot these points and join them with a smooth curve. The graph is shown in Figure 25.

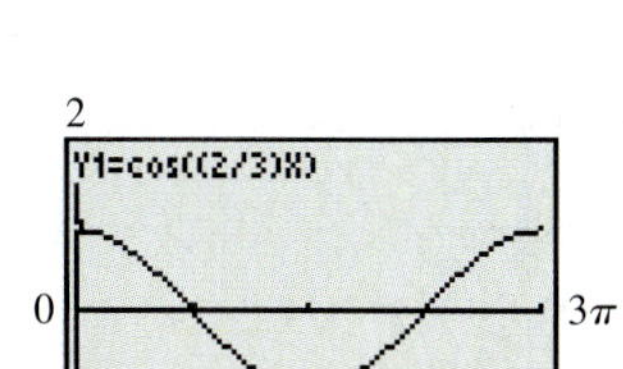

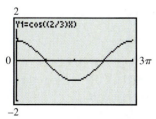

This screen shows a graph of the function in Example 3. By choosing Xscl $= \frac{3\pi}{4}$, the x-intercepts, maxima, and minima coincide with tick marks on the x-axis.

Now try Exercise 13.

NOTE Look at the middle row of the table in Example 3. The method of dividing the interval $\left[0, \frac{2\pi}{b}\right]$ into four equal parts will always give the values 0, $\frac{\pi}{2}$, π, $\frac{3\pi}{2}$, and 2π for this row, resulting in values of -1, 0, or 1 for the circular function. These lead to key points on the graph, which can then be easily sketched.

The method used in Examples 1–3 is summarized as follows.

Guidelines for Sketching Graphs of Sine and Cosine Functions

To graph $y = a \sin bx$ or $y = a \cos bx$, with $b > 0$, follow these steps.

Step 1 Find the period, $\frac{2\pi}{b}$. Start at 0 on the x-axis, and lay off a distance of $\frac{2\pi}{b}$.

Step 2 Divide the interval into four equal parts. (See the Note preceding Example 2.)

Step 3 Evaluate the function for each of the five x-values resulting from Step 2. The points will be maximum points, minimum points, and x-intercepts.

Step 4 Plot the points found in Step 3, and join them with a sinusoidal curve having amplitude $|a|$.

Step 5 Draw the graph over additional periods, to the right and to the left, as needed.

The function in Example 4 has both amplitude and period affected by the values of a and b.

EXAMPLE 4 Graphing $y = a \sin bx$

Graph $y = -2 \sin 3x$ over one period using the preceding guidelines.

Solution

Step 1 For this function, $b = 3$, so the period is $\frac{2\pi}{3}$. The function will be graphed over the interval $\left[0, \frac{2\pi}{3}\right]$.

Step 2 Divide the interval $\left[0, \frac{2\pi}{3}\right]$ into four equal parts to get the x-values 0, $\frac{\pi}{6}$, $\frac{\pi}{3}$, $\frac{\pi}{2}$, and $\frac{2\pi}{3}$.

Step 3 Make a table of values determined by the x-values from Step 2.

x	0	$\frac{\pi}{6}$	$\frac{\pi}{3}$	$\frac{\pi}{2}$	$\frac{2\pi}{3}$
$3x$	0	$\frac{\pi}{2}$	π	$\frac{3\pi}{2}$	2π
$\sin 3x$	0	1	0	-1	0
$-2 \sin 3x$	0	-2	0	2	0

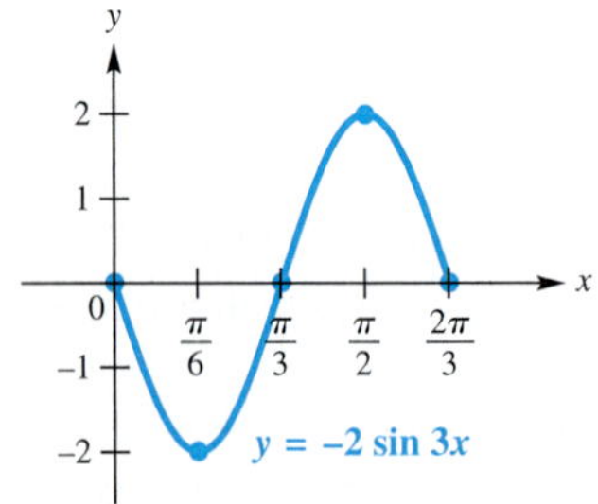

Figure 26

Step 4 Plot the points $(0, 0)$, $\left(\frac{\pi}{6}, -2\right)$, $\left(\frac{\pi}{3}, 0\right)$, $\left(\frac{\pi}{2}, 2\right)$, and $\left(\frac{2\pi}{3}, 0\right)$, and join them with a sinusoidal curve with amplitude 2. See Figure 26.

Step 5 The graph can be extended by repeating the cycle.

Notice that when a is negative, the graph of $y = a \sin bx$ is the reflection across the x-axis of the graph of $y = |a| \sin bx$.

Now try Exercise 19.

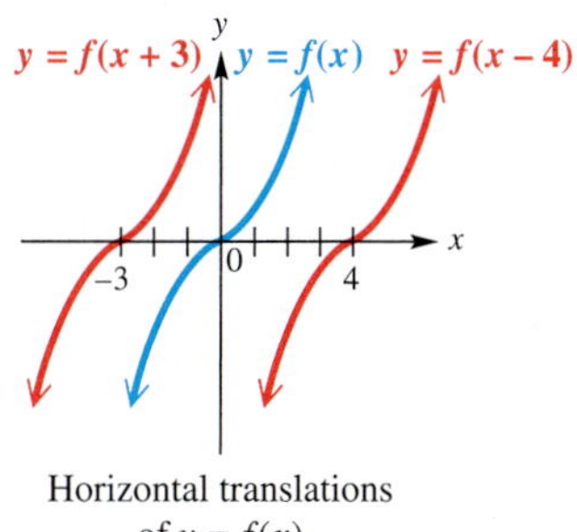

Figure 27

Translations In general, the graph of the function defined by $y = f(x - d)$ is translated *horizontally* when compared to the graph of $y = f(x)$. The translation is d units to the right if $d > 0$ and $|d|$ units to the left if $d < 0$. See Figure 27. With trigonometric functions, a horizontal translation is called a **phase shift.** In the function $y = f(x - d)$, the expression $x - d$ is called the **argument.**

In Example 5, we give two methods that can be used to sketch the graph of a circular function involving a phase shift.

EXAMPLE 5 Graphing $y = \sin(x - d)$

Graph $y = \sin\left(x - \frac{\pi}{3}\right)$.

Solution *Method 1* For the argument $x - \frac{\pi}{3}$ to result in all possible values throughout one period, it must take on all values between 0 and 2π, inclusive. Therefore, to find an interval of one period, we solve the three-part inequality

$$0 \le x - \frac{\pi}{3} \le 2\pi$$

$$\frac{\pi}{3} \le x \le \frac{7\pi}{3}. \qquad \text{Add } \tfrac{\pi}{3} \text{ to each part.}$$

Divide the interval $\left[\frac{\pi}{3}, \frac{7\pi}{3}\right]$ into four equal parts to get the following x-values.

$$\frac{\pi}{3}, \quad \frac{5\pi}{6}, \quad \frac{4\pi}{3}, \quad \frac{11\pi}{6}, \quad \frac{7\pi}{3}$$

A table of values using these x-values follows.

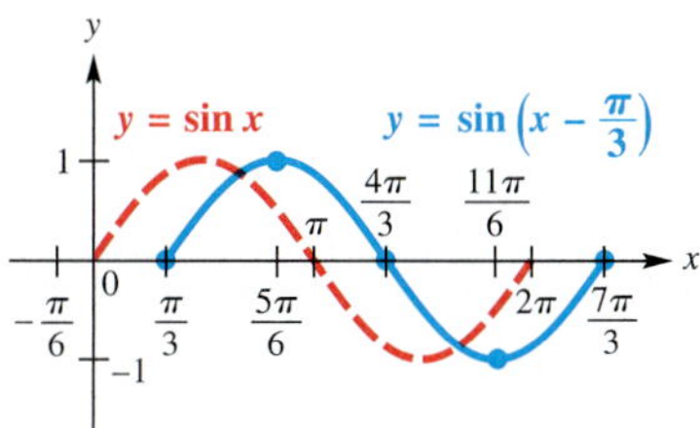

Figure 28

x	$\frac{\pi}{3}$	$\frac{5\pi}{6}$	$\frac{4\pi}{3}$	$\frac{11\pi}{6}$	$\frac{7\pi}{3}$
$x - \frac{\pi}{3}$	0	$\frac{\pi}{2}$	π	$\frac{3\pi}{2}$	2π
$\sin\left(x - \frac{\pi}{3}\right)$	0	1	0	−1	0

We join the corresponding points to get the graph shown in Figure 28. The period is 2π, and the amplitude is 1.

Method 2 We can also graph $y = \sin\left(x - \frac{\pi}{3}\right)$ using a horizontal translation. The argument $x - \frac{\pi}{3}$ indicates that the graph will be translated $\frac{\pi}{3}$ units to the *right* (the phase shift) as compared to the graph of $y = \sin x$. In Figure 28 we show the graph of $y = \sin x$ as a dashed curve, and the graph of $y = \sin\left(x - \frac{\pi}{3}\right)$ as a solid curve. Therefore, to graph a function using this method, first graph the basic circular function, and then graph the desired function by using the appropriate translation.

The graph can be extended through additional periods by repeating this portion of the graph over and over, as necessary.

Now try Exercise 37.

The graph of a function of the form $y = c + f(x)$ is translated *vertically* as compared with the graph of $y = f(x)$. See Figure 29. The translation is c units up if $c > 0$ and $|c|$ units down if $c < 0$.

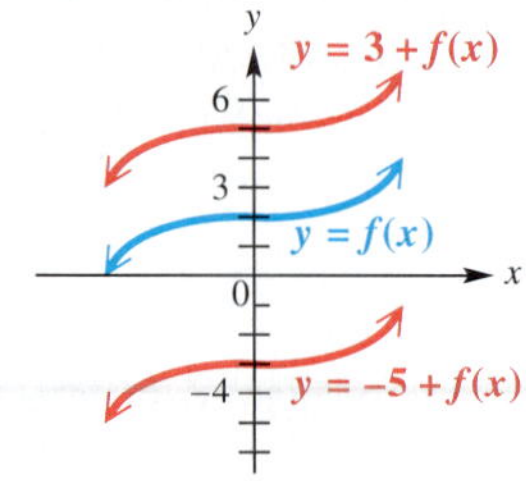

Vertical translations of $y = f(x)$

Figure 29

EXAMPLE 6 Graphing $y = c + a \cos bx$

Graph $y = 3 - 2 \cos 3x$.

Solution The values of y will be 3 greater than the corresponding values of y in $y = -2 \cos 3x$. This means that the graph of $y = 3 - 2 \cos 3x$ is the same as the graph of $y = -2 \cos 3x$, vertically translated 3 units up. Since the period of $y = -2 \cos 3x$ is $\frac{2\pi}{3}$, the key points have x-values

$$0, \quad \frac{\pi}{6}, \quad \frac{\pi}{3}, \quad \frac{\pi}{2}, \quad \frac{2\pi}{3}.$$

Use these x-values to make a table of points.

x	0	$\frac{\pi}{6}$	$\frac{\pi}{3}$	$\frac{\pi}{2}$	$\frac{2\pi}{3}$
$\cos 3x$	1	0	−1	0	1
$2 \cos 3x$	2	0	−2	0	2
$3 - 2 \cos 3x$	1	3	5	3	1

The key points are shown on the graph in Figure 30, along with more of the graph, sketched using the fact that the function is periodic.

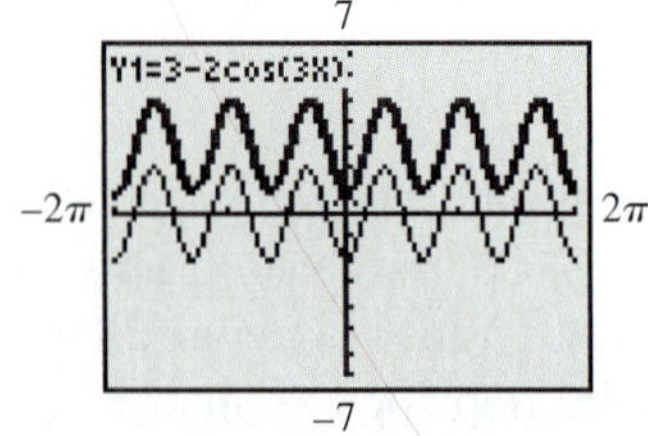

The function in Example 6 is shown using the thick graph style. Notice also the thin graph style for $y = -2 \cos 3x$.

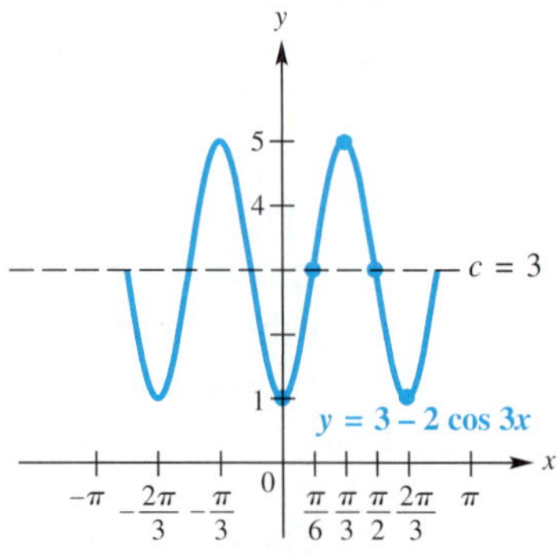

Figure 30

Now try Exercise 45.

Combinations of Translations A function of the form

$$y = c + a \sin b(x - d) \qquad \text{or} \qquad y = c + a \cos b(x - d), \qquad b > 0,$$

can be graphed according to the following guidelines.

Further Guidelines for Sketching Graphs of Sine and Cosine Functions

Method 1 Follow these steps.

Step 1 Find an interval whose length is one period $\frac{2\pi}{b}$ by solving the three-part inequality $0 \le b(x - d) \le 2\pi$.

Step 2 Divide the interval into four equal parts.

Step 3 Evaluate the function for each of the five x-values resulting from Step 2. The points will be maximum points, minimum points, and points that intersect the line $y = c$ ("middle" points of the wave).

Step 4 Plot the points found in Step 3, and join them with a sinusoidal curve having amplitude $|a|$.

Step 5 Draw the graph over additional periods, to the right and to the left, as needed.

Method 2 First graph the basic circular function. The amplitude of the function is $|a|$, and the period is $\frac{2\pi}{b}$. Then use translations to graph the desired function. The vertical translation is c units up if $c > 0$ and $|c|$ units down if $c < 0$. The horizontal translation (phase shift) is d units to the right if $d > 0$ and $|d|$ units to the left if $d < 0$.

EXAMPLE 7 Graphing $y = c + a \sin b(x - d)$

Graph $y = -1 + 2\sin(4x + \pi)$.

Solution We use Method 1. First write the expression in the form $c + a \sin b(x - d)$ by rewriting $4x + \pi$ as $4\left(x + \frac{\pi}{4}\right)$:

$$y = -1 + 2\sin\left[4\left(x + \frac{\pi}{4}\right)\right]. \qquad \text{Rewrite } 4x + \pi \text{ as } 4\left(x + \tfrac{\pi}{4}\right).$$

Step 1 Find an interval whose length is one period.

$$0 \le 4\left(x + \frac{\pi}{4}\right) \le 2\pi$$

$$0 \le x + \frac{\pi}{4} \le \frac{\pi}{2} \qquad \text{Divide by 4.}$$

$$-\frac{\pi}{4} \le x \le \frac{\pi}{4} \qquad \text{Subtract } \tfrac{\pi}{4}.$$

Step 2 Divide the interval $\left[-\frac{\pi}{4}, \frac{\pi}{4}\right]$ into four equal parts to get the x-values

$$-\frac{\pi}{4}, \quad -\frac{\pi}{8}, \quad 0, \quad \frac{\pi}{8}, \quad \frac{\pi}{4}.$$

Step 3 Make a table of values.

x	$-\frac{\pi}{4}$	$-\frac{\pi}{8}$	0	$\frac{\pi}{8}$	$\frac{\pi}{4}$
$x + \frac{\pi}{4}$	0	$\frac{\pi}{8}$	$\frac{\pi}{4}$	$\frac{3\pi}{8}$	$\frac{\pi}{2}$
$4\left(x + \frac{\pi}{4}\right)$	0	$\frac{\pi}{2}$	π	$\frac{3\pi}{2}$	2π
$\sin 4\left(x + \frac{\pi}{4}\right)$	0	1	0	-1	0
$2 \sin 4\left(x + \frac{\pi}{4}\right)$	0	2	0	-2	0
$-1 + 2 \sin(4x + \pi)$	-1	1	-1	-3	-1

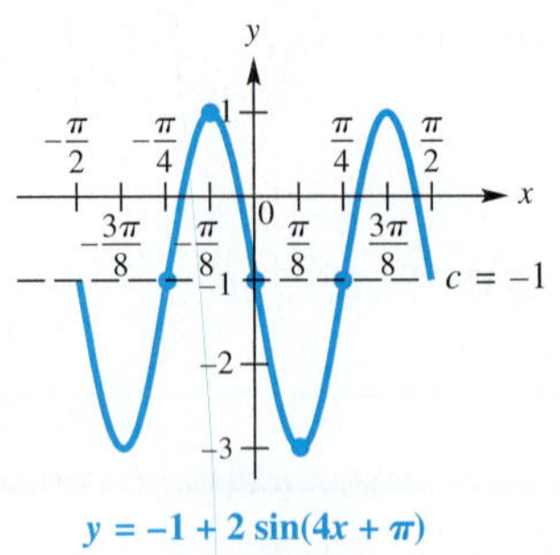

Figure 31

Steps 4 and 5 Plot the points found in the table and join them with a sinusoidal curve. Figure 31 shows the graph, extended to the right and left to include two full periods.

Now try Exercise 49.

Determining a Trigonometric Model Using Curve Fitting A sinusoidal function is often a good approximation of a set of real data points.

EXAMPLE 8 Modeling Temperature with a Sine Function

The maximum average monthly temperature in New Orleans is 82°F and the minimum is 54°F. The table shows the average monthly temperatures. The scatter diagram for a 2-year interval in Figure 32 strongly suggests that the temperatures can be modeled with a sine curve.

Month	°F	Month	°F
Jan	54	July	82
Feb	55	Aug	81
Mar	61	Sept	77
Apr	69	Oct	71
May	73	Nov	59
June	79	Dec	55

Source: Miller, A., J. Thompson, and R. Peterson, *Elements of Meteorology, 4th Edition,* Charles E. Merrill Publishing Co., 1983.

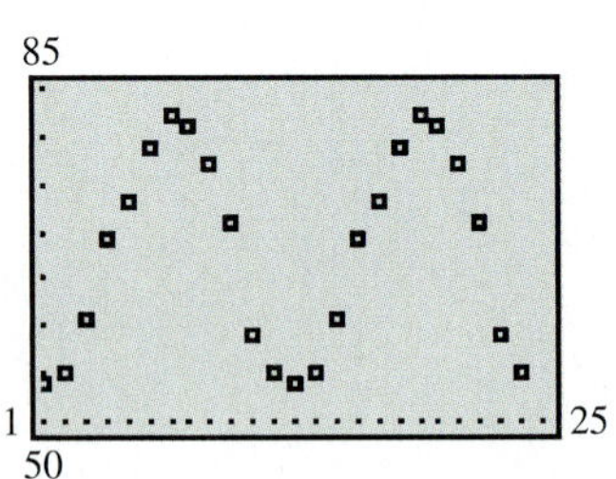

Figure 32

(a) Using only the maximum and minimum temperatures, determine a function of the form $f(x) = a \sin[b(x - d)] + c$, where a, b, c, and d are constants, that models the average monthly temperature in New Orleans. Let x represent the month, with January corresponding to $x = 1$.

(b) On the same coordinate axes, graph f for a two-year period together with the actual data values found in the table.

(c) Use the *sine regression* feature of a graphing calculator to determine a second model for these data.

Solution

(a) We use the maximum and minimum average monthly temperatures to find the amplitude a.

$$a = \frac{82 - 54}{2} = 14$$

The average of the maximum and minimum temperatures is a good choice for c. The average is

$$\frac{82 + 54}{2} = 68.$$

Since the coldest month is January, when $x = 1$, and the hottest month is July, when $x = 7$, we should choose d to be about 4. We experiment with values just greater than 4 to find d. Trial and error using a calculator leads to $d = 4.2$. Since temperatures repeat every 12 months, b is $\frac{2\pi}{12} = \frac{\pi}{6}$. Thus,

$$f(x) = a \sin[b(x - d)] + c = 14 \sin\left[\frac{\pi}{6}(x - 4.2)\right] + 68.$$

(b) Figure 33 shows the data points and the graph of $y = 14 \sin \frac{\pi}{6}x + 68$ for comparison. The horizontal translation of the model is fairly obvious here.

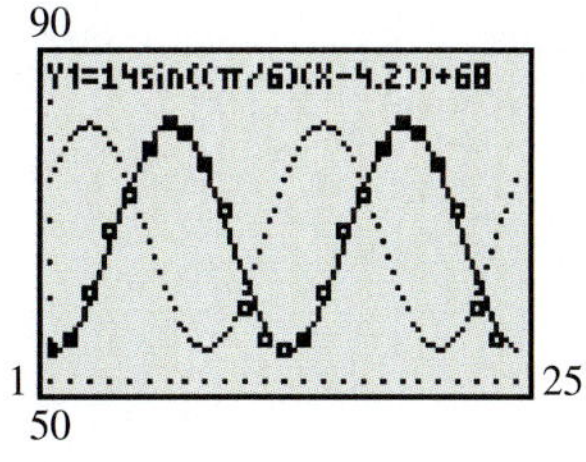

Figure 33

```
SinReg
 y=a*sin(bx+c)+d
 a=14.39
 b=.52
 c=-2.15
 d=67.99
```

Values are rounded to the nearest hundredth.

(a)

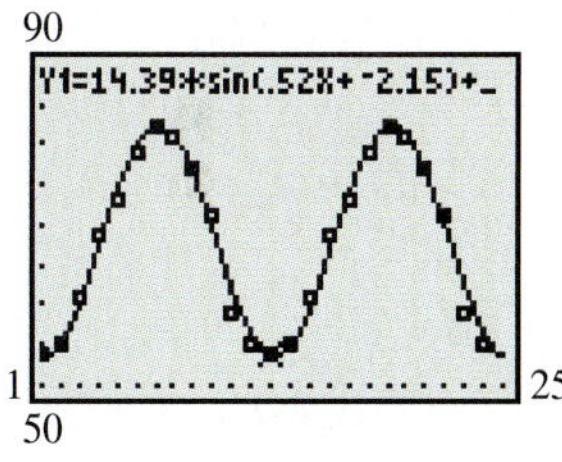

(b)

Figure 34

(c) We used the given data for a two-year period to produce the model described in Figure 34(a). Figure 34(b) shows its graph along with the data points.

Now try Exercise 73.

6.3 Exercises

Concept Check *In Exercises 1–4, match each function with its graph.*

1. $y = -\sin x$

2. $y = -\cos x$

3. $y = \sin 2x$

4. $y = 2 \cos x$

A. 
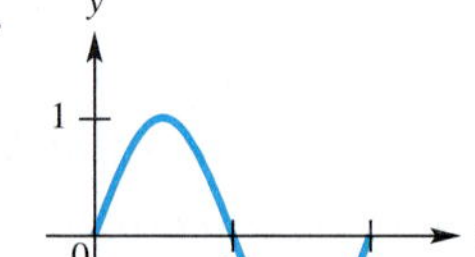

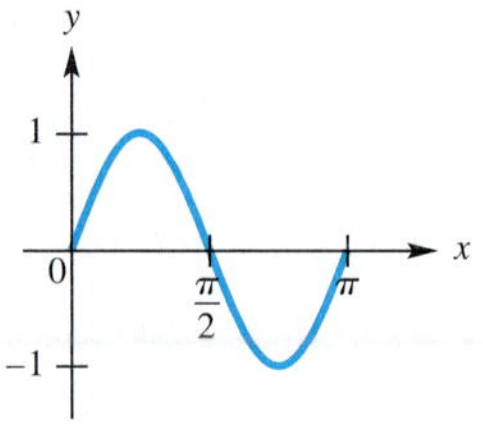

B. 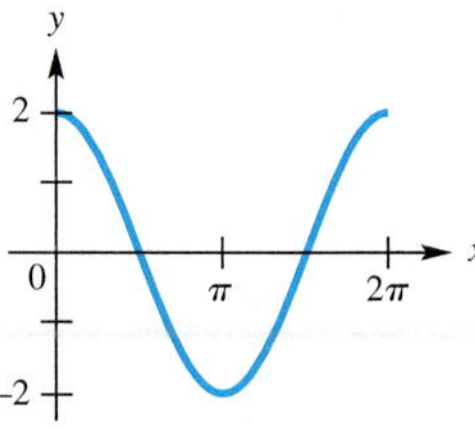

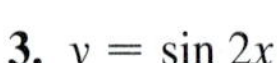

C. 
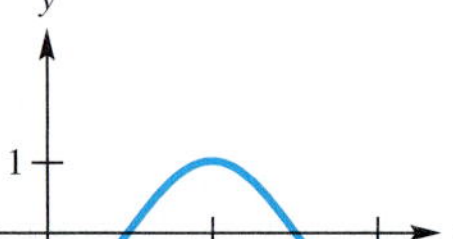

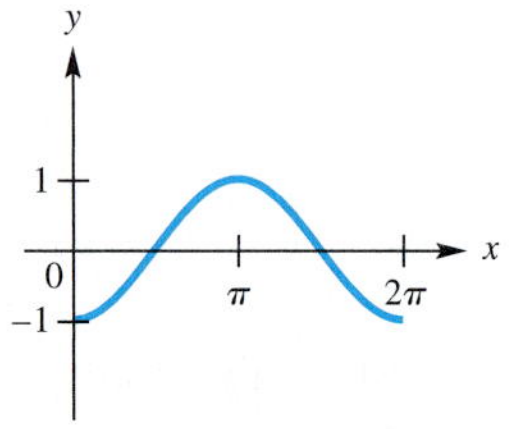

D. 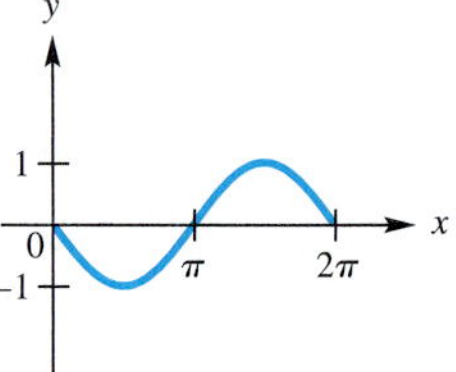

Graph each function over the interval $[-2\pi, 2\pi]$. Give the amplitude. See Example 1.

5. $y = 2 \cos x$

6. $y = 3 \sin x$

7. $y = \frac{2}{3} \sin x$

8. $y = \frac{3}{4} \cos x$

9. $y = -\cos x$

10. $y = -\sin x$

11. $y = -2 \sin x$

12. $y = -3 \cos x$

Graph each function over a two-period interval. Give the period and amplitude. See Examples 2–4.

13. $y = \sin \frac{1}{2}x$

14. $y = \sin \frac{2}{3}x$

15. $y = \cos 2x$

16. $y = \cos \frac{3}{4}x$

17. $y = 2 \sin \frac{1}{4}x$

18. $y = 3 \sin 2x$

19. $y = -2 \cos 3x$

20. $y = -5 \cos 2x$

Concept Check *In Exercises 21 and 22, give the equation of a sine function having the given graph.*

21.

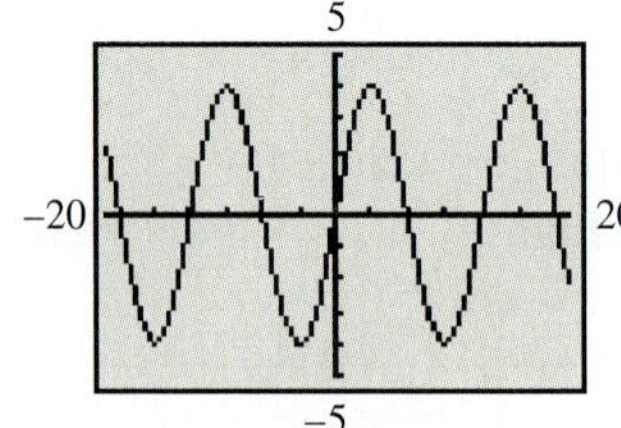

WINDOW
Xmin=-20
Xmax=20
Xscl=3.1415926...
Ymin=-5
Ymax=5
Yscl=1
Xres=1

22.

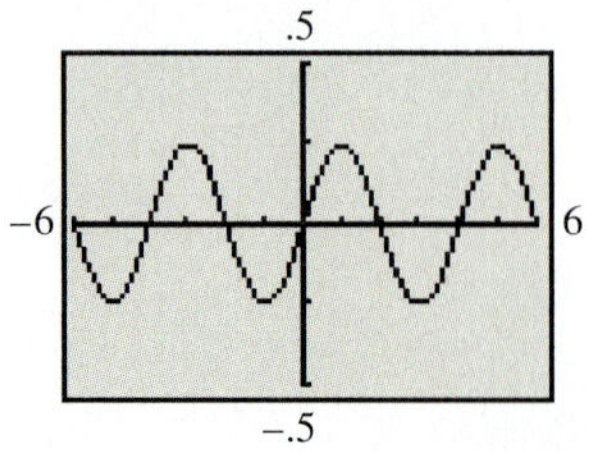

.5
–6
6
–.5

WINDOW
Xmin=-6
Xmax=6
Xscl=1
Ymin=-.5
Ymax=.5
Yscl=.25
Xres=1

Concept Check *Match each function in Column I with the appropriate description in Column II.*

I	II
23. $y = 3\sin(2x - 4)$	**A.** amplitude $= 2$, period $= \frac{\pi}{2}$, phase shift $= \frac{3}{4}$
24. $y = 2\sin(3x - 4)$	**B.** amplitude $= 3$, period $= \pi$, phase shift $= 2$
25. $y = 4\sin(3x - 2)$	**C.** amplitude $= 4$, period $= \frac{2\pi}{3}$, phase shift $= \frac{2}{3}$
26. $y = 2\sin(4x - 3)$	**D.** amplitude $= 2$, period $= \frac{2\pi}{3}$, phase shift $= \frac{4}{3}$

Concept Check *Match each function with its graph.*

27. $y = \sin\left(x - \frac{\pi}{4}\right)$

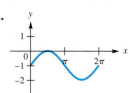

28. $y = \sin\left(x + \frac{\pi}{4}\right)$

A.

B.

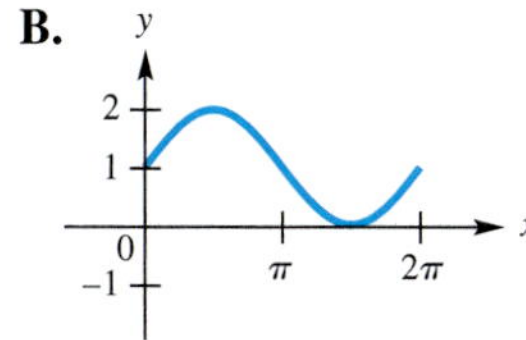

29. $y = 1 + \sin x$

30. $y = -1 + \sin x$

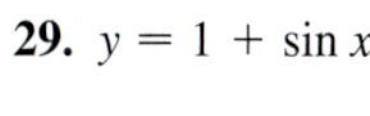

C.

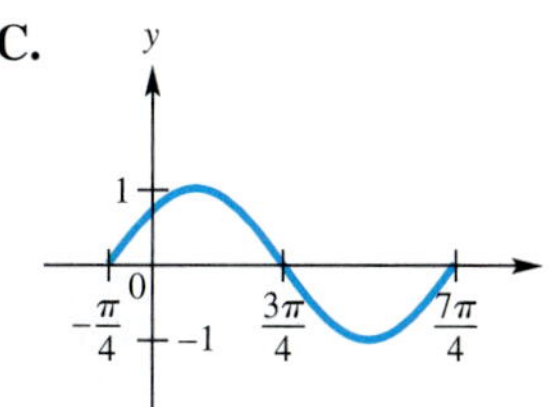

D.

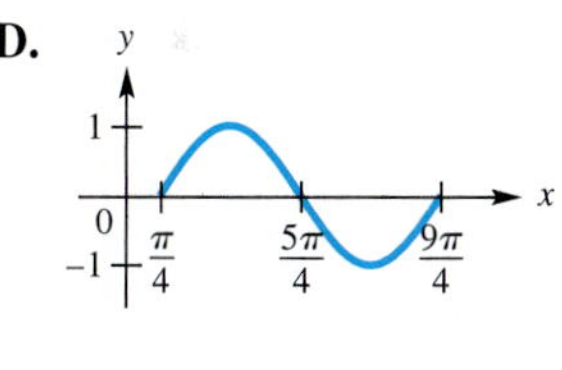

Find the amplitude, the period, any vertical translation, and any phase shift of the graph of each function. See Examples 5–7.

31. $y = 2\sin(x - \pi)$

32. $y = \frac{2}{3}\sin\left(x + \frac{\pi}{2}\right)$

33. $y = 4\cos\left(\frac{x}{2} + \frac{\pi}{2}\right)$

34. $y = \frac{1}{2}\sin\left(\frac{x}{2} + \pi\right)$

35. $y = 2 - \sin\left(3x - \frac{\pi}{5}\right)$

36. $y = -1 + \frac{1}{2}\cos(2x - 3\pi)$

Graph each function over a two-period interval. See Example 5.

37. $y = \sin\left(x - \frac{\pi}{4}\right)$

38. $y = \cos\left(x - \frac{\pi}{3}\right)$

39. $y = 2\cos\left(x - \frac{\pi}{3}\right)$

40. $y = 3\sin\left(x - \frac{3\pi}{2}\right)$

Graph each function over a one-period interval.

41. $y = -4\sin(2x - \pi)$

42. $y = 3\cos(4x + \pi)$

43. $y = \frac{1}{2}\cos\left(\frac{1}{2}x - \frac{\pi}{4}\right)$

44. $y = -\frac{1}{4}\sin\left(\frac{3}{4}x + \frac{\pi}{8}\right)$

Graph each function over a two-period interval. See Example 6.

45. $y = -1 - 2\cos 5x$

46. $y = 1 - \frac{2}{3}\sin\frac{3}{4}x$

47. $y = 1 - 2 \cos \frac{1}{2}x$

48. $y = -3 + 3 \sin \frac{1}{2}x$

Graph each function over a one-period interval. See Example 7.

49. $y = -3 + 2 \sin\left(x + \frac{\pi}{2}\right)$

50. $y = 4 - 3 \cos(x - \pi)$

51. $y = \frac{1}{2} + \sin 2\left(x + \frac{\pi}{4}\right)$

52. $y = -\frac{5}{2} + \cos 3\left(x - \frac{\pi}{6}\right)$

Concept Check *In Exercises 53 and 54, find the equation of a sine function having the given graph.*

53. (*Note:* Xscl $= \frac{\pi}{4}$.)

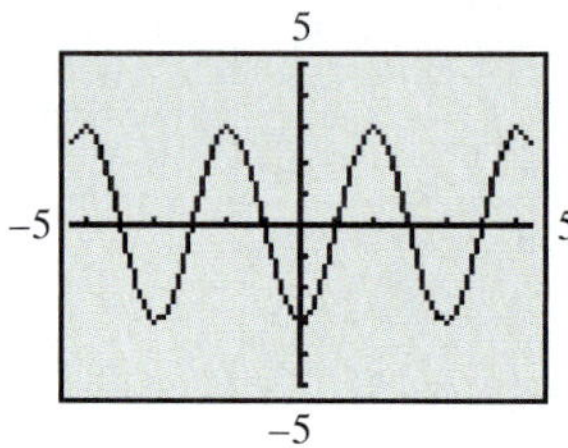

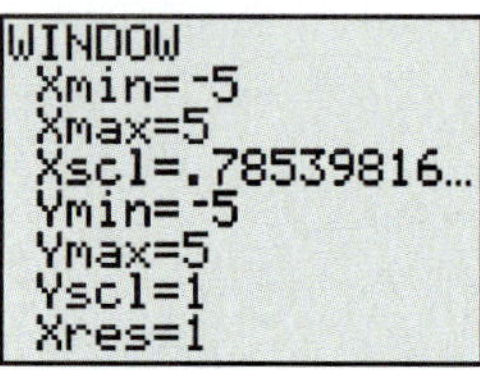

```
WINDOW
 Xmin=-5
 Xmax=5
 Xscl=.78539816…
 Ymin=-5
 Ymax=5
 Yscl=1
 Xres=1
```

54. (*Note:* Yscl $= \pi$.)

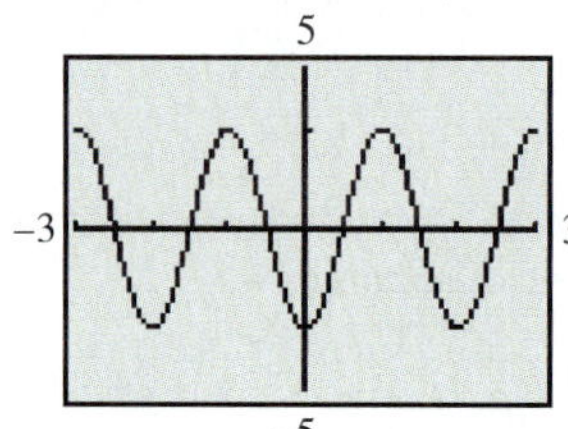

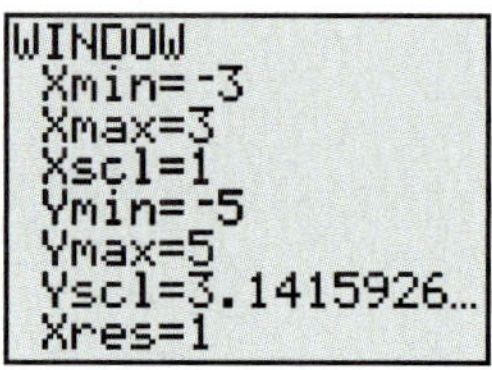

```
WINDOW
 Xmin=-3
 Xmax=3
 Xscl=1
 Ymin=-5
 Ymax=5
 Yscl=3.1415926…
 Xres=1
```

(Modeling) *Solve each problem.*

55. ***Average Annual Temperature*** Scientists believe that the average annual temperature in a given location is periodic. The average temperature at a given place during a given season fluctuates as time goes on, from colder to warmer, and back to colder. The graph shows an idealized description of the temperature (in °F) for the last few thousand years of a location at the same latitude as Anchorage, Alaska.

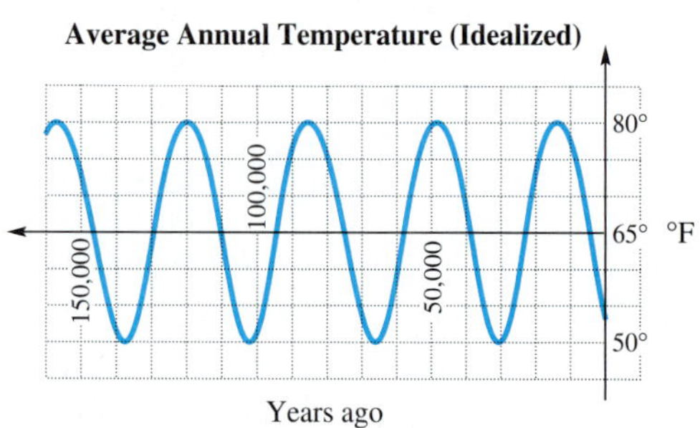

(a) Find the highest and lowest temperatures recorded.

(b) Use these two numbers to find the amplitude.

(c) Find the period of the function.

(d) What is the trend of the temperature now?

56. *Blood Pressure Variation* The graph gives the variation in blood pressure for a typical person. Systolic and diastolic pressures are the upper and lower limits of the periodic changes in pressure that produce the pulse. The length of time between peaks is called the period of the pulse.

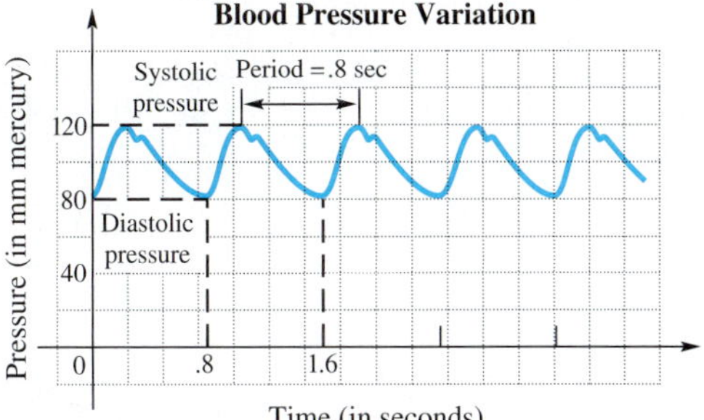

(a) Find the amplitude of the graph.
(b) Find the pulse rate (the number of pulse beats in 1 min) for this person.

57. *Activity of a Nocturnal Animal* Many of the activities of living organisms are periodic. For example, the graph below shows the time that a certain nocturnal animal begins its evening activity.

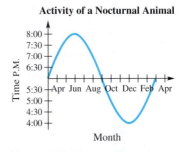

Activity of a Nocturnal Animal

(a) Find the amplitude of this graph. **(b)** Find the period.

58. *Position of a Moving Arm* The figure shows schematic diagrams of a rhythmically moving arm. The upper arm RO rotates back and forth about the point R; the position of the arm is measured by the angle y between the actual position and the downward vertical position. (*Source:* De Sapio, Rodolfo, *Calculus for the Life Sciences.* Copyright © 1978 by W. H. Freeman and Company. Reprinted by permission.)

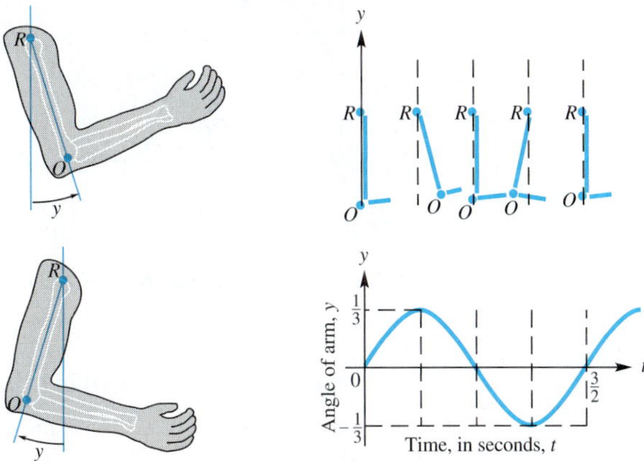

This graph shows the relationship between angle y and time t in seconds.

(a) Find an equation of the form $y = a \sin kt$ for the graph shown.
(b) How long does it take for a complete movement of the arm?

Tides for Kahului Harbor *The chart shows the tides for Kahului Harbor (on the island of Maui, Hawaii). To identify high and low tides and times for other Maui areas, the following adjustments must be made.*

Hana:	High, +40 min, +.1 ft;	Makena:	High, +1:21, −.5 ft;
	Low, +18 min, −.2 ft		Low, +1:09, −.2 ft
Maalaea:	High, +1:52, −.1 ft;	Lahaina:	High, +1:18, −.2 ft;
	Low, +1:19, −.2 ft		Low, +1:01, −.1 ft

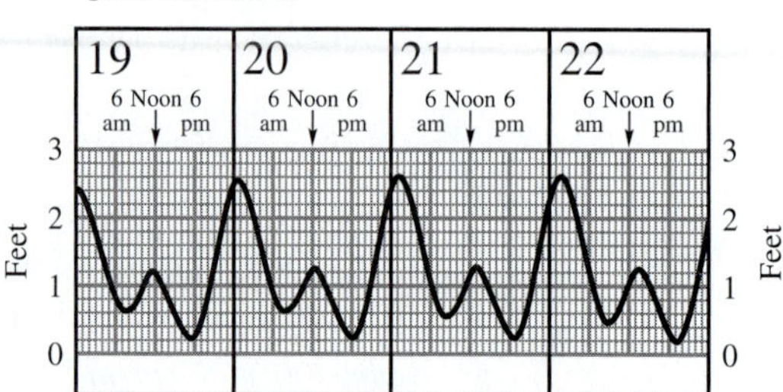

Source: Maui News. Original chart prepared by Edward K. Noda and Associates.

Use the graph to work Exercises 59–64.

59. The graph is an example of a periodic function. What is the period (in hours)?

60. What is the amplitude?

61. At what time on January 20 was low tide at Kahului? What was the height?

62. Repeat Exercise 61 for Maalaea.

63. At what time on January 22 was high tide at Kahului? What was the height?

64. Repeat Exercise 63 for Lahaina.

Musical Sound Waves *Pure sounds produce single sine waves on an oscilloscope. Find the amplitude and period of each sine wave graph in Exercises 65 and 66. On the vertical scale, each square represents .5; on the horizontal scale, each square represents* $30°$ *or* $\frac{\pi}{6}$.

65.

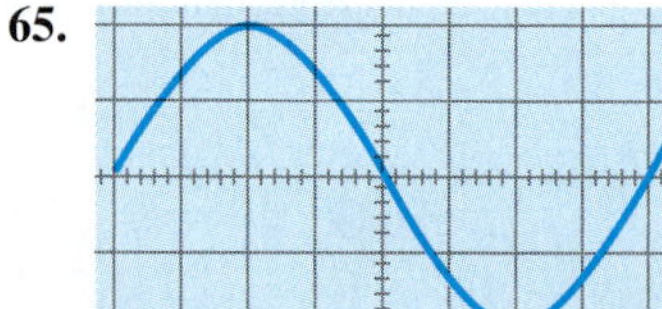

66.

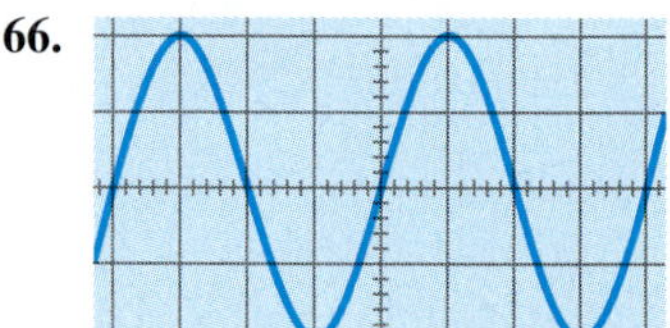

(Modeling) *Solve each problem.*

67. ***Voltage of an Electrical Circuit*** The voltage E in an electrical circuit is modeled by

$$E = 5 \cos 120\pi t,$$

where t is time measured in seconds.

(a) Find the amplitude and the period.

(b) How many cycles are completed in 1 sec? (The number of cycles (periods) completed in 1 sec is the *frequency* of the function.)

(c) Find E when $t = 0, .03, .06, .09, .12$.

(d) Graph E for $0 \leq t \leq \frac{1}{30}$.

68. ***Voltage of an Electrical Circuit*** For another electrical circuit, the voltage E is modeled by

$$E = 3.8 \cos 40\pi t,$$

where t is time measured in seconds.

(a) Find the amplitude and the period.
(b) Find the frequency. See Exercise 67(b).
(c) Find E when $t = .02, .04, .08, .12, .14$.
(d) Graph one period of E.

69. ***Atmospheric Carbon Dioxide*** At Mauna Loa, Hawaii, atmospheric carbon dioxide levels in parts per million (ppm) have been measured regularly since 1958. The function defined by

$$L(x) = .022x^2 + .55x + 316 + 3.5 \sin(2\pi x)$$

can be used to model these levels, where x is in years and $x = 0$ corresponds to 1960. (*Source:* Nilsson, A., *Greenhouse Earth,* John Wiley & Sons, 1992.)

(a) Graph L in the window $[15, 35]$ by $[325, 365]$.
(b) When do the seasonal maximum and minimum carbon dioxide levels occur?
(c) L is the sum of a quadratic function and a sine function. What is the significance of each of these functions? Discuss what physical phenomena may be responsible for each function.

70. ***Atmospheric Carbon Dioxide*** Refer to Exercise 69. The carbon dioxide content in the atmosphere at Barrow, Alaska, in parts per million (ppm) can be modeled using the function defined by

$$C(x) = .04x^2 + .6x + 330 + 7.5 \sin(2\pi x),$$

where $x = 0$ corresponds to 1970. (*Source:* Zeilik, M. and S. Gregory, *Introductory Astronomy and Astrophysics,* Brooks/Cole, 1998.)

(a) Graph C in the window $[5, 25]$ by $[320, 380]$.
(b) Discuss possible reasons why the amplitude of the oscillations in the graph of C is larger than the amplitude of the oscillations in the graph of L in Exercise 69, which models Hawaii.
(c) Define a new function C that is valid if x represents the actual year, where $1970 \le x \le 1995$.

71. ***Temperature in Fairbanks*** The temperature in Fairbanks is modeled by

$$T(x) = 37 \sin\left[\frac{2\pi}{365}(x - 101)\right] + 25,$$

where $T(x)$ is the temperature in degrees Fahrenheit on day x, with $x = 1$ corresponding to January 1 and $x = 365$ corresponding to December 31. Use a calculator to estimate the temperature on the following days. (*Source:* Lando, B. and C. Lando, "Is the Graph of Temperature Variation a Sine Curve?", *The Mathematics Teacher,* 70, September 1977.)

(a) March 1 (day 60) **(b)** April 1 (day 91) **(c)** Day 150
(d) June 15 **(e)** September 1 **(f)** October 31

72. ***Fluctuation in the Solar Constant*** The *solar constant* S is the amount of energy per unit area that reaches Earth's atmosphere from the sun. It is equal to 1367 watts per square meter but varies slightly throughout the seasons. This fluctuation ΔS in S can be calculated using the formula

$$\Delta S = .034S \sin\left[\frac{2\pi(82.5 - N)}{365.25}\right].$$

In this formula, N is the day number covering a four-year period, where $N = 1$ corresponds to January 1 of a leap year and $N = 1461$ corresponds to December 31 of the fourth year. (*Source:* Winter, C., R. Sizmann, and Vant-Hunt (Editors), *Solar Power Plants,* Springer-Verlag, 1991.)

(a) Calculate ΔS for $N = 80$, which is the spring equinox in the first year.
(b) Calculate ΔS for $N = 1268$, which is the summer solstice in the fourth year.
(c) What is the maximum value of ΔS?
(d) Find a value for N where ΔS is equal to 0.

(Modeling) Solve each problem. See Example 8.

73. ***Average Monthly Temperature*** The average monthly temperature (in °F) in Vancouver, Canada, is shown in the table.

Month	°F	Month	°F
Jan	36	July	64
Feb	39	Aug	63
Mar	43	Sept	57
Apr	48	Oct	50
May	55	Nov	43
June	59	Dec	39

Source: Miller, A. and J. Thompson, *Elements of Meteorology, 4th Edition,* Charles E. Merrill Publishing Co., 1983.

(a) Plot the average monthly temperature over a two-year period letting $x = 1$ correspond to the month of January during the first year. Do the data seem to indicate a translated sine graph?
(b) The highest average monthly temperature is 64°F in July, and the lowest average monthly temperature is 36°F in January. Their average is 50°F. Graph the data together with the line $y = 50$. What does this line represent with regard to temperature in Vancouver?
(c) Approximate the amplitude, period, and phase shift of the translated sine wave.
(d) Determine a function of the form $f(x) = a \sin b(x - d) + c$, where a, b, c, and d are constants, that models the data.
(e) Graph f together with the data on the same coordinate axes. How well does f model the given data?
(f) Use the sine regression capability of a graphing calculator to find the equation of a sine curve that fits these data.

74. ***Average Monthly Temperature*** The average monthly temperature (in °F) in Phoenix, Arizona, is shown in the table.

Month	°F	Month	°F
Jan	51	July	90
Feb	55	Aug	90
Mar	63	Sept	84
Apr	67	Oct	71
May	77	Nov	59
June	86	Dec	52

Source: Miller, A. and J. Thompson, *Elements of Meteorology, 4th Edition,* Charles E. Merrill Publishing Co., 1983.

(a) Predict the average yearly temperature and compare it to the actual value of 70°F.
(b) Plot the average monthly temperature over a two-year period by letting $x = 1$ correspond to January of the first year.
(c) Determine a function of the form $f(x) = a \cos b(x - d) + c$, where a, b, c, and d are constants, that models the data.
(d) Graph f together with the data on the same coordinate axes. How well does f model the data?
(e) Use the sine regression capability of a graphing calculator to find the equation of a sine curve that fits these data.

6.4 Graphs of the Other Circular Functions

Graphs of the Cosecant and Secant Functions ▪ Graphs of the Tangent and Cotangent Functions ▪ Addition of Ordinates

Graphs of the Cosecant and Secant Functions Since cosecant values are reciprocals of the corresponding sine values, the period of the function $y = \csc x$ is 2π, the same as for $y = \sin x$. When $\sin x = 1$, the value of $\csc x$ is also 1, and when $0 < \sin x < 1$, then $\csc x > 1$. Also, if $-1 < \sin x < 0$, then $\csc x < -1$. (Verify these statements.) As $|x|$ approaches 0, $|\sin x|$ approaches 0, and $|\csc x|$ gets larger and larger. The graph of $y = \csc x$ approaches the vertical line $x = 0$ but never touches it, so the line $x = 0$ is a *vertical asymptote.* In fact, the lines $x = n\pi$, where n is any integer, are all vertical asymptotes.

Using this information and plotting a few points shows that the graph takes the shape of the solid curve shown in Figure 35. To show how the two graphs are related, the graph of $y = \sin x$ is shown as a dashed curve.

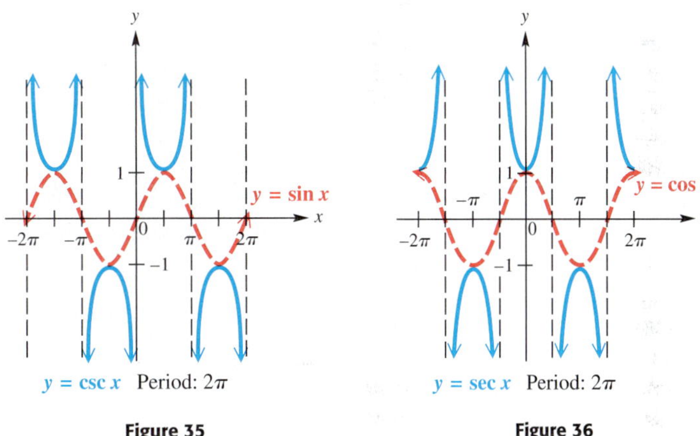

Figure 35

Figure 36

A similar analysis for the secant leads to the solid curve shown in Figure 36. The dashed curve, $y = \cos x$, is shown so that the relationship between these two reciprocal functions can be seen.

Typically, calculators do not have keys for the cosecant and secant functions. To graph $y = \csc x$ with a graphing calculator, use the fact that

$$\csc x = \frac{1}{\sin x}.$$

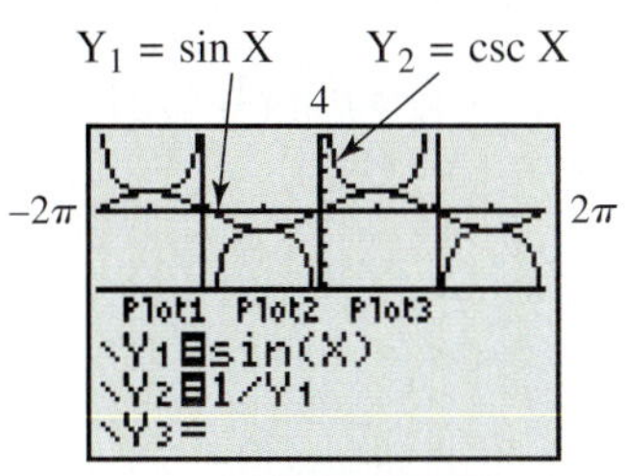

Trig window; connected mode

Figure 37

The graphs of $Y_1 = \sin X$ and $Y_2 = \csc X$ are shown in Figure 37. The calculator is in split screen and connected modes. Similarly, the secant function is graphed by using the identity

$$\sec x = \frac{1}{\cos x},$$

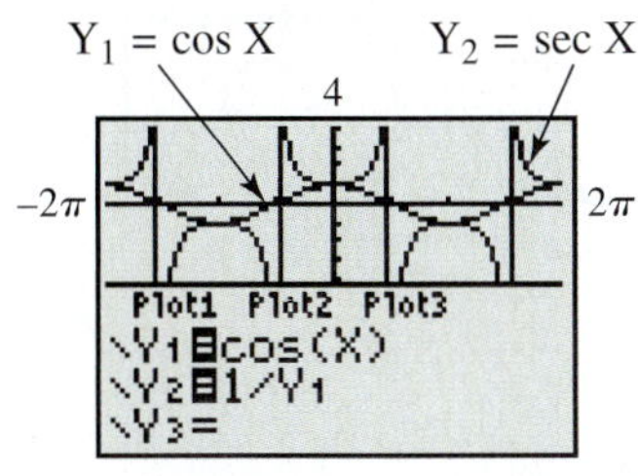

Trig window; connected mode

Figure 38

as shown in Figure 38.

Using dot mode for graphing will eliminate the vertical lines that appear in Figures 37 and 38. While they suggest asymptotes and are sometimes called *pseudo-asymptotes,* they are not actually parts of the graphs. See Figure 39 on the next page, for example. ■

COSECANT FUNCTION $f(x) = \csc x$

Domain: $\{x \mid x \neq n\pi$, where n is an integer$\}$ Range: $(-\infty, -1] \cup [1, \infty)$

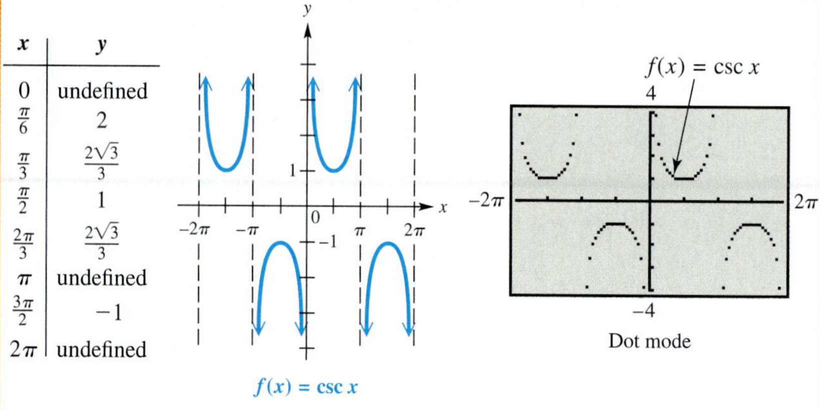

x	y
0	undefined
$\frac{\pi}{6}$	2
$\frac{\pi}{3}$	$\frac{2\sqrt{3}}{3}$
$\frac{\pi}{2}$	1
$\frac{2\pi}{3}$	$\frac{2\sqrt{3}}{3}$
π	undefined
$\frac{3\pi}{2}$	-1
2π	undefined

Figure 39

- The graph is discontinuous at values of x of the form $x = n\pi$ and has vertical asymptotes at these values.
- There are no x-intercepts.
- Its period is 2π.
- Its graph has no amplitude, since there are no maximum or minimum values.
- The graph is symmetric with respect to the origin, so the function is an odd function. For all x in the domain, $\csc(-x) = -\csc x$.

SECANT FUNCTION $f(x) = \sec x$

Domain: $\left\{x \mid x \neq (2n + 1)\frac{\pi}{2}\right.$, where n is an integer$\left.\right\}$ Range: $(-\infty, -1] \cup [1, \infty)$

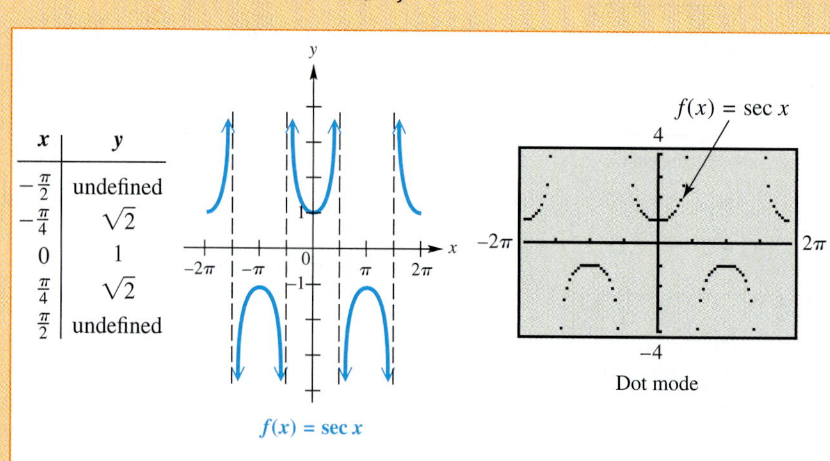

x	y
$-\frac{\pi}{2}$	undefined
$-\frac{\pi}{4}$	$\sqrt{2}$
0	1
$\frac{\pi}{4}$	$\sqrt{2}$
$\frac{\pi}{2}$	undefined

Figure 40

(continued)

- The graph is discontinuous at values of x of the form $x = (2n + 1)\frac{\pi}{2}$ and has vertical asymptotes at these values.
- There are no x-intercepts.
- Its period is 2π.
- Its graph has no amplitude, since there are no maximum or minimum values.
- The graph is symmetric with respect to the y-axis, so the function is an even function. For all x in the domain, $\sec(-x) = \sec x$.

In the previous section, we gave guidelines for sketching graphs of sine and cosine functions. We now present similar guidelines for graphing cosecant and secant functions.

Guidelines for Sketching Graphs of Cosecant and Secant Functions

To graph $\boldsymbol{y = a \csc bx}$ or $\boldsymbol{y = a \sec bx}$, with $b > 0$, follow these steps.

Step 1 Graph the corresponding reciprocal function as a guide, using a dashed curve.

To Graph	Use as a Guide
$y = a \csc bx$	$y = a \sin bx$
$y = a \sec bx$	$y = a \cos bx$

Step 2 Sketch the vertical asymptotes. They will have equations of the form $x = k$, where k is an x-intercept of the graph of the guide function.

Step 3 Sketch the graph of the desired function by drawing the typical U-shaped branches between the adjacent asymptotes. The branches will be above the graph of the guide function when the guide function values are positive and below the graph of the guide function when the guide function values are negative. The graph will resemble those in Figures 39 and 40 in the function boxes on the previous page.

Like graphs of the sine and cosine functions, graphs of the secant and cosecant functions may be translated vertically and horizontally. The period of both basic functions is 2π.

EXAMPLE 1 Graphing $y = a \sec bx$

Graph $y = 2 \sec \frac{1}{2}x$.

Solution

Step 1 This function involves the secant, so the corresponding reciprocal function will involve the cosine. The guide function to graph is

$$y = 2 \cos \frac{1}{2}x.$$

Using the guidelines of Section 6.3, we find that this guide function has amplitude 2 and one period of the graph lies along the interval that satisfies the inequality

$$0 \leq \frac{1}{2}x \leq 2\pi, \qquad \text{or} \qquad [0, 4\pi]. \quad \text{(Section 1.7)}$$

Dividing this interval into four equal parts gives the key points

$$(0, 2), \quad (\pi, 0), \quad (2\pi, -2), \quad (3\pi, 0), \quad (4\pi, 2),$$

which are joined with a smooth dashed curve to indicate that this graph is only a guide. An additional period is graphed as seen in Figure 41(a).

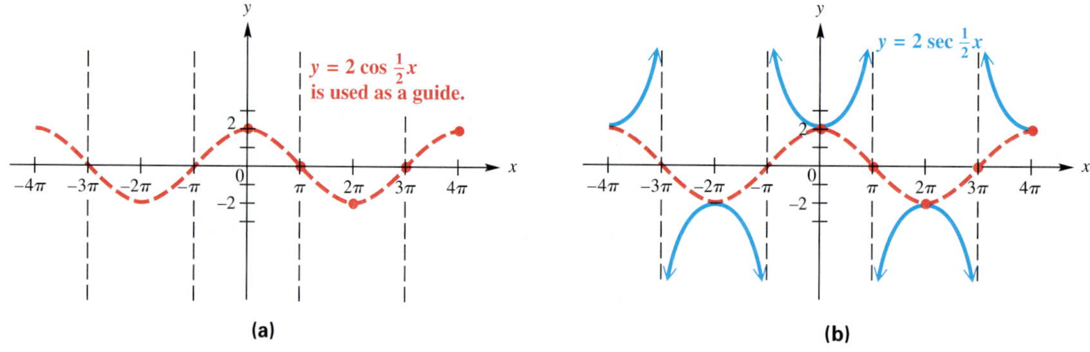

Figure 41

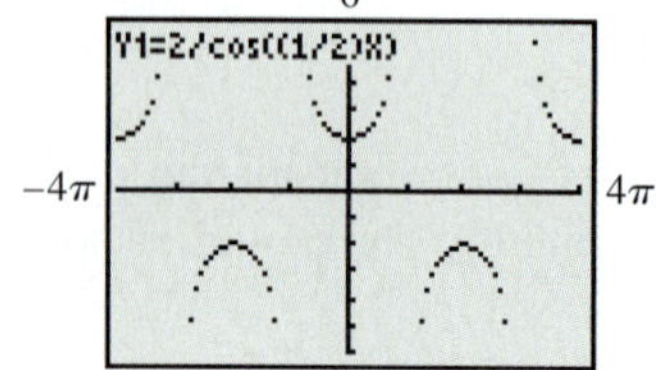

Dot mode

This is a calculator graph of the function in Example 1.

Step 2 Sketch the vertical asymptotes. These occur at x-values for which the guide function equals 0, such as

$$x = -3\pi, \qquad x = -\pi, \qquad x = \pi, \qquad x = 3\pi.$$

See Figure 41(a).

Step 3 Sketch the graph of $y = 2 \sec \frac{1}{2}x$ by drawing the typical U-shaped branches, approaching the asymptotes. See Figure 41(b).

Now try Exercise 7.

EXAMPLE 2 Graphing $y = a \csc(x - d)$

Graph $y = \frac{3}{2}\csc\left(x - \frac{\pi}{2}\right)$.

Solution

Step 1 Use the guidelines of Section 6.3 to graph the corresponding reciprocal function

$$y = \frac{3}{2}\sin\left(x - \frac{\pi}{2}\right),$$

shown as a red dashed curve in Figure 42.

Step 2 Sketch the vertical asymptotes through the x-intercepts of the graph of $y = \frac{3}{2}\sin\left(x - \frac{\pi}{2}\right)$. These have the form $x = (2n + 1)\frac{\pi}{2}$, where n is an integer. See the black dashed lines in Figure 42.

Step 3 Sketch the graph of $y = \frac{3}{2}\csc\left(x - \frac{\pi}{2}\right)$ by drawing the typical U-shaped branches between adjacent asymptotes. See the solid blue graph in Figure 42.

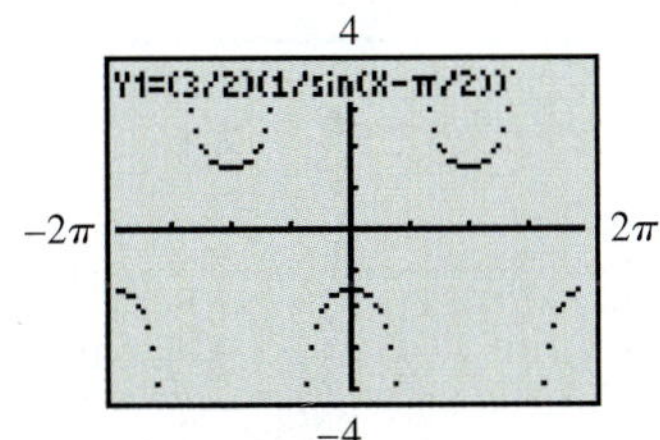

Dot mode

This is a calculator graph of the function in Example 2.

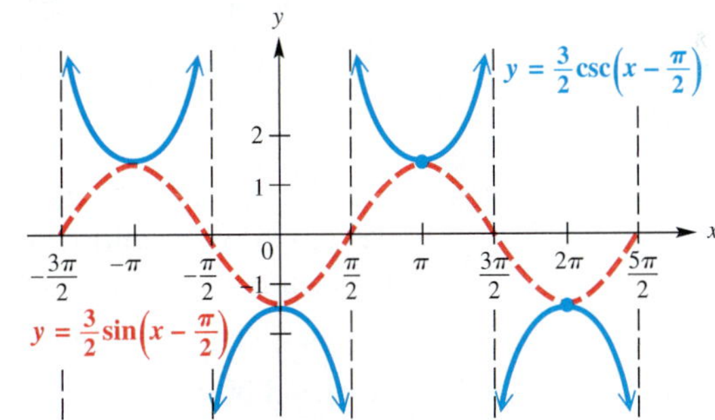

Figure 42

Now try Exercise 9.

Graphs of the Tangent and Cotangent Functions

Unlike the four functions whose graphs we studied previously, the tangent function has period π. Because $\tan x = \frac{\sin x}{\cos x}$, tangent values are 0 when sine values are 0, and undefined when cosine values are 0. As x-values go from $-\frac{\pi}{2}$ to $\frac{\pi}{2}$, tangent values go from $-\infty$ to ∞ and increase throughout the interval. Those same values are repeated as x goes from $\frac{\pi}{2}$ to $\frac{3\pi}{2}$, $\frac{3\pi}{2}$ to $\frac{5\pi}{2}$, and so on. The graph of $y = \tan x$ from $-\pi$ to $\frac{3\pi}{2}$ is shown in Figure 43.

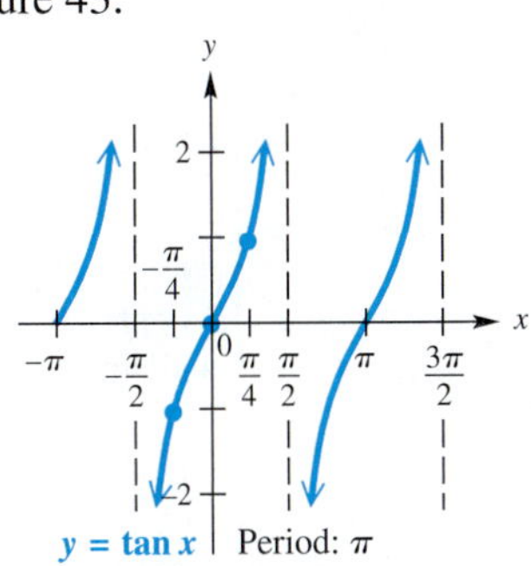

Figure 43

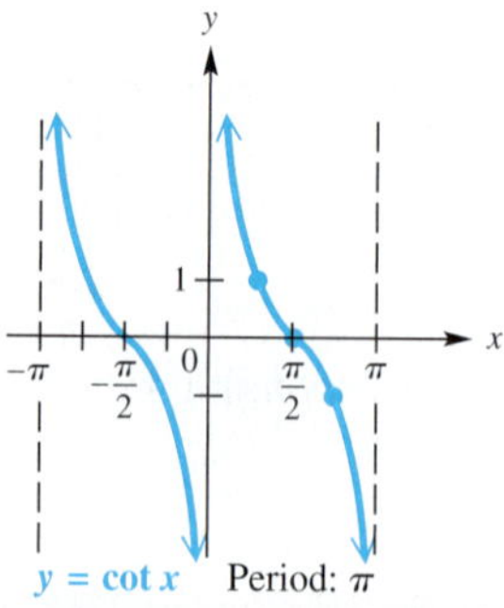

Figure 44

The cotangent function also has period π. Cotangent values are 0 when cosine values are 0, and undefined when sine values are 0. (Verify this also.) As x-values go from 0 to π, cotangent values go from ∞ to $-\infty$ and decrease throughout the interval. Those same values are repeated as x goes from π to 2π, 2π to 3π, and so on. The graph of $y = \cot x$ from $-\pi$ to π is shown in Figure 44.

TANGENT FUNCTION $f(x) = \tan x$

Domain: $\{x \mid x \neq (2n + 1)\frac{\pi}{2}$, where n is an integer$\}$ Range: $(-\infty, \infty)$

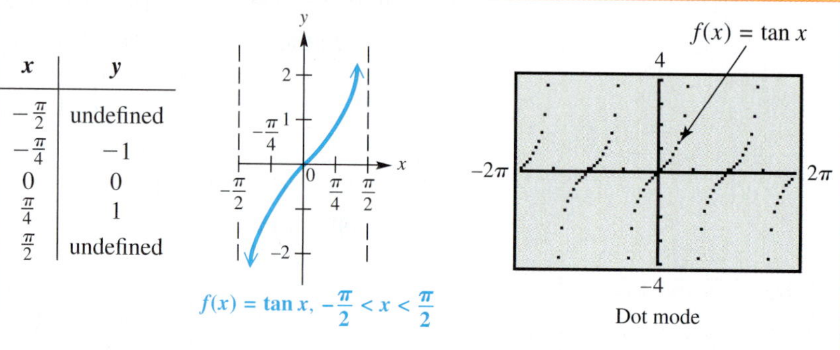

x	y
$-\frac{\pi}{2}$	undefined
$-\frac{\pi}{4}$	-1
0	0
$\frac{\pi}{4}$	1
$\frac{\pi}{2}$	undefined

Figure 45

- The graph is discontinuous at values of x of the form $x = (2n + 1)\frac{\pi}{2}$ and has vertical asymptotes at these values.
- Its x-intercepts are of the form $x = n\pi$.
- Its period is π.
- Its graph has no amplitude, since there are no minimum or maximum values.
- The graph is symmetric with respect to the origin, so the function is an odd function. For all x in the domain, $\tan(-x) = -\tan x$.

COTANGENT FUNCTION $f(x) = \cot x$

Domain: $\{x \mid x \neq n\pi$, where n is an integer$\}$ Range: $(-\infty, \infty)$

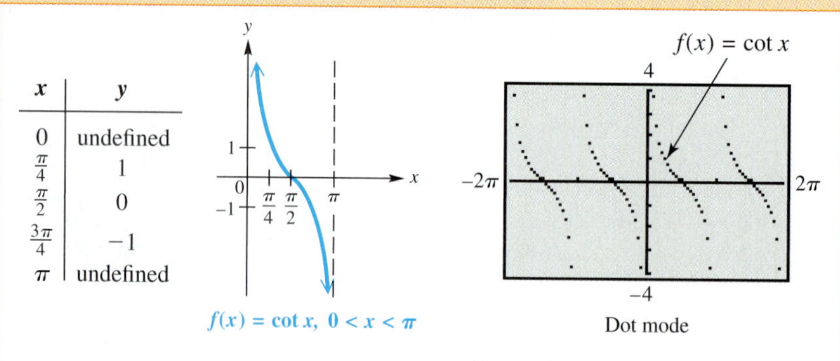

x	y
0	undefined
$\frac{\pi}{4}$	1
$\frac{\pi}{2}$	0
$\frac{3\pi}{4}$	-1
π	undefined

Figure 46

(continued)

- The graph is discontinuous at values of x of the form $x = n\pi$ and has vertical asymptotes at these values.
- Its x-intercepts are of the form $x = (2n + 1)\frac{\pi}{2}$.
- Its period is π.
- Its graph has no amplitude, since there are no minimum or maximum values.
- The graph is symmetric with respect to the origin, so the function is an odd function. For all x in the domain, $\cot(-x) = -\cot x$.

The tangent function can be graphed directly with a graphing calculator, using the tangent key. To graph the cotangent function, however, we must use one of the identities $\cot x = \frac{1}{\tan x}$ or $\cot x = \frac{\cos x}{\sin x}$ since graphing calculators generally do not have cotangent keys. ■

Guidelines for Sketching Graphs of Tangent and Cotangent Functions

To graph $\boldsymbol{y = a \tan bx}$ or $\boldsymbol{y = a \cot bx}$, with $b > 0$, follow these steps.

Step 1 Determine the period, $\frac{\pi}{b}$. To locate two adjacent vertical asymptotes, solve the following equations for x:

For $y = a \tan bx$: $\quad bx = -\dfrac{\pi}{2}$ and $bx = \dfrac{\pi}{2}$.

For $y = a \cot bx$: $\quad bx = 0$ and $bx = \pi$.

Step 2 Sketch the two vertical asymptotes found in Step 1.

Step 3 Divide the interval formed by the vertical asymptotes into four equal parts.

Step 4 Evaluate the function for the first-quarter point, midpoint, and third-quarter point, using the x-values found in Step 3.

Step 5 Join the points with a smooth curve, approaching the vertical asymptotes. Indicate additional asymptotes and periods of the graph as necessary.

EXAMPLE 3 Graphing $y = \tan bx$

Graph $y = \tan 2x$.

Solution

Step 1 The period of this function is $\frac{\pi}{2}$. To locate two adjacent vertical asymptotes, solve $2x = -\frac{\pi}{2}$ and $2x = \frac{\pi}{2}$ (since this is a tangent function). The two asymptotes have equations $x = -\frac{\pi}{4}$ and $x = \frac{\pi}{4}$.

Step 2 Sketch the two vertical asymptotes $x = \pm\frac{\pi}{4}$, as shown in Figure 47 on the next page.

Step 3 Divide the interval $\left(-\frac{\pi}{4}, \frac{\pi}{4}\right)$ into four equal parts. This gives the following key x-values.

first-quarter value: $-\frac{\pi}{8}$, middle value: 0, third-quarter value: $\frac{\pi}{8}$

Step 4 Evaluate the function for the x-values found in Step 3.

x	$-\frac{\pi}{8}$	0	$\frac{\pi}{8}$
$2x$	$-\frac{\pi}{4}$	0	$\frac{\pi}{4}$
$\tan 2x$	-1	0	1

$y = \tan 2x$ Period: $\frac{\pi}{2}$

Figure 47

Step 5 Join these points with a smooth curve, approaching the vertical asymptotes. See Figure 47. Another period has been graphed, one half period to the left and one half period to the right.

Now try Exercise 21.

EXAMPLE 4 Graphing $y = a \tan bx$

Graph $y = -3 \tan \frac{1}{2}x$.

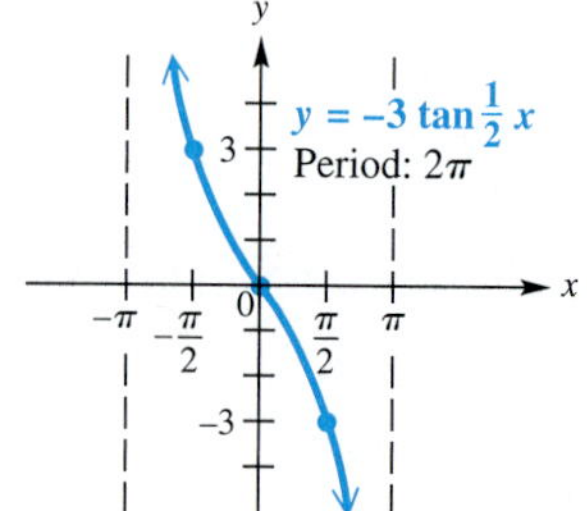

Figure 48

Solution The period is $\frac{\pi}{\frac{1}{2}} = 2\pi$. Adjacent asymptotes are at $x = -\pi$ and $x = \pi$. Dividing the interval $(-\pi, \pi)$ into four equal parts gives key x-values of $-\frac{\pi}{2}$, 0, and $\frac{\pi}{2}$. Evaluating the function at these x-values gives the key points.

$$\left(-\frac{\pi}{2}, 3\right), \quad (0, 0), \quad \left(\frac{\pi}{2}, -3\right)$$

By plotting these points and joining them with a smooth curve, we obtain the graph shown in Figure 48. Because the coefficient -3 is negative, the graph is reflected across the x-axis compared to the graph of $y = 3 \tan \frac{1}{2}x$.

Now try Exercise 29.

NOTE The function defined by $y = -3 \tan \frac{1}{2}x$ in Example 4, graphed in Figure 48, has a graph that compares to the graph of $y = \tan x$ as follows.

1. The period is larger because $b = \frac{1}{2}$, and $\frac{1}{2} < 1$.
2. The graph is "stretched" because $a = -3$, and $|-3| > 1$.
3. Each branch of the graph goes down from left to right (that is, the function decreases) between each pair of adjacent asymptotes because $a = -3 < 0$. When $a < 0$, the graph is reflected across the x-axis compared to the graph of $y = |a| \tan bx$.

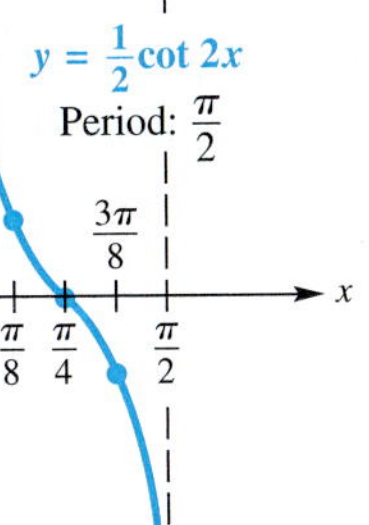

Figure 49

EXAMPLE 5 Graphing $y = a \cot bx$

Graph $y = \frac{1}{2} \cot 2x$.

Solution Because this function involves the cotangent, we can locate two adjacent asymptotes by solving the equations $2x = 0$ and $2x = \pi$. The lines $x = 0$ (the y-axis) and $x = \frac{\pi}{2}$ are two such asymptotes. Divide the interval $\left(0, \frac{\pi}{2}\right)$ into four equal parts, getting key x-values of $\frac{\pi}{8}, \frac{\pi}{4}$, and $\frac{3\pi}{8}$. Evaluating the function at these x-values gives the following key points.

$$\left(\frac{\pi}{8}, \frac{1}{2}\right), \quad \left(\frac{\pi}{4}, 0\right), \quad \left(\frac{3\pi}{8}, -\frac{1}{2}\right)$$

Joining these points with a smooth curve approaching the asymptotes gives the graph shown in Figure 49.

Now try Exercise 31.

Like the other circular functions, the graphs of the tangent and cotangent functions may be translated horizontally and vertically.

EXAMPLE 6 Graphing a Tangent Function with a Vertical Translation

Graph $y = 2 + \tan x$.

Analytic Solution

Every value of y for this function will be 2 units more than the corresponding value of y in $y = \tan x$, causing the graph of $y = 2 + \tan x$ to be translated 2 units up compared with the graph of $y = \tan x$. See Figure 50.

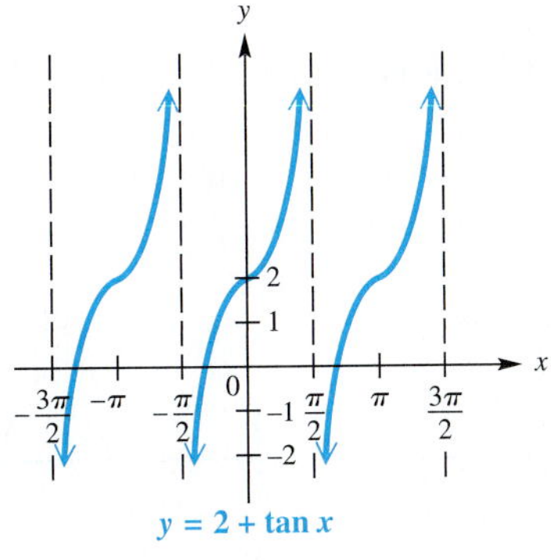

Figure 50

Graphing Calculator Solution

To see the vertical translation, observe the coordinates displayed at the bottoms of the screens in Figures 51 and 52. For $X = \frac{\pi}{4} \approx .78539816$,

$$Y_1 = \tan X = 1,$$

while for the same X-value,

$$Y_2 = 2 + \tan X = 2 + 1 = 3.$$

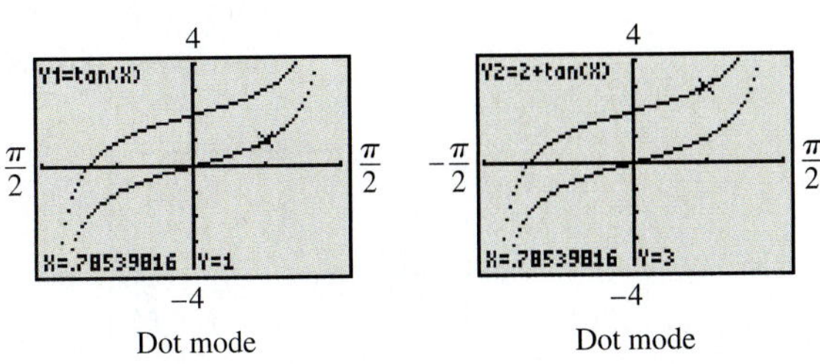

Figure 51 **Figure 52**

Now try Exercise 37.

EXAMPLE 7 Graphing a Cotangent Function with Vertical and Horizontal Translations

Graph $y = -2 - \cot\left(x - \frac{\pi}{4}\right)$.

Solution Here $b = 1$, so the period is π. The graph will be translated down 2 units (because $c = -2$), reflected across the x-axis (because of the negative sign in front of the cotangent), and will have a phase shift (horizontal translation) $\frac{\pi}{4}$ unit to the right $\left(\text{because of the argument } \left(x - \frac{\pi}{4}\right)\right)$. To locate adjacent asymptotes, since this function involves the cotangent, we solve the following equations:

$$x - \frac{\pi}{4} = 0, \quad \text{so } x = \frac{\pi}{4} \qquad \text{and} \qquad x - \frac{\pi}{4} = \pi, \quad \text{so } x = \frac{5\pi}{4}.$$

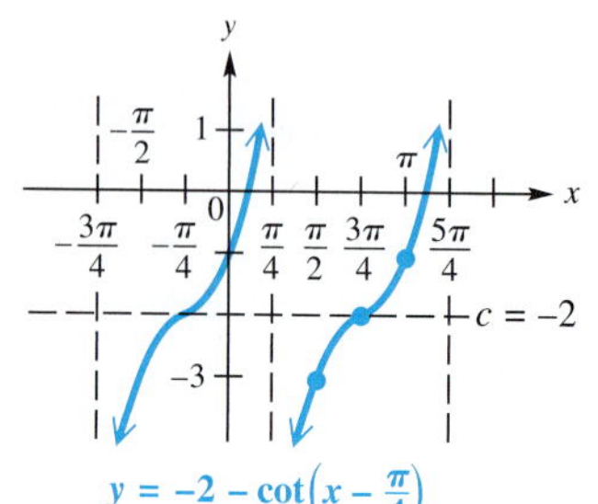

Figure 53

Dividing the interval $\left(\frac{\pi}{4}, \frac{5\pi}{4}\right)$ into four equal parts and evaluating the function at the three key x-values within the interval gives these points.

$$\left(\frac{\pi}{2}, -3\right), \quad \left(\frac{3\pi}{4}, -2\right), \quad (\pi, -1)$$

Join these points with a smooth curve. This period of the graph, along with the one in the domain interval $\left(-\frac{3\pi}{4}, \frac{\pi}{4}\right)$, is shown in Figure 53.

Now try Exercise 45.

Addition of Ordinates New functions can be formed by adding or subtracting other functions. A function formed by combining two other functions, such as

$$y = \cos x + \sin x,$$

has historically been graphed using a method known as *addition of ordinates.* (The x-value of a point is sometimes called its *abscissa,* while its y-value is called its *ordinate.*) To apply this method to this function, we graph the functions $y = \cos x$ and $y = \sin x$. Then, for selected values of x, we add $\cos x$ and $\sin x$, and plot the points $(x, \cos x + \sin x)$. Joining the resulting points with a sinusoidal curve gives the graph of the desired function. While this method illustrates some valuable concepts involving the arithmetic of functions, it is time-consuming.

With graphing calculators, this technique is easily illustrated. Let $Y_1 = \cos X$, $Y_2 = \sin X$, and $Y_3 = Y_1 + Y_2$. Figure 54 shows the result when Y_1 and Y_2 are graphed in thin graph style, and $Y_3 = \cos X + \sin X$ is graphed in thick graph style. Notice that for $X = \frac{\pi}{6} \approx .52359878$, $Y_1 + Y_2 = Y_3$.

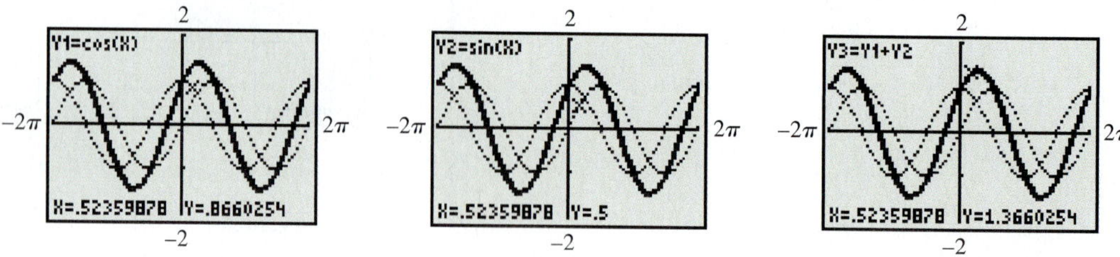

Figure 54

Now try Exercise 61.

6.4 Exercises

Concept Check *In Exercises 1–6, match each function with its graph from choices A–F.*

1. $y = -\csc x$ **2.** $y = -\sec x$ **3.** $y = -\tan x$

4. $y = -\cot x$ **5.** $y = \tan\left(x - \frac{\pi}{4}\right)$ **6.** $y = \cot\left(x - \frac{\pi}{4}\right)$

A.

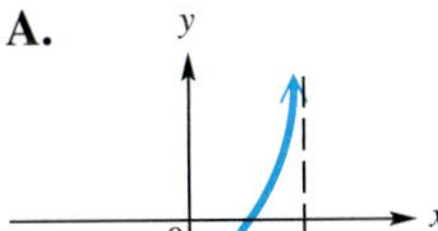

B.

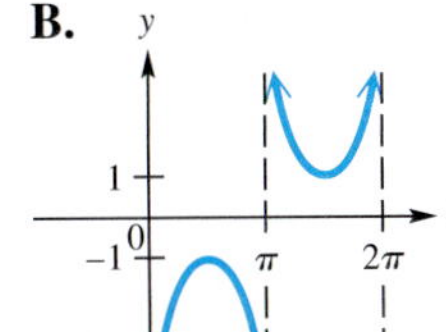

C.

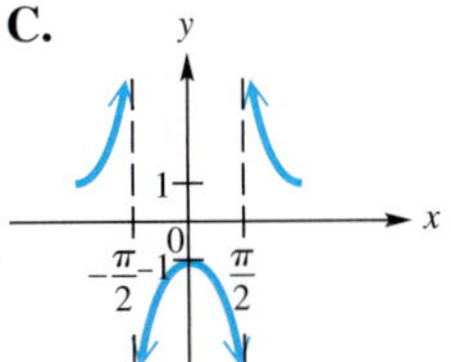

D.

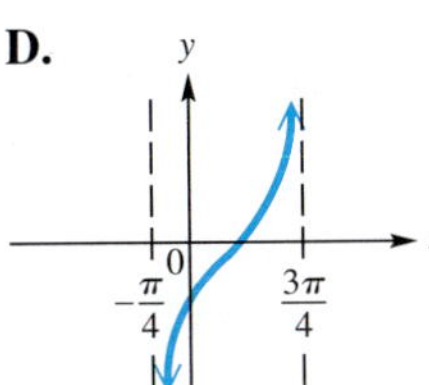

E.

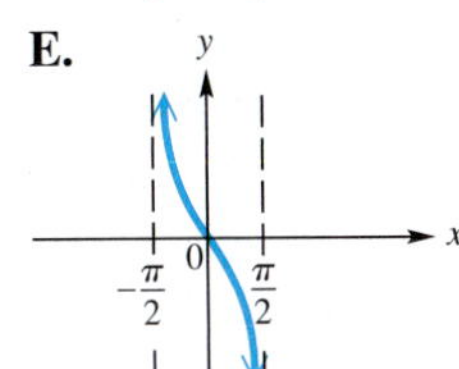

F.

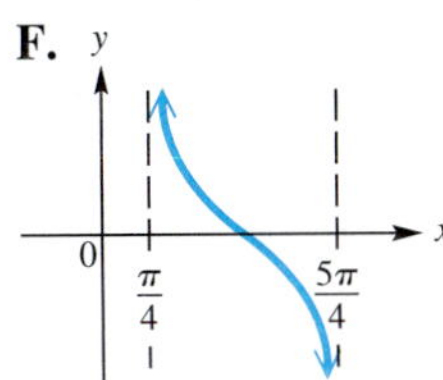

Graph each function over a one-period interval. See Examples 1 and 2.

7. $y = 3 \sec \frac{1}{4}x$ **8.** $y = -2 \sec \frac{1}{2}x$ **9.** $y = -\frac{1}{2} \csc\left(x + \frac{\pi}{2}\right)$

10. $y = \frac{1}{2} \csc\left(x - \frac{\pi}{2}\right)$ **11.** $y = \csc\left(x - \frac{\pi}{4}\right)$ **12.** $y = \sec\left(x + \frac{3\pi}{4}\right)$

13. $y = \sec\left(x + \frac{\pi}{4}\right)$ **14.** $y = \csc\left(x + \frac{\pi}{3}\right)$ **15.** $y = \sec\left(\frac{1}{2}x + \frac{\pi}{3}\right)$

16. $y = \csc\left(\frac{1}{2}x - \frac{\pi}{4}\right)$ **17.** $y = 2 + 3 \sec(2x - \pi)$

18. $y = 1 - 2 \csc\left(x + \frac{\pi}{2}\right)$ **19.** $y = 1 - \frac{1}{2} \csc\left(x - \frac{3\pi}{4}\right)$

20. $y = 2 + \frac{1}{4} \sec\left(\frac{1}{2}x - \pi\right)$

Graph each function over a one-period interval. See Examples 3–5.

21. $y = \tan 4x$ **22.** $y = \tan \frac{1}{2}x$ **23.** $y = 2 \tan x$

24. $y = 2 \cot x$ **25.** $y = 2 \tan \frac{1}{4}x$ **26.** $y = \frac{1}{2} \cot x$

27. $y = \cot 3x$ **28.** $y = -\cot \frac{1}{2}x$ **29.** $y = -2 \tan \frac{1}{4}x$

30. $y = 3 \tan \frac{1}{2}x$ **31.** $y = \frac{1}{2} \cot 4x$ **32.** $y = -\frac{1}{2} \cot 2x$

Graph each function over a two-period interval. See Examples 6 and 7.

33. $y = \tan(2x - \pi)$ **34.** $y = \tan\left(\frac{x}{2} + \pi\right)$ **35.** $y = \cot\left(3x + \frac{\pi}{4}\right)$

36. $y = \cot\left(2x - \frac{3\pi}{2}\right)$ **37.** $y = 1 + \tan x$ **38.** $y = 1 - \tan x$

39. $y = 1 - \cot x$ **40.** $y = -2 - \cot x$

41. $y = -1 + 2 \tan x$ **42.** $y = 3 + \frac{1}{2} \tan x$

43. $y = -1 + \frac{1}{2} \cot(2x - 3\pi)$ **44.** $y = -2 + 3 \tan(4x + \pi)$

45. $y = 1 - 2 \cot 2\left(x + \frac{\pi}{2}\right)$ **46.** $y = \frac{2}{3} \tan\left(\frac{3}{4}x - \pi\right) - 2$

Concept Check In Exercises 47–52, tell whether each statement is true *or* false. *If false, tell why.*

47. The smallest positive number k for which $x = k$ is an asymptote for the tangent function is $\frac{\pi}{2}$.

48. The smallest positive number k for which $x = k$ is an asymptote for the cotangent function is $\frac{\pi}{2}$.

49. The tangent and secant functions are undefined for the same values.

50. The secant and cosecant functions are undefined for the same values.

51. The graph of $y = \tan x$ in Figure 45 suggests that $\tan(-x) = \tan x$ for all x in the domain of $\tan x$.

52. The graph of $y = \sec x$ in Figure 40 suggests that $\sec(-x) = \sec x$ for all x in the domain of $\sec x$.

53. *Concept Check* If c is any number, then how many solutions does the equation $c = \tan x$ have in the interval $(-2\pi, 2\pi]$?

54. *Concept Check* If c is any number such that $-1 < c < 1$, then how many solutions does the equation $c = \sec x$ have over the entire domain of the secant function?

55. Consider the function defined by $f(x) = -4 \tan(2x + \pi)$. What is the domain of f? What is its range?

56. Consider the function defined by $g(x) = -2 \csc(4x + \pi)$. What is the domain of g? What is its range?

(Modeling) Solve each problem.

57. ***Distance of a Rotating Beacon*** A rotating beacon is located at point A next to a long wall. (See the figure.) The beacon is 4 m from the wall. The distance d is given by

$$d = 4 \tan 2\pi t,$$

where t is time measured in seconds since the beacon started rotating. (When $t = 0$, the beacon is aimed at point R. When the beacon is aimed to the right of R, the value of d is positive; d is negative if the beacon is aimed to the left of R.) Find d for each time.

(a) $t = 0$
(b) $t = .4$
(c) $t = .8$
(d) $t = 1.2$
(e) Why is .25 a meaningless value for t?

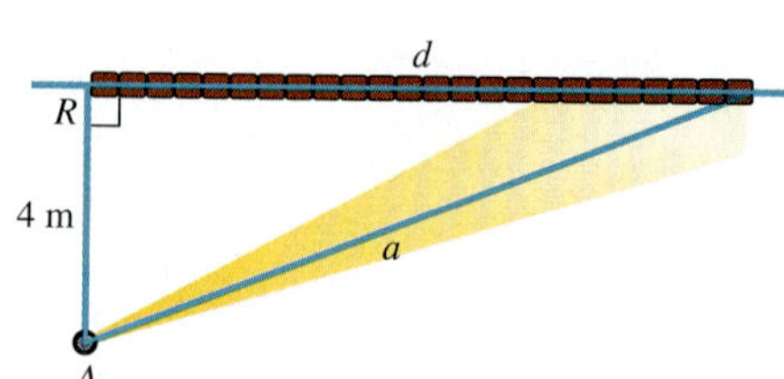

58. ***Distance of a Rotating Beacon*** In the figure for Exercise 57, the distance a is given by

$$a = 4|\sec 2\pi t|.$$

Find a for each time.

(a) $t = 0$ **(b)** $t = .86$ **(c)** $t = 1.24$

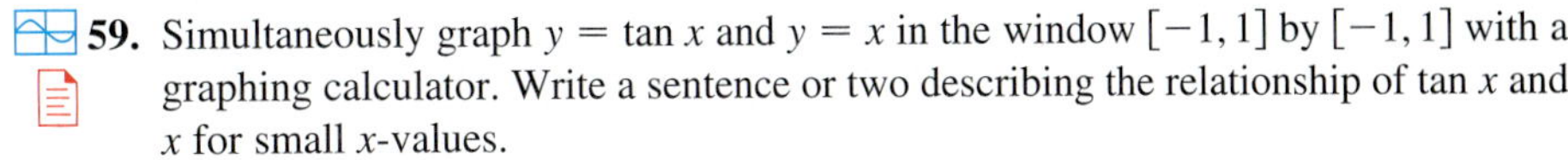

59. Simultaneously graph $y = \tan x$ and $y = x$ in the window $[-1, 1]$ by $[-1, 1]$ with a graphing calculator. Write a sentence or two describing the relationship of $\tan x$ and x for small x-values.

60. Between each pair of successive asymptotes, a portion of the graph of $y = \sec x$ or $y = \csc x$ resembles a parabola. Can each of these portions actually be a parabola? Explain.

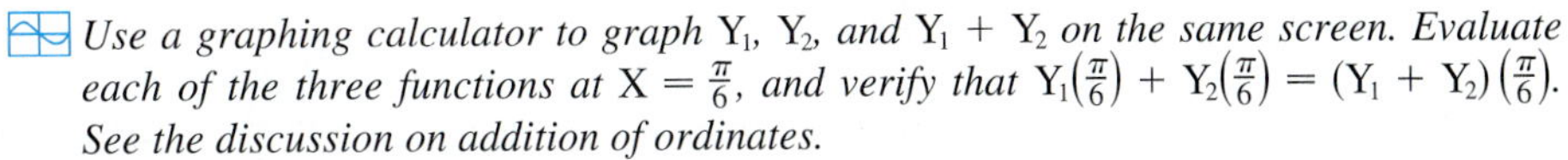

Use a graphing calculator to graph Y_1, Y_2, *and* $Y_1 + Y_2$ *on the same screen. Evaluate each of the three functions at* $X = \frac{\pi}{6}$, *and verify that* $Y_1\left(\frac{\pi}{6}\right) + Y_2\left(\frac{\pi}{6}\right) = (Y_1 + Y_2)\left(\frac{\pi}{6}\right)$. *See the discussion on addition of ordinates.*

61. $Y_1 = \sin X, \quad Y_2 = \sin 2X$

62. $Y_1 = \cos X, \quad Y_2 = \sec X$

Relating Concepts

For individual or collaborative investigation

(Exercises 63–68)

Consider the function defined by $y = -2 - \cot\left(x - \frac{\pi}{4}\right)$ *from Example 7.* ***Work these exercises in order.***

63. What is the smallest positive number for which $y = \cot x$ is undefined?

64. Let k represent the number you found in Exercise 63. Set $x - \frac{\pi}{4}$ equal to k, and solve to find a positive number for which $\cot\left(x - \frac{\pi}{4}\right)$ is undefined.

65. Based on your answer in Exercise 64 and the fact that the cotangent function has period π, give the general form of the equations of the asymptotes of the graph of $y = -2 - \cot\left(x - \frac{\pi}{4}\right)$. Let n represent any integer.

66. Use the capabilities of your calculator to find the smallest positive x-intercept of the graph of this function.

67. Use the fact that the period of this function is π to find the next positive x-intercept.

68. Give the solution set of the equation $-2 - \cot\left(x - \frac{\pi}{4}\right) = 0$ over all real numbers. Let n represent any integer.

Summary Exercises on Graphing Circular Functions

These summary exercises provide practice with the various graphing techniques presented in this chapter. Graph each function over a one-period interval.

1. $y = 2 \sin \pi x$

2. $y = 4 \cos 1.5x$

3. $y = -2 + .5 \cos \frac{\pi}{4} x$

4. $y = 3 \sec \frac{\pi x}{2}$

5. $y = -4 \csc .5x$

6. $y = 3 \tan\left(\frac{\pi x}{2} + \pi\right)$

Graph each function over a two-period interval.

7. $y = -5 \sin \frac{x}{3}$

8. $y = 10 \cos\left(\frac{x}{4} + \frac{\pi}{2}\right)$

9. $y = 3 - 4 \sin(2.5x + \pi)$

10. $y = 2 - \sec[\pi(x - 3)]$

6.5 Harmonic Motion

Simple Harmonic Motion ▪ **Damped Oscillatory Motion**

Simple Harmonic Motion In part A of Figure 55, a spring with a weight attached to its free end is in equilibrium (or rest) position. If the weight is pulled down a units and released (part B of the figure), the spring's elasticity causes the weight to rise a units ($a > 0$) above the equilibrium position, as seen in part C, and then oscillate about the equilibrium position. If friction is neglected, this oscillatory motion is described mathematically by a sinusoid. Other applications of this type of motion include sound, electric current, and electromagnetic waves.

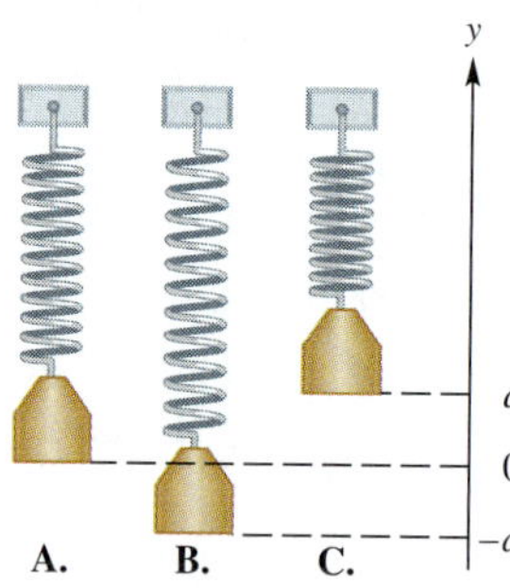

Figure 55

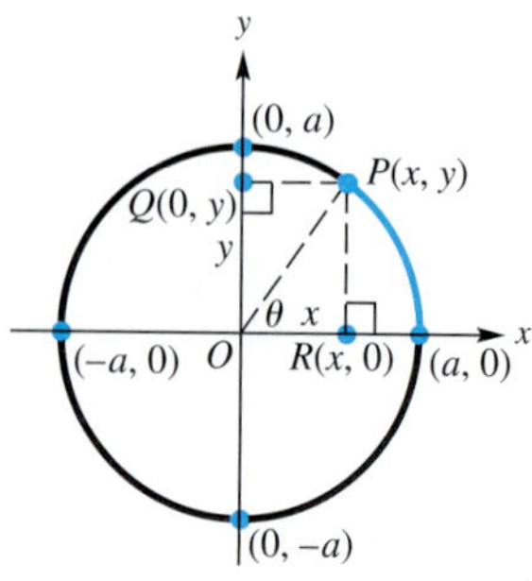

Figure 56

To develop a general equation for such motion, consider Figure 56. Suppose the point $P(x, y)$ moves around the circle counterclockwise at a uniform angular speed ω. Assume that at time $t = 0$, P is at $(a, 0)$. The angle swept out by ray OP at time t is given by $\theta = \omega t$. The coordinates of point P at time t are

$$x = a \cos \theta = a \cos \omega t \qquad \text{and} \qquad y = a \sin \theta = a \sin \omega t.$$

As P moves around the circle from the point $(a, 0)$, the point $Q(0, y)$ oscillates back and forth along the y-axis between the points $(0, a)$ and $(0, -a)$. Similarly, the point $R(x, 0)$ oscillates back and forth between $(a, 0)$ and $(-a, 0)$. This oscillatory motion is called **simple harmonic motion.**

The amplitude of the motion is $|a|$, and the period is $\frac{2\pi}{\omega}$. The moving points P and Q or P and R complete one oscillation or cycle per period. The number of cycles per unit of time, called the **frequency,** is the reciprocal of the period, $\frac{\omega}{2\pi}$, where $\omega > 0$.

Simple Harmonic Motion

The position of a point oscillating about an equilibrium position at time t is modeled by either

$$s(t) = a \cos \omega t \qquad \text{or} \qquad s(t) = a \sin \omega t,$$

where a and ω are constants, with $\omega > 0$. The amplitude of the motion is $|a|$, the period is $\frac{2\pi}{\omega}$, and the frequency is $\frac{\omega}{2\pi}$.

EXAMPLE 1 Modeling the Motion of a Spring

Suppose that an object is attached to a coiled spring such as the one in Figure 55. It is pulled down a distance of 5 in. from its equilibrium position, and then released. The time for one complete oscillation is 4 sec.

(a) Give an equation that models the position of the object at time t.

(b) Determine the position at $t = 1.5$ sec.

(c) Find the frequency.

Solution

(a) When the object is released at $t = 0$, the distance of the object from the equilibrium position is 5 in. below equilibrium. If $s(t)$ is to model the motion, then $s(0)$ must equal -5. We use

$$s(t) = a \cos \omega t,$$

with $a = -5$. We choose the cosine function because $\cos \omega(0) = \cos 0 = 1$, and $-5 \cdot 1 = -5$. (Had we chosen the sine function, a phase shift would have been required.) The period is 4, so

$$\frac{2\pi}{\omega} = 4, \qquad \text{or} \qquad \omega = \frac{\pi}{2}.$$ Solve for ω. (Section 1.1)

Thus, the motion is modeled by

$$s(t) = -5 \cos \frac{\pi}{2} t.$$

(b) After 1.5 sec, the position is

$$s(1.5) = -5 \cos\left[\frac{\pi}{2}(1.5)\right] \approx 3.54 \text{ in.}$$

Since $3.54 > 0$, the object is above the equilibrium position.

(c) The frequency is the reciprocal of the period, or $\frac{1}{4}$.

Now try Exercise 9.

EXAMPLE 2 Analyzing Harmonic Motion

Suppose that an object oscillates according to the model

$$s(t) = 8 \sin 3t,$$

where t is in seconds and $s(t)$ is in feet. Analyze the motion.

Solution The motion is harmonic because the model is of the form $s(t) = a \sin \omega t$. Because $a = 8$, the object oscillates 8 ft in either direction from its starting point. The period $\frac{2\pi}{3} \approx 2.1$ is the time, in seconds, it takes for one complete oscillation. The frequency is the reciprocal of the period, so the object completes $\frac{3}{2\pi} \approx .48$ oscillation per sec.

Now try Exercise 15.

Damped Oscillatory Motion In the example of the stretched spring, we disregard the effect of friction. Friction causes the amplitude of the motion to diminish gradually until the weight comes to rest. In this situation, we say that the motion has been *damped* by the force of friction. Most oscillatory motions are damped, and the decrease in amplitude follows the pattern of exponential decay. A typical example of **damped oscillatory motion** is provided by the function defined by

$$s(t) = e^{-t} \sin t.$$

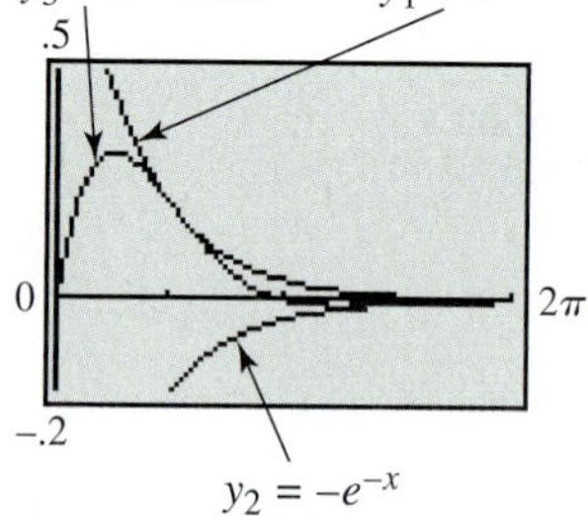

Figure 57

Figure 57 shows how the graph of $y_3 = e^{-x} \sin x$ is bounded above by the graph of $y_1 = e^{-x}$ and below by the graph of $y_2 = -e^{-x}$. The damped motion curve dips below the x-axis at $x = \pi$ but stays above the graph of y_2. Figure 58 shows a traditional graph of $s(t) = e^{-t} \sin t$, along with the graph of $y = \sin t$.

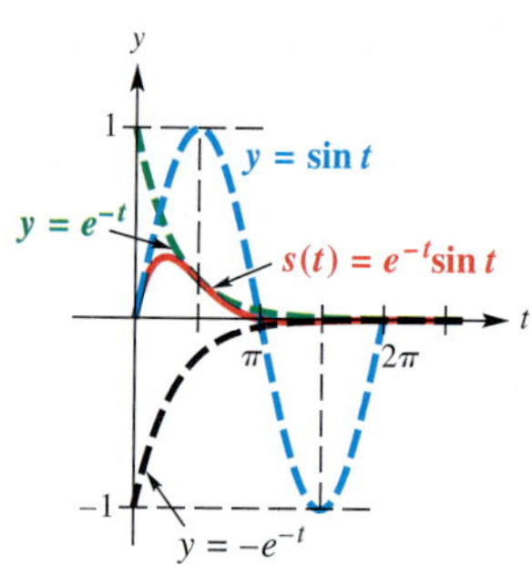

Figure 58

Shock absorbers are put on an automobile in order to damp oscillatory motion. Instead of oscillating up and down for a long while after hitting a bump or pothole, the oscillations of the car are quickly damped out for a smoother ride.

Now try Exercise 21.

6.5 Exercises

(Modeling) Springs Suppose that a weight on a spring has initial position $s(0)$ and period P.

(a) *Find a function s given by $s(t) = a \cos \omega t$ that models the displacement of the weight.*

(b) *Evaluate $s(1)$. Is the weight moving upward, downward, or neither when $t = 1$? Support your results graphically or numerically.*

1. $s(0) = 2$ in.; $P = .5$ sec

2. $s(0) = 5$ in.; $P = 1.5$ sec

3. $s(0) = -3$ in.; $P = .8$ sec

4. $s(0) = -4$ in.; $P = 1.2$ sec

(Modeling) Music A note on the piano has given frequency F. Suppose the maximum displacement at the center of the piano wire is given by $s(0)$. Find constants a and ω so that the equation $s(t) = a \cos \omega t$ models this displacement. Graph s in the viewing window $[0, .05]$ by $[-.3, .3]$.

5. $F = 27.5$; $s(0) = .21$

6. $F = 110$; $s(0) = .11$

7. $F = 55$; $s(0) = .14$

8. $F = 220$; $s(0) = .06$

(Modeling) Solve each problem. See Examples 1 and 2.

9. *Spring* An object is attached to a coiled spring, as in Figure 55. It is pulled down a distance of 4 units from its equilibrium position, and then released. The time for one complete oscillation is 3 sec.

(a) Give an equation that models the position of the object at time t.
(b) Determine the position at $t = 1.25$ sec.
(c) Find the frequency.

10. *Spring* Repeat Exercise 9, but assume that the object is pulled down 6 units and the time for one complete oscillation is 4 sec.

11. *Particle Movement* Write the equation and then determine the amplitude, period, and frequency of the simple harmonic motion of a particle moving uniformly around a circle of radius 2 units, with angular speed

(a) 2 radians per sec **(b)** 4 radians per sec.

12. *Pendulum* What are the period P and frequency T of oscillation of a pendulum of length $\frac{1}{2}$ ft? $\left(\textit{Hint: } P = 2\pi\sqrt{\frac{L}{32}}\right.$, where L is the length of the pendulum in feet and P is in seconds.$\left.\right)$

13. *Pendulum* In Exercise 12, how long should the pendulum be to have period 1 sec?

14. *Spring* The formula for the up and down motion of a weight on a spring is given by

$$s(t) = a \sin \sqrt{\frac{k}{m}}\,t.$$

If the spring constant k is 4, what mass m must be used to produce a period of 1 sec?

15. *Spring* The height attained by a weight attached to a spring set in motion is

$$s(t) = -4 \cos 8\pi t$$

inches after t seconds.

(a) Find the maximum height that the weight rises above the equilibrium position of $y = 0$.
(b) When does the weight first reach its maximum height, if $t \geq 0$?
(c) What are the frequency and period?

16. *Spring* (See Exercise 14.) A spring with spring constant and a 1-unit mass m attached to it is stretched and then allowed to come to rest.

(a) If the spring is stretched $\frac{1}{2}$ ft and released, what are the amplitude, period, and frequency of the resulting oscillatory motion?

(b) What is the equation of the motion?

17. *Spring* The position of a weight attached to a spring is

$$s(t) = -5 \cos 4\pi t$$

inches after t seconds.

(a) What is the maximum height that the weight rises above the equilibrium position?

(b) What are the frequency and period?

(c) When does the weight first reach its maximum height?

(d) Calculate and interpret $s(1.3)$.

18. *Spring* The position of a weight attached to a spring is

$$s(t) = -4 \cos 10t$$

inches after t seconds.

(a) What is the maximum height that the weight rises above the equilibrium position?

(b) What are the frequency and period?

(c) When does the weight first reach its maximum height?

(d) Calculate and interpret $s(1.466)$.

19. *Spring* A weight attached to a spring is pulled down 3 in. below the equilibrium position.

(a) Assuming that the frequency is $\frac{6}{\pi}$ cycles per sec, determine a model that gives the position of the weight at time t seconds.

(b) What is the period?

20. *Spring* A weight attached to a spring is pulled down 2 in. below the equilibrium position.

(a) Assuming that the period is $\frac{1}{3}$ sec, determine a model that gives the position of the weight at time t seconds.

(b) What is the frequency?

Use a graphing calculator to graph $y_1 = e^{-t} \sin t$, $y_2 = e^{-t}$, and $y_3 = -e^{-t}$ in the viewing window $[0, \pi]$ by $[-.5, .5]$.

21. Find the t-intercepts of the graph of y_1. Explain the relationship of these intercepts with the x-intercepts of the graph of $y = \sin x$.

22. Find any points of intersection of y_1 and y_2 or y_1 and y_3. How are these points related to the graph of $y = \sin x$?

Chapter 6 Summary

KEY TERMS

6.1 radian
sector of a circle
6.2 unit circle
circular functions
angular speed ω
linear speed v
6.3 periodic function
period
sine wave (sinusoid)
amplitude
phase shift
argument
6.4 addition of ordinates
6.5 simple harmonic motion
frequency
damped oscillatory motion

QUICK REVIEW

CONCEPTS	EXAMPLES

6.1 Radian Measure

An angle with its vertex at the center of a circle that intercepts an arc on the circle equal in length to the radius of the circle has a measure of **1 radian.**

Degree/Radian Relationship $\quad 180° = \pi$ **radians**

Converting Between Degrees and Radians

1. Multiply a degree measure by $\frac{\pi}{180}$ radian and simplify to convert to radians.
2. Multiply a radian measure by $\frac{180°}{\pi}$ and simplify to convert to degrees.

Convert 135° to radians.

$$135° = 135\left(\frac{\pi}{180}\text{ radian}\right) = \frac{3\pi}{4}\text{ radians}$$

Convert $-\frac{5\pi}{3}$ radians to degrees.

$$-\frac{5\pi}{3}\text{ radians} = -\frac{5\pi}{3}\left(\frac{180°}{\pi}\right) = -300°$$

Arc Length

The length s of the arc intercepted on a circle of radius r by a central angle of measure θ radians is given by the product of the radius and the radian measure of the angle, or

$$s = r\theta, \qquad \theta \text{ in radians.}$$

In the figure, $s = r\theta$ so

$$\theta = \frac{s}{r} = \frac{3}{4}\text{ radian.}$$

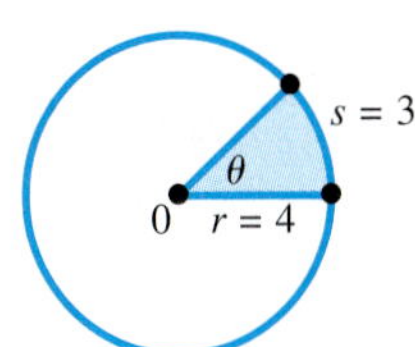

Area of a Sector

The area of a sector of a circle of radius r and central angle θ is given by

$$A = \frac{1}{2}r^2\theta, \qquad \theta \text{ in radians.}$$

The area of the sector in the figure is

$$A = \frac{1}{2}(4)^2\left(\frac{3}{4}\right) = 6\text{ square units.}$$

CONCEPTS | **EXAMPLES**

6.2 The Unit Circle and Circular Functions

Circular Functions

Start at the point $(1, 0)$ on the unit circle $x^2 + y^2 = 1$ and lay off an arc of length $|s|$ along the circle, going counterclockwise if s is positive, and clockwise if s is negative. Let the endpoint of the arc be at the point (x, y). The six circular functions of s are defined as follows. (Assume that no denominators are 0.)

$$\sin s = y \qquad \cos s = x \qquad \tan s = \frac{y}{x}$$

$$\csc s = \frac{1}{y} \qquad \sec s = \frac{1}{x} \qquad \cot s = \frac{x}{y}$$

Use the unit circle to find each value.

$$\sin\frac{5\pi}{6} = \frac{1}{2}$$

$$\cos\frac{3\pi}{2} = 0$$

$$\tan\frac{\pi}{4} = \frac{\frac{\sqrt{2}}{2}}{\frac{\sqrt{2}}{2}} = 1$$

$$\csc\frac{7\pi}{4} = \frac{1}{-\frac{\sqrt{2}}{2}} = -\sqrt{2}$$

$$\sec\frac{7\pi}{6} = \frac{1}{-\frac{\sqrt{3}}{2}} = -\frac{2\sqrt{3}}{3}$$

$$\cot\frac{\pi}{3} = \frac{\frac{1}{2}}{\frac{\sqrt{3}}{2}} = \frac{\sqrt{3}}{3}$$

The Unit Circle

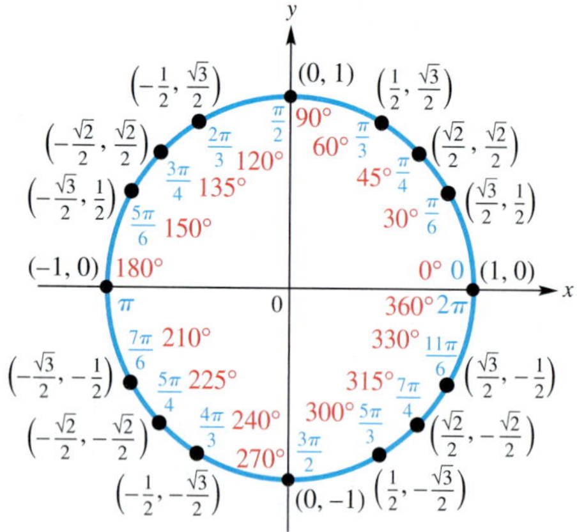

Unit circle $x^2 + y^2 = 1$

Formulas for Angular and Linear Speed

Angular Speed	Linear Speed
$\omega = \frac{\theta}{t}$	$v = \frac{s}{t}$
(ω in radians per unit time, θ in radians)	$v = \frac{r\theta}{t}$
	$v = r\omega$

A belt runs a pulley of radius 8 in. at 60 revolutions per min. Find the angular speed ω in radians per minute, and the linear speed v of the belt in inches per minute.

$$\omega = 60(2\pi) = 120\pi \text{ radians per min}$$

$$v = r\omega = 8(120\pi) = 960\pi \text{ in. per min}$$

CONCEPTS	EXAMPLES

6.3 Graphs of the Sine and Cosine Functions

Sine and Cosine Functions

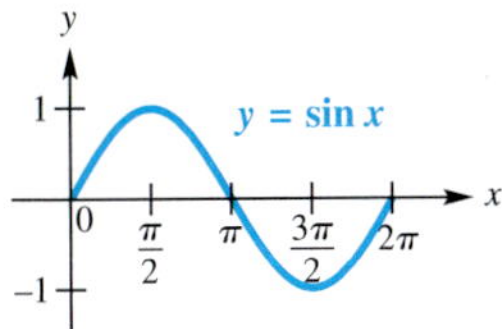

Domain: $(-\infty, \infty)$
Range: $[-1, 1]$
Amplitude: 1
Period: 2π

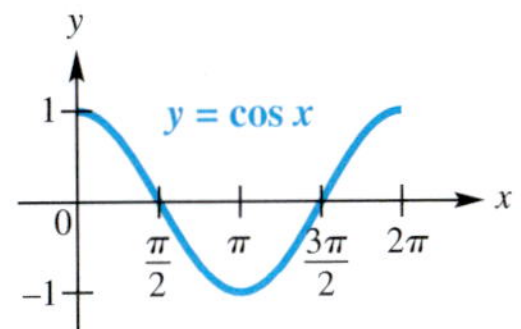

Domain: $(-\infty, \infty)$
Range: $[-1, 1]$
Amplitude: 1
Period: 2π

The graph of

$$y = c + a \sin b(x - d) \quad \text{or} \quad y = c + a \cos b(x - d),$$

$b > 0$, has

1. amplitude $|a|$,
2. period $\frac{2\pi}{b}$,
3. vertical translation c units up if $c > 0$ or $|c|$ units down if $c < 0$, and
4. phase shift d units to the right if $d > 0$ or $|d|$ units to the left if $d < 0$.

See pages 562 and 565 for a summary of graphing techniques.

Graph $y = \sin 3x$.

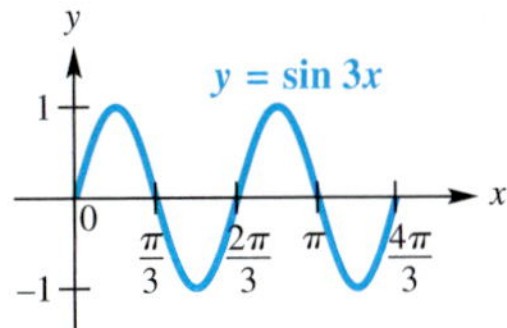

period: $\frac{2\pi}{3}$ amplitude: 1
domain: $(-\infty, \infty)$ range: $[-1, 1]$

Graph $y = -2 \cos x$.

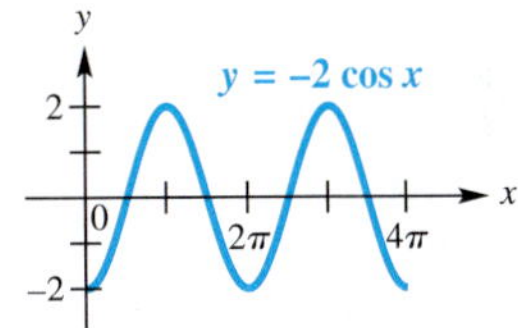

period: 2π amplitude: 2
domain: $(-\infty, \infty)$ range: $[-2, 2]$

6.4 Graphs of the Other Circular Functions

Cosecant and Secant Functions

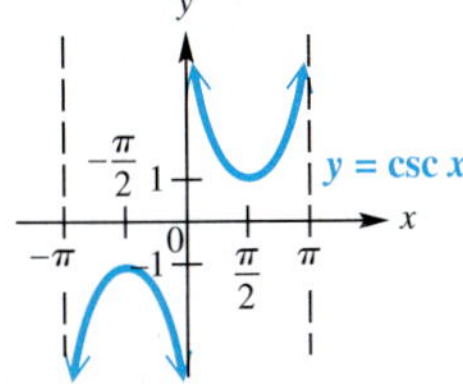

Domain: $\{x \mid x \neq n\pi,$ where n is an integer$\}$
Range: $(-\infty, -1] \cup [1, \infty)$
Period: 2π

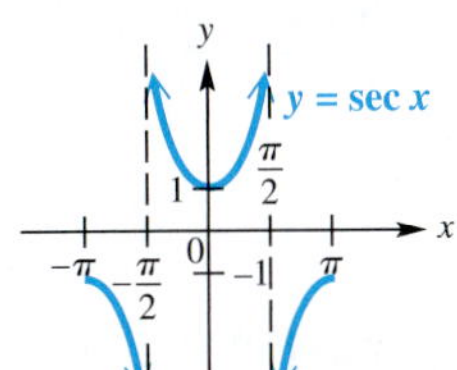

Domain: $\{x \mid x \neq (2n + 1)\frac{\pi}{2},$ where n is an integer$\}$
Range: $(-\infty, -1] \cup [1, \infty)$
Period: 2π

See page 577 for a summary of graphing techniques.

Graph one period of $y = \sec\left(x + \frac{\pi}{4}\right)$.

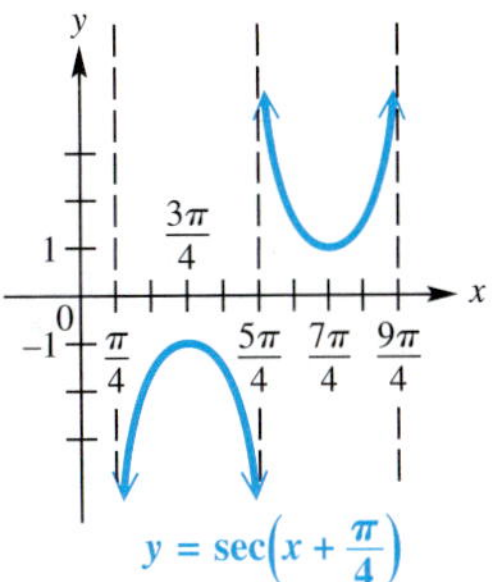

period: 2π
phase shift: $-\frac{\pi}{4}$
domain: $\{x \mid x \neq \frac{\pi}{4} + n\pi,$ where n is an integer$\}$
range: $(-\infty, -1] \cup [1, \infty)$

(continued)

CONCEPTS	EXAMPLES
Tangent and Cotangent Functions	Graph one period of $y = 2 \tan x$.

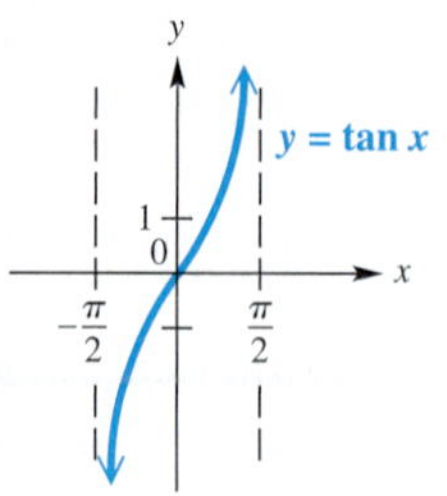

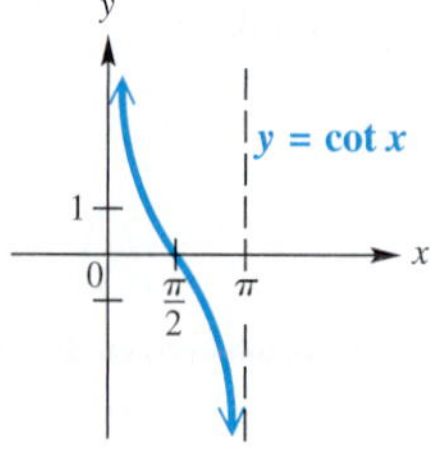

Domain: $\{x \mid x \neq (2n + 1)\frac{\pi}{2},$ where n is an integer$\}$
Range: $(-\infty, \infty)$
Period: π

Domain: $\{x \mid x \neq n\pi,$ where n is an integer$\}$
Range: $(-\infty, \infty)$
Period: π

See page 581 for a summary of graphing techniques.

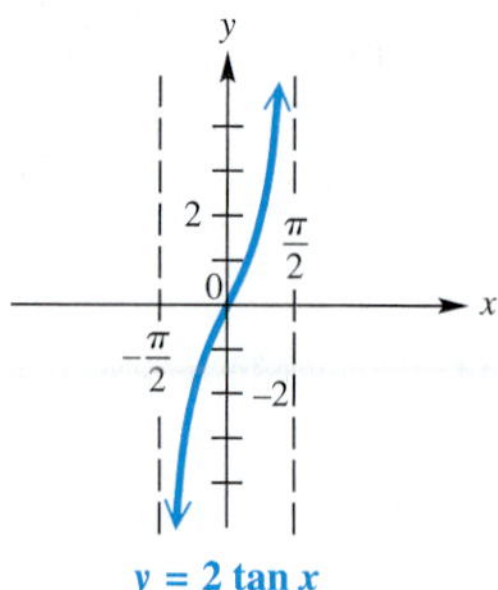

period: π
domain: $\{x \mid x \neq (2n + 1)\frac{\pi}{2},$ where n is an integer$\}$
range: $(-\infty, \infty)$

6.5 Harmonic Motion

Simple Harmonic Motion
The position of a point oscillating about an equilibrium position at time t is modeled by either

$$s(t) = a \cos \omega t \quad \text{or} \quad s(t) = a \sin \omega t,$$

where a and ω are constants, with $\omega > 0$. The amplitude of the motion is $|a|$, the period is $\frac{2\pi}{\omega}$, and the frequency is $\frac{\omega}{2\pi}$.

A spring oscillates according to

$$s(t) = -5 \cos 6t,$$

where t is in seconds and $s(t)$ is in inches.

$$\text{amplitude} = |-5| = 5$$

$$\text{period} = \frac{2\pi}{6} = \frac{\pi}{3}$$

$$\text{frequency} = \frac{3}{\pi}$$

Chapter 6 Review Exercises

1. Consider each angle in standard position having the given radian measure. In what quadrant does the terminal side lie?

   (a) 3 (b) 4 (c) −2 (d) 7

2. Which is larger—an angle of 1° or an angle of 1 radian? Discuss and justify your answer.

Convert each degree measure to radians. Leave answers as multiples of π.

3. 120°
4. 800°

Convert each radian measure to degrees.

5. $\frac{5\pi}{4}$
6. $-\frac{6\pi}{5}$

7. *Railroad Engineering* The term *grade* has several different meanings in construction work. Some engineers use the term grade to represent $\frac{1}{100}$ of a right angle and express grade as a percent. For instance, an angle of $.9°$ would be referred to as a 1% grade. (*Source:* Hay, W., *Railroad Engineering,* John Wiley & Sons, 1982.)
   (a) By what number should you multiply a grade to convert it to radians?
   (b) In a rapid-transit rail system, the maximum grade allowed between two stations is 3.5%. Express this angle in degrees and radians.

8. *Concept Check* Suppose the tip of the minute hand of a clock is 2 in. from the center of the clock. For each of the following durations, determine the distance traveled by the tip of the minute hand.
   (a) 20 min
   (b) 3 hr

Solve each problem. Use a calculator as necessary.

9. *Arc Length* The radius of a circle is 15.2 cm. Find the length of an arc of the circle intercepted by a central angle of $\frac{3\pi}{4}$ radians.

10. *Area of a Sector* A central angle of $\frac{7\pi}{4}$ radians forms a sector of a circle. Find the area of the sector if the radius of the circle is 28.69 in.

11. *Height of a Tree* A tree 2000 yd away subtends an angle of $1° \, 10'$. Find the height of the tree to two significant digits.

12. *Rotation of a Seesaw* The seesaw at a playground is 12 ft long. Through what angle does the board rotate when a child rises 3 ft along the circular arc?

Consider the figure here for Exercises 13 and 14.

13. What is the measure of θ in radians?

14. What is the area of the sector?

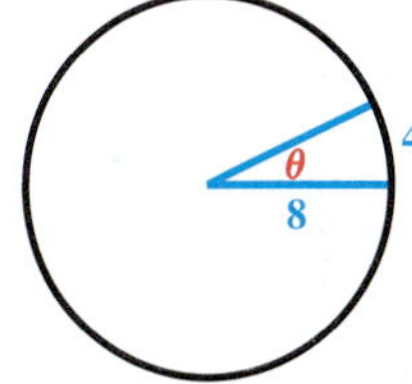

Solve each problem.

15. Find the time t if $\theta = \frac{5\pi}{12}$ radians and $\omega = \frac{8\pi}{9}$ radians per sec.

16. Find angle θ if $t = 12$ sec and $\omega = 9$ radians per sec.

17. *Linear Speed of a Flywheel* Find the linear speed of a point on the edge of a flywheel of radius 7 m if the flywheel is rotating 90 times per sec.

18. *Angular Speed of a Ferris Wheel* A Ferris wheel has radius 25 ft. If it takes 30 sec for the wheel to turn $\frac{5\pi}{6}$ radians, what is the angular speed of the wheel?

19. ***Concept Check*** Consider the area-of-a-sector formula $A = \frac{1}{2}r^2\theta$. What well-known formula corresponds to the special case $\theta = 2\pi$?

Find the exact function value. Do not use a calculator.

20. $\cos \frac{2\pi}{3}$ **21.** $\tan\left(-\frac{7\pi}{3}\right)$ **22.** $\csc\left(-\frac{11\pi}{6}\right)$

Use a calculator to find an approximation for each circular function value. Be sure your calculator is set in radian mode.

23. $\cos(-.2443)$ **24.** $\cot 3.0543$

25. Approximate the value of s in the interval $\left[0, \frac{\pi}{2}\right]$ if $\sin s = .4924$.

Find the exact value of s in the given interval that has the given circular function value. Do not use a calculator.

26. $\left[\frac{\pi}{2}, \pi\right]$; $\tan s = -\sqrt{3}$ **27.** $\left[\pi, \frac{3\pi}{2}\right]$; $\sec s = -\frac{2\sqrt{3}}{3}$

(Modeling) *Solve each problem.*

28. ***Phase Angle of the Moon*** Because the moon orbits Earth, we observe different phases of the moon during the period of a month. In the figure, t is called the *phase angle*.

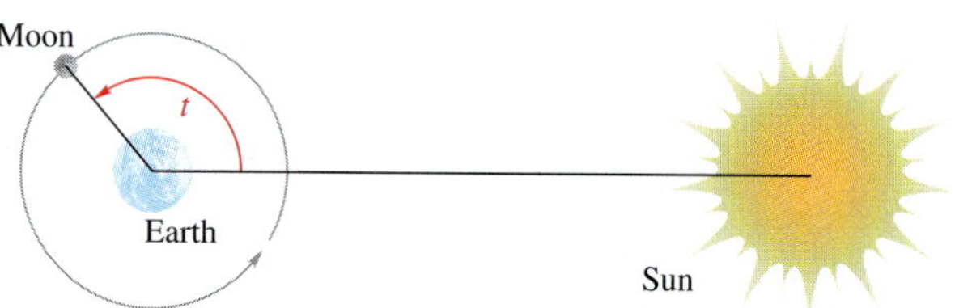

The *phase F* of the moon is computed by

$$F(t) = \frac{1}{2}(1 - \cos t),$$

and gives the fraction of the moon's face that is illuminated by the sun. (*Source:* Duffet-Smith, P., *Practical Astronomy with Your Calculator,* Cambridge University Press, 1988.) Evaluate each expression and interpret the result.

(a) $F(0)$ **(b)** $F\left(\frac{\pi}{2}\right)$ **(c)** $F(\pi)$ **(d)** $F\left(\frac{3\pi}{2}\right)$

29. ***Atmospheric Effect on Sunlight*** The shortest path for the sun's rays through Earth's atmosphere occurs when the sun is directly overhead. Disregarding the curvature of Earth, as the sun moves lower on the horizon, the distance that sunlight passes through the atmosphere increases by a factor of $\csc \theta$, where θ is the angle of elevation of the sun. This increased distance reduces both the intensity of the sun and the amount of ultraviolet light that reaches Earth's surface. See the figure at the top of the next page. (*Source:* Winter, C., R. Sizmann, and Vant-Hunt (Editors), *Solar Power Plants,* Springer-Verlag, 1991.)

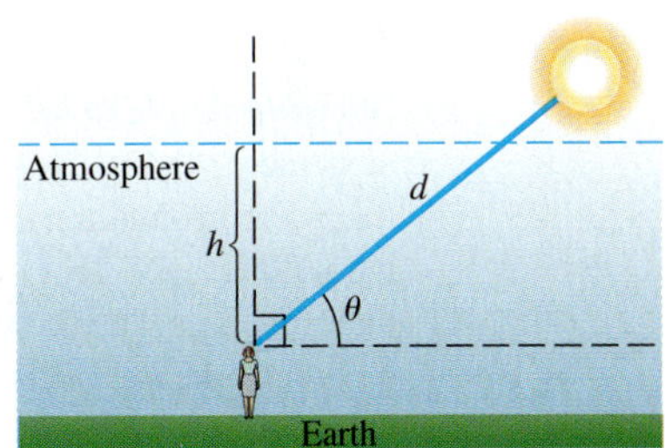

(a) Verify that $d = h \csc \theta$.

(b) Determine θ when $d = 2h$.

(c) The atmosphere filters out the ultraviolet light that causes skin to burn. Compare the difference between sunbathing when $\theta = \frac{\pi}{2}$ and when $\theta = \frac{\pi}{3}$. Which measure gives less ultraviolet light?

30. *Concept Check* Which one of the following is true about the graph of $y = 4 \sin 2x$?

A. It has amplitude 2 and period $\frac{\pi}{2}$. **B.** It has amplitude 4 and period π.
C. Its range is $[0, 4]$. **D.** Its range is $[-4, 0]$.

31. *Concept Check* Which one of the following is false about the graph of $y = -3 \cos \frac{1}{2}x$?

A. Its range is $[-3, 3]$. **B.** Its domain is $(-\infty, \infty)$.
C. Its amplitude is 3, and its period is 4π.
D. Its amplitude is 3, and its period is π.

For each function, give the amplitude, period, vertical translation, and phase shift, as applicable.

32. $y = 2 \sin x$ **33.** $y = \tan 3x$ **34.** $y = -\frac{1}{2} \cos 3x$

35. $y = 2 \sin 5x$ **36.** $y = 1 + 2 \sin \frac{1}{4}x$ **37.** $y = 3 - \frac{1}{4} \cos \frac{2}{3}x$

38. $y = 3 \cos\left(x + \frac{\pi}{2}\right)$ **39.** $y = -\sin\left(x - \frac{3\pi}{4}\right)$ **40.** $y = \frac{1}{2} \csc\left(2x - \frac{\pi}{4}\right)$

Concept Check *Identify the circular function that satisfies each description.*

41. period is π, x-intercepts are of the form $n\pi$, where n is an integer

42. period is 2π, graph passes through the origin

43. period is 2π, graph passes through the point $\left(\frac{\pi}{2}, 0\right)$

Graph each function over a one-period interval.

44. $y = 3 \cos 2x$ **45.** $y = \frac{1}{2} \cot 3x$ **46.** $y = \cos\left(x - \frac{\pi}{4}\right)$

47. $y = \tan\left(x - \frac{\pi}{2}\right)$ **48.** $y = 1 + 2 \cos 3x$ **49.** $y = -1 - 3 \sin 2x$

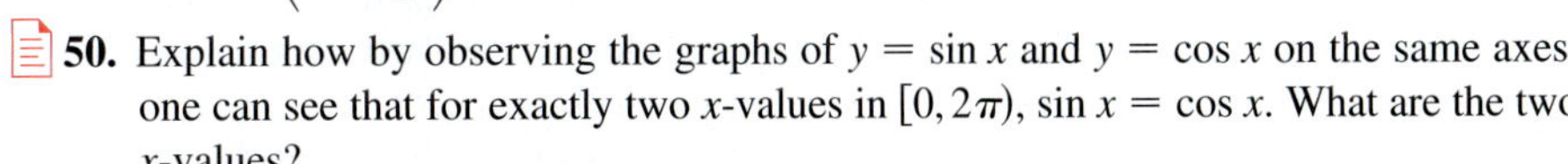

50. Explain how by observing the graphs of $y = \sin x$ and $y = \cos x$ on the same axes, one can see that for exactly two x-values in $[0, 2\pi)$, $\sin x = \cos x$. What are the two x-values?

Solve each problem.

51. ***Viewing Angle to an Object*** Let a person whose eyes are h_1 feet from the ground stand d feet from an object h_2 feet tall, where $h_2 > h_1$. Let θ be the angle of elevation to the top of the object. See the figure.

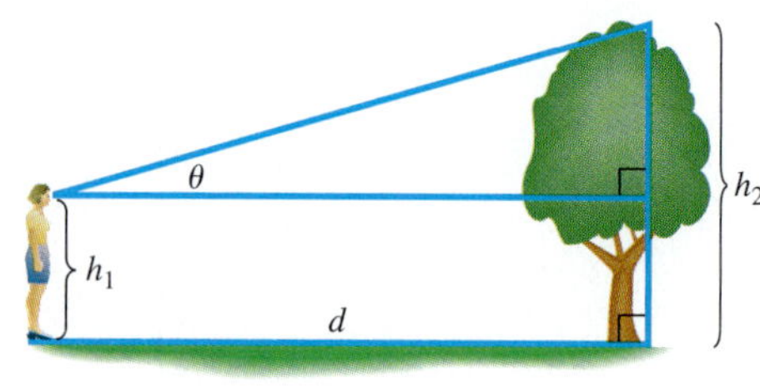

(a) Show that $d = (h_2 - h_1)\cot\theta$.

(b) Let $h_2 = 55$ and $h_1 = 5$. Graph d for the interval $0 < \theta \le \frac{\pi}{2}$.

52. ***(Modeling) Tides*** The figure shows a function f that models the tides in feet at Clearwater Beach, Florida, x hours after midnight starting on August 26, 1998. (*Source:* Pentcheff, D., *WWW Tide and Current Predictor.*)

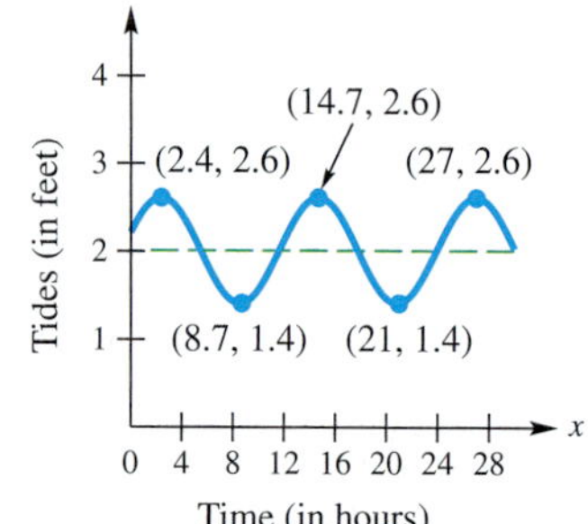

(a) Find the time between high tides.

(b) What is the difference in water levels between high tide and low tide?

(c) The tides can be modeled by

$$f(x) = .6\cos[.511(x - 2.4)] + 2.$$

Estimate the tides when $x = 10$.

53. ***(Modeling) Maximum Temperatures*** The maximum afternoon temperature in a given city might be modeled by

$$t = 60 - 30\cos\frac{x\pi}{6},$$

where t represents the maximum afternoon temperature in month x, with $x = 0$ representing January, $x = 1$ representing February, and so on. Find the maximum afternoon temperature for each month.

(a) January **(b)** April **(c)** May
(d) June **(e)** August **(f)** October

54. ***(Modeling) Average Monthly Temperature*** The average monthly temperature (in °F) in Chicago, Illinois, is shown in the table.

Month	°F	Month	°F
Jan	25	July	74
Feb	28	Aug	75
Mar	36	Sept	66
Apr	48	Oct	55
May	61	Nov	39
June	72	Dec	28

Source: Miller, A., J. Thompson, and R. Peterson, *Elements of Meteorology, 4th Edition,* Charles E. Merrill Publishing Co., 1983.

(a) Plot the average monthly temperature over a two-year period. Let $x = 1$ correspond to January of the first year.

(b) Determine a model function of the form $f(x) = a\sin b(x - d) + c$, where a, b, c, and d are constants.

(c) Explain the significance of each constant.

(d) Graph f together with the data on the same coordinate axes. How well does f model the data?

(e) Use the sine regression capability of a graphing calculator to find the equation of a sine curve that fits these data.

55. ***(Modeling) Pollution Trends*** The amount of pollution in the air fluctuates with the seasons. It is lower after heavy spring rains and higher after periods of little rain. In addition to this seasonal fluctuation, the long-term trend is upward. An idealized graph of this situation is shown in the figure on the next page.

55. *(continued)* Circular functions can be used to model the fluctuating part of the pollution levels, and exponential functions can be used to model long-term growth. The pollution level in a certain area might be given by

$$y = 7(1 - \cos 2\pi x)(x + 10) + 100e^{.2x},$$

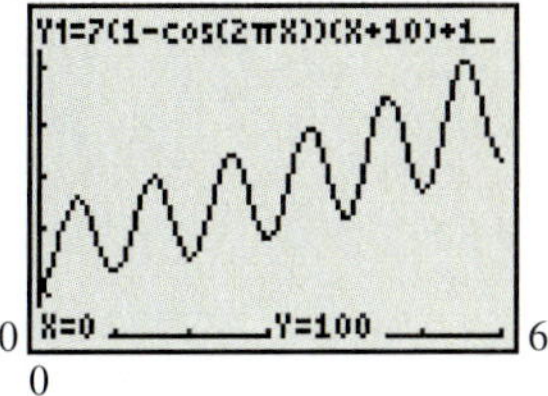

where x is the time in years, with $x = 0$ representing January 1 of the base year. July 1 of the same year would be represented by $x = .5$, October 1 of the following year would be represented by $x = 1.75$, and so on. Find the pollution levels on each date.

(a) January 1, base year
(b) July 1, base year
(c) January 1, following year
(d) July 1, following year

An object in simple harmonic motion has position function s inches from an initial point, where t is the time in seconds. Find the amplitude, period, and frequency.

56. $s(t) = 3 \cos 2t$

57. $s(t) = 4 \sin \pi t$

58. In Exercise 56, what does the period represent? What does the amplitude represent?

59. In Exercise 57, what does the frequency represent? Find the position of the object from the initial point at 1.5 sec, 2 sec, and 3.25 sec.

Chapter 6 Test

Convert each degree measure to radians.

1. 120°

2. −45°

3. 5° (to the nearest hundredth)

Convert each radian measure to degrees.

4. $\frac{3\pi}{4}$

5. $-\frac{7\pi}{6}$

6. 4 (to the nearest hundredth)

7. A central angle of a circle with radius 150 cm cuts off an arc of 200 cm. Find each measure.

(a) the radian measure of the angle
(b) the area of a sector with that central angle

8. ***Rotation of Gas Gauge Arrow*** The arrow on a car's gasoline gauge is $\frac{1}{2}$ in. long. See the figure. Through what angle does the arrow rotate when it moves 1 in. on the gauge?

9. *Angular and Linear Speed of a Point* Suppose that point P is on a circle with radius 10 cm, and ray OP is rotating with angular speed $\frac{\pi}{18}$ radian per sec.

(a) Find the angle generated by P in 6 sec.
(b) Find the distance traveled by P along the circle in 6 sec.
(c) Find the linear speed of P.

Find each circular function value.

10. $\sin \frac{3\pi}{4}$ **11.** $\cos\left(-\frac{7\pi}{6}\right)$ **12.** $\tan \frac{3\pi}{2}$ **13.** $\sec \frac{2\pi}{3}$

14. (a) Use a calculator to approximate s in the interval $\left[0, \frac{\pi}{2}\right]$, if $\sin s = .8258$.

(b) Find the exact value of s in the interval $\left[0, \frac{\pi}{2}\right]$, if $\cos s = \frac{1}{2}$.

15. Consider the function defined by $y = 3 - 6\sin\left(2x + \frac{\pi}{2}\right)$.

(a) What is its period?
(b) What is the amplitude of its graph?
(c) What is its range?
(d) What is the y-intercept of its graph?
(e) What is its phase shift?

Graph each function over a two-period interval. Identify asymptotes when applicable.

16. $y = -\cos 2x$ **17.** $y = -\csc 2x$

18. $y = \tan\left(x - \frac{\pi}{2}\right)$ **19.** $y = -1 + 2\sin(x + \pi)$

20. $y = -2 - \cot\left(x - \frac{\pi}{2}\right)$

21. *(Modeling) Average Monthly Temperature* The average monthly temperature (in °F) in Austin, Texas, can be modeled using the circular function defined by

$$f(x) = 17.5 \sin\left[\frac{\pi}{6}(x - 4)\right] + 67.5,$$

where x is the month and $x = 1$ corresponds to January. (*Source:* Miller, A., J. Thompson, and R. Peterson, *Elements of Meteorology, 4th Edition,* Charles E. Merrill Publishing Co., 1983.)

(a) Graph f in the window $[1, 25]$ by $[45, 90]$.
(b) Determine the amplitude, period, phase shift, and vertical translation of f.
(c) What is the average monthly temperature for the month of December?
(d) Determine the maximum and minimum average monthly temperatures and the months when they occur.
(e) What would be an approximation for the average *yearly* temperature in Austin? How is this related to the vertical translation of the sine function in the formula for f?

22. *(Modeling) Spring* The height of a weight attached to a spring is

$$s(t) = -4\cos 8\pi t$$

inches after t seconds.

(a) Find the maximum height that the weight rises above the equilibrium position of $s(t) = 0$.
(b) When does the weight first reach its maximum height, if $t \geq 0$?
(c) What are the frequency and period?

Chapter 6 Quantitative Reasoning

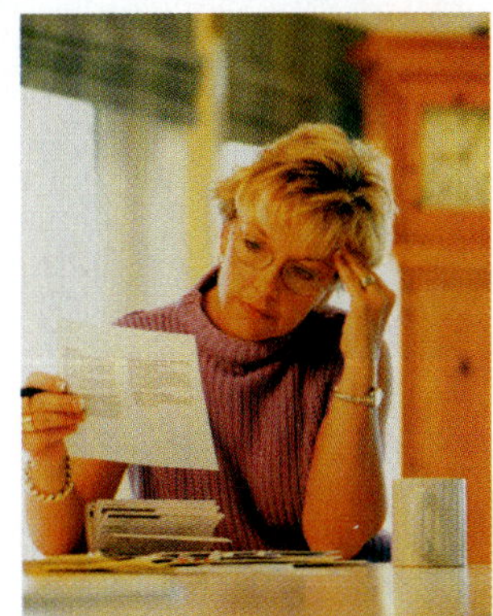

Does the fact that average monthly temperatures are periodic affect your utility bills?

In an article entitled "I Found Sinusoids in My Gas Bill" (*Mathematics Teacher*, January 2000), Cathy G. Schloemer presents the following graph that accompanied her gas bill.

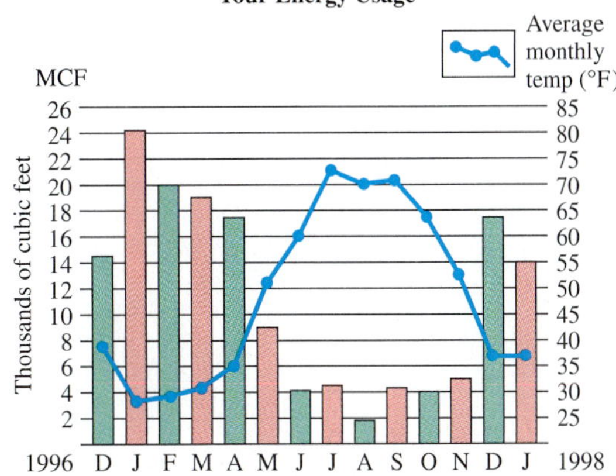

Notice that two sinusoids are suggested here: one for the behavior of the average monthly temperature and another for gas use in MCF (thousands of cubic feet).

1. If January 1997 is represented by $x = 1$, the data of estimated ordered pairs (month, temperature) is given in the list shown on the two graphing calculator screens below.

L1	L2	L3 1
1	28	------
2	29	
3	31	
4	35	
5	51	
6	60	
7	73	

L1(1)=1

L1	L2	L3 1
7	73	
8	70	
9	71	
10	64	
11	53	
12	37	

L1(13) =

Use the sine regression feature of a graphing calculator to find a sine function that fits these data points. Then make a scatter diagram, and graph the function.

2. If January 1997 is again represented by $x = 1$, the data of estimated ordered pairs (month, gas use in MCF) is given in the list shown on the two graphing calculator screens below.

L1	L2	L3 1
1	24.2	------
2	20	
3	18.8	
4	17.5	
5	9.2	
6	4.2	
7	4.8	

L1(1)=1

L1	L2	L3 1
7	4.8	
8	1.8	
9	4.8	
10	4	
11	5	
12	17.5	

L1(13) =

Use the sine regression feature of a graphing calculator to find a sine function that fits these data points. Then make a scatter diagram, and graph the function.

3. Answer the question posed at the top of the page, in the form of a short paragraph.

7

Trigonometric Identities and Equations

In 1831 Michael Faraday discovered that when a wire passes by a magnet, a small electric current is produced in the wire. Now we generate massive amounts of electricity by simultaneously rotating thousands of wires near large electromagnets. Because electric current alternates its direction on electrical wires, it is modeled accurately by either the sine or the cosine function.

We give many examples of applications of the trigonometric functions to electricity and other phenomena in the examples and exercises in this chapter, including a model of the wattage consumption of a toaster in Section 7.4, Example 5.

7.1 Fundamental Identities

Negative-Angle Identities ▪ Fundamental Identities ▪ Using the Fundamental Identities

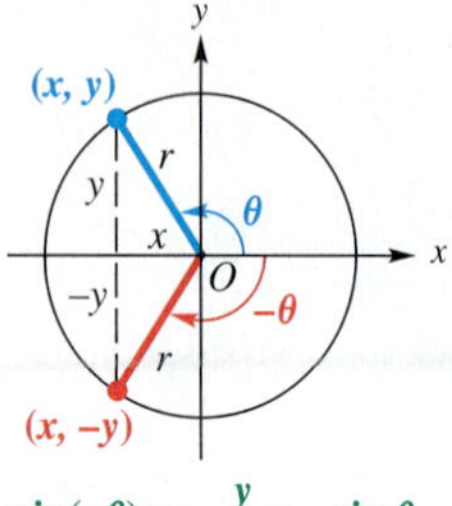

$\sin(-\theta) = -\frac{y}{r} = -\sin\theta$

Figure 1

Negative-Angle Identities

As suggested by the circle shown in Figure 1, an angle θ having the point (x, y) on its terminal side has a corresponding angle $-\theta$ with the point $(x, -y)$ on its terminal side. From the definition of sine,

$$\sin(-\theta) = \frac{-y}{r} \quad \text{and} \quad \sin\theta = \frac{y}{r}, \quad \text{(Section 5.2)}$$

so $\sin(-\theta)$ and $\sin\theta$ are negatives of each other, or

$$\sin(-\theta) = -\sin\theta.$$

Figure 1 shows an angle θ in quadrant II, but the same result holds for θ in any quadrant. Also, by definition,

$$\cos(-\theta) = \frac{x}{r} \quad \text{and} \quad \cos\theta = \frac{x}{r}, \quad \text{(Section 5.2)}$$

so

$$\cos(-\theta) = \cos\theta.$$

We can use these identities for $\sin(-\theta)$ and $\cos(-\theta)$ to find $\tan(-\theta)$ in terms of $\tan\theta$:

$$\tan(-\theta) = \frac{\sin(-\theta)}{\cos(-\theta)} = \frac{-\sin\theta}{\cos\theta} = -\frac{\sin\theta}{\cos\theta}$$

$$\tan(-\theta) = -\tan\theta.$$

Similar reasoning gives the following identities.

$$\csc(-\theta) = -\csc\theta, \qquad \sec(-\theta) = \sec\theta, \qquad \cot(-\theta) = -\cot\theta$$

This group of identities is known as the **negative-angle** or **negative-number identities.**

Fundamental Identities

In Chapter 5 we used the definitions of the trigonometric functions to derive the reciprocal, quotient, and Pythagorean identities. Together with the negative-angle identities, these are called the **fundamental identities.**

Fundamental Identities

Reciprocal Identities

$$\cot\theta = \frac{1}{\tan\theta} \qquad \sec\theta = \frac{1}{\cos\theta} \qquad \csc\theta = \frac{1}{\sin\theta}$$

Quotient Identities

$$\tan\theta = \frac{\sin\theta}{\cos\theta} \qquad \cot\theta = \frac{\cos\theta}{\sin\theta}$$

(continued)

Pythagorean Identities

$$\sin^2 \theta + \cos^2 \theta = 1 \qquad \tan^2 \theta + 1 = \sec^2 \theta \qquad 1 + \cot^2 \theta = \csc^2 \theta$$

Negative-Angle Identities

$$\sin(-\theta) = -\sin \theta \qquad \cos(-\theta) = \cos \theta \qquad \tan(-\theta) = -\tan \theta$$

$$\csc(-\theta) = -\csc \theta \qquad \sec(-\theta) = \sec \theta \qquad \cot(-\theta) = -\cot \theta$$

NOTE The most commonly recognized forms of the fundamental identities are given above. Throughout this chapter you must also recognize alternative forms of these identities. For example, two other forms of $\sin^2 \theta + \cos^2 \theta = 1$ are

$$\sin^2 \theta = 1 - \cos^2 \theta \quad \text{and} \quad \cos^2 \theta = 1 - \sin^2 \theta.$$

Using the Fundamental Identities One way we use these identities is to find the values of other trigonometric functions from the value of a given trigonometric function. Although we could find such values using a right triangle, this is a good way to practice using the fundamental identities.

EXAMPLE 1 Finding Trigonometric Function Values Given One Value and the Quadrant

If $\tan \theta = -\frac{5}{3}$ and θ is in quadrant II, find each function value.

(a) $\sec \theta$ **(b)** $\sin \theta$ **(c)** $\cot(-\theta)$

Solution

(a) Look for an identity that relates tangent and secant.

$$\tan^2 \theta + 1 = \sec^2 \theta \qquad \text{Pythagorean identity}$$

$$\left(-\frac{5}{3}\right)^2 + 1 = \sec^2 \theta \qquad \tan \theta = -\tfrac{5}{3}$$

$$\frac{25}{9} + 1 = \sec^2 \theta$$

$$\frac{34}{9} = \sec^2 \theta \qquad \text{Combine terms.}$$

$$-\sqrt{\frac{34}{9}} = \sec \theta \qquad \text{Take the negative square root. (Section 1.4)}$$

$$-\frac{\sqrt{34}}{3} = \sec \theta \qquad \text{Simplify the radical. (Section R.7)}$$

We chose the negative square root since $\sec \theta$ is negative in quadrant II.

(b)

$$\tan \theta = \frac{\sin \theta}{\cos \theta} \qquad \text{Quotient identity}$$

$$\cos \theta \tan \theta = \sin \theta \qquad \text{Multiply by } \cos \theta.$$

$$\left(\frac{1}{\sec \theta}\right)\tan \theta = \sin \theta \qquad \text{Reciprocal identity}$$

$$\left(-\frac{3\sqrt{34}}{34}\right)\left(-\frac{5}{3}\right) = \sin \theta \qquad \text{From part (a), } \frac{1}{\sec \theta} = -\frac{3}{\sqrt{34}} = -\frac{3\sqrt{34}}{34};\ \tan \theta = -\frac{5}{3}$$

$$\sin \theta = \frac{5\sqrt{34}}{34}$$

(c)

$$\cot(-\theta) = \frac{1}{\tan(-\theta)} \qquad \text{Reciprocal identity}$$

$$\cot(-\theta) = \frac{1}{-\tan \theta} \qquad \text{Negative-angle identity}$$

$$\cot(-\theta) = \frac{1}{-\left(-\frac{5}{3}\right)} = \frac{3}{5} \qquad \tan \theta = -\frac{5}{3};\ \text{Simplify. (Section R.5)}$$

Now try Exercises 5, 7, and 9.

CAUTION To avoid a common error, when taking the square root, be sure to choose the sign based on the quadrant of θ and the function being evaluated.

Any trigonometric function of a number or angle can be expressed in terms of any other function.

EXAMPLE 2 Expressing One Function in Terms of Another

Express $\cos x$ in terms of $\tan x$.

Solution Since $\sec x$ is related to both $\cos x$ and $\tan x$ by identities, start with $1 + \tan^2 x = \sec^2 x$.

$$\frac{1}{1 + \tan^2 x} = \frac{1}{\sec^2 x} \qquad \text{Take reciprocals.}$$

$$\frac{1}{1 + \tan^2 x} = \cos^2 x \qquad \text{Reciprocal identity}$$

$$\pm\sqrt{\frac{1}{1 + \tan^2 x}} = \cos x \qquad \text{Take square roots.}$$

$$\cos x = \frac{\pm 1}{\sqrt{1 + \tan^2 x}} \qquad \text{Quotient rule (Section R.7); rewrite.}$$

$$\cos x = \frac{\pm\sqrt{1 + \tan^2 x}}{1 + \tan^2 x} \qquad \text{Rationalize the denominator. (Section R.7)}$$

Choose the + sign or the − sign, depending on the quadrant of x.

Now try Exercise 43.

Plot1 Plot2 Plot3
\Y1=(sin(X))²+(cos(X))²
\Y2=1
\Y3=
\Y4=
\Y5=
\Y6=

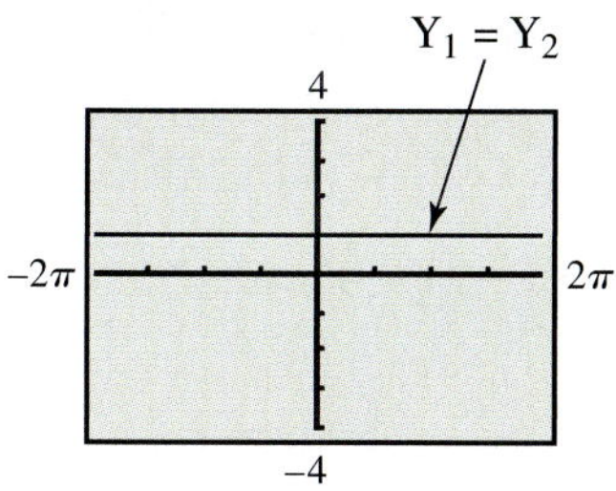

Figure 2

We can use a graphing calculator to decide whether two functions are identical. See Figure 2, which supports the identity $\sin^2 x + \cos^2 x = 1$. With an identity, you should see no difference in the two graphs. ■

All other trigonometric functions can easily be expressed in terms of $\sin\theta$ and/or $\cos\theta$. We often make such substitutions in an expression to simplify it.

EXAMPLE 3 Rewriting an Expression in Terms of Sine and Cosine

Write $\tan\theta + \cot\theta$ in terms of $\sin\theta$ and $\cos\theta$, and then simplify the expression.

Solution

$$\tan\theta + \cot\theta = \frac{\sin\theta}{\cos\theta} + \frac{\cos\theta}{\sin\theta} \qquad \text{Quotient identities}$$

$$= \frac{\sin^2\theta}{\cos\theta\sin\theta} + \frac{\cos^2\theta}{\cos\theta\sin\theta} \qquad \text{Write each fraction with the LCD. (Section R.5)}$$

$$= \frac{\sin^2\theta + \cos^2\theta}{\cos\theta\sin\theta} \qquad \text{Add fractions.}$$

$$\tan\theta + \cot\theta = \frac{1}{\cos\theta\sin\theta} \qquad \text{Pythagorean identity}$$

Now try Exercise 55.

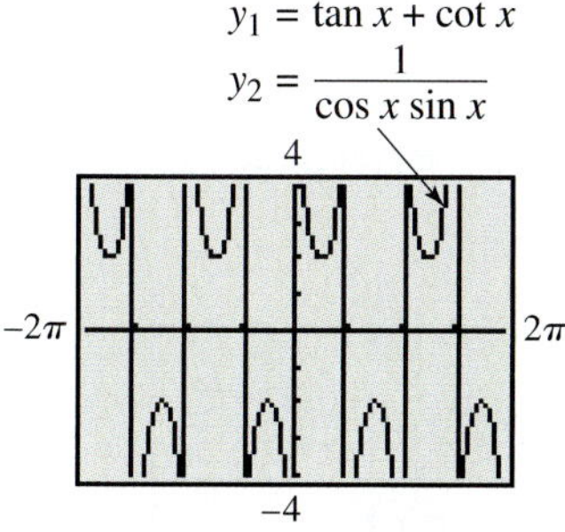

The graph supports the result in Example 3. The graphs of y_1 and y_2 appear to be identical.

CAUTION When working with trigonometric expressions and identities, be sure to write the argument of the function. For example, we would *not* write $\sin^2 + \cos^2 = 1$; an argument such as θ is necessary in this identity.

7.1 Exercises

Concept Check *Fill in the blanks.*

1. If $\tan x = 2.6$, then $\tan(-x) =$ ______.

2. If $\cos x = -.65$, then $\cos(-x) =$ ______.

3. If $\tan x = 1.6$, then $\cot x =$ ______.

4. If $\cos x = .8$ and $\sin x = .6$, then $\tan(-x) =$ ______.

Find $\sin s$. *See Example 1.*

5. $\cos s = \frac{3}{4}$, s in quadrant I

6. $\cot s = -\frac{1}{3}$, s in quadrant IV

7. $\cos(-s) = \frac{\sqrt{5}}{5}$, $\tan s < 0$

8. $\tan s = -\frac{\sqrt{7}}{2}$, $\sec s > 0$

9. $\sec s = \frac{11}{4}$, $\tan s < 0$

10. $\csc s = -\frac{8}{5}$

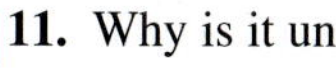

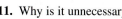

11. Why is it unnecessary to give the quadrant of s in Exercise 10?

Relating Concepts

For individual or collaborative investigation

(Exercises 12–17)

A function is called an even function if $f(-x) = f(x)$ for all x in the domain of f. A function is called an odd function if $f(-x) = -f(x)$ for all x in the domain of f. ***Work Exercises 12–17 in order,*** *to see the connection between the negative-angle identities and even and odd functions.*

12. Complete the statement: $\sin(-x) =$ ______.

13. Is the function defined by $f(x) = \sin x$ even or odd?

14. Complete the statement: $\cos(-x) =$ ______.

15. Is the function defined by $f(x) = \cos x$ even or odd?

16. Complete the statement: $\tan(-x) =$ ______.

17. Is the function defined by $f(x) = \tan x$ even or odd?

Concept Check For each graph, determine whether $f(-x) = f(x)$ or $f(-x) = -f(x)$ is true.

18.

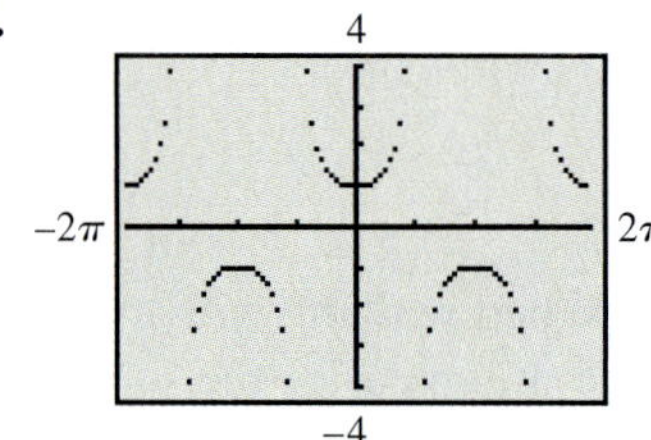

19.

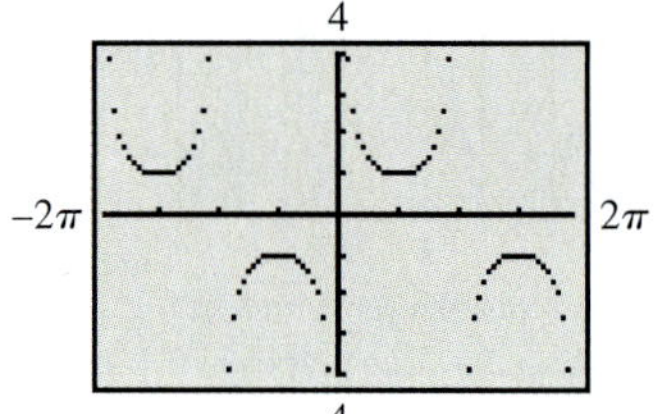

20.

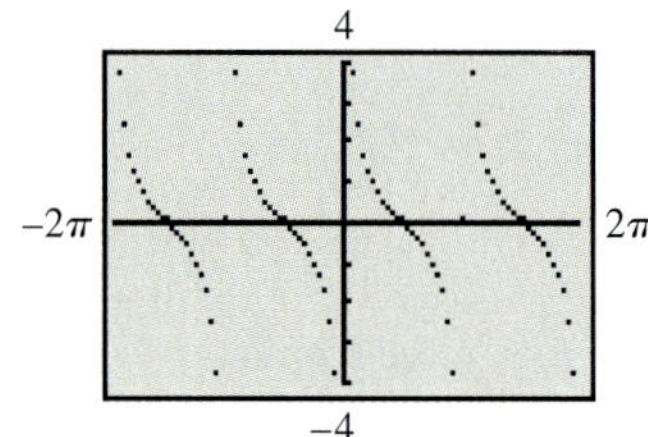

Find the remaining five trigonometric functions of θ. See Example 1.

21. $\sin \theta = \frac{2}{3}$, θ in quadrant II

22. $\cos \theta = \frac{1}{5}$, θ in quadrant I

23. $\tan \theta = -\frac{1}{4}$, θ in quadrant IV

24. $\csc \theta = -\frac{5}{2}$, θ in quadrant III

25. $\cot \theta = \frac{4}{3}$, $\sin \theta > 0$

26. $\sin \theta = -\frac{4}{5}$, $\cos \theta < 0$

27. $\sec \theta = \frac{4}{3}$, $\sin \theta < 0$

28. $\cos \theta = -\frac{1}{4}$, $\sin \theta > 0$

Concept Check *For each expression in Column I, choose the expression from Column II that completes an identity.*

I	II
29. $\dfrac{\cos x}{\sin x} =$ ______	**A.** $\sin^2 x + \cos^2 x$
30. $\tan x =$ ______	**B.** $\cot x$
31. $\cos(-x) =$ ______	**C.** $\sec^2 x$
32. $\tan^2 x + 1 =$ ______	**D.** $\dfrac{\sin x}{\cos x}$
33. $1 =$ ______	**E.** $\cos x$

Concept Check *For each expression in Column I, choose the expression from Column II that completes an identity. You may have to rewrite one or both expressions.*

I	II
34. $-\tan x \cos x =$ ______	**A.** $\dfrac{\sin^2 x}{\cos^2 x}$
35. $\sec^2 x - 1 =$ ______	**B.** $\dfrac{1}{\sec^2 x}$
36. $\dfrac{\sec x}{\csc x} =$ ______	**C.** $\sin(-x)$
37. $1 + \sin^2 x =$ ______	**D.** $\csc^2 x - \cot^2 x + \sin^2 x$
38. $\cos^2 x =$ ______	**E.** $\tan x$

39. A student writes "$1 + \cot^2 = \csc^2$." Comment on this student's work.

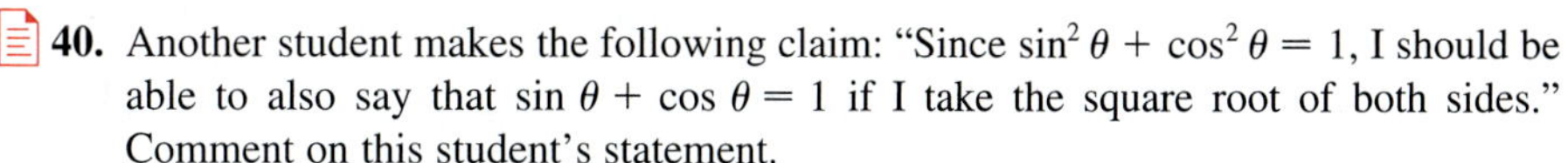

40. Another student makes the following claim: "Since $\sin^2 \theta + \cos^2 \theta = 1$, I should be able to also say that $\sin \theta + \cos \theta = 1$ if I take the square root of both sides." Comment on this student's statement.

41. ***Concept Check*** Suppose that $\cos \theta = \frac{x}{x+1}$. Find $\sin \theta$.

42. ***Concept Check*** Find $\tan \alpha$ if $\sec \alpha = \frac{p+4}{p}$.

Write the first trigonometric function in terms of the second trigonometric function. See Example 2.

43. $\sin x$; $\cos x$

44. $\cot x$; $\sin x$

45. $\tan x$; $\sec x$

46. $\cot x$; $\csc x$

47. $\csc x$; $\cos x$

48. $\sec x$; $\sin x$

Write each expression in terms of sine and cosine, and simplify it. See Example 3.

49. $\cot \theta \sin \theta$

50. $\sec \theta \cot \theta \sin \theta$

51. $\cos \theta \csc \theta$

52. $\cot^2 \theta(1 + \tan^2 \theta)$

53. $\sin^2 \theta(\csc^2 \theta - 1)$

54. $(\sec \theta - 1)(\sec \theta + 1)$

55. $(1 - \cos \theta)(1 + \sec \theta)$

56. $\dfrac{\cos \theta + \sin \theta}{\sin \theta}$

57. $\dfrac{\cos^2 \theta - \sin^2 \theta}{\sin \theta \cos \theta}$

58. $\dfrac{1 - \sin^2 \theta}{1 + \cot^2 \theta}$

59. $\sec \theta - \cos \theta$

60. $(\sec \theta + \csc \theta)(\cos \theta - \sin \theta)$

61. $\sin \theta(\csc \theta - \sin \theta)$

62. $\dfrac{1 + \tan^2 \theta}{1 + \cot^2 \theta}$

63. $\sin^2 \theta + \tan^2 \theta + \cos^2 \theta$

64. $\dfrac{\tan(-\theta)}{\sec \theta}$

65. *Concept Check* Let $\cos x = \frac{1}{5}$. Find all possible values for $\frac{\sec x - \tan x}{\sin x}$.

66. *Concept Check* Let $\csc x = -3$. Find all possible values for $\frac{\sin x + \cos x}{\sec x}$.

Relating Concepts

For individual or collaborative investigation

(Exercises 67–71)

In Chapter 6 we graphed functions defined by

$$y = c + a \cdot f[b(x - d)]$$

with the assumption that $b > 0$. To see what happens when $b < 0$, ***work Exercises 67–71 in order.***

67. Use a negative-angle identity to write $y = \sin(-2x)$ as a function of $2x$.

68. How does your answer to Exercise 67 relate to $y = \sin(2x)$?

69. Use a negative-angle identity to write $y = \cos(-4x)$ as a function of $4x$.

70. How does your answer to Exercise 69 relate to $y = \cos(4x)$?

71. Use your results from Exercises 67–70 to rewrite the following with a positive value of b.

(a) $\sin(-4x)$ **(b)** $\cos(-2x)$ **(c)** $-5 \sin(-3x)$

Use a graphing calculator to decide whether each equation is an identity. (Hint: In Exercise 76, graph the function of x for a few different values of y (in radians).)

72. $\cos 2x = 1 - 2 \sin^2 x$

73. $2 \sin s = \sin 2s$

74. $\sin x = \sqrt{1 - \cos^2 x}$

75. $\cos 2x = \cos^2 x - \sin^2 x$

76. $\cos(x - y) = \cos x - \cos y$

7.2 Verifying Trigonometric Identities

Verifying Identities by Working with One Side ▪ Verifying Identities by Working with Both Sides

Recall that an identity is an equation that is satisfied for all meaningful replacements of the variable. One of the skills required for more advanced work in mathematics, especially in calculus, is the ability to use identities to write expressions in alternative forms. We develop this skill by using the fundamental identities to verify that a trigonometric equation is an identity (for those values of the variable for which it is defined). Here are some hints to help you get started.

Looking Ahead to Calculus

Trigonometric identities are used in calculus to simplify trigonometric expressions, determine derivatives of trigonometric functions, and change the form of some integrals.

Hints for Verifying Identities

1. Learn the fundamental identities given in the last section. Whenever you see either side of a fundamental identity, the other side should come to mind. Also, be aware of equivalent forms of the fundamental identities. For example, $\sin^2 \theta = 1 - \cos^2 \theta$ is an alternative form of the identity $\sin^2 \theta + \cos^2 \theta = 1$.
2. Try to rewrite the more complicated side of the equation so that it is identical to the simpler side.
3. It is sometimes helpful to express all trigonometric functions in the equation in terms of sine and cosine and then simplify the result.
4. Usually, any factoring or indicated algebraic operations should be performed. For example, the expression $\sin^2 x + 2 \sin x + 1$ can be factored as $(\sin x + 1)^2$. The sum or difference of two trigonometric expressions, such as $\frac{1}{\sin \theta} + \frac{1}{\cos \theta}$, can be added or subtracted in the same way as any other rational expression.

$$\frac{1}{\sin \theta} + \frac{1}{\cos \theta} = \frac{\cos \theta}{\sin \theta \cos \theta} + \frac{\sin \theta}{\sin \theta \cos \theta}$$
$$= \frac{\cos \theta + \sin \theta}{\sin \theta \cos \theta}$$

5. As you select substitutions, keep in mind the side you are not changing, because it represents your goal. For example, to verify the identity

$$\tan^2 x + 1 = \frac{1}{\cos^2 x},$$

try to think of an identity that relates $\tan x$ to $\cos x$. In this case, since $\sec x = \frac{1}{\cos x}$ and $\sec^2 x = \tan^2 x + 1$, the secant function is the best link between the two sides.
6. If an expression contains $1 + \sin x$, multiplying both numerator and denominator by $1 - \sin x$ would give $1 - \sin^2 x$, which could be replaced with $\cos^2 x$. Similar results for $1 - \sin x$, $1 + \cos x$, and $1 - \cos x$ may be useful.

CAUTION ***Verifying identities is not the same as solving equations.*** Techniques used in solving equations, such as adding the same terms to both sides, or multiplying both sides by the same term, should not be used when working with identities since you are starting with a statement (to be verified) that may not be true.

Verifying Identities by Working with One Side To avoid the temptation to use algebraic properties of equations to verify identities, *work with only one side and rewrite it to match the other side,* as shown in Examples 1–4.

EXAMPLE 1 Verifying an Identity (Working with One Side)

Verify that the following equation is an identity.

$$\cot s + 1 = \csc s(\cos s + \sin s)$$

Solution We use the fundamental identities from Section 7.1 to rewrite one side of the equation so that it is identical to the other side. Since the right side is more complicated, we work with it, using the third hint to change all functions to sine or cosine.

Steps	Reasons
Right side of given equation: $\csc s(\cos s + \sin s) = \frac{1}{\sin s}(\cos s + \sin s)$	$\csc s = \frac{1}{\sin s}$
$= \frac{\cos s}{\sin s} + \frac{\sin s}{\sin s}$	Distributive property (Section R.1)
$= \cot s + 1$ (Left side of given equation)	$\frac{\cos s}{\sin s} = \cot s$; $\frac{\sin s}{\sin s} = 1$

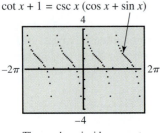

For $s = x$,
$\cot x + 1 = \csc x(\cos x + \sin x)$

The graphs coincide, supporting the conclusion in Example 1.

The given equation is an identity since the right side equals the left side.

Now try Exercise 33.

EXAMPLE 2 Verifying an Identity (Working with One Side)

Verify that the following equation is an identity.

$$\tan^2 x(1 + \cot^2 x) = \frac{1}{1 - \sin^2 x}$$

Solution We work with the more complicated left side, as suggested in the second hint. Again, we use the fundamental identities from Section 7.1.

$$\tan^2 x(1 + \cot^2 x) = \tan^2 x + \tan^2 x \cot^2 x \qquad \text{Distributive property}$$

$$= \tan^2 x + \tan^2 x \cdot \frac{1}{\tan^2 x} \qquad \cot^2 x = \frac{1}{\tan^2 x}$$

$$= \tan^2 x + 1 \qquad \tan^2 x \cdot \frac{1}{\tan^2 x} = 1$$

$$= \sec^2 x \qquad \tan^2 x + 1 = \sec^2 x$$

$$= \frac{1}{\cos^2 x} \qquad \sec^2 x = \frac{1}{\cos^2 x}$$

$$= \frac{1}{1 - \sin^2 x} \qquad \cos^2 x = 1 - \sin^2 x$$

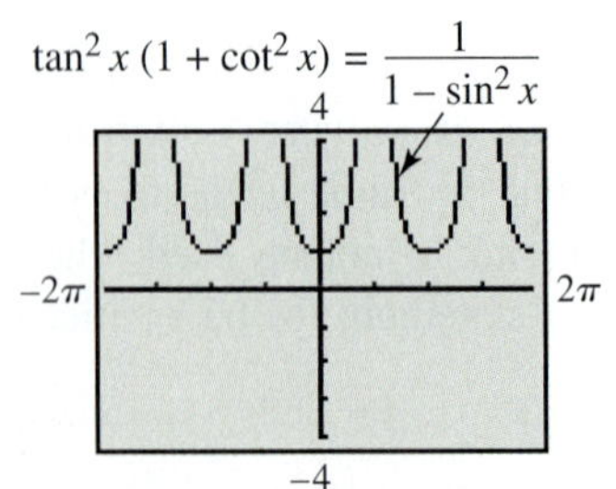

The screen supports the conclusion in Example 2.

Since the left side is identical to the right side, the given equation is an identity.

Now try Exercise 37.

EXAMPLE 3 Verifying an Identity (Working with One Side)

Verify that the following equation is an identity.

$$\frac{\tan t - \cot t}{\sin t \cos t} = \sec^2 t - \csc^2 t$$

Solution We transform the more complicated left side to match the right side.

$$\frac{\tan t - \cot t}{\sin t \cos t} = \frac{\tan t}{\sin t \cos t} - \frac{\cot t}{\sin t \cos t} \qquad \frac{a-b}{c} = \frac{a}{c} - \frac{b}{c} \text{ (Section R.5)}$$

$$= \tan t \cdot \frac{1}{\sin t \cos t} - \cot t \cdot \frac{1}{\sin t \cos t} \qquad \frac{a}{b} = a \cdot \frac{1}{b}$$

$$= \frac{\sin t}{\cos t} \cdot \frac{1}{\sin t \cos t} - \frac{\cos t}{\sin t} \cdot \frac{1}{\sin t \cos t} \qquad \tan t = \frac{\sin t}{\cos t};\ \cot t = \frac{\cos t}{\sin t}$$

$$= \frac{1}{\cos^2 t} - \frac{1}{\sin^2 t}$$

$$= \sec^2 t - \csc^2 t \qquad \frac{1}{\cos^2 t} = \sec^2 t;\ \frac{1}{\sin^2 t} = \csc^2 t$$

The third hint about writing all trigonometric functions in terms of sine and cosine was used in the third line of the solution.

Now try Exercise 41.

EXAMPLE 4 Verifying an Identity (Working with One Side)

Verify that the following equation is an identity.

$$\frac{\cos x}{1 - \sin x} = \frac{1 + \sin x}{\cos x}$$

Solution We work on the right side, using the last hint in the list given earlier to multiply numerator and denominator on the right by $1 - \sin x$.

$$\frac{1 + \sin x}{\cos x} = \frac{(1 + \sin x)(1 - \sin x)}{\cos x(1 - \sin x)} \qquad \text{Multiply by 1. (Section R.1)}$$

$$= \frac{1 - \sin^2 x}{\cos x(1 - \sin x)} \qquad (x+y)(x-y) = x^2 - y^2 \text{ (Section R.3)}$$

$$= \frac{\cos^2 x}{\cos x(1 - \sin x)} \qquad 1 - \sin^2 x = \cos^2 x$$

$$= \frac{\cos x}{1 - \sin x} \qquad \text{Lowest terms (Section R.5)}$$

Now try Exercise 47.

Verifying Identities by Working with Both Sides If both sides of an identity appear to be equally complex, the identity can be verified by working independently on the left side and on the right side, until each side is changed into some common third result. *Each step, on each side, must be reversible.*

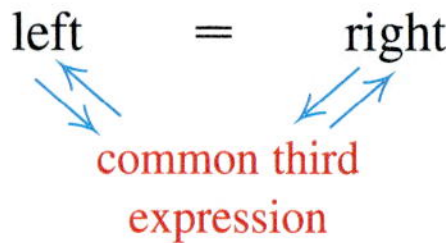

With all steps reversible, the procedure is as shown in the margin. The left side leads to a common third expression, which leads back to the right side. This procedure is just a shortcut for the procedure used in Examples 1–4: one side is changed into the other side, but by going through an intermediate step.

EXAMPLE 5 Verifying an Identity (Working with Both Sides)

Verify that the following equation is an identity.

$$\frac{\sec \alpha + \tan \alpha}{\sec \alpha - \tan \alpha} = \frac{1 + 2 \sin \alpha + \sin^2 \alpha}{\cos^2 \alpha}$$

Solution Both sides appear equally complex, so we verify the identity by changing each side into a common third expression. We work first on the left, multiplying numerator and denominator by $\cos \alpha$.

$$\frac{\sec \alpha + \tan \alpha}{\sec \alpha - \tan \alpha} = \frac{(\sec \alpha + \tan \alpha) \cos \alpha}{(\sec \alpha - \tan \alpha) \cos \alpha} \qquad \text{Multiply by 1.}$$

$$= \frac{\sec \alpha \cos \alpha + \tan \alpha \cos \alpha}{\sec \alpha \cos \alpha - \tan \alpha \cos \alpha} \qquad \text{Distributive property}$$

$$= \frac{1 + \tan \alpha \cos \alpha}{1 - \tan \alpha \cos \alpha} \qquad \sec \alpha \cos \alpha = 1$$

$$= \frac{1 + \frac{\sin \alpha}{\cos \alpha} \cdot \cos \alpha}{1 - \frac{\sin \alpha}{\cos \alpha} \cdot \cos \alpha} \qquad \tan \alpha = \frac{\sin \alpha}{\cos \alpha}$$

$$= \frac{1 + \sin \alpha}{1 - \sin \alpha}$$

On the right side of the original equation, begin by factoring.

$$\frac{1 + 2 \sin \alpha + \sin^2 \alpha}{\cos^2 \alpha} = \frac{(1 + \sin \alpha)^2}{\cos^2 \alpha} \qquad x^2 + 2xy + y^2 = (x + y)^2 \text{ (Section R.4)}$$

$$= \frac{(1 + \sin \alpha)^2}{1 - \sin^2 \alpha} \qquad \cos^2 \alpha = 1 - \sin^2 \alpha$$

$$= \frac{(1 + \sin \alpha)^2}{(1 + \sin \alpha)(1 - \sin \alpha)} \qquad \text{Factor } 1 - \sin^2 \alpha.$$

$$= \frac{1 + \sin \alpha}{1 - \sin \alpha} \qquad \text{Lowest terms}$$

We have shown that

$$\underbrace{\frac{\sec \alpha + \tan \alpha}{\sec \alpha - \tan \alpha}}_{\text{Left side of given equation}} = \underbrace{\frac{1 + \sin \alpha}{1 - \sin \alpha}}_{\text{Common third expression}} = \underbrace{\frac{1 + 2 \sin \alpha + \sin^2 \alpha}{\cos^2 \alpha}}_{\text{Right side of given equation}},$$

verifying that the given equation is an identity.

Now try Exercise 51.

CAUTION Use the method of Example 5 *only* if the steps are reversible.

There are usually several ways to verify a given identity. For instance, another way to begin verifying the identity in Example 5 is to work on the left as follows.

$$\frac{\sec \alpha + \tan \alpha}{\sec \alpha - \tan \alpha} = \frac{\frac{1}{\cos \alpha} + \frac{\sin \alpha}{\cos \alpha}}{\frac{1}{\cos \alpha} - \frac{\sin \alpha}{\cos \alpha}}$$ Fundamental identities (Section 7.1)

$$= \frac{\frac{1 + \sin \alpha}{\cos \alpha}}{\frac{1 - \sin \alpha}{\cos \alpha}}$$ Add and subtract fractions. (Section R.5)

$$= \frac{1 + \sin \alpha}{1 - \sin \alpha}$$ Simplify the complex fraction. (Section R.5)

Compare this with the result shown in Example 5 for the right side to see that the two sides indeed agree.

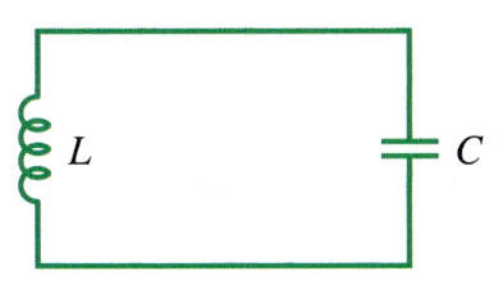

An Inductor and a Capacitor

Figure 3

EXAMPLE 6 Applying a Pythagorean Identity to Radios

Tuners in radios select a radio station by adjusting the frequency. A tuner may contain an inductor L and a capacitor C, as illustrated in Figure 3. The energy stored in the inductor at time t is given by

$$L(t) = k \sin^2(2\pi Ft)$$

and the energy stored in the capacitor is given by

$$C(t) = k \cos^2(2\pi Ft),$$

where F is the frequency of the radio station and k is a constant. The total energy E in the circuit is given by

$$E(t) = L(t) + C(t).$$

Show that E is a constant function. (*Source:* Weidner, R. and R. Sells, *Elementary Classical Physics,* Vol. 2, Allyn & Bacon, 1973.)

Solution

$$E(t) = L(t) + C(t)$$ Given equation

$$= k \sin^2(2\pi Ft) + k \cos^2(2\pi Ft)$$ Substitute.

$$= k[\sin^2(2\pi Ft) + \cos^2(2\pi Ft)]$$ Factor. (Section R.4)

$$= k(1)$$ $\sin^2 \theta + \cos^2 \theta = 1$ (Here $\theta = 2\pi Ft$).

$$= k$$

Since k is constant, $E(t)$ is a constant function.

Now try Exercise 85.

7.2 Exercises

Perform each indicated operation and simplify the result.

1. $\cot \theta + \dfrac{1}{\cot \theta}$
2. $\dfrac{\sec x}{\csc x} + \dfrac{\csc x}{\sec x}$
3. $\tan s(\cot s + \csc s)$
4. $\cos \beta(\sec \beta + \csc \beta)$
5. $\dfrac{1}{\csc^2 \theta} + \dfrac{1}{\sec^2 \theta}$
6. $\dfrac{1}{\sin \alpha - 1} - \dfrac{1}{\sin \alpha + 1}$
7. $\dfrac{\cos x}{\sec x} + \dfrac{\sin x}{\csc x}$
8. $\dfrac{\cos \theta}{\sin \theta} + \dfrac{\sin \theta}{1 + \cos \theta}$
9. $(1 + \sin t)^2 + \cos^2 t$
10. $(1 + \tan s)^2 - 2 \tan s$
11. $\dfrac{1}{1 + \cos x} - \dfrac{1}{1 - \cos x}$
12. $(\sin \alpha - \cos \alpha)^2$

Factor each trigonometric expression.

13. $\sin^2 \theta - 1$
14. $\sec^2 \theta - 1$
15. $(\sin x + 1)^2 - (\sin x - 1)^2$
16. $(\tan x + \cot x)^2 - (\tan x - \cot x)^2$
17. $2 \sin^2 x + 3 \sin x + 1$
18. $4 \tan^2 \beta + \tan \beta - 3$
19. $\cos^4 x + 2 \cos^2 x + 1$
20. $\cot^4 x + 3 \cot^2 x + 2$
21. $\sin^3 x - \cos^3 x$
22. $\sin^3 \alpha + \cos^3 \alpha$

Each expression simplifies to a constant, a single function, or a power of a function. Use fundamental identities to simplify each expression.

23. $\tan \theta \cos \theta$
24. $\cot \alpha \sin \alpha$
25. $\sec r \cos r$
26. $\cot t \tan t$
27. $\dfrac{\sin \beta \tan \beta}{\cos \beta}$
28. $\dfrac{\csc \theta \sec \theta}{\cot \theta}$
29. $\sec^2 x - 1$
30. $\csc^2 t - 1$
31. $\dfrac{\sin^2 x}{\cos^2 x} + \sin x \csc x$
32. $\dfrac{1}{\tan^2 \alpha} + \cot \alpha \tan \alpha$

In Exercises 33–68, verify that each trigonometric equation is an identity. See Examples 1–5.

33. $\dfrac{\cot \theta}{\csc \theta} = \cos \theta$
34. $\dfrac{\tan \alpha}{\sec \alpha} = \sin \alpha$
35. $\dfrac{1 - \sin^2 \beta}{\cos \beta} = \cos \beta$
36. $\dfrac{\tan^2 \alpha + 1}{\sec \alpha} = \sec \alpha$
37. $\cos^2 \theta(\tan^2 \theta + 1) = 1$
38. $\sin^2 \beta(1 + \cot^2 \beta) = 1$
39. $\cot s + \tan s = \sec s \csc s$
40. $\sin^2 \alpha + \tan^2 \alpha + \cos^2 \alpha = \sec^2 \alpha$
41. $\dfrac{\cos \alpha}{\sec \alpha} + \dfrac{\sin \alpha}{\csc \alpha} = \sec^2 \alpha - \tan^2 \alpha$
42. $\dfrac{\sin^2 \theta}{\cos \theta} = \sec \theta - \cos \theta$
43. $\sin^4 \theta - \cos^4 \theta = 2 \sin^2 \theta - 1$
44. $\dfrac{\cos \theta}{\sin \theta \cot \theta} = 1$
45. $(1 - \cos^2 \alpha)(1 + \cos^2 \alpha) = 2 \sin^2 \alpha - \sin^4 \alpha$
46. $\tan^2 \alpha \sin^2 \alpha = \tan^2 \alpha + \cos^2 \alpha - 1$

47. $\dfrac{\cos\theta + 1}{\tan^2\theta} = \dfrac{\cos\theta}{\sec\theta - 1}$

48. $\dfrac{(\sec\theta - \tan\theta)^2 + 1}{\sec\theta\csc\theta - \tan\theta\csc\theta} = 2\tan\theta$

49. $\dfrac{1}{1 - \sin\theta} + \dfrac{1}{1 + \sin\theta} = 2\sec^2\theta$

50. $\dfrac{1}{\sec\alpha - \tan\alpha} = \sec\alpha + \tan\alpha$

51. $\dfrac{\tan s}{1 + \cos s} + \dfrac{\sin s}{1 - \cos s} = \cot s + \sec s\csc s$

52. $\dfrac{1 - \cos x}{1 + \cos x} = (\cot x - \csc x)^2$

53. $\dfrac{\cot\alpha + 1}{\cot\alpha - 1} = \dfrac{1 + \tan\alpha}{1 - \tan\alpha}$

54. $\dfrac{1}{\tan\alpha - \sec\alpha} + \dfrac{1}{\tan\alpha + \sec\alpha} = -2\tan\alpha$

55. $\sin^2\alpha\sec^2\alpha + \sin^2\alpha\csc^2\alpha = \sec^2\alpha$

56. $\dfrac{\csc\theta + \cot\theta}{\tan\theta + \sin\theta} = \cot\theta\csc\theta$

57. $\sec^4 x - \sec^2 x = \tan^4 x + \tan^2 x$

58. $\dfrac{1 - \sin\theta}{1 + \sin\theta} = \sec^2\theta - 2\sec\theta\tan\theta + \tan^2\theta$

59. $\sin\theta + \cos\theta = \dfrac{\sin\theta}{1 - \dfrac{\cos\theta}{\sin\theta}} + \dfrac{\cos\theta}{1 - \dfrac{\sin\theta}{\cos\theta}}$

60. $\dfrac{\sin\theta}{1 - \cos\theta} - \dfrac{\sin\theta\cos\theta}{1 + \cos\theta} = \csc\theta(1 + \cos^2\theta)$

61. $\dfrac{\sec^4 s - \tan^4 s}{\sec^2 s + \tan^2 s} = \sec^2 s - \tan^2 s$

62. $\dfrac{\cot^2 t - 1}{1 + \cot^2 t} = 1 - 2\sin^2 t$

63. $\dfrac{\tan^2 t - 1}{\sec^2 t} = \dfrac{\tan t - \cot t}{\tan t + \cot t}$

64. $(1 + \sin x + \cos x)^2 = 2(1 + \sin x)(1 + \cos x)$

65. $\dfrac{1 + \cos x}{1 - \cos x} - \dfrac{1 - \cos x}{1 + \cos x} = 4\cot x\csc x$

66. $(\sec\alpha - \tan\alpha)^2 = \dfrac{1 - \sin\alpha}{1 + \sin\alpha}$

67. $(\sec\alpha + \csc\alpha)(\cos\alpha - \sin\alpha) = \cot\alpha - \tan\alpha$

68. $\dfrac{\sin^4\alpha - \cos^4\alpha}{\sin^2\alpha - \cos^2\alpha} = 1$

69. A student claims that the equation $\cos\theta + \sin\theta = 1$ is an identity, since by letting $\theta = 90°$ $\left(\text{or } \frac{\pi}{2} \text{ radians}\right)$ we get $0 + 1 = 1$, a true statement. Comment on this student's reasoning.

70. An equation that is an identity has an infinite number of solutions. If an equation has an infinite number of solutions, is it necessarily an identity? Explain.

Concept Check *Graph each expression and conjecture an identity. Then verify your conjecture algebraically.*

71. $(\sec \theta + \tan \theta)(1 - \sin \theta)$

72. $(\csc \theta + \cot \theta)(\sec \theta - 1)$

73. $\dfrac{\cos \theta + 1}{\sin \theta + \tan \theta}$

74. $\tan \theta \sin \theta + \cos \theta$

Graph the expressions on each side of the equals sign to determine whether the equation might be an identity. (Note: Use a domain whose length is at least 2π.) If the equation looks like an identity, prove it algebraically. See Example 1.

75. $\dfrac{2 + 5 \cos s}{\sin s} = 2 \csc s + 5 \cot s$

76. $1 + \cot^2 s = \dfrac{\sec^2 s}{\sec^2 s - 1}$

77. $\dfrac{\tan s - \cot s}{\tan s + \cot s} = 2 \sin^2 s$

78. $\dfrac{1}{1 + \sin s} + \dfrac{1}{1 - \sin s} = \sec^2 s$

By substituting a number for s or t, show that the equation is not an identity.

79. $\sin(\csc s) = 1$

80. $\sqrt{\cos^2 s} = \cos s$

81. $\csc t = \sqrt{1 + \cot^2 t}$

82. $\cos t = \sqrt{1 - \sin^2 t}$

83. When is $\sin x = \sqrt{1 - \cos^2 x}$ a true statement?

(Modeling) *Work each problem.*

84. ***Intensity of a Lamp*** According to Lambert's law, the intensity of light from a single source on a flat surface at point P is given by

$$I = k \cos^2 \theta,$$

where k is a constant. (*Source:* Winter, C., *Solar Power Plants,* Springer-Verlag, 1991.)

(a) Write I in terms of the sine function.

(b) Explain why the maximum value of I occurs when $\theta = 0$.

85. ***Oscillating Spring*** The distance or displacement y of a weight attached to an oscillating spring from its natural position is modeled by

$$y = 4 \cos(2\pi t),$$

where t is time in seconds. Potential energy is the energy of position and is given by

$$P = ky^2,$$

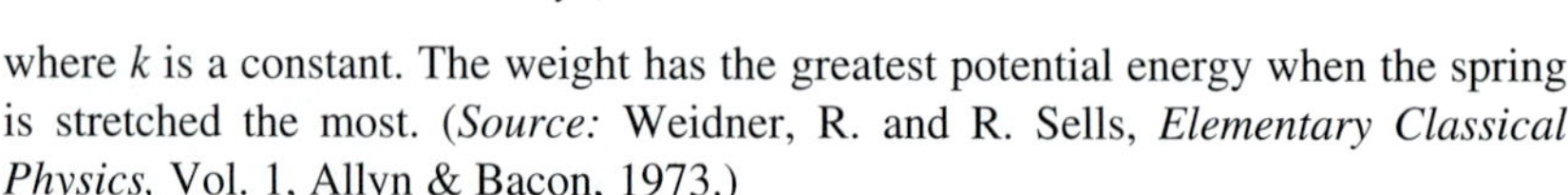

where k is a constant. The weight has the greatest potential energy when the spring is stretched the most. (*Source:* Weidner, R. and R. Sells, *Elementary Classical Physics,* Vol. 1, Allyn & Bacon, 1973.)

(a) Write an expression for P that involves the cosine function.

(b) Use a fundamental identity to write P in terms of $\sin(2\pi t)$.

86. ***Radio Tuners*** Refer to Example 6. Let the energy stored in the inductor be given by

$$L(t) = 3 \cos^2(6{,}000{,}000t)$$

and the energy in the capacitor be given by

$$C(t) = 3 \sin^2(6{,}000{,}000t),$$

where t is time in seconds. The total energy E in the circuit is given by $E(t) = L(t) + C(t)$.

(a) Graph L, C, and E in the window $[0, 10^{-6}]$ by $[-1, 4]$, with Xscl $= 10^{-7}$ and Yscl $= 1$. Interpret the graph.

(b) Make a table of values for L, C, and E starting at $t = 0$, incrementing by 10^{-7}. Interpret your results.

(c) Use a fundamental identity to derive a simplified expression for $E(t)$.

7.3 Sum and Difference Identities

Cosine Sum and Difference Identities ▪ Cofunction Identities ▪ Sine and Tangent Sum and Difference Identities

Cosine Sum and Difference Identities Several examples presented earlier should have convinced you by now that $\cos(A - B)$ *does not equal* $\cos A - \cos B$. For example, if $A = \frac{\pi}{2}$ and $B = 0$, then

$$\cos(A - B) = \cos\left(\frac{\pi}{2} - 0\right) = \cos\frac{\pi}{2} = 0,$$

while $$\cos A - \cos B = \cos\frac{\pi}{2} - \cos 0 = 0 - 1 = -1.$$

We can now derive a formula for $\cos(A - B)$. We start by locating angles A and B in standard position on a unit circle, with $B < A$. Let S and Q be the points where the terminal sides of angles A and B, respectively, intersect the circle. Locate point R on the unit circle so that angle POR equals the difference $A - B$. See Figure 4.

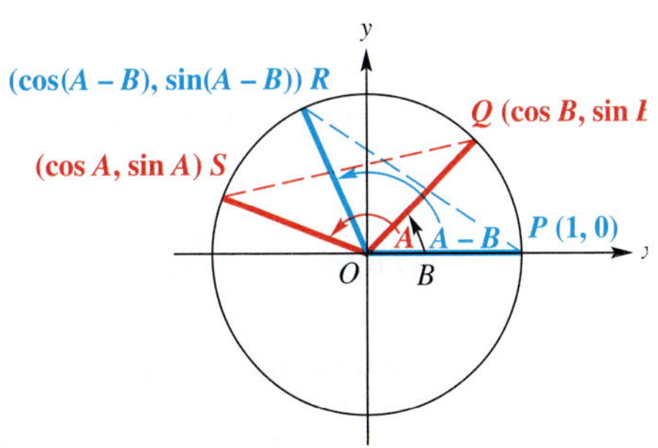

Figure 4

Point Q is on the unit circle, so by the work with circular functions in Chapter 6, the x-coordinate of Q is the cosine of angle B, while the y-coordinate of Q is the sine of angle B.

$$Q \text{ has coordinates } (\cos B, \sin B).$$

In the same way,

$$S \text{ has coordinates } (\cos A, \sin A),$$

and $$R \text{ has coordinates } (\cos(A - B), \sin(A - B)).$$

Angle SOQ also equals $A - B$. Since the central angles SOQ and POR are equal, chords PR and SQ are equal. By the distance formula, since $PR = SQ$,

$$\sqrt{[\cos(A - B) - 1]^2 + [\sin(A - B) - 0]^2} = \sqrt{(\cos A - \cos B)^2 + (\sin A - \sin B)^2}. \quad \text{(Section 2.1)}$$

Squaring both sides and clearing parentheses gives

$$\cos^2(A - B) - 2\cos(A - B) + 1 + \sin^2(A - B) = \cos^2 A - 2\cos A \cos B + \cos^2 B + \sin^2 A - 2\sin A \sin B + \sin^2 B.$$

Since $\sin^2 x + \cos^2 x = 1$ for any value of x, we can rewrite the equation as

$$2 - 2\cos(A - B) = 2 - 2\cos A \cos B - 2\sin A \sin B$$

$$\cos(A - B) = \cos A \cos B + \sin A \sin B. \quad \text{Subtract 2; divide by } -2.$$

Although Figure 4 shows angles A and B in the second and first quadrants, respectively, this result is the same for any values of these angles.

To find a similar expression for $\cos(A + B)$, rewrite $A + B$ as $A - (-B)$ and use the identity for $\cos(A - B)$.

$$\cos(A + B) = \cos[A - (-B)]$$

$$= \cos A \cos(-B) + \sin A \sin(-B) \quad \text{Cosine difference identity}$$

$$= \cos A \cos B + \sin A(-\sin B) \quad \text{Negative-angle identities (Section 7.1)}$$

$$\cos(A + B) = \cos A \cos B - \sin A \sin B$$

Cosine of a Sum or Difference

$$\cos(A - B) = \cos A \cos B + \sin A \sin B$$

$$\cos(A + B) = \cos A \cos B - \sin A \sin B$$

These identities are important in calculus and useful in certain applications. Although a calculator can be used to find an approximation for cos 15°, for example, the method shown below can be applied to get an exact value, as well as to practice using the sum and difference identities.

EXAMPLE 1 Finding Exact Cosine Function Values

Find the *exact* value of each expression.

(a) $\cos 15°$ **(b)** $\cos \frac{5\pi}{12}$ **(c)** $\cos 87° \cos 93° - \sin 87° \sin 93°$

Solution

(a) To find cos 15°, we write 15° as the sum or difference of two angles with known function values. Since we know the exact trigonometric function values of 45° and 30°, we write 15° as $45° - 30°$. (We could also use $60° - 45°$.) Then we use the identity for the cosine of the difference of two angles.

$$\begin{aligned}\cos 15° &= \cos(45° - 30°)\\ &= \cos 45° \cos 30° + \sin 45° \sin 30° &&\text{Cosine difference identity}\\ &= \frac{\sqrt{2}}{2} \cdot \frac{\sqrt{3}}{2} + \frac{\sqrt{2}}{2} \cdot \frac{1}{2} &&\text{(Section 5.3)}\\ &= \frac{\sqrt{6} + \sqrt{2}}{4}\end{aligned}$$

(b) $$\begin{aligned}\cos \frac{5\pi}{12} &= \cos\left(\frac{\pi}{6} + \frac{\pi}{4}\right) &&\frac{\pi}{6} = \frac{2\pi}{12};\ \frac{\pi}{4} = \frac{3\pi}{12}\\ &= \cos \frac{\pi}{6} \cos \frac{\pi}{4} - \sin \frac{\pi}{6} \sin \frac{\pi}{4} &&\text{Cosine sum identity}\\ &= \frac{\sqrt{3}}{2} \cdot \frac{\sqrt{2}}{2} - \frac{1}{2} \cdot \frac{\sqrt{2}}{2} &&\text{(Section 6.2)}\\ &= \frac{\sqrt{6} - \sqrt{2}}{4}\end{aligned}$$

The screen supports the solution in Example 1(b) by showing that $\cos \frac{5\pi}{12} = \frac{\sqrt{6} - \sqrt{2}}{4}$.

(c) $$\begin{aligned}\cos 87° \cos 93° - \sin 87° \sin 93° &= \cos(87° + 93°) &&\text{Cosine sum identity}\\ &= \cos 180°\\ &= -1 &&\text{(Section 5.2)}\end{aligned}$$

Now try Exercises 7, 9, and 11.

Cofunction Identities We can use the identity for the cosine of the difference of two angles and the fundamental identities to derive **cofunction identities.**

Cofunction Identities

$\cos(90° - \theta) = \sin \theta$ $\cot(90° - \theta) = \tan \theta$

$\sin(90° - \theta) = \cos \theta$ $\sec(90° - \theta) = \csc \theta$

$\tan(90° - \theta) = \cot \theta$ $\csc(90° - \theta) = \sec \theta$

Similar identities can be obtained for a real number domain by replacing $90°$ with $\frac{\pi}{2}$.

Substituting $90°$ for A and θ for B in the identity for $\cos(A - B)$ gives

$$\begin{aligned}\cos(90° - \theta) &= \cos 90° \cos \theta + \sin 90° \sin \theta\\ &= 0 \cdot \cos \theta + 1 \cdot \sin \theta\\ &= \sin \theta.\end{aligned}$$

This result is true for *any* value of θ since the identity for $\cos(A - B)$ is true for any values of A and B.

EXAMPLE 2 Using Cofunction Identities to Find θ

Find an angle θ that satisfies each of the following.

(a) $\cot \theta = \tan 25°$ **(b)** $\sin \theta = \cos(-30°)$ **(c)** $\csc \frac{3\pi}{4} = \sec \theta$

Solution

(a) Since tangent and cotangent are cofunctions, $\tan(90° - \theta) = \cot \theta$.

$$\begin{aligned} \cot \theta &= \tan 25° \\ \tan(90° - \theta) &= \tan 25° && \text{Cofunction identity} \\ 90° - \theta &= 25° && \text{Set angle measures equal.} \\ \theta &= 65° \end{aligned}$$

(b)

$$\begin{aligned} \sin \theta &= \cos(-30°) \\ \cos(90° - \theta) &= \cos(-30°) && \text{Cofunction identity} \\ 90° - \theta &= -30° \\ \theta &= 120° \end{aligned}$$

(c)

$$\begin{aligned} \csc \frac{3\pi}{4} &= \sec \theta \\ \sec\left(\frac{\pi}{2} - \frac{3\pi}{4}\right) &= \sec \theta && \text{Cofunction identity} \\ \sec\left(-\frac{\pi}{4}\right) &= \sec \theta && \text{Combine terms.} \\ -\frac{\pi}{4} &= \theta \end{aligned}$$

Now try Exercises 25 and 27.

NOTE Because trigonometric (circular) functions are periodic, the solutions in Example 2 are not unique. We give only one of infinitely many possibilities.

If one of the angles A or B in the identities for $\cos(A + B)$ and $\cos(A - B)$ is a quadrantal angle, then the identity allows us to write the expression in terms of a single function of A or B.

EXAMPLE 3 Reducing $\cos(A - B)$ to a Function of a Single Variable

Write $\cos(180° - \theta)$ as a trigonometric function of θ.

Solution Use the difference identity. Replace A with $180°$ and B with θ.

$$\begin{aligned} \cos(180° - \theta) &= \cos 180° \cos \theta + \sin 180° \sin \theta \\ &= (-1) \cos \theta + (0) \sin \theta && \text{(Section 5.2)} \\ &= -\cos \theta \end{aligned}$$

Now try Exercise 39.

Sine and Tangent Sum and Difference Identities We can use the cosine sum and difference identities to derive similar identities for sine and tangent. Since $\sin\theta = \cos(90° - \theta)$, we replace θ with $A + B$ to get

$$\begin{aligned}\sin(A+B) &= \cos[90° - (A+B)] && \text{Cofunction identity}\\ &= \cos[(90° - A) - B]\\ &= \cos(90° - A)\cos B + \sin(90° - A)\sin B && \text{Cosine difference identity}\\ \sin(A+B) &= \sin A\cos B + \cos A\sin B. && \text{Cofunction identities}\end{aligned}$$

Now we write $\sin(A - B)$ as $\sin[A + (-B)]$ and use the identity for $\sin(A + B)$.

$$\begin{aligned}\sin(A-B) &= \sin[A + (-B)]\\ &= \sin A\cos(-B) + \cos A\sin(-B) && \text{Sine sum identity}\\ \sin(A-B) &= \sin A\cos B - \cos A\sin B && \text{Negative-angle identities}\end{aligned}$$

Sine of a Sum or Difference

$$\sin(A+B) = \sin A\cos B + \cos A\sin B$$

$$\sin(A-B) = \sin A\cos B - \cos A\sin B$$

To derive the identity for $\tan(A + B)$, we start with

$$\begin{aligned}\tan(A+B) &= \frac{\sin(A+B)}{\cos(A+B)} && \text{Fundamental identity (Section 7.1)}\\ &= \frac{\sin A\cos B + \cos A\sin B}{\cos A\cos B - \sin A\sin B}. && \text{Sum identities}\end{aligned}$$

We express this result in terms of the tangent function by multiplying both numerator and denominator by $\frac{1}{\cos A\cos B}$.

$$\begin{aligned}\tan(A+B) &= \frac{\dfrac{\sin A\cos B + \cos A\sin B}{1}}{\dfrac{\cos A\cos B - \sin A\sin B}{1}}\cdot\frac{\dfrac{1}{\cos A\cos B}}{\dfrac{1}{\cos A\cos B}} && \text{Simplify the complex fraction. (Section R.5)}\\ &= \frac{\dfrac{\sin A\cos B}{\cos A\cos B} + \dfrac{\cos A\sin B}{\cos A\cos B}}{\dfrac{\cos A\cos B}{\cos A\cos B} - \dfrac{\sin A\sin B}{\cos A\cos B}} && \text{Multiply numerators; multiply denominators.}\\ &= \frac{\dfrac{\sin A}{\cos A} + \dfrac{\sin B}{\cos B}}{1 - \dfrac{\sin A}{\cos A}\cdot\dfrac{\sin B}{\cos B}}\\ \tan(A+B) &= \frac{\tan A + \tan B}{1 - \tan A\tan B} && \tan\theta = \frac{\sin\theta}{\cos\theta}\end{aligned}$$

Replacing B with $-B$ and using the fact that $\tan(-B) = -\tan B$ gives the identity for the tangent of the difference of two angles.

Tangent of a Sum or Difference

$$\tan(A+B)=\frac{\tan A+\tan B}{1-\tan A\tan B} \qquad \tan(A-B)=\frac{\tan A-\tan B}{1+\tan A\tan B}$$

EXAMPLE 4 Finding Exact Sine and Tangent Function Values

Find the *exact* value of each expression.

(a) $\sin 75°$ **(b)** $\tan\frac{7\pi}{12}$ **(c)** $\sin 40°\cos 160° - \cos 40°\sin 160°$

Solution

(a)
$$\begin{aligned}\sin 75° &= \sin(45° + 30°)\\ &= \sin 45°\cos 30° + \cos 45°\sin 30° && \text{Sine sum identity}\\ &= \frac{\sqrt{2}}{2}\cdot\frac{\sqrt{3}}{2}+\frac{\sqrt{2}}{2}\cdot\frac{1}{2} && \text{(Section 5.3)}\\ &= \frac{\sqrt{6}+\sqrt{2}}{4}\end{aligned}$$

(b)
$$\begin{aligned}\tan\frac{7\pi}{12} &= \tan\left(\frac{\pi}{3}+\frac{\pi}{4}\right)\\ &= \frac{\tan\frac{\pi}{3}+\tan\frac{\pi}{4}}{1-\tan\frac{\pi}{3}\tan\frac{\pi}{4}} && \text{Tangent sum identity}\\ &= \frac{\sqrt{3}+1}{1-\sqrt{3}\cdot 1} && \text{(Section 6.2)}\\ &= \frac{\sqrt{3}+1}{1-\sqrt{3}}\cdot\frac{1+\sqrt{3}}{1+\sqrt{3}} && \text{Rationalize the denominator. (Section R.7)}\\ &= \frac{\sqrt{3}+3+1+\sqrt{3}}{1-3} && \text{Multiply. (Section R.7)}\\ &= \frac{4+2\sqrt{3}}{-2} && \text{Combine terms.}\\ &= \frac{2(2+\sqrt{3})}{-2} && \text{Factor out 2. (Section R.5)}\\ &= -2-\sqrt{3} && \text{Lowest terms}\end{aligned}$$

(c)
$$\begin{aligned}\sin 40°\cos 160° - \cos 40°\sin 160° &= \sin(40° - 160°) && \text{Sine difference identity}\\ &= \sin(-120°)\\ &= -\sin 120° && \text{Negative-angle identity}\\ &= -\frac{\sqrt{3}}{2} && \text{(Section 5.3)}\end{aligned}$$

Now try Exercises 29, 31, and 35.

EXAMPLE 5 Writing Functions as Expressions Involving Functions of θ

Write each function as an expression involving functions of θ.

(a) $\sin(30° + \theta)$ **(b)** $\tan(45° - \theta)$ **(c)** $\sin(180° + \theta)$

Solution

(a) Using the identity for $\sin(A + B)$,

$$\sin(30° + \theta) = \sin 30° \cos\theta + \cos 30° \sin\theta$$
$$= \frac{1}{2}\cos\theta + \frac{\sqrt{3}}{2}\sin\theta.$$

(b) $\tan(45° - \theta) = \dfrac{\tan 45° - \tan\theta}{1 + \tan 45° \tan\theta} = \dfrac{1 - \tan\theta}{1 + \tan\theta}$

(c) $\sin(180° + \theta) = \sin 180° \cos\theta + \cos 180° \sin\theta$

$$= 0 \cdot \cos\theta + (-1)\sin\theta$$
$$= -\sin\theta$$

Now try Exercises 43 and 47.

EXAMPLE 6 Finding Function Values and the Quadrant of $A + B$

Suppose that A and B are angles in standard position, with $\sin A = \frac{4}{5}$, $\frac{\pi}{2} < A < \pi$, and $\cos B = -\frac{5}{13}$, $\pi < B < \frac{3\pi}{2}$. Find each of the following.

(a) $\sin(A + B)$ **(b)** $\tan(A + B)$ **(c)** the quadrant of $A + B$

Solution

(a) The identity for $\sin(A + B)$ requires $\sin A$, $\cos A$, $\sin B$, and $\cos B$. We are given values of $\sin A$ and $\cos B$. We must find values of $\cos A$ and $\sin B$.

$$\sin^2 A + \cos^2 A = 1 \qquad \text{Fundamental identity}$$

$$\frac{16}{25} + \cos^2 A = 1 \qquad \sin A = \tfrac{4}{5}$$

$$\cos^2 A = \frac{9}{25} \qquad \text{Subtract } \tfrac{16}{25}.$$

$$\cos A = -\frac{3}{5} \qquad \text{Since } A \text{ is in quadrant II, } \cos A < 0.$$

In the same way, $\sin B = -\frac{12}{13}$. Now use the formula for $\sin(A + B)$.

$$\sin(A + B) = \frac{4}{5}\left(-\frac{5}{13}\right) + \left(-\frac{3}{5}\right)\left(-\frac{12}{13}\right)$$
$$= -\frac{20}{65} + \frac{36}{65} = \frac{16}{65}$$

(b) To find $\tan(A + B)$, first use the values of sine and cosine from part (a) to get $\tan A = -\frac{4}{3}$ and $\tan B = \frac{12}{5}$.

$$\tan(A + B) = \frac{-\frac{4}{3} + \frac{12}{5}}{1 - \left(-\frac{4}{3}\right)\left(\frac{12}{5}\right)} = \frac{\frac{16}{15}}{1 + \frac{48}{15}} = \frac{\frac{16}{15}}{\frac{63}{15}} = \frac{16}{63}$$

(c) From parts (a) and (b), $\sin(A + B) = \frac{16}{65}$ and $\tan(A + B) = \frac{16}{63}$, both positive. Therefore, $A + B$ must be in quadrant I, since it is the only quadrant in which both sine and tangent are positive.

Now try Exercise 51.

EXAMPLE 7 Applying the Cosine Difference Identity to Voltage

Common household electric current is called *alternating current* because the current alternates direction within the wires. The voltage V in a typical 115-volt outlet can be expressed by the function $V(t) = 163 \sin \omega t$, where ω is the angular speed (in radians per second) of the rotating generator at the electrical plant and t is time measured in seconds. (*Source:* Bell, D., *Fundamentals of Electric Circuits,* Fourth Edition, Prentice-Hall, 1988.)

(a) It is essential for electric generators to rotate at precisely 60 cycles per sec so household appliances and computers will function properly. Determine ω for these electric generators.

(b) Graph V in the window $[0, .05]$ by $[-200, 200]$.

(c) Determine a value of ϕ so that the graph of $V(t) = 163 \cos(\omega t - \phi)$ is the same as the graph of $V(t) = 163 \sin \omega t$.

Solution

(a) Each cycle is 2π radians at 60 cycles per sec, so the angular speed is $\omega = 60(2\pi) = 120\pi$ radians per sec.

(b) $V(t) = 163 \sin \omega t = 163 \sin 120\pi t$. Because the amplitude of the function is 163 (from Section 6.3), $[-200, 200]$ is an appropriate interval for the range, as shown in Figure 5.

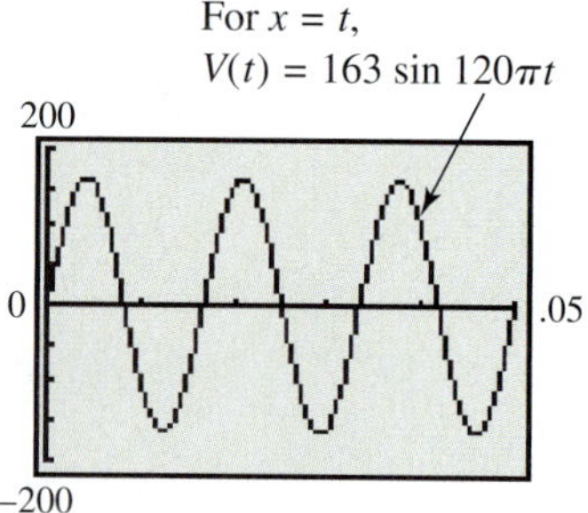

Figure 5

(c) Using the negative-angle identity for cosine and a cofunction identity,

$$\cos\left(x - \frac{\pi}{2}\right) = \cos\left[-\left(\frac{\pi}{2} - x\right)\right] = \cos\left(\frac{\pi}{2} - x\right) = \sin x.$$

Therefore, if $\phi = \frac{\pi}{2}$, then

$$V(t) = 163 \cos\left(\omega t - \frac{\pi}{2}\right) = 163 \sin \omega t.$$

Now try Exercise 81.

7.3 Exercises

Concept Check *Match each expression in Column I with the correct expression in Column II to form an identity.*

I	II
1. $\cos(x + y) =$ ______	**A.** $\cos x \cos y + \sin x \sin y$
2. $\cos(x - y) =$ ______	**B.** $\sin x \sin y - \cos x \cos y$
3. $\sin(x + y) =$ ______	**C.** $\sin x \cos y + \cos x \sin y$
4. $\sin(x - y) =$ ______	**D.** $\sin x \cos y - \cos x \sin y$
	E. $\cos x \sin y - \sin x \cos y$
	F. $\cos x \cos y - \sin x \sin y$

Use identities to find each exact value. (Do not use a calculator.) See Example 1.

5. $\cos 75°$

6. $\cos(-15°)$

7. $\cos 105°$
(*Hint:* $105° = 60° + 45°$)

8. $\cos(-105°)$
(*Hint:* $-105° = -60° + (-45°)$)

9. $\cos \frac{7\pi}{12}$

10. $\cos\left(-\frac{\pi}{12}\right)$

11. $\cos 40° \cos 50° - \sin 40° \sin 50°$

12. $\cos \frac{7\pi}{9} \cos \frac{2\pi}{9} - \sin \frac{7\pi}{9} \sin \frac{2\pi}{9}$

Write each function value in terms of the cofunction of a complementary angle. See Example 2.

13. $\tan 87°$

14. $\sin 15°$

15. $\cos \frac{\pi}{12}$

16. $\sin \frac{2\pi}{5}$

17. $\sin \frac{5\pi}{8}$

18. $\cot \frac{9\pi}{10}$

19. $\sec 146° 42'$

20. $\tan 174° 3'$

Use the cofunction identities to fill in each blank with the appropriate trigonometric function name. See Example 2.

21. $\cot \frac{\pi}{3} =$ ______ $\frac{\pi}{6}$

22. $\sin \frac{2\pi}{3} =$ ______ $\left(-\frac{\pi}{6}\right)$

23. ______ $33° = \sin 57°$

24. ______ $72° = \cot 18°$

Find an angle θ that makes each statement true. See Example 2.

25. $\tan \theta = \cot(45° + 2\theta)$

26. $\sin \theta = \cos(2\theta - 10°)$

27. $\sin(3\theta - 15°) = \cos(\theta + 25°)$

28. $\cot(\theta - 10°) = \tan(2\theta + 20°)$

Use identities to find the exact value of each of the following. See Example 4.

29. $\sin \frac{5\pi}{12}$

30. $\tan \frac{5\pi}{12}$

31. $\tan \frac{\pi}{12}$

32. $\sin \frac{\pi}{12}$

33. $\sin\left(-\frac{7\pi}{12}\right)$

34. $\tan\left(-\frac{7\pi}{12}\right)$

35. $\sin 76° \cos 31° - \cos 76° \sin 31°$

36. $\sin 40° \cos 50° + \cos 40° \sin 50°$

37. $\dfrac{\tan 80° + \tan 55°}{1 - \tan 80° \tan 55°}$

38. $\dfrac{\tan 80° - \tan(-55°)}{1 + \tan 80° \tan(-55°)}$

Use identities to write each expression as a function of x or θ. See Examples 3 and 5.

39. $\cos(90° - \theta)$

40. $\cos(180° - \theta)$

41. $\cos\left(x - \frac{3\pi}{2}\right)$

42. $\cos\left(\frac{\pi}{2} + x\right)$

43. $\sin(45° + \theta)$

44. $\sin(180° - \theta)$

45. $\sin\left(\frac{5\pi}{6} - x\right)$

46. $\sin\left(\frac{\pi}{4} - x\right)$

47. $\tan(60° + \theta)$

48. $\tan(\theta - 30°)$

49. $\tan\left(x + \frac{\pi}{6}\right)$

50. $\tan\left(\frac{\pi}{4} + x\right)$

Use the given information to find **(a)** $\cos(s + t)$, **(b)** $\sin(s - t)$, **(c)** $\tan(s + t)$, *and* **(d)** *the quadrant of $s + t$. See Example 6.*

51. $\cos s = \frac{3}{5}$ and $\sin t = \frac{5}{13}$, s and t in quadrant I

52. $\cos s = -\frac{1}{5}$ and $\sin t = \frac{3}{5}$, s and t in quadrant II

53. $\sin s = \frac{2}{3}$ and $\sin t = -\frac{1}{3}$, s in quadrant II and t in quadrant IV

54. $\sin s = \frac{3}{5}$ and $\sin t = -\frac{12}{13}$, s in quadrant I and t in quadrant III

55. $\cos s = -\frac{8}{17}$ and $\cos t = -\frac{3}{5}$, s and t in quadrant III

56. $\cos s = -\frac{15}{17}$ and $\sin t = \frac{4}{5}$, s in quadrant II and t in quadrant I

Relating Concepts

For individual or collaborative investigation
(Exercises 57–60)

The identities for $\cos(A + B)$ *and* $\cos(A - B)$ *can be used to find exact values of expressions like* $\cos 195°$ *and* $\cos 255°$, *where the angle is not in the first quadrant.* ***Work Exercises 57–60 in order,*** *to see how this is done.*

57. By writing 195° as 180° + 15°, use the identity for $\cos(A + B)$ to express $\cos 195°$ as $-\cos 15°$.

58. Use the identity for $\cos(A - B)$ to find $-\cos 15°$.

59. By the results of Exercises 57 and 58, $\cos 195° =$ ________.

60. Find each exact value using the method shown in Exercises 57–59.

(a) $\cos 255°$ **(b)** $\cos \frac{11\pi}{12}$

Find each exact value. Use the technique developed in Relating Concepts Exercises 57–60.

61. $\sin 165°$

62. $\tan 165°$

63. $\sin 255°$

64. $\tan 285°$

65. $\tan \frac{11\pi}{12}$

66. $\sin\left(-\frac{13\pi}{12}\right)$

67. Use the identity $\cos(90° - \theta) = \sin\theta$, and replace θ with $90° - A$, to derive the identity $\cos A = \sin(90° - A)$.

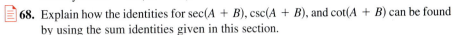

68. Explain how the identities for $\sec(A + B)$, $\csc(A + B)$, and $\cot(A + B)$ can be found by using the sum identities given in this section.

69. Why is it not possible to use a method similar to that of Example 5(c) to find a formula for $\tan(270° - \theta)$?

70. *Concept Check* Show that if A, B, and C are angles of a triangle, then $\sin(A + B + C) = 0$.

Graph each expression and use the graph to conjecture an identity. Then verify your conjecture algebraically.

71. $\sin\left(\frac{\pi}{2} + x\right)$

72. $\frac{1 + \tan x}{1 - \tan x}$

Verify that each equation is an identity.

73. $\sin(x + y) + \sin(x - y) = 2\sin x\cos y$

74. $\tan(x - y) - \tan(y - x) = \frac{2(\tan x - \tan y)}{1 + \tan x\tan y}$

75. $\frac{\cos(\alpha - \beta)}{\cos\alpha\sin\beta} = \tan\alpha + \cot\beta$

76. $\frac{\sin(s + t)}{\cos s\cos t} = \tan s + \tan t$

77. $\frac{\sin(x - y)}{\sin(x + y)} = \frac{\tan x - \tan y}{\tan x + \tan y}$

78. $\frac{\sin(s - t)}{\sin t} + \frac{\cos(s - t)}{\cos t} = \frac{\sin s}{\sin t\cos t}$

Exercises 79 and 80 refer to Example 7.

79. How many times does the current oscillate in .05 sec?

80. What are the maximum and minimum voltages in this outlet? Is the voltage always equal to 115 volts?

(Modeling) Solve each problem.

81. *Back Stress* If a person bends at the waist with a straight back making an angle of θ degrees with the horizontal, then the force F exerted on the back muscles can be modeled by the equation

$$F = \frac{.6W\sin(\theta + 90°)}{\sin 12°},$$

where W is the weight of the person. (*Source:* Metcalf, H., *Topics in Classical Biophysics,* Prentice-Hall, 1980.)

(a) Calculate F when $W = 170$ lb and $\theta = 30°$.
(b) Use an identity to show that F is approximately equal to $2.9W\cos\theta$.
(c) For what value of θ is F maximum?

82. *Back Stress* Refer to Exercise 81.

(a) Suppose a 200-lb person bends at the waist so that $\theta = 45°$. Estimate the force exerted on the person's back muscles.

(b) Approximate graphically the value of θ that results in the back muscles exerting a force of 400 lb.

83. *Sound Waves* Sound is a result of waves applying pressure to a person's eardrum. For a pure sound wave radiating outward in a spherical shape, the trigonometric function defined by

$$P = \frac{a}{r}\cos\left(\frac{2\pi r}{\lambda} - ct\right)$$

can be used to model the sound pressure at a radius of r feet from the source, where t is time in seconds, λ is length of the sound wave in feet, c is speed of sound in feet per second, and a is maximum sound pressure at the source measured in pounds per square foot. (*Source:* Beranek, L., *Noise and Vibration Control,* Institute of Noise Control Engineering, Washington, D.C., 1988.) Let $\lambda = 4.9$ ft and $c = 1026$ ft per sec.

(a) Let $a = .4$ lb per ft^2. Graph the sound pressure at distance $r = 10$ ft from its source in the window $[0, .05]$ by $[-.05, .05]$. Describe P at this distance.

(b) Now let $a = 3$ and $t = 10$. Graph the sound pressure in the window $[0, 20]$ by $[-2, 2]$. What happens to pressure P as radius r increases?

(c) Suppose a person stands at a radius r so that $r = n\lambda$, where n is a positive integer. Use the difference identity for cosine to simplify P in this situation.

84. *Voltage of a Circuit* When the two voltages

$$V_1 = 30 \sin 120\pi t \qquad \text{and} \qquad V_2 = 40 \cos 120\pi t$$

are applied to the same circuit, the resulting voltage V will be equal to their sum. (*Source:* Bell, D., *Fundamentals of Electric Circuits,* Second Edition, Reston Publishing Company, 1981.)

(a) Graph the sum in the window $[0, .05]$ by $[-60, 60]$.

(b) Use the graph to estimate values for a and ϕ so that $V = a \sin(120\pi t + \phi)$.

(c) Use identities to verify that your expression for V is valid.

7.4 Double-Angle Identities and Half-Angle Identities

Double-Angle Identities ▪ Product-to-Sum and Sum-to-Product Identities ▪ Half-Angle Identities

Double-Angle Identities When $A = B$ in the identities for the sum of two angles, these identities are called the **double-angle identities.** For example, to derive an expression for $\cos 2A$, we let $B = A$ in the identity $\cos(A + B) = \cos A \cos B - \sin A \sin B$.

$$\begin{aligned} \cos 2A &= \cos(A + A) \\ &= \cos A \cos A - \sin A \sin A \qquad \text{Cosine sum identity (Section 7.3)} \\ \cos 2A &= \cos^2 A - \sin^2 A \end{aligned}$$

Two other useful forms of this identity can be obtained by substituting either $\cos^2 A = 1 - \sin^2 A$ or $\sin^2 A = 1 - \cos^2 A$. Replace $\cos^2 A$ with the expression $1 - \sin^2 A$ to get

$$\begin{aligned} \cos 2A &= \cos^2 A - \sin^2 A \\ &= (1 - \sin^2 A) - \sin^2 A \quad \text{Fundamental identity (Section 7.1)} \\ \cos 2A &= 1 - 2\sin^2 A, \end{aligned}$$

or replace $\sin^2 A$ with $1 - \cos^2 A$ to get

$$\begin{aligned} \cos 2A &= \cos^2 A - \sin^2 A \\ &= \cos^2 A - (1 - \cos^2 A) \quad \text{Fundamental identity} \\ &= \cos^2 A - 1 + \cos^2 A \\ \cos 2A &= 2\cos^2 A - 1. \end{aligned}$$

We find $\sin 2A$ with the identity $\sin(A + B) = \sin A \cos B + \cos A \sin B$, letting $B = A$.

$$\begin{aligned} \sin 2A &= \sin(A + A) \\ &= \sin A \cos A + \cos A \sin A \quad \text{Sine sum identity} \\ \sin 2A &= 2 \sin A \cos A \end{aligned}$$

Using the identity for $\tan(A + B)$, we find $\tan 2A$.

$$\begin{aligned} \tan 2A &= \tan(A + A) \\ &= \frac{\tan A + \tan A}{1 - \tan A \tan A} \quad \text{Tangent sum identity} \\ \tan 2A &= \frac{2 \tan A}{1 - \tan^2 A} \end{aligned}$$

Looking Ahead to Calculus

The identities

$$\cos 2A = 1 - 2\sin^2 A$$

and $\cos 2A = 2\cos^2 A - 1$

can be rewritten as

$$\sin^2 A = \frac{1}{2}(1 - \cos 2A)$$

and $\cos^2 A = \frac{1}{2}(1 + \cos 2A)$.

These identities are used to integrate the functions $f(A) = \sin^2 A$ and $g(A) = \cos^2 A$.

Double-Angle Identities

$$\cos 2A = \cos^2 A - \sin^2 A \qquad \cos 2A = 1 - 2\sin^2 A$$

$$\cos 2A = 2\cos^2 A - 1 \qquad \sin 2A = 2 \sin A \cos A$$

$$\tan 2A = \frac{2 \tan A}{1 - \tan^2 A}$$

EXAMPLE 1 Finding Function Values of 2θ Given Information about θ

Given $\cos \theta = \frac{3}{5}$ and $\sin \theta < 0$, find $\sin 2\theta$, $\cos 2\theta$, and $\tan 2\theta$.

Solution To find $\sin 2\theta$, we must first find the value of $\sin \theta$.

$$\sin^2 \theta + \left(\frac{3}{5}\right)^2 = 1 \quad \sin^2 \theta + \cos^2 \theta = 1;\ \cos \theta = \tfrac{3}{5}$$

$$\sin^2 \theta = \frac{16}{25} \quad \text{Simplify.}$$

$$\sin \theta = -\frac{4}{5} \quad \text{Choose the negative square root since } \sin \theta < 0.$$

Using the double-angle identity for sine,

$$\sin 2\theta = 2 \sin \theta \cos \theta$$
$$= 2\left(-\frac{4}{5}\right)\left(\frac{3}{5}\right) = -\frac{24}{25}. \quad \sin \theta = -\tfrac{4}{5};\ \cos \theta = \tfrac{3}{5}$$

Now we find $\cos 2\theta$, using the first of the double-angle identities for cosine. (Any of the three forms may be used.)

$$\cos 2\theta = \cos^2 \theta - \sin^2 \theta = \frac{9}{25} - \frac{16}{25} = -\frac{7}{25}$$

The value of $\tan 2\theta$ can be found in either of two ways. We can use the double-angle identity and the fact that $\tan \theta = \frac{\sin \theta}{\cos \theta} = \frac{-\frac{4}{5}}{\frac{3}{5}} = -\frac{4}{3}$.

$$\tan 2\theta = \frac{2 \tan \theta}{1 - \tan^2 \theta} = \frac{2\left(-\frac{4}{3}\right)}{1 - \frac{16}{9}} = \frac{-\frac{8}{3}}{-\frac{7}{9}} = \frac{24}{7}$$

Simplify. (Section R.5)

Alternatively, we can find $\tan 2\theta$ by finding the quotient of $\sin 2\theta$ and $\cos 2\theta$.

$$\tan 2\theta = \frac{\sin 2\theta}{\cos 2\theta} = \frac{-\frac{24}{25}}{-\frac{7}{25}} = \frac{24}{7}$$

Now try Exercise 9.

EXAMPLE 2 Verifying a Double-Angle Identity

Verify that the following equation is an identity.

$$\cot x \sin 2x = 1 + \cos 2x$$

Solution We start by working on the left side, using the hint from Section 7.1 about writing all functions in terms of sine and cosine.

$$\cot x \sin 2x = \frac{\cos x}{\sin x} \cdot \sin 2x \qquad \text{Quotient identity}$$
$$= \frac{\cos x}{\sin x}(2 \sin x \cos x) \qquad \text{Double-angle identity}$$
$$= 2 \cos^2 x$$
$$= 1 + \cos 2x \qquad \cos 2x = 2\cos^2 x - 1, \text{ so } 2\cos^2 x = 1 + \cos 2x$$

The final step illustrates the importance of being able to recognize alternative forms of identities.

Now try Exercise 27.

EXAMPLE 3 Simplifying Expressions Using Double-Angle Identities

Simplify each expression.

(a) $\cos^2 7x - \sin^2 7x$ **(b)** $\sin 15° \cos 15°$

Solution

(a) This expression suggests one of the double-angle identities for cosine: $\cos 2A = \cos^2 A - \sin^2 A$. Substituting $7x$ for A gives

$$\cos^2 7x - \sin^2 7x = \cos 2(7x) = \cos 14x.$$

(b) If this expression were $2 \sin 15° \cos 15°$, we could apply the identity for $\sin 2A$ directly since $\sin 2A = 2 \sin A \cos A$. We can still apply the identity with $A = 15°$ by writing the multiplicative identity element 1 as $\frac{1}{2}(2)$.

$$\sin 15° \cos 15° = \frac{1}{2}(2) \sin 15° \cos 15° \quad \text{Multiply by 1 in the form } \tfrac{1}{2}(2).$$

$$= \frac{1}{2}(2 \sin 15° \cos 15°) \quad \text{Associative property (Section R.1)}$$

$$= \frac{1}{2} \sin(2 \cdot 15°) \quad 2 \sin A \cos A = \sin 2A, \text{ with } A = 15°$$

$$= \frac{1}{2} \sin 30°$$

$$= \frac{1}{2} \cdot \frac{1}{2} = \frac{1}{4} \quad \sin 30° = \tfrac{1}{2} \text{ (Section 5.3)}$$

Now try Exercises 13 and 15.

Identities involving larger multiples of the variable can be derived by repeated use of the double-angle identities and other identities.

EXAMPLE 4 Deriving a Multiple-Angle Identity

Write $\sin 3x$ in terms of $\sin x$.

Solution

$$\sin 3x = \sin(2x + x)$$

$$= \sin 2x \cos x + \cos 2x \sin x \quad \text{Sine sum identity (Section 7.3)}$$

$$= (2 \sin x \cos x) \cos x + (\cos^2 x - \sin^2 x) \sin x \quad \text{Double-angle identities}$$

$$= 2 \sin x \cos^2 x + \cos^2 x \sin x - \sin^3 x \quad \text{Multiply.}$$

$$= 2 \sin x(1 - \sin^2 x) + (1 - \sin^2 x) \sin x - \sin^3 x \quad \cos^2 x = 1 - \sin^2 x$$

$$= 2 \sin x - 2 \sin^3 x + \sin x - \sin^3 x - \sin^3 x \quad \text{Distributive property (Section R.1)}$$

$$= 3 \sin x - 4 \sin^3 x \quad \text{Combine terms.}$$

Now try Exercise 21.

The next example applies a multiple-angle identity to answer a question about electric current.

EXAMPLE 5 Determining Wattage Consumption

If a toaster is plugged into a common household outlet, the wattage consumed is not constant. Instead, it varies at a high frequency according to the model

$$W = \frac{V^2}{R},$$

where V is the voltage and R is a constant that measures the resistance of the toaster in ohms. (*Source:* Bell, D., *Fundamentals of Electric Circuits,* Fourth Edition, Prentice-Hall, 1998.) Graph the wattage W consumed by a typical toaster with $R = 15$ and $V = 163 \sin 120\pi t$ in the window $[0, .05]$ by $[-500, 2000]$. How many oscillations are there?

Solution Substituting the given values into the wattage equation gives

$$W = \frac{V^2}{R} = \frac{(163 \sin 120\pi t)^2}{15}.$$

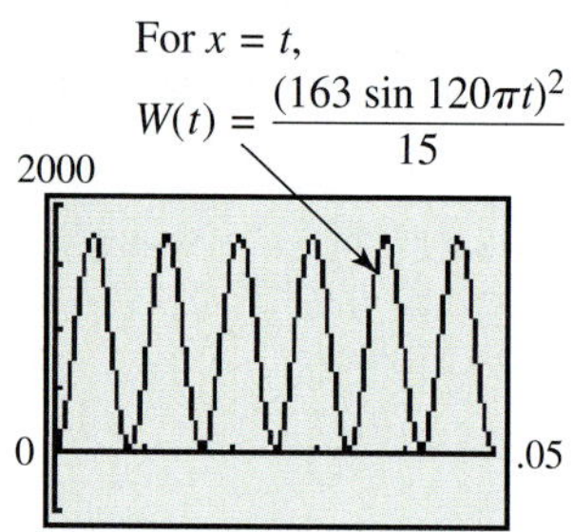

Figure 6

To determine the range of W, we note that $\sin 120\pi t$ has maximum value 1, so the expression for W has maximum value $\frac{163^2}{15} \approx 1771$. The minimum value is 0. The graph in Figure 6 shows that there are six oscillations.

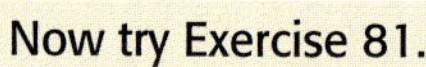

Now try Exercise 81.

Product-to-Sum and Sum-to-Product Identities Because they make it possible to rewrite a product as a sum, the identities for $\cos(A + B)$ and $\cos(A - B)$ are used to derive a group of identities useful in calculus.

Adding the identities for $\cos(A + B)$ and $\cos(A - B)$ gives

$$\begin{array}{r} \cos(A + B) = \cos A \cos B - \sin A \sin B \\ \cos(A - B) = \cos A \cos B + \sin A \sin B \\ \hline \cos(A + B) + \cos(A - B) = 2 \cos A \cos B \end{array}$$

or

$$\cos A \cos B = \frac{1}{2}[\cos(A + B) + \cos(A - B)].$$

Similarly, subtracting $\cos(A + B)$ from $\cos(A - B)$ gives

$$\sin A \sin B = \frac{1}{2}[\cos(A - B) - \cos(A + B)].$$

Using the identities for $\sin(A + B)$ and $\sin(A - B)$ in the same way, we get two more identities. Those and the previous ones are now summarized.

Looking Ahead to Calculus

The product-to-sum identities are used in calculus to find integrals of functions that are products of trigonometric functions. One classic calculus text includes the following example:

Evaluate $\int \cos 5x \cos 3x\,dx$.

The first solution line reads:

"We may write

$\cos 5x \cos 3x = \frac{1}{2}[\cos 8x + \cos 2x]$."

Product-to-Sum Identities

$$\mathbf{\cos A \cos B = \frac{1}{2}[\cos(A + B) + \cos(A - B)]}$$

$$\mathbf{\sin A \sin B = \frac{1}{2}[\cos(A - B) - \cos(A + B)]}$$

(continued)

$$\sin A \cos B = \frac{1}{2}[\sin(A + B) + \sin(A - B)]$$

$$\cos A \sin B = \frac{1}{2}[\sin(A + B) - \sin(A - B)]$$

EXAMPLE 6 Using a Product-to-Sum Identity

Write $\cos 2\theta \sin \theta$ as the sum or difference of two functions.

Solution Use the identity for $\cos A \sin B$, with $2\theta = A$ and $\theta = B$.

$$\begin{aligned}\cos 2\theta \sin \theta &= \frac{1}{2}[\sin(2\theta + \theta) - \sin(2\theta - \theta)] \\ &= \frac{1}{2}\sin 3\theta - \frac{1}{2}\sin \theta\end{aligned}$$

Now try Exercise 41.

From these new identities we can derive another group of identities that are used to write sums of trigonometric functions as products.

Sum-to-Product Identities

$$\sin A + \sin B = 2 \sin\left(\frac{A + B}{2}\right) \cos\left(\frac{A - B}{2}\right)$$

$$\sin A - \sin B = 2 \cos\left(\frac{A + B}{2}\right) \sin\left(\frac{A - B}{2}\right)$$

$$\cos A + \cos B = 2 \cos\left(\frac{A + B}{2}\right) \cos\left(\frac{A - B}{2}\right)$$

$$\cos A - \cos B = -2 \sin\left(\frac{A + B}{2}\right) \sin\left(\frac{A - B}{2}\right)$$

EXAMPLE 7 Using a Sum-to-Product Identity

Write $\sin 2\theta - \sin 4\theta$ as a product of two functions.

Solution Use the identity for $\sin A - \sin B$, with $2\theta = A$ and $4\theta = B$.

$$\begin{aligned}\sin 2\theta - \sin 4\theta &= 2 \cos\left(\frac{2\theta + 4\theta}{2}\right) \sin\left(\frac{2\theta - 4\theta}{2}\right) \\ &= 2 \cos \frac{6\theta}{2} \sin\left(\frac{-2\theta}{2}\right) \\ &= 2 \cos 3\theta \sin(-\theta) \\ &= -2 \cos 3\theta \sin \theta \quad \sin(-\theta) = -\sin \theta \text{ (Section 7.1)}\end{aligned}$$

Now try Exercise 45.

Half-Angle Identities From the alternative forms of the identity for $\cos 2A$, we derive three additional identities for $\sin \frac{A}{2}$, $\cos \frac{A}{2}$, and $\tan \frac{A}{2}$. These are known as **half-angle identities.**

To derive the identity for $\sin \frac{A}{2}$, start with the following double-angle identity for cosine and solve for $\sin x$.

$$\cos 2x = 1 - 2\sin^2 x$$

$$2\sin^2 x = 1 - \cos 2x \qquad \text{Add } 2\sin^2 x\text{; subtract } \cos 2x.$$

$$\sin x = \pm\sqrt{\frac{1 - \cos 2x}{2}} \qquad \text{Divide by 2; take square roots. (Section 1.4)}$$

$$\sin \frac{A}{2} = \pm\sqrt{\frac{1 - \cos A}{2}} \qquad \text{Let } 2x = A\text{, so } x = \tfrac{A}{2}\text{; substitute.}$$

The $\pm$ sign in this identity indicates that the appropriate sign is chosen depending on the quadrant of $\frac{A}{2}$. For example, if $\frac{A}{2}$ is a quadrant III angle, we choose the negative sign since the sine function is negative in quadrant III.

We derive the identity for $\cos \frac{A}{2}$ using the double-angle identity $\cos 2x = 2\cos^2 x - 1$.

$$1 + \cos 2x = 2\cos^2 x \qquad \text{Add 1.}$$

$$\cos^2 x = \frac{1 + \cos 2x}{2} \qquad \text{Rewrite; divide by 2.}$$

$$\cos x = \pm\sqrt{\frac{1 + \cos 2x}{2}} \qquad \text{Take square roots.}$$

$$\cos \frac{A}{2} = \pm\sqrt{\frac{1 + \cos A}{2}} \qquad \text{Replace } x \text{ with } \tfrac{A}{2}.$$

An identity for $\tan \frac{A}{2}$ comes from the identities for $\sin \frac{A}{2}$ and $\cos \frac{A}{2}$.

$$\tan \frac{A}{2} = \frac{\sin \frac{A}{2}}{\cos \frac{A}{2}} = \frac{\pm\sqrt{\frac{1 - \cos A}{2}}}{\pm\sqrt{\frac{1 + \cos A}{2}}} = \pm\sqrt{\frac{1 - \cos A}{1 + \cos A}}$$

We derive an alternative identity for $\tan \frac{A}{2}$ using double-angle identities.

$$\tan \frac{A}{2} = \frac{\sin \frac{A}{2}}{\cos \frac{A}{2}} = \frac{2\sin \frac{A}{2}\cos \frac{A}{2}}{2\cos^2 \frac{A}{2}} \qquad \text{Multiply by } 2\cos \tfrac{A}{2} \text{ in numerator and denominator.}$$

$$= \frac{\sin 2\left(\frac{A}{2}\right)}{1 + \cos 2\left(\frac{A}{2}\right)} \qquad \text{Double-angle identities}$$

$$\tan \frac{A}{2} = \frac{\sin A}{1 + \cos A}$$

From this identity for $\tan \frac{A}{2}$, we can also derive

$$\tan \frac{A}{2} = \frac{1 - \cos A}{\sin A}.$$

Half-Angle Identities

$$\cos\frac{A}{2} = \pm\sqrt{\frac{1+\cos A}{2}} \qquad \sin\frac{A}{2} = \pm\sqrt{\frac{1-\cos A}{2}}$$

$$\tan\frac{A}{2} = \pm\sqrt{\frac{1-\cos A}{1+\cos A}} \qquad \tan\frac{A}{2} = \frac{\sin A}{1+\cos A} \qquad \tan\frac{A}{2} = \frac{1-\cos A}{\sin A}$$

NOTE The last two identities for $\tan\frac{A}{2}$ do not require a sign choice. When using the other half-angle identities, select the plus or minus sign according to the quadrant in which $\frac{A}{2}$ terminates. For example, if an angle $A = 324°$, then $\frac{A}{2} = 162°$, which lies in quadrant II. In quadrant II, $\cos\frac{A}{2}$ and $\tan\frac{A}{2}$ are negative, while $\sin\frac{A}{2}$ is positive.

EXAMPLE 8 Using a Half-Angle Identity to Find an Exact Value

Find the exact value of cos 15° using the half-angle identity for cosine.

Solution

$$\cos 15° = \cos\frac{1}{2}(30°) = \sqrt{\frac{1+\cos 30°}{2}}$$

Choose the positive square root.

$$= \sqrt{\frac{1+\frac{\sqrt{3}}{2}}{2}} = \sqrt{\frac{\left(1+\frac{\sqrt{3}}{2}\right)\cdot 2}{2\cdot 2}} = \frac{\sqrt{2+\sqrt{3}}}{2}$$

Simplify the radicals. (Section R.7)

Now try Exercise 51.

EXAMPLE 9 Using a Half-Angle Identity to Find an Exact Value

Find the exact value of tan 22.5° using the identity $\tan\frac{A}{2} = \frac{\sin A}{1+\cos A}$.

Solution Since $22.5° = \frac{1}{2}(45°)$, replace A with 45°.

$$\tan 22.5° = \tan\frac{45°}{2} = \frac{\sin 45°}{1+\cos 45°} = \frac{\frac{\sqrt{2}}{2}}{1+\frac{\sqrt{2}}{2}}$$

Now multiply numerator and denominator by 2. Then rationalize the denominator.

$$\tan 22.5° = \frac{\sqrt{2}}{2+\sqrt{2}} = \frac{\sqrt{2}}{2+\sqrt{2}}\cdot\frac{2-\sqrt{2}}{2-\sqrt{2}} = \frac{2\sqrt{2}-2}{2}$$

$$= \frac{2(\sqrt{2}-1)}{2} = \sqrt{2}-1$$

Now try Exercise 53.

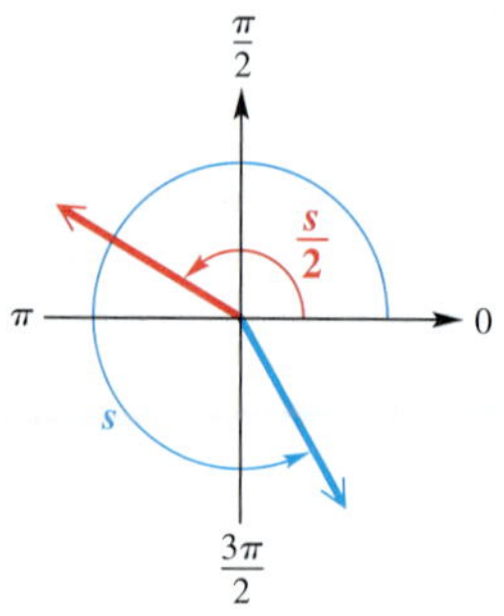

EXAMPLE 10 Finding Functions of $\frac{s}{2}$ Given Information about s

Given $\cos s = \frac{2}{3}$, with $\frac{3\pi}{2} < s < 2\pi$, find $\cos \frac{s}{2}$, $\sin \frac{s}{2}$, and $\tan \frac{s}{2}$.

Solution Since

$$\frac{3\pi}{2} < s < 2\pi$$

and

$$\frac{3\pi}{4} < \frac{s}{2} < \pi, \quad \text{Divide by 2. (Section 1.7)}$$

$\frac{s}{2}$ terminates in quadrant II. See Figure 7. In quadrant II, the values of $\cos \frac{s}{2}$ and $\tan \frac{s}{2}$ are negative and the value of $\sin \frac{s}{2}$ is positive. Now use the appropriate half-angle identities and simplify the radicals.

Figure 7

$$\sin \frac{s}{2} = \sqrt{\frac{1 - \frac{2}{3}}{2}} = \sqrt{\frac{1}{6}} = \frac{\sqrt{6}}{6}$$

$$\cos \frac{s}{2} = -\sqrt{\frac{1 + \frac{2}{3}}{2}} = -\sqrt{\frac{5}{6}} = -\frac{\sqrt{30}}{6}$$

$$\tan \frac{s}{2} = \frac{\sin \frac{s}{2}}{\cos \frac{s}{2}} = \frac{\frac{\sqrt{6}}{6}}{-\frac{\sqrt{30}}{6}} = -\frac{\sqrt{5}}{5}$$

Notice that it is not necessary to use a half-angle identity for $\tan \frac{s}{2}$ once we find $\sin \frac{s}{2}$ and $\cos \frac{s}{2}$. However, using this identity would provide an excellent check.

Now try Exercise 59.

EXAMPLE 11 Simplifying Expressions Using the Half-Angle Identities

Simplify each expression.

(a) $\pm\sqrt{\frac{1 + \cos 12x}{2}}$ **(b)** $\frac{1 - \cos 5\alpha}{\sin 5\alpha}$

Solution

(a) This matches part of the identity for $\cos \frac{A}{2}$.

$$\cos \frac{A}{2} = \pm\sqrt{\frac{1 + \cos A}{2}}$$

Replace A with $12x$ to get

$$\pm\sqrt{\frac{1 + \cos 12x}{2}} = \cos \frac{12x}{2} = \cos 6x.$$

(b) Use the third identity for $\tan \frac{A}{2}$ given earlier with $A = 5\alpha$ to get

$$\frac{1 - \cos 5\alpha}{\sin 5\alpha} = \tan \frac{5\alpha}{2}.$$

Now try Exercises 67 and 71.

7.4 Exercises

Use identities to find values of the sine and cosine functions for each angle measure. See Example 1.

1. θ, given $\cos 2\theta = \frac{3}{5}$ and θ terminates in quadrant I

2. θ, given $\cos 2\theta = \frac{3}{4}$ and θ terminates in quadrant III

3. x, given $\cos 2x = -\frac{5}{12}$ and $\frac{\pi}{2} < x < \pi$

4. x, given $\cos 2x = \frac{2}{3}$ and $\frac{\pi}{2} < x < \pi$

5. 2θ, given $\sin \theta = \frac{2}{5}$ and $\cos \theta < 0$

6. 2θ, given $\cos \theta = -\frac{12}{13}$ and $\sin \theta > 0$

7. $2x$, given $\tan x = 2$ and $\cos x > 0$

8. $2x$, given $\tan x = \frac{5}{3}$ and $\sin x < 0$

9. 2θ, given $\sin \theta = -\frac{\sqrt{5}}{7}$ and $\cos \theta > 0$

10. 2θ, given $\cos \theta = \frac{\sqrt{3}}{5}$ and $\sin \theta > 0$

Use an identity to write each expression as a single trigonometric function value or as a single number. See Example 3.

11. $\cos^2 15° - \sin^2 15°$

12. $\dfrac{2 \tan 15°}{1 - \tan^2 15°}$

13. $1 - 2 \sin^2 15°$

14. $1 - 2 \sin^2 22\frac{1}{2}°$

15. $2 \cos^2 67\frac{1}{2}° - 1$

16. $\cos^2 \frac{\pi}{8} - \frac{1}{2}$

17. $\dfrac{\tan 51°}{1 - \tan^2 51°}$

18. $\dfrac{\tan 34°}{2(1 - \tan^2 34°)}$

19. $\frac{1}{4} - \frac{1}{2} \sin^2 47.1°$

20. $\frac{1}{8} \sin 29.5° \cos 29.5°$

Express each function as a trigonometric function of x. See Example 4.

21. $\cos 3x$

22. $\sin 4x$

23. $\tan 3x$

24. $\cos 4x$

Graph each expression and use the graph to conjecture an identity. Then verify your conjecture algebraically.

25. $\cos^4 x - \sin^4 x$

26. $\dfrac{4 \tan x \cos^2 x - 2 \tan x}{1 - \tan^2 x}$

Verify that each equation is an identity. See Example 2.

27. $(\sin x + \cos x)^2 = \sin 2x + 1$

28. $\sec 2x = \dfrac{\sec^2 x + \sec^4 x}{2 + \sec^2 x - \sec^4 x}$

29. $\sin 4x = 4 \sin x \cos x \cos 2x$

30. $\dfrac{1 + \cos 2x}{\sin 2x} = \cot x$

31. $\dfrac{2 \cos 2x}{\sin 2x} = \cot x - \tan x$

32. $\sin 4x = 4 \sin x \cos x - 8 \sin^3 x \cos x$

33. $\sin 2x \cos 2x = \sin 2x - 4 \sin^3 x \cos x$

34. $\cos 2x = \dfrac{1 - \tan^2 x}{1 + \tan^2 x}$

35. $\tan x + \cot x = 2 \csc 2x$

36. $\dfrac{\cot x - \tan x}{\cot x + \tan x} = \cos 2x$

37. $\sec^2 \dfrac{x}{2} = \dfrac{2}{1 + \cos x}$

38. $\cot^2 \dfrac{x}{2} = \dfrac{(1 + \cos x)^2}{\sin^2 x}$

39. $\sin^2 \dfrac{x}{2} = \dfrac{\tan x - \sin x}{2 \tan x}$

40. $\dfrac{\sin 2x}{2 \sin x} = \cos^2 \dfrac{x}{2} - \sin^2 \dfrac{x}{2}$

Write each expression as a sum or difference of trigonometric functions. See Example 6.

41. $2 \sin 58° \cos 102°$

42. $2 \cos 85° \sin 140°$

43. $5 \cos 3x \cos 2x$

44. $\sin 4x \sin 5x$

Write each expression as a product of trigonometric functions. See Example 7.

45. $\cos 4x - \cos 2x$

46. $\cos 5x + \cos 8x$

47. $\sin 25° + \sin(-48°)$

48. $\sin 102° - \sin 95°$

49. $\cos 4x + \cos 8x$

50. $\sin 9x - \sin 3x$

Use a half-angle identity to find each exact value. See Examples 8 and 9.

51. $\sin 67.5°$

52. $\sin 195°$

53. $\cos 195°$

54. $\tan 195°$

55. $\cos 165°$

56. $\sin 165°$

57. Explain how you could use an identity of this section to find the exact value of $\sin 7.5°$. (*Hint:* $7.5 = \frac{1}{2}\left(\frac{1}{2}\right)(30)$.)

58. The identity $\tan \frac{A}{2} = \pm\sqrt{\frac{1 - \cos A}{1 + \cos A}}$ can be used to find $\tan 22.5° = \sqrt{3 - 2\sqrt{2}}$, and the identity $\tan \frac{A}{2} = \frac{\sin A}{1 + \cos A}$ can be used to find $\tan 22.5° = \sqrt{2} - 1$. Show that these answers are the same, without using a calculator. (*Hint:* If $a > 0$ and $b > 0$ and $a^2 = b^2$, then $a = b$.)

Find each of the following. See Example 10.

59. $\cos \dfrac{x}{2}$, given $\cos x = \dfrac{1}{4}$, with $0 < x < \dfrac{\pi}{2}$

60. $\sin \dfrac{x}{2}$, given $\cos x = -\dfrac{5}{8}$, with $\dfrac{\pi}{2} < x < \pi$

61. $\tan \dfrac{\theta}{2}$, given $\sin \theta = \dfrac{3}{5}$, with $90° < \theta < 180°$

62. $\cos \dfrac{\theta}{2}$, given $\sin \theta = -\dfrac{1}{5}$, with $180° < \theta < 270°$

63. $\sin \dfrac{x}{2}$, given $\tan x = 2$, with $0 < x < \dfrac{\pi}{2}$

64. $\cos \dfrac{x}{2}$, given $\cot x = -3$, with $\dfrac{\pi}{2} < x < \pi$

65. $\tan \dfrac{\theta}{2}$, given $\tan \theta = \dfrac{\sqrt{7}}{3}$, with $180° < \theta < 270°$

66. $\cot \dfrac{\theta}{2}$, given $\tan \theta = -\dfrac{\sqrt{5}}{2}$, with $90° < \theta < 180°$

Use an identity to write each expression with a single trigonometric function. See Example 11.

67. $\sqrt{\frac{1 - \cos 40°}{2}}$ **68.** $\sqrt{\frac{1 + \cos 76°}{2}}$ **69.** $\sqrt{\frac{1 - \cos 147°}{1 + \cos 147°}}$

70. $\sqrt{\frac{1 + \cos 165°}{1 - \cos 165°}}$ **71.** $\frac{1 - \cos 59.74°}{\sin 59.74°}$ **72.** $\frac{\sin 158.2°}{1 + \cos 158.2°}$

73. Use the identity $\tan \frac{A}{2} = \frac{\sin A}{1 + \cos A}$ to derive the equivalent identity $\tan \frac{A}{2} = \frac{1 - \cos A}{\sin A}$ by multiplying both the numerator and denominator by $1 - \cos A$.

74. Consider the expression $\tan\left(\frac{\pi}{2} + x\right)$.

(a) Why can't we use the identity for $\tan(A + B)$ to express it as a function of x alone?

(b) Use the identity $\tan \theta = \frac{\sin \theta}{\cos \theta}$ to rewrite the expression in terms of sine and cosine.

(c) Use the result of part (b) to show that $\tan\left(\frac{\pi}{2} + x\right) = -\cot x$.

(Modeling) Mach Number *An airplane flying faster than sound sends out sound waves that form a cone, as shown in the figure. The cone intersects the ground to form a* hyperbola. *As this hyperbola passes over a particular point on the ground, a sonic boom is heard at that point. If θ is the angle at the vertex of the cone, then*

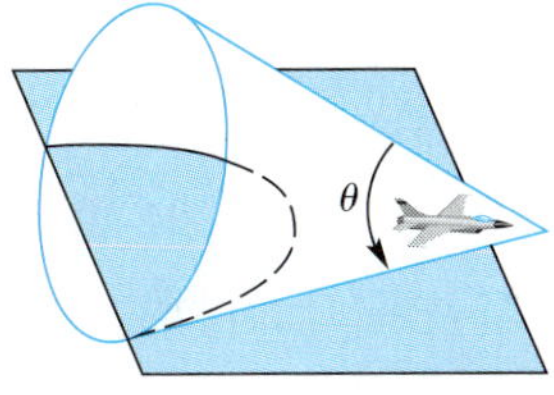

$$\sin \frac{\theta}{2} = \frac{1}{m},$$

where m is the Mach number for the speed of the plane. (We assume $m > 1$.) The Mach number is the ratio of the speed of the plane and the speed of sound. Thus, a speed of Mach 1.4 means that the plane is flying at 1.4 times the speed of sound. In Exercises 75–78, one of the values θ or m is given. Find the other value.

75. $m = \frac{3}{2}$ **76.** $m = \frac{5}{4}$ **77.** $\theta = 30°$ **78.** $\theta = 60°$

(Modeling) *Solve each problem. See Example 5.*

79. *Railroad Curves* In the United States, circular railroad curves are designated by the *degree of curvature,* the central angle subtended by a chord of 100 ft. See the figure. (*Source:* Hay, W. W., *Railroad Engineering,* John Wiley & Sons, 1982.)

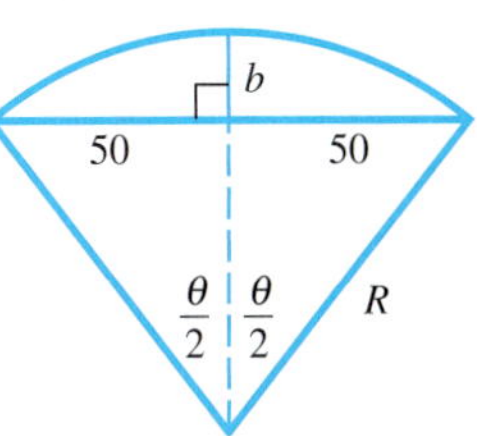

(a) Use the figure to write an expression for $\cos \frac{\theta}{2}$.

(b) Use the result of part (a) and the third half-angle identity for tangent to write an expression for $\tan \frac{\theta}{4}$.

(c) If $b = 12$, what is the measure of angle θ to the nearest degree?

80. *Distance Traveled by a Stone* The distance D of an object thrown (or propelled) from height h (feet) at angle θ with initial velocity v is modeled by the formula

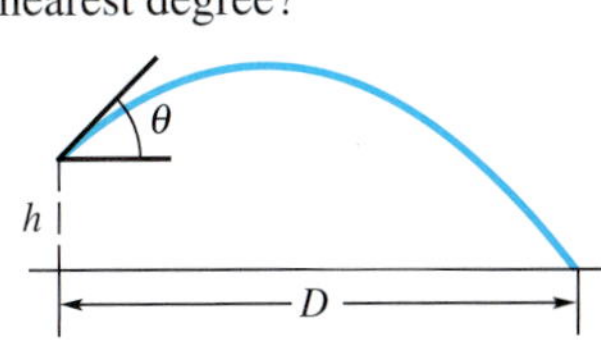

$$D = \frac{v^2 \sin \theta \cos \theta + v \cos \theta \sqrt{(v \sin \theta)^2 + 64h}}{32}.$$

See the figure. (*Source:* Kreighbaum, E. and K. Barthels, *Biomechanics,* Allyn & Bacon, 1996.) Also see the Chapter 5 Quantitative Reasoning.

(a) Find D when $h = 0$; that is, when the object is propelled from the ground.

(b) Suppose a car driving over loose gravel kicks up a small stone at a velocity of 36 ft per sec (about 25 mph) and an angle $\theta = 30°$. How far will the stone travel?

81. ***Wattage Consumption*** Refer to Example 5. Use an identity to determine values of a, c, and ω so that $W = a\cos(\omega t) + c$. Check your answer by graphing both expressions for W on the same coordinate axes.

82. ***Amperage, Wattage, and Voltage*** Amperage is a measure of the amount of electricity that is moving through a circuit, whereas voltage is a measure of the force pushing the electricity. The wattage W consumed by an electrical device can be determined by calculating the product of the amperage I and voltage V. (*Source:* Wilcox, G. and C. Hesselberth, *Electricity for Engineering Technology,* Allyn & Bacon, 1970.)

(a) A household circuit has voltage $V = 163\sin(120\pi t)$ when an incandescent lightbulb is turned on with amperage $I = 1.23\sin(120\pi t)$. Graph the wattage $W = VI$ consumed by the lightbulb in the window $[0, .05]$ by $[-50, 300]$.

(b) Determine the maximum and minimum wattages used by the lightbulb.

(c) Use identities to determine values for a, c, and ω so that $W = a\cos(\omega t) + c$.

(d) Check your answer by graphing both expressions for W on the same coordinate axes.

(e) Use the graph to estimate the average wattage used by the light. For how many watts do you think this incandescent lightbulb is rated?

Summary Exercises on Verifying Trigonometric Identities

These summary exercises provide practice with the various types of trigonometric identities presented in this chapter. Verify that each equation is an identity.

1. $\tan\theta + \cot\theta = \sec\theta\csc\theta$

2. $\csc\theta\cos^2\theta + \sin\theta = \csc\theta$

3. $\tan\frac{x}{2} = \csc x - \cot x$

4. $\sec(\pi - x) = -\sec x$

5. $\frac{\sin t}{1 + \cos t} = \frac{1 - \cos t}{\sin t}$

6. $\frac{1 - \sin t}{\cos t} = \frac{1}{\sec t + \tan t}$

7. $\sin 2\theta = \frac{2\tan\theta}{1 + \tan^2\theta}$

8. $\frac{2}{1 + \cos x} - \tan^2\frac{x}{2} = 1$

9. $\cot\theta - \tan\theta = \frac{2\cos^2\theta - 1}{\sin\theta\cos\theta}$

10. $\frac{1}{\sec t - 1} + \frac{1}{\sec t + 1} = 2\cot t\csc t$

11. $\frac{\sin(x + y)}{\cos(x - y)} = \frac{\cot x + \cot y}{1 + \cot x\cot y}$

12. $1 - \tan^2\frac{\theta}{2} = \frac{2\cos\theta}{1 + \cos\theta}$

13. $\frac{\sin\theta + \tan\theta}{1 + \cos\theta} = \tan\theta$

14. $\csc^4 x - \cot^4 x = \frac{1 + \cos^2 x}{1 - \cos^2 x}$

15. $\cos x = \frac{1 - \tan^2\frac{x}{2}}{1 + \tan^2\frac{x}{2}}$

16. $\cos 2x = \frac{2 - \sec^2 x}{\sec^2 x}$

17. $\frac{\tan^2 t + 1}{\tan t\csc^2 t} = \tan t$

18. $\frac{\sin s}{1 + \cos s} + \frac{1 + \cos s}{\sin s} = 2\csc s$

19. $\tan 4\theta = \frac{2\tan 2\theta}{2 - \sec^2 2\theta}$

20. $\tan\left(\frac{x}{2} + \frac{\pi}{4}\right) = \sec x + \tan x$

21. $\frac{\cot s - \tan s}{\cos s + \sin s} = \frac{\cos s - \sin s}{\sin s\cos s}$

22. $\frac{\tan\theta - \cot\theta}{\tan\theta + \cot\theta} = 1 - 2\cos^2\theta$

23. $\dfrac{\tan(x+y)-\tan y}{1+\tan(x+y)\tan y}=\tan x$

24. $2\cos^2\dfrac{x}{2}\tan x=\tan x+\sin x$

25. $\dfrac{\cos^4 x-\sin^4 x}{\cos^2 x}=1-\tan^2 x$

26. $\dfrac{\csc t+1}{\csc t-1}=(\sec t+\tan t)^2$

27. $\dfrac{2(\sin x-\sin^3 x)}{\cos x}=\sin 2x$

28. $\dfrac{1}{2}\cot\dfrac{x}{2}-\dfrac{1}{2}\tan\dfrac{x}{2}=\cot x$

7.5 Inverse Circular Functions

Inverse Functions ▪ Inverse Sine Function ▪ Inverse Cosine Function ▪ Inverse Tangent Function ▪ Remaining Inverse Circular Functions ▪ Inverse Function Values

Inverse Functions We first discussed inverse functions in Section 4.1. We give a quick review here for a pair of inverse functions f and f^{-1}.

1. If a function f is one-to-one, then f has an inverse function f^{-1}.
2. In a one-to-one function, each x-value corresponds to only one y-value and each y-value corresponds to only one x-value.
3. The domain of f is the range of f^{-1}, and the range of f is the domain of f^{-1}.
4. The graphs of f and f^{-1} are reflections of each other about the line $y = x$.
5. To find $f^{-1}(x)$ from $f(x)$, follow these steps.

   Step 1 Replace $f(x)$ with y and interchange x and y.

   Step 2 Solve for y.

   Step 3 Replace y with $f^{-1}(x)$.

In the remainder of this section, we use these facts to develop the inverse circular (trigonometric) functions.

Looking Ahead to Calculus

The inverse circular functions are used in calculus to solve certain types of related-rates problems and to integrate certain rational functions.

Inverse Sine Function From Figure 8 and the horizontal line test, we see that $y = \sin x$ does not define a one-to-one function. If we restrict the domain to the interval $\left[-\frac{\pi}{2},\frac{\pi}{2}\right]$, which is the part of the graph in Figure 8 shown in color, this restricted function is one-to-one and has an inverse function. The range of $y = \sin x$ is $[-1, 1]$, so the domain of the inverse function will be $[-1, 1]$, and its range will be $\left[-\frac{\pi}{2},\frac{\pi}{2}\right]$.

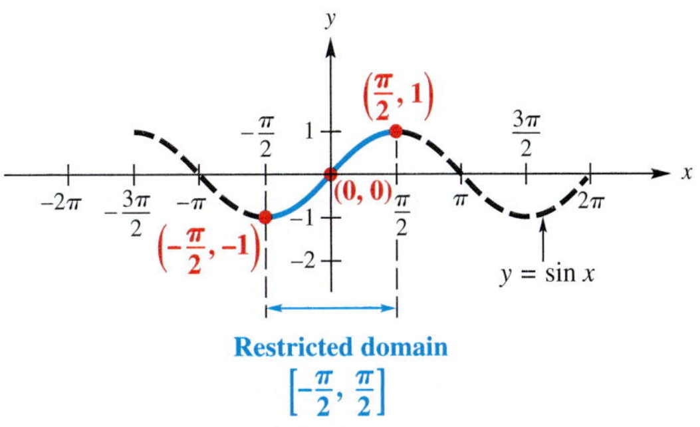

Figure 8

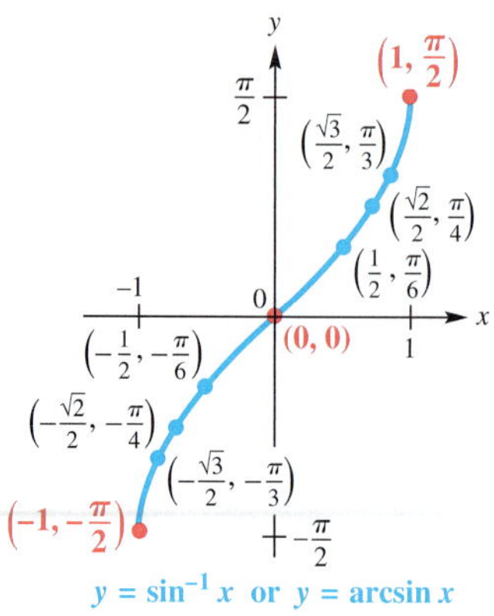

Figure 9

Reflecting the graph of $y = \sin x$ on the restricted domain across the line $y = x$ gives the graph of the inverse function, shown in Figure 9. Some key points are labeled on the graph. The equation of the inverse of $y = \sin x$ is found by interchanging x and y to get $x = \sin y$. This equation is solved for y by writing $y = \sin^{-1} x$ (read "inverse sine of x"). As Figure 9 shows, the domain of $y = \sin^{-1} x$ is $[-1, 1]$, while the restricted domain of $y = \sin x$, $\left[-\frac{\pi}{2}, \frac{\pi}{2}\right]$, is the range of $y = \sin^{-1} x$. An alternative notation for $\sin^{-1} x$ is $\arcsin x$.

Inverse Sine Function

$y = \sin^{-1} x$ **or** $y = \arcsin x$ means that $x = \sin y$, for $-\frac{\pi}{2} \leq y \leq \frac{\pi}{2}$.

We can think of $y = \sin^{-1} x$ or $y = \arcsin x$ as "y is the number in the interval $\left[-\frac{\pi}{2}, \frac{\pi}{2}\right]$ whose sine is x." Just as we evaluated $y = \log_2 4$ by writing it in exponential form as $2^y = 4$ (Section 4.3), we can write $y = \sin^{-1} x$ as $\sin y = x$ to evaluate it. We must pay close attention to the domain and range intervals.

EXAMPLE 1 Finding Inverse Sine Values

Find y in each equation.

(a) $y = \arcsin \frac{1}{2}$ **(b)** $y = \sin^{-1}(-1)$ **(c)** $y = \sin^{-1}(-2)$

Algebraic Solution

(a) The graph of the function defined by $y = \arcsin x$ (Figure 9) includes the point $\left(\frac{1}{2}, \frac{\pi}{6}\right)$. Thus,

$$\arcsin \frac{1}{2} = \frac{\pi}{6}.$$

Alternatively, we can think of $y = \arcsin \frac{1}{2}$ as "y is the number in $\left[-\frac{\pi}{2}, \frac{\pi}{2}\right]$ whose sine is $\frac{1}{2}$." Then we can write the given equation as $\sin y = \frac{1}{2}$. Since $\sin \frac{\pi}{6} = \frac{1}{2}$ and $\frac{\pi}{6}$ is in the range of the arcsine function, $y = \frac{\pi}{6}$.

(b) Writing the equation $y = \sin^{-1}(-1)$ in the form $\sin y = -1$ shows that $y = -\frac{\pi}{2}$. This can be verified by noticing that the point $\left(-1, -\frac{\pi}{2}\right)$ is on the graph of $y = \sin^{-1} x$.

(c) Because -2 is not in the domain of the inverse sine function, $\sin^{-1}(-2)$ does not exist.

Graphing Calculator Solution

To find these values with a graphing calculator, we graph $Y_1 = \sin^{-1} X$ and locate the points with X-values $\frac{1}{2}$ and -1. Figure 10(a) shows that when $X = \frac{1}{2}$, $Y = \frac{\pi}{6} \approx .52359878$. Similarly, Figure 10(b) shows that when $X = -1$, $Y = -\frac{\pi}{2} \approx -1.570796$.

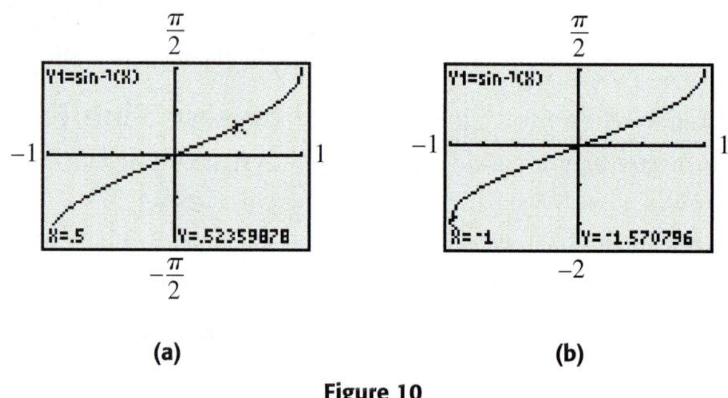

Figure 10

Since $\sin^{-1}(-2)$ does not exist, a calculator will give an error message for this input.

Now try Exercises 13 and 23.

CAUTION In Example 1(b), it is tempting to give the value of $\sin^{-1}(-1)$ as $\frac{3\pi}{2}$, since $\sin \frac{3\pi}{2} = -1$. Notice, however, that $\frac{3\pi}{2}$ is not in the range of the inverse sine function. Be certain that the number given for an inverse function value is in the range of the particular inverse function being considered.

Our observations about the inverse sine function from Figure 9 lead to the following generalizations.

INVERSE SINE FUNCTION

$y = \sin^{-1} x$ or $y = \arcsin x$

Domain: $[-1, 1]$ Range: $\left[-\frac{\pi}{2}, \frac{\pi}{2}\right]$

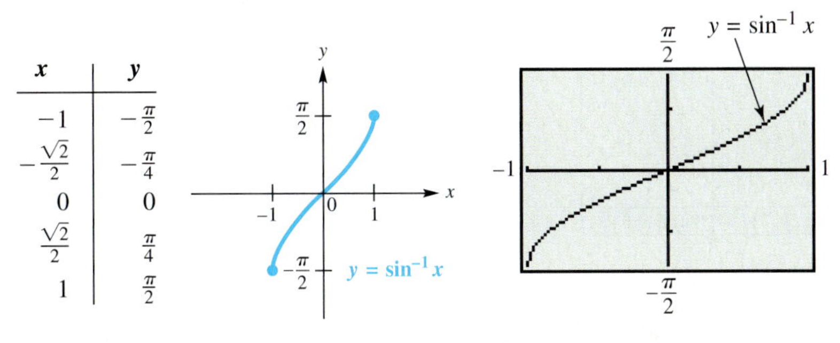

x	y
-1	$-\frac{\pi}{2}$
$-\frac{\sqrt{2}}{2}$	$-\frac{\pi}{4}$
0	0
$\frac{\sqrt{2}}{2}$	$\frac{\pi}{4}$
1	$\frac{\pi}{2}$

Figure 11

- The inverse sine function is increasing and continuous on its domain $[-1, 1]$.
- Its x-intercept is 0, and its y-intercept is 0.
- Its graph is symmetric with respect to the origin; it is an odd function.

Inverse Cosine Function The function $y = \cos^{-1} x$ (or $y = \arccos x$) is defined by restricting the domain of the function $y = \cos x$ to the interval $[0, \pi]$ as in Figure 12, and then interchanging the roles of x and y. The graph of $y = \cos^{-1} x$ is shown in Figure 13. Again, some key points are shown on the graph.

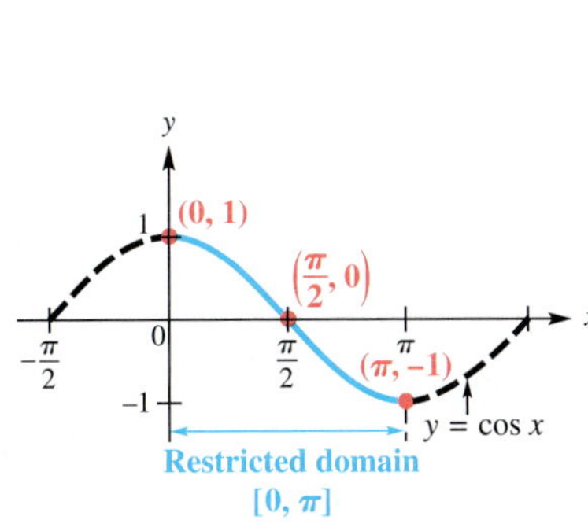

Figure 12

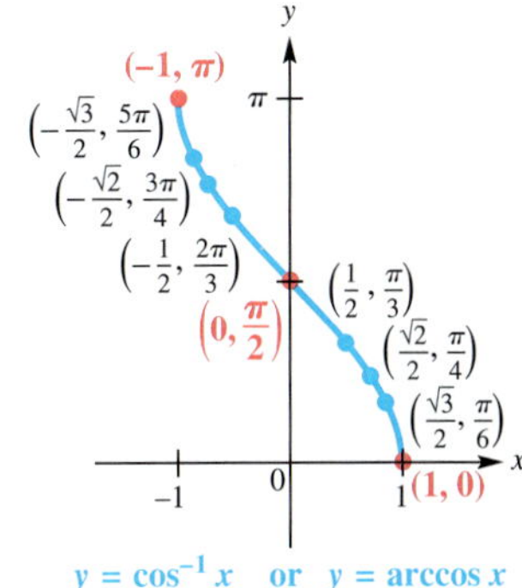

Figure 13

Inverse Cosine Function

$y = \cos^{-1} x$ or $y = \arccos x$ means that $x = \cos y$, for $0 \leq y \leq \pi$.

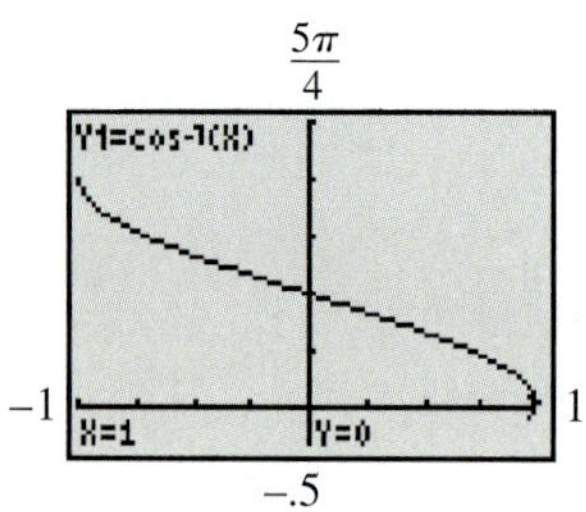

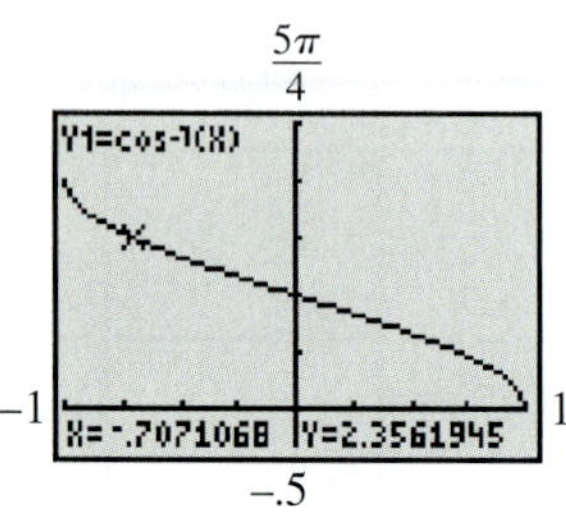

These screens support the results of Example 2, since $\frac{3\pi}{4} \approx 2.3561945$.

EXAMPLE 2 Finding Inverse Cosine Values

Find y in each equation.

(a) $y = \arccos 1$ **(b)** $y = \cos^{-1}\left(-\frac{\sqrt{2}}{2}\right)$

Solution

(a) Since the point $(1, 0)$ lies on the graph of $y = \arccos x$ in Figure 13 on the previous page, the value of y is 0. Alternatively, we can think of $y = \arccos 1$ as "y is the number in $[0, \pi]$ whose cosine is 1," or $\cos y = 1$. Then $y = 0$, since $\cos 0 = 1$ and 0 is in the range of the arccosine function.

(b) We must find the value of y that satisfies $\cos y = -\frac{\sqrt{2}}{2}$, where y is in the interval $[0, \pi]$, the range of the function $y = \cos^{-1} x$. The only value for y that satisfies these conditions is $\frac{3\pi}{4}$. Again, this can be verified from the graph in Figure 13.

Now try Exercises 15 and 21.

Our observations about the inverse cosine function lead to the following generalizations.

INVERSE COSINE FUNCTION
$y = \cos^{-1} x$ or $y = \arccos x$

Domain: $[-1, 1]$ Range: $[0, \pi]$

x	y
-1	π
$-\frac{\sqrt{2}}{2}$	$\frac{3\pi}{4}$
0	$\frac{\pi}{2}$
$\frac{\sqrt{2}}{2}$	$\frac{\pi}{4}$
1	0

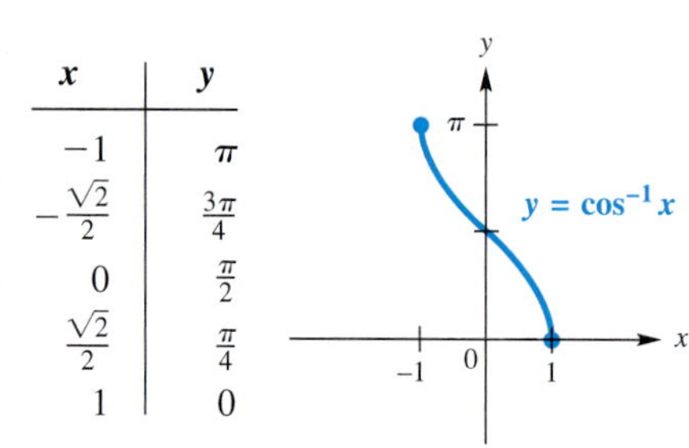

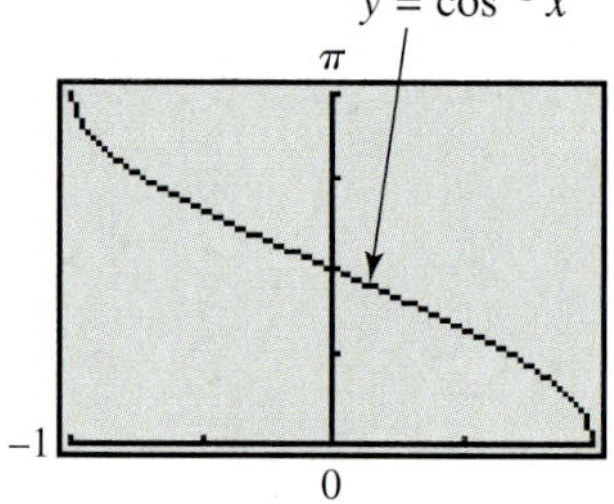

Figure 14

- The inverse cosine function is decreasing and continuous on its domain $[-1, 1]$.
- Its x-intercept is 1, and its y-intercept is $\frac{\pi}{2}$.
- Its graph is not symmetric with respect to the y-axis or the origin.

Inverse Tangent Function Restricting the domain of the function $y = \tan x$ to the open interval $\left(-\frac{\pi}{2}, \frac{\pi}{2}\right)$ yields a one-to-one function. By interchanging the roles of x and y, we obtain the inverse tangent function given by $y = \tan^{-1} x$ or $y = \arctan x$. Figure 15 shows the graph of the restricted tangent function. Figure 16 gives the graph of $y = \tan^{-1} x$.

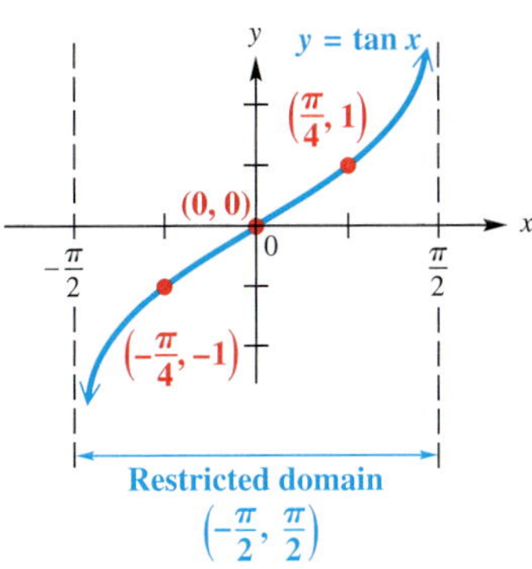

Figure 15

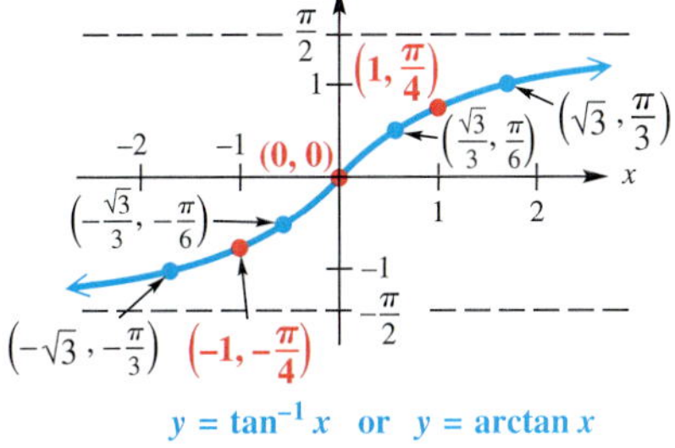

Figure 16

Inverse Tangent Function

$y = \tan^{-1} x$ or $y = \arctan x$ means that $x = \tan y$, for $-\frac{\pi}{2} < y < \frac{\pi}{2}$.

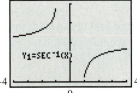

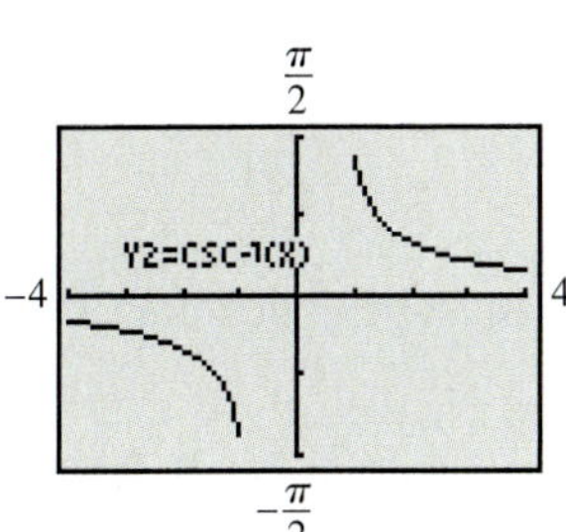

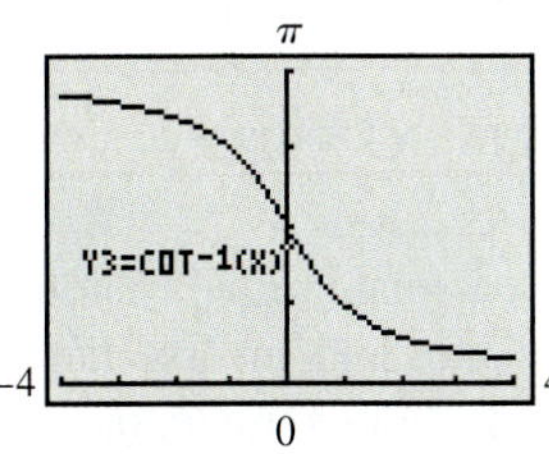

```
Plot1 Plot2 Plot3
\Y1■cos⁻¹(1/X)
\Y2■sin⁻¹(1/X)
\Y3■(tan⁻¹(1/X))*
(X≥0)+(tan⁻¹(1/X)
+π)*(X<0)
\Y4=
\Y5=
```

The first three screens show the graphs of the three remaining inverse circular functions. The last screen shows how they are defined.

Figure 18

INVERSE TANGENT FUNCTION
$y = \tan^{-1} x$ or $y = \arctan x$

Domain: $(-\infty, \infty)$ Range: $\left(-\frac{\pi}{2}, \frac{\pi}{2}\right)$

x	y
-1	$-\frac{\pi}{4}$
$-\frac{\sqrt{3}}{3}$	$-\frac{\pi}{6}$
0	0
$\frac{\sqrt{3}}{3}$	$\frac{\pi}{6}$
1	$\frac{\pi}{4}$

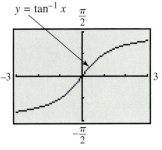

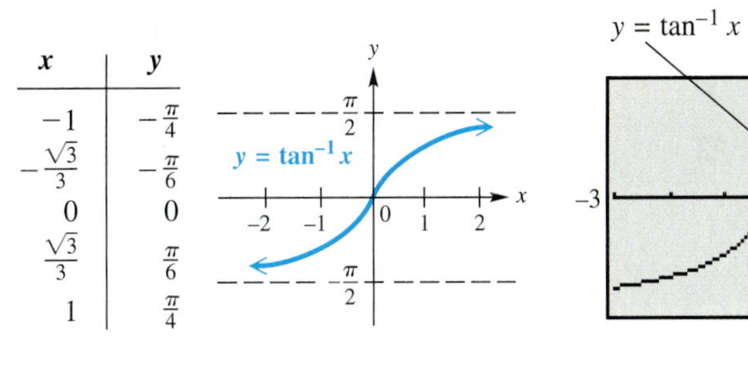

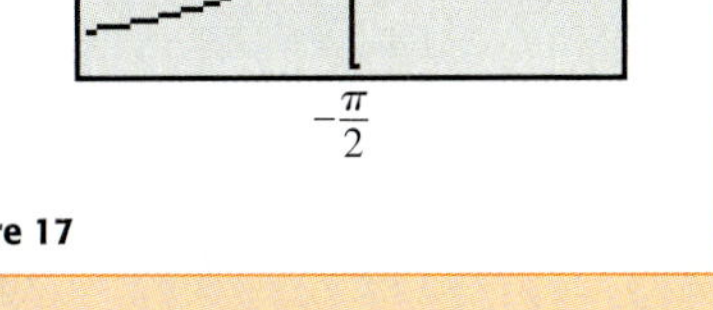

Figure 17

- The inverse tangent function is increasing and continuous on its domain $(-\infty, \infty)$.
- Its x-intercept is 0, and its y-intercept is 0.
- Its graph is symmetric with respect to the origin; it is an odd function.
- The lines $y = \frac{\pi}{2}$ and $y = -\frac{\pi}{2}$ are horizontal asymptotes.

Remaining Inverse Circular Functions The remaining three inverse trigonometric functions are defined similarly; their graphs are shown in Figure 18. All six inverse trigonometric functions with their domains and ranges are given in the table on the next page.

Inverse Function	Domain	Range	
		Interval	Quadrants of the Unit Circle
$y = \sin^{-1} x$	$[-1, 1]$	$\left[-\frac{\pi}{2}, \frac{\pi}{2}\right]$	I and IV
$y = \cos^{-1} x$	$[-1, 1]$	$[0, \pi]$	I and II
$y = \tan^{-1} x$	$(-\infty, \infty)$	$\left(-\frac{\pi}{2}, \frac{\pi}{2}\right)$	I and IV
$y = \cot^{-1} x$	$(-\infty, \infty)$	$(0, \pi)$	I and II
$y = \sec^{-1} x$	$(-\infty, -1] \cup [1, \infty)$	$[0, \pi], y \neq \frac{\pi}{2}$*	I and II
$y = \csc^{-1} x$	$(-\infty, -1] \cup [1, \infty)$	$\left[-\frac{\pi}{2}, \frac{\pi}{2}\right], y \neq 0$*	I and IV

Inverse Function Values The inverse circular functions are formally defined with real number ranges. However, there are times when it may be convenient to find degree-measured angles equivalent to these real number values. It is also often convenient to think in terms of the unit circle and choose the inverse function values based on the quadrants given in the preceding table.

EXAMPLE 3 Finding Inverse Function Values (Degree-Measured Angles)

Find the *degree measure* of θ in the following.

(a) $\theta = \arctan 1$ **(b)** $\theta = \sec^{-1} 2$

Solution

(a) Here θ must be in $(-90°, 90°)$, but since $1 > 0$, θ must be in quadrant I. The alternative statement, $\tan \theta = 1$, leads to $\theta = 45°$.

(b) Write the equation as $\sec \theta = 2$. For $\sec^{-1} x$, θ is in quadrant I or II. Because 2 is positive, θ is in quadrant I and $\theta = 60°$, since $\sec 60° = 2$. Note that $60°$ (the degree equivalent of $\frac{\pi}{3}$) is in the range of the inverse secant function.

Now try Exercises 33 and 39.

The inverse trigonometric function keys on a calculator give results in the proper quadrant for the inverse sine, inverse cosine, and inverse tangent functions, according to the definitions of these functions. For example, on a calculator, in degrees, $\sin^{-1} .5 = 30°$, $\sin^{-1}(-.5) = -30°$, $\tan^{-1}(-1) = -45°$, and $\cos^{-1}(-.5) = 120°$.

Finding $\cot^{-1} x$, $\sec^{-1} x$, and $\csc^{-1} x$ with a calculator is not as straightforward, because these functions must be expressed in terms of $\tan^{-1} x$, $\cos^{-1} x$, and $\sin^{-1} x$, respectively. If $y = \sec^{-1} x$, for example, then $\sec y = x$, which must be written as a cosine function as follows:

$$\text{If } \sec y = x, \text{ then } \frac{1}{\cos y} = x \quad \text{or} \quad \cos y = \frac{1}{x}, \quad \text{and} \quad y = \cos^{-1} \frac{1}{x}.$$

*The inverse secant and inverse cosecant functions are sometimes defined with different ranges. We use intervals that match their reciprocal functions (except for one missing point).

In summary, to find $\sec^{-1} x$, we find $\cos^{-1} \frac{1}{x}$. Similar statements apply to $\csc^{-1} x$ and $\cot^{-1} x$. There is one additional consideration with $\cot^{-1} x$. Since we take the inverse tangent of the reciprocal to find inverse cotangent, the calculator gives values of inverse cotangent with the same range as inverse tangent, $\left(-\frac{\pi}{2}, \frac{\pi}{2}\right)$, which is not the correct range for inverse cotangent. For inverse cotangent, the proper range must be considered and the results adjusted accordingly.

EXAMPLE 4 Finding Inverse Function Values with a Calculator

(a) Find y in radians if $y = \csc^{-1}(-3)$.

(b) Find θ in degrees if $\theta = \text{arccot}(-.3541)$.

Solution

Figure 19

(a) With the calculator in radian mode, enter $\csc^{-1}(-3)$ as $\sin^{-1}\left(-\frac{1}{3}\right)$ to get $y \approx -.3398369095$. See Figure 19.

(b) Set the calculator to degree mode. A calculator gives the inverse tangent value of a negative number as a quadrant IV angle. The restriction on the range of arccotangent implies that θ must be in quadrant II, so enter

$$\text{arccot}(-.3541) \quad \text{as} \quad \tan^{-1}\left(\frac{1}{-.3541}\right) + 180°.$$

As shown in Figure 19, $\theta \approx 109.4990544°$.

Now try Exercises 43 and 49.

EXAMPLE 5 Finding Function Values Using Definitions of the Trigonometric Functions

Evaluate each expression without using a calculator.

(a) $\sin\left(\tan^{-1} \frac{3}{2}\right)$ **(b)** $\tan\left(\cos^{-1}\left(-\frac{5}{13}\right)\right)$

Solution

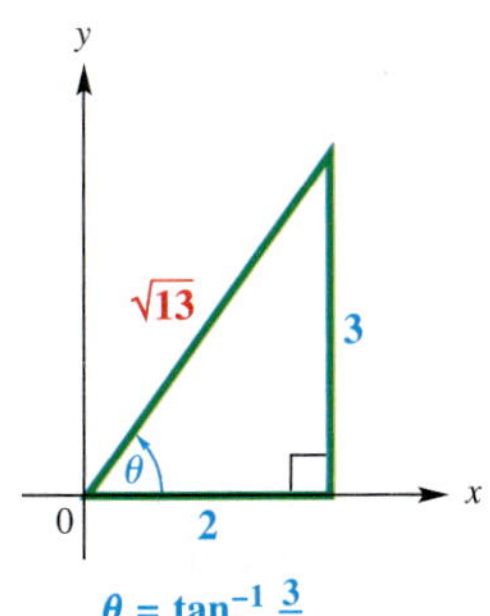

Figure 20

(a) Let $\theta = \tan^{-1} \frac{3}{2}$, so $\tan \theta = \frac{3}{2}$. The inverse tangent function yields values only in quadrants I and IV, and since $\frac{3}{2}$ is positive, θ is in quadrant I. Sketch θ in quadrant I, and label a triangle, as shown in Figure 20. By the Pythagorean theorem, the hypotenuse is $\sqrt{13}$. The value of sine is the quotient of the side opposite and the hypotenuse, so

$$\sin\left(\tan^{-1} \frac{3}{2}\right) = \sin \theta = \frac{3}{\sqrt{13}} = \frac{3\sqrt{13}}{13}. \quad \text{(Section 5.3)}$$

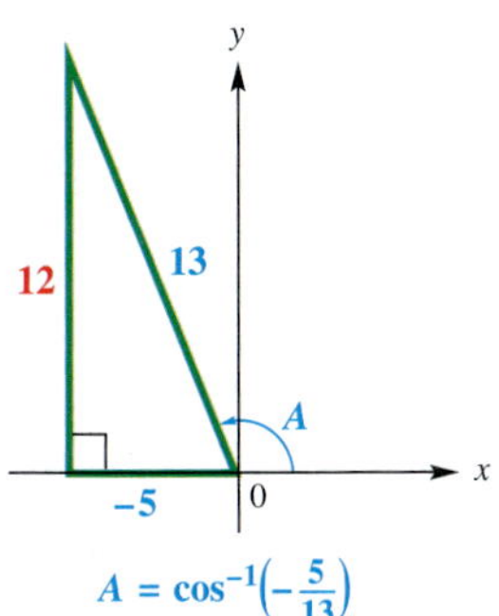

Figure 21

(b) Let $A = \cos^{-1}\left(-\frac{5}{13}\right)$. Then, $\cos A = -\frac{5}{13}$. Since $\cos^{-1} x$ for a negative value of x is in quadrant II, sketch A in quadrant II, as shown in Figure 21.

$$\tan\left(\cos^{-1}\left(-\frac{5}{13}\right)\right) = \tan A = -\frac{12}{5}$$

Now try Exercises 63 and 65.

EXAMPLE 6 Finding Function Values Using Identities

Evaluate each expression without using a calculator.

(a) $\cos\left(\arctan\sqrt{3} + \arcsin\dfrac{1}{3}\right)$ **(b)** $\tan\left(2\arcsin\dfrac{2}{5}\right)$

Solution

(a) Let $A = \arctan\sqrt{3}$ and $B = \arcsin\frac{1}{3}$, so $\tan A = \sqrt{3}$ and $\sin B = \frac{1}{3}$. Sketch both A and B in quadrant I, as shown in Figure 22. Now, use the cosine sum identity.

$$\cos(A + B) = \cos A \cos B - \sin A \sin B \quad \text{(Section 7.3)}$$

$$\cos\left(\arctan\sqrt{3} + \arcsin\frac{1}{3}\right) = \cos\left(\arctan\sqrt{3}\right)\cos\left(\arcsin\frac{1}{3}\right) - \sin\left(\arctan\sqrt{3}\right)\sin\left(\arcsin\frac{1}{3}\right) \quad (1)$$

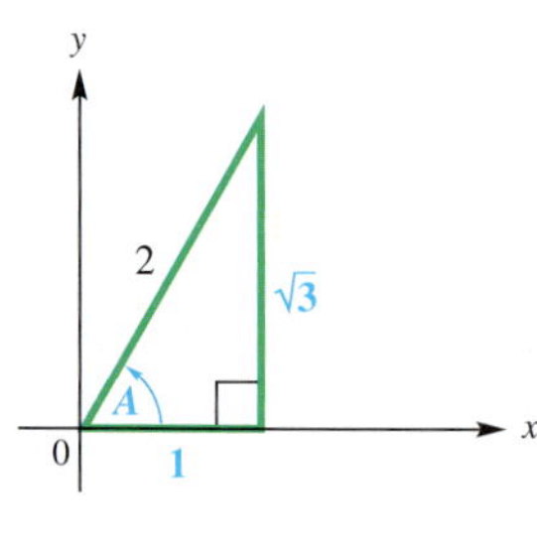

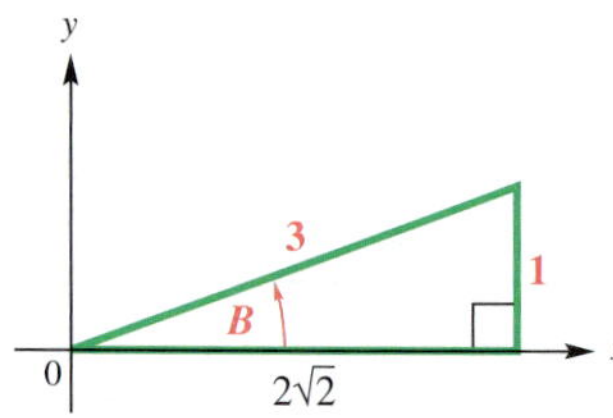

Figure 22

From Figure 22,

$$\cos\left(\arctan\sqrt{3}\right) = \cos A = \frac{1}{2}, \quad \cos\left(\arcsin\frac{1}{3}\right) = \cos B = \frac{2\sqrt{2}}{3},$$

$$\sin\left(\arctan\sqrt{3}\right) = \sin A = \frac{\sqrt{3}}{2}, \quad \sin\left(\arcsin\frac{1}{3}\right) = \sin B = \frac{1}{3}.$$

Substitute these values into equation (1) to get

$$\cos\left(\arctan\sqrt{3} + \arcsin\frac{1}{3}\right) = \frac{1}{2}\cdot\frac{2\sqrt{2}}{3} - \frac{\sqrt{3}}{2}\cdot\frac{1}{3} = \frac{2\sqrt{2} - \sqrt{3}}{6}.$$

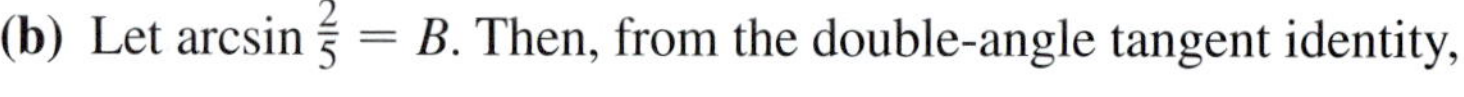

(b) Let $\arcsin\frac{2}{5} = B$. Then, from the double-angle tangent identity,

$$\tan\left(2\arcsin\frac{2}{5}\right) = \tan 2B = \frac{2\tan B}{1 - \tan^2 B}. \quad \text{(Section 7.4)}$$

Since $\arcsin\frac{2}{5} = B$, $\sin B = \frac{2}{5}$. Sketch a triangle in quadrant I, find the length of the third side, and then find $\tan B$. From the triangle in Figure 23, $\tan B = \frac{2}{\sqrt{21}}$, and

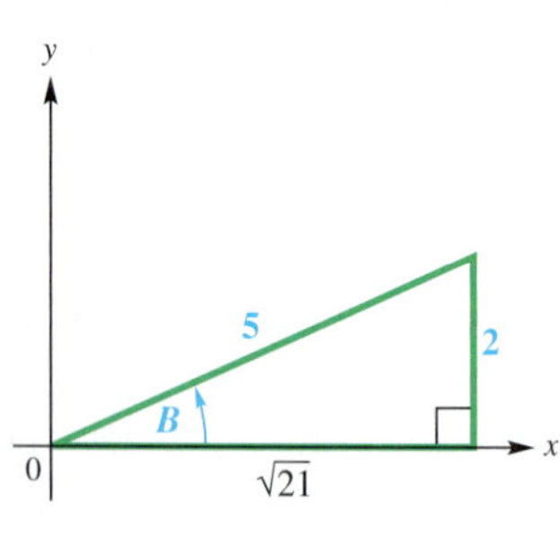

Figure 23

$$\tan\left(2\arcsin\frac{2}{5}\right) = \frac{2\left(\frac{2}{\sqrt{21}}\right)}{1 - \left(\frac{2}{\sqrt{21}}\right)^2} = \frac{\frac{4}{\sqrt{21}}}{1 - \frac{4}{21}} = \frac{4\sqrt{21}}{17}.$$

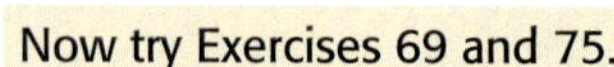

Now try Exercises 69 and 75.

While the work shown in Examples 5 and 6 does not rely on a calculator, we can support our algebraic work with one. By entering $\cos\left(\arctan\sqrt{3} + \arcsin\frac{1}{3}\right)$ from Example 6(a) into a calculator, we get the approximation .1827293862, the same approximation as when we enter $\frac{2\sqrt{2} - \sqrt{3}}{6}$ (the exact value we obtained algebraically). Similarly, we obtain the same approximation when we evaluate $\tan\left(2\arcsin\frac{2}{5}\right)$ and $\frac{4\sqrt{21}}{17}$, supporting our answer in Example 6(b).

EXAMPLE 7 Writing Function Values in Terms of *u*

Write each trigonometric expression as an algebraic expression in u.

(a) $\sin(\tan^{-1} u)$ **(b)** $\cos(2 \sin^{-1} u)$

Solution

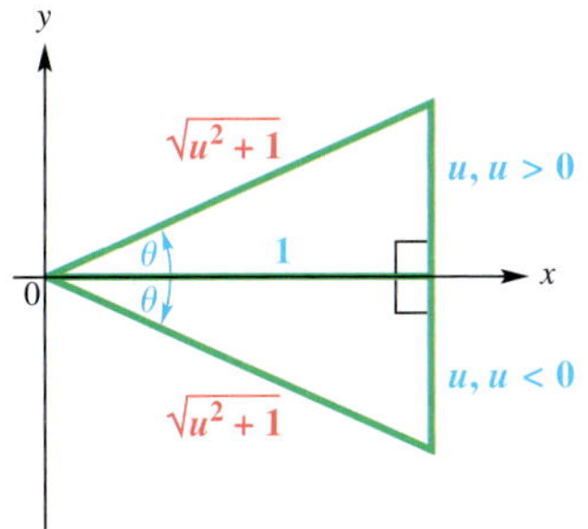

Figure 24

(a) Let $\theta = \tan^{-1} u$, so $\tan \theta = u$. Here, u may be positive or negative. Since $-\frac{\pi}{2} < \tan^{-1} u < \frac{\pi}{2}$, sketch θ in quadrants I and IV and label two triangles, as shown in Figure 24. Since sine is given by the quotient of the side opposite and the hypotenuse,

$$\sin(\tan^{-1} u) = \sin \theta = \frac{u}{\sqrt{u^2+1}} = \frac{u\sqrt{u^2+1}}{u^2+1}.$$

The result is positive when u is positive and negative when u is negative.

(b) Let $\theta = \sin^{-1} u$, so $\sin \theta = u$. To find $\cos 2\theta$, use the identity $\cos 2\theta = 1 - 2\sin^2 \theta$.

$$\cos(2 \sin^{-1} u) = \cos 2\theta = 1 - 2\sin^2 \theta = 1 - 2u^2$$

Now try Exercises 83 and 85.

EXAMPLE 8 Finding the Optimal Angle of Elevation of a Shot Put

The optimal angle of elevation θ a shot-putter should aim for to throw the greatest distance depends on the velocity v of the throw and the initial height h of the shot. See Figure 25. One model for θ that achieves this greatest distance is

$$\theta = \arcsin\left(\sqrt{\frac{v^2}{2v^2 + 64h}}\right).$$

(*Source:* Townend, M. Stewart, *Mathematics in Sport,* Chichester, Ellis Horwood Limited, 1984.)

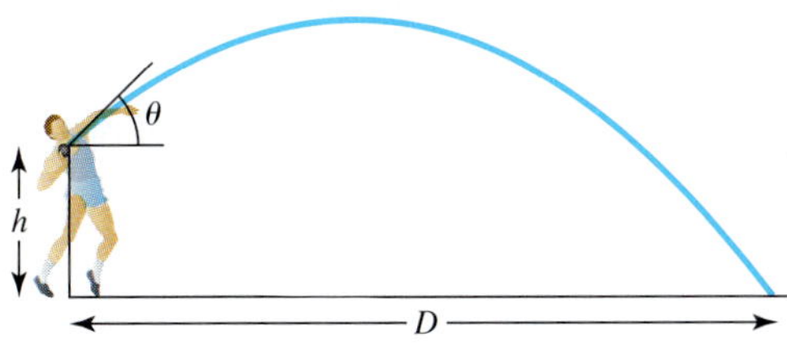

Figure 25

Suppose a shot-putter can consistently throw the steel ball with $h = 6.6$ ft and $v = 42$ ft per sec. At what angle should he throw the ball to maximize distance?

Solution To find this angle, substitute and use a calculator in degree mode.

$$\theta = \arcsin\left(\sqrt{\frac{42^2}{2(42^2) + 64(6.6)}}\right) \approx 41.9° \quad h = 6.6,\ v = 42$$

Now try Exercise 93.

7.5 Exercises

Concept Check *Complete each statement.*

1. For a function to have an inverse, it must be ______________.

2. The domain of $y = \arcsin x$ equals the ______________ of $y = \sin x$.

3. The range of $y = \cos^{-1} x$ equals the ______________ of $y = \cos x$.

4. The point $\left(\frac{\pi}{4}, 1\right)$ lies on the graph of $y = \tan x$. Therefore, the point ______________ lies on the graph of ______________.

5. If a function f has an inverse and $f(\pi) = -1$, then $f^{-1}(-1) =$ ______________.

6. How can the graph of f^{-1} be sketched if the graph of f is known?

Concept Check *In Exercises 7–10, write short answers.*

7. Consider the inverse sine function, defined by $y = \sin^{-1} x$ or $y = \arcsin x$.

(a) What is its domain? **(b)** What is its range?
(c) Is this function increasing or decreasing?
(d) Why is $\arcsin(-2)$ not defined?

8. Consider the inverse cosine function, defined by $y = \cos^{-1} x$ or $y = \arccos x$.

(a) What is its domain? **(b)** What is its range?
(c) Is this function increasing or decreasing?
(d) $\text{Arccos}\left(-\frac{1}{2}\right) = \frac{2\pi}{3}$. Why is $\arccos\left(-\frac{1}{2}\right)$ not equal to $-\frac{4\pi}{3}$?

9. Consider the inverse tangent function, defined by $y = \tan^{-1} x$ or $y = \arctan x$.

(a) What is its domain? **(b)** What is its range?
(c) Is this function increasing or decreasing?
(d) Is there any real number x for which $\arctan x$ is not defined? If so, what is it (or what are they)?

10. Give the domain and range of the three other inverse trigonometric functions, as defined in this section.

(a) inverse cosecant function **(b)** inverse secant function
(c) inverse cotangent function

11. *Concept Check* Is $\sec^{-1} a$ calculated as $\cos^{-1} \frac{1}{a}$ or as $\frac{1}{\cos^{-1} a}$?

12. *Concept Check* For positive values of a, $\cot^{-1} a$ is calculated as $\tan^{-1} \frac{1}{a}$. How is $\cot^{-1} a$ calculated for negative values of a?

Find the exact value of each real number y. Do not use a calculator. See Examples 1 and 2.

13. $y = \sin^{-1} 0$ **14.** $y = \tan^{-1} 1$ **15.** $y = \cos^{-1}(-1)$

16. $y = \arctan(-1)$ **17.** $y = \sin^{-1}(-1)$ **18.** $y = \cos^{-1} \frac{1}{2}$

19. $y = \arctan 0$ **20.** $y = \arcsin\left(-\frac{\sqrt{3}}{2}\right)$ **21.** $y = \arccos 0$

22. $y = \tan^{-1}(-1)$ **23.** $y = \sin^{-1} \frac{\sqrt{2}}{2}$ **24.** $y = \cos^{-1}\left(-\frac{1}{2}\right)$

25. $y = \arccos\left(-\frac{\sqrt{3}}{2}\right)$ **26.** $y = \arcsin\left(-\frac{\sqrt{2}}{2}\right)$ **27.** $y = \cot^{-1}(-1)$

28. $y = \sec^{-1}\left(-\sqrt{2}\right)$ **29.** $y = \csc^{-1}(-2)$ **30.** $y = \text{arccot}\left(-\sqrt{3}\right)$

31. $y = \text{arcsec}\, \frac{2\sqrt{3}}{3}$ **32.** $y = \csc^{-1} \sqrt{2}$

Give the degree measure of θ. Do not use a calculator. See Example 3.

33. $\theta = \arctan(-1)$ **34.** $\theta = \arccos\left(-\frac{1}{2}\right)$ **35.** $\theta = \arcsin\left(-\frac{\sqrt{3}}{2}\right)$

36. $\theta = \arcsin\left(-\frac{\sqrt{2}}{2}\right)$ **37.** $\theta = \cot^{-1}\left(-\frac{\sqrt{3}}{3}\right)$ **38.** $\theta = \csc^{-1}(-2)$

39. $\theta = \sec^{-1}(-2)$ **40.** $\theta = \csc^{-1}(-1)$

Use a calculator to give each value in decimal degrees. See Example 4.

41. $\theta = \sin^{-1}(-.13349122)$ **42.** $\theta = \cos^{-1}(-.13348816)$

43. $\theta = \arccos(-.39876459)$ **44.** $\theta = \arcsin .77900016$

45. $\theta = \csc^{-1} 1.9422833$ **46.** $\theta = \cot^{-1} 1.7670492$

Use a calculator to give each real number value. (Be sure the calculator is in radian mode.) See Example 4.

47. $y = \arctan 1.1111111$ **48.** $y = \arcsin .81926439$

49. $y = \cot^{-1}(-.92170128)$ **50.** $y = \sec^{-1}(-1.2871684)$

51. $y = \arcsin .92837781$ **52.** $y = \arccos .44624593$

Graph each inverse function as defined in the text.

53. $y = \cot^{-1} x$ **54.** $y = \csc^{-1} x$ **55.** $y = \sec^{-1} x$

56. $y = \operatorname{arccsc} 2x$ **57.** $y = \operatorname{arcsec} \frac{1}{2}x$

58. Explain why attempting to find $\sin^{-1} 1.003$ on your calculator will result in an error message.

59. Explain why you are able to find $\tan^{-1} 1.003$ on your calculator. Why is this situation different from the one described in Exercise 58?

Relating Concepts

For individual or collaborative investigation

(Exercises 60–62)*

60. Consider the function defined by $f(x) = 3x - 2$ and its inverse $f^{-1}(x) = \frac{x+2}{3}$. Simplify $f[f^{-1}(x)]$ and $f^{-1}[f(x)]$. What do you notice in each case? What would the graph look like in each case?

61. Use a graphing calculator to graph $y = \tan(\tan^{-1} x)$ in the standard viewing window, using radian mode. How does this compare to the graph you described in Exercise 60?

62. Use a graphing calculator to graph $y = \tan^{-1}(\tan x)$ in the standard viewing window, using radian and dot modes. Why does this graph not agree with the graph you found in Exercise 61?

*The authors wish to thank Carol Walker of Hinds Community College for making a suggestion on which these exercises are based.

Give the exact value of each expression without using a calculator. See Examples 5 and 6.

63. $\tan\left(\arccos \frac{3}{4}\right)$ **64.** $\sin\left(\arccos \frac{1}{4}\right)$ **65.** $\cos(\tan^{-1}(-2))$

66. $\sec\left(\sin^{-1}\left(-\frac{1}{5}\right)\right)$ **67.** $\sin\left(2 \tan^{-1} \frac{12}{5}\right)$ **68.** $\cos\left(2 \sin^{-1} \frac{1}{4}\right)$

69. $\cos\left(2 \arctan \frac{4}{3}\right)$ **70.** $\tan\left(2 \cos^{-1} \frac{1}{4}\right)$ **71.** $\sin\left(2 \cos^{-1} \frac{1}{5}\right)$

72. $\cos(2 \tan^{-1}(-2))$ **73.** $\sec(\sec^{-1} 2)$ **74.** $\csc(\csc^{-1} \sqrt{2})$

75. $\cos\left(\tan^{-1} \frac{5}{12} - \tan^{-1} \frac{3}{4}\right)$ **76.** $\cos\left(\sin^{-1} \frac{3}{5} + \cos^{-1} \frac{5}{13}\right)$

77. $\sin\left(\sin^{-1} \frac{1}{2} + \tan^{-1}(-3)\right)$ **78.** $\tan\left(\cos^{-1} \frac{\sqrt{3}}{2} - \sin^{-1}\left(-\frac{3}{5}\right)\right)$

Use a calculator to find each value. Give answers as real numbers.

79. $\cos(\tan^{-1} .5)$ **80.** $\sin(\cos^{-1} .25)$

81. $\tan(\arcsin .12251014)$ **82.** $\cot(\arccos .58236841)$

Write each expression as an algebraic (nontrigonometric) expression in u, u > 0. *See Example 7.*

83. $\sin(\arccos u)$ **84.** $\tan(\arccos u)$ **85.** $\cos(\arcsin u)$

86. $\cot(\arcsin u)$ **87.** $\sin\left(\sec^{-1} \frac{u}{2}\right)$ **88.** $\cos\left(\tan^{-1} \frac{3}{u}\right)$

89. $\tan\left(\sin^{-1} \frac{u}{\sqrt{u^2 + 2}}\right)$ **90.** $\sec\left(\cos^{-1} \frac{u}{\sqrt{u^2 + 5}}\right)$

91. $\sec\left(\operatorname{arccot} \frac{\sqrt{4 - u^2}}{u}\right)$ **92.** $\csc\left(\arctan \frac{\sqrt{9 - u^2}}{u}\right)$

(Modeling) Solve each problem.

93. *Angle of Elevation of a Shot Put* Refer to Example 8.

(a) What is the optimal angle when $h = 0$?

(b) Fix h at 6 ft and regard θ as a function of v. As v gets larger and larger, the graph approaches an asymptote. Find the equation of that asymptote.

94. *Angle of Elevation of a Plane* Suppose an airplane flying faster than sound goes directly over you. Assume that the plane is flying at a constant altitude. At the instant you feel the sonic boom from the plane, the angle of elevation to the plane is given by

$$\theta = 2 \arcsin \frac{1}{m},$$

where m is the *Mach number* of the plane's speed. (The Mach number is the ratio of the speed of the plane and the speed of sound.) Find θ to the nearest degree for each value of m.

(a) $m = 1.2$ **(b)** $m = 1.5$ **(c)** $m = 2$ **(d)** $m = 2.5$

95. *Observation of a Painting* A painting 1 m high and 3 m from the floor will cut off an angle θ to an observer, where

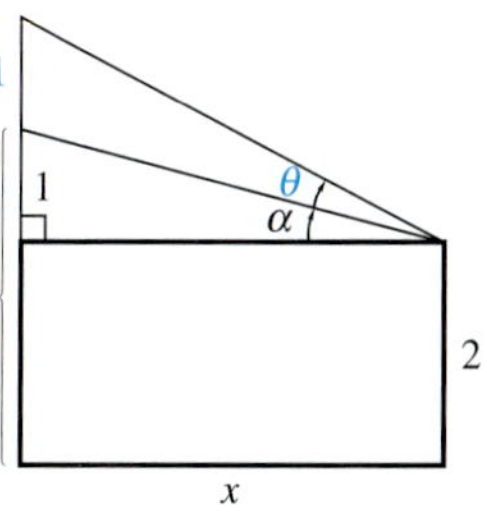

$$\theta = \tan^{-1}\left(\frac{x}{x^2 + 2}\right).$$

Assume that the observer is x meters from the wall where the painting is displayed and that the eyes of the observer are 2 m above the ground. (See the figure.) Find the value of θ for the following values of x. Round to the nearest degree.

(a) 1 **(b)** 2 **(c)** 3

(d) Derive the formula given above. (*Hint:* Use the identity for $\tan(\theta + \alpha)$. Use right triangles.)

(e) Graph the function for θ with a graphing calculator, and determine the distance that maximizes the angle.

(f) The idea in part (e) was first investigated in 1471 by the astronomer Regiomontanus. (*Source:* Maor, E., *Trigonometric Delights,* Princeton University Press, 1998.) If the bottom of the picture is a meters above eye level and the top of the picture is b meters above eye level, then the optimum value of x is $\sqrt{ab}$ meters. Use this result to find the exact answer to part (e).

96. *Landscaping Formula* A shrub is planted in a 100-ft-wide space between buildings measuring 75 ft and 150 ft tall. The location of the shrub determines how much sun it receives each day. Show that if θ is the angle in the figure and x is the distance of the shrub from the taller building, then the value of θ (in radians) is given by

$$\theta = \pi - \arctan\left(\frac{75}{100 - x}\right) - \arctan\left(\frac{150}{x}\right).$$

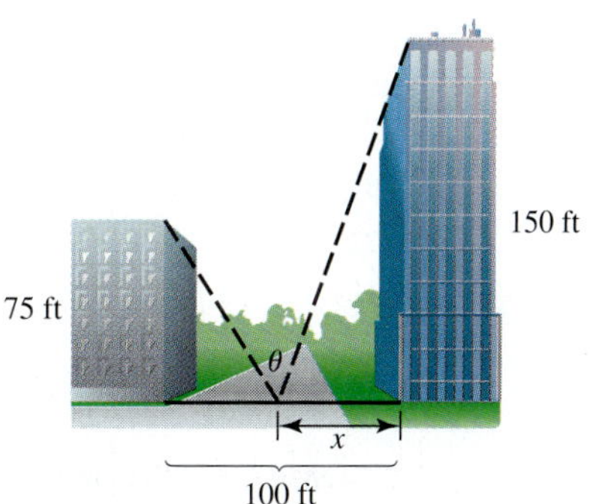

97. *Communications Satellite Coverage* The figure shows a stationary communications satellite positioned 20,000 mi above the equator. What percent of the equator can be seen from the satellite? The diameter of Earth is 7927 mi at the equator.

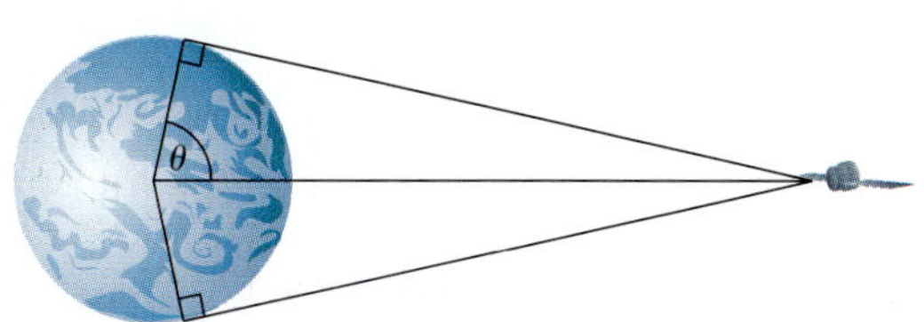

7.6 Trigonometric Equations

Solving by Linear Methods ▪ Solving by Factoring ▪ Solving by Quadratic Methods ▪ Solving by Using Trigonometric Identities ▪ Equations with Half-Angles ▪ Equations with Multiple Angles ▪ Applications

Looking Ahead to Calculus

There are many instances in calculus where it is necessary to solve trigonometric equations. Examples include solving related-rates problems and optimization problems.

Earlier in this chapter, we studied trigonometric equations that were identities. We now consider trigonometric equations that are *conditional;* that is, equations that are satisfied by some values but not others.

Solving by Linear Methods

Conditional equations with trigonometric (or circular) functions can usually be solved using algebraic methods and trigonometric identities.

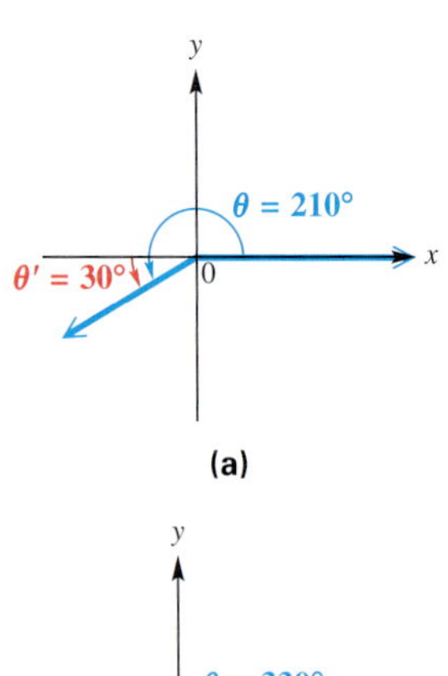

(a)

(b)

Figure 26

EXAMPLE 1 Solving a Trigonometric Equation by Linear Methods

Solve $2 \sin \theta - 1 = 0$ over the interval $[0°, 360°)$.

Solution Because $\sin \theta$ is the first power of a trigonometric function, we use the same method as we would to solve the linear equation $2x + 1 = 0$.

$$2 \sin \theta + 1 = 0$$

$$2 \sin \theta = -1 \quad \text{Subtract 1. (Section 1.1)}$$

$$\sin \theta = -\frac{1}{2} \quad \text{Divide by 2.}$$

To find values of θ that satisfy $\sin \theta = -\frac{1}{2}$, we observe that θ must be in either quadrant III or IV since the sine function is negative only in these two quadrants. Furthermore, the reference angle must be 30° since $\sin 30° = \frac{1}{2}$. The graphs in Figure 26 show the two possible values of θ, 210° and 330°. The solution set is $\{210°, 330°\}$.

Alternatively, we could determine the solutions by referring to Figure 11 in Section 6.2 on page 546.

Now try Exercise 11.

Solving by Factoring

EXAMPLE 2 Solving a Trigonometric Equation by Factoring

Solve $\sin x \tan x = \sin x$ over the interval $[0°, 360°)$.

Solution

$$\sin x \tan x = \sin x$$

$$\sin x \tan x - \sin x = 0 \quad \text{Subtract } \sin x.$$

$$\sin x(\tan x - 1) = 0 \quad \text{Factor. (Section R.4)}$$

$$\sin x = 0 \quad \text{or} \quad \tan x - 1 = 0 \quad \text{Zero-factor property (Section 1.4)}$$

$$\tan x = 1$$

$$x = 0° \quad \text{or} \quad x = 180° \qquad x = 45° \quad \text{or} \quad x = 225°$$

The solution set is $\{0°, 45°, 180°, 225°\}$.

Now try Exercise 31.

CAUTION There are four solutions in Example 2. Trying to solve the equation by dividing each side by $\sin x$ would lead to just $\tan x = 1$, which would give $x = 45°$ or $x = 225°$. The other two solutions would not appear. The missing solutions are the ones that make the divisor, $\sin x$, equal 0. For this reason, we avoid dividing by a variable expression.

Solving by Quadratic Methods In Section 1.6, we saw that an equation in the form $au^2 + bu + c = 0$, where u is an algebraic expression, is solved by quadratic methods. The expression u may also be a trigonometric function, as in the equation $\tan^2 x + \tan x - 2 = 0$.

EXAMPLE 3 Solving a Trigonometric Equation by Factoring

Solve $\tan^2 x + \tan x - 2 = 0$ over the interval $[0, 2\pi)$.

Solution This equation is quadratic in form and can be solved by factoring.

$$\tan^2 x + \tan x - 2 = 0$$

$$(\tan x - 1)(\tan x + 2) = 0 \quad \text{Factor.}$$

$$\tan x - 1 = 0 \quad \text{or} \quad \tan x + 2 = 0 \quad \text{Zero-factor property}$$

$$\tan x = 1 \quad \text{or} \quad \tan x = -2$$

The solutions for $\tan x = 1$ over the interval $[0, 2\pi)$ are $x = \frac{\pi}{4}$ and $x = \frac{5\pi}{4}$.

To solve $\tan x = -2$ over that interval, we use a scientific calculator set in *radian* mode. We find that $\tan^{-1}(-2) \approx -1.1071487$. This is a quadrant IV number, based on the range of the inverse tangent function. (Refer to Figure 11 in Section 6.2 on page 546.) However, since we want solutions over the interval $[0, 2\pi)$, we must first add π to -1.1071487, and then add 2π.

$$x \approx -1.1071487 + \pi \approx 2.0344439$$

$$x \approx -1.1071487 + 2\pi \approx 5.1760366$$

The solutions over the required interval form the solution set

$$\left\{ \underbrace{\frac{\pi}{4}, \quad \frac{5\pi}{4},}_{\text{Exact values}} \quad \underbrace{2.0, \quad 5.2}_{\text{Approximate values to the nearest tenth}} \right\}.$$

Now try Exercise 21.

EXAMPLE 4 Solving a Trigonometric Equation Using the Quadratic Formula

Find all solutions of $\cot x(\cot x + 3) = 1$.

Solution We multiply the factors on the left and subtract 1 to get the equation in standard quadratic form.

$$\cot^2 x + 3 \cot x - 1 = 0 \quad \text{(Section 1.4)}$$

Since this equation cannot be solved by factoring, we use the quadratic formula, with $a = 1$, $b = 3$, $c = -1$, and $\cot x$ as the variable.

$$\cot x = \frac{-3 \pm \sqrt{9+4}}{2} = \frac{-3 \pm \sqrt{13}}{2}$$ Quadratic formula with $a = 1, b = 3, c = -1$ **(Section 1.4)**

$\cot x \approx -3.302775638$ or $\cot x \approx .3027756377$ Use a calculator.

We cannot find inverse cotangent values directly on a calculator, so we use the fact that $\cot x = \frac{1}{\tan x}$, and take reciprocals to get

$$\tan x \approx -.3027756377 \quad \text{or} \quad \tan x \approx 3.302775638$$
$$x \approx -.2940013018 \quad \text{or} \quad x \approx 1.276795025.$$

To find *all* solutions, we add integer multiples of the period of the tangent function, which is π, to each solution found above. Thus, all solutions of the equation are written as

$-.2940013018 + n\pi$ and $1.276795025 + n\pi$, where n is any integer.*

Now try Exercise 43.

Solving by Using Trigonometric Identities Recall that squaring both sides of an equation, such as $\sqrt{x+4} = x + 2$, will yield all solutions but may also give extraneous values. (In this equation, 0 is a solution, while -3 is extraneous. Verify this.) The same situation may occur when trigonometric equations are solved in this manner.

EXAMPLE 5 Solving a Trigonometric Equation by Squaring

Solve $\tan x + \sqrt{3} = \sec x$ over the interval $[0, 2\pi)$.

Solution Since the tangent and secant functions are related by the identity $1 + \tan^2 x = \sec^2 x$, square both sides and express $\sec^2 x$ in terms of $\tan^2 x$.

$$\tan x + \sqrt{3} = \sec x$$

$$\tan^2 x + 2\sqrt{3}\tan x + 3 = \sec^2 x$$ $(x+y)^2 = x^2 + 2xy + y^2$ **(Section R.3)**

$$\tan^2 x + 2\sqrt{3}\tan x + 3 = 1 + \tan^2 x$$ Pythagorean identity **(Section 7.1)**

$$2\sqrt{3}\tan x = -2$$ Subtract $3 + \tan^2 x$.

$$\tan x = -\frac{1}{\sqrt{3}} = -\frac{\sqrt{3}}{3}$$ Divide by $2\sqrt{3}$; rationalize the denominator. **(Section R.7)**

The possible solutions are $\frac{5\pi}{6}$ and $\frac{11\pi}{6}$. Now check them. Try $\frac{5\pi}{6}$ first.

Left side: $\tan x + \sqrt{3} = \tan\frac{5\pi}{6} + \sqrt{3} = -\frac{\sqrt{3}}{3} + \sqrt{3} = \frac{2\sqrt{3}}{3}$ **(Section R.7)**

Right side: $\sec x = \sec\frac{5\pi}{6} = -\frac{2\sqrt{3}}{3}$ ← Not equal

*We usually give solutions of equations as solution sets, except when we ask for all solutions of a trigonometric equation.

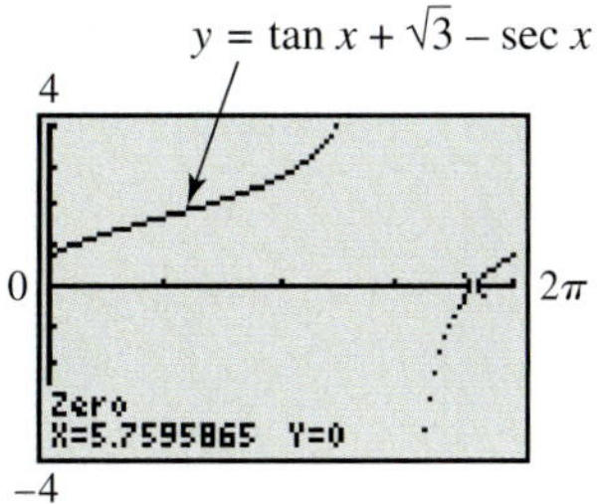

Dot mode; radian mode

The graph shows that on the interval $[0, 2\pi)$, the only x-intercept of the graph of $y = \tan x + \sqrt{3} - \sec x$ is 5.7595865, which is an approximation for $\frac{11\pi}{6}$, the solution found in Example 5.

The check shows that $\frac{5\pi}{6}$ is not a solution. Now check $\frac{11\pi}{6}$.

$$\text{Left side:} \quad \tan\frac{11\pi}{6} + \sqrt{3} = -\frac{\sqrt{3}}{3} + \sqrt{3} = \frac{2\sqrt{3}}{3}$$

$$\text{Right side:} \quad \sec\frac{11\pi}{6} = \frac{2\sqrt{3}}{3} \quad \leftarrow \text{Equal}$$

This solution satisfies the equation, so $\left\{\frac{11\pi}{6}\right\}$ is the solution set.

Now try Exercise 41.

The methods for solving trigonometric equations illustrated in the examples can be summarized as follows.

Solving a Trigonometric Equation

1. Decide whether the equation is linear or quadratic in form, so you can determine the solution method.
2. If only one trigonometric function is present, first solve the equation for that function.
3. If more than one trigonometric function is present, rearrange the equation so that one side equals 0. Then try to factor and set each factor equal to 0 to solve.
4. If the equation is quadratic in form, but not factorable, use the quadratic formula. Check that solutions are in the desired interval.
5. Try using identities to change the form of the equation. It may be helpful to square both sides of the equation first. If this is done, check for extraneous solutions.

Some trigonometric equations involve functions of half-angles or multiple angles.

Equations with Half-Angles

EXAMPLE 6 Solving an Equation Using a Half-Angle Identity

Solve $2 \sin \dfrac{x}{2} = 1$

(a) over the interval $[0, 2\pi)$, and **(b)** give all solutions.

Solution

(a) Write the interval $[0, 2\pi)$ as the inequality

$$0 \le x < 2\pi.$$

The corresponding interval for $\frac{x}{2}$ is

$$0 \le \frac{x}{2} < \pi. \quad \text{Divide by 2. (Section 1.7)}$$

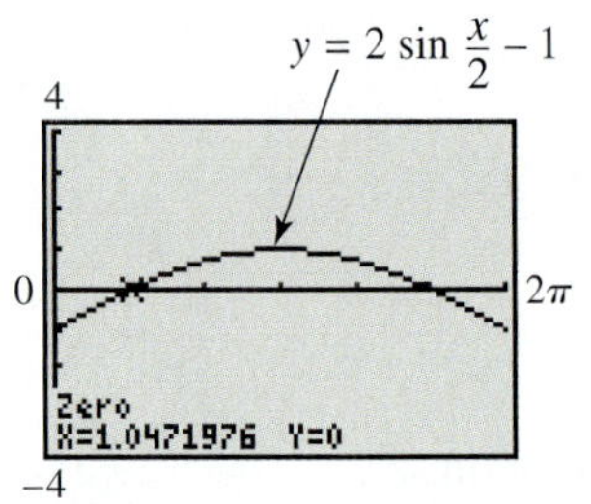

The x-intercepts are the solutions found in Example 6. Using Xscl $= \frac{\pi}{3}$ makes it possible to support the exact solutions by counting the tick marks from 0 on the graph.

To find all values of $\frac{x}{2}$ over the interval $[0, \pi)$ that satisfy the given equation, first solve for $\sin \frac{x}{2}$.

$$2 \sin \frac{x}{2} = 1$$

$$\sin \frac{x}{2} = \frac{1}{2} \quad \text{Divide by 2.}$$

The two numbers over the interval $[0, \pi)$ with sine value $\frac{1}{2}$ are $\frac{\pi}{6}$ and $\frac{5\pi}{6}$, so

$$\frac{x}{2} = \frac{\pi}{6} \quad \text{or} \quad \frac{x}{2} = \frac{5\pi}{6}$$

$$x = \frac{\pi}{3} \quad \text{or} \quad x = \frac{5\pi}{3}. \quad \text{Multiply by 2.}$$

The solution set over the given interval is $\left\{\frac{\pi}{3}, \frac{5\pi}{3}\right\}$.

(b) Since this is a sine function with period 4π, all solutions are given by the expressions

$$\frac{\pi}{3} + 4n\pi \quad \text{and} \quad \frac{5\pi}{3} + 4n\pi, \qquad \text{where } n \text{ is any integer.}$$

Now try Exercise 63.

Equations with Multiple Angles

EXAMPLE 7 Solving an Equation with a Double Angle

Solve $\cos 2x = \cos x$ over the interval $[0, 2\pi)$.

Solution First change $\cos 2x$ to a trigonometric function of x. Use the identity $\cos 2x = 2\cos^2 x - 1$ so the equation involves only $\cos x$. Then factor.

$$\cos 2x = \cos x$$

$$2\cos^2 x - 1 = \cos x \quad \text{Substitute; double-angle identity (Section 7.4)}$$

$$2\cos^2 x - \cos x - 1 = 0 \quad \text{Subtract } \cos x.$$

$$(2\cos x + 1)(\cos x - 1) = 0 \quad \text{Factor.}$$

$$2\cos x + 1 = 0 \quad \text{or} \quad \cos x - 1 = 0 \quad \text{Zero-factor property}$$

$$\cos x = -\frac{1}{2} \quad \text{or} \quad \cos x = 1$$

Over the required interval,

$$x = \frac{2\pi}{3} \quad \text{or} \quad x = \frac{4\pi}{3} \quad \text{or} \quad x = 0.$$

The solution set is $\left\{0, \frac{2\pi}{3}, \frac{4\pi}{3}\right\}$.

Now try Exercise 65.

CAUTION In the solution of Example 7, $\cos 2x$ cannot be changed to $\cos x$ by dividing by 2 since 2 is not a factor of $\cos 2x$.

$$\frac{\cos 2x}{2} \neq \cos x$$

The only way to change $\cos 2x$ to a trigonometric function of x is by using one of the identities for $\cos 2x$.

EXAMPLE 8 Solving an Equation Using a Multiple-Angle Identity

Solve $4 \sin \theta \cos \theta = \sqrt{3}$ over the interval $[0°, 360°)$.

Solution The identity $2 \sin \theta \cos \theta = \sin 2\theta$ is useful here.

$$4 \sin \theta \cos \theta = \sqrt{3}$$

$$2(2 \sin \theta \cos \theta) = \sqrt{3} \quad 4 = 2 \cdot 2$$

$$2 \sin 2\theta = \sqrt{3} \quad 2 \sin \theta \cos \theta = \sin 2\theta \text{ (Section 7.4)}$$

$$\sin 2\theta = \frac{\sqrt{3}}{2} \quad \text{Divide by 2.}$$

From the given interval $0° \leq \theta < 360°$, the interval for 2θ is $0° \leq 2\theta < 720°$. List all solutions over this interval.

$$2\theta = 60°, 120°, 420°, 480°$$

or

$$\theta = 30°, 60°, 210°, 240° \quad \text{Divide by 2.}$$

The final two solutions for 2θ were found by adding 360° to 60° and 120°, respectively, giving the solution set $\{30°, 60°, 210°, 240°\}$.

Now try Exercise 83.

Applications

Music is closely related to mathematics.

EXAMPLE 9 Describing a Musical Tone from a Graph

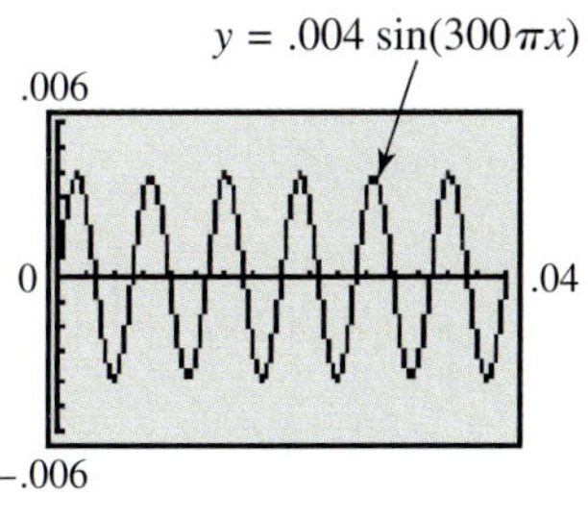

Figure 27

A basic component of music is a pure tone. The graph in Figure 27 models the sinusoidal pressure $y = P$ in pounds per square foot from a pure tone at time $x = t$ in seconds.

(a) The frequency of a pure tone is often measured in hertz. One hertz is equal to one cycle per second and is abbreviated Hz. What is the frequency f in hertz of the pure tone shown in the graph?

(b) The time for the tone to produce one complete cycle is called the *period.* Approximate the period T in seconds of the pure tone.

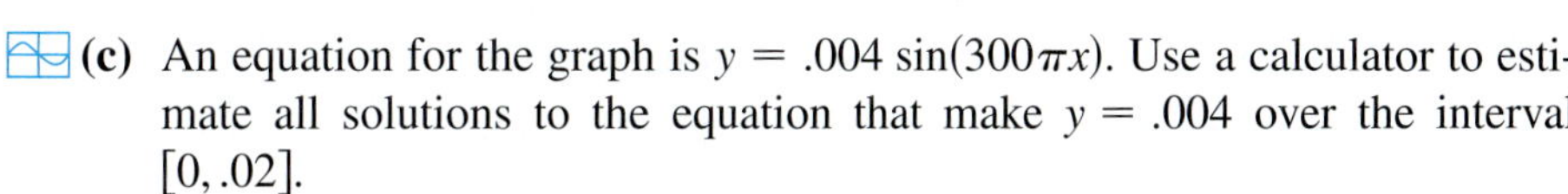

(c) An equation for the graph is $y = .004 \sin(300\pi x)$. Use a calculator to estimate all solutions to the equation that make $y = .004$ over the interval $[0, .02]$.

Solution

(a) From the graph in Figure 27 on the previous page, we see that there are 6 cycles in .04 sec. This is equivalent to $\frac{6}{.04} = 150$ cycles per sec. The pure tone has a frequency of $f = 150$ Hz.

(b) Six periods cover a time of .04 sec. One period would be equal to $T = \frac{.04}{6} = \frac{1}{150}$ or $.00\overline{6}$ sec.

(c) If we reproduce the graph in Figure 27 on a calculator as Y_1 and also graph a second function as $Y_2 = .004$, we can determine that the approximate values of x at the points of intersection of the graphs over the interval $[0, .02]$ are

$$.0017, \quad .0083, \quad \text{and} \quad .015.$$

The first value is shown in Figure 28.

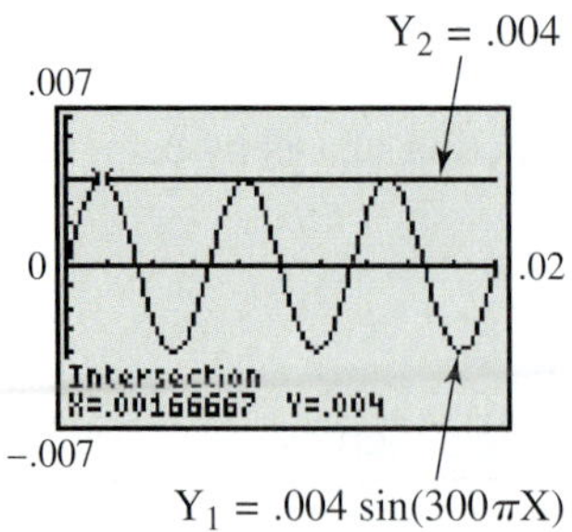

Figure 28

Now try Exercise 87.

A piano string can vibrate at more than one frequency when it is struck. It produces a complex wave that can mathematically be modeled by a sum of several pure tones. If a piano key with a frequency of f_1 is played, then the corresponding string will not only vibrate at f_1 but it will also vibrate at the higher frequencies of $2f_1, 3f_1, 4f_1, \ldots, nf_1, \ldots$. f_1 is called the **fundamental frequency** of the string, and higher frequencies are called the **upper harmonics.** The human ear will hear the sum of these frequencies as one complex tone. (*Source:* Roederer, J., *Introduction to the Physics and Psychophysics of Music,* Second Edition, Springer-Verlag, 1975.)

EXAMPLE 10 Analyzing Pressures of Upper Harmonics

Suppose that the A key above middle C is played. Its fundamental frequency is $f_1 = 440$ Hz, and its associated pressure is expressed as

$$P_1 = .002 \sin(880\pi t).$$

The string will also vibrate at

$$f_2 = 880, \quad f_3 = 1320, \quad f_4 = 1760, \quad f_5 = 2200, \ldots \text{Hz}.$$

The corresponding pressures of these upper harmonics are

$$P_2 = \frac{.002}{2}\sin(1760\pi t), \qquad P_3 = \frac{.002}{3}\sin(2640\pi t),$$

$$P_4 = \frac{.002}{4}\sin(3520\pi t), \qquad \text{and} \qquad P_5 = \frac{.002}{5}\sin(4400\pi t).$$

The graph of

$$P = P_1 + P_2 + P_3 + P_4 + P_5,$$

shown in Figure 29, is "saw-toothed."

(a) What is the maximum value of P?

(b) At what values of x does this maximum occur over the interval $[0, .01]$?

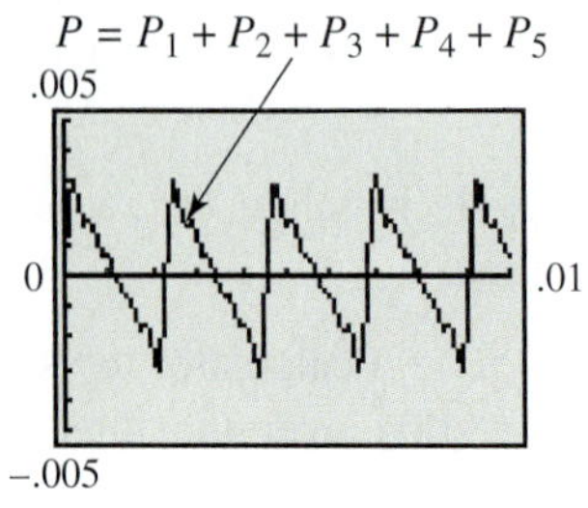

Figure 29

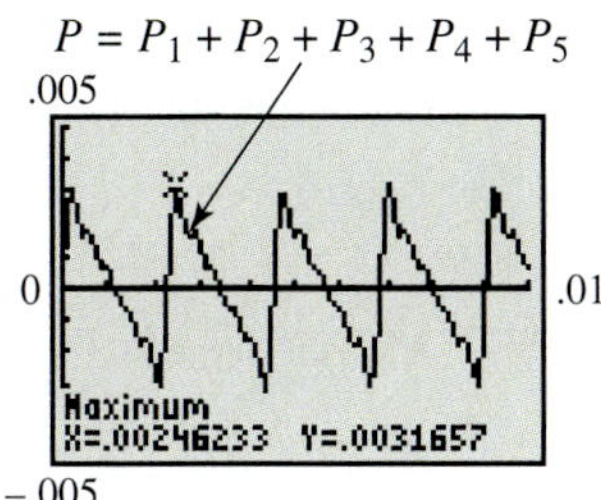

Figure 30

Solution

(a) A graphing calculator shows that the maximum value of P is approximately .00317. See Figure 30.

(b) The maximum occurs at $x \approx .000188, .00246, .00474, .00701$, and $.00928$. Figure 30 shows how the second value is found; the others are found similarly.

Now try Exercise 89.

7.6 Exercises

Concept Check *Refer to the summary box on solving a trigonometric equation. Decide on the appropriate technique to begin the solution of each equation. Do not solve the equation.*

1. $2 \cot x + 1 = -1$

2. $\sin x + 2 = 3$

3. $5 \sec^2 x = 6 \sec x$

4. $2 \cos^2 x - \cos x = 1$

5. $9 \sin^2 x - 5 \sin x = 1$

6. $\tan^2 x - 4 \tan x + 2 = 0$

7. $\tan x - \cot x = 0$

8. $\cos^2 x = \sin^2 x + 1$

Concept Check *Answer each question.*

9. Suppose you are solving a trigonometric equation for solutions over the interval $[0, 2\pi)$, and your work leads to $2x = \frac{2\pi}{3}, 2\pi, \frac{8\pi}{3}$. What are the corresponding values of x?

10. Suppose you are solving a trigonometric equation for solutions over the interval $[0°, 360°)$, and your work leads to $\frac{1}{3}\theta = 45°, 60°, 75°, 90°$. What are the corresponding values of θ?

Solve each equation for exact solutions over the interval $[0, 2\pi)$. See Examples 1–4.

11. $2 \cot x + 1 = -1$

12. $\sin x + 2 = 3$

13. $2 \sin x + 3 = 4$

14. $2 \sec x + 1 = \sec x + 3$

15. $\tan^2 x + 3 = 0$

16. $\sec^2 x + 2 = -1$

17. $(\cot x - 1)(\sqrt{3} \cot x + 1) = 0$

18. $(\csc x + 2)(\csc x - \sqrt{2}) = 0$

19. $\cos^2 x + 2 \cos x + 1 = 0$

20. $2 \cos^2 x - \sqrt{3} \cos x = 0$

21. $-2 \sin^2 x = 3 \sin x + 1$

22. $2 \cos^2 x - \cos x = 1$

Solve each equation for exact solutions over the interval $[0°, 360°)$. See Examples 2–5.

23. $(\cot \theta - \sqrt{3})(2 \sin \theta + \sqrt{3}) = 0$

24. $(\tan \theta - 1)(\cos \theta - 1) = 0$

25. $2 \sin \theta - 1 = \csc \theta$

26. $\tan \theta + 1 = \sqrt{3} + \sqrt{3} \cot \theta$

27. $\tan \theta - \cot \theta = 0$

28. $\cos^2 \theta = \sin^2 \theta + 1$

29. $\csc^2 \theta - 2 \cot \theta = 0$

30. $\sin^2 \theta \cos \theta = \cos \theta$

31. $2 \tan^2 \theta \sin \theta - \tan^2 \theta = 0$

32. $\sin^2 \theta \cos^2 \theta = 0$

33. $\sec^2 \theta \tan \theta = 2 \tan \theta$

34. $\cos^2 \theta - \sin^2 \theta = 0$

35. $9 \sin^2 \theta - 6 \sin \theta = 1$

36. $4 \cos^2 \theta + 4 \cos \theta = 1$

37. $\tan^2 \theta + 4 \tan \theta + 2 = 0$

38. $3 \cot^2 \theta - 3 \cot \theta - 1 = 0$

39. $\sin^2 \theta - 2 \sin \theta + 3 = 0$

40. $2 \cos^2 \theta + 2 \cos \theta - 1 = 0$

41. $\cot \theta + 2 \csc \theta = 3$

42. $2 \sin \theta = 1 - 2 \cos \theta$

Determine all solutions of each equation in radians (for x) or degrees (for θ) to the nearest tenth as appropriate. See Example 4.

43. $3 \sin^2 x - \sin x - 1 = 0$

44. $2 \cos^2 x + \cos x = 1$

45. $4 \cos^2 x - 1 = 0$

46. $2 \cos^2 x + 5 \cos x + 2 = 0$

47. $5 \sec^2 \theta = 6 \sec \theta$

48. $3 \sin^2 \theta - \sin \theta = 2$

49. $\dfrac{2 \tan \theta}{3 - \tan^2 \theta} = 1$

50. $\sec^2 \theta = 2 \tan \theta + 4$

The following equations cannot be solved by algebraic methods. Use a graphing calculator to find all solutions over the interval $[0, 2\pi)$. Express solutions to four decimal places.

51. $x^2 + \sin x - x^3 - \cos x = 0$

52. $x^3 - \cos^2 x = \dfrac{1}{2}x - 1$

53. Explain what is wrong with the following solution.
Solve $\tan 2\theta = 2$ over the interval $[0, 2\pi)$.

$$\tan 2\theta = 2$$

$$\frac{\tan 2\theta}{2} = \frac{2}{2}$$

$$\tan \theta = 1$$

$$\theta = \frac{\pi}{4} \quad \text{or} \quad \theta = \frac{5\pi}{4}$$

The solutions are $\frac{\pi}{4}$ and $\frac{5\pi}{4}$.

54. The equation $\cot \frac{x}{2} - \csc \frac{x}{2} - 1 = 0$ has no solution over the interval $[0, 2\pi)$. Using this information, what can be said about the graph of

$$y = \cot \frac{x}{2} - \csc \frac{x}{2} - 1$$

over this interval? Confirm your answer by actually graphing the function over the interval.

Solve each equation for exact solutions over the interval $[0, 2\pi)$. See Examples 6–8.

55. $\cos 2x = \dfrac{\sqrt{3}}{2}$

56. $\cos 2x = -\dfrac{1}{2}$

57. $\sin 3x = -1$

58. $\sin 3x = 0$

59. $3 \tan 3x = \sqrt{3}$

60. $\cot 3x = \sqrt{3}$

61. $\sqrt{2} \cos 2x = -1$

62. $2\sqrt{3} \sin 2x = \sqrt{3}$

63. $\sin \dfrac{x}{2} = \sqrt{2} - \sin \dfrac{x}{2}$

64. $\tan 4x = 0$

65. $\sin x = \sin 2x$

66. $\cos 2x - \cos x = 0$

67. $8 \sec^2 \dfrac{x}{2} = 4$

68. $\sin^2 \dfrac{x}{2} - 2 = 0$

69. $\sin \dfrac{x}{2} = \cos \dfrac{x}{2}$

70. $\sec \dfrac{x}{2} = \cos \dfrac{x}{2}$

Solve each equation in Exercises 71–78 for exact solutions over the interval $[0°, 360°)$. *In Exercises 79–86, give all exact solutions. See Examples 6–8.*

71. $\sqrt{2} \sin 3\theta - 1 = 0$ **72.** $-2 \cos 2\theta = \sqrt{3}$ **73.** $\cos \frac{\theta}{2} = 1$

74. $\sin \frac{\theta}{2} = 1$ **75.** $2\sqrt{3} \sin \frac{\theta}{2} = 3$ **76.** $2\sqrt{3} \cos \frac{\theta}{2} = -3$

77. $2 \sin \theta = 2 \cos 2\theta$ **78.** $\cos \theta - 1 = \cos 2\theta$ **79.** $1 - \sin \theta = \cos 2\theta$

80. $\sin 2\theta = 2 \cos^2 \theta$ **81.** $\csc^2 \frac{\theta}{2} = 2 \sec \theta$ **82.** $\cos \theta = \sin^2 \frac{\theta}{2}$

83. $2 - \sin 2\theta = 4 \sin 2\theta$ **84.** $4 \cos 2\theta = 8 \sin \theta \cos \theta$ **85.** $2 \cos^2 2\theta = 1 - \cos 2\theta$

86. $\sin \theta - \sin 2\theta = 0$

(Modeling) Solve each problem. See Examples 9 and 10.

87. ***Pressure on the Eardrum*** No musical instrument can generate a true pure tone. A pure tone has a unique, constant frequency and amplitude that sounds rather dull and uninteresting. The pressures caused by pure tones on the eardrum are sinusoidal. The change in pressure P in pounds per square foot on a person's eardrum from a pure tone at time t in seconds can be modeled using the equation

$$P = A \sin(2\pi f t + \phi),$$

where f is the frequency in cycles per second, and ϕ is the phase angle. When P is positive, there is an increase in pressure and the eardrum is pushed inward; when P is negative, there is a decrease in pressure and the eardrum is pushed outward. (*Source:* Roederer, J., *Introduction to the Physics and Psychophysics of Music,* Second Edition, Springer-Verlag, 1975.) A graph of the tone middle C is shown in the figure.

(a) Determine algebraically the values of t for which $P = 0$ over $[0, .005]$.

(b) From the graph and your answer in part (a), determine the interval for which $P \leq 0$ over $[0, .005]$.

(c) Would an eardrum hearing this tone be vibrating outward or inward when $P < 0$?

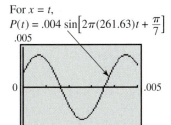

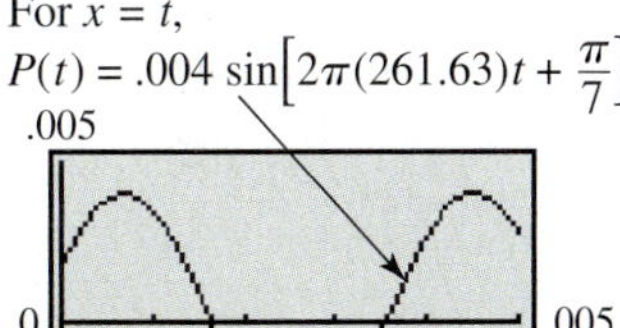

88. ***Hearing Beats in Music*** Musicians sometimes tune instruments by playing the same tone on two different instruments and listening for a phenomenon known as *beats*. Beats occur when two tones vary in frequency by only a few hertz. When the two instruments are in tune, the beats disappear. The ear hears beats because the pressure slowly rises and falls as a result of this slight variation in the frequency. This phenomenon can be seen using a graphing calculator. (*Source:* Pierce, J., *The Science of Musical Sound,* Scientific American Books, 1992.)

(a) Consider two tones with frequencies of 220 and 223 Hz and pressures $P_1 = .005 \sin 440\pi t$ and $P_2 = .005 \sin 446\pi t$, respectively. Graph the pressure $P = P_1 + P_2$ felt by an eardrum over the 1-sec interval $[.15, 1.15]$. How many beats are there in 1 sec?

(b) Repeat part (a) with frequencies of 220 and 216 Hz.

(c) Determine a simple way to find the number of beats per second if the frequency of each tone is given.

89. *Pressure of a Plucked String* If a string with a fundamental frequency of 110 Hz is plucked in the middle, it will vibrate at the odd harmonics of 110, 330, 550, . . . Hz but not at the even harmonics of 220, 440, 660, . . . Hz. The resulting pressure P caused by the string can be modeled by the equation

$$P = .003 \sin 220\pi t + \frac{.003}{3} \sin 660\pi t + \frac{.003}{5} \sin 1100\pi t + \frac{.003}{7} \sin 1540\pi t.$$

(*Source:* Benade, A., *Fundamentals of Musical Acoustics,* Dover Publications, 1990; Roederer, J., *Introduction to the Physics and Psychophysics of Music,* Second Edition, Springer-Verlag, 1975.)

(a) Graph P in the window $[0, .03]$ by $[-.005, .005]$.
(b) Use the graph to describe the shape of the sound wave that is produced.
(c) See Exercise 87. At lower frequencies, the inner ear will hear a tone only when the eardrum is moving outward. Determine the times over the interval $[0, .03]$ when this will occur.

90. *Hearing Difference Tones* Small speakers like those found in older radios and telephones often cannot vibrate slower than 200 Hz—yet 35 keys on a piano have frequencies below 200 Hz. When a musical instrument creates a tone of 110 Hz, it also creates tones at 220, 330, 440, 550, 660, . . . Hz. A small speaker cannot reproduce the 110-Hz vibration but it can reproduce the higher frequencies, which are called the upper harmonics. The low tones can still be heard because the speaker produces *difference tones* of the upper harmonics. The difference between consecutive frequencies is 110 Hz, and this difference tone will be heard by a listener. We can model this phenomenon using a graphing calculator. (*Source:* Benade, A., *Fundamentals of Musical Acoustics,* Dover Publications, 1990.)

(a) In the window $[0, .03]$ by $[-1, 1]$, graph the upper harmonics represented by the pressure

$$P = \frac{1}{2} \sin[2\pi(220)t] + \frac{1}{3} \sin[2\pi(330)t] + \frac{1}{4} \sin[2\pi(440)t].$$

(b) Estimate all t-coordinates where P is maximum.
(c) What does a person hear in addition to the frequencies of 220, 330, and 440 Hz?
(d) Graph the pressure produced by a speaker that can vibrate at 110 Hz and above.

91. *Daylight Hours in New Orleans* The seasonal variation in length of daylight can be modeled by a sine function. For example, the daily number of hours of daylight in New Orleans is given by

$$h = \frac{35}{3} + \frac{7}{3} \sin \frac{2\pi x}{365},$$

where x is the number of days after March 21 (disregarding leap year). (*Source:* Bushaw, Donald et al., *A Sourcebook of Applications of School Mathematics.* Copyright © 1980 by The Mathematical Association of America.)

(a) On what date will there be about 14 hr of daylight?
(b) What date has the least number of hours of daylight?
(c) When will there be about 10 hr of daylight?

(Modeling) Alternating Electric Current The study of alternating electric current requires the solutions of equations of the form

$$i = I_{\max} \sin 2\pi f t,$$

for time t in seconds, where i is instantaneous current in amperes, $I_{\max}$ is maximum current in amperes, and f is the number of cycles per second. (*Source:* Hannon, R. H., *Basic*

Technical Mathematics with Calculus, W. B. Saunders Company, 1978.) *Find the smallest positive value of t, given the following data.*

92. $i = 40, I_{max} = 100, f = 60$

93. $i = 50, I_{max} = 100, f = 120$

94. $i = I_{max}, f = 60$

95. $i = \frac{1}{2}I_{max}, f = 60$

(Modeling) *Solve each problem.*

96. ***Accident Reconstruction*** The model

$$.342D \cos \theta + h \cos^2 \theta = \frac{16D^2}{V_0^2}$$

is used to reconstruct accidents in which a vehicle vaults into the air after hitting an obstruction. V_0 is velocity in feet per second of the vehicle when it hits, D is distance (in feet) from the obstruction to the landing point, and h is the difference in height (in feet) between landing point and takeoff point. Angle θ is the takeoff angle, the angle between the horizontal and the path of the vehicle. Find θ to the nearest degree if $V_0 = 60$, $D = 80$, and $h = 2$.

97. ***Electromotive Force*** In an electric circuit, let

$$V = \cos 2\pi t$$

model the electromotive force in volts at t seconds. Find the smallest positive value of t where $0 \le t \le \frac{1}{2}$ for each value of V.

(a) $V = 0$ **(b)** $V = .5$ **(c)** $V = .25$

98. ***Voltage Induced by a Coil of Wire*** A coil of wire rotating in a magnetic field induces a voltage modeled by

$$e = 20 \sin\left(\frac{\pi t}{4} - \frac{\pi}{2}\right),$$

where t is time in seconds. Find the smallest positive time to produce each voltage.

(a) 0 **(b)** $10\sqrt{3}$

99. ***Movement of a Particle*** A particle moves along a straight line. The distance of the particle from the origin at time t is modeled by

$$s(t) = \sin t + 2 \cos t.$$

Find a value of t that satisfies each equation.

(a) $s(t) = \frac{2 + \sqrt{3}}{2}$ **(b)** $s(t) = \frac{3\sqrt{2}}{2}$

100. Explain what is wrong with the following solution for all x over the interval $[0, 2\pi)$ of the equation $\sin^2 x - \sin x = 0$.

$$\sin^2 x - \sin x = 0$$

$$\sin x - 1 = 0 \quad \text{Divide by } \sin x.$$

$$\sin x = 1 \quad \text{Add 1.}$$

$$x = \frac{\pi}{2}$$

The solution set is $\left\{\frac{\pi}{2}\right\}$.

7.7 Equations Involving Inverse Trigonometric Functions

Solving for x in Terms of y Using Inverse Functions ▪ Solving Inverse Trigonometric Equations

Until now, the equations in this chapter have involved trigonometric functions of angles or real numbers. Now we examine equations involving *inverse* trigonometric functions.

Solving for x in Terms of y Using Inverse Functions

EXAMPLE 1 Solving an Equation for a Variable Using Inverse Notation

Solve $y = 3 \cos 2x$ for x.

Solution We want $\cos 2x$ alone on one side of the equation so we can solve for $2x$, and then for x.

$$y = 3 \cos 2x$$

$$\frac{y}{3} = \cos 2x \qquad \text{Divide by 3.}$$

$$2x = \arccos \frac{y}{3} \qquad \text{Definition of arccosine (Section 7.5)}$$

$$x = \frac{1}{2} \arccos \frac{y}{3} \qquad \text{Multiply by } \tfrac{1}{2}.$$

Now try Exercise 7.

Solving Inverse Trigonometric Equations

EXAMPLE 2 Solving an Equation Involving an Inverse Trigonometric Function

Solve $2 \arcsin x = \pi$.

Solution First solve for $\arcsin x$, and then for x.

$$2 \arcsin x = \pi$$

$$\arcsin x = \frac{\pi}{2} \qquad \text{Divide by 2.}$$

$$x = \sin \frac{\pi}{2} \qquad \text{Definition of arcsine (Section 7.5)}$$

$$x = 1 \qquad \text{(Section 6.2)}$$

Verify that the solution satisfies the given equation. The solution set is $\{1\}$.

Now try Exercise 23.

EXAMPLE 3 Solving an Equation Involving Inverse Trigonometric Functions

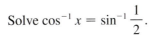

Solve $\cos^{-1} x = \sin^{-1} \dfrac{1}{2}$.

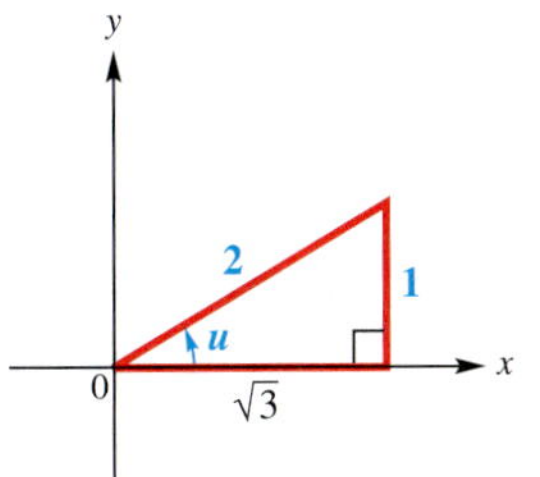

Figure 31

Solution Let $\sin^{-1} \frac{1}{2} = u$. Then $\sin u = \frac{1}{2}$ and for u in quadrant I, the equation becomes

$$\cos^{-1} x = u$$

$$\cos u = x. \quad \text{Alternative form}$$

Sketch a triangle and label it using the facts that u is in quadrant I and $\sin u = \frac{1}{2}$. See Figure 31. Since $x = \cos u$, $x = \frac{\sqrt{3}}{2}$, and the solution set is $\left\{\frac{\sqrt{3}}{2}\right\}$. *Check.*

EXAMPLE 4 Solving an Inverse Trigonometric Equation Using an Identity

Solve $\arcsin x - \arccos x = \dfrac{\pi}{6}$.

Solution Isolate one inverse function on one side of the equation.

$$\arcsin x - \arccos x = \frac{\pi}{6}$$

$$\arcsin x = \arccos x + \frac{\pi}{6} \qquad \text{Add } \arccos x. \quad (1)$$

$$\sin\left(\arccos x + \frac{\pi}{6}\right) = x \qquad \text{Definition of arcsine}$$

Let $u = \arccos x$, so $0 \leq u \leq \pi$ by definition.

$$\sin\left(u + \frac{\pi}{6}\right) = x \qquad \text{Substitute.} \quad (2)$$

$$\sin\left(u + \frac{\pi}{6}\right) = \sin u \cos \frac{\pi}{6} + \cos u \sin \frac{\pi}{6}$$

Sine sum identity **(Section 7.3)**

Substitute this result into equation (2) to get

$$\sin u \cos \frac{\pi}{6} + \cos u \sin \frac{\pi}{6} = x. \quad (3)$$

From equation (1) and by the definition of the arcsine function,

$$-\frac{\pi}{2} \leq \arccos x + \frac{\pi}{6} \leq \frac{\pi}{2}$$

$$-\frac{2\pi}{3} \leq \arccos x \leq \frac{\pi}{3}. \qquad \text{Subtract } \tfrac{\pi}{6}. \text{ (Section 1.7)}$$

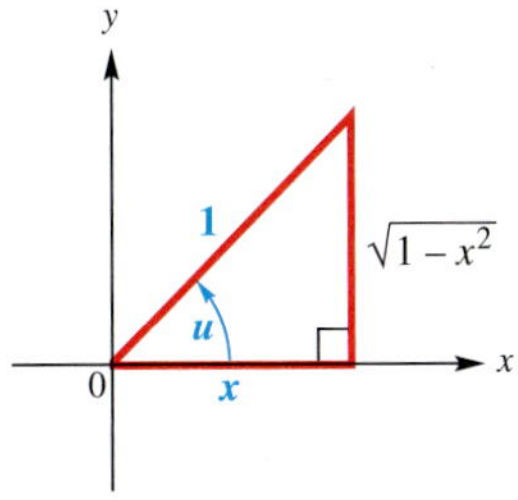

Figure 32

Since $0 \leq \arccos x \leq \pi$, we must have $0 \leq \arccos x \leq \frac{\pi}{3}$. Thus, $x > 0$, and we can sketch the triangle in Figure 32. From this triangle we find that $\sin u = \sqrt{1 - x^2}$.

Now substitute into equation (3) using $\sin u = \sqrt{1 - x^2}$, $\sin \frac{\pi}{6} = \frac{1}{2}$, $\cos \frac{\pi}{6} = \frac{\sqrt{3}}{2}$, and $\cos u = x$.

$$\sin u \cos \frac{\pi}{6} + \cos u \sin \frac{\pi}{6} = x \qquad (3)$$

$$\left(\sqrt{1 - x^2}\right)\frac{\sqrt{3}}{2} + x \cdot \frac{1}{2} = x$$

$$\left(\sqrt{1 - x^2}\right)\sqrt{3} + x = 2x \qquad \text{Multiply by 2.}$$

$$\left(\sqrt{3}\right)\sqrt{1 - x^2} = x \qquad \text{Subtract } x.$$

$$3(1 - x^2) = x^2 \qquad \text{Square both sides. (Section 1.6)}$$

$$3 - 3x^2 = x^2 \qquad \text{Distributive property (Section R.1)}$$

$$3 = 4x^2 \qquad \text{Add } 3x^2.$$

$$x = \sqrt{\frac{3}{4}} \qquad \text{Solve for } x\text{; Choose the positive square root because } x > 0.$$

$$x = \frac{\sqrt{3}}{2} \qquad \text{Quotient rule (Section R.7)}$$

To *check*, replace x with $\frac{\sqrt{3}}{2}$ in the original equation:

$$\arcsin \frac{\sqrt{3}}{2} - \arccos \frac{\sqrt{3}}{2} = \frac{\pi}{3} - \frac{\pi}{6} = \frac{\pi}{6},$$

as required. The solution set is $\left\{\frac{\sqrt{3}}{2}\right\}$.

Now try Exercise 31.

7.7 Exercises

Concept Check *Answer each question.*

1. Which one of the following equations has solution 0?

A. $\arctan 1 = x$ **B.** $\arccos 0 = x$ **C.** $\arcsin 0 = x$

2. Which one of the following equations has solution $\frac{\pi}{4}$?

A. $\arcsin \frac{\sqrt{2}}{2} = x$ **B.** $\arccos\left(-\frac{\sqrt{2}}{2}\right) = x$ **C.** $\arctan \frac{\sqrt{3}}{3} = x$

3. Which one of the following equations has solution $\frac{3\pi}{4}$?

A. $\arctan 1 = x$ **B.** $\arcsin \frac{\sqrt{2}}{2} = x$ **C.** $\arccos\left(-\frac{\sqrt{2}}{2}\right) = x$

4. Which one of the following equations has solution $-\frac{\pi}{6}$?

A. $\arctan \frac{\sqrt{3}}{3} = x$ **B.** $\arccos\left(-\frac{1}{2}\right) = x$ **C.** $\arcsin\left(-\frac{1}{2}\right) = x$

Solve each equation for x. See Example 1.

5. $y = 5 \cos x$

6. $4y = \sin x$

7. $2y = \cot 3x$

8. $6y = \frac{1}{2} \sec x$

9. $y = 3 \tan 2x$

10. $y = 3 \sin \frac{x}{2}$

11. $y = 6 \cos \frac{x}{4}$

12. $y = -\sin \frac{x}{3}$

13. $y = -2 \cos 5x$

14. $y = 3 \cot 5x$

15. $y = \cos(x + 3)$

16. $y = \tan(2x - 1)$

17. $y = \sin x - 2$

18. $y = \cot x + 1$

19. $y = 2 \sin x - 4$

20. $y = 4 + 3 \cos x$

21. Refer to Exercise 17. A student attempting to solve this equation wrote as the first step $y = \sin(x - 2)$, inserting parentheses as shown. Explain why this is incorrect.

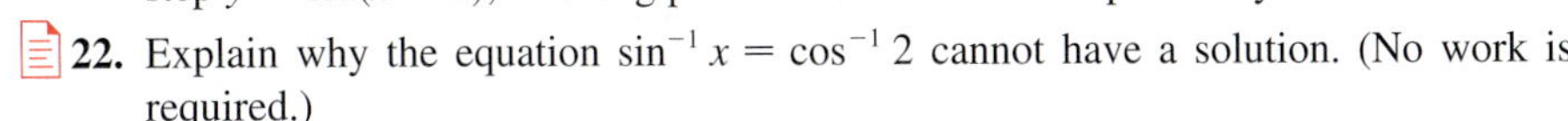

22. Explain why the equation $\sin^{-1} x = \cos^{-1} 2$ cannot have a solution. (No work is required.)

Solve each equation for exact solutions. See Examples 2 and 3.

23. $\frac{4}{3} \cos^{-1} \frac{y}{4} = \pi$

24. $4\pi + 4 \tan^{-1} y = \pi$

25. $2 \arccos\left(\frac{y - \pi}{3}\right) = 2\pi$

26. $\arccos\left(y - \frac{\pi}{3}\right) = \frac{\pi}{6}$

27. $\arcsin x = \arctan \frac{3}{4}$

28. $\arctan x = \arccos \frac{5}{13}$

29. $\cos^{-1} x = \sin^{-1} \frac{3}{5}$

30. $\cot^{-1} x = \tan^{-1} \frac{4}{3}$

Solve each equation for exact solutions. See Example 4.

31. $\sin^{-1} x - \tan^{-1} 1 = -\frac{\pi}{4}$

32. $\sin^{-1} x + \tan^{-1} \sqrt{3} = \frac{2\pi}{3}$

33. $\arccos x + 2 \arcsin \frac{\sqrt{3}}{2} = \pi$

34. $\arccos x + 2 \arcsin \frac{\sqrt{3}}{2} = \frac{\pi}{3}$

35. $\arcsin 2x + \arccos x = \frac{\pi}{6}$

36. $\arcsin 2x + \arcsin x = \frac{\pi}{2}$

37. $\cos^{-1} x + \tan^{-1} x = \frac{\pi}{2}$

38. $\sin^{-1} x + \tan^{-1} x = 0$

39. Provide graphical support for the solution in Example 4 by showing that the graph of $y = \arcsin x - \arccos x - \frac{\pi}{6}$ has x-intercept $\frac{\sqrt{3}}{2} \approx .8660254$.

40. Provide graphical support for the solution in Example 4 by showing that the x-coordinate of the point of intersection of the graphs of $Y_1 = \arcsin X - \arccos X$ and $Y_2 = \frac{\pi}{6}$ is $\frac{\sqrt{3}}{2} \approx .8660254$.

The following equations cannot be solved by algebraic methods. Use a graphing calculator to find all solutions over the interval $[0, 6]$. *Express solutions to as many decimal places as your calculator displays.*

41. $(\arctan x)^3 - x + 2 = 0$

42. $\pi \sin^{-1}(.2x) - 3 = -\sqrt{x}$

(Modeling) Solve each problem.

43. *Tone Heard by a Listener* When two sources located at different positions produce the same pure tone, the human ear will often hear one sound that is equal to the sum of the individual tones. Since the sources are at different locations, they will have different phase angles ϕ. If two speakers located at different positions produce pure tones $P_1 = A_1 \sin(2\pi ft + \phi_1)$ and $P_2 = A_2 \sin(2\pi ft + \phi_2)$, where $-\frac{\pi}{4} \leq \phi_1, \phi_2 \leq \frac{\pi}{4}$, then the resulting tone heard by a listener can be written as $P = A \sin(2\pi ft + \phi)$, where

$$A = \sqrt{(A_1 \cos \phi_1 + A_2 \cos \phi_2)^2 + (A_1 \sin \phi_1 + A_2 \sin \phi_2)^2}$$

$$\text{and} \quad \phi = \arctan\left(\frac{A_1 \sin \phi_1 + A_2 \sin \phi_2}{A_1 \cos \phi_1 + A_2 \cos \phi_2}\right).$$

(*Source:* Fletcher, N. and T. Rossing, *The Physics of Musical Instruments,* Second Edition, Springer-Verlag, 1998.)

(a) Calculate A and ϕ if $A_1 = .0012$, $\phi_1 = .052$, $A_2 = .004$, and $\phi_2 = .61$. Also find an expression for $P = A \sin(2\pi ft + \phi)$ if $f = 220$.

(b) Graph $Y_1 = P$ and $Y_2 = P_1 + P_2$ on the same coordinate axes over the interval $[0, .01]$. Are the two graphs the same?

44. *Tone Heard by a Listener* Repeat Exercise 43, with $A_1 = .0025$, $\phi_1 = \frac{\pi}{7}$, $A_2 = .001$, $\phi_2 = \frac{\pi}{6}$, and $f = 300$.

45. *Depth of Field* When a large-view camera is used to take a picture of an object that is not parallel to the film, the lens board should be tilted so that the planes containing the subject, the lens board, and the film intersect in a line. This gives the best "depth of field." See the figure. (*Source:* Bushaw, Donald et al., *A Sourcebook of Applications of School Mathematics.* Copyright © 1980 by The Mathematical Association of America.)

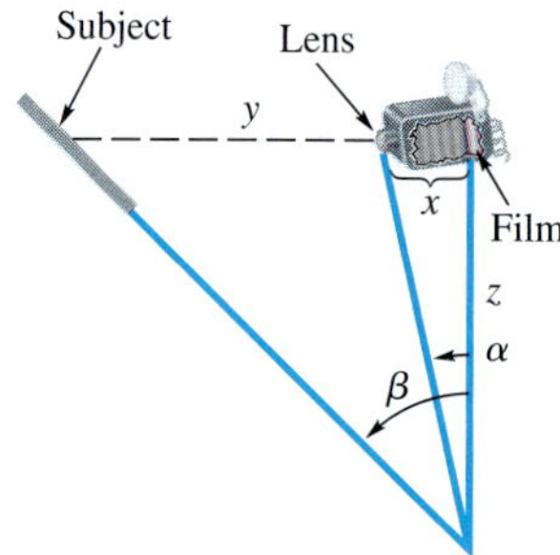

(a) Write two equations, one relating α, x, and z, and the other relating β, x, y, and z.

(b) Eliminate z from the equations in part (a) to get one equation relating α, β, x, and y.

(c) Solve the equation from part (b) for α.

(d) Solve the equation from part (b) for β.

46. *Programming Language for Inverse Functions* In Visual Basic, a widely used programming language for PCs, the only inverse trigonometric function available is arctangent. The other inverse trigonometric functions can be expressed in terms of arctangent as follows.

(a) Let $u = \arcsin x$. Solve the equation for x in terms of u.

(b) Use the result of part (a) to label the three sides of the triangle in the figure in terms of x.

(c) Use the triangle from part (b) to write an equation for $\tan u$ in terms of x.

(d) Solve the equation from part (c) for u.

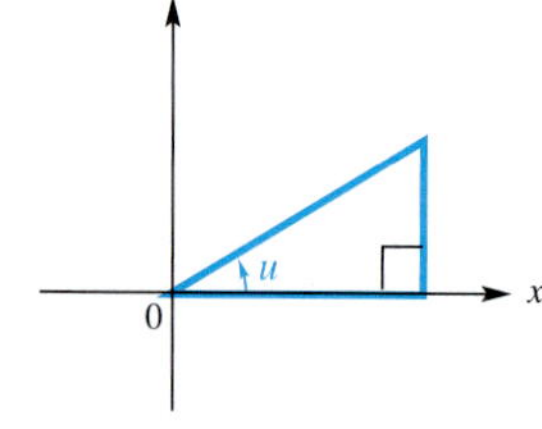

47. ***Alternating Electric Current*** In the study of alternating electric current, instantaneous voltage is modeled by

$$e = E_{max} \sin 2\pi f t,$$

where f is the number of cycles per second, E_{max} is the maximum voltage, and t is time in seconds.

(a) Solve the equation for t.

(b) Find the smallest positive value of t if $E_{max} = 12$, $e = 5$, and $f = 100$. Use a calculator.

48. ***Viewing Angle of an Observer*** While visiting a museum, Marsha Langlois views a painting that is 3 ft high and hangs 6 ft above the ground. See the figure. Assume her eyes are 5 ft above the ground, and let x be the distance from the spot where she is standing to the wall displaying the painting.

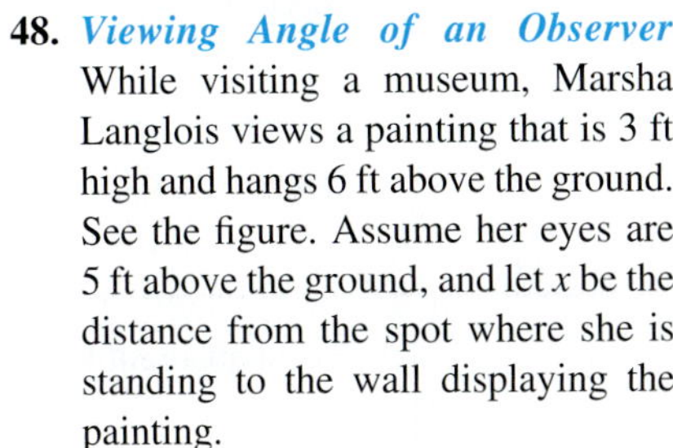

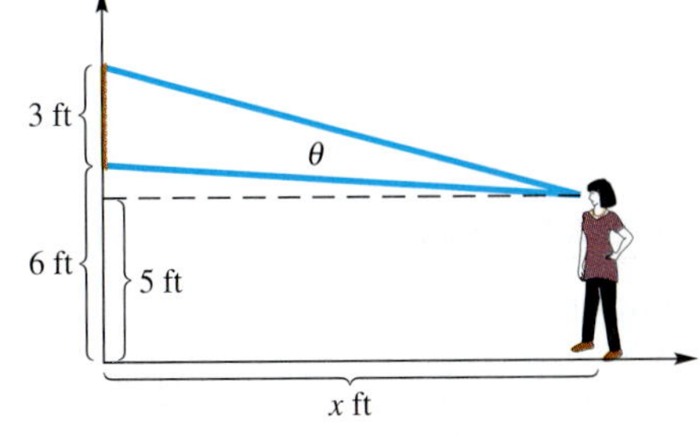

(a) Show that θ, the viewing angle subtended by the painting, is given by

$$\theta = \tan^{-1}\left(\frac{4}{x}\right) - \tan^{-1}\left(\frac{1}{x}\right).$$

(b) Find the value of x for each value of θ.

(i) $\theta = \frac{\pi}{6}$ **(ii)** $\theta = \frac{\pi}{8}$

(c) Find the value of θ for each value of x.

(i) $x = 4$ **(ii)** $x = 3$

49. ***Movement of an Arm*** In the exercises for Section 6.3 we found the equation

$$y = \frac{1}{3} \sin \frac{4\pi t}{3},$$

where t is time (in seconds) and y is the angle formed by a rhythmically moving arm.

(a) Solve the equation for t.

(b) At what time(s) does the arm form an angle of .3 radian?

50. The function $y = \sec^{-1} x$ is not found on graphing calculators. However, with some models it can be graphed as

$$y = \frac{\pi}{2} - ((x > 0) - (x < 0))\left(\frac{\pi}{2} - \tan^{-1}\left(\sqrt{(x^2 - 1)}\right)\right).$$

(This formula appears as Y_1 in the screen here.) Use the formula to obtain the graph of $y = \sec^{-1} x$ in the window $[-4, 4]$ by $[0, \pi]$.

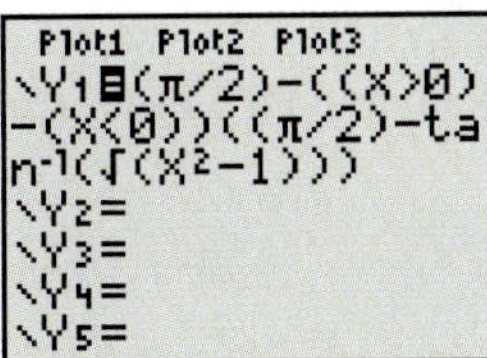

Chapter 7 Summary

NEW SYMBOLS

$\sin^{-1} x$ ($\arcsin x$)	inverse sine of x
$\cos^{-1} x$ ($\arccos x$)	inverse cosine of x
$\tan^{-1} x$ ($\arctan x$)	inverse tangent of x
$\cot^{-1} x$ ($\text{arccot}\, x$)	inverse cotangent of x
$\sec^{-1} x$ ($\text{arcsec}\, x$)	inverse secant of x
$\csc^{-1} x$ ($\text{arccsc}\, x$)	inverse cosecant of x

QUICK REVIEW

CONCEPTS | **EXAMPLES**

7.1 Fundamental Identities

Reciprocal Identities

$$\cot\theta = \frac{1}{\tan\theta} \qquad \sec\theta = \frac{1}{\cos\theta} \qquad \csc\theta = \frac{1}{\sin\theta}$$

Quotient Identities

$$\tan\theta = \frac{\sin\theta}{\cos\theta} \qquad \cot\theta = \frac{\cos\theta}{\sin\theta}$$

Pythagorean Identities

$$\sin^2\theta + \cos^2\theta = 1 \qquad \tan^2\theta + 1 = \sec^2\theta$$

$$1 + \cot^2\theta = \csc^2\theta$$

Negative-Angle Identities

$$\sin(-\theta) = -\sin\theta \qquad \cos(-\theta) = \cos\theta \qquad \tan(-\theta) = -\tan\theta$$

$$\csc(-\theta) = -\csc\theta \qquad \sec(-\theta) = \sec\theta \qquad \cot(-\theta) = -\cot\theta$$

If θ is in quadrant IV and $\sin\theta = -\frac{3}{5}$, find $\csc\theta$, $\cos\theta$, and $\sin(-\theta)$.

$$\csc\theta = \frac{1}{\sin\theta} = \frac{1}{-\frac{3}{5}} = -\frac{5}{3}$$

$$\sin^2\theta + \cos^2\theta = 1$$

$$\left(-\frac{3}{5}\right)^2 + \cos^2\theta = 1$$

$$\cos^2\theta = 1 - \frac{9}{25} = \frac{16}{25}$$

$$\cos\theta = +\sqrt{\frac{16}{25}} = \frac{4}{5}$$

$\cos\theta$ is positive in quadrant IV.

$$\sin(-\theta) = -\sin\theta = \frac{3}{5}$$

7.2 Verifying Trigonometric Identities

See the box titled Hints for Verifying Identities on page 613.

7.3 Sum and Difference Identities

Cofunction Identities

$$\cos(90^\circ - \theta) = \sin\theta \qquad \cot(90^\circ - \theta) = \tan\theta$$

$$\sin(90^\circ - \theta) = \cos\theta \qquad \sec(90^\circ - \theta) = \csc\theta$$

$$\tan(90^\circ - \theta) = \cot\theta \qquad \csc(90^\circ - \theta) = \sec\theta$$

Find a value of θ such that $\tan\theta = \cot 78^\circ$.

$$\tan\theta = \cot 78^\circ$$

$$\cot(90^\circ - \theta) = \cot 78^\circ$$

$$90^\circ - \theta = 78^\circ$$

$$\theta = 12^\circ$$

Sum and Difference Identities

$$\cos(A - B) = \cos A\cos B + \sin A\sin B$$

$$\cos(A + B) = \cos A\cos B - \sin A\sin B$$

$$\sin(A + B) = \sin A\cos B + \cos A\sin B$$

$$\sin(A - B) = \sin A\cos B - \cos A\sin B$$

Find the exact value of $\cos(-15^\circ)$.

$$\cos(-15^\circ) = \cos(30^\circ - 45^\circ)$$

$$= \cos 30^\circ\cos 45^\circ + \sin 30^\circ\sin 45^\circ$$

$$= \frac{\sqrt{3}}{2}\cdot\frac{\sqrt{2}}{2} + \frac{1}{2}\cdot\frac{\sqrt{2}}{2}$$

$$= \frac{\sqrt{6}}{4} + \frac{\sqrt{2}}{4} = \frac{\sqrt{6}+\sqrt{2}}{4}$$

CONCEPTS	EXAMPLES

Sum and Difference Identities

$$\tan(A + B) = \frac{\tan A + \tan B}{1 - \tan A \tan B}$$

$$\tan(A - B) = \frac{\tan A - \tan B}{1 + \tan A \tan B}$$

Write $\tan\left(\frac{\pi}{4} + \theta\right)$ in terms of $\tan\theta$.

$$\tan\left(\frac{\pi}{4} + \theta\right) = \frac{\tan\frac{\pi}{4} + \tan\theta}{1 - \tan\frac{\pi}{4}\tan\theta}$$

$$= \frac{1 + \tan\theta}{1 - \tan\theta} \qquad \tan\frac{\pi}{4} = 1$$

7.4 Double-Angle Identities and Half-Angle Identities

Double-Angle Identities

$$\cos 2A = \cos^2 A - \sin^2 A \qquad \cos 2A = 1 - 2\sin^2 A$$

$$\cos 2A = 2\cos^2 A - 1 \qquad \sin 2A = 2\sin A\cos A$$

$$\tan 2A = \frac{2\tan A}{1 - \tan^2 A}$$

Given $\cos\theta = -\frac{5}{13}$ and $\sin\theta > 0$, find $\sin 2\theta$.

Sketch a triangle in quadrant II and use it to find $\sin\theta$: $\sin\theta = \frac{12}{13}$.

$$\sin 2\theta = 2\sin\theta\cos\theta$$

$$= 2\left(\frac{12}{13}\right)\left(-\frac{5}{13}\right) = -\frac{120}{169}$$

Product-to-Sum Identities

$$\cos A\cos B = \frac{1}{2}[\cos(A + B) + \cos(A - B)]$$

$$\sin A\sin B = \frac{1}{2}[\cos(A - B) - \cos(A + B)]$$

$$\sin A\cos B = \frac{1}{2}[\sin(A + B) + \sin(A - B)]$$

$$\cos A\sin B = \frac{1}{2}[\sin(A + B) - \sin(A - B)]$$

Write $\sin(-\theta)\sin 2\theta$ as the difference of two functions.

$$\sin(-\theta)\sin 2\theta = \frac{1}{2}[\cos(-\theta - 2\theta) - \cos(-\theta + 2\theta)]$$

$$= \frac{1}{2}[\cos(-3\theta) - \cos\theta]$$

$$= \frac{1}{2}\cos(-3\theta) - \frac{1}{2}\cos\theta$$

$$= \frac{1}{2}\cos 3\theta - \frac{1}{2}\cos\theta$$

Sum-to-Product Identities

$$\sin A + \sin B = 2\sin\left(\frac{A + B}{2}\right)\cos\left(\frac{A - B}{2}\right)$$

$$\sin A - \sin B = 2\cos\left(\frac{A + B}{2}\right)\sin\left(\frac{A - B}{2}\right)$$

$$\cos A + \cos B = 2\cos\left(\frac{A + B}{2}\right)\cos\left(\frac{A - B}{2}\right)$$

$$\cos A - \cos B = -2\sin\left(\frac{A + B}{2}\right)\sin\left(\frac{A - B}{2}\right)$$

Write $\cos\theta + \cos 3\theta$ as a product of two functions.

$$\cos\theta + \cos 3\theta = 2\cos\left(\frac{\theta + 3\theta}{2}\right)\cos\left(\frac{\theta - 3\theta}{2}\right)$$

$$= 2\cos\left(\frac{4\theta}{2}\right)\cos\left(\frac{-2\theta}{2}\right)$$

$$= 2\cos 2\theta\cos(-\theta)$$

$$= 2\cos 2\theta\cos\theta$$

Half-Angle Identities

$$\cos\frac{A}{2} = \pm\sqrt{\frac{1 + \cos A}{2}} \qquad \sin\frac{A}{2} = \pm\sqrt{\frac{1 - \cos A}{2}}$$

$$\tan\frac{A}{2} = \pm\sqrt{\frac{1 - \cos A}{1 + \cos A}} \qquad \tan\frac{A}{2} = \frac{\sin A}{1 + \cos A}$$

$$\tan\frac{A}{2} = \frac{1 - \cos A}{\sin A}$$

(The sign is chosen based on the quadrant of $\frac{A}{2}$.)

Find the exact value of $\tan 67.5°$.

We choose the last form with $A = 135°$.

$$\tan 67.5° = \tan\frac{135°}{2} = \frac{1 - \cos 135°}{\sin 135°} = \frac{1 - \left(-\frac{\sqrt{2}}{2}\right)}{\frac{\sqrt{2}}{2}}$$

$$= \frac{1 + \frac{\sqrt{2}}{2}}{\frac{\sqrt{2}}{2}} \cdot \frac{2}{2} = \frac{2 + \sqrt{2}}{\sqrt{2}} \quad \text{or} \quad \sqrt{2} + 1$$

Rationalize the denominator; simplify.

CONCEPTS | **EXAMPLES**

7.5 Inverse Circular Functions

Inverse Function	Domain	Range: Interval	Range: Quadrants of the Unit Circle
$y = \sin^{-1} x$	$[-1, 1]$	$\left[-\frac{\pi}{2}, \frac{\pi}{2}\right]$	I and IV
$y = \cos^{-1} x$	$[-1, 1]$	$[0, \pi]$	I and II
$y = \tan^{-1} x$	$(-\infty, \infty)$	$\left(-\frac{\pi}{2}, \frac{\pi}{2}\right)$	I and IV
$y = \cot^{-1} x$	$(-\infty, \infty)$	$(0, \pi)$	I and II
$y = \sec^{-1} x$	$(-\infty, -1] \cup [1, \infty)$	$[0, \pi], y \neq \frac{\pi}{2}$	I and II
$y = \csc^{-1} x$	$(-\infty, -1] \cup [1, \infty)$	$\left[-\frac{\pi}{2}, \frac{\pi}{2}\right], y \neq 0$	I and IV

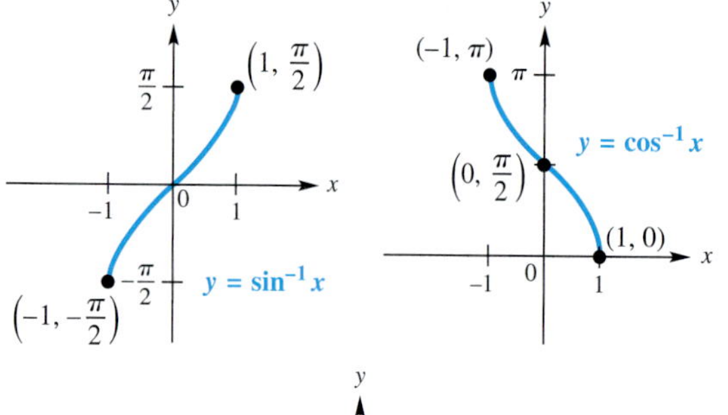

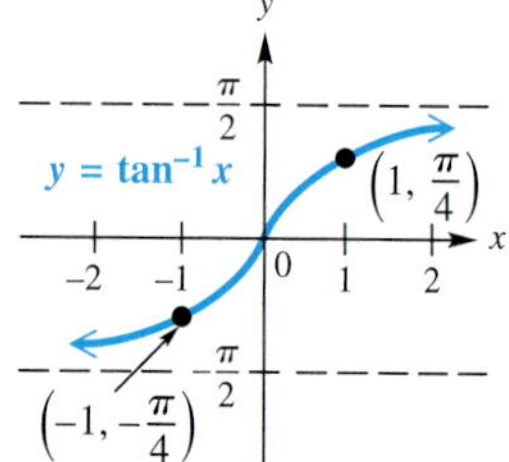

See page 649 for graphs of the other inverse circular (trigonometric) functions.

Evaluate $y = \cos^{-1} 0$.

Write $y = \cos^{-1} 0$ as $\cos y = 0$. Then $y = \frac{\pi}{2}$, because $\cos \frac{\pi}{2} = 0$ and $\frac{\pi}{2}$ is in the range of $\cos^{-1} x$.

Evaluate $\sin\left[\tan^{-1}\left(-\frac{3}{4}\right)\right]$.

Let $u = \tan^{-1}\left(-\frac{3}{4}\right)$. Then $\tan u = -\frac{3}{4}$. Since $\tan^{-1} x$ is negative in quadrant IV, sketch a triangle as shown.

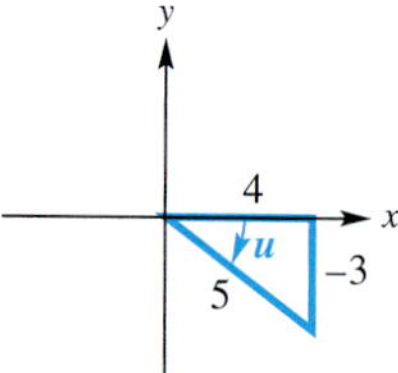

We want $\sin\left[\tan^{-1}\left(-\frac{3}{4}\right)\right] = \sin u$. From the triangle, $\sin u = -\frac{3}{5}$.

7.6 Trigonometric Equations

Solving a Trigonometric Equation

1. Decide whether the equation is linear or quadratic in form, so you can determine the solution method.
2. If only one trigonometric function is present, first solve the equation for that function.
3. If more than one trigonometric function is present, rearrange the equation so that one side equals 0. Then try to factor and set each factor equal to 0 to solve.
4. If the equation is quadratic in form, but not factorable, use the quadratic formula. Check that solutions are in the desired interval.
5. Try using identities to change the form of the equation. It may be helpful to square both sides of the equation first. If this is done, check for extraneous solutions.

Solve $\tan \theta + \sqrt{3} = 2\sqrt{3}$ over the interval $[0°, 360°)$. Use a linear method.

$$\tan \theta + \sqrt{3} = 2\sqrt{3}$$
$$\tan \theta = \sqrt{3}$$
$$\theta = 60°$$

Another solution over $[0°, 360°)$ is

$$\theta = 60° + 180° = 240°.$$

The solution set is $\{60°, 240°\}$.

Chapter 7 Review Exercises

Concept Check *For each expression in Column I, choose the expression from Column II that completes an identity.*

I		II	
1. $\sec x =$ ______	**2.** $\csc x =$ ______	**A.** $\dfrac{1}{\sin x}$	**B.** $\dfrac{1}{\cos x}$
3. $\tan x =$ ______	**4.** $\cot x =$ ______	**C.** $\dfrac{\sin x}{\cos x}$	**D.** $\dfrac{1}{\cot^2 x}$
5. $\tan^2 x =$ ______	**6.** $\sec^2 x =$ ______	**E.** $\dfrac{1}{\cos^2 x}$	**F.** $\dfrac{\cos x}{\sin x}$

Use identities to write each expression in terms of $\sin \theta$ *and* $\cos \theta$, *and simplify.*

7. $\sec^2 \theta - \tan^2 \theta$

8. $\dfrac{\cot \theta}{\sec \theta}$

9. $\tan^2 \theta(1 + \cot^2 \theta)$

10. $\csc \theta + \cot \theta$

11. Use the trigonometric identities to find $\sin x$, $\tan x$, and $\cot(-x)$, given $\cos x = \frac{3}{5}$ and x is in quadrant IV.

12. Given $\tan x = -\frac{5}{4}$, where $\frac{\pi}{2} < x < \pi$, use the trigonometric identities to find $\cot x$, $\csc x$, and $\sec x$.

Concept Check *For each expression in Column I, use an identity to choose an expression from Column II with the same value.*

I		II	
13. $\cos 210°$	**14.** $\sin 35°$	**A.** $\sin(-35°)$	**B.** $\cos 55°$
15. $\tan(-35°)$	**16.** $-\sin 35°$	**C.** $\sqrt{\dfrac{1 + \cos 150°}{2}}$	**D.** $2 \sin 150° \cos 150°$
17. $\cos 35°$	**18.** $\cos 75°$	**E.** $\cos 150° \cos 60° - \sin 150° \sin 60°$	
19. $\sin 75°$	**20.** $\sin 300°$	**F.** $\cot(-35°)$	
21. $\cos 300°$	**22.** $\cos(-55°)$	**G.** $\cos^2 150° - \sin^2 150°$	
		H. $\sin 15° \cos 60° + \cos 15° \sin 60°$	
		I. $\cos(-35°)$	**J.** $\cot 125°$

For each of the following, find $\sin(x + y)$, $\cos(x - y)$, $\tan(x + y)$, *and the quadrant of* $x + y$.

23. $\sin x = -\dfrac{1}{4}$, $\cos y = -\dfrac{4}{5}$, x and y in quadrant III

24. $\sin x = \dfrac{1}{10}$, $\cos y = \dfrac{4}{5}$, x in quadrant I, y in quadrant IV

Find each of the following.

25. $\cos \dfrac{\theta}{2}$, given $\cos \theta = -\dfrac{1}{2}$, with $90° < \theta < 180°$

26. $\sin y$, given $\cos 2y = -\dfrac{1}{3}$, with $\dfrac{\pi}{2} < y < \pi$

Graph each expression and use the graph to conjecture an identity. Then verify your conjecture algebraically.

27. $-\dfrac{\sin 2x + \sin x}{\cos 2x - \cos x}$

28. $\dfrac{1 - \cos 2x}{\sin 2x}$

Verify that each equation is an identity.

29. $\sin^2 x - \sin^2 y = \cos^2 y - \cos^2 x$

30. $2\cos^3 x - \cos x = \dfrac{\cos^2 x - \sin^2 x}{\sec x}$

31. $\dfrac{\sin^2 x}{2 - 2\cos x} = \cos^2 \dfrac{x}{2}$

32. $\dfrac{\sin 2x}{\sin x} = \dfrac{2}{\sec x}$

33. $2\cos A - \sec A = \cos A - \dfrac{\tan A}{\csc A}$

34. $\dfrac{2\tan B}{\sin 2B} = \sec^2 B$

35. $1 + \tan^2 \alpha = 2\tan \alpha \csc 2\alpha$

36. $\dfrac{2\cot x}{\tan 2x} = \csc^2 x - 2$

37. $\tan \theta \sin 2\theta = 2 - 2\cos^2 \theta$

38. $\csc A \sin 2A - \sec A = \cos 2A \sec A$

39. $2\tan x \csc 2x - \tan^2 x = 1$

40. $2\cos^2 \theta - 1 = \dfrac{1 - \tan^2 \theta}{1 + \tan^2 \theta}$

41. $\tan \theta \cos^2 \theta = \dfrac{2\tan \theta \cos^2 \theta - \tan \theta}{1 - \tan^2 \theta}$

42. $\sec^2 \alpha - 1 = \dfrac{\sec 2\alpha - 1}{\sec 2\alpha + 1}$

43. $2\cos^3 x - \cos x = \dfrac{\cos^2 x - \sin^2 x}{\sec x}$

44. $\sin^3 \theta = \sin \theta - \cos^2 \theta \sin \theta$

Give the exact real number value of y. Do not use a calculator.

45. $y = \sin^{-1} \dfrac{\sqrt{2}}{2}$

46. $y = \arccos\left(-\dfrac{1}{2}\right)$

47. $y = \tan^{-1}(-\sqrt{3})$

48. $y = \arcsin(-1)$

49. $y = \cos^{-1}\left(-\dfrac{\sqrt{2}}{2}\right)$

50. $y = \arctan \dfrac{\sqrt{3}}{3}$

51. $y = \sec^{-1}(-2)$

52. $y = \operatorname{arccsc} \dfrac{2\sqrt{3}}{3}$

53. $y = \operatorname{arccot}(-1)$

Give the degree measure of θ. Do not use a calculator.

54. $\theta = \arccos \dfrac{1}{2}$

55. $\theta = \arcsin\left(-\dfrac{\sqrt{3}}{2}\right)$

56. $\theta = \tan^{-1} 0$

Use a calculator to give the degree measure of θ.

57. $\theta = \arctan 1.7804675$

58. $\theta = \sin^{-1}(-.66045320)$

59. $\theta = \cos^{-1} .80396577$

60. $\theta = \cot^{-1} 4.5046388$

61. $\theta = \operatorname{arcsec} 3.4723155$

62. $\theta = \csc^{-1} 7.4890096$

Evaluate the following without using a calculator.

63. $\cos(\arccos(-1))$

64. $\sin\left(\arcsin\left(-\dfrac{\sqrt{3}}{2}\right)\right)$

65. $\arccos\left(\cos \dfrac{3\pi}{4}\right)$

66. $\operatorname{arcsec}(\sec \pi)$

67. $\tan^{-1}\left(\tan \dfrac{\pi}{4}\right)$

68. $\cos^{-1}(\cos 0)$

69. $\sin\left(\arccos \dfrac{3}{4}\right)$

70. $\cos(\arctan 3)$

71. $\cos(\csc^{-1}(-2))$

72. $\sec\left(2\sin^{-1}\left(-\frac{1}{3}\right)\right)$

73. $\tan\left(\arcsin\frac{3}{5} + \arccos\frac{5}{7}\right)$

Write each of the following as an algebraic (nontrigonometric) expression in u, $u > 0$.

74. $\cos\left(\arctan\frac{u}{\sqrt{1-u^2}}\right)$

75. $\tan\left(\operatorname{arcsec}\frac{\sqrt{u^2+1}}{u}\right)$

Solve each equation for solutions over the interval $[0, 2\pi)$.

76. $\sin^2 x = 1$

77. $2\tan x - 1 = 0$

78. $3\sin^2 x - 5\sin x + 2 = 0$

79. $\tan x = \cot x$

80. $\sec^4 2x = 4$

81. $\tan^2 2x - 1 = 0$

Give all solutions for each equation.

82. $\sec\frac{x}{2} = \cos\frac{x}{2}$

83. $\cos 2x + \cos x = 0$

84. $4\sin x\cos x = \sqrt{3}$

Solve each equation for solutions over the interval $[0°, 360°)$. If necessary, express solutions to the nearest tenth of a degree.

85. $\sin^2\theta + 3\sin\theta + 2 = 0$

86. $2\tan^2\theta = \tan\theta + 1$

87. $\sin 2\theta = \cos 2\theta + 1$

88. $2\sin 2\theta = 1$

89. $3\cos^2\theta + 2\cos\theta - 1 = 0$

90. $5\cot^2\theta - \cot\theta - 2 = 0$

Solve each equation in Exercises 91–97 for x.

91. $4y = 2\sin x$

92. $y = 3\cos\frac{x}{2}$

93. $2y = \tan(3x + 2)$

94. $5y = 4\sin x - 3$

95. $\frac{4}{3}\arctan\frac{x}{2} = \pi$

96. $\arccos x = \arcsin\frac{2}{7}$

97. $\arccos x + \arctan 1 = \frac{11\pi}{12}$

98. Solve $d = 550 + 450\cos\left(\frac{\pi}{50}t\right)$ for t in terms of d.

(Modeling) Solve each problem.

99. ***Viewing Angle of an Observer*** A 10-ft-wide chalkboard is situated 5 ft from the left wall of a classroom. See the figure. A student sitting next to the wall x feet from the front of the classroom has a viewing angle of θ radians.

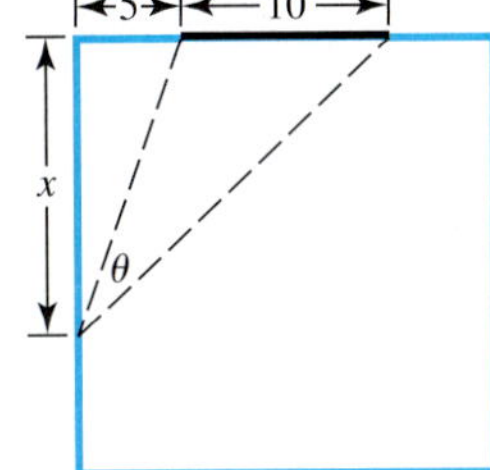

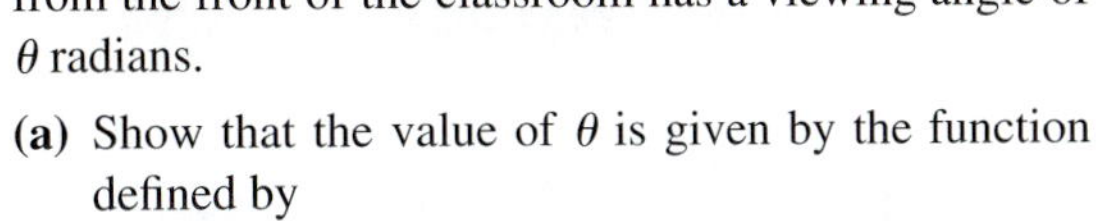

(a) Show that the value of θ is given by the function defined by

$$f(x) = \arctan\left(\frac{15}{x}\right) - \arctan\left(\frac{5}{x}\right).$$

(b) Graph $f(x)$ with a graphing calculator to estimate the value of x that maximizes the viewing angle.

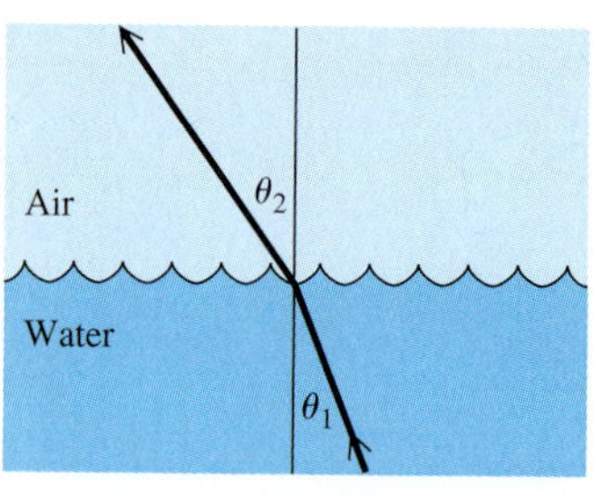

100. ***Snell's Law*** Recall Snell's law from Exercises 102 and 103 of Section 5.3:

$$\frac{c_1}{c_2} = \frac{\sin \theta_1}{\sin \theta_2},$$

where c_1 is the speed of light in one medium, c_2 is the speed of light in a second medium, and θ_1 and θ_2 are the angles shown in the figure. Suppose a light is shining up through water into the air as in the figure. As θ_1 increases, θ_2 approaches 90°, at which point no light will emerge from the water. Assume the ratio $\frac{c_1}{c_2}$ in this case is .752. For what value of θ_1 does $\theta_2 = 90°$? This value of θ_1 is called the *critical angle* for water.

101. ***Snell's Law*** Refer to Exercise 100. What happens when θ_1 is greater than the critical angle?

102. ***British Nautical Mile*** The British nautical mile is defined as the length of a minute of arc of a meridian. Since Earth is flat at its poles, the nautical mile, in feet, is given by

$$L = 6077 - 31 \cos 2\theta,$$

A nautical mile is the length on any of the meridians cut by a central angle of measure 1 minute.

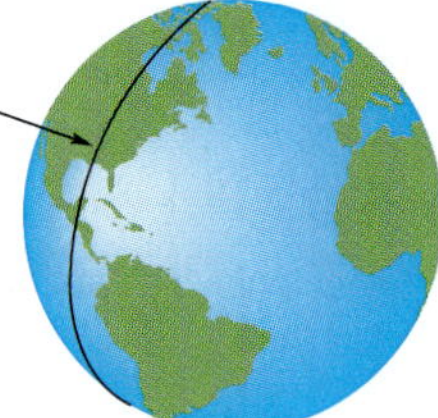

where θ is the latitude in degrees. See the figure. (*Source:* Bushaw, Donald et al., *A Sourcebook of Applications of School Mathematics.* Copyright © 1980 by The Mathematical Association of America.)

(a) Find the latitude between 0° and 90° at which the nautical mile is 6074 ft.

(b) At what latitude between 0° and 180° is the nautical mile 6108 ft?

(c) In the United States, the nautical mile is defined everywhere as 6080.2 ft. At what latitude between 0° and 90° does this agree with the British nautical mile?

103. The function $y = \csc^{-1} x$ is not found on graphing calculators. However, with some models it can be graphed as

$$y = ((x > 0) - (x < 0))\left(\frac{\pi}{2} - \tan^{-1}\left(\sqrt{(x^2 - 1)}\right)\right).$$

```
Plot1 Plot2 Plot3
\Y1■((X>0)-(X<0)
)*(π/2-tan⁻¹(√(X²
-1)))
\Y2=
\Y3=
\Y4=
\Y5=
```

(This formula appears as Y_1 in the screen here.) Use the formula to obtain the graph of $y = \csc^{-1} x$ in the window $[-4, 4]$ by $\left[-\frac{\pi}{2}, \frac{\pi}{2}\right]$.

104. **(a)** Use the graph of $y = \sin^{-1} x$ to approximate $\sin^{-1} .4$.

(b) Use the inverse sine key of a graphing calculator to approximate $\sin^{-1} .4$.

Chapter 7 Test

1. Given $\tan x = -\frac{5}{6}$, $\frac{3\pi}{2} < x < 2\pi$, use trigonometric identities to find $\sin x$ and $\cos x$.
2. Express $\tan^2 x - \sec^2 x$ in terms of $\sin x$ and $\cos x$, and simplify.
3. Find $\sin(x + y)$, $\cos(x - y)$, and $\tan(x + y)$, if $\sin x = -\frac{1}{3}$, $\cos y = -\frac{2}{5}$, x is in quadrant III, and y is in quadrant II.
4. Use a half-angle identity to find $\sin(-22.5°)$.

Graph each expression and use the graph to conjecture an identity. Then verify your conjecture algebraically.

5. $\sec x - \sin x \tan x$

6. $\cot \frac{x}{2} - \cot x$

Verify that each equation is an identity.

7. $\sec^2 B = \frac{1}{1 - \sin^2 B}$

8. $\cos 2A = \frac{\cot A - \tan A}{\csc A \sec A}$

9. Use an identity to write each expression as a trigonometric function of θ alone.

(a) $\cos(270° - \theta)$

(b) $\sin(\pi + \theta)$

10. ***Voltage*** The voltage in common household current is expressed as $V = 163 \sin \omega t$, where ω is the angular speed (in radians per second) of the generator at the electrical plant and t is time (in seconds).

(a) Use an identity to express V in terms of cosine.

(b) If $\omega = 120\pi$, what is the maximum voltage? Give the smallest positive value of t when the maximum voltage occurs.

11. Graph $y = \sin^{-1} x$, and indicate the coordinates of three points on the graph. Give the domain and range.

12. Find the exact value of y for each equation.

(a) $y = \arccos\left(-\frac{1}{2}\right)$

(b) $y = \sin^{-1}\left(-\frac{\sqrt{3}}{2}\right)$

(c) $y = \tan^{-1} 0$

(d) $y = \text{arcsec}(-2)$

13. Find each exact value.

(a) $\cos\left(\arcsin \frac{2}{3}\right)$

(b) $\sin\left(2 \cos^{-1} \frac{1}{3}\right)$

14. Write $\tan(\arcsin u)$ as an algebraic (nontrigonometric) expression in u, $u > 0$.

15. Solve $\sin^2 \theta = \cos^2 \theta + 1$ for solutions over the interval $[0°, 360°)$.

16. Solve $\csc^2 \theta - 2 \cot \theta = 4$ for solutions over the interval $[0°, 360°)$. Express approximate solutions to the nearest tenth of a degree.

17. Solve $\cos x = \cos 2x$ for solutions over the interval $[0, 2\pi)$.

18. Solve $2\sqrt{3} \sin \frac{x}{2} = 3$, giving all solutions in radians.

19. Solve each equation for x.

(a) $y = \cos 3x$

(b) $\arcsin x = \arctan \frac{4}{3}$

20. ***(Modeling) Movement of a Runner's Arm*** A runner's arm swings rhythmically according to the model

$$y = \frac{\pi}{8} \cos\left[\pi\left(t - \frac{1}{3}\right)\right],$$

where y represents the angle between the actual position of the upper arm and the downward vertical position and t represents time in seconds. At what times over the interval $[0, 3)$ is the angle y equal to 0?

Chapter 7 Quantitative Reasoning

How can we determine the amount of oil in a submerged storage tank?

The level of oil in a storage tank buried in the ground can be found in much the same way a dipstick is used to determine the oil level in an automobile crankcase. The person in the figure on the left has lowered a calibrated rod into an oil storage tank. When the rod is removed, the reading on the rod can be used with the dimensions of the storage tank to calculate the amount of oil in the tank.

Suppose the ends of the cylindrical storage tank in the figure are circles of radius 3 ft and the cylinder is 20 ft long. Determine the volume of oil in the tank if the rod shows a depth of 2 ft. (*Hint:* The volume will be 20 times the area of the shaded segment of the circle shown in the figure on the right.)

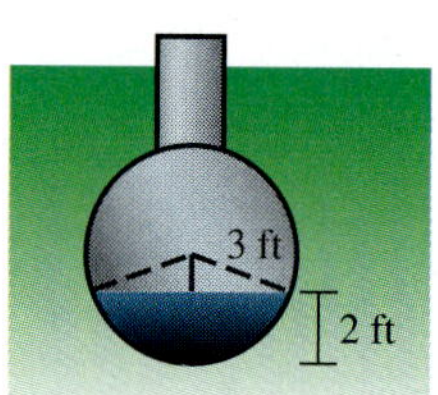

8 Applications of Trigonometry

High-resolution computer graphics and complex numbers make it possible to produce beautiful shapes called *fractals*. Benoit B. Mandelbrot first used the term *fractal* in 1975. At its basic level, a fractal is a unique, enchanting geometric figure with an endless self-similarity property. A fractal image repeats itself infinitely with ever-decreasing dimensions. If you look at smaller and smaller portions of a fractal image, you will continue to see the whole—much like looking into two parallel mirrors that are facing each other.

The above fractal, called *Newton's basins of attraction for the cube roots of unity,* is discussed in Exercise 50, Section 8.6. (*Source:* Crownover, R., *Introduction to Fractals and Chaos,* Jones and Bartlett, 1995; Lauwerier, H., *Fractals,* Princeton University Press, 1991. Reprinted with permission.)

8.1 The Law of Sines

Congruency and Oblique Triangles ▪ Derivation of the Law of Sines ▪ Applications ▪ Ambiguous Case ▪ Area of a Triangle

Until now, our work with triangles has been limited to right triangles. However, the concepts developed in Chapter 5 can be extended to *all* triangles. Every triangle has three sides and three angles. If any three of the six measures of a triangle are known (provided at least one measure is the length of a side), then the other three measures can be found. This is called *solving the triangle.*

Congruency and Oblique Triangles The following axioms from geometry allow us to prove that two triangles are congruent (that is, their corresponding sides and angles are equal).

Congruence Axioms

Side-Angle-Side (SAS)	If two sides and the included angle of one triangle are equal, respectively, to two sides and the included angle of a second triangle, then the triangles are congruent.
Angle-Side-Angle (ASA)	If two angles and the included side of one triangle are equal, respectively, to two angles and the included side of a second triangle, then the triangles are congruent.
Side-Side-Side (SSS)	If three sides of one triangle are equal, respectively, to three sides of a second triangle, then the triangles are congruent.

Keep in mind that whenever SAS, ASA, or SSS is given, the triangle is unique. We continue to label triangles as we did earlier with right triangles: side a opposite angle A, side b opposite angle B, and side c opposite angle C.

A triangle that is not a right triangle is called an **oblique triangle.** The measures of the three sides and the three angles of a triangle can be found if at least one side and any other two measures are known. There are four possible cases.

Data Required for Solving Oblique Triangles

Case 1	One side and two angles are known (SAA or ASA).
Case 2	Two sides and one angle not included between the two sides are known (SSA). This case may lead to more than one triangle.
Case 3	Two sides and the angle included between the two sides are known (SAS).
Case 4	Three sides are known (SSS).

NOTE If we know three angles of a triangle, we cannot find unique side lengths since AAA assures us only of similarity, not congruence. For example, there are infinitely many triangles ABC with $A = 35°$, $B = 65°$, and $C = 80°$.

Cases 1 and 2, discussed in this section, require the *law of sines*. Cases 3 and 4, discussed in the next section, require the *law of cosines*.

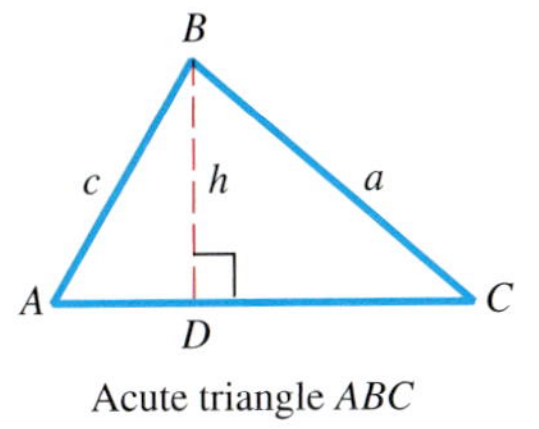

Acute triangle ABC

(a)

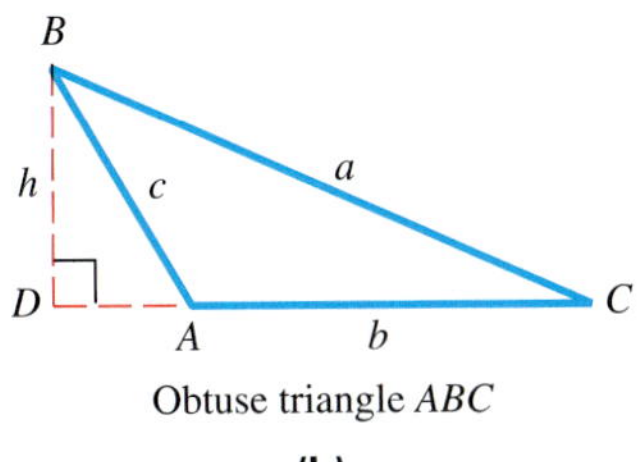

Obtuse triangle ABC

(b)

Figure 1

Derivation of the Law of Sines To derive the law of sines, we start with an oblique triangle, such as the *acute triangle* in Figure 1(a) or the *obtuse triangle* in Figure 1(b). This discussion applies to both triangles. First, construct the perpendicular from B to side AC (or its extension). Let h be the length of this perpendicular. Then c is the hypotenuse of right triangle ADB, and a is the hypotenuse of right triangle BDC.

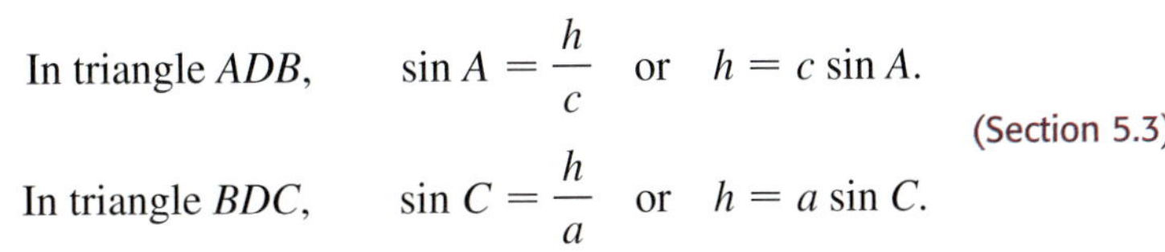

In triangle ADB, $\quad \sin A = \frac{h}{c} \quad$ or $\quad h = c \sin A$.

(Section 5.3)

In triangle BDC, $\quad \sin C = \frac{h}{a} \quad$ or $\quad h = a \sin C$.

Since $h = c \sin A$ and $h = a \sin C$,

$$a \sin C = c \sin A$$

$$\frac{a}{\sin A} = \frac{c}{\sin C}. \quad \text{Divide each side by } \sin A \sin C.$$

In a similar way, by constructing perpendicular lines from the other vertices, it can be shown that

$$\frac{a}{\sin A} = \frac{b}{\sin B} \quad \text{and} \quad \frac{b}{\sin B} = \frac{c}{\sin C}.$$

This discussion proves the following theorem.

Law of Sines

In any triangle ABC, with sides a, b, and c,

$$\frac{a}{\sin A} = \frac{b}{\sin B}, \quad \frac{a}{\sin A} = \frac{c}{\sin C}, \quad \text{and} \quad \frac{b}{\sin B} = \frac{c}{\sin C}.$$

This can be written in compact form as

$$\frac{a}{\sin A} = \frac{b}{\sin B} = \frac{c}{\sin C}.$$

That is, according to the law of sines, the lengths of the sides in a triangle are proportional to the sines of the measures of the angles opposite them.

When solving for an angle, we use an alternative form of the law of sines,

$$\frac{\sin A}{a} = \frac{\sin B}{b} = \frac{\sin C}{c}.$$

NOTE When using the law of sines, a good strategy is to select an equation so that the unknown variable is in the numerator and all other variables are known.

Applications If two angles and one side of a triangle are known (Case 1, SAA or ASA), then the law of sines can be used to solve the triangle.

EXAMPLE 1 Using the Law of Sines to Solve a Triangle (SAA)

Solve triangle ABC if $A = 32.0°$, $B = 81.8°$, and $a = 42.9$ cm.

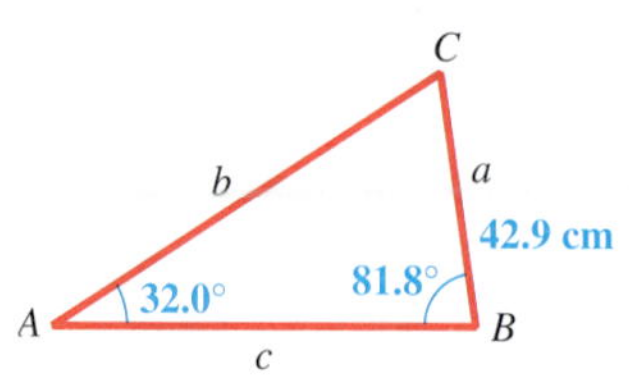

Figure 2

Solution Start by drawing a triangle, roughly to scale, and labeling the given parts as in Figure 2. Since the values of A, B, and a are known, use the form of the law of sines that involves these variables, and then solve for b.

$$\frac{a}{\sin A} = \frac{b}{\sin B} \qquad \text{Law of sines}$$

$$\frac{42.9}{\sin 32.0°} = \frac{b}{\sin 81.8°} \qquad \text{Substitute the given values.}$$

$$b = \frac{42.9 \sin 81.8°}{\sin 32.0°} \qquad \text{Multiply by } \sin 81.8°\text{; rewrite.}$$

$$b \approx 80.1 \text{ cm} \qquad \text{Approximate with a calculator.}$$

To find C, use the fact that the sum of the angles of any triangle is 180°.

$$A + B + C = 180°$$

$$C = 180° - A - B$$

$$C = 180° - 32.0° - 81.8° = 66.2°$$

Use the law of sines to find c. (Why does the Pythagorean theorem not apply?)

$$\frac{a}{\sin A} = \frac{c}{\sin C}$$

$$\frac{42.9}{\sin 32.0°} = \frac{c}{\sin 66.2°}$$

$$c = \frac{42.9 \sin 66.2°}{\sin 32.0°}$$

$$c \approx 74.1 \text{ cm}$$

Now try Exercise 7.

CAUTION When solving oblique triangles such as the one in Example 1, be sure to label a sketch carefully to help set up the correct equation.

EXAMPLE 2 Using the Law of Sines in an Application (ASA)

Jerry Keefe wishes to measure the distance across the Big Muddy River. See Figure 3. He determines that $C = 112.90°$, $A = 31.10°$, and $b = 347.6$ ft. Find the distance a across the river.

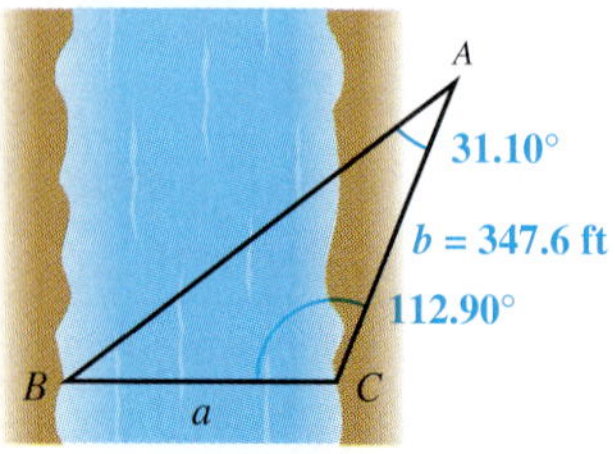

Figure 3

Solution To use the law of sines, one side and the angle opposite it must be known. Since b is the only side whose length is given, angle B must be found before the law of sines can be used.

$$B = 180° - A - C$$

$$B = 180° - 31.10° - 112.90° = 36.00°$$

Now use the form of the law of sines involving A, B, and b to find a.

$$\frac{a}{\sin A} = \frac{b}{\sin B} \quad \text{Law of sines}$$

$$\frac{a}{\sin 31.10°} = \frac{347.6}{\sin 36.00°} \quad \text{Substitute.}$$

$$a = \frac{347.6 \sin 31.10°}{\sin 36.00°} \quad \text{Multiply by } \sin 31.10°.$$

$$a \approx 305.5 \text{ ft} \quad \text{Use a calculator.}$$

Now try Exercise 47.

The next example involves the concept of bearing, first discussed in Chapter 5.

EXAMPLE 3 Using the Law of Sines in an Application (ASA)

Two ranger stations are on an east-west line 110 mi apart. A forest fire is located on a bearing of N 42° E from the western station at A and a bearing of N 15° E from the eastern station at B. How far is the fire from the western station? (See Figure 4.)

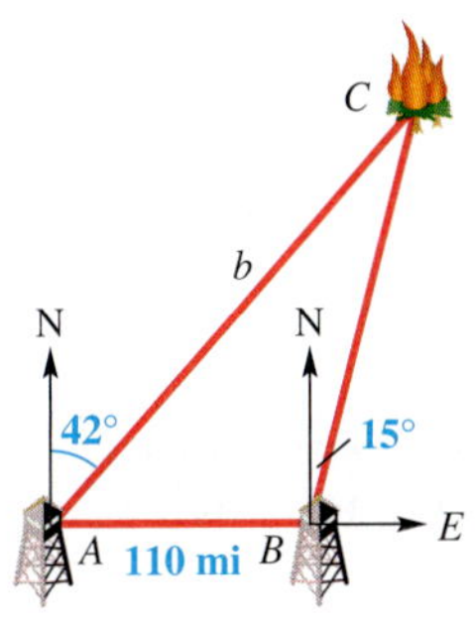

Figure 4

Solution Figure 4 shows the two stations at points A and B and the fire at point C. Angle $BAC = 90° - 42° = 48°$, the obtuse angle at B equals $90° + 15° = 105°$, and the third angle, C, equals $180° - 105° - 48° = 27°$. We use the law of sines to find side b.

$$\frac{b}{\sin B} = \frac{c}{\sin C} \quad \text{Law of sines}$$

$$\frac{b}{\sin 105°} = \frac{110}{\sin 27°} \quad \text{Substitute.}$$

$$b = \frac{110 \sin 105°}{\sin 27°} \quad \text{Multiply by } \sin 105°.$$

$$b \approx 234 \text{ mi}$$

Now try Exercise 49.

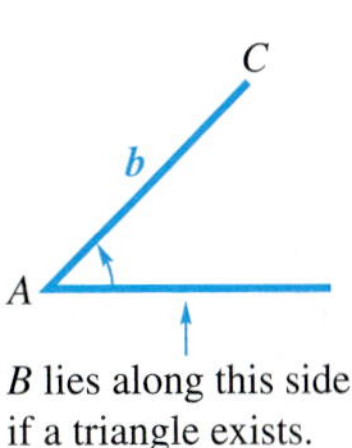

B lies along this side if a triangle exists.

Ambiguous Case If we are given the lengths of two sides and the angle opposite one of them (Case 2, SSA), it is possible that zero, one, or two such triangles exist. (Recall that there is no SSA congruence theorem.) To illustrate, suppose we know the measure of acute angle A of triangle ABC, the length of side a, and the length of side b. We draw angle A having adjacent side of length b, as shown in the margin. Now we must draw the side of length a opposite angle A. The table on the next page shows possible outcomes. This situation (SSA) is called the **ambiguous case** of the law of sines.

If angle A is acute, there are four possible outcomes.

Number of Possible Triangles	Sketch	Conditions Necessary for Case to Hold
0	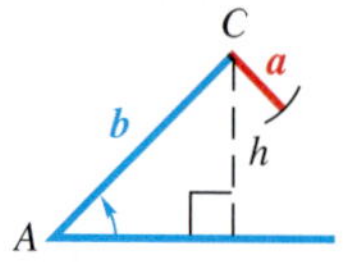	$\sin B > 1$, $a < h < b$
1	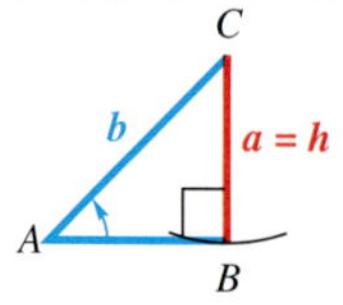	$\sin B = 1$, $a = h < b$
1	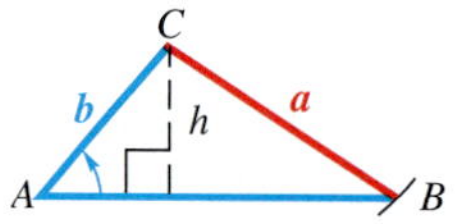	$0 < \sin B < 1$, $a \geq b$
2	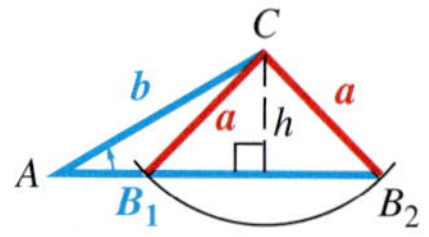	$0 < \sin B_2 < 1$, $h < a < b$

If angle A is obtuse, there are two possible outcomes.

Number of Possible Triangles	Sketch	Conditions Necessary for Case to Hold
0	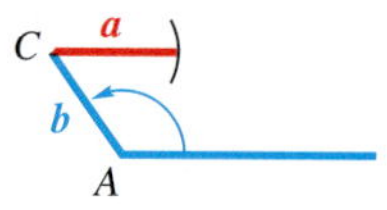	$\sin B \geq 1$, $a \leq b$
1	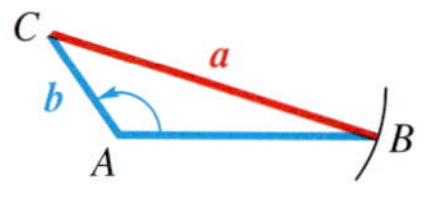	$0 < \sin B < 1$, $a > b$

We can apply the law of sines to the values of a, b, and A and use some basic properties of geometry and trigonometry to determine which situation applies. The following basic facts should be kept in mind.

1. For any angle θ of a triangle, $0 < \sin \theta \leq 1$. If $\sin \theta = 1$, then $\theta = 90°$ and the triangle is a right triangle.
2. $\sin \theta = \sin(180° - \theta)$ (That is, supplementary angles have the same sine value.)
3. The smallest angle is opposite the shortest side, the largest angle is opposite the longest side, and the middle-valued angle is opposite the intermediate side (assuming the triangle has sides that are all of different lengths).

EXAMPLE 4 Solving the Ambiguous Case (No Such Triangle)

Solve triangle ABC if $B = 55° 40'$, $b = 8.94$ m, and $a = 25.1$ m.

Solution Since we are given B, b, and a, we can use the law of sines to find A.

$$\frac{\sin A}{a} = \frac{\sin B}{b} \qquad \text{Law of sines (alternative form)}$$

$$\frac{\sin A}{25.1} = \frac{\sin 55° 40'}{8.94} \qquad \text{Substitute the given values.}$$

$$\sin A = \frac{25.1 \sin 55° 40'}{8.94} \qquad \text{Multiply by 25.1.}$$

$$\sin A \approx 2.3184379 \qquad \text{Use a calculator.}$$

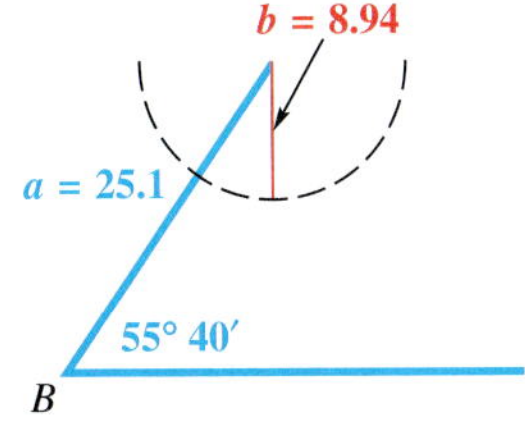

Figure 5

Since $\sin A$ cannot be greater than 1, there can be no such angle A and thus, no triangle with the given information. An attempt to sketch such a triangle leads to the situation shown in Figure 5.

Now try Exercise 29.

NOTE In the ambiguous case we are given SSA; that is, two sides and an angle opposite one of the sides. For example, suppose that b, c, and angle C are given. This situation represents the ambiguous case because angle C is opposite side c.

EXAMPLE 5 Solving the Ambiguous Case (Two Triangles)

Solve triangle ABC if $A = 55.3°$, $a = 22.8$ ft, and $b = 24.9$ ft.

Solution To begin, use the law of sines to find angle B.

$$\frac{\sin A}{a} = \frac{\sin B}{b}$$

$$\frac{\sin 55.3°}{22.8} = \frac{\sin B}{24.9}$$

$$\sin B = \frac{24.9 \sin 55.3°}{22.8}$$

$$\sin B \approx .8978678$$

There are two angles B between 0° and 180° that satisfy this condition. Since $\sin B \approx .8978678$, to the nearest tenth one value of B is

$$B_1 = 63.9°. \qquad \text{Use the inverse sine function. (Section 7.5)}$$

Supplementary angles have the same sine value, so another *possible* value of B is

$$B_2 = 180° - 63.9° = 116.1°.$$

To see if $B_2 = 116.1°$ is a valid possibility, add 116.1° to the measure of A, 55.3°. Since $116.1° + 55.3° = 171.4°$, and this sum is less than 180°, it is a valid angle measure for this triangle.

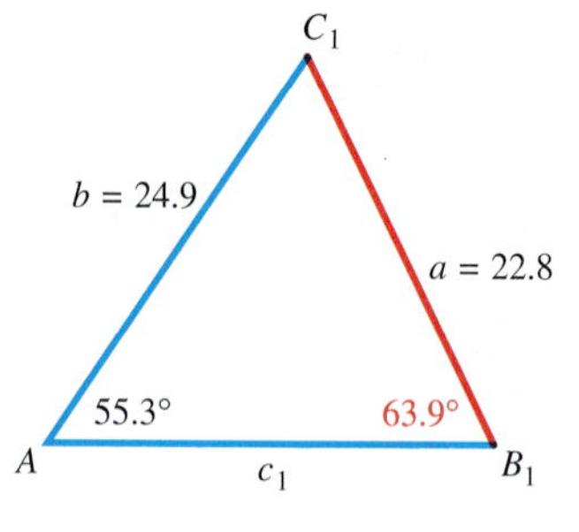

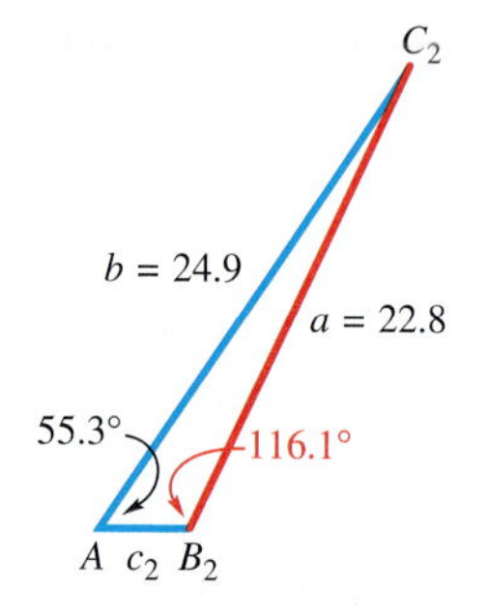

Figure 6

Now separately solve triangles AB_1C_1 and AB_2C_2 shown in Figure 6. Begin with AB_1C_1. Find C_1 first.

$$C_1 = 180° - A - B_1$$
$$C_1 = 180° - 55.3° - 63.9°$$
$$C_1 = 60.8°$$

Now, use the law of sines to find c_1.

$$\frac{a}{\sin A} = \frac{c_1}{\sin C_1}$$
$$\frac{22.8}{\sin 55.3°} = \frac{c_1}{\sin 60.8°}$$
$$c_1 = \frac{22.8 \sin 60.8°}{\sin 55.3°}$$
$$c_1 \approx 24.2 \text{ ft}$$

To solve triangle AB_2C_2, first find C_2.

$$C_2 = 180° - A - B_2$$
$$C_2 = 180° - 55.3° - 116.1°$$
$$C_2 = 8.6°$$

By the law of sines,

$$\frac{a}{\sin A} = \frac{c_2}{\sin C_2}$$
$$\frac{22.8}{\sin 55.3°} = \frac{c_2}{\sin 8.6°}$$
$$c_2 = \frac{22.8 \sin 8.6°}{\sin 55.3°}$$
$$c_2 \approx 4.15 \text{ ft.}$$

Now try Exercise 37.

The ambiguous case results in zero, one, or two triangles. The following guidelines can be used to determine how many triangles there are.

Number of Triangles Satisfying the Ambiguous Case (SSA)

Let sides a and b and angle A be given in triangle ABC. (The law of sines can be used to calculate the value of $\sin B$.)

1. If applying the law of sines results in an equation having $\sin B > 1$, then *no triangle* satisfies the given conditions.
2. If $\sin B = 1$, then *one triangle* satisfies the given conditions and $B = 90°$.
3. If $0 < \sin B < 1$, then either *one or two triangles* satisfy the given conditions.
   (a) If $\sin B = k$, then let $B_1 = \sin^{-1} k$ and use B_1 for B in the first triangle.
   (b) Let $B_2 = 180° - B_1$. If $A + B_2 < 180°$, then a second triangle exists. In this case, use B_2 for B in the second triangle.

NOTE The preceding guidelines can be applied whenever two sides and an angle opposite one of the sides are given.

EXAMPLE 6 Solving the Ambiguous Case (One Triangle)

Solve triangle ABC, given $A = 43.5°$, $a = 10.7$ in., and $c = 7.2$ in.

Solution To find angle C, use an alternative form of the law of sines.

$$\frac{\sin C}{c} = \frac{\sin A}{a}$$

$$\frac{\sin C}{7.2} = \frac{\sin 43.5°}{10.7}$$

$$\sin C = \frac{7.2 \sin 43.5°}{10.7} \approx .46319186$$

$$C \approx 27.6° \quad \text{Use the inverse sine function.}$$

There is another angle C that has sine value .46319186; it is $C = 180° - 27.6° = 152.4°$. However, notice in the given information that $c < a$, meaning that in the triangle, angle C must have measure *less than* angle A. Notice also that when we add this obtuse value to the given angle $A = 43.5°$, we obtain

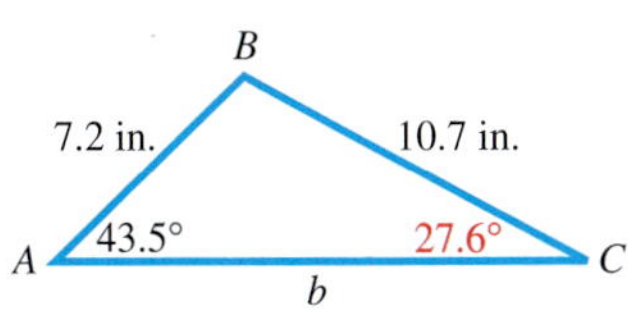

Figure 7

$$152.4° + 43.5° = 195.9°,$$

which is greater than 180°. So either of these approaches shows that there can be only one triangle. See Figure 7. Then

$$B = 180° - 27.6° - 43.5° = 108.9°,$$

and we can find side b with the law of sines.

$$\frac{b}{\sin B} = \frac{a}{\sin A}$$

$$\frac{b}{\sin 108.9°} = \frac{10.7}{\sin 43.5°}$$

$$b = \frac{10.7 \sin 108.9°}{\sin 43.5°}$$

$$b \approx 14.7 \text{ in.}$$

Now try Exercise 33.

EXAMPLE 7 Analyzing Data Involving an Obtuse Angle

Without using the law of sines, explain why $A = 104°$, $a = 26.8$ m, and $b = 31.3$ m cannot be valid for a triangle ABC.

Solution Since A is an obtuse angle, it is the largest angle and so the largest side of the triangle must be a. However, we are given $b > a$; thus, $B > A$, which is impossible if A is obtuse. Therefore, no such triangle ABC exists.

Now try Exercise 43.

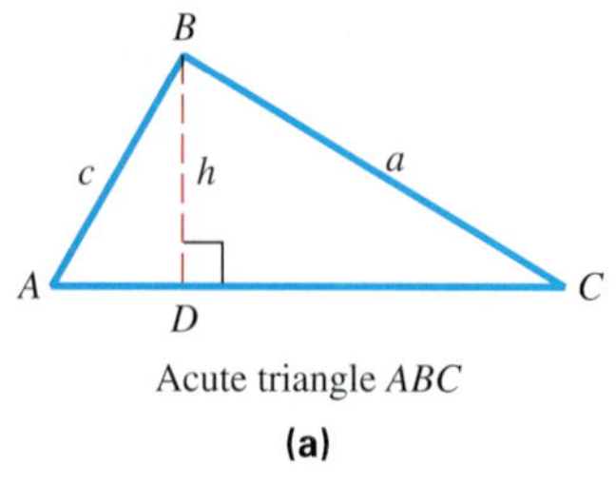

Acute triangle ABC

(a)

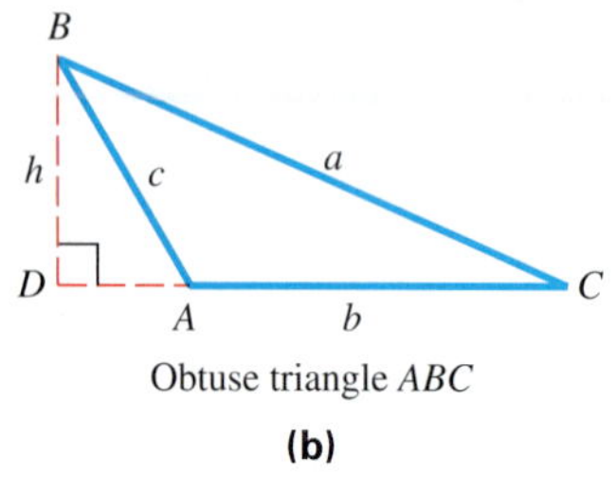

Obtuse triangle ABC

(b)

Figure 8

Area of a Triangle The method used to derive the law of sines can also be used to derive a formula to find the area of a triangle. A familiar formula for the area of a triangle is $\mathcal{A} = \frac{1}{2}bh$, where $\mathcal{A}$ represents area, b base, and h height. This formula cannot always be used easily since in practice, h is often unknown. To find another formula, refer to acute triangle ABC in Figure 8(a) or obtuse triangle ABC in Figure 8(b).

A perpendicular has been drawn from B to the base of the triangle (or the extension of the base). This perpendicular forms two right triangles. Using triangle ABD,

$$\sin A = \frac{h}{c} \quad \text{or} \quad h = c \sin A.$$

Substituting into the formula $\mathcal{A} = \frac{1}{2}bh$,

$$\mathcal{A} = \frac{1}{2}b(c \sin A) \quad \text{or} \quad \mathcal{A} = \frac{1}{2}bc \sin A.$$

Any other pair of sides and the angle between them could have been used.

Area of a Triangle (SAS)

In any triangle ABC, the area $\mathcal{A}$ is given by the following formulas:

$$\mathcal{A} = \frac{1}{2}bc \sin A, \quad \mathcal{A} = \frac{1}{2}ab \sin C, \quad \text{and} \quad \mathcal{A} = \frac{1}{2}ac \sin B.$$

That is, the area is half the product of the lengths of two sides and the sine of the angle included between them.

NOTE If the included angle measures 90°, its sine is 1, and the formula becomes the familiar $\mathcal{A} = \frac{1}{2}bh$.

EXAMPLE 8 Finding the Area of a Triangle (SAS)

Find the area of triangle ABC in Figure 9.

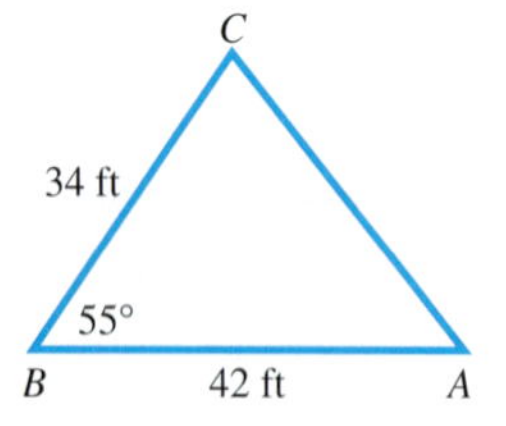

Figure 9

Solution We are given $B = 55°$, $a = 34$ ft, and $c = 42$ ft, so

$$\mathcal{A} = \frac{1}{2}ac \sin B = \frac{1}{2}(34)(42) \sin 55° \approx 585 \text{ ft}^2.$$

Now try Exercise 65.

EXAMPLE 9 Finding the Area of a Triangle (ASA)

Find the area of triangle ABC if $A = 24° 40'$, $b = 27.3$ cm, and $C = 52° 40'$.

Solution Before the area formula can be used, we must find either a or c. Since the sum of the measures of the angles of any triangle is 180°,

$$B = 180° - 24° 40' - 52° 40' = 102° 40'.$$

We can use the law of sines to find a.

$$\frac{a}{\sin A} = \frac{b}{\sin B}$$

$$\frac{a}{\sin 24° 40'} = \frac{27.3}{\sin 102° 40'}$$

$$a = \frac{27.3 \sin 24° 40'}{\sin 102° 40'}$$

$$a \approx 11.7 \text{ cm}$$

Now, we find the area.

$$\mathcal{A} = \frac{1}{2}ab \sin C = \frac{1}{2}(11.7)(27.3) \sin 52° 40' \approx 127 \text{ cm}^2$$

Now try Exercise 71.

8.1 Exercises

1. *Concept Check* Consider the oblique triangle ABC. Which one of the following proportions is *not* valid?

A. $\dfrac{a}{b} = \dfrac{\sin A}{\sin B}$ **B.** $\dfrac{a}{\sin A} = \dfrac{b}{\sin B}$

C. $\dfrac{\sin A}{a} = \dfrac{b}{\sin B}$ **D.** $\dfrac{\sin A}{a} = \dfrac{\sin B}{b}$

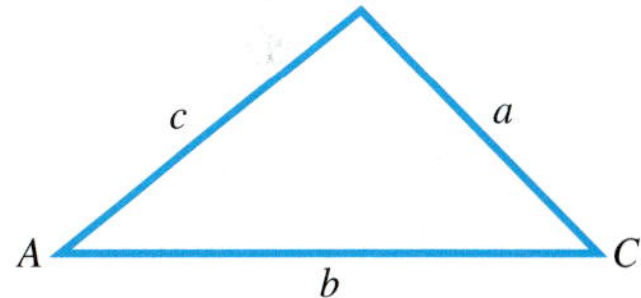

2. *Concept Check* Which two of the following situations do not provide sufficient information for solving a triangle by the law of sines?

A. You are given two angles and the side included between them.
B. You are given two angles and a side opposite one of them.
C. You are given two sides and the angle included between them.
D. You are given three sides.

Find the length of each side a. Do not use a calculator.

3.

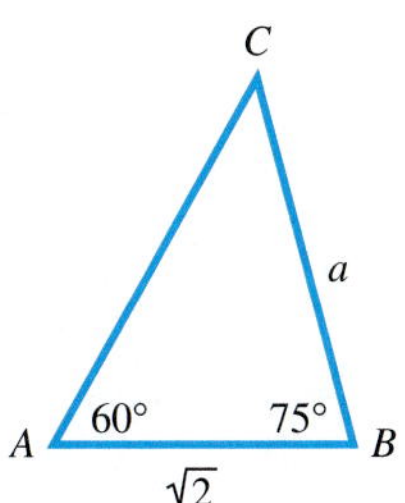

4.

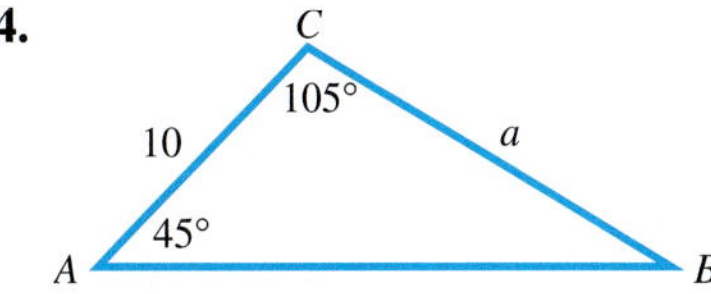

Determine the remaining sides and angles of each triangle ABC. See Example 1.

5.

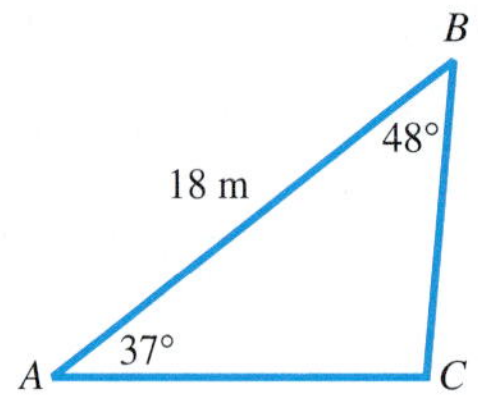

6.

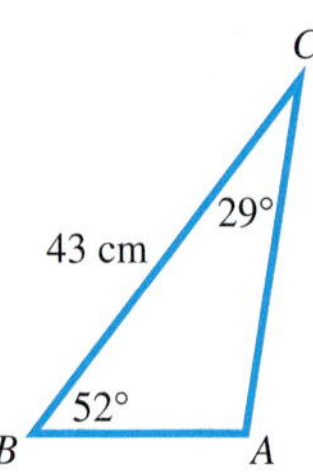

7.

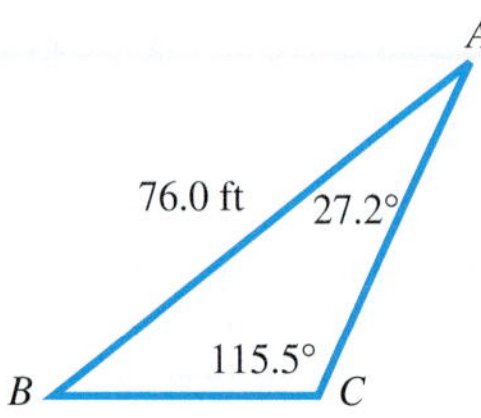

8.

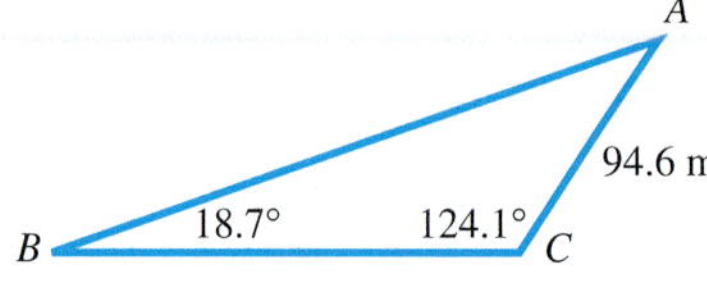

9. $A = 68.41°, B = 54.23°, a = 12.75$ ft

10. $C = 74.08°, B = 69.38°, c = 45.38$ m

11. $B = 20° \, 50', C = 103° \, 10', AC = 132$ ft

12. $A = 35.3°, B = 52.8°, AC = 675$ ft

13. $A = 39.70°, C = 30.35°, b = 39.74$ m

14. $C = 71.83°, B = 42.57°, a = 2.614$ cm

15. $B = 42.88°, C = 102.40°, b = 3974$ ft

16. $A = 18.75°, B = 51.53°, c = 2798$ yd

17. *Concept Check* Which one of the following sets of data does not determine a unique triangle?

A. $A = 40°, B = 60°, C = 80°$ **B.** $a = 5, b = 12, c = 13$
C. $a = 3, b = 7, C = 50°$ **D.** $a = 2, b = 2, c = 2$

18. *Concept Check* Which one of the following sets of data determines a unique triangle?

A. $A = 50°, B = 50°, C = 80°$ **B.** $a = 3, b = 5, c = 20$
C. $A = 40°, B = 20°, C = 30°$ **D.** $a = 7, b = 24, c = 25$

Concept Check *In each figure, a line of length h is to be drawn from the given point to the positive x-axis in order to form a triangle. For what value(s) of h can you draw the following?*

(a) *two triangles* **(b)** *exactly one triangle* **(c)** *no triangle*

19.

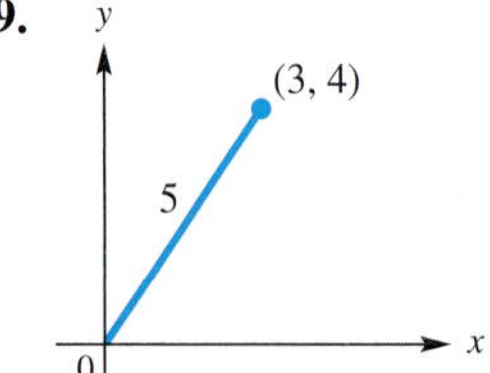

20.

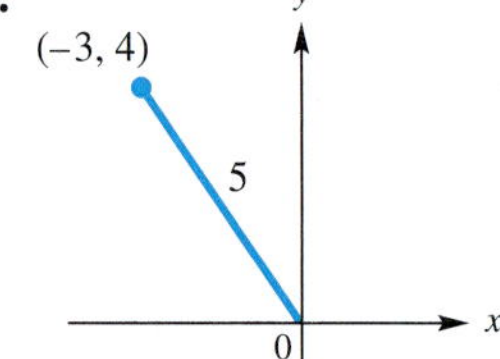

Determine the number of triangles ABC possible with the given parts. See Examples 4–7.

21. $a = 31, b = 26, B = 48°$

22. $a = 35, b = 30, A = 40°$

23. $a = 50, b = 61, A = 58°$

24. $B = 54°, c = 28, b = 23$

Find each angle B. Do not use a calculator.

25.

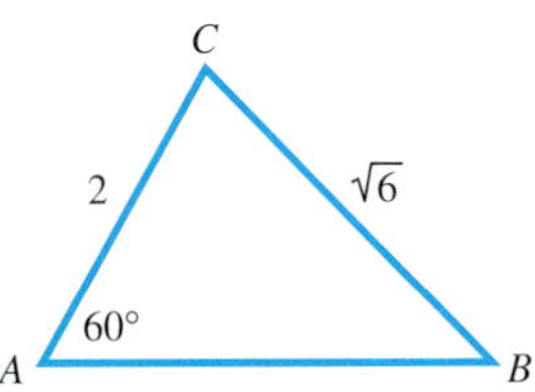

26.

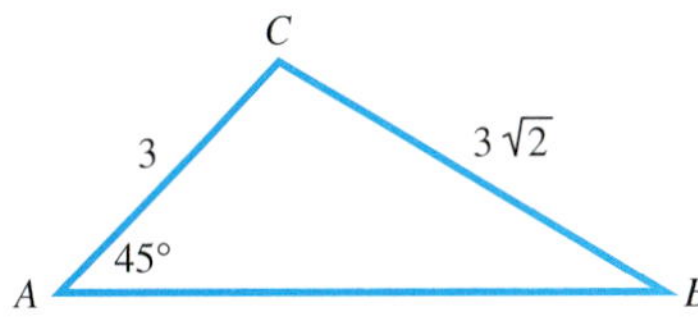

Find the unknown angles in triangle ABC for each triangle that exists. See Examples 4–6.

27. $A = 29.7°$, $b = 41.5$ ft, $a = 27.2$ ft

28. $B = 48.2°$, $a = 890$ cm, $b = 697$ cm

29. $B = 74.3°$, $a = 859$ m, $b = 783$ m

30. $C = 82.2°$, $a = 10.9$ km, $c = 7.62$ km

31. $A = 142.13°$, $b = 5.432$ ft, $a = 7.297$ ft

32. $B = 113.72°$, $a = 189.6$ yd, $b = 243.8$ yd

Solve each triangle ABC that exists. See Examples 4–6.

33. $A = 42.5°$, $a = 15.6$ ft, $b = 8.14$ ft

34. $C = 52.3°$, $a = 32.5$ yd, $c = 59.8$ yd

35. $B = 72.2°$, $b = 78.3$ m, $c = 145$ m

36. $C = 68.5°$, $c = 258$ cm, $b = 386$ cm

37. $A = 38° 40'$, $a = 9.72$ km, $b = 11.8$ km

38. $C = 29° 50'$, $a = 8.61$ m, $c = 5.21$ m

39. $B = 39.68°$, $a = 29.81$ m, $b = 23.76$ m

40. $A = 51.20°$, $c = 7986$ cm, $a = 7208$ cm

41. Apply the law of sines to the following: $a = \sqrt{5}$, $c = 2\sqrt{5}$, $A = 30°$. What is the value of sin C? What is the measure of C? Based on its angle measures, what kind of triangle is triangle ABC?

42. Explain the condition that must exist to determine that there is no triangle satisfying the given values of a, b, and B, once the value of sin B is found.

43. Without using the law of sines, explain why no triangle ABC exists satisfying $A = 103° 20'$, $a = 14.6$ ft, $b = 20.4$ ft.

44. Apply the law of sines to the data given in Example 7. Describe in your own words what happens when you try to find the measure of angle B using a calculator.

45. ***Property Survey*** A surveyor reported the following data about a piece of property: "The property is triangular in shape, with dimensions as shown in the figure." Use the law of sines to see whether such a piece of property could exist.

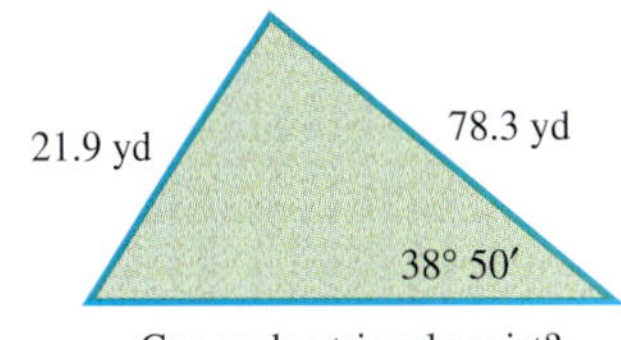

Can such a triangle exist?

46. ***Property Survey*** The surveyor tries again: "A second triangular piece of property has dimensions as shown." This time it turns out that the surveyor did not consider every possible case. Use the law of sines to show why.

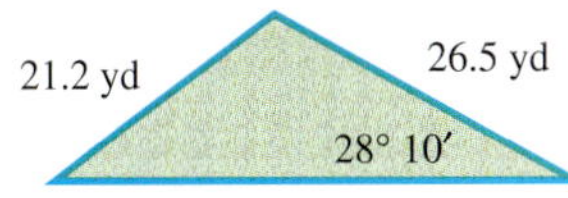

Solve each problem. See Examples 2 and 3.

47. ***Distance Across a River*** To find the distance AB across a river, a distance $BC = 354$ m is laid off on one side of the river. It is found that $B = 112° 10'$ and $C = 15° 20'$. Find AB. See the figure.

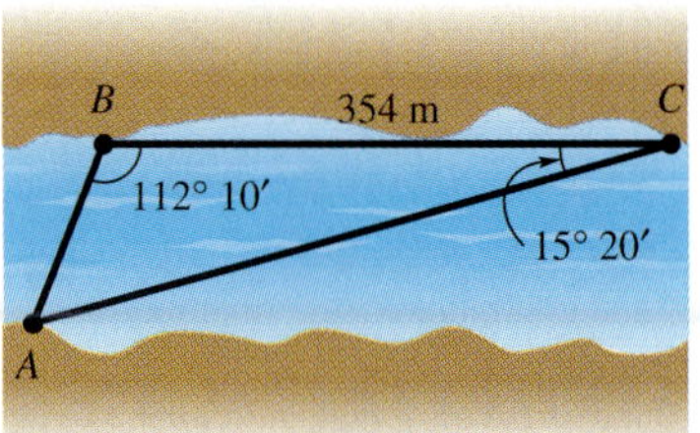

48. ***Distance Across a Canyon*** To determine the distance RS across a deep canyon, Rhonda lays off a distance $TR = 582$ yd. She then finds that $T = 32° 50'$ and $R = 102° 20'$. Find RS.

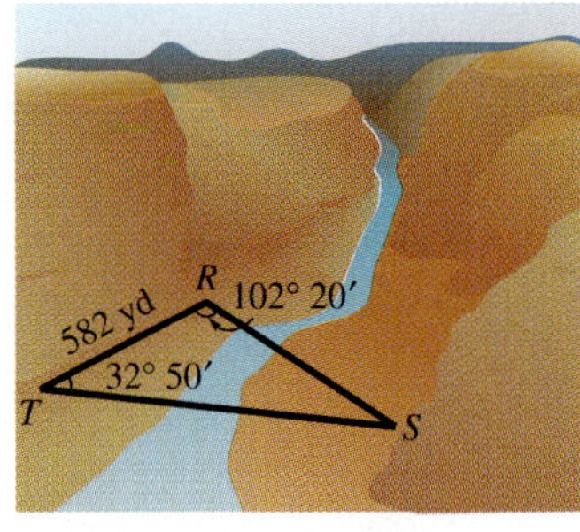

49. ***Distance a Ship Travels*** A ship is sailing due north. At a certain point the bearing of a lighthouse 12.5 km distant is N 38.8° E. Later on, the captain notices that the bearing of the lighthouse has become S 44.2° E. How far did the ship travel between the two observations of the lighthouse?

50. ***Distance Between Radio Direction Finders*** Radio direction finders are placed at points A and B, which are 3.46 mi apart on an east-west line, with A west of B. From A the bearing of a certain radio transmitter is 47.7°, and from B the bearing is 302.5°. Find the distance of the transmitter from A.

51. ***Measurement of a Folding Chair*** A folding chair is to have a seat 12.0 in. deep with angles as shown in the figure. How far down from the seat should the crossing legs be joined? (Find x in the figure.)

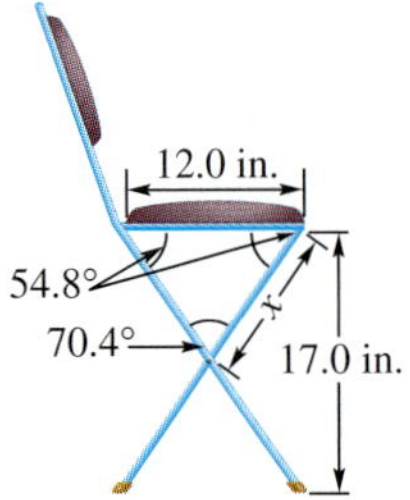

52. ***Distance Across a River*** Mark notices that the bearing of a tree on the opposite bank of a river flowing north is 115.45°. Lisa is on the same bank as Mark, but 428.3 m away. She notices that the bearing of the tree is 45.47°. The two banks are parallel. What is the distance across the river?

53. ***Angle Formed by Radii of Gears*** Three gears are arranged as shown in the figure. Find angle θ.

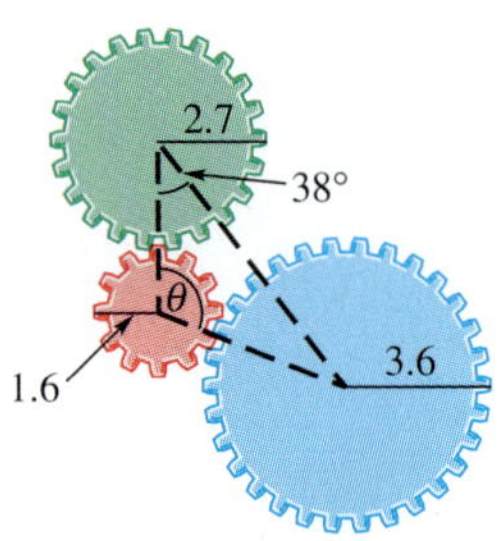

54. ***Distance Between Atoms*** Three atoms with atomic radii of 2.0, 3.0, and 4.5 are arranged as in the figure. Find the distance between the centers of atoms *A* and *C*.

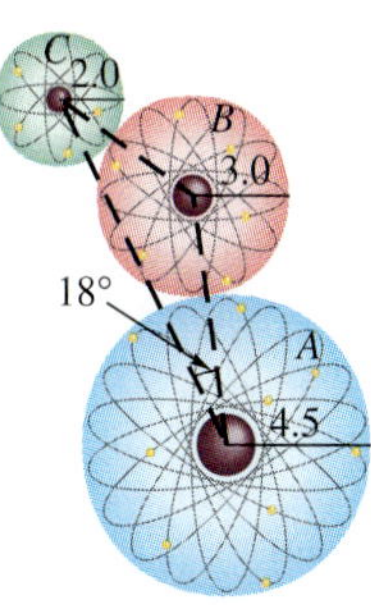

55. ***Distance Between a Ship and a Lighthouse*** The bearing of a lighthouse from a ship was found to be N 37° E. After the ship sailed 2.5 mi due south, the new bearing was N 25° E. Find the distance between the ship and the lighthouse at each location.

56. ***Height of a Balloon*** A balloonist is directly above a straight road 1.5 mi long that joins two villages. She finds that the town closer to her is at an angle of depression of 35°, and the farther town is at an angle of depression of 31°. How high above the ground is the balloon?

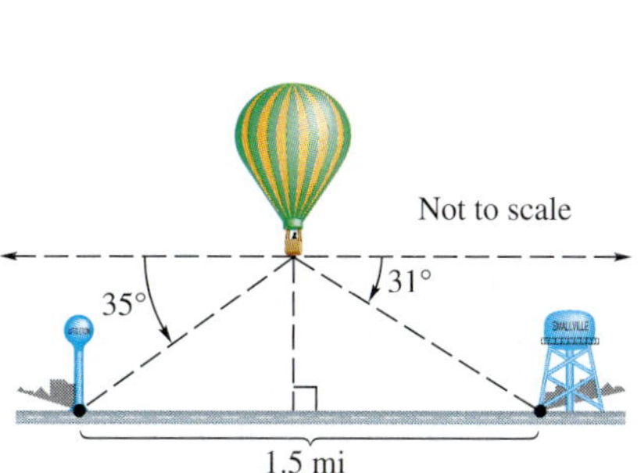

57. ***Distance Between a Pin and a Rod*** A slider crank mechanism is shown in the figure. Find the distance between the wrist pin *W* and the connecting rod center *C*.

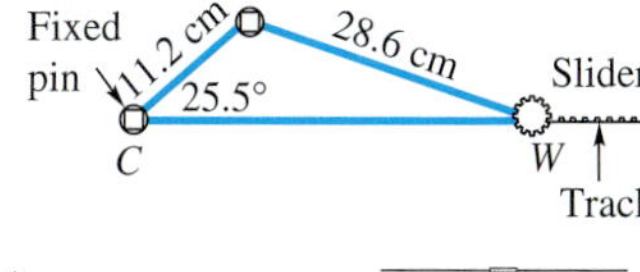

58. ***Height of a Helicopter*** A helicopter is sighted at the same time by two ground observers who are 3 mi apart on the same side of the helicopter. (See the figure.) They report the angles of elevation as 20.5° and 27.8°. How high is the helicopter?

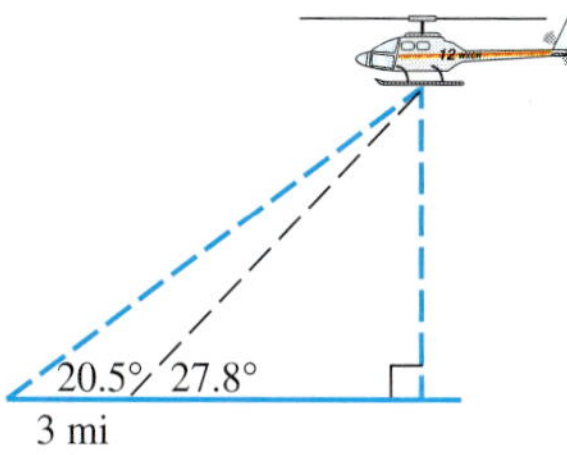

59. ***Distance from a Rocket to a Radar Station*** A rocket tracking facility has two radar stations T_1 and T_2, placed 1.73 km apart, that lock onto the rocket and continuously transmit the angles of elevation to a computer. Find the distance to the rocket from T_1 at the moment when the angles of elevation are 28.1° and 79.5°, as shown in the figure.

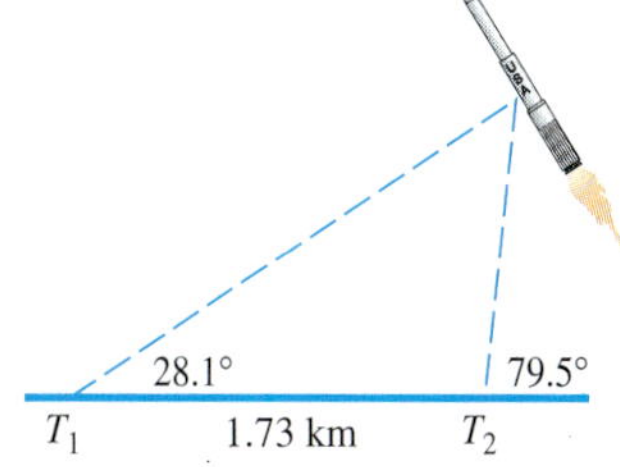

60. *Height of a Clock Tower* A surveyor standing 48.0 m from the base of a building measures the angle to the top of the building and finds it to be 37.4°. The surveyor then measures the angle to the top of a clock tower on the building, finding that it is 45.6°. Find the height of the clock tower.

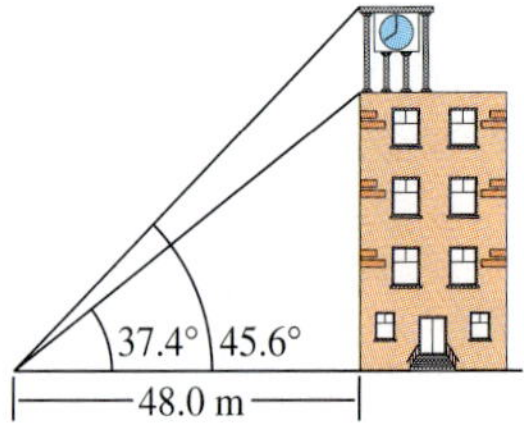

61. *Path of a Satellite* A satellite is traveling in a circular orbit 1600 km above Earth. It will pass directly over a tracking station at noon. The satellite takes 2 hr to make a complete orbit. Assume that the radius of Earth is 6400 km. The tracking antenna is aimed 30° above the horizon. See the figure. At what time (before noon) will the satellite pass through the beam of the antenna? (*Source: Space Mathematics* by B. Kastner, Ph.D. Copyright © 1985 by the National Aeronautics and Space Administration. Courtesy of NASA.)

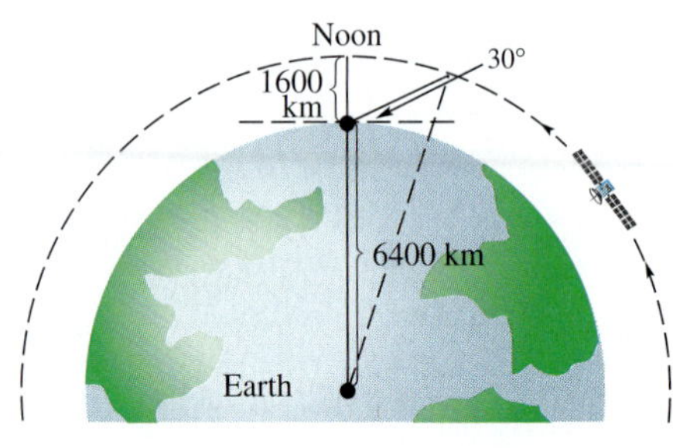

62. *Distance to the Moon* Since the moon is a relatively close celestial object, its distance can be measured directly by taking two different photographs at precisely the same time from two different locations. The moon will have a different angle of elevation at each location. On April 29, 1976, at 11:35 A.M., the lunar angles of elevation during a partial solar eclipse at Bochum in upper Germany and at Donaueschingen in lower Germany were measured as 52.6997° and 52.7430°, respectively. The two cities are 398 km apart. Calculate the distance to the moon from Bochum on this day, and compare it with the actual value of 406,000 km. Disregard the curvature of Earth in this calculation. (*Source:* Scholosser, W., T. Schmidt-Kaler, and E. Milone, *Challenges of Astronomy,* Springer-Verlag, 1991.)

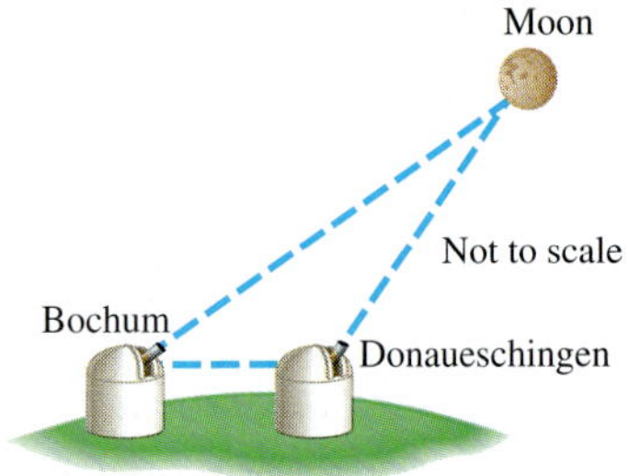

Find the area of each triangle using the formula $\mathcal{A} = \frac{1}{2}bh$, and then verify that the formula $\mathcal{A} = \frac{1}{2}ab \sin C$ gives the same result.

63.

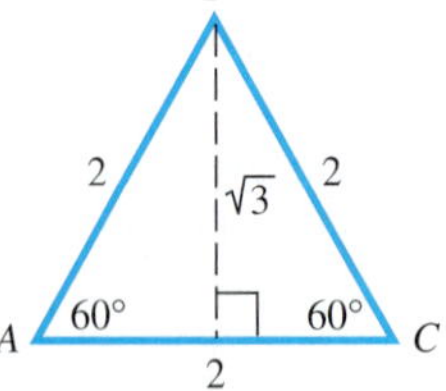

64.

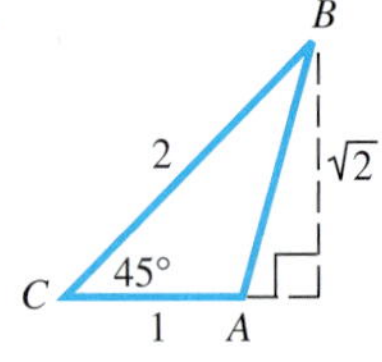

Find the area of each triangle ABC. See Examples 8 and 9.

65. $A = 42.5°$, $b = 13.6$ m, $c = 10.1$ m

66. $C = 72.2°$, $b = 43.8$ ft, $a = 35.1$ ft

67. $B = 124.5°$, $a = 30.4$ cm, $c = 28.4$ cm

68. $C = 142.7°$, $a = 21.9$ km, $b = 24.6$ km

69. $A = 56.80°$, $b = 32.67$ in., $c = 52.89$ in.

70. $A = 34.97°$, $b = 35.29$ m, $c = 28.67$ m

71. $A = 30.50°$, $b = 13.00$ cm, $C = 112.60°$

72. $A = 59.80°$, $b = 15.00$ m, $C = 53.10°$

Solve each problem.

73. ***Area of a Metal Plate*** A painter is going to apply a special coating to a triangular metal plate on a new building. Two sides measure 16.1 m and 15.2 m. She knows that the angle between these sides is 125°. What is the area of the surface she plans to cover with the coating?

74. ***Area of a Triangular Lot*** A real estate agent wants to find the area of a triangular lot. A surveyor takes measurements and finds that two sides are 52.1 m and 21.3 m, and the angle between them is 42.2°. What is the area of the triangular lot?

75. ***Triangle Inscribed in a Circle*** For a triangle inscribed in a circle of radius r, each of the law of sines ratios $\frac{a}{\sin A}$, $\frac{b}{\sin B}$, and $\frac{c}{\sin C}$ have value $2r$. The circle in the figure has diameter 1. What are the values of a, b, and c? (*Note:* This result provides an alternative way to define the sine function for angles between 0° and 180°. It was used nearly 2000 yr ago by the mathematician Ptolemy to construct one of the earliest trigonometric tables.)

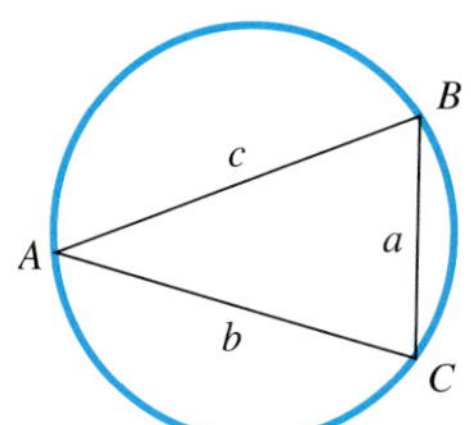

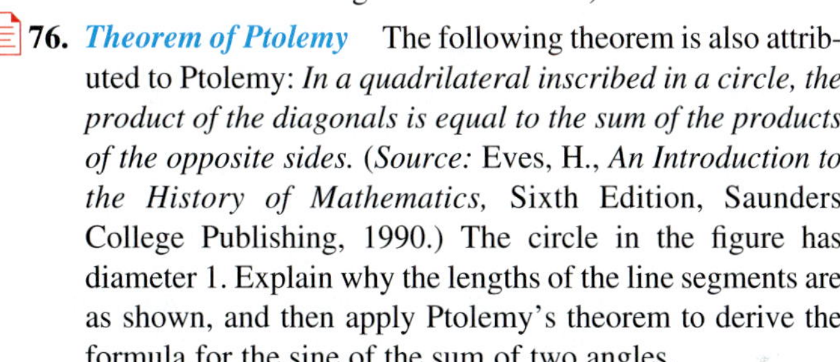
76. ***Theorem of Ptolemy*** The following theorem is also attributed to Ptolemy: *In a quadrilateral inscribed in a circle, the product of the diagonals is equal to the sum of the products of the opposite sides.* (*Source:* Eves, H., *An Introduction to the History of Mathematics*, Sixth Edition, Saunders College Publishing, 1990.) The circle in the figure has diameter 1. Explain why the lengths of the line segments are as shown, and then apply Ptolemy's theorem to derive the formula for the sine of the sum of two angles.

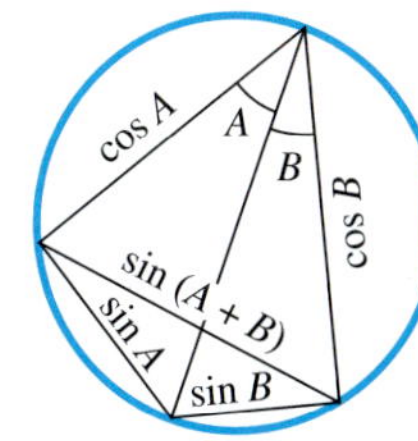

8.2 The Law of Cosines

Derivation of the Law of Cosines ▪ Applications ▪ Heron's Formula for the Area of a Triangle

As mentioned in Section 8.1, if we are given two sides and the included angle (Case 3) or three sides (Case 4) of a triangle, then a unique triangle is determined. These are the SAS and SSS cases, respectively. Both cases require using the *law of cosines*.

Remember the following property of triangles when applying the law of cosines to solve a triangle.

Triangle Side Length Restriction

In any triangle, the sum of the lengths of any two sides must be greater than the length of the remaining side.

For example, it would be impossible to construct a triangle with sides of lengths 3, 4, and 10. See Figure 10.

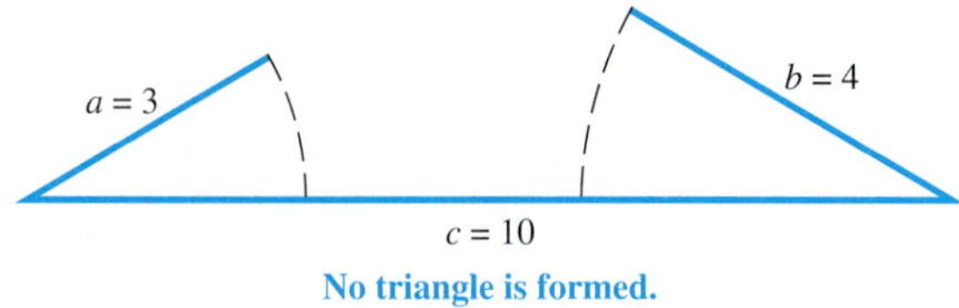

Figure 10

Derivation of the Law of Cosines To derive the law of cosines, let ABC be any oblique triangle. Choose a coordinate system so that vertex B is at the origin and side BC is along the positive x-axis. See Figure 11.

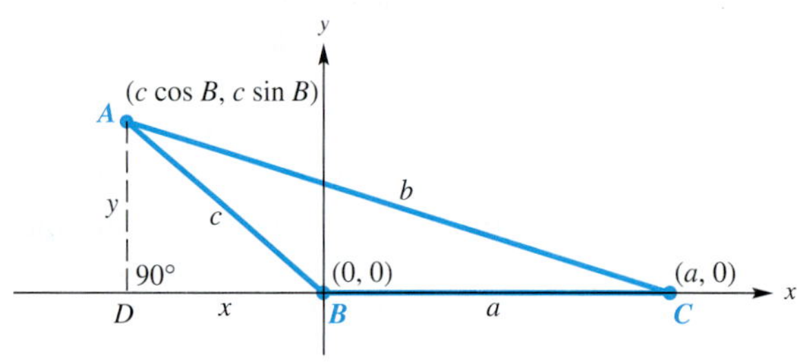

Figure 11

Let (x, y) be the coordinates of vertex A of the triangle. Verify that for angle B, whether obtuse or acute,

$$\sin B = \frac{y}{c} \quad \text{and} \quad \cos B = \frac{x}{c}. \quad \text{(Section 5.2)}$$

(Here x is negative when B is obtuse.) From these results

$$y = c \sin B \quad \text{and} \quad x = c \cos B,$$

so the coordinates of point A become $(c \cos B, c \sin B)$.

Point C has coordinates $(a, 0)$, and AC has length b. By the distance formula,

$$b = \sqrt{(c \cos B - a)^2 + (c \sin B)^2} \qquad \text{(Section 2.1)}$$

$$b^2 = (c \cos B - a)^2 + (c \sin B)^2 \qquad \text{Square both sides. (Section 1.6)}$$

$$= c^2 \cos^2 B - 2ac \cos B + a^2 + c^2 \sin^2 B \qquad \text{Multiply. (Section R.3)}$$

$$= a^2 + c^2(\cos^2 B + \sin^2 B) - 2ac \cos B \qquad \text{Properties of real numbers (Section R.1)}$$

$$= a^2 + c^2(1) - 2ac \cos B \qquad \text{Fundamental identity (Section 7.1)}$$

$$b^2 = a^2 + c^2 - 2ac \cos B.$$

This result is one form of the law of cosines. In our work, we could just as easily have placed A or C at the origin. This would have given the same result, but with the variables rearranged.

Law of Cosines

In any triangle ABC, with sides a, b, and c,

$$a^2 = b^2 + c^2 - 2bc \cos A,$$
$$b^2 = a^2 + c^2 - 2ac \cos B,$$
$$c^2 = a^2 + b^2 - 2ab \cos C.$$

That is, according to the law of cosines, the square of a side of a triangle is equal to the sum of the squares of the other two sides, minus twice the product of those two sides and the cosine of the angle included between them.

NOTE If we let $C = 90°$ in the third form of the law of cosines, we have $\cos C = \cos 90° = 0$, and the formula becomes $c^2 = a^2 + b^2$, the Pythagorean theorem. The Pythagorean theorem is a special case of the law of cosines.

Applications

EXAMPLE 1 Using the Law of Cosines in an Application (SAS)

A surveyor wishes to find the distance between two inaccessible points A and B on opposite sides of a lake. While standing at point C, she finds that $AC = 259$ m, $BC = 423$ m, and angle ACB measures $132° 40'$. Find the distance AB. See Figure 12.

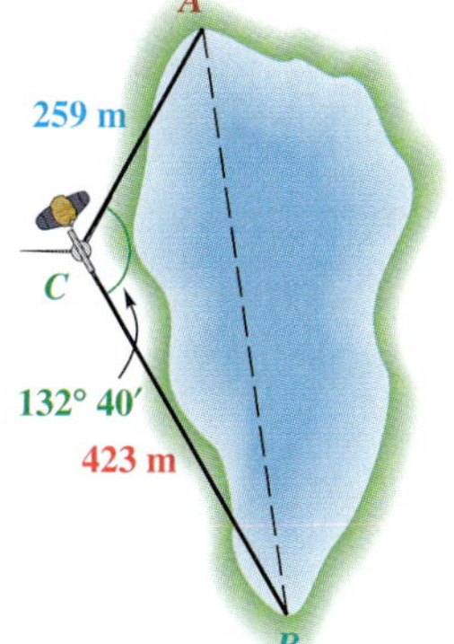

Figure 12

Solution The law of cosines can be used here since we know the lengths of two sides of the triangle and the measure of the included angle.

$$AB^2 = 259^2 + 423^2 - 2(259)(423) \cos 132° 40'$$

$AB^2 \approx 394{,}510.6$ Use a calculator.

$AB \approx 628$ Take the square root of each side. (Section 1.4)

The distance between the points is approximately 628 m.

Now try Exercise 39.

EXAMPLE 2 Using the Law of Cosines to Solve a Triangle (SAS)

Solve triangle ABC if $A = 42.3°$, $b = 12.9$ m, and $c = 15.4$ m. See Figure 13.

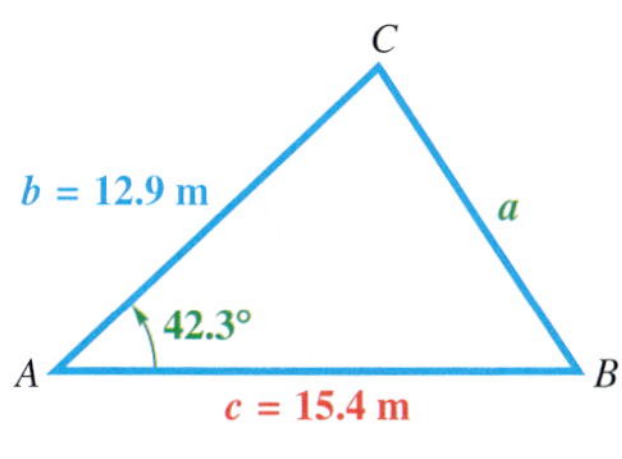

Figure 13

Solution We start by finding a with the law of cosines.

$$a^2 = b^2 + c^2 - 2bc \cos A$$
$$a^2 = 12.9^2 + 15.4^2 - 2(12.9)(15.4) \cos 42.3°$$
$$a^2 \approx 109.7$$
$$a \approx 10.47 \text{ m}$$

We now find the measures of angles B and C. Of the two remaining angles, B must be the smaller since it is opposite the shorter of the two sides b and c.

Therefore, B cannot be obtuse.

$$\frac{\sin A}{a} = \frac{\sin B}{b} \qquad \text{Law of sines (Section 8.1)}$$

$$\frac{\sin 42.3°}{10.47} = \frac{\sin B}{12.9}$$

$$\sin B = \frac{12.9 \sin 42.3°}{10.47} \qquad \text{Multiply by 10.47; rewrite.}$$

$$B \approx 56.0° \qquad \text{Use the inverse sine function. (Section 7.5)}$$

The easiest way to find C is to subtract the measures of A and B from 180°.

$$C = 180° - 42.3° - 56.0° = 81.7°$$

Now try Exercise 19.

CAUTION Had we chosen to use the law of sines to find C rather than B in Example 2, we would not have known whether C equals 81.7° or its supplement, 98.3°.

EXAMPLE 3 Using the Law of Cosines to Solve a Triangle (SSS)

Solve triangle ABC if $a = 9.47$ ft, $b = 15.9$ ft, and $c = 21.1$ ft.

Solution Given the lengths of three sides of the triangle, we can use the law of cosines to solve for any angle of the triangle. We solve for C, the largest angle. We will know that C is obtuse if $\cos C < 0$.

$$c^2 = a^2 + b^2 - 2ab \cos C \qquad \text{Law of cosines}$$

$$\cos C = \frac{a^2 + b^2 - c^2}{2ab} \qquad \text{Solve for } \cos C.$$

$$\cos C = \frac{9.47^2 + 15.9^2 - 21.1^2}{2(9.47)(15.9)} \qquad \text{Substitute.}$$

$$\cos C \approx -.34109402 \qquad \text{Use a calculator.}$$

$$C \approx 109.9° \qquad \text{Use the inverse cosine function. (Section 7.5)}$$

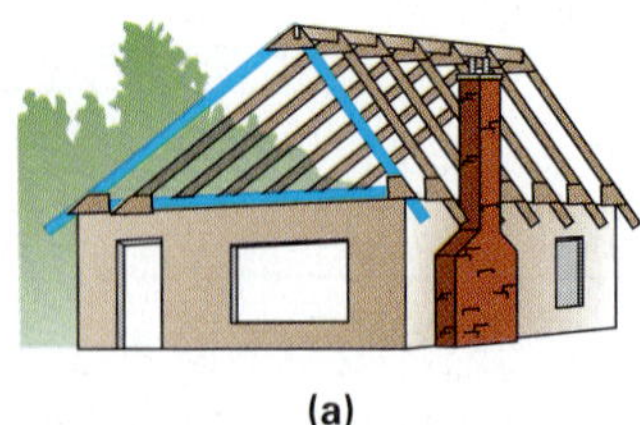

(a)

We can use either the law of sines or the law of cosines to find $B \approx 45.1°$. (Verify this.) Since $A = 180° - B - C$, we obtain $A \approx 25.0°$.

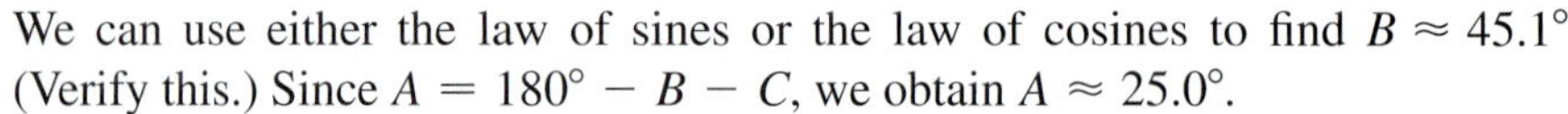

Now try Exercise 23.

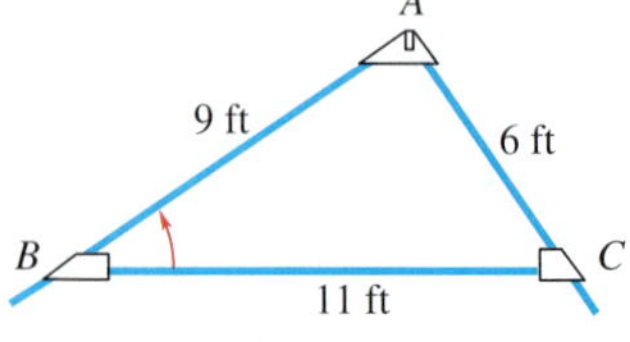

(b)

Figure 14

Trusses are frequently used to support roofs on buildings, as illustrated in Figure 14(a). The simplest type of roof truss is a triangle, as shown in Figure 14(b). One basic task when constructing a roof truss is to cut the ends of the rafters so that the roof has the correct slope. (*Source:* Riley, W., L. Sturges, and D. Morris, *Statics and Mechanics of Materials,* John Wiley and Sons, 1995.)

EXAMPLE 4 Designing a Roof Truss (SSS)

Find angle B for the truss shown in Figure 14(b).

Solution Let $a = 11$, $b = 6$, and $c = 9$ in the law of cosines.

$$b^2 = a^2 + c^2 - 2ac \cos B \quad \text{Law of cosines}$$

$$\cos B = \frac{a^2 + c^2 - b^2}{2ac} \quad \text{Solve for } \cos B.$$

$$\cos B = \frac{11^2 + 9^2 - 6^2}{2(11)(9)} \quad \text{Substitute.}$$

$$\cos B \approx .8384 \quad \text{Use a calculator.}$$

$$B \approx 33° \quad \text{Use the inverse cosine function.}$$

Now try Exercise 45.

Four possible cases can occur when solving an oblique triangle. These cases are summarized in the following table, along with a suggested procedure for solving in each case. There are other procedures that will work, but we give the one that is usually most efficient. In all four cases, it is assumed that the given information actually produces a triangle.

Oblique Triangle	Suggested Procedure for Solving
Case 1: One side and two angles are known. (SAA or ASA)	*Step 1* Find the remaining angle using the angle sum formula ($A + B + C = 180°$). *Step 2* Find the remaining sides using the law of sines.
Case 2: Two sides and one angle (not included between the two sides) are known. (SSA)	*This is the ambiguous case; there may be no triangle, one triangle, or two triangles.* *Step 1* Find an angle using the law of sines. *Step 2* Find the remaining angle using the angle sum formula. *Step 3* Find the remaining side using the law of sines. *If two triangles exist, repeat Steps 2 and 3.*
Case 3: Two sides and the included angle are known. (SAS)	*Step 1* Find the third side using the law of cosines. *Step 2* Find the smaller of the two remaining angles using the law of sines. *Step 3* Find the remaining angle using the angle sum formula.
Case 4: Three sides are known. (SSS)	*Step 1* Find the largest angle using the law of cosines. *Step 2* Find either remaining angle using the law of sines. *Step 3* Find the remaining angle using the angle sum formula.

Heron of Alexandria

Heron's Formula for the Area of a Triangle The law of cosines can be used to derive a formula for the area of a triangle given the lengths of the three sides. This formula is known as **Heron's formula,** named after the Greek mathematician Heron of Alexandria, who lived around A.D. 75. It is found in his work *Metrica*. Heron's formula can be used for the case SSS.

Heron's Area Formula (SSS)

If a triangle has sides of lengths a, b, and c, with **semiperimeter**

$$s = \frac{1}{2}(a + b + c),$$

then the area of the triangle is

$$\mathcal{A} = \sqrt{s(s - a)(s - b)(s - c)}.$$

That is, according to Heron's formula, the area of a triangle is the square root of the product of four factors: (1) the semiperimeter, (2) the semiperimeter minus the first side, (3) the semiperimeter minus the second side, and (4) the semiperimeter minus the third side.

EXAMPLE 5 Using Heron's Formula to Find an Area (SSS)

The distance "as the crow flies" from Los Angeles to New York is 2451 mi, from New York to Montreal is 331 mi, and from Montreal to Los Angeles is 2427 mi. What is the area of the triangular region having these three cities as vertices? (Ignore the curvature of Earth.)

Solution In Figure 15 we let $a = 2451$, $b = 331$, and $c = 2427$.

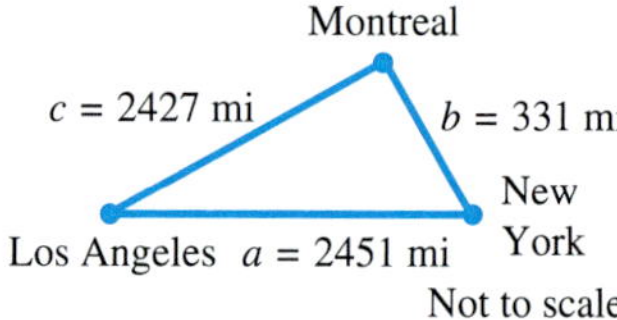

Figure 15

The semiperimeter s is

$$s = \frac{1}{2}(2451 + 331 + 2427) = 2604.5.$$

Using Heron's formula, the area $\mathcal{A}$ is

$$\mathcal{A} = \sqrt{s(s - a)(s - b)(s - c)}$$
$$\mathcal{A} = \sqrt{2604.5(2604.5 - 2451)(2604.5 - 331)(2604.5 - 2427)}$$
$$\mathcal{A} \approx 401{,}700 \text{ mi}^2.$$

Now try Exercise 73.

CONNECTIONS We have introduced two new formulas for the area of a triangle in this chapter. You should now be able to find the area $\mathcal{A}$ of a triangle using one of three formulas:

(a) $\mathcal{A} = \frac{1}{2}bh$

(b) $\mathcal{A} = \frac{1}{2}ab \sin C \quad \left(\text{or } \mathcal{A} = \frac{1}{2}ac \sin B \text{ or } \mathcal{A} = \frac{1}{2}bc \sin A\right)$

(c) $\mathcal{A} = \sqrt{s(s-a)(s-b)(s-c)}$.

Another area formula can be used when the coordinates of the vertices of a triangle are given. If the vertices are the ordered pairs (x_1, y_1), (x_2, y_2), and (x_3, y_3), then

$$\mathcal{A} = \frac{1}{2}\left|(x_1y_2 - y_1x_2 + x_2y_3 - y_2x_3 + x_3y_1 - y_3x_1)\right|.$$

For Discussion or Writing

Consider triangle PQR with vertices $P(2, 5)$, $Q(-1, 3)$, and $R(4, 0)$. (*Hint:* Draw a sketch first.)

1. Find the area of the triangle using the new formula just introduced.
2. Find the area of the triangle using (c) above. Use the distance formula to find the lengths of the three sides.
3. Find the area of the triangle using (b) above. First use the law of cosines to find the measure of an angle.

8.2 Exercises

Concept Check *Assume triangle ABC has standard labeling and complete the following.*

(a) *Determine whether* SAA, ASA, SSA, SAS, *or* SSS *is given.*

(b) *Decide whether the law of sines or the law of cosines should be used to begin solving the triangle.*

1. a, b, and C

2. A, C, and c

3. a, b, and A

4. a, b, and c

5. A, B, and c

6. a, c, and A

7. a, B, and C

8. b, c, and A

Find the length of the remaining side of each triangle. Do not use a calculator.

9.

3

60°

8

10.

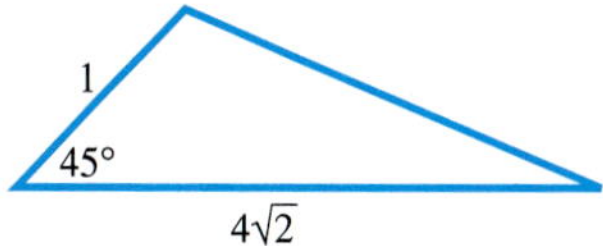

Find the value of θ in each triangle. Do not use a calculator.

11.

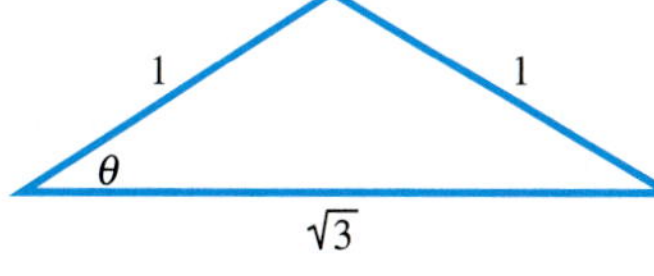

12.

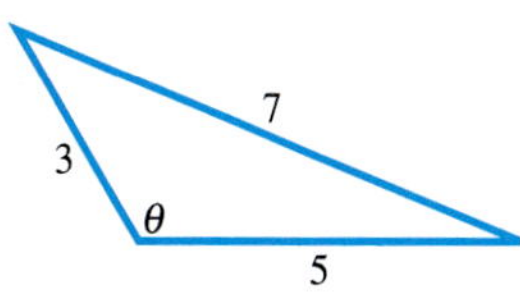

Solve each triangle. Approximate values to the nearest tenth.

13.

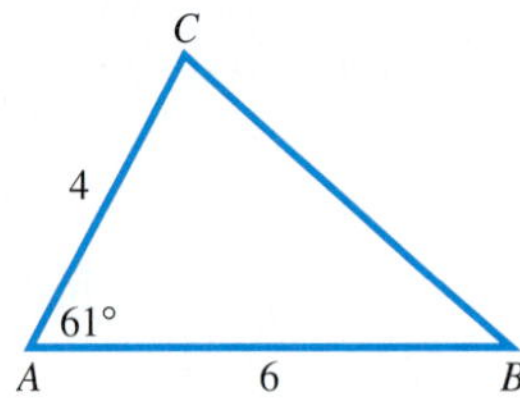

14.

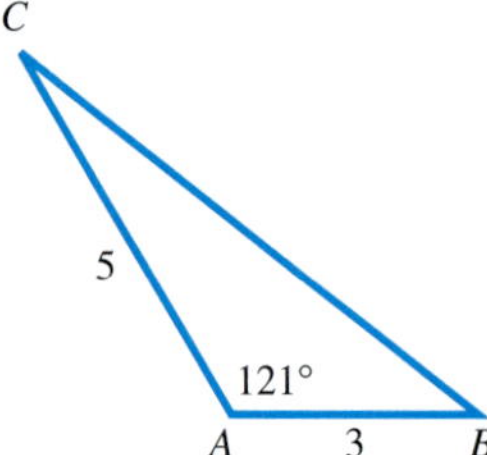

15.

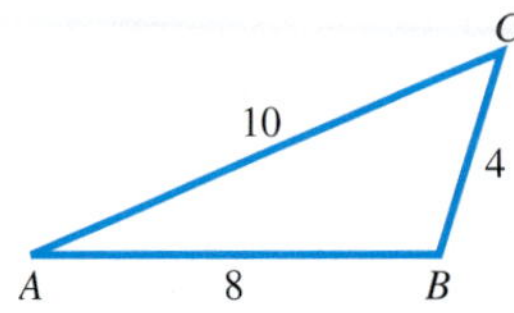

16.

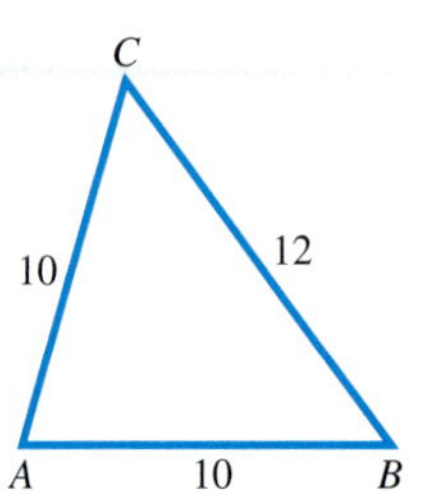

17.

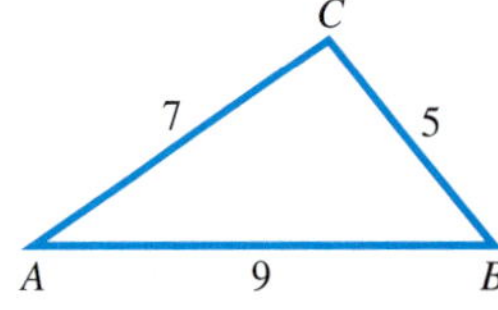

18.

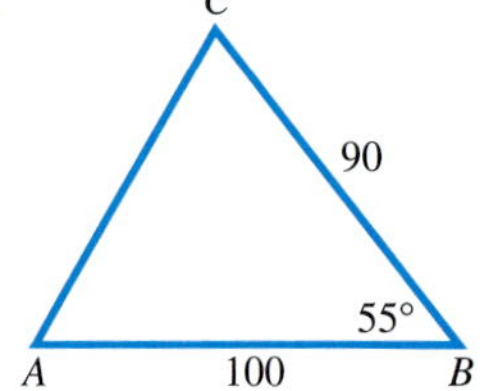

Solve each triangle. See Examples 2 and 3.

19. $A = 41.4°$, $b = 2.78$ yd, $c = 3.92$ yd

20. $C = 28.3°$, $b = 5.71$ in., $a = 4.21$ in.

21. $C = 45.6°$, $b = 8.94$ m, $a = 7.23$ m

22. $A = 67.3°$, $b = 37.9$ km, $c = 40.8$ km

23. $a = 9.3$ cm, $b = 5.7$ cm, $c = 8.2$ cm

24. $a = 28$ ft, $b = 47$ ft, $c = 58$ ft

25. $a = 42.9$ m, $b = 37.6$ m, $c = 62.7$ m

26. $a = 189$ yd, $b = 214$ yd, $c = 325$ yd

27. $AB = 1240$ ft, $AC = 876$ ft, $BC = 965$ ft

28. $AB = 298$ m, $AC = 421$ m, $BC = 324$ m

29. $A = 80° 40'$, $b = 143$ cm, $c = 89.6$ cm

30. $C = 72° 40'$, $a = 327$ ft, $b = 251$ ft

31. $B = 74.80°$, $a = 8.919$ in., $c = 6.427$ in.

32. $C = 59.70°$, $a = 3.725$ mi, $b = 4.698$ mi

33. $A = 112.8°$, $b = 6.28$ m, $c = 12.2$ m

34. $B = 168.2°$, $a = 15.1$ cm, $c = 19.2$ cm

35. $a = 3.0$ ft, $b = 5.0$ ft, $c = 6.0$ ft

36. $a = 4.0$ ft, $b = 5.0$ ft, $c = 8.0$ ft

37. Refer to Figure 10. If you attempt to find any angle of a triangle with the values $a = 3$, $b = 4$, and $c = 10$ with the law of cosines, what happens?

38. "The shortest distance between two points is a straight line." Explain how this relates to the geometric property that states that the sum of the lengths of any two sides of a triangle must be greater than the length of the remaining side.

Solve each problem. See Examples 1–4.

39. ***Distance Across a Lake*** Points A and B are on opposite sides of Lake Yankee. From a third point, C, the angle between the lines of sight to A and B is 46.3°. If AC is 350 m long and BC is 286 m long, find AB.

40. ***Diagonals of a Parallelogram*** The sides of a parallelogram are 4.0 cm and 6.0 cm. One angle is 58° while another is 122°. Find the lengths of the diagonals of the parallelogram.

41. ***Flight Distance*** Airports A and B are 450 km apart, on an east-west line. Tom flies in a northeast direction from A to airport C. From C he flies 359 km on a bearing of 128° 40′ to B. How far is C from A?

42. ***Distance Between Two Ships*** Two ships leave a harbor together, traveling on courses that have an angle of 135° 40′ between them. If they each travel 402 mi, how far apart are they?

43. ***Distance Between a Ship and a Rock*** A ship is sailing east. At one point, the bearing of a submerged rock is 45° 20′. After the ship has sailed 15.2 mi, the bearing of the rock has become 308° 40′. Find the distance of the ship from the rock at the later point.

44. ***Distance Between a Ship and a Submarine*** From an airplane flying over the ocean, the angle of depression to a submarine lying under the surface is 24° 10′. At the same moment, the angle of depression from the airplane to a battleship is 17° 30′. (See the figure.) The distance from the airplane to the battleship is 5120 ft. Find the distance between the battleship and the submarine. (Assume the airplane, submarine, and battleship are in a vertical plane.)

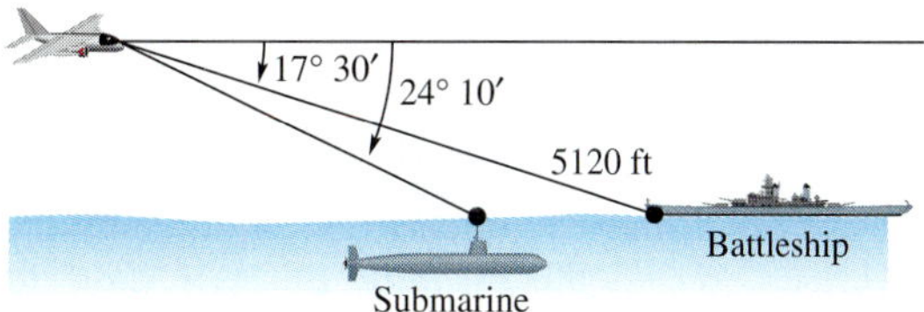

45. ***Truss Construction*** A triangular truss is shown in the figure. Find angle θ.

16 ft

13 ft

θ

20 ft

46. ***Distance Between Points on a Crane*** A crane with a counterweight is shown in the figure. Find the horizontal distance between points A and B.

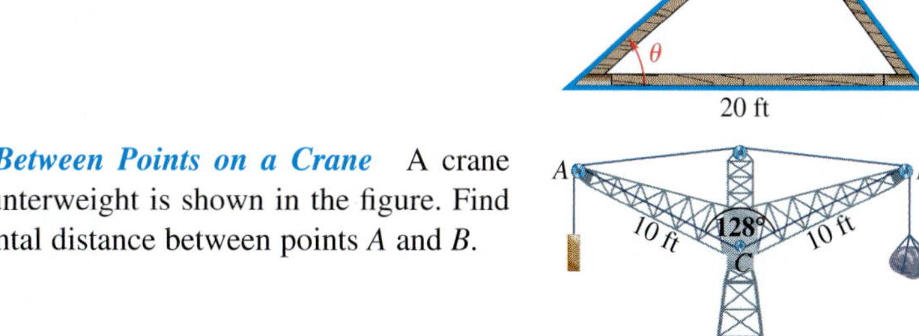

47. ***Distance Between a Beam and Cables*** A weight is supported by cables attached to both ends of a balance beam, as shown in the figure. What angles are formed between the beam and the cables?

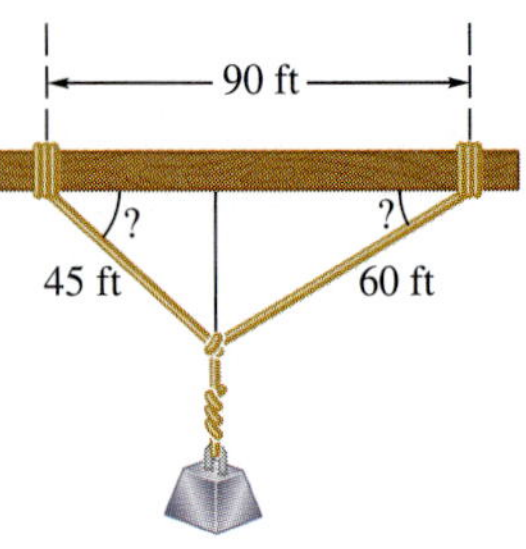

48. ***Length of a Tunnel*** To measure the distance through a mountain for a proposed tunnel, a point C is chosen that can be reached from each end of the tunnel. (See the figure.) If $AC = 3800$ m, $BC = 2900$ m, and angle $C = 110°$, find the length of the tunnel.

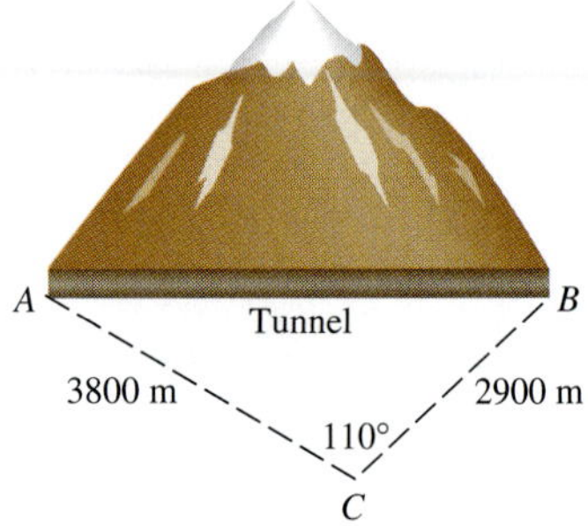

49. ***Distance on a Baseball Diamond*** A baseball diamond is a square, 90.0 ft on a side, with home plate and the three bases as vertices. The pitcher's rubber is 60.5 ft from home plate. Find the distance from the pitcher's rubber to each of the bases.

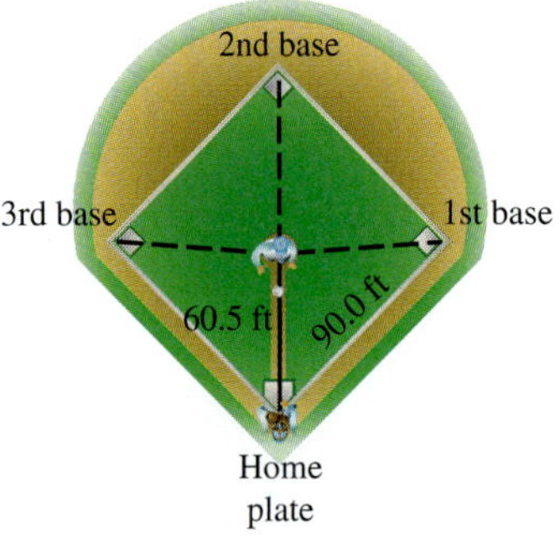

50. ***Distance Between Ends of the Vietnam Memorial*** The Vietnam Veterans' Memorial in Washington, D.C., is V-shaped with equal sides of length 246.75 ft. The angle between these sides measures 125° 12′. Find the distance between the ends of the two sides. (*Source:* Pamphlet obtained at Vietnam Veterans' Memorial.)

51. ***Distance Between a Ship and a Point*** Starting at point A, a ship sails 18.5 km on a bearing of 189°, then turns and sails 47.8 km on a bearing of 317°. Find the distance of the ship from point A.

52. ***Distance Between Two Factories*** Two factories blow their whistles at exactly 5:00. A man hears the two blasts at 3 sec and 6 sec after 5:00, respectively. The angle between his lines of sight to the two factories is 42.2°. If sound travels 344 m per sec, how far apart are the factories?

53. *Bearing of One Town to Another* Two towns 21 mi apart are separated by a dense forest. (See the figure.) To travel from town A to town B, a person must go 17 mi on a bearing of 325°, then turn and continue for 9 mi to reach town B. Find the bearing of B from A.

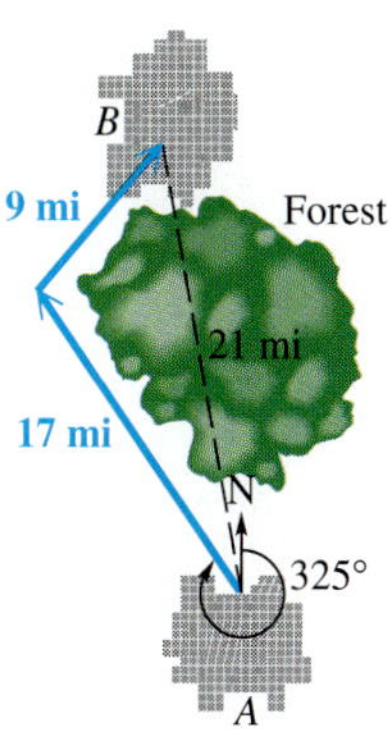

54. *Distance of a Plane* An airplane flies 180 mi from point X at a bearing of 125°, and then turns and flies at a bearing of 230° for 100 mi. How far is the plane from point X?

55. *Measurement Using Triangulation* Surveyors are often confronted with obstacles, such as trees, when measuring the boundary of a lot. One technique used to obtain an accurate measurement is the so-called *triangulation method.* In this technique, a triangle is constructed around the obstacle and one angle and two sides of the triangle are measured. Use this technique to find the length of the property line (the straight line between the two markers) in the figure. (*Source:* Kavanagh, B., *Surveying Principles and Applications,* Sixth Edition, Prentice-Hall, 2003.)

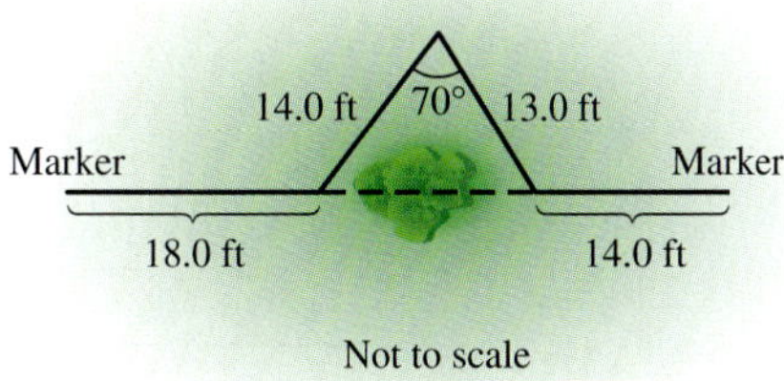

56. *Path of a Ship* A ship sailing due east in the North Atlantic has been warned to change course to avoid icebergs. The captain turns and sails on a bearing of 62°, then changes course again to a bearing of 115° until the ship reaches its original course. (See the figure.) How much farther did the ship have to travel to avoid the icebergs?

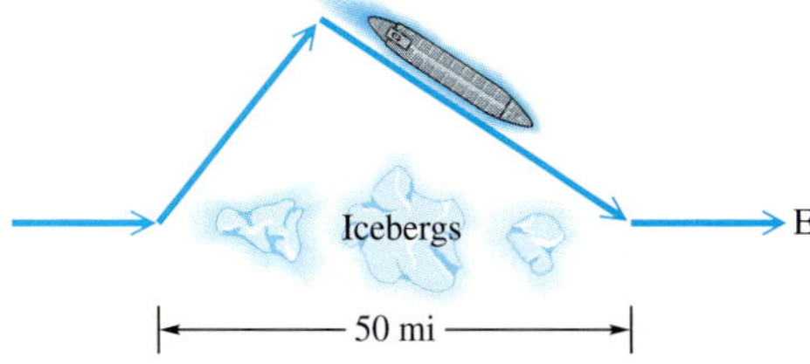

57. *Angle in a Parallelogram* A parallelogram has sides of length 25.9 cm and 32.5 cm. The longer diagonal has length 57.8 cm. Find the angle opposite the longer diagonal.

58. *Distance Between an Airplane and a Mountain* A person in a plane flying straight north observes a mountain at a bearing of 24.1°. At that time, the plane is 7.92 km from the mountain. A short time later, the bearing to the mountain becomes 32.7°. How far is the airplane from the mountain when the second bearing is taken?

59. *Layout for a Playhouse* The layout for a child's playhouse has the dimensions given in the figure. Find x.

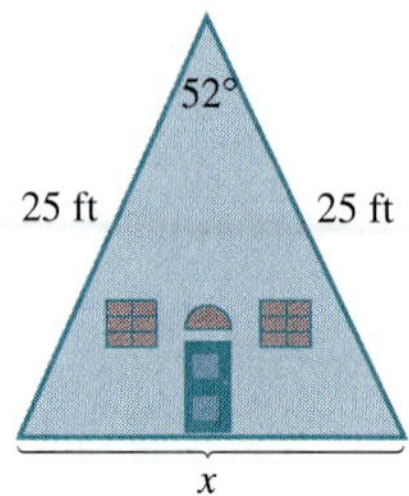

60. *Distance Between Two Towns* To find the distance between two small towns, an electronic distance measuring (EDM) instrument is placed on a hill from which both towns are visible. The distance to each town from the EDM and the angle between the two lines of sight are measured. (See the figure.) Find the distance between the towns.

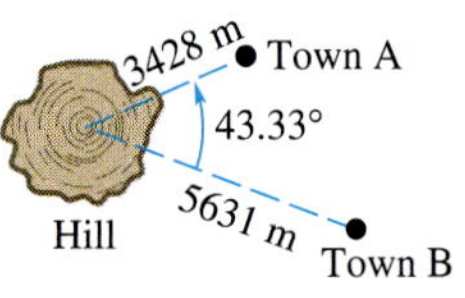

Find the measure of each angle θ to two decimal places.

61.

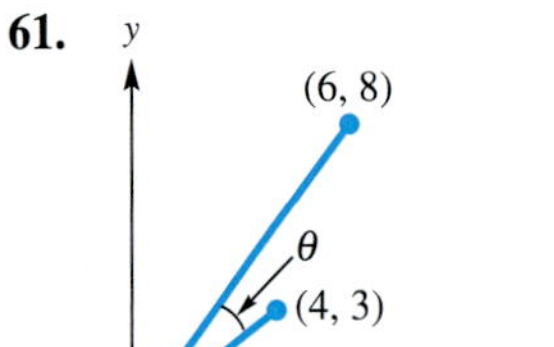

62.

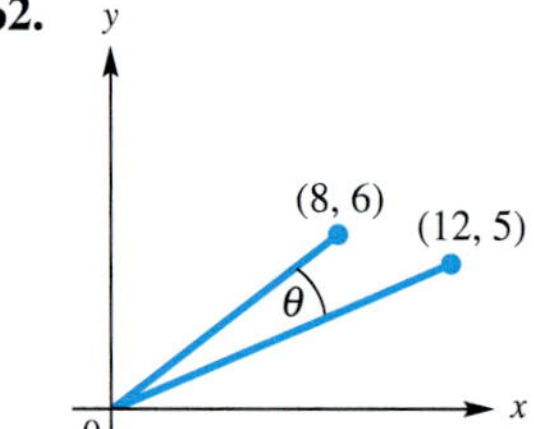

Find the exact area of each triangle using the formula $\mathcal{A} = \frac{1}{2}bh$, and then verify that Heron's formula gives the same result.

63.

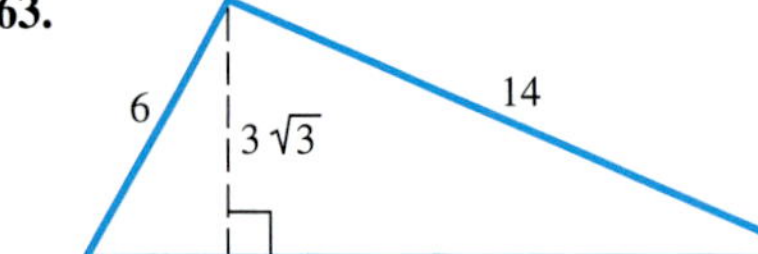

64.

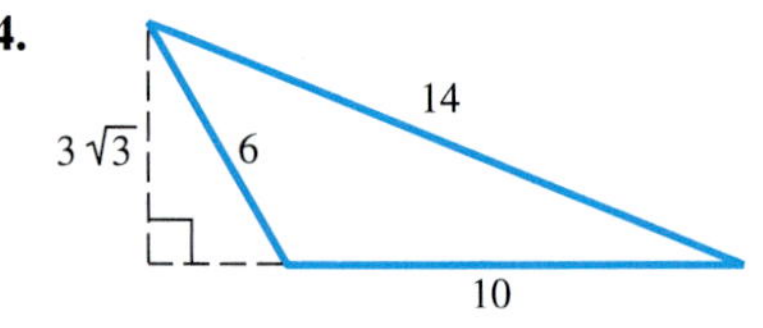

Find the area of each triangle ABC. See Example 5.

65. $a = 12$ m, $b = 16$ m, $c = 25$ m

66. $a = 22$ in., $b = 45$ in., $c = 31$ in.

67. $a = 154$ cm, $b = 179$ cm, $c = 183$ cm

68. $a = 25.4$ yd, $b = 38.2$ yd, $c = 19.8$ yd

69. $a = 76.3$ ft, $b = 109$ ft, $c = 98.8$ ft

70. $a = 15.89$ in., $b = 21.74$ in., $c = 10.92$ in.

Volcano Movement To help predict eruptions from the volcano Mauna Loa on the island of Hawaii, scientists keep track of the volcano's movement by using a "super triangle" with vertices on the three volcanoes shown on the map at the right. (For example, in one year, Mauna Loa moved 6 in., a result of increasing internal pressure.) Refer to the map to work Exercises 71 and 72.

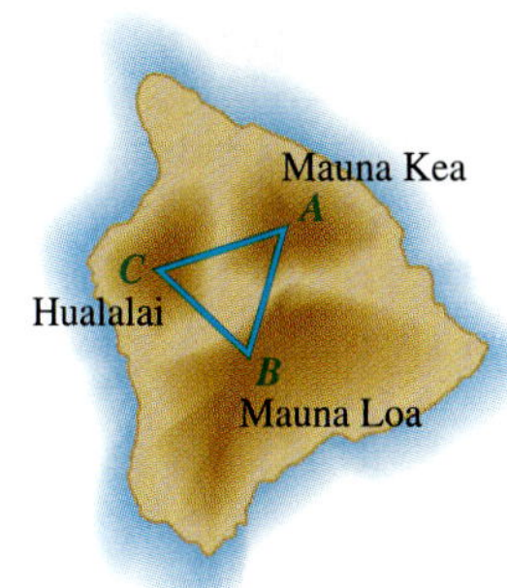

71. $AB = 22.47928$ mi, $AC = 28.14276$ mi, $A = 58.56989°$; find BC

72. $AB = 22.47928$ mi, $BC = 25.24983$ mi, $A = 58.56989°$; find B

Solve each problem. See Example 5.

73. *Area of the Bermuda Triangle* Find the area of the Bermuda Triangle if the sides of the triangle have approximate lengths 850 mi, 925 mi, and 1300 mi.

74. *Required Amount of Paint* A painter needs to cover a triangular region 75 m by 68 m by 85 m. A can of paint covers 75 m^2 of area. How many cans (to the next higher number of cans) will be needed?

75. *Perfect Triangles* A *perfect triangle* is a triangle whose sides have whole number lengths and whose area is numerically equal to its perimeter. Show that the triangle with sides of length 9, 10, and 17 is perfect.

76. *Heron Triangles* A *Heron triangle* is a triangle having integer sides and area. Show that each of the following is a Heron triangle.

(a) $a = 11, b = 13, c = 20$ **(b)** $a = 13, b = 14, c = 15$
(c) $a = 7, b = 15, c = 20$

77. Consider triangle ABC shown here.

(a) Use the law of sines to find candidates for the value of angle C.
(b) Rework part (a) using the law of cosines.
(c) Why is the law of cosines a better method in this case?

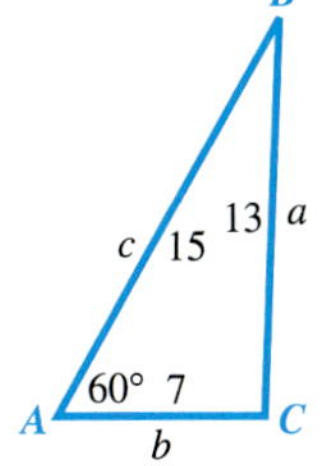

78. Show that the measure of angle A is twice the measure of angle B. (*Hint:* Use the law of cosines to find $\cos A$ and $\cos B$, and then show that $\cos A = 2\cos^2 B - 1$.)

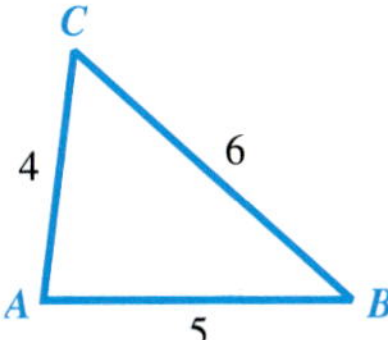

79. Let point D on side AB of triangle ABC be such that CD bisects angle C. Show that $\frac{AD}{DB} = \frac{b}{a}$.

80. In addition to the law of sines and the law of cosines, there is a **law of tangents.** In any triangle ABC,

$$\frac{\tan \frac{1}{2}(A - B)}{\tan \frac{1}{2}(A + B)} = \frac{a - b}{a + b}.$$

Verify this law for the triangle ABC with $a = 2$, $b = 2\sqrt{3}$, $A = 30°$, and $B = 60°$.

8.3 Vectors, Operations, and the Dot Product

Basic Terminology ▪ Algebraic Interpretation of Vectors ▪ Operations with Vectors ▪ Dot Product and the Angle Between Vectors

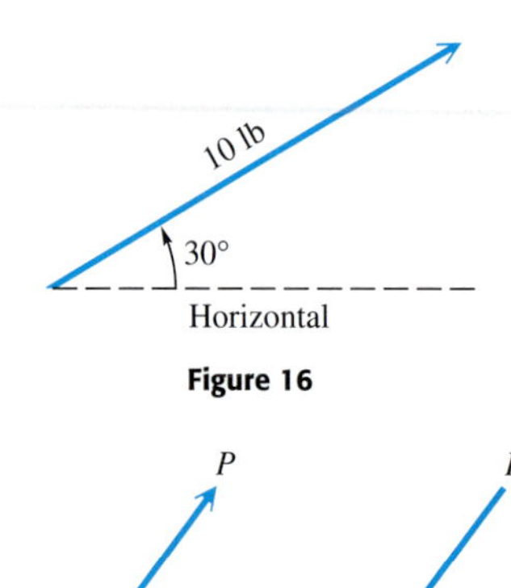

Figure 16

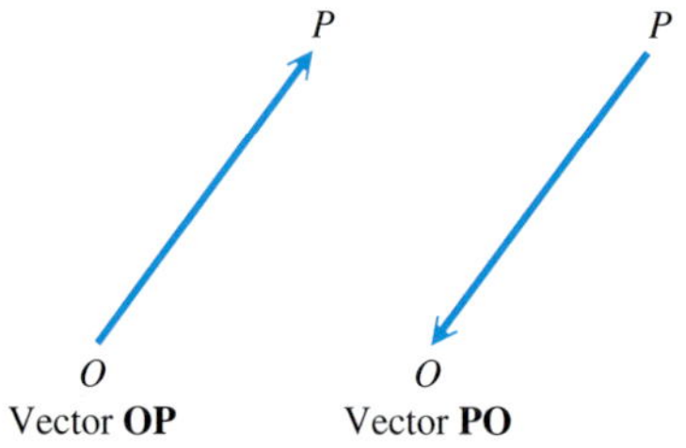

Vector **OP** Vector **PO**

Figure 17

Basic Terminology

Many quantities involve magnitudes, such as 45 lb or 60 mph. These quantities are called **scalars** and can be represented by real numbers. Other quantities, called **vector quantities,** involve both magnitude and direction. Typical vector quantities are velocity, acceleration, and force. For example, traveling 50 mph *east* represents a vector quantity.

A vector quantity is often represented with a directed line segment (a segment that uses an arrowhead to indicate direction), called a **vector.** The length of the vector represents the **magnitude** of the vector quantity. The direction of the vector, indicated by the arrowhead, represents the direction of the quantity. For example, the vector in Figure 16 represents a force of 10 lb applied at an angle of 30° from the horizontal.

The symbol for a vector is often printed in boldface type. When writing vectors by hand, it is customary to use an arrow over the letter or letters. Thus **OP** and $\overrightarrow{OP}$ both represent the vector **OP**. Vectors may be named with either one lowercase or uppercase letter, or two uppercase letters. When two letters are used, the first indicates the **initial point** and the second indicates the **terminal point** of the vector. Knowing these points gives the direction of the vector. For example, vectors **OP** and **PO** in Figure 17 are not the same vector. They have the same magnitude, but *opposite* directions. The magnitude of vector **OP** is written |**OP**|.

Two vectors are equal if and only if they have the same direction and the same magnitude. In Figure 18, vectors **A** and **B** are equal, as are vectors **C** and **D**. As Figure 18 shows, equal vectors need not coincide, but they must be parallel. Vectors **A** and **E** are unequal because they do not have the same direction, while **A** ≠ **F** because they have different magnitudes.

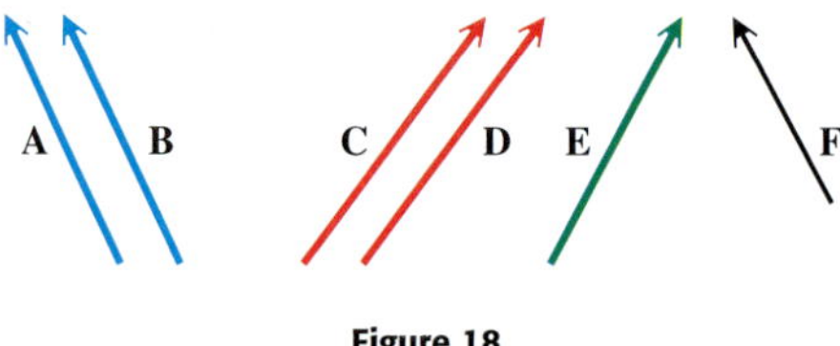

Figure 18

Now try Exercise 1.

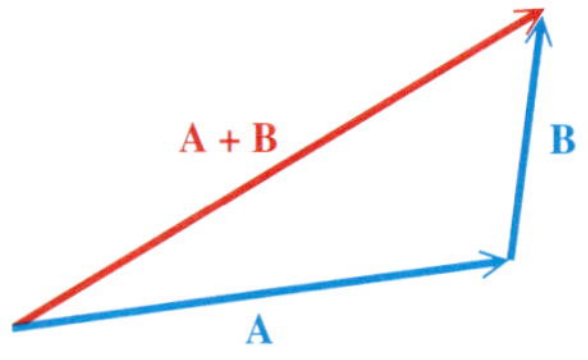

Figure 19

To find the sum of two vectors **A** and **B**, we place the initial point of vector **B** at the terminal point of vector **A**, as shown in Figure 19. The vector with the same initial point as **A** and the same terminal point as **B** is the sum **A** + **B**. The sum of two vectors is also a vector.

Another way to find the sum of two vectors is to use the **parallelogram rule.** Place vectors **A** and **B** so that their initial points coincide. Then, complete a parallelogram that has **A** and **B** as two sides. The diagonal of the parallelogram with the same initial points as **A** and **B** is the sum **A** + **B** found by the definition. Compare Figures 19 and 20. Parallelograms can be used to show that vector **B** + **A** is the same as vector **A** + **B**, or that **A** + **B** = **B** + **A**, so vector addition is commutative. The vector sum **A** + **B** is called the **resultant** of vectors **A** and **B**.

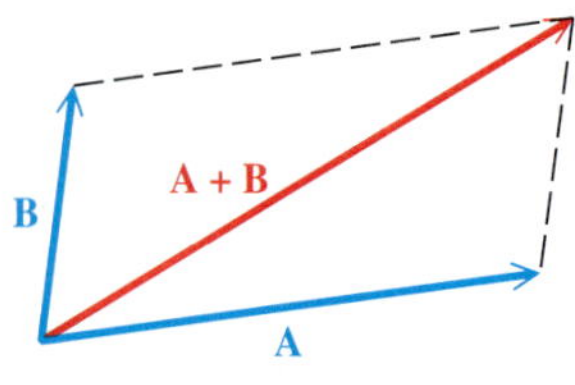

Figure 20

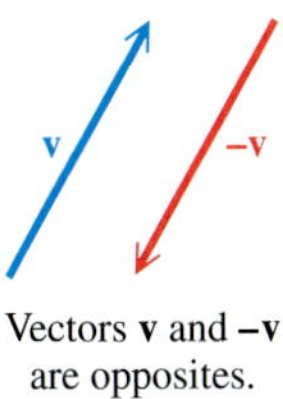

Vectors **v** and **−v** are opposites.

Figure 21

For every vector **v** there is a vector $-\mathbf{v}$ that has the same magnitude as **v** but opposite direction. Vector $-\mathbf{v}$ is called the **opposite** of **v**. (See Figure 21.) The sum of **v** and $-\mathbf{v}$ has magnitude 0 and is called the **zero vector.** As with real numbers, to subtract vector **B** from vector **A**, find the vector sum $\mathbf{A} + (-\mathbf{B})$. (See Figure 22.)

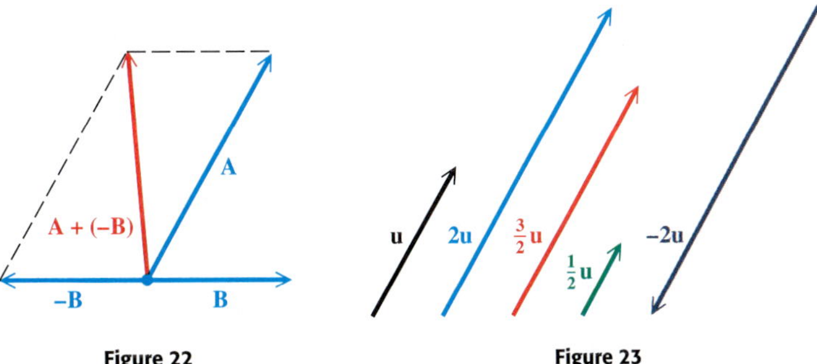

Figure 22 **Figure 23**

The **scalar product** of a real number (or scalar) k and a vector **u** is the vector $k \cdot \mathbf{u}$, which has magnitude $|k|$ times the magnitude of **u**. As suggested by Figure 23, the vector $k \cdot \mathbf{u}$ has the same direction as **u** if $k > 0$, and opposite direction if $k < 0$.

Now try Exercises 5, 7, 9, and 11.

Algebraic Interpretation of Vectors

We now consider vectors in a rectangular coordinate system. A vector with its initial point at the origin is called a **position vector.** A position vector **u** with its endpoint at the point (a, b) is written $\langle a, b \rangle$, so

$$\mathbf{u} = \langle a, b \rangle.$$

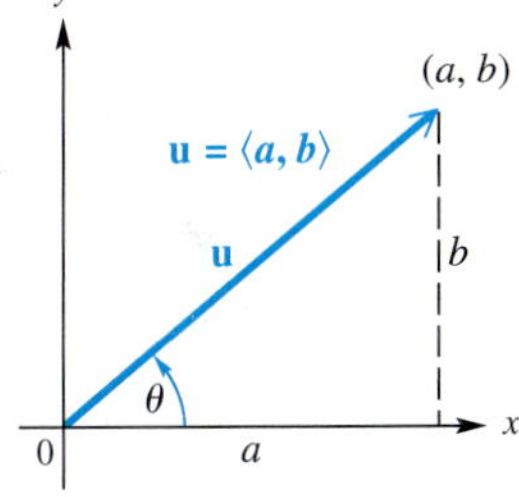

Figure 24

This means that every vector in the real plane corresponds to an ordered pair of real numbers. Thus, geometrically a vector is a directed line segment; algebraically, it is an ordered pair. The numbers a and b are the **horizontal component** and **vertical component,** respectively, of vector **u**. Figure 24 shows the vector $\mathbf{u} = \langle a, b \rangle$. The positive angle between the x-axis and a position vector is the **direction angle** for the vector. In Figure 24, θ is the direction angle for vector **u**.

From Figure 24, we can see that the magnitude and direction of a vector are related to its horizontal and vertical components.

Looking Ahead to Calculus

In addition to two-dimensional vectors in a plane, calculus courses introduce three-dimensional vectors in space. The magnitude of the two-dimensional vector $\langle a, b \rangle$ is given by $\sqrt{a^2 + b^2}$. If we extend this to the three-dimensional vector $\langle a, b, c \rangle$, the expression becomes $\sqrt{a^2 + b^2 + c^2}$. Similar extensions are made for other concepts.

Magnitude and Direction Angle of a Vector $\langle a, b \rangle$

The magnitude (length) of vector $\mathbf{u} = \langle a, b \rangle$ is given by

$$|\mathbf{u}| = \sqrt{a^2 + b^2}.$$

The direction angle θ satisfies $\tan \theta = \frac{b}{a}$, where $a \neq 0$.

EXAMPLE 1 Finding Magnitude and Direction Angle

Find the magnitude and direction angle for $\mathbf{u} = \langle 3, -2 \rangle$.

Algebraic Solution

The magnitude is $|\mathbf{u}| = \sqrt{3^2 + (-2)^2} = \sqrt{13}$. To find the direction angle θ, start with $\tan \theta = \frac{b}{a} = \frac{-2}{3} = -\frac{2}{3}$. Vector $\mathbf{u}$ has a positive horizontal component and a negative vertical component, placing the position vector in quadrant IV. A calculator gives $\tan^{-1}\left(-\frac{2}{3}\right) \approx -33.7°$. Adding 360° yields the direction angle $\theta = 326.3°$. See Figure 25.

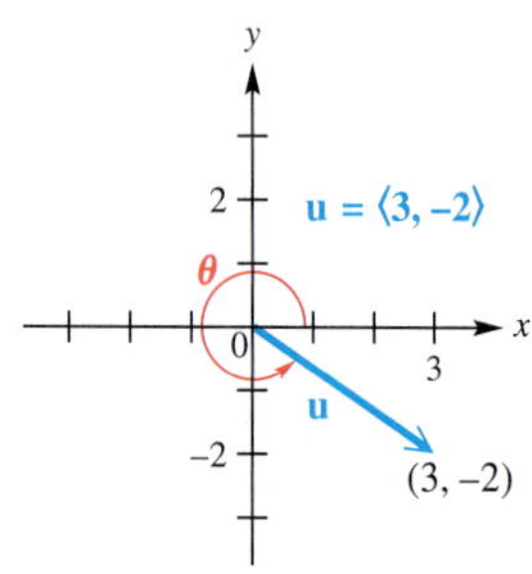

Figure 25

Graphing Calculator Solution

A calculator returns the magnitude and direction angle, given the horizontal and vertical components. An approximation for $\sqrt{13}$ is given, and the direction angle has a measure with smallest possible absolute value. We must add 360° to the value of θ to obtain the positive direction angle. See Figure 26.

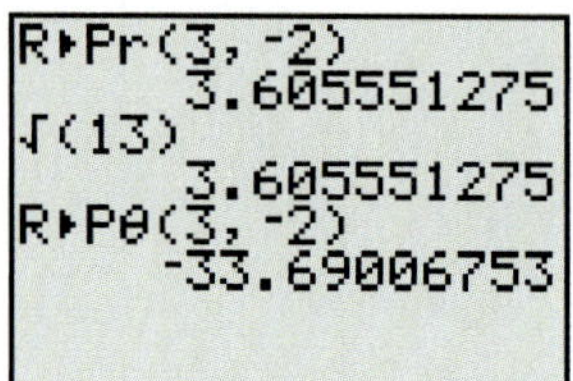

Figure 26

For more information, see your owner's manual or the graphing calculator manual that accompanies this text.

Now try Exercise 33.

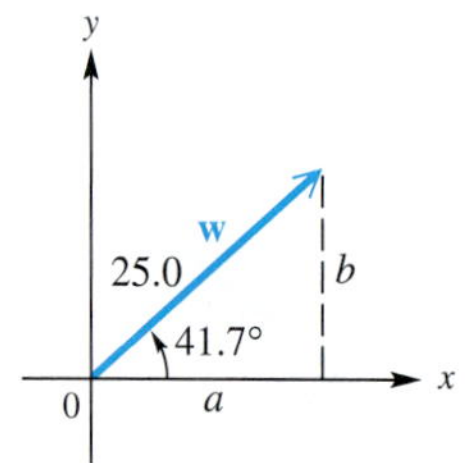

Figure 27

Horizontal and Vertical Components

The horizontal and vertical components, respectively, of a vector $\mathbf{u}$ having magnitude $|\mathbf{u}|$ and direction angle θ are given by

$$a = |\mathbf{u}| \cos \theta \quad \text{and} \quad b = |\mathbf{u}| \sin \theta.$$

That is, $\mathbf{u} = \langle a, b \rangle = \langle |\mathbf{u}| \cos \theta, |\mathbf{u}| \sin \theta \rangle$.

EXAMPLE 2 Finding Horizontal and Vertical Components

Vector $\mathbf{w}$ in Figure 27 has magnitude 25.0 and direction angle 41.7°. Find the horizontal and vertical components.

Algebraic Solution

Use the two formulas in the box, with $|\mathbf{w}| = 25.0$ and $\theta = 41.7°$.

$a = 25.0 \cos 41.7°$	$b = 25.0 \sin 41.7°$
$a \approx 18.7$	$b \approx 16.6$

Therefore, $\mathbf{w} = \langle 18.7, 16.6 \rangle$. The horizontal component is 18.7, and the vertical component is 16.6 (rounded to the nearest tenth).

Graphing Calculator Solution

See Figure 28. The results support the algebraic solution.

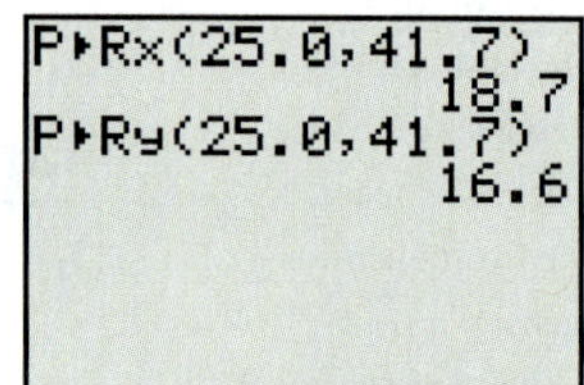

Figure 28

Now try Exercise 37.

EXAMPLE 3 Writing Vectors in the Form $\langle a, b\rangle$

Write each vector in Figure 29 in the form $\langle a, b\rangle$.

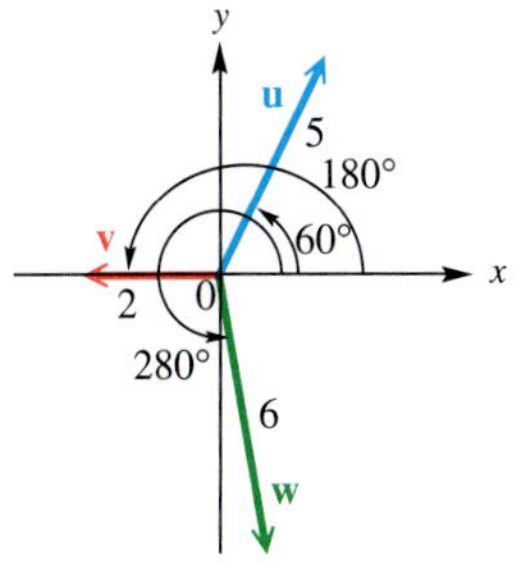

Solution

$$\mathbf{u} = \langle 5\cos 60°, 5\sin 60°\rangle = \left\langle 5\cdot\frac{1}{2}, 5\cdot\frac{\sqrt{3}}{2}\right\rangle = \left\langle \frac{5}{2}, \frac{5\sqrt{3}}{2}\right\rangle$$

$$\mathbf{v} = \langle 2\cos 180°, 2\sin 180°\rangle = \langle 2(-1), 2(0)\rangle = \langle -2, 0\rangle$$

$$\mathbf{w} = \langle 6\cos 280°, 6\sin 280°\rangle \approx \langle 1.0419, -5.9088\rangle \quad \text{Use a calculator.}$$

Figure 29

Now try Exercise 43.

The following properties of parallelograms are helpful when studying vectors.

1. A parallelogram is a quadrilateral whose opposite sides are parallel.
2. The opposite sides and opposite angles of a parallelogram are equal, and adjacent angles of a parallelogram are supplementary.
3. The diagonals of a parallelogram bisect each other, but do not necessarily bisect the angles of the parallelogram.

EXAMPLE 4 Finding the Magnitude of a Resultant

Two forces of 15 and 22 newtons act on a point in the plane. (A *newton* is a unit of force that equals .225 lb.) If the angle between the forces is 100°, find the magnitude of the resultant force.

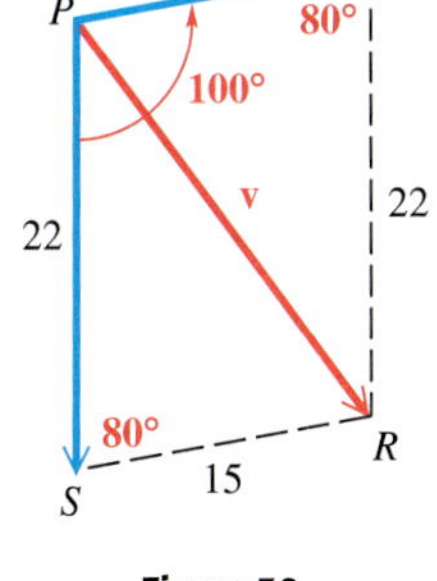

Figure 30

Solution As shown in Figure 30, a parallelogram that has the forces as adjacent sides can be formed. The angles of the parallelogram adjacent to angle P measure 80°, since adjacent angles of a parallelogram are supplementary. Opposite sides of the parallelogram are equal in length. The resultant force divides the parallelogram into two triangles. Use the law of cosines with either triangle.

$$|\mathbf{v}|^2 = 15^2 + 22^2 - 2(15)(22)\cos 80° \quad \text{Law of cosines (Section 8.2)}$$

$$\approx 225 + 484 - 115$$

$$|\mathbf{v}|^2 \approx 594$$

$$|\mathbf{v}| \approx 24 \quad \text{Take square roots. (Section 1.4)}$$

To the nearest unit, the magnitude of the resultant force is 24 newtons.

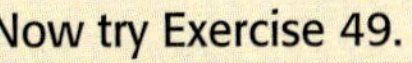

Operations with Vectors In Figure 31, $\mathbf{m} = \langle a, b\rangle$, $\mathbf{n} = \langle c, d\rangle$, and $\mathbf{p} = \langle a + c, b + d\rangle$. Using geometry, we can show that the endpoints of the three vectors and the origin form a parallelogram. Since a diagonal of this parallelogram gives the resultant of $\mathbf{m}$ and $\mathbf{n}$, we have $\mathbf{p} = \mathbf{m} + \mathbf{n}$ or

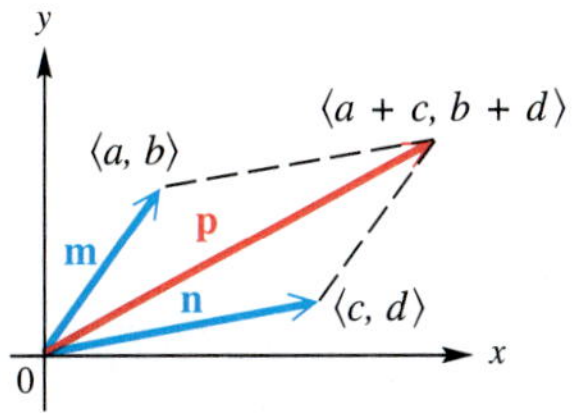

$$\langle a + c, b + d\rangle = \langle a, b\rangle + \langle c, d\rangle.$$

Figure 31

Similarly, we could verify the following vector operations.

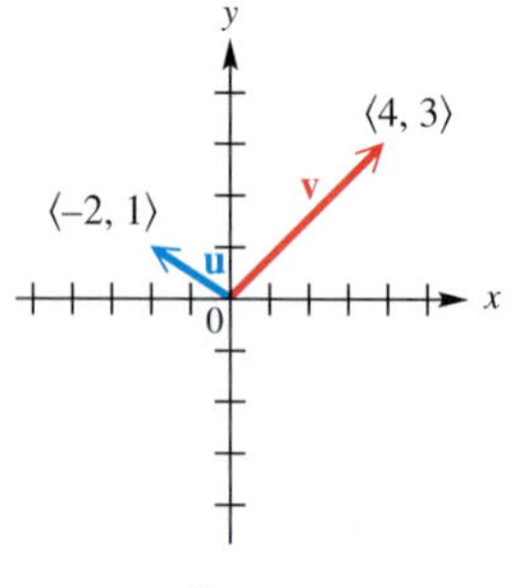

Figure 32

Vector Operations

For any real numbers a, b, c, d, and k,

$$\langle a, b\rangle + \langle c, d\rangle = \langle a + c, b + d\rangle$$
$$k \cdot \langle a, b\rangle = \langle ka, kb\rangle.$$
$$\text{If } \mathbf{a} = \langle a_1, a_2\rangle, \text{ then } -\mathbf{a} = \langle -a_1, -a_2\rangle.$$
$$\langle a, b\rangle - \langle c, d\rangle = \langle a, b\rangle + (-\langle c, d\rangle) = \langle a - c, b - d\rangle$$

EXAMPLE 5 Performing Vector Operations

Let $\mathbf{u} = \langle -2, 1\rangle$ and $\mathbf{v} = \langle 4, 3\rangle$. (See Figure 32.) Find the following: **(a)** $\mathbf{u} + \mathbf{v}$, **(b)** $-2\mathbf{u}$, **(c)** $4\mathbf{u} - 3\mathbf{v}$.

Algebraic Solution

(a) $\mathbf{u} + \mathbf{v} = \langle -2, 1\rangle + \langle 4, 3\rangle$
$= \langle -2 + 4, 1 + 3\rangle$
$= \langle 2, 4\rangle$

(b) $-2\mathbf{u} = -2 \cdot \langle -2, 1\rangle$
$= \langle -2(-2), -2(1)\rangle$
$= \langle 4, -2\rangle$

(c) $4\mathbf{u} - 3\mathbf{v} = 4 \cdot \langle -2, 1\rangle - 3 \cdot \langle 4, 3\rangle$
$= \langle -8, 4\rangle - \langle 12, 9\rangle$
$= \langle -8 - 12, 4 - 9\rangle$
$= \langle -20, -5\rangle$

Graphing Calculator Solution

Vector arithmetic can be performed with a graphing calculator, as shown in Figure 33.

```
{-2,1}+{4,3}
                {2 4}
-2{-2,1}
               {4 -2}
4{-2,1}-3{4,3}
             {-20 -5}
```

Figure 33

Now try Exercises 59, 61, and 63.

A **unit vector** is a vector that has magnitude 1. Two very useful unit vectors are defined as follows and shown in Figure 34(a).

$$\mathbf{i} = \langle 1, 0\rangle \qquad \mathbf{j} = \langle 0, 1\rangle$$

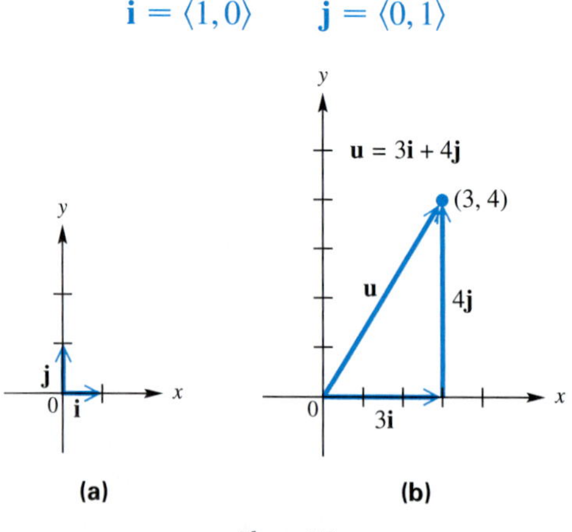

Figure 34

With the unit vectors **i** and **j**, we can express any other vector $\langle a, b\rangle$ in the form $a\mathbf{i} + b\mathbf{j}$, as shown in Figure 34(b), where $\langle 3, 4\rangle = 3\mathbf{i} + 4\mathbf{j}$. The vector operations previously given can be restated, using $a\mathbf{i} + b\mathbf{j}$ notation.

i, j Form for Vectors

If $\mathbf{v} = \langle a, b \rangle$, then $\mathbf{v} = a\mathbf{i} + b\mathbf{j}$.

Dot Product and the Angle Between Vectors The *dot product* of two vectors is a real number, not a vector. It is also known as the *inner product.* Dot products are used to determine the angle between two vectors, derive geometric theorems, and solve physics problems.

Dot Product

The **dot product** of the two vectors $\mathbf{u} = \langle a, b \rangle$ and $\mathbf{v} = \langle c, d \rangle$ is denoted $\mathbf{u} \cdot \mathbf{v}$, read "**u** dot **v**," and given by

$$\mathbf{u} \cdot \mathbf{v} = ac + bd.$$

That is, the dot product of two vectors is the sum of the product of their first components and the product of their second components.

EXAMPLE 6 Finding Dot Products

Find each dot product.

(a) $\langle 2, 3 \rangle \cdot \langle 4, -1 \rangle$

(b) $\langle 6, 4 \rangle \cdot \langle -2, 3 \rangle$

Solution

(a) $\langle 2, 3 \rangle \cdot \langle 4, -1 \rangle = 2(4) + 3(-1) = 5$

(b) $\langle 6, 4 \rangle \cdot \langle -2, 3 \rangle = 6(-2) + 4(3) = 0$

Now try Exercise 71.

The following properties of dot products are easily verified.

Properties of the Dot Product

For all vectors **u**, **v**, and **w** and real numbers k,

(a) $\mathbf{u} \cdot \mathbf{v} = \mathbf{v} \cdot \mathbf{u}$

(b) $\mathbf{u} \cdot (\mathbf{v} + \mathbf{w}) = \mathbf{u} \cdot \mathbf{v} + \mathbf{u} \cdot \mathbf{w}$

(c) $(\mathbf{u} + \mathbf{v}) \cdot \mathbf{w} = \mathbf{u} \cdot \mathbf{w} + \mathbf{v} \cdot \mathbf{w}$

(d) $(k\mathbf{u}) \cdot \mathbf{v} = k(\mathbf{u} \cdot \mathbf{v}) = \mathbf{u} \cdot (k\mathbf{v})$

(e) $\mathbf{0} \cdot \mathbf{u} = 0$

(f) $\mathbf{u} \cdot \mathbf{u} = |\mathbf{u}|^2$.

For example, to prove the first part of (d), we let $\mathbf{u} = \langle a, b \rangle$ and $\mathbf{v} = \langle c, d \rangle$. Then,

$$\begin{aligned}(k\mathbf{u}) \cdot \mathbf{v} &= (k\langle a, b \rangle) \cdot \langle c, d \rangle = \langle ka, kb \rangle \cdot \langle c, d \rangle \\ &= kac + kbd = k(ac + bd) \\ &= k(\langle a, b \rangle \cdot \langle c, d \rangle) = k(\mathbf{u} \cdot \mathbf{v}).\end{aligned}$$

The proofs of the remaining properties are similar.

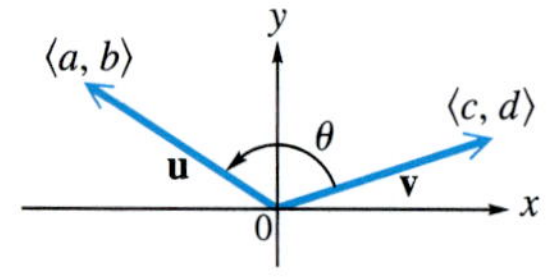

Figure 35

The dot product of two vectors can be positive, 0, or negative. A geometric interpretation of the dot product explains when each of these cases occurs. This interpretation involves the angle between the two vectors. Consider the vectors $\mathbf{u} = \langle a, b\rangle$ and $\mathbf{v} = \langle c, d\rangle$, as shown in Figure 35. The **angle $\boldsymbol{\theta}$ between u and v** is defined to be the angle having the two vectors as its sides for which $0° \le \theta \le 180°$. The following theorem relates the dot product to the angle between the vectors. Its proof is outlined in Exercise 32 in Section 8.4.

Geometric Interpretation of Dot Product

If θ is the angle between the two nonzero vectors $\mathbf{u}$ and $\mathbf{v}$, where $0° \le \theta \le 180°$, then

$$\mathbf{u} \cdot \mathbf{v} = |\mathbf{u}||\mathbf{v}| \cos \theta \quad \text{or, equivalently,} \quad \cos \theta = \frac{\mathbf{u} \cdot \mathbf{v}}{|\mathbf{u}||\mathbf{v}|}.$$

EXAMPLE 7 Finding the Angle Between Two Vectors

Find the angle θ between the two vectors $\mathbf{u} = \langle 3, 4\rangle$ and $\mathbf{v} = \langle 2, 1\rangle$.

Solution By the geometric interpretation,

$$\begin{aligned} \cos \theta &= \frac{\mathbf{u} \cdot \mathbf{v}}{|\mathbf{u}||\mathbf{v}|} = \frac{\langle 3, 4\rangle \cdot \langle 2, 1\rangle}{|\langle 3, 4\rangle||\langle 2, 1\rangle|} \\ &= \frac{3(2) + 4(1)}{\sqrt{9 + 16}\sqrt{4 + 1}} \\ &= \frac{10}{5\sqrt{5}} \approx .894427191. \end{aligned}$$

Therefore, $\theta \approx \cos^{-1} .894427191 \approx 26.57°$. (Section 7.5)

Now try Exercise 77.

For angles θ between 0° and 180°, $\cos \theta$ is positive, 0, or negative when θ is less than, equal to, or greater than 90°, respectively. Therefore, the dot product is positive, 0, or negative according to this table.

Dot Product	Angle Between Vectors
Positive	Acute
0	Right
Negative	Obtuse

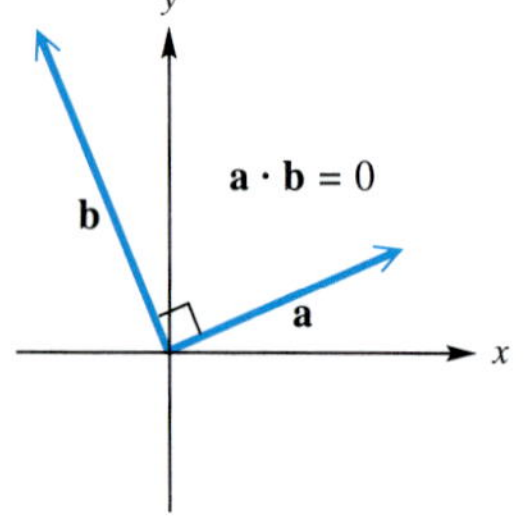

Figure 36

NOTE If $\mathbf{a} \cdot \mathbf{b} = 0$ for two nonzero vectors $\mathbf{a}$ and $\mathbf{b}$, then $\cos \theta = 0$ and $\theta = 90°$. Thus, **a** and **b** are perpendicular or **orthogonal vectors.** See Figure 36.

Now try Exercises 87 and 89.

8.3 Exercises

Concept Check Exercises 1–4 refer to the vectors **m**–**t** at the right.

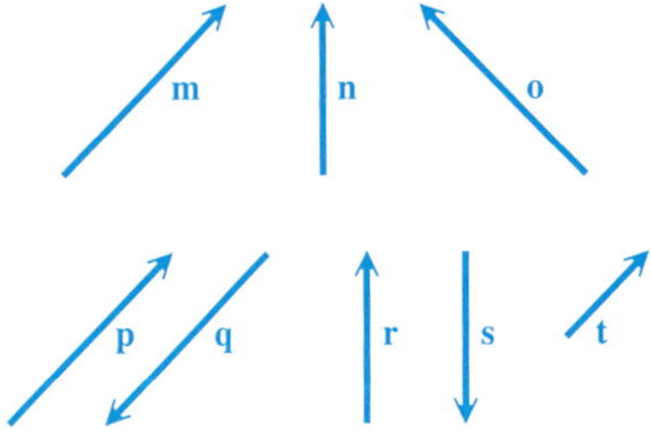

1. Name all pairs of vectors that appear to be equal.
2. Name all pairs of vectors that are opposites.
3. Name all pairs of vectors where the first is a scalar multiple of the other, with the scalar positive.
4. Name all pairs of vectors where the first is a scalar multiple of the other, with the scalar negative.

Concept Check Refer to vectors **a**–**h** below. Make a copy or a sketch of each vector, and then draw a sketch to represent each vector in Exercises 5–16. For example, find **a** + **e** by placing **a** and **e** so that their initial points coincide. Then use the parallelogram rule to find the resultant, as shown in the figure on the right.

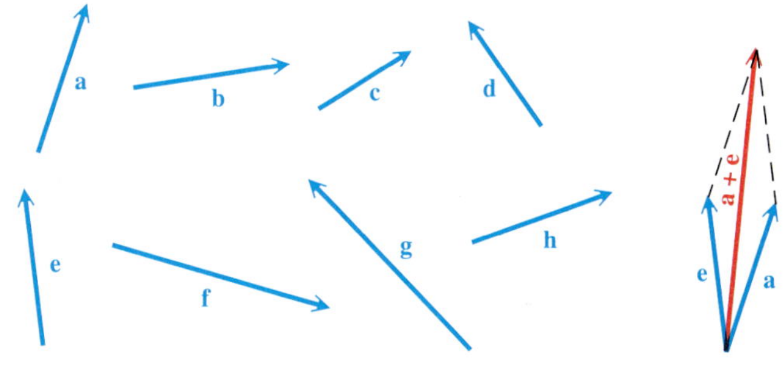

5. −**b**
6. −**g**
7. 3**a**
8. 2**h**
9. **a** + **b**
10. **h** + **g**
11. **a** − **c**
12. **d** − **e**
13. **a** + (**b** + **c**)
14. (**a** + **b**) + **c**
15. **c** + **d**
16. **d** + **c**
17. From the results of Exercises 13 and 14, does it appear that vector addition is associative?
18. From the results of Exercises 15 and 16, does it appear that vector addition is commutative?

In Exercises 19–24, use the figure to find each vector: **(a) a + b (b) a − b (c) −a**. *Use* $\langle x, y\rangle$ *notation.*

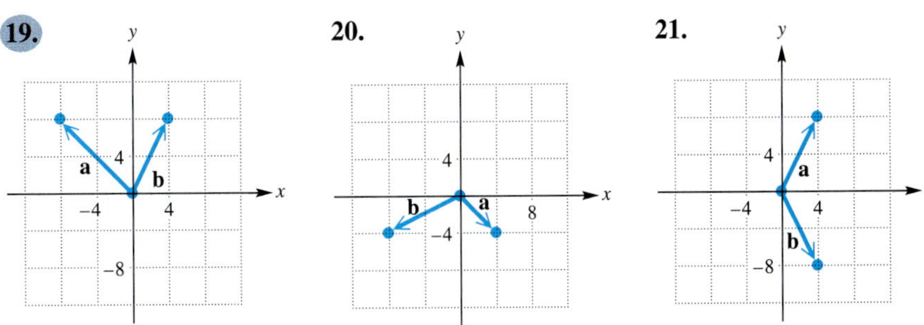

22.

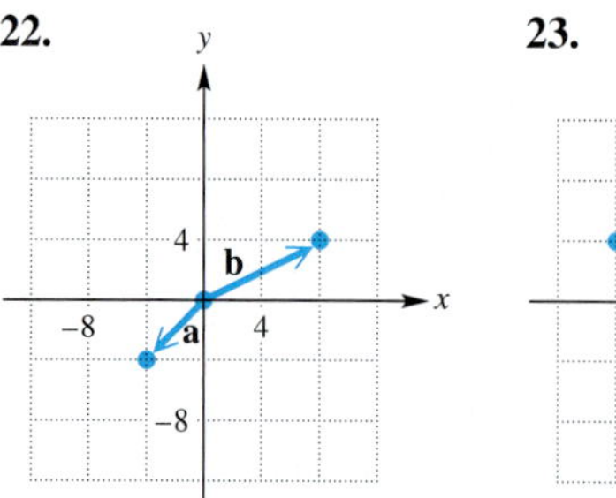

23.

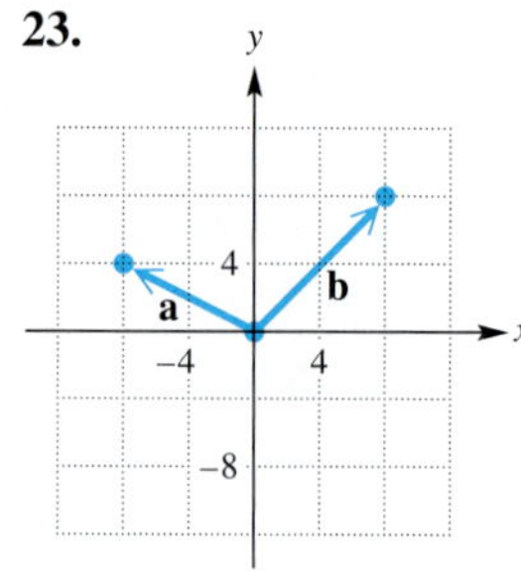

24.

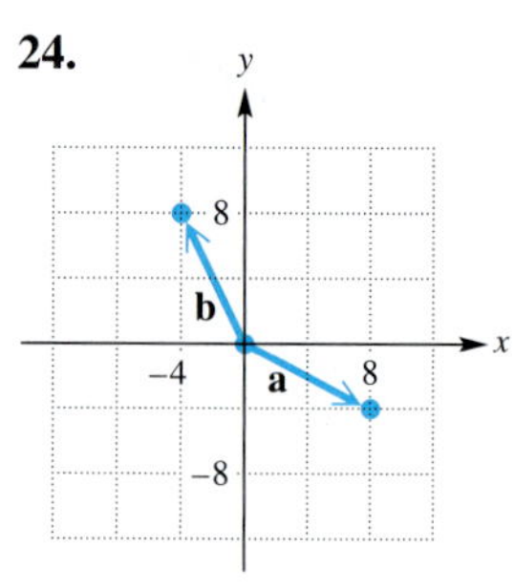

Given vectors **a** *and* **b**, *find:* **(a)** 2**a** **(b)** 2**a** + 3**b** **(c)** **b** − 3**a**.

25. $\mathbf{a} = 2\mathbf{i}, \mathbf{b} = \mathbf{i} + \mathbf{j}$

26. $\mathbf{a} = -\mathbf{i} + 2\mathbf{j}, \mathbf{b} = \mathbf{i} - \mathbf{j}$

27. $\mathbf{a} = \langle -1, 2 \rangle, \mathbf{b} = \langle 3, 0 \rangle$

28. $\mathbf{a} = \langle -2, -1 \rangle, \mathbf{b} = \langle -3, 2 \rangle$

For each pair of vectors **u** *and* **w** *with angle* θ *between them, sketch the resultant.*

29. $|\mathbf{u}| = 12, |\mathbf{w}| = 20, \theta = 27°$

30. $|\mathbf{u}| = 8, |\mathbf{w}| = 12, \theta = 20°$

31. $|\mathbf{u}| = 20, |\mathbf{w}| = 30, \theta = 30°$

32. $|\mathbf{u}| = 50, |\mathbf{w}| = 70, \theta = 40°$

Find the magnitude and direction angle for **u**. *See Example 1.*

33. $\langle 15, -8 \rangle$

34. $\langle -7, 24 \rangle$

35. $\langle -4, 4\sqrt{3} \rangle$

36. $\langle 8\sqrt{2}, -8\sqrt{2} \rangle$

For each of the following, vector **v** *has the given magnitude and direction. Find the magnitudes of the horizontal and vertical components of* **v**, *if* α *is the angle of inclination of* **v** *from the horizontal. See Example 2.*

37. $\alpha = 20°, |\mathbf{v}| = 50$

38. $\alpha = 50°, |\mathbf{v}| = 26$

39. $\alpha = 35° 50', |\mathbf{v}| = 47.8$

40. $\alpha = 27° 30', |\mathbf{v}| = 15.4$

41. $\alpha = 128.5°, |\mathbf{v}| = 198$

42. $\alpha = 146.3°, |\mathbf{v}| = 238$

Write each vector in the form $\langle a, b \rangle$. *See Example 3.*

43.

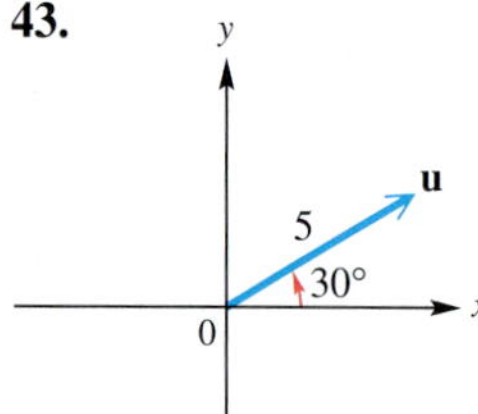

44.

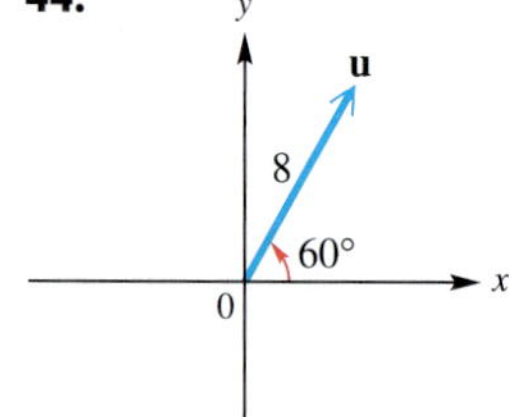

45.

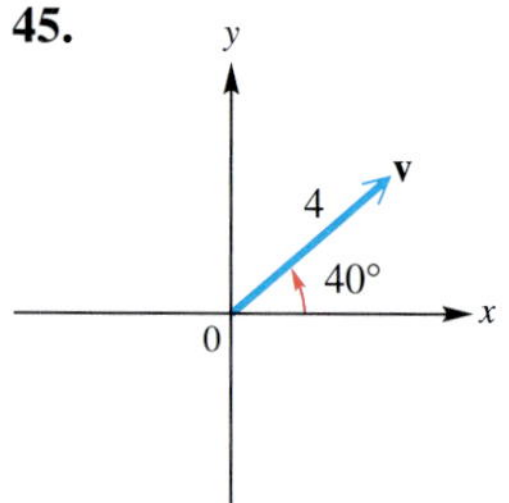

46.

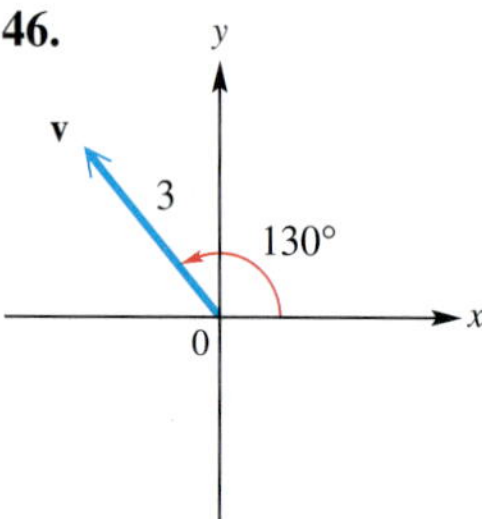

47.

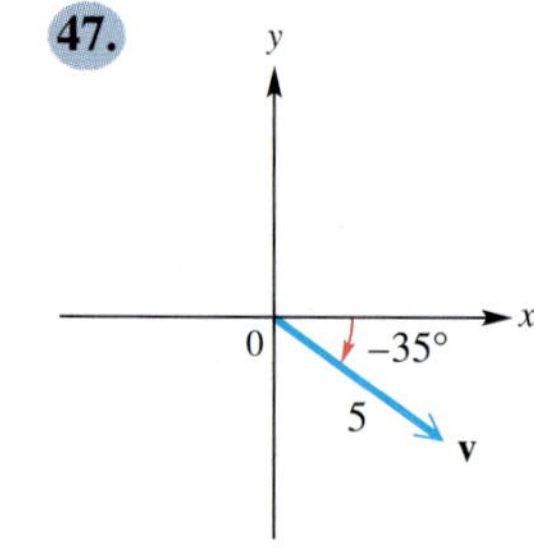

48.

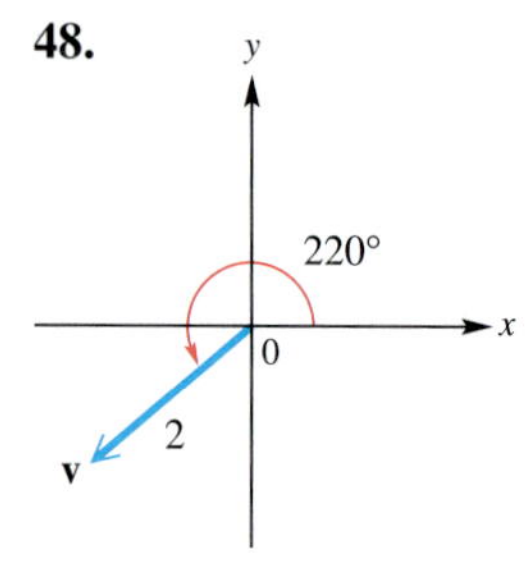

Two forces act at a point in the plane. The angle between the two forces is given. Find the magnitude of the resultant force. See Example 4.

49. forces of 250 and 450 newtons, forming an angle of 85°

50. forces of 19 and 32 newtons, forming an angle of 118°

51. forces of 116 and 139 lb, forming an angle of 140° 50′

52. forces of 37.8 and 53.7 lb, forming an angle of 68.5°

Use the parallelogram rule to find the magnitude of the resultant force for the two forces shown in each figure. Round answers to the nearest tenth.

53.

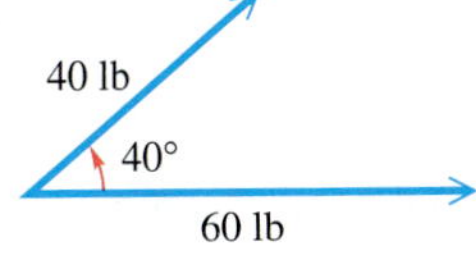

54.

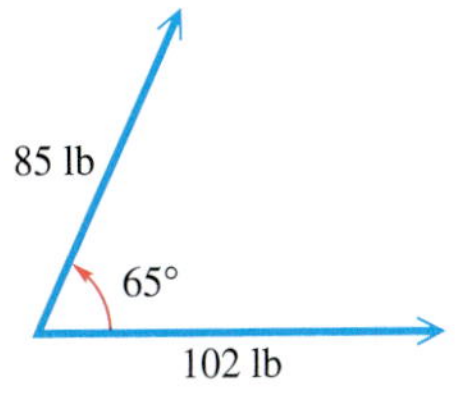

55.

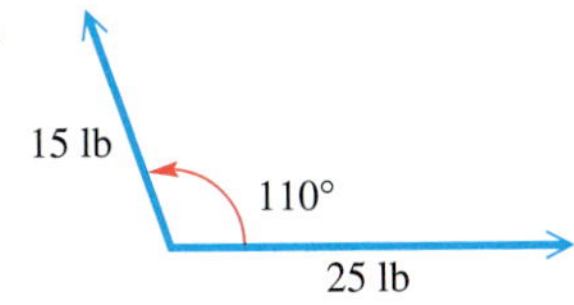

56.

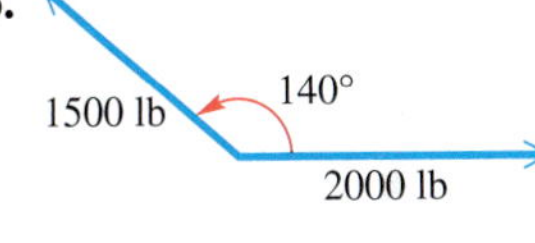

57. ***Concept Check*** If $\mathbf{u} = \langle a, b \rangle$ and $\mathbf{v} = \langle c, d \rangle$, what is $\mathbf{u} + \mathbf{v}$?

58. Explain how addition of vectors is similar to addition of complex numbers.

Given $\mathbf{u} = \langle -2, 5 \rangle$ *and* $\mathbf{v} = \langle 4, 3 \rangle$, *find the following. See Example 5.*

59. $\mathbf{u} + \mathbf{v}$ **60.** $\mathbf{u} - \mathbf{v}$ **61.** $-4\mathbf{u}$ **62.** $-5\mathbf{v}$

63. $3\mathbf{u} - 6\mathbf{v}$ **64.** $-2\mathbf{u} + 4\mathbf{v}$ **65.** $\mathbf{u} + \mathbf{v} - 3\mathbf{u}$ **66.** $2\mathbf{u} + \mathbf{v} - 6\mathbf{v}$

Write each vector in the form $a\mathbf{i} + b\mathbf{j}$. *See Figure 34(b).*

67. $\langle -5, 8 \rangle$ **68.** $\langle 6, -3 \rangle$ **69.** $\langle 2, 0 \rangle$ **70.** $\langle 0, -4 \rangle$

Find the dot product for each pair of vectors. See Example 6.

71. $\langle 6, -1 \rangle, \langle 2, 5 \rangle$ **72.** $\langle -3, 8 \rangle, \langle 7, -5 \rangle$ **73.** $\langle 2, -3 \rangle, \langle 6, 5 \rangle$

74. $\langle 1, 2 \rangle, \langle 3, -1 \rangle$ **75.** $4\mathbf{i}, 5\mathbf{i} - 9\mathbf{j}$ **76.** $2\mathbf{i} + 4\mathbf{j}, -\mathbf{j}$

Find the angle between each pair of vectors. See Example 7.

77. $\langle 2, 1 \rangle, \langle -3, 1 \rangle$ **78.** $\langle 1, 7 \rangle, \langle 1, 1 \rangle$ **79.** $\langle 1, 2 \rangle, \langle -6, 3 \rangle$

80. $\langle 4, 0 \rangle, \langle 2, 2 \rangle$ **81.** $3\mathbf{i} + 4\mathbf{j}, \mathbf{j}$ **82.** $-5\mathbf{i} + 12\mathbf{j}, 3\mathbf{i} + 2\mathbf{j}$

Let $\mathbf{u} = \langle -2, 1 \rangle$, $\mathbf{v} = \langle 3, 4 \rangle$, *and* $\mathbf{w} = \langle -5, 12 \rangle$. *Evaluate each expression.*

83. $(3\mathbf{u}) \cdot \mathbf{v}$ **84.** $\mathbf{u} \cdot (\mathbf{v} - \mathbf{w})$ **85.** $\mathbf{u} \cdot \mathbf{v} - \mathbf{u} \cdot \mathbf{w}$ **86.** $\mathbf{u} \cdot (3\mathbf{v})$

Determine whether each pair of vectors is orthogonal. See Figure 36.

87. $\langle 1, 2 \rangle, \langle -6, 3 \rangle$ **88.** $\langle 3, 4 \rangle, \langle 6, 8 \rangle$

89. $\langle 1, 0 \rangle, \langle \sqrt{2}, 0 \rangle$ **90.** $\langle 1, 1 \rangle, \langle 1, -1 \rangle$

91. $\sqrt{5}\mathbf{i} - 2\mathbf{j}, -5\mathbf{i} + 2\sqrt{5}\mathbf{j}$ **92.** $-4\mathbf{i} + 3\mathbf{j}, 8\mathbf{i} - 6\mathbf{j}$

Relating Concepts

For individual or collaborative investigation
(Exercises 93–98)

Consider the two vectors **u** *and* **v** *shown. Assume all values are exact.* ***Work Exercises 93–98 in order.***

93. Use trigonometry alone (without using vector notation) to find the magnitude and direction angle of **u** + **v**. Use the law of cosines and the law of sines in your work.

94. Find the horizontal and vertical components of **u**, using your calculator.

95. Find the horizontal and vertical components of **v**, using your calculator.

96. Find the horizontal and vertical components of **u** + **v** by adding the results you obtained in Exercises 94 and 95.

97. Use your calculator to find the magnitude and direction angle of the vector **u** + **v**.

98. Compare your answers in Exercises 93 and 97. What do you notice? Which method of solution do you prefer?

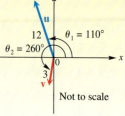

y
u
12
$\theta_1 = 110°$
$\theta_2 = 260°$
0
x
3
v
Not to scale

8.4 Applications of Vectors

The Equilibrant ▪ Incline Applications ▪ Navigation Applications

The Equilibrant The previous section covered methods for finding the resultant of two vectors. Sometimes it is necessary to find a vector that will counterbalance the resultant. This opposite vector is called the **equilibrant;** that is, the equilibrant of vector **u** is the vector −**u**.

EXAMPLE 1 Finding the Magnitude and Direction of an Equilibrant

Find the magnitude of the equilibrant of forces of 48 newtons and 60 newtons acting on a point A, if the angle between the forces is 50°. Then find the angle between the equilibrant and the 48-newton force.

Solution

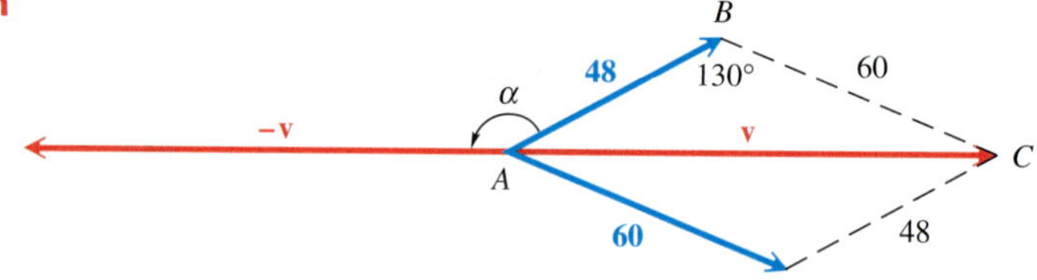

Figure 37

In Figure 37, the equilibrant is −**v**. The magnitude of **v**, and hence of −**v**, is found by using triangle ABC and the law of cosines.

$$|\mathbf{v}|^2 = 48^2 + 60^2 - 2(48)(60)\cos 130° \quad \text{Law of cosines (Section 8.2)}$$

$$|\mathbf{v}|^2 \approx 9606.5$$

$$|\mathbf{v}| \approx 98 \text{ newtons} \quad \text{Two significant digits}$$

The required angle, labeled α in Figure 37, can be found by subtracting angle CAB from $180°$. Use the law of sines to find angle CAB.

$$\frac{98}{\sin 130°} = \frac{60}{\sin CAB} \qquad \text{Law of sines (Section 8.1)}$$

$$\sin CAB \approx .46900680$$

$$CAB \approx 28° \qquad \text{Use the inverse sine function. (Section 7.5)}$$

Finally, $\alpha \approx 180° - 28° = 152°$.

Now try Exercise 1.

Incline Applications

We can use vectors to solve incline problems.

EXAMPLE 2 Finding a Required Force

Find the force required to keep a 50-lb wagon from sliding down a ramp inclined at $20°$ to the horizontal. (Assume there is no friction.)

Solution In Figure 38, the vertical 50-lb force **BA** represents the force of gravity. It is the sum of vectors **BC** and **−AC**. The vector **BC** represents the force with which the weight pushes against the ramp. Vector **BF** represents the force that would pull the weight up the ramp. Since vectors **BF** and **AC** are equal, $|\mathbf{AC}|$ gives the magnitude of the required force.

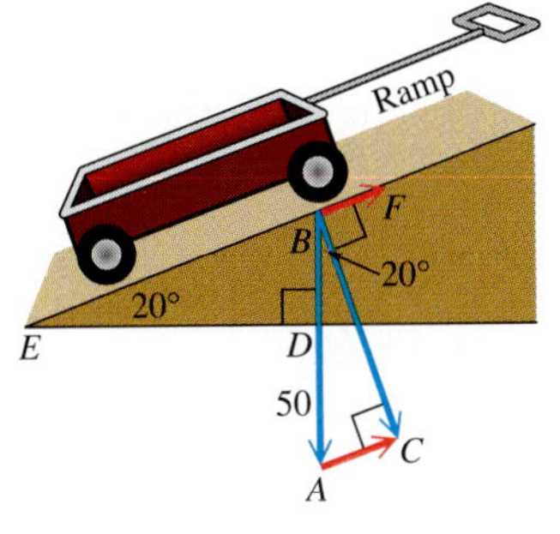

Figure 38

Vectors **BF** and **AC** are parallel, so angle EBD equals angle A. Since angle BDE and angle C are right angles, triangles CBA and DEB have two corresponding angles equal and so are similar triangles. Therefore, angle ABC equals angle E, which is $20°$. From right triangle ABC,

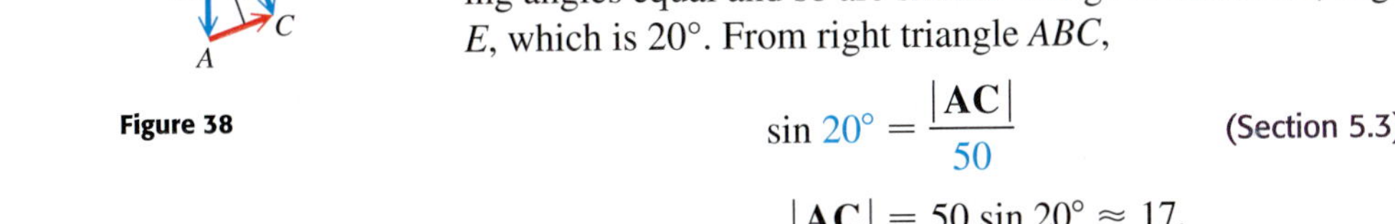

$$\sin 20° = \frac{|\mathbf{AC}|}{50} \qquad \text{(Section 5.3)}$$

$$|\mathbf{AC}| = 50 \sin 20° \approx 17.$$

Approximately a 17-lb force will keep the wagon from sliding down the ramp.

Now try Exercise 9.

EXAMPLE 3 Finding an Incline Angle

A force of 16.0 lb is required to hold a 40.0 lb lawn mower on an incline. What angle does the incline make with the horizontal?

Solution Figure 39 illustrates the situation. Consider right triangle ABC. Angle B equals angle θ, the magnitude of vector **BA** represents the weight of the mower, and vector **AC** equals vector **BE**, which represents the force required to hold the mower on the incline. From the figure,

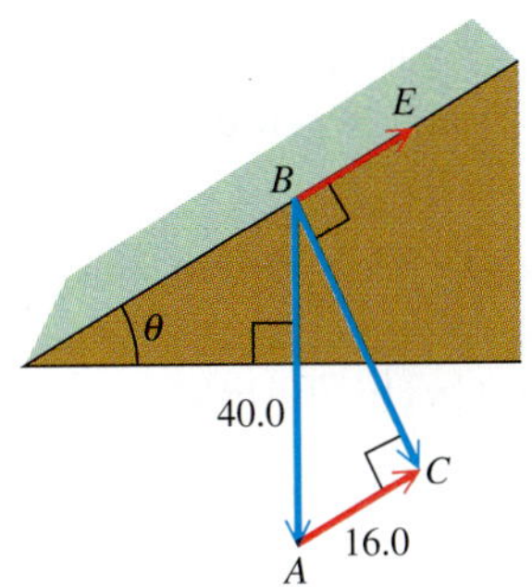

Figure 39

$$\sin B = \frac{16}{40} = .4$$

$$B \approx 23.5782°. \qquad \text{Use the inverse sine function.}$$

Therefore, the hill makes an angle of about $23.6°$ with the horizontal.

Now try Exercise 11.

Navigation Applications Problems involving bearing (defined in Section 5.4) can also be worked with vectors.

EXAMPLE 4 Applying Vectors to a Navigation Problem

A ship leaves port on a bearing of 28.0° and travels 8.20 mi. The ship then turns due east and travels 4.30 mi. How far is the ship from port? What is its bearing from port?

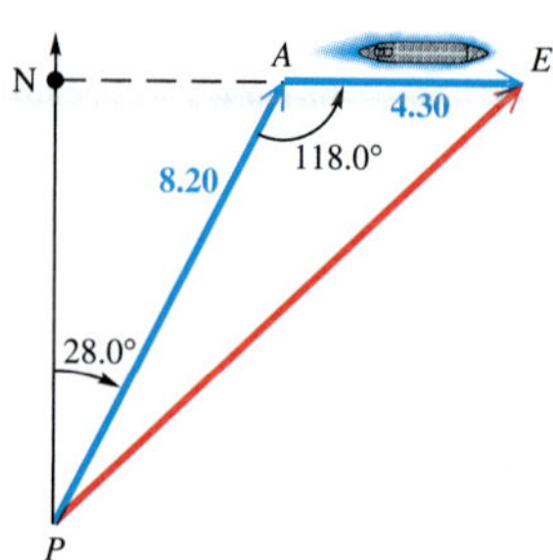

Figure 40

Solution In Figure 40, vectors **PA** and **AE** represent the ship's path. The magnitude and bearing of the resultant **PE** can be found as follows. Triangle *PNA* is a right triangle, so angle $NAP = 90° - 28.0° = 62.0°$. Then angle $PAE = 180° - 62.0° = 118.0°$. Use the law of cosines to find $|\mathbf{PE}|$, the magnitude of vector **PE**.

$$|\mathbf{PE}|^2 = 8.20^2 + 4.30^2 - 2(8.20)(4.30)\cos 118.0°$$
$$|\mathbf{PE}|^2 \approx 118.84$$
$$|\mathbf{PE}| \approx 10.9$$

The ship is about 10.9 mi from port.

To find the bearing of the ship from port, first find angle *APE*. Use the law of sines.

$$\frac{\sin APE}{4.30} = \frac{\sin 118.0°}{10.9}$$
$$\sin APE = \frac{4.30 \sin 118.0°}{10.9}$$
$$APE \approx 20.4° \quad \text{Use the inverse sine function.}$$

Now add 20.4° to 28.0° to find that the bearing is 48.4°.

Now try Exercise 15.

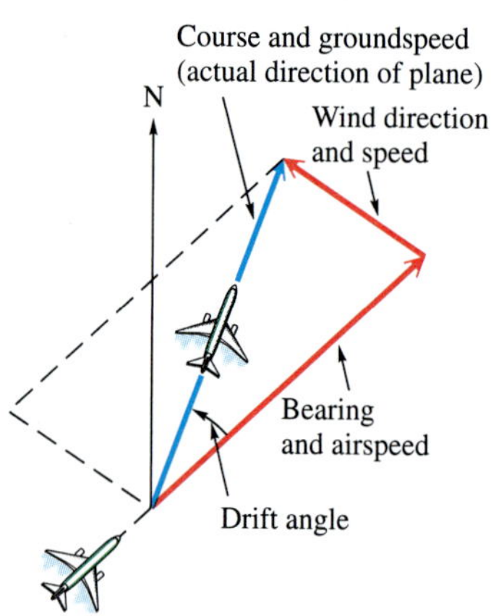

Figure 41

In air navigation, the **airspeed** of a plane is its speed relative to the air, while the **groundspeed** is its speed relative to the ground. Because of wind, these two speeds are usually different. The groundspeed of the plane is represented by the vector sum of the airspeed and windspeed vectors. See Figure 41.

EXAMPLE 5 Applying Vectors to a Navigation Problem

A plane with an airspeed of 192 mph is headed on a bearing of 121°. A north wind is blowing (from north to south) at 15.9 mph. Find the groundspeed and the actual bearing of the plane.

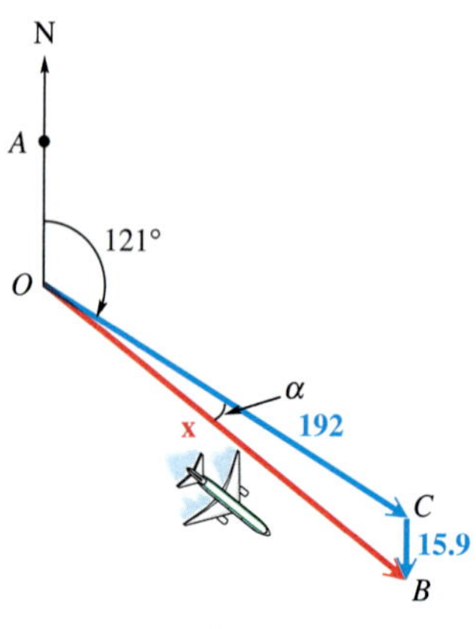

Figure 42

Solution In Figure 42, the groundspeed is represented by $|\mathbf{x}|$. We must find angle α to determine the bearing, which will be $121° + \alpha$. From Figure 42, angle *BCO* equals angle *AOC*, which equals 121°. Find $|\mathbf{x}|$ by the law of cosines.

$$|\mathbf{x}|^2 = 192^2 + 15.9^2 - 2(192)(15.9)\cos 121°$$
$$|\mathbf{x}|^2 \approx 40{,}261$$
$$|\mathbf{x}| \approx 200.7 \quad \text{or} \quad \text{about 201 mph}$$

Now find α by using the law of sines. Use the value of $|\mathbf{x}|$ before rounding.

$$\frac{\sin \alpha}{15.9} = \frac{\sin 121^\circ}{200.7}$$
$$\sin \alpha \approx .0679$$
$$\alpha \approx 3.89^\circ$$

To the nearest degree, α is 4°. The groundspeed is about 201 mph on a bearing of $121^\circ + 4^\circ = 125^\circ$.

Now try Exercise 23.

8.4 Exercises

Solve each problem. See Examples 1–5.

1. ***Direction and Magnitude of an Equilibrant*** Two tugboats are pulling a disabled speedboat into port with forces of 1240 lb and 1480 lb. The angle between these forces is 28.2°. Find the direction and magnitude of the equilibrant.

2. ***Direction and Magnitude of an Equilibrant*** Two rescue vessels are pulling a broken-down motorboat toward a boathouse with forces of 840 lb and 960 lb. The angle between these forces is 24.5°. Find the direction and magnitude of the equilibrant.

3. ***Angle Between Forces*** Two forces of 692 newtons and 423 newtons act at a point. The resultant force is 786 newtons. Find the angle between the forces.

4. ***Angle Between Forces*** Two forces of 128 lb and 253 lb act at a point. The equilibrant force is 320 lb. Find the angle between the forces.

5. ***Magnitudes of Forces*** A force of 176 lb makes an angle of 78° 50′ with a second force. The resultant of the two forces makes an angle of 41° 10′ with the first force. Find the magnitudes of the second force and of the resultant.

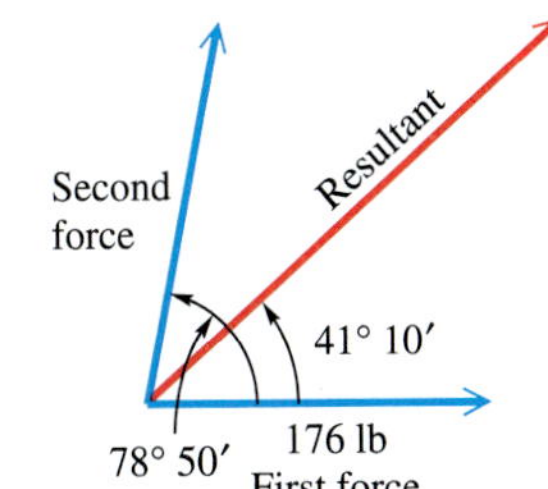

6. ***Magnitudes of Forces*** A force of 28.7 lb makes an angle of 42° 10′ with a second force. The resultant of the two forces makes an angle of 32° 40′ with the first force. Find the magnitudes of the second force and of the resultant.

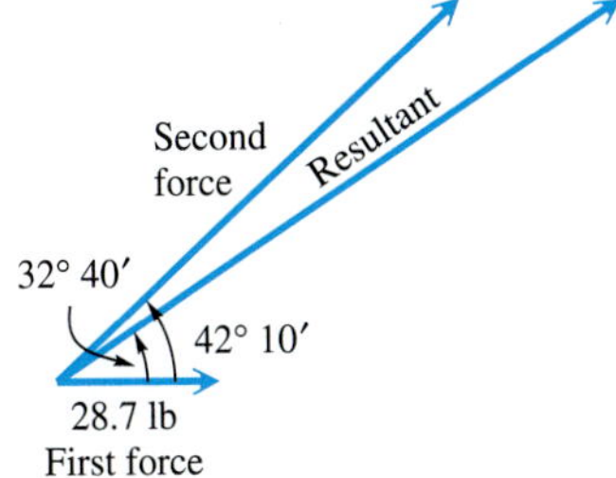

7. ***Angle of a Hill Slope*** A force of 25 lb is required to hold an 80-lb crate on a hill. What angle does the hill make with the horizontal?

8. ***Force Needed to Keep a Car Parked*** Find the force required to keep a 3000-lb car parked on a hill that makes an angle of 15° with the horizontal.

9. ***Force Needed for a Monolith*** To build the pyramids in Egypt, it is believed that giant causeways were constructed to transport the building materials to the site. One such causeway is said to have been 3000 ft long, with a slope of about 2.3°. How much force would be required to hold a 60-ton monolith on this causeway?

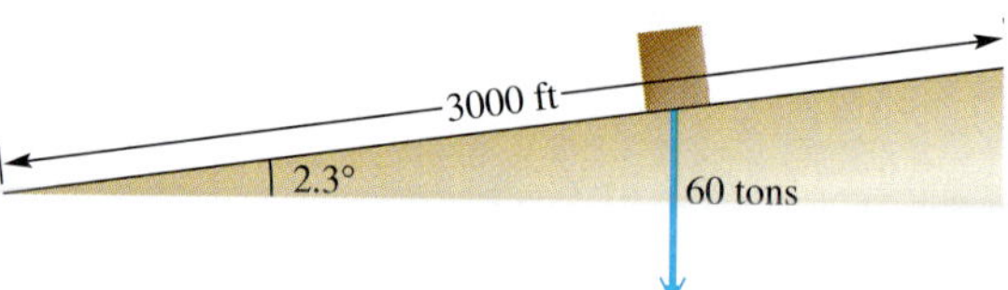

10. ***Weight of a Box*** Two people are carrying a box. One person exerts a force of 150 lb at an angle of 62.4° with the horizontal. The other person exerts a force of 114 lb at an angle of 54.9°. Find the weight of the box.

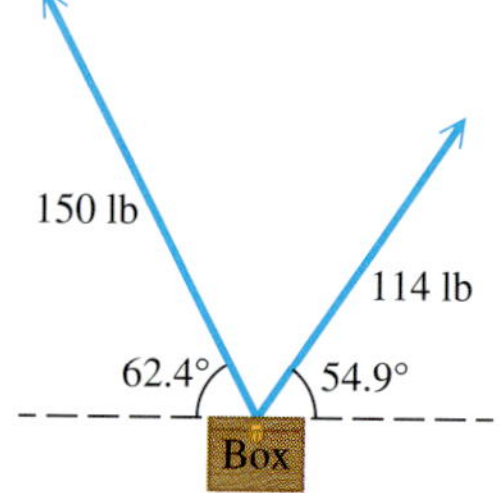

11. ***Incline Angle*** A force of 18 lb is required to hold a 60-lb stump grinder on an incline. What angle does the incline make with the horizontal?

12. ***Incline Angle*** A force of 30 lb is required to hold an 80-lb pressure washer on an incline. What angle does the incline make with the horizontal?

13. ***Weight of a Crate and Tension of a Rope*** A crate is supported by two ropes. One rope makes an angle of 46° 20′ with the horizontal and has a tension of 89.6 lb on it. The other rope is horizontal. Find the weight of the crate and the tension in the horizontal rope.

14. ***Angles Between Forces*** Three forces acting at a point are in equilibrium. The forces are 980 lb, 760 lb, and 1220 lb. Find the angles between the directions of the forces. (*Hint:* Arrange the forces to form the sides of a triangle.)

15. ***Distance and Bearing of a Ship*** A ship leaves port on a bearing of 34.0° and travels 10.4 mi. The ship then turns due east and travels 4.6 mi. How far is the ship from port, and what is its bearing from port?

16. ***Distance and Bearing of a Luxury Liner*** A luxury liner leaves port on a bearing of 110.0° and travels 8.8 mi. It then turns due west and travels 2.4 mi. How far is the liner from port, and what is its bearing from port?

17. *Distance of a Ship from Its Starting Point* Starting at point A, a ship sails 18.5 km on a bearing of 189°, then turns and sails 47.8 km on a bearing of 317°. Find the distance of the ship from point A.

18. *Distance of a Ship from Its Starting Point* Starting at point X, a ship sails 15.5 km on a bearing of 200°, then turns and sails 2.4 km on a bearing of 320°. Find the distance of the ship from point X.

19. *Distance and Direction of a Motorboat* A motorboat sets out in the direction N 80° E. The speed of the boat in still water is 20.0 mph. If the current is flowing directly south, and the actual direction of the motorboat is due east, find the speed of the current and the actual speed of the motorboat.

20. *Path Traveled by a Plane* The aircraft carrier *Tallahassee* is traveling at sea on a steady course with a bearing of 30° at 32 mph. Patrol planes on the carrier have enough fuel for 2.6 hr of flight when traveling at a speed of 520 mph. One of the pilots takes off on a bearing of 338° and then turns and heads in a straight line, so as to be able to catch the carrier and land on the deck at the exact instant that his fuel runs out. If the pilot left at 2 P.M., at what time did he turn to head for the carrier?

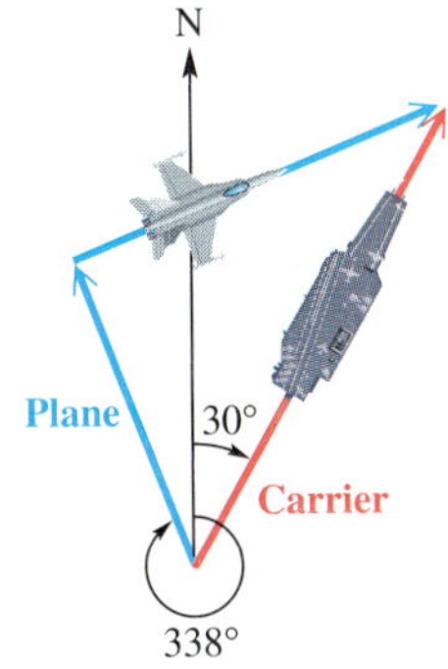

21. *Bearing and Groundspeed of a Plane* An airline route from San Francisco to Honolulu is on a bearing of 233.0°. A jet flying at 450 mph on that bearing runs into a wind blowing at 39.0 mph from a direction of 114.0°. Find the resulting bearing and groundspeed of the plane.

22. *Movement of a Motorboat* Suppose you would like to cross a 132-ft-wide river in a motorboat. Assume that the motorboat can travel at 7 mph relative to the water and that the current is flowing west at the rate of 3 mph. The bearing θ is chosen so that the motorboat will land at a point exactly across from the starting point.

    (a) At what speed will the motorboat be traveling relative to the banks?

    (b) How long will it take for the motorboat to make the crossing?

    (c) What is the measure of angle θ?

23. *Airspeed and Groundspeed* A pilot wants to fly on a bearing of 74.9°. By flying due east, he finds that a 42-mph wind, blowing from the south, puts him on course. Find the airspeed and the groundspeed.

24. *Bearing of a Plane* A plane flies 650 mph on a bearing of 175.3°. A 25-mph wind, from a direction of 266.6°, blows against the plane. Find the resulting bearing of the plane.

25. *Bearing and Groundspeed of a Plane* A pilot is flying at 190 mph. He wants his flight path to be on a bearing of 64° 30′. A wind is blowing from the south at 35.0 mph. Find the bearing he should fly, and find the plane's groundspeed.

26. *Bearing and Groundspeed of a Plane* A pilot is flying at 168 mph. She wants her flight path to be on a bearing of 57° 40′. A wind is blowing from the south at 27.1 mph. Find the bearing the pilot should fly, and find the plane's groundspeed.

27. *Bearing and Airspeed of a Plane* What bearing and airspeed are required for a plane to fly 400 mi due north in 2.5 hr if the wind is blowing from a direction of 328° at 11 mph?

28. *Groundspeed and Bearing of a Plane* A plane is headed due south with an airspeed of 192 mph. A wind from a direction of 78° is blowing at 23 mph. Find the groundspeed and resulting bearing of the plane.

29. *Groundspeed and Bearing of a Plane* An airplane is headed on a bearing of 174° at an airspeed of 240 km per hr. A 30 km per hr wind is blowing from a direction of 245°. Find the groundspeed and resulting bearing of the plane.

30. *Velocity of a Star* The space velocity $\mathbf{v}$ of a star relative to the sun can be expressed as the resultant vector of two perpendicular vectors—the radial velocity $\mathbf{v}_r$ and the tangential velocity $\mathbf{v}_t$, where $\mathbf{v} = \mathbf{v}_r + \mathbf{v}_t$. If a star is located near the sun and its space velocity is large, then its motion across the sky will also be large. Barnard's Star is a relatively close star with a distance of 35 trillion mi from the sun. It moves across the sky through an angle of 10.34″ per year, which is the largest motion of any known star. Its radial velocity is $\mathbf{v}_r = 67$ mi per sec toward the sun. (*Source:* Zeilik, M., S. Gregory, and E. Smith, *Introductory Astronomy and Astrophysics,* Second Edition, Saunders College Publishing, 1998; Acker, A. and C. Jaschek, *Astronomical Methods and Calculations,* John Wiley & Sons, 1986.)

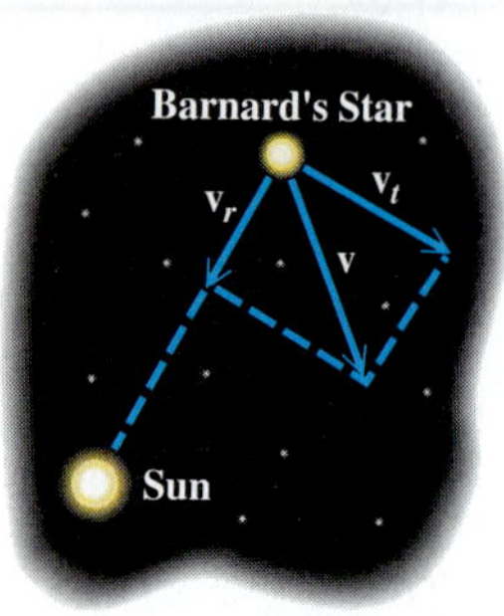

Not to scale

(a) Approximate the tangential velocity $\mathbf{v}_t$ of Barnard's Star. (*Hint:* Use the arc length formula $s = r\theta$.)

(b) Compute the magnitude of $\mathbf{v}$.

31. *(Modeling) Measuring Rainfall* Suppose that vector $\mathbf{R}$ models the amount of rainfall in inches and the direction it falls, and vector $\mathbf{A}$ models the area in square inches and orientation of the opening of a rain gauge, as illustrated in the figure. The total volume V of water collected in the rain gauge is given by $V = |\mathbf{R} \cdot \mathbf{A}|$. This formula calculates the volume of water collected even if the wind is blowing the rain in a slanted direction or the rain gauge is not exactly vertical. Let $\mathbf{R} = \mathbf{i} - 2\mathbf{j}$ and $\mathbf{A} = .5\mathbf{i} + \mathbf{j}$.

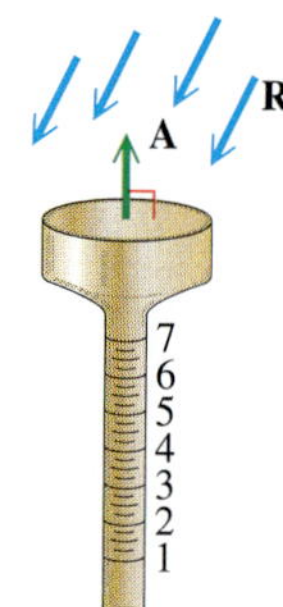

(a) Find $|\mathbf{R}|$ and $|\mathbf{A}|$. Interpret your results.

(b) Calculate V and interpret this result.

(c) For the rain gauge to collect the maximum amount of water, what should be true about vectors $\mathbf{R}$ and $\mathbf{A}$?

32. *The Dot Product* In the figure below $\mathbf{a} = \langle a_1, a_2 \rangle$, $\mathbf{b} = \langle b_1, b_2 \rangle$, and $\mathbf{a} - \mathbf{b} = \langle a_1 - b_1, a_2 - b_2 \rangle$. Apply the law of cosines to the triangle and derive the equation $\mathbf{a} \cdot \mathbf{b} = |\mathbf{a}||\mathbf{b}| \cos \theta$.

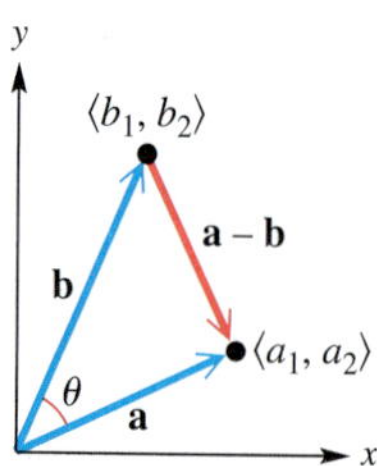

Summary Exercises on Applications of Trigonometry and Vectors

These summary exercises provide practice with applications that involve solving triangles and using vectors.

1. ***Wires Supporting a Flagpole*** A flagpole stands vertically on a hillside that makes an angle of 20° with the horizontal. Two supporting wires are attached as shown in the figure. What are the lengths of the supporting wires?

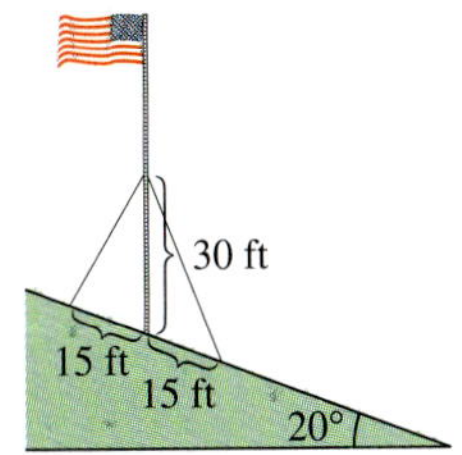

2. ***Walking Dogs on Leashes*** While Michael is walking his two dogs, Duke and Prince, they reach a corner and must wait for a WALK sign. Michael is holding the two leashes in the same hand, and the dogs are pulling on their leashes at the angles and forces shown in the figure. Find the magnitude of the force Michael must apply to restrain the dogs.

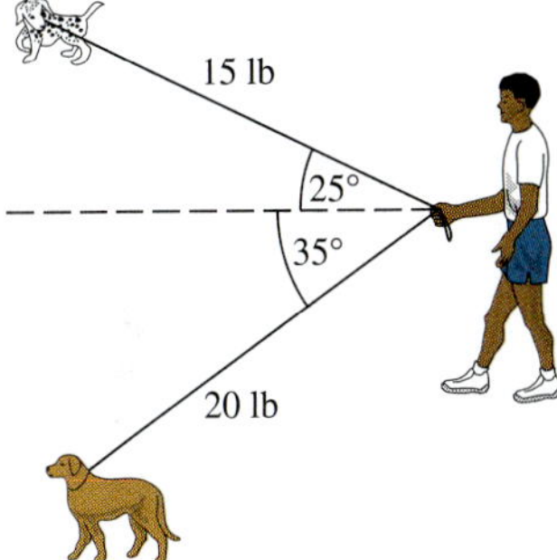

3. ***Distance Between Two Lighthouses*** Two lighthouses are located on a north-south line. From lighthouse A, the bearing of a ship 3742 m away is 129° 43′. From lighthouse B, the bearing of a ship is 39° 43′. Find the distance between the lighthouses.

4. ***Hot-Air Balloon*** A hot-air balloon is rising straight up at the speed of 15 ft per sec. Then a wind starts blowing horizontally at 5 ft per sec. What will the new speed of the balloon be and what angle with the horizontal will the balloon's path make?

5. ***Playing on a Swing*** Mary is playing with her daughter Brittany on a swing. Starting from rest, Mary pulls the swing through an angle of 40° and holds it briefly before releasing the swing. If Brittany weighs 50 lb, what horizontal force must Mary apply while holding the swing?

6. ***Height of an Airplane*** Two observation points A and B are 950 ft apart. From these points the angles of elevation of an airplane are 52° and 57°. (See the figure.) Find the height of the airplane.

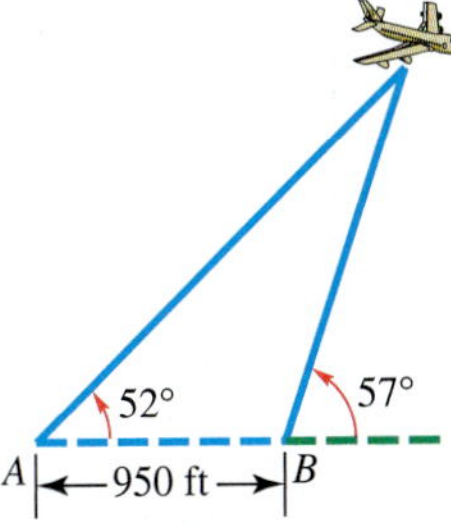

7. ***Ground Distances Measured by Aerial Photography*** The distance shown in an aerial photograph is determined by both the focal length of the lens and the tilt of the camera from the perpendicular to the ground. A camera lens with a 12-in. focal length has an angular coverage of 60°. If an aerial photograph is taken with this camera tilted 35° at an altitude of 5000 ft, calculate the distance d in miles that will be shown in this photograph. (See the figure.) (*Source:* Brooks, R., and D. Johannes, *Photoarchaeology,* Dioscorides Press, 1990; Moffitt, F., *Photogrammetry,* International Textbook Company, 1967.)

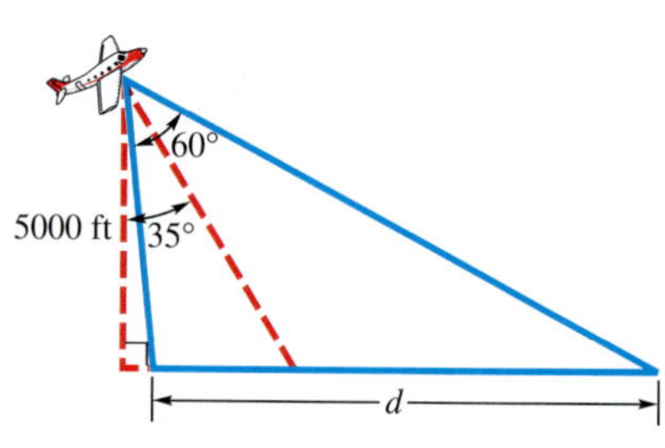

8. ***Wind and Vectors*** A wind can be described by $\mathbf{v} = 6\mathbf{i} + 8\mathbf{j}$, where vector $\mathbf{j}$ points north and represents a south wind of 1 mph.

(a) What is the speed of the wind? (b) Find $3\mathbf{v}$. Interpret the result.

(c) Interpret the wind if it switches to $\mathbf{u} = -8\mathbf{i} + 8\mathbf{j}$.

8.5 Trigonometric (Polar) Form of Complex Numbers; Products and Quotients

The Complex Plane and Vector Representation ▪ Trigonometric (Polar) Form ▪ Products of Complex Numbers in Trigonometric Form ▪ Quotients of Complex Numbers in Trigonometric Form

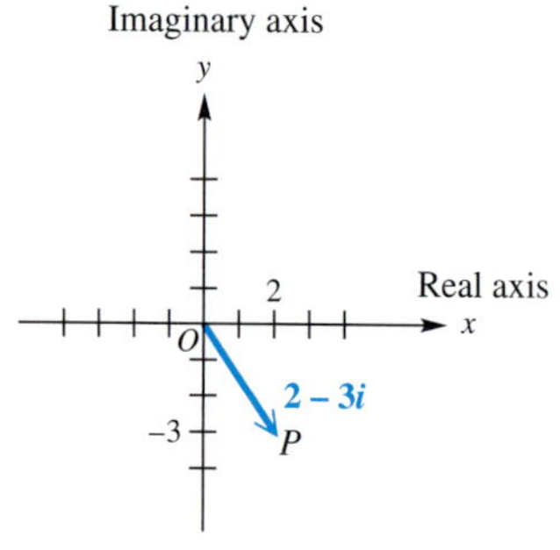

Figure 43

The Complex Plane and Vector Representation

Unlike real numbers, complex numbers cannot be ordered. One way to organize and illustrate them is by using a graph. To graph a complex number such as $2 - 3i$, we modify the familiar coordinate system by calling the horizontal axis the **real axis** and the vertical axis the **imaginary axis.** Then complex numbers can be graphed in this **complex plane,** as shown in Figure 43. Each complex number $a + bi$ determines a unique position vector with initial point $(0, 0)$ and terminal point (a, b). This shows the close connection between vectors and complex numbers.

NOTE This geometric representation is the reason that $a + bi$ is called the **rectangular form** of a complex number. (*Rectangular form* is also called *standard form.*)

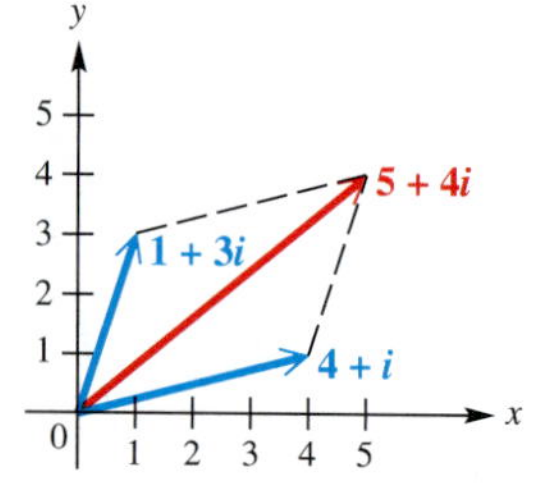

Figure 44

Recall that the sum of the two complex numbers $4 + i$ and $1 + 3i$ is

$$(4 + i) + (1 + 3i) = 5 + 4i. \quad \text{(Section 1.3)}$$

Graphically, the sum of two complex numbers is represented by the vector that is the resultant of the vectors corresponding to the two numbers, as shown in Figure 44.

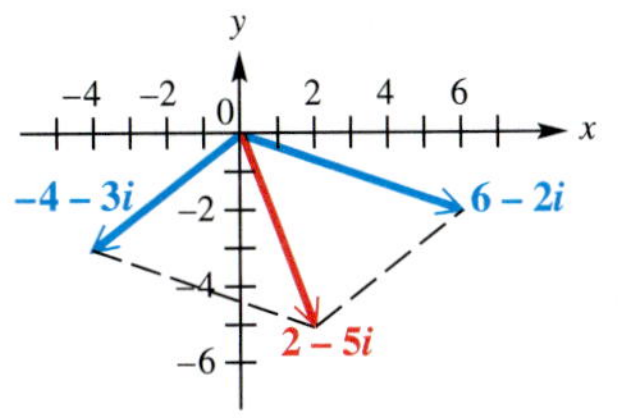

Figure 45

EXAMPLE 1 Expressing the Sum of Complex Numbers Graphically

Find the sum of $6 - 2i$ and $-4 - 3i$. Graph both complex numbers and their resultant.

Solution The sum is found by adding the two numbers.

$$(6 - 2i) + (-4 - 3i) = 2 - 5i$$

The graphs are shown in Figure 45.

Now try Exercises 1 and 11.

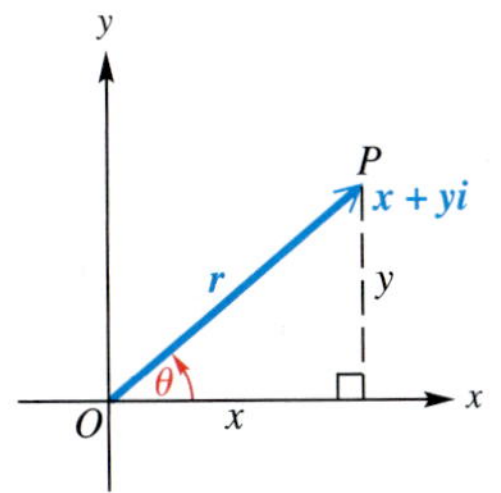

Figure 46

Trigonometric (Polar) Form Figure 46 shows the complex number $x + yi$ that corresponds to a vector **OP** with direction angle θ and magnitude r. The following relationships among x, y, r, and θ can be verified from Figure 46.

Relationships Among x, y, r, and θ

$$x = r \cos \theta \qquad y = r \sin \theta$$

$$r = \sqrt{x^2 + y^2} \qquad \tan \theta = \frac{y}{x}, \quad \text{if } x \neq 0$$

Substituting $x = r \cos \theta$ and $y = r \sin \theta$ from these relationships into $x + yi$ gives

$$\begin{aligned} x + yi &= r \cos \theta + (r \sin \theta)i \\ &= r(\cos \theta + i \sin \theta). \end{aligned}$$

Trigonometric (Polar) Form of a Complex Number

The expression

$$r(\cos \theta + i \sin \theta)$$

is called the **trigonometric form** (or **polar form**) of the complex number $x + yi$. The expression $\cos \theta + i \sin \theta$ is sometimes abbreviated cis θ. Using this notation,

$$r(\cos \theta + i \sin \theta) \text{ is written } r \text{ cis } \theta.$$

The number r is the **absolute value** (or **modulus**) of $x + yi$, and θ is the **argument** of $x + yi$. In this section we choose the value of θ in the interval $[0°, 360°)$. However, any angle coterminal with θ also could serve as the argument.

EXAMPLE 2 Converting from Trigonometric Form to Rectangular Form

Express $2(\cos 300° + i \sin 300°)$ in rectangular form.

Algebraic Solution

From Chapter 5, $\cos 300° = \frac{1}{2}$ and $\sin 300° = -\frac{\sqrt{3}}{2}$, so

$$2(\cos 300° + i \sin 300°) = 2\left(\frac{1}{2} - i\frac{\sqrt{3}}{2}\right)$$
$$= 1 - i\sqrt{3}.$$

Notice that the real part is positive and the imaginary part is negative; this is consistent with 300° being a quadrant IV angle.

Graphing Calculator Solution

Figure 47 confirms the algebraic solution.

```
2(cos(300)+isin(
300))
  1-1.732050808i
-√(3)
      -1.732050808
```

The imaginary part is an approximation for $-\sqrt{3}$.

Figure 47

Now try Exercise 19.

To convert from rectangular form to trigonometric form, we use the following procedure.

Converting from Rectangular Form to Trigonometric Form

Step 1 Sketch a graph of the number $x + yi$ in the complex plane.

Step 2 Find r by using the equation $r = \sqrt{x^2 + y^2}$.

Step 3 Find θ by using the equation $\tan \theta = \frac{y}{x}$, $x \neq 0$, choosing the quadrant indicated in Step 1.

CAUTION Errors often occur in Step 3. Be sure to choose the correct quadrant for θ by referring to the graph sketched in Step 1.

EXAMPLE 3 Converting from Rectangular Form to Trigonometric Form

Write each complex number in trigonometric form.

(a) $-\sqrt{3} + i$ **(b)** $-3i$

Solution

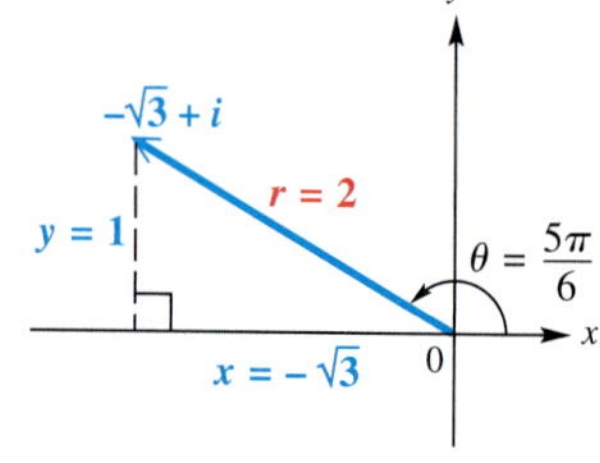

Figure 48

(a) We start by sketching the graph of $-\sqrt{3} + i$ in the complex plane, as shown in Figure 48. Since $x = -\sqrt{3}$ and $y = 1$,

$$r = \sqrt{x^2 + y^2} = \sqrt{(-\sqrt{3})^2 + 1^2} = \sqrt{3 + 1} = 2,$$

and $$\tan \theta = \frac{y}{x} = \frac{1}{-\sqrt{3}} = -\frac{\sqrt{3}}{3}.$$ Rationalize the denominator. (Section R.7)

Since $\tan \theta = -\frac{\sqrt{3}}{3}$, the reference angle for θ in radians is $\frac{\pi}{6}$. From the graph, we see that θ is in quadrant II, so $\theta = \pi - \frac{\pi}{6} = \frac{5\pi}{6}$. Therefore,

$$-\sqrt{3} + i = 2\left(\cos \frac{5\pi}{6} + i \sin \frac{5\pi}{6}\right) = 2 \text{ cis } \frac{5\pi}{6}.$$

ANGLE
1:°
2:'
3:r
4:▸DMS
5:R▸Pr(
6:R▸Pθ(
7↓P▸Rx(

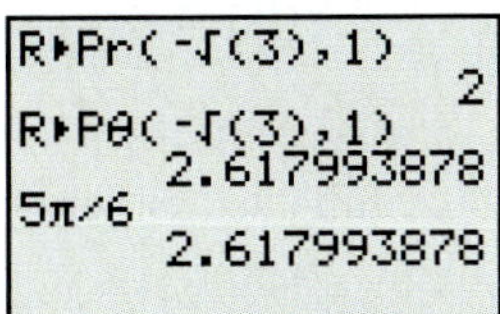

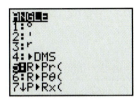

Choices 5 and 6 in the top screen show how to convert from rectangular (x, y) form to trigonometric form. The calculator is in radian mode. The results agree with our algebraic results in Example 3(a).

(b) The sketch of $-3i$ is shown in Figure 49.

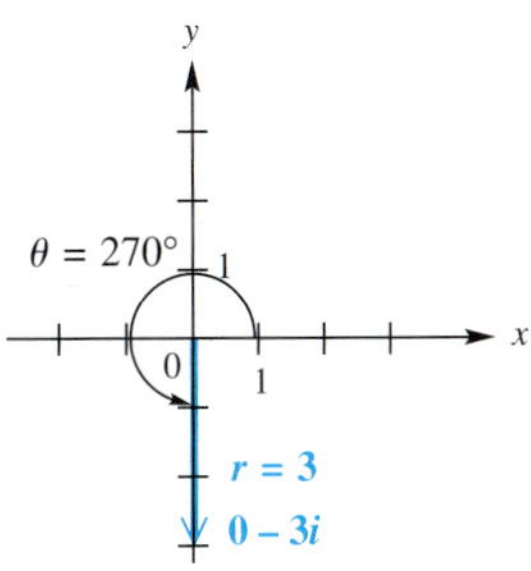

Figure 49

Since $-3i = 0 - 3i$, we have $x = 0$ and $y = -3$ and

$$r = \sqrt{0^2 + (-3)^2} = \sqrt{0 + 9} = \sqrt{9} = 3.$$

We cannot find θ by using $\tan\theta = \frac{y}{x}$, because $x = 0$. From the graph, a value for θ is 270°. In trigonometric form,

$$-3i = 3(\cos 270° + i \sin 270°) = 3 \text{ cis } 270°.$$

Now try Exercises 25 and 31.

NOTE In Example 3, we gave answers in both forms: $r(\cos\theta + i\sin\theta)$ and $r \text{ cis } \theta$. These forms will be used interchangeably from now on.

EXAMPLE 4 Converting Between Trigonometric and Rectangular Forms Using Calculator Approximations

Write each complex number in its alternative form, using calculator approximations as necessary.

(a) $6(\cos 115° + i \sin 115°)$ **(b)** $5 - 4i$

Solution

(a) Since 115° does not have a special angle as a reference angle, we cannot find exact values for cos 115° and sin 115°. Use a calculator set in degree mode to find $\cos 115° \approx -.4226182617$ and $\sin 115° \approx .906307787$. Therefore, in rectangular form,

$$\begin{aligned} 6(\cos 115° + i \sin 115°) &\approx 6(-.4226182617 + .906307787i) \\ &= -2.53570957 + 5.437846722i. \end{aligned}$$

(b) A sketch of $5 - 4i$ shows that θ must be in quadrant IV. See Figure 50. Here $r = \sqrt{5^2 + (-4)^2} = \sqrt{41}$ and $\tan\theta = -\frac{4}{5}$. Use a calculator to find that one measure of θ is $-38.66°$. In order to express θ in the interval $[0, 360°)$, we find $\theta = 360° - 38.66° = 321.34°$. Use these results to get

$$5 - 4i = \sqrt{41} \text{ cis } 321.34°.$$

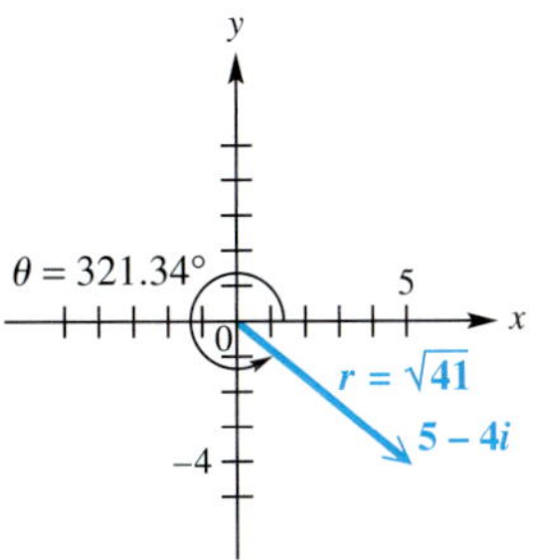

Figure 50

Now try Exercises 33 and 37.

Source: Figure from Crownover, R., *Introduction to Fractals and Chaos.* Copyright © 1995. Boston: Jones and Bartlett Publishers. Reprinted with permission.

Figure 51

We can apply complex numbers to the study of *fractals*. An example of a fractal is the *Mandelbrot set* shown in Figure 51.

EXAMPLE 5 Deciding Whether a Complex Number Is in the Julia Set

The fractal called the *Julia set* is shown in Figure 52. To determine if a complex number $z = a + bi$ is in this Julia set, perform the following sequence of calculations. Repeatedly compute the values of $z^2 - 1$, $(z^2 - 1)^2 - 1$, $[(z^2 - 1)^2 - 1]^2 - 1, \ldots.$ If the absolute values of any of the resulting complex numbers exceed 2, then the complex number z is not in the Julia set. Otherwise z is part of this set and the point (a, b) should be shaded in the graph.

Source: Figure from Crownover, R., *Introduction to Fractals and Chaos.* Copyright © 1995. Boston: Jones and Bartlett Publishers. Reprinted with permission.

Figure 52

Determine whether each number belongs to the Julia set.

(a) $z = 0 + 0i$ **(b)** $z = 1 + 1i$

Solution

(a) Here

$$\begin{aligned} z &= 0 + 0i = 0, \\ z^2 - 1 &= 0^2 - 1 = -1, \\ (z^2 - 1)^2 - 1 &= (-1)^2 - 1 = 0, \\ [(z^2 - 1)^2 - 1]^2 - 1 &= 0^2 - 1 = -1, \end{aligned}$$

and so on. We see that the calculations repeat as $0, -1, 0, -1$, and so on. The absolute values are either 0 or 1, which do not exceed 2, so $0 + 0i$ is in the Julia set and the point $(0, 0)$ is part of the graph.

(b) We have $z^2 - 1 = (1 + i)^2 - 1 = (1 + 2i + i^2) - 1 = -1 + 2i$. The absolute value is $\sqrt{(-1)^2 + 2^2} = \sqrt{5}$. Since $\sqrt{5}$ is greater than 2, $1 + 1i$ is not in the Julia set and $(1, 1)$ is not part of the graph.

Now try Exercise 43.

Products of Complex Numbers in Trigonometric Form Using the FOIL method to multiply complex numbers in rectangular form, we find the product of $1 + i\sqrt{3}$ and $-2\sqrt{3} + 2i$ as follows.

$$\begin{aligned} (1 + i\sqrt{3})(-2\sqrt{3} + 2i) &= -2\sqrt{3} + 2i - 2i(3) + 2i^2\sqrt{3} \quad \text{(Section 1.3)} \\ &= -2\sqrt{3} + 2i - 6i - 2\sqrt{3} \\ &= -4\sqrt{3} - 4i \end{aligned}$$

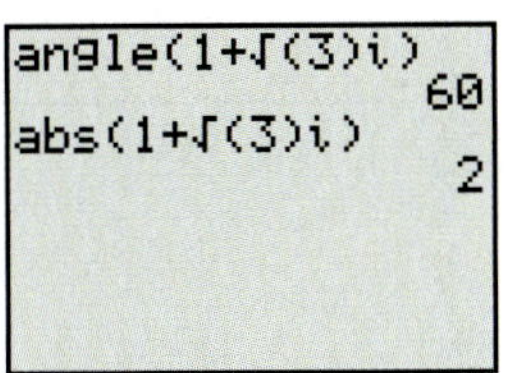

With the calculator in complex and degree modes, the MATH menu can be used to find the angle and the magnitude (absolute value) of the vector that corresponds to a given complex number.

We can also find this same product by first converting the complex numbers $1 + i\sqrt{3}$ and $-2\sqrt{3} + 2i$ to trigonometric form. Using the method explained earlier in this section,

$$1 + i\sqrt{3} = 2(\cos 60° + i \sin 60°)$$

and

$$-2\sqrt{3} + 2i = 4(\cos 150° + i \sin 150°).$$

If we multiply the trigonometric forms, and if we use identities for the cosine and the sine of the sum of two angles, then the result is

$$\begin{aligned}
&[2(\cos 60° + i \sin 60°)][4(\cos 150° + i \sin 150°)] \\
&= 2 \cdot 4(\cos 60° \cdot \cos 150° + i \sin 60° \cdot \cos 150° \\
&\quad + i \cos 60° \cdot \sin 150° + i^2 \sin 60° \cdot \sin 150°) \quad \text{(Section 7.3)} \\
&= 8[(\cos 60° \cdot \cos 150° - \sin 60° \cdot \sin 150°) \\
&\quad + i(\sin 60° \cdot \cos 150° + \cos 60° \cdot \sin 150°)] \\
&= 8[\cos(60° + 150°) + i \sin(60° + 150°)] \\
&= 8(\cos 210° + i \sin 210°).
\end{aligned}$$

The absolute value of the product, 8, is equal to the product of the absolute values of the factors, $2 \cdot 4$, and the argument of the product, 210°, is equal to the sum of the arguments of the factors, $60° + 150°$.

As we would expect, the product obtained when multiplying by the first method is the rectangular form of the product obtained when multiplying by the second method.

$$\begin{aligned}
8(\cos 210° + i \sin 210°) &= 8\left(-\frac{\sqrt{3}}{2} - \frac{1}{2}i\right) \\
&= -4\sqrt{3} - 4i
\end{aligned}$$

We can generalize this work in the following *product theorem.*

Product Theorem

If $r_1(\cos \theta_1 + i \sin \theta_1)$ and $r_2(\cos \theta_2 + i \sin \theta_2)$ are any two complex numbers, then

$$\begin{aligned}
&[r_1(\cos \theta_1 + i \sin \theta_1)] \cdot [r_2(\cos \theta_2 + i \sin \theta_2)] \\
&= r_1 r_2[\cos(\theta_1 + \theta_2) + i \sin(\theta_1 + \theta_2)].
\end{aligned}$$

In compact form, this is written

$$(r_1 \operatorname{cis} \theta_1)(r_2 \operatorname{cis} \theta_2) = r_1 r_2 \operatorname{cis}(\theta_1 + \theta_2).$$

That is, to multiply complex numbers in trigonometric form, multiply their absolute values and add their arguments.

EXAMPLE 6 Using the Product Theorem

Find the product of $3(\cos 45° + i \sin 45°)$ and $2(\cos 135° + i \sin 135°)$.

Solution

$$[3(\cos 45° + i \sin 45°)][2(\cos 135° + i \sin 135°)]$$
$$= 3 \cdot 2[\cos(45° + 135°) + i \sin(45° + 135°)] \quad \text{Product theorem}$$
$$= 6(\cos 180° + i \sin 180°) \quad \text{Multiply and add.}$$
$$= 6(-1 + i \cdot 0) = 6(-1) = -6$$

Now try Exercise 45.

Quotients of Complex Numbers in Trigonometric Form The rectangular form of the quotient of the complex numbers $1 + i\sqrt{3}$ and $-2\sqrt{3} + 2i$ is

$$\frac{1 + i\sqrt{3}}{-2\sqrt{3} + 2i} = \frac{(1 + i\sqrt{3})(-2\sqrt{3} - 2i)}{(-2\sqrt{3} + 2i)(-2\sqrt{3} - 2i)} \quad \text{Multiply by the conjugate of the denominator. (Section 1.3)}$$

$$= \frac{-2\sqrt{3} - 2i - 6i - 2i^2\sqrt{3}}{12 - 4i^2} \quad \text{FOIL; } (x + y)(x - y) = x^2 - y^2$$

$$= \frac{-8i}{16} = -\frac{1}{2}i. \quad \text{Simplify.}$$

Writing $1 + i\sqrt{3}$, $-2\sqrt{3} + 2i$, and $-\frac{1}{2}i$ in trigonometric form gives

$$1 + i\sqrt{3} = 2(\cos 60° + i \sin 60°),$$
$$-2\sqrt{3} + 2i = 4(\cos 150° + i \sin 150°),$$

and

$$-\frac{1}{2}i = \frac{1}{2}[\cos(-90°) + i \sin(-90°)].$$

The absolute value of the quotient, $\frac{1}{2}$, is the quotient of the two absolute values, $\frac{2}{4} = \frac{1}{2}$. The argument of the quotient, $-90°$, is the difference of the two arguments, $60° - 150° = -90°$. Generalizing from this example leads to the *quotient theorem.*

Quotient Theorem

If $r_1(\cos \theta_1 + i \sin \theta_1)$ and $r_2(\cos \theta_2 + i \sin \theta_2)$ are any two complex numbers, where $r_2(\cos \theta_2 + i \sin \theta_2) \neq 0$, then

$$\frac{r_1(\cos \theta_1 + i \sin \theta_1)}{r_2(\cos \theta_2 + i \sin \theta_2)} = \frac{r_1}{r_2}[\cos(\theta_1 - \theta_2) + i \sin(\theta_1 - \theta_2)].$$

In compact form, this is written

$$\frac{r_1 \operatorname{cis} \theta_1}{r_2 \operatorname{cis} \theta_2} = \frac{r_1}{r_2} \operatorname{cis}(\theta_1 - \theta_2).$$

That is, to divide complex numbers in trigonometric form, divide their absolute values and subtract their arguments.

EXAMPLE 7 Using the Quotient Theorem

Find the quotient $\frac{10 \text{ cis}(-60°)}{5 \text{ cis } 150°}$. Write the result in rectangular form.

Solution

$$\begin{aligned}\frac{10 \text{ cis}(-60°)}{5 \text{ cis } 150°} &= \frac{10}{5} \text{cis}(-60° - 150°) && \text{Quotient theorem}\\ &= 2 \text{ cis}(-210°) && \text{Divide and subtract.}\\ &= 2[\cos(-210°) + i \sin(-210°)] && \text{Rewrite.}\\ &= 2\left[-\frac{\sqrt{3}}{2} + i\left(\frac{1}{2}\right)\right] && \cos(-210°) = -\tfrac{\sqrt{3}}{2};\ \sin(-210°) = \tfrac{1}{2} \text{ (Section 5.3)}\\ &= -\sqrt{3} + i && \text{Rectangular form}\end{aligned}$$

Now try Exercise 55.

8.5 Exercises

Graph each complex number. See Example 1.

1. $-3 + 2i$
2. $6 - 5i$
3. $\sqrt{2} + \sqrt{2}i$
4. $2 - 2i\sqrt{3}$
5. $-4i$
6. $3i$
7. -8
8. 2

Concept Check *Give the rectangular form of the complex number represented in each graph.*

9.

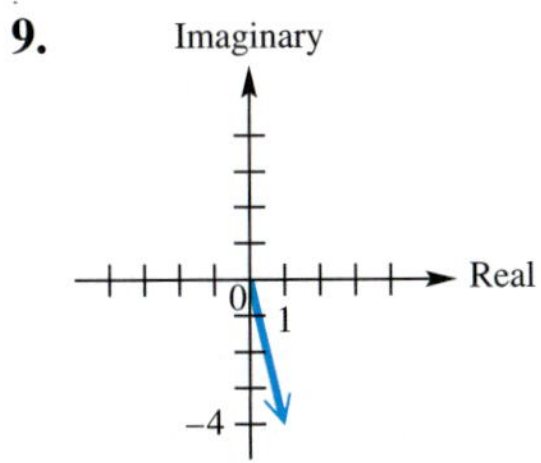

10.

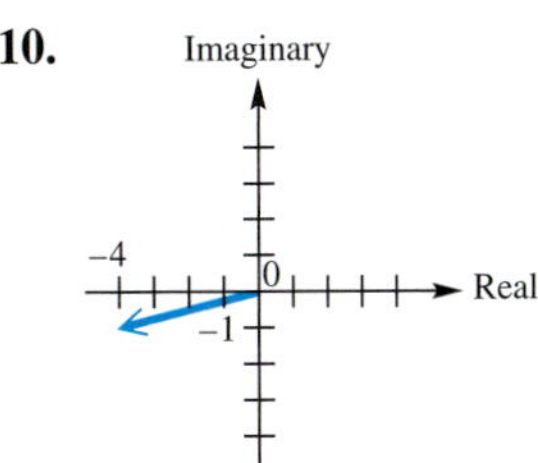

Find the sum of each pair of complex numbers. See Example 1.

11. $5 - 6i, -2 + 3i$
12. $7 - 3i, -4 + 3i$
13. $-3, 3i$
14. $6, -2i$
15. $7 + 6i, 3i$
16. $-5 - 8i, -1$

Write each complex number in rectangular form. See Example 2.

17. $10(\cos 90° + i \sin 90°)$
18. $8(\cos 270° + i \sin 270°)$
19. $4(\cos 240° + i \sin 240°)$
20. $2(\cos 330° + i \sin 330°)$
21. $3 \text{ cis } 150°$
22. $6 \text{ cis } 135°$
23. $\sqrt{2} \text{ cis } 180°$
24. $\sqrt{3} \text{ cis } 315°$

Write each complex number in trigonometric form $r(\cos\theta + i\sin\theta)$, with θ in the interval $[0°, 360°)$. See Example 3.

25. $\sqrt{3} - i$ **26.** $4\sqrt{3} + 4i$ **27.** $-5 - 5i$ **28.** $-\sqrt{2} + i\sqrt{2}$

29. $2 + 2i$ **30.** $-\frac{3\sqrt{3}}{2} - \frac{3}{2}i$ **31.** $5i$ **32.** -4

Perform each conversion, using a calculator as necessary. See Example 4.

	Rectangular Form	Trigonometric Form
33.	______________	$3(\cos 250° + i\sin 250°)$
34.	$-4 + i$	______________
35.	$12i$	______________
36.	______________	$3 \text{ cis } 180°$
37.	$3 + 5i$	______________
38.	______________	$\text{cis } 110.5°$

Concept Check *The complex number z, where $z = x + yi$, can be graphed in the plane as (x, y). Describe the graphs of all complex numbers z satisfying the conditions in Exercises 39–42.*

39. The absolute value of z is 1.

40. The real and imaginary parts of z are equal.

41. The real part of z is 1.

42. The imaginary part of z is 1.

Julia Set *Refer to Example 5 to solve Exercises 43 and 44.*

43. Is $z = -.2i$ in the Julia set?

44. The graph of the Julia set in Figure 52 appears to be symmetric with respect to both the x-axis and y-axis. Complete the following to show that this is true.

(a) Show that complex conjugates have the same absolute value.

(b) Compute $z_1^2 - 1$ and $z_2^2 - 1$, where $z_1 = a + bi$ and $z_2 = a - bi$.

(c) Discuss why if (a, b) is in the Julia set then so is $(a, -b)$.

(d) Conclude that the graph of the Julia set must be symmetric with respect to the x-axis.

(e) Using a similar argument, show that the Julia set must also be symmetric with respect to the y-axis.

Find each product and write it in rectangular form. See Example 6.

45. $[2(\cos 45° + i\sin 45°)][2(\cos 225° + i\sin 225°)]$

46. $[8(\cos 300° + i\sin 300°)][5(\cos 120° + i\sin 120°)]$

47. $[4(\cos 60° + i\sin 60°)][6(\cos 330° + i\sin 330°)]$

48. $[8(\cos 210° + i\sin 210°)][2(\cos 330° + i\sin 330°)]$

49. $(5 \text{ cis } 90°)(3 \text{ cis } 45°)$

50. $(6 \text{ cis } 120°)[5 \text{ cis}(-30°)]$

51. $(\sqrt{3} \text{ cis } 45°)(\sqrt{3} \text{ cis } 225°)$

52. $(\sqrt{2} \text{ cis } 300°)(\sqrt{2} \text{ cis } 270°)$

Find each quotient and write it in rectangular form. In Exercises 57–60, first convert the numerator and the denominator to trigonometric form. See Example 7.

53. $\frac{10(\cos 225° + i\sin 225°)}{5(\cos 45° + i\sin 45°)}$

54. $\frac{16(\cos 300° + i\sin 300°)}{8(\cos 60° + i\sin 60°)}$

55. $\frac{3 \text{ cis } 305°}{9 \text{ cis } 65°}$

56. $\frac{12 \text{ cis } 293°}{6 \text{ cis } 23°}$

57. $\frac{-i}{1 + i}$

58. $\frac{1}{2 - 2i}$

59. $\frac{2\sqrt{6} - 2i\sqrt{2}}{\sqrt{2} - i\sqrt{6}}$

60. $\frac{4 + 4i}{2 - 2i}$

Use a calculator to perform the indicated operations. Give answers in rectangular form.

61. $[2.5(\cos 35° + i \sin 35°)][3.0(\cos 50° + i \sin 50°)]$

62. $[4.6(\cos 12° + i \sin 12°)][2.0(\cos 13° + i \sin 13°)]$

63. $(12 \text{ cis } 18.5°)(3 \text{ cis } 12.5°)$

64. $(4 \text{ cis } 19.25°)(7 \text{ cis } 41.75°)$

65. $\frac{45(\cos 127° + i \sin 127°)}{22.5(\cos 43° + i \sin 43°)}$

66. $\frac{30(\cos 130° + i \sin 130°)}{10(\cos 21° + i \sin 21°)}$

Relating Concepts

For individual or collaborative investigation

(Exercises 67–73)

Consider the complex numbers $w = -1 + i$ and $z = -1 - i$. ***Work Exercises 67–73 in order.***

67. Multiply w and z using their rectangular forms and the FOIL method. Leave the product in rectangular form.

68. Find the trigonometric forms of w and z.

69. Multiply w and z using their trigonometric forms and the method described in this section.

70. Use the result of Exercise 69 to find the rectangular form of wz. How does this compare to your result in Exercise 67?

71. Find the quotient $\frac{w}{z}$ using their rectangular forms and multiplying both the numerator and the denominator by the conjugate of the denominator. Leave the quotient in rectangular form.

72. Use the trigonometric forms of w and z, found in Exercise 68, to divide w by z using the method described in this section.

73. Use the result of Exercise 72 to find the rectangular form of $\frac{w}{z}$. How does this compare to your result in Exercise 71?

74. Notice that $(r \text{ cis } \theta)^2 = (r \text{ cis } \theta)(r \text{ cis } \theta) = r^2 \text{ cis}(\theta + \theta) = r^2 \text{ cis } 2\theta$. State in your own words how we can square a complex number in trigonometric form. (In the next section, we will develop this idea more fully.)

(Modeling) *Solve each problem.*

75. ***Electrical Current*** The alternating current in an electric inductor is $I = \frac{E}{Z}$ amperes, where E is voltage and $Z = R + X_L i$ is impedance. If $E = 8(\cos 20° + i \sin 20°)$, $R = 6$, and $X_L = 3$, find the current. Give the answer in rectangular form, with real and imaginary parts to the nearest hundredth.

76. ***Electrical Current*** The current I in a circuit with voltage E, resistance R, capacitive reactance X_c, and inductive reactance X_L is

$$I = \frac{E}{R + (X_L - X_c)i}.$$

Find I if $E = 12(\cos 25° + i \sin 25°)$, $R = 3$, $X_L = 4$, and $X_c = 6$. Give the answer in rectangular form, with real and imaginary parts to the nearest tenth.

(Modeling) Impedance *In the parallel electrical circuit shown in the figure, the impedance Z can be calculated using the equation*

$$Z = \frac{1}{\frac{1}{Z_1} + \frac{1}{Z_2}},$$

where Z_1 and Z_2 are the impedances for the branches of the circuit.

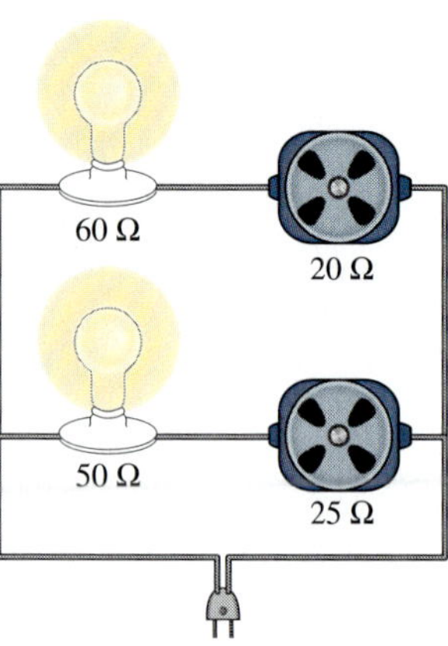

77. If $Z_1 = 50 + 25i$ and $Z_2 = 60 + 20i$, calculate Z.

78. Determine the phase angle θ for the value of Z found in Exercise 77.

8.6 De Moivre's Theorem; Powers and Roots of Complex Numbers

Powers of Complex Numbers (De Moivre's Theorem) ▪ **Roots of Complex Numbers**

Powers of Complex Numbers (De Moivre's Theorem) In the previous section, we studied the product theorem for complex numbers in trigonometric form. Because raising a number to a positive integer power is a repeated application of the product rule, it would seem likely that a theorem for finding powers of complex numbers exists. This is indeed the case. For example, the square of the complex number $r(\cos\theta + i\sin\theta)$ is

$$\begin{aligned}[r(\cos\theta + i\sin\theta)]^2 &= [r(\cos\theta + i\sin\theta)][r(\cos\theta + i\sin\theta)]\\ &= r \cdot r[\cos(\theta+\theta) + i\sin(\theta+\theta)]\\ &= r^2(\cos 2\theta + i\sin 2\theta).\end{aligned}$$

In the same way,

$$[r(\cos\theta + i\sin\theta)]^3 = r^3(\cos 3\theta + i\sin 3\theta).$$

These results suggest the following theorem for positive integer values of n. Although the theorem is stated and can be proved for all n, we use it only for positive integer values of n and their reciprocals.

Abraham De Moivre (1667–1754)

De Moivre's Theorem

If $r(\cos\theta + i\sin\theta)$ is a complex number, and if n is any real number, then

$$[r(\cos\theta + i\sin\theta)]^n = r^n(\cos n\theta + i\sin n\theta).$$

In compact form, this is written

$$[r \operatorname{cis}\theta]^n = r^n(\operatorname{cis} n\theta).$$

This theorem is named after the French expatriate friend of Isaac Newton, Abraham De Moivre, although he never explicitly stated it.

EXAMPLE 1 Finding a Power of a Complex Number

Find $\left(1 + i\sqrt{3}\right)^8$ and express the result in rectangular form.

Solution First convert $1 + i\sqrt{3}$ into trigonometric form.

$$1 + i\sqrt{3} = 2(\cos 60° + i \sin 60°) \quad \text{(Section 8.5)}$$

Now, apply De Moivre's theorem.

$$\begin{aligned}
\left(1 + i\sqrt{3}\right)^8 &= [2(\cos 60° + i \sin 60°)]^8 \\
&= 2^8[\cos(8 \cdot 60°) + i \sin(8 \cdot 60°)] && \text{De Moivre's theorem} \\
&= 256(\cos 480° + i \sin 480°) \\
&= 256(\cos 120° + i \sin 120°) && \text{480° and 120° are coterminal. (Section 5.1)} \\
&= 256\left(-\frac{1}{2} + i\frac{\sqrt{3}}{2}\right) && \cos 120° = -\tfrac{1}{2};\ \sin 120° = \tfrac{\sqrt{3}}{2} \text{ (Section 5.3)} \\
&= -128 + 128i\sqrt{3} && \text{Rectangular form}
\end{aligned}$$

Now try Exercise 7.

Roots of Complex Numbers Every nonzero complex number has exactly n distinct complex nth roots. De Moivre's theorem can be extended to find all nth roots of a complex number.

*n*th Root

For a positive integer n, the complex number $a + bi$ is an ***n*th root** of the complex number $x + yi$ if

$$(a + bi)^n = x + yi.$$

To find the three complex cube roots of $8(\cos 135° + i \sin 135°)$, for example, look for a complex number, say $r(\cos \alpha + i \sin \alpha)$, that will satisfy

$$[r(\cos \alpha + i \sin \alpha)]^3 = 8(\cos 135° + i \sin 135°).$$

By De Moivre's theorem, this equation becomes

$$r^3(\cos 3\alpha + i \sin 3\alpha) = 8(\cos 135° + i \sin 135°).$$

Set $r^3 = 8$ and $\cos 3\alpha + i \sin 3\alpha = \cos 135° + i \sin 135°$, to satisfy this equation. The first of these conditions implies that $r = 2$, and the second implies that

$$\cos 3\alpha = \cos 135° \quad \text{and} \quad \sin 3\alpha = \sin 135°.$$

For these equations to be satisfied, 3α must represent an angle that is coterminal with 135°. Therefore, we must have

$$3\alpha = 135° + 360° \cdot k, \quad k \text{ any integer}$$

or

$$\alpha = \frac{135° + 360° \cdot k}{3}, \quad k \text{ any integer.}$$

Now, let k take on the integer values 0, 1, and 2.

If $k = 0$, then $\alpha = \dfrac{135° + 0°}{3} = 45°$.

If $k = 1$, then $\alpha = \dfrac{135° + 360°}{3} = \dfrac{495°}{3} = 165°$.

If $k = 2$, then $\alpha = \dfrac{135° + 720°}{3} = \dfrac{855°}{3} = 285°$.

In the same way, $\alpha = 405°$ when $k = 3$. But note that $405° = 45° + 360°$ so $\sin 405° = \sin 45°$ and $\cos 405° = \cos 45°$. Similarly, if $k = 4$, $\alpha = 525°$, which has the same sine and cosine values as 165°. To continue with larger values of k would just be repeating solutions already found. Therefore, all of the cube roots (three of them) can be found by letting $k = 0, 1,$ or 2.

When $k = 0$, the root is $2(\cos 45° + i \sin 45°)$.

When $k = 1$, the root is $2(\cos 165° + i \sin 165°)$.

When $k = 2$, the root is $2(\cos 285° + i \sin 285°)$.

In summary, we see that $2(\cos 45° + i \sin 45°)$, $2(\cos 165° + i \sin 165°)$, and $2(\cos 285° + i \sin 285°)$ are the three cube roots of $8(\cos 135° + i \sin 135°)$.

Generalizing our results, we state the following theorem.

*n*th Root Theorem

If n is any positive integer, r is a positive real number, and θ is in degrees, then the nonzero complex number $r(\cos \theta + i \sin \theta)$ has exactly n distinct nth roots, given by

$$\sqrt[n]{r}(\cos \alpha + i \sin \alpha) \quad \text{or} \quad \sqrt[n]{r} \operatorname{cis} \alpha,$$

where

$$\alpha = \frac{\theta + 360° \cdot k}{n} \quad \text{or} \quad \alpha = \frac{\theta}{n} + \frac{360° \cdot k}{n}, \qquad k = 0, 1, 2, \ldots, n - 1.$$

NOTE In the statement of the nth root theorem, if θ is in radians, then

$$\alpha = \frac{\theta + 2\pi k}{n} \quad \text{or} \quad \alpha = \frac{\theta}{n} + \frac{2\pi k}{n}.$$

EXAMPLE 2 Finding Complex Roots

Find the two square roots of $4i$. Write the roots in rectangular form.

Solution First write $4i$ in trigonometric form as

$$4i = 4\left(\cos \frac{\pi}{2} + i \sin \frac{\pi}{2}\right).$$

Here $r = 4$ and $\theta = \frac{\pi}{2}$. The square roots have absolute value $\sqrt{4} = 2$ and arguments as follows.

$$\alpha = \frac{\frac{\pi}{2}}{2} + \frac{2\pi k}{2} = \frac{\pi}{4} + \pi k$$

Since there are two square roots, let $k = 0$ and 1.

$$\text{If } k = 0, \text{ then} \qquad \alpha = \frac{\pi}{4} + \pi \cdot 0 = \frac{\pi}{4}.$$

$$\text{If } k = 1, \text{ then} \qquad \alpha = \frac{\pi}{4} + \pi \cdot 1 = \frac{5\pi}{4}.$$

Using these values for α, the square roots are $2 \text{ cis } \frac{\pi}{4}$ and $2 \text{ cis } \frac{5\pi}{4}$, which can be written in rectangular form as

$$\sqrt{2} + i\sqrt{2} \qquad \text{and} \qquad -\sqrt{2} - i\sqrt{2}.$$

```
(√(2)+i√(2))²
                4i
(-√(2)-i√(2))²
                4i
```

This screen confirms the result of Example 2.

Now try Exercise 17(a).

EXAMPLE 3 Finding Complex Roots

Find all fourth roots of $-8 + 8i\sqrt{3}$. Write the roots in rectangular form.

Solution First write $-8 + 8i\sqrt{3}$ in trigonometric form as

$$-8 + 8i\sqrt{3} = 16 \text{ cis } 120°.$$

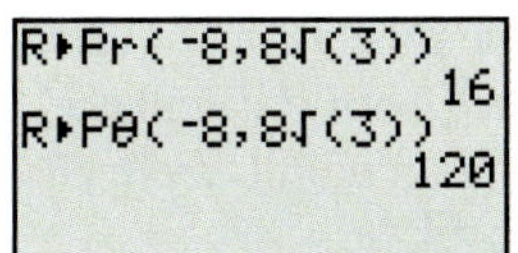

Degree mode

This screen shows how a calculator finds r and θ for the number in Example 3.

Here $r = 16$ and $\theta = 120°$. The fourth roots of this number have absolute value $\sqrt[4]{16} = 2$ and arguments as follows.

$$\alpha = \frac{120°}{4} + \frac{360° \cdot k}{4} = 30° + 90° \cdot k$$

Since there are four fourth roots, let $k = 0$, 1, 2, and 3.

$$\text{If } k = 0, \text{ then} \qquad \alpha = 30° + 90° \cdot 0 = 30°.$$

$$\text{If } k = 1, \text{ then} \qquad \alpha = 30° + 90° \cdot 1 = 120°.$$

$$\text{If } k = 2, \text{ then} \qquad \alpha = 30° + 90° \cdot 2 = 210°.$$

$$\text{If } k = 3, \text{ then} \qquad \alpha = 30° + 90° \cdot 3 = 300°.$$

Using these angles, the fourth roots are

$$2 \text{ cis } 30°, \qquad 2 \text{ cis } 120°, \qquad 2 \text{ cis } 210°, \qquad \text{and} \qquad 2 \text{ cis } 300°.$$

These four roots can be written in rectangular form as

$$\sqrt{3} + i, \qquad -1 + i\sqrt{3}, \qquad -\sqrt{3} - i, \qquad \text{and} \qquad 1 - i\sqrt{3}.$$

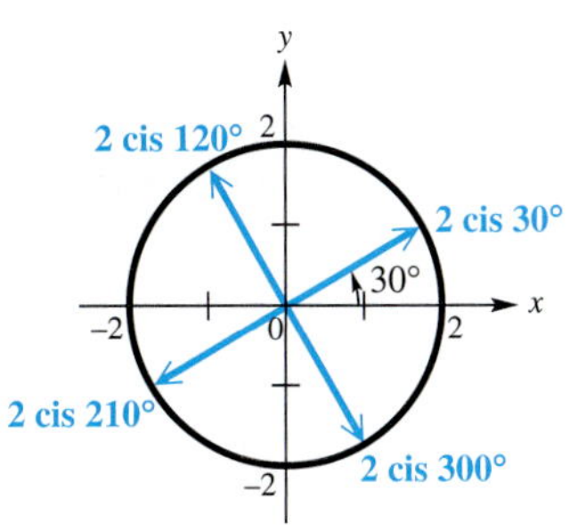

Figure 53

The graphs of these roots are all on a circle that has center at the origin and radius 2, as shown in Figure 53. Notice that the roots are equally spaced about the circle, 90° apart.

Now try Exercises 23(a) and (b).

EXAMPLE 4 Solving an Equation by Finding Complex Roots

Find all complex number solutions of $x^5 - 1 = 0$. Graph them as vectors in the complex plane.

Solution Write the equation as

$$x^5 - 1 = 0 \qquad \text{or} \qquad x^5 = 1.$$

While there is only one real number solution, 1, there are five complex number solutions. To find these solutions, first write 1 in trigonometric form as

$$1 = 1 + 0i = 1(\cos 0° + i \sin 0°).$$

The absolute value of the fifth roots is $\sqrt[5]{1} = 1$, and the arguments are given by

$$0° + 72° \cdot k, \qquad k = 0, 1, 2, 3, \text{ and } 4.$$

By using these arguments, the fifth roots are

$$1(\cos 0° + i \sin 0°), \qquad k = 0$$
$$1(\cos 72° + i \sin 72°), \qquad k = 1$$
$$1(\cos 144° + i \sin 144°), \qquad k = 2$$
$$1(\cos 216° + i \sin 216°), \qquad k = 3$$

and

$$1(\cos 288° + i \sin 288°). \qquad k = 4$$

The solution set of the equation can be written as {cis 0°, cis 72°, cis 144°, cis 216°, cis 288°}. The first of these roots equals 1; the others cannot easily be expressed in rectangular form but can be approximated with a calculator. The tips of the arrows representing the five fifth roots all lie on a unit circle and are equally spaced around it every 72°, as shown in Figure 54.

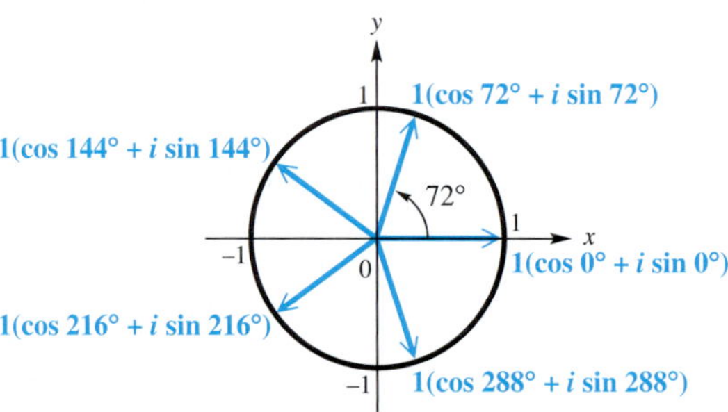

Figure 54

Now try Exercise 35.

8.6 Exercises

Find each power. Write each answer in rectangular form. See Example 1.

1. $[3(\cos 30° + i \sin 30°)]^3$

2. $[2(\cos 135° + i \sin 135°)]^4$

3. $(\cos 45° + i \sin 45°)^8$

4. $[2(\cos 120° + i \sin 120°)]^3$

5. $[3 \text{ cis } 100°]^3$

6. $[3 \text{ cis } 40°]^3$

7. $(\sqrt{3} + i)^5$

8. $(2\sqrt{2} - 2i\sqrt{2})^6$

9. $(2 - 2i\sqrt{3})^4$

10. $\left(\frac{\sqrt{2}}{2} - \frac{\sqrt{2}}{2}i\right)^8$

11. $(-2 - 2i)^5$

12. $(-1 + i)^7$

In Exercises 13–24, (a) find all cube roots of each complex number. Leave answers in trigonometric form. (b) Graph each cube root as a vector in the complex plane. See Examples 2 and 3.

13. $\cos 0° + i \sin 0°$

14. $\cos 90° + i \sin 90°$

15. $8 \text{ cis } 60°$

16. $27 \text{ cis } 300°$

17. $-8i$

18. $27i$

19. -64

20. 27

21. $1 + i\sqrt{3}$

22. $2 - 2i\sqrt{3}$

23. $-2\sqrt{3} + 2i$

24. $\sqrt{3} - i$

Find and graph all specified roots of 1.

25. second (square) **26.** fourth **27.** sixth **28.** eighth

Find and graph all specified roots of i.

29. second (square) **30.** fourth

Find all complex number solutions of each equation. Leave answers in trigonometric form. See Example 4.

31. $x^3 - 1 = 0$

32. $x^3 + 1 = 0$

33. $x^3 + i = 0$

34. $x^4 + i = 0$

35. $x^3 - 8 = 0$

36. $x^3 + 27 = 0$

37. $x^4 + 1 = 0$

38. $x^4 + 16 = 0$

39. $x^4 - i = 0$

40. $x^5 - i = 0$

41. $x^3 - (4 + 4i\sqrt{3}) = 0$

42. $x^4 - (8 + 8i\sqrt{3}) = 0$

43. Solve the equation $x^3 - 1 = 0$ by factoring the left side as the difference of two cubes and setting each factor equal to 0. Apply the quadratic formula as needed. Then compare your solutions to those of Exercise 31.

44. Solve the equation $x^3 + 27 = 0$ by factoring the left side as the sum of two cubes and setting each factor equal to 0. Apply the quadratic formula as needed. Then compare your solutions to those of Exercise 36.

Relating Concepts

For individual or collaborative investigation

(Exercises 45–48)

Earlier we derived identities, or formulas, for $\cos 2\theta$ *and* $\sin 2\theta$*. These identities can also be derived using De Moivre's theorem.* ***Work Exercises 45–48 in order,*** *to see how this is done.*

45. De Moivre's theorem states that $(\cos \theta + i \sin \theta)^2 =$ ____________.

46. Expand the left side of the equation in Exercise 45 as a binomial and collect terms to write the left side in the form $a + bi$.

47. Use the result of Exercise 46 to obtain the double-angle formula for cosine.

48. Repeat Exercise 47, but find the double-angle formula for sine.

Solve each problem.

49. ***Mandelbrot Set*** The fractal called the *Mandelbrot set* is shown in the figure. To determine if a complex number $z = a + bi$ is in this set, perform the following sequence of calculations. Repeatedly compute

$$z, \quad z^2 + z, \quad (z^2 + z)^2 + z,$$
$$[(z^2 + z)^2 + z]^2 + z, \ldots.$$

Source: Figure from Crownover, R., *Introduction to Fractals and Chaos.* Copyright © 1995. Boston: Jones and Bartlett Publishers. Reprinted with permission.

In a manner analogous to the Julia set, the complex number z does not belong to the Mandelbrot set if any of the resulting absolute values exceed 2. Otherwise z is in the set and the point (a, b) should be shaded in the graph. Determine whether or not the following numbers belong to the Mandelbrot set. (*Source:* Lauwerier, H., *Fractals,* Princeton University Press, 1991.)

(a) $z = 0 + 0i$ **(b)** $z = 1 - 1i$ **(c)** $z = -.5i$

50. ***Basins of Attraction*** The fractal shown in the figure is the solution to Cayley's problem of determining the basins of attraction for the cube roots of unity. The three cube roots of unity are

$$w_1 = 1, \quad w_2 = -\frac{1}{2} + \frac{\sqrt{3}}{2}i,$$

and $$w_3 = -\frac{1}{2} - \frac{\sqrt{3}}{2}i.$$

This fractal can be generated by repeatedly evaluating the function defined by $f(z) = \dfrac{2z^3 + 1}{3z^2}$, where z is a complex number. One begins by picking $z_1 = a + bi$ and then successively computing $z_2 = f(z_1)$, $z_3 = f(z_2)$, $z_4 = f(z_3), \ldots$. If the resulting values of $f(z)$ approach w_1, color the pixel at (a, b) red. If it approaches w_2, color it blue, and if it approaches w_3, color it yellow. If this process continues for a large number of different z_1, the fractal in the figure will appear. Determine the appropriate color of the pixel for each value of z_1. (*Source:* Crownover, R., *Introduction to Fractals and Chaos,* Jones and Bartlett Publishers, 1995.)

(a) $z_1 = i$ **(b)** $z_1 = 2 + i$ **(c)** $z_1 = -1 - i$

51. The screens here illustrate how a pentagon can be graphed using a graphing calculator. Note that a pentagon has five sides, and the T-step is $\frac{360}{5} = 72$. The display at the bottom of the graph screen indicates that one fifth root of 1 is $1 + 0i = 1$. Use this technique to find all fifth roots of 1, and express the real and imaginary parts in decimal form.

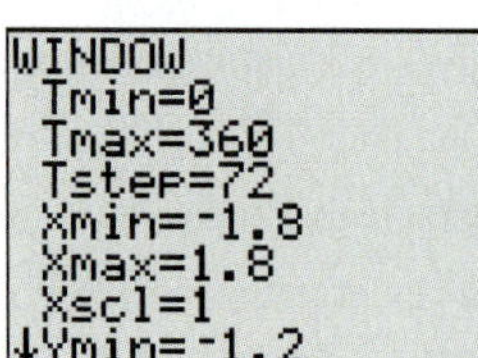


```
WINDOW
↑Tstep=72
 Xmin=-1.8
 Xmax=1.8
 Xscl=1
 Ymin=-1.2
 Ymax=1.2
 Yscl=1
```

This is a continuation of the previous screen.

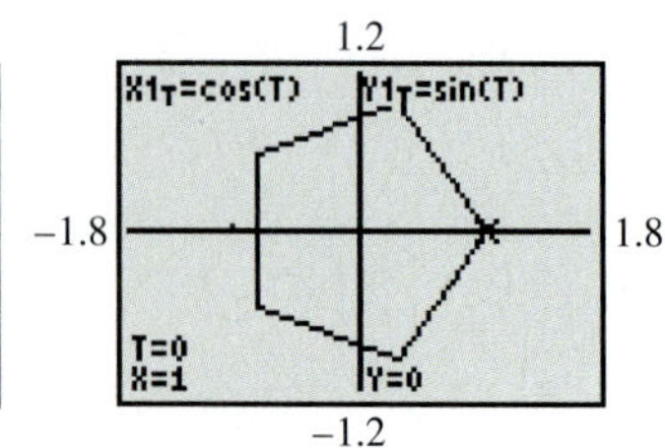

The calculator is in parametric, degree, and connected graph modes.

52. Use the method of Exercise 51 to find the first three of the ten 10th roots of 1.

53. One of the three cube roots of a complex number is $2 + 2\sqrt{3}i$. Determine the rectangular form of its other two cube roots.

Use a calculator to find all solutions of each equation in rectangular form.

54. $x^3 + 4 - 5i = 0$

55. $x^5 + 2 + 3i = 0$

56. *Concept Check* How many complex 64th roots does 1 have? How many are real? How many are not?

57. *Concept Check* True or false: Every real number must have two distinct real square roots.

58. *Concept Check* True or false: Some real numbers have three real cube roots.

59. Show that if z is an nth root of 1, then so is $\frac{1}{z}$.

60. Explain why a real number can have only one real cube root.

61. Explain why the n nth roots of 1 are equally spaced around the unit circle.

62. Refer to Figure 54. A regular pentagon can be created by joining the tips of the arrows. Explain how you can use this principle to create a regular octagon.

8.7 Polar Equations and Graphs

Polar Coordinate System ▪ **Graphs of Polar Equations** ▪ **Converting from Polar to Rectangular Equations** ▪ **Classifying Polar Equations**

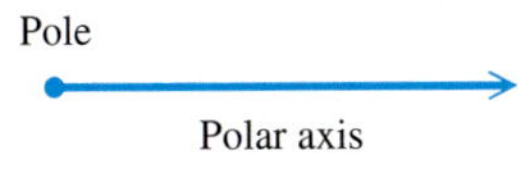

Figure 55

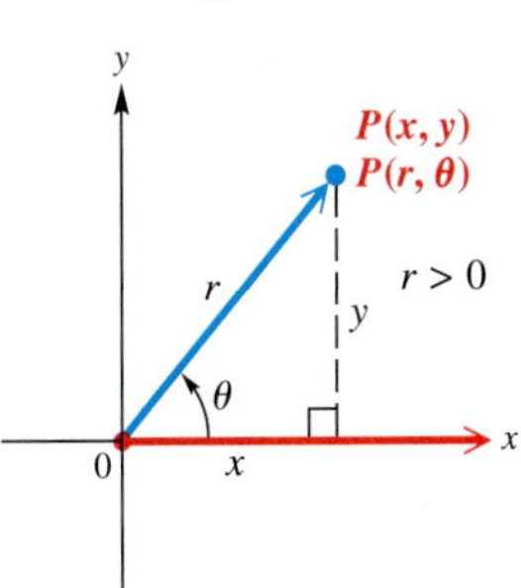

Figure 56

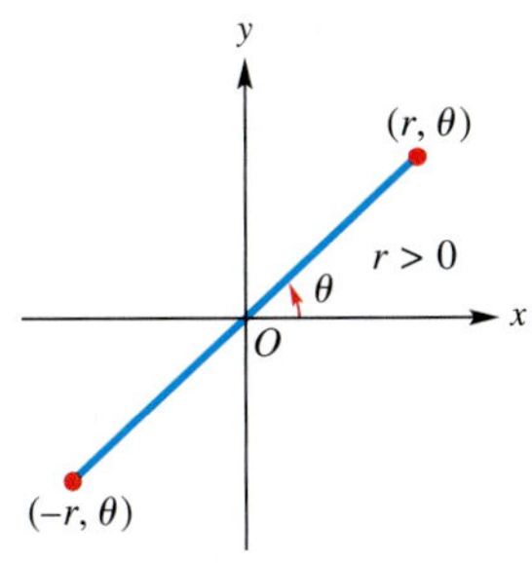

Figure 57

Polar Coordinate System

We have been using the rectangular coordinate system to graph equations. The **polar coordinate system** is based on a point, called the **pole,** and a ray, called the **polar axis.** The polar axis is usually drawn in the direction of the positive x-axis, as shown in Figure 55.

In Figure 56, the pole has been placed at the origin of a rectangular coordinate system so that the polar axis coincides with the positive x-axis. Point P has rectangular coordinates (x, y). Point P can also be located by giving the directed angle θ from the positive x-axis to ray OP and the directed distance r from the pole to point P. The ordered pair (r, θ) gives the **polar coordinates** of point P. If $r > 0$, then point P lies on the terminal side of θ and if $r < 0$, then point P lies on the ray pointing in the opposite direction of the terminal side of θ, a distance $|r|$ from the pole. See Figure 57.

Using Figure 56 and trigonometry, we can establish the following relationships between rectangular and polar coordinates.

Rectangular and Polar Coordinates

If a point has rectangular coordinates (x, y) and polar coordinates (r, θ), then these coordinates are related as follows.

$$x = r\cos\theta \qquad y = r\sin\theta$$

$$r^2 = x^2 + y^2 \qquad \tan\theta = \frac{y}{x}, \quad \text{if } x \neq 0$$

EXAMPLE 1 Plotting Points with Polar Coordinates

Plot each point by hand in the polar coordinate system. Then, determine the rectangular coordinates of each point.

(a) $P(2, 30°)$ **(b)** $Q\left(-4, \frac{2\pi}{3}\right)$ **(c)** $R\left(5, -\frac{\pi}{4}\right)$

Solution

(a) In this case, $r = 2$ and $\theta = 30°$, so the point P is located 2 units from the origin in the positive direction on a ray making a $30°$ angle with the polar axis, as shown in Figure 58.

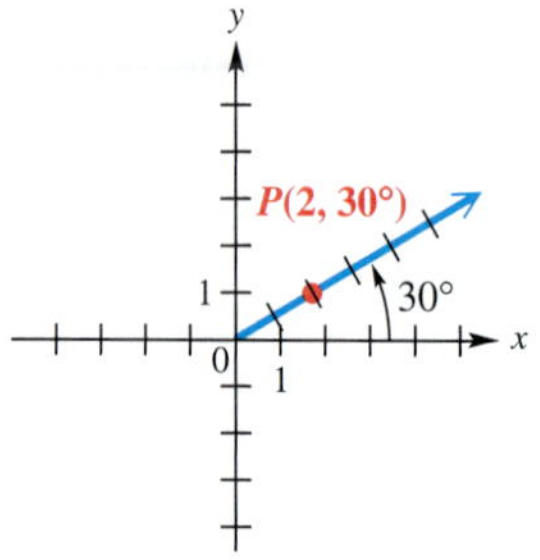

Using the conversion equations, we find the rectangular coordinates as follows.

$$x = r\cos\theta \qquad y = r\sin\theta$$
$$x = 2\cos 30° \qquad y = 2\sin 30°$$
$$x = 2\left(\frac{\sqrt{3}}{2}\right) = \sqrt{3} \qquad y = 2\left(\frac{1}{2}\right) = 1 \quad \text{(Section 5.3)}$$

Figure 58

The rectangular coordinates are $(\sqrt{3}, 1)$.

(b) Since r is negative, Q is 4 units in the opposite direction from the pole on an extension of the $\frac{2\pi}{3}$ ray. See Figure 59. The rectangular coordinates are

$$x = -4\cos\frac{2\pi}{3} = -4\left(-\frac{1}{2}\right) = 2 \quad \text{and} \quad y = -4\sin\frac{2\pi}{3} = -4\left(\frac{\sqrt{3}}{2}\right) = -2\sqrt{3}.$$

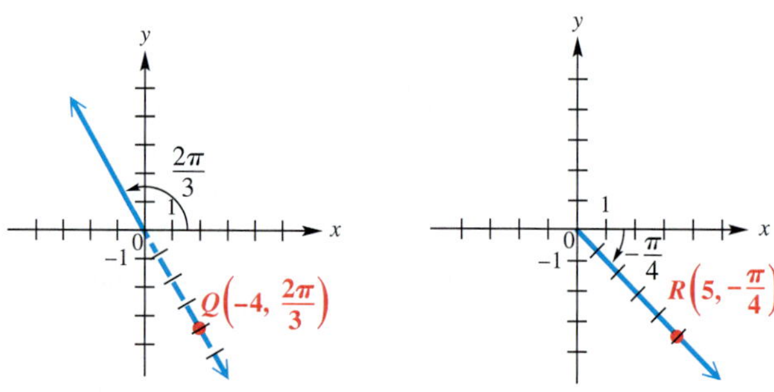

Figure 59 Figure 60

Looking Ahead to Calculus

Techniques studied in calculus associated with derivatives and integrals provide methods of finding slopes of tangent lines to polar curves, areas bounded by such curves, and lengths of their arcs.

(c) Point R is shown in Figure 60. Since θ is negative, the angle is measured in the clockwise direction. Furthermore, we have

$$x = 5\cos\left(-\frac{\pi}{4}\right) = \frac{5\sqrt{2}}{2} \quad \text{and} \quad y = 5\sin\left(-\frac{\pi}{4}\right) = -\frac{5\sqrt{2}}{2}.$$

Now try Exercises 3(a), (c), 5(a), (c), and 11(a), (c).

One important difference between rectangular coordinates and polar coordinates is that while a given point in the plane can have only one pair of rectangular coordinates, this same point can have an infinite number of pairs of polar coordinates. For example, $(2, 30°)$ locates the same point as $(2, 390°)$ or $(2, -330°)$ or $(-2, 210°)$.

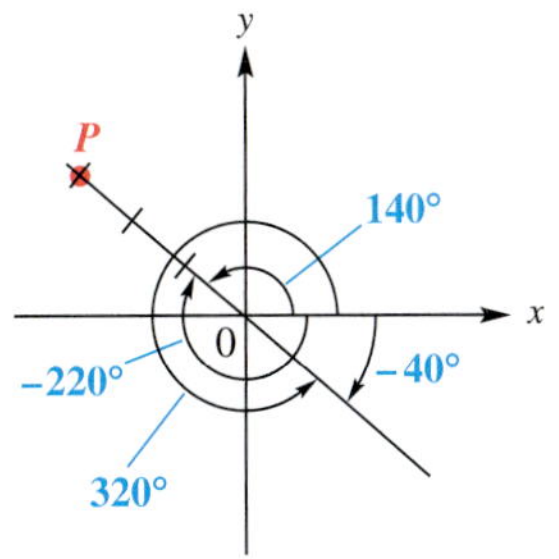

Figure 61

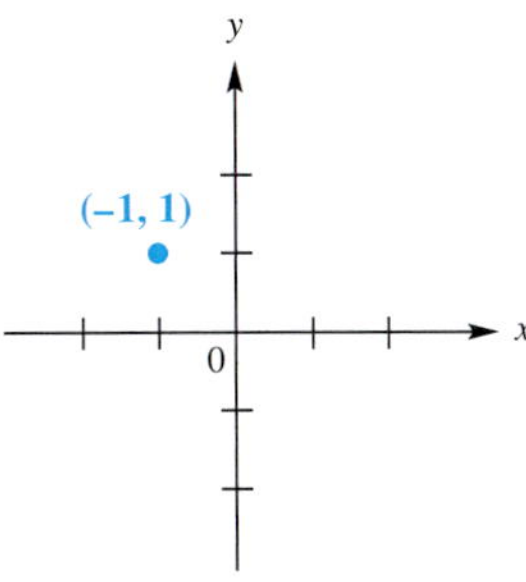

Figure 62

EXAMPLE 2 Giving Alternative Forms for Coordinates of a Point

(a) Give three other pairs of polar coordinates for the point $P(3, 140°)$.

(b) Determine two pairs of polar coordinates for the point with rectangular coordinates $(-1, 1)$.

Solution

(a) Three pairs that could be used for the point are $(3, -220°)$, $(-3, 320°)$, and $(-3, -40°)$. See Figure 61.

(b) As shown in Figure 62, the point $(-1, 1)$ lies in the second quadrant. Since $\tan \theta = \frac{1}{-1} = -1$, one possible value for θ is $135°$. Also,

$$r = \sqrt{x^2 + y^2} = \sqrt{(-1)^2 + 1^2} = \sqrt{2}.$$

Two pairs of polar coordinates are $\left(\sqrt{2}, 135°\right)$ and $\left(-\sqrt{2}, 315°\right)$.

Now try Exercises 3(b), 5(b), 11(b), and 13.

Graphs of Polar Equations The graphs that we have studied so far have used rectangular coordinates. Equations in x and y are called **rectangular** (or **Cartesian**) **equations.** We now examine how equations in polar coordinates are graphed. An equation where r and θ are the variables is a **polar equation.**

$$r = 3 \sin \theta, \qquad r = 2 + \cos \theta, \qquad r = \theta \qquad \text{Polar equations}$$

Lines and circles are two of the most common types of graphs studied in rectangular coordinates. While their rectangular forms are the ones most often encountered, they can also be defined in terms of polar coordinates. The polar equation of the line $ax + by = c$ can be derived as follows.

$$ax + by = c \qquad \text{Rectangular equation}$$

$$a(r \cos \theta) + b(r \sin \theta) = c \qquad \text{Convert from rectangular to polar coordinates.}$$

$$r(a \cos \theta + b \sin \theta) = c \qquad \text{Factor out } r. \text{ (Section R.4)}$$

$$r = \frac{c}{a \cos \theta + b \sin \theta}$$

For the circle $x^2 + y^2 = a^2$, we have

$$x^2 + y^2 = a^2$$

$$\left(\sqrt{x^2 + y^2}\right)^2 = a^2 \qquad \left(\sqrt{k}\right)^2 = k \text{ (Section R.7)}$$

$$r^2 = a^2 \qquad \sqrt{x^2 + y^2} = r$$

$$r = \pm a. \qquad r \text{ can be negative in polar coordinates.}$$

EXAMPLE 3 Examining Polar and Rectangular Equations of Lines and Circles

For each rectangular equation, give its equivalent polar equation and sketch its graph.

(a) $y = x - 3$ **(b)** $x^2 + y^2 = 4$

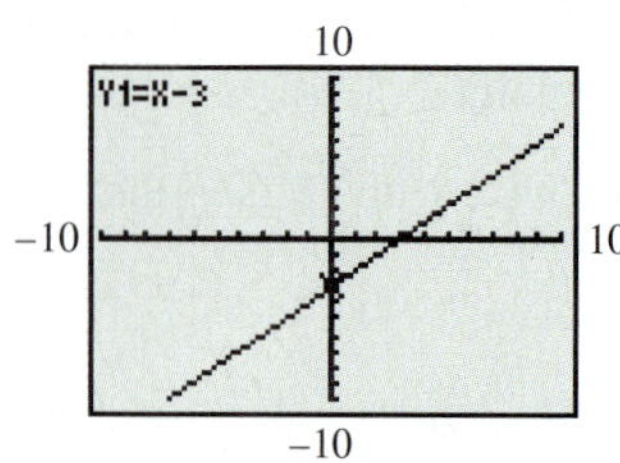

Function graphing mode

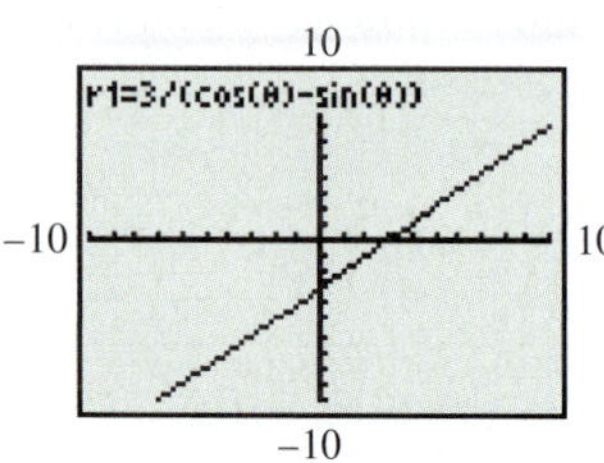

Polar graphing mode
Compare with Figure 63.

Solution

(a) This is the equation of a line. Rewrite $y = x - 3$ as

$$x - y = 3. \quad \text{Standard form } ax + by = c \text{ (Section 2.3)}$$

Using the general form for the polar equation of a line,

$$r = \frac{c}{a \cos \theta + b \sin \theta}$$

$$r = \frac{3}{1 \cdot \cos \theta + (-1) \cdot \sin \theta} \quad a = 1, b = -1, c = 3$$

$$r = \frac{3}{\cos \theta - \sin \theta}.$$

Its graph is shown in Figure 63.

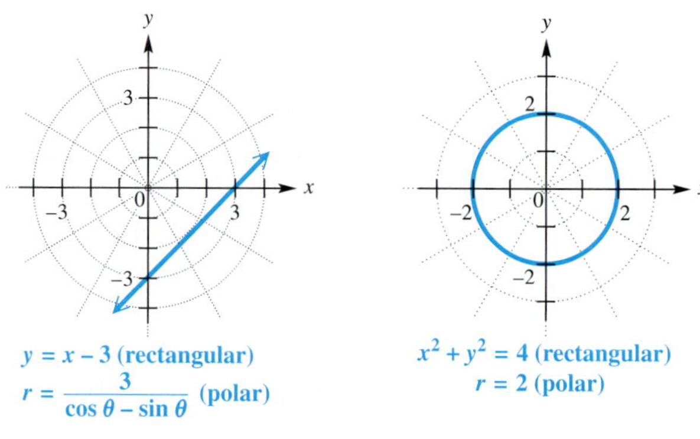

Figure 63 **Figure 64**

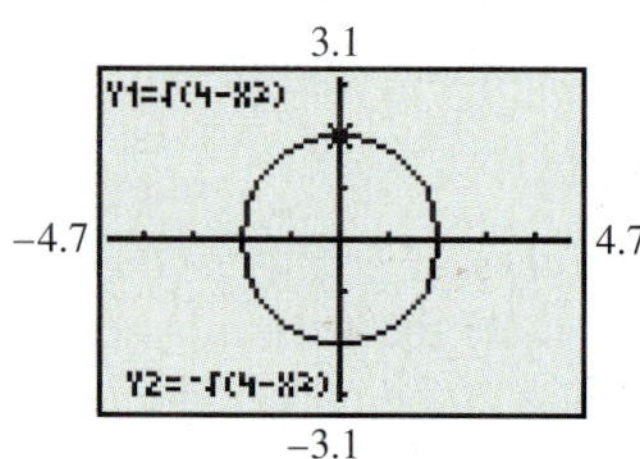

Function graphing mode

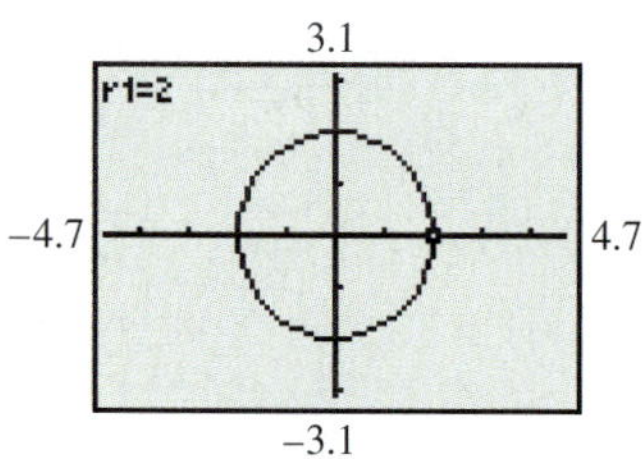

Polar graphing mode
Compare with Figure 64.

(b) The graph of $x^2 + y^2 = 4$ is a circle with center at the origin and radius 2.

$$x^2 + y^2 = 4 \quad \text{(Section 2.1)}$$

$$\left(\sqrt{x^2 + y^2}\right)^2 = (\pm 2)^2$$

$$\sqrt{x^2 + y^2} = 2 \quad \text{or} \quad \sqrt{x^2 + y^2} = -2$$

$$r = 2 \quad \text{or} \quad r = -2$$

(In polar coordinates, we may have $r < 0$.) The graphs of $r = 2$ and $r = -2$ coincide. The resulting graph is shown in Figure 64.

Now try Exercises 23 and 25.

To graph polar equations, evaluate r for various values of θ until a pattern appears, and then join the points with a smooth curve.

A graphing calculator can be used to graph an equation in the form $r = f(\theta)$. (See the margin screens next to Example 3.) Refer to your owner's manual to see how your model handles polar graphs. As always, you must set an appropriate window and choose the correct angle mode (radians or degrees). You will need to decide on maximum and minimum values of θ. Keep in mind the periods of the functions, so a complete set of ordered pairs is generated. ■

EXAMPLE 4 Graphing a Polar Equation (Cardioid)

Graph $r = 1 + \cos\theta$.

Algebraic Solution

To graph this equation, find some ordered pairs (as in the table) and then connect the points in order—from $(2, 0°)$ to $(1.9, 30°)$ to $(1.7, 45°)$ and so on. The graph is shown in Figure 65. This curve is called a **cardioid** because of its heart shape.

θ	$\cos\theta$	$r = 1 + \cos\theta$	θ	$\cos\theta$	$r = 1 + \cos\theta$
0°	1	2	135°	−.7	.3
30°	.9	1.9	150°	−.9	.1
45°	.7	1.7	180°	−1	0
60°	.5	1.5	270°	0	1
90°	0	1	315°	.7	1.7
120°	−.5	.5	330°	.9	1.9

Once the pattern of values of r becomes clear, it is not necessary to find more ordered pairs. That is why we stopped with the ordered pair $(1.9, 330°)$ in the table. Notice that the curve has been graphed on a **polar grid.**

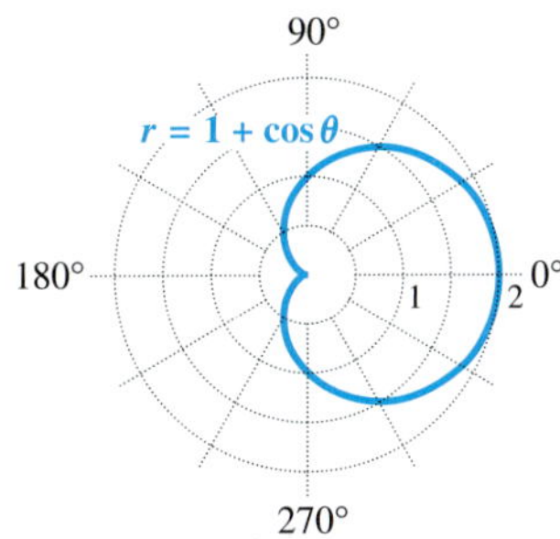

Figure 65

Graphing Calculator Solution

We choose degree mode and graph values of θ in the interval $[0°, 360°]$. The screens in Figure 66(a) show the choices needed to generate the graph in Figure 66(b).

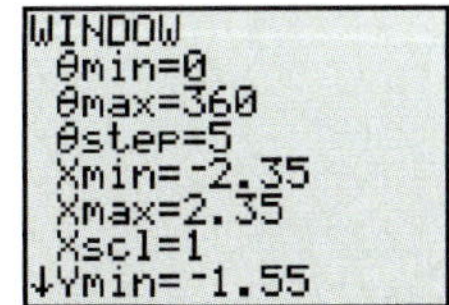

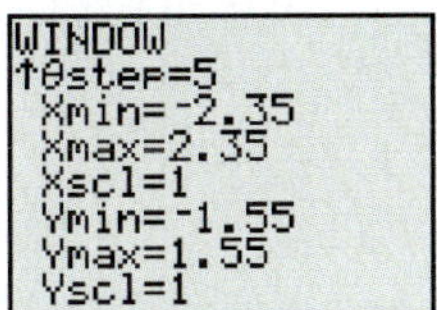

This is a continuation of the previous screen.

(a)

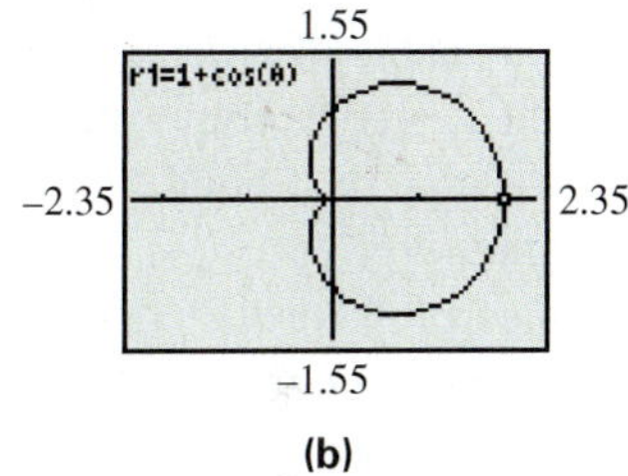

(b)

Figure 66

Now try Exercise 41.

EXAMPLE 5 Graphing a Polar Equation (Rose)

Graph $r = 3\cos 2\theta$.

Solution Because of the argument 2θ, the graph requires a larger number of points than when the argument is just θ. A few ordered pairs are given in the table. You should complete the table similarly through the first 180° so that 2θ has values up to 360°.

θ	0°	15°	30°	45°	60°	75°	90°
2θ	0°	30°	60°	90°	120°	150°	180°
$\cos 2\theta$	1	.9	.5	0	−.5	−.9	−1
$r = 3\cos 2\theta$	3	2.6	1.5	0	−1.5	−2.6	−3

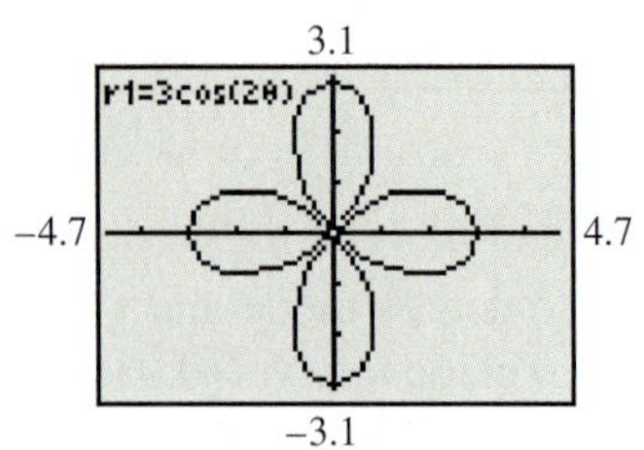

This is a calculator graph of the rose in Example 5.

Plotting the points from the table on the previous page in order gives the graph, called a **four-leaved rose.** Notice in Figure 67 how the graph is developed with a continuous curve, beginning with the upper half of the right horizontal leaf and ending with the lower half of that leaf. As the graph is traced, the curve goes through the pole four times.

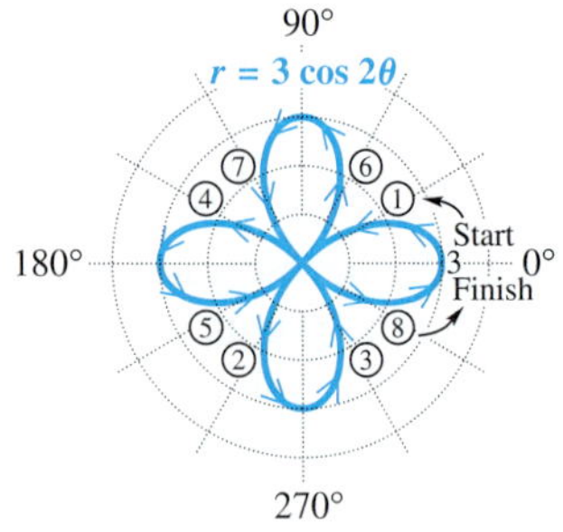

Figure 67

Now try Exercise 45.

The equation $r = 3 \cos 2\theta$ in Example 5 has a graph that belongs to a family of curves called **roses.** The graphs of $\boldsymbol{r = a \sin n\theta}$ and $\boldsymbol{r = a \cos n\theta}$ are roses, with n leaves if n is odd, and $2n$ leaves if n is even. The value of a determines the length of the leaves.

EXAMPLE 6 Graphing a Polar Equation (Lemniscate)

Graph $r^2 = \cos 2\theta$.

Algebraic Solution

Complete a table of ordered pairs, and sketch the graph, as in Figure 68. The point $(-1, 0°)$, with r negative, may be plotted as $(1, 180°)$. Also, $(-.7, 30°)$ may be plotted as $(.7, 210°)$, and so on. This curve is called a **lemniscate.**

Values of θ for $45° < \theta < 135°$ are not included in the table because the corresponding values of $\cos 2\theta$ are negative (quadrants II and III) and so do not have real square roots. Values of θ larger than $180°$ give 2θ larger than $360°$ and would repeat the points already found.

$\boldsymbol{\theta}$	0°	30°	45°	135°	150°	180°
$\boldsymbol{2\theta}$	0°	60°	90°	270°	300°	360°
$\boldsymbol{\cos 2\theta}$	1	.5	0	0	.5	1
$\boldsymbol{r = \pm\sqrt{\cos 2\theta}}$	±1	±.7	0	0	±.7	±1

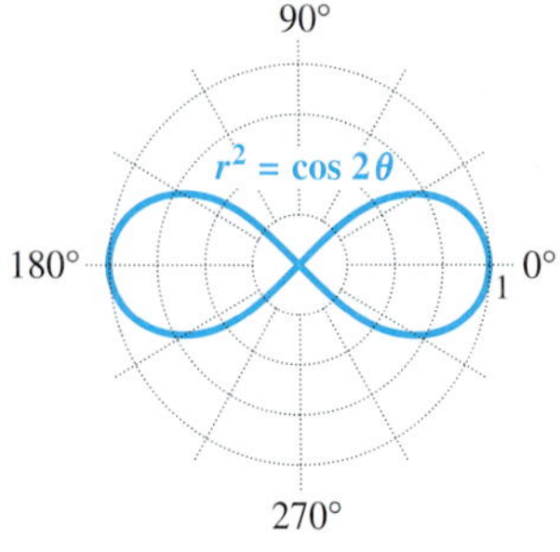

Figure 68

Graphing Calculator Solution

To graph $r^2 = \cos 2\theta$ with a graphing calculator, let

$$r_1 = \sqrt{\cos 2\theta}$$

and $$r_2 = -\sqrt{\cos 2\theta}.$$

See Figures 69(a) and (b).

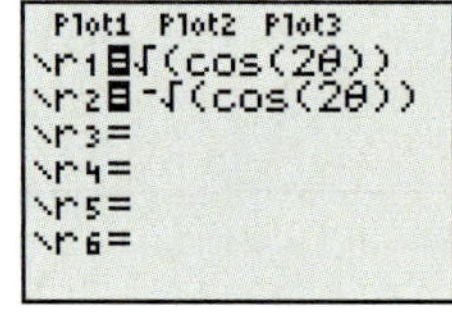

(a)

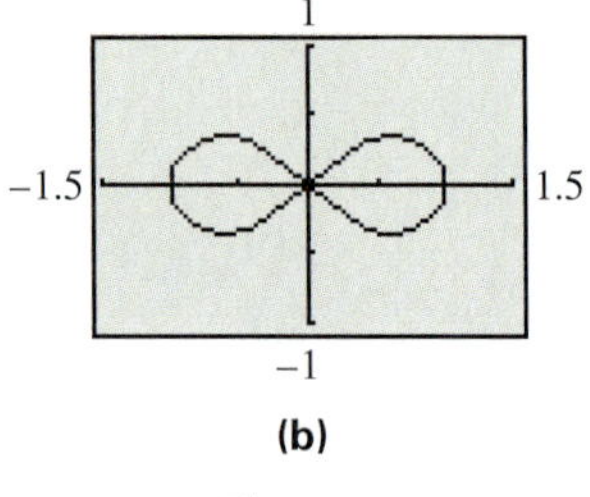

(b)

Figure 69

Now try Exercise 47.

EXAMPLE 7 Graphing a Polar Equation (Spiral of Archimedes)

Graph $r = 2\theta$ (θ measured in radians).

Solution Some ordered pairs are shown in the table. Since $r = 2\theta$, rather than a trigonometric function of θ, we must also consider negative values of θ. Radian measures have been rounded. These and other ordered pairs lead to the graph in Figure 70, called a **spiral of Archimedes.**

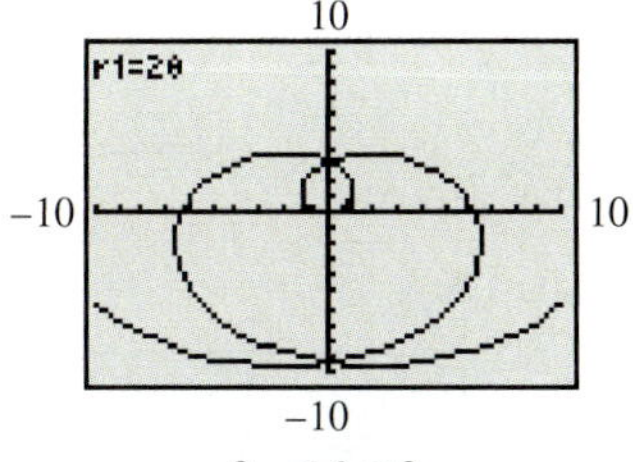

$-2\pi \le \theta \le 2\pi$

More of the spiral can be seen in this calculator graph of the spiral in Example 7 than is shown in Figure 70.

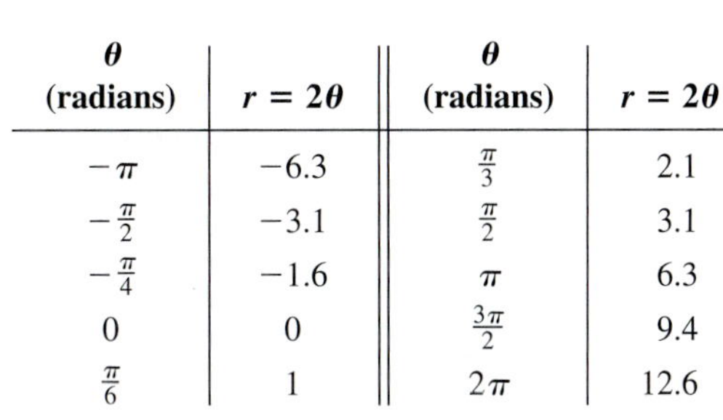

θ (radians)	$r = 2\theta$	θ (radians)	$r = 2\theta$
$-\pi$	-6.3	$\frac{\pi}{3}$	2.1
$-\frac{\pi}{2}$	-3.1	$\frac{\pi}{2}$	3.1
$-\frac{\pi}{4}$	-1.6	π	6.3
0	0	$\frac{3\pi}{2}$	9.4
$\frac{\pi}{6}$	1	2π	12.6

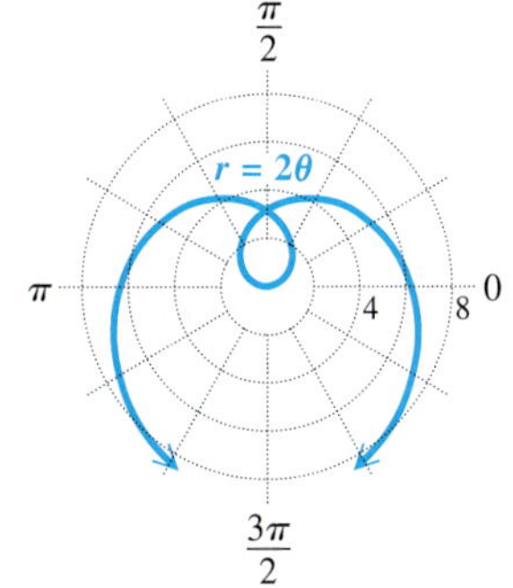

Figure 70

Now try Exercise 63.

Converting from Polar to Rectangular Equations

We can transform polar equations to rectangular equations, as shown in the following example.

EXAMPLE 8 Converting a Polar Equation to a Rectangular Equation

Convert the equation to rectangular coordinates, and graph.

$$r = \frac{4}{1 + \sin\theta}$$

Solution

$$r = \frac{4}{1 + \sin\theta} \qquad \text{Polar equation}$$

$$r + r\sin\theta = 4 \qquad \text{Multiply by } 1 + \sin\theta.$$

$$\sqrt{x^2 + y^2} + y = 4 \qquad \text{Let } r = \sqrt{x^2 + y^2} \text{ and } r\sin\theta = y.$$

$$\sqrt{x^2 + y^2} = 4 - y \qquad \text{Subtract } y.$$

$$x^2 + y^2 = (4 - y)^2 \qquad \text{Square each side. (Section 1.6)}$$

$$x^2 + y^2 = 16 - 8y + y^2 \qquad \text{Expand the right side. (Section R.3)}$$

$$x^2 = -8y + 16 \qquad \text{Subtract } y^2.$$

$$x^2 = -8(y - 2) \qquad \text{Rectangular equation}$$

The final equation represents a parabola and can be graphed using rectangular coordinates. See Figure 71.

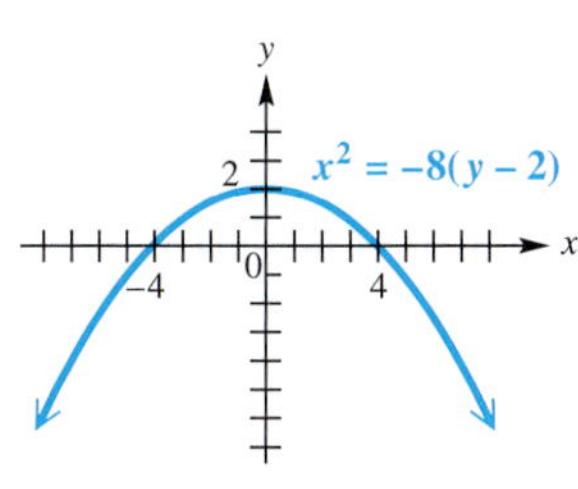

Figure 71

Now try Exercise 55.

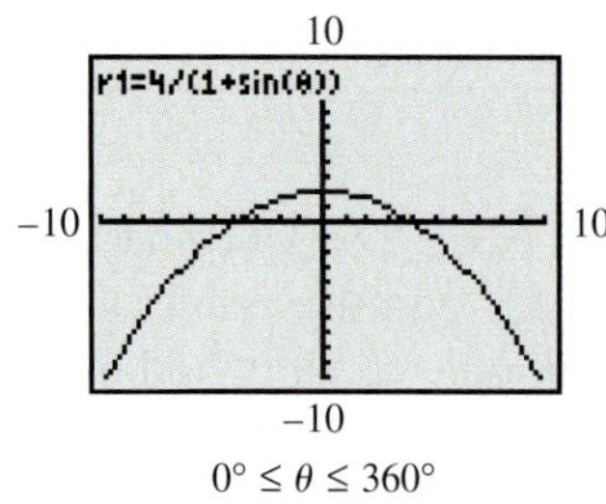

$0° \le \theta \le 360°$

Figure 72

The conversion in Example 8 is not necessary when using a graphing calculator. Figure 72 shows the graph of $r = \frac{4}{1 + \sin \theta}$, graphed directly using polar mode. ■

Classifying Polar Equations The following table summarizes some common polar graphs and forms of their equations. (In addition to circles, lemniscates, and roses, we include **limaçons**. Cardioids are a special case of limaçons, where $\left|\frac{a}{b}\right| = 1$.)

Circles and Lemniscates

Circles

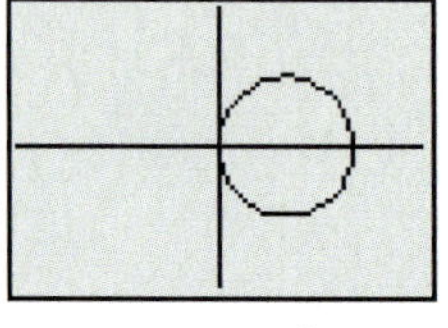
$r = a \cos \theta$

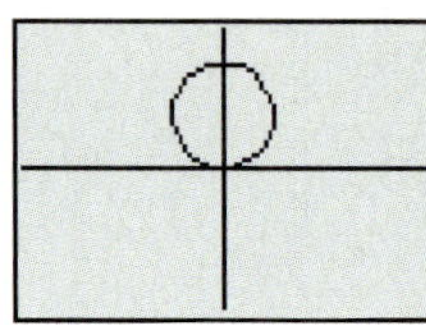
$r = a \sin \theta$

Lemniscates

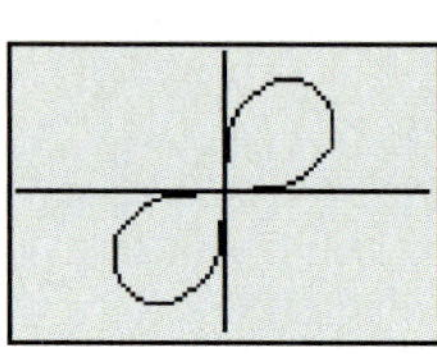
$r^2 = a^2 \sin 2\theta$

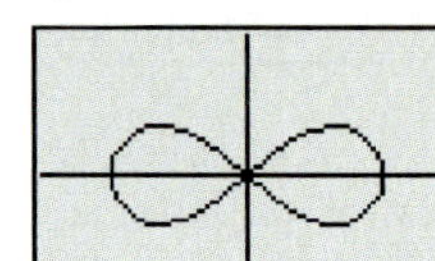
$r^2 = a^2 \cos 2\theta$

Limaçons

$r = a \pm b \sin \theta$ or $r = a \pm b \cos \theta$

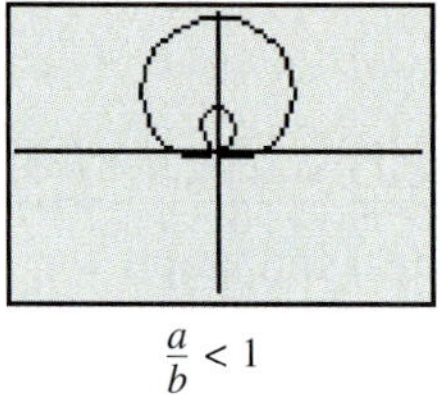
$\frac{a}{b} < 1$

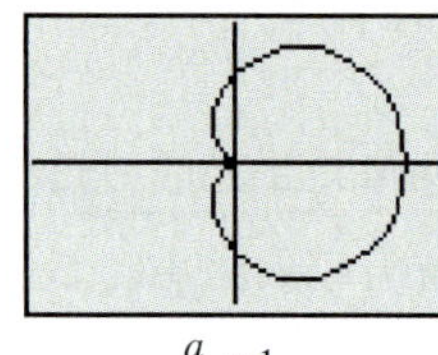
$\frac{a}{b} = 1$

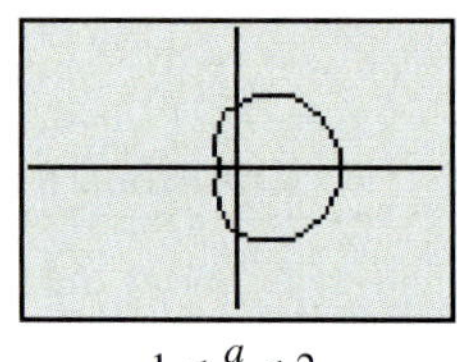
$1 < \frac{a}{b} < 2$

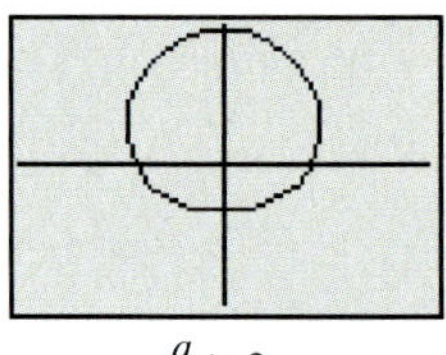
$\frac{a}{b} \ge 2$

Rose Curves

$2n$ leaves if n is even, $n \ge 2$

$n = 2$

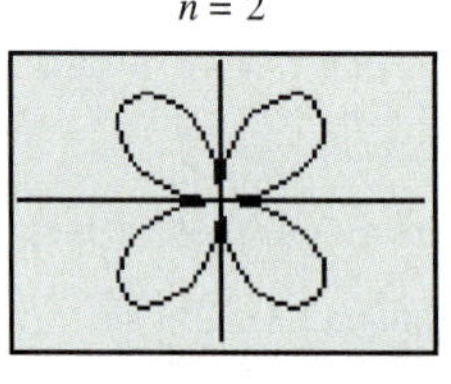
$r = a \sin n\theta$

$n = 4$

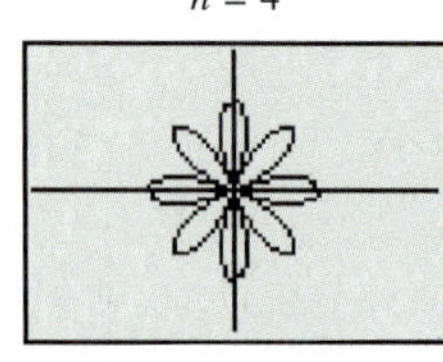
$r = a \cos n\theta$

n leaves if n is odd

$n = 3$

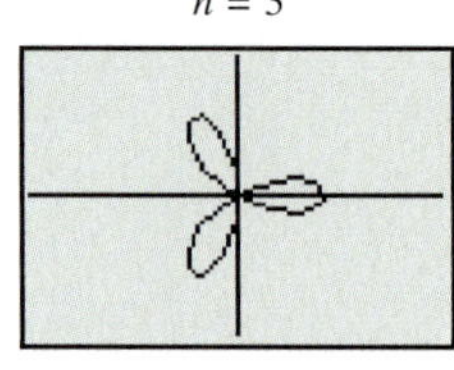
$r = a \cos n\theta$

$n = 5$

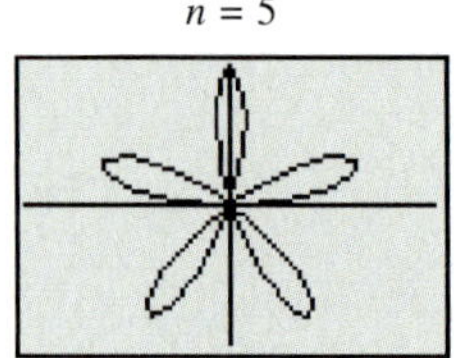
$r = a \sin n\theta$

8.7 Exercises

1. *Concept Check* For each point given in polar coordinates, state the quadrant in which the point lies if it is graphed in a rectangular coordinate system.
   (a) $(5, 135°)$ **(b)** $(2, 60°)$ **(c)** $(6, -30°)$ **(d)** $(4.6, 213°)$

2. *Concept Check* For each point given in polar coordinates, state the axis on which the point lies if it is graphed in a rectangular coordinate system. Also state whether it is on the positive portion or the negative portion of the axis. (For example, $(5, 0°)$ lies on the positive x-axis.)
   (a) $(7, 360°)$ **(b)** $(4, 180°)$ **(c)** $(2, -90°)$ **(d)** $(8, 450°)$

For each pair of polar coordinates, **(a)** *plot the point,* **(b)** *give two other pairs of polar coordinates for the point, and* **(c)** *give the rectangular coordinates for the point. See Examples 1 and 2.*

3. $(1, 45°)$
4. $(3, 120°)$
5. $(-2, 135°)$
6. $(-4, 30°)$
7. $(5, -60°)$
8. $(2, -45°)$
9. $(-3, -210°)$
10. $(-1, -120°)$
11. $\left(3, \frac{5\pi}{3}\right)$
12. $\left(4, \frac{3\pi}{2}\right)$

For each pair of rectangular coordinates, **(a)** *plot the point and* **(b)** *give two pairs of polar coordinates for the point, where $0° \le \theta < 360°$. See Example 2(b).*

13. $(1, -1)$
14. $(1, 1)$
15. $(0, 3)$
16. $(0, -3)$
17. $(\sqrt{2}, \sqrt{2})$
18. $(-\sqrt{2}, \sqrt{2})$
19. $\left(\frac{\sqrt{3}}{2}, \frac{3}{2}\right)$
20. $\left(-\frac{\sqrt{3}}{2}, -\frac{1}{2}\right)$
21. $(3, 0)$
22. $(-2, 0)$

For each rectangular equation, give its equivalent polar equation and sketch its graph. See Example 3.

23. $x - y = 4$
24. $x + y = -7$
25. $x^2 + y^2 = 16$
26. $x^2 + y^2 = 9$
27. $2x + y = 5$
28. $3x - 2y = 6$

Relating Concepts

For individual or collaborative investigation

(Exercises 29–36)

In rectangular coordinates, the graph of $ax + by = c$ is a horizontal line if $a = 0$ or a vertical line if $b = 0$. **Work Exercises 29–36 in order,** *to determine the general forms of polar equations for horizontal and vertical lines.*

29. Begin with the equation $y = k$, whose graph is a horizontal line. Make a trigonometric substitution for y using r and θ.
30. Solve the equation in Exercise 29 for r.
31. Rewrite the equation in Exercise 30 using the appropriate reciprocal function.
32. Sketch the graph of $r = 3 \csc \theta$. What is the corresponding rectangular equation?
33. Begin with the equation $x = k$, whose graph is a vertical line. Make a trigonometric substitution for x using r and θ.

(continued)

34. Solve the equation in Exercise 33 for r.

35. Rewrite the equation in Exercise 34 using the appropriate reciprocal function.

36. Sketch the graph of $r = 3 \sec \theta$. What is the corresponding rectangular equation?

Concept Check *Match the polar graphs in Column II with their corresponding equations in Column I.*

I

37. $r = 3$

38. $r = \cos 3\theta$

39. $r = \cos 2\theta$

40. $r = \dfrac{2}{\cos \theta + \sin \theta}$

II

A.

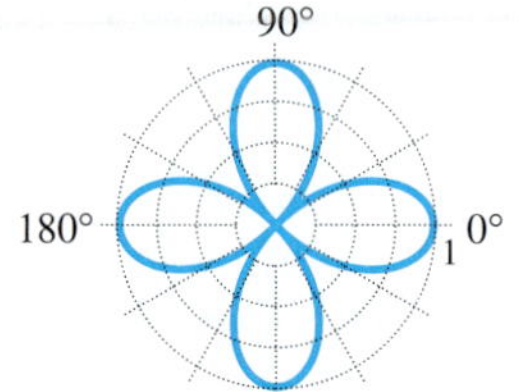

B.

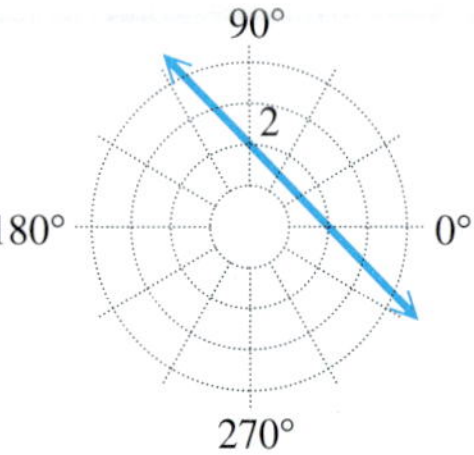

C.

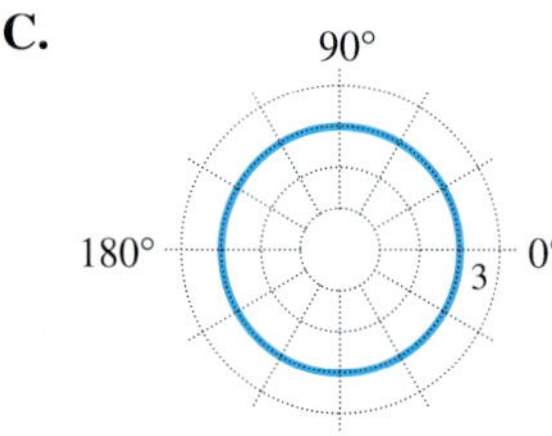

D.

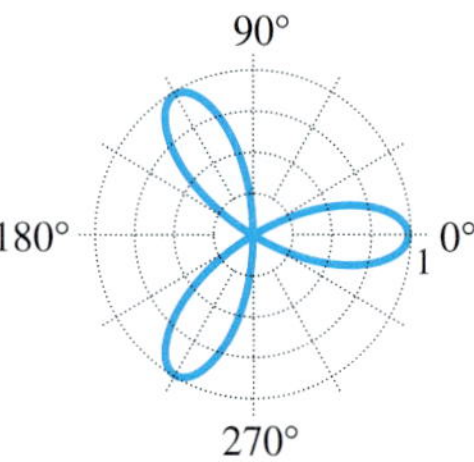

Give a complete graph of each polar equation. In Exercises 41–50, also identify the type of polar graph. See Examples 4–6.

41. $r = 2 + 2 \cos \theta$

42. $r = 8 + 6 \cos \theta$

43. $r = 3 + \cos \theta$

44. $r = 2 - \cos \theta$

45. $r = 4 \cos 2\theta$

46. $r = 3 \cos 5\theta$

47. $r^2 = 4 \cos 2\theta$

48. $r^2 = 4 \sin 2\theta$

49. $r = 4 - 4 \cos \theta$

50. $r = 6 - 3 \cos \theta$

51. $r = 2 \sin \theta \tan \theta$
(This is a *cissoid*.)

52. $r = \dfrac{\cos 2\theta}{\cos \theta}$
(This is a *cissoid* with a loop.)

For each equation, find an equivalent equation in rectangular coordinates and graph. See Example 8.

53. $r = 2 \sin \theta$

54. $r = 2 \cos \theta$

55. $r = \dfrac{2}{1 - \cos \theta}$

56. $r = \dfrac{3}{1 - \sin \theta}$

57. $r = -2 \cos \theta - 2 \sin \theta$

58. $r = \dfrac{3}{4 \cos \theta - \sin \theta}$

59. $r = 2 \sec \theta$

60. $r = -5 \csc \theta$

61. $r = \dfrac{2}{\cos \theta + \sin \theta}$

62. $r = \dfrac{2}{2 \cos \theta + \sin \theta}$

63. Graph $r = \theta$, a spiral of Archimedes. (See Example 7.) Use both positive and nonpositive values for θ.

64. Use a graphing calculator window of $[-1250, 1250]$ by $[-1250, 1250]$, in degree mode, to graph more of $r = 2\theta$ (a spiral of Archimedes) than what is shown in Figure 70 and the accompanying margin screen. Use $-1250° \leq \theta \leq 1250°$.

65. Find the polar equation of the line that passes through the points $(1, 0°)$ and $(2, 90°)$.

66. Explain how to plot a point (r, θ) in polar coordinates, if $r < 0$.

Concept Check *The polar graphs in this section exhibit symmetry. Visualize an xy-plane superimposed on the polar coordinate system, with the pole at the origin and the polar axis on the positive x-axis. Then a polar graph may be symmetric with respect to the x-axis (the polar axis), the y-axis (the line $\theta = \frac{\pi}{2}$), or the origin (the pole). Use this information to work Exercises 67 and 68.*

67. Complete the missing ordered pairs in the graphs below.

(a)

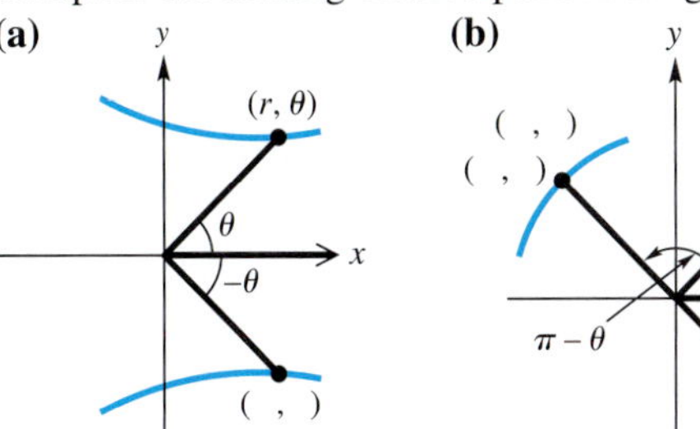

(b)

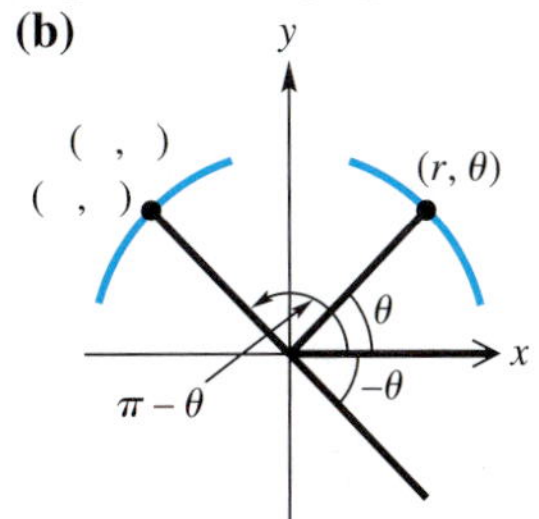

(c)

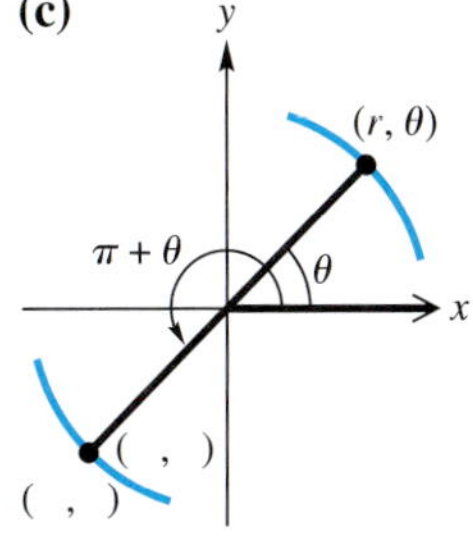

68. Based on your results in Exercise 67, fill in the blanks with the correct responses.

(a) The graph of $r = f(\theta)$ is symmetric with respect to the polar axis if substitution of __________ for θ leads to an equivalent equation.

(b) The graph of $r = f(\theta)$ is symmetric with respect to the vertical line $\theta = \frac{\pi}{2}$ if substitution of __________ for θ leads to an equivalent equation.

(c) Alternatively, the graph of $r = f(\theta)$ is symmetric with respect to the vertical line $\theta = \frac{\pi}{2}$ if substitution of __________ for r and __________ for θ leads to an equivalent equation.

(d) The graph of $r = f(\theta)$ is symmetric with respect to the pole if substitution of __________ for r leads to an equivalent equation.

(e) Alternatively, the graph of $r = f(\theta)$ is symmetric with respect to the pole if substitution of __________ for θ leads to an equivalent equation.

(f) In general, the completed statements in parts (a)–(e) mean that the graphs of polar equations of the form $r = a \pm b \cos\theta$ (where a may be 0) are symmetric with respect to __________.

(g) In general, the completed statements in parts (a)–(e) mean that the graphs of polar equations of the form $r = a \pm b \sin\theta$ (where a may be 0) are symmetric with respect to __________.

The graph of $r = a\theta$ in polar coordinates is an example of the spiral of Archimedes. With your calculator set to radian mode, use the given value of a and interval of θ to graph the spiral in the window specified.

69. $a = 1, 0 \leq \theta \leq 4\pi, [-15, 15]$ by $[-15, 15]$

70. $a = 2, -4\pi \leq \theta \leq 4\pi, [-30, 30]$ by $[-30, 30]$

71. $a = 1.5, -4\pi \leq \theta \leq 4\pi, [-20, 20]$ by $[-20, 20]$

72. $a = -1, 0 \leq \theta \leq 12\pi, [-40, 40]$ by $[-40, 40]$

Find the polar coordinates of the points of intersection of the given curves for the specified interval of θ.

73. $r = 4 \sin \theta$, $r = 1 + 2 \sin \theta$; $0 \leq \theta < 2\pi$

74. $r = 3$, $r = 2 + 2 \cos \theta$; $0° \leq \theta < 360°$

75. $r = 2 + \sin \theta$, $r = 2 + \cos \theta$; $0 \leq \theta < 2\pi$

76. $r = \sin 2\theta$, $r = \sqrt{2} \cos \theta$; $0 \leq \theta < \pi$

(Modeling) *Solve each problem.*

77. ***Planetary Orbits*** The polar equation

$$r = \frac{a(1 - e^2)}{1 + e \cos \theta}$$

can be used to graph the orbits of the planets, where a is the average distance in astronomical units from the sun and e is a constant called the *eccentricity*. The sun will be located at the pole. The table lists a and e for the planets.

Planet	a	e
Mercury	.39	.206
Venus	.78	.007
Earth	1.00	.017
Mars	1.52	.093
Jupiter	5.20	.048
Saturn	9.54	.056
Uranus	19.20	.047
Neptune	30.10	.009
Pluto	39.40	.249

Source: Karttunen, H., P. Kröger, H. Oja, M. Putannen, and K. Donners (Editors), *Fundamental Astronomy, 4th edition,* Springer-Verlag, 2003; Zeilik, M., S. Gregory, and E. Smith, *Introductory Astronomy and Astrophysics,* Saunders College Publishers, 1992.

(a) Graph the orbits of the four closest planets on the same polar grid. Choose a viewing window that results in a graph with nearly circular orbits.

(b) Plot the orbits of Earth, Jupiter, Uranus, and Pluto on the same polar grid. How does Earth's distance from the sun compare to the other planets' distances from the sun?

(c) Use graphing to determine whether or not Pluto is always the farthest planet from the sun.

78. ***Radio Towers and Broadcasting Patterns*** Many times radio stations do not broadcast in all directions with the same intensity. To avoid interference with an existing station to the north, a new station may be licensed to broadcast only east and west. To create an east-west signal, two radio towers are sometimes used, as illustrated in the figure. Locations where the radio signal is received correspond to the interior of the curve defined by $r^2 = 40{,}000 \cos 2\theta$, where the polar axis (or positive x-axis) points east.

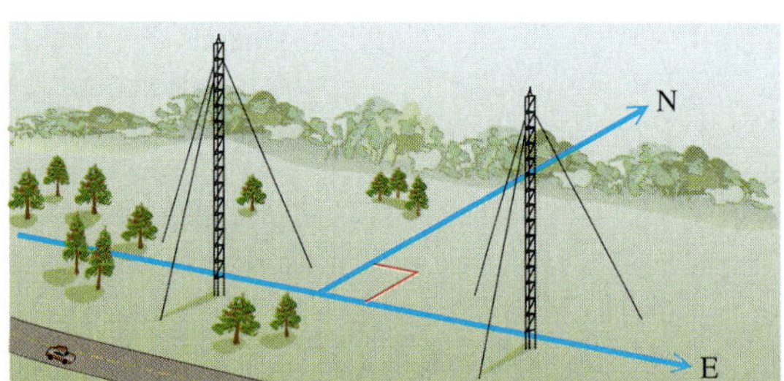

(a) Graph $r^2 = 40{,}000 \cos 2\theta$ for $0° \leq \theta \leq 360°$, where units are in miles. Assuming the radio towers are located near the pole, use the graph to describe the regions where the signal can be received and where the signal cannot be received.

(b) Suppose a radio signal pattern is given by $r^2 = 22{,}500 \sin 2\theta$. Graph this pattern and interpret the results.

8.8 Parametric Equations, Graphs, and Applications

Basic Concepts ▪ **Parametric Graphs and Their Rectangular Equivalents** ▪ **The Cycloid** ▪ **Applications of Parametric Equations**

Basic Concepts

Throughout this text, we have graphed sets of ordered pairs of real numbers that correspond to a function of the form $y = f(x)$ or $r = g(\theta)$. Another way to determine a set of ordered pairs involves two functions f and g defined by $x = f(t)$ and $y = g(t)$, where t is a real number in some interval I. Each value of t leads to a corresponding x-value and a corresponding y-value, and thus to an ordered pair (x, y).

Parametric Equations of a Plane Curve

A **plane curve** is a set of points (x, y) such that $x = f(t)$, $y = g(t)$, and f and g are both defined on an interval I. The equations $x = f(t)$ and $y = g(t)$ are **parametric equations** with **parameter** t.

Graphing calculators are capable of graphing plane curves defined by parametric equations. The calculator must be set in parametric mode, and the window requires intervals for the parameter t, as well as for x and y. ■

```
WINDOW
 Tmin=-3
 Tmax=3
 Tstep=.05
 Xmin=-2
 Xmax=10
 Xscl=1
↓Ymin=-4
```

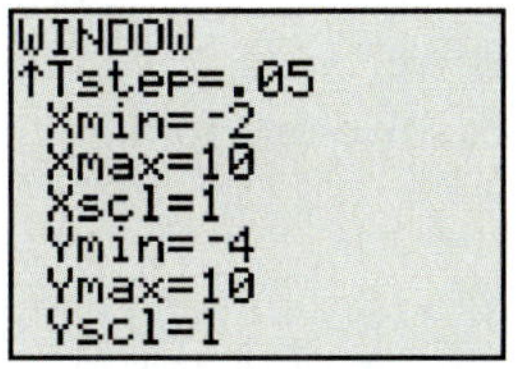

```
WINDOW
↑Tstep=.05
 Xmin=-2
 Xmax=10
 Xscl=1
 Ymin=-4
 Ymax=10
 Yscl=1
```

This is a continuation of the previous screen.

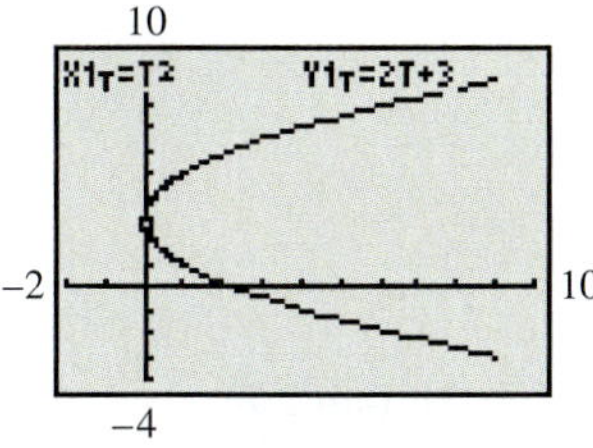

For the equation in Example 1, we make the choices seen in the first two screens to obtain the graph shown in the final screen.

Parametric Graphs and Their Rectangular Equivalents

EXAMPLE 1 Graphing a Plane Curve Defined Parametrically

Let $x = t^2$ and $y = 2t + 3$, for t in $[-3, 3]$. Graph the set of ordered pairs (x, y).

Solution Make a table of corresponding values of t, x, and y over the domain of t. Then plot the points as shown in Figure 73. The graph is a portion of a parabola with horizontal axis $y = 3$. The arrowheads indicate the direction the curve traces as t increases.

t	x	y
-3	9	-3
-2	4	-1
-1	1	1
0	0	3
1	1	5
2	4	7
3	9	9

$x = t^2$, $y = 2t + 3$ for t in $[-3, 3]$; (9, 9); (9, −3)

Figure 73

Now try Exercise 5(a).

EXAMPLE 2 Finding an Equivalent Rectangular Equation

Find a rectangular equation for the plane curve of Example 1 defined as follows.

$$x = t^2, \quad y = 2t + 3, \qquad \text{for } t \text{ in } [-3, 3]$$

Solution To eliminate the parameter t, solve either equation for t. Here, only the second equation, $y = 2t + 3$, leads to a unique solution for t, so we choose it.

$$y = 2t + 3 \qquad \text{Solve for } t. \text{ (Section 1.1)}$$
$$2t = y - 3$$
$$t = \frac{y-3}{2}$$

Now substitute this result into the first equation to get

$$x = t^2 = \left(\frac{y-3}{2}\right)^2 = \frac{(y-3)^2}{4} \qquad \text{or} \qquad 4x = (y-3)^2.$$

This is the equation of a horizontal parabola opening to the right, which agrees with the graph given in Figure 73. Because t is in $[-3, 3]$, x is in $[0, 9]$ and y is in $[-3, 9]$. The rectangular equation must be given with its restricted domain as

$$4x = (y-3)^2, \qquad \text{for } x \text{ in } [0, 9].$$

Now try Exercise 5(b).

Trigonometric functions are often used to define a plane curve parametrically.

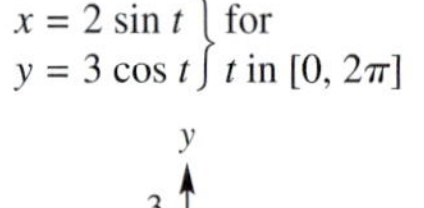

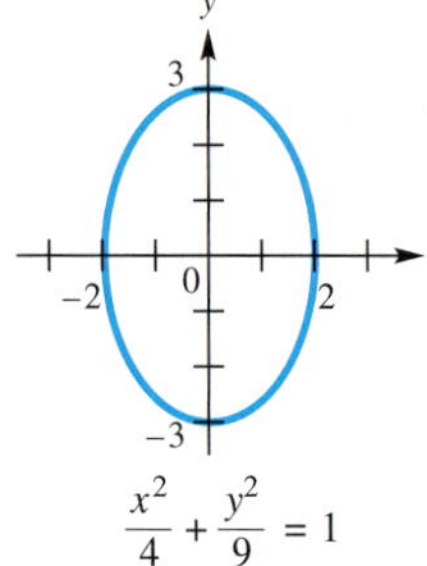

Figure 74

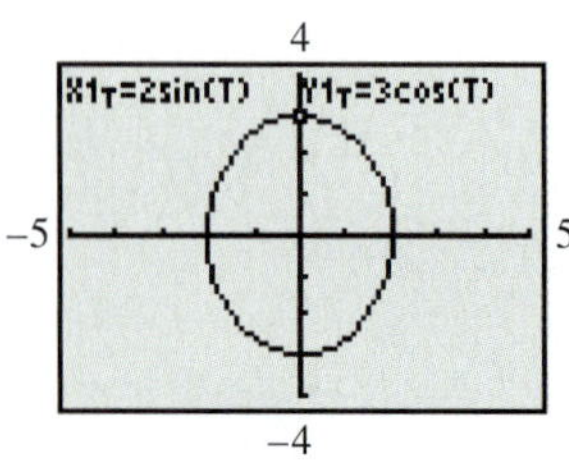

This is a calculator graph of the plane curve in Example 3.

EXAMPLE 3 Graphing a Plane Curve Defined Parametrically

Graph the plane curve defined by $x = 2 \sin t$, $y = 3 \cos t$, for t in $[0, 2\pi]$.

Solution To convert to a rectangular equation, it is not productive here to solve either equation for t. Instead, we use the fact that $\sin^2 t + \cos^2 t = 1$ to apply another approach. Square both sides of each equation; solve one for $\sin^2 t$, the other for $\cos^2 t$.

$x = 2 \sin t$	$y = 3 \cos t$	Given equations
$x^2 = 4 \sin^2 t$	$y^2 = 9 \cos^2 t$	Square both sides. (Section 1.6)
$\frac{x^2}{4} = \sin^2 t$	$\frac{y^2}{9} = \cos^2 t$	Divide.

Now add corresponding sides of the two equations.

$$\frac{x^2}{4} + \frac{y^2}{9} = \sin^2 t + \cos^2 t$$

$$\frac{x^2}{4} + \frac{y^2}{9} = 1 \qquad \sin^2 t + \cos^2 t = 1 \text{ (Section 5.2)}$$

This is the equation of an *ellipse,* as shown in Figure 74. (Ellipses are covered in more detail in Chapter 10.)

Now try Exercise 27.

Parametric representations of a curve are not unique. In fact, there are infinitely many parametric representations of a given curve. If the curve can be described by a rectangular equation $y = f(x)$, with domain X, then one simple parametric representation is

$$x = t, \quad y = f(t), \qquad \text{for } t \text{ in } X.$$

EXAMPLE 4 Finding Alternative Parametric Equation Forms

Give two parametric representations for the equation of the parabola

$$y = (x - 2)^2 + 1.$$

Solution The simplest choice is to let

$$x = t, \quad y = (t - 2)^2 + 1, \qquad \text{for } t \text{ in } (-\infty, \infty).$$

Another choice, which leads to a simpler equation for y, is

$$x = t + 2, \quad y = t^2 + 1, \qquad \text{for } t \text{ in } (-\infty, \infty).$$

Now try Exercise 29.

NOTE Sometimes trigonometric functions are desirable. One choice in Example 4 might be $x = 2 + \tan t$, $y = \sec^2 t$, for t in $\left(-\frac{\pi}{2}, \frac{\pi}{2}\right)$.

Looking Ahead to Calculus

The cycloid is a special case of a curve traced out by a point at a given distance from the center of a circle as the circle rolls along a straight line. Such a curve is called a *trochoid.* It is just one of several parametrically defined curves studied in calculus. *Bezier curves* are used in manufacturing, and *Conchoids of Nicodemes* are so named because the shape of their outer branches resembles a conch shell. Other examples are *hypocycloids, epicycloids, the witch of Agnesi, swallowtail catastrophe curves,* and *Lissajou figures.* (*Source:* Stewart, J., *Calculus*, Fifth Edition, Brooks/Cole Publishing Co., 2003.)

The Cycloid The path traced by a fixed point on the circumference of a circle rolling along a line is called a *cycloid.* A **cycloid** is defined by

$$x = at - a \sin t, \quad y = a - a \cos t, \qquad \text{for } t \text{ in } (-\infty, \infty).$$

EXAMPLE 5 Graphing a Cycloid

Graph the cycloid $x = t - \sin t$, $y = 1 - \cos t$, for t in $[0, 2\pi]$.

Algebraic Solution

There is no simple way to find a rectangular equation for the cycloid from its parametric equations. Instead, begin with a table of values.

t	0	$\frac{\pi}{4}$	$\frac{\pi}{2}$	π	$\frac{3\pi}{2}$	2π
x	0	.08	.6	π	5.7	2π
y	0	.3	1	2	1	0

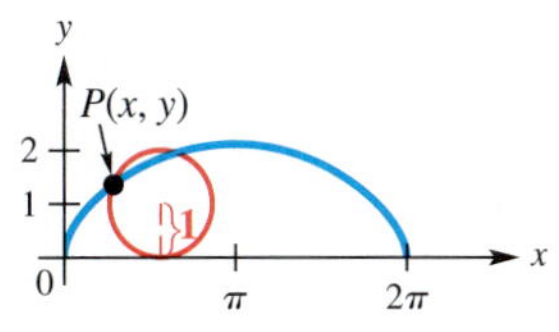

Figure 75

Plotting the ordered pairs (x, y) from the table of values leads to the portion of the graph in Figure 75 from 0 to 2π.

Graphing Calculator Solution

It is easier to graph a cycloid with a graphing calculator in parametric mode than with traditional methods. See Figure 76.

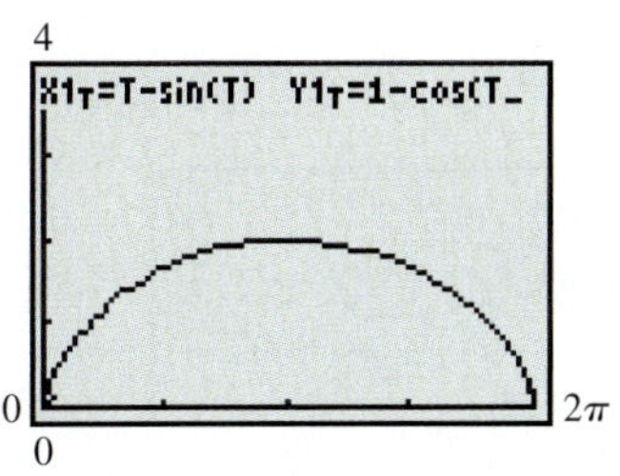

Figure 76

Now try Exercise 33.

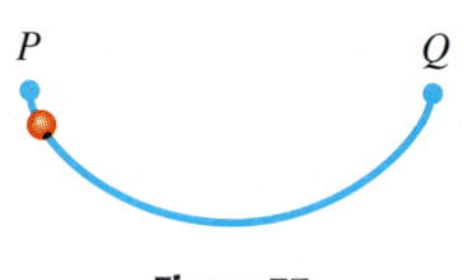

Figure 77

The cycloid has an interesting physical property. If a flexible cord or wire goes through points P and Q as in Figure 77, and a bead is allowed to slide due to the force of gravity without friction along this path from P to Q, the path that requires the shortest time takes the shape of the graph of an inverted cycloid.

Looking Ahead to Calculus

At any time t, the velocity of an object is given by the vector $\mathbf{v} = \langle f'(t), g'(t) \rangle$. The object's speed at time t is

$$|\mathbf{v}| = \sqrt{(f'(t))^2 + (g'(t))^2}.$$

Applications of Parametric Equations Parametric equations are used to simulate motion. If a ball is thrown with a velocity of v feet per second at an angle θ with the horizontal, its flight can be modeled by the parametric equations

$$x = (v \cos \theta)t \quad \text{and} \quad y = (v \sin \theta)t - 16t^2 + h,$$

where t is in seconds and h is the ball's initial height in feet above the ground. The term $-16t^2$ occurs because gravity is pulling downward. See Figure 78. These equations ignore air resistance.

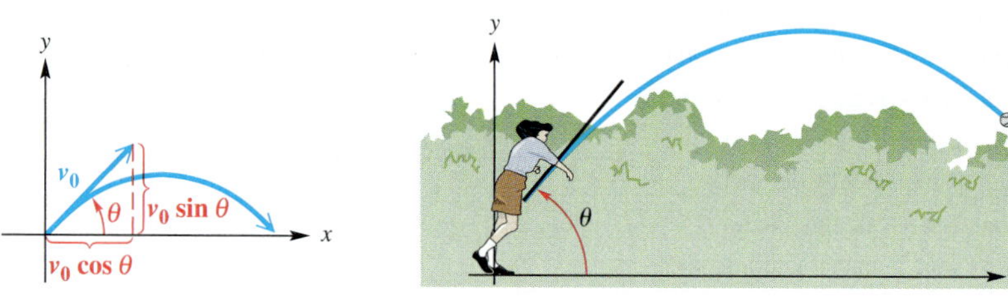

Figure 78

EXAMPLE 6 Simulating Motion with Parametric Equations

Three golf balls are hit simultaneously into the air at 132 ft per sec (90 mph) at angles of 30°, 50°, and 70° with the horizontal.

(a) Assuming the ground is level, determine graphically which ball travels the farthest. Estimate this distance.

(b) Which ball reaches the greatest height? Estimate this height.

Solution

(a) The three sets of parametric equations determined by the three golf balls are as follows since $h = 0$.

$$x_1 = (132 \cos 30°)t, \qquad y_1 = (132 \sin 30°)t - 16t^2$$
$$x_2 = (132 \cos 50°)t, \qquad y_2 = (132 \sin 50°)t - 16t^2$$
$$x_3 = (132 \cos 70°)t, \qquad y_3 = (132 \sin 70°)t - 16t^2$$

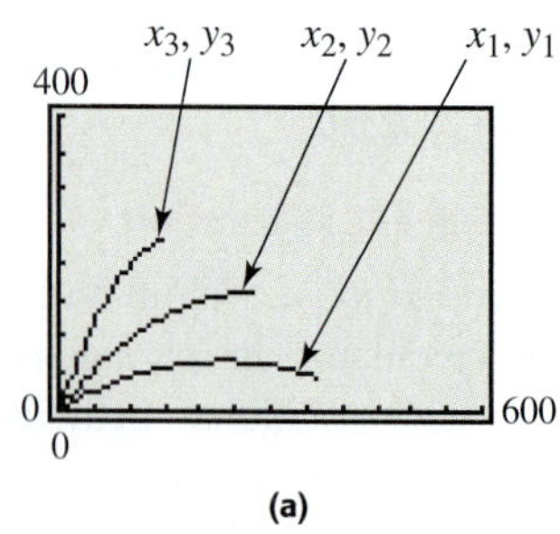

(a)

The graphs of the three sets of parametric equations are shown in Figure 79(a), where $0 \leq t \leq 9$. We used a graphing calculator in simultaneous mode so that we can view all three balls in flight at the same time. From the graph in Figure 79(b), we can see that the ball hit at 50° travels the farthest distance. Using the TRACE feature, we estimate this distance to be about 540 ft.

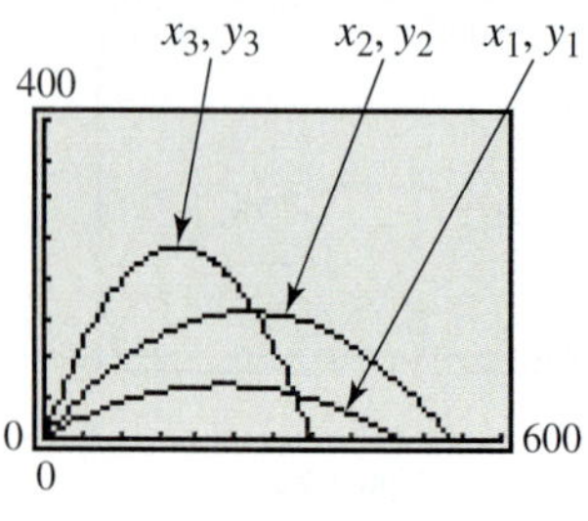

(b)

Figure 79

(b) Again, use the TRACE feature to find that the ball hit at 70° reaches the greatest height, about 240 ft.

Now try Exercise 39.

EXAMPLE 7 Examining Parametric Equations of Flight

A small rocket is launched from a table that is 3.36 ft above the ground. Its initial velocity is 64 ft per sec, and it is launched at an angle of 30° with respect to the ground. Find the rectangular equation that models its path. What type of path does the rocket follow?

Solution The path of the rocket is defined by the parametric equations

$$x = (64 \cos 30°)t \qquad \text{and} \qquad y = (64 \sin 30°)t - 16t^2 + 3.36$$

or, equivalently,

$$x = 32\sqrt{3}t \qquad \text{and} \qquad y = -16t^2 + 32t + 3.36.$$

From $x = 32\sqrt{3}t$, we obtain

$$t = \frac{x}{32\sqrt{3}}.$$

Substituting into the other parametric equation for t yields

$$y = -16\left(\frac{x}{32\sqrt{3}}\right)^2 + 32\left(\frac{x}{32\sqrt{3}}\right) + 3.36.$$

Simplifying, we find that the rectangular equation is

$$y = -\frac{1}{192}x^2 + \frac{\sqrt{3}}{3}x + 3.36.$$

Because this equation defines a parabola, the rocket follows a parabolic path.

Now try Exercise 43(a).

EXAMPLE 8 Analyzing the Path of a Projectile

Determine the total flight time and the horizontal distance traveled by the rocket in Example 7.

Algebraic Solution

The equation $y = -16t^2 + 32t + 3.36$ tells the vertical position of the rocket at time t. We need to determine those values of t for which $y = 0$ since these values correspond to the rocket at ground level. This yields

$$0 = -16t^2 + 32t + 3.36.$$

Using the quadratic formula, the solutions are $t = -.1$ or $t = 2.1$. Since t represents time, $t = -.1$ is an unacceptable answer. Therefore, the flight time is 2.1 sec.

The rocket was in the air for 2.1 sec, so we can use $t = 2.1$ and the parametric equation that models the horizontal position, $x = 32\sqrt{3}t$, to obtain

$$x = 32\sqrt{3}(2.1) \approx 116.4 \text{ ft}.$$

Graphing Calculator Solution

Figure 80 shows that when T = 2.1, the horizontal distance X covered is approximately 116.4 ft, which agrees with the algebraic solution.

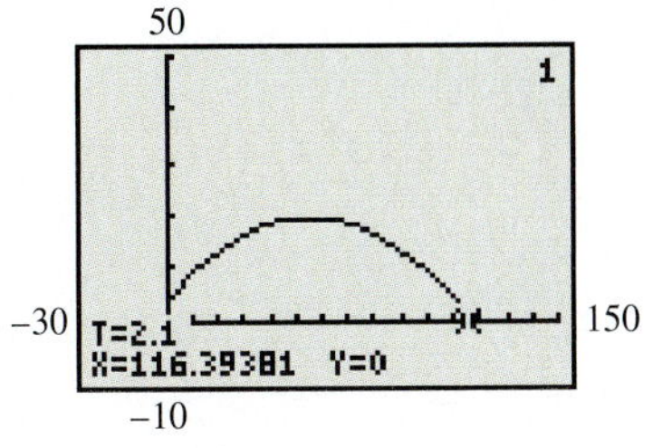

Figure 80

Now try Exercise 43(b).

8.8 Exercises

Concept Check *Match the ordered pair from Column II with the pair of parametric equations in Column I on whose graph the point lies. In each case, consider the given value of t.*

I	II
1. $x = 3t + 6, y = -2t + 4;\quad t = 2$	**A.** $(5, 25)$
2. $x = \cos t, y = \sin t;\quad t = \frac{\pi}{4}$	**B.** $(7, 2)$
3. $x = t, y = t^2;\quad t = 5$	**C.** $(12, 0)$
4. $x = t^2 + 3, y = t^2 - 2;\quad t = 2$	**D.** $\left(\frac{\sqrt{2}}{2}, \frac{\sqrt{2}}{2}\right)$

*For each plane curve **(a)** graph the curve, and **(b)** find a rectangular equation for the curve. See Examples 1 and 2.*

5. $x = t + 2, y = t^2,\quad$ for t in $[-1, 1]$

6. $x = 2t, y = t + 1,\quad$ for t in $[-2, 3]$

7. $x = \sqrt{t}, y = 3t - 4,\quad$ for t in $[0, 4]$

8. $x = t^2, y = \sqrt{t},\quad$ for t in $[0, 4]$

9. $x = t^3 + 1, y = t^3 - 1,\quad$ for t in $(-\infty, \infty)$

10. $x = 2t - 1, y = t^2 + 2,\quad$ for t in $(-\infty, \infty)$

11. $x = 2 \sin t, y = 2 \cos t,\quad$ for t in $[0, 2\pi]$

12. $x = \sqrt{5} \sin t, y = \sqrt{3} \cos t,\quad$ for t in $[0, 2\pi]$

13. $x = 3 \tan t, y = 2 \sec t,\quad$ for t in $\left(-\frac{\pi}{2}, \frac{\pi}{2}\right)$

14. $x = \cot t, y = \csc t,\quad$ for t in $(0, \pi)$

15. $x = \sin t, y = \csc t,\quad$ for t in $(0, \pi)$

16. $x = \tan t, y = \cot t,\quad$ for t in $\left(0, \frac{\pi}{2}\right)$

17. $x = t, y = \sqrt{t^2 + 2},\quad$ for t in $(-\infty, \infty)$

18. $x = \sqrt{t}, y = t^2 - 1,\quad$ for t in $[0, \infty)$

19. $x = 2 + \sin t, y = 1 + \cos t,\quad$ for t in $[0, 2\pi]$

20. $x = 1 + 2 \sin t, y = 2 + 3 \cos t,\quad$ for t in $[0, 2\pi]$

21. $x = t + 2, y = \frac{1}{t + 2},\quad$ for $t \neq -2$

22. $x = t - 3, y = \frac{2}{t - 3},\quad$ for $t \neq 3$

23. $x = t + 2, y = t - 4,\quad$ for t in $(-\infty, \infty)$

24. $x = t^2 + 2, y = t^2 - 4,\quad$ for t in $(-\infty, \infty)$

Graph each plane curve defined by the parametric equations for t in $[0, 2\pi]$. Then find a rectangular equation for the plane curve. See Example 3.

25. $x = 3 \cos t, y = 3 \sin t$

26. $x = 2 \cos t, y = 2 \sin t$

27. $x = 3 \sin t, y = 2 \cos t$

28. $x = 4 \sin t, y = 3 \cos t$

Give two parametric representations for the equation of each parabola. See Example 4.

29. $y = (x + 3)^2 - 1$

30. $y = (x + 4)^2 + 2$

31. $y = x^2 - 2x + 3$

32. $y = x^2 - 4x + 6$

Graph each cycloid defined by the given equations for t in the specified interval. See Example 5.

33. $x = 2t - 2 \sin t, y = 2 - 2 \cos t,\quad$ for t in $[0, 8\pi]$

34. $x = t - \sin t, y = 1 - \cos t,\quad$ for t in $[0, 4\pi]$

Lissajous Figures *The screen shown here is an example of a Lissajous figure. Lissajous figures occur in electronics and may be used to find the frequency of an unknown voltage. Graph each Lissajous figure for t in* $[0, 6.5]$ *in the window* $[-6, 6]$ *by* $[-4, 4]$.

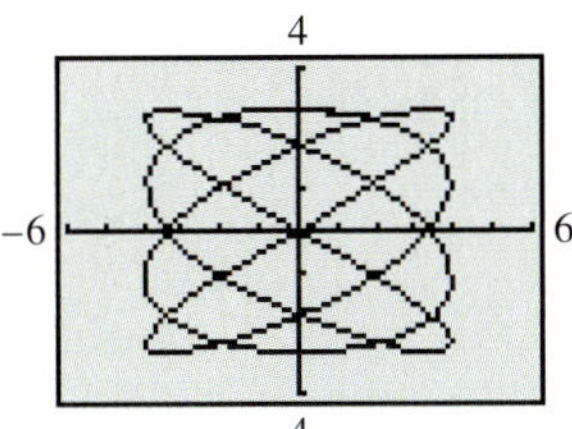

35. $x = 2 \cos t,\ y = 3 \sin 2t$

36. $x = 3 \cos 2t,\ y = 3 \sin 3t$

37. $x = 3 \sin 4t,\ y = 3 \cos 3t$

38. $x = 4 \sin 4t,\ y = 3 \sin 5t$

(Modeling) *In Exercises 39–42, do the following.*

(a) *Determine the parametric equations that model the path of the projectile.*
(b) *Determine the rectangular equation that models the path of the projectile.*
(c) *Determine how long the projectile is in flight and the horizontal distance covered.*

See Examples 6–8.

39. *Flight of a Model Rocket* A model rocket is launched from the ground with velocity 48 ft per sec at an angle of 60° with respect to the ground.

40. *Flight of a Golf Ball* Tiger is playing golf. He hits a golf ball from the ground at an angle of 60° with respect to the ground at velocity 150 ft per sec.

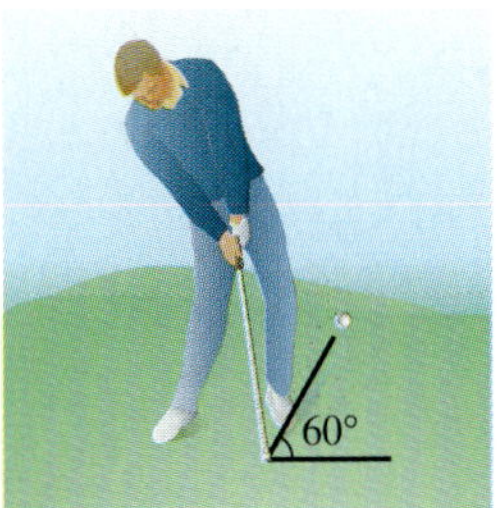

41. *Flight of a Softball* Sally hits a softball when it is 2 ft above the ground. The ball leaves her bat at an angle of 20° with respect to the ground at velocity 88 ft per sec.

42. *Flight of a Baseball* Luis hits a baseball when it is 2.5 ft above the ground. The ball leaves his bat at an angle of 29° from the horizontal with velocity 136 ft per sec.

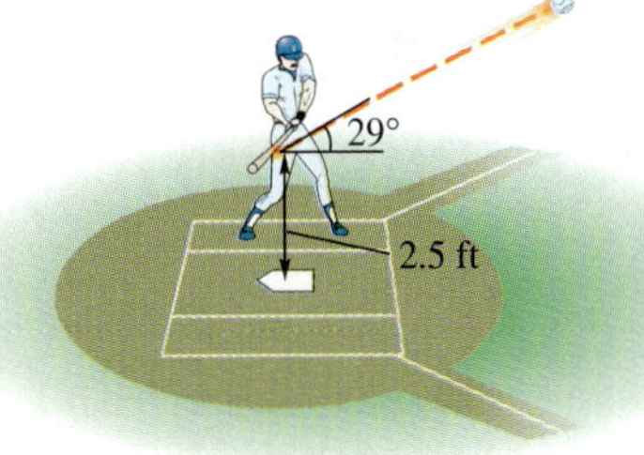

(Modeling) *Solve each problem. See Examples 7 and 8.*

43. *Path of a Rocket* A small rocket is launched from the top of an 8-ft ladder. Its initial velocity is 128 ft per sec, and it is launched at an angle of 60° with respect to the ground.

(a) Find the rectangular equation that models its path. What type of path does the rocket follow?

(b) Determine the total flight time and the horizontal distance the rocket travels.

44. *Simulating Gravity on the Moon* If an object is thrown on the moon, then the parametric equations of flight are

$$x = (v \cos \theta)t \quad \text{and} \quad y = (v \sin \theta)t - 2.66t^2 + h.$$

Estimate the distance that a golf ball hit at 88 ft per sec (60 mph) at an angle of 45° with the horizontal travels on the moon if the moon's surface is level.

45. *Flight of a Baseball* A baseball is hit from a height of 3 ft at a 60° angle above the horizontal. Its initial velocity is 64 ft per sec.

(a) Write parametric equations that model the flight of the baseball.
(b) Determine the horizontal distance traveled by the ball in the air. Assume that the ground is level.
(c) What is the maximum height of the baseball? At that time, how far has the ball traveled horizontally?
(d) Would the ball clear a 5-ft-high fence that is 100 ft from the batter?

(Modeling) Path of a Projectile In Exercises 46 and 47, a projectile has been launched from the ground with initial velocity 88 ft per sec. You are supplied with the parametric equations modeling the path of the projectile.

(a) *Graph the parametric equations.*
(b) *Approximate θ, the angle the projectile makes with the horizontal at launch, to the nearest tenth of a degree.*
(c) *Based on your answer to part (b), write parametric equations for the projectile using the cosine and sine functions.*

46. $x = 82.69295063t$, $y = -16t^2 + 30.09777261t$

47. $x = 56.56530965t$, $y = -16t^2 + 67.41191099t$

48. Give two parametric representations of the line through the point (x_1, y_1) with slope m.

49. Give two parametric representations of the parabola $y = a(x - h)^2 + k$.

50. Give a parametric representation of the rectangular equation $\frac{x^2}{a^2} - \frac{y^2}{b^2} = 1$.

51. Give a parametric representation of the rectangular equation $\frac{x^2}{a^2} + \frac{y^2}{b^2} = 1$.

52. The spiral of Archimedes has polar equation $r = a\theta$, where $r^2 = x^2 + y^2$. Show that a parametric representation of the spiral of Archimedes is

$$x = a\theta \cos \theta, \quad y = a\theta \sin \theta, \quad \text{for } \theta \text{ in } (-\infty, \infty).$$

53. Show that the hyperbolic spiral $r\theta = a$, where $r^2 = x^2 + y^2$, is given parametrically by

$$x = \frac{a \cos \theta}{\theta}, \quad y = \frac{a \sin \theta}{\theta}, \quad \text{for } \theta \text{ in } (-\infty, 0) \cup (0, \infty).$$

54. The parametric equations $x = \cos t$, $y = \sin t$, for t in $[0, 2\pi]$ and the parametric equations $x = \cos t$, $y = -\sin t$, for t in $[0, 2\pi]$ both have the unit circle as their graph. However, in one case the circle is traced out clockwise (as t moves from 0 to 2π) and in the other case the circle is traced out counterclockwise. For which pair of equations is the circle traced out in the clockwise direction?

Concept Check Consider the parametric equations $x = f(t)$, $y = g(t)$, for t in $[a, b]$, with $c > 0$, $d > 0$.

55. How is the graph affected if the equation $x = f(t)$ is replaced by $x = c + f(t)$?

56. How is the graph affected if the equation $y = g(t)$ is replaced by $y = d + g(t)$?

Chapter 8 Summary

KEY TERMS

8.1 Side-Angle-Side (SAS)
Angle-Side-Angle (ASA)
Side-Side-Side (SSS)
oblique triangle
Side-Angle-Angle (SAA)
ambiguous case

8.3 scalar
vector quantity
vector
magnitude
initial point
terminal point
parallelogram rule
resultant
opposite (of a vector)
zero vector
scalar product
position vector
horizontal component
vertical component
direction angle
unit vector
dot product
angle between two vectors
orthogonal vectors

8.4 equilibrant
airspeed
groundspeed

8.5 real axis
imaginary axis
complex plane
rectangular form of a complex number
trigonometric (polar) form of a complex number
absolute value (modulus)
argument

8.6 *n*th root of a complex number

8.7 polar coordinate system
pole
polar axis
polar coordinates
rectangular (Cartesian) equation
polar equation
cardioid
rose curve (four-leaved rose)
lemniscate
spiral of Archimedes
limaçon

8.8 parametric equations of a plane curve
parameter
cycloid

NEW SYMBOLS

OP or $\overrightarrow{\mathbf{OP}}$	vector **OP**
$\lvert\mathbf{OP}\rvert$	magnitude of vector **OP**
$\langle a, b\rangle$	position vector
i, **j**	unit vectors

QUICK REVIEW

CONCEPTS	EXAMPLES

8.1 The Law of Sines

Law of Sines

In any triangle *ABC*, with sides *a*, *b*, and *c*,

$$\frac{a}{\sin A} = \frac{b}{\sin B}, \quad \frac{a}{\sin A} = \frac{c}{\sin C}, \quad \text{and} \quad \frac{b}{\sin B} = \frac{c}{\sin C}.$$

In triangle *ABC*, find *c* if $A = 44°$, $C = 62°$, and $a = 12.00$ units. Then find its area.

$$\frac{a}{\sin A} = \frac{c}{\sin C}$$

$$\frac{12.00}{\sin 44°} = \frac{c}{\sin 62°}$$

$$c = \frac{12.00 \sin 62°}{\sin 44°} \approx 15.25 \text{ units}$$

Area of a Triangle

In any triangle *ABC*, the area is half the product of the lengths of two sides and the sine of the angle between them.

$$\mathcal{A} = \frac{1}{2}bc \sin A, \quad \mathcal{A} = \frac{1}{2}ab \sin C, \quad \mathcal{A} = \frac{1}{2}ac \sin B$$

$$\mathcal{A} = \frac{1}{2}ac \sin B = \frac{1}{2}(12.00)(15.25) \sin 74° \qquad B = 180° - 44° - 62°$$

$$\approx 87.96 \text{ sq units}$$

CONCEPTS | **EXAMPLES**

8.2 The Law of Cosines

Law of Cosines

In any triangle ABC, with sides a, b, and c,

$$a^2 = b^2 + c^2 - 2bc \cos A$$
$$b^2 = a^2 + c^2 - 2ac \cos B$$
$$c^2 = a^2 + b^2 - 2ab \cos C.$$

In triangle ABC, find C if $a = 11$ units, $b = 13$ units, and $c = 20$ units. Then find its area.

$$c^2 = a^2 + b^2 - 2ab \cos C$$
$$20^2 = 11^2 + 13^2 - 2(11)(13) \cos C$$
$$400 = 121 + 169 - 286 \cos C$$
$$\frac{400 - 121 - 169}{-286} = \cos C$$
$$C = \cos^{-1}\left(\frac{400 - 121 - 169}{-286}\right)$$
$$C \approx 112.6°$$

Heron's Area Formula

If a triangle has sides of lengths a, b, and c, with semiperimeter

$$s = \frac{1}{2}(a + b + c),$$

then the area of the triangle is

$$\mathcal{A} = \sqrt{s(s - a)(s - b)(s - c)}.$$

The semiperimeter s is

$$s = \frac{1}{2}(11 + 13 + 20) = 22,$$

so

$$\mathcal{A} = \sqrt{22(22 - 11)(22 - 13)(22 - 20)} = 66 \text{ sq units.}$$

8.3 Vectors, Operations, and the Dot Product

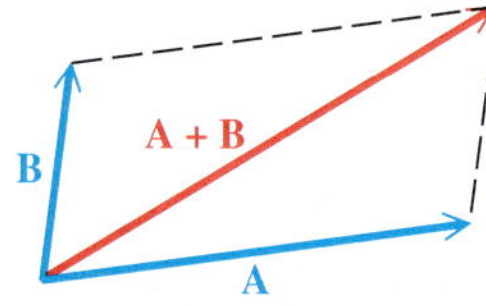

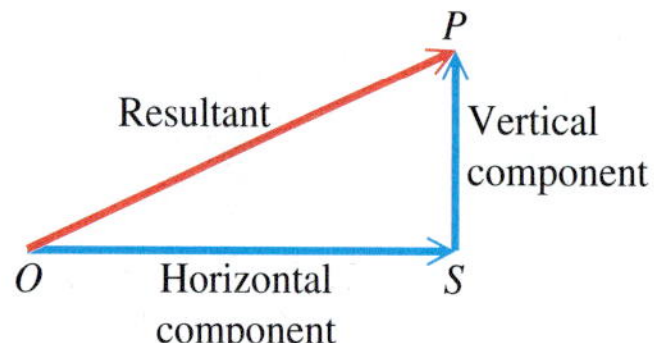

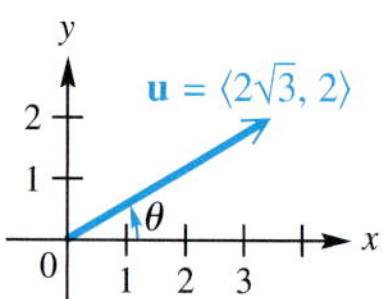

Magnitude and Direction Angle of a Vector

The magnitude (length) of vector $\mathbf{u} = \langle a, b \rangle$ is given by $|\mathbf{u}| = \sqrt{a^2 + b^2}$.

The direction angle θ satisfies $\tan \theta = \frac{b}{a}$, where $a \neq 0$.

$$|\mathbf{u}| = \sqrt{(2\sqrt{3})^2 + 2^2} = \sqrt{16} = 4$$

Since $\tan \theta = \frac{2}{2\sqrt{3}} = \frac{\sqrt{3}}{3}$, it follows that $\theta = 30°$.

Vector Operations

For any real numbers a, b, c, d, and k,

$$\langle a, b \rangle + \langle c, d \rangle = \langle a + c, b + d \rangle$$
$$k \cdot \langle a, b \rangle = \langle ka, kb \rangle.$$
$$\text{If } \mathbf{a} = \langle a_1, a_2 \rangle, \text{ then } -\mathbf{a} = \langle -a_1, -a_2 \rangle.$$
$$\langle a, b \rangle - \langle c, d \rangle = \langle a, b \rangle + (-\langle c, d \rangle) = \langle a - c, b - d \rangle$$

$$\langle 4, 6 \rangle + \langle -8, 3 \rangle = \langle -4, 9 \rangle$$
$$5\langle -2, 1 \rangle = \langle -10, 5 \rangle$$
$$-\langle -9, 6 \rangle = \langle 9, -6 \rangle$$
$$\langle 4, 6 \rangle - \langle -8, 3 \rangle = \langle 12, 3 \rangle$$

If $\mathbf{u} = \langle x, y \rangle$ has direction angle θ, then

$$\mathbf{u} = \langle |\mathbf{u}| \cos \theta, |\mathbf{u}| \sin \theta \rangle.$$

For $\mathbf{u}$ defined above,

$$\mathbf{u} = 2\sqrt{3}\mathbf{i} + 2\mathbf{j}.$$

i, j Form for Vectors

If $\mathbf{v} = \langle a, b \rangle$, then $\mathbf{v} = a\mathbf{i} + b\mathbf{j}$, where $\mathbf{i} = \langle 1, 0 \rangle$ and $\mathbf{j} = \langle 0, 1 \rangle$.

CONCEPTS	EXAMPLES

Dot Product

The dot product of the two vectors $\mathbf{u} = \langle a, b \rangle$ and $\mathbf{v} = \langle c, d \rangle$, denoted $\mathbf{u} \cdot \mathbf{v}$, is given by

$$\mathbf{u} \cdot \mathbf{v} = ac + bd.$$

If θ is the angle between $\mathbf{u}$ and $\mathbf{v}$, where $0° \leq \theta \leq 180°$, then

$$\mathbf{u} \cdot \mathbf{v} = |\mathbf{u}||\mathbf{v}| \cos \theta \quad \text{or} \quad \cos \theta = \frac{\mathbf{u} \cdot \mathbf{v}}{|\mathbf{u}||\mathbf{v}|}.$$

Examples:

$$\langle 2, 1 \rangle \cdot \langle 5, -2 \rangle = 2 \cdot 5 + 1(-2) = 8$$

Find the angle θ between $\mathbf{u} = \langle 3, 1 \rangle$ and $\mathbf{v} = \langle 2, -3 \rangle$.

$$\cos \theta = \frac{\langle 3, 1 \rangle \cdot \langle 2, -3 \rangle}{\sqrt{3^2 + 1^2} \cdot \sqrt{2^2 + (-3)^2}}$$

$$\cos \theta = \frac{6 + (-3)}{\sqrt{10} \cdot \sqrt{13}}$$

$$\cos \theta = \frac{3}{\sqrt{130}}$$

$$\theta \approx 74.7°$$

8.5 Trigonometric (Polar) Form of Complex Numbers; Products and Quotients

Trigonometric (Polar) Form of Complex Numbers

If the complex number $x + yi$ corresponds to the vector with direction angle θ and magnitude r, then

$$x = r \cos \theta \qquad y = r \sin \theta$$

$$r = \sqrt{x^2 + y^2} \qquad \tan \theta = \frac{y}{x}, \quad \text{if } x \neq 0.$$

The expression

$$r(\cos \theta + i \sin \theta) \quad \text{or} \quad r \text{ cis } \theta$$

is the trigonometric form (or polar form) of $x + yi$.

Examples:

Write $2(\cos 60° + i \sin 60°)$ in rectangular form.

$$2(\cos 60° + i \sin 60°) = 2\left(\frac{1}{2} + i \cdot \frac{\sqrt{3}}{2}\right) = 1 + \sqrt{3}i$$

Write $-\sqrt{2} + i\sqrt{2}$ in trigonometric form.

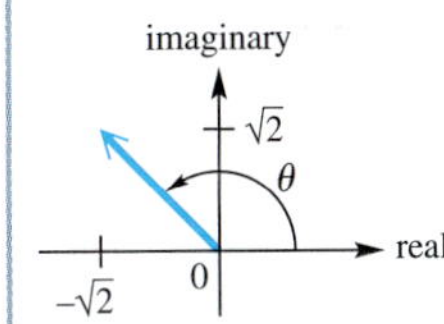

$$r = \sqrt{(-\sqrt{2})^2 + (\sqrt{2})^2} = 2$$

$\tan \theta = -1$ and θ is in quadrant II, so $\theta = 180° - 45° = 135°$. Therefore,

$$-\sqrt{2} + i\sqrt{2} = 2 \text{ cis } 135°.$$

Product and Quotient Theorems

For any two complex numbers $r_1(\cos \theta_1 + i \sin \theta_1)$ and $r_2(\cos \theta_2 + i \sin \theta_2)$,

$$[r_1(\cos \theta_1 + i \sin \theta_1)] \cdot [r_2(\cos \theta_2 + i \sin \theta_2)] = r_1 r_2[\cos(\theta_1 + \theta_2) + i \sin(\theta_1 + \theta_2)]$$

and

$$\frac{r_1(\cos \theta_1 + i \sin \theta_1)}{r_2(\cos \theta_2 + i \sin \theta_2)} = \frac{r_1}{r_2}[\cos(\theta_1 - \theta_2) + i \sin(\theta_1 - \theta_2)],$$

where $r_2 \text{ cis } \theta_2 \neq 0$.

Examples:

If $z_1 = 4(\cos 135° + i \sin 135°)$

and $z_2 = 2(\cos 45° + i \sin 45°)$, then

$$z_1 z_2 = 8(\cos 180° + i \sin 180°) = 8(-1 + i \cdot 0) = -8,$$

and

$$\frac{z_1}{z_2} = 2(\cos 90° + i \sin 90°) = 2(0 + i \cdot 1) = 2i.$$

CONCEPTS | **EXAMPLES**

8.6 De Moivre's Theorem; Powers and Roots of Complex Numbers

De Moivre's Theorem

$$[r(\cos \theta + i \sin \theta)]^n = r^n(\cos n\theta + i \sin n\theta)$$

***n*th Root Theorem**

If n is any positive integer, r is a positive real number, and θ is in degrees, then the nonzero complex number $r(\cos \theta + i \sin \theta)$ has exactly n distinct nth roots, given by

$$\sqrt[n]{r}\,(\cos \alpha + i \sin \alpha), \quad \text{or} \quad \sqrt[n]{r} \operatorname{cis} \alpha,$$

where

$$\alpha = \frac{\theta + 360° \cdot k}{n} \quad \text{or} \quad \alpha = \frac{\theta}{n} + \frac{360° \cdot k}{n},$$

$k = 0, 1, 2, \dots, n - 1.$

Let $z = 4(\cos 180° + i \sin 180°)$. Find z^3 and the square roots of z.

$$\begin{aligned} z^3 &= 4^3(\cos 3 \cdot 180° + i \sin 3 \cdot 180°) \\ &= 64(\cos 540° + i \sin 540°) \\ &= 64(-1 + i \cdot 0) \\ &= -64 \end{aligned}$$

For the given z, $r = 4$ and $\theta = 180°$. Its square roots are

$$\sqrt{4}\left(\cos \frac{180°}{2} + i \sin \frac{180°}{2}\right) = 2(0 + i \cdot 1) = 2i$$

and

$$\sqrt{4}\left(\cos \frac{180° + 360°}{2} + i \sin \frac{180° + 360°}{2}\right) = 2(0 + i(-1)) = -2i.$$

8.7 Polar Equations and Graphs

Rectangular and Polar Coordinates

The following relationships hold between the point (x, y) in the rectangular coordinate plane and the same point (r, θ) in the polar coordinate plane.

$$x = r \cos \theta \qquad y = r \sin \theta$$

$$r^2 = x^2 + y^2 \qquad \tan \theta = \frac{y}{x}, \quad \text{if } x \neq 0$$

Find the rectangular coordinates for the point $(5, 60°)$ in polar coordinates.

$$x = 5 \cos 60° = 5\left(\frac{1}{2}\right) = \frac{5}{2}$$

$$y = 5 \sin 60° = 5\left(\frac{\sqrt{3}}{2}\right) = \frac{5\sqrt{3}}{2}$$

The rectangular coordinates are $\left(\frac{5}{2}, \frac{5\sqrt{3}}{2}\right)$.

Find polar coordinates for $(-1, -1)$ in rectangular coordinates.

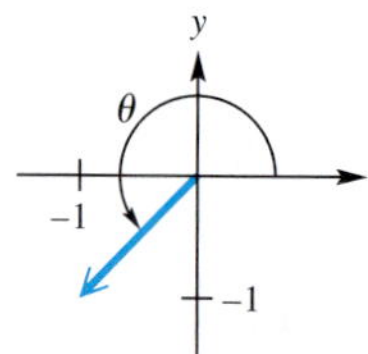

$$r = \sqrt{(-1)^2 + (-1)^2} = \sqrt{2}$$

$\tan \theta = 1$ and θ is in quadrant III, so $\theta = 225°$.

One pair of polar coordinates for $(-1, -1)$ is $(\sqrt{2}, 225°)$.

Polar Graphs

Examples of polar graphs include lines, circles, limaçons, lemniscates, roses, and spirals. (See page 756.)

Graph $r = 4 \cos 2\theta$.

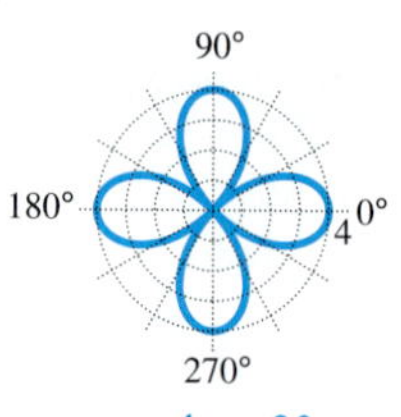

$r = 4 \cos 2\theta$

CONCEPTS	EXAMPLES

8.8 Parametric Equations, Graphs, and Applications

Plane Curve

A **plane curve** is a set of points (x, y) such that $x = f(t)$, $y = g(t)$, and f and g are both defined on an interval I. The equations $x = f(t)$ and $y = g(t)$ are **parametric equations** with **parameter t.**

Graph $x = 2 - \sin t$, $y = \cos t - 1$, for $0 \leq t \leq 2\pi$.

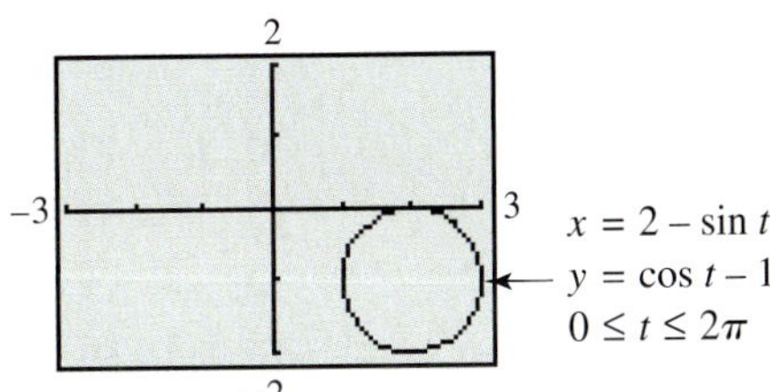

Flight of an Object

If an object has an initial velocity v, initial height h, and travels so that its initial angle of elevation is θ, then its flight after t seconds is modeled by the parametric equations

$$x = (v \cos \theta)t \quad \text{and} \quad y = (v \sin \theta)t - 16t^2 + h.$$

Joe kicks a football from the ground at an angle of 45° with a velocity of 48 ft per sec. Give the parametric equations that model its path and the distance it travels before hitting the ground.

$$x = (48 \cos 45°)t = 24\sqrt{2}t$$
$$y = (48 \sin 45°)t - 16t^2 = 24\sqrt{2}t - 16t^2$$

When the ball hits the ground, $y = 0$.

$$24\sqrt{2}t - 16t^2 = 0$$
$$8t\left(3\sqrt{2} - 2t\right) = 0$$
$$t = 0 \quad \text{or} \quad t = \frac{3\sqrt{2}}{2}$$

(Reject)

The distance it travels is $x = 24\sqrt{2}\left(\frac{3\sqrt{2}}{2}\right) = 72$ ft.

Chapter 8 Review Exercises

Use the law of sines to find the indicated part of each triangle ABC.

1. $C = 74.2°$, $c = 96.3$ m, $B = 39.5°$; find b

2. $A = 129.7°$, $a = 127$ ft, $b = 69.8$ ft; find B

3. $C = 51.3°$, $c = 68.3$ m, $b = 58.2$ m; find B

4. $a = 165$ m, $A = 100.2°$, $B = 25.0°$; find b

5. $B = 39° \, 50'$, $b = 268$ m, $a = 340$ m; find A

6. $C = 79° \, 20'$, $c = 97.4$ mm, $a = 75.3$ mm; find A

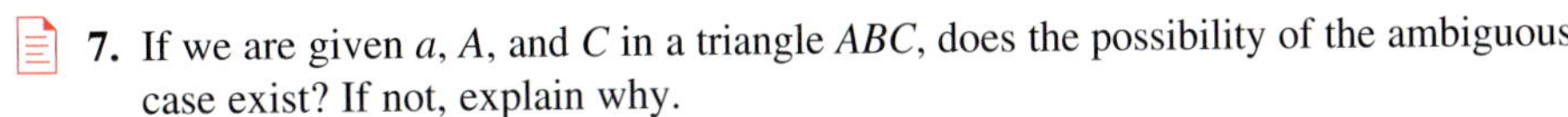

7. If we are given a, A, and C in a triangle ABC, does the possibility of the ambiguous case exist? If not, explain why.

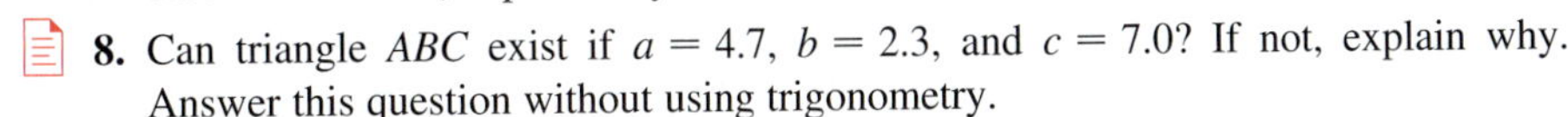

8. Can triangle ABC exist if $a = 4.7$, $b = 2.3$, and $c = 7.0$? If not, explain why. Answer this question without using trigonometry.

9. Given $a = 10$ and $B = 30°$, determine the values of b for which A has

(a) exactly one value **(b)** two possible values **(c)** no value.

10. Given $a = 10$ and $B = 150°$, determine the values of b for which A has

(a) exactly one value **(b)** two possible values **(c)** no value.

Use the law of cosines to find the indicated part of each triangle ABC.

11. $a = 86.14$ in., $b = 253.2$ in., $c = 241.9$ in.; find A

12. $B = 120.7°$, $a = 127$ ft, $c = 69.8$ ft; find b

13. $A = 51° 20'$, $c = 68.3$ m, $b = 58.2$ m; find a

14. $a = 14.8$ m, $b = 19.7$ m, $c = 31.8$ m; find B

15. $A = 60°$, $b = 5$ cm, $c = 21$ cm; find a

16. $a = 13$ ft, $b = 17$ ft, $c = 8$ ft; find A

Solve each triangle ABC having the given information.

17. $A = 25.2°$, $a = 6.92$ yd, $b = 4.82$ yd

18. $A = 61.7°$, $a = 78.9$ m, $b = 86.4$ m

19. $a = 27.6$ cm, $b = 19.8$ cm, $C = 42° 30'$

20. $a = 94.6$ yd, $b = 123$ yd, $c = 109$ yd

Find the area of each triangle ABC with the given information.

21. $b = 840.6$ m, $c = 715.9$ m, $A = 149.3°$

22. $a = 6.90$ ft, $b = 10.2$ ft, $C = 35° 10'$

23. $a = .913$ km, $b = .816$ km, $c = .582$ km

24. $a = 43$ m, $b = 32$ m, $c = 51$ m

Solve each problem.

25. ***Distance Across a Canyon*** To measure the distance AB across a canyon for a power line, a surveyor measures angles B and C and the distance BC, as shown in the figure. What is the distance from A to B?

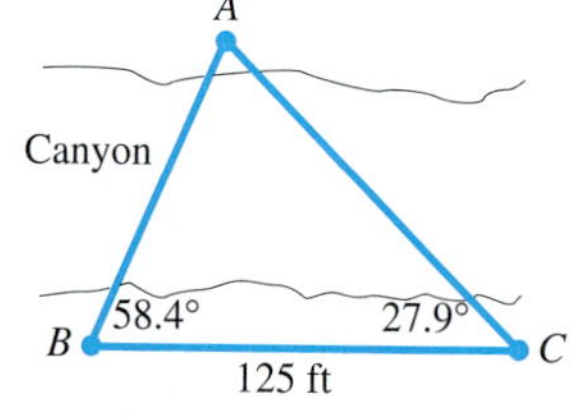

26. ***Length of a Brace*** A banner on an 8.0-ft pole is to be mounted on a building at an angle of 115°, as shown in the figure. Find the length of the brace.

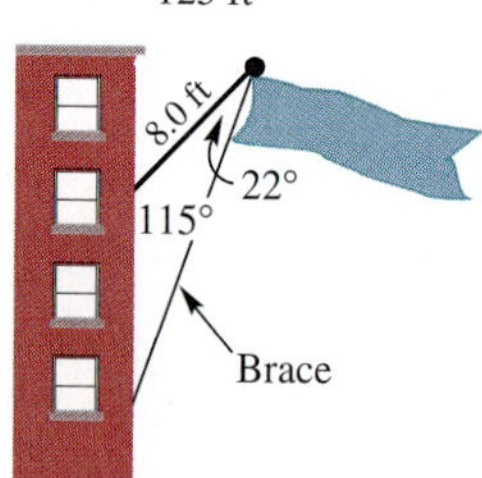

27. ***Height of a Tree*** A tree leans at an angle of 8.0° from the vertical. From a point 7.0 m from the bottom of the tree, the angle of elevation to the top of the tree is 68°. How tall is the tree?

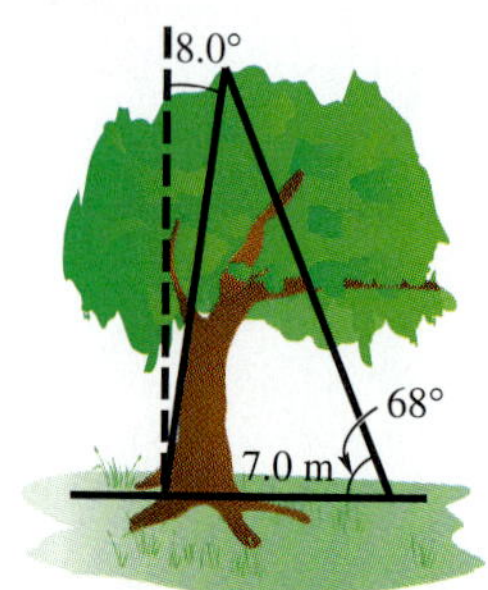

28. *Hanging Sculpture* A hanging sculpture in an art gallery is to be hung with two wires of lengths 15.0 ft and 12.2 ft so that the angle between them is 70.3°. How far apart should the ends of the wire be placed on the ceiling?

29. *Height of a Tree* A hill makes an angle of 14.3° with the horizontal. From the base of the hill, the angle of elevation to the top of a tree on top of the hill is 27.2°. The distance along the hill from the base to the tree is 212 ft. Find the height of the tree.

30. *Pipeline Position* A pipeline is to run between points A and B, which are separated by a protected wetlands area. To avoid the wetlands, the pipe will run from point A to C and then to B. The distances involved are $AB = 150$ km, $AC = 102$ km, and $BC = 135$ km. What angle should be used at point C?

31. *Distance Between Two Boats* Two boats leave a dock together. Each travels in a straight line. The angle between their courses measures 54° 10′. One boat travels 36.2 km per hr, and the other travels 45.6 km per hr. How far apart will they be after 3 hr?

32. *Distance from a Ship to a Lighthouse* A ship sailing parallel to shore sights a lighthouse at an angle of 30° from its direction of travel. After the ship travels 2.0 mi farther, the angle has increased to 55°. At that time, how far is the ship from the lighthouse?

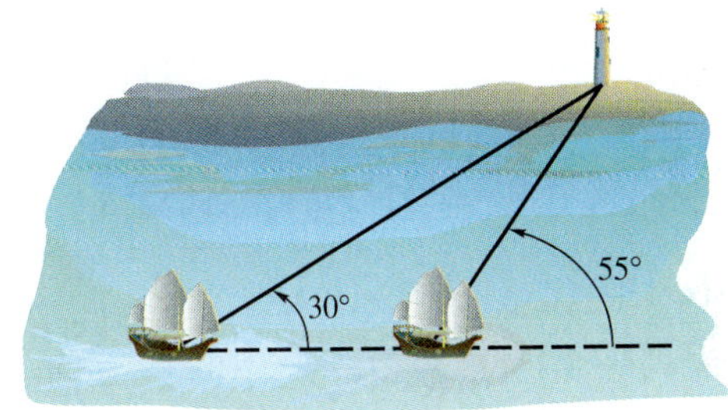

33. *Area of a Triangle* Find the area of the triangle shown in the figure using Heron's area formula.

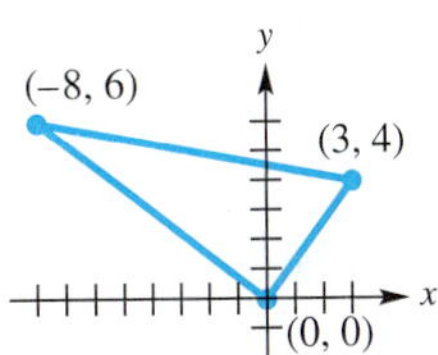

34. *Area of a Quadrilateral* A lot has the shape of the quadrilateral in the figure. What is its area?

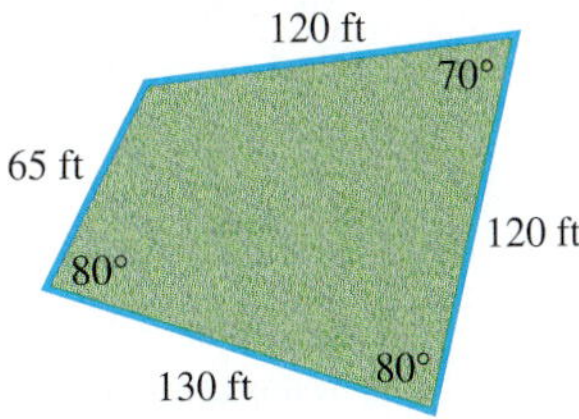

In Exercises 35 and 36, use the given vectors to sketch the following.

35. $\mathbf{a} - \mathbf{b}$

36. $\mathbf{a} + 3\mathbf{c}$

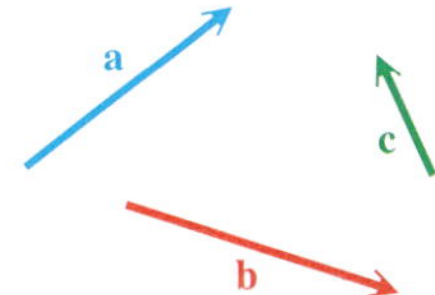

37. *Concept Check* Decide whether each statement is true or false.

(a) Opposite angles of a parallelogram are equal.

(b) A diagonal of a parallelogram must bisect two angles of the parallelogram.

Given two forces and the angle between them, find the magnitude of the resultant force.

38. forces of 142 and 215 newtons, forming an angle of 112°

39.

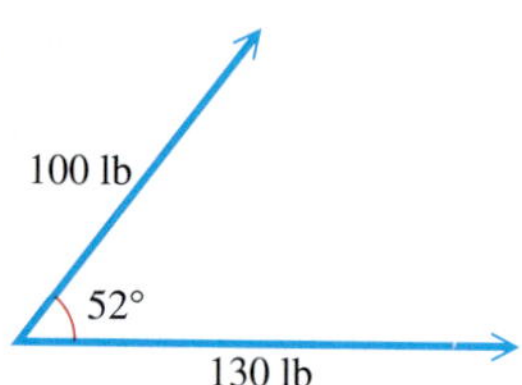

Vector **v** *has the given magnitude and direction angle. Find the magnitudes of the horizontal and vertical components of* **v**.

40. $|\mathbf{v}| = 50$, $\theta = 45°$ (Give exact values.)

41. $|\mathbf{v}| = 964$, $\theta = 154° \, 20'$

Find the magnitude and direction angle for **u** *rounded to the nearest tenth.*

42. $\mathbf{u} = \langle 21, -20 \rangle$

43. $\mathbf{u} = \langle -9, 12 \rangle$

44. Let $\mathbf{v} = 2\mathbf{i} - \mathbf{j}$ and $\mathbf{u} = -3\mathbf{i} + 2\mathbf{j}$. Express each in terms of **i** and **j**.

(a) $2\mathbf{v} + \mathbf{u}$ **(b)** $2\mathbf{v}$ **(c)** $\mathbf{v} - 3\mathbf{u}$

45. Let $\mathbf{a} = \langle 3, -2 \rangle$ and $\mathbf{b} = \langle -1, 3 \rangle$. Find $\mathbf{a} \cdot \mathbf{b}$ and the angle between **a** and **b**, rounded to the nearest tenth of a degree.

Find the vector of magnitude 1 *having the same direction angle as the given vector.*

46. $\mathbf{u} = \langle -4, 3 \rangle$

47. $\mathbf{u} = \langle 5, 12 \rangle$

Solve each problem.

48. ***Force Placed on a Barge*** One rope pulls a barge directly east with a force of 100 newtons. Another rope pulls the barge to the northeast with a force of 200 newtons. Find the resultant force acting on the barge and the angle between the resultant and the first rope.

49. ***Weight of a Sled and Passenger*** Paula and Steve are pulling their daughter Jessie on a sled. Steve pulls with a force of 18 lb at an angle of 10°. Paula pulls with a force of 12 lb at an angle of 15°. Find the magnitude of the resultant force on Jessie and the sled.

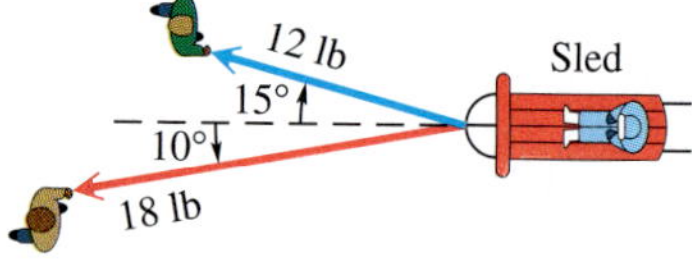

50. ***Angle of a Hill*** A 186-lb force just keeps a 2800-lb car from rolling down a hill. What angle does the hill make with the horizontal?

51. ***Direction and Speed of a Plane*** A plane has an airspeed of 520 mph. The pilot wishes to fly on a bearing of 310°. A wind of 37 mph is blowing from a bearing of 212°. What direction should the pilot fly, and what will be her actual speed?

52. ***Speed and Direction of a Boat*** A boat travels 15 km per hr in still water. The boat is traveling across a large river, on a bearing of 130°. The current in the river, coming from the west, has a speed of 7 km per hr. Find the resulting speed of the boat and its resulting direction of travel.

53. ***Control Points*** To obtain accurate aerial photographs, ground control must determine the coordinates of *control points* located on the ground that can be identified in the photographs. Using these known control points, the orientation and scale of each photograph can be found. Then, unknown positions and distances can easily be determined. The figure shows three consecutive control points A, B, and C.

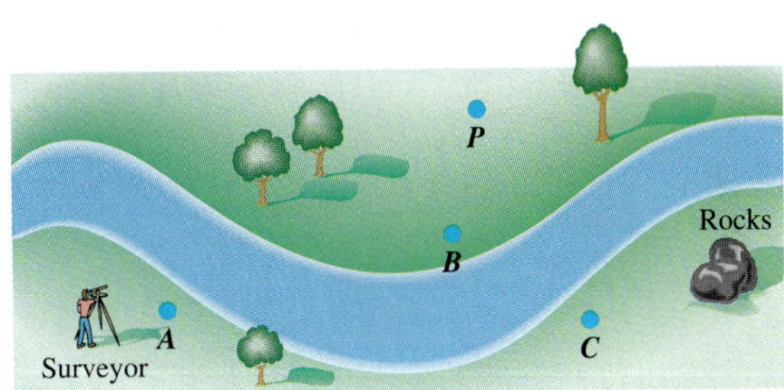

A surveyor measures a baseline distance of 92.13 ft from B to an arbitrary point P. Angles BAP and BCP are found to be $2° 22' 47''$ and $5° 13' 11''$, respectively. Then, angles APB and CPB are determined to be $63° 4' 25''$ and $74° 19' 49''$, respectively. Determine the distance between control points A and B and between B and C. (*Source:* Moffitt, F. and E. Mikhail, *Photogrammetry,* Third Edition, Harper & Row, 1980.)

Perform each operation. Write answers in rectangular form.

54. $[5(\cos 90° + i \sin 90°)][6(\cos 180° + i \sin 180°)]$

55. $[3 \text{ cis } 135°][2 \text{ cis } 105°]$

56. $\dfrac{2(\cos 60° + i \sin 60°)}{8(\cos 300° + i \sin 300°)}$

57. $\dfrac{4 \text{ cis } 270°}{2 \text{ cis } 90°}$

58. $(2 - 2i)^5$

59. $(\cos 100° + i \sin 100°)^6$

60. *Concept Check* The vector representing a real number will lie on the ______-axis in the complex plane.

Graph each complex number as a vector.

61. $5i$

62. $-4 + 2i$

In Exercises 63–69, perform each conversion.

	Rectangular Form	Trigonometric Form
63.	$-2 + 2i$	______
64.	______	$3(\cos 90° + i \sin 90°)$
65.	______	$2(\cos 225° + i \sin 225°)$
66.	$-4 + 4i\sqrt{3}$	______
67.	$1 - i$	______
68.	______	$4 \text{ cis } 240°$
69.	$-4i$	______

Concept Check The complex number z, where $z = x + yi$, can be graphed in the plane as (x, y). Describe the graphs of all complex numbers z satisfying the conditions in Exercises 70 and 71.

70. The absolute value of z is 2.

71. The imaginary part of z is the negative of the real part of z.

Find all roots as indicated. Express them in trigonometric form.

72. the fifth roots of $-2 + 2i$

73. the cube roots of $1 - i$

74. *Concept Check* How many real fifth roots does -32 have?

75. *Concept Check* How many real sixth roots does -64 have?

Solve each equation. Leave answers in trigonometric form.

76. $x^3 + 125 = 0$

77. $x^4 + 16 = 0$

78. Convert $(-1, \sqrt{3})$ to polar coordinates, with $0° \le \theta < 360°$ and $r > 0$.

79. Convert $(5, 315°)$ to rectangular coordinates.

80. *Concept Check* What will the graph of $r = k$ be, for $k > 0$?

Identify and graph each polar equation for θ in $[0°, 360°)$.

81. $r = 4 \cos \theta$

82. $r = -1 + \cos \theta$

83. $r = 2 \sin 4\theta$

84. $r = \dfrac{2}{2 \cos \theta - \sin \theta}$

Find an equivalent equation in rectangular coordinates.

85. $r = \dfrac{3}{1 + \cos \theta}$

86. $r = \sin \theta + \cos \theta$

87. $r = 2$

Find an equivalent equation in polar coordinates.

88. $y = x$

89. $y = x^2$

In Exercises 90–93, find a polar equation having the given graph.

90.

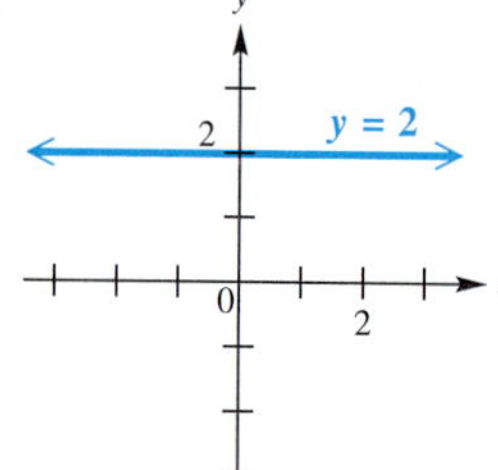

91.

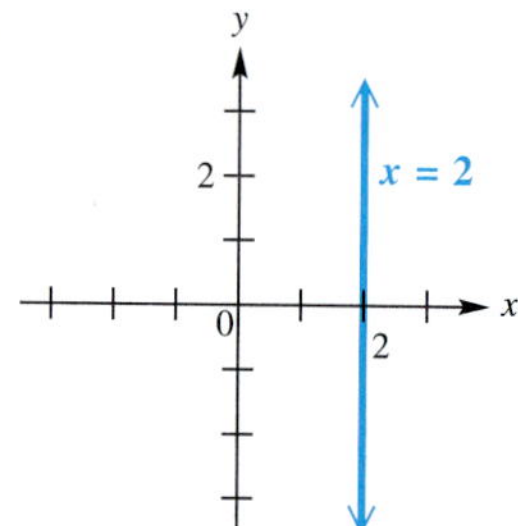

92.

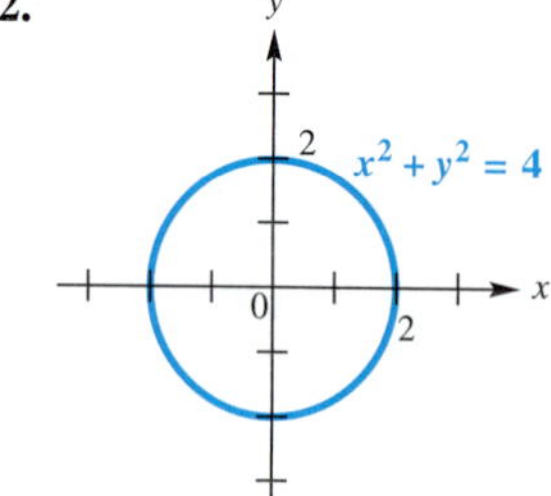

93.

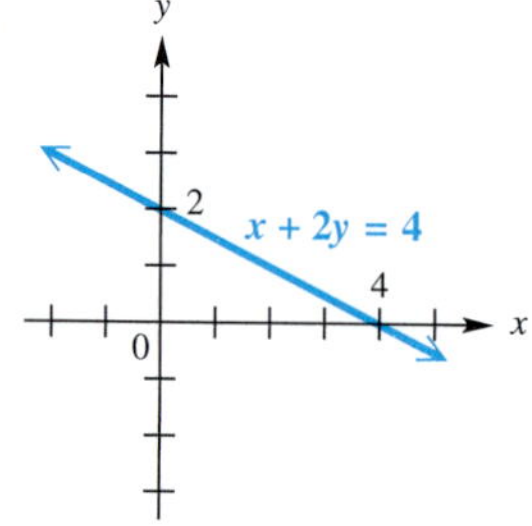

94. Show that the distance between (r_1, θ_1) and (r_2, θ_2) in polar coordinates is given by

$$d = \sqrt{r_1^2 + r_2^2 - 2r_1r_2\cos(\theta_1 - \theta_2)}.$$

95. Graph the plane curve defined by the parametric equations $x = t + \cos t$, $y = \sin t$, for t in $[0, 2\pi]$.

Find a rectangular equation for each plane curve with the given parametric equations.

96. $x = 3t + 2, y = t - 1,$ for t in $[-5, 5]$

97. $x = \sqrt{t - 1}, y = \sqrt{t},$ for t in $[1, \infty)$

98. $x = t^2 + 5, y = \dfrac{1}{t^2 + 1},$ for t in $(-\infty, \infty)$

99. $x = 5 \tan t, y = 3 \sec t,$ for t in $\left(-\dfrac{\pi}{2}, \dfrac{\pi}{2}\right)$

100. $x = \cos 2t, y = \sin t,$ for t in $(-\pi, \pi)$

101. Find a pair of parametric equations whose graph is the circle having center $(3, 4)$ and passing through the origin.

102. *Mandelbrot Set* Follow the steps in Exercise 44 of Section 8.5 to show that the graph of the Mandelbrot set in Exercise 49 of Section 8.6 is symmetric with respect to the x-axis.

103. *Flight of a Baseball* Manny hits a baseball when it is 3.2 ft above the ground. It leaves the bat with velocity 118 ft per sec at an angle of 27° with respect to the ground.

(a) Determine the parametric equations that model the path of the baseball.

(b) Determine the rectangular equation that models the path of the baseball.

(c) Determine how long the projectile is in flight and the horizontal distance covered.

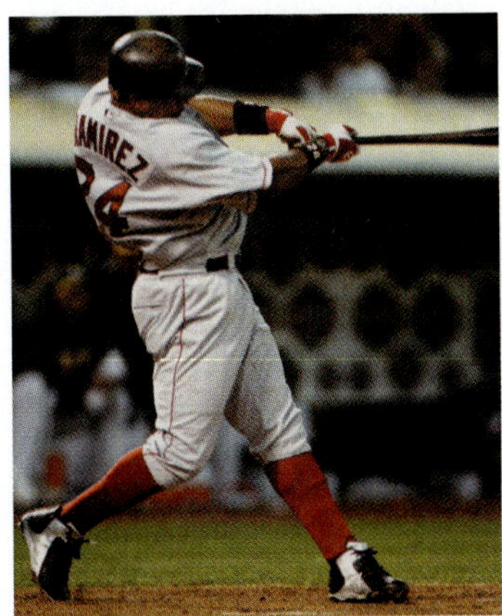

Chapter 8 Test

Find the indicated part of each triangle ABC.

1. $A = 25.2°, a = 6.92$ yd, $b = 4.82$ yd; find C

2. $C = 118°, b = 130$ km, $a = 75$ km; find c

3. $a = 17.3$ ft, $b = 22.6$ ft, $c = 29.8$ ft; find B

4. Find the area of triangle ABC in Exercise 2.

5. Given $a = 10$ and $B = 150°$ in triangle ABC, determine the values of b for which A has

(a) exactly one value **(b)** two possible values **(c)** no value.

6. Find the area of the triangle having sides of lengths 22, 26, and 40.

7. Find the magnitude and the direction angle for the vector shown in the figure.

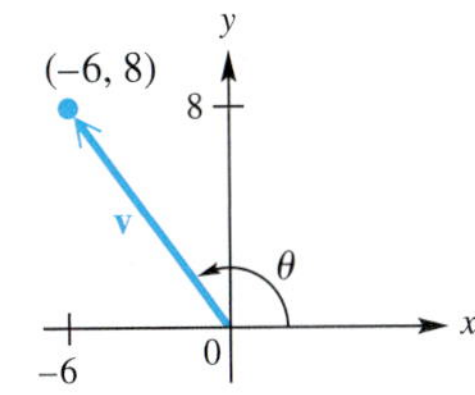

8. For the vectors $\mathbf{u} = \langle -1, 3 \rangle$ and $\mathbf{v} = \langle 2, -6 \rangle$, find each of the following.

(a) $\mathbf{u} + \mathbf{v}$ **(b)** $-3\mathbf{v}$ **(c)** $\mathbf{u} \cdot \mathbf{v}$ **(d)** $|\mathbf{u}|$

Solve each problem.

9. ***Height of a Balloon*** The angles of elevation of a balloon from two points A and B on level ground are $24° \, 50'$ and $47° \, 20'$, respectively. As shown in the figure, points A, B, and C are in the same vertical plane and points A and B are 8.4 mi apart. Approximate the height of the balloon above the ground to the nearest tenth of a mile.

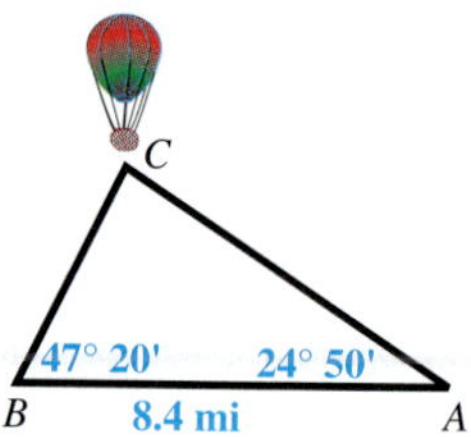

10. ***Horizontal and Vertical Components*** Find the horizontal and vertical components of the vector with magnitude 569 that is inclined 127.5° from the horizontal. Give your answer in the form $\langle a, b \rangle$.

11. ***Radio Direction Finders*** Radio direction finders are placed at points A and B, which are 3.46 mi apart on an east-west line, with A west of B. From A, the bearing of a certain illegal pirate radio transmitter is 48°, and from B the bearing is 302°. Find the distance between the transmitter and A to the nearest hundredth of a mile.

12. Write each complex number in trigonometric (polar) form, where $0° \le \theta < 360°$.

(a) $3i$ **(b)** $1 + 2i$ **(c)** $-1 - \sqrt{3}i$

13. Write each complex number in rectangular form.

(a) $3(\cos 30° + i \sin 30°)$ **(b)** $4 \text{ cis } 40°$ **(c)** $3(\cos 90° + i \sin 90°)$

14. For the complex numbers $w = 8(\cos 40° + i \sin 40°)$ and $z = 2(\cos 10° + i \sin 10°)$, find each of the following in the form specified.

(a) wz (trigonometric form) **(b)** $\frac{w}{z}$ (rectangular form) **(c)** z^3 (rectangular form)

15. Find the four complex fourth roots of $-16i$. Express them in trigonometric form.

16. Convert the given rectangular coordinates to polar coordinates. Give two pairs of polar coordinates for each point.

(a) $(0, 5)$ **(b)** $(-2, -2)$

17. Convert the given polar coordinates to rectangular coordinates.

(a) $(3, 315°)$ **(b)** $(-4, 90°)$

Identify and graph each polar equation for θ in $[0°, 360°)$.

18. $r = 1 - \cos \theta$ **19.** $r = 3 \cos 3\theta$

20. Convert the polar equation to a rectangular equation, and sketch its graph.

(a) $r = \dfrac{4}{2 \sin \theta - \cos \theta}$ **(b)** $r = 6$

Graph each pair of parametric equations.

21. $x = 4t - 3, y = t^2$, for t in $[-3, 4]$

22. $x = 2 \cos 2t, y = 2 \sin 2t$, for t in $[0, 2\pi]$

Chapter 8 Quantitative Reasoning

Just how much does the U.S. flag "show its colors"?

The flag of the United States includes the colors red, white, and blue. Which color is predominant? Clearly the answer is either red or white. (It can be shown that only 18.73% of the total area is blue.) (*Source: Slicing Pizzas, Racing Turtles, and Further Adventures in Applied Mathematics,* Banks, R., Princeton University Press, 1999.)

1. Let R denote the radius of the circumscribing circle of a five-pointed star appearing on the American flag. The star can be decomposed into ten congruent triangles. In the figure, r is the radius of the circumscribing circle of the pentagon in the interior of the star. Show that the area of a star is

$$A = \left[5\frac{\sin A \sin B}{\sin(A + B)}\right]R^2.$$

(*Hint:* $\sin C = \sin[180° - (A + B)] = \sin(A + B)$.)

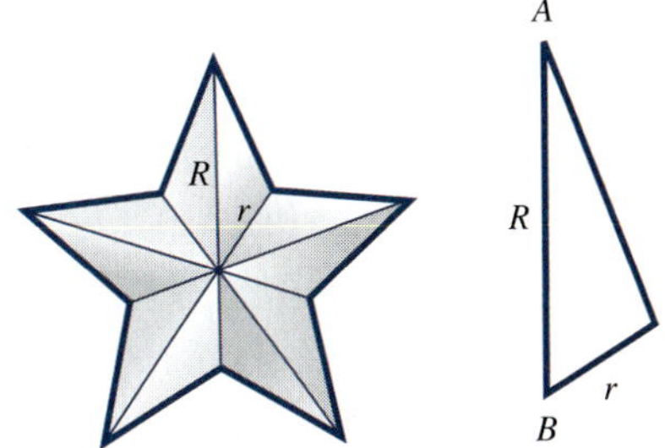

2. Angles A and B have values 18° and 36°, respectively. Express the area of a star in terms of its radius, R.

3. To determine whether red or white is predominant, we must know the measurements of the flag. Consider a flag of width 10 in., length 19 in., length of each upper stripe 11.4 in., and radius R of the circumscribing circle of each star .308 in. The thirteen stripes consist of six matching pairs of red and white stripes and one additional red, upper stripe. Therefore, we must compare the area of a red, upper stripe with the total area of the 50 white stars.

(a) Compute the area of the red, upper stripe.
(b) Compute the total area of the 50 white stars.
(c) Which color occupies the greatest area on the flag?

9

Systems and Matrices

To make predictions and forecasts about the future, professionals in many fields attempt to determine relationships between different factors. When quantities are interrelated, *systems of equations* in several variables are used to describe their relationship. As early as 4000 B.C. in Mesopotamia, people were able to solve up to 10 equations having 10 variables. In 1940 John Atanasoff, a physicist from Iowa State University, needed to solve a system of equations containing 29 equations and 29 variables. This led to the invention of the first fully electronic digital computer, dubbed ABC for Atanasoff-Berry Computer. (*Source: The Gazette,* Jan. 1, 1999.)

Systems can be used to predict future populations. In Exercises 59–62 of Section 9.2, we do this for the pronghorn antelope.

9.1 Systems of Linear Equations

Linear Systems ▪ Substitution Method ▪ Elimination Method ▪ Special Systems ▪ Applying Systems of Equations ▪ Solving Linear Systems with Three Unknowns (Variables) ▪ Using Systems of Equations to Model Data

Linear Systems The definition of a linear equation given in Chapter 1 can be extended to more variables; any equation of the form

$$a_1x_1 + a_2x_2 + \cdots + a_nx_n = b,$$

for real numbers $a_1, a_2, \ldots, a_n$ (not all of which are 0) and b, is a **linear equation** or a **first-degree equation in n unknowns.**

A set of equations is called a **system of equations.** The **solutions** of a system of equations must satisfy every equation in the system. If all the equations in a system are linear, the system is a **system of linear equations,** or a **linear system.**

The solution set of a linear equation in two unknowns (or variables) is an infinite set of ordered pairs. Since the graph of such an equation is a straight line, there are three possibilities for the solution set of a system of two linear equations in two unknowns, as shown in Figure 1. The possible graphs of a linear system in two unknowns are as follows.

1. The graphs of the two equations intersect in a single point. The coordinates of this point give the only solution of the system. In this case, the system is **consistent.** This is the most common case. See Figure 1(a).
2. The graphs are distinct parallel lines. In this case, the system is said to be **inconsistent** and the equations are **independent.** That is, there is no solution common to both equations. See Figure 1(b).
3. The graphs are the same line. In this case, the equations are said to be **dependent,** and any solution of one equation is also a solution of the other. Thus, there are infinitely many solutions. See Figure 1(c).

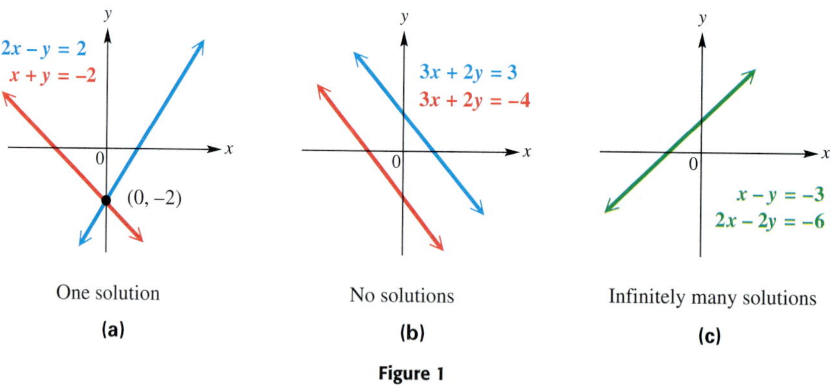

Figure 1

In this section, we introduce two algebraic methods for solving systems: *substitution* and *elimination.*

Substitution Method In a system of two equations with two variables, the **substitution method** involves using one equation to find an expression for one variable in terms of the other, then substituting into the other equation of the system.

EXAMPLE 1 Solving a System by Substitution

Solve the system.

$$3x + 2y = 11 \quad (1)$$
$$-x + y = 3 \quad (2)$$

Solution Begin by solving one of the equations for one of the variables. We solve equation (2) for y.

$$-x + y = 3 \quad (2)$$
$$y = x + 3 \quad \text{Add } x. \text{ (Section 1.1)} \quad (3)$$

Now replace y with $x + 3$ in equation (1), and solve for x.

$$3x + 2y = 11 \quad (1)$$
$$3x + 2(x + 3) = 11 \quad \text{Let } y = x + 3 \text{ in (1).}$$
$$3x + 2x + 6 = 11 \quad \text{Distributive property (Section R.1)}$$
$$5x + 6 = 11 \quad \text{Combine terms.}$$
$$5x = 5 \quad \text{Subtract 6.}$$
$$x = 1 \quad \text{Divide by 5.}$$

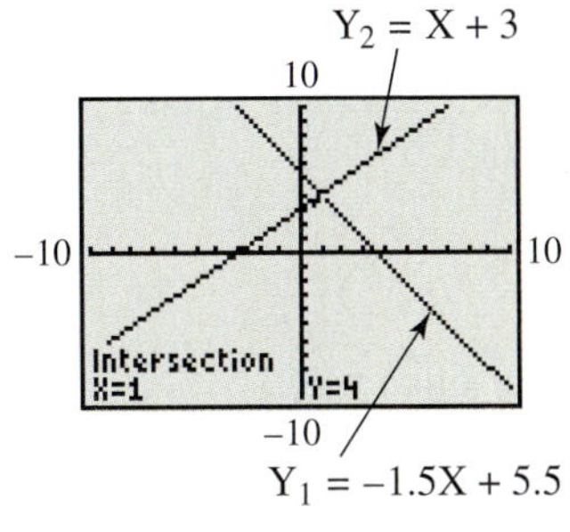

To solve the system in Example 1 graphically, solve both equations for y:

$3x + 2y = 11$ leads to

$$Y_1 = -1.5X + 5.5.$$

$-x + y = 3$ leads to

$$Y_2 = X + 3.$$

Graph both Y_1 and Y_2 in the standard window to find that their point of intersection is $(1, 4)$.

Replace x with 1 in equation (3) to obtain $y = 1 + 3 = 4$. The solution of the system is the ordered pair $(1, 4)$. Check this solution in *both* equations (1) and (2).

Check:

$3x + 2y = 11$	(1)	$-x + y = 3$	(2)
$3(1) + 2(4) = 11$	?	$-1 + 4 = 3$	?
$11 = 11$	True	$3 = 3$	True

Both check; the solution set is $\{(1, 4)\}$.

Now try Exercise 7.

Elimination Method Another way to solve a system of two equations, called the **elimination method,** uses multiplication and addition to eliminate a variable from one equation. To eliminate a variable, the coefficients of that variable in the two equations must be additive inverses. To achieve this, we use properties of algebra to change the system to an **equivalent system,** one with the same solution set. The three transformations that produce an equivalent system are listed here.

Transformations of a Linear System

1. Interchange any two equations of the system.
2. Multiply or divide any equation of the system by a nonzero real number.
3. Replace any equation of the system by the sum of that equation and a multiple of another equation in the system.

EXAMPLE 2 Solving a System by Elimination

Solve the system.

$$3x - 4y = 1 \quad (1)$$
$$2x + 3y = 12 \quad (2)$$

Solution One way to eliminate a variable is to use the second transformation and multiply both sides of equation (2) by -3, giving the equivalent system

$$3x - 4y = 1 \quad (1)$$
$$-6x - 9y = -36. \quad \text{Multiply (2) by } -3. \quad (3)$$

Now multiply both sides of equation (1) by 2, and use the third transformation to add the result to equation (3), eliminating x. Solve the result for y.

$$\begin{aligned} 6x - 8y &= 2 && \text{Multiply (1) by 2.} \\ -6x - 9y &= -36 && (3) \\ \hline -17y &= -34 && \text{Add.} \\ y &= 2 && \text{Solve for } y. \end{aligned}$$

Substitute 2 for y in either of the original equations and solve for x.

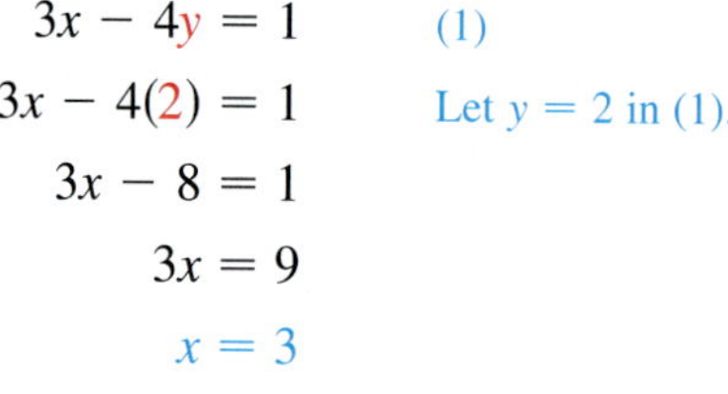

$$\begin{aligned} 3x - 4y &= 1 && (1) \\ 3x - 4(2) &= 1 && \text{Let } y = 2 \text{ in (1).} \\ 3x - 8 &= 1 \\ 3x &= 9 \\ x &= 3 \end{aligned}$$

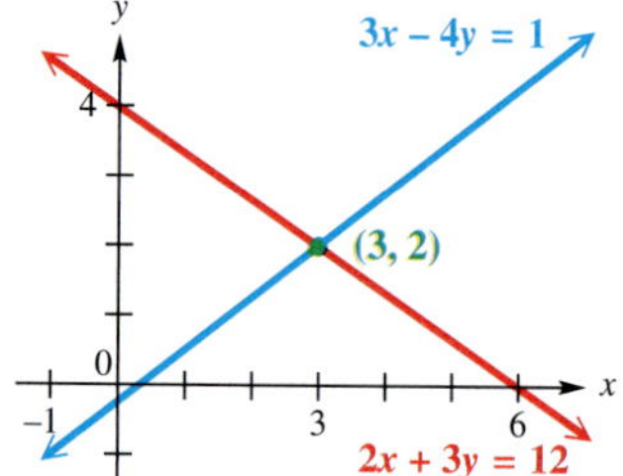

Figure 2

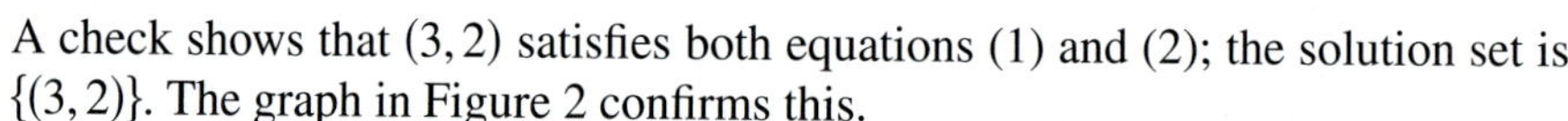

A check shows that $(3, 2)$ satisfies both equations (1) and (2); the solution set is $\{(3, 2)\}$. The graph in Figure 2 confirms this.

Now try Exercise 21.

Special Systems The systems in Examples 1 and 2 were both consistent, having a single solution. This is not always the case.

EXAMPLE 3 Solving an Inconsistent System

Solve the system.

$$3x - 2y = 4 \quad (1)$$
$$-6x + 4y = 7 \quad (2)$$

Solution To eliminate the variable x, multiply both sides of equation (1) by 2.

$$\begin{aligned} 6x - 4y &= 8 && \text{Multiply (1) by 2.} \\ -6x + 4y &= 7 && (2) \\ \hline 0 &= 15 && \text{False} \end{aligned}$$

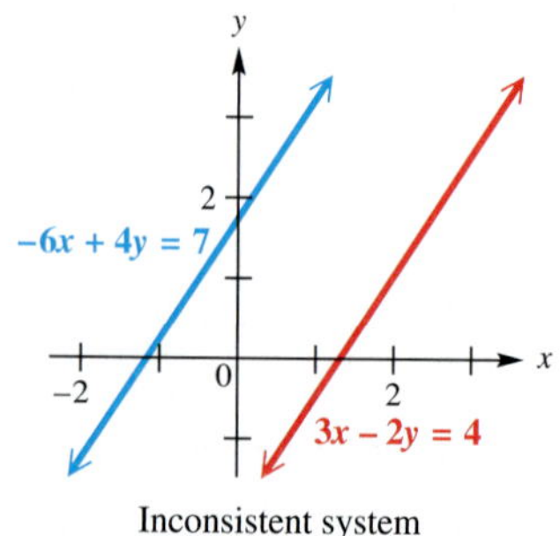

Inconsistent system

Figure 3

Since $0 = 15$ is false, the system is inconsistent and has no solution. As suggested by Figure 3, this means that the graphs of the equations of the system never intersect. (The lines are parallel.) The solution set is $\emptyset$, the empty set.

Now try Exercise 31.

EXAMPLE 4 Solving a System with Infinitely Many Solutions

Solve the system.

$$\begin{aligned} 8x - 2y &= -4 \quad (1) \\ -4x + y &= 2 \quad (2) \end{aligned}$$

Algebraic Solution

Divide both sides of equation (1) by 2, and add the result to equation (2).

$$\begin{aligned} 4x - y &= -2 \quad \text{Divide (1) by 2.} \\ -4x + y &= 2 \quad (2) \\ \hline 0 &= 0 \quad \text{True} \end{aligned}$$

The result, $0 = 0$, is a true statement, which indicates that the equations of the original system are equivalent. Any ordered pair (x, y) that satisfies either equation will satisfy the system. From equation (2),

$$\begin{aligned} -4x + y &= 2 \quad (2) \\ y &= 2 + 4x. \end{aligned}$$

The solutions of the system can be written in the form of a set of ordered pairs $(x, 2 + 4x)$, for any real number x. Some ordered pairs in the solution set are $(0, 2 + 4 \cdot 0) = (0, 2)$, $(1, 2 + 4 \cdot 1) = (1, 6)$, $(3, 14)$, and $(-2, -6)$. As shown in Figure 4, the equations of the original system are dependent and lead to the same straight line graph. The solution set is written $\{(x, 2 + 4x)\}$.

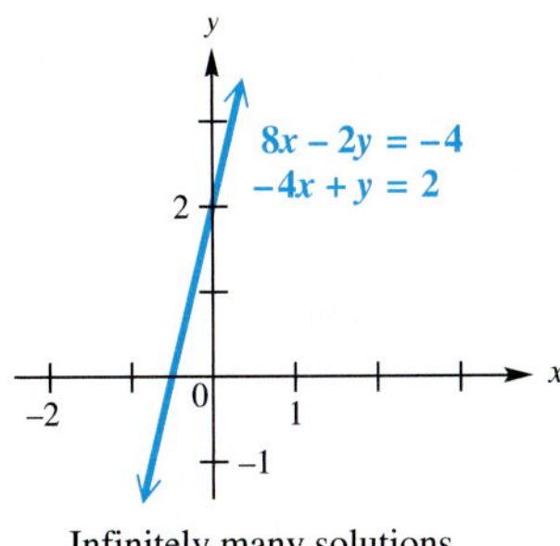

Figure 4

Graphing Calculator Solution

Solving the equations for y gives

$$Y_1 = \frac{8X + 4}{2}$$

and $$Y_2 = 2 + 4X.$$

The graphs of the two equations coincide, as seen in the top screen in Figure 5. The table indicates that $Y_1 = Y_2$ for arbitrarily selected values of X, providing another way to show that the two equations lead to the same graph.

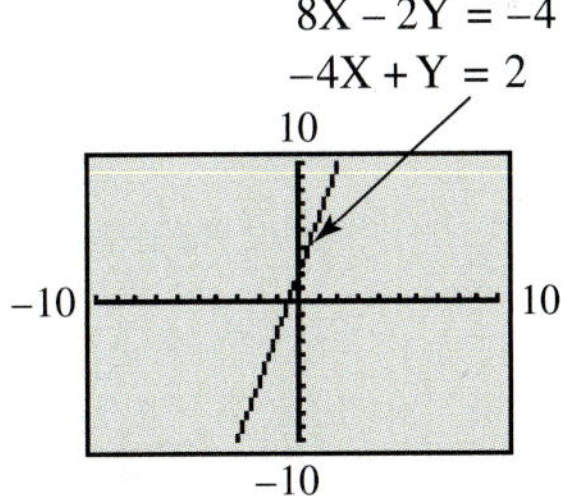

X	Y1	Y2
0	2	2
1	6	6
3	14	14
-2	-6	-6

Y1=(8X+4)/2

Figure 5

Refer to the algebraic solution to see how the solution set can be written using an arbitrary variable.

Now try Exercise 33.

NOTE In the algebraic solution for Example 4, we wrote the solution set with the variable x arbitrary. We could write the solution set with y arbitrary: $\left\{\left(\frac{y-2}{4}, y\right)\right\}$. By selecting values for y and solving for x in this ordered pair, we can find individual solutions. Verify again that $(0, 2)$ is a solution by letting $y = 2$ and solving for x to obtain $\frac{2-2}{4} = 0$.

Applying Systems of Equations Many applied problems involve more than one unknown quantity. Although some problems with two unknowns can be solved using just one variable, it is often easier to use two variables. To solve a problem with two unknowns, we must write two equations that relate the unknown quantities. The system formed by the pair of equations can then be solved using the methods of this chapter.

The following steps, based on the six-step problem-solving method first introduced in Chapter 1, give a strategy for solving applied problems using more than one variable.

Solving an Applied Problem by Writing a System of Equations

Step 1 **Read** the problem carefully until you understand what is given and what is to be found.

Step 2 **Assign variables** to represent the unknown values, using diagrams or tables as needed. *Write down* what each variable represents.

Step 3 **Write a system of equations** that relates the unknowns.

Step 4 **Solve** the system of equations.

Step 5 **State the answer** to the problem. Does it seem reasonable?

Step 6 **Check** the answer in the words of the original problem.

EXAMPLE 5 Using a Linear System to Solve an Application

Title IX legislation prohibits sex discrimination in sports programs. In 1997 the national average spent on two varsity athletes, one female and one male, was \$6050 for Division I-A schools. However, average expenditures for a male athlete exceeded those for a female athlete by \$3900. Determine how much was spent per varsity athlete for each gender. (*Source: USA Today.*)

Solution

Step 1 **Read** the problem. We must find the amount spent per varsity athlete for both male and female athletes.

Step 2 **Assign variables.** Let x represent average expenditures per male athlete and y represent average expenditures per female athlete.

Step 3 **Write a system of equations.** Since the average amount spent on one female and one male athlete was \$6050, one equation is

$$\frac{x + y}{2} = 6050.$$

The expenditures for a male athlete exceeded those for a female athlete by \$3900. Thus, $x - y = 3900$, which gives the system of equations

$$\frac{x + y}{2} = 6050 \quad (1)$$

$$x - y = 3900. \quad (2)$$

Step 4 **Solve** the system. To eliminate y, multiply equation (1) by 2 and then add the resulting equations.

$$\begin{aligned} x + y &= 12{,}100 && \text{Multiply (1) by 2.} \\ x - y &= 3\,900 && (2) \\ \hline 2x &= 16{,}000 && \text{Add.} \\ x &= 8000 && \text{Solve for } x. \end{aligned}$$

To find y, substitute 8000 for x in equation (2).

$$\begin{aligned} 8000 - y &= 3900 && \text{Let } x = 8000 \text{ in (2).} \\ -y &= -4100 && \text{Subtract 8000.} \\ y &= 4100 && \text{Multiply by } -1. \end{aligned}$$

Step 5 **State the answer.** The average expenditure for each male was \$8000 and for each female was \$4100.

Step 6 **Check.** The average of \$8000 and \$4100 is

$$\frac{\$8000 + \$4100}{2} = \$6050.$$

Also, \$8000 is \$3900 more than \$4100, as required. The answer checks.

Now try Exercise 89.

Solving Linear Systems with Three Unknowns (Variables) Earlier, we saw that the graph of a linear equation in two unknowns is a straight line. The graph of a linear equation in three unknowns requires a three-dimensional coordinate system. The three number lines are placed at right angles. The graph of a linear equation in three unknowns is a plane. Some possible intersections of planes representing three equations in three variables are shown in Figure 6.

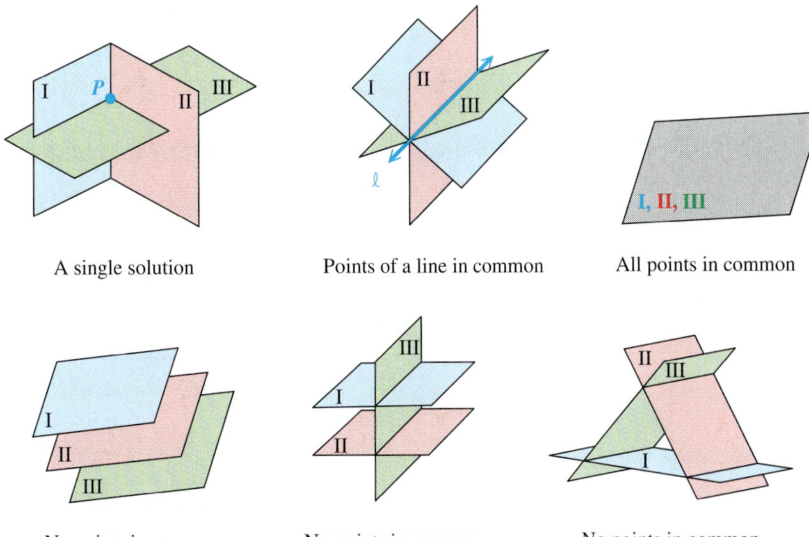

Figure 6

To solve a linear system with three unknowns, first eliminate a variable from any two of the equations. Then eliminate the *same variable* from a different pair of equations. Eliminate a second variable using the resulting two equations in two variables to get an equation with just one variable whose value you can now determine. Find the values of the remaining variables by substitution. Solutions of the system are written as **ordered triples.**

EXAMPLE 6 Solving a System of Three Equations with Three Variables

Solve the system.

$$\begin{aligned} 3x + 9y + 6z &= 3 \quad (1)\\ 2x + y - z &= 2 \quad (2)\\ x + y + z &= 2 \quad (3) \end{aligned}$$

Solution Eliminate z by adding equations (2) and (3) to get

$$3x + 2y = 4. \quad (4)$$

To eliminate z from another pair of equations, multiply both sides of equation (2) by 6 and add the result to equation (1).

$$\begin{aligned} 12x + 6y - 6z &= 12 \quad \text{Multiply (2) by 6.}\\ 3x + 9y + 6z &= 3 \quad (1)\\ \hline 15x + 15y \qquad &= 15 \quad (5) \end{aligned}$$

To eliminate x from equations (4) and (5), multiply both sides of equation (4) by -5 and add the result to equation (5). Solve the resulting equation for y.

$$\begin{aligned} -15x - 10y &= -20 \quad \text{Multiply (4) by } -5.\\ 15x + 15y &= 15 \quad (5)\\ \hline 5y &= -5\\ y &= -1 \end{aligned}$$

Using $y = -1$, find x from equation (4) by substitution.

$$\begin{aligned} 3x + 2(-1) &= 4 \quad \text{(4) with } y = -1\\ x &= 2 \end{aligned}$$

Substitute 2 for x and -1 for y in equation (3) to find z.

$$\begin{aligned} 2 + (-1) + z &= 2 \quad (3)\\ z &= 1 \end{aligned}$$

Verify that the ordered triple $(2, -1, 1)$ satisfies all three equations in the *original* system. The solution set is $\{(2, -1, 1)\}$.

Now try Exercise 47.

CAUTION Be careful not to end up with two equations that still have *three* variables. Remember to *eliminate the same variable* from each pair of equations.

EXAMPLE 7 Solving a System of Two Equations with Three Variables

Solve the system.

$$x + 2y + z = 4 \quad (1)$$
$$3x - y - 4z = -9 \quad (2)$$

Solution Geometrically, the solution is the intersection of the two planes given by equations (1) and (2). The intersection of two different nonparallel planes is a line. Thus there will be an infinite number of ordered triples in the solution set, representing the points on the line of intersection.

To eliminate x, multiply both sides of equation (1) by -3 and add the result to equation (2). (Either y or z could have been eliminated instead.)

$$-3x - 6y - 3z = -12 \quad \text{Multiply (1) by } -3.$$
$$3x - y - 4z = -9 \quad (2)$$
$$-7y - 7z = -21 \quad (3)$$

Now solve equation (3) for z.

$$-7y - 7z = -21 \quad (3)$$
$$-7z = 7y - 21$$
$$z = -y + 3$$

This gives z in terms of y. Express x also in terms of y by solving equation (1) for x and substituting $-y + 3$ for z in the result.

$$x + 2y + z = 4 \quad (1)$$
$$x = -2y - z + 4 \quad \text{Solve for } x.$$
$$x = -2y - (-y + 3) + 4 \quad \text{Substitute } -y + 3 \text{ for } z.$$
$$x = -y + 1 \quad \text{Simplify.}$$

The system has an infinite number of solutions. For any value of y, the value of z is $-y + 3$ and the value of x is $-y + 1$. For example, if $y = 1$, then $x = -1 + 1 = 0$ and $z = -1 + 3 = 2$, giving the solution $(0, 1, 2)$. Verify that another solution is $(-1, 2, 1)$.

With y arbitrary, the solution set is of the form $\{(-y + 1, y, -y + 3)\}$.

Now try Exercise 59.

NOTE Had we solved equation (3) in Example 7 for y instead of z, the solution would have had a different form but would have led to the same set of solutions. In that case we would have z arbitrary, and the solution set would be of the form $\{(-2 + z, 3 - z, z)\}$. By choosing $z = 2$, one solution would be $(0, 1, 2)$, which was found above.

Using Systems of Equations to Model Data Applications with three unknowns usually require solving a system of three equations. We can find the equation of a parabola in the form

$$y = ax^2 + bx + c \quad \text{(Section 3.1)}$$

by solving a system of three equations with three variables.

EXAMPLE 8 Using Curve Fitting to Find an Equation Through Three Points

Find the equation of the parabola $y = ax^2 + bx + c$ that passes through the points $(2, 4)$, $(-1, 1)$, and $(-2, 5)$.

Solution Since the three points lie on the graph of the equation $y = ax^2 + bx + c$, they must satisfy the equation. Substituting each ordered pair into the equation gives three equations with three variables.

$$4 = a(2)^2 + b(2) + c \quad \text{or} \quad 4 = 4a + 2b + c \qquad (1)$$
$$1 = a(-1)^2 + b(-1) + c \quad \text{or} \quad 1 = a - b + c \qquad (2)$$
$$5 = a(-2)^2 + b(-2) + c \quad \text{or} \quad 5 = 4a - 2b + c \qquad (3)$$

To solve this system, first eliminate c using equations (1) and (2).

$$\begin{aligned} 4 &= 4a + 2b + c && (1) \\ -1 &= -a + b - c && \text{Multiply (2) by } -1. \\ \hline 3 &= 3a + 3b && (4) \end{aligned}$$

Now, use equations (2) and (3) to eliminate the same variable, c.

$$\begin{aligned} 1 &= a - b + c && (2) \\ -5 &= -4a + 2b - c && \text{Multiply (3) by } -1. \\ \hline -4 &= -3a + b && (5) \end{aligned}$$

Solve the system of equations (4) and (5) in two variables by eliminating a.

$$\begin{aligned} 3 &= 3a + 3b && (4) \\ -4 &= -3a + b && (5) \\ \hline -1 &= 4b && \text{Add.} \\ -\frac{1}{4} &= b \end{aligned}$$

Find a by substituting $-\frac{1}{4}$ for b in equation (4).

$$\begin{aligned} 1 &= a + b && \text{Equation (4) divided by 3} \\ 1 &= a - \frac{1}{4} && \text{Let } b = -\tfrac{1}{4}. \\ \frac{5}{4} &= a \end{aligned}$$

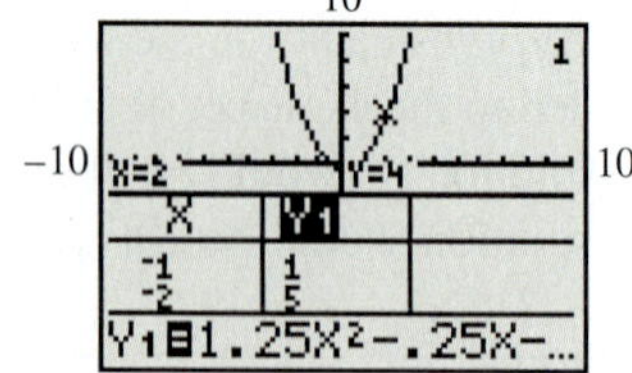

This graph/table screen shows that the points (2, 4), (–1, 1), and (–2, 5) lie on the graph of $Y_1 = 1.25X^2 - .25X - .5$. This supports the result of Example 8.

Finally, find c by substituting $a = \frac{5}{4}$ and $b = -\frac{1}{4}$ in equation (2).

$$\begin{aligned} 1 &= a - b + c && (2) \\ 1 &= \frac{5}{4} - \left(-\frac{1}{4}\right) + c && \text{Let } a = \tfrac{5}{4},\ b = -\tfrac{1}{4}. \\ 1 &= \frac{6}{4} + c \\ -\frac{1}{2} &= c \end{aligned}$$

The required equation is $y = \frac{5}{4}x^2 - \frac{1}{4}x - \frac{1}{2}$.

Now try Exercise 77.

EXAMPLE 9 Solving an Application Using a System of Three Equations

An animal feed is made from three ingredients: corn, soybeans, and cottonseed. One unit of each ingredient provides units of protein, fat, and fiber as shown in the table. How many units of each ingredient should be used to make a feed that contains 22 units of protein, 28 units of fat, and 18 units of fiber?

	Corn	Soybeans	Cottonseed	Total
Protein	.25	.4	.2	22
Fat	.4	.2	.3	28
Fiber	.3	.2	.1	18

Solution

Step 1 **Read** the problem. We must determine the number of units of corn, soybeans, and cottonseed.

Step 2 **Assign variables.** Let x represent the number of units of corn, y the number of units of soybeans, and z the number of units of cottonseed.

Step 3 **Write a system of equations.** Since the total amount of protein is to be 22 units, the first row of the table yields

$$.25x + .4y + .2z = 22. \quad (1)$$

Also, for the 28 units of fat,

$$.4x + .2y + .3z = 28, \quad (2)$$

and, for the 18 units of fiber,

$$.3x + .2y + .1z = 18. \quad (3)$$

Multiply equation (1) on both sides by 100, and equations (2) and (3) by 10 to get the equivalent system

$$\begin{aligned} 25x + 40y + 20z &= 2200 \quad (4) \\ 4x + 2y + 3z &= 280 \quad (5) \\ 3x + 2y + z &= 180. \quad (6) \end{aligned}$$

Step 4 **Solve** the system. Using the methods described earlier in this section, we find that $x = 40$, $y = 15$, and $z = 30$.

Step 5 **State the answer.** The feed should contain 40 units of corn, 15 units of soybeans, and 30 units of cottonseed to fulfill the given requirements.

Step 6 **Check.** Show that the ordered triple $(40, 15, 30)$ satisfies the system formed by equations (1), (2), and (3).

Now try Exercise 95.

NOTE Notice how the table in Example 9 is used to set up the equations of the system. The coefficients in each equation are read from left to right. This idea is extended in the next section, where we introduce solution of systems by matrices.

9.1 Exercises

Effects of NAFTA *The North American Free Trade Agreement (NAFTA) made the United States, Mexico, and Canada the largest free-trade zone in the world. The figure shows the projected levels of migration with and without NAFTA.*

Mexico–U.S. Migration with and without NAFTA

Number of migrants (in millions)

4
3
2
1

Without NAFTA

With NAFTA

1994 2000 2005 2010 2015

Year

Source: *Population Bulletin*, June 1999.

1. In what year after 1994 do both projections produce the same level of migration?
2. Refer to Exercise 1. In the year that the two projections produced the same level of migration, what was that level?
3. Express as an ordered pair the solution of the system containing the graphs of the two projections.
4. Use the terms *increasing* and *decreasing* to describe the trends for the "With NAFTA" graph.
5. If equations of the form $y = f(t)$ were determined that modeled either of the two graphs, then the variable t would represent ____________ and the variable y would represent ____________.

6. Explain why each graph is that of a function.

Solve each system by substitution. See Example 1.

7. $\begin{aligned} 4x + 3y &= -13 \\ -x + y &= 5 \end{aligned}$

8. $\begin{aligned} 3x + 4y &= 4 \\ x - y &= 13 \end{aligned}$

9. $\begin{aligned} x - 5y &= 8 \\ x &= 6y \end{aligned}$

10. $\begin{aligned} 6x - y &= 5 \\ y &= 11x \end{aligned}$

11. $\begin{aligned} 8x - 10y &= -22 \\ 3x + y &= 6 \end{aligned}$

12. $\begin{aligned} 4x - 5y &= -11 \\ 2x + y &= 5 \end{aligned}$

13. $\begin{aligned} 7x - y &= -10 \\ 3y - x &= 10 \end{aligned}$

14. $\begin{aligned} 4x + 5y &= 7 \\ 9y &= 31 + 2x \end{aligned}$

15. $\begin{aligned} -2x &= 6y + 18 \\ -29 &= 5y - 3x \end{aligned}$

16. $\begin{aligned} 3x - 7y &= 15 \\ 3x + 7y &= 15 \end{aligned}$

17. $\begin{aligned} 3y &= 5x + 6 \\ x + y &= 2 \end{aligned}$

18. $\begin{aligned} 4y &= 2x - 4 \\ x - y &= 4 \end{aligned}$

Solve each system by elimination. In Exercises 27–30, first clear denominators. See Example 2.

19. $\begin{aligned} 3x - y &= -4 \\ x + 3y &= 12 \end{aligned}$

20. $\begin{aligned} 4x + y &= -23 \\ x - 2y &= -17 \end{aligned}$

21. $\begin{aligned} 2x - 3y &= -7 \\ 5x + 4y &= 17 \end{aligned}$

22. $\begin{aligned} 4x + 3y &= -1 \\ 2x + 5y &= 3 \end{aligned}$

23. $\begin{aligned} 5x + 7y &= 6 \\ 10x - 3y &= 46 \end{aligned}$

24. $\begin{aligned} 12x - 5y &= 9 \\ 3x - 8y &= -18 \end{aligned}$

25. $\begin{aligned} 6x + 7y + 2 &= 0 \\ 7x - 6y - 26 &= 0 \end{aligned}$

26. $\begin{aligned} 5x + 4y + 2 &= 0 \\ 4x - 5y - 23 &= 0 \end{aligned}$

27. $\begin{aligned} \frac{x}{2} + \frac{y}{3} &= 4 \\ \frac{3x}{2} + \frac{3y}{2} &= 15 \end{aligned}$

28. $\begin{aligned} \frac{3x}{2} + \frac{y}{2} &= -2 \\ \frac{x}{2} + \frac{y}{2} &= 0 \end{aligned}$

29. $\begin{aligned} \frac{2x - 1}{3} + \frac{y + 2}{4} &= 4 \\ \frac{x + 3}{2} - \frac{x - y}{3} &= 3 \end{aligned}$

30. $\begin{aligned} \frac{x + 6}{5} + \frac{2y - x}{10} &= 1 \\ \frac{x + 2}{4} + \frac{3y + 2}{5} &= -3 \end{aligned}$

Solve each system. State whether it is inconsistent or has infinitely many solutions. If the system has infinitely many solutions, write the solution set with y arbitrary. See Examples 3 and 4.

31. $9x - 5y = 1$
$-18x + 10y = 1$

32. $3x + 2y = 5$
$6x + 4y = 8$

33. $4x - y = 9$
$-8x + 2y = -18$

34. $3x + 5y = -2$
$9x + 15y = -6$

35. $5x - 5y - 3 = 0$
$x - y - 12 = 0$

36. $2x - 3y - 7 = 0$
$-4x + 6y - 14 = 0$

37. $7x + 2y = 6$
$14x + 4y = 12$

38. $2x - 8y = 4$
$x - 4y = 2$

39. **Concept Check** Only one of the following screens gives the correct graphical solution of the system in Exercise 12. Which one is it? (*Hint:* Solve for y first in each equation and use the slope-intercept forms to help you answer the question.)

A. 

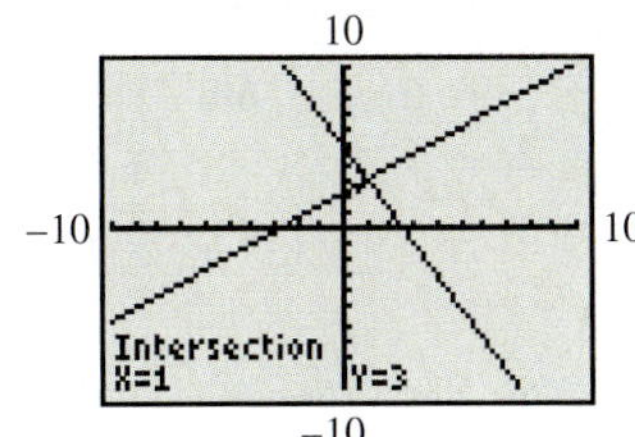

B.

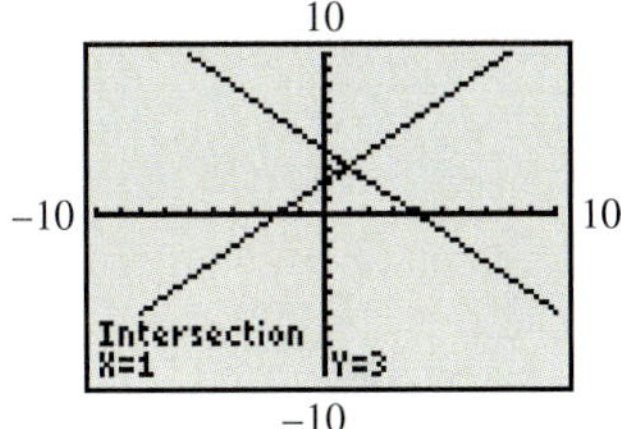

C.

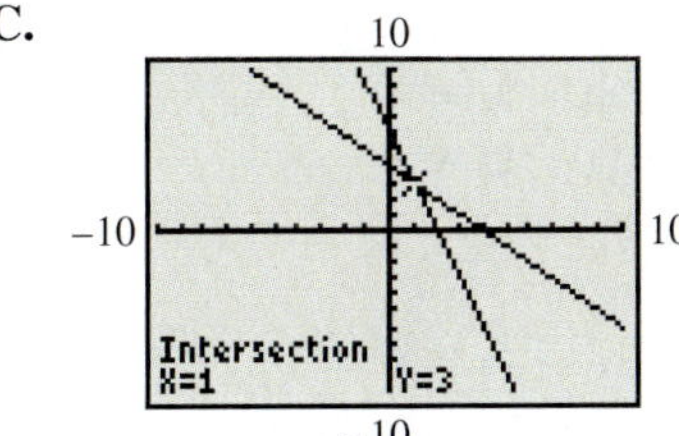

40. **Concept Check** If you attempt to solve a linear system with infinitely many solutions using a graphing calculator, how will the lines appear on the screen?

Use a graphing calculator to solve each system. Express solutions with approximations to the nearest thousandth.

41. $\sqrt{3}x - y = 5$
$100x + y = 9$

42. $\frac{11}{3}x + y = .5$
$.6x - y = 3$

43. $.2x + \sqrt{2}y = 1$
$\sqrt{5}x + .7y = 1$

44. $\sqrt{7}x + \sqrt{2}y - 3 = 0$
$\sqrt{6}x - y - \sqrt{3} = 0$

45. **Concept Check** For what value(s) of k will the following system of linear equations have no solution? infinitely many solutions?

$$x - 2y = 3$$
$$-2x + 4y = k$$

46. Explain how one can determine whether a linear system has no solution or infinitely many solutions when using the substitution or elimination method.

Solve each system. See Example 6.

47. $x + y + z = 2$
$2x + y - z = 5$
$x - y + z = -2$

48. $2x + y + z = 9$
$-x - y + z = 1$
$3x - y + z = 9$

49. $x + 3y + 4z = 14$
$2x - 3y + 2z = 10$
$3x - y + z = 9$

50. $\begin{aligned} 4x - y + 3z &= -2 \\ 3x + 5y - z &= 15 \\ -2x + y + 4z &= 14 \end{aligned}$

51. $\begin{aligned} x + 4y - z &= 6 \\ 2x - y + z &= 3 \\ 3x + 2y + 3z &= 16 \end{aligned}$

52. $\begin{aligned} 4x - 3y + z &= 9 \\ 3x + 2y - 2z &= 4 \\ x - y + 3z &= 5 \end{aligned}$

53. $\begin{aligned} x - 3y - 2z &= -3 \\ 3x + 2y - z &= 12 \\ -x - y + 4z &= 3 \end{aligned}$

54. $\begin{aligned} x + y + z &= 3 \\ 3x - 3y - 4z &= -1 \\ x + y + 3z &= 11 \end{aligned}$

55. $\begin{aligned} 2x + 6y - z &= 6 \\ 4x - 3y + 5z &= -5 \\ 6x + 9y - 2z &= 11 \end{aligned}$

56. $\begin{aligned} 8x - 3y + 6z &= -2 \\ 4x + 9y + 4z &= 18 \\ 12x - 3y + 8z &= -2 \end{aligned}$

57. $\begin{aligned} 2x - 3y + 2z - 3 &= 0 \\ 4x + 8y + z - 2 &= 0 \\ -x - 7y + 3z - 14 &= 0 \end{aligned}$

58. $\begin{aligned} -x + 2y - z - 1 &= 0 \\ -x - y - z + 2 &= 0 \\ x - y + 2z - 2 &= 0 \end{aligned}$

Solve each system in terms of the arbitrary variable x. See Example 7.

59. $\begin{aligned} x - 2y + 3z &= 6 \\ 2x - y + 2z &= 5 \end{aligned}$

60. $\begin{aligned} 3x - 2y + z &= 15 \\ x + 4y - z &= 11 \end{aligned}$

61. $\begin{aligned} 5x - 4y + z &= 9 \\ x + y &= 15 \end{aligned}$

62. $\begin{aligned} 3x - 5y - 4z &= -7 \\ y - z &= -13 \end{aligned}$

63. $\begin{aligned} 3x + 4y - z &= 13 \\ x + y + 2z &= 15 \end{aligned}$

64. $\begin{aligned} x - y + z &= -6 \\ 4x + y + z &= 7 \end{aligned}$

Solve each system. State whether it is inconsistent or has infinitely many solutions. If the system has infinitely many solutions, write the solution set with z arbitrary. See Examples 3, 4, 6, and 7.

65. $\begin{aligned} 3x + 5y - z &= -2 \\ 4x - y + 2z &= 1 \\ -6x - 10y + 2z &= 0 \end{aligned}$

66. $\begin{aligned} 3x + y + 3z &= 1 \\ x + 2y - z &= 2 \\ 2x - y + 4z &= 4 \end{aligned}$

67. $\begin{aligned} 5x - 4y + z &= 0 \\ x + y &= 0 \\ -10x + 8y - 2z &= 0 \end{aligned}$

68. $\begin{aligned} 2x + y - 3z &= 0 \\ 4x + 2y - 6z &= 0 \\ x - y + z &= 0 \end{aligned}$

Solve each system. (Hint: In Exercises 69–72, let $\frac{1}{x} = t$ and $\frac{1}{y} = u$.)

69. $\begin{aligned} \frac{2}{x} + \frac{1}{y} &= \frac{3}{2} \\ \frac{3}{x} - \frac{1}{y} &= 1 \end{aligned}$

70. $\begin{aligned} \frac{1}{x} + \frac{3}{y} &= \frac{16}{5} \\ \frac{5}{x} + \frac{4}{y} &= 5 \end{aligned}$

71. $\begin{aligned} \frac{2}{x} + \frac{1}{y} &= 11 \\ \frac{3}{x} - \frac{5}{y} &= 10 \end{aligned}$

72. $\begin{aligned} \frac{2}{x} + \frac{3}{y} &= 18 \\ \frac{4}{x} - \frac{5}{y} &= -8 \end{aligned}$

73. $\begin{aligned} \frac{2}{x} + \frac{3}{y} - \frac{2}{z} &= -1 \\ \frac{8}{x} - \frac{12}{y} + \frac{5}{z} &= 5 \\ \frac{6}{x} + \frac{3}{y} - \frac{1}{z} &= 1 \end{aligned}$

74. $\begin{aligned} -\frac{5}{x} + \frac{4}{y} + \frac{3}{z} &= 2 \\ \frac{10}{x} + \frac{3}{y} - \frac{6}{z} &= 7 \\ \frac{5}{x} + \frac{2}{y} - \frac{9}{z} &= 6 \end{aligned}$

75. *Concept Check* Consider the linear equation in three variables $x + y + z = 4$. Find a pair of linear equations in three variables that, when considered together with the given equation, will form a system having **(a)** exactly one solution, **(b)** no solution, **(c)** infinitely many solutions.

76. *Concept Check* Using your immediate surroundings:

(a) Give an example of three planes that intersect in a single point.
(b) Give an example of three planes that intersect in a line.

Curve Fitting Use a system of equations to solve each problem. See Example 8.

77. Find the equation of the parabola $y = ax^2 + bx + c$ that passes through the points $(2, 3)$, $(-1, 0)$, and $(-2, 2)$.

78. Find the equation of the line $y = ax + b$ that passes through the points $(-2, 1)$ and $(-1, -2)$.

79. Find the equation of the line through the given points.

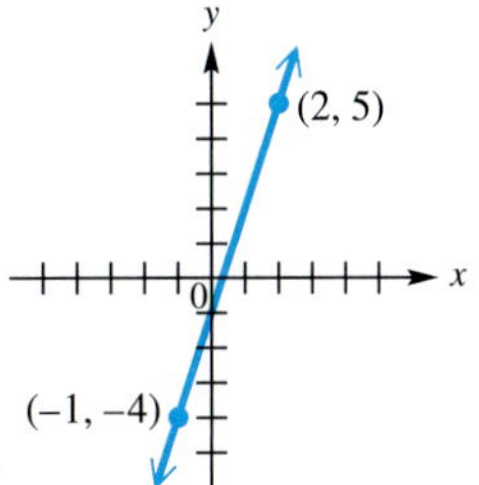

80. Find the equation of the parabola through the given points.

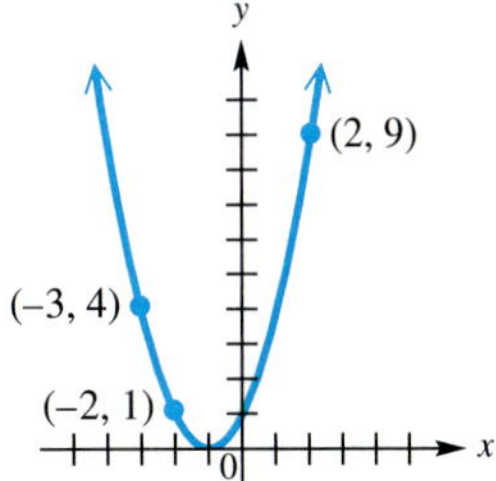

81. *Curve Fitting* Find the equation of the parabola. Three views of the same curve are given.

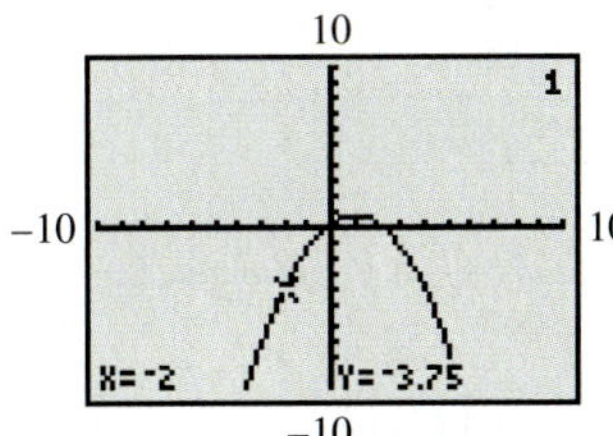

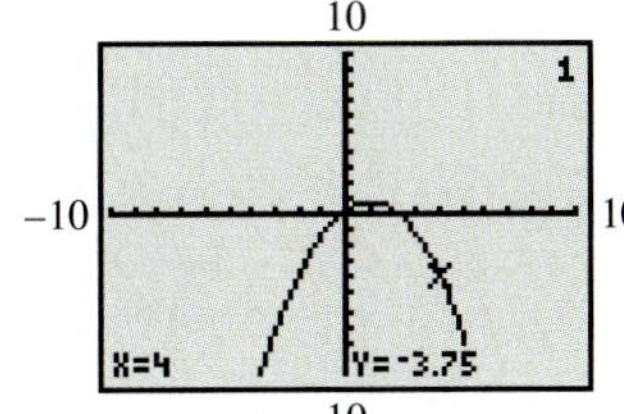

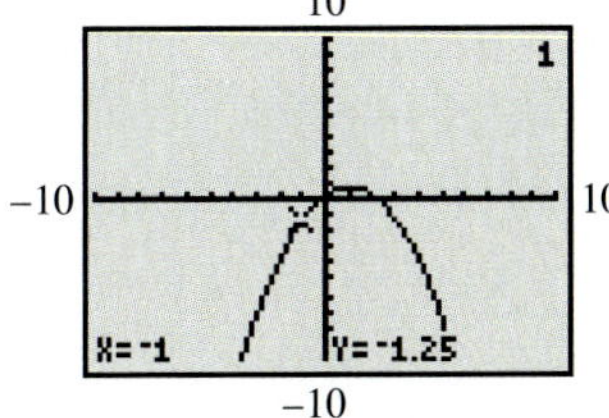

82. *Curve Fitting* The table was generated using a function defined by $Y_1 = aX^2 + bX + c$. Use any three points from the table to find the equation that defines the function.

X	Y1
-3	5.48
-2	2.9
-1	1.26
0	.56
1	.8
2	1.98
3	4.1

X=-3

Curve Fitting *Given three noncollinear points, there is one and only one circle that passes through them. Knowing that the equation of a circle may be written in the form*

$$x^2 + y^2 + ax + by + c = 0, \quad \text{(Section 2.1)}$$

find the equation of the circle passing through the given points.

83. $(-1, 3)$, $(6, 2)$, and $(-2, -4)$

84. $(-1, 5)$, $(6, 6)$, and $(7, -1)$

85. $(2, 1)$, $(-1, 0)$, and $(3, 3)$

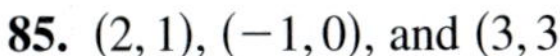

86.

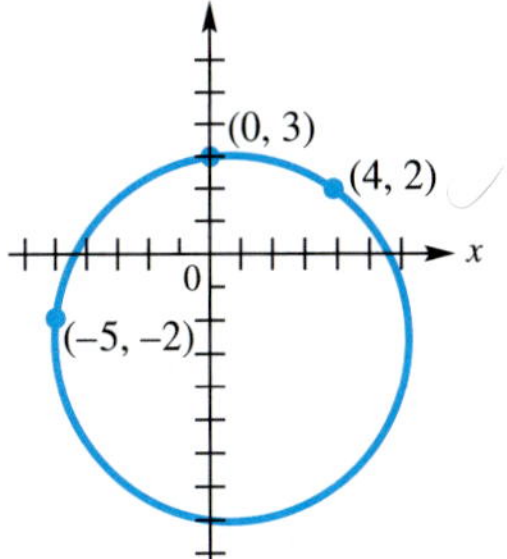

(Modeling) Use the method of Example 8 to work Exercises 87 and 88.

87. *Atmospheric Carbon Dioxide* Carbon dioxide concentrations (in parts per million) have been measured at Mauna Loa, Hawaii, over the past 45 yr. This concentration has increased quadratically. The table lists January readings for three years.

Year	CO_2
1962	318
1982	341
2002	371

Source: U.S. Department of Energy; Carbon Dioxide Information Analysis Center.

(a) If the quadratic relationship between the carbon dioxide concentration C and the year t is expressed as $C = at^2 + bt + c$, where $t = 0$ corresponds to 1962, use a system of linear equations to determine the constants a, b, and c, and give the equation.

(b) Predict the year when the amount of carbon dioxide in the atmosphere will double from its 1962 level.

88. *Aircraft Speed and Altitude* For certain aircraft there exists a quadratic relationship between an airplane's maximum speed S (in knots) and its ceiling C or highest altitude possible (in thousands of feet). The table lists three airplanes that conform to this relationship.

Airplane	Max Speed (S)	Ceiling (C)
Hawkeye	320	33
Corsair	600	40
Tomcat	1283	50

Source: Sanders, D., *Statistics: A First Course,* Sixth Edition, McGraw Hill, 2000.

(a) If the quadratic relationship between C and S is written as $C = aS^2 + bS + c$, use a linear system of equations to determine the constants a, b and c. Give the equation.

(b) A new aircraft of this type has a ceiling of 45,000 ft. Predict its top speed.

Solve each problem. See Examples 5 and 9.

89. *Budget Deficits* For fiscal years 2003 and 2004, California and New York were the two states in the United States with the largest budget deficits. The deficit for California was $1 billion less than three times that of New York. Together, the deficits of the two states totaled $47 billion. What were the deficits of each state? (*Source:* "States in a Sorry State," *Business Week*, March 17, 2003.)

90. *Fan Cost Index* The Fan Cost Index (FCI) is a measure of how much it will cost a family of four to attend a professional sports event. In a recent year, the FCI prices for Major League Baseball and the National Football League totaled $311.03. The FCI for baseball was $105.87 less than that of football. What were the FCIs for these sports? (*Source:* Team Marketing Report, Chicago.)

91. *Mixing Water* A sparkling-water distributor wants to make up 300 gal of sparkling water to sell for \$6.00 per gallon. She wishes to mix three grades of water selling for \$9.00, \$3.00, and \$4.50 per gallon, respectively. She must use twice as much of the \$4.50 water as the \$3.00 water. How many gallons of each should she use?

92. *Mixing Glue* A glue company needs to make some glue that it can sell for \$120 per barrel. It wants to use 150 barrels of glue worth \$100 per barrel, along with some glue worth \$150 per barrel, and some glue worth \$190 per barrel. It must use the same number of barrels of \$150 and \$190 glue. How much of the \$150 and \$190 glue will be needed? How many barrels of \$120 glue will be produced?

93. *Triangle Dimensions* The perimeter of a triangle is 59 in. The longest side is 11 in. longer than the medium side, and the medium side is 3 in. more than the shortest side. Find the length of each side of the triangle.

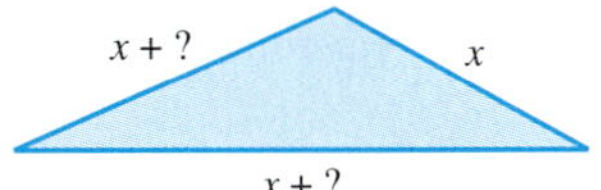

94. *Triangle Dimensions* The sum of the measures of the angles of any triangle is 180°. In a certain triangle, the largest angle measures 55° less than twice the medium angle, and the smallest angle measures 25° less than the medium angle. Find the measures of each of the three angles.

95. *Investment Decisions* Pat Summers wins \$200,000 in the Louisiana state lottery. He invests part of the money in real estate with an annual return of 3% and another part in a money market account at 2.5% interest. He invests the rest, which amounts to \$80,000 less than the sum of the other two parts, in certificates of deposit that pay 1.5%. If the total annual interest on the money is \$4900, how much was invested at each rate?

	Amount Invested	Rate (as a decimal)	Annual Interest
Real Estate		.03	
Money Market		.025	
CDs		.015	

96. *Investment Decisions* Jane Hooker invests \$40,000 received in an inheritance in three parts. With one part she buys mutual funds that offer a return of 2% per year. The second part, which amounts to twice the first, is used to buy government bonds paying 2.5% per year. She puts the rest of the money into a savings account that pays 1.25% annual interest. During the first year, the total interest is \$825. How much did she invest at each rate?

	Amount Invested	Rate (as a decimal)	Annual Interest
Mutual Funds		.02	
Government Bonds		.025	
Savings Account		.0125	

Relating Concepts

For individual or collaborative investigation
(Exercises 97–102)

Supply and Demand In many applications of economics, as the price of an item goes up, demand for the item goes down and supply of the item goes up. The price where supply and demand are equal is called the *equilibrium price*, and the resulting supply or demand is called the *equilibrium supply* or *equilibrium demand*. Suppose the supply of a product is related to its price by the equation

$$p = \frac{2}{3}q,$$

where p is in dollars and q is supply in appropriate units. (Here, q stands for quantity.) Furthermore, suppose demand and price for the same product are related by

$$p = -\frac{1}{3}q + 18,$$

where p is price and q is demand. The system formed by these two equations has solution (18, 12), as seen in the graph. Use this information to ***work Exercises 97–102 in order.***

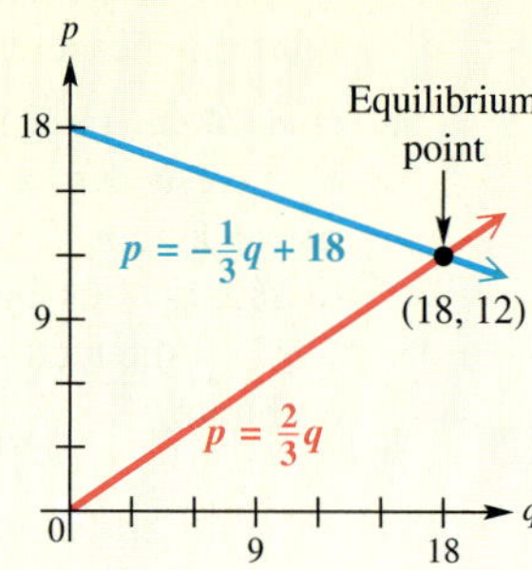

97. Suppose the demand and price for a certain model of electric can opener are related by $p = 16 - \frac{5}{4}q$, where p is price, in dollars, and q is demand, in appropriate units. Find the price when the demand is at each level.

(a) 0 units **(b)** 4 units **(c)** 8 units

98. Find the demand for the electric can opener at each price.

(a) \$6 **(b)** \$11 **(c)** \$16

99. Graph $p = 16 - \frac{5}{4}q$.

100. Suppose the price and supply of the can opener are related by $p = \frac{3}{4}q$, where q represents the supply and p the price. Find the supply at each price.

(a) \$0 **(b)** \$10 **(c)** \$20

101. Graph $p = \frac{3}{4}q$ on the same axes used for Exercise 99.

102. Use the result of Exercise 101 to find the equilibrium price and the equilibrium demand.

103. Solve the system of equations (4), (5), and (6) in Example 9:

$$\begin{aligned} 25x + 40y + 20z &= 2200 \\ 4x + 2y + 3z &= 280 \\ 3x + 2y + z &= 180. \end{aligned}$$

104. Check your solution in Exercise 103, showing that it satisfies all three equations.

9.2 Matrix Solution of Linear Systems

The Gauss-Jordan Method ▪ **Special Systems**

$$\begin{bmatrix} 2 & 3 & 7 \\ 5 & -1 & 10 \end{bmatrix} \quad \text{Matrix}$$

Since systems of linear equations occur in so many practical situations, computer methods have been developed for efficiently solving linear systems. Computer solutions of linear systems depend on the idea of a **matrix** (plural **matrices**), a rectangular array of numbers enclosed in brackets. Each number is called an **element** of the matrix.

The Gauss-Jordan Method Matrices in general are discussed in more detail later in this chapter. In this section, we develop a method for solving linear systems using matrices. As an example, start with a system and write the coefficients of the variables and the constants as a matrix, called the **augmented matrix** of the system.

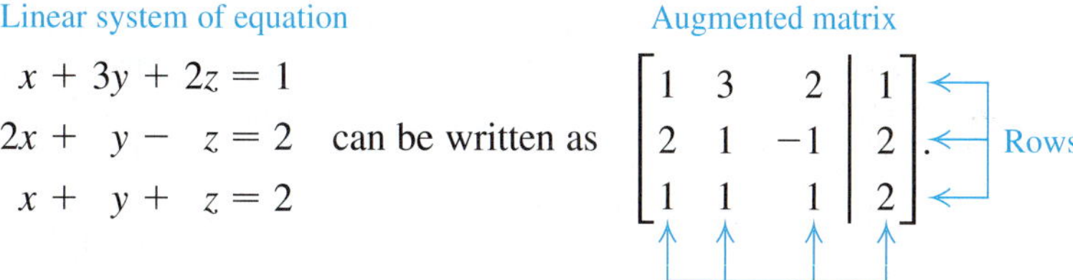

$$\begin{aligned} x + 3y + 2z &= 1 \\ 2x + y - z &= 2 \\ x + y + z &= 2 \end{aligned} \quad \text{can be written as} \quad \left[\begin{array}{ccc|c} 1 & 3 & 2 & 1 \\ 2 & 1 & -1 & 2 \\ 1 & 1 & 1 & 2 \end{array}\right].$$

The vertical line, which is optional, separates the coefficients from the constants. Because this matrix has 3 rows (horizontal) and 4 columns (vertical), we say its **size*** is 3×4 (read "three by four"). To refer to a number in the matrix, use its row and column numbers. For example, the number 3 is in the first row, second column.

We can treat the rows of this matrix just like the equations of the corresponding system of linear equations. Since the augmented matrix is nothing more than a shorthand form of the system, any transformation of the matrix that results in an equivalent system of equations can be performed.

Matrix Row Transformations

For any augmented matrix of a system of linear equations, the following row transformations will result in the matrix of an equivalent system.

1. Interchange any two rows.
2. Multiply or divide the elements of any row by a nonzero real number.
3. Replace any row of the matrix by the sum of the elements of that row and a multiple of the elements of another row.

These transformations are just restatements in matrix form of the transformations of systems discussed in the previous section. From now on, when referring to the third transformation, "a multiple of the elements of a row" will be abbreviated as "a multiple of a row."

*Other terms used to describe the size of a matrix are *order* and *dimension.*

Before using matrices to solve a linear system, the system must be arranged in the proper form, with variable terms on the left side of the equation and constant terms on the right. The variable terms must be in the same order in each of the equations.

The **Gauss-Jordan method** is a systematic technique for applying matrix row transformations in an attempt to reduce a matrix to *diagonal form,* with 1s along the diagonal, such as

$$\left[\begin{array}{cc|c} 1 & 0 & a \\ 0 & 1 & b \end{array}\right] \quad \text{or} \quad \left[\begin{array}{ccc|c} 1 & 0 & 0 & a \\ 0 & 1 & 0 & b \\ 0 & 0 & 1 & c \end{array}\right],$$

from which the solutions are easily obtained. This form is also called *reduced-row echelon form.*

Using the Gauss-Jordan Method to Put a Matrix into Diagonal Form

Step 1 Obtain 1 as the first element of the first column.
Step 2 Use the first row to transform the remaining entries in the first column to 0.
Step 3 Obtain 1 as the second entry in the second column.
Step 4 Use the second row to transform the remaining entries in the second column to 0.
Step 5 Continue in this manner as far as possible.

NOTE The Gauss-Jordan method proceeds *column by column,* from left to right. When you are working with a particular column, no row operation should undo the form of a preceding column.

EXAMPLE 1 Using the Gauss-Jordan Method

Solve the system.

$$\begin{aligned} 3x - 4y &= 1 \\ 5x + 2y &= 19 \end{aligned}$$

Solution Both equations are in the same form, with variable terms in the same order on the left, and constant terms on the right.

$$\left[\begin{array}{cc|c} 3 & -4 & 1 \\ 5 & 2 & 19 \end{array}\right] \quad \text{Write the augmented matrix.}$$

The goal is to transform the augmented matrix into one in which the value of the variables will be easy to see. That is, since each column in the matrix represents the coefficients of one variable, the augmented matrix should be transformed so that it is of the form

$$\left[\begin{array}{cc|c} 1 & 0 & k \\ 0 & 1 & j \end{array}\right],$$

for real numbers k and j. Once the augmented matrix is in this form, the matrix can be rewritten as a linear system to get

$$\begin{aligned} x &= k \\ y &= j. \end{aligned}$$

It is best to work in columns beginning in each column with the element that is to become 1. In the augmented matrix

$$\left[\begin{array}{cc|c} 3 & -4 & 1 \\ 5 & 2 & 19 \end{array}\right],$$

3 is in the first row, first column position. Use transformation 2, multiplying each entry in the first row by $\frac{1}{3}$ to get 1 in this position. (This step is abbreviated as $\frac{1}{3}$ R1.)

$$\left[\begin{array}{cc|c} 1 & -\frac{4}{3} & \frac{1}{3} \\ 5 & 2 & 19 \end{array}\right] \quad \tfrac{1}{3}\text{R1}$$

Introduce 0 in the second row, first column by multiplying each element of the first row by -5 and adding the result to the corresponding element in the second row, using transformation 3.

$$\left[\begin{array}{cc|c} 1 & -\frac{4}{3} & \frac{1}{3} \\ 0 & \frac{26}{3} & \frac{52}{3} \end{array}\right] \quad -5\text{R1} + \text{R2}$$

Obtain 1 in the second row, second column by multiplying each element of the second row by $\frac{3}{26}$, using transformation 2.

$$\left[\begin{array}{cc|c} 1 & -\frac{4}{3} & \frac{1}{3} \\ 0 & 1 & 2 \end{array}\right] \quad \tfrac{3}{26}\text{R2}$$

Finally, get 0 in the first row, second column by multiplying each element of the second row by $\frac{4}{3}$ and adding the result to the corresponding element in the first row.

$$\left[\begin{array}{cc|c} 1 & 0 & 3 \\ 0 & 1 & 2 \end{array}\right] \quad \tfrac{4}{3}\text{R2} + \text{R1}$$

This last matrix corresponds to the system

$$\begin{aligned} x &= 3 \\ y &= 2 \end{aligned}$$

that has solution set $\{(3, 2)\}$. We can read this solution directly from the third column of the final matrix. Check the solution in both equations of the *original* system.

Now try Exercise 17.

A linear system with three equations is solved in a similar way. Row transformations are used to get 1s down the diagonal from left to right and 0s above and below each 1.

EXAMPLE 2 Using the Gauss-Jordan Method

Solve the system.

$$\begin{aligned} x - y + 5z &= -6 \\ 3x + 3y - z &= 10 \\ x + 3y + 2z &= 5 \end{aligned}$$

Solution

$$\left[\begin{array}{rrr|r} 1 & -1 & 5 & -6 \\ 3 & 3 & -1 & 10 \\ 1 & 3 & 2 & 5 \end{array}\right] \quad \text{Write the augmented matrix.}$$

There is already a 1 in the first row, first column. Introduce 0 in the second row of the first column by multiplying each element in the first row by -3 and adding the result to the corresponding element in the second row.

$$\left[\begin{array}{rrr|r} 1 & -1 & 5 & -6 \\ 0 & 6 & -16 & 28 \\ 1 & 3 & 2 & 5 \end{array}\right] \quad -3\text{R1} + \text{R2}$$

To change the third element in the first column to 0, multiply each element of the first row by -1, and add the result to the corresponding element of the third row.

$$\left[\begin{array}{rrr|r} 1 & -1 & 5 & -6 \\ 0 & 6 & -16 & 28 \\ 0 & 4 & -3 & 11 \end{array}\right] \quad -1\text{R1} + \text{R3}$$

Use the same procedure to transform the second and third columns. For both of these columns, perform the additional step of getting 1 in the appropriate position of each column. Do this by multiplying the elements of the row by the reciprocal of the number in that position.

$$\left[\begin{array}{rrr|r} 1 & -1 & 5 & -6 \\ 0 & 1 & -\frac{8}{3} & \frac{14}{3} \\ 0 & 4 & -3 & 11 \end{array}\right] \quad \tfrac{1}{6}\text{R2}$$

$$\left[\begin{array}{rrr|r} 1 & 0 & \frac{7}{3} & -\frac{4}{3} \\ 0 & 1 & -\frac{8}{3} & \frac{14}{3} \\ 0 & 4 & -3 & 11 \end{array}\right] \quad \text{R2} + \text{R1}$$

$$\left[\begin{array}{rrr|r} 1 & 0 & \frac{7}{3} & -\frac{4}{3} \\ 0 & 1 & -\frac{8}{3} & \frac{14}{3} \\ 0 & 0 & \frac{23}{3} & -\frac{23}{3} \end{array}\right] \quad -4\text{R2} + \text{R3}$$

$$\left[\begin{array}{rrr|r} 1 & 0 & \frac{7}{3} & -\frac{4}{3} \\ 0 & 1 & -\frac{8}{3} & \frac{14}{3} \\ 0 & 0 & 1 & -1 \end{array}\right] \quad \tfrac{3}{23}\text{R3}$$

$$\left[\begin{array}{rrr|r} 1 & 0 & 0 & 1 \\ 0 & 1 & -\frac{8}{3} & \frac{14}{3} \\ 0 & 0 & 1 & -1 \end{array}\right] \quad -\tfrac{7}{3}\text{R3} + \text{R1}$$

$$\left[\begin{array}{ccc|c} 1 & 0 & 0 & 1 \\ 0 & 1 & 0 & 2 \\ 0 & 0 & 1 & -1 \end{array}\right] \quad \tfrac{8}{3}\text{R3} + \text{R2}$$

The linear system associated with this final matrix is

$$x = 1$$
$$y = 2$$
$$z = -1.$$

The solution set is $\{(1, 2, -1)\}$. Check the solution in the original system.

Now try Exercise 23.

A graphing calculator with matrix capability can perform row operations. See Figure 7(a). The screens in Figures 7(b) and (c) show typical entries for the matrices in the second and third steps of the solution in Example 2. The entire Gauss-Jordan method can be carried out in one step with the rref (reduced-row echelon form) command, as shown in Figure 7(d).

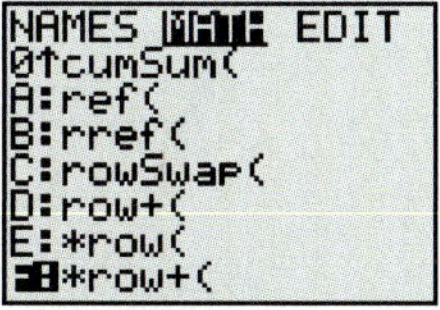

This typical menu shows various options for matrix row transformations in choices C–F.

(a)

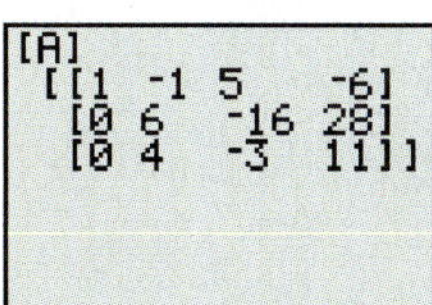

(b)

This is the matrix that results when row 2 is multiplied by $\frac{1}{6}$.

(c)

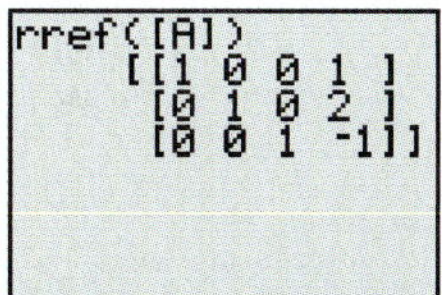

This screen shows the final matrix in the algebraic solution, found by using the rref command.

(d)

Figure 7

Special Systems The next two examples show how to recognize inconsistent systems or systems with infinitely many solutions when solving such systems using row transformations.

EXAMPLE 3 Solving an Inconsistent System

Use the Gauss-Jordan method to solve the system.

$$x + y = 2$$
$$2x + 2y = 5$$

Solution

$$\left[\begin{array}{cc|c} 1 & 1 & 2 \\ 2 & 2 & 5 \end{array}\right] \quad \text{Write the augmented matrix.}$$

$$\left[\begin{array}{cc|c} 1 & 1 & 2 \\ 0 & 0 & 1 \end{array}\right] \quad -2\text{R1} + \text{R2}$$

The next step would be to get 1 in the second row, second column. Because of the 0 there, it is impossible to go further. Since the second row corresponds to the equation $0x + 0y = 1$, which has no solution, the system is inconsistent and the solution set is $\emptyset$.

Now try Exercise 21.

EXAMPLE 4 Solving a System with Infinitely Many Solutions

Use the Gauss-Jordan method to solve the system.

$$\begin{aligned} 2x - 5y + 3z &= 1 \\ x - 2y - 2z &= 8 \end{aligned}$$

Solution Recall from the previous section that a system with two equations in three variables usually has an infinite number of solutions. We can use the Gauss-Jordan method to give the solution with z arbitrary.

$$\left[\begin{array}{rrr|r} 2 & -5 & 3 & 1 \\ 1 & -2 & -2 & 8 \end{array}\right] \quad \text{Write the augmented matrix.}$$

$$\left[\begin{array}{rrr|r} 1 & -2 & -2 & 8 \\ 2 & -5 & 3 & 1 \end{array}\right] \quad \text{Interchange rows to get 1 in the first row, first column position.}$$

$$\left[\begin{array}{rrr|r} 1 & -2 & -2 & 8 \\ 0 & -1 & 7 & -15 \end{array}\right] \quad -2\text{R1} + \text{R2}$$

$$\left[\begin{array}{rrr|r} 1 & -2 & -2 & 8 \\ 0 & 1 & -7 & 15 \end{array}\right] \quad -1\text{R2}$$

$$\left[\begin{array}{rrr|r} 1 & 0 & -16 & 38 \\ 0 & 1 & -7 & 15 \end{array}\right] \quad 2\text{R2} + \text{R1}$$

It is not possible to go further with the Gauss-Jordan method. The equations that correspond to the final matrix are

$$x - 16z = 38 \quad \text{and} \quad y - 7z = 15.$$

Solve these equations for x and y, respectively.

$$\begin{aligned} x - 16z &= 38 \\ x &= 16z + 38 \end{aligned} \qquad \Bigg| \qquad \begin{aligned} y - 7z &= 15 \\ y &= 7z + 15 \end{aligned} \quad \text{(Section 1.1)}$$

The solution set, written with z arbitrary, is $\{(16z + 38, 7z + 15, z)\}$.

Now try Exercise 35.

Summary of Possible Cases

When matrix methods are used to solve a system of linear equations and the resulting matrix is written in diagonal form:

1. If the number of rows with nonzero elements to the left of the vertical line is equal to the number of variables in the system, then the system has a single solution. See Examples 1 and 2.
2. If one of the rows has the form $[0\ 0 \cdots 0 \,|\, a]$ with $a \neq 0$, then the system has no solution. See Example 3.
3. If there are fewer rows in the matrix containing nonzero elements than the number of variables, then the system has either no solution or infinitely many solutions. If there are infinitely many solutions, give the solutions in terms of one or more arbitrary variables. See Example 4.

CONNECTIONS The Gaussian reduction method (which is similar to the Gauss-Jordan method) is named after the mathematician Carl F. Gauss. In 1811, Gauss published a paper showing how he determined the orbit of the asteroid Pallas. Over the years he had kept data on his numerous observations. Each observation produced a linear equation in six unknowns. A typical equation was

$$.79363x + 143.66y + .39493z + .95929u - .18856v + .17387w$$
$$= 183.93.$$

Eventually he had twelve equations of this type. Certainly a method was needed to simplify the computation of a simultaneous solution. The method he developed was the reduction of the system from a rectangular to a triangular one, hence the Gaussian reduction. However, Gauss did not use matrices to carry out the computation.

When computers are programmed to solve large linear systems involved in applications like designing aircraft or electrical circuits, they frequently use an algorithm that is similar to the Gauss-Jordan method presented here. Solving a linear system with n equations and n variables requires the computer to perform a total of

$$T(n) = \frac{2}{3}n^3 + \frac{3}{2}n^2 - \frac{7}{6}n$$

arithmetic calculations (additions, subtractions, multiplications, and divisions).*

For Discussion or Writing

1. Compute T for $n = 3, 6, 10, 29, 100, 200, 400, 1000, 5000, 10{,}000, 100{,}000$ and write the results in a table.
2. John Atanasoff wanted to solve a 29×29 linear system of equations. How many arithmetic operations would this have required? Is this too many to do by hand?
3. If the number of equations and variables is doubled, does the number of arithmetic operations double?
4. A Cray-T90 supercomputer can execute up to 60 billion arithmetic operations per second. How many hours would be required to solve a linear system with 100,000 variables?

**Source:* Burden, R. and J. Faires, *Numerical Analysis,* Sixth Edition, Brooks/Cole Publishing Company 1997.

9.2 Exercises

Use the third row transformation to change each matrix as indicated. See Example 1.

1. $\begin{bmatrix} 2 & 4 \\ 4 & 7 \end{bmatrix}$; -2 times row 1 added to row 2

2. $\begin{bmatrix} -1 & 4 \\ 7 & 0 \end{bmatrix}$; 7 times row 1 added to row 2

3. $\begin{bmatrix} 1 & 5 & 6 \\ -2 & 3 & -1 \\ 4 & 7 & 0 \end{bmatrix}$; 2 times row 1 added to row 2

4. $\begin{bmatrix} 2 & 5 & 6 \\ 4 & -1 & 2 \\ 3 & 7 & 1 \end{bmatrix}$; -6 times row 3 added to row 1

Concept Check *Write the augmented matrix for each system and give its size. Do not solve the system.*

5. $\begin{aligned} 2x + 3y &= 11 \\ x + 2y &= 8 \end{aligned}$

6. $\begin{aligned} 3x + 5y &= -13 \\ 2x + 3y &= -9 \end{aligned}$

7. $\begin{aligned} 2x + y + z - 3 &= 0 \\ 3x - 4y + 2z + 7 &= 0 \\ x + y + z - 2 &= 0 \end{aligned}$

8. $\begin{aligned} 4x - 2y + 3z - 4 &= 0 \\ 3x + 5y + z - 7 &= 0 \\ 5x - y + 4z - 7 &= 0 \end{aligned}$

Concept Check *Write the system of equations associated with each augmented matrix. Do not solve.*

9. $\left[\begin{array}{ccc|c} 3 & 2 & 1 & 1 \\ 0 & 2 & 4 & 22 \\ -1 & -2 & 3 & 15 \end{array}\right]$

10. $\left[\begin{array}{ccc|c} 2 & 1 & 3 & 12 \\ 4 & -3 & 0 & 10 \\ 5 & 0 & -4 & -11 \end{array}\right]$

11. $\left[\begin{array}{ccc|c} 1 & 0 & 0 & 2 \\ 0 & 1 & 0 & 3 \\ 0 & 0 & 1 & -2 \end{array}\right]$

12. $\left[\begin{array}{ccc|c} 1 & 0 & 0 & 4 \\ 0 & 1 & 0 & 2 \\ 0 & 0 & 1 & 3 \end{array}\right]$

13.
```
[A]
   [[1 1 0  3 ]
    [0 2 1  -4]
    [1 0 -1 5 ]]
```

14.
```
[B]
   [[2 0  1  9]
    [0 -1 -1 5]
    [3 1  0  8]]
```

Use the Gauss-Jordan method to solve each system of equations. For systems in three variables with infinitely many solutions, give the solution with z arbitrary; for any such equations with four variables, let w be the arbitrary variable. See Examples 1–4.

15. $\begin{aligned} x + y &= 5 \\ x - y &= -1 \end{aligned}$

16. $\begin{aligned} x + 2y &= 5 \\ 2x + y &= -2 \end{aligned}$

17. $\begin{aligned} 3x + 2y &= -9 \\ 2x - 5y &= -6 \end{aligned}$

18. $\begin{aligned} 2x - 3y &= 10 \\ 2x + 2y &= 5 \end{aligned}$

19. $\begin{aligned} 6x + y - 5 &= 0 \\ 5x + y - 3 &= 0 \end{aligned}$

20. $\begin{aligned} 2x - 5y - 10 &= 0 \\ 3x + y - 15 &= 0 \end{aligned}$

21. $\begin{aligned} 2x - y &= 6 \\ 4x - 2y &= 0 \end{aligned}$

22. $\begin{aligned} 3x - 2y &= 1 \\ 6x - 4y &= -1 \end{aligned}$

23. $\begin{aligned} x + y - 5z &= -18 \\ 3x - 3y + z &= 6 \\ x + 3y - 2z &= -13 \end{aligned}$

24. $\begin{aligned} -x + 2y + 6z &= 2 \\ 3x + 2y + 6z &= 6 \\ x + 4y - 3z &= 1 \end{aligned}$

25. $\begin{aligned} x + y - z &= 6 \\ 2x - y + z &= -9 \\ x - 2y + 3z &= 1 \end{aligned}$

26. $\begin{aligned} x + 3y - 6z &= 7 \\ 2x - y + z &= 1 \\ x + 2y + 2z &= -1 \end{aligned}$

27. $\begin{aligned} x - z &= -3 \\ y + z &= 9 \\ x + z &= 7 \end{aligned}$

28. $\begin{aligned} -x + y &= -1 \\ y - z &= 6 \\ x + z &= -1 \end{aligned}$

29. $\begin{aligned} y &= -2x - 2z + 1 \\ x &= -2y - z + 2 \\ z &= x - y \end{aligned}$

30. $\begin{aligned} x + y &= 1 \\ 2x - z &= 0 \\ y + 2z &= -2 \end{aligned}$

31. $\begin{aligned} 2x - y + 3z &= 0 \\ x + 2y - z &= 5 \\ 2y + z &= 1 \end{aligned}$

32. $\begin{aligned} 4x + 2y - 3z &= 6 \\ x - 4y + z &= -4 \\ -x + 2z &= 2 \end{aligned}$

33. $\begin{aligned} 3x + 5y - z + 2 &= 0 \\ 4x - y + 2z - 1 &= 0 \\ -6x - 10y + 2z &= 0 \end{aligned}$

34. $\begin{aligned} 3x + y + 3z &= 1 \\ x + 2y - z &= 2 \\ 2x - y + 4z &= 4 \end{aligned}$

35. $\begin{aligned} x - 8y + z &= 4 \\ 3x - y + 2z &= -1 \end{aligned}$

36. $\begin{aligned} 5x - 3y + z &= 1 \\ 2x + y - z &= 4 \end{aligned}$

37. $\begin{aligned} x - y + 2z + w &= 4 \\ y + z &= 3 \\ z - w &= 2 \\ x - y &= 0 \end{aligned}$

38. $\begin{aligned} x + 2y + z - 3w &= 7 \\ y + z &= 0 \\ x - w &= 4 \\ -x + y &= -3 \end{aligned}$

39. $\begin{aligned} x + 3y - 2z - w &= 9 \\ 4x + y + z + 2w &= 2 \\ -3x - y + z - w &= -5 \\ x - y - 3z - 2w &= 2 \end{aligned}$

40. $\begin{aligned} 2x + y - z + 3w &= 0 \\ 3x - 2y + z - 4w &= -24 \\ x + y - z + w &= 2 \\ x - y + 2z - 5w &= -16 \end{aligned}$

Solve each system using a graphing calculator capable of performing row operations. Give solutions with values correct to the nearest thousandth.

41. $\begin{aligned} .3x + 2.7y - \sqrt{2}z &= 3 \\ \sqrt{7}x - 20y + 12z &= -2 \\ 4x + \sqrt{3}y - 1.2z &= \frac{3}{4} \end{aligned}$

42. $\begin{aligned} \sqrt{5}x - 1.2y + z &= -3 \\ \frac{1}{2}x - 3y + 4z &= \frac{4}{3} \\ 4x + 7y - 9z &= \sqrt{2} \end{aligned}$

Graph each system of three equations together on the same axes and determine the number of solutions (exactly one, none, or infinitely many). If there is exactly one solution, estimate the solution. Then confirm your answer by solving the system with the Gauss-Jordan method.

43. $\begin{aligned} 2x + 3y &= 5 \\ -3x + 5y &= 22 \\ 2x + y &= -1 \end{aligned}$

44. $\begin{aligned} 3x - 2y &= 3 \\ -2x + 4y &= 14 \\ x + y &= 11 \end{aligned}$

For each equation, determine the constants A and B that make the equation an identity. (Hint: Combine terms on the right, and set coefficients of corresponding terms in the numerators equal.)

45. $\frac{1}{(x-1)(x+1)} = \frac{A}{x-1} + \frac{B}{x+1}$

46. $\frac{x+4}{x^2} = \frac{A}{x} + \frac{B}{x^2}$

47. $\frac{x}{(x-a)(x+a)} = \frac{A}{x-a} + \frac{B}{x+a}$

48. $\frac{2x}{(x+2)(x-1)} = \frac{A}{x+2} + \frac{B}{x-1}$

Solve each problem using matrices.

49. ***Mixing Acid Solutions*** A chemist has two prepared acid solutions, one of which is 2% acid by volume, another 7% acid. How many cubic centimeters of each should the chemist mix together to obtain 40 cm^3 of a 3.2% acid solution?

50. *Financing an Expansion* To get the necessary funds for a planned expansion, a small company took out three loans totaling \$25,000. The company was able to borrow some of the money at 8% interest. It borrowed \$2000 more than one-half the amount of the 8% loan at 10%, and the rest at 9%. The total annual interest was \$2220. How much did the company borrow at each rate?

Amount Invested	Rate (in %)	Annual Interest
x	8	
y	10	
z	9	
		2220

51. *Financing an Expansion* In Exercise 50, suppose we drop the condition that the amount borrowed at 10% is \$2000 more than one-half the amount borrowed at 8%. How is the solution changed?

52. *Financing an Expansion* Suppose the company in Exercise 50 can borrow only \$6000 at 9%. Is a solution possible that still meets the given conditions? Explain.

53. *Planning a Diet* A hospital dietician is planning a special diet for a certain patient. The total amount per meal of food groups A, B, and C must equal 400 g. The diet should include one-third as much of group A as of group B, and the sum of the amounts of group A and group C should equal twice the amount of group B. How many grams of each food group should be included? (Give answers to the nearest tenth.)

				Total
Food Group	A	B	C	
Grams/Meal	x			400

54. *Planning a Diet* In Exercise 53, suppose that, in addition to the conditions given there, foods A and B cost 2¢ per gram and food C costs 3¢ per gram, and that a meal must cost \$8. Is a solution possible? Explain.

(Modeling) Age Distribution in the United States *The age distribution in the United States has been steadily shifting. As people live longer, a larger percent of the population is 65 or over and a smaller percent is in younger age brackets. Use matrices to solve the problems in Exercises 55 and 56. Let* $x = 0$ *represent* 2000 *and* $x = 50$ *represent* 2050. *Express percents in decimal form.*

55. In 2000, 12.4% of the population was 65 or older. By 2050, this percent is expected to be 20.3%. The percent of the population age 25–34 in 2000 was 14.2%. That age group is expected to include 12.5% of the population in 2050. (*Source:* U.S. Bureau of the Census.)

(a) Assuming these population changes are linear, use the data for the 65 or over age group to write a linear equation. Then do the same for the 25–34 age group. (Use three significant figures.)

(b) Solve the system of linear equations from part (a). In what year will the two age groups include the same percent of the population? What is that percent?

56. In 2000, 16.0% of the U.S. population was age 35–44. This percent is expected to decrease to 12.3% in 2050. (*Source:* U.S. Bureau of the Census.)

(a) Write a linear equation representing this population change. (Use three significant figures.)

(b) Solve the system containing the equation from part (a) and the equation from Exercise 55 for the 65 or older age group. Give the year and percent when these two age groups will include the same percent of the population.

57. *(Modeling) Athlete's Weight and Height* The relationship between a professional basketball player's height H (in inches) and weight W (in pounds) was modeled using two different samples of players. The resulting equations that modeled each sample were

$$W = 7.46H - 374 \quad \text{and} \quad W = 7.93H - 405.$$

(a) Use each equation to predict the weight of a 6 ft 11 in. professional basketball player.

(b) According to each model, what change in weight is associated with a 1-in. increase in height?

(c) Determine the weight and height where the two models agree.

58. *(Modeling) Traffic Congestion* At rush hours, substantial traffic congestion is encountered at the traffic intersections shown in the figure. (All streets are one-way.) The city wishes to improve the signals at these corners to speed the flow of traffic. The traffic engineers first gather data. As the figure shows, 700 cars per hour come down M Street to intersection A, and 300 cars per hour come to intersection A on 10th Street. A total of x_1 of these cars leave A on M Street, while x_4 cars leave A on 10th Street. The number of cars entering A must equal the number leaving, so

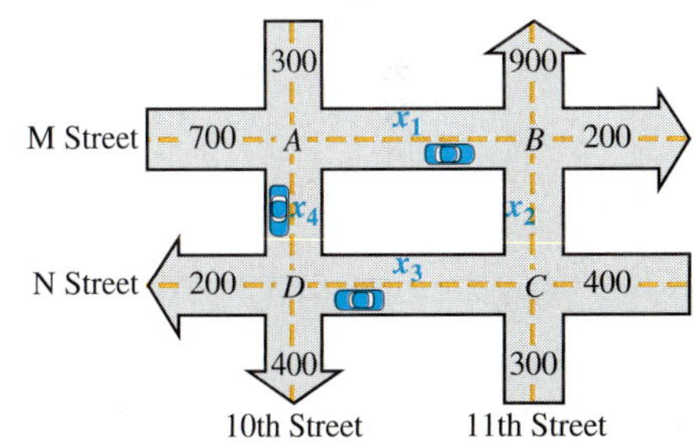

$$x_1 + x_4 = 700 + 300$$
$$x_1 + x_4 = 1000.$$

For intersection B, x_1 cars enter B on M Street, and x_2 cars enter B on 11th Street. The figure shows that 900 cars leave B on 11th, while 200 leave on M. We have

$$x_1 + x_2 = 900 + 200$$
$$x_1 + x_2 = 1100.$$

At intersection C, 400 cars enter on N Street and 300 on 11th Street, while x_2 leave on 11th Street and x_3 leave on N Street. This gives

$$x_2 + x_3 = 400 + 300$$
$$x_2 + x_3 = 700.$$

Finally, intersection D has x_3 cars entering on N and x_4 entering on 10th. There are 400 cars leaving D on 10th and 200 leaving on N.

(a) Set up an equation for intersection D.

(b) Use the four equations to write an augmented matrix, and then reduce by 1s on the diagonal to obtain 0s below. This is called *triangular form.*

(c) Since you got a row of all 0s, the system of equations does not have a unique solution. Write three equations, corresponding to the three nonzero rows of the matrix. Solve each of the equations for x_4.

(d) One of your equations should have been $x_4 = 1000 - x_1$. What is the largest possible value of x_1 so that x_4 is not negative?

(e) Another equation should have been $x_4 = x_2 - 100$. Find the smallest possible value of x_2 so that x_4 is not negative.

(f) Find the largest possible values of x_3 and x_4 so that neither variable is negative.

(g) Use the results of parts (a)–(f) to give a solution for the problem in which all the equations are satisfied and all variables are nonnegative. Is the solution unique?

Relating Concepts

For individual or collaborative investigation
(Exercises 59–62)

(Modeling) Number of Fawns *To model the spring fawn count F from the adult pronghorn population A, the precipitation P, and the severity of the winter W, environmentalists have used the equation*

$$F = a + bA + cP + dW,$$

where the coefficients a, b, c, and d are constants that must be determined before using the equation. (Winter severity is scaled between 1 and 5, with 1 being mild and 5 being severe.) ***Work Exercises 59–62 in order.*** *(Source:* Brase, C. and C. Brase, *Understandable Statistics,* D.C. Heath and Company, 1995; Bureau of Land Management.*)*

59. Substitute the values for F, A, P, and W from the table for Years 1–4 into the equation $F = a + bA + cP + dW$ and obtain four linear equations involving a, b, c, and d.

Year	Fawns	Adults	Precip. (in inches)	Winter Severity
1	239	871	11.5	3
2	234	847	12.2	2
3	192	685	10.6	5
4	343	969	14.2	1
5	?	960	12.6	3

60. Write an augmented matrix representing the system in Exercise 59, and solve for a, b, c, and d.

61. Write the equation for F using the values found in Exercise 60 for the coefficients.

62. Use the information in the table to predict the spring fawn count in Year 5. (Compare this with the actual count of 320.)

9.3 Determinant Solution of Linear Systems

Determinants ▪ Cofactors ▪ Evaluating $n \times n$ Determinants ▪ Cramer's Rule

Determinants

Every $n \times n$ matrix A is associated with a real number called the **determinant** of A, written $|A|$. The determinant of a 2×2 matrix is defined as follows.

Looking Ahead to Calculus

Determinants are used in calculus to find *vector cross products,* which are used to study the effect of forces in the plane or in space.

Determinant of a 2 × 2 Matrix

If $A = \begin{bmatrix} a_{11} & a_{12} \\ a_{21} & a_{22} \end{bmatrix}$, then $|A| = \begin{vmatrix} a_{11} & a_{12} \\ a_{21} & a_{22} \end{vmatrix} = a_{11}a_{22} - a_{21}a_{12}$.

NOTE Matrices are enclosed with square brackets, while determinants are denoted with vertical bars. Also, a matrix is an *array* of numbers, but its determinant is a *single* number.

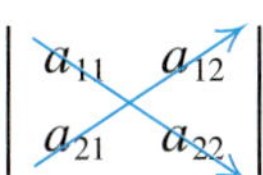

The arrows in the diagram in the margin will remind you which products to find when evaluating a 2×2 determinant.

EXAMPLE 1 Evaluating a 2 × 2 Determinant

Let $A = \begin{bmatrix} -3 & 4 \\ 6 & 8 \end{bmatrix}$. Find $|A|$.

Algebraic Solution

Use the definition with

$$a_{11} = -3, \quad a_{12} = 4, \quad a_{21} = 6, \quad a_{22} = 8.$$

$$|A| = \underset{a_{11}}{-3} \cdot \underset{a_{22}}{8} - \underset{a_{21}}{6} \cdot \underset{a_{12}}{4}$$

$$= -24 - 24$$

$$= -48$$

Graphing Calculator Solution

We can define a matrix and then use the capability of a graphing calculator to find the determinant of the matrix. In the screen in Figure 8, the symbol det([A]) represents the determinant of [A].

```
[A]
        [[-3  4]
         [6   8]]
det([A])
            -48
```

Figure 8

Now try Exercise 1.

The determinant of a 3×3 matrix A is defined as follows.

Determinant of a 3 × 3 Matrix

If $A = \begin{bmatrix} a_{11} & a_{12} & a_{13} \\ a_{21} & a_{22} & a_{23} \\ a_{31} & a_{32} & a_{33} \end{bmatrix}$, then

$$|A| = \begin{vmatrix} a_{11} & a_{12} & a_{13} \\ a_{21} & a_{22} & a_{23} \\ a_{31} & a_{32} & a_{33} \end{vmatrix} = (a_{11}a_{22}a_{33} + a_{12}a_{23}a_{31} + a_{13}a_{21}a_{32}) - (a_{31}a_{22}a_{13} + a_{32}a_{23}a_{11} + a_{33}a_{21}a_{12}).$$

The terms on the right side of the equation in the definition of $|A|$ can be rearranged to get

$$\begin{vmatrix} a_{11} & a_{12} & a_{13} \\ a_{21} & a_{22} & a_{23} \\ a_{31} & a_{32} & a_{33} \end{vmatrix} = a_{11}(a_{22}a_{33} - a_{32}a_{23}) - a_{21}(a_{12}a_{33} - a_{32}a_{13}) + a_{31}(a_{12}a_{23} - a_{22}a_{13}).$$

Each quantity in parentheses represents the determinant of a 2×2 matrix that is the part of the 3×3 matrix remaining when the row and column of the multiplier are eliminated, as shown below.

$$a_{11}(a_{22}a_{33} - a_{32}a_{23}) \qquad \begin{bmatrix} a_{11} & a_{12} & a_{13} \\ a_{21} & a_{22} & a_{23} \\ a_{31} & a_{32} & a_{33} \end{bmatrix}$$

$$a_{21}(a_{12}a_{33} - a_{32}a_{13}) \qquad \begin{bmatrix} a_{11} & a_{12} & a_{13} \\ a_{21} & a_{22} & a_{23} \\ a_{31} & a_{32} & a_{33} \end{bmatrix}$$

$$a_{31}(a_{12}a_{23} - a_{22}a_{13}) \qquad \begin{bmatrix} a_{11} & a_{12} & a_{13} \\ a_{21} & a_{22} & a_{23} \\ a_{31} & a_{32} & a_{33} \end{bmatrix}$$

Cofactors The determinant of each 2×2 matrix above is called the **minor** of the associated element in the 3×3 matrix. The symbol M_{ij} represents the minor that results when row i and column j are eliminated. The following list gives some of the minors from the matrix above.

Element	Minor	Element	Minor
a_{11}	$M_{11} = \begin{vmatrix} a_{22} & a_{23} \\ a_{32} & a_{33} \end{vmatrix}$	a_{22}	$M_{22} = \begin{vmatrix} a_{11} & a_{13} \\ a_{31} & a_{33} \end{vmatrix}$
a_{21}	$M_{21} = \begin{vmatrix} a_{12} & a_{13} \\ a_{32} & a_{33} \end{vmatrix}$	a_{23}	$M_{23} = \begin{vmatrix} a_{11} & a_{12} \\ a_{31} & a_{32} \end{vmatrix}$
a_{31}	$M_{31} = \begin{vmatrix} a_{12} & a_{13} \\ a_{22} & a_{23} \end{vmatrix}$	a_{33}	$M_{33} = \begin{vmatrix} a_{11} & a_{12} \\ a_{21} & a_{22} \end{vmatrix}$

In a 4×4 matrix, the minors are determinants of 3×3 matrices. Similarly, an $n \times n$ matrix has minors that are determinants of $(n - 1) \times (n - 1)$ matrices.

To find the determinant of a 3×3 or larger matrix, first choose any row or column. Then the minor of each element in that row or column must be multiplied by $+1$ or -1, depending on whether the sum of the row number and column number is even or odd. The product of a minor and the number $+1$ or -1 is called a *cofactor*.

Cofactor

Let M_{ij} be the minor for element a_{ij} in an $n \times n$ matrix. The **cofactor** of a_{ij}, written A_{ij}, is

$$\mathbf{A_{ij} = (-1)^{i+j} \cdot M_{ij}.}$$

EXAMPLE 2 Finding Cofactors of Elements

Find the cofactor of each of the following elements of the matrix

$$\begin{bmatrix} 6 & 2 & 4 \\ 8 & 9 & 3 \\ 1 & 2 & 0 \end{bmatrix}.$$

(a) 6 **(b)** 3 **(c)** 8

Solution

(a) Since 6 is in the first row, first column of the matrix, $i = 1$ and $j = 1$ so $M_{11} = \begin{vmatrix} 9 & 3 \\ 2 & 0 \end{vmatrix} = -6$. The cofactor is

$$(-1)^{1+1}(-6) = 1(-6) = -6.$$

(b) Here $i = 2$ and $j = 3$, so $M_{23} = \begin{vmatrix} 6 & 2 \\ 1 & 2 \end{vmatrix} = 10$. The cofactor is

$$(-1)^{2+3}(10) = -1(10) = -10.$$

(c) We have $i = 2$ and $j = 1$, and $M_{21} = \begin{vmatrix} 2 & 4 \\ 2 & 0 \end{vmatrix} = -8$. The cofactor is

$$(-1)^{2+1}(-8) = -1(-8) = 8.$$

Now try Exercise 11.

Evaluating $n \times n$ Determinants The determinant of a 3×3 or larger matrix is found as follows.

Finding the Determinant of a Matrix

Multiply each element in any row or column of the matrix by its cofactor. The sum of these products gives the value of the determinant.

The process of forming this sum of products is called **expansion by a given row or column.**

EXAMPLE 3 Evaluating a 3 × 3 Determinant

Evaluate $\begin{vmatrix} 2 & -3 & -2 \\ -1 & -4 & -3 \\ -1 & 0 & 2 \end{vmatrix}$, expanding by the second column.

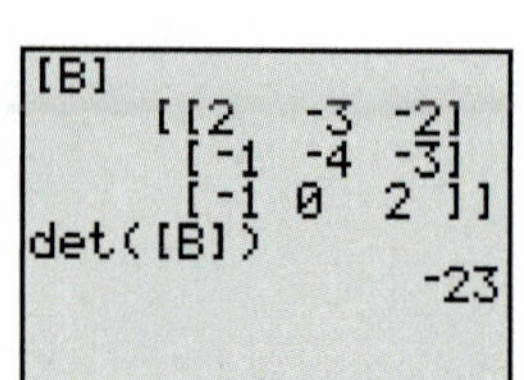

The matrix in Example 3 can be entered as [B] and its determinant found using a graphing calculator, as shown in the screen above.

Solution To find this determinant, first find the minors of each element in the second column.

$$M_{12} = \begin{vmatrix} -1 & -3 \\ -1 & 2 \end{vmatrix} = -1(2) - (-1)(-3) = -5$$

$$M_{22} = \begin{vmatrix} 2 & -2 \\ -1 & 2 \end{vmatrix} = 2(2) - (-1)(-2) = 2$$

$$M_{32} = \begin{vmatrix} 2 & -2 \\ -1 & -3 \end{vmatrix} = 2(-3) - (-1)(-2) = -8$$

Now find the cofactor of each element of these minors.

$$A_{12} = (-1)^{1+2} \cdot M_{12} = (-1)^3 \cdot (-5) = -1(-5) = 5$$
$$A_{22} = (-1)^{2+2} \cdot M_{22} = (-1)^4 \cdot 2 = 1 \cdot 2 = 2$$
$$A_{32} = (-1)^{3+2} \cdot M_{32} = (-1)^5 \cdot (-8) = -1(-8) = 8$$

Find the determinant by multiplying each cofactor by its corresponding element in the matrix and finding the sum of these products.

$$\begin{vmatrix} 2 & -3 & -2 \\ -1 & -4 & -3 \\ -1 & 0 & 2 \end{vmatrix} = a_{12} \cdot A_{12} + a_{22} \cdot A_{22} + a_{32} \cdot A_{32}$$

$$= -3(5) + (-4)2 + 0(8)$$
$$= -15 + (-8) + 0 = -23$$

Now try Exercise 15.

In Example 3, we would have found the same answer using any row or column of the matrix. One reason we used column 2 is that it contains a 0 element, so it was not really necessary to calculate M_{32} and A_{32}.

Instead of calculating $(-1)^{i+j}$ for a given element, the sign checkerboard in the margin can be used. The signs alternate for each row and column, beginning with + in the first row, first column position. If we expand a 3 × 3 matrix about row 3, for example, the first minor would have a + sign associated with it, the second minor a − sign, and the third minor a + sign. This array of signs can be extended for determinants of 4 × 4 and larger matrices.

For 3 × 3 matrices

+	−	+
−	+	−
+	−	+

Cramer's Rule Determinants can be used to solve a linear system in the form

$$a_1x + b_1y = c_1 \quad (1)$$
$$a_2x + b_2y = c_2 \quad (2)$$

by elimination as follows.

$$\begin{aligned} a_1b_2x + b_1b_2y &= c_1b_2 && \text{Multiply (1) by } b_2. \\ -a_2b_1x - b_1b_2y &= -c_2b_1 && \text{Multiply (2) by } -b_1. \\ \hline (a_1b_2 - a_2b_1)x \qquad &= c_1b_2 - c_2b_1 && \text{Add.} \\ x &= \frac{c_1b_2 - c_2b_1}{a_1b_2 - a_2b_1}, && \text{if } a_1b_2 - a_2b_1 \neq 0. \end{aligned}$$

Similarly,

$$\begin{aligned} -a_1a_2x - a_2b_1y &= -a_2c_1 && \text{Multiply (1) by } -a_2. \\ a_1a_2x + a_1b_2y &= a_1c_2 && \text{Multiply (2) by } a_1. \\ \hline (a_1b_2 - a_2b_1)y &= a_1c_2 - a_2c_1 && \text{Add.} \\ y &= \frac{a_1c_2 - a_2c_1}{a_1b_2 - a_2b_1}, && \text{if } a_1b_2 - a_2b_1 \neq 0. \end{aligned}$$

Both numerators and the common denominator of these values for x and y can be written as determinants, since

$$c_1b_2 - c_2b_1 = \begin{vmatrix} c_1 & b_1 \\ c_2 & b_2 \end{vmatrix}, \quad a_1c_2 - a_2c_1 = \begin{vmatrix} a_1 & c_1 \\ a_2 & c_2 \end{vmatrix}, \quad \text{and} \quad a_1b_2 - a_2b_1 = \begin{vmatrix} a_1 & b_1 \\ a_2 & b_2 \end{vmatrix}.$$

Using these determinants, the solutions for x and y become

$$x = \frac{\begin{vmatrix} c_1 & b_1 \\ c_2 & b_2 \end{vmatrix}}{\begin{vmatrix} a_1 & b_1 \\ a_2 & b_2 \end{vmatrix}} \quad \text{and} \quad y = \frac{\begin{vmatrix} a_1 & c_1 \\ a_2 & c_2 \end{vmatrix}}{\begin{vmatrix} a_1 & b_1 \\ a_2 & b_2 \end{vmatrix}}, \quad \text{if} \begin{vmatrix} a_1 & b_1 \\ a_2 & b_2 \end{vmatrix} \neq 0.$$

We denote the three determinants in the solution as

$$\begin{vmatrix} a_1 & b_1 \\ a_2 & b_2 \end{vmatrix} = D, \quad \begin{vmatrix} c_1 & b_1 \\ c_2 & b_2 \end{vmatrix} = D_x, \quad \text{and} \quad \begin{vmatrix} a_1 & c_1 \\ a_2 & c_2 \end{vmatrix} = D_y.$$

NOTE The elements of D are the four coefficients of the variables in the given system. The elements of D_x are obtained by replacing the coefficients of x in D by the respective constants, and the elements of D_y are obtained by replacing the coefficients of y in D by the respective constants.

These results are summarized as **Cramer's rule.**

Cramer's Rule for Two Equations in Two Variables

Given the system

$$\begin{aligned} a_1x + b_1y &= c_1 \\ a_2x + b_2y &= c_2, \end{aligned}$$

if $D \neq 0$, then the system has the unique solution

$$\boldsymbol{x = \frac{D_x}{D}} \quad \text{and} \quad \boldsymbol{y = \frac{D_y}{D}},$$

where $D = \begin{vmatrix} a_1 & b_1 \\ a_2 & b_2 \end{vmatrix}$, $\quad D_x = \begin{vmatrix} c_1 & b_1 \\ c_2 & b_2 \end{vmatrix}$, $\quad$ and $D_y = \begin{vmatrix} a_1 & c_1 \\ a_2 & c_2 \end{vmatrix}$.

CAUTION As indicated in the preceding box, Cramer's rule does not apply if $D = 0$. When $D = 0$, the system is inconsistent or has infinitely many solutions. For this reason, evaluate D first.

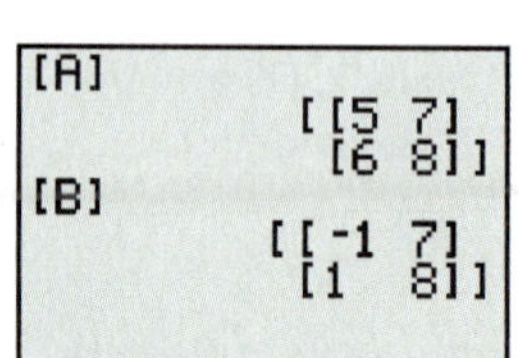

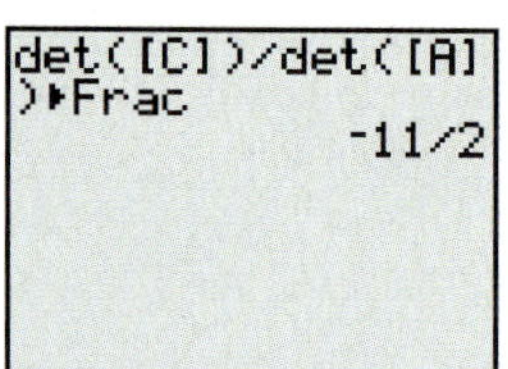

Because graphing calculators can evaluate determinants, they can also be used to apply Cramer's rule to solve a system of linear equations. The screens above support the result in Example 4.

EXAMPLE 4 Applying Cramer's Rule to a 2 × 2 System

Use Cramer's rule to solve the system.

$$5x + 7y = -1$$
$$6x + 8y = 1$$

Solution By Cramer's rule, $x = \frac{D_x}{D}$ and $y = \frac{D_y}{D}$. Find D first, since if $D = 0$, Cramer's rule does not apply. If $D \neq 0$, then find D_x and D_y.

$$D = \begin{vmatrix} 5 & 7 \\ 6 & 8 \end{vmatrix} = 5(8) - 6(7) = -2$$

$$D_x = \begin{vmatrix} -1 & 7 \\ 1 & 8 \end{vmatrix} = -1(8) - 1(7) = -15$$

$$D_y = \begin{vmatrix} 5 & -1 \\ 6 & 1 \end{vmatrix} = 5(1) - 6(-1) = 11$$

By Cramer's rule,

$$x = \frac{D_x}{D} = \frac{-15}{-2} = \frac{15}{2} \quad \text{and} \quad y = \frac{D_y}{D} = \frac{11}{-2} = -\frac{11}{2}.$$

The solution set is $\left\{\left(\frac{15}{2}, -\frac{11}{2}\right)\right\}$, as can be verified by substituting in the given system.

Now try Exercise 63.

By much the same method as used earlier for two equations, Cramer's rule can be generalized to a system of n linear equations with n variables.

General Form of Cramer's Rule

Let an $n \times n$ system have linear equations of the form

$$a_1x_1 + a_2x_2 + a_3x_3 + \cdots + a_nx_n = b.$$

Define D as the determinant of the $n \times n$ matrix of all coefficients of the variables. Define D_{x_1} as the determinant obtained from D by replacing the entries in column 1 of D with the constants of the system. Define D_{x_i} as the determinant obtained from D by replacing the entries in column i with the constants of the system. If $D \neq 0$, the unique solution of the system is

$$x_1 = \frac{D_{x_1}}{D}, \quad x_2 = \frac{D_{x_2}}{D}, \quad x_3 = \frac{D_{x_3}}{D}, \ldots, \quad x_n = \frac{D_{x_n}}{D}.$$

EXAMPLE 5 Applying Cramer's Rule to a 3 × 3 System

Use Cramer's rule to solve the system.

$$\begin{aligned} x + y - z + 2 &= 0 \\ 2x - y + z + 5 &= 0 \\ x - 2y + 3z - 4 &= 0 \end{aligned}$$

Solution

$$\begin{aligned} x + y - z &= -2 \\ 2x - y + z &= -5 \quad \text{Rewrite the system.} \\ x - 2y + 3z &= 4 \end{aligned}$$

Verify that the required determinants are

$$D = \begin{vmatrix} 1 & 1 & -1 \\ 2 & -1 & 1 \\ 1 & -2 & 3 \end{vmatrix} = -3, \qquad D_x = \begin{vmatrix} -2 & 1 & -1 \\ -5 & -1 & 1 \\ 4 & -2 & 3 \end{vmatrix} = 7,$$

$$D_y = \begin{vmatrix} 1 & -2 & -1 \\ 2 & -5 & 1 \\ 1 & 4 & 3 \end{vmatrix} = -22, \qquad D_z = \begin{vmatrix} 1 & 1 & -2 \\ 2 & -1 & -5 \\ 1 & -2 & 4 \end{vmatrix} = -21.$$

Thus, $\quad x = \dfrac{D_x}{D} = \dfrac{7}{-3} = -\dfrac{7}{3}, \qquad y = \dfrac{D_y}{D} = \dfrac{-22}{-3} = \dfrac{22}{3},$

and $\quad z = \dfrac{D_z}{D} = \dfrac{-21}{-3} = 7,$

so the solution set is $\left\{\left(-\frac{7}{3}, \frac{22}{3}, 7\right)\right\}$.

Now try Exercise 73.

CAUTION As shown in Example 5, each equation in the system must be written in the form $ax + by + cz + \cdots = k$ before using Cramer's rule.

EXAMPLE 6 Showing That Cramer's Rule Does Not Apply

Show that Cramer's rule does not apply to the following system.

$$\begin{aligned} 2x - 3y + 4z &= 10 \\ 6x - 9y + 12z &= 24 \\ x + 2y - 3z &= 5 \end{aligned}$$

Solution We need to show that $D = 0$. Expanding about column 1 gives

$$D = \begin{vmatrix} 2 & -3 & 4 \\ 6 & -9 & 12 \\ 1 & 2 & -3 \end{vmatrix} = 2\begin{vmatrix} -9 & 12 \\ 2 & -3 \end{vmatrix} - 6\begin{vmatrix} -3 & 4 \\ 2 & -3 \end{vmatrix} + 1\begin{vmatrix} -3 & 4 \\ -9 & 12 \end{vmatrix}$$

$$= 2(3) - 6(1) + 1(0) = 0.$$

Since $D = 0$, Cramer's rule does not apply.

Now try Exercise 77.

NOTE When $D = 0$, as in Example 6, the system is either inconsistent or has infinitely many solutions. Use the elimination method to tell which is the case. Verify that the system in Example 6 is inconsistent, so the solution set is $\emptyset$.

9.3 Exercises

Find the value of each determinant. See Example 1.

1. $\begin{vmatrix} -5 & 9 \\ 4 & -1 \end{vmatrix}$
2. $\begin{vmatrix} -1 & 3 \\ -2 & 9 \end{vmatrix}$
3. $\begin{vmatrix} -1 & -2 \\ 5 & 3 \end{vmatrix}$
4. $\begin{vmatrix} 6 & -4 \\ 0 & -1 \end{vmatrix}$
5. $\begin{vmatrix} 9 & 3 \\ -3 & -1 \end{vmatrix}$
6. $\begin{vmatrix} 0 & 2 \\ 1 & 5 \end{vmatrix}$
7. $\begin{vmatrix} 3 & 4 \\ 5 & -2 \end{vmatrix}$
8. $\begin{vmatrix} -9 & 7 \\ 2 & 6 \end{vmatrix}$
9. $\begin{vmatrix} -7 & 0 \\ 3 & 0 \end{vmatrix}$
10. *Concept Check* Refer to Exercise 9. Make a conjecture about the value of the determinant of a matrix that has a column of 0s.

Find the cofactor of each element in the second row for each determinant. See Example 2.

11. $\begin{vmatrix} -2 & 0 & 1 \\ 1 & 2 & 0 \\ 4 & 2 & 1 \end{vmatrix}$
12. $\begin{vmatrix} 1 & -1 & 2 \\ 1 & 0 & 2 \\ 0 & -3 & 1 \end{vmatrix}$
13. $\begin{vmatrix} 1 & 2 & -1 \\ 2 & 3 & -2 \\ -1 & 4 & 1 \end{vmatrix}$
14. $\begin{vmatrix} 2 & -1 & 4 \\ 3 & 0 & 1 \\ -2 & 1 & 4 \end{vmatrix}$

Find the value of each determinant. See Example 3.

15. $\begin{vmatrix} 4 & -7 & 8 \\ 2 & 1 & 3 \\ -6 & 3 & 0 \end{vmatrix}$
16. $\begin{vmatrix} 8 & -2 & -4 \\ 7 & 0 & 3 \\ 5 & -1 & 2 \end{vmatrix}$
17. $\begin{vmatrix} 1 & 2 & 0 \\ -1 & 2 & -1 \\ 0 & 1 & 4 \end{vmatrix}$
18. $\begin{vmatrix} 2 & 1 & -1 \\ 4 & 7 & -2 \\ 2 & 4 & 0 \end{vmatrix}$
19. $\begin{vmatrix} 10 & 2 & 1 \\ -1 & 4 & 3 \\ -3 & 8 & 10 \end{vmatrix}$
20. $\begin{vmatrix} 7 & -1 & 1 \\ 1 & -7 & 2 \\ -2 & 1 & 1 \end{vmatrix}$
21. $\begin{vmatrix} 1 & -2 & 3 \\ 0 & 0 & 0 \\ 1 & 10 & -12 \end{vmatrix}$
22. $\begin{vmatrix} 2 & 3 & 0 \\ 1 & 9 & 0 \\ -1 & -2 & 0 \end{vmatrix}$
23. $\begin{vmatrix} 3 & 3 & -1 \\ 2 & 6 & 0 \\ -6 & -6 & 2 \end{vmatrix}$
24. $\begin{vmatrix} 5 & -3 & 2 \\ -5 & 3 & -2 \\ 1 & 0 & 1 \end{vmatrix}$
25. $\begin{vmatrix} 1 & 0 & 0 \\ 0 & 1 & 0 \\ 0 & 0 & 1 \end{vmatrix}$
26. $\begin{vmatrix} 1 & 0 & 0 \\ 0 & -1 & 0 \\ 1 & 0 & 1 \end{vmatrix}$
27. $\begin{vmatrix} -2 & 0 & 1 \\ 0 & 1 & 0 \\ 0 & 0 & -1 \end{vmatrix}$
28. $\begin{vmatrix} 0 & 0 & -1 \\ -1 & 0 & 1 \\ 0 & -1 & 0 \end{vmatrix}$
29. $\begin{vmatrix} .4 & -.8 & .6 \\ .3 & .9 & .7 \\ 3.1 & 4.1 & -2.8 \end{vmatrix}$
30. $\begin{vmatrix} -.3 & -.1 & .9 \\ 2.5 & 4.9 & -3.2 \\ -.1 & .4 & .8 \end{vmatrix}$

Relating Concepts

For individual or collaborative investigation
(Exercises 31–34)

An example of a determinant equation is

$$\begin{vmatrix} x & 2 & 1 \\ -1 & x & 4 \\ -2 & 0 & 5 \end{vmatrix} = 45.$$

This equation can be solved by finding an expression in x for the determinant and then solving the resulting equation. **Work Exercises 31–34 in order,** *to see how to solve this equation.*

31. Use one of the methods described in this section to write the determinant as a polynomial in x.

32. Replace the determinant with the expression you found in Exercise 31. What kind of equation is this (based on the degree of the polynomial)?

33. Solve the equation found in Exercise 32.

34. Verify that when the solutions are substituted for x in the original determinant, the equation is satisfied.

Solve each equation for x. Refer to Relating Concepts Exercises 31–34.

35. $\begin{vmatrix} 5 & x \\ -3 & 2 \end{vmatrix} = 6$

36. $\begin{vmatrix} -.5 & 2 \\ x & x \end{vmatrix} = 0$

37. $\begin{vmatrix} x & 3 \\ x & x \end{vmatrix} = 4$

38. $\begin{vmatrix} 2x & x \\ 11 & x \end{vmatrix} = 6$

39. $\begin{vmatrix} -2 & 0 & 1 \\ -1 & 3 & x \\ 5 & -2 & 0 \end{vmatrix} = 3$

40. $\begin{vmatrix} 4 & 3 & 0 \\ 2 & 0 & 1 \\ -3 & x & -1 \end{vmatrix} = 5$

41. $\begin{vmatrix} 5 & 3x & -3 \\ 0 & 2 & -1 \\ 4 & -1 & x \end{vmatrix} = -7$

42. $\begin{vmatrix} 2x & 1 & -1 \\ 0 & 4 & x \\ 3 & 0 & 2 \end{vmatrix} = x$

Area of a Triangle *A triangle with vertices at* (x_1, y_1), (x_2, y_2), *and* (x_3, y_3), *as shown in the figure, has area equal to the absolute value of D, where*

$$D = \frac{1}{2}\begin{vmatrix} x_1 & y_1 & 1 \\ x_2 & y_2 & 1 \\ x_3 & y_3 & 1 \end{vmatrix}.$$

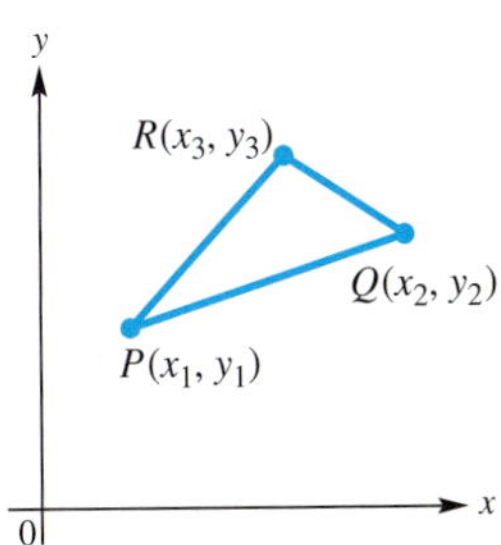

Find the area of each triangle having vertices at P, Q, and R.

43. $P(0, 0)$, $Q(0, 2)$, $R(1, 4)$

44. $P(0, 1)$, $Q(2, 0)$, $R(1, 5)$

45. $P(2, 5)$, $Q(-1, 3)$, $R(4, 0)$

46. $P(2, -2)$, $Q(0, 0)$, $R(-3, -4)$

47. ***Area of a Triangle*** Find the area of a triangular lot whose vertices have coordinates in feet of (101.3, 52.7), (117.2, 253.9), and (313.1, 301.6). (*Source:* Al-Khafaji, A. and J. Tooley, *Numerical Methods in Engineering Practice,* Holt, Rinehart, and Winston, 1995.)

48. Let $A = \begin{bmatrix} a_{11} & a_{12} & a_{13} \\ a_{21} & a_{22} & a_{23} \\ a_{31} & a_{32} & a_{33} \end{bmatrix}$. Find $|A|$ by expansion about row 3 of the matrix. Show that your result is really equal to $|A|$ as given in the definition of the determinant of a 3×3 matrix at the beginning of this section.

The following theorems are true for square matrices of any size.

Determinant Theorems

1. If every element in a row (or column) of matrix A is 0, then $|A| = 0$.
2. If the rows of matrix A are the corresponding columns of matrix B, then $|B| = |A|$.
3. If any two rows (or columns) of matrix A are interchanged to form matrix B, then $|B| = -|A|$.
4. Suppose matrix B is formed by multiplying every element of a row (or column) of matrix A by the real number k. Then $|B| = k \cdot |A|$.
5. If two rows (or columns) of a matrix A are identical, then $|A| = 0$.
6. Changing a row (or column) of a matrix by adding to it a constant times another row (or column) does not change the determinant of the matrix.

Use the determinant theorems to find the value of each determinant.

49. $\begin{vmatrix} 1 & 0 & 0 \\ 1 & 0 & 1 \\ 3 & 0 & 0 \end{vmatrix}$

50. $\begin{vmatrix} -1 & 2 & 4 \\ 4 & -8 & -16 \\ 3 & 0 & 5 \end{vmatrix}$

51. $\begin{vmatrix} 6 & 8 & -12 \\ -1 & 0 & 2 \\ 4 & 0 & -8 \end{vmatrix}$

52. $\begin{vmatrix} 4 & 8 & 0 \\ -1 & -2 & 1 \\ 2 & 4 & 3 \end{vmatrix}$

53. $\begin{vmatrix} -4 & 1 & 4 \\ 2 & 0 & 1 \\ 0 & 2 & 4 \end{vmatrix}$

54. $\begin{vmatrix} 6 & 3 & 2 \\ 1 & 0 & 2 \\ 5 & 7 & 3 \end{vmatrix}$

Evaluate each determinant.

55. $\begin{vmatrix} 3 & -6 & 5 & -1 \\ 0 & 2 & -1 & 3 \\ -6 & 4 & 2 & 0 \\ -7 & 3 & 1 & 1 \end{vmatrix}$

56. $\begin{vmatrix} 4 & 5 & -1 & -1 \\ 2 & -3 & 1 & 0 \\ -5 & 1 & 3 & 9 \\ 0 & -2 & 1 & 5 \end{vmatrix}$

57. $\begin{vmatrix} 4 & 0 & 0 & 2 \\ -1 & 0 & 3 & 0 \\ 2 & 4 & 0 & 1 \\ 0 & 0 & 1 & 2 \end{vmatrix}$

58. $\begin{vmatrix} -2 & 0 & 4 & 2 \\ 3 & 6 & 0 & 4 \\ 0 & 0 & 0 & 3 \\ 9 & 0 & 2 & -1 \end{vmatrix}$

59. *Concept Check* For the system below, match each determinant in (a)–(d) with its equivalent from choices A–D.

$$\begin{aligned} 4x + 3y - 2z &= 1 \\ 7x - 4y + 3z &= 2 \\ -2x + y - 8z &= 0 \end{aligned}$$

(a) D **(b)** D_x **(c)** D_y **(d)** D_z

A. $\begin{vmatrix} 1 & 3 & -2 \\ 2 & -4 & 3 \\ 0 & 1 & -8 \end{vmatrix}$ **B.** $\begin{vmatrix} 4 & 3 & 1 \\ 7 & -4 & 2 \\ -2 & 1 & 0 \end{vmatrix}$

C. $\begin{vmatrix} 4 & 1 & -2 \\ 7 & 2 & 3 \\ -2 & 0 & -8 \end{vmatrix}$ **D.** $\begin{vmatrix} 4 & 3 & -2 \\ 7 & -4 & 3 \\ -2 & 1 & -8 \end{vmatrix}$

60. *Concept Check* For the following system, $D = -43$, $D_x = -43$, $D_y = 0$, and $D_z = 43$. What is the solution set of the system?

$$\begin{aligned} x + 3y - 6z &= 7 \\ 2x - y + z &= 1 \\ x + 2y + 2z &= -1 \end{aligned}$$

Use Cramer's rule to solve each system of equations. If $D = 0$, use another method to determine the solution set. See Examples 4–6.

61. $x + y = 4$, $2x - y = 2$

62. $3x + 2y = -4$, $2x - y = -5$

63. $4x + 3y = -7$, $2x + 3y = -11$

64. $4x - y = 0$, $2x + 3y = 14$

65. $5x + 4y = 10$, $3x - 7y = 6$

66. $3x + 2y = -4$, $5x - y = 2$

67. $1.5x + 3y = 5$, $2x + 4y = 3$

68. $12x + 8y = 3$, $15x + 10y = 9$

69. $3x + 2y = 4$, $6x + 4y = 8$

70. $4x + 3y = 9$, $12x + 9y = 27$

71. $\frac{1}{2}x + \frac{1}{3}y = 2$, $\frac{3}{2}x - \frac{1}{2}y = -12$

72. $-\frac{3}{4}x + \frac{2}{3}y = 16$, $\frac{5}{2}x + \frac{1}{2}y = -37$

73. $2x - y + 4z + 2 = 0$, $3x + 2y - z + 3 = 0$, $x + 4y + 2z - 17 = 0$

74. $x + y + z - 4 = 0$, $2x - y + 3z - 4 = 0$, $4x + 2y - z + 15 = 0$

75. $4x - 3y + z = -1$, $5x + 7y + 2z = -2$, $3x - 5y - z = 1$

76. $2x - 3y + z = 8$, $-x - 5y + z = -4$, $3x - 5y + 2z = 12$

77. $x + 2y + 3z = 4$, $4x + 3y + 2z = 1$, $-x - 2y - 3z = 0$

78. $2x - y + 3z = 1$, $-2x + y - 3z = 2$, $5x - y + z = 2$

79. $-2x - 2y + 3z = 4$, $5x + 7y - z = 2$, $2x + 2y - 3z = -4$

80. $3x - 2y + 4z = 1$, $4x + y - 5z = 2$, $-6x + 4y - 8z = -2$

81. $5x - y = -4$, $3x + 2z = 4$, $4y + 3z = 22$

82. $3x + 5y = -7$, $2x + 7z = 2$, $4y + 3z = -8$

83. $x + 2y = 10$, $3x + 4z = 7$, $-y - z = 1$

84. $5x - 2y = 3$, $4y + z = 8$, $x + 2z = 4$

(Modeling) *Use one of the methods described in this section to solve the systems in Exercises 85 and 86.*

85. *Roof Trusses* Linear systems occur in the design of roof trusses for new homes and buildings. The simplest type of roof truss is a triangle. The truss shown in the figure is used to frame roofs of small buildings. If a 100-pound force is applied at the peak of the truss, then the forces or weights W_1 and W_2 exerted parallel to each rafter of the truss are determined by the following linear system of equations.

$$\frac{\sqrt{3}}{2}(W_1 + W_2) = 100$$

$$W_1 - W_2 = 0$$

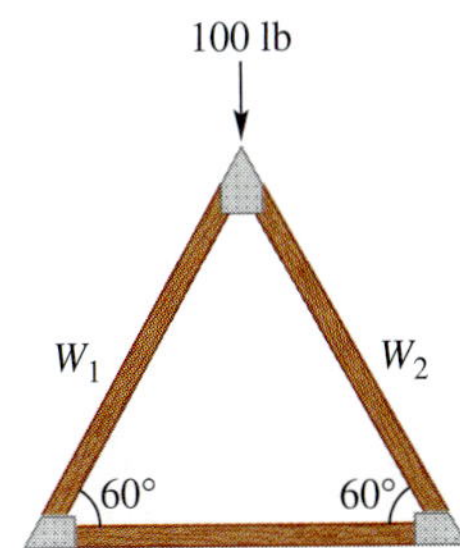

Solve the system to find W_1 and W_2. (*Source:* Hibbeler, R., *Structural Analysis,* 4th ed. © 1995. Reprinted by permission of Pearson Education, Inc. Upper Saddle River, NJ.)

86. *Roof Trusses* (Refer to Exercise 85.) Use the following system of equations to determine the forces or weights W_1 and W_2 exerted on each rafter for the truss shown in the figure.

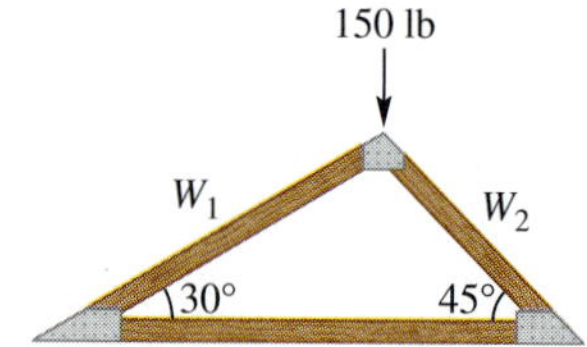

$$W_1 + \sqrt{2}W_2 = 300$$
$$\sqrt{3}W_1 - \sqrt{2}W_2 = 0$$

Solve each system for x and y using Cramer's rule. Assume a and b are nonzero constants.

87. $bx + y = a^2$
$ax + y = b^2$

88. $ax + by = \frac{b}{a}$
$x + y = \frac{1}{b}$

89. $b^2x + a^2y = b^2$
$ax + by = a$

90. $x + by = b$
$ax + y = a$

9.4 Partial Fractions

Decomposition of Rational Expressions ▪ Distinct Linear Factors ▪ Repeated Linear Factors ▪ Distinct Linear and Quadratic Factors ▪ Repeated Quadratic Factors

Add rational expressions

$$\frac{2}{x+1} + \frac{3}{x} = \frac{5x+3}{x(x+1)}$$

Partial fraction decomposition

Decomposition of Rational Expressions

The sums of rational expressions are found by combining two or more rational expressions into one rational expression. Here, the reverse process is considered: Given one rational expression, express it as the sum of two or more rational expressions. A special type of sum of rational expressions is called the **partial fraction decomposition;** each term in the sum is called a **partial fraction.** The technique of decomposing a rational expression into partial fractions is useful in calculus and other areas of mathematics.

To form a partial fraction decomposition of a rational expression, follow these steps.

Looking Ahead to Calculus

In calculus, partial fraction decomposition provides a powerful technique for determining integrals of rational functions.

Partial Fraction Decomposition of $\frac{f(x)}{g(x)}$

Step 1 If $\frac{f(x)}{g(x)}$ is not a proper fraction (a fraction with the numerator of lower degree than the denominator), divide $f(x)$ by $g(x)$. For example,

$$\frac{x^4 - 3x^3 + x^2 + 5x}{x^2 + 3} = x^2 - 3x - 2 + \frac{14x + 6}{x^2 + 3}.$$

Then apply the following steps to the remainder, which is a proper fraction.

Step 2 Factor the denominator $g(x)$ completely into factors of the form $(ax + b)^m$ or $(cx^2 + dx + e)^n$, where $cx^2 + dx + e$ is irreducible and m and n are positive integers.

(continued)

Step 3 **(a)** For each distinct linear factor $(ax + b)$, the decomposition must include the term $\dfrac{A}{ax + b}$.

(b) For each repeated linear factor $(ax + b)^m$, the decomposition must include the terms

$$\frac{A_1}{ax + b} + \frac{A_2}{(ax + b)^2} + \cdots + \frac{A_m}{(ax + b)^m}.$$

Step 4 **(a)** For each distinct quadratic factor $(cx^2 + dx + e)$, the decomposition must include the term $\dfrac{Bx + C}{cx^2 + dx + e}$.

(b) For each repeated quadratic factor $(cx^2 + dx + e)^n$, the decomposition must include the terms

$$\frac{B_1x + C_1}{cx^2 + dx + e} + \frac{B_2x + C_2}{(cx^2 + dx + e)^2} + \cdots + \frac{B_nx + C_n}{(cx^2 + dx + e)^n}.$$

Step 5 Use algebraic techniques to solve for the constants in the numerators of the decomposition.

To find the constants in Step 5, the goal is to get a system of equations with as many equations as there are unknowns in the numerators. One method for getting these equations is to substitute values for x on both sides of the rational equation formed in Step 3 or 4.

Distinct Linear Factors

EXAMPLE 1 Finding a Partial Fraction Decomposition

Find the partial fraction decomposition of $\dfrac{2x^4 - 8x^2 + 5x - 2}{x^3 - 4x}$.

Solution The given fraction is not a proper fraction; the numerator has greater degree than the denominator. Perform the division.

$$\begin{array}{r} 2x \phantom{{}- 8x^2 + 5x - 2} \\ x^3 - 4x \overline{)2x^4 - 8x^2 + 5x - 2} \\ \underline{2x^4 - 8x^2 \phantom{{}+ 5x - 2}} \\ 5x - 2 \end{array}$$ (Section R.3)

The quotient is $\dfrac{2x^4 - 8x^2 + 5x - 2}{x^3 - 4x} = 2x + \dfrac{5x - 2}{x^3 - 4x}$. Now, work with the remainder fraction. Factor the denominator as $x^3 - 4x = x(x + 2)(x - 2)$. Since the factors are distinct linear factors, use Step 3(a) to write the decomposition as

$$\frac{5x - 2}{x^3 - 4x} = \frac{A}{x} + \frac{B}{x + 2} + \frac{C}{x - 2}, \qquad (1)$$

where A, B, and C are constants that need to be found. Multiply both sides of equation (1) by $x(x+2)(x-2)$ to obtain

$$5x - 2 = A(x+2)(x-2) + Bx(x-2) + Cx(x+2). \quad (2)$$

Equation (1) is an identity since both sides represent the same rational expression. Thus, equation (2) is also an identity. Equation (1) holds for all values of x except 0, -2, and 2. However, equation (2) holds for all values of x. In particular, substituting 0 for x in equation (2) gives $-2 = -4A$, so $A = \frac{1}{2}$. Similarly, choosing $x = -2$ gives $-12 = 8B$, so $B = -\frac{3}{2}$. Finally, choosing $x = 2$ gives $8 = 8C$, so $C = 1$. The remainder rational expression can be written as the following sum of partial fractions:

$$\frac{5x-2}{x^3-4x} = \frac{1}{2x} + \frac{-3}{2(x+2)} + \frac{1}{x-2},$$

and the given rational expression can be written as

$$\frac{2x^4 - 8x^2 + 5x - 2}{x^3 - 4x} = 2x + \frac{1}{2x} + \frac{-3}{2(x+2)} + \frac{1}{x-2}.$$

Check the work by combining the terms on the right.

Now try Exercise 15.

Repeated Linear Factors

EXAMPLE 2 Finding a Partial Fraction Decomposition

Find the partial fraction decomposition of $\frac{2x}{(x-1)^3}$.

Solution This is a proper fraction. The denominator is already factored with repeated linear factors. Write the decomposition as shown, by using Step 3(b).

$$\frac{2x}{(x-1)^3} = \frac{A}{x-1} + \frac{B}{(x-1)^2} + \frac{C}{(x-1)^3}$$

Clear the denominators by multiplying both sides of this equation by $(x-1)^3$.

$$2x = A(x-1)^2 + B(x-1) + C \quad \text{(Section R.5)}$$

Substituting 1 for x leads to $C = 2$, so

$$2x = A(x-1)^2 + B(x-1) + 2. \quad (1)$$

The only root has been substituted, and values for A and B still need to be found. However, *any* number can be substituted for x. For example, when we choose $x = -1$ (because it is easy to substitute), equation (1) becomes

$$-2 = 4A - 2B + 2$$
$$-4 = 4A - 2B$$
$$-2 = 2A - B. \quad (2)$$

Substituting 0 for x in equation (1) gives

$$0 = A - B + 2$$
$$2 = -A + B. \quad (3)$$

Now, solve the system of equations (2) and (3) to get $A = 0$ and $B = 2$. The partial fraction decomposition is

$$\frac{2x}{(x-1)^3} = \frac{2}{(x-1)^2} + \frac{2}{(x-1)^3}.$$

We needed three substitutions because there were three constants to evaluate, A, B, and C. To check this result, we could combine the terms on the right.

Now try Exercise 11.

Distinct Linear and Quadratic Factors

EXAMPLE 3 Finding a Partial Fraction Decomposition

Find the partial fraction decomposition of $\dfrac{x^2 + 3x - 1}{(x+1)(x^2+2)}$.

Solution This denominator has distinct linear and quadratic factors, where neither is repeated. Since $x^2 + 2$ cannot be factored, it is irreducible. The partial fraction decomposition is

$$\frac{x^2 + 3x - 1}{(x+1)(x^2+2)} = \frac{A}{x+1} + \frac{Bx + C}{x^2 + 2}.$$

Multiply both sides by $(x+1)(x^2+2)$ to get

$$x^2 + 3x - 1 = A(x^2 + 2) + (Bx + C)(x + 1). \quad (1)$$

First, substitute -1 for x to get

$$(-1)^2 + 3(-1) - 1 = A[(-1)^2 + 2] + 0$$
$$-3 = 3A$$
$$A = -1.$$

Replace A with -1 in equation (1) and substitute any value for x. If $x = 0$, then

$$0^2 + 3(0) - 1 = -1(0^2 + 2) + (B \cdot 0 + C)(0 + 1)$$
$$-1 = -2 + C$$
$$C = 1.$$

Now, letting $A = -1$ and $C = 1$, substitute again in equation (1), using another value for x. If $x = 1$, then

$$3 = -3 + (B + 1)(2)$$
$$6 = 2B + 2$$
$$B = 2.$$

Using $A = -1$, $B = 2$, and $C = 1$, the partial fraction decomposition is

$$\frac{x^2 + 3x - 1}{(x+1)(x^2+2)} = \frac{-1}{x+1} + \frac{2x + 1}{x^2 + 2}.$$

Again, this work can be checked by combining terms on the right.

Now try Exercise 21.

For fractions with denominators that have quadratic factors, another method is often more convenient. The system of equations is formed by equating coefficients of like terms on both sides of the partial fraction decomposition. For instance, in Example 3, after both sides were multiplied by the common denominator, the equation was

$$x^2 + 3x - 1 = A(x^2 + 2) + (Bx + C)(x + 1).$$

Multiplying on the right and collecting like terms, we have

$$x^2 + 3x - 1 = Ax^2 + 2A + Bx^2 + Bx + Cx + C$$
$$x^2 + 3x - 1 = (A + B)x^2 + (B + C)x + (C + 2A).$$

Now, equating the coefficients of like powers of x gives the three equations

$$1 = A + B$$ (Section 9.1)
$$3 = B + C$$
$$-1 = C + 2A.$$

Solving this system of equations for A, B, and C would give the partial fraction decomposition. The next example uses a combination of the two methods.

Repeated Quadratic Factors

EXAMPLE 4 Finding a Partial Fraction Decomposition

Find the partial fraction decomposition of $\dfrac{2x}{(x^2 + 1)^2(x - 1)}$.

Solution This expression has both a linear factor and a repeated quadratic factor. By Steps 3(a) and 4(b) from the beginning of this section,

$$\frac{2x}{(x^2 + 1)^2(x - 1)} = \frac{Ax + B}{x^2 + 1} + \frac{Cx + D}{(x^2 + 1)^2} + \frac{E}{x - 1}.$$

Multiplying both sides by $(x^2 + 1)^2(x - 1)$ leads to

$$2x = (Ax + B)(x^2 + 1)(x - 1) + (Cx + D)(x - 1) + E(x^2 + 1)^2. \quad (1)$$

If $x = 1$, then equation (1) reduces to $2 = 4E$, or $E = \frac{1}{2}$. Substituting $\frac{1}{2}$ for E in equation (1) and combining terms on the right gives

$$2x = \left(A + \frac{1}{2}\right)x^4 + (-A + B)x^3 + (A - B + C + 1)x^2 +$$
$$(-A + B + D - C)x + \left(-B - D + \frac{1}{2}\right). \quad (2)$$

To get additional equations involving the unknowns, equate the coefficients of like powers of x on the two sides of equation (2). Setting corresponding coefficients of x^4 equal, $0 = A + \frac{1}{2}$ or $A = -\frac{1}{2}$. From the corresponding coefficients of x^3, $0 = -A + B$. Since $A = -\frac{1}{2}$, $B = -\frac{1}{2}$. Using the coefficients of x^2, $0 = A - B + C + 1$. Since $A = -\frac{1}{2}$ and $B = -\frac{1}{2}$, $C = -1$. Finally, from the coefficients of x, $2 = -A + B + D - C$. Substituting for A, B, and C gives

$D = 1$. With $A = -\frac{1}{2}$, $B = -\frac{1}{2}$, $C = -1$, $D = 1$, and $E = \frac{1}{2}$, the given fraction has the partial fraction decomposition

$$\frac{2x}{(x^2+1)^2(x-1)} = \frac{-\frac{1}{2}x - \frac{1}{2}}{x^2+1} + \frac{-x+1}{(x^2+1)^2} + \frac{\frac{1}{2}}{x-1}$$

or $$\frac{2x}{(x^2+1)^2(x-1)} = \frac{-(x+1)}{2(x^2+1)} + \frac{-x+1}{(x^2+1)^2} + \frac{1}{2(x-1)}.$$

Simplify complex fractions. (Section R.5)

Now try Exercise 25.

In summary, to solve for the constants in the numerators of a partial fraction decomposition, use either of the following methods or a combination of the two.

Decomposition into Partial Fractions

Method 1 For Linear Factors

1. Multiply both sides by the common denominator.
2. Substitute the zero of each factor in the resulting equation. For repeated linear factors, substitute as many other numbers as necessary to find all the constants in the numerators. The number of substitutions required will equal the number of constants $A, B, \ldots$.

Method 2 For Quadratic Factors

1. Multiply both sides by the common denominator.
2. Collect like terms on the right side of the resulting equation.
3. Equate the coefficients of like terms to get a system of equations.
4. Solve the system to find the constants in the numerators.

9.4 Exercises

Find the partial fraction decomposition for each rational expression. See Examples 1–4.

1. $\frac{5}{3x(2x+1)}$ **2.** $\frac{3x-1}{x(x+1)}$ **3.** $\frac{4x+2}{(x+2)(2x-1)}$

4. $\frac{x+2}{(x+1)(x-1)}$ **5.** $\frac{x}{x^2+4x-5}$ **6.** $\frac{5x-3}{(x+1)(x-3)}$

7. $\frac{2x}{(x+1)(x+2)^2}$ **8.** $\frac{2}{x^2(x+3)}$ **9.** $\frac{4}{x(1-x)}$

10. $\frac{4x^2-4x^3}{x^2(1-x)}$ **11.** $\frac{2x+1}{(x+2)^3}$ **12.** $\frac{4x^2-x-15}{x(x+1)(x-1)}$

13. $\frac{x^2}{x^2+2x+1}$ **14.** $\frac{3}{x^2+4x+3}$

15. $\dfrac{2x^5 + 3x^4 - 3x^3 - 2x^2 + x}{2x^2 + 5x + 2}$

16. $\dfrac{6x^5 + 7x^4 - x^2 + 2x}{3x^2 + 2x - 1}$

17. $\dfrac{x^3 + 4}{9x^3 - 4x}$

18. $\dfrac{x^3 + 2}{x^3 - 3x^2 + 2x}$

19. $\dfrac{-3}{x^2(x^2 + 5)}$

20. $\dfrac{2x + 1}{(x + 1)(x^2 + 2)}$

21. $\dfrac{3x - 2}{(x + 4)(3x^2 + 1)}$

22. $\dfrac{3}{x(x + 1)(x^2 + 1)}$

23. $\dfrac{1}{x(2x + 1)(3x^2 + 4)}$

24. $\dfrac{x^4 + 1}{x(x^2 + 1)^2}$

25. $\dfrac{3x - 1}{x(2x^2 + 1)^2}$

26. $\dfrac{3x^4 + x^3 + 5x^2 - x + 4}{(x - 1)(x^2 + 1)^2}$

27. $\dfrac{-x^4 - 8x^2 + 3x - 10}{(x + 2)(x^2 + 4)^2}$

28. $\dfrac{x^2}{x^4 - 1}$

29. $\dfrac{5x^5 + 10x^4 - 15x^3 + 4x^2 + 13x - 9}{x^3 + 2x^2 - 3x}$

30. $\dfrac{3x^6 + 3x^4 + 3x}{x^4 + x^2}$

Determine whether each partial fraction decomposition is correct by graphing the left side and the right side of the equation on the same coordinate axes and observing whether the graphs coincide.

31. $\dfrac{4x^2 - 3x - 4}{x^3 + x^2 - 2x} = \dfrac{2}{x} + \dfrac{-1}{x - 1} + \dfrac{3}{x + 2}$

32. $\dfrac{1}{(x - 1)(x + 2)} = \dfrac{1}{x - 1} - \dfrac{1}{x + 2}$

33. $\dfrac{x^3 - 2x}{(x^2 + 2x + 2)^2} = \dfrac{x - 2}{x^2 + 2x + 2} + \dfrac{2}{(x^2 + 2x + 2)^2}$

34. $\dfrac{2x + 4}{x^2(x - 2)} = \dfrac{-2}{x} + \dfrac{-2}{x^2} + \dfrac{2}{x - 2}$

9.5 Nonlinear Systems of Equations

Solving Nonlinear Systems with Real Solutions ▪ Solving Nonlinear Systems with Nonreal Complex Solutions ▪ Applying Nonlinear Systems

Solving Nonlinear Systems with Real Solutions A system of equations in which at least one equation is *not* linear is called a **nonlinear system.** The substitution method works well for solving many such systems, particularly when one of the equations is linear, as in the next example.

EXAMPLE 1 Solving a Nonlinear System by Substitution

Solve the system.

$$x^2 - y = 4 \quad (1)$$
$$x + y = -2 \quad (2)$$

Algebraic Solution

When one of the equations in a nonlinear system is linear, it is usually best to begin by solving the linear equation for one of the variables.

$y = -2 - x$ Solve equation (2) for y.

Substitute this result for y in equation (1).

$x^2 - (-2 - x) = 4$ (Section 9.1)

$x^2 + 2 + x = 4$ Distributive property (Section R.1)

$x^2 + x - 2 = 0$ Standard form

$(x + 2)(x - 1) = 0$ Factor. (Section R.4)

$x + 2 = 0$ or $x - 1 = 0$ Zero-factor property (Section 1.4)

$x = -2$ or $x = 1$

Substituting -2 for x in equation (2) gives $y = 0$. If $x = 1$, then $y = -3$. The solution set of the given system is $\{(-2, 0), (1, -3)\}$. A graph of the system is shown in Figure 9.

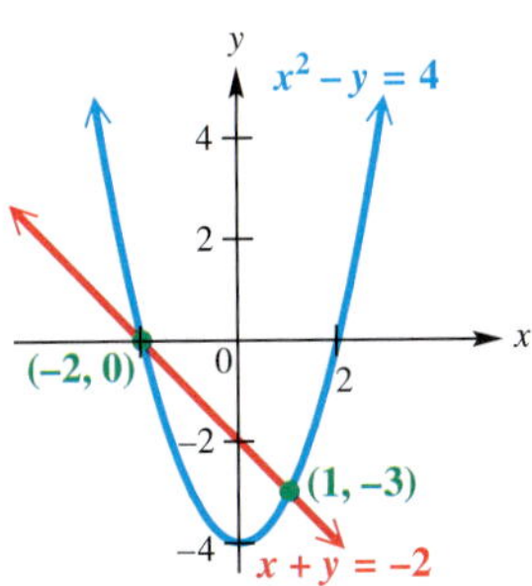

Figure 9

Graphing Calculator Solution

Solve each equation for y and graph them in the same viewing window. We obtain

$$Y_1 = X^2 - 4$$

and

$$Y_2 = -X - 2.$$

The screens in Figure 10, which indicate that the points of intersection are $(-2, 0)$ and $(1, -3)$, support the solution found algebraically.

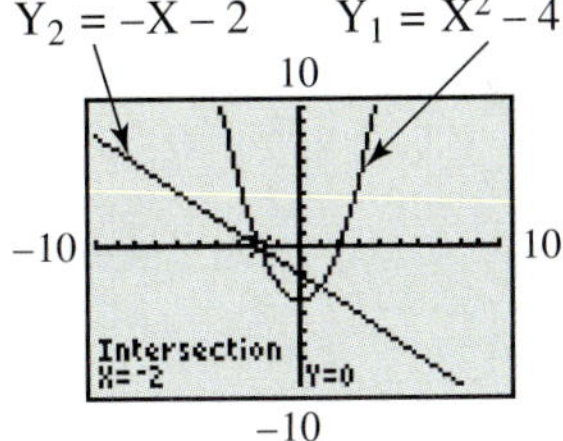

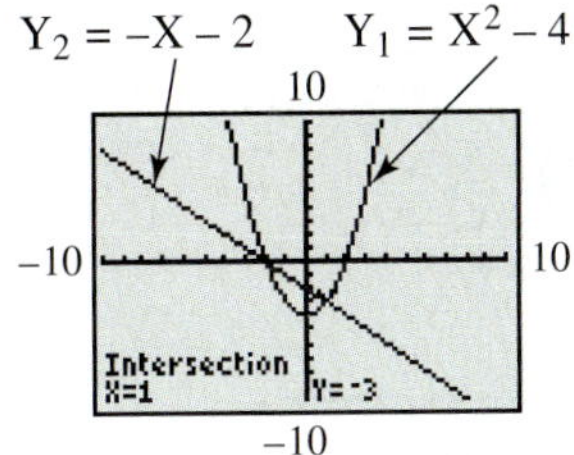

Figure 10

Now try Exercise 9.

CAUTION If we had solved for x in equation (2) to begin the algebraic solution in Example 1, we would find $y = 0$ or $y = -3$. Substituting $y = 0$ into equation (1) gives $x^2 = 4$, so $x = 2$ or $x = -2$, leading to the ordered pairs $(2, 0)$ and $(-2, 0)$. The ordered pair $(2, 0)$ does not satisfy equation (2), however. This shows the *necessity* of checking by substituting all potential solutions into each equation of the system.

Looking Ahead to Calculus

In calculus, finding the maximum and minimum points for a function of several variables usually requires the solution of a nonlinear system of equations.

Visualizing the types of graphs involved in a nonlinear system helps predict the possible numbers of ordered pairs of real numbers that may be in the solution set of the system. (Graphs of some nonlinear equations were discussed in Chapter 2.) For example, a line and a parabola may have 0, 1, or 2 points of intersection, as shown in Figure 11.

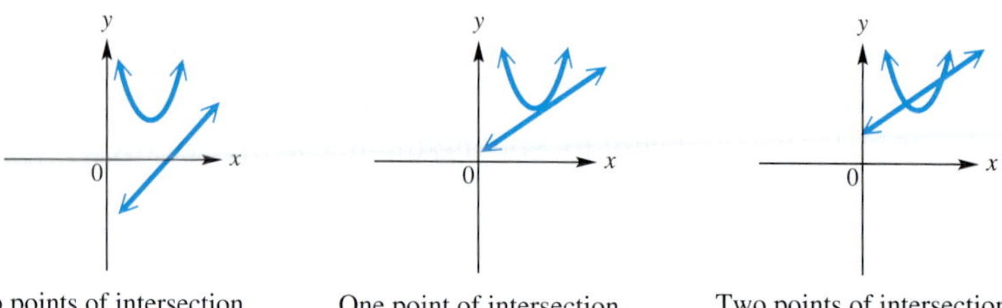

No points of intersection | One point of intersection | Two points of intersection

Figure 11

Nonlinear systems where both variables are squared in both equations are best solved by elimination, as shown in the next example.

EXAMPLE 2 Solving a Nonlinear System by Elimination

Solve the system.

$$x^2 + y^2 = 4 \quad (1)$$
$$2x^2 - y^2 = 8 \quad (2)$$

Solution The graph of equation (1) is a circle and, as we will learn in the next chapter, the graph of equation (2) is a *hyperbola*. These graphs may intersect in 0, 1, 2, 3, or 4 points.

Add the two equations to eliminate y^2.

$$\begin{aligned} x^2 + y^2 &= 4 && (1) \\ 2x^2 - y^2 &= 8 && (2) \\ \hline 3x^2 \qquad &= 12 && \text{Add. (Section 9.1)} \\ x^2 &= 4 && \text{Divide by 3.} \\ x &= \pm 2 && \text{Square root property (Section 1.4)} \end{aligned}$$

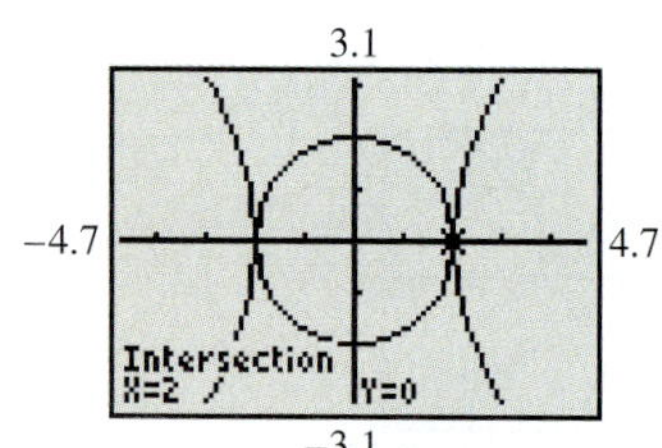

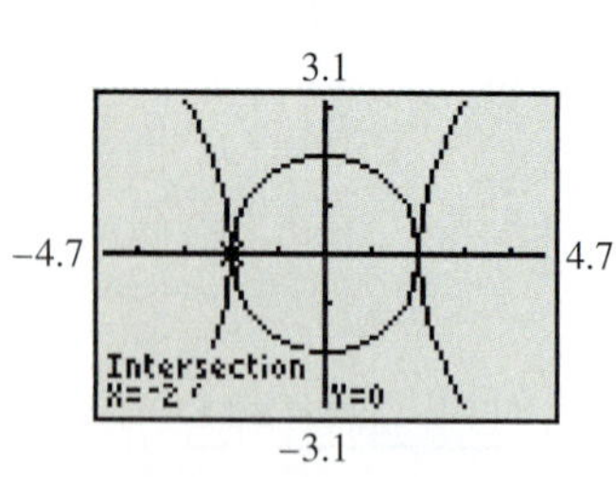

To solve the system in Example 2 graphically, solve equation (1) for y to get

$$y = \pm\sqrt{4 - x^2}.$$

Similarly, equation (2) yields

$$y = \pm\sqrt{-8 + 2x^2}.$$

Graph these *four* functions to find the two solutions.

Find y by substituting back into equation (1).

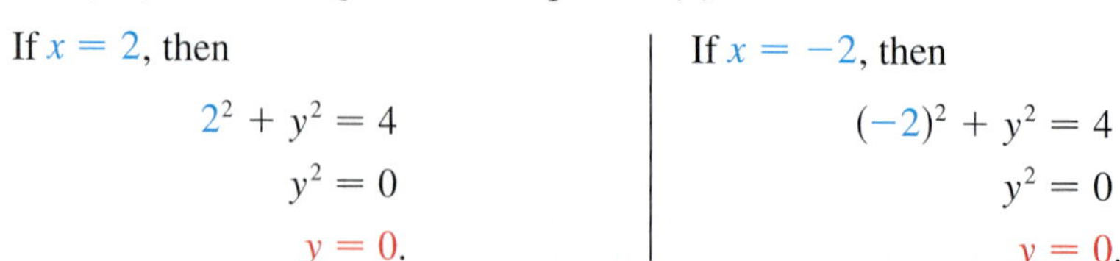

If $x = 2$, then

$$\begin{aligned} 2^2 + y^2 &= 4 \\ y^2 &= 0 \\ y &= 0. \end{aligned}$$

If $x = -2$, then

$$\begin{aligned} (-2)^2 + y^2 &= 4 \\ y^2 &= 0 \\ y &= 0. \end{aligned}$$

The solutions of the given system are $(2, 0)$ and $(-2, 0)$, so the solution set is $\{(2, 0), (-2, 0)\}$.

Now try Exercise 17.

NOTE The elimination method works with the system in Example 2 since the system can be thought of as a system of linear equations where the variables are x^2 and y^2. In other words, the system is *linear in* x^2 *and* y^2. To see this, substitute u for x^2 and v for y^2. The resulting system is linear in u and v.

Sometimes a combination of the elimination method and the substitution method is effective in solving a system, as illustrated in Example 3.

EXAMPLE 3 Solving a Nonlinear System by a Combination of Methods

Solve the system.

$$\begin{aligned} x^2 + 3xy + y^2 &= 22 \quad (1) \\ x^2 - xy + y^2 &= 6 \quad (2) \end{aligned}$$

Solution

$$\begin{aligned} x^2 + 3xy + y^2 &= 22 && (1) \\ -x^2 + xy - y^2 &= -6 && \text{Multiply (2) by } -1. \\ \hline 4xy &= 16 && \text{Add.} \quad (3) \\ y &= \frac{4}{x} && \text{Solve for } y \ (x \neq 0). \quad (4) \end{aligned}$$

Now substitute $\frac{4}{x}$ for y in either equation (1) or (2). We use equation (2).

$$\begin{aligned} x^2 - x\left(\frac{4}{x}\right) + \left(\frac{4}{x}\right)^2 &= 6 && \text{Let } y = \tfrac{4}{x} \text{ in (2).} \\ x^2 - 4 + \frac{16}{x^2} &= 6 && \text{Multiply and square.} \\ x^4 - 4x^2 + 16 &= 6x^2 && \text{Multiply by } x^2 \text{ to clear fractions. (Section 1.6)} \\ x^4 - 10x^2 + 16 &= 0 && \text{Subtract } 6x^2. \\ (x^2 - 2)(x^2 - 8) &= 0 && \text{Factor.} \end{aligned}$$

$$\begin{aligned} x^2 = 2 \quad &\text{or} \quad x^2 = 8 && \text{Zero-factor property} \\ x = \pm\sqrt{2} \quad &\text{or} \quad x = \pm 2\sqrt{2} && \text{Square root property; } \pm\sqrt{8} = \pm 2\sqrt{2} \text{ (Section R.7)} \end{aligned}$$

Substitute these x-values into equation (4) to find corresponding values of y.

If $x = \sqrt{2}$, then	If $x = -\sqrt{2}$, then	If $x = 2\sqrt{2}$, then	If $x = -2\sqrt{2}$, then
$y = \frac{4}{\sqrt{2}} = 2\sqrt{2}$.	$y = \frac{4}{-\sqrt{2}} = -2\sqrt{2}$.	$y = \frac{4}{2\sqrt{2}} = \sqrt{2}$.	$y = \frac{4}{-2\sqrt{2}} = -\sqrt{2}$.

The solution set of the system is

$$\{(\sqrt{2}, 2\sqrt{2}), (-\sqrt{2}, -2\sqrt{2}), (2\sqrt{2}, \sqrt{2}), (-2\sqrt{2}, -\sqrt{2})\}.$$

Verify these solutions by substitution in the original system.

Now try Exercise 35.

EXAMPLE 4 Solving a Nonlinear System with an Absolute Value Equation

Solve the system.

$$x^2 + y^2 = 16 \quad (1)$$
$$|x| + y = 4 \quad (2)$$

Solution Use the substitution method. Solving equation (2) for $|x|$ gives

$$|x| = 4 - y. \quad (3)$$

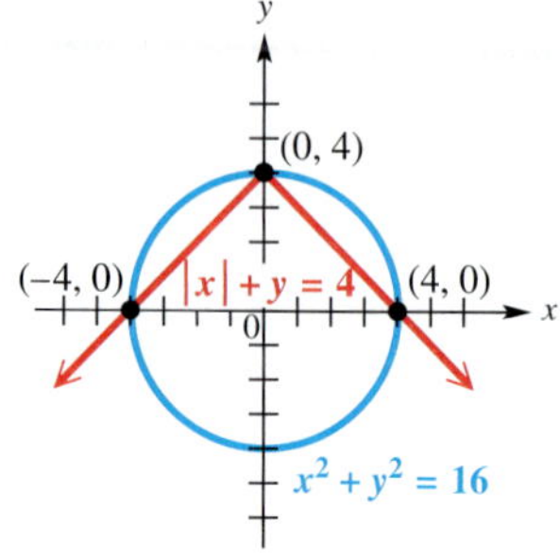

Since $|x| \geq 0$ for all x, $4 - y \geq 0$ and thus $y \leq 4$. In equation (1), the first term is x^2, which is the same as $|x|^2$. Therefore,

$$(4 - y)^2 + y^2 = 16$$
$$(16 - 8y + y^2) + y^2 = 16 \quad \text{Square the binomial. (Section R.3)}$$
$$2y^2 - 8y = 0 \quad \text{Combine terms.}$$
$$2y(y - 4) = 0 \quad \text{Factor.}$$
$$y = 0 \quad \text{or} \quad y = 4. \quad \text{Zero-factor property}$$

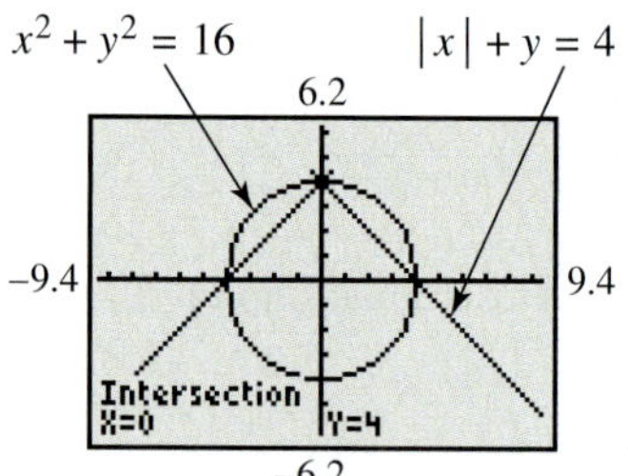

Figure 12

To solve for the corresponding values of x, use either equation (1) or (2). We use equation (1).

If $y = 0$, then

$$x^2 + 0^2 = 16$$
$$x^2 = 16$$
$$x = \pm 4.$$

If $y = 4$, then

$$x^2 + 4^2 = 16$$
$$x^2 = 0$$
$$x = 0.$$

The solution set, $\{(4, 0), (-4, 0), (0, 4)\}$, includes the points of intersection shown in Figure 12. Check the solutions in the original system.

Now try Exercise 39.

Solving Nonlinear Systems with Nonreal Complex Solutions Nonlinear systems sometimes have solutions with nonreal complex numbers in the ordered pairs.

EXAMPLE 5 Solving a Nonlinear System with Nonreal Complex Numbers in Its Solutions

Solve the system.

$$x^2 + y^2 = 5 \quad (1)$$
$$4x^2 + 3y^2 = 11 \quad (2)$$

Solution

$$-3x^2 - 3y^2 = -15 \quad \text{Multiply (1) by } -3.$$
$$4x^2 + 3y^2 = 11 \quad (2)$$
$$x^2 = -4 \quad \text{Add.}$$
$$x = \pm\sqrt{-4} \quad \text{Square root property}$$
$$x = \pm 2i \quad \text{(Section 1.3)}$$

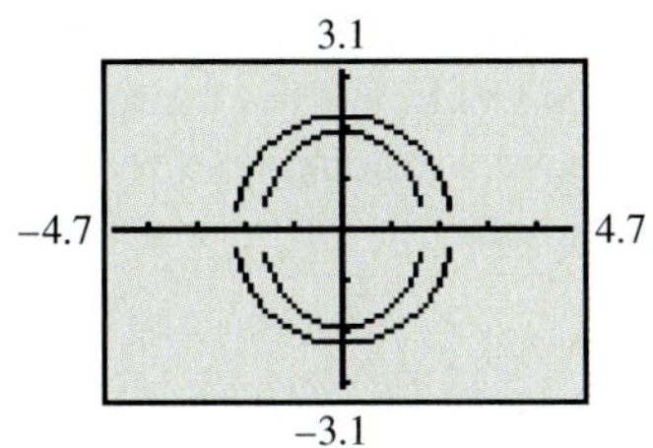

The graphs of the two equations in Example 5 do not intersect, as seen here. The graphs are obtained by graphing

$$y = \pm\sqrt{5 - x^2}$$

and

$$y = \pm\sqrt{\frac{11 - 4x^2}{3}}.$$

To find the corresponding values of y, substitute into equation (1).

If $x = 2i$, then

$$\begin{aligned}(2i)^2 + y^2 &= 5\\ -4 + y^2 &= 5\\ y^2 &= 9\\ y &= \pm 3.\end{aligned}$$

If $x = -2i$, then

$$\begin{aligned}(-2i)^2 + y^2 &= 5\\ -4 + y^2 &= 5\\ y^2 &= 9\\ y &= \pm 3.\end{aligned}$$

Checking the solutions in the given system shows that the solution set is

$$\{(2i, 3), (2i, -3), (-2i, 3), (-2i, -3)\}.$$

Now try Exercise 37.

Applying Nonlinear Systems Some applications require solving nonlinear systems of equations.

EXAMPLE 6 Using a Nonlinear System to Find the Dimensions of a Box

A box with an open top has a square base and four sides of equal height. The volume of the box is 75 in.^{3}, and the surface area is 85 in.^{2}. What are the dimensions of the box?

Solution

Step 1 **Read** the problem. We must find the dimensions (width, length, and height) of the box.

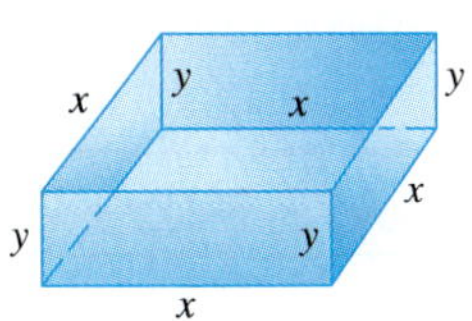

Figure 13

Step 2 **Assign variables.** Let x represent the length and width of the square base, and let y represent the height. See Figure 13.

Step 3 **Write a system of equations.** Use the formula for the volume of a box, $V = LWH$, to write one equation.

$$x^2y = 75$$

The surface consists of the base, whose area is x^2, and four sides, each having area xy. The total surface area is 85 in.^{2}, so a second equation is

$$x^2 + 4xy = 85.$$

The system to solve is

$$\begin{aligned}x^2y &= 75 && (1)\\ x^2 + 4xy &= 85. && (2)\end{aligned}$$

Step 4 **Solve** the system. Solve equation (1) for y to get $y = \frac{75}{x^2}$, and substitute into equation (2).

$$\begin{aligned}x^2 + 4x\left(\frac{75}{x^2}\right) &= 85 && \text{Let } y = \tfrac{75}{x^2} \text{ in (2).}\\ x^2 + \frac{300}{x} &= 85 && \text{Multiply.}\\ x^3 + 300 &= 85x && \text{Multiply by } x,\ x \neq 0.\\ x^3 - 85x + 300 &= 0 && \text{Subtract } 85x.\end{aligned}$$

We are restricted to positive values for x, and considering the nature of the problem, any solution should be relatively small. By the rational zeros theorem, factors of 300 are the only possible rational solutions. Using synthetic division, we see that 5 is a solution.

$$\begin{array}{r|rrrr} 5 & 1 & 0 & -85 & 300 \\ & & 5 & 25 & -300 \\ \hline & 1 & 5 & -60 & 0 \end{array}$$ (Section 3.2)

Coefficients of a quadratic polynomial factor

Therefore, one value of x is 5 and $y = \frac{75}{5^2} = 3$. We must now solve

$$x^2 + 5x - 60 = 0$$

for any other possible positive solutions. Using the quadratic formula, the positive solution is

$$x = \frac{-5 + \sqrt{5^2 - 4(1)(-60)}}{2(1)} \approx 5.639.$$ $a = 1, b = 5, c = -60$ (Section 1.4)

This value of x leads to $y \approx 2.359$.

Step 5 **State the answer.** There are two possible answers.

First answer: length = width = 5 in.; height = 3 in.

Second answer: length = width $\approx$ 5.639 in.; height $\approx$ 2.359 in.

Step 6 **Check.** The check is left for Exercise 61.

Now try Exercises 59 and 61.

9.5 Exercises

In Exercises 1–6 a nonlinear system is given, along with the graphs of both equations in the system. Verify that the points of intersection specified on the graph are solutions of the system by substituting directly into both equations.

1. $x^2 = y - 1$
$y = 3x + 5$

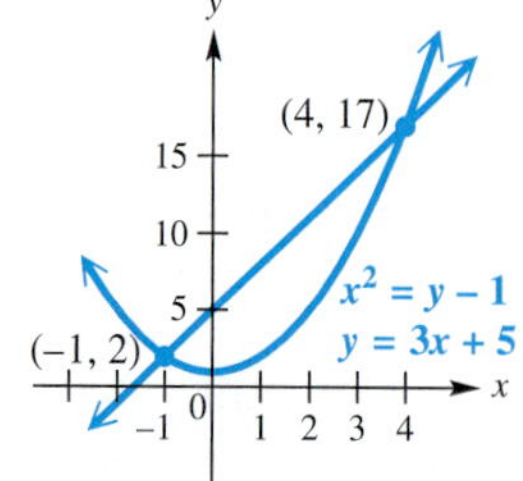

2. $2x^2 = 3y + 23$
$y = 2x - 5$

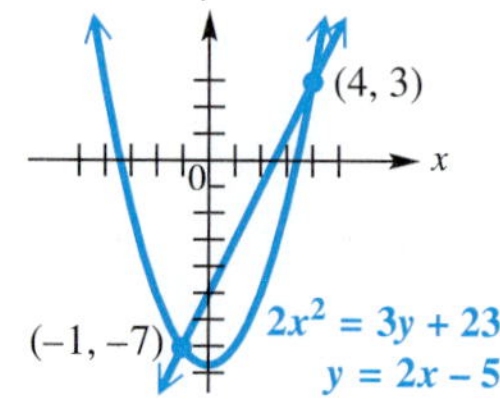

3. $x^2 + y^2 = 5$
$-3x + 4y = 2$

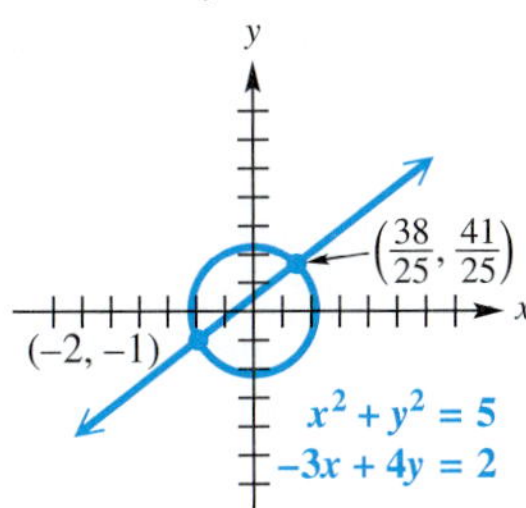

4. $x + y = -3$
$x^2 + y^2 = 45$

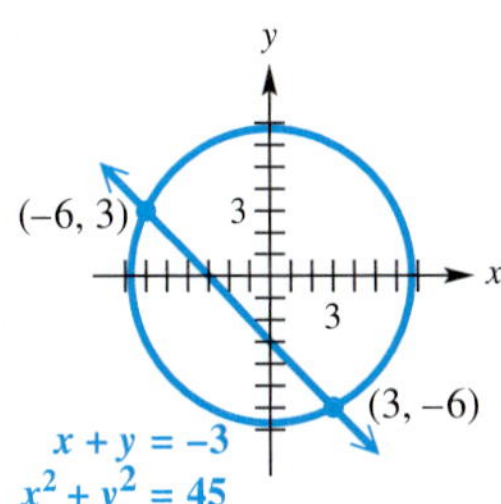

5. $y = \log x$
$x^2 - y^2 = 4$

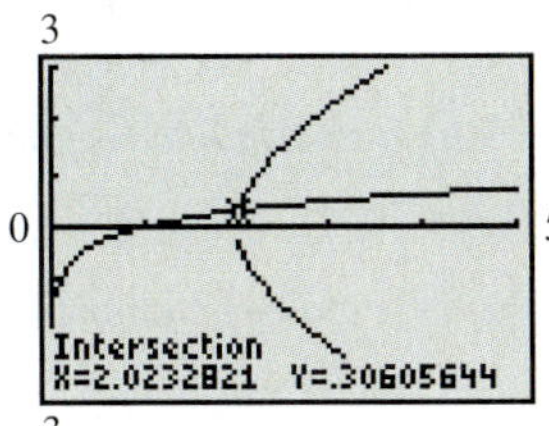

6. $y = \ln(2x + 3)$
$y = \sqrt{2 - .5x^2}$

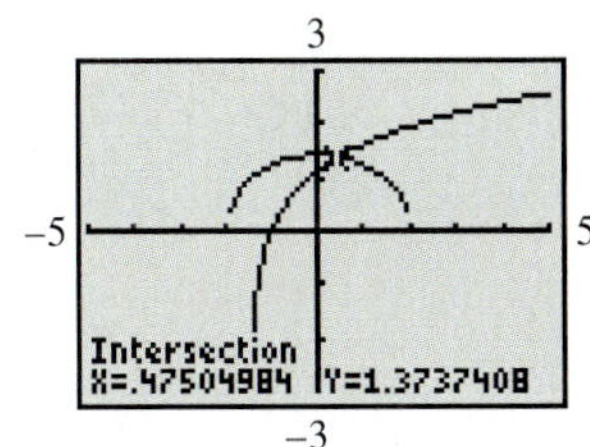

7. *Concept Check* In Example 1, we solved the following system. How can we tell before doing any work that this system cannot have more than two solutions?

$$x^2 - y = 4$$
$$x + y = -2$$

8. *Concept Check* In Example 5, there were four solutions to the system, but there were no points of intersection of the graphs. Will the number of solutions of a system ever be less than the number of points of intersection of the graphs of the equations of the system?

Give all solutions of each nonlinear system of equations, including those with nonreal complex components. See Examples 1–5.

9. $x^2 - y = 0$
$x + y = 2$

10. $x^2 + y = 2$
$x - y = 0$

11. $y = x^2 - 2x + 1$
$x - 3y = -1$

12. $y = x^2 + 6x + 9$
$x + 2y = -2$

13. $y = x^2 + 4x$
$2x - y = -8$

14. $y = 6x + x^2$
$3x - 2y = 10$

15. $3x^2 + 2y^2 = 5$
$x - y = -2$

16. $x^2 + y^2 = 5$
$-3x + 4y = 2$

17. $x^2 + y^2 = 8$
$x^2 - y^2 = 0$

18. $x^2 + y^2 = 10$
$2x^2 - y^2 = 17$

19. $5x^2 - y^2 = 0$
$3x^2 + 4y^2 = 0$

20. $x^2 + y^2 = 4$
$2x^2 - 3y^2 = -12$

21. $3x^2 + y^2 = 3$
$4x^2 + 5y^2 = 26$

22. $x^2 + 2y^2 = 9$
$3x^2 - 4y^2 = 27$

23. $2x^2 + 3y^2 = 5$
$3x^2 - 4y^2 = -1$

24. $3x^2 + 5y^2 = 17$
$2x^2 - 3y^2 = 5$

25. $2x^2 + 2y^2 = 20$
$4x^2 + 4y^2 = 30$

26. $x^2 + y^2 = 4$
$5x^2 + 5y^2 = 28$

27. $2x^2 - 3y^2 = 8$
$6x^2 + 5y^2 = 24$

28. $5x^2 - 2y^2 = 25$
$10x^2 + y^2 = 50$

29. $xy = -15$
$4x + 3y = 3$

30. $xy = 8$
$3x + 2y = -16$

31. $2xy + 1 = 0$
$x + 16y = 2$

32. $-5xy + 2 = 0$
$x - 15y = 5$

33. $x^2 + 4y^2 = 25$
$xy = 6$

34. $5x^2 - 2y^2 = 6$
$xy = 2$

35. $x^2 - xy + y^2 = 5$
$2x^2 + xy - y^2 = 10$

36. $3x^2 + xy + 3y^2 = 7$
$x^2 + y^2 = 2$

37. $x^2 + 2xy - y^2 = 14$
$x^2 - y^2 = -16$

38. $3x^2 + 2xy - y^2 = 9$
$x^2 - xy + y^2 = 9$

39. $x = |y|$
$x^2 + y^2 = 18$

40. $2x + |y| = 4$
$x^2 + y^2 = 5$

41. $2x^2 - y^2 = 4$
$|x| = |y|$

42. $x^2 + y^2 = 9$
$|x| = |y|$

Many nonlinear systems cannot be solved algebraically, so graphical analysis is the only way to determine the solutions of such systems. Use a graphing calculator to solve each nonlinear system. Give x- and y-coordinates to the nearest hundredth.

43. $y = \log(x + 5)$
$y = x^2$

44. $y = 5^x$
$xy = 1$

45. $y = e^{x+1}$
$2x + y = 3$

46. $y = \sqrt[3]{x - 4}$
$x^2 + y^2 = 6$

Solve each problem using a system of equations in two variables. See Example 6.

47. *Unknown Numbers* Find two numbers whose sum is 17 and whose product is 42.

48. *Unknown Numbers* Find two numbers whose sum is 10 and whose squares differ by 20.

49. *Unknown Numbers* Find two numbers whose squares have a sum of 100 and a difference of 28.

50. *Unknown Numbers* Find two numbers whose squares have a sum of 194 and a difference of 144.

51. *Unknown Numbers* Find two numbers whose ratio is 9 to 2 and whose product is 162.

52. *Unknown Numbers* Find two numbers whose ratio is 4 to 3 and are such that the sum of their squares is 100.

53. *Triangle Dimensions* The longest side of a right triangle is 13 m in length. One of the other sides is 7 m longer than the shortest side. Find the lengths of the two shorter sides of the triangle.

54. *Triangle Dimensions* The longest side of a right triangle is 29 ft in length. One of the other two sides is 1 ft longer than the shortest side. Find the lengths of the two shorter sides of the triangle.

55. Does the straight line $3x - 2y = 9$ intersect the circle $x^2 + y^2 = 25$? (*Hint:* To find out, solve the system formed by these two equations.)

56. For what value(s) of b will the line $x + 2y = b$ touch the circle $x^2 + y^2 = 9$ in only one point?

57. Find the equation of the line passing through the points of intersection of the graphs of $y = x^2$ and $x^2 + y^2 = 90$.

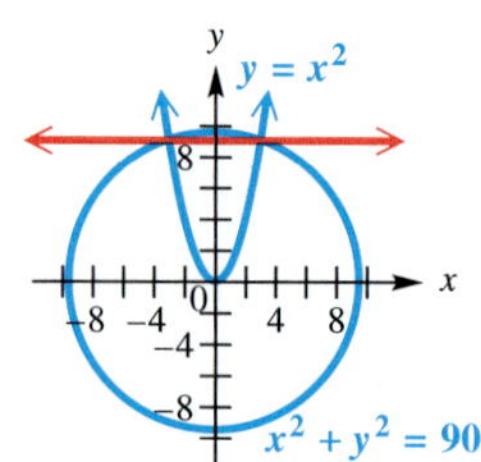

58. Suppose you are given the equations of two circles that are known to intersect in exactly two points. Explain how you would find the equation of the only chord common to these circles.

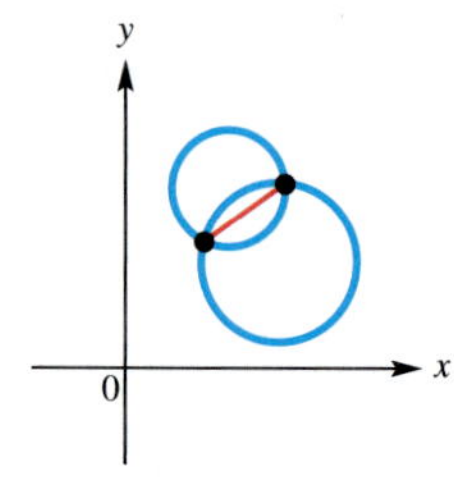

59. *Dimensions of a Box* A box with an open top has a square base and four sides of equal height. The volume of the box is 360 ft^3. The height is 4 ft greater than both the length and the width. If the surface area is 276 ft^2, what are the dimensions of the box?

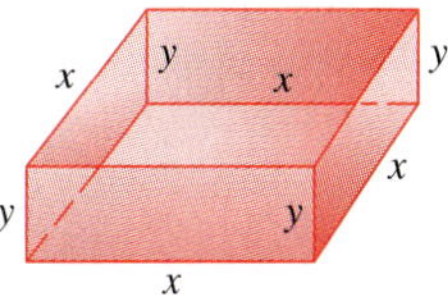

60. *Dimensions of a Cylinder* Find the radius and height of an open-ended cylinder with volume 50 in.^{3} and lateral surface area 65 in.^{2}.

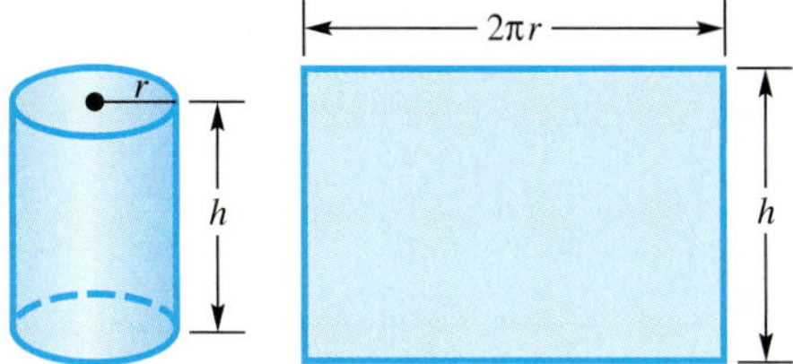

61. Check the two answers in Example 6.

62. *(Modeling) Equilibrium Demand and Price* Let the supply and demand equations for a certain commodity be

$$\text{supply: } p = \frac{2000}{2000 - q}$$

and

$$\text{demand: } p = \frac{7000 - 3q}{2q}.$$

(a) Find the equilibrium demand. **(b)** Find the equilibrium price (in dollars).

63. *(Modeling) Equilibrium Demand and Price* Let the supply and demand equations for a certain commodity be

$$\text{supply: } p = \sqrt{.1q + 9} - 2$$

and

$$\text{demand: } p = \sqrt{25 - .1q}.$$

(a) Find the equilibrium demand. **(b)** Find the equilibrium price (in dollars).

64. *(Modeling) Circuit Gain* In electronics, circuit gain is modeled by

$$G = \frac{Bt}{R + R_t},$$

where R is the value of a resistor, t is temperature, R_t is the value of R at temperature t, and B is a constant. The sensitivity of the circuit to temperature is modeled by

$$S = \frac{BR}{(R + R_t)^2}.$$

If $B = 3.7$ and t is 90 K (Kelvin), find the values of R and R_t that will make $G = .4$ and $S = .001$.

65. *(Modeling) Atmospheric Carbon Emissions* The emissions of carbon into the atmosphere from 1950–1995 are modeled in the graph for both Western Europe and Eastern Europe together with the former USSR. This carbon combines with oxygen to form carbon dioxide, which is believed to contribute to the greenhouse effect.

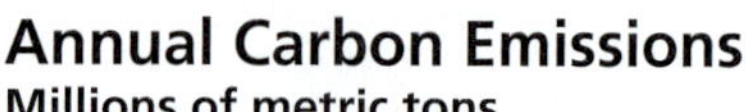

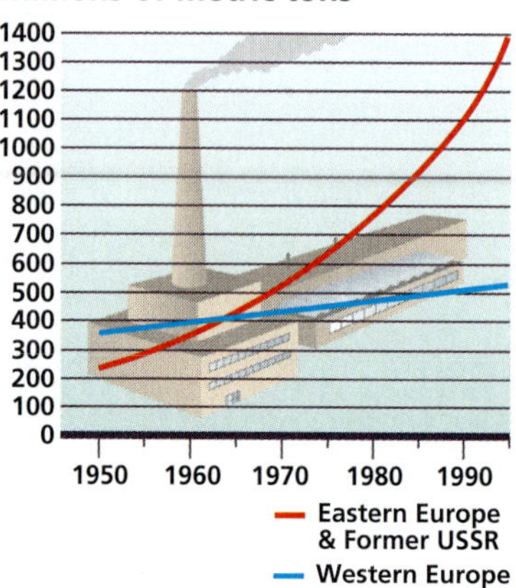

Source: Rosenberg, N. (editor), *Greenhouse Warming: Abatement and Adaptation,* Resources for the Future, Washington, D.C.

(a) Interpret this graph. How are emissions changing with time?

(b) Use the graph to estimate the year and the amount when the carbon emissions were equal.

(c) The equations below model carbon emissions in Western Europe and in Eastern Europe/former USSR. Use these equations to determine the year and emissions levels when $W = E$.

$$W = 375(1.008)^{(t-1950)} \quad \text{Western Europe}$$
$$E = 260(1.038)^{(t-1950)} \quad \text{Eastern Europe/former USSR}$$

Relating Concepts

For individual or collaborative investigation
(Exercises 66–71)

Consider the nonlinear system $\begin{array}{l} y = |x-1| \\ y = x^2 - 4 \end{array}$. ***Work Exercises 66–71 in order,*** *to see how concepts from previous chapters relate to the graphs and the solutions of this system.*

66. How is the graph of $y = |x - 1|$ obtained by transforming the graph of $y = |x|$?

67. How is the graph of $y = x^2 - 4$ obtained by transforming the graph of $y = x^2$?

68. Use the definition of absolute value to write $y = |x - 1|$ as a piecewise-defined function.

69. Write two quadratic equations that will be used to solve the system. (*Hint:* Set both parts of the piecewise-defined function in Exercise 68 equal to $x^2 - 4$.)

70. Use the quadratic formula to solve both equations from Exercise 69. Pay close attention to the restriction on x.

71. Use the values of x found in Exercise 70 to find the solution set of the system.

Summary Exercises on Systems of Equations

This chapter has introduced methods for solving systems of equations, including substitution and elimination, and matrix methods such as the Gauss-Jordan method and Cramer's rule. Use each method at least once when solving the systems below. Include solutions with nonreal complex number components. For systems with infinitely many solutions, write the solution set using an arbitrary variable.

1. $2x + 5y = 4$
$3x - 2y = -13$

2. $x - 3y = 7$
$-3x + 4y = -1$

3. $2x^2 + y^2 = 5$
$3x^2 + 2y^2 = 10$

4. $2x - 3y = -2$
$x + y = -16$
$3x - 2y + z = 7$

5. $6x - y = 5$
$xy = 4$

6. $4x + 2z = -12$
$x + y - 3z = 13$
$-3x + y - 2z = 13$

7. $x + 2y + z = 5$
$y + 3z = 9$

8. $x - 4 = 3y$
$x^2 + y^2 = 8$

9. $3x + 6y - 9z = 1$
$2x + 4y - 6z = 1$
$3x + 4y + 5z = 0$

10. $x + 2y + z = 0$
$x + 2y - z = 6$
$2x - y = -9$

11. $x^2 + y^2 = 4$
$y = x + 6$

12. $x + 5y = -23$
$4y - 3z = -29$
$2x + 5z = 19$

13. $y + 1 = x^2 + 2x$
$y + 2x = 4$

14. $3x + 6z = -3$
$y - z = 3$
$2x + 4z = -1$

15. $2x + 3y + 4z = 3$
$-4x + 2y - 6z = 2$
$4x + 3z = 0$

16. $\frac{3}{x} + \frac{4}{y} = 4$
$\frac{1}{x} + \frac{2}{y} = \frac{2}{3}$

17. $-5x + 2y + z = 5$
$-3x - 2y - z = 3$
$-x + 6y = 1$

18. $x + 5y + 3z = 9$
$2x + 9y + 7z = 5$

19. $2x^2 + y^2 = 9$
$3x - 2y = -6$

20. $2x - 4y - 6 = 0$
$-x + 2y + 3 = 0$

21. $x + y - z = 0$
$2y - z = 1$
$2x + 3y - 4z = -4$

22. $2y = 3x - x^2$
$x + 2y = 12$

23. $4x - z = -6$
$\frac{3}{5}y + \frac{1}{2}z = 0$
$\frac{1}{3}x + \frac{2}{3}z = -5$

24. $x - y + 3z = 3$
$-2x + 3y - 11z = -4$
$x - 2y + 8z = 6$

25. $x^2 + 3y^2 = 28$
$y - x = -2$

26. $5x - 2z = 8$
$4y + 3z = -9$
$\frac{1}{2}x + \frac{2}{3}y = -1$

27. $2x^2 + 3y^2 = 20$
$x^2 + 4y^2 = 5$

28. $x + y + z = -1$
$2x + 3y + 2z = 3$
$2x + y + 2z = -7$

29. $x + 2z = 9$
$y + z = 1$
$3x - 2y = 9$

30. $x^2 - y^2 = 15$
$x - 2y = 2$

31. $-x + y = -1$
$x + z = 4$
$6x - 3y + 2z = 10$

32. $2x - y - 2z = -1$
$-x + 2y + 13z = 12$
$3x + 9z = 6$

33. $xy = -3$
$x + y = -2$

34. $-3x + 2z = 1$
$4x + y - 2z = -6$
$x + y + 4z = 3$

35. $y = x^2 + 6x + 9$
$x + y = 3$

36. $x^2 + y^2 = 9$
$2x - y = 3$

37. $2x - 3y = -2$
$x + y - 4z = -16$
$3x - 2y + z = 7$

38. $3x - y = -2$
$y + 5z = -4$
$-2x + 3y - z = -8$

9.6 Systems of Inequalities and Linear Programming

Solving Linear Inequalities ▪ Solving Systems of Inequalities ▪ Linear Programming

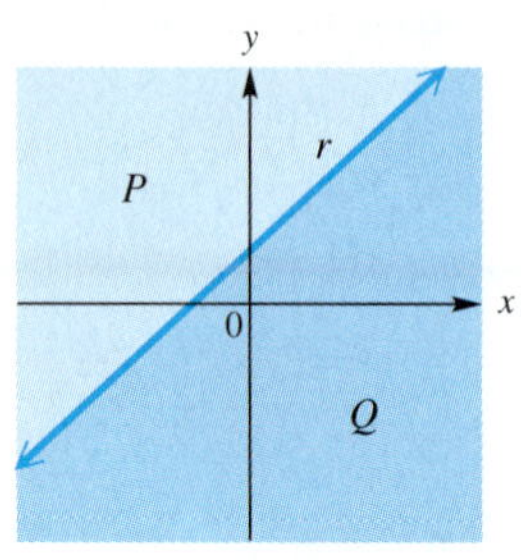

Figure 14

Many mathematical descriptions of real situations are best expressed as inequalities rather than equations. For example, a firm might be able to use a machine *no more* than 12 hr per day, while production of *at least* 500 cases of a certain product might be required to meet a contract. The simplest way to see the solution of an inequality in two variables is to draw its graph.

A line divides a plane into three sets of points: the points of the line itself and the points belonging to the two regions determined by the line. Each of these two regions is called a **half-plane.** In Figure 14, line r divides the plane into three different sets of points: line r, half-plane P, and half-plane Q. The points on r belong neither to P nor to Q. Line r is the **boundary** of each half-plane.

Solving Linear Inequalities

A **linear inequality in two variables** is an inequality of the form

$$Ax + By \leq C,$$

where A, B, and C are real numbers, with A and B not both equal to 0. (The symbol $\leq$ could be replaced with $\geq$, $<$, or $>$.) The graph of a linear inequality is a half-plane, perhaps with its boundary. For example, to graph the linear inequality

$$3x - 2y \leq 6,$$

first graph the boundary, $3x - 2y = 6$, as shown in Figure 15.

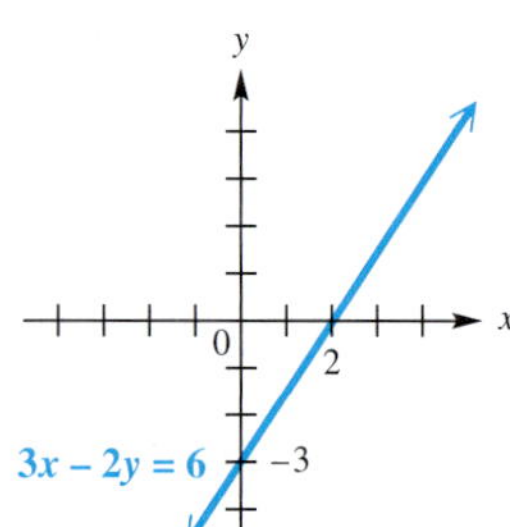

Figure 15

Since the points of the line $3x - 2y = 6$ satisfy $3x - 2y \leq 6$, this line is part of the solution set. To decide which half-plane (the one above the line $3x - 2y = 6$ or the one below the line) is part of the solution set, solve the original inequality for y.

$$3x - 2y \leq 6$$

$$-2y \leq -3x + 6 \quad \text{Subtract } 3x.$$

$$y \geq \frac{3}{2}x - 3 \quad \text{Divide by } -2\text{; change} \leq \text{ to } \geq. \text{ (Section 1.7)}$$

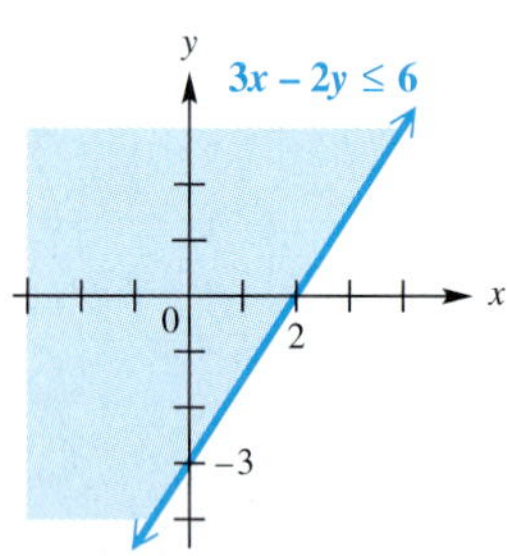

Figure 16

For a particular value of x, the inequality will be satisfied by all values of y that are *greater than* or equal to $\frac{3}{2}x - 3$. Thus, the solution set contains the half-plane *above* the line, as shown in Figure 16.

CAUTION A linear inequality must be in slope-intercept form (solved for y) to tell from a $<$ or $>$ symbol whether to shade the lower or upper half-plane. In Figure 16, the upper half-plane is shaded, even though the inequality is $3x - 2y \leq 6$ (with a $<$ symbol) in standard form. Only when we write the inequality as $y \geq \frac{3}{2}x - 3$ (slope-intercept form) does the $>$ symbol indicate to shade the upper half-plane.

EXAMPLE 1 Graphing a Linear Inequality

Graph $x + 4y > 4$.

Solution The boundary is the straight line $x + 4y = 4$. Since points on this line do not satisfy $x + 4y > 4$, it is customary to make the line dashed, as in Figure 17. To decide which half-plane represents the solution set, solve for y.

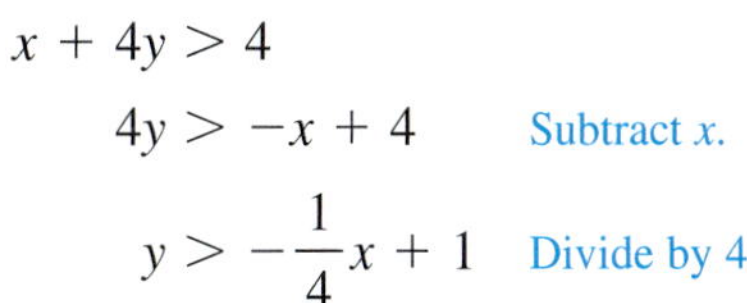

$$x + 4y > 4$$
$$4y > -x + 4 \quad \text{Subtract } x.$$
$$y > -\frac{1}{4}x + 1 \quad \text{Divide by 4.}$$

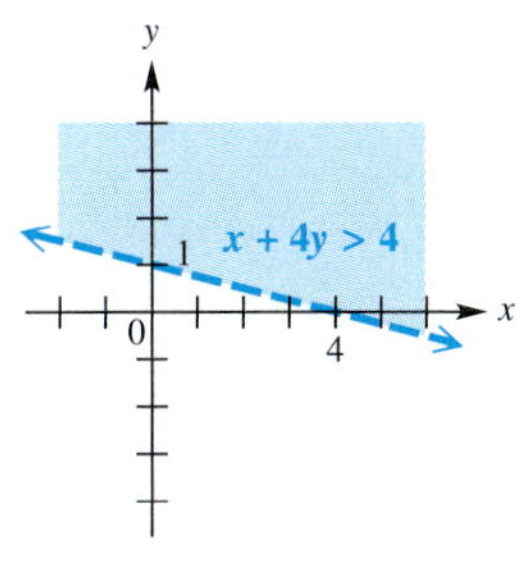

Figure 17

Since y is *greater than* $-\frac{1}{4}x + 1$, the graph of the solution set is the half-plane *above* the boundary, as shown in Figure 17.

Alternatively, or as a check, choose a point not on the boundary line and substitute into the inequality. The point $(0, 0)$ is a good choice if it does not lie on the boundary, since the substitution is easily done.

$$x + 4y > 4 \quad \text{Original inequality}$$
$$0 + 4(0) > 4 \quad \text{Use } (0, 0) \text{ as a test point.}$$
$$0 > 4 \quad \text{False}$$

Since the point $(0, 0)$ is below the boundary, the points that satisfy the inequality must be above the boundary, which agrees with the result above.

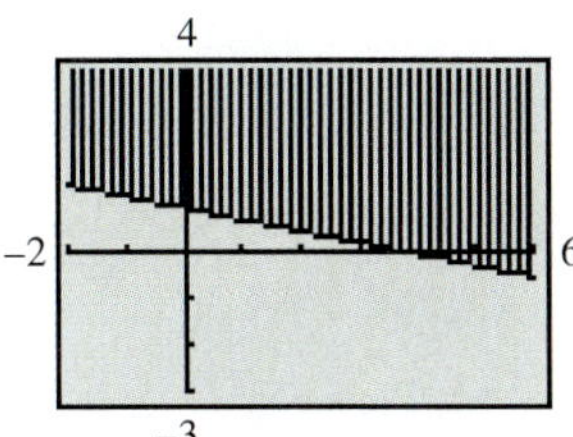

To graph the inequality in Example 1 using a graphing calculator, solve for y, and then direct the calculator to shade *above* the boundary line, $y = -\frac{1}{4}x + 1$. The calculator graph does not distinguish between solid boundary lines and dashed boundary lines. We must understand the mathematics to interpret a calculator graph correctly.

Now try Exercise 5.

The methods used to graph inequalities in two variables are summarized here. (When the symbol is $\leq$ or $\geq$, include the boundary.)

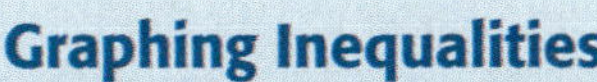

Graphing Inequalities

1. For a function f, the graph of $y < f(x)$ consists of all the points that are *below* the graph of $y = f(x)$; the graph of $y > f(x)$ consists of all the points that are *above* the graph of $y = f(x)$.
2. If the inequality is not or cannot be solved for y, choose a test point not on the boundary. If the test point satisfies the inequality, the graph includes all points on the same side of the boundary as the test point. Otherwise, the graph includes all points on the other side of the boundary.

Solving Systems of Inequalities The solution set of a **system of inequalities,** such as

$$x > 6 - 2y$$
$$x^2 < 2y,$$

is the intersection of the solution sets of its members. We find this intersection by graphing the solution sets of all inequalities on the same coordinate axes and identifying, by shading, the region common to all graphs.

EXAMPLE 2 Graphing Systems of Inequalities

Graph the solution set of each system.

(a) $x > 6 - 2y$
$x^2 < 2y$

(b) $|x| \leq 3$
$y \leq 0$
$y \geq |x| + 1$

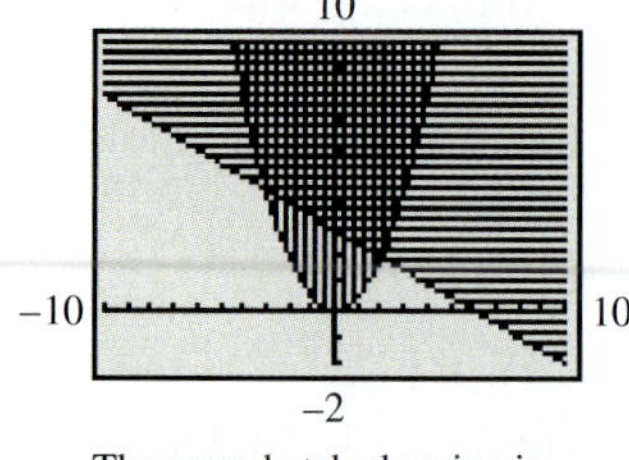

The cross-hatched region is the solution set of the system in Example 2(a).

Solution

(a) Figures 18(a) and (b) show the graphs of $x > 6 - 2y$ and $x^2 < 2y$. The methods of the first section in this chapter can be used to show that the boundaries intersect at the points $(2, 2)$ and $\left(-3, \frac{9}{2}\right)$. The solution set of the system is shown in Figure 18(c). Since the points on the boundaries of $x > 6 - 2y$ and $x^2 < 2y$ do not belong to the graph of the solution set, the boundaries are dashed.

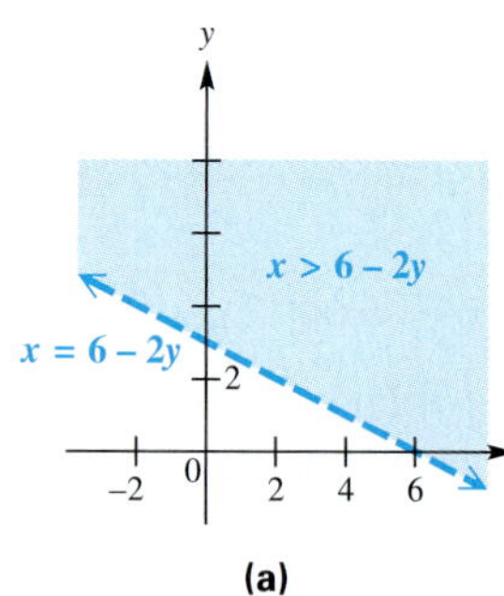

(a)

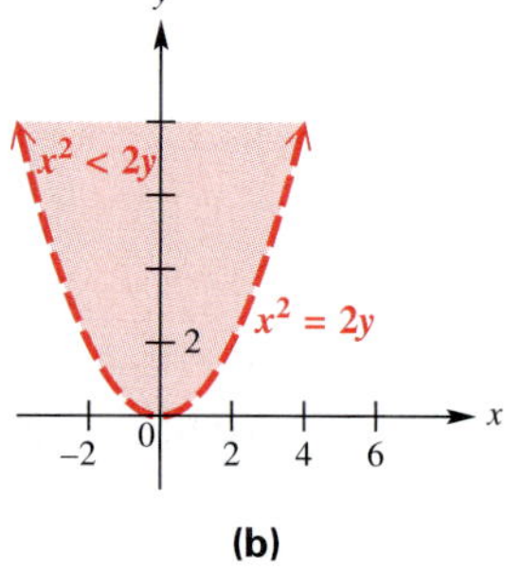

(b)

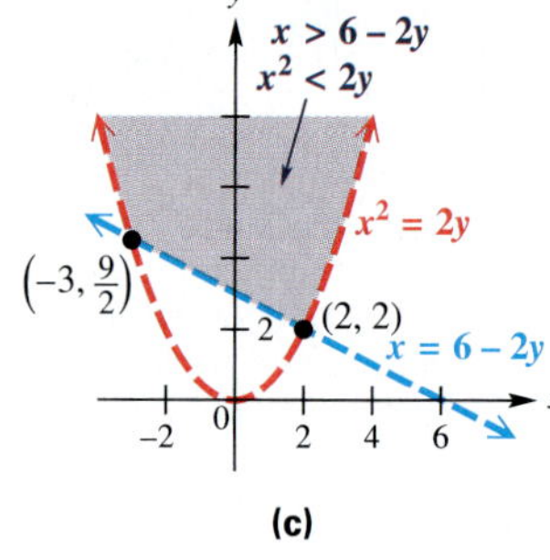

(c)

Figure 18

(b) Writing $|x| \leq 3$ as $-3 \leq x \leq 3$ shows that this inequality is satisfied by points in the region between and including $x = -3$ and $x = 3$. See Figure 19(a). The set of points that satisfies $y \leq 0$ includes the points below or on the x-axis. See Figure 19(b). Graph $y = |x| + 1$ and use a test point to verify that the solutions of $y \geq |x| + 1$ are on or above the boundary. See Figure 19(c). Since the solution sets of $y \leq 0$ and $y \geq |x| + 1$ shown in Figures 19(b) and (c) have no points in common, the solution set of the system is $\emptyset$.

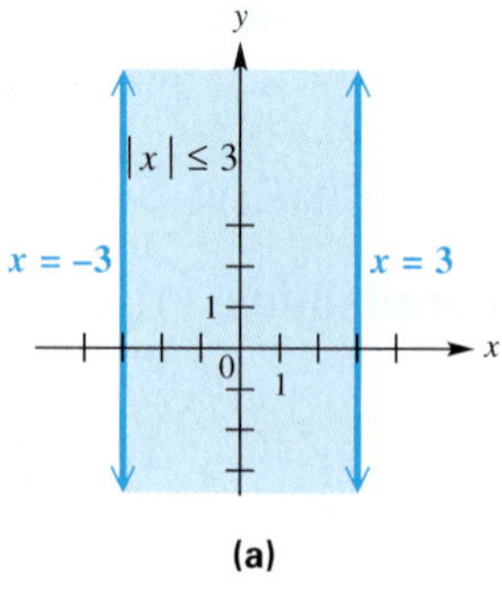

(a)

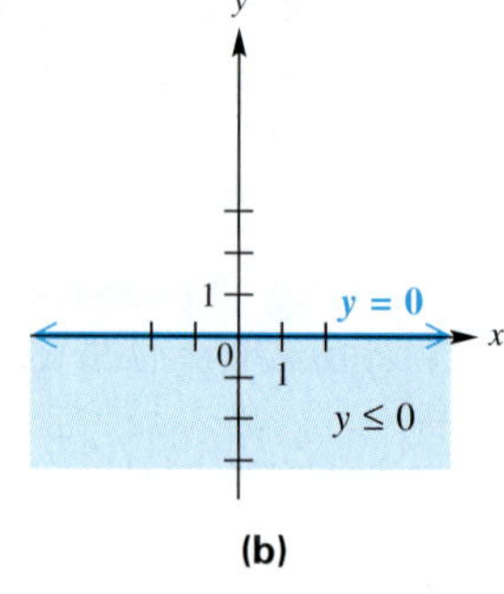

(b)

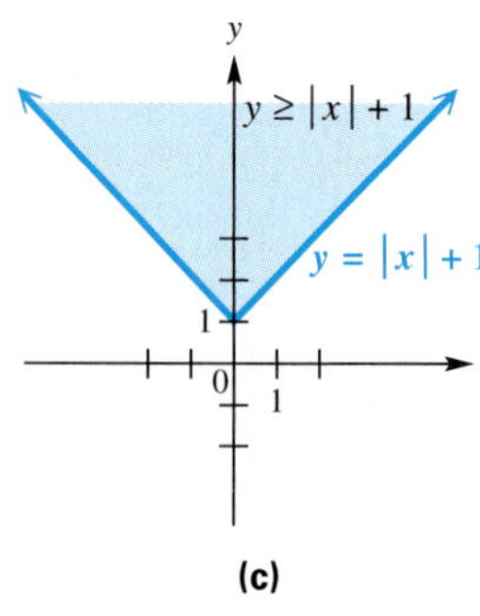

(c)

Figure 19

Now try Exercises 35, 41, and 43.

NOTE While we gave three graphs in the solutions of Example 2, in practice we usually give only a final graph showing the solution set of the system.

Linear Programming An important application of mathematics to business and social science is called *linear programming.* We use **linear programming** to find an optimum value, for example, minimum cost or maximum profit. It was first developed to solve problems in allocating supplies for the U.S. Air Force during World War II.

EXAMPLE 3 Finding a Maximum Profit Model

The Charlson Company makes two products—VCRs and DVD players. Each VCR gives a profit of \$30, while each DVD player produces \$70 profit. The company must manufacture at least 10 VCRs per day to satisfy one of its customers, but no more than 50 because of production problems. The number of DVD players produced cannot exceed 60 per day, and the number of VCRs cannot exceed the number of DVD players. How many of each should the company manufacture to obtain maximum profit?

Solution First we translate the statement of the problem into symbols.

Let $x =$ number of VCRs to be produced daily,

and $y =$ number of DVD players to be produced daily.

The company must produce at least 10 VCRs (10 or more), so

$$x \geq 10.$$

Since no more than 50 VCRs may be produced,

$$x \leq 50.$$

No more than 60 DVD players may be made in one day, so

$$y \leq 60.$$

The number of VCRs may not exceed the number of DVD players translates as

$$x \leq y.$$

The numbers of VCRs and of DVD players cannot be negative, so

$$x \geq 0 \qquad \text{and} \qquad y \geq 0.$$

These restrictions, or **constraints,** form the system of inequalities

$$x \geq 10, \quad x \leq 50, \quad y \leq 60, \quad x \leq y, \quad x \geq 0, \quad y \geq 0.$$

Each VCR gives a profit of \$30, so the daily profit from production of x VCRs is $30x$ dollars. Also, the profit from production of y DVD players will be $70y$ dollars per day. Total daily profit is thus

$$\text{profit} = 30x + 70y.$$

This equation defines the function to be maximized, called the **objective function.**

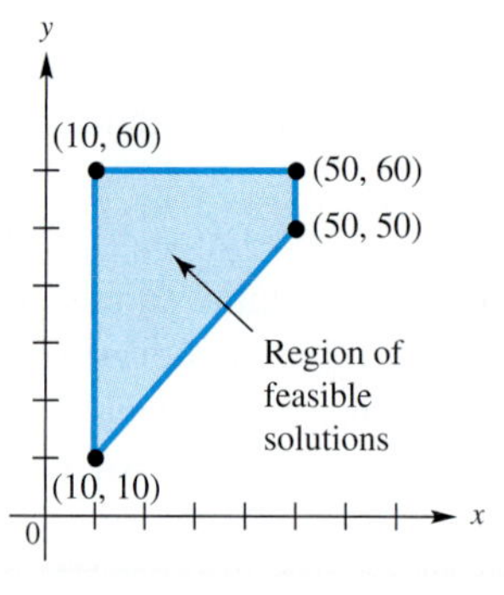

Figure 20

To find the maximum possible profit that the company can make, subject to these constraints, we sketch the graph of each constraint. The only feasible values of x and y are those that satisfy all constraints—that is, the values that lie in the intersection of the graphs of the constraints. The intersection is shown in Figure 20. Any point lying inside the shaded region or on the boundary in Figure 20 satisfies the restrictions as to the number of VCRs and DVD players that may be produced. (For practical purposes, however, only points with integer coefficients are useful.) This region is called the **region of feasible solutions.** The **vertices** (singular **vertex**) or **corner points** of the region of feasible solutions have coordinates

$$(10, 10), \quad (10, 60), \quad (50, 50), \quad \text{and} \quad (50, 60).$$

We must find the value of the objective function $30x + 70y$ for each vertex. We want the vertex that produces the maximum possible value of $30x + 70y$.

$$(10, 10): \quad 30(10) + 70(10) = 1000$$

$$(10, 60): \quad 30(10) + 70(60) = 4500$$

$$(50, 50): \quad 30(50) + 70(50) = 5000$$

$$(50, 60): \quad 30(50) + 70(60) = 5700 \longleftarrow \text{Maximum}$$

The maximum profit is obtained when 50 VCRs and 60 DVD players are produced each day. This maximum profit will be $30(50) + 70(60) = 5700$ dollars per day.

Now try Exercise 73.

To justify the procedure used in Example 3, consider the following. The Charlson Company needed to find values of x and y in the shaded region of Figure 20 that produce the maximum profit—that is, the maximum value of $30x + 70y$. To locate the point (x, y) that gives the maximum profit, add to the graph of Figure 20 lines corresponding to arbitrarily chosen profits of \$0, \$1000, \$3000, and \$7000.

$$30x + 70y = 0$$

$$30x + 70y = 1000$$

$$30x + 70y = 3000$$

$$30x + 70y = 7000$$

For instance, each point on the line $30x + 70y = 3000$ corresponds to production values that yield a profit of \$3000.

Figure 21 shows the region of feasible solutions together with these lines. The lines are parallel, and the higher the line, the greater the profit. The line $30x + 70y = 7000$ yields the greatest profit but does not contain any points of the region of feasible solutions. To find the feasible solution of greatest profit, lower the line $30x + 70y = 7000$ until it contains a feasible solution—that is, until it just touches the region of feasible solutions. This occurs at point A, a vertex of the region. See Figure 22.

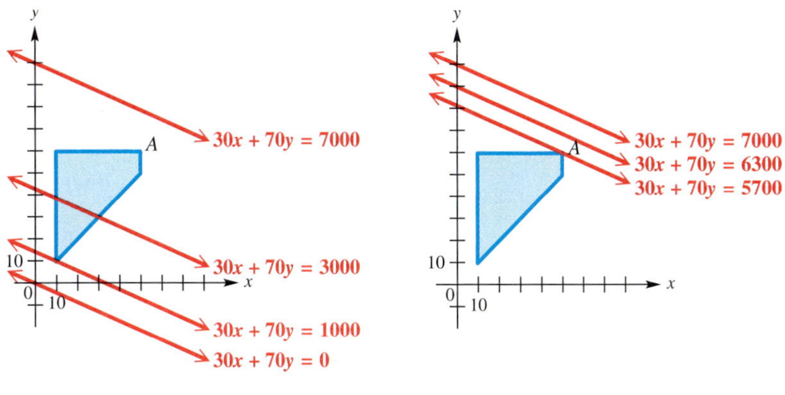

Figure 21 **Figure 22**

The result observed in Figure 22 holds for *every* linear programming problem.

Fundamental Theorem of Linear Programming

If an optimal value for a linear programming problem exists, it occurs at a vertex of the region of feasible solutions.

Solving a Linear Programming Problem

Step 1 Write the objective function and all necessary constraints.

Step 2 Graph the region of feasible solutions.

Step 3 Identify all vertices or corner points.

Step 4 Find the value of the objective function at each vertex.

Step 5 The solution is given by the vertex producing the optimal value of the objective function.

EXAMPLE 4 Finding a Minimum Cost Model

Robin takes vitamin pills each day. She wants at least 16 units of Vitamin A, at least 5 units of Vitamin B_1, and at least 20 units of Vitamin C. She can choose between red pills, costing 10¢ each, that contain 8 units of A, 1 of B_1, and 2 of C; and blue pills, costing 20¢ each, that contain 2 units of A, 1 of B_1, and 7 of C. How many of each pill should she buy to minimize her cost and yet fulfill her daily requirements?

Solution

Step 1 Let x represent the number of red pills to buy, and let y represent the number of blue pills to buy. Then the cost in pennies per day is given by

$$\text{cost} = 10x + 20y.$$

Since Robin buys x of the 10¢ pills and y of the 20¢ pills, she gets 8 units of Vitamin A from each red pill and 2 units of Vitamin A from each blue pill. Altogether she gets $8x + 2y$ units of A per day. Since she wants at least 16 units,

$$8x + 2y \geq 16.$$

Each red pill and each blue pill supplies 1 unit of Vitamin B_1. Robin wants at least 5 units per day, so

$$x + y \geq 5.$$

For Vitamin C, the inequality is

$$2x + 7y \geq 20.$$

Also, $x \geq 0$ and $y \geq 0$, since Robin cannot buy negative numbers of the pills.

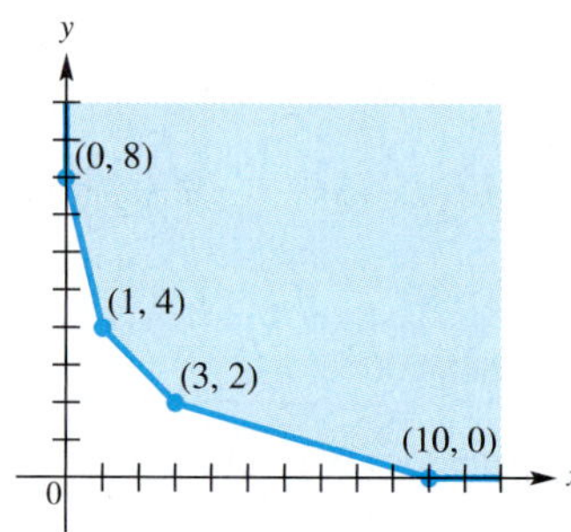

Figure 23

Step 2 The intersection of the graphs of

$$8x + 2y \geq 16, \quad x + y \geq 5, \quad 2x + 7y \geq 20,$$
$$x \geq 0, \quad \text{and} \quad y \geq 0$$

is given in Figure 23.

Step 3 The vertices are $(0, 8)$, $(1, 4)$, $(3, 2)$, and $(10, 0)$.

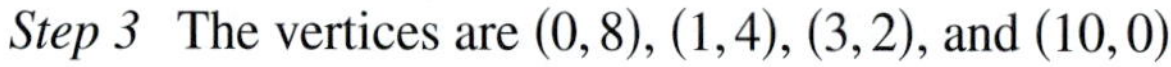

Steps 4 and 5 We find that the minimum cost occurs at $(3, 2)$. See the table.

Point	Cost = $10x + 20y$
$(0, 8)$	$10(0) + 20(8) = 160$
$(1, 4)$	$10(1) + 20(4) = 90$
$(3, 2)$	$10(3) + 20(2) = 70$ ← Minimum
$(10, 0)$	$10(10) + 20(0) = 100$

Robin's best choice is to buy 3 red pills and 2 blue pills, for a total cost of 70¢ per day. She receives just the minimum amounts of Vitamins B_1 and C, and an excess of Vitamin A.

Now try Exercises 67 and 77.

9.6 Exercises

Graph each inequality. See Example 1.

1. $x + 2y \leq 6$

2. $x - y \geq 2$

3. $2x + 3y \geq 4$

4. $4y - 3x < 5$

5. $3x - 5y > 6$

6. $x < 3 + 2y$

7. $5x \leq 4y - 2$

8. $2x > 3 - 4y$

9. $x \leq 3$

10. $y \leq -2$

11. $y < 3x^2 + 2$

12. $y \leq x^2 - 4$

13. $y > (x - 1)^2 + 2$

14. $y > 2(x + 3)^2 - 1$

15. $x^2 + (y + 3)^2 \leq 16$

16. $(x - 4)^2 + (y + 3)^2 \leq 9$

17. In your own words, explain how to determine whether the boundary of the graph of an inequality is solid or dashed.

18. When graphing $y \leq 3x - 6$, would you shade above or below the line $y = 3x - 6$? Explain your answer.

Concept Check *Work each problem.*

19. For $Ax + By \geq C$, if $B > 0$, would you shade above or below the line?

20. For $Ax + By \geq C$, if $B < 0$, would you shade above or below the line?

21. Which one of the following is a description of the graph of the inequality

$$(x - 5)^2 + (y - 2)^2 < 4?$$

A. the region inside a circle with center $(-5, -2)$ and radius 2
B. the region inside a circle with center $(5, 2)$ and radius 2
C. the region inside a circle with center $(-5, -2)$ and radius 4
D. the region outside a circle with center $(5, 2)$ and radius 4

22. Find a linear inequality in two variables whose graph does not intersect the graph of $y \geq 3x + 5$.

Concept Check *In Exercises 23–26, match each inequality with the appropriate calculator graph. Do not use your calculator; instead, use your knowledge of the concepts involved in graphing inequalities.*

23. $y \leq 3x - 6$

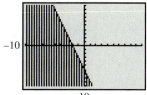

24. $y \geq 3x - 6$

25. $y \leq -3x - 6$

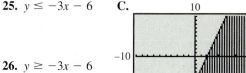

26. $y \geq -3x - 6$

A.

B.

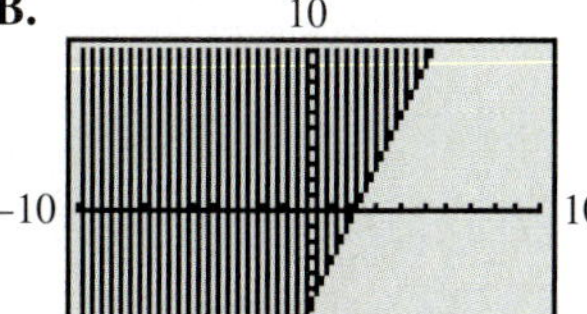

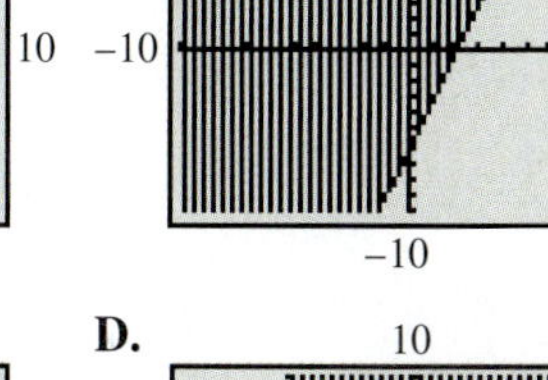

C.

D.

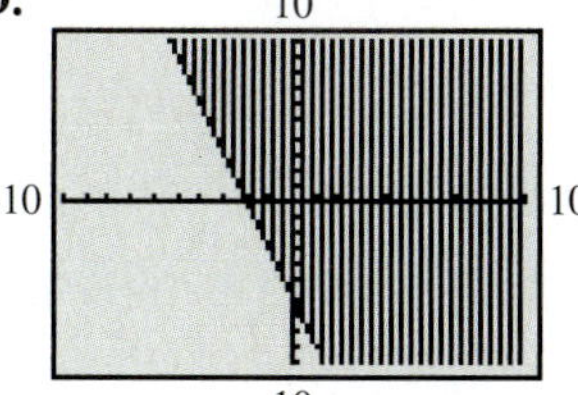

Graph the solution set of each system of inequalities. See Example 2.

27. $x + y \geq 0$
$2x - y \geq 3$

28. $x + y \leq 4$
$x - 2y \geq 6$

29. $2x + y > 2$
$x - 3y < 6$

30. $4x + 3y < 12$
$y + 4x > -4$

31. $3x + 5y \leq 15$
$x - 3y \geq 9$

32. $y \leq x$
$x^2 + y^2 < 1$

33. $4x - 3y \leq 12$
$y \leq x^2$

34. $y \leq -x^2$
$y \geq x^2 - 6$

35. $x + 2y \leq 4$
$y \geq x^2 - 1$

36. $x + y \leq 9$
$x \leq -y^2$

37. $y \leq (x + 2)^2$
$y \geq -2x^2$

38. $x - y < 1$
$-1 < y < 1$

39. $x + y \leq 36$
$-4 \leq x \leq 4$

40. $y \geq (x - 2)^2 + 3$
$y \leq -(x - 1)^2 + 6$

41. $y \geq x^2 + 4x + 4$
$y < -x^2$

42. $x \geq 0$
$x + y \leq 4$
$2x + y \leq 5$

43. $3x - 2y \geq 6$
$x + y \leq -5$
$y \leq 4$

44. $-2 < x < 3$
$-1 \leq y \leq 5$
$2x + y < 6$

45. $-2 < x < 2$
$y > 1$
$x - y > 0$

46. $x + y \leq 4$
$x - y \leq 5$
$4x + y \leq -4$

47. $x \leq 4$
$x \geq 0$
$y \geq 0$
$x + 2y \geq 2$

48. $2y + x \geq -5$
$y \leq 3 + x$
$x \leq 0$
$y \leq 0$

49. $2x + 3y \leq 12$
$2x + 3y > -6$
$3x + y < 4$
$x \geq 0$
$y \geq 0$

50. $y \geq 3^x$
$y \geq 2$

51. $y \leq \left(\frac{1}{2}\right)^x$
$y \geq 4$

52. $\ln x - y \geq 1$
$x^2 - 2x - y \leq 1$

53. $y \leq \log x$
$y \geq |x - 2|$

54. $e^{-x} - y \leq 1$
$x - 2y \geq 4$

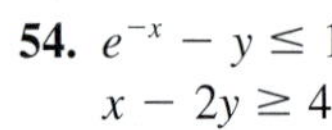

Concept Check *In Exercises 55–58, match each system of inequalities with the appropriate calculator graph. Do not use your calculator; instead, use your knowledge of the concepts involved in graphing systems of inequalities.*

55. $y \geq x$
$y \leq 2x - 3$

56. $y \geq x^2$
$y < 5$

57. $x^2 + y^2 \leq 16$
$y \geq 0$

58. $y \leq x$
$y \geq 2x - 3$

A.

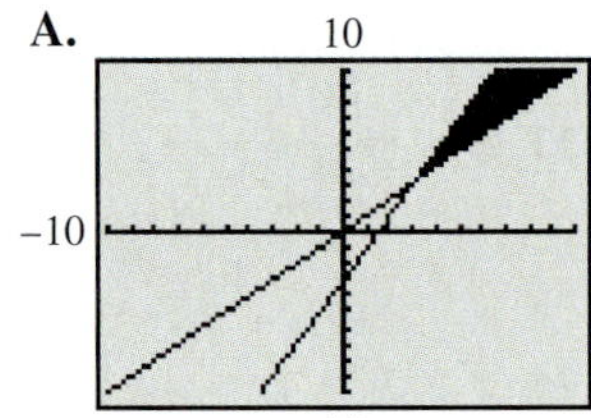

B.

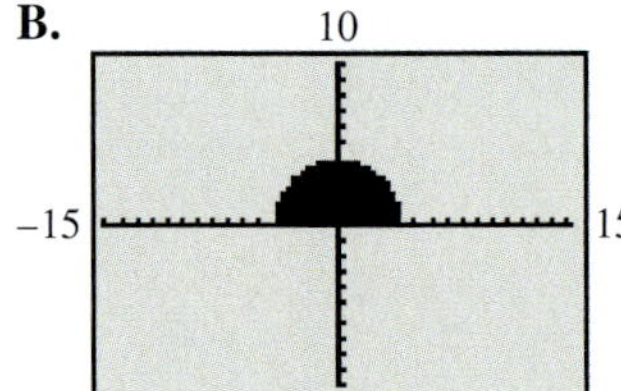

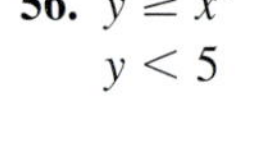

C.

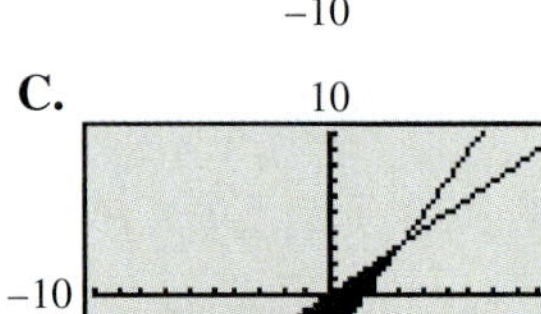

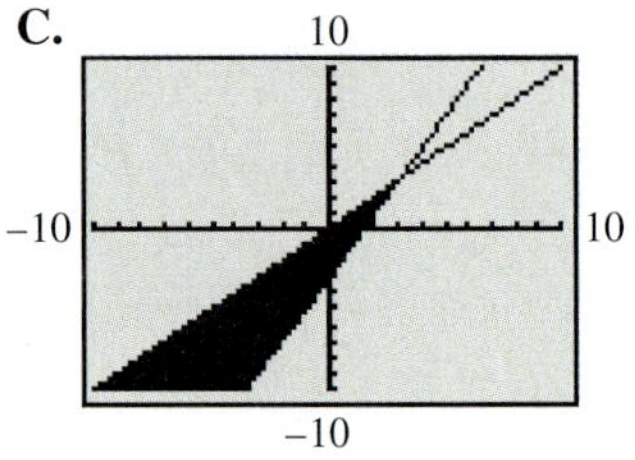

D.

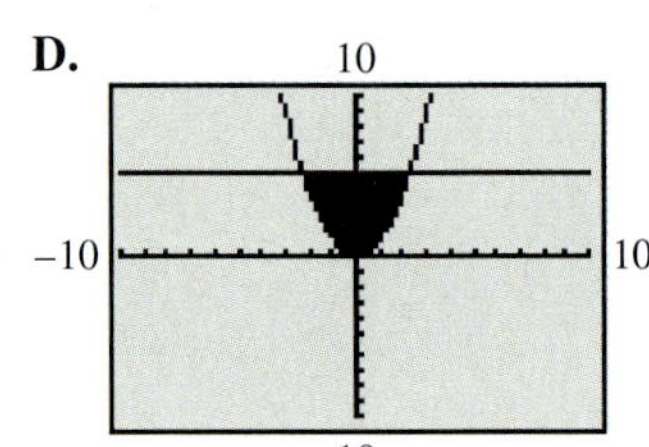

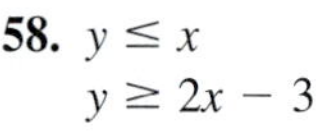

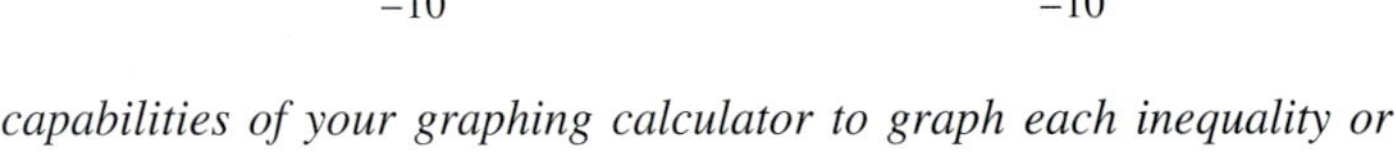

Use the shading capabilities of your graphing calculator to graph each inequality or system of inequalities.

59. $3x + 2y \geq 6$

60. $y \leq x^2 + 5$

61. $x + y \geq 2$
$x + y \leq 6$

62. $y \geq |x + 2|$
$y \leq 6$

63. ***Concept Check*** Find a system of linear inequalities for which the graph is the region in the first quadrant between and inclusive of the pair of lines $x + 2y - 8 = 0$ and $x + 2y = 12$.

64. ***Cost of Vitamins*** The figure shows the region of feasible solutions for the vitamin problem of Example 4 and the straight line graph of all combinations of red and blue pills for which the cost is 40 cents.

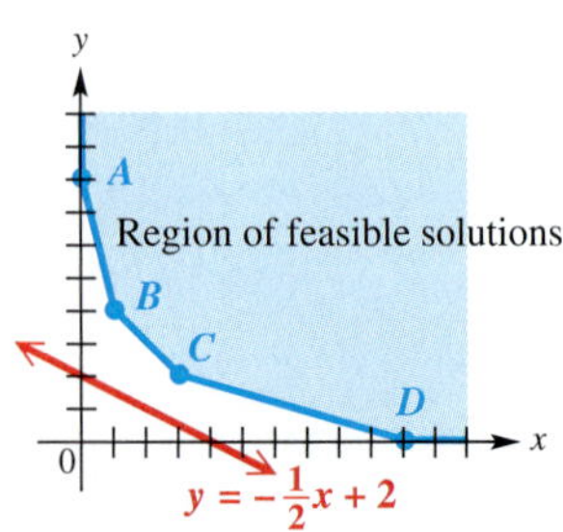

(a) The cost function is $10x + 20y$. Give the linear equation (in slope-intercept form) of the line of constant cost c.

(b) As c increases, does the line of constant cost move up or down?

(c) By inspection, find the vertex of the region of feasible solutions that gives the optimal solution.

The graphs show regions of feasible solutions. Find the maximum and minimum values of each objective function. See Examples 3 and 4.

65. objective function $= 3x + 5y$

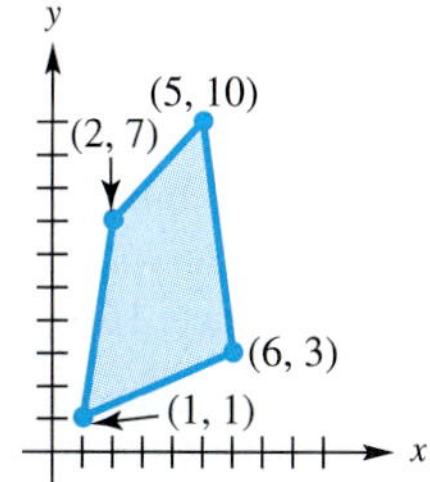

66. objective function $= 6x + y$

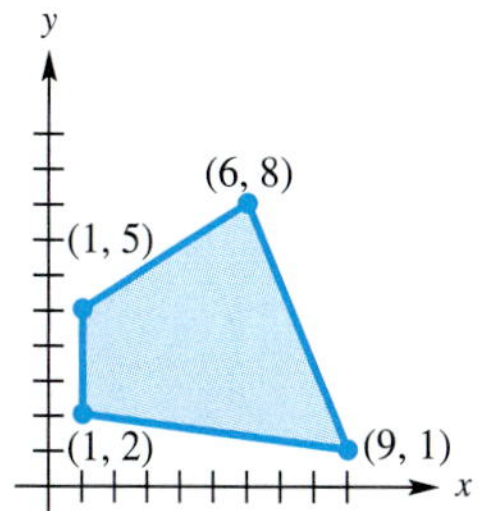

Find the maximum and minimum values of each objective function over the region of feasible solutions shown at the right. See Examples 3 and 4.

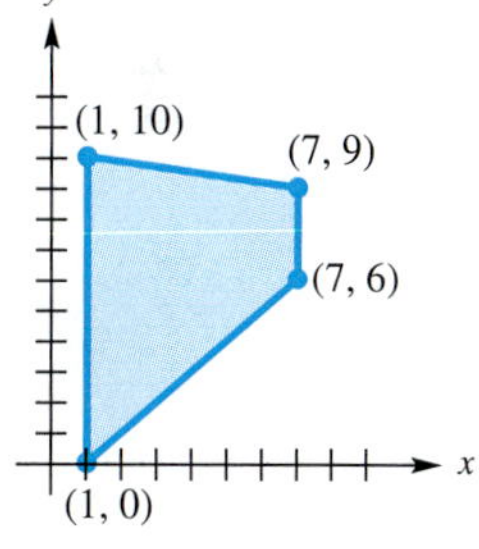

67. objective function $= 3x + 5y$

68. objective function $= 5x + 5y$

69. objective function $= 10y$

70. objective function $= 3x - y$

Write a system of inequalities for each problem and then graph the region of feasible solutions of the system. See Examples 3 and 4.

71. ***Vitamin Requirements*** Jane Lukas was given the following advice. She should supplement her daily diet with at least 6000 USP units of Vitamin A, at least 195 mg of Vitamin C, and at least 600 USP units of Vitamin D. She also finds that Mason's Pharmacy carries Brand X and Brand Y vitamins. Each Brand X pill contains 3000 USP units of A, 45 mg of C, and 75 USP units of D, while the Brand Y pills contain 1000 USP units of A, 50 mg of C, and 200 USP units of D.

72. ***Shipping Requirements*** The California Almond Growers have 2400 boxes of almonds to be shipped from their plant in Sacramento to Des Moines and San Antonio. The Des Moines market needs at least 1000 boxes, while the San Antonio market must have at least 800 boxes.

Solve each linear programming problem. See Examples 3 and 4.

73. ***Aid to Disaster Victims*** An agency wants to ship food and clothing to hurricane victims in Mexico. Commercial carriers have volunteered to transport the packages, provided they fit in the available cargo space. Each 20-ft^3 box of food weighs 40 lb and each 30-ft^3 box of clothing weighs 10 lb. The total weight cannot exceed 16,000 lb, and the total volume must be at most 18,000 ft^3. Each carton of food will feed 10 people, while each carton of clothing will help 8 people.

(a) How many cartons of food and clothing should be sent to maximize the number of people helped?

(b) What is the maximum number helped?

74. *Aid to Disaster Victims* Earthquake victims in Japan need medical supplies and bottled water. Each medical kit measures 1 ft^3 and weighs 10 lb. Each container of water is also 1 ft^3 but weighs 20 lb. The plane can only carry 80,000 lb with a total volume of 6000 ft^3. Each medical kit will aid 4 people, while each container of water will serve 10 people.

(a) How many of each should be sent to maximize the number of people helped?
(b) If each medical kit could aid 6 people instead of 4, how would the results from part (a) change?

75. *Storage Capacity* An office manager wants to buy some filing cabinets. He knows that cabinet A costs \$10 each, requires 6 ft^2 of floor space, and holds 8 ft^3 of files. Cabinet B costs \$20 each, requires 8 ft^2 of floor space, and holds 12 ft^3. He can spend no more than \$140 due to budget limitations, and his office has room for no more than 72 ft^2 of cabinets. He wants to maximize storage capacity within the limits imposed by funds and space. How many of each type of cabinet should he buy?

76. *Gasoline Revenues* The manufacturing process requires that oil refineries manufacture at least 2 gal of gasoline for each gallon of fuel oil. To meet the winter demand for fuel oil, at least 3 million gal per day must be produced. The demand for gasoline is no more than 6.4 million gal per day. If the price of gasoline is \$1.90 per gal and the price of fuel oil is \$1.50 per gal, how much of each should be produced to maximize revenue?

77. *Diet Requirements* Theo, who is dieting, requires two food supplements, I and II. He can get these supplements from two different products, *A* and *B*, as shown in the table. Theo's physician has recommended that he include at least 15 g of each supplement in his daily diet. If product *A* costs 25¢ per serving and product *B* costs 40¢ per serving, how can he satisfy his requirements most economically?

Supplement (g/serving)	I	II
Product *A*	3	2
Product *B*	2	4

78. *Profit from Televisions* Seall Manufacturing Company makes television monitors. It produces a bargain monitor that sells for \$100 profit and a deluxe monitor that sells for \$150 profit. On the assembly line the bargain monitor requires 3 hr, while the deluxe monitor takes 5 hr. The cabinet shop spends 1 hr on the cabinet for the bargain monitor and 3 hr on the cabinet for the deluxe monitor. Both models require 2 hr of time for testing and packing. On a particular production run, the Seall Company has available 3900 work hours on the assembly line, 2100 work hours in the cabinet shop, and 2200 work hours in the testing and packing department. How many of each model should it produce to make the maximum profit? What is the maximum profit?

9.7 Properties of Matrices

We used matrix notation to solve a system of linear equations earlier in this chapter. In this section and the next, we discuss algebraic properties of matrices.

Basic Definitions It is customary to use capital letters to name matrices. Also, subscript notation is often used to name elements of a matrix, as in the following matrix A.

$$A = \begin{bmatrix} a_{11} & a_{12} & a_{13} & \cdots & a_{1n} \\ a_{21} & a_{22} & a_{23} & \cdots & a_{2n} \\ a_{31} & a_{32} & a_{33} & \cdots & a_{3n} \\ \vdots & \vdots & \vdots & & \vdots \\ a_{m1} & a_{m2} & a_{m3} & \cdots & a_{mn} \end{bmatrix}$$

With this notation, the first row, first column element is a_{11} (read "a-sub-one-one"); the second row, third column element is a_{23}; and in general, the ith row, jth column element is a_{ij}.

Certain matrices have special names: an $n \times n$ matrix is a **square matrix** because the number of rows is equal to the number of columns. A matrix with just one row is a **row matrix,** and a matrix with just one column is a **column matrix.**

Two matrices are equal if they are the same size and if corresponding elements, position by position, are equal. Using this definition, the matrices

$$\begin{bmatrix} 2 & 1 \\ 3 & -5 \end{bmatrix} \quad \text{and} \quad \begin{bmatrix} 1 & 2 \\ -5 & 3 \end{bmatrix}$$

are *not* equal (even though they contain the same elements and are the same size), since the corresponding elements differ.

EXAMPLE 1 Finding Values to Make Two Matrices Equal

Find the values of the variables for which each statement is true, if possible.

(a) $\begin{bmatrix} 2 & 1 \\ p & q \end{bmatrix} = \begin{bmatrix} x & y \\ -1 & 0 \end{bmatrix}$ **(b)** $\begin{bmatrix} x \\ y \end{bmatrix} = \begin{bmatrix} 1 \\ 4 \\ 0 \end{bmatrix}$

Solution

(a) From the definition of equality given above, the only way that the statement can be true is if $2 = x$, $1 = y$, $p = -1$, and $q = 0$.

(b) This statement can never be true since the two matrices are different sizes. (One is 2×1 and the other is 3×1.)

Now try Exercises 1 and 7.

Adding Matrices Addition of matrices is defined as follows.

Addition of Matrices

To add two matrices of the same size, add corresponding elements. Only matrices of the same size can be added.

It can be shown that matrix addition satisfies the commutative, associative, closure, identity, and inverse properties. (See Exercises 87 and 88.)

EXAMPLE 2 Adding Matrices

Find each sum if possible.

(a) $\begin{bmatrix} 5 & -6 \\ 8 & 9 \end{bmatrix} + \begin{bmatrix} -4 & 6 \\ 8 & -3 \end{bmatrix}$ **(b)** $\begin{bmatrix} 2 \\ 5 \\ 8 \end{bmatrix} + \begin{bmatrix} -6 \\ 3 \\ 12 \end{bmatrix}$

(c) $A + B$, if $A = \begin{bmatrix} 5 & 8 \\ 6 & 2 \end{bmatrix}$ and $B = \begin{bmatrix} 3 & 9 & 1 \\ 4 & 2 & 5 \end{bmatrix}$

Algebraic Solution

(a) $$\begin{bmatrix} 5 & -6 \\ 8 & 9 \end{bmatrix} + \begin{bmatrix} -4 & 6 \\ 8 & -3 \end{bmatrix} = \begin{bmatrix} 5 + (-4) & -6 + 6 \\ 8 + 8 & 9 + (-3) \end{bmatrix} = \begin{bmatrix} 1 & 0 \\ 16 & 6 \end{bmatrix}$$

(b) $$\begin{bmatrix} 2 \\ 5 \\ 8 \end{bmatrix} + \begin{bmatrix} -6 \\ 3 \\ 12 \end{bmatrix} = \begin{bmatrix} -4 \\ 8 \\ 20 \end{bmatrix}$$

(c) The matrices

$$A = \begin{bmatrix} 5 & 8 \\ 6 & 2 \end{bmatrix}$$

and $$B = \begin{bmatrix} 3 & 9 & 1 \\ 4 & 2 & 5 \end{bmatrix}$$

have different sizes so A and B cannot be added; the sum $A + B$ does not exist.

Graphing Calculator Solution

(a) Figure 24(b) shows the sum of matrices A and B as defined in Figure 24(a).

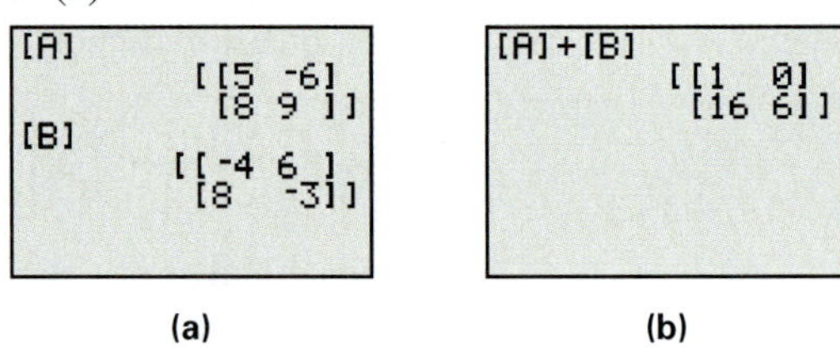

(a) (b)

Figure 24

(b) The screen in Figure 25 shows how the sum of two column matrices entered directly on the home screen is displayed.

(c) A graphing calculator will return an ERROR message if it is directed to perform an operation on matrices that is not possible due to dimension (size) mismatch. See Figure 26.

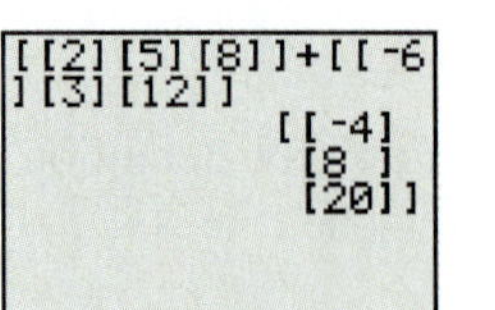

Figure 25

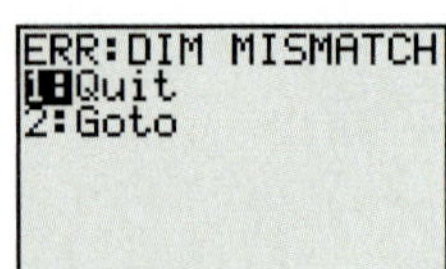

Figure 26

Now try Exercises 21, 23, and 25.

Special Matrices A matrix containing only zero elements is called a **zero matrix.** A zero matrix can be written with any size.

$$1 \times 3 \text{ zero matrix} \qquad O = \begin{bmatrix} 0 & 0 & 0 \end{bmatrix} \qquad\qquad 2 \times 3 \text{ zero matrix} \qquad O = \begin{bmatrix} 0 & 0 & 0 \\ 0 & 0 & 0 \end{bmatrix}$$

By the additive inverse property, each real number has an additive inverse: if a is a real number, then there is a real number $-a$ such that

$$a + (-a) = 0 \quad \text{and} \quad -a + a = 0. \quad \text{(Section R.1)}$$

Given matrix A, there is a matrix $-A$ such that $A + (-A) = O$. The matrix $-A$ has as elements the additive inverses of the elements of A. (Remember, each element of A is a real number and therefore has an additive inverse.) For example, if

$$A = \begin{bmatrix} -5 & 2 & -1 \\ 3 & 4 & -6 \end{bmatrix}, \quad \text{then} \quad -A = \begin{bmatrix} 5 & -2 & 1 \\ -3 & -4 & 6 \end{bmatrix}.$$

To check, test that $A + (-A)$ equals the zero matrix, O.

$$A + (-A) = \begin{bmatrix} -5 & 2 & -1 \\ 3 & 4 & -6 \end{bmatrix} + \begin{bmatrix} 5 & -2 & 1 \\ -3 & -4 & 6 \end{bmatrix} = \begin{bmatrix} 0 & 0 & 0 \\ 0 & 0 & 0 \end{bmatrix} = O$$

Matrix $-A$ is called the **additive inverse,** or **negative,** of matrix A. Every matrix has an additive inverse.

Subtracting Matrices The real number b is subtracted from the real number a, written $a - b$, by adding a and the additive inverse of b. That is,

$$a - b = a + (-b).$$

The same definition applies to subtraction of matrices.

Subtraction of Matrices

If A and B are two matrices of the same size, then

$$\mathbf{A - B = A + (-B).}$$

In practice, the difference of two matrices of the same size is found by subtracting corresponding elements.

EXAMPLE 3 Subtracting Matrices

Find each difference if possible.

(a) $\begin{bmatrix} -5 & 6 \\ 2 & 4 \end{bmatrix} - \begin{bmatrix} -3 & 2 \\ 5 & -8 \end{bmatrix}$ **(b)** $\begin{bmatrix} 8 & 6 & -4 \end{bmatrix} - \begin{bmatrix} 3 & 5 & -8 \end{bmatrix}$

(c) $A - B$, if $A = \begin{bmatrix} -2 & 5 \\ 0 & 1 \end{bmatrix}$ and $B = \begin{bmatrix} 3 \\ 5 \end{bmatrix}$

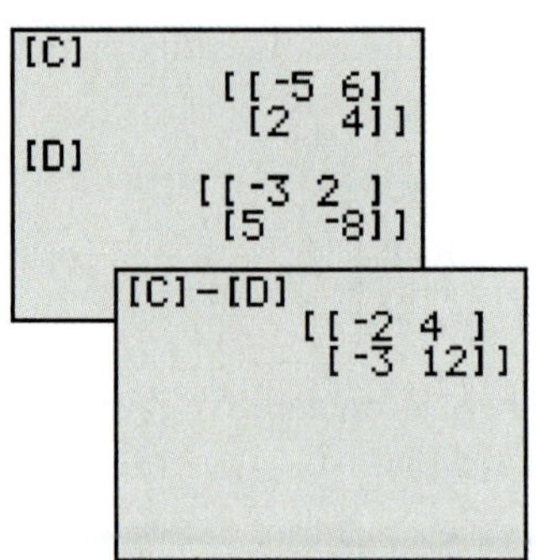

These screens support the result in Example 3(a).

Solution

(a) $\begin{bmatrix} -5 & 6 \\ 2 & 4 \end{bmatrix} - \begin{bmatrix} -3 & 2 \\ 5 & -8 \end{bmatrix} = \begin{bmatrix} -5-(-3) & 6-2 \\ 2-5 & 4-(-8) \end{bmatrix}$

$$= \begin{bmatrix} -2 & 4 \\ -3 & 12 \end{bmatrix}$$

(b) $[8 \quad 6 \quad -4] - [3 \quad 5 \quad -8] = [5 \quad 1 \quad 4]$

(c) The matrices

$$A = \begin{bmatrix} -2 & 5 \\ 0 & 1 \end{bmatrix} \quad \text{and} \quad B = \begin{bmatrix} 3 \\ 5 \end{bmatrix}$$

have different sizes and cannot be subtracted, so the difference $A - B$ does not exist.

Now try Exercises 27, 29, and 31.

Multiplying Matrices In work with matrices, a real number is called a **scalar** to distinguish it from a matrix. The product of a scalar k and a matrix X is the matrix kX, each of whose elements is k times the corresponding element of X.

```
[A]
        [[2 -3]
         [0 4 ]]
5[A]
        [[10 -15]
         [0  20 ]]
```

```
[B]
        [[20 36 ]
         [12 -16]]
(3/4)[B]
        [[15 27 ]
         [9  -12]]
```

These screens support the results in Example 4.

EXAMPLE 4 Multiplying Matrices by Scalars

Find each product.

(a) $5\begin{bmatrix} 2 & -3 \\ 0 & 4 \end{bmatrix}$ **(b)** $\frac{3}{4}\begin{bmatrix} 20 & 36 \\ 12 & -16 \end{bmatrix}$

Solution

(a) $5\begin{bmatrix} 2 & -3 \\ 0 & 4 \end{bmatrix} = \begin{bmatrix} 5(2) & 5(-3) \\ 5(0) & 5(4) \end{bmatrix}$ Multiply each element of the matrix by the scalar, 5.

$$= \begin{bmatrix} 10 & -15 \\ 0 & 20 \end{bmatrix}$$

(b) $\frac{3}{4}\begin{bmatrix} 20 & 36 \\ 12 & -16 \end{bmatrix} = \begin{bmatrix} 15 & 27 \\ 9 & -12 \end{bmatrix}$

Now try Exercises 37 and 39.

The proofs of the following properties of scalar multiplication are left for Exercises 91–94.

Properties of Scalar Multiplication

If A and B are matrices of the same size and c and d are scalars, then

$$(c + d)A = cA + dA \qquad c(A)d = cd(A)$$
$$c(A + B) = cA + cB \qquad (cd)A = c(dA).$$

We have seen how to multiply a real number (scalar) and a matrix. To find the product of two matrices, such as

$$A = \begin{bmatrix} -3 & 4 & 2 \\ 5 & 0 & 4 \end{bmatrix} \quad \text{and} \quad B = \begin{bmatrix} -6 & 4 \\ 2 & 3 \\ 3 & -2 \end{bmatrix},$$

first locate *row* 1 of A and *column* 1 of B, shown shaded below.

$$A = \begin{bmatrix} -3 & 4 & 2 \\ 5 & 0 & 4 \end{bmatrix} \qquad B = \begin{bmatrix} -6 & 4 \\ 2 & 3 \\ 3 & -2 \end{bmatrix}$$

Multiply corresponding elements, and find the sum of the products.

$$-3(-6) + 4(2) + 2(3) = 32$$

This result is the element for row 1, column 1 of the product matrix.

Now use *row* 1 of A and *column* 2 of B to determine the element in row 1, column 2 of the product matrix.

$$\begin{bmatrix} -3 & 4 & 2 \\ 5 & 0 & 4 \end{bmatrix}\begin{bmatrix} -6 & 4 \\ 2 & 3 \\ 3 & -2 \end{bmatrix} \qquad -3(4) + 4(3) + 2(-2) = -4$$

Next, use *row* 2 of A and *column* 1 of B; this will give the row 2, column 1 entry of the product matrix.

$$\begin{bmatrix} -3 & 4 & 2 \\ 5 & 0 & 4 \end{bmatrix}\begin{bmatrix} -6 & 4 \\ 2 & 3 \\ 3 & -2 \end{bmatrix} \qquad 5(-6) + 0(2) + 4(3) = -18$$

Finally, use *row* 2 of A and *column* 2 of B to find the entry for row 2, column 2 of the product matrix.

$$\begin{bmatrix} -3 & 4 & 2 \\ 5 & 0 & 4 \end{bmatrix}\begin{bmatrix} -6 & 4 \\ 2 & 3 \\ 3 & -2 \end{bmatrix} \qquad 5(4) + 0(3) + 4(-2) = 12$$

The product matrix can now be written.

$$\begin{bmatrix} -3 & 4 & 2 \\ 5 & 0 & 4 \end{bmatrix}\begin{bmatrix} -6 & 4 \\ 2 & 3 \\ 3 & -2 \end{bmatrix} = \begin{bmatrix} 32 & -4 \\ -18 & 12 \end{bmatrix}$$

As seen here, the product of a 2×3 matrix and a 3×2 matrix is a 2×2 matrix.

By definition, the product AB of an $m \times n$ matrix A and an $n \times p$ matrix B is found as follows. Multiply each element of the first row of A by the corresponding element of the first column of B. The sum of these n products is the first row, first column element of AB. Also, the sum of the products found by multiplying the elements of the first row of A times the corresponding elements of the second column of B gives the first row, second column element of AB, and so on.

To find the ith row, jth column element of AB, multiply each element in the ith row of A by the corresponding element in the jth column of B. (Note the shaded areas in the matrices below.) The sum of these products will give the row i, column j element of AB.

$$A = \begin{bmatrix} a_{11} & a_{12} & a_{13} & \cdots & a_{1n} \\ a_{21} & a_{22} & a_{23} & \cdots & a_{2n} \\ \vdots & & & & \\ a_{i1} & a_{i2} & a_{i3} & \cdots & a_{in} \\ \vdots & & & & \\ a_{m1} & a_{m2} & a_{m3} & \cdots & a_{mn} \end{bmatrix} \qquad B = \begin{bmatrix} b_{11} & b_{12} & \cdots & b_{1j} & \cdots & b_{1p} \\ b_{21} & b_{22} & \cdots & b_{2j} & \cdots & b_{2p} \\ \vdots & & & & & \\ b_{n1} & b_{n2} & \cdots & b_{nj} & \cdots & b_{np} \end{bmatrix}$$

Matrix Multiplication

The number of columns of an $m \times n$ matrix A is the same as the number of rows of an $n \times p$ matrix B (i.e., both n). The element c_{ij} of the product matrix $C = AB$ is found as follows:

$$c_{ij} = a_{i1}b_{1j} + a_{i2}b_{2j} + \cdots + a_{in}b_{nj}.$$

Matrix AB will be an $m \times p$ matrix.

EXAMPLE 5 Deciding Whether Two Matrices Can Be Multiplied

Suppose A is a 3×2 matrix, while B is a 2×4 matrix.

(a) Can the product AB be calculated?

(b) If AB can be calculated, what size is it?

(c) Can BA be calculated? **(d)** If BA can be calculated, what size is it?

Solution

(a) The following diagram shows that AB can be calculated, because the number of columns of A is equal to the number of rows of B. (Both are 2.)

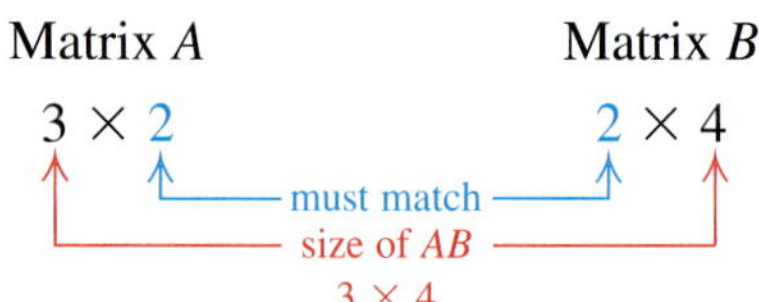

(b) As indicated in the diagram above, the product AB is a 3×4 matrix.

(c) The diagram below shows that BA cannot be calculated.

(d) The product BA cannot be calculated, because B has 4 columns and A has only 3 rows.

Now try Exercises 45 and 47.

EXAMPLE 6 Multiplying Matrices

Let $A = \begin{bmatrix} 1 & -3 \\ 7 & 2 \end{bmatrix}$ and $B = \begin{bmatrix} 1 & 0 & -1 & 2 \\ 3 & 1 & 4 & -1 \end{bmatrix}$. Find each product, if possible.

(a) AB **(b)** BA

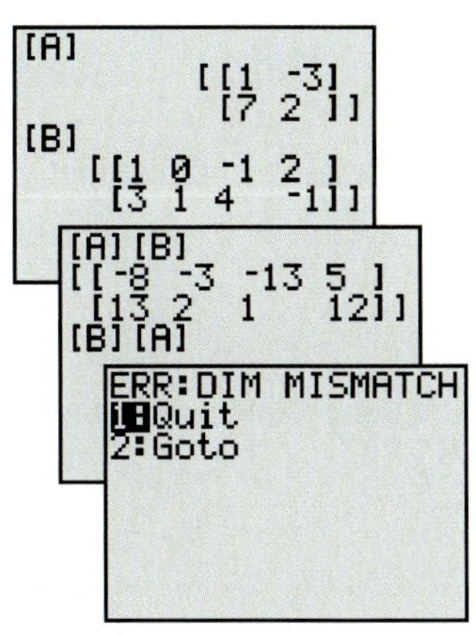

The three screens here illustrate matrix multiplication using a graphing calculator. The product in the middle screen and the error message in the final screen support the results in Example 6.

Solution

(a) First decide whether AB can be found. Since A is 2×2 and B is 2×4, the product can be found and will be a 2×4 matrix.

$$AB = \begin{bmatrix} 1 & -3 \\ 7 & 2 \end{bmatrix} \begin{bmatrix} 1 & 0 & -1 & 2 \\ 3 & 1 & 4 & -1 \end{bmatrix}$$

$$= \begin{bmatrix} 1(1) + (-3)3 & 1(0) + (-3)1 & 1(-1) + (-3)4 & 1(2) + (-3)(-1) \\ 7(1) + 2(3) & 7(0) + 2(1) & 7(-1) + 2(4) & 7(2) + 2(-1) \end{bmatrix}$$

Use the definition of matrix multiplication.

$$= \begin{bmatrix} -8 & -3 & -13 & 5 \\ 13 & 2 & 1 & 12 \end{bmatrix}$$

Perform the operations.

(b) Since B is a 2×4 matrix, and A is a 2×2 matrix, the number of columns of B (4) does not equal the number of rows of A (2). Therefore, the product BA cannot be found.

Now try Exercises 65 and 69.

EXAMPLE 7 Multiplying Square Matrices in Different Orders

Let $A = \begin{bmatrix} 1 & 3 \\ -2 & 5 \end{bmatrix}$ and $B = \begin{bmatrix} -2 & 7 \\ 0 & 2 \end{bmatrix}$. Find each product.

(a) AB **(b)** BA

Solution

(a) $$AB = \begin{bmatrix} 1 & 3 \\ -2 & 5 \end{bmatrix} \begin{bmatrix} -2 & 7 \\ 0 & 2 \end{bmatrix}$$

$$= \begin{bmatrix} 1(-2) + 3(0) & 1(7) + 3(2) \\ -2(-2) + 5(0) & -2(7) + 5(2) \end{bmatrix} = \begin{bmatrix} -2 & 13 \\ 4 & -4 \end{bmatrix}$$

(b) $$BA = \begin{bmatrix} -2 & 7 \\ 0 & 2 \end{bmatrix} \begin{bmatrix} 1 & 3 \\ -2 & 5 \end{bmatrix}$$

$$= \begin{bmatrix} -2(1) + 7(-2) & -2(3) + 7(5) \\ 0(1) + 2(-2) & 0(3) + 2(5) \end{bmatrix} = \begin{bmatrix} -16 & 29 \\ -4 & 10 \end{bmatrix}$$

Note that $AB \neq BA$.

Now try Exercise 75.

Examples 5 and 6 showed that the order in which two matrices are to be multiplied may determine whether their product can be found. Example 7 showed that even when both products AB and BA can be found, they may not be equal.

In general, if A and B are matrices, then $AB \neq BA$. ***Matrix multiplication is not commutative.***

Matrix multiplication does, however, satisfy the associative and distributive properties.

Properties of Matrix Multiplication

If A, B, and C are matrices such that all the following products and sums exist, then

$$(AB)C = A(BC), \quad A(B + C) = AB + AC, \quad (B + C)A = BA + CA.$$

For proofs of these results for the special cases when A, B, and C are square matrices, see Exercises 89 and 90. The identity and inverse properties for matrix multiplication are discussed in the next section.

Applying Matrix Algebra

EXAMPLE 8 Using Matrix Multiplication to Model Plans for a Subdivision

A contractor builds three kinds of houses, models A, B, and C, with a choice of two styles, colonial or ranch. Matrix P below shows the number of each kind of house the contractor is planning to build for a new 100-home subdivision. The amounts for each of the main materials used depend on the style of the house. These amounts are shown in matrix Q, while matrix R gives the cost in dollars for each kind of material. Concrete is measured here in cubic yards, lumber in 1000 board feet, brick in 1000s, and shingles in 100 square feet.

$$\begin{array}{c} \\ \text{Model A} \\ \text{Model B} \\ \text{Model C} \end{array} \begin{array}{c} \begin{array}{cc} \text{Colonial} & \text{Ranch} \end{array} \\ \begin{bmatrix} 0 & 30 \\ 10 & 20 \\ 20 & 20 \end{bmatrix} \end{array} = P$$

$$\begin{array}{c} \\ \text{Colonial} \\ \text{Ranch} \end{array} \begin{array}{c} \begin{array}{cccc} \text{Concrete} & \text{Lumber} & \text{Brick} & \text{Shingles} \end{array} \\ \begin{bmatrix} 10 & 2 & 0 & 2 \\ 50 & 1 & 20 & 2 \end{bmatrix} \end{array} = Q \qquad \begin{array}{c} \text{Concrete} \\ \text{Lumber} \\ \text{Brick} \\ \text{Shingles} \end{array} \begin{array}{c} \text{Cost per Unit} \\ \begin{bmatrix} 20 \\ 180 \\ 60 \\ 25 \end{bmatrix} \end{array} = R$$

(a) What is the total cost of materials for all houses of each model?

(b) How much of each of the four kinds of material must be ordered?

(c) What is the total cost of the materials?

Solution

(a) To find the materials cost for each model, first find matrix PQ, which will show the total amount of each material needed for all houses of each model.

$$PQ = \begin{bmatrix} 0 & 30 \\ 10 & 20 \\ 20 & 20 \end{bmatrix} \begin{bmatrix} 10 & 2 & 0 & 2 \\ 50 & 1 & 20 & 2 \end{bmatrix} = \begin{array}{c} \begin{array}{cccc} \text{Concrete} & \text{Lumber} & \text{Brick} & \text{Shingles} \end{array} \\ \begin{bmatrix} 1500 & 30 & 600 & 60 \\ 1100 & 40 & 400 & 60 \\ 1200 & 60 & 400 & 80 \end{bmatrix} \end{array} \begin{array}{l} \text{Model A} \\ \text{Model B} \\ \text{Model C} \end{array}$$

Multiplying PQ and the cost matrix R gives the total cost of materials for each model.

$$(PQ)R = \begin{bmatrix} 1500 & 30 & 600 & 60 \\ 1100 & 40 & 400 & 60 \\ 1200 & 60 & 400 & 80 \end{bmatrix} \begin{bmatrix} 20 \\ 180 \\ 60 \\ 25 \end{bmatrix} = \begin{array}{c} \text{Cost} \\ \begin{bmatrix} 72{,}900 \\ 54{,}700 \\ 60{,}800 \end{bmatrix} \end{array} \begin{array}{l} \text{Model A} \\ \text{Model B} \\ \text{Model C} \end{array}$$

(b) To find how much of each kind of material to order, refer to the columns of matrix PQ. The sums of the elements of the columns will give a matrix whose elements represent the total amounts of each material needed for the subdivision. Call this matrix T and write it as a row matrix.

$$T = [3800 \quad 130 \quad 1400 \quad 200]$$

(c) The total cost of all the materials is given by the product of matrix T, the total amounts matrix, and matrix R, the cost matrix. To multiply these and get a 1×1 matrix, representing the total cost, requires multiplying a 1×4 matrix and a 4×1 matrix. This is why in part (b) a row matrix was written rather than a column matrix. The total materials cost is given by TR, so

$$TR = [3800 \quad 130 \quad 1400 \quad 200] \begin{bmatrix} 20 \\ 180 \\ 60 \\ 25 \end{bmatrix} = [188{,}400].$$

The total cost of materials is \$188,400.

Now try Exercise 81.

To help keep track of the quantities a matrix represents, let matrix P, from Example 8, represent models/styles, matrix Q represent styles/materials, and matrix R represent materials/cost. In each case the meaning of the rows is written first and that of the columns second. When the product PQ was found in Example 8, the rows of the matrix represented models and the columns represented materials. Therefore, the matrix product PQ represents models/materials. The common quantity, styles, in both P and Q was eliminated in the product PQ. Do you see that the product $(PQ)R$ represents models/cost?

In practical problems this notation helps to identify the order in which two matrices should be multiplied so that the results are meaningful. In Example 8(c), either product RT or product TR could have been found. However, since T represents subdivisions/materials and R represents materials/cost, only TR gave the required matrix representing subdivisions/cost.

9.7 Exercises

Find the values of the variables, for which each statement is true, if possible. See Examples 1 and 2.

1. $\begin{bmatrix} -3 & a \\ b & 5 \end{bmatrix} = \begin{bmatrix} c & 0 \\ 4 & d \end{bmatrix}$

2. $\begin{bmatrix} w & x \\ 8 & -12 \end{bmatrix} = \begin{bmatrix} 9 & 17 \\ y & z \end{bmatrix}$

3. $\begin{bmatrix} x+2 & y-6 \\ z-3 & w+5 \end{bmatrix} = \begin{bmatrix} -2 & 8 \\ 0 & 3 \end{bmatrix}$

4. $\begin{bmatrix} 6 & a+3 \\ b+2 & 9 \end{bmatrix} = \begin{bmatrix} c-3 & 4 \\ -2 & d-4 \end{bmatrix}$

5. $\begin{bmatrix} 0 & 5 & x \\ -1 & 3 & y+2 \\ 4 & 1 & z \end{bmatrix} = \begin{bmatrix} 0 & w+3 & 6 \\ -1 & 3 & 0 \\ 4 & 1 & 8 \end{bmatrix}$

6. $\begin{bmatrix} 5 & x-4 & 9 \\ 2 & -3 & 8 \\ 6 & 0 & 5 \end{bmatrix} = \begin{bmatrix} y+3 & 2 & 9 \\ z+4 & -3 & 8 \\ 6 & 0 & w \end{bmatrix}$

7. $[x \quad y \quad z] = [21 \quad 6]$

8. $\begin{bmatrix} p \\ q \\ r \end{bmatrix} = \begin{bmatrix} 3 \\ -9 \end{bmatrix}$

9. $\begin{bmatrix} -7+z & 4r & 8s \\ 6p & 2 & 5 \end{bmatrix} + \begin{bmatrix} -9 & 8r & 3 \\ 2 & 5 & 4 \end{bmatrix} = \begin{bmatrix} 2 & 36 & 27 \\ 20 & 7 & 12a \end{bmatrix}$

10. $\begin{bmatrix} a+2 & 3z+1 & 5m \\ 8k & 0 & 3 \end{bmatrix} + \begin{bmatrix} 3a & 2z & 5m \\ 2k & 5 & 6 \end{bmatrix} = \begin{bmatrix} 10 & -14 & 80 \\ 10 & 5 & 9 \end{bmatrix}$

11. *Concept Check* A 3×8 matrix has _____ columns and _____ rows.

12. *Concept Check* A square matrix with 2 rows has _____ columns.

Concept Check Find the size of each matrix. Identify any square, column, or row matrices.

13. $\begin{bmatrix} -4 & 8 \\ 2 & 3 \end{bmatrix}$

14. $\begin{bmatrix} -9 & 6 & 2 \\ 4 & 1 & 8 \end{bmatrix}$

15. $\begin{bmatrix} -6 & 8 & 0 & 0 \\ 4 & 1 & 9 & 2 \\ 3 & -5 & 7 & 1 \end{bmatrix}$

16. $[8 \quad -2 \quad 4 \quad 6 \quad 3]$

17. $\begin{bmatrix} 2 \\ 4 \end{bmatrix}$

18. $[-9]$

19. Your friend missed the lecture on adding matrices. In your own words, explain to him how to add two matrices.

20. Explain to a friend in your own words how to subtract two matrices.

Perform each operation when possible. See Examples 2 and 3.

21. $\begin{bmatrix} -4 & 3 \\ 12 & -6 \end{bmatrix} + \begin{bmatrix} 2 & -8 \\ 5 & 10 \end{bmatrix}$

22. $\begin{bmatrix} 9 & 4 \\ -8 & 2 \end{bmatrix} + \begin{bmatrix} -3 & 2 \\ -4 & 7 \end{bmatrix}$

23. $\begin{bmatrix} 6 & -9 & 2 \\ 4 & 1 & 3 \end{bmatrix} + \begin{bmatrix} -8 & 2 & 5 \\ 6 & -3 & 4 \end{bmatrix}$

24. $\begin{bmatrix} 4 & -3 \\ 7 & 2 \\ -6 & 8 \end{bmatrix} + \begin{bmatrix} 9 & -10 \\ 0 & 5 \\ -1 & 6 \end{bmatrix}$

25. $[2 \quad 4 \quad 6] + \begin{bmatrix} -2 \\ -4 \\ -6 \end{bmatrix}$

26. $\begin{bmatrix} 3 \\ 1 \\ 0 \end{bmatrix} + \begin{bmatrix} 2 \\ -6 \end{bmatrix}$

27. $\begin{bmatrix} -6 & 8 \\ 0 & 0 \end{bmatrix} - \begin{bmatrix} 0 & 0 \\ -4 & -2 \end{bmatrix}$

28. $\begin{bmatrix} 11 & 0 \\ -4 & 0 \end{bmatrix} - \begin{bmatrix} 0 & 12 \\ 0 & -14 \end{bmatrix}$

29. $\begin{bmatrix} 12 \\ -1 \\ 3 \end{bmatrix} - \begin{bmatrix} 8 \\ 4 \\ -1 \end{bmatrix}$

30. $[10 \quad -4 \quad 6] - [-2 \quad 5 \quad 3]$

31. $[-4 \quad 3] - [5 \quad 8 \quad 2]$

32. $[4 \quad 6] - \begin{bmatrix} 2 \\ 3 \end{bmatrix}$

33. $\begin{bmatrix} 1 & -4 \\ 2 & -3 \\ -8 & 4 \end{bmatrix} - \begin{bmatrix} -6 & 9 \\ -2 & 5 \\ -7 & -12 \end{bmatrix}$

34. $\begin{bmatrix} 2 & -8 & 7 \\ 4 & 3 & -6 \end{bmatrix} - \begin{bmatrix} -1 & 0 & 4 \\ -5 & -5 & 2 \end{bmatrix}$

35. $\begin{bmatrix} 3x + y & x - 2y & 2x \\ 5x & 3y & x + y \end{bmatrix} + \begin{bmatrix} 2x & 3y & 5x + y \\ 3x + 2y & x & 2x \end{bmatrix}$

36. $\begin{bmatrix} 4k - 8y \\ 6z - 3x \\ 2k + 5a \\ -4m + 2n \end{bmatrix} - \begin{bmatrix} 5k + 6y \\ 2z + 5x \\ 4k + 6a \\ 4m - 2n \end{bmatrix}$

Let $A = \begin{bmatrix} -2 & 4 \\ 0 & 3 \end{bmatrix}$ *and* $B = \begin{bmatrix} -6 & 2 \\ 4 & 0 \end{bmatrix}$. *Find each of the following. See Example 4.*

37. $2A$

38. $-3B$

39. $\frac{3}{2}B$

40. $-1.5A$

41. $2A - B$

42. $-2A + 4B$

43. $-A + \frac{1}{2}B$

44. $\frac{3}{4}A - B$

Suppose that matrix A has size 2×3, *B has size* 3×5, *and C has size* 5×2. *Decide whether the given product can be calculated. If it can, determine its size. See Example 5.*

45. AB

46. CA

47. BA

48. AC

49. BC

50. CB

Find each matrix product when possible. See Examples 5–7.

51. $\begin{bmatrix} 1 & 2 \\ 3 & 4 \end{bmatrix}\begin{bmatrix} -1 \\ 7 \end{bmatrix}$

52. $\begin{bmatrix} -1 & 5 \\ 7 & 0 \end{bmatrix}\begin{bmatrix} 6 \\ 2 \end{bmatrix}$

53. $\begin{bmatrix} 3 & -4 & 1 \\ 5 & 0 & 2 \end{bmatrix}\begin{bmatrix} -1 \\ 4 \\ 2 \end{bmatrix}$

54. $\begin{bmatrix} -6 & 3 & 5 \\ 2 & 9 & 1 \end{bmatrix}\begin{bmatrix} -2 \\ 0 \\ 3 \end{bmatrix}$

55. $\begin{bmatrix} 5 & 2 \\ -1 & 4 \end{bmatrix}\begin{bmatrix} 3 & -2 \\ 1 & 0 \end{bmatrix}$

56. $\begin{bmatrix} -4 & 0 \\ 1 & 3 \end{bmatrix}\begin{bmatrix} -2 & 4 \\ 0 & 1 \end{bmatrix}$

57. $\begin{bmatrix} 2 & 2 & -1 \\ 3 & 0 & 1 \end{bmatrix}\begin{bmatrix} 0 & 2 \\ -1 & 4 \\ 0 & 2 \end{bmatrix}$

58. $\begin{bmatrix} -9 & 2 & 1 \\ 3 & 0 & 0 \end{bmatrix}\begin{bmatrix} 2 \\ -1 \\ 4 \end{bmatrix}$

59. $\begin{bmatrix} -3 & 0 & 2 & 1 \\ 4 & 0 & 2 & 6 \end{bmatrix}\begin{bmatrix} -4 & 2 \\ 0 & 1 \end{bmatrix}$

60. $\begin{bmatrix} -1 & 2 & 4 & 1 \\ 0 & 2 & -3 & 5 \end{bmatrix}\begin{bmatrix} 1 & 2 & 4 \\ -2 & 5 & 1 \end{bmatrix}$

61. $[-2 \quad 4 \quad 1]\begin{bmatrix} 3 & -2 & 4 \\ 2 & 1 & 0 \\ 0 & -1 & 4 \end{bmatrix}$

62. $[0 \quad 3 \quad -4]\begin{bmatrix} -2 & 6 & 3 \\ 0 & 4 & 2 \\ -1 & 1 & 4 \end{bmatrix}$

63. $\begin{bmatrix} -2 & -3 & -4 \\ 2 & -1 & 0 \\ 4 & -2 & 3 \end{bmatrix}\begin{bmatrix} 0 & 1 & 4 \\ 1 & 2 & -1 \\ 3 & 2 & -2 \end{bmatrix}$

64. $\begin{bmatrix} -1 & 2 & 0 \\ 0 & 3 & 2 \\ 0 & 1 & 4 \end{bmatrix}\begin{bmatrix} 2 & -1 & 2 \\ 0 & 2 & 1 \\ 3 & 0 & -1 \end{bmatrix}$

Given $A = \begin{bmatrix} 4 & -2 \\ 3 & 1 \end{bmatrix}$, $B = \begin{bmatrix} 5 & 1 \\ 0 & -2 \\ 3 & 7 \end{bmatrix}$, *and* $C = \begin{bmatrix} -5 & 4 & 1 \\ 0 & 3 & 6 \end{bmatrix}$, *find each product when possible. See Examples 5–7.*

65. BA **66.** AC **67.** BC **68.** CB

69. AB **70.** CA **71.** A^2 **72.** A^3 (*Hint:* $A^3 = A^2 \cdot A$)

73. ***Concept Check*** Compare the answers to Exercises 65 and 69, 67, and 68, and 66 and 70. How do they show that matrix multiplication is not commutative?

74. ***Concept Check*** For any matrices P and Q, what must be true for both PQ and QP to exist?

For each pair of matrices A and B, find **(a)** *AB and* **(b)** *BA. See Example 7.*

75. $A = \begin{bmatrix} 3 & 4 \\ -2 & 1 \end{bmatrix}, B = \begin{bmatrix} 6 & 0 \\ 5 & -2 \end{bmatrix}$

76. $A = \begin{bmatrix} 0 & -5 \\ -4 & 2 \end{bmatrix}, B = \begin{bmatrix} 3 & -1 \\ -5 & 4 \end{bmatrix}$

77. $A = \begin{bmatrix} 0 & 1 & -1 \\ 0 & 1 & 0 \\ 0 & 0 & 1 \end{bmatrix}, B = \begin{bmatrix} 1 & 0 & 0 \\ 0 & 1 & 0 \\ 0 & 0 & 1 \end{bmatrix}$

78. $A = \begin{bmatrix} -1 & 0 & 1 \\ 0 & 1 & 1 \\ -1 & -1 & 0 \end{bmatrix}, B = \begin{bmatrix} 0 & 0 & 1 \\ 0 & 1 & 0 \\ 1 & 0 & 0 \end{bmatrix}$

79. ***Concept Check*** In Exercise 77, $AB = A$ and $BA = A$. For this pair of matrices, B acts the same way for matrix multiplication as the number ______ acts for multiplication of real numbers.

80. ***Concept Check*** For $A = \begin{bmatrix} a & b \\ c & d \end{bmatrix}$ and $B = \begin{bmatrix} 1 & 0 \\ 0 & 1 \end{bmatrix}$, find AB and BA. What do you notice? Matrix B acts as the multiplicative ______ element for 2×2 square matrices.

Solve each problem. See Example 8.

81. ***Income from Yogurt*** Yagel's Yogurt sells three types of yogurt: nonfat, regular, and super creamy, at three locations. Location I sells 50 gal of nonfat, 100 gal of regular, and 30 gal of super creamy each day. Location II sells 10 gal of nonfat, and Location III sells 60 gal of nonfat each day. Daily sales of regular yogurt are 90 gal at Location II and 120 gal at Location III. At Location II, 50 gal of super creamy are sold each day, and 40 gal of super creamy are sold each day at Location III.

(a) Write a 3×3 matrix that shows the sales figures for the three locations, with the rows representing the three locations.

(b) The incomes per gallon for nonfat, regular, and super creamy are \$12, \$10, and \$15, respectively. Write a 1×3 or 3×1 matrix displaying the incomes.

(c) Find a matrix product that gives the daily income at each of the three locations.

(d) What is Yagel's Yogurt's total daily income from the three locations?

82. ***Purchasing Costs*** The Bread Box, a small neighborhood bakery, sells four main items: sweet rolls, bread, cakes, and pies. The amount of each ingredient (in cups, except for eggs) required for these items is given by matrix A.

	Eggs	Flour	Sugar	Shortening	Milk
Rolls (doz)	1	4	$\frac{1}{4}$	$\frac{1}{4}$	1
Bread (loaf)	0	3	0	$\frac{1}{4}$	0
Cake	4	3	2	1	1
Pie (crust)	0	1	0	$\frac{1}{3}$	0

$= A$

The cost (in cents) for each ingredient when purchased in large lots or small lots is given by matrix B.

$$\begin{array}{l} \\ \\ \text{Eggs} \\ \text{Flour} \\ \text{Sugar} \\ \text{Shortening} \\ \text{Milk} \end{array} \begin{array}{c} \text{Cost} \\ \begin{array}{cc} \text{Large Lot} & \text{Small Lot} \end{array} \\ \begin{bmatrix} 5 & 5 \\ 8 & 10 \\ 10 & 12 \\ 12 & 15 \\ 5 & 6 \end{bmatrix} \end{array} = B$$

(a) Use matrix multiplication to find a matrix giving the comparative cost per bakery item for the two purchase options.

(b) Suppose a day's orders consist of 20 dozen sweet rolls, 200 loaves of bread, 50 cakes, and 60 pies. Write the orders as a 1×4 matrix and, using matrix multiplication, write as a matrix the amount of each ingredient needed to fill the day's orders.

(c) Use matrix multiplication to find a matrix giving the costs under the two purchase options to fill the day's orders.

83. *(Modeling) Northern Spotted Owl Population* Several years ago, mathematical ecologists created a model to analyze population dynamics of the endangered northern spotted owl in the Pacific Northwest. The ecologists divided the female owl population into three categories: juvenile (up to 1 yr old), subadult (1 to 2 yr old), and adult (over 2 yr old). They concluded that the change in the makeup of the northern spotted owl population in successive years could be described by the matrix equation

$$\begin{bmatrix} j_{n+1} \\ s_{n+1} \\ a_{n+1} \end{bmatrix} = \begin{bmatrix} 0 & 0 & .33 \\ .18 & 0 & 0 \\ 0 & .71 & .94 \end{bmatrix} \begin{bmatrix} j_n \\ s_n \\ a_n \end{bmatrix}.$$

The numbers in the column matrices give the numbers of females in the three age groups after n years and $n + 1$ years. Multiplying the matrices yields

$j_{n+1} = .33a_n$ Each year 33 juvenile females are born for each 100 adult females.

$s_{n+1} = .18j_n$ Each year 18% of the juvenile females survive to become subadults.

$a_{n+1} = .71s_n + .94a_n.$ Each year 71% of the subadults survive to become adults and 94% of the adults survive.

(*Source:* Lamberson, R. H., R. McKelvey, B. R. Noon, and C. Voss, "A Dynamic Analysis of Northern Spotted Owl Viability in a Fragmented Forest Landscape," *Conservation Biology,* Vol. 6, No. 4, December, 1992, pp. 505–512.)

(a) Suppose there are currently 3000 female northern spotted owls made up of 690 juveniles, 210 subadults, and 2100 adults. Use the matrix equation above to determine the total number of female owls for each of the next 5 yr.

(b) Using advanced techniques from linear algebra, we can show that in the long run,

$$\begin{bmatrix} j_{n+1} \\ s_{n+1} \\ a_{n+1} \end{bmatrix} \approx .98359 \begin{bmatrix} j_n \\ s_n \\ a_n \end{bmatrix}.$$

What can we conclude about the long-term fate of the northern spotted owl?

(c) In the model, the main impediment to the survival of the northern spotted owl is the number .18 in the second row of the 3×3 matrix. This number is low for two reasons. The first year of life is precarious for most animals living in the wild. In addition, juvenile owls must eventually leave the nest and establish their own territory. If much of the forest near their original home has been cleared, then they are vulnerable to predators while searching for a new home. Suppose that due to better forest management, the number .18 can be increased to .3. Rework part (a) under this new assumption.

84. *(Modeling) Predator-Prey Relationship* In certain parts of the Rocky Mountains, deer provide the main food source for mountain lions. When the deer population is large, the mountain lions thrive. However, a large mountain lion population drives down the size of the deer population. Suppose the fluctuations of the two populations from year to year can be modeled with the matrix equation

$$\begin{bmatrix} m_{n+1} \\ d_{n+1} \end{bmatrix} = \begin{bmatrix} .51 & .4 \\ -.05 & 1.05 \end{bmatrix} \begin{bmatrix} m_n \\ d_n \end{bmatrix}.$$

The numbers in the column matrices give the numbers of animals in the two populations after n years and $n + 1$ years, where the number of deer is measured in hundreds.

(a) Give the equation for d_{n+1} obtained from the second row of the square matrix. Use this equation to determine the rate the deer population will grow from year to year if there are no mountain lions.

(b) Suppose we start with a mountain lion population of 2000 and a deer population of 500,000 (that is, 5000 hundred deer). How large would each population be after 1 yr? 2 yr?

(c) Consider part (b) but change the initial mountain lion population to 4000. Show that the populations would each grow at a steady annual rate of 1.01.

85. *Northern Spotted Owl Population* Refer to Exercise 83(b). Show that the number .98359 is an approximate zero of the polynomial represented by

$$\begin{vmatrix} -x & 0 & .33 \\ .18 & -x & 0 \\ 0 & .71 & .94 - x \end{vmatrix}.$$

86. *Predator-Prey Relationship* Refer to Exercise 84(c). Show that the number 1.01 is a zero of the polynomial represented by

$$\begin{vmatrix} .51 - x & .4 \\ -.05 & 1.05 - x \end{vmatrix}.$$

For Exercises 87–94, let

$$A = \begin{bmatrix} a_{11} & a_{12} \\ a_{21} & a_{22} \end{bmatrix}, \quad B = \begin{bmatrix} b_{11} & b_{12} \\ b_{21} & b_{22} \end{bmatrix}, \quad \text{and} \quad C = \begin{bmatrix} c_{11} & c_{12} \\ c_{21} & c_{22} \end{bmatrix},$$

where all the elements are real numbers. Use these matrices to show that each statement is true for 2×2 matrices.

87. $A + B = B + A$ (commutative property)

88. $A + (B + C) = (A + B) + C$ (associative property)

89. $(AB)C = A(BC)$ (associative property)

90. $A(B + C) = AB + AC$ (distributive property)

91. $c(A + B) = cA + cB$, for any real number c.

92. $(c + d)A = cA + dA$, for any real numbers c and d.

93. $c(A)d = (cd)A$, for any real numbers c and d.

94. $(cd)A = c(dA)$, for any real numbers c and d.

9.8 Matrix Inverses

Identity Matrices ▪ **Multiplicative Inverses** ▪ **Solving Systems Using Inverse Matrices**

In the previous section, we saw several parallels between the set of real numbers and the set of matrices. Another similarity is that both sets have identity and inverse elements for multiplication.

Identity Matrices

By the identity property for real numbers,

$$a \cdot 1 = a \quad \text{and} \quad 1 \cdot a = a \qquad \text{(Section R.1)}$$

for any real number a. If there is to be a multiplicative **identity matrix** I, such that

$$AI = A \quad \text{and} \quad IA = A,$$

for any matrix A, then A and I must be square matrices of the same size.

2 × 2 Identity Matrix

If I_2 represents the 2×2 identity matrix, then

$$I_2 = \begin{bmatrix} 1 & 0 \\ 0 & 1 \end{bmatrix}.$$

To verify that I_2 is the 2×2 identity matrix, we must show that $AI = A$ and $IA = A$ for any 2×2 matrix A. Let

$$A = \begin{bmatrix} x & y \\ z & w \end{bmatrix}.$$

Then

$$AI = \begin{bmatrix} x & y \\ z & w \end{bmatrix}\begin{bmatrix} 1 & 0 \\ 0 & 1 \end{bmatrix} = \begin{bmatrix} x \cdot 1 + y \cdot 0 & x \cdot 0 + y \cdot 1 \\ z \cdot 1 + w \cdot 0 & z \cdot 0 + w \cdot 1 \end{bmatrix} = \begin{bmatrix} x & y \\ z & w \end{bmatrix} = A,$$

and

$$IA = \begin{bmatrix} 1 & 0 \\ 0 & 1 \end{bmatrix}\begin{bmatrix} x & y \\ z & w \end{bmatrix} = \begin{bmatrix} 1 \cdot x + 0 \cdot z & 1 \cdot y + 0 \cdot w \\ 0 \cdot x + 1 \cdot z & 0 \cdot y + 1 \cdot w \end{bmatrix} = \begin{bmatrix} x & y \\ z & w \end{bmatrix} = A.$$

Generalizing from this example, there is an $n \times n$ identity matrix having 1s on the main diagonal and 0s elsewhere.

n × *n* Identity Matrix

The $n \times n$ identity matrix is I_n, where

$$I_n = \begin{bmatrix} 1 & 0 & \cdots & 0 \\ 0 & 1 & \cdots & 0 \\ \vdots & \vdots & a_{ij} & \vdots \\ 0 & 0 & \cdots & 1 \end{bmatrix}.$$

The element $a_{ij} = 1$ when $i = j$ (the diagonal elements) and $a_{ij} = 0$ otherwise.

EXAMPLE 1 Verifying the Identity Property of I_3

Let $A = \begin{bmatrix} -2 & 4 & 0 \\ 3 & 5 & 9 \\ 0 & 8 & -6 \end{bmatrix}$. Give the 3×3 identity matrix I_3 and show that $AI_3 = A$.

Algebraic Solution

The 3×3 identity matrix is

$$I_3 = \begin{bmatrix} 1 & 0 & 0 \\ 0 & 1 & 0 \\ 0 & 0 & 1 \end{bmatrix}.$$

By the definition of matrix multiplication,

$$AI_3 = \begin{bmatrix} -2 & 4 & 0 \\ 3 & 5 & 9 \\ 0 & 8 & -6 \end{bmatrix} \begin{bmatrix} 1 & 0 & 0 \\ 0 & 1 & 0 \\ 0 & 0 & 1 \end{bmatrix}$$

$$= \begin{bmatrix} -2 & 4 & 0 \\ 3 & 5 & 9 \\ 0 & 8 & -6 \end{bmatrix} = A.$$

Graphing Calculator Solution

The calculator screen in Figure 27(a) shows the identity matrix for $n = 3$. The screens in Figures 27(b) and (c) support the algebraic result.

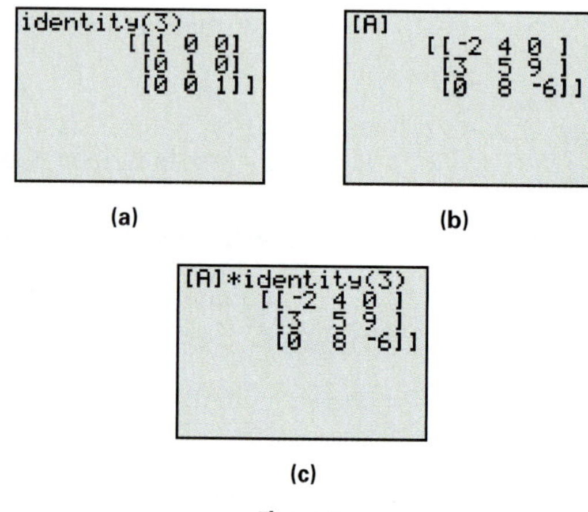

(a) (b) (c)

Figure 27

Now try Exercise 1.

Multiplicative Inverses

For every nonzero real number a, there is a multiplicative inverse $\frac{1}{a}$ such that

$$a \cdot \frac{1}{a} = 1 \quad \text{and} \quad \frac{1}{a} \cdot a = 1. \quad \text{(Section R.1)}$$

(Recall: $\frac{1}{a}$ is also written a^{-1}.) In a similar way, if A is an $n \times n$ matrix, then its **multiplicative inverse,** written A^{-1}, must satisfy both

$$AA^{-1} = I_n \quad \text{and} \quad A^{-1}A = I_n.$$

This means that only a square matrix can have a multiplicative inverse.

CAUTION Although $a^{-1} = \frac{1}{a}$ for any nonzero real number a, if A is a matrix,

$$A^{-1} \neq \frac{1}{A}.$$

In fact, $\frac{1}{A}$ has no meaning, since 1 is a *number* and A is a *matrix*.

To find the matrix A^{-1}, we use row transformations, introduced earlier in this chapter. As an example, we find the inverse of

$$A = \begin{bmatrix} 2 & 4 \\ 1 & -1 \end{bmatrix}.$$

Let the unknown inverse matrix be

$$A^{-1} = \begin{bmatrix} x & y \\ z & w \end{bmatrix}.$$

By the definition of matrix inverse, $AA^{-1} = I_2$, or

$$AA^{-1} = \begin{bmatrix} 2 & 4 \\ 1 & -1 \end{bmatrix}\begin{bmatrix} x & y \\ z & w \end{bmatrix} = \begin{bmatrix} 1 & 0 \\ 0 & 1 \end{bmatrix}.$$

By matrix multiplication,

$$\begin{bmatrix} 2x + 4z & 2y + 4w \\ x - z & y - w \end{bmatrix} = \begin{bmatrix} 1 & 0 \\ 0 & 1 \end{bmatrix}.$$

Setting corresponding elements equal gives the system of equations

$$\begin{aligned} 2x + 4z &= 1 \quad (1) \\ 2y + 4w &= 0 \quad (2) \\ x - z &= 0 \quad (3) \\ y - w &= 1. \quad (4) \end{aligned}$$

Since equations (1) and (3) involve only x and z, while equations (2) and (4) involve only y and w, these four equations lead to two systems of equations,

$$\begin{aligned} 2x + 4z &= 1 \\ x - z &= 0 \end{aligned} \quad \text{and} \quad \begin{aligned} 2y + 4w &= 0 \\ y - w &= 1. \end{aligned}$$

Writing the two systems as augmented matrices gives

$$\left[\begin{array}{cc|c} 2 & 4 & 1 \\ 1 & -1 & 0 \end{array}\right] \quad \text{and} \quad \left[\begin{array}{cc|c} 2 & 4 & 0 \\ 1 & -1 & 1 \end{array}\right].$$

Each of these systems can be solved by the Gauss-Jordan method. However, since the elements to the left of the vertical bar are identical, the two systems can be combined into one matrix,

$$\left[\begin{array}{cc|cc} 2 & 4 & 1 & 0 \\ 1 & -1 & 0 & 1 \end{array}\right],$$

and solved simultaneously using matrix row transformations. We need to change the numbers on the left of the vertical bar to the 2×2 identity matrix.

$$\left[\begin{array}{cc|cc} 1 & -1 & 0 & 1 \\ 2 & 4 & 1 & 0 \end{array}\right]$$ Interchange R1 and R2 to get 1 in the upper left corner. (Section 9.2)

$$\left[\begin{array}{cc|cc} 1 & -1 & 0 & 1 \\ 0 & 6 & 1 & -2 \end{array}\right]$$ $-2\text{R1} + \text{R2}$

$$\left[\begin{array}{cc|cc} 1 & -1 & 0 & 1 \\ 0 & 1 & \frac{1}{6} & -\frac{1}{3} \end{array}\right]$$ $\frac{1}{6}\text{R2}$

$$\left[\begin{array}{cc|cc} 1 & 0 & \frac{1}{6} & \frac{2}{3} \\ 0 & 1 & \frac{1}{6} & -\frac{1}{3} \end{array}\right]$$ $\text{R2} + \text{R1}$

The numbers in the first column to the right of the vertical bar in the final matrix give the values of x and z. The second column gives the values of y and w. That is,

$$\left[\begin{array}{cc|cc} 1 & 0 & x & y \\ 0 & 1 & z & w \end{array}\right] = \left[\begin{array}{cc|cc} 1 & 0 & \frac{1}{6} & \frac{2}{3} \\ 0 & 1 & \frac{1}{6} & -\frac{1}{3} \end{array}\right]$$

so that
$$A^{-1} = \begin{bmatrix} x & y \\ z & w \end{bmatrix} = \begin{bmatrix} \frac{1}{6} & \frac{2}{3} \\ \frac{1}{6} & -\frac{1}{3} \end{bmatrix}.$$

To check, multiply A by A^{-1}. The result should be I_2.

$$AA^{-1} = \begin{bmatrix} 2 & 4 \\ 1 & -1 \end{bmatrix}\begin{bmatrix} \frac{1}{6} & \frac{2}{3} \\ \frac{1}{6} & -\frac{1}{3} \end{bmatrix} = \begin{bmatrix} \frac{1}{3} + \frac{2}{3} & \frac{4}{3} - \frac{4}{3} \\ \frac{1}{6} - \frac{1}{6} & \frac{2}{3} + \frac{1}{3} \end{bmatrix}$$
$$= \begin{bmatrix} 1 & 0 \\ 0 & 1 \end{bmatrix} = I_2$$

Thus,
$$A^{-1} = \begin{bmatrix} \frac{1}{6} & \frac{2}{3} \\ \frac{1}{6} & -\frac{1}{3} \end{bmatrix}.$$

The process for finding the multiplicative inverse A^{-1} for any $n \times n$ matrix A that has an inverse is summarized below.

Finding an Inverse Matrix

To obtain A^{-1} for any $n \times n$ matrix A for which A^{-1} exists, follow these steps.

Step 1 Form the augmented matrix $[A \mid I_n]$, where I_n is the $n \times n$ identity matrix.

Step 2 Perform row transformations on $[A \mid I_n]$ to obtain a matrix of the form $[I_n \mid B]$.

Step 3 Matrix B is A^{-1}.

NOTE To confirm that two $n \times n$ matrices A and B are inverses of each other, it is sufficient to show that $AB = I_n$. It is not necessary to show also that $BA = I_n$.

EXAMPLE 2 Finding the Inverse of a 3 × 3 Matrix

Find A^{-1} if $A = \begin{bmatrix} 1 & 0 & 1 \\ 2 & -2 & -1 \\ 3 & 0 & 0 \end{bmatrix}$.

Solution Use row transformations as follows.

Step 1 Write the augmented matrix $[A \mid I_3]$.

$$\left[\begin{array}{ccc|ccc} 1 & 0 & 1 & 1 & 0 & 0 \\ 2 & -2 & -1 & 0 & 1 & 0 \\ 3 & 0 & 0 & 0 & 0 & 1 \end{array}\right]$$

Step 2 Since 1 is already in the upper left-hand corner as desired, begin by using the row transformation that will result in 0 for the first element in the second row. Multiply the elements of the first row by -2, and add the result to the second row.

$$\left[\begin{array}{ccc|ccc} 1 & 0 & 1 & 1 & 0 & 0 \\ 0 & -2 & -3 & -2 & 1 & 0 \\ 3 & 0 & 0 & 0 & 0 & 1 \end{array}\right] \quad -2R1 + R2$$

To get 0 for the first element in the third row, multiply the elements of the first row by -3 and add to the third row.

$$\left[\begin{array}{ccc|ccc} 1 & 0 & 1 & 1 & 0 & 0 \\ 0 & -2 & -3 & -2 & 1 & 0 \\ 0 & 0 & -3 & -3 & 0 & 1 \end{array}\right] \quad -3R1 + R3$$

To get 1 for the second element in the second row, multiply the elements of the second row by $-\frac{1}{2}$.

$$\left[\begin{array}{ccc|ccc} 1 & 0 & 1 & 1 & 0 & 0 \\ 0 & 1 & \frac{3}{2} & 1 & -\frac{1}{2} & 0 \\ 0 & 0 & -3 & -3 & 0 & 1 \end{array}\right] \quad -\tfrac{1}{2}R2$$

To get 1 for the third element in the third row, multiply the elements of the third row by $-\frac{1}{3}$.

$$\left[\begin{array}{ccc|ccc} 1 & 0 & 1 & 1 & 0 & 0 \\ 0 & 1 & \frac{3}{2} & 1 & -\frac{1}{2} & 0 \\ 0 & 0 & 1 & 1 & 0 & -\frac{1}{3} \end{array}\right] \quad -\tfrac{1}{3}R3$$

To get 0 for the third element in the first row, multiply the elements of the third row by -1 and add to the first row.

$$\left[\begin{array}{ccc|ccc} 1 & 0 & 0 & 0 & 0 & \frac{1}{3} \\ 0 & 1 & \frac{3}{2} & 1 & -\frac{1}{2} & 0 \\ 0 & 0 & 1 & 1 & 0 & -\frac{1}{3} \end{array}\right] \quad -1R3 + R1$$

To get 0 for the third element in the second row, multiply the elements of the third row by $-\frac{3}{2}$ and add to the second row.

$$\left[\begin{array}{ccc|ccc} 1 & 0 & 0 & 0 & 0 & \frac{1}{3} \\ 0 & 1 & 0 & -\frac{1}{2} & -\frac{1}{2} & \frac{1}{2} \\ 0 & 0 & 1 & 1 & 0 & -\frac{1}{3} \end{array}\right] \quad -\tfrac{3}{2}R3 + R2$$

Step 3 The last transformation shows that the inverse is

$$A^{-1} = \begin{bmatrix} 0 & 0 & \frac{1}{3} \\ -\frac{1}{2} & -\frac{1}{2} & \frac{1}{2} \\ 1 & 0 & -\frac{1}{3} \end{bmatrix}.$$

Confirm this by forming the product $A^{-1}A$ or AA^{-1}, each of which should equal the matrix I_3.

Now try Exercises 11 and 19.

```
[A]
    [[1 0  1 ]
     [2 -2 -1]
     [3 0  0 ]]
```

```
[A]-1▸Frac
...0     0    1/3 ...
...-1/2 -1/2  1/2 ...
...1     0   -1/3...
```

A graphing calculator can be used to find the inverse of a matrix. The screens support the result in Example 2. The elements of the inverse are expressed as fractions, so it is easier to compare with the inverse matrix found in the example.

As illustrated by the examples, the most efficient order for the transformations in Step 2 is to make the changes column by column from left to right, so that for each column the required 1 is the result of the first change. Next, perform the steps that obtain the 0s in that column. Then proceed to the next column.

EXAMPLE 3 Identifying a Matrix with No Inverse

Find A^{-1} if possible, given that $A = \begin{bmatrix} 2 & -4 \\ 1 & -2 \end{bmatrix}$.

Algebraic Solution

Using row transformations to change the first column of the augmented matrix

$$\left[\begin{array}{cc|cc} 2 & -4 & 1 & 0 \\ 1 & -2 & 0 & 1 \end{array}\right]$$

results in the following matrices:

$$\left[\begin{array}{cc|cc} 1 & -2 & \frac{1}{2} & 0 \\ 1 & -2 & 0 & 1 \end{array}\right] \quad \text{and} \quad \left[\begin{array}{cc|cc} 1 & -2 & \frac{1}{2} & 0 \\ 0 & 0 & -\frac{1}{2} & 1 \end{array}\right].$$

(We multiplied the elements in row one by $\frac{1}{2}$ in the first step, and in the second step we added the negative of row one to row two.) At this point, the matrix should be changed so that the second row, second element will be 1. Since that element is now 0, there is no way to complete the desired transformation, so A^{-1} does not exist for this matrix A. Just as there is no multiplicative inverse for the real number 0, not every matrix has a multiplicative inverse. Matrix A is an example of such a matrix.

Graphing Calculator Solution

If the inverse of a matrix does not exist, the matrix is called *singular*, as shown in Figure 28 for matrix [A]. This occurs when the determinant of the matrix is 0. (See Exercises 27–32.)

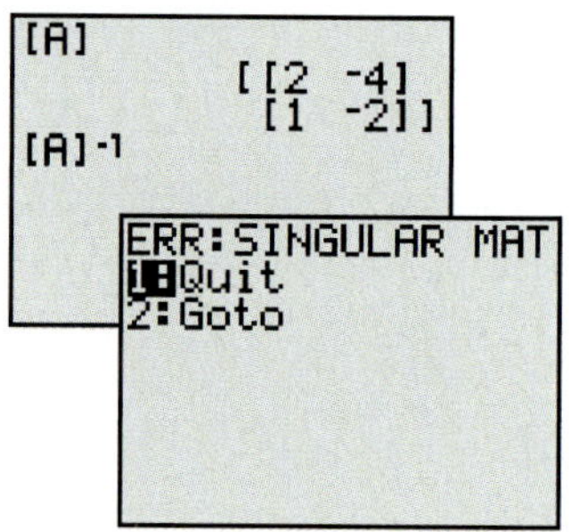

Figure 28

Now try Exercise 15.

If the inverse of a matrix exists, it is unique. That is, any given square matrix has no more than one inverse. The proof of this is left as Exercise 64.

Solving Systems Using Inverse Matrices Matrix inverses can be used to solve square linear systems of equations. (A square system has the same number of equations as variables.) For example, given the linear system

$$\begin{aligned} a_{11}x + a_{12}y + a_{13}z &= b_1 \\ a_{21}x + a_{22}y + a_{23}z &= b_2 \\ a_{31}x + a_{32}y + a_{33}z &= b_3, \end{aligned}$$

the definition of matrix multiplication can be used to rewrite the system as

$$\begin{bmatrix} a_{11} & a_{12} & a_{13} \\ a_{21} & a_{22} & a_{23} \\ a_{31} & a_{32} & a_{33} \end{bmatrix} \cdot \begin{bmatrix} x \\ y \\ z \end{bmatrix} = \begin{bmatrix} b_1 \\ b_2 \\ b_3 \end{bmatrix}. \quad (1)$$

(To see this, multiply the matrices on the left.)

$$\text{If} \quad A = \begin{bmatrix} a_{11} & a_{12} & a_{13} \\ a_{21} & a_{22} & a_{23} \\ a_{31} & a_{32} & a_{33} \end{bmatrix}, \quad X = \begin{bmatrix} x \\ y \\ z \end{bmatrix}, \quad \text{and} \quad B = \begin{bmatrix} b_1 \\ b_2 \\ b_3 \end{bmatrix},$$

then the system given in (1) becomes $AX = B$. If A^{-1} exists, then both sides of $AX = B$ can be multiplied on the left to get

$$A^{-1}(AX) = A^{-1}B$$

$$(A^{-1}A)X = A^{-1}B \quad \text{Associative property}$$

$$I_3X = A^{-1}B \quad \text{Inverse property}$$

$$X = A^{-1}B. \quad \text{Identity property}$$

Matrix $A^{-1}B$ gives the solution of the system.

Solution of the Matrix Equation $AX = B$

If A is an $n \times n$ matrix with inverse A^{-1}, X is an $n \times 1$ matrix of variables, and B is an $n \times 1$ matrix, then the matrix equation

$$\mathbf{AX = B}$$

has the solution

$$\mathbf{X = A^{-1}B.}$$

This method of using matrix inverses to solve systems of equations is useful when the inverse is already known or when many systems of the form $AX = B$ must be solved and only B changes.

EXAMPLE 4 Solving Systems of Equations Using Matrix Inverses

Use the inverse of the coefficient matrix to solve each system.

(a) $2x - 3y = 4$
$x + 5y = 2$

(b) $x + \quad z = -1$
$2x - 2y - z = 5$
$3x \quad = 6$

Algebraic Solution

(a) To represent the system as a matrix equation, use one matrix for the coefficients, one for the variables, and one for the constants.

$$A = \begin{bmatrix} 2 & -3 \\ 1 & 5 \end{bmatrix}, \quad X = \begin{bmatrix} x \\ y \end{bmatrix}, \quad \text{and} \quad B = \begin{bmatrix} 4 \\ 2 \end{bmatrix}$$

The system can then be written in matrix form $AX = B$, since

$$AX = \begin{bmatrix} 2 & -3 \\ 1 & 5 \end{bmatrix}\begin{bmatrix} x \\ y \end{bmatrix} = \begin{bmatrix} 2x - 3y \\ x + 5y \end{bmatrix} = \begin{bmatrix} 4 \\ 2 \end{bmatrix} = B.$$

To solve the system, first find A^{-1}. Then find $A^{-1}B$.

$$A^{-1} = \begin{bmatrix} \frac{5}{13} & \frac{3}{13} \\ -\frac{1}{13} & \frac{2}{13} \end{bmatrix}$$

$$A^{-1}B = \begin{bmatrix} \frac{5}{13} & \frac{3}{13} \\ -\frac{1}{13} & \frac{2}{13} \end{bmatrix}\begin{bmatrix} 4 \\ 2 \end{bmatrix} = \begin{bmatrix} 2 \\ 0 \end{bmatrix}$$

Since $X = A^{-1}B$, $\quad X = \begin{bmatrix} x \\ y \end{bmatrix} = \begin{bmatrix} 2 \\ 0 \end{bmatrix}.$

The final matrix shows that the solution set of the system is $\{(2, 0)\}$.

Graphing Calculator Solution

(a) Enter [A] and [B], and then find the product $[A]^{-1}[B]$ as shown in Figure 29.

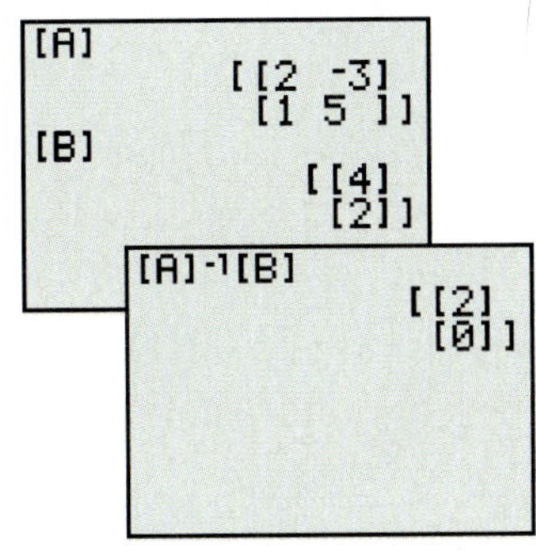

Figure 29

The column matrix indicates the solution set, $\{(2, 0)\}$.

(continued)

(b) The coefficient matrix A for this system is

$$A = \begin{bmatrix} 1 & 0 & 1 \\ 2 & -2 & -1 \\ 3 & 0 & 0 \end{bmatrix},$$

and its inverse A^{-1} was found in Example 2. Let

$$X = \begin{bmatrix} x \\ y \\ z \end{bmatrix} \quad \text{and} \quad B = \begin{bmatrix} -1 \\ 5 \\ 6 \end{bmatrix}.$$

Since $X = A^{-1}B$, we have

$$\begin{bmatrix} x \\ y \\ z \end{bmatrix} = \underbrace{\begin{bmatrix} 0 & 0 & \frac{1}{3} \\ -\frac{1}{2} & -\frac{1}{2} & \frac{1}{2} \\ 1 & 0 & -\frac{1}{3} \end{bmatrix}}_{A^{-1} \text{ from Example 2}} \begin{bmatrix} -1 \\ 5 \\ 6 \end{bmatrix}$$

$$= \begin{bmatrix} 2 \\ 1 \\ -3 \end{bmatrix}.$$

The solution set is $\{(2, 1, -3)\}$.

(b) Figure 30 shows the coefficient matrix [A] and the column matrix of constants [B]. Note that when $[A]^{-1}[B]$ is determined, it is not even necessary to display $[A]^{-1}$; as long as the inverse exists, the product will be computed.

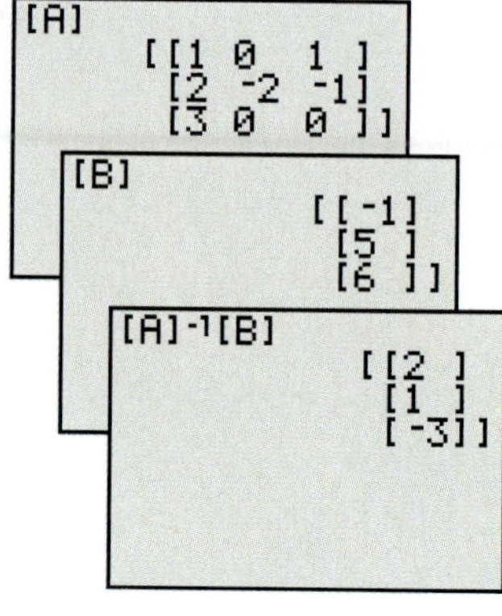

Figure 30

Now try Exercises 35 and 45.

9.8 Exercises

1. Show that for

$$A = \begin{bmatrix} -2 & 4 & 0 \\ 3 & 5 & 9 \\ 0 & 8 & -6 \end{bmatrix} \quad \text{and} \quad I_3 = \begin{bmatrix} 1 & 0 & 0 \\ 0 & 1 & 0 \\ 0 & 0 & 1 \end{bmatrix},$$

$I_3A = A$. (This result, along with that of Example 1, illustrates that the commutative property holds when one of the matrices is an identity matrix.)

2. Let $A = \begin{bmatrix} a & b \\ c & d \end{bmatrix}$ and $I_2 = \begin{bmatrix} 1 & 0 \\ 0 & 1 \end{bmatrix}$. Show that $AI_2 = I_2A = A$, thus proving that I_2 is the identity element for matrix multiplication for 2×2 square matrices.

Decide whether or not the given matrices are inverses of each other. (Hint: Check to see if their products are the identity matrix I_n.)

3. $\begin{bmatrix} 5 & 7 \\ 2 & 3 \end{bmatrix}$ and $\begin{bmatrix} 3 & -7 \\ -2 & 5 \end{bmatrix}$

4. $\begin{bmatrix} 2 & 3 \\ 1 & 1 \end{bmatrix}$ and $\begin{bmatrix} -1 & 3 \\ 1 & -2 \end{bmatrix}$

5. $\begin{bmatrix} -1 & 2 \\ 3 & -5 \end{bmatrix}$ and $\begin{bmatrix} -5 & -2 \\ -3 & -1 \end{bmatrix}$

6. $\begin{bmatrix} 2 & 1 \\ 3 & 2 \end{bmatrix}$ and $\begin{bmatrix} 2 & 1 \\ -3 & 2 \end{bmatrix}$

7. $\begin{bmatrix} 0 & 1 & 0 \\ 0 & 0 & -2 \\ 1 & -1 & 0 \end{bmatrix}$ and $\begin{bmatrix} 1 & 0 & 1 \\ 1 & 0 & 0 \\ 0 & -1 & 0 \end{bmatrix}$ **8.** $\begin{bmatrix} 1 & 2 & 0 \\ 0 & 1 & 0 \\ 0 & 1 & 0 \end{bmatrix}$ and $\begin{bmatrix} 1 & -2 & 0 \\ 0 & 1 & 0 \\ 0 & -1 & 1 \end{bmatrix}$

9. $\begin{bmatrix} -1 & -1 & -1 \\ 4 & 5 & 0 \\ 0 & 1 & -3 \end{bmatrix}$ and $\begin{bmatrix} 15 & 4 & -5 \\ -12 & -3 & 4 \\ -4 & -1 & 1 \end{bmatrix}$

10. $\begin{bmatrix} 1 & 3 & 3 \\ 1 & 4 & 3 \\ 1 & 3 & 4 \end{bmatrix}$ and $\begin{bmatrix} 7 & -3 & -3 \\ -1 & 1 & 0 \\ -1 & 0 & 1 \end{bmatrix}$

Find the inverse, if it exists, for each matrix. See Examples 2 and 3.

11. $\begin{bmatrix} -1 & 2 \\ -2 & -1 \end{bmatrix}$ **12.** $\begin{bmatrix} 1 & -1 \\ 2 & 0 \end{bmatrix}$ **13.** $\begin{bmatrix} -1 & -2 \\ 3 & 4 \end{bmatrix}$

14. $\begin{bmatrix} 3 & -1 \\ -5 & 2 \end{bmatrix}$ **15.** $\begin{bmatrix} 5 & 10 \\ -3 & -6 \end{bmatrix}$ **16.** $\begin{bmatrix} -6 & 4 \\ -3 & 2 \end{bmatrix}$

17. $\begin{bmatrix} 1 & 0 & 1 \\ 0 & -1 & 0 \\ 2 & 1 & 1 \end{bmatrix}$ **18.** $\begin{bmatrix} 1 & 0 & 0 \\ 0 & -1 & 0 \\ 1 & 0 & 1 \end{bmatrix}$ **19.** $\begin{bmatrix} 1 & 3 & 3 \\ 1 & 4 & 3 \\ 1 & 3 & 4 \end{bmatrix}$

20. $\begin{bmatrix} -2 & 2 & 4 \\ -3 & 4 & 5 \\ 1 & 0 & 2 \end{bmatrix}$ **21.** $\begin{bmatrix} 2 & 2 & -4 \\ 2 & 6 & 0 \\ -3 & -3 & 5 \end{bmatrix}$ **22.** $\begin{bmatrix} 2 & 4 & 6 \\ -1 & -4 & -3 \\ 0 & 1 & -1 \end{bmatrix}$

23. $\begin{bmatrix} 1 & 1 & 0 & 2 \\ 2 & -1 & 1 & -1 \\ 3 & 3 & 2 & -2 \\ 1 & 2 & 1 & 0 \end{bmatrix}$ **24.** $\begin{bmatrix} 1 & -2 & 3 & 0 \\ 0 & 1 & -1 & 1 \\ -2 & 2 & -2 & 4 \\ 0 & 2 & -3 & 1 \end{bmatrix}$

Each graphing calculator screen shows A^{-1} for some matrix A. Find each matrix A. (Hint: $(A^{-1})^{-1} = A$.)

25.
```
[A]-1
        [[5  -9]
         [-1 2 ]]
```

26.
```
[A]-1►Frac
  [[2/3 -1/3 0]
   [1/3 -5/3 1]
   [1/3 1/3  0]]
```

Relating Concepts

For individual or collaborative investigation

(Exercises 27–32)

It can be shown that the inverse of matrix $A = \begin{bmatrix} a & b \\ c & d \end{bmatrix}$ *is*

$$A^{-1} = \begin{bmatrix} \dfrac{d}{ad - bc} & \dfrac{-b}{ad - bc} \\ \dfrac{-c}{ad - bc} & \dfrac{a}{ad - bc} \end{bmatrix}.$$

(continued)

Work Exercises 27–32 in order, *to discover connections between the material in this section and a topic studied earlier in this chapter.*

27. With respect to the matrix $A = \begin{bmatrix} a & b \\ c & d \end{bmatrix}$, what do we call $ad - bc$?

28. Refer to A^{-1} as given on the previous page, and write it using determinant notation.

29. Write A^{-1} using scalar multiplication, where the scalar is $\frac{1}{|A|}$.

30. Explain in your own words how the inverse of matrix A can be found using a determinant.

31. Use the method described here to find the inverse of $A = \begin{bmatrix} 4 & 2 \\ 7 & 3 \end{bmatrix}$.

32. Complete the following statement: The inverse of a 2×2 matrix A does not exist if the determinant of A has value ______. (*Hint:* Look at the denominators in A^{-1} as given earlier.)

Solve each system by using the inverse of the coefficient matrix. See Example 4.

33. $-x + y = 1$
$2x - y = 1$

34. $x + y = 5$
$x - y = -1$

35. $2x - y = -8$
$3x + y = -2$

36. $x + 3y = -12$
$2x - y = 11$

37. $2x + 3y = -10$
$3x + 4y = -12$

38. $2x - 3y = 10$
$2x + 2y = 5$

39. $6x + 9y = 3$
$-8x + 3y = 6$

40. $5x - 3y = 0$
$10x + 6y = -4$

41. $.2x + .3y = -1.9$
$.7x - .2y = 4.6$

42. *Concept Check* Show that the matrix inverse method cannot be used to solve each system.

(a) $7x - 2y = 3$
$14x - 4y = 1$

(b) $x - 2y + 3z = 4$
$2x - 4y + 6z = 8$
$3x - 6y + 9z = 14$

Solve each system by using the inverse of the coefficient matrix. For Exercises 45–50, the inverses were found in Exercises 19–24. See Example 4.

43. $x + y + z = 6$
$2x + 3y - z = 7$
$3x - y - z = 6$

44. $2x + 5y + 2z = 9$
$4x - 7y - 3z = 7$
$3x - 8y - 2z = 9$

45. $x + 3y + 3z = 1$
$x + 4y + 3z = 0$
$x + 3y + 4z = -1$

46. $-2x + 2y + 4z = 3$
$-3x + 4y + 5z = 1$
$x + 2z = 2$

47. $2x + 2y - 4z = 12$
$2x + 6y = 16$
$-3x - 3y + 5z = -20$

48. $2x + 4y + 6z = 4$
$-x - 4y - 3z = 8$
$y - z = -4$

49. $x + y + 2w = 3$
$2x - y + z - w = 3$
$3x + 3y + 2z - 2w = 5$
$x + 2y + z = 3$

50. $x - 2y + 3z = 1$
$y - z + w = -1$
$-2x + 2y - 2z + 4w = 2$
$2y - 3z + w = -3$

(Modeling) *Solve each problem.*

51. *Plate-Glass Sales* The amount of plate-glass sales S (in millions of dollars) can be affected by the number of new building contracts B issued (in millions) and automobiles A produced (in millions). A plate-glass company in California wants to forecast future sales by using the past three years of sales. The totals for the three years are given in the table. To describe the relationship between these variables, the equation

S	A	B
602.7	5.543	37.14
656.7	6.933	41.30
778.5	7.638	45.62

$$S = a + bA + cB$$

was used, where the coefficients a, b, and c are constants that must be determined before the equation can be used. (*Source:* Makridakis, S. and S. Wheelwright, *Forecasting Methods for Management,* John Wiley & Sons, 1989.)

(a) Substitute the values for S, A, and B for each year from the table into the equation $S = a + bA + cB$, and obtain three linear equations involving a, b, and c.

(b) Use a graphing calculator to solve this linear system for a, b, and c. Use matrix inverse methods.

(c) Write the equation for S using these values for the coefficients.

(d) For the next year it is estimated that $A = 7.752$ and $B = 47.38$. Predict S. (The actual value for S was 877.6.)

(e) It is predicted that in 6 yr $A = 8.9$ and $B = 66.25$. Find the value of S in this situation and discuss its validity.

52. *Tire Sales* The number of automobile tire sales is dependent on several variables. In one study the relationship between annual tire sales S (in thousands of dollars), automobile registrations R (in millions), and personal disposable income I (in millions of dollars) was investigated. The results for three years are given in the table. To describe the relationship between these variables, we can use the equation

S	R	I
10,170	112.9	307.5
15,305	132.9	621.63
21,289	159.2	1937.13

$$S = a + bR + cI,$$

where the coefficients a, b, and c are constants that must be determined before the equation can be used. (*Source:* Jarrett, J., *Business Forecasting Methods,* Basil Blackwell, Ltd., 1991.)

(a) Substitute the values for S, R, and I for each year from the table into the equation $S = a + bR + cI$, and obtain three linear equations involving a, b, and c.

(b) Use a graphing calculator to solve this linear system for a, b, and c. Use matrix inverse methods.

(c) Write the equation for S using these values for the coefficients.

(d) If $R = 117.6$ and $I = 310.73$, predict S. (The actual value for S was 11,314.)

(e) If $R = 143.8$ and $I = 829.06$, predict S. (The actual value for S was 18,481.)

53. *Social Security Numbers* In Exercise 81 of Section 3.4, construction of your own personal Social Security polynomial was discussed. It is also possible to find a polynomial that goes through a given set of points in the plane by a process called *polynomial interpolation.* Recall that three points define a second-degree polynomial, four points define a third-degree polynomial, and so on. The only restriction on the points, since polynomials define functions, is that no two distinct points can have the same x-coordinate.

(continued)

53. *(continued)* Using the same SSN (539–58–0954) as in that exercise, we can find an eighth degree polynomial that lies on the nine points with x-coordinates 1 through 9 and y-coordinates that are digits of the SSN: (1, 5), (2, 3), (3, 9), . . . , (9, 4). This is done by writing a system of nine equations with nine variables, which is then solved by the inverse matrix method. The graph of this polynomial is shown. Find such a polynomial using your own SSN.

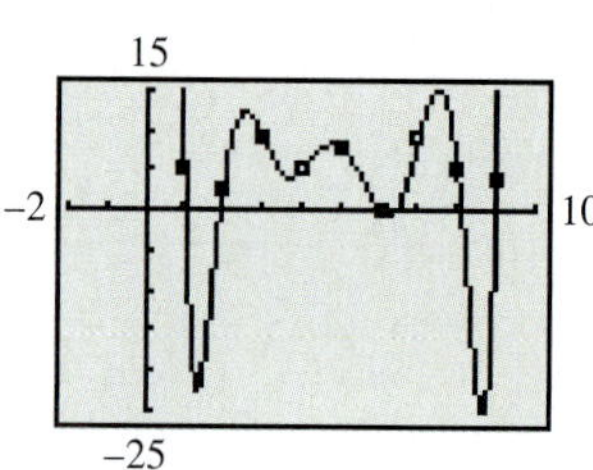

Use a graphing calculator to find the inverse of each matrix. Give as many decimal places as the calculator shows. See Example 2.

54. $\begin{bmatrix} \sqrt{2} & .5 \\ -17 & \frac{1}{2} \end{bmatrix}$

55. $\begin{bmatrix} \frac{2}{3} & .7 \\ 22 & \sqrt{3} \end{bmatrix}$

56. $\begin{bmatrix} 1.4 & .5 & .59 \\ .84 & 1.36 & .62 \\ .56 & .47 & 1.3 \end{bmatrix}$

57. $\begin{bmatrix} \frac{1}{2} & \frac{1}{4} & \frac{1}{3} \\ 0 & \frac{1}{4} & \frac{1}{3} \\ \frac{1}{2} & \frac{1}{2} & \frac{1}{3} \end{bmatrix}$

Use a graphing calculator and the method of matrix inverses to solve each system. Give as many decimal places as the calculator shows. See Example 4.

58. $\begin{aligned} x - \sqrt{2}y &= 2.6 \\ .75x + \quad y &= -7 \end{aligned}$

59. $\begin{aligned} 2.1x + \quad y &= \sqrt{5} \\ \sqrt{2}x - 2y &= 5 \end{aligned}$

60. $\begin{aligned} \pi x + ey + \sqrt{2}z &= 1 \\ ex + \pi y + \sqrt{2}z &= 2 \\ \sqrt{2}x + ey + \quad \pi z &= 3 \end{aligned}$

61. $\begin{aligned} (\log 2)x + \quad (\ln 3)y + (\ln 4)z &= 1 \\ (\ln 3)x + (\log 2)y + (\ln 8)z &= 5 \\ (\log 12)x + \quad (\ln 4)y + (\ln 8)z &= 9 \end{aligned}$

Let $A = \begin{bmatrix} a & b \\ c & d \end{bmatrix}$*, and let O be the 2×2 zero matrix. Show that the statements in Exercises 62 and 63 are true.*

62. $A \cdot O = O \cdot A = O$

63. For square matrices A and B of the same size, if $AB = O$ and if A^{-1} exists, then $B = O$.

64. Prove that any square matrix has no more than one inverse.

65. Give an example of two matrices A and B, where $(AB)^{-1} \neq A^{-1}B^{-1}$.

66. Suppose A and B are matrices, where A^{-1}, B^{-1}, and AB all exist. Show that $(AB)^{-1} = B^{-1}A^{-1}$.

67. Let $A = \begin{bmatrix} a & 0 & 0 \\ 0 & b & 0 \\ 0 & 0 & c \end{bmatrix}$, where a, b, and c are nonzero real numbers. Find A^{-1}.

68. Let $A = \begin{bmatrix} 1 & 0 & 0 \\ 0 & 0 & -1 \\ 0 & 1 & -1 \end{bmatrix}$. Show that $A^3 = I_3$, and use this result to find the inverse of A.

69. What are the inverses of I_n, $-A$ (in terms of A), and kA (k a scalar)?

70. Discuss the similarities and differences between solving the linear equation $ax = b$ and solving the matrix equation $AX = B$.

Chapter 9 Summary

KEY TERMS

9.1 linear equation (first-degree equation) in n unknowns
system of equations
solutions of a system of equations
system of linear equations (linear system)
consistent system
inconsistent system
dependent equations
equivalent systems
ordered triple

9.2 matrix (matrices)
element (of a matrix)
augmented matrix
size (of a matrix)

9.3 determinant
minor
cofactor
expansion by a row or column
Cramer's rule

9.4 partial fraction decomposition
partial fraction

9.5 nonlinear system

9.6 half-plane
boundary
linear inequality in two variables
system of inequalities
linear programming
constraints
objective function
region of feasible solutions
vertex (corner point)

9.7 square matrix
row matrix
column matrix
zero matrix
additive inverse (negative) of a matrix
scalar

9.8 identity matrix
multiplicative inverse (of a matrix)

NEW SYMBOLS

(a, b, c)	ordered triple
$[A]$	matrix A (graphing calculator symbolism)
$\lvert A \rvert$	determinant of matrix A
a_{ij}	the element in row i, column j, of a matrix
D, D_x, D_y	determinants used in Cramer's rule
I_2, I_3	identity matrices
A^{-1}	multiplicative inverse of matrix A

QUICK REVIEW

CONCEPTS	EXAMPLES

9.1 Systems of Linear Equations

Transformations of a Linear System

1. Interchange any two equations of the system.
2. Multiply or divide any equation of the system by a nonzero real number.
3. Replace any equation of the system by the sum of that equation and a multiple of another equation in the system.

Systems may be solved by substitution, elimination, or a combination of the two methods.

Substitution Method

Use one equation to find an expression for one variable in terms of the other, then substitute into the other equation of the system.

Solve by substitution.

$$4x - y = 7 \quad (1)$$
$$3x + 2y = 30 \quad (2)$$

Solve for y in equation (1).

$$y = 4x - 7$$

Substitute $4x - 7$ for y in equation (2), and solve for x.

$$3x + 2(4x - 7) = 30$$
$$3x + 8x - 14 = 30$$
$$11x - 14 = 30$$
$$11x = 44$$
$$x = 4$$

Substitute 4 for x in the equation $y = 4x - 7$ to find that $y = 9$. The solution set is $\{(4, 9)\}$.

CONCEPTS	EXAMPLES

Elimination Method
Use multiplication and addition to eliminate a variable from one equation. To eliminate a variable, the coefficients of that variable in the equations must be additive inverses.

Solve the system.

$$\begin{aligned} x + 2y - z &= 6 \quad (1)\\ x + y + z &= 6 \quad (2)\\ 2x + y - z &= 7 \quad (3) \end{aligned}$$

Add equations (1) and (2); z is eliminated and the result is $2x + 3y = 12$.

Eliminate z again by adding equations (2) and (3) to get $3x + 2y = 13$. Now solve the system

$$\begin{aligned} 2x + 3y &= 12 \quad (4)\\ 3x + 2y &= 13. \quad (5) \end{aligned}$$

$$\begin{aligned} -6x - 9y &= -36 \quad \text{Multiply (4) by } -3.\\ 6x + 4y &= 26 \quad \text{Multiply (5) by 2.}\\ \hline -5y &= -10 \quad \text{Add.}\\ y &= 2 \quad x \text{ is eliminated.} \end{aligned}$$

Let $y = 2$ in equation (4).

$$\begin{aligned} 2x + 3(2) &= 12 \quad (4) \text{ with } y = 2\\ 2x + 6 &= 12\\ 2x &= 6\\ x &= 3 \end{aligned}$$

Let $y = 2$ and $x = 3$ in any of the original equations to find $z = 1$. The solution set is $\{(3, 2, 1)\}$.

9.2 Matrix Solution of Linear Systems

Matrix Row Transformations
For any augmented matrix of a system of linear equations, the following row transformations will result in the matrix of an equivalent system.

1. Interchange any two rows.
2. Multiply or divide the elements of any row by a nonzero real number.
3. Replace any row of the matrix by the sum of the elements of that row and a multiple of the elements of another row.

Gauss-Jordan Method
The Gauss-Jordan method is a systematic technique for applying matrix row transformations in an attempt to reduce a matrix to diagonal form, with 1s along the diagonal.

Solve the system.

$$\begin{aligned} x + 3y &= 7\\ 2x + y &= 4 \end{aligned}$$

$$\left[\begin{array}{cc|c} 1 & 3 & 7\\ 2 & 1 & 4 \end{array}\right] \quad \text{Augmented matrix}$$

$$\left[\begin{array}{cc|c} 1 & 3 & 7\\ 0 & -5 & -10 \end{array}\right] \quad -2\text{R1} + \text{R2}$$

$$\left[\begin{array}{cc|c} 1 & 3 & 7\\ 0 & 1 & 2 \end{array}\right] \quad -\tfrac{1}{5}\text{R2}$$

$$\left[\begin{array}{cc|c} 1 & 0 & 1\\ 0 & 1 & 2 \end{array}\right] \quad -3\text{R2} + \text{R1}$$

This leads to

$$\begin{aligned} x &= 1\\ y &= 2, \end{aligned}$$

and the solution set is $\{(1, 2)\}$.

CONCEPTS | **EXAMPLES**

9.3 Determinant Solution of Linear Systems

Determinant of a 2 × 2 Matrix

If $A = \begin{bmatrix} a_{11} & a_{12} \\ a_{21} & a_{22} \end{bmatrix}$, then

$$|A| = \begin{vmatrix} a_{11} & a_{12} \\ a_{21} & a_{22} \end{vmatrix} = a_{11}a_{22} - a_{21}a_{12}.$$

Evaluate.

$$\begin{vmatrix} 3 & 5 \\ -2 & 6 \end{vmatrix} = 3(6) - (-2)5 = 28$$

Determinant of a 3 × 3 Matrix

If $A = \begin{bmatrix} a_{11} & a_{12} & a_{13} \\ a_{21} & a_{22} & a_{23} \\ a_{31} & a_{32} & a_{33} \end{bmatrix}$, then

$$|A| = \begin{vmatrix} a_{11} & a_{12} & a_{13} \\ a_{21} & a_{22} & a_{23} \\ a_{31} & a_{32} & a_{33} \end{vmatrix} = (a_{11}a_{22}a_{33} + a_{12}a_{23}a_{31} + a_{13}a_{21}a_{32}) - (a_{31}a_{22}a_{13} + a_{32}a_{23}a_{11} + a_{33}a_{21}a_{12}).$$

In practice, we usually evaluate determinants by expansion by minors.

Evaluate by expanding about the second column.

$$\begin{vmatrix} 2 & -3 & -2 \\ -1 & -4 & -3 \\ -1 & 0 & 2 \end{vmatrix} = -(-3)\begin{vmatrix} -1 & -3 \\ -1 & 2 \end{vmatrix} + (-4)\begin{vmatrix} 2 & -2 \\ -1 & 2 \end{vmatrix} - 0\begin{vmatrix} 2 & -2 \\ -1 & -3 \end{vmatrix}$$

$$= 3(-5) - 4(2) - 0(-8)$$

$$= -15 - 8 + 0$$

$$= -23$$

Cramer's Rule for Two Equations in Two Variables

Given the system

$$a_1x + b_1y = c_1$$
$$a_2x + b_2y = c_2,$$

if $D \neq 0$, then the system has the unique solution

$$x = \frac{D_x}{D} \quad \text{and} \quad y = \frac{D_y}{D},$$

where $D = \begin{vmatrix} a_1 & b_1 \\ a_2 & b_2 \end{vmatrix}$, $D_x = \begin{vmatrix} c_1 & b_1 \\ c_2 & b_2 \end{vmatrix}$, and $D_y = \begin{vmatrix} a_1 & c_1 \\ a_2 & c_2 \end{vmatrix}$.

Solve using Cramer's rule.

$$x - 2y = -1$$
$$2x + 5y = 16$$

$$x = \frac{\begin{vmatrix} -1 & -2 \\ 16 & 5 \end{vmatrix}}{\begin{vmatrix} 1 & -2 \\ 2 & 5 \end{vmatrix}} = \frac{-5 + 32}{5 + 4} = \frac{27}{9} = 3$$

$$y = \frac{\begin{vmatrix} 1 & -1 \\ 2 & 16 \end{vmatrix}}{\begin{vmatrix} 1 & -2 \\ 2 & 5 \end{vmatrix}} = \frac{16 + 2}{5 + 4} = \frac{18}{9} = 2$$

The solution set is $\{(3, 2)\}$.

General Form of Cramer's Rule

Let an $n \times n$ system have linear equations of the form

$$a_1x_1 + a_2x_2 + a_3x_3 + \cdots + a_nx_n = b.$$

Define D as the determinant of the $n \times n$ matrix of coefficients of the variables. Define D_{x_1} as the determinant obtained from D by replacing the entries in column 1 of D with the constants of the system. Define D_{x_i} as the determinant obtained from D by replacing the entries in column i with the constants of the system. If $D \neq 0$, the unique solution of the system is

$$x_1 = \frac{D_{x_1}}{D}, \quad x_2 = \frac{D_{x_2}}{D}, \quad x_3 = \frac{D_{x_3}}{D}, \quad \ldots, \quad x_n = \frac{D_{x_n}}{D}.$$

Solve using Cramer's rule.

$$3x + 2y + z = -5$$
$$x - y + 3z = -5$$
$$2x + 3y + z = 0$$

Using the method of expansion by minors, it can be shown that $D_x = 45$, $D_y = -30$, $D_z = 0$, and $D = -15$. Therefore,

$$x = \frac{D_x}{D} = \frac{45}{-15} = -3, \quad y = \frac{D_y}{D} = \frac{-30}{-15} = 2,$$

$$z = \frac{D_z}{D} = \frac{0}{-15} = 0.$$

The solution set is $\{(-3, 2, 0)\}$.

CONCEPTS	EXAMPLES

9.4 Partial Fractions

To solve for the constants in the numerators of a partial fraction decomposition, use either of the following methods or a combination of the two.

Method 1 For Linear Factors

1. Multiply both sides by the common denominator.
2. Substitute the zero of each factor in the resulting equation. For repeated linear factors, substitute as many other numbers as necessary to find all the constants in the numerators. The number of substitutions required will equal the number of constants $A, B, \ldots$.

Method 2 For Quadratic Factors

1. Multiply both sides by the common denominator.
2. Collect like terms on the right side of the resulting equation.
3. Equate the coefficients of like terms to get a system of equations.
4. Solve the system to find the constants in the numerators.

Decompose $\frac{9}{2x^2 + 9x + 9}$ into partial fractions.

$$\frac{9}{2x^2 + 9x + 9} = \frac{9}{(2x + 3)(x + 3)}$$

$$\frac{9}{(2x + 3)(x + 3)} = \frac{A}{2x + 3} + \frac{B}{x + 3}$$

Multiply by $(2x + 3)(x + 3)$.

$$9 = A(x + 3) + B(2x + 3)$$

$$9 = Ax + 3A + 2Bx + 3B$$

$$9 = (A + 2B)x + (3A + 3B)$$

Now solve the system

$$A + 2B = 0$$

$$3A + 3B = 9$$

to obtain $A = 6$ and $B = -3$. Therefore,

$$\frac{9}{2x^2 + 9x + 9} = \frac{6}{2x + 3} + \frac{-3}{x + 3}.$$

9.5 Nonlinear Systems of Equations

Solving a Nonlinear System

A nonlinear system can be solved by the substitution method, the elimination method, or a combination of the two.

Solve the system.

$$x^2 + 2xy - y^2 = 14 \quad (1)$$

$$x^2 - y^2 = -16 \quad (2)$$

$$x^2 + 2xy - y^2 = 14$$

$$-x^2 \qquad + y^2 = 16 \quad \text{Multiply (2) by } -1.$$

$$2xy \qquad = 30 \quad \text{Add; eliminate } x^2 \text{ and } y^2.$$

$$xy = 15$$

Solve for y to obtain $y = \frac{15}{x}$; substitute into equation (2).

$$x^2 - \left(\frac{15}{x}\right)^2 = -16 \quad (2)$$

$$x^2 - \frac{225}{x^2} = -16$$

$$x^4 + 16x^2 - 225 = 0 \quad \text{Multiply by } x^2\text{; add } 16x^2.$$

$$(x^2 - 9)(x^2 + 25) = 0 \quad \text{Factor.}$$

$$x = \pm 3 \quad \text{or} \quad x = \pm 5i \quad \text{Zero-factor property}$$

Find corresponding y-values to obtain the solution set

$$\{(3, 5), (-3, -5), (5i, -3i), (-5i, 3i)\}.$$

CONCEPTS	EXAMPLES

9.6 Systems of Inequalities and Linear Programming

Graphing Inequalities

1. For a function f, the graph of $y < f(x)$ consists of all the points that are *below* the graph of $y = f(x)$; the graph of $y > f(x)$ consists of all the points that are *above* the graph of $y = f(x)$.
2. If the inequality is not or cannot be solved for y, choose a test point not on the boundary. If the test point satisfies the inequality, the graph includes all points on the same side of the boundary as the test point. Otherwise, the graph includes all points on the other side of the boundary.

Solving Systems of Inequalities

To solve a system of inequalities, graph all inequalities on the same axes, and find the intersection of their solution sets.

Graph $y \geq x^2 - 2x + 3$.

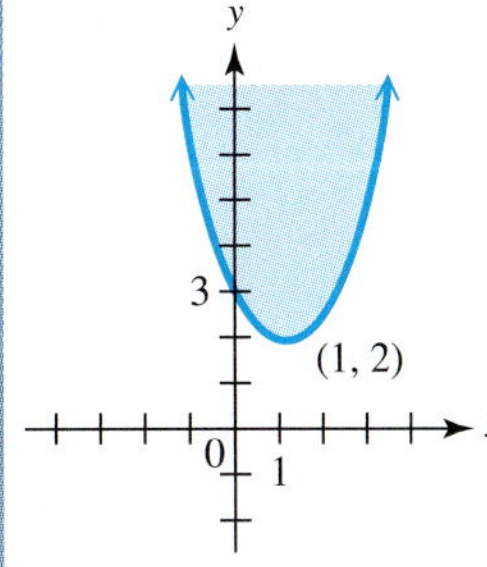

Graph the solution set of the system

$$3x - 5y > -15$$
$$x^2 + y^2 \leq 25.$$

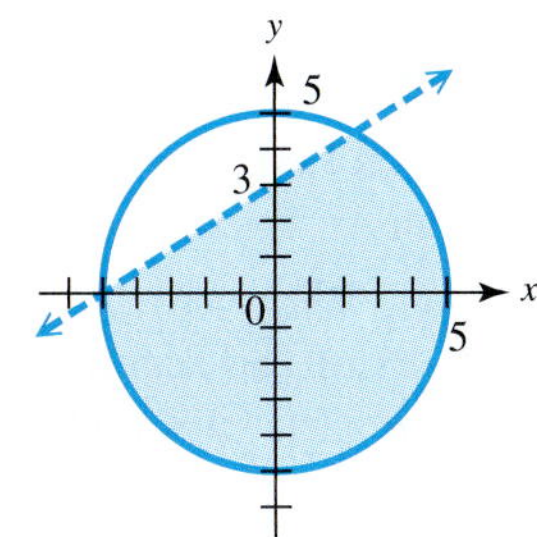

Fundamental Theorem of Linear Programming

The optimum value for a linear programming problem occurs at a vertex of the region of feasible solutions.

Solving a Linear Programming Problem

Write the objective function and all constraints, graph the region of feasible solutions, identify all vertex points (corner points), and find the value of the objective function at each vertex point. Choose the required maximum or minimum value accordingly.

The region of feasible solutions for

$$x + 2y \leq 14$$
$$3x + 4y \leq 36$$
$$x \geq 0$$
$$y \geq 0$$

is given in the figure. Maximize the objective function $8x + 12y$.

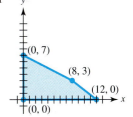

Vertex Point	Value of $8x + 12y$
(0, 0)	0
(0, 7)	84
(8, 3)	100 ← Maximum
(12, 0)	96

The objective function is maximized for $x = 8$ and $y = 3$.

9.7 Properties of Matrices

Addition and Subtraction of Matrices

To add (subtract) matrices of the same size, add (subtract) the corresponding elements.

Find the sum or difference.

$$\begin{bmatrix} 2 & 3 & -1 \\ 0 & 4 & 9 \end{bmatrix} + \begin{bmatrix} -8 & 12 & 1 \\ 5 & 3 & -3 \end{bmatrix} = \begin{bmatrix} -6 & 15 & 0 \\ 5 & 7 & 6 \end{bmatrix}$$

$$\begin{bmatrix} 5 & -1 \\ -8 & 8 \end{bmatrix} - \begin{bmatrix} -2 & 4 \\ 3 & -6 \end{bmatrix} = \begin{bmatrix} 7 & -5 \\ -11 & 14 \end{bmatrix}$$

Scalar Multiplication

To multiply a matrix by a scalar, multiply each element of the matrix by the scalar.

Find the scalar product.

$$3\begin{bmatrix} 6 & 2 \\ 1 & -2 \\ 0 & 8 \end{bmatrix} = \begin{bmatrix} 18 & 6 \\ 3 & -6 \\ 0 & 24 \end{bmatrix}$$

(continued)

CONCEPTS	EXAMPLES
Multiplication of Matrices The product AB of an $m \times n$ matrix A and an $n \times p$ matrix B is found as follows. To get the ith row, jth column element of the $m \times p$ matrix AB, multiply each element in the ith row of A by the corresponding element in the jth column of B. The sum of these products will give the element of row i, column j of AB.	Find the matrix product. $\begin{bmatrix} 1 & -2 & 3 \\ 5 & 0 & 4 \\ -8 & 7 & -7 \end{bmatrix} \begin{bmatrix} 1 \\ -2 \\ 3 \end{bmatrix} = \begin{bmatrix} 14 \\ 17 \\ -43 \end{bmatrix}$

9.8 Matrix Inverses

Finding an Inverse Matrix To obtain A^{-1} for any $n \times n$ matrix A for which A^{-1} exists, follow these steps. *Step 1* Form the augmented matrix $[A \mid I_n]$, where I_n is the $n \times n$ identity matrix. *Step 2* Perform row transformations on $[A \mid I_n]$ to get a matrix of the form $[I_n \mid B]$. *Step 3* Matrix B is A^{-1}.	Find A^{-1} if $A = \begin{bmatrix} 5 & 2 \\ 2 & 1 \end{bmatrix}$. $\left[\begin{array}{cc\|cc} 5 & 2 & 1 & 0 \\ 2 & 1 & 0 & 1 \end{array}\right]$ $\left[\begin{array}{cc\|cc} 1 & 0 & 1 & -2 \\ 2 & 1 & 0 & 1 \end{array}\right]$ $-2R2 + R1$ $\left[\begin{array}{cc\|cc} 1 & 0 & 1 & -2 \\ 0 & 1 & -2 & 5 \end{array}\right]$ $-2R1 + R2$ $\underbrace{}_{I_2}$ $\underbrace{}_{A^{-1}}$ Therefore, $A^{-1} = \begin{bmatrix} 1 & -2 \\ -2 & 5 \end{bmatrix}$.

Chapter 9 Review Exercises

Use the substitution or elimination method to solve each linear system. Identify any inconsistent systems or systems with infinitely many solutions. If the system has infinitely many solutions, write the solution set with y arbitrary.

1. $\begin{aligned} 2x + 6y &= 6 \\ 5x + 9y &= 9 \end{aligned}$

2. $\begin{aligned} 3x - 5y &= 7 \\ 2x + 3y &= 30 \end{aligned}$

3. $\begin{aligned} x + 5y &= 9 \\ 2x + 10y &= 18 \end{aligned}$

4. $\begin{aligned} \frac{1}{6}x + \frac{1}{3}y &= 8 \\ \frac{1}{4}x + \frac{1}{2}y &= 12 \end{aligned}$

5. $\begin{aligned} y &= -x + 3 \\ 2x + 2y &= 1 \end{aligned}$

6. $\begin{aligned} .2x + .5y &= 6 \\ .4x + y &= 9 \end{aligned}$

7. $\begin{aligned} 3x - 2y &= 0 \\ 9x + 8y &= 7 \end{aligned}$

8. $\begin{aligned} 6x + 10y &= -11 \\ 9x + 6y &= -3 \end{aligned}$

9. $\begin{aligned} 2x - 5y + 3z &= -1 \\ x + 4y - 2z &= 9 \\ -x + 2y + 4z &= 5 \end{aligned}$

10. $\begin{aligned} 4x + 3y + z &= -8 \\ 3x + y - z &= -6 \\ x + y + 2z &= -2 \end{aligned}$

11. $\begin{aligned} 5x - y &= 26 \\ 4y + 3z &= -4 \\ 3x + 3z &= 15 \end{aligned}$

12. $\begin{aligned} x + z &= 2 \\ 2y - z &= 2 \\ -x + 2y &= -4 \end{aligned}$

13. *Concept Check* Create your own inconsistent system of two equations.

14. *Concept Check* Create your own system of two equations with infinitely many solutions.

Write a system of linear equations, and then use the system to solve the problem.

15. *Meal Planning* A cup of uncooked rice contains 15 g of protein and 810 calories. A cup of uncooked soybeans contains 22.5 g of protein and 270 calories. How many cups of each should be used for a meal containing 9.5 g of protein and 324 calories?

16. *Order Quantities* A company sells recordable CDs for 80¢ each and play-only CDs for 60¢ each. The company receives $76 for an order of 100 CDs. However, the customer neglected to specify how many of each type to send. Determine the number of each type of CD that should be sent.

17. *Indian Weavers* The Waputi Indians make woven blankets, rugs, and skirts. Each blanket requires 24 hr for spinning the yarn, 4 hr for dyeing the yarn, and 15 hr for weaving. Rugs require 30, 5, and 18 hr and skirts 12, 3, and 9 hr, respectively. If there are 306, 59, and 201 hr available for spinning, dyeing, and weaving, respectively, how many of each item can be made? (*Hint:* Simplify the equations you write, if possible, before solving the system.)

18. *(Modeling) Populations of Minorities in the United States* The current and estimated resident populations (in percent) of blacks and Hispanics in the United States for the years 1990–2050 are modeled by the linear functions defined below.

$$y = .0515x + 12.3 \quad \text{Blacks}$$
$$y = .255x + 9.01 \quad \text{Hispanics}$$

In each case, x represents the number of years since 1990. (*Source:* U.S. Bureau of the Census, *U.S. Census of Population.*)

(a) Solve the system to find the year when these population percents will be equal.
(b) What percent of the U.S. resident population will be black or Hispanic in the year found in part (a)?
(c) Use a calculator graph of the system to support your algebraic solution.
(d) Which population is increasing more rapidly? (*Hint:* Consider the slopes of the lines.)

19. *(Modeling) Heart Rate* In a study, a group of athletes was exercised to exhaustion. Let x represent an athlete's heart rate 5 sec after stopping exercise and y this rate after 10 sec. It was found that the maximum heart rate H for these athletes satisfied the two equations

$$H = .491x + .468y + 11.2$$
$$H = -.981x + 1.872y + 26.4.$$

If an athlete had maximum heart rate $H = 180$, determine x and y graphically. Interpret your answer. (*Source:* Thomas, V., *Science and Sport,* Faber and Faber, 1970.)

20. *(Modeling) Equilibrium Supply and Demand* Let the supply and demand equations for units of banana smoothies be

$$\text{supply: } p = \frac{3}{2}q \qquad \text{and} \qquad \text{demand: } p = 81 - \frac{3}{4}q.$$

(a) Graph these equations on the same axes.
(b) Find the equilibrium demand.
(c) Find the equilibrium price.

21. *Curve Fitting* Find the equation of the vertical parabola that passes through the points shown in the table.

X	Y1
1	-2.3
2	-1.3
3	4.5
4	15.1
5	30.5
6	50.7
7	75.7

X=1

Find solutions for each system with the specified arbitrary variable.

22. $\begin{aligned} 2x - 6y + 4z &= 5 \\ 5x + y - 3z &= 1 \end{aligned}$; z

23. $\begin{aligned} 3x - 4y + z &= 2 \\ 2x + y &= 1 \end{aligned}$; x

Use the Gauss-Jordan method to solve each system.

24. $\begin{aligned} 2x + 3y &= 10 \\ -3x + y &= 18 \end{aligned}$

25. $\begin{aligned} 5x + 2y &= -10 \\ 3x - 5y &= -6 \end{aligned}$

26. $\begin{aligned} 3x + y &= -7 \\ x - y &= -5 \end{aligned}$

27. $\begin{aligned} 2x - y + 4z &= -1 \\ -3x + 5y - z &= 5 \\ 2x + 3y + 2z &= 3 \end{aligned}$

28. $\begin{aligned} x - z &= -3 \\ y + z &= 6 \\ 2x - 3z &= -9 \end{aligned}$

Solve each problem by writing a system of equations and then solving it using the Gauss-Jordan method.

29. ***Mixing Teas*** Three kinds of tea worth \$4.60, \$5.75, and \$6.50 per lb are to be mixed to get 20 lb of tea worth \$5.25 per lb. The amount of \$4.60 tea used is to be equal to the total amount of the other two kinds together. How many pounds of each tea should be used?

30. ***Mixing Solutions*** A 5% solution of a drug is to be mixed with some 15% solution and some 10% solution to get 20 ml of 8% solution. The amount of 5% solution used must be 2 ml more than the sum of the other two solutions. How many milliliters of each solution should be used?

31. ***(Modeling) College Enrollments*** In the past half-decade, college enrollments have shifted from being male-dominated to being female-dominated. During the period 1960–2000 both male and female enrollments grew, but female enrollments grew at a greater rate. If $x = 0$ represents 1960 and $x = 40$ represents 2000, the enrollments (in millions) are closely modeled by the following system.

$$y = .11x + 2.9 \quad \text{Males}$$
$$y = .19x + 1.4 \quad \text{Females}$$

Solve the system to find the year in which male and female enrollments were the same. What was the total enrollment at that time? (*Source:* U.S. Bureau of the Census.)

Evaluate each determinant.

32. $\begin{vmatrix} -2 & 4 \\ 0 & 3 \end{vmatrix}$

33. $\begin{vmatrix} -1 & 8 \\ 2 & 9 \end{vmatrix}$

34. $\begin{vmatrix} x & 4x \\ 2x & 8x \end{vmatrix}$

35. $\begin{vmatrix} -1 & 2 & 3 \\ 4 & 0 & 3 \\ 5 & -1 & 2 \end{vmatrix}$

36. $\begin{vmatrix} -2 & 4 & 1 \\ 3 & 0 & 2 \\ -1 & 0 & 3 \end{vmatrix}$

Solve each determinant equation.

37. $\begin{vmatrix} 3x & 7 \\ -x & 4 \end{vmatrix} = 8$

38. $\begin{vmatrix} 6x & 2 & 0 \\ 1 & 5 & 3 \\ x & 2 & -1 \end{vmatrix} = 2x$

Solve each system by Cramer's rule if possible. Identify any inconsistent systems or systems with infinitely many solutions. (Use another method if Cramer's rule cannot be used.)

39. $\begin{aligned} 3x + 7y &= 2 \\ 5x - y &= -22 \end{aligned}$

40. $\begin{aligned} 3x + y &= -1 \\ 5x + 4y &= 10 \end{aligned}$

41. $\begin{aligned} 6x + y &= -3 \\ 12x + 2y &= 1 \end{aligned}$

42. $\begin{aligned} 3x + 2y + z &= 2 \\ 4x - y + 3z &= -16 \\ x + 3y - z &= 12 \end{aligned}$

43. $\begin{aligned} x + y &= -1 \\ 2y + z &= 5 \\ 3x - 2z &= -28 \end{aligned}$

44. $\begin{aligned} 5x - 2y - z &= 8 \\ -5x + 2y + z &= -8 \\ x - 4y - 2z &= 0 \end{aligned}$

Find the partial fraction decomposition for each rational expression.

45. $\dfrac{2}{3x^2 - 5x + 2}$

46. $\dfrac{11 - 2x}{x^2 - 8x + 16}$

47. $\dfrac{5 - 2x}{(x^2 + 2)(x - 1)}$

48. $\dfrac{x^3 + 2x^2 - 3}{x^4 - 4x^2 + 4}$

Solve each nonlinear system.

49. $y = 2x + 10$
$x^2 + y = 13$

50. $x^2 = 2y - 3$
$x + y = 3$

51. $x^2 + y^2 = 17$
$2x^2 - y^2 = 31$

52. $2x^2 + 3y^2 = 30$
$x^2 + y^2 = 13$

53. $xy = -10$
$x + 2y = 1$

54. $xy + 2 = 0$
$y - x = 3$

55. $x^2 + 2xy + y^2 = 4$
$x - 3y = -2$

56. $x^2 + 2xy = 15 + 2x$
$xy - 3x + 3 = 0$

57. Find all values of b so that the straight line $3x - y = b$ touches the circle $x^2 + y^2 = 25$ at only one point.

58. Do the circle $x^2 + y^2 = 144$ and the line $x + 2y = 8$ have any points in common? If so, what are they?

Graph the solution set of each system of inequalities.

59. $x + y \le 6$
$2x - y \ge 3$

60. $y \le \frac{1}{3}x - 2$
$y^2 \le 16 - x^2$

61. Find $x \ge 0$ and $y \ge 0$ such that

$$3x + 2y \le 12$$
$$5x + y \ge 5$$

and $2x + 4y$ is maximized.

62. Find $x \ge 0$ and $y \ge 0$ such that

$$y \le -3x + 5$$
$$y \ge 4x - 3$$

and $7x + 14y$ is maximized.

Solve each linear programming problem.

63. *Cost of Nutrients* Certain laboratory animals must have at least 30 g of protein and at least 20 g of fat per feeding period. These nutrients come from food A, which costs 18 cents per unit and supplies 2 g of protein and 4 g of fat; and food B, which costs 12 cents per unit and has 6 g of protein and 2 g of fat. Food B is bought under a long-term contract requiring that at least 2 units of B be used per serving. How much of each food must be bought to produce the minimum cost per serving? What is the minimum cost?

64. *Profit from Farm Animals* A 4-H member raises only geese and pigs. She wants to raise no more than 16 animals, including no more than 10 geese. She spends \$5 to raise a goose and \$15 to raise a pig, and she has \$180 available for this project. Each goose produces \$6 in profit, and each pig produces \$20 in profit. How many of each animal should she raise to maximize her profit? What is her maximum profit?

Find the value of each variable.

65. $\begin{bmatrix} 5 & x+2 \\ -6y & z \end{bmatrix} = \begin{bmatrix} a & 3x-1 \\ 5y & 9 \end{bmatrix}$

66. $\begin{bmatrix} -6+k & 2 & a+3 \\ -2+m & 3p & 2r \end{bmatrix} + \begin{bmatrix} 3-2k & 5 & 7 \\ 5 & 8p & 5r \end{bmatrix} = \begin{bmatrix} 5 & y & 6a \\ 2m & 11 & -35 \end{bmatrix}$

Perform each operation when possible.

67. $\begin{bmatrix} 3 \\ 2 \\ 5 \end{bmatrix} - \begin{bmatrix} 8 \\ -4 \\ 6 \end{bmatrix} + \begin{bmatrix} 1 \\ 0 \\ 2 \end{bmatrix}$

68. $4\begin{bmatrix} 3 & -4 & 2 \\ 5 & -1 & 6 \end{bmatrix} + \begin{bmatrix} -3 & 2 & 5 \\ 1 & 0 & 4 \end{bmatrix}$

69. $\begin{bmatrix} 2 & 5 & 8 \\ 1 & 9 & 2 \end{bmatrix} - \begin{bmatrix} 3 & 4 \\ 7 & 1 \end{bmatrix}$

70. $\begin{bmatrix} -3 & 4 \\ 2 & 8 \end{bmatrix}\begin{bmatrix} -1 & 0 \\ 2 & 5 \end{bmatrix}$

71. $\begin{bmatrix} -1 & 0 \\ 2 & 5 \end{bmatrix}\begin{bmatrix} -3 & 4 \\ 2 & 8 \end{bmatrix}$

72. $\begin{bmatrix} 1 & 2 \\ 3 & 0 \\ -6 & 5 \end{bmatrix}\begin{bmatrix} 4 & 8 \\ -1 & 2 \end{bmatrix}$

73. $\begin{bmatrix} 3 & 2 & -1 \\ 4 & 0 & 6 \end{bmatrix}\begin{bmatrix} -2 & 0 \\ 0 & 2 \\ 3 & 1 \end{bmatrix}$

74. $\begin{bmatrix} 1 & -2 & 4 & 2 \\ 0 & 1 & -1 & 8 \end{bmatrix}\begin{bmatrix} -1 \\ 2 \\ 0 \\ 1 \end{bmatrix}$

75. $\begin{bmatrix} -2 & 5 & 5 \\ 0 & 1 & 4 \\ 3 & -4 & -1 \end{bmatrix}\begin{bmatrix} 1 & 0 & -1 \\ -1 & 0 & 0 \\ 1 & 1 & -1 \end{bmatrix}$

Find the inverse of each matrix that has an inverse.

76. $\begin{bmatrix} -4 & 2 \\ 0 & 3 \end{bmatrix}$ **77.** $\begin{bmatrix} 2 & 1 \\ 5 & 3 \end{bmatrix}$ **78.** $\begin{bmatrix} 2 & 3 & 5 \\ -2 & -3 & -5 \\ 1 & 4 & 2 \end{bmatrix}$ **79.** $\begin{bmatrix} 2 & -1 & 0 \\ 1 & 0 & 1 \\ 1 & -2 & 0 \end{bmatrix}$

Use the method of matrix inverses to solve each system.

80. $2x + y = 5$
$3x - 2y = 4$

81. $3x + 2y + z = -5$
$x - y + 3z = -5$
$2x + 3y + z = 0$

82. $x + y + z = 1$
$2x - y = -2$
$3y + z = 2$

Chapter 9 Test

Use substitution or elimination to solve each system. Identify any system that is inconsistent or has infinitely many solutions. If a system has infinitely many solutions, express the solution set with y arbitrary.

1. $3x - y = 9$
$x + 2y = 10$

2. $6x + 9y = -21$
$4x + 6y = -14$

3. $\frac{1}{4}x - \frac{1}{3}y = -\frac{5}{12}$
$\frac{1}{10}x + \frac{1}{5}y = \frac{1}{2}$

4. $x - 2y = 4$
$-2x + 4y = 6$

5. $2x + y + z = 3$
$x + 2y - z = 3$
$3x - y + z = 5$

Use the Gauss-Jordan method to solve each system.

6. $3x - 2y = 13$
$4x - y = 19$

7. $3x - 4y + 2z = 15$
$2x - y + z = 13$
$x + 2y - z = 5$

8. *Curve Fitting* Find the equation that defines the parabola shown.

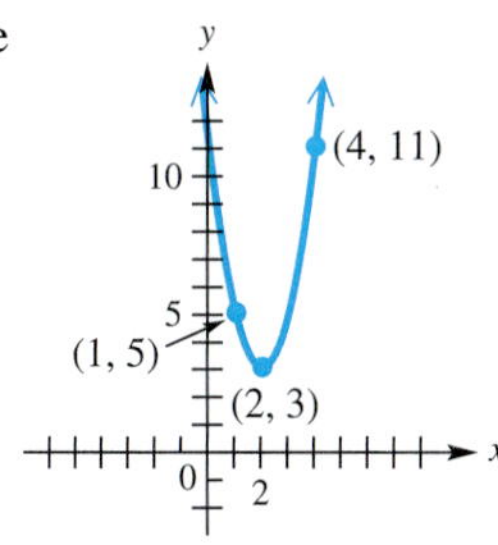

9. *Ordering Supplies* A knitting shop orders yarn from three suppliers in Toronto, Montreal, and Ottawa. One month the shop ordered a total of 100 units of yarn from these suppliers. The delivery costs were \$80, \$50, and \$65 per unit for the orders from Toronto, Montreal, and Ottawa, respectively, with total delivery costs of \$5990. The shop ordered the same amount from Toronto and Ottawa. How many units were ordered from each supplier?

Evaluate each determinant.

10. $\begin{vmatrix} 6 & 8 \\ 2 & -7 \end{vmatrix}$

11. $\begin{vmatrix} 2 & 0 & 8 \\ -1 & 7 & 9 \\ 12 & 5 & -3 \end{vmatrix}$

Solve each system by Cramer's rule.

12. $\begin{aligned} 2x - 3y &= -33 \\ 4x + 5y &= 11 \end{aligned}$

13. $\begin{aligned} x + y - z &= -4 \\ 2x - 3y - z &= 5 \\ x + 2y + 2z &= 3 \end{aligned}$

14. Find the partial fraction decomposition of $\dfrac{x + 2}{x^3 + 2x^2 + x}$.

15. *Concept Check* If a system of two nonlinear equations contains one equation whose graph is a circle and another equation whose graph is a line, can the system have exactly one solution? If so, draw a sketch to indicate this situation.

Solve each nonlinear system of equations.

16. $\begin{aligned} 2x^2 + y^2 &= 6 \\ x^2 - 4y^2 &= -15 \end{aligned}$

17. $\begin{aligned} x^2 + y^2 &= 25 \\ x + y &= 7 \end{aligned}$

18. *Unknown Numbers* Find two numbers such that their sum is -1 and the sum of their squares is 61.

19. Graph the solution set of

$$\begin{aligned} x - 3y &\geq 6 \\ y^2 &\leq 16 - x^2. \end{aligned}$$

20. Find $x \geq 0$ and $y \geq 0$ such that

$$\begin{aligned} x + 2y &\leq 24 \\ 3x + 4y &\leq 60 \end{aligned}$$

and $2x + 3y$ is maximized.

21. *Jewelry Profits* The J. J. Gravois Company designs and sells two types of rings: the VIP and the SST. The company can produce up to 24 rings each day using up to 60 total hours of labor. It takes 3 hr to make one VIP ring, and 2 hr to make one SST ring. How many of each type of ring should be made daily in order to maximize the company's profit, if the profit on one VIP ring is \$30 and the profit on one SST ring is \$40? What is the maximum profit?

22. Find the value of each variable in the equation $\begin{bmatrix} 5 & x + 6 \\ 0 & 4 \end{bmatrix} = \begin{bmatrix} y - 2 & 4 - x \\ 0 & w + 7 \end{bmatrix}$.

Perform each operation when possible.

23. $3\begin{bmatrix} 2 & 3 \\ 1 & -4 \\ 5 & 9 \end{bmatrix} - \begin{bmatrix} -2 & 6 \\ 3 & -1 \\ 0 & 8 \end{bmatrix}$

24. $\begin{bmatrix} 1 \\ 2 \end{bmatrix} + \begin{bmatrix} 4 \\ -6 \end{bmatrix} + \begin{bmatrix} 2 & 8 \\ -7 & 5 \end{bmatrix}$

25. $\begin{bmatrix} 2 & 1 & -3 \\ 4 & 0 & 5 \end{bmatrix}\begin{bmatrix} 1 & 3 \\ 2 & 4 \\ 3 & -2 \end{bmatrix}$

26. $\begin{bmatrix} 2 & -4 \\ 3 & 5 \end{bmatrix}\begin{bmatrix} 4 \\ 2 \\ 7 \end{bmatrix}$

27. *Concept Check* Which of the following properties does not apply to multiplication of matrices?

A. commutative **B.** associative **C.** distributive **D.** identity

Find the inverse, if it exists, of each matrix.

28. $\begin{bmatrix} -8 & 5 \\ 3 & -2 \end{bmatrix}$

29. $\begin{bmatrix} 4 & 12 \\ 2 & 6 \end{bmatrix}$

30. $\begin{bmatrix} 1 & 3 & 4 \\ 2 & 7 & 8 \\ -2 & -5 & -7 \end{bmatrix}$

Use matrix inverses to solve each system.

31. $2x + y = -6$
$3x - y = -29$

32. $x + y = 5$
$y - 2z = 23$
$x + 3z = -27$

Chapter 9 Quantitative Reasoning

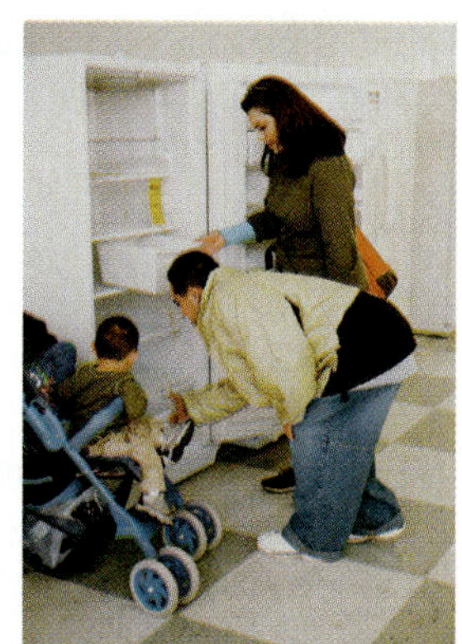

Do consumers use common sense?

Imagine two refrigerators in the appliance section of a department store. One sells for \$700 and uses \$85 worth of electricity a year. The other is \$100 more expensive but costs only \$25 a year to run. Given that either refrigerator should last at least 10 yr without repair, consumers would overwhelmingly buy the second model, right?

Well, not exactly. Many studies by economists have shown that in a wide range of decisions about money—from paying taxes to buying major appliances—consumers consistently make decisions that defy common sense. In some cases—as in the refrigerator example—this means that people are generally unwilling to pay a little more money up front to save a lot of money in the long run. (*Source:* Gladwell, Malcolm, "Consumers Often Defy Common Sense," Copyright © 1990, *The Washington Post*. Reprinted with permission.)

We can apply the concepts of the solution of a linear system to this situation. Over a 10-yr period, one refrigerator will cost \$700 + 10(\$85) = \$1550, while the other will cost \$800 + 10(\$25) = \$1050, a difference of \$500.

1. In how many years will the costs for the two refrigerators be equal? (*Hint:* Solve the system $y = 800 + 25x$, $y = 700 + 85x$ for x.)
2. Suppose you win a lottery and you can take either \$1400 in a year or \$1000 now. Suppose also that you can invest the \$1000 now at 6% interest. Decide which is a better deal and by how much.

10 Analytic Geometry

In this chapter, we study a group of curves known as *conic sections.* One conic section, the *ellipse,* has a special reflecting property responsible for "whispering galleries." In a whispering gallery, a person whispering at a certain point in the room can be heard clearly at another point across the room.

The Old House Chamber of the U.S. Capitol, now called Statuary Hall, is a whispering gallery. History has it that John Quincy Adams, whose desk was positioned at exactly the right point beneath the ellipsoidal ceiling, often pretended to sleep at his desk as he listened to political opponents whispering strategies in an area across the room. *(Source: We, the People, The Story of the United States Capitol,* 1991.)

In Section 10.2, we investigate this reflective property of ellipses.

10.1 Parabolas

Conic Sections ▪ Horizontal Parabolas ▪ Geometric Definition and Equations of Parabolas ▪ An Application of Parabolas

Conic Sections *Parabolas, circles, ellipses,* and *hyperbolas* form a group of curves known as **conic sections,** because they are the results of intersecting a cone with a plane. Figure 1 illustrates these curves. We studied circles and some parabolas (those that open up or down, that is, vertical parabolas) in Chapters 2 and 3.

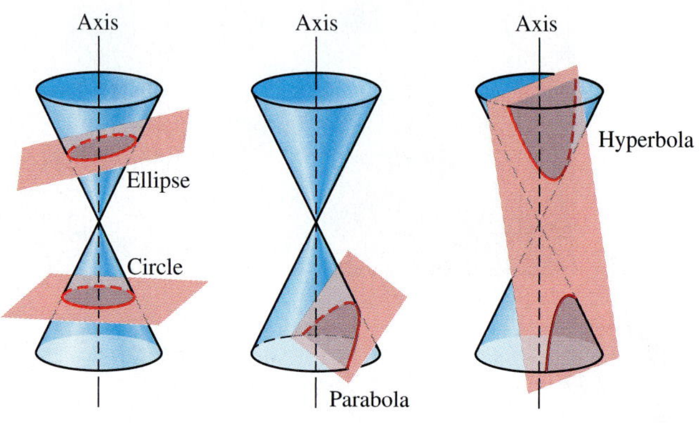

Figure 1

Horizontal Parabolas From Chapter 3, we know that the graph of the equation

$$y = a(x - h)^2 + k \quad \text{(Section 3.1)}$$

is a parabola with vertex (h, k) and the vertical line $x = h$ as axis. If we subtract k from each side, this equation becomes

$$y - k = a(x - h)^2. \quad \text{Subtract } k. \quad (1)$$

The equation

$$x - h = a(y - k)^2 \quad \text{Interchange the roles of } x - h \text{ and } y - k. \quad (2)$$

also has a parabola as its graph. While the graph of $y - k = a(x - h)^2$ has a *vertical* axis, the graph of $x - h = a(y - k)^2$ has a *horizontal* axis. The graph of the first equation is the graph of a function (specifically a quadratic function), while the graph of the second equation is not. (Why?)

Parabola with Horizontal Axis

The parabola with vertex (h, k) and the horizontal line $y = k$ as axis has an equation of the form

$$\mathbf{x - h = a(y - k)^2.}$$

The parabola opens to the right if $a > 0$ and to the left if $a < 0$.

NOTE When the vertex (h, k) is $(0, 0)$ and $a = 1$ in

$$y - k = a(x - h)^2 \quad (1)$$

and

$$x - h = a(y - k)^2, \quad (2)$$

the equations $y = x^2$ and $x = y^2$ result. The graphs of these equations are shown in Figure 2. Notice that the graphs are mirror images of each other with respect to the line $y = x$.

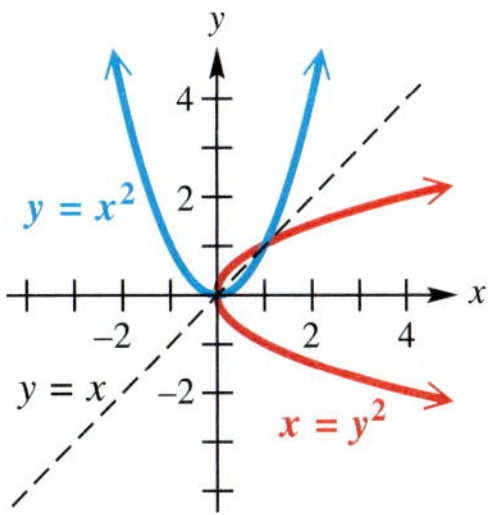

Figure 2

EXAMPLE 1 Graphing a Parabola with Horizontal Axis

Graph $x + 3 = (y - 2)^2$. Give the domain and range.

Solution The graph of $x + 3 = (y - 2)^2$ or $x - (-3) = (y - 2)^2$ has vertex $(-3, 2)$ and opens to the right because $a = 1 > 0$. Plotting a few additional points gives the graph shown in Figure 3. Note that the graph is symmetric about its axis, $y = 2$. The domain is $[-3, \infty)$, and the range is $(-\infty, \infty)$.

x	y
−3	2
−2	3
−2	1
1	4
1	0

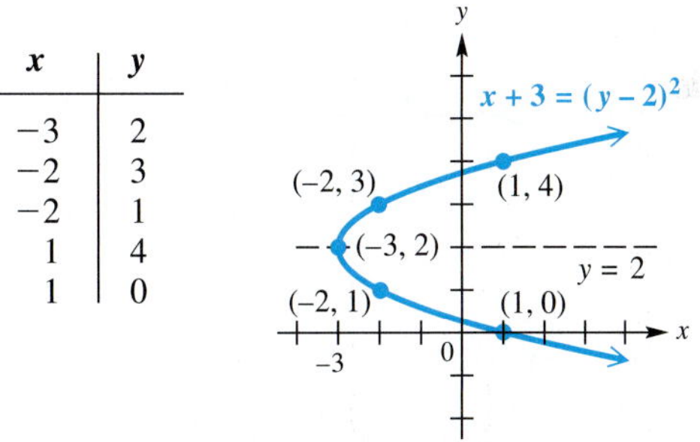

Figure 3

Now try Exercise 7.

When an equation of a horizontal parabola is given in the form

$$x = ay^2 + by + c,$$

completing the square on y allows us to write the equation in the form

$$x - h = a(y - k)^2$$

and more easily find the vertex, as shown in Example 2 on the next page.

EXAMPLE 2 Graphing a Parabola with Horizontal Axis

Graph $x = 2y^2 + 6y + 5$. Give the domain and range.

Algebraic Solution

$$x = 2y^2 + 6y + 5$$

$$x = 2(y^2 + 3y \quad) + 5 \quad \text{Factor out 2.}$$

$$x = 2\left(y^2 + 3y + \frac{9}{4} - \frac{9}{4}\right) + 5$$

Complete the square; $\left[\frac{1}{2}(3)\right]^2 = \frac{9}{4}$. (Section 1.4)

$$x = 2\left(y^2 + 3y + \frac{9}{4}\right) + 2\left(-\frac{9}{4}\right) + 5$$

Distributive property (Section R.1)

$$x = 2\left(y + \frac{3}{2}\right)^2 + \frac{1}{2}$$

Factor (Section R.4); simplify.

$$x - \frac{1}{2} = 2\left(y + \frac{3}{2}\right)^2 \quad \text{Subtract } \tfrac{1}{2}. \ (*)$$

The vertex of the parabola is $\left(\frac{1}{2}, -\frac{3}{2}\right)$. The axis is the horizontal line $y = k$, or $y = -\frac{3}{2}$. Using the vertex and the axis and plotting a few additional points gives the graph in Figure 4. The domain is $\left[\frac{1}{2}, \infty\right)$, and the range is $(-\infty, \infty)$.

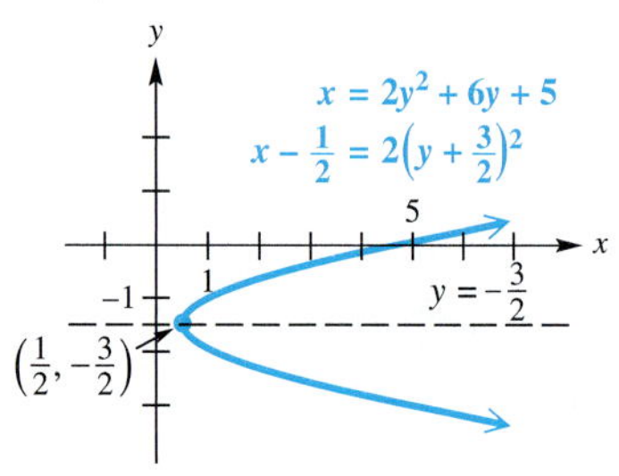

Figure 4

Graphing Calculator Solution

Since a horizontal parabola is *not* the graph of a function, to graph it using a graphing calculator we must write two equations by solving for y.

$$x - \frac{1}{2} = 2\left(y + \frac{3}{2}\right)^2 \quad (*) \text{ from algebraic solution}$$

$$x - .5 = 2(y + 1.5)^2 \quad \text{Write with decimals.}$$

$$\frac{x - .5}{2} = (y + 1.5)^2 \quad \text{Divide by 2.}$$

$$\pm\sqrt{\frac{x - .5}{2}} = y + 1.5 \quad \text{Take square roots on both sides. (Section 1.4)}$$

$$y = -1.5 \pm \sqrt{\frac{x - .5}{2}} \quad \text{Subtract 1.5; rewrite.}$$

Figure 5 shows the graphs of the two functions defined in the final equation. Their union is the graph of $x = 2y^2 + 6y + 5$.

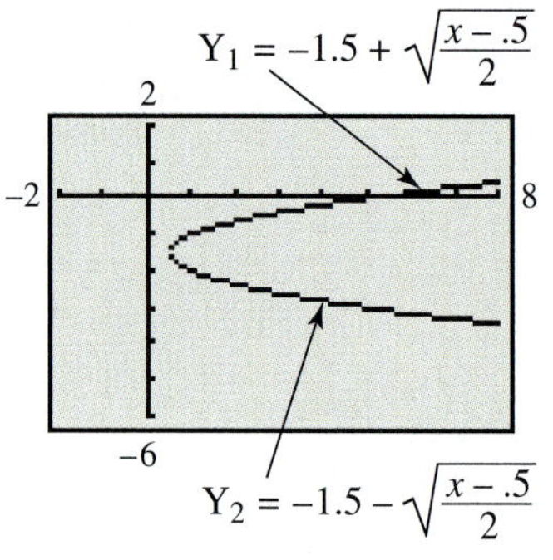

Figure 5

Now try Exercise 15.

Geometric Definition and Equations of Parabolas We can also develop the equation of a parabola from the geometric definition of a parabola as a set of points.

Parabola

A **parabola** is the set of points in a plane equidistant from a fixed point and a fixed line. The fixed point is called the **focus,** and the fixed line is called the **directrix** of the parabola.

As shown in Figure 6, the axis of symmetry of a parabola passes through the focus and is perpendicular to the directrix. The vertex is the midpoint of the line segment joining the focus and directrix on the axis.

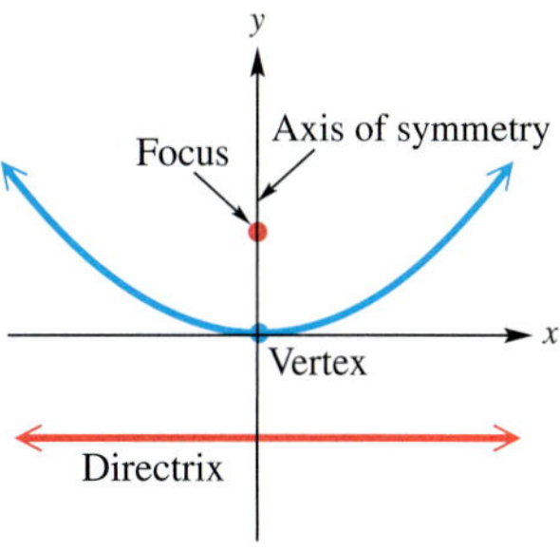

Figure 6

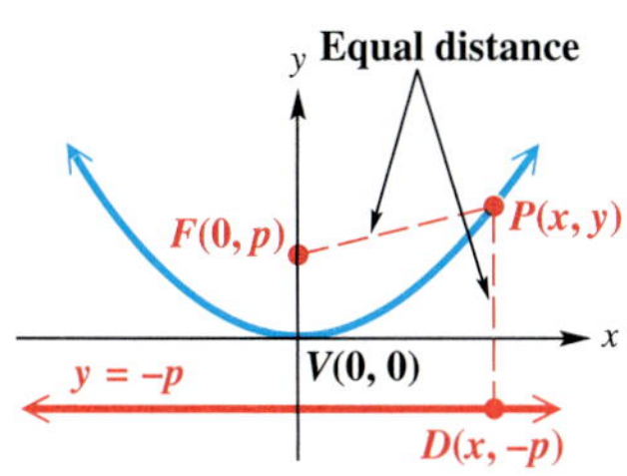

$d(P, F) = d(P, D)$
for all P on the parabola.

Figure 7

We can find an equation of a parabola from the preceding definition. Let the directrix be the line $y = -p$ and the focus be the point F with coordinates $(0, p)$, as shown in Figure 7. To find the equation of the set of points that are the same distance from the line $y = -p$ and the point $(0, p)$, choose one such point P and give it coordinates (x, y). Since $d(P, F)$ and $d(P, D)$ must be equal, using the distance formula gives

$$d(P, F) = d(P, D)$$

$$\sqrt{(x-0)^2 + (y-p)^2} = \sqrt{(x-x)^2 + (y-(-p))^2}$$ Distance formula **(Section 2.1)**

$$\sqrt{x^2 + (y-p)^2} = \sqrt{(y+p)^2}$$

$$x^2 + y^2 - 2yp + p^2 = y^2 + 2yp + p^2$$ Square both sides **(Section 1.6)**; multiply. **(Section R.3)**

$$x^2 = 4py.$$ Simplify.

This discussion is summarized as follows.

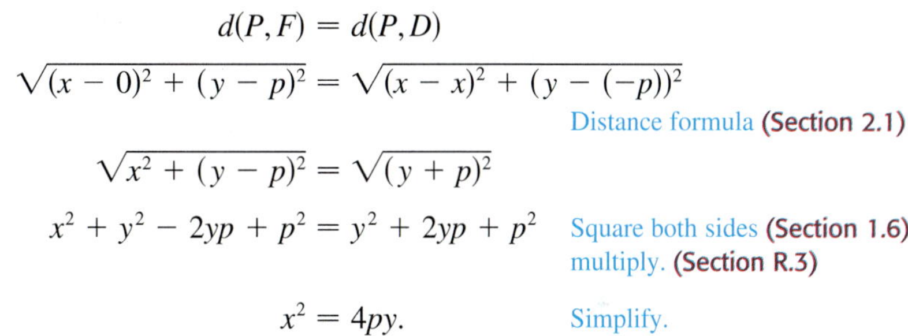

Parabola with Vertical Axis and Vertex (0, 0)

The parabola with focus $(0, p)$ and directrix $y = -p$ has equation

$$\mathbf{x^2 = 4py.}$$

The parabola has vertical axis $x = 0$ and opens up if $p > 0$ or down if $p < 0$.

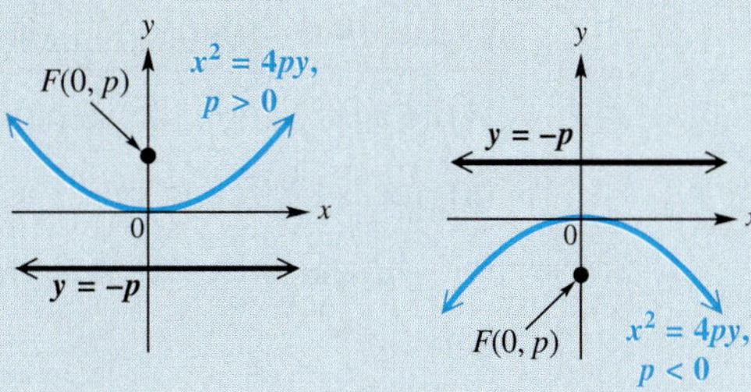

If the directrix is the line $x = -p$ and the focus is $(p, 0)$, using the definition of a parabola and the distance formula leads to the equation of a parabola with a horizontal axis. (See Exercise 59.)

Parabola with Horizontal Axis and Vertex (0,0)

The parabola with focus $(p, 0)$ and directrix $x = -p$ has equation

$$y^2 = 4px.$$

The parabola has horizontal axis $y = 0$ and opens to the right if $p > 0$ or to the left if $p < 0$.

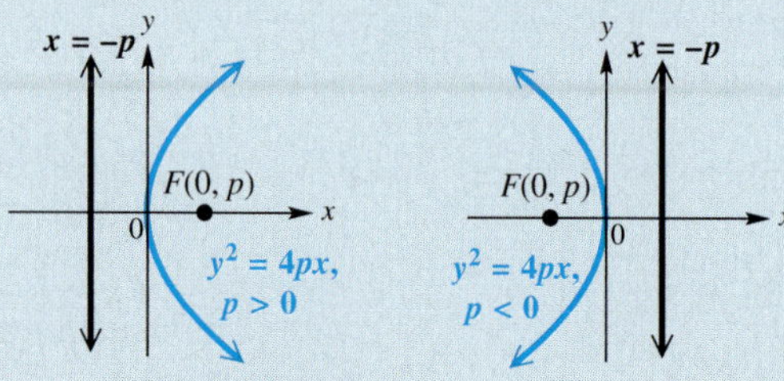

EXAMPLE 3 Determining Information about Parabolas from Their Equations

Find the focus, directrix, vertex, and axis of each parabola. Then use this information to graph the parabola.

(a) $x^2 = 8y$ **(b)** $y^2 = -28x$

Solution

(a) The equation $x^2 = 8y$ has the form $x^2 = 4py$, so $4p = 8$, from which $p = 2$. Since the x-term is squared, the parabola is vertical, with focus $(0, p) = (0, 2)$ and directrix $y = -2$. The vertex is $(0, 0)$, and the axis of the parabola is the y-axis. See Figure 8.

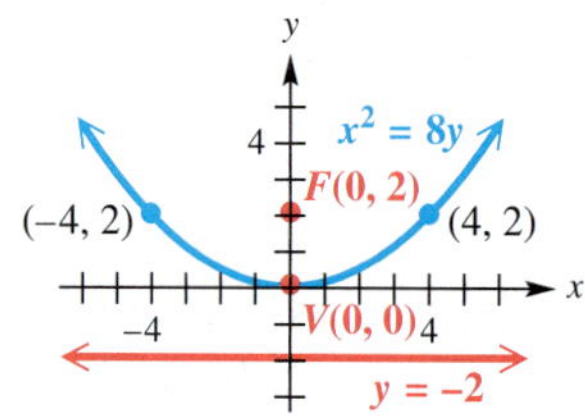

Figure 8

(b) The equation $y^2 = -28x$ has the form $y^2 = 4px$, with $4p = -28$, so $p = -7$. The parabola is horizontal, with focus $(-7, 0)$, directrix $x = 7$, vertex $(0, 0)$, and x-axis as axis of the parabola. Since p is negative, the graph opens to the left, as shown in Figure 9.

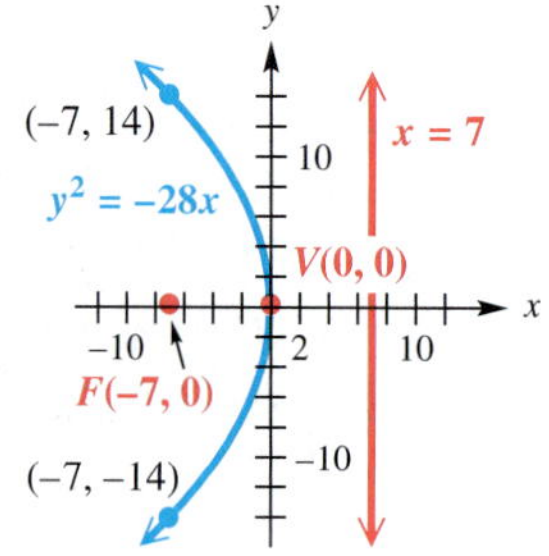

Figure 9

Now try Exercises 19 and 23.

EXAMPLE 4 Writing Equations of Parabolas

Write an equation for each parabola.

(a) focus $\left(\frac{2}{3}, 0\right)$ and vertex at the origin

(b) vertical axis, vertex at the origin, through the point $(-2, 12)$

Solution

(a) Since the focus $\left(\frac{2}{3}, 0\right)$ and the vertex $(0, 0)$ are both on the x-axis, the parabola is horizontal. It opens to the right because $p = \frac{2}{3}$ is positive. See Figure 10. The equation, which will have the form $y^2 = 4px$, is

$$y^2 = 4\left(\frac{2}{3}\right)x \quad \text{or} \quad y^2 = \frac{8}{3}x.$$

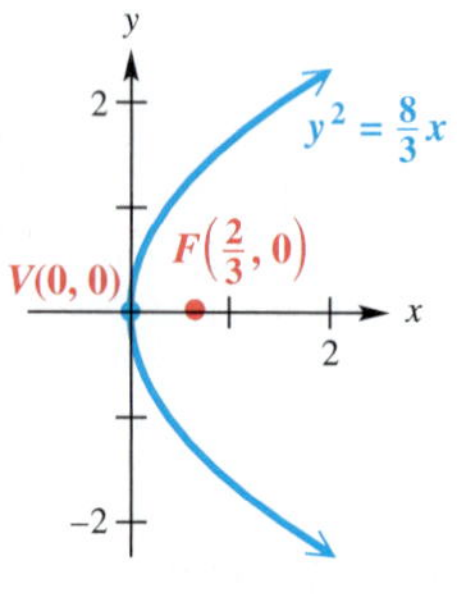

Figure 10

(b) The parabola will have an equation of the form $x^2 = 4py$ because the axis is vertical and the vertex is $(0, 0)$. Since the point $(-2, 12)$ is on the graph, it must satisfy the equation.

$$x^2 = 4py$$

$$(-2)^2 = 4p(12) \qquad \text{Let } x = -2 \text{ and } y = 12.$$

$$4 = 48p$$

$$p = \frac{1}{12}$$

Thus,

$$x^2 = 4py$$

$$x^2 = 4\left(\frac{1}{12}\right)y,$$

which gives the equation

$$x^2 = \frac{1}{3}y \qquad \text{or} \qquad y = 3x^2.$$

Now try Exercises 31 and 35.

The equations $x^2 = 4py$ and $y^2 = 4px$ can be extended to parabolas having vertex (h, k) by replacing x and y with $x - h$ and $y - k$, respectively.

Equation Forms for Translated Parabolas

A parabola with vertex (h, k) has an equation of the form

$$(x - h)^2 = 4p(y - k) \qquad \text{Vertical axis}$$

or

$$(y - k)^2 = 4p(x - h), \qquad \text{Horizontal axis}$$

where the focus is distance $|p|$ from the vertex.

EXAMPLE 5 Writing an Equation of a Parabola

Write an equation for the parabola with vertex $(1, 3)$ and focus $(-1, 3)$, and graph it. Give the domain and range.

Solution Since the focus is to the left of the vertex, the axis is horizontal and the parabola opens to the left. See Figure 11. The distance between the vertex and the focus is $1 - (-1)$ or 2, so $p = -2$ (since the parabola opens to the left). The equation of the parabola is

$$(y - k)^2 = 4p(x - h) \qquad \text{Parabola with horizontal axis}$$

$$(y - 3)^2 = 4(-2)(x - 1) \qquad \text{Substitute for } p, h, \text{ and } k.$$

$$(y - 3)^2 = -8(x - 1).$$

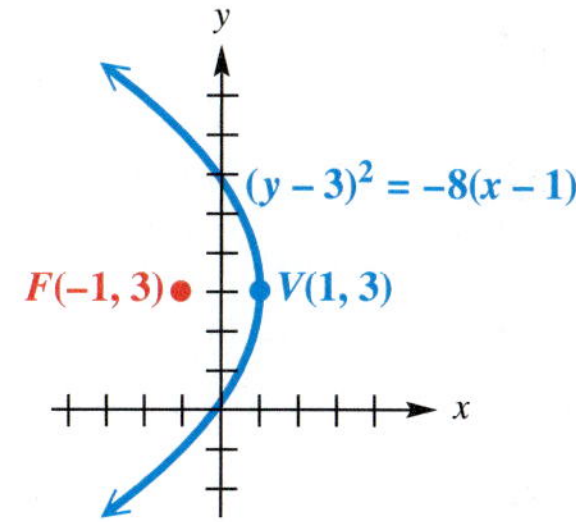

Figure 11

The domain is $(-\infty, 1]$, and the range is $(-\infty, \infty)$.

Now try Exercise 41.

CONNECTIONS Parabolas have a special reflecting property that makes them useful in the design of telescopes, radar equipment, auto headlights, and solar furnaces. When a ray of light or a sound wave traveling parallel to the axis of a parabolic shape bounces off the parabola, it passes through the focus. For example, in the solar furnace shown in the figure, a parabolic mirror collects light at the focus and thereby generates intense heat at that point. The reflecting property can be used in reverse. If a light source is placed at the focus, then the reflected light rays will be directed straight ahead. This is why the reflector in a car headlight is parabolic.

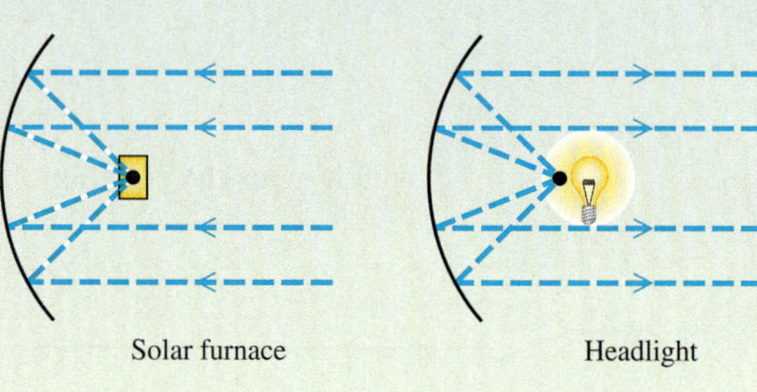

An Application of Parabolas

EXAMPLE 6 Modeling the Reflective Property of Parabolas

The Parkes radio telescope has a parabolic dish shape with diameter 210 ft and depth 32 ft. Because of this parabolic shape, distant rays hitting the dish will be reflected directly toward the focus. A cross section of the dish is shown in Figure 12. (*Source:* Mar, J. and H. Liebowitz, *Structure Technology for Large Radio and Radar Telescope Systems,* The MIT Press, Massachusetts Institute of Technology, 1969.)

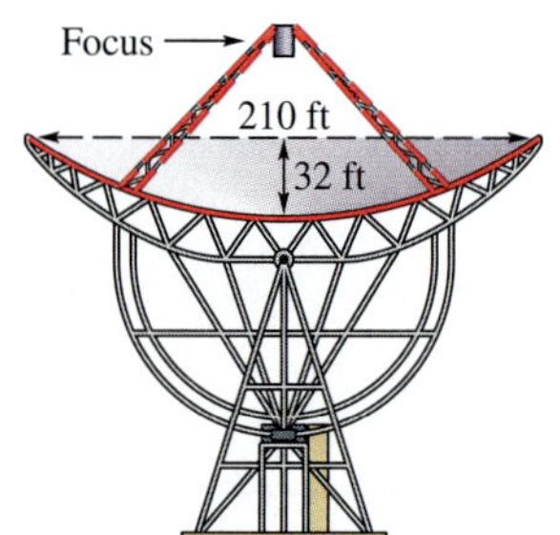

Figure 12

(a) Determine an equation that models this cross section by placing the vertex at the origin with the parabola opening up.

(b) The receiver must be placed at the focus of the parabola. How far from the vertex of the parabolic dish should the receiver be located?

Solution

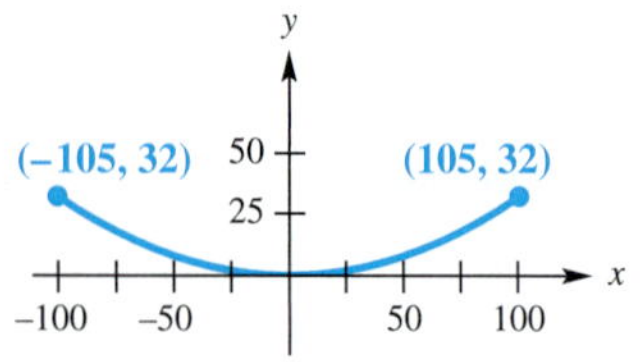

Figure 13

(a) Locate the vertex at the origin as shown in Figure 13. The form of the parabola is $x^2 = 4py$. The parabola must pass through the point $\left(\frac{210}{2}, 32\right) = (105, 32)$. Thus,

$$x^2 = 4py \quad \text{Vertical parabola}$$

$$(105)^2 = 4p(32) \quad \text{Let } x = 105 \text{ and } y = 32.$$

$$11{,}025 = 128p$$

$$p = \frac{11{,}025}{128}. \quad \text{Solve for } p.$$

The cross section can be modeled by the equation

$$x^2 = 4py$$
$$x^2 = 4\left(\frac{11{,}025}{128}\right)y \quad \text{Substitute for } p.$$
$$x^2 = \frac{11{,}025}{32}y.$$

(b) The distance between the vertex and the focus is p. In part (a), we found $p = \frac{11{,}025}{128} \approx 86.1$, so the receiver should be located at $(0, 86.1)$ or 86.1 ft above the vertex.

Now try Exercise 51.

10.1 Exercises

1. *Concept Check* Match each equation of a parabola in Column I with its description in Column II.

I	II
(a) $y + 2 = (x - 4)^2$	**A.** vertex $(2, -4)$; opens down
(b) $y + 4 = (x - 2)^2$	**B.** vertex $(2, -4)$; opens up
(c) $y + 2 = -(x - 4)^2$	**C.** vertex $(4, -2)$; opens down
(d) $y + 4 = -(x - 2)^2$	**D.** vertex $(4, -2)$; opens up
(e) $x + 2 = (y - 4)^2$	**E.** vertex $(-2, 4)$; opens left
(f) $x + 4 = (y - 2)^2$	**F.** vertex $(-2, 4)$; opens right
(g) $x + 2 = -(y - 4)^2$	**G.** vertex $(-4, 2)$; opens left
(h) $x + 4 = -(y - 2)^2$	**H.** vertex $(-4, 2)$; opens right

2. *Concept Check* Match each equation of a parabola in Column I with the appropriate description in Column II.

I	II
(a) $y = 2x^2 - 3x - 9$	**A.** opens right
(b) $y = -3x^2 + 4x + 2$	**B.** opens up
(c) $x = 2y^2 - 3y - 9$	**C.** opens left
(d) $x = -3y^2 + 4y + 2$	**D.** opens down

Graph each horizontal parabola, and give the domain and range. See Examples 1 and 2.

3. $x = -y^2$
4. $x = y^2 + 2$
5. $x = (y - 3)^2$
6. $x = (y + 1)^2$
7. $x - 2 = (y - 4)^2$
8. $x + 1 = (y + 2)^2$
9. $x - 2 = -3(y - 1)^2$
10. $x = -2(y + 3)^2$
11. $x - 4 = \frac{1}{2}(y - 1)^2$
12. $x = -\frac{1}{3}(y - 3)^2 + 3$
13. $x = y^2 + 4y + 2$
14. $x = 2y^2 - 4y + 6$
15. $x = -4y^2 - 4y + 3$
16. $x = -2y^2 + 2y - 3$
17. $2x = y^2 - 4y + 6$
18. $x + 3y^2 + 18y + 22 = 0$

Give the focus, directrix, and axis for each parabola. See Example 3.

19. $x^2 = 24y$
20. $x^2 = \frac{1}{8}y$
21. $y = -4x^2$

22. $9y = x^2$
23. $y^2 = -4x$
24. $y^2 = -16x$
25. $x = -32y^2$
26. $x = 16y^2$
27. $(y - 3)^2 = 12(x - 1)$
28. $(x + 2)^2 = 20y$
29. $(x - 7)^2 = 16(y + 5)$
30. $(y - 2)^2 = 24(x - 3)$

Write an equation for each parabola with vertex at the origin. See Example 4.

31. focus $(5, 0)$
32. focus $\left(-\frac{1}{2}, 0\right)$
33. focus $\left(0, \frac{1}{4}\right)$
34. focus $\left(0, -\frac{1}{3}\right)$
35. through $\left(\sqrt{3}, 3\right)$, opening up
36. through $\left(2, -2\sqrt{2}\right)$, opening right
37. through $(3, 2)$, symmetric with respect to the x-axis
38. through $(2, -4)$, symmetric with respect to the y-axis

Write an equation for each parabola. See Example 5.

39. vertex $(4, 3)$, focus $(4, 5)$
40. vertex $(-2, 1)$, focus $(-2, -3)$
41. vertex $(-5, 6)$, focus $(2, 6)$
42. vertex $(1, 2)$, focus $(4, 2)$

Determine the two equations necessary to graph each horizontal parabola using a graphing calculator, and graph it in the viewing window specified. See Example 2.

43. $x = 3y^2 + 6y - 4$; $[-10, 2]$ by $[-4, 4]$
44. $x = -2y^2 + 4y + 3$; $[-10, 6]$ by $[-4, 4]$
45. $x + 2 = -(y + 1)^2$; $[-10, 2]$ by $[-4, 4]$
46. $x - 5 = 2(y - 2)^2$; $[-2, 12]$ by $[-2, 6]$

Relating Concepts

For individual or collaborative investigation
(Exercises 47–50)

Curve Fitting *Given three noncollinear points, the equation of a horizontal parabola joining them can be found by solving a system of equations. The parabola will have an equation of the form $x = ay^2 + by + c$.* ***Work Exercises 47–50 in order,*** *to find the equation of the horizontal parabola containing $(-5, 1)$, $(-14, -2)$, and $(-10, 2)$.*

47. Write three equations in a, b, and c, by substituting the given values of x and y into the equation $x = ay^2 + by + c$.

48. Solve the system of three equations determined in Exercise 47.

49. Does the horizontal parabola open to the left or to the right? Why?

50. Write the equation of the horizontal parabola.

Solve each problem. See Example 6.

51. ***(Modeling) Radio Telescope Design*** The U.S. Naval Research Laboratory designed a giant radio telescope that had diameter 300 ft and maximum depth 44 ft. (*Source:* Mar, J., and H. Liebowitz, *Structure Technology for Large Radio and Radar Telescope Systems,* The MIT Press, 1969.)

(a) Find the equation of a parabola that models the cross section of the dish if the vertex is placed at the origin and the parabola opens up.

(b) The receiver must be placed at the focus of the parabola. How far from the vertex should the receiver be located?

52. *(Modeling) Radio Telescope Design* Suppose the telescope in Exercise 51 had diameter 400 ft and maximum depth 50 ft.

(a) Find the equation of this parabola.

(b) The receiver must be placed at the focus of the parabola. How far from the vertex should the receiver be located?

53. *Parabolic Arch* An arch in the shape of a parabola has the dimensions shown in the figure. How wide is the arch 9 ft up?

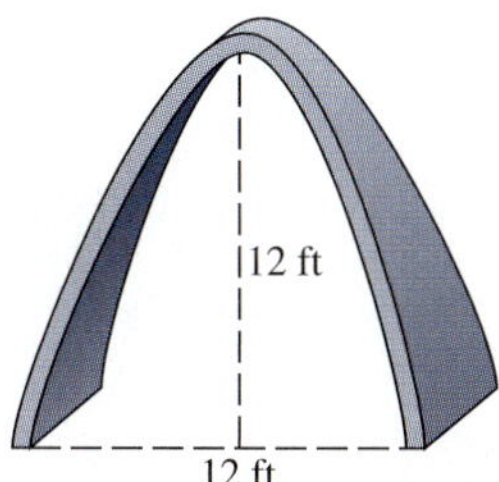

54. *Height of Bridge Cable Supports* The cable in the center portion of a bridge is supported as shown in the figure to form a parabola. The center vertical cable is 10 ft high, the supports are 210 ft high, and the distance between the two supports is 400 ft. Find the height of the remaining vertical cables, if the vertical cables are evenly spaced. (Ignore the width of the supports and cables.)

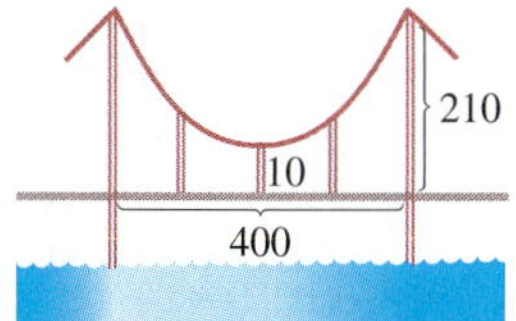

55. *(Modeling) Path of a Cannon Shell* About 400 yr ago, the physicist Galileo observed that certain projectiles follow a parabolic path. For instance, if a cannon fires a shell at a 45° angle with a speed of v feet per second, then the path of the shell (see the figure on the top) is modeled by the equation

$$y = x - \frac{32}{v^2}x^2.$$

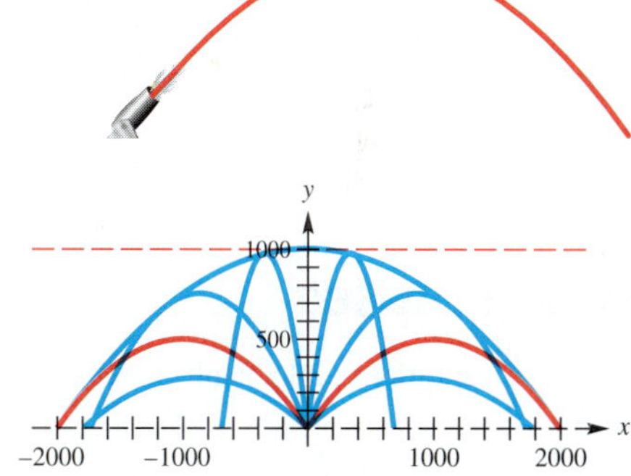

The figure on the bottom shows the paths of shells all fired at the same speed but at different angles. The greatest distance is achieved with a 45° angle. The outline, called the *envelope,* of this family of curves is another parabola with the cannon as focus. The horizontal line through the vertex of the envelope parabola is a directrix for all the other parabolas. Suppose all the shells are fired at a speed of 252.982 ft per sec.

(a) What is the greatest distance that a shell can be fired?

(b) What is the equation of the envelope parabola?

(c) Can a shell reach a helicopter 1500 ft due east of the cannon flying at a height of 450 ft?

56. *(Modeling) Path of an Object* When an object moves under the influence of a constant force (without air resistance), its path is parabolic. This would occur if a ball is thrown near the surface of a planet or other celestial object. Suppose two balls are simultaneously thrown upward at a 45° angle on two different planets. If their initial velocities are both 30 mph, then their xy-coordinates in feet at time x in seconds can be modeled by the equation

$$y = x - \frac{g}{1922}x^2,$$

where g is the acceleration due to gravity. The value of g will vary depending on the mass and size of the planet. (*Source:* Zeilik, M. and S. Gregory, *Introductory Astronomy and Astrophysics,* Fourth Edition, Brooks/Cole, 1998.)

(a) For Earth $g = 32.2$, while on Mars $g = 12.6$. Find the two equations, and graph on the same screen of a graphing calculator the paths of the two balls thrown on Earth and Mars. Use the window $[0, 180]$ by $[0, 120]$. (*Hint:* If possible, set the mode on your graphing calculator to simultaneous.)

(b) Determine the difference in the horizontal distances traveled by the two balls.

57. ***(Modeling) Path of an Object*** (Refer to Exercise 56.) Suppose the two balls are now thrown upward at a 60° angle on Mars and the moon. If their initial velocity is 60 mph, then their xy-coordinates in feet at time x in seconds can be modeled by the equation

$$y = \frac{19}{11}x - \frac{g}{3872}x^2.$$

(*Source:* Zeilik, M. and S. Gregory, *Introductory Astronomy and Astrophysics,* Fourth Edition, Brooks/Cole, 1998.)

(a) Graph on the same coordinate axes the paths of the balls if $g = 5.2$ for the moon. Use the window $[0, 1500]$ by $[0, 1000]$.

(b) Determine the maximum height of each ball to the nearest foot.

58. Explain how you can tell, just by looking at the equation of a parabola, whether it has a horizontal or a vertical axis.

59. Prove that the parabola with focus $(p, 0)$ and directrix $x = -p$ has the equation $y^2 = 4px$.

60. Write a short paragraph on the appearances of parabolic shapes in your surroundings.

10.2 Ellipses

Equations and Graphs of Ellipses ▪ **Translated Ellipses** ▪ **Eccentricity** ▪ **Applications of Ellipses**

Equations and Graphs of Ellipses

Like the parabola, the ellipse is defined as a set of points.

Ellipse

An **ellipse** is the set of all points in a plane the sum of whose distances from two fixed points is constant. Each fixed point is called a **focus** (plural, **foci**) of the ellipse.

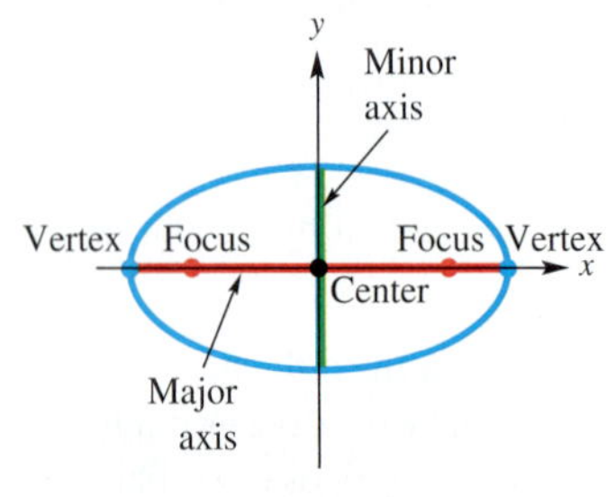

Figure 14

As shown in Figure 14, an ellipse has two axes of symmetry, the **major axis** (the longer one) and the **minor axis** (the shorter one). The foci are always located on the major axis. The midpoint of the major axis is the **center** of the ellipse, and the endpoints of the major axis are the **vertices** of the ellipse. The graph of an ellipse is not the graph of a function. (Why?)

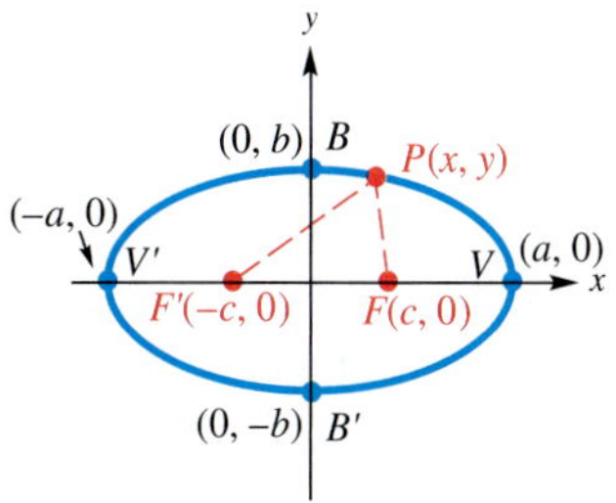

Figure 15

The ellipse in Figure 15 has its center at the origin, foci $F(c, 0)$ and $F'(-c, 0)$, and vertices $V(a, 0)$ and $V'(-a, 0)$. From Figure 15, the distance from V to F is $a - c$ and the distance from V to F' is $a + c$. The sum of these distances is $2a$. Since V is on the ellipse, this sum is the constant referred to in the definition of an ellipse. Thus, for any point $P(x, y)$ on the ellipse,

$$d(P, F) + d(P, F') = 2a.$$

By the distance formula,

$$d(P, F) = \sqrt{(x - c)^2 + y^2},$$

and $$d(P, F') = \sqrt{[x - (-c)]^2 + y^2} = \sqrt{(x + c)^2 + y^2}.$$

Thus,

$$\sqrt{(x - c)^2 + y^2} + \sqrt{(x + c)^2 + y^2} = 2a$$

$$\sqrt{(x - c)^2 + y^2} = 2a - \sqrt{(x + c)^2 + y^2}$$

Isolate $\sqrt{(x - c)^2 + y^2}$.

$$(x - c)^2 + y^2 = 4a^2 - 4a\sqrt{(x + c)^2 + y^2} + (x + c)^2 + y^2$$

Square both sides. **(Sections R.3, 1.6)**

$$x^2 - 2cx + c^2 + y^2 = 4a^2 - 4a\sqrt{(x + c)^2 + y^2} + x^2 + 2cx + c^2 + y^2$$

Square $x - c$; square $x + c$.

$$4a\sqrt{(x + c)^2 + y^2} = 4a^2 + 4cx \qquad \text{Isolate } 4a\sqrt{(x + c)^2 + y^2}.$$

$$a\sqrt{(x + c)^2 + y^2} = a^2 + cx \qquad \text{Divide by 4.}$$

$$a^2(x^2 + 2cx + c^2 + y^2) = a^4 + 2ca^2x + c^2x^2$$

Square both sides.

$$a^2x^2 + 2ca^2x + a^2c^2 + a^2y^2 = a^4 + 2ca^2x + c^2x^2$$

Distributive property **(Section R.1)**

$$a^2x^2 + a^2c^2 + a^2y^2 = a^4 + c^2x^2 \qquad \text{Subtract } 2ca^2x.$$

$$a^2x^2 - c^2x^2 + a^2y^2 = a^4 - a^2c^2 \qquad \text{Rearrange terms.}$$

$$(a^2 - c^2)x^2 + a^2y^2 = a^2(a^2 - c^2) \qquad \text{Factor. } \textbf{(Section R.4)}$$

$$\frac{x^2}{a^2} + \frac{y^2}{a^2 - c^2} = 1. \qquad (*) \qquad \text{Divide by } a^2(a^2 - c^2).$$

Since $B(0, b)$ is on the ellipse in Figure 15, we have

$$d(B, F) + d(B, F') = 2a$$

$$\sqrt{(-c)^2 + b^2} + \sqrt{c^2 + b^2} = 2a$$

$$2\sqrt{c^2 + b^2} = 2a \qquad \text{Combine terms.}$$

$$\sqrt{c^2 + b^2} = a \qquad \text{Divide by 2.}$$

$$c^2 + b^2 = a^2 \qquad \text{Square both sides.}$$

$$b^2 = a^2 - c^2. \qquad \text{Subtract } c^2.$$

Replacing $a^2 - c^2$ with b^2 in equation (*) gives

$$\frac{x^2}{a^2} + \frac{y^2}{b^2} = 1,$$

the standard form of the equation of an ellipse centered at the origin with foci on the x-axis. If the vertices and foci were on the y-axis, an almost identical derivation could be used to get the standard form

$$\frac{x^2}{b^2} + \frac{y^2}{a^2} = 1.$$

Looking Ahead to Calculus

Methods of calculus can be used to solve problems involving ellipses. For example, differentiation is used to find the slope of the tangent line at a point on the ellipse, and integration is used to find the length of any arc of the ellipse.

Standard Forms of Equations for Ellipses

The ellipse with center at the origin and equation

$$\frac{x^2}{a^2} + \frac{y^2}{b^2} = 1 \quad (a > b)$$

has vertices $(\pm a, 0)$, endpoints of the minor axis $(0, \pm b)$, and foci $(\pm c, 0)$, where $c^2 = a^2 - b^2$.

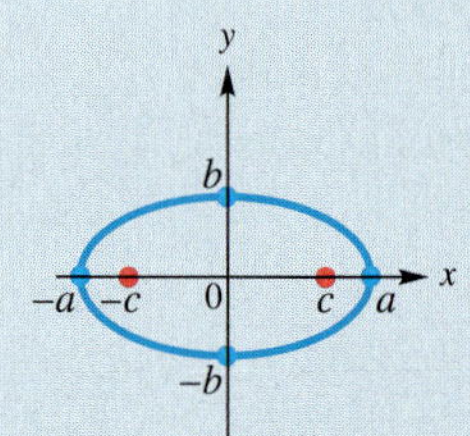

Major axis on x-axis

The ellipse with center at the origin and equation

$$\frac{x^2}{b^2} + \frac{y^2}{a^2} = 1 \quad (a > b)$$

has vertices $(0, \pm a)$, endpoints of the minor axis $(\pm b, 0)$, and foci $(0, \pm c)$, where $c^2 = a^2 - b^2$.

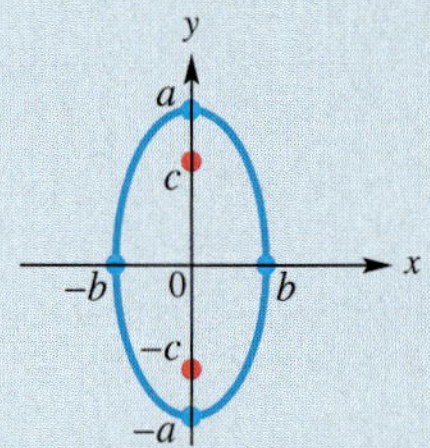

Major axis on y-axis

Do not be confused by the two standard forms—in one case a^2 is associated with x^2; in the other case a^2 is associated with y^2. In practice it is necessary only to find the intercepts of the graph—if the positive x-intercept is larger than the positive y-intercept, then the major axis is horizontal; otherwise, it is vertical. When using the relationship $a^2 - c^2 = b^2$, or $a^2 - b^2 = c^2$, choose a^2 and b^2 so that $a^2 > b^2$.

EXAMPLE 1 Graphing Ellipses Centered at the Origin

Graph each ellipse, and find the coordinates of the foci. Give the domain and range.

(a) $4x^2 + 9y^2 = 36$ **(b)** $4x^2 + y^2 = 64$

Solution

(a) Divide each side of $4x^2 + 9y^2 = 36$ by 36 to get

$$\frac{x^2}{9} + \frac{y^2}{4} = 1. \quad \text{Standard form of an ellipse}$$

Thus, the x-intercepts are ± 3, and the y-intercepts are ± 2. The graph of the ellipse is shown in Figure 16.

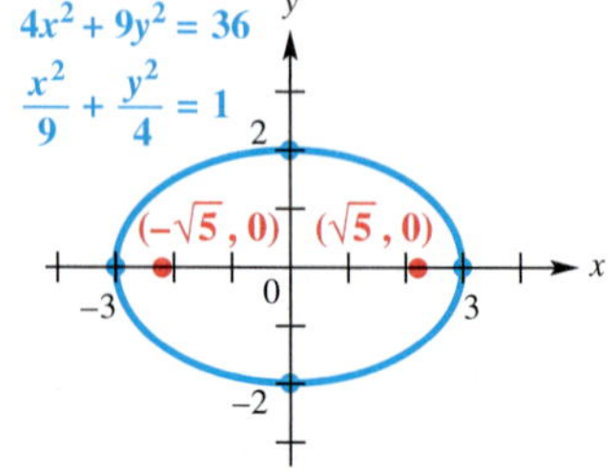

Figure 16

Since $9 > 4$, we find the foci of the ellipse by letting $a^2 = 9$ and $b^2 = 4$ in $c^2 = a^2 - b^2$.

$$c^2 = 9 - 4 = 5, \qquad \text{so} \qquad c = \sqrt{5}.$$

(By definition, $c > 0$. See Figure 15 on page 593.) The major axis is along the x-axis, so the foci have coordinates $\left(-\sqrt{5}, 0\right)$ and $\left(\sqrt{5}, 0\right)$. The domain of this relation is $[-3, 3]$, and the range is $[-2, 2]$.

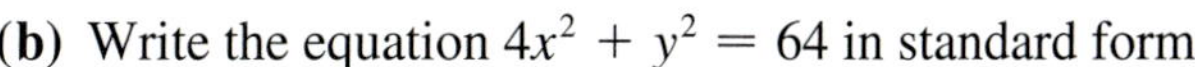

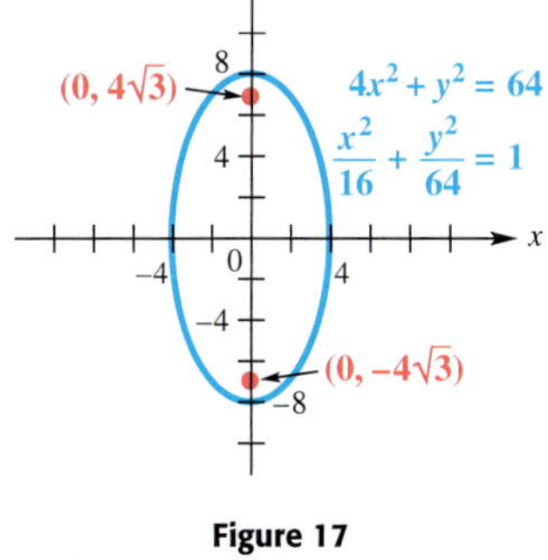

Figure 17

(b) Write the equation $4x^2 + y^2 = 64$ in standard form.

$$\frac{x^2}{16} + \frac{y^2}{64} = 1 \quad \text{Divide by 64.}$$

The x-intercepts are ± 4; the y-intercepts are ± 8. See the graph in Figure 17. Here $64 > 16$, so $a^2 = 64$ and $b^2 = 16$. Thus,

$$c^2 = 64 - 16 = 48, \qquad \text{so} \qquad c = \sqrt{48} = 4\sqrt{3}. \quad \text{(Section R.7)}$$

The major axis is on the y-axis, so the coordinates of the foci are $\left(0, -4\sqrt{3}\right)$ and $\left(0, 4\sqrt{3}\right)$. The domain of the relation is $[-4, 4]$; the range is $[-8, 8]$.

Now try Exercises 7 and 9.

$Y_1 = 2\sqrt{1 - \frac{x^2}{9}}$

$Y_2 = -2\sqrt{1 - \frac{x^2}{9}}$

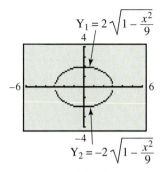

Since $Y_2 = -Y_1$, $-Y_1$ can be entered to get the portion of the graph below the x-axis.

Figure 18

The graph of an ellipse is *not* the graph of a function. To graph the ellipse in Example 1(a) with a graphing calculator, solve for y in $4x^2 + 9y^2 = 36$ to get equations of the two functions

$$y = 2\sqrt{1 - \frac{x^2}{9}} \qquad \text{and} \qquad y = -2\sqrt{1 - \frac{x^2}{9}}.$$

See Figure 18. ■

EXAMPLE 2 Writing the Equation of an Ellipse

Write the equation of the ellipse having center at the origin, foci at $(0, 3)$ and $(0, -3)$, and major axis of length 8 units.

Solution Since the major axis is 8 units long, $2a = 8$ and $a = 4$. To find b^2, use the relationship $a^2 - b^2 = c^2$, with $a = 4$ and $c = 3$.

$$a^2 - b^2 = c^2$$
$$4^2 - b^2 = 3^2 \quad \text{Substitute for } a \text{ and } c.$$
$$16 - b^2 = 9$$
$$b^2 = 7$$

Since the foci are on the y-axis, we use the larger intercept, a, to find the denominator for y^2, giving the equation in standard form as

$$\frac{x^2}{7} + \frac{y^2}{16} = 1.$$

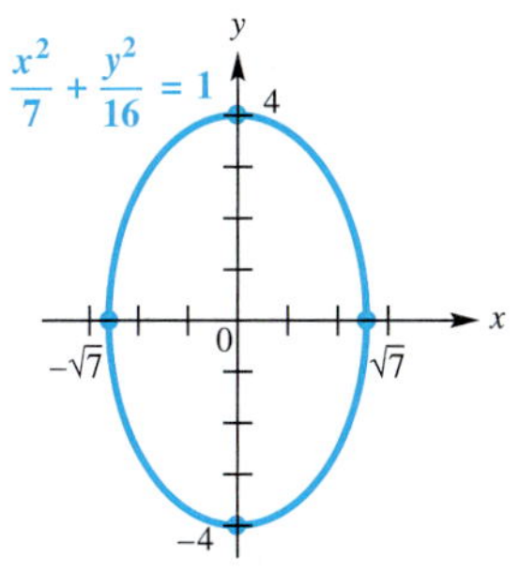

Figure 19

A graph of this ellipse is shown in Figure 19. The domain of this relation is $\left[-\sqrt{7}, \sqrt{7}\right]$, and the range is $[-4, 4]$.

Now try Exercise 17.

EXAMPLE 3 Graphing a Half-Ellipse

Graph $\frac{y}{4} = \sqrt{1 - \frac{x^2}{25}}$. Give the domain and range.

Solution Square both sides to get

$$\frac{y^2}{16} = 1 - \frac{x^2}{25} \quad \text{or} \quad \frac{x^2}{25} + \frac{y^2}{16} = 1,$$

the equation of an ellipse with x-intercepts ± 5 and y-intercepts ± 4. Since

$$\sqrt{1 - \frac{x^2}{25}} \geq 0,$$

the only possible values of y are those making $\frac{y}{4} \geq 0$, giving the half-ellipse shown in Figure 20. The half-ellipse in Figure 20 is the graph of a function. The domain is the interval $[-5, 5]$, and the range is $[0, 4]$.

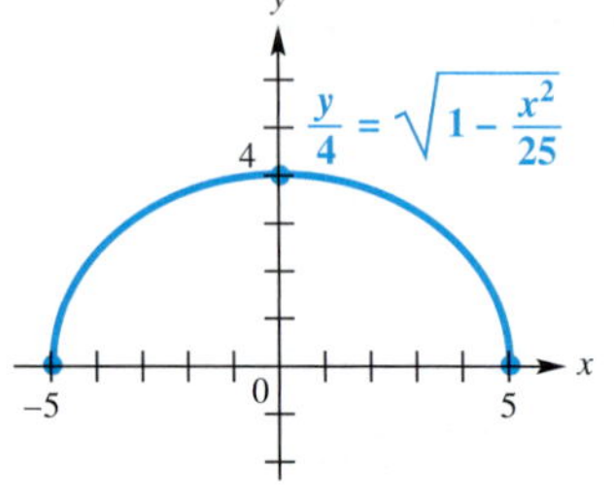

Figure 20

Now try Exercise 29.

Translated Ellipses Just as a circle need not have its center at the origin, an ellipse may also have its center translated away from the origin.

Ellipse Centered at (h, k)

An ellipse centered at (h, k) with horizontal major axis of length $2a$ has equation

$$\frac{(x-h)^2}{a^2} + \frac{(y-k)^2}{b^2} = 1.$$

There is a similar result for ellipses having a vertical major axis.

This result can be proven from the definition of an ellipse.

EXAMPLE 4 Graphing an Ellipse Translated Away from the Origin

Graph $\frac{(x-2)^2}{9} + \frac{(y+1)^2}{16} = 1$. Give the domain and range.

Solution The graph of this equation is an ellipse centered at $(2, -1)$. As mentioned earlier, ellipses always have $a > b$. For this ellipse, $a = 4$ and $b = 3$. Since $a = 4$ is associated with y^2, the vertices of the ellipse are on the vertical line through $(2, -1)$. Find the vertices by locating two points on the vertical line through $(2, -1)$, one 4 units up from $(2, -1)$ and one 4 units down. The vertices are $(2, 3)$ and $(2, -5)$. Two other points on the ellipse are on the horizontal line through $(2, -1)$, one 3 units to the right and one 3 units to the left.

The graph is shown in Figure 21. The domain is $[-1, 5]$, and the range is $[-5, 3]$.

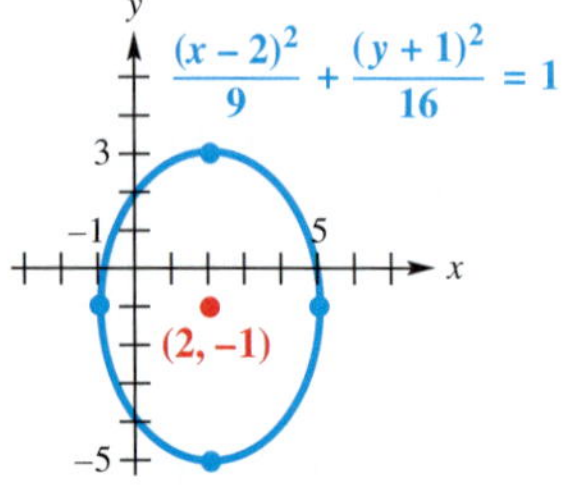

Figure 21

Now try Exercise 13.

NOTE As suggested by the graphs in this section, an ellipse is symmetric with respect to its major axis, its minor axis, and its center. Also, if $a = b$ in the equation of an ellipse, then the ellipse is a circle.

Eccentricity The ellipse is the third conic section (or *conic*) we have studied. (The circle and the parabola were the first two.) The fourth conic section, the hyperbola, will be introduced in the next section. All conics can be characterized by one general definition.

Conic

A **conic** is the set of all points $P(x, y)$ in a plane such that the ratio of the distance from P to a fixed point and the distance from P to a fixed line is constant.

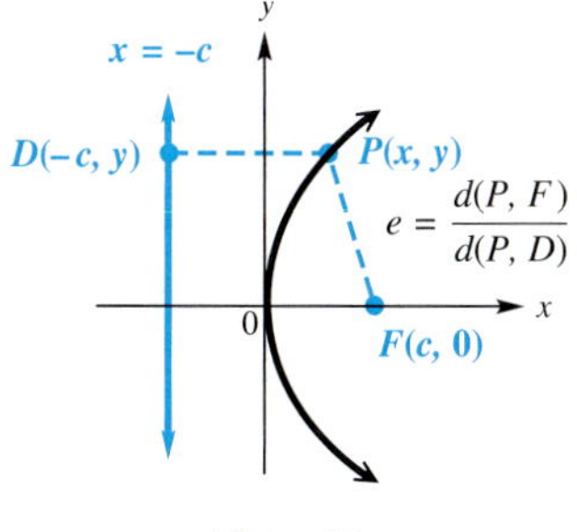

Figure 22

As with parabolas, the fixed line is the directrix, and the fixed point is a focus. In Figure 22, the focus is $F(c, 0)$, and the directrix is the line $x = -c$. The constant ratio is called the **eccentricity** of the conic, written e. (This is not the same e as the base of natural logarithms.)

If the conic is a parabola, then by definition, the distances $d(P, F)$ and $d(P, D)$ in Figure 22 are equal. Thus, every parabola has eccentricity 1.

For the ellipse, eccentricity is a measure of its "roundness." The constant ratio in the definition is $e = \frac{c}{a}$, where (as before) c is the distance from the center of the figure to a focus, and a is the distance from the center to a vertex. By the definition of an ellipse, $a^2 > b^2$ and $c = \sqrt{a^2 - b^2}$. Thus, for the ellipse,

$$0 < c < a$$

$$0 < \frac{c}{a} < 1 \quad \text{Divide by } a.$$

$$0 < e < 1.$$

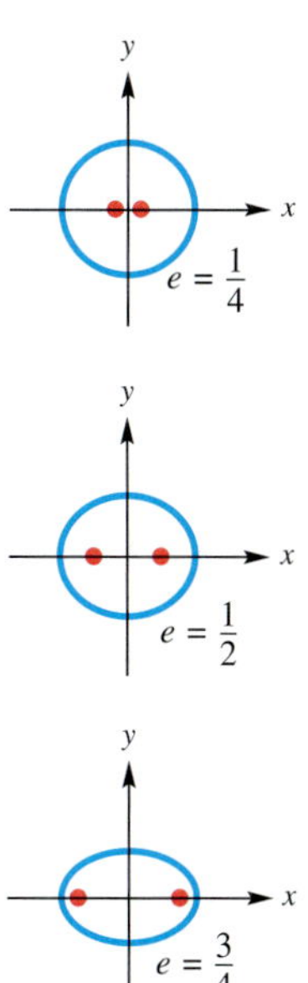

Figure 23

If a is constant, letting c approach 0 would force the ratio $\frac{c}{a}$ to approach 0, which also forces b to approach a (so that $\sqrt{a^2 - b^2} = c$ would approach 0). Since b determines the endpoints of the minor axis, this means that the lengths of the major and minor axes are almost the same, producing an ellipse very close in shape to a circle when e is very close to 0. In a similar manner, if e approaches 1, then b will approach 0.

The path of Earth around the sun is an ellipse that is very nearly circular. In fact, for this ellipse, $e \approx .017$. On the other hand, the path of Halley's comet is a very flat ellipse, with $e \approx .97$. Figure 23 compares ellipses with different eccentricities. The locations of the foci are shown in each case.

The equation of a circle can be written

$$(x - h)^2 + (y - k)^2 = r^2$$

$$\frac{(x - h)^2}{r^2} + \frac{(y - k)^2}{r^2} = 1.$$

In a circle, the foci coincide with the center, so $a = b$, $c = \sqrt{a^2 - b^2} = 0$, and thus $e = 0$.

EXAMPLE 5 Finding Eccentricity from Equations of Ellipses

Find the eccentricity of each ellipse.

(a) $\frac{x^2}{9} + \frac{y^2}{16} = 1$ **(b)** $5x^2 + 10y^2 = 50$

Solution

(a) Since $16 > 9$, $a^2 = 16$, which gives $a = 4$. Also,

$$c = \sqrt{a^2 - b^2} = \sqrt{16 - 9} = \sqrt{7}.$$

Finally, $e = \frac{c}{a} = \frac{\sqrt{7}}{4} \approx .66.$

(b) Divide by 50 to obtain $\frac{x^2}{10} + \frac{y^2}{5} = 1$. Here, $a^2 = 10$, with $a = \sqrt{10}$. Now, find c.

$$c = \sqrt{10 - 5} = \sqrt{5} \quad \text{and} \quad e = \frac{\sqrt{5}}{\sqrt{10}} = \frac{1}{\sqrt{2}} \approx .71$$

Now try Exercises 37 and 39.

Applications of Ellipses

EXAMPLE 6 Applying the Equation of an Ellipse to the Orbit of a Planet

The orbit of the planet Mars is an ellipse with the sun at one focus. The eccentricity of the ellipse is .0935, and the closest distance that Mars comes to the sun is 128.5 million mi. (*Source: The World Almanac and Book of Facts.*) Find the maximum distance of Mars from the sun.

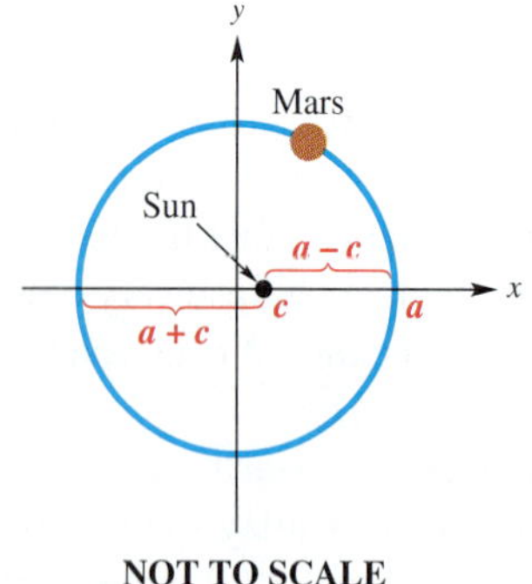

NOT TO SCALE

Figure 24

Solution Figure 24 shows the orbit of Mars with the origin at the center of the ellipse and the sun at one focus. Mars is closest to the sun when Mars is at the right endpoint of the major axis and farthest from the sun when Mars is at the left endpoint. Therefore, the smallest distance is $a - c$, and the greatest distance is $a + c$.

Since $a - c = 128.5$, $c = a - 128.5$. Using $e = \frac{c}{a}$,

$$\frac{a - 128.5}{a} = .0935$$

$$a - 128.5 = .0935a \quad \text{Multiply by } a.$$

$$.9065a = 128.5 \quad \text{Subtract } .0935a\text{; add } 128.5.$$

$$a \approx 141.8. \quad \text{Divide by } .9065.$$

Then $c = 141.8 - 128.5 = 13.3$

and $a + c = 141.8 + 13.3 = 155.1.$

The maximum distance of Mars from the sun is about 155.1 million mi.

Now try Exercise 45.

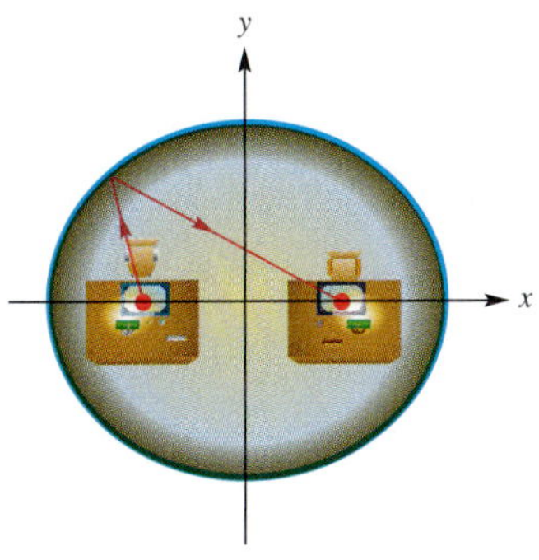

Aerial view of Old House Chamber

Figure 25

When a ray of light or sound emanating from one focus of an ellipse bounces off the ellipse, it passes through the other focus. See Figure 25. As mentioned in the chapter introduction, this reflecting property is responsible for whispering galleries. John Quincy Adams was able to listen in on his opponents' conversations because his desk was positioned at one of the foci beneath the ellipsoidal ceiling and his opponents were located across the room at the other focus.

A lithotripter is a machine used to crush kidney stones using shock waves. The patient is placed in an elliptical tub with the kidney stone at one focus of the ellipse. A beam is projected from the other focus to the tub so that it reflects to hit the kidney stone. See Figures 26 and 27.

EXAMPLE 7 Modeling the Reflective Property of Ellipses

If a lithotripter is based on the ellipse

$$\frac{x^2}{36} + \frac{y^2}{27} = 1,$$

determine how many units both the kidney stone and the source of the beam must be placed from the center of the ellipse.

Solution The kidney stone and the source of the beam must be placed at the foci, $(c, 0)$ and $(-c, 0)$. Here $a^2 = 36$ and $b^2 = 27$, so

$$c = \sqrt{a^2 - b^2} = \sqrt{36 - 27} = \sqrt{9} = 3.$$

Thus, the foci are $(3, 0)$ and $(-3, 0)$, so the kidney stone and the source both must be placed on a line 3 units from the center. See Figure 27.

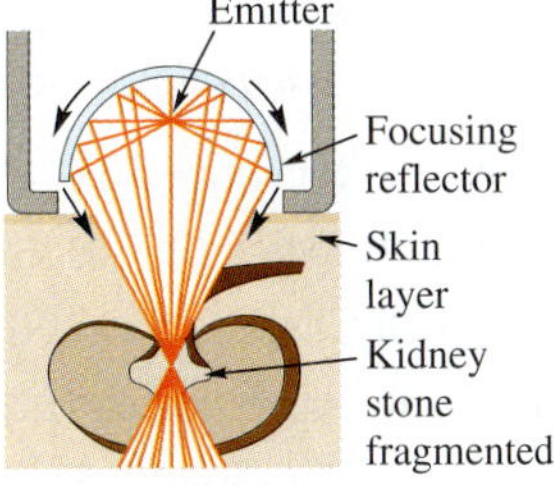

The top of an ellipse is illustrated in this depiction of how a lithotripter crushes kidney stones.*

Figure 26

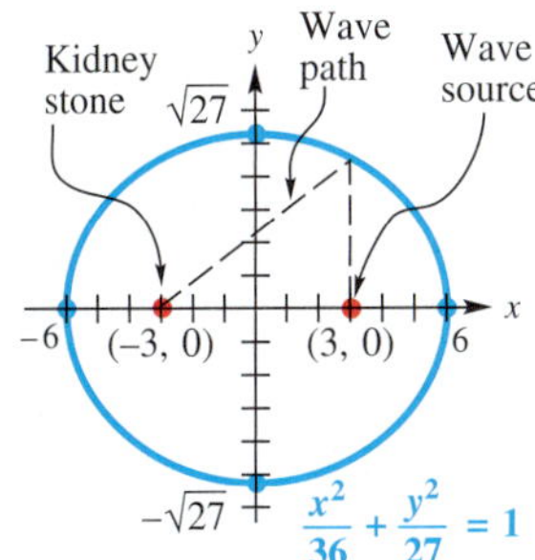

Figure 27

Now try Exercise 49.

*Source: Adapted drawing of an ellipse in illustration of a lithotripter. The American Medical Association, *Encyclopedia of Medicine,* 1989.

10.2 Exercises

1. *Concept Check* Match each equation of an ellipse in Column I with the appropriate description in Column II.

I	II
(a) $36x^2 + 9y^2 = 324$	A. x-intercepts ± 3; y-intercepts ± 6
(b) $9x^2 + 36y^2 = 324$	B. x-intercepts ± 4; y-intercepts ± 5
(c) $\frac{x^2}{25} = 1 - \frac{y^2}{16}$	C. x-intercepts ± 6; y-intercepts ± 3
(d) $\frac{x^2}{16} = 1 - \frac{y^2}{25}$	D. x-intercepts ± 5; y-intercepts ± 4

2. *Concept Check* Determine whether or not each equation is that of an ellipse. If it is not, state the kind of graph the equation has.

(a) $x^2 + 4y^2 = 4$ (b) $x^2 + y^2 = 4$ (c) $x^2 + y = 4$ (d) $\frac{x}{4} + \frac{y}{25} = 1$

Graph each ellipse. Identify the domain, range, center, vertices, endpoints of the minor axis, and the foci in each figure. See Examples 1 and 4.

3. $\frac{x^2}{25} + \frac{y^2}{9} = 1$
4. $\frac{x^2}{16} + \frac{y^2}{25} = 1$
5. $\frac{x^2}{9} + y^2 = 1$
6. $\frac{x^2}{36} + \frac{y^2}{16} = 1$
7. $9x^2 + y^2 = 81$
8. $4x^2 + 16y^2 = 64$
9. $4x^2 + 25y^2 = 100$
10. $4x^2 + y^2 = 16$
11. $\frac{(x-2)^2}{25} + \frac{(y-1)^2}{4} = 1$
12. $\frac{(x+2)^2}{16} + \frac{(y+1)^2}{9} = 1$
13. $\frac{(x+3)^2}{16} + \frac{(y-2)^2}{36} = 1$
14. $\frac{(x-1)^2}{9} + \frac{(y+3)^2}{25} = 1$

Write an equation for each ellipse. See Example 2.

15. x-intercepts ± 5; foci at $(-3, 0)$, $(3, 0)$
16. y-intercepts ± 4; foci at $(0, -1)$, $(0, 1)$
17. major axis with length 6; foci at $(0, 2)$, $(0, -2)$
18. minor axis with length 4; foci at $(-5, 0)$, $(5, 0)$
19. center at $(5, 2)$; minor axis vertical, with length 8; $c = 3$
20. center at $(-3, 6)$; major axis vertical, with length 10; $c = 2$
21. vertices at $(4, 9)$, $(4, 1)$; minor axis with length 6
22. foci at $(-3, -3)$, $(7, -3)$; the point $(2, 1)$ on ellipse
23. foci at $(0, -3)$, $(0, 3)$; the point $(8, 3)$ on ellipse
24. foci at $(-4, 0)$, $(4, 0)$; sum of distances from foci to point on ellipse is 9 (*Hint:* Consider one of the vertices.)
25. foci at $(0, 4)$, $(0, -4)$; sum of distances from foci to point on ellipse is 10
26. eccentricity $\frac{1}{2}$; vertices at $(-4, 0)$, $(4, 0)$
27. eccentricity $\frac{3}{4}$; foci at $(0, -2)$, $(0, 2)$
28. eccentricity $\frac{2}{3}$; foci at $(0, -9)$, $(0, 9)$

Graph each equation. Give the domain and range. Identify any that are graphs of functions. See Example 3.

29. $\dfrac{y}{2} = \sqrt{1 - \dfrac{x^2}{25}}$

30. $\dfrac{x}{4} = \sqrt{1 - \dfrac{y^2}{9}}$

31. $x = -\sqrt{1 - \dfrac{y^2}{64}}$

32. $y = -\sqrt{1 - \dfrac{x^2}{100}}$

Determine the two equations necessary to graph each ellipse with a graphing calculator, and graph it in the viewing window indicated. See Figure 18.

33. $\dfrac{x^2}{16} + \dfrac{y^2}{4} = 1$; $[-4.7, 4.7]$ by $[-3.1, 3.1]$

34. $\dfrac{x^2}{4} + \dfrac{y^2}{25} = 1$; $[-9.4, 9.4]$ by $[-6.2, 6.2]$

35. $\dfrac{(x-3)^2}{25} + \dfrac{y^2}{9} = 1$; $[-9.4, 9.4]$ by $[-6.2, 6.2]$

36. $\dfrac{x^2}{36} + \dfrac{(y+4)^2}{4} = 1$; $[-9.4, 9.4]$ by $[-6.2, 6.2]$

Find the eccentricity of each ellipse. If necessary, round to the nearest hundredth. See Example 5.

37. $\dfrac{x^2}{3} + \dfrac{y^2}{4} = 1$

38. $\dfrac{x^2}{8} + \dfrac{y^2}{4} = 1$

39. $4x^2 + 7y^2 = 28$

40. $x^2 + 25y^2 = 25$

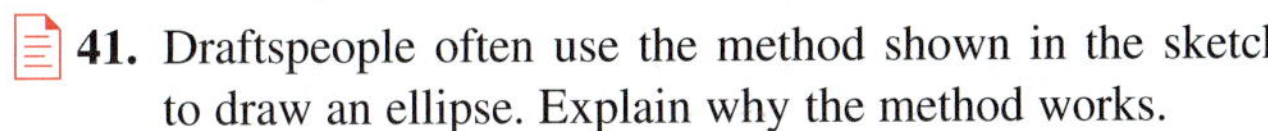

41. Draftspeople often use the method shown in the sketch to draw an ellipse. Explain why the method works.

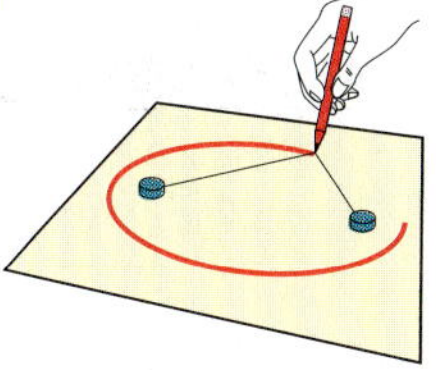

42. Explain how the method of Exercise 41 can be modified to draw a circle.

Solve each problem. See Examples 6 and 7.

43. ***Height of an Overpass*** A one-way road passes under an overpass in the shape of half an ellipse, 15 ft high at the center and 20 ft wide. Assuming a truck is 12 ft wide, what is the tallest truck that can pass under the overpass?

NOT TO SCALE

44. ***Height and Width of an Overpass*** An arch has the shape of half an ellipse. The equation of the ellipse is $100x^2 + 324y^2 = 32{,}400$, where x and y are in meters.

(a) How high is the center of the arch?

(b) How wide is the arch across the bottom?

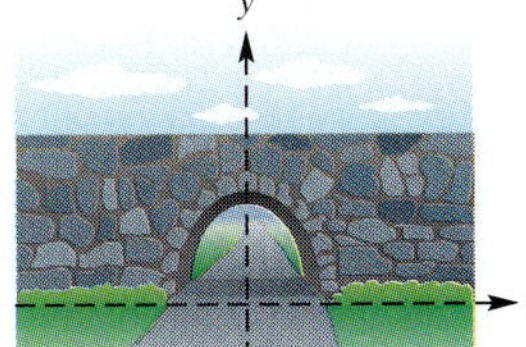

NOT TO SCALE

45. *Orbit of Halley's Comet* The famous Halley's comet last passed by Earth in February 1986 and will next return in 2062. It has an elliptical orbit of eccentricity .9673 with the sun at one focus. The greatest distance of the comet from the sun is 3281 million mi. (*Source: The World Almanac and Book of Facts.*) Find the least distance between Halley's comet and the sun.

46. *(Modeling) Orbit of a Satellite* The coordinates in miles for the orbit of the artificial satellite Explorer VII can be modeled by the equation

$$\frac{x^2}{a^2} + \frac{y^2}{b^2} = 1,$$

where $a = 4465$ and $b = 4462$. Earth's center is located at one focus of the elliptical orbit. (*Source:* Loh, W., *Dynamics and Thermodynamics of Planetary Entry,* Prentice-Hall, 1963; Thomson, W., *Introduction to Space Dynamics,* John Wiley & Sons, 1961.)

(a) Graph both the orbit of Explorer VII and of Earth on the same coordinate axes if the average radius of Earth is 3960 mi. Use the window $[-6750, 6750]$ by $[-4500, 4500]$.

(b) Determine the maximum and minimum heights of the satellite above Earth's surface.

47. *(Modeling) Orbits of Planets* Neptune and Pluto both have elliptical orbits with the sun at one focus. Neptune's orbit has $a = 30.1$ astronomical units (AU) with an eccentricity of $e = .009$, whereas Pluto's orbit has $a = 39.4$ and $e = .249$. (*Source:* Zeilik, M., S. Gregory, and E. Smith, *Introductory Astronomy and Astrophysics,* Fourth Edition, Saunders College Publishers, 1998.)

(a) Position the sun at the origin and determine equations that model each orbit.

(b) Graph both equations on the same coordinate axes. Use the window $[-60, 60]$ by $[-40, 40]$.

48. *(Modeling) The Roman Colisseum*

(a) The Roman Colisseum is an ellipse with major axis 620 ft and minor axis 513 ft. Find the distance between the foci of this ellipse.

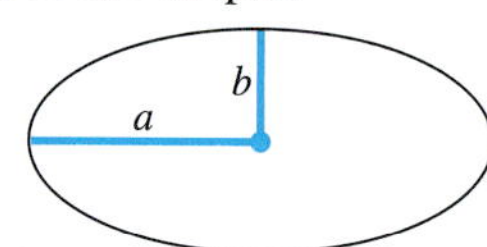

(b) A formula for the approximate circumference of an ellipse is

$$C \approx 2\pi\sqrt{\frac{a^2 + b^2}{2}},$$

where a and b are the lengths shown in the figure. Use this formula to find the approximate circumference of the Roman Colisseum.

49. *Design of a Lithotripter* Suppose a lithotripter is based on the ellipse with equation

$$\frac{x^2}{36} + \frac{y^2}{9} = 1.$$

How far from the center of the ellipse must the kidney stone and the source of the beam be placed? Give the exact answer.

50. *Design of a Lithotripter* Rework Exercise 49 if the equation of the ellipse is $9x^2 + 4y^2 = 36$.

10.3 Hyperbolas

Equations and Graphs of Hyperbolas ▪ **Translated Hyperbolas** ▪ **Eccentricity**

Equations and Graphs of Hyperbolas

An ellipse was defined as the set of all points in a plane the sum of whose distances from two fixed points is a constant. A *hyperbola* is defined similarly.

Hyperbola

A **hyperbola** is the set of all points in a plane such that the absolute value of the difference of the distances from two fixed points is constant. The two fixed points are called the **foci** of the hyperbola.

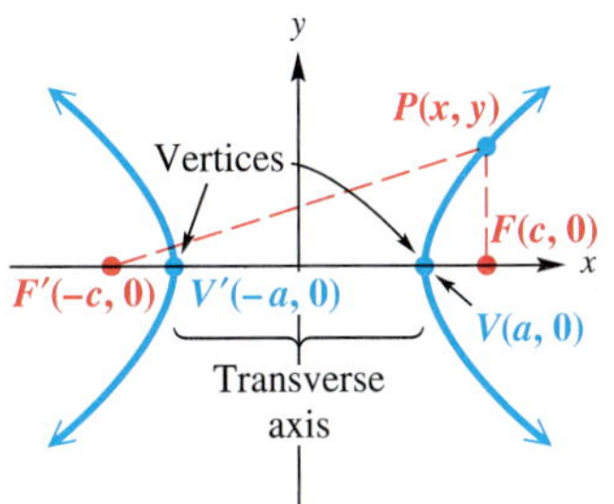

Figure 28

Suppose a hyperbola has center at the origin and foci at $F'(-c, 0)$ and $F(c, 0)$. See Figure 28. The midpoint of the segment $F'F$ is the center of the hyperbola and the points $V'(-a, 0)$ and $V(a, 0)$ are the **vertices** of the hyperbola. The line segment $V'V$ is the **transverse axis** of the hyperbola.

As with the ellipse,

$$d(V, F') - d(V, F) = (c + a) - (c - a) = 2a,$$

so the constant in the definition is $2a$, and

$$|d(P, F') - d(P, F)| = 2a$$

for any point $P(x, y)$ on the hyperbola. The distance formula and algebraic manipulation similar to that used for finding an equation for an ellipse (see Exercise 60) produce the result

$$\frac{x^2}{a^2} - \frac{y^2}{c^2 - a^2} = 1.$$

Letting $b^2 = c^2 - a^2$ gives

$$\frac{x^2}{a^2} - \frac{y^2}{b^2} = 1$$

as an equation of the hyperbola in Figure 28. Letting $y = 0$ shows that the x-intercepts are $\pm a$. If $x = 0$, the equation becomes

$$\frac{0^2}{a^2} - \frac{y^2}{b^2} = 1 \qquad \text{Let } x = 0.$$

$$-\frac{y^2}{b^2} = 1$$

$$y^2 = -b^2,$$

which has no real number solutions, showing that this hyperbola has no y-intercepts. Again, see Figure 28.

Starting with the equation for a hyperbola and solving for y, we get

$$\frac{x^2}{a^2} - \frac{y^2}{b^2} = 1$$

$$\frac{x^2}{a^2} - 1 = \frac{y^2}{b^2} \qquad \text{Subtract 1; add } \tfrac{y^2}{b^2}.$$

$$\frac{x^2 - a^2}{a^2} = \frac{y^2}{b^2} \qquad \text{Write the left side as a single fraction.}$$

$$y = \pm\frac{b}{a}\sqrt{x^2 - a^2}. \qquad \text{Take square roots on both sides (Section 1.6); multiply by } b.$$

If x^2 is very large in comparison to a^2, the difference $x^2 - a^2$ would be very close to x^2. If this happens, then the points satisfying the final equation above would be very close to one of the lines

$$y = \pm\frac{b}{a}x.$$

Thus, as $|x|$ gets larger and larger, the points of the hyperbola $\frac{x^2}{a^2} - \frac{y^2}{b^2} = 1$ get closer and closer to the lines $y = \pm\frac{b}{a}x$. These lines, called **asymptotes** of the hyperbola, are useful when sketching the graph.

EXAMPLE 1 Using Asymptotes to Graph a Hyperbola

Graph $\frac{x^2}{25} - \frac{y^2}{49} = 1$. Sketch the asymptotes, and find the coordinates of the vertices and foci. Give the domain and range.

Algebraic Solution

For this hyperbola, $a = 5$ and $b = 7$. With these values,

$$y = \pm\frac{b}{a}x \quad \text{becomes} \quad y = \pm\frac{7}{5}x. \qquad \text{Asymptotes}$$

If we choose $x = 5$, then $y = \pm 7$. Choosing $x = -5$ also gives $y = \pm 7$. These four points—$(5, 7)$, $(5, -7)$, $(-5, 7)$, and $(-5, -7)$—are the corners of the rectangle shown in Figure 29.

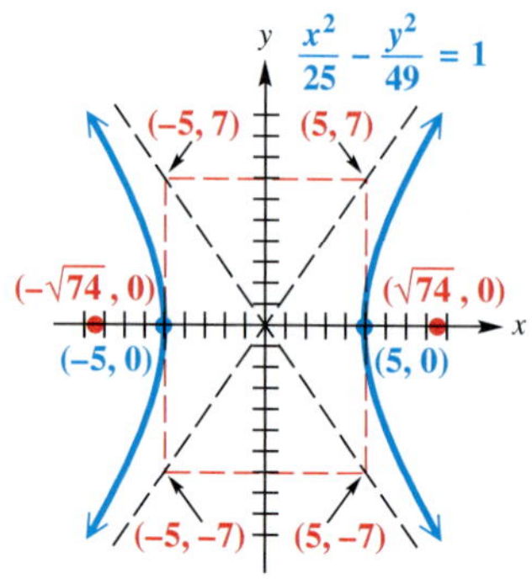

Figure 29

Graphing Calculator Solution

The graph of a hyperbola is not the graph of a function. We must solve for y in $\frac{x^2}{25} - \frac{y^2}{49} = 1$ to get equations of the two functions

$$y = \pm\frac{7}{5}\sqrt{x^2 - 25}.$$

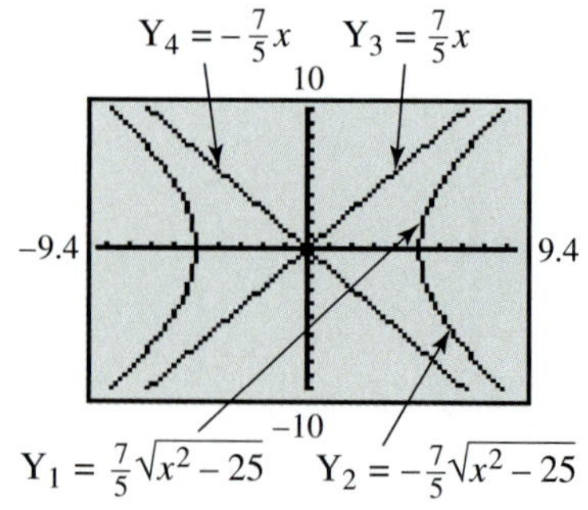

Figure 30

The extended diagonals of this rectangle, called the **fundamental rectangle,** are the asymptotes of the hyperbola. Since $a = 5$, the vertices of the hyperbola are $(5, 0)$ and $(-5, 0)$, as shown in Figure 29. We find the foci by letting

$$c^2 = a^2 + b^2 = 25 + 49 = 74, \quad \text{so} \quad c = \sqrt{74}.$$

Therefore, the foci are $(\sqrt{74}, 0)$ and $(-\sqrt{74}, 0)$. The domain is $(-\infty, -5] \cup [5, \infty)$, and the range is $(-\infty, \infty)$.

The graph of Y_1 is the upper portion of each branch of the hyperbola shown in Figure 30, and the graph of Y_2 is the lower portion of each branch. We could enter $Y_2 = -Y_1$ to get the part of the graph below the x-axis.

The asymptotes are also shown. We can use tracing to observe how the branches of the hyperbola approach the asymptotes.

Now try Exercise 5.

While $a > b$ for an ellipse, examples would show that for hyperbolas, it is possible that $a > b$, $a < b$, or $a = b$. If the foci of a hyperbola are on the y-axis, the equation of the hyperbola has the form

$$\frac{y^2}{a^2} - \frac{x^2}{b^2} = 1, \quad \text{with asymptotes} \quad y = \pm\frac{a}{b}x.$$

CAUTION If the foci of a hyperbola are on the x-axis, the asymptotes have equations $y = \pm\frac{b}{a}x$, while foci on the y-axis lead to asymptotes $y = \pm\frac{a}{b}x$. To avoid errors, write the equation of the hyperbola in either the form

$$\frac{x^2}{a^2} - \frac{y^2}{b^2} = 1 \quad \text{or} \quad \frac{y^2}{a^2} - \frac{x^2}{b^2} = 1,$$

and replace 1 with 0. Solving the resulting equation for y produces the proper equations for the asymptotes.

The basic information on hyperbolas is summarized as follows.

Standard Forms of Equations for Hyperbolas

The hyperbola with center at the origin and equation

$$\frac{x^2}{a^2} - \frac{y^2}{b^2} = 1$$

has vertices $(\pm a, 0)$, asymptotes $y = \pm\frac{b}{a}x$, and foci $(\pm c, 0)$, where $c^2 = a^2 + b^2$.

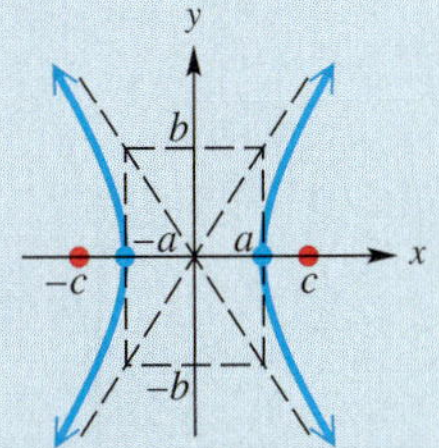

Transverse axis on x-axis

The hyperbola with center at the origin and equation

$$\frac{y^2}{a^2} - \frac{x^2}{b^2} = 1$$

has vertices $(0, \pm a)$, asymptotes $y = \pm\frac{a}{b}x$, and foci $(0, \pm c)$, where $c^2 = a^2 + b^2$.

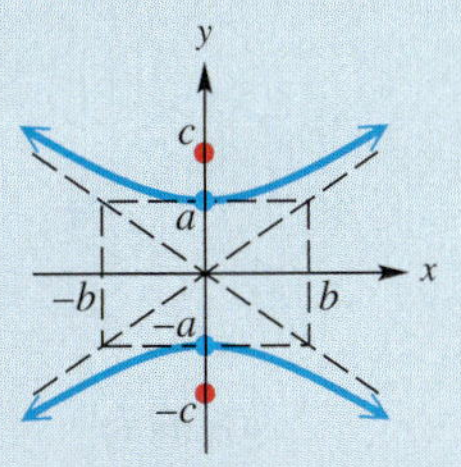

Transverse axis on y-axis

EXAMPLE 2 Graphing a Hyperbola

Graph $25y^2 - 4x^2 = 100$. Give the domain and range.

Solution $\frac{y^2}{4} - \frac{x^2}{25} = 1$ Divide by 100; standard form.

This hyperbola is centered at the origin, has foci on the y-axis, and has vertices $(0, 2)$ and $(0, -2)$. To find the equations of the asymptotes, replace 1 with 0.

$$\frac{y^2}{4} - \frac{x^2}{25} = 0$$

$$\frac{y^2}{4} = \frac{x^2}{25} \quad \text{Add } \tfrac{x^2}{25}.$$

$$y^2 = \frac{4x^2}{25} \quad \text{Multiply by 4.}$$

$$y = \pm\frac{2}{5}x \quad \text{Square root property (Section 1.4)}$$

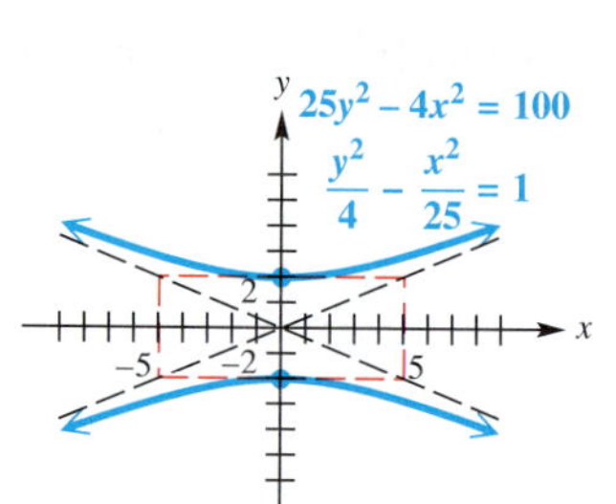

Figure 31

To graph the asymptotes, use the points $(5, 2)$, $(5, -2)$, $(-5, 2)$, and $(-5, -2)$ to determine the fundamental rectangle. The diagonals of this rectangle are the asymptotes for the graph, as shown in Figure 31. The domain of the relation is $(-\infty, \infty)$, and the range is $(-\infty, -2] \cup [2, \infty)$.

Now try Exercise 13.

Translated Hyperbolas In the graph of each hyperbola considered so far, the center is the origin and the asymptotes pass through the origin. This feature holds in general; the asymptotes of *any* hyperbola pass through the center of the hyperbola. Like an ellipse, a hyperbola can have its center translated away from the origin.

EXAMPLE 3 Graphing a Hyperbola Translated Away from the Origin

Graph $\frac{(y+2)^2}{9} - \frac{(x+3)^2}{4} = 1$. Give the domain and range.

Solution This equation represents a hyperbola centered at $(-3, -2)$. For this vertical hyperbola, $a = 3$ and $b = 2$. The x-values of the vertices are -3. Locate the y-values of the vertices by taking the y-value of the center, -2, and adding and subtracting 3. Thus, the vertices are $(-3, 1)$ and $(-3, -5)$. The asymptotes have slopes $\pm\frac{3}{2}$ and pass through the center $(-3, -2)$. The equations of the asymptotes,

$$y = -2 \pm \frac{3}{2}(x + 3),$$

can be found either by using the point-slope form of the equation of a line or by replacing 1 with 0 in the equation of the hyperbola as was done in Example 2. The graph is shown in Figure 32. The domain of the relation is $(-\infty, \infty)$, and the range is $(-\infty, -5] \cup [1, \infty)$.

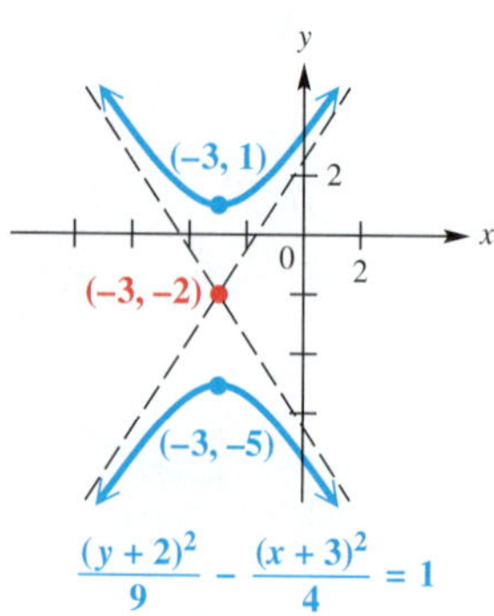

Figure 32

Now try Exercise 17.

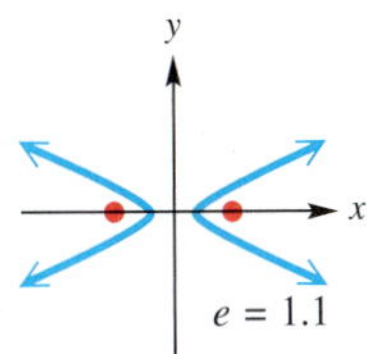

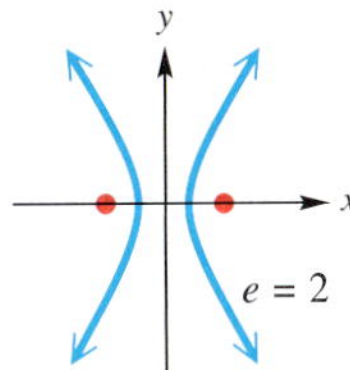

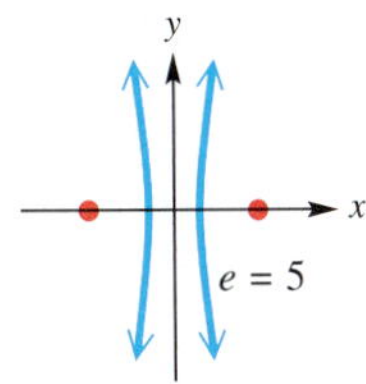

Figure 33

Eccentricity If we apply the definition of eccentricity from the previous section to the hyperbola, we get

$$e = \frac{\sqrt{a^2 + b^2}}{a} = \frac{c}{a}.$$

Since $c > a$, we have $e > 1$. Narrow hyperbolas have e near 1, and wide hyperbolas have large e. See Figure 33.

EXAMPLE 4 Finding Eccentricity from the Equation of a Hyperbola

Find the eccentricity of the hyperbola $\frac{x^2}{9} - \frac{y^2}{4} = 1$.

Solution Here $a^2 = 9$; thus, $a = 3$, $c = \sqrt{9 + 4} = \sqrt{13}$, and

$$e = \frac{c}{a} = \frac{\sqrt{13}}{3} \approx 1.2.$$

Now try Exercise 27.

EXAMPLE 5 Finding the Equation of a Hyperbola

Find the equation of the hyperbola with eccentricity 2 and foci at $(-9, 5)$ and $(-3, 5)$.

Solution Since the foci have the same y-coordinate, the line through them, and therefore the hyperbola, is horizontal. The center of the hyperbola is halfway between the two foci at $(-6, 5)$. The distance from each focus to the center is $c = 3$. Since $e = \frac{c}{a}$, $a = \frac{c}{e} = \frac{3}{2}$ and $a^2 = \frac{9}{4}$. Thus,

$$b^2 = c^2 - a^2 = 9 - \frac{9}{4} = \frac{27}{4}.$$

The equation of the hyperbola is

$$\frac{(x + 6)^2}{\frac{9}{4}} - \frac{(y - 5)^2}{\frac{27}{4}} = 1 \quad \text{or} \quad \frac{4(x + 6)^2}{9} - \frac{4(y - 5)^2}{27} = 1.$$

Simplify complex fractions. **(Section R.5)**

Now try Exercise 39.

The following chart summarizes our discussion of eccentricity in this chapter.

Conic	Eccentricity
Parabola	$e = 1$
Circle	$e = 0$
Ellipse	$e = \frac{c}{a}$ and $0 < e < 1$
Hyperbola	$e = \frac{c}{a}$ and $e > 1$

CONNECTIONS Ships and planes often use a location-finding system called LORAN. With this system, a radio transmitter at M in the figure sends out a series of pulses. When each pulse is received at transmitter S, it then sends out a pulse. A ship at P receives pulses from both M and S. A receiver on the ship measures the difference in the arrival times of the pulses. The navigator then consults a special map showing hyperbolas that correspond to the differences in arrival times (which give the distances d_1 and d_2 in the figure). In this way the ship can be located as lying on a branch of a particular hyperbola.

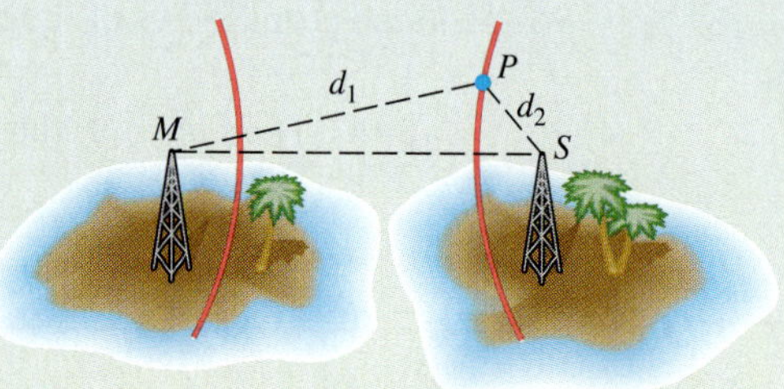

For Discussion or Writing

Suppose in the figure $d_1 = 80$ mi, $d_2 = 30$ mi, and the distance between the transmitters is 100 mi. Use the definition of a hyperbola to find an equation of the hyperbola on which the ship is located.

10.3 Exercises

Concept Check *Match each equation with the correct graph.*

1. $\dfrac{x^2}{25} + \dfrac{y^2}{9} = 1$

2. $\dfrac{x^2}{9} + \dfrac{y^2}{25} = 1$

3. 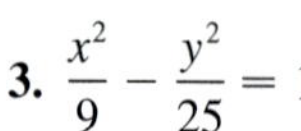 $\dfrac{x^2}{9} - \dfrac{y^2}{25} = 1$

4. 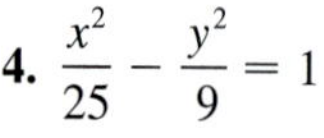 $\dfrac{x^2}{25} - \dfrac{y^2}{9} = 1$

A.

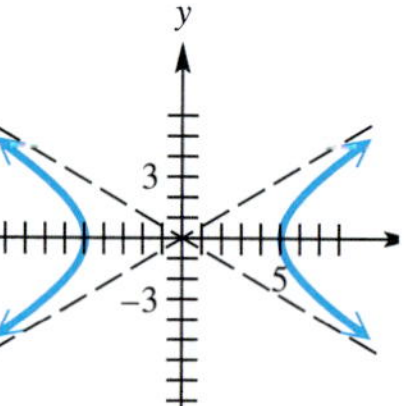

B.

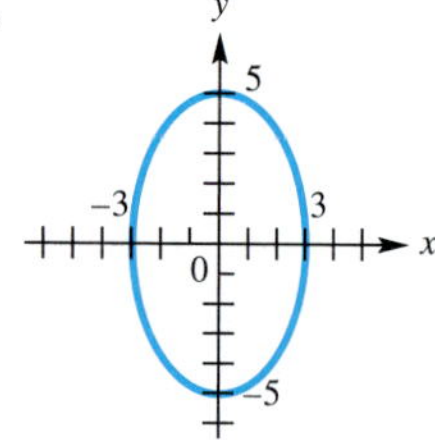

C.

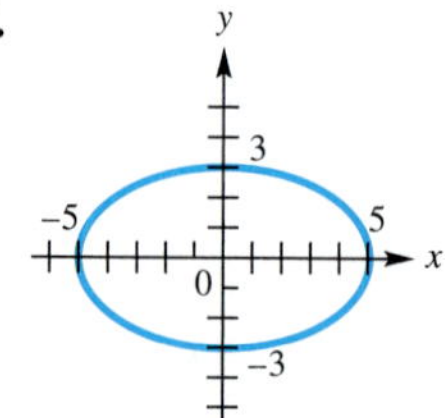

D.

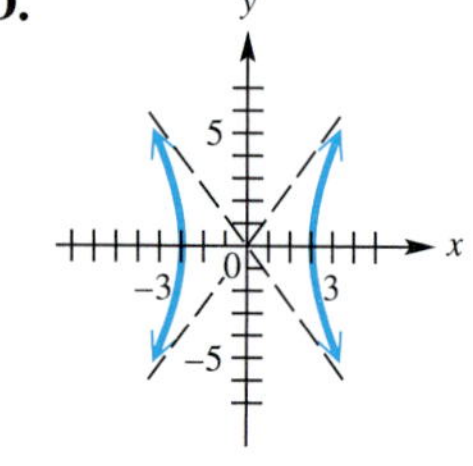

Graph each hyperbola. Give the domain, range, center, vertices, foci, and equations of the asymptotes for each figure. See Examples 1–3.

5. $\frac{x^2}{16} - \frac{y^2}{9} = 1$

6. $\frac{x^2}{25} - \frac{y^2}{144} = 1$

7. $\frac{y^2}{25} - \frac{x^2}{49} = 1$

8. $\frac{y^2}{64} - \frac{x^2}{4} = 1$

9. $x^2 - y^2 = 9$

10. $x^2 - 4y^2 = 64$

11. $9x^2 - 25y^2 = 225$

12. $y^2 - 4x^2 = 16$

13. $4y^2 - 25x^2 = 100$

14. $\frac{x^2}{4} - y^2 = 4$

15. $9x^2 - 4y^2 = 1$

16. $25y^2 - 9x^2 = 1$

17. $\frac{(y-7)^2}{36} - \frac{(x-4)^2}{64} = 1$

18. $\frac{(x+6)^2}{144} - \frac{(y+4)^2}{81} = 1$

19. $\frac{(x+3)^2}{16} - \frac{(y-2)^2}{9} = 1$

20. $\frac{(y+5)^2}{4} - \frac{(x-1)^2}{16} = 1$

21. $16(x+5)^2 - (y-3)^2 = 1$

22. $4(x+9)^2 - 25(y+6)^2 = 100$

Graph each equation. Give the domain and range. Identify any that are graphs of functions. See Example 3 in the previous section.

23. $\frac{y}{3} = \sqrt{1 + \frac{x^2}{16}}$

24. $\frac{x}{3} = -\sqrt{1 + \frac{y^2}{25}}$

25. $5x = -\sqrt{1 + 4y^2}$

26. $3y = \sqrt{4x^2 - 16}$

Find the eccentricity to the nearest tenth of each hyperbola. See Example 4.

27. $\frac{x^2}{4} - \frac{y^2}{4} = 1$

28. $\frac{x^2}{18} - \frac{y^2}{2} = 1$

29. $8y^2 - 4x^2 = 8$

30. $16y^2 - 4x^2 = 32$

Write an equation for each hyperbola. See Examples 4 and 5.

31. x-intercepts ± 4; foci at $(-5, 0)$, $(5, 0)$

32. y-intercepts ± 9; foci at $(0, -15)$, $(0, 15)$

33. vertices at $(0, 6)$, $(0, -6)$; asymptotes $y = \pm\frac{1}{2}x$

34. vertices at $(-10, 0)$, $(10, 0)$; asymptotes $y = \pm 5x$

35. vertices at $(-3, 0)$, $(3, 0)$; passing through $(6, 1)$

36. vertices at $(0, 5)$, $(0, -5)$; passing through $(3, 10)$

37. foci at $(0, \sqrt{13})$, $(0, -\sqrt{13})$; asymptotes $y = \pm 5x$

38. foci at $(-3\sqrt{5}, 0)$, $(3\sqrt{5}, 0)$; asymptotes $y = \pm 2x$

39. vertices at $(4, 5)$, $(4, 1)$; asymptotes $y = 3 \pm 7(x - 4)$

40. vertices at $(5, -2)$, $(1, -2)$; asymptotes $y = -2 \pm \frac{3}{2}(x - 3)$

41. center at $(1, -2)$; focus at $(4, -2)$; vertex at $(3, -2)$

42. center at $(9, -7)$; focus at $(9, 3)$; vertex at $(9, -1)$

43. eccentricity 3; center at $(0, 0)$; vertex at $(0, 7)$

44. center at $(8, 7)$; focus at $(13, 7)$; eccentricity $\frac{5}{3}$

45. vertices at $(-2, 10)$, $(-2, 2)$; eccentricity $\frac{5}{4}$

46. foci at $(9, 2)$, $(-11, 2)$; eccentricity $\frac{25}{9}$

Determine the two equations necessary to graph each hyperbola with a graphing calculator, and graph it in the viewing window indicated. See Example 1.

47. $\frac{x^2}{4} - \frac{y^2}{16} = 1$; $[-9.4, 9.4]$ by $[-10, 10]$

48. $\frac{x^2}{25} - \frac{y^2}{49} = 1$; $[-9.4, 9.4]$ by $[-10, 10]$

49. $4y^2 - 36x^2 = 144$; $[-10, 10]$ by $[-15, 15]$

50. $y^2 - 9x^2 = 9$; $[-10, 10]$ by $[-10, 10]$

Relating Concepts

For individual or collaborative investigation

(Exercises 51–56)

The graph of $\frac{x^2}{4} - y^2 = 1$ is a hyperbola. We know that the graph of this hyperbola approaches its asymptotes as $|x|$ gets larger and larger. ***Work Exercises 51–56 in order,*** *to see the relationship between the hyperbola and one of its asymptotes.*

51. Solve $\frac{x^2}{4} - y^2 = 1$ for y, and choose the positive square root.

52. Find the equation of the asymptote with positive slope.

53. Use a calculator to evaluate the y-coordinate of the point where $x = 50$ on the graph of the portion of the hyperbola represented by the equation obtained in Exercise 51. Round your answer to the nearest hundredth.

54. Find the y-coordinate of the point where $x = 50$ on the graph of the asymptote found in Exercise 52.

55. Compare your results in Exercises 53 and 54. How do they support the following statement?

> When $x = 50$, the graph of the function defined by the equation found in Exercise 51 lies *below* the graph of the asymptote found in Exercise 52.

56. What do you think will happen if we choose x-values larger than 50?

Solve each problem.

57. ***(Modeling) Atomic Structure*** In 1911 Ernest Rutherford discovered the basic structure of the atom by "shooting" positively charged alpha particles with a speed of 10^7 m per sec at a piece of gold foil 6×10^{-7} m thick. Only a small percentage of the alpha particles struck a gold nucleus head-on and were deflected directly back toward their source. The rest of the particles often followed a hyperbolic trajectory because they were repelled by positively charged gold nuclei. As a result of this famous experiment, Rutherford proposed that the atom was composed of mostly empty space with a small and dense nucleus. The figure shows an alpha particle A initially approaching a gold nucleus N and being deflected at an angle $\theta = 90°$. N is located at a focus of the hyperbola, and the trajectory of A passes through a vertex of the hyperbola. (*Source:* Semat, H. and J. Albright, *Introduction to Atomic and Nuclear Physics,* Fifth Edition, International Thomson Publishing, 1972.)

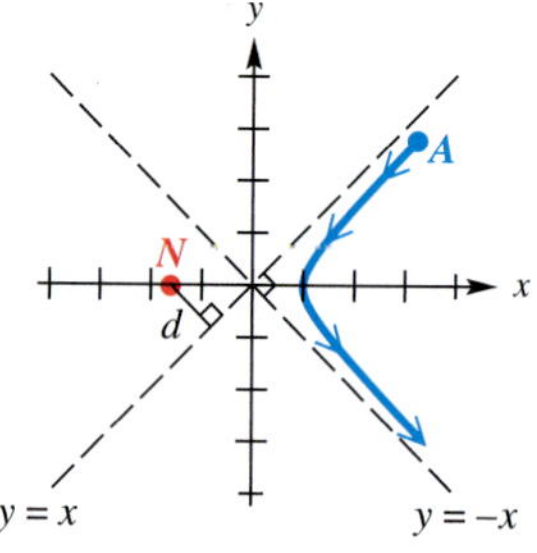

(a) Determine the equation of the trajectory of the alpha particle if $d = 5 \times 10^{-14}$ m.

(b) What was the minimum distance between the centers of the alpha particle and the gold nucleus?

58. ***(Modeling) Design of a Sports Complex*** Two buildings in a sports complex are shaped and positioned like a portion of the branches of the hyperbola

$$400x^2 - 625y^2 = 250{,}000,$$

where x and y are in meters.

(a) How far apart are the buildings at their closest point?

(b) Find the distance d in the figure.

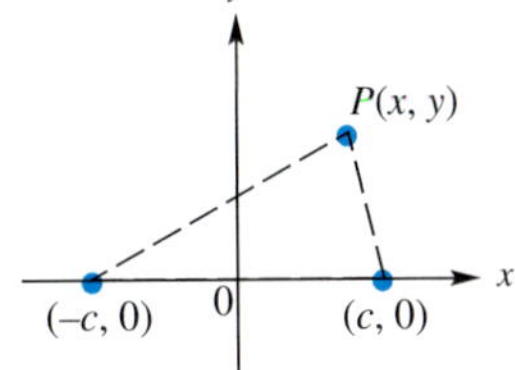

59. ***Sound Detection*** Microphones are placed at points $(-c, 0)$ and $(c, 0)$. An explosion occurs at point $P(x, y)$ having positive x-coordinate. See the figure. The sound is detected at the closer microphone t seconds before being detected at the farther microphone. Assume that sound travels at a speed of 330 m per sec, and show that P must be on the hyperbola

$$\frac{x^2}{330^2t^2} - \frac{y^2}{4c^2 - 330^2t^2} = \frac{1}{4}.$$

60. Suppose a hyperbola has center at the origin, foci at $F'(-c, 0)$ and $F(c, 0)$, and $|d(P, F') - d(P, F)| = 2a$. Let $b^2 = c^2 - a^2$, and show that an equation of the hyperbola is

$$\frac{x^2}{a^2} - \frac{y^2}{b^2} = 1.$$

10.4 Summary of the Conic Sections

Characteristics ▪ Identifying Conic Sections ▪ Geometric Definition of Conic Sections

Characteristics The graphs of parabolas, circles, ellipses, and hyperbolas are called conic sections since each graph can be obtained by cutting a cone with a plane, as suggested by Figure 1 at the beginning of the chapter. All conic sections of the types presented in this chapter have equations of the form

$$Ax^2 + Cy^2 + Dx + Ey + F = 0,$$

where either A or C must be nonzero.

The special characteristics of the equation of each conic section presented earlier are summarized below.

Conic Section	Characteristic	Example
Parabola	Either $A = 0$ or $C = 0$, but not both.	$x^2 = y + 4$ $(y - 2)^2 = -(x + 3)$
Circle	$A = C \neq 0$	$x^2 + y^2 = 16$
Ellipse	$A \neq C, AC > 0$	$\frac{x^2}{16} + \frac{y^2}{25} = 1$
Hyperbola	$AC < 0$	$x^2 - y^2 = 1$

The following chart summarizes our work with conic sections.

Equation	Graph	Description	Identification
$y - k = a(x - h)^2$	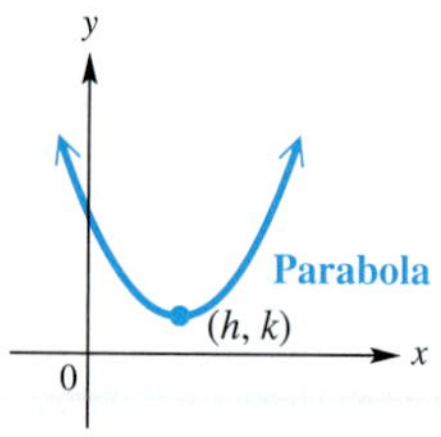	Opens up if $a > 0$, down if $a < 0$. Vertex is (h, k).	x^2-term y is not squared.
$x - h = a(y - k)^2$	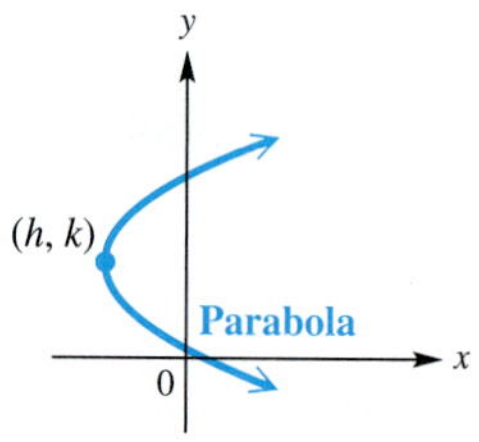	Opens to the right if $a > 0$, to the left if $a < 0$. Vertex is (h, k).	y^2-term x is not squared.
$(x - h)^2 + (y - k)^2 = r^2$	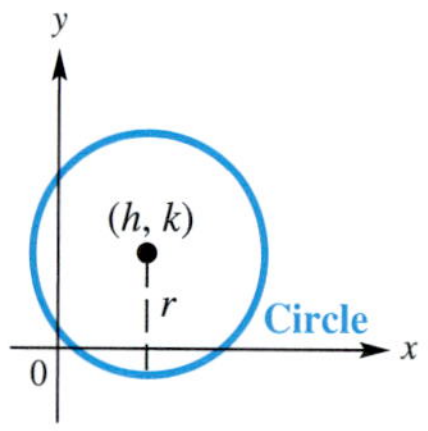	Center is (h, k), radius is r.	x^2- and y^2-terms have the same positive coefficient.
$\frac{x^2}{a^2} + \frac{y^2}{b^2} = 1 \quad (a > b)$	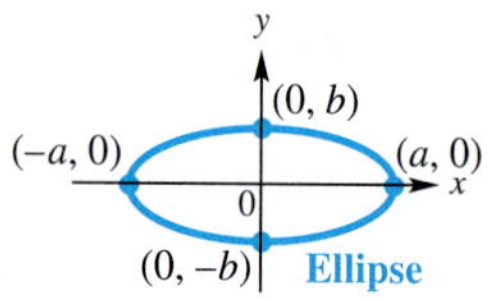	x-intercepts are a and $-a$. y-intercepts are b and $-b$.	x^2- and y^2-terms have different positive coefficients.
$\frac{x^2}{b^2} + \frac{y^2}{a^2} = 1 \quad (a > b)$	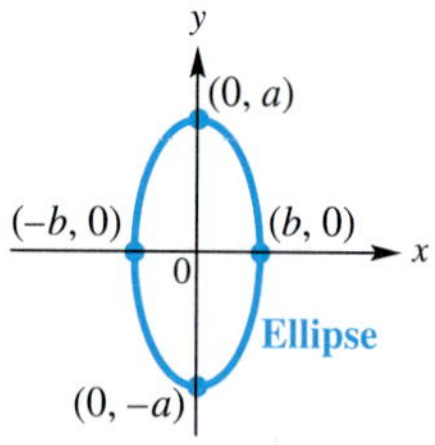	x-intercepts are b and $-b$. y-intercepts are a and $-a$.	x^2- and y^2-terms have different positive coefficients.
$\frac{x^2}{a^2} - \frac{y^2}{b^2} = 1$	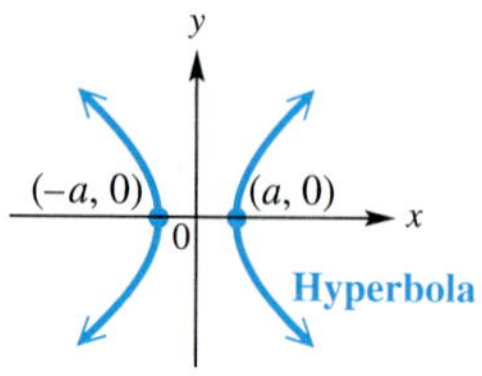	x-intercepts are a and $-a$. Asymptotes are found from (a, b), $(a, -b)$, $(-a, -b)$, and $(-a, b)$.	x^2-term has a positive coefficient. y^2-term has a negative coefficient.

Equation	Graph	Description	Identification
$\dfrac{y^2}{a^2} - \dfrac{x^2}{b^2} = 1$	y, (0, a), x, 0, (0, −a), **Hyperbola**	y-intercepts are a and $-a$. Asymptotes are found from (b, a), $(b, -a)$, $(-b, -a)$, and $(-b, a)$.	y^2-term has a positive coefficient. x^2-term has a negative coefficient.

Now try Exercises 1, 5, 7, and 11.

Identifying Conic Sections To recognize the type of graph that a given conic section has, we sometimes need to transform the equation into a more familiar form.

EXAMPLE 1 Determining Types of Conic Sections from Equations

Determine the type of conic section represented by each equation, and graph it.

(a) $x^2 = 25 + 5y^2$ **(b)** $x^2 - 8x + y^2 + 10y = -41$

(c) $4x^2 - 16x + 9y^2 + 54y = -61$ **(d)** $x^2 - 6x + 8y - 7 = 0$

Solution

(a)

$$x^2 = 25 + 5y^2$$

$$x^2 - 5y^2 = 25 \qquad \text{Subtract } 5y^2.$$

$$\frac{x^2}{25} - \frac{y^2}{5} = 1 \qquad \text{Divide by 25.}$$

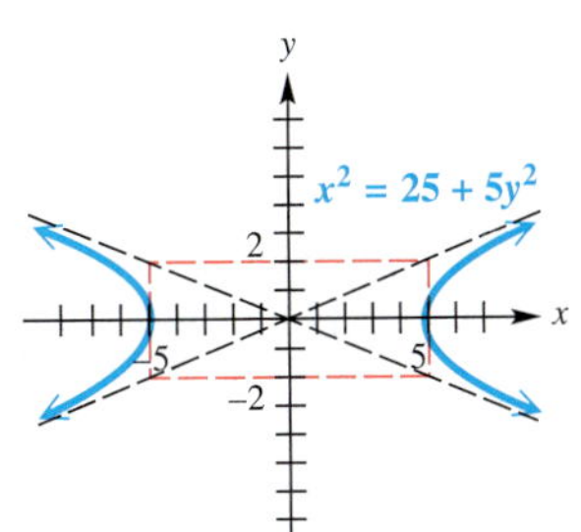

Figure 34

The equation represents a hyperbola centered at the origin, with asymptotes

$$\frac{x^2}{25} - \frac{y^2}{5} = 0, \quad \text{or} \quad y = \pm\frac{\sqrt{5}}{5}x.$$

The x-intercepts are ± 5; the graph is shown in Figure 34.

(b)

$$x^2 - 8x + y^2 + 10y = -41$$

$$(x^2 - 8x + 16 - 16) + (y^2 + 10y + 25 - 25) = -41$$

Complete the square on both x and y. (Section 1.4)

$$(x^2 - 8x + 16) - 16 + (y^2 + 10y + 25) - 25 = -41$$

Regroup terms.

$$(x - 4)^2 + (y + 5)^2 = -41 + 16 + 25$$

Factor (Section R.4); add 16 and 25.

$$(x - 4)^2 + (y + 5)^2 = 0$$

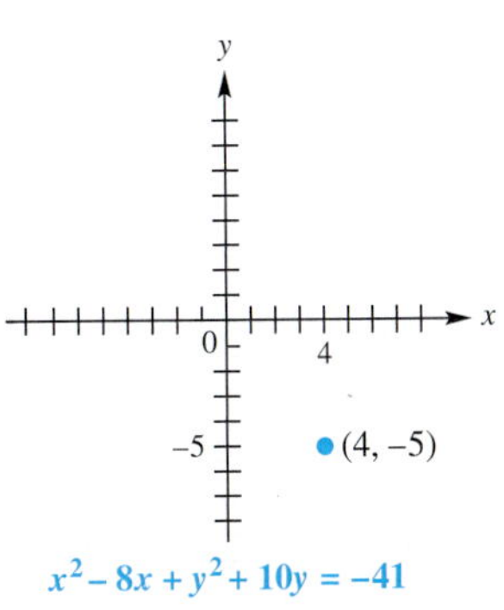

Figure 35

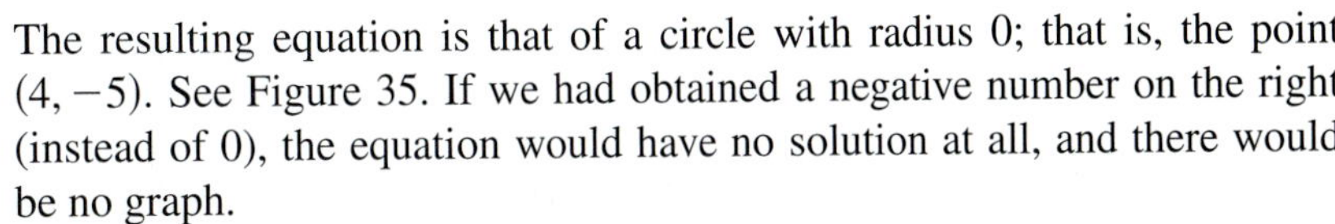

The resulting equation is that of a circle with radius 0; that is, the point $(4, -5)$. See Figure 35. If we had obtained a negative number on the right (instead of 0), the equation would have no solution at all, and there would be no graph.

(c) In $4x^2 - 16x + 9y^2 + 54y = -61$, the coefficients of the x^2- and y^2-terms are unequal and both positive, so the equation might represent an ellipse but not a circle. (It might also represent a single point or no points at all.)

$$4x^2 - 16x + 9y^2 + 54y = -61$$

$$4(x^2 - 4x \qquad) + 9(y^2 + 6y \qquad) = -61 \quad \text{Factor out 4; factor out 9.}$$

$$4(x^2 - 4x + 4 - 4) + 9(y^2 + 6y + 9 - 9) = -61 \quad \text{Complete the square.}$$

$$4(x^2 - 4x + 4) - 16 + 9(y^2 + 6y + 9) - 81 = -61 \quad \text{Distributive property (Section R.1)}$$

$$4(x - 2)^2 + 9(y + 3)^2 = 36 \quad \text{Factor; add 97.}$$

$$\frac{(x - 2)^2}{9} + \frac{(y + 3)^2}{4} = 1 \quad \text{Divide by 36.}$$

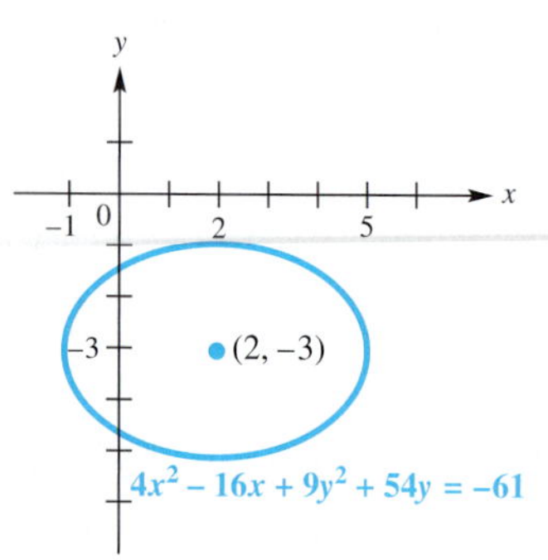

Figure 36

This equation represents an ellipse having center $(2, -3)$ and graph as shown in Figure 36.

(d) Since only one variable in $x^2 - 6x + 8y - 7 = 0$ is squared (x, and not y), the equation represents a parabola. Get the term with y (the variable that is not squared) alone on one side.

$$x^2 - 6x + 8y - 7 = 0$$

$$8y = -x^2 + 6x + 7 \quad \text{Isolate the } y\text{-term.}$$

$$8y = -(x^2 - 6x \qquad) + 7 \quad \text{Regroup terms; factor out } -1.$$

$$8y = -(x^2 - 6x + 9 - 9) + 7 \quad \text{Complete the square.}$$

$$8y = -(x^2 - 6x + 9) + 9 + 7 \quad \text{Distributive property; } -(-9) = +9$$

$$8y = -(x - 3)^2 + 16 \quad \text{Factor; add.}$$

$$y = -\frac{1}{8}(x - 3)^2 + 2 \quad \text{Multiply by } \tfrac{1}{8}.$$

$$y - 2 = -\frac{1}{8}(x - 3)^2 \quad \text{Subtract 2.}$$

The parabola has vertex $(3, 2)$ and opens down, as shown in the graph in Figure 37.

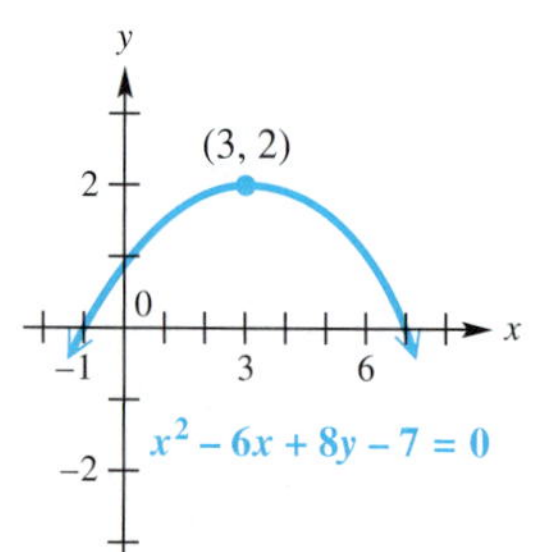

Figure 37

Now try Exercises 17, 29, 33, and 35.

The next example is designed to serve as a warning about a common error.

EXAMPLE 2 Determining the Type of Conic Section from Its Equation

Identify the graph of $4y^2 - 16y - 9x^2 + 18x = -43$.

Solution

$$\begin{aligned}
4y^2 - 16y - 9x^2 + 18x &= -43 \\
4(y^2 - 4y \qquad) - 9(x^2 - 2x \qquad) &= -43 && \text{Factor out 4; factor out } -9. \\
4(y^2 - 4y + 4 - 4) - 9(x^2 - 2x + 1 - 1) &= -43 && \text{Complete the square.} \\
4(y^2 - 4y + 4) - 16 - 9(x^2 - 2x + 1) + 9 &= -43 && \text{Distributive property} \\
4(y-2)^2 - 9(x-1)^2 &= -36 && \text{Factor; add 16 and subtract 9.}
\end{aligned}$$

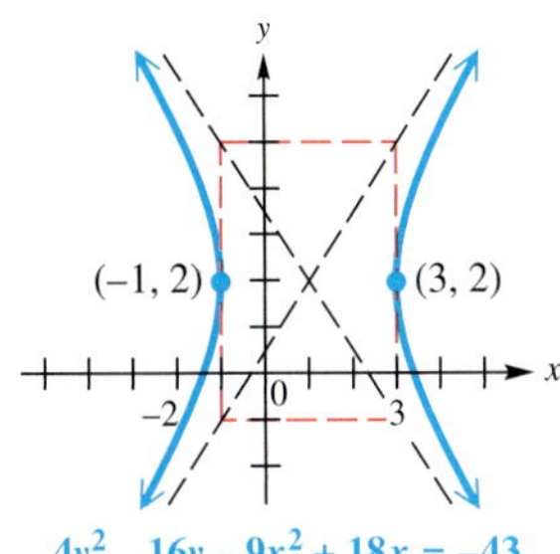

Figure 38

Because of the -36, we might think that this equation does not have a graph. However, the minus sign in the middle on the left shows that the graph is that of a hyperbola.

$$\frac{(x-1)^2}{4} - \frac{(y-2)^2}{9} = 1 \quad \text{Divide by } -36\text{; rearrange terms.}$$

This hyperbola has center $(1, 2)$. The graph is shown in Figure 38.

Now try Exercise 31.

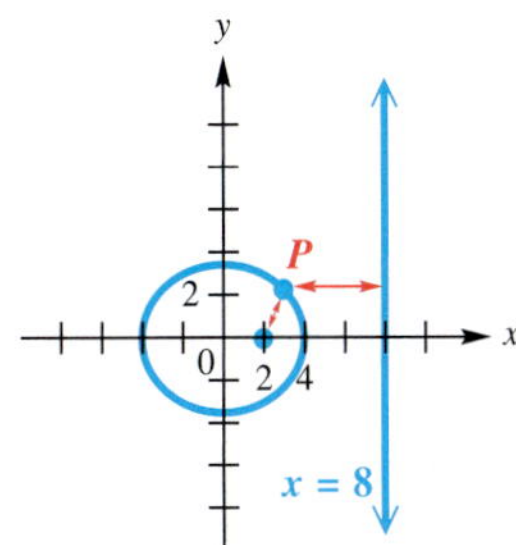

Figure 39

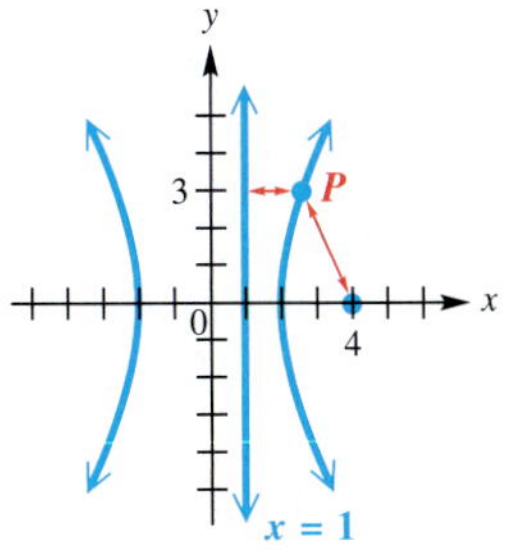

Figure 40

Geometric Definition of Conic Sections In the first section of this chapter, a parabola was defined as the set of points in a plane equidistant from a fixed point (focus) and a fixed line (directrix). A parabola has eccentricity 1. This definition can be generalized to apply to the ellipse and the hyperbola. Figure 39 shows an ellipse with $a = 4$, $c = 2$, and $e = \frac{1}{2}$. The line $x = 8$ is shown also. For any point P on the ellipse,

$$[\text{distance of } P \text{ from the focus}] = \frac{1}{2}[\text{distance of } P \text{ from the line}].$$

Figure 40 shows a hyperbola with $a = 2$, $c = 4$, and $e = 2$, along with the line $x = 1$. For any point P on the hyperbola,

$$[\text{distance of } P \text{ from the focus}] = 2[\text{distance of } P \text{ from the line}].$$

The following geometric definition applies to all conic sections except circles, which have $e = 0$.

Geometric Definition of a Conic Section

Given a fixed point F (focus), a fixed line L (directrix), and a positive number e, the set of all points P in the plane such that

$$[\text{distance of } P \text{ from } F] = e \cdot [\text{distance of } P \text{ from } L]$$

is a conic section of eccentricity e. The conic section is a parabola when $e = 1$, an ellipse when $0 < e < 1$, and a hyperbola when $e > 1$.

Now try Exercise 51.

10.4 Exercises

The equation of a conic section is given in a familiar form. Identify the type of graph that each equation has, without actually graphing. See the summary chart in this section.

1. $x^2 + y^2 = 144$

2. $(x - 2)^2 + (y + 3)^2 = 25$

3. $y = 2x^2 + 3x - 4$

4. $x = 3y^2 + 5y - 6$

5. $x - 1 = -3(y - 4)^2$

6. $\frac{x^2}{25} + \frac{y^2}{36} = 1$

7. $\frac{x^2}{49} + \frac{y^2}{100} = 1$

8. $x^2 - y^2 = 1$

9. $\frac{x^2}{4} - \frac{y^2}{16} = 1$

10. $\frac{(x + 2)^2}{9} + \frac{(y - 4)^2}{16} = 1$

11. $\frac{x^2}{25} - \frac{y^2}{25} = 1$

12. $y + 7 = 4(x + 3)^2$

For each equation that has a graph, identify the type of graph. It may be necessary to transform the equation. See Examples 1 and 2.

13. $\frac{x^2}{4} = 1 - \frac{y^2}{9}$

14. $\frac{x^2}{4} = 1 + \frac{y^2}{9}$

15. $\frac{x^2}{4} + \frac{y^2}{4} = 1$

16. $\frac{x^2}{4} + \frac{y^2}{4} = -1$

17. $x^2 = 25 + y^2$

18. $x^2 = 25 - y^2$

19. $9x^2 + 36y^2 = 36$

20. $x^2 = 4y - 8$

21. $\frac{(x + 3)^2}{16} + \frac{(y - 2)^2}{16} = 1$

22. $\frac{(x - 4)^2}{8} + \frac{(y + 1)^2}{2} = 0$

23. $y^2 - 4y = x + 4$

24. $11 - 3x = 2y^2 - 8y$

25. $(x + 7)^2 + (y - 5)^2 + 4 = 0$

26. $4(x - 3)^2 + 3(y + 4)^2 = 0$

27. $3x^2 + 6x + 3y^2 - 12y = 12$

28. $2x^2 - 8x + 2y^2 + 20y = 12$

29. $x^2 - 6x + y = 0$

30. $x - 4y^2 - 8y = 0$

31. $-4x^2 + 8x + y^2 + 6y = -6$

32. $x^2 + 2x = -4y$

33. $4x^2 - 8x + 9y^2 + 54y = -84$

34. $3x^2 + 12x + 3y^2 = -11$

35. $4x^2 - 24x + 5y^2 + 10y + 41 = 0$

36. $6x^2 - 12x + 6y^2 - 18y + 25 = 0$

37. Identify the type of conic section consisting of the set of all points in the plane for which the sum of the distances from the points $(5, 0)$ and $(-5, 0)$ is 14.

38. Identify the type of conic section consisting of the set of all points in the plane for which the absolute value of the difference of the distances from the points $(3, 0)$ and $(-3, 0)$ is 2.

39. Identify the type of conic section consisting of the set of all points in the plane for which the distance from the point $(3, 0)$ is one and one-half times the distance from the line $x = \frac{4}{3}$.

40. Identify the type of conic section consisting of the set of all points in the plane for which the distance from the point $(2, 0)$ is one-third of the distance from the line $x = 10$.

Find the eccentricity of each conic section. The point shown on the x-axis is a focus and the line shown is a directrix.

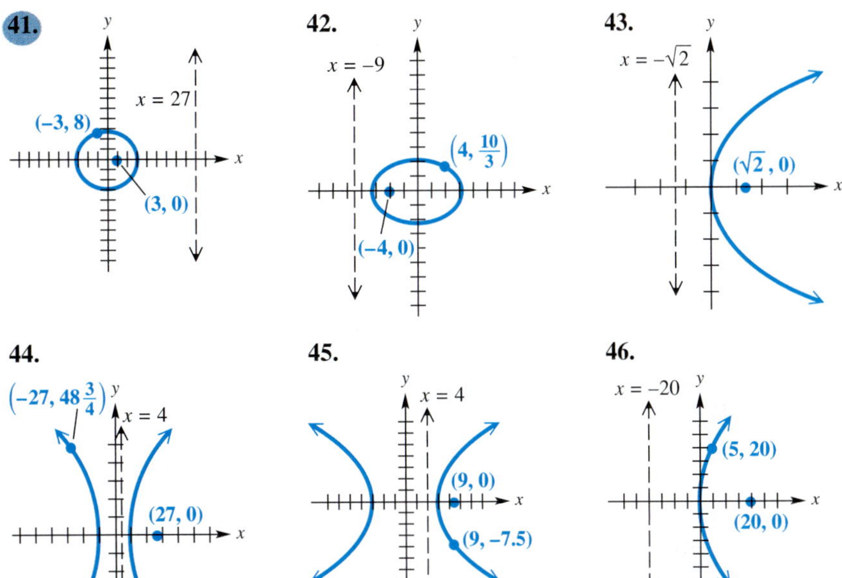

Satellite Trajectory *When a satellite is near Earth, its orbital trajectory may trace out a hyperbola, a parabola, or an ellipse. The type of trajectory depends on the satellite's velocity V in meters per second. It will be hyperbolic if* $V > \frac{k}{\sqrt{D}}$, *parabolic if* $V = \frac{k}{\sqrt{D}}$, *and elliptical if* $V < \frac{k}{\sqrt{D}}$, *where* $k = 2.82 \times 10^7$ *is a constant and D is the distance in meters from the satellite to the center of Earth. Use this information in Exercises 47–49. (Source:* Loh, W., *Dynamics and Thermodynamics of Planetary Entry,* Prentice-Hall, 1963; Thomson, W., *Introduction to Space Dynamics,* John Wiley & Sons, 1961.*)*

47. When the artificial satellite Explorer IV was at a maximum distance D of 42.5×10^6 m from Earth's center, it had a velocity V of 2090 m per sec. Determine the shape of its trajectory.

48. If a satellite is scheduled to leave Earth's gravitational influence, its velocity must be increased so that its trajectory changes from elliptical to hyperbolic. Determine the minimum increase in velocity necessary for Explorer IV to escape Earth's gravitational influence when $D = 42.5 \times 10^6$ m.

49. Explain why it is easier to change a satellite's trajectory from an ellipse to a hyperbola when D is maximum rather than minimum.

50. If $Ax^2 + Cy^2 + Dx + Ey + F = 0$ is the general equation of an ellipse, find its center point by completing the square.

51. Graph the ellipse $\frac{x^2}{16} + \frac{y^2}{12} = 1$ with a graphing calculator. Trace to find the coordinates of several points on the ellipse. For each of these points P, verify that

$$[\text{distance of } P \text{ from } (2,0)] = \frac{1}{2}[\text{distance of } P \text{ from the line } x = 8].$$

52. Graph the hyperbola $\frac{x^2}{4} - \frac{y^2}{12} = 1$ with a graphing calculator. Trace to find the coordinates of several points on the hyperbola. For each of these points P, verify that

$$[\text{distance of } P \text{ from } (4,0)] = 2[\text{distance of } P \text{ from the line } x = 1].$$

Chapter 10 Summary

KEY TERMS

10.1 conic sections, parabola, focus, directrix

10.2 ellipse, foci, major axis, minor axis, center, vertices, conic, eccentricity

10.3 hyperbola, transverse axis, asymptotes, fundamental rectangle

NEW SYMBOLS

e eccentricity

QUICK REVIEW

CONCEPTS | **EXAMPLES**

10.1 Parabolas

Parabola with Vertical Axis and Vertex (0, 0)
The parabola with focus $(0, p)$ and directrix $y = -p$ has equation $\boldsymbol{x^2 = 4py}$. The parabola has vertical axis $x = 0$ and opens up if $p > 0$ or down if $p < 0$.

Parabola with Horizontal Axis and Vertex (0, 0)
The parabola with focus $(p, 0)$ and directrix $x = -p$ has equation $\boldsymbol{y^2 = 4px}$. The parabola has horizontal axis $y = 0$ and opens to the right if $p > 0$ or to the left if $p < 0$.

Equation Forms for Translated Parabolas
A parabola with vertex (h, k) has an equation of the form

$$(x - h)^2 = 4p(y - k) \qquad \text{Vertical axis}$$

or

$$(y - k)^2 = 4p(x - h), \qquad \text{Horizontal axis}$$

where the focus is distance p or $-p$ from the vertex.

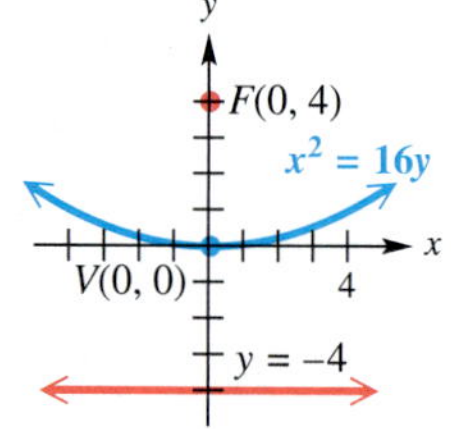

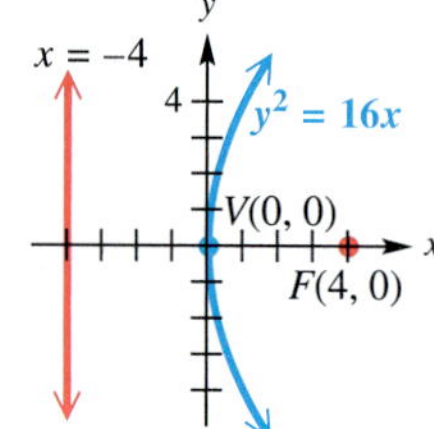

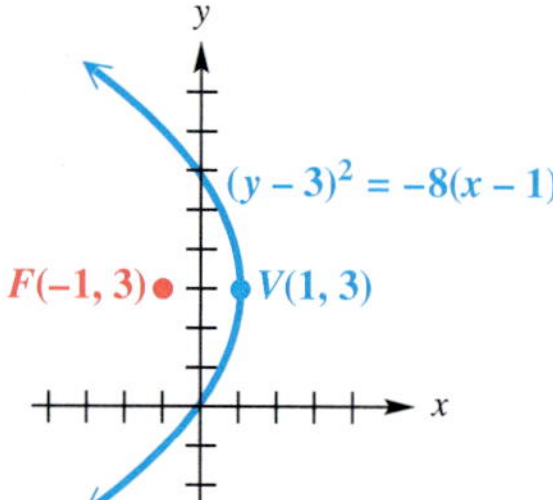

10.2 Ellipses

Standard Forms of Equations for Ellipses
The ellipse with center at the origin and equation

$$\frac{x^2}{a^2} + \frac{y^2}{b^2} = 1 \qquad (a > b)$$

has vertices $(\pm a, 0)$, endpoints of the minor axis $(0, \pm b)$, and foci $(\pm c, 0)$, where $c^2 = a^2 - b^2$.

The ellipse with center at the origin and equation

$$\frac{x^2}{b^2} + \frac{y^2}{a^2} = 1 \qquad (a > b)$$

has vertices $(0, \pm a)$, endpoints of the minor axis $(\pm b, 0)$, and foci $(0, \pm c)$, where $c^2 = a^2 - b^2$.

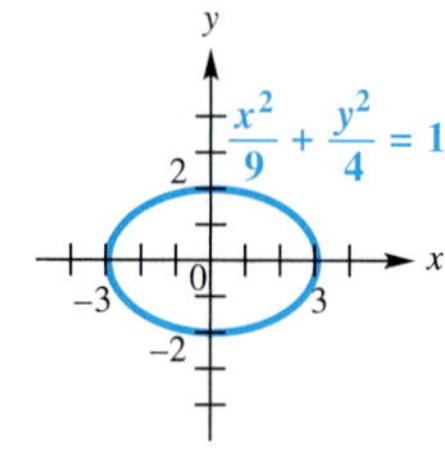

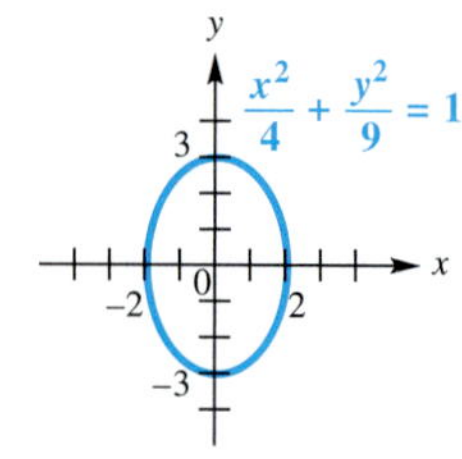

CONCEPTS	EXAMPLES

Translated Ellipses

The preceding equations can be extended to ellipses having center (h, k) by replacing x and y with $x - h$ and $y - k$, respectively.

10.3 Hyperbolas

Standard Forms of Equations for Hyperbolas

The hyperbola with center at the origin and equation

$$\frac{x^2}{a^2} - \frac{y^2}{b^2} = 1$$

has vertices $(\pm a, 0)$, asymptotes $y = \pm \frac{b}{a}x$, and foci $(\pm c, 0)$, where $c^2 = a^2 + b^2$.

The hyperbola with center at the origin and equation

$$\frac{y^2}{a^2} - \frac{x^2}{b^2} = 1$$

has vertices $(0, \pm a)$, asymptotes $y = \pm \frac{a}{b}x$, and foci $(0, \pm c)$, where $c^2 = a^2 + b^2$.

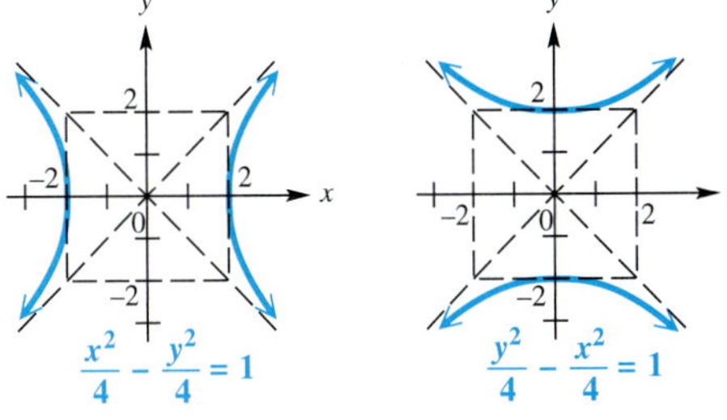

Translated Hyperbolas

The preceding equations can be extended to hyperbolas having center (h, k) by replacing x and y with $x - h$ and $y - k$, respectively.

10.4 Summary of the Conic Sections

Conic sections in this chapter have equations that can be written in the form

$$Ax^2 + Cy^2 + Dx + Ey + F = 0.$$

Conic Section	Characteristic	Example
Parabola	Either $A = 0$ or $C = 0$, but not both.	$y = x^2$ $x = 3y^2 + 2y - 4$
Circle	$A = C \neq 0$	$x^2 + y^2 = 16$
Ellipse	$A \neq C, AC > 0$	$\frac{x^2}{16} + \frac{y^2}{25} = 1$
Hyperbola	$AC < 0$	$x^2 - y^2 = 1$

See the summary chart on pages 922 and 923.

Chapter 10 Review Exercises

Graph each parabola. In Exercises 1–4, give the domain, range, vertex, and axis. In Exercises 5–8, give the domain, range, focus, directrix, and axis.

1. $x = 4(y - 5)^2 + 2$

2. $x = -(y + 1)^2 - 7$

3. $x = 5y^2 - 5y + 3$

4. $x = 2y^2 - 4y + 1$

5. $y^2 = -\frac{2}{3}x$

6. $y^2 = 2x$

7. $3x^2 = y$

8. $x^2 + 2y = 0$

Write an equation for each parabola with vertex at the origin.

9. focus $(4, 0)$

10. focus $(0, -3)$

11. through $(-3, 4)$, opening up

12. through $(2, 5)$, opening right

An equation of a conic section is given. Identify the type of conic section. It may be necessary to transform the equation into a more familiar form.

13. $y^2 + 9x^2 = 9$

14. $9x^2 - 16y^2 = 144$

15. $3y^2 - 5x^2 = 30$

16. $y^2 + x = 4$

17. $4x^2 - y = 0$

18. $x^2 + y^2 = 25$

19. $4x^2 - 8x + 9y^2 + 36y = -4$

20. $9x^2 - 18x - 4y^2 - 16y - 43 = 0$

Concept Check *Match each equation with its calculator graph. In all cases except choice B,* Xscl = Yscl = 1.

21. $4x^2 + y^2 = 36$

22. $x = 2y^2 + 3$

23. $(x - 2)^2 + (y + 3)^2 = 36$

24. $\frac{x^2}{36} + \frac{y^2}{9} = 1$

25. $(y - 1)^2 - (x - 2)^2 = 36$

26. $y^2 = 36 + 4x^2$

A.

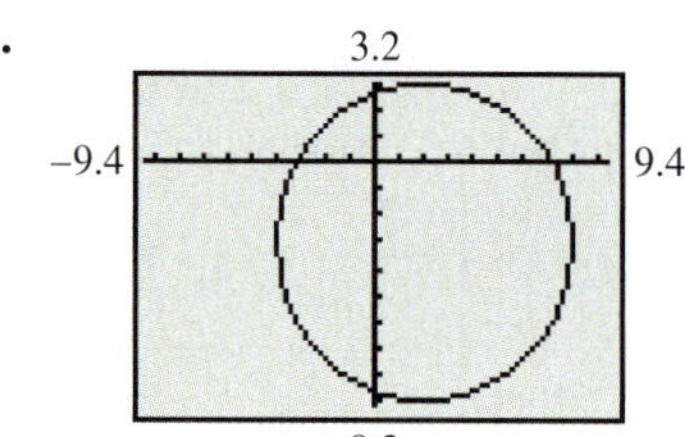

B.

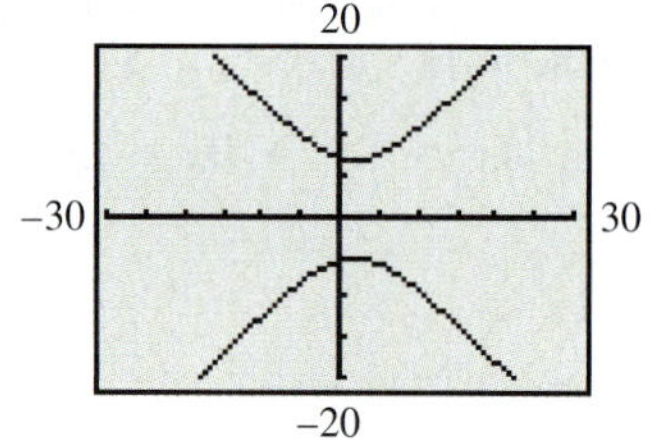

In this screen, Xscl = Yscl = 5.

C.

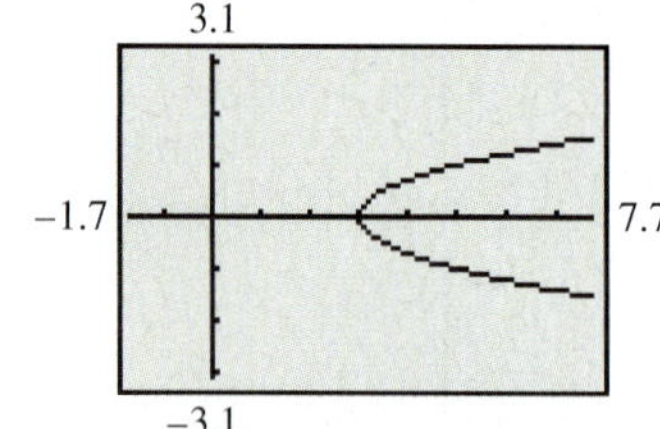

D.

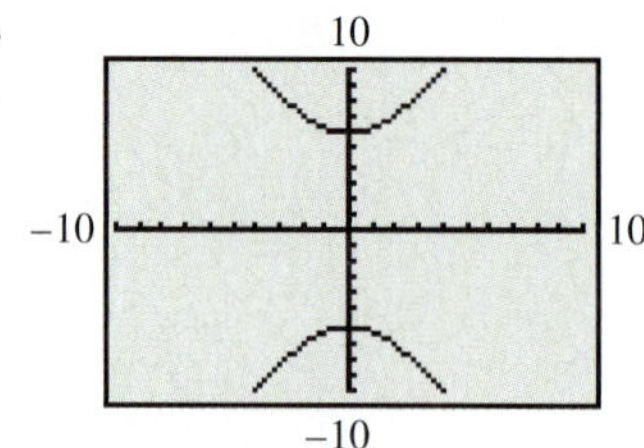

E.

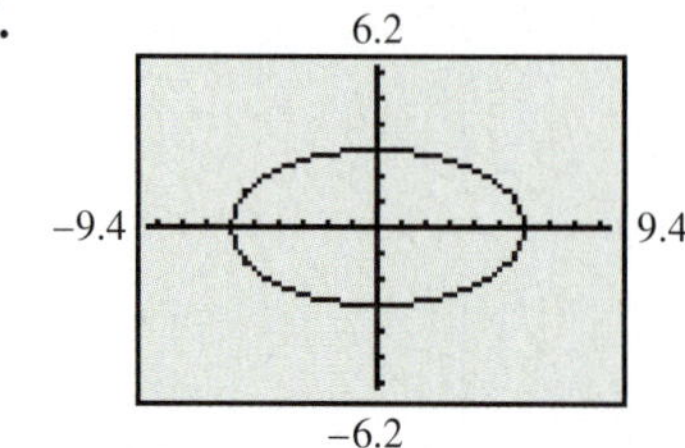

F.

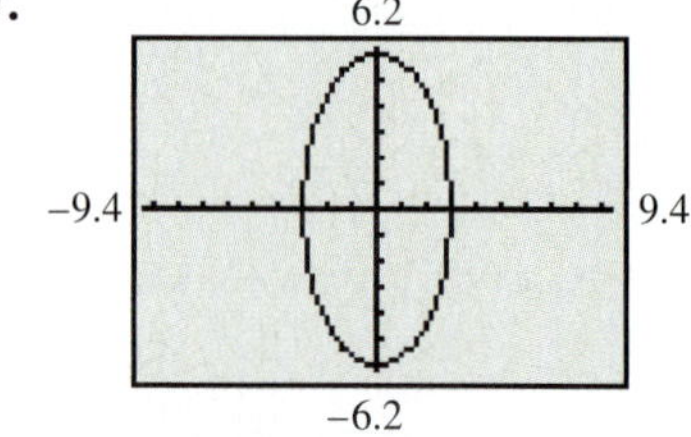

Graph each relation and identify the graph. Give the domain, range, coordinates of the vertices for each ellipse or hyperbola, and equations of the asymptotes for each hyperbola. Give the domain and range for each circle.

27. $\frac{x^2}{4} + \frac{y^2}{9} = 1$

28. $\frac{x^2}{16} + \frac{y^2}{4} = 1$

29. $\frac{x^2}{64} - \frac{y^2}{36} = 1$

30. $\frac{y^2}{25} - \frac{x^2}{9} = 1$

31. $\frac{(x+1)^2}{16} + \frac{(y-1)^2}{16} = 1$

32. $(x-3)^2 + (y+2)^2 = 9$

33. $4x^2 + 9y^2 = 36$

34. $x^2 = 16 + y^2$

35. $\frac{(x-3)^2}{4} + (y+1)^2 = 1$

36. $\frac{(x-2)^2}{9} + \frac{(y+3)^2}{4} = 1$

37. $\frac{(y+2)^2}{4} - \frac{(x+3)^2}{9} = 1$

38. $\frac{(x+1)^2}{16} - \frac{(y-2)^2}{4} = 1$

Graph each relation. Give the domain and range, and identify any that are graphs of functions.

39. $\frac{x}{3} = -\sqrt{1 - \frac{y^2}{16}}$

40. $x = -\sqrt{1 - \frac{y^2}{36}}$

41. $y = -\sqrt{1 + x^2}$

42. $y = -\sqrt{1 - \frac{x^2}{25}}$

Write an equation for each conic section with center at the origin.

43. ellipse; vertex at $(0, 4)$, focus at $(0, 2)$

44. ellipse; x-intercept 6, focus at $(-2, 0)$

45. hyperbola; focus at $(0, -5)$, transverse axis of length 8

46. hyperbola; y-intercept -2, passing through $(2, 3)$

Write an equation for each conic section satisfying the given conditions.

47. parabola with focus at $(3, 2)$ and directrix $x = -3$

48. parabola with vertex at $(-3, 2)$ and y-intercepts 5 and -1

49. ellipse with foci at $(-2, 0)$ and $(2, 0)$ and major axis of length 10

50. ellipse with foci at $(0, 3)$ and $(0, -3)$ and vertex at $(0, 7)$

51. hyperbola with x-intercepts ± 3 and foci at $(-5, 0)$, $(5, 0)$

52. hyperbola with foci at $(0, 12)$, $(0, -12)$ and asymptotes $y = \pm x$

53. Find the equation of the ellipse consisting of all points in the plane the sum of whose distances from $(0, 0)$ and $(4, 0)$ is 8.

54. Find the equation of the hyperbola consisting of all points in the plane for which the absolute value of the difference of the distances from $(0, 0)$ and $(0, 4)$ is 2.

55. Calculator graphs are shown in Figures A–D. Arrange the figures in order so that the first in the list has the smallest eccentricity and the rest have eccentricities in increasing order.

A.

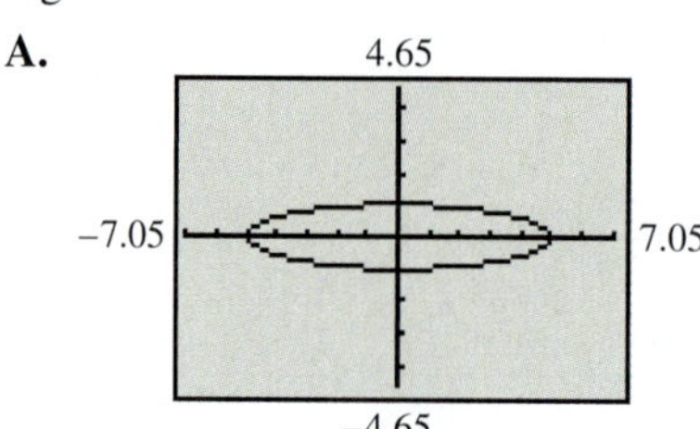

B.

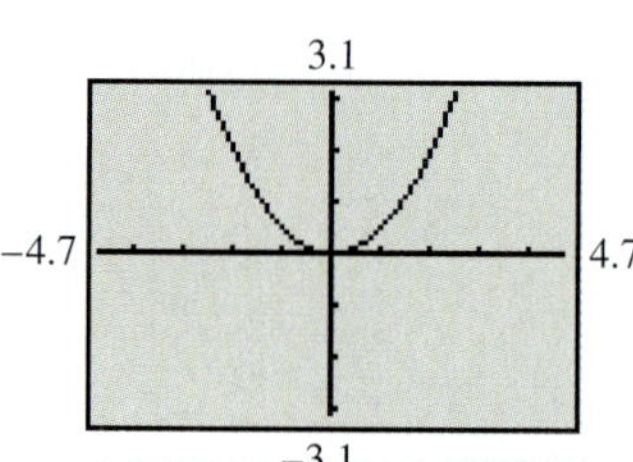

C.

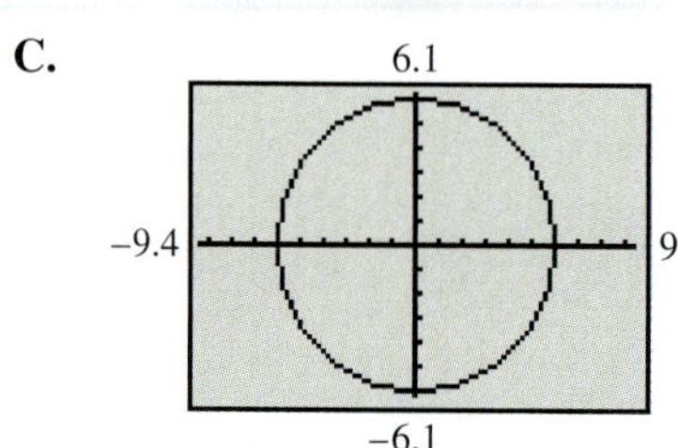

D.

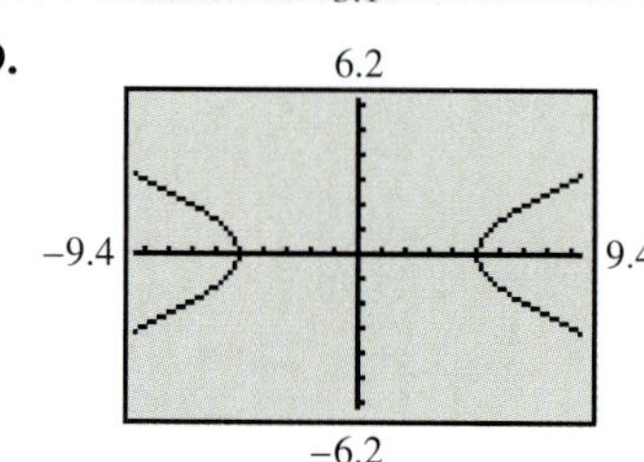

56. *Orbit of Venus* The orbit of Venus is an ellipse with the sun at one of the foci. The eccentricity of the orbit is $e = .006775$, and the major axis has length 134.5 million mi. (*Source: The World Almanac and Book of Facts.*) Find the least and greatest distances of Venus from the sun.

57. *Orbit of a Comet* The comet Swift-Tuttle has an elliptical orbit of eccentricity $e = .964$, with the sun at one of the foci. Find the equation of the comet given that the closest it comes to the sun is 89 million mi.

58. Find the equation of the hyperbola consisting of all points P in the plane for which the absolute value of the difference of the distances of P from $(-5, 0)$ and $(5, 0)$ is 8. Then graph the hyperbola with a graphing calculator and trace to find the coordinates of several points on the graph of the hyperbola. For each of these points, verify that the absolute value of the differences of the distances is indeed 8.

Chapter 10 Test

Graph each parabola. Give the domain, range, vertex, and axis.

1. $y = -x^2 + 6x$

2. $x = 4y^2 + 8y$

3. Give the coordinates of the focus and the equation of the directrix for the parabola with equation $x = 8y^2$.

4. Write an equation for the parabola with vertex $(2, 3)$, passing through the point $(-18, 1)$, and opening to the left.

5. Explain how to determine just by looking at the equation whether a parabola has a vertical or a horizontal axis, and whether it opens up, down, to the left, or to the right.

Graph each ellipse. Give the domain and range.

6. $\dfrac{(x-8)^2}{100}+\dfrac{(y-5)^2}{49}=1$

7. $16x^2+4y^2=64$

8. Graph $y=-\sqrt{1-\dfrac{x^2}{36}}$. Tell whether the graph is that of a function.

9. Write an equation for the ellipse centered at the origin having horizontal major axis with length 6 and minor axis with length 4.

10. ***Height of the Arch of a Bridge*** An arch of a bridge has the shape of the top half of an ellipse. The arch is 40 ft wide and 12 ft high at the center. Find the equation of the complete ellipse. Find the height of the arch 10 ft from the center of the bottom.

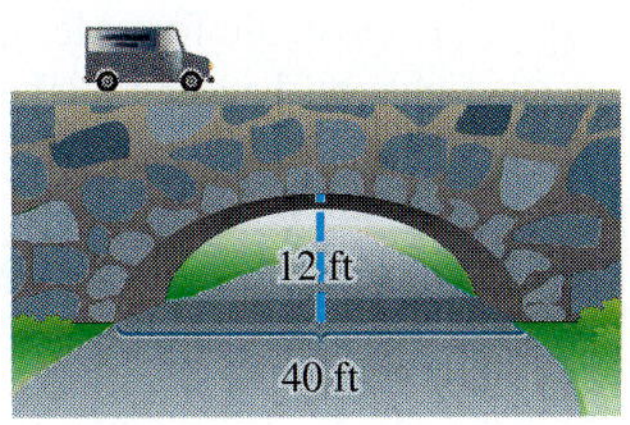

Graph each hyperbola. Give the domain, range, and equations of the asymptotes.

11. $\dfrac{x^2}{4}-\dfrac{y^2}{4}=1$

12. $9x^2-4y^2=36$

13. Find the equation of the hyperbola with y-intercepts ± 5 and foci at $(0,-6)$ and $(0,6)$.

Identify the type of graph, if any, defined by each equation.

14. $x^2+8x+y^2-4y+2=0$

15. $5x^2+10x-2y^2-12y-23=0$

16. $3x^2+10y^2-30=0$

17. $x^2-4y=0$

18. $(x+9)^2+(y-3)^2=0$

19. $x^2+4x+y^2-6y+30=0$

20. The screen shown here gives the graph of

$$\frac{x^2}{25}-\frac{y^2}{49}=1$$

as generated by a graphing calculator. What two functions Y_1 and Y_2 were used to obtain the graph?

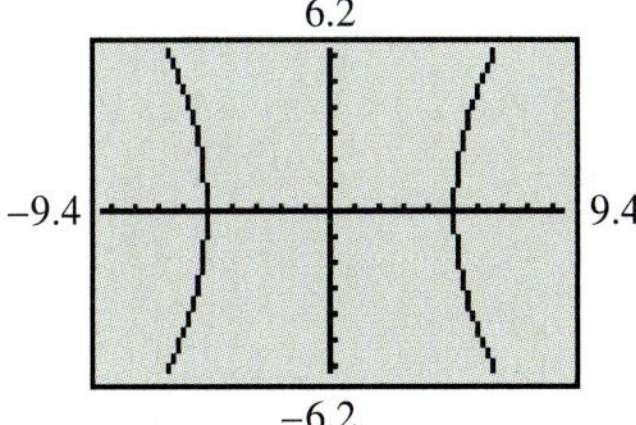

Chapter 10 Quantitative Reasoning

Does a rugby player need to know algebra?

A rugby field is similar to a modern football field with the exception that the goalpost, which is 18.5 ft wide, is located on the goal line instead of at the back of the endzone. The rugby equivalent of a touchdown, called a *try,* is scored by touching the ball down beyond the goal line. After a try is scored, the scoring team can earn extra points by kicking the ball through the goalposts. The ball must be placed somewhere on the line perpendicular to the goal line and passing through the point where the try was scored. See the figure below on the left. If that line passes through the goalposts, then the kicker should place the ball at whatever distance he is most comfortable. If the line passes outside the goalposts, then the player might choose the point on the line where angle θ in the figure on the left is as large as possible. The problem of determining this optimal point is similar to a problem posed in 1471 by the astronomer Regiomontanus. (*Source:* Maor, E., *Trigonometric Delights,* Princeton University Press, 1998.)

The figure on the right below shows a vertical line segment AB, where A and B are a and b units above the horizontal axis, respectively. If point P is located on the axis at a distance of x units from point Q, then angle θ is greatest when $x = \sqrt{ab}$.

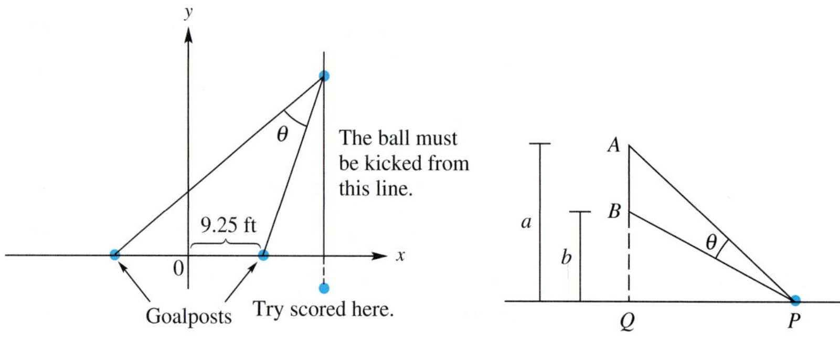

1. Use the result from Regiomontanus' problem to show that when the line is outside the goalposts the optimal location to kick the rugby ball lies on the hyperbola $x^2 - y^2 = 9.25^2$.

2. If the line on which the ball must be kicked is 10 feet to the right of the goalpost, how far from the goal line should the ball be placed to maximize angle θ?

3. Rugby players find it easier to kick the ball from the hyperbola's asymptote. When the line on which the ball must be kicked is 10 ft to the right of the goalpost, how far will this point differ from the exact optimal location?

11 Further Topics in Algebra

Amazing as it may seem, the male honeybee hatches from an unfertilized egg, while the female hatches from a fertilized one. As a result, the "family tree" of a male honeybee exhibits an interesting pattern. If we start with a male honeybee and count the number of bees in successive generations, we obtain the sequence of numbers

$$1, 1, 2, 3, 5, 8, 13, 21, 34, 55, \ldots,$$

known as the *Fibonacci sequence.* This fascinating sequence has countless interesting properties and appears in many places in nature.

In Section 11.1, we further investigate the Fibonacci sequence as we study *sequences* and sums of terms of sequences, known as *series.*

11.1 Sequences and Series

Sequences ▪ **Series and Summation Notation** ▪ **Summation Properties**

Sequences

A *sequence* is a function that computes an ordered list. For example, the average person in the United States uses 100 gallons of water each day. The function defined by $f(n) = 100n$ generates the terms of the sequence

$$100, 200, 300, 400, 500, 600, 700, \ldots,$$

when $n = 1, 2, 3, 4, 5, 6, 7, \ldots$. This function represents the gallons of water used by the average person after n days.

A second example of a sequence involves investing money. If \$100 is deposited into a savings account paying 5% interest compounded annually, then the function defined by $g(n) = 100(1.05)^n$ calculates the account balance after n years. The terms of the sequence are

$$g(1), g(2), g(3), g(4), g(5), g(6), g(7), \ldots,$$

and can be approximated as

$$105, 110.25, 115.76, 121.55, 127.63, 134.01, 140.71, \ldots.$$

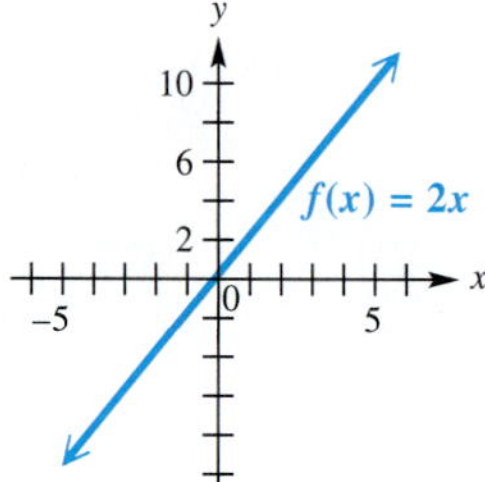

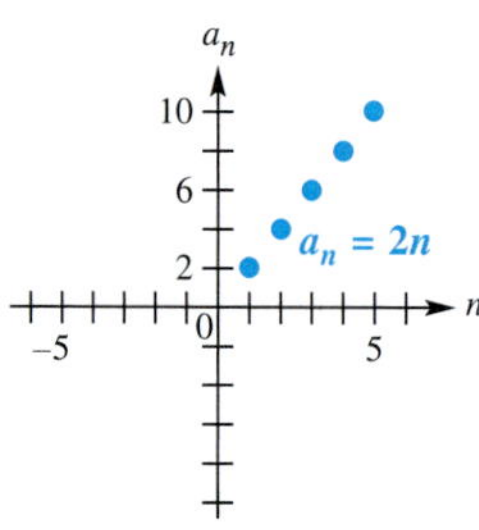

Figure 1

Sequence

A **finite sequence** is a function that has a set of natural numbers of the form $\{1, 2, 3, \ldots, n\}$ as its domain. An **infinite sequence** has the set of natural numbers as its domain.

For example, the sequence of natural-number multiples of 2,

$$2, 4, 6, 8, 10, 12, 14, \ldots, \quad \text{is infinite,}$$

but the sequence of days in June,

$$1, 2, 3, 4, \ldots, 29, 30, \quad \text{is finite.}$$

Instead of using $f(x)$ notation to indicate a sequence, it is customary to use a_n, where $a_n = f(n)$. The letter n is used instead of x as a reminder that n represents a *natural number*. The elements in the range of a sequence, called the **terms** of the sequence, are $a_1, a_2, a_3, \ldots$. The elements of both the domain and the range of a sequence are *ordered*. The first term is found by letting $n = 1$, the second term is found by letting $n = 2$, and so on. The **general term, or *n*th term,** of the sequence is a_n.

Figure 1 shows graphs of $f(x) = 2x$ and $a_n = 2n$. Notice that $f(x)$ is a continuous function, and a_n is discontinuous. To graph a_n, we plot points of the form $(n, 2n)$ for $n = 1, 2, 3, \ldots$.

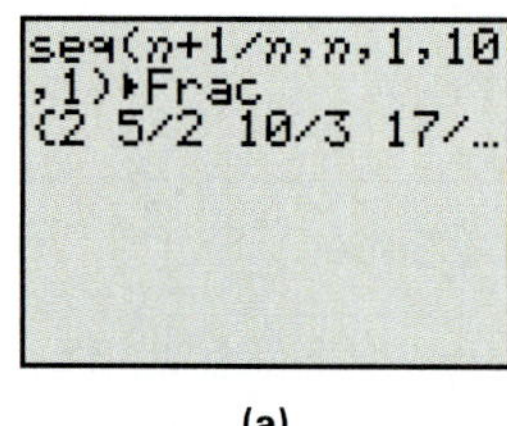

(a)

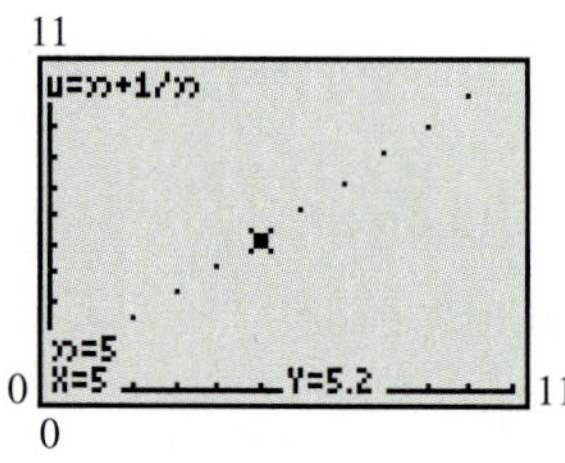

The fifth term is 5.2.

(b)

Figure 2

A graphing calculator can list the terms in a sequence. Using sequence mode to list the first 10 terms of the sequence with general term $a_n = n + \frac{1}{n}$ produces the result shown in Figure 2(a). Additional terms of the sequence can be seen by scrolling to the right. Sequences can also be graphed in sequence mode. Figure 2(b) shows a calculator screen with the graph of $a_n = n + \frac{1}{n}$. Notice that for $n = 5$, the term is $5 + \frac{1}{5} = 5.2$. ■

EXAMPLE 1 Finding Terms of Sequences

Write the first five terms for each sequence.

(a) $a_n = \dfrac{n+1}{n+2}$ **(b)** $a_n = (-1)^n \cdot n$ **(c)** $a_n = \dfrac{(-1)^n}{2^n}$

Solution

(a) Replacing n in $a_n = \frac{n+1}{n+2}$, with 1, 2, 3, 4, and 5 gives

$$n = 1: \quad a_1 = \frac{1+1}{1+2} = \frac{2}{3}$$

$$n = 2: \quad a_2 = \frac{2+1}{2+2} = \frac{3}{4}$$

$$n = 3: \quad a_3 = \frac{3+1}{3+2} = \frac{4}{5}$$

$$n = 4: \quad a_4 = \frac{4+1}{4+2} = \frac{5}{6}$$

$$n = 5: \quad a_5 = \frac{5+1}{5+2} = \frac{6}{7}.$$

(b) Replace n in $a_n = (-1)^n \cdot n$ with 1, 2, 3, 4, and 5 to obtain

$$n = 1: \quad a_1 = (-1)^1 \cdot 1 = -1$$

$$n = 2: \quad a_2 = (-1)^2 \cdot 2 = 2$$

$$n = 3: \quad a_3 = (-1)^3 \cdot 3 = -3$$

$$n = 4: \quad a_4 = (-1)^4 \cdot 4 = 4$$

$$n = 5: \quad a_5 = (-1)^5 \cdot 5 = -5.$$

(c) For $a_n = \frac{(-1)^n}{2^n}$, we have

$$a_1 = -\frac{1}{2}, \quad a_2 = \frac{1}{4}, \quad a_3 = -\frac{1}{8}, \quad a_4 = \frac{1}{16}, \quad \text{and} \quad a_5 = -\frac{1}{32}.$$

Now try Exercises 1, 7, and 9.

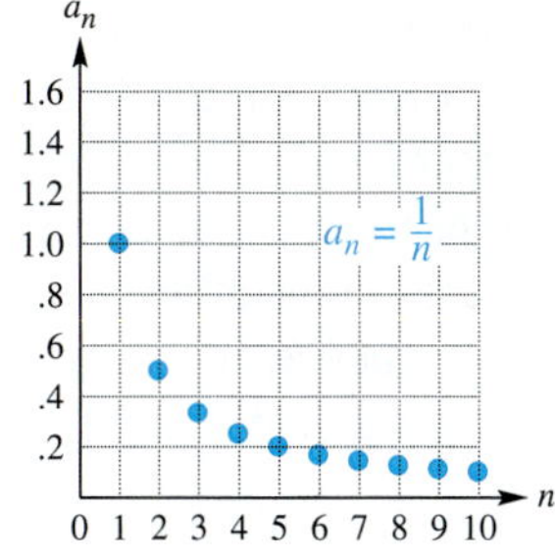

Figure 3

If the terms of an infinite sequence get closer and closer to some real number, the sequence is said to be **convergent** and to **converge** to that real number. For example, the sequence defined by $a_n = \frac{1}{n}$ approaches 0 as n becomes large. Thus, a_n is a convergent sequence that converges to 0. A graph of this sequence for $n = 1, 2, 3, \ldots, 10$ is shown in Figure 3. The terms of a_n approach the horizontal axis.

A sequence that does not converge to any number is **divergent.** The terms of the sequence $a_n = n^2$ are

$$1, 4, 9, 16, 25, 36, 49, 64, 81, \ldots.$$

This sequence is divergent because as n becomes large, the values of a_n do not approach a fixed number; rather, they increase without bound.

Some sequences are defined by a **recursive definition,** one in which each term after the first term or first few terms is defined as an expression involving the previous term or terms. On the other hand, the sequences in Example 1 were defined *explicitly,* with a formula for a_n that does not depend on a previous term.

EXAMPLE 2 Using a Recursion Formula

Find the first four terms of each sequence.

(a) $a_1 = 4$
$a_n = 2 \cdot a_{n-1} + 1$, if $n > 1$

(b) $a_1 = 2$
$a_n = a_{n-1} + n - 1$, if $n > 1$

Solution

(a) This is a recursive definition. We know $a_1 = 4$. Since $a_n = 2 \cdot a_{n-1} + 1$,

$$\begin{aligned} a_1 &= 4 \\ a_2 &= 2 \cdot a_1 + 1 = 2 \cdot 4 + 1 = 9 \\ a_3 &= 2 \cdot a_2 + 1 = 2 \cdot 9 + 1 = 19 \\ a_4 &= 2 \cdot a_3 + 1 = 2 \cdot 19 + 1 = 39. \end{aligned}$$

(b) In this recursive definition, $a_1 = 2$ and $a_n = a_{n-1} + n - 1$.

$$\begin{aligned} a_1 &= 2 \\ a_2 &= a_1 + 2 - 1 = 2 + 1 = 3 \\ a_3 &= a_2 + 3 - 1 = 3 + 2 = 5 \\ a_4 &= a_3 + 4 - 1 = 5 + 3 = 8 \end{aligned}$$

Now try Exercises 21 and 25.

Leonardo of Pisa (Fibonacci) (1170–1250)

CONNECTIONS One of the most famous sequences in mathematics is the **Fibonacci sequence,**

$$1, 1, 2, 3, 5, 8, 13, 21, 34, 55, \ldots,$$

named for the Italian mathematician Leonardo of Pisa, who was also known as Fibonacci. The Fibonacci sequence is found in numerous places in nature. For example, male honeybees hatch from eggs that have not been fertilized, so a male bee has only one parent, a female. On the other hand, female honeybees hatch from fertilized eggs, so a female has two parents, one male and one female. The number of ancestors in consecutive generations of bees follows the Fibonacci sequence. Successive terms in the sequence also appear in plants, such as in the spirals of the daisy head, pineapple, and pine cone.

For Discussion or Writing

1. Try to discover the pattern in the Fibonacci sequence.
2. Using the description given above, write a recursive definition that calculates the number of ancestors of a male bee in each generation.

EXAMPLE 3 Modeling Insect Population Growth

Frequently the population of a particular insect does not continue to grow indefinitely. Instead, its population grows rapidly at first, and then levels off because of competition for limited resources. In one study, the behavior of the winter moth was modeled with a sequence similar to the following, where a_n represents the population density in thousands per acre during year n. (*Source:* Varley, G. and G. Gradwell, "Population models for the winter moth," Symposium of the Royal Entomological Society of London, 4.)

$$a_1 = 1$$
$$a_n = 2.85a_{n-1} - .19a_{n-1}^{\,2}, \quad \text{for } n \geq 2$$

(a) Give a table of values for $n = 1, 2, 3, \ldots, 10$.

(b) Graph the sequence. Describe what happens to the population density.

Solution

n	$u(n)$
1	1
2	2.66
3	6.2366
4	10.384
5	9.1069
6	10.197
7	9.3056

n=1

Figure 4

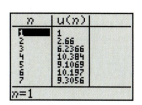

(a) Evaluate $a_1, a_2, a_3, \ldots, a_{10}$ recursively. Since $a_1 = 1$,

$$a_2 = 2.85a_1 - .19a_1^{\,2} = 2.85(1) - .19(1)^2 = 2.66,$$

and $\quad a_3 = 2.85a_2 - .19a_2^{\,2} = 2.85(2.66) - .19(2.66)^2 \approx 6.24.$

Approximate values for $n = 1, 2, 3, \ldots, 10$ are shown in the table. Figure 4 shows the computation of the sequence, denoted by $u(n)$ rather than a_n, using a calculator.

n	1	2	3	4	5	6	7	8	9	10
a_n	1	2.66	6.24	10.4	9.11	10.2	9.31	10.1	9.43	9.98

(b) The graph of a sequence is a set of discrete points. Plot the points (1, 1), (2, 2.66), (3, 6.24), …, (10, 9.98), as shown in Figure 5(a). At first, the insect population increases rapidly, and then oscillates about the line $y = 9.7$. (See the Note on the next page.) The oscillations become smaller as n increases, indicating that the population density may stabilize near 9.7 thousand per acre. In Figure 5(b), the first 20 terms have been plotted with a calculator.

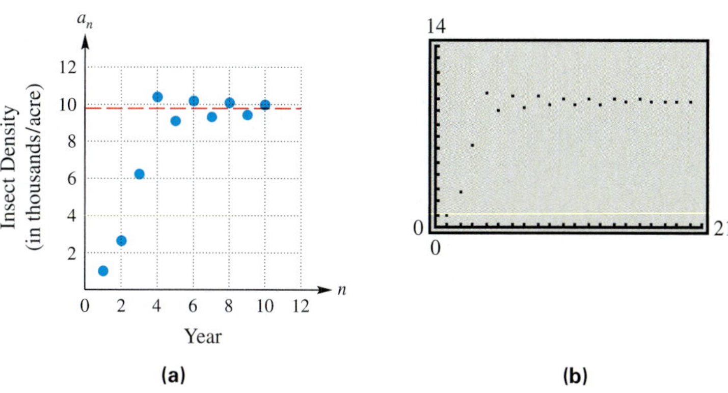

(a) (b)

Figure 5

Now try Exercise 81.

NOTE In Example 3, the insect population stabilizes near the value $k = 9.7$ thousand. This value of k can be found by solving the quadratic equation $k = 2.85k - .19k^2$. Why is this true?

Series and Summation Notation Suppose a person has a starting salary of \$30,000 and receives a \$2000 raise each year. Then,

$$30{,}000, \quad 32{,}000, \quad 34{,}000, \quad 36{,}000, \quad 38{,}000$$

are terms of the sequence that describe this person's salaries over a 5-year period. The total earned is given by the *finite series*

$$30{,}000 + 32{,}000 + 34{,}000 + 36{,}000 + 38{,}000,$$

whose sum is \$170,000. Any sequence can be used to define a *series.* For example, the infinite sequence

$$1, \frac{1}{3}, \frac{1}{9}, \frac{1}{27}, \frac{1}{81}, \frac{1}{243}, \dots$$

defines the terms of the *infinite series*

$$1 + \frac{1}{3} + \frac{1}{9} + \frac{1}{27} + \frac{1}{81} + \frac{1}{243} + \cdots.$$

Looking Ahead to Calculus

A series converges if the sequence of partial sums $S_1, S_2, S_3, \dots$ converges. For example, it can be shown that

$$1 + \frac{1}{2} + \frac{1}{3} + \frac{1}{4} + \dots \text{ diverges,}$$

while

$$1 - \frac{1}{2} + \frac{1}{3} - \frac{1}{4} + \dots \text{ converges.}$$

If a sequence has terms $a_1, a_2, a_3, \dots$, then S_n is defined as the sum of the first n terms. That is,

$$S_n = a_1 + a_2 + a_3 + \cdots + a_n.$$

The sum of the terms of a sequence, called a **series,** is written using **summation notation.** The symbol Σ, the Greek capital letter *sigma,* is used to indicate a sum.

Series

A **finite series** is an expression of the form

$$S_n = a_1 + a_2 + a_3 + \cdots + a_n = \sum_{i=1}^{n} a_i,$$

and an **infinite series** is an expression of the form

$$S_\infty = a_1 + a_2 + a_3 + \cdots + a_n + \cdots = \sum_{i=1}^{\infty} a_i.$$

The letter i is called the **index of summation.**

CAUTION Do not confuse this use of i with the use of i to represent the imaginary unit. Other letters may be used for the index of summation, such as k and j.

EXAMPLE 4 Using Summation Notation

Evaluate the series $\sum_{k=1}^{6} (2^k + 1)$.

Algebraic Solution

Write each of the six terms, then evaluate the sum.

$$\begin{aligned}\sum_{k=1}^{6} (2^k + 1) &= (2^1 + 1) + (2^2 + 1) + (2^3 + 1) \\ &\quad + (2^4 + 1) + (2^5 + 1) + (2^6 + 1) \\ &= (2 + 1) + (4 + 1) + (8 + 1) \\ &\quad + (16 + 1) + (32 + 1) + (64 + 1) \\ &= 3 + 5 + 9 + 17 + 33 + 65 \\ &= 132\end{aligned}$$

Graphing Calculator Solution

A graphing calculator can store the sequence into a list, L_1, and then compute the sum of the six terms in the list. See Figure 6.

```
seq(2^K+1,K,1,6)
→L1
{3 5 9 17 33 65}
sum(L1)
                132
```

Figure 6

Now try Exercise 41.

EXAMPLE 5 Using Summation Notation with Subscripts

Write the terms for each series. Evaluate each sum, if possible.

(a) $\sum_{j=3}^{6} a_j$ **(b)** $\sum_{i=1}^{3} (6x_i - 2)$, if $x_1 = 2$, $x_2 = 4$, and $x_3 = 6$

(c) $\sum_{i=1}^{4} f(x_i)\Delta x$, if $f(x) = x^2$, $x_1 = 0$, $x_2 = 2$, $x_3 = 4$, $x_4 = 6$, and $\Delta x = 2$

Solution

(a) $\sum_{j=3}^{6} a_j = a_3 + a_4 + a_5 + a_6$

(b) Let $i = 1, 2,$ and 3, respectively, to obtain

$$\begin{aligned}\sum_{i=1}^{3} (6x_i - 2) &= (6x_1 - 2) + (6x_2 - 2) + (6x_3 - 2) \\ &= (6 \cdot 2 - 2) + (6 \cdot 4 - 2) + (6 \cdot 6 - 2) \quad \text{Substitute the given values for } x_1, x_2, \text{ and } x_3. \\ &= 10 + 22 + 34 \\ &= 66.\end{aligned}$$

(c)

$$\begin{aligned}\sum_{i=1}^{4} f(x_i)\Delta x &= f(x_1)\Delta x + f(x_2)\Delta x + f(x_3)\Delta x + f(x_4)\Delta x \\ &= x_1^2\Delta x + x_2^2\Delta x + x_3^2\Delta x + x_4^2\Delta x \\ &= 0^2(2) + 2^2(2) + 4^2(2) + 6^2(2) \quad f(x) = x^2;\ \Delta x = 2 \\ &= 0 + 8 + 32 + 72 \\ &= 112\end{aligned}$$

Now try Exercises 47, 49, and 57.

Looking Ahead to Calculus

Summation notation is used in calculus to describe the area under a curve, the volume of a figure rotated about an axis, and in many other applications, as well as in the definition of integral. In the definition of the definite integral, Σ is replaced with an elongated S:

$$\int_a^b f(x)\,dx = \lim_{n\to\infty} \sum_{i=1}^{n} f(x_i)\,\Delta x_i.$$

In some cases, the definite integral can be interpreted as the sum of the areas of rectangles.

Summation Properties Properties of summation provide useful shortcuts for evaluating series.

Summation Properties

If $a_1, a_2, a_3, \ldots, a_n$ and $b_1, b_2, b_3, \ldots, b_n$ are two sequences, and c is a constant, then for every positive integer n,

(a) $\sum_{i=1}^{n} c = nc$ (b) $\sum_{i=1}^{n} ca_i = c\sum_{i=1}^{n} a_i$

(c) $\sum_{i=1}^{n} (a_i + b_i) = \sum_{i=1}^{n} a_i + \sum_{i=1}^{n} b_i$ (d) $\sum_{i=1}^{n} (a_i - b_i) = \sum_{i=1}^{n} a_i - \sum_{i=1}^{n} b_i$.

To prove Property (a), expand the series to obtain

$$c + c + c + c + \cdots + c,$$

where there are n terms of c, so the sum is nc.

Property (c) also can be proved by first expanding the series:

$$\sum_{i=1}^{n} (a_i + b_i) = (a_1 + b_1) + (a_2 + b_2) + \cdots + (a_n + b_n)$$

$$= (a_1 + a_2 + \cdots + a_n) + (b_1 + b_2 + \cdots + b_n)$$

Commutative and associative properties (Section R.1)

$$= \sum_{i=1}^{n} a_i + \sum_{i=1}^{n} b_i.$$

Proofs of the other two properties are similar. The following results can be proved by mathematical induction. (See Section 5 of this chapter.)

Summation Rules

$$\sum_{i=1}^{n} i = 1 + 2 + \cdots + n = \frac{n(n+1)}{2}$$

$$\sum_{i=1}^{n} i^2 = 1^2 + 2^2 + \cdots + n^2 = \frac{n(n+1)(2n+1)}{6}$$

$$\sum_{i=1}^{n} i^3 = 1^3 + 2^3 + \cdots + n^3 = \frac{n^2(n+1)^2}{4}$$

EXAMPLE 6 Using the Summation Properties

Use the summation properties to find each sum.

(a) $\sum_{i=1}^{40} 5$ (b) $\sum_{i=1}^{22} 2i$ (c) $\sum_{i=1}^{14} (2i^2 - 3)$

Solution

(a) $\sum_{i=1}^{40} 5 = 40(5) = 200$ Property (a) with $n = 40$ and $c = 5$

(b) $\sum_{i=1}^{22} 2i = 2\sum_{i=1}^{22} i$ — Property (b) with $c = 2$ and $a_i = i$

$= 2 \cdot \frac{22(22+1)}{2}$ — Summation rules

$= 506$ — Simplify.

(c) $\sum_{i=1}^{14} (2i^2 - 3) = \sum_{i=1}^{14} 2i^2 - \sum_{i=1}^{14} 3$ — Property (d) with $a_i = 2i^2$ and $b_i = 3$

$= 2\sum_{i=1}^{14} i^2 - \sum_{i=1}^{14} 3$ — Property (b) with $c = 2$ and $a_i = i^2$

$= 2 \cdot \frac{14(14+1)(2 \cdot 14 + 1)}{6} - 14(3)$ — Summation rules; Property (a)

$= 1988$ — Simplify.

Now try Exercises 61, 63, and 65.

EXAMPLE 7 Using the Summation Properties

Evaluate $\sum_{i=1}^{6} (i^2 + 3i + 5)$.

Solution

$\sum_{i=1}^{6} (i^2 + 3i + 5) = \sum_{i=1}^{6} i^2 + \sum_{i=1}^{6} 3i + \sum_{i=1}^{6} 5$ — Property (c)

$= \sum_{i=1}^{6} i^2 + 3\sum_{i=1}^{6} i + \sum_{i=1}^{6} 5$ — Property (b)

$= \sum_{i=1}^{6} i^2 + 3\sum_{i=1}^{6} i + 6(5)$ — Property (a)

$= \frac{6(6+1)(2 \cdot 6 + 1)}{6} + 3\left[\frac{6(6+1)}{2}\right] + 6(5)$ — Summation rules

$= 91 + 3(21) + 6(5)$

$= 184$

Now try Exercise 67.

11.1 Exercises

Write the first five terms of each sequence. See Example 1.

1. $a_n = 4n + 10$

2. $a_n = 6n - 3$

3. $a_n = 2^{n-1}$

4. $a_n = -3^n$

5. $a_n = \left(\frac{1}{3}\right)^n (n - 1)$

6. $a_n = (-2)^n(n)$

7. $a_n = (-1)^n(2n)$

8. $a_n = (-1)^{n-1}(n + 1)$

9. $a_n = \frac{4n - 1}{n^2 + 2}$

10. $a_n = \frac{n^2 - 1}{n^2 + 1}$

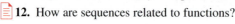

11. Your friend does not understand what is meant by the nth term or general term of a sequence. How would you explain this idea?

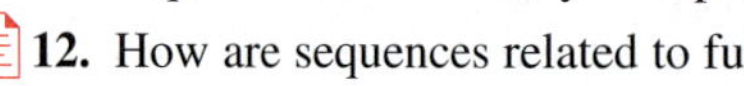

12. How are sequences related to functions?

Decide whether each sequence is finite *or* infinite.

13. The sequence of days of the week

14. The sequence of dates in the month of November

15. 1, 2, 3, 4

16. $-1, -2, -3, -4$

17. $1, 2, 3, 4, \ldots$

18. $-1, -2, -3, -4, \ldots$

19. $a_1 = 3$
$a_n = 3 \cdot a_{n-1}$, if $2 \leq n \leq 10$

20. $a_1 = 1$
$a_2 = 3$
$a_n = a_{n-1} + a_{n-2}$, if $n \geq 3$

Find the first four terms of each sequence. See Example 2.

21. $a_1 = -2$
$a_n = a_{n-1} + 3$, if $n > 1$

22. $a_1 = -1$
$a_n = a_{n-1} - 4$, if $n > 1$

23. $a_1 = 1$
$a_2 = 1$
$a_n = a_{n-1} + a_{n-2}$, if $n \geq 3$
(This is the Fibonacci sequence.)

24. $a_1 = 2$
$a_2 = 5$
$a_n = a_{n-1} + a_{n-2}$, if $n \geq 3$

25. $a_1 = 2$
$a_n = n \cdot a_{n-1}$, if $n > 1$

26. $a_1 = -3$
$a_n = 2n \cdot a_{n-1}$, if $n > 1$

Evaluate each series. See Example 4.

27. $\sum_{i=1}^{5} (2i + 1)$

28. $\sum_{i=1}^{6} (3i - 2)$

29. $\sum_{j=1}^{4} \frac{1}{j}$

30. $\sum_{i=1}^{5} (i + 1)^{-1}$

31. $\sum_{i=1}^{4} i^i$

32. $\sum_{k=1}^{4} (k + 1)^2$

33. $\sum_{k=1}^{6} (-1)^k \cdot k$

34. $\sum_{i=1}^{7} (-1)^{i+1} \cdot i^2$

35. $\sum_{i=2}^{5} (6 - 3i)$

36. $\sum_{i=3}^{7} (5i + 2)$

37. $\sum_{i=-2}^{3} 2(3)^i$

38. $\sum_{i=-1}^{2} 5(2)^i$

39. $\sum_{i=-1}^{5} (i^2 - 2i)$

40. $\sum_{i=3}^{6} (2i^2 + 1)$

41. $\sum_{i=1}^{5} (3^i - 4)$

42. $\sum_{i=1}^{4} [(-2)^i - 3]$

Use a graphing calculator to evaluate each series. See Example 4.

43. $\sum_{i=1}^{10} (4i^2 - 5)$

44. $\sum_{i=1}^{10} (i^3 - 6)$

45. $\sum_{j=3}^{9} (3j - j^2)$

46. $\sum_{k=5}^{10} (k^2 - 4k + 7)$

Write the terms for each series. Evaluate the sum, given that $x_1 = -2$, $x_2 = -1$, $x_3 = 0$, $x_4 = 1$, and $x_5 = 2$. See Examples 5(a) and 5(b).

47. $\sum_{i=1}^{5} x_i$

48. $\sum_{i=1}^{5} -x_i$

49. $\sum_{i=1}^{5} (2x_i + 3)$

50. $\sum_{i=1}^{4} x_i^2$ **51.** $\sum_{i=1}^{3} (3x_i - x_i^2)$ **52.** $\sum_{i=1}^{3} (x_i^2 + 1)$

53. $\sum_{i=2}^{5} \frac{x_i + 1}{x_i + 2}$ **54.** $\sum_{i=1}^{5} \frac{x_i}{x_i + 3}$

Write the terms of $\sum_{i=1}^{4} f(x_i)\Delta x$, *with* $x_1 = 0$, $x_2 = 2$, $x_3 = 4$, $x_4 = 6$, *and* $\Delta x = .5$, *for each function. Evaluate the sum. See Example 5(c).*

55. $f(x) = 4x - 7$ **56.** $f(x) = 6 + 2x$ **57.** $f(x) = 2x^2$

58. $f(x) = x^2 - 1$ **59.** $f(x) = \frac{-2}{x + 1}$ **60.** $f(x) = \frac{5}{2x - 1}$

Use the summation properties and rules to evaluate each series. See Examples 6 and 7.

61. $\sum_{i=1}^{100} 6$ **62.** $\sum_{i=1}^{20} 5i$ **63.** $\sum_{i=1}^{15} i^2$ **64.** $\sum_{i=1}^{50} 2i^3$

65. $\sum_{i=1}^{5} (5i + 3)$ **66.** $\sum_{i=1}^{5} (8i - 1)$ **67.** $\sum_{i=1}^{5} (4i^2 - 2i + 6)$

68. $\sum_{i=1}^{6} (2 + i - i^2)$ **69.** $\sum_{i=1}^{4} (3i^3 + 2i - 4)$ **70.** $\sum_{i=1}^{6} (i^2 + 2i^3)$

Concept Check *Use summation notation to write each series.**

71. $\frac{1}{3(1)} + \frac{1}{3(2)} + \frac{1}{3(3)} + \cdots + \frac{1}{3(9)}$ **72.** $\frac{5}{1 + 1} + \frac{5}{1 + 2} + \frac{5}{1 + 3} + \cdots + \frac{5}{1 + 15}$

73. $1 - \frac{1}{2} + \frac{1}{4} - \frac{1}{8} + \cdots - \frac{1}{128}$ **74.** $1 - \frac{1}{4} + \frac{1}{9} - \frac{1}{16} + \cdots - \frac{1}{400}$

Use the sequence feature of a graphing calculator to graph the first ten terms of each sequence as defined. Use the graph to make a conjecture as to whether the sequence converges or diverges. If you think it converges, determine the number to which it converges.

75. $a_n = \frac{n + 4}{2n}$ **76.** $a_n = \frac{1 + 4n}{2n}$ **77.** $a_n = 2e^n$

78. $a_n = n(n + 2)$ **79.** $a_n = \left(1 + \frac{1}{n}\right)^n$ **80.** $a_n = (1 + n)^{1/n}$

Solve each problem involving sequences and series. See Example 3.

81. ***(Modeling) Insect Population*** Suppose an insect population density in thousands per acre during year n can be modeled by the recursively defined sequence

$$a_1 = 8$$
$$a_n = 2.9a_{n-1} - .2a_{n-1}^2, \quad \text{for } n > 1.$$

(a) Find the population for $n = 1, 2, 3$.

(b) Graph the sequence for $n = 1, 2, 3, \ldots, 20$. Use the window $[0, 21]$ by $[0, 14]$. Interpret the graph.

*These exercises were suggested by Joe Lloyd Harris, Gulf Coast Community College.

82. ***Male Bee Ancestors*** As mentioned in the chapter introduction and in the Connections box in this section, one of the most famous sequences in mathematics is the Fibonacci sequence,

$$1, 1, 2, 3, 5, 8, 13, 21, 34, 55, \ldots .$$

(Also see Exercise 23.) Recall that male honeybees hatch from eggs that have not been fertilized, so a male bee has only one parent, a female. On the other hand, female honeybees hatch from fertilized eggs, so a female has two parents, one male and one female. The number of ancestors in consecutive generations of bees follows the Fibonacci sequence. Draw a tree showing the number of ancestors of a male bee in each generation following the description given above.

83. ***(Modeling) Bacteria Growth*** If certain bacteria are cultured in a medium with sufficient nutrients, they will double in size and then divide every 40 min. Let N_1 be the initial number of bacteria cells, N_2 the number after 40 min, N_3 the number after 80 min, and N_j the number after $40(j - 1)$ min. (*Source:* Hoppensteadt, F. and C. Peskin, *Mathematics in Medicine and the Life Sciences,* Springer-Verlag, 1992.)

(a) Write N_{j+1} in terms of N_j for $j \geq 1$.
(b) Determine the number of bacteria after 2 hr if $N_1 = 230$.
(c) Graph the sequence N_j for $j = 1, 2, 3, \ldots, 7$ where $N_1 = 230$. Use the window $[0, 10]$ by $[0, 15{,}000]$.
(d) Describe the growth of these bacteria when there are unlimited nutrients.

84. ***(Modeling) Verhulst's Model for Bacteria Growth*** Refer to Exercise 83. If the bacteria are not cultured in a medium with sufficient nutrients, competition will ensue and growth will slow. According to Verhulst's model, the number of bacteria N_j at time $40(j - 1)$ in minutes can be determined by the sequence

$$N_{j+1} = \left[\frac{2}{1 + \frac{N_j}{K}}\right] N_j,$$

where K is a constant and $j \geq 1$. (*Source:* Hoppensteadt, F. and C. Peskin, *Mathematics in Medicine and the Life Sciences,* Springer-Verlag, 1992.)

(a) If $N_1 = 230$ and $K = 5000$, make a table of N_j for $j = 1, 2, 3, \ldots, 20$. Round values in the table to the nearest integer.
(b) Graph the sequence N_j for $j = 1, 2, 3, \ldots, 20$. Use the window $[0, 20]$ by $[0, 6000]$.
(c) Describe the growth of these bacteria when there are limited nutrients.
(d) Make a conjecture as to why K is called the *saturation constant.* Test your conjecture by changing the value of K in the given formula.

85. ***(Modeling) Consumer Credit Growth*** The table gives total outstanding consumer credit, in billions of dollars, for the years 1995–2002. If we let $x = 1$ represent 1995, $x = 2$ represent 1996, and so on, this data can be represented as a sequence whose terms are 1096, 1185, and so on.

(a) Use a graphing calculator to find a linear function that models the data.
(b) Repeat part (a) for a quadratic function.
(c) Repeat part (a) for a cubic function.
(d) Use the results of parts (a)–(c) to predict consumer debt, in billions of dollars, for 2003.

Year	Credit in Billions
1995	1096
1996	1185
1997	1243
1998	1317
1999	1416
2000	1560
2001	1667
2002	1726

Source: Federal Reserve Board.

86. ***Estimating*** π Find the sum of the first six terms of the series

$$\frac{\pi^4}{90} = \frac{1}{1^4} + \frac{1}{2^4} + \frac{1}{3^4} + \frac{1}{4^4} + \frac{1}{5^4} + \cdots + \frac{1}{n^4} + \cdots.$$

Then multiply the result by 90, and take the fourth root. This will provide an estimate of π. Compare your answer to the actual decimal approximation of π.

87. ***Estimating Powers of e*** The series

$$e^a \approx 1 + a + \frac{a^2}{2!} + \frac{a^3}{3!} + \cdots + \frac{a^n}{n!},$$

where $n! = 1 \cdot 2 \cdot 3 \cdot 4 \cdot \cdots \cdot n$, can be used to estimate the value of e^a for any real number a. Use the first eight terms of this series to approximate each expression. Compare this estimate with the actual value.

(a) e **(b)** e^{-1}

88. ***Estimating Square Roots*** The recursively defined sequence

$$a_1 = k$$

$$a_n = \frac{1}{2}\left(a_{n-1} + \frac{k}{a_{n-1}}\right), \quad \text{if } n > 1$$

can be used to compute $\sqrt{k}$ for any positive number k. This sequence was known to Sumerian mathematicians 4000 yr ago, and it is still used today. Use this sequence to approximate the given square root by finding a_6. Compare your result with the actual value. (*Source:* Heinz-Otto, P., *Chaos and Fractals,* Springer-Verlag, 1993.)

(a) $\sqrt{2}$ **(b)** $\sqrt{11}$

11.2 Arithmetic Sequences and Series

Arithmetic Sequences ▪ Arithmetic Series

Arithmetic Sequences A sequence in which each term after the first is obtained by adding a fixed number to the previous term is an **arithmetic sequence** (or **arithmetic progression**). The fixed number that is added is the **common difference.** The sequence

$$5, 9, 13, 17, 21, \ldots$$

is an arithmetic sequence since each term after the first is obtained by adding 4 to the previous term. That is,

$$9 = 5 + 4$$
$$13 = 9 + 4$$
$$17 = 13 + 4$$
$$21 = 17 + 4,$$

and so on. The common difference is 4.

If the common difference of an arithmetic sequence is d, then by the definition of an arithmetic sequence,

$$d = a_{n+1} - a_n,$$

for every positive integer n in the domain of the sequence.

EXAMPLE 1 Finding the Common Difference

Find the common difference, d, for the arithmetic sequence

$$-9, -7, -5, -3, -1, \ldots.$$

Solution We find d by choosing any two adjacent terms and subtracting the first from the second. Choosing -7 and -5 gives

$$d = -5 - (-7) = 2.$$

Choosing -9 and -7 would give $d = -7 - (-9) = 2$, the same result.

Now try Exercise 1.

If a_1 and d are known, then all the terms of an arithmetic sequence can be found.

EXAMPLE 2 Finding Terms Given a_1 and d

Write the first five terms for each arithmetic sequence.

(a) The first term is 7, and the common difference is -3.

(b) $a_1 = -12, d = 5$

Solution

(a) Here $a_1 = 7$ and $d = -3$, so

$$a_2 = 7 + (-3) = 4$$
$$a_3 = 4 + (-3) = 1$$
$$a_4 = 1 + (-3) = -2$$
$$a_5 = -2 + (-3) = -5.$$

(b) Starting with a_1, add d to each term to get the next term.

$$a_1 = -12$$
$$a_2 = -12 + d = -12 + 5 = -7$$
$$a_3 = -7 + d = -7 + 5 = -2$$
$$a_4 = -2 + d = -2 + 5 = 3$$
$$a_5 = 3 + d = 3 + 5 = 8$$

Now try Exercises 7 and 9.

If a_1 is the first term of an arithmetic sequence and d is the common difference, then the terms of the sequence are given by

$$a_1 = a_1$$
$$a_2 = a_1 + d$$
$$a_3 = a_2 + d = a_1 + d + d = a_1 + 2d$$
$$a_4 = a_3 + d = a_1 + 2d + d = a_1 + 3d$$
$$a_5 = a_1 + 4d$$
$$a_6 = a_1 + 5d,$$

and, by this pattern,

$$a_n = a_1 + (n - 1)d.$$

This result can be proven by mathematical induction. (See Section 5 of this chapter.)

*n*th Term of an Arithmetic Sequence

In an arithmetic sequence with first term a_1 and common difference d, the nth term, a_n, is given by

$$\mathbf{a_n = a_1 + (n - 1)d.}$$

EXAMPLE 3 Finding Terms of an Arithmetic Sequence

Find a_{13} and a_n for the arithmetic sequence $-3, 1, 5, 9, \ldots$.

Solution Here $a_1 = -3$ and $d = 1 - (-3) = 4$. To find a_{13}, substitute 13 for n in the formula for the nth term.

$$a_n = a_1 + (n - 1)d$$

$$a_{13} = a_1 + (13 - 1)d$$

$$a_{13} = -3 + (12)4 \qquad \text{Let } a_1 = -3, d = 4.$$

$$a_{13} = -3 + 48$$

$$a_{13} = 45$$

Find a_n by substituting values for a_1 and d in the formula for a_n.

$$a_n = -3 + (n - 1) \cdot 4$$

$$a_n = -3 + 4n - 4 \qquad \text{Distributive property (Section R.1)}$$

$$a_n = 4n - 7$$

Now try Exercise 13.

EXAMPLE 4 Finding Terms of an Arithmetic Sequence

Find a_{18} and a_n for the arithmetic sequence having $a_2 = 9$ and $a_3 = 15$.

Solution Find d first; $d = a_3 - a_2 = 15 - 9 = 6$.

Since $$a_2 = a_1 + d,$$

$$9 = a_1 + 6 \qquad \text{Let } a_2 = 9, d = 6.$$

$$a_1 = 3.$$

Then, $$a_{18} = 3 + (18 - 1)6 \qquad \text{Formula for } a_n;\ a_1 = 3, n = 18, d = 6$$

$$a_{18} = 105,$$

and $$a_n = 3 + (n - 1)6$$

$$a_n = 3 + 6n - 6 \qquad \text{Distributive property}$$

$$a_n = 6n - 3.$$

Now try Exercise 17.

EXAMPLE 5 Finding the First Term of an Arithmetic Sequence

Suppose that an arithmetic sequence has $a_8 = -16$ and $a_{16} = -40$. Find a_1.

Solution Since $a_{16} = a_8 + 8d$, it follows that

$$8d = a_{16} - a_8 = -40 - (-16) = -24,$$

and so $d = -3$. To find a_1, use the equation $a_8 = a_1 + 7d$.

$$-16 = a_1 + 7d \quad \text{Let } a_8 = -16.$$

$$-16 = a_1 + 7(-3) \quad \text{Let } d = -3.$$

$$a_1 = 5$$

Now try Exercise 21.

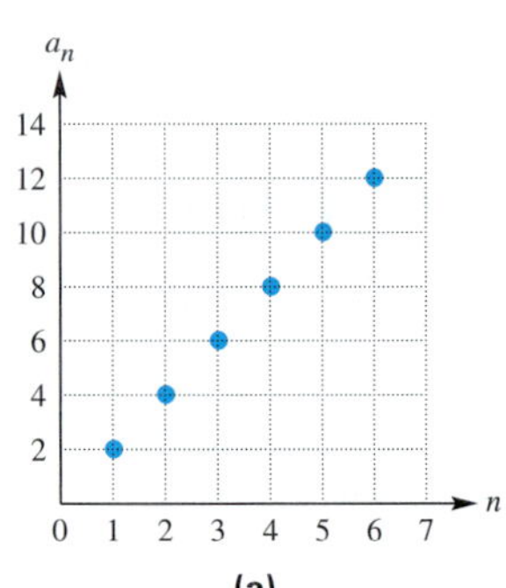

(a)

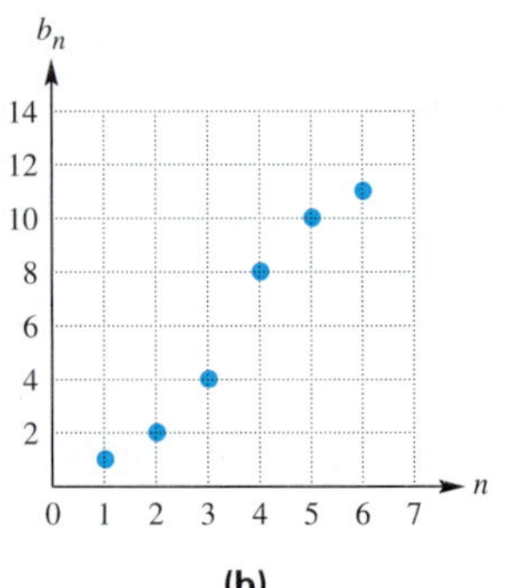

(b)

Figure 7

The graph of any sequence is a scatter diagram. To determine the characteristics of the graph of an arithmetic sequence, start by rewriting the formula for the nth term.

$$\begin{aligned} a_n &= a_1 + (n - 1)d && \text{Formula for the } n\text{th term} \\ &= a_1 + nd - d && \text{Distributive property} \\ &= dn + (a_1 - d) && \text{Commutative and associative properties (Section R.1)} \\ &= dn + c && \text{Let } c = a_1 - d. \end{aligned}$$

The points in the graph of an arithmetic sequence are determined by $f(n) = dn + c$, where n is a natural number. Thus, the points in the graph of f must lie on the *line*

$$y = dx + c. \quad \text{(Section 2.4)}$$

(d: slope; c: y-intercept)

For example, the sequence a_n shown in Figure 7(a) is an arithmetic sequence because the points that comprise its graph are collinear (lie on a line). The slope determined by these points is 2, so the common difference d equals 2. On the other hand, the sequence b_n shown in Figure 7(b) is not an arithmetic sequence because the points are not collinear.

EXAMPLE 6 Finding the *n*th Term from a Graph

Find a formula for the nth term of the sequence a_n shown in Figure 8. What are the domain and range of this sequence?

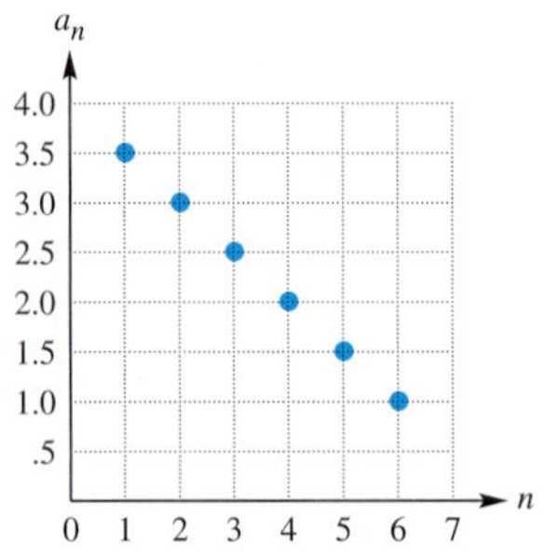

Figure 8

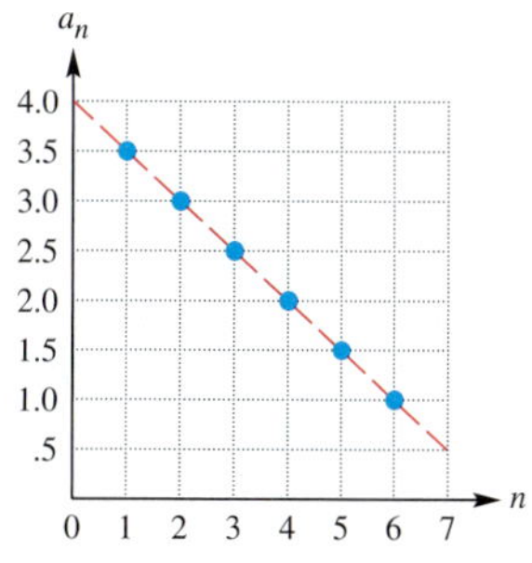

Figure 9

Solution The points in Figure 8 lie on a line, so the sequence is arithmetic. The equation of the dashed line shown in Figure 9 is $y = -.5x + 4$, so the nth term of this sequence is determined by

$$a_n = -.5n + 4.$$

The sequence is comprised of the points

$$(1, 3.5), (2, 3), (3, 2.5), (4, 2), (5, 1.5), (6, 1).$$

Thus, the domain of the sequence is given by $\{1, 2, 3, 4, 5, 6\}$, and the range is given by $\{3.5, 3, 2.5, 2, 1.5, 1\}$.

Now try Exercise 25.

Arithmetic Series The sum of the terms of an arithmetic sequence is an **arithmetic series.** To illustrate, suppose that a person borrows \$3000 and agrees to pay \$100 per month plus interest of 1% per month on the unpaid balance until the loan is paid off. The first month, \$100 is paid to reduce the loan, plus interest of $(.01)3000 = 30$ dollars. The second month, another \$100 is paid toward the loan, and $(.01)2900 = 29$ dollars is paid for interest. Since the loan is reduced by \$100 each month, interest payments decrease by $(.01)100 = 1$ dollar each month, forming the arithmetic sequence

$$30, 29, 28, \ldots, 3, 2, 1.$$

The total amount of interest paid is given by the sum of the terms of this sequence. Now we develop a formula to find this sum without adding all 30 numbers directly. Since the sequence is arithmetic, we can write the sum of the first n terms as

$$S_n = a_1 + [a_1 + d] + [a_1 + 2d] + \cdots + [a_1 + (n - 1)d].$$

We used the formula for the general term in the last expression. Now we write the same sum in reverse order, beginning with a_n and *subtracting d.*

$$S_n = a_n + [a_n - d] + [a_n - 2d] + \cdots + [a_n - (n - 1)d]$$

Adding respective sides of these two equations term by term, we obtain

$$S_n + S_n = (a_1 + a_n) + (a_1 + a_n) + \cdots + (a_1 + a_n)$$

or

$$2S_n = n(a_1 + a_n),$$

since there are n terms of $a_1 + a_n$ on the right. Now solve for S_n to get

$$S_n = \frac{n}{2}(a_1 + a_n).$$

Using the formula $a_n = a_1 + (n - 1)d$, we can also write this result for S_n as

$$S_n = \frac{n}{2}[a_1 + a_1 + (n - 1)d]$$

or

$$S_n = \frac{n}{2}[2a_1 + (n - 1)d],$$

which is an alternative formula for the sum of the first n terms of an arithmetic sequence.

A summary of this work follows.

Sum of the First *n* Terms of an Arithmetic Sequence

If an arithmetic sequence has first term a_1 and common difference d, then the sum of the first n terms is given by

$$S_n = \frac{n}{2}(a_1 + a_n) \qquad \text{or} \qquad S_n = \frac{n}{2}[2a_1 + (n - 1)d].$$

The first formula is used when the first and last terms are known; otherwise the second formula is used.

In the sequence of interest payments discussed earlier, $n = 30$, $a_1 = 30$, and $a_n = 1$. Choosing the first formula,

$$S_n = \frac{n}{2}(a_1 + a_n),$$

gives

$$S_{30} = \frac{30}{2}(30 + 1) = 15(31) = 465,$$

so a total of \$465 interest will be paid over the 30 months.

EXAMPLE 7 Using the Sum Formulas

(a) Evaluate S_{12} for the arithmetic sequence $-9, -5, -1, 3, 7, \ldots$.

(b) Use a formula for S_n to evaluate the sum of the first 60 positive integers.

Solution

(a) We want the sum of the first 12 terms. Using $a_1 = -9$, $n = 12$, and $d = 4$ in the second formula,

$$S_n = \frac{n}{2}[2a_1 + (n - 1)d],$$

gives

$$S_{12} = \frac{12}{2}[2(-9) + 11(4)] = 156.$$

(b) The first 60 positive integers form the arithmetic sequence $1, 2, 3, 4, \ldots, 60$. Thus, $n = 60$, $a_1 = 1$, and $a_{60} = 60$, so we use the first formula in the preceding box to find the sum.

$$S_n = \frac{n}{2}(a_1 + a_n)$$

$$S_{60} = \frac{60}{2}(1 + 60) = 1830$$

Now try Exercises 31 and 39.

EXAMPLE 8 Using the Sum Formulas

The sum of the first 17 terms of an arithmetic sequence is 187. If $a_{17} = -13$, find a_1 and d.

Solution

$$S_{17} = \frac{17}{2}(a_1 + a_{17}) \quad \text{Use the first formula for } S_n, \text{ with } n = 17.$$

$$187 = \frac{17}{2}(a_1 - 13) \quad \text{Let } S_{17} = 187,\ a_{17} = -13.$$

$$22 = a_1 - 13 \quad \text{Multiply by } \tfrac{2}{17}.$$

$$a_1 = 35$$

Since $a_{17} = a_1 + (17 - 1)d$,

$$-13 = 35 + 16d \quad \text{Let } a_{17} = -13,\ a_1 = 35.$$

$$-48 = 16d$$

$$d = -3.$$

Now try Exercise 45.

Any sum of the form $\sum_{i=1}^{n}(di + c)$, where d and c are real numbers, represents the sum of the terms of an arithmetic sequence having first term $a_1 = d(1) + c = d + c$ and common difference d. These sums can be evaluated using the formulas in this section.

EXAMPLE 9 Using Summation Notation

Evaluate each sum.

(a) $\sum_{i=1}^{10}(4i + 8)$ **(b)** $\sum_{k=3}^{9}(4 - 3k)$

Solution

(a) This sum contains the first 10 terms of the arithmetic sequence having

$$a_1 = 4 \cdot 1 + 8 = 12, \quad \text{First term}$$

and

$$a_{10} = 4 \cdot 10 + 8 = 48. \quad \text{Last term}$$

Thus, $$\sum_{i=1}^{10}(4i + 8) = S_{10} = \frac{10}{2}(12 + 48) = 5(60) = 300.$$

(b) The first few terms are

$$[4 - 3(3)] + [4 - 3(4)] + [4 - 3(5)] + \cdots = -5 + (-8) + (-11) + \cdots.$$

Thus, $a_1 = -5$ and $d = -3$. If the sequence started with $k = 1$, there would be nine terms. Since it starts at 3, two of those terms are missing, so there are seven terms and $n = 7$. Use the second formula for S_n.

$$\sum_{k=3}^{9}(4 - 3k) = \frac{7}{2}[2(-5) + 6(-3)] = -98$$

Now try Exercises 51 and 53.

```
sum(seq(4I+8,I,1
,10,1))
             300
sum(seq(4-3K,K,3
,9,1))
             -98
```

As shown in the previous section, a graphing calculator will give the sum of a sequence without having to first store the sequence. The screen here illustrates this method for the sequences in Example 9.

11.2 Exercises

Find the common difference d for each arithmetic sequence. See Example 1.

1. 2, 5, 8, 11, . . .
2. 4, 10, 16, 22, . . .
3. 3, −2, −7, −12, . . .
4. −8, −12, −16, −20, . . .
5. $x + 3y, 2x + 5y, 3x + 7y, \ldots$
6. $t^2 + q, -4t^2 + 2q, -9t^2 + 3q, \ldots$

Write the first five terms of each arithmetic sequence. See Example 2.

7. The first term is 8, and the common difference is 6.
8. The first term is −2, and the common difference is 12.
9. $a_1 = 5, d = -2$
10. $a_1 = 4, d = 3$
11. $a_3 = 10, d = -2$
12. $a_1 = 3 - \sqrt{2}, a_2 = 3$

Find a_8 and a_n for each arithmetic sequence. See Examples 3 and 4.

13. 5, 7, 9, . . .
14. −3, −7, −11, . . .
15. $a_1 = 5, a_4 = 15$
16. $a_1 = -4, a_5 = 16$
17. $a_{10} = 6, a_{12} = 15$
18. $a_{15} = 8, a_{17} = 2$
19. $a_1 = x, a_2 = x + 3$
20. $a_2 = y + 1, d = -3$

Find a_1 for each arithmetic sequence. See Examples 5 and 8.

21. $a_5 = 27, a_{15} = 87$
22. $a_{12} = 60, a_{20} = 84$
23. $S_{16} = -160, a_{16} = -25$
24. $S_{28} = 2926, a_{28} = 199$

Find a formula for the nth term of the finite arithmetic sequence shown in each graph. Then state the domain and range of the sequence. See Example 6.

25.
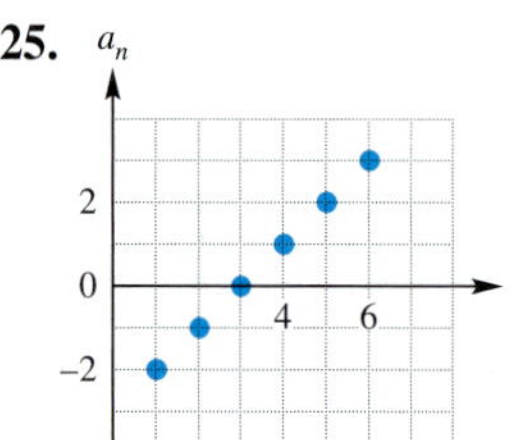
26.
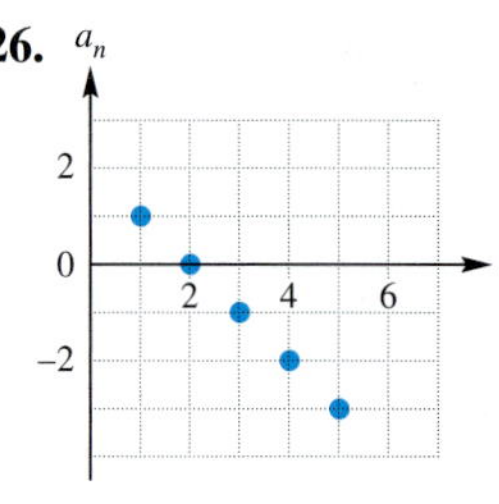
27.
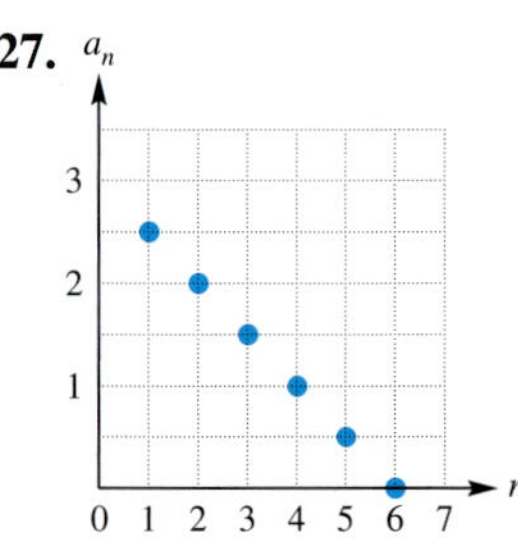
28.
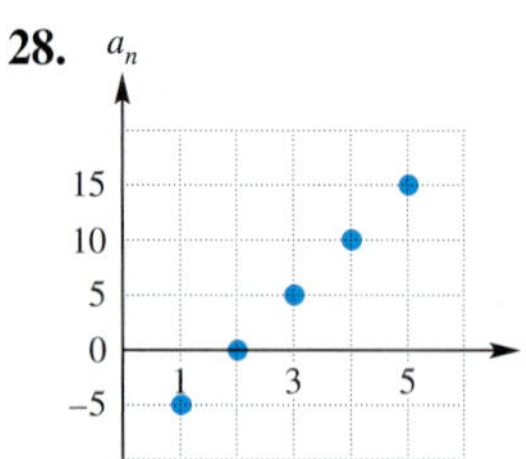
29.
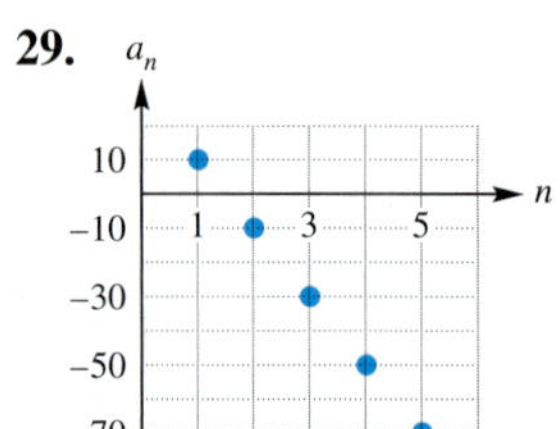
30.
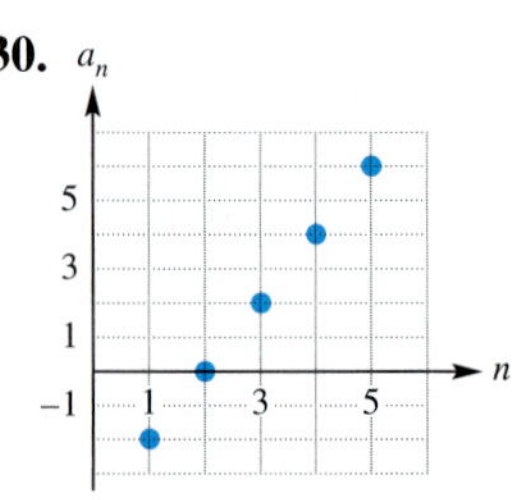

Evaluate S_{10}, the sum of the first ten terms, of each arithmetic sequence. See Example 7(a).

31. 8, 11, 14, . . .
32. −9, −5, −1, . . .
33. 5, 9, 13, . . .
34. 8, 6, 4, . . .
35. $a_2 = 9, a_4 = 13$
36. $a_3 = 5, a_4 = 8$
37. $a_1 = 10, a_{10} = 5.5$
38. $a_1 = -8, a_{10} = -1.25$

Find each sum as described. See Example 7(b).

39. the sum of the first 80 positive integers

40. the sum of the first 120 positive integers

41. the sum of the first 50 positive odd integers

42. the sum of the first 90 positive odd integers

43. the sum of the first 60 positive even integers

44. the sum of the first 70 positive even integers

Find a_1 and d for each arithmetic series. See Example 8.

45. $S_{20} = 1090, a_{20} = 102$

46. $S_{31} = 5580, a_{31} = 360$

47. $S_{12} = -108, a_{12} = -19$

48. $S_{25} = 650, a_{25} = 62$

Evaluate each sum. See Example 9.

49. $\sum_{i=1}^{3} (i + 4)$

50. $\sum_{i=1}^{5} (i - 8)$

51. $\sum_{j=1}^{10} (2j + 3)$

52. $\sum_{j=1}^{15} (5j - 9)$

53. $\sum_{i=4}^{12} (-5 - 8i)$

54. $\sum_{k=5}^{19} (-3 - 4k)$

55. $\sum_{i=1}^{1000} i$

56. $\sum_{k=1}^{2000} k$

Relating Concepts

For individual or collaborative investigation

(Exercises 57–60)

Let $f(x) = mx + b$. ***Work Exercises 57–60 in order.***

57. Find $f(1), f(2)$, and $f(3)$.

58. Consider the sequence $f(1), f(2), f(3), \ldots$. Is it an arithmetic sequence?

59. If the sequence is arithmetic, what is the common difference?

60. What is a_n for the sequence described in Exercise 58?

Use the sum and sequence features of a graphing calculator to evaluate the sum of the first ten terms of each arithmetic series with a_n defined as shown. In Exercises 63 and 64, round to the nearest thousandth.

61. $a_n = 4.2n + 9.73$

62. $a_n = 8.42n + 36.18$

63. $a_n = \sqrt{8}n + \sqrt{3}$

64. $a_n = -\sqrt[3]{4}n + \sqrt{7}$

Solve each problem.

65. ***Integer Sum*** Find the sum of all the integers from 51 to 71.

66. ***Integer Sum*** Find the sum of all the integers from -8 to 30.

67. ***Clock Chimes*** If a clock strikes the proper number of chimes each hour on the hour, how many times will it chime in a month of 30 days?

68. ***Telephone Pole Stack*** A stack of telephone poles has 30 in the bottom row, 29 in the next, and so on, with one pole in the top row. How many poles are in the stack?

69. ***Population Growth*** Five years ago, the population of a city was 49,000. Each year the zoning commission permits an increase of 580 in the population. What will the maximum population be 5 yr from now?

70. *Slide Supports* A super slide of uniform slope is to be built on a level piece of land. There are to be 20 equally spaced vertical supports, with the longest support 15 m long and the shortest 2 m long. Find the total length of all the supports.

71. *Rungs of a Ladder* How much material would be needed for the rungs of a ladder of 31 rungs, if the rungs taper uniformly from 18 in. to 28 in.?

72. *(Modeling) Children's Growth Pattern* The normal growth pattern for children age 3–11 follows that of an arithmetic sequence. An increase in height of about 6 cm per year is expected. Thus, 6 would be the common difference of the sequence. For example, a child who measures 96 cm at age 3 would have his expected height in subsequent years represented by the sequence 102, 108, 114, 120, 126, 132, 138, 144. Each term differs from the adjacent terms by the common difference, 6.

(a) If a child measures 98.2 cm at age 3 and 109.8 cm at age 5, what would be the common difference of the arithmetic sequence describing her yearly height?

(b) What would we expect her height to be at age 8?

73. Suppose that $a_1, a_2, a_3, \ldots$ and $b_1, b_2, b_3, \ldots$ are both arithmetic sequences. Let $d_n = a_n + c \cdot b_n$, for any real number c and every positive integer n. Show that $d_1, d_2, d_3, \ldots$ is an arithmetic sequence.

74. *Concept Check* Suppose that $a_1, a_2, a_3, a_4, a_5, \ldots$ is an arithmetic sequence. Is $a_1, a_3, a_5, \ldots$ an arithmetic sequence?

75. Explain why the sequence log 2, log 4, log 8, log 16, . . . is arithmetic.

11.3 Geometric Sequences and Series

Geometric Sequences ▪ **Geometric Series** ▪ **Infinite Geometric Series** ▪ **Annuities**

Geometric Sequences Suppose you agreed to work for 1¢ the first day, 2¢ the second day, 4¢ the third day, 8¢ the fourth day, and so on, with your wages doubling each day. How much will you earn on day 20, after working 5 days a week for a month? How much will you have earned altogether in 20 days? These questions will be answered in this section.

A **geometric sequence** (or **geometric progression**) is a sequence in which each term after the first is obtained by multiplying the preceding term by a fixed nonzero real number, called the **common ratio.** The sequence discussed above,

$$1, 2, 4, 8, 16, \ldots$$

is an example of a geometric sequence in which the first term is 1 and the common ratio is 2.

Notice that if we divide any term after the first term by the preceding term, we obtain the common ratio $r = 2$.

$$\frac{a_2}{a_1} = \frac{2}{1} = 2; \quad \frac{a_3}{a_2} = \frac{4}{2} = 2; \quad \frac{a_4}{a_3} = \frac{8}{4} = 2; \quad \frac{a_5}{a_4} = \frac{16}{8} = 2$$

If the common ratio of a geometric sequence is r, then by the definition of a geometric sequence,

$$r = \frac{a_{n+1}}{a_n},$$

for every positive integer n. Therefore, we find the common ratio by choosing any term except the first and dividing it by the preceding term.

In the geometric sequence

$$2, 8, 32, 128, \ldots,$$

$r = 4$. Notice that

$$\begin{aligned} 8 &= 2 \cdot 4 \\ 32 &= 8 \cdot 4 = (2 \cdot 4) \cdot 4 = 2 \cdot 4^2 \\ 128 &= 32 \cdot 4 = (2 \cdot 4^2) \cdot 4 = 2 \cdot 4^3. \end{aligned}$$

To generalize this, assume that a geometric sequence has first term a_1 and common ratio r. The second term is $a_2 = a_1 r$, the third is $a_3 = a_2 r = (a_1 r)r = a_1 r^2$, and so on. Following this pattern, the nth term is $a_n = a_1 r^{n-1}$. Again, this result can be proven by mathematical induction. (See Section 5 of this chapter.)

*n*th Term of a Geometric Sequence

In a geometric sequence with first term a_1 and common ratio r, the nth term, a_n, is given by

$$a_n = a_1 r^{n-1}.$$

EXAMPLE 1 Finding the *n*th Term of a Geometric Sequence

Use the formula for the nth term of a geometric sequence to answer the first question posed at the beginning of this section. How much will be earned on day 20 if daily wages follow the sequence 1, 2, 4, 8, 16, ... cents?

Solution To answer the question, let $a_1 = 1$ and $r = 2$, and find a_{20}.

$$a_{20} = a_1 r^{19} = 1(2)^{19} = 524{,}288 \text{ cents} \quad \text{or} \quad \$5242.88$$

Now try Exercise 1(a).

EXAMPLE 2 Finding Terms of a Geometric Sequence

Find a_5 and a_n for the geometric sequence 4, 12, 36, 108, ...

Solution The first term, a_1, is 4. Find r by choosing any term after the first and dividing it by the preceding term. For example,

$$r = \frac{36}{12} = 3.$$

Since $a_4 = 108$, $a_5 = 3 \cdot 108 = 324$. We could also find the fifth term by using the formula for a_n, $a_n = a_1 r^{n-1}$, and replacing n with 5, r with 3, and a_1 with 4.

$$a_5 = 4 \cdot (3)^{5-1} = 4 \cdot 3^4 = 324$$

By the formula, $\quad a_n = 4 \cdot 3^{n-1}.$

Now try Exercise 11.

EXAMPLE 3 Finding Terms of a Geometric Sequence

Find r and a_1 for the geometric sequence with third term 20 and sixth term 160.

Solution Use the formula for the nth term of a geometric sequence:

$$\text{for } n = 3, \quad a_3 = a_1 r^2 = 20;$$
$$\text{for } n = 6, \quad a_6 = a_1 r^5 = 160.$$

Since $a_1 r^2 = 20$, $a_1 = \frac{20}{r^2}$. Substitute this value for a_1 in the second equation.

$$a_1 r^5 = 160$$
$$\left(\frac{20}{r^2}\right) r^5 = 160$$
$$20r^3 = 160$$
$$r^3 = 8$$
$$r = 2$$

Since $a_1 r^2 = 20$ and $r = 2$,

$$a_1(2)^2 = 20$$
$$4a_1 = 20$$
$$a_1 = 5.$$

Now try Exercise 17.

EXAMPLE 4 Modeling a Population of Fruit Flies

A population of fruit flies is growing in such a way that each generation is 1.5 times as large as the last generation. Suppose there were 100 insects in the first generation. How many would there be in the fourth generation?

Solution Write the population of each generation as a geometric sequence with a_1 as the first-generation population, a_2 the second-generation population, and so on. Then the fourth-generation population is a_4. Using the formula for a_n, with $n = 4$, $r = 1.5$, and $a_1 = 100$, gives

$$a_4 = a_1 r^3 = 100(1.5)^3 = 100(3.375) = 337.5.$$

In the fourth generation, the population will number about 338 insects.

Now try Exercise 61.

Geometric Series A **geometric series** is the sum of the terms of a geometric sequence. In applications, it may be necessary to find the sum of the terms of such a sequence. For example, a scientist might want to know the total number of insects in four generations of the population discussed in Example 4. This population would equal $a_1 + a_2 + a_3 + a_4$, or

$$100 + 100(1.5) + 100(1.5)^2 + 100(1.5)^3 = 812.5 \approx 813 \text{ insects.}$$

To find a formula for the sum of the first n terms of a geometric sequence, S_n, first write the sum as

$$S_n = a_1 + a_2 + a_3 + \cdots + a_n$$

or

$$S_n = a_1 + a_1r + a_1r^2 + \cdots + a_1r^{n-1}. \quad (1)$$

If $r = 1$, then $S_n = na_1$, which is a correct formula for this case. If $r \neq 1$, then multiply both sides of equation (1) by r to obtain

$$rS_n = a_1r + a_1r^2 + a_1r^3 + \cdots + a_1r^n. \quad (2)$$

If equation (2) is subtracted from equation (1),

$$\begin{aligned} S_n &= a_1 + a_1r + a_1r^2 + \cdots + a_1r^{n-1} & &(1)\\ rS_n &= \phantom{a_1 + {}} a_1r + a_1r^2 + \cdots + a_1r^{n-1} + a_1r^n & &(2)\\ \hline S_n - rS_n &= a_1 - a_1r^n & &\text{Subtract.}\\ S_n(1-r) &= a_1(1-r^n) & &\text{Factor. (Section R.4)}\\ S_n &= \frac{a_1(1-r^n)}{1-r}, \quad \text{where } r \neq 1. & &\text{Divide by } 1-r. \text{ (Section 1.1)} \end{aligned}$$

Sum of the First n Terms of a Geometric Sequence

If a geometric sequence has first term a_1 and common ratio r, then the sum of the first n terms is given by

$$S_n = \frac{a_1(1-r^n)}{1-r}, \quad \text{where } r \neq 1.$$

We can use a geometric series to find the total fruit fly population in Example 4 over the four-generation period. With $n = 4$, $a_1 = 100$, and $r = 1.5$,

$$S_4 = \frac{100(1-1.5^4)}{1-1.5} = \frac{100(1-5.0625)}{-.5} = 812.5 \approx 813 \text{ insects},$$

which agrees with our previous result.

EXAMPLE 5 Finding the Sum of the First n Terms

At the beginning of this section, we posed the following question: How much will you have earned altogether after 20 days? Answer this question.

Solution We must find the total amount earned in 20 days with daily wages of 1, 2, 4, 8, ... cents. Since $a_1 = 1$ and $r = 2$,

$$S_{20} = \frac{1(1-2^{20})}{1-2} = \frac{1-1{,}048{,}576}{-1} = 1{,}048{,}575 \text{ cents}, \quad \text{or} \quad \$10{,}485.75.$$

Not bad for 20 days of work!

Now try Exercise 1(b).

EXAMPLE 6 Finding the Sum of the First *n* Terms

Find $\sum_{i=1}^{6} 2 \cdot 3^i$.

Solution This series is the sum of the first six terms of a geometric sequence having $a_1 = 2 \cdot 3^1 = 6$ and $r = 3$. Using the formula for S_n,

$$S_6 = \frac{6(1 - 3^6)}{1 - 3} = \frac{6(1 - 729)}{-2} = \frac{6(-728)}{-2} = 2184.$$

Now try Exercise 29.

Infinite Geometric Series We extend our discussion of sums of sequences to include infinite geometric sequences such as

$$2, 1, \frac{1}{2}, \frac{1}{4}, \frac{1}{8}, \frac{1}{16}, \ldots,$$

with first term 2 and common ratio $\frac{1}{2}$. Using the formula for S_n gives the following sequence of sums.

$$S_1 = 2, \quad S_2 = 3, \quad S_3 = \frac{7}{2}, \quad S_4 = \frac{15}{4}, \quad S_5 = \frac{31}{8}, \quad S_6 = \frac{63}{16}, \ldots$$

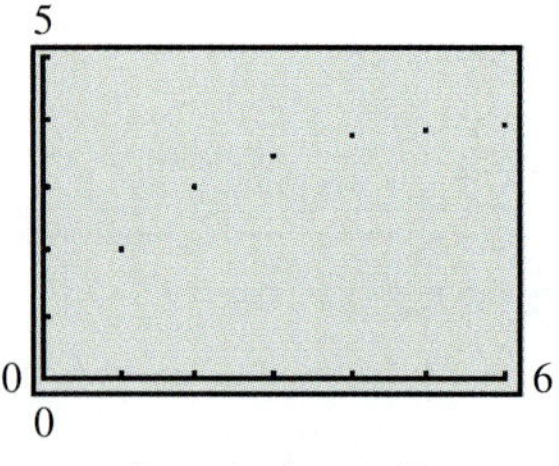

As n gets larger, S_n approaches 4.

Figure 10

The formula for S_n can also be written as $S_n = 4 - \left(\frac{1}{2}\right)^{n-2}$. As this formula and Figure 10 suggest, these sums seem to be getting closer and closer to the number 4. For no value of n is $S_n = 4$. However, if n is large enough, then S_n is as close to 4 as desired. As mentioned earlier, we say the sequence converges to 4. This is expressed as

$$\lim_{n \to \infty} S_n = 4.$$

(Read: "the limit of S_n as n increases without bound is 4.") Since $\lim_{n \to \infty} S_n = 4$, the number 4 is called the *sum of the terms* of the infinite geometric sequence

$$2, 1, \frac{1}{2}, \frac{1}{4}, \ldots$$

and

$$2 + 1 + \frac{1}{2} + \frac{1}{4} + \cdots = 4.$$

Looking Ahead to Calculus

In the discussion of

$$\lim_{n \to \infty} S_n = 4,$$

we used the phrases "large enough" and "as close as desired." This description is made more precise in a standard calculus course.

EXAMPLE 7 Summing the Terms of an Infinite Geometric Series

Evaluate $1 + \frac{1}{3} + \frac{1}{9} + \frac{1}{27} + \cdots$.

Solution Use the formula for the sum of the first n terms of a geometric sequence to obtain

$$S_1 = 1, \quad S_2 = \frac{4}{3}, \quad S_3 = \frac{13}{9}, \quad S_4 = \frac{40}{27},$$

and, in general,

$$S_n = \frac{1\left[1 - \left(\frac{1}{3}\right)^n\right]}{1 - \frac{1}{3}}. \quad \text{Let } a_1 = 1, r = \tfrac{1}{3}.$$

The table shows the value of $\left(\frac{1}{3}\right)^n$ for larger and larger values of n.

n	1	10	100	200
$\left(\frac{1}{3}\right)^n$	$\frac{1}{3}$	1.69×10^{-5}	1.94×10^{-48}	3.76×10^{-96}

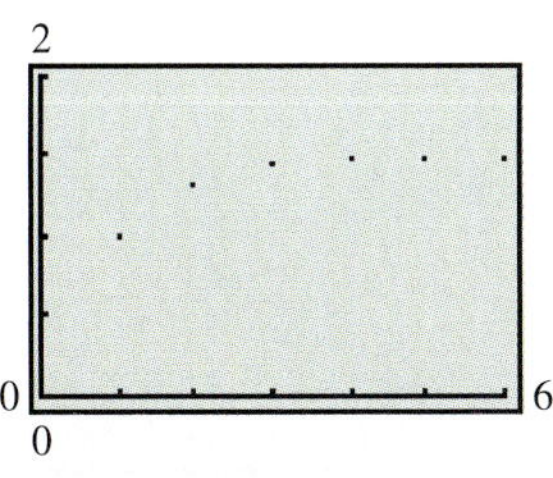

This graph of the first six values of S_n in Example 7 shows its value approaching $\frac{3}{2}$. (The y-scale here is $\frac{1}{2}$.)

As n gets larger and larger, $\left(\frac{1}{3}\right)^n$ gets closer and closer to 0. That is,

$$\lim_{n\to\infty}\left(\frac{1}{3}\right)^n = 0,$$

making it reasonable that

$$\lim_{n\to\infty} S_n = \lim_{n\to\infty}\frac{1\left[1-\left(\frac{1}{3}\right)^n\right]}{1-\frac{1}{3}} = \frac{1(1-0)}{1-\frac{1}{3}} = \frac{1}{\frac{2}{3}} = \frac{3}{2}.$$

Simplify the complex fraction. **(Section R.5)**

Hence,

$$1 + \frac{1}{3} + \frac{1}{9} + \frac{1}{27} + \cdots = \frac{3}{2}.$$

Now try Exercise 39.

If a geometric series has first term a_1 and common ratio r, then

$$S_n = \frac{a_1(1-r^n)}{1-r} \qquad (r \neq 1)$$

for every positive integer n. If $-1 < r < 1$, then $\lim_{n\to\infty} r^n = 0$, and

$$\lim_{n\to\infty} S_n = \frac{a_1(1-0)}{1-r} = \frac{a_1}{1-r}.$$

This quotient, $\frac{a_1}{1-r}$, is called the **sum of the terms of an infinite geometric sequence.** The limit $\lim_{n\to\infty} S_n$ is often expressed as S_∞ or $\sum_{i=1}^{\infty} a_i$.

Sum of the Terms of an Infinite Geometric Sequence

The sum of the terms of an infinite geometric sequence with first term a_1 and common ratio r, where $-1 < r < 1$, is given by

$$S_\infty = \frac{a_1}{1-r}.$$

If $|r| > 1$, then the terms get larger and larger in absolute value, so there is no limit as $n \to \infty$. Hence the terms of the sequence will not have a sum.

EXAMPLE 8 Finding the Sum of the Terms of an Infinite Geometric Sequence

Find each sum.

(a) $\sum_{i=1}^{\infty}\left(-\frac{3}{4}\right)\left(-\frac{1}{2}\right)^{i-1}$ **(b)** $\sum_{i=1}^{\infty}\left(\frac{3}{5}\right)^{i}$

Solution

(a) Here, $a_1 = -\frac{3}{4}$ and $r = -\frac{1}{2}$. Since $-1 < r < 1$, the preceding formula applies.

$$S_\infty = \frac{a_1}{1 - r} = \frac{-\frac{3}{4}}{1 - \left(-\frac{1}{2}\right)} = -\frac{1}{2}$$

(b) $\sum_{i=1}^{\infty}\left(\frac{3}{5}\right)^{i} = \frac{\frac{3}{5}}{1 - \frac{3}{5}} = \frac{3}{2}$ $\quad a_1 = \frac{3}{5}, r = \frac{3}{5}$

Now try Exercises 45 and 47.

Looking Ahead to Calculus

In calculus, functions are sometimes defined in terms of infinite series. Here are three functions we studied earlier in the text defined that way.

$$e^x = \frac{x^0}{0!} + \frac{x^1}{1!} + \frac{x^2}{2!} + \frac{x^3}{3!} + \cdots$$

$$\ln(1 + x) = x - \frac{x^2}{2} + \frac{x^3}{3} - \cdots$$

for x in $(-1, 1)$

$$\frac{1}{1 + x} = 1 - x + x^2 - x^3 + \cdots$$

for x in $(-1, 1)$

Annuities A sequence of equal payments made after equal periods of time, such as car payments or house payments, is called an **annuity.** If the payments are accumulated in an account (with no withdrawals), the sum of the payments and interest on the payments is called the **future value** of the annuity.

EXAMPLE 9 Finding the Future Value of an Annuity

To save money for a trip, Taylor Wells deposited \$1000 at the *end* of each year for 4 yr in an account paying 3% interest, compounded annually. Find the future value of this annuity.

Solution To find the future value, we use the formula for interest compounded annually,

$$A = P(1 + r)^t. \quad \text{(Section 4.2)}$$

The first payment earns interest for 3 yr, the second payment for 2 yr, and the third payment for 1 yr. The last payment earns no interest. The total amount is

$$1000(1.03)^3 + 1000(1.03)^2 + 1000(1.03) + 1000.$$

This is the sum of the terms of a geometric sequence with first term (starting at the end of the sum as written above) $a_1 = 1000$ and common ratio $r = 1.03$. Using the formula for S_4, the sum of four terms, gives

$$S_4 = \frac{1000[1 - (1.03)^4]}{1 - 1.03} \approx 4183.63.$$

The future value of the annuity is \$4183.63.

Now try Exercise 71.

The general formula for the future value of an annuity can be stated as follows. (See Exercise 79.)

Future Value of an Annuity

The formula for the future value of an annuity is given by

$$S = R\left[\frac{(1+i)^n - 1}{i}\right],$$

where S is future value, R is payment at the end of each period, i is interest rate per period, and n is number of periods.

11.3 Exercises

Refer to the first sentence of this section, which describes a method of payment for a job. Determine **(a)** *how much you will earn on the day indicated and* **(b)** *how much you will have earned altogether after your wages are paid on the day indicated. See Examples 1 and 5.*

1. day 10 **2.** day 12 **3.** day 15 **4.** day 18

Find a_5 and a_n for each geometric sequence. See Example 2.

5. $a_1 = 5, r = -2$ **6.** $a_1 = 8, r = -5$ **7.** $a_2 = -4, r = 3$

8. $a_3 = -2, r = 4$ **9.** $a_4 = 243, r = -3$ **10.** $a_4 = 18, r = 2$

11. $-4, -12, -36, -108, \ldots$ **12.** $-2, 6, -18, 54, \ldots$

13. $\frac{4}{5}, 2, 5, \frac{25}{2}, \ldots$ **14.** $\frac{1}{2}, \frac{2}{3}, \frac{8}{9}, \frac{32}{27}, \ldots$

15. $10, -5, \frac{5}{2}, -\frac{5}{4}, \ldots$ **16.** $3, -\frac{9}{4}, \frac{27}{16}, -\frac{81}{64}, \ldots$

Find a_1 and r for each geometric sequence. See Example 3.

17. $a_2 = -6, a_7 = -192$ **18.** $a_3 = 5, a_8 = \frac{1}{625}$

19. $a_3 = 50, a_7 = .005$ **20.** $a_4 = -\frac{1}{4}, a_9 = -\frac{1}{128}$

Use the formula for S_n to find the sum of the first five terms of each geometric sequence. In Exercises 25 and 26, round to the nearest hundredth. See Example 5.

21. $2, 8, 32, 128, \ldots$ **22.** $4, 16, 64, 256, \ldots$

23. $18, -9, \frac{9}{2}, -\frac{9}{4}, \ldots$ **24.** $12, -4, \frac{4}{3}, -\frac{4}{9}, \ldots$

25. $a_1 = 8.423, r = 2.859$ **26.** $a_1 = -3.772, r = -1.553$

Find each sum. See Example 6.

27. $\sum_{i=1}^{5} 3^i$ **28.** $\sum_{i=1}^{4} (-2)^i$ **29.** $\sum_{j=1}^{6} 48\left(\frac{1}{2}\right)^j$

30. $\sum_{j=1}^{5} 243\left(\frac{2}{3}\right)^j$ **31.** $\sum_{k=4}^{10} 2^k$ **32.** $\sum_{k=3}^{9} (-3)^k$

33. *Concept Check* Under what conditions does the sum of an infinite geometric series exist?

34. The number .999. . . can be written as the sum of the terms of an infinite geometric sequence: $.9 + .09 + .009 + \cdots$. Here we have $a_1 = .9$ and $r = .1$. Use the formula for S_∞ to find this sum. Does your intuition indicate that your answer is correct?

Find r for each infinite geometric sequence. Identify any whose sum does not converge.

35. 12, 24, 48, 96, . . .

36. 625, 125, 25, 5, . . .

37. $-48, -24, -12, -6, \ldots$

38. $2, -10, 50, -250, \ldots$

39. Use $\lim_{n\to\infty} S_n$ to show that $2 + 1 + \frac{1}{2} + \frac{1}{4} + \cdots$ converges to 4. See Example 7.

40. In Example 7, we determined that $1 + \frac{1}{3} + \frac{1}{9} + \frac{1}{27} + \cdots$ converges to $\frac{3}{2}$ using an argument involving limits. Use the formula for the sum of the terms of an infinite geometric sequence to obtain the same result.

Find each sum that converges. See Example 8.

41. $18 + 6 + 2 + \dfrac{2}{3} + \cdots$

42. $100 + 10 + 1 + \cdots$

43. $\dfrac{1}{4} - \dfrac{1}{6} + \dfrac{1}{9} - \dfrac{2}{27} + \cdots$

44. $\dfrac{4}{3} + \dfrac{2}{3} + \dfrac{1}{3} + \cdots$

45. $\displaystyle\sum_{i=1}^{\infty} 3\left(\frac{1}{4}\right)^{i-1}$

46. $\displaystyle\sum_{i=1}^{\infty} 5\left(-\frac{1}{4}\right)^{i-1}$

47. $\displaystyle\sum_{k=1}^{\infty} (.3)^k$

48. $\displaystyle\sum_{k=1}^{\infty} 10^{-k}$

Relating Concepts

For individual or collaborative investigation
(Exercises 49–52)

Let $g(x) = ab^x$. ***Work Exercises 49–52 in order.***

49. Find $g(1)$, $g(2)$, and $g(3)$.

50. Consider the sequence $g(1), g(2), g(3), \ldots$. Is it a geometric sequence? If so, what is the common ratio?

51. What is the general term of the sequence in Exercise 50?

52. Explain how geometric sequences are related to exponential functions.

Use the sum and sequence features of a graphing calculator to evaluate each sum. Round to the nearest thousandth.

53. $\displaystyle\sum_{i=1}^{10} (1.4)^i$

54. $\displaystyle\sum_{j=1}^{6} -(3.6)^j$

55. $\displaystyle\sum_{j=3}^{8} 2(.4)^j$

56. $\displaystyle\sum_{i=4}^{9} 3(.25)^i$

Solve each problem. See Examples 1–8.

57. *(Modeling) Investment for Retirement* According to T. Rowe Price Associates, a person with a moderate investment strategy and n years to retirement should have accumulated savings of a_n percent of his or her annual salary. The geometric sequence defined by

$$a_n = 1276(.916)^n$$

gives the appropriate percent for each year n.

(a) Find a_1 and r. Round a_1 to the nearest whole number.
(b) Find and interpret the terms a_{10} and a_{20}. Round to the nearest whole number.

58. ***(Modeling) Investment for Retirement*** Refer to Exercise 57. For someone who has a conservative investment strategy with n years to retirement, the geometric sequence is

$$a_n = 1278(.935)^n.$$

(*Source:* T. Rowe Price Associates.)

(a) Repeat part (a) of Exercise 57. **(b)** Repeat part (b) of Exercise 57.
(c) Why are the answers in parts (a) and (b) larger than those in Exercise 57?

59. ***(Modeling) Bacterial Growth*** The strain of bacteria described in Exercise 83 in the first section of this chapter will double in size and then divide every 40 min. Let a_1 be the initial number of bacteria cells, a_2 the number after 40 min, and a_n the number after $40(n - 1)$ min. (*Source:* Hoppensteadt, F. and C. Peskin, *Mathematics in Medicine and the Life Sciences,* Springer-Verlag, 1992.)

(a) Write a formula for the nth term a_n of the geometric sequence

$$a_1, a_2, a_3, \ldots, a_n, \ldots.$$

(b) Determine the first n where $a_n > 1{,}000{,}000$ if $a_1 = 100$.
(c) How long does it take for the number of bacteria to exceed one million?

60. ***Photo Processing*** The final step in processing a black-and-white photographic print is to immerse the print in a chemical fixer. The print is then washed in running water. Under certain conditions, 98% of the fixer in a print will be removed with 15 min of washing. How much of the original fixer would be left after 1 hr of washing?

61. ***(Modeling) Fruit Flies Population*** A population of fruit flies is growing in such a way that each generation is 1.25 times as large as the last generation. Suppose there were 200 insects in the first generation. How many would there be in the fifth generation?

62. ***Depreciation*** Each year a machine loses 20% of the value it had at the beginning of the year. Find the value of the machine at the end of 6 yr if it cost $100,000 new.

63. ***Sugar Processing*** A sugar factory receives an order for 1000 units of sugar. The production manager thus orders production of 1000 units of sugar. He forgets, however, that the production of sugar requires some sugar (to prime the machines, for example), and so he ends up with only 900 units of sugar. He then orders an additional 100 units, and receives only 90 units. A further order for 10 units produces 9 units. Finally seeing he is wrong, the manager decides to try mathematics. He views the production process as an infinite geometric progression with $a_1 = 1000$ and $r = .1$. Using this, find the number of units of sugar that he should have ordered originally.

64. ***Height of a Dropped Ball*** Mitzi drops a ball from a height of 10 m and notices that on each bounce the ball returns to about $\frac{3}{4}$ of its previous height. About how far will the ball travel before it comes to rest? (*Hint:* Consider the sum of two sequences.)

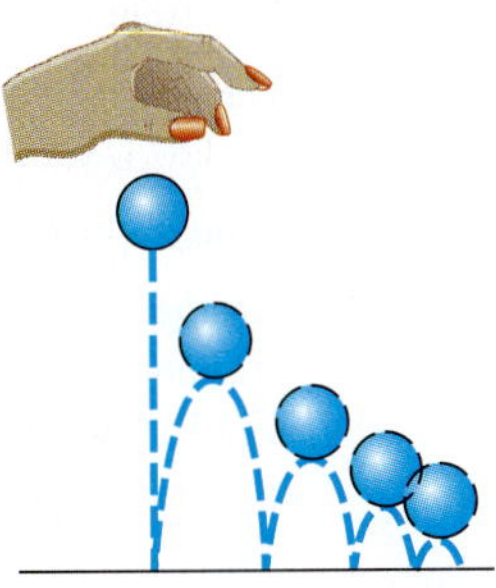

65. *Number of Ancestors* Each person has two parents, four grandparents, eight great-grandparents, and so on. What is the total number of ancestors a person has, going back five generations? ten generations?

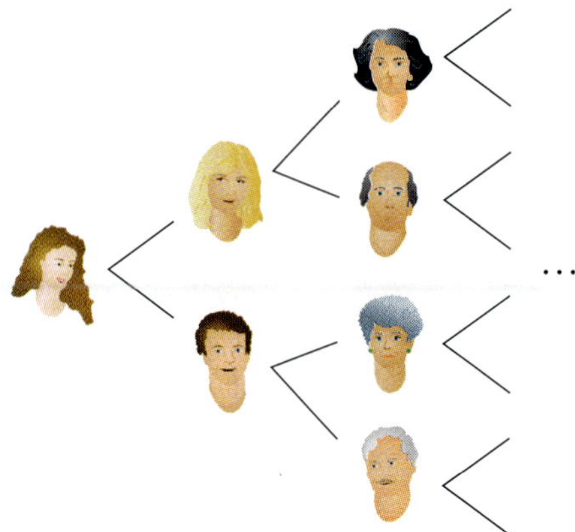

66. *Drug Dosage* Certain medical conditions are treated with a fixed dose of a drug administered at regular intervals. Suppose a person is given 2 mg of a drug each day and that during each 24-hr period, the body utilizes 40% of the amount of drug that was present at the beginning of the period.

(a) Show that the amount of the drug present in the body at the end of n days is $\sum_{i=1}^{n} 2(.6)^i$.

(b) What will be the approximate quantity of the drug in the body at the end of each day after the treatment has been administered for a long period of time?

67. *Side Length of a Triangle* A sequence of equilateral triangles is constructed. The first triangle has sides 2 m in length. To get the second triangle, midpoints of the sides of the original triangle are connected. What is the length of each side of the eighth such triangle? See the figure.

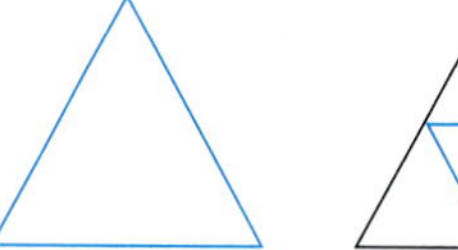
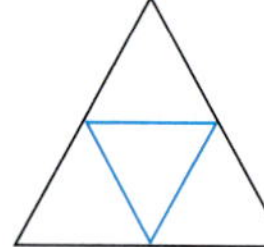
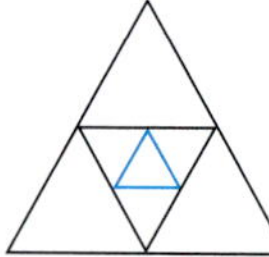
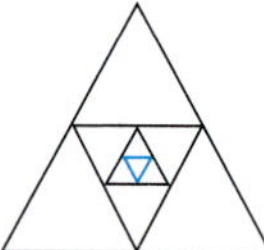

68. *Perimeter and Area of Triangles* In Exercise 67, if the process could be continued indefinitely, what would be the total perimeter of all the triangles? What would be the total area of all the triangles, disregarding the overlapping?

69. *Salaries* You are offered a 6-week summer job and are asked to select one of the following salary options.

Option 1: \$5000 for the first day with a \$10,000 raise each day for the remaining 29 days (that is, \$15,000 for day 2, \$25,000 for day 3, and so on)

Option 2: 1¢ for the first day with the pay doubled each day (that is, 2¢ for day 2, 4¢ for day 3, and so on)

Which option would you choose?

70. *Number of Ancestors* Suppose a genealogical Web site allows you to identify all your ancestors that lived during the last 300 yr. Assuming that each generation spans about 25 yr, guess the number of ancestors that would be found during the 12 generations. Then use the formula for a geometric series to find the correct value.

Future Value of an Annuity *Find the future value of each annuity. See Example 9.*

71. payments of \$1000 at the end of each year for 9 yr at 3% interest compounded annually

72. payments of \$800 at the end of each year for 12 yr at 2% interest compounded annually

73. payments of \$2430 at the end of each year for 10 yr at 1% interest compounded annually

74. payments of \$1500 at the end of each year for 6 yr at .5% interest compounded annually

75. Refer to Exercise 73. Use the answer and recursion to find the balance after 11 yr.

76. Refer to Exercise 74. Use the answer and recursion to find the balance after 7 yr.

77. ***Individual Retirement Account*** Starting on his fortieth birthday, Michael Branson deposits \$2000 per year in an Individual Retirement Account until age 65 (last payment at age 64). Find the total amount in the account if he had a guaranteed interest rate of 3% compounded annually.

78. ***Retirement Savings*** To save for retirement, Karla Harby put \$3000 at the end of each year into an ordinary annuity for 20 yr at 2.5% annual interest. At the end of the 20 yr, what was the amount of the annuity?

79. Show that the formula for future value of an annuity gives the correct answer when compared to the solution in Example 9.

80. The screen here shows how the TI-83 Plus calculator computes the future value of the annuity described in Example 9. Use a calculator with this capability to support your answers in Exercises 71–78.

81. ***Concept Check*** Let $a_1, a_2, a_3, \ldots$ and $b_1, b_2, b_3, \ldots$ be geometric sequences. Let $d_n = c \cdot a_n \cdot b_n$ for a fixed real number c and every positive integer n. Show that $d_1, d_2, d_3, \ldots$ is a geometric sequence.

82. Explain why the sequence log 6, log 36, log 1296, log 1,679,616, . . . is geometric.

Summary Exercises on Sequences and Series

Use the following guidelines in Exercises 1–14.

Given a sequence $a_1, a_2, a_3, a_4, a_5, \ldots,$

If the differences $a_2 - a_1, a_3 - a_2, a_4 - a_3, a_5 - a_4, \ldots$ are all equal to the same number d, then the sequence is *arithmetic*, and d is the *common difference*.

If the ratios $\frac{a_2}{a_1}, \frac{a_3}{a_2}, \frac{a_4}{a_3}, \frac{a_5}{a_4}, \ldots$ are all equal to the same number r, then the sequence is *geometric*, and r is the *common ratio*.

In Exercises 1–8, determine whether the sequence is arithmetic, geometric, *or* neither. *If the sequence is arithmetic, give its common difference d. If the sequence is geometric, give its common ratio r.*

1. 2, 4, 8, 16, 32, . . .

2. 1, 4, 7, 10, 13, . . .

3. $3, \frac{1}{2}, -2, -\frac{9}{2}, -7, \ldots$

4. $1, -2, 3, -4, 5, \ldots$

5. $\frac{3}{4}, 1, \frac{4}{3}, \frac{16}{9}, \frac{64}{27}, \ldots$

6. $4, -12, 36, -108, 324, \ldots$

7. $\frac{1}{2}, \frac{1}{3}, \frac{1}{4}, \frac{1}{5}, \frac{1}{6}, \ldots$

8. $5, 2, -1, -4, -7, \ldots$

In Exercises 9–14, determine whether the given sequence is either arithmetic *or* geometric. *Then find* a_n *and* $\sum_{i=1}^{10} a_i$.

9. $3, 6, 12, 24, 48, \ldots$

10. $2, 6, 10, 14, 18, \ldots$

11. $4, \frac{5}{2}, 1, -\frac{1}{2}, -2, \ldots$

12. $\frac{3}{2}, 1, \frac{2}{3}, \frac{4}{9}, \frac{8}{27}, \ldots$

13. $3, -6, 12, -24, 48, \ldots$

14. $-5, -8, -11, -14, -17, \ldots$

Evaluate each sum, where possible. Identify any that diverge.

15. $\sum_{i=1}^{\infty} \frac{1}{3}(-2)^{i-1}$

16. $\sum_{j=1}^{4} 2\left(\frac{1}{10}\right)^{j-1}$

17. $\sum_{i=1}^{25} (4 - 6i)$

18. $\sum_{i=1}^{6} 3^i$

19. $\sum_{i=1}^{\infty} 4\left(-\frac{1}{2}\right)^i$

20. $\sum_{i=1}^{\infty} (3i - 2)$

21. $\sum_{j=1}^{12} (2j - 1)$

22. $\sum_{k=1}^{\infty} 3^{-k}$

11.4 The Binomial Theorem

A Binomial Expansion Pattern ▪ Pascal's Triangle ▪ *n*-Factorial ▪ Binomial Coefficients ▪ The Binomial Theorem ▪ *k*th Term of a Binomial Expansion

Looking Ahead to Calculus

Students taking calculus study the binomial series, which follows from Isaac Newton's extension to the case where the exponent is no longer a positive integer. His result led to a series for $(1 + x)^k$, where k is a real number and $|x| < 1$.

A Binomial Expansion Pattern In this section, we introduce a method for writing the expansion of expressions of the form $(x + y)^n$, where n is a natural number. The formula for writing the terms of the expansion of $(x + y)^n$ is called the *binomial theorem*. Some expansions of $(x + y)^n$, for various nonnegative integer values of n, are given below.

$$(x + y)^0 = 1$$
$$(x + y)^1 = x + y$$
$$(x + y)^2 = x^2 + 2xy + y^2$$
$$(x + y)^3 = x^3 + 3x^2y + 3xy^2 + y^3$$
$$(x + y)^4 = x^4 + 4x^3y + 6x^2y^2 + 4xy^3 + y^4$$
$$(x + y)^5 = x^5 + 5x^4y + 10x^3y^2 + 10x^2y^3 + 5xy^4 + y^5$$

Notice that after the special case $(x + y)^0 = 1$, each expansion begins with x raised to the same power as the binomial itself. That is, the expansion of $(x + y)^1$ has a first term of x^1, $(x + y)^2$ has a first term of x^2, $(x + y)^3$ has a first term of x^3, and so on. Also, the last term in each expansion is y to the same power as the binomial. Thus, the expansion of $(x + y)^n$ should begin with the term x^n and end with the term y^n.

Notice that the exponent on x decreases by 1 in each term after the first, while the exponent on y, beginning with y in the second term, increases by 1 in each succeeding term. That is, the *variables* in the terms of the expansion of $(x + y)^n$ have the following pattern.

$$x^n, x^{n-1}y, x^{n-2}y^2, x^{n-3}y^3, \ldots, xy^{n-1}, y^n$$

This pattern suggests that the sum of the exponents on x and y in each term is n. For example, the third term in the list above is $x^{n-2}y^2$, and the sum of the exponents is $n - 2 + 2 = n$.

Pascal's Triangle Now, examine the *coefficients* in the terms of the expansion of $(x + y)^n$. Writing the coefficients alone gives the following pattern.

Pascal's Triangle

											Row
					1						0
				1		1					1
			1		2		1				2
		1		3		3		1			3
	1		4		6		4		1		4
1		5		10		10		5		1	5

Blaise Pascal (1623–1662)

With the coefficients arranged in this way, each number in the triangle is the sum of the two numbers directly above it (one to the right and one to the left). For example, in row four, 1 is the sum of 1 (the only number above it), 4 is the sum of 1 and 3, 6 is the sum of 3 and 3, and so on. This triangular array of numbers is called **Pascal's triangle,** in honor of the seventeenth-century mathematician Blaise Pascal, one of the first to use it extensively.

To find the coefficients for $(x + y)^6$, we need to include row six in Pascal's triangle. Adding adjacent numbers, we find that row six is

$$1 \quad 6 \quad 15 \quad 20 \quad 15 \quad 6 \quad 1.$$

Using these coefficients, we obtain the expansion of $(x + y)^6$:

$$(x + y)^6 = x^6 + 6x^5y + 15x^4y^2 + 20x^3y^3 + 15x^2y^4 + 6xy^5 + y^6.$$

***n*-Factorial** Although it is possible to use Pascal's triangle to find the coefficients of $(x + y)^n$ for any positive integer n, this calculation becomes impractical for large values of n because of the need to write all the preceding rows. A more efficient way of finding these coefficients uses **factorial notation.** The number $n!$ (read "n-factorial") is defined as follows.

n-Factorial

For any positive integer n,

$$\mathbf{n! = n(n-1)(n-2)\cdots(3)(2)(1) \quad \text{and} \quad 0! = 1.}$$

For example,

$$5! = 5 \cdot 4 \cdot 3 \cdot 2 \cdot 1 = 120,$$

$$7! = 7 \cdot 6 \cdot 5 \cdot 4 \cdot 3 \cdot 2 \cdot 1 = 5040,$$

and

$$2! = 2 \cdot 1 = 2.$$

Binomial Coefficients Now look at the coefficients of the expansion

$$(x + y)^5 = x^5 + 5x^4y + 10x^3y^2 + 10x^2y^3 + 5xy^4 + y^5.$$

The coefficient of the second term, $5x^4y$, is 5, and the exponents on the variables are 4 and 1. Note that

$$5 = \frac{5!}{4!1!}.$$

The coefficient of the third term is 10, with exponents of 3 and 2, and

$$10 = \frac{5!}{3!2!}.$$

The last term (the sixth term) can be written as $y^5 = 1x^0y^5$, with coefficient 1, and exponents of 0 and 5. Since $0! = 1$,

$$1 = \frac{5!}{0!5!}.$$

Generalizing from these examples, the coefficient for the term of the expansion of $(x + y)^n$ in which the variable part is x^ry^{n-r} (where $r \leq n$) is

$$\frac{n!}{r!(n-r)!}.$$

This number, called a **binomial coefficient,** is often symbolized $\binom{n}{r}$ or ${}_nC_r$ (read "n choose r").

Binomial Coefficient

For nonnegative integers n and r, with $r \leq n$,

$$ {}_nC_r = \binom{n}{r} = \frac{n!}{r!(n-r)!}.$$

The binomial coefficients are numbers from Pascal's triangle. For example, $\binom{3}{0}$ is the first number in row three, and $\binom{7}{4}$ is the fifth number in the row seven.

Graphing calculators are capable of finding binomial coefficients. A calculator with a 10-digit display will give exact values for $n!$ for $n \leq 13$ and approximate values of $n!$ for $14 \leq n \leq 69$. ■

EXAMPLE 1 Evaluating Binomial Coefficients

Evaluate each binomial coefficient.

(a) $\binom{6}{2}$ **(b)** $\binom{8}{0}$ **(c)** $\binom{10}{10}$ **(d)** ${}_{12}C_{10}$

Algebraic Solution

(a) $\binom{6}{2} = \frac{6!}{2!(6-2)!} = \frac{6!}{2!4!}$

$= \frac{6 \cdot 5 \cdot 4 \cdot 3 \cdot 2 \cdot 1}{2 \cdot 1 \cdot 4 \cdot 3 \cdot 2 \cdot 1} = 15$

(b) $\binom{8}{0} = \frac{8!}{0!(8-0)!} = \frac{8!}{0!8!} = \frac{8!}{1 \cdot 8!} = 1$

(c) $\binom{10}{10} = \frac{10!}{10!(10-10)!} = \frac{10!}{10!0!} = 1$

(d) ${}_{12}C_{10} = \frac{12!}{10!(12-10)!} = \frac{12!}{10!2!} = 66$

Graphing Calculator Solution

Graphing calculators calculate binomial coefficients using the notation ${}_nC_r$. This function is often found in MATH mode. Figure 11 shows the values of the binomial coefficients found in parts (a), (b), and (c).

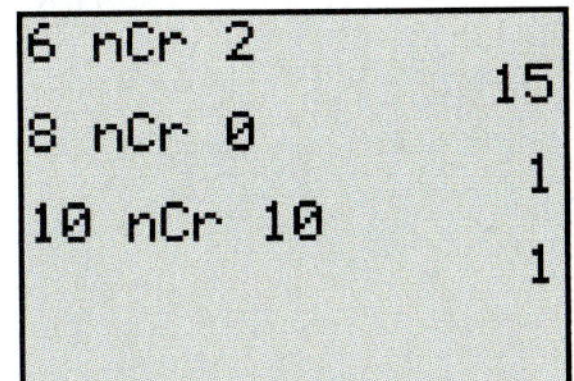

Figure 11

Now try Exercises 5, 9, and 17.

Refer again to Pascal's triangle. Notice the symmetry in each row. This suggests that binomial coefficients should have the same property. That is,

$$\binom{n}{r} = \binom{n}{n-r}.$$

This is true, since

$$\binom{n}{r} = \frac{n!}{r!(n-r)!} \quad \text{and} \quad \binom{n}{n-r} = \frac{n!}{(n-r)!r!}.$$

The Binomial Theorem Our observations about the expansion of $(x + y)^n$ are summarized as follows.

1. There are $n + 1$ terms in the expansion.
2. The first term is x^n, and the last term is y^n.
3. In each succeeding term, the exponent on x decreases by 1 and the exponent on y increases by 1.
4. The sum of the exponents on x and y in any term is n.
5. The coefficient of the term with $x^r y^{n-r}$ or $x^{n-r} y^r$ is $\binom{n}{r}$.

These observations about the expansion of $(x + y)^n$ for any positive integer value of n suggest the **binomial theorem.** A proof of the binomial theorem using mathematical induction is given in the next section.

Looking Ahead to Calculus

The binomial theorem is used to show that the derivative of $f(x) = x^n$ is given by the function $f'(x) = nx^{n-1}$. This fact is used extensively in calculus.

Binomial Theorem

For any positive integer n and any complex numbers x and y,

$$(x + y)^n = x^n + \binom{n}{1}x^{n-1}y + \binom{n}{2}x^{n-2}y^2 + \binom{n}{3}x^{n-3}y^3 + \cdots + \binom{n}{r}x^{n-r}y^r + \cdots + \binom{n}{n-1}xy^{n-1} + y^n.$$

NOTE The binomial theorem looks much more manageable written as a series. The theorem can be summarized as

$$(x + y)^n = \sum_{r=0}^{n} \binom{n}{r} x^{n-r}y^r.$$

EXAMPLE 2 Applying the Binomial Theorem

Write the binomial expansion of $(x + y)^9$.

Solution Using the binomial theorem,

$$(x + y)^9 = x^9 + \binom{9}{1}x^8y + \binom{9}{2}x^7y^2 + \binom{9}{3}x^6y^3 + \binom{9}{4}x^5y^4 + \binom{9}{5}x^4y^5 + \binom{9}{6}x^3y^6 + \binom{9}{7}x^2y^7 + \binom{9}{8}xy^8 + y^9.$$

Now evaluate each of the binomial coefficients.

$$\begin{aligned}(x + y)^9 &= x^9 + \frac{9!}{1!8!}x^8y + \frac{9!}{2!7!}x^7y^2 + \frac{9!}{3!6!}x^6y^3 + \frac{9!}{4!5!}x^5y^4 + \frac{9!}{5!4!}x^4y^5 \\ &\quad + \frac{9!}{6!3!}x^3y^6 + \frac{9!}{7!2!}x^2y^7 + \frac{9!}{8!1!}xy^8 + y^9 \\ &= x^9 + 9x^8y + 36x^7y^2 + 84x^6y^3 + 126x^5y^4 + 126x^4y^5 \\ &\quad + 84x^3y^6 + 36x^2y^7 + 9xy^8 + y^9\end{aligned}$$

Now try Exercise 21.

EXAMPLE 3 Applying the Binomial Theorem

Expand $\left(a - \frac{b}{2}\right)^5$.

Solution Write the binomial as follows.

$$\left(a - \frac{b}{2}\right)^5 = \left(a + \left(-\frac{b}{2}\right)\right)^5$$

Now use the binomial theorem with $x = a$, $y = -\frac{b}{2}$, and $n = 5$ to obtain

$$\left(a - \frac{b}{2}\right)^5 = a^5 + \binom{5}{1}a^4\left(-\frac{b}{2}\right) + \binom{5}{2}a^3\left(-\frac{b}{2}\right)^2 + \binom{5}{3}a^2\left(-\frac{b}{2}\right)^3 + \binom{5}{4}a\left(-\frac{b}{2}\right)^4 + \left(-\frac{b}{2}\right)^5$$

$$= a^5 + 5a^4\left(-\frac{b}{2}\right) + 10a^3\left(-\frac{b}{2}\right)^2 + 10a^2\left(-\frac{b}{2}\right)^3 + 5a\left(-\frac{b}{2}\right)^4 + \left(-\frac{b}{2}\right)^5$$

$$= a^5 - \frac{5}{2}a^4b + \frac{5}{2}a^3b^2 - \frac{5}{4}a^2b^3 + \frac{5}{16}ab^4 - \frac{1}{32}b^5.$$

Now try Exercise 33.

NOTE As Example 3 illustrates, any expansion of the *difference* of two terms has alternating signs.

EXAMPLE 4 Applying the Binomial Theorem

Expand $\left(\frac{3}{m^2} - 2\sqrt{m}\right)^4$. (Assume $m > 0$.)

Solution By the binomial theorem,

$$\left(\frac{3}{m^2} - 2\sqrt{m}\right)^4 = \left(\frac{3}{m^2}\right)^4 + \binom{4}{1}\left(\frac{3}{m^2}\right)^3(-2\sqrt{m})^1 + \binom{4}{2}\left(\frac{3}{m^2}\right)^2(-2\sqrt{m})^2$$

$$+ \binom{4}{3}\left(\frac{3}{m^2}\right)^1(-2\sqrt{m})^3 + (-2\sqrt{m})^4$$

$$= \frac{81}{m^8} + 4\left(\frac{27}{m^6}\right)(-2m^{1/2}) + 6\left(\frac{9}{m^4}\right)(4m)$$

$$+ 4\left(\frac{3}{m^2}\right)(-8m^{3/2}) + 16m^2 \qquad \sqrt{m} = m^{1/2} \text{ (Section R.7)}$$

$$= \frac{81}{m^8} - \frac{216}{m^{11/2}} + \frac{216}{m^3} - \frac{96}{m^{1/2}} + 16m^2.$$

Now try Exercise 35.

kth Term of a Binomial Expansion Earlier in this section, we wrote the binomial theorem in summation notation as $\sum_{r=0}^{n}\binom{n}{r}x^{n-r}y^r$, which gives the form of each term. We can use this form to write any particular term of a binomial expansion without writing out the entire expansion.

For example, to find the tenth term of $(x + y)^n$, where $n \geq 9$, first notice that in the tenth term y is raised to the ninth power (since y has the power 1 in the second term, the power 2 in the third term, and so on). Because the exponents on x and y in any term must have a sum of n, the exponent on x in the tenth term is $n - 9$. Thus, the tenth term of the expansion is

$$\binom{n}{9}x^{n-9}y^9 = \frac{n!}{9!(n-9)!}x^{n-9}y^9.$$

We give this result in the following theorem.

*k*th Term of the Binomial Expansion

The kth term of the binomial expansion of $(x + y)^n$, where $n \geq k - 1$, is

$$\binom{n}{k-1} x^{n-(k-1)} y^{k-1}.$$

To find the kth term of the binomial expansion, use the following steps.

Step 1 Find $k - 1$. This is the exponent on the second term of the binomial.

Step 2 Subtract the exponent found in Step 1 from n to get the exponent on the first term of the binomial.

Step 3 Determine the coefficient by using the exponents found in the first two steps and n.

EXAMPLE 5 Finding a Particular Term of a Binomial Expansion

Find the seventh term of $(a + 2b)^{10}$.

Solution In the seventh term, $2b$ has an exponent of 6 while a has an exponent of $10 - 6$, or 4. The seventh term is

$$\binom{10}{6} a^4 (2b)^6 = 210a^4(64b^6) = 13{,}440a^4b^6.$$

Now try Exercise 37.

11.4 Exercises

Evaluate each expression, if possible. See Example 1.

1. $\dfrac{6!}{3!\,3!}$ **2.** $\dfrac{5!}{2!\,3!}$ **3.** $\dfrac{7!}{3!\,4!}$ **4.** $\dfrac{8!}{5!\,3!}$

5. $\dbinom{8}{3}$ **6.** $\dbinom{7}{4}$ **7.** $\dbinom{10}{8}$ **8.** $\dbinom{9}{6}$

9. $\dbinom{13}{13}$ **10.** $\dbinom{12}{12}$ **11.** $\dbinom{n}{n-1}$ **12.** $\dbinom{n}{n-2}$

13. ${}_8C_3$ **14.** ${}_9C_7$ **15.** ${}_{100}C_2$ **16.** ${}_{20}C_{15}$

17. ${}_4C_0$ **18.** ${}_4C_1$

19. *Concept Check* What are the first and last terms in the expansion of $(2x + 3y)^4$?

20. Determine the binomial coefficient for the fifth term in the expansion of $(x + y)^8$.

Write the binomial expansion for each expression. See Examples 2–4.

21. $(x + y)^6$ **22.** $(m + n)^4$ **23.** $(p - q)^5$

24. $(a - b)^7$ **25.** $(r^2 + s)^5$ **26.** $(m + n^2)^4$

27. $(p + 2q)^4$
28. $(3r - s)^6$
29. $(7p + 2q)^4$
30. $(4a - 5b)^5$
31. $(3x - 2y)^6$
32. $(7k - 9j)^4$
33. $\left(\frac{m}{2} - 1\right)^6$
34. $\left(3 + \frac{y}{3}\right)^5$
35. $\left(\sqrt{2}r + \frac{1}{m}\right)^4$
36. $\left(\frac{1}{k} - \sqrt{3}p\right)^3$

Write the indicated term of each binomial expansion. See Example 5.

37. Sixth term of $(4h - j)^8$
38. Eighth term of $(2c - 3d)^{14}$
39. Fifteenth term of $(a^2 + b)^{22}$
40. Twelfth term of $(2x + y^2)^{16}$
41. Fifteenth term of $(x - y^3)^{20}$
42. Tenth term of $(a^3 + 3b)^{11}$

Concept Check *Work Exercises 43–46.*

43. Find the middle term of $(3x^7 + 2y^3)^8$.
44. Find the two middle terms of $(-2m^{-1} + 3n^{-2})^{11}$.
45. Find the value of n for which the coefficients of the fifth and eighth terms in the expansion of $(x + y)^n$ are the same.
46. Find the term(s) in the expansion of $\left(3 + \sqrt{x}\right)^{11}$ that contains x^4.

Relating Concepts

For individual or collaborative investigation

(Exercises 47–50)

In this section, we saw how the factorial of a positive integer n can be computed as a product: $n! = 1 \cdot 2 \cdot 3 \cdot \cdots \cdot n$. Calculators and computers can evaluate factorials very quickly. Before the days of technology, mathematicians developed a formula, called ***Stirling's formula,*** *for approximating large factorials. Interestingly enough, the formula involves the irrational numbers π and e.*

$$n! \approx \sqrt{2\pi n} \cdot n^n \cdot e^{-n}$$

As an example, the exact value of 5! is 120, and Stirling's formula gives the approximation as 118.019168 with a graphing calculator. This is "off" by less than 2, an error of only 1.65%. ***Work Exercises 47–50 in order.***

47. Use a calculator to find the exact value of 10! and its approximation, using Stirling's formula.
48. Subtract the smaller value from the larger value in Exercise 47. Divide it by 10! and convert to a percent. What is the percent error?
49. Repeat Exercises 47 and 48 for $n = 12$.
50. Repeat Exercises 47 and 48 for $n = 13$. What seems to happen as n gets larger?

In later courses, it is shown that

$$(1 + x)^n = 1 + nx + \frac{n(n-1)}{2!}x^2 + \frac{n(n-1)(n-2)}{3!}x^3 + \cdots$$

for any real number n (not just positive integer values) and any real number x, where $|x| < 1$. Use this series to approximate the given number to the nearest thousandth.

51. $(1.02)^{-3}$
52. $(1.04)^{-5}$
53. $(1.01)^{1.5}$
54. $(1.03)^2$

11.5 Mathematical Induction

Proof by Mathematical Induction ▪ Proving Statements ▪ Generalized Principle of Mathematical Induction ▪ Proof of the Binomial Theorem

Proof by Mathematical Induction Many results in mathematics are claimed true for every positive integer. Any of these results could be checked for $n = 1$, $n = 2$, $n = 3$, and so on, but since the set of positive integers is infinite it would be impossible to check every possible case. For example, let S_n represent the statement that the sum of the first n positive integers is $\frac{n(n+1)}{2}$.

$$S_n:\quad 1 + 2 + 3 + \cdots + n = \frac{n(n+1)}{2}$$

The truth of this statement is easily verified for the first few values of n:

If $n = 1$, then S_1 is	$1 = \frac{1(1+1)}{2}$,	which is true.
If $n = 2$, then S_2 is	$1 + 2 = \frac{2(2+1)}{2}$,	which is true.
If $n = 3$, then S_3 is	$1 + 2 + 3 = \frac{3(3+1)}{2}$,	which is true.
If $n = 4$, then S_4 is	$1 + 2 + 3 + 4 = \frac{4(4+1)}{2}$,	which is true.

Continuing in this way for any amount of time would still not prove that S_n is true for *every* positive integer value of n. To prove that such statements are true for every positive integer value of n, the following principle is often used.

Principle of Mathematical Induction

Let S_n be a statement concerning the positive integer n. Suppose that

1. S_1 is true;
2. for any positive integer k, $k \le n$, if S_k is true, then S_{k+1} is also true.

Then S_n is true for every positive integer value of n.

A proof by mathematical induction can be explained as follows. By assumption (1) above, the statement is true when $n = 1$. By assumption (2) above, the fact that the statement is true for $n = 1$ implies that it is true for $n = 1 + 1 = 2$. Using (2) again, the statement is thus true for $2 + 1 = 3$, for $3 + 1 = 4$, for $4 + 1 = 5$, and so on. Continuing in this way shows that the statement must be true for *every* positive integer.

Figure 12

The situation is similar to that of a number of dominoes lined up as shown in Figure 12. If the first domino is pushed over, it pushes the next, which pushes the next, and so on until all are down.

Another example of the principle of mathematical induction might be an infinite ladder. Suppose the rungs are spaced so that whenever you are on a rung, you know you can move to the next rung. Then *if* you can get to the first rung, you can go as high up the ladder as you wish.

Two separate steps are required for a proof by mathematical induction.

Proof by Mathematical Induction

Step 1 Prove that the statement is true for $n = 1$.

Step 2 Show that, for any positive integer k, $k \leq n$, if S_k is true, then S_{k+1} is also true.

Proving Statements Mathematical induction is used in the next example to prove the statement S_n mentioned at the beginning of this section.

EXAMPLE 1 Proving an Equality Statement

Let S_n represent the statement

$$1 + 2 + 3 + \cdots + n = \frac{n(n + 1)}{2}.$$

Prove that S_n is true for every positive integer n.

Solution The proof by mathematical induction is as follows.

Step 1 Show that the statement is true when $n = 1$. If $n = 1$, S_1 becomes

$$1 = \frac{1(1 + 1)}{2},$$

which is true.

Step 2 Show that S_k implies S_{k+1}, where S_k is the statement

$$1 + 2 + 3 + \cdots + k = \frac{k(k + 1)}{2},$$

and S_{k+1} is the statement

$$1 + 2 + 3 + \cdots + k + (k + 1) = \frac{(k + 1)[(k + 1) + 1]}{2}.$$

Start with S_k and assume it is a true statement.

$$1 + 2 + 3 + \cdots + k = \frac{k(k + 1)}{2}$$

Add $k + 1$ to both sides of this equation to obtain S_{k+1}.

$$1 + 2 + 3 + \cdots + k + (k + 1) = \frac{k(k + 1)}{2} + (k + 1)$$

$$= (k + 1)\left(\frac{k}{2} + 1\right) \qquad \text{Factor out } k + 1. \text{ (Section R.4)}$$

$$= (k + 1)\left(\frac{k + 2}{2}\right) \qquad \text{Add inside the parentheses. (Section R.5)}$$

$$1 + 2 + 3 + \cdots + k + (k + 1) = \frac{(k + 1)[(k + 1) + 1]}{2} \qquad \text{Multiply; } k + 2 = (k + 1) + 1.$$

This final result is the statement for $n = k + 1$; it has been shown that if S_k is true, then S_{k+1} is also true.

The two steps required for a proof by mathematical induction have been completed, so the statement S_n is true for every positive integer value of n.

Now try Exercise 1.

CAUTION Notice that the left side of the statement S_n always includes *all* the terms up to the nth term, as well as the nth term.

EXAMPLE 2 Proving an Inequality Statement

Prove: If x is a real number between 0 and 1, then for every positive integer n,

$$0 < x^n < 1.$$

Solution

Step 1 Here S_1 is the statement

$$\text{if } 0 < x < 1, \text{ then } 0 < x^1 < 1,$$

which is true.

Step 2 S_k is the statement

$$\text{if } 0 < x < 1, \text{ then } 0 < x^k < 1.$$

To show that this implies that S_{k+1} is true, multiply all three parts of $0 < x^k < 1$ by x to get

$$x \cdot 0 < x \cdot x^k < x \cdot 1. \quad \text{(Section 1.7)}$$

(Here the fact that $0 < x$ is used.) Simplify to obtain

$$0 < x^{k+1} < x.$$

Since $x < 1$,

$$0 < x^{k+1} < x < 1$$

and thus $$0 < x^{k+1} < 1.$$

This work shows that if S_k is true, then S_{k+1} is true.

Since both steps for a proof by mathematical induction have been completed, the given statement is true for every positive integer n.

Now try Exercise 23.

Generalized Principle of Mathematical Induction Some statements S_n are not true for the first few values of n, but are true for all values of n that are greater than or equal to some fixed integer j. The following slightly generalized form of the principle of mathematical induction takes care of these cases.

Generalized Principle of Mathematical Induction

Let S_n be a statement concerning the positive integer n. Let j be a fixed positive integer. Suppose that

Step 1 S_j is true;

Step 2 for any positive integer k, $k \geq j$, S_k implies S_{k+1}.

Then S_n is true for all positive integers n, where $n \geq j$.

EXAMPLE 3 Using the Generalized Principle

Let S_n represent the statement $2^n > 2n + 1$. Show that S_n is true for all values of n such that $n \geq 3$.

Solution (Check that S_n is false for $n = 1$ and $n = 2$.)

Step 1 Show that S_n is true for $n = 3$. If $n = 3$, then S_n is

$$2^3 > 2 \cdot 3 + 1,$$

or

$$8 > 7.$$

Thus, S_3 is true.

Step 2 Now show that S_k implies S_{k+1}, where $k \geq 3$, and where

$$S_k \text{ is } 2^k > 2k + 1,$$

and

$$S_{k+1} \text{ is } 2^{k+1} > 2(k + 1) + 1.$$

Multiply both sides of $2^k > 2k + 1$ by 2, obtaining

$$2 \cdot 2^k > 2(2k + 1)$$
$$2^{k+1} > 4k + 2.$$

Rewrite $4k + 2$ as $2k + 2 + 2k = 2(k + 1) + 2k$.

$$2^{k+1} > 2(k + 1) + 2k \quad (1)$$

Since k is a positive integer greater than 3,

$$2k > 1. \quad (2)$$

Adding $2(k + 1)$ to both sides of inequality (2) gives

$$2(k + 1) + 2k > 2(k + 1) + 1. \quad (3)$$

From inequalities (1) and (3),

$$2^{k+1} > 2(k + 1) + 2k > 2(k + 1) + 1,$$

or

$$2^{k+1} > 2(k + 1) + 1, \quad \text{as required.}$$

Thus, S_k implies S_{k+1}, and this, together with the fact that S_3 is true, shows that S_n is true for every positive integer value of n greater than or equal to 3.

Now try Exercise 21.

Proof of the Binomial Theorem The binomial theorem can be proved by mathematical induction. That is, for any positive integer n and any complex numbers x and y,

$$(x+y)^n = x^n + \binom{n}{1}x^{n-1}y + \binom{n}{2}x^{n-2}y^2 + \binom{n}{3}x^{n-3}y^3 + \cdots + \binom{n}{r}x^{n-r}y^r + \cdots + \binom{n}{n-1}xy^{n-1} + y^n. \quad (1)$$

(Section 11.4)

Proof Let S_n be statement (1). Begin by verifying S_n for $n = 1$,

$$S_1: \quad (x+y)^1 = x^1 + y^1,$$

which is true.

Now assume that S_n is true for the positive integer k. Statement S_k becomes

$$S_k: \quad (x+y)^k = x^k + \frac{k!}{1!(k-1)!}x^{k-1}y + \frac{k!}{2!(k-2)!}x^{k-2}y^2 + \cdots + \frac{k!}{(k-1)!1!}xy^{k-1} + y^k. \quad (2)$$

Definition of the binomial coefficient **(Section 11.4)**

Multiply both sides of equation (2) by $x + y$.

$$\begin{aligned}(x+y)^k \cdot (x+y) &= x(x+y)^k + y(x+y)^k \\ &= \left[x \cdot x^k + \frac{k!}{1!(k-1)!}x^k y + \frac{k!}{2!(k-2)!}x^{k-1}y^2 + \cdots + \frac{k!}{(k-1)!1!}x^2y^{k-1} + xy^k\right] \\ &\quad + \left[x^k \cdot y + \frac{k!}{1!(k-1)!}x^{k-1}y^2 + \cdots + \frac{k!}{(k-1)!1!}xy^k + y \cdot y^k\right]\end{aligned}$$

Distributive property **(Section R.1)**

Rearrange terms to get

$$(x+y)^{k+1} = x^{k+1} + \left[\frac{k!}{1!(k-1)!} + 1\right]x^k y + \left[\frac{k!}{2!(k-2)!} + \frac{k!}{1!(k-1)!}\right]x^{k-1}y^2 + \cdots + \left[1 + \frac{k!}{(k-1)!1!}\right]xy^k + y^{k+1}. \quad (3)$$

The first expression in brackets in equation (3) simplifies to $\binom{k+1}{1}$. To see this, note that

$$\binom{k+1}{1} = \frac{(k+1)(k)(k-1)(k-2)\cdots 1}{1 \cdot (k)(k-1)(k-2)\cdots 1} = k+1.$$

Also, $$\frac{k!}{1!(k-1)!} + 1 = \frac{k(k-1)!}{1(k-1)!} + 1 = k+1.$$

The second expression becomes $\binom{k+1}{2}$, the last $\binom{k+1}{k}$, and so on. The result of equation (3) is just equation (2) with every k replaced by $k + 1$. Thus, the truth of S_n when $n = k$ implies the truth of S_n for $n = k + 1$, which completes the proof of the theorem by mathematical induction.

11.5 Exercises

Write out in full and verify the statements S_1, S_2, S_3, S_4, and S_5 for the following. Then use mathematical induction to prove that each statement is true for every positive integer n. See Example 1.

1. $1 + 3 + 5 + \cdots + (2n - 1) = n^2$

2. $2 + 4 + 6 + \cdots + 2n = n(n + 1)$

Assume that n is a positive integer. Use mathematical induction to prove each statement S by following these steps. See Example 1.

(a) *Verify the statement for $n = 1$.*
(b) *Write the statement for $n = k$.*
(c) *Write the statement for $n = k + 1$.*
(d) *Assume the statement is true for $n = k$. Use algebra to change the statement in part (b) to the statement in part (c).*
(e) *Write a conclusion based on Steps (a)–(d).*

3. $3 + 6 + 9 + \cdots + 3n = \dfrac{3n(n + 1)}{2}$

4. $5 + 10 + 15 + \cdots + 5n = \dfrac{5n(n + 1)}{2}$

5. $2 + 4 + 8 + \cdots + 2^n = 2^{n+1} - 2$

6. $3 + 3^2 + 3^3 + \cdots + 3^n = \dfrac{3(3^n - 1)}{2}$

7. $1^2 + 2^2 + 3^2 + \cdots + n^2 = \dfrac{n(n + 1)(2n + 1)}{6}$

8. $1^3 + 2^3 + 3^3 + \cdots + n^3 = \dfrac{n^2(n + 1)^2}{4}$

9. $5 \cdot 6 + 5 \cdot 6^2 + 5 \cdot 6^3 + \cdots + 5 \cdot 6^n = 6(6^n - 1)$

10. $7 \cdot 8 + 7 \cdot 8^2 + 7 \cdot 8^3 + \cdots + 7 \cdot 8^n = 8(8^n - 1)$

11. $\dfrac{1}{1 \cdot 2} + \dfrac{1}{2 \cdot 3} + \dfrac{1}{3 \cdot 4} + \cdots + \dfrac{1}{n(n + 1)} = \dfrac{n}{n + 1}$

12. $\dfrac{1}{1 \cdot 4} + \dfrac{1}{4 \cdot 7} + \dfrac{1}{7 \cdot 10} + \cdots + \dfrac{1}{(3n - 2)(3n + 1)} = \dfrac{n}{3n + 1}$

13. $\dfrac{1}{2} + \dfrac{1}{2^2} + \dfrac{1}{2^3} + \cdots + \dfrac{1}{2^n} = 1 - \dfrac{1}{2^n}$

14. $\dfrac{4}{5} + \dfrac{4}{5^2} + \dfrac{4}{5^3} + \cdots + \dfrac{4}{5^n} = 1 - \dfrac{1}{5^n}$

Find all natural number values for n for which the given statement is false.

15. $2^n > 2n$

16. $3^n > 2n + 1$

17. $2^n > n^2$

18. $n! > 2n$

Prove each statement by mathematical induction. See Examples 2 and 3.

19. $(a^m)^n = a^{mn}$ (Assume a and m are constant.)

20. $(ab)^n = a^n b^n$ (Assume a and b are constant.)

21. $2^n > 2n$, if $n \geq 3$

22. $3^n > 2n + 1$, if $n \geq 2$

23. If $a > 1$, then $a^n > 1$.

24. If $a > 1$, then $a^n > a^{n-1}$.

25. If $0 < a < 1$, then $a^n < a^{n-1}$.

26. $2^n > n^2$, for $n \geq 5$

27. If $n \geq 4$, then $n! > 2^n$, where $n! = n(n-1)(n-2)\cdots(3)(2)(1)$.

28. $4^n > n^4$, for $n \geq 5$

Solve each problem.

29. ***Number of Handshakes*** Suppose that each of the n ($n \geq 2$) people in a room shakes hands with everyone else, but not with himself. Show that the number of handshakes is $\frac{n^2 - n}{2}$.

30. ***Sides of a Polygon*** The series of sketches below starts with an equilateral triangle having sides of length 1. In the following steps, equilateral triangles are constructed on each side of the preceding figure. The length of the sides of each new triangle is $\frac{1}{3}$ the length of the sides of the preceding triangles. Develop a formula for the number of sides of the nth figure. Use mathematical induction to prove your answer.

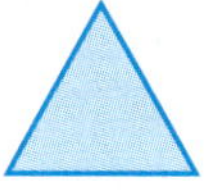

31. ***Perimeter*** Find the perimeter of the nth figure in Exercise 30.

32. ***Area*** Show that the area of the nth figure in Exercise 30 is

$$\sqrt{3}\left[\frac{2}{5} - \frac{3}{20}\left(\frac{4}{9}\right)^{n-1}\right].$$

33. ***Tower of Hanoi*** A pile of n rings, each ring smaller than the one below it, is on a peg. Two other pegs are attached to a board with this peg. In the game called the *Tower of Hanoi* puzzle, all the rings must be moved to a different peg, with only one ring moved at a time, and with no ring ever placed on top of a smaller ring. Find the least number of moves that would be required. Prove your result with mathematical induction.

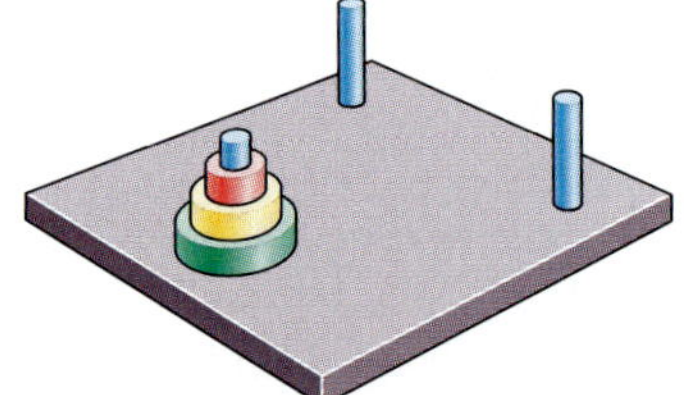

11.6 Counting Theory

Fundamental Principle of Counting ▪ Permutations ▪ Combinations ▪ Distinguishing Between Permutations and Combinations

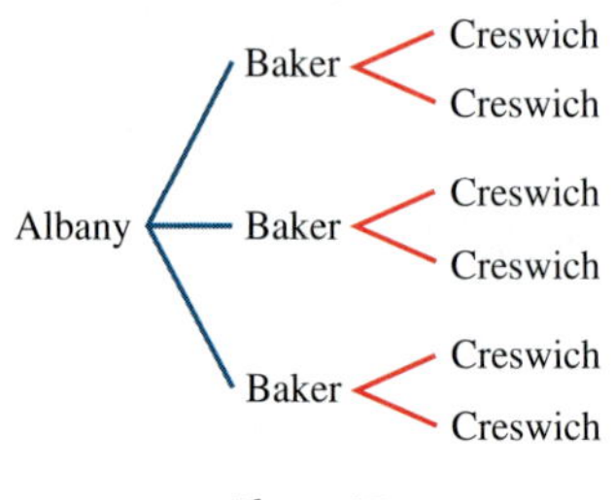

Figure 13

Fundamental Principle of Counting

If there are 3 roads from Albany to Baker and 2 roads from Baker to Creswich, in how many ways can one travel from Albany to Creswich by way of Baker? For each of the 3 roads from Albany to Baker, there are 2 different roads from Baker to Creswich. Hence, there are $3 \cdot 2 = 6$ different ways to make the trip, as shown in the **tree diagram** in Figure 13.

In this situation, each choice of road is an example of an *event.* Two events are **independent events** if neither influences the outcome of the other. The opening example illustrates the fundamental principle of counting with independent events.

Fundamental Principle of Counting

If n independent events occur, with

m_1 ways for event 1 to occur,

m_2 ways for event 2 to occur,

$\vdots$

and m_n ways for event n to occur,

then there are

$$m_1 \cdot m_2 \cdot \cdots \cdot m_n$$

different ways for all n events to occur.

EXAMPLE 1 Using the Fundamental Principle of Counting

A restaurant offers a choice of 3 salads, 5 main dishes, and 2 desserts. Use the fundamental principle of counting to find the number of different 3-course meals that can be selected.

Solution Three events are involved: selecting a salad, selecting a main dish, and selecting a dessert. The first event can occur in 3 ways, the second event can occur in 5 ways, and the third event can occur in 2 ways; thus there are

$$3 \cdot 5 \cdot 2 = 30 \text{ possible meals.}$$

Now try Exercise 23.

EXAMPLE 2 Using the Fundamental Principle of Counting

A teacher has 5 different books that he wishes to arrange in a row. How many different arrangements are possible?

Solution Five events are involved: selecting a book for the first spot, selecting a book for the second spot, and so on. For the first spot the teacher has 5 choices. After a choice has been made, the teacher has 4 choices for the second spot. Continuing in this manner, there are 3 choices for the third spot, 2 for the fourth spot, and 1 for the fifth spot. By the fundamental principle of counting, there are

$$5 \cdot 4 \cdot 3 \cdot 2 \cdot 1 = 120 \text{ different arrangements.}$$

Now try Exercise 27.

In using the fundamental principle of counting, products such as $5 \cdot 4 \cdot 3 \cdot 2 \cdot 1$ occur often. We use the symbol $n!$ (read "n-factorial"), for any counting number n, as follows.

$$n! = n(n-1)(n-2)\cdots(3)(2)(1) \quad \text{(Section 11.4)}$$

Thus, $5 \cdot 4 \cdot 3 \cdot 2 \cdot 1$ is written $5!$ and $3 \cdot 2 \cdot 1$ is written $3!$. By the definition of $n!$, $n[(n-1)!] = n!$ for all natural numbers $n \geq 2$. It is convenient to have this relation hold also for $n = 1$, so, by definition,

$$0! = 1. \quad \text{(Section 11.4)}$$

EXAMPLE 3 Arranging r of n Items ($r < n$)

Suppose the teacher in Example 2 wishes to place only 3 of the 5 books in a row. How many arrangements of 3 books are possible?

Solution The teacher still has 5 ways to fill the first spot, 4 ways to fill the second spot, and 3 ways to fill the third. Since only 3 books will be used, there are only 3 spots to be filled (3 events) instead of 5, with

$$5 \cdot 4 \cdot 3 = 60 \text{ arrangements.}$$

Now try Exercise 33.

Permutations Since each ordering of three books is considered a different *arrangement,* the number 60 in the preceding example is called the number of *permutations* of 5 things taken 3 at a time, written $P(5, 3) = 60$. The number of ways of arranging 5 elements from a set of 5 elements, written $P(5, 5) = 120$, was found in Example 2.

A **permutation** of n elements taken r at a time is one of the *arrangements* of r elements from a set of n elements. Generalizing from the examples above, the number of permutations of n elements taken r at a time, denoted by $P(n, r)$, is

$$\begin{aligned} P(n, r) &= n(n-1)(n-2)\cdots(n-r+1) \\ &= \frac{n(n-1)(n-2)\cdots(n-r+1)(n-r)(n-r-1)\cdots(2)(1)}{(n-r)(n-r-1)\cdots(2)(1)} \\ &= \frac{n!}{(n-r)!}. \end{aligned}$$

Permutations of n Elements Taken r at a Time

If $P(n, r)$ denotes the number of permutations of n elements taken r at a time, with $r \le n$, then

$$P(n, r) = \frac{n!}{(n-r)!}.$$

Alternative notations for $P(n, r)$ are P_r^n and ${}_nP_r$.

EXAMPLE 4 Using the Permutations Formula

Find each permutation.

(a) The number of permutations of the letters L, M, and N

(b) The number of permutations of 2 of the letters L, M, and N

Solution

(a) By the formula for $P(n, r)$, with $n = 3$ and $r = 3$,

$$P(3, 3) = \frac{3!}{(3-3)!} = \frac{3!}{0!} = \frac{3!}{1} = 3 \cdot 2 \cdot 1 = 6.$$

As shown in the tree diagram in Figure 14, the 6 permutations are

LMN, LNM, MLN, MNL, NLM, NML.

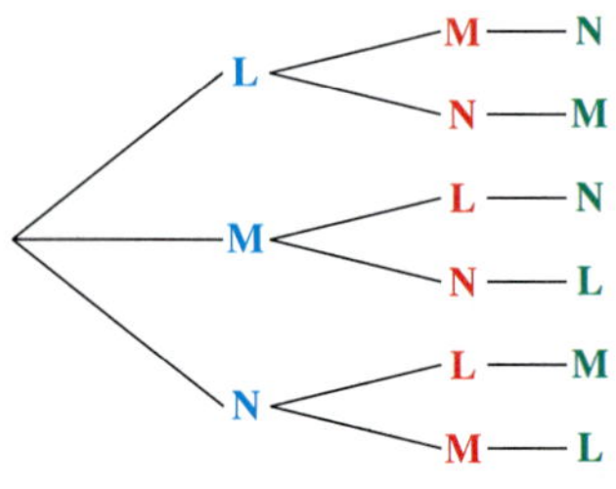

Figure 14

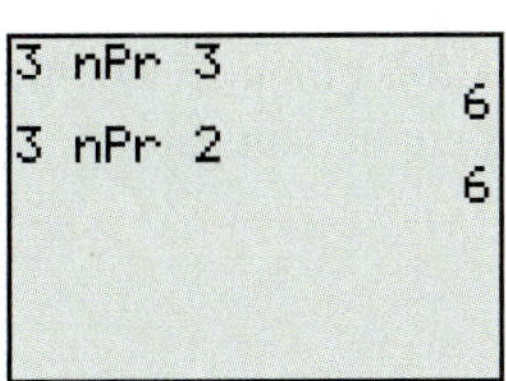

This screen shows how the TI-83 Plus calculates $P(3, 3)$ and $P(3, 2)$. See Example 4.

(b) Find $P(3, 2)$.

$$P(3,2) = \frac{3!}{(3-2)!} = \frac{3!}{1!} = \frac{3!}{1} = 6$$

This result is the same as the answer in part (a). After the first two choices are made, the third is already determined since only one letter is left.

Now try Exercise 37.

EXAMPLE 5 Using the Permutations Formula

Suppose 8 people enter an event in a swim meet. In how many ways could the gold, silver, and bronze medals be awarded?

Solution Using the fundamental principle of counting, there are 3 events, giving $8 \cdot 7 \cdot 6 = 336$ choices. We can also use the formula for $P(n, r)$ to get the same result.

$$P(8,3) = \frac{8!}{5!} = \frac{8 \cdot 7 \cdot 6 \cdot 5 \cdot 4 \cdot 3 \cdot 2 \cdot 1}{5 \cdot 4 \cdot 3 \cdot 2 \cdot 1}$$
$$= 8 \cdot 7 \cdot 6 = 336$$

Now try Exercise 35.

EXAMPLE 6 Using the Permutations Formula

In how many ways can 6 students be seated in a row of 6 desks?

Solution Use $P(n, r)$ with $n = 6$ and $r = 6$ to get

$$P(6,6) = 6! = 6 \cdot 5 \cdot 4 \cdot 3 \cdot 2 \cdot 1 = 720.$$

Now try Exercise 31.

Combinations In Example 3 we saw that there are 60 ways that a teacher can arrange 3 of 5 different books in a row. That is, there are 60 permutations of 5 things taken 3 at a time. Suppose now that the teacher does not wish to arrange the books in a row, but rather wishes to choose, without regard to order, any 3 of the 5 books to donate to a book sale to raise money for the school. In how many ways can the teacher do this?

The number 60 counts all possible *arrangements* of 3 books chosen from 5. The following 6 arrangements, however, would all lead to the same set of 3 books being given to the book sale.

mystery-biography-textbook	biography-textbook-mystery
mystery-textbook-biography	textbook-biography-mystery
biography-mystery-textbook	textbook-mystery-biography

The list shows 6 different *arrangements* of 3 books but only one *set* of 3 books. A subset of items selected *without regard to order* is called a **combination.** The number of combinations of 5 things taken 3 at a time is written $\binom{5}{3}$, ${}_5C_3$, or $C(5, 3)$.

NOTE This combinations notation also represents the binomial coefficient defined in Section 4 of this chapter. That is, binomial coefficients are the combinations of n elements chosen r at a time.

To evaluate $\binom{5}{3}$ or $C(5, 3)$, start with the $5 \cdot 4 \cdot 3$ *permutations* of 5 things taken 3 at a time. Since order does not matter, and each subset of 3 items from the set of 5 items can have its elements rearranged in $3 \cdot 2 \cdot 1 = 3!$ ways, we find $\binom{5}{3}$ by dividing the number of permutations by $3!$, or

$$\binom{5}{3} = \frac{5 \cdot 4 \cdot 3}{3!} = \frac{5 \cdot 4 \cdot 3}{3 \cdot 2 \cdot 1} = 10.$$

The teacher can choose 3 books for the book sale in 10 ways.

Generalizing this discussion gives the following formula for the number of combinations of n elements taken r at a time:

$$C(n, r) = \binom{n}{r} = \frac{P(n, r)}{r!}.$$

A more useful version of this formula is found as follows.

$$C(n, r) = \binom{n}{r} = \frac{P(n, r)}{r!} = \frac{n!}{(n - r)!} \cdot \frac{1}{r!} = \frac{n!}{(n - r)!\,r!}$$

This version is most useful for calculation and is the one we used earlier to calculate binomial coefficients.

Combinations of *n* Elements Taken *r* at a Time

If $C(n, r)$ or $\binom{n}{r}$ represents the number of combinations of n elements taken r at a time, with $r \leq n$, then

$$C(n,r) = \binom{n}{r} = \frac{n!}{(n - r)!r!}.$$

NOTE The formula for $C(n, r)$ given above is equivalent to the binomial coefficient formula given in Section 4 of this chapter.

EXAMPLE 7 Using the Combinations Formula

How many different committees of 3 people can be chosen from a group of 8 people?

Solution Since a committee is an unordered set, use combinations.

$$C(8, 3) = \binom{8}{3} = \frac{8!}{5!\,3!} = \frac{8 \cdot 7 \cdot 6 \cdot 5 \cdot 4 \cdot 3 \cdot 2 \cdot 1}{5 \cdot 4 \cdot 3 \cdot 2 \cdot 1 \cdot 3 \cdot 2 \cdot 1} = 56$$

Now try Exercise 39.

8 nCr 3
56

This screen shows how the TI-83 Plus calculates $C(8, 3)$. See Example 7.

EXAMPLE 8 Using the Combinations Formula

Three stockbrokers are to be selected from a group of 30 to work on a special project.

(a) In how many different ways can the stockbrokers be selected?

(b) In how many ways can the group of 3 be selected if a particular stockbroker must work on the project?

Solution

(a) Here we wish to know the number of 3-element combinations that can be formed from a set of 30 elements. (We want combinations, not permutations, since order within the group does not matter.)

$$C(30, 3) = \binom{30}{3} = \frac{30!}{27!3!} = 4060$$

There are 4060 ways to select the project group.

(b) Since 1 broker has already been selected for the project, the problem is reduced to selecting 2 more from the remaining 29 brokers.

$$C(29, 2) = \binom{29}{2} = \frac{29!}{27!2!} = 406$$

In this case, the project group can be selected in 406 ways.

Now try Exercise 41.

Distinguishing Between Permutations and Combinations Students often have difficulty determining whether to use permutations or combinations in solving problems. The following chart lists some of the similarities and differences between these two concepts.

Permutations	Combinations
Number of ways of selecting r items out of n items	
Repetitions are not allowed.	
Order is important.	Order is not important.
Arrangements of r items from a set of n items	Subsets of r items from a set of n items
$P(n, r) = \frac{n!}{(n-r)!}$	$C(n, r) = \binom{n}{r} = \frac{n!}{(n-r)!r!}$
Clue words: arrangement, schedule, order	Clue words: group, committee, sample, selection

EXAMPLE 9 Distinguishing Between Permutations and Combinations

Should permutations or combinations be used to solve each problem?

(a) How many 4-digit codes are possible if no digits are repeated?

(b) A sample of 3 light bulbs is randomly selected from a batch of 15 bulbs. How many different samples are possible?

(c) In a basketball tournament with 8 teams, how many games must be played so that each team plays every other team exactly once?

(d) In how many ways can 4 stockbrokers be assigned to 6 offices so that each broker has a private office?

Solution

(a) Since changing the order of the 4 digits results in a different code, permutations should be used.

(b) The order in which the 3 light bulbs are selected is not important. The sample is unchanged if the items are rearranged, so combinations should be used.

(c) Selection of 2 teams for a game is an *unordered* subset of 2 from the set of 8 teams. Use combinations.

(d) The office assignments are an *ordered* selection of 4 offices from the 6 offices. Exchanging the offices of any 2 brokers within a selection of 4 offices gives a different assignment, so permutations should be used.

Now try Exercise 21.

To illustrate the differences between permutations and combinations in another way, suppose we want to select 2 cans of soup from 4 cans on a shelf: noodle (N), bean (B), mushroom (M), and tomato (T). As shown in Figure 15(a), there are 12 ways to select 2 cans from the 4 cans if order matters (if noodle first and bean second is considered different from bean, then noodle, for example). On the other hand, if order is unimportant, then there are 6 ways to choose 2 cans of soup from the 4, as illustrated in Figure 15(b).

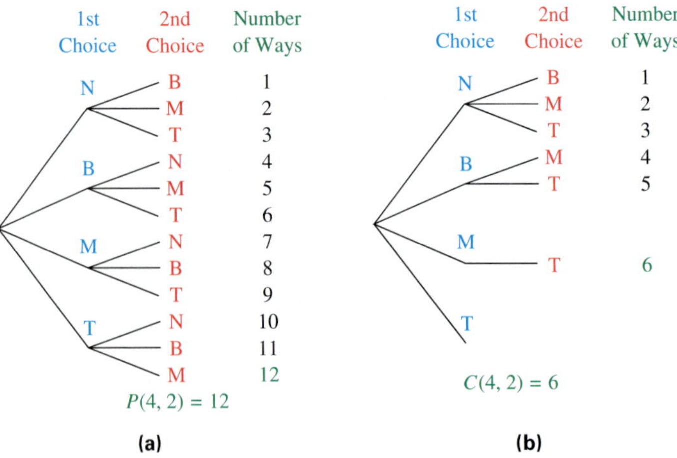

Figure 15

CAUTION Not all counting problems lend themselves to either permutations or combinations. Whenever a tree diagram or the fundamental principle of counting can be used directly, as in the soup example, use it.

11.6 Exercises

Evaluate each expression. See Examples 4–9.

1. $P(12, 8)$ **2.** $P(5, 5)$ **3.** $P(9, 2)$ **4.** $P(10, 9)$

5. $P(5, 1)$ **6.** $P(6, 0)$ **7.** $C(4, 2)$ **8.** $C(9, 3)$

9. $C(6, 0)$ **10.** $C(8, 1)$ **11.** $\binom{12}{4}$ **12.** $\binom{16}{3}$

Use a calculator to evaluate each expression. See Examples 4 and 7.

13. ${}_{20}P_5$ **14.** ${}_{100}P_5$ **15.** ${}_{15}P_8$ **16.** ${}_{32}P_4$

17. ${}_{20}C_5$ **18.** ${}_{100}C_5$ **19.** $\binom{15}{8}$ **20.** $\binom{32}{4}$

21. Decide whether the situation described involves a permutation or a combination of objects. See Example 9.

(a) a telephone number **(b)** a Social Security number
(c) a hand of cards in poker **(d)** a committee of politicians
(e) the "combination" on a combination lock
(f) a lottery choice of six numbers where order does not matter
(g) an automobile license plate

22. Explain the difference between a permutation and a combination. What should you look for in a problem to decide which is an appropriate method of solution?

Use the fundamental principle of counting or permutations to solve each problem. See Examples 1–6.

23. *Home Plan Choices* How many different types of homes are available if a builder offers a choice of 5 basic plans, 4 roof styles, and 2 exterior finishes?

24. *Auto Varieties* An auto manufacturer produces 7 models, each available in 6 different colors, with 4 different upholstery fabrics, and 5 interior colors. How many varieties of the auto are available?

25. *Radio-Station Call Letters* How many different 4-letter radio-station call letters can be made

(a) if the first letter must be K or W and no letter may be repeated?
(b) if repeats are allowed (but the first letter is K or W)?
(c) How many of the 4-letter call letters (starting with K or W) with no repeats end in R?

26. *Meal Choices* A menu offers a choice of 3 salads, 8 main dishes, and 5 desserts. How many different 3-course meals (salad, main dish, dessert) are possible?

27. *Arranging Blocks* Baby Finley wants to arrange 7 blocks in a row. How many different arrangements can he make?

28. *Names for a Baby* A couple has narrowed down the choice of a name for their new baby to 3 first names and 5 middle names. How many different first- and middle-name combinations are possible?

29. *License Plates* For many years, the state of California used 3 letters followed by 3 digits on its automobile license plates.
   (a) How many different license plates are possible with this arrangement?
   (b) When the state ran out of new plates, the order was reversed to 3 digits followed by 3 letters. How many additional plates were then possible?
   (c) When the plates described in part (b) were also used up, the state then issued plates with 1 letter followed by 3 digits and then 3 letters. How many plates does this scheme provide?

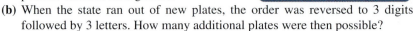

30. *Telephone Numbers* How many 7-digit telephone numbers are possible if the first digit cannot be 0 and
   (a) only odd digits may be used?
   (b) the telephone number must be a multiple of 10 (that is, it must end in 0)?
   (c) the telephone number must be a multiple of 100?
   (d) the first 3 digits are 481?
   (e) no repetitions are allowed?

31. *Seating People in a Row* In an experiment on social interaction, 8 people will sit in 8 seats in a row. In how many ways can this be done?

32. *Genetics Experiment* In how many ways can 7 of 10 monkeys be arranged in a row for a genetics experiment?

33. *Course Schedule Arrangement* A business school offers courses in keyboarding, spreadsheets, transcription, business English, technical writing, and accounting. In how many ways can a student arrange a schedule if 3 courses are taken?

34. *Course Schedule Arrangement* If your college offers 400 courses, 20 of which are in mathematics, and your counselor arranges your schedule of 4 courses by random selection, how many schedules are possible that do not include a math course?

35. *Club Officer Choices* In a club with 15 members, how many ways can a slate of 3 officers consisting of president, vice-president, and secretary/treasurer be chosen?

36. *Batting Orders* A baseball team has 20 players. How many 9-player batting orders are possible?

37. *Letter Arrangement* Consider the word TOUGH.
   (a) In how many ways can all the letters of the word TOUGH be arranged?
   (b) In how many ways can the first 3 letters of the word TOUGH be arranged?

38. *Basketball Positions* In how many ways can 5 players be assigned to the 5 positions on a basketball team, assuming that any player can play any position? In how many ways can 10 players be assigned to the 5 positions?

Solve each problem involving combinations. See Examples 7 and 8.

39. *Seminar Presenters* A banker's association has 30 members. If 4 members are selected at random to present a seminar, how many different groups of 4 are possible?

40. *Apple Samples* How many different samples of 3 apples can be drawn from a crate of 25 apples?

41. *Hamburger Choices* Howard's Hamburger Heaven sells hamburgers with cheese, relish, lettuce, tomato, mustard, or ketchup.

(a) How many different hamburgers can be made that use any 4 of the extras?

(b) How many different hamburgers can be made if one of the 4 extras must be cheese?

42. *Financial Planners* Three financial planners are to be selected from a group of 12 to participate in a special program. In how many ways can this be done? In how many ways can the group that will not participate be selected?

43. *Card Combinations* Five cards are marked with the numbers 1, 2, 3, 4, or 5, shuffled, and 2 cards are then drawn. How many different 2-card hands are possible?

44. *Marble Samples* If a bag contains 15 marbles, how many samples of 2 marbles can be drawn from it? How many samples of 4 marbles can be drawn?

45. *Marble Samples* In Exercise 44, if the bag contains 3 yellow, 4 white, and 8 blue marbles, how many samples of 2 can be drawn in which both marbles are blue?

46. *Apple Samples* In Exercise 40, if it is known that there are 5 rotten apples in the crate,

(a) how many samples of 3 could be drawn in which all 3 are rotten?

(b) how many samples of 3 could be drawn in which there are 1 rotten apple and 2 good apples?

47. *Convention Delegation Choices* A city council is composed of 5 liberals and 4 conservatives. Three members are to be selected randomly as delegates to a convention.

(a) How many delegations are possible?

(b) How many delegations could have all liberals?

(c) How many delegations could have 2 liberals and 1 conservative?

(d) If 1 member of the council serves as mayor, how many delegations are possible that include the mayor?

48. *Delegation Choices* Seven workers decide to send a delegation of 2 to their supervisor to discuss their grievances.

(a) How many different delegations are possible?

(b) If it is decided that a certain employee must be in the delegation, how many different delegations are possible?

(c) If there are 2 women and 5 men in the group, how many delegations would include at least 1 woman?

Use any or all of the methods described in this section to solve each problem. See Examples 1–9.

49. *Course Schedule* If Dwight Johnston has 8 courses to choose from, how many ways can he arrange his schedule if he must pick 4 of them?

50. *Pineapple Samples* How many samples of 3 pineapples can be drawn from a crate of 12?

51. *Soup Ingredients* Velma specializes in making different vegetable soups with carrots, celery, beans, peas, mushrooms, and potatoes. How many different soups can she make with any 4 ingredients?

52. *Secretary/Manager Assignments* From a pool of 7 secretaries, 3 are selected to be assigned to 3 managers, 1 secretary to each manager. In how many ways can this be done?

53. *Musical Chairs Seatings* In a game of musical chairs, 12 children will sit in 11 chairs. (1 will be left out.) How many seatings are possible?

54. *Plant Samples* In an experiment on plant hardiness, a researcher gathers 6 wheat plants, 3 barley plants, and 2 rye plants. She wishes to select 4 plants at random.

(a) In how many ways can this be done?

(b) In how many ways can this be done if exactly 2 wheat plants must be included?

55. *Committee Choices* In a club with 8 men and 11 women members, how many 5-member committees can be chosen that have the following?

(a) all men **(b)** all women

(c) 3 men and 2 women **(d)** no more than 3 women

56. *Committee Choices* From 10 names on a ballot, 4 will be elected to a political party committee. In how many ways can the committee of 4 be formed if each person will have a different responsibility?

57. *Combination Lock* A briefcase has 2 locks. The combination to each lock consists of a 3-digit number, where digits may be repeated. How many combinations are possible? (*Hint:* The word *combination* is a misnomer. Lock combinations are permutations where the arrangement of the numbers is important.)

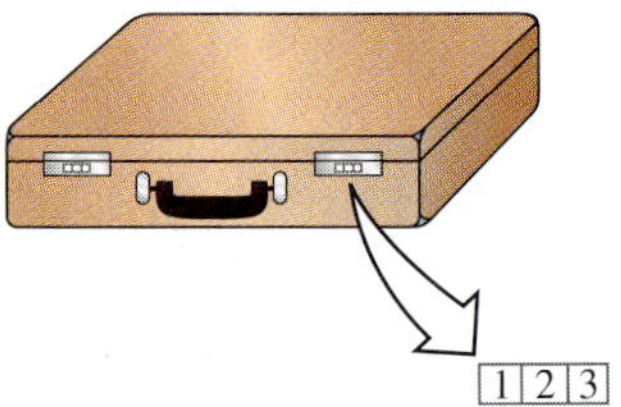

58. *Combination Lock* A typical combination for a padlock consists of 3 numbers from 0 to 39. Find the number of combinations that are possible with this type of lock, if a number may be repeated.

59. *Garage Door Openers* The code for some garage door openers consists of 12 electrical switches that can be set to either 0 or 1 by the owner. With this type of opener, how many codes are possible? (*Source:* Promax.)

60. *Lottery* To win the jackpot in a lottery game, a person must pick 3 numbers from 0 to 9 in the correct order. If a number can be repeated, how many ways are there to play the game?

61. *Keys* How many distinguishable ways can 4 keys be put on a circular key ring?

62. *Sitting at a Round Table* How many ways can 7 people sit at a round table? Assume that a different way means that at least 1 person is sitting next to someone different.

Prove each statement for positive integers n and r, with $r \leq n$. (Hint: Use the definitions of permutations and combinations.)

63. $P(n, n-1) = P(n, n)$ **64.** $P(n, 1) = n$ **65.** $P(n, 0) = 1$

66. $\binom{n}{n} = 1$ **67.** $\binom{n}{0} = 1$ **68.** $\binom{n}{n-1} = n$

69. $\binom{n}{n-r} = \binom{n}{r}$

70. Explain why the restriction $r \leq n$ is needed in the formula for $P(n, r)$.

Relating Concepts

For individual or collaborative investigation

(Exercises 71 and 72)

The value of n! can quickly become too large for most calculators to evaluate. To estimate n! for large values of n, we can use the property of logarithms that

$$\log(n!) = \log(1 \times 2 \times 3 \times \cdots \times n)$$
$$= \log 1 + \log 2 + \log 3 + \cdots + \log n.$$

Using a sum and sequence utility on a calculator, we can then determine r such that $n! \approx 10^r$ *since* $r = \log n!$. *For example, the screen illustrates that a calculator gives the same approximation of* 30! *using the factorial function and the formula just discussed. Use this technique to approximate each quantity in Exercises 71 and 72. Then, try to compute each value directly on your calculator.*

71. **(a)** 50! **(b)** 60! **(c)** 65!

72. **(a)** $P(47, 13)$ **(b)** $P(50, 4)$ **(c)** $P(29, 21)$

11.7 Basics of Probability

Basic Concepts ▪ Complements and Venn Diagrams ▪ Odds ▪ Union of Two Events ▪ Binomial Probability

Basic Concepts Consider an experiment that has one or more possible **outcomes,** each of which is equally likely to occur. For example, the experiment of tossing a fair coin has 2 equally likely possible outcomes: landing heads up (H) or landing tails up (T). Also, the experiment of rolling a fair die has 6 equally likely outcomes: landing so the face that is up shows 1, 2, 3, 4, 5, or 6 dots.

The set S of all possible outcomes of a given experiment is called the **sample space** for the experiment. (In this text, all sample spaces are finite.) One sample space for the experiment of tossing a coin could consist of the outcomes H and T. This sample space can be written as

$$S = \{H, T\}. \quad \text{Use set notation.}$$

Similarly, a sample space for the experiment of rolling a single die once is

$$S = \{1, 2, 3, 4, 5, 6\}.$$

Now try Exercises 1 and 5.

Any subset of the sample space is called an **event.** In the experiment with the die, for example, "the number showing is a 3" is an event, say E_1, such that $E_1 = \{3\}$. "The number showing is greater than 3" is also an event, say E_2, such that $E_2 = \{4, 5, 6\}$. To represent the number of outcomes that belong to event E, the notation $n(E)$ is used. Then $n(E_1) = 1$ and $n(E_2) = 3$.

The notation $P(E)$ is used for the *probability* of an event E. If the outcomes in the sample space for an experiment are equally likely, then the probability of event E occurring is found as follows.

Probability of Event E

In a sample space with equally likely outcomes, the **probability** of an event E, written $P(E)$, is the ratio of the number of outcomes in sample space S that belong to event E, $n(E)$, to the total number of outcomes in sample space S, $n(S)$. That is,

$$P(E) = \frac{n(E)}{n(S)}.$$

To use this definition to find the probability of the event E_1 in the die experiment, start with the sample space, $S = \{1, 2, 3, 4, 5, 6\}$, and the desired event, $E_1 = \{3\}$. Since $n(E_1) = 1$ and since there are 6 outcomes in the sample space,

$$P(E_1) = \frac{n(E_1)}{n(S)} = \frac{1}{6}.$$

EXAMPLE 1 Finding Probabilities of Events

A single die is rolled. Write each event in set notation and give the probability of the event.

(a) E_3: the number showing is even

(b) E_4: the number showing is greater than 4

(c) E_5: the number showing is less than 7

(d) E_6: the number showing is 7

Solution

(a) Since $E_3 = \{2, 4, 6\}$, $n(E_3) = 3$. As given earlier, $n(S) = 6$, so

$$P(E_3) = \frac{3}{6} = \frac{1}{2}.$$

(b) Again $n(S) = 6$. Event $E_4 = \{5, 6\}$, with $n(E_4) = 2$.

$$P(E_4) = \frac{2}{6} = \frac{1}{3}$$

(c) $E_5 = \{1, 2, 3, 4, 5, 6\}$ and $P(E_5) = \frac{6}{6} = 1.$

(d) $E_6 = \emptyset$ and $P(E_6) = \frac{0}{6} = 0.$

Now try Exercises 7 and 9.

In Example 1(c), $E_5 = S$. Therefore, the event E_5 is certain to occur every time the experiment is performed. An event that is certain to occur always has probability 1. In Example 1(d), $E_6 = \emptyset$ and $P(E_6) = 0$. The probability of an impossible event, such as E_6, is always 0, since none of the outcomes in the sample space satisfy the event. For any event E, $P(E)$ is between 0 and 1 inclusive.

Complements and Venn Diagrams The set of all outcomes in the sample space that do *not* belong to event E is called the **complement** of E, written E'. For example, in the experiment of drawing a single card from a standard deck of 52 cards, let E be the event "the card is an ace." Then E' is the event "the card is not an ace." From the definition of E', for an event E,

$$E \cup E' = S \quad \text{and} \quad E \cap E' = \emptyset.^*$$

NOTE A standard deck of 52 cards has four suits: hearts ♥, diamonds ♦, spades ♠, and clubs ♣, with thirteen cards of each suit. Each suit has a jack, a queen, and a king (sometimes called the "face cards"), an ace, and cards numbered from 2 to 10. The hearts and diamonds are red and the spades and clubs are black. We will refer to this standard deck of cards in this section.

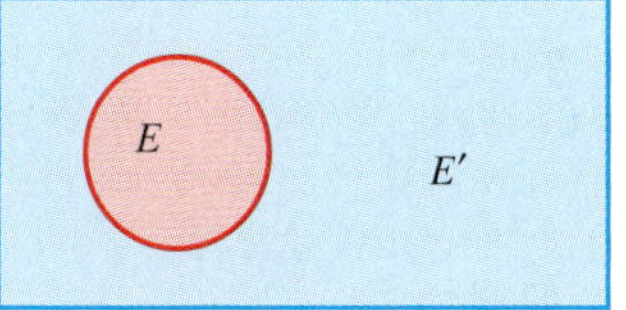

Figure 16

Probability concepts can be illustrated using **Venn diagrams,** as shown in Figure 16. The rectangle in Figure 16 represents the sample space in an experiment. The area inside the circle represents event E, while the area inside the rectangle, but outside the circle, represents event E'.

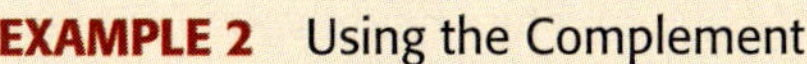

EXAMPLE 2 Using the Complement

In the experiment of drawing a card from a well-shuffled deck, find the probability of event E, the card is an ace, and event E'.

Solution Since there are 4 aces in the deck of 52 cards, $n(E) = 4$ and $n(S) = 52$. Therefore,

$$P(E) = \frac{n(E)}{n(S)} = \frac{4}{52} = \frac{1}{13}.$$

Of the 52 cards, 48 are not aces, so

$$P(E') = \frac{n(E')}{n(S)} = \frac{48}{52} = \frac{12}{13}.$$

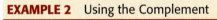

Now try Exercises 15(a)–(c).

In Example 2, $P(E) + P(E') = \frac{1}{13} + \frac{12}{13} = 1$. This is always true for any event E and its complement E'. That is,

$$P(E) + P(E') = 1.$$

This can be restated as

$$P(E) = 1 - P(E') \quad \text{or} \quad P(E') = 1 - P(E).$$

These two equations suggest an alternative way to compute the probability of an event. For example, if it is known that $P(E) = \frac{1}{10}$, then

$$P(E') = 1 - \frac{1}{10} = \frac{9}{10}.$$

*The **union** of two sets A and B is the set $A \cup B$ of all elements from either A or B, or both. The **intersection** of sets A and B, written $A \cap B$, includes all elements that belong to both sets.

Odds Sometimes probability statements are expressed in terms of odds, a comparison of $P(E)$ with $P(E')$. The **odds** in favor of an event E are expressed as the ratio of $P(E)$ to $P(E')$ or as the quotient $\frac{P(E)}{P(E')}$. For example, if the probability of rain can be established as $\frac{1}{3}$, the odds that it will rain are

$$P(\text{rain}) \text{ to } P(\text{no rain}) = \frac{1}{3} \text{ to } \frac{2}{3} = \frac{\frac{1}{3}}{\frac{2}{3}} = \frac{1}{2} \quad \text{or} \quad 1 \text{ to } 2.$$

On the other hand, the odds that it will not rain are 2 to 1 $\left(\text{or } \frac{2}{3} \text{ to } \frac{1}{3}\right)$. If the odds in favor of an event are, say, 3 to 5, then the probability of the event is $\frac{3}{8}$, and the probability of the complement of the event is $\frac{5}{8}$. If the odds favoring event E are m to n, then

$$P(E) = \frac{m}{m+n} \quad \text{and} \quad P(E') = \frac{n}{m+n}.$$

EXAMPLE 3 Finding Odds in Favor of an Event

A shirt is selected at random from a dark closet containing 6 blue shirts and 4 shirts that are not blue. Find the odds in favor of a blue shirt being selected.

Solution Let E represent "a blue shirt is selected." Then,

$$P(E) = \frac{6}{10} \quad \text{or} \quad \frac{3}{5} \quad \text{and} \quad P(E') = 1 - \frac{3}{5} = \frac{2}{5}.$$

Therefore, the odds in favor of a blue shirt being selected are

$$P(E) \text{ to } P(E') = \frac{3}{5} \text{ to } \frac{2}{5} = \frac{\frac{3}{5}}{\frac{2}{5}} = \frac{3}{2} \quad \text{or} \quad 3 \text{ to } 2.$$

(Section R.5)

Now try Exercises 15(d) and (e).

Union of Two Events Since events are sets, we can use set operations to find the union of two events. Suppose a fair die is rolled. Let H be the event "the result is a 3," and K the event "the result is an even number." From the results earlier in this section,

$$H = \{3\} \qquad K = \{2,4,6\} \qquad H \cup K = \{2,3,4,6\}$$

$$P(H) = \frac{1}{6} \qquad P(K) = \frac{3}{6} = \frac{1}{2} \qquad P(H \cup K) = \frac{4}{6} = \frac{2}{3}.$$

Notice that $P(H) + P(K) = P(H \cup K)$.

Before assuming that this relationship is true in general, consider another event G for this experiment, "the result is a 2."

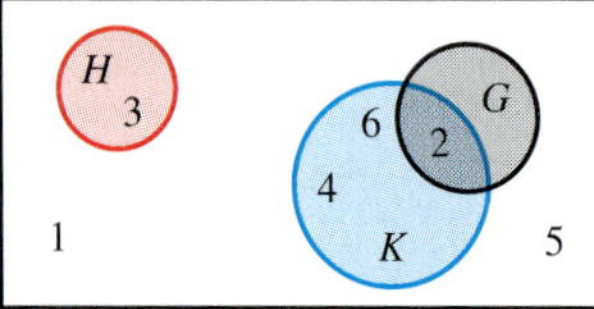

Figure 17

$$G = \{2\} \qquad K = \{2,4,6\} \qquad G \cup K = \{2,4,6\}$$

$$P(G) = \frac{1}{6} \qquad P(K) = \frac{3}{6} = \frac{1}{2} \qquad P(G \cup K) = \frac{3}{6} = \frac{1}{2}$$

In this case, $P(G) + P(K) \neq P(G \cup K)$. As Figure 17 suggests, the difference in the two examples above comes from the fact that events H and K cannot occur simultaneously. Such events are called **mutually exclusive events.** In fact, $H \cap K = \emptyset$, which is always true for mutually exclusive events. Events G and K, however, can occur simultaneously. Both are satisfied if the result of the roll is a 2, the element in their intersection ($G \cap K = \{2\}$). This example suggests the following property.

Probability of the Union of Two Events

For any events E and F,

$$P(E \text{ or } F) = P(E \cup F) = P(E) + P(F) - P(E \cap F).$$

EXAMPLE 4 Finding Probabilities of Unions

One card is drawn from a well-shuffled deck of 52 cards. What is the probability of the following outcomes?

(a) The card is an ace or a spade. **(b)** The card is a 3 or a king.

Solution

(a) The events "drawing an ace" and "drawing a spade" are not mutually exclusive since it is possible to draw the ace of spades, an outcome satisfying both events. The probability is

$$\begin{aligned} P(\text{ace or spade}) &= P(\text{ace}) + P(\text{spade}) - P(\text{ace and spade}) \\ &= \frac{4}{52} + \frac{13}{52} - \frac{1}{52} = \frac{16}{52} = \frac{4}{13}. \end{aligned}$$

(b) "Drawing a 3" and "drawing a king" are mutually exclusive events because it is impossible to draw one card that is both a 3 and a king.

$$\begin{aligned} P(3 \text{ or } K) &= P(3) + P(K) - P(3 \text{ and } K) \\ &= \frac{4}{52} + \frac{4}{52} - 0 = \frac{8}{52} = \frac{2}{13} \end{aligned}$$

Now try Exercises 17(a) and (b).

EXAMPLE 5 Finding Probabilities of Unions

Suppose two fair dice are rolled. Find each probability.

(a) The first die shows a 2, or the sum of the two dice is 6 or 7.

(b) The sum of the dots showing is at most 4.

Solution

(a) Think of the two dice as being distinguishable, one red and one green for example. (Actually, the sample space is the same even if they are not apparently distinguishable.) A sample space with equally likely outcomes is shown in Figure 18 on the next page, where $(1, 1)$ represents the event "the first die (red) shows a 1 and the second die (green) shows a 1," $(1, 2)$ represents "the first die shows a 1 and the second die shows a 2," and so on.

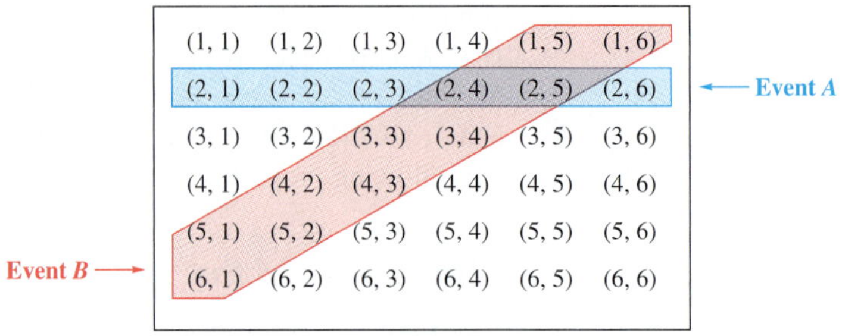

(1, 1)	(1, 2)	(1, 3)	(1, 4)	(1, 5)	(1, 6)
(2, 1)	(2, 2)	(2, 3)	(2, 4)	(2, 5)	(2, 6)
(3, 1)	(3, 2)	(3, 3)	(3, 4)	(3, 5)	(3, 6)
(4, 1)	(4, 2)	(4, 3)	(4, 4)	(4, 5)	(4, 6)
(5, 1)	(5, 2)	(5, 3)	(5, 4)	(5, 5)	(5, 6)
(6, 1)	(6, 2)	(6, 3)	(6, 4)	(6, 5)	(6, 6)

Event *A* (row 2) · Event *B* (diagonal band)

Figure 18

Let A represent the event "the first die shows a 2," and B represent the event "the sum of the two dice is 6 or 7." See Figure 18. Event A has 6 elements, event B has 11 elements, and the sample space has 36 elements. Thus,

$$P(A) = \frac{6}{36}, \qquad P(B) = \frac{11}{36}, \qquad \text{and} \qquad P(A \cap B) = \frac{2}{36},$$

so

$$\begin{aligned} P(A \cup B) &= P(A) + P(B) - P(A \cap B) \\ &= \frac{6}{36} + \frac{11}{36} - \frac{2}{36} = \frac{15}{36} = \frac{5}{12}. \end{aligned}$$

(b) "At most 4" can be written as "2 or 3 or 4." (A sum of 1 is meaningless here.) Since the events represented by "2," "3," or "4" are mutually exclusive,

$$P(\text{at most } 4) = P(2 \text{ or } 3 \text{ or } 4) = P(2) + P(3) + P(4). \quad (*)$$

The sample space for this experiment includes the 36 possible pairs of numbers shown in Figure 18. The pair $(1, 1)$ is the only one with a sum of 2, so $P(2) = \frac{1}{36}$. Also $P(3) = \frac{2}{36}$ since both $(1, 2)$ and $(2, 1)$ give a sum of 3. The pairs $(1, 3)$, $(2, 2)$, and $(3, 1)$ have a sum of 4, so $P(4) = \frac{3}{36}$. Substituting into equation (*) above gives

$$P(\text{at most } 4) = \frac{1}{36} + \frac{2}{36} + \frac{3}{36} = \frac{6}{36} = \frac{1}{6}.$$

Now try Exercise 17(c).

The properties of probability are summarized as follows.

Properties of Probability

For any events E and F:

1. $0 \le P(E) \le 1$

2. $P(\text{a certain event}) = 1$

3. $P(\text{an impossible event}) = 0$

4. $P(E') = 1 - P(E)$

5. $P(E \text{ or } F) = P(E \cup F) = P(E) + P(F) - P(E \cap F)$.

CAUTION When finding the probability of a union, remember to subtract the probability of the intersection from the sum of the probabilities of the individual events.

Binomial Probability A **binomial experiment** is an experiment that consists of repeated independent trials with only two outcomes in each trial, success or failure. Let the probability of success in one trial be p. Then the probability of failure is $1 - p$, and the probability of exactly r successes in n trials is given by

$$\binom{n}{r}p^r(1-p)^{n-r}.$$

This expression is equivalent to the general term of the binomial expansion given earlier. Thus the terms of the binomial expansion give the probabilities of exactly r successes in n trials, for $0 \le r \le n$, in a binomial experiment.

EXAMPLE 6 Finding Probabilities in a Binomial Experiment

An experiment consists of rolling a die 10 times. Find each probability.

(a) The probability that in exactly 4 of the rolls, the result is a 3.

(b) The probability that in exactly 9 of the rolls, the result is not a 3.

Algebraic Solution

(a) The probability p of a 3 on one roll is $\frac{1}{6}$. Here $n = 10$ and $r = 4$, so the required probability is

$$\binom{10}{4}\left(\frac{1}{6}\right)^4\left(1-\frac{1}{6}\right)^{10-4} = 210\left(\frac{1}{6}\right)^4\left(\frac{5}{6}\right)^6$$
$$\approx .054.$$

(b) The probability is

$$\binom{10}{9}\left(\frac{5}{6}\right)^9\left(\frac{1}{6}\right)^1 \approx .323.$$

Use $n = 10$, $r = 9$, and $p = 1 - \frac{1}{6} = \frac{5}{6}$.

Graphing Calculator Solution

Graphing calculators that have statistical distribution functions give binomial probabilities. Figure 19 shows the results for parts (a) and (b). The numbers in parentheses separated by commas represent n, p, and r, respectively.

```
binompdf(10,(1/6
),4)
        .0542658759
binompdf(10,(5/6
),9)
        .3230111658
```

Figure 19

Now try Exercise 35.

11.7 Exercises

Write a sample space with equally likely outcomes for each experiment.

1. A two-headed coin is tossed once.
2. Two ordinary coins are tossed.
3. Three ordinary coins are tossed.
4. Slips of paper marked with the numbers 1, 2, 3, 4, and 5 are placed in a box. After mixing well, two slips are drawn.
5. The spinner shown here is spun twice.

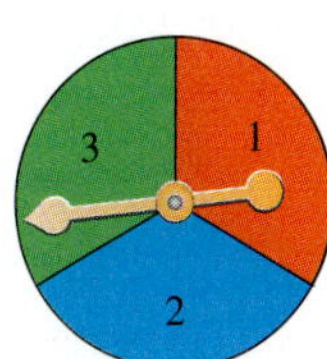

6. A die is rolled and then a coin is tossed.

Write each event in set notation and give the probability of the event. See Example 1.

7. In Exercise 1:

(a) the result of the toss is heads; **(b)** the result of the toss is tails.

8. In Exercise 2:

(a) both coins show the same face; **(b)** at least one coin turns up heads.

9. In Exercise 5:

(a) the result is a repeated number; **(b)** the second number is 1 or 3;
(c) the first number is even and the second number is odd.

10. In Exercise 4:

(a) both slips are marked with even numbers;
(b) both slips are marked with odd numbers;
(c) both slips are marked with the same number;
(d) one slip is marked with an odd number, the other with an even number.

11. A student gives the probability of an event in a problem as $\frac{6}{5}$. Explain why this answer must be incorrect.

12. ***Concept Check*** If the probability of an event is .857, what is the probability that the event will not occur?

13. ***Concept Check*** Associate each probability in parts (a)–(g) with one of the statements in A–F. Choices may be used more than once.

(a) $P(E) = -.1$ **(b)** $P(E) = .01$ **(c)** $P(E) = 1$ **(d)** $P(E) = 2$
(e) $P(E) = .99$ **(f)** $P(E) = 0$ **(g)** $P(E) = .5$

A. The event is certain to occur. **B.** The event cannot occur.
C. The event is very likely to occur. **D.** The event is very unlikely to occur.
E. The event is just as likely to occur as not occur.
F. The probability value is impossible.

Work each problem. See Examples 1–6.

14. ***Batting Average*** A baseball player with a batting average of .300 comes to bat. What are the odds in favor of the ball player getting a hit?

15. ***Drawing a Marble*** A marble is drawn at random from a box containing 3 yellow, 4 white, and 8 blue marbles. Find the probabilities in parts (a)–(c).

(a) A yellow marble is drawn. **(b)** A black marble is drawn.
(c) The marble is yellow or white.
(d) What are the odds in favor of drawing a yellow marble?
(e) What are the odds against drawing a blue marble?

16. ***Small Business Loan*** The probability that a bank with assets greater than or equal to \$30 billion will make a loan to a small business is .002. What are the odds against such a bank making a small business loan? (*Source: The Wall Street Journal* analysis of *CA1 Reports* filed with federal banking authorities.)

17. ***Dice Rolls*** Two dice are rolled. Find the probability of each event.

(a) The sum of the dots is at least 10.
(b) The sum of the dots is either 7 or at least 10.
(c) The sum of the dots is 2, or the dice both show the same number.

18. ***Languages Spoken in Hispanic Households*** In a recent survey of Hispanic households, 20.4% of the respondents said that only Spanish was spoken at home, 11.9% said that only English was spoken, and the remainder said that both Spanish and English were spoken. What are the odds that English is spoken in a randomly selected Hispanic household? (*Source:* American Demographics.)

19. *U.S. Population Origins* Projected Hispanic and non-Hispanic U.S. populations (in thousands) for the year 2025 are given in the table. (Other populations are all non-Hispanic.) Assume these projections are accurate. Find the probability that a U.S. resident selected at random in 2025 is the following.

(a) of Hispanic origin
(b) not White
(c) Indian (Native American) or Black
(d) What are the odds that a randomly selected U.S. resident is Asian?

Type	Number
Hispanic origin	58,930
White	209,117
Black	43,511
Indian (Native American)	2,744
Asian	20,748

Source: U.S. Bureau of the Census.

20. *U.S. Population by Region* The U.S. resident population by region (in millions) for selected years is given in the table. Find the probability that a U.S. resident selected at random satisfies the following.

(a) lived in the West in 1997
(b) lived in the Midwest in 1995
(c) lived in the Northeast or Midwest in 1997
(d) lived in the South or West in 1997
(e) What are the odds that a randomly selected U.S. resident in 2000 was not from the South?

Region	1995	1997	2000
Northeast	51.4	51.6	53.6
Midwest	61.8	62.5	64.4
South	91.8	94.2	100.2
West	57.7	59.4	63.2

Source: U.S. Bureau of the Census.

21. *State Lottery* One game in a state lottery requires you to pick 1 heart, 1 club, 1 diamond, and 1 spade, in that order, from the 13 cards in each suit. What is the probability of getting all four picks correct and winning $5000?

22. *State Lottery* If three of the four selections in Exercise 21 are correct, the player wins $200. Find the probability of this outcome.

23. *Male Life Table* The table is an abbreviated version of a *life table* used by the Office of the Chief Actuary of the Social Security Administration. (The actual table includes every age, not just every tenth age.) Theoretically, this table follows a group of 100,000 males at birth and gives the number still alive at each age.

Exact Age	Number of Lives	Exact Age	Number of Lives
0	100,000	60	82,963
10	98,924	70	66,172
20	98,233	80	39,291
30	96,735	90	10,537
40	94,558	100	468
50	90,757	110	1

Source: Office of the Actuary, Social Security Administration.

(a) What is the probability that a 40-year-old man will live 30 more years?
(b) What is the probability that a 40-year-old man will not live 30 more years?
(c) Consider a group of five 40-year-old men. What is the probability that exactly three of them survive to age 70? (*Hint:* The longevities of the individual men can be considered as independent trials.)

(d) Consider two 40-year-old men. What is the probability that at least one of them survives to age 70? (*Hint:* The probability that both survive is the product of the probabilities that each survives.)

24. ***Opinion Survey*** The management of a firm wishes to survey the opinions of its workers, classified as follows for the purpose of an interview:

30% have worked for the company 5 or more years,
28% are female,
65% contribute to a voluntary retirement plan, and 50% of the female workers contribute to the retirement plan.

Find each probability if a worker is selected at random.

(a) A male worker is selected.
(b) A worker is selected who has worked for the company less than 5 yr.
(c) A worker is selected who contributes to the retirement plan or is female.

25. ***Growth in Stock Value*** A financial analyst has determined the possibilities (and their probabilities) for the growth in value of a certain stock during the next year. (Assume these are the only possibilities.) See the table. For instance, the probability of a 5% growth is .15. If you invest \$10,000 in the stock, what is the probability that the stock will be worth at least \$11,400 by the end of the year?

Percent Growth	Probability
5	.15
8	.20
10	.35
14	.20
18	.10

26. ***Growth in Stock Value*** Refer to Exercise 25. Suppose the percents and probabilities in the table are estimates of annual growth during the next 3 yr. What is the probability that an investment of \$10,000 will grow in value to *at least* \$15,000 during the next 3 yr? (*Hint:* Use the formula for (annual) compound interest discussed in Section 4.2.)

College Student Smokers *The table gives the results of a survey of* 14,000 *college students who were cigarette smokers in a recent year.*

Number of Cigarettes Per Day	Less than 1	1 to 9	10 to 19	A pack of 20 or more
Percent (as a decimal)	.45	.24	.20	.11

Source: Harvard School of Public Health Study in the *Journal of AMA*.

Using the percents as probabilities, find the probability (to six decimal places) that, out of 10 *of these student smokers selected at random, the following were true.*

27. Four smoked less than 10 cigarettes per day.

28. Five smoked a pack or more per day.

29. Fewer than 2 smoked between 1 and 19 cigarettes per day.

30. No more than 3 smoked less than 1 cigarette per day.

College Applications *The table gives the results of a survey of* 282,549 *freshmen from the class of 2006 at* 437 *of the nation's baccalaureate colleges and universities.*

Number of Colleges Applied to	1	2 or 3	4–6	7 or more
Percent (as a decimal)	.20	.29	.37	.14

Source: Higher Education Research Institute, UCLA, 2002.

Using the percents as probabilities, find the probability of each event for a randomly selected student.

31. The student applied to fewer than 4 colleges.

32. The student applied to at least 2 colleges.

33. The student applied to more than 3 colleges.

34. The student applied to no colleges.

35. *Color-Blind Males* The probability that a male will be color-blind is .042. Find the probabilities (to six decimal places) that in a group of 53 men, the following are true.

(a) Exactly 5 are color-blind.
(b) No more than 5 are color-blind.
(c) At least 1 is color-blind.

36. The screens illustrate how the TABLE feature of a graphing calculator can be used to find the probabilities of having 0, 1, 2, 3, or 4 girls in a family of 4 children. (Note that 0 appears for X = 5 and X = 6. Why is this so?)

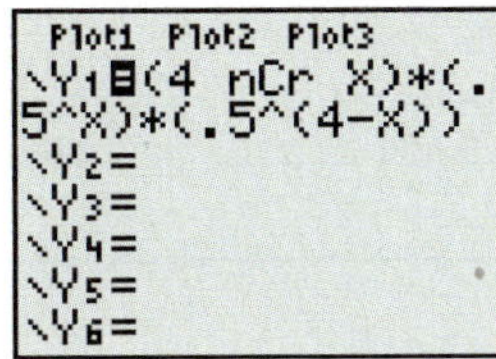

X	Y1
0	.0625
1	.25
2	.375
3	.25
4	.0625
5	0
6	0

Y1■(4 nCr X)*(.…

Use this approach to determine the following.

(a) Find the probabilities of having 0, 1, 2, or 3 boys in a family of 3 children.
(b) Find the probabilities of having 0, 1, 2, 3, 4, 5, or 6 girls in a family of 6 children.

37. *(Modeling) Spread of Disease* What will happen when an infectious disease is introduced into a family? Suppose a family has I infected members and S members who are not infected but are susceptible to contracting the disease. The probability P of exactly k people not contracting the disease during a 1-week period can be calculated by the formula

$$P = \binom{S}{k} q^k (1 - q)^{S-k},$$

where $q = (1 - p)^I$, and p is the probability that a susceptible person contracts the disease from an infected person. For example, if $p = .5$, then there is a 50% chance that a susceptible person exposed to 1 infected person for 1 week will contract the disease. (*Source:* Hoppensteadt, F. and C. Peskin, *Mathematics in Medicine and the Life Sciences,* Springer-Verlag, 1992.)

(a) Compute the probability P of 3 family members not becoming infected within 1 week if there are currently 2 infected and 4 susceptible members. Assume that $p = .1$. (*Hint:* To use the formula, first determine the values of k, I, S, and q.)
(b) A highly infectious disease can have $p = .5$. Repeat part (a) with this value of p.
(c) Determine the probability that everyone would become sick in a large family if initially, $I = 1$, $S = 9$, and $p = .5$. Discuss the results.

38. *(Modeling) Spread of Disease* (Refer to Exercise 37.) Suppose that in a family $I = 2$ and $S = 4$. If the probability P is .25 of there being $k = 2$ uninfected members after 1 week, estimate graphically the possible values of p. (*Hint:* Write P as a function of p.)

Chapter 11 Summary

KEY TERMS

11.1 finite sequence, infinite sequence, terms of a sequence, general term (nth term), convergent sequence, divergent sequence, recursive definition, Fibonacci sequence, series, summation notation, finite series, infinite series, index of summation

11.2 arithmetic sequence (arithmetic progression), common difference, arithmetic series

11.3 geometric sequence (geometric progression), common ratio, geometric series, annuity, future value of an annuity

11.4 Pascal's triangle, factorial notation, binomial coefficient, binomial theorem (general binomial expansion)

11.6 tree diagram, independent events, permutation, combination

11.7 outcome, sample space, event, probability, complement, Venn diagram, odds, mutually exclusive events, binomial experiment

NEW SYMBOLS

a_n	nth term of a sequence
$\sum_{i=1}^{n} a_i$	summation notation; sum of n terms
i	index of summation
S_n	sum of first n terms of a sequence
$\sum_{i=1}^{\infty} a_i$	sum of an infinite number of terms
$\lim_{n\to\infty} a_n$	limit of a_n as n gets larger and larger
$n!$	n-factorial
${}_nC_r$ or $\binom{n}{r}$	binomial coefficient (combinations of n elements taken r at a time)
$P(n, r)$	permutations of n elements taken r at a time
$C(n, r)$ or $\binom{n}{r}$	combinations of n elements taken r at a time
$n(E)$	number of outcomes that belong to event E
$P(E)$	probability of event E
E'	complement of event E

QUICK REVIEW

CONCEPTS	EXAMPLES
11.1 Sequences and Series	
Sequence **General Term a_n** **Series**	The sequence $1, \frac{1}{2}, \frac{1}{3}, \frac{1}{4}, \ldots, \frac{1}{n}$ has general term $a_n = \frac{1}{n}$. The corresponding series is the *sum* $$1 + \frac{1}{2} + \frac{1}{3} + \frac{1}{4} + \cdots + \frac{1}{n}.$$
Summation Properties If $a_1, a_2, a_3, \ldots, a_n$ and $b_1, b_2, b_3, \ldots, b_n$ are sequences and c is a constant, then for every positive integer n, (a) $\sum_{i=1}^{n} c = nc$	$$\sum_{i=1}^{6} 5 = 6 \cdot 5 = 30$$

CONCEPTS	EXAMPLES
(b) $\sum_{i=1}^{n} ca_i = c\sum_{i=1}^{n} a_i$	$\sum_{i=1}^{4} 3(2i+1) = 3\sum_{i=1}^{4}(2i+1)$ $= 3(3+5+7+9)$ $= 72$
(c) $\sum_{i=1}^{n} (a_i \pm b_i) = \sum_{i=1}^{n} a_i \pm \sum_{i=1}^{n} b_i.$	$\sum_{i=1}^{3} 5i + 6i^2 = \sum_{i=1}^{3} 5i + \sum_{i=1}^{3} 6i^2$ $= (5+10+15) + (6+24+54)$ $= 30 + 84 = 114$

11.2 Arithmetic Sequences and Series

Assume a_1 is the first term, a_n is the nth term, and d is the common difference.	The arithmetic sequence 2, 5, 8, 11, . . . has $a_1 = 2$.
Common Difference $\quad d = a_{n+1} - a_n$	$d = 5 - 2 = 3$ (Any two successive terms could have been used.)
nth Term $\quad a_n = a_1 + (n-1)d$	Suppose that $n = 10$. Then the 10th term is $a_{10} = 2 + (10-1)3$ $= 2 + 9 \cdot 3 = 29.$
Sum of the First n Terms $S_n = \frac{n}{2}(a_1 + a_n)$ or $S_n = \frac{n}{2}[2a_1 + (n-1)d]$	The sum of the first 10 terms is $S_{10} = \frac{10}{2}(a_1 + a_{10})$ $= 5(2+29) = 5(31) = 155$ or $\quad S_{10} = \frac{10}{2}[2(2) + (10-1)3]$ $= 5(4 + 9 \cdot 3)$ $= 5(4+27) = 5(31) = 155.$

11.3 Geometric Sequences and Series

Assume a_1 is the first term, a_n is the nth term, and r is the common ratio.	The geometric sequence 1, 2, 4, 8, . . . has $a_1 = 1$.
Common Ratio $\quad r = \frac{a_{n+1}}{a_n}$	$r = \frac{8}{4} = 2$ (Any two successive terms could have been used.)
nth Term $\quad a_n = a_1 r^{n-1}$	Suppose that $n = 6$. Then the sixth term is $a_6 = (1)(2)^{6-1} = 1(2)^5 = 32.$
Sum of the First n Terms $S_n = \frac{a_1(1-r^n)}{1-r} \quad (r \neq 1)$	The sum of the first six terms is $S_6 = \frac{1(1-2^6)}{1-2} = \frac{1-64}{-1} = 63.$

(continued)

CONCEPTS	EXAMPLES
Sum of the Terms of an Infinite Geometric Sequence with $\lvert r \rvert < 1$ $$S_\infty = \frac{a_1}{1-r}$$	The sum of the terms of the infinite geometric sequence $$\sum_{k=0}^{\infty}\left(\frac{1}{2}\right)^k = 1 + \frac{1}{2} + \frac{1}{4} + \cdots$$ is $$S_\infty = \frac{1}{1-\frac{1}{2}} = \frac{1}{\frac{1}{2}} = 2.$$

11.4 The Binomial Theorem

For any positive integer n, $$n! = n(n-1)(n-2)\cdots(3)(2)(1)$$ $$0! = 1.$$	$$4! = 4 \cdot 3 \cdot 2 \cdot 1 = 24$$
Binomial Coefficient For nonnegative integers n and r, with $r \le n$, $${}_nC_r = \binom{n}{r} = \frac{n!}{r!(n-r)!}.$$	$${}_5C_3 = \frac{5!}{3!(5-3)!} = \frac{5!}{3!2!} = \frac{5 \cdot 4 \cdot 3 \cdot 2 \cdot 1}{3 \cdot 2 \cdot 1 \cdot 2 \cdot 1} = 10$$
Binomial Theorem For any positive integer n and any complex numbers x and y, $$(x+y)^n = x^n + \binom{n}{1}x^{n-1}y + \binom{n}{2}x^{n-2}y^2 + \binom{n}{3}x^{n-3}y^3 + \cdots + \binom{n}{r}x^{n-r}y^r + \cdots + \binom{n}{n-1}xy^{n-1} + y^n.$$	$$(2m+3)^4 = (2m)^4 + \frac{4!}{3!1!}(2m)^3(3) + \frac{4!}{2!2!}(2m)^2(3)^2 + \frac{4!}{1!3!}(2m)(3)^3 + 3^4$$ $$= 2^4m^4 + 4(2)^3m^3(3) + 6(2)^2m^2(9) + 4(2m)(27) + 81$$ $$= 16m^4 + 12(8)m^3 + 54(4)m^2 + 216m + 81$$ $$= 16m^4 + 96m^3 + 216m^2 + 216m + 81$$
***k*th Term of the Binomial Expansion of $(x+y)^n$** $$\binom{n}{k-1}x^{n-(k-1)}y^{k-1} \quad (n \ge k-1)$$	The eighth term of $(a-2b)^{10}$ is $$\binom{10}{7}a^3(-2b)^7 = \frac{10!}{7!3!}a^3(-2)^7b^7$$ $$= 120(-128)a^3b^7$$ $$= -15{,}360a^3b^7.$$

11.5 Mathematical Induction

Principle of Mathematical Induction Let S_n be a statement concerning the positive integer n. Suppose that **1.** S_1 is true; **2.** for any positive integer k, $k \le n$, if S_k is true, then S_{k+1} is also true. Then S_n is true for every positive integer value of n.	See Examples 1 and 2 in Section 11.5. Example 3 in Section 11.5 illustrates the Generalized Principle of Mathematical Induction.

CONCEPTS	EXAMPLES

11.6 Counting Theory

Fundamental Principle of Counting

If n independent events occur, with

m_1 ways for event 1 to occur,

m_2 ways for event 2 to occur,

$\vdots$

and m_n ways for event n to occur,

then there are $m_1 \cdot m_2 \cdot \cdots \cdot m_n$ different ways for all n events to occur.

If there are 2 ways to choose a pair of socks and 5 ways to choose a pair of shoes, then there are $2 \cdot 5 = 10$ ways to choose socks and shoes.

Permutations Formula

If $P(n, r)$ denotes the number of permutations of n elements taken r at a time, with $r \leq n$, then

$$P(n,r) = \frac{n!}{(n-r)!}.$$

How many ways are there to arrange the letters of the word *triangle* using 5 letters at a time?

Here, $n = 8$ and $r = 5$, so the number of ways is

$$P(8,5) = \frac{8!}{(8-5)!} = \frac{8!}{3!} = 6720.$$

Combinations Formula

The number of combinations of n elements taken r at a time, with $r \leq n$, is

$$C(n,r) = \binom{n}{r} = \frac{n!}{(n-r)!r!}.$$

How many committees of 4 senators can be formed from a group of 9 senators?

Since the arrangement of senators does not matter, this is a combinations problem. The number of committees is

$$C(9,4) = \binom{9}{4} = \frac{9!}{5!4!} = 126.$$

11.7 Basics of Probability

Probability of an Event E

In a sample space S with equally likely outcomes, the probability of an event E is

$$P(E) = \frac{n(E)}{n(S)}.$$

A number is chosen at random from $S = \{1, 2, 3, 4, 5, 6\}$. What is the probability that the number is less than 3?

The event is $E = \{1, 2\}$; $n(S) = 6$ and $n(E) = 2$, so

$$P(E) = \frac{2}{6} = \frac{1}{3}.$$

Properties of Probability

For any events E and F:

1. $0 \leq P(E) \leq 1$

2. $P(\text{a certain event}) = 1$

3. $P(\text{an impossible event}) = 0$

4. $P(E') = 1 - P(E)$

5. $P(E \text{ or } F) = P(E \cup F)$
$= P(E) + P(F) - P(E \cap F)$.

What is the probability that the number is 3 or more?

This event is E'.

$$P(E') = 1 - \frac{1}{3} = \frac{2}{3}$$

Binomial Probability

If the probability of success in a binomial experiment is p, then the probability of r successes in n trials is

$$\binom{n}{r}p^r(1-p)^{n-r}.$$

An experiment consists of rolling a die 8 times. Find the probability that exactly 5 rolls result in a 2.

Here, we have $n = 8$, $r = 5$, and $p = \frac{1}{6}$.

$$\binom{8}{5}\left(\frac{1}{6}\right)^5\left(1 - \frac{1}{6}\right)^{8-5} = 56\left(\frac{1}{6}\right)^5\left(\frac{5}{6}\right)^3 \approx .00417$$

Chapter 11 Review Exercises

Write the first five terms of each sequence. State whether the sequence is arithmetic, geometric, *or* neither.

1. $a_n = \dfrac{n}{n+1}$

2. $a_n = (-2)^n$

3. $a_n = 2(n+3)$

4. $a_n = n(n+1)$

5. $a_1 = 5$
$a_n = a_{n-1} - 3$, if $n \geq 2$

In Exercises 6–9, write the first five terms of the sequence described.

6. arithmetic, $a_2 = 10, d = -2$

7. arithmetic, $a_3 = \pi, a_4 = 1$

8. geometric, $a_1 = 6, r = 2$

9. geometric, $a_1 = -5, a_2 = -1$

10. An arithmetic sequence has $a_5 = -3$ and $a_{15} = 17$. Find a_1 and a_n.

11. A geometric sequence has $a_1 = -8$ and $a_7 = -\frac{1}{8}$. Find a_4 and a_n.

Find a_8 for each arithmetic sequence.

12. $a_1 = 6, d = 2$

13. $a_1 = 6x - 9, a_2 = 5x + 1$

Find S_{12} for each arithmetic sequence.

14. $a_1 = 2, d = 3$

15. $a_2 = 6, d = 10$

Find a_5 for each geometric sequence.

16. $a_1 = -2, r = 3$

17. $a_3 = 4, r = \dfrac{1}{5}$

Find S_4 for each geometric sequence.

18. $a_1 = 3, r = 2$

19. $a_1 = -1, r = 3$

20. $\dfrac{3}{4}, -\dfrac{1}{2}, \dfrac{1}{3}, \ldots$

Evaluate each sum that exists.

21. $\sum_{i=1}^{7} (-1)^{i-1}$

22. $\sum_{i=1}^{5} (i^2 + i)$

23. $\sum_{i=1}^{4} \dfrac{i+1}{i}$

24. $\sum_{j=1}^{10} (3j - 4)$

25. $\sum_{j=1}^{2500} j$

26. $\sum_{i=1}^{5} 4 \cdot 2^i$

27. $\sum_{i=1}^{\infty} \left(\dfrac{4}{7}\right)^i$

28. $\sum_{i=1}^{\infty} -2\left(\dfrac{6}{5}\right)^i$

Evaluate each series that converges. If the series diverges, say so.

29. $24 + 8 + \dfrac{8}{3} + \dfrac{8}{9} + \cdots$

30. $-\dfrac{3}{4} + \dfrac{1}{2} - \dfrac{1}{3} + \dfrac{2}{9} - \cdots$

31. $\dfrac{1}{12} + \dfrac{1}{6} + \dfrac{1}{3} + \dfrac{2}{3} + \cdots$

32. $.9 + .09 + .009 + .0009 + \cdots$

Evaluate each sum where $x_1 = 0$, $x_2 = 1$, $x_3 = 2$, $x_4 = 3$, $x_5 = 4$, and $x_6 = 5$.

33. $\sum_{i=1}^{4} (x_i^2 - 6)$

34. $\sum_{i=1}^{6} f(x_i)\,\Delta x;\ f(x) = (x - 2)^3,\ \Delta x = .1$

Write each sum using summation notation.

35. $4 - 1 - 6 - \cdots - 66$

36. $10 + 14 + 18 + \cdots + 86$

37. $4 + 12 + 36 + \cdots + 972$

38. $\frac{5}{6} + \frac{6}{7} + \frac{7}{8} + \cdots + \frac{12}{13}$

Use the binomial theorem to expand each expression.

39. $(x + 2y)^4$

40. $(3z - 5w)^3$

41. $\left(3\sqrt{x} - \frac{1}{\sqrt{x}}\right)^5$

42. $(m^3 - m^{-2})^4$

Find the indicated term or terms for each expansion.

43. sixth term of $(4x - y)^8$

44. seventh term of $(m - 3n)^{14}$

45. first four terms of $(x + 2)^{12}$

46. last three terms of $(2a + 5b)^{16}$

47. Describe a proof by mathematical induction.

48. What kinds of statements are proved by mathematical induction? Give examples.

Use mathematical induction to prove that each statement is true for every positive integer n.

49. $1 + 3 + 5 + 7 + \cdots + (2n - 1) = n^2$

50. $2 + 6 + 10 + 14 + \cdots + (4n - 2) = 2n^2$

51. $2 + 2^2 + 2^3 + \cdots + 2^n = 2(2^n - 1)$

52. $1^3 + 3^3 + 5^3 + \cdots + (2n - 1)^3 = n^2(2n^2 - 1)$

Find the value of each expression.

53. $P(9, 2)$

54. $P(6, 0)$

55. $\binom{8}{3}$

56. $9!$

57. $C(10, 5)$

Solve each problem.

58. ***Wedding Plans*** Two people are planning their wedding. They can select from 2 different chapels, 4 soloists, 3 organists, and 2 ministers. How many different wedding arrangements are possible?

59. ***Couch Styles*** Bob Schiffer, who is furnishing his apartment, wants to buy a new couch. He can select from 5 different styles, each available in 3 different fabrics, with 6 color choices. How many different couches are available?

60. ***Summer Job Assignments*** Four students are to be assigned to 4 different summer jobs. Each student is qualified for all 4 jobs. In how many ways can the jobs be assigned?

61. ***Conference Delegations*** A student body council consists of a president, vice-president, secretary/treasurer, and 3 representatives at large. Three members are to be selected to attend a conference.

(a) How many different such delegations are possible?

(b) How many are possible if the president must attend?

62. ***Tournament Outcomes*** Nine football teams are competing for first-, second-, and third-place titles in a statewide tournament. In how many ways can the winners be determined?

63. ***License Plates*** How many different license plates can be formed with a letter followed by 3 digits and then 3 letters? How many such license plates have no repeats?

64. ***Racetrack Bets*** Most racetracks have "compound" bets on 2 or more horses. An *exacta* is a bet in which the first and second finishers in a race are specified in order. A *quinella* is a bet on the first 2 finishers in a race, with order not specified.

(a) In a field of 9 horses, how many different exacta bets can be placed?
(b) How many different quinella bets can be placed in a field of 9 horses?

65. ***Drawing a Marble*** A marble is drawn at random from a box containing 4 green, 5 black, and 6 white marbles. Find the following probabilities.

(a) A green marble is drawn. **(b)** A marble that is not black is drawn.
(c) A blue marble is drawn.
(d) What are the odds in favor of drawing a marble that is not white?

66. ***Drawing a Card*** A card is drawn from a standard deck of 52 cards. Find the probability of each of the following events.

(a) a black king **(b)** a face card or an ace
(c) an ace or a diamond **(d)** a card that is not a diamond
(e) What are the odds in favor of drawing an ace?

67. ***Political Orientation*** The table describes the political orientation of college freshmen in the class of 2006, as determined from a survey of 282,200 freshmen.

Political Orientation	Number of Freshmen (in thousands)
Far left	7.06
Liberal	71.48
Middle of the road	143.5
Conservative	56.51
Far right	3.673
Total	282.2

Source: Higher Education Research Institute, UCLA, 2002.

(a) What is the probability that a randomly selected student from the class is in the conservative group?
(b) What is the probability that a randomly selected student from the class is on the far left or the far right politically?
(c) What is the probability of a randomly selected student from the class not being politically middle of the road?

68. ***Defective Toaster Ovens*** A sample shipment of 5 toaster ovens is chosen. The probability of exactly 0, 1, 2, 3, 4, or 5 toaster ovens being defective is given in the table.

Number Defective	0	1	2	3	4	5
Probability	.31	.25	.18	.12	.08	.06

Find the probability that the given number of toaster ovens are defective.

(a) no more than 3 **(b)** at least 2 **(c)** more than 5

69. *Rolling a Die* A die is rolled 12 times. Find the probability (to three decimal places) that exactly 2 of the rolls result in a 5.

70. *Tossing a Coin* A coin is tossed 10 times. Find the probability (to three decimal places) that exactly 4 of the tosses result in a tail.

Chapter 11 Test

Write the first five terms of each sequence. State whether the sequence is arithmetic, geometric, *or* neither.

1. $a_n = (-1)^n(n^2 + 2)$

2. $a_n = -3\left(\frac{1}{2}\right)^n$

3. $a_1 = 2,\ a_2 = 3,\ a_n = a_{n-1} + 2a_{n-2},\quad \text{for } n \geq 3$

4. A certain arithmetic sequence has $a_1 = 1$ and $a_3 = 25$. Find a_5.

5. A certain geometric sequence has $a_1 = 81$ and $r = -\frac{2}{3}$. Find a_6.

Find the sum of the first ten terms of each series.

6. arithmetic, $a_1 = -43,\ d = 12$

7. geometric, $a_1 = 5,\ r = -2$

Evaluate each sum that exists.

8. $\sum_{i=1}^{30} (5i + 2)$

9. $\sum_{i=1}^{5} (-3 \cdot 2^i)$

10. $\sum_{i=1}^{\infty} (2^i) \cdot 4$

11. $\sum_{i=1}^{\infty} 54\left(\frac{2}{9}\right)^i$

Use the binomial theorem to expand each expression.

12. $(x + y)^6$

13. $(2x - 3y)^4$

14. Find the third term in the expansion of $(w - 2y)^6$.

Evaluate each expression.

15. $8!$

16. $C(10, 2)$

17. $\binom{7}{3}$

18. $P(11, 3)$

19. Use mathematical induction to prove that for all positive integers n,

$$1 + 7 + 13 + \cdots + (6n - 5) = n(3n - 2).$$

Solve each problem.

20. *Athletic Shoe Styles* A sports-shoe manufacturer makes athletic shoes in 4 different styles. Each style comes in 3 different colors, and each color comes in 2 different shades. How many different types of shoes can be made?

21. *Seminar Attendees* A mortgage company has 10 loan officers: 1 black, 2 Asian, and the rest white. In how many ways can 3 of these officers be selected to attend a seminar? How many ways are there if the black officer and exactly 1 Asian officer must be included?

22. *Project Workers* Refer to Exercise 21. If 4 of the loan officers are women and 6 are men, in how many ways can 2 women and 2 men be selected to work on a special project?

23. Write a few sentences to a friend explaining how to determine when to use permutations and when to use combinations in an applied problem.

24. *Drawing Cards* A card is drawn from a standard deck of 52 cards. Find the probability that each of the following is drawn.

(a) a red three
(b) a card that is not a face card
(c) a king or a spade
(d) What are the odds in favor of drawing a face card?

25. *Defective Transistors* A sample of 4 light bulbs is chosen. The probability of exactly 0, 1, 2, 3, or 4 light bulbs being defective is given in the table. Find the probability that at most 2 are defective.

Number Defective	0	1	2	3	4
Probability	.19	.43	.30	.07	.01

26. *Rolling a Die* Find the probability (to three decimal places) of obtaining 5 on exactly two of six rolls of a single die.

Chapter 11 Quantitative Reasoning

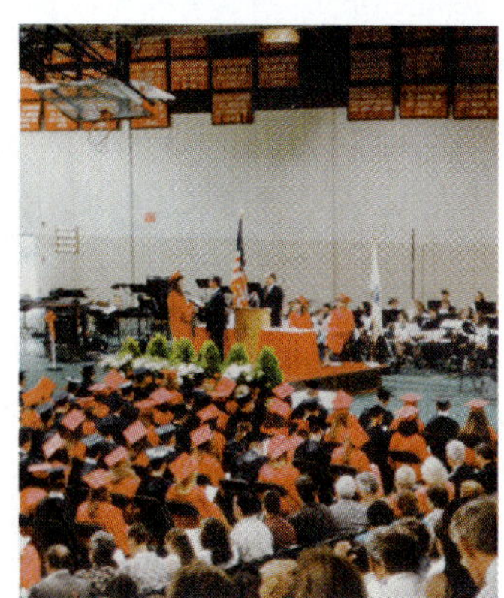

What is the value of a college education?

A high school graduate must decide whether the cost of investing in a college education will be worth it in the long run versus jumping immediately into the job market. Suppose you have estimated the cost of a 4-yr college education, including tuition and living expenses, at $130,000. You have also located the following statistics of annual earnings.

> In 2000, the median annual earnings of a person with 4 yr of college was $43,368, and the median annual earnings of a high school graduate with no college attendance was $26,364. The annual median earnings of the two groups have risen at rates of about $994 and $534 per year, respectively. (*Source:* U.S. Bureau of Labor Statistics.)

Now, do the math. Assume the average 18-yr-old high school graduate in 2000 will work until age 65, earning the median amount throughout those years. (Of course, such a person will receive less than the median earnings in the beginning and more than the median earnings in later years. These differences should balance out to produce a reasonable approximation of lifetime earnings.)

1. How much will a person earning the median amount earn until retirement if he or she joins the work force immediately after high school graduation without going to college?
2. How much will a person earning the median amount earn until retirement if he or she attends college for 4 yr and then joins the work force?
3. How much more will a person earning the median amount who attends 4 yr of college earn over his or her lifetime? Is the $130,000 cost worth it?

Appendix

A Polar Form of Conic Sections

Up to this point, we have worked with equations of conic sections in rectangular form. If the focus of a conic section is at the pole, the polar form of its equation is

$$r = \frac{ep}{1 \pm e \cdot f(\theta)},$$

where f is either the sine or cosine function.

Polar Forms of Conic Sections

A polar equation of the form

$$r = \frac{ep}{1 \pm e \cos \theta} \quad \text{or} \quad r = \frac{ep}{1 \pm e \sin \theta}$$

has a conic section as its graph. The eccentricity is e (where $e > 0$), and $|p|$ is the distance between the pole (focus) and the directrix.

We can verify that $r = \frac{ep}{1 + e \cos \theta}$ does indeed satisfy the definition of a conic section. Consider Figure 1, where the directrix is vertical and $p > 0$ units to the right of the focus $F(0, 0°)$. Let $P(r, \theta)$ be a point on the graph. Then the distance between P and the directrix is

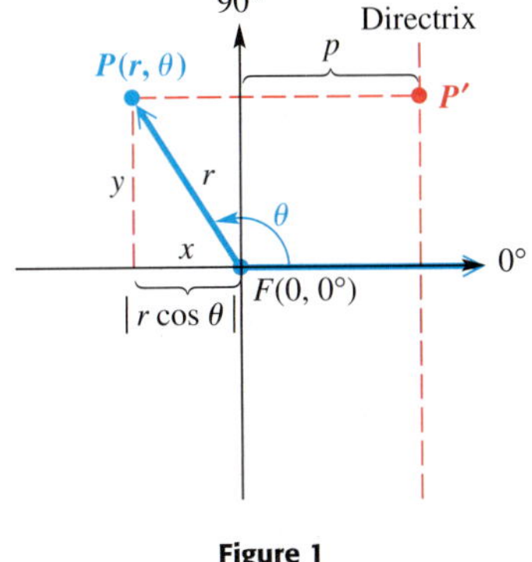

Figure 1

$$
\begin{aligned}
PP' &= |p - x| \\
&= |p - r \cos \theta| && x = r \cos \theta \text{ (Section 8.5)} \\
&= \left| p - \left(\frac{ep}{1 + e \cos \theta} \right) \cos \theta \right| && \text{Use the equation for } r. \\
&= \left| \frac{p(1 + e \cos \theta) - ep \cos \theta}{1 + e \cos \theta} \right| && \text{Write with a common denominator. (Section R.5)} \\
&= \left| \frac{p + ep \cos \theta - ep \cos \theta}{1 + e \cos \theta} \right| && \text{Distributive property (Section R.1)} \\
PP' &= \left| \frac{p}{1 + e \cos \theta} \right|.
\end{aligned}
$$

Since
$$r = \frac{ep}{1 + e\cos\theta},$$
we can multiply each side by $\frac{1}{e}$ to obtain
$$\frac{r}{e} = \frac{p}{1 + e\cos\theta}.$$
We substitute $\frac{r}{e}$ for the expression in the absolute value bars for PP'.
$$PP' = \left|\frac{p}{1 + e\cos\theta}\right| = \left|\frac{r}{e}\right| = \frac{|r|}{|e|} = \frac{|r|}{e}$$
The distance between the pole and P is $PF = |r|$, so the ratio of PF to PP' is
$$\frac{PF}{PP'} = \frac{|r|}{\frac{|r|}{e}} = e.$$
Simplify the complex fraction. **(Section R.5)**

Thus, by the definition, the graph has eccentricity e and must be a conic.

In the preceding discussion, we assumed a vertical directrix to the right of the pole. There are three other possible situations, and all four are summarized in the table.

If the equation is:	then the directrix is:
$r = \frac{ep}{1 + e\cos\theta}$	*vertical*, p units to the *right* of the pole.
$r = \frac{ep}{1 - e\cos\theta}$	*vertical*, p units to the *left* of the pole.
$r = \frac{ep}{1 + e\sin\theta}$	*horizontal*, p units *above* the pole.
$r = \frac{ep}{1 - e\sin\theta}$	*horizontal*, p units *below* the pole.

EXAMPLE 1 Graphing a Conic Section with Equation in Polar Form

Graph $r = \frac{8}{4 + 4\sin\theta}$.

Algebraic Solution

Divide both numerator and denominator by 4 to get
$$r = \frac{2}{1 + \sin\theta}.$$
Based on the preceding table, this is the equation of a conic with $ep = 2$ and $e = 1$. Thus, $p = 2$. Since $e = 1$, the graph is a parabola. The focus is at the pole, and the directrix is horizontal, 2 units *above* the pole. The vertex must have polar coordinates $(1, 90°)$. Letting $\theta = 0°$ and $\theta = 180°$ gives the additional points $(2, 0°)$ and $(2, 180°)$. See Figure 2 on the next page.

Graphing Calculator Solution

Enter
$$r_1 = \frac{8}{4 + 4\sin\theta},$$
with the calculator in polar and degree modes. The first two screens in Figure 3 on the next page show the window settings, and the third screen shows the graph. Notice that the point $(1, 90°)$ is indicated at the bottom of the third screen.

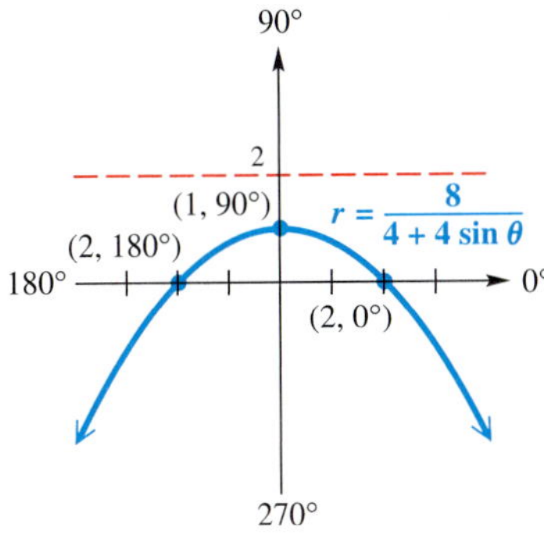

Figure 2

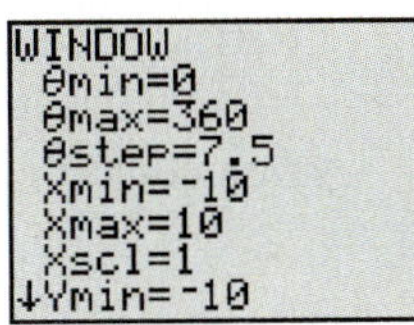

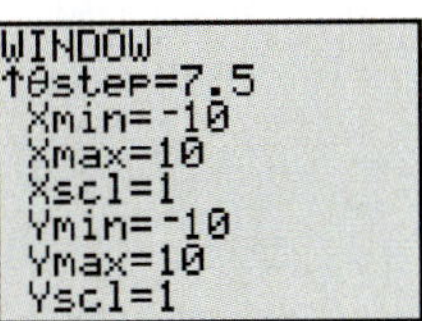

This is a continuation of the screen to the left.

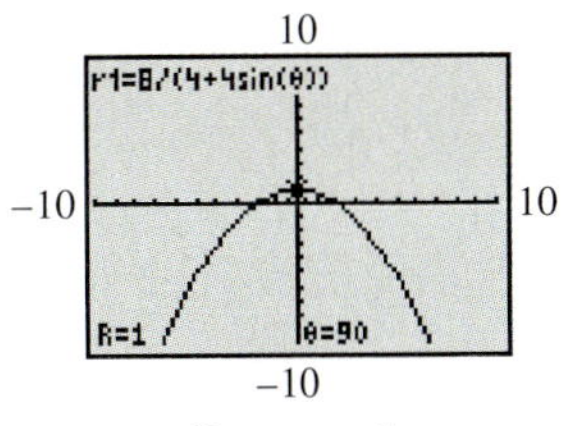

Degree mode

Figure 3

Now try Exercise 1.

EXAMPLE 2 Finding a Polar Equation

Find the polar equation of a parabola with focus at the pole and vertical directrix 3 units to the left of the pole.

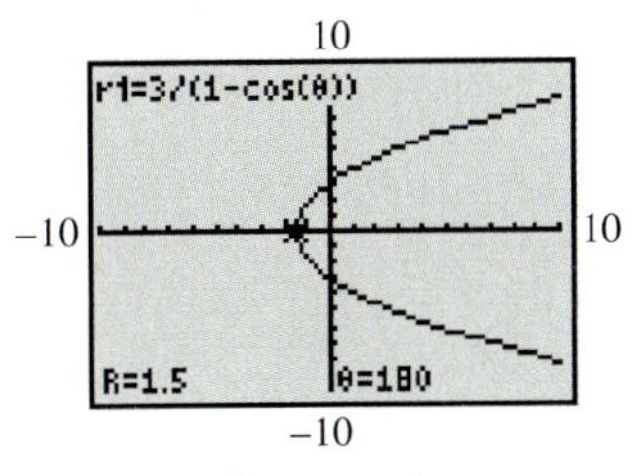

Degree mode

Figure 4

Solution The eccentricity e must be 1, p must equal 3, and the equation must be of the form

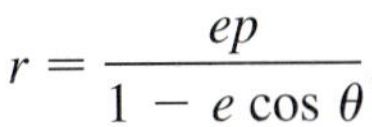

$$r = \frac{ep}{1 - e\cos\theta}.$$

Thus, we have
$$r = \frac{1 \cdot 3}{1 - 1\cos\theta} = \frac{3}{1 - \cos\theta}.$$

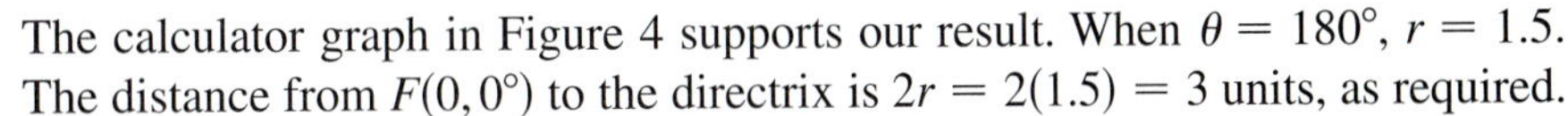

The calculator graph in Figure 4 supports our result. When $\theta = 180°$, $r = 1.5$. The distance from $F(0, 0°)$ to the directrix is $2r = 2(1.5) = 3$ units, as required.

Now try Exercise 13.

EXAMPLE 3 Identifying and Converting from Polar Form to Rectangular Form

Identify the type of conic represented by $r = \frac{8}{2 - \cos\theta}$. Then convert the equation to rectangular form.

Solution To identify the type of conic, we divide both the numerator and the denominator on the right side by 2 to obtain

$$r = \frac{4}{1 - \frac{1}{2}\cos\theta}.$$

From the table, we see that this is a conic that has a vertical directrix, with $e = \frac{1}{2}$; thus, it is an ellipse. To convert to rectangular form, we start with the given equation.

$$r = \frac{8}{2 - \cos\theta}$$

$r(2 - \cos\theta) = 8$	Multiply by $2 - \cos\theta$.
$2r - r\cos\theta = 8$	Distributive property
$2r = r\cos\theta + 8$	Add $r\cos\theta$ to each side.
$(2r)^2 = (r\cos\theta + 8)^2$	Square each side. **(Section 1.6)**
$(2r)^2 = (x + 8)^2$	$r\cos\theta = x$
$4r^2 = x^2 + 16x + 64$	Multiply. **(Section R.3)**
$4(x^2 + y^2) = x^2 + 16x + 64$	$r^2 = x^2 + y^2$ **(Section 8.5)**
$4x^2 + 4y^2 = x^2 + 16x + 64$	Distributive property
$3x^2 + 4y^2 - 16x - 64 = 0$	Standard form **(Section 10.2)**

The coefficients of x^2 and y^2 are both positive and are not equal, further supporting our assertion that the graph is an ellipse.

Now try Exercise 21.

Appendix A Exercises

Graph each conic whose equation is given in polar form. See Example 1.

1. $r = \dfrac{6}{3 + 3\sin\theta}$

2. $r = \dfrac{10}{5 + 5\sin\theta}$

3. $r = \dfrac{-4}{6 + 2\cos\theta}$

4. $r = \dfrac{-8}{4 + 2\cos\theta}$

5. $r = \dfrac{2}{2 - 4\sin\theta}$

6. $r = \dfrac{6}{2 - 4\sin\theta}$

7. $r = \dfrac{4}{2 - 4\cos\theta}$

8. $r = \dfrac{6}{2 - 4\cos\theta}$

9. $r = \dfrac{-1}{1 + 2\sin\theta}$

10. $r = \dfrac{-1}{1 - 2\sin\theta}$

11. $r = \dfrac{-1}{2 + \cos\theta}$

12. $r = \dfrac{-1}{2 - \cos\theta}$

Find a polar equation of the parabola with focus at the pole, satisfying the given conditions. See Example 2.

13. Vertical directrix 3 units to the right of the pole

14. Vertical directrix 4 units to the left of the pole

15. Horizontal directrix 5 units below the pole

16. Horizontal directrix 6 units above the pole

Find a polar equation for the conic with focus at the pole, satisfying the given conditions. Also identify the type of conic represented.

17. $e = \frac{4}{5}$; vertical directrix 5 units to the right of the pole

18. $e = \frac{2}{3}$; vertical directrix 6 units to the left of the pole

19. $e = \frac{5}{4}$; horizontal directrix 8 units below the pole

20. $e = \frac{3}{2}$; horizontal directrix 4 units above the pole

Identify the type of conic represented, and convert the equation to rectangular form. See Example 3.

21. $r = \dfrac{6}{3 - \cos\theta}$

22. $r = \dfrac{8}{4 - \cos\theta}$

23. $r = \dfrac{-2}{1 + 2\cos\theta}$

24. $r = \dfrac{-3}{1 + 3\cos\theta}$

25. $r = \dfrac{-6}{4 + 2\sin\theta}$

26. $r = \dfrac{-12}{6 + 3\sin\theta}$

27. $r = \dfrac{10}{2 - 2\sin\theta}$

28. $r = \dfrac{12}{4 - 4\sin\theta}$

B Rotation of Axes

Derivation of Rotation Equations ▪ Applying a Rotation Equation

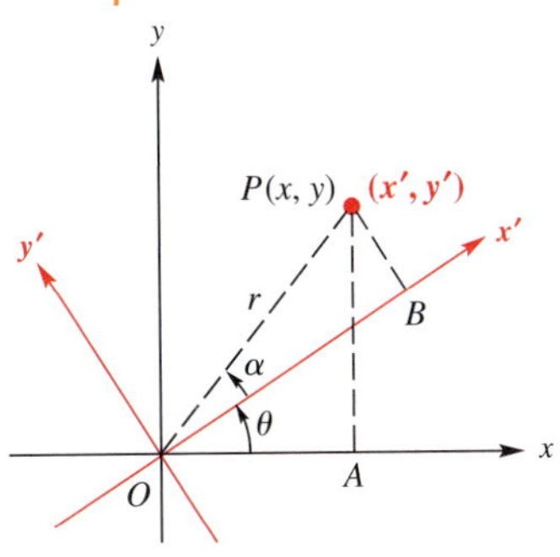

Figure 1

Derivation of Rotation Equations If we begin with an xy-coordinate system having origin O and rotate the axes about O through an angle θ, the new coordinate system is called a **rotation** of the xy-system. Trigonometric identities can be used to obtain equations for converting the coordinates of a point from the xy-system to the rotated $x'y'$-system. Let P be any point other than the origin, with coordinates (x, y) in the xy-system and (x', y') in the $x'y'$-system. See Figure 1. Let $OP = r$, and let α represent the angle made by OP and the x'-axis. As shown in Figure 1,

$$\cos(\theta + \alpha) = \frac{OA}{r} = \frac{x}{r}, \qquad \sin(\theta + \alpha) = \frac{AP}{r} = \frac{y}{r},$$

$$\cos\alpha = \frac{OB}{r} = \frac{x'}{r}, \qquad \sin\alpha = \frac{PB}{r} = \frac{y'}{r}.$$

(Section 5.2)

These four statements can be written as

$$x = r\cos(\theta + \alpha), \qquad y = r\sin(\theta + \alpha), \qquad x' = r\cos\alpha, \qquad y' = r\sin\alpha.$$

Looking Ahead to Calculus

Rotation of axes is a topic traditionally covered in calculus texts, in conjunction with parametric equations and polar coordinates. The coverage in calculus is typically the same as that seen in this section.

Using the trigonometric identity for the cosine of the sum of two angles gives

$$\begin{aligned} x &= r\cos(\theta + \alpha) \\ &= r(\cos\theta\cos\alpha - \sin\theta\sin\alpha) && \text{(Section 7.3)} \\ &= (r\cos\alpha)\cos\theta - (r\sin\alpha)\sin\theta && \text{Distributive property (Section R.1)} \\ &= x'\cos\theta - y'\sin\theta. && \text{Substitute.} \end{aligned}$$

In the same way, by using the identity for the sine of the sum of two angles, $y = x'\sin\theta + y'\cos\theta$. This proves the following result.

Rotation Equations

If the rectangular coordinate axes are rotated about the origin through an angle θ, and if the coordinates of a point P are (x, y) and (x', y') with respect to the xy-system and the $x'y'$-system, respectively, then the **rotation equations** are

$$\mathbf{x = x'\cos\theta - y'\sin\theta} \quad \text{and} \quad \mathbf{y = x'\sin\theta + y'\cos\theta.}$$

Applying a Rotation Equation

EXAMPLE 1 Finding an Equation after a Rotation

The equation of a curve is $x^2 + y^2 + 2xy + 2\sqrt{2}x - 2\sqrt{2}y = 0$. Find the resulting equation if the axes are rotated 45°. Graph the equation.

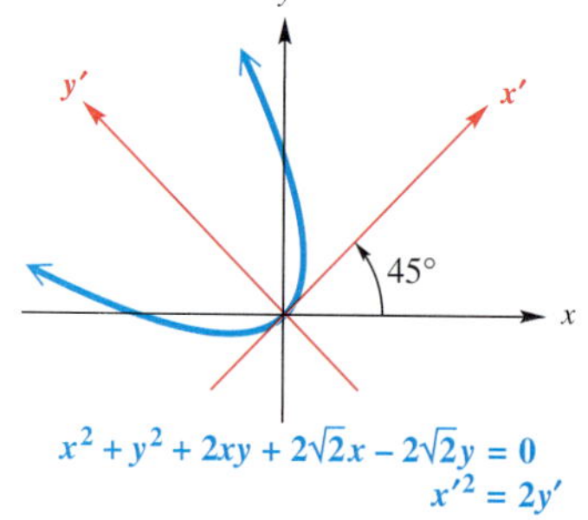

Figure 2

Solution If $\theta = 45°$, then $\sin\theta = \frac{\sqrt{2}}{2}$ and $\cos\theta = \frac{\sqrt{2}}{2}$, and the rotation equations become

$$x = \frac{\sqrt{2}}{2}x' - \frac{\sqrt{2}}{2}y' \quad \text{and} \quad y = \frac{\sqrt{2}}{2}x' + \frac{\sqrt{2}}{2}y'.$$

Substituting these values into the given equation yields

$$x^2 + y^2 + 2xy + 2\sqrt{2}x - 2\sqrt{2}y = 0$$

$$\left[\frac{\sqrt{2}}{2}x' - \frac{\sqrt{2}}{2}y'\right]^2 + \left[\frac{\sqrt{2}}{2}x' + \frac{\sqrt{2}}{2}y'\right]^2 + 2\left[\frac{\sqrt{2}}{2}x' - \frac{\sqrt{2}}{2}y'\right]\left[\frac{\sqrt{2}}{2}x' + \frac{\sqrt{2}}{2}y'\right]$$
$$+ 2\sqrt{2}\left[\frac{\sqrt{2}}{2}x' - \frac{\sqrt{2}}{2}y'\right] - 2\sqrt{2}\left[\frac{\sqrt{2}}{2}x' + \frac{\sqrt{2}}{2}y'\right] = 0.$$

Expanding these terms yields

$$\frac{1}{2}x'^2 - x'y' + \frac{1}{2}y'^2 + \frac{1}{2}x'^2 + x'y' + \frac{1}{2}y'^2 + x'^2 - y'^2 + 2x' - 2y' - 2x' - 2y' = 0.$$

Collecting terms gives

$$\begin{aligned} 2x'^2 - 4y' &= 0 \\ x'^2 - 2y' &= 0 && \text{Divide by 2.} \end{aligned}$$

or, finally $\qquad x'^2 = 2y',\quad$ (Section 10.1)

the equation of a parabola. The graph is shown in Figure 2.

Now try Exercise 13.

We have graphed equations written in the general form

$$Ax^2 + Cy^2 + Dx + Ey + F = 0. \quad \text{(Section 10.4)}$$

To graph an equation that has an xy-term by hand, it is necessary to find an appropriate **angle of rotation** to eliminate the xy-term. The necessary angle of rotation can be determined by using the following result. The proof is quite lengthy and is not presented here.

Angle of Rotation

The xy-term is removed from the general equation

$$\boldsymbol{Ax^2 + Bxy + Cy^2 + Dx + Ey + F = 0}$$

by a rotation of the axes through an angle θ, $0° < \theta < 90°$, where

$$\cot 2\theta = \frac{A - C}{B}.$$

Now try Exercise 7.

To find the rotation equations, first find $\sin\theta$ and $\cos\theta$. Example 2 illustrates a way to obtain $\sin\theta$ and $\cos\theta$ from $\cot 2\theta$ without first identifying angle θ.

EXAMPLE 2 Rotating and Graphing

Remove the xy-term from $52x^2 - 72xy + 73y^2 = 200$ by performing a suitable rotation, and graph the equation.

Solution Here $A = 52$, $B = -72$, and $C = 73$. By substitution,

$$\cot 2\theta = \frac{52 - 73}{-72} = \frac{-21}{-72} = \frac{7}{24}.$$

To find $\sin\theta$ and $\cos\theta$, use the trigonometric identities

$$\sin\theta = \sqrt{\frac{1 - \cos 2\theta}{2}} \quad \text{and} \quad \cos\theta = \sqrt{\frac{1 + \cos 2\theta}{2}}. \quad \text{(Section 7.4)}$$

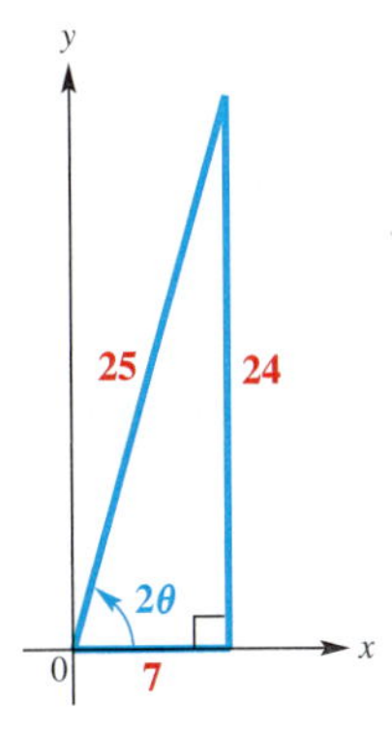

Figure 3

Sketch a right triangle and label it as in Figure 3, to see that $\cos 2\theta = \frac{7}{25}$. (Recall from Section 5.2 that in the two quadrants for which we are concerned, cosine and cotangent have the same sign.) Then

$$\sin\theta = \sqrt{\frac{1 - \frac{7}{25}}{2}} = \sqrt{\frac{9}{25}} = \frac{3}{5} \quad \text{and} \quad \cos\theta = \sqrt{\frac{1 + \frac{7}{25}}{2}} = \sqrt{\frac{16}{25}} = \frac{4}{5}.$$

Use these values for $\sin\theta$ and $\cos\theta$ to obtain

$$x = \frac{4}{5}x' - \frac{3}{5}y' \quad \text{and} \quad y = \frac{3}{5}x' + \frac{4}{5}y'.$$

Substituting these expressions for x and y into the original equation yields

$$52\left[\frac{4}{5}x' - \frac{3}{5}y'\right]^2 - 72\left[\frac{4}{5}x' - \frac{3}{5}y'\right]\left[\frac{3}{5}x' + \frac{4}{5}y'\right] + 73\left[\frac{3}{5}x' + \frac{4}{5}y'\right]^2 = 200$$

$$52\left[\frac{16}{25}x'^2 - \frac{24}{25}x'y' + \frac{9}{25}y'^2\right] - 72\left[\frac{12}{25}x'^2 + \frac{7}{25}x'y' - \frac{12}{25}y'^2\right] + 73\left[\frac{9}{25}x'^2 + \frac{24}{25}x'y' + \frac{16}{25}y'^2\right] = 200$$

$$25x'^2 + 100y'^2 = 200 \quad \text{Combine terms.}$$

$$\frac{x'^2}{8} + \frac{y'^2}{2} = 1. \quad \text{Divide by 200.}$$

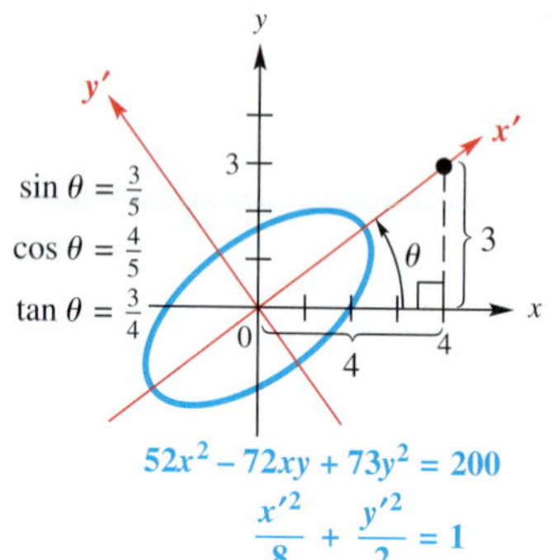

Figure 4

This is an equation of an ellipse having x'-intercepts $\pm 2\sqrt{2}$ and y'-intercepts $\pm\sqrt{2}$. The graph is shown in Figure 4. To find θ, use the fact that

$$\frac{\sin\theta}{\cos\theta} = \frac{\frac{3}{5}}{\frac{4}{5}} = \frac{3}{4} = \tan\theta, \quad \text{(Section 7.1)}$$

from which $\theta \approx 37°$.

Now try Exercise 17.

The following summary enables us to use the general equation to decide on the type of graph to expect.

Equation of a Conic with an *xy*-Term

If the general second-degree equation

$$Ax^2 + Bxy + Cy^2 + Dx + Ey + F = 0$$

has a graph, it will be one of the following:

(a) a circle or an ellipse (or a point) if $B^2 - 4AC < 0$;

(b) a parabola (or one line or two parallel lines) if $B^2 - 4AC = 0$;

(c) a hyperbola (or two intersecting lines) if $B^2 - 4AC > 0$;

(d) a straight line if $A = B = C = 0$, and $D \neq 0$ or $E \neq 0$.

Now try Exercise 1.

Appendix B Exercises

Concept Check *Use the summary in this section to predict the type of graph of each second-degree equation.*

1. $4x^2 + 3y^2 + 2xy - 5x = 8$

2. $x^2 + 2xy - 3y^2 + 2y = 12$

3. $2x^2 + 3xy - 4y^2 = 0$

4. $x^2 - 2xy + y^2 + 4x - 8y = 0$

5. $4x^2 + 4xy + y^2 + 15 = 0$

6. $-x^2 + 2xy - y^2 + 16 = 0$

Find the angle of rotation θ that will remove the xy-term in each equation.

7. $2x^2 + \sqrt{3}xy + y^2 + x = 5$

8. $4\sqrt{3}x^2 + xy + 3\sqrt{3}y^2 = 10$

9. $3x^2 + \sqrt{3}xy + 4y^2 + 2x - 3y = 12$

10. $4x^2 + 2xy + 2y^2 + x - 7 = 0$

11. $x^2 - 4xy + 5y^2 = 18$

12. $3\sqrt{3}x^2 - 2xy + \sqrt{3}y^2 = 25$

Use the given angle of rotation to remove the xy-term and graph each equation. See Example 1.

13. $x^2 - xy + y^2 = 6;\ \theta = 45°$

14. $2x^2 - xy + 2y^2 = 25;\ \theta = 45°$

15. $8x^2 - 4xy + 5y^2 = 36;\ \sin\theta = \dfrac{2}{\sqrt{5}}$

16. $5y^2 + 12xy = 10;\ \sin\theta = \dfrac{3}{\sqrt{13}}$

Remove the xy-term from each equation by performing a suitable rotation. Graph each equation. See Example 2.

17. $3x^2 - 2xy + 3y^2 = 8$

18. $x^2 + xy + y^2 = 3$

19. $x^2 - 4xy + y^2 = -5$

20. $x^2 + 2xy + y^2 + 4\sqrt{2}x - 4\sqrt{2}y = 0$

21. $7x^2 + 6\sqrt{3}xy + 13y^2 = 64$

22. $7x^2 + 2\sqrt{3}xy + 5y^2 = 24$

23. $3x^2 - 2\sqrt{3}xy + y^2 - 2x - 2\sqrt{3}y = 0$

24. $2x^2 + 2\sqrt{3}xy + 4y^2 = 5$

In each equation, remove the xy-term by rotation. Then translate the axes and sketch the graph.

25. $x^2 + 3xy + y^2 - 5\sqrt{2}y = 15$

26. $x^2 - \sqrt{3}xy + 2\sqrt{3}x - 3y - 3 = 0$

27. $4x^2 + 4xy + y^2 - 24x + 38y - 19 = 0$

28. $12x^2 + 24xy + 19y^2 - 12x - 40y + 31 = 0$

29. $16x^2 + 24xy + 9y^2 - 130x + 90y = 0$

30. $9x^2 - 6xy + y^2 - 12\sqrt{10}x - 36\sqrt{10}y = 0$

31. Look at the box titled "Angle of Rotation." Why is no rotation applicable if the value of B is 0?

32. Look at the equation involving $\cot 2\theta$ in the box titled "Angle of Rotation." Why must the angle of rotation be 45° if the coefficients of x^2 and y^2 are equal, and $B \neq 0$?

C Geometry Formulas

Square

Perimeter: $P = 4s$

Area: $A = s^2$

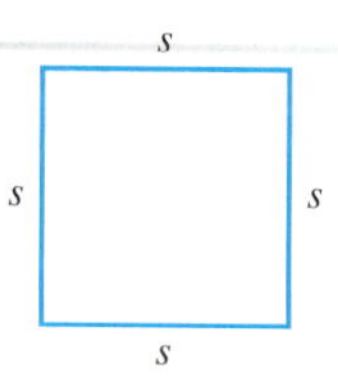

Rectangle

Perimeter: $P = 2L + 2W$

Area: $A = LW$

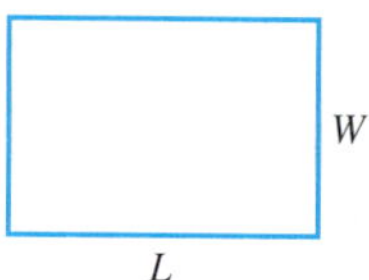

Triangle

Perimeter: $P = a + b + c$

Area: $A = \frac{1}{2}bh$

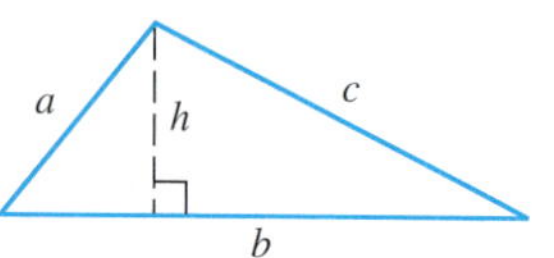

Parallelogram

Perimeter: $P = 2a + 2b$

Area: $A = bh$

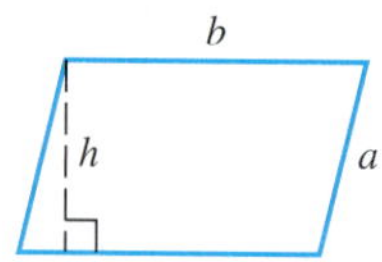

Trapezoid

Perimeter: $P = a + b + c + B$

Area: $A = \frac{1}{2}(B + b)h$

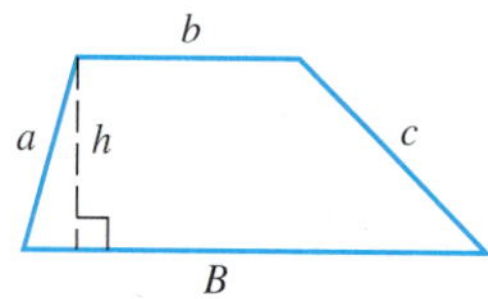

Circle

Diameter: $d = 2r$

Circumference: $C = 2\pi r = \pi d$

Area: $A = \pi r^2$

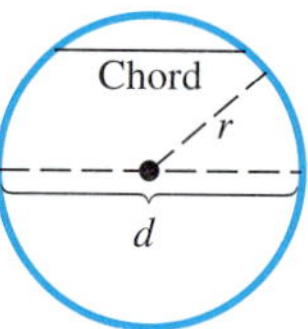

Cube

Volume: $V = e^3$

Surface area: $S = 6e^2$

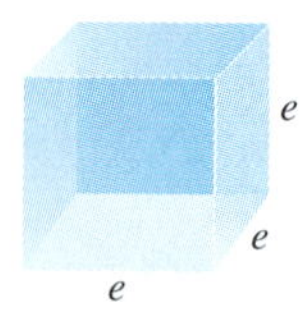

Rectangular Solid

Volume: $V = LWH$

Surface area: $S = 2HW + 2LW + 2LH$

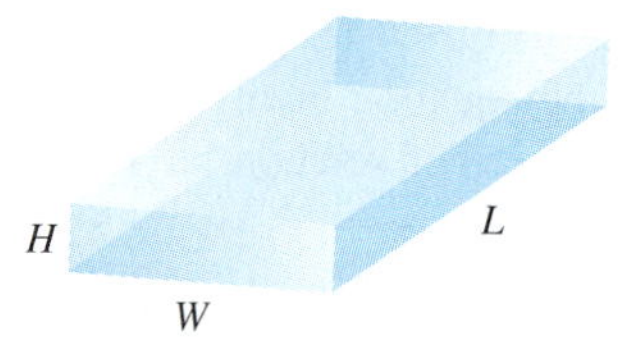

Sphere

Volume: $V = \frac{4}{3}\pi r^3$

Surface area: $S = 4\pi r^2$

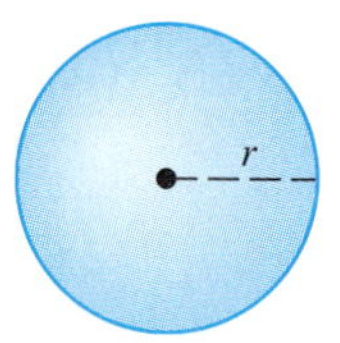

Cone

Volume: $V = \frac{1}{3}\pi r^2 h$

Surface area: $S = \pi r\sqrt{r^2 + h^2}$ (excludes the base)

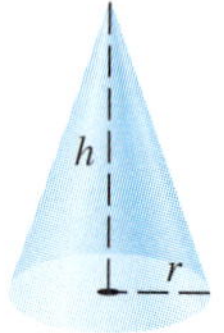

Right Circular Cylinder

Volume: $V = \pi r^2 h$

Surface area: $S = 2\pi rh + 2\pi r^2$ (includes top and bottom)

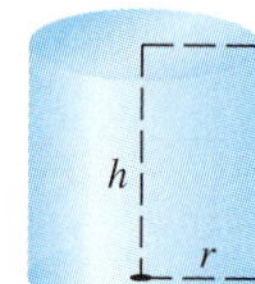

Right Pyramid

Volume: $V = \frac{1}{3}Bh$

B = area of the base

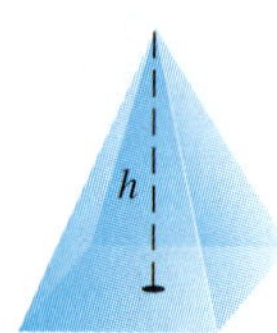

Glossary

A

absolute value The distance on the number line from a number to 0 is called the absolute value of that number. (Section R.2)

acute angle An acute angle is an angle measuring between $0°$ and $90°$. (Section 5.1)

addition of ordinates Addition of ordinates is a method for graphing a function that is the sum of two other functions by adding the y-values of the two functions at selected x-values. (Section 6.4)

additive inverse (negative) For each real number a, the number $-a$ is the additive inverse or negative of a. (Section R.1)

additive inverse (negative) of a matrix When two matrices are added and a zero matrix results, the matrices are additive inverses (negatives) of each other. (Section 9.7)

airspeed In air navigation, the airspeed of a plane is its speed relative to the air. (Section 8.4)

algebraic expression An algebraic expression is the result of adding, subtracting, multiplying, dividing (except by 0), or raising to powers or taking roots on any combination of variables or constants. (Section R.3)

amplitude The amplitude of a periodic function is half the difference between the maximum and minimum values of the function. (Section 6.3)

angle An angle is formed by rotating a ray around its endpoint. (Section 5.1)

angle of depression The angle of depression from point X to point Y (below X) is the acute angle formed by ray XY and a horizontal ray with endpoint at X. (Section 5.4)

angle of elevation The angle of elevation from point X to point Y (above X) is the acute angle formed by ray XY and a horizontal ray with endpoint at X. (Section 5.4)

angle of rotation The angle through which the axes of an xy-coordinate system are rotated to obtain a new coordinate system is called the angle of rotation. (Appendix B)

Angle-Side-Angle (ASA) The Angle-Side-Angle (ASA) congruence axiom states that if two angles and the included side of one triangle are equal, respectively, to two angles and the included side of a second triangle, then the triangles are congruent. (Section 8.1)

angle in standard position An angle is in standard position if its vertex is at the origin and its initial side is along the positive x-axis. (Section 5.1)

angle between two vectors The angle between two vectors is defined to be the angle θ, for $0° \leq \theta \leq 180°$, having the two vectors as its sides. (Section 8.3)

angular speed ω Angular speed ω (omega) measures the speed of rotation and is defined by $\omega = \frac{\theta}{t}$, where θ is the angle of rotation in radians and t is time. (Section 6.2)

annuity An annuity is a sequence of equal payments made at equal periods of time. (Section 11.3)

argument of a complex number When a complex number $x + yi$ is written in trigonometric (or polar) form as $r(\cos \theta + i \sin \theta)$, the angle θ is called the argument of the complex number. (Section 8.5)

argument of a function The argument of a function is the expression containing the independent variable of the function. For example, in the function $y = f(x - d)$, the expression $x - d$ is the argument. (Section 6.3)

arithmetic sequence (arithmetic progression) An arithmetic sequence is a sequence in which each term after the first is obtained by adding a fixed number to the previous term. (Section 11.2)

arithmetic series An arithmetic series is the sum of the terms of an arithmetic sequence. (Section 11.2)

asymptotes of a hyperbola The asymptotes of a hyperbola are two lines that the hyperbola approaches but never touches or intersects. (Section 10.3)

augmented matrix An augmented matrix is a matrix whose elements are the coefficients of the variables and the constants of a system of equations. An augmented matrix is often written with a vertical bar that separates the coefficients of the variables from the constants. (Section 9.2)

average rate of change The average rate of change is an interpretation of slope that applies to many linear models. The slope of a line gives the average rate of change in y per unit of change in x, where the value of y depends on the value of x. (Section 2.3)

axis The line of symmetry for a parabola is called the axis of the parabola. (Sections 3.1, 10.1)

B

base The base is the number that is a repeated factor in exponential notation. In the expression a^n, a is the base. (Section R.1)

bearing Bearing is used to identify angles in navigation. One method for expressing bearing uses a single angle, with bearing measured in a clockwise direction from due north. A second method for expressing bearing starts with a north-south line and uses an acute angle to show the direction, either east or west, from this line. (Section 5.4)

binomial A binomial is a polynomial containing exactly two terms. (Section R.3)

binomial coefficient For nonnegative integers n and r, with $r \leq n$, the binomial coefficient is the value of $\frac{n!}{r!(n-r)!}$. Binomial coefficients are used in calculating the terms of a binomial expansion. (Section 11.4)

binomial experiment In probability, an experiment that consists of repeated independent trials with only two outcomes in each trial, success or failure, is called a binomial experiment. (Section 11.7)

binomial theorem (general binomial expansion) The binomial theorem is a formula used to expand a binomial raised to a power. (Section 11.4)

boundary A line that separates a plane into two half-planes is called the boundary of each half-plane. (Section 9.6)

break-even point The break-even point is the point where the revenue from selling a product is equal to the cost of producing it. (Section 1.7)

C

cardioid A cardioid is a heart-shaped curve that is the graph of a polar equation of the form $r = a \pm b \sin \theta$ or $r = a \pm b \cos \theta$, where $\left|\frac{a}{b}\right| = 1$. (Section 8.7)

center of a circle The center of a circle is the given point that is a given distance from all points on the circle. (Section 2.1)

center of an ellipse The center of an ellipse is the midpoint of the major axis. (Section 10.2)

center of a hyperbola The center of a hyperbola is the midpoint of the transverse axis. (Section 10.3)

change in x The change in x is the horizontal difference (the difference in x-coordinates) between two points on a line. (Section 2.3)

change in y The change in y is the vertical difference (the difference in y-coordinates) between two points on a line. (Section 2.3)

circle A circle is the set of all points in a plane that lie a given distance from a given point. (Section 2.1)

circular functions The trigonometric functions of arc lengths, or real numbers, are called circular functions. (Section 6.2)

closed interval A closed interval is an interval that includes both of its endpoints. (Section 1.7)

coefficient (numerical coefficient) The real number in a term of an algebraic expression is called a coefficient. (Section R.3)

cofactor The product of a minor of an element of a square matrix and the number $+1$ (if the sum of the row number and column number is even) or -1 (if the sum of the row number and column number is odd) is called a cofactor. (Section 9.3)

collinear Points are collinear if they lie on the same line. (Section 2.1)

column matrix A matrix with just one column is called a column matrix. (Section 9.7)

combination A subset of items selected *without regard to order* is called a combination. (Section 11.6)

combined variation Variation in which one variable depends on more than one other variable is called combined variation. (Section 3.6)

common difference In an arithmetic sequence, the fixed number that is added to each term to get the next term is called the common difference. (Section 11.2)

common logarithm A base 10 logarithm is called a common logarithm. (Section 4.4)

common ratio In a geometric sequence, the fixed number by which each term is multiplied to get the next term is called the common ratio. (Section 11.3)

complement of an event In probability, the set of all outcomes in a sample space that do *not* belong to an event E is called the complement of E, written E'. (Section 11.7)

complement of a set The set of all elements in the universal set that do not belong to set A is the complement of A, written A'. (Section 11.7)

complementary angles Two positive angles are complementary angles if the sum of their measures is 90°. (Section 5.1)

completing the square The process of adding to a binomial the number that makes it a perfect square trinomial is called completing the square. (Section 1.4)

complex conjugates Two complex numbers that differ only in the sign of their imaginary parts are called complex conjugates. (Section 1.3)

complex fraction A complex fraction is a quotient of two rational expressions. (Section R.5)

complex number A complex number is a number of the form $a + bi$, where a and b are real numbers and $i = \sqrt{-1}$. (Section 1.3)

complex plane The complex plane is a two-dimensional representation of the complex numbers in which the horizontal axis is the real axis and the vertical axis is the imaginary axis. (Section 8.5)

composite function (composition) If f and g are functions, then the composite function, or composition, of g and f is defined by $(g \circ f)(x) = g[f(x)]$. (Section 2.7)

compound amount In an investment paying compound interest, the compound amount is the balance *after* interest has been earned. (The compound amount is sometimes called the *future value.*) (Section 4.2)

compound interest In compound interest, interest is paid both on the principal and previously earned interest. (Section 4.2)

conditional equation An equation that is satisfied by some numbers but not by others is called a conditional equation. (Section 1.1)

conic (conic section) (geometric definition) A conic is the set of all points $P(x, y)$ in a plane such that the ratio of the distance from P to a fixed point and the distance from P to a fixed line is constant. (Section 10.2)

conic sections (conics) When a plane intersects a double cone at different angles, the figures formed by the intersections are called conic sections. (Section 10.1)

conjugates The expressions $a - b$ and $a + b$ are called conjugates. (Section R.7)

consistent system A consistent system is a system of equations with at least one solution. (Section 9.1)

constant function A function f is constant on an interval I if, for every x_1 and x_2 in I, $f(x_1) = f(x_2)$. (Section 2.2)

constant of variation In a variation equation, such as $y = kx$, $y = \frac{k}{x}$, or $y = kx^n z^m$, the real number k is called the constant of variation. (Section 3.6)

constraints In linear programming, the inequalities that represent restrictions on a particular situation are called the constraints. (Section 9.6)

continuous compounding As the frequency of compounding increases, compound interest approaches a limit, called continuous compounding. (Section 4.2)

continuous function A function is continuous over an interval of its domain if its hand-drawn graph over that interval can be sketched without lifting the pencil from the paper. (Section 2.5)

contradiction An equation that has no solution is called a contradiction. (Section 1.1)

convergent sequence An infinite sequence is convergent if its terms get closer and closer to some real number. (Section 11.1)

coordinate (on a number line) A number that corresponds to a particular point on a number line is called the coordinate of the point. (Section R.2)

coordinate plane (*xy*-plane) The plane into which the rectangular coordinate system is introduced is called the coordinate plane, or *xy*-plane. (Section 2.1)

coordinates (in the *xy*-plane) The coordinates of a point in the *xy*-plane are the numbers in the ordered pair that correspond to that point. (Section 2.1)

coordinate system (on a number line) The correspondence between points on a line and the real numbers is called a coordinate system. (Section R.2)

cosecant Let $P(x, y)$ be a point other than the origin on the terminal side of an angle θ in standard position. Let $r = \sqrt{x^2 + y^2}$ represent the distance from the origin to P. Then the cosecant function is defined by $\csc \theta = \frac{r}{y}$ $(y \neq 0)$. (Section 5.2)

cosine Let $P(x, y)$ be a point other than the origin on the terminal side of an angle θ in standard position. Let $r = \sqrt{x^2 + y^2}$ represent the distance from the origin to P. Then the cosine function is defined by $\cos \theta = \frac{x}{r}$. (Section 5.2)

cotangent Let $P(x, y)$ be a point other than the origin on the terminal side of an angle θ in standard position. Let $r = \sqrt{x^2 + y^2}$ represent the distance from the origin to P. Then the cotangent function is defined by $\cot \theta = \frac{x}{y}$ $(y \neq 0)$. (Section 5.2)

coterminal angles Two angles that have the same initial side and the same terminal side, but different amounts of rotation, are called coterminal angles. The measures of coterminal angles differ by a multiple of 360°. (Section 5.1)

Cramer's rule Cramer's rule uses determinants to solve systems of linear equations. (Section 9.3)

cycloid A cycloid is a curve that represents the path traced by a fixed point on the circumference of a circle rolling along a line. (Section 8.8)

D

damped oscillatory motion Damped oscillatory motion is oscillatory motion that has been slowed down (damped) by the force of friction. Friction causes the amplitude of the motion to diminish gradually until the weight comes to rest. (Section 6.5)

decreasing function A function f is decreasing on an interval I if, whenever $x_1 < x_2$ in I, $f(x_1) > f(x_2)$. (Section 2.2)

degree The degree is the most common unit of measure for angles. One degree, written 1°, represents $\frac{1}{360}$ of a rotation. (Section 5.1)

degree of a polynomial The greatest degree of any term in a polynomial is called the degree of the polynomial. (Section R.3)

degree of a term The degree of a term is the sum of the exponents on the variables in the term. (Section R.3)

dependent equations Two equations are dependent if any solution of one equation is also a solution of the other. (Section 9.1)

dependent variable If the value of the variable y depends on the value of the variable x, then y is called the dependent variable. (Section 2.2)

determinant Every $n \times n$ matrix A is associated with a single real number called the determinant of A, written $|A|$. (Section 9.3)

difference quotient If f is a function and h is a positive number, then the expression $\frac{f(x + h) - f(x)}{h}$ is called the difference quotient. (Section 2.7)

direction angle The positive angle between the x-axis and a position vector is the direction angle for the vector. (Section 8.3)

directrix A directrix is a fixed line that, together with a focus, is used to determine the points that form a conic section. (Sections 10.1, 10.4)

discriminant The quantity under the radical in the quadratic formula, $b^2 - 4ac$, is called the discriminant. (Section 1.4)

divergent sequence If an infinite sequence does not converge to some number, then it is called a divergent sequence. (Section 11.1)

division algorithm The division algorithm is a method for dividing polynomials that is similar to that used for long division of whole numbers. The division algorithm can be stated as follows: Let $f(x)$ and $g(x)$ be polynomials with $g(x)$ of lower degree than $f(x)$ and $g(x)$ of degree one or more. There exist unique polynomials $q(x)$ and $r(x)$ such that $f(x) = g(x) \cdot q(x) + r(x)$, where either $r(x) = 0$ or the degree of $r(x)$ is less than the degree of $g(x)$. (Section 3.2)

domain of a rational expression The domain of a rational expression is the set of real numbers for which the expression is defined. (Section R.5)

domain of a relation In a relation, the set of all values of the independent variable (x) is called the domain. (Section 2.2)

dominating term In a polynomial function, the dominating term is the term of greatest degree. (Section 3.4)

dot product The dot product of two vectors is the sum of the product of their first components and the product of their second components. The dot product of the two vectors $\mathbf{u} = \langle a, b \rangle$ and $\mathbf{v} = \langle c, d \rangle$ is denoted $\mathbf{u} \cdot \mathbf{v}$ and given by $\mathbf{u} \cdot \mathbf{v} = ac + bd$. (Section 8.3)

doubling time The amount of time that it takes for a quantity that grows exponentially to become twice its initial amount is called its doubling time. (Section 4.6)

E

eccentricity The eccentricity of a parabola, ellipse, or hyperbola is the fixed ratio of the distance from a point P to a focus compared to the distance from the same point to the directrix. The eccentricity of a circle is 0. (Section 10.2)

element of a matrix Each number in a matrix is called an element of the matrix. (Section 9.2)

ellipse An ellipse is the set of all points in a plane the sum of whose distances from two fixed points is constant. (Section 10.2)

empty set (null set) The empty set or null set, written $\emptyset$ or $\{\ \}$, is the set containing no elements. (Section 1.1)

end behavior The end behavior of a graph of a polynomial function describes what happens to the values of y as x gets larger and larger in absolute value. (Section 3.4)

equation An equation is a statement that two expressions are equal. (Section 1.1)

equilibrant The opposite vector of the resultant of two vectors is called the equilibrant. (Section 8.4)

equivalent equations Equations with the same solution set are called equivalent equations. (Section 1.1)

equivalent systems Equivalent systems are systems of equations that have the same solution set. (Section 9.1)

even function A function f is called an even function if $f(-x) = f(x)$ for all x in the domain of f. (Section 2.6)

event In probability, any subset of the sample space is called an event. (Section 11.7)

exact number A number that represents the result of counting, or a number that results from theoretical work and is not the result of a measurement, is an exact number. (Section 5.4)

expansion by a row or column Expansion by a row or column is a method for evaluating a determinant of a 3×3 or larger matrix. This process involves multiplying each element of any row or column of the matrix by its cofactor and then adding these products. (Section 9.3)

exponent An exponent is a number that indicates how many times a factor is repeated. In the expression a^n, n is the exponent. (Section R.1)

exponential equation An exponential equation is an equation with a variable in an exponent. (Section 4.2)

exponential function If $a > 0$ and $a \neq 1$, then $f(x) = a^x$ defines the exponential function with base a. (Section 4.2)

exponential growth or decay function An exponential growth or decay function models a situation in which a quantity changes at a rate proportional to the amount present. Such a function is defined by an equation of the form $y = y_0 e^{kt}$, where y_0 is the amount or number present at time $t = 0$ and k is a constant. (Section 4.6)

F

factor When two numbers are multiplied, each number is called a factor of the product. (Section R.1)

factored completely A polynomial is factored completely when it is written as a product of prime polynomials. (Section R.4)

factored form A polynomial is written in factored form when it is written as a product of prime polynomials. (Section R.4)

factorial notation Factorial notation is a compact way of writing a product of consecutive natural numbers. The symbol $n!$, read "n-factorial," is defined as follows: For any positive integer n, $n! = n(n-1)(n-2)\cdots(3)(2)(1)$, and $0! = 1$. (Section 11.4)

factoring The process of finding polynomials whose product equals a given polynomial is called factoring. (Section R.4)

factoring by grouping Factoring by grouping is a method of grouping the terms of a polynomial in such a way that the polynomial can be factored even though its greatest common factor is 1. (Section R.4)

factor theorem The factor theorem states that the polynomial $x - k$ is a factor of the polynomial $f(x)$ if and only if $f(k) = 0$. (Section 3.3)

Fibonacci sequence The Fibonacci sequence is the sequence $1, 1, 2, 3, 5, 8, 13, \ldots$. In this sequence, each term starting with the third term is the sum of the previous two terms. (Section 11.1)

finite sequence A sequence is a finite sequence if its domain is the set $\{1, 2, 3, \ldots, n\}$, where n is a natural number. (Section 11.1)

finite series A finite series is an expression of the form $S_n = a_1 + a_2 + a_3 + \cdots + a_n = \sum_{i=1}^{n} a_i$. (Section 11.1)

foci (singular, **focus**) Foci are fixed points used to determine the points that form a parabola, an ellipse, or a hyperbola. A parabola has one focus, while an ellipse or a hyperbola has two foci. (Sections 10.1, 10.2, 10.3)

four-leaved rose A four-leaved rose is a curve that is the graph of a polar equation of the form $r = a \sin 2\theta$ or $r = a \cos 2\theta$. (Section 8.7)

frequency In simple harmonic motion, the frequency is the number of cycles per unit of time, or the reciprocal of the period. (Section 6.5)

function A function is a relation (set of ordered pairs) in which, for each value of the first component of the ordered pairs, there is *exactly one* value of the second component. (Section 2.2)

function notation Function notation $f(x)$ (read "f of x") represents the y-value of the function f for the indicated x-value. (Section 2.2)

fundamental rectangle The fundamental rectangle is used as a guide in sketching the graph of a hyperbola. The extended diagonals of this rectangle are the asymptotes of the hyperbola. (Section 10.3)

future value In an investment paying compound interest, the future value is the balance *after* interest has been earned. (The future value is sometimes called the *compound amount.*) (Section 4.2)

future value of an annuity If the payments on an annuity are accumulated in an account (with no withdrawals), then the sum of the payments and interest on the payments is called the future value of the annuity. (Section 11.3)

G

general term (*n*th term) In the sequence $a_1, a_2, a_3, \ldots$, the general term (or *n*th term) is a_n. (Section 11.1)

geometric sequence (geometric progression) A geometric sequence is a sequence in which each term after the first is obtained by multiplying the preceding term by a constant nonzero real number. (Section 11.3)

geometric series A geometric series is the sum of the terms of a geometric sequence. (Section 11.3)

graph of an equation The graph of an equation is the set of all points that correspond to all of the ordered pairs that satisfy the equation. (Section 2.1)

groundspeed In air navigation, the groundspeed of a plane is its speed relative to the ground. (Section 8.4)

H

half-life The amount of time that it takes for a quantity that decays exponentially to become half its initial amount is called its half-life. (Section 4.6)

half-plane A line separates a plane into two regions, each of which is called a half-plane. (Section 9.6)

horizontal asymptote A horizontal line that a graph approaches as $|x|$ gets larger and larger without bound is called a horizontal asymptote. The line $y = b$ is a horizontal asymptote if $y \to b$ as $|x| \to \infty$. (Section 3.5)

horizontal component When a vector **u** is expressed as an ordered pair in the form $\mathbf{u} = \langle a, b \rangle$, the number a is the horizontal component of the vector. (Section 8.3)

hyperbola A hyperbola is the set of all points in a plane such that the absolute value of the difference of the distances from two fixed points is constant. (Section 10.3)

I

identity An equation satisfied by every number that is a meaningful replacement for the variable is called an identity. (Section 1.1)

identity element for addition (additive identity) The number 0 is the identity element for addition, or the additive identity. (Section R.1)

identity element for multiplication (multiplicative identity) The number 1 is the identity element for multiplication, or the multiplicative identity. (Section R.1)

identity matrix (multiplicative identity matrix) An identity matrix is an $n \times n$ matrix with 1s on the main diagonal and 0s everywhere else. (Section 9.8)

imaginary axis In the complex plane, the vertical axis is called the imaginary axis. (Section 8.5)

imaginary part In the complex number $a + bi$, b is called the imaginary part. (Section 1.3)

imaginary unit The number i, defined by $i^2 = -1$ $\left(\text{so } i = \sqrt{-1}\right)$, is called the imaginary unit. (Section 1.3)

inconsistent system An inconsistent system is a system of equations with no solution. (Section 9.1)

increasing function A function f is increasing on an interval I if, whenever $x_1 < x_2$ in I, $f(x_1) < f(x_2)$. (Section 2.2)

independent events Two events are independent events if neither influences the outcomes of the other. (Section 11.6)

independent variable If the value of the variable y depends on the value of the variable x, then x is called the independent variable. (Section 2.2)

index of a radical In a radical of the form $\sqrt[n]{a}$, n is called the index. (Section R.7)

index of summation When using summation notation, such as $\sum_{i=1}^{n} a_i$, the letter i is called the index of summation. Other letters may also be used, such as j and k. (Section 11.1)

inequality An inequality says that one expression is greater than, greater than or equal to, less than, or less than or equal to, another. (Section 1.7)

infinite sequence A sequence is an infinite sequence if its domain is the set of *all* natural numbers. (Section 11.1)

infinite series An infinite series is an expression of the form $S_\infty = a_1 + a_2 + a_3 + \cdots + a_n + \cdots = \sum_{i=1}^{\infty} a_i$. (Section 11.1)

initial point When two letters are used to name a vector, the first letter indicates the initial (starting) point of the vector. (Section 8.3)

initial side When a ray is rotated around its endpoint to form an angle, the ray in its initial position is called the initial side of the angle. (Section 5.1)

integers The set of integers is $\{\ldots -3, -2, -1, 0, 1, 2, 3, \ldots\}$. (Section R.1)

intersection The intersection of sets A and B, written $A \cap B$, is the set of elements that belong to both A and B. (Sections 1.7, 11.7)

interval An interval is a portion of the real number line, which may or may not include its endpoint(s). (Section 1.7)

interval notation Interval notation is a simplified notation for writing intervals. It uses parentheses and brackets to show whether the endpoints are included. (Section 1.7)

inverse function Let f be a one-to-one function. Then g is the inverse function of f if $(f \circ g)(x) = x$ for every x in the domain of g, and $(g \circ f)(x) = x$ for every x in the domain of f. (Section 4.1)

irrational numbers Real numbers that cannot be represented as fractions of integers are called irrational numbers. (Section R.1)

L

leading coefficient In a polynomial function of degree n, the leading coefficient is a_n, that is, the coefficient of the highest-degree term. (Section 3.1)

lemniscate A lemniscate is a figure-eight-shaped curve that is the graph of a polar equation of the form $r^2 = a^2 \sin 2\theta$ or $r^2 = a^2 \cos 2\theta$. (Section 8.7)

like radicals Radicals with the same radicand and the same index are called like radicals. (Section R.7)

like terms Terms with the same variables each raised to the same powers are called like terms. (Section R.3)

line Two distinct points A and B determine a line called line AB. (Section 5.1)

line segment (segment) Line segment AB (or segment AB) is the portion of line AB between A and B, including A and B themselves. (Section 5.1)

linear cost function A linear cost function is a linear function that has the form $C(x) = mx + b$, where x represents the number of items produced, m represents the variable cost per item, and b represents the fixed cost. (Section 2.3)

linear equation (first-degree equation) in *n* unknowns Any equation of the form $a_1x_1 + a_2x_2 + \cdots + a_nx_n = b$, for real numbers $a_1, a_2, \ldots, a_n$ (not all of which are 0) and b, is a linear equation or a first-degree equation in n unknowns. (Section 9.1)

linear equation (first-degree equation) in one variable A linear equation in one variable is an equation that can be written in the form $ax + b = 0$, where a and b are real numbers with $a \neq 0$. (Section 1.1)

linear function A function f is a linear function if $f(x) = ax + b$ for real numbers a and b. (Section 2.3)

linear inequality in one variable A linear inequality in one variable is an inequality that can be written in the form $ax + b > 0$, where a and b are real numbers with $a \neq 0$. (Any of the symbols $<$, $\geq$, or $\leq$ may also be used.) (Section 1.7)

linear inequality in two variables A linear inequality in two variables is an inequality of the form $Ax + By \leq C$, where A, B, and C are real numbers with A and B not both equal to 0. (Any of the symbols $\geq$, $<$, or $>$ may also be used.) (Section 9.6)

linear programming Linear programming, an application of mathematics to business and social science, is a method for finding an optimum value, such as minimum cost or maximum profit. (Section 9.6)

linear speed *v* The linear speed v measures the distance traveled per unit of time. (Section 6.2)

literal equation A literal equation is an equation that relates two or more variables. (Section 1.1)

logarithm A logarithm is an exponent; $\log_a x$ is the power to which the base a must be raised to obtain x. (Section 4.3)

logarithmic equation A logarithmic equation is an equation with a logarithm in at least one term. (Section 4.3)

logarithmic function If $a > 0$, $a \neq 1$, and $x > 0$, then $f(x) = \log_a x$ defines the logarithmic function with base a. (Section 4.3)

lowest terms A rational expression is in lowest terms when the greatest common factor of its numerator and its denominator is 1. (Section R.5)

M

magnitude The length of a vector represents the magnitude of the vector quantity. (Section 8.3)

major axis The major axis of an ellipse is its longer axis of symmetry. (Section 10.2)

mathematical induction Mathematical induction is a method for proving that a statement S_n is true for every positive integer n. (Section 11.5)

mathematical model A mathematical model is an equation (or inequality) that describes the relationship between two or more quantities. (Section 1.2)

matrix (plural, **matrices**) A matrix is a rectangular array of numbers enclosed in brackets. (Section 9.2)

minor In an $n \times n$ matrix with $n \geq 3$, the minor of a particular element is the determinant of the $(n - 1) \times (n - 1)$ matrix that results when the row and column that contain the chosen element are eliminated. (Section 9.3)

minor axis The minor axis of an ellipse is its shorter axis of symmetry. (Section 10.2)

minute One minute, written $1'$, is $\frac{1}{60}$ of a degree. (Section 5.1)

modulus (absolute value) of a complex number When a complex number $x + yi$ is written in trigonometric (or polar) form as $r(\cos \theta + i \sin \theta)$, the number r is called the modulus or absolute value of the complex number. (Section 8.5)

monomial A monomial is a polynomial containing only one term. (Section R.3)

multiplicative inverse (reciprocal) For each nonzero real number a, the number $\frac{1}{a}$ is called the multiplicative inverse or reciprocal of a. (Section R.1)

multiplicative inverse of a matrix If A is an $n \times n$ matrix, then its multiplicative inverse, written A^{-1}, must satisfy both $AA^{-1} = I_n$ and $A^{-1}A = I_n$. (Section 9.8)

mutually exclusive events In probability, two events that cannot occur simultaneously are called mutually exclusive events. (Section 11.7)

N

natural logarithm A logarithm to base e is called a natural logarithm. (Section 4.4)

natural numbers (counting numbers) The natural numbers or counting numbers form the set of numbers $\{1, 2, 3, 4, \ldots\}$. (Section R.1)

negative angle A negative angle is an angle that is formed by clockwise rotation around its endpoint. (Section 5.1)

nonlinear system A system of equations in which at least one equation is *not* linear is called a nonlinear system. (Section 9.5)

nonstrict inequality An inequality in which the symbol is either $\leq$ or $\geq$ is called a nonstrict inequality. (Section 1.7)

***n*th root of a complex number** For a positive integer n, the complex number $a + bi$ is an nth root of the complex number $x + yi$ if $(a + bi)^n = x + yi$. (Section 8.6)

O

objective function In linear programming, the function to be maximized or minimized is called the objective function. (Section 9.6)

oblique asymptote A nonvertical, nonhorizontal line that a graph approaches as $|x|$ gets larger and larger without bound is called an oblique asymptote. (Section 3.5)

oblique triangle A triangle that is not a right triangle is called an oblique triangle. (Section 8.1)

obtuse angle An obtuse angle is an angle measuring more than 90° but less than 180°. (Section 5.1)

odd function A function f is called an odd function if $f(-x) = -f(x)$ for all x in the domain of f. (Section 2.6)

odds The odds in favor of an event are expressed as the ratio of the probability of the event to the probability of the complement of the event. (Section 11.7)

one-to-one function In a one-to-one function, each x-value corresponds to only one y-value, and each y-value corresponds to only one x-value. (Section 4.1)

open interval An open interval is an interval that does not include its endpoint(s). (Section 1.7)

opposite of a vector The opposite of a vector **v** is a vector **−v** that has the same magnitude as **v** but opposite direction. (Section 8.3)

ordered pair An ordered pair consists of two components, written inside parentheses, in which the order of the components is important. Ordered pairs are used to identify points in the coordinate plane. (Section 2.1)

ordered triple An ordered triple consists of three components, written inside parentheses, in which the order of the components is important. Ordered triples are used to identify points in space. (Section 9.1)

origin The point of intersection of the x-axis and the y-axis of a rectangular coordinate system is called the origin. (Section 2.1)

orthogonal vectors Orthogonal vectors are vectors that are perpendicular, that is, the angle between the two vectors is 90°. (Section 8.3)

outcome In probability, a possible result of each trial in an experiment is called an outcome of the experiment. (Section 11.7)

P

parabola A parabola is a curve that is the graph of a quadratic function. (Sections 2.5, 3.1)

parabola (geometric definition) A parabola is the set of all points in a plane equidistant from a fixed point (called the focus) and a fixed line (called the directrix). (Section 10.1)

parallelogram rule The parallelogram rule is a way to find the sum of two vectors. If the two vectors are placed so that their initial points coincide and a parallelogram is completed that has these two vectors as two of its sides, then the diagonal vector of the parallelogram that has the same initial point as the two vectors is their sum. (Section 8.3)

parameter A parameter is a variable in terms of which two or more other variables are expressed. In a pair of parametric equations $x = f(t)$ and $y = g(t)$, the variable t is the parameter. (Section 8.8)

parametric equations of a plane curve A pair of equations $x = f(t)$ and $y = g(t)$ are parametric equations of a plane curve. (Section 8.8)

partial fraction Each term in a partial fraction decomposition is called a partial fraction. (Section 9.4)

partial fraction decomposition When one rational expression is expressed as the sum of two or more rational expressions, the sum is called the partial fraction decomposition. (Section 9.4)

Pascal's triangle Pascal's triangle is a triangular array of numbers that is helpful in expanding binomials.The numbers in the triangle are the binomial coefficients. (Section 11.4)

period For a periodic function such that $f(x) = f(x + np)$, the smallest possible positive value of p is the period of the function. (Section 6.3)

periodic function A periodic function is a function f such that $f(x) = f(x + np)$, for every real number x in the domain of f, every integer n, and some positive real number p. (Section 6.3)

permutation A permutation of n elements taken r at a time is one of the *arrangements* of r elements from a set of n elements. (Section 11.6)

phase shift With trigonometric functions, a horizontal translation is called a phase shift. (Section 6.3)

piecewise-defined function A piecewise-defined function is a function that is defined by different rules over different parts of its domain. (Section 2.5)

plane curve A plane curve is a set of points (x, y) such that $x = f(t)$ and $y = g(t)$, and f and g are both defined on an interval I. (Section 8.8)

point-slope form The point-slope form of the equation of the line with slope m through the point (x_1, y_1) is $y - y_1 = m(x - x_1)$. (Section 2.4)

polar axis The polar axis is a specific ray in the polar coordinate system that has the pole as its endpoint. The polar axis is usually drawn in the direction of the positive x-axis. (Section 8.7)

polar coordinates In the polar coordinate system, the ordered pair (r, θ) gives polar coordinates of point P, where r is the directed distance from the pole to P and θ is the directed angle from the positive x-axis to ray OP. (Section 8.7)

polar coordinate system The polar coordinate system is a coordinate system based on a point (the pole) and a ray (the polar axis). (Section 8.7)

polar equation A polar equation is an equation that uses polar coordinates. The variables are r and θ. (Section 8.7)

pole The pole is the single fixed point in the polar coordinate system that is the endpoint of the polar axis. The pole is usually placed at the origin of a rectangular coordinate system. (Section 8.7)

polynomial A polynomial is a term or a finite sum of terms, with only positive or zero integer exponents permitted on the variables. (Section R.3)

polynomial function of degree *n* A polynomial function of degree n, where n is a nonnegative integer, is a function defined by an expression of the form $f(x) = a_n x^n + a_{n-1}x^{n-1} + \cdots + a_1 x + a_0$, where $a_n, a_{n-1}, \ldots, a_1$, and a_0 are real numbers, with $a_n \neq 0$. (Section 3.1)

position vector A vector with its initial point at the origin is called a position vector. (Section 8.3)

positive angle A positive angle is an angle that is formed by counterclockwise rotation around its endpoint. (Section 5.1)

power (exponential expression, exponential) An expression of the form a^n is called a power, an exponential expression, or an exponential. (Section R.1)

present value In an investment paying compound interest, the principal is sometimes called the present value. (Section 4.2)

prime polynomial A polynomial with variable terms that cannot be written as a product of two polynomials of lower degree is a prime polynomial. (Section R.4)

principal *n*th root For even values of n (square roots, fourth roots, and so on), when a is positive, there are two real nth roots, one positive and one negative. In such cases, the notation $\sqrt[n]{a}$ represents the positive root, or principal nth root. (Section R.7)

probability of an event In a sample space with equally likely outcomes, the probability of an event is the ratio of the number of outcomes in the sample space that belong to the event to the number of outcomes in the sample space. (Section 11.7)

pure imaginary number A complex number $a + bi$ in which $a = 0$ and $b \neq 0$ is called a pure imaginary number. (Section 1.3)

Pythagorean theorem The Pythagorean theorem states that in a right triangle, the sum of the squares of the lengths of the legs is equal to the square of the length of the hypotenuse. (Section 1.5)

Q

quadrantal angle A quadrantal angle is an angle that, when placed in standard position, has its terminal side along the x-axis or the y-axis. (Section 5.1)

quadrants The quadrants are the four regions into which the x-axis and y-axis divide the coordinate plane. (Section 2.1)

quadratic equation An equation that can be written in the form $ax^2 + bx + c = 0$, where a, b, and c are real numbers with $a \neq 0$, is a quadratic equation. (Section 1.4)

quadratic in form An equation is said to be quadratic in form if it can be written as $au^2 + bu + c = 0$, where $a \neq 0$ and u is some algebraic expression. (Section 1.6)

quadratic formula The quadratic formula $x = \frac{-b \pm \sqrt{b^2 - 4ac}}{2a}$ is a general formula that can be used to solve any quadratic equation. (Section 1.4)

quadratic function A function f is a quadratic function if $f(x) = ax^2 + bx + c$, where a, b, and c are real numbers, with $a \neq 0$. (Section 3.1)

quadratic inequality A quadratic inequality is an inequality that can be written in the form $ax^2 + bx + c < 0$ for real numbers a, b, and c with $a \neq 0$. (The symbol $<$ can be replaced with $>$, $\leq$, or $\geq$.) (Section 1.7)

R

radian A radian is a unit of measure for angles. An angle with its vertex at the center of a circle that intercepts an arc on the circle equal in length to the radius of the circle has a measure of 1 radian. (Section 6.1)

radicand The number or expression under a radical sign is called the radicand. (Section R.7)

radius The radius of a circle is the given distance between the center and any point on the circle. (Section 2.1)

range In a relation, the set of all values of the dependent variable (y) is called the range. (Section 2.2)

rational equation A rational equation is an equation that has a rational expression for one or more of its terms. (Section 1.6)

rational expression The quotient of two polynomials P and Q, with $Q \neq 0$, is called a rational expression. (Section R.5)

rational function A function f of the form $f(x) = \frac{p(x)}{q(x)}$, where $p(x)$ and $q(x)$ are polynomials, with $q(x) \neq 0$, is called a rational function. (Section 3.5)

rational inequality A rational inequality is an inequality in which one or both sides contain rational expressions. (Section 1.7)

rationalizing the denominator Rationalizing the denominator is a way of simplifying a radical expression so that there are no radicals in the denominator. (Section R.7)

rational numbers The rational numbers are the set of numbers $\frac{p}{q}$, where p and q are integers and $q \neq 0$. (Section R.1)

ray The portion of line AB that starts at A and continues through B, and on past B, is called ray AB. (Section 5.1)

real axis In the complex plane, the horizontal axis is called the real axis. (Section 8.5)

real numbers The set of all numbers that correspond to points on a number line is called the real numbers. (Section R.1)

real part In the complex number $a + bi$, a is called the real part. (Section 1.3)

reciprocal function The function defined by $f(x) = \frac{1}{x}$ is called the reciprocal function. (Section 3.5)

rectangular (Cartesian) coordinate system The x-axis and y-axis together make up a rectangular coordinate system. (Section 2.1)

rectangular (Cartesian) equation A rectangular or Cartesian equation is an equation that uses rectangular coordinates. If it is an equation in two variables, the variables are x and y. (Section 8.7)

rectangular form (standard form) of a complex number The rectangular form or standard form of a complex number is $a + bi$, where a and b are real numbers. (Section 8.5)

recursive definition A sequence is defined by a recursive definition if each term after the first term or first few terms is defined as an expression involving the previous term or terms. (Section 11.1)

reference angle The reference angle for an angle θ, written θ', is the positive acute angle made by the terminal side of angle θ and the x-axis. (Section 5.3)

region of feasible solutions In linear programming, the region of feasible solutions is the region of the graph that satisfies all of the constraints. (Section 9.6)

relation A relation is a set of ordered pairs. (Section 2.2)

resultant If **A** and **B** are vectors, the vector sum **A** + **B** is called the resultant of vectors **A** and **B**. (Section 8.3)

right angle A right angle is an angle measuring exactly 90°. (Section 5.1)

root (solution) of an equation A zero of $f(x)$ is called a root or solution of the equation $f(x) = 0$. (Section 3.2)

rose curve A rose curve is a member of a family of curves that resemble flowers and that is the graph of an equation of the form $r = a \sin n\theta$ or $r = a \cos n\theta$. (Section 8.7)

rotation of axes If the axes in an xy-coordinate system having origin O are rotated about O through an angle θ, the new coordinate system is called a rotation of the xy-system. (Appendix B)

row matrix A matrix with just one row is called a row matrix. (Section 9.7)

S

sample space In probability, the set of all possible outcomes of a given experiment is called the sample space. (Section 11.7)

scalar In work with matrices or vectors, a real number is called a scalar to distinguish it from a matrix or a vector. (Sections 8.3, 9.7)

scalar product The scalar product of a real number (or scalar) k and a vector **u** is the vector $k \cdot \mathbf{u}$, which has magnitude $|k|$ times the magnitude of **u**. (Section 8.3)

scatter diagram A scatter diagram is a graph of specific ordered pairs of data. (Section 2.4)

secant Let $P(x, y)$ be a point other than the origin on the terminal side of an angle θ in standard position. Let $r = \sqrt{x^2 + y^2}$ represent the distance from the origin to P. Then the secant function is defined by $\sec \theta = \frac{r}{x}$ $(x \neq 0)$. (Section 5.2)

second One second, written 1″, is $\frac{1}{60}$ of a minute. (Section 5.1)

sector of a circle A sector of a circle is the portion of the interior of a circle intercepted by a central angle. (Section 6.1)

sequence A sequence is a function that has a set of natural numbers of the form $\{1, 2, 3, \ldots, n\}$ or $\{1, 2, 3, \ldots, n, \ldots\}$ as its domain. (Section 11.1)

series A series is the sum of the terms of a sequence. (Section 11.1)

Side-Angle-Side (SAS) The Side-Angle-Side (SAS) congruence axiom states that if two sides and the included angle of one triangle are equal, respectively, to two sides and the included angle of a second triangle, then the triangles are congruent. (Section 8.1)

Side-Side-Side (SSS) The Side-Side-Side (SSS) congruence axiom states that if three sides of one triangle are equal, respectively, to three sides of a second triangle, then the triangles are congruent. (Section 8.1)

significant digit A significant digit is a digit obtained by actual measurement. (Section 5.4)

simple harmonic motion Simple harmonic motion is oscillatory motion about an equilibrium position. If friction is neglected, this motion can be described by a sinusoid. (Section 6.5)

simple interest In simple interest, interest is paid only on the principal, not on previously earned interest. (Section 1.1)

sine Let $P(x, y)$ be a point other than the origin on the terminal side of an angle θ in standard position. Let $r = \sqrt{x^2 + y^2}$ represent the distance from the origin to P. Then the sine function is defined by $\sin \theta = \frac{y}{r}$. (Section 5.2)

sine wave (sinusoid) The graph of a sine function is called a sine wave or sinusoid. (Section 6.3)

size (order, dimension) of a matrix The size of a matrix indicates the number of rows and columns that the matrix has, with the number of rows given first. (Section 9.2)

slope The slope of a nonvertical line is the ratio of the change in y to the change in x. (Section 2.3)

slope-intercept form The slope-intercept form of the equation of the line with slope m and y-intercept b is $y = mx + b$. (Section 2.4)

solution (root) A solution or root of an equation is a number that makes the equation a true statement. (Section 1.1)

solution set The solution set of an equation is the set of all numbers that satisfy the equation. (Section 1.1)

solutions of a system of equations The solutions of a system of equations must satisfy every equation in the system. (Section 9.1)

spiral of Archimedes A spiral of Archimedes is an infinite curve that is the graph of a polar equation of the form $r = n\theta$. (Section 8.7)

square matrix An $n \times n$ matrix, that is, a matrix with the same number of columns as rows, is called a square matrix. (Section 9.7)

standard form of a complex number A complex number written in the form $a + bi$ (or $a + ib$) is in standard form. (Section 1.3)

standard form of a linear equation The form $Ax + By = C$ is called the standard form of a linear equation. (Section 2.3)

step function A step function is a function whose graph looks like a series of steps. (Section 2.5)

straight angle A straight angle is an angle measuring exactly 180°. (Section 5.1)

strict inequality An inequality in which the symbol is either $<$ or $>$ is called a strict inequality. (Section 1.7)

summation notation Summation notation is a compact way of writing a series using the general term of the corresponding sequence and the symbol Σ, the Greek capital letter sigma. (Section 11.1)

supplementary angles Two positive angles are supplementary angles if the sum of their measures is 180°. (Section 5.1)

synthetic division Synthetic division is a shortcut method of dividing a polynomial by a binomial of the form $x - k$. (Section 3.2)

system of equations A set of equations that are considered at the same time is called a system of equations. (Section 9.1)

system of inequalities A set of inequalities that are considered at the same time is called a system of inequalities. (Section 9.6)

system of linear equations (linear system) If all the equations in a system are linear, then the system is a system of linear equations, or a linear system. (Section 9.1)

T

tangent Let $P(x, y)$ be a point other than the origin on the terminal side of an angle θ in standard position. Let $r = \sqrt{x^2 + y^2}$ represent the distance from the origin to P. Then the tangent function is defined by $\tan \theta = \frac{y}{x}$ $(x \neq 0)$. (Section 5.2)

term The product of a real number and one or more real variables raised to powers is called a term. (Section R.3)

terminal point When two letters are used to name a vector, the second letter indicates the terminal (ending) point of the vector. (Section 8.3)

terminal side When a ray is rotated around its endpoint to form an angle, the ray in its location after rotation is called the terminal side of the angle. (Section 5.1)

terms of a sequence The elements in the range of a sequence are called the terms of the sequence. (Section 11.1)

transverse axis The line segment that connects the vertices of a hyperbola is called the transverse axis of the hyperbola. (Section 10.3)

tree diagram A tree diagram is a diagram with branches that is used to systematically list all the outcomes of a counting situation or probability experiment. (Section 11.6)

trinomial A trinomial is a polynomial containing exactly three terms. (Section R.3)

turning points The points on the graph of a function where the function changes from increasing to decreasing or from decreasing to increasing are called turning points. (Section 3.4).

U

union The union of sets A and B, written $A \cup B$, is the set of all elements that belong to set A or set B (or both). (Sections 1.7, 11.7)

unit circle The unit circle is the circle with center at the origin and radius 1. (Section 6.2)

unit vector A unit vector is a vector that has magnitude 1. Two useful unit vectors are $\mathbf{i} = \langle 1, 0 \rangle$ and $\mathbf{j} = \langle 0, 1 \rangle$. (Section 8.3)

V

varies directly (directly proportional to) y varies directly as x, or y is directly proportional to x, if there exists a nonzero real number k such that $y = kx$. (Section 3.6)

varies inversely (inversely proportional to) y varies inversely as x, or y is inversely proportional to x, if there exists a nonzero real number k such that $y = \frac{k}{x}$. (Section 3.6)

varies jointly In joint variation, a variable depends on the product of two or more other variables. If m and n are real numbers, then y varies jointly as the nth power of x and the mth power of z if there exists a nonzero real number k such that $y = kx^n z^m$. (Section 3.6)

vector A vector is a directed line segment that represents a vector quantity. (Section 8.3)

vector quantities Quantities that involve both magnitude and direction are called vector quantities. (Section 8.3)

Venn diagram A Venn diagram is a diagram used to illustrate relationships among sets or probability concepts. (Section 11.7)

vertex (corner point) In linear programming, a vertex or corner point is one of the vertices of the region of feasible solutions. (Section 9.6)

vertex of an angle The vertex of an angle is the endpoint of the ray that is rotated to form the angle. (Section 5.1)

vertex of a parabola The vertex of a parabola is the point where the axis of symmetry intersects the parabola. This is the turning point of the parabola. (Sections 2.5, 3.1)

vertical asymptote A vertical line that a graph approaches, but never touches or intersects, is called a vertical asymptote. The line $x = a$ is a vertical asymptote if $|f(x)| \to \infty$ as $x \to a$. (Section 3.5)

vertical component When a vector $\mathbf{u}$ is expressed as an ordered pair in the form $\mathbf{u} = \langle a, b \rangle$, the number b is the vertical component of the vector. (Section 8.3)

vertices of an ellipse The vertices of an ellipse are the endpoints of the major axis. (Section 10.2)

vertices of a hyperbola The vertices of a hyperbola are the two points on the hyperbola that are closest to the center. (Section 10.3)

W

whole numbers The set of whole numbers $\{0, 1, 2, 3, 4, \ldots\}$ is formed by combining the set of natural numbers and the number 0. (Section R.1)

X

x-axis The horizontal number line in a rectangular coordinate system is called the x-axis. (Section 2.1)

x-intercept An x-intercept is the x-value of a point where the graph of an equation intersects the x-axis. (Section 2.1)

Y

y-axis The vertical number line in a rectangular coordinate system is called the y-axis. (Section 2.1)

y-intercept A y-intercept is the y-value of a point where the graph of an equation intersects the y-axis. (Section 2.1)

Z

zero-factor property The zero-factor property states that if the product of two (or more) complex numbers is 0, then at least one of the numbers must be 0. (Section 1.4)

zero matrix A matrix containing only zero elements is called a zero matrix. (Section 9.7)

zero of multiplicity n A polynomial function has a zero k of multiplicity n if the zero occurs n times, that is, the polynomial has n factors of $x - k$. (Section 3.3)

zero polynomial The function defined by $f(x) = 0$ is called the zero polynomial. (Section 3.1)

zero of a polynomial function A zero of a polynomial function f is a number k such that $f(k) = 0$. (Section 3.2)

zero vector The zero vector is the vector with magnitude 0. (Section 8.3)

Solutions to Selected Exercises

CHAPTER R REVIEW OF BASIC CONCEPTS

R.1 Exercises *(page 11)*

61. No; in general $a - b \neq b - a$. *Examples:*

$a = 15, b = 0$: $a - b = 15 - 0 = 15$, but
$b - a = 0 - 15 = -15$.
$a = 12, b = 7$: $a - b = 12 - 7 = 5$, but
$b - a = 7 - 12 = -5$.

$a = -6, b = 4$: $a - b = -6 - 4 = -10$, but
$b - a = 4 - (-6) = 10$.

$a = -18, b = -3$: $a - b = -18 - (-3) = -15$, but
$b - a = -3 - (-18) = 15$.

75. Assuming the Earth's orbit is circular, the length of its orbit is the circumference of a circle with radius 92,960,000 mi.

$$C = 2\pi r \approx 2(3.14)(92{,}960{,}000)$$
$$\approx 583{,}788{,}800,$$

so the Earth travels approximately 583,788,800 mi per yr. Convert this speed to miles per hour.

$$\frac{583{,}788{,}800 \dfrac{\text{mi}}{\text{yr}}}{365.26 \dfrac{\text{days}}{\text{yr}}} \approx 1{,}598{,}283 \frac{\text{mi}}{\text{day}}$$

$$\frac{1{,}598{,}283 \dfrac{\text{mi}}{\text{day}}}{24 \dfrac{\text{hr}}{\text{day}}} \approx 67{,}000 \text{ mph}$$

85. The process in your head should be like the following:

$$72 \cdot 17 + 28 \cdot 17 = 17(72 + 28)$$
$$= 17(100)$$
$$= 1700.$$

R.2 Exercises *(page 22)*

33. $|2k - 8|$, if $k < 4$

If $k < 4$, then $2k < 8$, and $2k - 8 < 0$, so

$$|2k - 8| = -(2k - 8)$$
$$= -2k + 8$$
$$= 8 - 2k.$$

37. $|3 + x^2|$

If x is any real number, $x^2 \geq 0$, so $3 + x^2 > 0$. Therefore, $|3 + x^2| = 3 + x^2$.

65. $\dfrac{x^3}{y} > 0$

The quotient of two numbers is positive if they have the same sign (both positive or both negative). The sign of x^3 is the same as the sign of x. Therefore, $\dfrac{x^3}{y} > 0$ if x and y have the same sign.

R.3 Exercises *(page 32)*

11.
$$-(4m^3n^0)^2 = -[4^2(m^3)^2(n^0)^2] \quad \text{Power Rule 2}$$
$$= -(4^2)m^6n^0 \quad \text{Power Rule 1}$$
$$= -(4^2)m^6 \cdot 1 \quad \text{Zero exponent}$$
$$= -4^2m^6 \quad \text{or} \quad -16m^6$$

33.
$$(6m^4 - 3m^2 + m) - (2m^3 + 5m^2 + 4m) + (m^2 - m)$$
$$= (6m^4 - 3m^2 + m) + (-2m^3 - 5m^2 - 4m) + (m^2 - m)$$
$$= 6m^4 - 2m^3 + (-3 - 5 + 1)m^2 + (1 - 4 - 1)m$$
$$= 6m^4 - 2m^3 - 7m^2 - 4m$$

53. $[(2p - 3) + q]^2$

$$= (2p - 3)^2 + 2(2p - 3)(q) + q^2$$
Square of a binomial, treating $(2p - 3)$ as one term
$$= (2p)^2 - 2(2p)(3) + 3^2 + 2(2p - 3)(q) + q^2$$
Square the binomial $(2p - 3)$.
$$= 4p^2 - 12p + 9 + 4pq - 6q + q^2$$

73.
$$\frac{-4x^7 - 14x^6 + 10x^4 - 14x^2}{-2x^2}$$
$$= \frac{-4x^7}{-2x^2} + \frac{-14x^6}{-2x^2} + \frac{10x^4}{-2x^2} + \frac{-14x^2}{-2x^2}$$
$$= 2x^5 + 7x^4 - 5x^2 + 7$$

R.4 Exercises *(page 43)*

33.
$$24a^4 + 10a^3b - 4a^2b^2$$
$$= 2a^2(12a^2 + 5ab - 2b^2) \quad \text{Factor out the GCF, } 2a^2.$$
$$= 2a^2(4a - b)(3a + 2b) \quad \text{Factor the trinomial.}$$

41. $(a - 3b)^2 - 6(a - 3b) + 9$

$= [(a - 3b) - 3]^2$ Factor the perfect square trinomial.

$= (a - 3b - 3)^2$

61. $27 - (m + 2n)^3$

$= 3^3 - (m + 2n)^3$ Write as a difference of cubes.

$= [3 - (m + 2n)][3^2 + 3(m + 2n) + (m + 2n)^2]$

Factor the difference of cubes.

$= (3 - m - 2n)(9 + 3m + 6n + m^2 + 4mn + 4n^2)$

Distributive property; square the binomial $(m + 2n)$.

75. $9(a - 4)^2 + 30(a - 4) + 25$

$= 9t^2 + 30t + 25$ Replace $a - 4$ with t.

$= (3t)^2 + 2(3t)(5) + 5^2$

$= (3t + 5)^2$ Factor the perfect square trinomial.

$= [3(a - 4) + 5]^2$ Replace t with $a - 4$.

$= (3a - 12 + 5)^2$

$= (3a - 7)^2$

R.5 Exercises *(page 52)*

19. $\dfrac{8m^2 + 6m - 9}{16m^2 - 9} = \dfrac{(2m + 3)(4m - 3)}{(4m + 3)(4m - 3)}$ Factor.

$= \dfrac{2m + 3}{4m + 3}$ Lowest terms

35. $\dfrac{x^3 + y^3}{x^3 - y^3} \cdot \dfrac{x^2 - y^2}{x^2 + 2xy + y^2}$

$= \dfrac{(x + y)(x^2 - xy + y^2)}{(x - y)(x^2 + xy + y^2)} \cdot \dfrac{(x + y)(x - y)}{(x + y)(x + y)}$ Factor.

$= \dfrac{x^2 - xy + y^2}{x^2 + xy + y^2}$ Lowest terms

55. $\dfrac{4}{x + 1} + \dfrac{1}{x^2 - x + 1} - \dfrac{12}{x^3 + 1}$

$= \dfrac{4}{x + 1} + \dfrac{1}{x^2 - x + 1} - \dfrac{12}{(x + 1)(x^2 - x + 1)}$

Factor the sum of cubes.

$= \dfrac{4(x^2 - x + 1)}{(x + 1)(x^2 - x + 1)} + \dfrac{1(x + 1)}{(x + 1)(x^2 - x + 1)}$

$- \dfrac{12}{(x + 1)(x^2 - x + 1)}$

Write each fraction with the common denominator.

$= \dfrac{4(x^2 - x + 1) + 1(x + 1) - 12}{(x + 1)(x^2 - x + 1)}$

Add and subtract numerators.

$= \dfrac{4x^2 - 4x + 4 + x + 1 - 12}{(x + 1)(x^2 - x + 1)}$

Distributive property

$= \dfrac{4x^2 - 3x - 7}{(x + 1)(x^2 - x + 1)}$ Combine like terms.

$= \dfrac{(4x - 7)(x + 1)}{(x + 1)(x^2 - x + 1)}$ Factor the numerator.

$= \dfrac{4x - 7}{x^2 - x + 1}$ Lowest terms

67. $\dfrac{\frac{1}{x + h} - \frac{1}{x}}{h}$

Multiply both numerator and denominator by the LCD of all the fractions, $x(x + h)$.

$$\frac{\frac{1}{x + h} - \frac{1}{x}}{h} = \frac{x(x + h)\left(\frac{1}{x + h} - \frac{1}{x}\right)}{x(x + h)(h)}$$

$$= \frac{x(x + h)\left(\frac{1}{x + h}\right) - x(x + h)\left(\frac{1}{x}\right)}{x(x + h)(h)}$$

Distributive property

$$= \frac{x - (x + h)}{x(x + h)(h)}$$

$$= \frac{-h}{x(x + h)(h)}$$

$$= \frac{-1}{x(x + h)}$$ Lowest terms

R.6 Exercises *(page 62)*

41. $\left(-\dfrac{64}{27}\right)^{1/3} = -\dfrac{4}{3}$ because $\left(-\dfrac{4}{3}\right)^3 = -\dfrac{64}{27}$.

69. $\dfrac{p^{1/5}p^{7/10}p^{1/2}}{(p^3)^{-1/5}} = \dfrac{p^{1/5+7/10+1/2}}{p^{-3/5}}$ Product rule; power rule 1

$= \dfrac{p^{2/10+7/10+5/10}}{p^{-6/10}}$ Write all fractions with the LCD, 10.

$= \dfrac{p^{14/10}}{p^{-6/10}}$

$= p^{(14/10)-(-6/10)}$ Quotient rule

$= p^{20/10}$

$= p^2$

79. $(r^{1/2} - r^{-1/2})^2 = (r^{1/2})^2 - 2(r^{1/2})(r^{-1/2}) + (r^{-1/2})^2$ Square of a binomial

$= r - 2r^0 + r^{-1}$ Power rule 1; product rule

$= r - 2 \cdot 1 + r^{-1}$ Zero exponent

$= r - 2 + r^{-1}$ or $r - 2 + \dfrac{1}{r}$ Negative exponent

93. $\dfrac{x - 9y^{-1}}{(x - 3y^{-1})(x + 3y^{-1})}$

$= \dfrac{x - \frac{9}{y}}{\left(x - \frac{3}{y}\right)\left(x + \frac{3}{y}\right)}$ Definition of negative exponent

$= \dfrac{x - \frac{9}{y}}{x^2 - \frac{9}{y^2}}$ Multiply in the denominator.

$= \dfrac{y^2\left(x - \frac{9}{y}\right)}{y^2\left(x^2 - \frac{9}{y^2}\right)}$ Multiply numerator and denominator by the LCD, y^2.

$= \dfrac{y^2x - 9y}{y^2x^2 - 9}$ Distributive property

$= \dfrac{y(xy - 9)}{x^2y^2 - 9}$ Factor numerator.

R.7 Exercises *(page 73)*

53. $\sqrt[4]{\dfrac{g^3h^5}{9r^6}} = \dfrac{\sqrt[4]{g^3h^5}}{\sqrt[4]{9r^6}}$ Quotient rule

$= \dfrac{\sqrt[4]{h^4(g^3h)}}{\sqrt[4]{r^4(9r^2)}}$ Factor out perfect fourth powers.

$= \dfrac{h\sqrt[4]{g^3h}}{r\sqrt[4]{9r^2}}$ Remove all perfect fourth powers from the radicals.

$= \dfrac{h\sqrt[4]{g^3h}}{r\sqrt[4]{9r^2}} \cdot \dfrac{\sqrt[4]{9r^2}}{\sqrt[4]{9r^2}}$ Rationalize the denominator.

$= \dfrac{h\sqrt[4]{9g^3hr^2}}{r\sqrt[4]{81r^4}}$

$= \dfrac{h\sqrt[4]{9g^3hr^2}}{3r^2}$

69. $\left(\sqrt[3]{11} - 1\right)\left(\sqrt[3]{11^2} + \sqrt[3]{11} + 1\right)$

This product has the pattern

$$(x - y)(x^2 + xy + y^2) = x^3 - y^3,$$

the difference of cubes. Thus,

$\left(\sqrt[3]{11} - 1\right)\left(\sqrt[3]{11^2} + \sqrt[3]{11} + 1\right) = \left(\sqrt[3]{11}\right)^3 - 1^3$

$= 11 - 1$

$= 10.$

71. $\left(\sqrt{3} + \sqrt{8}\right)^2 = \left(\sqrt{3}\right)^2 + 2\left(\sqrt{3}\right)\left(\sqrt{8}\right) + \left(\sqrt{8}\right)^2$ Square of a binomial

$= 3 + 2\sqrt{24} + 8$

$= 3 + 2\sqrt{4 \cdot 6} + 8$

$= 3 + 2\left(2\sqrt{6}\right) + 8$ Product rule

$= 11 + 4\sqrt{6}$

81. $\dfrac{-4}{\sqrt[3]{3}} + \dfrac{1}{\sqrt[3]{24}} - \dfrac{2}{\sqrt[3]{81}} = \dfrac{-4}{\sqrt[3]{3}} + \dfrac{1}{\sqrt[3]{8 \cdot 3}} - \dfrac{2}{\sqrt[3]{27 \cdot 3}}$

$= \dfrac{-4}{\sqrt[3]{3}} + \dfrac{1}{2\sqrt[3]{3}} - \dfrac{2}{3\sqrt[3]{3}}$ Simplify radicals.

$= \dfrac{-4 \cdot 6}{\sqrt[3]{3} \cdot 6} + \dfrac{1 \cdot 3}{2\sqrt[3]{3} \cdot 3} - \dfrac{2 \cdot 2}{3\sqrt[3]{3} \cdot 2}$ Write fractions with common denominator, $6\sqrt[3]{3}$.

$= \dfrac{-24}{6\sqrt[3]{3}} + \dfrac{3}{6\sqrt[3]{3}} - \dfrac{4}{6\sqrt[3]{3}}$

$= \dfrac{-25}{6\sqrt[3]{3}}$

$= \dfrac{-25}{6\sqrt[3]{3}} \cdot \dfrac{\sqrt[3]{3^2}}{\sqrt[3]{3^2}}$ Rationalize the denominator.

$= \dfrac{-25\sqrt[3]{9}}{6\sqrt[3]{27}}$

$= \dfrac{-25\sqrt[3]{9}}{6 \cdot 3}$

$= \dfrac{-25\sqrt[3]{9}}{18}$

85. $\dfrac{\sqrt{7} - 1}{2\sqrt{7} + 4\sqrt{2}}$

$= \dfrac{\sqrt{7} - 1}{2\sqrt{7} + 4\sqrt{2}} \cdot \dfrac{2\sqrt{7} - 4\sqrt{2}}{2\sqrt{7} - 4\sqrt{2}}$

Multiply numerator and denominator by the conjugate of the denominator.

$= \dfrac{\left(\sqrt{7} - 1\right)\left(2\sqrt{7} - 4\sqrt{2}\right)}{\left(2\sqrt{7} + 4\sqrt{2}\right)\left(2\sqrt{7} - 4\sqrt{2}\right)}$

Multiply numerators; multiply denominators.

$= \dfrac{\sqrt{7} \cdot 2\sqrt{7} - \sqrt{7} \cdot 4\sqrt{2} - 1 \cdot 2\sqrt{7} + 1 \cdot 4\sqrt{2}}{\left(2\sqrt{7}\right)^2 - \left(4\sqrt{2}\right)^2}$

Use FOIL in the numerator; product of the sum and difference of two terms in the denominator.

$= \dfrac{2 \cdot 7 - 4\sqrt{14} - 2\sqrt{7} + 4\sqrt{2}}{4 \cdot 7 - 16 \cdot 2}$

$= \dfrac{14 - 4\sqrt{14} - 2\sqrt{7} + 4\sqrt{2}}{-4}$

(continued)

$$= \frac{-2(-7 + 2\sqrt{14} + \sqrt{7} - 2\sqrt{2})}{-2 \cdot 2}$$

Factor the numerator and denominator.

$$= \frac{-7 + 2\sqrt{14} + \sqrt{7} - 2\sqrt{2}}{2}$$

CHAPTER 1 EQUATIONS AND INEQUALITIES

1.1 Exercises *(page 90)*

25. $.5x + \frac{4}{3}x = x + 10$

$\frac{1}{2}x + \frac{4}{3}x = x + 10$ Change decimal to fraction.

$6\left(\frac{1}{2}x + \frac{4}{3}x\right) = 6(x + 10)$ Multiply by the LCD, 6.

$3x + 8x = 6x + 60$ Distributive property

$11x = 6x + 60$ Combine terms.

$5x = 60$ Subtract $6x$.

$x = 12$ Divide by 5.

Solution set: $\{12\}$

33. $.3(x + 2) - .5(x + 2) = -.2x - .4$

$10[.3(x + 2) - .5(x + 2)] = 10(-.2x - .4)$ Multiply by 10 to clear decimals.

$3(x + 2) - 5(x + 2) = -2x - 4$ Distributive property

$3x + 6 - 5x - 10 = -2x - 4$ Distributive property

$-2x - 4 = -2x - 4$ Combine terms.

$0 = 0$ Add $2x$; add 4.

$0 = 0$ is a true statement, so the equation is an identity. Solution set: {all real numbers}

47. $S = 2lw + 2wh + 2hl$, for h

$S - 2lw = 2wh + 2hl$ Isolate terms with h on one side of the equation.

$S - 2lw = h(2w + 2l)$ Factor out h.

$\frac{S - 2lw}{2w + 2l} = h$ or $h = \frac{S - 2lw}{2w + 2l}$ Divide by $2w + 2l$.

57. $3x = (2x - 1)(m + 4)$, for x

$3x = 2xm + 8x - m - 4$ FOIL

$m + 4 = 2xm + 5x$ Add $m + 4$; subtract $3x$.

$m + 4 = x(2m + 5)$ Factor out x.

$\frac{m + 4}{2m + 5} = x$ or $x = \frac{m + 4}{2m + 5}$ Divide by $2m + 5$.

1.2 Exercises *(page 100)*

17. Let h = the height of the box. Use the formula for the surface area of a rectangular box.

$S = 2lw + 2wh + 2hl$

$496 = 2 \cdot 18 \cdot 8 + 2 \cdot 8 \cdot h + 2 \cdot h \cdot 18$ Let $S = 496$, $l = 18$, $w = 8$.

$496 = 288 + 16h + 36h$

$496 = 288 + 52h$

$208 = 52h$

$4 = h$

The height of the box is 4 ft.

21. Let x = David's biking speed.

Then $x + 4.5$ = David's driving speed.

Summarize the given information in a table, using the equation $d = rt$. Because the speeds are given in miles per hour, the times must be changed from minutes to hours.

	r	t	d
Driving	$x + 4.5$	$\frac{1}{3}$	$\frac{1}{3}(x + 4.5)$
Biking	x	$\frac{3}{4}$	$\frac{3}{4}x$

Distance driving = Distance biking

$\frac{1}{3}(x + 4.5) = \frac{3}{4}x$

$12\left(\frac{1}{3}(x + 4.5)\right) = 12\left(\frac{3}{4}x\right)$ Multiply by the LCD, 12.

$4(x + 4.5) = 9x$

$4x + 18 = 9x$ Distributive property

$18 = 5x$ Subtract $4x$.

$\frac{18}{5} = x$ Divide by 5.

Now find the distance.

$$d = rt = \frac{3}{4}x = \frac{3}{4}\left(\frac{18}{5}\right) = \frac{27}{10} = 2.7$$

David travels 2.7 mi to work.

33. Let x = the number of hours to fill the pool with both pipes open.

	Rate	Time	Part of the Job Accomplished
Inlet Pipe	$\frac{1}{5}$	x	$\frac{1}{5}x$
Outlet Pipe	$\frac{1}{8}$	x	$\frac{1}{8}x$

Filling the pool is 1 whole job, but because the outlet pipe empties the pool, its contribution should be subtracted from the contribution of the inlet pipe.

$$\frac{1}{5}x - \frac{1}{8}x = 1$$

$$40\left(\frac{1}{5}x - \frac{1}{8}x\right) = 40 \cdot 1 \quad \text{Multiply by the LCD, 40.}$$

$$8x - 5x = 40 \quad \text{Distributive property}$$

$$3x = 40 \quad \text{Combine terms.}$$

$$x = \frac{40}{3} = 13\frac{1}{3} \quad \text{Divide by 3.}$$

It took $13\frac{1}{3}$ hr to fill the pool.

39. Let x = the number of milliliters of water to be added.

Strength	Milliliters of Solution	Milliliters of Salt
6%	8	.06(8)
0%	x	$0(x)$
4%	$8 + x$	$.04(8 + x)$

The number of milliliters of salt in the 6% solution plus the number of milliliters of salt in the water (0% solution) must equal the number of milliliters of salt in the 4% solution, so

$$.06(8) + 0(x) = .04(8 + x).$$

$$.48 = .32 + .04x$$

$$.16 = .04x$$

$$4 = x$$

To reduce the saline concentration to 4%, 4 mL of water should be added.

49. **(a)** $V = lwh = (10 \text{ ft})(10 \text{ ft})(8 \text{ ft}) = 800 \text{ ft}^3$

(b) Area of panel $= (8 \text{ ft})(4 \text{ ft}) = 32 \text{ ft}^2$

$$32 \text{ ft}^2\left(\frac{3365\ \mu\text{g}}{\text{ft}^2}\right) = 107{,}680\ \mu\text{g}$$

(c) $F = 107{,}680x$

(d) $107{,}680x = 33(800)$

$$x = \frac{33(800)}{107{,}680} \approx .25$$

It will take approximately .25 day, or 6 hr.

1.3 Exercises *(page 113)*

49. $-i - 2 - (6 - 4i) - (5 - 2i)$

$= (-2 - 6 - 5) + [-1 - (-4) - (-2)]i$

$= -13 + 5i$

67. $(2 + i)(2 - i)(4 + 3i)$

$= [(2 + i)(2 - i)](4 + 3i)$ Associative property

$= (2^2 - i^2)(4 + 3i)$ Product of the sum and difference of two terms

$= [4 - (-1)](4 + 3i)$ $i^2 = -1$

$= 5(4 + 3i)$

$= 20 + 15i$ Distributive property

79. $\dfrac{1}{i^{-11}} = i^{11}$

$= i^8 \cdot i^3$

$= (i^4)^2 \cdot i^3$

$= 1(-i)$

$= -i$

95. $\left(\dfrac{\sqrt{2}}{2} + \dfrac{\sqrt{2}}{2}i\right)^2 = \left(\dfrac{\sqrt{2}}{2}\right)^2 + 2 \cdot \dfrac{\sqrt{2}}{2} \cdot \dfrac{\sqrt{2}}{2}i + \left(\dfrac{\sqrt{2}}{2}i\right)^2$

Square of a binomial

$= \dfrac{2}{4} + 2 \cdot \dfrac{2}{4}i + \dfrac{2}{4}i^2$

$= \dfrac{1}{2} + i + \dfrac{1}{2}i^2$

$= \dfrac{1}{2} + i + \dfrac{1}{2}(-1)$ $i^2 = -1$

$= \dfrac{1}{2} + i - \dfrac{1}{2}$

$= i$

Thus, $\dfrac{\sqrt{2}}{2} + \dfrac{\sqrt{2}}{2}i$ is a square root of i.

1.4 Exercises *(page 123)*

17. $-4x^2 + x = -3$

$0 = 4x^2 - x - 3$ Standard form

$0 = (4x + 3)(x - 1)$ Factor.

$4x + 3 = 0$ or $x - 1 = 0$ Zero-factor property

$x = -\dfrac{3}{4}$ or $x = 1$

Solution set: $\left\{-\dfrac{3}{4}, 1\right\}$

53. $\dfrac{1}{2}x^2 + \dfrac{1}{4}x - 3 = 0$

$4\left(\dfrac{1}{2}x^2 + \dfrac{1}{4}x - 3\right) = 4 \cdot 0$ Multiply by the LCD, 4.

$2x^2 + x - 12 = 0$ Distributive property; standard form

$x = \dfrac{-b \pm \sqrt{b^2 - 4ac}}{2a}$

Quadratic formula

$x = \dfrac{-1 \pm \sqrt{1^2 - 4(2)(-12)}}{2(2)}$

$a = 2, b = 1, c = -12$

$x = \dfrac{-1 \pm \sqrt{97}}{4}$

Solution set: $\left\{\dfrac{-1 \pm \sqrt{97}}{4}\right\}$

69. $4x^2 - 2xy + 3y^2 = 2$

(a) Solve for x in terms of y.

$4x^2 - 2yx + 3y^2 - 2 = 0$ Standard form

$4x^2 - (2y)x + (3y^2 - 2) = 0$

$a = 4, b = -2y, c = 3y^2 - 2$

$$x = \frac{-b \pm \sqrt{b^2 - 4ac}}{2a}$$

$$= \frac{-(-2y) \pm \sqrt{(-2y)^2 - 4(4)(3y^2 - 2)}}{2(4)}$$

$$= \frac{2y \pm \sqrt{4y^2 - 16(3y^2 - 2)}}{8}$$

$$= \frac{2y \pm \sqrt{4y^2 - 48y^2 + 32}}{8}$$

$$= \frac{2y \pm \sqrt{32 - 44y^2}}{8}$$

$$= \frac{2y \pm \sqrt{4(8 - 11y^2)}}{8}$$

$$= \frac{2y \pm 2\sqrt{8 - 11y^2}}{8}$$

$$= \frac{2(y \pm \sqrt{8 - 11y^2})}{2(4)}$$

$$x = \frac{y \pm \sqrt{8 - 11y^2}}{4}$$

(b) Solve for y in terms of x.

$3y^2 - 2xy + 4x^2 - 2 = 0$

$3y^2 - (2x)y + (4x^2 - 2) = 0$

$a = 3, b = -2x, c = 4x^2 - 2$

$$y = \frac{-b \pm \sqrt{b^2 - 4ac}}{2a}$$

$$= \frac{-(-2x) \pm \sqrt{(-2x)^2 - 4(3)(4x^2 - 2)}}{2(3)}$$

$$= \frac{2x \pm \sqrt{4x^2 - 12(4x^2 - 2)}}{6}$$

$$= \frac{2x \pm \sqrt{4x^2 - 48x^2 + 24}}{6}$$

$$= \frac{2x \pm \sqrt{24 - 44x^2}}{6}$$

$$= \frac{2x \pm \sqrt{4(6 - 11x^2)}}{6}$$

$$= \frac{2x \pm 2\sqrt{6 - 11x^2}}{6}$$

$$= \frac{2(x \pm \sqrt{6 - 11x^2})}{2(3)}$$

$$y = \frac{x \pm \sqrt{6 - 11x^2}}{3}$$

83. $x = 4$ or $x = 5$

$x - 4 = 0$ or $x - 5 = 0$ Zero-factor property

$(x - 4)(x - 5) = 0$

$x^2 - 9x + 20 = 0$

$a = 1, b = -9, c = 20$

(Any nonzero constant multiple of these numbers will also work.)

1.5 Exercises *(page 130)*

13. $S = 2\pi rh + 2\pi r^2$ Surface area of a cylinder

$8\pi = 2\pi r \cdot 3 + 2\pi r^2$ Let $S = 8\pi, h = 3$.

$8\pi = 6\pi r + 2\pi r^2$

$0 = 2\pi r^2 + 6\pi r - 8\pi$

$0 = 2\pi(r^2 + 3r - 4)$

$0 = 2\pi(r + 4)(r - 1)$

$r + 4 = 0$ or $r - 1 = 0$

$r = -4$ or $r = 1$

Because r represents the radius of a cylinder, -4 is not reasonable. The radius is 1 ft.

19. Let r = radius of circle.

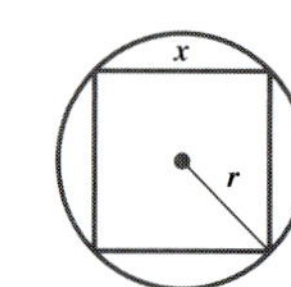

Let x = length of side of square.

From the figure, the radius is $\frac{1}{2}$ the length of the diagonal of the square.

$a^2 + b^2 = c^2$ Pythagorean theorem

$x^2 + x^2 = (2r)^2$ Let $a = x, b = x, c = 2r$.

$2x^2 = 4r^2$

$x^2 = 2r^2$

$800 = 2r^2$ Area of square $= x^2 = 800$

$400 = r^2$

$r = \pm\sqrt{400} = \pm 20$

Because r represents the radius of a circle, -20 is not reasonable. The radius is 20 ft.

43. Let x = number of passengers in excess of 75.
Then $225 - 5x$ = the cost per passenger (in dollars)
and $75 + x$ = the number of passengers.

(Cost per passenger) (Number of passengers) = Revenue

$(225 - 5x)(75 + x) = 16{,}000$

$16{,}875 - 150x - 5x^2 = 16{,}000$

$0 = 5x^2 + 150x - 875$

$0 = x^2 + 30x - 175$

$0 = (x + 35)(x - 5)$

$x + 35 = 0$ or $x - 5 = 0$

$x = -35$ or $x = 5$

The negative solution is not meaningful. Since there are 5 passengers in excess of 75, the total number of passengers is 80.

1.6 Exercises *(page 144)*

13. $$\frac{4}{x^2+x-6} - \frac{1}{x^2-4} = \frac{2}{x^2+5x+6}$$

$$\frac{4}{(x+3)(x-2)} - \frac{1}{(x+2)(x-2)} = \frac{2}{(x+3)(x+2)}$$

Factor denominators.

$$(x+3)(x-2)(x+2)\left(\frac{4}{(x+3)(x-2)} - \frac{1}{(x+2)(x-2)}\right)$$
$$= (x+3)(x-2)(x+2)\left(\frac{2}{(x+3)(x+2)}\right)$$

Multiply by the LCD, $(x+3)(x-2)(x+2)$, where $x \neq -3, x \neq 2, x \neq -2$.

$$4(x+2) - 1(x+3) = 2(x-2)$$

Simplify on both sides.

$$4x + 8 - x - 3 = 2x - 4$$

Distributive property

$$3x + 5 = 2x - 4$$

Combine terms.

$$x = -9$$

Subtract $2x$; subtract 5.

The restrictions $x \neq -3, x \neq 2, x \neq -2$ do not affect the result.

Solution set: $\{-9\}$

45. $$\sqrt{2\sqrt{7x+2}} = \sqrt{3x+2}$$

$$\left(\sqrt{2\sqrt{7x+2}}\right)^2 = \left(\sqrt{3x+2}\right)^2$$ Square both sides.

$$2\sqrt{7x+2} = 3x+2$$

$$\left(2\sqrt{7x+2}\right)^2 = (3x+2)^2$$

Square both sides again.

$$4(7x+2) = 9x^2 + 12x + 4$$

Square the binomial on the right.

$$28x + 8 = 9x^2 + 12x + 4$$
$$0 = 9x^2 - 16x - 4$$
$$0 = (9x+2)(x-2)$$

$9x + 2 = 0$ or $x - 2 = 0$

$x = -\frac{2}{9}$ or $x = 2$

Check both proposed solutions by substituting first $-\frac{2}{9}$ and then 2 in the *original* equation. These checks will verify that both of these numbers are solutions.

Solution set: $\left\{-\frac{2}{9}, 2\right\}$

77. $x^{-2/3} + x^{-1/3} - 6 = 0$

Since $(x^{-1/3})^2 = x^{-2/3}$, let $u = x^{-1/3}$.

$u^2 + u - 6 = 0$ Substitute.

$(u+3)(u-2) = 0$ Factor.

$u + 3 = 0$ or $u - 2 = 0$

$u = -3$ or $u = 2$

Now replace u with $x^{-1/3}$.

$x^{-1/3} = -3$ or $x^{-1/3} = 2$

$(x^{-1/3})^{-3} = (-3)^{-3}$ or $(x^{-1/3})^{-3} = 2^{-3}$

Raise both sides of each equation to the -3 power.

$x = \frac{1}{(-3)^3}$ or $x = \frac{1}{2^3}$

$x = -\frac{1}{27}$ or $x = \frac{1}{8}$

A check will show that both $-\frac{1}{27}$ and $\frac{1}{8}$ satisfy the original equation.

Solution set: $\left\{-\frac{1}{27}, \frac{1}{8}\right\}$

87. $m^{3/4} + n^{3/4} = 1$, for m

$$m^{3/4} = 1 - n^{3/4}$$

$$(m^{3/4})^{4/3} = (1 - n^{3/4})^{4/3}$$ Raise both sides to the $\frac{4}{3}$ power.

$$m = (1 - n^{3/4})^{4/3}$$

1.7 Exercises *(page 156)*

51. $x^2 - 2x \leq 1$

Step 1 Solve the corresponding quadratic equation.

$$x^2 - 2x = 1$$
$$x^2 - 2x - 1 = 0$$

The trinomial $x^2 - 2x - 1$ cannot be factored, so solve this equation with the quadratic formula.

$$x = \frac{-b \pm \sqrt{b^2 - 4ac}}{2a}$$

$$x = \frac{-(-2) \pm \sqrt{(-2)^2 - 4(1)(-1)}}{2(1)}$$

$a = 1, b = -2, c = -1$

Simplify to obtain $x = 1 \pm \sqrt{2}$, that is, $x = 1 - \sqrt{2}$ or $x = 1 + \sqrt{2}$.

Step 2 Identify the intervals determined by the solutions of the equation. The values $1 - \sqrt{2}$ and $1 + \sqrt{2}$ divide a number line into three intervals: $(-\infty, 1 - \sqrt{2})$, $(1 - \sqrt{2}, 1 + \sqrt{2})$, and $(1 + \sqrt{2}, \infty)$. Use solid circles on $1 - \sqrt{2}$ and $1 + \sqrt{2}$ because the inequality symbol is $\leq$. (Note that $1 - \sqrt{2} \approx -.4$ and $1 + \sqrt{2} \approx 2.4$.)

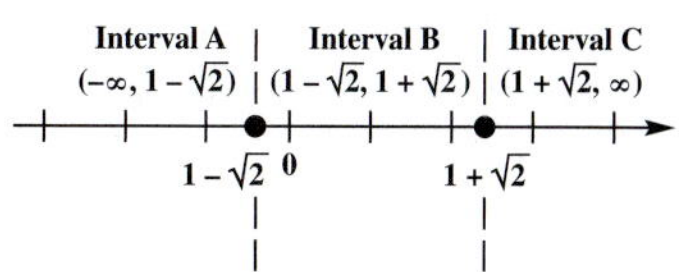

Step 3 Use a test value from each interval to determine which intervals form the solution set.

(continued)

Interval	Test Value	Is $x^2 - 2x \leq 1$ True or False?
A: $\left(-\infty, 1 - \sqrt{2}\right)$	-1	$(-1)^2 - 2(-1) \leq 1$? $3 \leq 1$ False
B: $\left(1 - \sqrt{2}, 1 + \sqrt{2}\right)$	0	$0^2 - 2(0) \leq 1$? $0 \leq 1$ True
C: $\left(1 + \sqrt{2}, \infty\right)$	3	$3^2 - 2(3) \leq 1$? $3 \leq 1$ False

Only Interval B makes the inequality true. Both endpoints are included because the given inequality is a nonstrict inequality.

Solution set: $\left[1 - \sqrt{2}, 1 + \sqrt{2}\right]$

61. $4x - x^3 \geq 0$

Step 1 $4x - x^3 = 0$ Corresponding equation

$x(4 - x^2) = 0$ Factor out the GCF, x.

$x(2 - x)(2 + x) = 0$

Factor the difference of squares.

$x = 0$ or $2 - x = 0$ or $2 + x = 0$

Zero-factor property

$x = 0$ or $x = 2$ or $x = -2$

Step 2 The values -2, 0, and 2 divide the number line into four intervals.

Interval A: $(-\infty, -2)$; Interval B: $(-2, 0)$;

Interval C: $(0, 2)$; Interval D: $(2, \infty)$

Step 3

Interval	Test Value	Is $4x - x^3 \geq 0$ True or False?
A: $(-\infty, -2)$	-3	$4(-3) - (-3)^3 \geq 0$? $15 \geq 0$ True
B: $(-2, 0)$	-1	$4(-1) - (-1)^3 \geq 0$? $-3 \geq 0$ False
C: $(0, 2)$	1	$4(1) - 1^3 \geq 0$? $3 \geq 0$ True
D: $(2, \infty)$	3	$4(3) - 3^3 \geq 0$? $-15 \geq 0$ False

Intervals A and C make the inequality true. The endpoints -2, 0, and 2 are all included because the given inequality is a nonstrict inequality.

Solution set: $(-\infty, -2] \cup [0, 2]$

79. $\dfrac{7}{x+2} \geq \dfrac{1}{x+2}$

Step 1 Rewrite the inequality so that 0 is on one side and there is a single fraction on the other side.

$$\frac{7}{x+2} - \frac{1}{x+2} \geq 0$$

$$\frac{6}{x+2} \geq 0$$

Step 2 Determine the values that will cause either the numerator or the denominator to equal 0.

Since $6 \neq 0$, the numerator is never equal to 0.

The denominator is equal to 0 when $x + 2 = 0$, or $x = -2$.

-2 divides a number line into two intervals, $(-\infty, -2)$ and $(-2, \infty)$.

Use an open circle on -2 because it makes the denominator 0.

Interval A $(-\infty, -2)$ | Interval B $(-2, \infty)$

-2 0

Step 3 Use a test value from each interval to determine which intervals form the solution set.

Interval	Test Value	Is $\frac{7}{x+2} \geq \frac{1}{x+2}$ True or False?
A: $(-\infty, -2)$	-3	$\frac{7}{-3+2} \geq \frac{1}{-3+2}$? $-7 \geq -1$ False
B: $(-2, \infty)$	0	$\frac{7}{0+2} \geq \frac{1}{0+2}$? $\frac{7}{2} \geq \frac{1}{2}$ True

Interval B satisfies the original inequality. The endpoint -2 is not included because it makes the denominator 0.

Solution set: $(-2, \infty)$

87. $\dfrac{2x - 3}{x^2 + 1} \geq 0$

The inequality already has 0 on one side, so set the numerator and denominator equal to 0. If $2x - 3 = 0$, then $x = \frac{3}{2}$. $x^2 + 1 = 0$ has no real solutions. $\frac{3}{2}$ divides a number line into two intervals, $\left(-\infty, \frac{3}{2}\right)$ and $\left(\frac{3}{2}, \infty\right)$.

Interval	Test Value	Is $\frac{2x-3}{x^2+1} \geq 0$ True or False?
A: $\left(-\infty, \frac{3}{2}\right)$	0	$\frac{2(0) - 3}{0^2 + 1} \geq 0$? $-3 \geq 0$ False
B: $\left(\frac{3}{2}, \infty\right)$	2	$\frac{2(2) - 3}{2^2 + 1} \geq 0$? $\frac{1}{5} \geq 0$ True

Interval B satisfies the original inequality, along with the endpoint $\frac{3}{2}$.

Solution set: $\left[\dfrac{3}{2}, \infty\right)$

1.8 Exercises *(page 164)*

33. $4|x-3| > 12$

$|x-3| > 3$ — Divide by 4.

$x-3 < -3$ or $x-3 > 3$ — Property 4

$x < 0$ or $x > 6$ — Add 3.

Solution set: $(-\infty, 0) \cup (6, \infty)$

47. $\left|5x + \frac{1}{2}\right| - 2 < 5$

$\left|5x + \frac{1}{2}\right| < 7$ — Add 2.

$-7 < 5x + \frac{1}{2} < 7$ — Property 3

$2(-7) < 2\left(5x + \frac{1}{2}\right) < 2(7)$ — Multiply each part by 2.

$-14 < 10x + 1 < 14$ — Distributive property

$-15 < 10x < 13$ — Subtract 1 from each part.

$\frac{-15}{10} < x < \frac{13}{10}$ — Divide each part by 10.

$-\frac{3}{2} < x < \frac{13}{10}$ — Lowest terms

Solution set: $\left(-\frac{3}{2}, \frac{13}{10}\right)$

59. $|2x+1| \le 0$

Since the absolute value of a number is always nonnegative, $|2x+1| < 0$ is never true, so $|2x+1| \le 0$ is only true when $|2x+1| = 0$. Solve this equation.

$$|2x+1| = 0$$
$$2x+1 = 0$$
$$x = -\frac{1}{2}$$

Solution set: $\left\{-\frac{1}{2}\right\}$

69. $|x^2+1| = |2x|$

$x^2+1 = 2x$ or $x^2+1 = -2x$ — Property 2

$x^2-2x+1 = 0$ or $x^2+2x+1 = 0$

$(x-1)^2 = 0$ or $(x+1)^2 = 0$

$x-1 = 0$ or $x+1 = 0$

$x = 1$ or $x = -1$

Solution set: $\{-1, 1\}$

CHAPTER 2 GRAPHS AND FUNCTIONS

2.1 Exercises *(page 192)*

15. $P(3\sqrt{2}, 4\sqrt{5}), Q(\sqrt{2}, -\sqrt{5})$

(a) $d(P,Q) = \sqrt{(\sqrt{2} - 3\sqrt{2})^2 + (-\sqrt{5} - 4\sqrt{5})^2}$

Let $x_1 = 3\sqrt{2}, y_1 = 4\sqrt{5}, x_2 = \sqrt{2}, y_2 = -\sqrt{5}$.

$= \sqrt{(-2\sqrt{2})^2 + (-5\sqrt{5})^2}$

$= \sqrt{8 + 125}$

$= \sqrt{133}$

(b) The midpoint of the segment PQ has coordinates

$$\left(\frac{3\sqrt{2}+\sqrt{2}}{2}, \frac{4\sqrt{5}+(-\sqrt{5})}{2}\right) = \left(2\sqrt{2}, \frac{3\sqrt{5}}{2}\right).$$

71. midpoint (5, 8), endpoint (13, 10)

$\frac{13+x}{2} = 5$ and $\frac{10+y}{2} = 8$

$13 + x = 10$ and $10 + y = 16$

$x = -3$ and $y = 6$

The other endpoint has coordinates $(-3, 6)$.

85. Let $P(x, y)$ be a point whose distance from $A(1, 0)$ is $\sqrt{10}$ and whose distance from $B(5, 4)$ is also $\sqrt{10}$.

$d(P, A) = \sqrt{10}$, so

$\sqrt{(1-x)^2 + (0-y)^2} = \sqrt{10}$ — Distance formula

$(1-x)^2 + y^2 = 10.$ — Square both sides.

$d(P, B) = \sqrt{10}$, so

$\sqrt{(5-x)^2 + (4-y)^2} = \sqrt{10}$

$(5-x)^2 + (4-y)^2 = 10.$

Thus,

$$(1-x)^2 + y^2 = (5-x)^2 + (4-y)^2$$
$$1 - 2x + x^2 + y^2 = 25 - 10x + x^2 + 16 - 8y + y^2$$
$$1 - 2x = 41 - 10x - 8y$$
$$8y = 40 - 8x$$
$$y = 5 - x.$$

Substitute $5 - x$ for y in the equation $(1-x)^2 + y^2 = 10$ and solve for x.

$$(1-x)^2 + (5-x)^2 = 10$$
$$1 - 2x + x^2 + 25 - 10x + x^2 = 10$$
$$2x^2 - 12x + 26 = 10$$
$$2x^2 - 12x + 16 = 0$$
$$x^2 - 6x + 8 = 0$$
$$(x-2)(x-4) = 0$$

$x - 2 = 0$ or $x - 4 = 0$

$x = 2$ or $x = 4$

To find the corresponding values of y, substitute in the equation $y = 5 - x$.

If $x = 2$, then $y = 5 - 2 = 3$.

If $x = 4$, then $y = 5 - 4 = 1$.

The points satisfying the given conditions are (2, 3) and (4, 1).

2.3 Exercises *(page 221)*

49. $5x - 2y = 10$

Find two ordered pairs that are solutions of the equation. If $x = 0$, then $5(0) - 2y = 10$, or $-2y = 10$, so $y = -5$. If $y = 0$, then $5x - 2(0) = 10$, or $5x = 10$, so $x = 2$. Thus, the two ordered pairs are $(0, -5)$ and $(2, 0)$. The slope is

$$m = \frac{\text{rise}}{\text{run}} = \frac{0 - (-5)}{2 - 0} = \frac{5}{2}.$$

Plot the points $(0, -5)$ and $(2, 0)$ and draw a line through them.

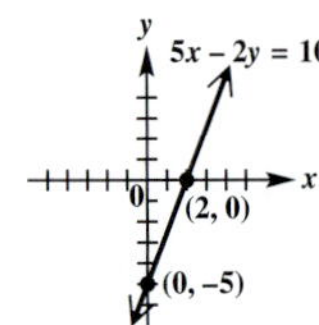

89. fixed cost = \$1650; variable cost = \$400; price of item = \$305

(a) $C(x) = 400x + 1650$ $\quad m = 400, b = 1650$

(b) $R(x) = 305x$

(c) $P(x) = R(x) - C(x)$

$= 305x - (400x + 1650)$

$P(x) = -95x - 1650$

(d)
$$C(x) = R(x)$$
$$400x + 1650 = 305x$$
$$95x = -1650$$
$$x \approx -17.4$$

This result indicates a negative "break-even point," but the number of units produced must be a positive number. A calculator graph of the lines $y = 400x + 1650$ and $y = 305x$ on the same screen or solving the inequality $305x < 400x + 1650$ will show that $R(x) < C(x)$ for all positive values of x (in fact whenever x is greater than about -17.4). Do not produce the product since it is impossible to make a profit.

2.4 Exercises *(page 236)*

13. Since the x-intercept is 3 and the y-intercept is -2, the line passes through the points $(3, 0)$ and $(0, -2)$. Use these points to find the slope.

$$m = \frac{-2 - 0}{0 - 3} = \frac{-2}{-3} = \frac{2}{3}$$

The slope is $\frac{2}{3}$ and the y-intercept is -2, so the equation of the line in slope-intercept form is

$$y = \frac{2}{3}x - 2.$$

43. (a) Find the slope of the line $3y + 2x = 6$.

$$3y + 2x = 6$$
$$3y = -2x + 6$$
$$y = -\frac{2}{3}x + 2$$

Thus, $m = -\frac{2}{3}$. A line parallel to $3y + 2x = 6$ will also have slope $-\frac{2}{3}$. Using the points $(4, -1)$ and $(k, 2)$ and the definition of slope,

$$\frac{2 - (-1)}{k - 4} = -\frac{2}{3}.$$

Solve this equation for k.

$$\frac{3}{k - 4} = -\frac{2}{3}$$
$$3(k - 4)\left(\frac{3}{k - 4}\right) = 3(k - 4)\left(-\frac{2}{3}\right)$$
$$9 = -2(k - 4)$$
$$9 = -2k + 8$$
$$2k = -1$$
$$k = -\frac{1}{2}$$

(b) Find the slope of the line $2y - 5x = 1$.

$$2y - 5x = 1$$
$$2y = 5x + 1$$
$$y = \frac{5}{2}x + \frac{1}{2}$$

Thus, $m = \frac{5}{2}$. A line perpendicular to $2y - 5x = 1$ will have slope $-\frac{2}{5}$ since $\frac{5}{2}\left(-\frac{2}{5}\right) = 1$. Using the points $(4, -1)$ and $(k, 2)$ and the definition of slope,

$$\frac{2 - (-1)}{k - 4} = -\frac{2}{5}.$$

Solve this equation for k.

$$\frac{3}{k - 4} = -\frac{2}{5}$$
$$5(k - 4)\left(\frac{3}{k - 4}\right) = 5(k - 4)\left(-\frac{2}{5}\right)$$
$$15 = -2(k - 4)$$
$$15 = -2k + 8$$
$$2k = -7$$
$$k = -\frac{7}{2}$$

69. $A(-1, 4)$, $B(-2, -1)$, $C(1, 14)$

For A and B, $m = \dfrac{-1 - 4}{-2 - (-1)} = \dfrac{-5}{-1} = 5.$

For B and C, $m = \dfrac{14 - (-1)}{1 - (-2)} = \dfrac{15}{3} = 5.$

For A and C, $m = \dfrac{14 - 4}{1 - (-1)} = \dfrac{10}{2} = 5.$

All three slopes are the same, so by Exercise 67, the three points are collinear.

2.6 Exercises *(page 264)*

61. $f(x) = 2x + 5$

Translate the graph of $f(x)$ up 2 units to obtain the graph of

$$t(x) = (2x + 5) + 2 = 2x + 7.$$

Now translate the graph of $t(x)$ left 3 units to obtain the graph of

$$\begin{aligned} g(x) &= 2(x + 3) + 7 \\ &= 2x + 6 + 7 \\ &= 2x + 13. \end{aligned}$$

(Note that if the original graph is first translated left 3 units and then up 2 units, the final result will be the same.)

73. **(a)** Choose any value of x. Find the corresponding values of y on both graphs and compare these values. For example, choose $x = 2$. From the graph, $f(2) = 1$ and $g(2) = 3$. For any value of x, the y-value for $g(x)$ is 2 greater than the y-value for $f(x)$, so the graph of $g(x)$ is a vertical translation of the graph of $f(x)$ up 2 units. Therefore, $g(x) = f(x) + 2$, that is, $c = 2$.

(b) Choose any value of y. Find the corresponding values of x on both graphs and compare these values. For example, choose $y = 3$. From the graph, $f(6) = 3$ and $g(2) = 3$. For any value of y, the x-value for $g(x)$ is 4 less than the x-value for $f(x)$, so the graph of $g(x)$ is a horizontal translation of the graph of $f(x)$ to the left 4 units. Therefore, $g(x) = f(x + 4)$, that is, $c = 4$.

2.7 Exercises *(page 276)*

25. **(a)** From the graph, $f(-1) = 0$ and $g(-1) = 3$, so

$$\begin{aligned} (f + g)(-1) &= f(-1) + g(-1) \\ &= 0 + 3 = 3. \end{aligned}$$

(b) From the graph, $f(-2) = -1$ and $g(-2) = 4$, so

$$\begin{aligned} (f - g)(-2) &= f(-2) - g(-2) \\ &= -1 - 4 = -5. \end{aligned}$$

(c) From the graph, $f(0) = 1$ and $g(0) = 2$, so

$$\begin{aligned} (fg)(0) &= f(0) \cdot g(0) \\ &= 1 \cdot 2 = 2. \end{aligned}$$

(d) From the graph, $f(2) = 3$ and $g(2) = 0$, so

$$\left(\frac{f}{g}\right)(2) = \frac{f(2)}{g(2)} = \frac{3}{0},$$

which is undefined.

39. $f(x) = x^2 - 4$

(a) $$\begin{aligned} f(x + h) &= (x + h)^2 - 4 \\ &= x^2 + 2xh + h^2 - 4 \end{aligned}$$

(b) $$\begin{aligned} f(x + h) - f(x) &= (x^2 + 2xh + h^2 - 4) - (x^2 - 4) \\ &= x^2 + 2xh + h^2 - 4 - x^2 + 4 \\ &= 2xh + h^2 \end{aligned}$$

(c) $$\begin{aligned} \frac{f(x + h) - f(x)}{h} &= \frac{2xh + h^2}{h} \\ &= \frac{h(2x + h)}{h} \\ &= 2x + h \end{aligned}$$

63. $f(x) = 9x^2 - 11x$, $g(x) = 2\sqrt{x + 2}$

$$\begin{aligned} (f \circ g)(x) &= f[g(x)] \\ &= f(2\sqrt{x + 2}) \\ &= 9(2\sqrt{x + 2})^2 - 11(2\sqrt{x + 2}) \\ &= 9[4(x + 2)] - 22\sqrt{x + 2} \\ &= 9(4x + 8) - 22\sqrt{x + 2} \\ &= 36x + 72 - 22\sqrt{x + 2} \end{aligned}$$

$$\begin{aligned} (g \circ f)(x) &= g[f(x)] \\ &= g(9x^2 - 11x) \\ &= 2\sqrt{(9x^2 - 11x) + 2} \\ &= 2\sqrt{9x^2 - 11x + 2} \end{aligned}$$

65. $g[f(2)] = g(1) = 2$; $g[f(3)] = g(2) = 5$

Since $g[f(1)] = 7$ and $f(1) = 3$, $g(3) = 7$.

Completed table:

x	$f(x)$	$g(x)$	$g[f(x)]$
1	3	2	7
2	1	5	2
3	2	7	5

CHAPTER 3 POLYNOMIAL AND RATIONAL FUNCTIONS

3.1 Exercises *(page 303)*

3. $f(x) = -2(x + 3)^2 + 2$

(a) domain: $(-\infty, \infty)$

range: $(-\infty, 2]$

(b) vertex: $(h, k) = (-3, 2)$

(c) axis: $x = -3$

(d) $$\begin{aligned} y &= -2(0 + 3)^2 + 2 \quad \text{Let } x = 0. \\ &= -16 \end{aligned}$$

y-intercept: -16

(continued)

(e) $0 = -2(x + 3)^2 + 2$ Let $f(x) = 0$.

$2(x + 3)^2 = 2$

$(x + 3)^2 = 1$

$x + 3 = \pm\sqrt{1}$

$x + 3 = 1$ or $x + 3 = -1$

$x = -2$ or $x = -4$

x-intercepts: $-4, -2$

59. $y = \dfrac{-16x^2}{.434v^2} + 1.15x + 8$

(a) $10 = \dfrac{-16(15)^2}{.434v^2} + 1.15(15) + 8$ Let $y = 10$, $x = 15$.

$10 = \dfrac{-3600}{.434v^2} + 17.25 + 8$

$\dfrac{3600}{.434v^2} = 15.25$

$3600 = 6.6185v^2$

$v^2 = \dfrac{3600}{6.6185}$

$v = \pm\sqrt{\dfrac{3600}{6.6185}} \approx \pm 23.32$

Because v represents velocity, only the positive square root is meaningful. The basketball should have an initial velocity of 23.32 ft per sec.

(b) $y = \dfrac{-16x^2}{.434(23.32)^2} + 1.15x + 8$

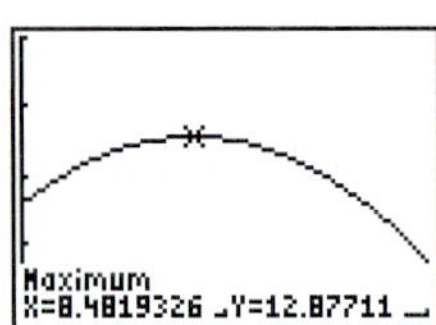

Graph this function in an appropriate window, such as [0, 20] by [0, 20], with X scale = 5, Y scale = 5. Use the calculator to find the vertex of the parabola, which is the maximum point. The y-coordinate of this point is approximately 12.88, so the maximum height of the basketball is about 12.88 ft.

71. $y = x^2 - 10x + c$

An x-intercept occurs where $y = 0$, or

$$0 = x^2 - 10x + c.$$

There will be exactly one x-intercept if this equation has exactly one solution, or the discriminant is 0.

$$b^2 - 4ac = 0$$

$$(-10)^2 - 4(1)c = 0$$

$$100 = 4c$$

$$c = 25$$

3.2 Exercises *(page 319)*

7. $\dfrac{x^5 + 3x^4 + 2x^3 + 2x^2 + 3x + 1}{x + 2}$

$x + 2 = x - (-2)$

$$\begin{array}{r|rrrrrr} -2 & 1 & 3 & 2 & 2 & 3 & 1 \\ & & -2 & -2 & 0 & -4 & 2 \\ \hline & 1 & 1 & 0 & 2 & -1 & 3 \end{array}$$

In the last row of the synthetic division, all numbers except the last one give the coefficients of the quotient, and the last number gives the remainder. The quotient is $1x^4 + 1x^3 + 0x^2 + 2x - 1$ or $x^4 + x^3 + 2x - 1$, and the remainder is 3. Thus,

$$\frac{x^5 + 3x^4 + 2x^3 + 2x^2 + 3x + 1}{x + 2} = x^4 + x^3 + 2x - 1 + \frac{3}{x + 2}.$$

19. $f(x) = 2x^3 + x^2 + x - 8;\quad k = -1$

$$\begin{array}{r|rrrr} -1 & 2 & 1 & 1 & -8 \\ & & -2 & 1 & -2 \\ \hline & 2 & -1 & 2 & -10 \end{array}$$

$f(x) = (x + 1)(2x^2 - x + 2) + (-10)$

$= (x + 1)(2x^2 - x + 2) - 10$

27. $f(x) = x^2 + 5x + 6;\quad k = -2$

$$\begin{array}{r|rrr} -2 & 1 & 5 & 6 \\ & & -2 & -6 \\ \hline & 1 & 3 & 0 \end{array}$$

The last number in the bottom row of the synthetic division gives the remainder, 0. Therefore, by the remainder theorem, $f(-2) = 0$.

3.3 Exercises *(page 329)*

23. $f(x) = x^3 + (7 - 3i)x^2 + (12 - 21i)x - 36i;\quad k = 3i$

$$\begin{array}{r|rrrr} 3i & 1 & 7 - 3i & 12 - 21i & -36i \\ & & 3i & 21i & 36i \\ \hline & 1 & 7 & 12 & 0 \end{array}$$

The quotient is $x^2 + 7x + 12$, so

$f(x) = (x - 3i)(x^2 + 7x + 12)$

$= (x - 3i)(x + 4)(x + 3)$.

45. $f(x) = 3(x - 2)(x + 3)(x^2 - 1)$

$0 = 3(x - 2)(x + 3)(x^2 - 1)$ Let $f(x) = 0$.

$0 = 3(x - 2)(x + 3)(x - 1)(x + 1)$ Factor the difference of squares.

$x - 2 = 0$ or $x + 3 = 0$ or $x - 1 = 0$ or $x + 1 = 0$ Zero-factor property

$x = 2$ or $x = -3$ or $x = 1$ or $x = -1$

The zeros are 2, -3, 1, and -1, all of multiplicity 1.

51. Zeros of -2, 1, and 0; $f(-1) = -1$

The factors of $f(x)$ are $x - (-2) = x + 2$, $x - 1$, and $x - 0 = x$.

$$\begin{aligned} f(x) &= a(x + 2)(x - 1)(x) \\ f(-1) &= a(-1 + 2)(-1 - 1)(-1) = -1 \\ a(1)(-2)(-1) &= -1 \\ 2a &= -1 \\ a &= -\frac{1}{2} \end{aligned}$$

Therefore,

$$\begin{aligned} f(x) &= -\frac{1}{2}(x + 2)(x - 1)(x) \\ &= -\frac{1}{2}(x^2 + x - 2)(x) \\ &= -\frac{1}{2}(x^3 + x^2 - 2x) \\ f(x) &= -\frac{1}{2}x^3 - \frac{1}{2}x^2 + x. \end{aligned}$$

3.4 Exercises *(page 342)*

37. $f(x) = x^3 + 5x^2 - x - 5$

$= x^2(x + 5) - 1(x + 5)$

$= (x + 5)(x^2 - 1)$ Factor by grouping.

$= (x + 5)(x + 1)(x - 1)$ Factor the difference of squares.

Find the real zeros of f. (All three zeros of this function are real.)

$x + 5 = 0$ or $x + 1 = 0$ or $x - 1 = 0$ Zero-factor property

$x = -5$ or $x = -1$ or $x = 1$

The zeros of f are -5, -1, and 1, so the x-intercepts of the graph are also -5, -1, and 1. Plot the points

$(-5, 0)$, $(-1, 0)$, and $(1, 0)$.

Now find the y-intercept.

$$f(0) = 0^3 + 5 \cdot 0^2 - 0 - 5 = -5,$$

so the y-intercept is -5. Plot the point $(0, -5)$.

The x-intercepts divide the x-axis into four intervals:

$(-\infty, -5)$, $(-5, -1)$, $(-1, 1)$, $(1, \infty)$.

Test a point in each interval to find the sign of $f(x)$ in each interval. See the table.

Interval	Test Point	Value of $f(x)$	Sign of $f(x)$	Graph Above or Below x-axis
$(-\infty, -5)$	-6	-35	Negative	Below
$(-5, -1)$	-2	9	Positive	Above
$(-1, 1)$	0	-5	Negative	Below
$(1, \infty)$	2	21	Positive	Above

Plot the points $(-6, -35)$, $(-2, 9)$, and $(2, 21)$. Note that $(0, -5)$ was already plotted when the y-intercept was found. Connect these test points, the zeros (or x-intercepts), and the y-intercept with a smooth curve to obtain the graph.

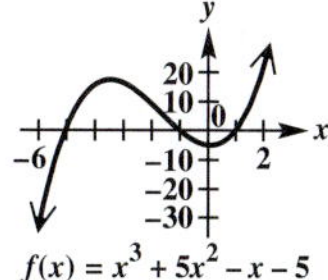

75. $f(x) = x^3 + 4x^2 - 8x - 8$; $[-3.8, -3]$

Graph this function in a window that will produce a comprehensive graph, such as $[-10, 10]$ by $[-50, 50]$, with X scale = 1, Y scale = 10.

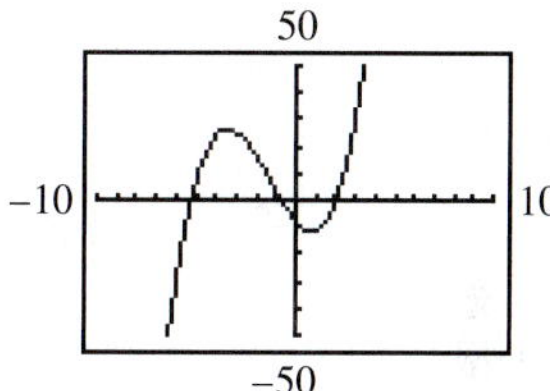

From this graph, we can see that the turning point in the interval $[-3.8, -3]$ is a maximum. Use "maximum" from the CALC menu to approximate the coordinates of this turning point.

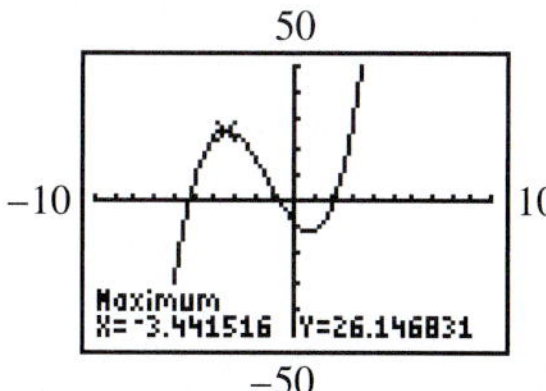

To the nearest hundredth, the turning point in the interval $[-3.8, -3]$ is $(-3.44, 26.15)$.

93. Use the following volume formulas:

$$V_{\text{cylinder}} = \pi r^2 h$$

$$V_{\text{hemisphere}} = \frac{1}{2}V_{\text{sphere}} = \frac{1}{2}\left(\frac{4}{3}\pi r^3\right) = \frac{2}{3}\pi r^3$$

(continued)

$$\pi r^2 h + 2\left(\frac{2}{3}\pi r^3\right) = \text{Total volume of tank}$$

$$\pi x^2(12) + \frac{4}{3}\pi x^3 = 144\pi \quad \text{Let } V = 144\pi,\ h = 12, \text{ and } r = x.$$

$$\frac{4}{3}\pi x^3 + 12\pi x^2 - 144\pi = 0$$

$$\frac{4}{3}x^3 + 12x^2 - 144 = 0 \quad \text{Divide by } \pi.$$

$$4x^3 + 36x^2 - 432 = 0 \quad \text{Multiply by 3.}$$

$$x^3 + 9x^2 - 108 = 0 \quad \text{Divide by 4.}$$

Synthetic division or graphing $f(x) = x^3 + 9x^2 - 108$ with a graphing calculator will show that this function has zeros of -6 (multiplicity 2) and 3. Because x represents the radius of the hemispheres, a negative solution for the equation $x^3 + 9x^2 - 108 = 0$ is not meaningful. In order to get a volume of 144π ft^3, a radius of 3 ft should be used.

3.5 Exercises *(page 362)*

61. $f(x) = \dfrac{1}{x^2 + 1}$

Because $x^2 + 1 = 0$ has no real solutions, the denominator is never 0, so there are no vertical asymptotes.

As $|x| \to \infty$, $y \to 0$, so the line $y = 0$ (the x-axis) is the horizontal asymptote.

$f(0) = \dfrac{1}{0^2 + 1} = 1$, so the y-intercept is 1.

The equation $f(x) = 0$ has no solutions (notice that the numerator has no zeros), so there are no x-intercepts.

$f(-x) = \dfrac{1}{(-x)^2 + 1} = \dfrac{1}{x^2 + 1} = f(x)$, so the graph is symmetric with respect to the y-axis.

Use a table of values to obtain several additional points on the graph. Use the asymptote, y-intercepts, and these additional points (which reflect the symmetry of the graph) to sketch the graph of the function.

x	y
$\pm.5$	.8
±1	.5
±2	.2

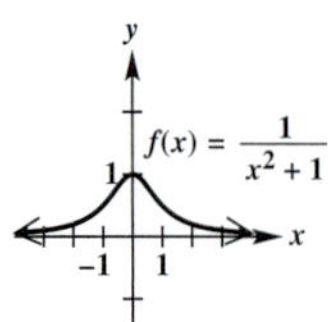

65. $f(x) = \dfrac{x^2 + 2x}{2x - 1}$

Step 1 Find any vertical asymptotes.

$$2x - 1 = 0$$

$$x = \frac{1}{2}$$

Step 2 Find any horizontal or oblique asymptotes. Because the numerator has degree exactly one more than the denominator, there is an oblique asymptote. Because $2x - 1$ is not of the form $x - a$, use polynomial long division rather than synthetic division to find the equation of this asymptote.

$$\begin{array}{r} \frac{1}{2}x + \frac{5}{4} \\ 2x - 1\overline{)x^2 + 2x + 0} \\ \underline{x^2 - \frac{1}{2}x} \\ \frac{5}{2}x + 0 \\ \underline{\frac{5}{2}x - \frac{5}{4}} \\ \frac{5}{4} \end{array}$$

Disregard the remainder. The equation of the oblique asymptote is $y = \frac{1}{2}x + \frac{5}{4}$.

Step 3 Find the y-intercept.

$$f(0) = \frac{0^2 + 2(0)}{2(0) - 1} = \frac{0}{-1} = 0,$$

so the y-intercept is 0.

Step 4 Find the x-intercepts, if any, by finding the zeros of the numerator.

$$x^2 + 2x = 0$$

$$x(x + 2) = 0$$

$$x = 0 \quad \text{or} \quad x = -2$$

There are two x-intercepts, -2 and 0.

Step 5 The graph does not intersect its oblique asymptote because the equation

$$\frac{x^2 + 2x}{2x - 1} = \frac{1}{2}x + \frac{5}{4},$$

which is equivalent to

$$x^2 + 2x = x^2 + 2x - \frac{5}{4},$$

has no solution.

Step 6

x	y
-3	$-\frac{3}{7} \approx -.43$
-1	$\frac{1}{3} \approx .33$
$\frac{1}{4} = .25$	$-\frac{9}{8} = -1.125$
1	3
3	3

Use the asymptotes, the intercepts, and these additional points to sketch the graph.

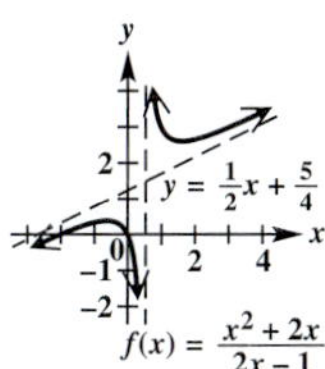

69. The graph has one vertical asymptote, $x = 2$, so $x - 2$ is a factor of the denominator of the rational expression. There is a point of discontinuity ("hole") in the graph at $x = -2$, so there is a factor of $x + 2$ in both numerator and denominator. There is one x-intercept, 3, so 3 is a zero of the numerator, which means that $x - 3$ is a factor of the numerator. Putting all of this information together, we have a possible function for the graph:

$$f(x) = \frac{(x - 3)(x + 2)}{(x - 2)(x + 2)} \quad \text{or} \quad f(x) = \frac{x^2 - x - 6}{x^2 - 4}.$$

Note: From the second form of the function, we can see that the graph of this function has a horizontal asymptote of $y = 1$, which is consistent with the given graph.

3.6 Exercises *(page 372)*

19. *Step 1* Write the general relationship among the variables as an equation. Use the constant k.

$$a = \frac{kmn^2}{y^3}$$

Step 2 Substitute $a = 9$, $m = 4$, $n = 9$, and $y = 3$ to find k.

$$9 = \frac{k \cdot 4 \cdot 9^2}{3^3}$$

$$9 = 12k$$

$$k = \frac{3}{4}$$

Step 3 Substitute this value of k into the equation from Step 1, obtaining a specific formula.

$$a = \frac{3}{4} \cdot \frac{mn^2}{y^3}$$

$$a = \frac{3mn^2}{4y^3}$$

Step 4 Substitute $m = 6$, $n = 2$, and $y = 5$ and solve for a.

$$a = \frac{3mn^2}{4y^3} = \frac{3 \cdot 6 \cdot 2^2}{4 \cdot 5^3} = \frac{18}{125}$$

35. *Step 1* Let F represent the force of the wind, A represent the area of the surface, and v represent the velocity of the wind.

$$F = kAv^2$$

Step 2 $50 = k\left(\frac{1}{2}\right)(40)^2$ Let $F = 50$, $A = \frac{1}{2}$, $v = 40$.

$$50 = 800k$$

$$k = \frac{50}{800} = \frac{1}{16}$$

Step 3 $F = \frac{1}{16}Av^2$

Step 4 $F = \frac{1}{16} \cdot 2 \cdot 80^2 = 800$

The force of the wind would be 800 lb.

43. Let t represent the Kelvin temperature.

$$R = kt^4$$

$213.73 = k \cdot 293^4$ Let $R = 213.73$, $t = 293$.

$$k = \frac{213.73}{293^4} \approx 2.9 \times 10^{-8}$$

Thus, $R = (2.9 \times 10^{-8})t^4$.

If $t = 335$, then

$$R = (2.9 \times 10^{-8})(335^4) \approx 365.24.$$

CHAPTER 4 EXPONENTIAL AND LOGARITHMIC FUNCTIONS

4.1 Exercises *(page 398)*

41. $f(x) = \frac{2}{x + 6}$, $\quad g(x) = \frac{6x + 2}{x}$

$$(f \circ g) = f[g(x)]$$

$$= f\left(\frac{6x + 2}{x}\right)$$

$$= \frac{2}{\frac{6x + 2}{x} + 6}$$

$$= \frac{2}{\frac{6x + 2 + 6x}{x}}$$

$$= \frac{2}{1} \cdot \frac{x}{12x + 2}$$

$$= \frac{2x}{12x + 2}$$

$$= \frac{2x}{2(6x + 1)}$$

$$= \frac{x}{6x + 1} \neq x$$

Since $(f \circ g)(x) \neq x$, the functions are not inverses. It is not necessary to check $(g \circ f)(x)$.

59. $f(x) = \dfrac{1}{x-3}$

(a)

$$y = \frac{1}{x-3} \qquad y = f(x)$$

$$x = \frac{1}{y-3} \qquad \text{Interchange } x \text{ and } y.$$

$$x(y-3) = 1 \qquad \text{Solve for } y.$$

$$xy - 3x = 1$$

$$xy = 1 + 3x$$

$$y = \frac{1+3x}{x}$$

$$f^{-1}(x) = \frac{1+3x}{x} \qquad \text{Replace } y \text{ with } f^{-1}(x),$$

(b) 

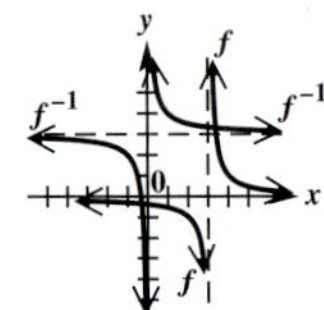

(c) For both f and f^{-1}, the domain contains all real numbers except those for which the denominator equals 0.

Domain of f = range of $f^{-1} = (-\infty, 3) \cup (3, \infty)$

Domain of f^{-1} = range of $f = (-\infty, 0) \cup (0, \infty)$

4.2 Exercises *(page 414)*

7. $g(x) = \left(\dfrac{1}{4}\right)^x$

$$g(-2) = \left(\frac{1}{4}\right)^{-2} = 4^2 = 16$$

55. $\left(\dfrac{1}{e}\right)^{-x} = \left(\dfrac{1}{e^2}\right)^{x+1}$

$$(e^{-1})^{-x} = (e^{-2})^{x+1} \qquad \text{Definition of negative exponent}$$

$$e^x = e^{-2(x+1)} \qquad (a^m)^n = a^{mn}$$

$$e^x = e^{-2x-2} \qquad \text{Distributive property}$$

$$x = -2x - 2 \qquad \text{Property (b)}$$

$$3x = -2 \qquad \text{Add } 2x.$$

$$x = -\frac{2}{3} \qquad \text{Divide by 3.}$$

Solution set: $\left\{-\dfrac{2}{3}\right\}$

57. $(\sqrt{2})^{x+4} = 4^x$

$$(2^{1/2})^{x+4} = (2^2)^x \qquad \text{Definition of } a^{1/n}\text{; write both sides as powers of a common base.}$$

$$2^{(1/2)(x+4)} = 2^{2x} \qquad (a^m)^n = a^{mn}$$

$$2^{(1/2)x+2} = 2^{2x} \qquad \text{Distributive property}$$

$$\frac{1}{2}x + 2 = 2x \qquad \text{Property (b)}$$

$$2 = \frac{3}{2}x \qquad \text{Subtract } \tfrac{1}{2}x.$$

$$\frac{4}{3} = x \qquad \text{Multiply by } \tfrac{2}{3}.$$

Solution set: $\left\{\dfrac{4}{3}\right\}$

59. $\dfrac{1}{27} = b^{-3}$

$$3^{-3} = b^{-3}$$

$$b = 3$$

Solution set: $\{3\}$

Alternate solution:

$$\frac{1}{27} = b^{-3}$$

$$\frac{1}{27} = \frac{1}{b^3}$$

$$27 = b^3$$

$$b = \sqrt[3]{27}$$

$$b = 3$$

Solution set: $\{3\}$

4.3 Exercises *(page 427)*

23. $\log_x 25 = -2$

$$x^{-2} = 25 \qquad \text{Write in exponential form.}$$

$$(x^{-2})^{-1/2} = 25^{-1/2} \qquad \text{Raise both sides to the same power.}$$

$$x = \frac{1}{25^{1/2}}$$

$$x = \frac{1}{5}$$

Solution set: $\left\{\dfrac{1}{5}\right\}$

71. $-\dfrac{2}{3}\log_5(5m^2) + \dfrac{1}{2}\log_5(25m^2)$

$$= \log_5(5m^2)^{-2/3} + \log_5(25m^2)^{1/2} \qquad \text{Power property}$$

$$= \log_5[(5m^2)^{-2/3} \cdot (25m^2)^{1/2}] \qquad \text{Product property}$$

$$= \log_5(5^{-2/3}m^{-4/3} \cdot 5m) \qquad \text{Properties of exponents}$$

$$= \log_5(5^{-2/3} \cdot 5^1 \cdot m^{-4/3} \cdot m^1)$$

$$= \log_5(5^{1/3} \cdot m^{-1/3}) \qquad a^m \cdot a^n = a^{m+n}$$

$$= \log_5 \frac{5^{1/3}}{m^{1/3}} \quad \text{or} \quad \log_5 \sqrt[3]{\frac{5}{m}}$$

77. $\log_{10}\sqrt{30} = \log_{10} 30^{1/2} \qquad$ Definition of $a^{1/n}$

$$= \frac{1}{2}\log_{10} 30 \qquad \text{Power property}$$

$= \frac{1}{2}\log_{10}(10 \cdot 3)$

$= \frac{1}{2}(\log_{10} 10 + \log_{10} 3)$ Product property

$= \frac{1}{2}(1 + .4771)$ $\log_a a = 1$; Substitute.

$= \frac{1}{2}(1.4771) \approx .7386$

4.4 Exercises *(page 438)*

39. $\log_{\sqrt{13}} 12 = \frac{\ln 12}{\ln \sqrt{13}}$ Change-of-base theorem

$= \frac{\ln 12}{\ln 13^{1/2}}$ Definition of $a^{1/n}$

$= \frac{\ln 12}{.5 \ln 13}$ Power property

≈ 1.9376 Use a calculator; round answer to four decimal places.

The required logarithm can also be found by entering $\ln \sqrt{13}$ into the calculator directly:

$\log_{\sqrt{13}} 12 = \frac{\ln 12}{\ln \sqrt{13}}$ Change-of-base theorem

≈ 1.9376 Use a calculator; round answer to four decimal places.

45. $\ln(b^4\sqrt{a}) = \ln(b^4 a^{1/2})$ Definition of $a^{1/n}$

$= \ln b^4 + \ln a^{1/2}$ Product property

$= 4 \ln b + \frac{1}{2}\ln a$ Power property

$= 4v + \frac{1}{2}u$ Substitute v for $\ln b$ and u for $\ln a$.

4.5 Exercises *(page 448)*

27. $\log(x + 25) = 1 + \log(2x - 7)$

$\log(x + 25) - \log(2x - 7) = 1$

$\log \frac{x + 25}{2x - 7} = 1$ Quotient property

$\frac{x + 25}{2x - 7} = 10^1$ Write in exponential form; $\log x = \log_{10} x$.

$x + 25 = 10(2x - 7)$ Multiply by $2x - 7$.

$x + 25 = 20x - 70$ Distributive property

$95 = 19x$ Add 70; subtract x.

$5 = x$ Divide by 19.

Solution set: $\{5\}$

39. $\log_2(\log_2 x) = 1$

$\log_2 x = 2^1$ Write in exponential form.

$x = 2^2$ Write in exponential form.

$x = 4$ Evaluate.

Solution set: $\{4\}$

41. $\log x^2 = (\log x)^2$

$2 \log x = (\log x)^2$ Power property

$(\log x)^2 - 2 \log x = 0$ Subtract $2 \log x$; rewrite.

$\log x(\log x - 2) = 0$ Factor.

$\log x = 0$ or $\log x - 2 = 0$ Zero-factor property

$\log x = 2$

$x = 10^0$ or $x = 10^2$ Write in exponential form.

$x = 1$ or $x = 100$

Solution set: $\{1, 100\}$

47. $p = a + \frac{k}{\ln x}$ for x

$p - a = \frac{k}{\ln x}$ Subtract a.

$(\ln x)(p - a) = k$ Multiply by $\ln x$.

$\ln x = \frac{k}{p - a}$ Divide by $p - a$.

$x = e^{k/(p-a)}$ Change to exponential form.

4.6 Exercises *(page 458)*

23. $A = Pe^{rt}$ Continuous compounding formula

$3P = Pe^{.05t}$ Let $A = 3P$ and $r = .05$.

$3 = e^{.05t}$ Divide by P.

$\ln 3 = \ln e^{.05t}$ Take logarithms on both sides.

$\ln 3 = .05t$ $\ln e^x = x$

$\frac{\ln 3}{.05} = t$ Divide by .05.

$t \approx 21.97$ Use a calculator.

It will take about 21.97 yr for the investment to triple.

31. $f(t) = 200(.90)^{t-1}$

Find t when $f(t) = 50$.

$50 = 200(.90)^{t-1}$ Let $f(t) = 50$.

$.25 = (.90)^{t-1}$ Divide by 200.

$\ln .25 = \ln[(.90)^{t-1}]$ Take logarithms on both sides.

$\ln .25 = (t - 1)\ln .90$ Power property

$\frac{\ln .25}{\ln .90} = t - 1$ Divide by ln .90.

(continued)

$$\frac{\ln .25}{\ln .90} + 1 = t \qquad \text{Add 1.}$$

$$t \approx 14.2 \qquad \text{Use a calculator.}$$

The initial dose will reach a level of 50 mg in about 14.2 hr.

CHAPTER 5 TRIGONOMETRIC FUNCTIONS

5.1 Exercises *(page 478)*

29. $90° - 72° 58' 11''$

$$\begin{array}{r} 89° 59' 60'' \\ -72° 58' 11'' \\ \hline 17° 01' 49'' \end{array} \qquad \text{Write } 90° \text{ as } 89° 59' 60''.$$

Thus, $90° - 72° 58' 11'' = 17° 1' 49''$.

77. 600 rotations per min

$= \frac{600}{60}$ rotations per sec

$= 10$ rotations per sec

$= 5$ rotations per $\frac{1}{2}$ sec

$= 5(360°)$ per $\frac{1}{2}$ sec

$= 1800°$ per $\frac{1}{2}$ sec

A point on the edge of the tire will move $1800°$ in $\frac{1}{2}$ sec.

5.2 Exercises *(page 490)*

55. $\tan(3\theta - 4°) = \frac{1}{\cot(5\theta - 8°)}$ Given equation

$\tan(3\theta - 4°) = \tan(5\theta - 8°)$ Reciprocal identity

The second equation above will be true if $3\theta - 4° = 5\theta - 8°$, so solving this equation will give a value (but not the only value) for which the given equation is true.

$$3\theta - 4° = 5\theta - 8°$$
$$4° = 2\theta$$
$$\theta = 2°$$

67. $\tan 30° = \frac{\sin 30°}{\cos 30°}$ Quotient identity

$\cos 30° < 1$, so $\tan 30° = \frac{\sin 30°}{\cos 30°} > \sin 30°$; that is, $\tan 30°$ is greater than $\sin 30°$.

87. Given $\tan \theta = -\frac{15}{8}$, with θ in quadrant II

Draw θ in standard position in quadrant II. (See the figure at the top of the next column.) Because $\tan \theta = \frac{y}{x}$ and θ is in quadrant II, we can use the values $y = 15$ and $x = -8$ for a point on its terminal side.

$$r = \sqrt{x^2 + y^2} = \sqrt{(-8)^2 + 15^2} = \sqrt{64 + 225}$$
$$= \sqrt{289} = 17$$

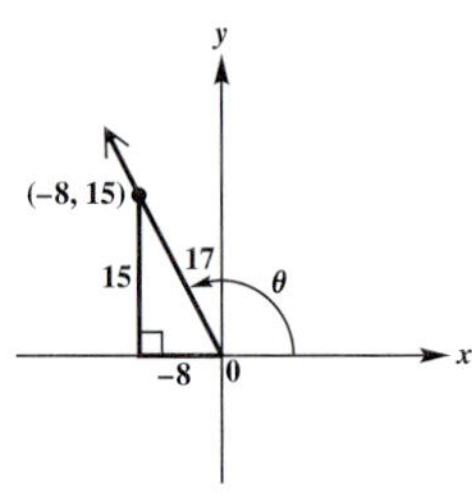

Use the values of x, y, and r and the definitions of the trigonometric functions to find the six trigonometric function values for θ.

$$\sin \theta = \frac{y}{r} = \frac{15}{17} \qquad \csc \theta = \frac{r}{y} = \frac{17}{15}$$

$$\cos \theta = \frac{x}{r} = \frac{-8}{17} = -\frac{8}{17} \qquad \sec \theta = \frac{r}{x} = \frac{17}{-8} = -\frac{17}{8}$$

$$\tan \theta = \frac{y}{x} = \frac{15}{-8} = -\frac{15}{8} \qquad \cot \theta = \frac{x}{y} = \frac{-8}{15} = -\frac{8}{15}$$

5.3 Exercises *(page 502)*

33. One point on the line $y = \sqrt{3}x$ is the origin, $(0, 0)$. Let (x, y) be any other point on this line. Then, by the definition of slope, $m = \frac{y - 0}{x - 0} = \frac{y}{x} = \sqrt{3}$, but also, by the definition of tangent, $\tan \theta = \frac{y}{x}$. Thus, $\tan \theta = \sqrt{3}$. Because $\tan 60° = \sqrt{3}$, the line $y = \sqrt{3}x$ makes a $60°$ angle with the positive x-axis. (See Exercise 30.)

35. Apply the relationships between the lengths of the sides of a $30°$–$60°$ right triangle first to the triangle on the left to find the values of x and y, and then to the triangle on the right to find the values of z and w. In a $30°$–$60°$ right triangle, the side opposite the $30°$ angle is $\frac{1}{2}$ the length of the hypotenuse. The longer leg is $\sqrt{3}$ times the shorter leg.

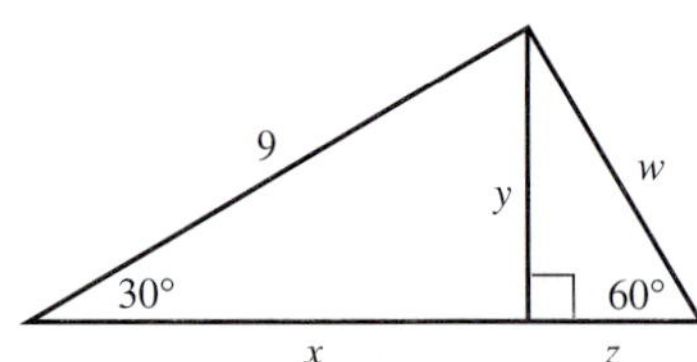

Thus,

$$y = \frac{1}{2}(9) = \frac{9}{2} \quad \text{and} \quad x = y\sqrt{3} = \frac{9\sqrt{3}}{2}.$$

$y = z\sqrt{3}$, so

$$z = \frac{y}{\sqrt{3}} = \frac{\frac{9}{2}}{\sqrt{3}} = \frac{9}{2\sqrt{3}} \cdot \frac{\sqrt{3}}{\sqrt{3}} = \frac{9\sqrt{3}}{6} = \frac{3\sqrt{3}}{2},$$

and $$w = 2z = 2\left(\frac{3\sqrt{3}}{2}\right) = 3\sqrt{3}.$$

61. To find the reference angle for $-300°$, sketch this angle in standard position.

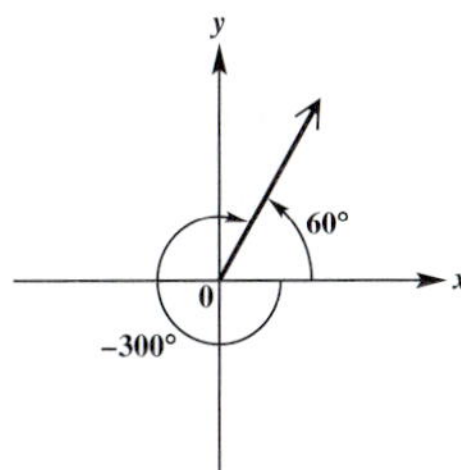

The reference angle for $-300°$ is $-300° + 360° = 60°$. Because $-300°$ is in quadrant I, the values of all its trigonometric functions will be positive, so these values will be identical to the trigonometric function values for 60°. (See the Function Values of Special Angles table that follows Example 2 in Section 5.3.)

$$\sin(-300°) = \frac{\sqrt{3}}{2} \qquad \csc(-300°) = \frac{2\sqrt{3}}{3}$$

$$\cos(-300°) = \frac{1}{2} \qquad \sec(-300°) = 2$$

$$\tan(-300°) = \sqrt{3} \qquad \cot(-300°) = \frac{\sqrt{3}}{3}$$

5.4 Exercises *(page 516)*

19. Solve the right triangle with $B = 73.00°$, $b = 128$ in., and $C = 90°$.

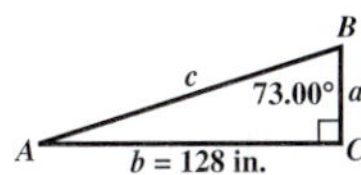

$$A = 90° - 73.00° = 17.00°$$

$$\tan 73.00° = \frac{128}{a}$$

$$a = \frac{128}{\tan 73.00°} \approx 39.1 \text{ in.} \quad \text{Three significant digits}$$

$$\sin 73.00° = \frac{128}{c}$$

$$c = \frac{128}{\sin 73.00°} \approx 134 \text{ in.} \quad \text{Three significant digits}$$

31. Let x represent the horizontal distance between the two buildings and y represent the height of the portion of the building across the street that is higher than the window.

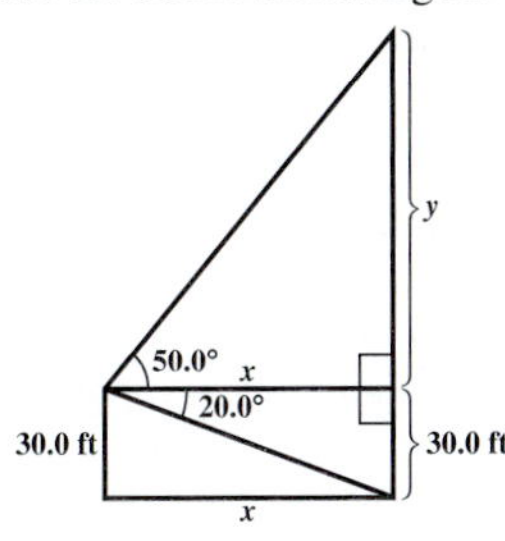

$$\tan 20.0° = \frac{30.0}{x}$$

$$x = \frac{30.0}{\tan 20.0°} \approx 82.4$$

$$\tan 50.0° = \frac{y}{x}$$

$$y = x \tan 50.0° = \left(\frac{30.0}{\tan 20.0°}\right) \tan 50.0° \approx 98.2$$

$$\text{height} = y + 30.0 = \left(\frac{30.0}{\tan 20.0°}\right) \tan 50.0° + 30.0 \approx 128$$

Three significant digits

The height of the building across the street is about 128 ft.

37. Let h represent the height of the tower.

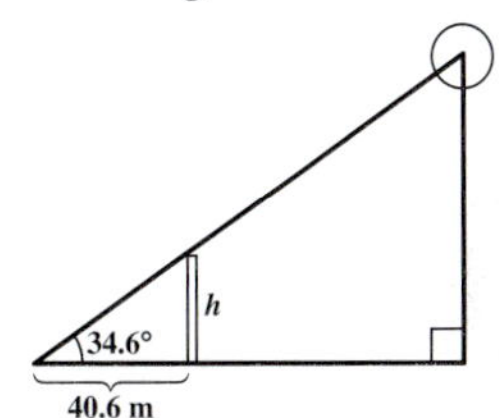

$$\tan 34.6° = \frac{h}{40.6}$$

$$h = 40.6 \tan 34.6° \approx 28.0$$

Three significant digits

The height of the tower is about 28.0 m.

CHAPTER 6 THE CIRCULAR FUNCTIONS AND THEIR GRAPHS

6.1 Exercises *(page 539)*

41. $\cos\left(-\frac{\pi}{6}\right) = \cos(-30°)$ Convert to degrees.

$= \cos(-30° + 360°)$ Add 360° to get a coterminal angle between 0° and 360°.

$= \cos 330°$

$= +\cos 30°$ Cosine is positive in quadrant IV; 30° is the reference angle for 330°.

$= \frac{\sqrt{3}}{2}$

65. For the large gear and pedal,

$$s = r\theta = 4.72\pi. \quad 180° = \pi \text{ radians}$$

Thus, the chain moves 4.72π in. Find the angle through which the small gear rotates.

$$\theta = \frac{s}{r} = \frac{4.72\pi}{1.38} \approx 3.42\pi$$

(continued)

The angle θ for the wheel and for the small gear are the same, so for the wheel,

$$s = r\theta = 13.6(3.42\pi) \approx 146 \text{ in.}$$

The bicycle will move about 146 in.

85. **(a)**

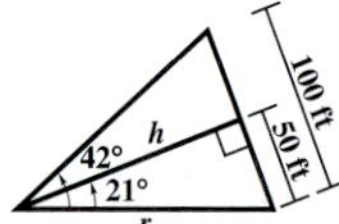

The triangle formed by the sides of the central angle and the chord is isosceles. Therefore, the bisector of the central angle is also the perpendicular bisector of the chord and divides the larger triangle into two congruent right triangles.

$$\sin 21° = \frac{50}{r}$$

$$r = \frac{50}{\sin 21°} \approx 140 \text{ ft}$$

The radius of the curve is about 140 ft.

(b) $r = \dfrac{50}{\sin 21°}$; $\theta = 42°$

$$42° = 42\left(\frac{\pi}{180} \text{ radian}\right) = \frac{7\pi}{30} \text{ radian}$$

$$s = r\theta = \frac{50}{\sin 21°} \cdot \frac{7\pi}{30} = \frac{35\pi}{3 \sin 21°} \approx 102 \text{ ft}$$

The length of the arc determined by the 100-ft chord is about 102 ft.

(c) The portion of the circle bounded by the arc and the 100-ft chord is the shaded region in the figure below.

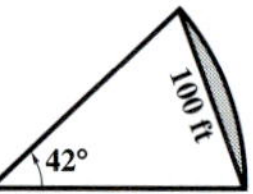

The area of the portion of the circle can be found by subtracting the area of the triangle from the area of the sector. From the figure in part (a),

$$\tan 21° = \frac{50}{h}, \quad \text{so} \quad h = \frac{50}{\tan 21°}.$$

$$A_{\text{sector}} = \frac{1}{2}r^2\theta$$

$$= \frac{1}{2}\left(\frac{50}{\sin 21°}\right)^2\left(\frac{7\pi}{30}\right) \quad \text{From part (b), } 42° = \tfrac{7\pi}{30}.$$

$$\approx 7135 \text{ ft}^2$$

$$A_{\text{triangle}} = \frac{1}{2}bh = \frac{1}{2}(100)\left(\frac{50}{\tan 21°}\right)$$

$$\approx 6513 \text{ ft}^2$$

$$A_{\text{portion}} = A_{\text{sector}} - A_{\text{triangle}} \approx 7135 \text{ ft}^2 - 6513 \text{ ft}^2$$

$$= 622 \text{ ft}^2$$

The area of the portion is about 622 ft^2.

87. Use the Pythagorean theorem to find the hypotenuse of the right triangle, which is also the radius of the sector of the circle.

$$r^2 = 30^2 + 40^2 = 900 + 1600 = 2500$$

$$r = \sqrt{2500} = 50$$

$$A_{\text{triangle}} = \frac{1}{2}bh = \frac{1}{2}(30)(40)$$

$$= 600 \text{ yd}^2$$

$$A_{\text{sector}} = \frac{1}{2}r^2\theta$$

$$= \frac{1}{2}(50)^2 \cdot \frac{\pi}{3} \quad 60° = \tfrac{\pi}{3}$$

$$= \frac{1250\pi}{3} \text{ yd}^2$$

$$\text{Total area} = A_{\text{triangle}} + A_{\text{sector}} = 600 \text{ yd}^2 + \frac{1250\pi}{3} \text{ yd}^2$$

$$\approx 1900 \text{ yd}^2$$

6.2 Exercises *(page 552)*

39. cos 2

$\frac{\pi}{2} \approx 1.57$ and $\pi \approx 3.14$, so $\frac{\pi}{2} < 2 < \pi$. Thus, an angle of 2 radians is in quadrant II. (The figure for Exercises 35–38 also shows that 2 radians is in quadrant II.) Because values of the cosine function are negative in quadrant II, cos 2 is negative.

57. $\left[\pi, \dfrac{3\pi}{2}\right]$; $\tan s = \sqrt{3}$

Recall that $\tan \frac{\pi}{3} = \sqrt{3}$ and in quadrant III tan s is positive. Therefore,

$$\tan\left(\pi + \frac{\pi}{3}\right) = \tan\frac{4\pi}{3} = \sqrt{3},$$

and thus, $s = \frac{4\pi}{3}$.

83. The hour hand of a clock moves through an angle of 2π radians (one complete revolution) in 12 hr, so

$$\omega = \frac{\theta}{t} = \frac{2\pi}{12} = \frac{\pi}{6} \text{ radian per hr.}$$

91. At 200 revolutions per min, the bicycle tire is moving $200(2\pi) = 400\pi$ radians per min. This is the angular velocity ω. The linear velocity of the bicycle is

$$v = r\omega = 13(400\pi) = 5200\pi \text{ in. per min.}$$

Convert this velocity to miles per hour.

$$v = \frac{5200\pi \text{ in.}}{\text{min}} \cdot \frac{60 \text{ min}}{\text{hr}} \cdot \frac{1 \text{ ft}}{12 \text{ in.}} \cdot \frac{1 \text{ mi}}{5280 \text{ ft}} \approx 15.5 \text{ mph}$$

6.3 Exercises *(page 568)*

51. $y = \frac{1}{2} + \sin 2\left(x + \frac{\pi}{4}\right)$

This equation has the form $y = c + a \sin b(x - d)$ with $c = \frac{1}{2}$, $a = 1$, $b = 2$, and $d = -\frac{\pi}{4}$. Start with the graph of $y = \sin x$ and modify it to take into account the amplitude, period, and translations required to obtain the desired graph.

Amplitude: $|a| = 1$

Period: $\frac{2\pi}{b} = \frac{2\pi}{2} = \pi$

Vertical translation: $\frac{1}{2}$ unit up

Phase shift (horizontal translation): $\frac{\pi}{4}$ units to the left

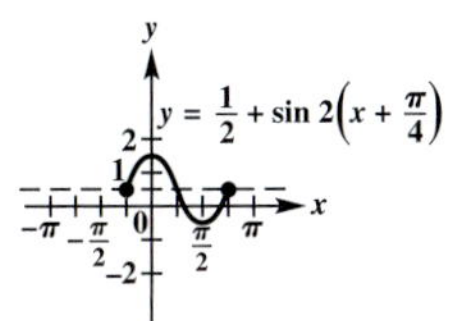

6.4 Exercises *(page 585)*

43. $y = -1 + \frac{1}{2}\cot(2x - 3\pi)$

$y = -1 + \frac{1}{2}\cot 2\left(x - \frac{3\pi}{2}\right)$ Rewrite $2x - 3\pi$ as $2\left(x - \frac{3\pi}{2}\right)$.

Period: $\frac{\pi}{b} = \frac{\pi}{2}$

Vertical translation: 1 unit down

Phase shift (horizontal translation): $\frac{3\pi}{2}$ units to the right

Because the function is to be graphed over a two-period interval, locate three adjacent vertical asymptotes. Because asymptotes of the graph of $y = \cot x$ occur at multiples of π, the following equations can be solved to locate asymptotes:

$$2\left(x - \frac{3\pi}{2}\right) = -2\pi, \quad 2\left(x - \frac{3\pi}{2}\right) = -\pi, \quad \text{and} \quad 2\left(x - \frac{3\pi}{2}\right) = 0.$$

Solve each of these equations.

$$2\left(x - \frac{3\pi}{2}\right) = -2\pi$$
$$x - \frac{3\pi}{2} = -\pi$$
$$x = -\pi + \frac{3\pi}{2} = \frac{\pi}{2}$$

$$2\left(x - \frac{3\pi}{2}\right) = -\pi$$
$$x - \frac{3\pi}{2} = -\frac{\pi}{2}$$
$$x = -\frac{\pi}{2} + \frac{3\pi}{2} = \frac{2\pi}{2} = \pi$$

$$2\left(x - \frac{3\pi}{2}\right) = 0$$
$$x - \frac{3\pi}{2} = 0$$
$$x = \frac{3\pi}{2}$$

Divide the interval $\left(\frac{\pi}{2}, \pi\right)$ into four equal parts to obtain the following key x-values:

first-quarter value: $\frac{5\pi}{8}$; middle value: $\frac{3\pi}{4}$;

third-quarter value: $\frac{7\pi}{8}$.

Evaluating the given function at these three key x-values gives the following points:

$$\left(\frac{5\pi}{8}, -\frac{1}{2}\right), \quad \left(\frac{3\pi}{4}, -1\right), \quad \left(\frac{7\pi}{8}, -\frac{3}{2}\right).$$

Connect these points with a smooth curve and continue the graph to approach the asymptotes $x = \frac{\pi}{2}$ and $x = \pi$ to complete one period of the graph. Sketch an identical curve between the asymptotes $x = \pi$ and $x = \frac{3\pi}{2}$ to complete a second period of the graph.

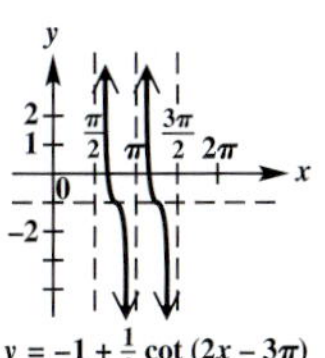

6.5 Exercises *(page 591)*

19. (a) We will use a model of the form $s(t) = a \cos \omega t$ with $a = -3$. Since

$$s(0) = -3\cos(0\omega) = -3\cos 0 = -3 \cdot 1 = -3,$$

using a cosine function rather than a sine function will avoid the need for a phase shift.

Since the frequency $= \frac{6}{\pi}$ cycles per sec, by definition,

$$\frac{6}{\pi} = \frac{\omega}{2\pi}$$
$$\pi\omega = 12\pi \quad \text{Cross products}$$
$$\omega = 12. \quad \text{Divide by } \pi.$$

(continued)

Therefore, a model for the position of the weight at time t seconds is

$$s(t) = -3 \cos 12t.$$

(b) Period $= \dfrac{1}{\frac{6}{\pi}} = \dfrac{\pi}{6}$ sec

CHAPTER 7 TRIGONOMETRIC IDENTITIES AND EQUATIONS

7.1 Exercises *(page 609)*

25. $\cot \theta = \dfrac{4}{3}$, $\sin \theta > 0$

Since $\cot \theta > 0$ and $\sin \theta > 0$, θ is in quadrant I, so all the function values are positive.

$$\tan \theta = \frac{1}{\cot \theta} = \frac{1}{\frac{4}{3}} = \frac{3}{4}$$

$$\sec^2 \theta = \tan^2 \theta + 1 \qquad \text{Pythagorean identity}$$

$$= \left(\frac{3}{4}\right)^2 + 1 = \frac{9}{16} + \frac{16}{16} = \frac{25}{16}$$

$$\sec \theta = \sqrt{\frac{25}{16}} = \frac{5}{4} \qquad \sec \theta > 0$$

$$\cos \theta = \frac{1}{\sec \theta} = \frac{1}{\frac{5}{4}} = \frac{4}{5}$$

$$\sin^2 \theta = 1 - \cos^2 \theta \qquad \text{Alternative form of Pythagorean identity}$$

$$= 1 - \left(\frac{4}{5}\right)^2 = \frac{9}{25}$$

$$\sin \theta = \sqrt{\frac{9}{25}} = \frac{3}{5} \qquad \sin \theta > 0$$

$$\csc \theta = \frac{1}{\sin \theta} = \frac{1}{\frac{3}{5}} = \frac{5}{3}$$

Thus, $\sin \theta = \frac{3}{5}$, $\cos \theta = \frac{4}{5}$, $\tan \theta = \frac{3}{4}$, $\sec \theta = \frac{5}{4}$, and $\csc \theta = \frac{5}{3}$.

47. $\csc x = \dfrac{1}{\sin x}$ Alternative form of reciprocal identity

$$= \frac{1}{\pm\sqrt{1 - \cos^2 x}} \qquad \text{Alternative form of Pythagorean identity}$$

$$= \frac{\pm 1}{\sqrt{1 - \cos^2 x}}$$

$$= \frac{\pm 1}{\sqrt{1 - \cos^2 x}} \cdot \frac{\sqrt{1 - \cos^2 x}}{\sqrt{1 - \cos^2 x}} \qquad \text{Rationalize the denominator.}$$

$$= \frac{\pm\sqrt{1 - \cos^2 x}}{1 - \cos^2 x}$$

59. $\sec \theta - \cos \theta = \dfrac{1}{\cos \theta} - \cos \theta$

$$= \frac{1}{\cos \theta} - \frac{\cos^2 \theta}{\cos \theta} \qquad \text{Get a common denominator.}$$

$$= \frac{1 - \cos^2 \theta}{\cos \theta} \qquad \text{Subtract fractions.}$$

$$= \frac{\sin^2 \theta}{\cos \theta} \qquad 1 - \cos^2 \theta = \sin^2 \theta$$

$$= \frac{\sin \theta}{\cos \theta}, \sin \theta = \tan \theta \sin \theta$$

65. Since $\cos x = \frac{1}{5}$, which is positive, x is in quadrant I or quadrant IV.

$$\sin x = \pm\sqrt{1 - \cos^2 x} = \pm\sqrt{1 - \left(\frac{1}{5}\right)^2}$$

$$= \pm\sqrt{\frac{24}{25}} = \pm\frac{2\sqrt{6}}{5}$$

$$\tan x = \frac{\sin x}{\cos x} = \frac{\pm\frac{2\sqrt{6}}{5}}{\frac{1}{5}} = \pm 2\sqrt{6}$$

$$\sec x = \frac{1}{\cos x} = \frac{1}{\frac{1}{5}} = 5$$

Quadrant I:

$$\frac{\sec x - \tan x}{\sin x} = \frac{5 - 2\sqrt{6}}{\frac{2\sqrt{6}}{5}} = \frac{5(5 - 2\sqrt{6})}{2\sqrt{6}}$$

$$= \frac{25 - 10\sqrt{6}}{2\sqrt{6}} \cdot \frac{\sqrt{6}}{\sqrt{6}} = \frac{25\sqrt{6} - 60}{12}$$

Quadrant IV:

$$\frac{\sec x - \tan x}{\sin x} = \frac{5 - (-2\sqrt{6})}{-\frac{2\sqrt{6}}{5}} = \frac{5(5 + 2\sqrt{6})}{-2\sqrt{6}}$$

$$= \frac{25 + 10\sqrt{6}}{-2\sqrt{6}} \cdot \frac{-\sqrt{6}}{-\sqrt{6}} = \frac{-25\sqrt{6} - 60}{12}$$

7.2 Exercises *(page 618)*

11. $\dfrac{1}{1 + \cos x} - \dfrac{1}{1 - \cos x} = \dfrac{1(1 - \cos x) - 1(1 + \cos x)}{(1 + \cos x)(1 - \cos x)}$

$$= \frac{1 - \cos x - 1 - \cos x}{1 - \cos^2 x}$$

$$= \frac{-2 \cos x}{\sin^2 x} = -\frac{2 \cos x}{\sin^2 x}$$

or $$-\frac{2 \cos x}{\sin^2 x} = -2\left(\frac{\cos x}{\sin x}\right)\left(\frac{1}{\sin x}\right)$$

$$= -2 \cot x \csc x$$

15. $(\sin x + 1)^2 - (\sin x - 1)^2$

$$= (\sin^2 x + 2 \sin x + 1) - (\sin^2 x - 2 \sin x + 1)$$

Square the binomials.

$$= \sin^2 x + 2 \sin x + 1 - \sin^2 x + 2 \sin x - 1$$

$$= 4 \sin x$$

63. Verify that $\dfrac{\tan^2 t - 1}{\sec^2 t} = \dfrac{\tan t - \cot t}{\tan t + \cot t}$ is an identity.

Work with the right side.

$$\frac{\tan t - \cot t}{\tan t + \cot t} = \frac{\tan t - \frac{1}{\tan t}}{\tan t + \frac{1}{\tan t}} \qquad \cot t = \tfrac{1}{\tan t}$$

$$= \frac{\tan t}{\tan t}\left(\frac{\tan t - \frac{1}{\tan t}}{\tan t + \frac{1}{\tan t}}\right)$$

Multiply numerator and denominator of the complex fraction by the LCD, $\tan t$.

$$= \frac{\tan^2 t - 1}{\tan^2 t + 1} \qquad \text{Distributive property}$$

$$= \frac{\tan^2 t - 1}{\sec^2 t} \qquad \tan^2 t + 1 = \sec^2 t$$

79. Show that $\sin(\csc s) = 1$ is not an identity.

We need find only one value for which the statement is false. Let $s = 2$. Use a calculator to find that $\sin(\csc 2) \approx .891094$, which is not equal to 1. Thus, $\sin(\csc s) = 1$ is not true for *all* real numbers s, so it is not an identity.

7.3 Exercises *(page 629)*

55. $\cos s = -\dfrac{8}{17}$ and $\cos t = -\dfrac{3}{5}$, s and t in quadrant III

In order to substitute into sum and difference identities, we need to find the values of $\sin s$ and $\sin t$, and also the values of $\tan s$ and $\tan t$. Because s and t are both in quadrant III, the values of $\sin s$ and $\sin t$ will be negative, while $\tan s$ and $\tan t$ will be positive.

$$\sin s = -\sqrt{1 - \cos^2 s} = -\sqrt{1 - \left(-\frac{8}{17}\right)^2}$$

$$= -\sqrt{\frac{225}{289}} = -\frac{15}{17}$$

$$\sin t = -\sqrt{1 - \cos^2 t} = -\sqrt{1 - \left(-\frac{3}{5}\right)^2}$$

$$= -\sqrt{\frac{16}{25}} = -\frac{4}{5}$$

$$\tan s = \frac{\sin s}{\cos s} = \frac{-\frac{15}{17}}{-\frac{8}{17}} = \frac{15}{8}$$

$$\tan t = \frac{\sin t}{\cos t} = \frac{-\frac{4}{5}}{-\frac{3}{5}} = \frac{4}{3}$$

(a) $\cos(s + t) = \cos s \cos t - \sin s \sin t$

$$= \left(-\frac{8}{17}\right)\left(-\frac{3}{5}\right) - \left(-\frac{15}{17}\right)\left(-\frac{4}{5}\right)$$

$$= \frac{24}{85} - \frac{60}{85} = -\frac{36}{85}$$

(b) $\sin(s - t) = \sin s \cos t - \cos s \sin t$

$$= \left(-\frac{15}{17}\right)\left(-\frac{3}{5}\right) - \left(-\frac{8}{17}\right)\left(-\frac{4}{5}\right)$$

$$= \frac{45}{85} - \frac{32}{85} = \frac{13}{85}$$

(c) $\tan(s + t) = \dfrac{\tan s + \tan t}{1 - \tan s \tan t} = \dfrac{\frac{15}{8} + \frac{4}{3}}{1 - \left(\frac{15}{8}\right)\left(\frac{4}{3}\right)}$

$$= \frac{\frac{45}{24} + \frac{32}{24}}{1 - \frac{60}{24}} = \frac{\frac{77}{24}}{-\frac{36}{24}} = -\frac{77}{36}$$

(d) From parts (a) and (c), $\cos(s + t) < 0$ and $\tan(s + t) < 0$. The only quadrant in which the values of both the cosine and the tangent are negative is quadrant II, so $s + t$ is in quadrant II.

65. $\tan \dfrac{11\pi}{12} = \tan\left(\pi - \dfrac{\pi}{12}\right)$

$$= \frac{\tan \pi - \tan \frac{\pi}{12}}{1 + \tan \pi \tan \frac{\pi}{12}} \qquad \text{Tangent difference identity}$$

$$= \frac{0 - \tan \frac{\pi}{12}}{1 + 0 \cdot \tan \frac{\pi}{12}} \qquad \tan \pi = 0$$

$$= -\tan \frac{\pi}{12}$$

Now use a difference identity to find $\tan \frac{\pi}{12}$.

$$\tan \frac{\pi}{12} = \tan\left(\frac{\pi}{4} - \frac{\pi}{6}\right)$$

$$= \frac{\tan \frac{\pi}{4} - \tan \frac{\pi}{6}}{1 + \tan \frac{\pi}{4} \cdot \tan \frac{\pi}{6}}$$

$$= \frac{1 - \frac{\sqrt{3}}{3}}{1 + 1 \cdot \frac{\sqrt{3}}{3}} \qquad \tan \tfrac{\pi}{4} = 1;\ \tan \tfrac{\pi}{6} = \tfrac{\sqrt{3}}{3}$$

$$= \frac{\frac{3 - \sqrt{3}}{3}}{\frac{3 + \sqrt{3}}{3}} \qquad \text{Add and subtract with the common denominator.}$$

$$= \frac{3 - \sqrt{3}}{3 + \sqrt{3}} \qquad \text{Multiply numerator and denominator by 3.}$$

$$= \frac{3 - \sqrt{3}}{3 + \sqrt{3}} \cdot \frac{3 - \sqrt{3}}{3 - \sqrt{3}} \qquad \text{Rationalize the denominator.}$$

$$= \frac{9 - 6\sqrt{3} + 3}{9 - 3}$$

Square of a binomial in numerator; product of sum and difference of two terms in denominator

$$= \frac{12 - 6\sqrt{3}}{6}$$

$$= \frac{6(2 - \sqrt{3})}{6} \qquad \text{Factor the numerator.}$$

$$= 2 - \sqrt{3} \qquad \text{Lowest terms}$$

(continued)

Thus,

$$\tan\frac{11\pi}{12} = -\tan\frac{\pi}{12} = -(2-\sqrt{3}) = -2+\sqrt{3}.$$

77. Verify that $\dfrac{\sin(x-y)}{\sin(x+y)} = \dfrac{\tan x - \tan y}{\tan x + \tan y}$ is an identity.

Work with the left side.

$$\frac{\sin(x-y)}{\sin(x+y)} = \frac{\sin x\cos y - \cos x\sin y}{\sin x\cos y + \cos x\sin y}$$

Sine sum and difference identities

$$= \frac{\dfrac{\sin x\cos y}{\cos x\cos y} - \dfrac{\cos x\sin y}{\cos x\cos y}}{\dfrac{\sin x\cos y}{\cos x\cos y} + \dfrac{\cos x\sin y}{\cos x\cos y}}$$ Divide numerator and denominator by $\cos x\cos y$.

$$= \frac{\dfrac{\sin x}{\cos x}\cdot 1 - 1\cdot\dfrac{\sin y}{\cos y}}{\dfrac{\sin x}{\cos x}\cdot 1 + 1\cdot\dfrac{\sin y}{\cos y}}$$

$$= \frac{\tan x - \tan y}{\tan x + \tan y}$$ Tangent quotient identity

7.4 Exercises *(page 641)*

19. $\dfrac{1}{4} - \dfrac{1}{2}\sin^2 47.1° = \dfrac{1}{4}(1 - 2\sin^2 47.1°)$ Factor out $\frac{1}{4}$.

$$= \frac{1}{4}\cos 2(47.1°)$$

$\cos 2A = 1 - 2\sin^2 A$

$$= \frac{1}{4}\cos 94.2°$$

23. $\tan 3x = \tan(2x + x)$

$$= \frac{\tan 2x + \tan x}{1 - \tan 2x\tan x}$$ Tangent sum identity

$$= \frac{\dfrac{2\tan x}{1-\tan^2 x} + \tan x}{1 - \dfrac{2\tan x}{1-\tan^2 x}\cdot\tan x}$$ Tangent double-angle identity

$$= \frac{\dfrac{2\tan x + (1-\tan^2 x)\tan x}{1-\tan^2 x}}{\dfrac{1-\tan^2 x - 2\tan^2 x}{1-\tan^2 x}}$$ Add and subtract using the common denominator.

$$= \frac{2\tan x + \tan x - \tan^3 x}{1 - \tan^2 x - 2\tan^2 x}$$

Multiply numerator and denominator by $1 - \tan^2 x$.

$$= \frac{3\tan x - \tan^3 x}{1 - 3\tan^2 x}$$ Combine terms.

29. Verify that $\sin 4x = 4\sin x\cos x\cos 2x$ is an identity.
Work with the left side.

$\sin 4x = \sin 2(2x)$

$= 2\sin 2x\cos 2x$ Sine double-angle identity

$= 2(2\sin x\cos x)\cos 2x$ Sine double-angle identity

$= 4\sin x\cos x\cos 2x$

37. Verify that $\sec^2\dfrac{x}{2} = \dfrac{2}{1+\cos x}$ is an identity.

Work with the left side.

$$\sec^2\frac{x}{2} = \frac{1}{\cos^2\frac{x}{2}}$$ Reciprocal identity

$$= \frac{1}{\left(\pm\sqrt{\dfrac{1+\cos x}{2}}\right)^2}$$ Cosine half-angle identity

$$= \frac{1}{\dfrac{1+\cos x}{2}}$$

$$= \frac{2}{1+\cos x}$$

7.5 Exercises *(page 654)*

71. $\sin\left(2\cos^{-1}\dfrac{1}{5}\right)$

Let $\theta = \cos^{-1}\frac{1}{5}$, so $\cos\theta = \frac{1}{5}$. The inverse cosine function yields values only in quadrants I and II, and since $\frac{1}{5}$ is positive, θ is in quadrant I. Sketch θ in quadrant I, and label the sides of a right triangle. By the Pythagorean theorem, the length of the side opposite θ will be $\sqrt{5^2 - 1^2} = \sqrt{24} = 2\sqrt{6}$.

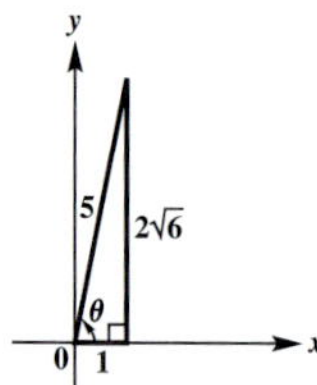

From the figure, $\sin\theta = \dfrac{2\sqrt{6}}{5}$.

Then,

$$\sin\left(2\cos^{-1}\frac{1}{5}\right) = \sin 2\theta$$

$$= 2\sin\theta\cos\theta$$ Sine double-angle identity

$$= 2\left(\frac{2\sqrt{6}}{5}\right)\left(\frac{1}{5}\right) = \frac{4\sqrt{6}}{25}.$$

7.6 Exercises *(page 665)*

15. $\tan^2 x + 3 = 0$

$\tan^2 x = -3$

The square of a real number cannot be negative, so this equation has no solution. Solution set: $\emptyset$

25.
$$2 \sin \theta - 1 = \csc \theta$$
$$2 \sin \theta - 1 = \frac{1}{\sin \theta} \quad \text{Reciprocal identity}$$
$$2 \sin^2 \theta - \sin \theta = 1 \quad \text{Multiply by } \sin \theta.$$
$$2 \sin^2 \theta - \sin \theta - 1 = 0 \quad \text{Subtract 1.}$$
$$(2 \sin \theta + 1)(\sin \theta - 1) = 0 \quad \text{Factor.}$$
$$2 \sin \theta + 1 = 0 \quad \text{or} \quad \sin \theta - 1 = 0 \quad \text{Zero-factor property}$$
$$\sin \theta = -\frac{1}{2} \quad \text{or} \quad \sin \theta = 1$$

Over the interval $[0°, 360°)$, the equation $\sin \theta = -\frac{1}{2}$ has two solutions, the angles in quadrants III and IV that have reference angle 30°. These are 210° and 330°. In the same interval, the only angle θ for which $\sin \theta = 1$ is 90°.

Solution set: {90°, 210°, 330°}

49.
$$\frac{2 \tan \theta}{3 - \tan^2 \theta} = 1$$
$$2 \tan \theta = 3 - \tan^2 \theta$$
$$\tan^2 \theta + 2 \tan \theta - 3 = 0$$
$$(\tan \theta - 1)(\tan \theta + 3) = 0$$
$$\tan \theta = 1 \quad \text{or} \quad \tan \theta = -3$$

Over the interval $[0°, 360°)$, the equation $\tan \theta = 1$ has two solutions, 45° and 225°. Over the same interval, the equation $\tan \theta = -3$ has two solutions that are approximately $-71.6° + 180° = 108.4°$ and $-71.6° + 360° = 288.4°$.

The period of the tangent function is 180°, so all solutions of the given equation are $45° + n \cdot 180°$ and $108.4° + n \cdot 180°$, where n is any integer.

77.
$$2 \sin \theta = 2 \cos 2\theta$$
$$\sin \theta = \cos 2\theta \quad \text{Divide by 2.}$$
$$\sin \theta = 1 - 2 \sin^2 \theta \quad \text{Cosine double-angle identity}$$
$$2 \sin^2 \theta + \sin \theta - 1 = 0$$
$$(2 \sin \theta - 1)(\sin \theta + 1) = 0$$
$$2 \sin \theta - 1 = 0 \quad \text{or} \quad \sin \theta + 1 = 0 \quad \text{Zero-factor property}$$
$$\sin \theta = \frac{1}{2} \quad \text{or} \quad \sin \theta = -1$$

Over the interval $[0°, 360°)$, the equation $\sin \theta = \frac{1}{2}$ has two solutions, 30° and 150°. Over the same interval, the equation $\sin \theta = -1$ has one solution, 270°.

Solution set: {30°, 150°, 270°}

7.7 Exercises *(page 672)*

33. $\arccos x + 2 \arcsin \frac{\sqrt{3}}{2} = \pi$

$$\arccos x = \pi - 2 \arcsin \frac{\sqrt{3}}{2}$$
$$\arccos x = \pi - 2\left(\frac{\pi}{3}\right) \quad \arcsin \tfrac{\sqrt{3}}{2} = \tfrac{\pi}{3}$$
$$\arccos x = \pi - \frac{2\pi}{3}$$
$$\arccos x = \frac{\pi}{3}$$
$$x = \cos \frac{\pi}{3} \quad \text{Definition of arccosine}$$
$$x = \frac{1}{2}$$

Solution set: $\left\{\frac{1}{2}\right\}$

37. $\cos^{-1} x + \tan^{-1} x = \frac{\pi}{2}$

$$\cos^{-1} x = \frac{\pi}{2} - \tan^{-1} x$$
$$x = \cos\left(\frac{\pi}{2} - \tan^{-1} x\right) \quad \text{Definition of } \cos^{-1} x$$
$$x = \cos \frac{\pi}{2} \cdot \cos(\tan^{-1} x) + \sin \frac{\pi}{2} \cdot \sin(\tan^{-1} x) \quad \text{Cosine difference identity}$$
$$x = 0 \cdot \cos(\tan^{-1} x) + 1 \cdot \sin(\tan^{-1} x) \quad \cos \tfrac{\pi}{2} = 0;\ \sin \tfrac{\pi}{2} = 1$$
$$x = \sin(\tan^{-1} x)$$

Let $u = \tan^{-1} x$, so $\tan u = x$.

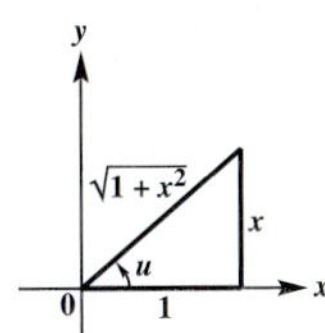

From the triangle, we find that $\sin u = \frac{x}{\sqrt{1 + x^2}}$, so the equation $x = \sin(\tan^{-1} x)$ becomes $x = \frac{x}{\sqrt{1 + x^2}}$. Solve this equation.

(continued)

$$x = \frac{x}{\sqrt{1+x^2}}$$

$x\sqrt{1+x^2} = x$ Multiply by $\sqrt{1+x^2}$.

$x\sqrt{1+x^2} - x = 0$

$x\left(\sqrt{1+x^2} - 1\right) = 0$ Factor.

$x = 0$ or $\sqrt{1+x^2} - 1 = 0$ Zero-factor property

$\sqrt{1+x^2} = 1$ Isolate the radical.

$1 + x^2 = 1$ Square both sides.

$x^2 = 0$

$x = 0$

Solution set: $\{0\}$

CHAPTER 8 APPLICATIONS OF TRIGONOMETRY

8.1 Exercises *(page 695)*

25. $\dfrac{\sin B}{b} = \dfrac{\sin A}{a}$ Alternative form of the law of sines

$\dfrac{\sin B}{2} = \dfrac{\sin 60°}{\sqrt{6}}$ Substitute values from the figure.

$\sin B = \dfrac{2 \sin 60°}{\sqrt{6}}$ Multiply by 2.

$\sin B = \dfrac{2 \cdot \frac{\sqrt{3}}{2}}{\sqrt{6}}$ $\sin 60° = \frac{\sqrt{3}}{2}$

$= \dfrac{\sqrt{3}}{\sqrt{6}} = \sqrt{\dfrac{1}{2}} = \dfrac{\sqrt{2}}{2}$

$B = 45°$ $\sin 45° = \frac{\sqrt{2}}{2}$

There is another angle between 0° and 180° whose sine is $\frac{\sqrt{2}}{2}$: $180° - 45° = 135°$. However, this is too large because $A = 60°$ and $60° + 135° = 195° > 180°$, so there is only one solution, $B = 45°$.

31. $A = 142.13°$, $b = 5.432$ ft, $a = 7.297$ ft

$\dfrac{\sin B}{b} = \dfrac{\sin A}{a}$ Alternative form of the law of sines

$\sin B = \dfrac{b \sin A}{a}$

$\sin B = \dfrac{5.432 \sin 142.13°}{7.297}$ Substitute given values.

$\sin B \approx .45697580$

$B \approx 27.19°$ Use the inverse sine function.

Because angle A is obtuse, angle B must be acute, so this is the only possible value for B and there is one triangle with the given measurements.

$C = 180° - A - B$ Sum of the angles of any triangle is 180°.

$= 180° - 142.13° - 27.19°$

$C = 10.68°$

Thus, $B \approx 27.19°$ and $C \approx 10.68°$.

53. We cannot find θ directly because the length of the side opposite angle θ is not given. Redraw the triangle shown in the figure and label the third angle as α.

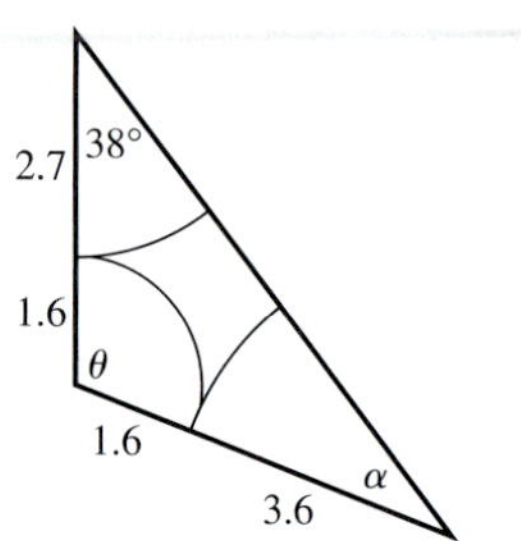

$\dfrac{\sin \alpha}{1.6 + 2.7} = \dfrac{\sin 38°}{1.6 + 3.6}$ Alternative form of the law of sines

$\dfrac{\sin \alpha}{4.3} = \dfrac{\sin 38°}{5.2}$

$\sin \alpha = \dfrac{4.3 \sin 38°}{5.2} \approx .50910468$

$\alpha \approx 31°$

Then $\theta = 180° - 38° - 31°$

$\theta \approx 111°$.

8.2 Exercises *(page 707)*

21. $C = 45.6°$, $b = 8.94$ m, $a = 7.23$ m

First find c.

$c^2 = a^2 + b^2 - 2ab \cos C$ Law of cosines

$c^2 = 7.23^2 + 8.94^2 - 2(7.23)(8.94) \cos 45.6°$ Substitute given values.

$c^2 \approx 41.7493$

$c \approx 6.46$

Find A next since angle A is smaller than angle B (because $a < b$), and thus angle A must be acute.

$\dfrac{\sin A}{a} = \dfrac{\sin C}{c}$ Alternative form of the law of sines

$\sin A = \dfrac{a \sin C}{c}$

$\sin A = \dfrac{7.23 \sin 45.6°}{6.46}$

$\sin A \approx .79963428$

$A \approx 53.1°$

Finally, find B.

$$B = 180° - C - A$$
$$= 180° - 45.6° - 53.1°$$
$$B = 81.3°$$

Thus, $c \approx 6.46$ m, $A \approx 53.1°$, and $B \approx 81.3°$.

41. Find AC, or b, in this figure.

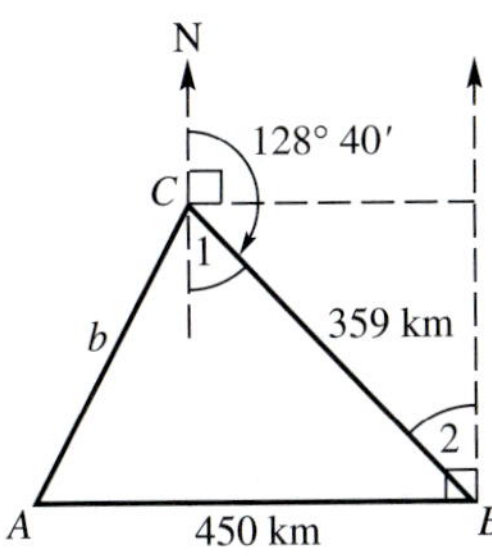

Angle 1 $= 180° - 128° 40' = 51° 20'$

Angles 1 and 2 are alternate interior angles formed when two parallel lines (the north lines) are cut by a transversal, line BC, so angle 2 = angle 1 $= 51° 20'$.

angle $ABC = 90° -$ angle 2 $= 90° - 51° 20' = 38° 40'$ Complementary angles

$$b^2 = a^2 + c^2 - 2ac \cos B \qquad \text{Law of cosines}$$

$$b^2 = 359^2 + 450^2 - 2(359)(450) \cos 38° 40' \qquad \text{Substitute values from the figure.}$$

$$b^2 \approx 79{,}106$$
$$b \approx 281$$

C is about 281 km from A.

8.3 Exercises *(page 721)*

19. Use the figure to find the components of **a** and **b**: $\mathbf{a} = \langle -8, 8 \rangle$ and $\mathbf{b} = \langle 4, 8 \rangle$.

(a) $\mathbf{a} + \mathbf{b} = \langle -8, 8 \rangle + \langle 4, 8 \rangle = \langle -8 + 4, 8 + 8 \rangle = \langle -4, 16 \rangle$

(b) $\mathbf{a} - \mathbf{b} = \langle -8, 8 \rangle - \langle 4, 8 \rangle = \langle -8 - 4, 8 - 8 \rangle$
$= \langle -12, 0 \rangle$

(c) $-\mathbf{a} = -\langle -8, 8 \rangle = \langle 8, -8 \rangle$

47. $\mathbf{v} = \langle a, b \rangle = \langle 5 \cos(-35°), 5 \sin(-35°) \rangle \approx \langle 4.0958, -2.8679 \rangle$

81. First write the given vectors in component form.

$$3\mathbf{i} + 4\mathbf{j} = \langle 3, 4 \rangle; \quad \mathbf{j} = \langle 0, 1 \rangle$$

$$\cos \theta = \frac{\langle 3, 4 \rangle \cdot \langle 0, 1 \rangle}{|\langle 3, 4 \rangle| \, |\langle 0, 1 \rangle|}$$

$$= \frac{3(0) + 4(1)}{\sqrt{9 + 16} \cdot \sqrt{0 + 1}}$$

$$= \frac{4}{5 \cdot 1} = \frac{4}{5} = .8$$

$$\theta = \cos^{-1} .8 \approx 36.87°$$

8.4 Exercises *(page 727)*

5. Use the parallelogram rule. In the figure, **x** represents the second force and **v** is the resultant.

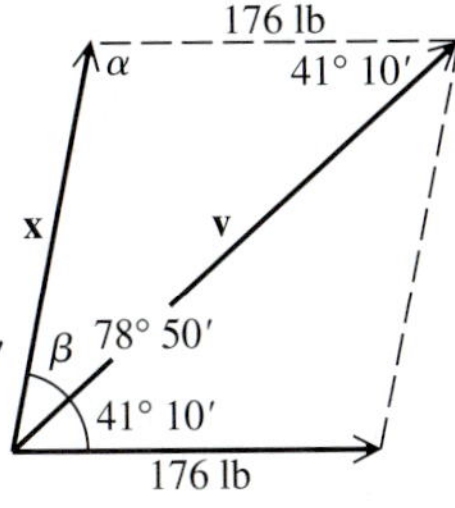

$\alpha = 180° - 78° 50' = 101° 10'$

$\beta = 78° 50' - 41° 10' = 37° 40'$

$$\frac{|\mathbf{x}|}{\sin 41° 10'} = \frac{176}{\sin 37° 40'} \qquad \text{Law of sines}$$

$$|\mathbf{x}| = \frac{176 \sin 41° 10'}{\sin 37° 40'} \approx 190$$

$$\frac{|\mathbf{v}|}{\sin \alpha} = \frac{176}{\sin \beta} \qquad \text{Law of sines}$$

$$|\mathbf{v}| = \frac{176 \sin 101° 10'}{\sin 37° 40'} \approx 283$$

Thus, the magnitude of the second force is about 190 lb and the magnitude of the resultant is about 283 lb.

27. Let **v** represent the airspeed vector.

The groundspeed is $\dfrac{400 \text{ mi}}{2.5 \text{ hr}} = 160$ mph.

angle $BAC = 328° - 180° = 148°$

$$|\mathbf{v}|^2 = 11^2 + 160^2 - 2(11)(160) \cos 148° \qquad \text{Law of cosines}$$

$$|\mathbf{v}|^2 \approx 28{,}706$$
$$|\mathbf{v}| \approx 170$$

The airspeed must be approximately 170 mph.

$$\frac{\sin B}{11} = \frac{\sin 148°}{170} \qquad \text{Law of sines}$$

$$\sin B = \frac{11 \sin 148°}{170} \approx .034289$$

$$B \approx 2°$$

The bearing must be approximately $360° - 2° = 358°$.

8.5 Exercises *(page 739)*

21. $3 \text{ cis } 150° = 3(\cos 150° + i \sin 150°)$

$$= 3\left(-\frac{\sqrt{3}}{2} + i \cdot \frac{1}{2}\right) \qquad \cos 150° = -\tfrac{\sqrt{3}}{2};\ \sin 150° = \tfrac{1}{2}$$

$$= -\frac{3\sqrt{3}}{2} + \frac{3}{2}i \qquad \text{Rectangular form}$$

57. $\dfrac{-i}{1 + i}$

Numerator: $-i = 0 - 1i$

$r = \sqrt{0^2 + (-1)^2} = 1$

$\theta = 270°$ since $\cos 270° = 0$ and $\sin 270° = -1$, so $-i = 1 \text{ cis } 270°$.

(continued)

Denominator: $1 + i = 1 + 1i$

$r = \sqrt{1^2 + 1^2} = \sqrt{2}$

$\tan\theta = \frac{y}{x} = \frac{1}{1} = 1$

Since x and y are both positive, θ is in quadrant I, so $\theta = \tan^{-1} 1 = 45°$. Thus, $1 + i = \sqrt{2}\text{ cis } 45°$.

$$\frac{-i}{1+i} = \frac{1 \text{ cis } 270°}{\sqrt{2}\text{ cis } 45°}$$

$$= \frac{1}{\sqrt{2}}\text{cis}(270° - 45°) \quad \text{Quotient theorem}$$

$$= \frac{\sqrt{2}}{2}\text{ cis } 225°$$

$$= \frac{\sqrt{2}}{2}(\cos 225° + i \sin 225°)$$

$$= \frac{\sqrt{2}}{2}\left(-\frac{\sqrt{2}}{2} - i \cdot \frac{\sqrt{2}}{2}\right) \quad \cos 225° = -\tfrac{\sqrt{2}}{2};\ \sin 225° = -\tfrac{\sqrt{2}}{2}$$

$$= -\frac{1}{2} - \frac{1}{2}i \quad \text{Rectangular form}$$

8.6 Exercises *(page 746)*

11. $(-2 - 2i)^5$

First write $-2 - 2i$ in trigonometric form.

$$r = \sqrt{(-2)^2 + (-2)^2} = \sqrt{8} = 2\sqrt{2}$$

$$\tan\theta = \frac{y}{x} = \frac{-2}{-2} = 1$$

Because x and y are both negative, θ is in quadrant III, so $\theta = 225°$.

$$-2 - 2i = 2\sqrt{2}(\cos 225° + i \sin 225°)$$

$$(-2 - 2i)^5 = \left[2\sqrt{2}(\cos 225° + i \sin 225°)\right]^5$$

$$= (2\sqrt{2})^5[\cos(5 \cdot 225°) + i \sin(5 \cdot 225°)]$$

De Moivre's theorem

$$= 32 \cdot 4\sqrt{2}(\cos 1125° + i \sin 1125°)$$

$$= 128\sqrt{2}(\cos 1125° + i \sin 1125°)$$

$$= 128\sqrt{2}(\cos 45° + i \sin 45°)$$

1125° and 45° are coterminal.

$$= 128\sqrt{2}\left(\frac{\sqrt{2}}{2} + i \cdot \frac{\sqrt{2}}{2}\right)$$

$\cos 45° = \frac{\sqrt{2}}{2}$; $\sin 45° = \frac{\sqrt{2}}{2}$

$$= 128 + 128i \quad \text{Rectangular form}$$

41. $x^3 - (4 + 4i\sqrt{3}) = 0$

$$x^3 = 4 + 4i\sqrt{3}$$

$$r = \sqrt{4^2 + (4\sqrt{3})^2} = \sqrt{16 + 48} = \sqrt{64} = 8$$

$$\tan\theta = \frac{4\sqrt{3}}{4} = \sqrt{3}$$

θ is in quadrant I, so $\theta = 60°$.

$$x^3 = 4 + 4i\sqrt{3}$$

$$x^3 = 8\left(\frac{1}{2} + i\frac{\sqrt{3}}{2}\right)$$

$$r^3(\cos 3\alpha + i \sin 3\alpha) = 8(\cos 60° + i \sin 60°)$$

$r^3 = 8$, so $r = 2$.

$$\alpha = \frac{60°}{3} + \frac{360° \cdot k}{3},\ k \text{ any integer} \quad n\text{th root theorem}$$

$\alpha = 20° + 120° \cdot k$, k any integer

If $k = 0$, then $\alpha = 20° + 0° = 20°$.

If $k = 1$, then $\alpha = 20° + 120° = 140°$.

If $k = 2$, then $\alpha = 20° + 240° = 260°$.

Solution set: $\{2(\cos 20° + i \sin 20°),$ $2(\cos 140° + i \sin 140°), 2(\cos 260° + i \sin 260°)\}$

8.7 Exercises *(page 757)*

53.

$r = 2\sin\theta$	
$r^2 = 2r\sin\theta$	Multiply by r.
$x^2 + y^2 = 2y$	$r^2 = x^2 + y^2$, $r\sin\theta = y$
$x^2 + y^2 - 2y = 0$	Subtract $2y$.
$x^2 + y^2 - 2y + 1 = 1$	Add 1 to complete the square on y.
$x^2 + (y - 1)^2 = 1$	Factor the perfect square trinomial.

The graph is a circle with center (0, 1) and radius 1.

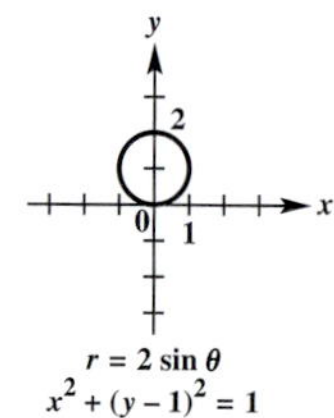

59.

$r = 2\sec\theta$	
$r = \frac{2}{\cos\theta}$	Reciprocal identity
$r\cos\theta = 2$	Multiply by $\cos\theta$.
$x = 2$	$r\cos\theta = x$

The graph is the vertical line through (2, 0).

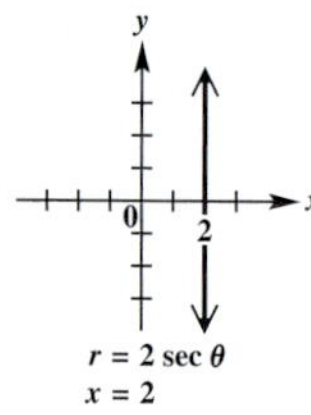

8.8 Exercises *(page 766)*

9. $x = t^3 + 1,\ y = t^3 - 1,\quad$ for t in $(-\infty, \infty)$

(a)

t	x	y
-2	-7	-9
-1	0	-2
0	1	-1
1	2	0
2	9	7
3	28	26

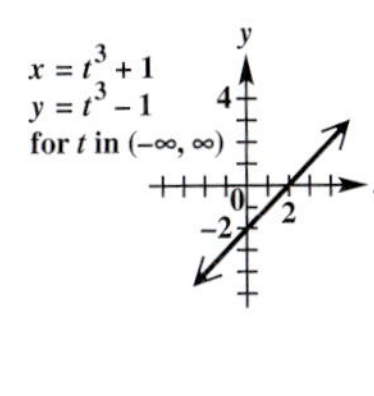

(b)
$$x = t^3 + 1$$
$$y = t^3 - 1$$
$$x - y = 2 \quad \text{Subtract equations to eliminate } t.$$
$$y = x - 2 \quad \text{Solve for } y.$$

The rectangular equation is $y = x - 2$, for x in $(-\infty, \infty)$. The graph is a line with slope 1 and y-intercept -2.

13. $x = 3 \tan t,\ y = 2 \sec t,\quad$ for t in $\left(-\frac{\pi}{2}, \frac{\pi}{2}\right)$

(a)

t	x	y
$-\frac{\pi}{3}$	$-3\sqrt{3} \approx -5.2$	4
$-\frac{\pi}{6}$	$-\sqrt{3} \approx -1.7$	$\frac{4\sqrt{3}}{3} \approx 2.3$
0	0	2
$\frac{\pi}{6}$	$\sqrt{3} \approx 1.7$	$\frac{4\sqrt{3}}{3} \approx 2.3$
$\frac{\pi}{3}$	$3\sqrt{3} \approx 5.2$	4

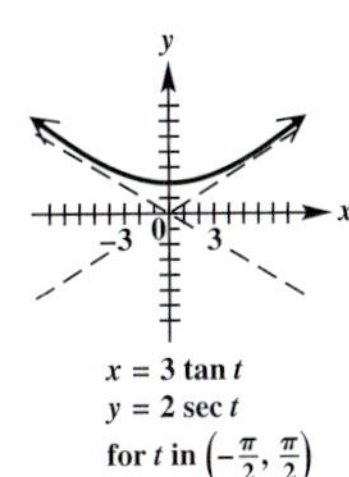

(b) $x = 3 \tan t$, so $\frac{x}{3} = \tan t$.

$y = 2 \sec t$, so $\frac{y}{2} = \sec t$.

$$1 + \tan^2 t = \sec^2 t \quad \text{Pythagorean identity}$$
$$1 + \left(\frac{x}{3}\right)^2 = \left(\frac{y}{2}\right)^2 \quad \text{Substitute expressions for } \tan t \text{ and } \sec t.$$
$$1 + \frac{x^2}{9} = \frac{y^2}{4}$$
$$y^2 = 4\left(1 + \frac{x^2}{9}\right) \quad \text{Multiply by 4.}$$
$$y = 2\sqrt{1 + \frac{x^2}{9}} \quad \text{Use the positive square root because } y > 0 \text{ in the given interval for } t.$$

The rectangular equation is $y = 2\sqrt{1 + \frac{x^2}{9}}$, for x in $(-\infty, \infty)$. The graph is the upper half of a hyperbola. (See Chapter 10.)

CHAPTER 9 SYSTEMS AND MATRICES

9.1 Exercises *(page 794)*

27. $\frac{x}{2} + \frac{y}{3} = 4 \quad (1)$

$\frac{3x}{2} + \frac{3y}{2} = 15 \quad (2)$

It is easier to work with equations that do not contain fractions. To clear fractions, multiply each equation by the LCD of all the fractions in the equation.

$$6\left(\frac{x}{2} + \frac{y}{3}\right) = 6(4) \quad \text{Multiply by the LCD, 6.}$$
$$2\left(\frac{3x}{2} + \frac{3y}{2}\right) = 2(15) \quad \text{Multiply by the LCD, 2.}$$

This gives the system

$$3x + 2y = 24 \quad (3)$$
$$3x + 3y = 30. \quad (4)$$

To solve this system by elimination, multiply equation (3) by -1 and add the result to equation (4), eliminating x.

$$-3x - 2y = -24$$
$$3x + 3y = 30$$
$$y = 6$$

Substitute 6 for y in equation (1).

$$\frac{x}{2} + \frac{6}{3} = 4 \quad \text{Let } y = 6.$$
$$\frac{x}{2} + 2 = 4$$
$$\frac{x}{2} = 2$$
$$x = 4$$

Solution set: $\{(4, 6)\}$

69. $\frac{2}{x} + \frac{1}{y} = \frac{3}{2} \quad (1)$

$\frac{3}{x} - \frac{1}{y} = 1 \quad (2)$

(continued)

Let and With these substitutions, the system becomes

$$2t + u = \frac{3}{2} \quad (3)$$

$$3t - u = 1. \quad (4)$$

Add these equations, eliminating u, and solve for t.

$$5t = \frac{5}{2}$$

$$t = \frac{1}{2} \quad \text{Multiply by } \tfrac{1}{5}.$$

Substitute $\frac{1}{2}$ for t in equation (3) and solve for u.

$$2\left(\frac{1}{2}\right) + u = \frac{3}{2} \quad \text{Let } t = \tfrac{1}{2}.$$

$$1 + u = \frac{3}{2}$$

$$u = \frac{1}{2}$$

Now find the values of x and y, the variables in the original system. Since $t = \frac{1}{x}$, $tx = 1$, and $x = \frac{1}{t}$. Likewise, $y = \frac{1}{u}$.

$$x = \frac{1}{t} = \frac{1}{\frac{1}{2}} = 2 \quad \text{Let } t = \tfrac{1}{2}.$$

$$y = \frac{1}{u} = \frac{1}{\frac{1}{2}} = 2 \quad \text{Let } u = \tfrac{1}{2}.$$

Solution set: $\{(2, 2)\}$

9.2 Exercises *(page 808)*

45. $$\frac{1}{(x-1)(x+1)} = \frac{A}{x-1} + \frac{B}{x+1}$$

$$\frac{1}{(x-1)(x+1)}$$

$$= \frac{A(x+1)}{(x-1)(x+1)} + \frac{B(x-1)}{(x-1)(x+1)} \quad \text{Write terms on the right with the LCD, } (x-1)(x+1).$$

$$= \frac{A(x+1) + B(x-1)}{(x-1)(x+1)} \quad \text{Combine terms on the right.}$$

$$1 = A(x+1) + B(x-1) \quad \text{Denominators are equal, so numerators must be equal.}$$

$$1 = Ax + A + Bx - B \quad \text{Distributive property}$$

$$1 = (Ax + Bx) + (A - B) \quad \text{Group like terms.}$$

$$1 = (A + B)x + (A - B) \quad \text{Factor } Ax + Bx.$$

Since $1 = 0x + 1$, we can equate the coefficients of like powers of x to obtain the system

$$A + B = 0$$

$$A - B = 1.$$

Solve this system by the Gauss-Jordan method.

$$\left[\begin{array}{cc|c} 1 & 1 & 0 \\ 1 & -1 & 1 \end{array}\right]$$

$$\left[\begin{array}{cc|c} 1 & 1 & 0 \\ 0 & -2 & 1 \end{array}\right] \quad -\text{R1} + \text{R2}$$

$$\left[\begin{array}{cc|c} 1 & 1 & 0 \\ 0 & 1 & -\frac{1}{2} \end{array}\right] \quad -\tfrac{1}{2}\text{R2}$$

$$\left[\begin{array}{cc|c} 1 & 0 & \frac{1}{2} \\ 0 & 1 & -\frac{1}{2} \end{array}\right] \quad -\text{R2} + \text{R1}$$

From the final matrix, $A = \frac{1}{2}$ and $B = -\frac{1}{2}$.

49. Let x = number of cubic centimeters of 2% solution and y = number of cubic centimeters of 7% solution. From the given information, we can write the system

$$x + y = 40$$

$$.02x + .07y = .032(40)$$

or

$$x + y = 40$$

$$.02x + .07y = 1.28.$$

Solve this system by the Gauss-Jordan method.

$$\left[\begin{array}{cc|c} 1 & 1 & 40 \\ .02 & .07 & 1.28 \end{array}\right]$$

$$\left[\begin{array}{cc|c} 1 & 1 & 40 \\ 0 & .05 & .48 \end{array}\right] \quad -.02\text{R1} + \text{R2}$$

$$\left[\begin{array}{cc|c} 1 & 1 & 40 \\ 0 & 1 & 9.6 \end{array}\right] \quad \tfrac{1}{.05}\text{R2 (or 20R2)}$$

$$\left[\begin{array}{cc|c} 1 & 0 & 30.4 \\ 0 & 1 & 9.6 \end{array}\right] \quad -\text{R2} + \text{R1}$$

From the final matrix, $x = 30.4$ and $y = 9.6$, so the chemist should mix 30.4 cm^3 of the 2% solution with 9.6 cm^3 of the 7% solution.

9.3 Exercises *(page 820)*

53. $$\begin{vmatrix} -4 & 1 & 4 \\ 2 & 0 & 1 \\ 0 & 2 & 4 \end{vmatrix} = \begin{vmatrix} 0 & 1 & 6 \\ 2 & 0 & 1 \\ 0 & 2 & 4 \end{vmatrix} \quad \text{Use determinant theorem 6; 2R2 + R1.}$$

$$= 0\begin{vmatrix} 0 & 1 \\ 2 & 4 \end{vmatrix} - 2\begin{vmatrix} 1 & 6 \\ 2 & 4 \end{vmatrix} + 0\begin{vmatrix} 1 & 6 \\ 0 & 1 \end{vmatrix} \quad \text{Expand about column 1.}$$

$$= 0 - 2(4 - 12) + 0$$

$$= -2(-8) = 16$$

67. $$1.5x + 3y = 5 \quad (1)$$

$$2x + 4y = 3 \quad (2)$$

$$D = \begin{vmatrix} 1.5 & 3 \\ 2 & 4 \end{vmatrix} = 1.5(4) - 2(3) = 6 - 6 = 0$$

Because $D = 0$, Cramer's rule does not apply. To determine whether the system is inconsistent or has infinitely many solutions, use the elimination method.

$$\begin{aligned} 6x + 12y &= 20 && \text{Multiply equation (1) by 4.} \\ -6x - 12y &= -9 && \text{Multiply equation (2) by } -3. \\ \hline 0 &= 11 && \text{False} \end{aligned}$$

The system is inconsistent.

Solution set: $\emptyset$

9.4 Exercises *(page 829)*

13. $\dfrac{x^2}{x^2 + 2x + 1}$

This is not a proper fraction; the numerator has degree greater than or equal to that of the denominator. Divide the numerator by the denominator.

$$\begin{array}{r} 1 \\ x^2 + 2x + 1 \overline{)\,x^2 \qquad\qquad} \\ \underline{x^2 + 2x + 1} \\ -2x - 1 \end{array}$$

Find the partial fraction decomposition for

$$\frac{-2x - 1}{x^2 + 2x + 1} = \frac{-2x - 1}{(x + 1)^2}.$$

$$\frac{-2x - 1}{(x + 1)^2} = \frac{A}{x + 1} + \frac{B}{(x + 1)^2} \quad \text{Factor the denominator of the given fraction.}$$

$$-2x - 1 = A(x + 1) + B \quad \text{Multiply by the LCD, } (x + 1)^2.$$

Use this equation to find the value of B.

$$\begin{aligned} -2(-1) - 1 &= A(-1 + 1) + B && \text{Let } x = -1. \\ 1 &= B \end{aligned}$$

Now use the same equation and the value of B to find the value of A.

$$\begin{aligned} -2(2) - 1 &= A(2 + 1) + 1 && \text{Let } x = 2 \text{ and } B = 1. \\ -5 &= 3A + 1 \\ -6 &= 3A \\ A &= -2 \end{aligned}$$

Thus,

$$\frac{x^2}{x^2 + 2x + 1} = 1 + \frac{-2}{x + 1} + \frac{1}{(x + 1)^2}.$$

23. $\dfrac{1}{x(2x + 1)(3x^2 + 4)}$

The denominator contains two linear factors and one quadratic factor. All factors are distinct, and $3x^2 + 4$ cannot be factored, so it is irreducible. The partial fraction decomposition is of the form

$$\frac{1}{x(2x + 1)(3x^2 + 4)} = \frac{A}{x} + \frac{B}{2x + 1} + \frac{Cx + D}{3x^2 + 4}.$$

We need to find the values of A, B, C, and D.

$$\begin{aligned} 1 &= A(2x + 1)(3x^2 + 4) + Bx(3x^2 + 4) \\ &\quad + (Cx + D)(x)(2x + 1) && \text{Multiply by the LCD, } x(2x + 1)(3x^2 + 4). \\ 1 &= A(6x^3 + 3x^2 + 8x + 4) + B(3x^3 + 4x) \\ &\quad + C(2x^3 + x^2) + D(2x^2 + x) && \text{Multiply on the right.} \\ 1 &= 6Ax^3 + 3Ax^2 + 8Ax + 4A + 3Bx^3 + 4Bx \\ &\quad + 2Cx^3 + Cx^2 + 2Dx^2 + Dx && \text{Distributive property} \\ 1 &= (6A + 3B + 2C)x^3 + (3A + C + 2D)x^2 \\ &\quad + (8A + 4B + D)x + 4A && \text{Collect like terms.} \end{aligned}$$

Since $1 = 0x^3 + 0x^2 + 0x + 1$, equating the coefficients of like powers of x produces the following system of equations:

$$\begin{aligned} 6A + 3B + 2C &= 0 \\ 3A + C + 2D &= 0 \\ 8A + 4B + D &= 0 \\ 4A &= 1. \end{aligned}$$

Any of the methods from this chapter can be used to solve this system. If the Gauss-Jordan method is used, begin by representing the system with the augmented matrix

$$\left[\begin{array}{cccc|c} 6 & 3 & 2 & 0 & 0 \\ 3 & 0 & 1 & 2 & 0 \\ 8 & 4 & 0 & 1 & 0 \\ 4 & 0 & 0 & 0 & 1 \end{array}\right].$$

After performing a series of row operations, we obtain the following final matrix:

$$\left[\begin{array}{cccc|c} 1 & 0 & 0 & 0 & \frac{1}{4} \\ 0 & 1 & 0 & 0 & -\frac{8}{19} \\ 0 & 0 & 1 & 0 & -\frac{9}{76} \\ 0 & 0 & 0 & 1 & -\frac{6}{19} \end{array}\right],$$

from which we read the values of the four variables: $A = \frac{1}{4}$, $B = -\frac{8}{19}$, $C = -\frac{9}{76}$, $D = -\frac{6}{19}$.

Substitute these values into the form given at the beginning of this solution for the partial fraction decomposition.

$$\begin{aligned} \frac{1}{x(2x + 1)(3x^2 + 4)} &= \frac{A}{x} + \frac{B}{2x + 1} + \frac{Cx + D}{3x^2 + 4} \\ \frac{1}{x(2x + 1)(3x^2 + 4)} &= \frac{\frac{1}{4}}{x} + \frac{-\frac{8}{19}}{2x + 1} + \frac{-\frac{9}{76}x - \frac{6}{19}}{3x^2 + 4} \\ &= \frac{\frac{1}{4}}{x} + \frac{-\frac{8}{19}}{2x + 1} + \frac{-\frac{9}{76}x - \frac{24}{76}}{3x^2 + 4} \end{aligned}$$

Get a common denominator for the numerator of the last term.

$$= \frac{1}{4x} + \frac{-8}{19(2x + 1)} + \frac{-9x - 24}{76(3x^2 + 4)}$$

Simplify the complex fractions.

9.5 Exercises *(page 836)*

51. Let x and y represent the two numbers.

$$\frac{x}{y} = \frac{9}{2} \qquad (1)$$

$$xy = 162 \qquad (2)$$

Solve equation (1) for x.

$$y\left(\frac{x}{y}\right) = y\left(\frac{9}{2}\right) \qquad \text{Multiply by } y.$$

$$x = \frac{9}{2}y$$

Substitute $\frac{9}{2}y$ for x in equation (2).

$$\left(\frac{9}{2}y\right)y = 162$$

$$\frac{9}{2}y^2 = 162$$

$$\frac{2}{9}\left(\frac{9}{2}y^2\right) = \frac{2}{9}(162) \qquad \text{Multiply by } \tfrac{2}{9}.$$

$$y^2 = 36$$

$$y = \pm 6$$

If $y = 6$, then $x = \frac{9}{2}(6) = 27$. If $y = -6$, then $x = \frac{9}{2}(-6) = -27$. The two numbers are either 27 and 6, or -27 and -6.

63. supply: $p = \sqrt{.1q + 9} - 2$

demand: $p = \sqrt{25 - .1q}$

(a) Equilibrium occurs when supply = demand, so solve the system formed by the supply and demand equations. This system can be solved by substitution. Substitute $\sqrt{.1q + 9} - 2$ for p in the demand equation and solve the resulting equation for q.

$$\sqrt{.1q + 9} - 2 = \sqrt{25 - .1q}$$

$$\left(\sqrt{.1q + 9} - 2\right)^2 = \left(\sqrt{25 - .1q}\right)^2$$

Square both sides.

$$.1q + 9 - 4\sqrt{.1q + 9} + 4 = 25 - .1q$$

$(x - y)^2 = x^2 - 2xy + y^2$

$$.2q - 12 = 4\sqrt{.1q + 9}$$

Combine like terms; isolate radical.

$$(.2q - 12)^2 = \left(4\sqrt{.1q + 9}\right)^2$$

Square both sides again.

$$.04q^2 - 4.8q + 144 = 16(.1q + 9)$$

$$.04q^2 - 4.8q + 144 = 1.6q + 144$$

Distributive property

$$.04q^2 - 6.4q = 0 \qquad \text{Combine like terms.}$$

$$.04q(q - 160) = 0 \qquad \text{Factor.}$$

$$q = 0 \quad \text{or} \quad q = 160 \qquad \text{Zero-factor property}$$

Disregard an equilibrium demand of 0. The equilibrium demand is 160 units.

(b) Substitute 160 for q in either equation and solve for p. Using the supply equation, we obtain

$$\begin{aligned} p &= \sqrt{.1q + 9} - 2 \\ &= \sqrt{.1(160) + 9} - 2 \qquad \text{Let } q = 160. \\ &= \sqrt{16 + 9} - 2 \\ &= \sqrt{25} - 2 \\ &= 5 - 2 = 3. \end{aligned}$$

The equilibrium price is \$3.

9.6 Exercises *(page 848)*

53. $y \le \log x$

$y \ge |x - 2|$

Graph $y = \log x$ as a solid curve because $y \le \log x$ is a nonstrict inequality.

(Recall that "$\log x$" means $\log_{10} x$.) This graph contains the points $(.1, -1)$ and $(1, 0)$ because $10^{-1} = .1$ and $10^0 = 1$. Use a calculator to approximate other points on the graph, such as (2, .30) and (4, .60). Because the symbol is $\le$, shade the region *below* the curve.

Now graph $y = |x - 2|$. Make this boundary solid because $y \ge |x - 2|$ is also a nonstrict inequality. This graph can be obtained by translating the graph of $y = |x|$ to the right 2 units. It contains the points (0, 2), (2, 0), and (4, 2). Because the symbol is $\ge$, shade the region *above* the absolute value graph.

The solution set of the system is the intersection (or overlap) of the two shaded regions, which is shown in the final graph.

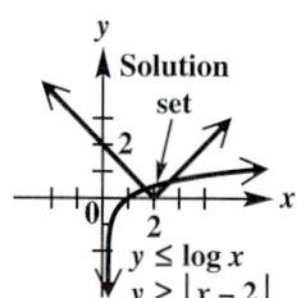

75. Let x = number of cabinet A

and y = number of cabinet B.

The cost constraint is

$$10x + 20y \le 140.$$

The space constraint is

$$6x + 8y \le 72.$$

Since the numbers of cabinets cannot be negative, we also have the constraints $x \ge 0$ and $y \ge 0$. We want to maximize the objective function,

$$\text{storage capacity} = 8x + 12y.$$

Find the region of feasible solutions by graphing the system of inequalities that is made up of the constraints.

To graph $10x + 20y \le 140$, draw the line with x-intercept 14 and y-intercept 7 as a solid line. Because the symbol is $\le$, shade the region *below* the line.

To graph $6x + 8y \le 72$, draw the line with x-intercept 12 and y-intercept 9 as a solid line. Because the symbol is $\le$, shade the region *below* the line.

The constraints $x \ge 0$ and $y \ge 0$ restrict the graph to the first quadrant. The graph of the feasible region is the intersection of the regions that are the graphs of the individual constraints.

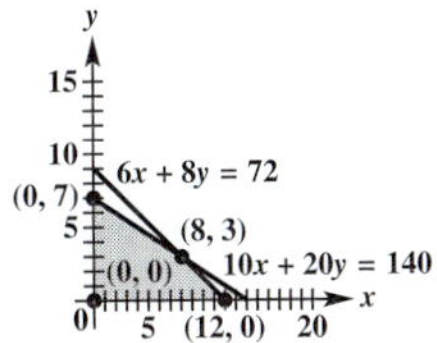

From the graph, observe that three of the vertices are (0, 0), (0, 7), and (12, 0). The fourth vertex is the intersection point of the lines $10x + 20y = 140$ and $6x + 8y = 72$. To find this point, solve the system

$$\begin{aligned} 10x + 20y &= 140 \\ 6x + 8y &= 72 \end{aligned}$$

to obtain the ordered pair (8, 3). Evaluate the objective function at each vertex.

Point	Storage Capacity = $8x + 12y$
(0, 0)	$8(0) + 12(0) = 0$
(0, 7)	$8(0) + 12(7) = 84$
(8, 3)	$8(8) + 12(3) = 100$
(12, 0)	$8(12) + 12(0) = 96$

The maximum value of $8x + 12y$ occurs at the vertex (8, 3), so the office manager should buy 8 of cabinet A and 3 of cabinet B for a total storage capacity of 100 ft^3.

9.8 Exercises *(page 874)*

23. $A = \begin{bmatrix} 1 & 1 & 0 & 2 \\ 2 & -1 & 1 & -1 \\ 3 & 3 & 2 & -2 \\ 1 & 2 & 1 & 0 \end{bmatrix}$

$$\left[\begin{array}{cccc|cccc} 1 & 1 & 0 & 2 & 1 & 0 & 0 & 0 \\ 2 & -1 & 1 & -1 & 0 & 1 & 0 & 0 \\ 3 & 3 & 2 & -2 & 0 & 0 & 1 & 0 \\ 1 & 2 & 1 & 0 & 0 & 0 & 0 & 1 \end{array}\right]$$ Write the augmented matrix $[A|I_4]$.

$$\left[\begin{array}{cccc|cccc} 1 & 1 & 0 & 2 & 1 & 0 & 0 & 0 \\ 0 & -3 & 1 & -5 & -2 & 1 & 0 & 0 \\ 0 & 0 & 2 & -8 & -3 & 0 & 1 & 0 \\ 0 & 1 & 1 & -2 & -1 & 0 & 0 & 1 \end{array}\right]$$ $-2\text{R1} + \text{R2}$; $-3\text{R1} + \text{R3}$; $-1\text{R1} + \text{R4}$

$$\left[\begin{array}{cccc|cccc} 1 & 1 & 0 & 2 & 1 & 0 & 0 & 0 \\ 0 & 1 & -\frac{1}{3} & \frac{5}{3} & \frac{2}{3} & -\frac{1}{3} & 0 & 0 \\ 0 & 0 & 2 & -8 & -3 & 0 & 1 & 0 \\ 0 & 1 & 1 & -2 & -1 & 0 & 0 & 1 \end{array}\right]$$ $-\frac{1}{3}\text{R2}$

$$\left[\begin{array}{cccc|cccc} 1 & 0 & \frac{1}{3} & \frac{1}{3} & \frac{1}{3} & \frac{1}{3} & 0 & 0 \\ 0 & 1 & -\frac{1}{3} & \frac{5}{3} & \frac{2}{3} & -\frac{1}{3} & 0 & 0 \\ 0 & 0 & 2 & -8 & -3 & 0 & 1 & 0 \\ 0 & 0 & \frac{4}{3} & -\frac{11}{3} & -\frac{5}{3} & \frac{1}{3} & 0 & 1 \end{array}\right]$$ $-1\text{R2} + \text{R1}$; $-1\text{R2} + \text{R4}$

$$\left[\begin{array}{cccc|cccc} 1 & 0 & \frac{1}{3} & \frac{1}{3} & \frac{1}{3} & \frac{1}{3} & 0 & 0 \\ 0 & 1 & -\frac{1}{3} & \frac{5}{3} & \frac{2}{3} & -\frac{1}{3} & 0 & 0 \\ 0 & 0 & 1 & -4 & -\frac{3}{2} & 0 & \frac{1}{2} & 0 \\ 0 & 0 & \frac{4}{3} & -\frac{11}{3} & -\frac{5}{3} & \frac{1}{3} & 0 & 1 \end{array}\right]$$ $\frac{1}{2}\text{R3}$

$$\left[\begin{array}{cccc|cccc} 1 & 0 & 0 & \frac{5}{3} & \frac{5}{6} & \frac{1}{3} & -\frac{1}{6} & 0 \\ 0 & 1 & 0 & \frac{1}{3} & \frac{1}{6} & -\frac{1}{3} & \frac{1}{6} & 0 \\ 0 & 0 & 1 & -4 & -\frac{3}{2} & 0 & \frac{1}{2} & 0 \\ 0 & 0 & 0 & \frac{5}{3} & \frac{1}{3} & \frac{1}{3} & -\frac{2}{3} & 1 \end{array}\right]$$ $-\frac{1}{3}\text{R3} + \text{R1}$; $\frac{1}{3}\text{R3} + \text{R2}$; $-\frac{4}{3}\text{R3} + \text{R4}$

$$\left[\begin{array}{cccc|cccc} 1 & 0 & 0 & \frac{5}{3} & \frac{5}{6} & \frac{1}{3} & -\frac{1}{6} & 0 \\ 0 & 1 & 0 & \frac{1}{3} & \frac{1}{6} & -\frac{1}{3} & \frac{1}{6} & 0 \\ 0 & 0 & 1 & -4 & -\frac{3}{2} & 0 & \frac{1}{2} & 0 \\ 0 & 0 & 0 & 1 & \frac{1}{5} & \frac{1}{5} & -\frac{2}{5} & \frac{3}{5} \end{array}\right]$$ $\frac{3}{5}\text{R4}$

$$\left[\begin{array}{cccc|cccc} 1 & 0 & 0 & 0 & \frac{1}{2} & 0 & \frac{1}{2} & -1 \\ 0 & 1 & 0 & 0 & \frac{1}{10} & -\frac{2}{5} & \frac{3}{10} & -\frac{1}{5} \\ 0 & 0 & 1 & 0 & -\frac{7}{10} & \frac{4}{5} & -\frac{11}{10} & \frac{12}{5} \\ 0 & 0 & 0 & 1 & \frac{1}{5} & \frac{1}{5} & -\frac{2}{5} & \frac{3}{5} \end{array}\right]$$ $-\frac{5}{3}\text{R4} + \text{R1}$; $-\frac{1}{3}\text{R4} + \text{R2}$; $4\text{R4} + \text{R3}$

$$A^{-1} = \begin{bmatrix} \frac{1}{2} & 0 & \frac{1}{2} & -1 \\ \frac{1}{10} & -\frac{2}{5} & \frac{3}{10} & -\frac{1}{5} \\ -\frac{7}{10} & \frac{4}{5} & -\frac{11}{10} & \frac{12}{5} \\ \frac{1}{5} & \frac{1}{5} & -\frac{2}{5} & \frac{3}{5} \end{bmatrix}$$

41. $.2x + .3y = -1.9$
$.7x - .2y = 4.6$

$$A = \begin{bmatrix} .2 & .3 \\ .7 & -.2 \end{bmatrix}, \quad X = \begin{bmatrix} x \\ y \end{bmatrix}, \quad B = \begin{bmatrix} -1.9 \\ 4.6 \end{bmatrix}$$

Find A^{-1}.

$$\left[\begin{array}{cc|cc} .2 & .3 & 1 & 0 \\ .7 & -.2 & 0 & 1 \end{array}\right]$$ Write the augmented matrix $[A|I_2]$.

$$\left[\begin{array}{cc|cc} 1 & 1.5 & 5 & 0 \\ .7 & -.2 & 0 & 1 \end{array}\right]$$ 5R1

$$\left[\begin{array}{cc|cc} 1 & 1.5 & 5 & 0 \\ 0 & -1.25 & -3.5 & 1 \end{array}\right]$$ $-.7\text{R1} + \text{R2}$

$$\left[\begin{array}{cc|cc} 1 & 1.5 & 5 & 0 \\ 0 & 1 & 2.8 & -.8 \end{array}\right]$$ $\frac{1}{-1.25}\text{R2}$ or $-.8\text{R2}$

$$\left[\begin{array}{cc|cc} 1 & 0 & .8 & 1.2 \\ 0 & 1 & 2.8 & -.8 \end{array}\right]$$ $-1.5\text{R2} + \text{R1}$

(continued)

$$A^{-1} = \begin{bmatrix} .8 & 1.2 \\ 2.8 & -.8 \end{bmatrix}$$

$$X = A^{-1}B = \begin{bmatrix} .8 & 1.2 \\ 2.8 & -.8 \end{bmatrix}\begin{bmatrix} -1.9 \\ 4.6 \end{bmatrix} = \begin{bmatrix} 4 \\ -9 \end{bmatrix}$$

Solution set: $\{(4, -9)\}$

CHAPTER 10 ANALYTIC GEOMETRY

10.1 Exercises *(page 899)*

29. $(x - 7)^2 = 16(y + 5)$

This equation can be rewritten as

$$(x - 7)^2 = 16[y - (-5)].$$

Thus, it has the form

$$(x - h)^2 = 4p(y - k),$$

with $h = 7$, $k = -5$, and $4p = 16$, so $p = 4$. The graph of the given equation is a parabola with vertical axis. The vertex (h, k) is $(7, -5)$. Because this parabola has a vertical axis and $p > 0$, the parabola opens up, so the focus is distance $p = 4$ units above the vertex. Thus, the focus is the point $(7, -1)$.

The directrix is a horizontal line $p = 4$ units below the vertex, so the directrix is the line $y = -9$. The axis is the vertical line through the vertex, so the equation of the axis is $x = 7$.

37. Through $(3, 2)$, symmetric with respect to the x-axis

This parabola has a horizontal axis (the x-axis) because of the symmetry and vertex $(0, 0)$, so its equation can be written in the form $y^2 = 4px$. Use this equation with the coordinates of the point $(3, 2)$ to find the value of p.

$$y^2 = 4px$$

$$2^2 = 4p \cdot 3 \quad \text{Let } x = 3 \text{ and } y = 2.$$

$$4 = 12p$$

$$p = \frac{1}{3}$$

Thus, the equation of the parabola is

$$y^2 = 4px = 4\left(\frac{1}{3}\right)x = \frac{4}{3}x.$$

53. Place the parabola that represents the arch on a coordinate system with the center of the bottom of the arch at the origin. Then the vertex will be at $(0, 12)$ and the points $(-6, 0)$ and $(6, 0)$ will also be on the parabola.

Because the axis of the parabola is the y-axis and the vertex is $(0, 12)$, the equation will have the form $x^2 = 4p(y - 12)$. Use the coordinates of the point $(6, 0)$ to find the value of p.

$$x^2 = 4p(y - 12)$$

$$6^2 = 4p(0 - 12) \quad \text{Let } x = 6 \text{ and } y = 0.$$

$$36 = -48p$$

$$p = -\frac{3}{4}$$

Thus, the equation of the parabola is

$$x^2 = 4\left(-\frac{3}{4}\right)(y - 12) \quad \text{or} \quad x^2 = -3(y - 12).$$

Now find the x-coordinate of a point whose y-coordinate is 9 and whose x-coordinate is positive.

$$x^2 = -3(y - 12)$$

$$x^2 = -3(9 - 12) \quad \text{Let } y = 9.$$

$$x^2 = 9$$

$$x = \sqrt{9} = 3 \quad x > 0$$

Using symmetry, the width of the arch 9 ft up is $2(3 \text{ ft}) = 6$ ft.

10.2 Exercises *(page 910)*

25. foci at $(0, 4)$, $(0, -4)$; sum of distances from foci to point on ellipse is 10

Center: $(0, 0)$

Distance from center to either focus $= 4 = c$

Sum of distances from foci to point on ellipse $= 2a = 10$, so $a = 5$.

Form of equation of ellipse with center at origin and vertical major axis:

$$\frac{x^2}{b^2} + \frac{y^2}{a^2} = 1$$

$c^2 = a^2 - b^2$, so $b^2 = a^2 - c^2$.

$$b^2 = 5^2 - 4^2 = 25 - 16 = 9$$

Equation of ellipse: $\dfrac{x^2}{9} + \dfrac{y^2}{25} = 1$

43. Place the half-ellipse that represents the overpass on a coordinate system with the center of the bottom of the overpass at the origin. If the complete ellipse were drawn, the center of the ellipse would also be at $(0, 0)$. Then the half-ellipse will include the points $(0, 15)$, $(-10, 0)$, and $(10, 0)$. Thus, for the complete ellipse, $a = 15$ and $b = 10$, and the equation would be

$$\frac{x^2}{b^2} + \frac{y^2}{a^2} = 1$$

$$\frac{x^2}{10^2} + \frac{y^2}{15^2} = 1 \quad \text{Let } a = 15 \text{ and } b = 10.$$

$$\frac{x^2}{100} + \frac{y^2}{225} = 1.$$

To find the equation of the half-ellipse, solve this equation for y and use the positive square root since the overpass is represented by the upper half of the ellipse.

$$\frac{y^2}{225} = 1 - \frac{x^2}{100}$$

$$y^2 = 225\left(1 - \frac{x^2}{100}\right)$$

$$y = \sqrt{225\left(1 - \frac{x^2}{100}\right)}$$

$$y = 15\sqrt{1 - \frac{x^2}{100}}$$

Find the y-coordinate of the point whose x-intercept is $\frac{1}{2}(12) = 6$.

$$y = 15\sqrt{1 - \frac{x^2}{100}} \quad \text{Equation of half-ellipse}$$

$$y = 15\sqrt{1 - \frac{6^2}{100}} \quad \text{Let } x = 6.$$

$$y = 15\sqrt{1 - \frac{36}{100}} = 15\sqrt{\frac{64}{100}} = 15(.8) = 12$$

The tallest truck that can pass under the overpass is 12 ft tall.

10.3 Exercises *(page 918)*

23. $\frac{y}{3} = \sqrt{1 + \frac{x^2}{16}}$

Square both sides to get

$$\frac{y^2}{9} = 1 + \frac{x^2}{16} \quad \text{or} \quad \frac{y^2}{9} - \frac{x^2}{16} = 1.$$

This is the equation of a hyperbola with center (0, 0), vertices (0, 3) and (0, −3), and asymptotes $y = \pm\frac{3}{4}x$. The original equation represents the top half of the hyperbola. The domain is $(-\infty, \infty)$, and the range is $[3, \infty)$. The vertical line test shows that this is the graph of a function.

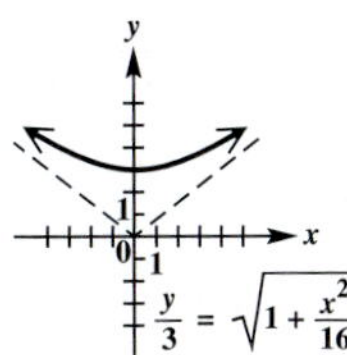

33. vertices at (0, 6) and (0, −6); asymptotes $y = \pm\frac{1}{2}x$

From the given vertices, the equation is of the form

$$\frac{y^2}{a^2} - \frac{x^2}{b^2} = 1$$

$$\frac{y^2}{6^2} - \frac{x^2}{b^2} = 1$$

$$\frac{y^2}{36} - \frac{x^2}{b^2} = 1.$$

The slopes of the asymptotes are $\pm\frac{1}{2}$. Use the positive slope to find the value of b.

$$\frac{a}{b} = \frac{1}{2}$$

$$\frac{6}{b} = \frac{1}{2} \quad \text{Let } a = 6.$$

$$b = 12$$

Use the values of a and b to write the equation of the hyperbola.

$$\frac{y^2}{6^2} - \frac{x^2}{12^2} = 1 \quad \text{Let } a = 6 \text{ and } b = 12.$$

$$\frac{y^2}{36} - \frac{x^2}{144} = 1$$

43. eccentricity 3; center at (0, 0); vertex at (0, 7)

Since the center and the given vertex lie on the y-axis, the equation is of the form

$$\frac{y^2}{a^2} - \frac{x^2}{b^2} = 1.$$

The distance between the center and a vertex is 7 units, so $a = 7$. Use the given eccentricity to find the value of c.

$$e = \frac{c}{a}$$

$$3 = \frac{c}{7} \quad \text{Let } e = 3 \text{ and } a = 7.$$

$$c = 21$$

Now find the value of b^2. Since $c^2 = a^2 + b^2$,

$$b^2 = c^2 - a^2 = 21^2 - 7^2 = 441 - 49 = 392.$$

The equation of the hyperbola is $\frac{y^2}{49} - \frac{x^2}{392} = 1$.

10.4 Exercises *(page 926)*

23.
$$y^2 - 4y = x + 4$$

$$y^2 - 4y + 4 = x + 4 + 4 \quad \text{Add 4 to complete the square.}$$

$$(y - 2)^2 = x + 8 \quad \text{Factor on the left; add on the right.}$$

The equation is of the form $x - h = a(y - k)^2$ with $a = 1$, $h = -8$, and $k = 2$, so the graph of the given equation is a parabola.

41. From the graph, the coordinates of P (a point on the graph) are (−3, 8), the coordinates of F (a focus) are (3, 0), and the equation of L (the directrix) is $x = 27$.

By the distance formula, the distance from P to F is

$$\sqrt{(x_2 - x_1)^2 + (y_2 - y_1)^2} = \sqrt{[3 - (-3)]^2 + (0 - 8)^2}$$
$$= \sqrt{6^2 + (-8)^2}$$
$$= \sqrt{36 + 64} = \sqrt{100} = 10.$$

(continued)

The distance between a point and a line is defined as the perpendicular distance, so the distance from P to L is $|27 - (-3)| = 30.$

$$e = \frac{\text{distance from } P \text{ to } F}{\text{distance from } P \text{ to } L} = \frac{10}{30} = \frac{1}{3}$$

CHAPTER 11 FURTHER TOPICS IN ALGEBRA

11.2 Exercises *(page 954)*

23. $S_{16} = -160,\ a_{16} = -25$

$$S_n = \frac{n}{2}(a_1 + a_n)$$

$$S_{16} = \frac{16}{2}(a_1 + a_{16}) \quad \text{Let } n = 16.$$

$$-160 = 8(a_1 - 25) \quad \text{Let } S_{16} = -160,\ a_{16} = -25.$$

$$-20 = a_1 - 25 \quad \text{Divide by 8.}$$

$$a_1 = 5 \quad \text{Add 25.}$$

43. The positive even integers form the arithmetic sequence 2, 4, 6, 8, ..., with $a_1 = 2$ and $d = 2$. Find the sum of the first 60 terms of this sequence.

$$S_n = \frac{n}{2}[2a_1 + (n-1)d]$$

$$S_{60} = \frac{60}{2}(2a_1 + 59d) \quad \text{Let } n = 60.$$

$$= \frac{60}{2}(2 \cdot 2 + 59 \cdot 2) \quad \text{Let } a_1 = 2,\ d = 2.$$

$$= 30(4 + 118)$$

$$= 30(122)$$

$$S_{60} = 3660$$

11.3 Exercises *(page 963)*

31. $\sum_{k=4}^{10} 2^k$

This series is the sum of the fourth through tenth terms of a geometric sequence with $a_1 = 2^1 = 2$ and $r = 2$. To find this sum, find the difference between the sum of the first ten terms and the sum of the first three terms.

$$\sum_{k=4}^{10} 2^k = \sum_{k=1}^{10} 2^k - \sum_{k=1}^{3} 2^k$$

$$= \frac{2(1 - 2^{10})}{1 - 2} - \frac{2(1 - 2^3)}{1 - 2}$$

$$= \frac{2(1 - 1024)}{-1} - \frac{2(1 - 8)}{-1}$$

$$= \frac{2(-1023)}{-1} - \frac{2(-7)}{-1}$$

$$= 2046 - 14$$

$$\sum_{k=4}^{10} 2^k = 2032$$

43. $\frac{1}{4} - \frac{1}{6} + \frac{1}{9} - \frac{2}{27} + \cdots$

For this infinite geometric series, $a_1 = \frac{1}{4}$ and

$$r = \frac{-\frac{1}{6}}{\frac{1}{4}} = -\frac{1}{6} \cdot \frac{4}{1} = -\frac{2}{3}.$$

Because $-1 < r < 1$, this series converges.

$$S_\infty = \frac{a_1}{1 - r} = \frac{\frac{1}{4}}{1 - \left(-\frac{2}{3}\right)} = \frac{\frac{1}{4}}{\frac{5}{3}} = \frac{1}{4} \cdot \frac{3}{5} = \frac{3}{20}$$

Let $a_1 = \frac{1}{4},\ r = -\frac{2}{3}.$

69. Option 1 is modeled by the arithmetic sequence $a_n = 5000 + 10{,}000(n - 1)$ with sum

$$S_n = \frac{n}{2}[2a_1 + (n-1)d]$$

$$S_{30} = \frac{30}{2}[2a_1 + 29d] \quad \text{Let } n = 30.$$

$$= \frac{30}{2}(2 \cdot 5000 + 29 \cdot 10{,}000) \quad \text{Let } a_1 = 5000 \text{ and } d = 10{,}000.$$

$$= 15(10{,}000 + 290{,}000)$$

$$S_{30} = 15(300{,}000) = 4{,}500{,}000.$$

Thus, Option 1 pays you a total of \$4,500,000.00.

Option 2 is modeled by the geometric sequence $a_n = .01(2)^{n-1}$ with sum

$$S_n = \frac{a_1(1 - r^n)}{1 - r}$$

$$S_{30} = \frac{.01(1 - 2^{30})}{1 - 2} = 10{,}737{,}418.23.$$

Let $n = 30,\ a_1 = .01,\ r = 2.$

Thus, Option 2 pays you a total of \$10,737,418.23.

You should choose Option 2.

11.4 Exercises *(page 974)*

11. $$\binom{n}{n-1} = \frac{n!}{(n-1)![n - (n-1)]!}$$

$$= \frac{n!}{(n-1)!1!} = \frac{n!}{(n-1)!}$$

$$= \frac{n(n-1)!}{(n-1)!} = n$$

41. Fifteenth term of $(x - y^3)^{20}$

Here, $n = 20$ and $k = 15$, so $k - 1 = 14$ and $n - (k - 1) = 6$. The fifteenth term of the expansion is

$$\binom{20}{14} x^6(-y^3)^{14} = 38{,}760x^6y^{42}.$$

43. Middle term of $(3x^7 + 2y^3)^8$

This expansion has nine terms, so the middle term is the fifth term. Here, $n = 8$ and $k = 5$, so $k - 1 = 4$ and $n - (k - 1) = 4$. The fifth term of the expansion is

$$\binom{8}{4}(3x^7)^4(2y^3)^4 = 70(81x^{28})(16y^{12}) = 90{,}720x^{28}y^{12}.$$

11.5 Exercises *(page 981)*

29. Let S_n represent the statement that the number of handshakes for n people is $\frac{n^2 - n}{2}$.

We need to prove that this statement is true for every positive integer $n \geq 2$, so we will use the generalized principle of mathematical induction.

Step 1 Show that the statement is true when $n = 2$. S_2 is the statement that for 2 people, the number of handshakes is

$$\frac{2^2 - 2}{1} = \frac{4 - 2}{2} = \frac{2}{2} = 1,$$

which is true.

Step 2 Show that S_k implies S_{k+1}, where S_k is the statement that the number of handshakes for k people is $\frac{k^2 - k}{2}$ and S_{k+1} is the statement that the number of handshakes for $(k + 1)$ people is $\frac{(k + 1)^2 - (k + 1)}{2}$.

Start with S_k and assume it is a true statement:

For k people, there are $\frac{k^2 - k}{2}$ handshakes.

If one more person enters the room that already contains k people, this person will shake hands once with each of the k people who were already in the room, so there will be k additional handshakes. Thus, the number of handshakes for $(k + 1)$ people is

$$\frac{k^2 - k}{2} + k = \frac{k^2 - k}{2} + \frac{2k}{2}$$

Get a common denominator.

$$= \frac{k^2 - k + 2k}{2}$$

Add rational expressions.

$$= \frac{k^2 + k}{2}$$

Combine like terms in the numerator.

$$= \frac{(k^2 + 2k + 1) - k - 1}{2}$$

Write k as $2k - k$; add 1 to complete the square; subtract 1.

$$= \frac{(k + 1)^2 - (k + 1)}{2}.$$

Factor out -1 in the numerator.

This work shows that if S_k is true, then S_{k+1} is true.

Since both steps for a proof by the generalized principle of mathematical induction have been completed, the given statement is true for every positive integer $n \geq 2$.

11.6 Exercises *(page 989)*

25. (a) There are two choices for the first letter, K and W.

The second letter can be any of the 26 letters of the alphabet except for the one chosen for the first letter, so there are 25 choices. The third letter can be any of the remaining 24 letters of the alphabet, so there are 24 choices. The fourth letter can be any of the remaining 23 letters of the alphabet, so there are 23 choices.

Therefore, by the fundamental principle of counting, the number of possible 4-letter radio-station call letters is

$$2 \cdot 25 \cdot 24 \cdot 23 = 27{,}600.$$

(b) There are two choices for the first letter.

Because repeats are allowed, there are 26 choices for each of the remaining three letters.

Therefore, by the fundamental principle of counting, the number of possible 4-letter radio-station call letters is

$$2 \cdot 26 \cdot 26 \cdot 26 = 35{,}152.$$

(c) There are two choices for the first letter.

There are 24 choices for the second letter since it cannot repeat the first letter and cannot be R. There are 23 choices for the third letter since it cannot repeat either of the first two letters and cannot be R. There is only one choice for the last letter since it must be R.

Therefore, by the fundamental principle of counting, the number of possible 4-letter radio-station call letters is

$$2 \cdot 24 \cdot 23 \cdot 1 = 1104.$$

55. (c) Choosing 3 men and 2 women involves two independent events, each of which involves combinations.

First, select the men. There are 8 men in the club, so the number of ways to do this is

$$C(8, 3) = \binom{8}{3} = \frac{8!}{5!3!} = 56.$$

(continued)

Now select the women. There are 11 women in the club, so the number of ways to do this is

$$C(11, 2) = \binom{11}{2} = \frac{11!}{9!2!} = 55.$$

To find the number of committees, use the fundamental principle of counting. The number of delegations with 3 men and 2 women is $56 \cdot 55 = 3080$.

61. Because the keys are arranged in a circle, there is no "first" key. The number of distinguishable arrangements is the number of ways to arrange the other three keys in relation to any one of the keys, which is

$$P(3, 3) = 3! = 6.$$

11.7 Exercises *(page 999)*

23. (a) A 40-yr-old man who lives 30 more yr would be 70 yr old.

Let E be the event "selected man will live to be 70"; then $n(E) = 66{,}172$. For this situation, the sample space S is the set of all 40-yr-old men, so $n(S) = 94{,}558$. Thus, the probability that a 40-yr-old man will live 30 more yr is

$$P(E) = \frac{n(E)}{n(S)} = \frac{66{,}172}{94{,}558} \approx .6998.$$

(b) Using the notation and result from part (a), the probability that a 40-yr-old man will not live 30 more yr is

$$P(E') = 1 - P(E) = 1 - .6998 = .3002.$$

(c) Use the notation and results from parts (a) and (b). In this binomial experiment, we call "a 40-yr-old man survives to age 70" a success. Then,

$$p = P(E) = .6998$$

and $$1 - p = P(E') = .3002.$$

There are 5 independent trials and we need the probability of 3 successes, so $n = 5$ and $r = 3$. The probability that exactly 3 of the 40-yr-old men survive to age 70 is

$$\binom{n}{r}p^r(1-p)^{n-r} = \binom{5}{3}(.6998)^3(.3002)^2$$
$$= 10(.6998)^3(.3002)^2$$
$$\approx .3088.$$

(d) Let F be the event "at least one man survives to age 70." The easiest way to find $P(F)$ is to first find the probability of the complementary event F': "neither man survives to age 70."

$$P(F') = P(E') \cdot P(E') = (.3002)^2 = .0901$$

Then,

$$P(F) = 1 - P(F') \approx 1 - .0901 = .9099.$$

29. P(smoked between 1 and 19)

$= P$(smoked 1 to 9, or smoked 10 to 19)

$= P$(smoked 1 to 9) $+ P$(smoked 10 to 19)

$= .24 + .20$

$= .44$

In this binomial experiment, call "smoked between 1 and 19" a success. Then $p = .44$ and $1 - p = .56$. "Fewer than 2" means 0 or 1, so

P(0 smoked between 1 and 19) $+ P$(1 smoked between 1 and 19)

$$= \binom{10}{0}(.44)^0(.56)^{10} + \binom{10}{1}(.44)^1(.56)^9$$
$$\approx .003033 + .023831$$
$$= .026864.$$

Answers to Selected Exercises

To The Student

In this section we provide the answers that we think most students will obtain when they work the exercises using the methods explained in the text. If your answer does not look exactly like the one given here, it is not necessarily wrong. In many cases there are equivalent forms of the answer. For example, if the answer section shows $\frac{3}{4}$ and your answer is .75, you have obtained the correct answer but written it in a different (yet equivalent) form. Unless the directions specify otherwise, .75 is just as valid an answer as $\frac{3}{4}$. In general, if your answer does not agree with the one given in the text, see whether it can be transformed into the other form. If it can, then it is the correct answer. If you still have doubts, talk with your instructor.

If you need further help with algebra, you may want to obtain a copy of the *Student's Solution Manual* that goes with this book. Your college bookstore either has this manual or can order it for you.

CHAPTER R REVIEW OF BASIC CONCEPTS

R.1 Exercises *(page 11)*

1. B, C, D, F **3.** D, F **5.** E, F **9.** Answers will vary. Three examples are $\frac{2}{3}$, $-\frac{4}{9}$, and $\frac{21}{2}$. **11.** 1, 3 **13.** $-6, -\frac{12}{4}$ (or -3), 0, 1, 3 **15.** -81 **17.** 81 **19.** -243 **21.** -162 **25.** -148 **27.** 23 **29.** 18 **31.** -12 **33.** $-\frac{25}{36}$ **35.** $-\frac{6}{7}$ **37.** 36 **39.** 36 **41.** $-\frac{1}{2}$ **43.** $-\frac{23}{20}$ **45.** $-\frac{13}{3}$ **47.** 92.9 **49.** 86.0 **51.** .031 **53.** .024; .023; Increased weight results in lower BACs. **55.** distributive **57.** inverse **59.** identity **63.** $(8 - 14)p = -6p$ **65.** $-3z + 3y$ **67.** $20z$ **69.** $m + 8$ **71.** $\frac{2}{3}y + \frac{4}{9}z - \frac{5}{3}$ **73.** 65.25 **75.** approximately 67,000 mph **77.** 930 **79.** 990 **81.** 31 ft **85.** 1700 **87.** 150

R.2 Exercises *(page 22)*

1. $-5, -4, -2, -\sqrt{3}, \sqrt{6}, \sqrt{8}, 3$ **3.** $\frac{3}{4}, \frac{7}{5}, \sqrt{2}, \frac{22}{15}, \frac{8}{5}$ **5.** $-|9|, -|-6|, |-8|$ **9.** false; $|5 - 7| = |7| - |5|$ **11.** true **13.** false; $|a - b| = |b| - |a|$ **15.** 9 **17.** $-\frac{4}{5}$ **19.** 8 **21.** 6 **23.** 4 **25.** -1 **27.** -5 **29.** $\pi - 3$ **31.** $3 - y$ **33.** $8 - 2k$ **35.** $y - x$ **37.** $3 + x^2$ **39.** property 2 **41.** property 3 **43.** property 5 **45.** property 1 **47.** 17,648 yd; No, it is not the same, because the sum of the absolute values is 17,660. **49.** 9 **51.** 47°F **53.** 22°F **55.** 3 **57.** 9 **59.** 13 **61.** x and y have the same sign. **63.** x and y have different signs. **65.** x and y have the same sign.

R.3 Exercises *(page 32)*

1. incorrect; $(mn)^2 = m^2n^2$ **3.** incorrect; $\left(\frac{k}{5}\right)^3 = \frac{k^3}{5^3}$ **5.** 9^8 **7.** $-16x^7$ **9.** 2^{10} **11.** -4^2m^6 or $-16m^6$ **13.** $\frac{r^{24}}{s^6}$ **15. (a)** B **(b)** C **(c)** B **(d)** C **19.** polynomial; degree 11; monomial **21.** polynomial; degree 6; binomial **23.** polynomial; degree 6; binomial **25.** polynomial; degree 6; trinomial **27.** not a polynomial **29.** $x^2 - x + 3$ **31.** $12y^2 + 4$ **33.** $6m^4 - 2m^3 - 7m^2 - 4m$ **35.** $28r^2 + r - 2$ **37.** $15x^4 - \frac{7}{3}x^3 - \frac{2}{9}x^2$ **39.** $12x^5 + 8x^4 - 20x^3 + 4x^2$ **41.** $-2z^3 + 7z^2 - 11z + 4$ **43.** $m^2 + mn - 2n^2 - 2km + 5kn - 3k^2$ **45.** $4m^2 - 9$ **47.** $16x^4 - 25y^2$ **49.** $16m^2 + 16mn + 4n^2$ **51.** $25r^2 - 30rt^2 + 9t^4$ **53.** $4p^2 - 12p + 9 + 4pq - 6q + q^2$ **55.** $9q^2 + 30q + 25 - p^2$ **57.** $9a^2 + 6ab + b^2 - 6a - 2b + 1$ **59.** $y^3 + 6y^2 + 12y + 8$

61. $q^4 - 8q^3 + 24q^2 - 32q + 16$ **63.** $p^3 - 7p^2 - p - 7$ **65.** $49m^2 - 4n^2$ **67.** $-14q^2 + 11q - 14$ **69.** $4p^2 - 16$ **71.** $11y^3 - 18y^2 + 4y$ **73.** $2x^5 + 7x^4 - 5x^2 + 7$ **75.** $-5x^2 + 8 + \frac{2}{x^2}$ **77.** $2m^2 + m - 2 + \frac{6}{3m + 2}$ **79.** $x^3 - x^2 - x + 4 + \frac{-17}{3x + 3}$ **81.** 9999 **82.** 3591 **83.** 10,404 **84.** 5041 **85. (a)** $(x + y)^2$ **(b)** $x^2 + 2xy + y^2$ **(d)** the special product for squaring a binomial **87. (a)** approximately 60,501,000 ft^3 **(b)** The shape becomes a rectangular box with a square base, with volume $V = b^2h$. **(c)** If we let $a = b$, then $V = \frac{1}{3}h(a^2 + ab + b^2)$ becomes $V = \frac{1}{3}h(b^2 + bb + b^2)$, which simplifies to $V = hb^2$. Yes, the Egyptian formula gives the same result. **89.** 6.2; .1 high **91.** 2.3; 0, exact **93.** 1,000,000 **95.** 32 **97.** Both expansions are equal to $-x^3 + 3x^2y - 3xy^2 + y^3$.

R.4 Exercises *(page 43)*

1. $12(m + 5)$ **3.** $8k(k^2 + 3)$ **5.** $xy(1 - 5y)$ **7.** $-2p^2q^4(2p + q)$ **9.** $4k^2m^3(1 + 2k^2 - 3m)$ **11.** $2(a + b)(1 + 2m)$ **13.** $(r + 3)(3r - 5)$ **15.** $(m - 1)(2m^2 - 7m + 7)$ **17.** $(2s + 3)(3t - 5)$ **19.** $(m^4 + 3)(2 - a)$ **21.** $(5z^2 - 2x)(4 + p)$ **23.** $(2a - 1)(3a - 4)$ **25.** $(3m + 2)(m + 4)$ **27.** $(3k - 2p)(2k + 3p)$ **29.** $(5a + 3b)(a - 2b)$ **31.** $x^2(3 - x)^2$ **33.** $2a^2(4a - b)(3a + 2b)$ **35.** $(3m - 2)^2$ **37.** $2(4a + 3b)^2$ **39.** $(2xy + 7)^2$ **41.** $(a - 3b - 3)^2$ **43. (a)** B **(b)** C **(c)** A **(d)** D **45.** $(3a + 4)(3a - 4)$ **47.** $(5s^2 + 3t)(5s^2 - 3t)$ **49.** $(a + b + 4)(a + b - 4)$ **51.** $(p^2 + 25)(p + 5)(p - 5)$ **53.** $(2 - a)(4 + 2a + a^2)$ **55.** $(5x - 3)(25x^2 + 15x + 9)$ **57.** $(3y^3 + 5z^2)(9y^6 - 15y^3z^2 + 25z^4)$ **59.** $r(r^2 + 18r + 108)$ **61.** $(3 - m - 2n)(9 + 3m + 6n + m^2 + 4mn + 4n^2)$ **63.** B **65.** $(x - 1)(x^2 + x + 1)(x + 1)(x^2 - x + 1)$ **66.** $(x - 1)(x + 1)(x^4 + x^2 + 1)$ **67.** $(x^2 - x + 1)(x^2 + x + 1)$ **68.** additive inverse property (0 in the form $x^2 - x^2$ was added on the right.); associative property of addition; factoring a perfect square trinomial; factoring a difference of squares; commutative property of addition **69.** They are the same. **70.** $(x^4 - x^2 + 1)(x^2 + x + 1)(x^2 - x + 1)$ **71.** $(m^2 - 5)(m^2 + 2)$ **73.** $9(7k - 3)(k + 1)$ **75.** $(3a - 7)^2$ **77.** $(2b + c + 4)(2b + c - 4)$ **79.** $(x + y)(x - 5)$ **81.** $(m - 2n)(p^4 + q)$ **83.** $(2z + 7)^2$ **85.** $(10x + 7y)(100x^2 - 70xy + 49y^2)$ **87.** $(5m^2 - 6)(25m^4 + 30m^2 + 36)$ **89.** $9(x + 2)(3x^2 + 4)$ **91.** $(4p - 1)(p + 1)$ **93.** prime **95.** $4xy$ **99.** ± 36 **101.** 9

R.5 Exercises *(page 52)*

1. $\{x \mid x \neq 6\}$ **3.** $\left\{x \mid x \neq -\frac{1}{2}, 1\right\}$ **5.** $\{x \mid x \neq -2, -3\}$ **7. (a)** $\frac{3}{4}$ **(b)** $\frac{1}{6}$ **8.** No; $\frac{1}{x} + \frac{1}{y} \neq \frac{1}{x + y}$. **9. (a)** $\frac{2}{15}$ **(b)** $-\frac{1}{2}$ **10.** No; $\frac{1}{x} - \frac{1}{y} \neq \frac{1}{x - y}$. **11.** $\frac{8}{9}$ **13.** $\frac{-3}{t + 5}$ **15.** $\frac{2x + 4}{x}$ **17.** $\frac{m - 2}{m + 3}$ **19.** $\frac{2m + 3}{4m + 3}$ **21.** $\frac{25p^2}{9}$ **23.** $\frac{2}{9}$ **25.** $\frac{5x}{y}$ **27.** $\frac{2a + 8}{a - 3}$ or $\frac{2(a + 4)}{a - 3}$ **29.** 1 **31.** $\frac{m + 6}{m + 3}$ **33.** $\frac{x + 2y}{4 - x}$ **35.** $\frac{x^2 - xy + y^2}{x^2 + xy + y^2}$ **37.** B, C **39.** $\frac{19}{6k}$ **41.** $\frac{137}{30m}$ **43.** $\frac{a - b}{a^2}$ **45.** $\frac{5 - 22x}{12x^2y}$ **47.** 3 **49.** $\frac{2x}{(x + z)(x - z)}$ **51.** $\frac{4}{a - 2}$ or $\frac{-4}{2 - a}$ **53.** $\frac{3x + y}{2x - y}$ or $\frac{-3x - y}{y - 2x}$ **55.** $\frac{4x - 7}{x^2 - x + 1}$ **57.** $\frac{2x^2 - 9x}{(x - 3)(x + 4)(x - 4)}$ **59.** $\frac{x + 1}{x - 1}$ **61.** $\frac{-1}{x + 1}$ **63.** $\frac{(2 - b)(1 + b)}{b(1 - b)}$ **65.** $\frac{m^3 - 4m - 1}{m - 2}$ **67.** $\frac{-1}{x(x + h)}$ **69.** 0 mi **71.** 20.1 (thousand dollars)

R.6 Exercises *(page 62)*

1. (a) B **(b)** D **(c)** B **(d)** D **3.** $\frac{1}{(-4)^3}$ or $-\frac{1}{64}$ **5.** $-\frac{1}{5^4}$ or $-\frac{1}{625}$ **7.** 3^2 or 9 **9.** $\frac{1}{16x^2}$ **11.** $\frac{4}{x^2}$ **13.** $-\frac{1}{a^3}$ **15.** 4^2 or 16 **17.** x^4 **19.** $\frac{1}{r^3}$ **21.** 6^6 **23.** $\frac{2r^3}{3}$ **25.** $\frac{4n^7}{3m^7}$ **27.** $-4r^6$ **29.** $\frac{5^4}{a^{10}}$ **31.** $\frac{p^4}{5}$ **33.** $\frac{1}{2pq}$

35. $\frac{4}{a^2}$ **37.** 13 **39.** 2 **41.** $-\frac{4}{3}$ **43.** not a real number **45.** **(a)** E **(b)** G **(c)** F **(d)** F **47.** 4 **49.** 1000 **51.** -27 **53.** $\frac{256}{81}$ **55.** 9 **57.** 4 **59.** y **61.** $k^{2/3}$ **63.** x^3y^8 **65.** $\frac{1}{x^{10/3}}$ **67.** $\frac{6}{m^{1/4}n^{3/4}}$ **69.** p^2 **71.** **(a)** approximately 250 sec **(b)** $\frac{1}{2^{1.5}} \approx .3536$ **73.** $y - 10y^2$ **75.** $-4k^{10/3} + 24k^{4/3}$ **77.** $x^2 - x$ **79.** $r - 2 + r^{-1}$ or $r - 2 + \frac{1}{r}$ **81.** $k^{-2}(4k + 1)$ **83.** $z^{-1/2}(9 + 2z)$ **85.** $p^{-7/4}(p - 2)$ **87.** $(p + 4)^{-3/2}(p^2 + 9p + 21)$ **89.** $b + a$ **91.** -1 **93.** $\frac{y(xy - 9)}{x^2y^2 - 9}$ **95.** 27,000 **97.** 27 **99.** 4 **101.** $\frac{1}{100}$

R.7 Exercises *(page 73)*

1. **(a)** F **(b)** H **(c)** G **(d)** C **3.** $\sqrt[3]{m^2}$ or $(\sqrt[3]{m})^2$ **5.** $\sqrt[3]{(2m + p)^2}$ or $(\sqrt[3]{2m + p})^2$ **7.** $k^{2/5}$ **9.** $-3 \cdot 5^{1/2}p^{3/2}$ **11.** A **13.** $x \geq 0$ **15.** 5 **17.** $5k^2|m|$ **19.** $|4x - y|$ **21.** 5 **23.** not a real number **25.** $3\sqrt[3]{3}$ **27.** $-2\sqrt[4]{2}$ **29.** $\sqrt{42pqr}$ **31.** $\sqrt[3]{14xy}$ **33.** $-\frac{3}{5}$ **35.** $-\frac{\sqrt[3]{5}}{2}$ **37.** $\frac{\sqrt[4]{m}}{n}$ **39.** -15 **41.** $32\sqrt[3]{2}$ **43.** $2x^2z^4\sqrt{2x}$ **45.** cannot be simplified further **47.** $\frac{\sqrt{6x}}{3x}$ **49.** $\frac{x^2y\sqrt{xy}}{z}$ **51.** $\frac{2\sqrt[3]{x}}{x}$ **53.** $\frac{h\sqrt[4]{9g^3hr^2}}{3r^2}$ **55.** $\sqrt{3}$ **57.** $\sqrt[3]{2}$ **59.** cannot be simplified further **61.** $12\sqrt{2x}$ **63.** $7\sqrt[3]{3}$ **65.** $3x\sqrt[4]{x^2y^3} - 2x^2\sqrt[4]{x^2y^3}$ **67.** -7 **69.** 10 **71.** $11 + 4\sqrt{6}$ **73.** $5\sqrt{6}$ **75.** $\frac{m\sqrt[3]{n^2}}{n}$ **77.** $\frac{x\sqrt[3]{2} - \sqrt[3]{5}}{x^3}$ **79.** $\frac{11\sqrt{2}}{8}$ **81.** $-\frac{25\sqrt[3]{9}}{18}$ **83.** $\frac{-\sqrt{21} - 7}{4}$ **85.** $\frac{-7 + 2\sqrt{14} + \sqrt{7} - 2\sqrt{2}}{2}$ **87.** $\frac{\sqrt{r}(3 + \sqrt{r})}{9 - r}$ **89.** $\frac{3m(2 - \sqrt{m + n})}{4 - m - n}$ **91.** 19.0°; The table gives 19°. **93.** 3 **95.** 2 **97.** 2 **99.** It gives six decimal places of accuracy.

Chapter R Review Exercises *(page 79)*

1. $-12, -6, -\sqrt{4}$ (or -2), 0, 6 **3.** whole number, integer, rational number, real number **5.** irrational number, real number **11.** commutative **13.** associative **15.** identity **17.** **(a)** 74°F **(b)** 84°F **19.** 32 **21.** $-\frac{37}{18}$ **23.** $-\frac{12}{5}$ **25.** -32 **27.** -13 **29.** $-|3 - (-2)|, -|-2|, |6 - 4|, |8 + 1|$ **31.** -3 **33.** $8 - \sqrt{8}$ **35.** $7q^3 - 9q^2 - 8q + 9$ **37.** $16y^2 + 42y - 49$ **39.** $9k^2 - 30km + 25m^2$ **41.** **(a)** 51 million **(b)** 52 million **(c)** The approximation is 1 million high. **43.** **(a)** 183 million **(b)** 183 million **(c)** They are the same. **45.** $6m^2 - 3m + 5$ **47.** $3b - 8 + \frac{2}{b^2 + 4}$ **49.** $3(z - 4)^2(3z - 11)$ **51.** $(z - 8k)(z + 2k)$ **53.** $6a^6(4a + 5b)(2a - 3b)$ **55.** $(7m^4 + 3n)(7m^4 - 3n)$ **57.** $3(9r - 10)(2r + 1)$ **59.** $(3x - 4)(9x - 34)$ **61.** $\frac{1}{2k^2(k - 1)}$ **63.** $\frac{x + 1}{x + 4}$ **65.** $\frac{(p + q)(p + 6q)^2}{5p}$ **67.** $\frac{2m}{m - 4}$ or $\frac{-2m}{4 - m}$ **69.** $\frac{q + p}{pq - 1}$ **71.** $\frac{1}{64}$ **73.** $\frac{16}{25}$ **75.** $-10z^8$ **77.** 1 **79.** $-8y^{11}p$ **81.** $\frac{1}{(p + q)^5}$ **83.** $-14r^{17/12}$ **85.** $y^{1/2}$ **87.** $10z^{7/3} - 4z^{1/3}$ **89.** $10\sqrt{2}$ **91.** $5\sqrt[4]{2}$ **93.** $-\frac{\sqrt[3]{50p}}{5p}$ **95.** $\sqrt[12]{m}$ **97.** 66 **99.** $-9m\sqrt{2m} + 5m\sqrt{m}$ or $m(-9\sqrt{2m} + 5\sqrt{m})$ **101.** $\frac{6(3 + \sqrt{2})}{7}$

In Exercises 103–113, we give only the corrected right-hand sides of the equations.

103. $x^3 + 5x$ **105.** m^6 **107.** $\frac{a}{2b}$ **109.** One possible answer is $\frac{\sqrt{a} - \sqrt{b}}{a - b}$. **111.** $4 - t - 1$ or $3 - t$ **113.** 5^2

Chapter R Test *(page 83)*

1. **(a)** $-13, -\frac{12}{4}$ (or -3), 0, $\sqrt{49}$ (or 7) **(b)** $-13, -\frac{12}{4}$ (or -3), 0, $\frac{3}{5}$, 5.9, $\sqrt{49}$ (or 7) **(c)** All are real numbers. **2.** 4

3. **(a)** associative **(b)** commutative **(c)** distributive **(d)** inverse **4.** 87.9 **5.** $11x^2 - x + 2$ **6.** $36r^2 - 60r + 25$

7. $3t^3 + 5t^2 + 2t + 8$ **8.** $2x^2 - x - 5 + \frac{3}{x-5}$ **9.** \$8401 **10.** \$8797 **11.** $(3x - 7)(2x - 1)$

12. $(x^2 + 4)(x + 2)(x - 2)$ **13.** $2m(4m + 3)(3m - 4)$ **14.** $(x - 2)(x^2 + 2x + 4)(y + 3)(y - 3)$

15. $\frac{x^4(x+1)}{3(x^2+1)}$ **16.** $\frac{x(4x+1)}{(x+2)(x+1)(2x-3)}$ **17.** $\frac{2a}{2a-3}$ or $\frac{-2a}{3-2a}$ **18.** $\frac{y}{y+2}$ **19.** $\frac{y}{x}$ **20.** $\frac{9}{16}$

21. $3x^2y^4\sqrt{2x}$ **22.** $2\sqrt{2x}$ **23.** $x - y$ **24.** $\frac{7(\sqrt{11}+\sqrt{7})}{2}$ **25.** approximately 2.1 sec

CHAPTER 1 EQUATIONS AND INEQUALITIES

1.1 Exercises *(page 90)*

1. true **3.** false **7.** B **9.** $\{-4\}$ **11.** $\{1\}$ **13.** $\left\{-\frac{2}{7}\right\}$ **15.** $\left\{-\frac{7}{8}\right\}$ **17.** $\{-1\}$ **19.** $\{10\}$ **21.** $\{75\}$

23. $\{0\}$ **25.** $\{12\}$ **27.** $\{50\}$ **29.** identity; {all real numbers} **31.** conditional equation; $\{0\}$

33. identity; {all real numbers} **35.** contradiction; $\emptyset$ **39.** $l = \frac{V}{wh}$ **41.** $c = P - a - b$

43. $B = \frac{2A - hb}{h}$ or $B = \frac{2A}{h} - b$ **45.** $h = \frac{S - 2\pi r^2}{2\pi r}$ or $h = \frac{S}{2\pi r} - r$ **47.** $h = \frac{S - 2lw}{2w + 2l}$

Answers in Exercises 49–57 exist in equivalent forms as well.

49. $x = -3a + b$ **51.** $x = \frac{3a + b}{3 - a}$ **53.** $x = \frac{3 - 3a}{a^2 - a - 1}$ **55.** $x = \frac{2a^2}{a^2 + 3}$ **57.** $x = \frac{m + 4}{2m + 5}$

59. **(a)** \$63 **(b)** \$1638 **61.** 68°F **63.** 15°C **65.** 37.8°C **67.** 463.9°C **69.** −14.9°C

Connections *(page 100)*

Step 1 compares to Polya's first step, Steps 2 and 3 compare to his second step, Step 4 compares to his third step, and Step 6 compares to his fourth step.

1.2 Exercises *(page 100)*

1. 20 mi **3.** \$8 **5.** A **7.** D **9.** 30 cm **11.** 7.6 cm **13.** 600 ft, 800 ft, 1000 ft

15. width: 3 mm; length: 7 mm **17.** 4 ft **19.** 50 mi **21.** 2.7 mi **23.** 15 min

25. 1 hr, 8 min, 12 sec; It is about $\frac{1}{2}$ the world record time. **27.** 35 km per hr **29.** $1\frac{7}{8}$ hr **31.** 78 hr **33.** $13\frac{1}{3}$ hr

35. $7\frac{1}{2}$ gal **37.** 2 L **39.** 4 mL **41.** short-term note: \$100,000; long-term note: \$140,000

43. \$10,000 at 2.5%; \$20,000 at 3% **45.** \$50,000 at 1.5%; \$90,000 at 4% **47.** **(a)** .0352 **(b)** approximately .015 or 1.5% **(c)** approximately 1 case **49.** **(a)** 800 ft^3 **(b)** 107,680 μg **(c)** $F = 107{,}680x$ **(d)** approximately .25 day, or 6 hr

51. **(a)** 16.8 million **(b)** 2009 **(c)** They are quite close. **(d)** 13.5 million

1.3 Exercises *(page 113)*

1. true **3.** true **5.** false; *Every* real number is a complex number. **7.** real, complex **9.** complex, pure imaginary

11. complex **13.** real, complex **15.** complex, pure imaginary **17.** $5i$ **19.** $i\sqrt{10}$ **21.** $12i\sqrt{2}$

23. $-3i\sqrt{2}$ **25.** -13 **27.** $-2\sqrt{6}$ **29.** $\sqrt{3}$ **31.** $i\sqrt{3}$ **33.** $\frac{1}{2}$ **35.** -2 **37.** $-3 + i\sqrt{6}$

39. $2 - 2i\sqrt{2}$ **41.** $\frac{1}{8} + \frac{\sqrt{2}}{8}i$ **43.** $7 - i$ **45.** 2 **47.** $1 - 10i$ **49.** $-13 + 5i$ **51.** $8 - i$ **53.** $-14 + 2i$ **55.** $5 - 12i$ **57.** $-8 - 6i$ **59.** 13 **61.** 7 **63.** $25i$ **65.** $12 + 9i$ **67.** $20 + 15i$ **69.** i **71.** -1 **73.** $-i$ **75.** 1 **77.** $-i$ **79.** $-i$ **83.** $2 - 2i$ **85.** $\frac{3}{5} - \frac{4}{5}i$ **87.** $-1 - 2i$ **89.** $-5i$ **91.** $-8i$ **93.** $-\frac{2}{3}i$ **97.** $4 + 6i$

1.4 Exercises *(page 123)*

1. G **3.** C **5.** H **7.** D **9.** D; $\left\{-\frac{1}{3}, 7\right\}$ **11.** C; $\{-4, 3\}$ **13.** $\{2, 3\}$ **15.** $\left\{-\frac{2}{5}, 1\right\}$ **17.** $\left\{-\frac{3}{4}, 1\right\}$ **19.** $\{\pm 4\}$ **21.** $\{\pm 3\sqrt{3}\}$ **23.** $\{\pm 4i\}$ **25.** $\left\{\frac{1 \pm 2\sqrt{3}}{3}\right\}$ **27.** $\{-5 \pm i\sqrt{3}\}$ **29.** $\left\{\frac{3}{5} \pm \frac{\sqrt{3}}{5}i\right\}$ **31.** $\{-3, -1\}$ **33.** $\left\{-\frac{7}{2}, 4\right\}$ **35.** $\{1 \pm \sqrt{3}\}$ **37.** $\left\{-\frac{5}{2}, 2\right\}$ **39.** $\left\{\frac{2 \pm \sqrt{10}}{2}\right\}$ **41.** $\left\{1 \pm \frac{\sqrt{3}}{2}i\right\}$ **43.** He is incorrect because $c = 0$. **45.** $\left\{\frac{1 \pm \sqrt{5}}{2}\right\}$ **47.** $\{3 \pm \sqrt{2}\}$ **49.** $\{1 \pm 2i\}$ **51.** $\left\{\frac{3}{2} \pm \frac{\sqrt{2}}{2}i\right\}$ **53.** $\left\{\frac{-1 \pm \sqrt{97}}{4}\right\}$ **55.** $\left\{\frac{-2 \pm \sqrt{10}}{2}\right\}$ **57.** $\left\{\frac{-3 \pm \sqrt{41}}{8}\right\}$ **59.** $\{2, -1 \pm i\sqrt{3}\}$ **61.** $\left\{-3, \frac{3}{2} \pm \frac{3\sqrt{3}}{2}i\right\}$ **63.** $t = \frac{\pm\sqrt{2sg}}{g}$ **65.** $v = \frac{\pm\sqrt{FrkM}}{kM}$ **67.** $t = \frac{v_0 \pm \sqrt{v_0^2 - 64h + 64s_0}}{32}$ **69.** **(a)** $x = \frac{y \pm \sqrt{8 - 11y^2}}{4}$ **(b)** $y = \frac{x \pm \sqrt{6 - 11x^2}}{3}$ **71.** 0; one rational solution (a double solution) **73.** 1; two distinct rational solutions **75.** 84; two distinct irrational solutions **77.** -23; two distinct nonreal complex solutions **79.** 2304; two distinct rational solutions

In Exercises 83 and 85, there are other answers.

83. $a = 1, b = -9, c = 20$ **85.** $a = 1, b = -2, c = -1$

Connections *(page 129)*

1. **(a)** 400 ft **(b)** 1600 ft; No, the second answer is $2^2 = 4$ times the first because the number of seconds is squared in the formula. **2.** Both formulas involve the number 16 times the square of the time. However, in the formula for the distance an object falls, 16 is positive, while in the formula for a propelled object, it is preceded by a negative sign. Also in the formula for a propelled object, the initial velocity and height affect the distance.

1.5 Exercises *(page 130)*

1. A **3.** D **5.** 100 yd by 400 yd **7.** 9 ft by 12 ft **9.** 20 in. by 30 in. **11.** 3.75 cm **13.** 1 ft **15.** 4 **17.** 5 ft **19.** 20 ft **21.** 16.4 ft **23.** 3000 yd **25.** **(a)** 1 sec, 5 sec **(b)** 6 sec **27.** **(a)** It will not reach 80 ft. **(b)** 2 sec **29.** **(a)** .19 sec, 10.92 sec **(b)** 11.32 sec **31.** **(a)** approximately 19.2 hr **(b)** 84.3 ppm (109.8 is not in the interval [50, 100].) **33.** **(a)** 10.3 hr **(b)** 722.2 ppm (857.8 is not in the interval [500, 800].) **35.** 23.93 million **37.** $80 - x$ **38.** $300 + 20x$ **39.** $(80 - x)(300 + 20x) = 24{,}000 + 1300x - 20x^2$ **40.** $20x^2 - 1300x + 11{,}000 = 0$ **41.** $\{10, 55\}$; Because of the restriction, only $x = 10$ is valid. The number of apartments rented is 70. **43.** 80

1.6 Exercises *(page 144)*

1. $-\frac{3}{2}, 6$ **3.** $2, -1$ **5.** 0 **7.** $\{-10\}$ **9.** $\emptyset$ **11.** $\emptyset$ **13.** $\{-9\}$ **15.** $\{-2\}$ **17.** $\emptyset$ **19.** $\left\{-\frac{5}{2}, \frac{1}{9}\right\}$ **21.** $\left\{\frac{3}{4}, 1\right\}$ **23.** $\{3, 5\}$ **25.** $\left\{-2, \frac{5}{4}\right\}$ **27.** $\{3\}$ **29.** $\{-1\}$ **31.** $\{5\}$ **33.** $\{9\}$ **35.** $\{9\}$ **37.** $\emptyset$ **39.** $\{\pm 2\}$ **41.** $\{0, 3\}$ **43.** $\{-2\}$ **45.** $\left\{-\frac{2}{9}, 2\right\}$ **47.** $\{4\}$ **49.** $\{-2\}$ **51.** $\left\{\frac{2}{5}, 1\right\}$ **53.** $\left\{\frac{3}{2}\right\}$ **55.** $\{31\}$

57. $\{-3, 1\}$ **59.** $\{-27, 3\}$ **61.** $\left\{\pm 1, \pm\frac{\sqrt{10}}{2}\right\}$ **63.** $\{\pm\sqrt{3}, \pm i\sqrt{5}\}$ **65.** $\left\{\frac{1}{4}, 1\right\}$ **67.** $\{0, 8\}$ **69.** $\{-63, 28\}$

71. $\{0, 31\}$ **73.** $\left\{\frac{-6 \pm 2\sqrt{3}}{3}, \frac{-4 \pm \sqrt{2}}{2}\right\}$ **75.** $\left\{-\frac{2}{7}, 5\right\}$ **77.** $\left\{-\frac{1}{27}, \frac{1}{8}\right\}$ **81.** $\{16\}$; $u = -3$ does not lead to a solution of the equation. **82.** $\{16\}$; 9 does not satisfy the equation. **84.** $\{4\}$ **85.** $h = \frac{d^2}{k^2}$ **87.** $m = (1 - n^{3/4})^{4/3}$

89. $e = \frac{Er}{R + r}$

Summary Exercises on Solving Equations *(page 146)*

1. $\{3\}$ **2.** $\{-1\}$ **3.** $\{-3 \pm 3\sqrt{2}\}$ **4.** $\{2, 6\}$ **5.** $\emptyset$ **6.** $\{-31\}$ **7.** $\{-6\}$ **8.** $\{6\}$ **9.** $\left\{\frac{1}{5} \pm \frac{2}{5}i\right\}$

10. $\{-2, 1\}$ **11.** $\left\{-\frac{1}{243}, \frac{1}{3125}\right\}$ **12.** $\{-1\}$ **13.** $\{\pm i, \pm 2\}$ **14.** $\{-2.4\}$ **15.** $\{4\}$ **16.** $\left\{\frac{1}{3} \pm i\frac{\sqrt{2}}{3}\right\}$

17. $\left\{\frac{15}{7}\right\}$ **18.** $\{4\}$ **19.** $\{3, 11\}$ **20.** $\{1\}$

1.7 Exercises *(page 156)*

1. F **3.** A **5.** I **7.** B **9.** E **13.** $(-\infty, 4]$;  **15.** $[-1, \infty)$; 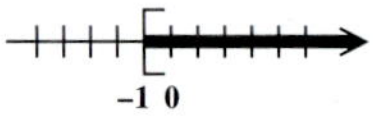

17. $(-\infty, 6]$;  **19.** $(-\infty, 4)$;  **21.** $\left[-\frac{11}{5}, \infty\right)$; 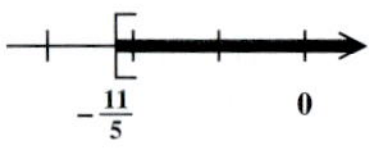

23. $\left(-\infty, \frac{48}{7}\right]$;  **25.** $(-5, 3)$; 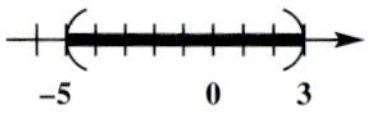 **27.** $[3, 6]$; 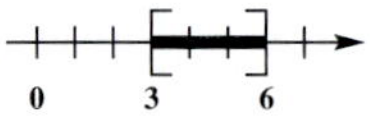

29. $(4, 6)$; 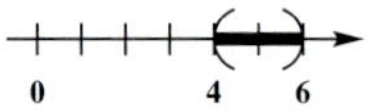 **31.** $[-9, 9]$; 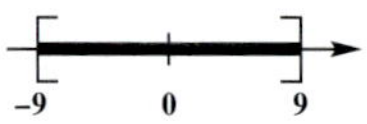 **33.** $(-16, 19]$; 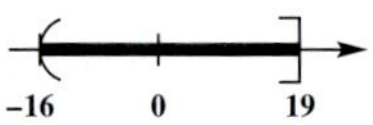 **35.** $[500, \infty)$

37. $[45, \infty)$ **39.** $(-\infty, -2) \cup (3, \infty)$ **41.** $\left[-\frac{3}{2}, 6\right]$ **43.** $(-\infty, -3] \cup [-1, \infty)$ **45.** $[-2, 3]$ **47.** $[-3, 3]$

49. $\left(\frac{-5 - \sqrt{33}}{2}, \frac{-5 + \sqrt{33}}{2}\right)$ **51.** $[1 - \sqrt{2}, 1 + \sqrt{2}]$ **53.** A **55.** $\left\{\frac{4}{3}, -2, -6\right\}$ **56.** 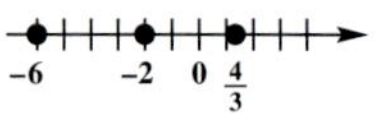

57. In the interval $(-\infty, -6)$, choose $x = -10$, for example. It satisfies the original inequality. In the interval $(-6, -2)$, choose $x = -4$, for example. It does not satisfy the inequality. In the interval $\left(-2, \frac{4}{3}\right)$, choose $x = 0$, for example. It satisfies the original inequality. In the interval $\left(\frac{4}{3}, \infty\right)$, choose $x = 4$, for example. It does not satisfy the original inequality.

58. 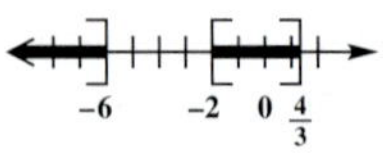 $(-\infty, -6] \cup \left[-2, \frac{4}{3}\right]$ **59.** $\left[-2, \frac{3}{2}\right] \cup [3, \infty)$ **61.** $(-\infty, -2] \cup [0, 2]$

63. $(-\infty, -1) \cup (-1, 3)$ **65.** $[-4, -3] \cup [3, \infty)$ **67.** $(-\infty, \infty)$ **69.** $(-5, 3]$ **71.** $(-\infty, -2)$

73. $(-\infty, 6) \cup \left[\frac{15}{2}, \infty\right)$ **75.** $(-\infty, 1) \cup \left(\frac{9}{5}, \infty\right)$ **77.** $\left(-\infty, -\frac{3}{2}\right) \cup \left[-\frac{1}{2}, \infty\right)$ **79.** $(-2, \infty)$

81. $\left(0, \frac{4}{11}\right) \cup \left(\frac{1}{2}, \infty\right)$ **83.** $(-\infty, -2] \cup (1, 2)$ **85.** $(-\infty, 5)$ **87.** $\left[\frac{3}{2}, \infty\right)$ **89.** $\left(\frac{5}{2}, \infty\right)$

91. $\left[-\frac{8}{3}, \frac{3}{2}\right] \cup (6, \infty)$ **93.** 1998 and 1999; 2001 and 2002; It agrees favorably, except for the year 2001. **95.** between (and inclusive of) 4 sec and 9.75 sec **97.** **(a)** $2.08 \times 10^{-5} \le \frac{R}{72} \le 8.33 \times 10^{-5}$ **(b)** between 5400 and 21,700

1.8 Exercises *(page 164)*

1. F **3.** D **5.** G **7.** C **9.** $\left\{-\frac{1}{3}, 1\right\}$ **11.** $\left\{\frac{2}{3}, \frac{8}{3}\right\}$ **13.** $\{-6, 14\}$ **15.** $\left\{\frac{5}{2}, \frac{7}{2}\right\}$ **17.** $\left\{-\frac{4}{3}, \frac{2}{9}\right\}$ **19.** $\left\{-\frac{7}{3}, -\frac{1}{7}\right\}$ **21.** $\{1\}$ **23.** $(-\infty, \infty)$ **27.** $(-4, -1)$ **29.** $(-\infty, -4] \cup [-1, \infty)$ **31.** $\left(-\frac{3}{2}, \frac{5}{2}\right)$ **33.** $(-\infty, 0) \cup (6, \infty)$ **35.** $\left(-\infty, -\frac{2}{3}\right) \cup (4, \infty)$ **37.** $\left[-\frac{2}{3}, 4\right]$ **39.** $\left[-1, -\frac{1}{2}\right]$ **41.** $\left\{-1, -\frac{1}{2}\right\}$ **43.** $\{2, 4\}$ **45.** $\left(-\frac{4}{3}, \frac{2}{3}\right)$ **47.** $\left(-\frac{3}{2}, \frac{13}{10}\right)$ **49.** $\left(-\infty, \frac{3}{2}\right] \cup \left[\frac{7}{2}, \infty\right)$ **51.** $(-\infty, \infty)$ **53.** $\emptyset$ **55.** $\left\{-\frac{5}{8}\right\}$ **57.** $\emptyset$ **59.** $\left\{-\frac{1}{2}\right\}$ **61.** $\left(-\infty, -\frac{2}{3}\right) \cup \left(-\frac{2}{3}, \infty\right)$ **63.** -6 or 6 **64.** $x^2 - x = 6$; $\{-2, 3\}$ **65.** $x^2 - x = -6$; $\left\{\frac{1}{2} \pm \frac{\sqrt{23}}{2}i\right\}$ **66.** $\left\{-2, 3, \frac{1}{2} \pm \frac{\sqrt{23}}{2}i\right\}$ **67.** $\left\{-\frac{1}{4}, 6\right\}$ **69.** $\{-1, 1\}$ **71.** $\emptyset$ **73.** $(-\infty, 4) \cup (4, \infty)$

In Exercises 75–81, the expression in absolute value bars may be replaced by its additive inverse. For example, in Exercise 75, $p - q$ may be written $q - p$. **75.** $|p - q| = 5$ **77.** $|m - 9| \le 8$ **79.** $|p - 9| < .0001$ **81.** $|r - 19| \ge 1$ **83.** $(.9996, 1.0004)$ **85.** $[6.5, 9.5]$ **87.** $|F - 730| \le 50$ **89.** $25.33 \le R_L \le 28.17$; $36.58 \le R_E \le 40.92$

Chapter 1 Review Exercises *(page 171)*

1. $\{6\}$ **3.** $\left\{-\frac{11}{3}\right\}$ **5.** $f = \frac{AB(p + 1)}{24}$ **7.** A, B **9.** 13 in. on each side **11.** $3\frac{3}{7}$ L **13.** 15 mph **15. (a)** $A = 36.525x$ **(b)** 2629.8 mg **17. (a)** \$3.40; It is quite close. They differ by just \$.05. **(b)** 1989 (rounded up); It is fairly close. 1990 would be exact. **19.** $10 - 3i$ **21.** $-8 + 13i$ **23.** $19 + 17i$ **25.** 146 **27.** $-30 - 40i$ **29.** $1 - 2i$ **31.** $-i$ **33.** i **35.** i **37.** $\{-7 \pm \sqrt{5}\}$ **39.** $\left\{-3, \frac{5}{2}\right\}$ **41.** $\left\{-\frac{3}{2}, 7\right\}$ **43.** $\{2 \pm \sqrt{6}\}$ **45.** $\left\{\frac{\sqrt{5} \pm 3}{2}\right\}$ **47.** D **49.** A **51.** 76; two distinct irrational solutions **53.** -124; two distinct nonreal complex solutions **55.** 0; one rational solution (a double solution) **57.** 6.25 sec and 7.5 sec **59.** $\frac{1}{2}$ ft **61.** 15,056 **63.** $\left\{\pm i, \pm\frac{1}{2}\right\}$ **65.** $\left\{-\frac{7}{24}\right\}$ **67.** $\emptyset$ **69.** $\left\{-\frac{7}{4}\right\}$ **71.** $\left\{-15, \frac{5}{2}\right\}$ **73.** $\{3\}$ **75.** $\{-2, -1\}$ **77.** $\emptyset$ **79.** $\{-4, 1\}$ **81.** $\{-1\}$ **83.** $\left(-\frac{7}{13}, \infty\right)$ **85.** $(-\infty, 1]$ **87.** $[4, 5]$ **89.** $[-4, 1]$ **91.** $\left(-\frac{2}{3}, \frac{5}{2}\right)$ **93.** $(-\infty, -4] \cup [0, 4]$ **95.** $(-\infty, -2) \cup (5, \infty)$ **97.** $(-2, 0)$ **99.** $(-3, 1) \cup [7, \infty)$ **101. (b)** 87.7 ppb **103. (a)** 20 sec **(b)** between 2 sec and 18 sec **107.** $W \ge 65$ **109.** $a \le 100{,}000$ **111.** $\{-11, 3\}$ **113.** $\left\{\frac{11}{27}, \frac{25}{27}\right\}$ **115.** $\left\{-\frac{2}{7}, \frac{4}{3}\right\}$ **117.** $[-6, -3]$ **119.** $\left(-\infty, -\frac{1}{7}\right) \cup (1, \infty)$ **121.** $\left\{-4, -\frac{2}{3}\right\}$ **123.** $(-\infty, \infty)$ **125.** $\{0, -4\}$ **127.** $|k - 3| = 12$ (or $|3 - k| = 12$) **129.** $|t - 4| \ge .01$ (or $|4 - t| \ge .01$)

Chapter 1 Test *(page 178)*

1. $\{0\}$ **2.** $\{-12\}$ **3.** $\left\{-\frac{1}{2}, \frac{7}{3}\right\}$ **4.** $\left\{\frac{-1 \pm 2\sqrt{2}}{3}\right\}$ **5.** $\left\{-\frac{1}{3} \pm \frac{\sqrt{5}}{3}i\right\}$ **6.** $\emptyset$ **7.** $\left\{-\frac{3}{4}\right\}$ **8.** $\{4\}$ **9.** $\{-3, 1\}$ **10.** $\{-2\}$ **11.** $\{\pm 1, \pm 4\}$ **12.** $\{-30, 5\}$ **13.** $\left\{-\frac{5}{2}, 1\right\}$ **14.** $\left\{-6, \frac{4}{3}\right\}$ **15.** $W = \frac{S - 2LH}{2H + 2L}$ **16. (a)** $5 - 8i$ **(b)** $-29 - 3i$ **(c)** $55 + 48i$ **(d)** $6 + i$ **17. (a)** -1 **(b)** i **(c)** i **18. (a)** $A = 806{,}400x$

18. **(b)** 24,192,000 gal **(c)** $P = 40.32x$; approximately 40 pools **(d)** approximately 24.8 days **19.** length: 100 m; width: 55 m **20.** cashews: $23\frac{1}{3}$ lb; walnuts: $11\frac{2}{3}$ lb **21.** 560 km per hr **22.** **(a)** 1 sec and 5 sec **(b)** 6 sec **23.** B **24.** $(-3, \infty)$ **25.** $[-10, 2]$ **26.** $(-\infty, -1] \cup \left[\frac{3}{2}, \infty\right)$ **27.** $(-\infty, 3) \cup (4, \infty)$ **28.** $(-2, 7)$ **29.** $(-\infty, -6] \cup [5, \infty)$ **30.** $\emptyset$

CHAPTER 2 GRAPHS AND FUNCTIONS

Connections *(page 192)*

1. Answers will vary. **2.** Latitude and longitude values pinpoint distances north or south of the equator and east or west of the prime meridian. Similarly on a Cartesian coordinate system, x- and y-coordinates give distances and directions from the y-axis and x-axis, respectively.

2.1 Exercises *(page 192)*

1. true **3.** false; The equation should be $x^2 + y^2 = 4$ to satisfy these conditions. **5.** any three of the following: $(2, -5)$, $(-1, 7)$, $(3, -9)$, $(5, -17)$, $(6, -21)$ **7.** any three of the following: (1993, 31), (1995, 35), (1997, 37), (1999, 35), (2001, 28) **9.** **(a)** $8\sqrt{2}$ **(b)** $(-9, -3)$ **11.** **(a)** $\sqrt{34}$ **(b)** $\left(\frac{11}{2}, \frac{7}{2}\right)$ **13.** **(a)** $\sqrt{202}$ **(b)** $\left(-\frac{5}{2}, -\frac{1}{2}\right)$ **15.** **(a)** $\sqrt{133}$ **(b)** $\left(2\sqrt{2}, \frac{3\sqrt{5}}{2}\right)$ **17.** yes **19.** no **21.** yes **23.** yes **25.** no **27.** no **29.** 74.2%; This is 1.1% less than the actual percent of 75.3. **31.** \$15,481

Other ordered pairs are possible in Exercises 33–43.

33. **(a)**

x	y
0	-2
4	0
2	-1

(b) $6y = 3x - 12$

35. **(a)**

x	y
0	$\frac{5}{3}$
$\frac{5}{2}$	0
4	-1

(b) $2x + 3y = 5$

37. **(a)**

x	y
0	0
1	1
-2	4

(b) $y = x^2$

39. **(a)**

x	y
3	0
4	1
7	2

(b) $y = \sqrt{x - 3}$

41. **(a)**

x	y
4	2
-2	4
0	2

(b) $y = |x - 2|$

43. **(a)**

x	y
0	0
-1	-1
2	8

(b) $y = x^3$

45. **(a)** $x^2 + y^2 = 36$ **(b)** $x^2 + y^2 = 36$

47. **(a)** $(x - 2)^2 + y^2 = 36$ **(b)** $(x - 2)^2 + y^2 = 36$

49. **(a)** $(x + 2)^2 + (y - 5)^2 = 16$ **(b)** $(x + 2)^2 + (y - 5)^2 = 16$

51. **(a)** $(x - 5)^2 + (y + 4)^2 = 49$ **(b)**

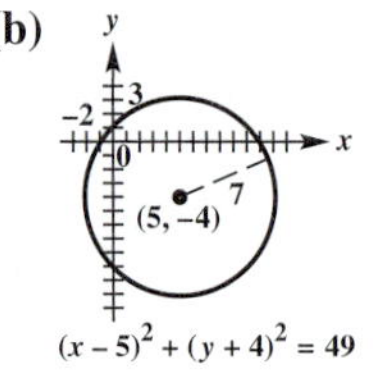

53. $(x - 3)^2 + (y - 2)^2 = 4$ **57.** yes; center: $(-3, -4)$; radius: 4 **59.** yes; center: $(2, -6)$; radius: 6 **61.** yes; center: $\left(-\frac{1}{2}, 2\right)$; radius: 3 **63.** no **65.** $(2, -3)$ **66.** $3\sqrt{5}$ **67.** $3\sqrt{5}$ **68.** $3\sqrt{5}$

69. $(x-2)^2+(y+3)^2=45$ **70.** $(x+2)^2+(y+1)^2=41$ **71.** $(-3,6)$ **73.** $(5,-4)$ **77.** $(4,0)$
79. III; I; IV; IV **81.** yes; no **83.** $(2+\sqrt{7}, 2+\sqrt{7}), (2-\sqrt{7}, 2-\sqrt{7})$ **85.** $(2,3)$ and $(4,1)$
87. $9+\sqrt{119}, 9-\sqrt{119}$

2.2 Exercises *(page 209)*

3. independent variable **5.** function **7.** not a function **9.** function **11.** not a function; domain: $\{0,1,2\}$; range: $\{-4,-1,0,1,4\}$ **13.** function; domain: $\{2,3,5,11,17\}$; range: $\{1,7,20\}$
15. function; domain: $\{1997, 1998, 1999, 2000\}$; range: $\{547{,}200{,}000,\ 528{,}500{,}000,\ 564{,}100{,}000,\ 587{,}100{,}000\}$
17. function; domain: $(-\infty,\infty)$; range: $(-\infty,\infty)$ **19.** not a function; domain: $[3,\infty)$; range: $(-\infty,\infty)$
21. not a function; domain: $[-4,4]$; range: $[-3,3]$ **23.** function; domain: $(-\infty,\infty)$; range: $[0,\infty)$
25. not a function; domain: $[0,\infty)$; range: $(-\infty,\infty)$ **27.** function; domain: $(-\infty,\infty)$; range: $(-\infty,\infty)$
29. not a function; domain: $(-\infty,\infty)$; range: $(-\infty,\infty)$ **31.** function; domain: $[0,\infty)$; range: $[0,\infty)$
33. function; domain: $(-\infty,0)\cup(0,\infty)$; range: $(-\infty,0)\cup(0,\infty)$ **35.** function; domain: $\left[-\frac{1}{2},\infty\right)$; range: $[0,\infty)$
37. function; domain: $(-\infty,9)\cup(9,\infty)$; range: $(-\infty,0)\cup(0,\infty)$ **39.** B **41.** 4 **43.** -11 **45.** $-3p+4$
47. $3x+4$ **49.** $-3x-2$ **51.** $-6m+13$ **53.** **(a)** 2 **(b)** 3 **55.** **(a)** 15 **(b)** 10 **57.** **(a)** 3 **(b)** -3
59. **(a)** $f(x)=-\frac{1}{3}x+4$ **(b)** 3 **61.** **(a)** $f(x)=-2x^2+3$ **(b)** -15 **63.** **(a)** $f(x)=\frac{4}{3}x-\frac{8}{3}$ **(b)** $\frac{4}{3}$ **67.** -4
69. **(a)** 0 **(b)** 4 **(c)** 2 **(d)** 4 **71.** **(a)** -3 **(b)** -2 **(c)** 0 **(d)** 2 **73.** **(a)** $[4,\infty)$ **(b)** $(-\infty,-1]$ **(c)** $[-1,4]$
75. **(a)** $(-\infty,4]$ **(b)** $[4,\infty)$ **(c)** none **77.** **(a)** none **(b)** $(-\infty,-2]$; $[3,\infty)$ **(c)** $(-2,3)$ **79.** **(a)** yes **(b)** $[0,24]$
(c) 1200 megawatts **(d)** at 17 hr or 5 P.M.; at 4 A.M. **(e)** $f(12)=2000$; At 12 noon, electricity use is 2000 megawatts.
(f) increasing from 4 A.M. to 5 P.M.; decreasing from midnight to 4 A.M. and from 5 P.M. to midnight
81. **(a)** about 12 noon to about 8 P.M. **(b)** from midnight until about 6 A.M. and after 10 P.M. **(c)** about 10 A.M. and 8:30 P.M.

2.3 Exercises *(page 221)*

1. B **3.** C **5.** A

In Exercises 7–23, we give the domain first and then the range.

7. $(-\infty,\infty)$; $(-\infty,\infty)$

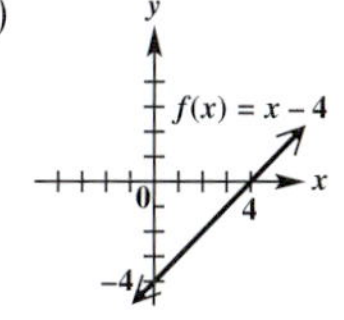

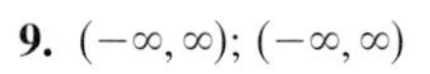
9. $(-\infty,\infty)$; $(-\infty,\infty)$

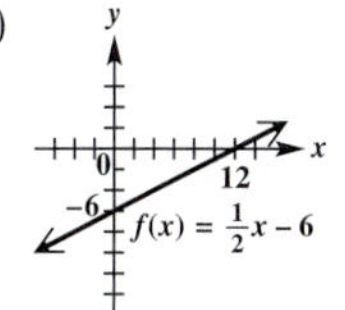

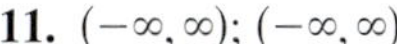
11. $(-\infty,\infty)$; $(-\infty,\infty)$

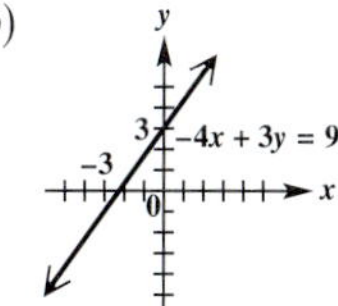

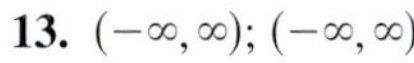
13. $(-\infty,\infty)$; $(-\infty,\infty)$

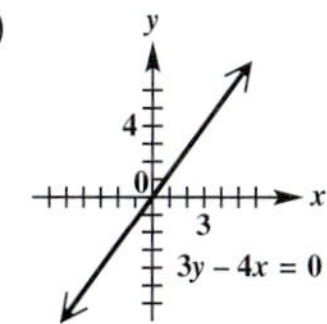

15. $(-\infty,\infty)$; $(-\infty,\infty)$

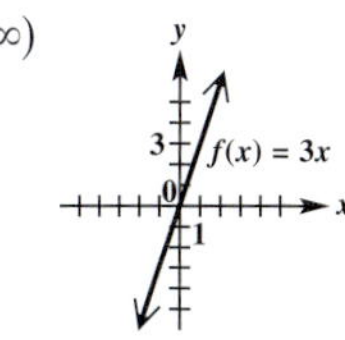

17. $(-\infty,\infty)$; $\{-4\}$; constant function

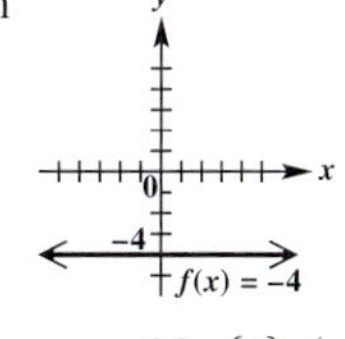

19. $\{3\}$; $(-\infty,\infty)$

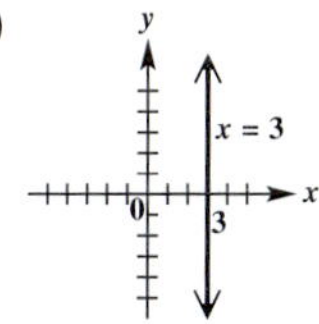

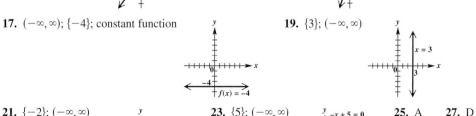

21. $\{-2\}$; $(-\infty,\infty)$

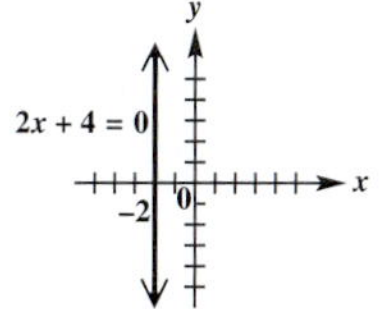

23. $\{5\}$; $(-\infty,\infty)$

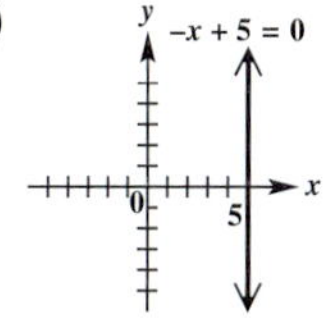
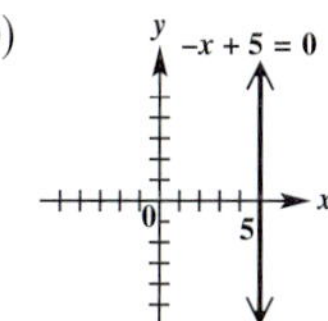

25. A **27.** D

29. 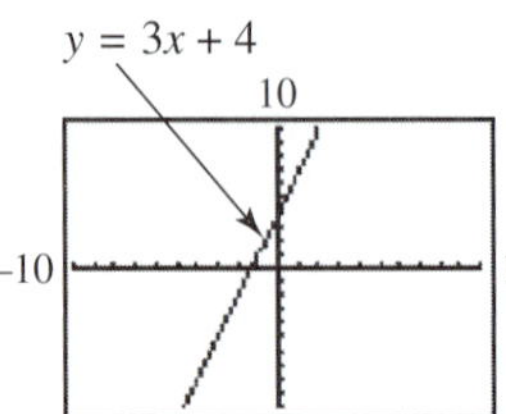

31. 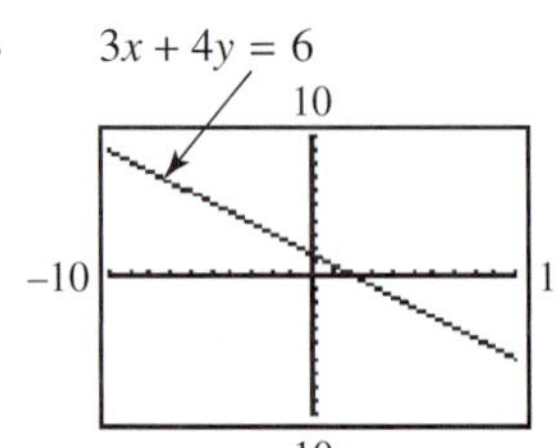

33. A, C, D, E **35.** $\frac{2}{5}$ **37.** 0

39. 0 **41.** undefined **45.** $m = 3$

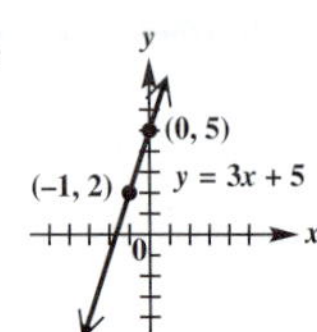

47. $m = -\frac{3}{2}$

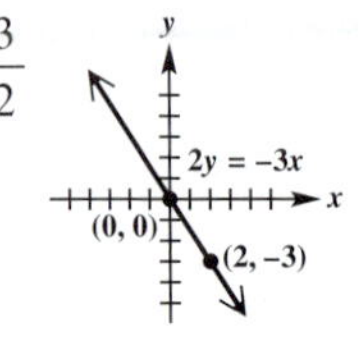

49. $m = \frac{5}{2}$

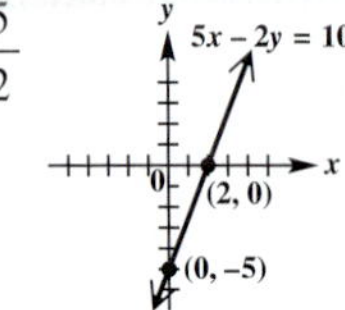

53. 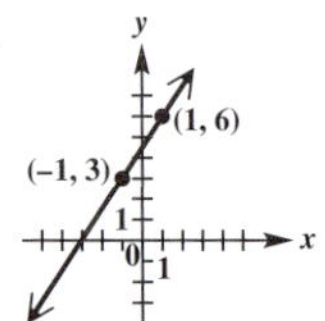

55. 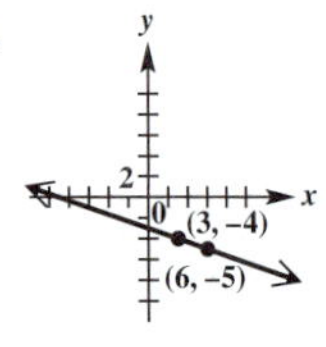

57. 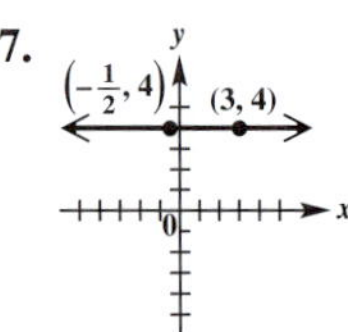

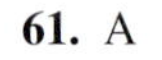
59. D **61.** A **63.** E

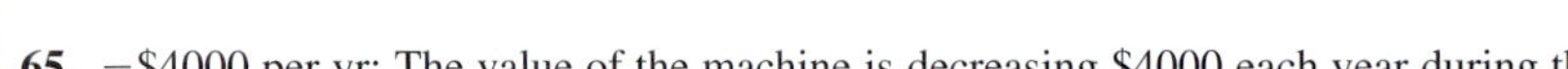
65. −\$4000 per yr; The value of the machine is decreasing \$4000 each year during these years.
67. 0% per yr (or no change); The percent of pay raise is not changing—it is 3% each year during these years.
69. **(b)** 189.5; This means that the average rate of change in the number of radio stations per year is an increase of 189.5.
71. **(a)** −1.7 million recipients per yr **(b)** The negative slope means the numbers of recipients *decreased* by 1.7 million each year.
73. .34% per yr; The percent of freshmen listing business as their probable field of study increased an average of .34% per yr from 1995 to 2000. **75.** 3.03 million per yr; Sales of DVD players increased an average of 3.03 million each year from 1997 to 2002. **76.** 3 **77.** 3 **78.** the same **79.** $\sqrt{10}$ **80.** $2\sqrt{10}$ **81.** $3\sqrt{10}$ **82.** The sum is $3\sqrt{10}$, which is equal to the answer in Exercise 81. **83.** B; C; A; C (The order of the last two may be reversed.) **84.** The midpoint is $(3, 3)$, which is the same as the middle entry in the table. **85.** 7.5 **87.** **(a)** $C(x) = 11x + 180$ **(b)** $R(x) = 20x$ **(c)** $P(x) = 9x - 180$ **(d)** 20 units; produce **89.** **(a)** $C(x) = 400x + 1650$ **(b)** $R(x) = 305x$ **(c)** $P(x) = -95x - 1650$ **(d)** $R(x) < C(x)$ for all positive x; don't produce, impossible to make a profit

2.4 Exercises *(page 236)*

1. D **3.** C **5.** $2x + y = 5$ **7.** $3x + 2y = -7$ **9.** $x = -8$ **11.** $y = \frac{1}{4}x + \frac{13}{4}$ **13.** $y = \frac{2}{3}x - 2$

15. $x = -6$ (cannot be written in slope-intercept form) **17.** $y = 5x + 15$ **19.** $y = -\frac{2}{3}x - \frac{4}{5}$ **21.** $y = \frac{3}{2}$

23. −2; does not; undefined; $\frac{1}{2}$; does not; 0 **25.** **(a)** B **(b)** D **(c)** A **(d)** C

27. slope: 3; y-intercept: −1 **29.** slope: 4; y-intercept: −7 **31.** slope: $-\frac{3}{4}$; y-intercept: 0

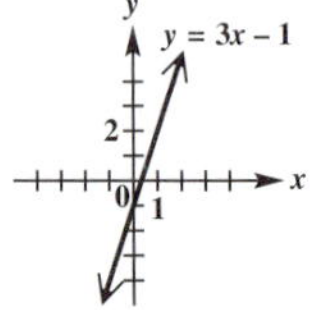

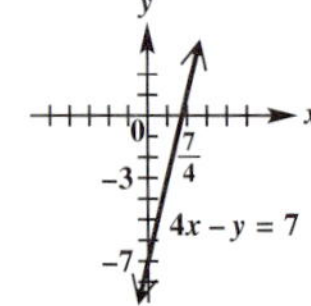

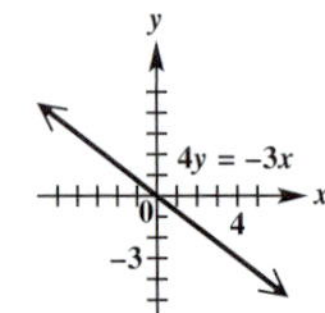

33. slope: $-\frac{1}{2}$; y-intercept: -2

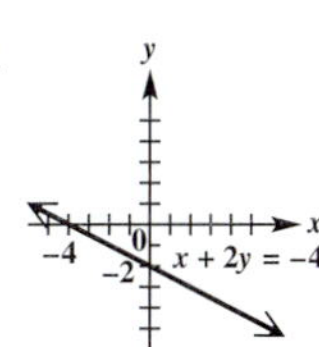

35. **(a)** $x + 3y = 11$ **(b)** $y = -\frac{1}{3}x + \frac{11}{3}$

37. **(a)** $5x - 3y = -13$ **(b)** $y = \frac{5}{3}x + \frac{13}{3}$

39. **(a)** $y = 1$ **(b)** $y = 1$

45. $y = .624x - 1185.98$; 59.5%; This figure is very close to the actual figure.

47. **(a)** $f(x) \approx 731.3x + 9340$

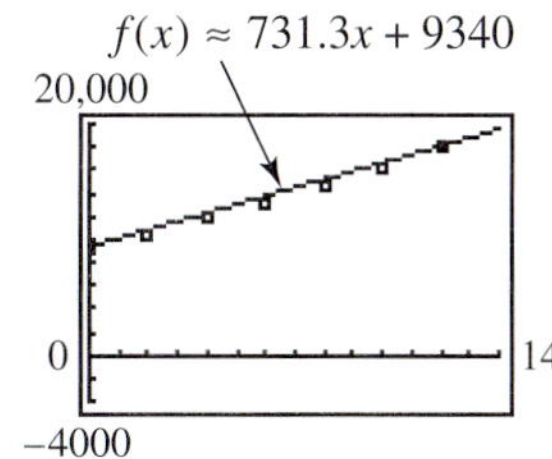

The average tuition increase is about \$731 per year for the period, because this is the slope of the line. **(b)** $f(5) \approx 12{,}996.5$; This is a fairly good approximation. **(c)** $f(x) \approx 730.14x + 8984.71$

49. **(a)** $F = \frac{9}{5}C + 32$ **(b)** $C = \frac{5}{9}(F - 32)$ **(c)** $-40°$ **51.** **(a)** $C = -.6797I + 8359$ **(b)** $-.6797$ **53.** **(a)** $y = 2x - 2$ **(b)** 1 **(c)** $\{1\}$ **55.** **(a)** $y = 2x - 8$ **(b)** 4 **(c)** $\{4\}$ **57.** **(a)** $\{12\}$ **58.** the Pythagorean theorem and its converse **59.** $\sqrt{x_1^2 + m_1^2x_1^2}$ **60.** $\sqrt{x_2^2 + m_2^2x_2^2}$ **61.** $\sqrt{(x_2 - x_1)^2 + (m_2x_2 - m_1x_1)^2}$ **63.** $-2x_1x_2(m_1m_2 + 1) = 0$ **64.** Since $x_1 \neq 0$, $x_2 \neq 0$, we have $m_1m_2 + 1 = 0$, implying that $m_1m_2 = -1$. **65.** If two nonvertical lines are perpendicular, then the product of the slopes of these lines is -1. **69.** yes

Summary Exercises on Graphs, Functions, and Equations *(page 241)*

1. **(a)** $\sqrt{65}$ **(b)** $\left(\frac{5}{2}, 1\right)$ **(c)** $y = 8x - 19$ **2.** **(a)** $\sqrt{29}$ **(b)** $\left(\frac{3}{2}, -1\right)$ **(c)** $y = -\frac{2}{5}x - \frac{2}{5}$ **3.** **(a)** 5 **(b)** $\left(\frac{1}{2}, 2\right)$ **(c)** $y = 2$ **4.** **(a)** $\sqrt{10}$ **(b)** $\left(\frac{3\sqrt{2}}{2}, 2\sqrt{2}\right)$ **(c)** $y = -2x + 5\sqrt{2}$ **5.** **(a)** 2 **(b)** $(5, 0)$ **(c)** $x = 5$ **6.** **(a)** $4\sqrt{2}$ **(b)** $(-1, -1)$ **(c)** $y = x$ **7.** **(a)** $4\sqrt{3}$ **(b)** $(4\sqrt{3}, 3\sqrt{5})$ **(c)** $y = 3\sqrt{5}$ **8.** **(a)** $\sqrt{34}$ **(b)** $\left(\frac{3}{2}, -\frac{3}{2}\right)$ **(c)** $y = \frac{5}{3}x - 4$

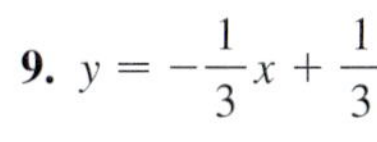

9. $y = -\frac{1}{3}x + \frac{1}{3}$

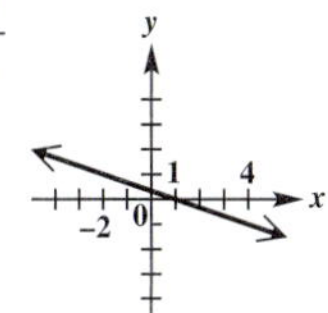

10. $y = 3$

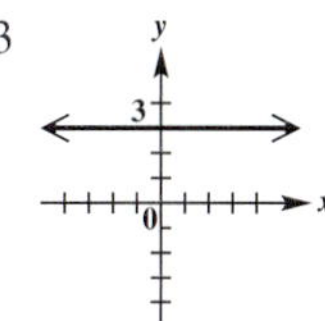

11. $(x - 2)^2 + (y + 1)^2 = 9$

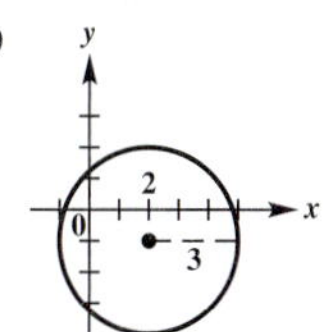

12. $x^2 + (y - 2)^2 = 4$

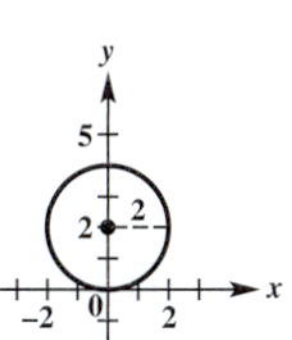

13. $y = -\frac{5}{6}x - \frac{5}{2}$

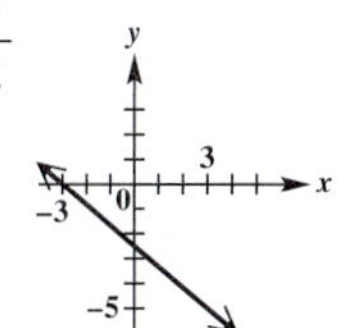

14. $y = -\frac{4}{3}x$

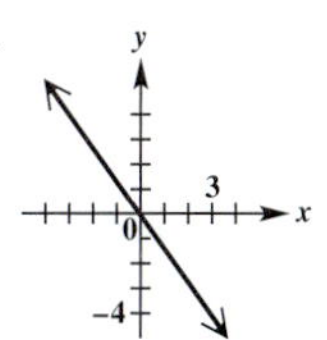

15. $y = -\frac{2}{3}x$

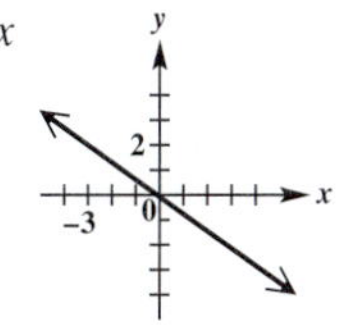

16. $x = -4$

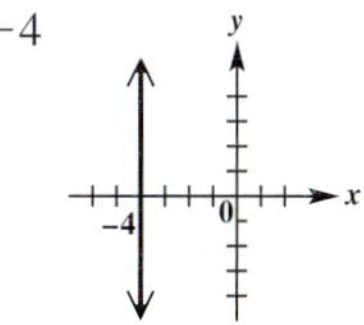

17. yes; center: $(2, -1)$; radius: 3 **18.** no

19. yes; center: $(6, 0)$; radius: 4 **20.** yes; center: $(-1, -8)$; radius: 2 **21.** no **22.** yes; center: $(0, 4)$; radius: 5 **23.** **(a)** domain: $(-\infty, \infty)$; range: $(-\infty, \infty)$ **(b)** $f(x) = \frac{3}{2}x - \frac{1}{2}$; $-\frac{7}{2}$ **24.** **(a)** domain: $(-\infty, \infty)$; range: $(-\infty, -1]$

24. **(b)** $f(x) = -3x^2 - 1$; -13 **25.** **(a)** domain: $(-\infty, \infty)$; range: $(-\infty, \infty)$ **(b)** $f(x) = \frac{1}{4}x + \frac{3}{2}$; 1 **26.** **(a)** domain: $[-5, \infty)$; range: $(-\infty, \infty)$ **(b)** y is not a function of x. **27.** **(a)** domain: $[-7, 3]$; range: $[-5, 5]$ **(b)** y is not a function of x. **28.** **(a)** domain: $(-\infty, \infty)$; range: $\left[-\frac{3}{2}, \infty\right)$ **(b)** $f(x) = \frac{1}{2}x^2 - \frac{3}{2}$; $\frac{1}{2}$

2.5 Exercises *(page 249)*

1. $(-\infty, \infty)$ **3.** $[0, \infty)$ **5.** $(-\infty, 1)$; $[1, \infty)$ **7.** E; $(-\infty, \infty)$ **9.** A; $(-\infty, \infty)$ **11.** F; $y = x$ **13.** H; no **15.** B; $\{\ldots, -3, -2, -1, 0, 1, 2, 3, \ldots\}$ **17.** **(a)** -10 **(b)** -2 **(c)** -1 **(d)** 2 **19.** **(a)** -3 **(b)** 1 **(c)** 0 **(d)** 9

21.

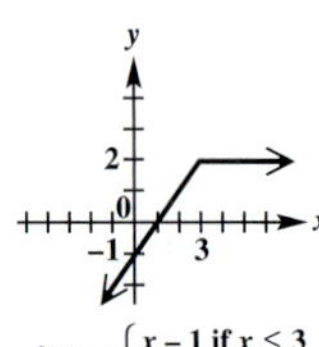

$f(x) = \begin{cases} x - 1 & \text{if } x \le 3 \\ 2 & \text{if } x > 3 \end{cases}$

23.

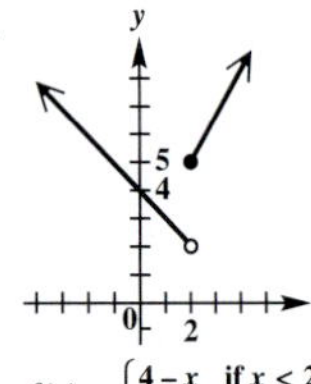

$f(x) = \begin{cases} 4 - x & \text{if } x < 2 \\ 1 + 2x & \text{if } x \ge 2 \end{cases}$

25.

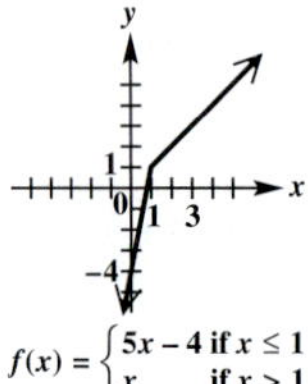

$f(x) = \begin{cases} 5x - 4 & \text{if } x \le 1 \\ x & \text{if } x > 1 \end{cases}$

27.

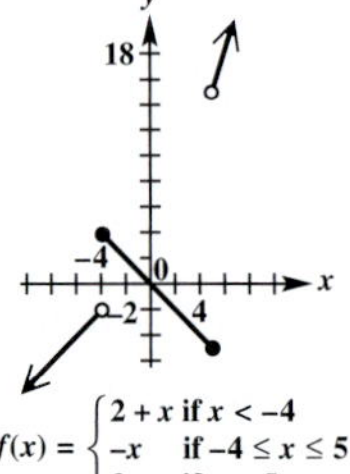

$f(x) = \begin{cases} 2 + x & \text{if } x < -4 \\ -x & \text{if } -4 \le x \le 5 \\ 3x & \text{if } x > 5 \end{cases}$

In Exercises 29 and 31 we give the rule, then the domain, and then the range.

29. $f(x) = \begin{cases} -1 & \text{if } x \le 0 \\ 1 & \text{if } x > 0 \end{cases}$; $(-\infty, \infty)$; $\{-1, 1\}$ **31.** $f(x) = \begin{cases} 2 & \text{if } x \le 0 \\ -1 & \text{if } x > 1 \end{cases}$; $(-\infty, 0] \cup (1, \infty)$; $\{-1, 2\}$

33. $(-\infty, \infty)$; $\{\ldots, -2, -1, 0, 1, 2, \ldots\}$

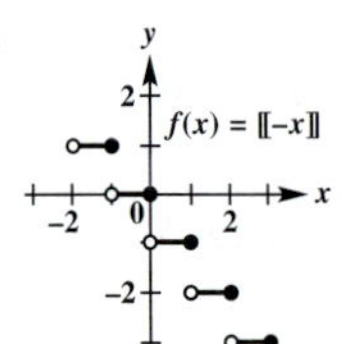

35. $(-\infty, \infty)$; $\{\ldots, -2, -1, 0, 1, 2, \ldots\}$

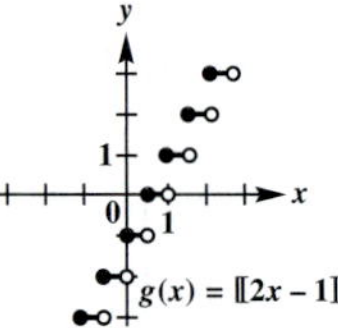

37.

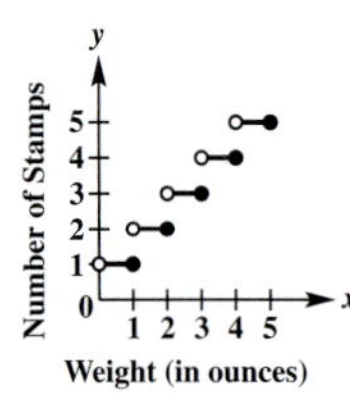

39. B **41.** D **43.** **(a)** for $[3, 6]$: $y = -\frac{1}{3}x + 74$; for $(6, 9]$: $y = -x + 78$

(b) $f(x) = \begin{cases} -\frac{1}{3}x + 74 & \text{if } 3 \le x \le 6 \\ -x + 78 & \text{if } 6 < x \le 9 \end{cases}$

45. **(a)** 140 **(b)** 220 **(c)** 220 **(d)** 220 **(e)** 140 **(f)** 60 **(g)** 60

(h)

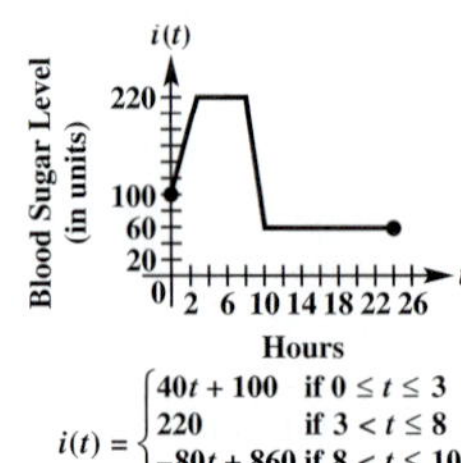

47. $y_1 = \sqrt{x}$; $y_2 = -\sqrt{x}$

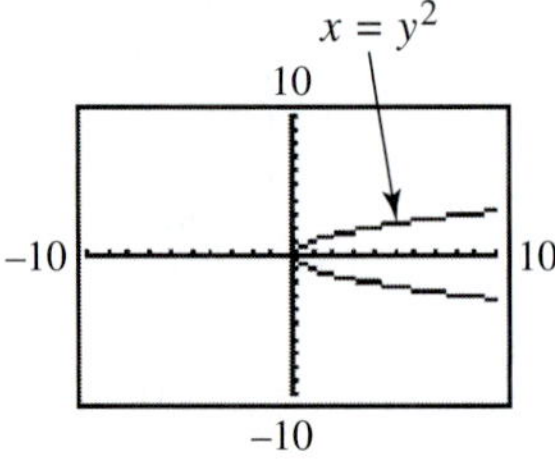

2.6 Exercises *(page 264)*

1. (a) B **(b)** D **(c)** E **(d)** A **(e)** C **3. (a)** B **(b)** A **(c)** G **(d)** C **(e)** F **(f)** D **(g)** H **(h)** E

5.

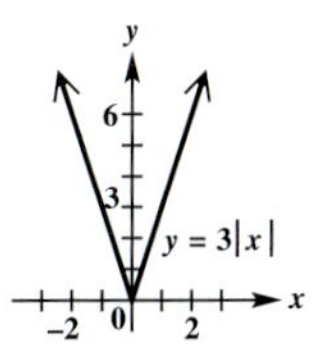

7.

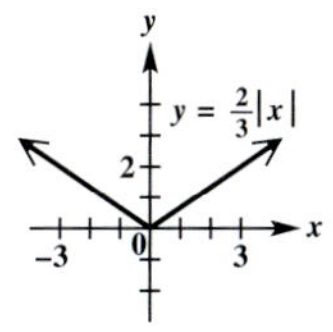

9.

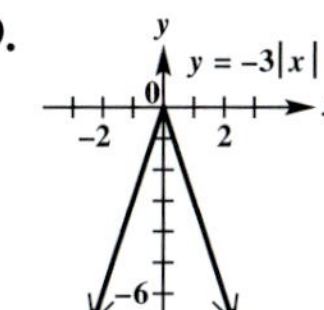

11.

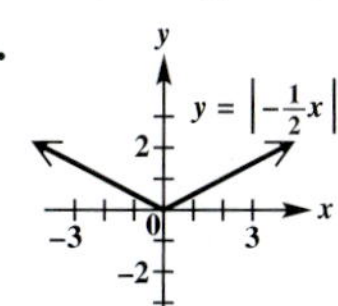

13. (a) $(8, 3)$ **(b)** $(8, 48)$

15.

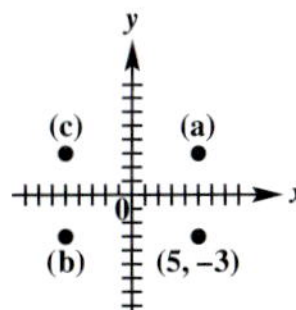

17.

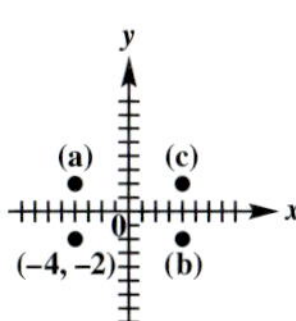

19. y-axis **21.** x-axis, y-axis, origin **23.** origin **25.** none of these

27. odd **29.** even **31.** neither **33.**

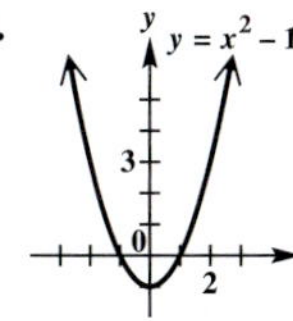

35.

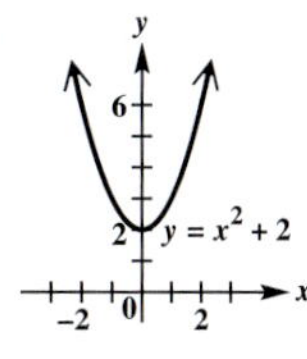

37.

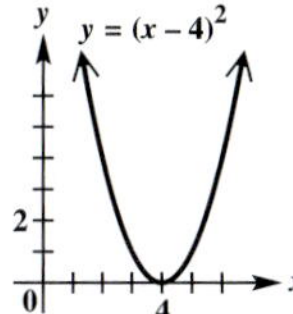

39.

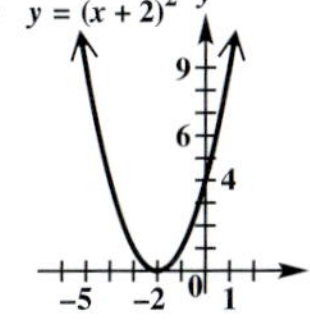

41.

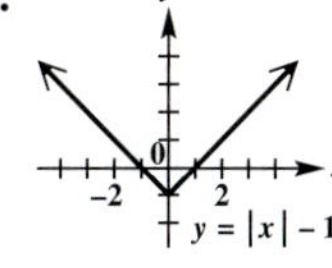

43.

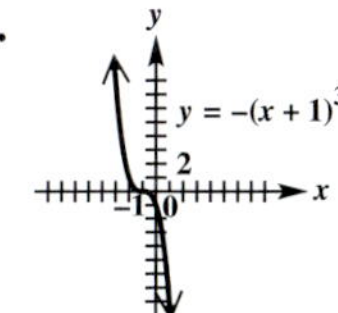

45.

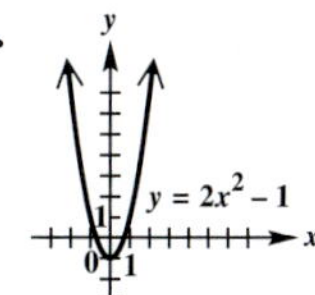

47.

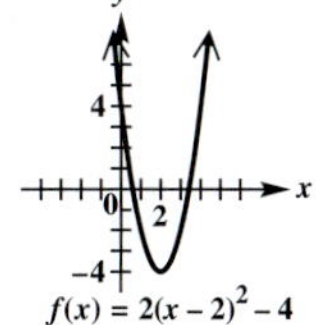

49. (a)

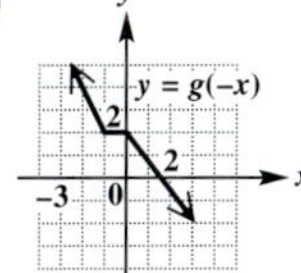

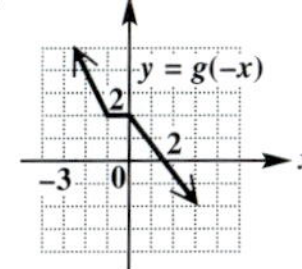

The graph of $g(x)$ is reflected across the y-axis.

(b)

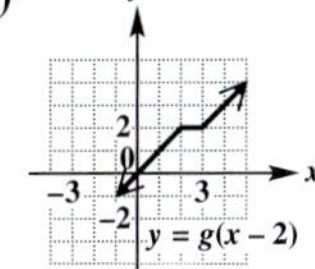

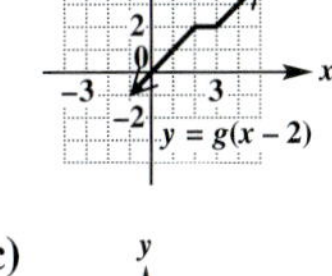

The graph of $g(x)$ is translated to the right 2 units.

(c)

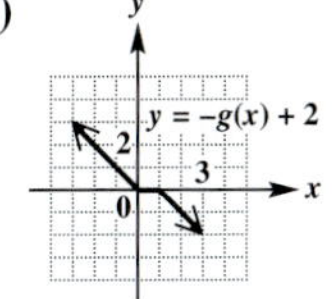

The graph of $g(x)$ is reflected across the x-axis and translated 2 units up.

51. It is the graph of $f(x) = |x|$ translated 1 unit to the left, reflected across the x-axis, and translated 3 units up. The equation is $y = -|x + 1| + 3$. **53.** It is the graph of $g(x) = \sqrt{x}$ translated 4 units to the left, stretched vertically by a factor of 2, and translated 4 units down. The equation is $y = 2\sqrt{x + 4} - 4$. **55.** $f(-3) = -6$ **57.** $f(9) = 6$ **59.** $f(-3) = -6$

61. $g(x) = 2x + 13$ **63. (a)**

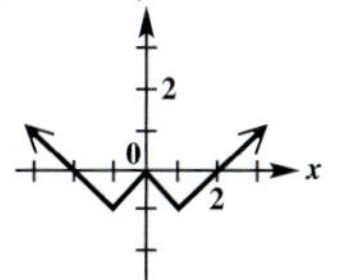

(b)

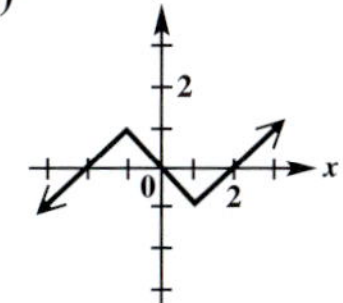

67. It is translated 6 units up.

68. $G(x) = x + 6$ **69.** It is translated 6 units up. **70.** It is translated 6 units to the right. **71.**

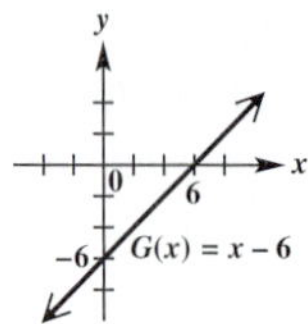

72. It is translated 6 units to the right. **73.** **(a)** 2 **(b)** 4

2.7 Exercises *(page 276)*

1. 61 **3.** 2016 **5.** $-\frac{7}{2}$ **7.** $5m^2 - 8m - 4$ **9.** $5x - 1$; $x + 9$; $6x^2 - 7x - 20$; $\frac{3x+4}{2x-5}$; all domains are $(-\infty, \infty)$ except for that of $\frac{f}{g}$, which is $\left(-\infty, \frac{5}{2}\right) \cup \left(\frac{5}{2}, \infty\right)$. **11.** $3x^2 - 4x + 3$; $x^2 - 2x - 3$; $2x^4 - 5x^3 + 9x^2 - 9x$; $\frac{2x^2 - 3x}{x^2 - x + 3}$; all domains are $(-\infty, \infty)$. **13.** $\sqrt{4x-1} + \frac{1}{x}$; $\sqrt{4x-1} - \frac{1}{x}$; $\frac{\sqrt{4x-1}}{x}$; $x\sqrt{4x-1}$; all domains are $\left[\frac{1}{4}, \infty\right)$.
15. 7.7; 11.8; 19.5 **17.** 1991–1996 **19.** 6; It represents the dollars in billions spent for general science in 2000.
21. space and other technologies; 1995–2000 **23.** **(a)** 2 **(b)** 4 **(c)** 0 **(d)** $-\frac{1}{3}$ **25.** **(a)** 3 **(b)** −5 **(c)** 2 **(d)** undefined
27. **(a)** 5 **(b)** 5 **(c)** 0 **(d)** undefined **29.**

x	$(f+g)(x)$	$(f-g)(x)$	$(fg)(x)$	$\left(\frac{f}{g}\right)(x)$
−2	6	−6	0	0
0	5	5	0	undefined
2	5	9	−14	−3.5
4	15	5	50	2

33. **(a)** $2 - x - h$ **(b)** $-h$ **(c)** −1 **35.** **(a)** $6x + 6h + 2$ **(b)** $6h$ **(c)** 6 **37.** **(a)** $-2x - 2h + 5$ **(b)** $-2h$ **(c)** −2
39. **(a)** $x^2 + 2xh + h^2 - 4$ **(b)** $2xh + h^2$ **(c)** $2x + h$ **41.** −5 **43.** 7 **45.** 6 **47.** −1 **49.** 1 **51.** 9
53. 1 **55.** $g(1) = 9$, and $f(9)$ cannot be determined from the table given. **57.** $-30x - 33$; $-30x + 52$
59. $4x^2 + 42x + 118$; $4x^2 + 2x + 13$ **61.** $\frac{2}{(2-x)^4}$; $(-\infty, 2) \cup (2, \infty)$; $2 - \frac{2}{x^4}$; $(-\infty, 0) \cup (0, \infty)$
63. $36x + 72 - 22\sqrt{x+2}$; $2\sqrt{9x^2 - 11x + 2}$ **65.**

x	$f(x)$	$g(x)$	$g[f(x)]$
1	3	2	7
2	1	5	2
3	2	7	5

In Exercises 73–77, we give only one of the many possible ways.
73. $g(x) = 6x - 2$, $f(x) = x^2$ **75.** $g(x) = x^2 - 1$, $f(x) = \sqrt{x}$ **77.** $g(x) = 6x$, $f(x) = \sqrt{x} + 12$
79. $(f \circ g)(x) = 63{,}360x$ computes the number of inches in x miles. **81.** **(a)** $A(2x) = \sqrt{3}x^2$ **(b)** $64\sqrt{3}$ square units
83. **(a)** $(A \circ r)(t) = 16\pi t^2$ **(b)** It defines the area of the leak in terms of the time t, in minutes. **(c)** 144π ft^2
85. **(a)** $N(x) = 100 - x$ **(b)** $G(x) = 20 + 5x$ **(c)** $C(x) = (100 - x)(20 + 5x)$ **(d)** \$9600

Chapter 2 Review Exercises *(page 285)*

1. $\sqrt{85}$; $\left(-\frac{1}{2}, 2\right)$ **3.** 5; $\left(-6, \frac{11}{2}\right)$ **5.** −7; −1; 8; 23 **7.** $(x+2)^2 + (y-3)^2 = 225$
9. $(x+8)^2 + (y-1)^2 = 289$ **11.** $(2, -3)$; 1 **13.** $\left(-\frac{7}{2}, -\frac{3}{2}\right)$; $\frac{3\sqrt{6}}{2}$ **15.** $3 + 2\sqrt{5}$; $3 - 2\sqrt{5}$
17. It is the single point $(4, -5)$. **19.** no; $[-6, 6]$; $[-6, 6]$ **21.** no; $(-\infty, \infty)$; $(-\infty, -1] \cup [1, \infty)$
23. no; $[0, \infty)$; $(-\infty, \infty)$ **25.** function of x **27.** function of x **29.** $(-\infty, 8) \cup (8, \infty)$ **31.** $[-7, 7]$ **33.** 0

35. -8.52 **37.** $-2k^2 + 3k - 6$ **39.**

41.
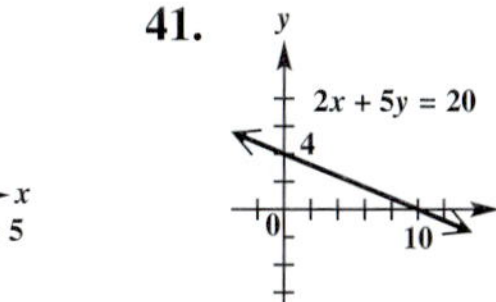
43.
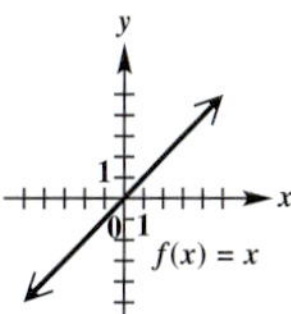

45.
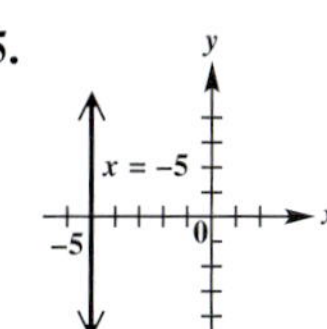
47.
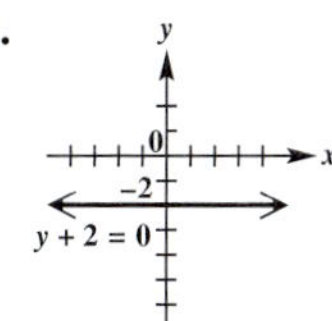
49.
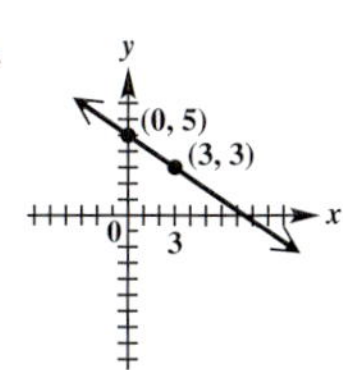
51. -2 **53.** 0 **55.** $-\frac{11}{2}$
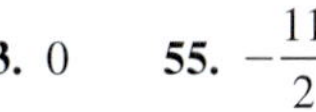

57. undefined **59.** Initially, the car is at home. After traveling 30 mph for 1 hr, the car is 30 mi away from home. During the second hour the car travels 20 mph until it is 50 mi away. During the third hour the car travels toward home at 30 mph until it is 20 mi away. During the fourth hour the car travels away from home at 40 mph until it is 60 mi away from home. During the last hour, the car travels 60 mi at 60 mph until it arrives home. **61. (a)** $y = 3.62x - 9.12$; The slope, 3.62, indicates that the number of e-filing taxpayers increased by 3.62% each year from 1996 to 2001. **(b)** 45.18%

63. $y = -2x + 1$ **65.** $y = 3x - 7$ **67.** $y = -10$ **69.** $x = -7$ **71.**

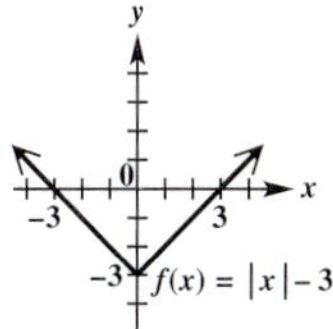

73.
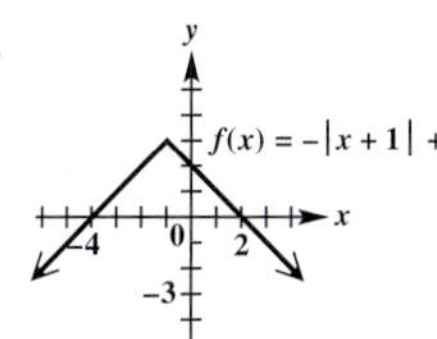
75.
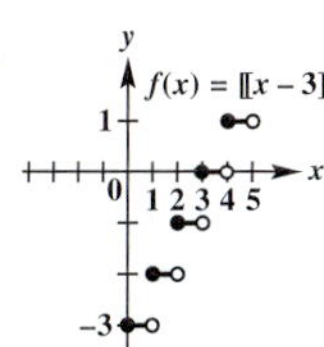
77.
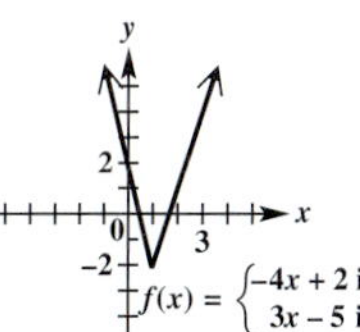
79.
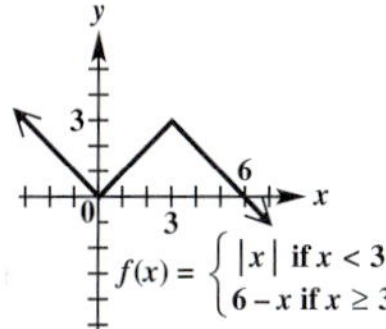

81. true **83.** false; For example, $f(x) = x^2$ is even, and $(2, 4)$ is on the graph but $(2, -4)$ is not. **85.** true **87.** x-axis **89.** y-axis **91.** none of these **93.** Reflect the graph of $f(x) = |x|$ across the x-axis. **95.** Translate the graph of $f(x) = |x|$ to the right 4 units and stretch vertically by a factor of 2. **97.** $y = -3x - 4$

99. (a)
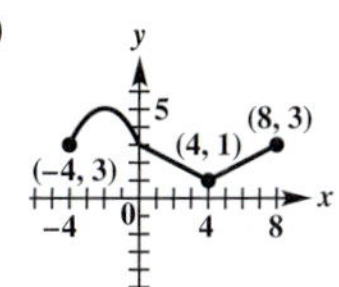
(b)
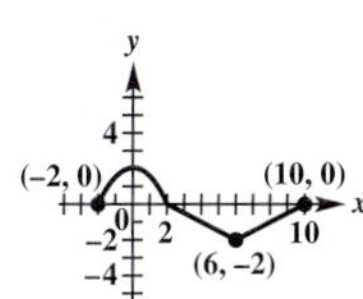
(c)
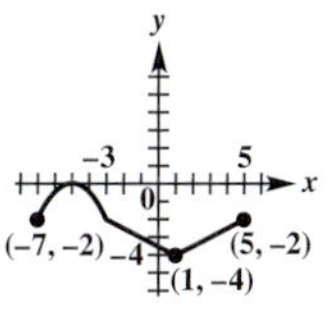
(d)
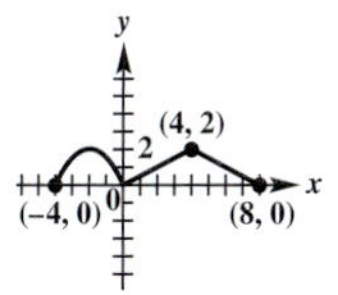

101. $3x^4 - 9x^3 - 16x^2 + 12x + 16$ **103.** 68 **105.** $-\frac{23}{4}$ **107.** $(-\infty, \infty)$ **109.** C and D **111.** $2x + h - 5$

113. $x - 2$ **115.** 1 **117.** 8 **119.** -6 **121.** 2 **123.** 1

125. $f(x) = 36x$; $g(x) = 1760x$; $(g \circ f)(x) = g[f(x)] = 1760(36x) = 63{,}360x$ **127.** $V(r) = \frac{4}{3}\pi(r + 3)^3 - \frac{4}{3}\pi r^3$

Chapter 2 Test *(page 290)*

1. (a) D **(b)** D **(c)** C **(d)** B **(e)** C **(f)** C **(g)** C **(h)** D **(i)** D **(j)** C **2.** $\frac{3}{5}$ **3.** $\sqrt{34}$ **4.** $\left(\frac{1}{2}, \frac{5}{2}\right)$

5. $3x - 5y = -11$ **6.** $f(x) = \frac{3}{5}x + \frac{11}{5}$ **7. (a)** not a function; domain: $[0, 4]$; range: $[-4, 4]$ **(b)** function; domain: $(-\infty, -1) \cup (-1, \infty)$; range: $(-\infty, 0) \cup (0, \infty)$; decreasing on $(-\infty, -1)$ and on $(-1, \infty)$ **8. (a)** $x = 5$ **(b)** $y = -3$

9. **(a)** $y = -3x + 9$ **(b)** $y = \frac{1}{3}x + \frac{7}{3}$ **10.** **(a)** $(-\infty, -3)$ **(b)** $(4, \infty)$ **(c)** $[-3, 4]$ **(d)** $(-\infty, -3)$; $[-3, 4]$; $(4, \infty)$ **(e)** $(-\infty, \infty)$ **(f)** $(-\infty, 2)$ **11.**

$y = |x - 2| - 1$

12.

$f(x) = [\![x + 1]\!]$

13.

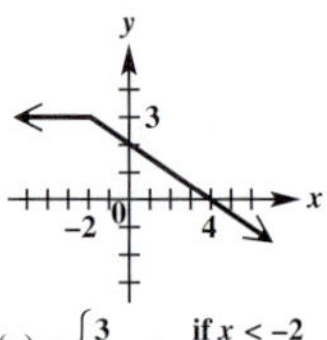

14. **(a)** $y = f(x) + 2$; (0, 2), (4, 2), (1, −1) **(b)** $y = f(x + 2)$; (−2, 0), (2, 0), (−1, −3) **(c)** $y = -f(x)$; (1, 3), (0, 0), (4, 0) **(d)**

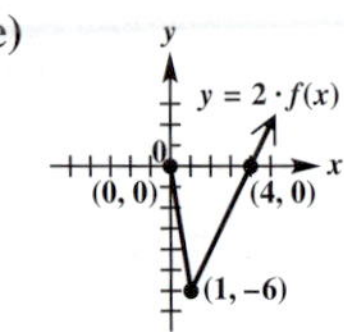

(e) $y = 2 \cdot f(x)$; (0, 0), (4, 0), (1, −6)

16. **(a)** yes **(b)** yes **(c)** yes **17.** **(a)** $2x^2 - x + 1$ **(b)** $\frac{2x^2 - 3x + 2}{-2x + 1}$ **(c)** $\left(-\infty, \frac{1}{2}\right) \cup \left(\frac{1}{2}, \infty\right)$ **(d)** $4x + 2h - 3$ **18.** **(a)** 0 **(b)** -12 **(c)** 1 **19.** \$2.75 **20.** **(a)** $C(x) = 3300 + 4.50x$ **(b)** $R(x) = 10.50x$ **(c)** $R(x) - C(x) = 6.00x - 3300$ **(d)** 551

CHAPTER 3 POLYNOMIAL AND RATIONAL FUNCTIONS

3.1 Exercises *(page 303)*

1. **(a)** domain: $(-\infty, \infty)$; range: $[-4, \infty)$ **(b)** $(-3, -4)$ **(c)** $x = -3$ **(d)** 5 **(e)** $-5, -1$ **3.** **(a)** domain: $(-\infty, \infty)$; range: $(-\infty, 2]$ **(b)** $(-3, 2)$ **(c)** $x = -3$ **(d)** -16 **(e)** $-4, -2$ **5.** B **7.** D **9.** **(e)** If the absolute value of the coefficient is greater than 1, it causes the graph to be stretched vertically, so it is narrower. If the absolute value of the coefficient is between 0 and 1, it causes the graph to shrink vertically, so it is broader.

11. **(e)** The graph of $(x - h)^2$ is translated h units to the right if h is positive and $|h|$ units to the left if h is negative.

13. vertex: $(2, 0)$; axis: $x = 2$; domain: $(-\infty, \infty)$; range: $[0, \infty)$

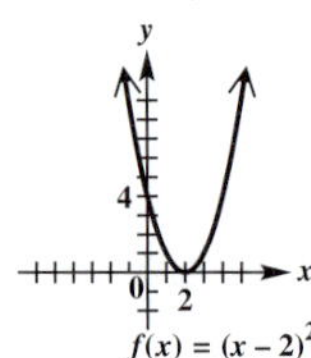

15. vertex: $(-3, -4)$; axis: $x = -3$; domain: $(-\infty, \infty)$; range: $[-4, \infty)$

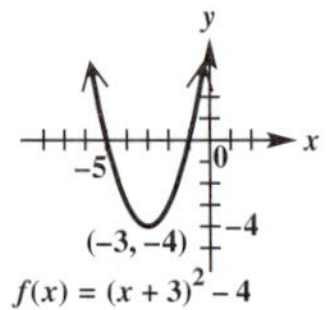

17. vertex: $(-1, -3)$; axis: $x = -1$; domain: $(-\infty, \infty)$; range: $(-\infty, -3]$

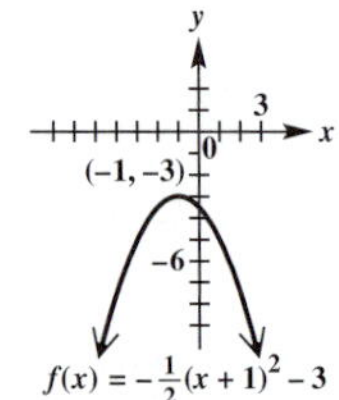

19. vertex: $(1, 2)$; axis: $x = 1$; domain: $(-\infty, \infty)$; range: $[2, \infty)$

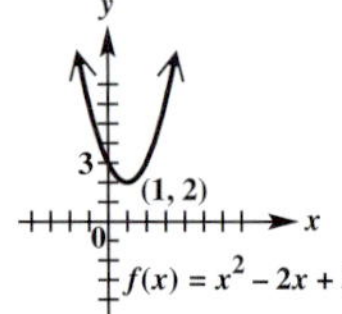

21. vertex: $(5, -4)$; axis: $x = 5$; domain: $(-\infty, \infty)$; range: $[-4, \infty)$

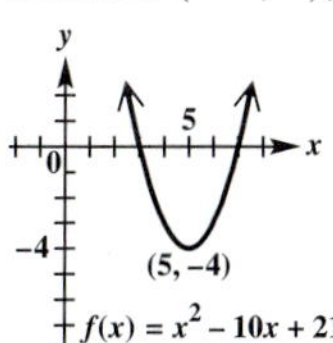

23. vertex: $(-3, 2)$; axis: $x = -3$; domain: $(-\infty, \infty)$; range: $(-\infty, 2]$

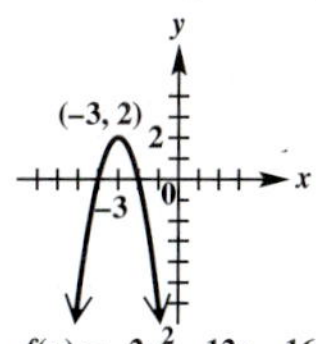

25. vertex: $(-3, 4)$; axis: $x = -3$; domain: $(-\infty, \infty)$; range: $(-\infty, 4]$

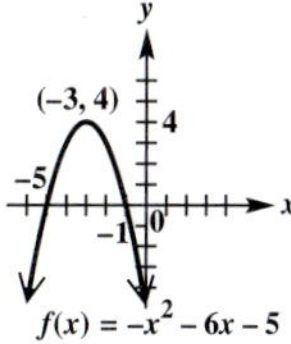

27. 3 **29.** none **31.** No, they are not equivalent.

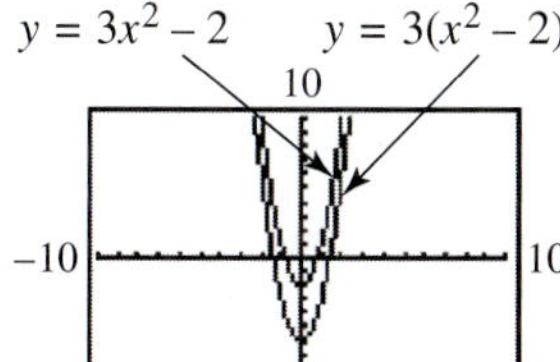

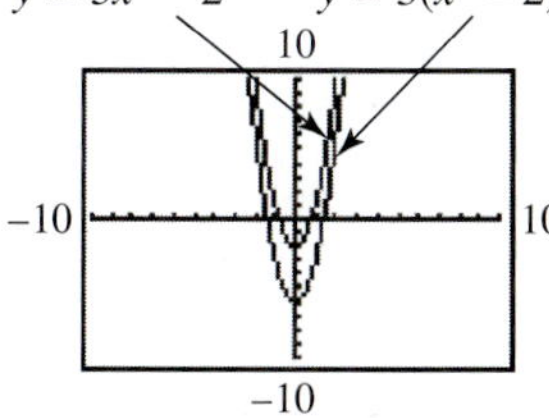

33. E **35.** D

37. C **39.** $f(x) = -2(x - 1)^2 + 4$ or $f(x) = -2x^2 + 4x + 2$

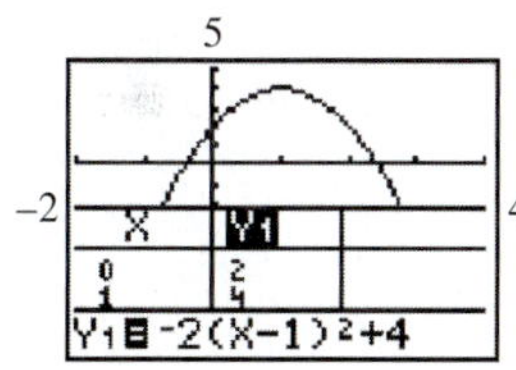

41. quadratic; negative

43. quadratic; positive **45.** linear; positive **47.** **(a)** $f(t) = -16t^2 + 200t + 50$ **(b)** 6.25 sec, 675 ft **(c)** between approximately 1.4 and 11.1 sec **(d)** approximately 12.75 sec **49.** The ball will not reach 400 ft because there are no real solutions of $-16t^2 + 150t = 400$. **51.** **(a)** $30 - x$ **(b)** $0 < x < 30$ **(c)** $f(x) = -x^2 + 30x$ **(d)** 15 and 15; The maximum product is 225. **(e)** 4 and 26 **53.** **(a)** $640 - 2x$ **(b)** $0 < x < 320$ **(c)** $A(x) = -2x^2 + 640x$ **(d)** between 57.04 ft and 85.17 ft or 234.83 ft and 262.96 ft **(e)** 160 ft by 320 ft; The maximum area is 51,200 ft^2. **55.** **(a)** $2x$ **(b)** length: $2x - 4$; width: $x - 4$; $x > 4$ **(c)** $V(x) = 4x^2 - 24x + 32$ **(d)** 8 in. by 20 in. **(e)** 13.0 in. to 14.2 in. **57.** **(a)** 3.5 ft **(b)** approximately .2 ft and 2.3 ft **(c)** 1.25 ft **(d)** approximately 3.78 ft **59.** **(a)** 23.32 ft per sec **(b)** 12.88 ft **61.** **(a)** 30.2 (percent) **(b)** It is not realistic to assume they will decrease, based on the trend seen in the period 1990–2000. **63.** 1992

65. **(a)**

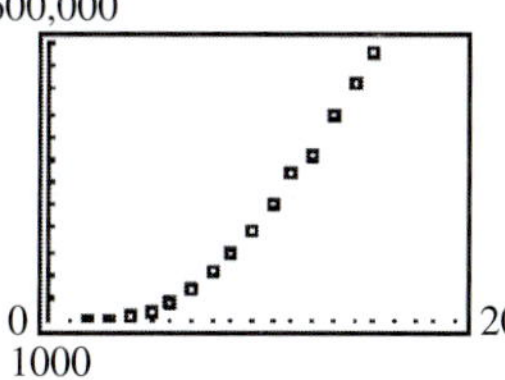

(c) $f(x) = 2974.76(x - 2)^2 + 1563$ (Other choices will lead to other models.)

(d)

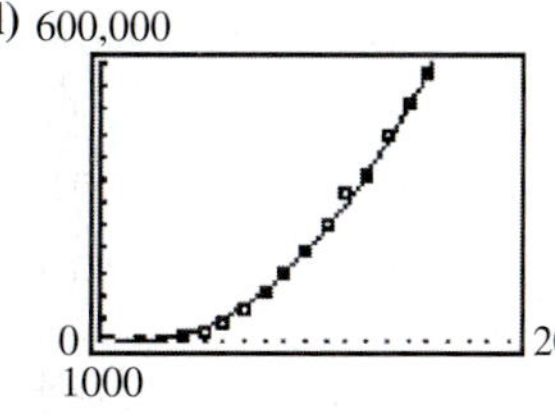

There is a relatively good fit. **(e)** 1999: 861,269; 2000: 965,385 **(f)** approximately 104,116

67. **(a)** 120.2

3 15
39.8

(b) $f(x) = .6(x - 4)^2 + 50$ **(c)** There is a good fit.

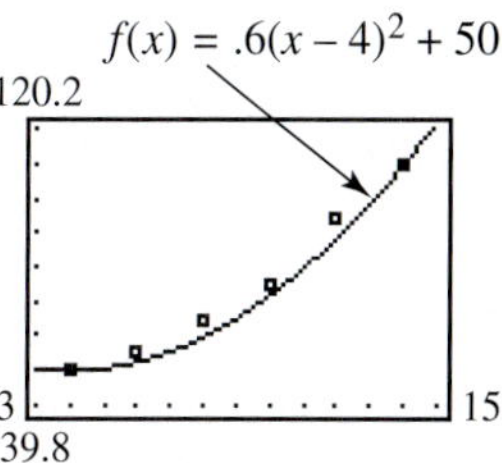

(d) $g(x) = .402x^2 - 1.175x + 48.343$ **(e)** $f(16) = 136.4$ thousand; $g(16) \approx 132.5$ thousand

69. **(a)**

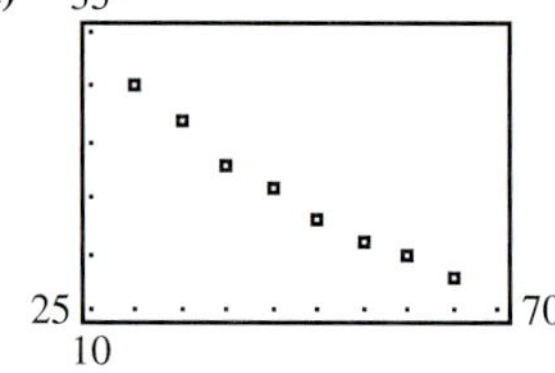

(b) $g(x) = .0074x^2 - 1.185x + 59.02$ models the data very well. **(c)** $g(70) \approx 12.33$ sec

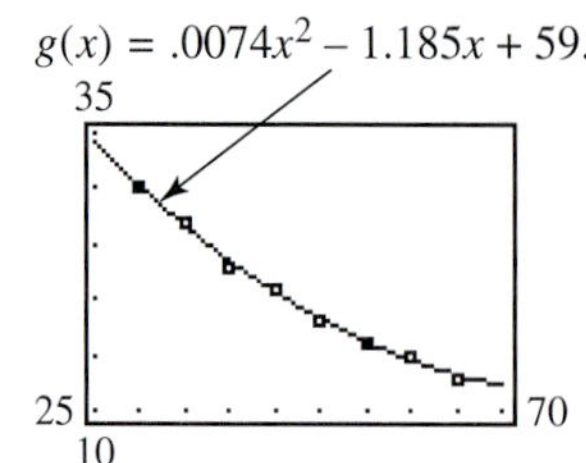

(d) about 39.1 mph **71.** $c = 25$ **73.** $f(x) = \frac{1}{2}x^2 - \frac{7}{2}x + 5$ **75.** 9 **(a)** $\sqrt{9} = 3$ **(b)** $\frac{1}{9}$ **77.** $(3, 6)$

81. The x-intercepts are -4 and 2. **82.** the open interval $(-4, 2)$ **83.** The graph of g is obtained by reflecting the graph of f across the x-axis.

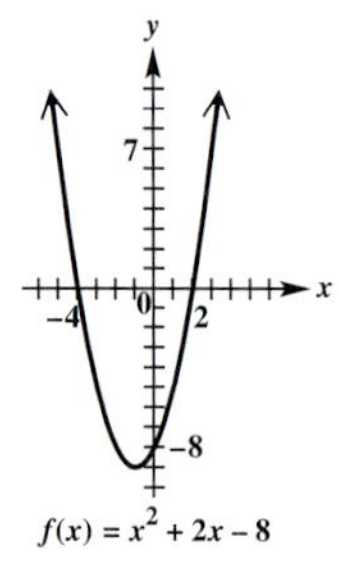

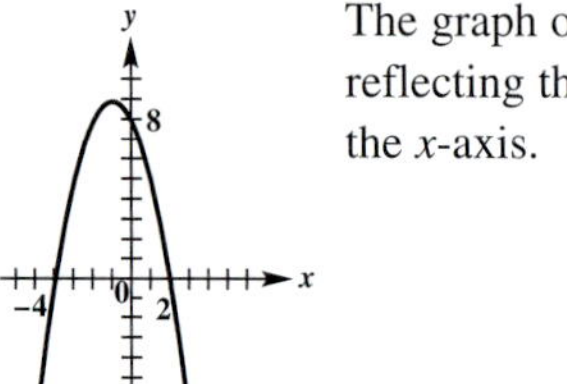

84. the open interval $(-4, 2)$ **85.** They are the same.

Connections *(page 318)*

1. -18 **2.** yes; no **3.** One example is $f(x) = x^2 + 1$.

3.2 Exercises *(page 319)*

1. $x^2 - 3x - 2 + \frac{-2}{x + 5}$ **3.** $4x^2 - 4x + 1 + \frac{-3}{x + 1}$ **5.** $x^3 + 2x + 1$ **7.** $x^4 + x^3 + 2x - 1 + \frac{3}{x + 2}$

9. $-9x^2 - 10x - 27 + \frac{-52}{x - 2}$ **11.** $\frac{1}{3}x^2 - \frac{1}{9}x + \frac{1}{x - \frac{1}{3}}$ **13.** $x^3 - x^2 - 6x$ **15.** $x^2 + x + 1$

17. $x^4 - x^3 + x^2 - x + 1$ **19.** $f(x) = (x + 1)(2x^2 - x + 2) - 10$ **21.** $f(x) = (x + 2)(-x^2 + 4x - 8) + 20$
23. $f(x) = (x - 3)(4x^3 + 9x^2 + 7x + 20) + 60$ **25.** $f(x) = (x + 1)(3x^3 + x^2 - 11x + 11) + 4$ **27.** 0 **29.** -1

31. -6 **33.** -5 **35.** 7 **37.** $-6 - i$ **39.** yes **41.** yes **43.** no; -9 **45.** yes **47.** yes **49.** no; $\frac{357}{125}$

51. yes **53.** no; $13 + 7i$ **55.** no; $-2 + 7i$ **57.** 1 **58.** It is equal to the real number because 1 is the identity element for multiplication. **59.** Add the coefficients of $f(x)$. **60.** $1 + (-4) + 9 + (-6) = 0$; The answers agree.
61. $f(-x) = -x^3 - 4x^2 - 9x - 6$; $f(-1) = -20$ **62.** Both are -20. To find $f(-1)$, add the coefficients of $f(-x)$.

Connections *(page 329)*

2

3.3 Exercises *(page 329)*

1. true **3.** false; -2 is a zero of multiplicity 4. **5.** yes **7.** no **9.** yes **11.** no **13.** yes **15.** yes
17. $f(x) = (x - 2)(2x - 5)(x + 3)$ **19.** $f(x) = (x + 3)(3x - 1)(2x - 1)$ **21.** $f(x) = (x + 4)(3x - 1)(2x + 1)$
23. $f(x) = (x - 3i)(x + 4)(x + 3)$ **25.** $f(x) = [x - (1 + i)](2x - 1)(x + 3)$ **27.** $f(x) = (x + 2)^2(x + 1)(x - 3)$
29. $-1 \pm i$ **31.** $3, 2 + i$ **33.** $i, \pm 2i$ **35.** **(a)** $\pm 1, \pm 2, \pm 5, \pm 10$ **(b)** $-1, -2, 5$ **(c)** $f(x) = (x + 1)(x + 2)(x - 5)$
37. **(a)** $\pm 1, \pm 2, \pm 3, \pm 5, \pm 6, \pm 10, \pm 15, \pm 30$ **(b)** $-5, -3, 2$ **(c)** $f(x) = (x + 5)(x + 3)(x - 2)$

39. **(a)** $\pm1, \pm2, \pm3, \pm4, \pm6, \pm12, \pm\frac{1}{2}, \pm\frac{3}{2}, \pm\frac{1}{3}, \pm\frac{2}{3}, \pm\frac{4}{3}, \pm\frac{1}{6}$ **(b)** $-4, -\frac{1}{3}, \frac{3}{2}$ **(c)** $f(x) = (x + 4)(3x + 1)(2x - 3)$

41. **(a)** $\pm1, \pm2, \pm3, \pm4, \pm6, \pm12, \pm\frac{1}{2}, \pm\frac{3}{2}, \pm\frac{1}{3}, \pm\frac{2}{3}, \pm\frac{4}{3}, \pm\frac{1}{4}, \pm\frac{3}{4}, \pm\frac{1}{6}, \pm\frac{1}{8}, \pm\frac{3}{8}, \pm\frac{1}{12}, \pm\frac{1}{24}$ **(b)** $-\frac{3}{2}, -\frac{2}{3}, \frac{1}{2}$ **(c)** $f(x) = 2(2x + 3)(3x + 2)(2x - 1)$ **43.** $0, \pm\frac{\sqrt{7}}{7}i$ **45.** $2, -3, 1, -1$ **47.** -2 (multiplicity 5), 1 (multiplicity 5), $1 - \sqrt{3}$ (multiplicity 2) **49.** $f(x) = -3x^3 + 6x^2 + 33x - 36$ **51.** $f(x) = -\frac{1}{2}x^3 - \frac{1}{2}x^2 + x$

53. $f(x) = \frac{1}{6}x^3 + \frac{3}{2}x^2 + \frac{9}{2}x + \frac{9}{2}$

In Exercises 55–71, we give only one possible answer.

55. $f(x) = x^2 - 10x + 26$ **57.** $f(x) = x^3 - 4x^2 + 6x - 4$ **59.** $f(x) = x^3 - 3x^2 + x + 1$

61. $f(x) = x^4 - 6x^3 + 10x^2 + 2x - 15$ **63.** $f(x) = x^3 - 8x^2 + 22x - 20$ **65.** $f(x) = x^4 - 4x^3 + 5x^2 - 2x - 2$

67. $f(x) = x^4 - 16x^3 + 98x^2 - 240x + 225$ **69.** $f(x) = x^5 - 12x^4 + 74x^3 - 248x^2 + 445x - 500$

71. $f(x) = x^4 - 6x^3 + 17x^2 - 28x + 20$ **73.** 2 or 0 positive; 1 negative **75.** 1 positive; 1 negative

77. 2 or 0 positive; 3 or 1 negative

3.4 Exercises *(page 342)*

1. A **3.** one **5.** B and D **7.** $f(x) = x(x + 5)^2(x - 3)$

9.

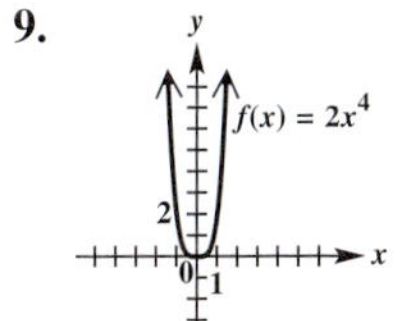

11.

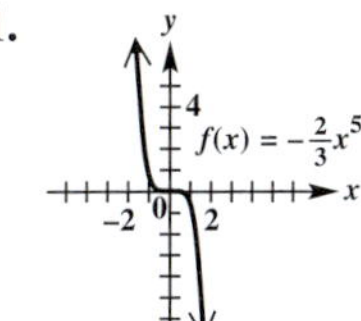

13.

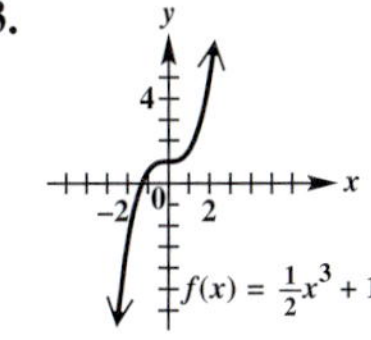

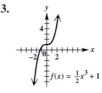

15.

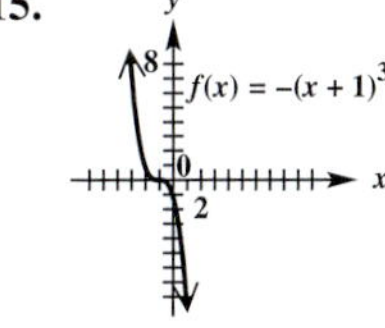

17.

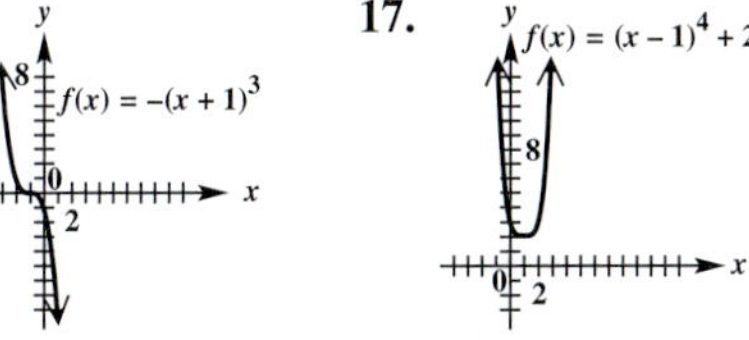

19.

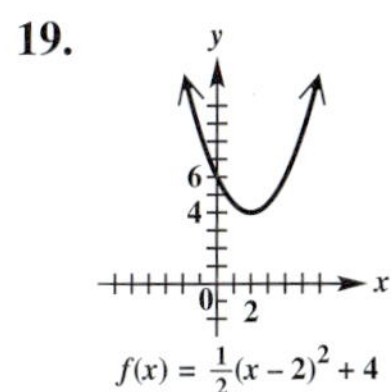

21. **23.**

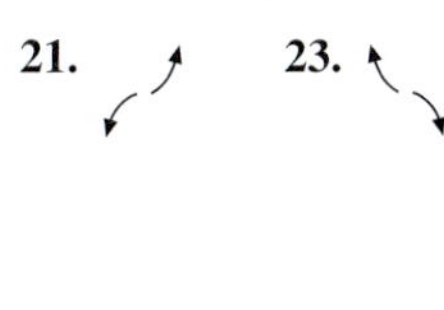

25. **27.**

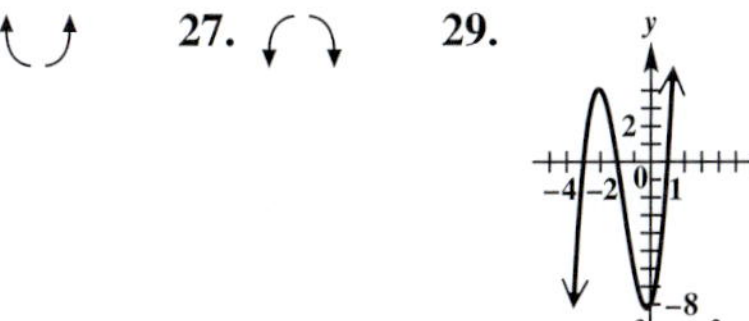

29. **31.**

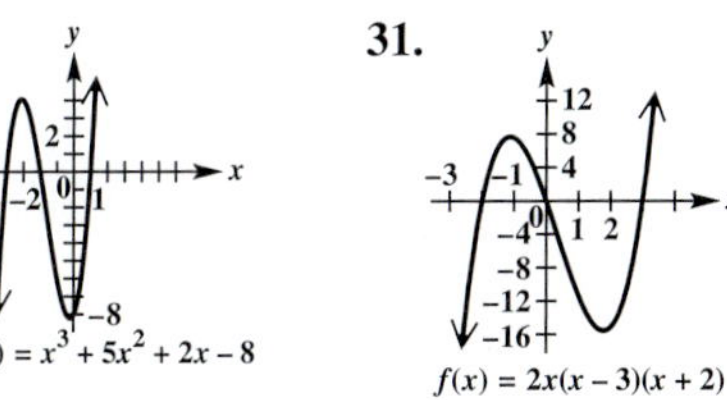

33.

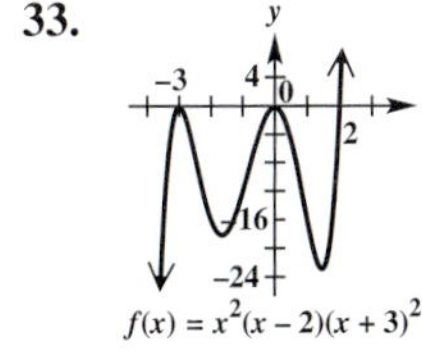

35.

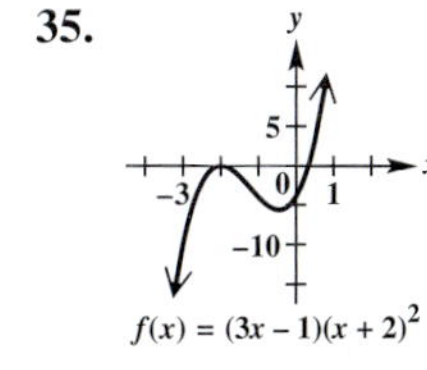

37.

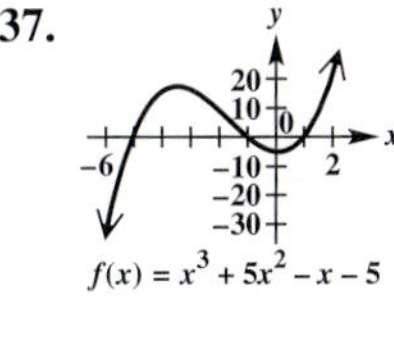

39.

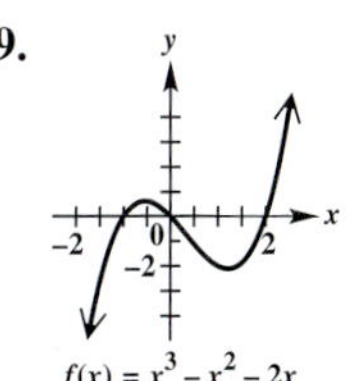

41.

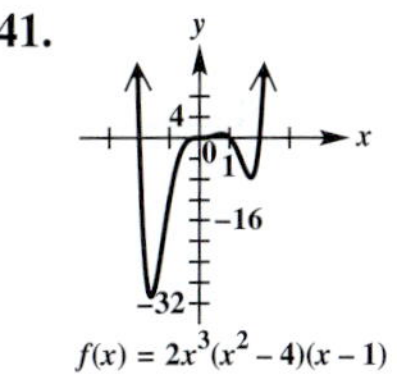

43. $f(2) = -2 < 0$; $f(3) = 1 > 0$ **45.** $f(0) = 7 > 0$; $f(1) = -1 < 0$ **47.** $f(1) = -6 < 0$; $f(2) = 16 > 0$

49. $f(3.2) = -3.8144 < 0$; $f(3.3) = 7.1891 > 0$ **51.** $f(-1) = -35 < 0$; $f(0) = 12 > 0$

61. $f(x) = \frac{1}{2}(x + 6)(x - 2)(x - 5)$ or $f(x) = \frac{1}{2}x^3 - \frac{1}{2}x^2 - 16x + 30$

63. $f(1.25) = -14.21875$ **65.** $f(1.25) = 29.046875$ **67.** 2.7807764 **69.** 1.543689 **71.** $-3.0, -1.4, 1.4$

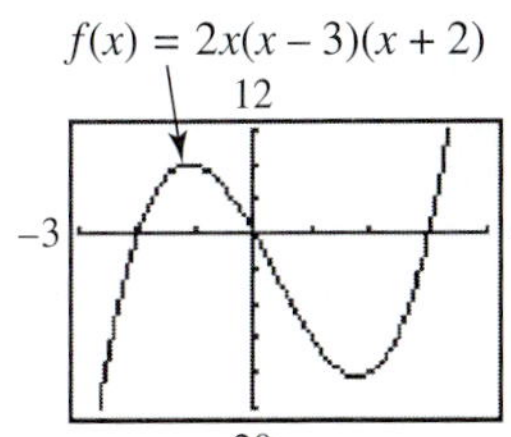

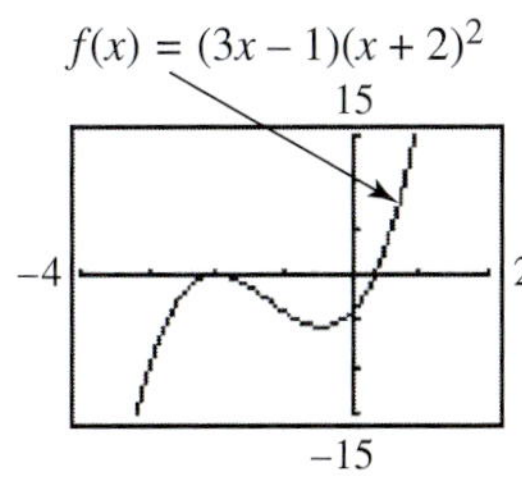

73. $-1.1, 1.2$ **75.** $(-3.44, 26.15)$ **77.** $(-.09, 1.05)$ **79.** $(-.20, -28.62)$ **81.** Answers will vary.

83.

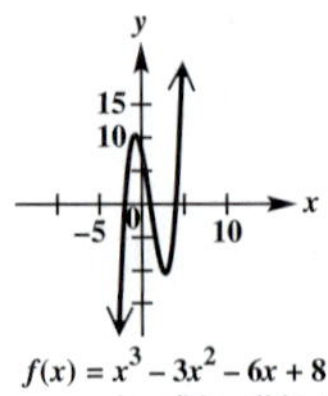

$f(x) = x^3 - 3x^2 - 6x + 8$
$= (x - 4)(x - 1)(x + 2)$

(a) $\{-2, 1, 4\}$
(b) $(-\infty, -2) \cup (1, 4)$
(c) $(-2, 1) \cup (4, \infty)$

84.

$f(x) = x^3 + 4x^2 - 11x - 30$
$= (x - 3)(x + 2)(x + 5)$

(a) $\{-5, -2, 3\}$
(b) $(-\infty, -5) \cup (-2, 3)$
(c) $(-5, -2) \cup (3, \infty)$

85.

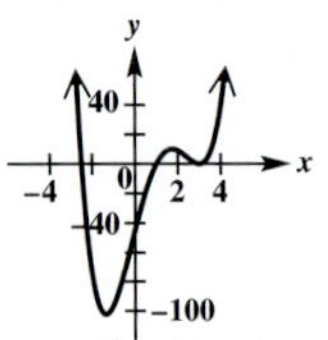

$f(x) = 2x^4 - 9x^3 - 5x^2 + 57x - 45$
$= (x - 3)^2(2x + 5)(x - 1)$

(a) $\{-2.5, 1, 3$ (multiplicity 2)$\}$
(b) $(-2.5, 1)$
(c) $(-\infty, -2.5) \cup (1, 3) \cup (3, \infty)$

86.

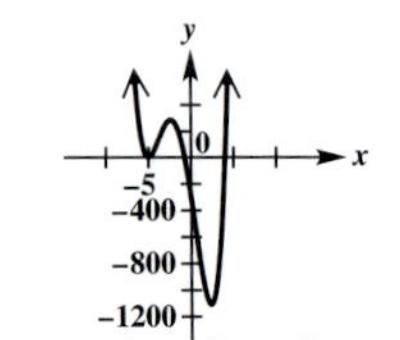

$f(x) = 4x^4 + 27x^3 - 42x^2 - 445x - 300$
$= (x + 5)^2(4x + 3)(x - 4)$

(a) $\{-5$ (multiplicity 2), $-.75, 4\}$
(b) $(-.75, 4)$
(c) $(-\infty, -5) \cup (-5, -.75) \cup (4, \infty)$

87.

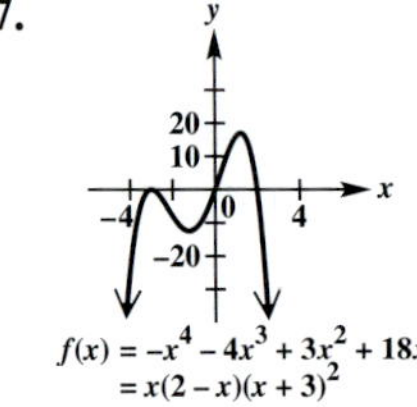

$f(x) = -x^4 - 4x^3 + 3x^2 + 18x$
$= x(2 - x)(x + 3)^2$

(a) $\{-3$ (multiplicity 2), $0, 2\}$
(b) $\{-3\} \cup [0, 2]$
(c) $(-\infty, 0] \cup [2, \infty)$

88.

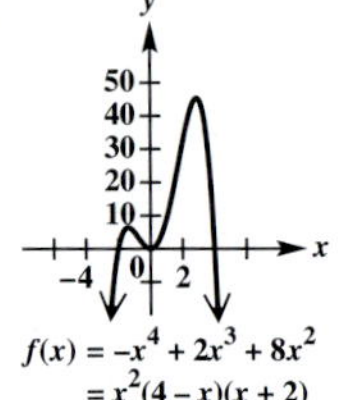

$f(x) = -x^4 + 2x^3 + 8x^2$
$= x^2(4 - x)(x + 2)$

(a) $\{-2, 0$ (multiplicity 2), $4\}$
(b) $[-2, 4]$
(c) $(-\infty, -2] \cup \{0\} \cup [4, \infty)$

89. **(a)** $0 < x < 6$ **(b)** $V(x) = x(18 - 2x)(12 - 2x)$ or $V(x) = 4x^3 - 60x^2 + 216x$ **(c)** $x \approx 2.35$; about 228.16 in.^{3} **(d)** $.42 < x < 5$ **91.** **(a)** $x - 1$; $(1, \infty)$ **(b)** $\sqrt{x^2 - (x - 1)^2}$ **(c)** $2x^3 - 5x^2 + 4x - 28{,}225 = 0$ **(d)** hypotenuse: 25 in.; legs: 24 in. and 7 in. **93.** 3 ft **95.** **(a)** about 7.13 cm; The ball floats partly above the surface. **(b)** The sphere is more dense than water and sinks below the surface. **(c)** 10 cm; The balloon is submerged with its top even with the surface.

97. **(a)**

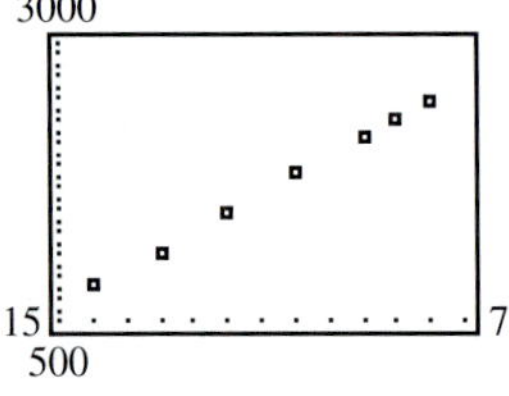

(b)

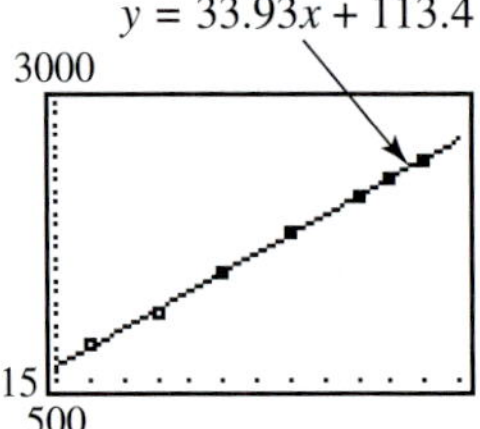

(c)

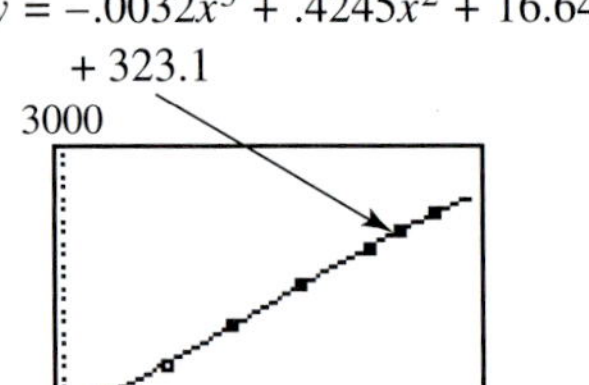

(d) linear: 1572 ft; cubic: 1570 ft **(e)** The cubic function is slightly better because only one data point is not on the curve.

99. B

Summary Exercises on Polynomial Functions, Zeros, and Graphs *(page 349)*

1. **(a)** positive zeros: 1; negative zeros: 3 or 1 **(b)** $\pm 1, \pm 2, \pm 3, \pm 6$ **(c)** $-3, -1$ (multiplicity 2), 2 **(d)** no other real zeros **(e)** no other complex zeros **(f)** $-3, -1, 2$ **(g)** -6 **(h)** $f(4) = 350$; $(4, 350)$ **(i)** ↖ ↗ **(j)**

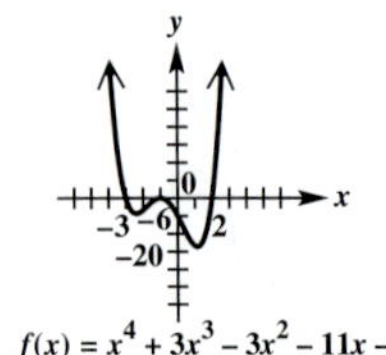

$f(x) = x^4 + 3x^3 - 3x^2 - 11x - 6$

2. **(a)** positive zeros: 3 or 1; negative zeros: 2 or 0 **(b)** $\pm 1, \pm 3, \pm 5, \pm 9, \pm 15, \pm 45, \pm \frac{1}{2}, \pm \frac{3}{2}, \pm \frac{5}{2}, \pm \frac{9}{2}, \pm \frac{15}{2}, \pm \frac{45}{2}$

(c) $-3, \frac{1}{2}, 5$ **(d)** $-\sqrt{3}, \sqrt{3}$ **(e)** no other complex zeros **(f)** $-3, \frac{1}{2}, 5, -\sqrt{3}, \sqrt{3}$ **(g)** 45 **(h)** $f(4) = 637; (4, 637)$ **(i)** ↖ ↘

(j)

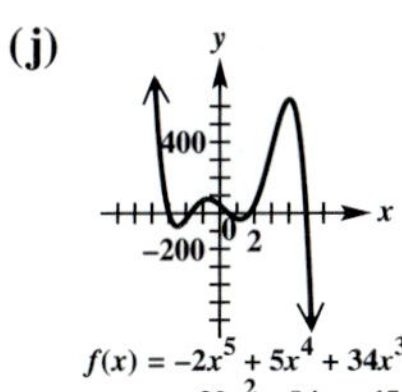

3. **(a)** positive zeros: 4, 2, or 0; negative zeros: 1 **(b)** $\pm 1, \pm 5, \pm \frac{1}{2}, \pm \frac{5}{2}$ **(c)** 5 **(d)** $-\frac{\sqrt{2}}{2}, \frac{\sqrt{2}}{2}$ **(e)** $-i, i$ **(f)** $-\frac{\sqrt{2}}{2}, \frac{\sqrt{2}}{2}, 5$

(g) 5 **(h)** $f(4) = -527; (4, -527)$ **(i)** ↙ ↗ **(j)**

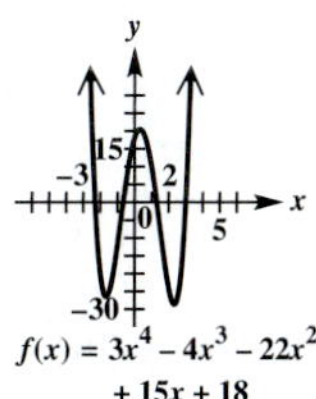

4. **(a)** positive zeros: 2 or 0; negative zeros: 2 or 0 **(b)** $\pm 1, \pm 2, \pm 3, \pm 6, \pm 9, \pm 18, \pm \frac{1}{3}, \pm \frac{2}{3}$ **(c)** $-\frac{2}{3}, 3$ **(d)** $\frac{-1 + \sqrt{13}}{2}$, $\frac{-1 - \sqrt{13}}{2}$ **(e)** no other complex zeros **(f)** $-\frac{2}{3}, 3, \frac{-1 \pm \sqrt{13}}{2}$ **(g)** 18 **(h)** $f(4) = 238; (4, 238)$ **(i)** ↖ ↗ **(j)**

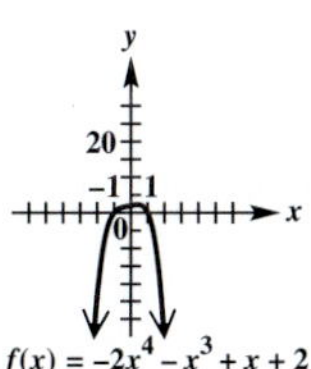

5. **(a)** positive zeros: 1; negative zeros: 3 or 1 **(b)** $\pm 1, \pm 2, \pm \frac{1}{2}$ **(c)** $-1, 1$ **(d)** no other real zeros

(e) $-\frac{1}{4} + i\frac{\sqrt{15}}{4}, -\frac{1}{4} - i\frac{\sqrt{15}}{4}$ **(f)** $-1, 1$ **(g)** 2 **(h)** $f(4) = -570; (4, -570)$ **(i)** ↙ ↘ **(j)**

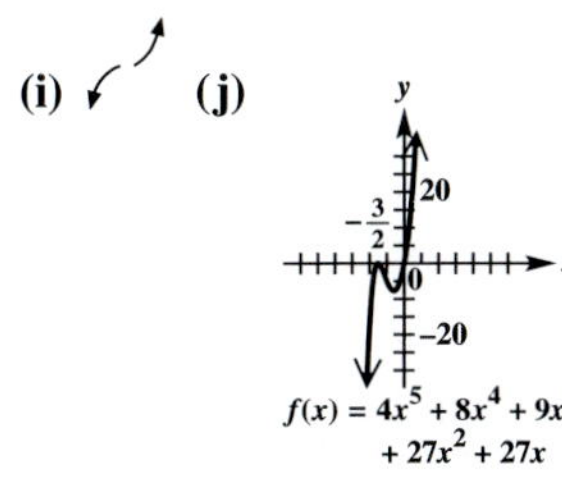

6. **(a)** positive zeros: 0; negative zeros: 4, 2, or 0 **(b)** $0, \pm 1, \pm 3, \pm 9, \pm 27, \pm \frac{1}{2}, \pm \frac{3}{2}, \pm \frac{9}{2}, \pm \frac{27}{2}, \pm \frac{1}{4}, \pm \frac{3}{4}, \pm \frac{9}{4}, \pm \frac{27}{4}$

(c) $0, -\frac{3}{2}$ (multiplicity 2) **(d)** no other real zeros **(e)** $\frac{1}{2} + i\frac{\sqrt{11}}{2}, \frac{1}{2} - i\frac{\sqrt{11}}{2}$ **(f)** $0, -\frac{3}{2}$ **(g)** 0 **(h)** $f(4) = 7260; (4, 7260)$

(i) ↙ ↗ **(j)**

7. **(a)** positive zeros: 1; negative zeros: 1 **(b)** $\pm 1, \pm 5, \pm \frac{1}{3}, \pm \frac{5}{3}$ **(c)** no rational zeros **(d)** $-\sqrt{5}, \sqrt{5}$ **(e)** $-i\frac{\sqrt{3}}{3}, i\frac{\sqrt{3}}{3}$ **(f)** $-\sqrt{5}, \sqrt{5}$ **(g)** -5 **(h)** $f(4) = 539$; $(4, 539)$ **(i)** **(j)**

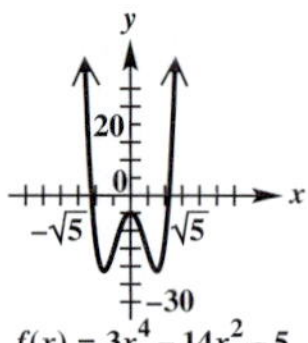

8. **(a)** positive zeros: 2 or 0; negative zeros: 3 or 1 **(b)** $\pm 1, \pm 3, \pm 9$ **(c)** $-3, -1$ (multiplicity 2), 1, 3 **(d)** no other real zeros **(e)** no other complex zeros **(f)** $-3, -1, 1, 3$ **(g)** -9 **(h)** $f(4) = -525$; $(4, -525)$ **(i)** **(j)**

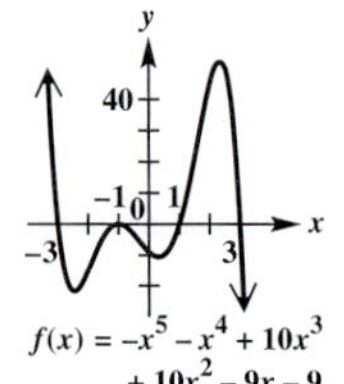

9. **(a)** positive zeros: 4, 2, or 0; negative zeros: 0 **(b)** $\pm 1, \pm 2, \pm 3, \pm 4, \pm 6, \pm 12, \pm \frac{1}{3}, \pm \frac{2}{3}, \pm \frac{4}{3}$ **(c)** $\frac{1}{3}$, 2 (multiplicity 2), 3 **(d)** no other real zeros **(e)** no other complex zeros **(f)** $\frac{1}{3}$, 2, 3 **(g)** -12 **(h)** $f(4) = -44$; $(4, -44)$ **(i)** **(j)**

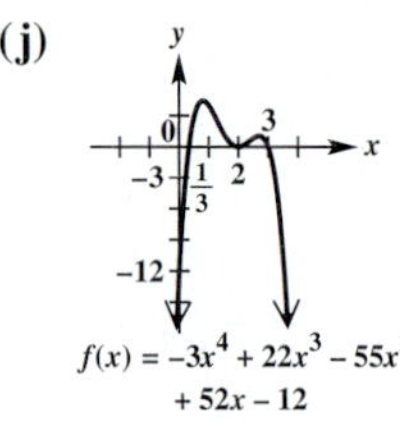

10. For the function in Exercise 2: ± 1.732; for the function in Exercise 3: $\pm .707$; for the function in Exercise 4: $-2.303, 1.303$; for the function in Exercise 7: ± 2.236

3.5 Exercises *(page 362)*

1. $(-\infty, 0) \cup (0, \infty)$; $(-\infty, 0) \cup (0, \infty)$ **3.** none; $(-\infty, 0) \cup (0, \infty)$; none **5.** $x = 3$; $y = 2$ **7.** even; symmetry with respect to the y-axis **9.** A, B, C **11.** A **13.** A **15.** A, C, D

17. To obtain the graph of f, stretch the graph of $y = \frac{1}{x}$ vertically by a factor of 2.

Domain: $(-\infty, 0) \cup (0, \infty)$; range: $(-\infty, 0) \cup (0, \infty)$

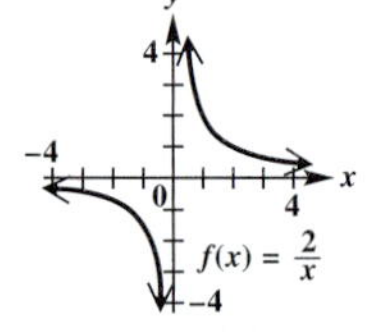

19. To obtain the graph of f, shift the graph of $y = \frac{1}{x}$ to the left 2 units.

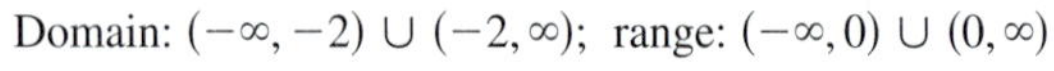
Domain: $(-\infty, -2) \cup (-2, \infty)$; range: $(-\infty, 0) \cup (0, \infty)$

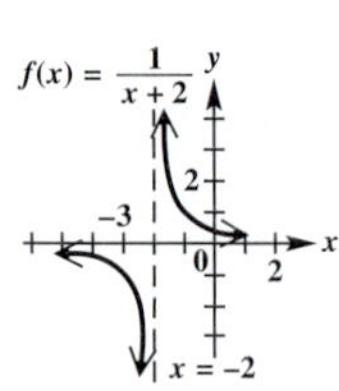

21. To obtain the graph of f, shift the graph of $y = \frac{1}{x}$ up 1 unit.

Domain: $(-\infty, 0) \cup (0, \infty)$; range: $(-\infty, 1) \cup (1, \infty)$

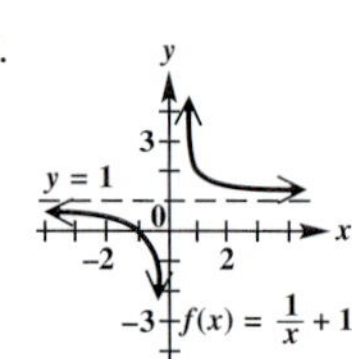

23. To obtain the graph of f, stretch the graph of $y = \frac{1}{x^2}$ vertically by a factor of 2, and reflect across the x-axis.
Domain: $(-\infty, 0) \cup (0, \infty)$; range: $(-\infty, 0)$

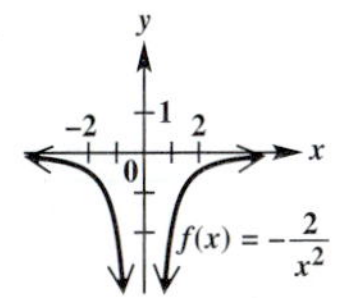

25. To obtain the graph of f, shift the graph of $y = \frac{1}{x^2}$ to the right 3 units.
Domain: $(-\infty, 3) \cup (3, \infty)$; range: $(0, \infty)$

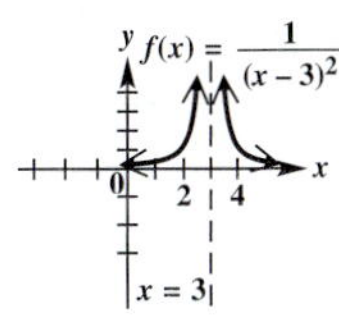

27. To obtain the graph of f, shift the graph of $y = \frac{1}{x^2}$ to the left 2 units, reflect across the x-axis, and shift 3 units down.
Domain: $(-\infty, -2) \cup (-2, \infty)$; range: $(-\infty, -3)$

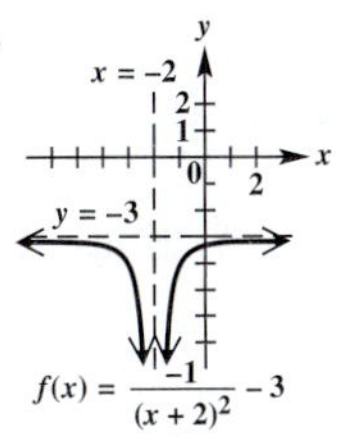

29. D **31.** G **33.** E **35.** F

In Exercises 37–45, V.A. represents vertical asymptote, H.A. represents horizontal asymptote, and O.A. represents oblique asymptote.

37. V.A.: $x = 5$; H.A.: $y = 0$ **39.** V.A.: $x = -\frac{1}{2}$; H.A.: $y = -\frac{3}{2}$ **41.** V.A.: $x = -3$; O.A.: $y = x - 3$

43. V.A.: $x = -2$, $x = \frac{5}{2}$; H.A.: $y = \frac{1}{2}$ **45.** V.A.: none; H.A.: $y = 1$ **47. (a)** $f(x) = \frac{2x - 5}{x - 3}$ **(b)** $\frac{5}{2}$
(c) horizontal asymptote: $y = 2$; vertical asymptote: $x = 3$ **49. (a)** $y = x + 1$ **(b)** at $x = 0$ and $x = 1$ **(c)** above

51.
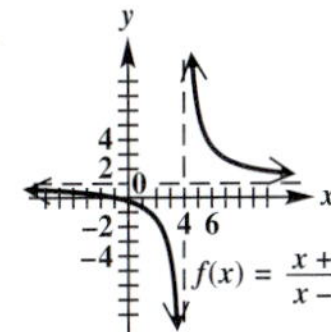

53.
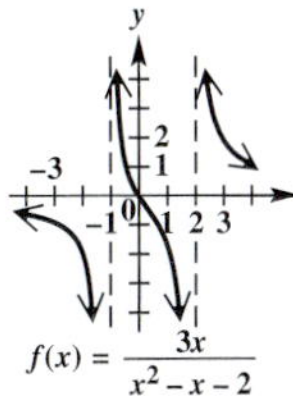

55.
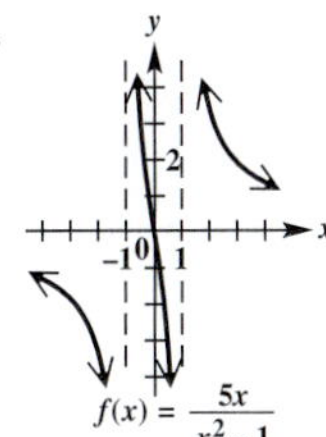

57.
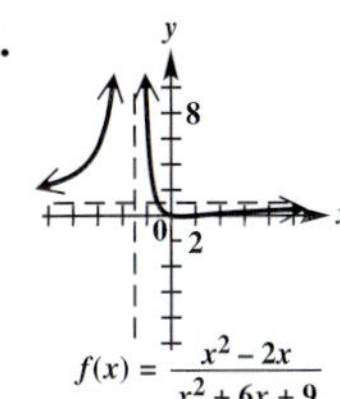

59.
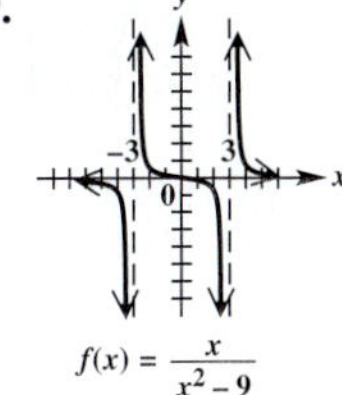

61.
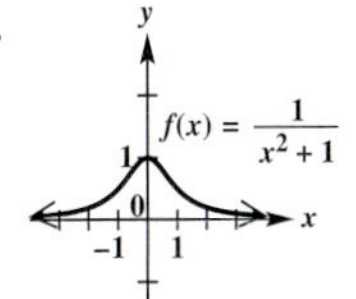

63.
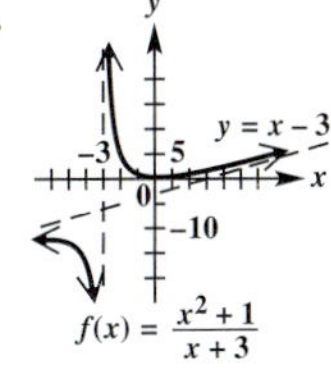

65.
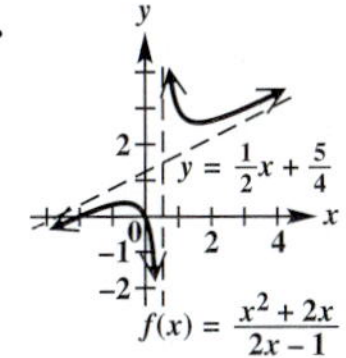

67.
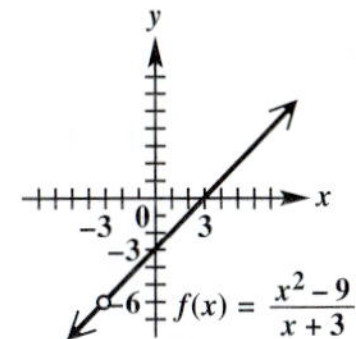

69. $f(x) = \frac{(x-3)(x+2)}{(x-2)(x+2)}$ or $f(x) = \frac{x^2 - x - 6}{x^2 - 4}$ **71.** $f(x) = \frac{x - 2}{x(x - 4)}$ or $f(x) = \frac{x - 2}{x^2 - 4x}$ **73.** Several answers are possible. One answer is $f(x) = \frac{(x-3)(x+1)}{(x-1)^2}$.

75. $f(1.25) = -.\overline{81}$

$f(x) = \dfrac{x+1}{x-4}$

6.2

−9.4 9.4

−6.2

77. $f(1.25) = 2.708\overline{3}$

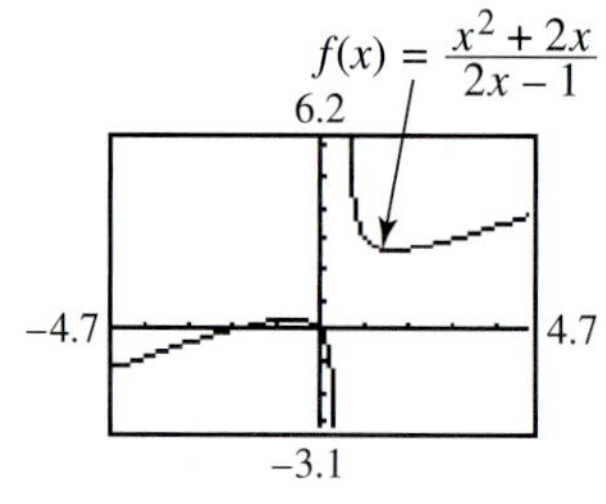

79. **(a)** 26 per min **(b)** 5 park attendants

For $r = x$,

$y = T(r) = \dfrac{2x-25}{2x^2-50x}$ $y = .5$

1

Intersection X=26.039936 Y=.5

25 40

0

81. **(a)** approximately 52.1 mph

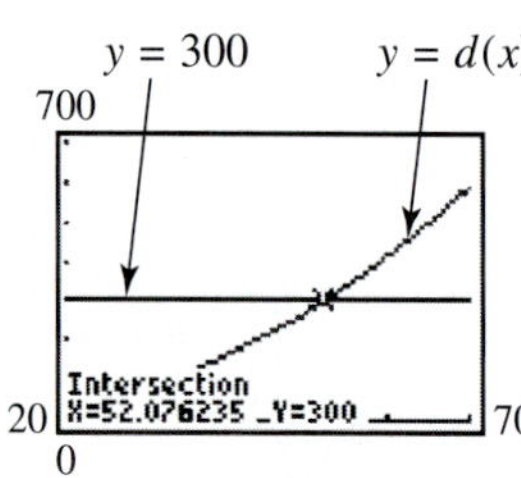

x	$d(x)$	x	$d(x)$
20	**34**	50	**273**
25	**56**	55	**340**
30	**85**	60	**415**
35	**121**	65	**499**
40	**164**	70	**591**
45	**215**		

83. All answers are given in tens of millions. **(a)** \$65.5 **(b)** \$64 **(c)** \$60 **(d)** \$40 **(e)** \$0

(f) $R(x) = \dfrac{80x - 8000}{x - 110}$

80

0 100

0

85. $y = 1$ **86.** $(x+4)(x+1)(x-3)(x-5)$

87. **(a)** $(x-1)(x-2)(x+2)(x-5)$ **(b)** $f(x) = \dfrac{(x+4)(x+1)(x-3)(x-5)}{(x-1)(x-2)(x+2)(x-5)}$

88. **(a)** $x - 5$ **(b)** 5 **89.** $-4, -1, 3$ **90.** -3 **91.** $x = 1, x = 2, x = -2$

92. $\left(\dfrac{7+\sqrt{241}}{6}, 1\right), \left(\dfrac{7-\sqrt{241}}{6}, 1\right)$

93.

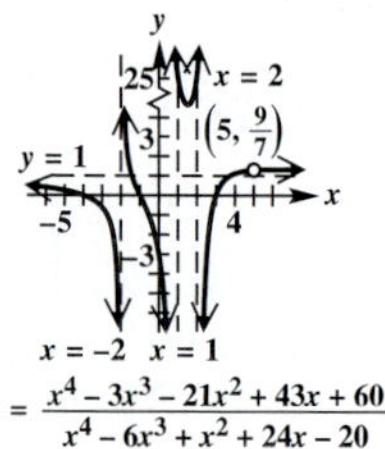

94. **(a)** $(-4, -2) \cup (-1, 1) \cup (2, 3)$ **(b)** $(-\infty, -4) \cup (-2, -1) \cup (1, 2) \cup (3, 5) \cup (5, \infty)$

3.6 Exercises *(page 372)*

1. The circumference of a circle varies directly as (or is proportional to) its radius. **3.** The average speed varies directly as (or is proportional to) the distance traveled and inversely as the time. **5.** The strength of a muscle varies directly as (or is proportional to) the cube of its length. **7.** C **9.** A **11.** -30 **13.** $\dfrac{220}{7}$ **15.** $\dfrac{5}{2}$ **17.** $\dfrac{32}{15}$ **19.** $\dfrac{18}{125}$

21. 69.08 in. **23.** 850 ohms **25.** 8 lb **27.** 16 in. **29.** 90 revolutions per minute **31.** .0444 ohm

33. \$1375 **35.** 800 lb **37.** $\dfrac{8}{9}$ metric ton **39.** $\dfrac{66\pi}{17}$ sec **41.** 21 **43.** 365.24 **45.** 92; undernourished

47. increases; decreases **49.** y is half as large as before. **51.** y is one-third as large as before.

53. p is $\dfrac{1}{32}$ as large as before.

Chapter 3 Review Exercises *(page 381)*

1. vertex: $(-4,-5)$; axis: $x=-4$; x-intercepts: $\dfrac{-12 \pm \sqrt{15}}{3}$; y-intercept: 43; domain: $(-\infty, \infty)$; range: $[-5, \infty)$

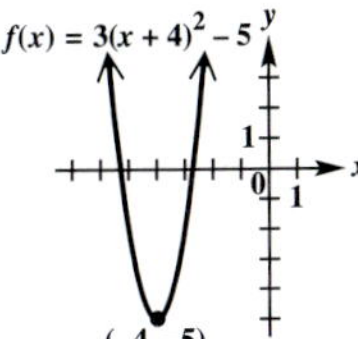

3. vertex: $(-2, 11)$; axis $x=-2$; x-intercepts: $\dfrac{-6 \pm \sqrt{33}}{3}$; y-intercept: -1; domain: $(-\infty, \infty)$; range: $(-\infty, 11]$

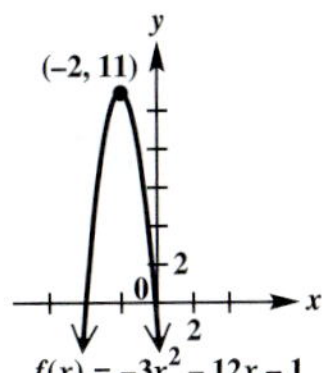

5. (h, k) **7.** $k \le 0$; $h \pm \sqrt{\dfrac{-k}{a}}$ **9.** 90 m by 45 m **11. (a)** 120 **(b)** 40 **(c)** 22 **(d)** 84 **(e)** 146 **(f)** minimum at $x = 8$ (August) **13.** Because the discriminant is 67.3033, a positive number, there are two x-intercepts.

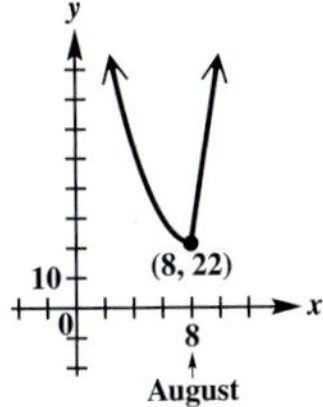

15. (a) the open interval $(-.52, 2.59)$ **(b)** $(-\infty, -.52) \cup (2.59, \infty)$ **17.** $x^2 + 4x + 1 + \dfrac{-7}{x-3}$

19. $2x^2 - 8x + 31 + \dfrac{-118}{x+4}$ **21.** $(x-2)(5x^2 + 7x + 16) + 26$ **23.** -1 **25.** 28 **27.** yes **29.** $7 - 2i$

In Exercises 31 and 33, other answers are possible.

31. $f(x) = x^3 - 10x^2 + 17x + 28$ **33.** $f(x) = x^4 - 5x^3 + 3x^2 + 15x - 18$ **35.** $\dfrac{1}{2}, -1, 5$

37. (a) $f(-1) = -10 < 0$; $f(0) = 2 > 0$ **(b)** $f(2) = -4 < 0$; $f(3) = 14 > 0$ **41.** yes

43. $f(x) = -2x^3 + 6x^2 + 12x - 16$ **45.** $1, -\dfrac{1}{2}, \pm 2i$ **47.** $\dfrac{13}{2}$

49. Any polynomial that can be factored into $a(x - b)^3$ works. One example is $f(x) = 2(x-1)^3$.

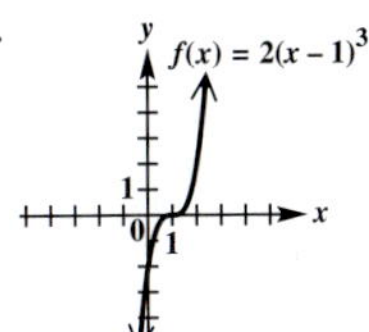

51. (a) $(-\infty, \infty)$ **(b)** $(-\infty, \infty)$ **(c)** $f(x) \to \infty$ as $x \to \infty$, $f(x) \to -\infty$ as $x \to -\infty$:

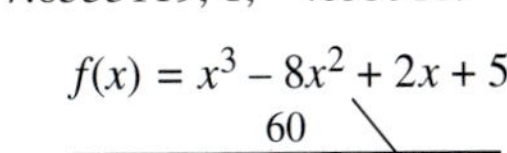

(d) at most 7 **(e)** at most 6

53. C **55.** E **57.** B **59.**

61.

63. 7.6533119, 1, $-.6533119$

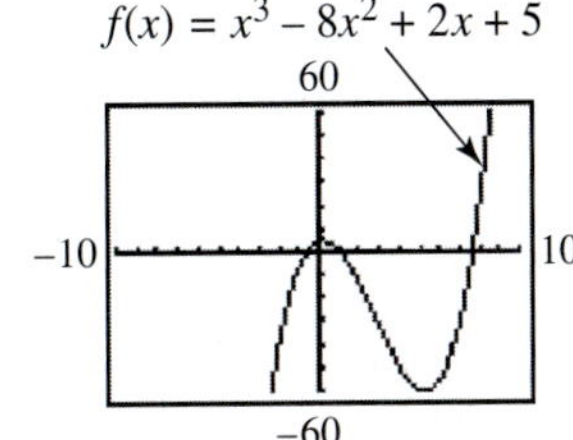

65. **(a)**

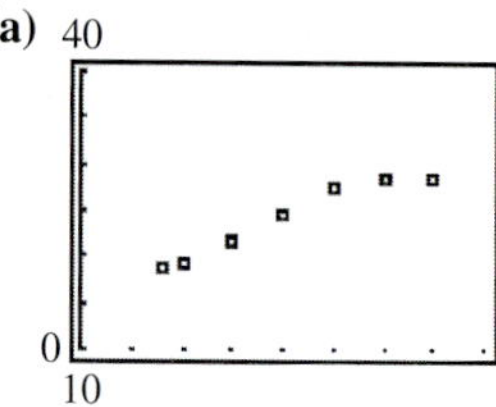

(b) $f(x) = -.011x^2 + .869x + 11.9$

(c) $f(x) = -.00087x^3 + .0456x^2 - .219x + 17.8$

(d)

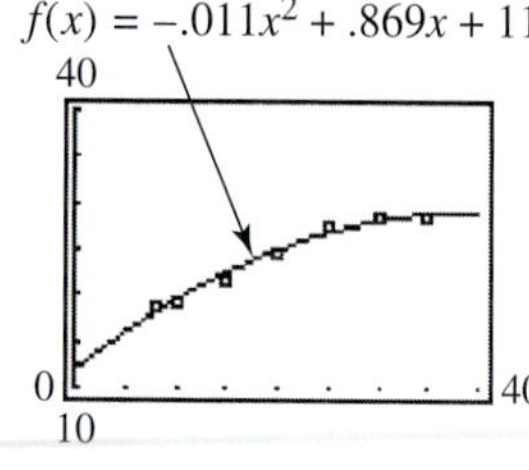

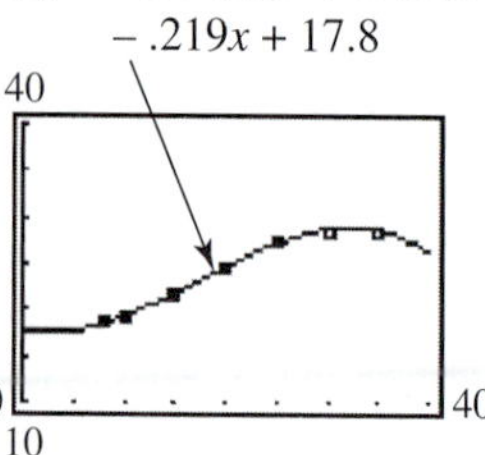

(e) Both functions approximate the data well. The quadratic function is probably better for prediction because it is unlikely that the percent of out-of-pocket spending would decrease after 2025 (as the cubic function shows) unless changes were made in Medicare law. **67.** 12 in. $\times$ 4 in. $\times$ 15 in.

69.

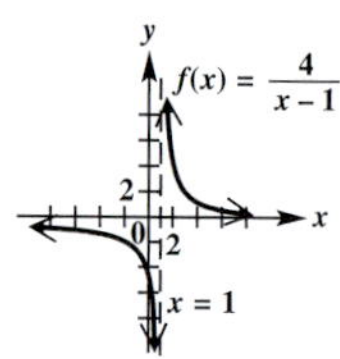

71.

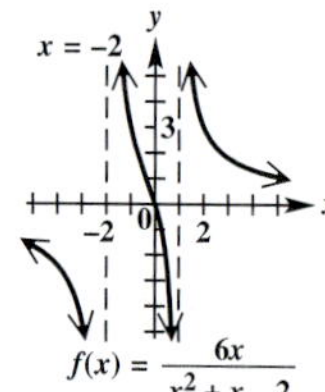

73.

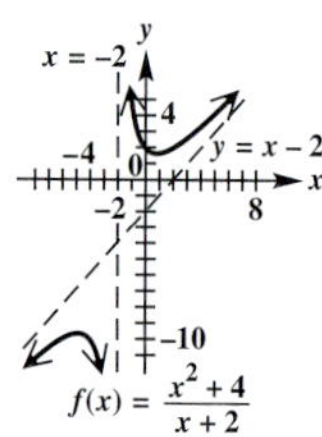

75.

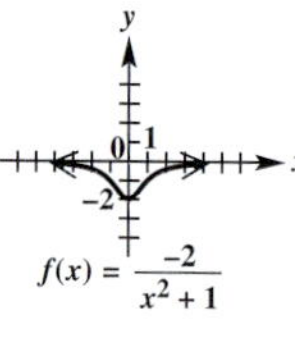

77. **(a)**

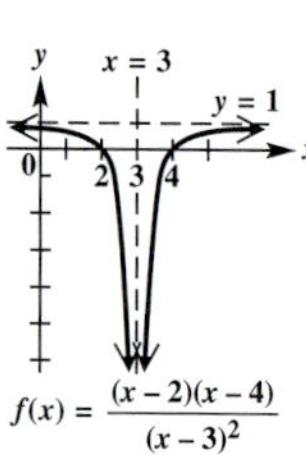

(b) One possibility is $f(x) = \dfrac{(x-2)(x-4)}{(x-3)^2}$. **79.** $f(x) = \dfrac{-3x+6}{x-1}$

81. **(a)**

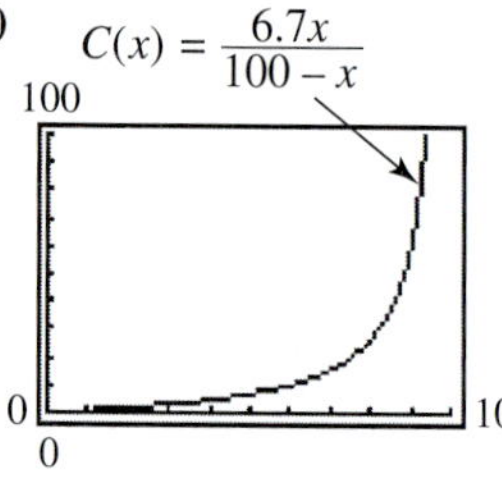

(b) approximately \$127.3 thousand **83.** 35 **85.** 3 **87.** $84\frac{3}{8}$ **89.** 7500 lb

Chapter 3 Test *(page 386)*

1.

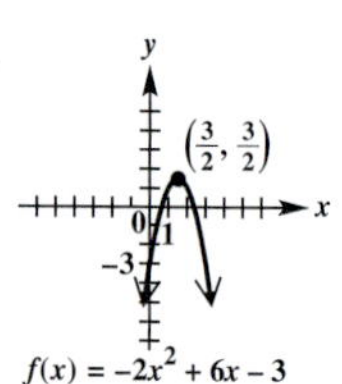

x-intercepts: $\dfrac{-3 \pm \sqrt{3}}{-2}\left(\text{or } \dfrac{3 \pm \sqrt{3}}{2}\right)$; y-intercept: -3; vertex: $\left(\dfrac{3}{2}, \dfrac{3}{2}\right)$; axis: $x = \dfrac{3}{2}$; domain: $(-\infty, \infty)$; range: $\left(-\infty, \dfrac{3}{2}\right]$ **2.** 8.1 million **3.** $3x^2 - 2x - 5 + \dfrac{16}{x+2}$ **4.** $2x^2 - x - 5 + \dfrac{3}{x-5}$ **5.** 53 **6.** It is a factor. The other factor is $6x^3 + 7x^2 - 14x - 8$. **7.** $-2, -3 - 2i, -3 + 2i$

8. $f(x) = 2x^4 - 2x^3 - 2x^2 - 2x - 4$ **9.** Because $f(x) > 0$ for all x, the graph never intersects or touches the x-axis, so $f(x)$ has no real zeros. **10.** **(a)** $f(1) = 5 > 0$; $f(2) = -1 < 0$ **(b)** 2 or 0 positive zeros; 1 negative zero **(c)** 4.0937635, 1.8370381, $-.9308016$

11. 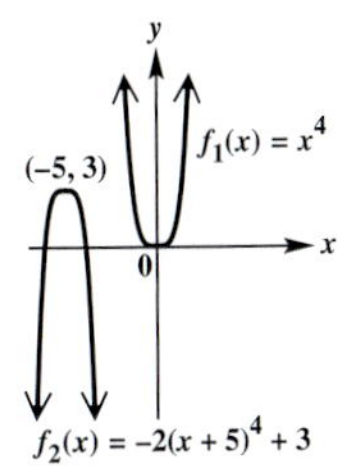

To obtain the graph of f_2, translate the graph of f_1 5 units to the left, stretch by a factor of 2, reflect across the x-axis, and translate 3 units up.

12. C **13.** 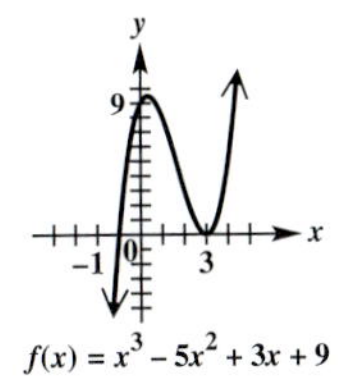

14. 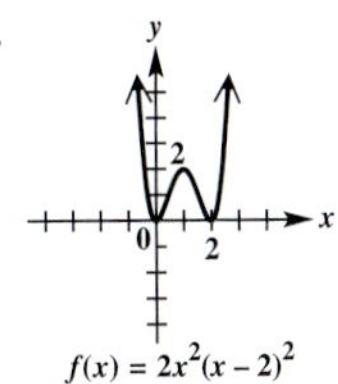

15. 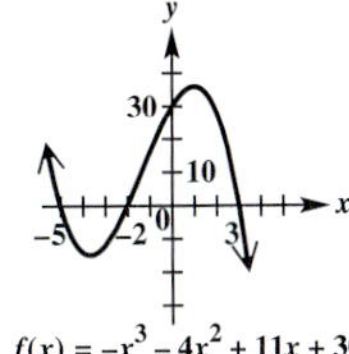

16. $f(x) = 2(x - 2)^2(x + 3)$ or $f(x) = 2x^3 - 2x^2 - 16x + 24$ **17.** **(a)** 270.08 **(b)** increasing from $t = 0$ to $t = 5.9$ and $t = 9.5$ to $t = 15$; decreasing from $t = 5.9$ to $t = 9.5$

18. 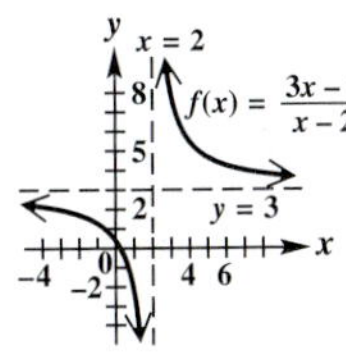

19. 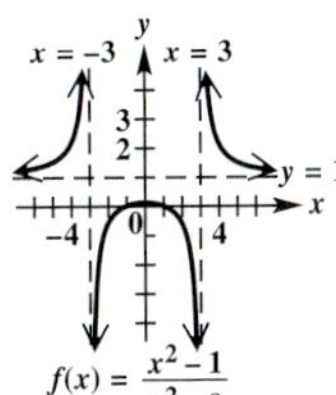

20. **(a)** $y = 2x + 3$ **(b)** $-2, \frac{3}{2}$ **(c)** 6 **(d)** $x = 1$

(e) 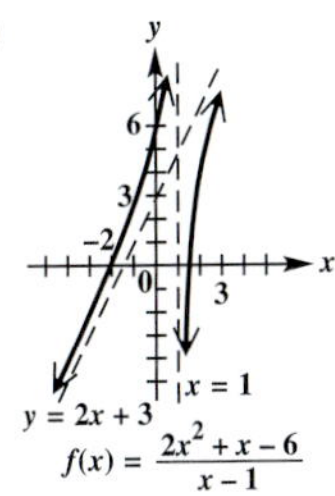

21. 60 **22.** $\frac{640}{9}$ kg

CHAPTER 4 EXPONENTIAL AND LOGARITHMIC FUNCTIONS

4.1 Exercises *(page 398)*

1. It is not one-to-one because both Illinois and Wisconsin are paired with the same range element, 40. **3.** one-to-one **5.** one-to-one **7.** not one-to-one **9.** not one-to-one **11.** not one-to-one **13.** one-to-one **15.** one-to-one **17.** one-to-one **19.** range; domain **21.** false **23.** −3 **27.** untying your shoelaces **29.** leaving a room **31.** rewinding a tape **33.** inverses **35.** not inverses **37.** inverses **39.** inverses **41.** not inverses **43.** inverses **45.** $\{(6, -3), (1, 2), (8, 5)\}$ **47.** not one-to-one

49. **(a)** $f^{-1}(x) = \frac{1}{3}x + \frac{4}{3}$ **(b)** **(c)** Domains and ranges of both f and f^{-1} are $(-\infty, \infty)$.

51. **(a)** $f^{-1}(x) = -\frac{1}{4}x + \frac{3}{4}$ **(b)**

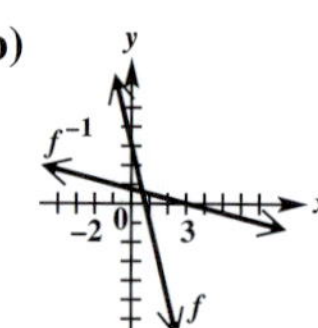

(c) Domains and ranges of both f and f^{-1} are $(-\infty, \infty)$.

53. **(a)** $f^{-1}(x) = \sqrt[3]{x - 1}$ **(b)**

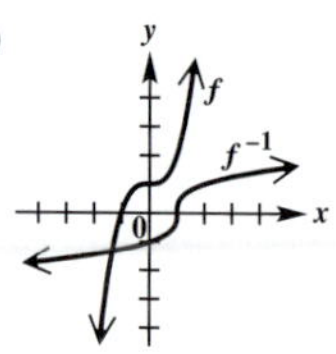

(c) Domains and ranges of both f and f^{-1} are $(-\infty, \infty)$.

55. not one-to-one

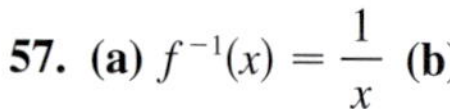

57. **(a)** $f^{-1}(x) = \frac{1}{x}$ **(b)**

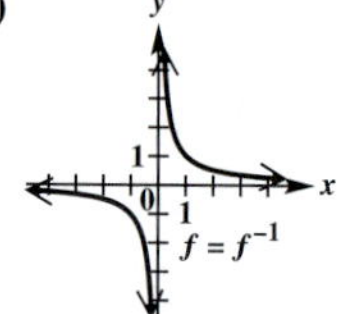

(c) Domains and ranges of both f and f^{-1} are $(-\infty, 0) \cup (0, \infty)$.

59. **(a)** $f^{-1}(x) = \frac{1 + 3x}{x}$ **(b)**

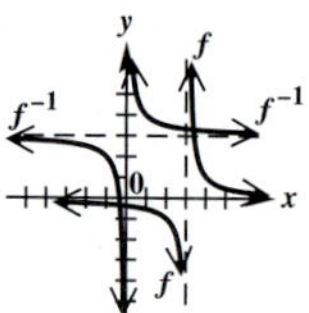

(c) Domain of f = range of $f^{-1} = (-\infty, 3) \cup (3, \infty)$; Domain of f^{-1} = range of $f = (-\infty, 0) \cup (0, \infty)$.

61. **(a)** $f^{-1}(x) = x^2 - 6, x \geq 0$ **(b)**

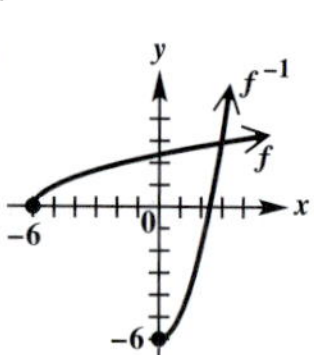

(c) Domain of f = range of $f^{-1} = [-6, \infty)$; Domain of f^{-1} = range of $f = [0, \infty)$.

63.

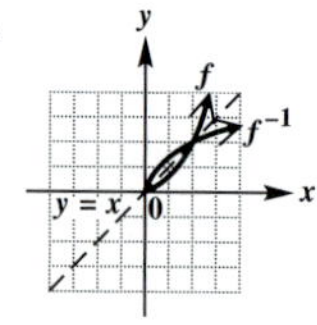

65.

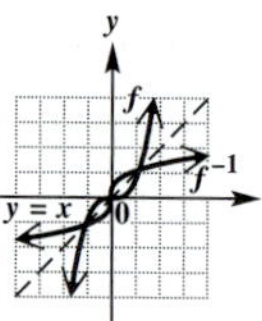

67.

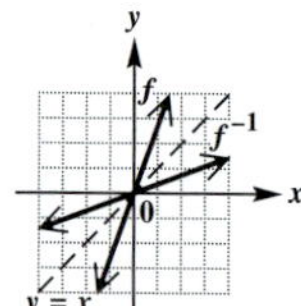

69. 4 **71.** 2

73. −2 **75.** It represents the cost, in dollars, of building 1000 cars. **77.** $\frac{1}{a}$ **79.** not one-to-one

81. one-to-one; $f^{-1}(x) = \frac{-5 - 3x}{x - 1}$ **83.** $f^{-1}(x) = \frac{x + 2}{3}$; MIGUEL HAS ARRIVED

85. 6858 124 2743 63 511 124 1727 4095; $f^{-1}(x) = \sqrt[3]{x + 1}$

Connections *(page 414)*

1. 2.717 (Calculator gives 2.718.) **2.** .9512 (Calculator gives .9512.) **3.** $\frac{x^6}{6 \cdot 5 \cdot 4 \cdot 3 \cdot 2 \cdot 1}$

4.2 Exercises *(page 414)*

1. 9 **3.** $\frac{1}{9}$ **5.** $\frac{1}{16}$ **7.** 16 **9.** 5.196 **11.** .039 **13.**

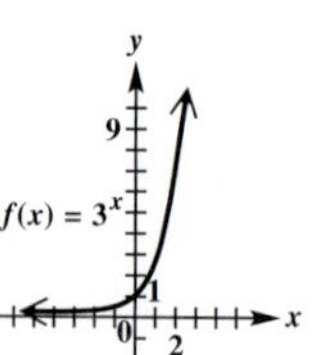

15.

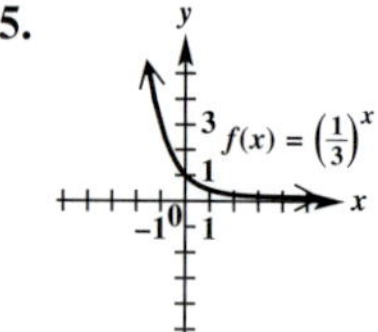

17. 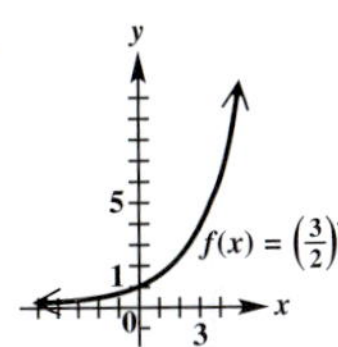

19. 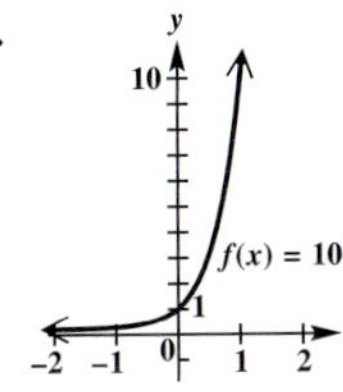

21. 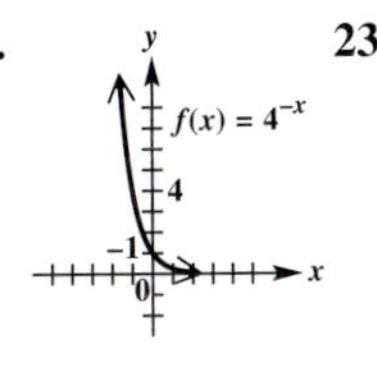

23. 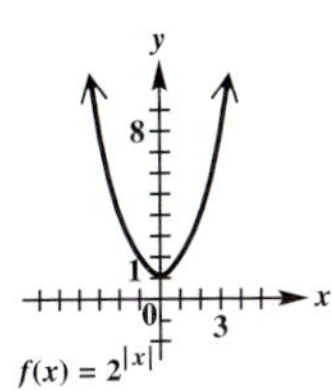

25. 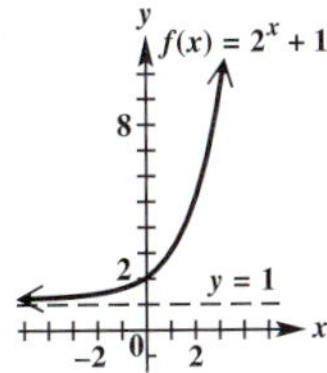

27. 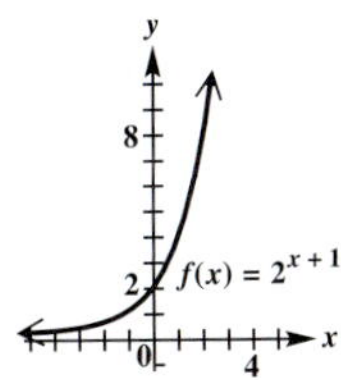

29. 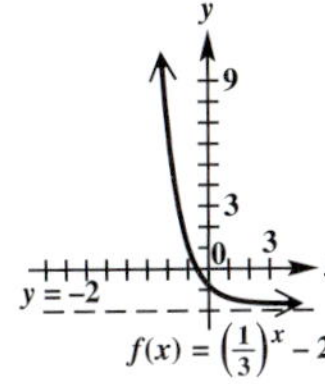

31. 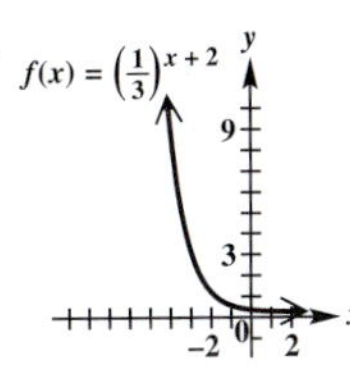

33. 2.3 **35.** .75 **37.** .31

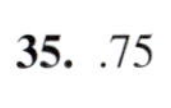

39. $f(x) = \dfrac{e^x - e^{-x}}{2}$

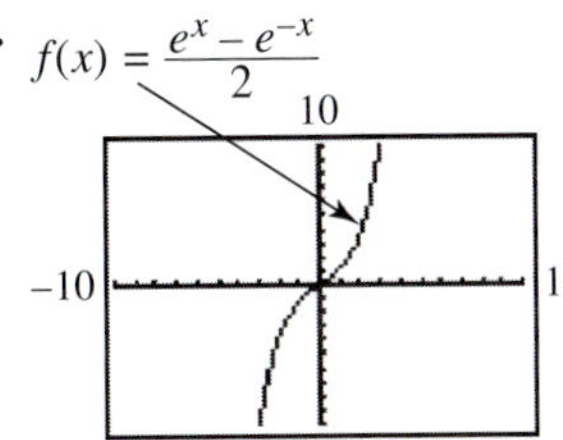

41. $f(x) = x \cdot 2^x$

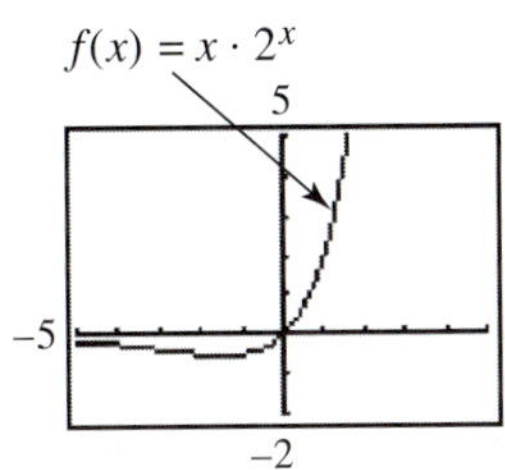

43. $\left\{\frac{1}{2}\right\}$ **45.** $\{-2\}$ **47.** $\{0\}$ **49.** $\left\{\frac{1}{2}\right\}$

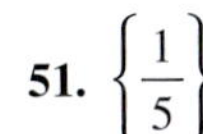

51. $\left\{\frac{1}{5}\right\}$ **53.** $\{-7\}$ **55.** $\left\{-\frac{2}{3}\right\}$ **57.** $\left\{\frac{4}{3}\right\}$ **59.** $\{3\}$ **61.** $\{-8, 8\}$ **63.** **(a)** \$13,891.16; \$4984.62 **(b)** \$13,968.24; \$5061.70 **65.** \$21,223.33 **67.** \$3528.81 **69.** 4.5% **71.** Bank A (even though it has the highest stated rate) **73.** **(a)**

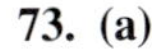

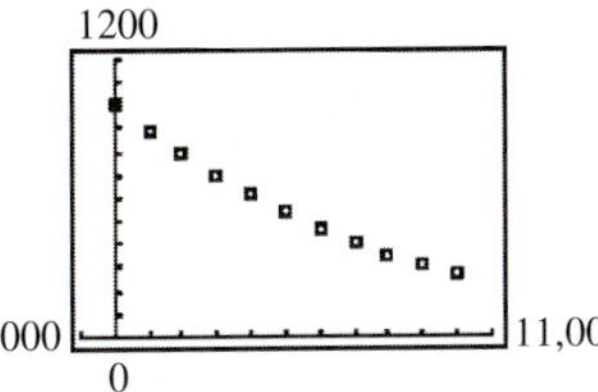

(b) exponential **(c)** $P(x) = 1013e^{-.0001341x}$

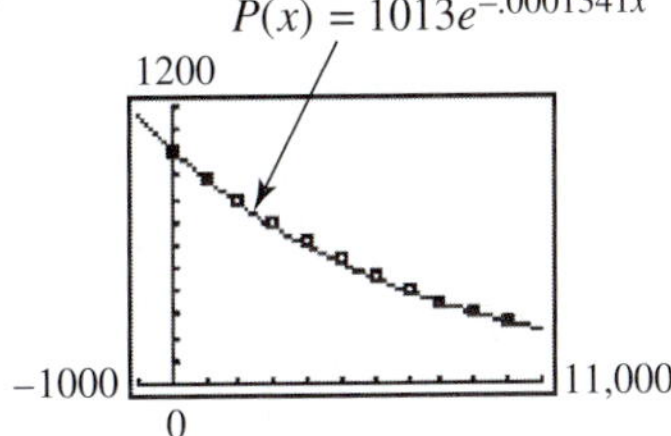

(d) $P(1500) \approx 828$ mb; $P(11{,}000) \approx 232$ mb **75.** **(a)** about 63,000 **(b)** about 42,000 **(c)** about 21,000 **77.** $\{.9\}$ **79.** $\{-.5, 1.3\}$ **83.** $f(x) = 2^x$ **85.** $f(t) = 27 \cdot 9^t$ **89.** yes; an inverse function **90.**

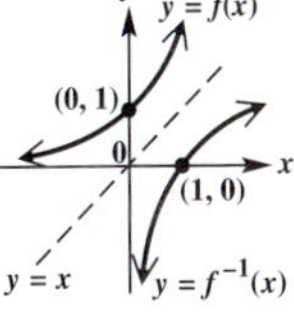

91. $x = a^y$ **92.** $x = 10^y$ **93.** $x = e^y$ **94.** (q, p)

Connections *(page 427)*

1.

$$\begin{array}{r} \log_{10} 458.3 \approx 2.661149857 \\ + \log_{10} 294.6 \approx 2.469232743 \\ \hline \approx 5.130382600 \\ 10^{5.130382600} \approx 135{,}015.18 \end{array}$$

A calculator gives $(458.3)(294.6) = 135{,}015.18$.

2. Answers will vary.

4.3 Exercises *(page 427)*

1. **(a)** C **(b)** A **(c)** E **(d)** B **(e)** F **(f)** D **3.** $\log_3 81 = 4$ **5.** $\log_{2/3} \frac{27}{8} = -3$ **7.** $6^2 = 36$ **9.** $(\sqrt{3})^8 = 81$

13. $\{-4\}$ **15.** $\{-3\}$ **17.** $\left\{\frac{1}{4}\right\}$ **19.** $\{8\}$ **21.** $\{9\}$ **23.** $\left\{\frac{1}{5}\right\}$ **25.** $\{64\}$ **27.** $\left\{\frac{2}{3}\right\}$ **29.** $\left\{\frac{1}{3}\right\}$

33. **35.** **37.**

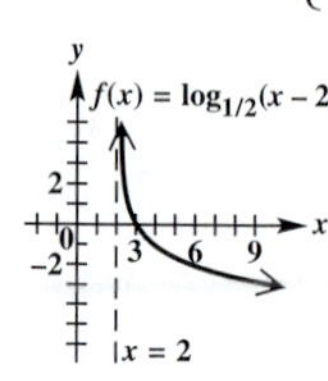

39. E **41.** B **43.** F

45.

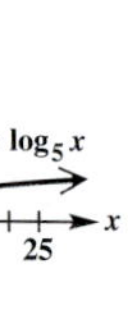

47.

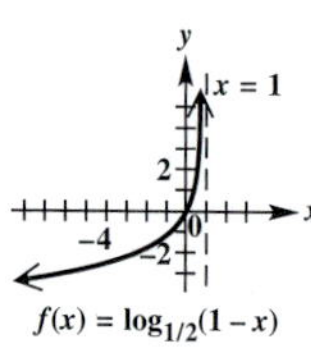

49.

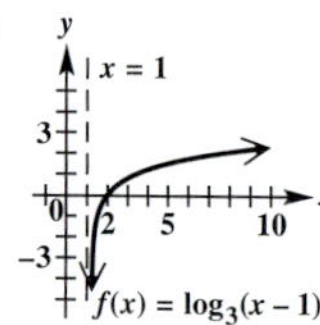

51.

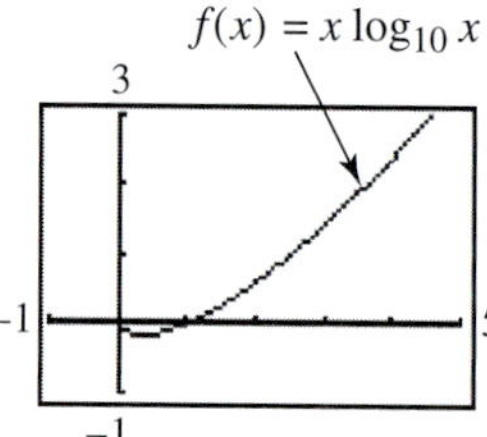

53. $\log_a x - \log_a y$ **54.** Since $\log_2 \frac{x}{4} = \log_2 x - \log_2 4$ by the quotient rule, the graph of $y = \log_2 \frac{x}{4}$ can be obtained by translating the graph of $y = \log_2 x$ down $\log_2 4 = 2$ units. **55.**

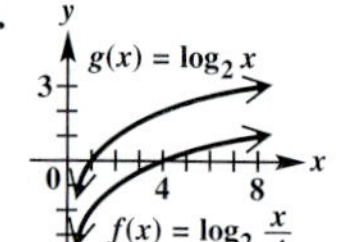

56. 0; 2; 2; 0; By the quotient rule, $\log_2 \frac{x}{4} = \log_2 x - \log_2 4$. Both sides should equal 0. Since $2 - 2 = 0$, they do. **57.** $\log_2 6 + \log_2 x - \log_2 y$

59. $1 + \frac{1}{2}\log_5 7 - \log_5 3$ **61.** cannot be simplified **63.** $\frac{1}{2}(\log_m 5 + 3 \log_m r - 5 \log_m z)$ **65.** $\log_a \frac{xy}{m}$

67. $\log_m \frac{a^2}{b^6}$ **69.** $\log_a [(z - 1)^2(3z + 2)]$ **71.** $\log_5 \frac{5^{1/3}}{m^{1/3}}$ or $\log_5 \sqrt[3]{\frac{5}{m}}$ **73.** .7781 **75.** .3522 **77.** .7386

79. **(a)**

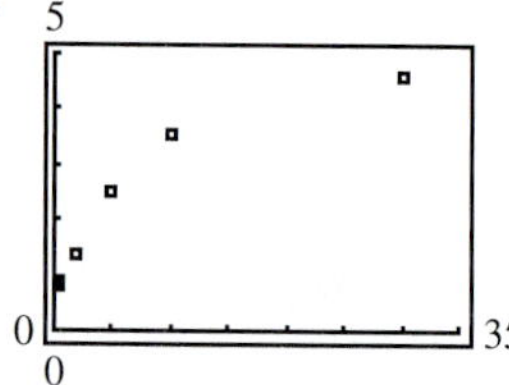

81. **(a)** -4 **(b)** 6 **85.** $\{.01, 2.38\}$

Summary Exercises on Inverse, Exponential, and Logarithmic Functions *(page 431)*

1. inverses **2.** not inverses **3.** inverses **4.** inverses **5.** **6.**

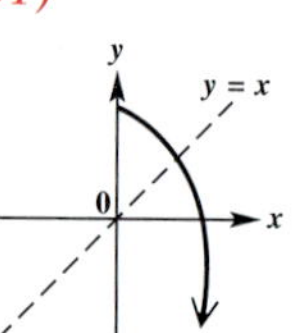

7. not one-to-one **8.** **9.** B **10.** D **11.** C **12.** A

13. The functions in Exercises 9 and 12 are inverses of one another. The functions in Exercises 10 and 11 are inverses of one another. **14.** $f^{-1}(x) = 5^x$ **15.** $f^{-1}(x) = \frac{x+6}{3}$; domains and ranges of both f and f^{-1} are $(-\infty, \infty)$.

16. $f^{-1}(x) = \sqrt[3]{\frac{x}{2}} - 1$; domains and ranges of both f and f^{-1} are $(-\infty, \infty)$. **17.** f is not one-to-one.

18. $f^{-1}(x) = \frac{5x+1}{2+3x}$; domain of f = range of $f^{-1} = \left(-\infty, \frac{5}{3}\right) \cup \left(\frac{5}{3}, \infty\right)$; domain of f^{-1} = range of $f = \left(-\infty, -\frac{2}{3}\right) \cup \left(-\frac{2}{3}, \infty\right)$ **19.** f is not one-to-one. **20.** $f^{-1}(x) = \sqrt{x^2+9}$, $x \ge 0$; domain of f = range of $f^{-1} = [3, \infty)$; domain of f^{-1} = range of $f = [0, \infty)$. **21.** $\log_{1/10} 1000 = -3$ **22.** $\log_a c = b$ **23.** $\log_{\sqrt{3}} 9 = 4$ **24.** $\log_4 \frac{1}{8} = -\frac{3}{2}$

25. $\log_2 32 = x$ **26.** $\log_{27} 81 = \frac{4}{3}$ **27.** $\{2\}$ **28.** $\{25\}$ **29.** $\{-2\}$ **30.** $\left\{\frac{3}{2}\right\}$ **31.** $\{5\}$ **32.** $\{-2\}$

33. $\{1\}$ **34.** $\left\{\frac{1}{9}\right\}$ **35.** $\left\{\frac{16}{3}\right\}$

4.4 Exercises *(page 438)*

1. increasing **3.** $f^{-1}(x) = \log_5 x$ **5.** natural; common **7.** There is no power of 2 that yields a result of 0. **9.** $\log 8 = .90308999$ **11.** 1.5563 **13.** -1.3768 **15.** 4.3010 **17.** 3.5835 **19.** -3.1701 **21.** 4.6931 **23.** 3.2 **25.** 1.8 **27.** 2.0×10^{-3} **29.** 1.6×10^{-5} **31.** poor fen **33.** rich fen **35.** 2.3219 **37.** $-.2537$

39. 1.9376 **41.** -1.4125 **43.** D **45.** $4v + \frac{1}{2}u$ **47.** $\frac{3}{2}u - \frac{5}{2}v$ **49.** **(a)** 3 **(b)** 5^2 or 25 **(c)** $\frac{1}{e}$

51. **(a)** 5 **(b)** ln 3 **(c)** 2 ln 3 or ln 9 **53.** domain: $(-\infty, 0) \cup (0, \infty)$; range: $(-\infty, \infty)$; symmetric with respect to the y-axis **55.** When $X \ge 4$, $4 - X \le 0$, and we cannot obtain a real value for the logarithm of a nonpositive number. **57.** $f(x) = 2 + \ln x$, so it is the graph of $f(x) = \ln x$ translated 2 units up. **59.** **(a)** 20 **(b)** 30 **(c)** 50 **(d)** 60 **(e)** about 3 decibels **61.** **(a)** 3 **(b)** 6 **(c)** 8 **63.** about $126{,}000{,}000 I_0$ **65.** about 70 million visitors; We must assume that the rate of increase continues to be logarithmic. **67.** **(a)** 2 **(b)** 2 **(c)** 2 **(d)** 1 **69.** 1 **71.** between 7°F and 11°F

73. **(a)**

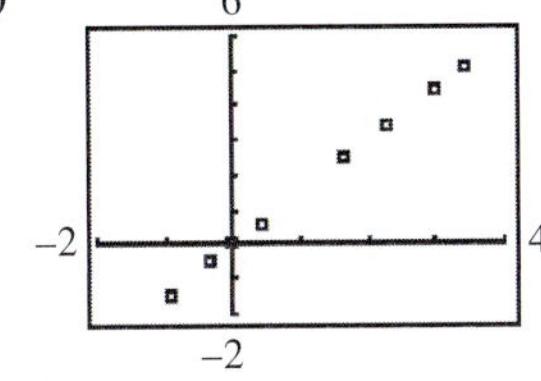

Let $x = \ln D$ and $y = \ln P$ for each planet. From the graph, the data appear to be linear.

(b) The points (0, 0) and (3.40, 5.10) determine the line $y = 1.5x$ or $\ln P = 1.5 \ln D$. (Answers will vary.) **(c)** $P \approx 248.3$ yr

4.5 Exercises *(page 448)*

1. $\log_7 19$; $\frac{\log 19}{\log 7}$; $\frac{\ln 19}{\ln 7}$ **3.** $\log_{1/2} 12$; $\frac{\log 12}{\log\left(\frac{1}{2}\right)}$; $\frac{\ln 12}{\ln\left(\frac{1}{2}\right)}$ **5.** $\{1.6309\}$ **7.** $\{-.0803\}$ **9.** $\{2.2694\}$ **11.** $\{2.3863\}$

13. $\{-.1227\}$ **15.** $\emptyset$ **17.** $\{2\}$ **19.** $\{140.0112\}$ **21.** $\left\{\frac{1}{3}\right\}$ **23.** $\emptyset$ **25.** $\{1\}$ **27.** $\{5\}$ **29.** $\{25\}$

31. {4} **33.** $\left\{\frac{9}{2}\right\}$ **35.** {−17.5314} **37.** {8} **39.** {4} **41.** {1, 100} **45.** $t = -\frac{2}{R}\ln\left(1 - \frac{RI}{E}\right)$

47. $x = e^{k/(p-a)}$ **49.** By the power rule for exponents, $(a^m)^n = a^{mn}$. **50.** $(e^x - 1)(e^x - 3) = 0$ **51.** {0, ln 3}

52. The graph intersects the x-axis at 0 and ln 3 ≈ 1.099.

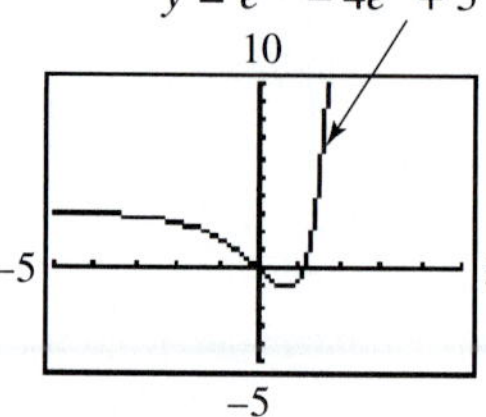

53. $(-\infty, 0) \cup (\ln 3, \infty)$

54. (0, ln 3) **55.** $f^{-1}(x) = \ln(x + 4) - 1$; domain: $(-4, \infty)$; range: $(-\infty, \infty)$ **57.** $(27, \infty)$ **59.** {1.52} **61.** {0}

63. {2.45, 5.66} **65.** during 2011 **67.** **(a)** about 24% **(b)** 1963 **69.** **(a)** $P(T) = 1 - e^{-.0034 - .0053T}$

(b) For $T = x$, $P(x) = 1 - e^{-.0034 - .0053x}$

1
0 1000
0

(c) $P(60) \approx .275$ or 27.5%. The reduction in carbon emissions from a tax of $60 per ton of carbon is 27.5%. **(d)** $T =$ $130.14 **71.** 2.6 yr **73.** 6.48%

4.6 Exercises *(page 458)*

1. B **3.** C **5.** **(a)** 440 g **(b)** 387 g **(c)** 264 g **(d)** 21.66 yr **7.** 1611.97 yr **9.** **(a)** 11% **(b)** 36% **(c)** 84%

11. about 9000 yr **13.** about 15,600 yr **15.** 6.25°C **17.** **(a)** $f(x) = .05(1.04)^{x-1950}$ **(b)** 4%

19. **(a)** 7% compounded quarterly **(b)** $800.32 **21.** about 27.73 yr **23.** about 21.97 yr **25.** **(a)** $P = 1$; $a \approx 1.01355$

(b) 1.14 billion **(c)** 2030 **27.** **(a)** 44.9 billion **(b)** 1992 **29.** **(a)** 11 **(b)** 12.6 **(c)** 18.0 **(d)**

For $L = y$ and $t = x$
$y = 9 + 2e^{.15x}$

30
0 10
0

(e) Living standards are increasing, but at a slow rate. **31.** about 14.2 hr **33.** about 22.7% **35.** **(a)** $S(1) \approx 45{,}200$; $S(3) \approx 37{,}000$ **(b)** $S(2) \approx 72{,}400$; $S(10) \approx 48{,}500$ **37.** about 18.3 yr **39.** about 34.7 yr

41. **(a)**

X	Y1
10	7.9832
20	31.582
30	46.965
40	49.645
50	49.96
60	49.996
70	50

X=10

The maximum height appears to be 50 ft.

(b)

$y = \frac{50}{1 + 47.5e^{-.22x}}$

60
10 70
0

The horizontal asymptote is $y = 50$. It tells us that this tree cannot grow taller than 50 ft.

(c) after about 19.4 yr

Chapter 4 Review Exercises *(page 467)*

1. not one-to-one **3.** one-to-one **5.** not one-to-one **7.** $f^{-1}(x) = \sqrt[3]{x+3}$ **9.** It represents the number of years after 2004 for the investment to reach \$50,000. **11.** one-to-one **13.** B **15.** C **17.** $\log_2 32 = 5$ **19.** $\log_{3/4} \frac{4}{3} = -1$ **21.** $\log_3 4$ **22.** 2 **23.** 3 **24.** It lies between 2 and 3. **25.** By the change-of-base theorem, $\log_3 16 = \frac{\log 16}{\log 3} = \frac{\ln 16}{\ln 3} \approx 2.523719014$. **26.** $-1; 0$ **27.** It lies between -1 and 0; $\log_5 .68 = \frac{\log .68}{\log 5} = \frac{\ln .68}{\ln 5} \approx -.2396255723$ **29.** $10^{.5378} \approx 3.45$ **31.** 3 **33.** $\log_3 m + \log_3 n - \log_3 5 - \log_3 r$ **35.** cannot be simplified **37.** -1.3862 **39.** 11.8776 **41.** 1.1592 **43.** $\left\{\frac{3}{2}\right\}$ **45.** $\{-.3138\}$ **47.** $\{4\}$ **49.** $\{3\}$ **51.** $\{5\}$ **53.** $n = a(e^{S/a} - 1)$ **55.** **(a)** about $200{,}000{,}000I_0$ **(b)** about $13{,}000{,}000I_0$ **(c)** The 1906 earthquake had a magnitude almost 16 times greater than the 1989 earthquake. **57.** 5.1% **59.** \$60,602.77 **61.** 17.3 yr **63.** 2007 **65.** **(a)** $\log_4(2x^2 - x) = \frac{\ln(2x^2 - x)}{\ln 4}$ **(b)**

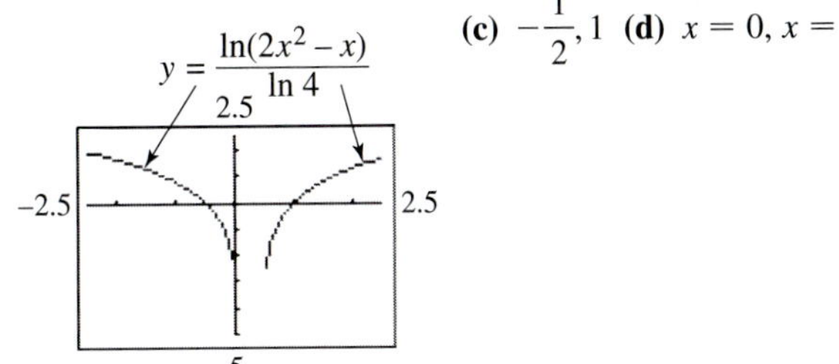

(c) $-\frac{1}{2}, 1$ **(d)** $x = 0, x = \frac{1}{2}$

Chapter 4 Test *(page 471)*

1. **(a)** $(-\infty, \infty); (-\infty, \infty)$ **(b)** The graph is a stretched translation of $y = \sqrt[3]{x}$, which passes the horizontal line test and is thus a one-to-one function. **(c)** $f^{-1}(x) = \frac{x^3 + 7}{2}$ **(d)** $(-\infty, \infty); (-\infty, \infty)$ **(e)**

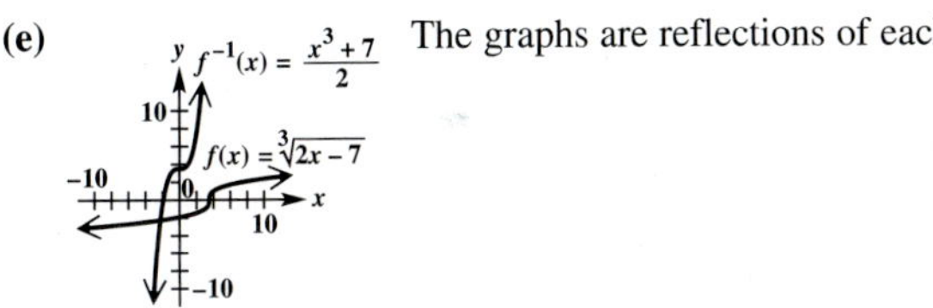

The graphs are reflections of each other across the line $y = x$. **2.** **(a)** B **(b)** A **(c)** C **(d)** D **3.** $\left\{\frac{1}{2}\right\}$ **4.** **(a)** $\log_4 8 = \frac{3}{2}$ **(b)** $8^{2/3} = 4$ **5.** They are inverses. **6.** $2 \log_7 x + \frac{1}{4} \log_7 y - 3 \log_7 z$ **7.** 2.3755 **8.** -3.0640 **9.** 1.1674 **10.** -2.6038 **11.** $\{5\}$ **12.** $\left\{\frac{5}{2}\right\}$ **13.** $\{2\}$ **14.** $\{4.7833\}$ **15.** $\{20.1246\}$ **17.** 10 sec **18.** **(a)** about 18.9 yr **(b)** about 18.8 yr **19.** **(a)** about 329.3 g **(b)** about 13.9 days **20.** near the end of 2015

CHAPTER 5 TRIGONOMETRIC FUNCTIONS

5.1 Exercises *(page 478)*

3. 45° **5.** **(a)** 60° **(b)** 150° **7.** **(a)** 45° **(b)** 135° **9.** **(a)** 36° **(b)** 126° **11.** 150° **13.** 70°; 110° **15.** 55°; 35° **17.** 80°; 100° **19.** $(90 - x)°$ **21.** $(x - 360)°$ **23.** 83° 59′ **25.** 23° 49′ **27.** 38° 32′ **29.** 17° 1′ 49″ **31.** 20.900° **33.** 91.598° **35.** 274.316° **37.** 31° 25′ 47″ **39.** 89° 54′ 1″ **41.** 178° 35′ 58″ **45.** 320°

47. 235° **49.** 179° **51.** 130° **53.** $30° + n \cdot 360°$ **55.** $135° + n \cdot 360°$ **57.** $-90° + n \cdot 360°$

Angles other than those given are possible in Exercises 61–67.

61.

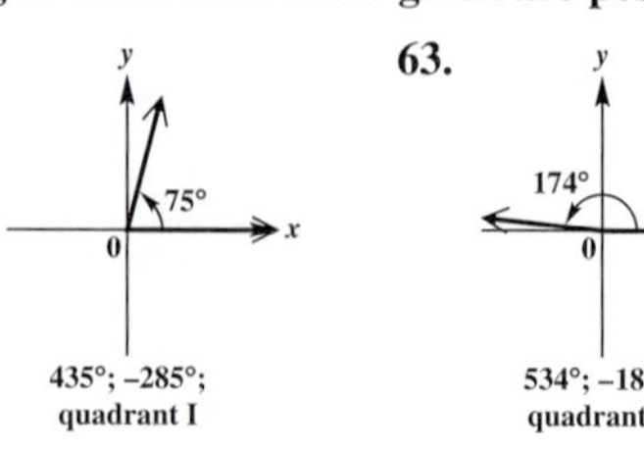

63.

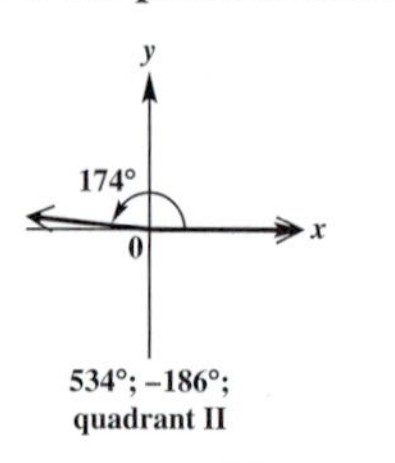

65.

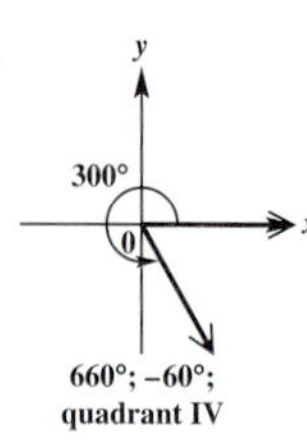

67.

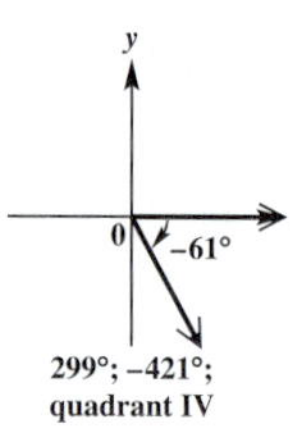

69. $3\sqrt{2}$

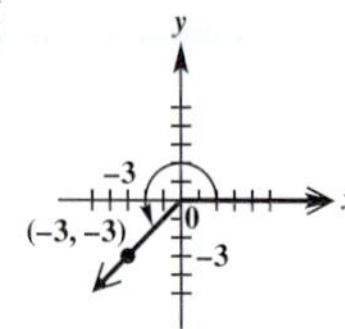

71. $\sqrt{34}$

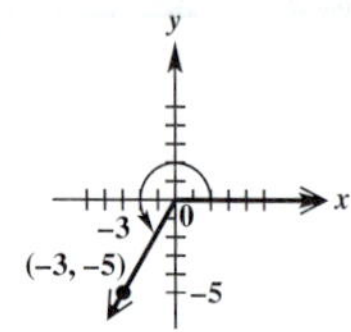

73. 4

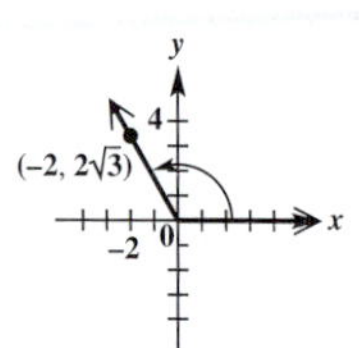
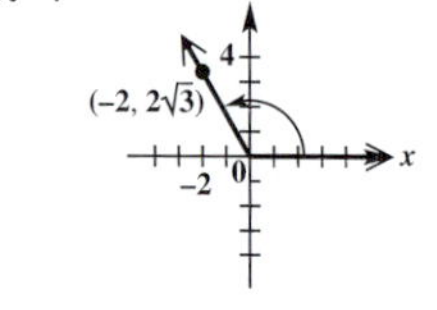

75. $\frac{3}{4}$ **77.** 1800°

79. 12.5 rotations per hr **81.** 4 sec

5.2 Exercises *(page 490)*

1.

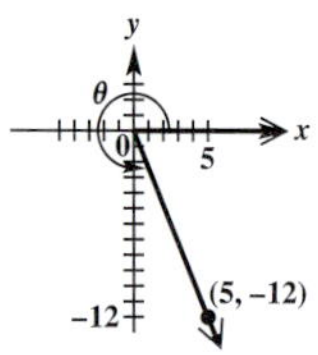

In Exercises 3–9 and 17–21, we give, in order, sine, cosine, tangent, cotangent, secant, and cosecant.

3. $\frac{4}{5}; -\frac{3}{5}; -\frac{4}{3}; -\frac{3}{4}; -\frac{5}{3}; \frac{5}{4}$ **5.** 1; 0; undefined; 0; undefined; 1 **7.** $\frac{\sqrt{3}}{2}; \frac{1}{2}; \sqrt{3}; \frac{\sqrt{3}}{3}; 2; \frac{2\sqrt{3}}{3}$

9. 0; -1; 0; undefined; -1; undefined **13.** negative **15.** negative

17. $-\frac{2\sqrt{5}}{5}; \frac{\sqrt{5}}{5}; -2; -\frac{1}{2}; \sqrt{5}; -\frac{\sqrt{5}}{2}$

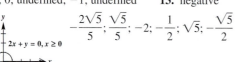

19. $\frac{6\sqrt{37}}{37}; -\frac{\sqrt{37}}{37}; -6; -\frac{1}{6}; -\sqrt{37}; \frac{\sqrt{37}}{6}$

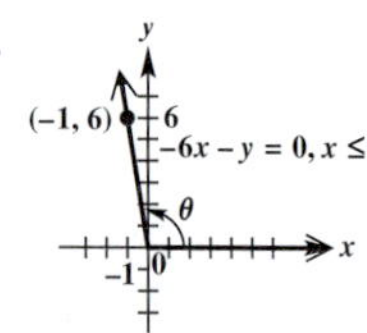

21. 1; 0; undefined; 0; undefined; 1 **23.** -7 **25.** 3 **27.** 1 **29.** 0 **31.** undefined **33.** They are equal.

35. They are equal. **37.** 40° **39.** 45° **41.** decrease; decrease **43.** -1; $\theta = 180°$ **45.** -5 **47.** $-\frac{3\sqrt{5}}{5}$

49. .10199657 **51.** The range of the cosine function is $[-1, 1]$, so $\cos \theta$ cannot equal $\frac{3}{2}$. **53.** $\sqrt{3}$ **55.** 2° **57.** II

59. I or III **61.** +; –; – **63.** –; +; – **65.** –; +; – **67.** tan 30° **69.** sec 33° **71.** impossible

73. possible **75.** possible **77.** impossible **79.** $\frac{\sqrt{15}}{4}$ **81.** $-\frac{4}{3}$ **83.** $-\frac{\sqrt{3}}{2}$ **85.** $-.56616682$

In Exercises 87–91, we give, in order, sine, cosine, tangent, cotangent, secant, and cosecant.

87. $\frac{15}{17}; -\frac{8}{17}; -\frac{15}{8}; -\frac{8}{15}; -\frac{17}{8}; \frac{17}{15}$ **89.** $-\frac{\sqrt{3}}{2}; -\frac{1}{2}; \sqrt{3}; \frac{\sqrt{3}}{3}; -2; -\frac{2\sqrt{3}}{3}$

91. $-.555762; .831342; -.668512; -1.49586; 1.20287; -1.79933$

95. false; for example, $\sin 30° + \cos 30° \approx .5 + .8660 = 1.3660 \neq 1$. **97.** 146 ft **99.** **(a)** $\tan \theta = \frac{y}{x}$ **(b)** $x = \frac{y}{\tan \theta}$

5.3 Exercises *(page 502)*

In Exercises 1 and 3, we give, in order, sine, cosine, and tangent.

1. $\frac{21}{29}; \frac{20}{29}; \frac{21}{20}$ **3.** $\frac{n}{p}; \frac{m}{p}; \frac{n}{m}$ **5.** C **7.** B **9.** E **11.** $\frac{\sqrt{3}}{3}$ **13.** $\frac{1}{2}$ **15.** $\frac{2\sqrt{3}}{3}$

17. $\sqrt{2}$ **19.** $\frac{\sqrt{2}}{2}$ **21.** **22.** 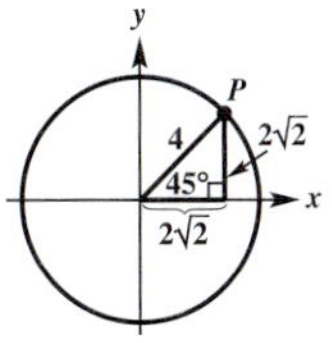 **23.** the legs; $(2\sqrt{2}, 2\sqrt{2})$ **24.** $(1, \sqrt{3})$

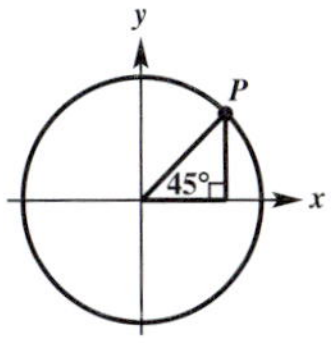

25. $\sin x$; $\tan x$ **27.** 60° **29.** $\left(\frac{\sqrt{2}}{2}, \frac{\sqrt{2}}{2}\right)$; 45° **31.** $y = \frac{\sqrt{3}}{3}x$ **33.** 60°

35. $x = \frac{9\sqrt{3}}{2}; y = \frac{9}{2}; z = \frac{3\sqrt{3}}{2}; w = 3\sqrt{3}$ **37.** $p = 15; r = 15\sqrt{2}; q = 5\sqrt{6}; t = 10\sqrt{6}$ **39.** $A = \frac{s^2}{2}$ **41.** F

43. B **45.** B **51.** $\frac{\sqrt{2}}{2}; \frac{\sqrt{2}}{2}; \sqrt{2}; \sqrt{2}$ **53.** $-\frac{1}{2}; -\frac{\sqrt{3}}{3}; -2$ **55.** $\frac{1}{2}; -\sqrt{3}; -\frac{2\sqrt{3}}{3}$ **57.** $\sqrt{3}; \frac{\sqrt{3}}{3}$

In Exercises 59–63, we give, in order, sine, cosine, tangent, cotangent, secant, and cosecant.

59. $-\frac{\sqrt{2}}{2}; \frac{\sqrt{2}}{2}; -1, -1; \sqrt{2}; -\sqrt{2}$ **61.** $\frac{\sqrt{3}}{2}; \frac{1}{2}; \sqrt{3}; \frac{\sqrt{3}}{3}; 2; \frac{2\sqrt{3}}{3}$ **63.** $\frac{1}{2}; \frac{\sqrt{3}}{2}; \frac{\sqrt{3}}{3}; \sqrt{3}; \frac{2\sqrt{3}}{3}; 2$

65. $-\frac{\sqrt{3}}{2}$ **67.** $\frac{\sqrt{3}}{2}$ **69.** true **71.** false; $\frac{1}{2} \neq \sqrt{3}$ **73.** true **75.** .6252427 **77.** 1.0273488

79. 15.055723 **81.** 1.4887142 **83.** .6743024 **85.** .9999905 **87.** .4327386 **89.** .2308682 **91.** 55.845496°

93. 38.491580° **95.** 30°; 330° **97.** 135°; 225° **99.** 45°; 315° **103.** **(a)** 19° **(b)** 50° **105.** 7.9°

107. approximately 78 mph **109.** 65.96 lb **111.** −2.87° **113.** **(a)** 703 ft **(b)** 1701 ft **(c)** R would decrease.

5.4 Exercises *(page 516)*

1. 20,385.5 to 20,386.5 **3.** 8958.5 to 8959.5 **7.** .05 **9.** $B = 53° 40'$; $a = 571$ m; $b = 777$ m

11. $M = 38.8°$; $n = 154$ m; $p = 198$ m **17.** $B = 62.00°$; $a = 8.17$ ft; $b = 15.4$ ft

19. $A = 17.00°$; $a = 39.1$ in.; $c = 134$ in. **21.** $c = 85.9$ yd; $A = 62° 50'$; $B = 27° 10'$

23. The angle of elevation from X to Y is 90° whenever Y is directly above X.

27. It should be shown as an angle measured from due north. **29.** 9.35 m **31.** 128 ft **33.** 26.92 in. **35.** 22°

37. 28.0 m **39.** 13.3 ft **41.** 146 m **43.** **(a)** 29,008 ft **(b)** shorter **45.** 220 mi **47.** 47 nautical mi

49. 130 mi **51.** 2.01 mi **53.** 147 m **55.** 2.47 km **57.** 10.8 ft **59.** **(a)** 323 ft **(b)** $R\left(1 - \cos \frac{\theta}{2}\right)$

61. **(a)** 23.4 ft **(b)** 48.3 ft **(c)** The faster the speed, the more land needs to be cleared inside the curve.

Chapter 5 Review Exercises *(page 527)*

1. complement: 55°; supplement: 145° **3.** $59.591\overline{6}°$ **5.** $270° + n \cdot 360°$

In Exercises 7–11, we give, in order, sine, cosine, tangent, cotangent, secant, and cosecant.

7. $-\frac{\sqrt{3}}{2}; \frac{1}{2}; -\sqrt{3}; -\frac{\sqrt{3}}{3}; 2; -\frac{2\sqrt{3}}{3}$ **9.** $-\frac{4}{5}; \frac{3}{5}; -\frac{4}{3}; -\frac{3}{4}; \frac{5}{3}; -\frac{5}{4}$ **11.** $\frac{\sqrt{2}}{2}; -\frac{\sqrt{2}}{2}; -1; -1; -\sqrt{2}; \sqrt{2}$

13.

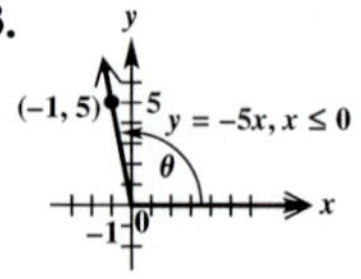

15. **(a)** impossible **(b)** possible **(c)** possible

In Exercises 17, 21, 25, and 27 we give, in order, sine, cosine, tangent, cotangent, secant, and cosecant.

17. $-\frac{\sqrt{39}}{8}; -\frac{5}{8}; \frac{\sqrt{39}}{5}; \frac{5\sqrt{39}}{39}; -\frac{8}{5}; -\frac{8\sqrt{39}}{39}$ **21.** $\frac{8}{17}, \frac{15}{17}, \frac{8}{15}, \frac{15}{8}, \frac{17}{15}, \frac{17}{8}$ **23.** Row 1: $\frac{1}{2}, \frac{\sqrt{3}}{2}, \frac{\sqrt{3}}{3}, \sqrt{3}, \frac{2\sqrt{3}}{3}, 2$; Row 2: $\frac{\sqrt{2}}{2}, \frac{\sqrt{2}}{2}, 1, 1, \sqrt{2}, \sqrt{2}$; Row 3: $\frac{\sqrt{3}}{2}, \frac{1}{2}, \sqrt{3}, \frac{\sqrt{3}}{3}, 2, \frac{2\sqrt{3}}{3}$ **25.** $-\frac{\sqrt{3}}{2}; \frac{1}{2}; -\sqrt{3}; -\frac{\sqrt{3}}{3}; 2; -\frac{2\sqrt{3}}{3}$

27. $-\frac{1}{2}; \frac{\sqrt{3}}{2}; -\frac{\sqrt{3}}{3}; -\sqrt{3}; \frac{2\sqrt{3}}{3}; -2$ **29.** -1.3563417 **31.** .20834446 **33.** D **35.** 41.635092°

39. $B = 31° 30'$; $a = 638$; $b = 391$ **41.** 73.7 ft **43.** 20.4 m **45.** 19,600 ft **47.** 110 km

Chapter 5 Test *(page 530)*

1. 203° **2.** **(a)** + **(b)** − **3.** III **4.** $\sin\theta = -\frac{5\sqrt{29}}{29}$; $\cos\theta = \frac{2\sqrt{29}}{29}$; $\tan\theta = -\frac{5}{2}$ **5.** 120°

6. $\sin\theta = -\frac{3}{5}$; $\tan\theta = -\frac{3}{4}$; $\cot\theta = -\frac{4}{3}$; $\sec\theta = \frac{5}{4}$; $\csc\theta = -\frac{5}{3}$ **7.** $x = 4$; $y = 4\sqrt{3}$; $z = 4\sqrt{2}$; $w = 8$

8. $-\sqrt{3}$ **9.** **(a)** false **(b)** false **10.** **(a)** .97939940 **(b)** -1.9056082 **(c)** 1.9362132 **11.** 221.8°; 318.2°

12. $B = 31° 30'$; $a = 638$; $b = 391$ **13.** 15.5 ft **14.** 92 km **15.** 448 m

CHAPTER 6 THE CIRCULAR FUNCTIONS AND THEIR GRAPHS

6.1 Exercises *(page 539)*

1. $\frac{\pi}{3}$ **3.** $\frac{5\pi}{6}$ **5.** $\frac{7\pi}{4}$ **7.** $-\frac{\pi}{4}$ **9.** .68 **11.** 2.43 **13.** 1.122 **15.** 1 **17.** 3 **21.** 60° **23.** 315°

25. 330° **27.** −30° **29.** 114.6° **31.** 99.7° **33.** −564.2° **35.** 1 **36.** $\sqrt{2}$ **37.** $-\frac{\sqrt{3}}{3}$ **38.** $\frac{1}{2}$

39. −1 **40.** $\frac{1}{2}$ **41.** $\frac{\sqrt{3}}{2}$ **42.** −1 **43.** 2π **45.** 8 **47.** 1 **49.** 25.8 cm **51.** 5.05 m

53. The length is doubled. **55.** 3500 km **57.** 5900 km **59.** 44° N **61.** **(a)** 11.6 in. **(b)** 37° 5′ **63.** 38.5°

65. 146 in. **67.** .20 km **69.** 6π **71.** 1.5 **73.** 1116.1 m^2 **75.** 706.9 ft^2 **77.** 114.0 cm^2 **79.** 1885.0 mi^2

81. 3.6 **83.** **(a)** $13\frac{1}{3}°$; $\frac{2\pi}{27}$ **(b)** 480 ft **(c)** $\frac{160}{9} \approx 17.8$ ft **(d)** approximately 672 ft^2 **85.** **(a)** 140 ft **(b)** 102 ft **(c)** 622 ft^2

87. 1900 yd^2 **89.** radius: 3947 mi; circumference: 24,800 mi **91.** The area is quadrupled.

Connections *(page 550)*

$PQ = y = \frac{y}{1} = \sin\theta$; $OQ = x = \frac{x}{1} = \cos\theta$; $AB = \frac{AB}{1} = \frac{AB}{AO} = \frac{y}{x}$ (by similar triangles) $= \tan\theta$

6.2 Exercises *(page 552)*

1. **(a)** 1 **(b)** 0 **(c)** undefined **3.** **(a)** 0 **(b)** 1 **(c)** 0 **5.** **(a)** 0 **(b)** −1 **(c)** 0 **7.** $-\frac{1}{2}$ **9.** −1 **11.** −2 **13.** $-\frac{1}{2}$

15. $\frac{\sqrt{2}}{2}$ **17.** $\frac{\sqrt{3}}{2}$ **19.** $\frac{2\sqrt{3}}{3}$ **21.** $-\frac{\sqrt{3}}{3}$ **23.** .5736 **25.** .4068 **27.** 1.2065 **29.** 14.3338

31. −1.0460 **33.** −3.8665 **35.** .7 **37.** 4 **39.** negative **41.** negative **43.** positive

45. $\sin\theta = \frac{\sqrt{2}}{2}$; $\cos\theta = \frac{\sqrt{2}}{2}$; $\tan\theta = 1$; $\cot\theta = 1$; $\sec\theta = \sqrt{2}$; $\csc\theta = \sqrt{2}$

47. $\sin\theta = -\frac{12}{13}$; $\cos\theta = \frac{5}{13}$; $\tan\theta = -\frac{12}{5}$; $\cot\theta = -\frac{5}{12}$; $\sec\theta = \frac{13}{5}$; $\csc\theta = -\frac{13}{12}$ **49.** .2095 **51.** 1.4426 **53.** .3887

55. $\frac{5\pi}{6}$ **57.** $\frac{4\pi}{3}$ **59.** $\frac{7\pi}{4}$ **61.** $(-.8011, .5985)$ **63.** $(.4385, -.8987)$ **65.** I **67.** II **69.** $\frac{3\pi}{32}$ radian per sec **71.** $\frac{6}{5}$ min **73.** .180311 radian per sec **75.** $\frac{9}{5}$ radians per sec **77.** 1.83333 radians per sec **79.** 18π cm **81.** 12 sec **83.** $\frac{\pi}{6}$ radian per hr **85.** $\frac{7\pi}{30}$ cm per min **87.** 1500π m per min **89.** 2π sec **91.** 15.5 mph **93.** **(a)** $\frac{2\pi}{365}$ radian **(b)** $\frac{\pi}{4380}$ radian per hr **(c)** 66,700 mph **95.** **(a)** .24 radian per sec **(b)** 3.11 cm per sec **97.** 3.73 cm **99.** 523.6 radians per sec

6.3 Exercises *(page 568)*

1. D **3.** A

5. 2 **7.** $\frac{2}{3}$ **9.** 1 **11.** 2

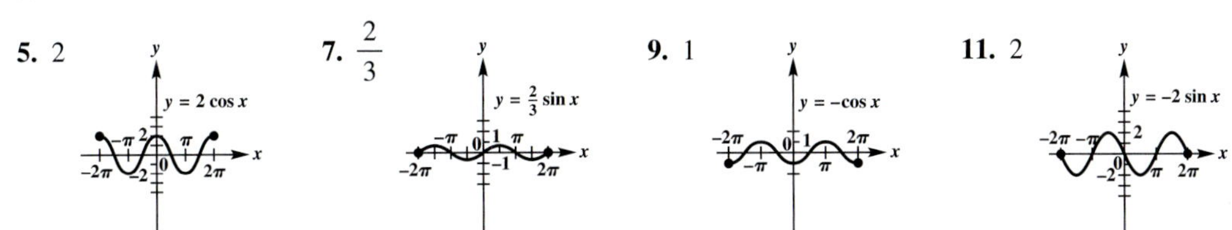

13. 4π; 1 **15.** π; 1 **17.** 8π; 2 **19.** $\frac{2\pi}{3}$; 2

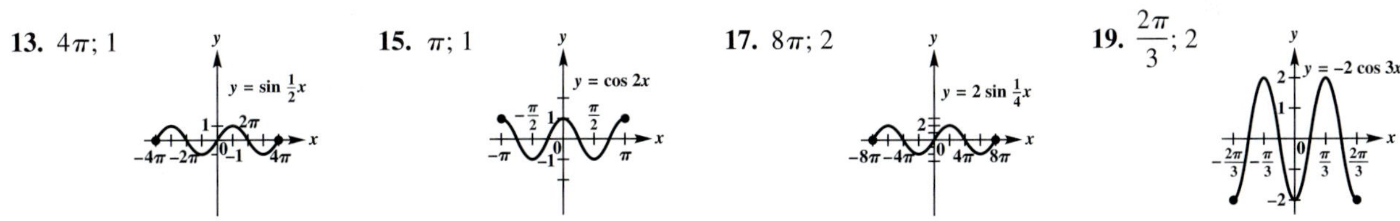

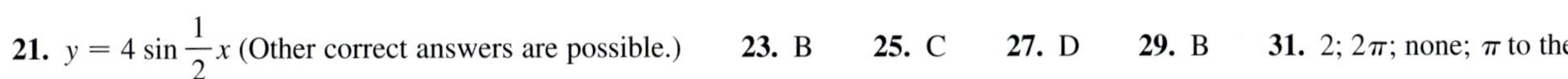

21. $y = 4 \sin \frac{1}{2}x$ (Other correct answers are possible.) **23.** B **25.** C **27.** D **29.** B **31.** 2; 2π; none; π to the right

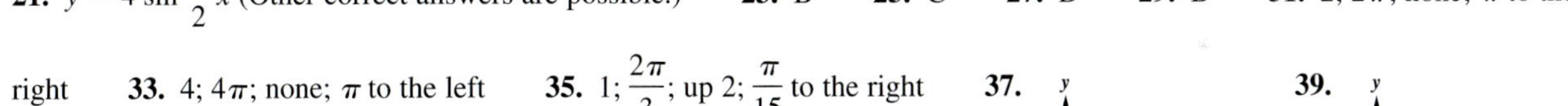

33. 4; 4π; none; π to the left **35.** 1; $\frac{2\pi}{3}$; up 2; $\frac{\pi}{15}$ to the right **37.** **39.**

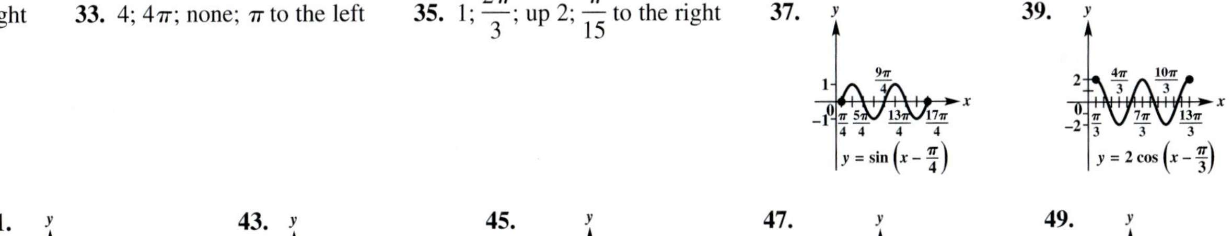

41. **43.** **45.** **47.** **49.**

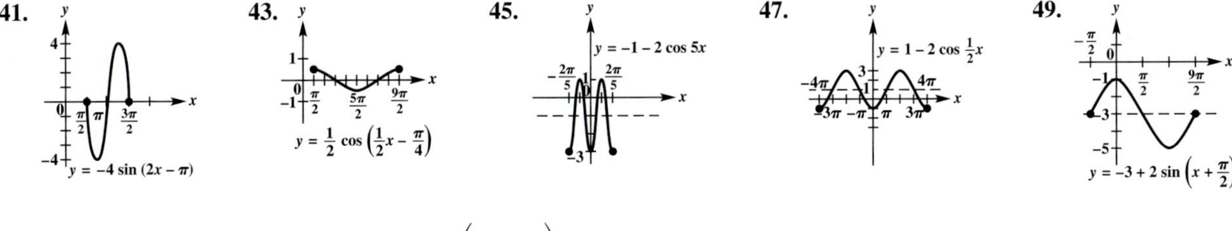

51.

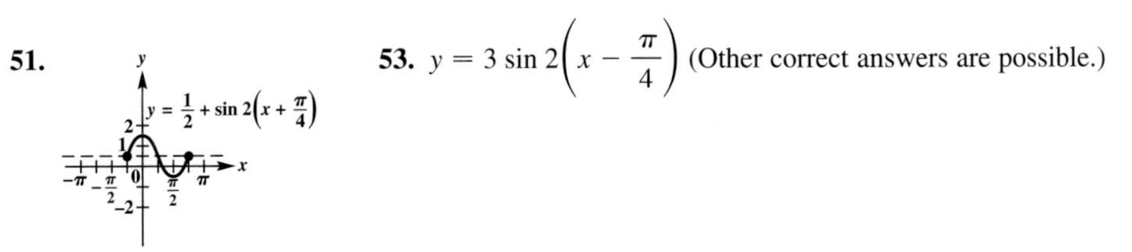

53. $y = 3 \sin 2\left(x - \frac{\pi}{4}\right)$ (Other correct answers are possible.)

55. **(a)** 80°; 50° **(b)** 15° **(c)** about 35,000 yr **(d)** downward **57.** **(a)** about 2 hr **(b)** 1 yr **59.** 24 hr **61.** approximately 6:00 P.M.; approximately .2 ft **63.** approximately 2:00 A.M.; approximately 2.6 ft **65.** 1; 240° or $\frac{4\pi}{3}$

67. **(a)** 5; $\frac{1}{60}$ **(b)** 60 **(c)** 5; 1.545; −4.045; −4.045; 1.545 **(d)**

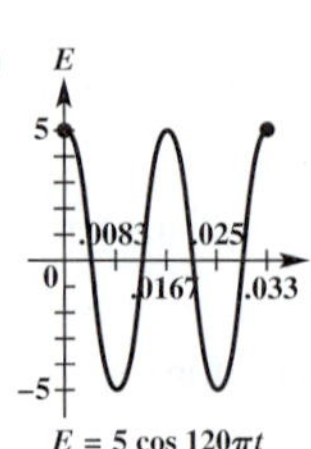

69. **(a)** $L(x) = .022x^2 + .55x + 316 + 3.5\sin(2\pi x)$ **(b)** maximums: $x = \frac{1}{4}, \frac{5}{4}, \frac{9}{4}, \ldots$; minimums: $x = \frac{3}{4}, \frac{7}{4}, \frac{11}{4}, \ldots$

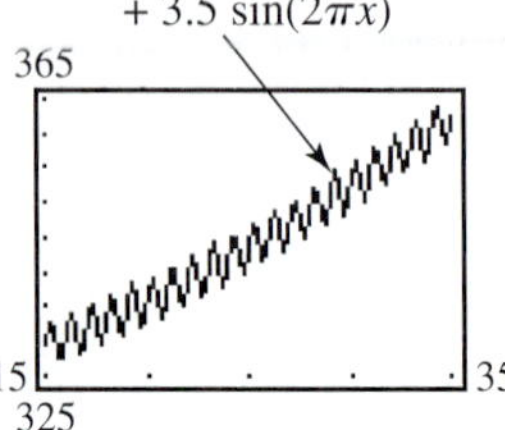

71. **(a)** 1° **(b)** 19° **(c)** 53° **(d)** 58° **(e)** 49° **(f)** 13°

73. **(a)** yes

70

1 25

30

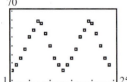

(b) It represents the average yearly temperature.

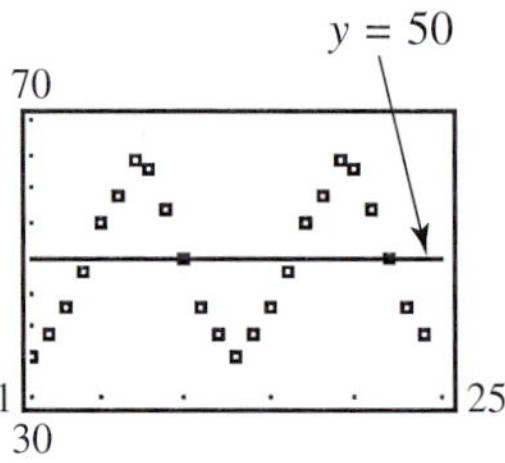

(c) 14; 12; 4.2 **(d)** $f(x) = 14\sin\left[\frac{\pi}{6}(x - 4.2)\right] + 50$

(e) The function gives an excellent model for the given data.

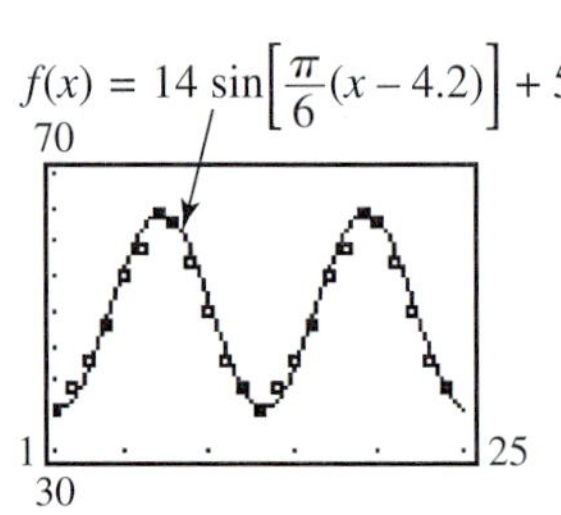

(f)

```
SinReg
 y=a*sin(bx+c)+d
 a=13.21
 b=.52
 c=-2.18
 d=49.68
```

TI-83 Plus fixed to the nearest hundredth.

6.4 Exercises *(page 585)*

1. B **3.** E **5.** D **7.** $y = 3\sec\frac{1}{4}x$

9.

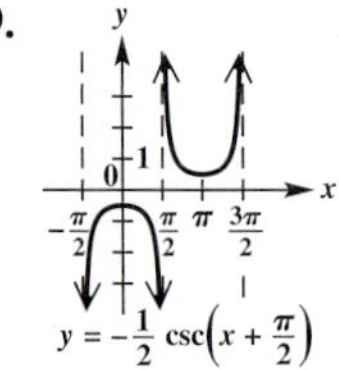

11.

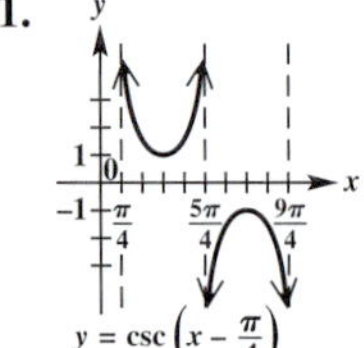

13.

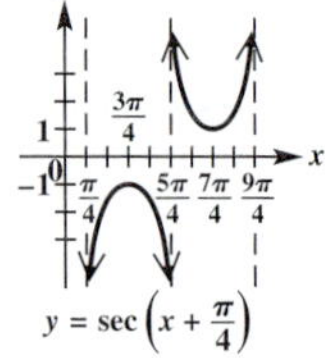

15. $y = \sec\left(\frac{1}{2}x + \frac{\pi}{3}\right)$

17.

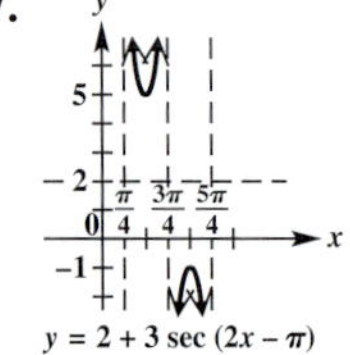

19.

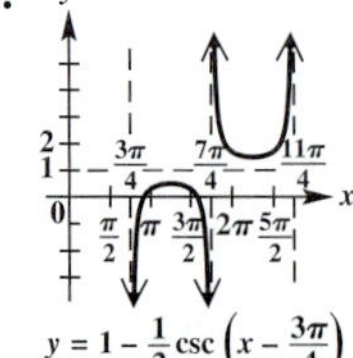

21.

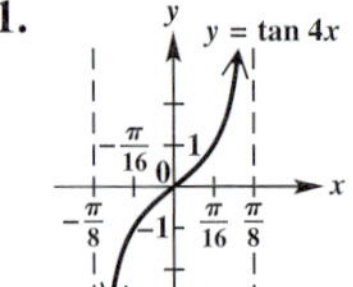

23.

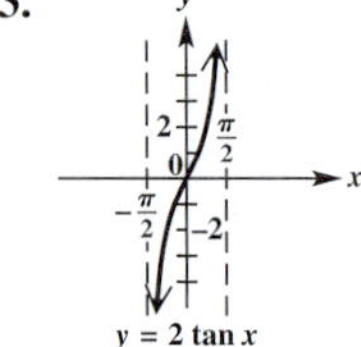

25.

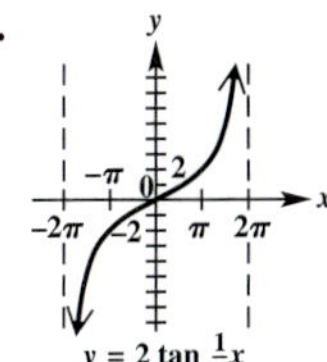

27.

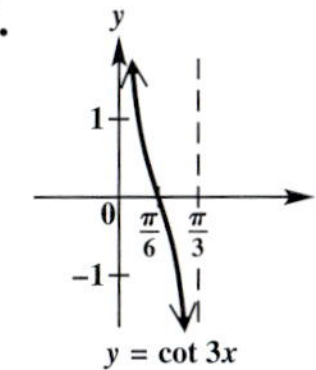

29.

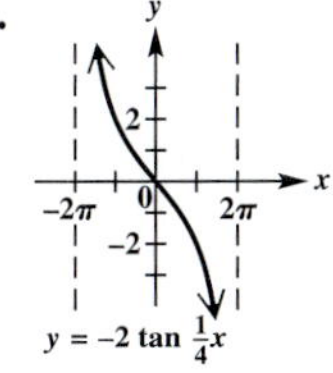

31.

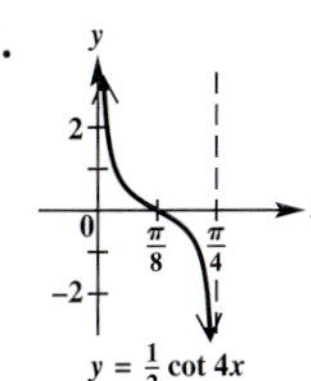

33.

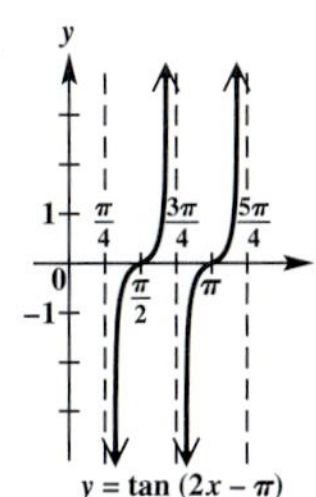

35.

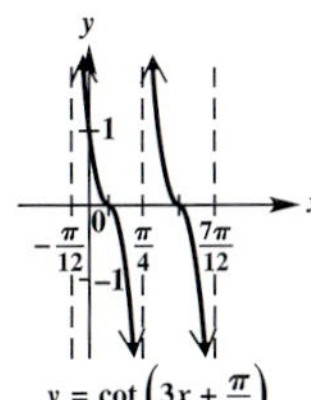

37.

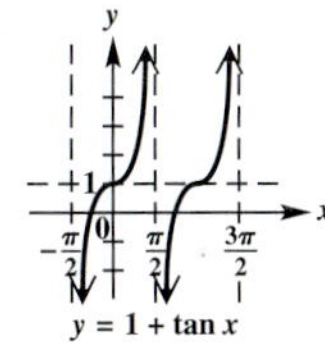

39.

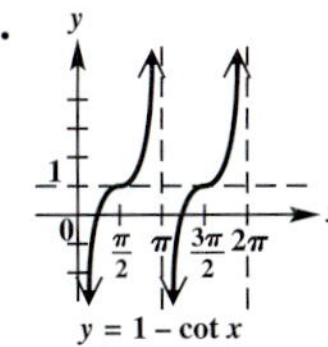

41.

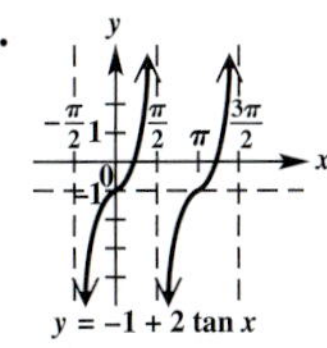

43.

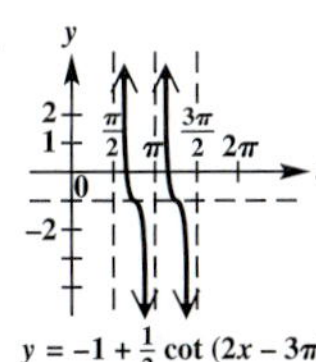

45.

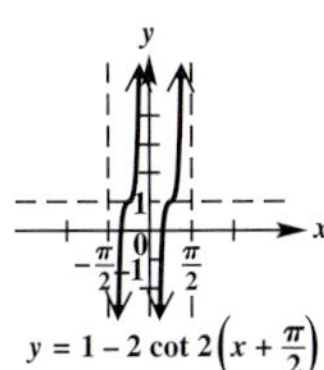
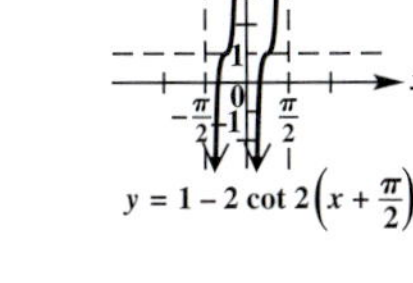

47. true **49.** true **51.** false; $\tan(-x) = -\tan x$ for all x in the domain. **53.** four

55. domain: $\left\{x \mid x \neq (2n + 1)\frac{\pi}{4}, \text{ where } n \text{ is an integer}\right\}$; range: $(-\infty, \infty)$ **57.** **(a)** 0 m **(b)** -2.9 m **(c)** -12.3 m **(d)** 12.3 m **(e)** It leads to $\tan \frac{\pi}{2}$, which is undefined.

61. We show the display for $Y_1 + Y_2$ at $X = \frac{\pi}{6}$. **63.** π **64.** $\frac{5\pi}{4}$ **65.** $y = \frac{5\pi}{4} + n\pi$

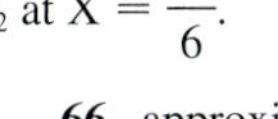

3; Y3=Y1+Y2; -2π; 2π; X=.52359878 Y=1.3660254; -3

66. approximately .3217505544

67. approximately 3.463343208 **68.** $\{x \mid x = .3217505544 + n\pi\}$

Summary Exercises on Graphing Circular Functions *(page 588)*

1. $y = 2 \sin \pi x$

2.

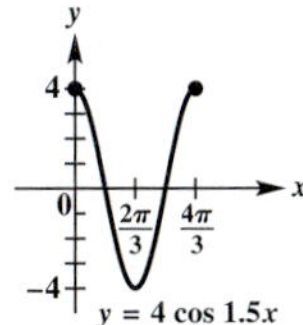

3.

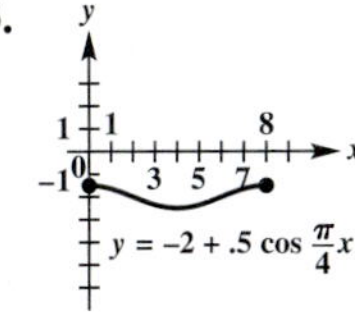

4.

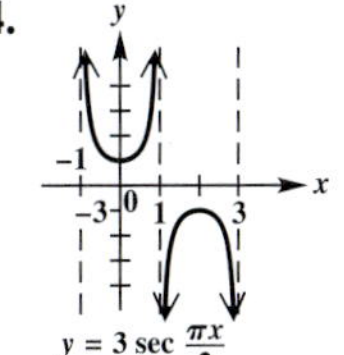

5.

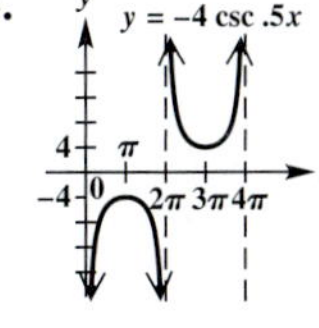

6.

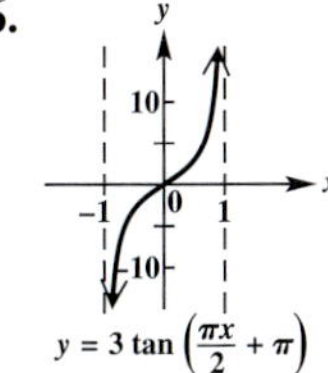

7.

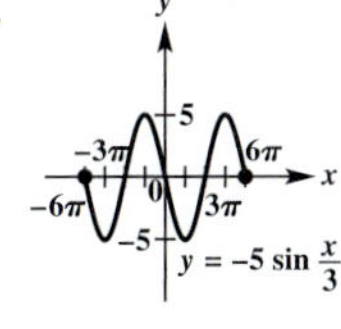

8.

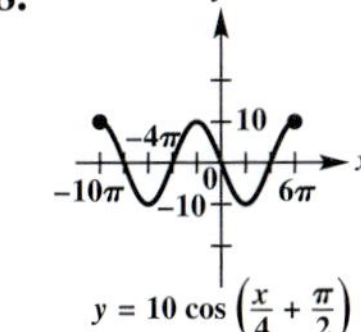

9.

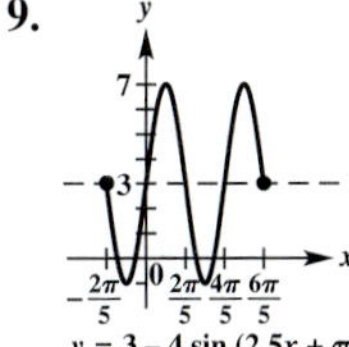

10.

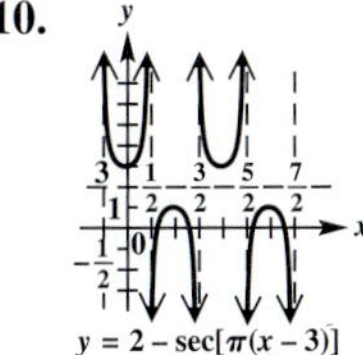

6.5 Exercises *(page 591)*

1. **(a)** $s(t) = 2 \cos 4\pi t$ **(b)** $s(1) = 2$; The weight is neither moving upward nor downward. At $t = 1$, the motion of the weight is changing from up to down. **3.** **(a)** $s(t) = -3 \cos 2.5\pi t$ **(b)** $s(1) = 0$; upward

5. $s(t) = .21 \cos 55\pi t$

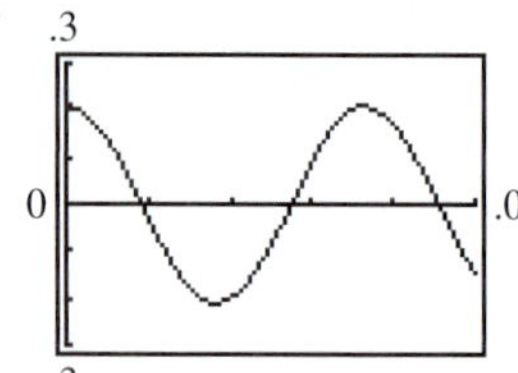

7. $s(t) = .14 \cos 110\pi t$

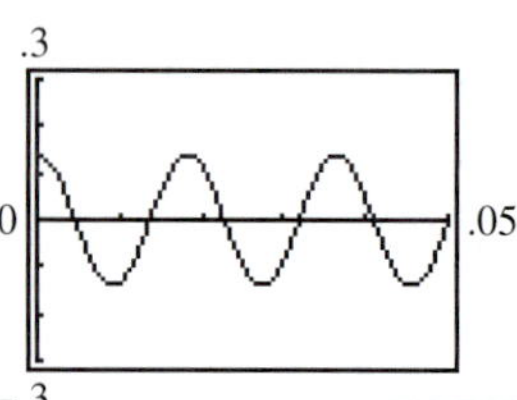

9. **(a)** $s(t) = -4 \cos \frac{2\pi}{3} t$ **(b)** 3.46 units **(c)** $\frac{1}{3}$ **11.** **(a)** $s(t) = 2 \sin 2t$; amplitude: 2; period: π; frequency: $\frac{1}{\pi}$ **(b)** $s(t) = 2 \sin 4t$; amplitude: 2; period: $\frac{\pi}{2}$; frequency: $\frac{2}{\pi}$ **13.** $\frac{8}{\pi^2}$ ft **15.** **(a)** 4 in. **(b)** after $\frac{1}{8}$ sec **(c)** 4 cycles per sec; $\frac{1}{4}$ sec **17.** **(a)** 5 in. **(b)** 2 cycles per sec; $\frac{1}{2}$ sec **(c)** after $\frac{1}{4}$ sec **(d)** approximately 4; After 1.3 sec, the weight is about 4 in. above the equilibrium position. **19.** **(a)** $s(t) = -3 \cos 12t$ **(b)** $\frac{\pi}{6}$ sec

21. 0; π; They are the same.

Chapter 6 Review Exercises *(page 596)*

1. **(a)** II **(b)** III **(c)** III **(d)** I **3.** $\frac{2\pi}{3}$ **5.** 225° **7.** **(a)** $\frac{\pi}{200}$ **(b)** 3.15°; .055 radian **9.** 35.8 cm **11.** 41 yd **13.** $\frac{1}{2}$ radian **15.** $\frac{15}{32}$ sec **17.** 1260π m per sec **19.** $A = \pi r^2$, the formula for the area of a circle **21.** $-\sqrt{3}$ **23.** .9703 **25.** .5148 **27.** $\frac{7\pi}{6}$ **29.** **(b)** $\frac{\pi}{6}$ **(c)** less ultraviolet light when $\theta = \frac{\pi}{3}$ **31.** D **33.** not applicable; $\frac{\pi}{3}$; none; none **35.** 2; $\frac{2\pi}{5}$; none; none **37.** $\frac{1}{4}$; 3π; 3 up; none **39.** 1; 2π; none; $\frac{3\pi}{4}$ to the right **41.** tangent

43. cosine **45.** **47.** **49.** **51.** **(b)**

53. **(a)** 30° **(b)** 60° **(c)** 75° **(d)** 86° **(e)** 86° **(f)** 60° **55.** **(a)** 100 **(b)** 258 **(c)** 122 **(d)** 296

57. amplitude: 4; period: 2; frequency: $\frac{1}{2}$ **59.** The frequency is the number of cycles in one unit of time; -4; 0; $-2\sqrt{2}$

Chapter 6 Test *(page 601)*

1. $\frac{2\pi}{3}$ **2.** $-\frac{\pi}{4}$ **3.** .09 **4.** 135° **5.** $-210°$ **6.** 229.18° **7.** **(a)** $\frac{4}{3}$ **(b)** 15,000 cm^2 **8.** 2 radians

9. **(a)** $\frac{\pi}{3}$ radians **(b)** $\frac{10\pi}{3}$ cm **(c)** $\frac{5\pi}{9}$ cm per sec **10.** $\frac{\sqrt{2}}{2}$ **11.** $-\frac{\sqrt{3}}{2}$ **12.** undefined **13.** -2

14. **(a)** .9716 **(b)** $\frac{\pi}{3}$ **15.** **(a)** π **(b)** 6 **(c)** $[-3, 9]$ **(d)** -3 **(e)** $\frac{\pi}{4}$ to the left $\left(\text{that is, } -\frac{\pi}{4}\right)$ **16.**

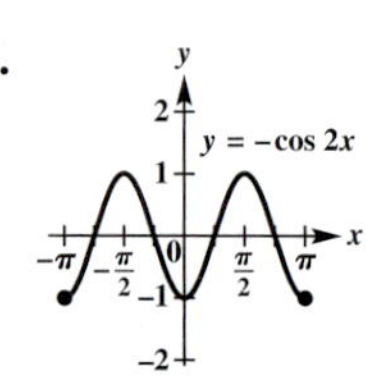

17.

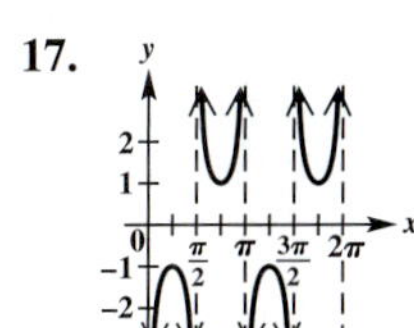

18.

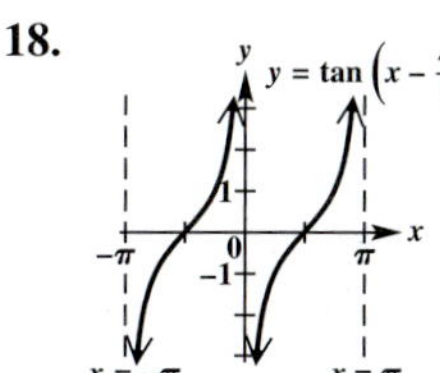

19.

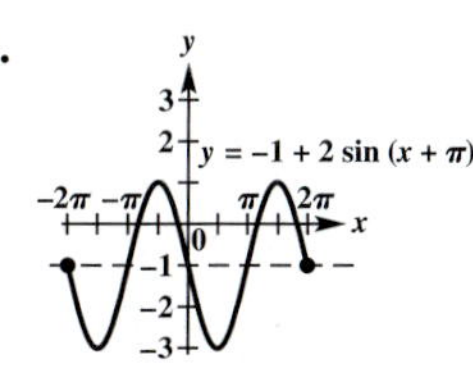

20.

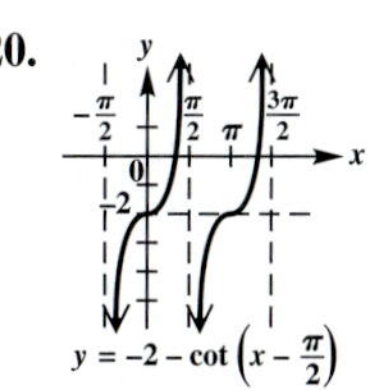

21. **(a)**

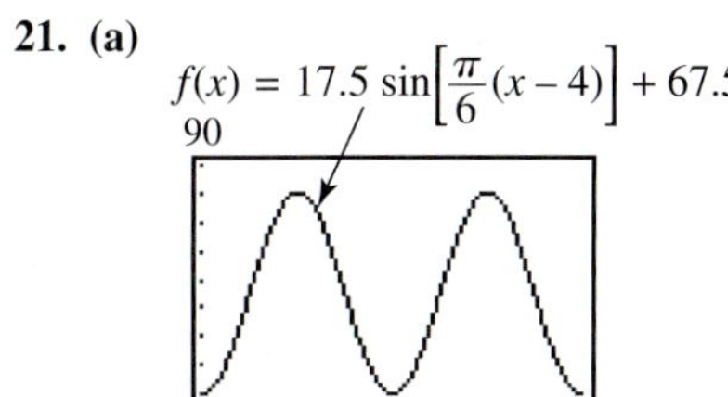

(b) 17.5; 12; 4 to the right; 67.5 up **(c)** approximately 52°F **(d)** 50°F in January; 85°F in July **(e)** approximately 67.5°; This is the vertical translation. **22.** **(a)** 4 in. **(b)** after $\frac{1}{8}$ sec **(c)** 4 cycles per sec; $\frac{1}{4}$ sec

CHAPTER 7 TRIGONOMETRIC IDENTITIES AND EQUATIONS

7.1 Exercises *(page 609)*

1. -2.6 **3.** $.625$ **5.** $\frac{\sqrt{7}}{4}$ **7.** $-\frac{2\sqrt{5}}{5}$ **9.** $-\frac{\sqrt{105}}{11}$ **12.** $-\sin x$ **13.** odd **14.** $\cos x$ **15.** even **16.** $-\tan x$ **17.** odd **19.** $f(-x) = -f(x)$

21. $\cos\theta = -\frac{\sqrt{5}}{3}$; $\tan\theta = -\frac{2\sqrt{5}}{5}$; $\cot\theta = -\frac{\sqrt{5}}{2}$; $\sec\theta = -\frac{3\sqrt{5}}{5}$; $\csc\theta = \frac{3}{2}$

23. $\sin\theta = -\frac{\sqrt{17}}{17}$; $\cos\theta = \frac{4\sqrt{17}}{17}$; $\cot\theta = -4$; $\sec\theta = \frac{\sqrt{17}}{4}$; $\csc\theta = -\sqrt{17}$

25. $\sin\theta = \frac{3}{5}$; $\cos\theta = \frac{4}{5}$; $\tan\theta = \frac{3}{4}$; $\sec\theta = \frac{5}{4}$; $\csc\theta = \frac{5}{3}$

27. $\sin\theta = -\frac{\sqrt{7}}{4}$; $\cos\theta = \frac{3}{4}$; $\tan\theta = -\frac{\sqrt{7}}{3}$; $\cot\theta = -\frac{3\sqrt{7}}{7}$; $\csc\theta = -\frac{4\sqrt{7}}{7}$ **29.** B **31.** E **33.** A **35.** A

37. D **41.** $\sin\theta = \frac{\pm\sqrt{2x+1}}{x+1}$ **43.** $\sin x = \pm\sqrt{1-\cos^2 x}$ **45.** $\tan x = \pm\sqrt{\sec^2 x - 1}$

47. $\csc x = \frac{\pm\sqrt{1-\cos^2 x}}{1-\cos^2 x}$ **49.** $\cos\theta$ **51.** $\cot\theta$ **53.** $\cos^2\theta$ **55.** $\sec\theta - \cos\theta$ **57.** $\cot\theta - \tan\theta$

59. $\tan\theta\sin\theta$ **61.** $\cos^2\theta$ **63.** $\sec^2\theta$ **65.** $\frac{25\sqrt{6}-60}{12}$; $\frac{-25\sqrt{6}-60}{12}$

67. $-\sin(2x)$ **68.** It is the negative of $\sin(2x)$. **69.** $\cos(4x)$ **70.** It is the same function. **71.** **(a)** $-\sin(4x)$ **(b)** $\cos(2x)$ **(c)** $5\sin(3x)$ **73.** not an identity **75.** identity

7.2 Exercises *(page 618)*

1. $\csc\theta\sec\theta$ or $\frac{1}{\sin\theta\cos\theta}$ **3.** $1 + \sec s$ **5.** 1 **7.** 1 **9.** $2 + 2\sin t$ **11.** $-\frac{2\cos x}{\sin^2 x}$ or $-2\cot x\csc x$

13. $(\sin\theta + 1)(\sin\theta - 1)$ **15.** $4\sin x$ **17.** $(2\sin x + 1)(\sin x + 1)$ **19.** $(\cos^2 x + 1)^2$

21. $(\sin x - \cos x)(1 + \sin x\cos x)$ **23.** $\sin\theta$ **25.** 1 **27.** $\tan^2\beta$ **29.** $\tan^2 x$ **31.** $\sec^2 x$

71. $(\sec\theta + \tan\theta)(1 - \sin\theta) = \cos\theta$ **73.** $\frac{\cos\theta + 1}{\sin\theta + \tan\theta} = \cot\theta$ **75.** identity **77.** not an identity

83. It is true when $\sin x \geq 0$. **85.** **(a)** $P = 16k\cos^2(2\pi t)$ **(b)** $P = 16k[1 - \sin^2(2\pi t)]$

7.3 Exercises *(page 629)*

1. F **3.** C **5.** $\dfrac{\sqrt{6}-\sqrt{2}}{4}$ **7.** $\dfrac{\sqrt{2}-\sqrt{6}}{4}$ **9.** $\dfrac{\sqrt{2}-\sqrt{6}}{4}$ **11.** 0 **13.** $\cot 3°$ **15.** $\sin \dfrac{5\pi}{12}$

17. $\cos\left(-\dfrac{\pi}{8}\right)$ **19.** $\csc(-56° 42')$ **21.** tan **23.** cos **25.** 15° **27.** 20° **29.** $\dfrac{\sqrt{6}+\sqrt{2}}{4}$ **31.** $2-\sqrt{3}$

33. $\dfrac{-\sqrt{6}-\sqrt{2}}{4}$ **35.** $\dfrac{\sqrt{2}}{2}$ **37.** -1 **39.** $\sin\theta$ **41.** $-\sin x$ **43.** $\dfrac{\sqrt{2}}{2}(\cos\theta + \sin\theta)$

45. $\dfrac{1}{2}(\cos x - \sqrt{3}\sin x)$ **47.** $\dfrac{\sqrt{3}+\tan\theta}{1-\sqrt{3}\tan\theta}$ **49.** $\dfrac{\sqrt{3}\tan x + 1}{\sqrt{3}-\tan x}$ **51.** **(a)** $\dfrac{16}{65}$ **(b)** $\dfrac{33}{65}$ **(c)** $\dfrac{63}{16}$ **(d)** I

53. **(a)** $\dfrac{-2\sqrt{10}+2}{9}$ **(b)** $\dfrac{4\sqrt{2}-\sqrt{5}}{9}$ **(c)** $\dfrac{-8\sqrt{5}-5\sqrt{2}}{20-2\sqrt{10}}$ **(d)** II **55.** **(a)** $-\dfrac{36}{85}$ **(b)** $\dfrac{13}{85}$ **(c)** $-\dfrac{77}{36}$ **(d)** II

58. $\dfrac{-\sqrt{6}-\sqrt{2}}{4}$ **59.** $\dfrac{-\sqrt{6}-\sqrt{2}}{4}$ **60.** **(a)** $\dfrac{\sqrt{2}-\sqrt{6}}{4}$ **(b)** $\dfrac{-\sqrt{6}-\sqrt{2}}{4}$ **61.** $\dfrac{\sqrt{6}-\sqrt{2}}{4}$ **63.** $\dfrac{-\sqrt{6}-\sqrt{2}}{4}$

65. $-2+\sqrt{3}$ **71.** $\sin\left(\dfrac{\pi}{2}+x\right) = \cos x$ **79.** 3 **81.** **(a)** 425 lb **(c)** 0°

83. **(a)** The pressure P is oscillating.

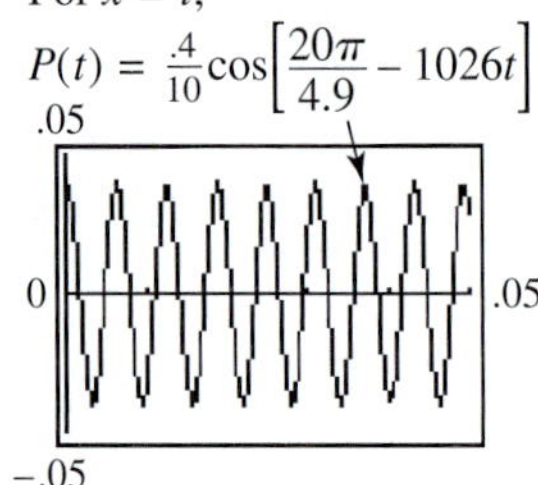

(b) The pressure oscillates and amplitude decreases as r increases.

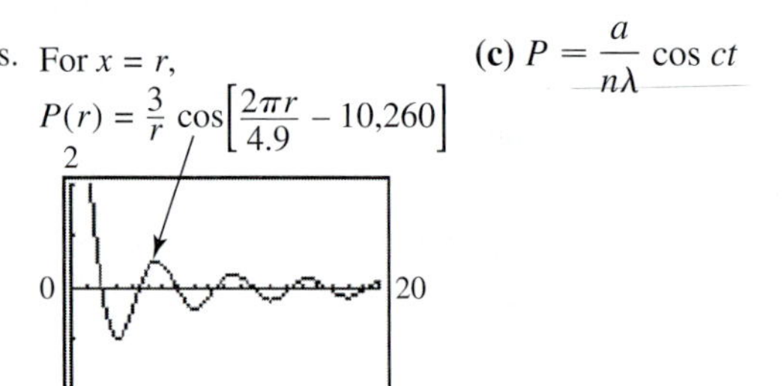

(c) $P = \dfrac{a}{n\lambda}\cos ct$

7.4 Exercises *(page 641)*

1. $\cos\theta = \dfrac{2\sqrt{5}}{5}$; $\sin\theta = \dfrac{\sqrt{5}}{5}$ **3.** $\cos x = -\dfrac{\sqrt{42}}{12}$; $\sin x = \dfrac{\sqrt{102}}{12}$ **5.** $\cos 2\theta = \dfrac{17}{25}$; $\sin 2\theta = -\dfrac{4\sqrt{21}}{25}$

7. $\cos 2x = -\dfrac{3}{5}$; $\sin 2x = \dfrac{4}{5}$ **9.** $\cos 2\theta = \dfrac{39}{49}$; $\sin 2\theta = -\dfrac{4\sqrt{55}}{49}$ **11.** $\dfrac{\sqrt{3}}{2}$ **13.** $\dfrac{\sqrt{3}}{2}$ **15.** $-\dfrac{\sqrt{2}}{2}$

17. $\dfrac{1}{2}\tan 102°$ **19.** $\dfrac{1}{4}\cos 94.2°$ **21.** $\cos 3x = 4\cos^3 x - 3\cos x$ **23.** $\tan 3x = \dfrac{3\tan x - \tan^3 x}{1 - 3\tan^2 x}$

25. $\cos^4 x - \sin^4 x = \cos 2x$ **41.** $\sin 160° - \sin 44°$ **43.** $\dfrac{5}{2}\cos 5x + \dfrac{5}{2}\cos x$ **45.** $-2\sin 3x \sin x$

47. $-2\sin 11.5° \cos 36.5°$ **49.** $2\cos 6x \cos 2x$ **51.** $\dfrac{\sqrt{2+\sqrt{2}}}{2}$ **53.** $-\dfrac{\sqrt{2+\sqrt{3}}}{2}$ **55.** $-\dfrac{\sqrt{2+\sqrt{3}}}{2}$

59. $\frac{\sqrt{10}}{4}$ **61.** 3 **63.** $\frac{\sqrt{50 - 10\sqrt{5}}}{10}$ **65.** $-\sqrt{7}$ **67.** sin 20° **69.** tan 73.5° **71.** tan 29.87° **75.** 84°

77. 3.9 **79.** **(a)** $\cos \frac{\theta}{2} = \frac{R - b}{R}$ **(b)** $\tan \frac{\theta}{4} = \frac{b}{50}$ **(c)** 54° **81.** $a = -885.6$, $c = 885.6$, $\omega = 240\pi$

7.5 Exercises *(page 654)*

1. one-to-one **3.** domain **5.** π **7.** **(a)** $[-1, 1]$ **(b)** $\left[-\frac{\pi}{2}, \frac{\pi}{2}\right]$ **(c)** increasing **(d)** -2 is not in the domain.

9. **(a)** $(-\infty, \infty)$ **(b)** $\left(-\frac{\pi}{2}, \frac{\pi}{2}\right)$ **(c)** increasing **(d)** no **11.** $\cos^{-1} \frac{1}{a}$ **13.** 0 **15.** π **17.** $-\frac{\pi}{2}$ **19.** 0 **21.** $\frac{\pi}{2}$

23. $\frac{\pi}{4}$ **25.** $\frac{5\pi}{6}$ **27.** $\frac{3\pi}{4}$ **29.** $-\frac{\pi}{6}$ **31.** $\frac{\pi}{6}$ **33.** $-45°$ **35.** $-60°$ **37.** 120° **39.** 120°

41. $-7.6713835°$ **43.** 113.500970° **45.** 30.987961° **47.** .83798122 **49.** 2.3154725 **51.** 1.1900238

53. **55.** **57.**

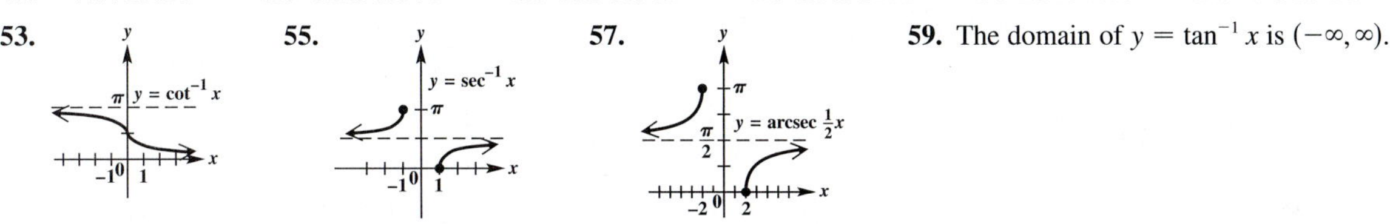

59. The domain of $y = \tan^{-1} x$ is $(-\infty, \infty)$.

60. In both cases, the result is x. In each case, the graph is a straight line bisecting quadrants I and III (i.e., the line $y = x$).

61. It is the graph of $y = x$.

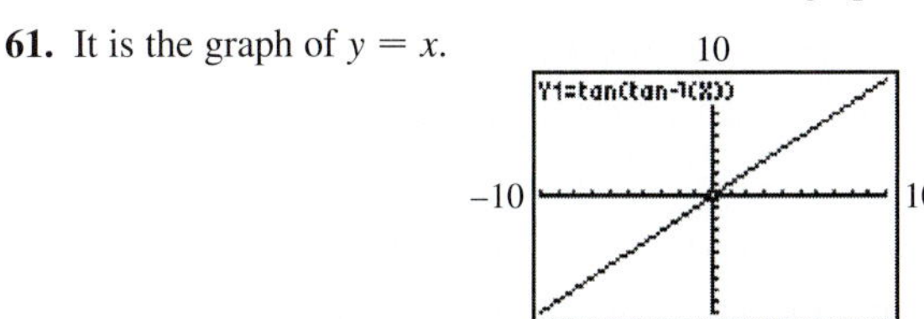

62. It does not agree because the range of the inverse tangent function is $\left(-\frac{\pi}{2}, \frac{\pi}{2}\right)$, not $(-\infty, \infty)$, as was the case in Exercise 61.

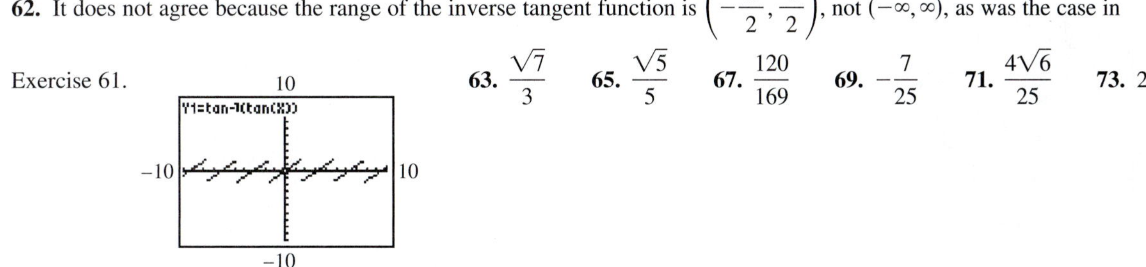

63. $\frac{\sqrt{7}}{3}$ **65.** $\frac{\sqrt{5}}{5}$ **67.** $\frac{120}{169}$ **69.** $-\frac{7}{25}$ **71.** $\frac{4\sqrt{6}}{25}$ **73.** 2

75. $\frac{63}{65}$ **77.** $\frac{\sqrt{10} - 3\sqrt{30}}{20}$ **79.** .894427191 **81.** .1234399811 **83.** $\sqrt{1 - u^2}$ **85.** $\sqrt{1 - u^2}$ **87.** $\frac{\sqrt{u^2 - 4}}{u}$

89. $\frac{u\sqrt{2}}{2}$ **91.** $\frac{2\sqrt{4 - u^2}}{4 - u^2}$ **93.** **(a)** 45° **(b)** $\theta = 45°$ **95.** **(a)** 18° **(b)** 18° **(c)** 15° **(e)** 1.4142151 m (Note: Due to the computational routine, there may be a discrepancy in the last few decimal places.) **(f)** $\sqrt{2}$

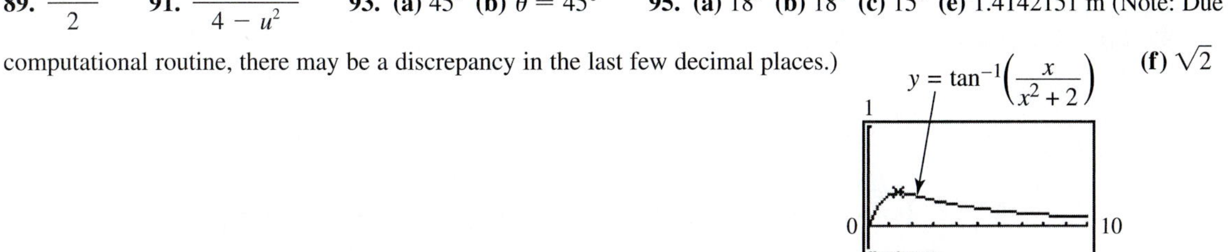

Radian mode

97. about 44.7%

7.6 Exercises *(page 665)*

1. Solve the linear equation for cot *x*. **3.** Solve the quadratic equation for sec *x* by factoring. **5.** Solve the quadratic equation for sin *x* using the quadratic formula. **7.** Use an identity to rewrite as an equation with one trigonometric function.

9. $\frac{\pi}{3}, \pi, \frac{4\pi}{3}$ **11.** $\left\{\frac{3\pi}{4}, \frac{7\pi}{4}\right\}$ **13.** $\left\{\frac{\pi}{6}, \frac{5\pi}{6}\right\}$ **15.** $\emptyset$ **17.** $\left\{\frac{\pi}{4}, \frac{2\pi}{3}, \frac{5\pi}{4}, \frac{5\pi}{3}\right\}$ **19.** $\{\pi\}$ **21.** $\left\{\frac{7\pi}{6}, \frac{3\pi}{2}, \frac{11\pi}{6}\right\}$

23. $\{30°, 210°, 240°, 300°\}$ **25.** $\{90°, 210°, 330°\}$ **27.** $\{45°, 135°, 225°, 315°\}$ **29.** $\{45°, 225°\}$

31. $\{0°, 30°, 150°, 180°\}$ **33.** $\{0°, 45°, 135°, 180°, 225°, 315°\}$ **35.** $\{53.6°, 126.4°, 187.9°, 352.1°\}$

37. $\{149.6°, 329.6°, 106.3°, 286.3°\}$ **39.** $\emptyset$ **41.** $\{57.7°, 159.2°\}$ **43.** $.9 + 2n\pi, 2.3 + 2n\pi, 3.6 + 2n\pi, 5.8 + 2n\pi,$ where *n* is any integer **45.** $\frac{\pi}{3} + 2n\pi, \frac{2\pi}{3} + 2n\pi, \frac{4\pi}{3} + 2n\pi, \frac{5\pi}{3} + 2n\pi,$ where *n* is any integer **47.** $33.6° + 360° n,$ $326.4° + 360° n,$ where *n* is any integer **49.** $45° + 180° n, 108.4° + 180° n,$ where *n* is any integer

51. $\{.6806, 1.4159\}$ **55.** $\left\{\frac{\pi}{12}, \frac{11\pi}{12}, \frac{13\pi}{12}, \frac{23\pi}{12}\right\}$ **57.** $\left\{\frac{\pi}{2}, \frac{7\pi}{6}, \frac{11\pi}{6}\right\}$ **59.** $\left\{\frac{\pi}{18}, \frac{7\pi}{18}, \frac{13\pi}{18}, \frac{19\pi}{18}, \frac{25\pi}{18}, \frac{31\pi}{18}\right\}$

61. $\left\{\frac{3\pi}{8}, \frac{5\pi}{8}, \frac{11\pi}{8}, \frac{13\pi}{8}\right\}$ **63.** $\left\{\frac{\pi}{2}, \frac{3\pi}{2}\right\}$ **65.** $\left\{0, \frac{\pi}{3}, \pi, \frac{5\pi}{3}\right\}$ **67.** $\emptyset$ **69.** $\left\{\frac{\pi}{2}\right\}$

71. $\{15°, 45°, 135°, 165°, 255°, 285°\}$ **73.** $\{0°\}$ **75.** $\{120°, 240°\}$ **77.** $\{30°, 150°, 270°\}$ **79.** $0° + 360° n,$ $30° + 360° n, 150° + 360° n, 180° + 360° n,$ where *n* is any integer **81.** $60° + 360° n, 300° + 360° n,$ where *n* is any integer **83.** $11.8° + 360° n, 78.2° + 360° n, 191.8° + 360° n, 258.2° + 360° n,$ where *n* is any integer

85. $30° + 360° n, 90° + 360° n, 150° + 360° n, 210° + 360° n, 270° + 360° n, 330° + 360° n,$ where *n* is any integer

87. (a) .00164 and .00355 **(b)** [.00164, .00355] **(c)** outward

89. (a) For $x = t$,
$P(t) = .003 \sin 220\pi t + \frac{.003}{3} \sin 660\pi t + \frac{.003}{5} \sin 1100\pi t + \frac{.003}{7} \sin 1540\pi t$

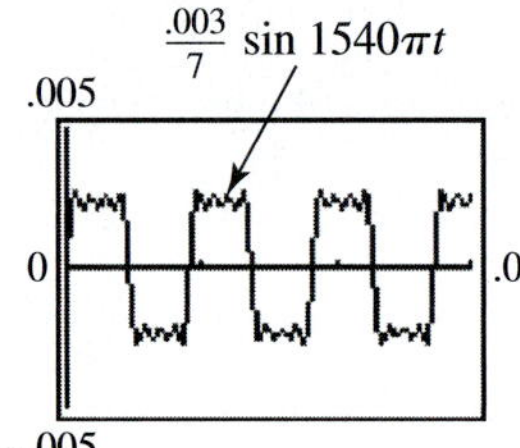

(b) The graph is periodic, and the wave has "jagged square" tops and bottoms.
(c) This will occur when *t* is in one of these intervals: (.0045, .0091), (.0136, .0182), (.0227, .0273).

91. (a) 91.3 days after March 21, on June 20 **(b)** 273.8 days after March 21, on December 19 **(c)** 228.7 days after March 21, on November 4, and again after 318.8 days, on February 2 **93.** .0007 sec **95.** .0014 sec **97. (a)** $\frac{1}{4}$ sec **(b)** $\frac{1}{6}$ sec **(c)** .21 sec **99. (a)** One such value is $\frac{\pi}{3}$. **(b)** One such value is $\frac{\pi}{4}$.

7.7 Exercises *(page 672)*

1. C **3.** C **5.** $x = \arccos \frac{y}{5}$ **7.** $x = \frac{1}{3} \operatorname{arccot} 2y$ **9.** $x = \frac{1}{2} \arctan \frac{y}{3}$ **11.** $x = 4 \arccos \frac{y}{6}$

13. $x = \frac{1}{5} \arccos\left(-\frac{y}{2}\right)$ **15.** $x = -3 + \arccos y$ **17.** $x = \arcsin(y + 2)$ **19.** $x = \arcsin\left(\frac{y + 4}{2}\right)$

23. $\{-2\sqrt{2}\}$ **25.** $\{\pi - 3\}$ **27.** $\left\{\frac{3}{5}\right\}$ **29.** $\left\{\frac{4}{5}\right\}$ **31.** $\{0\}$ **33.** $\left\{\frac{1}{2}\right\}$ **35.** $\left\{-\frac{1}{2}\right\}$ **37.** $\{0\}$

39. Y = arcsin X – arccos X – $\frac{\pi}{6}$

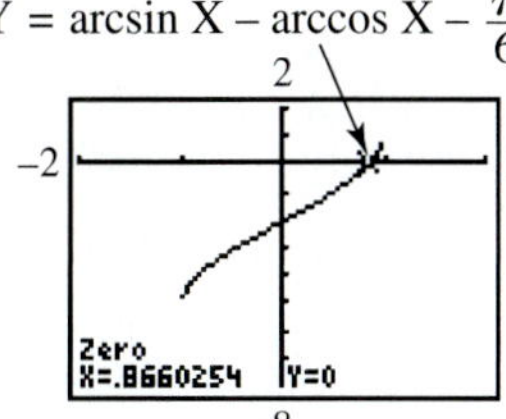

41. {4.4622037}

43. **(a)** $A \approx .00506$, $\phi \approx .484$; $P = .00506 \sin(440\pi t + .484)$

(b) The two graphs are the same. For $x = t$,
$P(t) = .00506 \sin(440\pi t + .484)$
$P_1(t) + P_2(t) = .0012 \sin(440\pi t + .052) + .004 \sin(440\pi t + .61)$

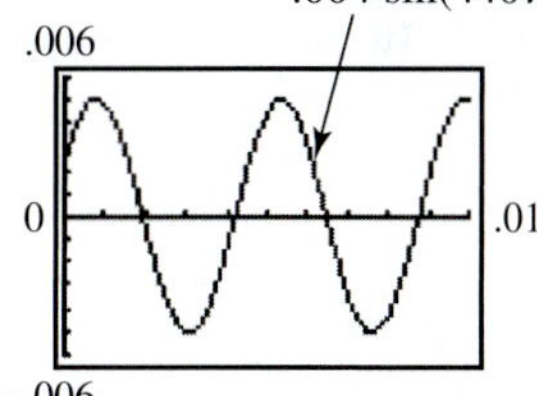

45. **(a)** $\tan\alpha = \frac{x}{z}$; $\tan\beta = \frac{x+y}{z}$ **(b)** $\frac{x}{\tan\alpha} = \frac{x+y}{\tan\beta}$ **(c)** $\alpha = \arctan\left(\frac{x\tan\beta}{x+y}\right)$ **(d)** $\beta = \arctan\left(\frac{(x+y)\tan\alpha}{x}\right)$

47. **(a)** $t = \frac{1}{2\pi f}\arcsin\frac{e}{E_{\max}}$ **(b)** .00068 sec **49.** **(a)** $t = \frac{3}{4\pi}\arcsin 3y$ **(b)** .27 sec

Chapter 7 Review Exercises *(page 679)*

1. B **3.** C **5.** D **7.** 1 **9.** $\frac{1}{\cos^2\theta}$ **11.** $\sin x = -\frac{4}{5}$; $\tan x = -\frac{4}{3}$; $\cot(-x) = \frac{3}{4}$ **13.** E **15.** J

17. I **19.** H **21.** G **23.** $\frac{4+3\sqrt{15}}{20}$; $\frac{4\sqrt{15}+3}{20}$; $\frac{4+3\sqrt{15}}{4\sqrt{15}-3}$; I **25.** $\frac{1}{2}$ **27.** $-\frac{\sin 2x + \sin x}{\cos 2x - \cos x} = \cot\frac{x}{2}$

45. $\frac{\pi}{4}$ **47.** $-\frac{\pi}{3}$ **49.** $\frac{3\pi}{4}$ **51.** $\frac{2\pi}{3}$ **53.** $\frac{3\pi}{4}$ **55.** $-60°$ **57.** 60.67924514° **59.** 36.4895081°

61. 73.26220613° **63.** -1 **65.** $\frac{3\pi}{4}$ **67.** $\frac{\pi}{4}$ **69.** $\frac{\sqrt{7}}{4}$ **71.** $\frac{\sqrt{3}}{2}$ **73.** $\frac{294+125\sqrt{6}}{92}$ **75.** $\frac{1}{u}$

77. {.463647609, 3.605240263} **79.** $\left\{\frac{\pi}{4}, \frac{3\pi}{4}, \frac{5\pi}{4}, \frac{7\pi}{4}\right\}$ **81.** $\left\{\frac{\pi}{8}, \frac{3\pi}{8}, \frac{5\pi}{8}, \frac{7\pi}{8}, \frac{9\pi}{8}, \frac{11\pi}{8}, \frac{13\pi}{8}, \frac{15\pi}{8}\right\}$

83. $\frac{\pi}{3} + 2n\pi$, $\pi + 2n\pi$, $\frac{5\pi}{3} + 2n\pi$, where n is any integer **85.** {270°} **87.** {45°, 90°, 225°, 270°}

89. {70.5°, 180°, 289.5°} **91.** $x = \arcsin 2y$ **93.** $x = \left(\frac{1}{3}\arctan 2y\right) - \frac{2}{3}$ **95.** $\emptyset$ **97.** $\left\{-\frac{1}{2}\right\}$

99. **(b)** 8.6602567 ft; There may be a discrepancy in the final digits.

$f(x) = \arctan\left(\frac{15}{x}\right) - \arctan\left(\frac{5}{x}\right)$

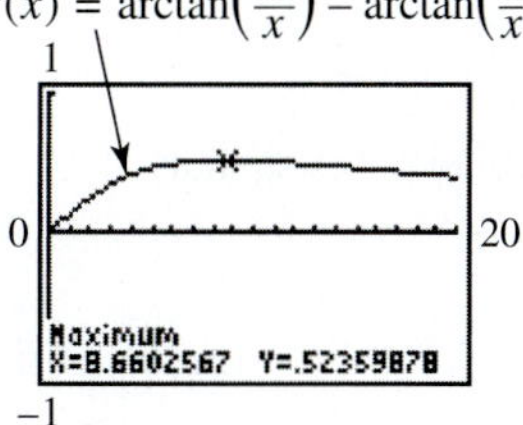

101. The light beam is completely underwater. **103.**

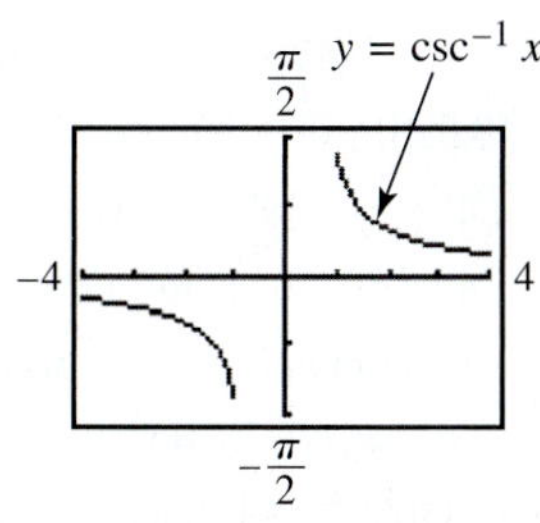

Radian mode

Chapter 7 Test (page 682)

1. $\sin x = -\frac{5\sqrt{61}}{61}$; $\cos x = \frac{6\sqrt{61}}{61}$ **2.** -1 **3.** $\sin(x+y) = \frac{2-2\sqrt{42}}{15}$; $\cos(x-y) = \frac{4\sqrt{2}-\sqrt{21}}{15}$; $\tan(x+y) = \frac{2\sqrt{2}-4\sqrt{21}}{8+\sqrt{42}}$ **4.** $\frac{-\sqrt{2-\sqrt{2}}}{2}$ **5.** $\sec x - \sin x \tan x = \cos x$ **6.** $\cot \frac{x}{2} - \cot x = \csc x$

9. **(a)** $-\sin\theta$ **(b)** $-\sin\theta$ **10.** **(a)** $V = 163 \cos\left(\frac{\pi}{2} - \omega t\right)$ **(b)** 163 volts; $\frac{1}{240}$ sec

11. $[-1, 1]$; $\left[-\frac{\pi}{2}, \frac{\pi}{2}\right]$

12. **(a)** $\frac{2\pi}{3}$ **(b)** $-\frac{\pi}{3}$ **(c)** 0 **(d)** $\frac{2\pi}{3}$ **13.** **(a)** $\frac{\sqrt{5}}{3}$ **(b)** $\frac{4\sqrt{2}}{9}$

14. $\frac{u\sqrt{1-u^2}}{1-u^2}$ **15.** $\{90°, 270°\}$ **16.** $\{18.4°, 135°, 198.4°, 315°\}$ **17.** $\left\{0, \frac{2\pi}{3}, \frac{4\pi}{3}\right\}$

18. $\frac{2\pi}{3} + 4n\pi$, $\frac{4\pi}{3} + 4n\pi$, where n is any integer **19.** **(a)** $x = \frac{1}{3} \arccos y$ **(b)** $\left\{\frac{4}{5}\right\}$ **20.** $\frac{5}{6}$ sec, $\frac{11}{6}$ sec, $\frac{17}{6}$ sec

CHAPTER 8 APPLICATIONS OF TRIGONOMETRY

8.1 Exercises (page 695)

1. C **3.** $\sqrt{3}$ **5.** $C = 95°$, $b = 13$ m, $a = 11$ m **7.** $B = 37.3°$, $a = 38.5$ ft, $b = 51.0$ ft **9.** $C = 57.36°$, $b = 11.13$ ft, $c = 11.55$ ft **11.** $A = 56°\ 00'$, $AB = 361$ ft, $BC = 308$ ft **13.** $B = 110.0°$, $a = 27.01$ m, $c = 21.37$ m **15.** $A = 34.72°$, $a = 3326$ ft, $c = 5704$ ft **17.** A **19.** **(a)** $4 < h < 5$ **(b)** $h = 4$ or $h > 5$ **(c)** $h < 4$ **21.** 2 **23.** 0 **25.** 45° **27.** $B_1 \approx 49.1°$, $C_1 \approx 101.2°$, $B_2 \approx 130.9°$, $C_2 \approx 19.4°$ **29.** no such triangle **31.** $B \approx 27.19°$, $C \approx 10.68°$ **33.** $B \approx 20.6°$, $C \approx 116.9°$, $c \approx 20.6$ ft **35.** no such triangle **37.** $B_1 \approx 49°\ 20'$, $C_1 \approx 92°\ 00'$, $c_1 \approx 15.5$ km, $B_2 \approx 130°\ 40'$, $C_2 \approx 10°\ 40'$, $c_2 \approx 2.88$ km **39.** $A_1 \approx 53.23°$, $C_1 \approx 87.09°$, $c_1 \approx 37.16$ m, $A_2 \approx 126.77°$, $C_2 \approx 13.55°$, $c_2 \approx 8.719$ m **41.** 1; 90°; a right triangle **45.** It cannot exist. **47.** 118 m **49.** 17.8 km **51.** 10.4 in. **53.** 111° **55.** first location: 5.1 mi; second location: 7.2 mi **57.** 38.3 cm **59.** 2.18 km **61.** around 11:55 A.M. **63.** $\sqrt{3}$ **65.** 46.4 m^2 **67.** 356 cm^2 **69.** 722.9 in.^{2} **71.** 65.94 cm^2 **73.** 100 m^2 **75.** $a = \sin A$, $b = \sin B$, $c = \sin C$

Connections (page 707)

1.–3. All three methods give the area as 9.5 sq units.

8.2 Exercises (page 707)

1. **(a)** SAS **(b)** law of cosines **3.** **(a)** SSA **(b)** law of sines **5.** **(a)** ASA **(b)** law of sines **7.** **(a)** ASA **(b)** law of sines **9.** 7 **11.** 30° **13.** $a \approx 5.4$, $B \approx 40.7°$, $C \approx 78.3°$ **15.** $A \approx 22.3°$, $B \approx 108.2°$, $C \approx 49.5°$ **17.** $A \approx 33.6°$, $B \approx 50.7°$, $C \approx 95.7°$ **19.** $a \approx 2.60$ yd, $B \approx 45.1°$, $C \approx 93.5°$ **21.** $c \approx 6.46$ m, $A \approx 53.1°$, $B \approx 81.3°$ **23.** $A \approx 82°$, $B \approx 37°$, $C \approx 61°$ **25.** $C \approx 102°\ 10'$, $B \approx 35°\ 50'$, $A \approx 42°\ 00'$ **27.** $C \approx 84°\ 30'$, $B \approx 44°\ 40'$, $A \approx 50°\ 50'$ **29.** $a \approx 156$ cm, $B \approx 64°\ 50'$, $C \approx 34°\ 30'$ **31.** $b \approx 9.529$ in., $A \approx 64.59°$, $C \approx 40.61°$ **33.** $a \approx 15.7$ m, $B \approx 21.6°$, $C \approx 45.6°$ **35.** $A \approx 30°$, $B \approx 56°$, $C \approx 94°$ **37.** The value of $\cos\theta$ will be greater than 1; your calculator will give you an error message (or a complex number) when using the inverse cosine function. **39.** 257 m **41.** 281 km **43.** 10.8 mi **45.** 40.5° **47.** 26.4° and 36.3° **49.** second base: 66.8 ft; first and third bases: 63.7 ft **51.** 39.2 km **53.** 350° **55.** approximately 47.5 ft **57.** 163.5° **59.** 22 ft

61. 16.26° **63.** $24\sqrt{3}$ **65.** 78 m^2 **67.** 12,600 cm^2 **69.** 3650 ft^2 **71.** 25.24983 mi **73.** 392,000 mi^2 **75.** Area and perimeter are both 36. **77.** **(a)** 87.8° and 92.2° both appear possible. **(b)** 92.2° **(c)** With the law of cosines we are required to find the inverse cosine of a negative number. Therefore, we know that angle C is greater than 90°.

8.3 Exercises *(page 721)*

1. **m** and **p**; **n** and **r** **3.** **m** and **p** equal 2**t**, or **t** is one-half **m** or **p**; also **m** = 1**p** and **n** = 1**r**

5. −**b**

7.

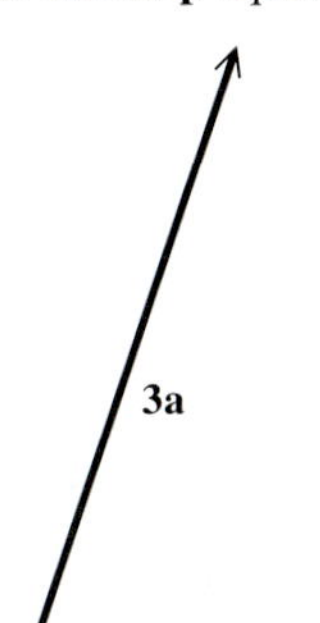

9.

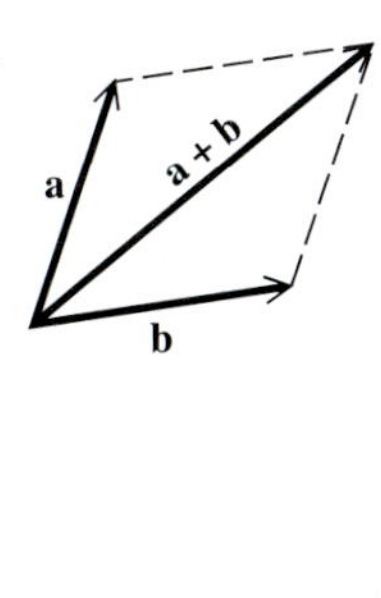

11.

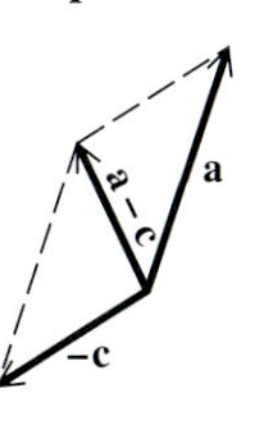

13.

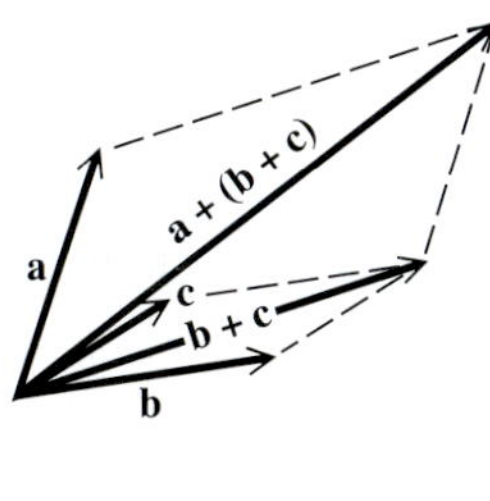

15. c + d, d, c

17. Yes, it appears that vector addition is associative (and this is true, in general).
19. **(a)** ⟨−4, 16⟩ **(b)** ⟨−12, 0⟩ **(c)** ⟨8, −8⟩ **21.** **(a)** ⟨8, 0⟩ **(b)** ⟨0, 16⟩ **(c)** ⟨−4, −8⟩
23. **(a)** ⟨0, 12⟩ **(b)** ⟨−16, −4⟩ **(c)** ⟨8, −4⟩ **25.** **(a)** 4**i** **(b)** 7**i** + 3**j** **(c)** −5**i** + **j**
27. **(a)** ⟨−2, 4⟩ **(b)** ⟨7, 4⟩ **(c)** ⟨6, −6⟩

29. 12, 27°, 20 **31.** 20, 30°, 30 **33.** 17; 331.9° **35.** 8; 120° **37.** 47, 17

39. 38.8, 28.0 **41.** 123, 155 **43.** $\left\langle \frac{5\sqrt{3}}{2}, \frac{5}{2} \right\rangle$ **45.** ⟨3.0642, 2.5712⟩ **47.** ⟨4.0958, −2.8679⟩ **49.** 530 newtons

51. 88.2 lb **53.** 94.2 lb **55.** 24.4 lb **57.** $\langle a + c, b + d \rangle$ **59.** ⟨2, 8⟩ **61.** ⟨8, −20⟩ **63.** ⟨−30, −3⟩ **65.** ⟨8, −7⟩ **67.** −5**i** + 8**j** **69.** 2**i** or 2**i** + 0**j** **71.** 7 **73.** −3 **75.** 20 **77.** 135° **79.** 90° **81.** 36.87° **83.** −6 **85.** −24 **87.** orthogonal **89.** not orthogonal **91.** not orthogonal **93.** magnitude: 9.5208; direction angle: 119.0647° **94.** ⟨−4.1042, 11.2763⟩ **95.** ⟨−.5209, −2.9544⟩ **96.** ⟨−4.6252, 8.3219⟩ **97.** magnitude: 9.5208; direction angle: 119.0647° **98.** They are the same. Preference of method is an individual choice.

8.4 Exercises *(page 727)*

1. 2640 lb at an angle of 167.2° with the 1480-lb force **3.** 93.9° **5.** 190 lb and 283 lb, respectively **7.** 18° **9.** 2.4 tons **11.** 17.5° **13.** weight: 64.8 lb; tension: 61.9 lb **15.** 13.5 mi; 50.4° **17.** 39.2 km **19.** current: 3.5 mph; motorboat: 19.7 mph **21.** bearing: 237°; groundspeed: 470 mph **23.** groundspeed: 161 mph; airspeed: 156 mph **25.** bearing: 74°; groundspeed: 202 mph **27.** bearing: 358°; airspeed: 170 mph **29.** groundspeed: 230 km per hr; bearing: 167° **31.** **(a)** $|\mathbf{R}| = \sqrt{5} \approx 2.2$, $|\mathbf{A}| = \sqrt{1.25} \approx 1.1$; About 2.2 in. of rain fell. The area of the opening of the rain gauge is about 1.1 in.^{2}. **(b)** $V = 1.5$; the volume of rain was 1.5 in.^{3}. **(c)** **R** and **A** should be parallel and point in opposite directions.

Summary Exercises on Applications of Trigonometry and Vectors *(page 731)*

1. 28.59 ft, 37.85 ft **2.** 30.4 lb **3.** 5856 m **4.** 15.8 ft per sec; 71.6° **5.** 32 lb **6.** 7200 ft **7.** 1.95 mi **8.** **(a)** 10 mph **(b)** 3**v** = 18**i** + 24**j**; This represents a 30 mph wind in the direction of **v**. **(c)** **u** represents a southeast wind of $\sqrt{128} \approx 11.3$ mph.

8.5 Exercises *(page 739)*

1.

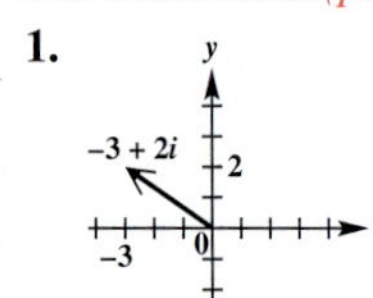

3. **5.** **7.**

9. $1 - 4i$ **11.** $3 - 3i$ **13.** $-3 + 3i$ **15.** $7 + 9i$ **17.** $10i$ **19.** $-2 - 2i\sqrt{3}$ **21.** $-\frac{3\sqrt{3}}{2} + \frac{3}{2}i$

23. $-\sqrt{2}$ **25.** $2(\cos 330° + i \sin 330°)$ **27.** $5\sqrt{2}(\cos 225° + i \sin 225°)$ **29.** $2\sqrt{2}(\cos 45° + i \sin 45°)$

31. $5(\cos 90° + i \sin 90°)$ **33.** $-1.0261 - 2.8191i$ **35.** $12(\cos 90° + i \sin 90°)$ **37.** $\sqrt{34}(\cos 59.04° + i \sin 59.04°)$

39. the circle of radius 1 centered at the origin **41.** the vertical line $x = 1$ **43.** yes **45.** $-4i$ **47.** $12\sqrt{3} + 12i$

49. $-\frac{15\sqrt{2}}{2} + \frac{15\sqrt{2}}{2}i$ **51.** $-3i$ **53.** -2 **55.** $-\frac{1}{6} - \frac{\sqrt{3}}{6}i$ **57.** $-\frac{1}{2} - \frac{1}{2}i$ **59.** $\sqrt{3} + i$

61. $.6537 + 7.4715i$ **63.** $30.8580 + 18.5414i$ **65.** $.2091 + 1.9890i$ **67.** 2 **68.** $w = \sqrt{2}$ cis 135°; $z = \sqrt{2}$ cis 225° **69.** 2 cis 0° **70.** 2; It is the same. **71.** $-i$ **72.** cis(−90°) **73.** $-i$; It is the same.

75. $1.18 - .14i$ **77.** approximately $27.43 + 11.5i$

8.6 Exercises *(page 746)*

1. $27i$ **3.** 1 **5.** $\frac{27}{2} - \frac{27\sqrt{3}}{2}i$ **7.** $-16\sqrt{3} + 16i$ **9.** $-128 + 128i\sqrt{3}$ **11.** $128 + 128i$

13. **(a)** $\cos 0° + i \sin 0°$, $\cos 120° + i \sin 120°$, $\cos 240° + i \sin 240°$ **(b)**

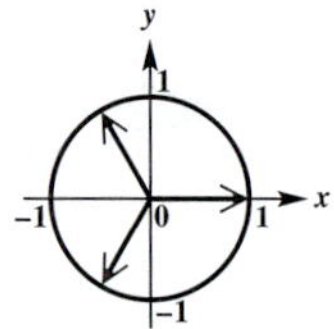

15. **(a)** 2 cis 20°, 2 cis 140°, 2 cis 260° **(b)**

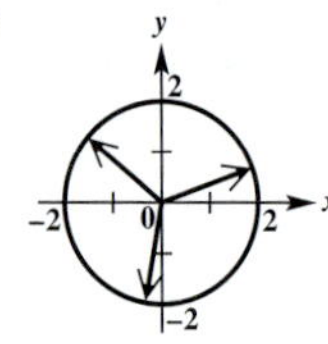

17. **(a)** $2(\cos 90° + i \sin 90°)$, $2(\cos 210° + i \sin 210°)$, $2(\cos 330° + i \sin 330°)$ **(b)**

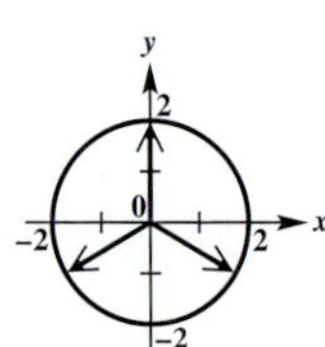

19. **(a)** $4(\cos 60° + i \sin 60°)$, $4(\cos 180° + i \sin 180°)$, $4(\cos 300° + i \sin 300°)$ **(b)**

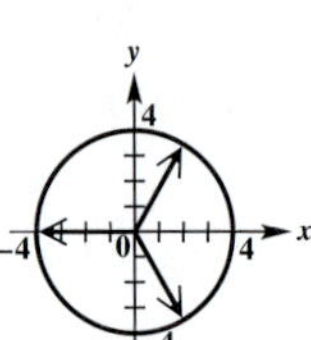

21. **(a)** $\sqrt[3]{2}(\cos 20° + i \sin 20°)$, $\sqrt[3]{2}(\cos 140° + i \sin 140°)$, $\sqrt[3]{2}(\cos 260° + i \sin 260°)$ **(b)**

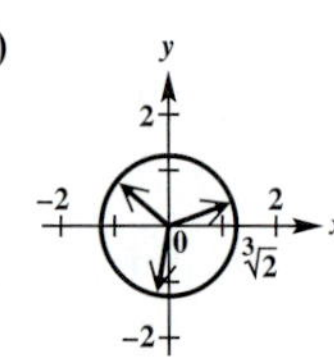

23. **(a)** $\sqrt[3]{4}(\cos 50° + i \sin 50°)$, $\sqrt[3]{4}(\cos 170° + i \sin 170°)$, $\sqrt[3]{4}(\cos 290° + i \sin 290°)$ **(b)**

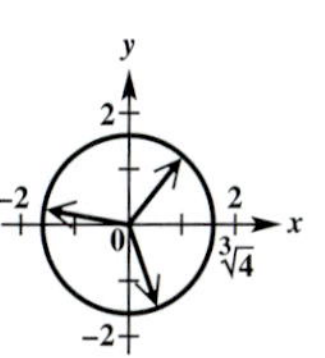

25.

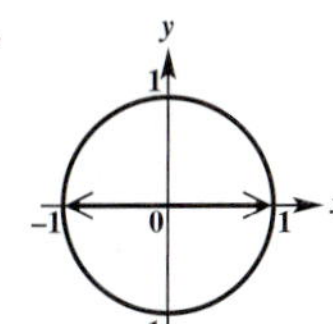

$\cos 0° + i \sin 0°$,
$\cos 180° + i \sin 180°$

27.

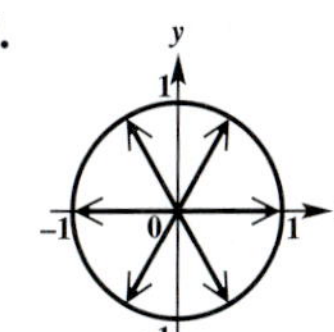

$\cos 0° + i \sin 0°$, $\cos 60° + i \sin 60°$,
$\cos 120° + i \sin 120°$, $\cos 180° + i \sin 180°$,
$\cos 240° + i \sin 240°$, $\cos 300° + i \sin 300°$

29.

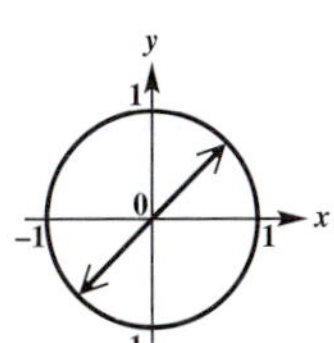

$\cos 45° + i \sin 45°$, $\cos 225° + i \sin 225°$

31. $\{\cos 0° + i \sin 0°, \cos 120° + i \sin 120°, \cos 240° + i \sin 240°\}$ **33.** $\{\cos 90° + i \sin 90°, \cos 210° + i \sin 210°, \cos 330° + i \sin 330°\}$ **35.** $\{2(\cos 0° + i \sin 0°), 2(\cos 120° + i \sin 120°), 2(\cos 240° + i \sin 240°)\}$ **37.** $\{\cos 45° + i \sin 45°, \cos 135° + i \sin 135°, \cos 225° + i \sin 225°, \cos 315° + i \sin 315°\}$ **39.** $\{\cos 22.5° + i \sin 22.5°, \cos 112.5° + i \sin 112.5°, \cos 202.5° + i \sin 202.5°, \cos 292.5° + i \sin 292.5°\}$ **41.** $\{2(\cos 20° + i \sin 20°), 2(\cos 140° + i \sin 140°), 2(\cos 260° + i \sin 260°)\}$ **43.** $1, -\frac{1}{2} + \frac{\sqrt{3}}{2}i, -\frac{1}{2} - \frac{\sqrt{3}}{2}i$ **45.** $\cos 2\theta + i \sin 2\theta$ **46.** $(\cos^2 \theta - \sin^2 \theta) + i(2 \cos \theta \sin \theta) = \cos 2\theta + i \sin 2\theta$ **47.** $\cos 2\theta = \cos^2 \theta - \sin^2 \theta$ **48.** $\sin 2\theta = 2 \cos \theta \sin \theta$ **49.** **(a)** yes **(b)** no **(c)** yes **51.** $1, .30901699 + .95105652i, -.809017 + .58778525i, -.809017 - .5877853i, .30901699 - .9510565i$ **53.** $-4, 2 - 2i\sqrt{3}$ **55.** $\{.87708 + .94922i, -.63173 + 1.1275i, -1.2675 - .25240i, -.15164 - 1.28347i, 1.1738 - .54083i\}$ **57.** false

8.7 Exercises *(page 757)*

1. **(a)** II **(b)** I **(c)** IV **(d)** III

Graphs for Exercises 3(a), 5(a), 7(a), 9(a), 11(a)

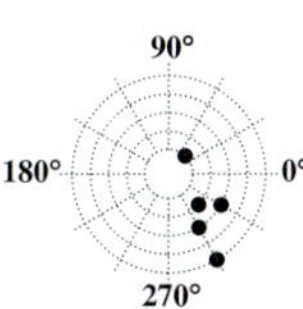

Answers may vary in Exercises 3(b)–11(b).

3. **(b)** $(1, 405°), (-1, 225°)$ **(c)** $\left(\frac{\sqrt{2}}{2}, \frac{\sqrt{2}}{2}\right)$ **5.** **(b)** $(-2, 495°), (2, 315°)$ **(c)** $\left(\sqrt{2}, -\sqrt{2}\right)$ **7.** **(b)** $(5, 300°), (-5, 120°)$ **(c)** $\left(\frac{5}{2}, -\frac{5\sqrt{3}}{2}\right)$ **9.** **(b)** $(-3, 150°), (3, -30°)$ **(c)** $\left(\frac{3\sqrt{3}}{2}, -\frac{3}{2}\right)$ **11.** **(b)** $\left(3, \frac{11\pi}{3}\right), \left(-3, \frac{2\pi}{3}\right)$ **(c)** $\left(\frac{3}{2}, -\frac{3\sqrt{3}}{2}\right)$

Graphs for Exercises 13(a), 15(a), 17(a), 19(a), 21(a)

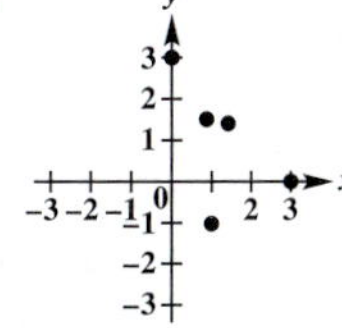

Answers may vary in Exercises 13(b)–21(b).

13. **(b)** $\left(\sqrt{2}, 315°\right), \left(-\sqrt{2}, 135°\right)$ **15.** **(b)** $(3, 90°), (-3, 270°)$ **17.** **(b)** $(2, 45°), (-2, 225°)$

19. **(b)** $(\sqrt{3}, 60°), (-\sqrt{3}, 240°)$ **21.** **(b)** $(3, 0°), (-3, 180°)$ **23.** $r = \dfrac{4}{\cos\theta - \sin\theta}$

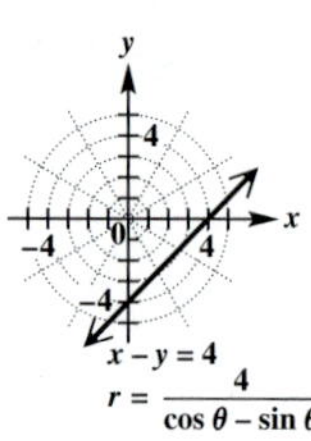

25. $r = 4$ or $r = -4$

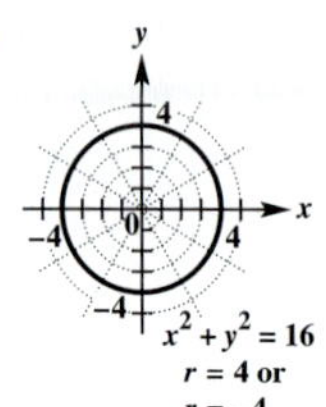

27. $r = \dfrac{5}{2\cos\theta + \sin\theta}$

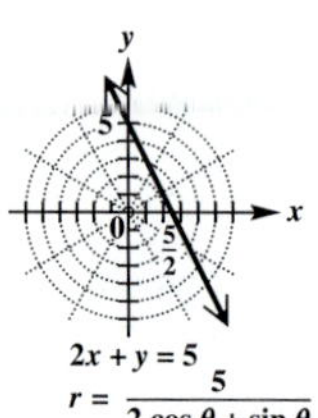

29. $r\sin\theta = k$ **30.** $r = \dfrac{k}{\sin\theta}$

31. $r = k\csc\theta$ **32.** $y = 3$

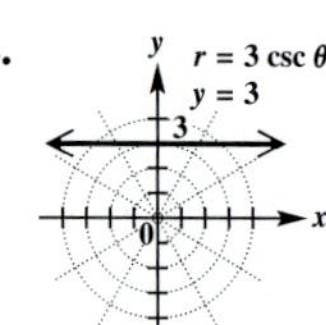

33. $r\cos\theta = k$ **34.** $r = \dfrac{k}{\cos\theta}$ **35.** $r = k\sec\theta$

36. $x = 3$

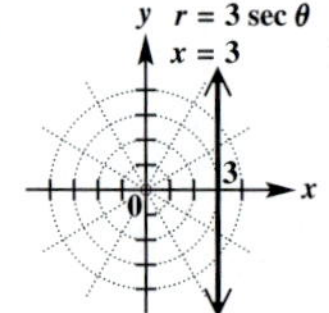

37. C **39.** A **41.** cardioid

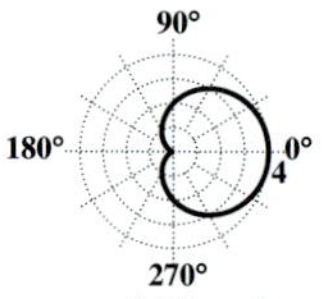

43. limaçon

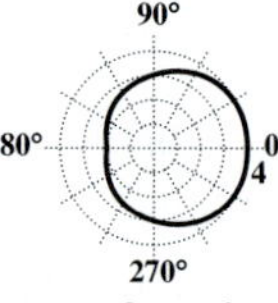

45. four-leaved rose

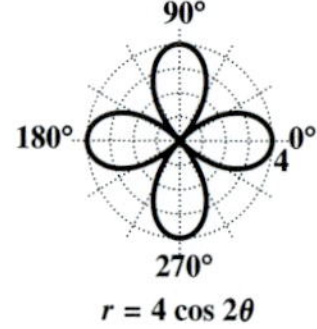

47. lemniscate

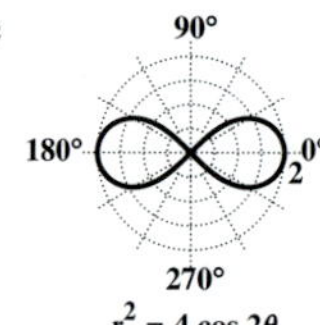

49. cardioid

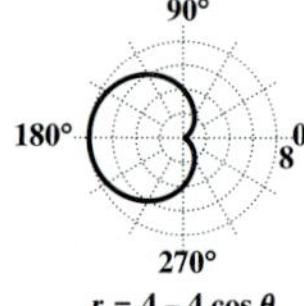

51.

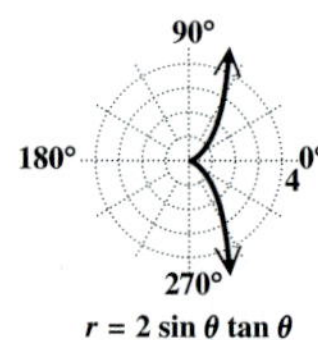

53. $x^2 + (y - 1)^2 = 1$

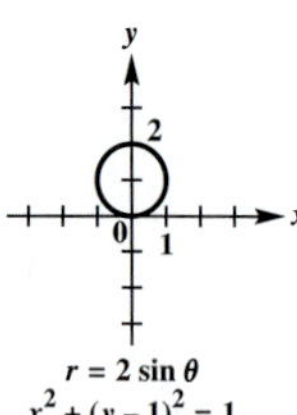

55. $y^2 = 4(x + 1)$

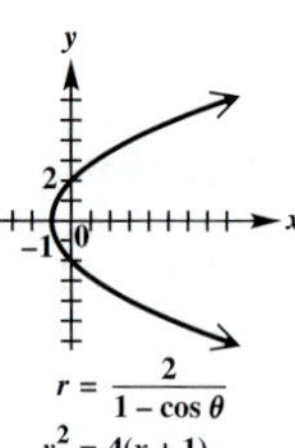

57. $(x + 1)^2 + (y + 1)^2 = 2$

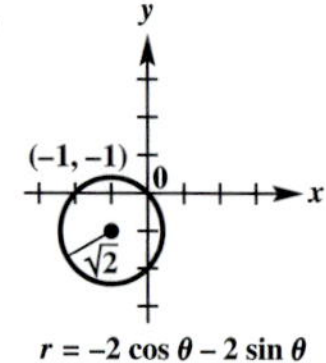

59. $x = 2$

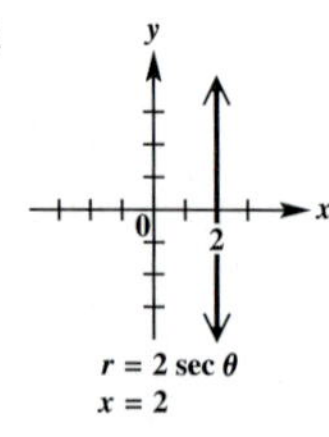

61. $x + y = 2$

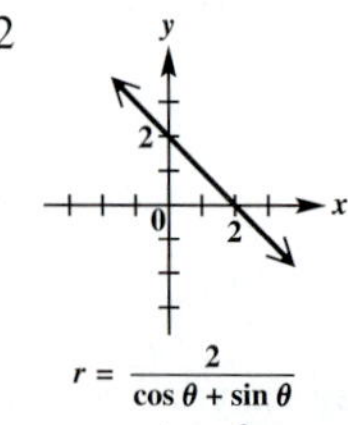

63.

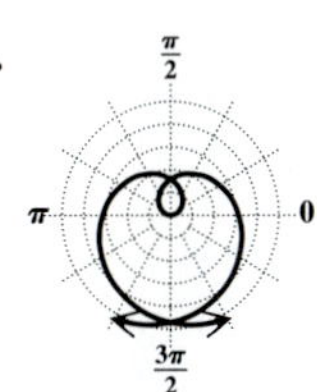
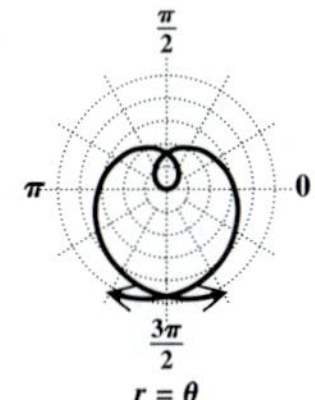

65. $r = \dfrac{2}{2\cos\theta + \sin\theta}$ **67. (a)** $(r, -\theta)$ **(b)** $(r, \pi - \theta)$ or $(-r, -\theta)$ **(c)** $(r, \pi + \theta)$ or $(-r, \theta)$

69.

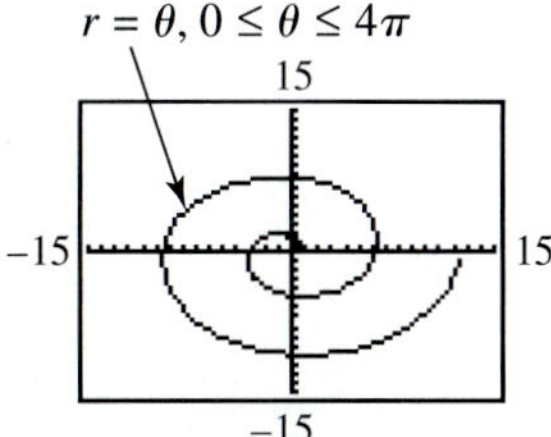

71.

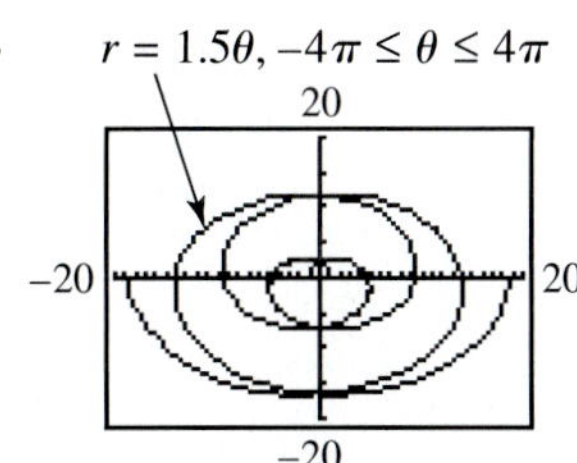

73. $\left(2, \dfrac{\pi}{6}\right), \left(2, \dfrac{5\pi}{6}\right)$

75. $\left(\dfrac{4 + \sqrt{2}}{2}, \dfrac{\pi}{4}\right), \left(\dfrac{4 - \sqrt{2}}{2}, \dfrac{5\pi}{4}\right)$

77. (a)

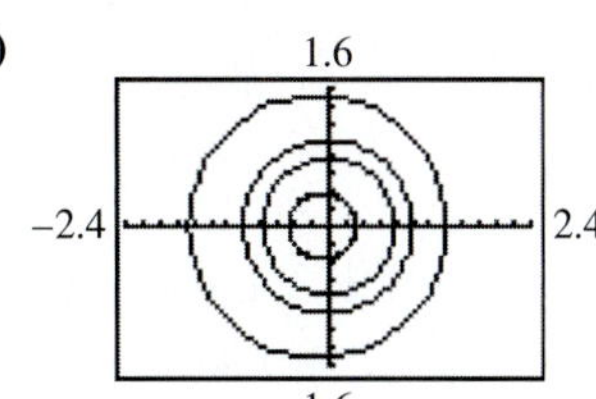

(b)

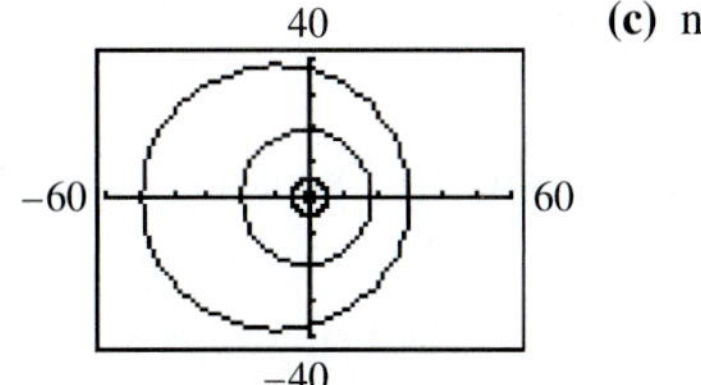

Earth is closest to the sun.

(c) no

8.8 Exercises *(page 766)*

1. C **3.** A **5. (a)**

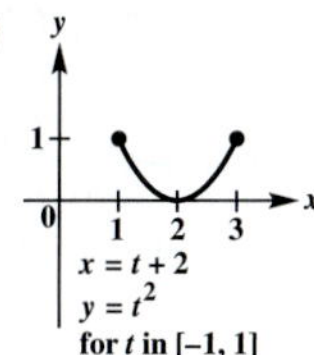
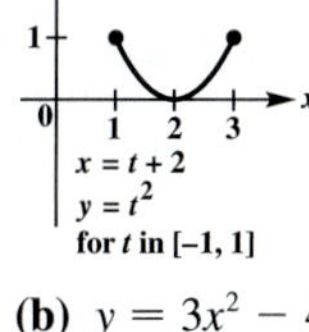

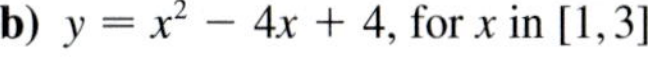

(b) $y = x^2 - 4x + 4$, for x in $[1, 3]$

7. (a)

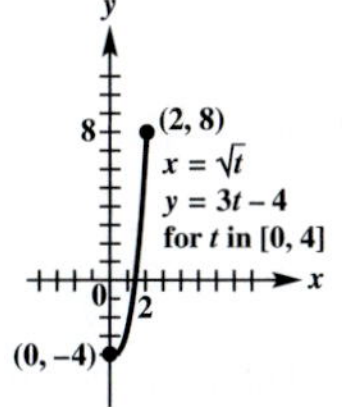

(b) $y = 3x^2 - 4$, for x in $[0, 2]$

9. (a)

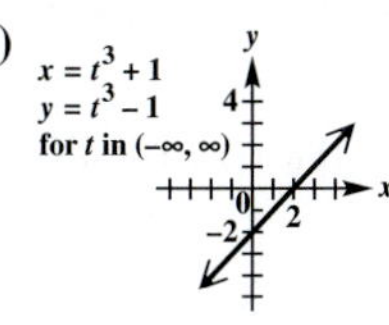
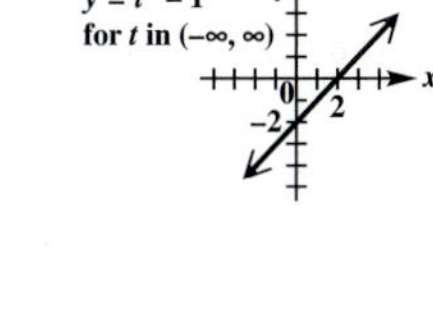

(b) $y = x - 2$, for x in $(-\infty, \infty)$

11. (a)

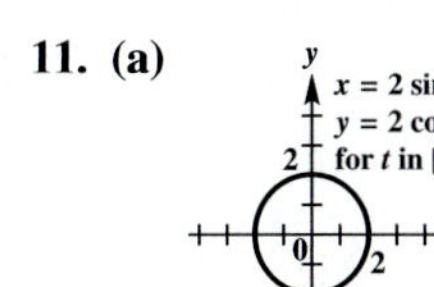
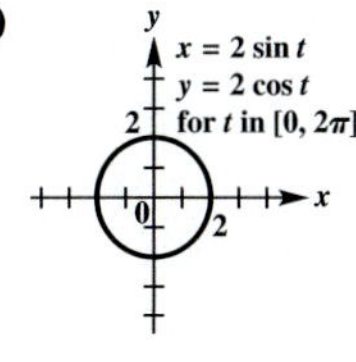

(b) $x^2 + y^2 = 4$, for x in $[-2, 2]$

13. (a)

(b) $y = 2\sqrt{1 + \dfrac{x^2}{9}}$, for x in $(-\infty, \infty)$

15. (a)

(b) $y = \dfrac{1}{x}$, for x in $(0, 1]$

17. (a)

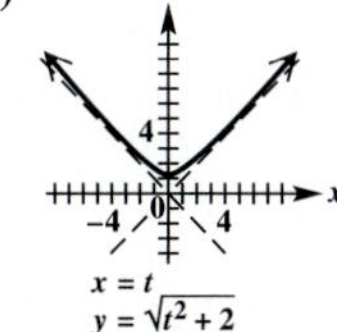

(b) $y = \sqrt{x^2 + 2}$, for x in $(-\infty, \infty)$

19. **(a)** $x = 2 + \sin t$, $y = 1 + \cos t$, for t in $[0, 2\pi]$ **(b)** $(x - 2)^2 + (y - 1)^2 = 1$, for x in $[1, 3]$

21. **(a)** $x = t + 2$, $y = \frac{1}{t + 2}$, for $t \neq -2$ **(b)** $y = \frac{1}{x}$, for x in $(-\infty, 0) \cup (0, \infty)$

23. **(a)** $x = t + 2$, $y = t - 4$, for t in $(-\infty, \infty)$ **(b)** $y = x - 6$, for x in $(-\infty, \infty)$

25. $x = 3 \cos t$, $y = 3 \sin t$, for t in $[0, 2\pi]$; $x^2 + y^2 = 9$

27. $x = 3 \sin t$, $y = 2 \cos t$, for t in $[0, 2\pi]$; $\frac{x^2}{9} + \frac{y^2}{4} = 1$

Answers may vary for Exercises 29 and 31.

29. $x = t$, $y = (t + 3)^2 - 1$, for t in $(-\infty, \infty)$; $x = t - 3$, $y = t^2 - 1$, for t in $(-\infty, \infty)$

31. $x = t$, $y = t^2 - 2t + 3$, for t in $(-\infty, \infty)$; $x = t + 1$, $y = t^2 + 2$, for t in $(-\infty, \infty)$

33. $x = 2t - 2 \sin t$, $y = 2 - 2 \cos t$, for t in $[0, 8\pi]$

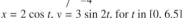

35. $x = 2 \cos t$, $y = 3 \sin 2t$, for t in $[0, 6.5]$

37.

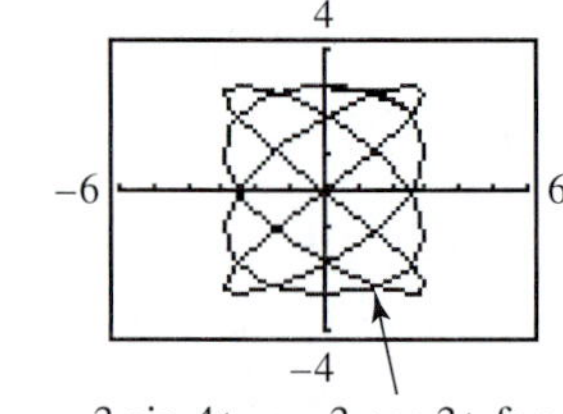

$x = 3 \sin 4t$, $y = 3 \cos 3t$, for t in $[0, 6.5]$

39. **(a)** $x = 24t$, $y = -16t^2 + 24\sqrt{3}t$ **(b)** $y = -\frac{1}{36}x^2 + \sqrt{3}x$ **(c)** 2.6 sec; 62 ft

41. **(a)** $x = (88 \cos 20°)t$, $y = 2 - 16t^2 + (88 \sin 20°)t$ **(b)** $y = 2 - \frac{x^2}{484 \cos^2 20°} + (\tan 20°)x$ **(c)** 1.9 sec; 161 ft

43. **(a)** $y = -\frac{1}{256}x^2 + \sqrt{3}x + 8$; parabolic path **(b)** approximately 7 sec; approximately 448 ft

45. **(a)** $x = 32t$, $y = 32\sqrt{3}t - 16t^2 + 3$ **(b)** about 112.6 ft **(c)** 51 ft maximum height; The ball had traveled horizontally about 55.4 ft. **(d)** yes

47. $x = 56.56530965t$
$y = -16t^2 + 67.41191099t$

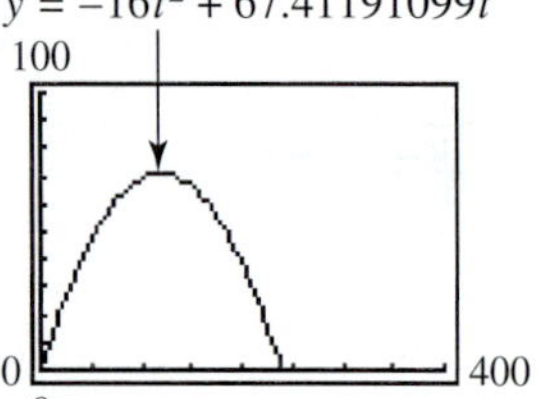

(b) 50.0° **(c)** $x = (88 \cos 50.0°)t$, $y = -16t^2 + (88 \sin 50.0°)t$

49. Many answers are possible; for example, $y = a(t - h)^2 + k$, $x = t$ and $y = at^2 + k$, $x = t + h$.

51. Many answers are possible; for example, $x = a \sin t$, $y = b \cos t$ and $x = t$, $y^2 = b^2\left(1 - \frac{t^2}{a^2}\right)$.

55. The graph is translated horizontally by c units.

Chapter 8 Review Exercises *(page 773)*

1. 63.7 m **3.** 41.7° **5.** 54° 20′ or 125° 40′ **9.** **(a)** $b = 5, b \geq 10$ **(b)** $5 < b < 10$ **(c)** $b < 5$ **11.** 19.87° or 19° 52′ **13.** 55.5 m **15.** 19 cm **17.** $B \approx 17.3°, C \approx 137.5°, c \approx 11.0$ yd **19.** $c \approx 18.7$ cm, $A \approx 91° 40', B \approx 45° 50'$ **21.** 153,600 m^2 **23.** .234 km^2 **25.** 58.6 ft **27.** 13 m **29.** 53.2 ft **31.** 115 km **33.** 25 **35.**

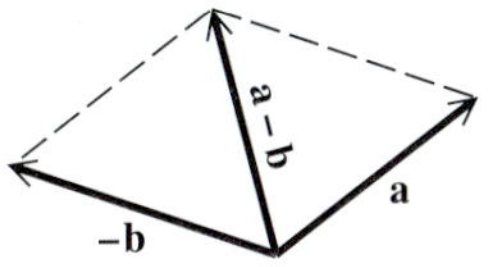

37. **(a)** true **(b)** false **39.** 207 lb **41.** 869; 418 **43.** 15; 126.9° **45.** −9; 142.1° **47.** $\left\langle \frac{5}{13}, \frac{12}{13} \right\rangle$ **49.** 29 lb **51.** bearing: 306°; speed: 524 mph **53.** $AB = 1978.28$ ft; $BC = 975.05$ ft **55.** $-3 - 3i\sqrt{3}$ **57.** −2 **59.** $-\frac{1}{2} - \frac{\sqrt{3}}{2}i$ **61.**

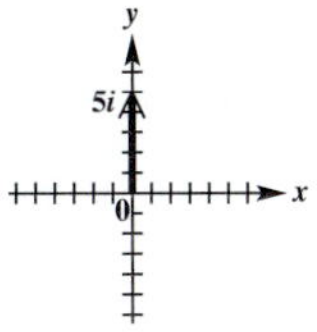

63. $2\sqrt{2}(\cos 135° + i \sin 135°)$ **65.** $-\sqrt{2} - i\sqrt{2}$ **67.** $\sqrt{2}(\cos 315° + i \sin 315°)$ **69.** $4(\cos 270° + i \sin 270°)$ **71.** the line $y = -x$ **73.** $\sqrt[6]{2}(\cos 105° + i \sin 105°), \sqrt[6]{2}(\cos 225° + i \sin 225°), \sqrt[6]{2}(\cos 345° + i \sin 345°)$ **75.** none **77.** $\{2(\cos 45° + i \sin 45°), 2(\cos 135° + i \sin 135°), 2(\cos 225° + i \sin 225°), 2(\cos 315° + i \sin 315°)\}$ **79.** $\left(\frac{5\sqrt{2}}{2}, -\frac{5\sqrt{2}}{2}\right)$ **81.** circle

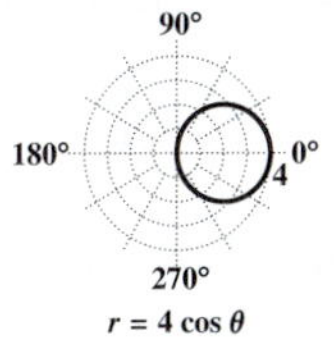

83. eight-leaved rose

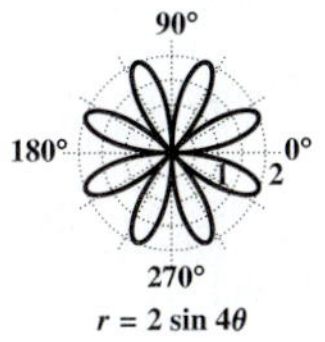

85. $y^2 = -6\left(x - \frac{3}{2}\right)$ or $y^2 + 6x - 9 = 0$ **87.** $x^2 + y^2 = 4$ **89.** $r = \tan\theta \sec\theta$ or $r = \frac{\tan\theta}{\cos\theta}$ **91.** $r = 2 \sec\theta$ or $r = \frac{2}{\cos\theta}$ **93.** $r = \frac{4}{\cos\theta + 2\sin\theta}$ **95.**

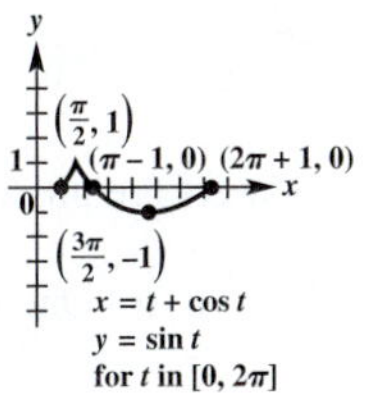

97. $y = \sqrt{x^2 + 1}$, for x in $[0, \infty)$ **99.** $y = 3\sqrt{1 + \frac{x^2}{25}}$, for x in $(-\infty, \infty)$ **101.** $x = 3 + 5\cos t, y = 4 + 5\sin t$, for t in $[0, 2\pi]$ **103.** **(a)** $x = (118 \cos 27°)t, y = 3.2 - 16t^2 + (118 \sin 27°)t$ **(b)** $y = 3.2 - \frac{4x^2}{3481 \cos^2 27°} + (\tan 27°)x$ **(c)** 3.4 sec; 358 ft

Chapter 8 Test *(page 779)*

1. 137.5° **2.** 180 km **3.** 49.0° **4.** 4300 km^2 **5.** **(a)** $b > 10$ **(b)** none **(c)** $b \leq 10$ **6.** 264 sq units **7.** $|\mathbf{v}| = 10; \theta \approx 126.9°$ **8.** **(a)** $\langle 1, -3 \rangle$ **(b)** $\langle -6, 18 \rangle$ **(c)** −20 **(d)** $\sqrt{10}$ **9.** 2.7 mi **10.** $\langle -346, 451 \rangle$ **11.** 1.91 mi **12.** **(a)** $3(\cos 90° + i \sin 90°)$ **(b)** $\sqrt{5}$ cis 63.43° **(c)** $2(\cos 240° + i \sin 240°)$ **13.** **(a)** $\frac{3\sqrt{3}}{2} + \frac{3}{2}i$ **(b)** $3.06 + 2.57i$ **(c)** $3i$ **14.** **(a)** $16(\cos 50° + i \sin 50°)$ **(b)** $2\sqrt{3} + 2i$ **(c)** $4\sqrt{3} + 4i$ **15.** 2 cis 67.5°, 2 cis 157.5°, 2 cis 247.5°, 2 cis 337.5°

16. Answers may vary. **(a)** $(5, 90°), (5, -270°)$ **(b)** $(2\sqrt{2}, 225°), (2\sqrt{2}, -135°)$ **17. (a)** $\left(\frac{3\sqrt{2}}{2}, -\frac{3\sqrt{2}}{2}\right)$ **(b)** $(0, -4)$

18. cardioid

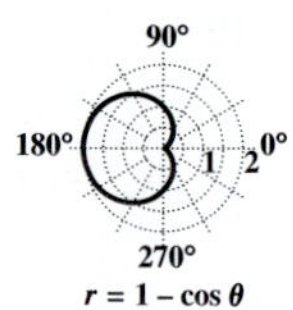

19. three-leaved rose

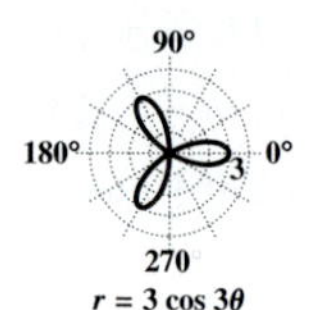

20. (a) $x - 2y = -4$

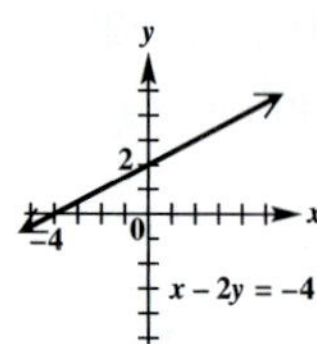

(b) $x^2 + y^2 = 36$

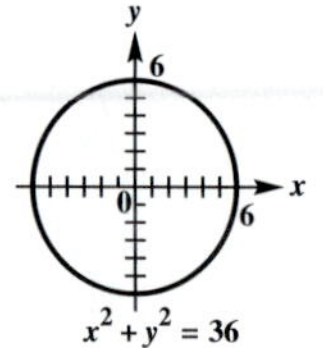

21.

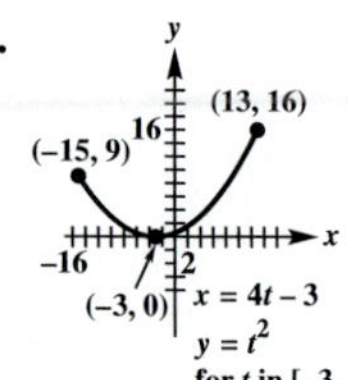

22.

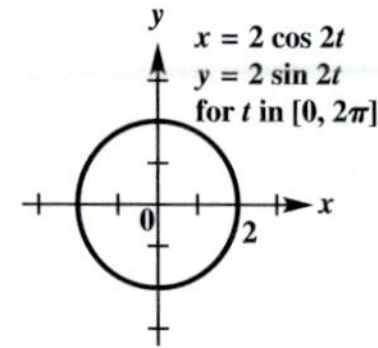

CHAPTER 9 SYSTEMS AND MATRICES

9.1 Exercises *(page 794)*

1. approximately 2002 **3.** (2002, 3.3 million) **5.** the year; number of migrants **7.** $\{(-4, 1)\}$ **9.** $\{(48, 8)\}$ **11.** $\{(1, 3)\}$ **13.** $\{(-1, 3)\}$ **15.** $\{(3, -4)\}$ **17.** $\{(0, 2)\}$ **19.** $\{(0, 4)\}$ **21.** $\{(1, 3)\}$ **23.** $\{(4, -2)\}$ **25.** $\{(2, -2)\}$ **27.** $\{(4, 6)\}$ **29.** $\{(5, 2)\}$ **31.** $\emptyset$; inconsistent system **33.** $\left\{\left(\frac{y+9}{4}, y\right)\right\}$; infinitely many solutions **35.** $\emptyset$; inconsistent system **37.** $\left\{\left(\frac{6-2y}{7}, y\right)\right\}$; infinitely many solutions **39.** A **41.** $\{(.138, -4.762)\}$ **43.** $\{(.236, .674)\}$ **45.** $k \neq -6$; $k = -6$ **47.** $\{(1, 2, -1)\}$ **49.** $\{(2, 0, 3)\}$ **51.** $\{(1, 2, 3)\}$ **53.** $\{(4, 1, 2)\}$ **55.** $\left\{\left(\frac{1}{2}, \frac{2}{3}, -1\right)\right\}$ **57.** $\{(-3, 1, 6)\}$ **59.** $\{(x, -4x + 3, -3x + 4)\}$ **61.** $\{(x, -x + 15, -9x + 69)\}$ **63.** $\left\{\left(x, \frac{41 - 7x}{9}, \frac{47 - x}{9}\right)\right\}$ **65.** $\emptyset$; inconsistent system **67.** $\left\{\left(-\frac{z}{9}, \frac{z}{9}, z\right)\right\}$; infinitely many solutions **69.** $\{(2, 2)\}$ **71.** $\left\{\left(\frac{1}{5}, 1\right)\right\}$ **73.** $\{(4, 6, 1)\}$ **75.** Other answers are possible. **(a)** One example is $x + 2y + z = 5$, $2x - y + 3z = 4$. **(b)** One example is $x + y + z = 5$, $2x - y + 3z = 4$. **(c)** One example is $2x + 2y + 2z = 8$, $2x - y + 3z = 4$. **77.** $y = \frac{3}{4}x^2 + \frac{1}{4}x - \frac{1}{2}$ **79.** $y = 3x - 1$ **81.** $y = -\frac{1}{2}x^2 + x + \frac{1}{4}$ **83.** $x^2 + y^2 - 4x + 2y - 20 = 0$ **85.** $x^2 + y^2 + x - 7y = 0$ **87. (a)** $a = \frac{7}{800}$, $b = \frac{39}{40}$, $c = 318$; $C = \frac{7}{800}t^2 + \frac{39}{40}t + 318$ **(b)** near the end of 2104 **89.** California: \$35 billion; New York: \$12 billion **91.** 120 gal of \$9.00; 60 gal of \$3.00; 120 gal of \$4.50 **93.** 28 in.; 17 in.; 14 in. **95.** \$100,000 at 3%; \$40,000 at 2.5%; \$60,000 at 1.5% **97. (a)** \$16 **(b)** \$11 **(c)** \$6 **98. (a)** 8 **(b)** 4 **(c)** 0 **99.** See the answer to Exercise 101. **100. (a)** 0 **(b)** $\frac{40}{3}$ **(c)** $\frac{80}{3}$

101.

p
16
(8, 6) $p = \frac{3}{4}q$
8
$p = 16 - \frac{5}{4}q$
0 2 6 10 q

102. price: \$6; demand: 8 **103.** $\{(40, 15, 30)\}$

45. $\{(10, -1, -2)\}$ **47.** $\{(11, -1, 2)\}$ **49.** $\{(1, 0, 2, 1)\}$

51. (a) $602.7 = a + 5.543b + 37.14c$ **(b)** $a \approx -490.547, b = -89, c = 42.71875$ **(c)** $S = -490.547 - 89A + 42.71875B$

$656.7 = a + 6.933b + 41.30c$

$778.5 = a + 7.638b + 45.62c$

(d) $S \approx 843.5$ **(e)** $S \approx 1547.5$ **53.** Answers will vary. **55.** $\begin{bmatrix} -.1215875322 & .0491390161 \\ 1.544369078 & -.046799063 \end{bmatrix}$

57. $\begin{bmatrix} 2 & -2 & 0 \\ -4 & 0 & 4 \\ 3 & 3 & -3 \end{bmatrix}$ **59.** $\{(1.68717058, -1.306990242)\}$ **61.** $\{(13.58736702, 3.929011993, -5.342780076)\}$

65. $A = \begin{bmatrix} 1 & 0 \\ 1 & 1 \end{bmatrix}, B = \begin{bmatrix} 1 & 1 \\ 0 & 1 \end{bmatrix}$ (Other answers are possible.) **67.** $\begin{bmatrix} \frac{1}{a} & 0 & 0 \\ 0 & \frac{1}{b} & 0 \\ 0 & 0 & \frac{1}{c} \end{bmatrix}$ **69.** $I_n, -A^{-1}, \dfrac{1}{k}A^{-1}$

Chapter 9 Review Exercises *(page 884)*

1. $\{(0, 1)\}$ **3.** $\{(9 - 5y, y)\}$; infinitely many solutions **5.** $\emptyset$; inconsistent system **7.** $\left\{\left(\dfrac{1}{3}, \dfrac{1}{2}\right)\right\}$ **9.** $\{(3, 2, 1)\}$

11. $\{(5, -1, 0)\}$ **13.** One possible answer is $\begin{matrix} x + y = 2 \\ x + y = 3 \end{matrix}$. **15.** $\dfrac{1}{3}$ cup of rice, $\dfrac{1}{5}$ cup of soybeans **17.** 5 blankets, 3 rugs, 8 skirts **19.** $x \approx 177.1, y \approx 174.9$; If an athlete's maximum heart rate is 180 beats per minute (bpm), then it will be about 177 bpm 5 sec after stopping and 175 bpm 10 sec after stopping. **21.** $Y_1 = 2.4X^2 - 6.2X + 1.5$

23. $\{(x, 1 - 2x, 6 - 11x)\}$ **25.** $\{(-2, 0)\}$ **27.** $\{(0, 1, 0)\}$ **29.** 10 lb of \$4.60 tea, 8 lb of \$5.75 tea, 2 lb of \$6.50 tea

31. 1979; approximately 10 million **33.** -25 **35.** -1 **37.** $\left\{\dfrac{8}{19}\right\}$ **39.** $\{(-4, 2)\}$ **41.** $\emptyset$; inconsistent system

43. $\{(14, -15, 35)\}$ **45.** $\dfrac{2}{x-1} - \dfrac{6}{3x-2}$ **47.** $\dfrac{1}{x-1} - \dfrac{x+3}{x^2+2}$ **49.** $\{(-3, 4), (1, 12)\}$

51. $\{(-4, 1), (-4, -1), (4, -1), (4, 1)\}$ **53.** $\left\{(5, -2), \left(-4, \dfrac{5}{2}\right)\right\}$ **55.** $\{(-2, 0), (1, 1)\}$ **57.** $b = \pm 5\sqrt{10}$

59.

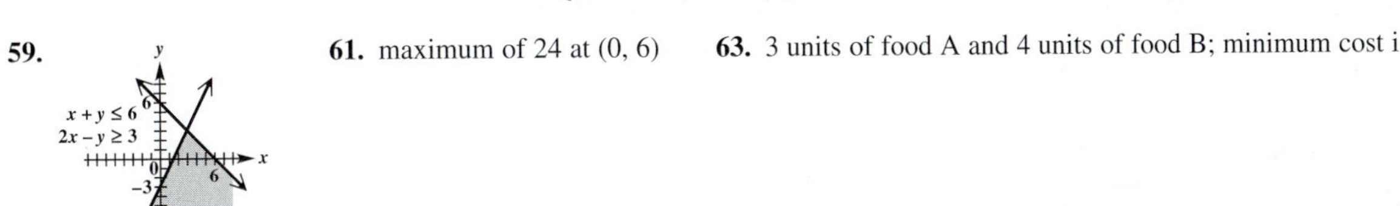

61. maximum of 24 at (0, 6) **63.** 3 units of food A and 4 units of food B; minimum cost is \$1.02 per serving. **65.** $a = 5, x = \dfrac{3}{2}, y = 0, z = 9$ **67.** $\begin{bmatrix} -4 \\ 6 \\ 1 \end{bmatrix}$ **69.** cannot be subtracted **71.** $\begin{bmatrix} 3 & -4 \\ 4 & 48 \end{bmatrix}$

73. $\begin{bmatrix} -9 & 3 \\ 10 & 6 \end{bmatrix}$ **75.** $\begin{bmatrix} -2 & 5 & -3 \\ 3 & 4 & -4 \\ 6 & -1 & -2 \end{bmatrix}$ **77.** $\begin{bmatrix} 3 & -1 \\ -5 & 2 \end{bmatrix}$ **79.** $\begin{bmatrix} \frac{2}{3} & 0 & -\frac{1}{3} \\ \frac{1}{3} & 0 & -\frac{2}{3} \\ -\frac{2}{3} & 1 & \frac{1}{3} \end{bmatrix}$ **81.** $\{(-3, 2, 0)\}$

Chapter 9 Test *(page 888)*

1. $\{(4, 3)\}$ **2.** $\left\{\left(\dfrac{-3y - 7}{2}, y\right)\right\}$; infinitely many solutions **3.** $\{(1, 2)\}$ **4.** $\emptyset$; inconsistent system **5.** $\{(2, 0, -1)\}$

6. $\{(5, 1)\}$ **7.** $\{(5, 3, 6)\}$ **8.** $y = 2x^2 - 8x + 11$ **9.** 22 units from Toronto, 56 units from Montreal, 22 units from Ottawa **10.** -58 **11.** -844 **12.** $\{(-6, 7)\}$ **13.** $\{(1, -2, 3)\}$ **14.** $\dfrac{2}{x} + \dfrac{-2}{x+1} + \dfrac{-1}{(x+1)^2}$

15. yes

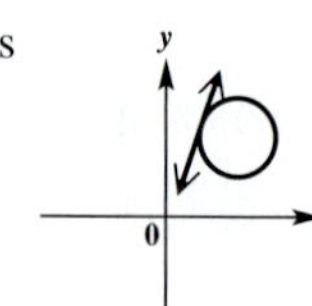

16. $\{(1, 2), (-1, 2), (1, -2), (-1, -2)\}$ **17.** $\{(3, 4), (4, 3)\}$ **18.** 5 and -6

19.

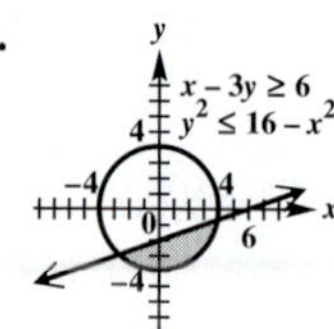

20. maximum of 42 at (12, 6) **21.** 0 VIP rings and 24 SST rings; maximum profit is \$960.

22. $x = -1$; $y = 7$; $w = -3$ **23.** $\begin{bmatrix} 8 & 3 \\ 0 & -11 \\ 15 & 19 \end{bmatrix}$ **24.** cannot be added **25.** $\begin{bmatrix} -5 & 16 \\ 19 & 2 \end{bmatrix}$

26. cannot be multiplied **27.** A **28.** $\begin{bmatrix} -2 & -5 \\ -3 & -8 \end{bmatrix}$ **29.** does not exist

30. $\begin{bmatrix} -9 & 1 & -4 \\ -2 & 1 & 0 \\ 4 & -1 & 1 \end{bmatrix}$ **31.** $\{(-7, 8)\}$ **32.** $\{(0, 5, -9)\}$

CHAPTER 10 ANALYTIC GEOMETRY

10.1 Exercises *(page 899)*

1. **(a)** D **(b)** B **(c)** C **(d)** A **(e)** F **(f)** H **(g)** E **(h)** G

In Exercises 3–17, we give the domain first, then the range.

3. $(-\infty, 0]$; $(-\infty, \infty)$ **5.** $[0, \infty)$; $(-\infty, \infty)$ **7.** $[2, \infty)$; $(-\infty, \infty)$ **9.** $(-\infty, 2]$; $(-\infty, \infty)$ **11.** $[4, \infty)$; $(-\infty, \infty)$

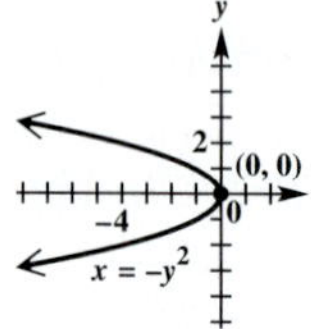

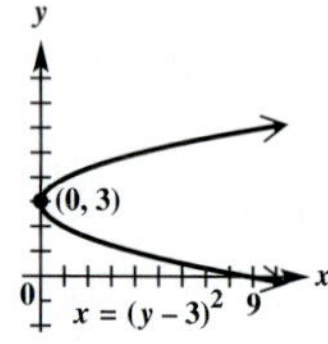

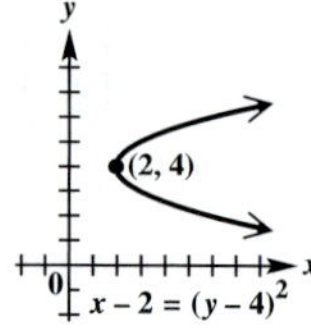

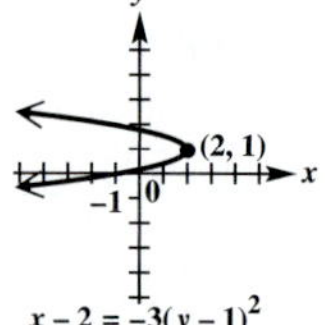

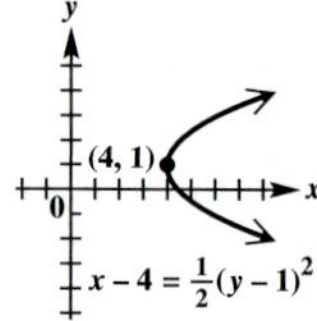

13. $[-2, \infty)$; $(-\infty, \infty)$ **15.** $(-\infty, 4]$; $(-\infty, \infty)$ **17.** $[1, \infty)$; $(-\infty, \infty)$ **19.** $(0, 6)$, $y = -6$, y-axis

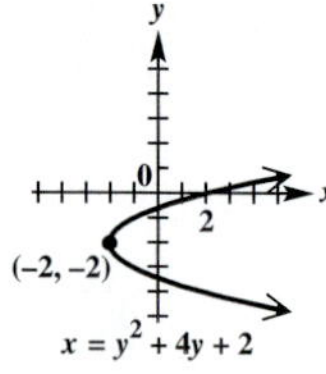

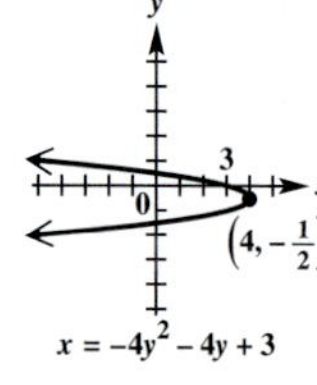

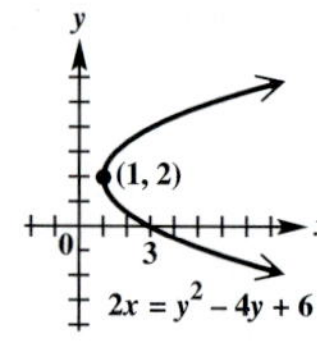

21. $\left(0, -\frac{1}{16}\right)$, $y = \frac{1}{16}$, y-axis **23.** $(-1, 0)$, $x = 1$, x-axis **25.** $\left(-\frac{1}{128}, 0\right)$, $x = \frac{1}{128}$, x-axis **27.** $(4, 3)$, $x = -2$, $y = 3$

29. $(7, -1)$, $y = -9$, $x = 7$ **31.** $y^2 = 20x$ **33.** $x^2 = y$ **35.** $x^2 = y$ **37.** $y^2 = \frac{4}{3}x$ **39.** $(x - 4)^2 = 8(y - 3)$

41. $(y - 6)^2 = 28(x + 5)$ **43.**

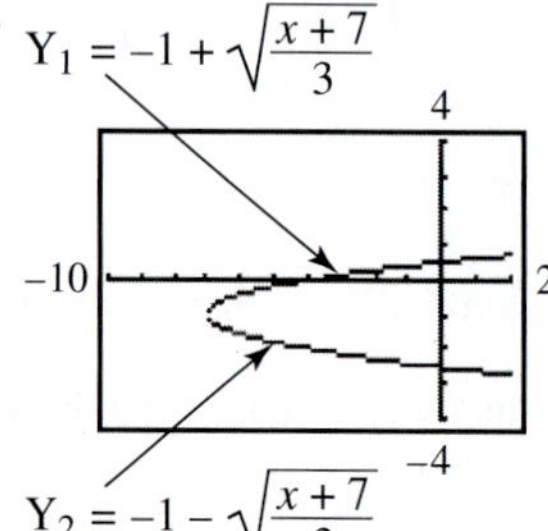

45.

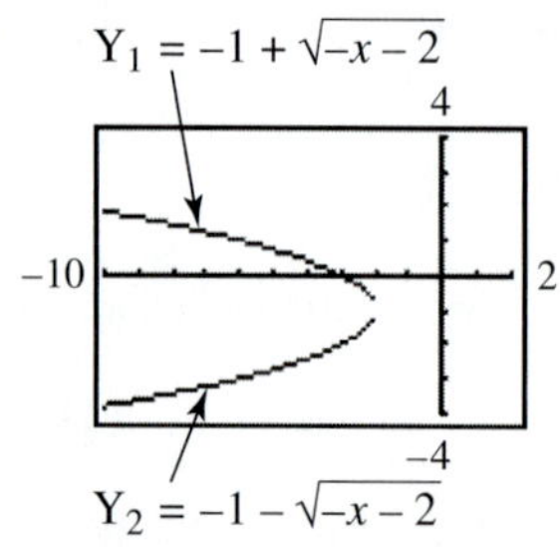

47. $a + b + c = -5$; $4a - 2b + c = -14$; $4a + 2b + c = -10$ **48.** $\{(-2, 1, -4)\}$ **49.** to the left, because $a = -2 < 0$

50. $x = -2y^2 + y - 4$ **51.** **(a)** $y = \frac{11}{5625}x^2$ **(b)** 127.8 ft **53.** 6 ft **55.** **(a)** 2000 ft **(b)** $y = 1000 - .00025x^2$ **(c)** no

57. **(a)**

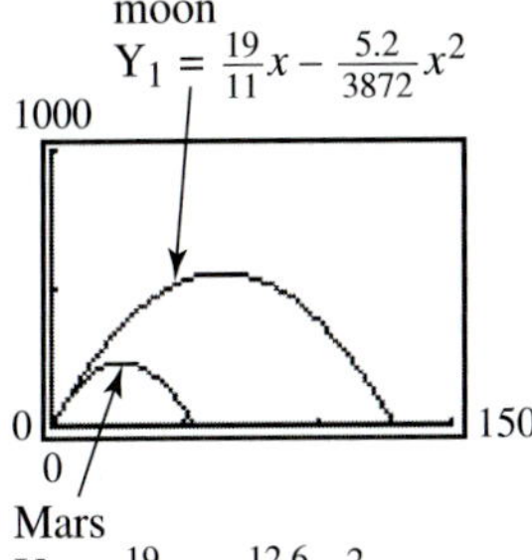

(b) Mars: approximately 229 ft; moon: approximately 555 ft

10.2 Exercises *(page 910)*

1. **(a)** A **(b)** C **(c)** D **(d)** B

3. $[-5, 5]$; $[-3, 3]$; $(0, 0)$; $(-5, 0), (5, 0)$; $(0, -3), (0, 3)$; $(-4, 0), (4, 0)$

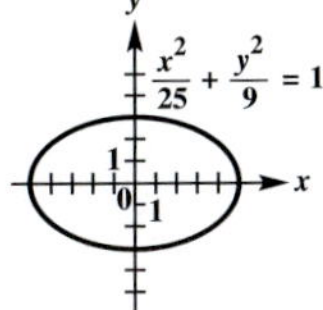

5. $[-3, 3]$; $[-1, 1]$; $(0, 0)$; $(-3, 0), (3, 0)$; $(0, -1), (0, 1)$; $(-2\sqrt{2}, 0), (2\sqrt{2}, 0)$

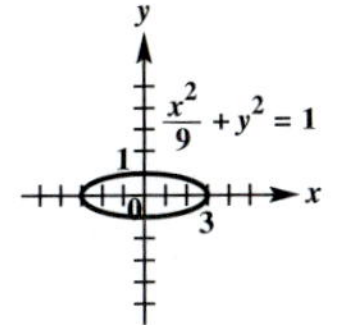

7. $[-3, 3]$; $[-9, 9]$; $(0, 0)$; $(0, -9), (0, 9)$; $(-3, 0), (3, 0)$; $(0, -6\sqrt{2}), (0, 6\sqrt{2})$

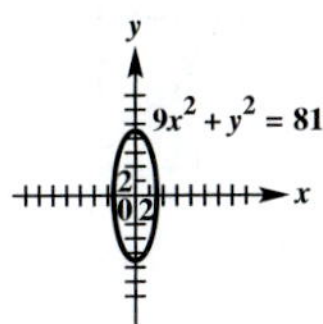

9. $[-5, 5]$; $[-2, 2]$; $(0, 0)$; $(-5, 0), (5, 0)$; $(0, -2), (0, 2)$; $(-\sqrt{21}, 0), (\sqrt{21}, 0)$

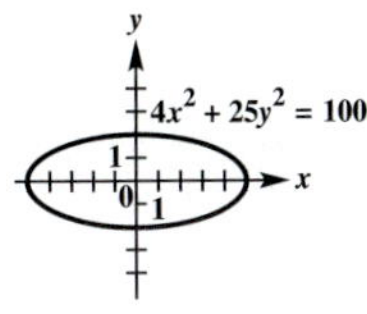

11. $[-3, 7]$; $[-1, 3]$; $(2, 1)$; $(-3, 1), (7, 1)$; $(2, -1), (2, 3)$; $(2 - \sqrt{21}, 1), (2 + \sqrt{21}, 1)$

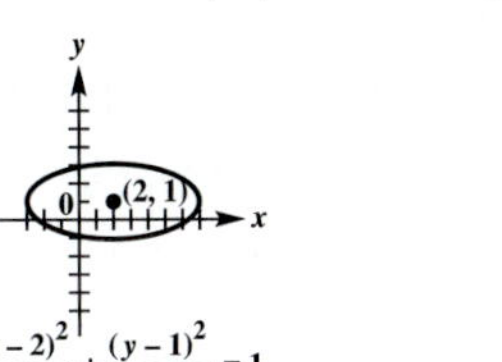

13. $[-7, 1]$; $[-4, 8]$; $(-3, 2)$; $(-3, -4), (-3, 8)$; $(-7, 2), (1, 2)$; $(-3, 2 - 2\sqrt{5}), (-3, 2 + 2\sqrt{5})$

15. $\frac{x^2}{25} + \frac{y^2}{16} = 1$ **17.** $\frac{x^2}{5} + \frac{y^2}{9} = 1$ **19.** $\frac{(x-5)^2}{25} + \frac{(y-2)^2}{16} = 1$ **21.** $\frac{(x-4)^2}{9} + \frac{(y-5)^2}{16} = 1$

23. $\frac{x^2}{72} + \frac{y^2}{81} = 1$ **25.** $\frac{x^2}{9} + \frac{y^2}{25} = 1$ **27.** $\frac{9x^2}{28} + \frac{9y^2}{64} = 1$

29. $[-5, 5]$; $[0, 2]$; function

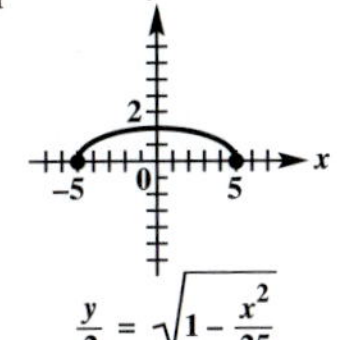

31. $[-1, 0]$; $[-8, 8]$

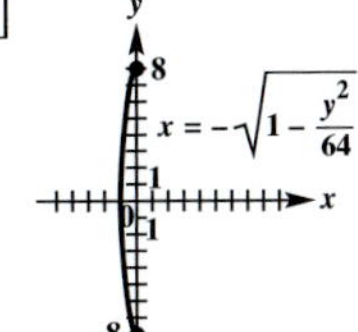

33.

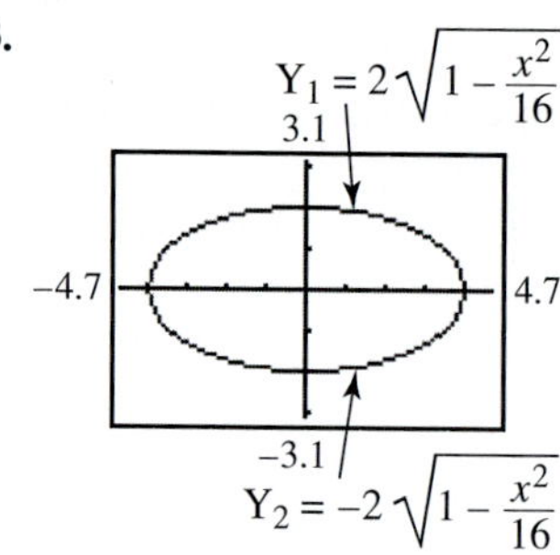

35.

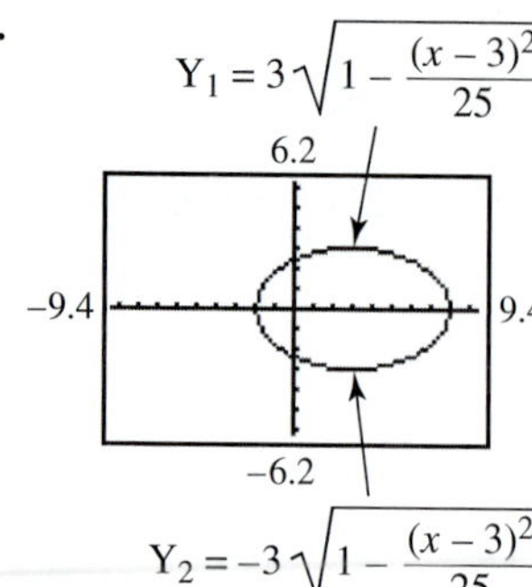

37. $\frac{1}{2}$ **39.** .65 **43.** 12 ft tall

45. approximately 55 million mi **47.** **(a)** Neptune: $\frac{(x - .2709)^2}{30.1^2} + \frac{y^2}{30.1^2} = 1$; Pluto: $\frac{(x - 9.8106)^2}{39.4^2} + \frac{y^2}{38.16^2} = 1$

(b)

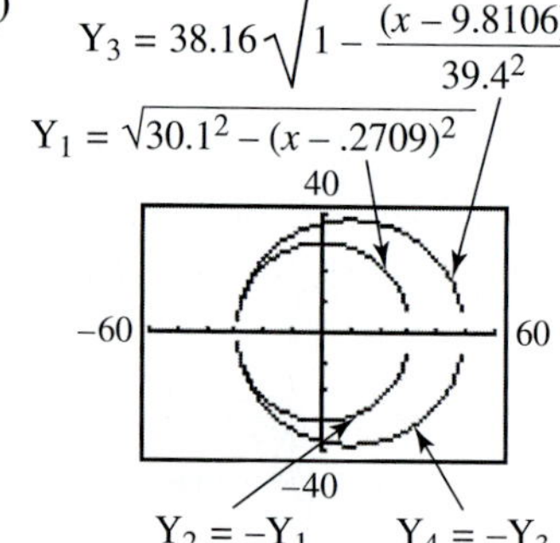

49. $3\sqrt{3}$ units

Connections *(page 918)*

$$\frac{x^2}{625} - \frac{y^2}{1875} = 1$$

10.3 Exercises *(page 918)*

1. C **3.** D **5.** $(-\infty, -4] \cup [4, \infty)$; $(-\infty, \infty)$; $(0, 0)$; $(-4, 0)$, $(4, 0)$; $(-5, 0)$, $(5, 0)$; $y = \pm\frac{3}{4}x$

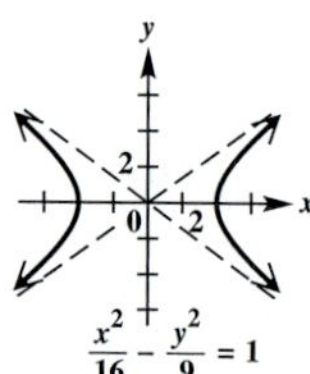

7. $(-\infty, \infty)$; $(-\infty, -5] \cup [5, \infty)$; $(0, 0)$; $(0, -5)$, $(0, 5)$; $(0, -\sqrt{74})$, $(0, \sqrt{74})$; $y = \pm\frac{5}{7}x$

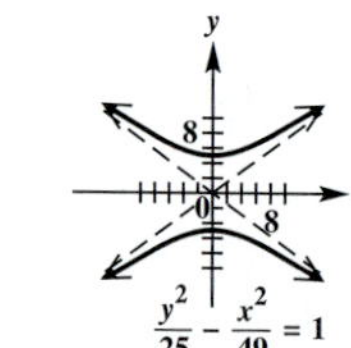

9. $(-\infty, -3] \cup [3, \infty)$; $(-\infty, \infty)$; $(0, 0)$; $(-3, 0)$, $(3, 0)$; $(-3\sqrt{2}, 0)$, $(3\sqrt{2}, 0)$; $y = \pm x$

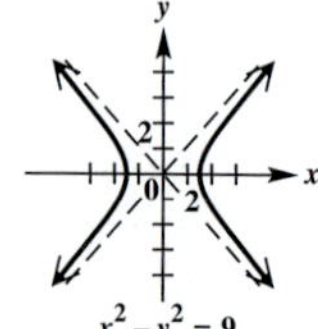

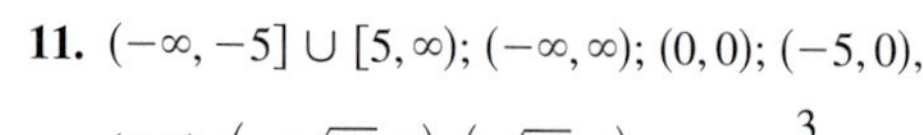
11. $(-\infty, -5] \cup [5, \infty)$; $(-\infty, \infty)$; $(0, 0)$; $(-5, 0)$, $(5, 0)$; $(-\sqrt{34}, 0)$, $(\sqrt{34}, 0)$; $y = \pm\frac{3}{5}x$

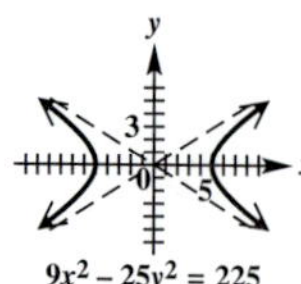

13. $(-\infty, \infty)$; $(-\infty, -5] \cup [5, \infty)$; $(0, 0)$; $(0, -5)$, $(0, 5)$; $(0, -\sqrt{29})$, $(0, \sqrt{29})$; $y = \pm\frac{5}{2}x$

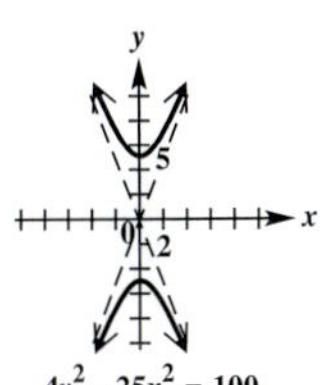

15. $\left(-\infty, -\frac{1}{3}\right] \cup \left[\frac{1}{3}, \infty\right)$; $(-\infty, \infty)$; $(0, 0)$; $\left(-\frac{1}{3}, 0\right)$, $\left(\frac{1}{3}, 0\right)$; $\left(-\frac{\sqrt{13}}{6}, 0\right)$, $\left(\frac{\sqrt{13}}{6}, 0\right)$; $y = \pm\frac{3}{2}x$

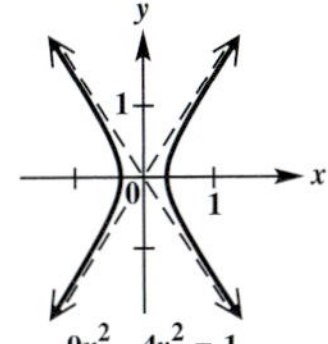

17. $(-\infty, \infty)$; $(-\infty, 1] \cup [13, \infty)$; $(4, 7)$; $(4, 1)$, $(4, 13)$; $(4, -3)$, $(4, 17)$; $y = 7 \pm \frac{3}{4}(x - 4)$

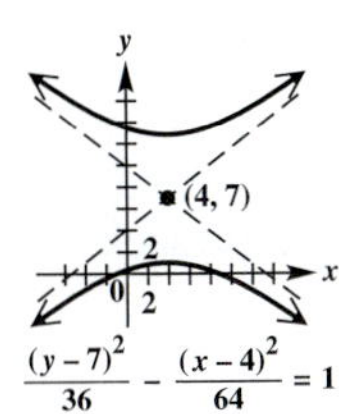

19. $(-\infty, -7] \cup [1, \infty)$; $(-\infty, \infty)$; $(-3, 2)$; $(-7, 2)$, $(1, 2)$; $(-8, 2)$, $(2, 2)$; $y = 2 \pm \frac{3}{4}(x + 3)$

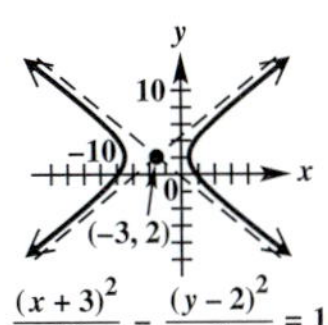

21. $\left(-\infty, -\frac{21}{4}\right] \cup \left[-\frac{19}{4}, \infty\right)$; $(-\infty, \infty)$; $(-5, 3)$; $\left(-\frac{21}{4}, 3\right)$, $\left(-\frac{19}{4}, 3\right)$; $\left(-5 - \frac{\sqrt{17}}{4}, 3\right)$, $\left(-5 + \frac{\sqrt{17}}{4}, 3\right)$; $y = 3 \pm 4(x + 5)$

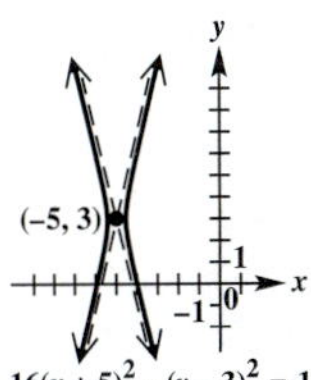

23. $(-\infty, \infty)$; $[3, \infty)$; function

25. $\left(-\infty, -\frac{1}{5}\right]$; $(-\infty, \infty)$

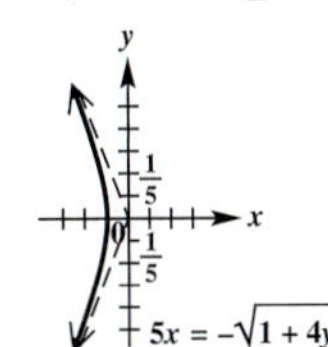

27. 1.4 **29.** 1.7 **31.** $\frac{x^2}{16} - \frac{y^2}{9} = 1$ **33.** $\frac{y^2}{36} - \frac{x^2}{144} = 1$ **35.** $\frac{x^2}{9} - 3y^2 = 1$ **37.** $\frac{2y^2}{25} - 2x^2 = 1$ **39.** $\frac{(y - 3)^2}{4} - \frac{49(x - 4)^2}{4} = 1$ **41.** $\frac{(x - 1)^2}{4} - \frac{(y + 2)^2}{5} = 1$ **43.** $\frac{y^2}{49} - \frac{x^2}{392} = 1$ **45.** $\frac{(y - 6)^2}{16} - \frac{(x + 2)^2}{9} = 1$

47.

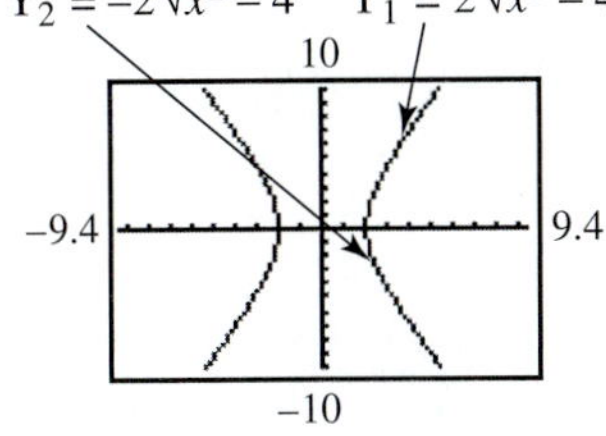

49.

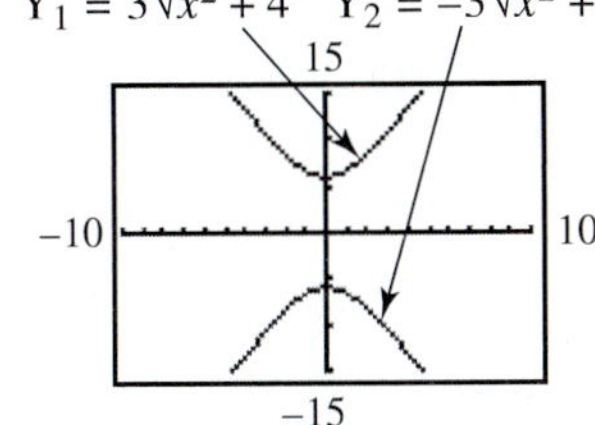

51. $y = \frac{1}{2}\sqrt{x^2 - 4}$ **52.** $y = \frac{1}{2}x$

53. $y \approx 24.98$ **54.** $y = 25$ **55.** Because $24.98 < 25$, the graph of $y = \frac{1}{2}\sqrt{x^2 - 4}$ lies below the graph of $y = \frac{1}{2}x$ when $x = 50$. **56.** The y-values on the hyperbola will approach the y-values on the asymptote. **57.** **(a)** $x = \sqrt{y^2 + 2.5 \times 10^{-27}}$ **(b)** approximately 1.2×10^{-13} m

10.4 Exercises *(page 926)*

1. circle **3.** parabola **5.** parabola **7.** ellipse **9.** hyperbola **11.** hyperbola **13.** ellipse **15.** circle **17.** hyperbola **19.** ellipse **21.** circle **23.** parabola **25.** no graph **27.** circle **29.** parabola **31.** hyperbola **33.** ellipse **35.** point **37.** ellipse **39.** hyperbola **41.** $\frac{1}{3}$ **43.** 1 **45.** 1.5 **47.** elliptical

Chapter 10 Review Exercises *(page 930)*

1. $[2, \infty)$; $(-\infty, \infty)$; $(2, 5)$; $y = 5$

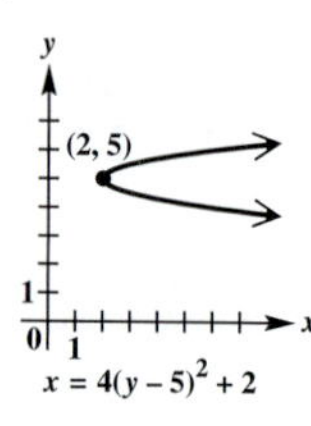

3. $\left[\frac{7}{4}, \infty\right)$; $(-\infty, \infty)$; $\left(\frac{7}{4}, \frac{1}{2}\right)$; $y = \frac{1}{2}$

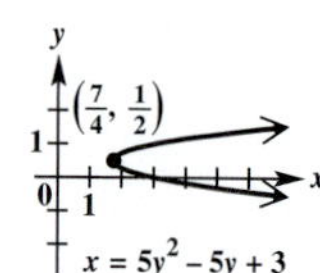

5. $(-\infty, 0]$; $(-\infty, \infty)$; $\left(-\frac{1}{6}, 0\right)$; $x = \frac{1}{6}$; x-axis

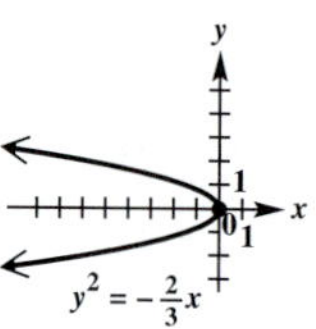

7. $(-\infty, \infty)$; $[0, \infty)$; $\left(0, \frac{1}{12}\right)$; $y = -\frac{1}{12}$; y-axis

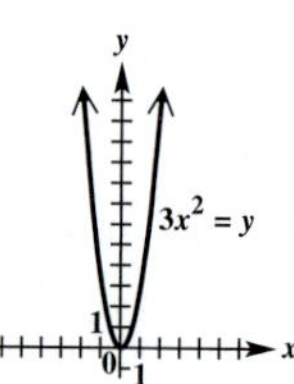

9. $y^2 = 16x$ **11.** $x^2 = \frac{9}{4}y$ **13.** ellipse **15.** hyperbola **17.** parabola **19.** ellipse **21.** F **23.** A **25.** B

27. ellipse; $[-2, 2]$; $[-3, 3]$; $(0, -3)$, $(0, 3)$

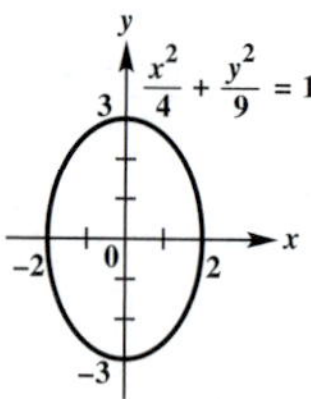

29. hyperbola; $(-\infty, -8] \cup [8, \infty)$; $(-\infty, \infty)$; $(-8, 0)$, $(8, 0)$; $y = \pm\frac{3}{4}x$

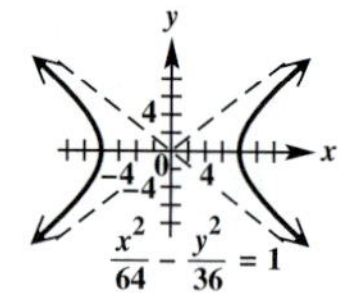

31. circle; $[-5, 3]$; $[-3, 5]$

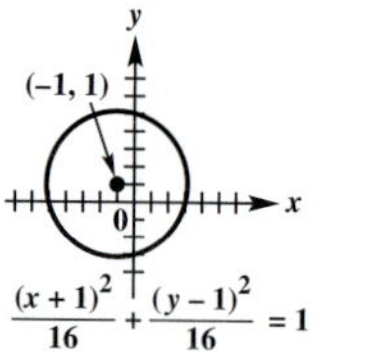

33. ellipse; $[-3, 3]$; $[-2, 2]$; $(-3, 0)$, $(3, 0)$

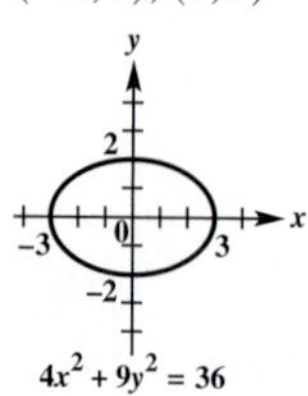

35. ellipse; $[1, 5]$; $[-2, 0]$; $(1, -1)$, $(5, -1)$

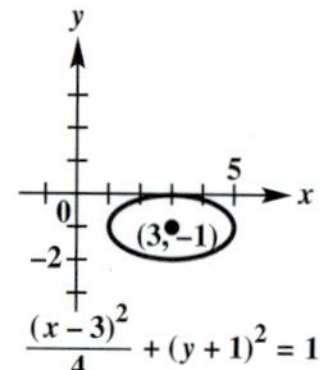

37. hyperbola; $(-\infty, \infty)$; $(-\infty, -4] \cup [0, \infty)$; $(-3, -4)$, $(-3, 0)$; $y = -2 \pm \frac{2}{3}(x + 3)$

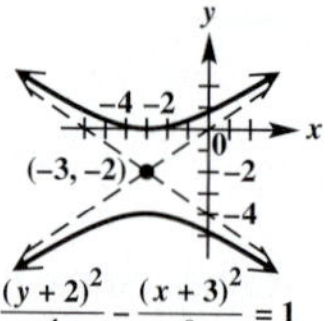

39. $[-3, 0]$; $[-4, 4]$

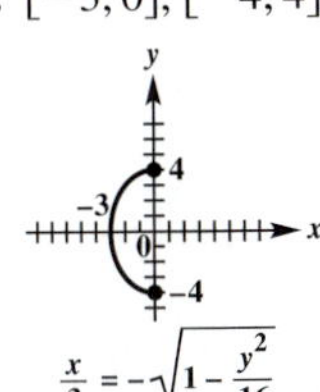

41. $(-\infty, \infty)$; $(-\infty, -1]$; function

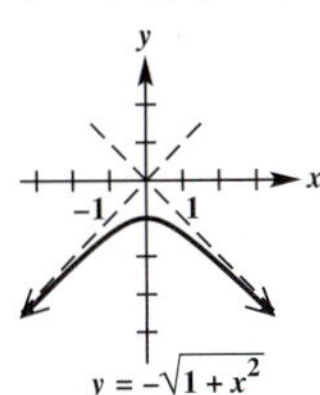

43. $\frac{x^2}{12} + \frac{y^2}{16} = 1$ **45.** $\frac{y^2}{16} - \frac{x^2}{9} = 1$ **47.** $(y - 2)^2 = 12x$ **49.** $\frac{x^2}{25} + \frac{y^2}{21} = 1$ **51.** $\frac{x^2}{9} - \frac{y^2}{16} = 1$

53. $\frac{(x - 2)^2}{16} + \frac{y^2}{12} = 1$ **55.** C, A, B, D **57.** $\frac{x^2}{6{,}111{,}883} + \frac{y^2}{432{,}135} = 1$

Chapter 10 Test *(page 932)*

1. $(-\infty, \infty)$; $(-\infty, 9]$; $(3, 9)$; $x = 3$

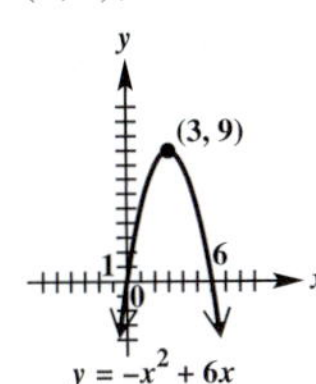

2. $[-4, \infty)$; $(-\infty, \infty)$; $(-4, -1)$; $y = -1$

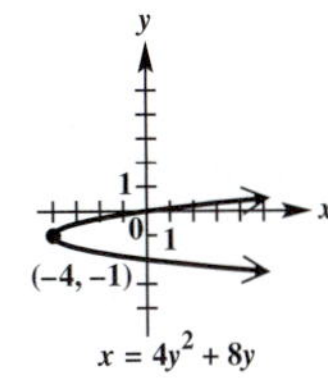

3. $\left(\frac{1}{32}, 0\right)$; $x = -\frac{1}{32}$ **4.** $x - 2 = -5(y - 3)^2$

6. $[-2, 18]$; $[-2, 12]$

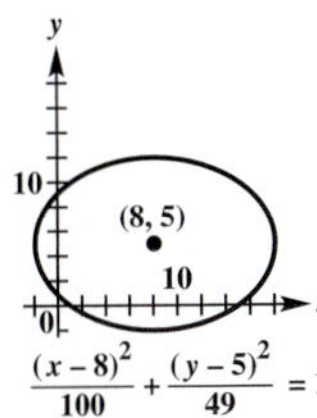

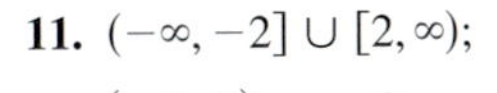

7. $[-2, 2]$; $[-4, 4]$

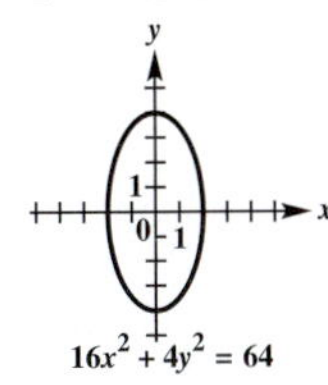

8. It is the graph of a function.

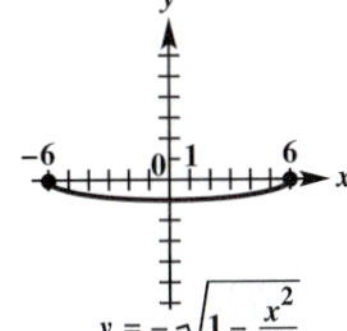

9. $\frac{x^2}{9} + \frac{y^2}{4} = 1$

10. $\frac{x^2}{400} + \frac{y^2}{144} = 1$; approximately 10.39 ft

11. $(-\infty, -2] \cup [2, \infty)$; $(-\infty, \infty)$; $y = \pm x$

$\frac{x^2}{4} - \frac{y^2}{4} = 1$

12. $(-\infty, -2] \cup [2, \infty)$; $(-\infty, \infty)$; $y = \pm\frac{3}{2}x$

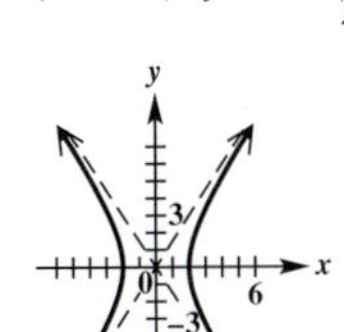

13. $\frac{y^2}{25} - \frac{x^2}{11} = 1$ **14.** circle **15.** hyperbola **16.** ellipse

17. parabola **18.** point **19.** no graph

20. $Y_1 = 7\sqrt{\frac{x^2}{25} - 1}$, $Y_2 = -7\sqrt{\frac{x^2}{25} - 1}$

CHAPTER 11 FURTHER TOPICS IN ALGEBRA

Connections *(page 938)*

1. After the first two terms, each of which is 1, the terms of the sequence are found by adding the two preceding terms. Thus, the third term is $1 + 1 = 2$, the fourth is $1 + 2 = 3$, and so on. **2.** (Assume the original male honeybee is in generation 1.) $a_1 = 1$, $a_2 = 1$, $a_n = a_{n-1} + a_{n-2}$, if $n \geq 3$

11.1 Exercises *(page 943)*

1. 14, 18, 22, 26, 30 **3.** 1, 2, 4, 8, 16 **5.** $0, \frac{1}{9}, \frac{2}{27}, \frac{1}{27}, \frac{4}{243}$ **7.** $-2, 4, -6, 8, -10$ **9.** $1, \frac{7}{6}, 1, \frac{5}{6}, \frac{19}{27}$

13. finite **15.** finite **17.** infinite **19.** finite **21.** $-2, 1, 4, 7$ **23.** 1, 1, 2, 3 **25.** 2, 4, 12, 48 **27.** 35

29. $\frac{25}{12}$ **31.** 288 **33.** 3 **35.** -18 **37.** $\frac{728}{9}$ **39.** 28 **41.** 343 **43.** 1490 **45.** -154

47. $-2 + (-1) + 0 + 1 + 2; 0$ **49.** $-1 + 1 + 3 + 5 + 7; 15$ **51.** $-10 - 4 + 0; -14$ **53.** $0 + \frac{1}{2} + \frac{2}{3} + \frac{3}{4}; \frac{23}{12}$

55. $-3.5 + .5 + 4.5 + 8.5; 10$ **57.** $0 + 4 + 16 + 36; 56$ **59.** $-1 - \frac{1}{3} - \frac{1}{5} - \frac{1}{7}; -\frac{176}{105}$ **61.** 600 **63.** 1240

65. 90 **67.** 220 **69.** 304

There are other acceptable forms of the answers in Exercises 71 and 73.

71. $\sum_{i=1}^{9} \frac{1}{3i}$ **73.** $\sum_{k=1}^{8} \left(-\frac{1}{2}\right)^{k-1}$ **75.** converges to $\frac{1}{2}$ **77.** diverges **79.** converges to $e \approx 2.71828$

81. **(a)** $a_1 = 8$ thousand per acre, $a_2 = 10.4$ thousand per acre, $a_3 = 8.528$ thousand per acre **(b)** The population density oscillates above and below 9.5 thousand per acre (approximately).

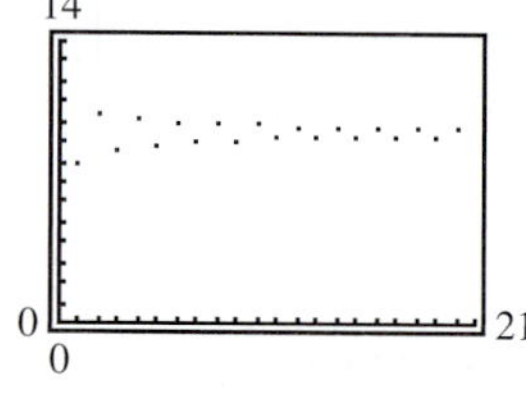

83. **(a)** $N_{j+1} = 2N_j$ for $j \geq 1$ **(b)** 1840 **(c)**

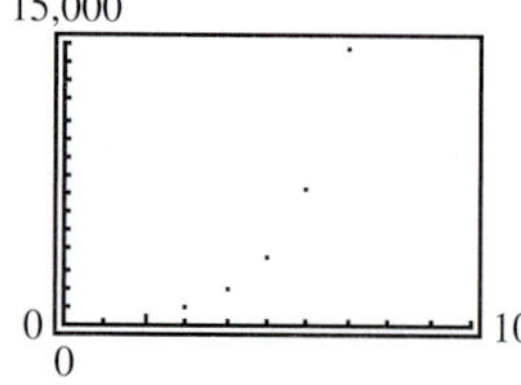

85. **(a)** $y = 93.69x + 979.6$ **(b)** $y = 3.167x^2 + 65.19x + 1027$ **(c)** $y = -1.303x^3 + 20.76x^2 - 1.916x + 1092$ **(d)** linear: 1823; quadratic: 1870; cubic: 1806 **87.** **(a)** 2.718254; $e \approx 2.718282$ **(b)** .367857; $e^{-1} \approx .367879$

11.2 Exercises *(page 954)*

1. 3 **3.** -5 **5.** $x + 2y$ **7.** 8, 14, 20, 26, 32 **9.** $5, 3, 1, -1, -3$ **11.** 14, 12, 10, 8, 6

13. $a_8 = 19; a_n = 3 + 2n$ **15.** $a_8 = \frac{85}{3}; a_n = \frac{5}{3} + \frac{10}{3}n$ **17.** $a_8 = -3; a_n = -39 + \frac{9}{2}n$

19. $a_8 = x + 21; a_n = x + 3n - 3$ **21.** 3 **23.** 5

In Exercises 25–29, *D* is the domain and *R* is the range.

25. $a_n = n - 3$; D: $\{1, 2, 3, 4, 5, 6\}$; R: $\{-2, -1, 0, 1, 2, 3\}$ **27.** $a_n = 3 - \frac{1}{2}n$; D: $\{1, 2, 3, 4, 5, 6\}$; R: $\{0, .5, 1, 1.5, 2, 2.5\}$

29. $a_n = 30 - 20n$; D: $\{1, 2, 3, 4, 5\}$; R: $\{-70, -50, -30, -10, 10\}$ **31.** 215 **33.** 230 **35.** 160 **37.** 77.5

39. 3240 **41.** 2500 **43.** 3660 **45.** $a_1 = 7, d = 5$ **47.** $a_1 = 1, d = -\frac{20}{11}$ **49.** 18 **51.** 140 **53.** -621

55. 500,500 **57.** $f(1) = m + b$; $f(2) = 2m + b$; $f(3) = 3m + b$ **58.** yes **59.** m **60.** $a_n = mn + b$

61. 328.3 **63.** 172.884 **65.** 1281 **67.** 4680 **69.** 54,800 **71.** 713 in.

11.3 Exercises *(page 963)*

1. **(a)** \$5.12 **(b)** \$10.23 **3.** **(a)** \$163.84 **(b)** \$327.67

In Exercises 5–15, there may be other ways to express a_n.

5. $a_5 = 80; a_n = 5(-2)^{n-1}$ **7.** $a_5 = -108; a_n = -\frac{4}{3}(3)^{n-1}$ **9.** $a_5 = -729; a_n = -9(-3)^{n-1}$

11. $a_5 = -324$; $a_n = -4(3)^{n-1}$ **13.** $a_5 = \frac{125}{4}$; $a_n = \frac{4}{5}\left(\frac{5}{2}\right)^{n-1}$ **15.** $a_5 = \frac{5}{8}$; $a_n = 10\left(-\frac{1}{2}\right)^{n-1}$ **17.** -3; 2

19. 5000; $\pm .1$ **21.** 682 **23.** $\frac{99}{8}$ **25.** 860.95 **27.** 363 **29.** $\frac{189}{4}$ **31.** 2032 **33.** The sum exists if $|r| < 1$.

35. 2; does not converge **37.** $\frac{1}{2}$ **41.** 27 **43.** $\frac{3}{20}$ **45.** 4 **47.** $\frac{3}{7}$ **49.** $g(1) = ab$; $g(2) = ab^2$; $g(3) = ab^3$

50. yes; The common ratio is b. **51.** $a_n = ab^n$ **53.** 97.739 **55.** .212 **57.** **(a)** $a_1 = 1169$; $r = .916$ **(b)** $a_{10} = 531$; $a_{20} = 221$; This means that a person who is 10 yr from retirement should have savings of 531% of his or her annual salary; a person 20 yr from retirement should have savings of 221% of his or her annual salary.

59. **(a)** $a_n = a_1 \cdot 2^{n-1}$ **(b)** 15 (rounded from 14.28) **(c)** 560 min or 9 hr, 20 min **61.** about 488 **63.** $\frac{10,000}{9}$ units

65. 62; 2046 **67.** $\frac{1}{64}$ m **69.** Option 2 pays better. **71.** \$10,159.11 **73.** \$25,423.18 **75.** \$28,107.41

77. \$72,918.53

Summary Exercises on Sequences and Series *(page 967)*

1. geometric; $r = 2$ **2.** arithmetic; $d = 3$ **3.** arithmetic; $d = -\frac{5}{2}$ **4.** neither **5.** geometric; $r = \frac{4}{3}$

6. geometric; $r = -3$ **7.** neither **8.** arithmetic; $d = -3$ **9.** geometric; $3(2)^{n-1}$; 3069

10. arithmetic; $-2 + 4n$; 200 **11.** arithmetic; $\frac{11}{2} - \frac{3}{2}n$; $-\frac{55}{2}$ **12.** geometric; $\frac{3}{2}\left(\frac{2}{3}\right)^{n-1}$ or $\left(\frac{2}{3}\right)^{n-2}$; $\frac{58,025}{13,122}$

13. geometric; $3(-2)^{n-1}$; -1023 **14.** arithmetic; $-2 - 3n$; -185 **15.** diverges **16.** $\frac{1111}{500}$ **17.** -1850

18. 1092 **19.** $-\frac{4}{3}$ **20.** diverges **21.** 144 **22.** $\frac{1}{2}$

11.4 Exercises *(page 974)*

1. 20 **3.** 35 **5.** 56 **7.** 45 **9.** 1 **11.** n **13.** 56 **15.** 4950 **17.** 1 **19.** $16x^4$; $81y^4$

21. $x^6 + 6x^5y + 15x^4y^2 + 20x^3y^3 + 15x^2y^4 + 6xy^5 + y^6$ **23.** $p^5 - 5p^4q + 10p^3q^2 - 10p^2q^3 + 5pq^4 - q^5$

25. $r^{10} + 5r^8s + 10r^6s^2 + 10r^4s^3 + 5r^2s^4 + s^5$ **27.** $p^4 + 8p^3q + 24p^2q^2 + 32pq^3 + 16q^4$

29. $2401p^4 + 2744p^3q + 1176p^2q^2 + 224pq^3 + 16q^4$

31. $729x^6 - 2916x^5y + 4860x^4y^2 - 4320x^3y^3 + 2160x^2y^4 - 576xy^5 + 64y^6$

33. $\frac{m^6}{64} - \frac{3m^5}{16} + \frac{15m^4}{16} - \frac{5m^3}{2} + \frac{15m^2}{4} - 3m + 1$ **35.** $4r^4 + \frac{8\sqrt{2}r^3}{m} + \frac{12r^2}{m^2} + \frac{4\sqrt{2}r}{m^3} + \frac{1}{m^4}$ **37.** $-3584h^3j^5$

39. $319,770a^{16}b^{14}$ **41.** $38,760x^6y^{42}$ **43.** $90,720x^{28}y^{12}$ **45.** 11 **47.** exact: 3,628,800; approximate: 3,598,695.619

48. about .830% **49.** exact: 479,001,600; approximate: 475,687,486.5; about .692% **50.** exact: 6,227,020,800; approximate: 6,187,239,475; about .639%; As n gets larger, the percent error decreases. **51.** .942 **53.** 1.015

11.5 Exercises *(page 981)*

1. S_1: $1 = 1^2$; S_2: $1 + 3 = 2^2$; S_3: $1 + 3 + 5 = 3^2$; S_4: $1 + 3 + 5 + 7 = 4^2$; S_5: $1 + 3 + 5 + 7 + 9 = 5^2$

Although we do not usually give proofs, the answers for Exercises 3 and 11 are given here.

3. **(a)** $3(1) = 3$ and $\frac{3(1)(1 + 1)}{2} = \frac{6}{2} = 3$, so S is true for $n = 1$. **(b)** $3 + 6 + 9 + \cdots + 3k = \frac{3(k)(k + 1)}{2}$

(c) $3 + 6 + 9 + \cdots + 3(k + 1) = \frac{3(k + 1)((k + 1) + 1)}{2}$ **(d)** Add $3(k + 1)$ to both sides of the equation in part (b). Simplify the expression on the right side to match the right side of the equation in part (c). **(e)** Since S is true for $n = 1$ and S is true for $n = k + 1$ when it is true for $n = k$, S is true for every positive integer n.

11. **(a)** $\frac{1}{1 \cdot 2} = \frac{1}{2}$ and $\frac{1}{1 + 1} = \frac{1}{2}$, so S is true for $n = 1$. **(b)** $\frac{1}{1 \cdot 2} + \frac{1}{2 \cdot 3} + \frac{1}{3 \cdot 4} + \cdots + \frac{1}{k(k + 1)} = \frac{k}{k + 1}$

(c) $\frac{1}{1 \cdot 2} + \frac{1}{2 \cdot 3} + \cdots + \frac{1}{(k+1)((k+1)+1)} = \frac{k+1}{(k+1)+1}$ **(d)** Add the last term on the left of the equation in part (c) to both sides of the equation in part (b). Simplify the right side until it matches the right side in part (c). **(e)** Since S is true for $n = 1$ and S is true for $n = k + 1$ when it is true for $n = k$, S is true for every positive integer n.

15. $n = 1$ or 2 **17.** $n = 2, 3,$ or 4

For Exercises 19–27, we show only the proof for Exercise 19.

19. (a) $(a^m)^1 = a^m$ and $a^{m(1)} = a^m$, so S is true for $n = 1$. **(b)** $(a^m)^k = a^{mk}$ **(c)** $(a^m)^{(k+1)} = a^{m(k+1)}$

(d) $(a^m)^k \cdot (a^m)^1 = a^{mk} \cdot (a^m)^1$

$(a^m)^{(k+1)} = a^{(mk+m)}$ Product rule for exponents

$(a^m)^{(k+1)} = a^{m(k+1)}$ Factor.

(e) Since S is true for $n = 1$ and S is true for $n = k + 1$ when it is true for $n = k$, S is true for every positive integer n.

31. $\frac{4^{n-1}}{3^{n-2}}$ or $3\left(\frac{4}{3}\right)^{n-1}$ **33.** $2^n - 1$

11.6 Exercises *(page 989)*

1. 19,958,400 **3.** 72 **5.** 5 **7.** 6 **9.** 1 **11.** 495 **13.** 1,860,480 **15.** 259,459,200 **17.** 15,504 **19.** 6435 **21. (a)** permutation **(b)** permutation **(c)** combination **(d)** combination **(e)** permutation **(f)** combination **(g)** permutation **23.** 40 **25. (a)** 27,600 **(b)** 35,152 **(c)** 1104 **27.** 5040 **29. (a)** 17,576,000 **(b)** 17,576,000 **(c)** 456,976,000 **31.** 40,320 **33.** 120 **35.** 2730 **37. (a)** 120 **(b)** 6 **39.** 27,405 **41. (a)** 15 **(b)** 10 **43.** 10 **45.** 28 **47. (a)** 84 **(b)** 10 **(c)** 40 **(d)** 28 **49.** 1680 **51.** 15 **53.** 479,001,600 **55. (a)** 56 **(b)** 462 **(c)** 3080 **(d)** 8526 **57.** 1,000,000 **59.** 4096 **61.** 6 **71. (a)** $3.04140932 \times 10^{64}$ **(b)** $8.320987113 \times 10^{81}$ **(c)** $8.247650592 \times 10^{90}$ **72. (a)** $8.759976613 \times 10^{20}$ **(b)** 5,527,200 **(c)** $2.19289732 \times 10^{26}$

11.7 Exercises *(page 999)*

1. $\{H\}$ **3.** $\{(H,H,H), (H,H,T), (H,T,H), (T,H,H), (H,T,T), (T,H,T), (T,T,H), (T,T,T)\}$

5. $\{(1,1), (1,2), (1,3), (2,1), (2,2), (2,3), (3,1), (3,2), (3,3)\}$

7. (a) $\{H\}$; 1 **(b)** $\emptyset$; 0 **9. (a)** $\{(1,1), (2,2), (3,3)\}$; $\frac{1}{3}$ **(b)** $\{(1,1), (1,3), (2,1), (2,3), (3,1), (3,3)\}$; $\frac{2}{3}$ **(c)** $\{(2,1), (2,3)\}$; $\frac{2}{9}$

13. (a) F **(b)** D **(c)** A **(d)** F **(e)** C **(f)** B **(g)** E **15. (a)** $\frac{1}{5}$ **(b)** 0 **(c)** $\frac{7}{15}$ **(d)** 1 to 4 **(e)** 7 to 8 **17. (a)** $\frac{1}{6}$ **(b)** $\frac{1}{3}$ **(c)** $\frac{1}{6}$ **19. (a)** .176 **(b)** .376 **(c)** .138 **(d)** 10,374 to 157,151 or about 1 to 15 **21.** $\frac{1}{28{,}561} \approx .000035$ **23. (a)** about .6998 **(b)** about .3002 **(c)** about .3088 **(d)** about .9099 **25.** .3 **27.** .042246 **29.** .026864 **31.** .49 **33.** .51 **35. (a)** .047822 **(b)** .976710 **(c)** .897110 **37. (a)** about .404 **(b)** about .047 **(c)** about .002

Chapter 11 Review Exercises *(page 1008)*

1. $\frac{1}{2}, \frac{2}{3}, \frac{3}{4}, \frac{4}{5}, \frac{5}{6}$; neither **3.** 8, 10, 12, 14, 16; arithmetic **5.** $5, 2, -1, -4, -7$; arithmetic

7. $3\pi - 2, 2\pi - 1, \pi, 1, -\pi + 2$ **9.** $-5, -1, -\frac{1}{5}, -\frac{1}{25}, -\frac{1}{125}$ **11.** ± 1; $-8\left(\frac{1}{2}\right)^{n-1} = -\left(\frac{1}{2}\right)^{n-4}$ or $-8\left(-\frac{1}{2}\right)^{n-1} = \left(-\frac{1}{2}\right)^{n-4}$ **13.** $-x + 61$ **15.** 612 **17.** $\frac{4}{25}$ **19.** -40 **21.** 1 **23.** $\frac{73}{12}$ **25.** 3,126,250

27. $\frac{4}{3}$ **29.** 36 **31.** diverges **33.** -10

In Exercises 35 and 37, other answers are possible.

35. $\sum_{i=1}^{15} (-5i + 9)$ **37.** $\sum_{i=1}^{6} 4(3)^{i-1}$ **39.** $x^4 + 8x^3y + 24x^2y^2 + 32xy^3 + 16y^4$

41. $243x^{5/2} - 405x^{3/2} + 270x^{1/2} - 90x^{-1/2} + 15x^{-3/2} - x^{-5/2}$ **43.** $-3584x^3y^5$ **45.** $x^{12} + 24x^{11} + 264x^{10} + 1760x^9$
53. 72 **55.** 56 **57.** 252 **59.** 90 **61.** **(a)** 20 **(b)** 10 **63.** 456,976,000; 258,336,000
65. **(a)** $\frac{4}{15}$ **(b)** $\frac{2}{3}$ **(c)** 0 **(d)** 3 to 2 **67.** **(a)** .2002 **(b)** .0380 **(c)** .4915 **69.** .296

Chapter 11 Test *(page 1011)*

1. $-3, 6, -11, 18, -27$; neither **2.** $-\frac{3}{2}, -\frac{3}{4}, -\frac{3}{8}, -\frac{3}{16}, -\frac{3}{32}$; geometric **3.** 2, 3, 7, 13, 27; neither **4.** 49
5. $-\frac{32}{3}$ **6.** 110 **7.** -1705 **8.** 2385 **9.** -186 **10.** does not exist **11.** $\frac{108}{7}$
12. $x^6 + 6x^5y + 15x^4y^2 + 20x^3y^3 + 15x^2y^4 + 6xy^5 + y^6$ **13.** $16x^4 - 96x^3y + 216x^2y^2 - 216xy^3 + 81y^4$ **14.** $60w^4y^2$
15. 40,320 **16.** 45 **17.** 35 **18.** 990 **20.** 24 **21.** 120; 14 **22.** 90 **24.** **(a)** $\frac{1}{26}$ **(b)** $\frac{10}{13}$ **(c)** $\frac{4}{13}$
(d) 3 to 10 **25.** .92 **26.** .201

Appendix A Exercises *(page 1016)*

1.

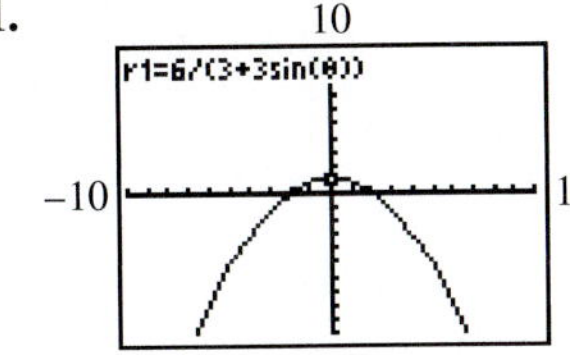

3.

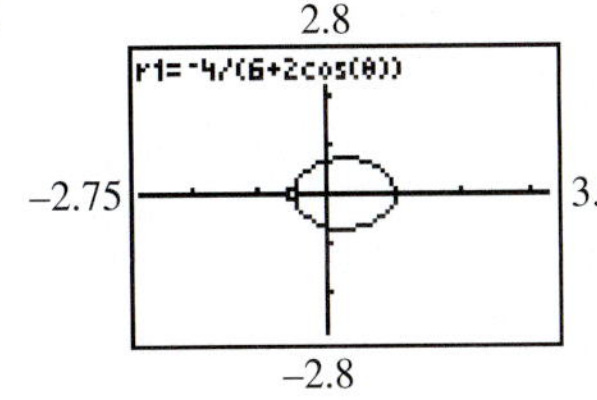

5.

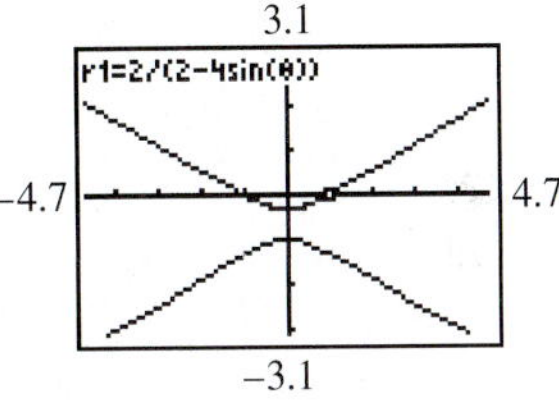

7.

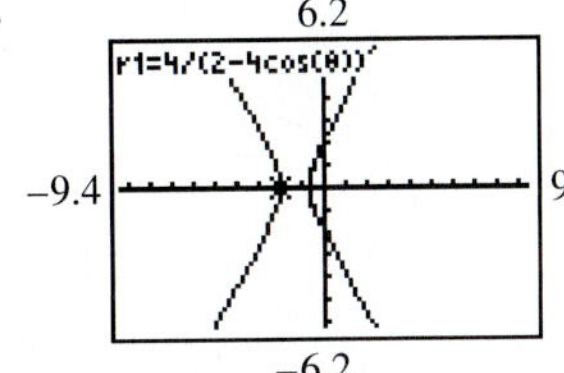

9.

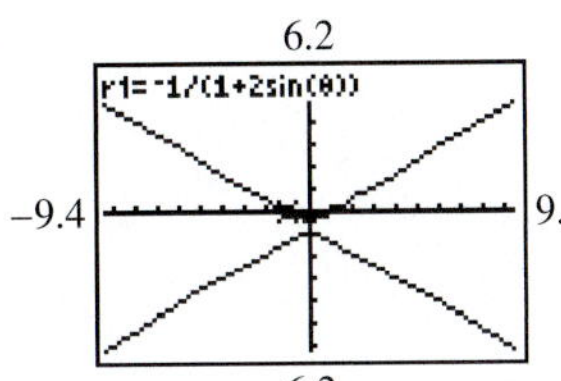

11.

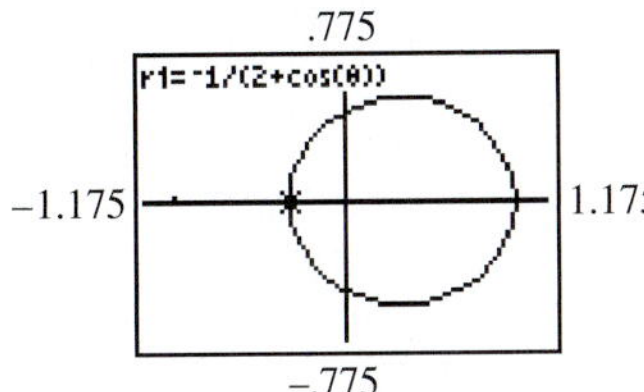

13. $r = \frac{3}{1 + \cos\theta}$ **15.** $r = \frac{5}{1 - \sin\theta}$ **17.** $r = \frac{20}{5 + 4\cos\theta}$; ellipse **19.** $r = \frac{40}{4 - 5\sin\theta}$; hyperbola
21. ellipse; $8x^2 + 9y^2 - 12x - 36 = 0$ **23.** hyperbola; $3x^2 - y^2 + 8x + 4 = 0$ **25.** ellipse; $4x^2 + 3y^2 - 6y - 9 = 0$
27. parabola; $x^2 - 10y - 25 = 0$

Appendix B Exercises *(page 1020)*

1. circle or ellipse or a point **3.** hyperbola or two intersecting lines **5.** parabola or one line or two parallel lines
7. 30° **9.** 60° **11.** 22.5°

13.

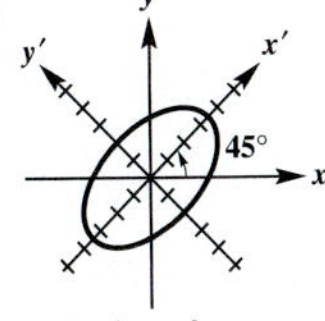

15.

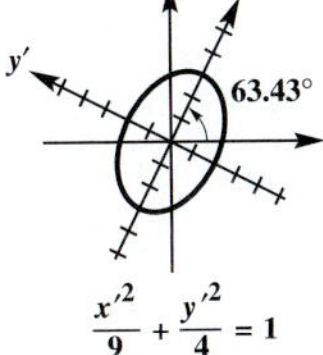

17.

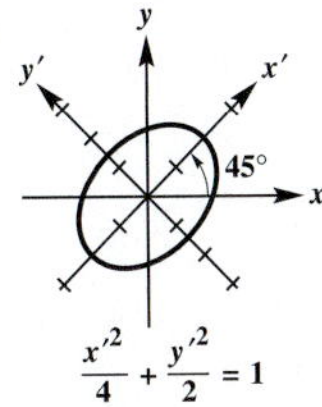

19.

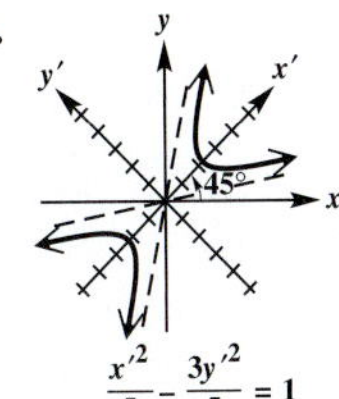

21.

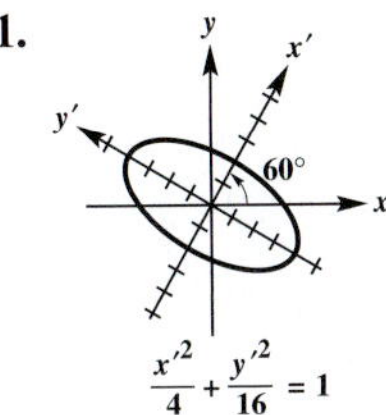

23.

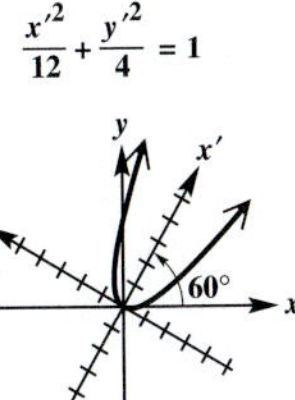

25.

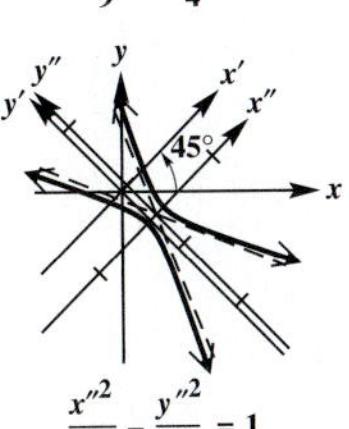

27.

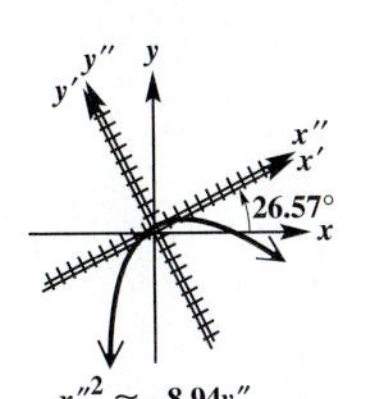

29.

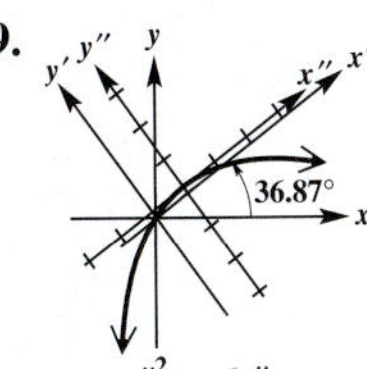

31. If $B = 0$, $\cot 2\theta$ is undefined. The graph may be translated but is not rotated.

Index of Applications

Economics

Education

Energy

Engineering

Entertainment

Environmental Issues

Finance

General Interest

Geology

Geometry

Government

Health and Medicine

Labor

Physics

Social Issues

Sports

Statistics and Probability

Transportation

U.S. and World Demographics

Weather

Index

G

H

O

P

W

X

Y

Z

Graphs of Functions

2.5 Identity Function

$f(x) = x$

Squaring Function

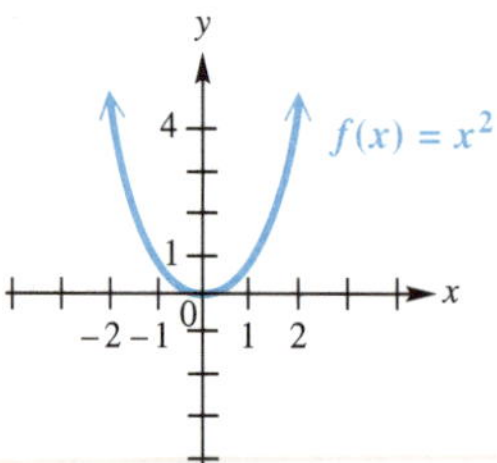

Cubing Function

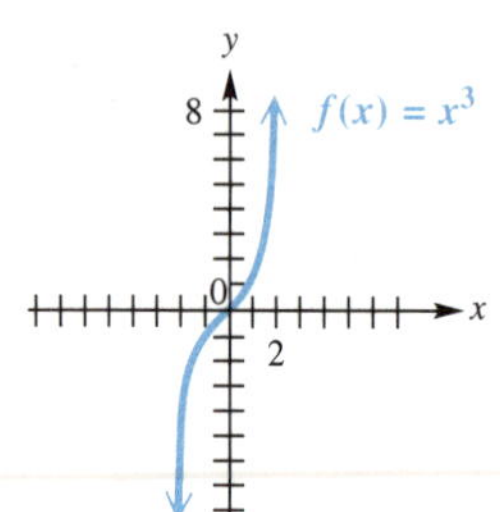

Square Root Function

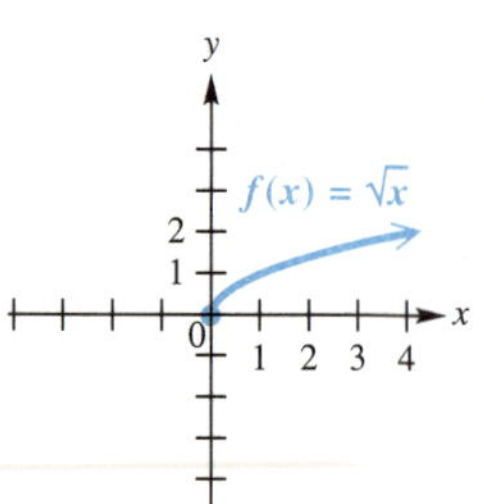

Cube Root Function

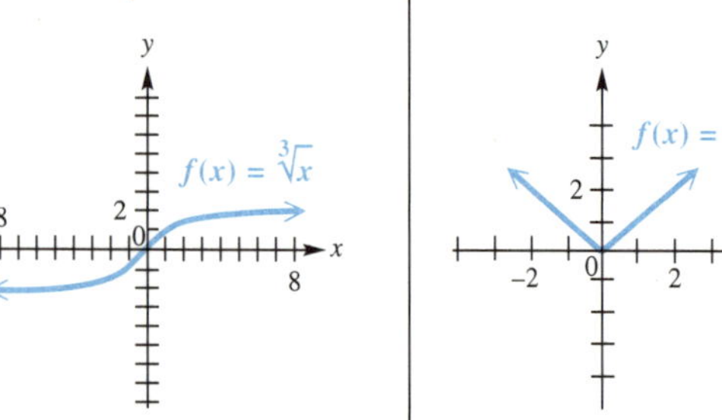

Absolute Value Function

$f(x) = |x|$

Greatest Integer Function

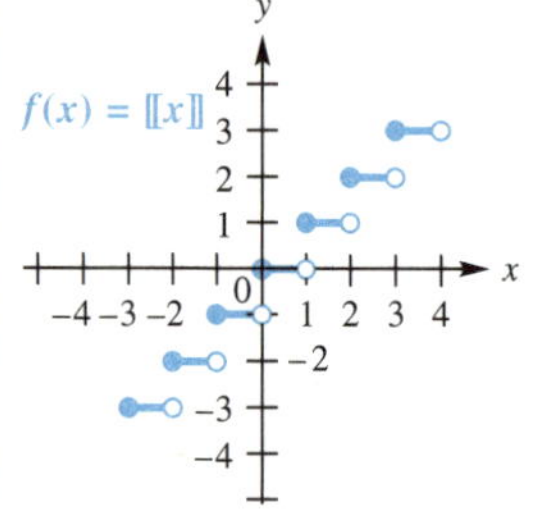

3.5 Reciprocal Function

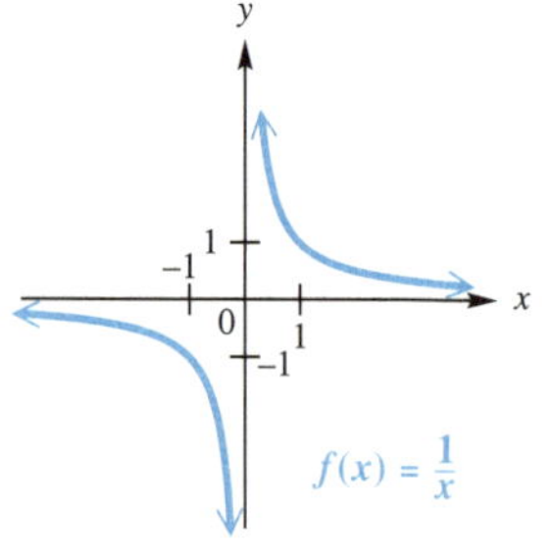

3.4 Polynomial Functions

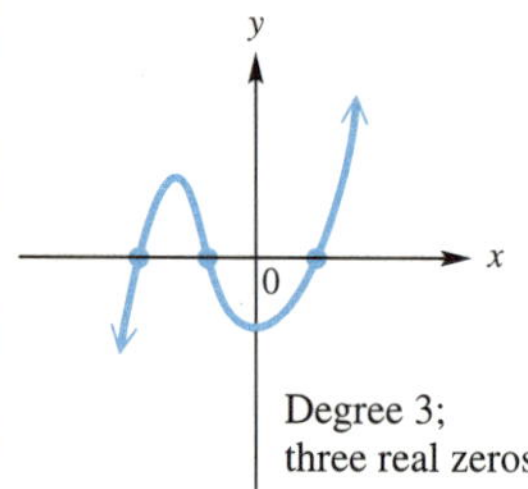

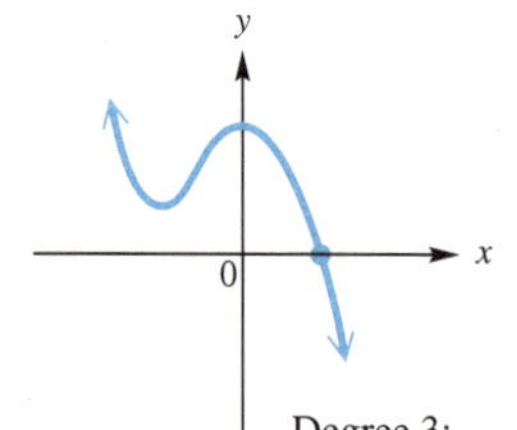

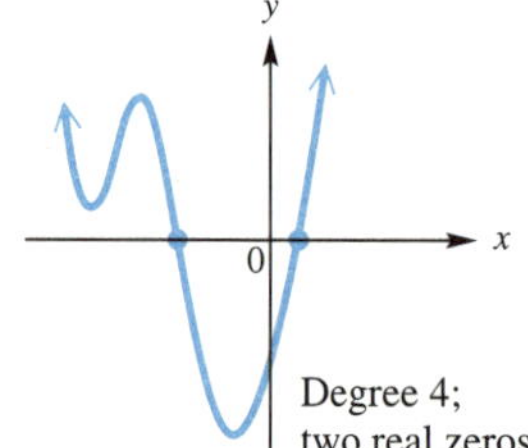

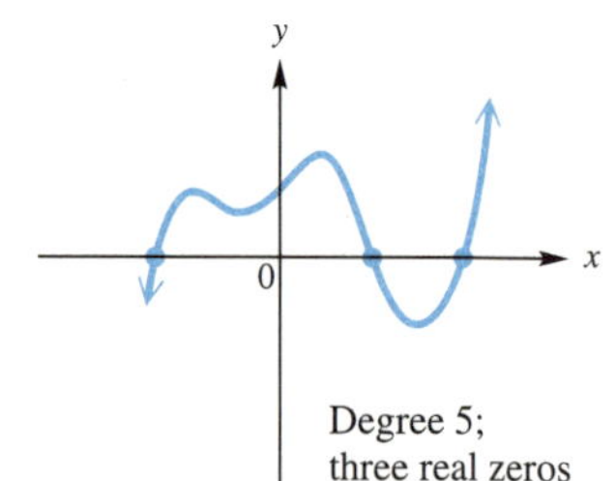

4.2 Exponential Functions

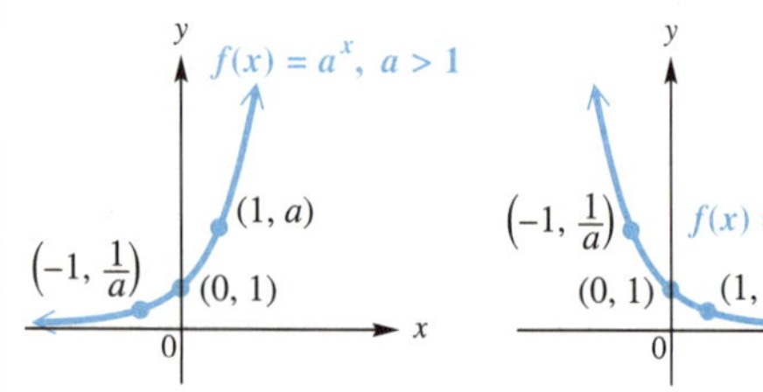

4.3 Logarithmic Functions

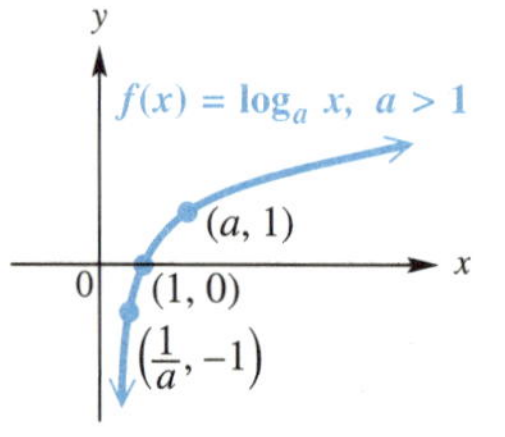

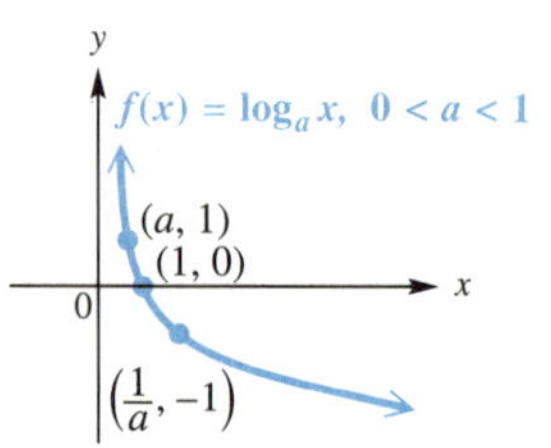

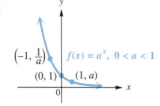